STUDY GUIDE AND SOLUTIONS MANUAL

Organic Chemistry: Principles and Mechanisms

Second Edition

Joel Karty

Joel Karty
ELON UNIVERSITY

Marie M. Melzer

W • W • NORTON & COMPANY • NEW YORK • LONDON

W. W. Norton & Company has been independent since its founding in 1923, when William Warder Norton and Mary D. Herter Norton first published lectures delivered at the People's Institute, the adult education division of New York City's Cooper Union. The firm soon expanded its program beyond the Institute, publishing books by celebrated academics from America and abroad. By midcentury, the two major pillars of Norton's publishing program—trade books and college texts—were firmly established. In the 1950s, the Norton family transferred control of the company to its employees, and today—with a staff of four hundred and a comparable number of trade, college, and professional titles published each year—W. W. Norton & Company stands as the largest and oldest publishing house owned wholly by its employees.

Printed in the United States of America.

Associate Media Editor: Arielle Holstein
Associate Managing Editor, College: Carla L. Talmadge
Production Manager: Eric Pier-Hocking
Assistant Media Editor: Doris Chiu
Project Editor: Holly Monteith

ISBN 978-0-393-65555-1

W. W. Norton & Company, Inc., 500 Fifth Avenue, New York, N.Y. 10110
www.wwnorton.com

W. W. Norton & Company Ltd., 15 Carlisle Street, London W1D 3BS
2 3 4 5 6 7 9 0

CONTENTS

PREFACE

We have written this *Study Guide and Solutions Manual* to be a valuable resource to help you excel in organic chemistry. It does more than just provide the answers to the Your Turn exercises and the Problems in each chapter (both the in-chapter and end-of-chapter problems); it can help you become a *better learner* and a *better problem solver*. Becoming a better learner and problem solver is important for your success in this course, because there is simply too much to memorize. Instead, your goals should be to *understand* the various concepts that underlie the chemistry and to become well practiced with the variety of ways those concepts are applied to arrive at the correct answers. This *Manual* and the textbook have a number of features to help you achieve such goals.

Think/Solve approach

Every solution in this *Manual* uses the Think/Solve approach, which is the same approach used in the Solved Problems throughout the textbook. In the Think part of every solution, we provide questions that relate to the particular problem at hand; when answered, those Think questions help guide you to the problem's solution. In the Solve part, we answer several of those questions and show how those answers are used to solve the problem.

Don't take the Think questions for granted. Before you read the Solve part of the solution, make an honest effort to answer the questions that we have posed. Doing so will provide two benefits. One benefit is that it will help enhance your approach to solving problems in general. Most organic chemistry problems are not solved algorithmically; there is no specific set of steps to memorize that can be used to solve every problem you encounter. To the contrary, solving organic chemistry problems requires you to bring together various concepts you've learned, often in widely different ways. The Think questions help you determine which concepts are relevant to the particular problem at hand and how you should consider bringing them together to solve it. The more you practice this vital step in problem solving, the better you will become at it.

The second benefit of spending time on the Think questions comes from realizing which ones you can answer easily and which ones you can't. Trouble answering a particular Think question could indicate that you have a gap in your understanding of a specific concept or idea. Take the time to review that concept or idea where it is discussed in the textbook. That way, you will be better poised to solve other problems that require you to apply the same concept (but perhaps in a different way).

Various levels of challenge

The Your Turn exercises, in-chapter problems, and end-of chapter problems are designed for you to attempt at different points in your learning process. Carry out the Your Turn exercises as you are reading the material in the textbook for the first time (even before the material is discussed in class). Those exercises are written to be relatively straightforward and to take little time, and they provide you quick feedback as to whether you "get" the concept at hand. Work through each in-chapter problem after you have had sufficient time to digest the relevant material that is being discussed and have gone through any related Solved Problems. Finally, challenge yourself with as many end-of-chapter problems as possible immediately after you have completed a chapter.

Don't put the end-of-chapter problems off to the days leading up to an exam. Instead, work on them regularly. Do your best to work through about 10 each day. That way, you will be able to internalize and learn from each problem much better. Importantly, if you get stuck on a particular problem, this will give you plenty of time to seek help with any questions or confusion you have, perhaps by rereading portions of the textbook, asking a study partner or teaching assistant, or visiting your professor. If you procrastinate on the end-of-chapter problems, you will find yourself rushing through them, you won't learn as much from them, and you will not have sufficient time to seek the proper help.

Organization of end-of-chapter problems

The end-of-chapter problems in the textbook are grouped together according to the sections of the chapter that deal with the material needed to solve the problems. Such an organization can help you focus your efforts efficiently. By using this *Manual* to check your answers, you can determine which topics and which types of problems deserve the most of your attention. Then, if you determine that you need more work in a particular area, you can quickly find a related problem to solve.

In the end-of-chapter problems, each chapter also has a group of integrated problems. These problems have an additional measure of challenge, because solving them requires integrating material from multiple sections of the same chapter or from multiple chapters. Being more open ended, these problems could therefore highlight a weakness in your ability to determine which of several concepts are appropriate to apply to the particular problem at hand, or they could highlight topics you may have forgotten from a previous chapter. Either way, difficulties with these kinds of problems could provide you with very useful feedback.

There are two ways you can use the end-of-chapter problems. One way is simply as practice. You will be expected to solve certain types of problems on exams, and by working through more end-of-chapter problems, you will become more familiar with the ways that certain types of problems might look. You will become more proficient in beginning and completing them. You will become more accurate and more efficient, and these are certainly things that will result in better performances on exams.

A much better way to use the end-of-chapter problems, however, is as *self-assessment*. After you have completed a chapter (along with the Your Turn exercises and the in-chapter problems), try to tackle each end-of-chapter problem with no assistance—that is, without having seen the solution and without consulting the relevant material in the textbook, in your notes, or with your study partners. If you successfully solve a problem in little time, take that as a message that you are relatively strong in that particular area. If you get the right answer but find that it took you a long time to solve the problem, then you need more practice. Take what you have learned from that problem and apply it to a similar problem. This time, challenge yourself to solve it faster. If, on the other hand, your answer to a problem is incorrect or if you needed help working through the problem (perhaps by reading the Think questions or by asking a study partner), then you may have a weakness in that particular area that needs to be addressed before the next exam. Take the time to strengthen that area of weakness. Reread the relevant portions of the textbook, discuss things with study partners, or visit your instructor. Then try another, similar problem unassisted, and repeat the process until you are no longer struggling.

We wish you all the best in your studying.

Joel Karty
Marie M. Melzer

CHAPTER 1 | Atomic and Molecular Structure

Your Turn Exercises

Your Turn 1.1

Think

When designating orbitals as 1*s*, 2*s*, 2*p*, and so on, what corresponds to the shell number? Which orbitals share that designation? Which electrons are in this shell?

Solve

The number appearing before the *s* and *p* designation specifies the shell number. The 2*s* and 2*p* orbitals are in the second shell, noted by the dashed box. Electrons (3)–(10) are in the second shell.

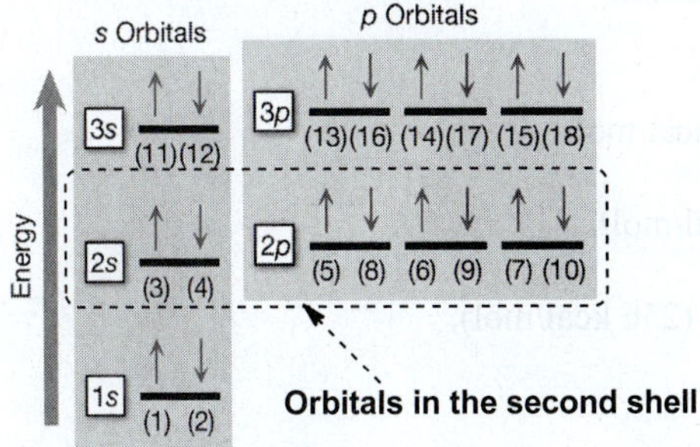

Your Turn 1.2

Think

How many total electrons does carbon have? What is the electron configuration for carbon? What distinguishes valence electrons from core electrons?

Solve

The valence electrons are in the highest energy shell, that is, the $n = 2$ shell. The core electrons are in all the other (lower energy) shells. Carbon has six total electrons: $1s^2 2s^2 2p^2$. Therefore, the valence electrons are in the second shell, $2s^2$ and $2p^2$, and the core electrons are in the first shell, $1s^2$ (noted in the figure by the circle and rectangle, respectively).

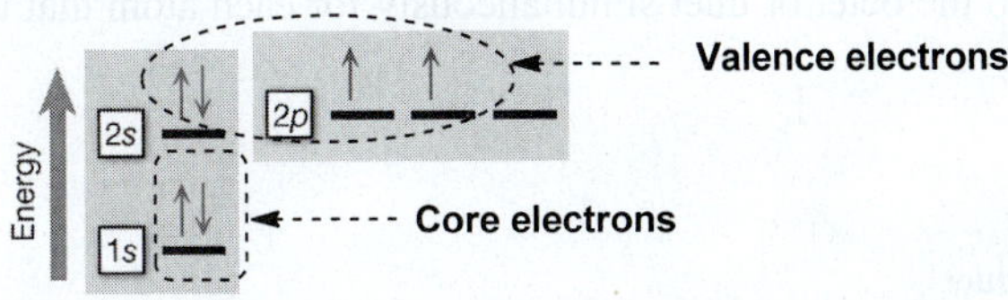

Your Turn 1.3

Think

What is the definition of bond energy? Looking at the plot of energy versus internuclear distance, what is the energy of the separate H atoms? What is the energy of the bonded H–H molecule? How does the bond energy correlate to these two values?

Solve

The bond energy represents the change in energy (on the vertical scale) needed to increase the bond length from its equilibrium value (the lowest energy point on the horizontal scale) to infinity. This can be calculated from the potential energy diagram (Fig. 1-10a). Using the energy difference of the separated H atoms (E = 450 kJ/mol) and bonded H–H molecule (E = 0 kJ/mol), a bond energy of 450 kJ/mol is calculated. See the figure on the next page.

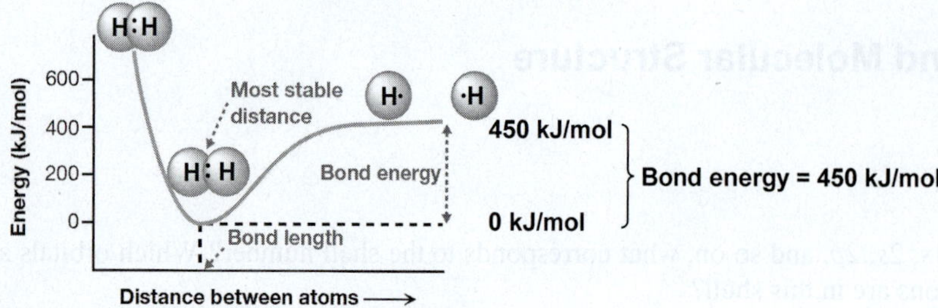

Your Turn 1.4

Think

Look at Tables 1-2 and 1-3. Does the strongest bond have the largest or smallest kJ/mol value? In general, are single bonds stronger or weaker compared to double and triple bonds?

Solve

(a) The strongest *single* bond has a value of 586 kJ/mol (140 kcal/mol).
(b) The strongest *single* bond is the Si–F bond.
(c) The weakest *single* bond has a value of 138 kJ/mol (33 kcal/mol).
(d) The weakest *single* bond is the O–O bond.
(e) The strongest bond of any type has a value of 1072 kJ/mol (256 kcal/mol).
(f) The strongest bond of any type is the C≡O triple bond.

Your Turn 1.5

Think

Lewis structures take into account only which type of electron? How are bonding and nonbonding electrons represented, and how are they counted? How do you determine whether a pair of electrons counts as part of an atom's octet or duet?

Solve

For the electrons to be counted around any atom, the lines (bonding) or dots (nonbonding) have to touch the atom. The C atom's octet (black dashed circle below) comprises the four sets of bonding electrons (lines) around the C atom. Each bond comprises two electrons shared between the atoms it touches. O's octet (black dashed oval below) comprises two sets of bonding electrons, one to the C atom and one to a H atom, and then two lone pairs. H's duet (gray dashed ovals) comprises the two shared electrons in the bond to either the C or O atom. You will notice that each bond contributes to the octet or duet simultaneously for each atom that touches the bond, hence the circles' overlap.

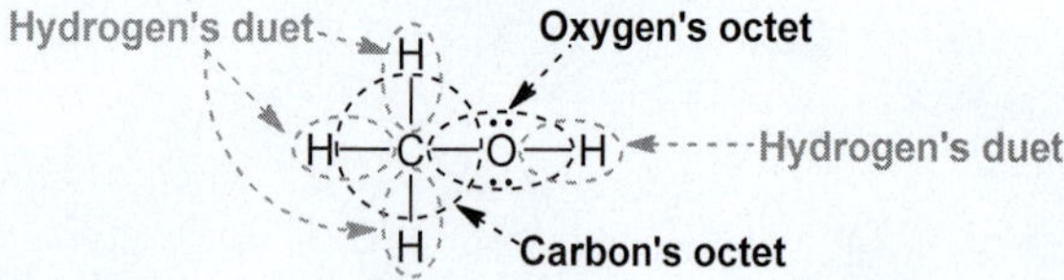

Your Turn 1.6

Think

Consult Figure 1-16 to determine the electronegativity value for boron (B) and hydrogen (H). How is a dipole arrow drawn? To which atom, more or less electronegative, does the arrow point? Which end of the dipole arrow receives the positive delta (δ^+) and negative delta (δ^-) symbol?

Solve

The electronegativity of B is 2.04 and of H is 2.20. This leads to the conclusion that H is *slightly* more electronegative than B. The difference in electronegativity is 0.16. The dipole arrow (⇸) points to the more electronegative atom (each H atom in this example). This leads to a partial negative charge on each H atom (δ^-) and a partial positive charge on the B atom (δ^+). See the figure on the next page.

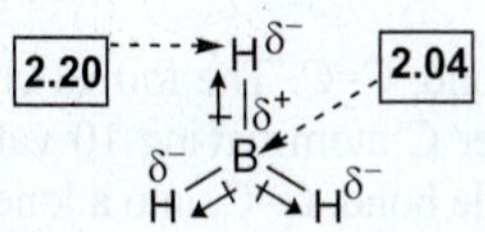

2.20 H$^{\delta-}$ 2.04
δ^+
B
δ^- H H$^{\delta-}$

Your Turn 1.7

Think

Consult Figure 1-16 to determine the electronegativity value for sodium (Na), chlorine (Cl), carbon (C), and hydrogen (H). To what does the *difference* in electronegativity correspond in terms of bond type (i.e., ionic vs. covalent)? Are any metals present? What type of bonding is likely to result?

Solve

For NaCl, the difference is 3.16 − 0.93 = 2.23. The electronegativity difference of 2.23 is large and correlates to an ionic bond. This means that Cl acquires the one extra electron needed to fill its octet from the less electronegative Na atom. This results in the formation of the chloride anion, Cl^-, and the sodium cation, Na^+. For CH_4, the difference is 2.55 − 2.20 = 0.35. This corresponds to a small difference in electronegativity. Therefore, the C–H bond is a covalent bond that is not very polar.

Your Turn 1.8

Think

What gives the atom its identity (protons or electrons)? What is the charge of a proton? What is the charge of an electron? How does the overall number of protons and electrons contribute to the charge on the atom? How many core electrons does carbon have?

Solve

The identity of an atom is determined by the number of protons in the nucleus. For a carbon nucleus, the number of protons is always six. Add the core electrons (always two for a carbon atom) to the number of valence electrons given to get the total number of electrons. The charge is determined by comparing the number of protons and electrons. If the number of protons and the number of electrons are equal, the charge is 0. It is +1 if there is one more proton than electron and −1 if there is one more electron than proton.

Number of Valence Electrons	Total Number of Electrons	Number of Protons	Charge
3	5	6	+1
4	6	6	0
5	7	6	−1

Your Turn 1.9

Think

What is the formal charge on each atom, determined from Solved Problem 1.16? How does the sum of formal charges compare to the total charge?

Solve

For HCO_2^-, the H, the C, and one O atom have a formal charge of 0, and the other O atom has a formal charge of −1. Therefore, the sum of formal charges is $0 + 0 + 0 + (-1) = -1$. The sum must always be the same as the charge of the molecule. In this case, both values are −1.

Your Turn 1.10

Think

Remember that the curved arrow illustrates electron movement necessary to change one resonance structure into another. In the structure on the left, which electrons move to form the double bond in the structure on the right? Which electrons move in the structure on the left to form the lone pair on the C atom in the structure on the right?

Solve

The first curved arrow below is used to convert the lone pair into a covalent double bond, C=C. The formal charge becomes more positive by 1. Without any other changes, this would lead to the center C atom having 10 valence electrons. The second arrow is used to convert a pair of electrons from the original double bond, C=C, into a lone pair on the other end's C atom, $:CH_2$. The formal charge on this C atom becomes more negative by 1.

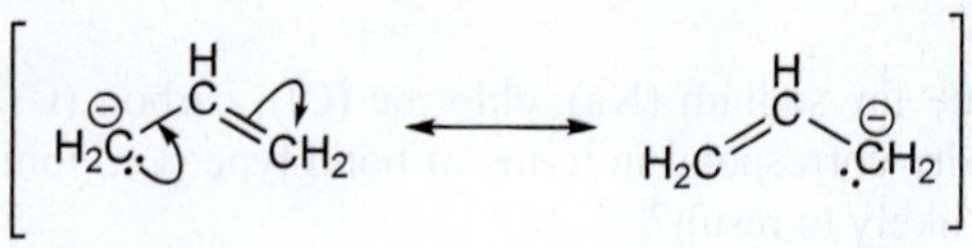

Your Turn 1.11

Think

Do any of the atoms have an unfilled valence shell? If so, is the atom with the unfilled valence shell adjacent to a double bond, a triple bond, or a lone pair? For **(a)**, how can you show via a curved arrow the electrons in the triple bond moving to form a double bond between the adjacent pair of atoms? For **(b)**, how can you show via a curved arrow the electrons in the lone pair moving to form a triple bond involving the same atom from which the lone pair came?

Solve

(a) The C^+ in the first structure has an unfilled valence shell and is adjacent to the triple bond. The curved arrow is used to convert the triple bond into a double bond involving the adjacent pair of C atoms. This leaves an unfilled valence on the left carbon.

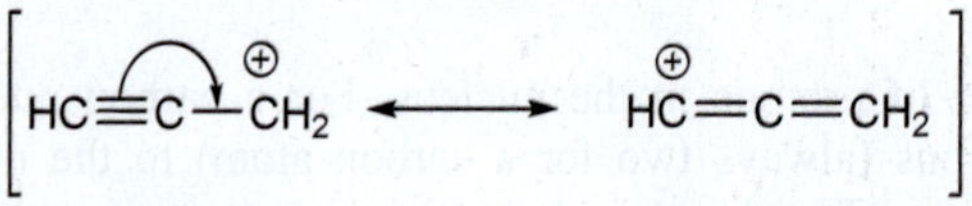

(b) The boron has an unfilled valence shell in the first structure and is adjacent to a lone pair. The curved arrow is used to convert the lone pair into a triple bond. This results in a formal charge of +1 on the nitrogen and of −1 on the boron.

Your Turn 1.12

Think

Are any lone pairs adjacent to a double or triple bond? Do any of the double bonds convert into lone pairs? How can you show via a curved arrow the electrons in the lone pair moving to form a double bond? How can you show via a curved arrow the electrons in a double bond moving to form a lone pair?

Solve

Recall that atoms do not move in resonance structures; only nonbonding electrons or electrons from double or triple bonds move. The first curved arrow is used to convert the lone pair on the fifth C atom into a covalent double bond, C=C. Without any other changes, this would lead to 10 valence electrons on the sixth C atom. The second arrow is used to convert a pair of electrons from the double bond, C=C, into a lone pair on the seventh C atom, :CH.

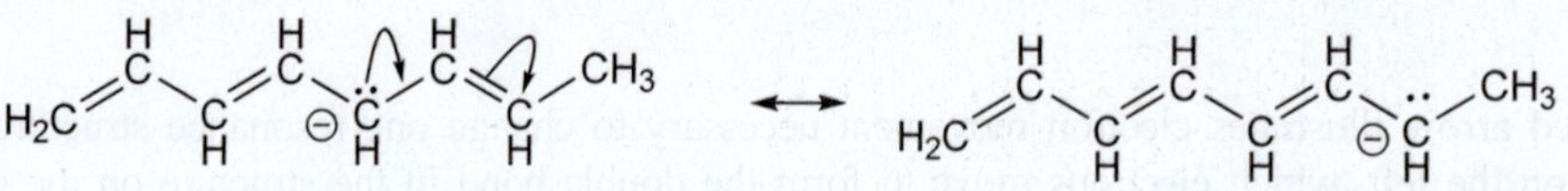

Your Turn 1.13

Think

What is implied at the intersection of every two or more lines or at the end of a drawn bond? How many carbon atoms are present in each structure shown?

Solve

A carbon atom is implied at the intersection of every two or more lines or at the end of a drawn bond. For the line structure shown below, going from left to right, there are two single bonds, then a triple bond, and then two more single bonds. Therefore, the carbon chain is C–C–C≡C–C–C. The compound has six carbon atoms, not four carbon atoms, and is hex-3-yne, $CH_3CH_2C{\equiv}CCH_2CH_3$.

Hex-3-yne → H_3C–H_2C–C≡C–CH_2–CH_3 Hex-3-yne

Your Turn 1.14

Think

Do you see any atoms other than carbon? Do you see any triple or double bonds? What atoms are attached by the double or triple bond? In the C=O bond, what atoms are bonded to either side of the carbon? What compound class does this functional group represent?

Solve

The C=O group, called the carbonyl group, has the only non-C/H atoms and the only nonsingle bond in both compounds. On either side of the C=O bond is a C atom. This is the ketone compound class.

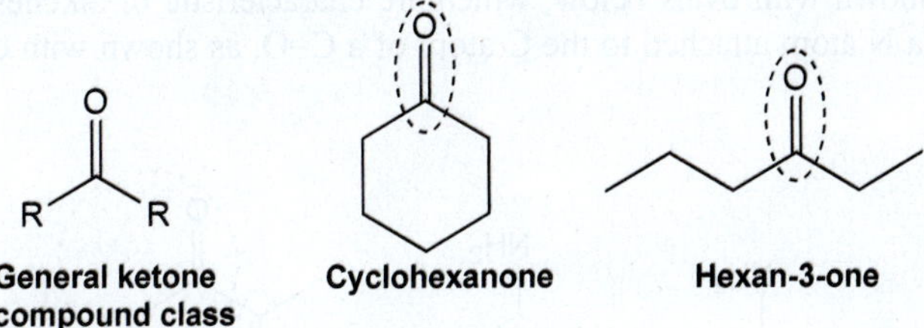

Your Turn 1.15

Think

Consult Table 1-7 and draw each of the four amino acids in the question. What functional groups from Table 1-6 appear in those amino acids? Look for atoms other than carbon and hydrogen and for the presence of multiple bonds.

Solve

Serine's side chain has an OH group attached to a C atom that has no other functional groups, which is characteristic of an alcohol. Aspartic acid's side chain has an OH group attached to a C=O group, which, together, is characteristic of a carboxylic acid. Lysine's side chain has a NH_2 group attached to a C atom that has no other functional groups, which is characteristic of an amine. Phenylalanine's side chain has a six-membered ring with alternating C=C bonds, which is characteristic of an arene.

Serine Aspartic acid Lysine Phenylalanine

Your Turn 1.16

Think

Are there any atoms other than C or H? Are there any bonds other than single bonds? Can you match these up with any functional groups in Table 1-6?

Solve

There are five OH (hydroxyl) groups attached to C atoms that are not part of any other functional group. Each of these is characteristic of an alcohol. There is also a C=O (carbonyl) group attached to a C on one side and to a H on the other, characteristic of an aldehyde.

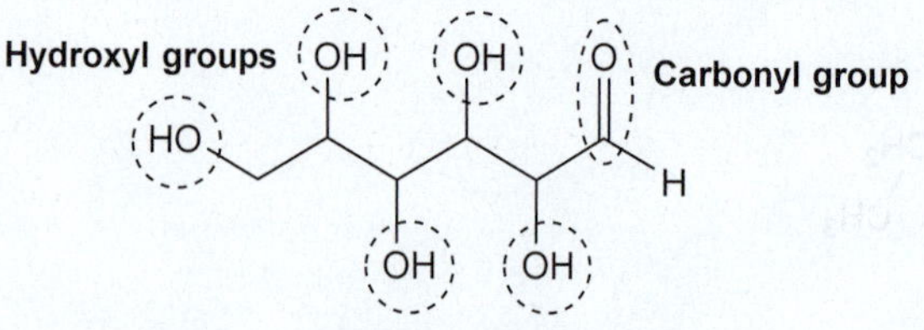

Hydroxyl groups — Carbonyl group

Glucose, $C_6H_{12}O_6$

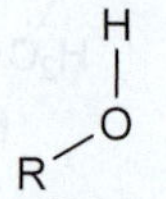

General alcohol compound class

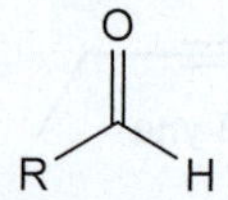

General aldehyde compound class

Your Turn 1.17

Think

Look at Table 1-6 if you need to look up the functional groups and compound classes. Where are the C=C bonds located in each nitrogenous base? Of what compound class is the C=C characteristic? What collection of bonds and atoms makes up the functional group characteristic of amides?

Solve

Each nitrogenous base has one C=C bond, as shown with ovals below, which are characteristic of alkenes. The functional group characteristic of an amide has a N atom attached to the C atom of a C=O, as shown with circles below.

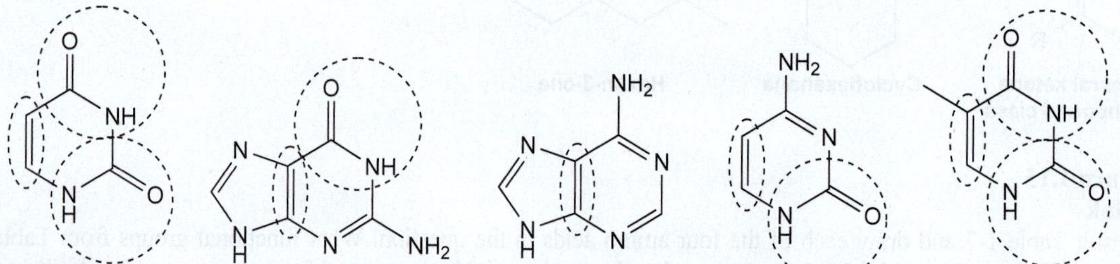

Uracil (U) — Guanine (G) — Adenine (A) — Cytosine (C) — Thymine (T)

In Chapter Problems

Problem 1.2

Think

How many protons are in the nucleus of a carbon atom? Of an oxygen atom? Does a cation have more protons than electrons, or vice versa? What does the charge of the species indicate about the number of protons compared to the number of electrons?

Solve

(a) The carbon nucleus always has six protons. A carbon anion with a −1 charge should have one more electron than it has protons; thus, this species has seven electrons.

(b) The oxygen nucleus always has eight protons. An oxygen cation with a +1 charge should have one more proton than it has electrons; thus, this species has seven electrons.

(c) An oxygen anion with a −1 charge should have one more electron than it has protons; thus, this species has nine electrons.

Problem 1.4

Think

How many total electrons are in an oxygen atom? What is the order in which the atomic orbitals should be filled (see Fig. 1-7)? What is the valence shell, and where do the core electrons reside?

Solve

There are eight total electrons ($Z = 8$ for O). The first two are placed in the $1s$ orbital, and the next two are placed in the $2s$ orbital, leaving four electrons for the three $2p$ orbitals. The electron configuration is $1s^2 2s^2 2p^4$. The valence shell is the second shell, so there are six valence electrons and two core electrons.

Problem 1.6

Think

How is bond length represented on each curve? Which curve represents a larger internuclear bond distance?

Solve

The bottom curve represents the longer bond. The minimum energy for the bottom curve occurs at a slightly longer bond length (farther to the right) than it does for the top curve. To see this, draw a line from the minimum on each curve to the x axis.

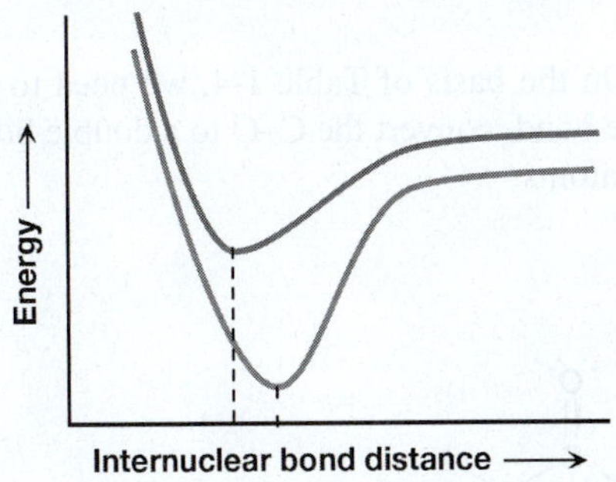

Problem 1.8

Think

Consider the steps for drawing Lewis structures. Which atoms must be bonded together?

Solve

First count the valence electrons for each atom and the total number of valence electrons for the compound C_2H_3N.

3 H atoms:	3 × 1 valence electrons	=	3 valence electrons
2 C atoms:	2 × 4 valence electrons	=	8 valence electrons
1 N atom:	1 × 5 valence electrons	=	5 valence electrons
	Total	=	16 valence electrons

A C atom is bonded to three H atoms and a second C atom. The second C atom is bonded to the N atom. This accounts for only 10 valence electrons. The remaining six electrons are added, and the N atom has an incomplete valence. The lone pairs on C are converted to a C≡N triple bond to give all atoms their octets.

Skeleton — 10 electrons → Unfilled valence — 16 electrons → All atoms have octets — 16 electrons

Problem 1.9

Think

Consider the steps for drawing Lewis structures. For which atoms is a deficient or expanded octet permissible? Which row in the periodic table is forbidden to exceed the octet?

Solve

The atoms that exceed the octet are circled below. Atoms in the second row, such as C, are strictly forbidden to exceed the octet. Atoms in the third row or below, including S and P, are allowed to have expanded octets.

(a) Acceptable

(b) Not acceptable; C's octet is exceeded

(c) Acceptable; S allowed to have expanded octet

(d) Acceptable; P allowed to have expanded octet

(e) Not acceptable; C's octet is exceeded

Problem 1.11

Think

Which atoms have an octet, and which atoms (other than H) do not? How many bonds and lone pairs are typical for each element? Refer to Table 1-4 for the common number of covalent bonds and lone pairs for H, C, N, and O.

Solve

The atoms not shown with an octet are in boldface on the left. On the basis of Table 1-4, we need to convert each C–C single bond and C–N single bond in the ring to a double bond, convert the C–O to a double bond, add a lone pair to the N atom, and add two lone pairs to each of the O atoms.

Problem 1.12

Think

Which atoms (other than H) need lone pairs or additional bonds to complete their octets? Which atom in each bond is more electronegative? To which atom does the bond dipole (↛) point? What does the length of the arrow indicate about the difference in electronegativity?

Solve

To complete octets, the O atom in **(i)** needs two lone pairs; the F atoms in **(ii)** need three lone pairs each and each C–C single bond needs to be converted to a double bond; and the N atom in **(iii)** needs one lone pair and double bonds need to alternate around the ring. The C–F bond dipole is the largest because the C and F have the greatest difference in electronegativity. The C–H bond dipole is the smallest. Bond dipoles point toward the more electronegative atom.

(i) (ii) (iii)

Problem 1.13

Think

What does an electrostatic potential map depict? Which color (red or blue) corresponds to a buildup of negative charge? Draw the bond dipoles and add the δ^+ and δ^- symbols. Match the electrostatic potential map to one of the four compounds.

Solve

The red region on an electrostatic potential map corresponds to excess negative charge (δ^-), whereas each blue region corresponds to excess positive charge (δ^+). The electrostatic potential map, therefore, is consistent with $H_2C{=}O$.

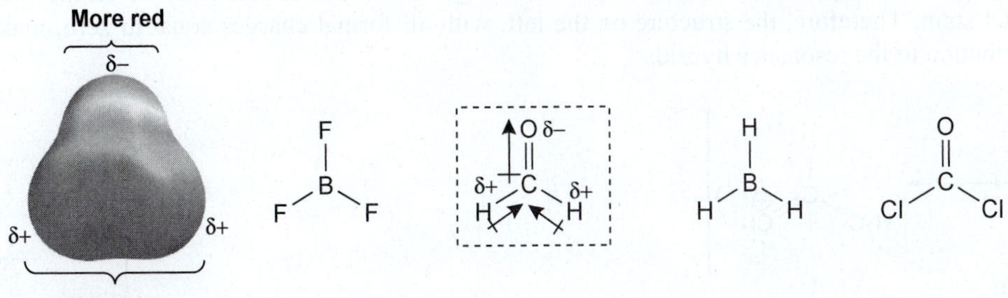

Problem 1.15

Think

Does the compound contain elements from both the left and right sides of the periodic table? Does the compound contain any recognizable polyatomic cations?

Solve

The first structure is ionic; its sodium cation Na^+ is easiest to spot, and its phenoxide anion has a formal charge of −1 on the O atom. The middle structure is covalent; all atoms have their "preferred" number of covalent bonds, and there are no recognizable polyatomic ions. The third structure contains a N atom that must have four bonds. This is a type of ammonium cation, $CH_3CH_2NH_3^+Cl^-$. The Cl^- anion is not covalently bonded to any other atom in this structure.

A Ionic | B Covalent | C Ionic

Problem 1.17

Think

How are lone pairs assigned when determining an atom's formal charge? How are electrons in covalent bonds assigned? How many valence electrons should an atom be assigned to have a formal charge of 0?

Solve

The figure below shows how valence electrons are assigned according to the formal charge method, in which each pair of bonding electrons is split evenly. The number of valence electrons assigned to the atom is then compared to the group number. All atoms have a formal charge of 0 in this compound.

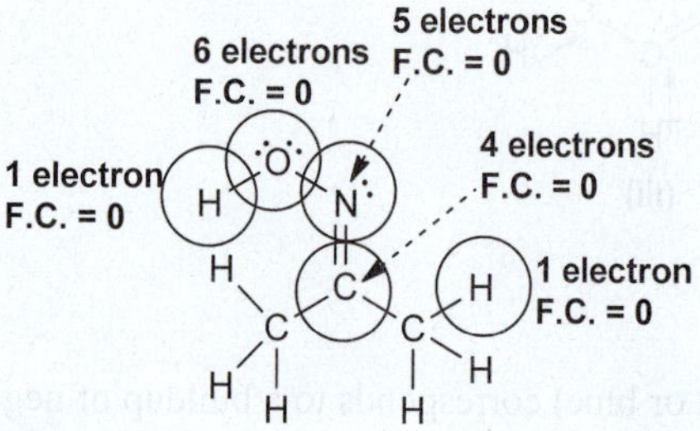

5 electrons F.C. = 0
6 electrons F.C. = 0
4 electrons F.C. = 0
1 electron F.C. = 0
1 electron F.C. = 0

Problem 1.19

Think

Are both resonance structures equivalent? Add in formal charges. Which structure has fewer atoms with formal charges other than zero?

Solve

The resonance structures are inequivalent. The structure on the right has a formal charge of −1 on the O atom and of +1 on the Cl atom. Therefore, the structure on the left, with all formal charges equal to zero, makes a much greater contribution to the resonance hybrid.

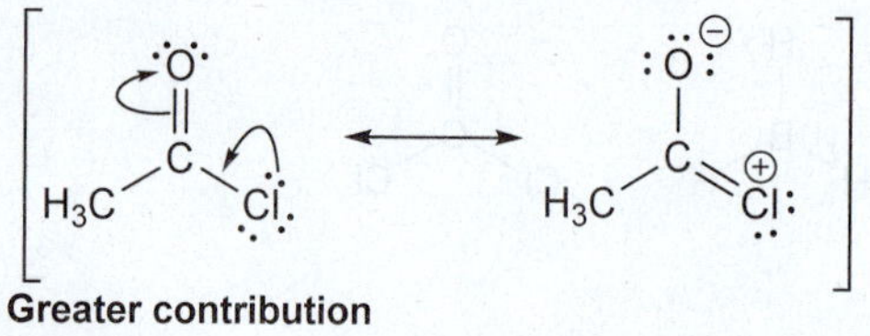

Greater contribution

Problem 1.21

Think

Which feature from Figure 1-27 does each of these species possess? How must the curved arrows be added? What happens to the respective formal charges?

Solve

(a) This species has a lone pair on an atom (N:) that is adjacent to a multiple bond (the C=C double bond), which is Feature 1 from Figure 1-27a. Two curved arrows are added, yielding the additional resonance structure shown below.

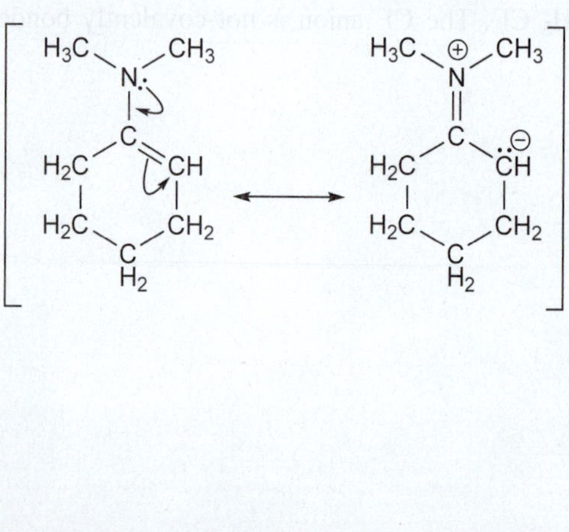

(b) With a ring of alternating single and double bonds (Feature 4 in Fig. 1-27d), a pair of electrons from each multiple bond can be shifted around the ring. This compound has nine C=C bonds and nine C–C bonds. Each C=C bond converts into a C–C bond, and vice versa. Therefore, nine curved arrows are needed to draw the new resonance structure.

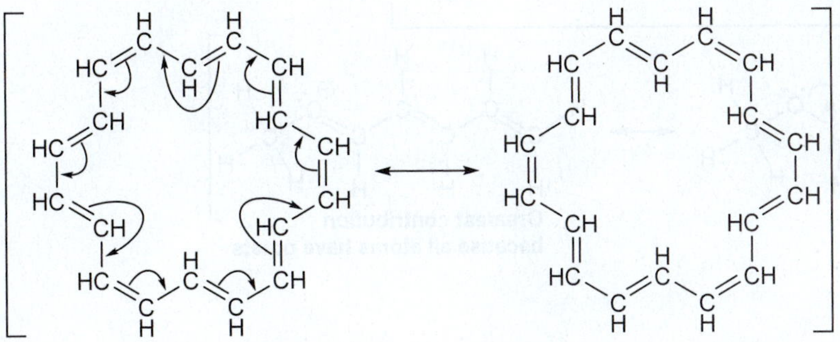

(c) This species has a lone pair on an atom (O:) that is adjacent to an atom lacking an octet (C^+), which is Feature 3 from Figure 1-27c. One curved arrow is added, yielding the additional resonance structure shown below.

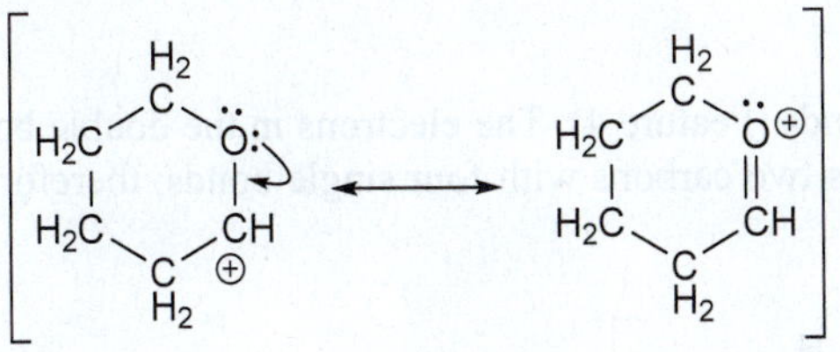

Problem 1.22

Think

Which feature from Figure 1-27 does each of these species possess? How must the curved arrows be added? What happens to the respective formal charges? After electron movement, does the new structure have a feature from Figure 1-27?

Solve

The resonance structures are shown below and on the next page for **(a)** and **(b)**.

(a) The carbocation has an unfilled valence shell and is adjacent to a double bond (Feature 2). A single curved arrow is used to show the movement of the electrons in the double bond. This leaves an unfilled valence on the C atom bearing the positive charge. There are a total of five resonance structures, as shown below.

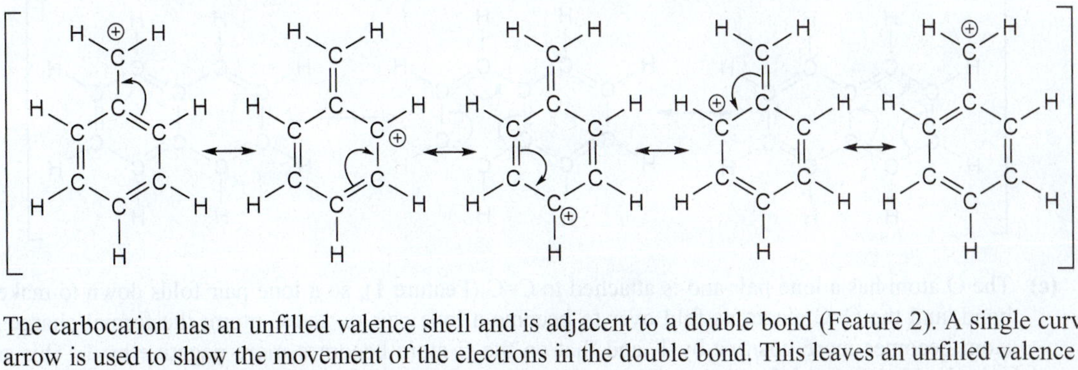

(b) The carbocation has an unfilled valence shell and is adjacent to a double bond (Feature 2). A single curved arrow is used to show the movement of the electrons in the double bond. This leaves an unfilled valence on the C atom bearing the positive charge in the second structure. Feature 2 appears in the second structure, too, so the same movement of electrons leads to the third structure. In the third structure, an atom with a lone pair is attached to an atom lacking an octet (Feature 3), so a single curved arrow is used to convert the lone pair into a double bond. There are four total resonance structures as shown on the next page.

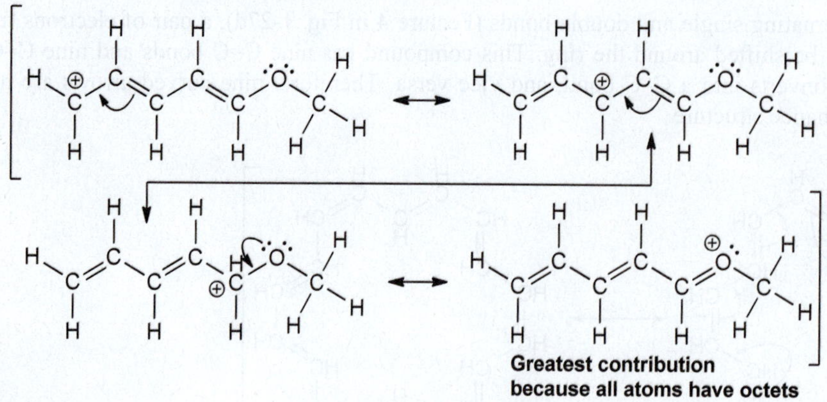

Greatest contribution because all atoms have octets

Problem 1.23

Think

Which feature from Figure 1-27 does each of these species possess? How must the curved arrows be added? What happens to the respective formal charges? After electron movement, does the new structure have a feature from Figure 1-27?

Solve

(a) There is a ring of alternating single and multiple bonds (Feature 4). The electrons in the double bond move around the ring on the left. The ring on the right has two carbons with four single bonds; therefore, it does not participate in resonance.

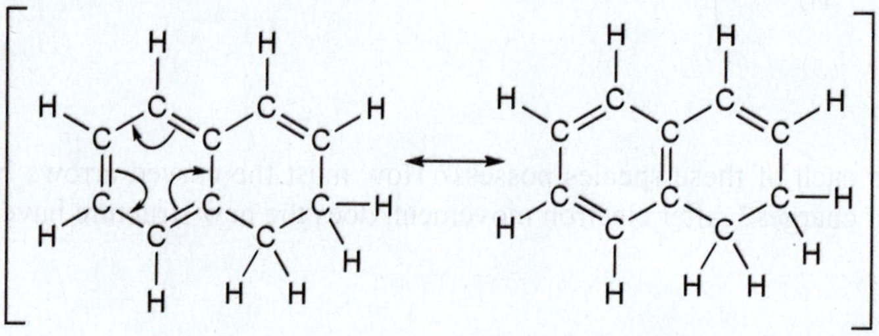

(b) The alternating double bonds are located around the ring on the left (Feature 4). Therefore, the electrons in the double bond can move around the entire ring. In the resulting structure, the ring on the right has alternating single and double bonds, so its electrons can be moved to generate a third resonance structure.

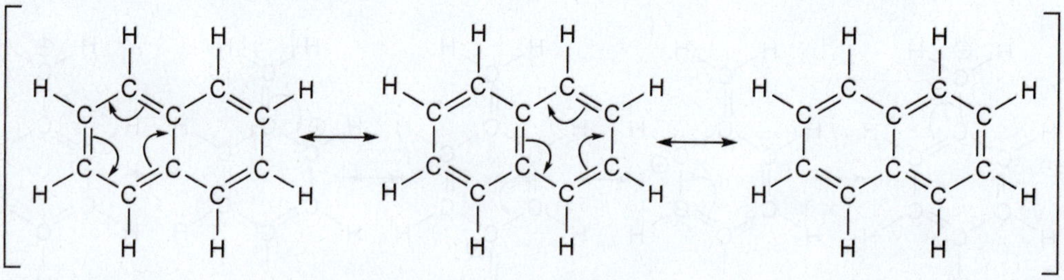

(c) The O atom has a lone pair and is attached to C=C (Feature 1), so a lone pair folds down to make an O=C bond, and the C=C electrons fold over to become a lone pair on the C atom; the formal charge on the O atom becomes more positive by 1, and that on the C atom becomes more negative by 1. This same lone pair–double bond and double bond–lone pair conversion occurs on each additional resonance structure until resonance is completed over the entirety of the ring as shown on the next page.

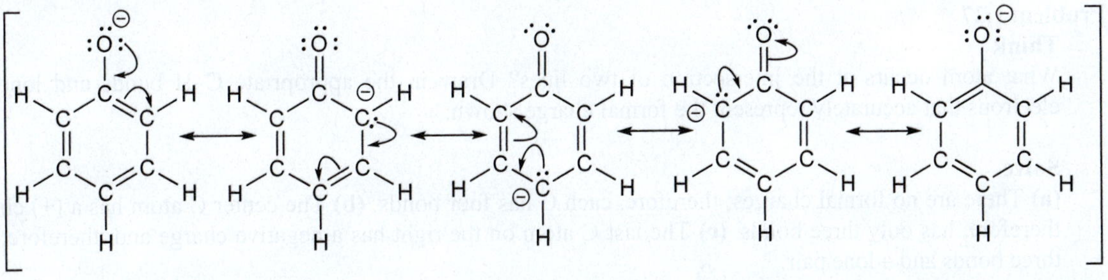

(d) The double bond on the right is adjacent to a C atom with an unfilled valence (Feature 2). Therefore, the double bond moves over one bond to the right. The same thing occurs from the second to the third structure.

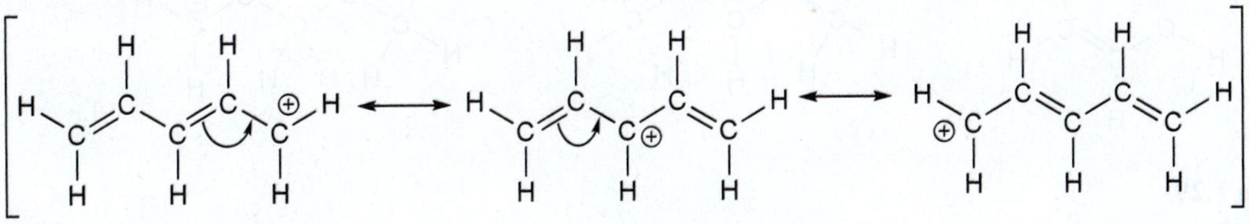

Problem 1.24

Think

Refer to Table 1-5 to review formal charges on atoms in specific bonding scenarios. Try to fill the valences on each atom and consider how the formal charge reflects the number of bonds and lone pairs.

Solve

The Lewis structures are completed below.

(a)	(b)	(c)	(d)	(e)
Fill in two lone pairs on the O atom and one lone pair on the N atom.	Fill in two lone pairs on the N atom to make N^-.	Already complete	Already complete	Fill in a lone pair on the C atom at far right to make C^-.

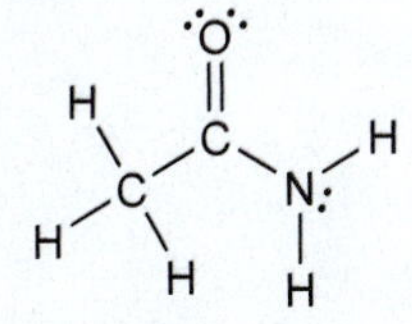

Acetamide

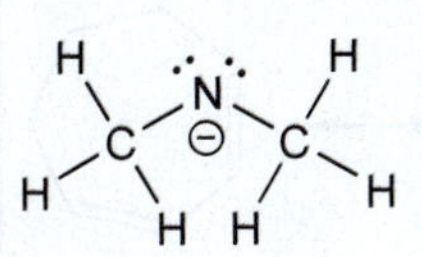

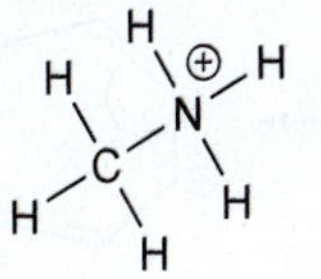

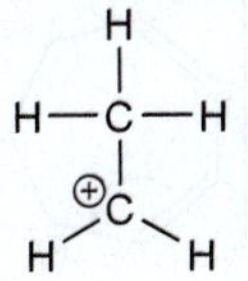

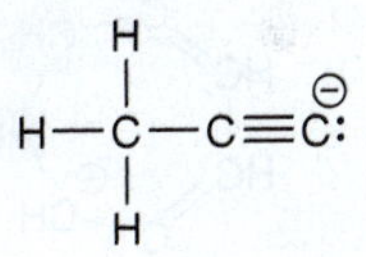

Problem 1.26

Think

Review the rules for drawing line structures. How are C atoms represented? Is it necessary to draw all lone pairs? What are the rules for drawing H atoms on C atoms? H atoms on heteroatoms?

Solve

The line structures are drawn below. C atoms are not drawn explicitly, nor are H atoms bonded to C atoms; all other atoms are drawn, as are their attached H atoms. Bonds between heteroatoms and H can be omitted, and so can lone pairs.

(a) Pyrrole

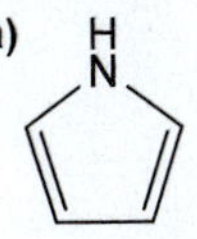

(b) Benzoic acid

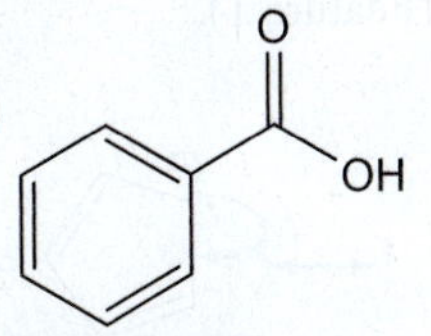

(c)

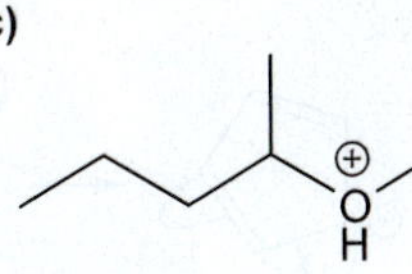

Problem 1.27

Think

What atom occurs at the intersection of two lines? Draw in the appropriate C–H bonds and lone pairs of electrons that accurately represent the formal charge shown.

Solve

(a) There are no formal charges; therefore, each C has four bonds. **(b)** The center C atom has a (+) charge and, therefore, has only three bonds. **(c)** The last C atom on the right has a negative charge and, therefore, has only three bonds and a lone pair.

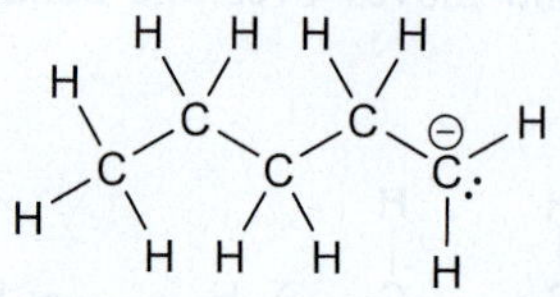

Problem 1.29

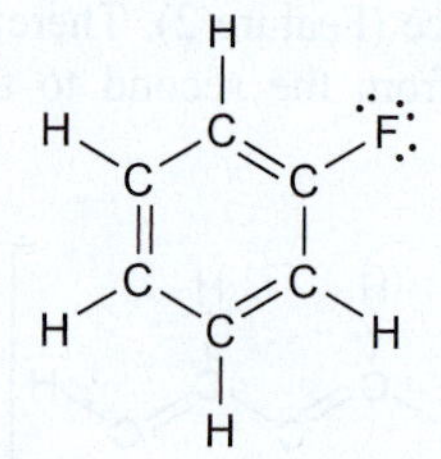

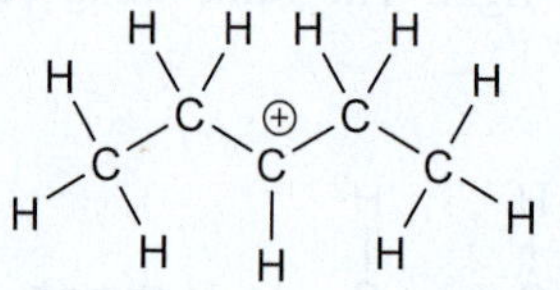

Think

What would the structures look like as complete Lewis structures (i.e., with all hydrogen atoms and lone pairs drawn in)? Which feature from Figure 1-27 does each of these species possess? How must the curved arrows be added? What happens to the respective formal charges? After electron movement, does the new structure have a feature from Figure 1-27?

Solve

(a) The two C atoms that have four single bonds and no formal charge (highlighted in gray below) do not participate in resonance. Therefore, resonance is only possible with the other five C atoms. The double bond on the bottom left is adjacent to a C atom with an unfilled valence (Feature 2), so one curved arrow is added to move electrons to achieve the second resonance structure. The second structure has the same feature, so the same type of electron movement leads to the third structure.

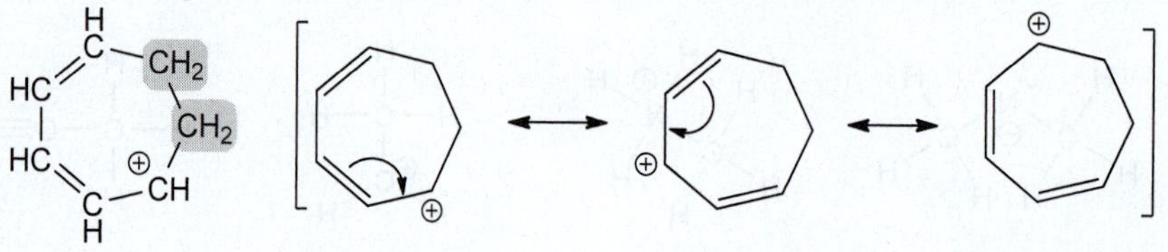

(b) No resonance structures are possible with this compound (without introducing formal charges), because two C atoms between the C=C and the C$^+$ have four bonds. These "block" the possibility of resonance.

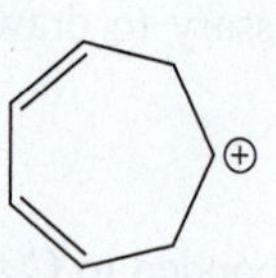

(c) All of the C atoms in the ring can participate in resonance. Five resonance structures are possible; each resonance structure transforms into the other when a lone pair converts into a double bond and a double bond converts into a lone pair on the adjacent atom (Feature 1).

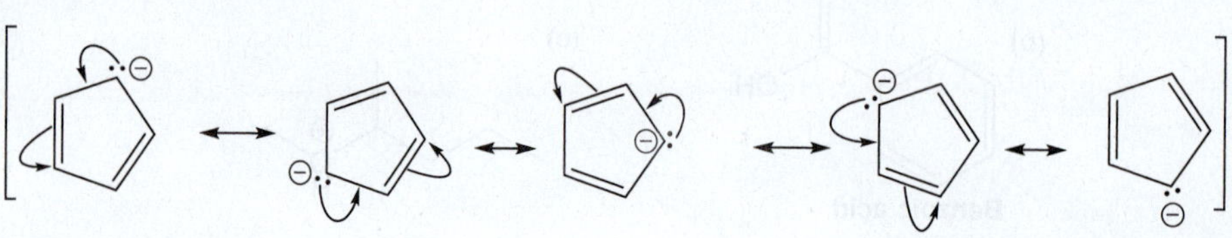

Problem 1.30

Think

What functional groups are made up of three or more atoms? Within those functional groups, can you find arrangements that involve fewer atoms, thus characterizing other functional groups?

Solve

The alternating ring of single and double bonds characteristic of an arene ring has what appears to be C=C groups that are characteristic of alkenes. The acetal contains two ether groups, the hemiacetal contains an ether and an OH, the ester contains an ether and carbonyl, and an amide contains a carbonyl and a singly bonded N that is characteristic of amines.

(Alkene) (Arene) (Alcohol) (Ether) (Ether) (Alcohol) (Ether) (Ether) (Amine)

Arene Phenol Acetal Hemiacetal Ester Amide

Problem 1.32

Think

Are the functional groups the same or different? What compound classes do those functional groups characterize? What impact will the ring have on the reactivity?

Solve

δ-Valerolactone (left) has the CO_2C functional group and is a cyclic ester, whereas pentanoic acid (right) has a CO_2H functional group and is a carboxylic acid. The presence of the six-membered ring does not cause a difference in reactivity, but since the functional groups are different, the chemical reactivity will be different, too.

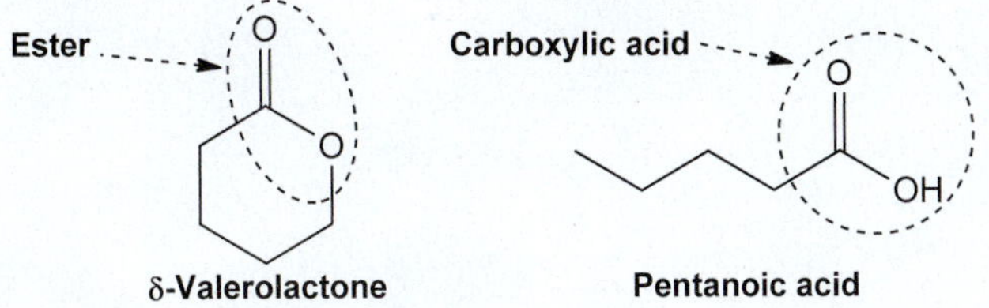

Problem 1.34

Think

Are any bonds present other than C–C and C–H single bonds? Are any *atoms* present other than carbon and hydrogen? Are any special rings present? What arrangements of bonds and atoms can you find that match with the entries in Table 1-6?

Solve

The functional groups are circled and the compound classes of which they are characteristic are labeled in the figure below.

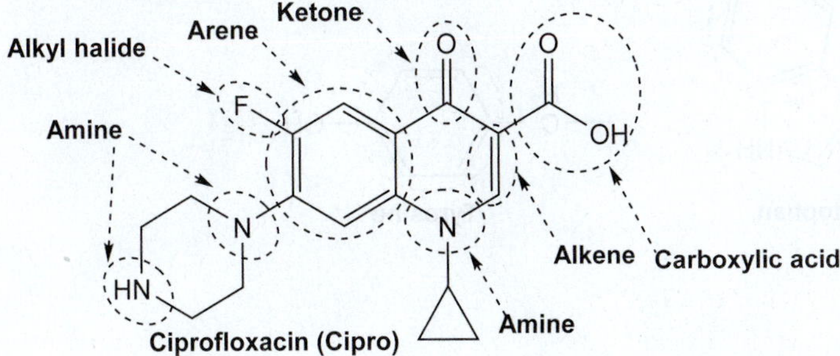

Problem 1.35

Think

Refer to Table 1-6 to review and identify the functional groups and compound classes present in the amino acid side groups shown in Table 1-7.

Solve

(a) OH group characteristic of an alcohol: serine, threonine, tyrosine

Serine Threonine Tyrosine

(b) O=C–N group characteristic of an amide: asparagine, glutamine

Asparagine Glutamine

(c) CO_2H group characteristic of a carboxylic acid: aspartic acid, glutamic acid

Aspartic acid Glutamic acid

(d) Singly bonded N characteristic of an amine: lysine

Lysine

(e) Alternating ring of single and double bonds characteristic of an arene: phenylalanine, tryptophan, tyrosine

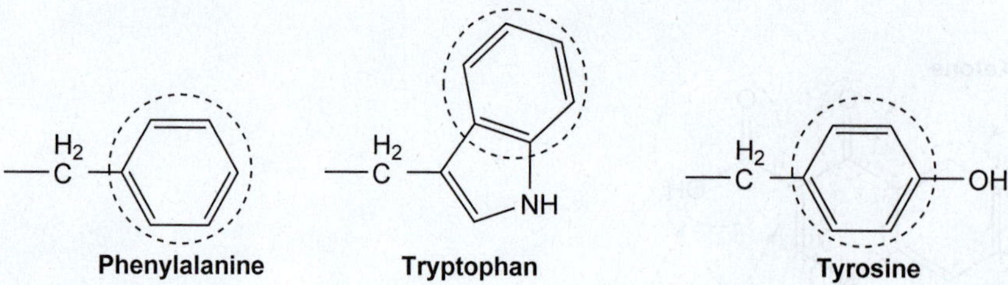

Problem 1.36

Think

What atoms are present in a carbohydrate? What is the general formula for a carbohydrate? What is the general formula for a monosaccharide? Write the formula for each compound and compare it to the definitions.

Solve

In a carbohydrate, carbon, hydrogen, and oxygen are present in a general formula of $C_xH_{2y}O_y$. A monosaccharide is a carbohydrate with one additional restriction to the chemical formula, $C_xH_{2x}O_x$. In the figures below, **A** and **C** are both carbohydrates and monosaccharides, **B** is neither, and **D** is a carbohydrate only.

A	**B**	**C**	**D**
Chemical formula: $C_5H_{10}O_5$	**Chemical formula:** $C_3H_4O_3$	**Chemical formula:** $C_6H_{12}O_6$	**Chemical formula:** $C_{12}H_{22}O_{11}$

Problem 1.37

Think

Refer to Figure 1-39. Which atoms compose the phosphate, sugar, and nitrogenous base? Refer to Figure 1-40 to identify the nitrogenous bases in DNA and RNA. What is the difference between a ribose and deoxyribose sugar?

Solve

(a) The portions are circled and labeled below.

(b) The first one can be part of DNA because the sugar unit is deoxyribose, given that it is missing an oxygen atom at the location indicated. The second one can be part of RNA because it has the OH group at that location.

(c) The nitrogenous bases are labeled. Notice that uracil is part of RNA, not DNA.

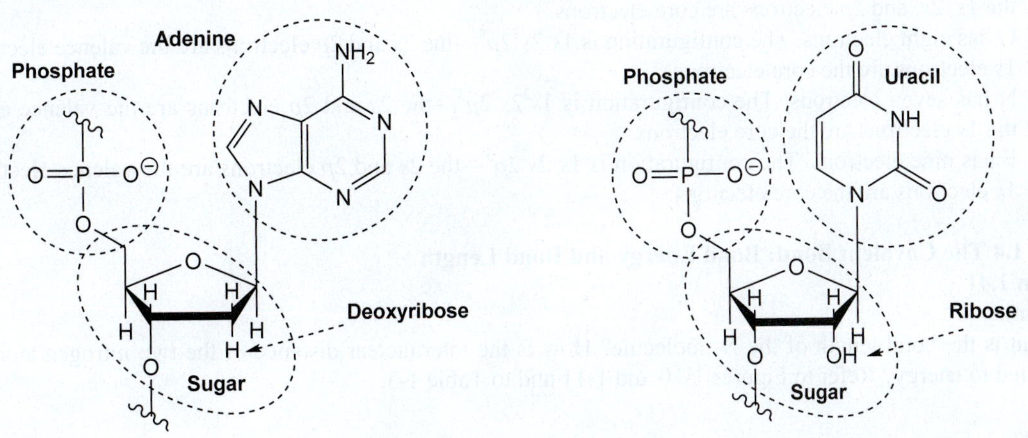

End of Chapter Problems
Section 1.3 Atomic Structure and Ground State Electron Configurations
Problem 1.38

Think
Refer to Figure 1-7 to view the orbital filling diagram. For each orbital, what designation characterizes the principal quantum number? How is the principal quantum number related to the potential energy?

Solve
The number that appears before the orbital type designates the principal quantum number (the shell number), and the potential energy increases with a greater principal quantum number. The 4*s* orbital has the largest principal quantum number, $n = 4$, so it possesses the most energy.

Problem 1.39

Think
Refer to Figure 1-7 to view the orbital filling diagram. For each orbital, what designation characterizes the principal quantum number? How is the principal quantum number related to the potential energy? For orbitals in the same shell, how do you determine energy differences?

Solve
(a) The 5*s* orbital possesses more potential energy than the 4*s* orbital. It has a higher principal quantum number, $n = 5$ versus $n = 4$ (fifth shell vs. fourth shell).
(b) The 5*d* orbital possesses more potential energy. When the principal quantum number (n) is the same for different orbitals, energy is dictated by type of orbital: $s < p < d$. In this case, for the two orbitals given that are in the $n = 5$ shell, energy increases in the order (lowest) $5p < 5d$ (highest).

Problem 1.40

Think
How many electrons does each element listed possess? What is the valence shell, and where do the core electrons reside? What distinguishes valence electrons from core electrons? Refer to Figure 1-7 to view the order in which electrons are filled.

Solve
(a) Al has 13 electrons. The configuration is $1s^22s^22p^63s^23p^1$—the 3*s* and 3*p* electrons are the valence electrons; the 1*s*, 2*s*, and 2*p* electrons are core electrons.
(b) S has 16 electrons. The configuration is $1s^22s^22p^63s^23p^4$—the 3*s* and 3*p* electrons are the valence electrons; the 1*s*, 2*s*, and 2*p* electrons are core electrons.
(c) O has eight electrons. The configuration is $1s^22s^22p^4$—the 2*s* and 2*p* electrons are the valence electrons; the 1*s* electrons are the core electrons.
(d) N has seven electrons. The configuration is $1s^22s^22p^3$—the 2*s* and 2*p* electrons are the valence electrons; the 1*s* electrons are the core electrons.
(e) F has nine electrons. The configuration is $1s^22s^22p^5$—the 2*s* and 2*p* electrons are the valence electrons; the 1*s* electrons are the core electrons.

Section 1.4 The Covalent Bond: Bond Energy and Bond Length
Problem 1.41

Think
What is the bond length of the N_2 molecule? How is the internuclear distance of the two nitrogen atoms in N_2 related to energy? Refer to Figures 1-10 and 1-11 and to Table 1-3.

Solve
The plot of energy as a function of the internuclear distance for two N atoms is shown on the next page. The bond length for a N_2 molecule is 110 pm. It takes energy to lengthen or shorten a covalent bond from its bond length. Therefore, values further from this optimal bond length are higher in energy.

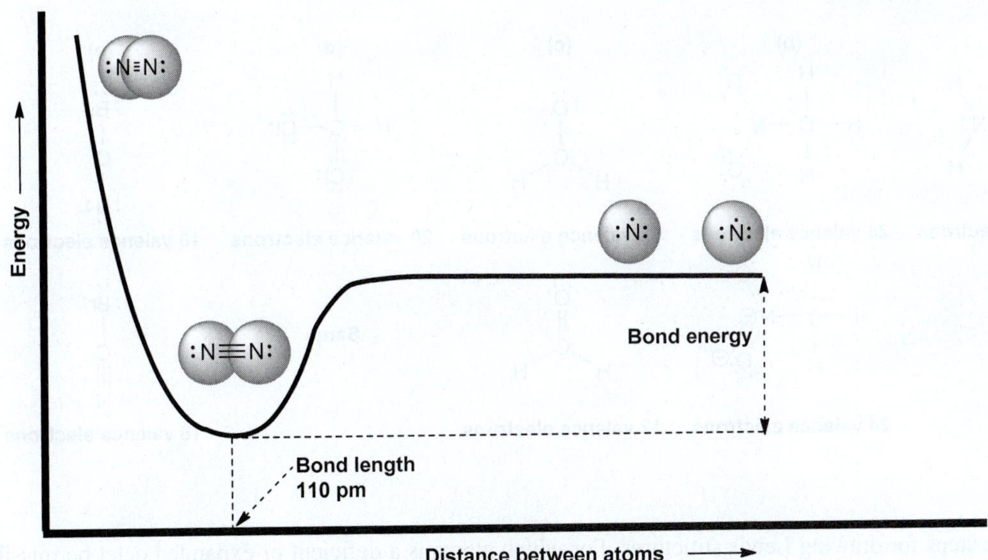

(a) The distances 50 pm and 75 pm are both below the bond length. It takes energy to shorten a covalent bond, and therefore, 50 pm has a higher potential energy.

(b) The distance 75 pm is below the bond length and 110 pm is the bond distance. At 110 pm, the energy is at a minimum, and therefore, 75 pm represents a higher energy.

(c) The distance 150 pm is above the bond length and 110 pm is the bond distance. At 110 pm, the energy is at a minimum, and therefore, 150 pm represents a higher energy.

(d) 150 pm and 160 pm are both above the bond length. It takes energy to lengthen a covalent bond, and therefore, 160 pm represents a higher potential energy.

Problem 1.42

Think

Refer to Table 1-3 and Figure 1-13. How does the strength of the bond change from single to double to triple? How does the length change?

Solve

From Table 1-3, we see that as the number of bonds increases between a pair of atoms, the bond energy increases and the bond length decreases. The HC≡N triple bond is the shortest and strongest bond.

Problem 1.43

Think

Refer to Table 1-2 for the average bond energies of single bonds. Does the strongest bond have the largest or smallest kJ/mol value?

Solve

HF has the strongest single bond (569 kJ/mol), and Cl_2 has the weakest single bond (243 kJ/mol).

Sections 1.5–1.8 Lewis Dot Structures, Polarity, and Ionic Bonds

Problem 1.44

Think

Consider the steps for drawing Lewis structures. Which atoms must be bonded together?

Solve

First count the total valence electrons, then write the skeleton of the molecule, distribute the remaining electrons as lone pairs in an attempt to fill each atom's valence shell, and convert lone pairs into double or triple bonds, as necessary. In the table on the next page, the first row shows the results before converting lone pairs to multiple bonds. The next row shows the completed Lewis structures.

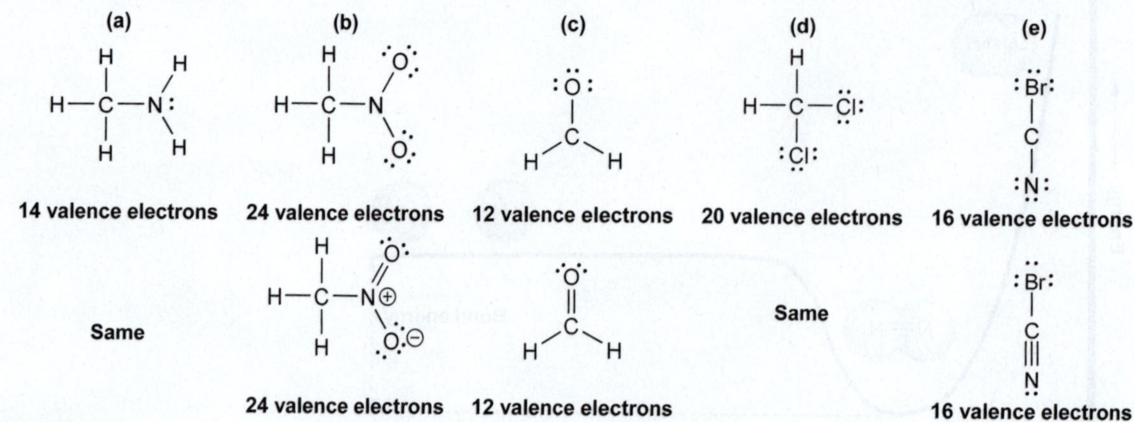

Problem 1.45

Think

Consider the steps for drawing Lewis structures. For which atoms is a deficient or expanded octet permissible? Which row in the periodic table is forbidden to exceed the octet?

Solve

The circled atoms below exceed the octet, but this is strictly forbidden for atoms in the second row.

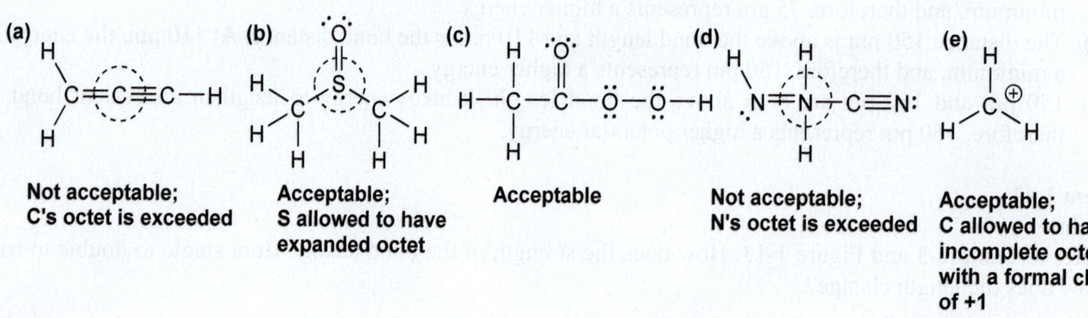

(a) Not acceptable; C's octet is exceeded

(b) Acceptable; S allowed to have expanded octet

(c) Acceptable

(d) Not acceptable; N's octet is exceeded

(e) Acceptable; C allowed to have an incomplete octet with a formal charge of +1

Problem 1.46

Think

Which atoms (other than hydrogen) have an octet, and which atoms don't? How many bonds and lone pairs are typical for each element? Refer to Table 1-4 for the common number of covalent bonds and lone pairs for H, C, N, and O.

Solve

The atoms without an octet are circled on the left. On the basis of Table 1-4, we need to do the following:

(a) Convert the terminal C–O single bond and the C–N single bond on the left to double bonds, and convert the terminal C–N single bond to a triple bond. Add a lone pair to each N, two lone pairs to each O, and three lone pairs to the F.

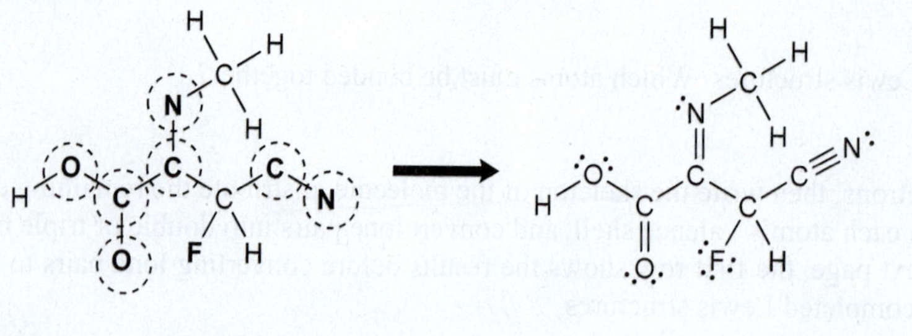

(b) Convert alternating C–C single bonds in the ring to double bonds and the terminal C–C bond outside the ring to a triple bond.

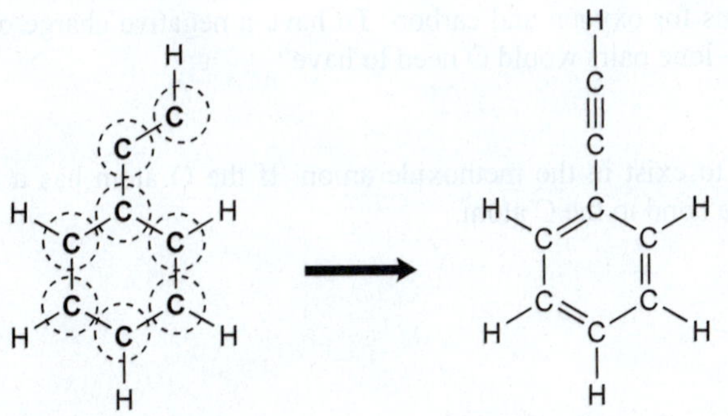

(c) Convert the second C–C single bond to a double bond and the N–O single bond to a double bond. Add a lone pair to the N and two lone pairs to the O.

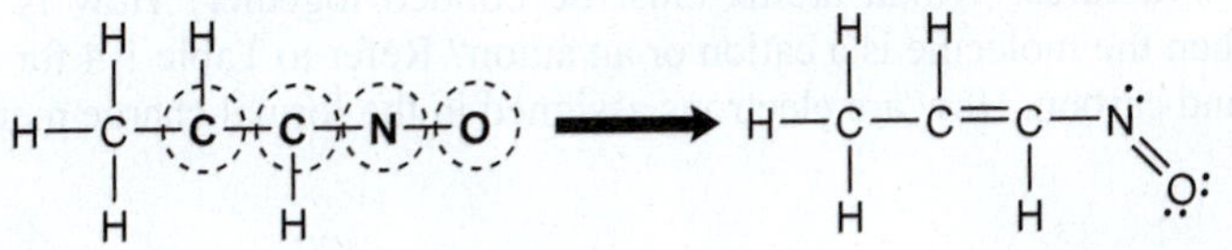

Problem 1.47

Think

Refer to Figure 1-16 to determine the electronegativity value for each C and the atoms that are in the C–X bond indicated. As the electronegativity of the atom bonded to C increases, what happens to the concentration of negative charge on C?

Solve

The greater the electronegativity of the atom bonded to C, the smaller the concentration of negative charge on C. F has the greatest electronegativity, giving C the smallest concentration of negative charge. Li has the smallest electronegativity, giving C the greatest concentration of negative charge.

$$\mathbf{CH_3–F < CH_3–OH < CH_3–NH_2 < CH_3–CH_3 < CH_3–MgBr < CH_3–Li}$$

Problem 1.48

Think

Does the compound contain elements from both the left (metals) and right (nonmetals) sides of the periodic table? Does the compound contain any recognizable polyatomic cations?

Solve

Items **(a)**, **(e)**, and **(f)** contain only nonmetals and have no recognizable polyatomic ions; they have only covalent bonds. Items **(b)**, **(c)**, **(d)**, **(g)**, **(h)**, and **(i)** are ionic compounds:

- **(b)** consists of Na^+ and Cl^-.
- **(c)** consists of Na^+ and HO^-.
- **(d)** consists of Na^+ and CH_3O^-.
- **(g)** consists of Li^+ and CH_3NH^-.
- **(h)** consists of K^+ and $CH_3CH_2CO_2^-$.
- **(i)** consists of $C_6H_5NH_3^+$ and Cl^-. Even though **(i)** has only nonmetals, a positive charge appears owing to N having four bonds and no lone pairs.

Section 1.9 Assigning Electrons to Atoms in Molecules: Formal Charge

Problem 1.49

Think

Refer to Table 1-4 for common bonding patterns for oxygen and carbon. To have a negative charge on the O, what type of bond must the C–O be? How many lone pairs would O need to have?

Solve

No, it is not possible for a C=O double bond to exist in the methoxide anion. If the O atom has a negative charge, it must have three lone pairs and a single bond to the C atom.

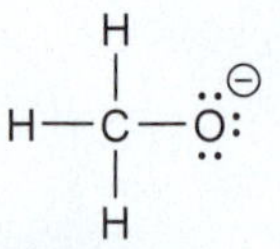

Problem 1.50

Think

Consider the steps for drawing Lewis structures. Which atoms must be bonded together? How is the total number of valence electrons changed when the molecule is a cation or an anion? Refer to Table 1-4 for common bonding patterns for oxygen, nitrogen, and carbon. How are electrons assigned in the formal charge method?

Solve

(a) C_2H_5 anion = 14 valence electrons. The negative charge is located on one of the C atoms. That C^- should have three bonds and one lone pair, and the other C should have four bonds.

(b) CH_3O cation = 12 valence electrons. The positive charge is located on the O, so it should have three bonds and one lone pair. The C should have four bonds: two single bonds and a double bond.

(c) CH_6N cation = 14 electrons. The positive charge is located on the N, so it should have four bonds, and the C should have four bonds.

(d) CH_5O cation = 14 valence electrons. The positive charge is located on the O, so it should have three bonds and one lone pair, and the C should have four bonds.

(e) C_3H_3 anion = 16 valence electrons. The negative charge is located on the last C, so it should have three bonds and one lone pair. The other two C atoms should each have four total bonds.

Note that **(a)** and **(e)** both have a carbon with a negative charge that bears three bonds and one lone pair. It does not matter that **(a)** has three single bonds and **(e)** has a triple bond.

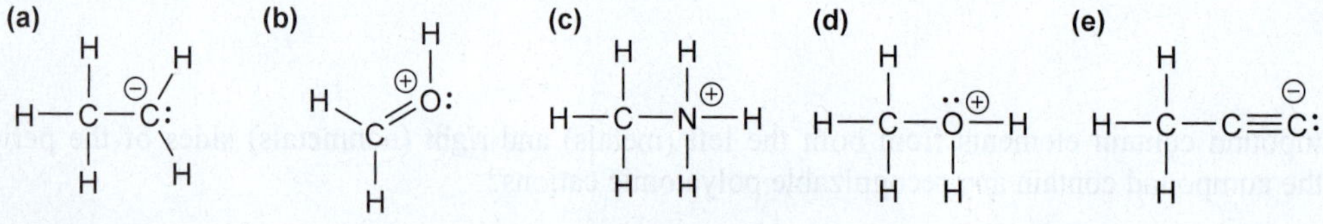

Problem 1.51

Think

When assigning valence electrons to atoms to determine formal charge, how are lone pairs assigned? How are electrons in covalent bonds assigned? How many valence electrons should each atom be assigned to have a formal charge of 0?

Solve

The figure on the next page shows the formal charges—each atom is assigned half of its bonding electrons and all of its lone pairs. To have a formal charge of 0, an atom must be assigned the same number of valence electrons as its group number. Negative charges result when there are additional electrons, and positive charges result when there are fewer electrons. Two atoms have a nonzero formal charge. The O with three lone pairs has a formal charge of −1, and the N that has a triple bond and a single bond has a formal charge of +1. Remember that the sum of the formal charges must equal the total charge on the compound—in this case, zero.

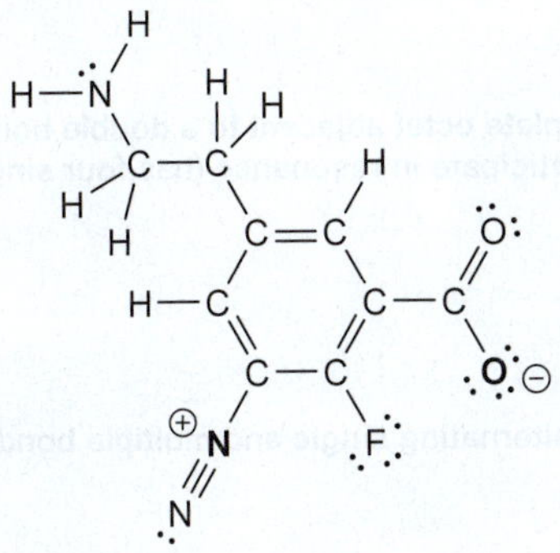

Problem 1.52

Think

Refer to Table 1-4 for common bonding patterns for O, N, and C. How many bonds and lone pairs does each atom possess with a −1 formal charge? How many bonds and lone pairs should the remaining atoms have if they are to be uncharged? Be sure to fill in the octet for the other nonhydrogen atoms in the molecule.

Solve

(a) A −1 formal charge on N requires two bonds and two lone pairs.
(b) A −1 formal charge on O requires one bond and three lone pairs.
(c) A −1 formal charge on C requires three bonds and one lone pair.

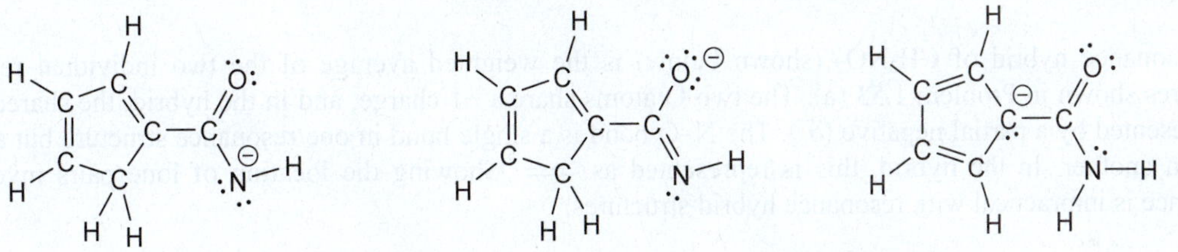

(a) −1 on N

(b) −1 on O

(c) −1 on C

Sections 1.10 and 1.11 Resonance Theory; Drawing All Resonance Structures

Problem 1.53

Think

Before you can draw resonance structures, you must begin with a complete Lewis structure or line structure. Which atoms are bonded together? How many total valence electrons must each species have? Do any of the completed molecules have a feature from Figure 1-27? How must the curved arrows be added for that feature? What happens to the respective formal charges when the electrons are moved? After electron movement, does the new structure have a feature from Figure 1-27?

Solve

The resonance contributors are shown below and on the next page. For **(e)**, resonance contributors are minor, because they have additional formal charges.

(a)

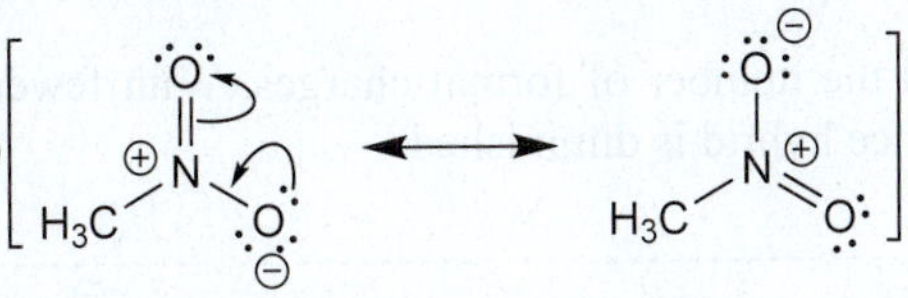

Feature 1: Lone pair adjacent to a double bond
CH_3 unable to participate in resonance (has four single bonds)

(b)

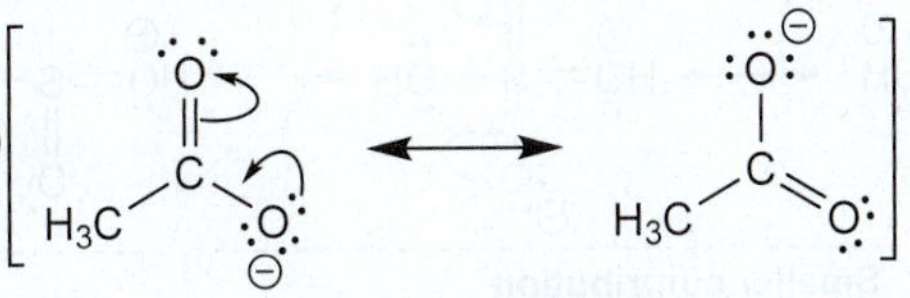

Feature 1: Lone pair adjacent to a double bond
CH_3 unable to participate in resonance (has four single bonds)

(c) 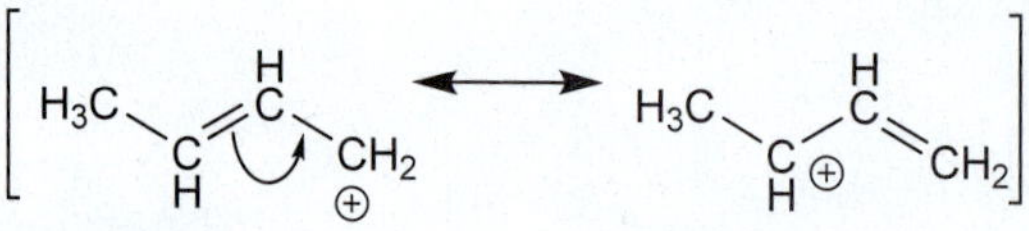 Feature 2: Incomplete octet adjacent to a double bond
CH_3 unable to participate in resonance (has four single bonds)

(d) 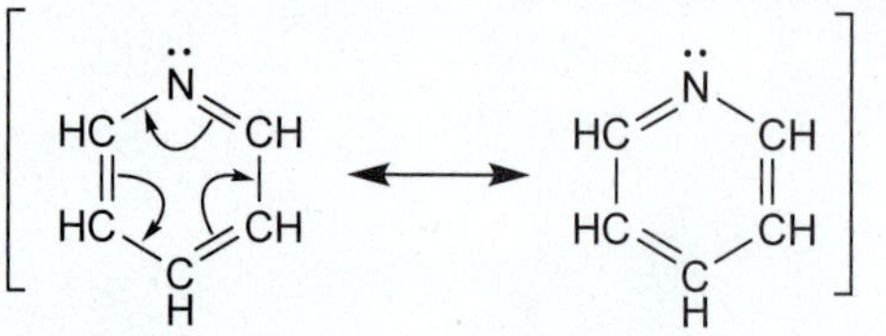 Feature 4: Ring of alternating single and multiple bonds

(e) 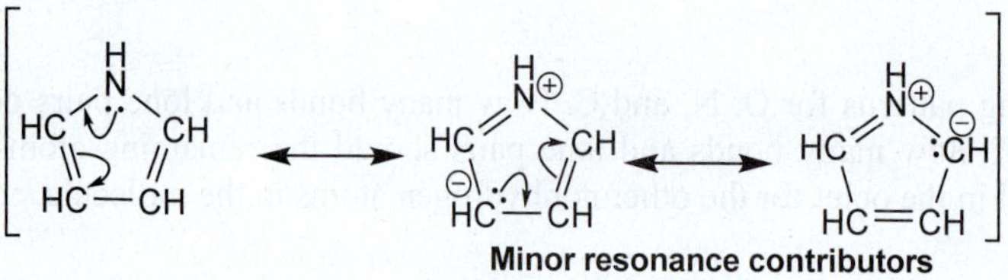 Feature 1: Lone pair adjacent to a double bond

Minor resonance contributors

Problem 1.54

Think

Refer to the individual resonance structures in Problem 1.53 (a). Where are the charges located? Which bonds change from one resonance structure to the next? How do you represent partial bonds and partial charges?

Solve

The resonance hybrid of CH_3NO_2 (shown below) is the weighted average of the two individual resonance structures shown in Problem 1.53 (a). The two O atoms share a −1 charge, and in the hybrid, the shared charge is represented by a partial negative (δ^-). The N–O bond is a single bond in one resonance structure but a double bond in another. In the hybrid, this is represented as ═══. Showing the location of lone pairs involved in resonance is impractical with resonance hybrid structures.

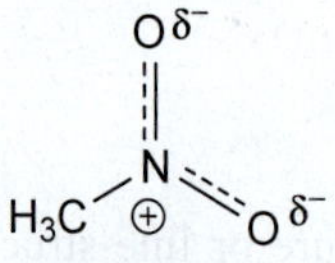

Hybrid

Problem 1.55

Think

How many bonds and lone pairs are typical of O and H? Are there any atoms allowed to have an expanded octet? Which feature from Figure 1-27 does each of these species possess? How must the curved arrows be added? What happens to the respective formal charges? After electron movement, does the new structure have a feature from Figure 1-27?

Solve

The resonance structures differ in the number of bonds and the number of formal charges. With fewer bonds and more formal charges, the contribution toward the resonance hybrid is diminished.

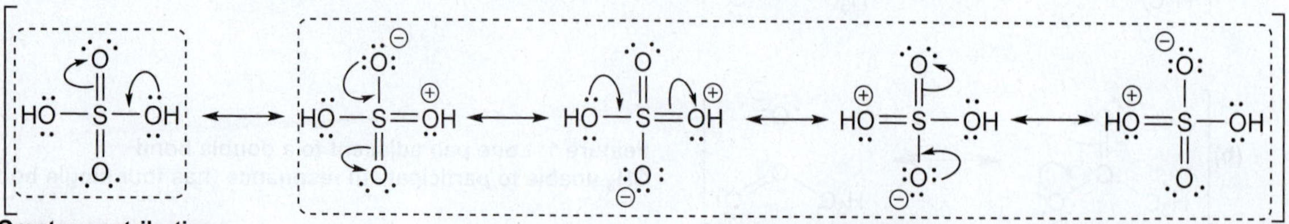

Greater contribution

Smaller contribution

Problem 1.56

Think

If there were another resonance structure, how would you add curved arrows to show electron movement? What would the resonance structure look like? Given the different bond lengths of the single versus double bonds, could all the atoms stay in the locations in which they began?

Solve

The molecule does not have a resonance structure, even though it has a ring of alternating single and double bonds. If we attempt to draw the resonance structure by shifting the electrons around the ring in the same way that we do for benzene, we do indeed arrive at a new structure (below). However, because the double bonds are shorter than the single bonds in this molecule, attempting to draw the resonance structure requires that the atoms also be moved. The reason that cyclobutadiene does not possess resonance structures will be discussed in Chapter 14.

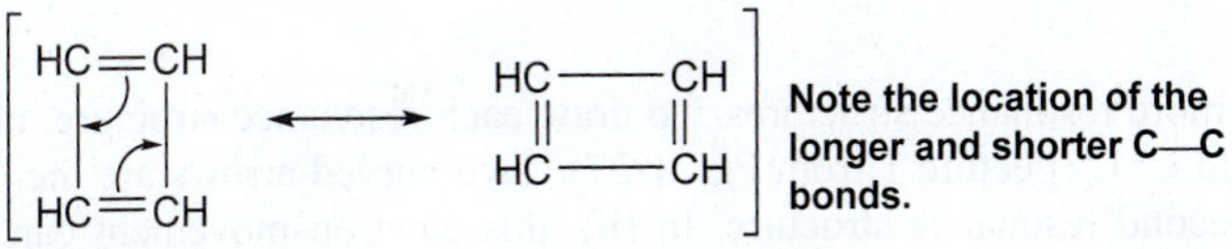

HC=CH / HC=CH ⟷ HC—CH / HC—CH

Note the location of the longer and shorter C—C bonds.

Problem 1.57

Think

Which feature from Figure 1-27 does each of these species possess? How must the curved arrows be added? What happens to the respective formal charges? After electron movement, does the new structure have a feature from Figure 1-27?

Solve

Structure **(a)** is more stable because it has more resonance structures. To draw each resonance structure, notice that C^+ lacks an octet and is attached to C=C (Feature 2 from Fig. 1-27), so one curved arrow is added to move electrons. This electron movement can be repeated three more times to produce a total of five resonance structures. A sixth resonance structure can be produced because the lone pair on oxygen can be involved in resonance without giving rise to additional formal charges (far right). This is not true for structure **(b)**, leaving it with only five total resonance structures (see next page).

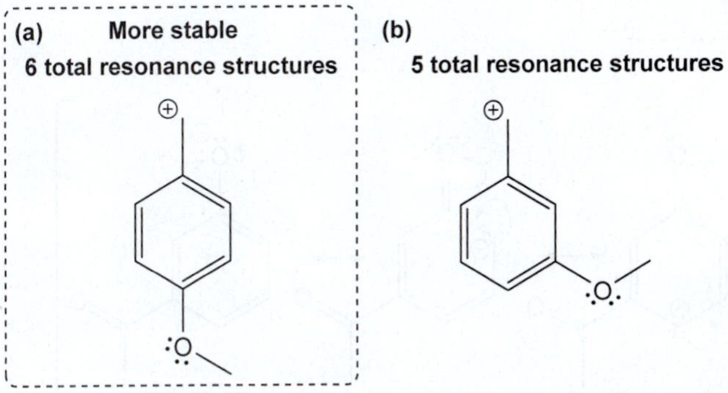

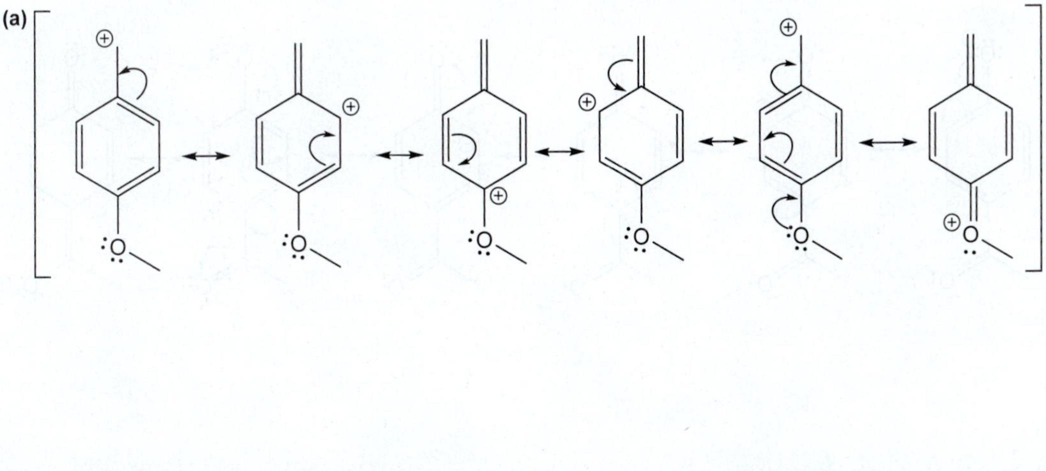

(a) **More stable**
6 total resonance structures

(b)
5 total resonance structures

(a)

(b)

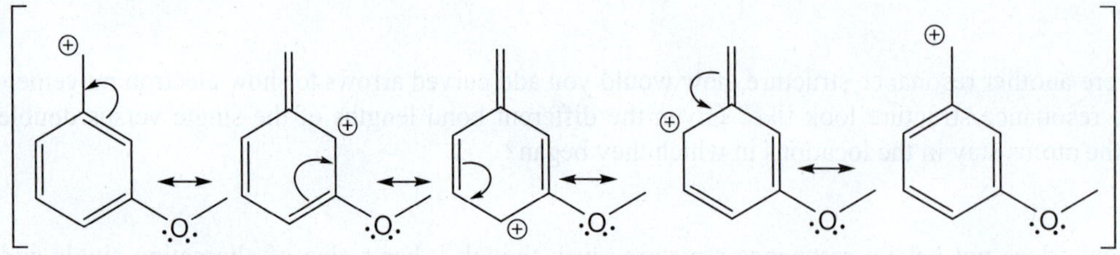

Problem 1.58

Think

Which feature from Figure 1-27 does each of these species possess? How must the curved arrows be added? What happens to the respective formal charges? After electron movement, does the new structure have a feature from Figure 1-27?

Solve

Structure **(b)** is more stable because it has more resonance structures. To draw each resonance structure, notice that the O^- has a lone pair and is adjacent to C=C (Feature 1 from Fig. 1-27). Two curved arrows are therefore added to move electrons to produce the second resonance structure. In **(b)**, this electron movement can take place three more times to produce five resonance structures. Then, the C=O can participate in resonance to produce a sixth resonance structure (far right). In structure **(a)**, the C=O double bond cannot participate in resonance, so only five resonance structures are possible.

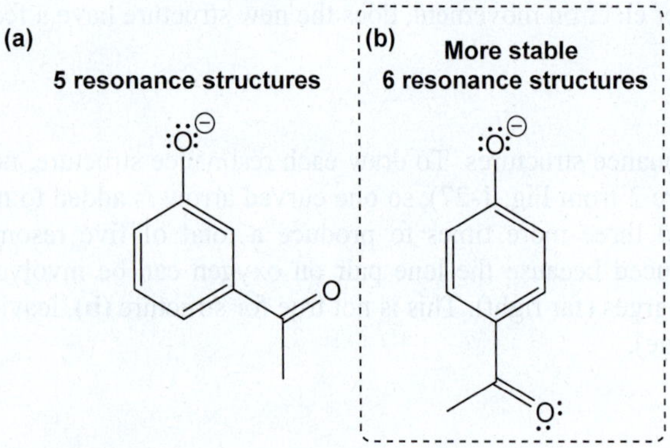

(a) 5 resonance structures

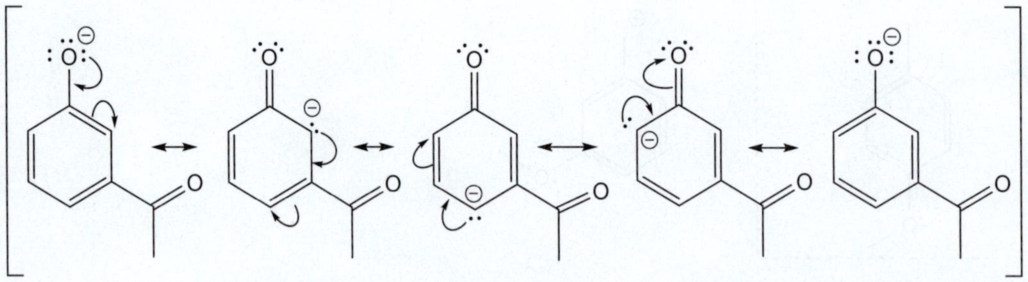

(b) More stable 6 resonance structures

(a)

(b)

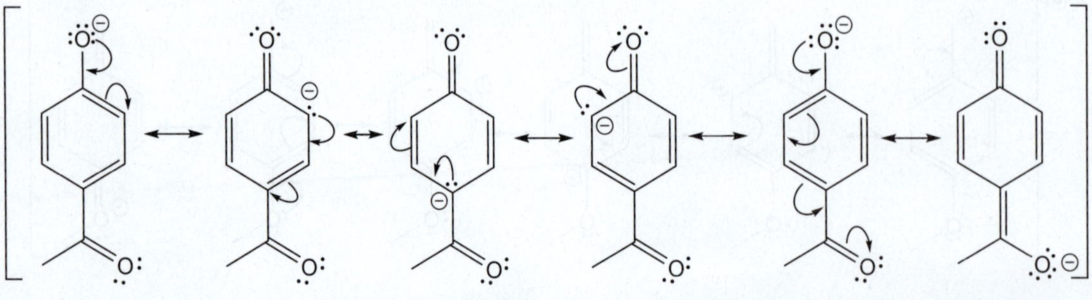

Section 1.12 Shorthand Notations

Problem 1.59

Think

Review the rules for drawing line structures. How are C atoms represented? Is it necessary to draw all lone pairs? What are the rules for drawing H atoms on C atoms? On heteroatoms? On H atoms bonded to heteroatoms?

Solve

The line structure is shown below. C atoms and CH bonds are not shown, nor are lone pairs.

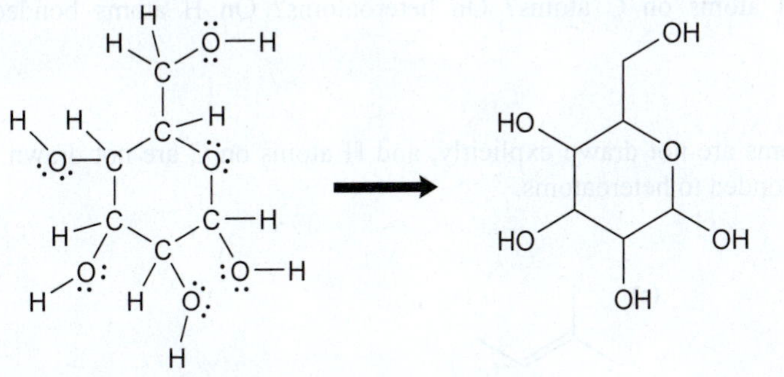

Glucose

Problem 1.60

Think

What atom occurs at the intersection of two lines? Draw in the appropriate C–H bonds and lone pairs of electrons that accurately represent the formal charges shown. Consider Table 1-5 when doing so.

Solve

To redraw the line structure into the complete Lewis structure, all of the C and H atoms and lone pairs of electrons must be explicitly shown (right figure). All C atoms are uncharged, so each should have a total of four bonds and no lone pairs. If a carbon has fewer than four bonds shown, add bonds to H.

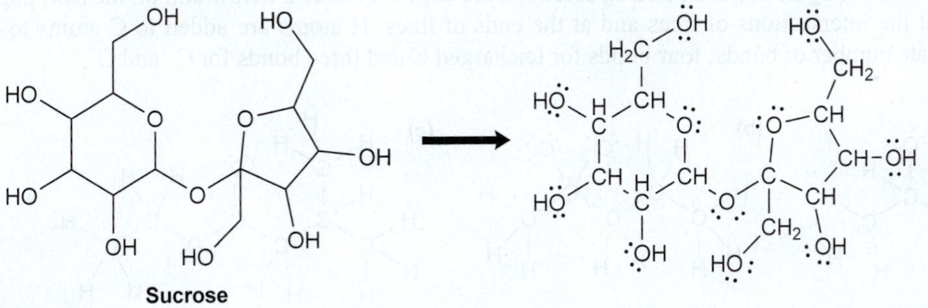

Sucrose

Problem 1.61

Think

Review the rules for drawing condensed formulas. How do you represent multiple H atoms bonded to the same atom? How are rings drawn? How are branches handled?

Solve

Condensed formulas are shown below and on the next page. Note that bonds to atoms making up rings must be shown explicitly. Parentheses are used to handle branching where more than two groups are attached to the C atom.

(a) $CH_3CH_2CH_2CH_3$

(b) $CH_3CH{=}CHCH_3$

(c) $CH_3C(CH_3){=}CHCH_3$

(d)

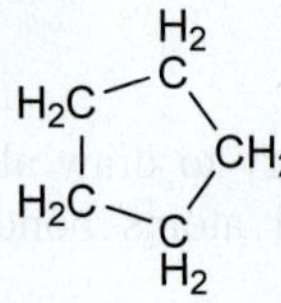

(e)

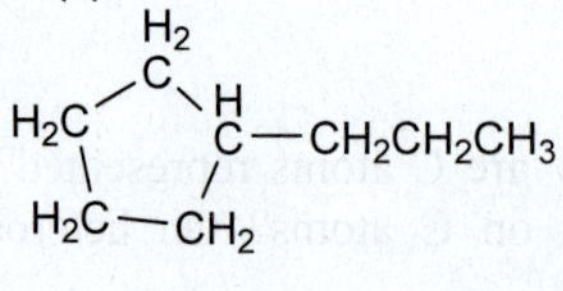

(f) $CH_3OCH_2CH_3$

Problem 1.62

Think

Review the rules for drawing line structures. How are C atoms represented? Is it necessary to draw all lone pairs? What are the rules for drawing H atoms on C atoms? On heteroatoms? On H atoms bonded to heteroatoms?

Solve

The line structures are drawn below. C atoms are not drawn explicitly, and H atoms on C are not drawn. All other atoms are drawn, including H atoms bonded to heteroatoms.

(a) (b) 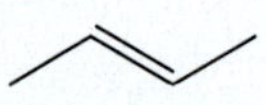 (c)

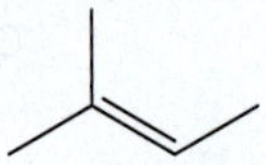

(d) 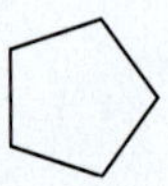 (e) 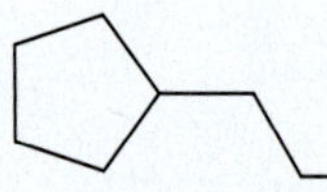 (f)

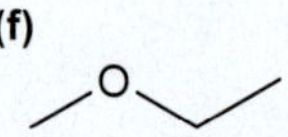

Problem 1.63

Think

What atom occurs at the intersection of two lines? At the end of a line? Draw in the appropriate C–H bonds and lone pairs of electrons that accurately represent the formal charges shown. Consider Table 1-5 when doing so.

Solve

All atoms, bonds, formal charges, and nonbonding electrons are explicitly drawn below and on the next page. C atoms are located at the intersections of lines and at the ends of lines. H atoms are added to C atoms to give each C the appropriate number of bonds: four bonds for uncharged C and three bonds for C^+ and C^-.

(a) (b) 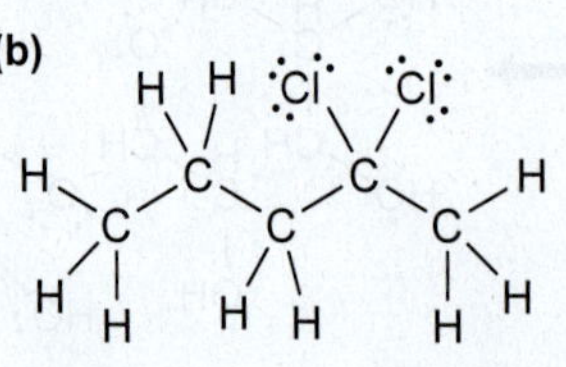 (c)

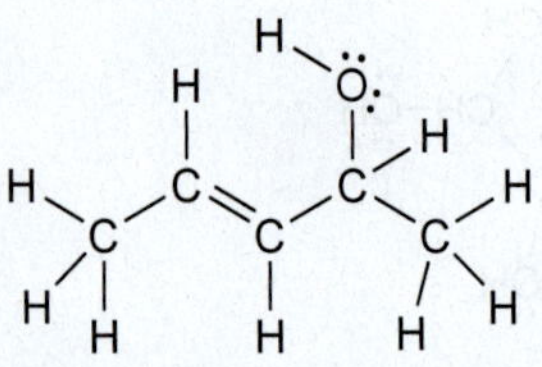

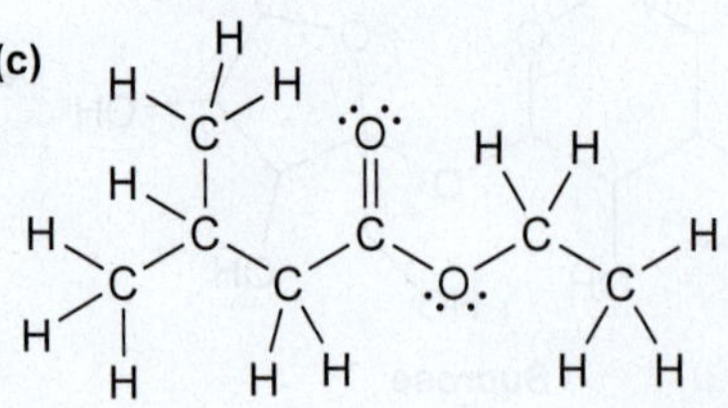

(d) 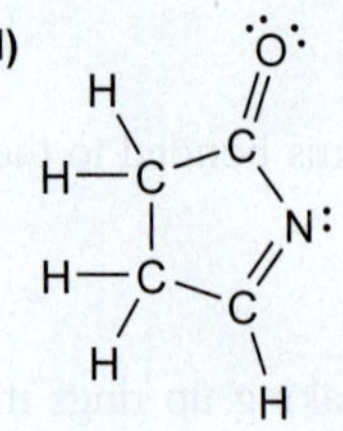 (e) 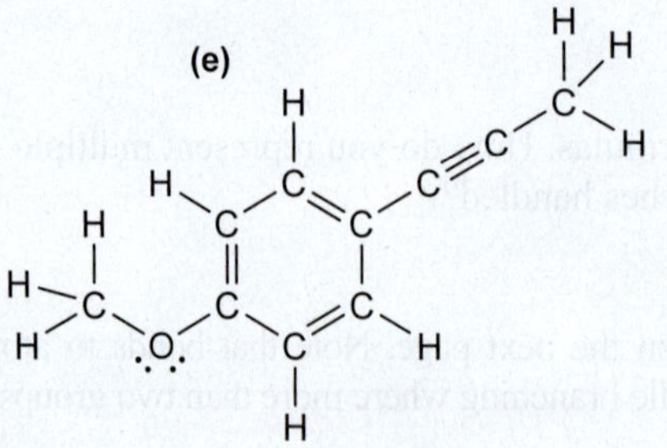 (f)

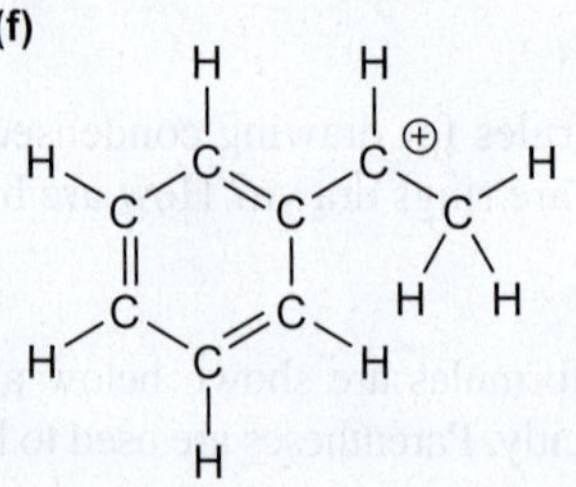

(g) (h) 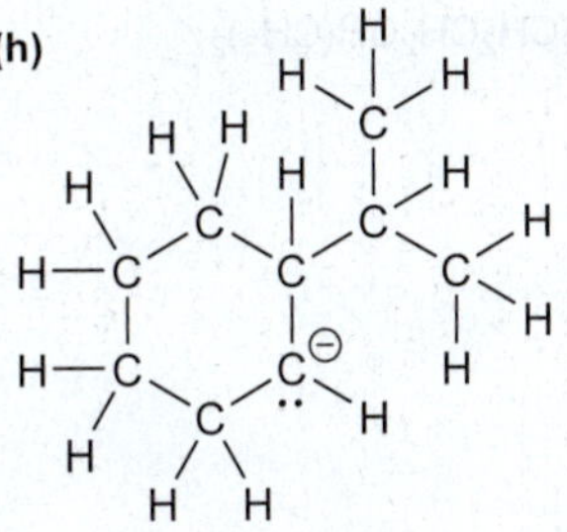 (i)

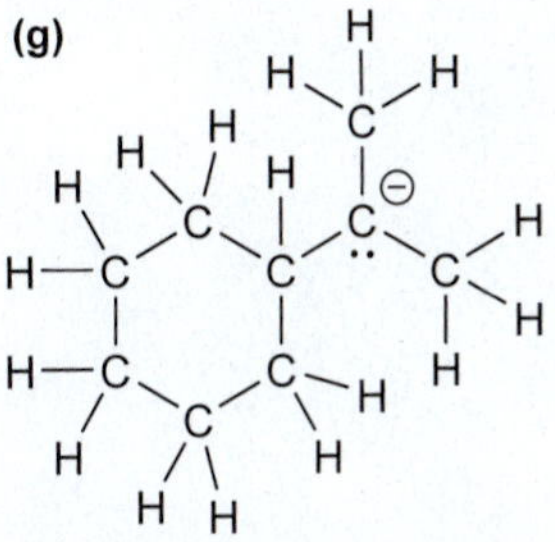

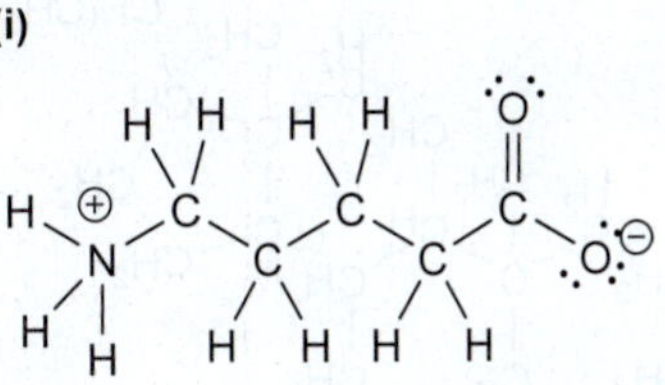

Problem 1.64

Think

Review the rules for drawing condensed formulas. How do you represent a H atom attached to a non-H atom? How are rings drawn? How are branches handled?

Solve

Condensed formulas are shown below. H atoms typically follow the atom to which they are bonded. Note that rings must be shown explicitly. The parentheses in **(a)**, **(g)**, and **(h)** are used to handle branching where the preceding C has three groups attached. The parentheses in **(i)** indicate multiple CH_2 groups attached linearly.

(a) $CH_3CH{=}CHCH(OH)CH_3$ **(b)** $CH_3CH_2CH_2CCl_2CH_3$ **(c)** $(CH_3)_2CHCH_2CO_2CH_2CH_3$

(d) (e) 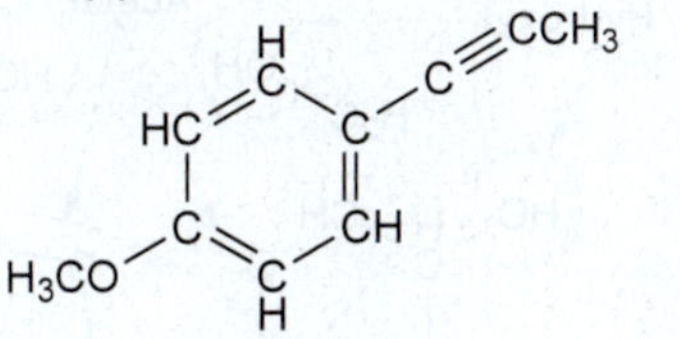 (f) 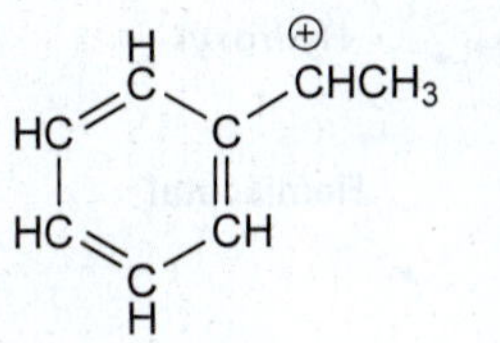 (g)

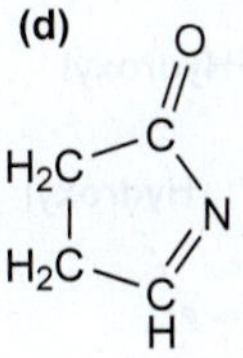

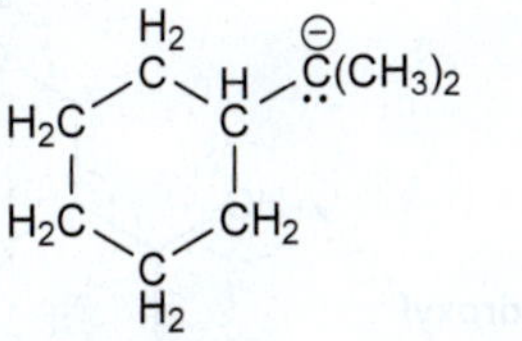

(h) 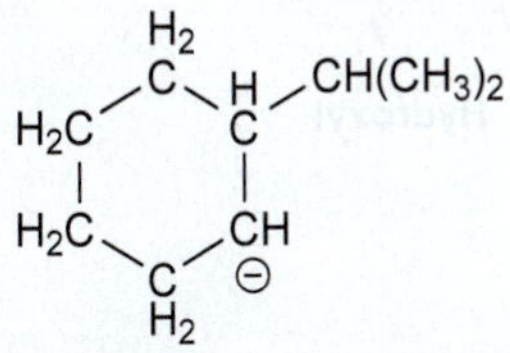 **(i)** $H_3\overset{\oplus}{N}(CH_2)_4CO_2^{\ominus}$

Problem 1.65

Think

Review the rules for drawing line structures and condensed formulas. For line structures, what atom occurs at the intersection of two lines? At the end of a line? Draw in the appropriate C–H bonds and lone pairs of electrons that accurately represent the formal charges shown. Consider Table 1-5 when doing so. In condensed formulas, how do you represent a H atom attached to a non-H atom? How are rings drawn? How are branches handled?

Solve

For the line structure given, C atoms are located at the intersections of lines and at the ends of lines. H atoms are added to C atoms to give each C the appropriate number of bonds—in this case, four bonds for each uncharged C. In a condensed formula, H atoms attached to a particular C are written after that C. All rings must have their bonds shown explicitly. Each instance of parentheses is used to handle branching where the preceding carbon is attached to three groups. Writing a condensed formula with all Cs and Hs explicitly drawn results in a very congested structure (see condensed formula on the next page). The line structure is less congested and simpler to draw, and it is easier to view all parts.

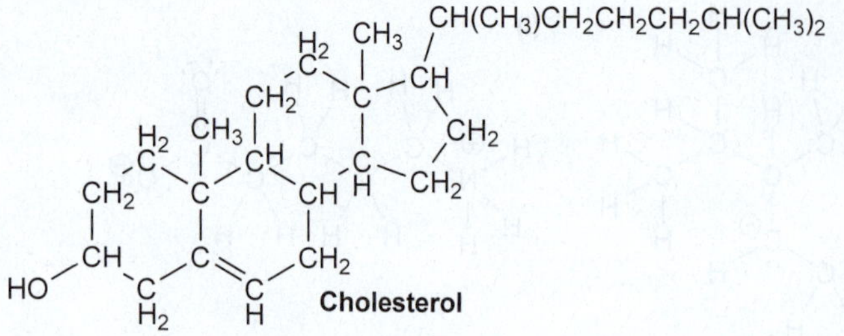

Cholesterol

Section 1.13 An Overview of Organic Compounds: Functional Groups

Problem 1.66

Think

Are any bonds present other than C–C and C–H single bonds? Are any *atoms* present other than carbon and hydrogen? Are any special rings present? What arrangements of bonds and atoms can you find that match with the entries in Table 1-6?

Solve

In glucose (Problem 1.59), there are four OH (hydroxyl) groups characteristic of an alcohol, and there is one hemiacetal. Notice that the hemiacetal is the result of fusing an OH and an ether together. In sucrose (Problem 1.60), there are eight OH groups and two acetals. Each acetal consists of a C atom with two separate C–O–C groups found in ethers.

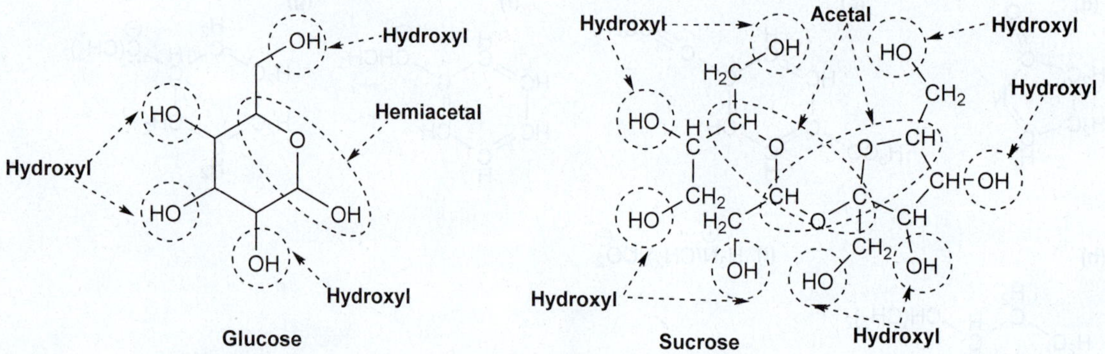

Glucose

Sucrose

Problem 1.67

Think

Refer to Table 1-6. Which group has a heteroatom between two singly bonded carbon groups? Where is sulfur located in the periodic table? How do groups in the same column behave chemically?

Solve

CH_3–S–CH_3 should most resemble the behavior of an ether (R–O–R), because S is located below O in the periodic table. Moreover, S and O each have two bonds and two lone pairs when uncharged.

Problem 1.68

Think

What functional group does ethanol have? Does that functional group match up with any functional groups that are present in molecules **A–E**? If so, which of the compounds **A–E** has a bonding arrangement of atoms that most closely resembles the bonding arrangement in that functional group?

Solve

The functional group in ethanol is −OH (hydroxyl), which is characteristic of all alcohols. **A** and **C** both have an −OH group, too, but **B** and **D** do not, so **B** and **D** are expected to behave differently. The −OH group in **C** is part of the larger CO_2H group, characteristic of carboxylic acids, so it is expected to behave differently. Therefore, ethanol should behave most like H_2O (**A**).

Problem 1.69

Think

Are any bonds present other than C–C and C–H single bonds? Are any *atoms* present other than carbon and hydrogen? Are any special rings present? Refer to Table 1-6 for common functional groups and compound classes.

Solve

The functional groups are circled below, and the compound classes of which they are characteristic are labeled.

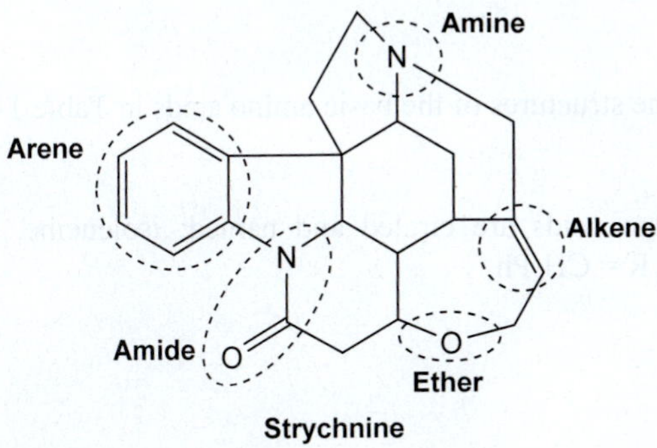

Problem 1.70

Think

Are any bonds present other than C–C and C–H single bonds? Are any *atoms* present other than carbon and hydrogen? Are any special rings present? Refer to Table 1-6 for common functional groups and compound classes.

Solve

The functional groups are circled below, and the compound classes of which they are characteristic are labeled.

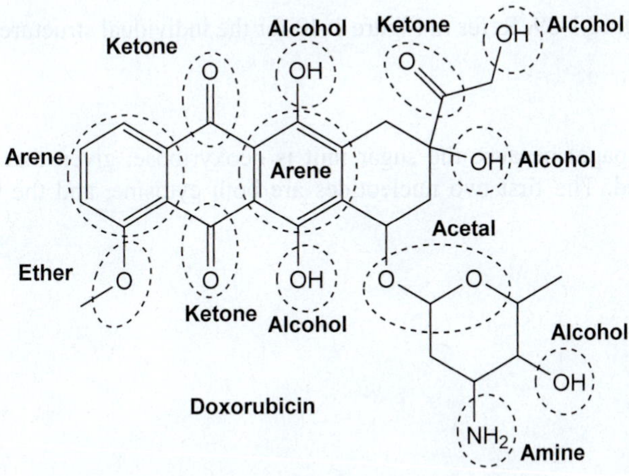

Section 1.14 The Organic Chemistry of Biomolecules

Problem 1.71

Think

What is the general Lewis structure for an amino acid? At what location in the general structure does each of the individual amino acids differ? What is the identity of the R group in alanine and aspartic acid?

Solve

The general amino acid structure (Fig. 1-34) is shown on the next page, along with the two Lewis structures for alanine (R = CH_3) and aspartic acid (R = CH_2CO_2H). The side groups are circled.

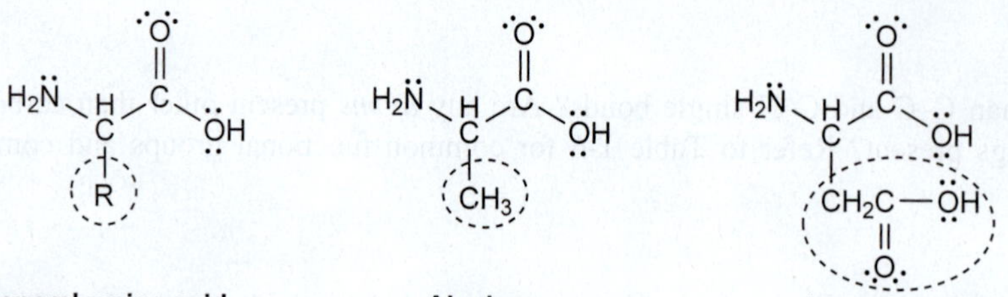

Problem 1.72

Think

What distinguishes one amino acid from another? Review the structures of the basic amino acids in Table 1-7.

Solve

In the tripeptide shown below, the three individual amino acids are circled and named: isoleucine, R = $CH_2(CH_3)CH_2CH_3$; serine, R = CH_2OH; and phenylalanine, R = CH_2Ph.

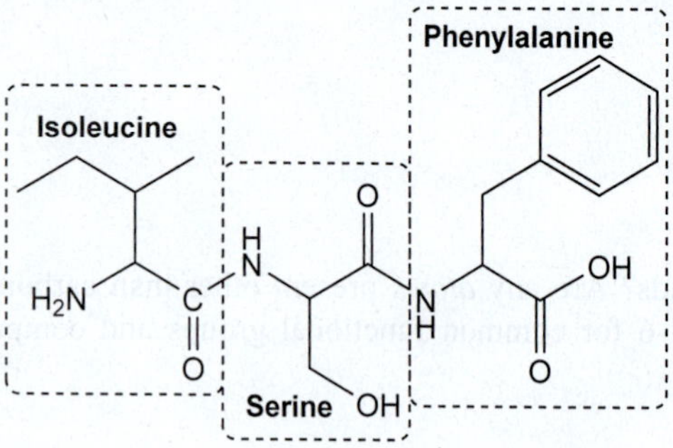

Problem 1.73

Think

What distinguishes RNA from DNA? Refer to Figure 1-39. Refer to Figure 1-40 for the individual structures for each nucleotide.

Solve

This is a segment of DNA (shown on the next page) because the sugar unit is deoxyribose, given that it is missing an oxygen atom at the location indicated. The first two nucleotides are both cytosine, and the third nucleotide is guanine.

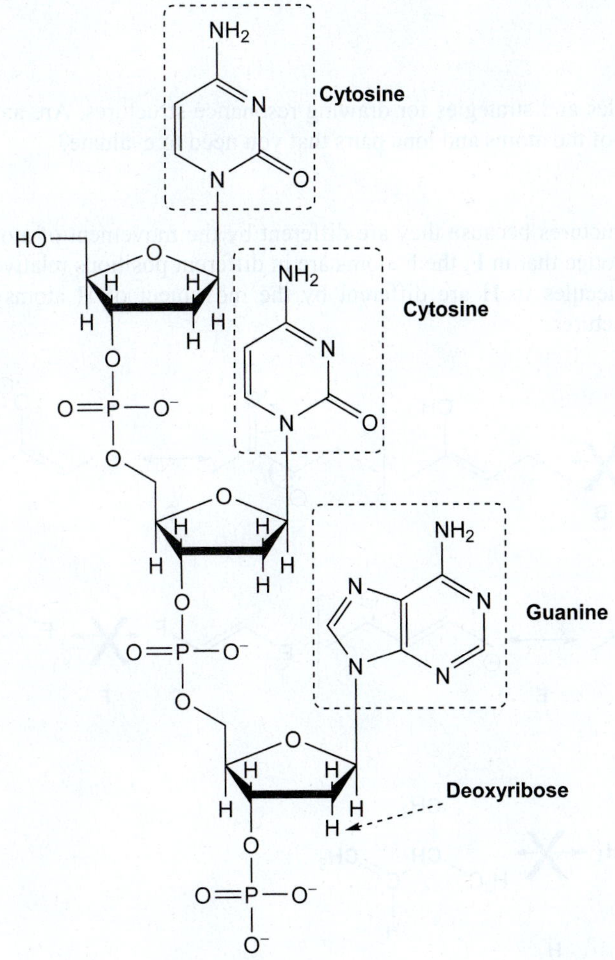

Problem 1.74

Think

What atoms are present in a carbohydrate? What is the general formula for a carbohydrate? What is the general formula for a monosaccharide? Write the formula for each compound and compare it to the definitions.

Solve

In a carbohydrate, carbon, hydrogen, and oxygen are present in a general formula of $C_xH_{2y}O_y$. A monosaccharide is a carbohydrate with one additional restriction to the chemical formula, $C_xH_{2x}O_x$. In the figures below, **(a)** and **(b)** are both carbohydrates and monosaccharides, and **(c)** is neither.

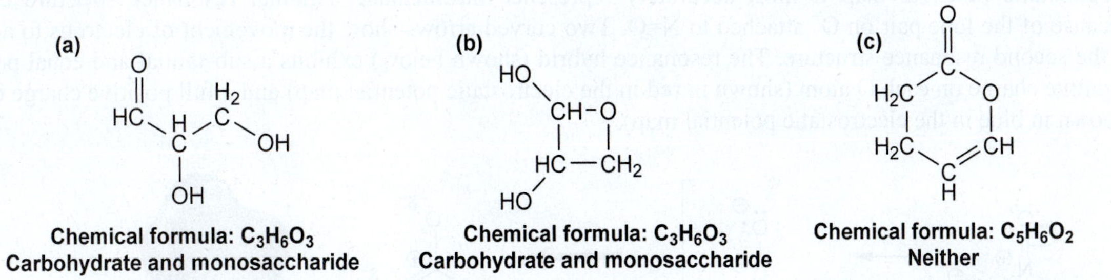

Integrated Problems

Problem 1.75

Think

Refer to Sections 1.10 and 1.11 to review the rules and strategies for drawing resonance structures. Are atoms allowed to move? Do the line structures show all of the atoms and lone pairs that you need to evaluate?

Solve

A, **B**, **F**, **G**, and **H** are not pairs of resonance structures because they are different by the movement of atoms, not just by the movement of valence electrons. Notice that in **F**, the F atoms are in different positions relative to the double bond. Also, notice that the two molecules in **H** are different by the movement of H atoms (in boldface). **C**, **D**, and **E** are pairs of resonance structures.

A B C

D E F

G H

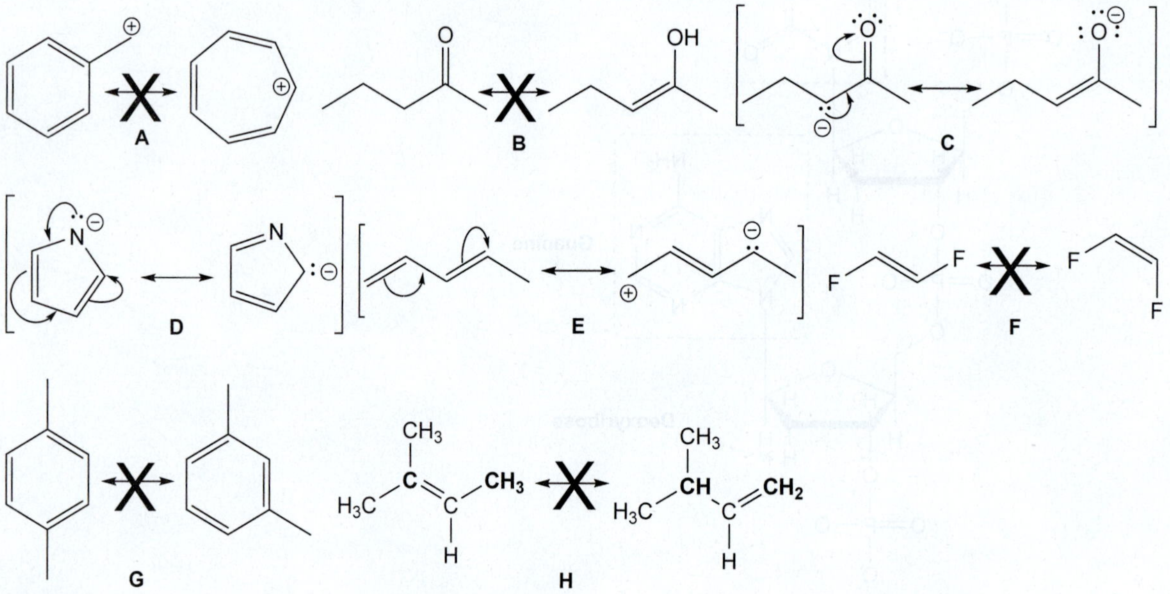

Problem 1.76

Think

Review the rules for drawing Lewis structures, and draw the Lewis structure for CH_3NO_2. Is there an atom with a lone pair attached to a multiple bond? If so, how many curved arrows must be drawn to indicate electron movement to arrive at another resonance structure? In the resonance hybrid, where are the partial positive and negative charges, δ^+ and δ^-? Can you match the hybrid to one of the electrostatic potential maps? Which color (red or blue) corresponds to a buildup of negative charge? Of positive charge?

Solve

Electrostatic potential map **C** most accurately represents nitromethane. Another resonance structure exists because of the lone pair on O^- attached to N=O. Two curved arrows show the movement of electrons to arrive at the second resonance structure. The resonance hybrid (shown below) exhibits a substantial and equal partial negative charge on each O atom (shown in red in the electrostatic potential map) and a full positive charge on N (shown in blue in the electrostatic potential map).

Resonance hybrid **C**

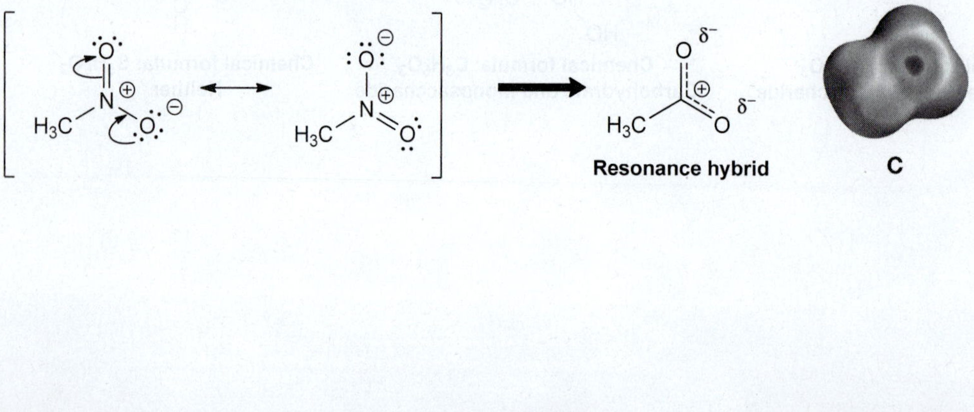

Problem 1.77

Think

Review the rules for drawing resonance structures, and draw the resonance hybrid for each compound below. In the structure you are given, is there a multiple bond attached to an atom with a lone pair? To an atom lacking an octet? Is there a ring of alternating single and multiple bonds? What does an electrostatic potential map depict? Which color (red or blue) corresponds to a buildup of negative charge? Match up the electrostatic potential map to one of the five compounds.

Solve

The electrostatic potential map indicates a concentration of negative charge shared among three different carbon atoms. This is consistent with **C**, which has three total resonance structures, as shown below. Each resonance structure has a C^- with a lone pair attached to C=C, so the curved arrows are drawn away from the lone pair.

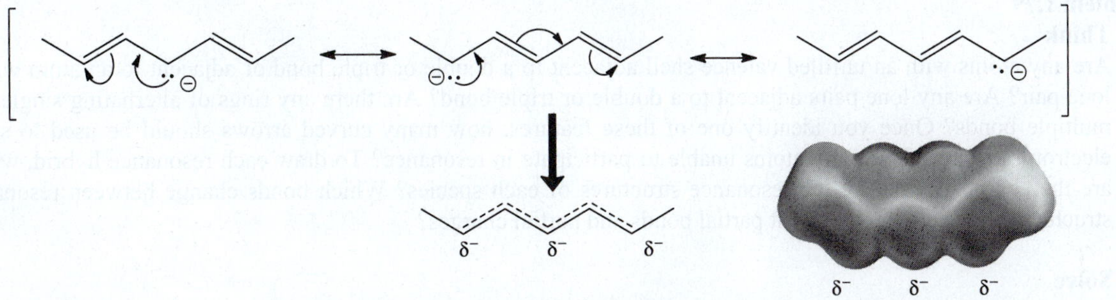

δ⁻ δ⁻ δ⁻

δ⁻ δ⁻ δ⁻

B and **D** have no additional resonance structures, so the negative charge is localized on just one C atom. **A** has three additional resonance structures, allowing the negative charge to be shared among four C atoms. **E** has one additional resonance structure, allowing the negative charge to be shared among two C atoms.

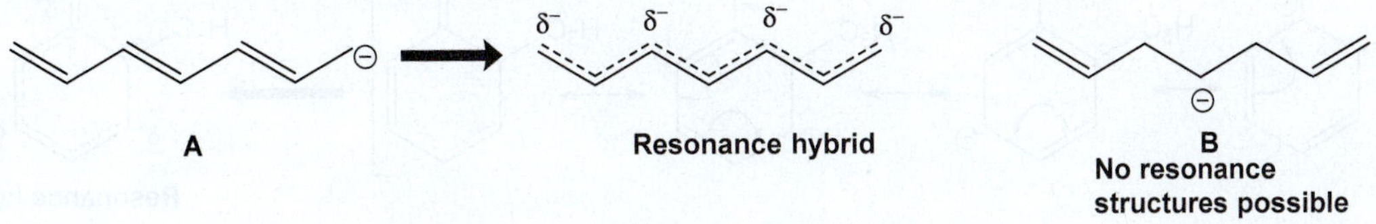

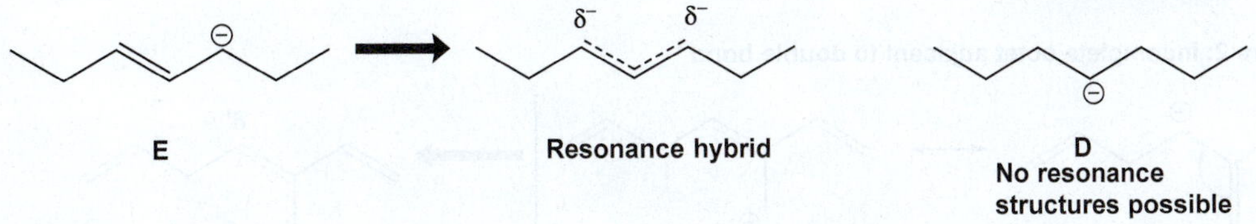

δ⁻ δ⁻ δ⁻ δ⁻

A **Resonance hybrid** **B** **No resonance structures possible**

δ⁻ δ⁻

E **Resonance hybrid** **D** **No resonance structures possible**

Problem 1.78

Think

Refer to Table 1-6 to review the functional groups and compound classes. Draw in the necessary functional groups first, and then fill in the rest of the formula. Be sure that all atoms (other than H) in your structures have octets and that there are nine H atoms.

Solve

(a) An amine is $R–NH_2$, R–NHR, or $R–NR_2$, and an alkyne has a C≡C bond.
(b) An alkene has a C=C bond.
(c) A nitrile has a C≡N bond. Two examples for each problem are shown on the next page.
Note: There are additional correct answers.

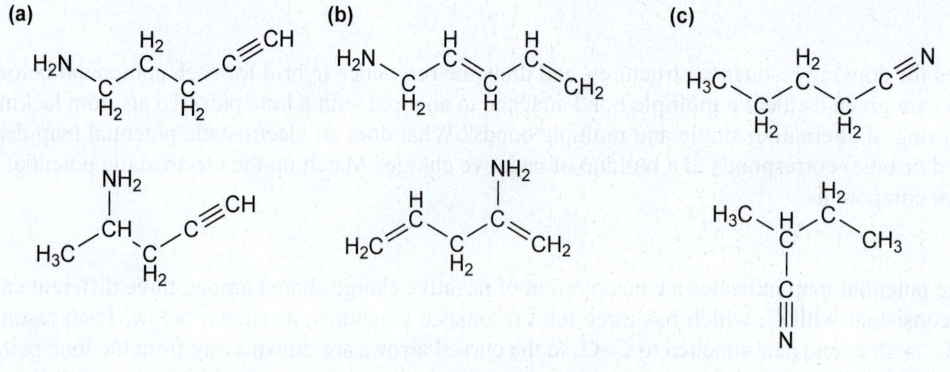

(a) (b) (c)

Problem 1.79

Think

Are any atoms with an unfilled valence shell adjacent to a double or triple bond or adjacent to an atom with a lone pair? Are any lone pairs adjacent to a double or triple bond? Are there any rings of alternating single and multiple bonds? Once you identify one of these features, how many curved arrows should be used to show electron movement? Are any atoms unable to participate in resonance? To draw each resonance hybrid, where are the charges located in the resonance structures of each species? Which bonds change between resonance structures? How do you represent partial bonds and partial charges?

Solve

Resonance structures and resonance hybrids for **(a)–(d)** are shown below and on the next page.

(a) Feature 2: Incomplete octet adjacent to double bond

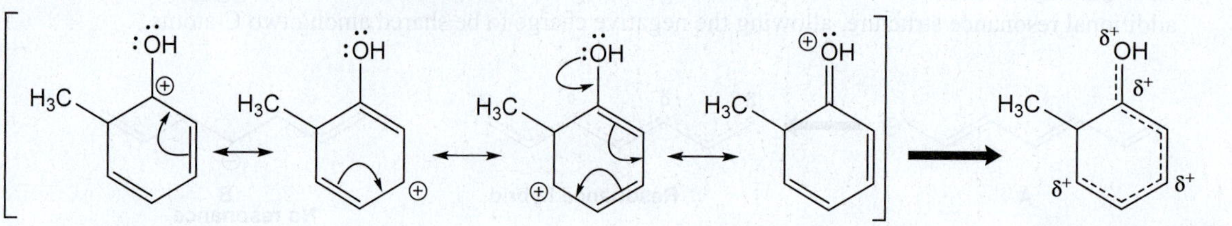

Resonance hybrid

(b) Feature 2: Incomplete octet adjacent to double bond

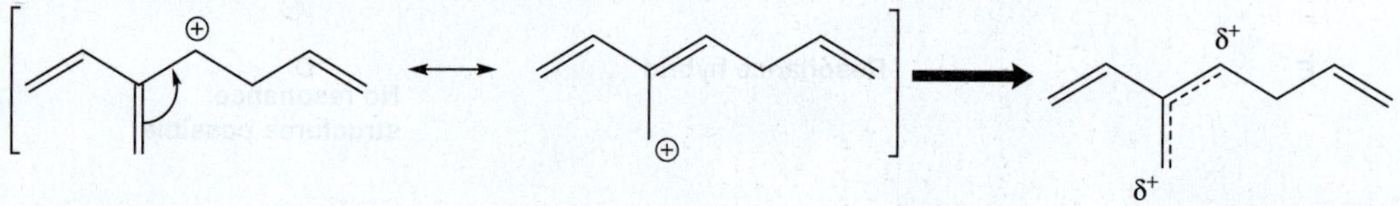

Resonance hybrid

(c) Feature 1: Lone pair adjacent to double bond

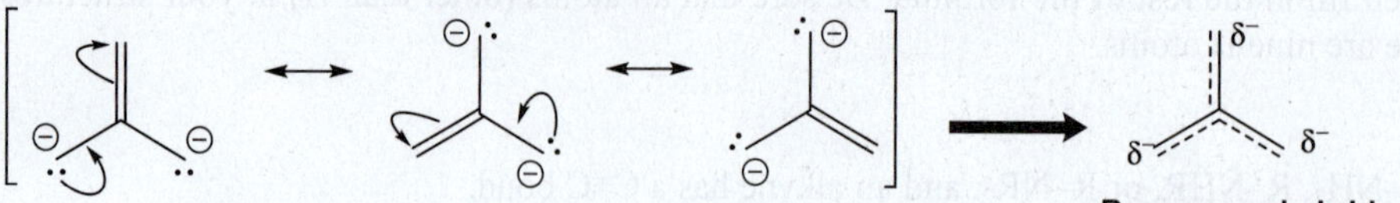

Resonance hybrid

(d) Feature 1: Lone pair adjacent to double bond

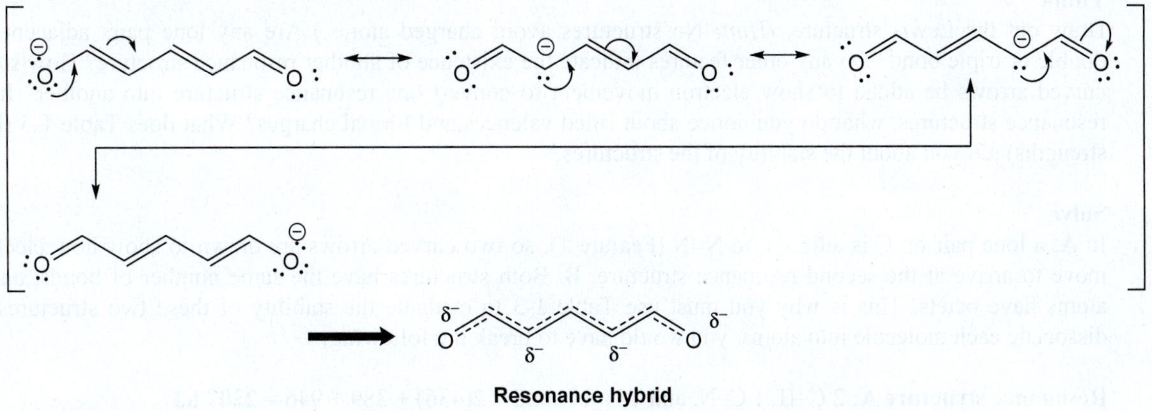

Problem 1.80

Think

Are any lone pairs adjacent to a double or triple bond? Do any atoms lack an octet adjacent to a double or triple bond? Any there any rings of alternating single and multiple bonds? After identifying one of these features, how many curved arrows should you add to show electron movement? Are any atoms unable to participate in resonance? To draw the resonance hybrid, where are the charges located? Which bonds change between resonance structures? How do you represent partial bonds and partial charges?

Solve

This species has an atom with an incomplete octet (C^+) adjacent to a double bond (Feature 2). One curved arrow is used to move electrons to arrive at the second resonance structure. This is repeated five more times, producing a total of seven resonance structures. Resonance is possible for all of the C atoms, except the two $-CH_3$ groups that already have four single bonds. In the resonance hybrid, all atoms that share the +1 charge receive a partial positive charge (δ^+). All bonds that appear as a single bond in one resonance structure and a double bond in another are represented as a single bond plus a partial bond in the hybrid (----).

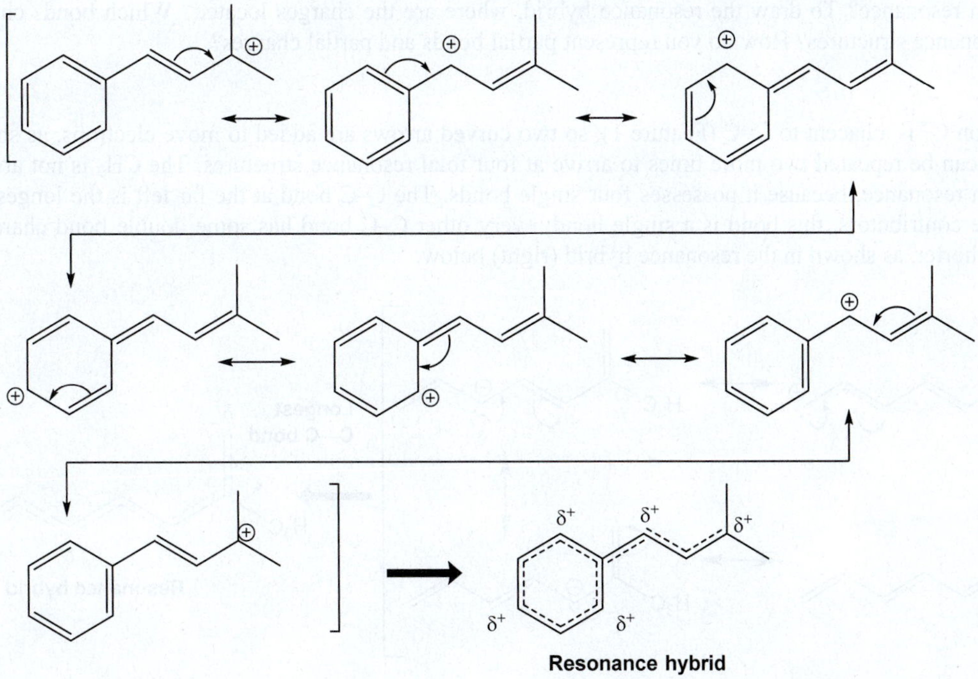

Problem 1.81

Think

Draw out the Lewis structure. *(Hint:* No structures avoid charged atoms.) Are any lone pairs adjacent to a double or triple bond? Do any other features indicate the existence of another resonance structure? How should curved arrows be added to show electron movement to convert one resonance structure into another? In the resonance structures, what do you notice about filled valences and formal charges? What does Table 1-3 (bond strengths) tell you about the stability of the structures?

Solve

In **A**, a lone pair on C is adjacent to N≡N (Feature 1), so two curved arrows are drawn to show how electrons move to arrive at the second resonance structure, **B**. Both structures have the same number of bonds, and all atoms have octets. This is why you must use Table 1-3 to evaluate the stability of these two structures. To dissociate each molecule into atoms, you would have to break the following:

Resonance structure A: 2 C–H, 1 C–N, and 1 N≡N bond = 2(436) + 389 + 946 = **2207 kJ** (larger bond energy)

Resonance structure B: 2 C–H, 1 C=N, and 1 N=N bond = 2(436) + 619 + 418 = **1909 kJ**

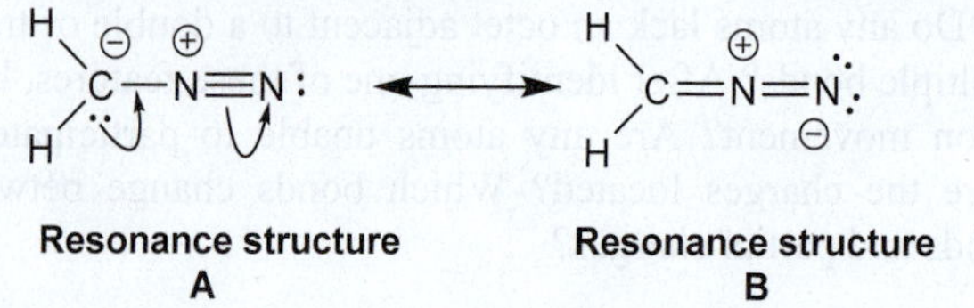

Resonance structure A

Resonance structure B

Thus, **A** is more stable and contributes more to the resonance hybrid.

Problem 1.82

Think

Are any lone pairs adjacent to a double or triple bond? Are any atoms lacking an octet adjacent to a double or triple bond? Are there any rings of alternating single and multiple bonds? After identifying one of these features, how many curved arrows should you add to show electron movement? Are any atoms unable to participate in resonance? To draw the resonance hybrid, where are the charges located? Which bonds change between resonance structures? How do you represent partial bonds and partial charges?

Solve

A lone pair on C^- is adjacent to C=C (Feature 1), so two curved arrows are added to move electrons, as shown below. This can be repeated two more times to arrive at four total resonance structures. The CH_3 is not able to participate in resonance, because it possesses four single bonds. The C–C bond at the far left is the longest. In all resonance contributors, this bond is a single bond; every other C–C bond has some double bond character and will be shorter, as shown in the resonance hybrid (right) below.

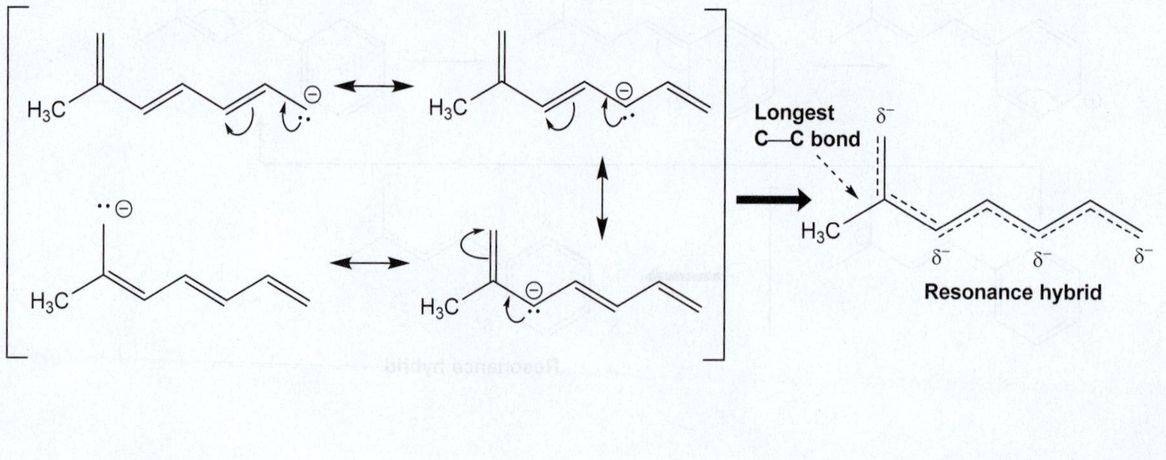

Longest C–C bond

Resonance hybrid

INTERCHAPTER A | Nomenclature: The Basic System for Naming Simple Organic Compounds—Alkanes, Haloalkanes, Nitroalkanes, Cycloalkanes, and Ethers

In Chapter Problems

Problem A.1

Think

What would the name of the alkane or cycloalkane be without the generic substituent? How many C atoms are present in the main chain?

Solve

(a) A two-carbon alkane with a substituent is a substituted *ethane*.

(b) A five-carbon alkane with a substituent is a substituted *pentane*.

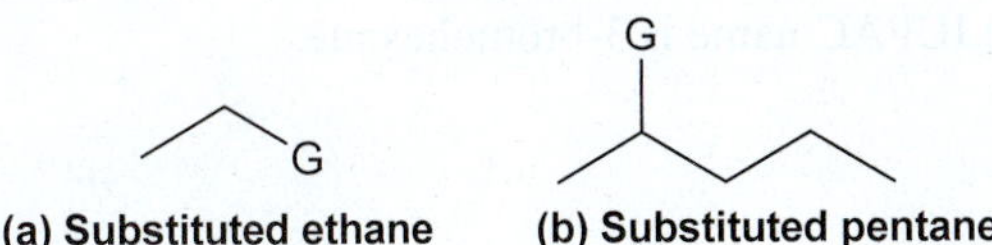

Problem A.2

Think

How many carbon atoms are in *propane*? How many are in *hexane*? How do you convert an alkane into a substituted alkane?

Solve

(a) *Propane* is a linear alkane with three carbon atoms. To generate a substituted propane, replace a H atom with a generic substituent, G.

(b) The substituted *hexane* is a linear alkane with six carbon atoms and a generic substituent. *Note:* Other substituted structures are possible.

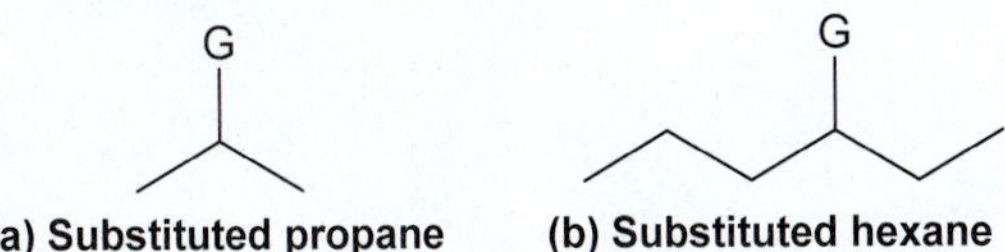

Problem A.4

Think

How many carbon atoms does the longest continuous carbon chain have? What is the corresponding root? What are the choices for assigning C1? Which of those choices would give the substituent group the lower locator number? How do you add the locator number and the name of the substituent to the IUPAC name?

Solve

The IUPAC names for molecules **(a)–(d)** are given below and on the next page. The main chain is numbered.

(a) The longest continuous carbon chain has five carbon atoms, so the root is *pentane*. The NO_2 group is a *nitro* substituent, making the molecule a *nitropentane*. C1 must be one of the carbon atoms at the end of the chain, so there are two possible numbering systems. In the first numbering system, the locator number for the NO_2 group is 5, whereas in the second numbering system, it is 1. Therefore, the second choice is correct, because the locator number is lower. The complete IUPAC name is 1-nitropentane.

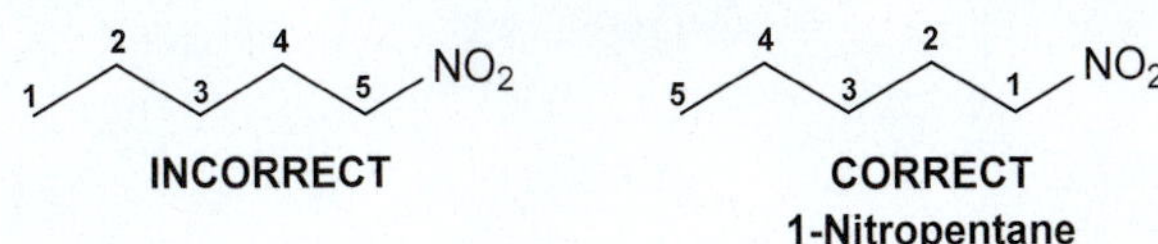

(b) The longest continuous carbon chain has six carbon atoms, so the root is *hexane*. The Cl is a *chloro* substituent, making this a *chlorohexane*. C1 must be one of the carbon atoms at the end of the chain, so there are two possible numbering systems. In the first numbering system, the locator number for the Cl group is 2, whereas in the second numbering system, it is 5. Therefore, the first choice is correct, because the locator number is lower. The complete IUPAC name is 2-chlorohexane.

CORRECT
2-Chlorohexane

INCORRECT

(c) The longest continuous carbon chain has six carbon atoms, so the root is *hexane*. The Br is a *bromo* substituent, making this a *bromohexane*. C1 must be one of the carbon atoms at the end of the chain, so there are two possible numbering systems. In the first numbering system, the locator number for the Br group is 4, whereas in the second numbering system, it is 3. Therefore, the second choice is correct, because the locator number is lower. The complete IUPAC name is 3-bromohexane.

INCORRECT

CORRECT
3-Bromohexane

(d) The longest continuous carbon chain has four carbon atoms, so the root is *butane*. The F is a *fluoro* substituent, making this a *fluorobutane*. C1 must be one of the carbon atoms at the end of the chain, so there are two possible numbering systems. In the first numbering system, the locator number for the F group is 4, whereas in the second numbering system, it is 1. Therefore, the second choice is correct, because the locator number is lower. The complete IUPAC name is 1-fluorobutane.

INCORRECT

CORRECT
1-Fluorobutane

Problem A.6

Think

How many carbon atoms does the longest continuous carbon chain have, and what is the corresponding root? On which end of the chain should numbering begin? What are the names of the substituents, and which substituent comes first alphabetically? How do you indicate the number of each kind of substituent, and what locator number is assigned to each substituent?

Solve

The IUPAC names for molecules **(a)–(d)** are given below and on the next page. The main chain is numbered.

(a) The longest continuous carbon chain has two carbon atoms, so the root is ethane. Numbering must begin on the right to give the first substituent encountered the smallest locator number, 1. There are two types of substituents: *chloro* and *fluoro*. *Chloro* comes first alphabetically, so it appears first in the IUPAC name. The prefix *di* is added to indicate two *fluoro* substituents, and the locator numbers 1,1 are added to indicate their positions along the chain. No prefix is necessary for the *chloro* because there is only one Cl atom. The IUPAC name, therefore, is 1-chloro-1,1-difluoroethane.

1-Chloro-1,1-difluoroethane

(b) The longest continuous carbon chain has four carbon atoms, so the root is *butane*. Numbering starts from the right to give each of the first three substituents a locator number of 1. If numbering were to begin at left, then the locator number for the second substituent would be 2. There are three types of substituents, *bromo*, *iodo*, and *nitro*, and these are in alphabetical order. The prefix *tri* is added to indicate three *bromo* substituents, and the locator numbers 1,3,4 are added to indicate their positions along the chain. The prefix *di* is added to indicate two *iodo* substituents, and the locator numbers 1,1 are added to indicate their positions along the chain. No prefix is necessary for the *nitro* because there is only one NO_2 group. The IUPAC name, therefore, is 1,3,4-tribromo-1,1-diiodo-2-nitrobutane.

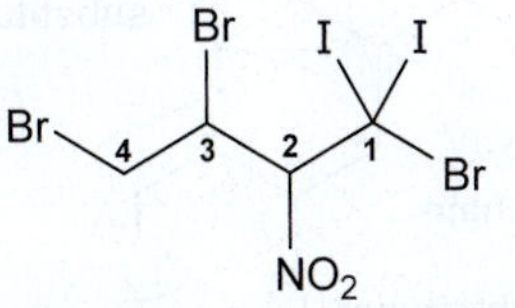

1,3,4-Tribromo-1,1-diiodo-2-nitrobutane

(c) The longest continuous carbon chain has three carbon atoms, so the root is *propane*. Numbering must begin on the left to give the three substituents encountered the smallest locator number, 1. If numbering were to begin at right, then the locator number for the second substituent would be 3. There are two types of substituents: *chloro* and *fluoro*. *Chloro* comes first alphabetically, so it appears first in the IUPAC name. The prefix *tri* is added to indicate three *chloro* substituents, and the locator numbers 1,1,1 are added to indicate their positions along the chain. No prefix is necessary for the *fluoro* because there is only one F atom. The IUPAC name, therefore, is 1,1,1-trichloro-3-fluoropropane.

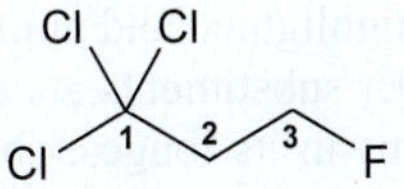

1,1,1-Trichloro-3-fluoropropane

(d) The longest continuous carbon chain has six carbon atoms, so the root is hexane. Numbering may begin on either the right or the left to give the first substituent encountered the smallest locator number, 2, the second 3, the third 4, and the fourth 5. The molecule is symmetrical. There is one type of substituent: nitro. The prefix tetra is added to indicate four nitro substituents, and the locator numbers 2,3,4,5 are added to indicate their positions along the chain. The IUPAC name, therefore, is 2,3,4,5-tetranitrohexane.

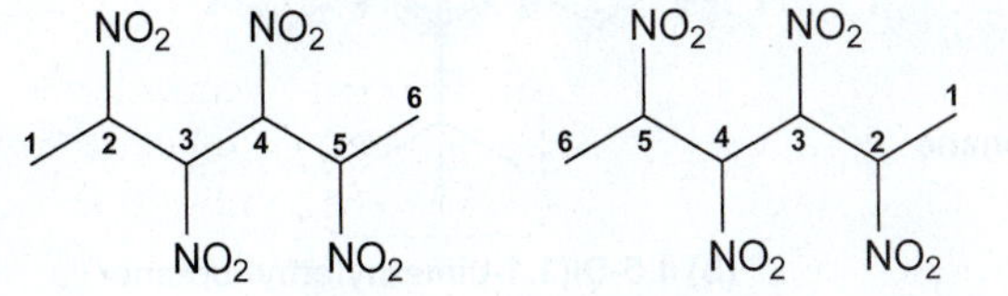

2,3,4,5-Tetranitrohexane

Problem A.7

Think

How many C atoms are in the longest continuous chain? Name the longest continuous chain. How many C atoms are in the longest chain of each alkyl substituent, and does that chain have substituents? How do you name the alkyl substituent, taking branching into account and using the *yl* suffix?

Solve

The IUPAC names for molecules **(a)** and **(b)** are given on the next page. The main chain is numbered.

(a) There are eight C atoms in the main chain, octane. C1 must be one of the carbon atoms at the end of the chain, so there are two possible numbering systems. Both numbering systems give 4 and 5 as the locator numbers for the first and second substituents. The correct numbering system is the one shown on the next page, because it gives the lower number to the alkyl group that comes first in alphabetical order. The first alkyl group is branched; it has two C atoms in the longest chain (*ethyl*) and a CH_3 group (*methyl*) attached to its C1, so it is 1-methylethyl. The second alkyl group has three C atoms in the longest chain, so it is propyl. The IUPAC name is 4-(1-methylethyl)-5-propyloctane.

(b) There are nine C atoms in the main chain, *nonane*. C1 must be one of the carbon atoms at the end of the chain, so there are two possible numbering systems. The correct numbering system is the one shown below, because it gives the lower number to the first substituent encountered, *fluoro*. The alkyl group is branched; it is 1,1-dimethylethyl. The IUPAC name is 3-fluoro-5-(1,1-dimethylethyl)nonane.

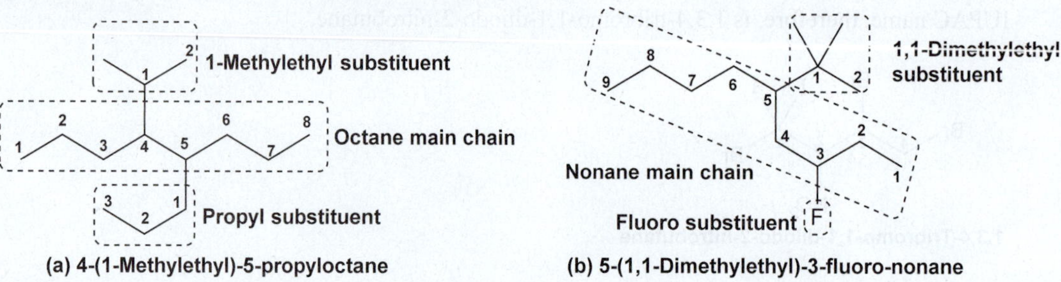

(a) 4-(1-Methylethyl)-5-propyloctane (b) 5-(1,1-Dimethylethyl)-3-fluoro-nonane

Problem A.8

Think

How many carbon atoms are in the longest continuous chain? What substituents are attached? How many of each kind? What do the parentheses indicate? Where does numbering begin for a chain? Add in the substituents at the designated locations.

Solve

Structures for IUPAC names **(a)** and **(b)** are drawn below. The main chain is highlighted and numbered.

(a) *Nonane* indicates a chain of nine C atoms. 3,3-Dibromo means that two Br substituents are attached at C3. (2-Methylpropyl) indicates a branched alkyl group that has three C atoms in its longest chain and a CH_3 group attached at its C2, and that branched alkyl group attaches to the main chain at C5.

(b) (*Dimethylethyl*) indicates a branched alkyl group that has two C atoms in its longest chain and two CH_3 groups attached to its C1. Two such branched alkyl groups are attached to the main chain at C4 and C5.

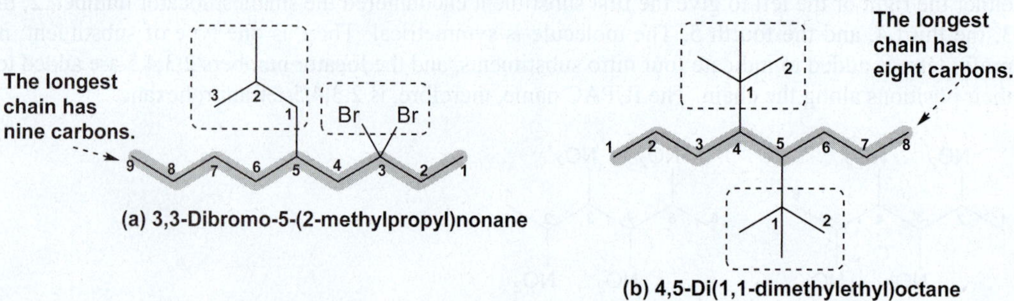

(a) 3,3-Dibromo-5-(2-methylpropyl)nonane (b) 4,5-Di(1,1-dimethylethyl)octane

Problem A.10

Think

How many carbon atoms are in the longest carbon chain? The largest carbon ring? Which one establishes the root? Which carbon atom gives the lowest locator number to the first substituent, to the second, and so on? In a ring, should numbering increase clockwise or counterclockwise to encounter the next substituent the earliest?

Solve

The IUPAC names for molecules **(a)–(d)** are given on the next page. The main chain or ring is numbered.

(a) The longest carbon chain has six C atoms, whereas the largest ring has three. Therefore, the chain establishes the root as *hexane*, and the ring is a cyclopropyl substituent located at C3.

(b) The longest carbon chain has three C atoms, whereas the largest ring has five. Therefore, the ring establishes the root as *cyclopentane*, and the chain is a (1-methylethyl) substituent located at C1.

(c) The longest carbon chain has three C atoms, whereas the largest ring has six. Therefore, the ring establishes the root as *cyclohexane*. The three-carbon substituent is a chain of three C atoms with a Br attached at its

C1, so it is a (1-bromopropyl) substituent. Numbering begins at the C that is attached to (1-bromopropyl) and proceeds counterclockwise to give the second substituent the lowest locator number, 2.

(d) The longest carbon chain has one C atom, and the largest ring has eight, so the root is *cyclooctane.* Numbering is chosen to give the lowest possible locator number to the first substituent, 1; the second substituent, 1; and the third substituent, 2.

(a) 3-Cyclopropylhexane **(b) 1-(1-Methylethyl)cyclopentane**

(c) 1-(1-Bromopropyl)-2-methyl-4-nitrocyclohexane **(d) 1,1,2,6,6-Pentamethylcyclooctane**

Problem A.11

Think

What is the longest continuous carbon chain or ring? How do you name an OR substituent as an alkoxy group? How do you add locator numbers to the main chain or ring to minimize the numbers?

Solve

The IUPAC names for molecules **(a)–(c)** are given below. The main chain or ring is numbered.

(a) The longest carbon chain has three C atoms. The substituent, CH_3CH_2O, has two C atoms, so it is an *ethoxy* group, and it is attached to C1 of the main chain.

(b) The longest carbon chain has two C atoms, whereas the largest ring has six, so the root is cyclohexane. An *ethoxy* group and a *nitro* group are substituents on the ring. Numbering could begin at the C atom attached to either substituent to give locator numbers of 1 and 3, but the *ethoxy* group has priority, because it is first alphabetically.

(c) The largest ring has five C atoms, making it a *cyclopentane.* The substituent is an *alkoxy* group with a cyclic ring of three C atoms attached to O, making it a *cyclopropoxy* group. Notice that no locator number is necessary in the name, because there is only one way to attach the substituent to the carbon ring.

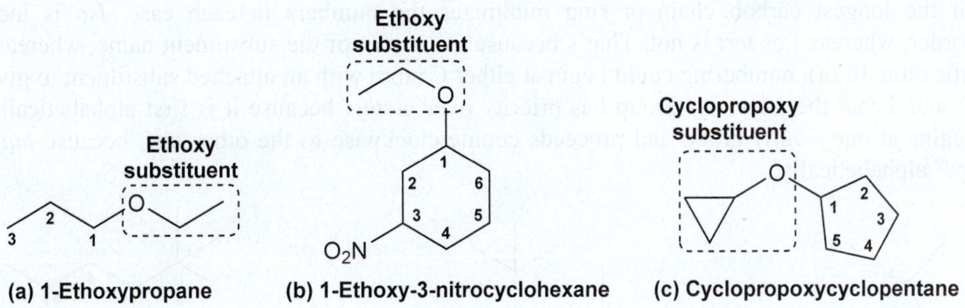

(a) 1-Ethoxypropane **(b) 1-Ethoxy-3-nitrocyclohexane** **(c) Cyclopropoxycyclopentane**

Problem A.12

Think

Based on the root, what is the longest continuous chain or ring of carbon atoms? Where does the numbering of the chain or ring begin? What substituents are present, and how many are there of each? Where are they located along the main chain or ring?

Solve

Structures for IUPAC names **(a)**–**(c)** are drawn below. The main chain or ring is numbered.

(a) Pentane has five C atoms in its chain. Diethoxy indicates two CH_3CH_2O groups, which are both attached to C3.

(b) Hexane has six C atoms in its chain. 1-Chloro indicates Cl attached to C1, and 5-methoxy indicates CH_3O attached to C5.

(c) Cyclopentane indicates a ring of five C atoms. Diethoxy indicates two CH_3CH_2O groups; one is attached to C1, and the other is attached to C2. A CH_3 group is also attached to C1.

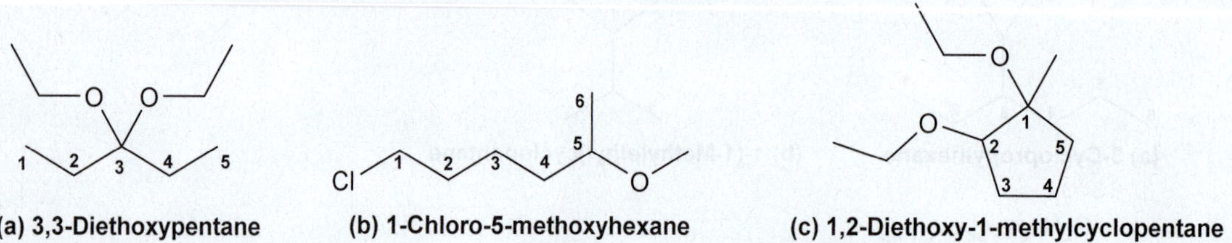

(a) 3,3-Diethoxypentane **(b) 1-Chloro-5-methoxyhexane** **(c) 1,2-Diethoxy-1-methylcyclopentane**

Problem A.13

Think

For each C atom in question, how many other C atoms are bonded directly to it?

Solve

Primary (1°) = one C atom; secondary (2°) = two C atoms; tertiary (3°) = three C atoms; and quaternary (4°) = four C atoms. Terminal (end) C atoms are always primary.

2° C
1° C
1° C
1° C
1° C
1° C

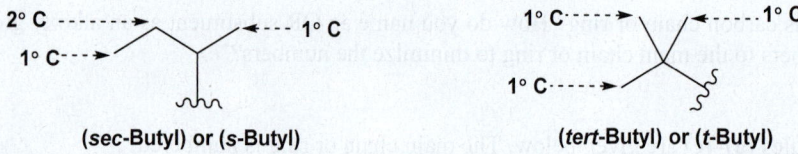

(*sec*-Butyl) or (*s*-Butyl) **(*tert*-Butyl) or (*t*-Butyl)**

Problem A.15

Think

Review the rules that you have learned thus far for naming alkanes and cycloalkanes. What is the longest continuous carbon chain or ring? What common substituents from Figure A-8 are attached to it? How is the longest chain or ring numbered? How do you put the pieces of the name together? Which parts of the substituents' names are considered when arranging them in alphabetical order, and which are not?

Solve

Trivial names for compounds **(a)**–**(c)** are given below. The main chain or ring is numbered. Notice how the numbering of the longest carbon chain or ring minimizes the numbers in each case. *Iso* is included in alphabetical order, whereas *t* or *tert* is not. That's because *iso* is part of the substituent name, whereas *tert* is a type of classification. In **(a)**, numbering could begin at either C atom with an attached substituent to give locator numbers of 1 and 2, but the *isopropyl* group has priority over *methyl* because it is first alphabetically. In **(c)**, numbering begins at one *t-butyl* group and proceeds counterclockwise to the other one, because *butyl* comes before "*methyl*" alphabetically.

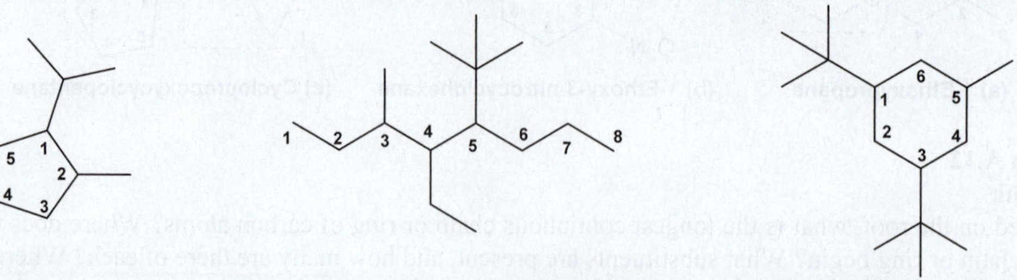

(a) 1-Isopropyl-2-methylcyclopentane **(b) 5-*t*-Butyl-4-ethyl-3-methyloctane** **(c) 1,3-Di-*t*-butyl-5-methylcyclohexane**

Problem A.16

Think

Use the trivial name to draw the structure for each compound. For the IUPAC name, how is the main chain or ring named? Are the halogen groups treated as the main chain or as a substituent? Where should numbering begin along the main chain or ring?

Solve

In the IUPAC rules, the halogens are treated as substituents. Note that **(a)** doesn't require a number locator, because there is only one way to number the ring.

(a) Iodocyclopentane (Cyclopentyl iodide)

(b) 1-Bromohexane (*n*-Hexyl bromide)

Problem A.17

Think

Use the trivial name to draw the structure for each compound. What alkyl groups are bonded to each side of the ether oxygen atom? For the IUPAC name, how is the main chain or ring named? How are *alkoxy* groups identified and named as substituents? Where should numbering begin along the main chain or ring?

Solve

Structures and IUPAC names for trivial names **(a)–(c)** are drawn below. The main chain or ring is numbered.

(a) A *t-butyl* and an *ethyl* group are attached to O. The longest chain has three C atoms, to which a *methyl* (CH_3) group and an *ethoxy* (OCH_2CH_3) group are attached at C2.

(b) A *cyclohexyl* group and a *methyl* group are attached to O. The six-carbon ring establishes the root as *cyclohexane*, to which a *methoxy* group is attached at C1. No locator number is required, because there is only one way to number the ring.

(c) A *sec-butyl* group and a *propyl* group are both attached to O. The longest carbon chain has four C atoms, to which a *propoxy* ($OCH_2CH_2CH_3$) group is attached at C2.

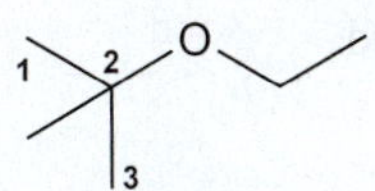

(a) 2-Ethoxy-2-methylpropane (*tert*-Butyl ethyl ether)

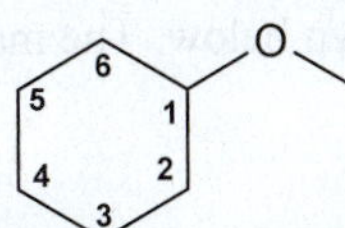

(b) Methoxycyclohexane (Cyclohexyl methyl ether)

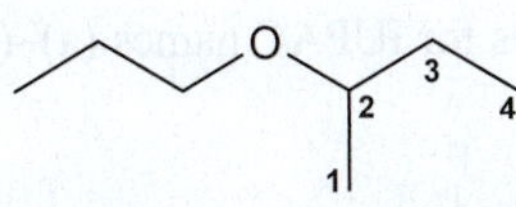

(c) 2-Propoxybutane (*sec*-Butyl propyl ether)

End of Chapter Problems

Section A.3 Haloalkanes and Nitroalkanes: Roots, Prefixes, and Locator Numbers

Problem A.18

Think

Review the rules that you have learned thus far for naming alkanes, haloalkanes, and nitroalkanes. What is the longest continuous carbon chain? What substituents are present, and how many are there of each? How is the longest chain numbered? How do you put the pieces of the name together?

Solve

Structures for IUPAC names **(a)–(d)** are drawn below. The main chain is numbered.

(a) 1,2,3-Tribromohexane

(b) 2,2,3,3,4-Pentachlorohexane

(c) 1,2-Dichloro-4-nitrohexane

(d) 1,2-Dichloro-3-nitropentane

Problem A.19

Think

Review the rules that you have learned thus far for naming alkanes, haloalkanes, and nitroalkanes. What is the longest continuous carbon chain? What substituents are present, and how many are there of each? How is the longest chain numbered? How do you put the pieces of the name together?

Solve

Structures for IUPAC names **(a)–(d)** are drawn below. The main chain is numbered.

(a) 2,2-Dichloro-3-fluorobutane

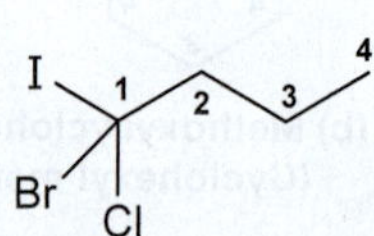

(b) 1-Bromo-1-chloro-1-iodobutane

(c) 2-Bromo-1,1-diiodohexane

(d) 3-Chloro-1,1,2,2-tetrafluoropentane

Problem A.20

Think

Review the rules that you have learned thus far for naming alkanes, haloalkanes, and nitroalkanes. What is the longest continuous carbon chain? What substituents are present, and how many are there of each? How is the longest chain numbered? How do you put the pieces of the name together?

Solve

Structures for IUPAC names **(a)–(d)** are drawn below. The main chain is numbered.

(a) 3-Bromo-2-nitropentane

(b) 2,2-Dichloro-4,4,5-trinitroheptane

(c) 1,2,3,4-Tetranitrobutane

(d) 6-Iodo-1,2-difluorohexane

Problem A.21

Think

How many carbon atoms are in the longest continuous chain of carbon atoms? How does that establish the root? Where should the numbering of the longest chain begin? What substituents are present, and how many are there of each? In what order should the substituents appear in the molecule's name?

Solve

The IUPAC names for molecules **(a)–(c)** are given below. The main chain is numbered.

(a) 2,3-Dibromobutane

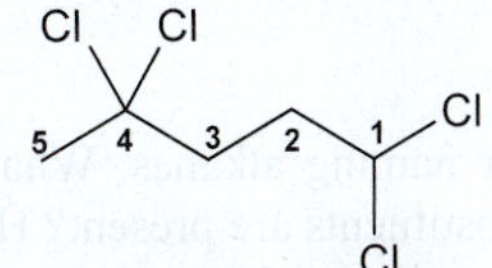

(b) 1,1,4,4-Tetrachloropentane

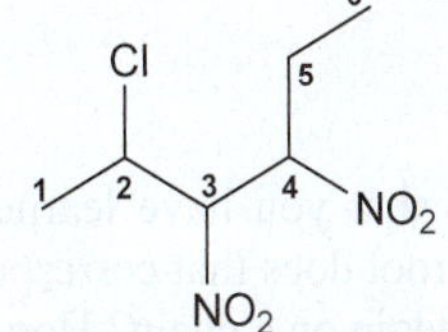

(c) 2-Chloro-3,4-dinitrohexane

Section A.4 Alkyl Substituents: Branched Alkanes and Substituted Branched Alkanes

Problem A.22

Think

How many carbon atoms are in the longest continuous chain? What substituents are attached? How many of each kind? Where does numbering begin for a chain? Add in the substituents at the designated locations.

Solve

Structures for IUPAC names **(a)–(d)** are drawn below. The main chain is numbered.

(a) 2-Methylhexane **(b) 3-Methylhexane** **(c) 2,3-Dimethylbutane** **(d) 2,2,3-Trimethylbutane**

Problem A.23

Think

How many carbon atoms are in the longest continuous chain? What substituents are attached? How many of each kind? Where does numbering begin for a chain? Add in the substituents at the designated locations.

Solve

Structures for IUPAC names **(a)–(c)** are drawn below. The main chain is numbered.

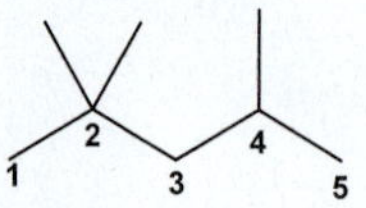

(a) 2,2,4-Trimethylpentane 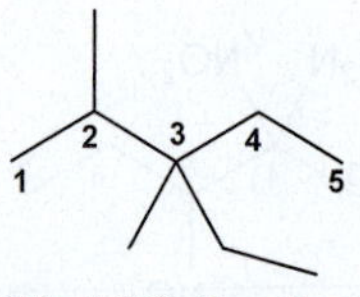 **(b) 3-Ethyl-2,3-dimethylpentane** 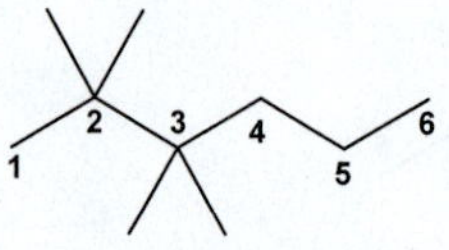 **(c) 2,2,3,3-Tetramethylhexane**

Problem A.24

Think

How many carbon atoms are in the longest continuous chain? What substituents are attached? How do you take into account branched alkyl groups? How many of each kind of substituent are there? Where does numbering begin for a chain? Add in the substituents at the designated locations.

Solve

Structures for IUPAC names **(a)** and **(b)** are drawn below. The main chain or ring is numbered. Parentheses are used to designate branched alkyl groups.

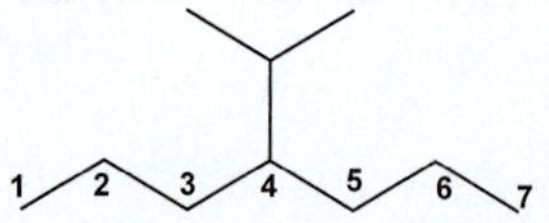

(a) 4-(1-Methylethyl)heptane 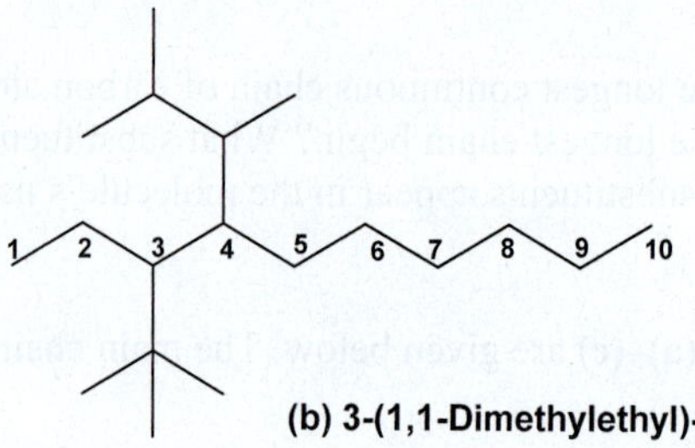 **(b) 3-(1,1-Dimethylethyl)-4-(1,2-dimethylpropyl)decane**

Problem A.25

Think

Review the rules that you have learned thus far for naming alkanes. What is the longest continuous carbon chain? To which root does that correspond? What substituents are present? How many of each are there? Where does numbering begin on a chain? How do you put the pieces of the name together?

Solve

The IUPAC names for molecules **(a)–(c)** are given below. The main chain is numbered.

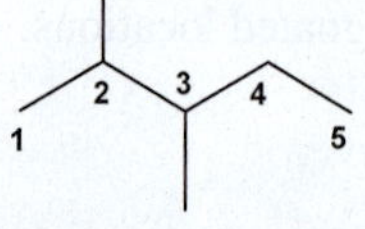

(a) 2,3-Dimethylpentane 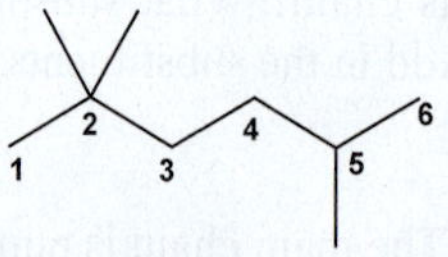 **(b) 2,2,5-Trimethylhexane** 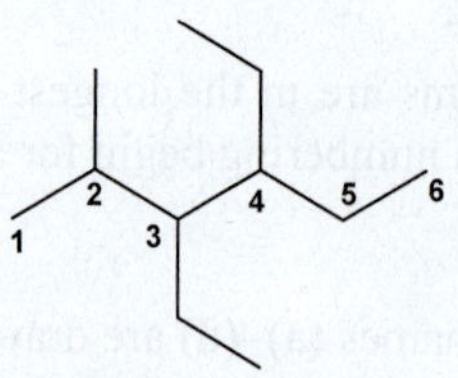 **(c) 3,4-Diethyl-2-methylhexane**

Problem A.26

Think

Review the rules that you have learned thus far for naming alkanes. What is the longest continuous carbon chain? To which root does that correspond? What substituents are present? How do you take into account branched alkyl groups? How many of each kind of substituent are there? Where does numbering begin on a chain? How do you put the pieces of the name together?

Solve

The IUPAC names for molecules **(a)–(d)** are given below. The main chain is numbered. Parentheses are used to designate branched alkyl groups.

(a) 2,3,4-Trimethylpentane **(b) 4-Ethyl-2-methyloctane** **(c) 2-Methyl-3-(1-methylethyl)heptane**

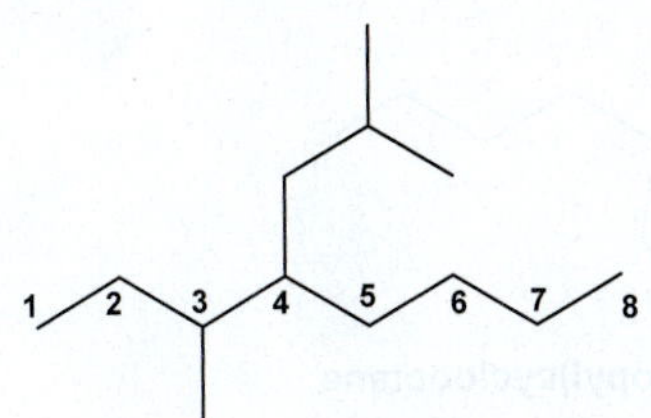

(d) 3-Methyl-4-(2-methylpropyl)octane

Section A.5 Cyclic Alkanes and Cyclic Alkyl Groups

Problem A.27

Think

How many carbon atoms are in the largest continuous ring? The longest chain? Which one establishes the root? What substituents are attached? How many of each kind? Where does numbering begin for a ring? Add in the substituents at the designated locations.

Solve

Structures for IUPAC names **(a)–(c)** are drawn below. The main ring is numbered.

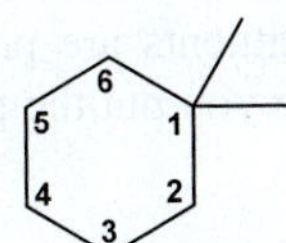

(a) 1,1-Dimethylcyclohexane

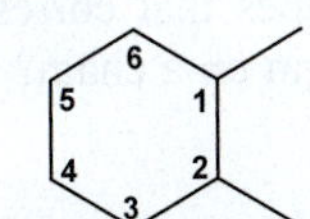

(b) 1,2-Dimethylcyclohexane

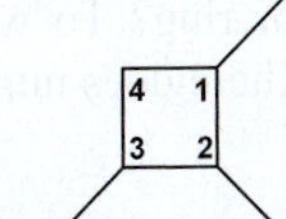

(c) 1,2,3-Trimethylcyclobutane

Problem A.28

Think

How many carbon atoms are in the largest continuous ring? The longest continuous chain? Which one establishes the root? What substituents are attached? How many of each kind? Where does numbering begin? Add in the substituents at the designated locations.

Solve

Structures for IUPAC names **(a)–(c)** are drawn below. The main chain or ring is numbered.

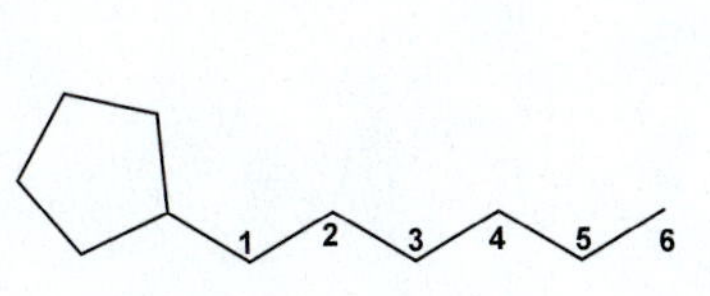

(a) 1-Cyclopentylhexane

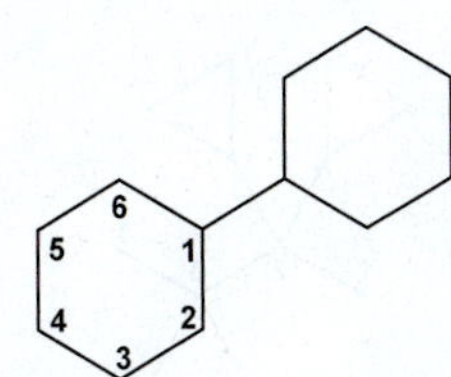

(b) Cyclohexylcyclohexane

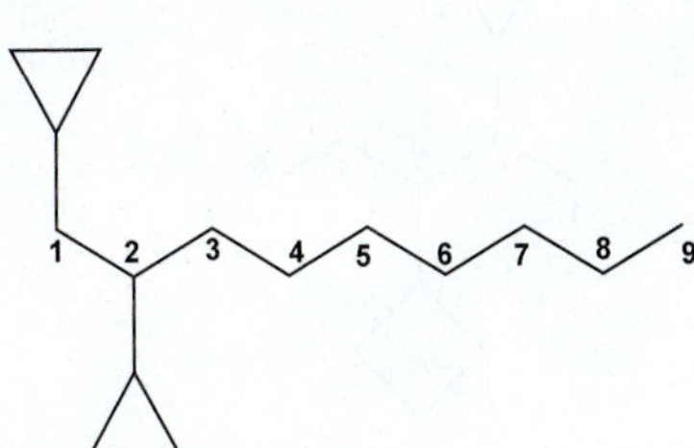

(c) 1,2-Dicyclopropylnonane

Problem A.29

Think

How many carbon atoms are in the longest continuous chain or ring? What substituents are attached? How do you take into account branched alkyl groups? How many of each kind of substituent are there? Where does numbering begin for a chain? For a ring? Add in the substituents at the designated locations.

Solve

Structures for IUPAC names **(a)–(c)** are drawn below. The main chain or ring is numbered. Parentheses are used to designate branched alkyl groups.

(a) 1-(1,1-Dimethylethyl)-2,4-diethylcyclohexane **(b) 1,4-Dibutyl-2-(1-methylpropyl)cyclooctane**

(c) 1,1-Dicyclopropyl-3-(1,1-dimethylethyl)cycloheptane

Problem A.30

Think

Review the rules that you have learned thus far for naming alkanes and cycloalkanes. What is the longest continuous carbon chain or ring? To which root does that correspond? What substituents are present? How many of each are there? Where does numbering begin on a chain? On a ring? How do you put the pieces of the name together?

Solve

The IUPAC names for molecules **(a)–(e)** are given below. The main chain or ring is numbered.

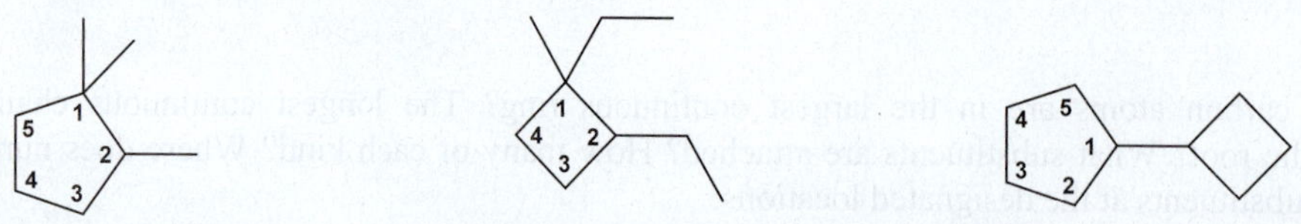

(a) 1,1,2-Trimethylcyclopentane **(b) 1,2-Diethyl-1-methylcyclobutane** **(c) Cyclobutylcyclopentane**

(d) 3-Cyclobutyl-2-cyclopropylpentane **(e) 2,2-Dicyclopropyl-3,3-dimethylbutane**

Problem A.31

Think

Review the rules that you have learned thus far for naming alkanes and cycloalkanes. What is the longest continuous carbon chain or ring? To which root does that correspond? What substituents are present? How do you take into account branched alkyl groups? How many of each kind of substituent are there? Where does numbering begin on a chain? On a ring? How do you put the pieces of the name together?

Solve

The IUPAC names for molecules **(a)–(c)** are given below. The main chain or ring is numbered. Parentheses are used to designate branched alkyl groups. *Note for* ***(b)***: Bis, rather than *di*, is the numeral term to indicate two of the same branched alkyl groups.

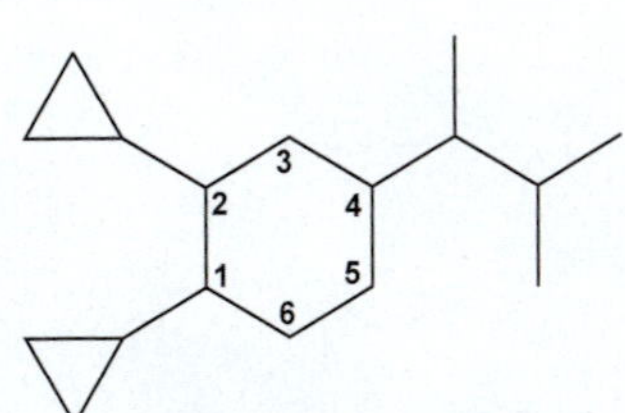

(a) 1,2-Dicyclopropyl-4-(1,2-dimethylpropyl)cyclohexane

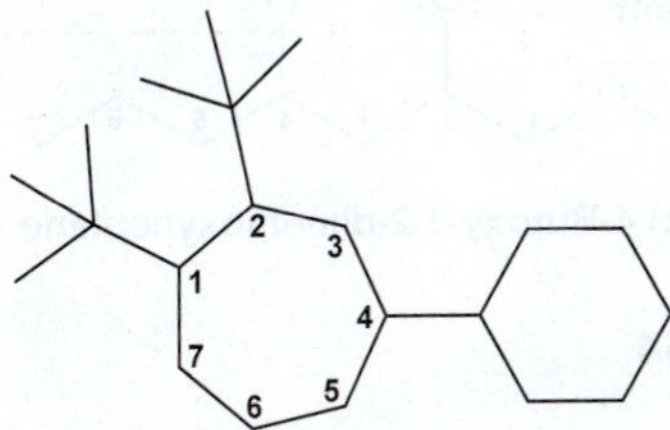

(b) 4-Cyclohexyl-1,2-bis(1,1-dimethylethyl)cycloheptane

(c) 2-Cyclobutyl-4-(2-methylpropyl)octane

Section A.6 Ethers and Alkoxy Groups

Problem A.32

Think

How many carbon atoms are in the longest continuous chain? How do you draw an alkoxy substituent? How many carbon atoms are in each alkoxy substituent?

Solve

Structures for IUPAC names **(a)–(c)** are drawn below. The main chain is numbered.

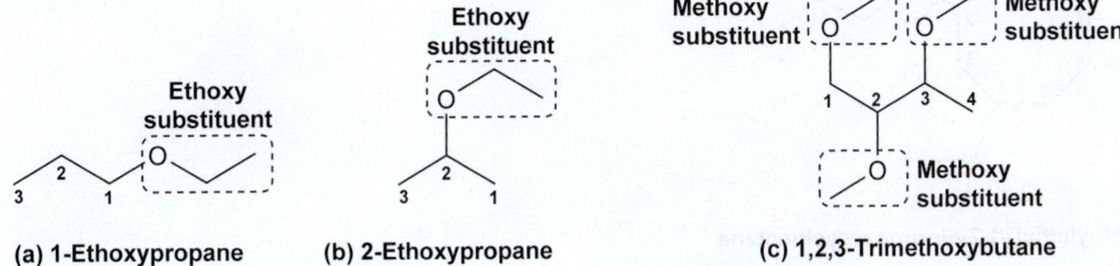

(a) 1-Ethoxypropane **(b) 2-Ethoxypropane** **(c) 1,2,3-Trimethoxybutane**

Problem A.33

Think

How many carbon atoms are in the longest continuous chain? How do you draw an alkoxy substituent? How many carbon atoms are in each alkoxy substituent?

Solve

Structures for IUPAC names **(a)–(c)** are drawn below. The main chain is numbered.

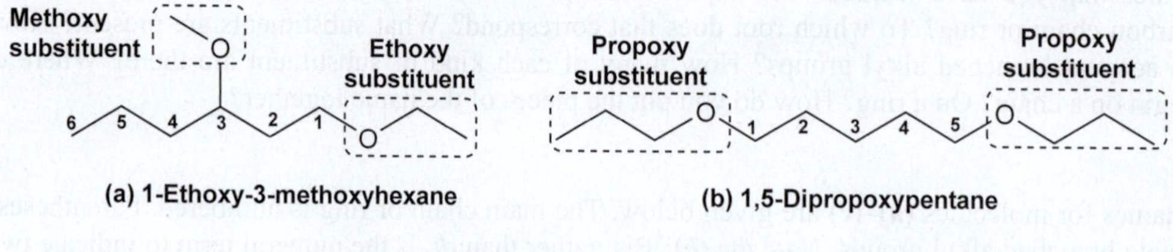

(a) 1-Ethoxy-3-methoxyhexane **(b) 1,5-Dipropoxypentane**

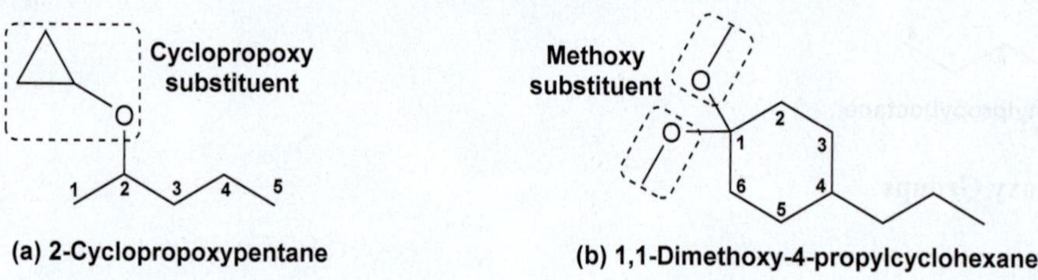

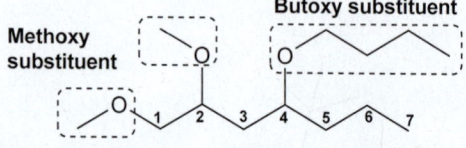

(c) 4-Butoxy-1,2-dimethoxyheptane

Problem A.34

Think

How many carbon atoms are in the longest continuous chain or ring? How do you draw an alkoxy substituent? How many carbon atoms are in each alkoxy substituent?

Solve

Structures for IUPAC names **(a)–(c)** are drawn below. The main chain or ring is numbered.

Cyclopropoxy substituent

(a) 2-Cyclopropoxypentane

Methoxy substituent

(b) 1,1-Dimethoxy-4-propylcyclohexane

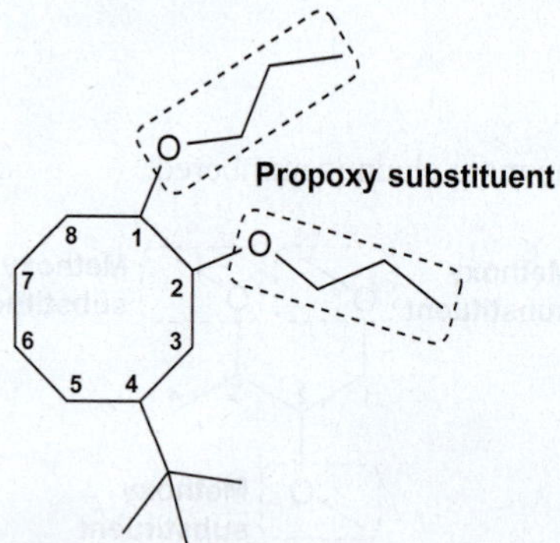

(c) 4-(1,1-Dimethylethyl)-1,2-dipropoxycyclooctane

Problem A.35

Think

How many carbon atoms are in the longest continuous chain? How do you name an OR substituent as an alkoxy group? How do you add number locators to the main chain to minimize the numbers?

Solve

The IUPAC names for molecules **(a)–(d)** are given below. The main chain is numbered.

(a) 1-Methoxybutane **(b) 1-Ethoxypropane** **(c) 1-Propoxypropane** **(d) 1-Methoxypentane**

Problem A.36

Think

How many carbon atoms are in the longest continuous chain? How do you name an OR substituent as an alkoxy group? How do you add number locators to the main chain or ring to minimize the numbers?

Solve

The IUPAC names for molecules **(a)–(d)** are given below. The main chain is numbered.

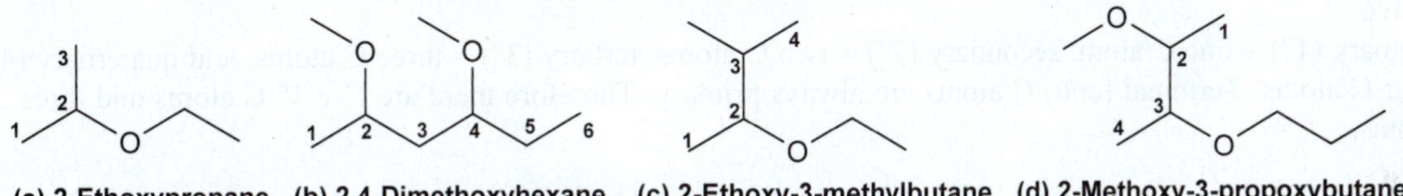

(a) 2-Ethoxypropane (b) 2,4-Dimethoxyhexane (c) 2-Ethoxy-3-methylbutane (d) 2-Methoxy-3-propoxybutane

Section A.7 Trivial Names or Common Names

Problem A.37

Think

Review the rules that you have learned thus far for naming alkanes, cycloalkanes, ethers, haloalkanes, and nitroalkanes. Are any common alkyl groups present? Do you use trivial names for these groups? What is the longest continuous carbon chain or ring? What other substituents are present, and how many are there of each? How is the longest chain or ring numbered? How do you put the pieces of the name together?

Solve

Structures for trivial names **(a)–(c)** are drawn below. The main chain or ring is numbered.

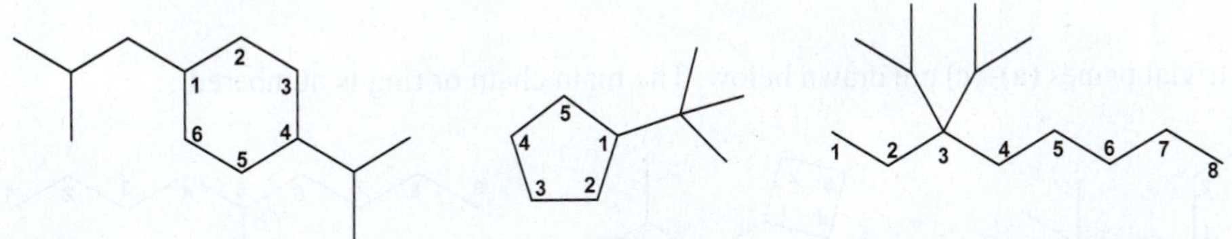

(a) 1-Isobutyl-4-isopropylcyclohexane (b) *tert*-Butylcyclopentane (c) 3,3-Diisopropyloctane

Problem A.38

Think

Draw the compound given the name listed. Using the name given, what carbon atom must have been numbered C1? Looking at the molecule, is there another way to number the main chain to give the carbon atoms with substituents lower numbers?

Solve

If the longest carbon chain (still six C atoms) is numbered as shown below, the lowest substituent now is on C2. The correct name is 3-ethyl-2,2-dimethylhexane.

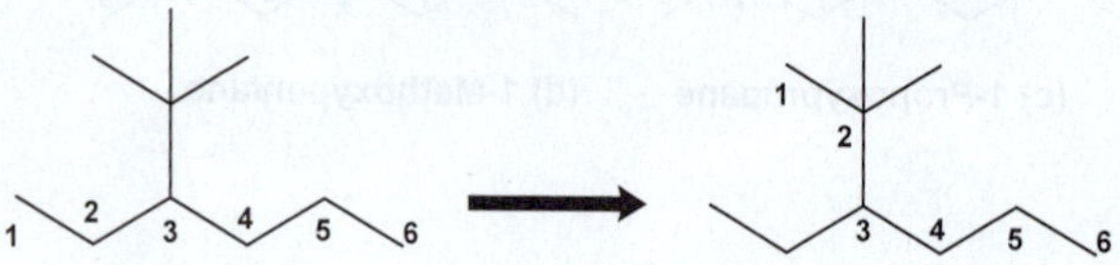

3-*tert*-Butylhexane **3-Ethyl-2,2-dimethylhexane**

Problem A.39

Think

For the structure of 3-*tert*-butylhexane, see Problem A.38. For each C atom in question, how many other C atoms are bonded directly to it?

Solve

Primary (1°) = one C atom, secondary (2°) = two C atoms, tertiary (3°) = three C atoms, and quaternary (4°) = four C atoms. Terminal (end) C atoms are always primary. Therefore there are five 1° C atoms and three 2° C atoms.

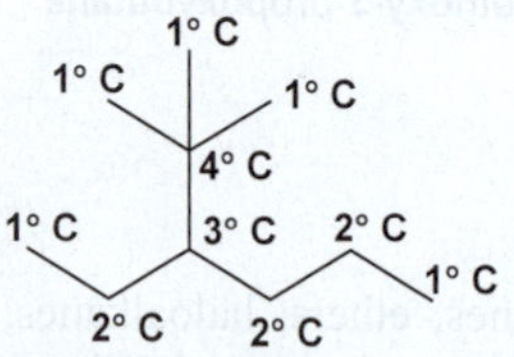

Problem A.40

Think

How many carbon atoms are in the longest continuous chain or ring? Based on the root name, what is the structure of the main chain or ring? On what carbon atom does numbering begin? What substituents are present? Do any of them have trivial names? Add in the substituents.

Solve

Structures for trivial names **(a)**–**(c)** are drawn below. The main chain or ring is numbered.

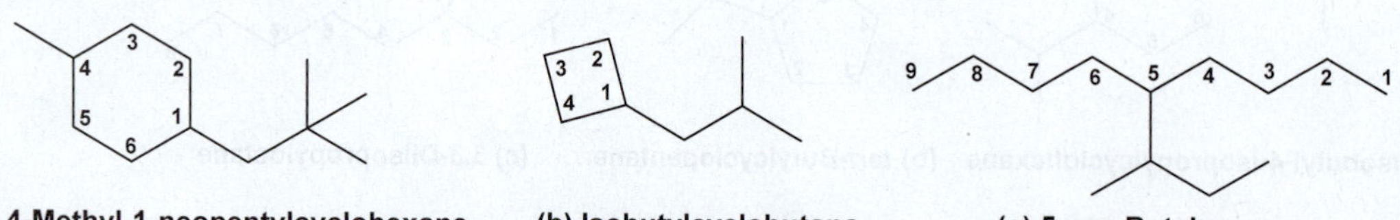

(a) 4-Methyl-1-neopentylcyclohexane **(b) Isobutylcyclobutane** **(c) 5-*sec*-Butylnonane**

Problem A.41

Think

Review the rules that you have learned thus far for naming alkanes and cycloalkanes. What is the longest continuous carbon chain or ring? What common substituents are attached to it? How is the longest chain or ring numbered? How do you put the pieces of the name together? Which parts of the substituents' names are considered when arranging them in alphabetical order, and which are not?

Solve

The IUPAC names for molecules **(a)–(c)** are given below. The main chain or ring is numbered.

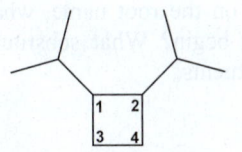

(a) 1,2-Diisopropylcyclobutane

(b) 4-*tert*-Butylheptane

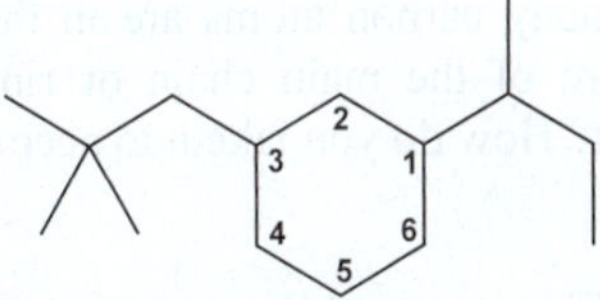

(c) 1-*sec*-Butyl-3-neopentylcyclohexane

Problem A.42

Think

Review the rules that you have learned thus far for naming alkanes, cycloalkanes, ethers, haloalkanes, and nitroalkanes. Are any common alkyl groups present? Do you use trivial names for these groups? What is the longest continuous carbon chain or ring? What substituents are present, and how many are there of each? How is the longest chain or ring numbered? How do you put the pieces of the name together?

Solve

Structures and IUPAC names for trivial names **(a)–(e)** are drawn below. The main chain or ring is numbered.

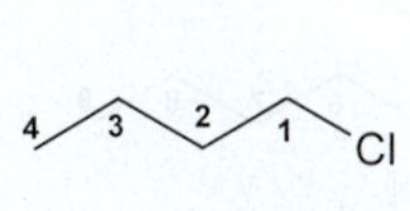

(a) 1-Chlorobutane
(*n*-Butyl chloride)

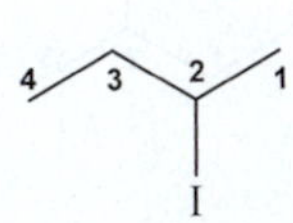

(b) 2-Iodobutane
(*sec*-Butyl iodide)

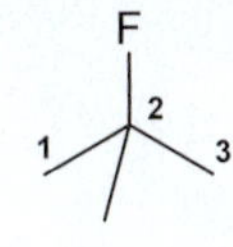

(c) 2-Fluoro-2-methylpropane
(*tert*-Butyl fluoride)

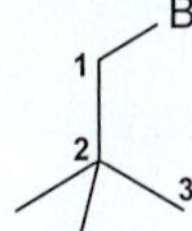

(d) 1-Bromo-2,2-dimethylpropane
(Neopentyl bromide)

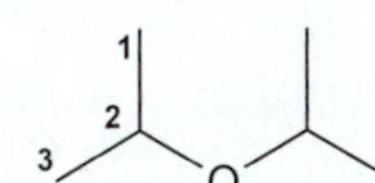

(e) 2-Isopropoxypropane
or 2-(1-Methylethoxy)propane
(Diisopropyl ether)

Problem A.43

Think

Draw the structure using the IUPAC name. Based on the root, how many carbon atoms are in the main chain or ring? What substituents are present, and how many are there of each? How should the main chain or ring be numbered? What functional group is present? What is the name of the alkyl group on each side of the ether oxygen atom? How do you put the pieces of the name together?

Solve

Structures and IUPAC names for trivial names **(a)** and **(b)** are drawn below. The main chain or ring is numbered.

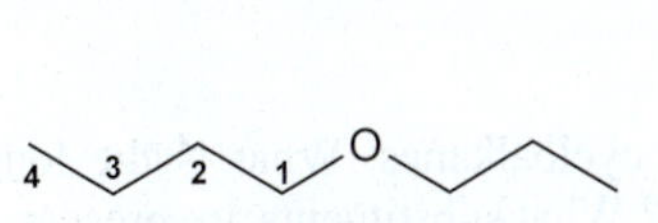

(a) 1-Propoxybutane
(Butyl propyl ether)

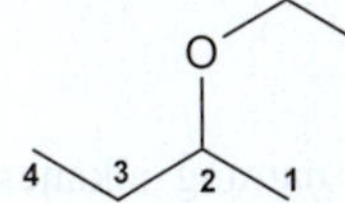

(b) 2-Ethoxybutane
(*sec*-Butyl ethyl ether)

Integrated Problems

Problem A.44

Think

How many carbon atoms are in the longest continuous chain or ring? Based on the root name, what is the structure of the main chain or ring? On what carbon atom does numbering begin? What substituents are present? How do you take into account branched alkyl groups? Add in the substituents.

Solve

Structures for IUPAC names **(a)–(h)** are drawn below. The main chain or ring is numbered.

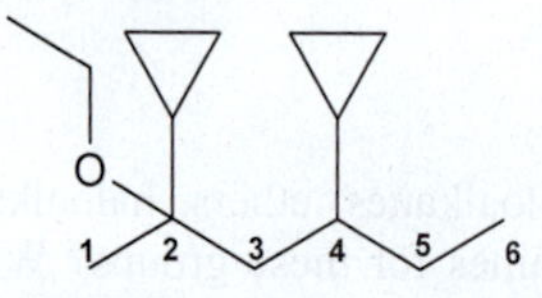

(a) 2,4-Dicyclopropyl-2-ethoxyhexane 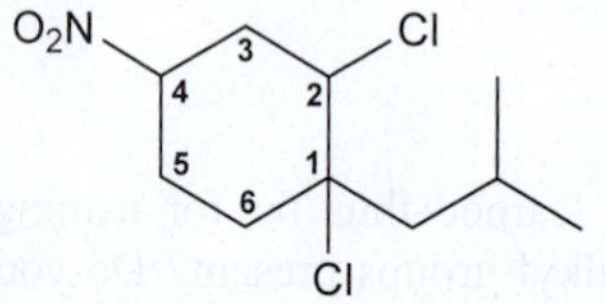 **(b) 1,2-Dichloro-1-(2-methylpropyl)-4-nitrocyclohexane**

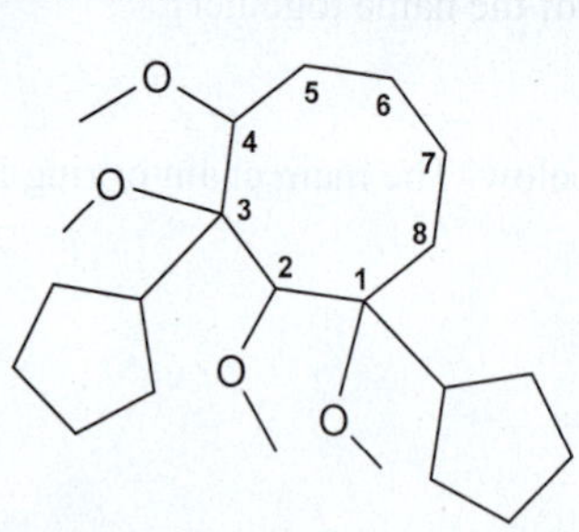

(c) 1,3-Dicyclopentyl-1,2,3,4-tetramethoxycyclooctane 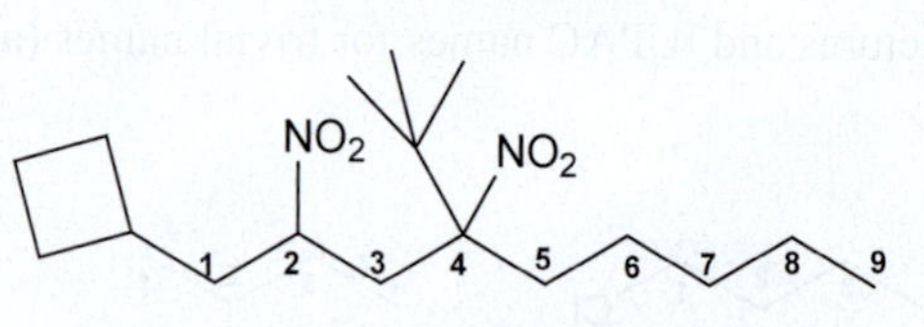 **(d) 1-Cyclobutyl-4-(1,1-dimethylethyl)-2,4-dinitrononane**

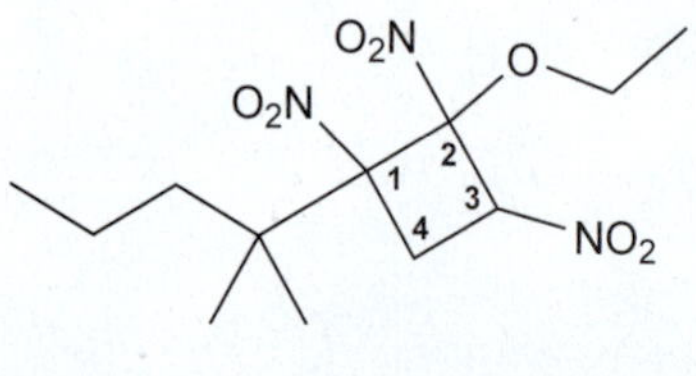

(e) 1-(1,1-Dimethylbutyl)-2-ethoxy-1,2,3-trinitrocyclobutane **(f) 1,2,4-Tricyclopropyl-1-(2,2-dichloropentyl)cyclohexane**

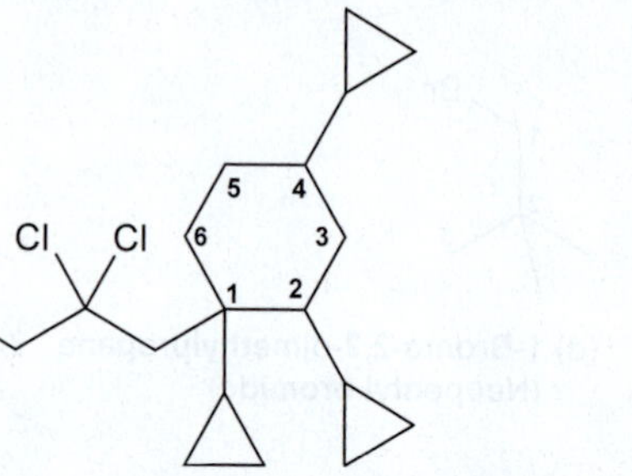

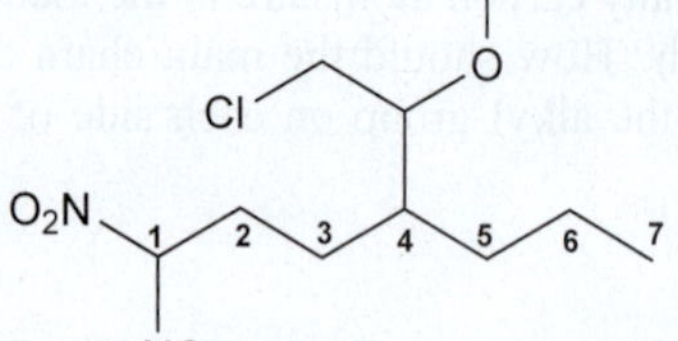

(g) 4-(2-Chloro-1-methoxyethyl)-1,1-dinitroheptane 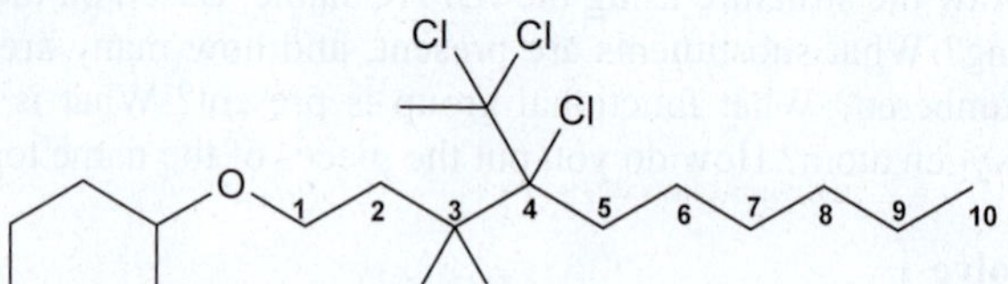 **(h) 3,3,4-Trichloro-1-cyclohexoxy-4-(1,1-dichloroethyl)decane**

Problem A.45

Think

Review the rules that you have learned thus far for naming alkanes and cycloalkanes. What is the longest continuous carbon chain or ring? How is the longest chain or ring numbered? What substituents are present, and how many are there of each? How do you take into account branched alkyl groups? How do you put the pieces of the name together? Which parts of the substituents' names are considered when arranging them in alphabetical order, and which are not?

Solve

The IUPAC names for molecules **(a)–(h)** are given below. The main chain or ring is numbered.

(a) 1-(1-Bromoethyl)-4-(1,1-dimethylethyl)cyclohexane

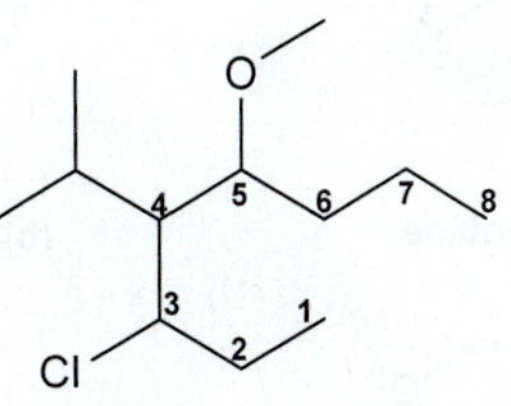

(b) 3-Chloro-5-methoxy-4-(1-methylethyl)octane

(c) Cyclopropoxycyclopentane

(d) 1-Chloro-2,4-diethoxy-1-(1-methylethyl)cycloheptane

(e) 4-(1,1-Dinitroethyl)-5-(methoxymethyl)octane

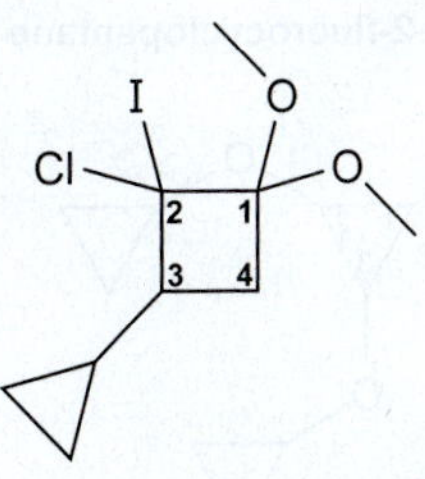

(f) 2-Chloro-3-cyclopropyl-2-iodo-1,1-dimethoxycyclobutane

(g) 1-methoxy-3-(1-methoxyethyl)-5-(3-methylbutyl)cyclohexane

(h) 3,3-Dibromo-5-cyclobutyl-4-(1-methoxyethyl)heptane

Problem A.46

Think

How many carbon atoms are in the longest continuous chain or ring of carbon atoms? How does that establish the root? Where should the numbering of the longest chain or ring begin? What substituents are present, and how many are there of each? In what order should the substituents appear in the molecule's name?

Solve

The IUPAC names for molecules **(a)–(j)** are given below. The main chain or ring is numbered.

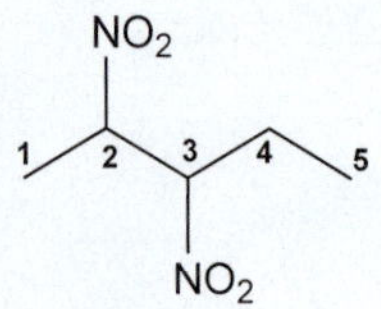

(a) 2,3-Dinitropentane

(b) 2,2-Diiodo-5-nitrohexane

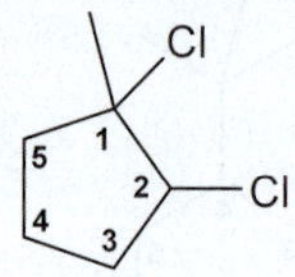

(c) 1,2-Dichloro-1-methylcyclopentane

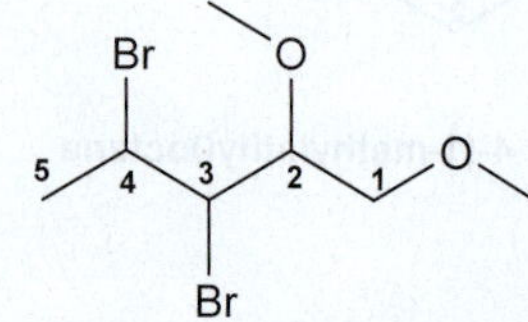

(d) 3,4-Dibromo-1,2-dimethoxypentane

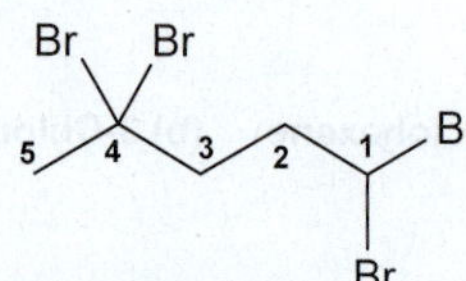

(e) 1,1,4,4-Tetrabromopentane

(f) 1,1,2-Trinitrocyclobutane

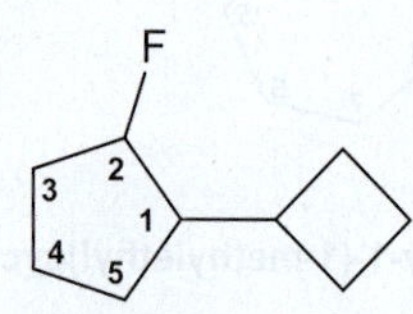

(g) 1-Cyclobutyl-2-fluorocyclopentane

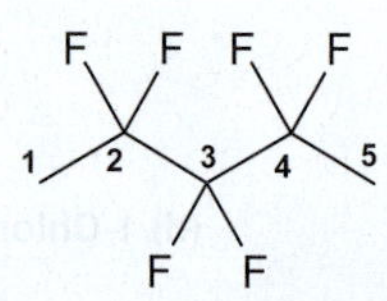

(h) 2,2,3,3,4,4-Hexafluoropentane

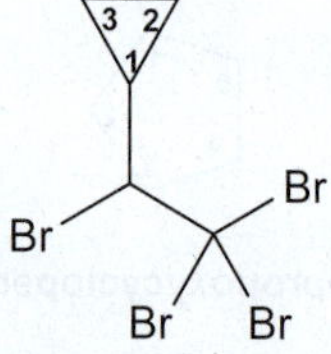

(i) (1,2,2,2-Tetrabromoethyl)cyclopropane

(j) 1,2,3-Tricyclopropoxycyclopropane

CHAPTER 2 | Three-Dimensional Geometry, Intermolecular Interactions, and Physical Properties

Your Turn Exercises

Your Turn 2.1

Think

In a Lewis structure, what constitutes a group of electrons? Consult Table 2-1.

Solve

In a Lewis structure, a group of electrons is a lone pair of electrons, a single bond, a double bond, or a triple bond.

(a) The triply bonded C in Figure 2-1a has one single bond and one triple bond, thus two groups of electrons.

(b) The central C in Figure 2-1b has two single bonds and one double bond, thus three groups of electrons.

(c) Each C atom in Figure 2-1c has four single bonds, thus four groups of electrons. *Note:* Only one C atom is circled, for clarity.

Your Turn 2.2

Think

In a Lewis structure, what constitutes a group of electrons? Consult Table 2-1.

Solve

In a Lewis structure, a group of electrons is a lone pair of electrons, a single bond, a double bond, or a triple bond. CH_3^+ has three single bonds and no lone pairs, thus three groups of electrons. CH_3^- has three single bonds and one lone pair, thus four groups of electrons.

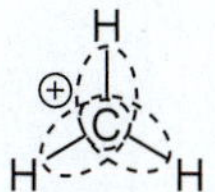

Your Turn 2.3

Think

Which N–H bonds are in the plane of the paper? Which N–H bonds come out of the plane of the paper? Which N–H bonds point away from you? How are each of these kinds of bonds represented in dash–wedge notation?

Solve

N–H bonds that are in the plane of the paper are represented by a flat line (—), N–H bonds that come out of the plane of the paper are represented by a wedge (▬), and N–H bonds that point away from you are represented by a dash (''''''''). The structure on the left has two bonds in the plane of the paper, one out of the plane, and one pointed away. The structure on the right has two horizontal bonds that are pointed away and two vertical bonds that are out of the plane of the paper.

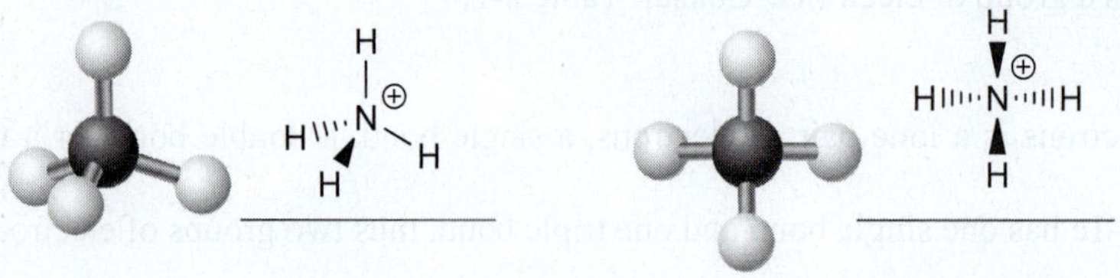

Your Turn 2.4

Think

Are the wedge and dash bonds or the flat lines in the plane of the paper? Which bonds are perpendicular to the plane of the paper? In which direction does the V open?

Solve

The flat lines are in the plane of the paper, and the V opens pointing to the top right corner (in this figure). The wedge and dash bonds are perpendicular to the plane of the paper, and the V opens to the bottom left (in this figure). See below.

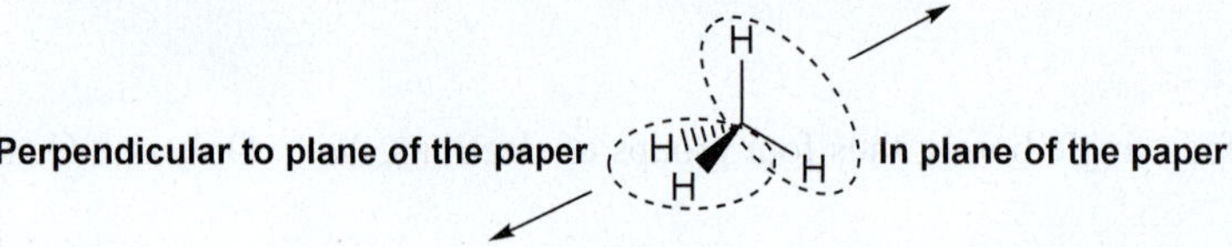

Your Turn 2.5

Think

In this structure, which bonds (C–C, C–H, or C–O) are in the plane of the paper? Which bonds come out of the plane of the paper? Which bonds point away? How are each of these kinds of bonds represented in dash–wedge notation?

Solve

By convention, the line structure drawings represent the carbon–carbon chain as a planar zigzag line. The four bonds in a tetrahedron can be thought of as two perpendicular Vs (Fig. 2-6). The carbon chain in more complex molecules is represented by alternating, flat Vs (i.e., ΛVΛV). At the point of each V is where the perpendicular V begins. In line structures, the C–H bond is not drawn. In butan-2-ol, the C–O bond points away from you, and this is the only bond where it is necessary to draw the dash–wedge notation.

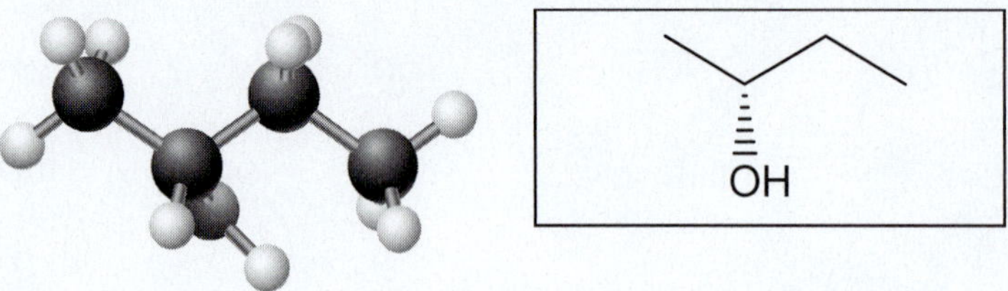

Your Turn 2.6

Think

Construct the molecular model and carry out the rotations indicated. What effect does flipping the molecule over 180° have on a bond in the plane of the paper? A bond pointed toward you? A bond pointed away from you? In the cyclopentane ring shown, where do bonds on the right point after a 180° flip?

Solve

A 180° turn does not have any effect on bonds in the plane of the paper (flat lines in both) and inverts wedges to dashes and dashes to wedges. A 180° turn about the vertical axis flips bonds on the right to bonds on the left, whereas a 180° turn about the horizontal axis flips bonds from top to bottom.

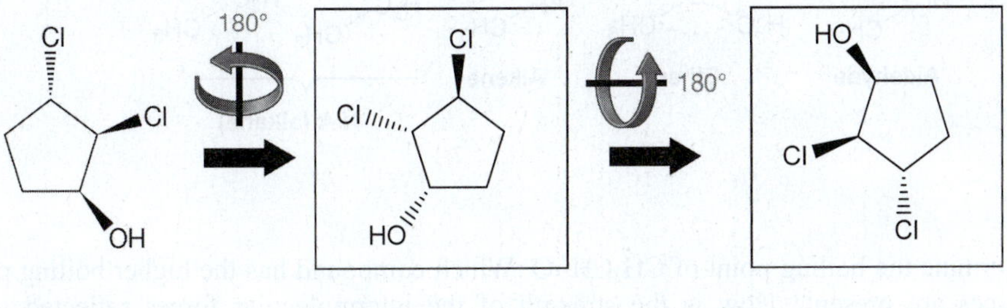

Your Turn 2.7

Think

Consult Figure 1-16 to determine the electronegativity value for beryllium and hydrogen. How is a dipole arrow drawn? To which atom, more or less electronegative, does the arrow point?

Solve

The electronegativity of Be is 1.57, and the electronegativity of H is 2.20 (see Fig. 1-16). H is more electronegative than Be, and the dipole arrow (+→) points to the more electronegative atom (each H atom in this example). BeH_2 is a linear nonpolar molecule.

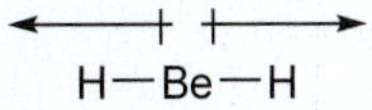

Your Turn 2.8

Think

Consult Figure 1-16 to determine the electronegativity value for carbon and fluorine. How is a dipole arrow drawn? To which atom, more or less electronegative, does the arrow point? Recall that dipole moments are vectors. In which direction does the net dipole point for each V (two C–F bonds)?

Solve

The electronegativity of C is 2.55, and the electronegativity of F is 3.98 (see Fig. 1-16). F is more electronegative than C, and the dipole arrow (+→) points to the more electronegative atom (each F in this example). The individual C–F bond dipoles can be added to yield a net dipole for each V that points from the C toward the center of the F atoms. The thin arrows indicate the bond dipoles. The thick arrows indicate the vector sum of each pair of bond dipoles. Notice that the thick arrows are equal in magnitude and point in opposite directions. The molecule has no net dipole moment and is, therefore, nonpolar.

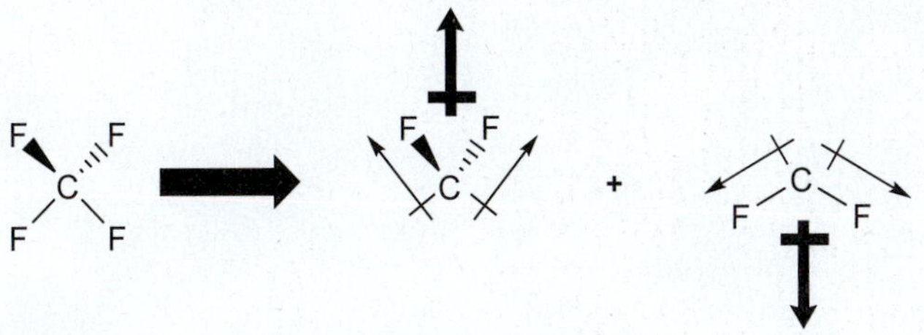

Your Turn 2.9

Think

Consult Table 1-6 for a list of functional group structures and compound class names. Look for multiple bonds and bonds other than C–C and C–H.

Solve

Functional groups are circled, and the compound classes are provided below.

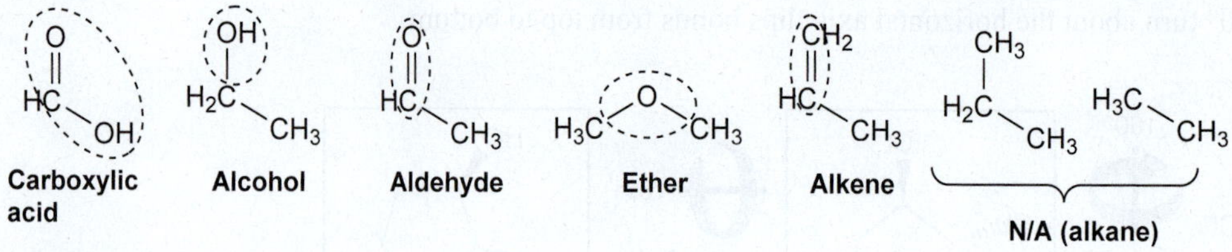

Carboxylic acid Alcohol Aldehyde Ether Alkene N/A (alkane)

Your Turn 2.10

Think

Consult Table 2-4 to determine the boiling point of $CH_3CH{=}O$. Which compound has the higher boiling point? What intermolecular forces are present? How is the strength of the intermolecular forces reflected in the physical property of boiling point?

Solve

Both CH_3CH_2F and $CH_3CH{=}O$ have permanent dipoles and, thus, have dipole–dipole interactions as the dominant intermolecular interaction. $CH_3CH{=}O$ has a higher boiling point than CH_3CH_2F (−20 °C vs. −37.1 °C). The stronger the attraction one molecule has to another molecule of itself, the higher the boiling point will be. $CH_3CH{=}O$, therefore, must have stronger dipole–dipole interactions and a greater dipole moment.

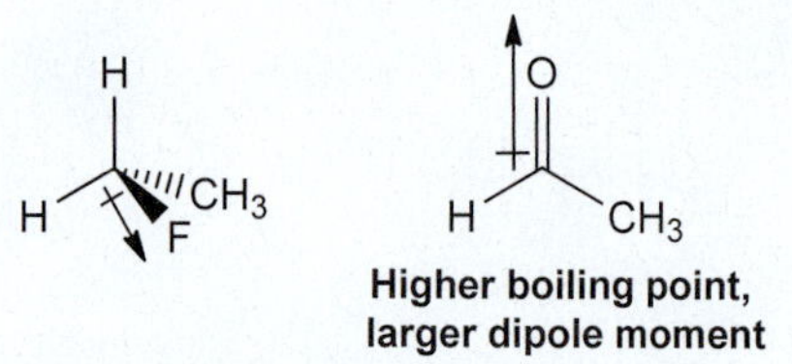

Higher boiling point, larger dipole moment

Your Turn 2.11

Think

Which atoms must be covalently bonded to a hydrogen atom for it to be considered a H-bond donor? Which atoms are considered H-bond acceptors? In a given H bond, how many donors and acceptors are there?

Solve

Any H bond consists of one H-bond donor and one H-bond acceptor. In the H bonds below, the donor is from the O–H covalent bond, and the acceptor is the O atom from the other molecule.

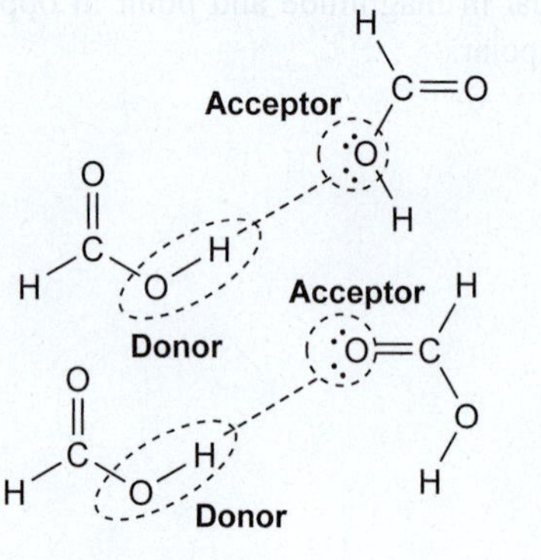

Acceptor

Donor

Acceptor

Donor

Your Turn 2.12

Think

Which atoms must be covalently bonded to a hydrogen atom for it to be considered a H-bond donor? Which atoms are considered H-bond acceptors?

Solve

All O atoms in both compounds are potential H-bond acceptors. The potential H-bond donors are the O–H hydrogen atoms.

Potential H-bond acceptor
Potential H-bond donor
Cholesterol
Potential H-bond acceptor
Potential H-bond donor
Octanoic acid (caprylic acid)

Your Turn 2.13

Think

Consult Table 2-5 for a list of the total number of electrons and the boiling points for the various straight-chain alkanes.

Solve

The graph below is a plot of boiling points against the total number of electrons. Positions of points in the graph are approximate, but a clear trend is indicated (a greater number of total electrons leads to an increase in boiling point for these nonbranching hydrocarbons). As the total number of electrons increases, so does the strength of the London dispersion forces.

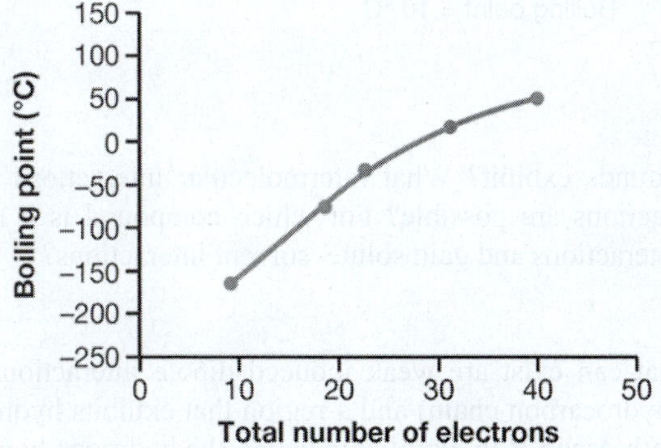

Your Turn 2.14

Think

Where do the molecules contact each other? How does the shape of the molecule affect the amount of surface area in contact with another molecule?

Solve

The molecules contact each other on the outside of the molecule. Pentane (left) has a greater contact surface area because of its long, linear shape (no branching). This leads to an increased boiling point.

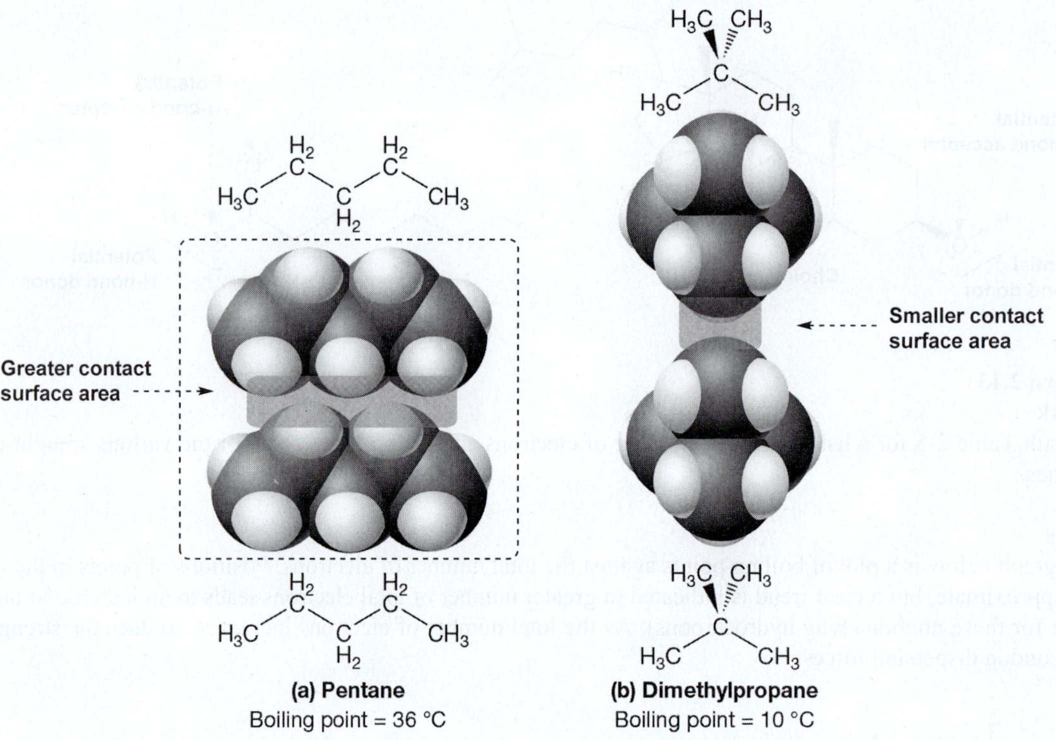

(a) Pentane
Boiling point = 36 °C

(b) Dimethylpropane
Boiling point = 10 °C

Your Turn 2.15

Think

What intermolecular interactions do each of the compounds exhibit? What intermolecular interactions does water exhibit? Upon mixing, what solute–solvent interactions are possible? For which compound is it more favorable to lose the solute–solute and solvent–solvent interactions and gain solute–solvent interactions?

Solve

Compound **A** is nonpolar, and the only interactions that can exist are weak induced dipole interactions, or dispersion forces. Compound **B** has a nonpolar region (hydrocarbon chain) and a region that exhibits hydrogen bonding (NH_2). Water exhibits hydrogen bonding. For both **A** and **B** to dissolve in water, the hydrogen bonding among water molecules must be disrupted. **A** will undergo dispersion forces with water, whereas **B** will undergo hydrogen bonding with water. Because **B** has the stronger interactions with water, **B** is more soluble in water than **A**.

NH_2

A

B
More soluble in H_2O

Your Turn 2.16

Think

What are the requirements for an ion–dipole interaction? Locate the partial charge on each polar molecule that will be attracted to the ion.

Solve

An ion–dipole interaction occurs between a molecule with a permanent dipole and an ion. In **(a)**, HCO_2^- is the anion and interacts with the δ^+ near H in water, and in **(b)**, Na^+ is the cation and interacts with the δ^- near O in water. The ion–dipole interactions are labeled below: **(a)** has six ion–dipole interactions, and **(b)** has six ion–dipole interactions.

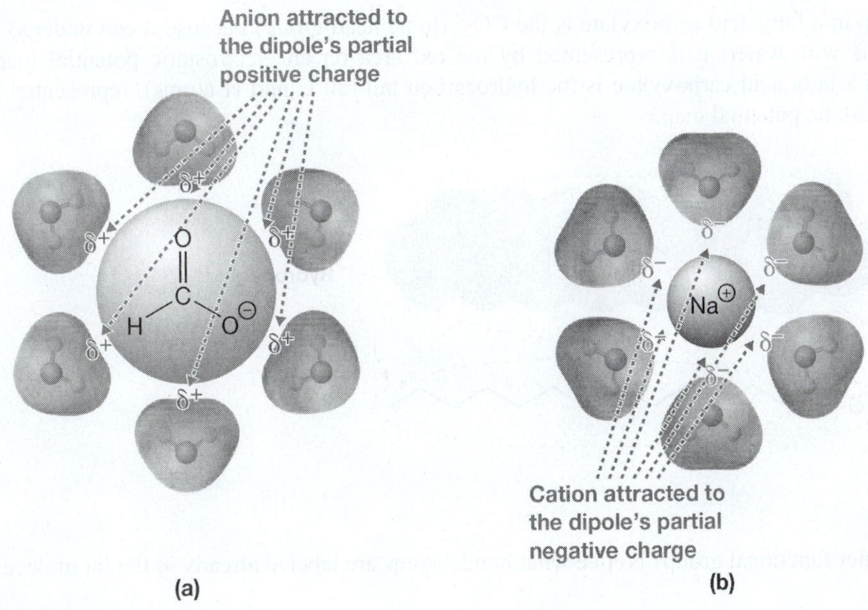

Your Turn 2.17

Think

For an H–D bond to be a hydrogen-bond donor, what must D be? Which end of the H–D covalent bond is responsible for the strong attraction to a hydrogen-bond acceptor: the partial positive or the partial negative?

Solve

A hydrogen-bond donor can be a H–O, H–N, or H–F bond. The partial positive end, namely, the H, is responsible for the strong attraction to the partial negative charge of a hydrogen-bond acceptor.

	POLAR PROTIC SOLVENTS		POLAR APROTIC SOLVENTS		
	Structure	Name	Structure	Name	
H-bond donors	H–O–H (δ^+ on H)	Water	H_3C–S(=O)–CH_3 (δ^+ on S)	Dimethyl sulfoxide (DMSO)	No H-bond donors
H-bond donor	H_3C–CH_2–O–H (δ^+ on H)	Ethanol	H_3C–C(=O)–CH_3 (δ^+ on C)	Propanone (Acetone)	
H-bond donor	H_3C–C(=O)–O–H (δ^+ on H)	Ethanoic acid (Acetic acid)	H–C(=O)–N(CH_3)–CH_3 (δ^+ on C)	*N,N*-Dimethyl-formamide (DMF)	

Your Turn 2.18

Think

What group in a fatty acid carboxylate is hydrophilic? What group in a fatty acid carboxylate is hydrophobic? From the electrostatic potential map, is the red region electron rich or electron poor?

Solve

The hydrophilic group in a fatty acid carboxylate is the CO_2^- (ionic head group) because it can undergo strong ion–dipole interactions with water; it is represented by the red area on an electrostatic potential map. The hydrophobic group in a fatty acid carboxylate is the hydrocarbon tail (all C and H atoms), represented by the blue area on an electrostatic potential map.

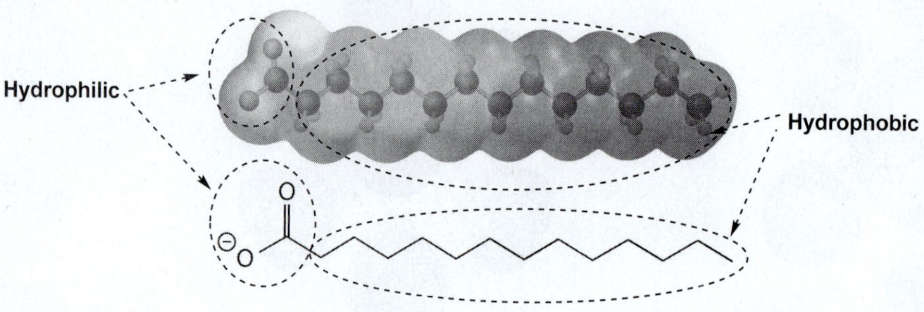

Your Turn 2.19

Think

What constitutes an ester functional group? Notice what bonds/atoms are labeled already in the fat molecule.

Solve

An ester consists of a carbonyl C=O adjacent to an ether linkage R–O. Triglyceride fat molecules have a total of three ester groups.

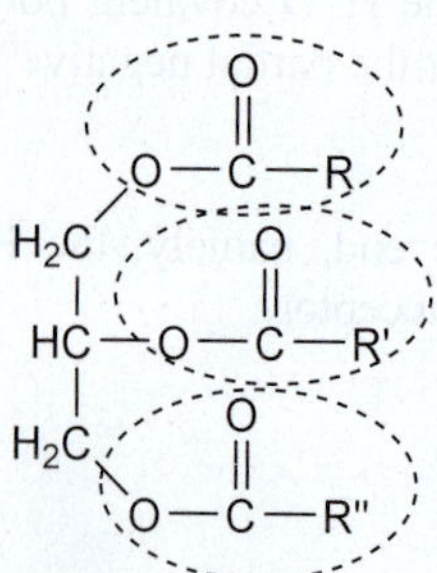

Your Turn 2.20

Think

What are the parts of a phospholipid? Are charged groups hydrophilic or hydrophobic? Are hydrocarbon chains hydrophilic or hydrophobic? Which part of the phospholipid makes up the interior of a lipid bilayer?

Solve

The two parts of a phospholipid are the glycerol backbone attached to fatty acids via ester linkages. The phosphate portion is ionic (hydrophilic), and the alkyl chains of the fatty acids are nonpolar (hydrophobic). The tails are hydrophobic, and to escape the aqueous environment of the cell, the hydrophobic tails associate together. The hydrophobic region of a lipid bilayer **(a)** and the cell membrane **(b)** are labeled in the dashed boxes in the figure on the next page.

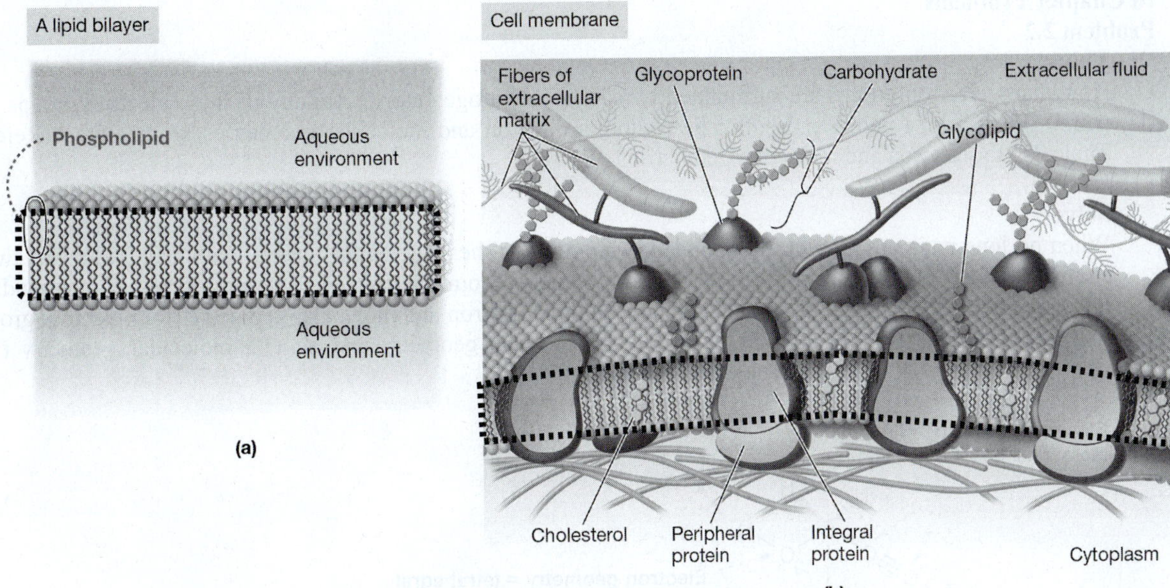
A lipid bilayer
Phospholipid
Aqueous environment
Aqueous environment
(a)
Cell membrane
Fibers of extracellular matrix
Glycoprotein
Carbohydrate
Extracellular fluid
Glycolipid
Cholesterol
Peripheral protein
Integral protein
Cytoplasm
(b)

In Chapter Problems

Problem 2.2

Think

How many electron groups are surrounded by each nonhydrogen atom? Are any of these electron groups lone pairs? When a lone pair is present, can an atom's electron and molecular geometries be the same? Refer to Table 2-3 for electron and molecular geometries.

Solve

When no lone pairs are present, the molecular geometry is the same as the electron geometry. The first two C atoms are linear in both electron geometry and molecular geometry (two electron groups: one single bond and one triple bond). The third C atom is tetrahedral in both electron and molecular geometry (four electron groups: four single bonds). The O atom is tetrahedral in electron geometry and bent in molecular geometry (four electron groups: two single bonds and two lone pairs).

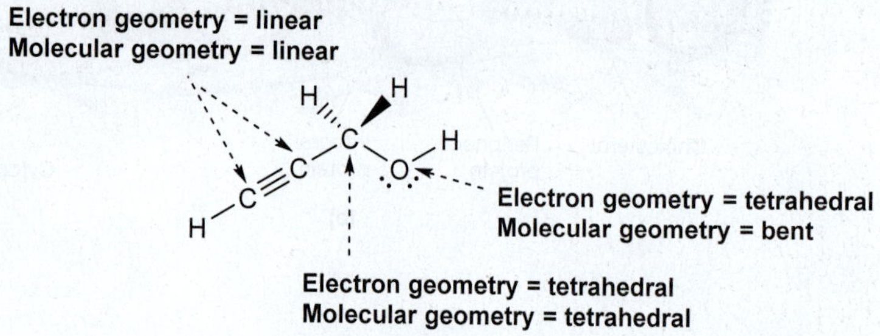

Electron geometry = linear
Molecular geometry = linear

Electron geometry = tetrahedral
Molecular geometry = bent

Electron geometry = tetrahedral
Molecular geometry = tetrahedral

Prop-2-yn-1-ol
(Propargyl alcohol)

Problem 2.3

Think

For each nonhydrogen atom, what is the molecular geometry? Review the rules for dash–wedge notation. How are linear carbon chains drawn in line structure notation? What atoms and types of bonds should be omitted in line structures?

Solve

In a line structure, a flat line (—) represents a bond in the plane of the paper, a wedge (◀) represents a bond that comes out of the plane of the paper, and a dash (''''''''''') represents a bond that points away from you. Remember to omit C atoms and H atoms that are bonded to C. See the structures below.

(a) The six C atoms of the ring are trigonal planar, so the ring is planar. The three C atoms of the chain are tetrahedral. The first Cl is attached to the carbon chain by a wedge, and the second Cl is attached also by a wedge.

(b) The four doubly bonded C atoms are trigonal planar, and the remaining one is tetrahedral. The ring, therefore, is essentially planar. The C atom of C≡N is linear. The substituent comes off the ring pointed toward you and, therefore, is attached by a wedge.

(c) The C atoms are all tetrahedral, and the C–C bonds in the linear carbon chain are drawn as a zigzag. The NH_2 group points away from you and is, therefore, attached to the carbon chain by a dash.

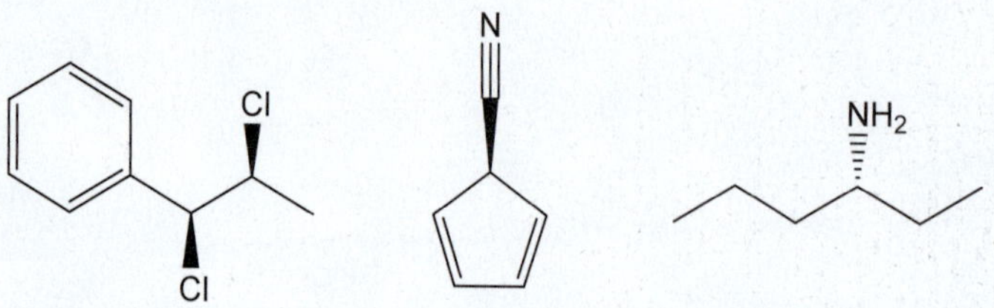

(a) **(b)** **(c)**

Problem 2.4

Think

How should the four bonds of a tetrahedral atom be represented? Refer to Figure 2-6. Should the two Vs be in the same plane or different planes? Should they open in the same direction or in opposite directions?

Solve

The four bonds that make up the tetrahedral geometry must consist of two perpendicular Vs that open in opposite directions. On the left, the V in the plane of the paper (solid oval) opens upward, but the V constructed from the two C–Br bonds (dashed oval) does not open downward. The problem is fixed in the structure on the right, where the V made from the C–Br bonds is perpendicular to the first V and opens downward.

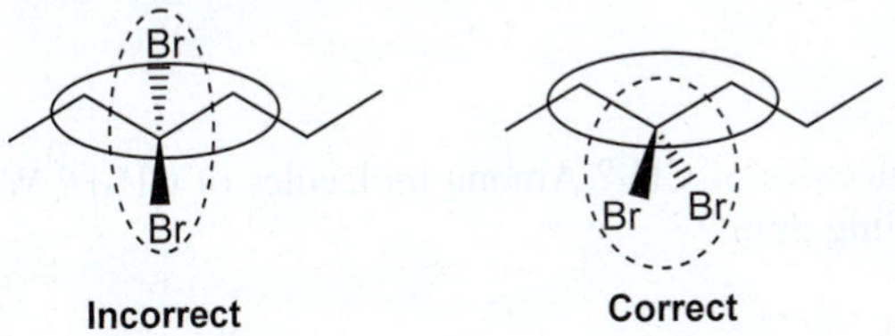

Problem 2.5

Think

What is the molecular geometry about each nonhydrogen atom? Consult Figure 1-16 to determine the electronegativity value for each atom. How is a bond dipole arrow drawn? To which atom, more or less electronegative, does each bond dipole arrow point? Do the bond dipoles perfectly cancel? If not, in which direction does the net dipole point?

Solve

Molecular geometry is very important in drawing the net dipole moment. The dipole arrow (+→) points to the more electronegative atom in each covalent bond. The net dipole moment is the sum of the individual bond dipoles. The thin arrows indicate the bond dipoles. The thick arrows indicate the vector sum of each set of bond dipoles. Structure **(c)** is the only one that is not polar, because it is symmetric—the two C=O bond dipoles point in opposite directions and, therefore, cancel out. From its Lewis structure, **(a)** appears to be nonpolar, but it is polar because of its tetrahedral geometry. You can decompose CBr_2Cl_2 into two Vs in the same way as you can decompose CH_2Cl_2. The resultant dipoles point in opposite directions, but with different magnitudes, so they *do not* cancel completely. Similarly, the bond dipoles in NH_3 **(b)** appear to cancel in the Lewis structure, but they do not, because NH_3 is pyramidal. Structure **(d)** is polar because the C=O bond dipoles do not point in opposite directions and, therefore, do not cancel. Structure **(e)** is polar because the C–H bonds are considered nonpolar and the only polar bond is the C–Li bond. The net dipole is similar to the dipole of the C–Li bond.

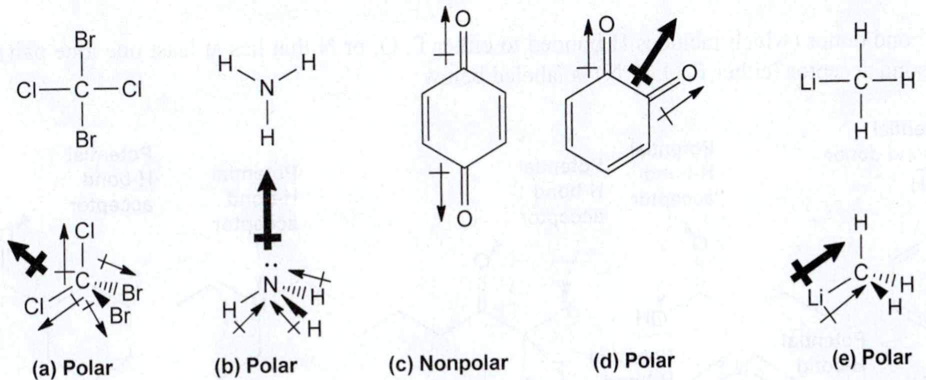

Problem 2.7

Think

What is the strongest intermolecular interaction in a sample of $NaOCH_2CH_3$? In a sample of $CH_3CH{=}O$? Which type of interaction is stronger? How does this affect the boiling point?

Solve

$NaOCH_2CH_3$ is an ionic compound made of Na^+ and $^-OCH_2CH_3$ ions, whereas $CH_3CH{=}O$ is polar covalent. Therefore, CH_3CH_2ONa is held together by ion–ion interactions, the strongest of the intermolecular forces, whereas $CH_3CH{=}O$ is not. The strongest intermolecular interaction between molecules of $CH_3CH{=}O$ is the dipole–dipole interaction. With stronger intermolecular interactions, more heat energy is needed for CH_3CH_2ONa to boil. Thus, CH_3CH_2ONa has a higher boiling point.

Problem 2.8

Think

What is the strongest intermolecular interaction among molecules of CH_4? Among molecules of CH_3F? Which type of interaction is stronger? How does this affect the boiling point?

Solve

CH_3F has a higher boiling point than CH_4 because CH_3F is a polar compound, whereas CH_4 is nonpolar. Thus, CH_3F molecules are held together by dipole–dipole interactions, whereas molecules of CH_4 are not.

Problem 2.9

Think

What are the types of intermolecular interactions that hold each species together? Are any ions present? Do any molecules have net dipoles? Can hydrogen bonds be formed?

Solve

In **A**, $(CH_3)_2CHNH^-$ and Li^+ are oppositely charged ions and are held together by ion–ion interactions. The pair of molecules in **B** are polar, so they are held together by dipole–dipole interactions. No hydrogen bonds can be formed in **B** because there are no F–H, O–H, or N–H covalent bonds. The molecules in **C** are polar, so they, too, are held together by dipole–dipole interactions. Furthermore, they undergo hydrogen bonding because of the N–H donor that exists in one molecule and the N acceptor that exists in the other. Of these interactions, the ion–ion interactions in **A** are the strongest, and the dipole–dipole interactions in **B** are the weakest.

Problem 2.10

Think

Which atoms must be covalently bonded to a hydrogen atom for it to be considered a hydrogen-bond (H-bond) donor? Which atoms are considered H-bond acceptors?

Solve

Each potential H-bond donor (which includes H bonded to either F, O, or N that has at least one lone pair) and each potential H-bond acceptor (either F, O, or N) is labeled below.

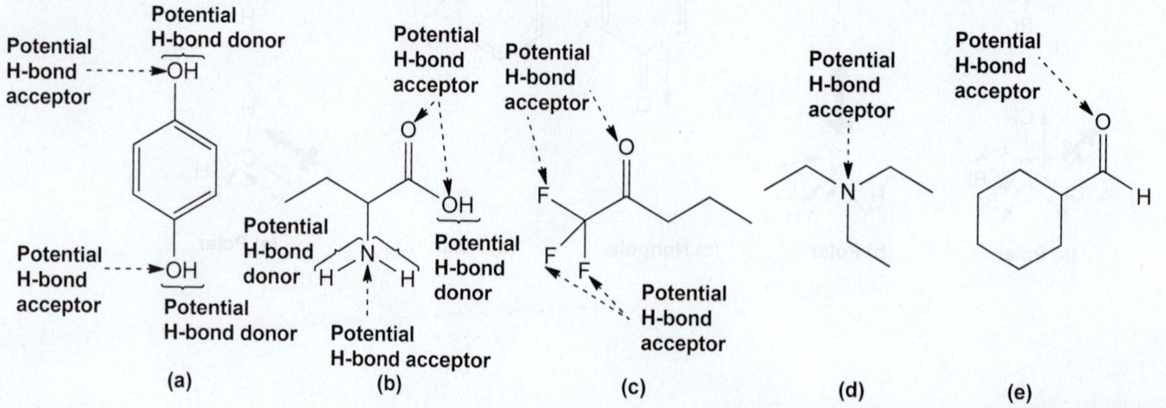

(a) Potential H-bond acceptor; Potential H-bond donor; Potential H-bond acceptor; Potential H-bond donor

(b) Potential H-bond acceptor; Potential H-bond donor; Potential H-bond donor; Potential H-bond acceptor

(c) Potential H-bond acceptor; Potential H-bond acceptor

(d) Potential H-bond acceptor

(e) Potential H-bond acceptor

Problem 2.11

Think

Which atoms must be covalently bonded to a hydrogen atom for it to be considered a H-bond donor? Which atoms are considered H-bond acceptors?

Solve

H-bond donors exist in covalent F–H, O–H, and N–H bonds. H-bond acceptors include F, O, and N atoms. Functional groups that possess at least one H-bond acceptor but no H-bond donors include the ones characteristic of alkyl fluorides, ethers, acetals, epoxides, nitriles, ketones, aldehydes, and esters. Also included are the functional groups in amines and amides if only C atoms are bonded to the N atom. Functional groups that contain at least one H-bond donor and one H-bond acceptor include the ones characteristic of alcohols, hemiacetals, and carboxylic acids. Also included are the functional groups in amines and amides if at least one H atom is bonded to N. Functional groups that contain no H-bond donors and no H-bond acceptors include the ones characteristic of alkyl halides other than fluoride, alkenes, alkynes, aryl groups, and thiols.

Problem 2.13

Think

What is the most important intermolecular interaction that will occur between two molecules of **C**? Between two molecules of **D**? Can the extent of hydrogen bonding be distinguished by the number of H-bond donors and acceptors? How would that affect the boiling points of **C** and **D**?

Solve

C has a higher boiling point than **D**. For both compounds, the most important intermolecular interaction present is hydrogen bonding. For compound **C**, a pair of molecules has a total of four potential H-bond acceptors (N atoms) and eight potential H-bond donors (N–H bonds). For compound **D**, a pair of molecules has a total of four potential H-bond acceptors (N atoms) and four potential H-bond donors (N–H bonds). Hydrogen bonding, therefore, is stronger in **C** than it is in **D**.

H_2N — NH_2 **C** **Higher boiling point** H_2N — N **D**

Problem 2.14

Think

Which atom, F or N, is more electronegative? How does that affect the concentration of negative charge on that atom? What is the effect on the strength of the hydrogen bond?

Solve

We should expect the F–H···F hydrogen bond to be stronger than the N–H···N hydrogen bond. Because F is more electronegative than N, the concentration of negative charge on F is greater than that on N, and the concentration of positive charge on H in a F–H bond is greater than that in a N–H bond. So the attraction between the opposite charges is stronger in the F–H···F hydrogen bond.

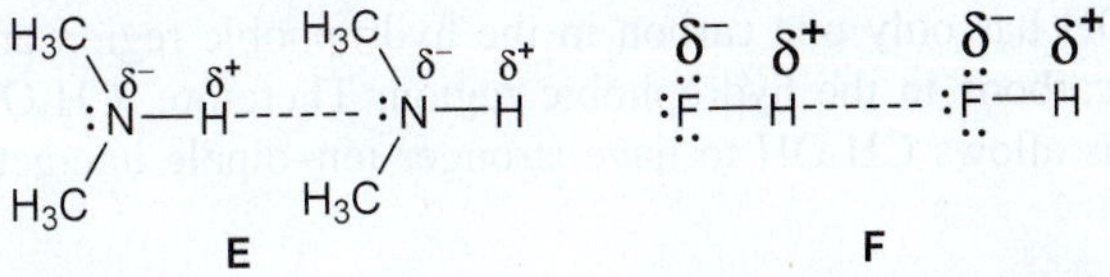

E **F** **Stronger hydrogen bond**

Problem 2.15

Think

What is the most important intermolecular interaction in each of the molecules in **(a)**? In each of the molecules in **(b)**? Do the molecules in **(a)** have the same number of electrons? Do they have the same contact surface area? How about the molecules in **(b)**?

Solve

For the two molecules in **(a)**, 2-methylpentane should have a higher boiling point than 2,2-dimethylbutane. Both molecules are nonpolar, so the most important intermolecular interaction in each sample is an induced dipole–induced dipole interaction. Both molecules have precisely the same number of electrons, but 2-methylpentane is less compact; therefore, it has more contact surface area, making the induced dipole–induced dipole interactions stronger. For the two molecules in **(b)**, 1,2-dimethylcyclopropane should have a higher boiling point than 1,1-dimethylcyclopropane for the same reason.

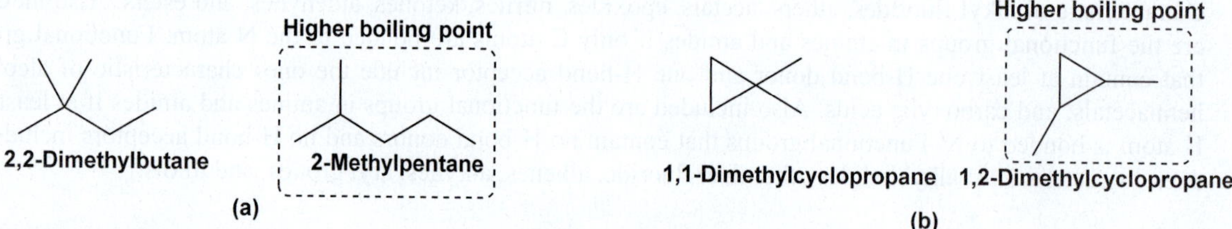

Problem 2.17

Think

What are the relative strengths of the intermolecular interactions in the pure substances that would be disrupted upon mixing? How do these compare to the strengths of the intermolecular interactions gained?

Solve

We should expect **C** to be more soluble than **D** in toluene. Because toluene is nonpolar, the interactions between each compound and toluene involve induced dipoles and so are quite weak. In dissolving, H bonding is destroyed in the first compound, whereas ion–ion interactions are destroyed in the second compound. Because ion–ion interactions are stronger than H bonding, it is more favorable for the second compound to remain intact (i.e., not to dissolve).

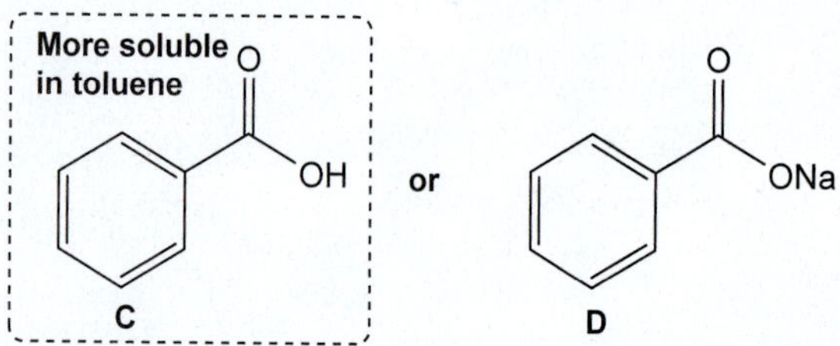

Problem 2.19

Think

How is the solubility of an ionic compound related to how strongly solvated the constituent ions are in solution? What is the structural difference between the two solvent molecules that makes one better than the other at solvating ions?

Solve

The more strongly solvated the ions are by the solvent, the more soluble the ionic compound is in that solvent. Therefore, CH_3OH is better at solvating ions. CH_3OH has only one carbon in the hydrophobic region and is more polar than $CH_3CH_2CH_2OH$, which has three carbons in the hydrophobic region. Therefore, CH_3OH is effectively more polar than $CH_3CH_2CH_2OH$, and this allows CH_3OH to have stronger ion–dipole interactions with the Na^+ and Cl^- ions.

Problem 2.20

Think

What are the intermolecular interactions in pure water that would be disrupted upon mixing with a solute? How do these compare to the intermolecular interactions gained between water and each functional group in Table 1-6? Which intermolecular interactions gained upon dissolving would be comparable to the ones lost from pure water?

Solve
Hydrogen bonding interactions among water molecules would be disrupted upon dissolving a solute. For a functional group to be considered hydrophilic, therefore, it should be able to undergo comparable intermolecular interactions with water, such as significant H bonding, strong dipole–dipole interactions, or ion–dipole interactions. Those include the ones characteristic of alkyl fluorides, alcohols, hemiacetals, nitriles, ketones, aldehydes, and carboxylic acids. They also include the ones in amines and amides if the N atom is bonded to at least one H atom. Functional groups are hydrophobic if such interactions with water are not possible. Those include the ones characteristic of alkenes, alkynes, aryl groups, alkyl halides other than fluorides, thiols, ethers, acetals, epoxides, and esters. They also include the ones characteristic of in amines and amides if N is bonded only to C.

Problem 2.22

Think
Do aldehydes possess potential H-bond donors, acceptors, or both? What about alcohols? Would that make an aldehyde group more or less hydrophilic than an alcohol group?

Solve
We should expect that the maximum number of carbon atoms a water-soluble aldehyde contains should be less than the maximum number for water-soluble alcohols. Whereas the OH group in alcohols has a potential H-bond donor and a potential acceptor that can participate in H bonding with water, an aldehyde has only a single potential H-bond acceptor and no donor. Thus, the H bonding that takes place between an aldehyde and water is less extensive than that between an alcohol and water, making an aldehyde less hydrophilic than a comparable alcohol.

Problem 2.24

Think
What intermolecular interactions are present in each compound? Are they of similar strength in each compound? What is the total number of electrons in each? How does the total number of electrons affect the boiling point?

Solve
Without concern for differences in polarizability, we would normally expect **A** to have a higher boiling point (b.p.) than **B**. Whereas **A** has a highly polar C=O bond that can give rise to relatively strong dipole–dipole interactions, **B** is nonpolar. However, a molecule of **B** has many more electrons (146) compared to **A** (56), meaning that **B** is much more polarizable. Therefore, the induced dipole–induced dipole interactions are much stronger in **B**, giving it a significantly higher boiling point (285 °C) than **A** (178 °C).

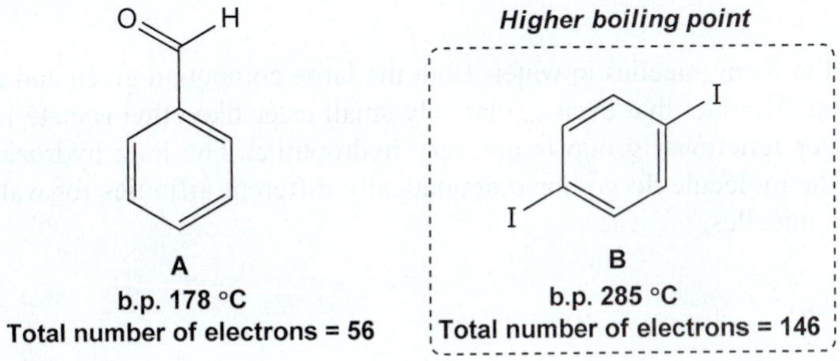

Problem 2.26

Think
What are the most important intermolecular interactions that exist in the isolated substance, and what are their relative strengths? What are the most important intermolecular interactions that would be lost from water? What are the most important intermolecular interactions that exist between each solute molecule and water? How does the solubility of the compounds in water compare with the solubility in hexane?

Solve

Hydrogen bonding is disrupted among water molecules when each solute is mixed in. Therefore, solutes that can undergo relatively strong intermolecular interactions with water will be soluble. The weaker the interaction is, the less soluble the compound will be. Water solubility increases in the order **D** < **A** < **E** < **B** < **C**:

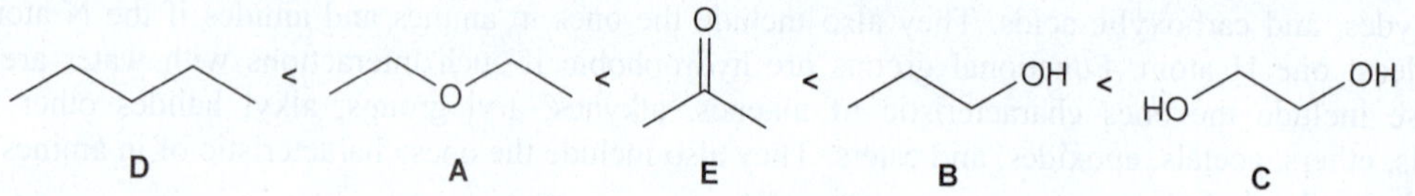

D < A < E < B < C

A, **B**, **C**, and **E** can all undergo H bonding with water. To distinguish the strengths of these interactions, consider the numbers of potential H-bond donors and acceptors. **C** has two potential donors and two potential acceptors; **B** has one potential donor and one potential acceptor; and both **A** and **E** have one potential acceptor only. This makes **C** the most soluble, followed by **B**. **E** is more polar than **A**, so it will undergo stronger dipole–dipole interactions with water, making **E** more soluble than **A**. **D** is the least soluble because, being nonpolar, it has the weakest interactions with water (induced dipoles). Notice that the order of solubility of these compounds in water is the reverse of the order in Solved Problem 2.25.

Problem 2.27

Think

What is the strongest intermolecular interaction in a sample of $KSCH_3$? Is ethanol a protic or aprotic solvent? What about acetone? How well can ethanol and acetone solvate K^+? How well can the two solvents solvate CH_3S^-?

Solve

$KSCH_3$ is an ionic compound made of K^+ and CH_3S^- ions, so they are held together by ion–ion interactions, the strongest of the intermolecular interactions. $KSCH_3$, therefore, will be highly soluble in solvents that can solvate ions well. Ethanol (CH_3CH_2OH) is a polar protic solvent, so it can solvate both K^+ and CH_3S^- ions very well. Acetone is a polar aprotic solvent, so it can solvate K^+ ions well, but it does not solvate CH_3S^- ions very well. Consequently, $KSCH_3$ will be more soluble in ethanol.

Problem 2.28

Think

Based on the solubility of ethyl acetate, is an ester functional group strongly or weakly hydrophilic? Is the long hydrocarbon chain of the compound shown hydrophilic or hydrophobic? Do the two ends of the molecule have dramatically different affinities for water?

Solve

We would not expect the given compound to form micelles in water. Both the large compound given and ethyl acetate have a single ester functional group. The fact that even a relatively small ester like ethyl acetate is not very soluble in water suggests that an ester functional group is not very hydrophilic. The long hydrocarbon chain is hydrophobic, so the two ends of the molecule do *not* have dramatically different affinities for water—something that would be necessary to form micelles.

Problem 2.29

Think

Does the compound have a very hydrophilic end? A very hydrophobic end? What is required for a compound to act as a soap?

Solve

We should not expect this compound to act as a soap. The two ends are ionic and, therefore, are very hydrophilic. The molecule, however, has no hydrophobic end. Although the hydrocarbon rings between the two ionic groups are hydrophobic, a particle of dirt or oil cannot dissolve that portion without at least partially dissolving an ionic group.

Problem 2.30

Think

What is the general structure for a triacylglycerol (Equation 2-4)? How does the fatty acid group RCO_2H combine with the HO group of glycerol to form triacylglycerol? What is the structure if all three R groups are from lauric acid? What is the structure if one group is from linoleic acid and the other two are from oleic acid?

Solve

The RCO_2H for each fatty acid combines with the HO of glycerol to form an ester. The general structure for a triacylglycerol becomes **(a)** if all three fatty acids are lauric acid and **(b)** if one fatty acid is linoleic acid and two are oleic acids, as shown below.

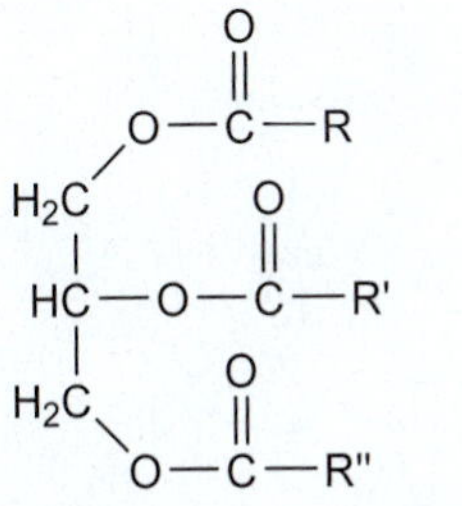

General triacylglycerol

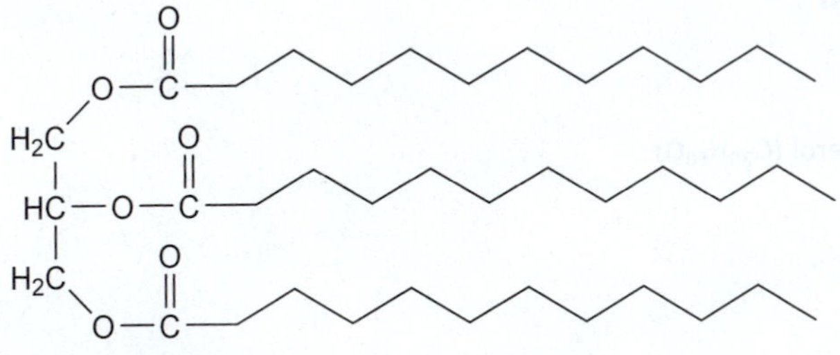

(a)
R = Lauric acid

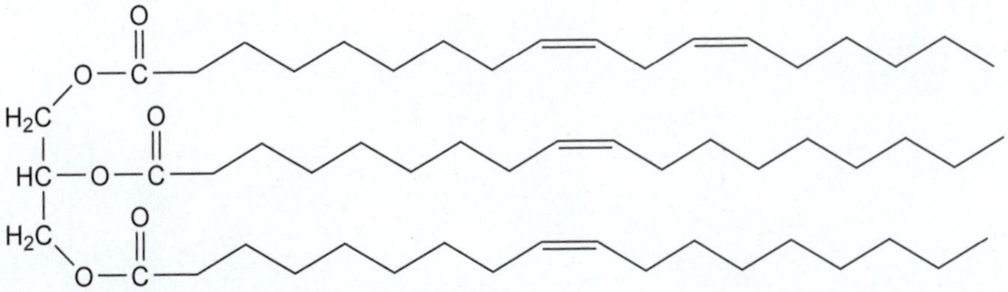

(b)
R = Linoleic acid
R', R" = Oleic acid

Problem 2.31

Think

What is the structure of an isoprene unit? Which groups of five C atoms can be assigned to their own isoprene units? Which groups of five C atoms cannot?

Solve

An isoprene unit has five carbon atoms consisting of a four-carbon chain, with the fifth carbon attached to C2; that is, it appears as follows:

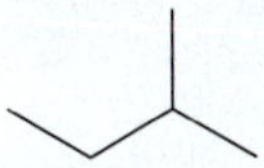

Lanosterol has 30 carbon atoms and theoretically could be assigned to six isoprene units if all of the carbon atoms could be assigned to distinct isoprene units. However, because two of the methyl groups shifted from their original locations, not all carbon atoms in lanosterol can be assigned unambiguously to their own isoprene units. The figure on the next page shows how 25 of the C atoms can be assigned to five different isoprene units (black highlight). The remaining five carbon atoms in the gray boxes cannot.

Lanosterol ($C_{30}H_{50}O$)

End of Chapter Problems

Sections 2.1–2.3 Valence Shell Electron Pair Repulsion (VSEPR) Theory; Dash–Wedge Notation; The Molecular Modeling Kit

Problem 2.32

Think

How many electron groups surround each indicated atom? Are any of these electron groups lone pairs? How does the number of electron groups relate to the electron geometries in Table 2-2? Refer to Table 2-3 for molecular geometries.

Solve

Bond angles and electron and molecular geometries are listed below. When no lone pairs are present, the molecular geometry is the same as the electron geometry. Lone pairs of electrons and bonds to H are added in to explicitly show each electron and molecular geometry.

Electron = tetrahedral
Molecular = tetrahedral
Bond angle ~109.5°

Electron = tetrahedral
Molecular = bent
Bond angle ~109.5°

Electron = tetrahedral
Molecular = bent
Bond angle ~109.5°

Electron = trigonal planar
Molecular = N/A
Bond angle = N/A

Electron = trigonal planar
Molecular = trigonal planar
Bond angle ~120°

Problem 2.33

Think

How many electron groups surround each indicated atom? Are any of these electron groups lone pairs? How does the number of electron groups relate to the electron geometries in Table 2-2? Refer to Table 2-3 for molecular geometries.

Solve

Bond angles and electron and molecular geometries are listed below. When no lone pairs are present, the molecular geometry is the same as the electron geometry. Lone pairs of electrons and bonds to H are added in to explicitly show each electron and molecular geometry.

Electron = tetrahedral
Molecular = bent
Bond angle ~109.5°

Electron = linear
Molecular = linear
Bond angle = 180°

Electron = trigonal planar
Molecular = trigonal planar
Bond angle ~ 120°

Electron = trigonal planar
Molecular = trigonal planar
Bond angle ~120°

Electron = tetrahedral
Molecular = tetrahedral
Bond angle ~ 109.5°

Problem 2.34

Think

How many electron groups surround each indicated atom? Are any of these electron groups lone pairs? How does the number of electron groups relate to the electron geometries in Table 2-2? Refer to Table 2-3 for molecular geometries.

Solve

Bond angles and electron and molecular geometries are listed below. When no lone pairs are present, the molecular geometry is the same as the electron geometry. Lone pairs of electrons and bonds to H are added in to explicitly show each electron and molecular geometry.

(a)

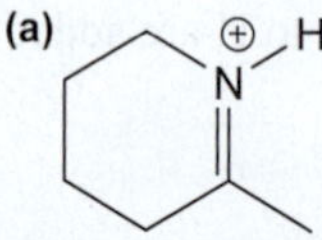

Electron = trigonal planar
Molecular = trigonal planar
Bond angle ~ 120°

(b)

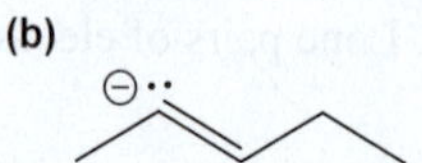

Electron = trigonal planar
Molecular = bent
Bond angle ~120°

(c)

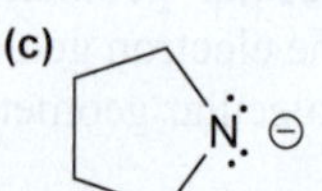

Electron = tetrahedral
Molecular = bent
Bond angle ~ 109.5°

(d)

CH_3
H_3C—N⊕—CH_3
CH_3

Electron = tetrahedral
Molecular = tetrahedral
Bond angle ~ 109.5°

(e)

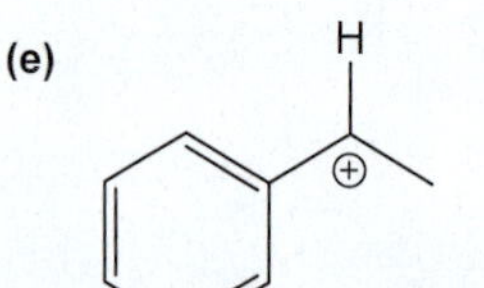

Electron = trigonal planar
Molecular = trigonal planar
Bond angle ~ 120°

(f)

Electron = linear
Molecular = N/A
Bond angle = N/A

Problem 2.35

Think

What is the ideal molecular geometry and angle for each carbon? What is the angle of each C–C bond in a three-membered ring? How does the difference between ideal angle and forced angle affect the angle strain?

Solve

In a three-carbon ring, the bond angles are forced to be 60° on average. For **A**, cyclopropane, each C atom's ideal geometry is tetrahedral and should have a bond angle of 109.5°. For **B**, cyclopropene, the C atom with four single bonds should have tetrahedral geometry and a bond angle of 109.5°; the ideal geometry of the two C atoms in the C=C bond is trigonal planar and should have a bond angle of 120°. The C atoms of the C=C bond, therefore, are forced to be farther away from their ideal geometry than are the tetrahedral C atoms, giving **B** more angle strain overall.

A

Ideal angle = 109.5°

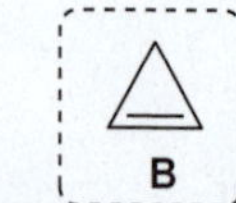

B

Ideal angle = 109.5° and 120°
Larger angle strain

Problem 2.36

Think

For each nonhydrogen atom, what is the molecular geometry? Review the rules for dash–wedge notation. Which bonds in each molecule are in the plane of the paper? Which ones are pointing toward you? Which ones are pointing away? How are carbon chains drawn in line notation?

Solve
A flat line (—) represents a bond in the plane of the paper, a wedge (◀) represents a bond that comes out of the plane of the paper, and a dash (''''''''') represents a bond that points away from you. See the structures below.

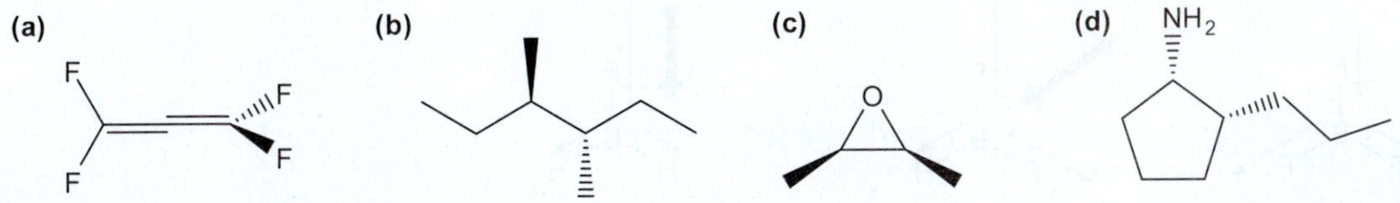

(a) (b) (c) (d) NH_2

F F F F O

Problem 2.37

Think
How does this rotation affect the wedge bonds? Dash bonds? Planar bonds? How does the rotation affect whether bonds point toward the top, bottom, left, or right of the figure? Be mindful of only performing the rotation indicated. Building a molecular model might be helpful.

Solve
(a) The 180° rotation indicated inverts all wedges to dash bonds, and vice versa. All bonds pointing toward the top of the figure end up pointing toward the bottom, and vice versa. The zigzag inverts all V to Λ and all Λ to V.

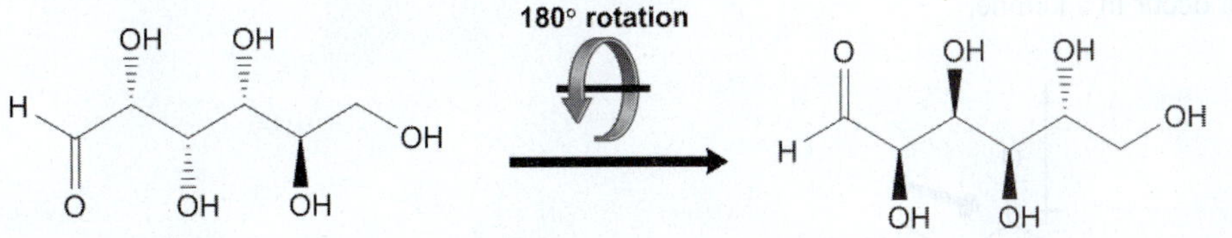

180° rotation

OH OH H OH O OH OH D-Glucose O OH OH OH H OH OH

(b) The 180° rotation indicated inverts all wedges to dash bonds, and vice versa. All groups on the left before the transformation end up on the right after the transformation.

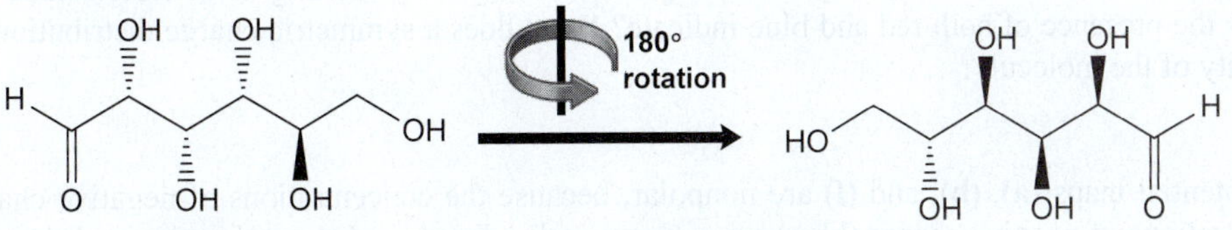

180° rotation

OH OH H OH O OH OH D-Glucose OH OH H HO OH OH O

Section 2.4 Net Molecular Dipoles and Dipole Moments
Problem 2.38

Think
What is the molecular geometry for each of the molecules? Which directions do the bond dipoles point along each bond? Draw in the dipole arrows for each polar bond. Consider the difference in electronegativity (Fig. 1-16) for each atom to determine the magnitude of each bond dipole. Do the bond dipoles cancel? How does the magnitude of the net dipole relate to the polarity of the molecule?

Solve
Polarity is as follows: (least) $BF_3 < BFH_2 < BF_2H$ (most). All these compounds are trigonal planar. Fluorine and hydrogen are both more electronegative than boron, so each B–F bond dipole points toward F, and each B–H bond dipole points toward H (bond dipoles are represented as thin arrows on the next page). Fluorine is more electronegative than hydrogen, however, so the B–F bond dipoles are larger. Bond dipoles in BF_3 cancel

entirely, as shown below, and, therefore, BF_3 is nonpolar. In BFH_2, the B–F bond dipole partially cancels with that from the two B–H bonds, resulting in a small net dipole (thick arrow) in the direction of the B–F bond. BF_2H has an even larger dipole moment because of the additional F.

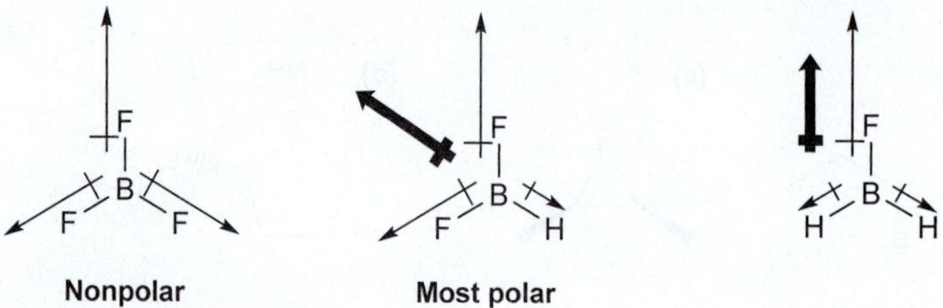

Problem 2.39

Think

In each compound shown below, where do the bond dipoles point? Do the additional bond dipoles in the ester point in the same direction as the C=O bond dipole? How does the net dipole result from the bond dipoles?

Solve

Oxygen is more electronegative than carbon, so each C=O and C–O bond dipole (thin arrows) points toward O. As we can see below, the C–O bond dipoles (not in boldface) in an ester do not point in the same direction as the C=O bond dipole, so they partially cancel with the C=O bond dipole to give the net dipole moment (thick arrows). The only significant bond dipole in a ketone is from the C=O bond. Therefore, cancellation of the bond dipole does not occur in a ketone.

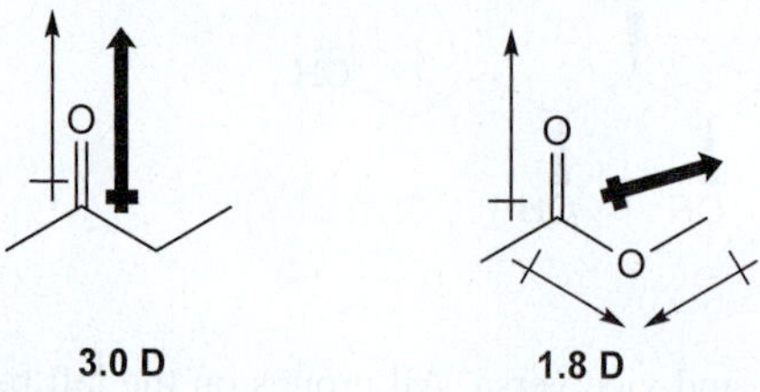

Problem 2.40

Think

What colors represent concentrations of negative and positive charge? By looking at an electrostatic potential map, what does the presence of both red and blue indicate? What does a symmetric charge distribution indicate about the polarity of the molecule?

Solve

Electrostatic potential maps **(a)**, **(b)**, and **(f)** are nonpolar, because the concentrations of negative charge (red) and the concentrations of positive charge (blue) are symmetrically distributed about the center of the molecule; thus, any bond dipoles must cancel completely. For **(c)**, **(d)**, and **(e)**, the net dipole points from the blue region (excess positive charge) to the red region (excess negative charge), as shown below.

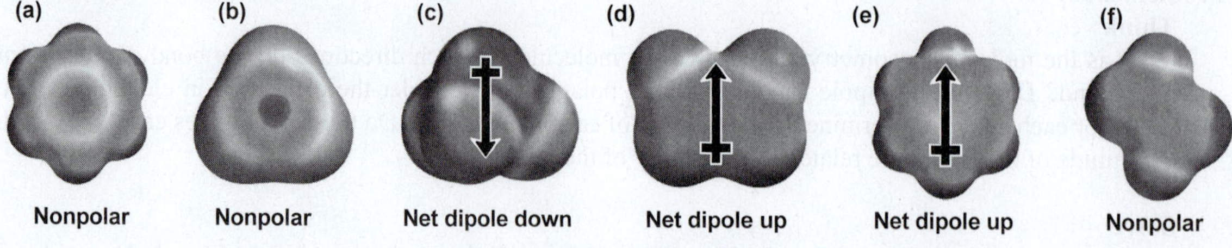

Problem 2.41

Think

What is the molecular geometry for each of the atoms? Draw in the dipole arrows for each polar bond. Consider the difference in electronegativity for each atom (Fig. 1-16) to determine the magnitude of the net dipole. How does the magnitude of the net dipole relate to the polarity of the molecule?

Solve

Molecules **(b)**, **(d)**, **(e)**, **(g)**, **(j)**, **(l)**, and **(n)** are polar. The significant bond dipoles are indicated by the thin arrows. The direction of the net dipole moment is indicated by the thick arrow. The nonpolar molecules **(a)**, **(c)**, **(h)**, **(i)**, **(k)**, and **(m)** all have polar bonds, but because of the symmetry of those bonds, the dipoles cancel, and no net dipole exists. In general, hydrocarbons like molecule **(f)** are considered nonpolar.

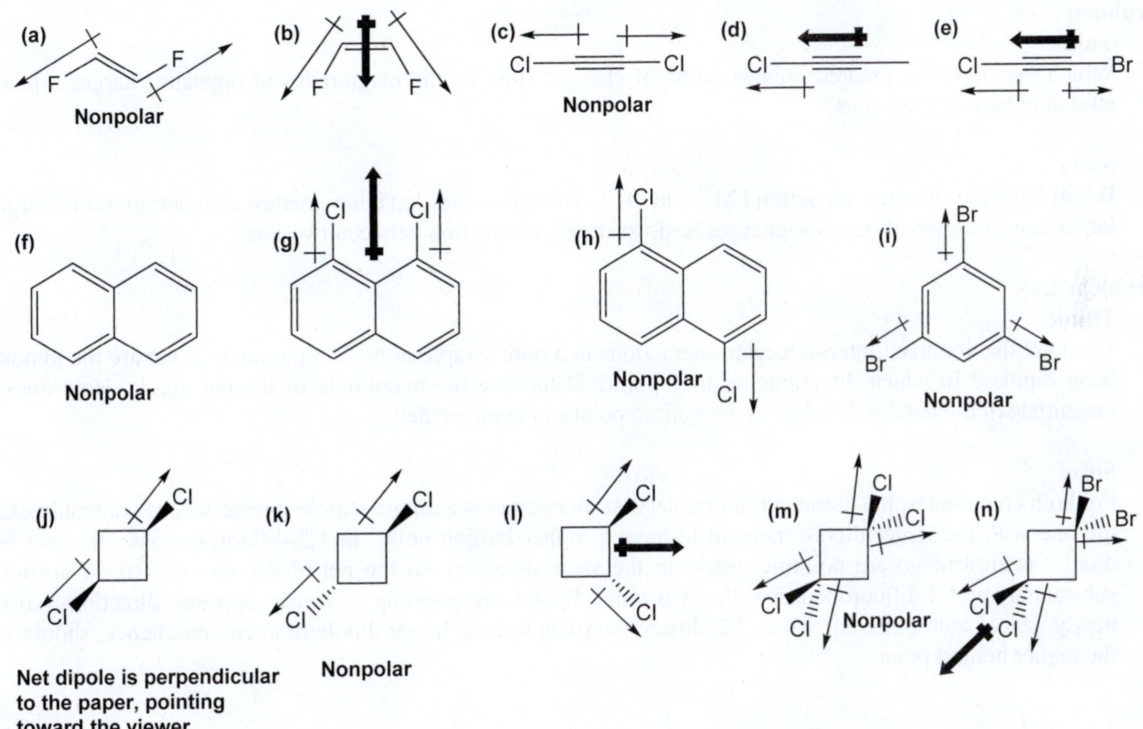

Problem 2.42

Think

What are possible ways you could draw CX_2Y_2 in a square planar geometry? Is there more than one way? Does the polarity differ in each drawing? How does this compare to a tetrahedral geometry?

Solve

If CX_2CY_2 were planar, there could be two different compounds, as shown below on the right. The first one would be polar, whereas the second one would be nonpolar. However, a tetrahedral geometry of such a compound, like CH_2Cl_2 (shown below on the left), is consistent with the fact that the compound is always found to be polar. This is because the net dipole from the V of the CCl_2 is not the same as that from the V of the CH_2.

Tetrahedral — Polar

Tetrahedral geometry

Square planar geometry

Square planar — Polar — Nonpolar

Section 2.6 Melting Points, Boiling Points, and Intermolecular Interactions

Problem 2.43

Think

In CH_3^+, what type of charge is concentrated on the carbon? Is it a partial or full charge? What type of charge would be attracted to the carbon? In the species that are given, are there partial or full charges?

Solve

CH_3^+ is a cation having a full positive charge concentrated on the C atom and would be attracted to full or partial negative charges. Both **(c)** Cl^- and **(d)** F^- anions will be attracted to CH_3^+ via ion–ion interactions. Both **(a)** H_2O and **(e)** $H_2C{=}O$ are polar, so they have a partial negative charge that will be attracted via ion–dipole interactions. **(b)** Na^+ has a positive charge and will thus be repelled by CH_3^+.

Problem 2.44

Think

Which ions have the greatest concentration of charge? How do the magnitudes of opposite charges affect the attraction between two ions?

Solve

B will have the strongest attraction (Al^{3+} and O^{2-}). These two ions have the greatest concentration of charge. A larger concentration of opposite charges leads to stronger attraction between two ions.

Problem 2.45

Think

What are the dominant intermolecular interactions in a pure sample of each compound? What are the important bond dipoles? In which directions do they point? Determine the magnitude of the net dipole. How does the magnitude of the net dipole relate to the boiling point of the molecule?

Solve

For both compounds, the dominant intermolecular interaction is a dipole–dipole interaction, so we would expect the one with the larger dipole moment to have a higher boiling point. In 1,2-difluorobenzene, the two bond dipoles (thin arrows) are pointing nearly in the same direction, so the net dipole moment (thick arrows) is substantial. In 1,3-difluorobenzene, the two bond dipoles are pointing in nearly opposite directions, so they nearly cancel each other out. Thus, 1,2-difluorobenzene has the larger dipole moment and, hence, should have the higher boiling point.

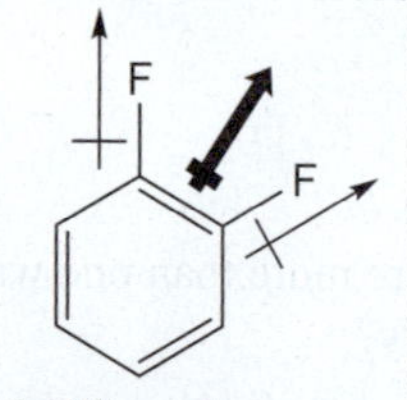

F F
1,2-Difluorobenzene

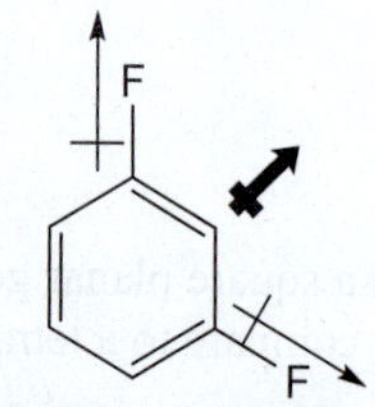

F F
1,3-Difluorobenzene

Problem 2.46

Think

Review the rules for condensed formulas (Section 1.12b). Which nonhydrogen atoms are bonded together? How can we add bonds and lone pairs to maximize the number of octets and also conform to the total charge of zero? How many total valence electrons must be accounted for in this compound? Which atoms must be covalently bonded to a hydrogen atom for it to be considered a hydrogen-bond (H-bond) donor? Which atoms are considered H-bond acceptors?

Solve

The Lewis structure for $(CH_3)_2CHCH(NH_2)CO_2H$ is shown on the next page. Each potential H-bond donor (which includes H bonded to either F, O, or N) is labeled using curly braces, and each potential H-bond acceptor (either F, O, or N) is labeled using a gray circle.

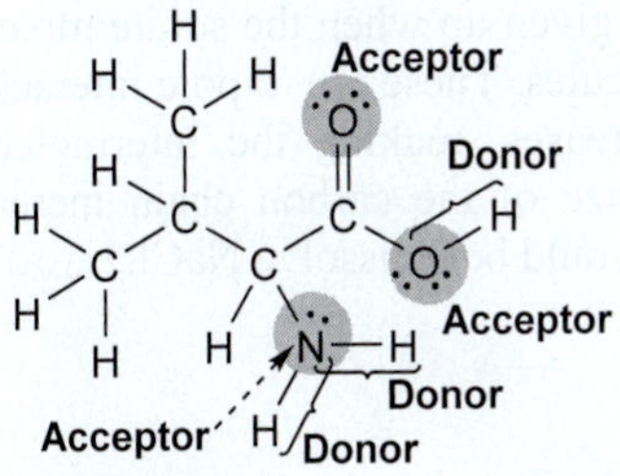

Problem 2.47

Think

Are any ions present? Any H-bond donors or acceptors? Any significant bond dipoles? Do the species have the same number of electrons? Do they have the same contact surface area?

Solve

(a) Being uncharged and nonpolar, the only intermolecular interaction that exists between two H_2 molecules or two He atoms is an induced dipole–induced dipole interaction.

(b) To have a higher boiling point, H_2 must be more polarizable than He. This should make sense, because even though both species have two electrons, the electrons in H_2 are distributed between two nuclei, so they occupy a greater region of space than electrons around a single He nucleus. Therefore, H_2 has a greater contact surface area than He.

Problem 2.48

Think

Given that CH_3O^- carries a full negative charge, what type of charge would Br_2 need to undergo attraction? Does it have that type of charge permanently? If not, how could it develop that kind of charge? Would I_2 or Br_2 develop that type of charge more easily?

Solve

(a) To be attracted to CH_3O^-, Br_2 would need to develop a partial positive charge. It does not have one initially but would develop one via an induced dipole when in the presence of CH_3O^-. The name of the attraction should be "ion–induced dipole interaction."

(b) The attraction between CH_3O^- and I_2 would be stronger, because I_2, containing more electrons than Br_2, is more polarizable.

Section 2.7 Solubility

Problem 2.49

Think

What is the dominant type of intermolecular interaction disrupted among water molecules when each organic solvent is mixed in? What intermolecular interactions exist between water and each organic solvent molecule? Which organic solvent has the strongest intermolecular interaction with water?

Solve

Strong hydrogen bonding is disrupted among water molecules when the organic solvents are mixed in. Only **B** and **C** will undergo hydrogen bonding with water, because they each have at least one O atom to act as a H-bond acceptor, so they will be more soluble than **A** or **D**. Molecule **C** ($CH_3OCH_2CH_2OCH_3$) has two acceptors, whereas **B** ($CH_3OCH_2CH_2CH_2CH_3$) has only one, so **C** would be more soluble.

Problem 2.50

Think

What are the relative strengths of the intermolecular interactions in the pure substances that would be disrupted upon mixing? How do these compare to the strengths of the intermolecular interactions gained when the solute is mixed in? Would a larger hydrocarbon portion in the solvent molecule lead to stronger or weaker intermolecular interactions with the solute?

Solve

H bonding between solvent molecules and ion–ion interactions in NaCl are given up when the solute mixes in. Ion–dipole interactions are gained between the solute ions and solvent molecules. These ion–dipole interactions are compromised as the size of the nonpolar hydrocarbon portion increases, making the intermolecular interactions weaker. Effectively, the alcohol becomes less polar as the size of the carbon chain increases. Therefore, **A** (CH_3OH), having the smallest nonpolar hydrocarbon portion, would best dissolve NaCl.

Problem 2.51

Think

What are the dominant intermolecular interactions between a pair of each molecule? Are ions present? Are strong bond dipoles present? Are there H-bond donors and acceptors present? How can you predict the extent of H bonding based on the numbers of potential donors and acceptors? How does the extent of H bonding affect boiling point (b.p.)?

Solve

The first two compounds can undergo H bonding because in a pair of each, there is at least one potential H-bond donor (NH) and one potential H-bond acceptor (N). For the third compound, a pair of molecules cannot undergo H bonding, because there are no H-bond donors. Therefore, the third compound has the lowest b.p. The first two compounds differ in the number of potential H-bond donor groups: two for the first and one for the second. Thus, H bonding is stronger for the first compound, giving it a higher b.p.

NH2	H N	N
2-Methylbutan-1-amine	***N*-Methylbutan-2-amine**	***N*-Ethyl-*N*-methylethan-1-amine**
b.p. = 97 °C	**b.p. = 84 °C**	**b.p. = 65 °C**
2 H-bond donors	**1 H-bond donor**	**Zero H-bond donors**

Section 2.8 Ranking Boiling Points and Solubilities of Structurally Similar Compounds

Problem 2.52

Think

What are the types of intermolecular interactions in a pure sample of each compound? Of those, what are the most important? Are any ions present? Any polar bonds? Any potential H-bond donors and acceptors? Are the net dipoles of the same magnitude? Do the compounds differ in the number of potential H-bond donors and acceptors? Do they differ in contact surface area? How does the strength of the intermolecular interactions affect the boiling point?

Solve

In order of increasing boiling point, **A < H < F < G < B < C < E < D**. The dominant interactions for each compound are shown on the next page. Induced dipole–induced dipole interactions are the weakest, giving rise to the lowest boiling points. Ion–ion interactions are the strongest, giving rise to the highest boiling points. The two nonpolar compounds **A** and **H** have the same number of electrons, so they are differentiated by their contact surface areas: A greater contact surface area results in a higher boiling point. The two polar compounds **F** and **G** are differentiated by their magnitude of the dipole moment; the ketone **G** has a larger dipole moment than the ether **F**. The H-bonding compounds **B, C,** and **E** are differentiated by the number of potential H-bond donors and acceptors in a pair of molecules: As the number of these features increases, so does the strength of H bonding, and, therefore, the boiling point increases.

Induced dipole–induced dipole interactions — Larger contact surface area — Dipole–dipole interactions

A < H < F < G <

Greater number of H-bonds

B < C < E < D

H bonding — Ion–ion interactions

Problem 2.53
Think
What are the types of intermolecular interactions between each pair of species? Are any ions present? Potential H-bond donors and acceptors? Significant net dipoles? How does the strength of the intermolecular interactions affect the boiling point?

Solve
The dominant intermolecular interactions (indicated below each molecule) establish the order of boiling points. Hydrogen bonding is the strongest of these interactions, giving rise to the highest boiling point. Dispersion forces are the weakest, giving rise to the lowest boiling point.

$CH_3CH_2CH_2CH_2CH_3$	<	$CH_3CH_2OCH_2CH_3$	<	$FCH_2CH_2OCH_2CH_3$	<	$CH_3CH(OH)CH_2CH_3$
B		**A**		**D**		**C**
Disperson forces only		One dipole		Two dipoles CF greater than CO		Hydrogen bonding

Problem 2.54
Think
What are the types of intermolecular interactions between each pair of species? Are any ions present? Potential H-bond donors and acceptors? Significant net dipoles? How does the strength of the intermolecular interactions affect the boiling point?

Solve
The dominant intermolecular interactions (indicated below each molecule) establish the order of boiling points. **A**, **B**, and **C** can all undergo hydrogen bonding, whereas **D** cannot. Therefore, **D** has the lowest boiling point. The extent of hydrogen bonding among **A**, **B**, and **C** is determined by the number of potential H-bond donors and acceptors. As the number increases, so does the boiling point.

$FCH_2CH_2CH_2CH_3$	<	$CH_3CH_2CH_2CH_2OH$	<	$CH_3CH_2CH(OH)CH_2OH$	<	$CH_3CH(OH)CH(OH)CH_2OH$
D		**B**		**A**		**C**
Dipole–dipole (No H bonding)		Hydrogen bonding 1 H-bond donor 1 H-bond acceptor		Hydrogen bonding 2 H-bond donors 2 H-bond acceptors		Hydrogen bonding 3 H-bond donors 3 H-bond acceptors

Section 2.9 Protic and Aprotic Solvents

Problem 2.55

Think

What do you think the word *protic* indicates? What types of intermolecular interactions should protic solvents be able to undergo with solute species?

Solve

For a solvent to be considered protic, it must possess a potential H-bond donor. This allows it to form strong ion–dipole interactions with dissolved ions, especially anions. Therefore, it must contain an O–H, N–H, or F–H bond. It cannot solely possess a H-bond acceptor. The word *protic* indicates that there is some kind of proton (H^+) character to the solvent. The H-bond donors are circled below.

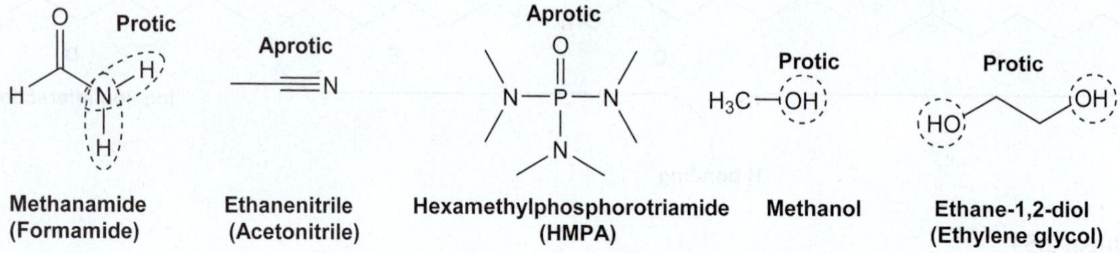

Protic — **Methanamide (Formamide)**

Aprotic — **Ethanenitrile (Acetonitrile)**

Aprotic — **Hexamethylphosphorotriamide (HMPA)**

Protic — **Methanol** (H_3C–OH)

Protic — **Ethane-1,2-diol (Ethylene glycol)**

Problem 2.56

Think

Is DMSO a protic or aprotic solvent? What is the dominant type of intermolecular interaction disrupted when NaCl mixes with DMSO? What intermolecular interactions exist between NaCl and each organic solvent molecule in Problem 2.55? How does the protic nature of a solvent affect the solubility of an ion?

Solve

DMSO is an aprotic solvent (Section 2.9, Fig. 2-25). Polar protic solvents solvate both cations and anions very strongly, whereas polar aprotic solvents solvate cations very strongly, but not anions. Therefore, the polar protic solvents in Problem 2.55 (formamide, methanol, and ethylene glycol) will solvate NaCl better, and NaCl will have a substantially greater solubility than it will in dimethyl sulfoxide.

Problem 2.57

Think

What are the dominant intermolecular interactions between NaCl and each solvent? Which ketone has a larger hydrocarbon portion? Would a larger hydrocarbon portion in the solvent molecule lead to stronger or weaker intermolecular interactions with the solute?

Solve

Ion–dipole interactions are the dominant intermolecular interactions between NaCl and each ketone solvent. Acetone $(CH_3)_2C{=}O$ has a smaller hydrocarbon portion surrounding the polar C=O bond than di-*tert*-butyl ketone ($[(CH_3)_3C]_2C{=}O$; dashed circles below). With smaller hydrocarbon groups, the Na^+ and Cl^- ions can approach closer to the partial charges of the dipole, allowing for stronger ion–dipole interactions. Therefore, NaCl will be more soluble in acetone than di-*tert*-butyl ketone.

Acetone

Di-*tert*-butyl ketone

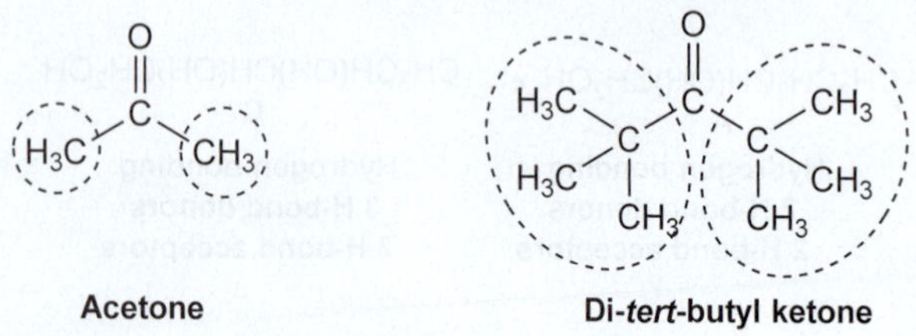

Section 2.10 Soaps and Detergents

Problem 2.58

Think

What is the charge on the carboxylate group? What intermolecular interactions are occurring? What difference does the +1 charge versus +2 charge have on the ability of the ions to participate in that kind of interaction with the carboxylate group in soap?

Solve

The attraction is due to ion–ion interactions involving a metal cation and a carboxylate anion (RCO_2^-). Ca^{2+} and Mg^{2+} both have a +2 charge and so will attract more strongly to the −1 charge on the carboxylate anion than will Na^+ or K^+.

Problem 2.59

Think

Draw out each structure. Consider the concentration of charge for each. Are resonance structures possible? If so, how does the number of resonance structures affect charge localization?

Solve

In a carboxylate anion (RCO_2^-), the negative charge is resonance delocalized over two O atoms, whereas in a sulfate anion ($ROSO_3^-$), the negative charge is resonance delocalized over three O atoms. Thus, in a sulfate anion, the concentration of negative charge is smaller than it is in a carboxylate anion, so its attraction to a positively charged ion is weaker.

Carboxylate hybrid

Alkyl sulfate hybrid

Problem 2.60

Think

What types of intermolecular interactions contribute to a functional group being hydrophilic? What types of intermolecular interactions contribute to a functional group being hydrophobic? Refer to Figure 2-27 as a guide for how the micelle is drawn. With water being the solvent, should the hydrophilic or hydrophobic portions of the molecules be on the outside of the micelle? What is the strongest intermolecular interaction between $CH_3(CH_2)_{15}N(CH_3)_3{}^+Cl^-$ and a molecule of water?

Solve

Functional groups are hydrophilic if they can interact strongly with water through H bonding, dipole–dipole interactions, or ion–dipole interactions. Functional groups are hydrophobic if such interactions with water are not possible (i.e., nonpolar). **(a)** The positively charged N establishes the hydrophilic head group, as shown below. **(b)** The depiction of a micelle is the same as in Figure 2-27, except the exterior of the micelle is the $N^+(CH_3)_3$ group. **(c)** Positively charged groups would be on the exterior of the micelle, which can undergo ion–dipole interactions with water.

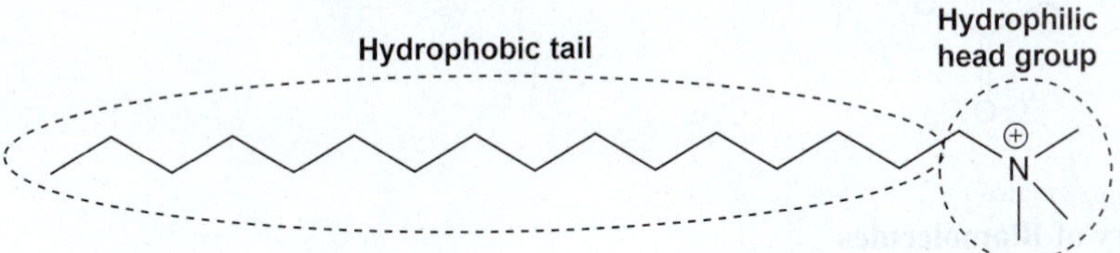

Problem 2.61

Think

What types of intermolecular interactions contribute to a functional group being hydrophilic? What types of intermolecular interactions contribute to a functional group being hydrophobic? Refer to Figure 2-27 as a guide for how the micelle is drawn. With water being the solvent, should the hydrophilic or hydrophobic portions of the molecules be on the outside of the micelle? Does the detergent molecule consist of ions? Does it have H-bond donors and acceptors?

Solve

Functional groups are hydrophilic if they can interact strongly with water through H bonding, dipole–dipole interactions, or ion–dipole interactions. Functional groups are hydrophobic if such interactions with water are not possible (i.e., nonpolar).

(a) The three OH groups, along with the ester group, are hydrophilic, as shown below.

(b) This is the same as Figure 2-27, with the OH groups on the outside.

(c) H-bonding interactions dominate between the OH groups of the head group and the OH groups of water.

(d) Because they do not consist of ions, these detergents have less of a tendency to attract metal cations, which would otherwise cause the detergent to precipitate out of solution.

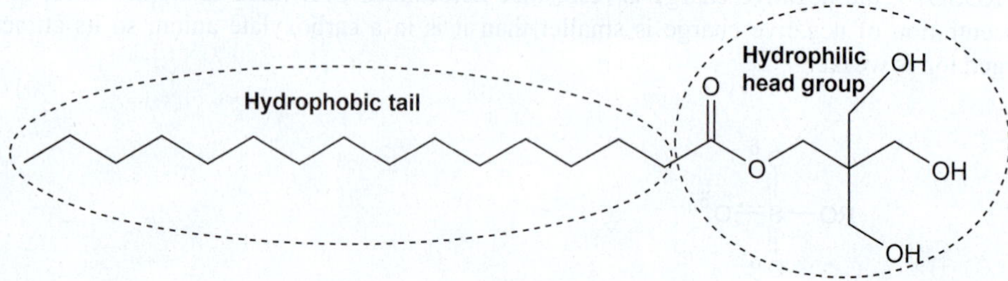

Problem 2.62

Think

In the depiction of the molecular species given in the problem, does the compound have a hydrophilic end? Hydrophobic end? What is required for a compound to act as a soap? If the chain is flexible, as described in the problem, can the molecule attain a shape in which one end is very hydrophilic and the other is very hydrophobic?

Solve

If the carbon chain were rigid, the compound could not be an effective soap, for the same reason as in Problem 2.29. However, because the chain is flexible, it can attain the geometry below, where one end is highly hydrophilic (the ionic end) and the other end is highly hydrophobic (the hydrocarbon portion). Thus, it can act as a soap.

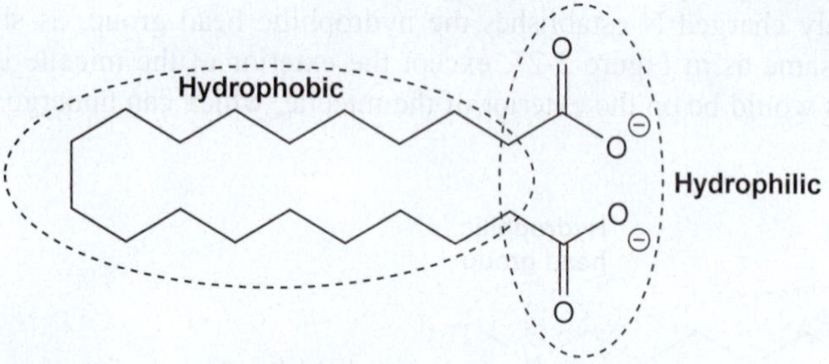

Section 2.11 The Organic Chemistry of Biomolecules

Problem 2.63

Think

What is the structure of an isoprene unit? Can you assign each C to an isoprene unit without having a C part of two isoprene units? How many isoprene units are present?

Solve

An isoprene unit has five carbon atoms, consisting of a four-carbon chain and a fifth carbon attached to C2; that is, it appears as follows:

The figure below shows how the C atoms can be assigned to different isoprene units. There are four isoprene units in all, making this compound a derivative of a diterpene.

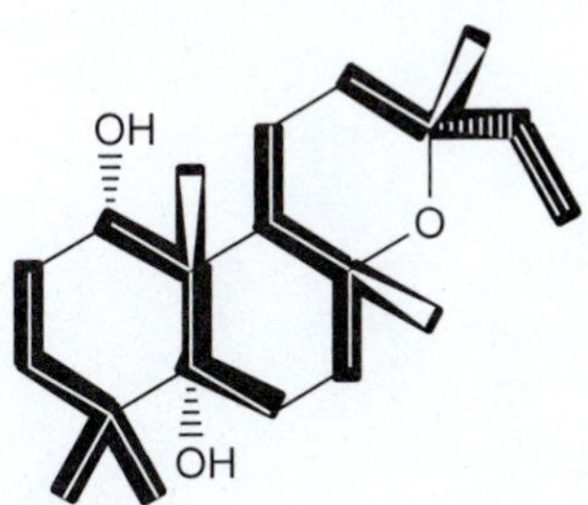

Problem 2.64

Think

What is the general structure of a fat or oil (Section 2.11a)? How does the structure change when you have a specific fatty acid group? Consult Table 2-8 for the structures of various fatty acids.

Solve

The generic structure for a fat or oil is shown below. A fat or oil contains three adjacent ester groups produced from fatty acids and the alcohol group of glycerol. Fats and oils are distinguished by the identities of the R groups, which are established by the fatty acids from which they come.

Typically long-chain hydrocarbon groups

An ester functional group

Glycerol

Fatty acids

+ 3 H_2O

Fat or oil (Triacylglycerol or triglyceride)

(a) Fat or oil from three molecules of stearic acid:

(b) Fat or oil from two molecules of oleic acid and one molecule of linolenic acid:

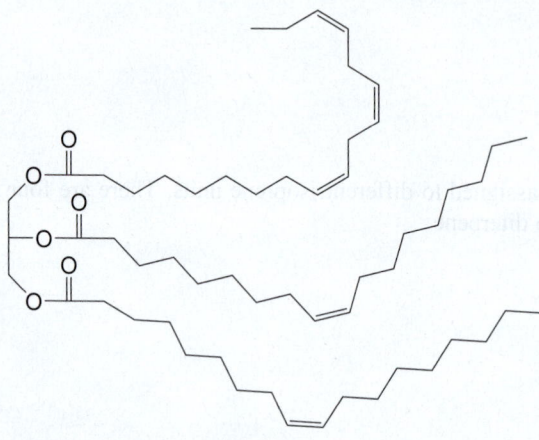

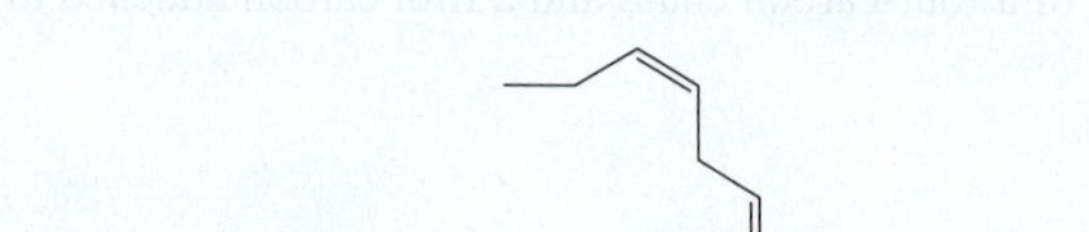

Problem 2.65

Think

Review Sections 2.11a, 2.11c, and 2.11d to study the basic structures of fatty acids, steroids, and waxes, respectively.

Solve

(a) This is a wax. Waxes are typically mixtures of compounds consisting mainly of long-chain esters and alkanes.

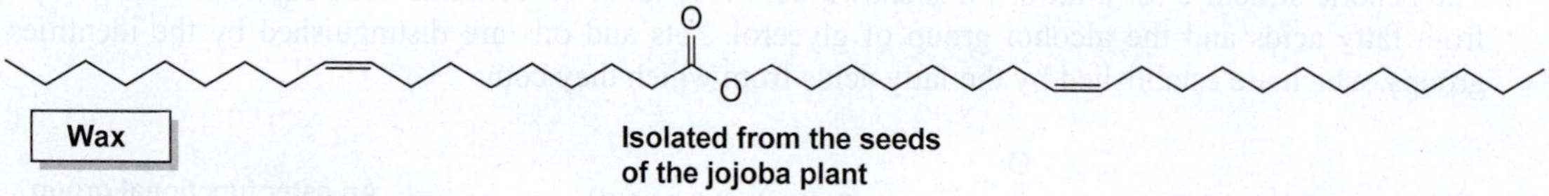

Wax

Isolated from the seeds of the jojoba plant

(b) This is a steroid. Steroids consist of three six-membered rings and one five-membered ring fused together.

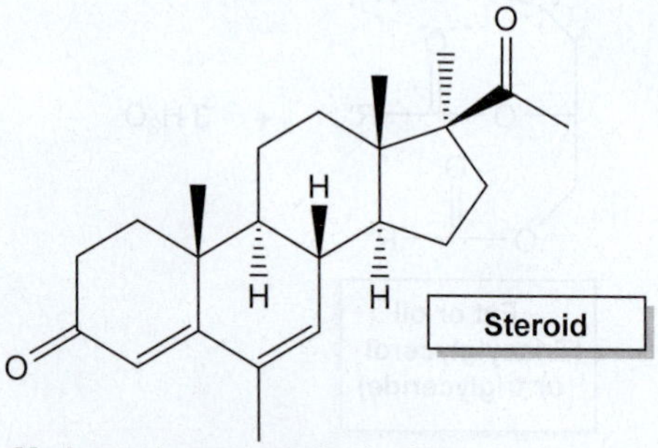

Steroid

Medrogestone, a synthetic drug

(c) This is a fatty acid. A fatty acid is a carboxylic acid with a long hydrophobic hydrocarbon tail that can include C=C bonds.

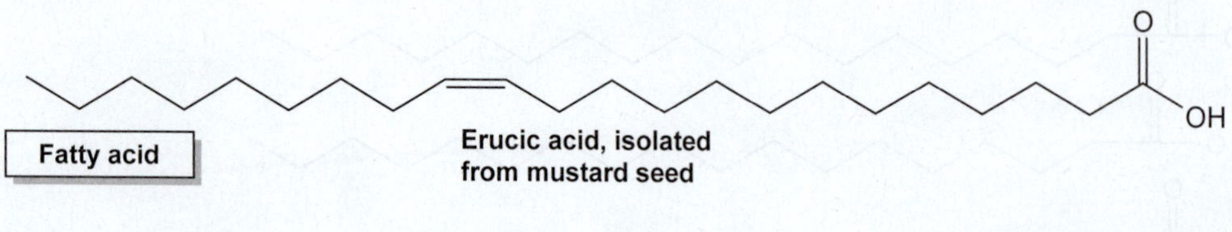

Fatty acid

Erucic acid, isolated from mustard seed

Integrated Problems

Problem 2.66

Think

What are the dominant intermolecular interactions in samples of the two pure compounds? Are they the same or different? If they are the same, how can the shapes of the molecules affect the strengths of those interactions? It may help to draw each pair of molecules undergoing that type of interaction.

Solve

The two molecules are made from exactly the same atoms (i.e., have the same formula), and each possesses both a H-bond donor and a H-bond acceptor. Thus, the dominant intermolecular interaction in each compound is H bonding. The higher boiling point of propan-1-ol indicates that it undergoes stronger H bonding than propan-2-ol does. The strength of H bonding is made weaker in propan-2-ol because of the bulkiness of each alkyl group (i.e., CH_3 group) bonded to the C atom to which the OH group is attached. That bulkiness makes it more difficult for the H-bond donor and acceptor to be close to each other to form the H bond. In propan-1-ol, the C atom bonded to the OH group is connected to only one alkyl group (i.e., a CH_2CH_3 group) and a H atom, which is very small.

Bulkiness of alkyl groups ***decreases*** **stability of H bond**

Problem 2.67

Think

What are the dominant intermolecular interactions in a pure sample of each compound? What are the important bond dipoles? In which directions do they point? Determine the magnitude of the net dipole. How does the magnitude of the net dipole relate to the boiling point of the molecule?

Solve

As shown below, for the small cyclic ethers, the two C–O bond dipoles (thin arrows) point in nearly the same directions and so add together to give a substantial net dipole (thick arrows). For these compounds, therefore, there are relatively strong dipole–dipole interactions. Small cyclic alkanes are nonpolar. Therefore, the significant dipole moments in the small cyclic ethers give rise to higher boiling points. For larger rings, the vectors of the two C–O bond dipoles point nearly at each other, allowing them to cancel more efficiently, making the larger rings less polar. Thus, the polarity of the larger cyclic ethers more closely resembles the polarity of the larger cyclic alkanes, and the boiling points of the larger cyclic ethers more closely resemble the boiling points of the larger cyclic alkanes.

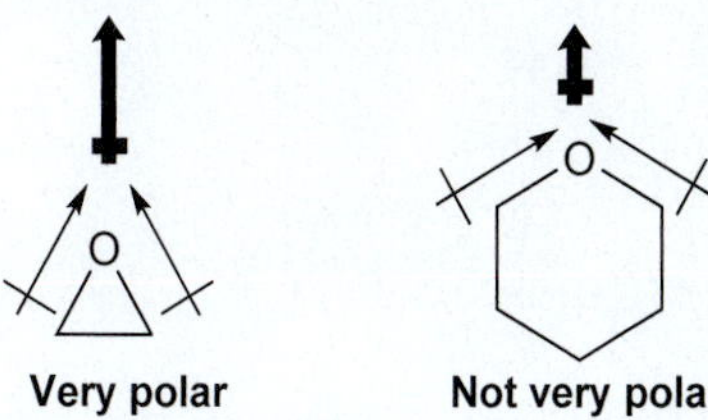

Very polar **Not very polar**

Problem 2.68

Think

What are the intermolecular interactions in a pair of **A** molecules? A pair of **B** molecules? How many H-bond donors and acceptors are there? Are the numbers of electrons significantly different? Consider the surface area of contact between each molecule.

Solve

Hydrogen bonding exists between a pair of each molecule, and the strength of hydrogen bonding is expected to be very similar because molecules **A** and **B** have the same number of H-bond donors and acceptors and are also structurally very similar. The difference must be in the hydrocarbon tails, which are highly nonpolar. Being nonpolar, those tails participate mainly in induced dipole–induced dipole interactions. Such interactions must be stronger with the saturated fatty acid because it has a higher melting point. Because the two molecules have nearly the same number of total electrons, the two molecules have about the same polarizability. Thus, the stronger interactions with the saturated fatty acid can be explained by a greater contact surface area.

H bond
Liquid
O
OH
O
OH
A

H bond
Solid
O
OH
O
OH
Stronger induced dipole–induced dipole interactions
B

Problem 2.69

Think

What are the major intermolecular interactions in each? Would you expect hexafluorobenzene to be more polarizable than benzene? To be more polarizable, what must the electrons be able to do easily? How do you think that the strong electronegativity of F would affect the electrons' ability to do this?

Solve

Because both compounds are nonpolar, induced dipole–induced dipole interactions govern their boiling points (b.p.). The fact that they have about the same boiling points means that they have about the same polarizability. We would normally expect hexafluorobenzene to have a greater polarizability (and hence a higher boiling point) because it has so many more electrons. However, F is very highly electronegative, which prevents its electrons from being moved around easily (i.e., F's high electronegativity decreases its polarizability).

Benzene
b.p. = 80 °C

F F F F F F
Hexafluorobenzene
b.p. = 81 °C

Problem 2.70

Think

How many electron groups are present about the O atom in H_2O? What is the electron geometry about the O in H_2O? What is the bond angle? How do you think the bond angle compares to the angle of the two H bonds?

Solve

The O atom has four electron groups and, therefore, a tetrahedral electron geometry, so its lone pairs are about 109.5° apart. Thus, the H bonds are about 109.5° apart.

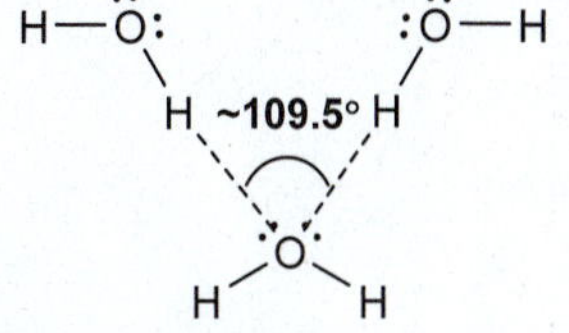

H—Ö:　　:Ö—H

H ~109.5° H

O

H　H

Problem 2.71

Think

What are the intermolecular interactions in each alcohol below? Are ions present? Are strong bond dipoles present? Are there H-bond donors and acceptors? How do you think the bulkiness of the alkyl groups affects the strength of the dominant intermolecular interaction?

Solve

For each compound, the dominant intermolecular interaction between a pair of molecules is H bonding, owing to the presence of a H-bond donor (OH) and a H-bond acceptor (O). The molecules have the same number of donors and acceptors, but with greater bulkiness from the alkyl groups surrounding the OH group, H bonding is more difficult because the H-bond donors and acceptors cannot approach each other as closely. With weaker H bonding come lower boiling points.

1 bulky alkyl group — OH — **Pentan-1-ol** — **b.p. = 136–138 °C**

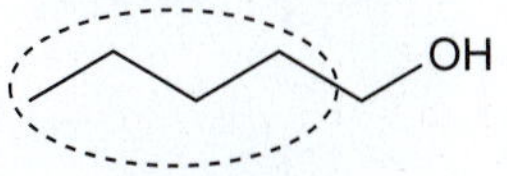

2 bulky alkyl groups — OH — **Pentan-3-ol** — **b.p. = 114–115 °C**

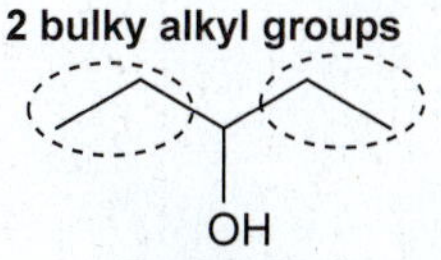

3 bulky alkyl groups — HO — **2-Methylbutan-2-ol** — **b.p. = 102 °C**

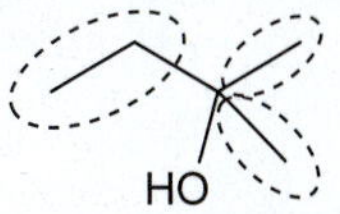

Problem 2.72

Think

What are the dominant intermolecular interactions in each alcohol below? How would the OH groups on the same molecule interact when they are in close proximity? What impact would that have on the molecules' ability to interact with other molecules?

Solve

For each compound, the dominant intermolecular interaction between a pair of molecules is H bonding. Because the OH groups in 1,2-dihydroxybenzene are so close together, they form an *internal* H bond (see below). In 1,3-dihydroxybenzene, this cannot happen, because the OH groups are too far apart. With the OH groups in 1,2-dihydroxybenzene tied up in internal H bonding, they are less available to H bond with other molecules. Therefore, the intermolecular forces between molecules in 1,2-dihydroxybenzene are less than in 1,3-dihydroxybenzene, giving 1,2-dihydroxybenzene a lower boiling point.

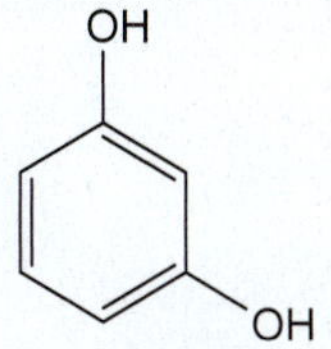

OH — OH — **b.p. = 281 °C** — **1,3-Dihydroxybenzene**

Internal H-bond — HO----H — O — **b.p. = 245 °C** — **1,2-Dihydroxybenzene**

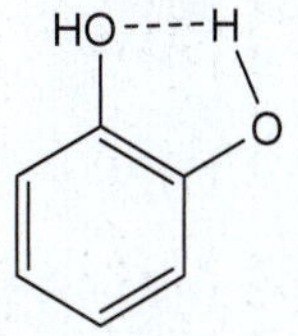

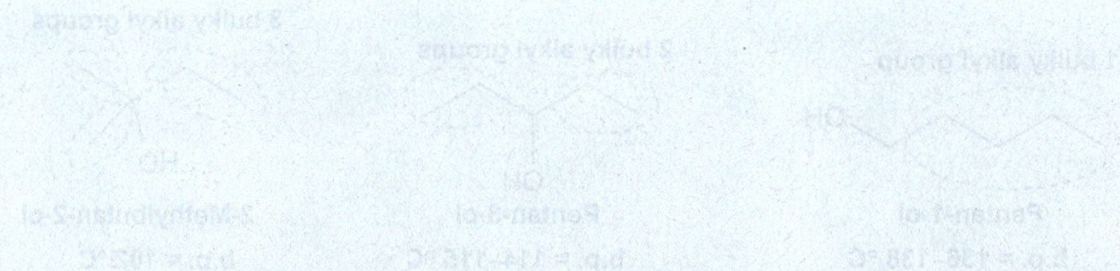

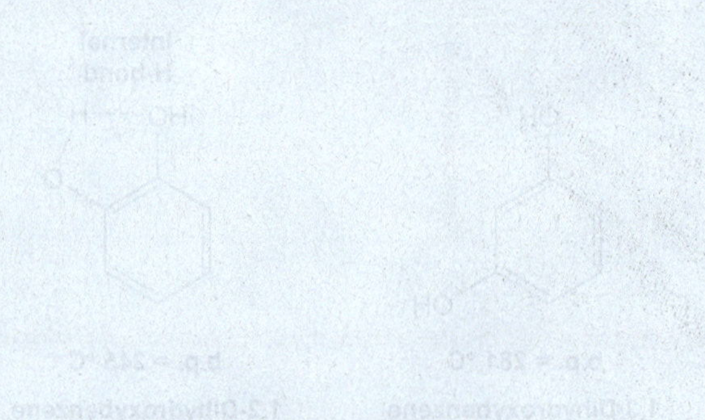

CHAPTER 3 | Orbital Interactions 1: Hybridization and Two-Center Molecular Orbitals

Your Turn Exercises

Your Turn 3.1

Think

What constitutes a nodal plane? How is electron density depicted in the representations of the p_x, p_y, and p_z?

Solve

A nodal plane is a plane in which the electron density is zero. In the depictions of the orbitals, electron density is high inside the surface shown and low outside. In a *p* orbital, the nodal plane passes through the nucleus between the two phases of the orbital. In looking at a Cartesian graph, the three nodes for the p_x, p_y, and p_z orbitals pass through the origin and are mutually perpendicular. The p_x nodal plane is the *yz* plane, the p_y nodal plane is the *xz* plane, and the p_z nodal plane is the *xy* plane. Therefore the nodal plane for the p_z orbital is parallel to the plane of the page.

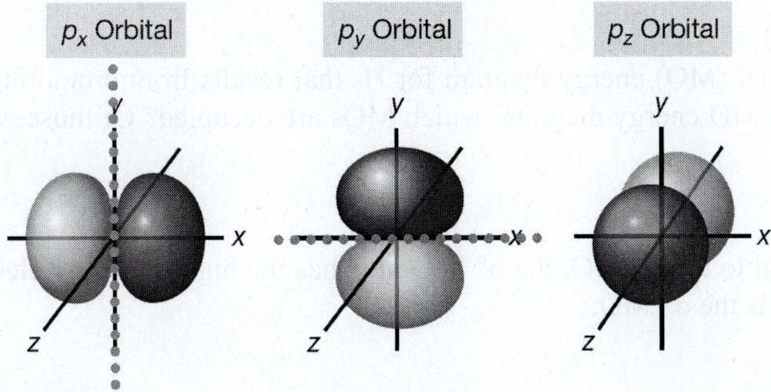

Your Turn 3.2

Think

If both waves are generated on the same side of the rope, what type of interference occurs when the waves meet? With that kind of interference, what happens to the displacement of the rope when the waves meet? If the waves are generated on opposite sides of the rope, what type of interference occurs when the waves meet? With that kind of interference, what happens to the displacement of the rope when the waves meet?

Solve

(a) Both initial waves are generated on the left side. This is analogous to two orbitals having the same phase. The waves propagate toward each other, and when the centers of the two waves meet, the result is a new wave that displaces the rope twice as much as each of the individual waves. This is **constructive interference**. Constructive interference results from addition of the two waves.

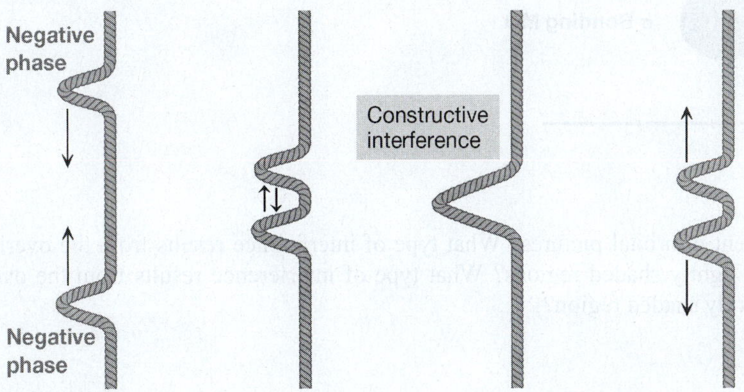

(b) The initial waves are generated on opposite sides of the rope with equal amplitudes. When the two waves meet in the middle, they cancel and yield zero displacement along the entire rope. This is **destructive interference**. Destructive interference results from subtraction of the two waves.

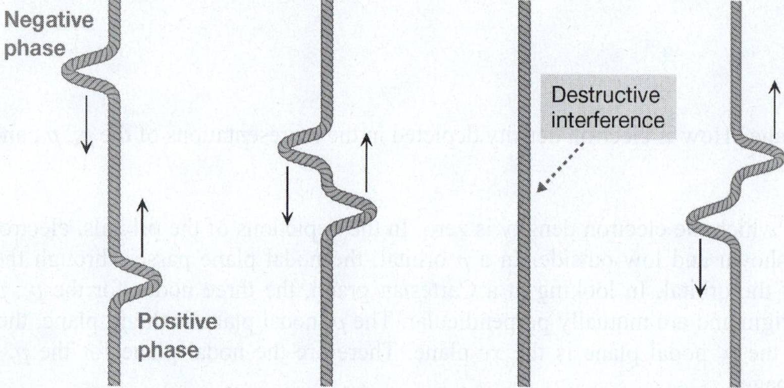

Your Turn 3.3

Think

Solved Problem 3.3 shows the molecular orbital (MO) energy diagram for H_2 that results from promoting one electron from the σ MO to the σ* MO. In that MO energy diagram, which MOs are occupied? Of those, which is the highest in energy?

Solve

When one electron from the σ MO is promoted to the σ* MO, the σ* MO now has the highest energy electron. Therefore, the highest occupied MO (HOMO) is the σ* MO.

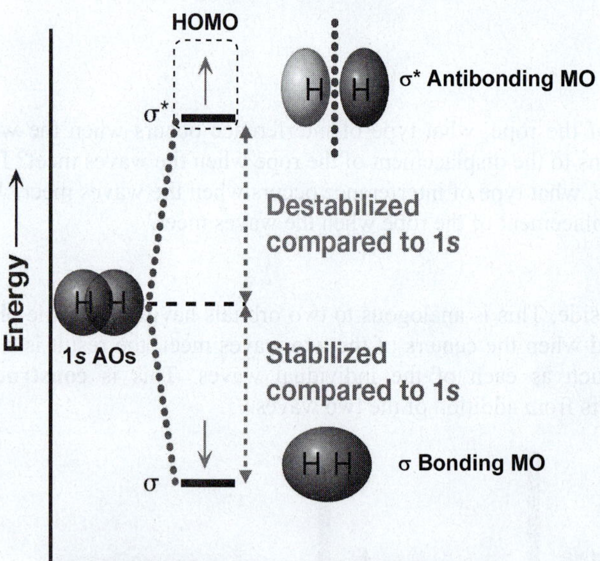

Your Turn 3.4

Think

What does the shading represent in orbital pictures? What type of interference results from the overlap of two darkly shaded regions or two lightly shaded regions? What type of interference results from the overlap of a lightly shaded region and a darkly shaded region?

Solve

Regions of constructive interference are those in which the lobes from the overlapping orbitals are in the same phase—in this case, both lobes are shaded darkly. Regions of destructive interference are those in which the lobes from overlapping orbitals are opposite in phase, that is, one is shaded darkly and the other is shaded lightly.

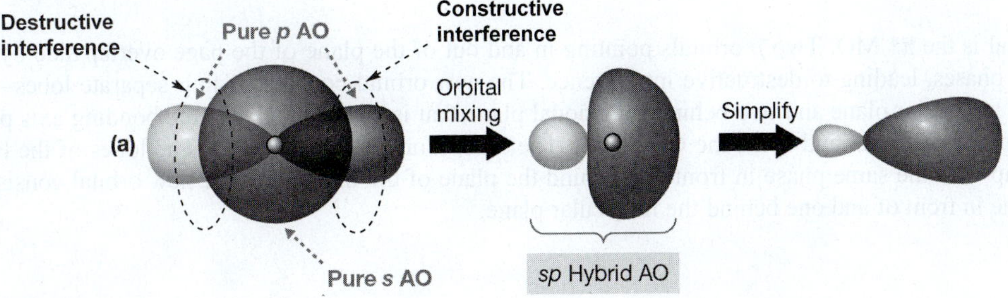

Your Turn 3.5

Think

What does the shading represent in orbital pictures? What type of interference results from the overlap of two darkly shaded regions or two lightly shaded regions? What type of interference results from the overlap of a lightly shaded region and a darkly shaded region?

Solve

Regions of constructive interference are those in which the lobes from the overlapping orbitals are in the same phase—in this case, both lobes are shaded. Regions of destructive interference are those in which the lobes from overlapping orbitals are opposite in phase, that is, one is shaded and the other is unshaded.

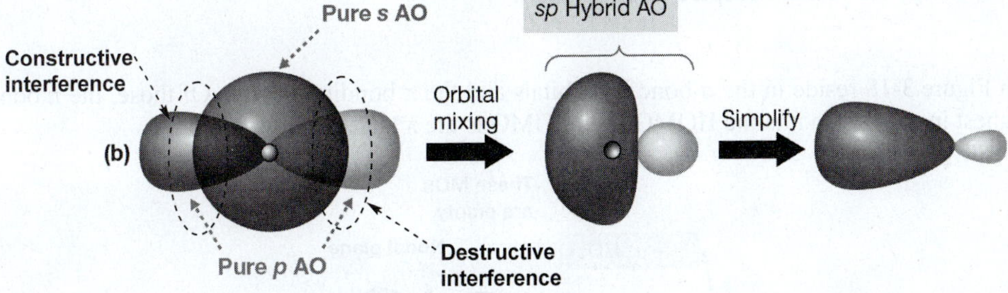

Your Turn 3.6

Think

Use Figure 3-11a and 3-11b a guide. How many hybrid orbitals are present? How many pure (unhybridized) *p* orbitals remain? In which plane do the hybrid orbitals lie? How is the unhybridized *p* orbital oriented relative to the plane of the hybrid orbitals?

Solve

Four orbitals are shown: three sp^2 hybrid orbitals (from the s, p_y, and p_z orbitals) and one unhybridized p_x orbital. The three sp^2 orbitals point 120° apart and are all in the yz plane. The unhybridized p_x orbital is perpendicular to the plane that contains the hybrid orbitals.

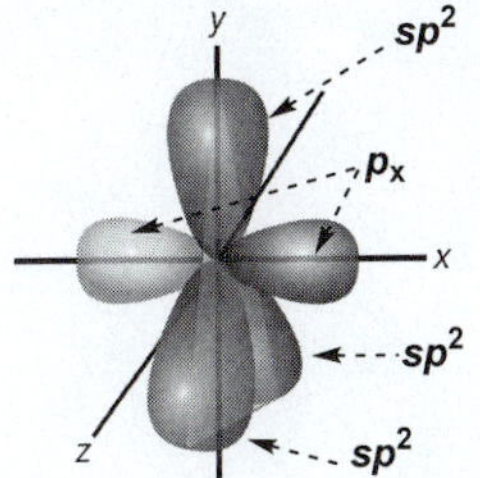

Your Turn 3.7

Think

What type of orbital overlap results in a π MO? What type of orbital overlap results in a π* MO? If the molecular plane of ethene is perpendicular to the molecular plane in Figure 3-18, what is the effect on the nodal plane perpendicular to the bonding axis?

Solve

The first orbital is the π* MO. Two *p* orbitals pointing in and out of the plane of the page overlap side by side with opposite phases, leading to destructive interference. The new orbital consists of four separate lobes—two in front of the molecular plane and two behind. The nodal plane that is perpendicular to the bonding axis points out of the plane of the paper and splits the C–C bond. The second orbital is the π MO. The lobes of the two *p* orbitals overlap with the same phase in front and behind the plane of the molecule. The new orbital consists of two lobes—one in front of and one behind the molecular plane.

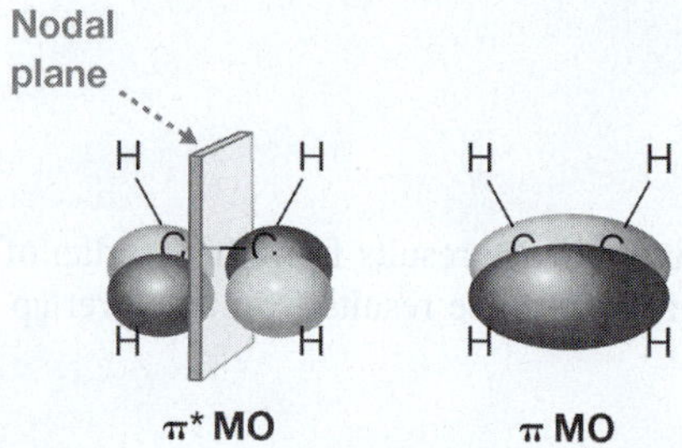

Your Turn 3.8

Think

Which orbitals contain electrons? Of those orbitals, which is the highest in energy (HOMO)? Which unfilled orbital is lowest in energy (lowest unoccupied MO, LUMO)?

Solve

The electrons in Figure 3-18 reside in the σ bonding orbitals and the π bonding orbital. Of those, the π bonding orbital is the highest in energy, so it is the HOMO. The LUMO is the π* antibonding orbital.

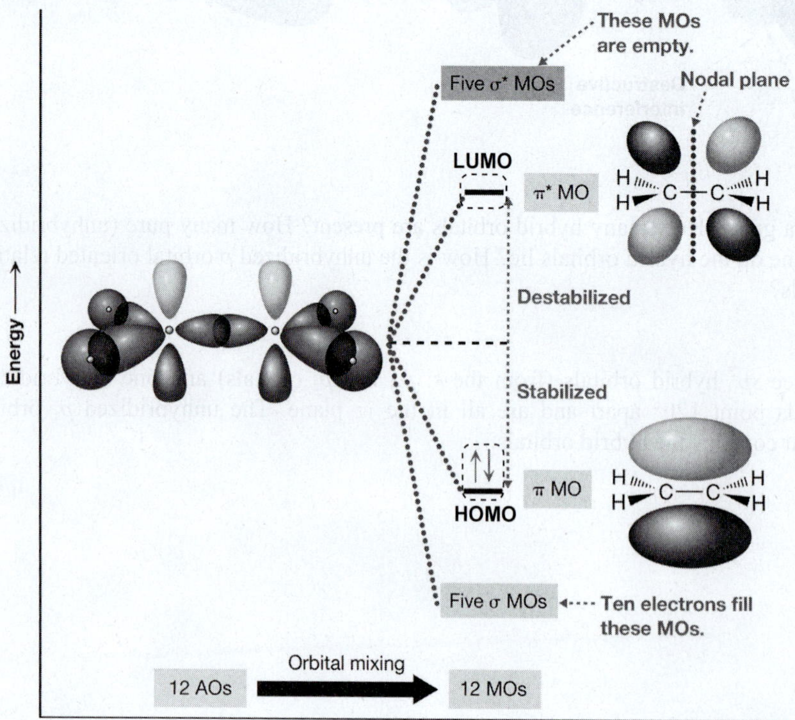

Your Turn 3.9

Think

Do σ interactions occur along the bonding axes or on either side of the bonding axes? Are the carbon σ interactions formed from pure *p* or from *sp* hybrid orbitals?

Solve

The σ interactions occur along the bonding axes. The C–C σ bond is formed from the overlap of two *sp* hybrid orbitals along the bonding axis, and the two C–H bonds are formed from the overlap of one *sp* hybrid orbital from carbon and one *s* orbital from hydrogen. In ethyne, there are three σ bonds.

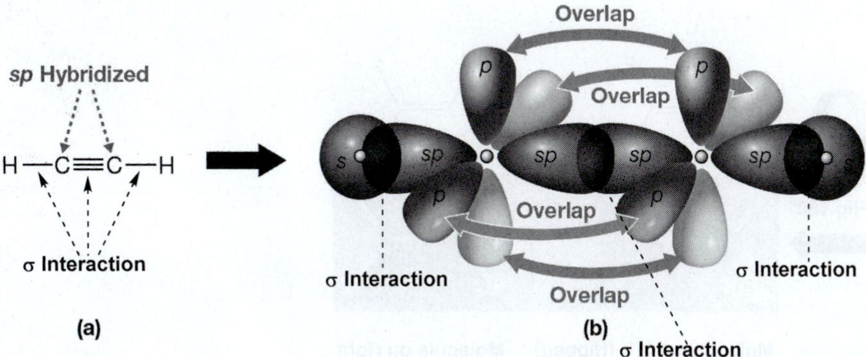

Your Turn 3.10

Think

Which orbitals contain electrons? Of those orbitals, which is the highest in energy (HOMO)? Which unfilled orbital is lowest in energy (LUMO)?

Solve

The electrons in Figure 3-24 reside in the σ bonding orbitals and the two π bonding orbitals. Of those, one of the π bonding orbitals is the highest in energy, making it the HOMO. The LUMO is one of the π* antibonding orbitals.

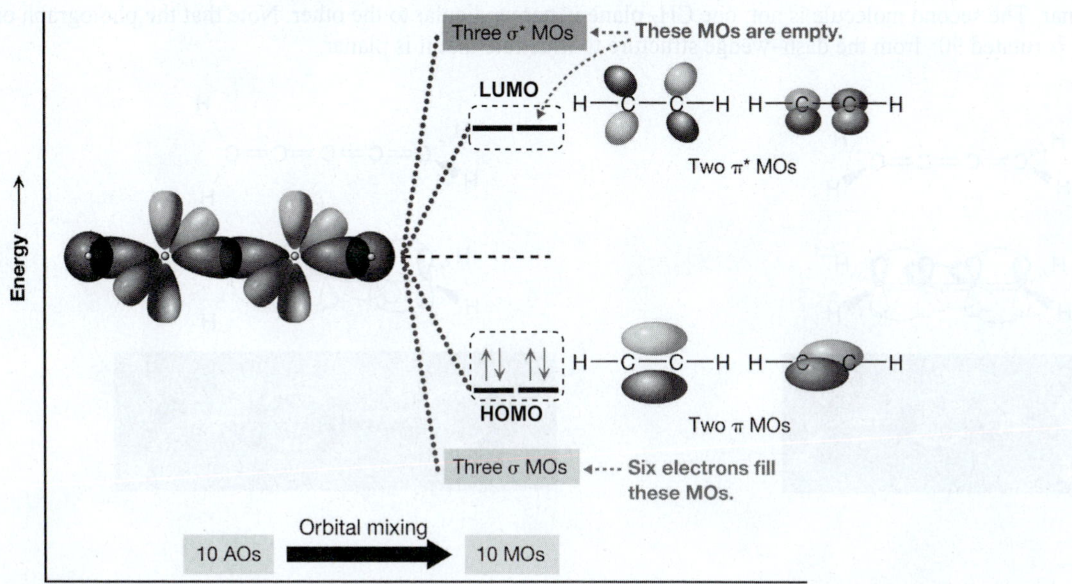

Your Turn 3.11

Think

In the orientations given for the molecules, do all atoms line up when the two molecules are brought together? Can you reorient one molecule so that all atoms in the two molecules line up? You may wish to construct the molecules using a model kit.

Solve

In the orientations given, all atoms do not line up perfectly when the two molecules are held together. However, if the molecule on the left is flipped 180°, as shown below, it appears to be identical to the molecule on the right, so they are, indeed, the same molecule. This happens because one of the C atoms in the double bond is bonded to two H atoms.

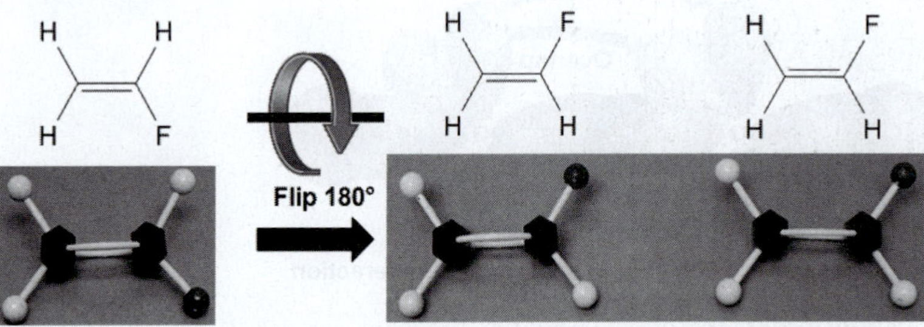

Your Turn 3.12

Think

A model kit is really helpful in understanding this problem. What is the hybridization of the second and third carbons in the first structure? What is the hybridization of the second, third, and fourth carbons in the second structure? Draw the orbitals that form the π bonds. What do you notice about the orientation of the π bonds on adjacent atoms? Look at Figure 3-31 for assistance in setting up your structure.

Solve

The internal carbons are *sp* hybridized. This means that there are two *p* orbitals that are unhybridized and are perpendicular. In a system like this, every π bond is perpendicular to the one next to it. The first molecule is planar. The second molecule is not; one CH_2 plane is perpendicular to the other. Note that the photograph on the left is rotated 90° from the dash–wedge structure to illustrate that it is planar.

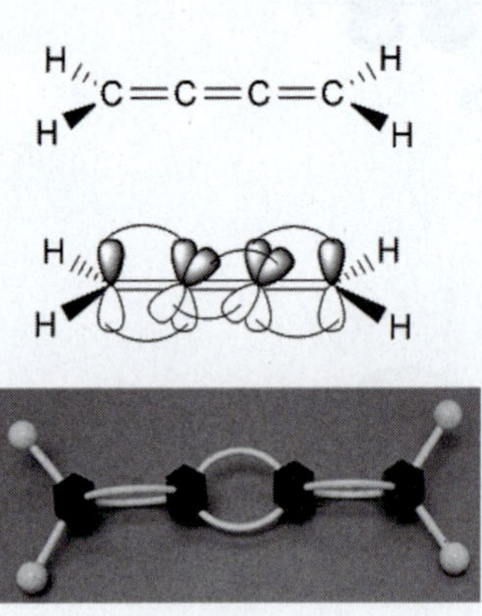

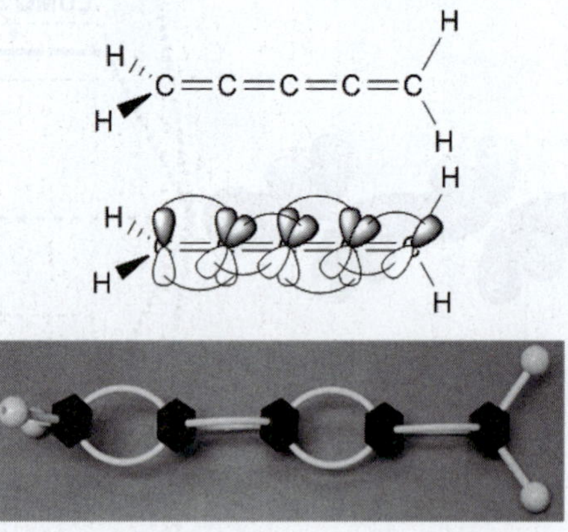

Your Turn 3.13

Think

How many orbitals are hybridized to form sp^3 orbitals? How many of those orbitals are *s* orbitals? What fraction does that correspond to? How many orbitals are hybridized to form sp^2 orbitals? How many of those are *s* orbitals? How many orbitals are hybridized to form *sp* orbitals? How many of those are *s* orbitals?

Solve

Four orbitals (*p*, *p*, *p*, and *s*) are hybridized to form sp^3 orbitals. Only one of those is an *s* orbital, giving a fraction of one-fourth, or 25%, *s* character. Three orbitals (*p*, *p*, and *s*) are hybridized to form sp^2 orbitals, only one of which is an *s* orbital. Thus, the *s* character is one-third, or 33.33%. Two orbitals (*p* and *s*) are hybridized to form *sp* orbitals. Thus, the *s* character is one-half, or 50%.

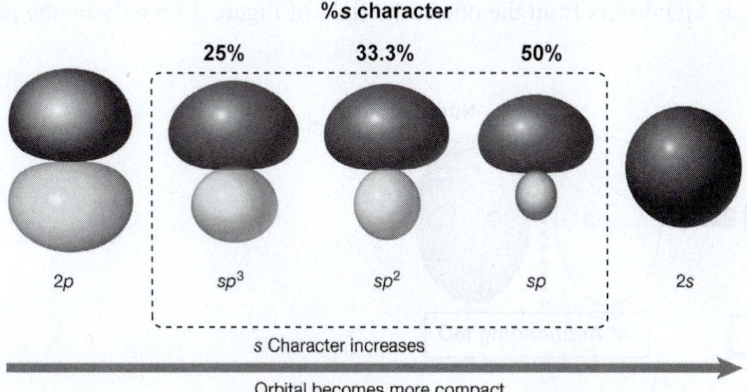

In Chapter Problems

Problem 3.2

Think

In the region in which the *s* orbitals overlap, are the phases the same or opposite? Will that lead to constructive interference or destructive interference? How does the resulting orbital shape compare to the one at the right of Figure 3-6b?

Solve

One orbital is shaded darkly and the other orbital is shaded lightly, indicating that they have opposite phases. Therefore, destructive interference will take place, and the result is an MO that has been diminished in the internuclear region and possesses a nodal plane between the two nuclei. This MO is designated by a sigma star, σ*, indicating an antibonding MO. This MO differs from the one at the right of Figure 3-6b only by the phases of the orbitals, so it is not unique.

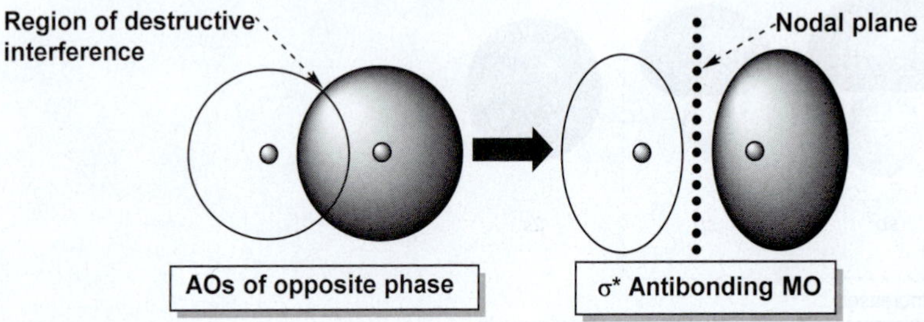

Problem 3.4

Think

How many electrons remain after one electron is removed? In which MO will that electron reside? Is an electron in the σ MO stabilized or destabilized compared to an electron in the 1*s* atomic orbital (AO)? Is the HOMO filled or unfilled?

Solve

The electron will be removed from the σ MO, leaving one electron in the lower-energy σ MO. In an isolated H atom and H^+ ion, the electron would be in a 1*s* orbital. Compared to the electron in the 1*s* H-atom AO, the one electron is stabilized in the σ MO for H_2^+, so H_2^+ is more stable than the isolated H atom and H^+ ion. However, H_2^+ is not as stable as the H_2 atom, because H_2 has one more electron stabilized in the σ MO.

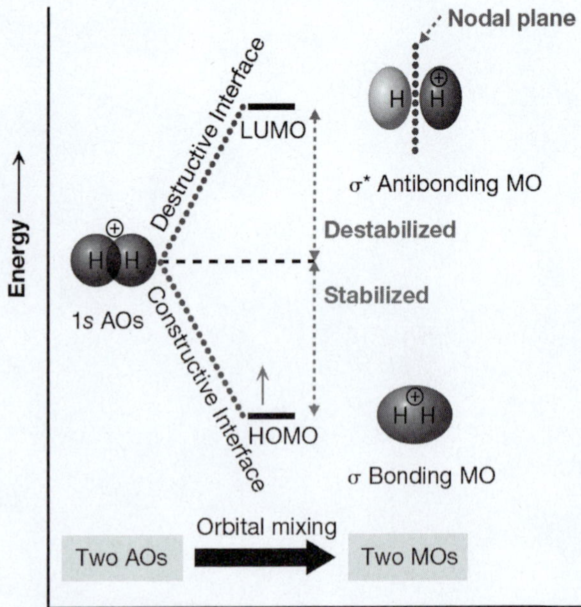

Problem 3.5

Think

Which regions of the orbitals overlap in the same phase? What type of interference results? Which regions of the orbitals overlap in opposite phases? What type of interference results? How does the shape of the resulting orbital compare to the one in Figure 3-9a? What is different about the two orbitals?

Solve

The *s* orbital overlaps in the same phase as the *p* orbital lobe on the right (both orbitals are shaded lightly there) and results in constructive interference. The *s* orbital overlaps in the opposite phase of the *p* orbital lobe on the left (one orbital is shaded lightly and the other is shaded darkly) and results in destructive interference. This is the same as Figure 3-9a, but the phases are inverted.

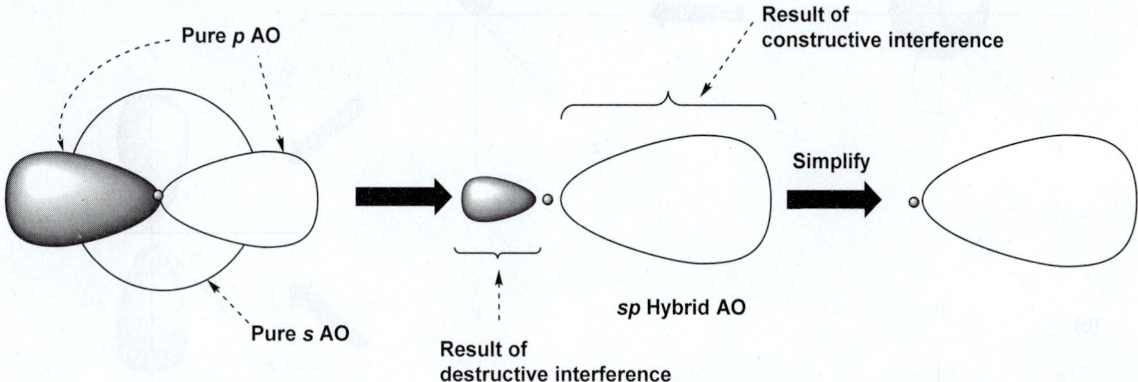

Problem 3.6

Think

Which regions of the orbitals overlap in the same phase? What type of interference results? Which regions of the orbitals overlap in opposite phases? What type of interference results? How does the shape of the resulting orbital compare to the one in Figure 3-9b? What is different about the two orbitals?

Solve

The *s* orbital overlaps in the same phase as the *p* orbital lobe on the left (both orbitals are shaded lightly there) and results in constructive interference. The *s* orbital overlaps in the opposite phase of the *p* orbital lobe on the right (one orbital is shaded darkly and the other is shaded lightly) and results in destructive interference. This is the same as Figure 3-9b, but the phases are inverted.

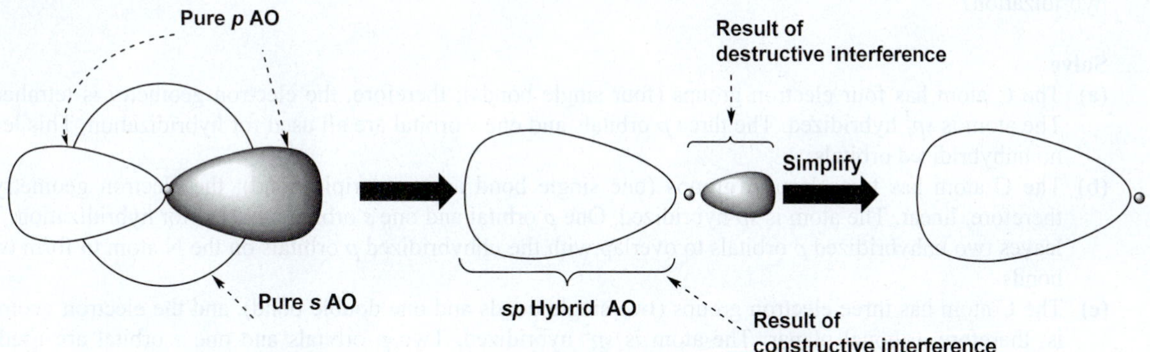

Problem 3.8

Think

Along which axis are these *sp* hybrid orbitals aligned? What orbitals, therefore, must interact? What is the phase of the large lobe of each hybrid orbital?

Solve

The *sp* hybrid orbitals are aligned along the *y* axis. Therefore, it must be the p_y orbital that has been used for hybridization, leaving the p_x and the p_z orbitals unhybridized. The large lobe is shaded darkly in both hybrid orbitals, so the *s* and p_y orbitals must both be shaded darkly where the overlap occurs. The respective interactions are as follows:

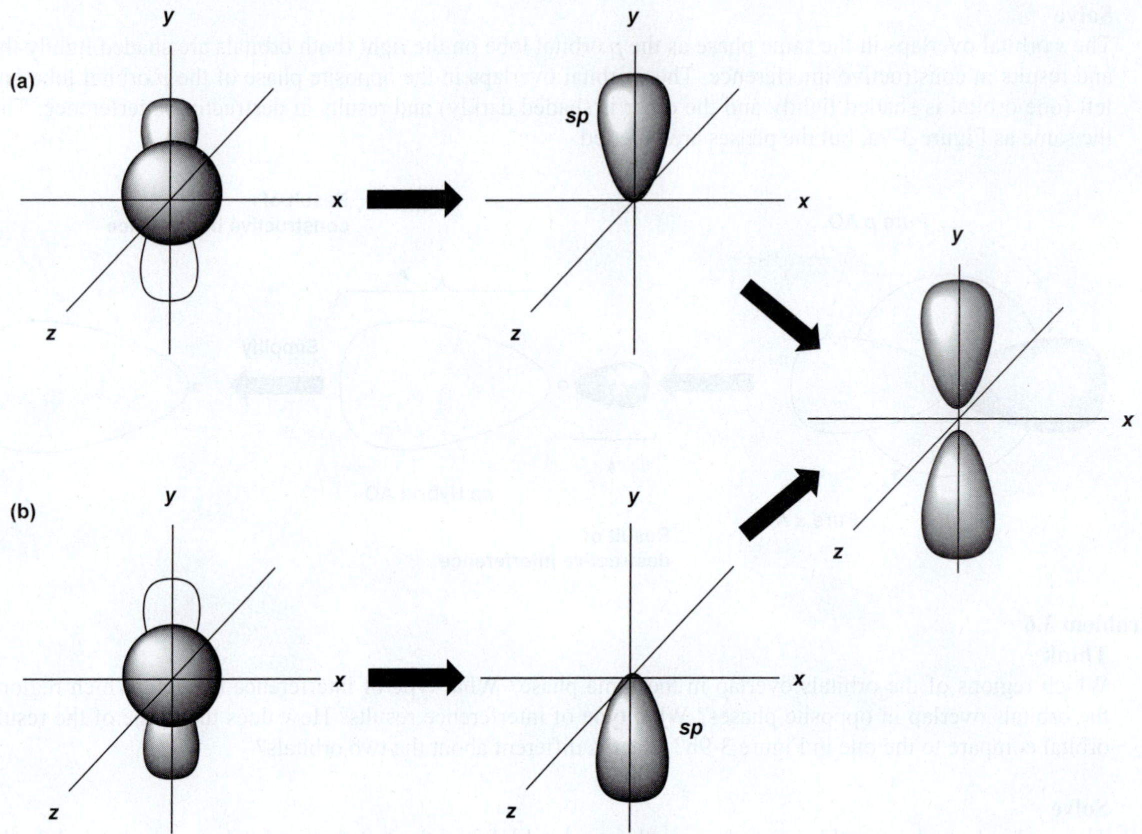

Problem 3.10

Think

What is the electron geometry about each atom? What is the relationship between electron geometry and hybridization?

Solve

(a) The C atom has four electron groups (four single bonds); therefore, the electron geometry is tetrahedral. The atom is sp^3 hybridized. The three *p* orbitals and one *s* orbital are all used for hybridization. This leaves no unhybridized orbitals.

(b) The C atom has two electron groups (one single bond and one triple bond); the electron geometry is, therefore, linear. The atom is *sp* hybridized. One *p* orbital and one *s* orbital are used for hybridization. This leaves two unhybridized *p* orbitals to overlap with the unhybridized *p* orbitals on the N atom to form two π bonds.

(c) The C atom has three electron groups (two single bonds and one double bond), and the electron geometry is, therefore, trigonal planar. The atom is sp^2 hybridized. Two *p* orbitals and one *s* orbital are used for hybridization. This leaves one unhybridized *p* orbital to overlap with one unhybridized *p* orbital on the O atom to form one π bond.

(d) The O atom has three electron groups (one double bond and two lone pairs). The atom is sp^2 hybridized. Two *p* orbitals and one *s* orbital are used for hybridization. This leaves one unhybridized *p* orbital to overlap with the unhybridized *p* orbital on the C atom to form one π bond.

(e) The C atom has four electron groups (four single bonds); therefore, the electron geometry is tetrahedral. The atom is sp^3 hybridized. The three *p* orbitals and one *s* orbital are all used for hybridization. This leaves no unhybridized orbitals.

Problem 3.11

Think

How many electron groups are around an *sp*-hybridized atom? An sp^2-hybridized atom? With the formula C_5H_8 that is given, can any of those electron groups be lone pairs? What types of bonds could those be for a carbon atom to have *sp* hybrid orbitals? To have sp^2 hybrid orbitals?

Solve

(a) Two electron groups must surround an *sp*-hybridized atom. For uncharged carbon atoms, those groups can be two double bonds or one single bond and one triple bond. Two examples are shown below. Other molecules are possible.

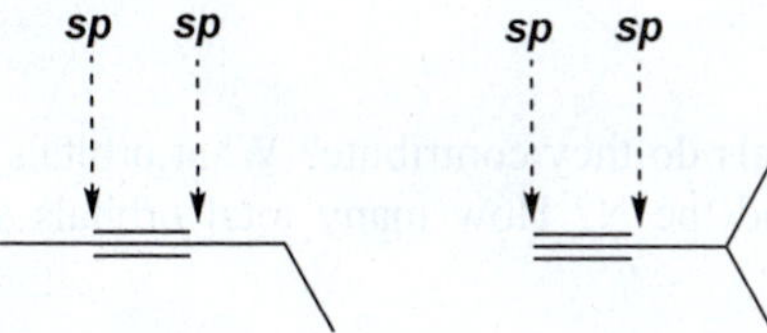

(b) Three electron groups must surround an sp^2-hybridized carbon. For an uncharged carbon atom, those groups can be a double bond and two single bonds. Two examples are shown below. Other molecules are possible.

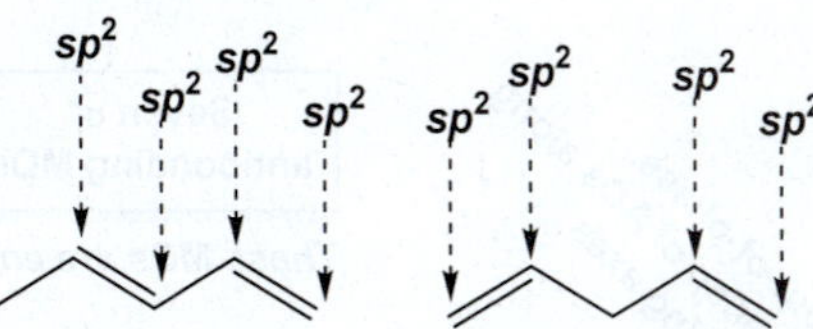

Problem 3.13

Think

What is the hybridization of N? What valence shell orbitals does it contribute? What orbitals do the H atoms contribute? How many total orbitals should be produced from AO mixing?

Solve

(a) The N atom is sp^3 hybridized, so it contributes four sp^3 hybrid orbitals in the valence shell. Each H atom contributes a 1*s* orbital. The overlap of AOs appears on the left in the diagram on the next page.

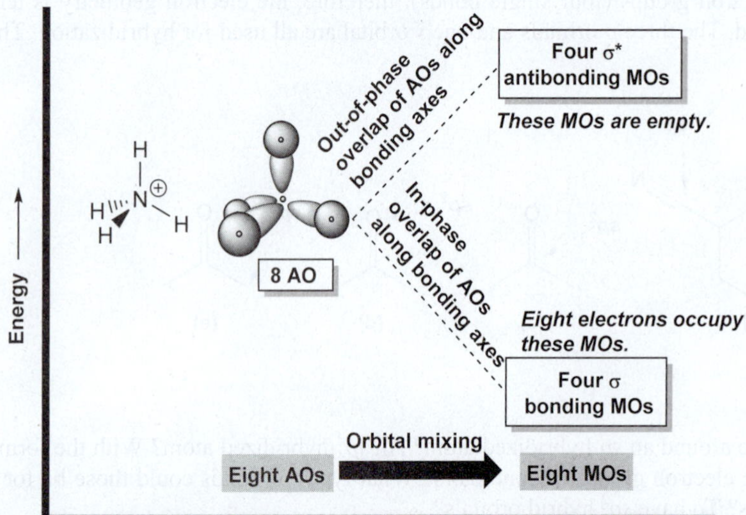

(b) The energies of the MOs appear on the right. The eight AOs that overlap do so via four σ interactions along the bonding axes, generating four σ and four σ* MOs. Eight valence electrons completely fill the σ MOs, leaving the σ* empty. This is the same orbital picture shown in Solved Problem 3.12, except that the C atom is replaced with a N^+.

Problem 3.14

Think

What is the hybridization of N and C? What valence shell orbitals do they contribute? What orbitals do the H atoms contribute? What type of bond exists between the C and the N? How many total orbitals should be produced from AO mixing?

Solve

(a) The N atom and the C atom are both sp^3 hybridized, and each contributes four sp^3 hybrid orbitals in the valence shell. Each H atom contributes a 1*s* orbital. The overlap of AOs appears on the left in the following diagram:

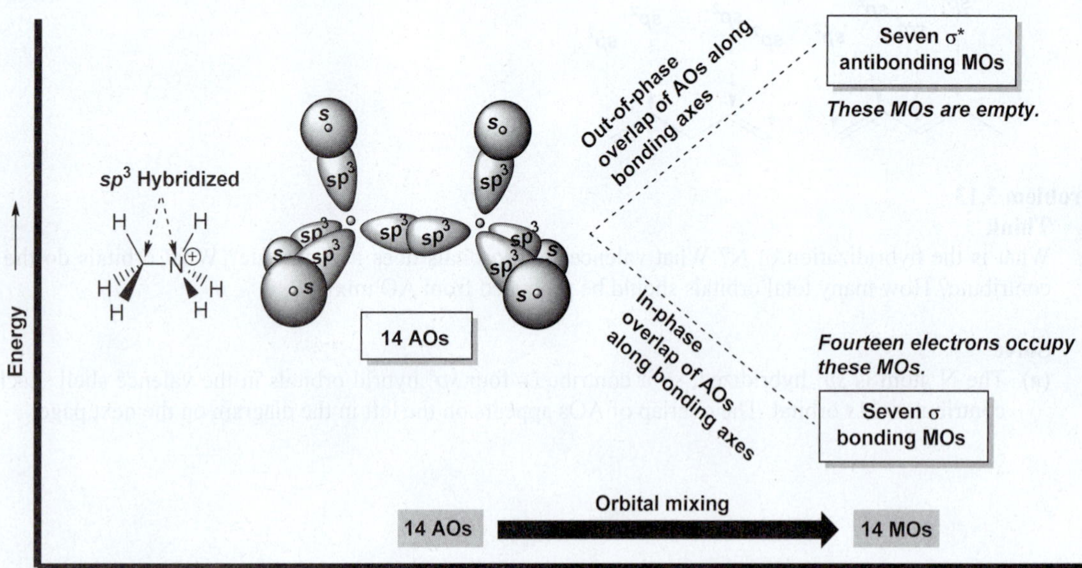

(b) The energies of the MOs appear on the right. The 14 AOs that overlap do so via seven σ interactions along the bonding axes, generating seven σ and seven σ* MOs. Fourteen valence electrons completely fill the σ MOs, leaving the σ* empty. This is nearly identical to the orbital energy diagram shown in Figure 3-14, except one of the C atoms is replaced by a N^+.

Problem 3.16

Think

How many double bonds are present? Of what is each double bond composed? How many single bonds are present? Of what is each single bond composed?

Solve

(a) There is one C=C double bond; it is composed of one σ bond and one π bond. Additionally, there are three C–C single bonds and 10 C–H single bonds. Each single bond is a σ bond. This gives a total of one π bond and 14 σ bonds.

(b) There are three C=C double bonds, and each is composed of one σ bond and one π bond, for a total of three σ bonds and three π bonds. Additionally, there are two C–C single bonds and eight C–H single bonds. Each single bond is a σ bond. This gives a total of three π bonds and 13 σ bonds.

(c) There are three C=C double bonds; each is composed of one σ bond and one π bond, for a total of three σ bonds and three π bonds. Additionally, there are eight C–C single bonds and 12 C–H single bonds. Each single bond is a σ bond. This gives a total of three π bonds and 23 σ bonds.

(d) There is one N=C double bond; it is composed of one σ bond and one π bond. Additionally, there are four C–C single bonds, one N–C single bond, one N–H single bond, and 13 C–H single bonds. Each single bond is a σ bond. This gives a total of one π bond and 20 σ bonds.

(a) 1 π, 14 σ (b) 3 π, 13 σ (c) 3 π, 23 σ (d) 1 π, 20 σ

Problem 3.17

Think

What is the hybridization of N and C? What valence shell orbitals does each contribute? What orbitals do the H atoms contribute? What type of bond exists between the C and the N? What types of orbitals are associated with lone pairs of electrons?

Solve

The C and N atoms are sp^2 hybridized, just as the C and O atoms are in Figure 3-21. The picture, therefore, is nearly identical to Figure 3-21. The only differences arise from (1) a N atom in place of an O atom and (2) a N–H bond in place of a lone pair on O. Consequently, we should expect one nonbonding MO instead of the two that appear in Figure 3-21.

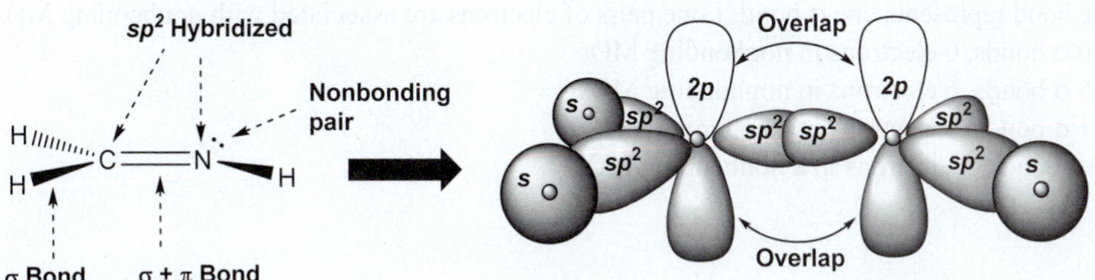

Problem 3.18

Think

How many double bonds are present? Of what is each double bond composed? How many single bonds are present? Of what is each single bond composed? How many lone pairs are present? What types of orbitals are associated with lone pairs of electrons?

Solve

Each double bond represents one σ bond and one π bond. Each single bond (including those to H) represents one σ bond. Each lone pair is associated with a nonbonding MO.

(a) 2 π bonds, 16 σ bonds, 4 electrons (2 pairs) in nonbonding MOs
(b) 2 π bonds, 16 σ bonds, 8 electrons (4 pairs) in nonbonding MOs
(c) 2 π bonds, 12 σ bonds, 4 electrons (2 pairs) in nonbonding MOs
(d) 1 π bond, 19 σ bonds, 2 electrons (1 pair) in nonbonding MOs
(e) 2 π bonds, 11 σ bonds, 8 electrons (4 pairs) in nonbonding MOs

Problem 3.19

Think

What is the hybridization of N and C? What valence shell orbitals does each contribute? What orbitals do the H atoms contribute? What type of bond exists between the C and the N?

Solve

The hybridizations of N and C are both *sp*, the same as the two C atoms in Figure 3-23. The orbital picture, therefore, is identical to that in Figure 3-23, except for (1) a N atom in place of the C atom and (2) a lone pair on N in place of a C–H bond. Without overlap between the N and H atoms, there is one fewer σ bond and one fewer σ* MO.

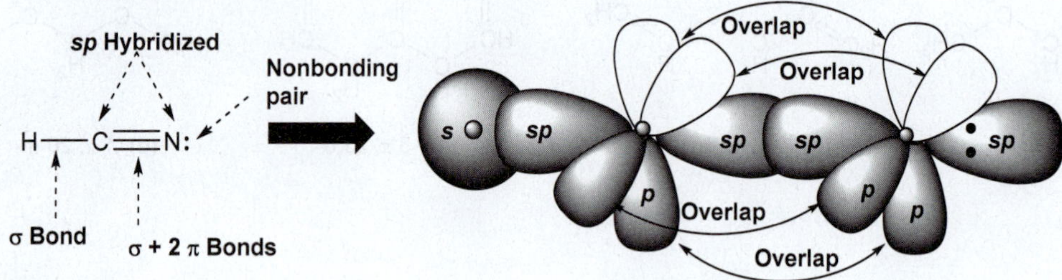

Problem 3.20

Think

How many triple bonds are present? Of what types of bonds is each triple bond composed? How many double bonds are present? Of what types of bonds is each double bond composed? How many single bonds are present? Of what type of bond is each single bond composed? How many lone pairs are present? What types of orbitals are associated with a lone pair of electrons?

Solve

Each triple bond represents two π bonds and one σ bond. Each double bond represents one π bond and one σ bond. Each single bond represents one σ bond. Lone pairs of electrons are associated with nonbonding MOs.

(a) 2 π bonds, 20 σ bonds, 0 electrons in nonbonding MOs
(b) 2 π bonds, 15 σ bonds, 6 electrons in nonbonding MOs
(c) 2 π bonds, 11 σ bonds, 2 electrons in a nonbonding MO
(d) 4 π bonds, 7 σ bonds, 2 electrons in a nonbonding MO

Problem 3.21

Think

What is the hybridization of each nonhydrogen atom? What is the molecular geometry about each nonhydrogen atom? Which geometries are planar? Are the planes of any adjacent atoms required to be parallel?

Solve

The atoms in the same plane (circled) are all connected to the sp^2-hybridized C atom or the sp^2-hybridized N atom. Atoms that are sp^2 hybridized have trigonal planar or bent molecular geometries and are always planar. When two sp^2-hybridized atoms are connected by a double bond, the π bond locks the electron groups on those atoms in the same plane. Therefore, atoms that are doubly bonded together and any atoms to which they are directly bonded must lie in the same plane.

(a) (b) (c)

Problem 3.22

Think

Are any ions present? Are polar bonds present? Are there any H-bond donors or acceptors? Do the bond dipoles completely cancel when they are summed?

Solve

No ions are present, so no ion–ion interactions are possible. There are no H-bond donors or acceptors, so no H bonding is present. Polar C–Cl bonds are present in both compounds. In *cis*-1,2-dichloroethane, those bond dipoles sum to give a net molecular dipole moment, making the molecule polar. A pair of molecules of *cis*-1,2-dichloroethane, therefore, has dipole–dipole interactions in addition to induced dipole–induced dipole interactions. In *trans*-1,2-dichloroethane, conversely, the bond dipoles cancel perfectly, making the molecule nonpolar. A pair of molecules of *trans*-1,2-dichloroethane, therefore, only has induced dipole–induced dipole interactions. The presence of dipole–dipole interactions gives the cis form a higher boiling point (b.p.) than the trans form.

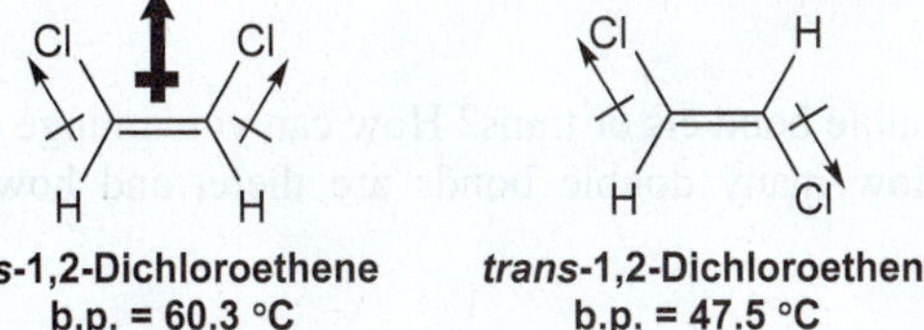

***cis*-1,2-Dichloroethene** **b.p. = 60.3 °C**

***trans*-1,2-Dichloroethene** **b.p. = 47.5 °C**

Problem 3.24

Think

Does the exchange of two groups attached to a doubly bonded C give rise to a different molecule? How can you tell? Perform an exchange of groups for each C=C for molecules **A**–**D** and compare molecules before and after the hypothetical rotation. Review Solved Problem 3.23.

Solve

(a) These molecules before and after the exchange of H and F are the same because both molecules have one Br atom on the same side as the H and the other Br atom on the same side as the F of the double bond. Therefore, two distinct configurations do *not* exist.

Exchange H and F

(b) These molecules before and after the exchange of Br and Cl are different because the first one has the Br atoms on opposite sides of the double bond, whereas the second one has the Br atoms on the same side. Therefore, two distinct configurations *do* exist.

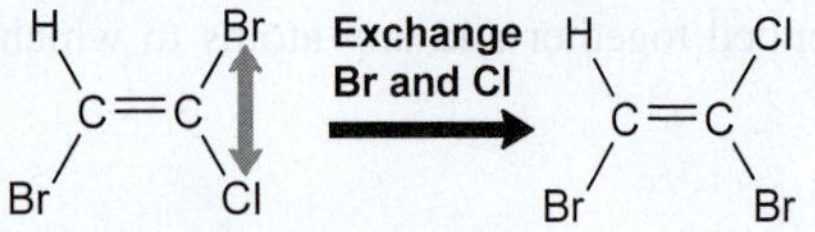

(c) These molecules before and after the exchange of the H and CH=CH_2 groups are the same because both molecules have two H atoms on the same side of the double bond and a H and CH=CH_2 on the same side of the double bond. Therefore, two configurations do *not* exist.

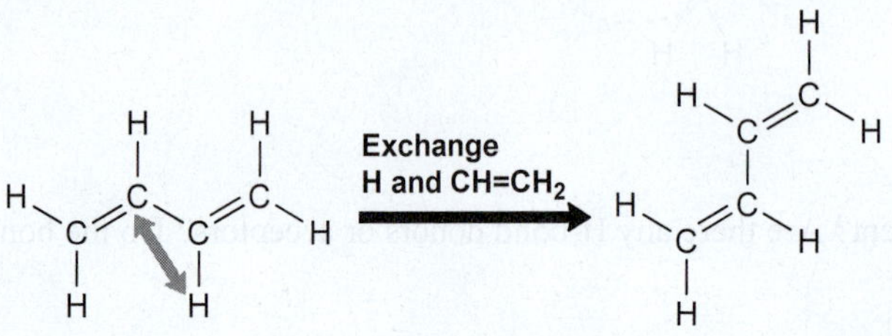

(d) These molecules before and after the CH_3 and lone pair are different because the first one has the CH_3 groups on opposite sides of the double bond, whereas the second one has the CH_3 groups on the same side. Therefore, two configurations *do* exist.

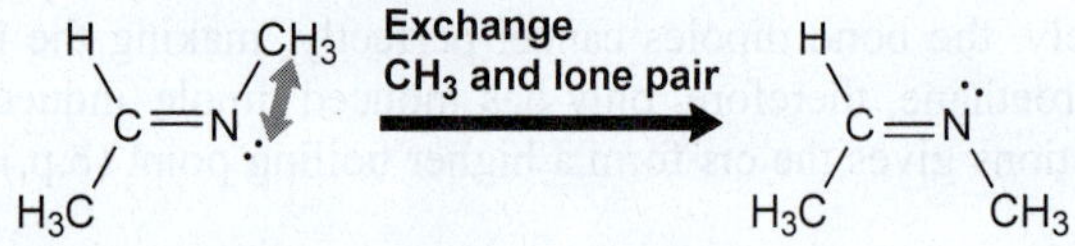

Problem 3.25

Think

What is the structure of α-linolenic acid? What makes a double bond cis or trans? How can you arrange each of these double bonds to be in the trans configuration? How many double bonds are there, and how many configurations are possible for each double bond?

Solve

The all-trans form is shown below. In each trans double bond, the non-H groups are attached to opposite sides of the C=C double bond. In all, because each double bond has two possible configurations (cis and trans), there are 2^3 (or eight) total possibilities.

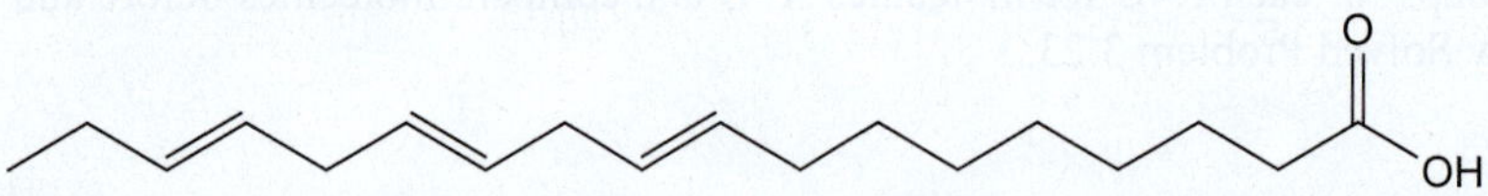

Problem 3.27

Think

What is the hybridization of each atom involved in the C–C bond? How does that affect the length and strength of bonds involving the atom?

Solve

(a) The molecule on the left has the shorter C–C single bond. In the molecule on the left, the C–C bond is between an sp^3-hybridized C atom and an sp-hybridized C atom, whereas in the second molecule, it is between an sp^3-hybridized C atom and an sp^2-hybridized C atom. With greater *s* character in the *sp*-hybridized C than in sp^2, the bond becomes shorter and stronger.

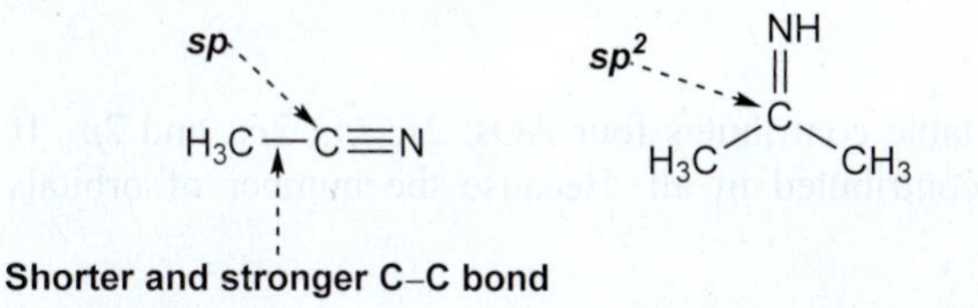

sp $H_3C—C≡N$ *sp²* $(H_3C)(CH_3)C{=}NH$

Shorter and stronger C–C bond

(b) The molecule on the left has the shorter carbon–nitrogen bond. In the molecule on the left, the carbon–nitrogen bond is a triple bond between an sp-hybridized C atom and an sp-hybridized N atom, whereas in the second molecule, the carbon–nitrogen bond is a double bond between an sp^2-hybridized C atom and an sp^2-hybridized N atom. Triple bonds are shorter than double bonds.

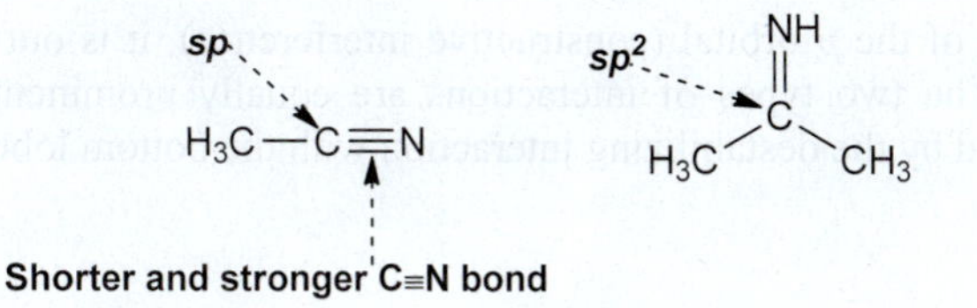

sp $H_3C—C≡N$ *sp²* $(H_3C)(CH_3)C{=}NH$

Shorter and stronger C≡N bond

End of Chapter Problems

Section 3.3 An Introduction to Molecular Orbital Theory and σ Bonds: An Example with H_2

Problem 3.28

Think

How many valence orbitals does each atom in the second row of the periodic table contribute? How many MOs can be produced by mixing together that number of AOs?

Solve

Each atom from the second row of the periodic table contributes four AOs: 2*s*, $2p_x$, $2p_y$, and $2p_z$. If there are three of these atoms, then 12 AOs would be contributed in all. Because the number of orbitals must be conserved, those 12 AOs would produce 12 MOs.

Problem 3.29

Think

Is there a region of constructive interference? Destructive interference? How does each type of interference affect the energy of the resulting orbital? Is one type of interference more prominent than the other?

Solve

Although the *s* orbital is in phase with one lobe of the *p* orbital (constructive interference), it is out of phase with the other lobe (destructive interference). The two types of interactions are equally prominent, so any stabilizing interaction with the top lobe is canceled by the destabilizing interaction with the bottom lobe.

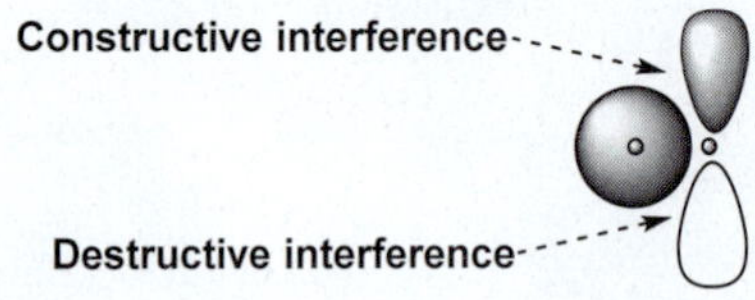

Problem 3.30

Think

How many MOs should result from the interaction of two AOs? How should those AOs mix together differently to give rise to different MOs? When the AOs are in phase, what type of MO results? When the AOs are out of phase, what type of MO results? Does the overlap occur along the bonding axis or on opposite sides of the bonding axis?

Solve

The two AOs should mix to form two MOs. They do so by mixing with different phase combinations of the AOs, as shown below. When the AOs are in phase in the region of overlap, constructive interference takes place and a bonding MO results; when they are out of phase in the overlap region, destructive interference takes place and an antibonding MO results. In both cases, the symmetry is σ, because overlap occurs *along* the bonding axis. Notice that in the antibonding MO, there is an additional node perpendicular to the bonding axis.

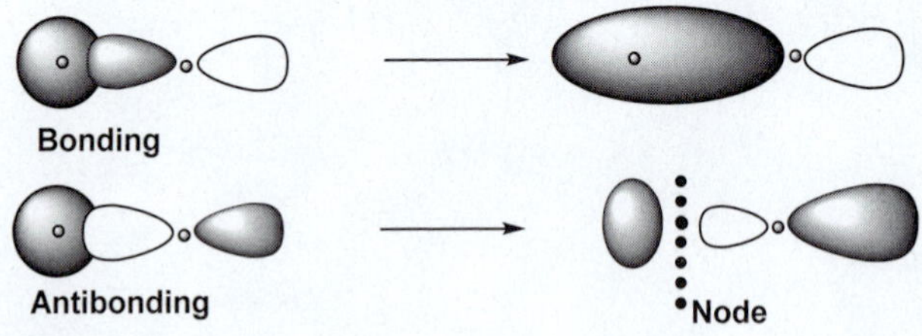

Problem 3.31

Think

How many MOs should result from the interaction of two AOs? How should those AOs mix together differently to give rise to different MOs? When the AOs are in phase, what type of MO results? When the AOs are out of phase, what type of MO results? Does the overlap occur along the bonding axis or on opposite sides of the bonding axis?

Solve

The two AOs should mix to form two MOs. They do so by mixing with different phase combinations of the AOs, as shown below. When the AOs are in phase in the region of overlap, constructive interference takes place and a bonding MO results; when they are out of phase in the overlap region, destructive interference takes place and an antibonding MO results. In both cases, the symmetry is σ, because overlap occurs *along* the bonding axis. Notice that in the antibonding MO, there is an additional node perpendicular to the bonding axis.

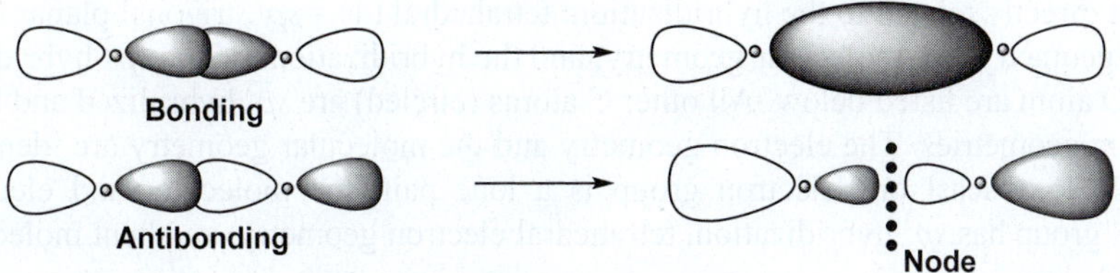

Section 3.4 Hybridized Atomic Orbitals and Geometry

Problem 3.32

Think

How many electron groups are on each atom? What is the electron geometry about each atom? What is the relationship between electron geometry and hybridization?

Solve

The number of electron groups is directly related to the hybridization: tetrahedral (4) = sp^3, trigonal planar (3) = sp^2, linear (2) = sp. The electron geometry (not the molecular geometry) and the hybridizations of all nonhydrogen atoms are listed below.

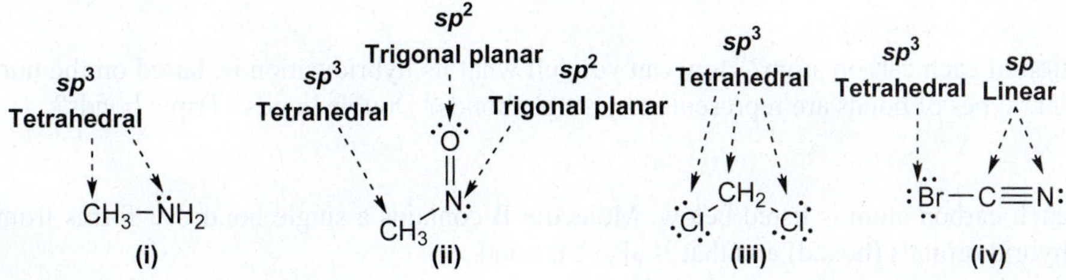

Problem 3.33

Think

How many electron groups are on each atom? What is the electron geometry about each atom? What is the relationship between electron geometry and hybridization?

Solve

The number of electron groups is directly related to the hybridization: tetrahedral (4) = sp^3, trigonal planar (3) = sp^2, linear (2) = sp. The electron geometry (not the molecular geometry) and the hybridizations of all nonhydrogen atoms are listed below.

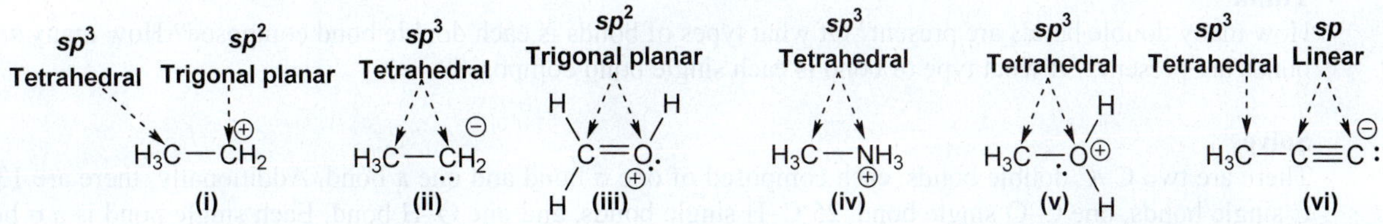

Problem 3.34

Think

How many electron groups are around each nonhydrogen atom? How does the number of groups correspond to electron geometry? What is the relationship between electron geometry and hybridization? Depending on the number of lone pairs around each atom, how does electron geometry relate to molecular geometry?

Solve

The number of electron groups is directly related to the hybridization: tetrahedral (4) = sp^3, trigonal planar (3) = sp^2, linear (2) = *sp*. The electron geometry, the molecular geometry, and the hybridizations of the sp^2-hybridized C atoms and the sp^3-hybridized O atom are listed below. All other C atoms (circled) are sp^3 hybridized and have tetrahedral molecular and electron geometries. The electron geometry and the molecular geometry are identical if all electron groups are bonds. If at least one electron group is a lone pair, the molecular and electron geometries are different. The OH group has sp^3 hybridization, tetrahedral electron geometry, and bent molecular geometry.

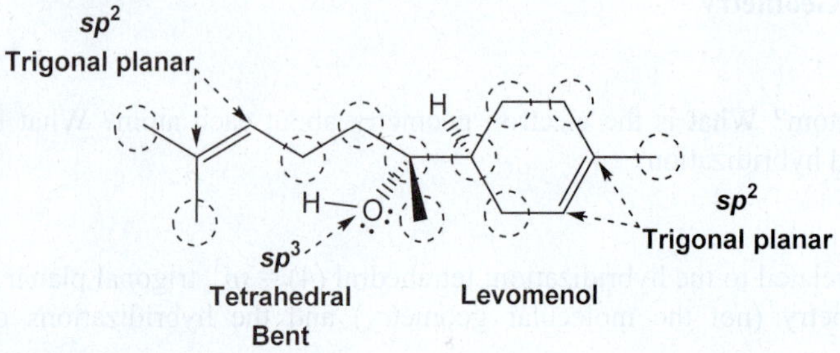

sp^2 Trigonal planar

sp^2 Trigonal planar

sp^3 Tetrahedral Bent

Levomenol

Problem 3.35

Think

What is the hybridization of each carbon atom? How can you tell what its hybridization is, based on the number of electron groups? What types of bonds are represented by single bonds? Double bonds? Triple bonds?

Solve

The hybridization of each carbon atom is listed below. Molecule **B** contains a single bond that forms from the overlap of *sp* and sp^2 hybrid orbitals (boxed) and that is also a σ bond.

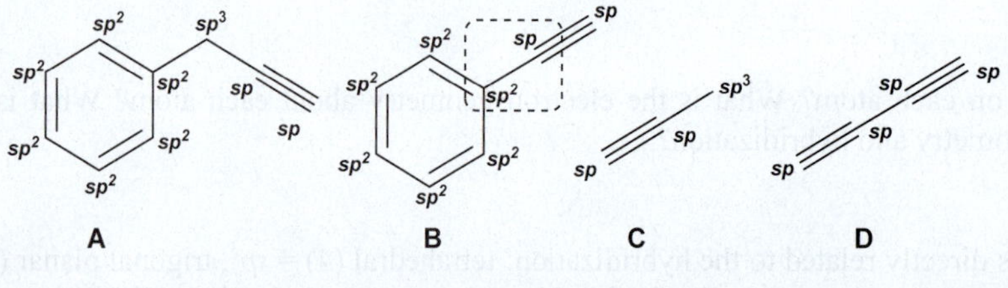

A B C D

Sections 3.5–3.8 An Introduction to π Bonds: An Example with Ethene ($H_2C{=}CH_2$)

Problem 3.36

Think

How many double bonds are present? Of what types of bonds is each double bond composed? How many single bonds are present? Of what type of bond is each single bond composed?

Solve

There are two C=C double bonds, each composed of one σ bond and one π bond. Additionally, there are 13 C–C single bonds, one C–O single bond, 25 C–H single bonds, and one O–H bond. Each single bond is a σ bond. This gives a total of two π bonds and 42 σ bonds.

Problem 3.37

Think

How many double bonds are present? What kinds of bonds make up each double bond? How many π bonds does that correspond to? How many electrons are in each π bond?

Solve

There are two double bonds; each double bond consists of one σ bond and one π bond, for a total of two π bonds. There are two electrons in each π bond for a total of four π electrons.

Problem 3.38

Think

How many electron groups are on each nonhydrogen atom? What is the electron geometry about each of those atoms? What is the relationship between electron geometry and hybridization? How many single/double/triple bonds are present? Of what types of bonds is each bond composed?

Solve

(a) All hybridizations are labeled on the figure below. The sp^3-hybridized C atoms each have four electron groups and a tetrahedral electron geometry, sp^2-hybridized C atoms each have three electron groups and a trigonal planar electron geometry, and sp-hybridized C atoms each have two electron groups and a linear electron geometry. The ROH oxygen is sp^3 hybridized, as it has four electron groups and a tetrahedral electron geometry. The O of the C=O bond is sp^2 hybridized, as it has three electron groups and a trigonal planar electron geometry.

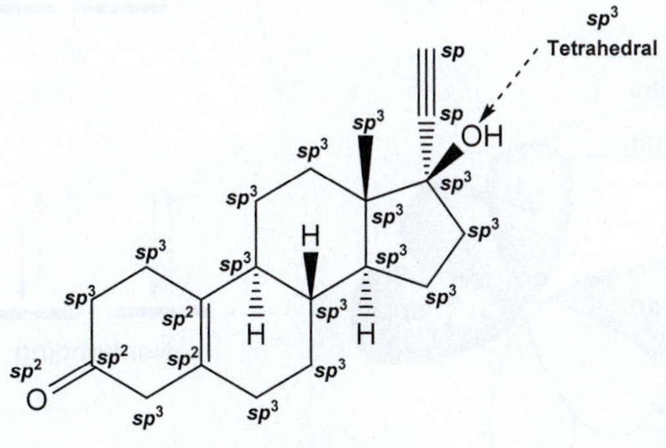

Norethynodrel

(b) There is one triple bond, composed of one σ bond and two π bonds. There are two double bonds, each composed of one σ bond and one π bond. There are 21 C–C single bonds, one C–O single bond, and 26 single bonds to H. In total, there are 51 σ bonds and four π bonds.

Problem 3.39

Think

How many double bonds are present? How many π bonds are in each double bond?

Solve

There are 11 double bonds present; each double bond consists of one σ bond and one π bond. All 11 π bonds are labeled on the figure below.

β-Carotene

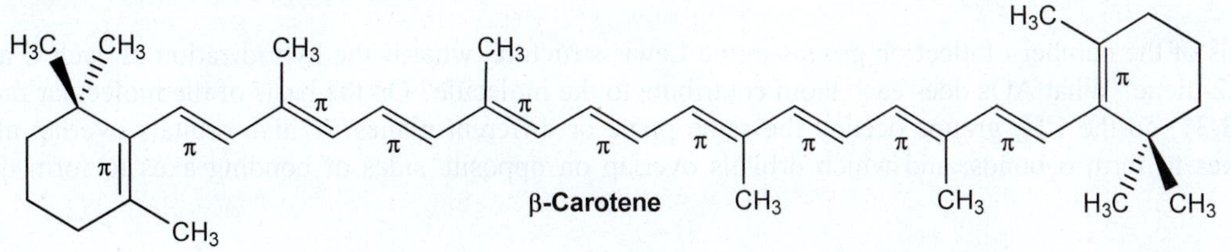

Problem 3.40

Think

What is the hybridization of C and O on the carbon monoxide molecule? How can you tell what its hybridization is, based on the number of electron groups? What AOs does each atom contribute to the molecule? Which orbitals overlap along bonding axes to form the σ bonds, and which orbitals overlap on opposite sides of bonding axes to form the two π bonds? Are there any nonbonding electrons?

Solve

The C and O atoms each have two electron groups in the Lewis structure and, therefore, are *sp* hybridized. Each atom contributes two *sp*-hybridized AOs and two unhybridized *p* AOs to the molecule. The orbital overlap picture is shown below. Two *sp* AOs overlap between the C and O atoms, leading to one σ MO and one σ* MO. There is overlap between adjacent p_y orbitals and between adjacent p_z orbitals. Each overlap involves two orbitals, giving rise to one π MO and one π* MO or to two MOs of π symmetry. In all, there are four MOs of π symmetry. The two *sp*-hybridized AOs on the outside of the molecule don't mix with other AOs, so they remain nonbonding. The 10 valence electrons are filled in the MOs from lowest to highest energy. The four nonbonding electrons represent the two lone pairs.

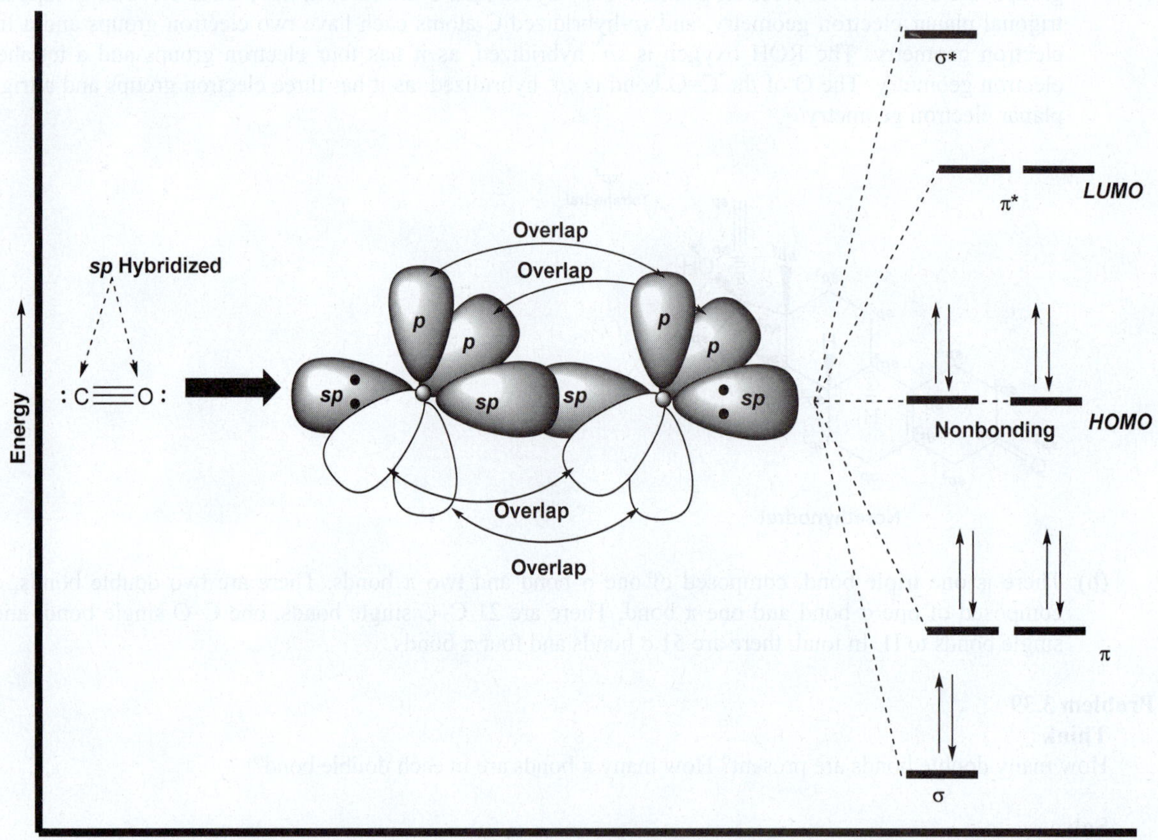

Problem 3.41

Think

On the basis of the number of electron groups in the Lewis structure, what is the hybridization of each C atom in propa-1,2-diene? What AOs does each atom contribute to the molecule? On the basis of the molecular model in Figure 3-31, do the CH_2 groups occupy the same plane or different planes? Which orbitals overlap along bonding axes to form σ bonds, and which orbitals overlap on opposite sides of bonding axes to form the π bonds?

Solve

The two end C atoms each have three electron groups in the Lewis structure and, therefore, are sp^2 hybridized. Each of these atoms contributes three sp^2 hybrid AOs and one unhybridized *p* AO to the molecule. The central C atom has two electron groups in the Lewis structure, making it *sp* hybridized. This atom contributes two *sp* hybrid AOs and two unhybridized *p* AOs to the molecule. Figure 3-31 shows that the two CH_2 groups are in perpendicular planes. The orbital picture and the energy diagram are shown below. Each terminal sp^2 hybrid AO overlaps with an *s* AO from a H atom to form one σ MO and one σ* MO. Each *sp* hybrid AO overlaps with an sp^2 hybrid AO to form one σ MO and one σ* MO. Two pairs of *p* AOs overlap in perpendicular planes, each pair forming one π MO and one π* MO. The 16 valence electrons are filled in the MOs from lowest energy to highest. From the energy diagram below, we can see that one of the π MOs is the HOMO and one of the π* MOs is the LUMO.

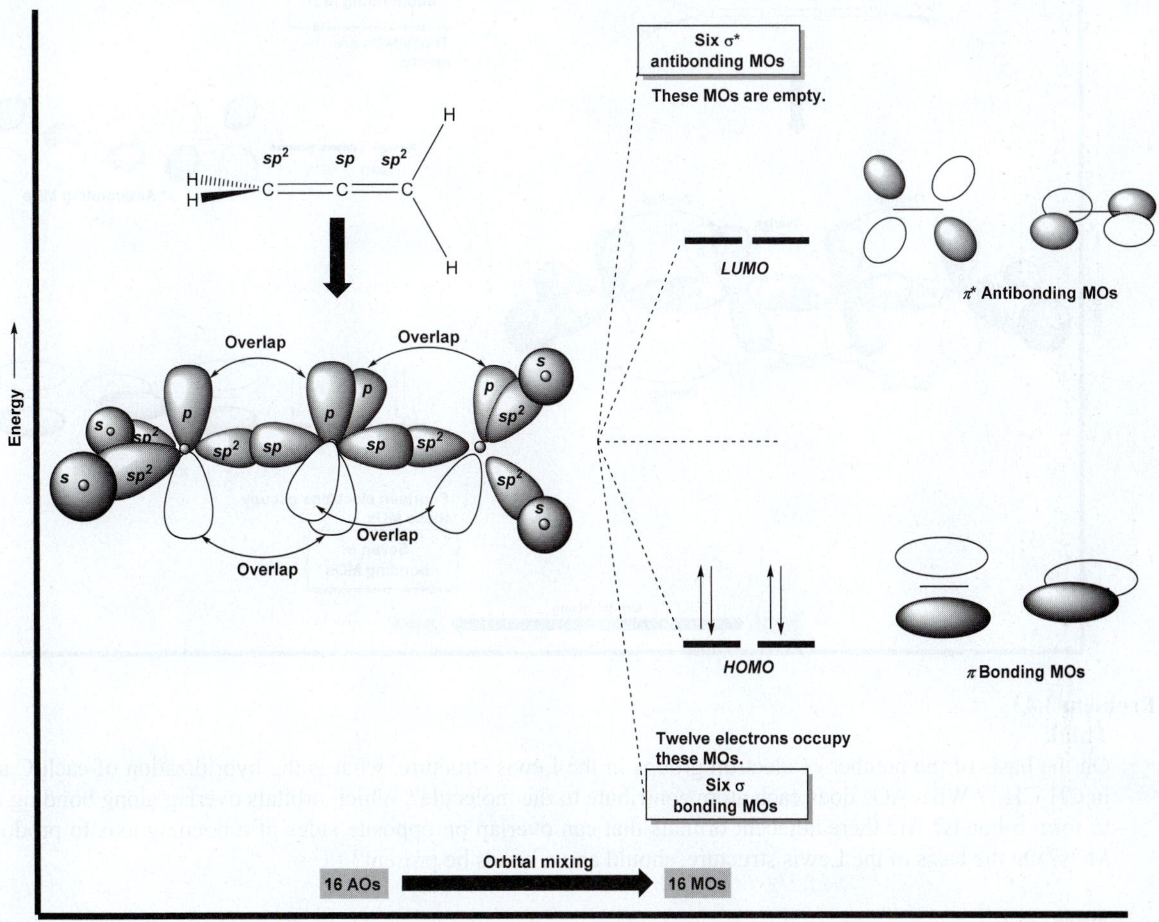

Problem 3.42

Think

On the basis of the number of electron groups in the Lewis structure, what is the hybridization of each C in buta-1,2,3-triene? What AOs does each atom contribute to the molecule? On the basis of Your Turn 3.12, are the CH_2 groups in the same plane or different planes? Which orbitals overlap along bonding axes to form σ bonds, and which orbitals overlap on opposite sides of bonding axes to form π bonds?

Solve

The two end C atoms each have three electron groups in the Lewis structure and are sp^2 hybridized, so they each contribute three sp^2 hybrid AOs and one unhybridized *p* AO. The two central C atoms each have two electron groups and are *sp* hybridized, so they each contribute two *sp* hybrid AOs and two unhybridized *p* AOs. According to Your Turn 3.12, the two CH_2 groups occupy the same plane. The orbital picture and the energy diagram are shown below. Each overlap of an *s* AO with an sp^2 AO results in one σ MO and one σ* MO. The same is true for each overlap of two hybrid AOs. Each overlap of adjacent *p* AOs results in one π MO and one π* MO. After filling in the 20 valence electrons, we can see that one of the π MOs is the HOMO and one of the π* MOs is the LUMO.

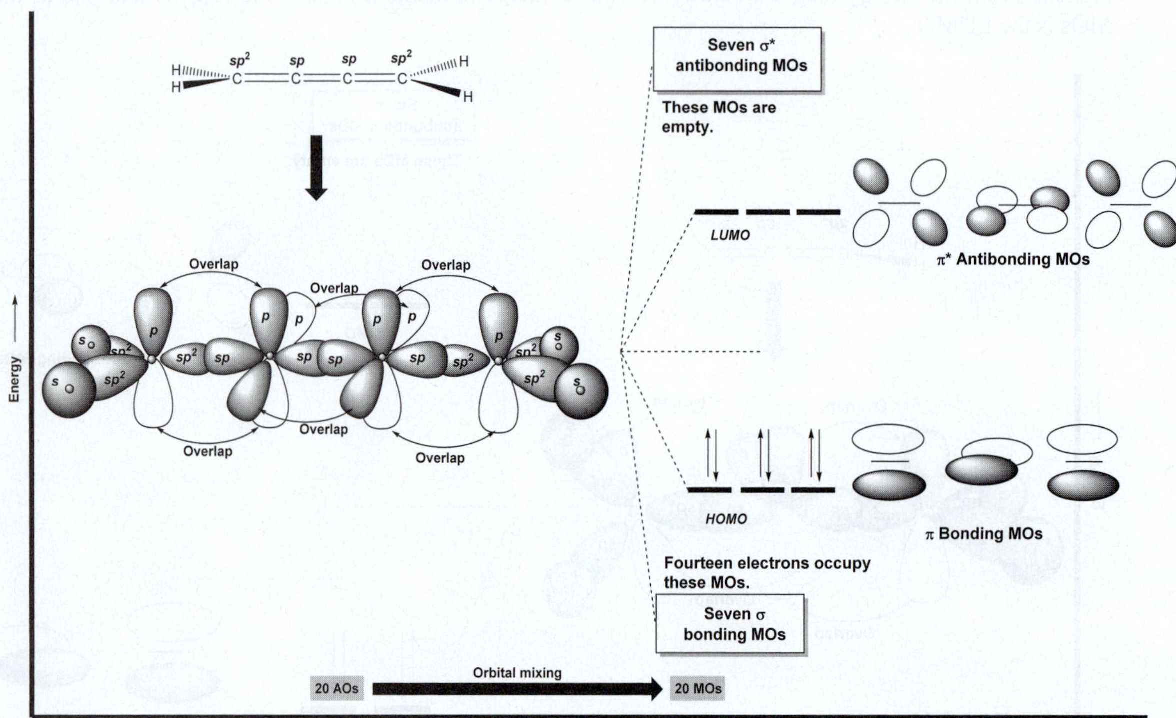

Problem 3.43

Think

On the basis of the number of electron groups in the Lewis structure, what is the hybridization of each C atom in $CH_3CH_2^+$? What AOs does each atom contribute to the molecule? Which orbitals overlap along bonding axes to form σ bonds? Are there adjacent orbitals that can overlap on opposite sides of a bonding axis to produce π MOs? On the basis of the Lewis structure, should any π bonds be present?

Solve

The C atom of the CH_3 has four electron groups in the Lewis structure and is sp^3 hybridized, whereas the C atom of the CH_2^+ has three electron groups and is sp^2 hybridized. Each *s* AO overlaps an sp^3 AO to form one σ MO and one σ* MO. The same is true for the overlap between the two hybrid AOs. The *p* orbital on the sp^2-hybridized C atom cannot overlap another *p* orbital and, therefore, remains unhybridized. The 12 valence electrons fill the six MOs that are lowest in energy. As we can see from the energy diagram on the next page, one of the σ MOs is the HOMO, and the unhybridized *p* orbital is the LUMO.

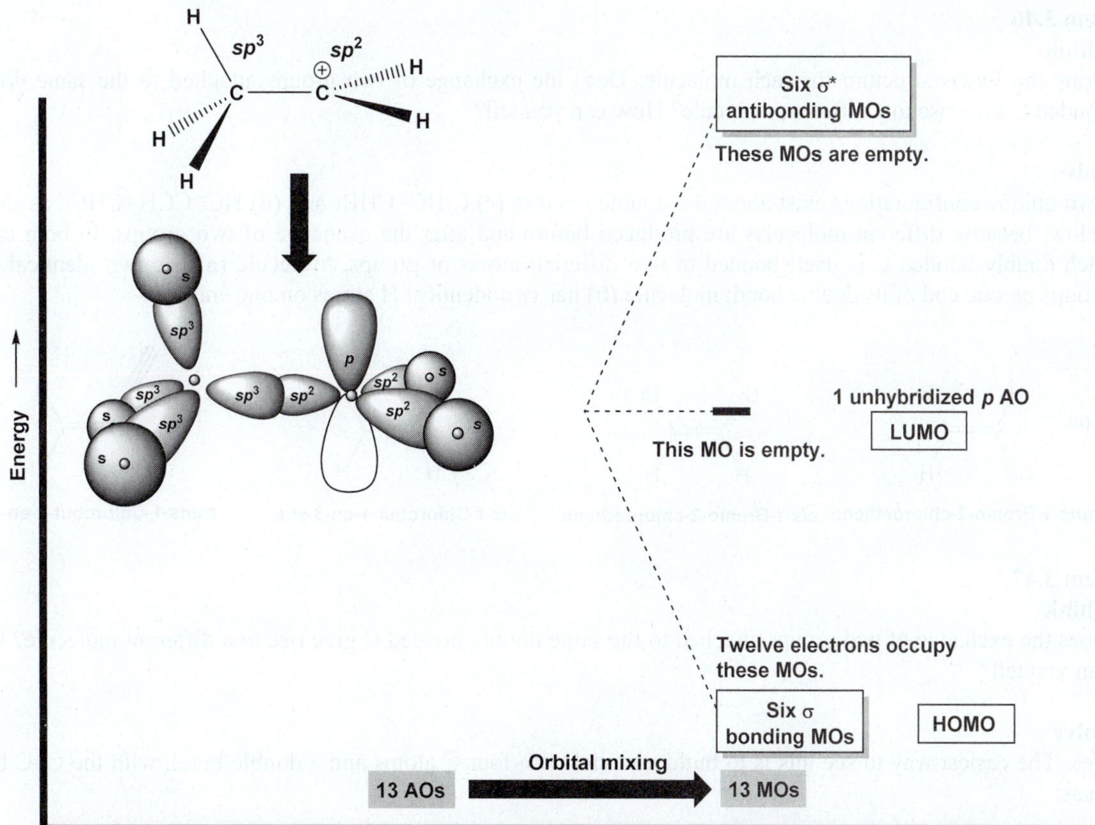

Problem 3.44

Think

Which MOs shown in Problem 3.43 have room for an additional electron? Of those, which MO is lowest in energy? After the electron is added, which occupied MO is highest in energy, and which is lowest in energy?

Solve

MO energy diagrams are filled from the lowest energy orbital (i.e., bottom to top). The additional electron is added to the lowest energy orbital available in the energy diagram shown in the solution to Problem 3.43, which is the nonbonding *p* orbital. After the electron is added, that nonbonding *p* orbital becomes the HOMO, and a σ* becomes the LUMO.

Section 3.9 Bond Rotation about Single and Double Bonds: Cis and Trans Configurations

Problem 3.45

Think

Where are the double bonds located? What is the relationship between the H atoms on each C atom of the C=C? Are they on the same side or different sides?

Solve

If the H atoms are on the same side of the double bond, the configuration is cis; if the H atoms are on opposite sides of the double bond, the configuration is trans. See below.

Problem 3.46

Think

Draw the Lewis structure for each molecule. Does the exchange of two groups attached to the same doubly bonded C give rise to a different molecule? How can you tell?

Solve

Two unique configurations exist about the double bond in **(c)** ClHC=CHBr and **(d)** HC≡CCH=CHCl, as shown below, because different molecules are produced before and after the exchange of two groups. In both cases, each doubly bonded C is itself bonded to two different atoms or groups. Molecule **(a)** has two identical CH_3 groups on one end of its double bond; molecule **(b)** has two identical H atoms on one end.

(c) ***trans*-1-Bromo-2-chloroethene** ***cis*-1-Bromo-2-chloroethene**

(d) ***cis*-1-Chlorobut-1-en-3-yne** ***trans*-1-Chlorobut-1-en-3-yne**

Problem 3.47

Think

Does the exchange of two groups attached to the same doubly bonded C give rise to a different molecule? How can you tell?

Solve

Yes. The easiest way to see this is to build a model with four C atoms and a double bond, with the C=C bond trans:

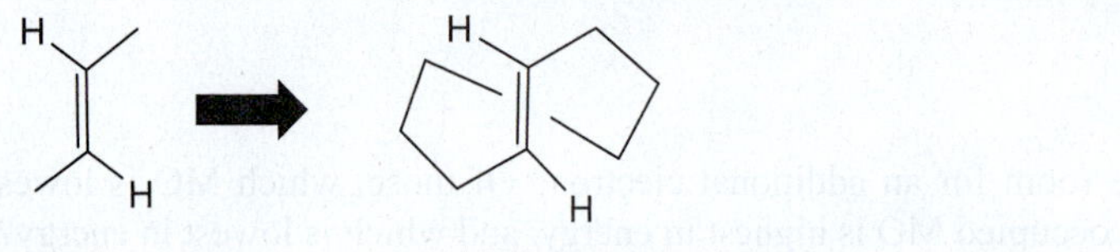

***trans*-Cyclooctene**

Then, using a bridge of four more C atoms, *make a connection behind* the double bond. The three-dimensional molecule forms a figure eight. This is different from the cis configuration, as shown below.

***cis*-Cyclooctene** ≠ ***trans*-Cyclooctene**

Problem 3.48

Think

What is the hybridization of each nonhydrogen atom? What is the molecular geometry about each nonhydrogen atom? What does a π bond require of the orientation of atoms connected by a double bond?

Solve

All atoms that are sp^2 hybridized (i.e., have three electron groups) have a trigonal planar geometry. The planes of sp^2-hybridized atoms that are connected by a double bond must be the same. Because of the alternating single and double bonds in these molecules, several atoms must lie in the same plane. The atoms that are required to be in the plane are highlighted (see figure on the next page). To see this more clearly, it would help to build models using your model kit.

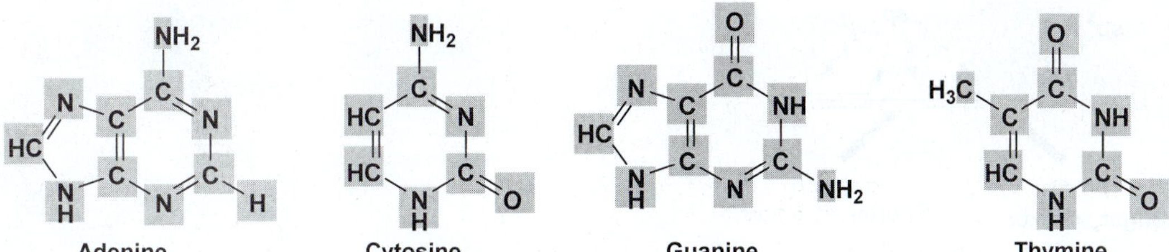

Problem 3.49

Think

What type of bond exists between carbons 2 and 3? Is free rotation permitted about this type of bond? What does this mean in regard to the atoms residing in the same plane?

Solve

A single bond exists between carbons 2 and 3, so the groups attached by that single bond are not locked into place. Single bonds can rotate; this means that the atoms around the single bond are not limited to being in the same plane.

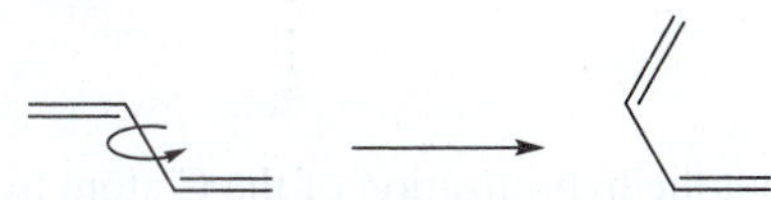

Rotation about single bond

Section 3.11 Hybridization, Bond Characteristics, and Effective Electronegativity

Problem 3.50

Think

Do sp^2 and sp^3 C atoms have the same effective electronegativities? If they are different, what does this mean in regard to the bond dipoles along the C–C bonds? Do those bond dipoles cancel, or do they result in a molecule with a net dipole? What is the strongest intermolecular interaction in a pair of each molecule?

Solve

The cis alkene is slightly polar because an sp^2 C atom has a greater effective electronegativity than an sp^3 C atom. This gives rise to bond dipoles (thin arrows), as shown below, and to the net dipole moment, indicated by the thick arrow. For the trans alkene, these bond dipoles cancel out completely, making the molecule nonpolar. Thus, the cis alkene can participate in dipole–dipole interactions, whereas the trans alkene cannot.

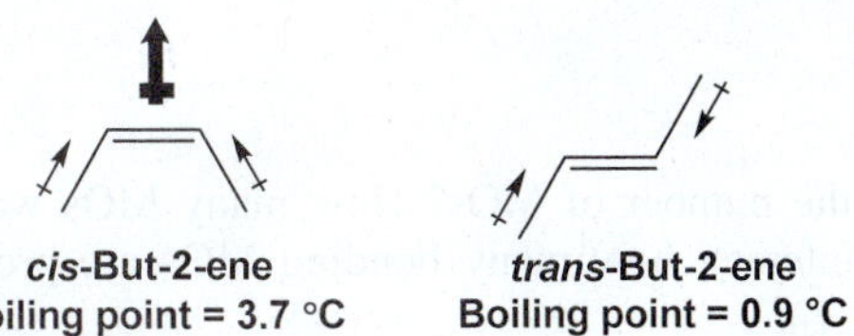

cis-But-2-ene
Boiling point = 3.7 °C

trans-But-2-ene
Boiling point = 0.9 °C

Problem 3.51

Think

What is the hybridization of each atom involved in each C–C bond? How does that affect bond length and strength? How does hybridization relate to effective electronegativity?

Solve

(a) and **(b)** The single bond on the left is between an sp^3-hybridized C atom and an *sp*-hybridized C atom, whereas that on the right is between two *sp*-hybridized C atoms. Because the sp^3 C atom has less *s* character in its hybrid orbitals, it tends to form longer and weaker bonds than an *sp*-hybridized C atom does. Thus, we should expect the single bond on the right to be shorter and stronger. See the figure on the next page.

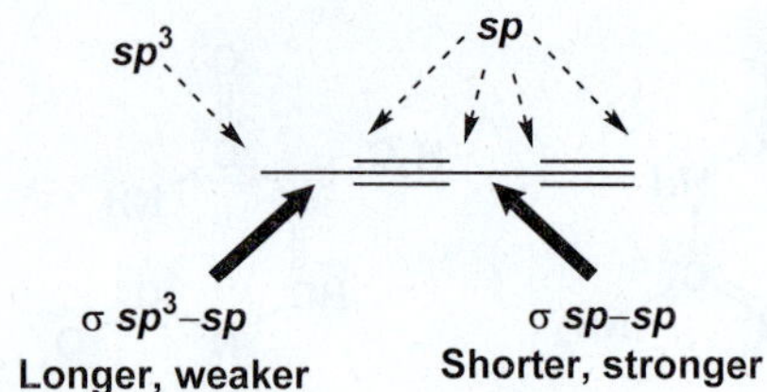

(c) Because an *sp*-hybridized C atom has a greater effective electronegativity than an sp^3-hybridized C atom, there will be a significant bond dipole in the C–C single bond on the left, pointing toward the *sp*-hybridized C atom (shown below). Thus, we should expect the net dipole to point to the right.

Problem 3.52

Think

What is the hybridization of each atom involved in each C–Cl bond? How does that affect bond length and strength? How does hybridization affect electronegativity?

Solve

(a) and **(b)** In molecules **A** and **B**, the Cl atom is sp^3 hybridized, but the hybridization of the C atom to which it is bonded is different in the two molecules. The C atom is sp^3 hybridized in molecule **A** but is *sp* hybridized in molecule **B**. Thus, the C atom in molecule **B** has more *s* character in its hybrid orbitals, giving rise to a shorter and stronger bond.

(c) With greater *s* character, the C atom in molecule **B** has a greater effective electronegativity and, therefore, draws electron density toward itself more than does the C atom in molecule **A**. Thus, the Cl atom in molecule **B** bears a lower concentration of negative charge, and the one in **A** bears a higher concentration.

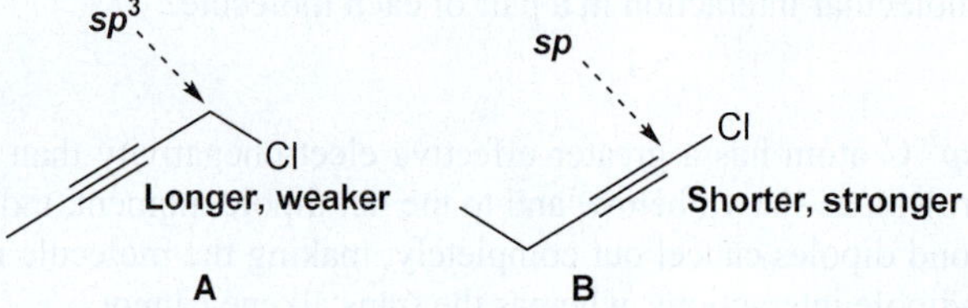

Integrated Problems

Problem 3.53

Think

What is the relationship between the number of AOs and the number of MOs? How many MOs would be produced in all? When two orbitals from different atoms interact, how many bonding MOs are produced? Antibonding MOs? Nonbonding MOs?

Solve

There are 12 MOs in all (see solution to Problem 3.28). Every time one AO mixes with another AO on a different atom, one bonding and one antibonding MO result. The problem describes four pairs of AOs interacting, which will result in four bonding MOs and four antibonding MOs. The remaining four MOs must be nonbonding.

Problem 3.54

Think

How many electron groups are present on each nonhydrogen atom? How does that relate to each atom's electron geometry and hybridization? Which bonds are not permitted to undergo free rotation? Does the exchange of two groups on the same doubly bonded C give rise to a different molecule? How can you tell? How does hybridization affect bond length and strength?

Solve

(a) As shown below, all atoms with two electron groups are *sp* hybridized. All atoms with three electron groups are sp^2 hybridized. All atoms with four electron groups are sp^3 hybridized. (As we will learn in Chapter 14, the singly bonded O atom is an exception in this case, because of the resonance involving its lone pairs and the attached C=O bond.)

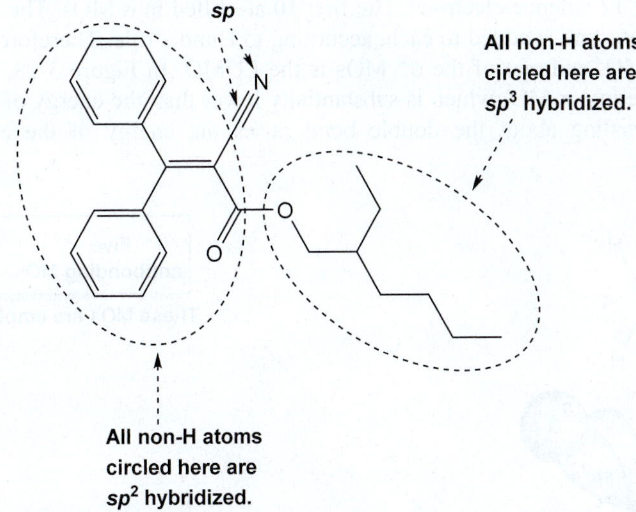

(b) The atoms included in the dashed lines must be in the same plane. Even though the N atom is not directly bonded to the C atoms of the double bond, it must be in the same plane, because the C atom to which it is bonded is linear and is also bonded to a C atom in a double bond.

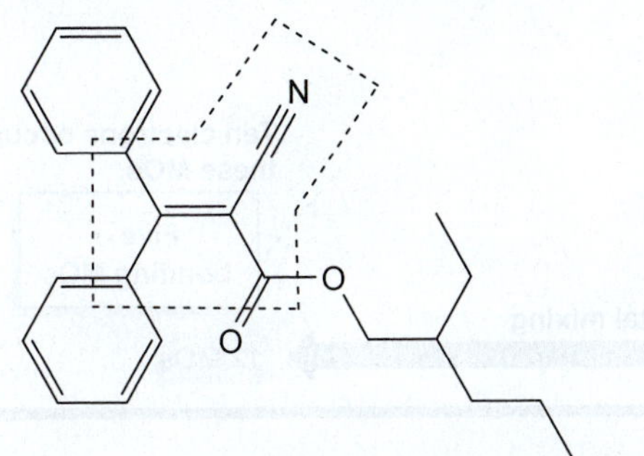

(c) Because two benzene rings are attached to one end of the double bond, cis and trans configurations are not possible. The exchange of two groups on one of the doubly bonded C atoms results in exactly the same structure.

(d) We should expect the single bond to the CN group to be shorter. The C atom of the CN group is *sp* hybridized and so has a greater *s* character than the C atom of the CO_2R group, which is sp^2 hybridized. With greater *s* character comes a shorter and stronger bond.

Problem 3.55

Think

On the basis of the number of electron groups in the Lewis structure, what is the hybridization of each carbon? What AOs does each atom contribute to the molecule? What orbitals overlap along bonding axes to form σ MOs? In twisted ethene, in which plane is each unhybridized *p* orbital? In this orientation, do you expect overlap between the two unhybridized *p* orbitals? Will that result in a π bond?

Solve

(a) The orbital overlap picture looks nearly the same as that for planar ethene (Fig. 3-17), except the one CH_2 group is perpendicular to the other. Thus, the unhybridized *p* orbitals are perpendicular to each other and do not interact.

(b) The energy diagram is shown below. The formation of σ MOs and σ* MOs is identical to that of planar ethene in Figure 3-18. Because the *p* orbitals do not interact, they remain as unhybridized *p* orbitals, or nonbonding MOs. There are a total of 12 valence electrons. The first 10 are filled in σ MOs. The last two are placed in the nonbonding *p* orbitals—one is added to each, according to Hund's rule. Therefore, one of the nonbonding *p* orbitals is the HOMO, and one of the σ* MOs is the LUMO. In Figure 3-18, the two electrons that are highest in energy are in a π MO, which is substantially lower than the energy of the two highest electrons here. Therefore, twisting about the double bond raises the energy of the electrons significantly, which precludes rotation.

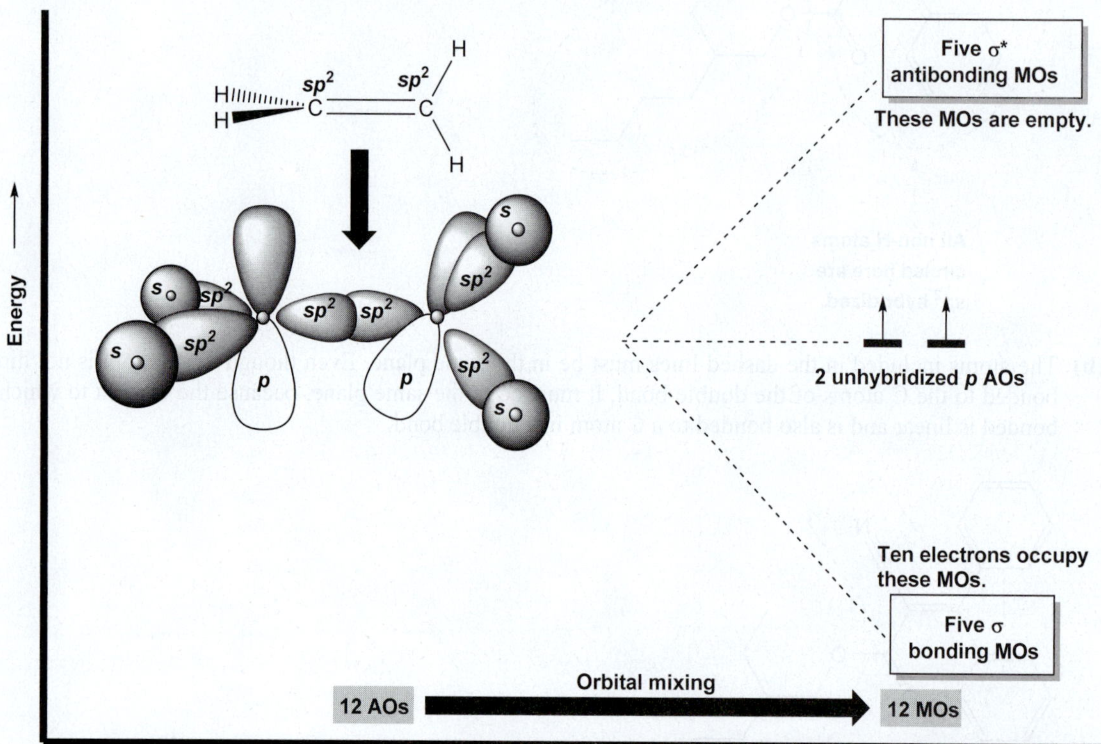

Problem 3.56

Think

Are there lone pairs adjacent to the double bond? Is resonance possible? If so, draw the resonance structure and consider the types of bonds and geometries suggested by it.

Solve

When the N atom with lone pairs is adjacent to a C=O group, a lone pair on N is able to participate in resonance. In the resonance structure, the N atom has three electron groups, suggesting a trigonal planar geometry and an sp^2 hybridization. The resonance structure also shows that the N and C atoms form a C=N double bond, which suggests that free rotation about this bond is not permitted.

An amide

INTERCHAPTER B | Naming Alkenes, Alkynes, and Benzene Derivatives

In Chapter Problems

Problem B.1

Think

What is the longest continuous carbon chain or largest carbon ring that contains the entire C=C or C≡C bond? What root name corresponds to the number of carbons in that chain or ring?

Solve

The root for each of the following molecules is highlighted in gray and labeled below the molecule. The root is determined by the longest continuous carbon chain or largest carbon ring that contains the entire C=C or C≡C bond. For **(b)**–**(d)**, the longest carbon chain has more carbon atoms than specified by the root, but notice that in these cases, the longest carbon chain or ring does not completely contain the C=C or C≡C group. For **(b)**, the largest ring has six C atoms; for **(c)**, the largest ring has five C atoms; and for **(d)**, the longest chain has seven C atoms.

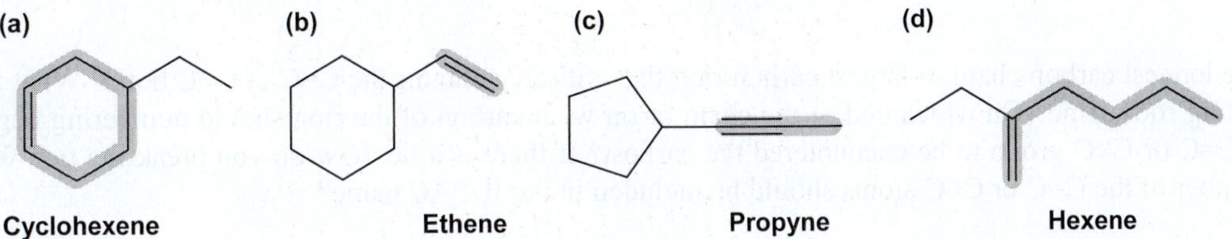

Problem B.2

Think

What is the longest continuous carbon chain or largest carbon ring that contains the entire C=C or C≡C bond? How can you assign the numbers to allow the C=C or C≡C bond to be encountered the earliest? How do you break a tie if the numbers are the same beginning from each end?

Solve

(a) The numbering begins at the left so that the C≡C bond is encountered earlier.

(b) The longest continuous chain that contains the C=C bond is seven C atoms. The longest continuous chain contains eight C atoms but does not contain the C=C bond. The numbering system begins at the top so that the C=C bond is encountered earlier.

(c) For a ring, the C=C bond must be C1 and C2. This could be achieved if the numbering system were to increase clockwise or counterclockwise around the ring, but counterclockwise is chosen to encounter one of the methyl groups the earliest.

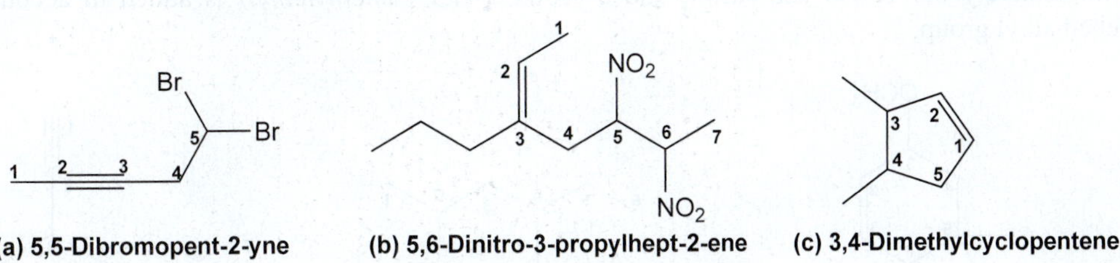

(a) 5,5-Dibromopent-2-yne **(b) 5,6-Dinitro-3-propylhept-2-ene** **(c) 3,4-Dimethylcyclopentene**

Problem B.4

Think

What is the longest carbon chain or largest carbon ring that entirely contains the C=C or C≡C bond? What is the corresponding root name? On which end of the chain should numbering begin to allow the C=C or C≡C group to be encountered the earliest? Which locator number of the C=C or C≡C atoms should be included in the IUPAC name?

Solve

The IUPAC names for molecules **(a)–(d)** in Problem B.2 are given in Problem B.2 and are also repeated here.

(a) The longest carbon chain that entirely contains the C≡C bond has five C atoms, so the root is *pentyne.* Numbering begins at the terminal carbon at the left to encounter the C≡C group the earliest, so the locator numbers for those two carbon atoms are C2 and C3. The lower of those two numbers, 2, is added to the root immediately before the *yne* suffix, and prefixes are added to account for the two *bromo* substituents. The name is 5,5-dibromopent-2-yne.

(b) The longest carbon chain that entirely contains the C=C bond has seven C atoms, so the root is *heptene.* Numbering begins at the terminal carbon at the top to encounter the C=C group the earliest, so the locator numbers for those two carbon atoms are C2 and C3. The lower of those two numbers, 2, is added to the root immediately before the *ene* suffix, and prefixes are added to account for the two *nitro* and one *propyl* substituents. The name is 5,6-dinitro-3-propylhept-2-ene.

(c) The largest carbon ring entirely contains the C=C bond and has five C atoms, so the root is *cyclopentene.* Numbering in rings must put the C=C at C1 and C2. The locator number is not included because it is understood to be 1. Prefixes are added to account for the two *methyl* substituents. The name is 3,4-dimethylcyclopentene.

Problem B.5

Think

What is the longest carbon chain or largest carbon ring that entirely contains the C=C or C≡C bond? What is the corresponding root name? On which end of the chain or on what carbon of the ring should numbering begin to allow the C=C or C≡C group to be encountered the earliest? If there is a tie, how do you break the tie? Which locator number of the C=C or C≡C atoms should be included in the IUPAC name?

Solve

The IUPAC names for molecules **(a)–(c)** are given below. The main chain or ring is numbered.

(a) The largest carbon ring that entirely contains the C=C bond has six C atoms, so the root is *cyclohexene.* Numbering in rings must put the C=C at C1 and C2. C1 is assigned to the bottom right carbon, and numbering increases counterclockwise to encounter the first substituent the earliest. The locator number is not included because it is understood to be 1. Prefixes are added to account for the *methoxy* and two *chloro* substituents.

(b) The longest carbon chain that entirely contains the C=C bond has six C atoms, so the root is *hexene.* Numbering from left or right will put the C=C at C3 and C4. Because there is a tie, numbering should begin at the right to arrive at the first substituent the earliest. The lower of those two numbers, 3, is added to the root immediately before the *ene* suffix. The locator number for the NO_2 group is C2 and for the CH_3 group is C4.

(c) The longest carbon chain that entirely contains the C≡C bond has seven C atoms, so the root is *heptyne.* Numbering begins at the terminal carbon at the top to encounter the C≡C group the earliest, so the locator numbers for those two carbon atoms are C1 and C2. The lower of those two numbers, 1, is added to the root immediately before the *yne* suffix, and a prefix *3-(1,1-Dimethylethyl)* is added to account for the branched alkyl group.

(a) 5,5-Dichloro-3-methoxycyclohexene

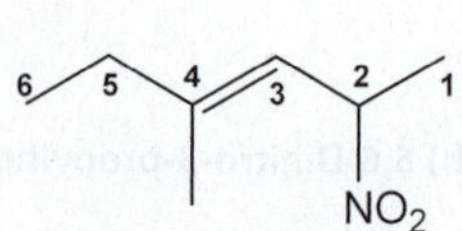

(b) 4-Methyl-2-nitrohex-3-ene

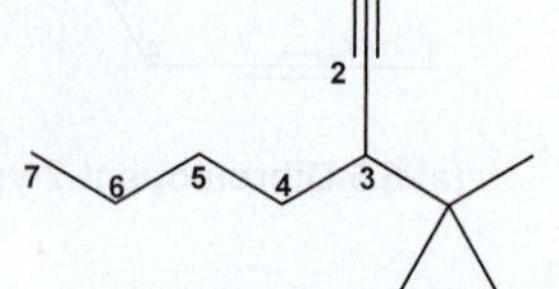

(c) 3-(1,1-Dimethylethyl)hept-1-yne

Problem B.7

Think

What is the longest carbon chain or largest carbon ring that entirely contains all of the C=C or C≡C groups? On which end of the chain should numbering begin so that the first C=C or C≡C group is encountered the earliest? How do you determine the set of locator numbers for the C=C or C≡C group? When do you add the letter *a* before the suffix?

Solve

The IUPAC names for molecules **(a)–(c)** are given below. The main chain or ring is numbered.

(a) There are two C≡C groups, and the longest carbon chain that contains both of them has seven C atoms. Because of the multiple C≡C groups, the letter *a* is added before the suffix, so this is a *heptadiyne*. Numbering begins at the bottom left terminal carbon because that gives the first C≡C group the lowest locator number (in this case, 1), so the set of locator numbers assigned to the C≡C groups is 1 and 5.

(b) There are two C=C groups, and the longest carbon chain that contains both of them has six C atoms. Because of the multiple C=C groups, the letter *a* is added before the suffix, so this is a *hexadiene*. Numbering at either end of the carbon chain would have the C=C bonds located at C2 and C4, so we choose numbering to begin at the left to arrive at the first substituent the earliest.

(c) There are two C=C groups, and the largest carbon ring that contains both of them has six C atoms. Because of the multiple C=C groups, the letter *a* is added before the suffix, so this is a *cyclohexadiene*. Numbering could begin at either C=C at the bottom C to give the first C=C group the lowest locator number (in this case, 1), so the set of locator numbers assigned to the C=C groups is 1 and 4.

(a) 7,7-Dichloro-4-methoxyhepta-1,5-diyne **(b) 4-(1,1-Dimethylethyl)-2-nitrohexa-2,4-diene**

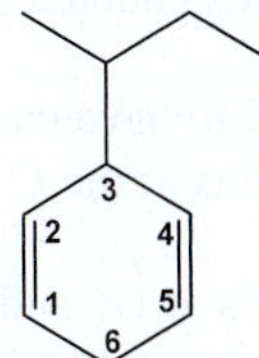

(c) 3-(1-Methylpropyl)cyclohexa-1,4-diene

Problem B.8

Think

How many carbon atoms are in the main chain or ring? Are there C=C double bonds present? C≡C triple bonds? On which carbon atoms are the multiple bonds located? Are there any substituents present? Which gets a lower number (higher priority), a double bond or a triple bond?

Solve

Structures for the given IUPAC names are drawn below. The main chain or ring is numbered. The C=C and C≡C groups are given equal priority, unless there is a tie between two numbering systems, and in the case of a tie, the C=C group is given priority (a lower number). The C=C groups come first in the name when a double and triple bond are both present. Note that the *e* is dropped on *diene* when there is more to follow for the name.

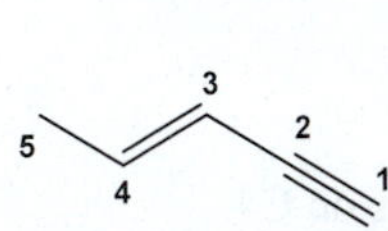

Pent-3-en-1-yne

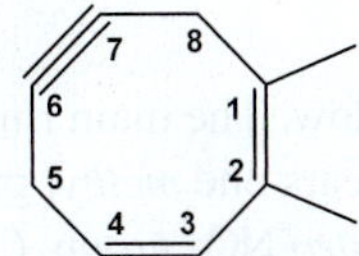

1,2-Dimethylcycloocta-1,3-dien-6-yne

Problem B.9

Think

How many C atoms are in the longest continuous carbon chain that contains the C=C or C≡C groups? Which functional group, C=C or C≡C, has the higher priority in the numbering system? How do you denote two C=C bonds? Which suffix comes last? What other substituents are present, and where are they located?

Solve

There are seven C atoms in the longest continuous carbon chain containing both C=C groups and the C≡C group. The C=C group is first in the chain and therefore gets the number C1. The *e* in the *ene* suffix is dropped if another suffix, in this case, *yne*, follows.

4-Methylhepta-1,3-dien-5-yne

Problem B.11

Think

Is there a six-membered ring of alternating C–C and C=C bonds? What is the root assigned in such cases? Where should C1 be to allow each successive substituent to be encountered the earliest? Should numbering proceed clockwise or counterclockwise?

Solve

The IUPAC names for molecules **(a)–(c)** are given below. Each of these compounds contains a six-membered ring of alternating C–C and C=C bonds, requiring the root *benzene*. The numbering system is then established by the substituents to encounter each successive substituent the earliest.

(a) Numbering begins at the C atom attached to the NO_2 group and proceeds counterclockwise to give the two *propyl* groups the lower numbers, C2 and C4.

(b) Numbering could begin at the C bonded to either the F or the bottom Cl to have the substituents attached to C1, C2, and C3. Alphabetical order breaks the tie, and numbering begins at the C atom attached to the Cl group and proceeds clockwise.

(c) Numbering could begin at either group to give the substituents numbers at C1 and C3. Alphabetical order breaks the tie; the *ethoxy* groups are assigned C1 and the *ethyl* groups C3.

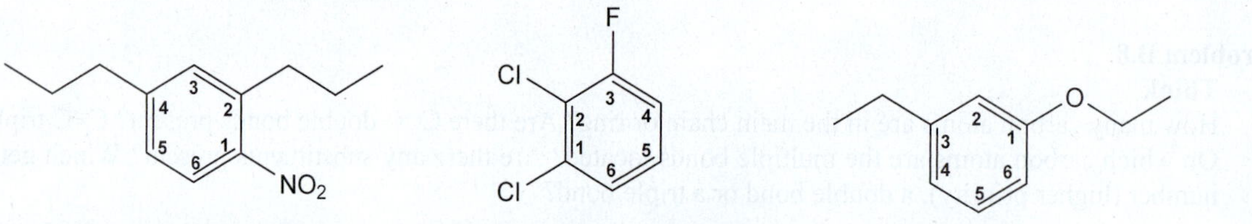

(a) 1-Nitro-2,4-dipropylbenzene **(b) 1,2-Dichloro-3-fluorobenzene** **(c) 1-Ethoxy-3-ethylbenzene**

Problem B.12

Think

What is the structure of benzene? How many substituents are present? What is the identity and location of each substituent?

Solve

Structures for the given IUPAC names are drawn below. The main ring is numbered.

- 1,2,3-Trimethylbenzene: C1, C2, and C3 each bears one *methyl* group.
- 4-Bromo-2-chloro-1-nitrobenzene: C1 has the *nitro* NO_2 group, Cl is on C2, and Br is on C4.

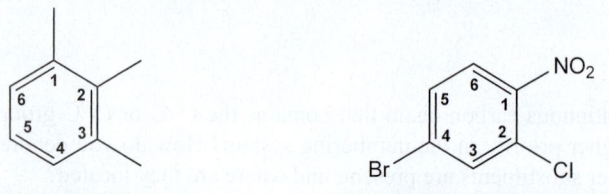

1,2,3-Trimethylbenzene **4-Bromo-2-chloro-1-nitrobenzene**

Problem B.13

Think

What structure corresponds to the root? What substituents are present, and how many are there of each? Which positioning, 1,2-, 1,3-, or 1,4-, correlates with *p*-, *m*-, and *o*- nomenclature?

Solve

Structures for the given IUPAC names are drawn below. The main ring is numbered. The *benzene* root corresponds to the six-membered rings below. Para, *p*-, is 1,4-; ortho, *o*-, is 1,2-; and meta, *m*-, is 1,3-positioning. For *m*-bromoethoxybenzene, the Br gets C1 because of alphabetical order rules.

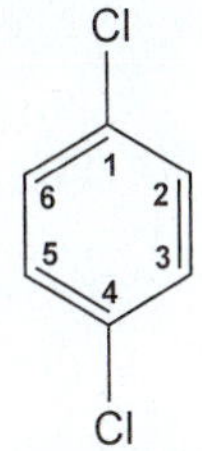

p-Dichlorobenzene
1,4-Dichlorobenzene

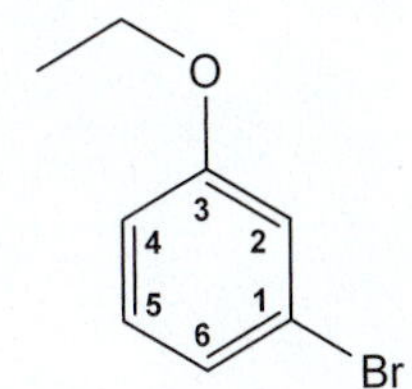

m-Bromoethoxybenzene
1-Bromo-3-ethoxybenzene

Problem B.14

Think

What does the *phenyl* (Ph) substituent look like? How many carbon atoms are in the main chain? Are any functional groups present? Which groups have the highest priority in establishing the numbering system of the carbon chain?

Solve

Structures for IUPAC names **(a)** and **(h)** are drawn below. The main chain is numbered. When benzene is not the root, the C_6H_5 becomes the *phenyl* substituent. In **(a)**, a C=C bond is located at C1 of a six-carbon chain and a *phenyl* group is located at C2. In **(b)**, a *phenyl* group is located at C1 and C5 of a five-carbon chain.

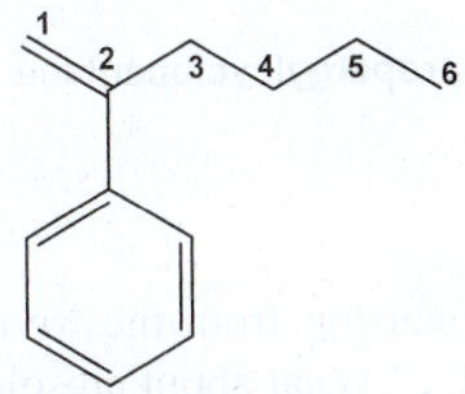

(a) 2-Phenyl-1-hexene

(b) 1,5-Diphenylpentane

Problem B.15

Think

Based on the number of carbon atoms, what is the name of the main carbon chain? What is the name of the Ph group, how many Ph groups are present, and on which C atoms are the Ph groups located?

Solve

The main carbon chain consists of two C atoms, so it is *ethane*. The Ph group is *phenyl*, and there are six Ph groups: three on C1 and three on C2.

Ph Ph
Ph—C₁—C₂—Ph
Ph Ph

1,1,1,2,2,2-Hexaphenylethane

Problem B.16

Think

On the basis of the number of carbon atoms, what is the name of the main carbon chain or ring? What is the suffix? What substituents are present, and what are their trivial names?

Solve

The main ring is a five-carbon ring with one C=C bond and is therefore named *cyclopentene*. The C=C must be C1 and C2, and therefore the substituents are located at C3 and C5. The trivial name for a $-CH{=}CH_2$ substituent is *vinyl*. Therefore, the name is 3,5-divinylcyclopentene.

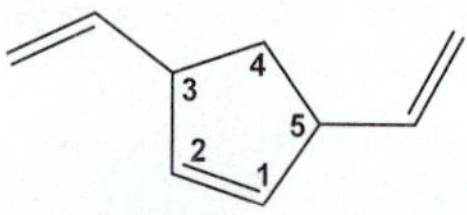

3,5-Divinylcyclopentene

Problem B.17

Think

Refer to Figure B-4 to familiarize yourself with the common names for substituents containing a C=C or C≡C group. On the basis of the root, how many carbon atoms make up the longest continuous carbon chain or ring? What substituents are present, how many are there of each, and where are they located?

Solve

Structures for IUPAC names **(a)–(c)** are drawn below. The main chain or ring is numbered where appropriate. *Vinyl* is the $-CH{=}CH_2$ substituent, *allyl* is the $-CH_2-CH{=}CH_2$ substituent, and *propargyl* is the $-CH_2-C{\equiv}CH$ substituent. In **(a)**, vinyl groups are located at C1 and C4 of the *cyclohexane* ring. In **(b)**, one *allyl* group and one *vinyl* group are attached to the same O atom to make an *ether*. In **(c)**, a *chloro* substituent is attached at C1 of a five-carbon ring, and a *propargyl* group is attached at C3.

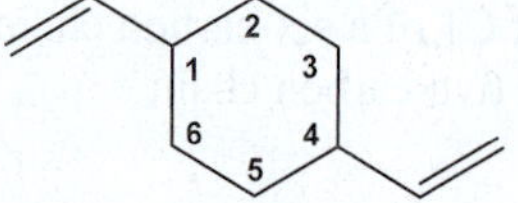

(a) 1,4-Divinylcyclohexane

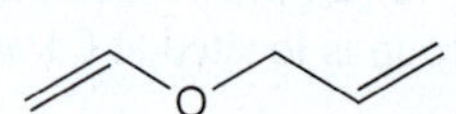

(b) Allyl vinyl ether

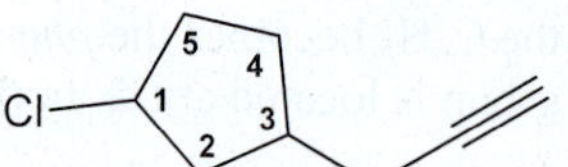

(c) 1-Chloro-3-propargylcyclopentane

Problem B.18

Think

Review the trivial names involving the benzene ring, and draw the structures starting from the trivial name. What substituent must be present in toluene, and which carbon atom is assigned C1? What about anisole?

Solve

Structures and IUPAC names for trivial names **(a)–(c)** are drawn below. The main chain or ring is numbered. In **(a)** and **(b)**, the root name is *toluene*, which is $C_6H_5-CH_3$, and the *methyl* group gets assigned to C1. In **(c)**, the root name is *anisole*, which is $C_6H_5-OCH_3$, and the *methoxy* group gets assigned to C1.

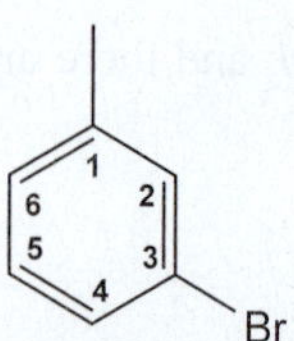

(a) *m*-Bromotoluene
3-Bromotoluene
3-Bromo-1-methylbenzene

(b) 2,5-Dinitrotoluene
1-Methyl-2,5-dinitrobenzene

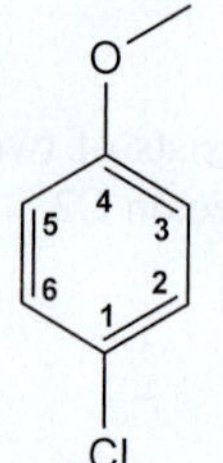

(c) *p*-Chloroanisole
1-Chloro-4-methoxybenzene

End of Chapter Problems

Section B.1 Alkenes, Alkynes, Cycloalkenes, and Cycloalkynes: Molecules with One C=C or C≡C

Problem B.19

Think

What would the root name of each molecule be without the double bond? Are any rings present? How should the suffix be changed to indicate the presence of a double bond? Are locator numbers necessary?

Solve

The IUPAC names for molecules **(a)–(c)** are given below. The main ring is numbered in **(b)** and **(c)**. The roots without the C=C bonds would be *ethane*, *cyclohexane*, and *cycloheptane*. Each name is derived by changing the corresponding alkane or cycloalkane name to an alkene or cycloalkene. No locator numbers are necessary for these molecules because the location of the double bond is implied between C1 and C2.

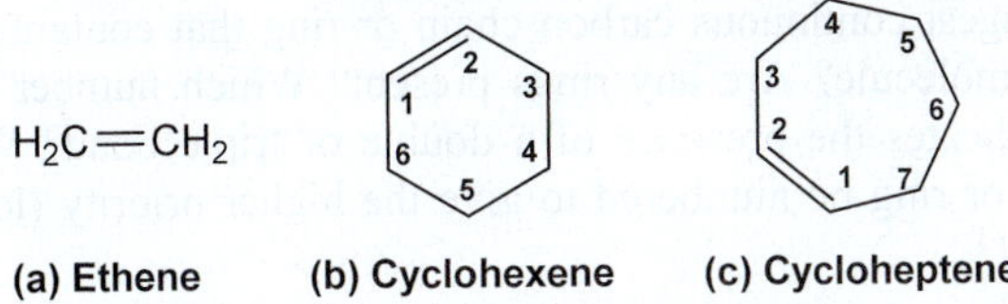

Problem B.20

Think

What would the root name of each molecule be without the double bond? Are any rings present? Which number C gets assigned to the double bond? What is the suffix for a C=C double bond? When do you need to include a number to indicate the location of the double bond?

Solve

The IUPAC names for molecules **(a)–(e)** are given below. The main chain is numbered. Without the C=C bonds, the roots for **(a)** and **(b)** would be *butane*, the root for **(c)** would be *pentane*, the root for **(d)** would be *octane*, and the root for **(e)** would be *cyclopropane*. Each molecule is named by replacing the *ane* suffix with *ene*. In **(a)** through **(d)**, the carbons of the chain are numbered beginning at the end that would allow the double bond to be encountered the earliest. In **(e)**, the C=C atoms are numbered C1 and C2 because they are in a ring. Locator numbers of double bonds are needed in the name when a different location of the double bond results in a different molecule or when the ring contains substituents.

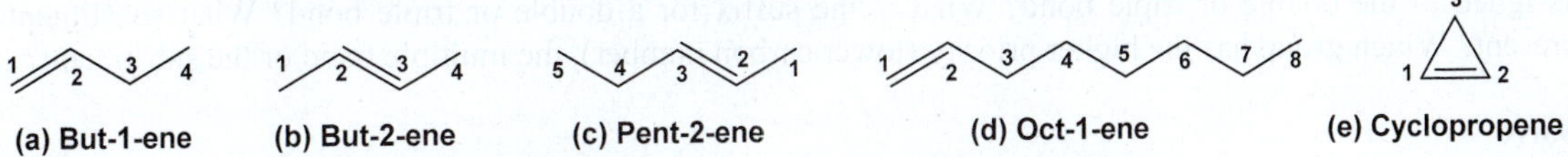

Problem B.21

Think

Is there a double or triple bond present? On the basis of the root, how many carbon atoms are in the longest continuous carbon chain or ring containing the C=C or C≡C bond? Where is the C=C or C≡C bond located?

Solve

Structures for IUPAC names **(a)–(e)** are drawn on the next page. The main chain or ring is numbered. In **(a)**, a double bond is located at C2 of a six-carbon chain. In **(b)**, a double bond is located at C3 of a six-carbon chain. In **(c)**, a double bond is located at C1 of a seven-carbon chain. In **(d)**, a triple bond is located at C2 of an eight-carbon chain. In **(e)**, a double bond is located in a seven-membered ring. Numbers for the multiple bond in a ring are not necessary when there are no substituents, as it is understood that the multiple bond is located at C1 and C2. Numbers are needed when a different location of the multiple bond results in a different molecule or the ring contains substituents. Numbers are required in the acyclic compounds to identify where along the chain the multiple bond is located.

(a) Hex-2-ene (b) Hex-3-ene (c) Hept-1-ene

(d) Oct-2-yne (e) Cycloheptene

Problem B.22

Think

On the basis of the number of carbon atoms in the longest continuous carbon chain or ring that contains the double or triple bond, what is the root name of each molecule? Are any rings present? Which number gets assigned to the double or triple bond? What suffix indicates the presence of a double or triple bond? What substituents are present? How should the carbon chain or ring be numbered to give the higher priority (lower carbon number) to the carbon atoms of the multiple bond?

Solve

The IUPAC names for molecules **(a)–(c)** are given below. The main chain is numbered. The double or triple bond has a higher priority than the substituents, so the carbon atoms involved in the multiple bond receive the lowest numbers.

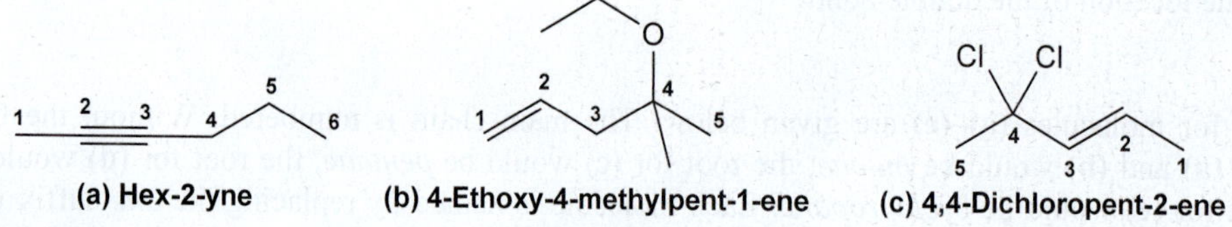

(a) Hex-2-yne (b) 4-Ethoxy-4-methylpent-1-ene (c) 4,4-Dichloropent-2-ene

Problem B.23

Think

On the basis of the number of carbon atoms in the longest continuous carbon chain or ring that contains the double or triple bond, what is the root name of each molecule? Are any rings present? Which number C gets assigned to the double or triple bond? What is the suffix for a double or triple bond? What substituents are present? Which group has the higher priority (lower carbon number), the multiple bond or the substituent?

Solve

The IUPAC names for molecules **(a)–(c)** are given below. The main ring is numbered. The double or triple bond is assigned the higher priority (or lower numbers). In a ring, this typically means that the carbon atoms in the multiple bond are assigned C1 and C2 (see the numbering below).

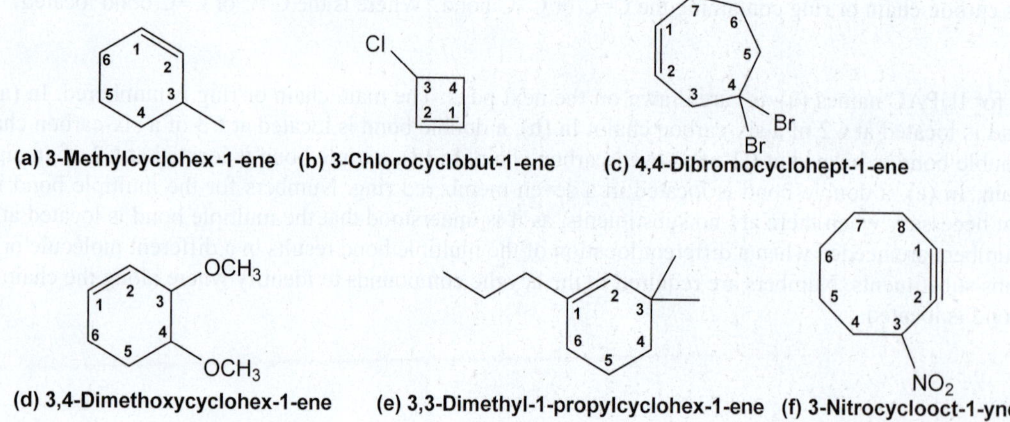

(a) 3-Methylcyclohex-1-ene (b) 3-Chlorocyclobut-1-ene (c) 4,4-Dibromocyclohept-1-ene

(d) 3,4-Dimethoxycyclohex-1-ene (e) 3,3-Dimethyl-1-propylcyclohex-1-ene (f) 3-Nitrocyclooct-1-yne

Problem B.24

Think

What is the longest continuous carbon chain that contains the multiple bond? Is the multiple bond part of the root, or is it a substituent? What substituents are attached to the main chain, and how many are there of each? How should the main chain be numbered to give the multiply bonded carbon atoms the lowest numbers?

Solve

The IUPAC names for molecules **(a)**–**(c)** are given below. The main chain is numbered. In each of the following examples, the multiple bond is part of the root and also part of the longest continuous carbon chain. Notice in **(b)** and **(c)** that the longest continuous carbon chain is not drawn entirely horizontally.

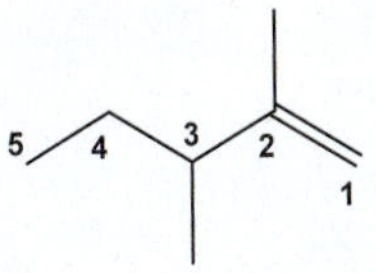

(a) 2,3-Dimethylpent-1-ene 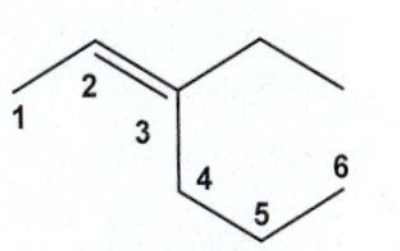 **(b) 3-Ethylhex-2-ene** 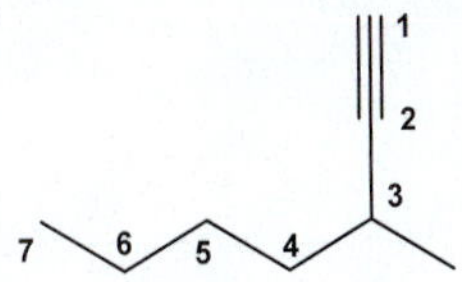 **(c) 3-Methylhept-1-yne**

Problem B.25

Think

What does the structure of the compound look like given the name provided? Does the longest continuous carbon chain contain the multiple bond? If not, how many carbon atoms are in the longest chain that contains the multiple bond? Is the multiple bond part of the root, or is it a substituent?

Solve

The structure of the molecule is shown below. The longest continuous carbon chain has six carbon atoms, but that chain does not entirely contain the double bond. The chain that contains the double bond has five carbon atoms. Thus, 2-propylbut-1-ene should be 2-ethylpent-1-ene.

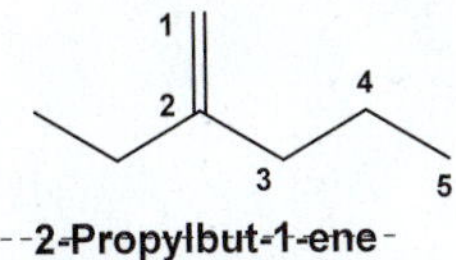

2-Propylbut-1-ene
2-Ethylpent-1-ene

Problem B.26

Think

On the basis of the root, how many carbon atoms make up the longest continuous carbon chain or ring? Are there any C=C or C≡C bonds present? If so, how many are there of each, and where are they located? What substituents are present? How many are there of each, and where are they located?

Solve

Structures for molecules **(a)**–**(h)** are drawn below and on the next page. The main chain or ring is numbered.

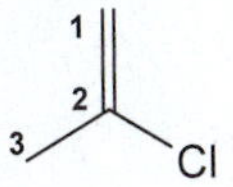

(a) 2-Chloropropene 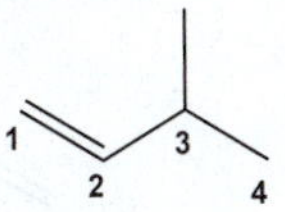 **(b) 3-Methylbut-1-ene** 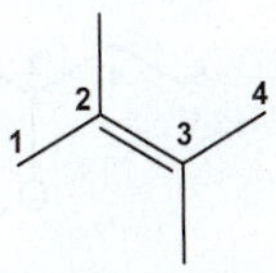 **(c) 2,3-Dimethyl-2-butene** 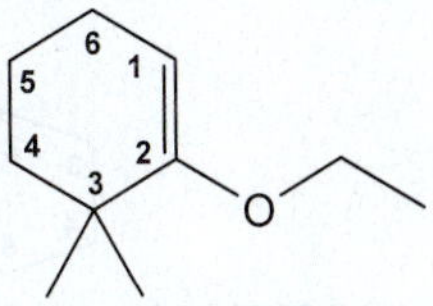 **(d) 2-Ethoxy-3,3-dimethylcyclohexene**

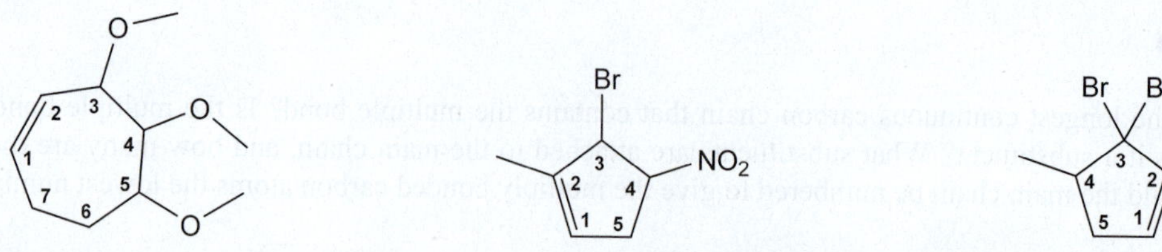

(e) 3,4,5-Trimethoxycycloheptene **(f) 3-Bromo-2-methyl-4-nitrocyclopentene** **(g) 3,3-Dibromo-4-methylcyclopentene**

(h) 4-Methyl-2-pentyne

Problem B.27

Think

Review the rules for naming alkenes and alkanes. What is the longest continuous carbon chain or ring that contains a multiple bond? What functional groups and substituents are present, and how many are there of each? How do you number the carbon atoms to give the highest-priority carbon atoms the lowest numbers? How do you name a C=C bond that is not entirely part of a chain or ring?

Solve

The IUPAC names for molecules **(a)–(i)** are given below. The main chain or ring is numbered. Numbering for chains begins at the end that allows the double or triple bond to be encountered the earliest. In **(g)**, the C=C bond is not entirely part of the ring or the chain, so the CH_2 group is named as a *methylene* substituent.

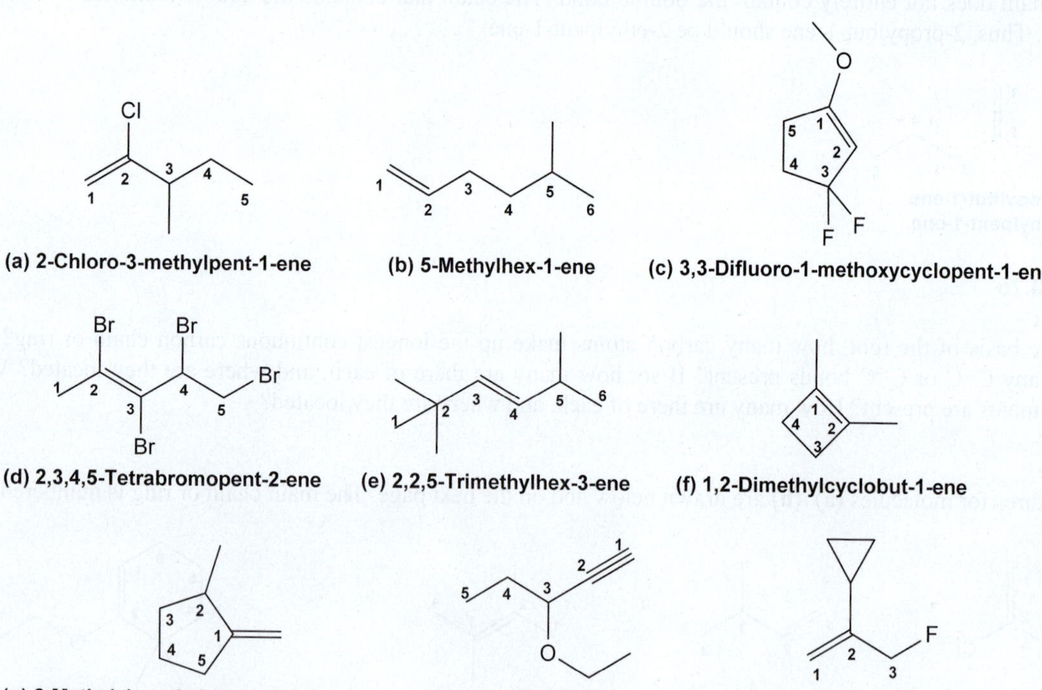

(a) 2-Chloro-3-methylpent-1-ene **(b) 5-Methylhex-1-ene** **(c) 3,3-Difluoro-1-methoxycyclopent-1-ene**

(d) 2,3,4,5-Tetrabromopent-2-ene **(e) 2,2,5-Trimethylhex-3-ene** **(f) 1,2-Dimethylcyclobut-1-ene**

(g) 2-Methyl-1-methylenecyclopentane **(h) 3-Ethoxypent-1-yne** **(i) 2-Cyclopropyl-3-fluoropropene**

Section B.2 Molecules with Multiple C=C or C≡C Bonds

Problem B.28

Think

How many carbons are in the ring? How many C=C double bonds are there, and where are the C=C bonds located? In the name given, the numbers are located at the beginning of the name. Where else can the C=C bond numbers be located later in the name?

Solve

The name indicates a six-carbon ring with C=C bonds located at C1 and C4, as shown below. The C=C bond numbers (1,4) can be located in the name after the six-carbon *cyclo* root, *cyclohexa*. Note that the *a* is added to *hex* (*hexa*) owing to the presence of *di* in *diene*.

Cyclohexa-1,4-diene

Problem B.29

Think

How many carbons are in the longest continuous carbon chain or ring? Are there double bonds present? If so, how many, and where are they located? Are they part of the main carbon chain, or are they substituents? What other substituents are present, and where are they located?

Solve

The two double bonds are assigned the higher priority (or lower numbers). When two double bonds are present, the name includes the number of the first C atom of each double bond and the suffix *diene*. Also, the letter *a* is added before *di*. See below.

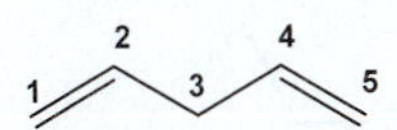

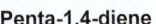

Penta-1,4-diene **Cyclopenta-1,3-diene**

Problem B.30

Think

How many C atoms are in the main chain? How many C=C bonds are present? On which C atoms are the C=C bonds located? How do you denote the location and number of C=C bonds? Do you need to add an *a* that appears before the suffix?

Solve

There are six C atoms in the main chain, which are part of three double bonds located at C1, C3, and C5. The three double bonds are denoted by *1,3,5-triene*. Therefore, the name is *hexa-1,3,5-triene*. Notice the extra *a* added after *hex* and before the suffix, owing to the presence of the three double bonds.

Hexa-1,3,5-triene

Problem B.31

Think

How many carbons are in the longest continuous carbon chain or ring? Are there double bonds present? If so, how many, and where are they located? Are they part of the main carbon chain, or are they substituents? What other substituents are present, and where are they located?

Solve

Structures for IUPAC names **(a)** and **(b)** are drawn below. The main chain is numbered. In **(a)**, the name indicates a six-membered carbon chain that has double bonds located at C1, C3, and C5, with a *methyl* group attached to C2. In **(b)**, the name indicates a six-carbon chain with double bonds located at C1 and C5 and two *methoxy* groups located at C1 and C6.

(a) 2-Methyl-1,3,5-hexatriene
2-Methylhexa-1,3,5-triene

(b) 1,6-Dimethoxyhexa-1,5-diene

Problem B.32

Think

What is the longest carbon chain or largest carbon ring that entirely contains all of the C=C or C≡C groups? For a carbon chain, on which end should numbering begin so that the first C=C or C≡C group is encountered the earliest? If there is a tie, how do you break the tie? How do you determine the set of locator numbers for the C=C or C≡C groups? When do you add the letter *a* before the suffix?

Solve

The IUPAC names for molecules **(a)–(g)** are given below. The main chain or ring is numbered. In **(b)**, there are two ways to number the carbons to have the double bonds begin at C1 and C3. The numbering system shown is chosen to also give the lowest number to the first substituent, Br. In **(e)** and **(f)**, the numbering system is chosen to give the lowest number to both the first and second double bonds. In all the names, the letter *a* is added before the suffix, because each has at least two double or triple bonds.

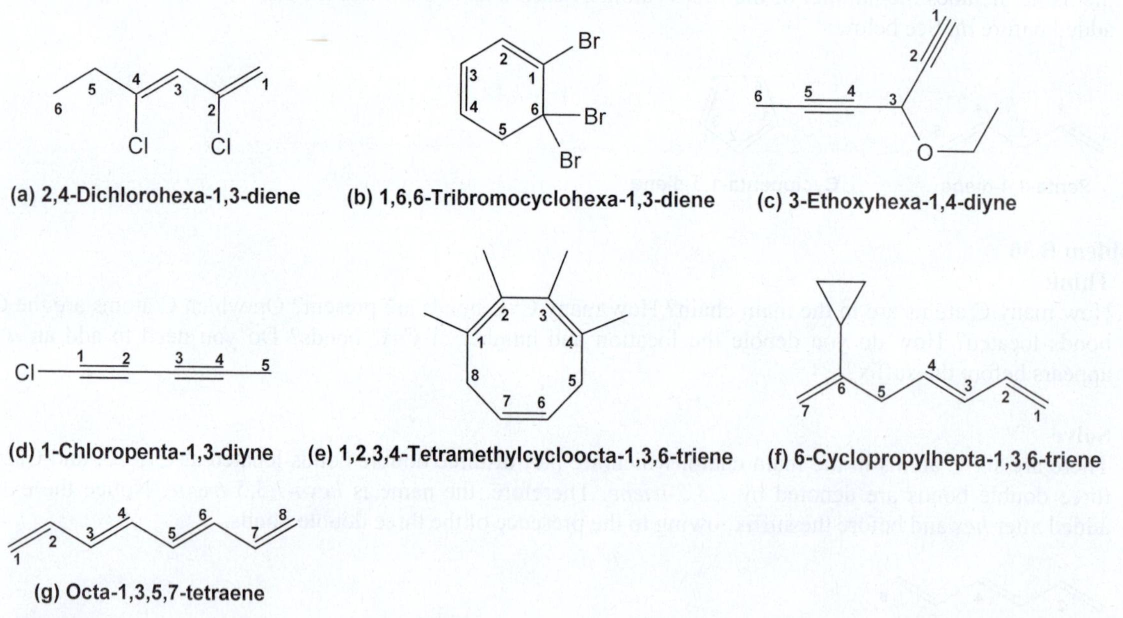

(a) 2,4-Dichlorohexa-1,3-diene **(b) 1,6,6-Tribromocyclohexa-1,3-diene** **(c) 3-Ethoxyhexa-1,4-diyne**

(d) 1-Chloropenta-1,3-diyne **(e) 1,2,3,4-Tetramethylcycloocta-1,3,6-triene** **(f) 6-Cyclopropylhepta-1,3,6-triene**

(g) Octa-1,3,5,7-tetraene

Problem B.33

Think

How many carbon atoms are in the main chain? Are C=C double bonds present? C≡C triple bonds? On which carbon atoms are the multiple bonds located? Are there any substituents present? Which gets a lower number (higher priority), a double bond or a triple bond?

Solve

The IUPAC names for molecules **(a)–(d)** are given below. The main chain or ring is numbered. Notice in **(a)** that the longest carbon chain has eight C atoms, but the root is *hept* because the longest carbon chain that contains both the double and triple bonds has seven C atoms. Similarly, in **(b)**, the longest carbon chain has seven C atoms, but the root is *hex*, because the longest carbon chain that contains both the double and triple bonds has six C atoms. In **(d)**, notice that the carbons are numbered to give both the double and triple bonds the lowest numbers.

(a) 4-Methyl-5-propylhept-5-en-1-yne

(b) 2-(1-Methylethyl)hex-1-en-3,5-diyne

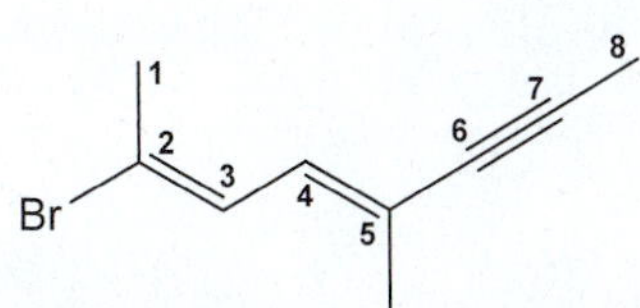

(c) 2-Bromo-5-methylocta-2,4-dien-6-yne

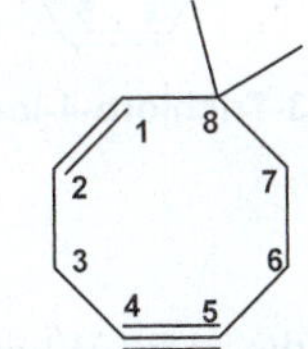

(d) 8,8-Dimethylcyclooct-1-en-4-yne

Section B.3 Benzene and Benzene Derivatives

Problem B.34

Think

What is the structure of benzene? What does the substituent look like in each case? Are any numbers necessary?

Solve

Structures for IUPAC names given are drawn below. Benzene is a six-membered carbon ring with alternating single and double bonds. No numbers are necessary, as there is only one substituent, and it is implied that the substituent is located at C1.

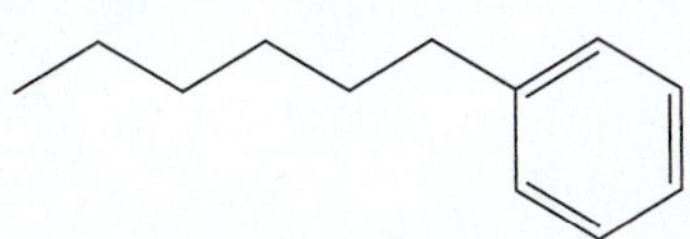

Hexylbenzene

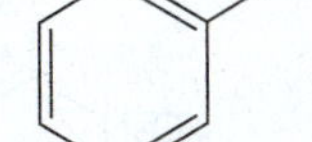

Bromobenzene

Problem B.35

Think

What is the name of the six-membered carbon ring with alternating single and double bonds? How many C atoms make up the chain of the substituent?

Solve

Benzene is the name of the six-membered carbon ring with alternating single and double bonds, and the substituent is a five-carbon alkyl chain. Thus, the name of the molecule is pentylbenzene.

Pentylbenzene

Problem B.36

Think

What is the root name of a molecule that has a six-carbon ring of alternating single and double bonds? What is the name of each substituent? How many of each substituent are present? How do you decide which carbon is C1 and whether numbering should increase clockwise or counterclockwise? Which substituent takes priority?

Solve

The IUPAC names for molecules **(a)** and **(b)** are given below. The main ring is numbered. In **(a)**, the carbon atoms are numbered to achieve the lowest set of numbers for the carbons that have substituents attached. In **(b)**, numbering could begin at the left or right carbons to have the substituents assigned to the 1, 2, 3, and 4 carbons. Halogens and alkyl groups have equal priorities in IUPAC naming rules, so to break the tie, C1 is assigned to the carbon bonded to Cl, because *chloro* comes before *iodo* alphabetically.

(a) 1,2-Dichloro-4-ethylbenzene **(b) 1,2,3-Trichloro-4-iodobenzene**

Problem B.37

Think

What structure corresponds to *benzene* as the root? What substituents are present, and how many are there of each? Where are they located? How do you determine the appropriate numbering system?

Solve

Structures for IUPAC names **(a)–(f)** are drawn below. The main ring is numbered. In each molecule, the root *benzene* corresponds to a six-carbon ring of alternating single and double bonds. C1 will have an attached substituent, and numbering increases in the direction that allows the next substituent to be encountered the earliest. In **(b)**, substituents would be attached to C1 and C2 regardless of whether C1 is attached to F or Cl. The tie is broken by giving priority to the Cl atom, because *chloro* comes before *fluoro* alphabetically. Similarly, in **(c)**, *iodo* is given priority over *nitro*, because it comes first alphabetically.

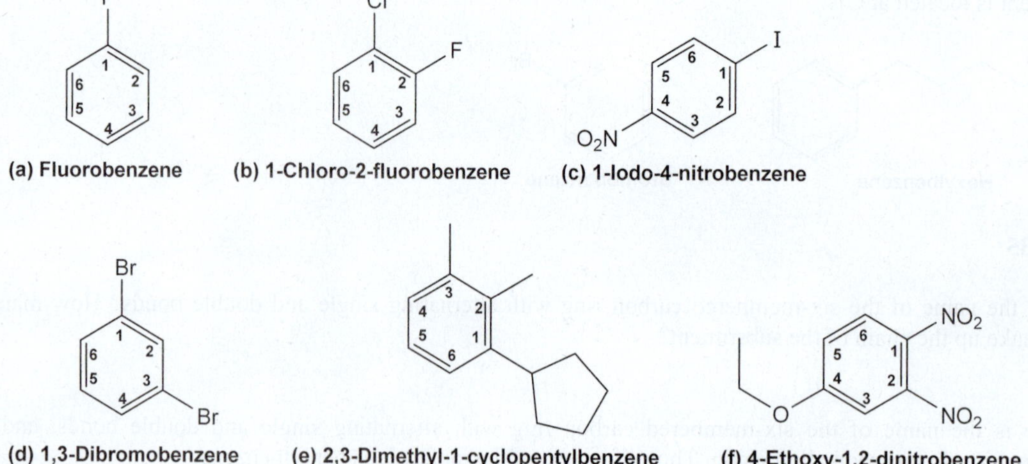

(a) Fluorobenzene **(b) 1-Chloro-2-fluorobenzene** **(c) 1-Iodo-4-nitrobenzene**

(d) 1,3-Dibromobenzene **(e) 2,3-Dimethyl-1-cyclopentylbenzene** **(f) 4-Ethoxy-1,2-dinitrobenzene**

Problem B.38

Think

What root is assigned to a molecule containing a six-carbon ring of alternating single and double bonds? What substituents are present? How many are there of each, and where are they located? How do you establish the numbering system? If a methyl group is attached to benzene, what does the root become? What types of molecules can be described using the *ortho*, *meta*, and *para* prefixes?

Solve

The IUPAC names for molecules **(a)–(g)** are given below. The main ring is numbered. Each molecule contains a six-carbon ring of alternating single and double bonds and receives the root *benzene*. Molecules **(c)** and **(d)** have a methyl group attached to the benzene ring, so the root becomes *toluene*, and C1 is assigned to the ring carbon that is attached to the methyl group. Only **(c)** and **(d)** can be named using the *ortho*, *meta*, or *para* system, because they are disubstituted benzenes.

(a) Iodobenzene

(b) Butylbenzene

(c) *m*-Chlorotoluene
3-Chlorotoluene
3-Chloro-1-methylbenzene

(d) *m*-Nitrotoluene
3-Nitrotoluene
1-Methyl-3-nitrobenzene

(e) 2-Bromo-1-ethoxy-4-nitrobenzene

(f) 1,2,4-Trichlorobenzene

(g) 1,3,5-Trichlorobenzene

Problem B.39

Think

Refer to Figures B-3 and B-4 to familiarize yourself with the common names for alkenes and alkynes and for substituents that contain a C=C or C≡C bond. On the basis of the root, how many carbon atoms make up the longest continuous carbon chain or ring? What substituents are present, how many are there of each, and where are they located?

Solve

Structures for IUPAC names **(a)–(f)** are drawn below. The main chain or ring is numbered, where appropriate. *Vinyl* is the –CH=CH_2 substituent, *allyl* is the –CH_2–CH=CH_2 substituent, and *acetylene* is –C≡C–.

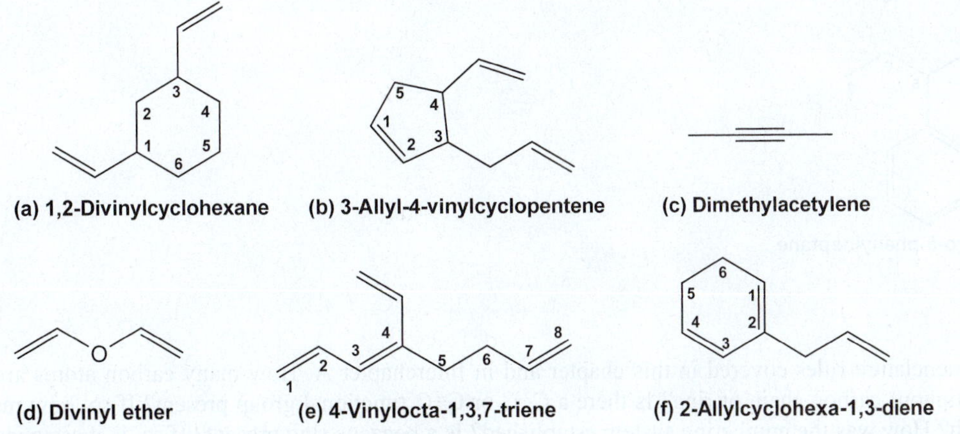

(a) 1,2-Divinylcyclohexane **(b) 3-Allyl-4-vinylcyclopentene** **(c) Dimethylacetylene**

(d) Divinyl ether **(e) 4-Vinylocta-1,3,7-triene** **(f) 2-Allylcyclohexa-1,3-diene**

Problem B.40

Think

Toluene represents a benzene ring with what attached group? *Anisole* represents a benzene ring with what attached group? What are the structures of the *phenyl* and *benzyl* substituents?

Solve

Structures for IUPAC names **(a)–(f)** are drawn below. The main chain or ring is numbered. *Toluene* is a benzene ring with an attached CH_3, whereas *anisole* is a benzene ring with an attached OCH_3. A C_6H_5 group is *phenyl*, and a $CH_2C_6H_5$ group is *benzyl.*

(a) 2-Fluorotoluene **(b) 4-Ethoxytoluene** **(c) 2-Ethoxyanisole** **(d) 1,3-Diphenylheptane**

(e) 4,4-Diphenyl-1-octene **(f) Benzylbenzene**

Problem B.41

Think

Is the C_6H_5 group the main chain or a substituent? Is the $C_6H_5CH_2$ group the main chain or a substituent? What are the names of the C_6H_5 and $C_6H_5CH_2$ groups? On the basis of the number of C atoms, what is the name of the main chain? Which C atom is assigned C1?

Solve

The main chain is the seven-membered carbon chain, so it is *heptane*. *Benzyl* is the $–CH_2C_6H_5$ substituent, and *phenyl* is the $–C_6H_5$ substituent. The numbering goes left to right to encounter the substituents the earliest.

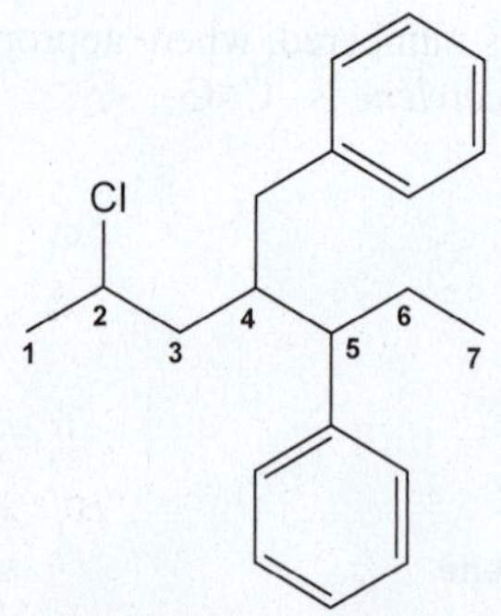

4-Benzyl-2-chloro-5-phenylheptane

Problem B.42

Think

Review the nomenclature rules covered in this chapter and in Interchapter A. How many carbon atoms are in the longest continuous carbon chain or ring? Is there a C=C or C≡C functional group present? If so, how many are there of each? How was the numbering system established? Is a benzene ring present? If so, is it assigned as the root, or is it treated as a substituent? Can any alkyl groups or complete molecules be assigned trivial names?

Solve

The trivial names for molecules **(a)–(d)** are given below. The main chain or ring is numbered.

(a) Styrene **(b) 3-Vinylhex-1-en-5-yne** **(c) 1-*tert*-Butyl-3-vinylbenzene** **(d) 4-Propargylcyclopent-1-ene**

Integrated Problems

Problem B.43

Think

Review the nomenclature rules covered in this chapter and in Interchapter A. How many carbon atoms are in the longest continuous carbon chain or ring? Is a C=C or C≡C functional group present? If so, how many are there of each? How is the numbering system established? Is a benzene ring present? If so, is it assigned as the root, or is it treated as a substituent?

Solve

The IUPAC names for molecules **(a)–(d)** are given below. The main chain or ring is numbered. Notice in **(a)** that the longest carbon chain has eight carbons but that the main chain has seven carbons, because that chain contains all three double bonds.

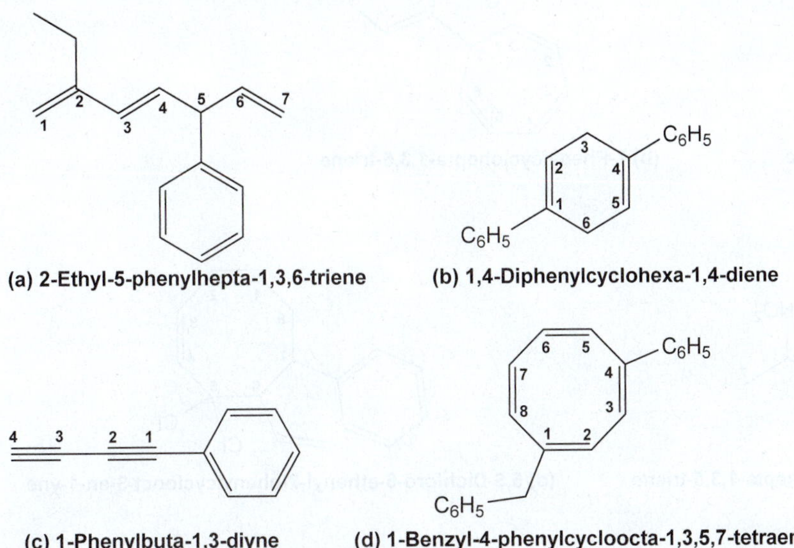

(a) 2-Ethyl-5-phenylhepta-1,3,6-triene **(b) 1,4-Diphenylcyclohexa-1,4-diene**

(c) 1-Phenylbuta-1,3-diyne **(d) 1-Benzyl-4-phenylcycloocta-1,3,5,7-tetraene**

Problem B.44

Think

Review the nomenclature rules covered in this chapter and in Interchapter A. How many carbon atoms are in the longest continuous carbon chain or ring? Is a C=C or C≡C functional group present? If so, how many are there of each? How is the numbering system established? Is a benzene ring present? If so, is it assigned as the root, or is it treated as a substituent?

Solve

The IUPAC names for molecules **(a)–(c)** are given below. The main chain or ring is numbered.

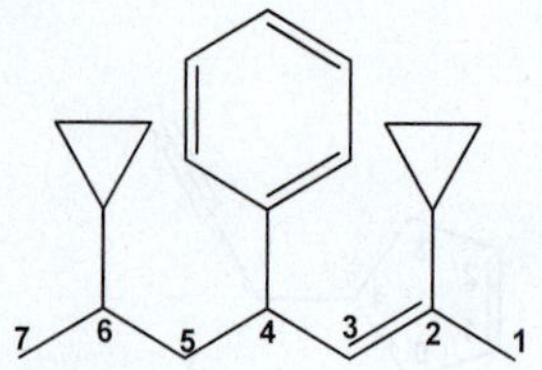

(a) 2,6-Dicyclopropyl-4-phenylhept-2-ene 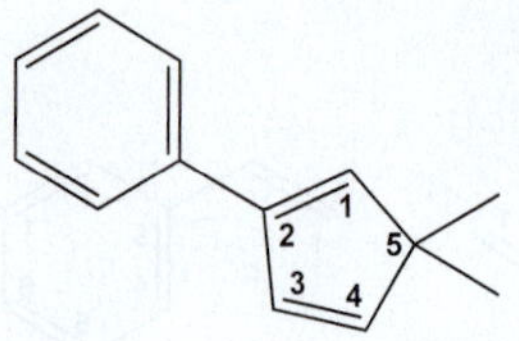 **(b) 5,5-Dimethyl-2-phenylcyclopenta-1,3-diene** 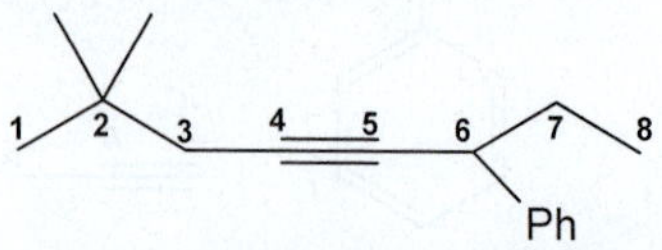 **(c) 2,2-Dimethyl-6-phenyloct-4-yne**

Problem B.45

Think

Review the nomenclature rules covered in this chapter and in Interchapter A. How many carbon atoms are in the longest continuous carbon chain or ring? Is a C=C or C≡C functional group present? If so, how many are there of each? How is the numbering system established? Is a benzene ring present? If so, is it assigned as the root, or is it treated as a substituent?

Solve

Structures for IUPAC names **(a)–(d)** are drawn below. The main chain or ring is numbered.

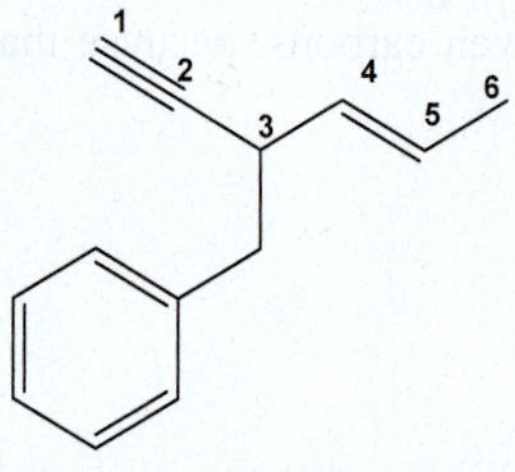

(a) 3-phenylmethylhex-4-en-1-yne 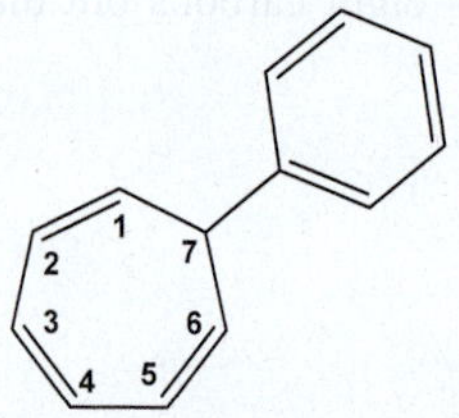 **(b) 7-Phenylcyclohepta-1,3,5-triene**

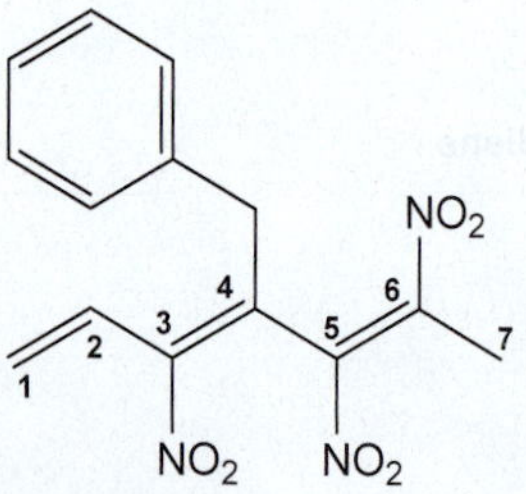

(c) 3,5,6-Trinitro-4-phenylmethylhepta-1,3,5-triene 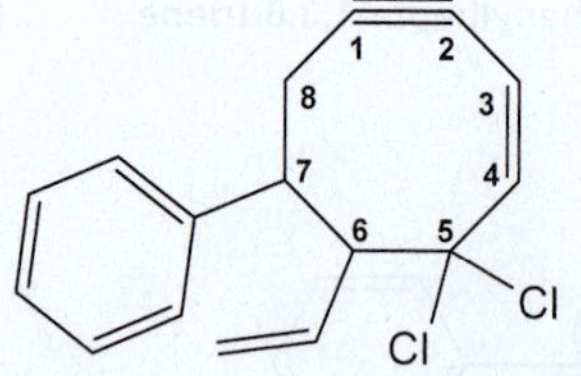 **(d) 5,5-Dichloro-6-ethenyl-7-phenylcyclooct-3-en-1-yne**

CHAPTER 4 | Isomerism 1: Conformers and Constitutional Isomers

Your Turn Exercises

Your Turn 4.1

Think

If you perform a 90° counterclockwise rotation of the molecule CX_3–CY_3 (as viewed from the top), which group is in the front? Which group is in the back? In which plane is the C–C bond? Do bonds intersecting at the dot represent bonds to the front C atom or back C atom? What about bonds drawn to the circle?

Solve

If the molecule is rotated 90° counterclockwise, the CX_3 group is in the front and the CY_3 group is in the back. The C–C bond is not visible in the Newman projection because it is perpendicular to the plane of the paper. The front C atom is depicted by a point, and the three other bonds to the front C atom converge at this point. The back C atom is depicted by a circle, and the three other bonds to the back C atom connect to the circle. Make sure to draw your three other bonds approximately equally spaced around the circle or point.

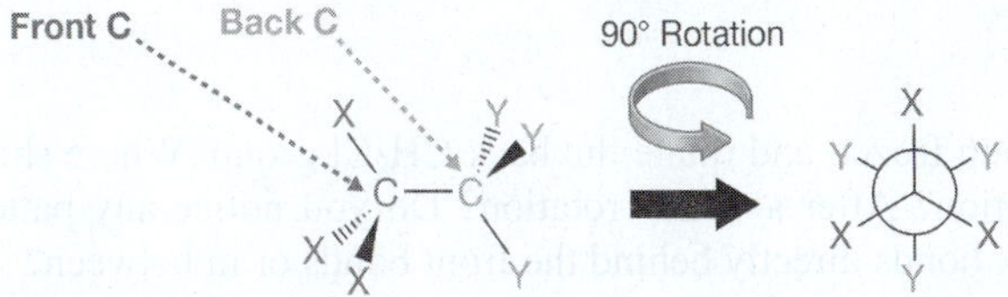

Your Turn 4.2

Think

Each carbon is sp^3 hybridized, and the CH_3 and CBr_3 groups all rotate freely. Once you have your model constructed, draw the C–C bond first and fill in with the other three bonds—wedge, dash, and planar. If you perform a 90° clockwise rotation of the molecule CH_3–CBr_3 (similar to what is shown in Your Turn 4.1), which group is in the front? Which group is in the back? In which plane is the C–C bond? How do you represent the three other bonds on the front C atom? The back C atom?

Solve

If the molecule in **(a)** drawn below is rotated 90° clockwise (as viewed from the top), the CBr_3 group is in the front and the CH_3 group is in the back; the Newman projection is drawn following the steps in Your Turn 4.1. You may have instead elected to draw the C–Br bonds on the left and the C–H bonds on the right, as in the molecule shown in **(b)** (next page). The Newman projection that results is 180° opposite of the first Newman projection. Both Newman projections are equivalent.

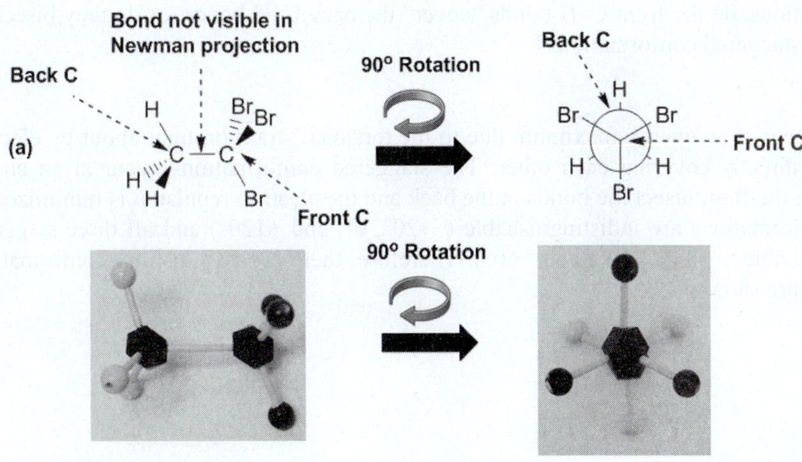

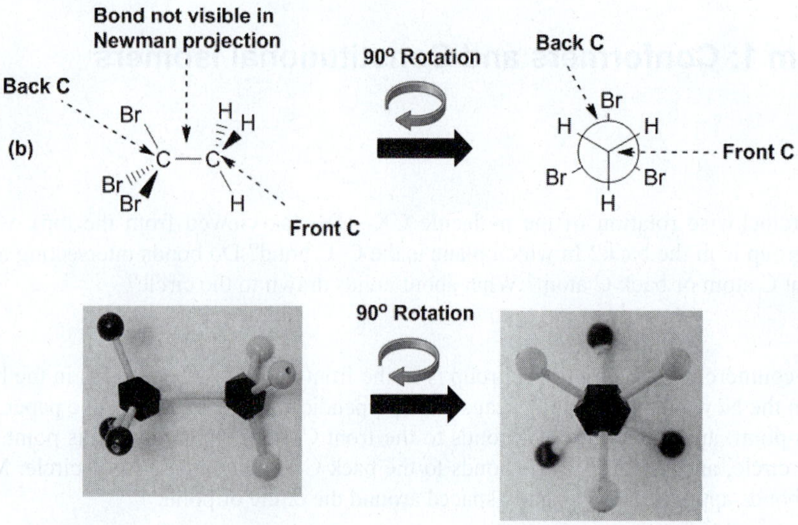

Your Turn 4.3

Think

Use Figure 4-3 as a guide. Keep the front CH_2Cl group frozen and rotate the back CH_2Cl group. Where should the Cl on the back C atom appear after a +60° rotation? After a +120° rotation? Do you notice any patterns forming in each successive 60° rotation? Are the back bonds directly behind the front bonds or in between?

Solve

The CH_2Cl group remains frozen in place on the front C atom as the back CH_2Cl is rotated. At 0°, the back bonds are directly behind the front bonds. Upon a +60° rotation clockwise, the back bonds are now in between the front bonds. Upon another +60° rotation clockwise (+120° from the start), the back bonds are again behind the front bonds.

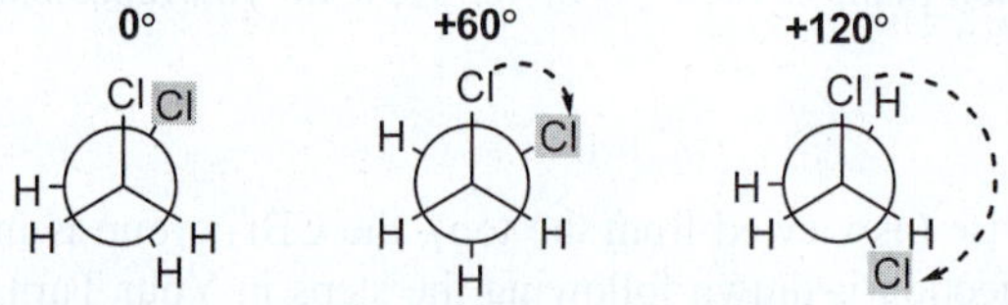

Your Turn 4.4

Think

Do eclipsed conformations pass through an energy maximum (higher energy) or an energy minimum (lower energy)? In eclipsed conformations, do the front C–H bonds "cover" the back C–H bonds, or do they bisect the back C–H bonds? What about staggered conformations?

Solve

The eclipsed conformations occur at an energy maximum due to the torsional strain brought about by electron repulsion between the bonds directly covering each other. The staggered conformations occur at an energy minimum because the bonds in the front bisect the bonds in the back and the electron repulsion is minimized. In this example, all eclipsed conformations are indistinguishable (−120°, 0°, and +120°) and all three staggered conformations are indistinguishable (±180°, −60°, and +60°). Therefore, there are two distinct conformations for ethane, as the following figure shows.

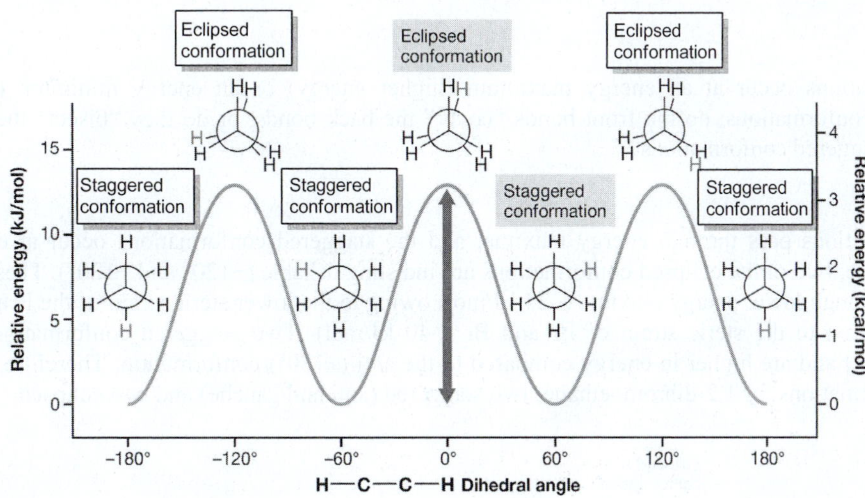

Your Turn 4.5

Think

After you construct your model of $H_3C–CH_3$, try to hold the CH_3 groups in place as you examine the molecule from different angles. When $\theta = 0°$, are the front and back C–H bonds at a maximum or minimum distance? To which conformation (staggered or eclipsed) does this correlate?

Solve

Your model should resemble the structures in Figure 4-5. The front and back C–H bonds are closer to each other in an eclipsed conformation and thus are *higher* in energy owing to electron repulsion among the front and back C–H bonds.

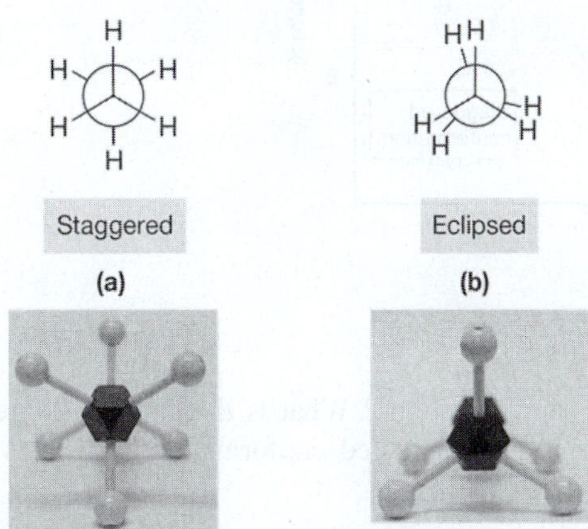

Your Turn 4.6

Think

Consult Figure 4-4. What does the thick red arrow represent? What is the energy value for the eclipsed conformations? What is the energy value for the staggered conformations? What is the difference in energy between these two conformations?

Solve

The thick red arrow represents the rotational energy barrier, which is the energy difference between the eclipsed and staggered conformations. The energy for the eclipsed conformations is about 13 kJ/mol, and the energy for the staggered conformation is 0 kJ/mol (according to Fig. 4-4). Thus, the energy difference is about 13 kJ/mol. This energy difference is the torsional strain of the ethane molecule in the eclipsed conformation, namely, the energy the molecule must acquire to pass through the less stable, higher energy conformation.

Your Turn 4.7

Think

Do eclipsed conformations occur at an energy maximum (higher energy) or an energy minimum (lower energy)? In eclipsed conformations, do the front bonds "cover" the back bonds, or do they "bisect" the back bonds? What about staggered conformations?

Solve

The eclipsed conformations pass through energy maxima, and the staggered conformations occur at energy minima. In this example, two of the eclipsed conformations are indistinguishable (−120° and +120°). These two conformations pass through lower energy maxima (~24 kJ/mol) owing to the lower steric strain of the H and Br atoms eclipsed compared to the steric strain of Br and Br (~40 kJ/mol). Two staggered conformations are gauche (−60° and +60°) and are higher in energy compared to the anti (±180°) conformation. Therefore, there are four distinct conformations for 1,2-dibromoethane: two staggered (anti and gauche) and two eclipsed.

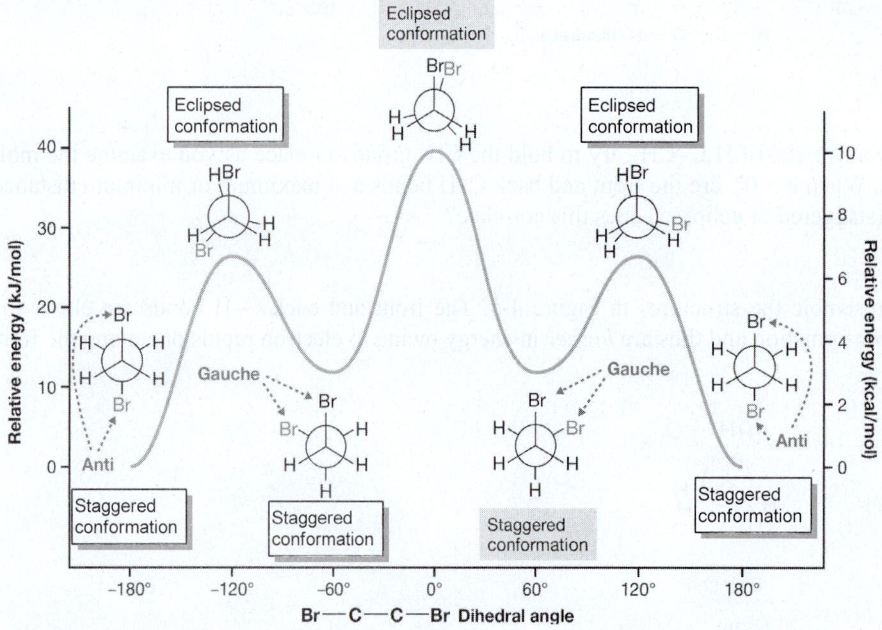

Your Turn 4.8

Think

What is the energy value for the two distinguishable eclipsed conformations? What is the energy difference between the two? What is the energy value for two distinguishable staggered conformations? What is the difference in energy between these two conformations?

Solve

The energy for the eclipsed conformation at 0° is about 40 kJ/mol (9.6 kcal/mol), and the energy for the eclipsed conformations at ±120° is about 24 kJ/mol (5.7 kcal/mol). Therefore, the eclipsed conformation at 0° is about 16 kJ/mol (3.8 kcal/mol) higher in energy than the other two. The energy for the gauche staggered conformation is about 12 kJ/mol (2.9 kcal/mol), and the energy for the anti staggered conformation is 0 kJ/mol (0 kcal/mol). Therefore, the anti staggered conformation at ±180° is about 12 kJ/mol (2.9 kcal/mol) lower than the other two (gauche).

Your Turn 4.9

Think

Which group is considered the bulky group (H or Cl)? In which conformation are the bulky groups 180° apart (anti)? In which conformation are the bulky groups 60° apart (gauche)?

Solve

Cl is a larger atom compared to H and, therefore, is considered the bulky group in the molecule. The conformation on the left has the two Cl groups 60° apart, so it is gauche. The conformation on the right has the two Cl groups 180° apart, so it is anti.

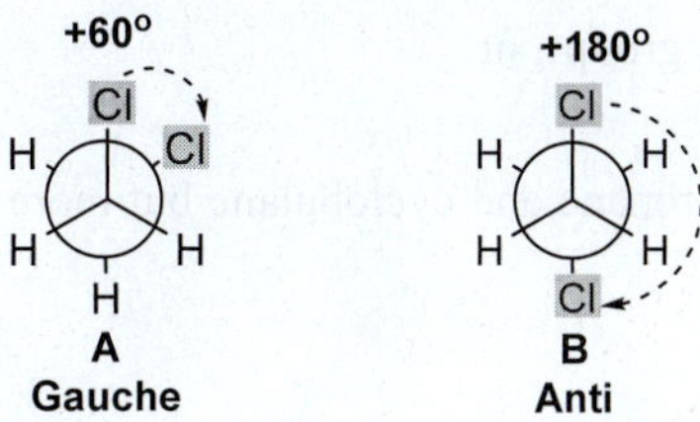

Your Turn 4.10

Think

At what energy is the anti staggered conformation (consult the *y* axis)? At what energy is the gauche staggered conformation? What conformation occurs at an energy maximum between the anti and gauche conformations? Draw the arrow from the anti staggered energy to the energy of that conformation at the energy maximum.

Solve

The anti conformation is located at 0 kJ/mol, and both gauche conformations are ~12 kJ/mol higher. The energy of the eclipsed conformation between the two is ~26 kJ/mol higher. The rotational energy barrier to go from anti to gauche is, therefore, the difference in energy between the eclipsed conformation and the anti conformation, indicated by the thick dotted arrow below, and is about 26 kJ/mol (6.2 kcal/mol).

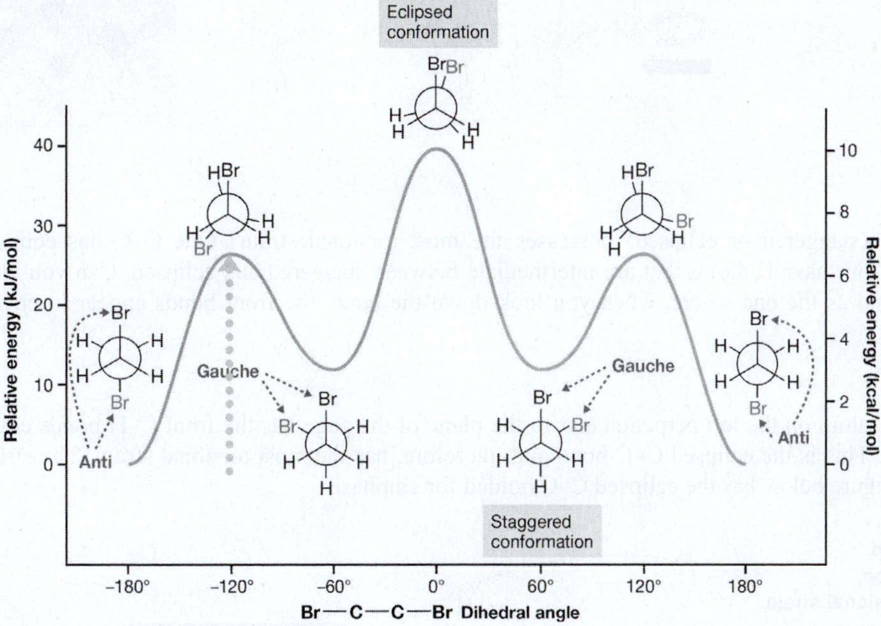

Your Turn 4.11

Think

Draw out each CH_2 on each point on the ring. The heat of combustion for cyclooctane is 4962.2 kJ/mol. How do you calculate the heat of combustion per CH_2? Once you calculate the heat of combustion ($\Delta H°$) per CH_2 for cyclooctane, subtract the heat of combustion per CH_2 for cyclohexane (consult Table 4-1) to obtain the ring strain per CH_2 group. Multiply the ring strain per CH_2 group by 8 to obtain the total ring strain for the molecule.

Solve

Eight CH_2 groups compose cyclooctane. The $\Delta H°$ per carbon in cyclooctane is $\frac{4962.2 \frac{kJ}{mol}}{8\ CH_2\ groups} = 620.3\ kJ/mol$.
Ring strain is the difference between $\Delta H°$ per CH_2 group in cyclooctane versus cyclohexane, or
$620.3 \frac{kJ}{mol} - 615.1 \frac{kJ}{mol} = 5.2 \frac{kJ}{mol}$.
Total ring strain is the ring strain per CH_2 group times the number of CH_2 groups, or
$5.2 \frac{kJ}{mol} \times 8\ CH_2\ groups = 41.6 \frac{kJ}{mol}$.
This value for total ring strain for cyclooctane is less than that of cyclopropane and cyclobutane but more than that of cyclopentane, cyclohexane, and cycloheptane.

Your Turn 4.12

Think

First, form your six-membered ring of tetrahedral carbons, and then add in two C–H bonds per C atom. How should the ring look in the chair conformation?

Solve

In the chair conformation, one CH_2 group points upward (head of the chair), four are in the middle (body of chair), and one points downward (foot of chair). Rotate the model so that the two C–C bonds are perpendicular to the plane of the paper—one involving a head CH_2 carbon and one involving the foot CH_2 carbon (bonds are in boldface for emphasis). Can you observe the staggered conformation about each C–C bond? It may help to draw a Newman projection for each side of the ring.

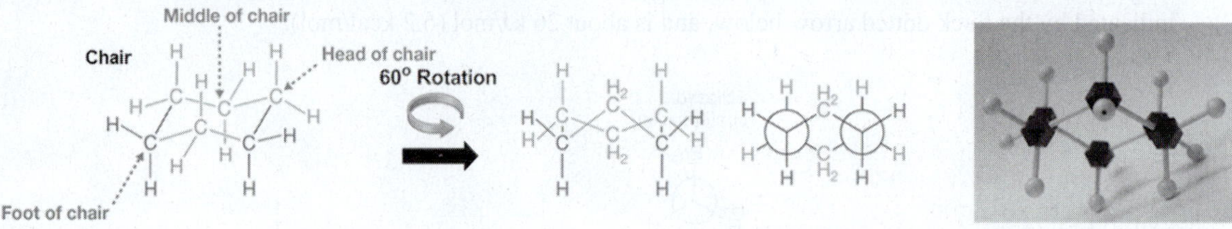

Your Turn 4.13

Think

Which conformation (staggered or eclipsed) possesses the most torsional strain? One C–C has eclipsed H atoms, and the other four have H atoms that are intermediate between staggered and eclipsed. Can you identify the eclipsed C–C bond as the one where, when you look down the bond, the front bonds appear to cover the back bonds?

Solve

In Figure 4-13, the C atom on the left perpendicular to the plane of the page has the front C–H bonds covering the back C–H bonds. This is the eclipsed C–C bond and, therefore, has the most torsional strain. The structure in the middle in the figure below has the eclipsed C–C bolded for emphasis.

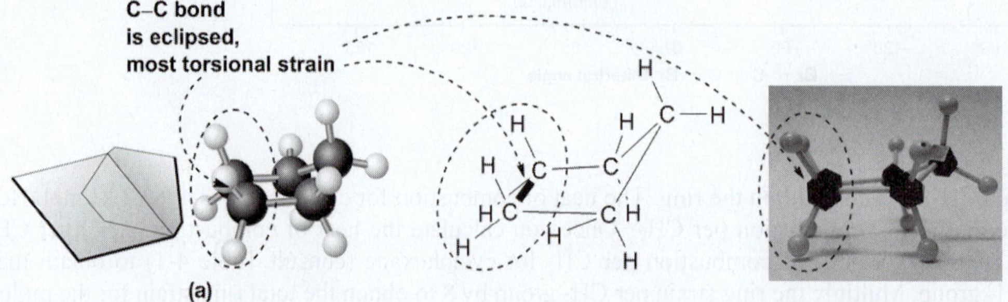

Your Turn 4.14

Think

What type of rotation is allowed in a ring structure? How does moving the orange C atom from its out-of-plane position into the plane and the green C atom from in the plane into an out-of-plane position affect the other C–C bonds? Which bonds are eclipsed before and after the rotation?

Solve

The bonds indicated below are the ones that undergo the most significant rotation. In this example, the two CH_2 groups on the left side of the molecule are the eclipsed groups in both conformations.

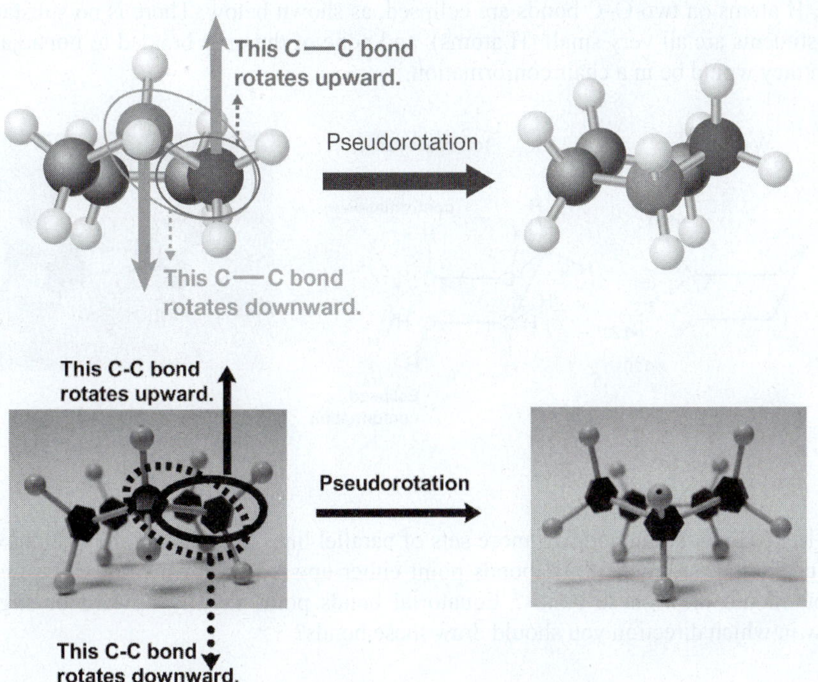

Your Turn 4.15

Think

When you build your cyclohexane model, make sure you include all C–C and C–H bonds. It might also be helpful to mark the H atoms in a blue or red marker to differentiate axial (all red initially) from equatorial (all blue initially) C–H bonds.

Solve

All of the C–C bonds undergo rotation as axial and equatorial groups change their positions. Every pair of adjacent C–C bonds undergoes rotation in the opposite direction. The "head" of the chair (C′ below) rotates to become the foot of the chair, and the foot of the chair (C″ highlighted below) rotates to become the head of the chair. All axial C–H bonds convert into equatorial, and vice versa. The axial H atoms are circled in the structure on the left, and the same H atoms that change into equatorial H atoms are circled on the right.

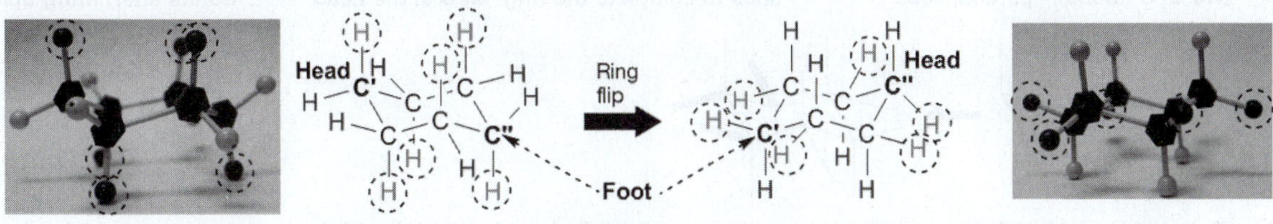

Your Turn 4.16

Think

What constitutes angle strain? On the basis of the drawing, attempt to identify any angles that deviate from the sp^3 109.5°. Are any of the C–H bonds eclipsed? If so, what type of strain results? Are any substituents not directly bonded to adjacent C atoms closer in this conformation than they otherwise could be?

Solve

Angle strain comes from the deviation between the actual bond angle and the ideal bond angle. All carbon atoms of cyclohexane are ideally 109.5° apart, but three of them in the half-chair conformation have C–C–C angles that are actually at ~120° angles and, therefore, provide substantial angle strain. Torsional strain comes from eclipsed conformations. H atoms on two C–C bonds are eclipsed, as shown below. There is no substantial steric strain, because the substituents are all very small (H atoms), and none of the ones bonded to nonadjacent C atoms are much closer than they would be in a chair conformation.

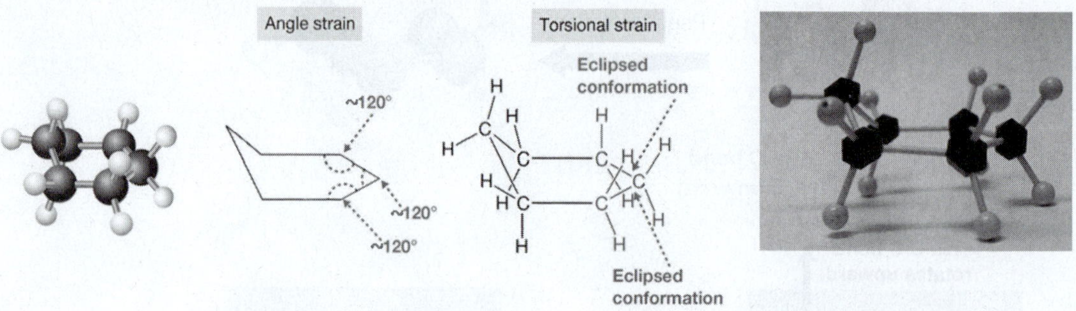
Angle strain
~120°
~120°
~120°
Torsional strain
Eclipsed conformation
Eclipsed conformation

Your Turn 4.17

Think

Which two bonds are drawn first? How do you add two more sets of parallel lines to complete the cyclohexane ring? Now the cyclohexane chair ring is drawn. Axial bonds point either upward or downward. How do you know in which direction you should draw these bonds? Equatorial bonds point slightly upward or slightly downward. How do you know in which direction you should draw these bonds?

Solve

Work with the models, referring to Figure 4-26. The progression of steps to draw this new chair conformation appears below, which is similar to Figure 4-26. Added bonds are in boldface for emphasis. Add the first axial bond to the C at the head of the chair so that it points upward. Now add in the remaining axial bonds, alternating up and down. Where there is an axial bond pointing upward, there must be an equatorial bond pointing slightly downward, and vice versa. These structures take time to draw accurately, so practice drawing the chair conformation of cyclohexane until you can draw it accurately.

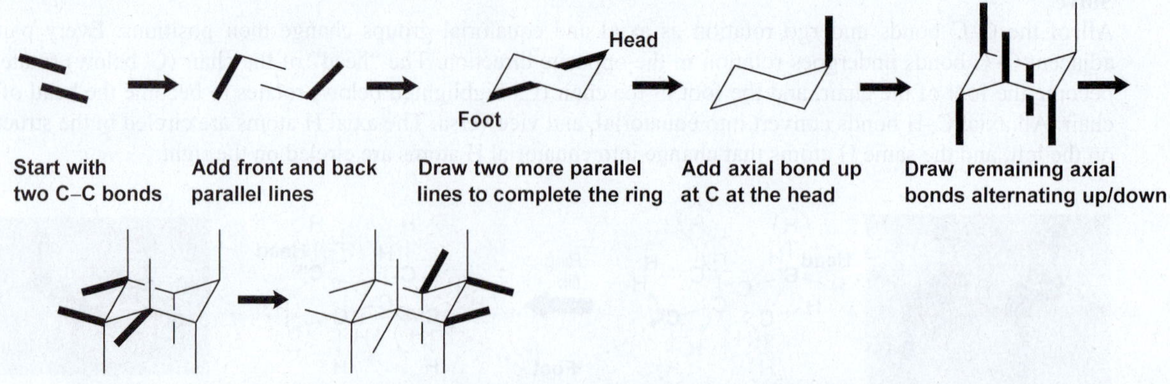
Head
Foot
Start with two C–C bonds
Add front and back parallel lines
Draw two more parallel lines to complete the ring
Add axial bond up at C at the head
Draw remaining axial bonds alternating up/down
Draw equatorial bonds alternating up/down; three on the left should point left and three on the right should point right

Your Turn 4.18

Think

The two bonds that are drawn first are drawn in which direction? Follow the steps outlined in Your Turn 4.17 to complete the ring and then draw in axial and equatorial C–H bonds.

Solve

The two bonds that are drawn first are drawn from top right to bottom left. These bonds serve as the framework upon which the rest of the structure is drawn. Adding in the front and back bonds, completing the ring, and drawing axial/equatorial C–H bonds follows the same steps as previously described in Your Turn 4.17.

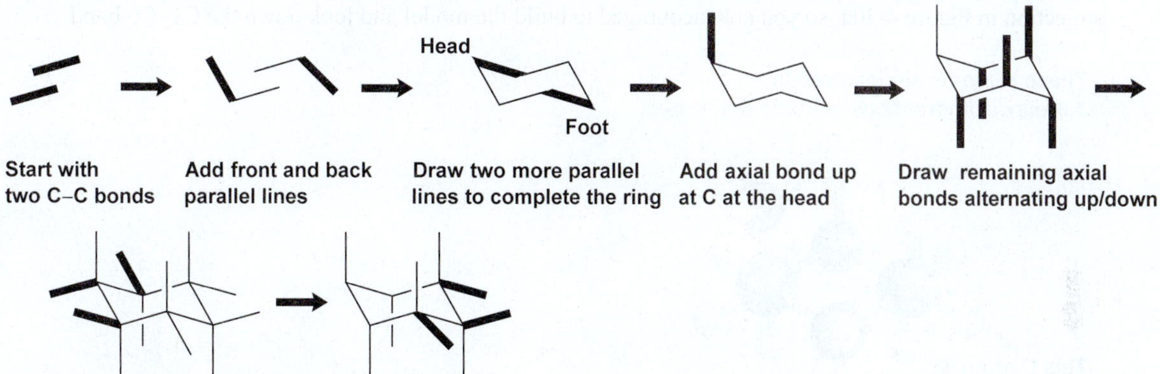
Head
Foot
Start with two C–C bonds
Add front and back parallel lines
Draw two more parallel lines to complete the ring
Add axial bond up at C at the head
Draw remaining axial bonds alternating up/down
Draw equatorial bonds alternating up/down; three on the left should point left and three on the right should point right

Your Turn 4.19

Think

Do axial C–H bonds lie almost in the plane defined by the ring or perpendicular to the plane? What about equatorial C–H bonds? How many total axial bonds should there be in each chair? How many equatorial bonds? What position is occupied by the CH_3 group in each of the conformations given?

Solve

Axial bonds are perpendicular to the plane defined by the ring, whereas equatorial bonds are closer to parallel. In each structure should be a total of six axial bonds and six equatorial bonds. Axial H atoms are labeled below. All unlabeled H atoms are equatorial. The conformation shown in **(a)** has five axial C–H bonds and six equatorial, and the conformation shown in **(b)** has six axial C–H bonds and five equatorial. The sixth axial bond in **(a)** and the sixth equatorial bond in **(b)** is the C–CH_3 bond. The CH_3 group is circled for clarity.

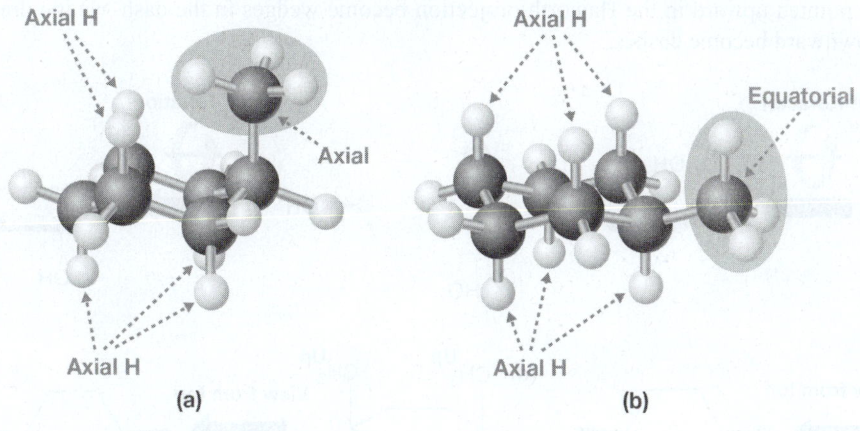
Axial H
Axial
Axial H
(a)
Axial H
Equatorial
Axial H
(b)

Axial H atoms are labeled. All other H atoms are equatorial.

Your Turn 4.20

Think

The H atoms on the ring involving 1,3-diaxial interactions in Figure 4-30a are on which C atoms? Count the C–CH_3 as C1 and then locate the 1,3-diaxial interactions. The CH_2 that is gauche to the CH_3 in Figure 4-30a is C3. This carbon is two C atoms away from the C–CH_3. Following this pattern, which other carbon number would have the CH_2 gauche to the CH_3?

Solve

The H atoms involved in 1,3-diaxial interactions are the C–H bonds on C3 and C5, as shown below. The other CH_2 that is gauche to the CH_3 is the one that contains C5. This is difficult to see if viewing only the Newman projection in Figure 4-30a, so you are encouraged to build the model and look down the C1–C6 bond.

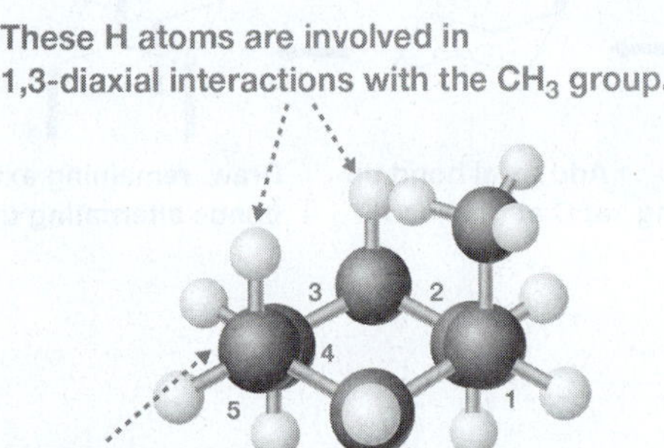
These H atoms are involved in
1,3-diaxial interactions with the CH3 group.
3
2
4
5
1
6
This C atom is
gauche to the
CH3 group.

Your Turn 4.21

Think

How is the cyclohexane ring represented in a Haworth projection? How do you transform a planar cyclohexane ring with dashes and wedges into a Haworth projection, and vice versa? In a Haworth projection, in which directions should the bonds to the substituents point? You are encouraged to build models of each molecule and look at them from different vantage points.

Solve

In both representations, the cyclohexane ring is depicted as planar and not as a chair conformation. In **(a)** and **(b)**, the ring in the plane of the page is rotated 90° to depict the Haworth projection. The wedge bonds in **(a)** and **(b)** are drawn up and the dash bonds are drawn down. To transform a Haworth projection into a dash–wedge drawing, rotate the plane of the ring to be in the plane of the page so that you view the Haworth projection from the top. Bonds that are pointed upward in the Haworth projection become wedges in the dash–wedge drawing, and bonds that point downward become dashes.

(a) HO OH — 90° Rotation → OH OH

(b) HO OH — 90° Rotation → OH OH

(c) Up CH_3, CH_3 Down — View from top →

(d) Up CH_3, Up CH_3 — View from top →

Your Turn 4.22

Think

Use your model kit to construct *trans*-1,3-dimethylcyclohexane exactly as shown in the structure on the left. In both structures given, there is one axial CH_3 and one equatorial CH_3. Without carrying out a chair flip, can you reorient the molecule to match up the axial and equatorial CH_3 groups with how they appear in the structure on the right?

Solve

To translate the conformation of *trans*-1,3-dimethylcyclohexane on the left into the "ring-flipped" conformation on the right, without flipping the chair, carry out the following rotations of the molecule.

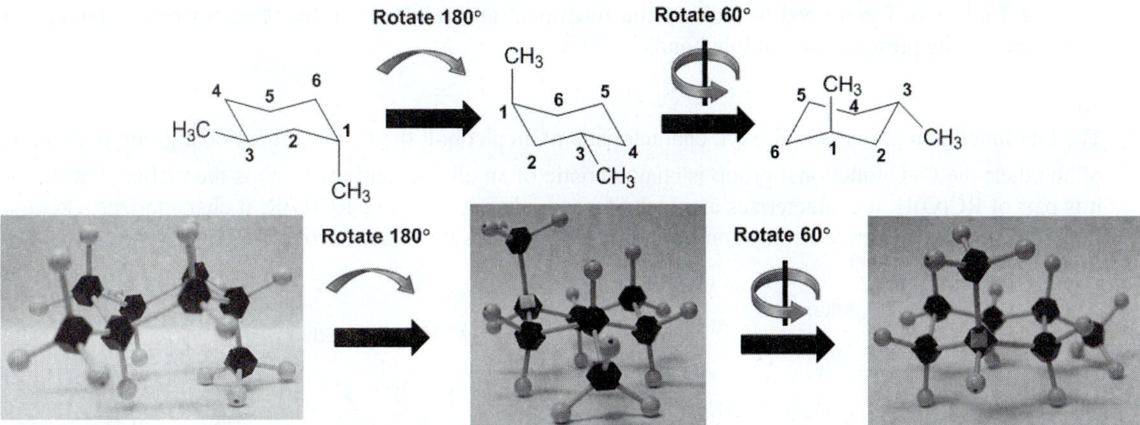

Your Turn 4.23

Think

Find the longest continuous chain of carbon atoms and number the chain beginning at one end so that the first carbon that has a substituent receives the lowest number. Is that carbon chain the same length as the ones in the previous two molecules? On what carbon atoms of that chain are the substituents located? Are these carbon atoms the same as in the previous two molecules?

Solve

The longest continuous chain of carbon atoms has six carbons. The first substituent is a CH_3 group located on C2. The second substituent is a CH_3 located on C4. This is the same connectivity as the previous two molecules. Recall that there is free rotation about single C–C bonds. These three molecules are *conformers*.

CH_3 groups on C2 and C4

Your Turn 4.24

Think

What cannot be present in a saturated molecule? Without those features present, can you add more hydrogen atoms to the nonhydrogen atoms given in each molecule?

Solve

Each saturated molecule must have only single bonds and no rings, allowing it to contain the maximum number of hydrogen atoms possible. The saturated molecules that correspond to each molecule given are shown on the next page. The formula changes from C_2H_4 to C_2H_6, C_2H_2 to C_2H_6, C_2H_4O to C_2H_6O, and C_2H_4O to C_2H_6O.

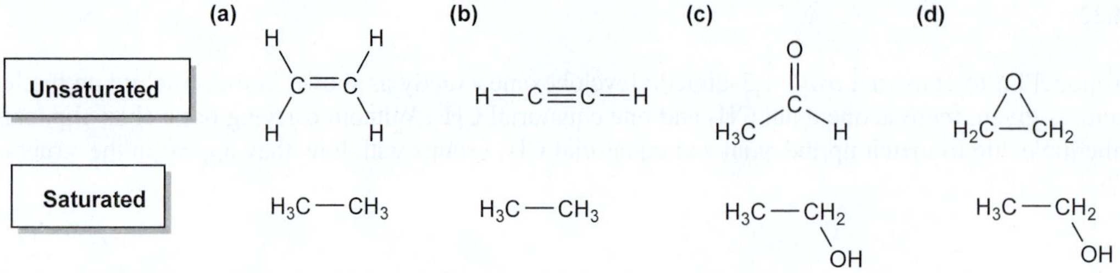

Your Turn 4.25

Think

Look at Table 1-6 if you need to look up the functional groups. Be mindful of atoms other than carbon and hydrogen and the presence of multiple bonds.

Solve

The OH functional group is hydroxyl, characteristic of an alcohol; the C–O–C functional group is characteristic of an ether; the C=C functional group is characteristic of an alkene; and the C=O is the carbonyl group—when it is part of RC(O)H, it characterizes an aldehyde, and when it is part of RC(O)R, it characterizes a ketone. The functional groups are circled below, and the compound classes they characterize are also given.

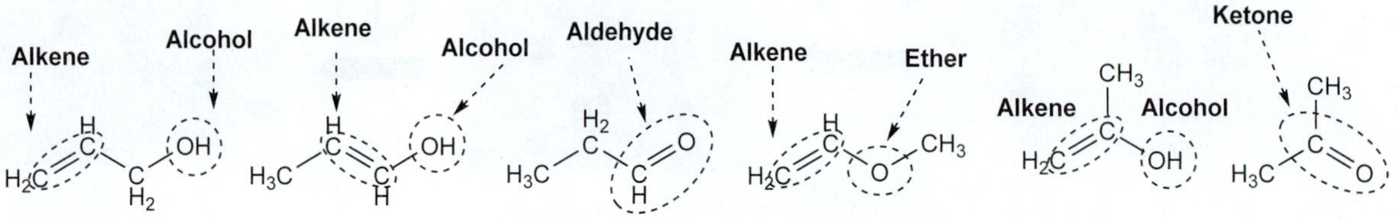

In Chapter Problems

Problem 4.2

Think

What atoms are connected by the bond not observable in the Newman projection? Which of those atoms is in front, and which is behind? What atoms or groups are attached to those atoms?

Solve

The bond not shown in each example is a C–C single bond. The C atom in front is represented by a point, and the three groups connected to the front C atom all attach at the point. The C atom in back is represented by a circle, and the three groups connected to the back C atom all attach to the circle. Each Lewis structure is drawn from a 90° counterclockwise rotation. The primes (′) below help visualize the rotation. Each rotation below is counterclockwise as viewed from above, but the rotation could instead be clockwise.

(a) The C atom in front is connected to two CH_3 groups and one H atom. The C atom in back is connected to two H atoms and a CH_3 group.

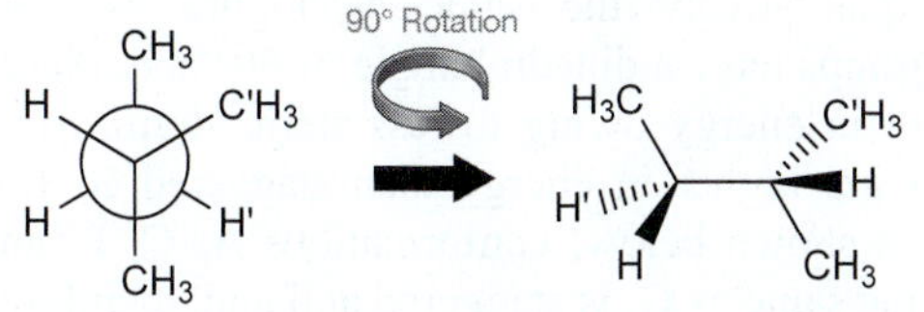

(b) The C atom in front is connected to three H atoms. The C atom in back is connected to three Cl atoms.

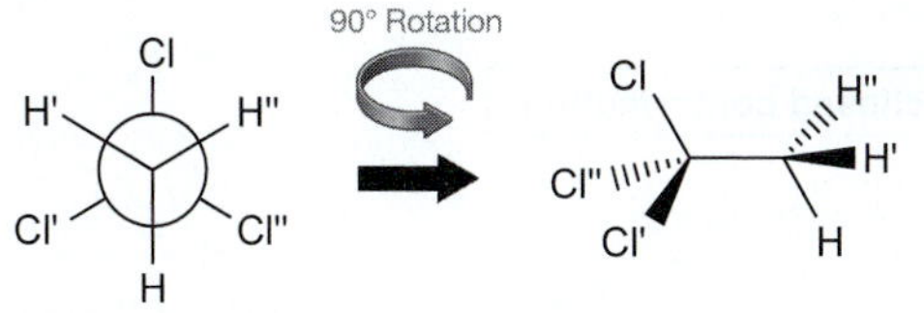

(c) The C atom in front is connected to two F atoms and one CF_3 group. The C atom in back is connected to three F atoms.

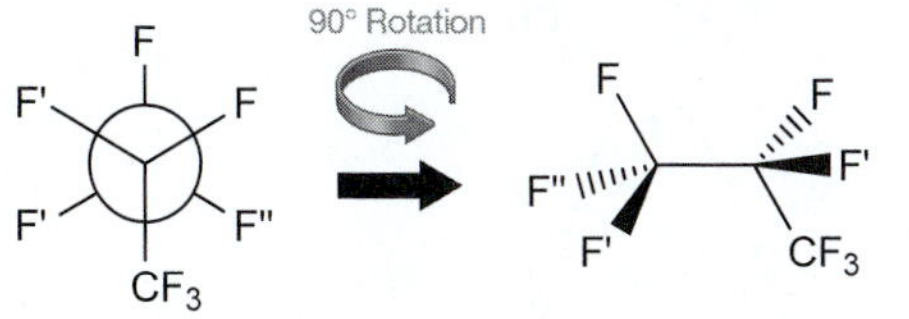

Problem 4.3

Think

What groups are connected by the bond not observable in the Newman projection? Which one is in back, and which is in front? Which group are you asked to rotate, and which should remain frozen in place? Using Figure 4-3 and Your Turn 4.3 as guides, what is the angle between each of the three groups on a C atom in a Newman projection? How many 60° rotations does it take for one group to get to the location of the next group? How many 60° rotations does it take to get all the way around the circle? You are encouraged to construct the molecule using a model kit and then to carry out the necessary manipulations on it.

Solve

Each bonding group on the same atom is 120° apart. It takes two 60° rotations to go from group to group, and it takes six 60° rotations to go all the way around the circle. See the rotations on the next page. The back group is locked in place, and thus, the **X** label will be pointed downward for the entire rotation sequence. The front group is rotating around the circle clockwise, and thus, the **Y** label will move 60° clockwise with each rotation. Both are in boldface for emphasis. *Note:* All rotations are performed clockwise.

–180° –120° –60° 0° +60° +120° +180°

Problem 4.4

Think

How is 1,2-dichloroethane related to 1,2-dibromoethane, whose conformational analysis is shown in Figure 4-7? How is steric strain related to the relative energy of the conformations? Which interactions (staggered or eclipsed) are higher in energy? Which staggered interactions, anti or gauche, are higher in energy?

Solve

1,2-Dichloroethane differs from 1,2-dibromoethane only by the exchange of the two Br atoms for two Cl atoms. Therefore, the conformational analysis should look qualitatively the same as Figure 4-7. Staggered conformations minimize the torsional strain because the groups have a dihedral angle of 60° and, therefore, are lower in energy. Staggered anti conformations are lowest in energy owing to less steric strain, followed by staggered gauche conformations. Eclipsed conformations are higher in energy than staggered conformations because of torsional strain. In the conformational analysis shown below, conformations **A**, **C**, **E**, and **G** are staggered and, therefore, are lower in energy. **A**, which is the same as **G**, is staggered anti and, therefore, has the *lowest* energy. **B**, **D**, and **F** are eclipsed and have the higher energy. **D** has the highest energy because the two larger Cl atoms are eclipsed and, therefore, have the largest steric strain.

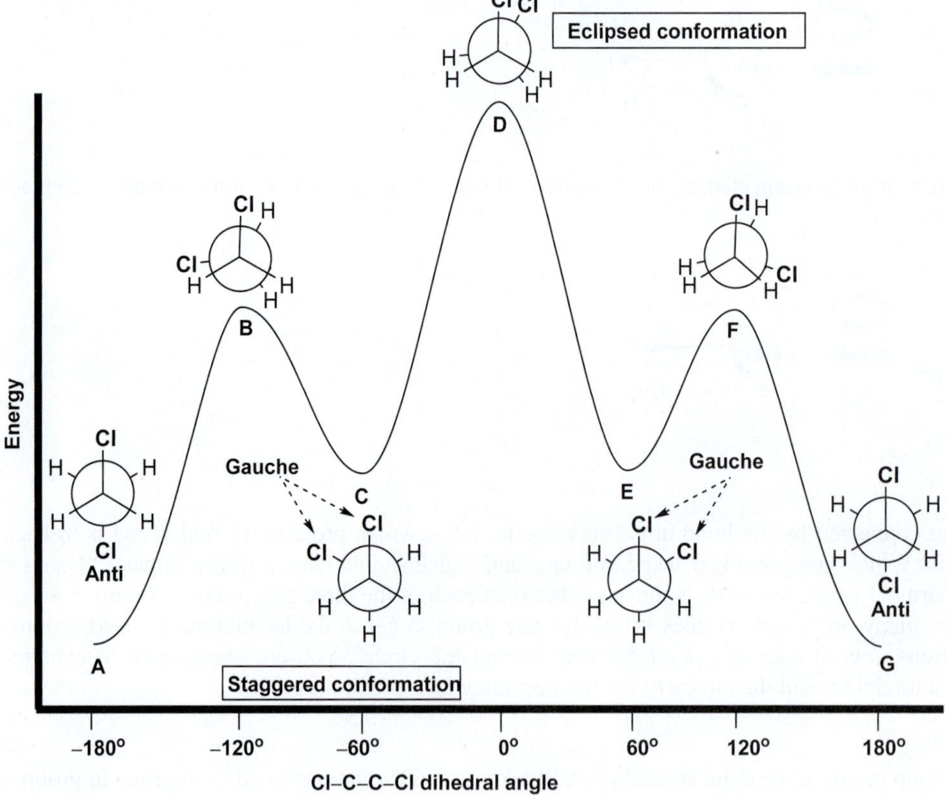

Problem 4.5

Think

Draw out each structure. Which compound has larger steric strain when rotated about the C–C bond? How does steric strain relate to the rotational energy barrier?

Solve

Because Br is much larger than F, steric strain in the eclipsed conformations in 1,2-dibromoethane is much greater than steric strain in the eclipsed conformations of 1,2-difluoroethane. Thus, the rotational energy barrier is larger in 1,2-dibromoethane. See the structures below.

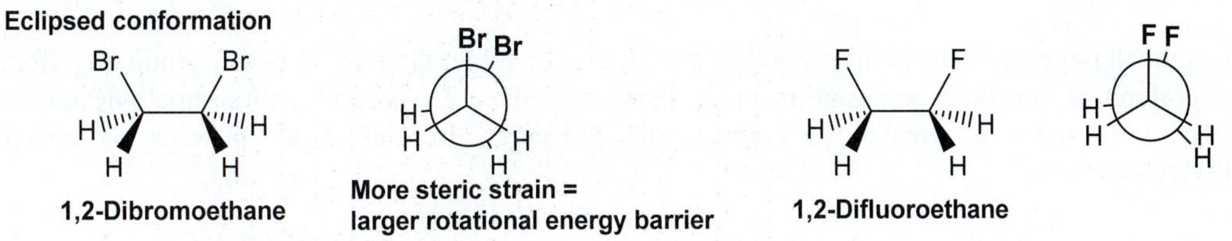

Eclipsed conformation

1,2-Dibromoethane

More steric strain = larger rotational energy barrier

1,2-Difluoroethane

Problem 4.7

Think

What are the molecular formulas of these molecules? What are the structural differences among them? How do those differences translate into heats of combustion? Which compound has the largest ring strain?

Solve

All four of these compounds have the molecular formula C_8H_{14} and contain only C–C and C–H single bonds. Differences in heats of combustion, therefore, will largely reflect differences in ring strain. The structure that contains two three-membered rings, **H**, is the most strained, so it will give off the most heat during combustion—it will have the greatest heat of combustion. Next comes the structure with the three- and four-membered rings, **G**, followed by the structure with two four-membered rings, **E**. The structure with the four- and five-membered rings, **F**, is the least strained. The order of heats of combustion is **F** < **E** < **G** < **H**. The ring numbers are listed below each structure.

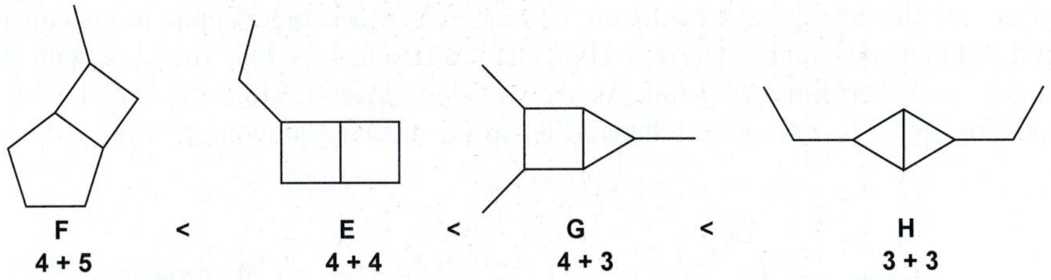

F < E < G < H

4 + 5, 4 + 4, 4 + 3, 3 + 3

Problem 4.9

Think

What is the difference, if any, in the sizes of hydrogen and deuterium (D)? What are the axial and equatorial interactions in each conformation? Are there any differences? How does the size of a substituent attached to a cyclohexane ring govern the preference for the axial versus equatorial position?

Solve

In the chair conformation on the left below, D is equatorial and H is axial. In the chair conformation on the right, D is axial and H is equatorial. Because D is an isotope of H, the electron clouds are essentially no different in size, so there will be essentially no difference in energy between the two conformations. Remember that the bulkier substituent has the stronger preference for equatorial. Thus, we should expect the two conformations to be present in equal abundance.

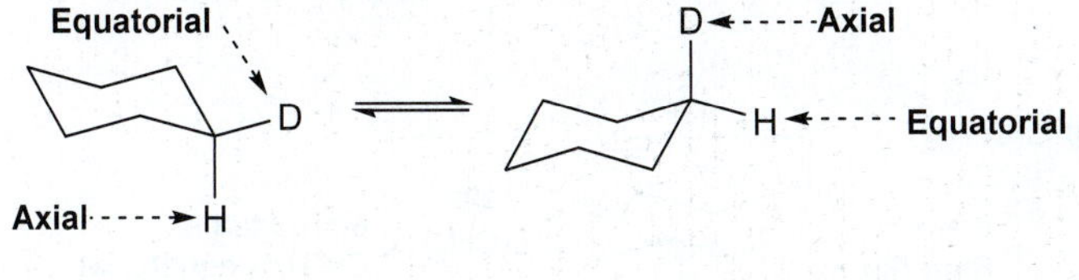

Equatorial, Axial

Problem 4.10

Think

What is the difference in size of I compared to F? Which group is bulkier, CI_3 or CF_3? Does the larger substituent have a greater preference for an axial or for an equatorial position?

Solve

The I atom is much larger than the F atom, making the CI_3 group bulkier than CF_3. The CI_3 group, therefore, has a stronger preference for the equatorial position than does the CF_3 group. Trifluoromethylcyclohexane, consequently, will have a greater percentage of molecules in the axial position compared to triiodomethylcylcohexane.

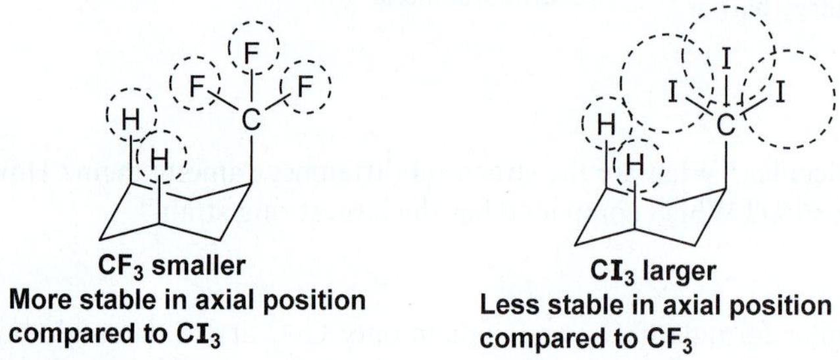

Problem 4.12

Think

Does one substituent on the ring require more room than the other? Which position, axial or equatorial, offers more room? Can both substituents achieve that position at the same time?

Solve

The two methyl (CH_3) groups are the largest substituents on the ring, and each is more stable in the equatorial position. Draw out *trans*-1,2-dimethylcyclohexane (in a Haworth projection or a line structure with dash–wedge) and then transfer to a chair. Perform a ring flip. As shown below, *trans*-1,2-dimethylcyclohexane can achieve equatorial positions for both CH_3 groups simultaneously, so the diequatorial conformation is the most stable conformation.

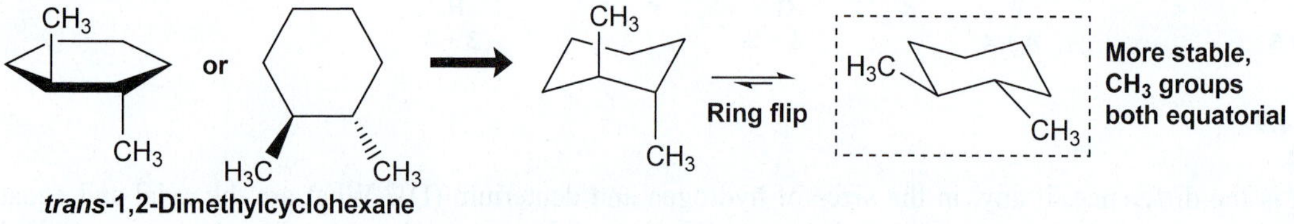

Problem 4.13

Think

Which substituent requires more room, methyl or *t*-butyl? Which position, axial or equatorial, offers more room? Can both substituents achieve that position at the same time?

Solve

Because the *t*-butyl group is so bulky, the ring is essentially "locked" in the equatorial position, which is the conformation shown below on the right. Two carbons away on the ring, the CH_3 group must be on the opposite side of the ring—if the *t*-butyl group is up, the methyl group must be down. The position occupied by CH_3, therefore, must be axial.

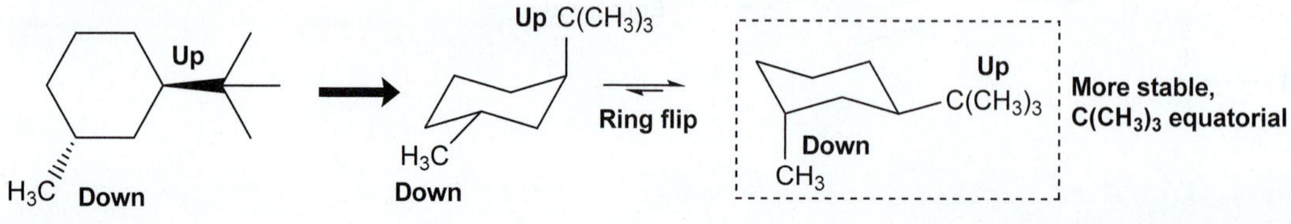

Problem 4.14

Think

If the two substituents are cis, are they both on the same side of the ring or on opposite sides? Which substituent requires more room, methyl or trichloromethyl? Which position, axial or equatorial, offers more room? Can both substituents achieve that position?

Solve

Both CH_3 and CCl_3 must be on the same side of the ring—for example, both pointed upward (as shown below). Only one of those positions can be equatorial, whereas the other is axial. Because CCl_3 is larger than CH_3, the compound is more stable with CCl_3 in the equatorial position, forcing CH_3 into the axial position, as shown below.

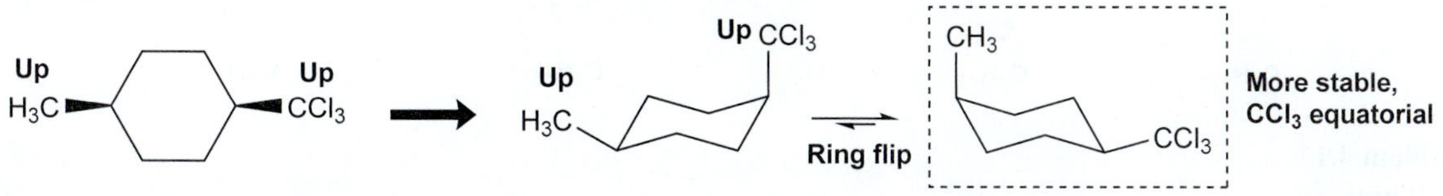

Up H_3C Up CCl_3 → Up H_3C Up CCl_3 ⇌ Ring flip CH_3 CCl_3 **More stable, CCl_3 equatorial**

Problem 4.16

Think

Do both compounds have the same molecular formula? Is the largest continuous chain or ring in each molecule the same? Are the C atoms involved in the double bonds assigned the same numbers in each molecule? In each molecule, are the substituents attached to the numbered C atoms in the same order?

Solve

Pairs **(a)**, **(b)**, and **(c)** have different chemical formulas and, therefore, are not isomers.

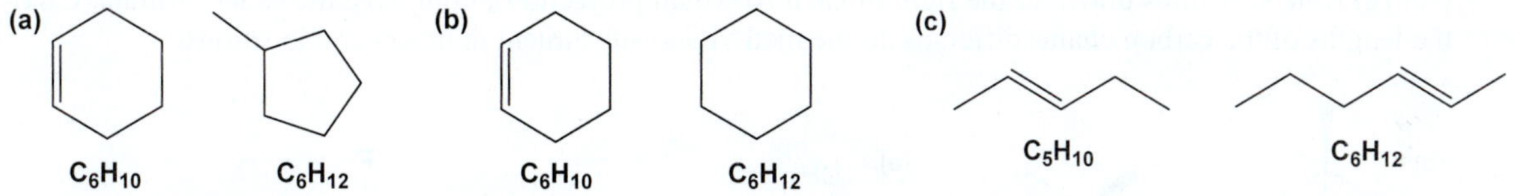

(a) C_6H_{10} C_6H_{12} (b) C_6H_{10} C_6H_{12} (c) C_5H_{10} C_6H_{12}

Pairs **(d)** and **(i)** have the same molecular formula but different connectivity, making them constitutional isomers. For pair **(d)**, both have the formula C_7H_{14}, but the methyl groups are on different C atoms. In the first molecule, the methyl groups are on C1 and C2; in the second molecule, the methyl groups are on C1 and C3. For pair **(i)**, both have the formula C_9H_{16}, but the propyl group is on a different C atom in each molecule, if we assign the doubly bonded C atoms C1 and C2. In the first molecule, the propyl group is on C3; in the second molecule, the propyl group is on C4.

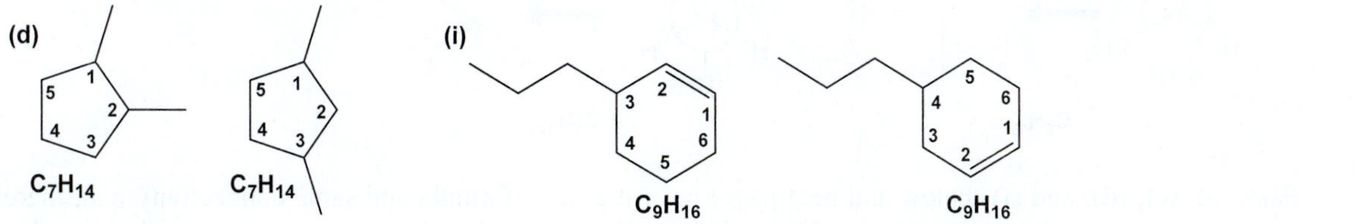

(d) C_7H_{14} C_7H_{14} (i) C_9H_{16} C_9H_{16}

Pairs **(e)**, **(f)**, **(g)**, and **(h)** (next page) have the same formula and the same connectivity and, therefore, are not constitutional isomers.

(e) C_9H_{20} C_9H_{20}

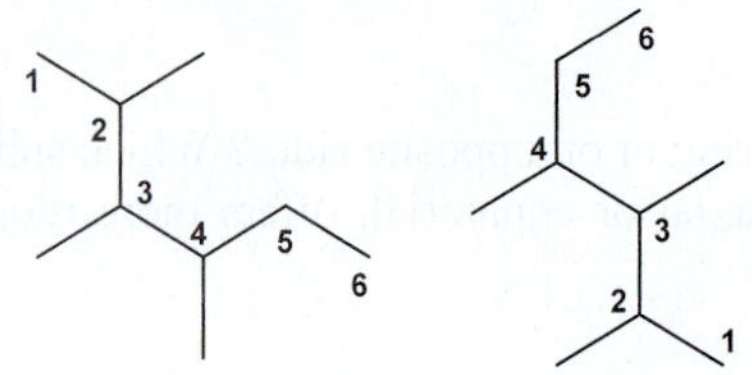

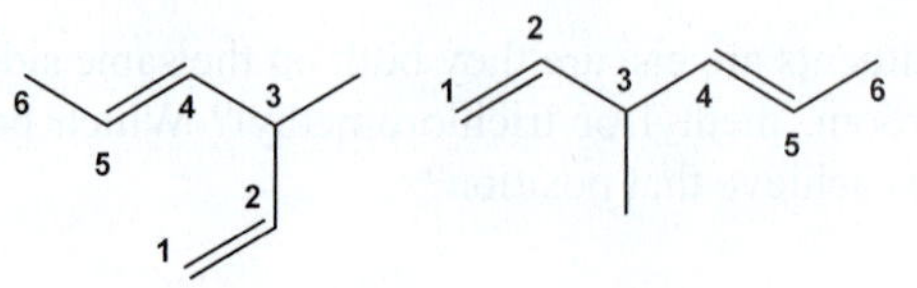

(f) C_7H_{12} C_7H_{12}

(g) C_6H_{12} C_6H_{12}

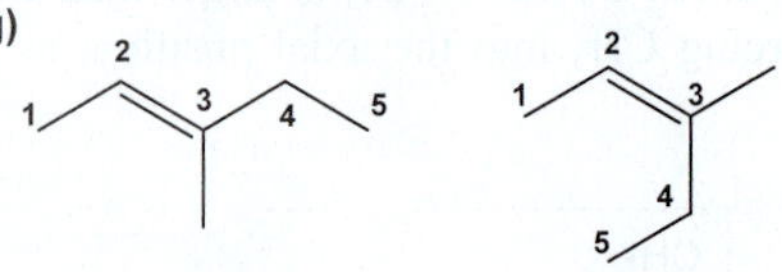

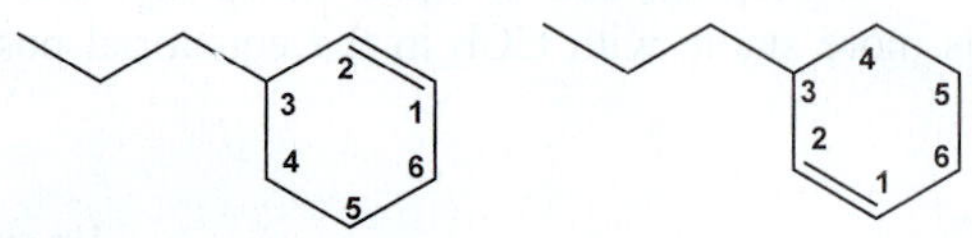

(h) C_9H_{16} C_9H_{16}

Problem 4.17

Think

Do both compounds have the same molecular formula? Is the largest continuous chain or ring in each molecule the same? Are the C atoms involved in the double bonds assigned the same numbers in each molecule? In each molecule, are the substituents attached to the numbered C atoms in the same order?

Solve

Pairs **(b)**, **(e)**, and **(g)** have the same molecular formula and different connectivities, and, therefore, are constitutional isomers. For pair **(b)**, both have the formula C_8H_{16}, but the methyl groups are on different C atoms. In the first molecule, the methyl groups are on C1 and C3, but in the second molecule, the methyl groups are on C1 and C2. For pair **(e)**, the methyl groups are on different carbon atoms—C3 or C4, respectively. For pair **(g)** (line structures drawn to the right of each Newman projection), both have the same formula, C_5H_{12}, but the lengths of the carbon chains differ, as do the methyl and substituent numbers and locations.

(b) C_8H_{16} C_8H_{16}

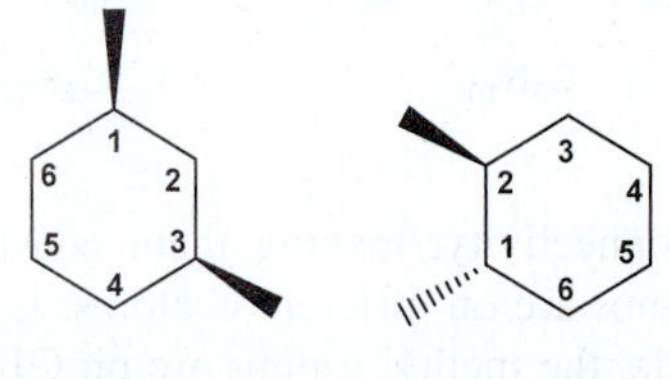

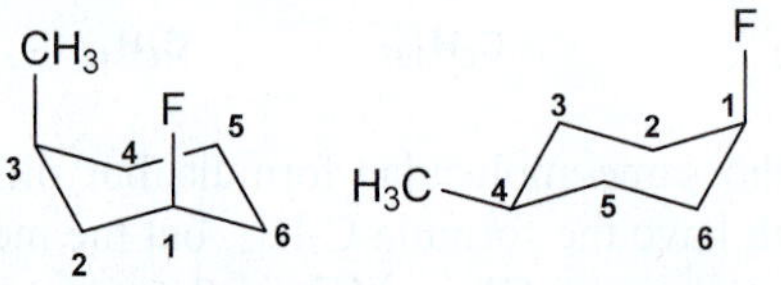

(e) $C_7H_{13}F$ $C_7H_{13}F$

(g) C_5H_{12} C_5H_{12}

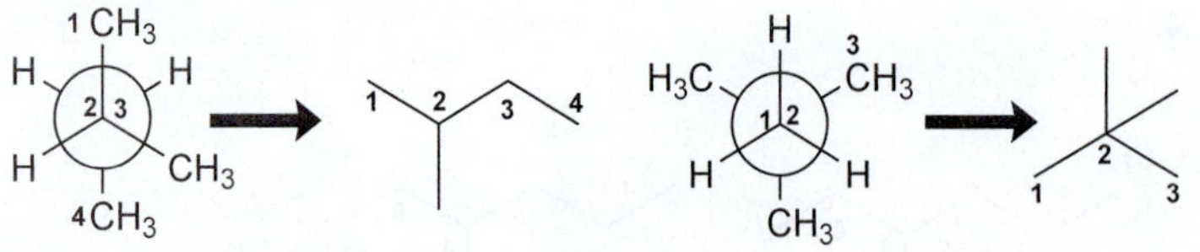

Pairs **(a)**, **(c)**, **(d)**, and **(f)** (below and next page) have the same formula and same connectivity and, therefore, are not constitutional isomers. See numbering below. Remember that differences in dash–wedge notation and differences in rotational angles about a single bond do *not* represent differences in connectivity.

(a) C_8H_{16} C_8H_{16}

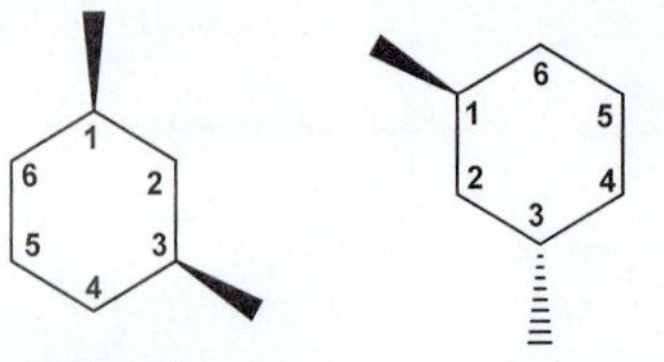

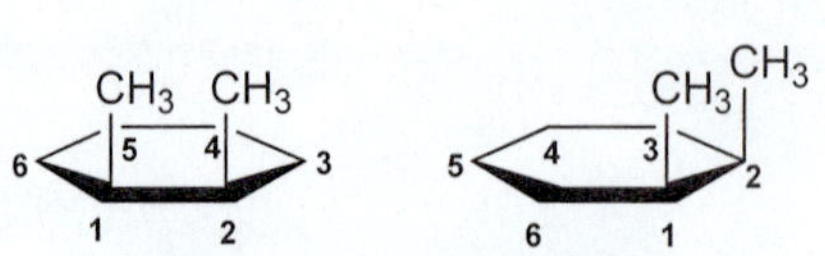

(c) C_8H_{16} C_8H_{16}

(d) $C_6H_{10}Cl_2$ $C_6H_{10}Cl_2$

(f) C_5H_{12} C_5H_{12}

Problem 4.19

Think

How many rings, double bonds, and triple bonds are present? How much does each contribute to the overall index of hydrogen deficiency (IHD)?

Solve

A double bond and a ring each contribute 1 to the IHD, and a triple bond contributes 2 to the IHD.

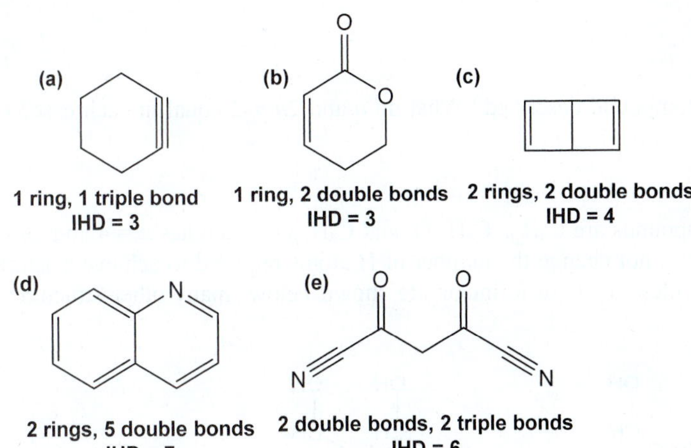

Problem 4.20

Think

In each compound class in Table 1-6, how many rings, double bonds, and triple bonds are present? How much does each of these features contribute to the overall IHD?

Solve

Compound classes that have an IHD of 1 are alkenes (one double bond), epoxides (one ring), ketones (one double bond), aldehydes (one double bond), carboxylic acids (one double bond), esters (one double bond), and amides (one double bond). Compound classes that have an IHD of 2 are alkynes and nitriles (one triple bond each). No compound classes in the table have an IHD of 3. Arenes have an IHD of 4.

IHD = 1

Alkene Epoxide Ketone Aldehyde Carboxylic acid Ester Amide

IHD = 2

Alkyne Nitrile

IHD = 4

Arene

Problem 4.22

Think

What is the formula for an analogous saturated compound? How many H atoms are missing from the formula given?

Solve

We can construct a (hypothetical) saturated molecule containing four C atoms, two N atoms, one O atom, and one F atom, as shown below. To be saturated, the molecule must have only single bonds and no rings. This molecule has 11 H atoms, so a compound with the formula $C_4H_7N_2OF$ is deficient by four H atoms, or two H_2 molecules. Thus, the IHD = 2.

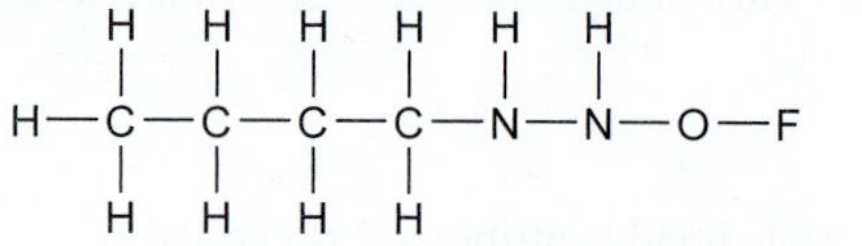

$C_4H_{11}N_2OF$

Problem 4.23

Think

What is the formula for each saturated compound described? What do n and $2n + 2$ equal in each case? How does $2n + 2$ compare to the number of H atoms?

Solve

The formulas for the three saturated compounds are C_4H_{10}, $C_4H_{10}O$, and $C_4H_{10}O_2$. Each has $n = 4$ and $2n + 2 = 10$. Therefore, each additional O atom does not change the number of H atoms required to achieve a saturated compound. Examples of saturated molecules for these formulas are shown below; many other structures are possible.

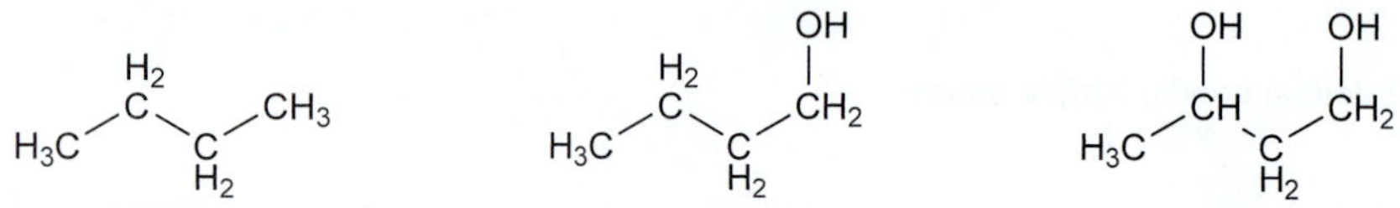

C_4H_{10} **$C_4H_{10}O$** **$C_4H_{10}O_2$**

Problem 4.24

Think

What is the formula for each saturated compound described? What do n and $2n + 2$ equal in each case? How does $2n + 2$ compare to the number of H atoms in each molecule? What is the effect of adding a N atom? How many bonds does an uncharged N have? What is the effect of adding a F atom? How many bonds does an uncharged F have?

Solve

Nitrogen. The formulas for the saturated compounds are C_4H_{10}, $C_4H_{11}N$, and $C_4H_{12}N_2$. In each case, $n = 4$ and $2n + 2 = 10$. With no N atoms, the number of H atoms is $2n + 2$; with one N atom, the number of H atoms is $(2n + 2) + 1$. With two N atoms, the number of H atoms is $(2n + 2) + 2$. Each additional N atom requires one additional H atom to achieve a saturated compound. Uncharged N atoms have three bonds. Two bonds are required to continue the chain, and the additional bond is made up by adding in a H atom. Examples of saturated molecules for these formulas are shown below; many other structures are possible.

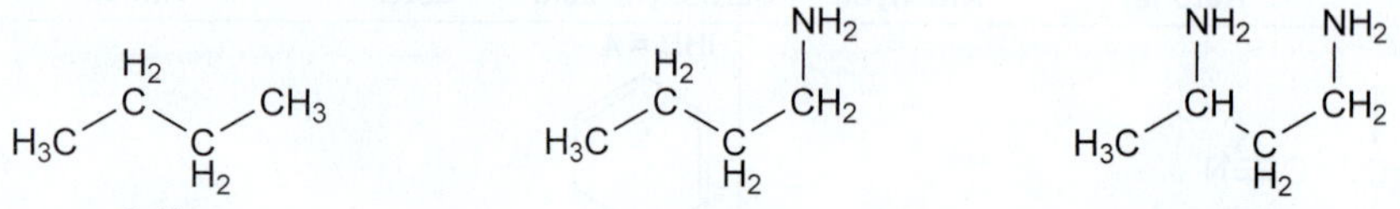

C_4H_{10} **$C_4H_{11}N$** **$C_4H_{12}N_2$**

Fluorine. The formulas for the saturated compounds are C_4H_{10}, C_4H_9F, and $C_4H_8F_2$. In each case, $n = 4$ and $2n + 2 = 10$. With no F atoms, the number of H atoms is $2n + 2$; with one F atom, the number of H atoms is $(2n + 2) - 1$. With two F atoms, the number of H atoms is $(2n + 2) - 2$. Each additional F atom requires one fewer H atom to achieve a saturated compound. An uncharged F atom has one bond and, therefore, replaces a H atom. This is the case for all halogen atoms. Examples of saturated molecules for these formulas are shown below; many other structures are possible.

$H_3C{-}CH_2{-}CH_2{-}CH_3$ — **C_4H_{10}**

$H_3C{-}CH_2{-}CH_2{-}CH_2F$ — **C_4H_9F**

$H_3C{-}CHF{-}CH_2{-}CH_2F$ — **$C_4H_8F_2$**

Problem 4.25

Think

What do the prefixes *aldo* and *keto* specify? What tells you the number of C atoms in each structure? If the molecule is a monosaccharide, how should the number of hydrogen and oxygen atoms compare to the number of carbon atoms? Refer to Figure 4-45 as a general reference to each acyclic sugar.

Solve

The *aldo* and *keto* prefixes specify whether the C=O group is at the end of a carbon chain and is of the form HC=O, as in aldehydes, or is internal and is of the form $R_2C{=}O$, as in ketones. The number of carbons in the chain is specified in the name, too: *tri* = 3; *tetr* = 4; *pent* = 5; *hex* = 6. For example, **(a)** aldotetrose has a terminal carbonyl, as we see in aldehydes, and has four C atoms total. See structures for **(a)–(e)** below. *Note:* There are many other possibilities for the dash–wedge bonds of the C–OH groups, but only one is drawn for each sugar below.

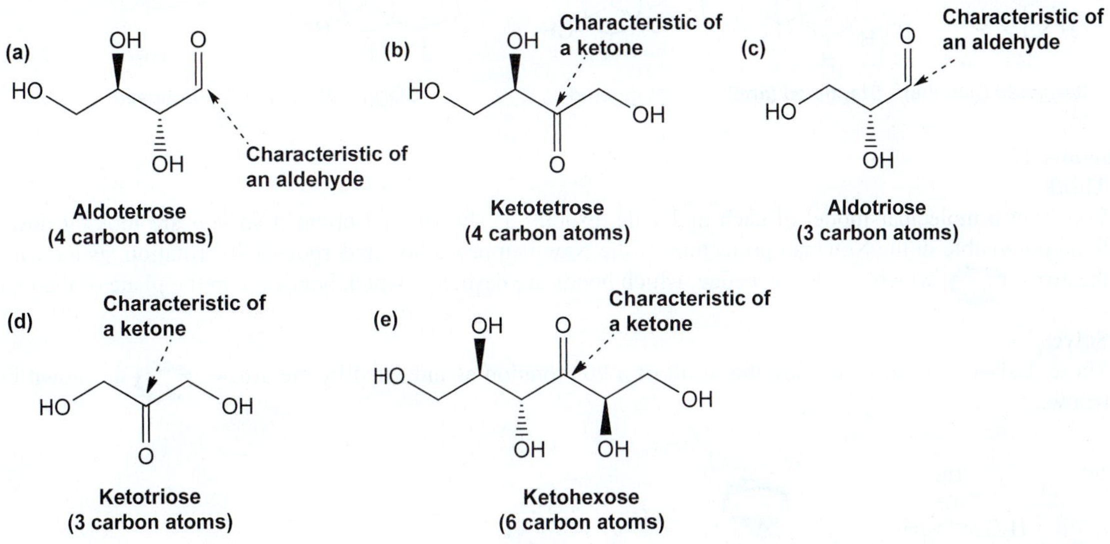

End of Chapter Problems
Section 4.2 Newman Projections
Problem 4.26

Think

Construct a molecular model of each molecule, exactly as shown, and orient it so you are looking down the bond indicated. How do you represent the front C atom in a Newman projection? How do you represent the back C atom? Should the C–C bond indicated be visible? How many atoms should be shown bonded to each C atom? As drawn, is the staggered or eclipsed conformation shown?

Solve

The front C atom is represented by a dot and the back C atom is represented by a circle. The C–C bond is not shown, as it is perpendicular to the plane of the paper and is not visible. The remaining three groups on the front C atom converge at the point, and the remaining three groups on the back C atom connect to the circle. These Newman projections result from looking down the structures in the problem from the left side, as shown by **(a)**. As you look from the left, the front C atom has H, H′, and Br, and the back C atom has CH_3, H, and H′. The flat bonds in the dash–wedge structure become up in the Newman projection, the wedges become bonds to the right, and dashes become bonds to the left. See the figure below.

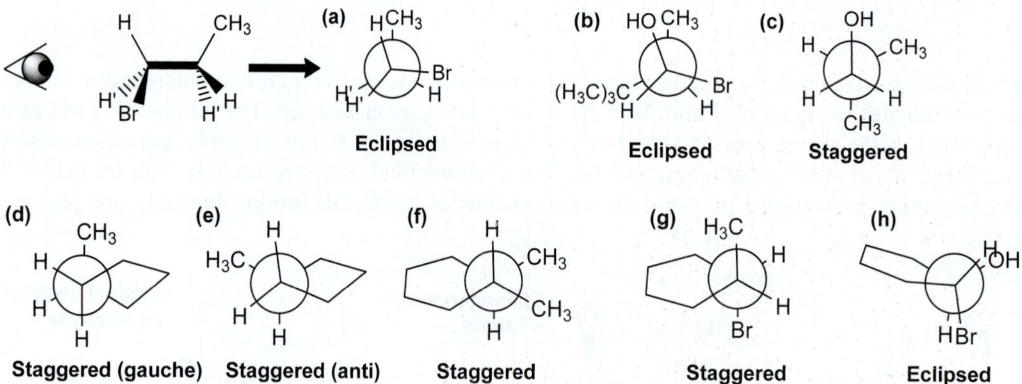

(d) Staggered (gauche)

(e) Staggered (anti)

(f) Staggered

(g) Staggered

(h) Eclipsed

Problem 4.27

Think

Construct a molecular model of each molecule, exactly as shown, and orient it so you are looking down the bond not visible in the Newman projection. If the Newman projection undergoes a 90° rotation, as indicated by the arrow, which bonds are wedge, which bonds are dash, and which bonds are in the plane of the paper?

Solve

These dash–wedge structures are the result of a 90° rotation as indicated by the arrow as shown by **(a)** below.

(a) Up, Dash, Wedge, Down — 90° Rotation

(b)

(c)

(d)

Problem 4.28

Think

Build a molecular model of cyclohexane and orient it so you are looking down the bonds indicated by the Newman projections. Can you identify the plane that is roughly defined by the cyclohexane ring? Which C–H bonds are straight up and down, perpendicular to the plane? Which atoms point outward from the center of the ring?

Solve

From the Newman projection, we can identify four axial H atoms (up and down) and four equatorial H atoms (point outward from the center of the ring). The other four H atoms that are labeled CH_2 are not distinguishable as either axial or equatorial. See the figure below.

Problem 4.29

Think

Construct a molecular model of each molecule, exactly as shown, and orient it so you are looking down the bond indicated. Looking down the C–C bond indicated, which C atom is in front, and which C atom is in the back? On the front C atom, are there groups pointing upward, downward, leftward, or rightward? On the back C atom, are there groups pointing upward, downward, leftward, or rightward? On which side is the ring connected? Which molecule minimizes steric strain?

Solve

As we can see in the Newman projections below, there is more steric strain in the cis compound than there is in the trans owing to the eclipsing of the CH_3 groups.

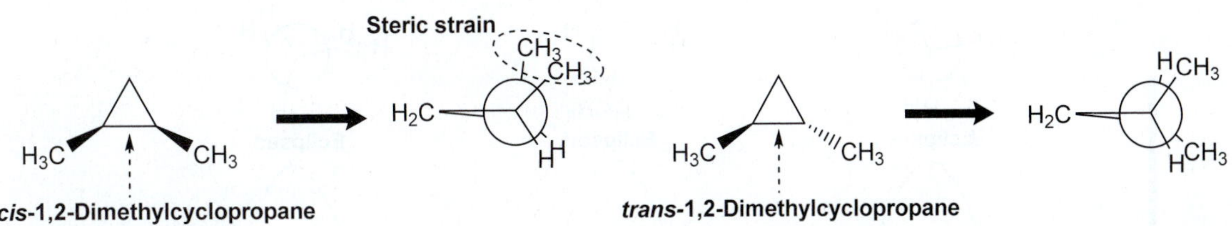

Section 4.3 Conformational Analysis

Problem 4.30

Think

Which interactions (staggered or eclipsed) are higher in energy? Which staggered interactions, anti or gauche, are higher in energy? Which group, CH_3, Cl, CBr_3, or H, takes up the most space?

Solve

Staggered anti conformations are lowest in energy (most stable), followed by staggered gauche conformations. Eclipsed conformations are higher in energy compared to staggered conformations because they have substantial torsional strain. Compounds **D**, **B**, and **C** are all eclipsed and, therefore, are the least stable. **D** is the least stable of these three, because the largest groups, CH_3 and CBr_3, have a dihedral angle of 0° and are closest in space. The staggered conformations are more stable. **F** is the most stable, because the largest groups, CH_3 and CBr_3, are anti. See figure on the next page.

Increasing stability

D < B < C < E < A < F

D B C E A F

E: CH_3–F anti; A: CH_3–Cl anti; F: CH_3–CBr_3 anti

Eclipsed (D, B, C) Staggered (E, A, F)

Problem 4.31

Think

Build a model of the molecule, look down the bond about which the rotation is to be performed, and carry out the 360° rotation. Can you identify each staggered and eclipsed conformation? In each conformation, is there any significant steric strain? If so, how is steric strain related to the relative energy of the conformations? Which conformations (staggered or eclipsed) are higher in energy? Which staggered conformations, anti or gauche, are higher in energy?

Solve

The conformational analysis is performed from the direction indicated below.

Front Back

The back group is locked in place and the front group rotates. Eclipsed conformations are higher than the staggered ones. All three staggered conformations are identical in energy, because the steric strain between the front and back substituents is identical in each, and all three eclipsed conformations are identical in energy for the same reason.

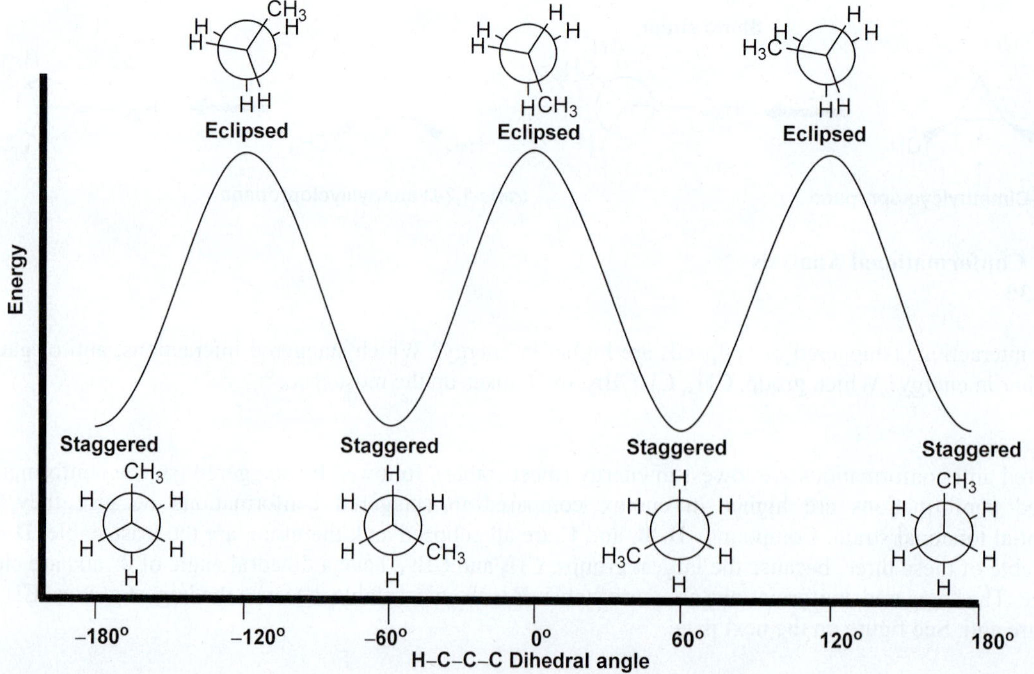

Problem 4.32

Think

Build a model of the molecule, look down the bond about which the rotation is to be performed, and carry out the 360° rotation. Can you identify each staggered and eclipsed conformation? In each conformation, is there any significant steric strain? If so, how is steric strain related to the relative energy of the conformations? Which conformations (staggered or eclipsed) are higher in energy? Which staggered conformations, anti or gauche, are higher in energy?

Solve

The conformational analysis is performed from the direction indicated below.

Front Back

The back group is locked in place, and the front group rotates. The plot is qualitatively identical to that for Br–CH_2–CH_2–Br (Fig. 4-7) because butane has CH_3 groups in place of Br atoms. The two gauche conformations at +60° and −60° are higher in energy than the anti conformation at ±180° owing to steric strain between the two CH_3 groups. The eclipsed conformation at 0° is higher in energy than those conformations at +120° and −120° because of steric strain between the two CH_3 groups.

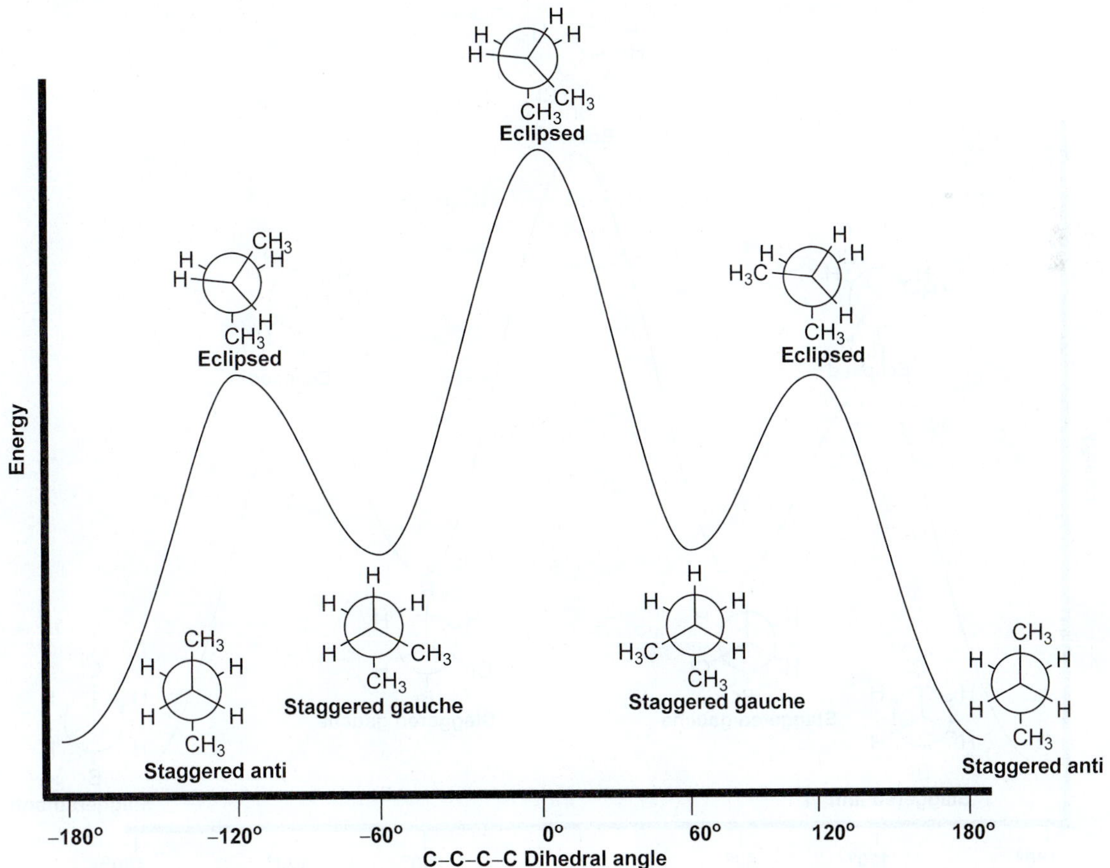

Problem 4.33

Think

Build a model of the molecule, look down the bond about which the rotation is to be performed, and carry out the 360° rotation. Can you identify each staggered and eclipsed conformation? In each conformation, is there any significant steric strain? If so, how is steric strain related to the relative energy of the conformations? Which conformations (staggered or eclipsed) are higher in energy? Which staggered conformations, anti or gauche, are higher in energy?

Solve

The conformational analysis is performed from the direction indicated below.

Front Back

The back group is locked in place, and the front group rotates. The plot is qualitatively identical to that for Br–CH_2–CH_2–Br (Fig. 4-7). The two gauche conformations at +60° and −60° are higher in energy than the anti conformation at ±180° owing to steric strain between the Br and Cl substituents. The eclipsed conformation at 0° is higher in energy than the conformations at +120° and −120° because of steric strain between the Br and Cl groups.

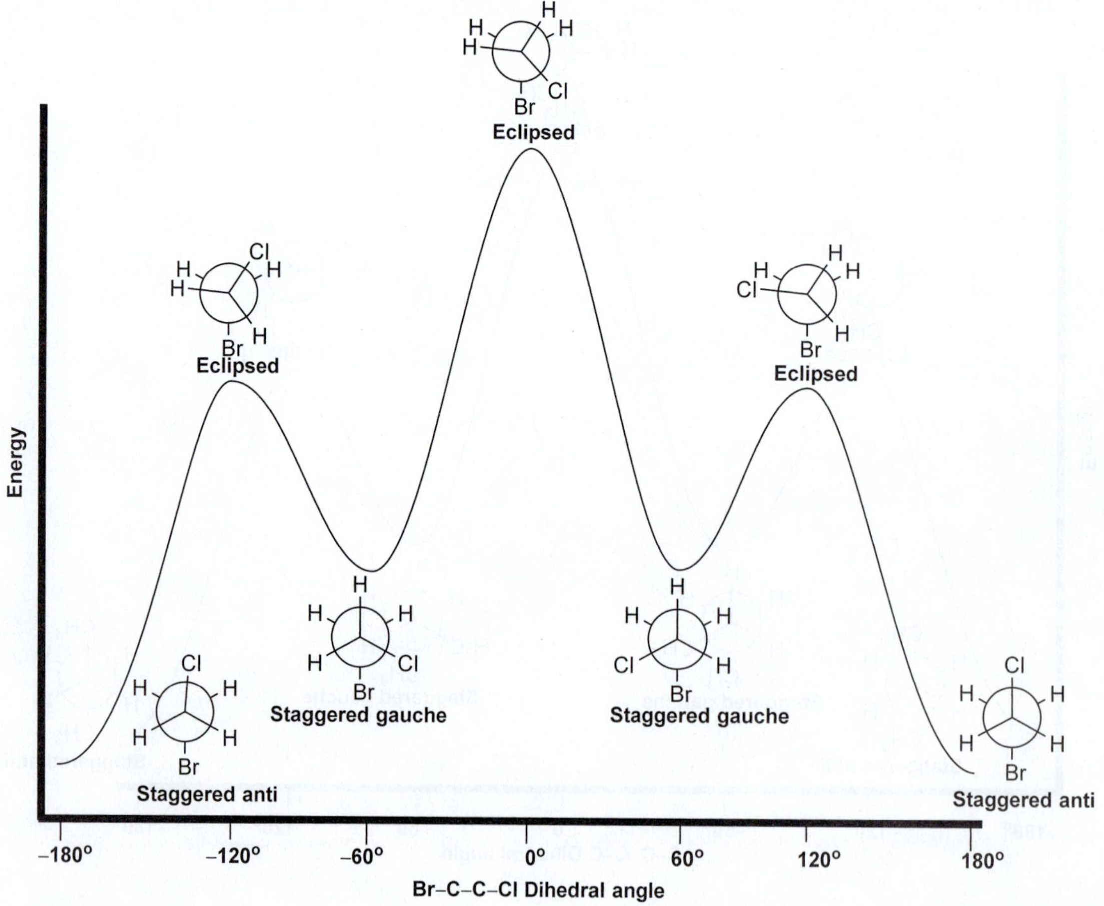

Problem 4.34

Think

Build a model of the molecule, look down the bond about which the rotation is to be performed, and carry out the 360° rotation. Can you identify each staggered and eclipsed conformation? In each conformation, is there any significant steric strain? If so, how is steric strain related to the relative energy of the conformations? Which conformations (staggered or eclipsed) are higher in energy? Which staggered conformations, anti or gauche, are higher in energy?

Solve

The conformational analysis is performed from the direction indicated below.

Front **Back**

The back group is locked in place, and the front group rotates. The eclipsed conformations are higher in energy than the staggered ones owing to torsional strain. The staggered conformation at −60° is higher in energy than the other two because both CH_3 groups on the front carbon atom are gauche to the CH_3 group on the back carbon atom. Each gauche interaction represents steric strain. The other two staggered conformations each have only one such gauche interaction. The eclipsed conformations at −120° and 0° are higher in energy than that at +120° because the two bulky CH_3 groups are eclipsed. In the eclipsed conformation at +120°, CH_3 groups are not eclipsed.

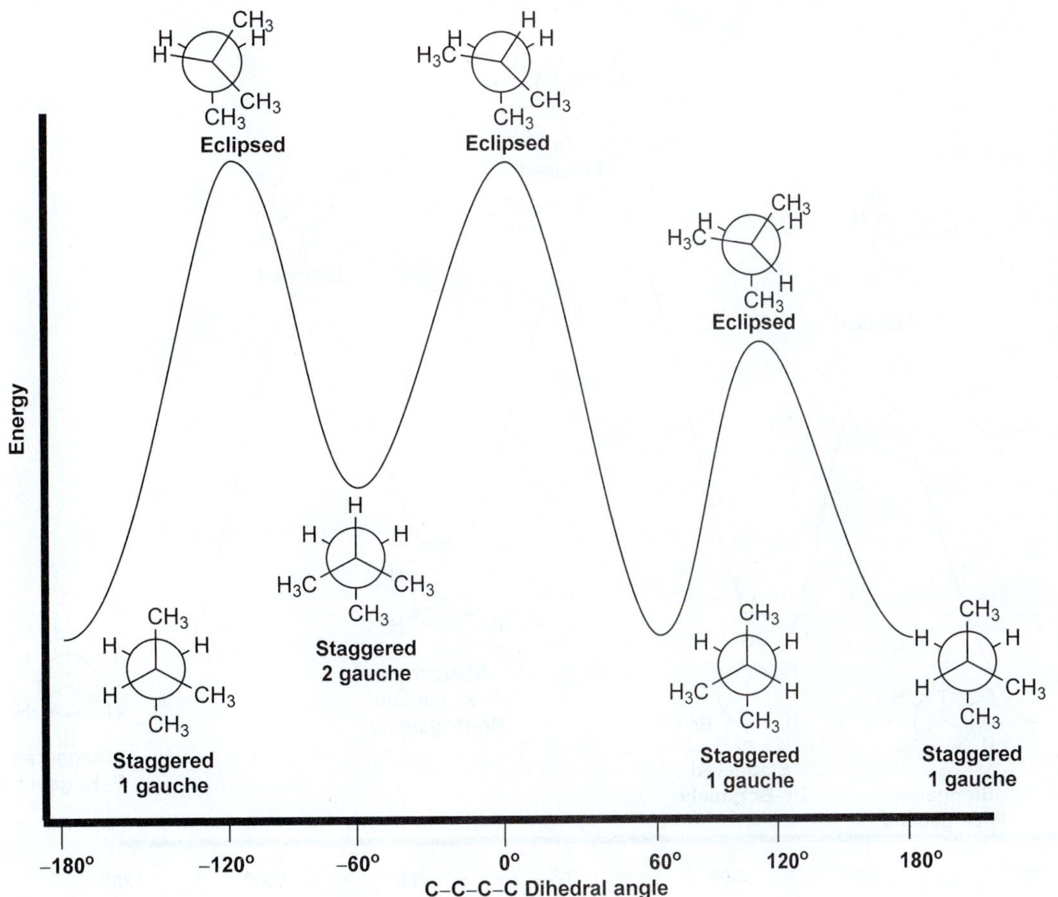

Problem 4.35

Think

Build a model of the molecule, look down the bond about which the rotation is to be performed, and carry out the 360° rotation. Can you identify each staggered and eclipsed conformation? In each conformation, is there any significant steric strain? If so, how is steric strain related to the relative energy of the conformations? Which conformations (staggered or eclipsed) are higher in energy? Which staggered conformations, anti or gauche, are higher in energy?

Solve

The conformational analysis is performed from the direction indicated below.

Front **Back**

The three staggered conformations are all at different energies, and the three eclipsed conformations are all at different energies. The staggered conformation at −60° is higher in energy than that at ±180° because the gauche interaction between two Br atoms is greater than that between a Br atom and a F atom. The staggered conformation at +60° is higher in energy than either of those because the Br atom on the back C atom is gauche to both the F and Br atoms on the front C atom. The lowest energy eclipsed conformation is at −120° because all of the halogen atoms are eclipsed only with a small H atom. That at +120° is higher in energy owing to the eclipsing interaction between F and Br atoms. That at 0° is highest in energy owing to eclipsing between the two Br atoms, and the Br atom is the largest of the substituents.

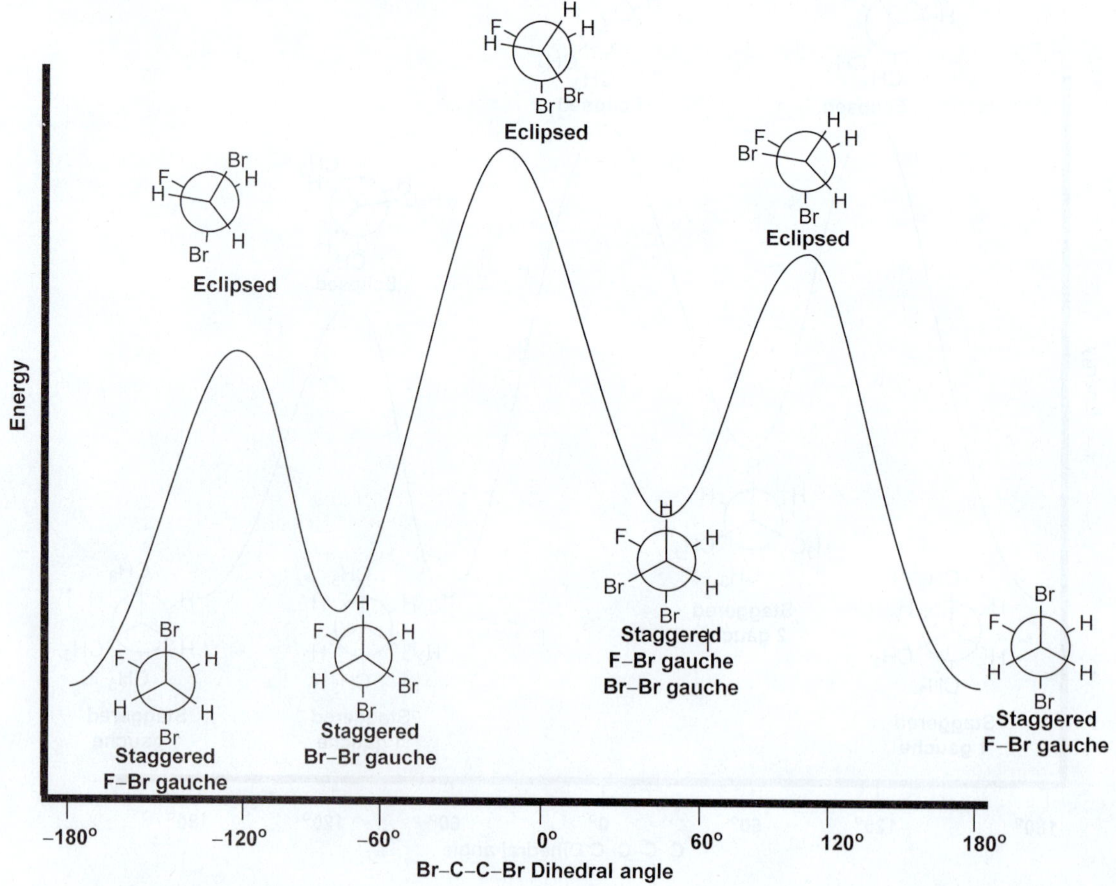

Sections 4.4–4.8 Ring Strain, Stable Conformations of Rings, and Chair Conformations of Monosubstituted Cyclohexanes

Problem 4.36

Think

How many CH_2 groups are present? The heat of combustion for cyclononane is 5586 kJ/mol. How do you calculate the heat of combustion per CH_2? How does that compare to the heat of combustion per CH_2 for cyclohexane (consult Table 4-1)? With that difference representing the ring strain per CH_2 group, how do you calculate the total ring strain? Compare this value to cycloheptane.

Solve

Nine CH_2 groups compose cyclononane. The $\Delta H°$ per carbon in cyclononane is $\frac{5586\frac{\text{kJ}}{\text{mol}}}{9\ \text{CH}_2\ \text{groups}} = 620.67\frac{\text{kJ}}{\text{mol}}$. Ring strain per CH_2 is the difference between $\Delta H°$ per CH_2 group in cyclononane versus cyclohexane, or $620.67\frac{\text{kJ}}{\text{mol}} - 615.1\frac{\text{kJ}}{\text{mol}} = 5.57\frac{\text{kJ}}{\text{mol}}$. Total ring strain is the ring strain per CH_2 group times the number of CH_2 groups, or $5.57\frac{\text{kJ}}{\text{mol}} \times 9\ \text{CH}_2\ \text{groups} = 50.13\frac{\text{kJ}}{\text{mol}}$. This value is more than that of cycloheptane, whose total ring strain is 26.6 kJ/mol.

Problem 4.37

Think

The molecular formula is the same for each structure. What would contribute to the difference in the heat of combustion? Which ring size is the most stable? Which is the least stable? How does stability relate to heat of combustion value?

Solve

All five of these compounds have the molecular formula C_9H_{16} and contain only C–C and C–H single bonds. Differences in heats of combustion, therefore, will largely reflect differences in ring strain. The structure that contains two three-membered rings, **B**, is the most strained, so it will give off the most heat during combustion—it will have the greatest heat of combustion. Next comes the structure with one three- and one four-membered ring, **D**, followed by the structure with two four-membered rings, **A**. The structure with one four- and one five-membered ring, **E**, is next, and the least strained structure is the compound with the two five-membered rings, **C**. The order of heats of combustion is **C** < **E** < **A** < **D** < **B**.

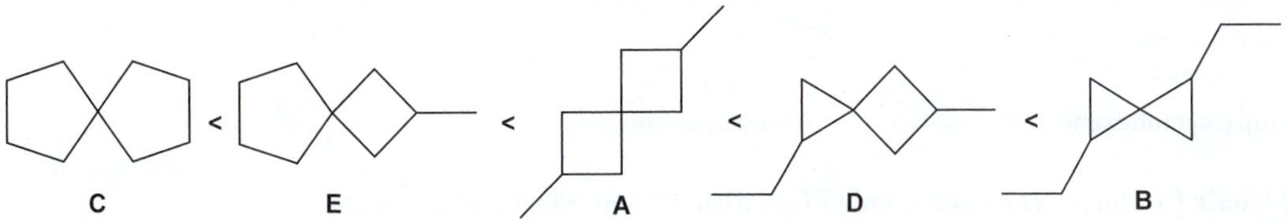

C < E < A < D < B

Problem 4.38

Think

Draw a chair for each ring. Can one ring occupy the axial position of the second ring? Equatorial? Can the second ring occupy the axial position of the first ring? Equatorial? How many combinations are possible?

Solve

The three conformations are shown below. Each ring can occupy the axial or equatorial position of the other ring. The most stable conformation is the one in which the bond connecting the two rings is equatorial with regard to each ring. The bond connecting the two rings is bolded for emphasis.

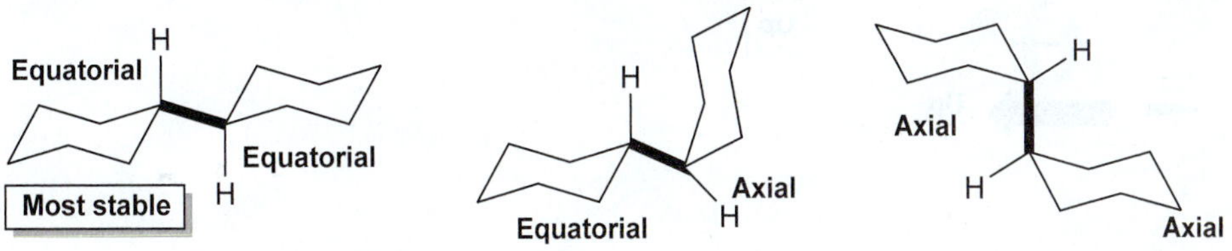

H
Equatorial
Equatorial
H
Most stable

H
Equatorial
Axial
H

H
Axial
H
Axial

Problem 4.39

Think

Can you draw both chair conformations of each compound? If a bonding pair of electrons on N occupies an axial position, what kind of position does the lone pair occupy? If a bonding pair of electrons on N occupies an equatorial position, what position does the lone pair occupy? Which position offers more room, an axial or an equatorial position?

Solve

In the molecule on the left, a lone pair occupies the equatorial position, forcing the H atom into the axial position. Because a chair conformation is more stable when the larger group is equatorial, this suggests that a lone pair occupies more room than a N–H bond. In the molecule on the right, the CH_3 group occupies the equatorial position, forcing the lone pair into the axial position. So a CH_3 group takes up more space than a lone pair.

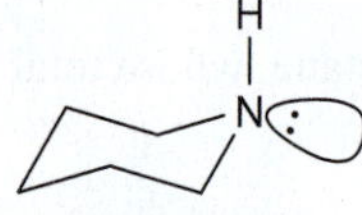

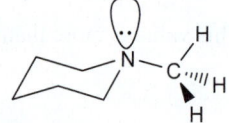

Problem 4.40

Think

Which group is bulkier, isopropyl or propyl? How does bulkiness affect a substituent's preference for equatorial versus axial?

Solve

The isopropyl group is bulkier. The C atom attached to the ring is attached to one other alkyl group in propyl and two other alkyl groups in isopropyl (see below). The isopropyl group, therefore, will have a greater preference for the equatorial position, meaning that the propyl group will be more likely than the isopropyl group to occupy the axial position.

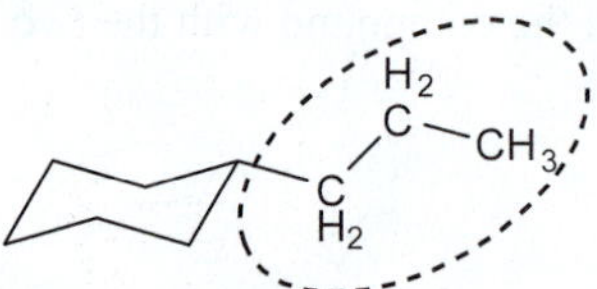

Isopropylcyclohexane

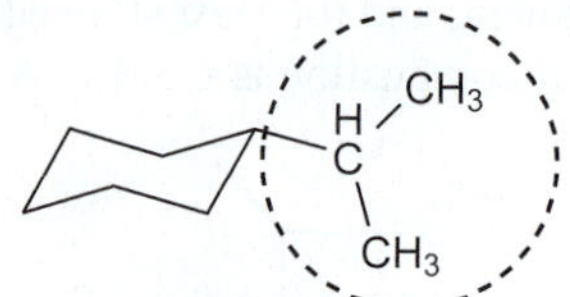

Propylcyclohexane

Section 4.9 Chair Conformations and Chair Flips with Two or More Substituents

Problem 4.41

Think

Which substituent attached to the cyclohexane ring requires more room? Which position, axial or equatorial, offers more room? Can both substituents achieve that position at the same time?

Solve

A *t*-butyl group [$C(CH_3)_3$] requires more room than a methyl (CH_3). The equatorial position offers more room.

(a) One CH_3 group will be axial, and the other CH_3 group will be equatorial. There is no way to achieve both CH_3 groups in the equatorial position.

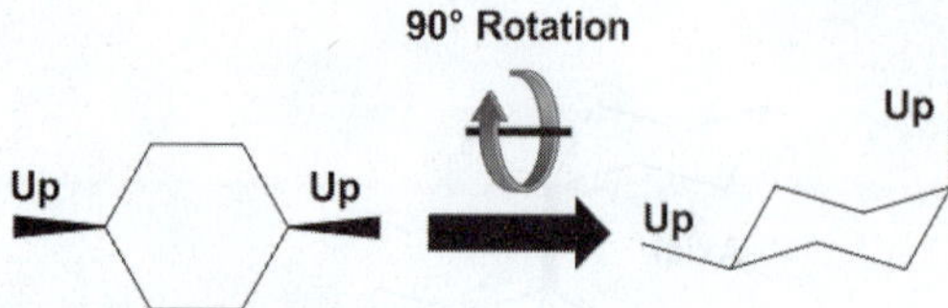

(b) Both CH_3 groups are able to be in the equatorial position.

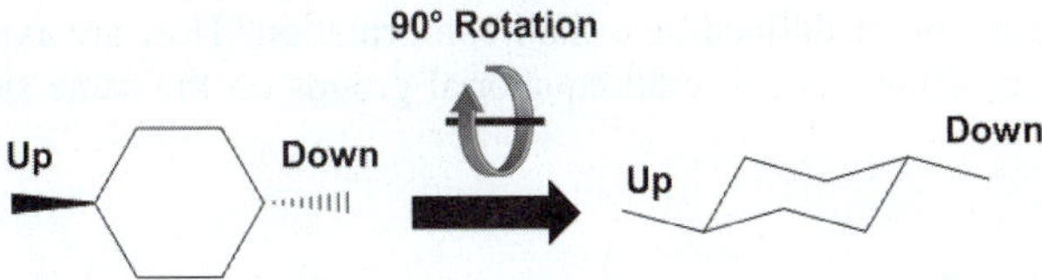

(c) Both the CH_3 and $C(CH_3)_3$ groups are able to be in the equatorial position.

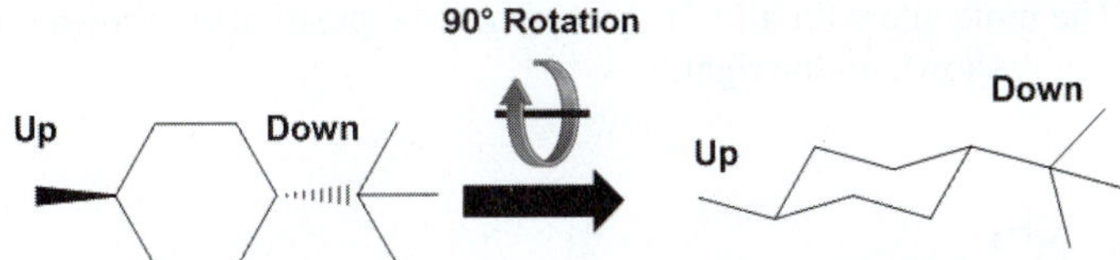

(d) Only one group is able to be in the equatorial position. The $C(CH_3)_3$ is larger, and therefore, the molecule will be more stable with the $C(CH_3)_3$ group in the equatorial position than the CH_3 group.

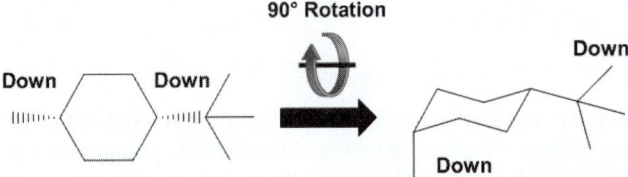

(e) Both the CH_3 and $C(CH_3)_3$ groups are able to be in the equatorial position.

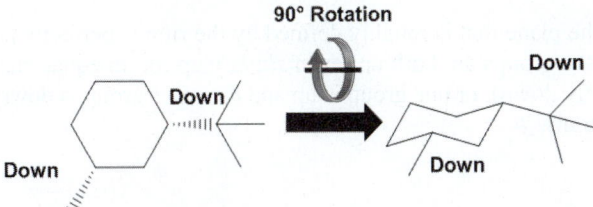

(f) Only one group is able to be in the equatorial position. The $C(CH_3)_3$ is larger, and therefore, the molecule will be more stable with the $C(CH_3)_3$ group in the equatorial position than the CH_3 group.

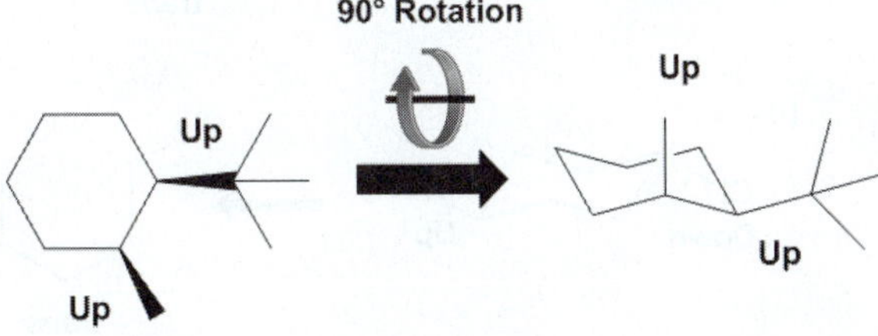

(g) Both the CH_3 and $C(CH_3)_3$ groups are able to be in the equatorial position.

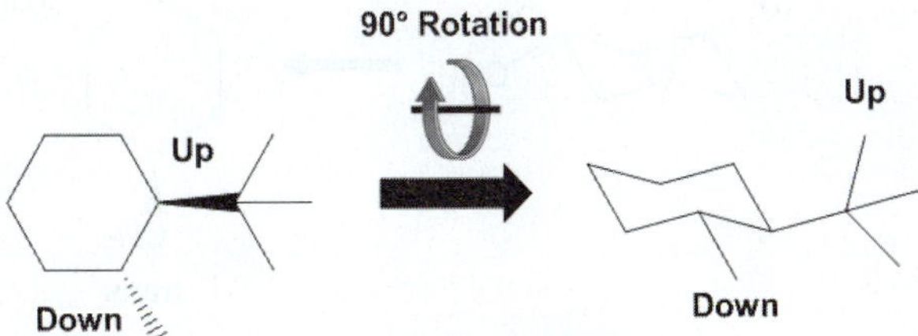

Problem 4.42

Think

How are equatorial groups oriented relative to the plane defined by a chair conformation? How are axial groups oriented? Are axial positions on the same side of the ring? Are all equatorial groups on the same side of the ring?

Solve

Axial positions are perpendicular to the plane defined by the cyclohexane chair, and going around the ring, they alternate sides of the plane. Equatorial positions point outward from the center of the ring and are oriented slightly upward or downward. They, too, alternate up and down. In both cases, each adjacent pair of CH_3 groups must be on opposite sides of the ring. The molecule with all CH_3 groups in axial positions is shown on the left. That in which the groups are all equatorial is shown on the right.

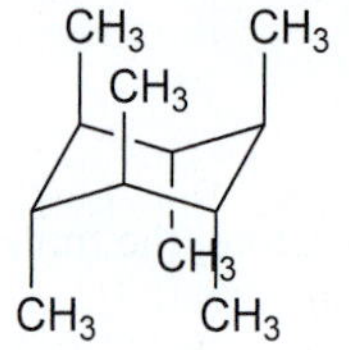

All axial

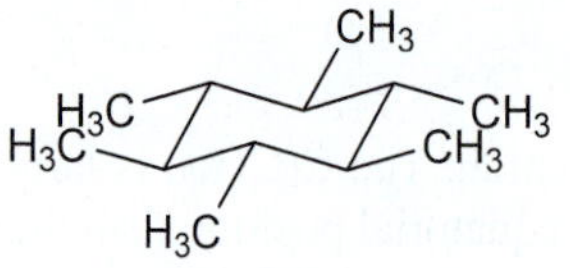

All equatorial

Problem 4.43

Think

In the chair conformations, can you identify the plane that is roughly defined by the cyclohexane chair? Relative to that plane, are the substituents on the same side of the plane or on opposite sides? If you are having difficulty working with the chair representation, transfer each chair conformation to a dash–wedge representation. For each group, consider if it is pointing upward or downward.

Solve

The way the chair conformations are drawn, the plane that is roughly defined by the ring is perfectly horizontal. The key is to notice if, relative to that plane, the groups are both up (even *slightly* up for an equatorial group is sufficient), the groups are both down (or slightly down), or one group is up and the other group is down. **(a)** cis; **(b)** trans; **(c)** cis; **(d)** trans; **(e)** cis; **(f)** trans.

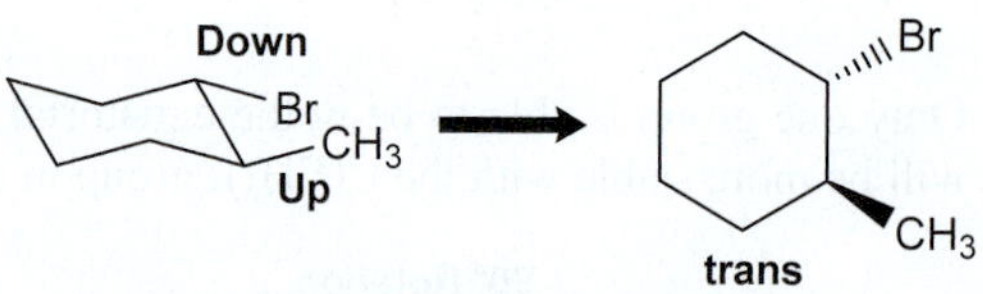

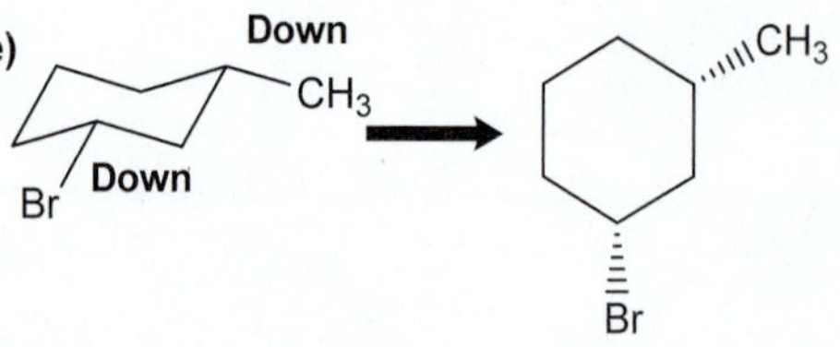

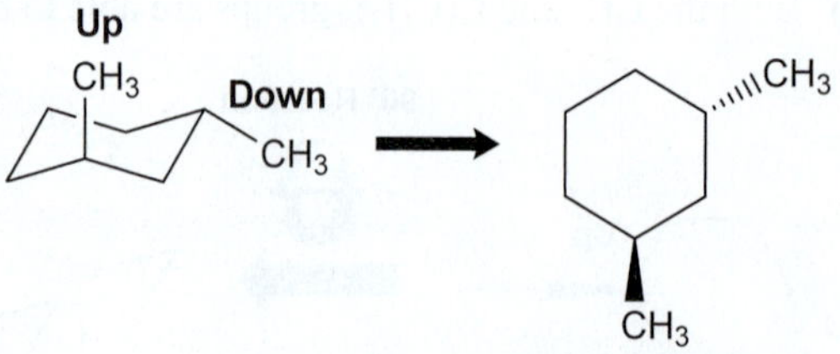

(a) Down, Down → cis

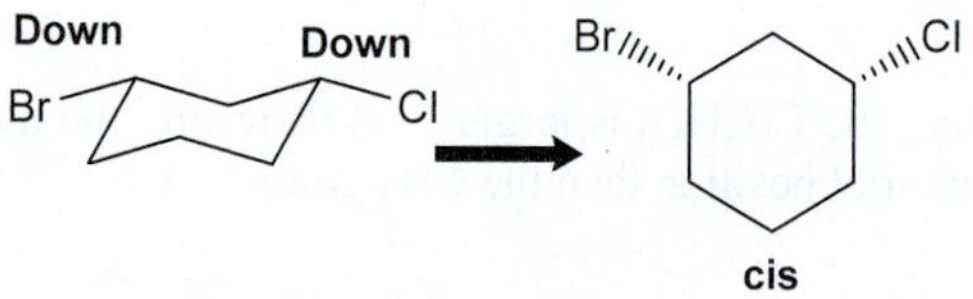

(b) Down, Up → trans

(c) Up, Up → cis

(d) Down, Up → trans

(e) Down, Down → cis

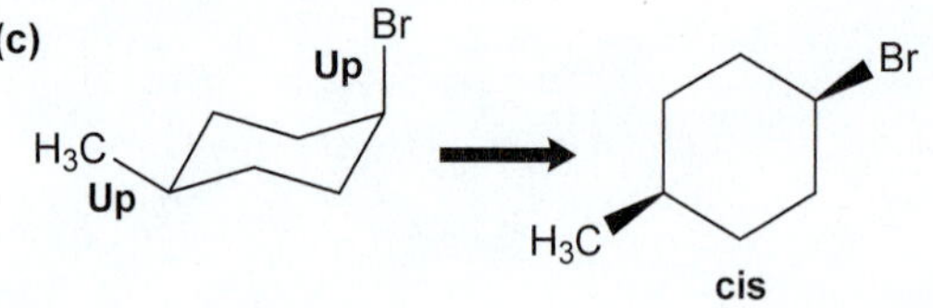

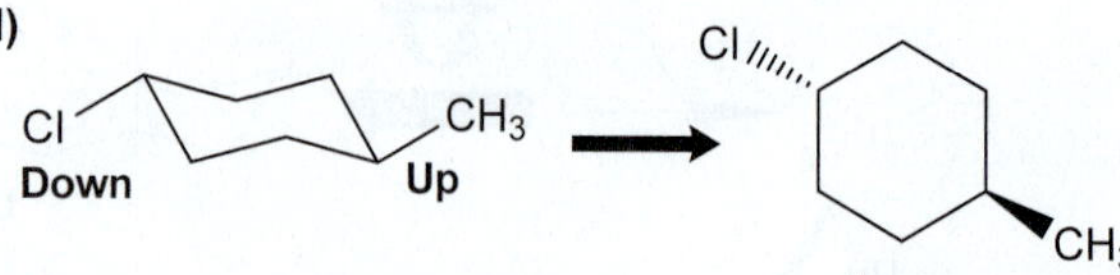

(f) Up, Down → trans

Problem 4.44

Think

Write the chair structure for the cis and trans of each molecule. It can help to draw each molecule in a dash–wedge line structure first and then, using a molecular model to help, transfer each structure to a chair. Which substituents require the most room? Which position, axial or equatorial, offers more room? Can both substituents achieve that position at the same time?

Solve

For each compound, both the cis and trans isomer forms are shown in chair structures, and a box is drawn around the more stable form. In each case, in the more stable compound, both CH_3 groups can attain equatorial positions; in the less stable form, one CH_3 group is forced into an axial position.

(a) CH_3 CH_3 CH_3 CH_3 **cis** CH_3 CH_3 CH_3 CH_3 **trans**

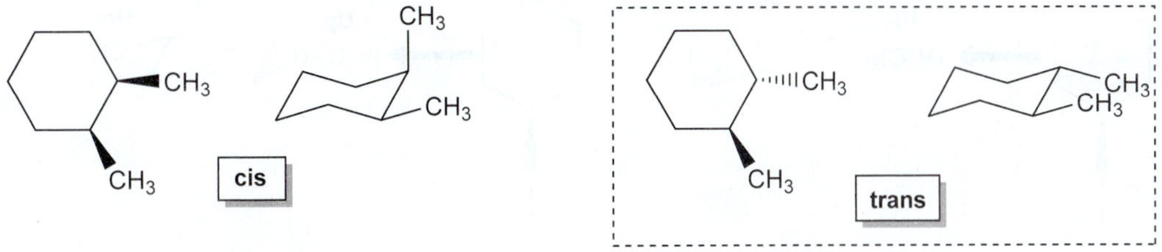

(b) CH_3 CH_3 CH_3 CH_3 **cis** CH_3 CH_3 CH_3 CH_3 **trans**

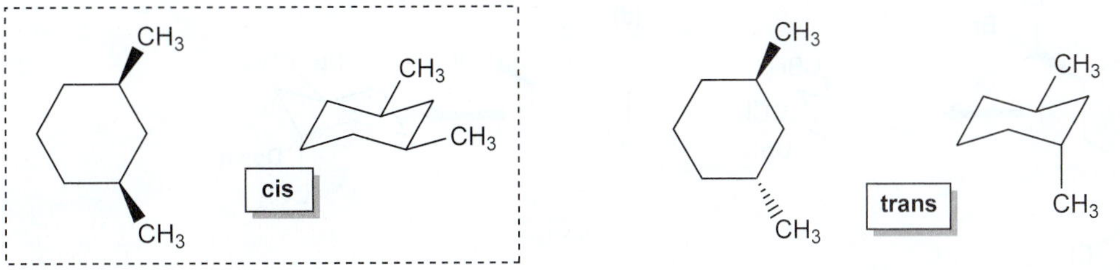

(c) H_3C CH_3 H_3C CH_3 **cis** H_3C CH_3 H_3C CH_3 **trans**

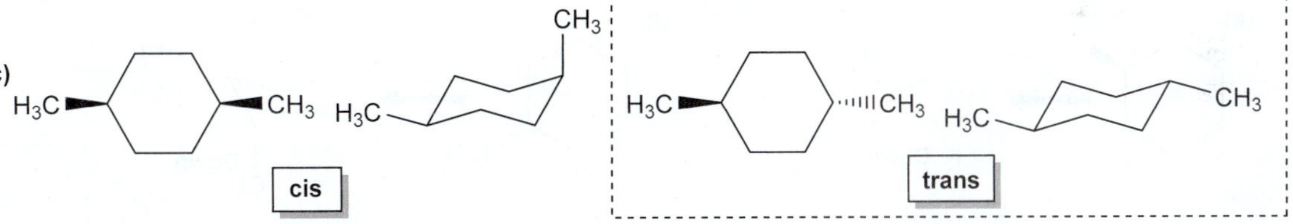

Problem 4.45

Think

To which bonds do cis and trans refer for this molecule? Build models of the two isomers, and try several chair flips in both rings. Draw the more stable chair conformation of each. In which position, axial or equatorial, is a bulky substituent more stable? Can both substituents occupy that position simultaneously?

Solve

Cis and trans refer to the bonds that are attached to the atoms that fuse the second ring to the first.

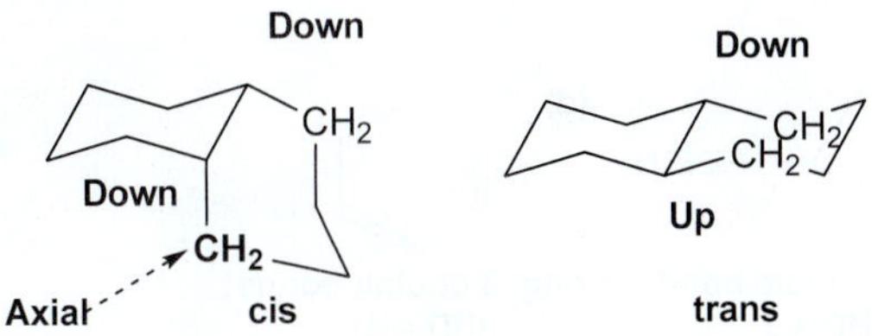

Down CH_2 Down CH_2 Axial cis

Down CH_2 CH_2 Up trans

In the cis isomer (left), a carbon in the right-hand ring cannot avoid being axial. In the trans isomer, the two CH_2 groups in the ring on the right occupy equatorial positions. Thus, the trans isomer is more stable.

Problem 4.46

Think

Which substituents require the most room? Which position, axial or equatorial, offers more room? Can both substituents achieve that position simultaneously?

Solve

The more stable conformation of each molecule is drawn below. Bulky substituents prefer to occupy the equatorial position. If both groups are not able to occupy an equatorial position simultaneously, the larger group will take priority over the smaller group.

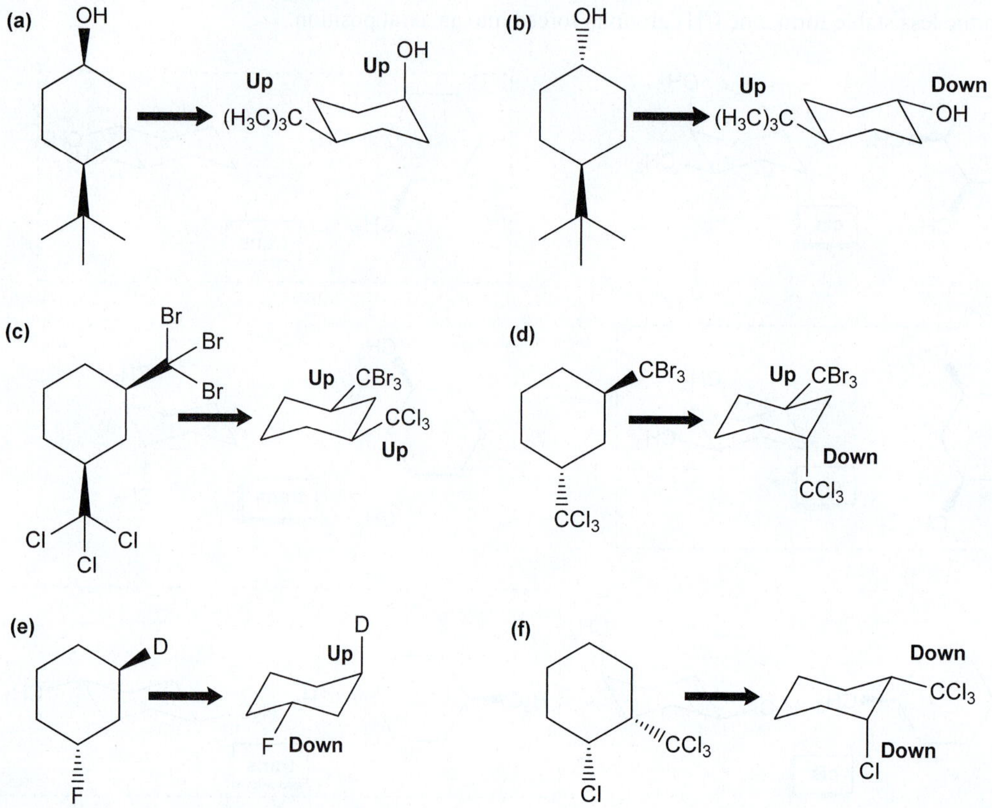

(a) OH, Up OH, Up $(H_3C)_3C$

(b) OH, Up $(H_3C)_3C$, Down OH

(c) Br, Br, Br, Cl, Cl, Cl; Up CBr_3, CCl_3 Up

(d) CBr_3, CCl_3; Up CBr_3, Down CCl_3

(e) D, F; D Up, F Down

(f) CCl_3, Cl; Down CCl_3, Down Cl

Section 4.12 Index of Hydrogen Deficiency

Problem 4.47

Think

How many rings, double bonds, and triple bonds are present? How much does each contribute to the overall IHD?

Solve

A double bond and a ring each contribute 1 to the IHD, and a triple bond contributes 2 to the IHD. See figure below and on the next page for IHD for compounds **(a)**–**(h)**.

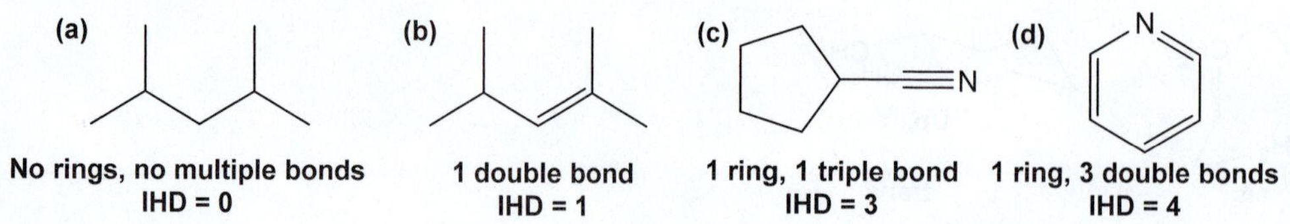

(a) No rings, no multiple bonds
IHD = 0

(b) 1 double bond
IHD = 1

(c) N; 1 ring, 1 triple bond
IHD = 3

(d) N; 1 ring, 3 double bonds
IHD = 4

(e) 1 ring IHD = 1

(f) 2 rings, 2 double bonds IHD = 4

(g) 2 rings, 3 double bonds IHD = 5

(h) 1 double bond, 2 triple bonds IHD = 5

Problem 4.48

Think

What is the formula for each saturated compound listed? For each IHD given, how many hydrogen atoms should be removed from the saturated compound?

Solve

For the nonhydrogen atoms given, the formula for a corresponding saturated molecule is provided. For each IHD, one H_2 molecule should be missing, so two H atoms need to be removed.

	Condition	Saturated Molecule	Formula of Saturated Molecule	IHD	Number of H Atoms to Remove	Formula of Described Molecule
(a)	4 C	$CH_3CH_2CH_2CH_3$	C_4H_{10}	0	0	C_4H_{10}
(b)	4 C	$CH_3CH_2CH_2CH_3$	C_4H_{10}	2	4	C_4H_6
(c)	3 C, 2 O	$CH_3CH_2CH_2OOH$	$C_3H_8O_2$	1	2	C_3H_6O
(d)	5 C, 2 Cl, 1 N	$CH_3(CH_2)_4Cl_2N$	$C_5H_{11}Cl_2N$	3	6	$C_5H_5Cl_2N$
(e)	1 C, 1 N	CH_3NH_2	CH_5N	2	4	CHN
(f)	6 C, 3 F , 2 N, 1 O	$CF_3(CH_2)_5NHNHOH$	$C_6H_{13}F_3N_2O$	4	8	$C_6H_5F_3N_2O$

Problem 4.49

Think

What is the formula for each saturated compound listed? Which atom does Si behave like? Which atom does S behave like? How does a negative charge affect the number of bonds and lone pairs on an atom? What is the difference in the saturated formula compared to the formula given? How many H atoms are missing from a saturated molecule does each IHD represent?

Solve

See the filled-in table below. Si behaves like C and should be added to the value of n in calculating $2n + 2$. S behaves like O, and therefore, there is no change to the number of H atoms. A negative charge decreases the number of H atoms, because it represents an additional pair of electrons existing as a lone pair rather than as a bond.

	Given Formula	Saturated Molecule	Formula of Saturated Molecule	Number of H Atoms Missing	IHD
(a)	C_6H_6	$CH_3(CH_2)_4CH_3$	C_6H_{14}	8	4
(b)	$C_6H_5NO_2$	$CH_3(CH_2)_5NHOOH$	$C_6H_{15}NO_2$	10	5
(c)	$C_8H_{13}F_2NO$	$CH_3(CH_2)_7N(F)O(F)$	$C_8H_{17}F_2NO$	4	2
(d)	$C_4H_{12}Si$	$CH_3(CH_2)_3SiH_3$	$C_4H_{12}Si$	0	0
(e)	$C_6H_5O^-$	$CH_3(CH_2)_5O^-$	$C_6H_{13}O^-$	8	4
(f)	$C_4H_6O_3S$	$CH_3(CH_2)_3OOOSH$	$C_4H_{10}O_3S$	4	2

Sections 4.11 and 4.13 Identifying and Drawing Constitutional Isomers

Problem 4.50

Think

Do both compounds have the same molecular formula? Is the largest continuous chain or ring in each molecule the same? Are the C atoms involved in the double bonds assigned the same numbers in each molecule? In each molecule, are the substituents attached to the numbered C atoms in the same order?

Solve

Pairs **(a)**, **(c)**, and **(e)** have the same formula and same connectivity and, therefore, are not constitutional isomers.

(a) C_6H_{10} C_6H_{10}

(c) C_7H_{14} C_7H_{14}

(e) $C_8H_{18}O$ $C_8H_{18}O$

Pairs **(b)**, **(d)**, **(f)**, and **(h)** have the same molecular formula but different connectivity, making them constitutional isomers. For pair **(b)**, both have the formula C_7H_{12}, but the methyl groups are on different C atoms. In the first molecule, the methyl group is on C1; in the second molecule, the methyl group is on C3. For pair **(d)**, both have the formula C_8H_{18}, but the methyl groups are on different C atoms. In the first molecule, the methyl groups are on C2 and C4; in the second molecule, the methyl groups are both on C3. For pair **(f)**, both have the formula C_6H_6, but the sizes of the rings and the locations of the double bonds are different. In the first molecule, the ring has six carbons and three double bonds. The second molecule is a bicylic molecule with two double bonds. For pair **(h)**, both have the formula C_5H_8O, but they have different functional groups. In the first molecule, there is a C=O outside of the ring. In the second molecule, there is a C=C in the ring and an OH bonded to the ring. This pair is considered to be tautomers (Section 7.9).

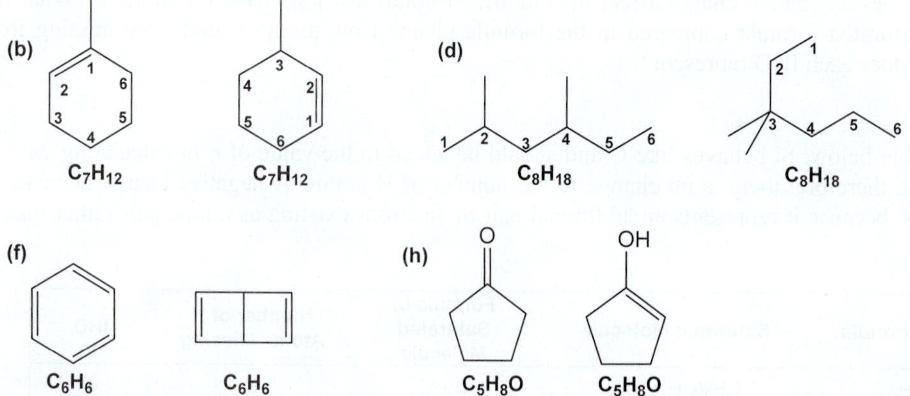

The members of pair **(g)** have different chemical formulas and, therefore, are not isomers.

(g) C_6H_6 C_6H_8

Problem 4.51

Think

Refer to Section 4.13 to follow the steps to draw all constitutional isomers for each formula.

Solve

The IHD of $C_5H_{11}Br$ is 0, so there are no rings, double bonds, or triple bonds. There are eight constitutional isomers in all, which are shown below.

5 C in a row

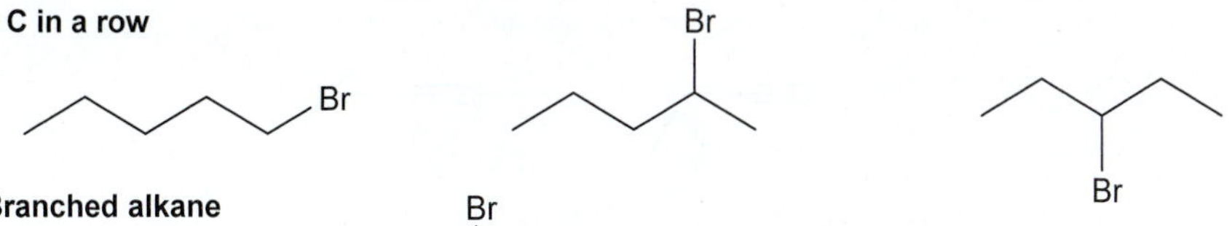

Branched alkane

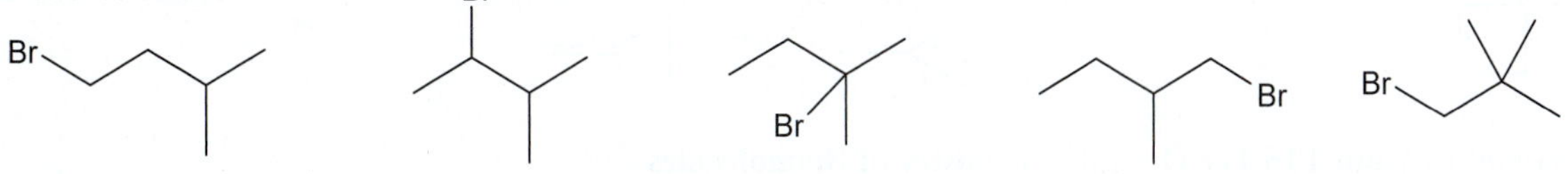

Problem 4.52

Think

Refer to Section 4.13 to follow the steps to draw all constitutional isomers for each formula.

Solve

An analogous saturated molecule is $CHF_2CH_2CH_2OH$, which has the formula $C_3H_6F_2O$, so the IHD for $C_3H_6F_2O$ = 0. Therefore, there should be no rings, double bonds, or triple bonds. There are nine possibilities with the backbone C–C–C–O and five possibilities with the backbone C–C(O)–C. The constitutional isomers are shown below.

Backbone C–C–C–O:

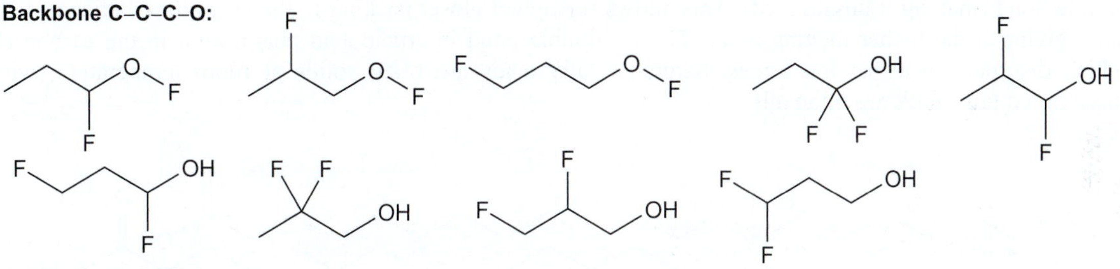

Backbone C–C(O)–C:

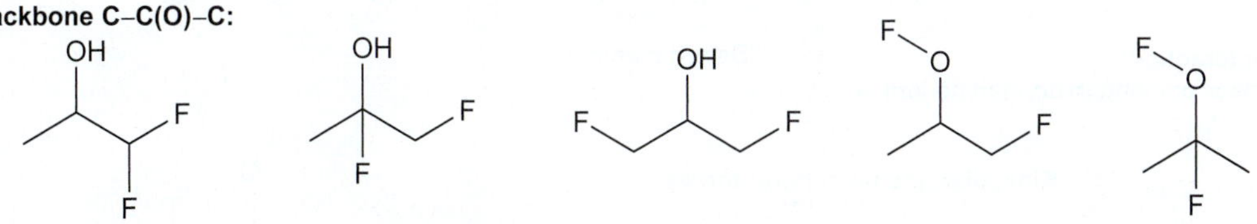

Problem 4.53

Think

Refer to Section 4.13 to follow the steps to draw all constitutional isomers for each formula.

Solve

The IHD of $C_3H_6F_2O$ = 0, just as in Problem 4.49. There can be no double bonds, no triple bonds, and no rings. The only backbone is C–C–O–C, and the six constitutional isomers are shown below.

Backbone C–C–O–C:

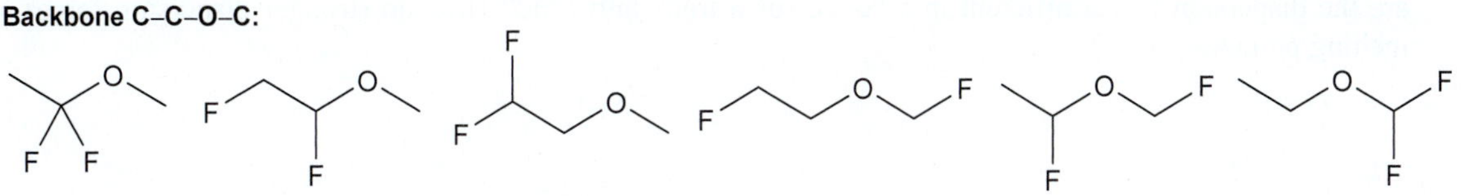

Problem 4.54

Think

Refer to Section 4.13 to follow the steps to draw all constitutional isomers for each formula.

Solve

An analogous saturated molecule is $CH_3CH_2CH_2CH_3$, whose formula is C_4H_{10}. The IHD of C_4H_6 = 2. Two rings; two double bonds, one of each; or one triple bond is possible. Four isomers have no ring, and four other isomers have one ring. Two isomers have a triple bond and one a bicyclic isomer.

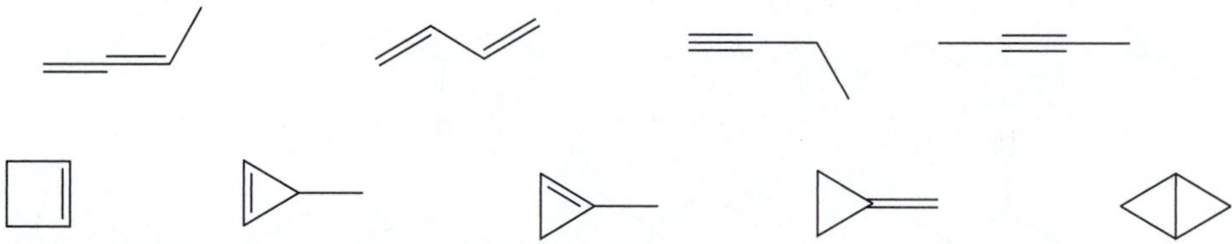

Sections 4.14 and 4.15 The Organic Chemistry of Biomolecules

Problem 4.55

Think

What intermolecular forces are available to each fatty acid? Which fatty acid has a fully saturated hydrophobic region? Which fatty acid can pack closer together in the crystalline lattice? How does tight packing affect melting point?

Solve

Both fatty acids have a long hydrophobic chain and a carboxylic acid end, so induced dipole–induced dipole and hydrogen bonding interactions are available to both. The difference between behenic acid and erucic acid is that behenic acid has all single bonds making up the carbon chain (it is saturated), whereas erucic acid has a cis double bond (making it unsaturated). This allows for tighter/closer packing in the crystalline lattice of behenic acid, giving it the higher melting point. The cis double bond in erucic acid puts a kink in the carbon chain, which disrupts the dispersion forces. Saturated fatty acids are often solids at room temperature, whereas unsaturated fatty acids are often oils.

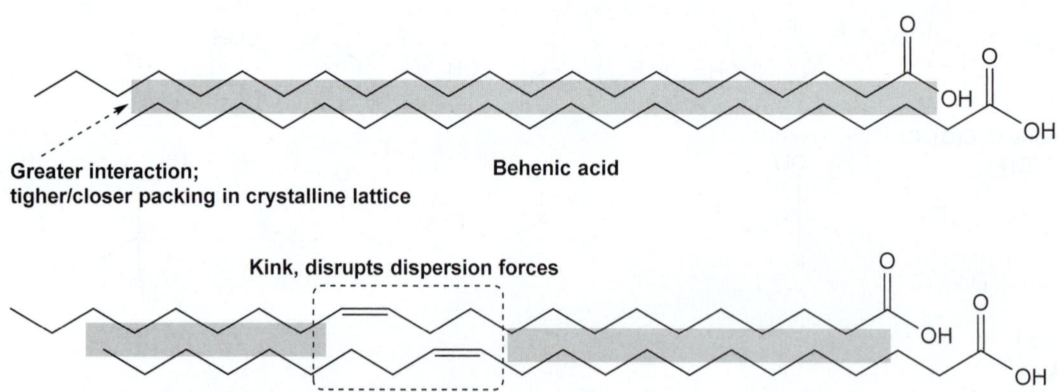

Problem 4.56

Think

What is the structure of the two fatty acids? What intermolecular forces are available to each fatty acid? How are the dispersion forces different in a cis versus a trans fatty acid? How do stronger dispersion forces affect melting point?

Solve

Both fatty acids have a long hydrophobic chain and a carboxylic acid end, so induced dipole–induced dipole and hydrogen bonding interactions are available to both. The difference between oleic acid and elaidic acid is the configuration of the C=C, cis versus trans. The cis double bond in oleic acid puts a kink in the carbon chain, which disrupts the dispersion forces, whereas elaidic acid is more like the saturated fatty acids. This allows for tighter/closer packing in the crystalline lattice of elaidic acid, giving it the higher melting point. Trans fatty acids are often solids at room temperature, whereas cis unsaturated fatty acids are often oils.

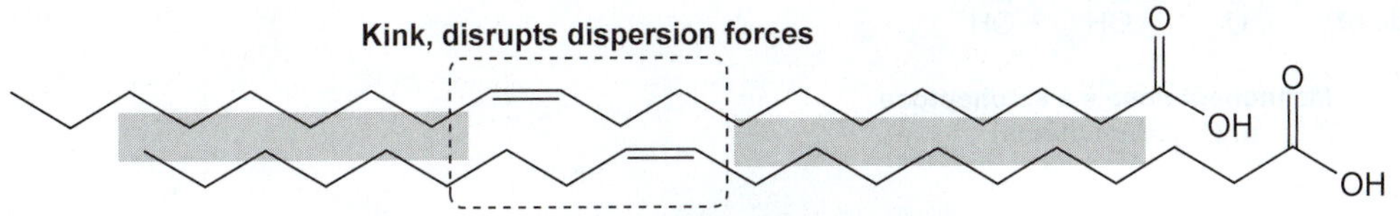

Kink, disrupts dispersion forces

Oleic acid

No kink, dispersion forces continuous

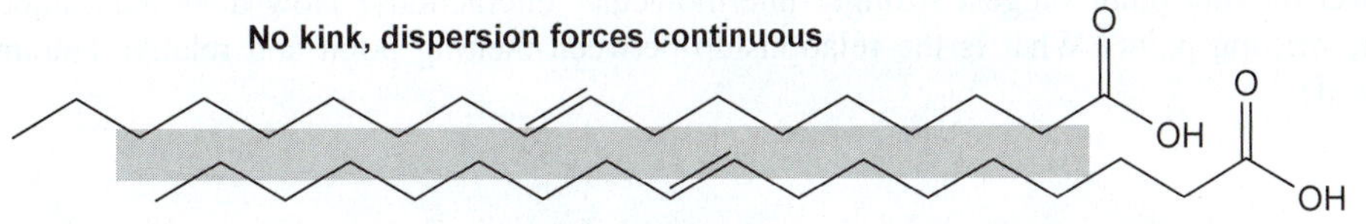

Elaidic acid

Problem 4.57

Think

How many C atoms are in a hexose sugar? What is the general formula for a monosaccharide (refer to Section 1.14b)? Where is the C=O group located (internal or terminal) for an aldohexose? For a ketohexose? How many OH groups are present?

Solve

The number of carbons in the chain is specified in the name: *hex* = 6. The *aldo* and *keto* prefixes specify whether the C=O group is part of a group characteristic of an aldehyde, requiring it to be at the end of the chain, or of a ketone, requiring it to be internal. The general formula for a six-carbon monosaccharide is $C_6H_{12}O_6$, where one O atom is part of a C=O group and the other five O atoms are part of OH groups. Each C atom can have only one OH group, and there has to be one OH group at the end of the carbon chain. Therefore, only one structure is possible for an aldohexose. Two constitutional isomers are possible for the ketohexose.

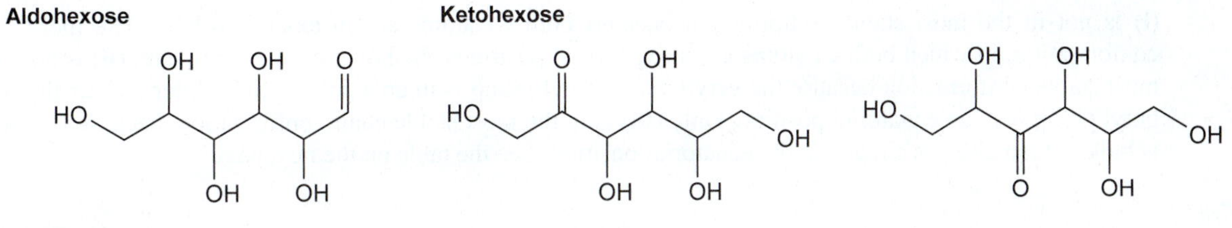

Aldohexose

Ketohexose

Problem 4.58

Think

Review the distinctions found in Section 4.14. How many C atoms are in mannoheptulose? Where is the C=O group located?

Solve

The C=O is internal and is therefore characteristic of a ketone and has the *keto* prefix. The number of carbons in the chain is seven, and therefore, it has the root *hept*. Therefore, mannoheptulose is classified as a ketoheptose.

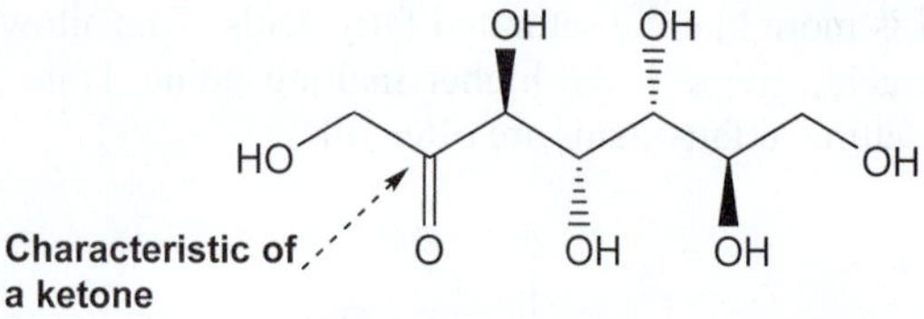

OH OH
HO OH
Characteristic of a ketone O OH OH

Mannoheptulose = a ketoheptose (7 carbon atoms)

Problem 4.59

Think

Does a higher or lower melting point suggest stronger intermolecular interactions? How does the number of C=C bonds affect the melting point? What is the relationship between melting point and relative amount of saturation in a fatty acid?

Solve

A higher melting point suggests stronger intermolecular interactions. Each cis C=C double bond in a fatty acid lowers the melting point and has a higher degree of unsaturation. Therefore, olive oil, with a lower melting point, has a higher degree of unsaturation compared to palm oil, with a higher melting point. The actual composition of both olive oil and palm oil varies, but olive oil consists of roughly 85% unsaturated fatty acids (~70% oleic and ~15% linoleic), whereas palm oil consists of roughly 46% unsaturated fatty acids (~37% oleic acid and ~9% linoleic).

Integrated Problems

Problem 4.60

Think

Build a molecular model of each molecule exactly as indicated in the Newman projections given. Which groups take up the most space? Are these groups axial or equatorial in the conformation given? Are those groups axial or equatorial after a chair flip? Is there more room in an axial or in an equatorial position? Once you have the molecular model in the more stable chair conformation, reorient it in space so you can use it as a guide to draw the dash–wedge and Haworth projections.

Solve

(i) is not in the most stable conformation because both Cl atoms are in axial positions. The more stable conformation, in which both Cl atoms are in equatorial positions, is shown on the next page. **(ii)** is not in the most stable conformation because the very bulky *t*-butyl group is in an axial position. After a chair flip, the *t*-butyl group is in an equatorial position. Only **(iii)** is in the most stable chair conformation, because CI_3, which is bulkier than CBr_3, is already in the equatorial position. See the table on the next page.

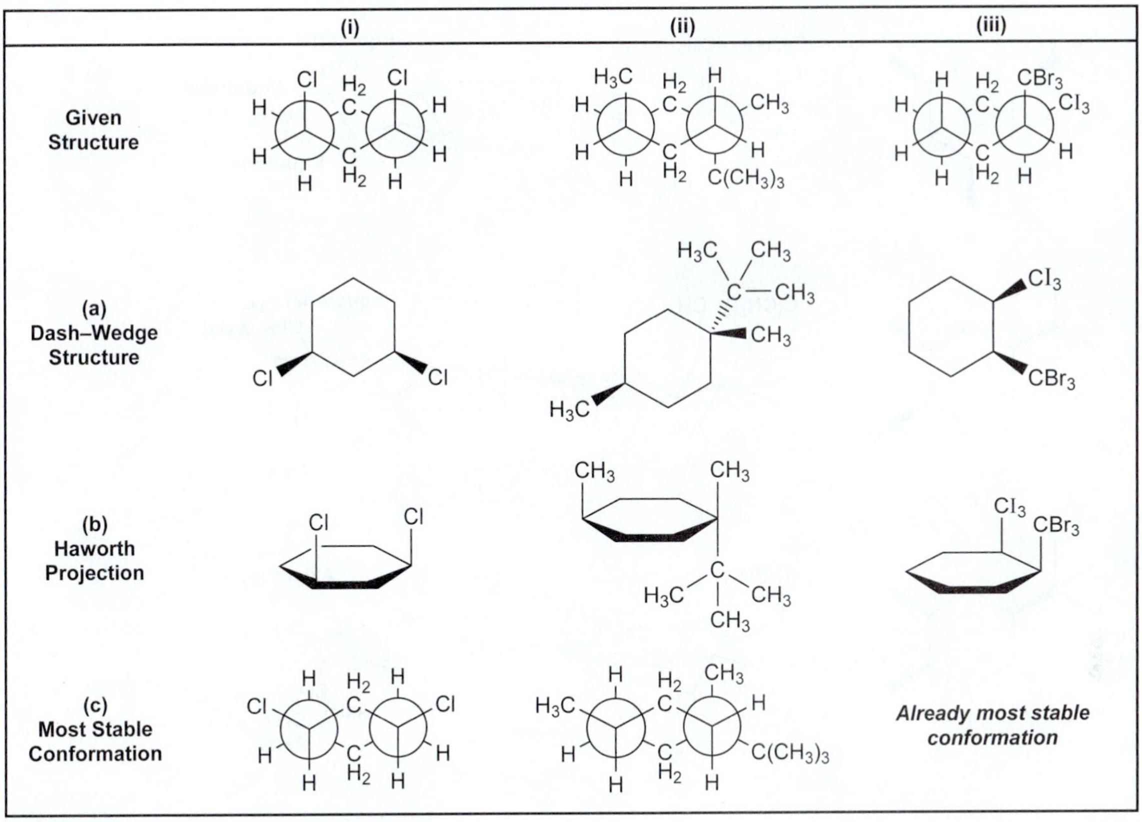

	(i)	(ii)	(iii)
Given Structure			
(a) Dash–Wedge Structure			
(b) Haworth Projection			
(c) Most Stable Conformation			***Already most stable conformation***

Problem 4.61

Think

The molecular formula is the same for each structure. What would contribute to the differences in the heat of combustion? Draw each structure in a chair conformation. Perform any necessary ring flip to get each structure in its most stable conformation. What is the order of stability for these four structures? How does stability relate to heat of combustion value?

Solve

The least stable isomer has the greatest heat of combustion. The *t*-butyl group is much larger than the CH_3 groups, so it must go equatorial in the most stable conformation of each structure. The remaining three CH_3 groups must follow the dash–wedge notation that is given. **B** has all four substituents equatorial and, therefore, is the most stable and has the smallest heat of combustion. Following the same logic, the order of heat of combustion from smallest to greatest is **B** < **A** < **C** < **D**. See structures below and on the next page.

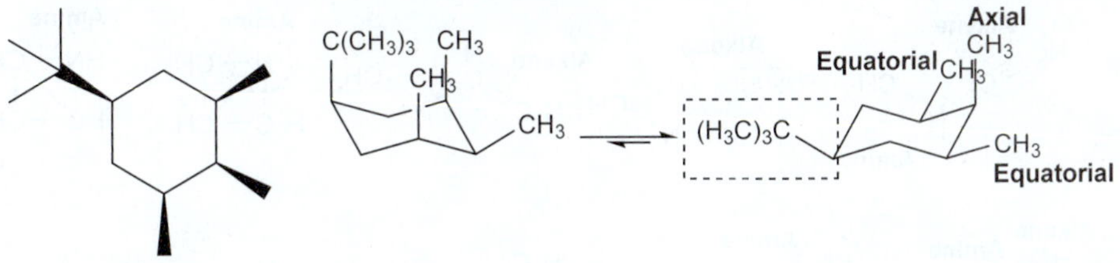

$C(CH_3)_3$ CH_3 CH_3 CH_3

Axial CH_3

Equatorial CH_3

$(H_3C)_3C$

CH_3 **Equatorial**

A

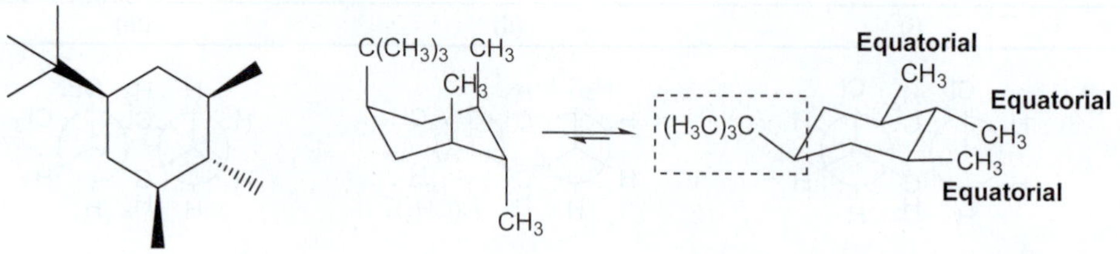

Equatorial
Equatorial
Equatorial

B

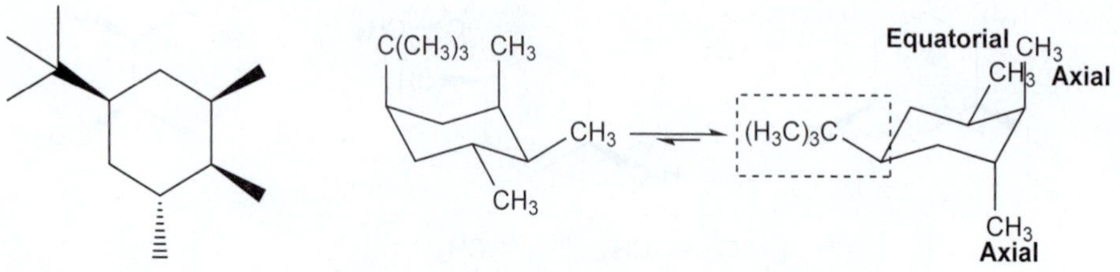

Equatorial
Axial
Axial

C

Axial
Axial
Axial

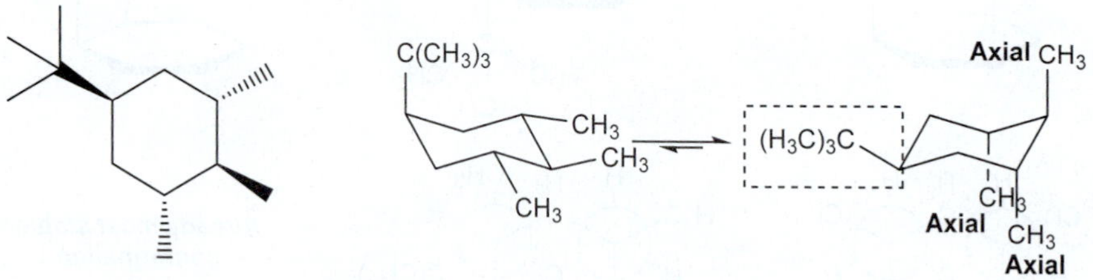

D

Problem 4.62

Think

Refer to Section 4.13 to follow the steps to draw all constitutional isomers for each formula.

Solve

An analogous saturated molecule is $CH_3CH_2CH_2NH_2$, which has a formula of C_3H_9N. Therefore, the formula C_3H_5N has an IHD of 2, so there are either two double bonds, two rings (one of each), or one triple bond. The functional groups are circled, and the compound classes of which they are characteristic are labeled in the figure below.

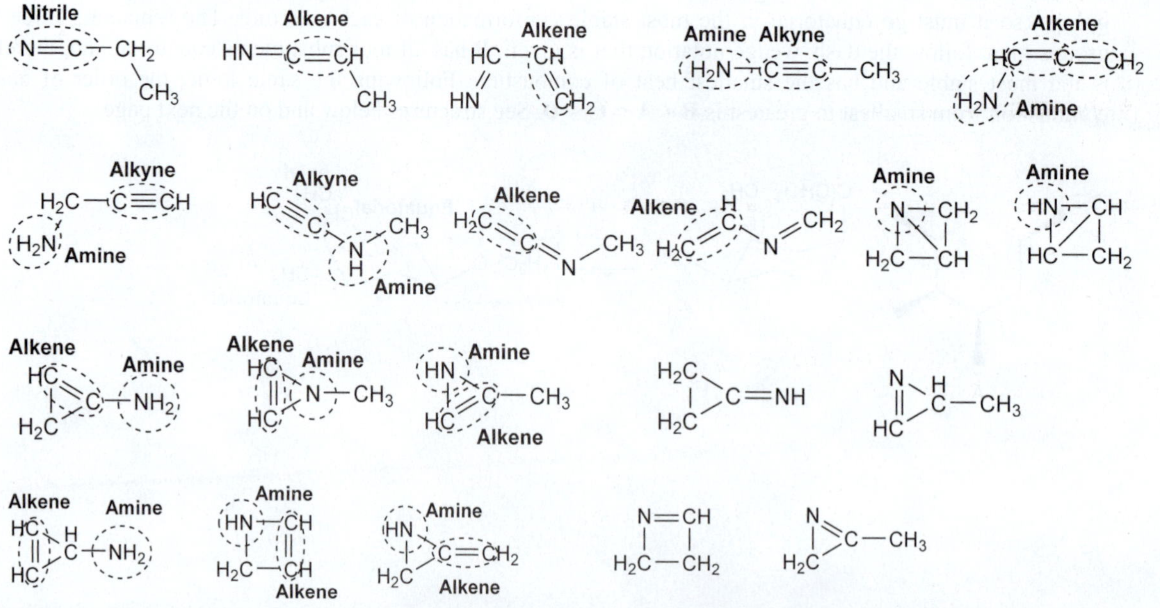

Problem 4.63

Think

Refer to Section 4.13 to follow the steps to draw all constitutional isomers for each formula. What is a benzene ring? How much of the IHD does a benzene ring account for?

Solve

An analogous saturated compound is $CH_3(CH_2)_7CH_3$. The IHD for C_9H_{12} is 4. A benzene ring has three formal double bonds and a ring, which accounts for the entire IHD of 4. So, aside from the benzene ring, there should be no double bonds, triple bonds, or rings.

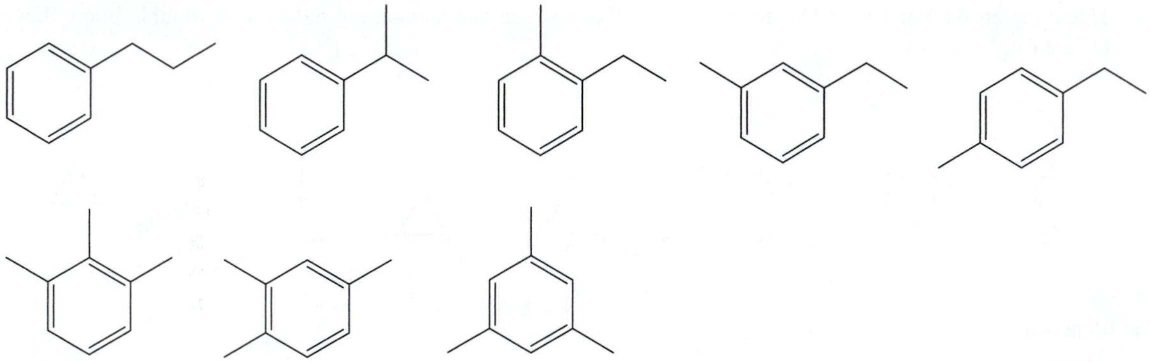

Problem 4.64

Think

Refer to Section 4.13 to follow the steps to draw all constitutional isomers for each formula. What functional group does a carboxylic acid have? How much of an IHD does that functional group account for? Where must that functional group be located—at the end of a chain, internal to the chain, or either?

Solve

An analogous saturated molecule is $CH_3(CH_2)_4OOH$, which has 12 H atoms. The formula $C_5H_8O_2$ has an IHD of 2. A carboxylic acid contains the CO_2H group, which consists of a C=O double bond and a C–OH. So the carboxylic acid C=O group accounts for an IHD of 1, leaving another IHD of 1 yet to be accounted for. This can be either a second double bond or a ring. A CO_2H carbon atom has only one available bond, so it must be located at the end of a chain. The 11 possible constitutional isomers are shown below.

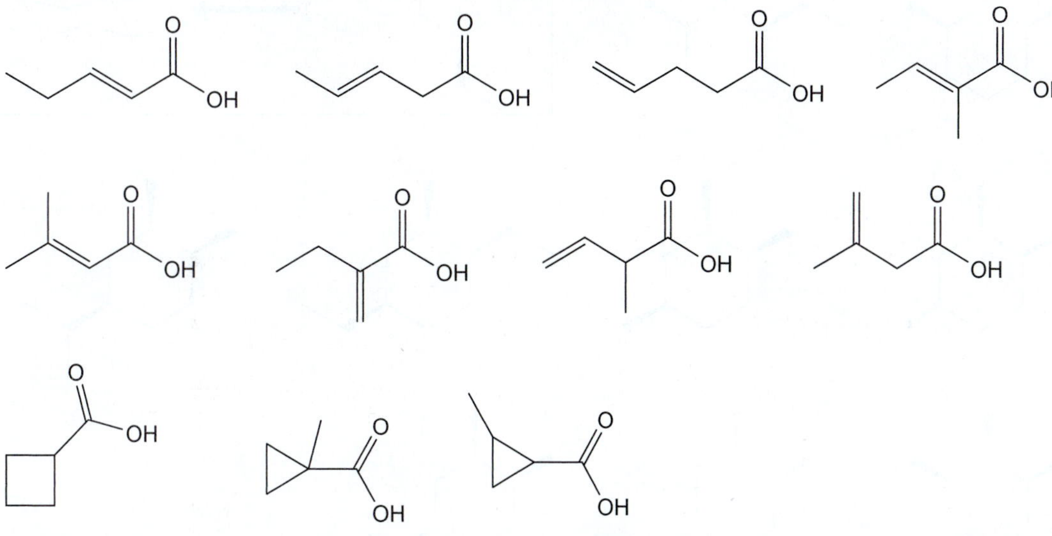

O OH O OH O OH O OH

O OH O OH O OH O OH

O OH O OH O OH

Problem 4.65

Think

Refer to Section 4.13 to follow the steps to draw all constitutional isomers for each formula. What functional group does a ketone have? How much of an IHD does that functional group account for? For a ketone, where is that functional group located—at the end of a chain, internal to the chain, or either?

Solve

An analogous saturated compound is $CH_3(CH_2)_4OH$, which has 12 H atoms. Compounds of the formula C_5H_8O have an IHD of 2. The functional group that makes up the ketone compound class is in the form C–C(O)–C, in which the C and O atoms are joined by a double bond (C=O). So the ketone C=O group accounts for an IHD of 1, leaving an IHD of 1 yet to be accounted for. Because the isomers cannot have a C=C double bond, they must have a ring.

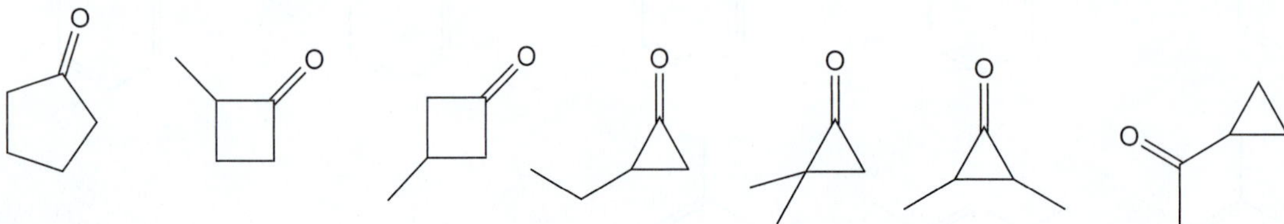

Problem 4.66

Think

Draw a dash–wedge line structure of each molecule, and then draw both chair conformations of each line structure. It will help to build a molecular model of each and carry out chair flips on the models. What positions—axial or equatorial—will each of the three substituents occupy in each constitutional isomer? Which position does each bulky substituent prefer? Can all bulky substituents occupy that position at the same time?

Solve

The various constitutional isomers are shown below. Because equatorial positions alternate up and down around a cyclohexane ring, only the isomer that is boxed can have all three of its substituents occupy equatorial positions at the same time. In each of the other isomers, at least one of the alkyl substituents will be forced to occupy an axial position.

Problem 4.67

Think

Which substituents require the most room? Which position, axial or equatorial, offers more room? Can all substituents achieve that position simultaneously? It will help to build a molecular model and carry out multiple chair flips on it.

Solve

The OH and CH_2OH groups attached to the ring all prefer the equatorial position. In the most stable conformation, all of the substituents occupy the equatorial position.

β-D-Glucopyranose

Problem 4.68

Think

What is the size of the ring in each compound? Which ring size has more strain associated with it?

Solve

β-D-Glucofuranose has a five-membered ring, whereas β-D-glucopyranose has a six-membered ring. Because there is greater ring strain with a five-membered ring, β-D-glucopyranose is more stable.

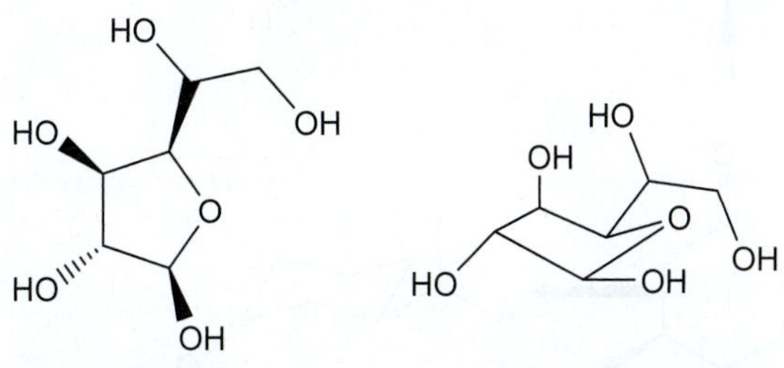

β-D-Glucofuranose

Problem 4.69

Think

What is the electron geometry for each sp^3-hybridized heteroatom? Is the chair conformation still possible? Which substituents require the most room? Which position, axial or equatorial, offers more room? Can all substituents achieve that position at the same time?

Solve

The electron geometry of each heteroatom is tetrahedral, and yes, the chair conformation is still possible. In the most stable conformation for **(a)**, the larger substituent, $-C(CH_3)_3$, occupies the equatorial position preferentially over the smaller substituent, $-CH_3$. In **(b)** and **(c)**, both substituents occupy the equatorial position. See the figure on the next page.

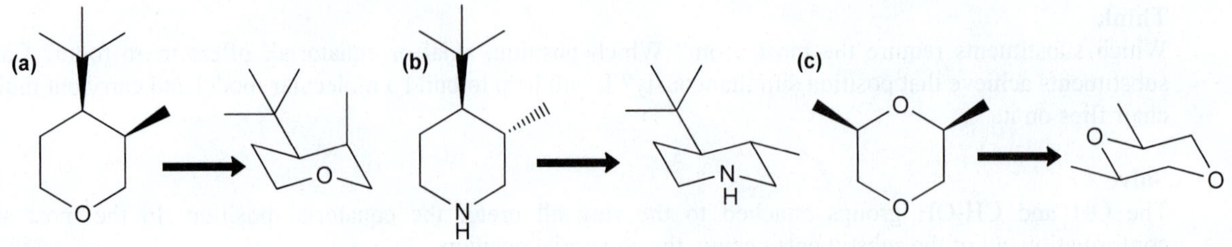

(a) (b) (c)

Problem 4.70

Think

What is the electron geometry for the sp^2-hybridized carbon? For the O atom that has two single bonds? Is the chair conformation still possible? Which substituents require the most room? Which position, axial or equatorial, offers more room? Can all substituents achieve that position?

Solve

The electron geometry for the sp^2-hybridized C is trigonal planar. Each ring, therefore, will still be able to attain the chair conformation, though with a geometry that is slightly altered from one composed entirely of sp^3-hybridized C atoms. To convince yourself that this is the case, you should build the molecules using your molecular modeling kit. In **(e)**, the sp^3-hybridized O can be treated just like an sp^3-hybridized C. The most stable conformations are shown below.

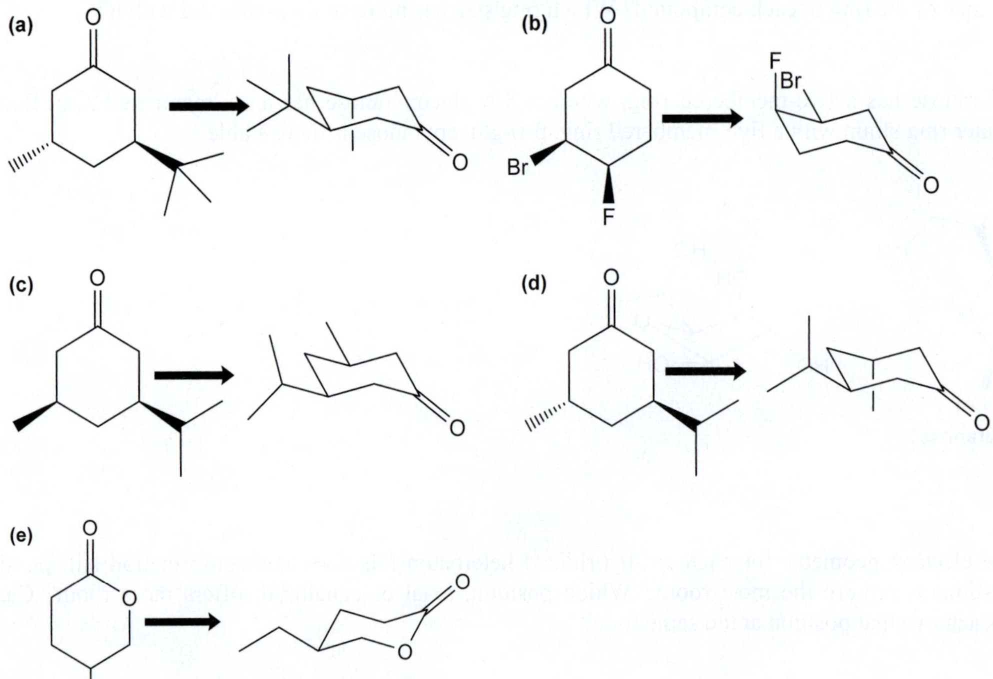

(a) (b)

(c) (d)

(e)

Problem 4.71

Think

The molecular formula is the same for each structure. What would then contribute to the difference in the heat of combustion? Do they have the same strain? Perform any necessary ring flips to get each structure in its most stable conformation. How does stability relate to heat of combustion value?

Solve

The least stable isomer has the greatest heat of combustion. β-D-Allopyranose has the greater heat of combustion owing to the presence of an axial OH (circled). This makes the structure less stable, and therefore, it has a greater heat of combustion.

β-D-Glucopyranose β-D-Allopyranose

Problem 4.72

Think

How does the size of the atom relate to the length of the bond? Which bond is longer? How does the length of the bond affect the 1,3-diaxial interactions?

Solve

I is much larger than Br. The C–I bond, therefore, will be longer than the C–Br bond. Even though the I atom is larger and you might expect the 1,3-diaxial repulsions to be stronger, the longer C–I bond allows the I atom to be farther from the axial H atoms. The two factors (larger I atom and longer C–I bond distance) essentially cancel each other out in this case.

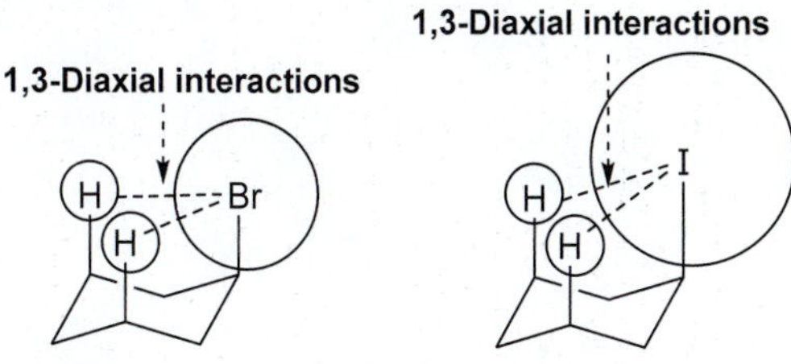

Problem 4.73

Think

What stabilizing interactions are available between the O and OH groups in the axial position but not available in the equatorial position? Draw out the structure in the chair conformation to visualize this interaction. How is this a stabilizing interaction?

Solve

Hydrogen bonding is available to OH (H-bond donor) and O (H-bond acceptor). In the equatorial position, this interaction is not possible because the donor and acceptors are too far apart. In the axial position, H bonding is available because the donor and acceptors are close enough.

H-bonding interactions

CHAPTER 5 | Isomerism 2: Chirality, Enantiomers, and Diastereomers

Your Turn Exercises

Your Turn 5.1

Think

First, construct your model given the orientation of the atoms in the structure on the right of Figure 5-2a. How can you orient this molecule to have the F atom on the left and pointing toward you and the Br atom on the left and pointing away from you? Draw the Cl and H atoms in the boxes and compare to the structure on the left. If two atoms are swapped, can it be the same molecule? If not, how are these molecules related?

Solve

Construct your model of the molecule on the right in Figure 5-2a so that the H atom is up, the Cl atom is to the left and parallel to the plane of the paper, the F atom is to the right and pointed toward you, and the Br atom is to the right and pointed away from you. Rotate the molecule 109.5°, as shown below, and make sure you keep the C–H and C–Cl bonds flat, the F atom pointed toward you, and the Br atom pointed away from you. You should see that the Cl atom is now up and the Br atom is to the left and pointed away from you. If you compare this molecule to the molecule on the left in Figure 5-2a (drawn below), you will note that the H and Cl atoms are swapped. This confirms that the two molecules are nonsuperimposable.

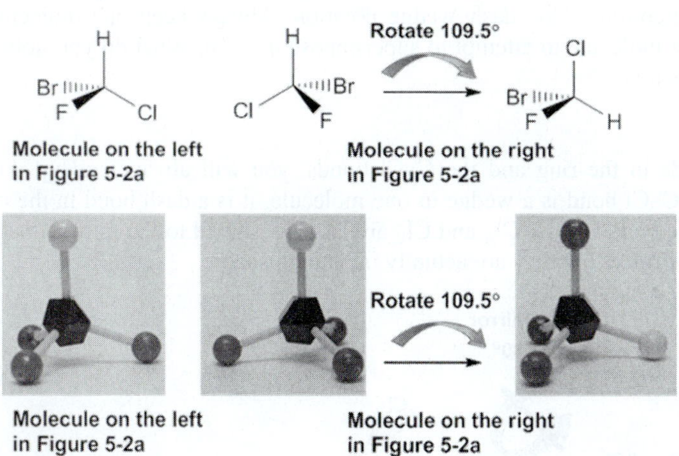

Your Turn 5.2

Think

When constructing the two molecules, pay particular attention to dash–wedge notation. When a carbon atom has four different groups, such as CHBrClF in Your Turn 5.1, is the mirror image superimposable? What is the effect of having two of the same groups (CH_2BrCl)? Do you anticipate that the mirror images are or are not superimposable?

Solve

Construct your models given the orientation of the atoms in Figure 5-3a. Rotate the mirror image molecule 109.5° as shown on the next page, and make sure you keep both C–H bonds flat, the F atom pointed toward you, and the Cl atom pointed away from you. You should see that one H atom in the resulting orientation (on the next page on the right) is up, the other H atom is pointed to the left and down, the F atom is to the right and pointed toward you, and the Cl atom is to the right and pointed away from you. If you compare this molecule to the molecule on the right in Figure 5-3a (i.e., the original molecule) drawn on the next page, you will notice that the two are identical. This illustrates that CH_2FCl has a mirror image that is the same as itself. *Note:* H_A and H_B are labels assigned to the mirror image to help you visualize the transformations performed, but they are actually indistinguishable.

Rotate 109.5°

Original molecule on the right in Figure 5-3a

Mirror image molecule on the left in Figure 5-3a

Same as molecule on right in Figure 5-3a (original)

Rotate 109.5°

Original molecule on the right in Figure 5-3a

Mirror image molecule on the left in Figure 5-3a

Same as molecule on right in Figure 5-3a (original)

Your Turn 5.3

Think

Consult Solved Problem 5.6 to build your model of the original molecule and the mirror image. When you construct the molecules, pay particular attention to the dash–wedge notation. Always keep one molecule in place as you perform rotations of the other molecule to attempt to superimpose the two. What do you notice as you attempt to line up the C–C and C–Cl bonds?

Solve

As you attempt to line up the C–C bonds in the ring and the C–Cl bonds, you will always end up with a mismatch of the C–Cl bonds. Where one C–Cl bond is a wedge in one molecule, it is a dash bond in the other molecule. No orientation lines up all of the bonds. *Note:* Cl_A and Cl_B are labels assigned to the mirror image to help you visualize the transformations performed, but they are actually indistinguishable.

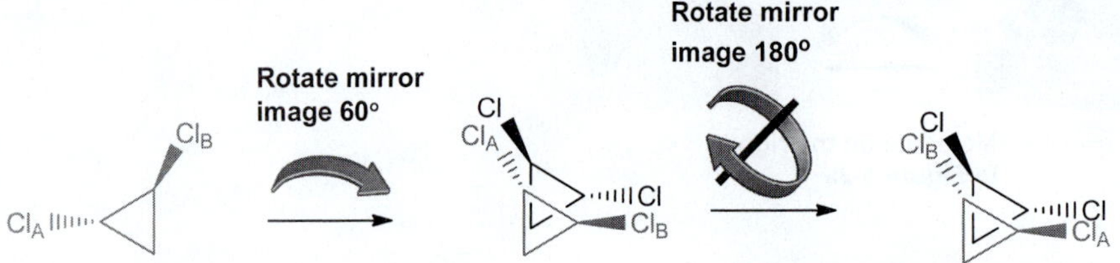

Your Turn 5.4

Think

In Solved Problem 5.6, is the mirror image of the red molecule shown? Are the two structures superimposable? Does that make the structures enantiomers?

Solve

The molecule is chiral. The red and black molecules in Solved Problem 5.6 are mirror images of each other. From Solved Problem 5.6, we have already verified that the two molecules are not the same and are, therefore, enantiomers. Because each molecule has an enantiomer, each molecule is chiral.

Your Turn 5.5

Think

Can you divide the molecule in half so that one half is the mirror image of the other half? Are molecules that possess a plane of symmetry chiral or achiral? Is the molecule in Solved Problem 5.6 chiral or achiral?

Solve

Molecules that possess a plane of symmetry are achiral. In Solved Problem 5.6, we showed that *trans*-1,2-dichlorocyclopropane is chiral. Therefore, *trans*-1,2-dichlorocyclopropane does not possess a plane of symmetry. The C–Cl_A dash bond is pointing away from you, and the C–Cl_B bond is pointing toward you; therefore, they are not reflections through a mirror that divides the molecule in half.

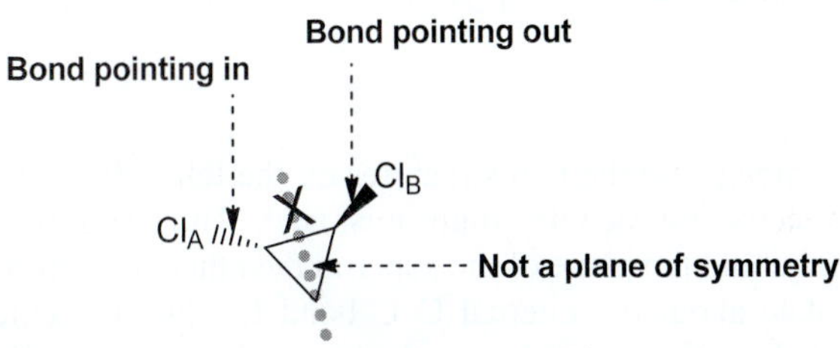

trans-1,2-Dichlorocyclopropane

Your Turn 5.6

Think

In which direction (toward you or away from you) are the C–Cl bonds oriented? Are they oriented in the same or in the opposite direction? Are molecules that possess a plane of symmetry chiral or achiral?

Solve

Both C–Cl bonds in *cis*-1,2-dichlorocyclopropane are pointing out of the plane of the paper. Thus, when dividing the molecule in half, as shown below, one Cl is the reflection of the other, so there is a plane of symmetry (indicated by the dotted line). The molecule is therefore achiral.

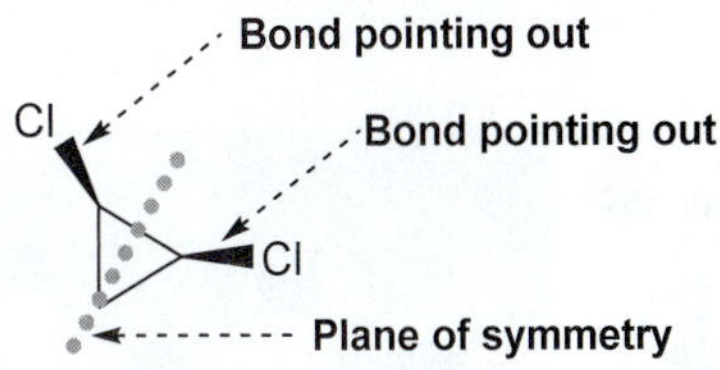

cis-1,2-Dichlorocyclopropane

Your Turn 5.7

Think

Which atoms are tetrahedral (*sp*^{3} hybridized)? For each *sp*^{3} atom, list the groups that are bonded to it. If you have four different groups, what does that mean? If you end up with two or more of the same group, what does that mean? For the ring structure, how do you look at the groups attached to an atom of the ring to determine if they are the same or different?

Solve

The chiral centers are marked below with an asterisk (*). Each chiral center has four different groups bonded to it. In butan-2-ol, the four groups are H, CH_3, CH_2CH_3, and OH; in the 2-bromo-1,1-dimethylcyclobutane, the groups are H, Br, and two groups that are part of the ring. The groups that make up the ring are different because, when we divide the ring in half such that the dividing line goes through the chiral center and the center of the ring, the two halves of the ring are different. One half of the ring has two CH_3 groups, and the other half doesn't.

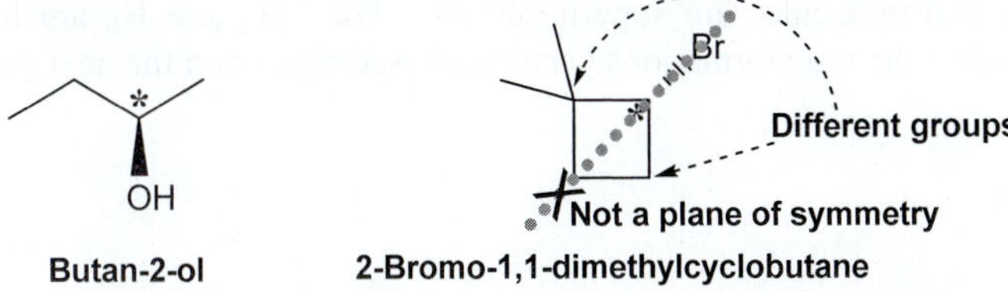

Butan-2-ol

2-Bromo-1,1-dimethylcyclobutane

Your Turn 5.8

Think

Draw the molecule of butan-2-ol that is given and its mirror image; build both molecules, paying special attention to the dash–wedge notation; and orient the molecules so their carbon chains line up. When they do, do all atoms from the two molecules line up? Does rotating about any of the C–C bonds help? Are four different groups attached to any of the carbon atoms? How many chiral centers are present? Can you draw any planes of symmetry?

Solve

The molecule of butan-2-ol that is given and its mirror image are the two structures on the left. After rotation, the zigzag orientations of the carbon chains line up perfectly, but the OH groups mismatch: The OH group is in front of the plane of the paper in one molecule, but it is behind the plane of the paper in the other. Therefore, the two mirror images are not superimposable. If you rotate about the internal C–C bond for one molecule, its carbon chain will no longer be zigzag, so this will not help with superimposing the two molecules. As with any molecule containing exactly one chiral center, the molecule has no planes of symmetry and must be chiral.

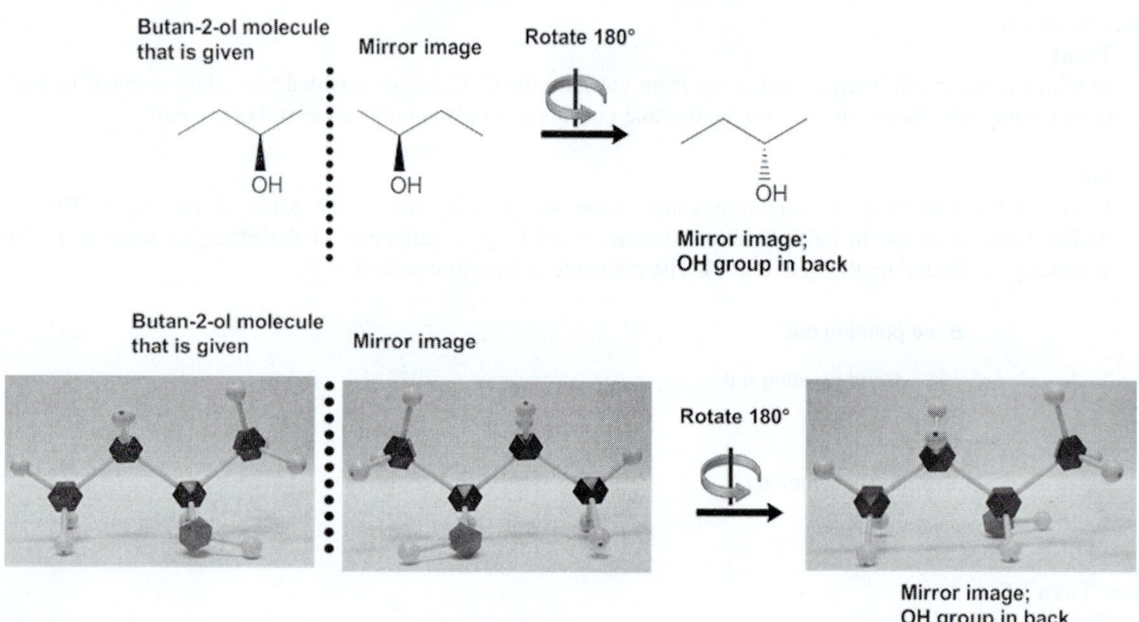

Your Turn 5.9

Think

In a Newman projection, which bond is not visible? When you build your models, make sure that the directions of the C–Br bonds on the front and back carbons agree with the Newman projections. Without rotating about the C–C bond, do the molecules line up perfectly? Do they line up perfectly after rotation about the C–C bond?

Solve

Once you have your models built, rotate them 90° so that the C–C bond that was once pointing toward you is now flat. Then, keep the original molecule as your reference point and attempt to superimpose the mirror image. The CH_2Br group on the right is in the same orientation in both molecules. But the CH_2Br groups on the left mismatch. Then, rotate the CH_2Br group on the left in the mirror image molecule 120° about the C–C bond to move the Br atom to point away. Now the two molecules are superimposable. *Note:* H_A and H_B are labels assigned to the mirror image to help you visualize the transformations performed. See figures on the next page.

Rotate 90°

Original molecule

Rotate 90°

Rotate about bond

Mirror image

Mirror image rotated to return to original

Your Turn 5.10

Think

When you construct the molecule and the mirror image in Figure 5-14a, make sure that the chair conformation is exactly as shown, and pay particular attention to whether each C–F bond is axial or equatorial. Attempt to superimpose the two molecules by lining up each atom in one molecule with the same atom in the other molecule. Are you able to do so? If you ring flip the mirror image, are you now able to superimpose the two molecules?

Solve

If you attempt to superimpose the two molecules, you will not be able to do so, as shown in the first row of graphics below. Ring flipping the mirror will now lead to a conformation that is superimposable on the original structure. After you carry out the ring flip on the mirror image, rotate the molecule around 120°, as shown in the second row of graphics below, and then you will be able to superimpose the two molecules. *Note:* F_A and F_B are labels assigned to the mirror image to help you visualize the transformations performed.

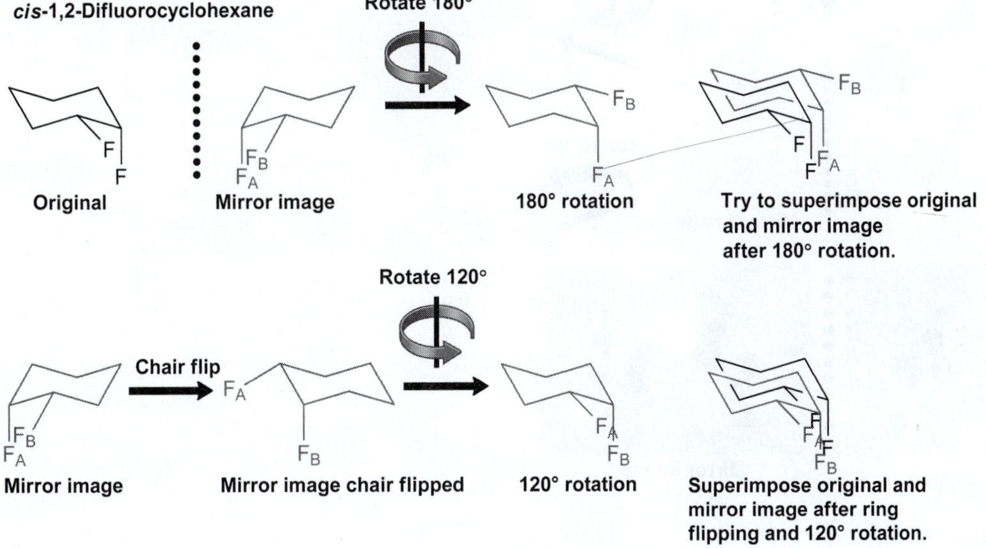

Your Turn 5.11

Think

When you construct the molecule and the mirror image, make sure that the chair conformation is exactly as shown, and pay particular attention to whether the C–F bonds shown in the figure are axial or equatorial. Would a ring flip aid your attempt to superimpose the two molecules?

Solve

Consult the steps in Figure 5-14 for assistance in your attempt to superimpose the two molecules. Every time you attempt to line up one of the C–F bonds in one molecule with a C–F bond in the other molecule, you will notice that the other C–F bonds will mismatch. This illustrates that the two mirror images of *trans*-1,2-difluorocyclohexane are *nonsuperimposable*. Ring flipping the two molecules turns the 1,2-diequatorial into the 1,2-diaxial conformation. The two molecules are still *nonsuperimposable*. *Note:* F_A and F_B are labels assigned to the mirror image to help you visualize the transformations performed.

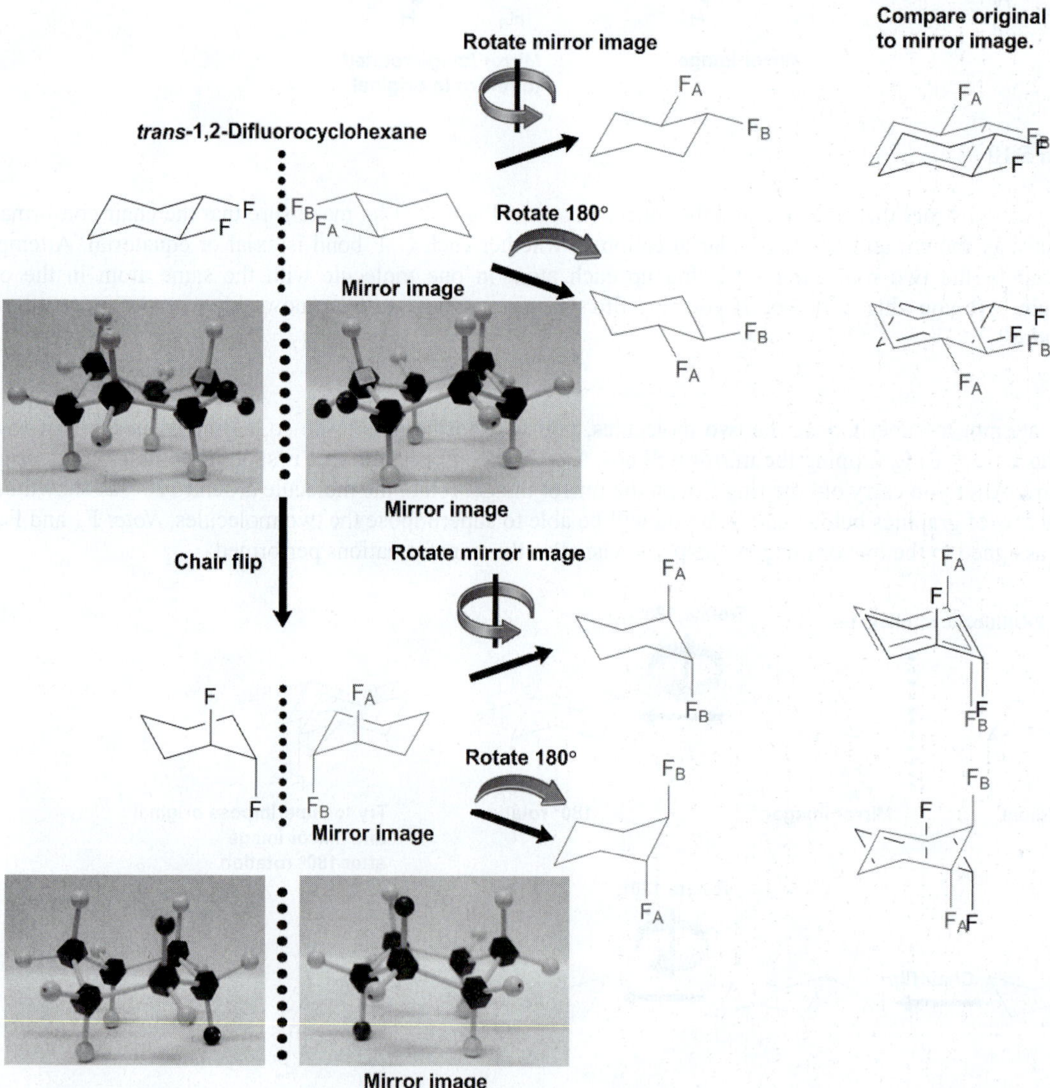

Your Turn 5.12

Think

As drawn, do the structures appear to be superimposable? Is there another orientation you can consider to convert a wedge bond into a dash bond? What would the resulting structure look like if you were to rotate about the C–C bond where the OH and Cl are bonded?

Solve

If the two structures are compared as given, they do not appear to be superimposable. Whereas Cl is in back in the molecule on the left, **A**, it is in front in the molecule on the right, **B**. If the left molecule is rotated 180° about the C–C bond, we arrive at conformation **A′**. In that new conformation, the OH atom is in the back and the Cl atom is in the front (like in molecule **B**). However, you will notice that the carbon chain has been rotated and is not superimposable on **B**. If you instead flip molecule **A** horizontally, you will obtain orientation **A″**. Comparison to molecule **B** once again shows that the two molecules are not superimposable: Whereas OH is in front in **A″**, it is in back in the molecule **A**.

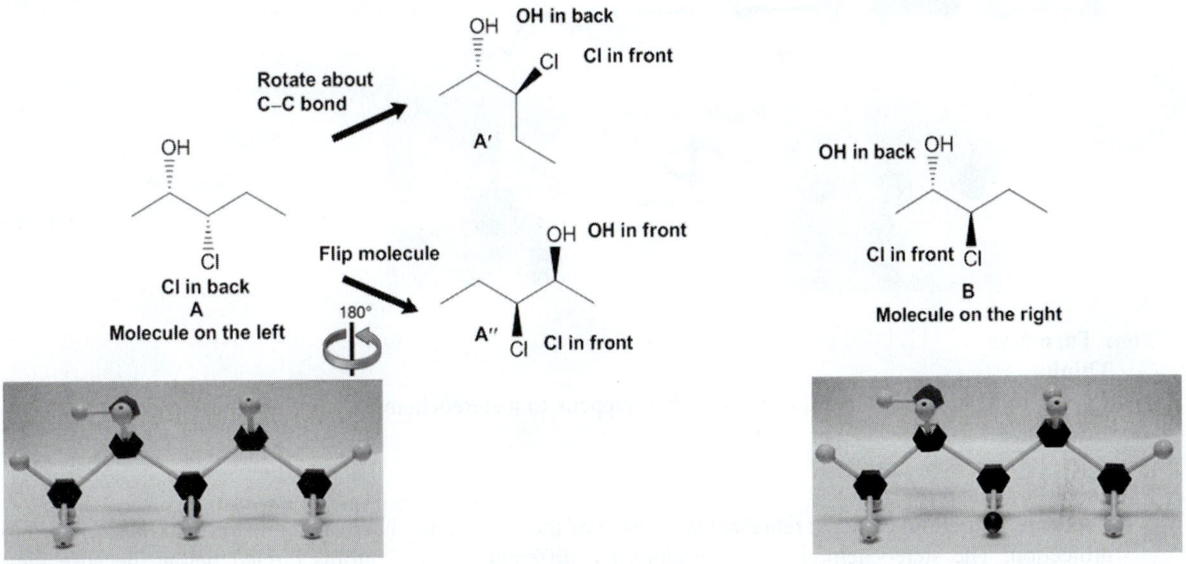

OH in back
Cl in front
Rotate about C–C bond
A′
Cl in back
A
Molecule on the left
Flip molecule
180°
OH in front
A″
Cl in front
OH in back
Cl in front
B
Molecule on the right

Your Turn 5.13

Think

What constitutes a chiral center for a carbon atom? For each sp^3 carbon atom, list the four groups that are attached. If you have four different groups, what does that mean? If you end up with two or more of the same group, what does that mean?

Solve

The three chiral centers are marked below with an asterisk (*). Each chiral center has four different groups bonded to it. In 2,3-dibromo-4-methylhexane, C2 is bonded to CH_3, H, Br, and CHBrR; C3 is bonded to Br, $CHBrCH_3$, H, and $CH(CH_3)R$; C4 is bonded to CH_3, H, CH_2CH_3, and CHBrR. The remaining C atoms are each bonded to at least two H atoms and are therefore not chiral centers.

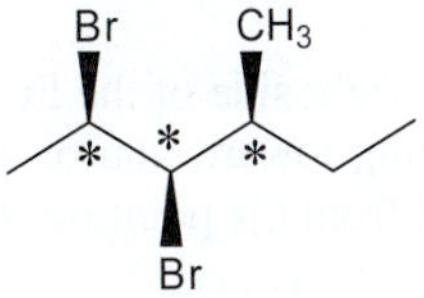

Your Turn 5.14

Think

In a Fischer projection, are the horizontal bonds pointed toward you or away from you, wedge or dash, respectively? What about the vertical bonds? After you build the molecule, paying particular attention to the dash–wedge notation, orient the molecule so the C–W and C–Y bonds are parallel to the plane of the paper. Does group X point toward you or away from you? Group Z?

Solve
Horizontal bonds in a Fischer projection point toward you (wedge) and vertical bonds point away from you (dash). The W and Y groups both go from dash to planar bonds by rotating the molecule 90°. After that rotation, the X group points to the left and toward you (wedge), whereas the Z group points to the left and away from you (dash).

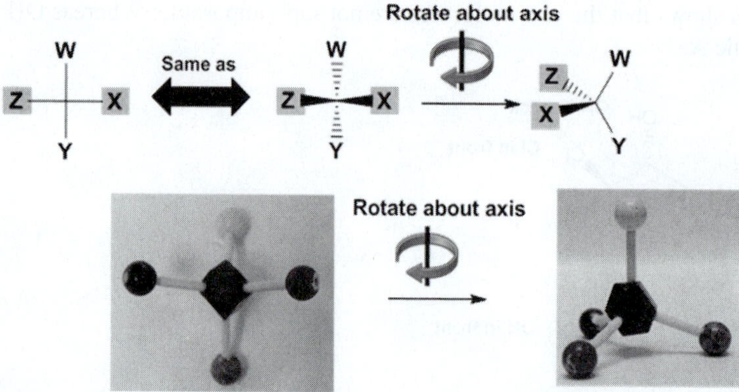

Your Turn 5.15

Think
Which of the atoms are chiral centers? What happens to a stereochemical configuration when two substituents are exchanged?

Solve
There are four chiral carbons, represented by each of the intersecting horizontal and vertical lines in the Fischer projection. The stereochemical configurations are different at the C atoms circled below, because they are related by the exchange of two substituents—in this case, H and OH.

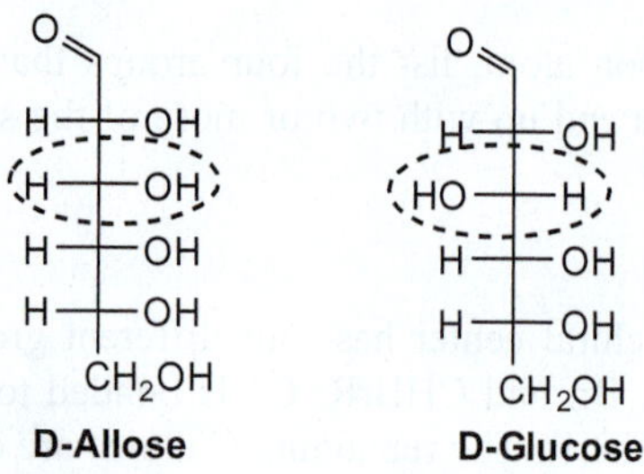

Your Turn 5.16

Think
In the Fischer projection in Figure 5-23b, are the C2 and C3 C–OH bonds on the left or right side of the Fischer projection? In the zigzag structure in Figure 5-23a, is the C–OH bond from C2 pointing toward you or away from you? What about the C–OH bond from C3? In which direction is the C–OH bond from C4 pointing in the zigzag structure and Fischer projection?

Solve
In the Fischer projection in Figure 5-23b, the C2 C–OH bond is on the left and the C3 C–OH bond is on the right (see figure on the next page). They are on the same side of the carbon chain's plane (both wedge) in the zigzag structure in Figure 5-23a but on opposite sides in the Fischer projection. In the Fischer projection in Figure 5-23b, the C4 C–OH bond is also on the right. In the zigzag structure, the C4 C–OH bond is on the opposite side of the carbon chain's plane (dash) compared to the C2 and C3 C–OH bonds.

The pattern is that bonds on the same side of the carbon chain's plane in a zigzag structure are on opposite sides in a Fischer projection (C2 and C3 C–OH bonds) and that bonds on opposite sides of the carbon chain's plane in a zigzag structure are on the same side of the plane in a Fischer projection.

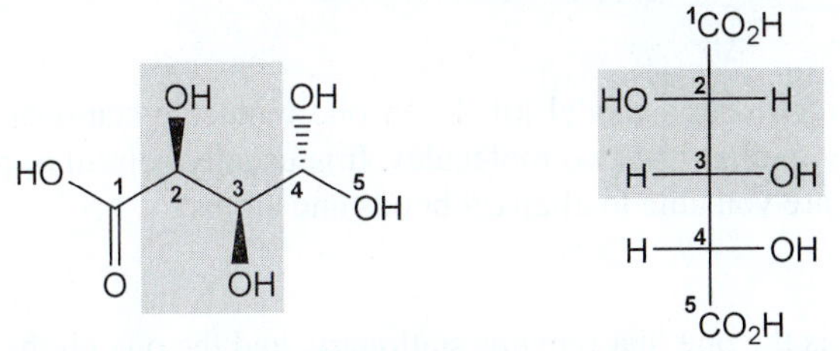

Same side of plane zigzag structure = Opposite side of Fischer projection

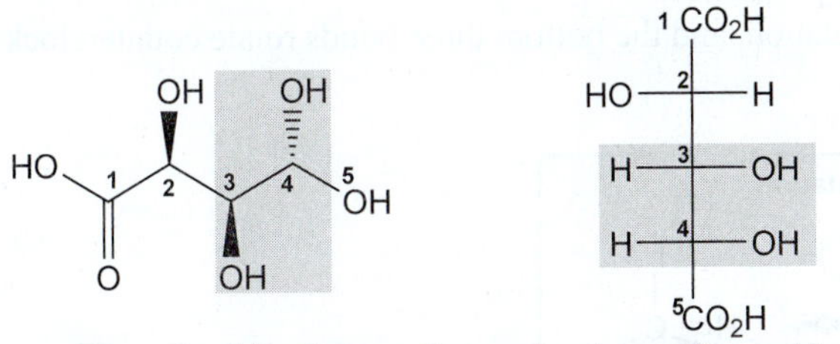

Opposite side of plane zigzag structure = Same side of Fischer projection

Your Turn 5.17

Think

Which enantiomer, **A** or **B**, is in excess? How can you tell from the solution percentages in the mixture the extent by which that enantiomer is in excess? What percentage of the solution is optically inactive? What percentage is racemic?

Solve

For a solution mixture that is 30% **A** and 70% **B**, the **B** enantiomer is in excess. All of enantiomer **A**, therefore, can be combined with 30% of the solution that consists of **B**, making 30% + 30% = 60% of the solution racemic. Thus, the remaining 40% of the total solution is excess **B**, so the solution has a 40% enantiomeric excess of **B**.

In Chapter Problems

Problem 5.1

Think

It might be helpful to build a model of each molecule using a model kit. Keep one molecule stationary and orient the other molecule every way imaginable to try to align the two molecules. It is usually helpful to pick a point of reference. When you perform your rotations, are you able to align *all* bonds and atoms?

Solve

In all of the solutions below, the molecule on the left is the one that remains stationary, and the one on the right is reoriented in space (when necessary) in an attempt to superimpose the two molecules. Prime symbols (′) are used to label atoms to help you visualize the rotations performed.

(a) Superimposable. The C–H″ bond is the axis of rotation, and the bottom three bonds rotate counterclockwise 120° as viewed from the top.

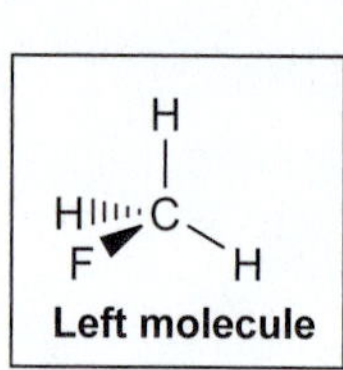

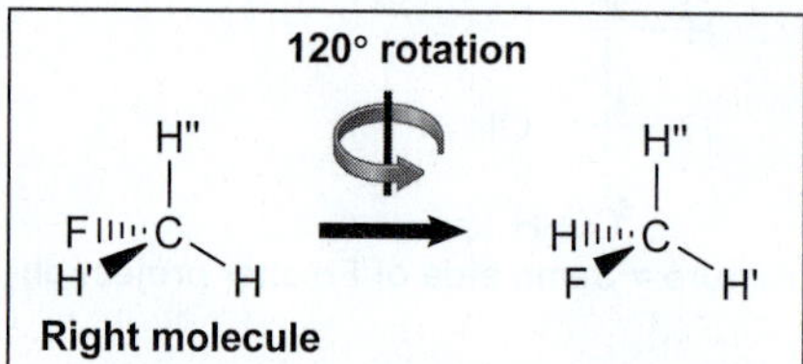

(b) Superimposable. The molecule is rotated 180° out of the plane of the paper.

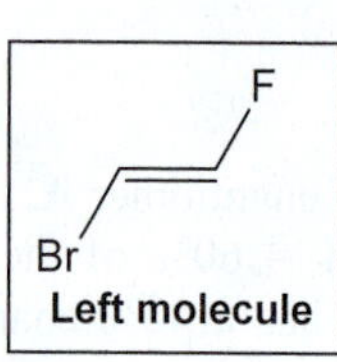

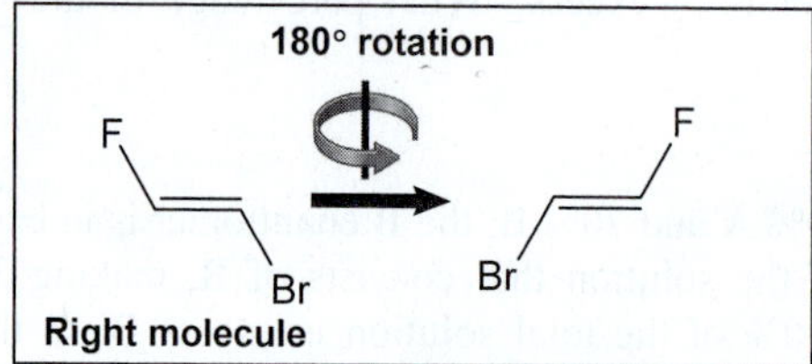

(c) Nonsuperimposable. Free rotation about a C=C bond is not permitted. The molecule on the left is the trans isomer, and the molecule on the right is the cis isomer.

(d) Superimposable. The molecule is rotated 180° out of the plane of the paper.

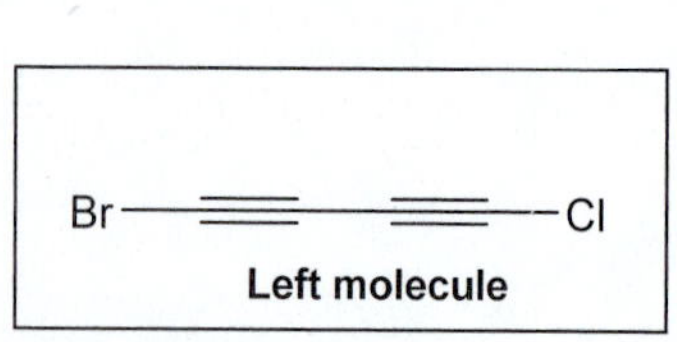

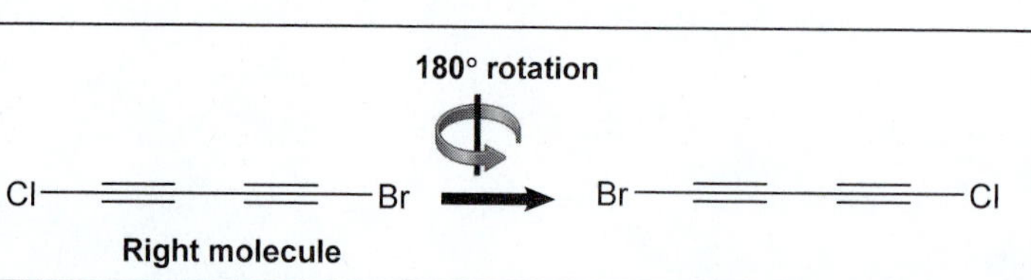

(e) Superimposable. A 180° rotation is performed about the diagonal on the cyclobutane ring, as shown. The dash C–Cl bonds convert to wedge C–Cl bonds, and thus the molecules are shown to be superimposable.

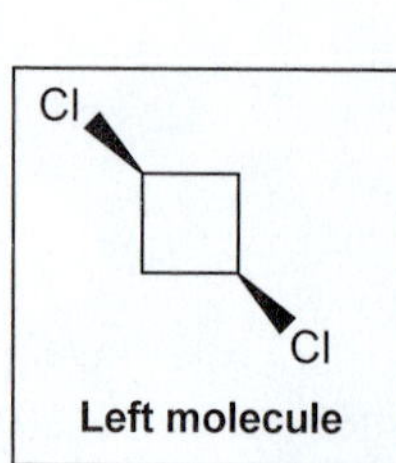

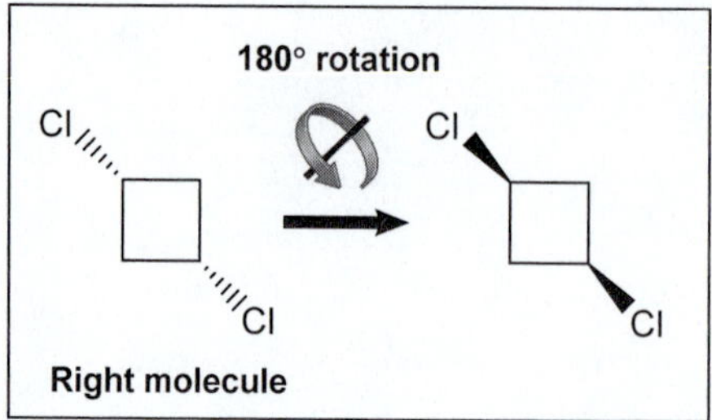

(f) Nonsuperimposable. Full rotation about single bonds in a ring is not permitted. The molecule on the left has the C–Cl bonds both wedge (cis), and the molecule on the right has one wedge and one dash (trans).

(g) Nonsuperimposable. Both molecules have one wedge and one dash C–Cl bond. However, if you attempt to superimpose the two molecules, you will see it is not possible: Where there is a dash bond in the first molecule, there is a wedge bond in the second, and vice versa. A 180° rotation of the second molecule does not change the situation, as shown below.

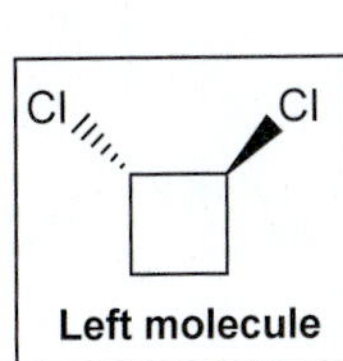

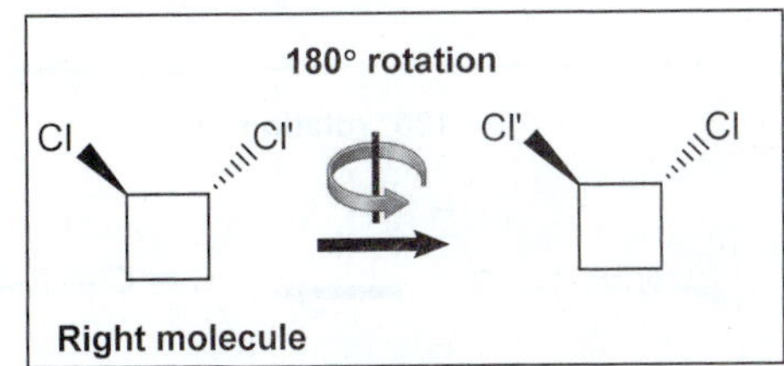

(h) Nonsuperimposable. Full rotation about single bonds in a ring is not permitted. The molecule on the left has the C–Cl bonds both wedge (cis), and the molecule on the right has one wedge and one dash (trans).

Problem 5.3

Think

For each molecule, draw a mirror next to the molecule. For each atom or group in the original molecule, where should its mirror image be with respect to the mirror? How should dash–wedge notation be treated in the mirror image?

Solve

Mirrors are represented by the dotted lines below, and the mirror images are drawn to the right of each mirror. If a bond is a wedge in the original, it will be a wedge in the mirror, and if a bond is a dash in the original, it will be a dash in the mirror. Make sure that each atom has a mirror image that is the same distance from the mirror, at the same position along the mirror, and has the same position relative to the plane of the paper.

(a) (b) (c) (d)

Problem 5.4

It might be helpful to build a model of each molecule and its mirror image using a model kit. Keep one molecule stationary, and try to align the other molecule to make it superimposable. It is usually helpful to pick a point of reference. When you perform your molecule and bond rotations, are you able to align *all* bonds and atoms?

Solve

Each molecule and mirror image from Problem 5.3 is repeated below. In all of the solutions below, the molecule on the left is the one that remains stationary, and the one on the right (the mirror image) has rotations performed on it in an attempt to superimpose the two molecules.

(a) Nonsuperimposable. From the 180° rotation shown below on the right molecule, it is evident that the two molecules are nonsuperimposable. The flat C–H and C–F bonds align, but the wedge and dash C–H and C–F bonds do not align.

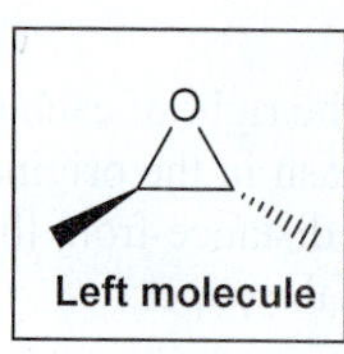

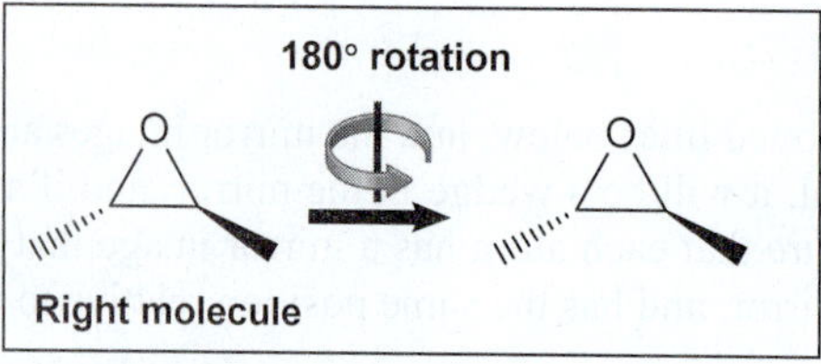

(b) Superimposable. It is evident from the mirror image that the two molecules are superimposable. No rotations are necessary.

(c) Nonsuperimposable. Both molecules have one wedge and one dash C–CH_3 bond. However, if you attempt to superimpose the two molecules, you will see it is not possible, because the dash–wedge notation does not agree. The rotation of the mirror image molecule, shown below, still does not allow the dash–wedge notation to agree.

180° rotation

Left molecule

Right molecule

(d) Superimposable. In the mirror image molecule, keep the front C of the Newman projection and its bonds frozen in place, because they are already superimposable. Rotate the back CHF_2 group 120° clockwise, as shown below, to make the two molecules entirely superimposable.

120° rotation about C–C bond

Left molecule

Right molecule

Problem 5.5

Think

Can you draw the mirror image of the molecule on the right? How do you represent a mirror? In a mirror image, how is dash–wedge notation treated? How is each atom in the mirror image drawn with respect to the mirror? Does the mirror image of the molecule on the right appear to be the same as or different from the molecule on the left?

Solve

A mirror can be represented by a dotted line adjacent to the molecule. Dash–wedge notation is conserved in the mirror image. Make sure that each atom has a mirror image that is the same distance from the mirror, at the same position along the mirror, and has the same position relative to the plane of the paper.

(a) The two molecules are not mirror images. The molecule on the left has two wedge C–Cl bonds, and the molecule on the right has one wedge and one dash C–Cl bond. The mirror image drawn is not the same.

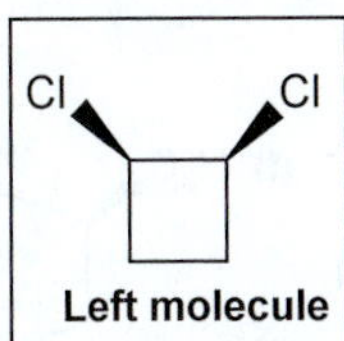

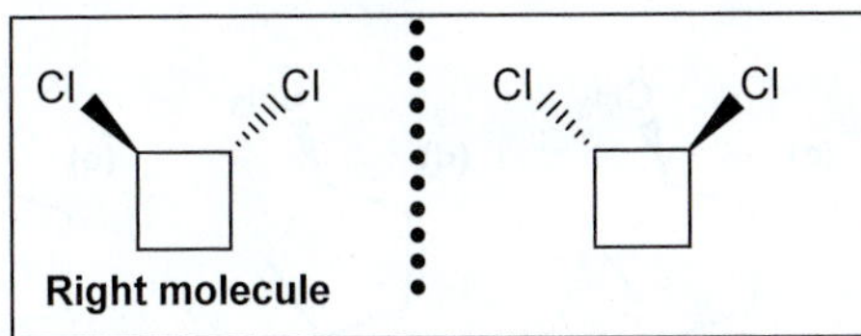

(b) The two molecules are mirror images.

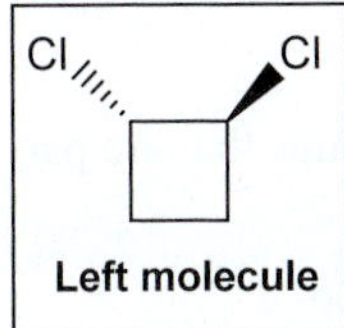

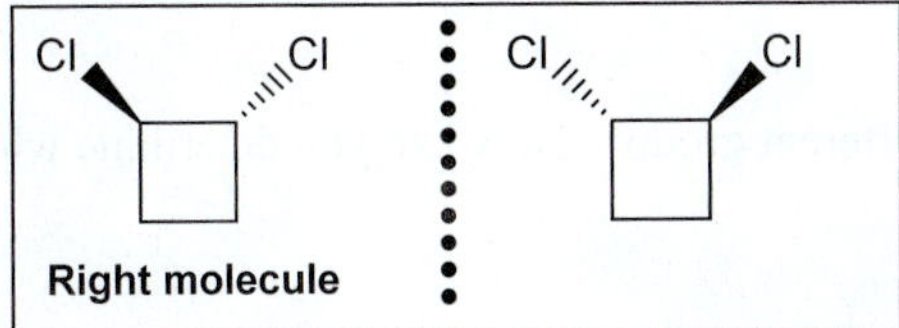

(c) The two molecules are not mirror images. The two molecules originally drawn are actually the same molecule, just different by rotation of the back CH_2F group about the C–C bond. When the mirror image for the right molecule is drawn, it is evident that the front group with the CH_3, Cl, and H groups will not align with the left molecule.

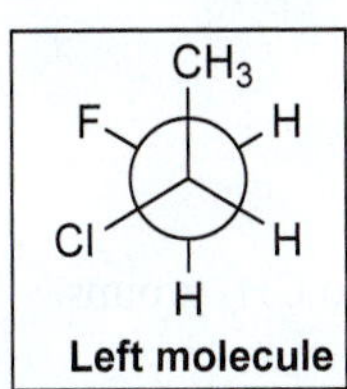

Problem 5.7

Think

Can you draw the mirror image of the molecule? Is the mirror image different from the original molecule, or are the two superimposable? It may help to construct models of the molecule and its mirror image and orient them in every way imaginable.

Solve

trans-1,2-Dichloroethene is not chiral, because *all the atoms lie in the same plane.* Below we show the mirror image, and after it is rotated 180°, we can see that it is the same as the original molecule.

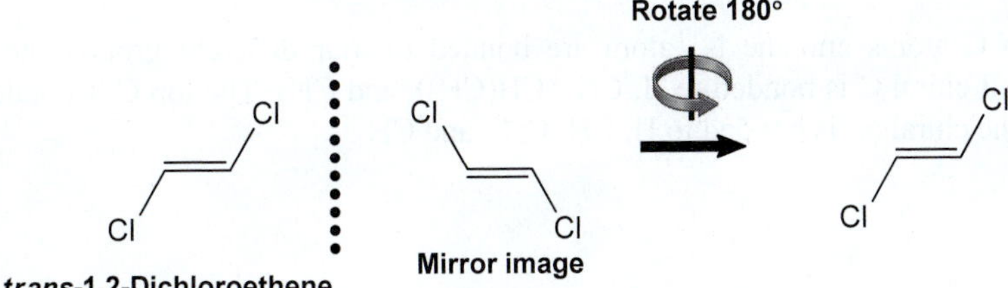

Problem 5.8

Think

Is there a way to bisect the molecule such that one half of the molecule is the mirror image of the other half? If so, draw in the dotted line that represents this internal plane of symmetry.

Solve

Molecule **(a)**, **(c)**, and **(f)** each has a plane of symmetry, as shown below. The other molecules do not.

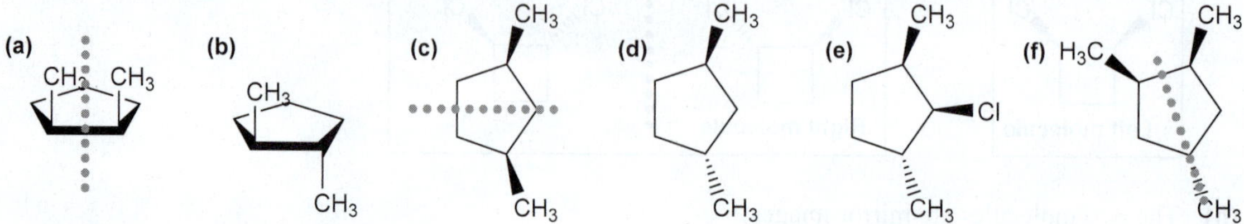

Problem 5.10

Think

Which atoms are bonded to four different groups? How can you determine whether groups that are part of a ring are different from each other?

Solve

For **(a)–(e)**, chiral centers are indicated by an asterisk.

(a) One chiral center. The second C atom from the left is not a chiral center, because it is bonded to two CH_3 groups (one is explicitly labeled, and the other is not).

CH_3 * CH_3

(b) Zero chiral centers. C2 and C4 are not chiral centers, because each is bonded to two CH_3 groups.

CH_3 CH_3

(c) Zero chiral centers. The central C atom is not a chiral center, because it is bonded to two benzene rings.

CH_3

(d) Three chiral centers. The two C atoms and the N^+ atom are bonded to four different groups and are, therefore, chiral centers. The left chiral C is bonded to H, CH_3, $CH(CH_3)$, and CH_2. The top C is bonded to H, CH_3, $CH(CH_3)$, and NH^+. The chiral N^+ is bonded to H, CH, CH_3, and CH_2.

CH_3 H_3C * * ⊕ *NH CH_3

(e) Three chiral centers. No plane of symmetry exists, so the C atom bonded to the CH_3 group pointing away from us *is* a chiral center. Alternatively, notice that the C atom is bonded to four different groups: H, CH_3, $CH(CH_3)R$, and $CH(CH_2Cl)R$.

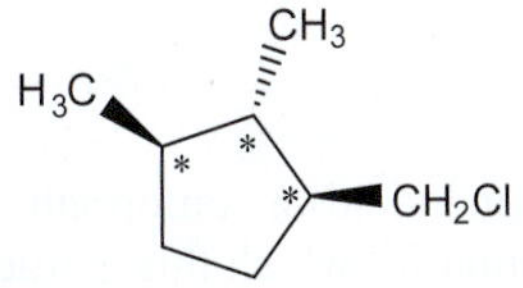

H3C
CH3
CH2Cl

Problem 5.11

Think

How many chiral centers are in each molecule? Do any of the molecules possess an internal mirror plane (i.e., a plane of symmetry)?

Solve

To be meso, a compound must contain at least two chiral centers and have a plane of symmetry. Compound **(a)** has no plane of symmetry and is chiral, so it is not meso. Compound **(b)** has a plane of symmetry and is achiral, and it contains two chiral centers, so it is meso. Compound **(c)** does not contain any chiral centers, so it is not meso, even though it has a plane of symmetry. Compound **(d)** does not have a plane of symmetry, so it is chiral and, therefore, cannot be meso. Compound **(e)** has a plane of symmetry and is achiral (which we can see after rotation about the C–C bond), and it contains two chiral centers, so it is meso.

Br Cl Br Br Cl Cl
(a) (b) *meso* (c)

Plane of symmetry
(d) OH OH Rotate about C–C bond OH OH No plane of symmetry
(e) OH *meso* OH Rotate about C–C bond OH OH

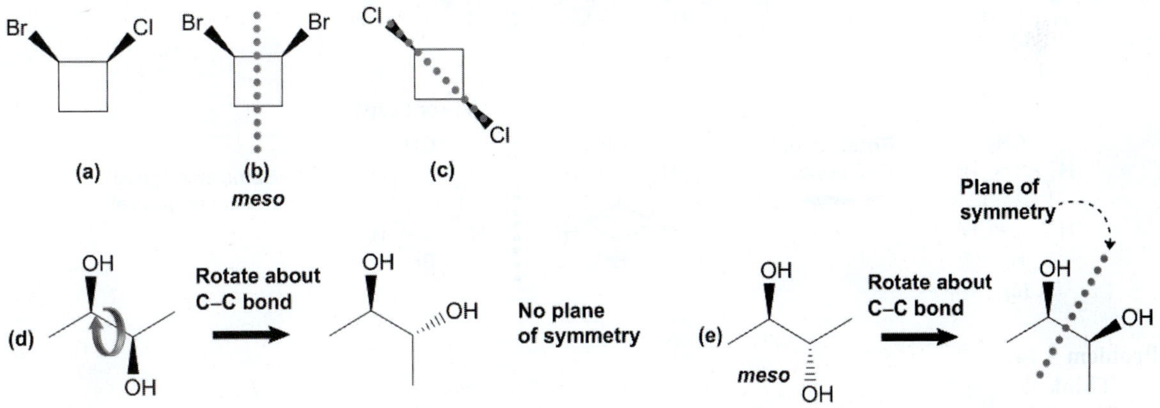

Problem 5.12

Think

Use Figure 5-12 as a guide to exchange W and Y. Are you able to align all four bonds in the resulting structure with the appropriate ones in the mirror image of the original molecule?

Solve

Groups W and Y are exchanged, and the resulting molecule (gray) is superimposable on the blue structure from Figure 15-12. To view how the two are superimposable, two rotations are necessary (see below).

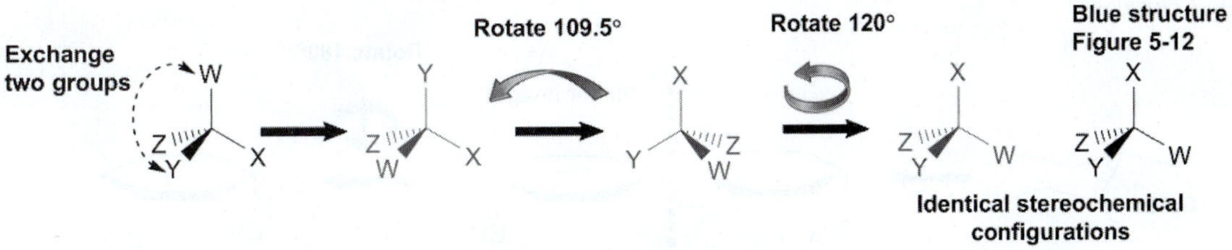

Exchange two groups
Rotate 109.5°
Rotate 120°
Blue structure Figure 5-12
Identical stereochemical configurations

Problem 5.13

Think

Can you draw the mirror image of the molecule? Is the mirror image different from the original molecule, or are the two superimposable? It may help to construct models of the molecule and its mirror image and orient them in every way imaginable. Make sure to try rotations about single bonds, too.

Solve

Molecules **(a)** and **(c)** are achiral. For each molecule, rotation about a single bond yields a conformation whose mirror image is the same as itself (see below). For **(b)**, there is not a conformation for which this is true, so it is chiral.

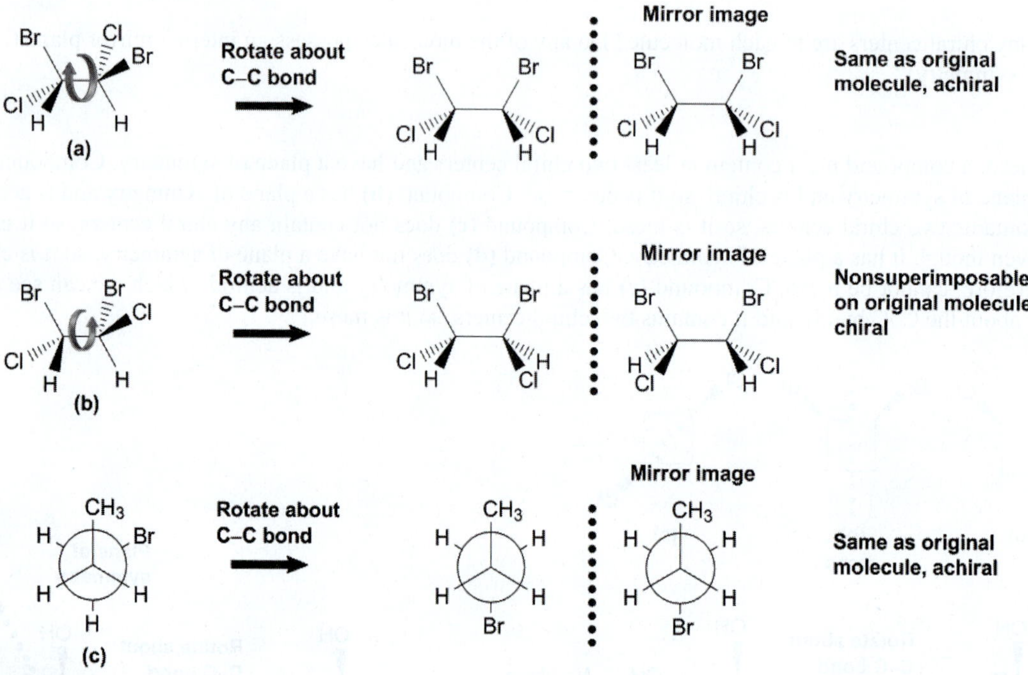

Problem 5.14

Think

Transform each chair to a Haworth projection. Sometimes it is easier to see mirror images in this type of structure. Does the molecule have an enantiomer, or is its mirror image superimposable?

Solve

We can more clearly identify chirality in cyclohexane rings by working with Haworth projections instead of chair representations. As we can see below and on the next page, the mirror image of **(b)** is exactly the same as the original molecule without rotating either molecule. The mirror images of **(a)**, **(d)**, and **(e)** are the same after rotation of the resulting molecules. Therefore, molecules **(a)**, **(b)**, **(d)**, and **(e)** are achiral. The mirror image of **(c)** is not superimposable on the original molecule, so it is chiral.

(a) Achiral

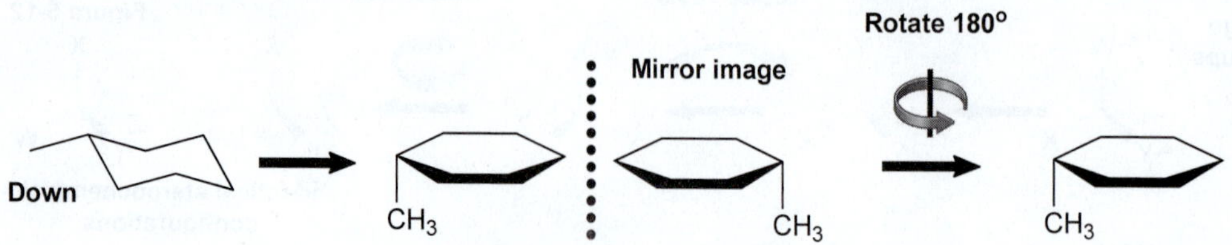

(b) Achiral

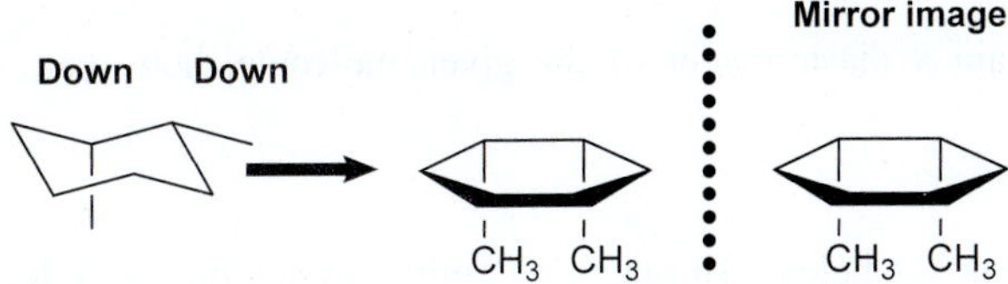

(c) Chiral

Mirror image

Up

Down

CH_3 CH_3 CH_3 CH_3

(d) Achiral

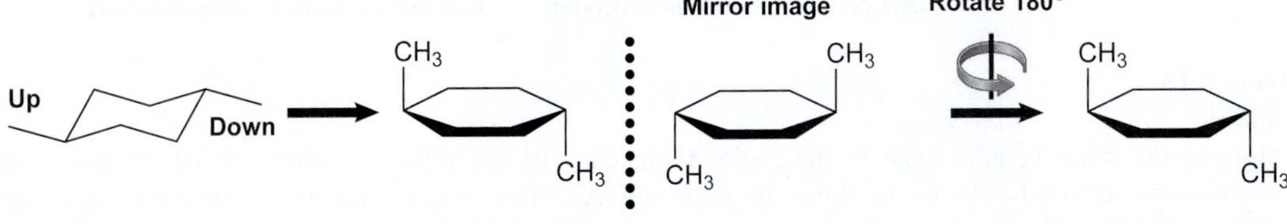

(e) Achiral

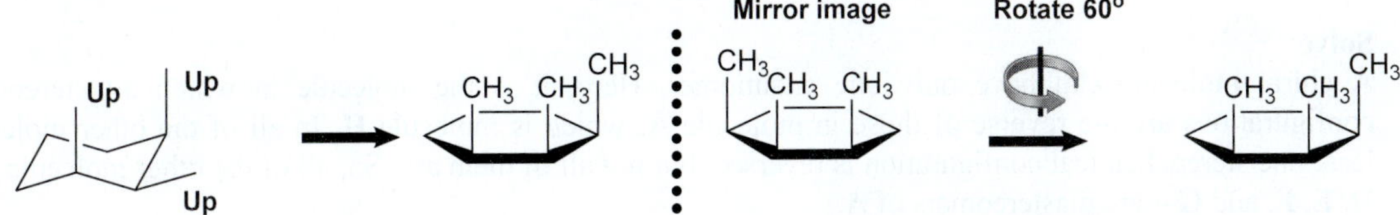

Problem 5.15

Think

Which C and N atoms are bonded to four different groups? How can you determine whether groups that are part of a ring are different from each other?

Solve

Molecules **(a)**, **(c)**, and **(d)** have no chiral centers. In **(a)**, each planar C atom is bonded to only three groups total. The tetrahedral C atoms are each bonded to at least two H atoms. The N atom is uncharged and has a lone pair of electrons, so it undergoes nitrogen inversion. In **(c)**, all planar C atoms are bonded to only three groups total, and the same is true of the N atom. The four tetrahedral C atoms are each bonded to at lease two H atoms. In **(d)**, each tetrahedral C atom is bonded to at least two H atoms. The N atom is bonded to two of the same ethyl groups. There are two chiral centers in molecule **(b)**. The nitrogen atom is a chiral center because four different groups are attached to it: CH_3, CH_2CH_3, $–CH_2–$(ring), and $–CH(CH_3)–$(ring). The four groups on the chiral carbon are H, CH_3, N, and the ring.

(a) Br N

(b) * * N ⊕

(c) ⊕ N

(d) ⊕ N C_6H_5

Problem 5.17

Think

Which atoms are the chiral centers? To obtain a diastereomer of the given molecule, how many of those configurations can be reversed?

Solve

There are three chiral centers, indicated by asterisks below. To obtain two other diastereomers, at least one of the configurations must be reversed from the original molecule, but not all of them. Three additional diastereomers are given below.

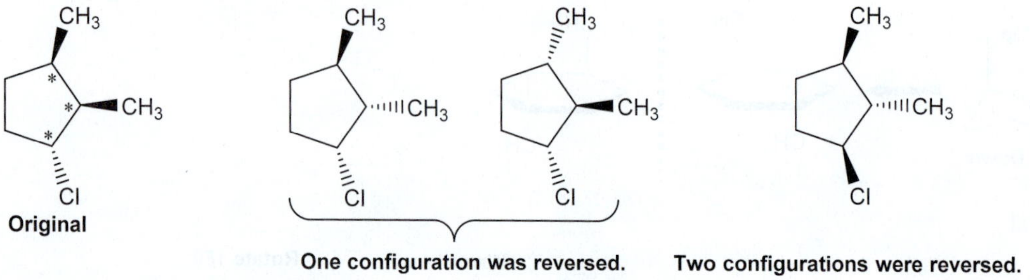

Original

One configuration was reversed. **Two configurations were reversed.**

Problem 5.18

Think

How many chiral centers exist in molecule **A**? In each of the following structures **B–H**, how many chiral centers are inverted relative to those in molecule **A**? How many have to be inverted to be considered enantiomers? How many configurations can be inverted to be considered diastereomers? How many enantiomers are possible for a chiral molecule?

Solve

A chiral molecule can have only one enantiomer. Here, it is the molecule in which all stereochemical configurations are the reverse of those in molecule **A**, which is molecule **H**. In all of the other molecules, at least one stereochemical configuration is reversed, but not all of them are. So, all of the other molecules—**B**, **C**, **D**, **E**, **F**, and **G**—are diastereomers of **A**.

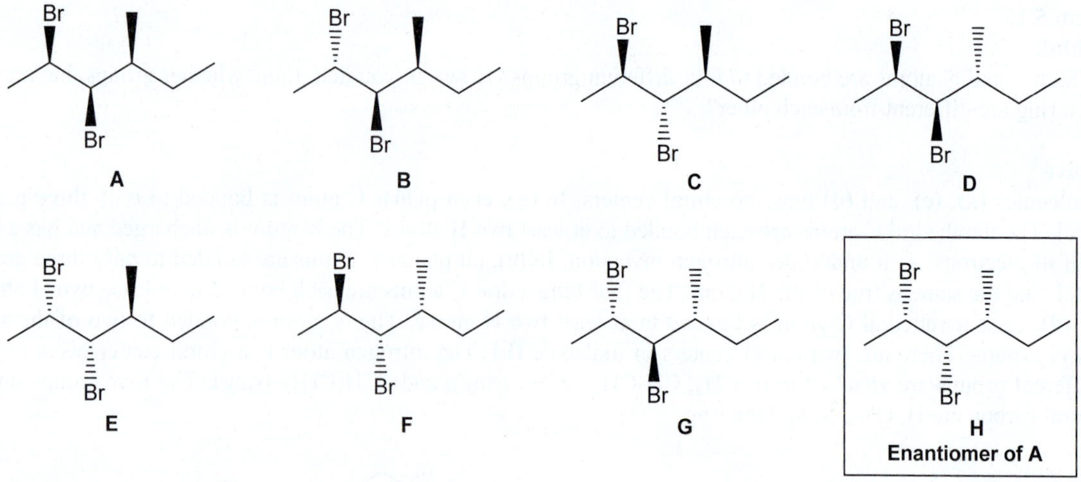

A **B** **C** **D**

E **F** **G** **H**

Enantiomer of A

Problem 5.20

Think

How many chiral centers are present? How can we change each chiral center to convert from one configurational isomer to another? How does the formula 2^n assist in determining the total number of stereoisomers? Are any of the isomers meso?

Solve

(a) There are two chiral centers, thus the maximum number of configurational isomers is $2^n = 2^2 = 4$. To obtain all possible configurations, systematically reverse the configurations at the different chiral centers. Notice that **A** and **B** both possess a plane of symmetry and are exactly the same (rotate **B** 180° to visualize this); they are meso. Therefore, **three** configurational isomers are possible.

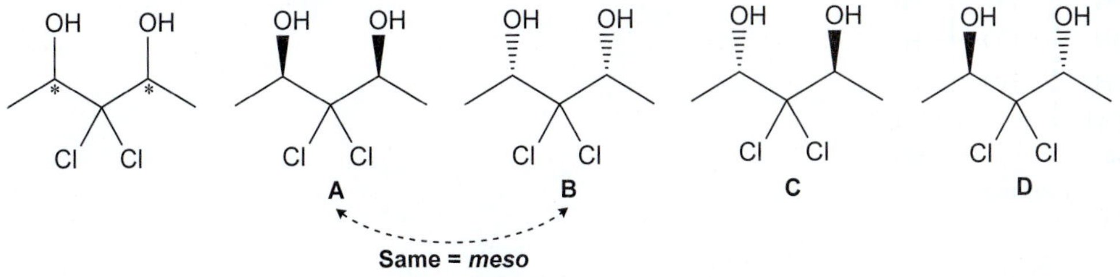

A **B** **C** **D**

Same = *meso*

(b) There are two chiral centers: one at C2 and one at C4. C3 is not bonded to four different groups. All of the permutations of reversing the configurations at just those chiral centers are shown below. There are four of them, which does equal $2^n = 2^2 = 4$. Notice that **A** and **B** are not the same in this example, because the C3 C–OH bond is denoted as a wedge and a 180° rotation would not interconvert these two molecules.

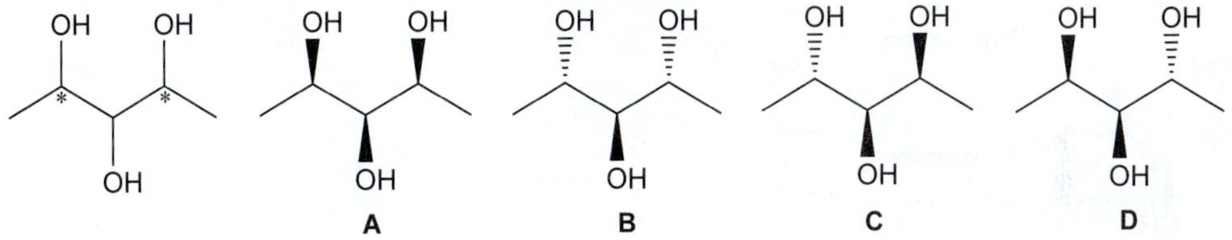

A **B** **C** **D**

Problem 5.21

Think

How many chiral centers do D-allose and L-glucose have? How many chiral centers are reversed between the two? How many must be reversed for the molecules to be enantiomers? How many can be reversed for the molecules to be diastereomers?

Solve

They cannot be enantiomers of each other, because any molecule can have only one mirror image, and D- and L-glucose are enantiomers. Because D-allose and L-glucose have the same connectivity but are not the same, they must be diastereomers. This is consistent with the fact that three of the four chiral centers are reversed, as seen below.

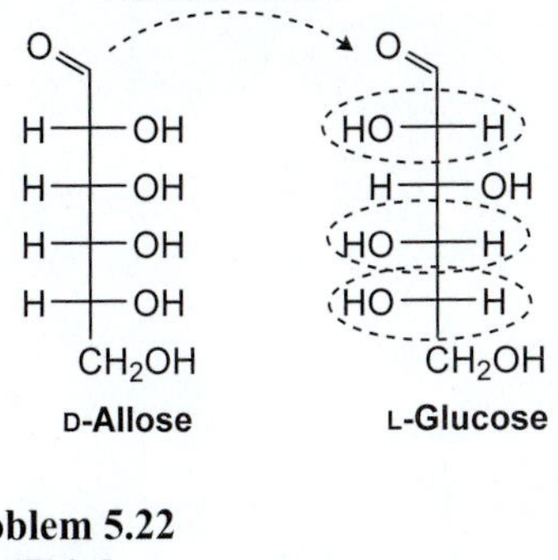

Diastereomers

D-Allose **L-Glucose**

Problem 5.22

Think

How many chiral centers are present? How many chiral centers must be reversed to draw the enantiomer? How is the mirror image of a chiral molecule related to the original molecule?

Solve

To obtain the enantiomer of D-allose, all four chiral centers are reversed, and/or you can draw the mirror image.

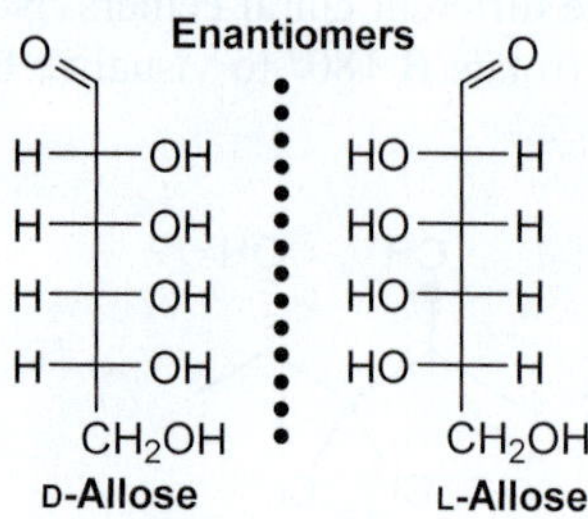

Problem 5.23

Think

How many stereoisomers are present? Review the steps for converting a zigzag structure to a Fischer projection.

Solve

(a) There are three chiral centers: C3, C4, and C5. Draw in the Fischer framework to denote the presence of chiral centers at these three C atoms, but temporarily leave the substituents on the horizontal bonds blank. C1 should be on top.

Top carbon

Use a model kit to build a molecular model of the molecule, paying special attention to the dash–wedge notation. Orient the molecule vertically so that C1 is on top and C6 is on the bottom, shown on the left below. In that orientation, the horizontal bonds of C3 and C5 point toward you, so add the H and OH substituents to C3 and C5 of the Fischer projection, as they appear in the model. Then, flip the model over as on the right below. The horizontal bonds of C4 point toward you, so add the H and OH substituents to C4 of the Fischer projection, as they appear in the model.

(b) The same methodology is followed.

Problem 5.24

Think

How many stereoisomers are present? Review the steps for converting a Fischer projection to a zigzag structure.

Solve

(a) There are four chiral centers: C2, C3, C4, and C5. Construct a molecular model with a six-carbon chain so that the chain is zigzag. Hold the molecule so that the zigzag is vertical. With the horizontal bonds pointed toward you on C2 and C4, add the H and Cl atoms as they appear in the Fischer projection. This is shown on the left below. Then, flip the molecule over so that the horizontal bonds on C3 and C5 are pointed toward you, and add the H and Cl atoms as they appear in the Fischer projection.

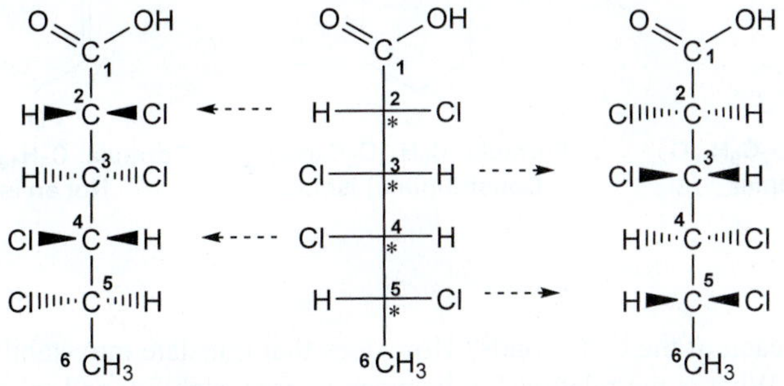

Once all the substituents have been added, you can reorient the completed model so the zigzag is in the plane of the paper, as shown below.

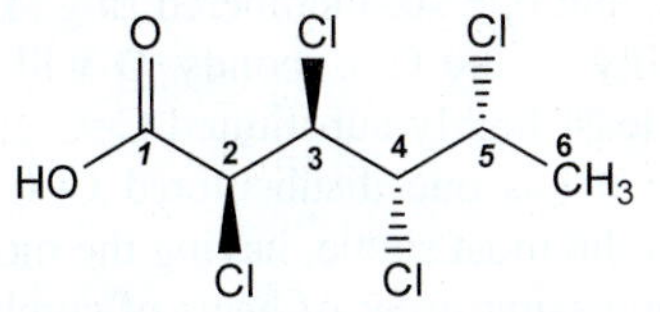

View from the side

(b) The same methodology is followed.

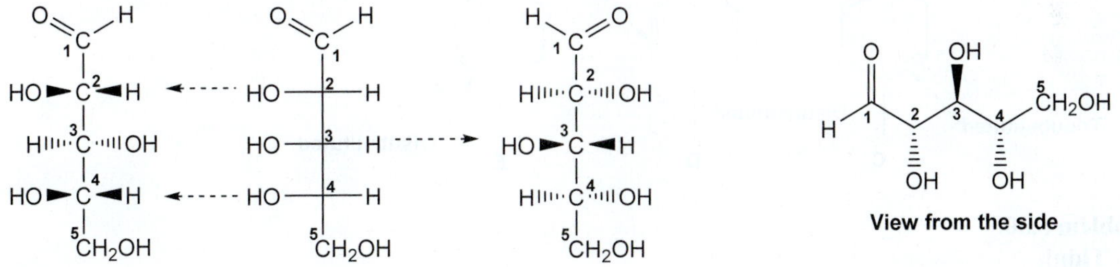

View from the side

Problem 5.26

Think

How is each molecule—**A** through **E**—related to **Y**? Are they constitutional isomers, enantiomers, diastereomers, unrelated, or the same molecule? How do these relationships translate into relative behavior? Explain.

Solve

The acidities will be different unless the molecules are enantiomers or identical molecules. Because molecule **E** is its enantiomer, it will have exactly the same acidity. All of the remaining molecules will have acidities different from the molecule given. The relationships are given on the next page.

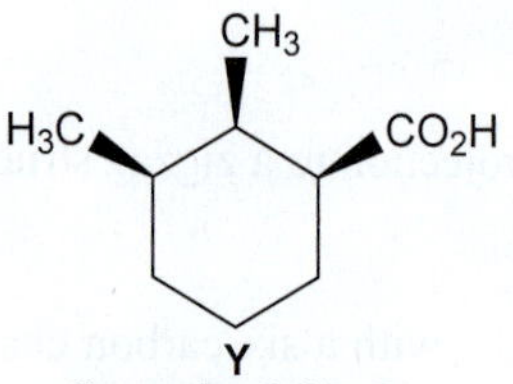

Y
Formula: $C_9H_{16}O_2$

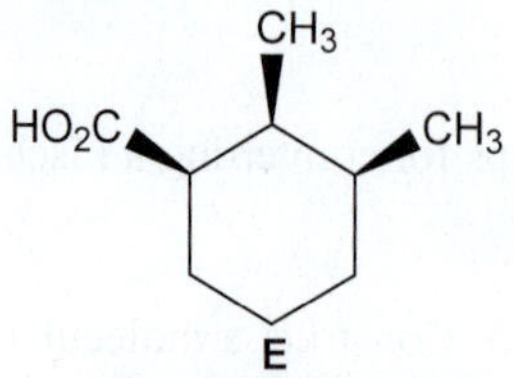

E
Formula: $C_9H_{16}O_2$
Enantiomer of Y

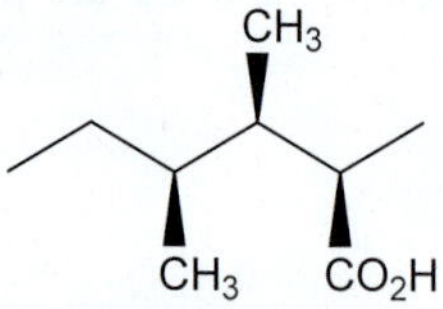

A
Formula: $C_9H_{18}O_2C_9H_{18}O_2$
Not an isomer

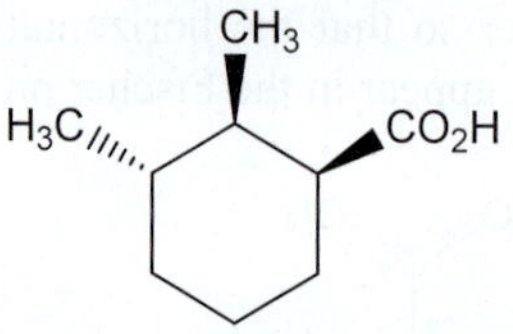

B
Formula: $C_9H_{16}O_2C_9H_{16}O_2$
Diastereomer

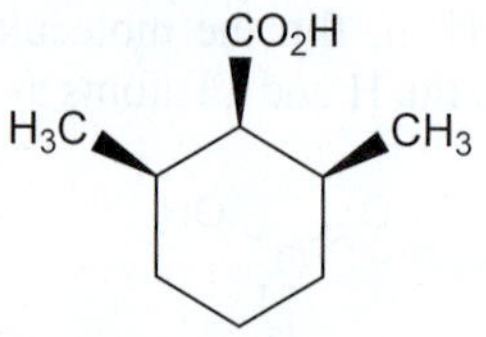

C
Formula: $C_9H_{16}O_2C_9H_{16}O_2$
Constitutional isomer

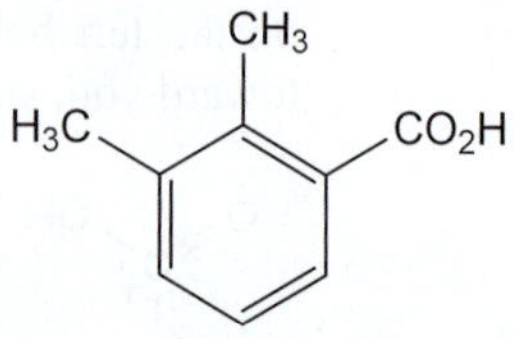

D
Formula: $C_9H_{10}O_2C_9H_{10}O_2$
Not an isomer

Problem 5.28

Think

What is the degree of alkyl substitution on each of the C=C bonds? How does that translate into stability? How many of each kind of C=C bond are there? What is the relationship between relative stability and relative heats of combustion?

Solve

All three have the same molecular formula (C_8H_{12}), two double bonds, and one six-membered ring. Therefore, the difference in heats of combustion will be due solely to the stability of the C=C bonds. **D** will have the greatest heat of combustion, because it is the least stable, having the least highly substituted C=C groups. Its C=C groups are both disubstituted. **E** will be in the middle, because it has one disubstituted C=C and one trisubstituted C=C. **C** will have the smallest heat of combustion, as it is the most stable, having the most highly substituted C=C groups. Its C=C groups are both trisubstituted. The increasing order of heats of combustion is as follows: **C** < **E** < **D**.

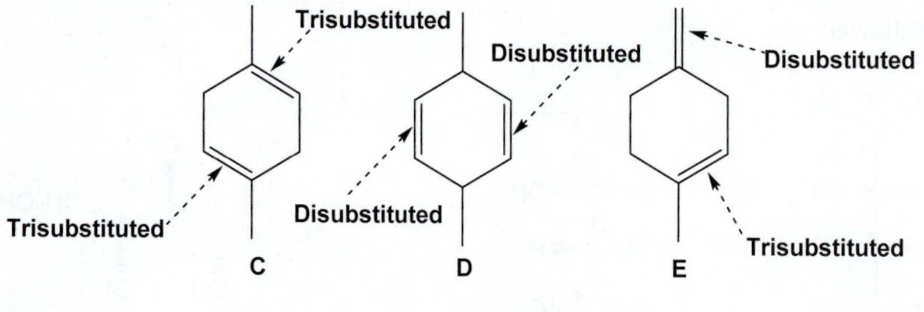

Trisubstituted
Trisubstituted
C

Disubstituted
Disubstituted
D

Disubstituted
Trisubstituted
E

Problem 5.30

Think

For which variable in Equation 5-2 are we solving? Are the units for concentration correct? Are the units for the cell length correct?

Solve

We are asked to solve for the measured angle of rotation α. The units for concentration are correct (g/mL), but the units for the cell length need to be converted from cm to dm: 10.0 cm $\times \frac{1\text{ dm}}{10\text{ cm}}$ = 1.00 dm.

$$\alpha = ([\alpha]_{\text{D}}^{20})(l)(c) = (+223°)(1.00\text{ dm})\left(0.00300\frac{\text{g}}{\text{mL}}\right) = +0.669°$$

Problem 5.32

Think

What is the sign of the specific rotation of the mixture of the two enantiomers? Which enantiomer is in excess? How can you tell from the specific rotation of the mixture the extent by which that enantiomer is in excess? What percentage of the solution is optically inactive?

Solve

The specific rotation of the mixture of the two enantiomers is +12°. This means that the (+) enantiomer is in excess.

$$\text{The \% enantiomeric excess} = \frac{\text{(Specific rotation of mixture)}}{\text{(Specific rotation of pure enantiomer)}} \times 100 = \frac{+12^\circ}{+49^\circ} \times 100 = 24\%.$$

Thus 76% is optically inactive, meaning that half of this percentage is the (+) enantiomer and the other half is the (−) enantiomer. Thus, 38% is the (−) enantiomer and 38% + 24% = 62% is the (+) enantiomer.

Problem 5.33

Think

How many stereochemical configurations have to be different for two sugars to be considered epimers? How do you know which configuration(s) to change? Consult Figure 5-32 for names and structures of the D family of aldoses.

Solve

Epimers differ only by the stereochemical configuration at one carbon atom. Reverse the configuration at the carbon atom indicated.

(a) The C2 epimer of D-glucose

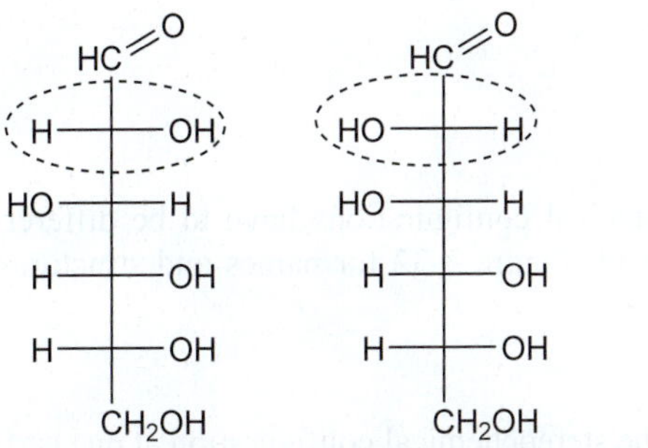

D-Glucose | **C2 epimer of D-glucose; D-mannose**

(b) The C3 epimer of D-talose

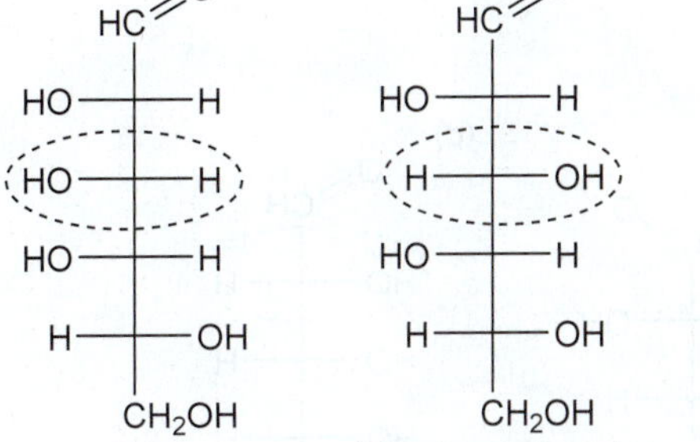

D-Talose | **C3 epimer of D-talose; D-idose**

(c) The C4 epimer of D-talose

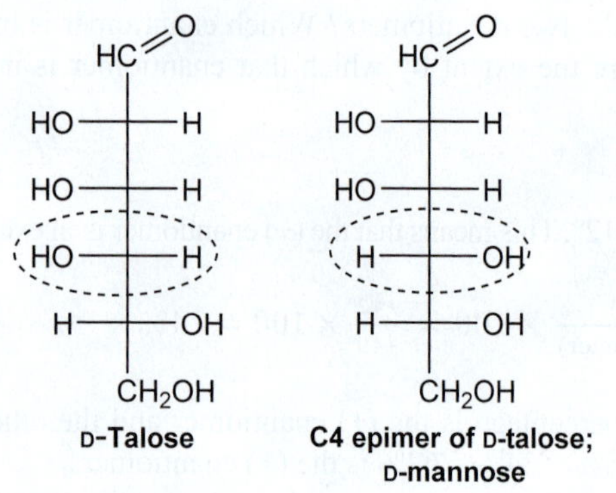

D-Talose

C4 epimer of D-talose; D-mannose

(d) The C3 epimer of D-xylose

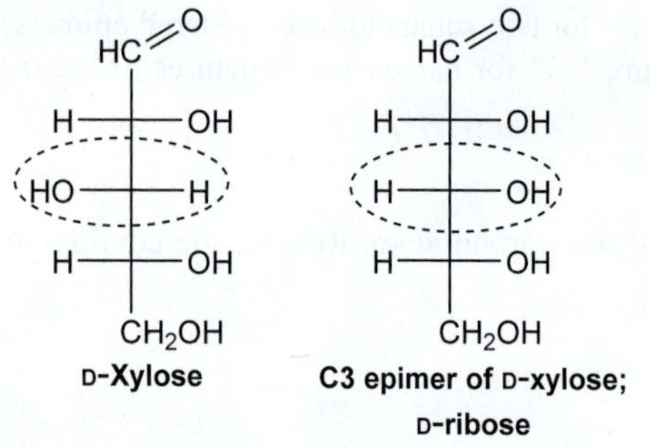

D-Xylose

C3 epimer of D-xylose; D-ribose

Problem 5.34

Think

How are the D and L sugars related? How many stereochemical configurations have to be different for two sugars to be considered enantiomers? To be epimers? Consult Figure 5-32 for names and structures of the D family of aldoses.

Solve

D and L structures are enantiomers. Epimers differ only by the stereochemical configuration at one carbon atom.

(a) L-Mannose is the enantiomer of D-mannose, so reverse the configuration at all four chiral centers.

(b) L-Arabinose is the enantiomer of D-arabinose, so reverse the configuration at all three chiral centers.

(c) L-Threose is the enantiomer of D-threose, so reverse the configuration at both of the chiral centers.

(d) The C2 epimer of L-arabinose should differ from L-arabinose only by reversing the configuration at C2 of L-arabinose.

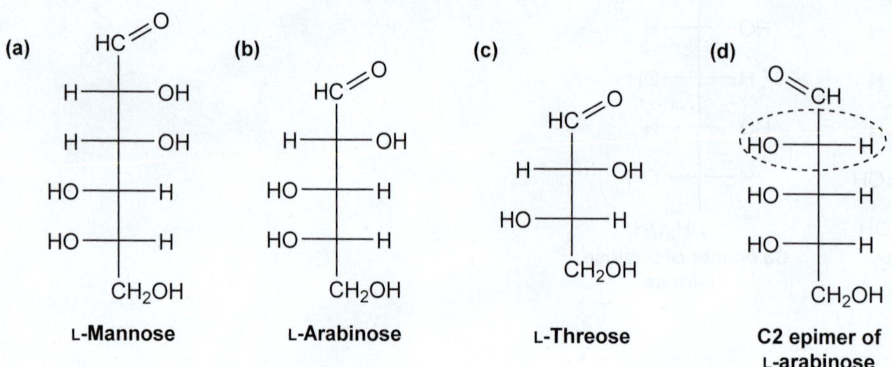

End of Chapter Problems

Sections 5.2–5.4 Enantiomers, Chirality, and Stereocenters

Problem 5.35

Think

Which of these objects possess a plane of symmetry? How can you recognize a plane of symmetry? How is the presence of a plane of symmetry related to an object being chiral or achiral?

Solve

A plane of symmetry exists if you can divide the object in half such that one half is the mirror image of the other.

Chiral objects (no plane of symmetry): **(b)**, **(d)**, **(e)**, **(h)**, and **(i)**
Achiral objects (plane of symmetry): **(a)**, **(c)**, **(f)**, **(g)**, and **(j)**

Problem 5.36

Think

How do you represent the C–C bond in a Newman projection? What is the dihedral angle for the Cl groups for the anti and gauche conformations? For a plane of symmetry to exist, what must be true of atoms on either side of the plane?

Solve

Newman projections for 1,2-dichloroethane are shown below. Only the anti conformation has a plane of symmetry, as shown. The others do not. That plane is not a plane of symmetry in either of the gauche conformations because the Cl atom that does not lie in the plane does not have a reflection on the other side of the plane. Where another Cl would need to appear, an H atom appears instead.

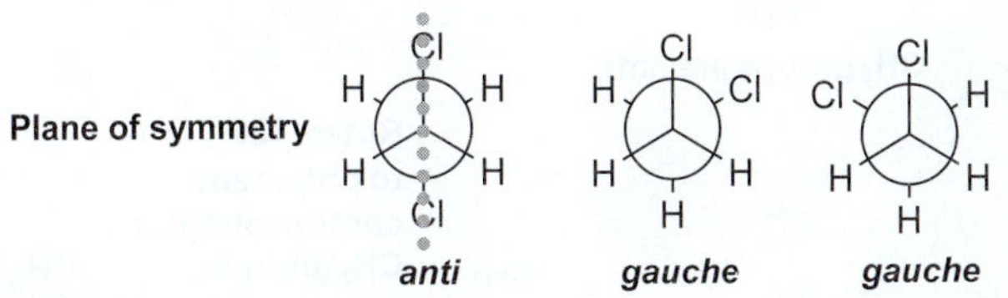

Plane of symmetry

Cl, H, H, H, H, Cl — *anti*

Cl, H, Cl, H, H, H — *gauche*

Cl, Cl, H, H, H, H — *gauche*

Problem 5.37

Think

Can you identify the chiral center in each molecule? Do the groups attached to the chiral center in one molecule superimpose on the same groups of the chiral center in the other molecule? Or are the chiral centers related by the exchange of two groups or by being mirror images of each other? It would help to build each molecule using your molecular modeling kit and then try to superimpose each molecule on the one given.

Solve

(a) Opposite configuration. This molecule is the mirror image of the original molecule, so the configuration of its chiral center must be the mirror image of the chiral center in the given molecule.

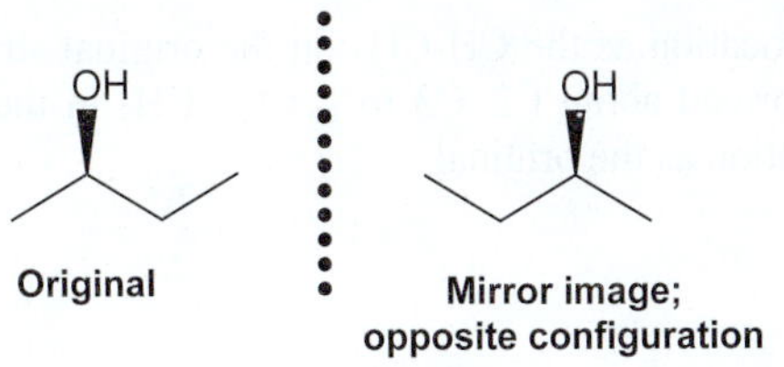

OH — **Original**

OH — **Mirror image; opposite configuration**

(b) Same configuration. A 180° rotation shows that the two molecules are the same.

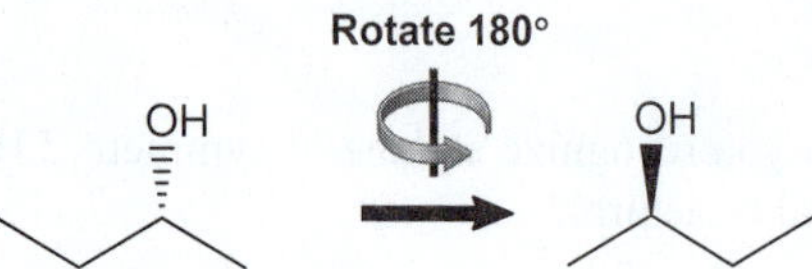

(c) Opposite configuration. The OH and CH_3 groups were exchanged.

H CH_3
HO
Exchange two groups
H OH
H_3C

(d) and **(e)** The original structure is translated to a Newman projection, with the back C atom being the chiral center. The CH_3 groups are anti to each other, and **(d)** and **(e)** both need to have a 120° rotation performed to obtain the same CH_3 anti conformation as the original. In **(d)**, rotating the back carbon yields the original molecule, so the configuration of the chiral center is the same in both. In **(e)**, rotating the back carbon 120° shows that the OH and CH_3 groups attached to the chiral center were switched, so **(e)** has the opposite configuration.

H OH
Back
Front
Original
CH_3, HO, H, H, H, CH_3
CH_3 groups are anti.

(d) Same
CH_3, H, CH_3, H, H, OH
Rotate 120° to obtain anti conformation for CH_3 groups.
CH_3, HO, H, H, H, CH_3

(e) Opposite
CH_3, HO, CH_3, H, H, H
Rotate 120° to obtain anti conformation for CH_3 groups.
CH_3, H, OH, H, H, CH_3

(f) and **(g)** The original is structure is converted to a Fischer projection to make direct comparisons.

H OH
Original
CH_3
HO—H
CH_2CH_3

(f) The CH_2CH_3 is held in place because it is in the same location as the CH_2CH_3 on the original structure; a counterclockwise 120° rotation of the molecule is performed about C2–C3 to get the CH_3 in the vertical position. It is now evident that **(f)** has the same configuration as the original.

Rotate 120° about C2-C3

OH
H—2—1CH_3
4
3CH_2CH_3
(f)

1CH_3
HO—2—H
3$CH_2$4CH_3
Same configuration

(g) None of the four groups is in the same location compared to the original structure. Thus, first perform a 120° clockwise rotation to get the CH_2CH_3 in the same location as the original ethyl group. Then perform a counterclockwise 120° rotation about C2–C3 to move the OH to the left. It is now evident that the two structures have opposite configurations because they differ by the exchange of the CH_3 and H groups.

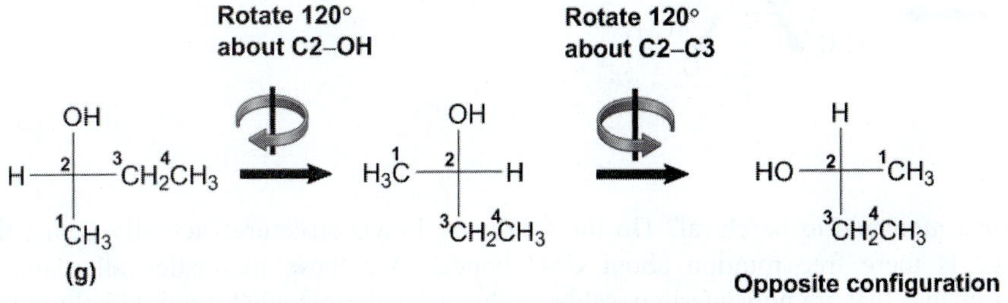

Rotate 120° about C2–OH

Rotate 120° about C2–C3

(g)

Opposite configuration

(h) In both the original and molecule **(h)**, the CH_2CH_3 group is flat and pointed down to the right. Rotate **(h)** about the C2–C3 bond to get the flat bond on the bottom left. It is evident that the configurations are the same.

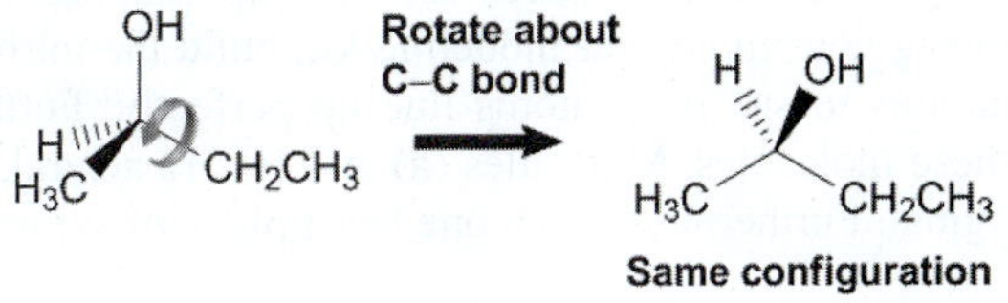

Rotate about C–C bond

Same configuration

Problems 5.38/5.39

Think

Is the mirror image of each molecule superimposable on the original molecule? Which C and N atoms are bonded to four different groups? Which molecules possess a plane of symmetry, and which do not? What is the minimum number of chiral centers a meso compound must have? Can a meso compound be chiral?

Solve

Molecules **(b)**, **(e)**, **(f)**, and **(h)** are chiral, because their mirror images are not superimposable on the original. Furthermore, none of them has a plane of symmetry. Molecules **(a)**, **(c)**, **(d)**, and **(g)** are achiral because their mirror images are superimposable on the original. Furthermore, they each have a plane of symmetry. The plane of symmetry in **(a)** is the plane of the paper. Molecule **(c)** has a plane of symmetry, because the N undergoes nitrogen inversion. During inversion, the N is temporarily planar, making the plane of symmetry the plane of the paper. The plane of symmetry in **(d)** is the plane of the paper. Molecule **(g)** has a plane of symmetry after rotation of the C–C bond, as shown on the next page. Chiral centers are identified with an asterisk below. Each chiral center is bonded to four different groups. Only **(g)** is meso, because it has two tetrahedral chiral centers but a plane of symmetry, making it achiral.

(a) 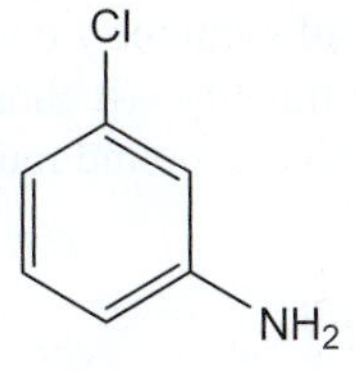 Achiral

(b) 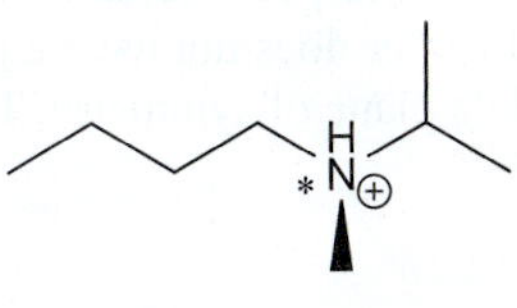 Chiral

(c) 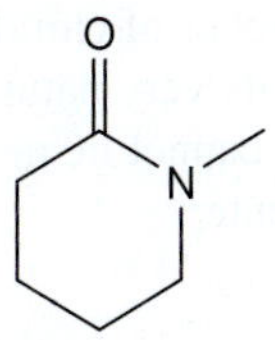 Achiral

(d) 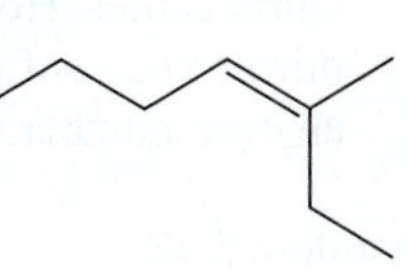 Achiral

(e) 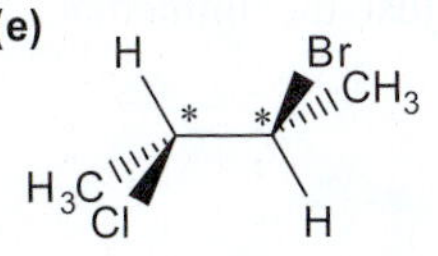 Chiral

(f) 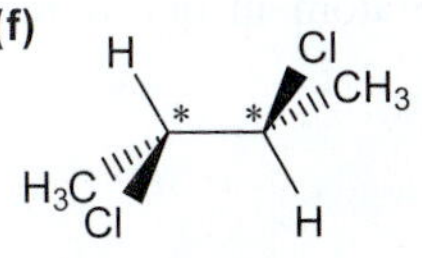 Chiral

(g) 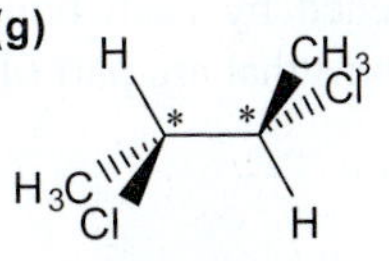 Achiral/meso

(h) 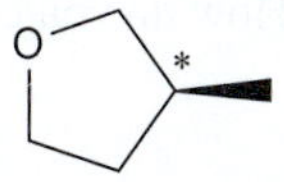 Chiral

Alternative view of **(g)** to view the plane of symmetry:

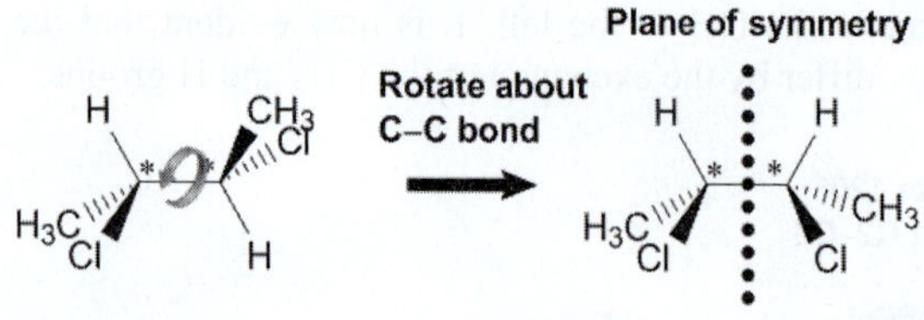

Problem 5.40

Think

What does it mean for a molecule to be chiral? Do the following Lewis structures actually depict the three-dimensional geometry? Is there free rotation about C=C bonds? Are these molecules all planar? Which molecules have mirror images that are nonsuperimposable on the original molecule? It would help to build each molecule using your molecular modeling kit.

Solve

In each of these molecules, the plane of one trigonal planar C is perpendicular to the plane of the other (shown below). Molecules **(b)**, **(c)**, and **(d)** are chiral, because their mirror images are not superimposable on the original. To see that explicitly, build each molecule using your molecular modeling kit, build the mirror image of it, and then hold them together in various orientations to see if all atoms line up perfectly. Furthermore, notice that there is no plane of symmetry in any of these molecules. Molecules **(a)** and **(e)** are achiral, because each one's mirror image is superimposable on the original. Furthermore, each one has a plane of symmetry that lies in the plane of the paper (not shown).

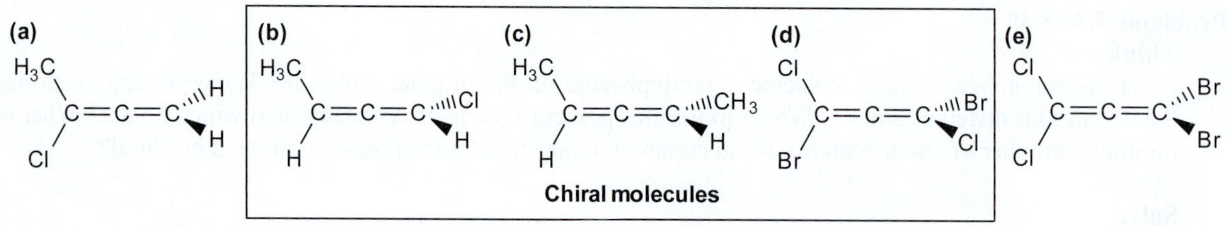

Problem 5.41

Think

With an odd number of chiral centers, is it possible for a molecule to possess a plane of symmetry? Can one chiral center be the mirror image of a second? Can a chiral center be the mirror image of itself?

Solve

No, it is not possible for a meso compound to contain exactly three chiral centers. To be meso, the compound must possess a plane of symmetry, making it achiral. To have a plane of symmetry, each chiral center on one side of the plane of symmetry must have a corresponding chiral center on the other side of that plane of symmetry. With an odd number of chiral centers, this is impossible, unless the plane of symmetry contains one chiral center. However, by its very nature, a chiral center does not have a plane of symmetry—it cannot be the mirror image of itself. So it cannot lie in a molecule's plane of symmetry. Thus, a meso compound must contain an even number of chiral centers.

Problem 5.42

Think

Which atoms are bonded to four different groups? To identify a chiral center, make sure to consider the entire rest of the molecule that is attached by each bond to the atom in question, not just the immediate atoms attached. How do you consider atoms that are part of a ring?

Solve

When considering atoms that are part of a ring, remember to divide the ring in half, where the dividing line goes through the atom in question and the center of the ring, and determine whether the two halves of the ring are the same. The chiral centers are indicated by an asterisk below.

(a) 4 chiral centers

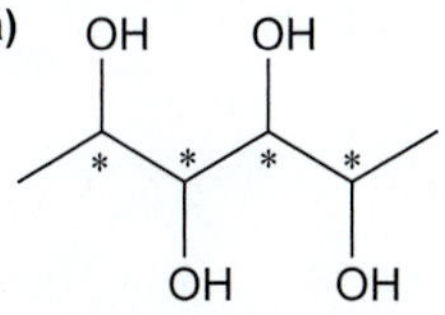

(b) 2 chiral centers

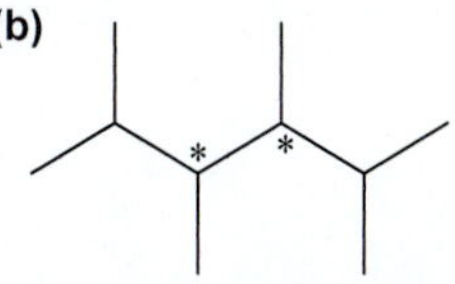

(c) 0 chiral centers

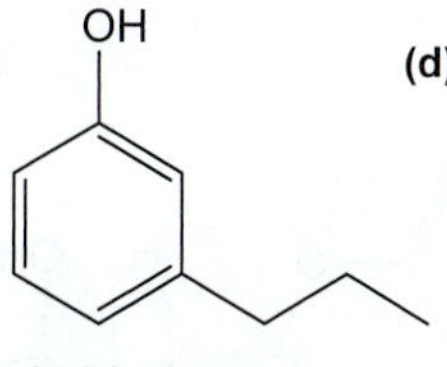

(d) 1 chiral center

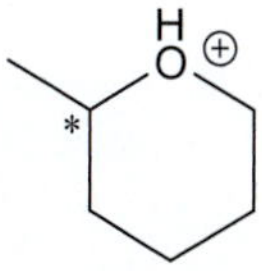

(e) 1 chiral center

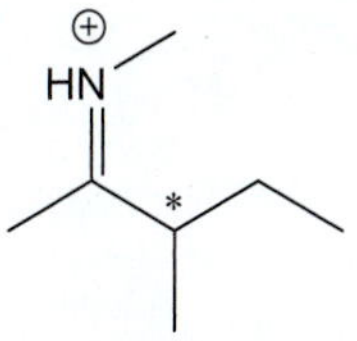

(f) 2 chiral centers

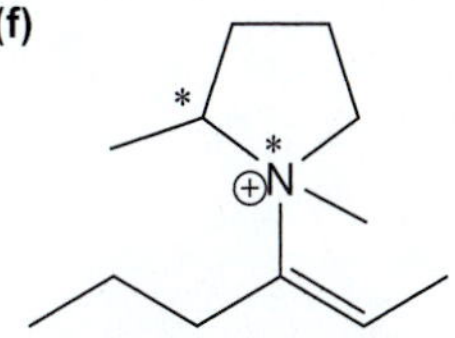

(g) 0 chiral centers

Problem 5.43

Think

Which atoms are bonded to four different groups? Remember that, when identifying groups on a potential chiral center, look at the entire rest of the molecule that is attached by each bond to the atom in question, not just the immediate atoms that are attached. How do you consider atoms that are part of a ring?

Solve

When considering atoms that are part of a ring, remember to divide the ring in half, where the dividing line goes through the atom in question and the center of the ring, and determine whether the two halves of the ring are the same. Aldosterone has seven chiral centers, shown below using an asterisk.

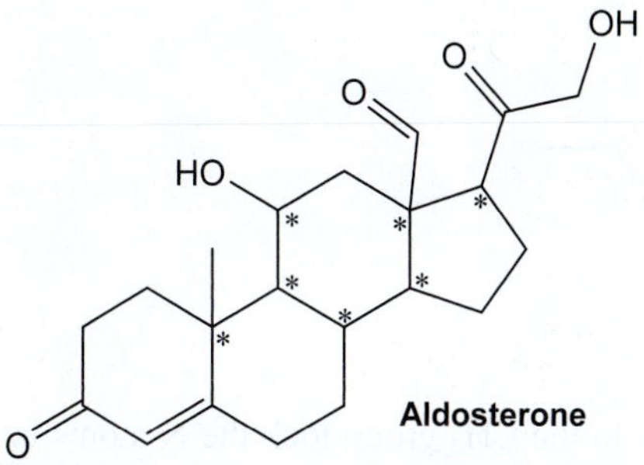

Aldosterone

Problem 5.44

Think

Which atoms are bonded to four different groups? Remember that, when identifying groups on a potential chiral center, look at the entire rest of the molecule that is attached by each bond to the atom in question, not just the immediate atoms that are attached. How do you consider atoms that are part of a ring? How do you consider N atoms?

Solve

When considering atoms that are part of a ring, remember to divide the ring in half, where the dividing line goes through the atom in question and the center of the ring, and determine whether the two halves of the ring are the same. The N atom is not a chiral center because of nitrogen inversion. Taxol has 11 chiral centers, shown below using an asterisk.

Taxol

Problem 5.45

Think

What is the relationship between any two enantiomers? How does the limited rotation in a ring affect the ability to interconvert the two mirror images? For nitrogen inversion to take place, the N atom is temporarily rehybridized to sp^2. What do you notice about the strain in Tröger's base if N were to be sp^2 hybridized?

Solve

(a) The enantiomer is the mirror image, as shown below.

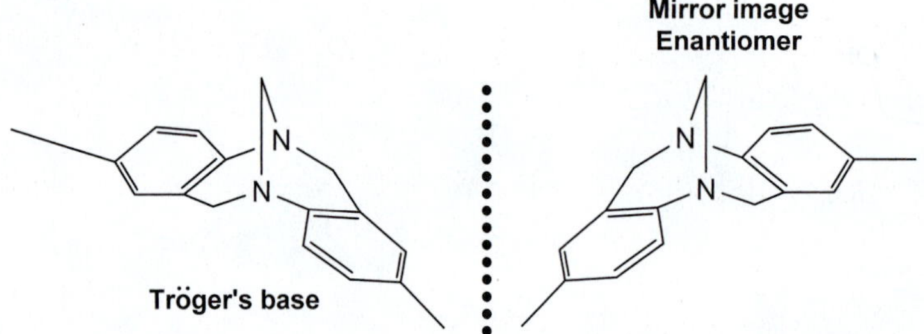

(b) Nitrogen inversion doesn't take place, because the N–C bonds to the CH_2 group lock the N atoms in their respective configurations. To undergo nitrogen inversion, N would need to temporarily attain a planar geometry, which would severely strain the ring system.

Sections 5.5–5.7 Diastereomers and Fischer Projections

Problem 5.46

Think

Which C atoms are bonded to four different groups? Remember that, when identifying groups on a potential chiral center, look at the entire rest of the molecule that is attached by each bond to the atom in question, not just the immediate atoms that are attached. What is the formula for determining the maximum number of chiral centers? What can you do to a chiral center to produce a new stereoisomer? How can you determine if one of the stereoisomers is meso?

Solve

There are six chiral centers, indicated by an asterisk below. Each configurational isomer (stereoisomer) can be generated by various ways of exchanging two groups attached to a chiral center, thus reversing the configuration. Because no plane of symmetry can exist in the molecule, we are guaranteed that none of the stereoisomers generated will be meso, so each permutation in which a configuration is reversed at one of those chiral centers will yield a new stereoisomer. Therefore, there are $2^6 = 64$ different configurational isomers for this molecule (stereoisomers are not shown).

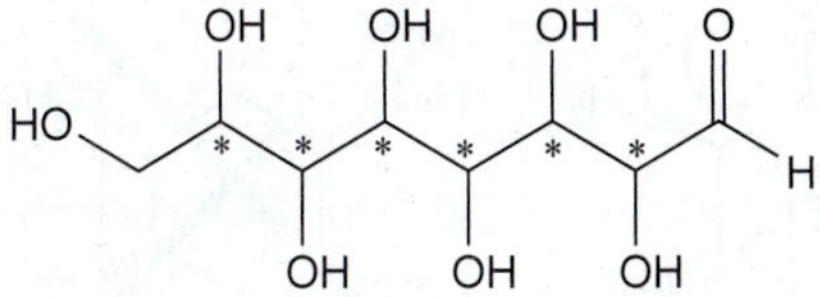

HO OH OH OH O H OH OH OH

Problem 5.47

Think

How many C atoms are bonded to four different groups? What is the formula to determine the maximum number of stereoisomers? Are any planes of symmetry present? Systematically draw all isomers by reversing the configuration at each chiral center.

Solve

There are four chiral centers, so there are at most $2^4 = 16$ possible configurational isomers. Those possibilities are shown below and on the next page. We obtain them by reversing configurations systematically at each of the four chiral centers. We can begin by having the substituents at each chiral center on the ring pointing toward us, as shown in **(a)** below. In **(b)–(e)**, one wedge bond was converted to dash. In **(f)–(k)**, two wedge bonds were converted to dashes. In **(l)–(o)**, three wedge bonds were converted to dashes. And in **(p)**, all four wedge bonds were converted to dashes. However, there are a number of redundancies, which are crossed out. Configurations **(p)** and **(a)** are redundant. Configurations **(m)** and **(b)** are redundant. Configurations **(l)** and **(c)** are redundant. Configurations **(o)** and **(d)** are redundant. Configurations **(n)** and **(e)** are redundant. Configurations **(h)** and **(f)** are redundant. For each redundancy, you can flip the molecule over, turning it about the C=O bond. To see this, it would help to build the molecules using your molecular modeling kit. So in all, there are 10 different configurational isomers. This number is lower than the maximum number of 16 owing to planes of symmetry. Notice that molecule **(a)** and **(f)** each has a plane of symmetry (shown below), making it achiral.

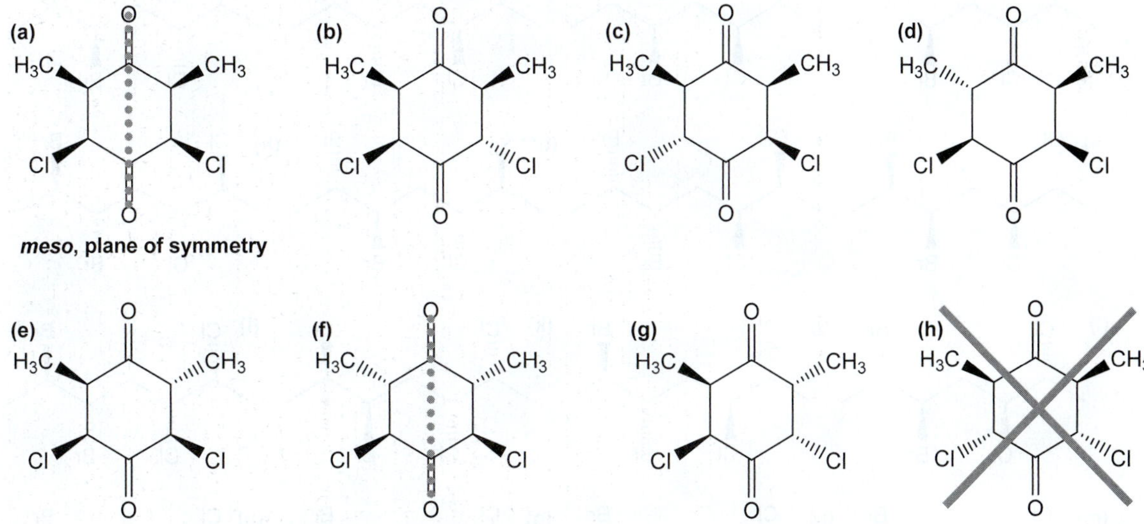

(a) O H_3C CH_3 Cl Cl O

***meso*, plane of symmetry**

(b) O H_3C CH_3 Cl Cl O

(c) O H_3C CH_3 Cl Cl O

(d) O H_3C CH_3 Cl Cl O

(e) O H_3C CH_3 Cl Cl O

(f) O H_3C CH_3 Cl Cl O

***meso*, plane of symmetry**

(g) O H_3C CH_3 Cl Cl O

(h) O H_3C CH_3 Cl Cl O

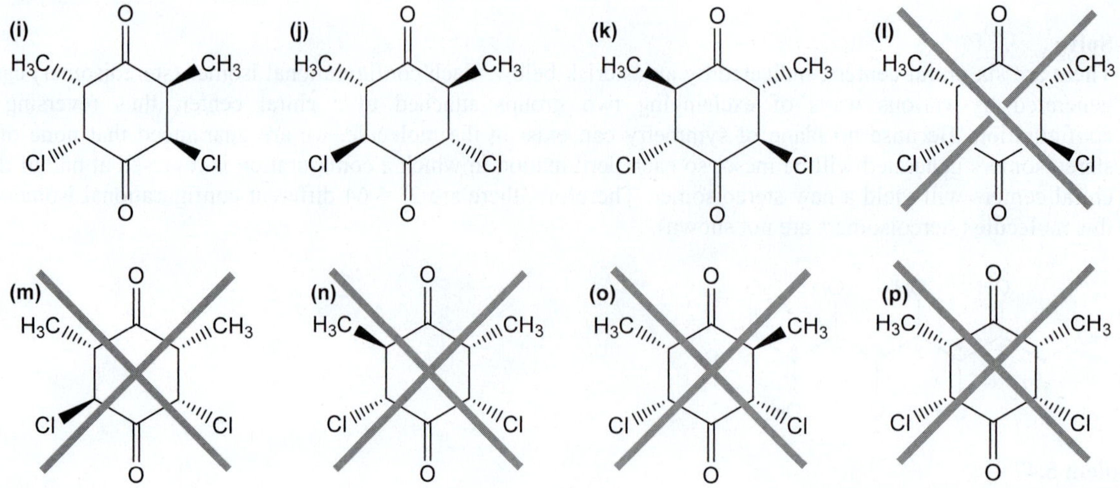

Problem 5.48

Think

How many C atoms are bonded to four different groups? What is the formula to determine the maximum number of stereoisomers? Are any planes of symmetry present? Systematically draw all isomers by reversing the configuration at each chiral center.

Solve

There are four chiral centers, so there are up to $2^4 = 16$ possible configurational isomers. Those possibilities are shown below. We obtain them by reversing configurations systematically (exchanging two groups attached) at each of the four chiral centers. See all 16 below. We can begin by assigning wedges to each bond to a halogen, as shown in **(a)** below. In **(b)–(e)**, one wedge bond was converted to dash. In **(f)–(k)**, two wedge bonds were converted to dashes. In **(l)–(o)**, three wedge bonds were converted to dashes. And in **(p)**, all four wedge bonds were converted to dashes. None of these molecules are redundant. Had there been a plane of symmetry, we would have expected the actual number of configurational isomers to be lower than the maximum number of 16.

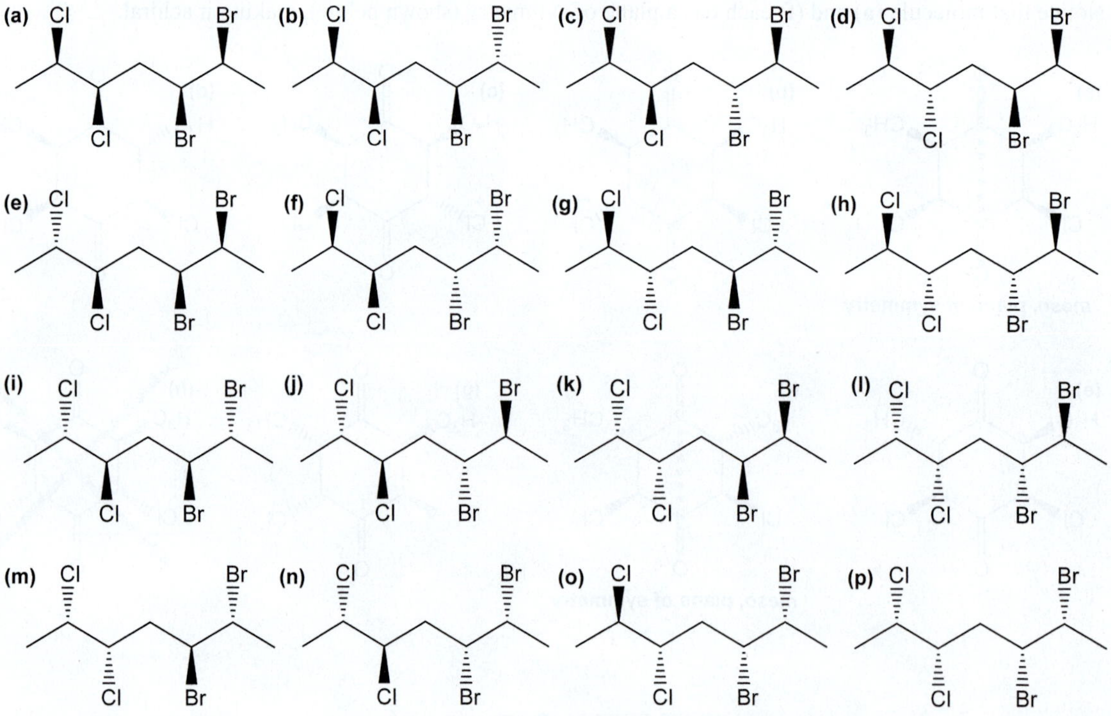

Problems 5.49/5.50

Think

What does it mean for a molecule to be chiral? Can you determine that using molecular models? Are chiral centers present? Do any of the molecules have a plane of symmetry? What is the minimum number of chiral centers a meso compound must have? Can a meso compound be chiral?

Solve

Molecules **(a)**, **(b)**, and **(d)** are chiral. To be chiral, a molecule and its mirror image must be nonsuperimposable. One way to test this is to build each molecule and its mirror image using a molecular modeling kit and then see if you can get every atom in one molecule to line up perfectly with every atom of the other. Alternatively, you can consider numbers of chiral centers and planes of symmetry. Molecule **(a)** is chiral because it has exactly one chiral center. Molecule **(b)** does not have a plane of symmetry. Molecule **(c)** is achiral owing to the plane of symmetry shown. Molecule **(d)** does not have a plane of symmetry. Chiral centers are indicated below using an asterisk. Only **(c)** is meso, as it has two chiral centers and a plane of symmetry, making it achiral.

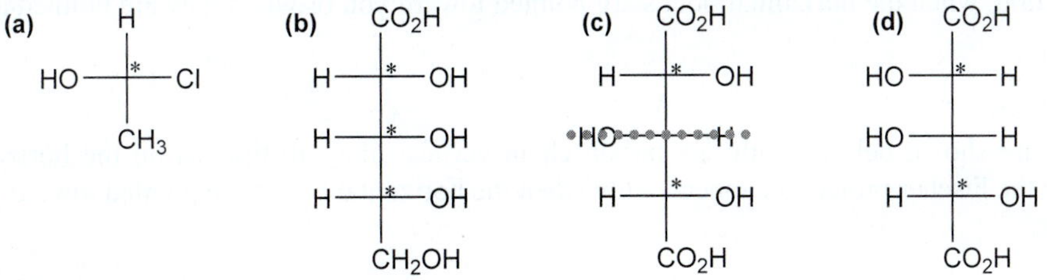

(a) H; HO–C*–Cl; CH_3

(b) CO_2H; H–C*–OH; H–C*–OH; H–C*–OH; CH_2OH

(c) CO_2H; H–C*–OH; HO–C–H; H–C*–OH; CO_2H

(d) CO_2H; HO–C*–H; HO–C–H; H–C*–OH; CO_2H

Problem 5.51

Think

How many chiral centers are present? Review the steps for converting a Fischer projection to a zigzag structure.

Solve

Zigzag structures are shown below and on the next page. Each Fischer projection is first converted to dash–wedge notation by adding substituents to horizontal bonds as they appear in the Fischer projection, but only when the horizontal bonds are pointing toward you.

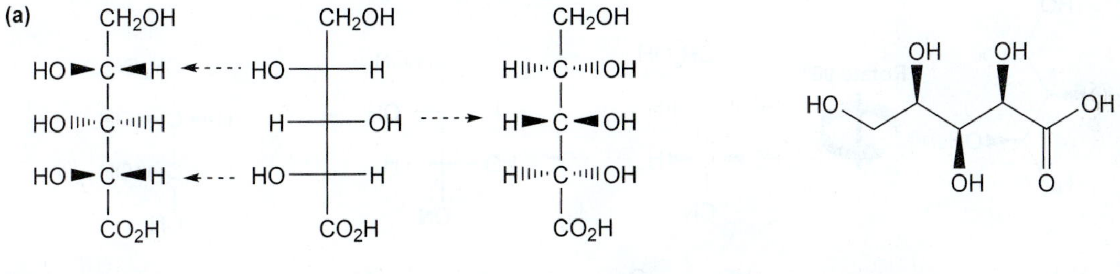

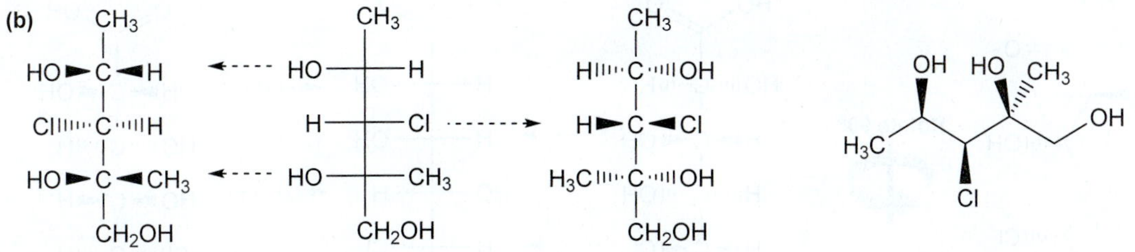

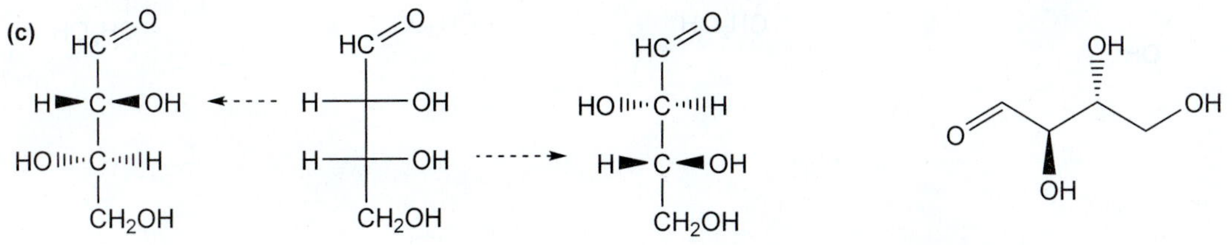

(a) CH_2OH; HO–C–H; H–C–OH; HO–C–H; CO_2H

(b) CH_3; HO–C–H; H–C–Cl; HO–C–CH_3; CH_2OH

(c) HC=O; H–C–OH; H–C–OH; CH_2OH

(d)

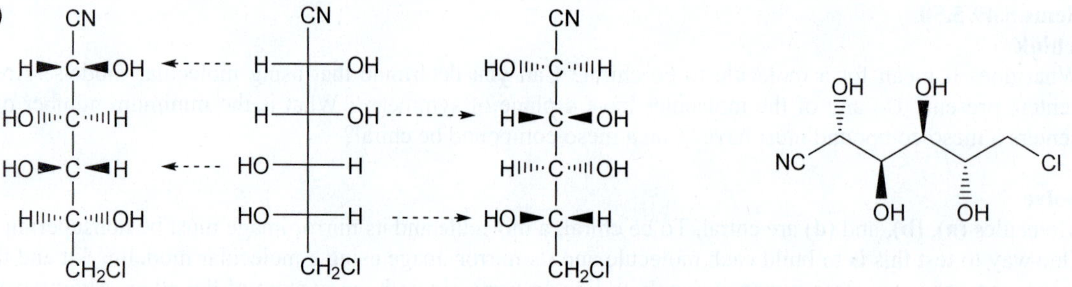

Problem 5.52

Think

How many chiral centers are present? Review the steps for converting a zigzag structure to a Fischer projection. With the carbon chain oriented vertically, when should the substituents on the horizontal bonds in a molecule match the Fischer projection: when the horizontal bonds are pointed toward you or when they are pointed away from you?

Solve

The Fischer projections are shown below. With the carbon chain vertical, the substituents on the horizontal bonds should match how the Fischer projection appears only when the horizontal bonds are pointed toward you.

(a)

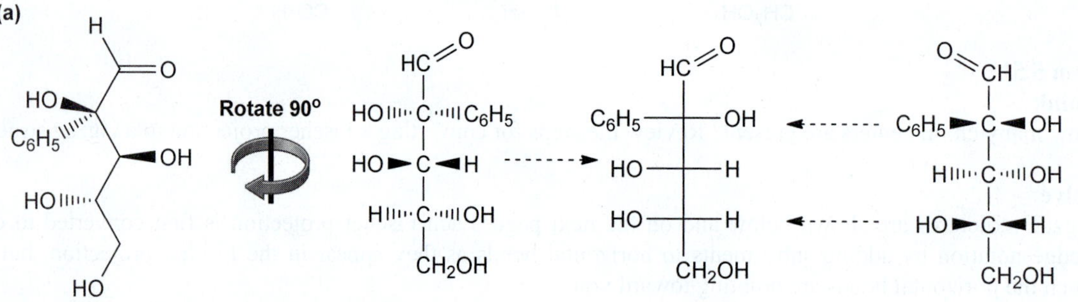

Rotate 90°

(b)

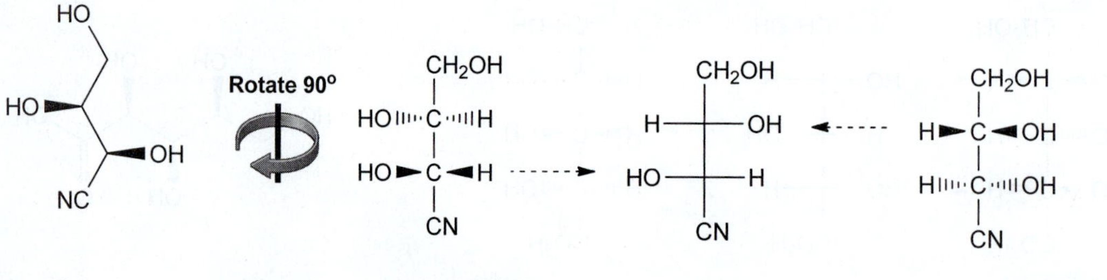

Rotate 90°

(c)

Rotate 90°

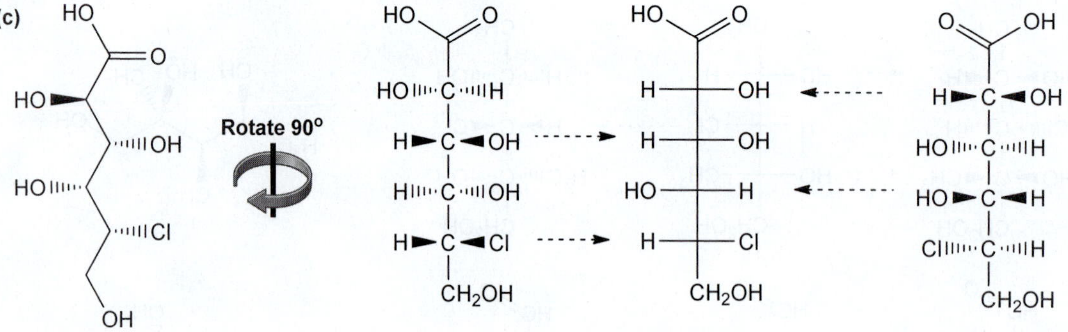

Sections 5.8–5.10 Physical and Chemical Properties of Isomers

Problem 5.53

Think

What type of isomer relationship exists between each pair? What is true about the physical properties of enantiomers, diastereomers, and constitutional isomers?

Solve

Enantiomers must have the same physical properties in an achiral environment. Diastereomers and constitutional isomers must have different physical properties. To determine the relationships between the molecules, apply the flow chart in Figure 5-1.

	Structure	Relationship	Boiling Point
(a)		Diastereomers	Different
(b)	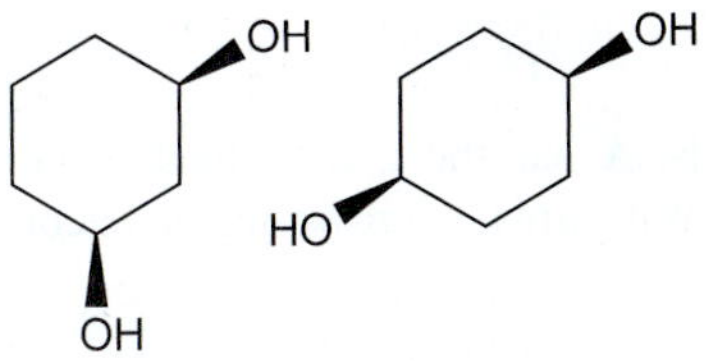 OH, OH; OH, HO	Constitutional isomers	Different
(c)	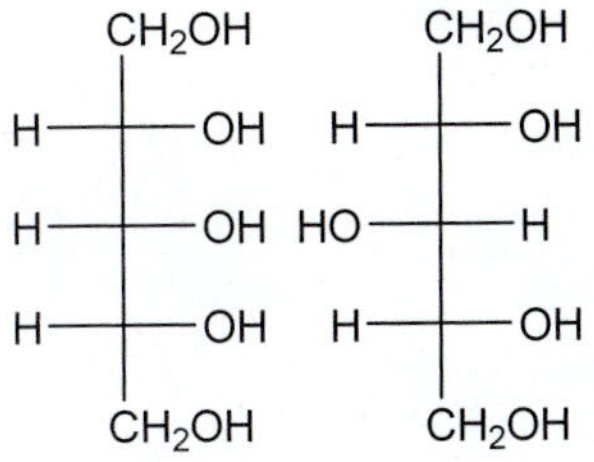 CH_2OH; H—OH; H—OH; H—OH; CH_2OH CH_2OH; H—OH; HO—H; H—OH; CH_2OH	Diastereomers	Different
(d)	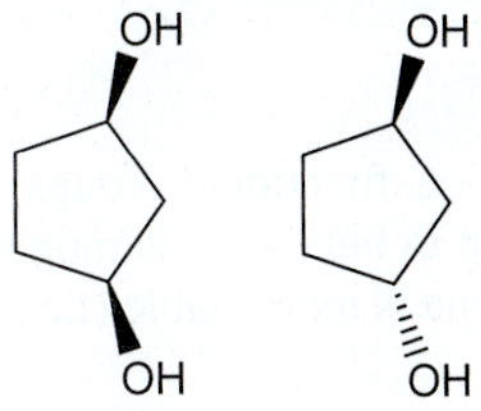 OH, OH; OH, OH	Diastereomers	Different
(e)	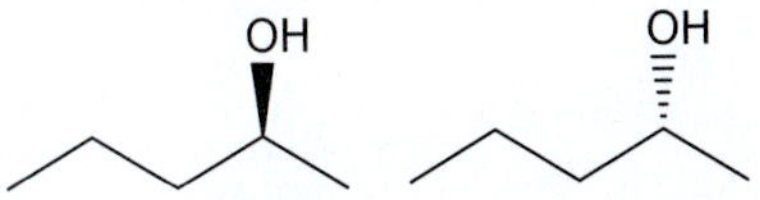 OH; OH	Enantiomers	Same

Problem 5.54

Think

Which of the C=C bonds is most highly alkyl substituted? Most stable? What is the relationship between relative stability and relative heats of combustion?

Solve

All three have the same molecular formula (C_8H_{14}), one double bond, and one six-membered ring. Therefore, the differences in heats of combustion will be due solely to the stability of the C=C bonds. The tetrasubstituted alkene is the most stable, so it has the smallest heat of combustion. The disubstituted alkene is the least stable, so it has the greatest heat of combustion. The order of increasing heat of combustion is **B** < **A** < **C**.

B < A < C

B
Tetrasubstituted double bond

A
Trisubstituted double bond

C
Disubstituted double bond

Problem 5.55

Think

What is the difference in stability of the two alkenes. Is one more highly substituted? What about cis/trans relationships? Does one have more steric strain?

Solve

Both molecules are disubstituted alkenes, and both are cis. **A** has the greater heat of combustion, however, owing to steric crowding between the isopropyl groups. With greater crowding, it becomes less stable (i.e., contains more energy).

H_3C CH_3 CH_3 H_3C

A

Problem 5.56

Think

Is the monosubstituted or disubstituted C≡C more stable? What is the relationship between alkyne stability and heat of combustion?

Solve

Just as the greater alkyl substitution provides more stability to C=C functional groups, a greater degree of alkyl substitution stabilizes the C≡C functional group. The C≡C group in but-1-yne is monosubstituted, whereas that in but-2-yne is disubstituted, as shown below. Therefore, but-2-yne is more stable (i.e., lower in energy) and has the smaller (less negative) heat of combustion.

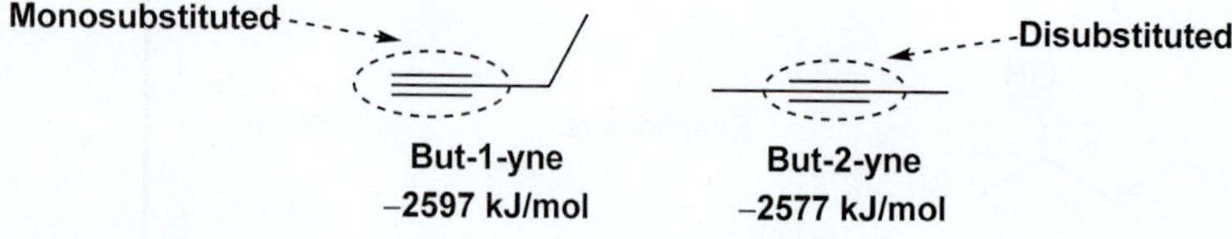

Problem 5.57

Think

Which C atom in **A** is the chiral center? Draw the two enantiomers. What does salt **C** look like with one enantiomer of **A**? With the other enantiomer of **A**? What is the isomeric relationship between these two salts? What is true of physical and chemical properties of enantiomer pairs? Of diastereomer pairs?

Solve

(a) The dash–wedge notation is shown below for the chiral carbons for both acid enantiomers of **A** and for the pure base enantiomer **B**. The salt products are also shown below.

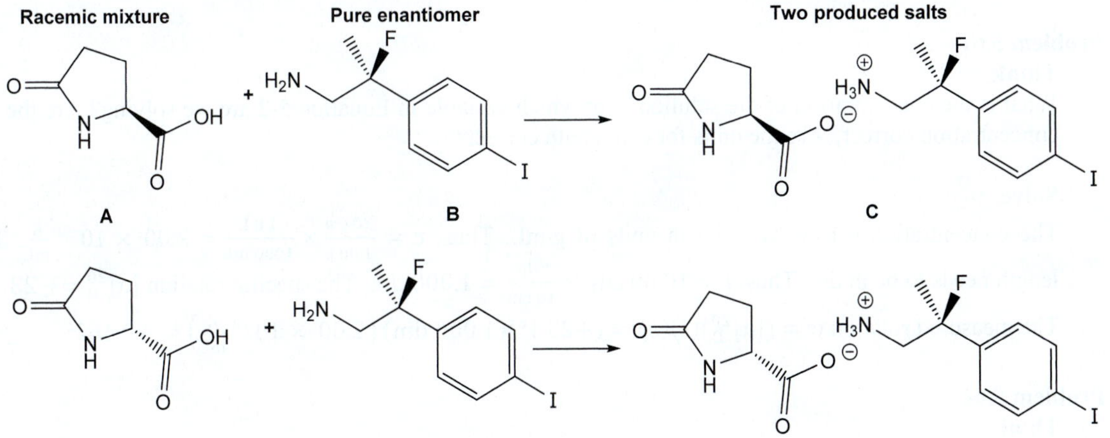

Racemic mixture | Pure enantiomer | Two produced salts

A | B | C

(b) The two salts have exactly the same connectivity. The cations of each salt have the same configuration, but the anions have an opposite configuration and are enantiomers. In other words, in the two products **C**, one side is the mirror image of the other (the part from **A**), and the other part is identical in both compounds (the part from **B**). This means that the two salts **C** cannot be mirror images overall, so they are diastereomers. Diastereomers have different physical and chemical properties, which is why they can be separated readily.

Section 5.11 Optical Activity

Problem 5.58

Think

Review Problem 5.48 to view the configurational isomers. Which molecules are chiral? How can you tell? Do any molecules have a plane of symmetry? How does chirality relate to optical activity?

Solve

Molecules are optically active if they are chiral. For a molecule to be chiral, its mirror image must not be superimposable on the original. To help determine chirality, consider using a model kit to build each molecule and its mirror image and try to align the two. Alternatively, you can apply the plane of symmetry test: Molecules that have a plane of symmetry must be achiral. For the 16 configurational isomers in Problem 5.48 (see solution to Problem 5.48 to review structures), there are no planes of symmetry; all 16 configurational isomers are optically active.

Problem 5.59

Think

What is true about the magnitude and sign of the specific rotation of enantiomers?

Solve

Enantiomers have optical rotations of the same magnitude but opposite sign. Thus, if taxol has a specific rotation of −49°, its enantiomer will have a specific rotation of +49°.

Problem 5.60

Think

Which enantiomer (+) or (−) is in excess? Which way does that enantiomer rotate plane-polarized light? What percentage of the mixture is racemic? What is the remaining percentage of the mixture?

Solve

In a mixture that is 60% (+) enantiomer and 40% (−) enantiomer, the (+) enantiomer is in excess, so optical rotation will be in the (+) direction, or clockwise. The amount of the mixture that is racemic is 80%, or 40% (+) and 40% (−), which leaves a 20% excess of the (+) enantiomer.

Problem 5.61

Think

What is the concentration of the solution? For which variable in Equation 5-2 are we solving? Are the units for concentration correct? Are the units for cell length correct?

Solve

The concentration (c) needs to be in units of g/mL. Thus, $c = \frac{20.0\text{ g}}{1.00\text{ L}} \times \frac{1.0\text{ L}}{1000\text{ mL}} = 2.00 \times 10^{-2}\frac{\text{g}}{\text{mL}}$. The path length needs to be in dm. Thus, $l = 10.00\text{ cm} \times \frac{1\text{ dm}}{10\text{ cm}} = 1.000\text{ dm}$. The specific rotation $[\alpha]_D^{20} = +23.1°$.

The measured rotation $\alpha = ([\alpha]_D^{20})(l)(c) = (+23.1°)(1.000\text{ dm})\left(2.00 \times 10^{-2}\frac{\text{g}}{\text{mL}}\right) = +0.462°$.

Problem 5.62

Think

If the enantiomeric excess is 84%, what percentage is left as racemic? What makes up a racemic mixture? What is the percentage of *R* and *S* enantiomers? What percentage of the solution is optically active?

Solve

If the enantiomeric excess is 84%, 16% is left. This leftover percentage is racemic and is made up of equal amounts of *R* and *S* enantiomers (8% each). Therefore, the *S* enantiomer is 84% + 8% = 92% and the *R* enantiomer is 8%. The enantiomeric excess is the only optically active part of the solution. Thus, 84% is optically active. The specific rotation, $[\alpha]_D$, therefore, is equal to 84% of +25.0°. Thus, $[\alpha]_D = +21°$.

Sections 5.12–5.14 Organic Chemistry of Biomolecules

Problem 5.63

Think

Consult Table 1-7 to review the structures of the amino acids. Which α carbon does not have four different groups?

Solve

Amino acids have NH_2, CO_2H, H, and R bonded to the same carbon atom—the α carbon. In the case of glycine, the R group is simply H, so the α carbon of glycine (shown below) has two H atoms and, therefore, is not a chiral center. Glycine, therefore, is achiral.

$H_2N{-}CH_2{-}C({=}O){-}OH$

Problem 5.64

Think

How many carbon atoms does a D-aldoheptose have? How many does a D-aldohexose have (see Fig. 5-32)? How many additional H–C–OH groups are present in a D-aldoheptose compared to a D-aldohexose? What does each additional H–C–OH group do to the number of possible configurational isomers?

Solve

A D-aldoheptose has seven C atoms, whereas a D-aldohexose has six. Therefore, a D-aldoheptose has one additional H–C–OH group. This introduces a new chiral center, which can have either of two stereochemical configurations. Following the pattern in Figure 5-32, each additional H–C–OH group doubles the possible number of D sugars. D-Aldotriose has one, D-aldotetrose has two, D-aldopentose has four, D-aldohexose has eight, and finally, D-aldoheptose has 16. The Fischer projections of one possible D-aldoheptose (D-mannoheptulose) and its enantiomer are drawn on the next page.

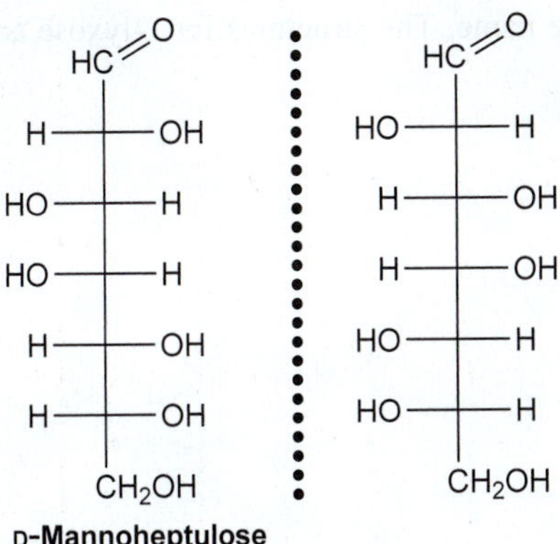

D-Mannoheptulose

Problem 5.65

Think

Refer to Figure 5-32 for the structure of the four D-aldopentoses. How many stereochemical configurations have to be different for two sugars to be considered epimers? How do you know which configuration(s) to change?

Solve

Epimers differ only by the stereochemical configuration at one carbon atom. Reverse the configuration at C2 for C2 epimers or C3 for C3 epimers. D-(−)-Ribose and D-(−)-arabinose are C2 epimers. D-(+)-Xylose and D-(+)-lyxose are C2 epimers. D-(−)-Ribose and D-(+)-xylose are C3 epimers. D-(−)-Arabinose and D-(+)-lyxose are C3 epimers.

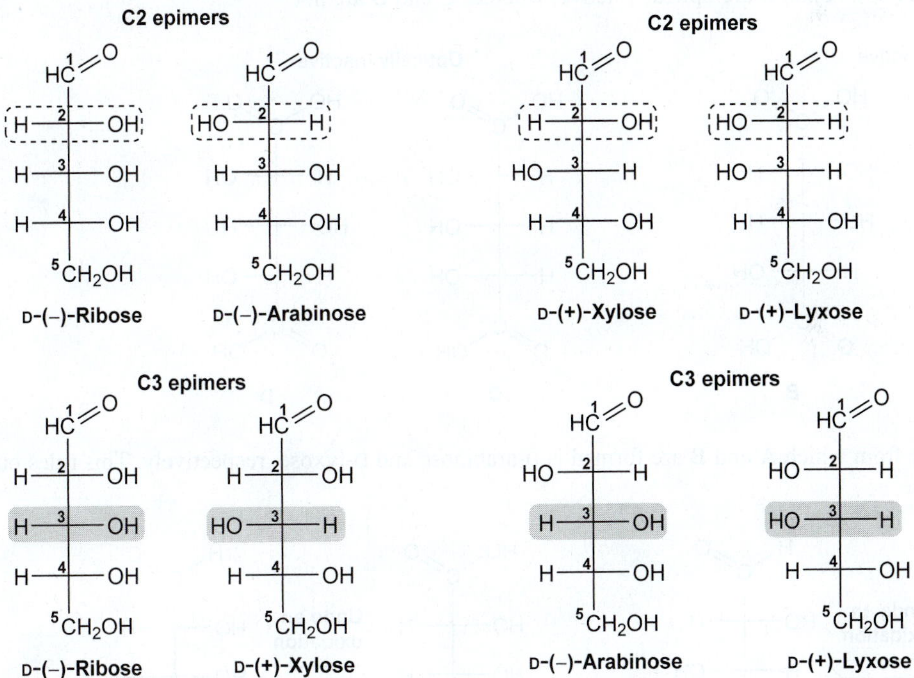

Problem 5.66

Think

What is the structure of D-lyxose? D-Talose? What is the relationship between D and L sugars?

Solve

The enantiomer of a D sugar is designated as an L sugar of the same name. The structures for L-lyxose and L-talose are drawn below.

(a) L-Lyxose

(b) L-Talose

Problem 5.67

Think

What is an aldaric acid? Which five-carbon aldaric acids are optically active? From which D-aldopentoses could these optically active aldaric acids be made? Which D-aldohexoses could have undergone a Wohl degradation to make those D-aldopentoses?

Solve

An aldaric acid [HOOC–(CHOH)$_n$–COOH] is a type of sugar acid in which the terminal HC=O and CH_2OH are replaced by COOH groups. In this example, $n = 3$, and all aldaric acids that can be made from D-aldopentoses are drawn below (**A**–**D**). **A** and **B** are optically active, whereas **C** and **D** are not.

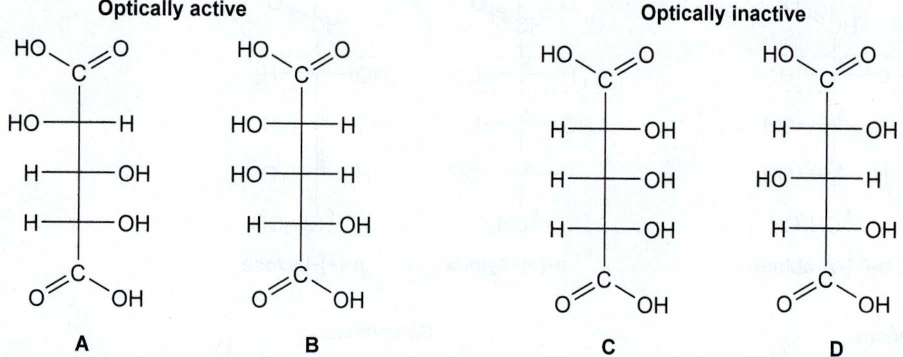

The D-aldopentose from which **A** and **B** are formed is D-arabinose and D-lyxose, respectively. This rules out D-ribose and D-xylose.

A — Undo an oxidation ⟹ L-Arabinose

B — Undo an oxidation ⟹ L-Lyxose

The D-aldohexoses that could have undergone a Wohl degradation to form D-arabinose are D-glucose and D-mannose. The D-aldohexoses that could have undergone a Wohl degradation to form D-lyxose are D-galactose and D-talose. This rules out D-allose, D-altrose, D-gulose, and D-iodose. The Wohl degradation and following oxidation for D-glucose are shown below.

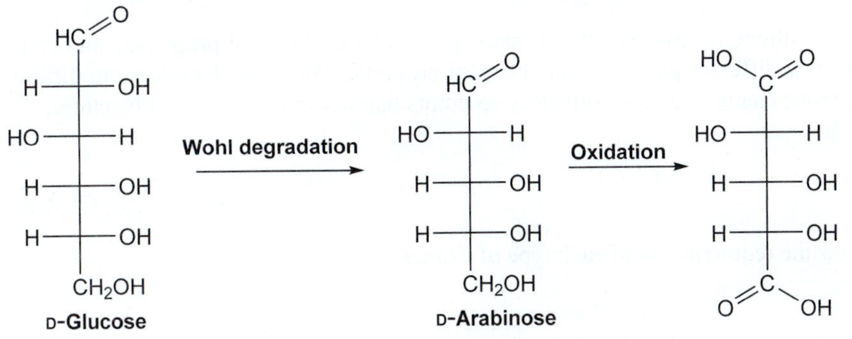

Problem 5.68

Think

What is the product when the C1 of D-glucose is converted from a CH=O to a CH_2OH and when the C6 is converted from CH_2OH to CH=O? What about for D-mannose? Can any rotations in the plane of the paper convert the resulting structures into the original compound?

Solve

The structures are shown below for the conversion of C1 from a CH=O to a CH_2OH and for the conversion of C6 from a CH_2OH to CH=O for both D-glucose and D-mannose. A rotation of 180° in the plane of the paper (to get the CH=O at the top and the CH_2OH at the bottom) shows that the new product for D-mannose is in fact D-mannose. By contrast, D-glucose results in the formation of L-gulose.

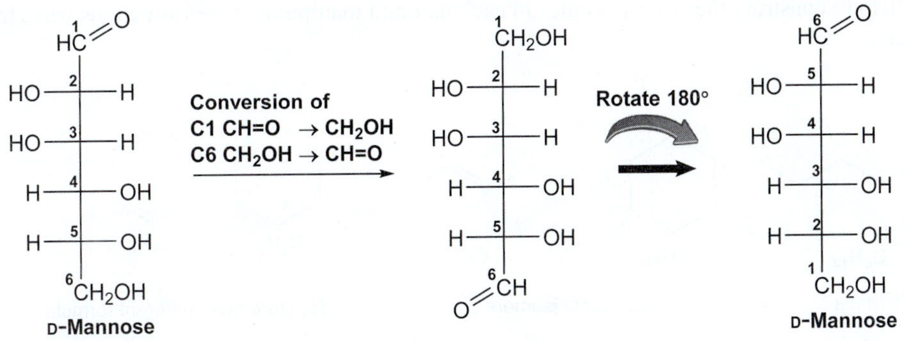

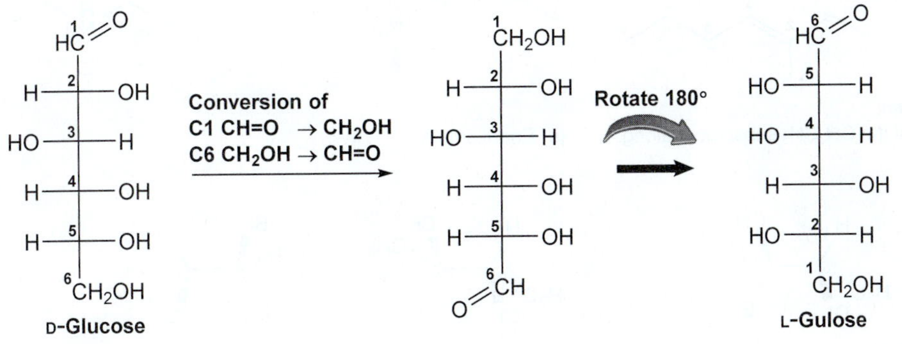

Problem 5.69

Think

How are the properties of an enantiomer affected in an achiral versus in a chiral environment? If the odor is different for the (+)-carvone and (−)-carvone enantiomers, what does that suggest about the environment?

Solve

In an achiral environment, enantiomers have exactly the same physical and chemical properties, and in a chiral environment, enantiomers have different physical and chemical properties. Because the odors are different for the (+)-carvone and (−)-carvone enantiomers, the olfactory receptors that detect carvone must be chiral.

Integrated Problems

Problem 5.70

Think

Consult Figure 5-1 to review the requirements of each type of isomer.

Solve

If **A** and **B** are isomers, it is understood that they have the same molecular formula but are different in some way.

(a) They could be enantiomers, diastereomers, or constitutional isomers, because they have the same index of hydrogen deficiency (IHD). No other information is given.
(b) They are constitutional isomers only because they will have different connectivity.
(c) They are constitutional isomers only because they will have different connectivity.
(d) They are enantiomers, diastereomers, or constitutional isomers.
(e) They could be diastereomers or constitutional isomers, because they cannot be enantiomers (mirror images) if one has a plane of symmetry and the other does not.

Problem 5.71

Think

Consult Figure 5-1 to review the requirements of each type of isomer. To see the relationships between the molecules, it would help to construct the two molecules in each pair and manipulate them in various ways to see if and how they differ.

Solve

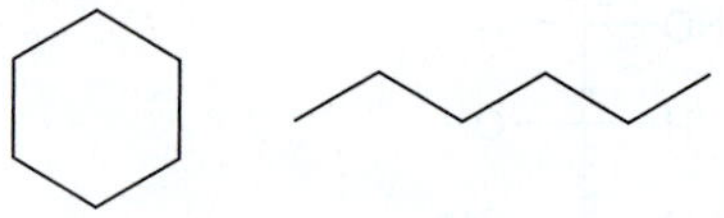

C_6H_{12} C_6H_{14}

(a) Unrelated, different formula

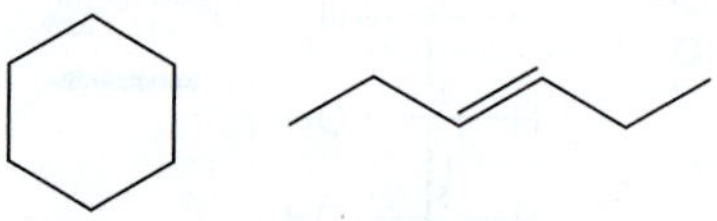

C_6H_{12} C_6H_{12}

(b) Constitutional isomers

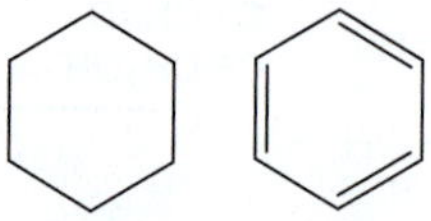

C_6H_{12} C_6H_6

(c) Unrelated, different formula

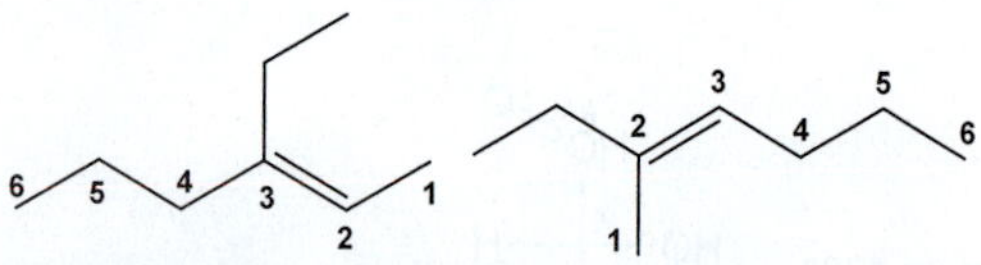

(d) Constitutional isomers
Different locations of the CH_2CH_3 substituent along the carbon chain

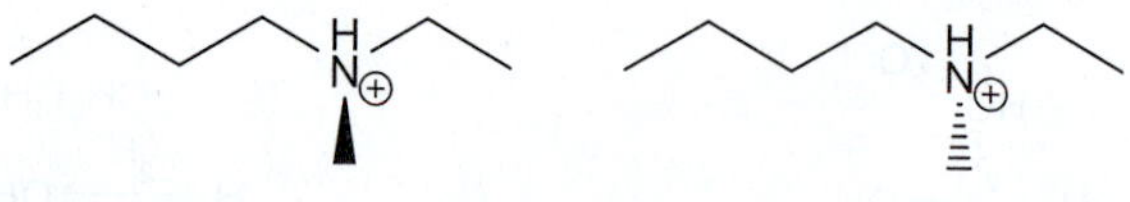

(e) Enantiomers

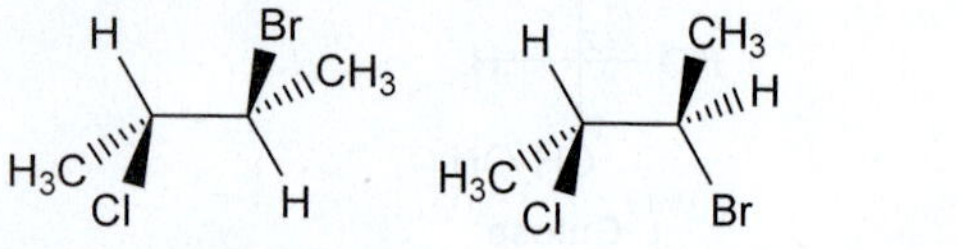

(f) Conformational isomers
Different by the rotation of the C atom in the plane on the right

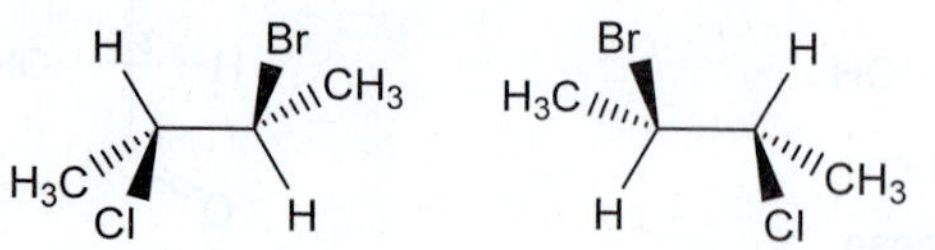

(g) Enantiomers

(h) Diastereomers

Problem 5.72

Think

Consult Figure 5-1 to review the requirements of each type of isomer. To see the relationships between the molecules, it would help to construct the two molecules in each pair and manipulate them in various ways to see if they differ and how they differ.

Solve

(a) Same. Flip the molecule on the right vertically.

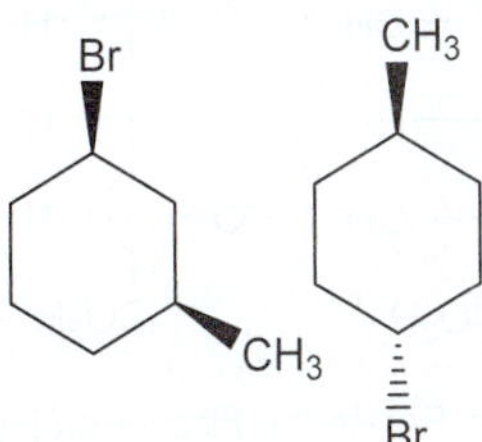

(b) Constitutional isomers. Different locations of the two substituents along the ring

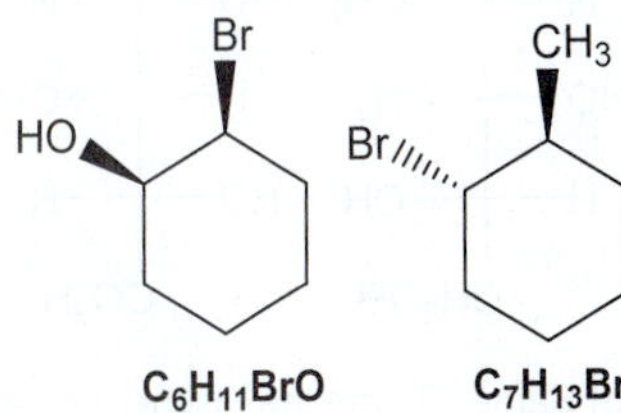

(c) Unrelated, different formulas

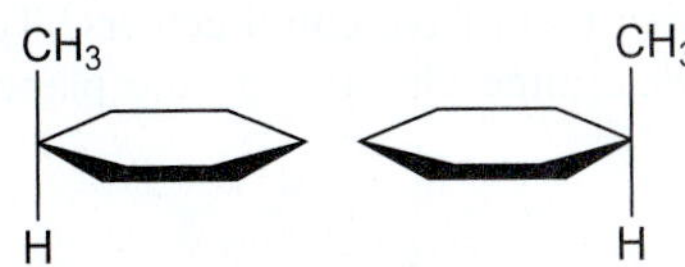

(d) Same. Rotate the molecule on the right horizontally.

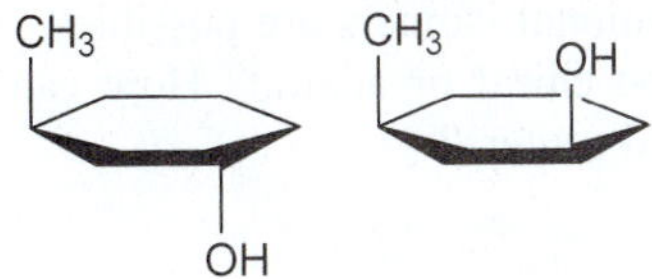

(e) Diastereomers. Different by the configuration of one out of two chiral centers

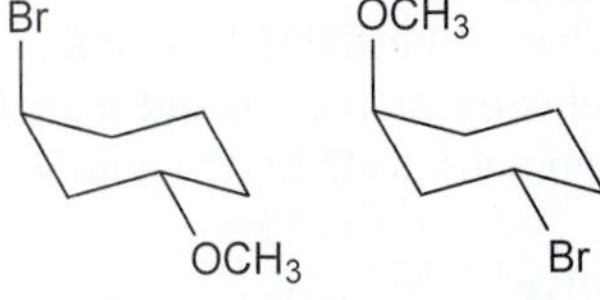

(f) Conformers. Different by a chair flip and then rotation of the molecule on the right 180° about the horizontal axis

Problem 5.73

Think

Consult Figure 5-1 to review the requirements of each type of isomer. To see the relationships between the molecules, it would help to construct the two molecules in each pair and manipulate them in various ways to see if and how they differ.

Solve

(a) Same. Rotate the entire molecule on the right about the C–C bond.

(b) Same. Rotate the Fischer projection on the right by 180°, which leaves the configuration unchanged.

(c) Enantiomers. Different by a 90° rotation of the Fischer projection, which reverses the configuration

(d) Enantiomers. Mirror images of each other, but not superimposable

(e) Enantiomers. Mirror images of each other, but not superimposable

(f) Diastereomers. Different by the inversion of one chiral center out of three

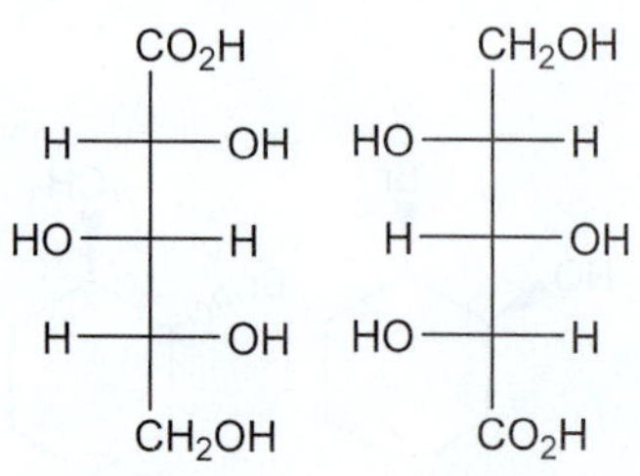

(g) Same. Rotate the Fischer projection on the right by 180°, which leaves the configurations the same.

(h) Same. Rotate the Fischer projection on the right by 180°, which leaves the configurations the same.

Problem 5.74

Think

What is the IHD? How many constitutional isomers are possible? Do any of them have chiral centers? To be optically active, should a molecule be chiral or achiral? How can you determine chirality by the plane-of-symmetry test? By the number of chiral centers?

Solve

The IHD is 0; thus, there are no rings and no multiple bonds. All isomers are shown below. To be optically active, a molecule must be chiral. The two optically active isomers are shown with dash–wedge notation. Notice that those are the only isomers that have a chiral center.

Problem 5.75

Think

Consult Figure 5-1 to review the requirements of each type of isomer. To help see the relationships between the molecules, it would help to construct the two molecules in each pair and manipulate them in various ways to see if and how they differ.

Solve

(a) Constitutional isomers have different connectivity. Replacing any two H atoms on *different* C atoms will yield constitutional isomers. Two examples are shown below.

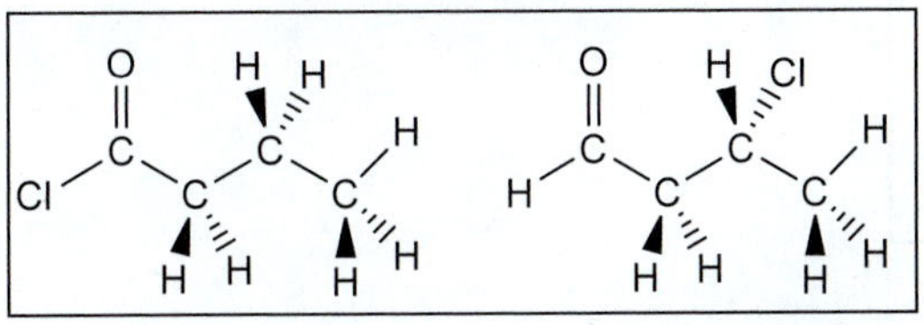

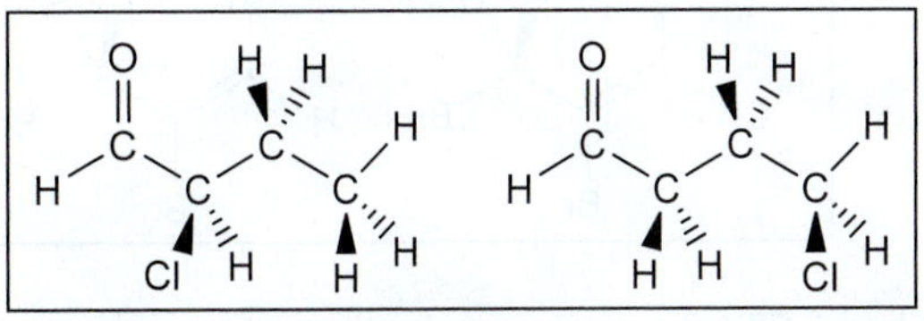

(b) Replacing H atoms on the same C atom of a CH_2 group will yield enantiomers. Examples are shown below.

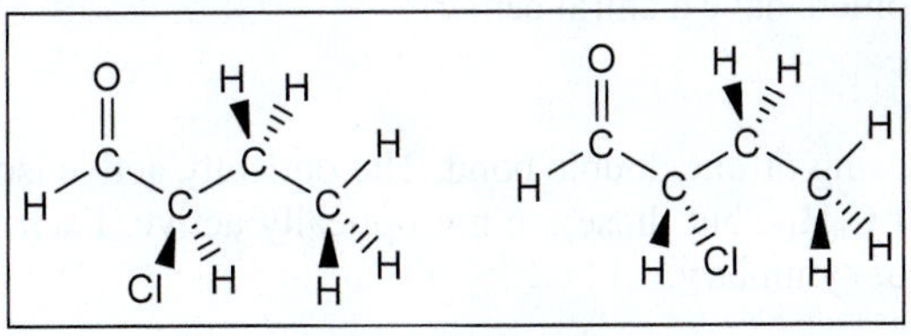

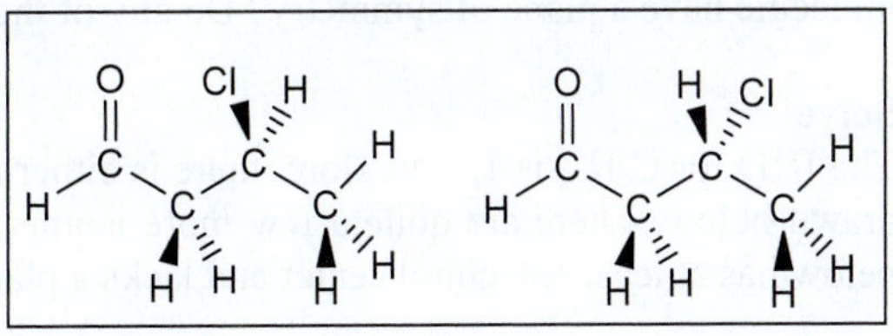

(c) Replacing H atoms on the CH_3 group will yield conformational isomers because the CH_3 carbon does not become a chiral center after the replacement. These two structures are the same.

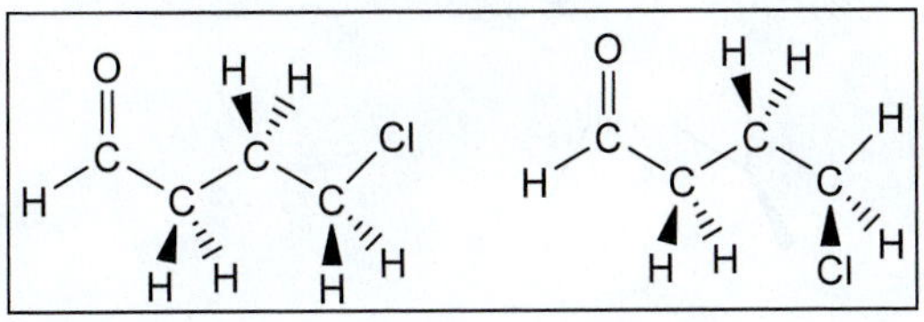

Problem 5.76

Think

Consult Figure 5-1 to review the requirements of each type of isomer. To see the relationships between the molecules, it would help to construct the two molecules in each pair and manipulate them in various ways to see if and how they differ.

Solve

(a) Replacing any pair of H atoms that are not on the same C atom will yield constitutional isomers. Examples are shown below.

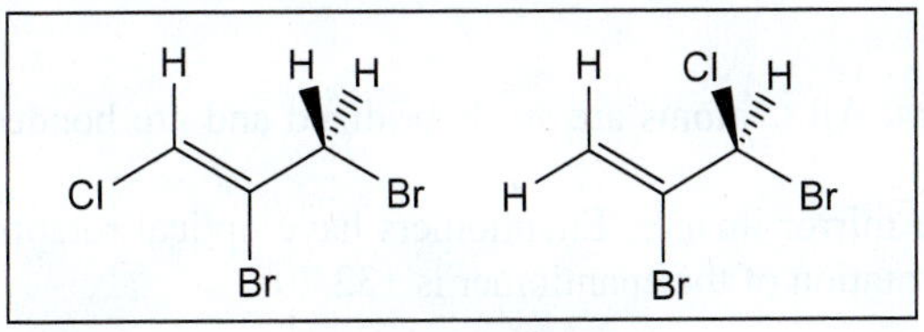

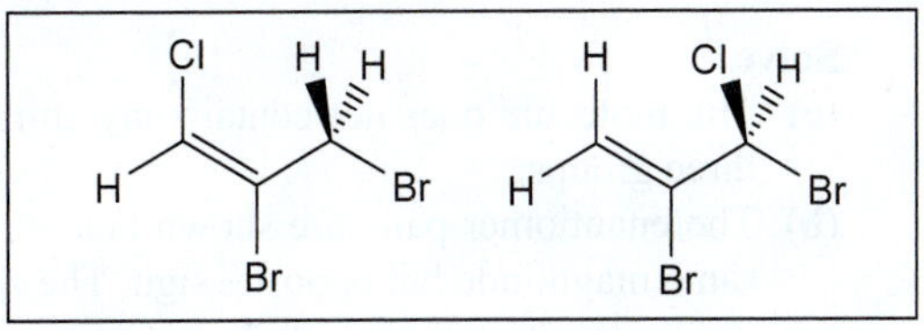

(b) Replacing the pair of H atoms only on the tetrahedral CH_2 group will yield enantiomers, as shown below.

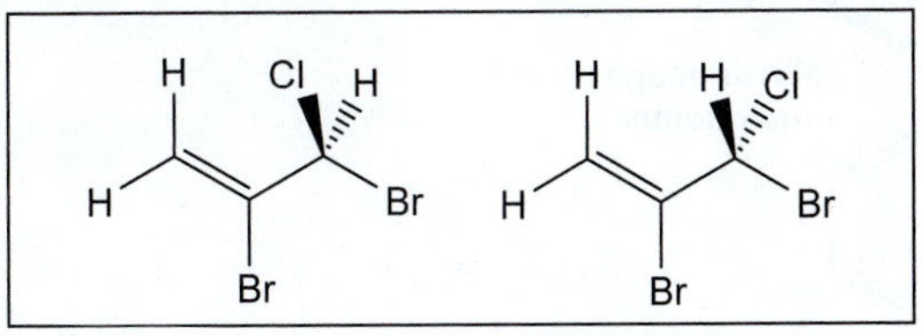

(c) Replacing the H atoms attached to the same C atom of the C=C bond will produce diastereomers. More precisely, this will produce *Z*/*E* isomers.

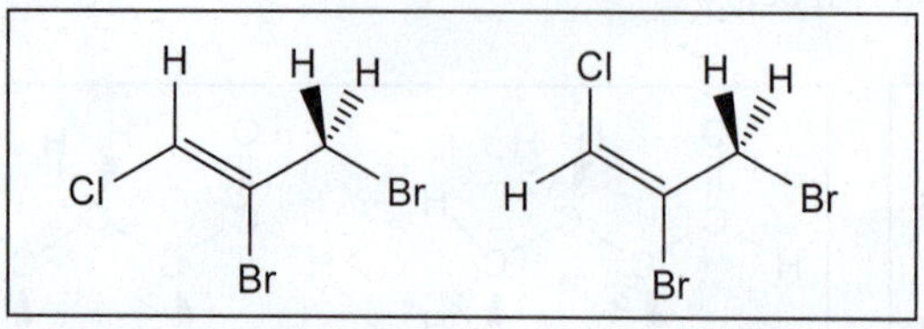

Problem 5.77

Think

What is the IHD for C_6H_{12}? For a compound to be optically active, must it be chiral or achiral? Should the molecule have a plane of symmetry? Do any of the isomers have a chiral center?

Solve

The IHD for C_6H_{12} is 1. Therefore, there is either one ring or one double bond. The optically active isomers are drawn below. There are quite a few more isomers of C_6H_{12}, but those are not optically active. Each molecule below has at least one chiral center and lacks a plane of symmetry.

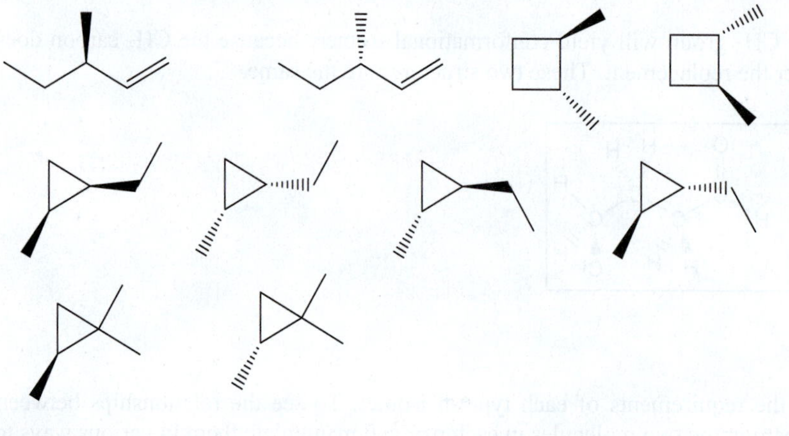

Problem 5.78

Think

Are any of the atoms bonded to four different groups? Remember, when identifying groups on a potential chiral center, to look at the entire rest of the molecule that is attached, not only at the immediate atoms that are attached. What is the hybridization of each C atom? If 1,1′-bi-2-naphthol is chiral, what must be true about its mirror image? What is true about the magnitude and sign of the specific rotation of enantiomers?

Solve

(a) The molecule does not contain any chiral centers. All C atoms are sp^2 hybridized and are bonded to only three groups.

(b) The enantiomer pairs are shown below; they are mirror images. Enantiomers have optical rotations of the same magnitude but opposite sign. The specific rotation of the enantiomer is +32.70°.

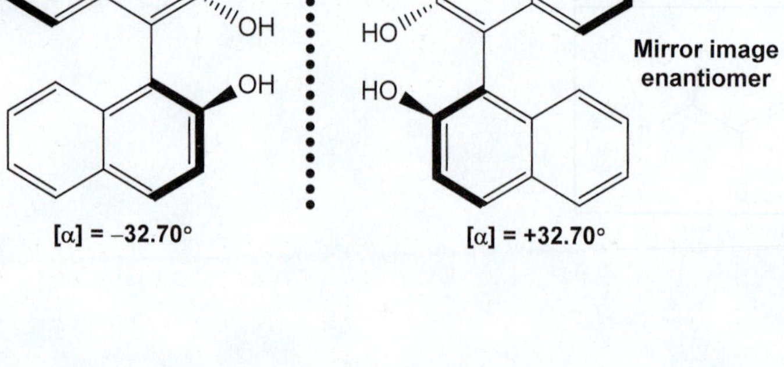

Mirror image enantiomer

[α] = –32.70°

[α] = +32.70°

(c) The two molecules are also related by the rotation about the C–C single bond that connects each pair of fused rings, as shown below. Thus, they are conformational isomers, too. They do not interconvert readily, however, because during that rotation, excessive steric hindrance results from the H atoms and the OH groups crashing into each other, as indicated below.

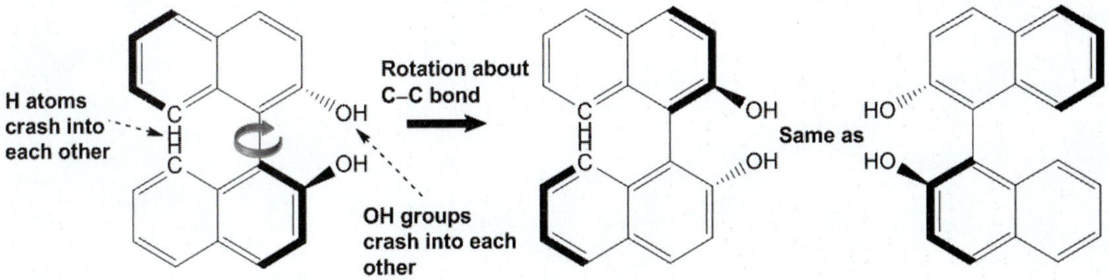

H atoms crash into each other

Rotation about C–C bond

OH groups crash into each other

Same as

Problem 5.79

Think

It might help to build a model of this compound and its mirror image. For a plane of symmetry to exist in a molecule, what must be true of the molecule on either side of that plane? If a plane of symmetry were to bisect the ring, could the C–Cl bonds reflect into each other if one were wedge and the other were dash? What about the C–Br bonds?

Solve

(a) Attempts at drawing a plane of symmetry are shown below. *None* of the attempts below shows a plane of symmetry. In situations where the dash–wedge notation reflects across the plane appropriately, the atoms do not, and vice versa.

(b) Shown below is the molecule's mirror image. If a 180° rotation is performed as shown, the molecule is seen to be exactly the same as the original. Note that the primes (′) are used to help in visualizing the rotation.

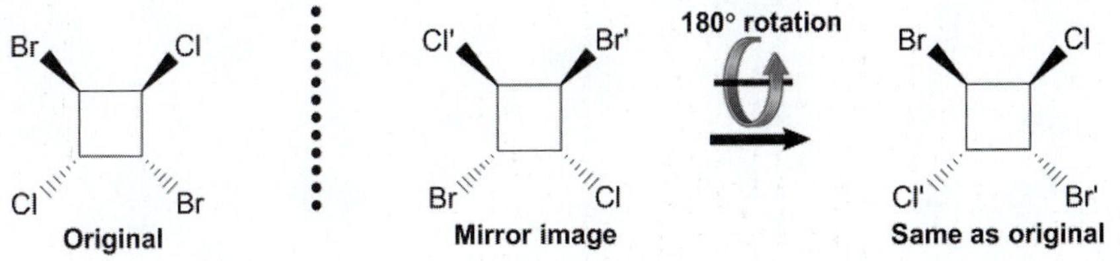

Original

Mirror image

180° rotation

Same as original

(c) The point of symmetry is in the center of the molecule, as shown below. To reflect an atom through the point of symmetry, draw a line from the atom to the point. The reflection of the atom is located along the same line on the other side of the point. Both atoms are found the same distance from the point. The Cl atoms, for example, are reflections through the point of symmetry, and so, too, are the Br atoms. Notice that a reflection through a point of symmetry reverses the dash–wedge notation. The Haworth projection (right) shows this for the Cl atoms.

Point of symmetry

Point of symmetry

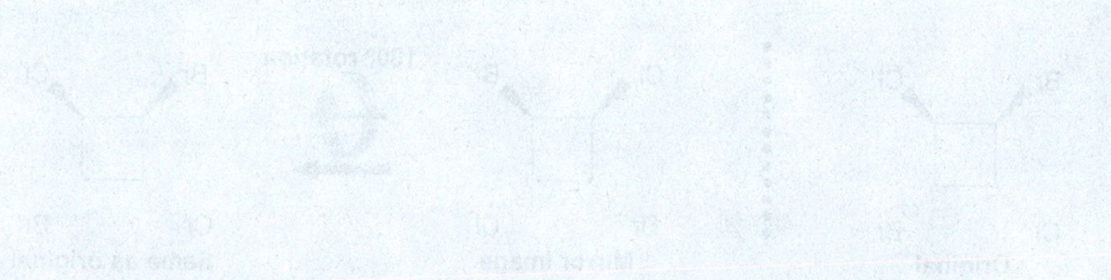

INTERCHAPTER C | Stereochemistry in Nomenclature: *R* and *S* Configurations about Asymmetric Carbons and *Z* and *E* Configurations about Double Bonds

In Chapter Problems

Problem C.1

Think

What are the atomic numbers for each atom in the pair? Which is greater? How does atomic number affect the relative priority? If the atomic number is the same, which atom has the larger mass?

Solve

(a) The atomic number for F is 9 and for O is 8; therefore F has a higher priority.

(b) The atomic number for P is 15 and for F is 9; therefore P has a higher priority.

(c) The atomic number for ^{13}C and ^{12}C is 6. ^{13}C has a mass of 13, and ^{12}C has a mass of 12; therefore ^{13}C has a higher priority.

Problem C.2

Think

Which atoms establish priority? Which atoms have the highest atomic number? Does the highest atomic number get the highest (first) or lowest (fourth) priority?

Solve

The atoms at the point of attachment establish priority, so for the CH_3 group, it is the C atom. The highest atomic number is assigned the highest priority. The atoms are arranged below from *highest* to *lowest* priority.

(a) Br, Cl, F, CH_3 **(b)** I, Br, CH_3, H **(c)** F, O, N, CH_3

Problem C.3

Think

Which bond is not shown on the asymmetric carbon? Is that bond dash or wedge for each molecule? What are the priority assignments for each of the four groups? Is the lowest-priority group pointed toward or away from you? Are the substituents arranged clockwise (*R*) or counterclockwise (*S*)? Review the nomenclature rules from Interchapter A to complete the IUPAC name.

Solve

Complete IUPAC names for molecules **(a)–(c)** are given below. In each compound, a dash C–H bond is not shown. In each molecule, H is the fourth-priority substituent and is pointing away.

(a) For the asymmetric carbon, the atoms at the point of attachment are C, H, N, and O. Because these atoms are all different, their atomic numbers establish the priorities of the substituents as follows: O > N > C > H. The configuration is *S*.

(b) For the asymmetric carbon, the atoms at the point of attachment are Cl, H, N, and O. Because these atoms are all different, their atomic numbers establish the priorities of the substituents as follows: Cl > O > N > H. The configuration is *S*.

(c) For the asymmetric carbon, the atoms at the point of attachment are F, H, O, and C. Because these atoms are all different, their atomic numbers establish the priorities of the substituents as follows: F > O > C > H. The configuration is *R*.

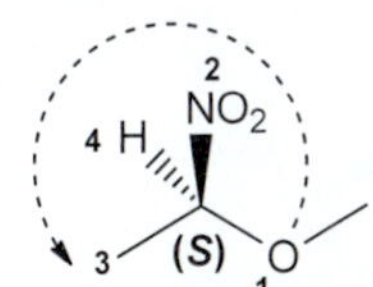

(a) (*S*)-1-Methoxy-1-nitroethane **(b) (*S*)-Chloromethoxynitromethane** **(c) (*R*)-1-Fluoro-1,4-dimethoxybutane**

Problem C.4

Think

Draw the molecule that corresponds to each name without drawing the configuration. Where is the asymmetric C? What are the first through fourth priority group assignments? How are the group priorities determined according to the atoms at the points of attachment? Draw the molecule in such a way that the first through fourth arrangement matches the configuration given.

Solve

In these examples, the substituents are distinguished by the atoms at their points of attachment, with the highest atomic number receiving the highest priority. The order of priority from highest to lowest is labeled: F > O > N > C > H.

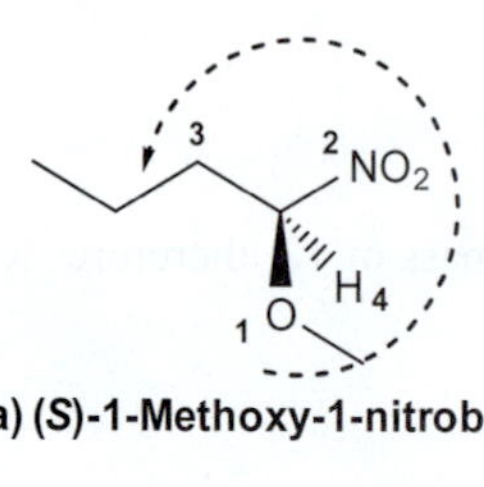

(a) (*S*)-1-Methoxy-1-nitrobutane

(b) (*R*)-1-Methoxy-1-nitrobutane

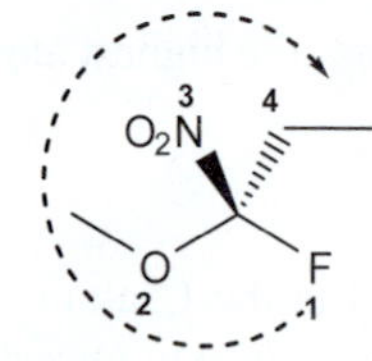

(c) (*R*)-1-Fluoro-1-methoxy-1-nitropropane

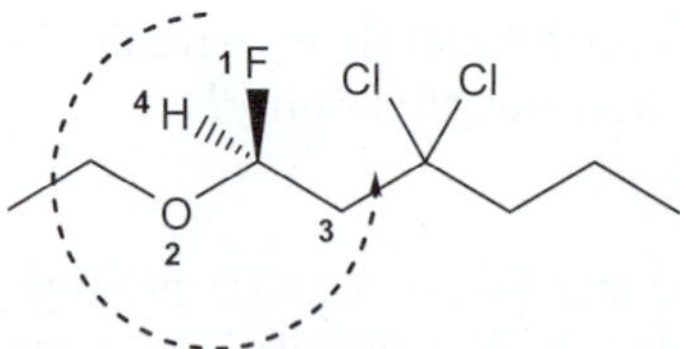

(d) (*S*)-3,3-Dichloro-1-ethoxy-1-fluorohexane

Problem C.5

Think

What are the priority assignments for each of the four groups? Which weighs more: ^{12}C or ^{13}C or ^{14}C, ^{16}O or ^{18}O? What effect does mass have on priority labeling? Is the lowest-priority group pointed toward or away from you? Are the substituents arranged clockwise (*R*) or counterclockwise (*S*)?

Solve

(a) Because of their different atomic numbers, F has priority over ^{13}C and ^{12}C, which have priority over H. ^{13}C has priority over ^{12}C owing to a greater atomic mass. The priorities of the substituents are as follows: F > ^{13}C > ^{12}C > H. The configuration is *R*.

(b) NH_2 has priority over $^{14}CH_3$, because N has the higher atomic number. D and H have the same atomic number, but D has priority over H, because it has a higher atomic mass. Number 4 is neither in the front nor in the back; therefore you need to rotate the molecule to get the number 4 H in the back. The priorities of the substituents are as follows: N > ^{14}C > D > H. The configuration is *R*.

(c) ^{18}OH has priority over $^{16}OCH_2CH_3$. Both of those have priority over C, which has priority over H. The lowest-priority group (H) is in the back, and the first to third groups are arranged clockwise. The priorities of the substituents are as follows: ^{18}O > ^{16}O > C > H. The configuration is *R*.

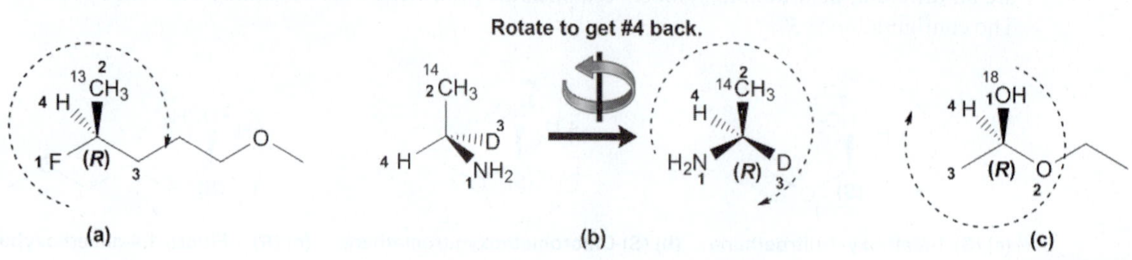

Problem C.6

Think

What is the atom at the point of attachment in each group? If it is the same, how do you break the tie? What is the set of atoms one bond away from the point of attachment? For each substituent, compare the sets of atoms one bond away from the point of attachment. What are their priorities?

Solve

(a) The first atom in each set is a C and therefore is tied. To break the tie, proceed to the second tiebreaker and examine the atoms one bond away from the point of attachment. For $CH_2CH_2CH_3$, the set of atoms is {C, H, H}, and for $CH(CH_3)_2$, the set of atoms is {C, C, H}. C has a higher priority than H, and therefore $CH(CH_3)_2$ has a higher priority; $CH(CH_3)_2 > CH_2CH_2CH_3$.

(b) The first atom in each set is an O and therefore is tied. To break the tie, proceed to the second tiebreaker and examine the atoms one bond away from the point of attachment. For OH, the next atom is H, and for OCH_3, the next atom is C. C has a higher priority than H, and therefore OCH_3 has a higher priority; $OCH_3 > OH$.

Problem C.8

Think

What are the priority assignments for each of the four groups attached to the asymmetric carbon? Are the priorities distinguished by the atoms at the points of attachment? In determining the priorities of those substituents, what distinctions can be made using the first tiebreaker? The second? The third? Is the lowest-priority group pointed toward or away from you? Are the first through third substituents arranged clockwise (*R*) or counterclockwise (*S*)? Review the rules for IUPAC nomenclature from Interchapters A and B to establish the name of the molecule without considering stereochemistry. Once you determine the stereochemical configuration, how do you incorporate that designation into the IUPAC name?

Solve

The numbers listed are the first through fourth priority numbers, not the nomenclature carbon chain numbers. The IUPAC names for molecules **(a)–(d)** are given on the next page.

(a) H has the lowest priority, because the atomic number of H is lower than that of C. The other three substituents are attached by C, and to break the tie, we look at the sets of atoms one bond away from the points of attachment. CH_3 has lower priority than either CH_2CH_3 or $CH_2CH_2CH_3$, because {C, H, H} beats {H, H, H}. $CH_3CH_2CH_2$ and CH_3CH_2 are still tied, however, but looking at the sets of atoms two bonds away, $CH_3CH_2CH_2$ has higher priority than CH_3CH_2, because {C, H, H} beats {H, H, H}. The configuration is *S*.

(b) Cl has first priority, and H has fourth priority, because of the atomic numbers of those atoms. The other two substituents are both attached by C, and to break the tie, we look at the sets of atoms one bond away from the points of attachment. CH_3OCH_2 has higher priority, because {O, H, H} beats {C, H, H}. So the configuration is *S*.

(c) F has first priority, and H has fourth priority, because of the atomic numbers of those atoms. The other two substituents are attached by C, and to break the tie, we look at the sets of atoms one bond away from the points of attachment. NO_2CH_2 has higher priority, because {N, H, H} beats {C, H, H}. The counterclockwise arrangement of the first-, second-, and third-priority groups makes it initially appear that the configuration is *S*, but the fourth-priority group is in front instead of behind, so you need to reverse the arrangement, and the configuration is *R*.

(d) F has first priority, and H has fourth priority, because of their atomic numbers. The other two substituents are attached by C, and to break the tie, we look at the sets of atoms one bond away from the points of attachment. $BrCH_2$ has higher priority, because {Br, H, H} beats {C, H, H}. The counterclockwise arrangement of the first-, second-, and third-priority groups makes it initially appear that the configuration is *S*, but the fourth-priority group is in front instead of behind, so you need to reverse the arrangement, and the configuration is *R*.

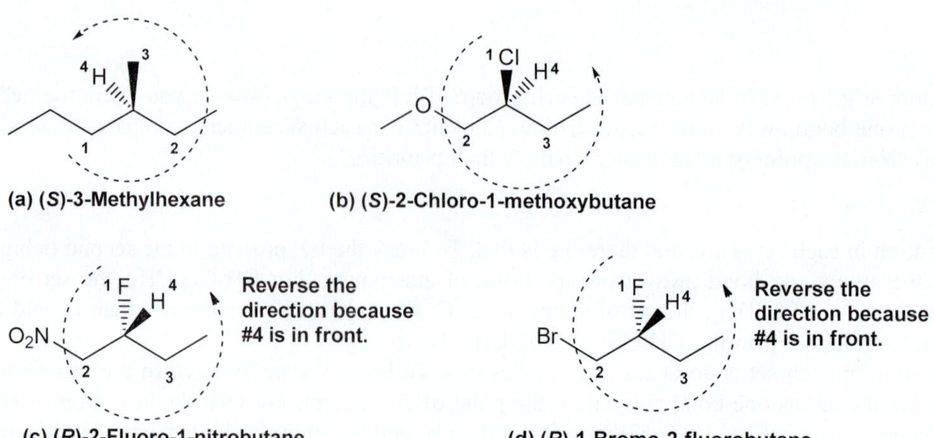
Reverse the direction because #4 is in front.
Reverse the direction because #4 is in front.

(a) (*S*)-3-Methylhexane (b) (*S*)-2-Chloro-1-methoxybutane

(c) (*R*)-2-Fluoro-1-nitrobutane (d) (*R*)-1-Bromo-2-fluorobutane

Problem C.9

Think

Draw the molecule that corresponds to each name at first without drawing the dash–wedge notation. Where is the asymmetric C? What are the first through fourth priority group assignments? Are substituents attached by the same type of atom or different atoms? If there is a tie at the points of attachment, how do you break the tie looking at atoms one bond farther away from the points of attachment? Complete the molecule by adding dash–wedge notation in such a way that the first through fourth arrangement matches the configuration given.

Solve

In these examples, the order of priority from highest to lowest is labeled.

(a) Group 2 beats Group 3 because, one bond away from the points of attachment, {C, H, H} beats {H, H, H}.
(b) Group 2 beats Group 3 because, one bond away from the points of attachment, {F, F, F} beats {C, H, H}.
(c) Group 2 beats Group 3 because, one bond away from the points of attachment, {C, H, H} beats {H, H, H}.
(d) Group 1 beats Group 2 because, two bonds away from the points of attachment, {C, H, H} beats {H, H, H}.
Group 3 beats Group 4 because, one bond away from the points of attachment, {C, H, H} beats {H, H, H}.

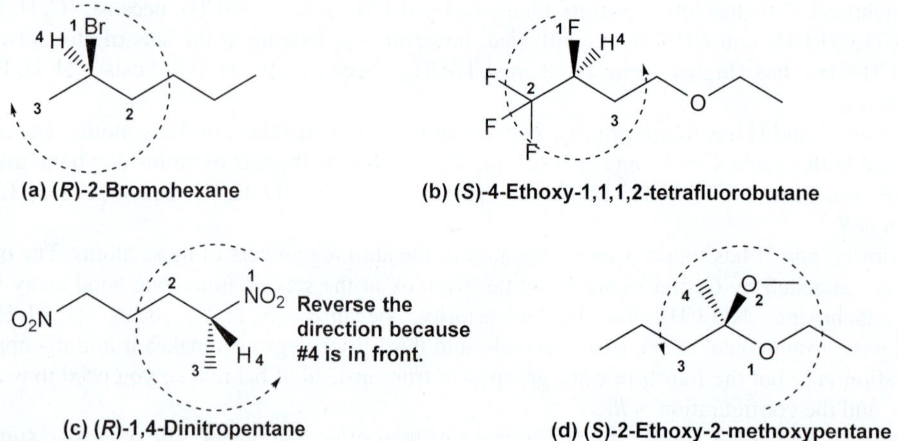
Reverse the direction because #4 is in front.

(a) (*R*)-2-Bromohexane (b) (*S*)-4-Ethoxy-1,1,1,2-tetrafluorobutane

(c) (*R*)-1,4-Dinitropentane (d) (*S*)-2-Ethoxy-2-methoxypentane

Problem C.10

Think

How do you treat the atoms in a double or triple bond? It might be helpful to draw out single bond representations for each multiple bond. Do the groups differ at the points of attachment? By the sets of atoms one bond farther away from the points of attachment?

Solve

Each of the multiple bonds is replaced by a single bond, and the configuration for each asymmetric C is labeled.

(a) *S* configuration. Group 2 beats Group 3 because, one bond away from the points of attachment, {C, C, H} beats {C, H, H}.

(b) *R* configuration. Groups 1, 2, and 3 have that order because, one bond away from the points of attachment, {C, C, C} beats {C, C, H}, which beats {H, H, H}.

(c) *R* configuration. Group 2 beats Group 3 because, one bond away from the points of attachment, {C, C, H} beats {C, H, H}.

(d) *S* configuration. Groups 1, 2, and 3 have that order because, one bond away from the points of attachment, {N, N, N} beats {C, C, C}, which beats {H, H, H}.

(e) *R* configuration. Group 2 beats Group 3 because, one bond away from the points of attachment, {C, C, H} beats {C, H, H}.

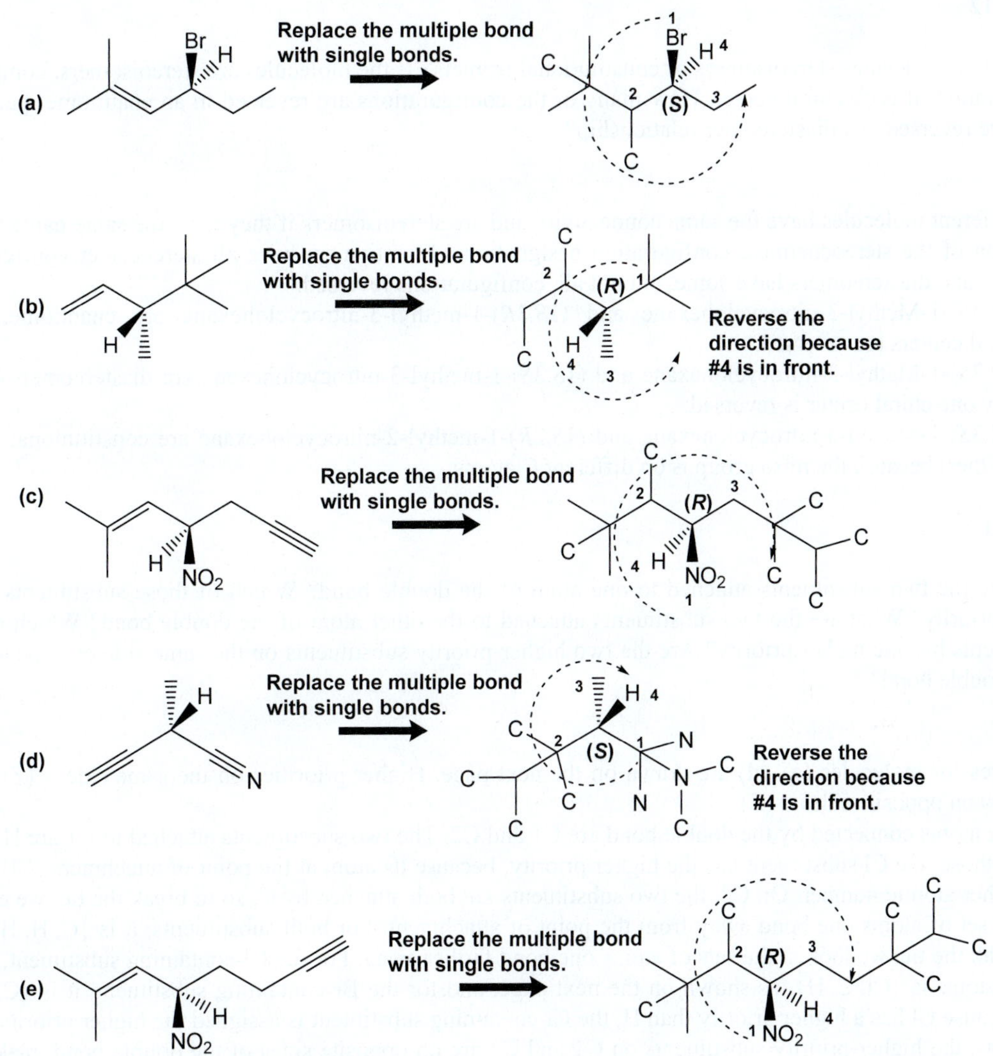

Problem C.11

Think

What are the priority assignments for each of the four groups attached to each asymmetric carbon? Is the lowest-priority group pointed toward you (horizontal bond) or away from you (vertical bond)? Are the first through third substituents arranged clockwise (*R*) or counterclockwise (*S*) when the lowest-priority substituent points away? Review the rules for IUPAC nomenclature from Interchapters A and B.

Solve

If fourth priority is on a horizontal bond, it points toward you instead of away. The configuration, therefore, is the reverse of what it appears to be. The IUPAC names for molecules **(a)–(c)** are given below.

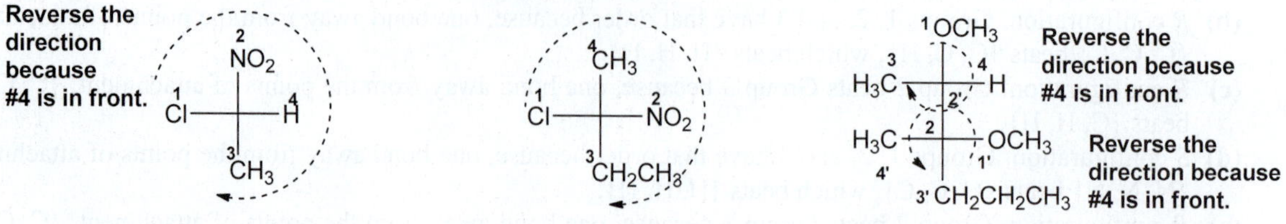

Reverse the direction because #4 is in front.

(a) (*S*)-1-Chloro-1-nitroethane **(b) (*R*)-2-Chloro-2-nitrobutane** **(c) (2*S*,3*R*)-2,3-Dimethoxy-3-methylhexane**

Problem C.12

Think

Are the two molecules stereoisomers or constitutional isomers? If the molecules are stereoisomers, compare the configuration of each chiral center. How many of the configurations are reversed in an enantiomer pair? How many are reversed in a diastereomer relationship?

Solve

Two different molecules have the same connectivity and are stereoisomers if they have the same name with the exception of the stereochemical configuration designations. Enantiomers have all stereocenter configurations reversed, and diastereomers have some, but not all, configurations reversed.

(a) (1*R*,3*S*)-1-Methyl-3-nitrocyclohexane and (1*S*,3*R*)-1-methyl-3-nitrocyclohexane are enantiomers; both chiral centers are reversed.

(b) (1*R*,3*S*)-1-Methyl-3-nitrocyclohexane and (1*S*,3*S*)-1-methyl-3-nitrocyclohexane are diastereomers because only one chiral center is reversed.

(c) (1*R*,3*S*)-1-Methyl-3-nitrocyclohexane and (1*S*,2*R*)-1-methyl-2-nitrocyclohexane are constitutional isomers (neither) because the nitro group is on different C atoms.

Problem C.14

Think

What are the two substituents attached to one atom of the double bond? Which of those substituents has the higher priority? What are the two substituents attached to the other atom of the double bond? Which of those substituents has the higher priority? Are the two higher-priority substituents on the same side or opposite sides of the double bond?

Solve

Structures for molecules **(a)–(d)** are drawn on the next page. Higher priorities on the same side = (*Z*); higher priorities on opposite sides = (*E*).

(a) The atoms connected by the double bond are C1 and C2. The two substituents attached to C1 are H and Cl. Of those, the Cl substituent has the higher priority, because its atom at the point of attachment (Cl) has the higher atomic number. On C2, the two substituents are both attached by C, so to break the tie, we examine the set of atoms one bond away from the point of attachment. For both substituents, it is {C, H, H}, so to break the tie, we look at the sets of atoms one bond farther away. For the Cl-containing substituent, that set of atoms is {Cl, C, H}, as shown on the next page, and for the Br-containing substituent, it is {C, H, H}. Because Cl has a higher priority than H, the Cl-containing substituent is assigned the higher priority on C2. Thus, the higher-priority substituents on C1 and C2 are on opposite sides of the double bond, making the configuration *E*.

(b) The atoms connected by the double bond are C1 and C2. The two substituents attached to C1 are H and Cl. Of those, the Cl substituent has the higher priority, because it has the higher atomic number. On C2, the two substituents are both attached by C, so to break the tie, we examine the set of atoms one bond away. For the ring-containing substituent, that set of atoms is {C, C, H}, as shown on the next page, and for the F-containing substituent, it is {C, H, H}. Because C has a higher priority than H, the ring-containing substituent is assigned the higher priority on C2. Thus, the higher-priority substituents on C1 and C2 are on the same side of the double bond, making the configuration *Z*.

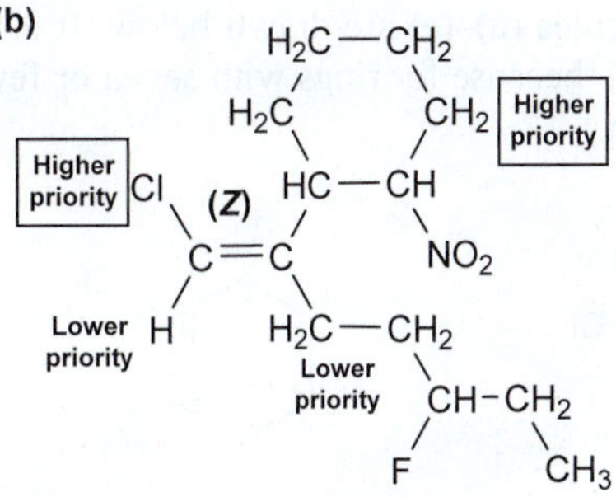

Problem C.15

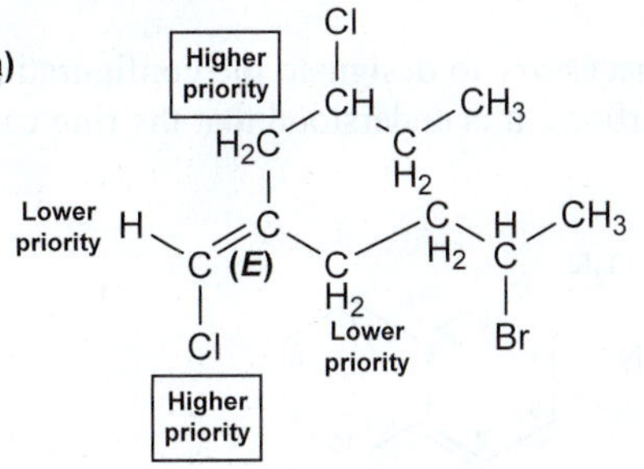

Think

How many C=C bonds require specifying the *E*/*Z* configuration? For each of those C=C bonds, assign higher and lower priority for each group attached at each end. Are the higher-priority groups on the same side or opposite sides of the double bond? Review the rules for IUPAC nomenclature from Interchapters A and B.

Solve

Higher priorities on the same side = (*Z*); higher priorities on opposite sides = (*E*). The IUPAC names for molecules **(a)–(c)** are given below

(a) The numbering is assigned left to right to give the double bonds the lower numbers. The double bond at C2 and C3 is assigned *E*, because the higher-priority groups (CH_3 and CH=CBrR) are on opposite sides. The double bond at C4 and C5 is assigned *Z*, because the higher-priority groups (CH=CHR and Br) are on the same side.

(b) The numbering is assigned left to right to give the double bonds the lower numbers. The double bond at C1 and C2 is neither *E* nor *Z* because C1 has two H atoms. The double bond at C3 and C4 is assigned *Z*, because the higher-priority groups [Cl and $C(CH_3)_3$] are on the same side.

(c) The numbering is assigned left to right to give the triple bond the lower number. The double bond at C4 and C5 is assigned *E*, because the higher-priority groups (NO_2 and phenyl) are on opposite sides. The double bond at C6 and C7 is assigned *E*, because the higher-priority groups [CPh=C(NO_2)R and OCH_3] are on opposite sides.

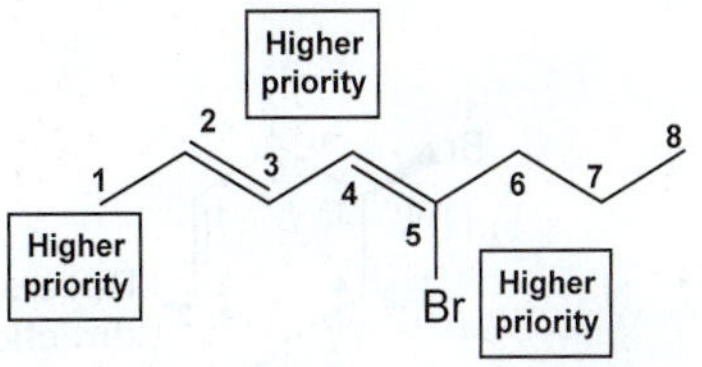

(a) (2*E*,4*Z*)-5-Bromoocta-2,4-diene

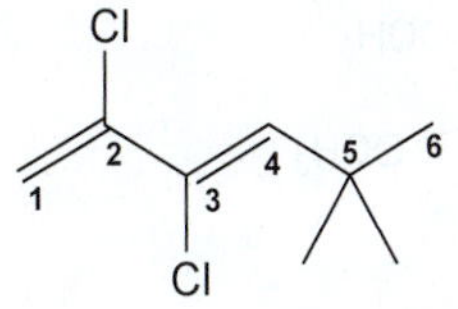

(b) (*Z*)-2,3-Dichloro-5,5-dimethylhexa-1,3-diene

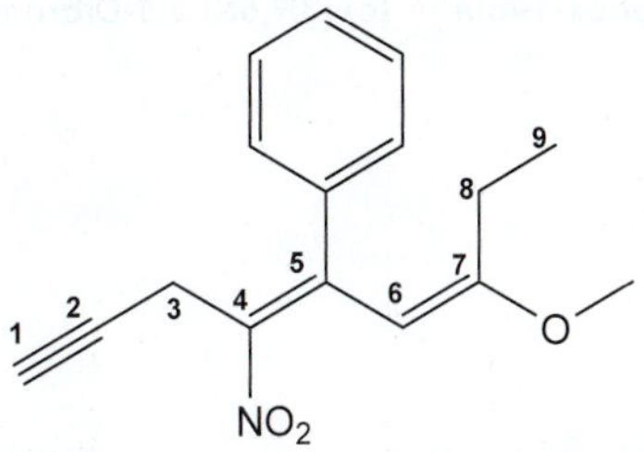

(c) (4*E*,6*E*)-7-Methoxy-4-nitro-5-phenylnona-4,6-dien-1-yne

Problem C.16

Think

Draw the molecule that corresponds to each name, temporarily ignoring stereochemical configurations. Then consider the rules for *R*/*S* and *E*/*Z* configurations. When is it necessary to designate the configuration at the C=C in a ring?

Solve

Structures for molecules **(a)–(e)** are drawn below. It is only necessary to designate the configuration about the C=C in molecule **(c)**, because for rings with seven or fewer carbons, it is understood that the ring carbons are in the cis configuration.

(a) 3,3-Dichlorocyclohexene **(b) 1-Chlorocyclohexene** **(c) (*E*)-4,4-Dinitrocyclodecene**

(d) (*R*)-3-Fluorocycloheptene **(e) Cyclohepta-1,3-diene**

Problem C.17

Think

What are the priority assignments for each of the four groups attached to the different chiral centers? Is the lowest-priority group pointed toward you or away from you? Are the first through third substituents arranged clockwise (*R*) or counterclockwise (*S*)? Which alkene bonds require specifying the *E*/*Z* configuration? For each of them, assign higher and lower priority for each group on both sides of the double bond. Are the higher-priority groups on the same side or opposite sides? Review the rules for IUPAC nomenclature from Interchapters A and B.

Solve

The IUPAC names for molecules **(a)–(c)** are given below The largest ring is numbered, and the first through fourth priorities are labeled using a number with a prime (′) or double prime (″).

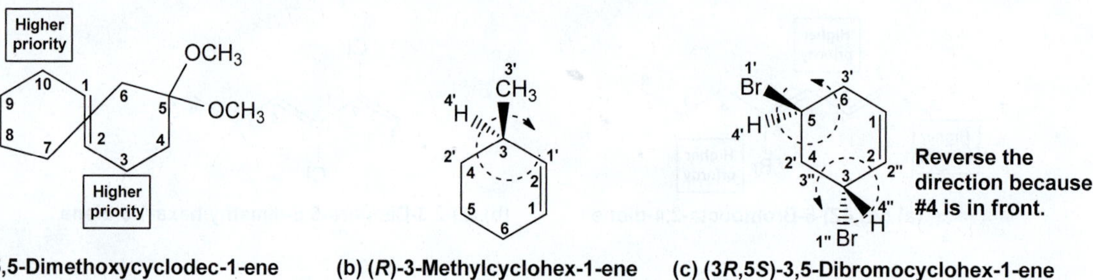

(a) (*E*)-5,5-Dimethoxycyclodec-1-ene **(b) (*R*)-3-Methylcyclohex-1-ene** **(c) (3*R*,5*S*)-3,5-Dibromocyclohex-1-ene**

End of Chapter Problems

Section C.1 *R* and *S* Configurations

Problem C.18

Think

How do you treat the atoms in a double or triple bond? It might be helpful to draw out single bond representations for each multiple bond. Do the groups differ at the points of attachment? By the sets of atoms one bond farther away from the points of attachment?

Solve

Each of the multiple bonds is replaced by a single bond, and the configuration for each chiral center is labeled.

(a) *S* configuration. Group 2 beats Group 3 because, one bond away from the points of attachment, {C, C, C} beats {H, H, H}.

(b) *S* configuration. Group 2 beats Group 3 because, one bond away from the points of attachment, {O, O, C} beats {O, H, H}.

(c) *R* configuration. Group 2 beats Group 3 because, one bond away from the points of attachment, {N, N, C} beats {C, H, H}.

(d) *R* configuration. Group 2 beats Group 3 because, one bond away from the points of attachment, {C, C, C} beats {C, H, H}.

(a) Replace the multiple bond with single bonds.

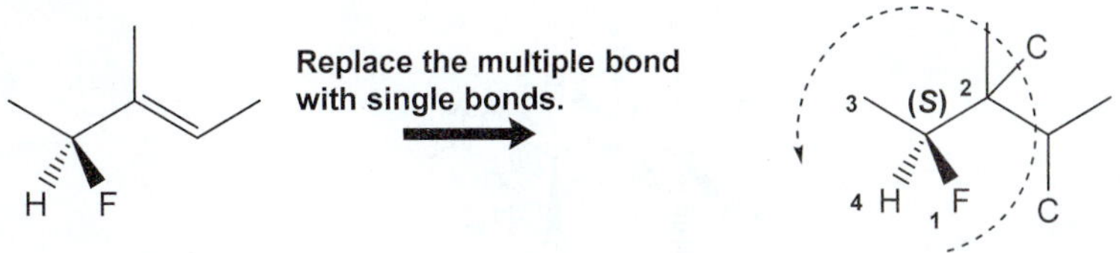

(b) Replace the multiple bond with single bonds.

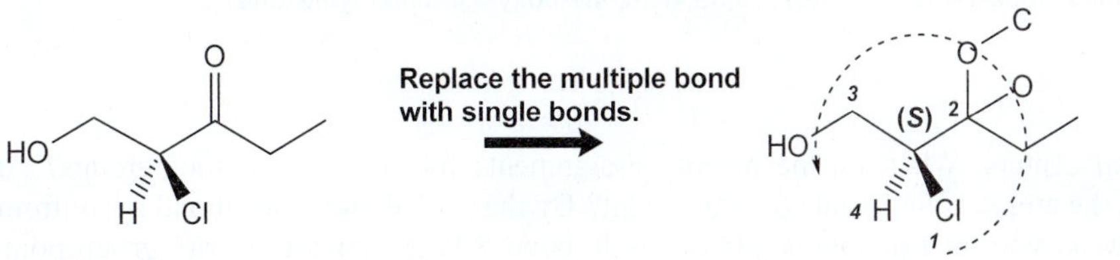

(c) Replace the multiple bond with single bonds.

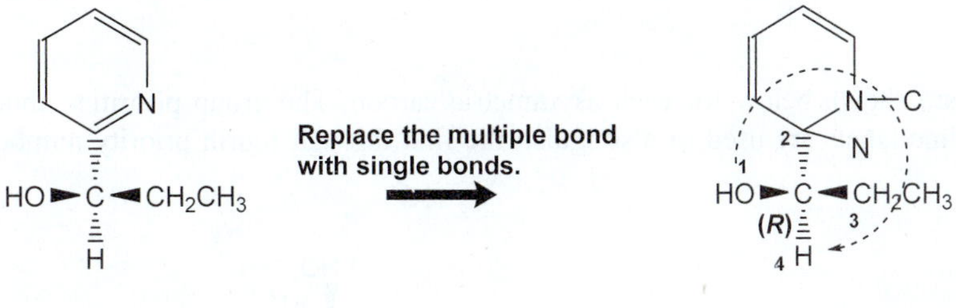

(d) Replace the multiple bond with single bonds.

Reverse the direction because #4 is in front.

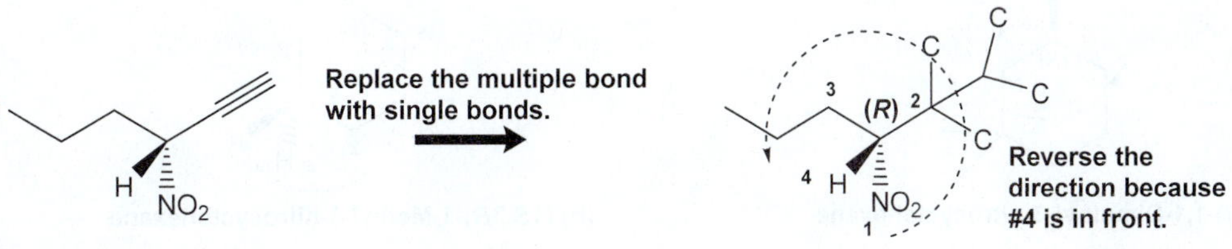

Problem C.19

Think

Identify all chiral centers. What are the priority assignments for the four groups in each chiral center? Are the groups distinguished by the atoms at the points of attachment? By the sets of atoms one bond away from the points of attachment? How do you treat double bonds as single bonds? Is the lowest-priority group pointed toward or away from you? Are the first through third substituents arranged clockwise (*R*) or counterclockwise (*S*)?

Solve

R and *S* configurations are shown below for each asymmetric carbon. The group priorities about each chiral center are numbered. Prime labels are used to distinguish the first through fourth priority numbering for each chiral center. In **(a)**, **(b)**, and **(d)**, the ties at the points of attachment are broken one bond away from the points of attachment. In the chiral center on the left in **(c)**, Group 1 beats Group 2, because the sets of atoms one bond away for the alkene and alkane are both {C, C, H}. The sets of atoms two bonds away for both the alkene and alkyl group are {C, H, H}. There are no atoms another bond away for the alkene, but the next atoms out for the alkyl group are {H, H, H}. Therefore, the alkyl group wins.

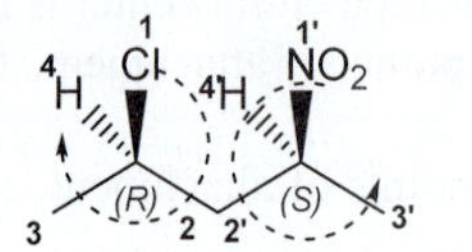

(a) (2*R*,4*S*)-2-Chloro-4-nitropentane

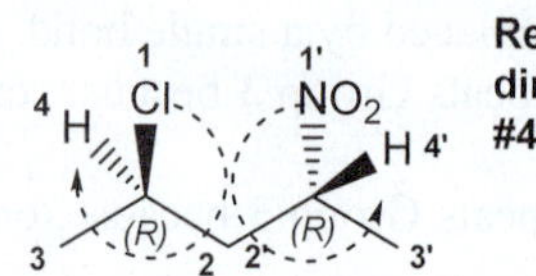

Reverse the direction because #4 is in front.

(b) (2*R*,4*R*)-2-Chloro-4-nitropentane

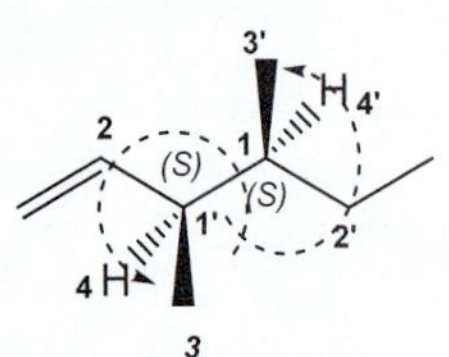

(c) (3S,4*S*)-3,4-Dimethylhex-1-ene

Reverse the direction because #4 is in front.

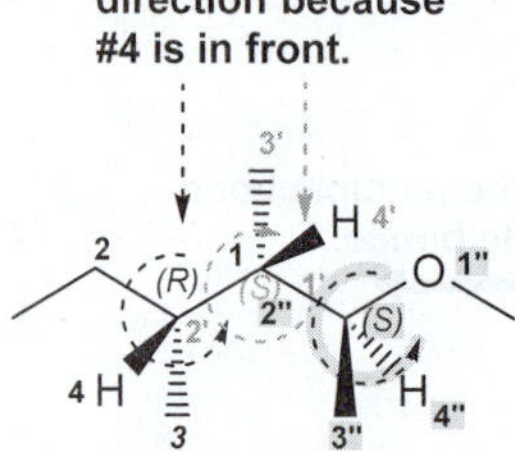

(d) (2S,3*S*,4R)-2-Methoxy-3,4-dimethylhexane

Problem C.20

Think

Identify all chiral centers. What are the priority assignments for each of the four groups? Are the groups distinguished by the atoms at the points of attachment? By the sets of atoms one bond away from the points of attachment? How do you treat double bonds as single bonds? Is the lowest-priority group pointed toward or away from you? Are the first through third substituents arranged clockwise (*R*) or counterclockwise (*S*)?

Solve

R and *S* configurations are shown below for each asymmetric carbon. The group priorities about each chiral center are numbered. Prime labels are used to distinguish the first through fourth priority numbering for each chiral center.

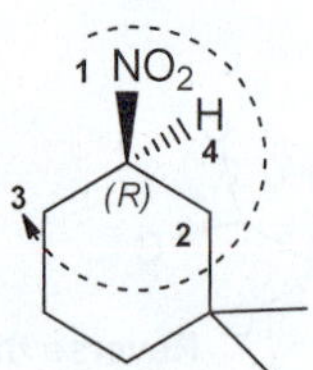

(a) (*R*)-1,1-Dimethyl-3-nitrocyclohexane

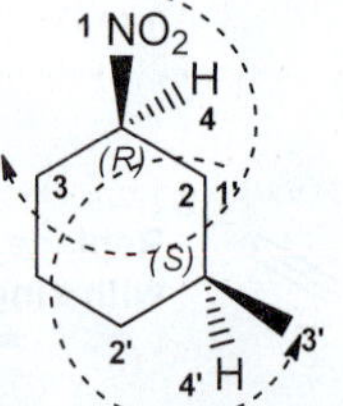

(b) (1*S*,3*R*)-1-Methyl-3-nitrocyclohexane

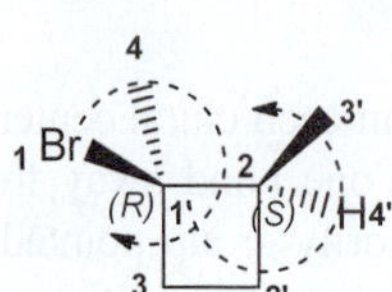

(c) (1*R*,2*S*)-1-Bromo-1,2-dimethylcyclobutane

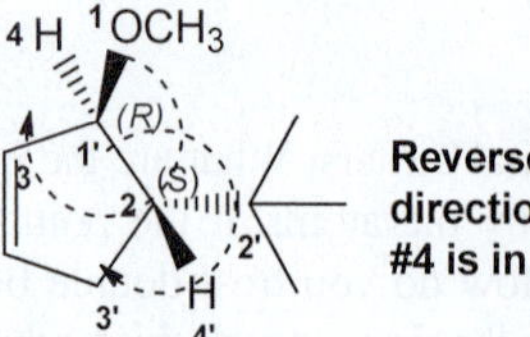

Reverse the direction because #4 is in front.

(d) (3*R*,4*S*)-4-(*tert*-Butyl)-3-methoxycyclopent-1-ene

Problem C.21

Think

How many of the configurations are reversed in an enantiomer pair? How many are reversed in a diastereomer relationship?

Solve

Enantiomers have all chiral center configurations reversed, and diastereomers have some, but not all, configurations reversed. For a molecule with (2*R*, 3*S*, 5*R*) configuration, the enantiomer would have (2*S*, 3*R*, 5*S*). Two examples of diastereomers of the original compound are (2*R*, 3*S*, 5*S*) and (2*R*, 3*R*, 5*S*). Other diastereomers exist as well.

Problem C.22

Think

Where are the *R* and *S* designations located in the given name? Is there another acceptable location to place the *R* and *S* designations?

Solve

The *R* and *S* designations can be placed in the IUPAC name in two ways. Each designation can be placed immediately before the first number used to locate a substituent on the corresponding asymmetric carbon, which is the way it is presented in the given name. The other method is to have all *R* and *S* designations placed together at the front of the name. The name given in this problem, (*R*)-1-chloro-(*R*)-4-methyl-(*S*)-2-nitrocyclohexane, can therefore also be written with all of the *R* and *S* designations placed together, (1*R*,2*R*,4*S*)-1-chloro-4-methyl-2-nitrocyclohexane.

Section C.2 *E* and *Z* Configurations

Problem C.23

Think

Of the two groups attached to one C atom of the C=C double bond, which has higher priority? Are the groups distinguished by the atoms at the points of attachment or by sets of atoms farther away from the points of attachment? Which group has higher priority on the other C atom of the C=C double bond? Are the higher-priority groups on the same side or opposite sides?

Solve

Higher priorities on the same side = (*Z*); higher priorities on opposite sides = (*E*).

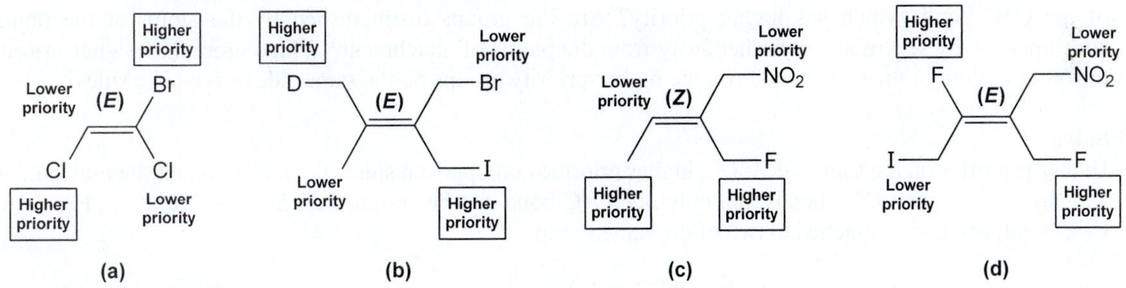

Problem C.24

Think

Of the two groups attached to one C atom of the C=C double bond, which has higher priority? Are the groups distinguished by the atoms at the points of attachment or by sets of atoms farther away from the points of attachment? Which group has higher priority on the other C atom of the C=C double bond? Are the higher-priority groups on the same side or opposite sides?

Solve

Higher priorities on the same side = (*Z*); higher priorities on opposite sides = (*E*).

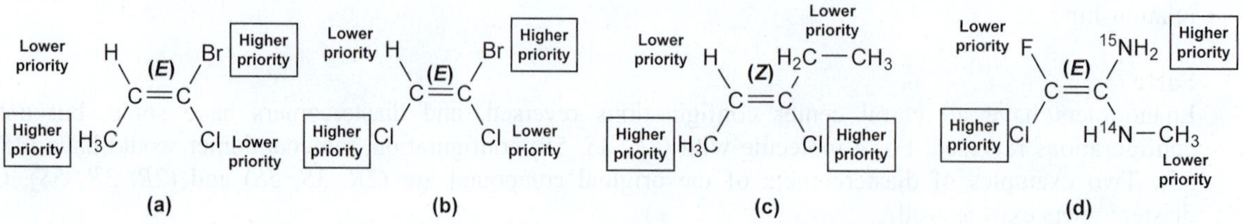

Problem C.25

Think

Of the two groups attached to one C atom of the C=C double bond, which has higher priority? Are the groups distinguished by the atoms at the points of attachment or by sets of atoms farther away from the points of attachment? Which group has higher priority on the other C atom of the C=C double bond? Are the higher-priority groups on the same side or opposite sides?

Solve

Higher priorities on the same side = (*Z*); higher priorities on opposite sides = (*E*). In **(a)**, the triple bond beats the pentyl group. After replacing the triple bond with three single bonds, the set of atoms one bond away from the point of attachment is {C, C, C} for the triple bond and {C, H, H} for the pentyl group.

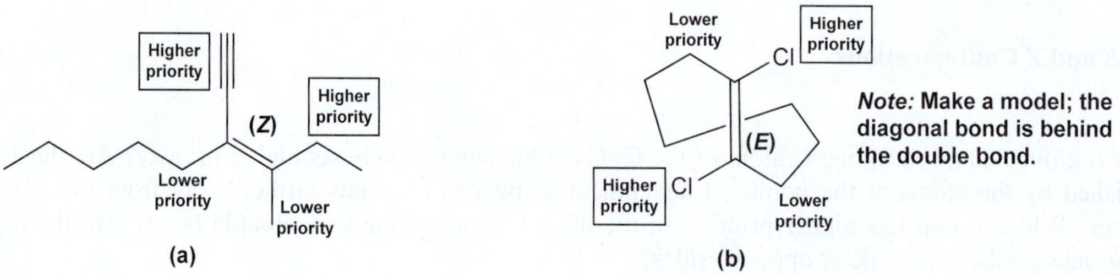

Note: **Make a model; the diagonal bond is behind the double bond.**

Problem C.26

Think

Which C=C bonds can have *E*/*Z* designations? For each one that does, of the two groups attached to one C atom of the C=C bond, which has higher priority? Are the groups distinguished by the atoms at the points of attachment or by sets of atoms farther away from the points of attachment? Which group has higher priority on the other C atom of the C=C bond? Are the higher-priority groups on the same side or opposite sides?

Solve

Higher priorities on the same side = (*Z*); higher priorities on opposite sides = (*E*). Notice that the molecule in **(a)** and **(b)** each has two C=C bonds, but only one C=C bond can be assigned an *E*/*Z* configuration. For the other C=C bond, one end is attached to two of the same group.

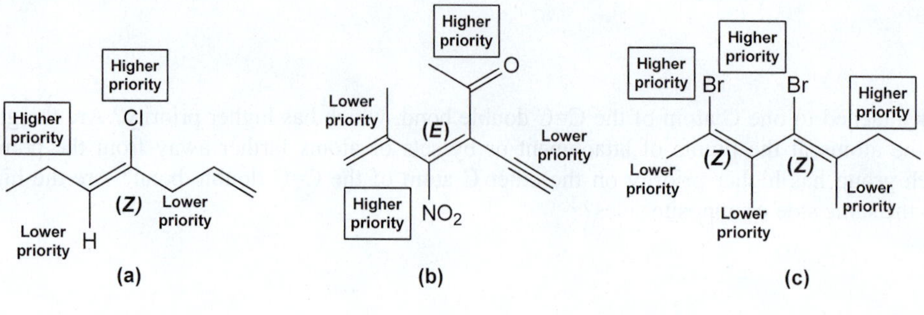

Integrated Problems

Problem C.27

Think

Which atoms are chiral centers? For each one, what are the priority assignments for each of the four groups? Is the lowest-priority group pointed toward you or away from you? Are the substituents arranged clockwise (*R*) or counterclockwise (*S*)? Review the rules for IUPAC nomenclature from Interchapters A and B.

Solve

IUPAC names for molecules **(a)–(e)** are given below. Group priorities about each chiral center are numbered.

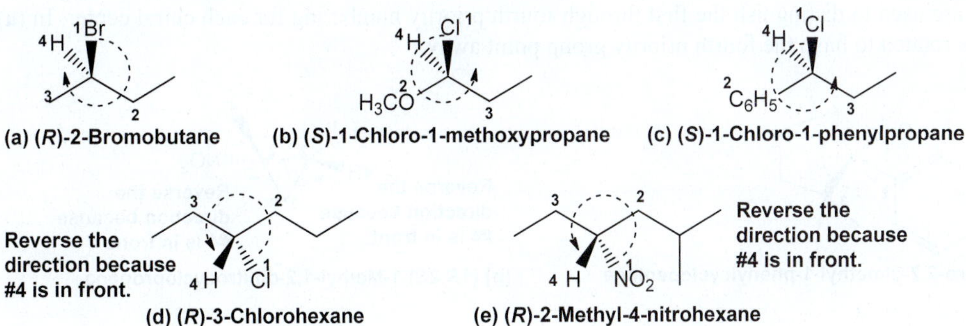

Reverse the direction because #4 is in front.
Reverse the direction because #4 is in front.

(a) (*R*)-2-Bromobutane (b) (*S*)-1-Chloro-1-methoxypropane (c) (*S*)-1-Chloro-1-phenylpropane

(d) (*R*)-3-Chlorohexane (e) (*R*)-2-Methyl-4-nitrohexane

Problem C.28

Think

First draw the molecules temporarily ignoring dash–wedge notation. Identify all chiral centers. What are the priority assignments for the four groups on each chiral center? Are the groups distinguished by the atoms at the points of attachment? By the sets of atoms one bond away from the points of attachment? How do you treat double bonds as single bonds? Are the first through third substituents arranged clockwise or counterclockwise? How can you add dash–wedge notation to the bonds to give the *R* or *S* configuration? Review the rules for IUPAC nomenclature from Interchapters A and B.

Solve

Structures for molecules **(a)–(c)** are drawn below. *R* and *S* configurations are shown for each asymmetric carbon. Prime labels are used to distinguish the first through fourth priority numbering for each chiral center. All ties at the points of attachment are broken with the sets of atoms one bond away from the points of attachment, with the exception of the middle chiral center in **(c)**. For that chiral center, groups 2′ and 3′ have {Cl, C, H} one bond away from the points of attachment, resulting in a tie. Two bonds away from the points of attachment, the group on the left has {C, C, C}, whereas the one on the right has {H, H, H}.

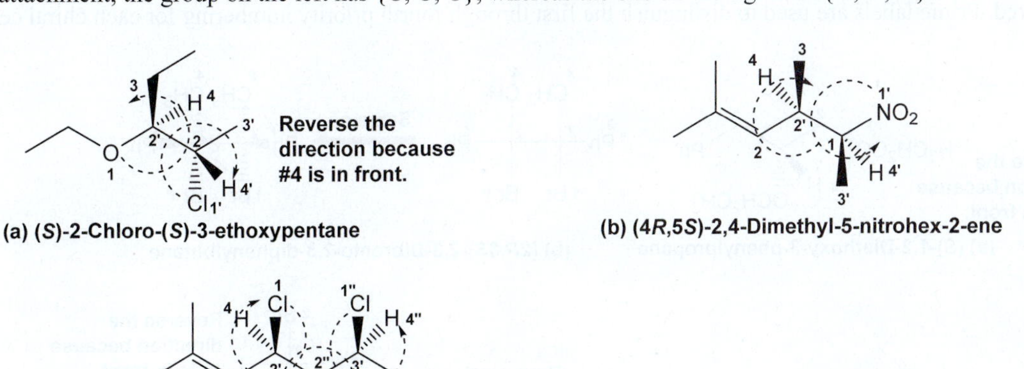

Reverse the direction because #4 is in front.

(a) (*S*)-2-Chloro-(*S*)-3-ethoxypentane (b) (4*R*,5*S*)-2,4-Dimethyl-5-nitrohex-2-ene

(c) (4*R*,5*R*,6*S*)-4,5,6-Trichloro-2-methyl-3-phenylhept-2-ene

Problem C.29

Think

First draw the molecule that corresponds to each name without drawing dash–wedge notation. Where are the asymmetric C atoms? For each one, what are the first through fourth priority group assignments? Are the groups distinguished by the atoms at the points of attachment? By the sets of atoms one bond away from the points of attachment? Add dash–wedge notation in such a way that the first through fourth arrangement matches the configuration given. Review the rules for IUPAC nomenclature from Interchapters A and B.

Solve

Structures for molecules **(a)–(d)** are drawn below. The group priorities about each chiral center are numbered. Prime labels are used to distinguish the first through fourth priority numbering for each chiral center. In **(a)**, the molecule was rotated to have the fourth priority group point away.

(a) (*S*)-1-Chloro-2,2-dimethyl-1-phenylcyclopentane

Reverse the direction because #4 is in front.

Reverse the direction because #4 is in front.

(b) (1*R*,2*S*)-1-Methyl-1,2-dinitrocyclopropane

Reverse the direction because #4 is in front.

(c) (*R*)-4-Ethoxycyclohexene

Reverse the direction because #4 is in front.

Reverse the direction because #4 is in front.

(d) (3*S*,4*S*)-3-Chloro-4-fluoro-2-methylhepta-1,6-diene

Problem C.30

Think

Which atoms are chiral centers? For each one, what are the priority assignments for each of the four groups? Is the lowest-priority group pointed toward you or away from you? Which bonds in a Fischer projection are pointed toward you, and which are pointed away? Are the substituents arranged clockwise (*R*) or counterclockwise (*S*)? Review the rules for IUPAC nomenclature from Interchapters A and B.

Solve

The IUPAC names for molecules **(a)–(d)** are given below. The group priorities about each chiral center are numbered. Prime labels are used to distinguish the first through fourth priority numbering for each chiral center.

Reverse the direction because #4 is in front.

(a) (*S*)-1,2-Diethoxy-3-phenylpropane

Same as

(b) (2*R*,3*S*)-2,3-Dibromo-2,3-diphenylbutane

(c) (1*S*,2*R*)-1-Chloro-2-methylcyclopentane

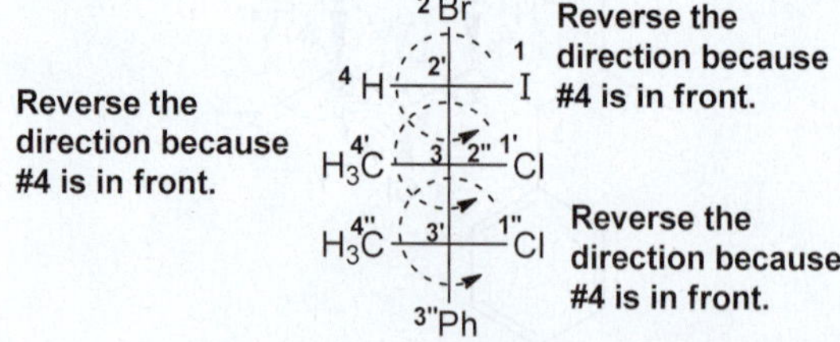

2 Br
4 H
I
1
H3C
Cl
H3C
Cl
3"Ph
Reverse the direction because #4 is in front.
Reverse the direction because #4 is in front.
Reverse the direction because #4 is in front.

(d) (1*R*,2*R*,3*R*)-1-Bromo-2,3-dichloro-1-iodo-2-methyl-3-phenylbutane

Problem C.31

Think

First draw the molecule that corresponds to each name without including dash–wedge notation. Where are the asymmetric C atoms? What are the first through fourth priority group assignments? Add the dash–wedge notation in such a way that the first through fourth arrangement matches the configuration given. Review the rules for IUPAC nomenclature from Interchapters A and B.

Solve

Structures for molecules **(a)–(f)** are drawn below. The group priorities about each chiral center are numbered.

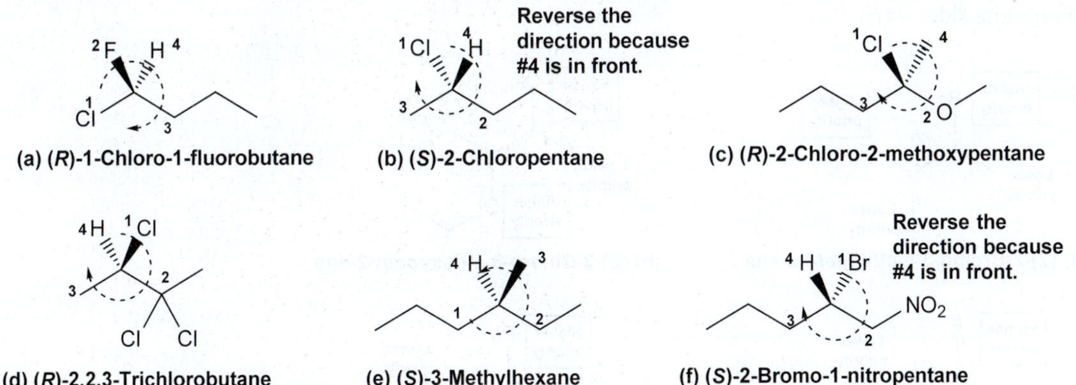
Reverse the direction because #4 is in front.
(a) (R)-1-Chloro-1-fluorobutane
(b) (S)-2-Chloropentane
(c) (R)-2-Chloro-2-methoxypentane
Reverse the direction because #4 is in front.
(d) (R)-2,2,3-Trichlorobutane
(e) (S)-3-Methylhexane
(f) (S)-2-Bromo-1-nitropentane

Problem C.32

Think

First draw the molecule that corresponds to each name without including dash–wedge notation. Where are the asymmetric C atoms? What are the first through fourth priority group assignments? Add the dash–wedge notation in such a way that the first through fourth arrangement matches the configuration given. Review the rules for IUPAC nomenclature from Interchapters A and B.

Solve

Structures for molecules **(a)–(e)** are drawn below. The group priorities about each chiral center are numbered. Prime labels are used to distinguish the first through fourth priority numbering for each chiral center. The main chain or ring is numbered.

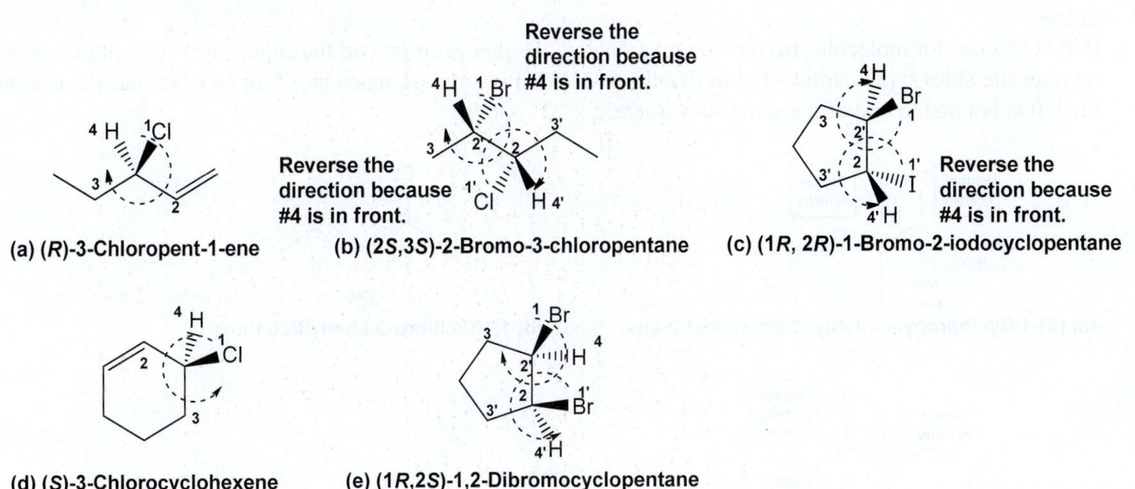
Reverse the direction because #4 is in front.
Reverse the direction because #4 is in front.
Reverse the direction because #4 is in front.
(a) (R)-3-Chloropent-1-ene
(b) (2S,3S)-2-Bromo-3-chloropentane
(c) (1R, 2R)-1-Bromo-2-iodocyclopentane
(d) (S)-3-Chlorocyclohexene
(e) (1R,2S)-1,2-Dibromocyclopentane

Problem C.33

Think

Of the two groups attached to one C atom of the C=C double bond, which has higher priority? Are the groups distinguished by the atoms at the points of attachment or by sets of atoms farther away from the points of attachment? Which group has higher priority on the other C atom of the C=C double bond? Are the higher-priority groups on the same side or opposite sides? Review the rules for IUPAC nomenclature from Interchapters A and B.

Solve

IUPAC names for molecules **(a)–(d)** are given below. Higher priorities on the same side = (*Z*); higher priorities on opposite sides = (*E*).

(a) (*Z*)-2-Bromo-3-methylpent-2-ene

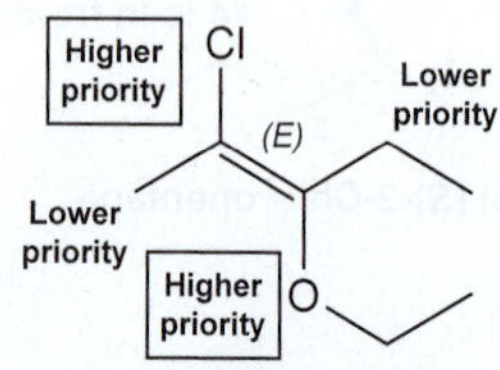

(b) (*E*)-2-Chloro-3-ethoxypent-2-ene

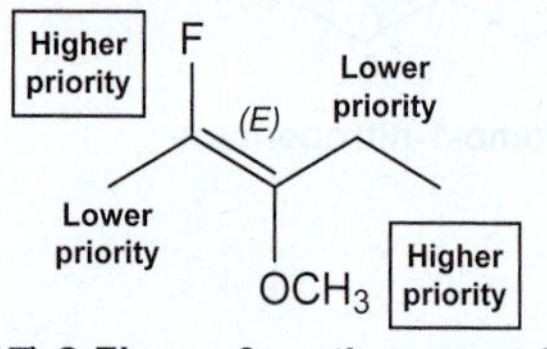

(c) (*E*)-2-Fluoro-3-methoxypent-2-ene

(d) (*E*)-3-Chloro-2-phenylpent-2-ene

Problem C.34

Think

Of the two groups attached to one C atom of the C=C double bond, which has higher priority? Are the groups distinguished by the atoms at the points of attachment or by sets of atoms farther away from the points of attachment? Which group has higher priority on the other C atom of the C=C double bond? Are the higher-priority groups on the same side or opposite sides? Review the rules for IUPAC nomenclature from Interchapters A and B.

Solve

IUPAC names for molecules **(a)–(c)** are given below. Higher priorities on the same side = (*Z*); higher priorities on opposite sides = (*E*). Notice that in **(b)**, the C=C bond is not designated as (*Z*) or (*E*), because the C atom on the left is bonded to two of the same substituent.

(a) (*Z*)-1-Cyclopropyl-2-methyl-3-phenylbut-2-ene

(b) 1,1-Dichloro-2-phenylbut-1-ene

(c) (*E*)-5,5-Dimethyl-2,3-diphenylhex-2-ene

Problem C.35

Think

Draw the molecule that corresponds to each name without drawing the configuration. Consider the rules for *R*/*S* and *E*/*Z* configurations.

Solve

Structures for molecules **(a)–(f)** are drawn below. Higher priorities on the same side = (*Z*); higher priorities on opposite sides = (*E*).

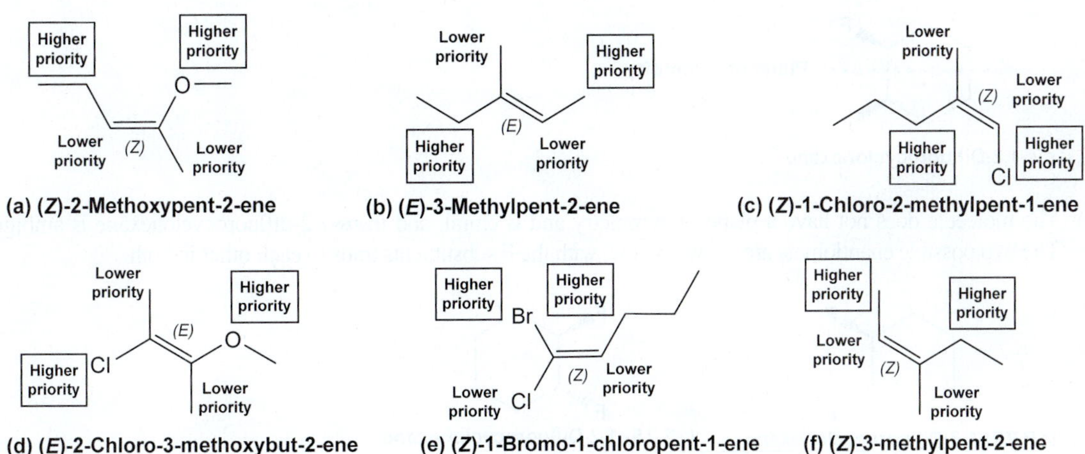

Problem C.36

Think

Draw the molecule that corresponds to each name without drawing the configuration. Consider the rules for *R*/*S* and *E*/*Z* configurations.

Solve

Structures for molecules **(a)–(d)** are drawn below. Higher priorities on the same side = (*Z*); higher priorities on opposite sides = (*E*). *Note:* Molecule **(c)** is drawn twice to show the priorities for each C=C bond.

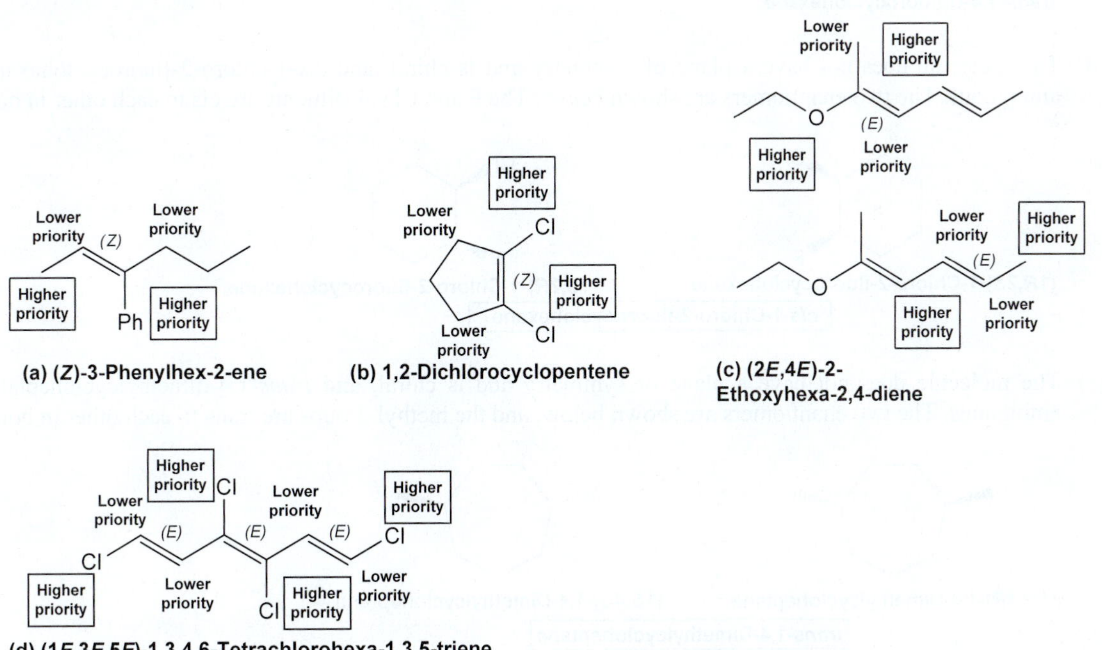

Problem C.37

Think

Which molecules have a plane of symmetry? Identify the chiral and achiral molecules. For the molecules that are chiral, are the cis/trans relationships the same in each enantiomer?

Solve

(a) The molecule has a plane of symmetry and, therefore, is achiral. There are no enantiomers for this molecule, so *cis* is an unambiguous way to describe the molecule.

F

Plane of symmetry

F

***cis*-1,2-Difluorocyclohexane**

(b) The molecule does not have a plane of symmetry and is chiral, and *trans*-1,2-difluorocyclohexane is ambiguous. The two possible enantiomers are shown below, with the F substituents trans to each other in both.

F F

F F

(1*R*,2*R*)-1,2-Difluorocyclohexane **(1*S*,2*S*)-1,2-Difluorocyclohexane**

***trans*-1,2-Difluorocyclohexane**

(c) The molecule has a plane of symmetry and, therefore, is achiral. There are no enantiomers for this molecule, so *trans* is an unambiguous way to describe the molecule.

F

Plane of symmetry

F

***trans*-1,4-Difluorocyclohexane**

(d) The molecule does not have a plane of symmetry and is chiral, and *cis*-1-chloro-2-fluorocyclohexane is ambiguous. The two enantiomers are shown below. The F and Cl substituents are cis to each other in both.

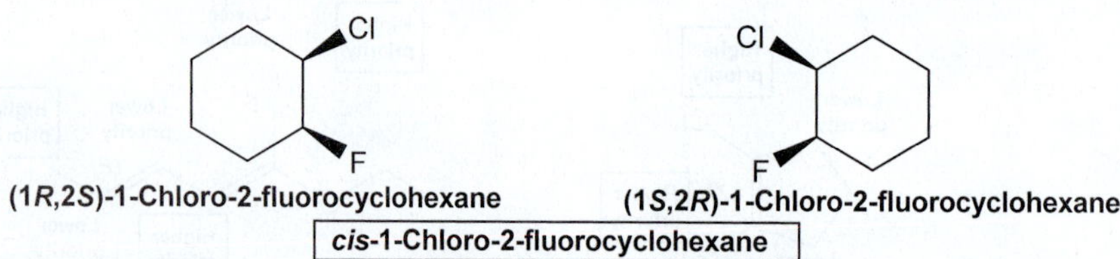

(1*R*,2*S*)-1-Chloro-2-fluorocyclohexane **(1*S*,2*R*)-1-Chloro-2-fluorocyclohexane**

***cis*-1-Chloro-2-fluorocyclohexane**

(e) The molecule does not have a plane of symmetry and is chiral, and *trans*-1,4-dimethylcycloheptane is ambiguous. The two enantiomers are shown below, and the methyl groups are trans to each other in both.

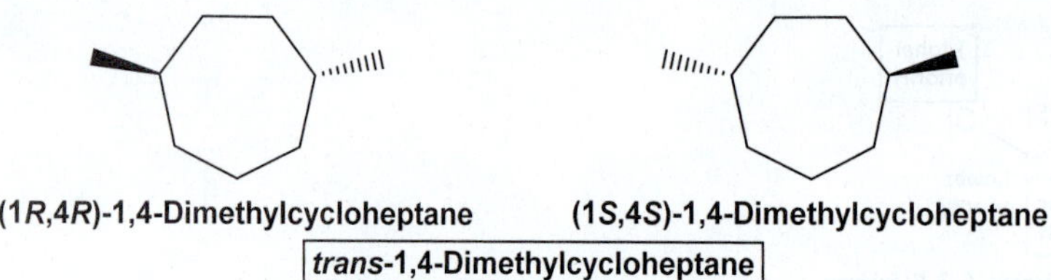

(1*R*,4*R*)-1,4-Dimethylcycloheptane **(1*S*,4*S*)-1,4-Dimethylcycloheptane**

***trans*-1,4-Dimethylcycloheptane**

CHAPTER 6 | The Proton Transfer Reaction: An Introduction to Mechanisms, Thermodynamics, and Charge Stability

Your Turn Exercises

Your Turn 6.1

Think

What does the double-barbed curved arrow represent? Is bond breaking shown at the tail or head of the arrow? Is bond making shown at the tail or head of the arrow? In comparing the reactants and products, what bonds were broken? What bonds were formed?

Solve

Each double-barbed curved arrow represents the movement of two valence electrons. The head of the first arrow shows bond making (HO–H), and the tail of the second arrow shows bond breaking (H–Cl). The electrons in the newly formed HO–H bond were initially the lone pair on the hydroxide oxygen. The electrons in the broken H–Cl bond end up as an additional lone pair on Cl. Each pair of electrons involved in the bond making/breaking steps is circled with a dashed oval.

Bond formation

Bond breaking

HO^{-} + H–Cl ⟶ HO–H + Cl^{-}

Your Turn 6.2

Think

In comparing the reactants and products, what bonds were broken? What bonds were formed? To indicate the conversion of a lone pair into a bond, where should the curved arrow originate? Where should it point? To indicate the conversion of a bond into a lone pair, where should the curved arrow originate? Where should it point? How does the magnitude of the K_{eq} inform you of the extent of product formation at equilibrium?

Solve

The lone pair on the O atom of the alcohol is used to form a bond to the H atom of the carboxylic acid, and the H–O bond in the carboxylic acid is broken, with the electrons from that bond ending up as an additional lone pair on the carboxylic acid's O. Therefore, one curved arrow is drawn from a lone pair on the alcohol O atom to the H atom on the carboxylic acid, and a second curved arrow originates from the center of the carboxylic acid's O–H bond and points to the carboxylic acid's O atom. The larger equilibrium constant favors more product formation, as K_{eq} = [products]/[reactants]. In this case, it is the second reaction that forms more products, because the $K_{eq} = 4.0 \times 10^{-3}$ is five orders of magnitude larger than that of the first reaction with $K_{eq} = 7.1 \times 10^{-8}$.

H_3C–C(=O)–O–H (Acid) + HO–CH_2–CH_3 (Base) ⇌ H_3C–C(=O)–O^{-} + H_2O^{+}–CH_2–CH_3 $K_{eq} = 7.1 \times 10^{-8}$

F_3C–C(=O)–O–H (Acid) + HO–CH_2–CH_3 (Base) ⇌ F_3C–C(=O)–O^{-} + H_2O^{+}–CH_2–CH_3 $K_{eq} = 4.0 \times 10^{-3}$

Your Turn 6.3

Think

Consult Table 6-1. Is the stronger acid the one with the more negative or more positive pK_a? What does the difference in pK_a values between two acids tell you?

Solve

The acid with the more negative pK_a is the stronger acid. Therefore, HCl (pK_a = −7) is a stronger acid than H_3O^+ (pK_a = −1.7). The difference in pK_a values is −1.7 − (−7) = 5.3, which corresponds to a difference in acid strength of >10^5. Thus, HCl is >100,000 times stronger an acid than H_3O^+.

Acid is stronger than H_3O^+

This side is heavily favored.

H_2O: + H—Cl: ⇌ $H_2O^{\oplus}$—H + :Cl:$^{\ominus}$

pK_a = −7 pK_a = −1.7

Your Turn 6.4

Think

Consult Table 6-1. Is the stronger acid the one with the more negative or more positive pK_a? What does the difference in pK_a values between two acids tell you?

Solve

The acid with the more negative pK_a is the stronger acid. Therefore, H_2O (pK_a = 15.7) is a stronger acid than $(CH_3)_2NH$ (pK_a = 38). The difference in pK_a values is 38 − 15.7 = 22.3, which corresponds to a difference in acid strength of $10^{22.3}$. Thus, H_2O is 2.0×10^{22} times stronger as an acid than $(CH_3)_2NH$. Moreover, HO^- is a weaker base than $(CH_3)_2N:^-$ because the stronger the acid is, the weaker the conjugate base is.

Base is stronger than $HO^{\ominus}$

$(H_3C)_2N:^{\ominus}$ + H—OH ⇌ $(H_3C)_2$N—H + :OH$^{\ominus}$

pK_a = 15.7 pK_a = 38

This side is heavily favored.

Your Turn 6.5

Think

Consult Table 6-1. Is the stronger acid the one with the more negative or more positive pK_a? For diethyl ether to be a suitable solvent, should it react with the solute? Should the proton transfer given favor the reactant side or the product side?

Solve

The pK_a values are written below. The reactant side of the reaction is favored because the stronger acid is on the product side. This is not the same side that is favored in Equation 6-11. Therefore, diethyl ether is a suitable solvent for $(CH_3)_2N^-$ because the equilibrium lies to the left, indicating that diethyl ether is relatively inert in the presence of $(CH_3)_2N^-$.

$(H_3C)_2N:^{\ominus}$ + H_3C–CH(H)–O–CH_2–CH_3 ⇌ $(H_3C)_2$N—H + H_3C–$\overset{\ominus}{C}$H–O–CH_2–CH_3

This side is favored.

pK_a = ~45 pK_a = 38

Your Turn 6.6

Think

Consult Figure 6-1. What is the pH at the pK_a? At what pH is the acid ~100% dissociated? How many pH units above the pK_a is this value? At what pH is the acid ~0% dissociated (~100% associated)? How many pH units below the pK_a is this value?

Solve

The acid is nearly 100% dissociated at around two pH units above the pK_a, or pH = 7. It is nearly 100% associated around two pH units below the pK_a, or pH = 3.

Your Turn 6.7

Think

Consult Figure 6-1 as a guide. What is the pH when an acid whose pK_a = 9 is dissociated 50%? How many pH units above the acid's pK_a of 9 must the solution be to cause the acid to dissociate nearly 100%? How many pH units below the acid's pK_a of 9 must the solution be to cause the acid to dissociate roughly 0%? How does the increase in pK_a by four units affect the appearance of the graph?

Solve

The pH curve shifts four units higher so that 50% dissociation takes place at pH = pK_a = 9. The acid is nearly 100% dissociated at around two pH units above the pK_a, or pH = 11. It is nearly 100% associated around two pH units below the pK_a, or pH = 7.

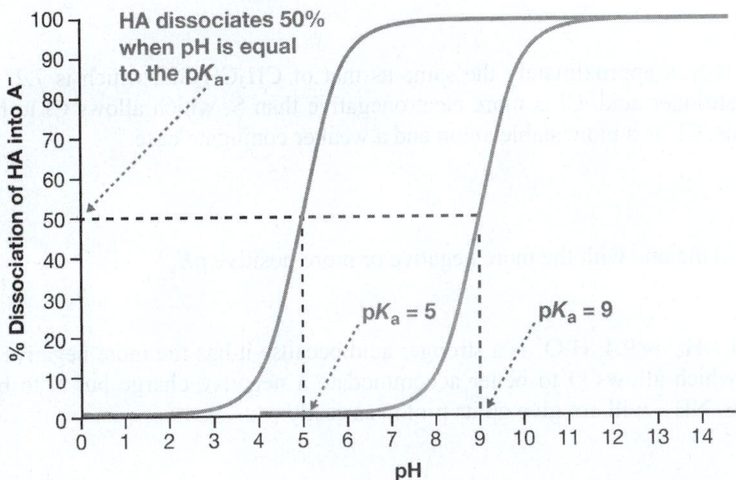

Your Turn 6.8

Think

As the reaction coordinate increases (going from left to right in the figure), which bonds are broken? Which bonds are formed? How does the distance between the atoms correlate to bond breaking and bond forming?

Solve

As the reaction coordinate increases, the distance between Cl and H **decreases**, because as the reaction proceeds, the H–Cl bond is forming; the distance between the O and H **increases** because the O–H bond is breaking.

Your Turn 6.9

Think

Draw an arrow to represent the difference between the free energy of products and reactants (ΔG°_{rxn}). Draw an arrow to represent the difference between the free energy of the reactants and the transition state ($\Delta G^{\circ\ddagger}$). Compare the length of the arrow you drew to the length of the arrow in Figure 6-2a. Which one is longer? What does that mean for the value of $\Delta G^{\circ\ddagger}$?

Solve
The free energy quantities are indicated in the diagram below. The $\Delta G^{\circ\ddagger}$ is larger in Figure 6-2b compared to the reaction in Figure 6-2a.

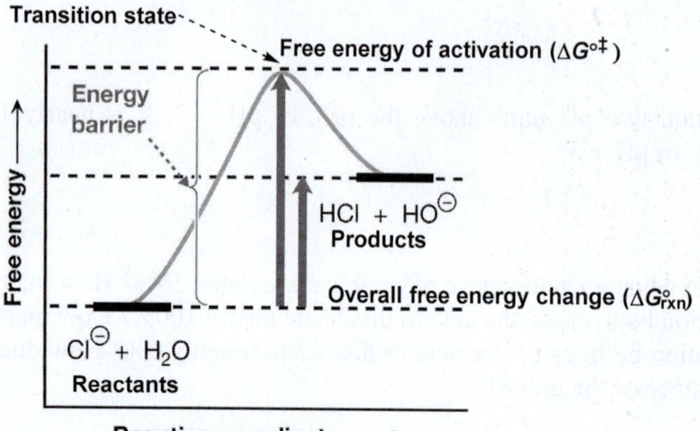

Your Turn 6.10

Think
Consult Table 6-1. Is the stronger acid the one with the more negative or more positive pK_a?

Solve
The pK_a of HCl is −7, and that of H_2S is approximately the same as that of CH_3CH_2SH, which is 7.2. HCl, having the more negative pK_a, is a stronger acid. Cl is more electronegative than S, which allows Cl to better accommodate a negative charge. Thus, Cl^- is a more stable anion and a weaker conjugate base.

Your Turn 6.11

Think
Consult Table 6-1. Is the stronger acid the one with the more negative or more positive pK_a?

Solve
The pK_a of H_3O^+ is −1.7, and that of NH_4^+ is 9.4. H_3O^+ is a stronger acid because it has the more negative pK_a. O is more electronegative than N, which allows O to better accommodate a negative charge but N to better accommodate a positive charge. Thus, NH_4^+ will not give up its proton as easily.

Your Turn 6.12

Think
Consult Table 6-1. Is the stronger acid the one with the more negative or more positive pK_a?

Solve
The pK_a of $H_3C–CH_3$ is about 50, that of $H_2C=CH_2$ is about 44, and that of HC≡CH is about 25. The acid strength goes in the following increasing order: C_2H_6 (sp^3) < C_2H_4 (sp^2) < C_2H_2 (sp). The difference in acid strength is due to the stability of the negative charge that develops on the conjugate base. Charge stability depends, in turn, on the effective electronegativity of the C atom, which depends on its hybridization: $sp^3 < sp^2 < sp$.

Your Turn 6.13

Think
Consult Table 6-1. Is the stronger acid the one with the more negative or more positive pK_a?

Solve
The pK_a of ethanoic acid (acetic acid) is 4.75. That of ethanol is 16. Ethanoic acid, having the lower pK_a, is a stronger acid owing to resonance delocalization of the negative charge in the conjugate base.

Your Turn 6.14

Think

Remember that the curved arrow illustrates electron movement necessary to change one resonance structure into another. In the structure on the left, which electrons move to form the double bond in the structure on the right? Which electrons move in the structure on the left to form the lone pair on the O atom in the structure on the right? How do you draw the average of the two structures (i.e., the hybrid)? How do you represent partial charges?

Solve

In the structure on the left, a lone pair from O^- is used to make the C=O double bond in the structure on the right. This requires a curved arrow from the lone pair to the center of the initial C–O bond. Without any other changes, this would lead to 10 valence electrons on the C atom. Therefore, in addition, a bonding pair of electrons from the C=O double bond on the left is converted to an additional lone pair on the O atom that is initially uncharged. This requires a second curved arrow to be drawn from the center of the C=O double bond to the O atom at the top. Because each O has a −1 charge in one structure and a 0 charge in the other, the hybrid charge is represented by a partial negative charge symbol, δ−. Because each C–O bond is a single bond in one structure and a double bond in the other, they are both represented as a bond intermediate between a single and double bond.

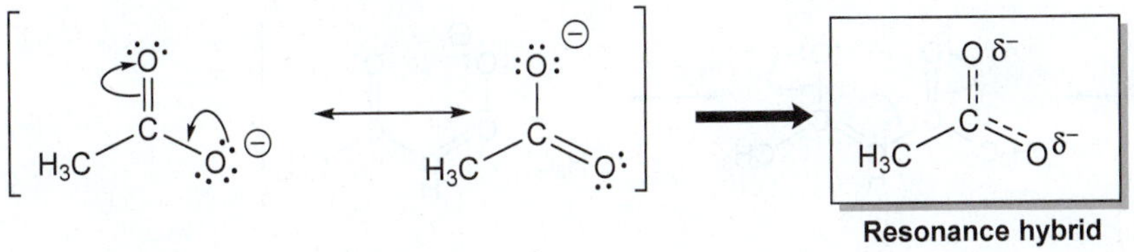

Resonance hybrid

Your Turn 6.15

Think

What structural feature in HSO_4^- indicates that there should be another resonance structure—a lone pair attached to a double/triple bond, an atom lacking an octet attached to a double/triple bond, or a ring of alternating single and double bonds? With that structural feature, how many curved arrows do you need to convert one resonance structure into another? When a lone pair is converted into a bonding pair (and vice versa), how does that change the formal charge?

Solve

In HSO_4^-, the negatively charged O has three lone pairs and is attached to a S=O bond. Therefore, two curved arrows are required to arrive at the next resonance structure: one from a lone pair on O^- that points to the center of the S–O bond, and one from the center of a S=O bond that points to the corresponding O. This is done twice to obtain the two additional resonance structures, which are shown below. You may have chosen to draw the third resonance structure second. This does not matter, as long as your arrow movement is consistent with the structure that results. The resonance hybrid is somewhat of an average of all three resonance structures. Three O atoms have a partial negative charge, and three S–O bonds are intermediate between single and double bonds.

Resonance hybrid

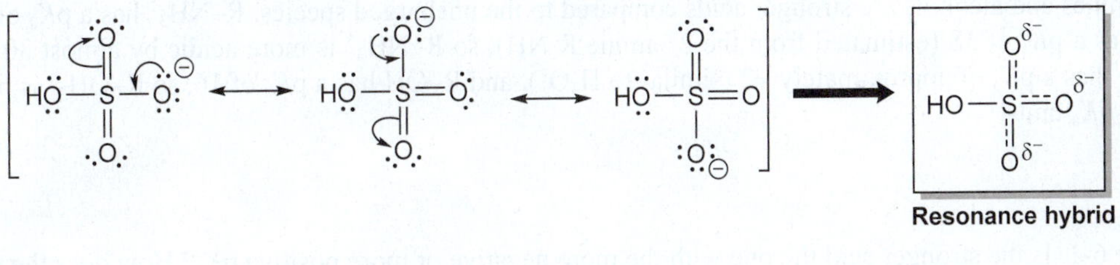

Your Turn 6.16

Think

Are there any lone pairs adjacent to a double or triple bond? Do any of the double bonds convert to lone pairs? How can you show via a curved arrow the electrons in the lone pair moving to form a double bond? How can you show via a curved arrow the electrons in a double bond moving to form a lone pair?

Solve

Recall that atoms do not move on going from one resonance structure to another. Only nonbonding electrons or electrons from multiple bonds are moved. *Top structure:* The first curved arrow is used to convert the lone pair on the fifth C atom into a covalent double bond, C=C. Without any other changes, this would lead to 10 valence electrons on the fourth C atom. The second arrow is used to convert a pair of electrons from the double bond, C=O, into a lone pair on the O. *Bottom structure:* The first curved arrow is used to convert the lone pair on the third C atom into a covalent double bond, C=C. Without any other changes, this would lead to 10 valence electrons on the fourth C atom. The second arrow is used to convert a pair of electrons from the double bond, C=O, into a lone pair on the O atom. This yields the resonance structure in the middle. A third resonance structure can be produced from the structure on the left by repeating the electron movement on the left side of the ion instead of the right. Alternatively, to arrive at that third resonance structure beginning from the one in the middle, the three curved arrows shown are required.

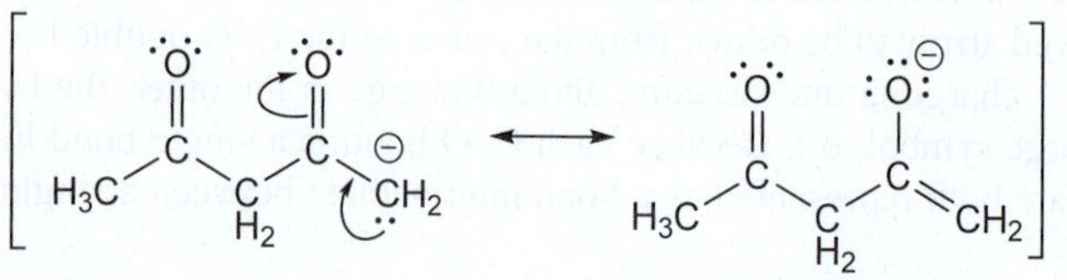

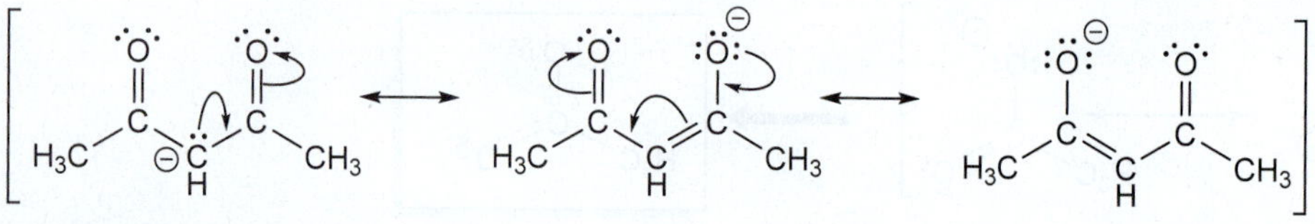

Your Turn 6.17

Think

Is the CH_3 group electron donating or electron withdrawing relative to the H atom? In which direction does the arrow point?

Solve

The CH_3 group is electron donating compared to the H atom, so an arrow is drawn away from the CH_3 group, toward the N^+.

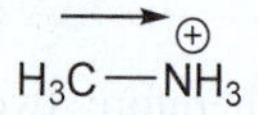

Your Turn 6.18

Think

Consult Table 6-1. Is the stronger acid the one with the more negative or more positive pK_a? Which species (charge or uncharged) do you expect to be the stronger acid? How is this reflected in the pK_a values?

Solve

Protonated amines and alcohols are stronger acids compared to the uncharged species. R–NH_3^+ has a pK_a of 9.4 and R–NH_2 has a pK_a of 38 (estimated from the 2° amine R_2NH), so R–NH_3^+ is more acidic by almost 30 pK_a units. R–OH_2^+ has a pK_a of approximately −2 (similar to H_3O^+), and R–OH has a pK_a of 16, so R–OH_2^+ is more acidic by ~18 pK_a units.

Your Turn 6.19

Think

Consult Table 6-1. Is the stronger acid the one with the more negative or more positive pK_a? How does the type of atom to which the acidic proton is bound affect the strength of the acid? How does hybridization of the C atom bound to the acidic proton affect the strength of the acid?

Solve

R–OH (pK_a = 16), R–NH_2 (pK_a = 38, estimated from the 2° amine R_2NH), R–CH_3 (pK_a = 50), R–C≡CH (pK_a =25). Therefore, R–OH is >20 pK_a units more acidic than R–NH_2, and R–C≡CH is about 25 pK_a units more acidic than R–CH_3. O, N, and C are in the same row and are about the same size. O is more electronegative than N, which is more electronegative than C. This allows O to accommodate negative charge more easily, and the conjugate base is stabilized. For comparison of the two CH species, the effective electronegativity depends on hybridization of the C atom ($sp^3 < sp^2 < sp$). Therefore, the conjugate base of R–C≡CH (*sp*-hybridized C) is more stable, and the acid is stronger.

Your Turn 6.20

Think

Are there any lone pairs adjacent to a double or triple bond? Can any of the double bonds convert into lone pairs? How can you show via a curved arrow the electrons in the lone pair moving to form a double bond? How can you show via a curved arrow the electrons in a double bond moving to form a lone pair? How do you draw the average of the four structures (hybrid)? How do you represent partial charges?

Solve

In each resonance structure, there is an atom with a lone pair attached to a double bond, so two curved arrows are necessary to convert from one to the other. In each case, the first arrow moves a lone pair to form a double bond, and the second arrow shows the double bond converting into a lone pair. When bonding electrons convert to nonbonding electrons, the formal charge decreases by 1. When nonbonding electrons convert to bonding electrons, the formal charge becomes more positive by 1. The hybrid structure shows the partial negative charge on each of the four atoms that bears a full negative charge on one of the resonance structures. You will note that the negative charge skips every other atom and goes around the entire ring. Also, the C–N and each C–C bond in the hybrid are intermediates between a single and double bond.

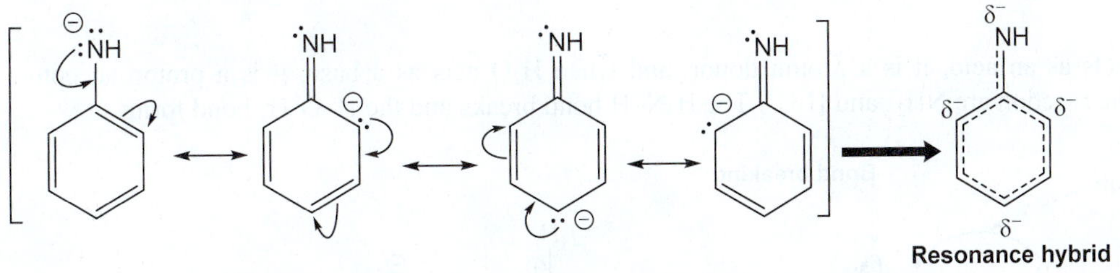

NH NH NH NH δ− NH δ− δ− δ−

Resonance hybrid

In Chapter Problems

Problem 6.2

Think

What does it mean to be an acid? A base? What are the important electrons to keep track of during the course of the reaction? What bonds are broken? What bonds are formed? What does a curved arrow represent? How many curved arrows are needed?

Solve

In the reverse reaction, hydroxide acts as the base (proton acceptor) and ammonium acts as the acid (proton donor). The H–OH bond forms and the $H–^{+}NH_3$ bond breaks. A curved arrow represents electron movement to show the breaking of the $H–^{+}NH_3$ bond and formation of the H–OH bond. The curved arrow notation is as follows:

Bond formation

Bond breaking

$$HO^- + H{-}NH_3^+ \longrightarrow H{-}O{-}H + H{-}NH_2$$

Problem 6.3

Think

What does it mean to be an acid? A base? What are the important electrons to keep track of during the course of the reaction? What bonds are broken? What bonds are formed? What does a curved arrow represent? How many curved arrows are needed? How can you derive the products by moving the electrons according to the curved arrows?

Solve

When NH_3 acts as an acid, it is a proton donor, and when H_2O acts as a base, it is a proton acceptor. The products of the reaction are NH_2^- and H_3O^+. The $H_2N–H$ bond breaks and the $H–O^{+}H_2$ bond forms.

Bond formation

Bond breaking

$$H{-}O{-}H + H{-}NH_2 \longrightarrow H_3O^+ + {}^{-}NH_2$$

Problem 6.5

Think

Is the stronger acid the one with the more positive pK_a or the lower pK_a? What is the difference between pK_a values, and how can that difference be used to calculate the difference in their acid strengths?

Solve

The compound with the lower pK_a value, phenol (pK_a = 10.00), is the stronger acid. The difference in pK_a values is $10.26 - 10.00 = 0.26$, which corresponds to a difference in acid strength of $10^{0.26} = 1.8$. Thus, phenol is 1.8 times stronger as an acid than 4-methylphenol.

OH

Phenol
pK_a = 10.00

OH

4-Methylphenol
pK_a = 10.26

Problem 6.6

Think

What is the relationship between the strength of an acid and the strength of its conjugate base? What is the conjugate acid of Cl^-? Of $C_6H_5O^-$? Consulting Table 6-1, what is the pK_a of each of those acids? Which acid is stronger?

Solve

The strength of a base decreases as the strength of its conjugate acid increases. Thus, the stronger the acid is, the weaker the conjugate base is, and vice versa. Phenoxide, $C_6H_5O^-$, is a stronger base than chloride, Cl^-, because its conjugate acid, phenol (C_6H_5OH; pK_a = 10.0), is weaker than the conjugate acid of Cl^-, which is HCl (pK_a = −7). Because the difference in pK_a is 17, phenoxide is 10^{17} times stronger as a base compared to chloride.

Problem 6.8

Think

What are the products of the reaction? What acid is present on each side of the reaction? Which one is stronger? What is the difference in their pK_a values, and how does that value relate to the difference in acid strengths?

Solve

The stronger acid (lower pK_a) is on the reactant side, as shown below, so the product side is favored. The difference in the two pK_a values is 18 (= 38 − 20). Therefore, the right side of the reaction is favored by 10^{18}. The equilibrium arrow ⇌ is drawn to show that the reaction favors the products and the equilibrium lies to the right.

pK_a = 20

pK_a = 38

Problem 6.9

Think

For the solvent not to react with the solute, should the solvent be a weaker or stronger acid than the carbanion's conjugate acid, HC≡CH? What is the pK_a of HC≡CH? What is the pK_a of each solvent given?

Solve

The solvent must be a weaker acid than the conjugate acid of HC≡C:⁻ so the solvent will not be deprotonated. The pK_a value, therefore, must be more positive (weaker acid) than that of HC≡CH, which is ~25. The only solvents that qualify are **(d)** and **(e)**, whose pK_a values are 35 and 45, respectively. The pK_a value of **(a)**, **(b)**, and **(c)** is 15.7, 16, and 17, respectively.

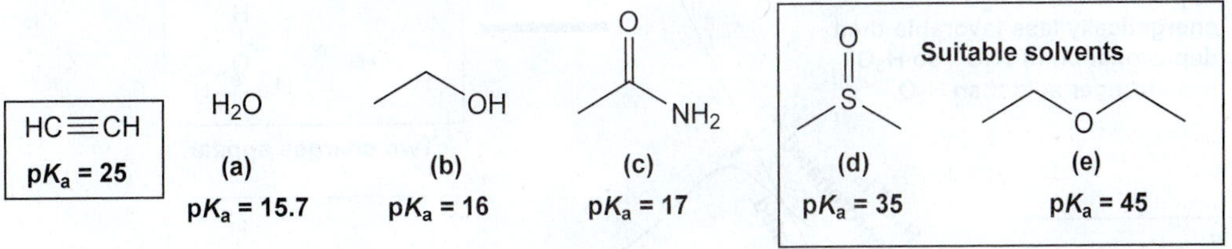

Problem 6.11

Think

On what functional group does each acidic H atom appear? What molecule(s) in Table 6-1 have the same functional group? Are there any nearby electronegative atoms or adjacent double bonds?

Solve

(a) C–H adjacent to C=O, which is characteristic of a ketone; p$K_a \approx 20$
(b) C–H adjacent to C=O, which is characteristic of an aldehyde; p$K_a \approx 20$
(c) C–H adjacent to –O–, which is characteristic of an ether; p$K_a \approx 45$
(d) H on *sp*^{3}-hybridized O^+, similar to H_3O^+; p$K_a \approx -1.7$
(e) H–O, which is characteristic of an alcohol; p$K_a \approx 16$

(a) p$K_a \approx 20$ (b) p$K_a \approx 20$ (c) p$K_a \gg 45$ (d) p$K_a \approx -1.7$ (e) p$K_a \approx 16$

Problem 6.12

Think

What are the products of each reaction? Is there better charge stabilization on the reactant side or the product side? What is the pK_a of each acid? What is the relationship between acid strength and the free energy change on loss of a proton?

Solve

The reactions and corresponding pK_a values are as follows:

pK_a = 15.7 ⇌ pK_a = –1.7 — Two additional charges

pK_a = –1.7 ⇌ pK_a = –1.7 — No additional charges

Creating a figure similar to Figure 6-6, we obtain the following free energy diagram. Notice that with water as an acid, two additional charges are created, so the products are higher in energy than the reactants. With H_3O^+ as the acid, on the other hand, no additional charges are created. Moreover, the reactants and products are the same species and have the same energy. The latter is more energetically favorable, therefore, making H_3O^+ a stronger acid, consistent with the much lower pK_a of H_3O^+ than of H_2O.

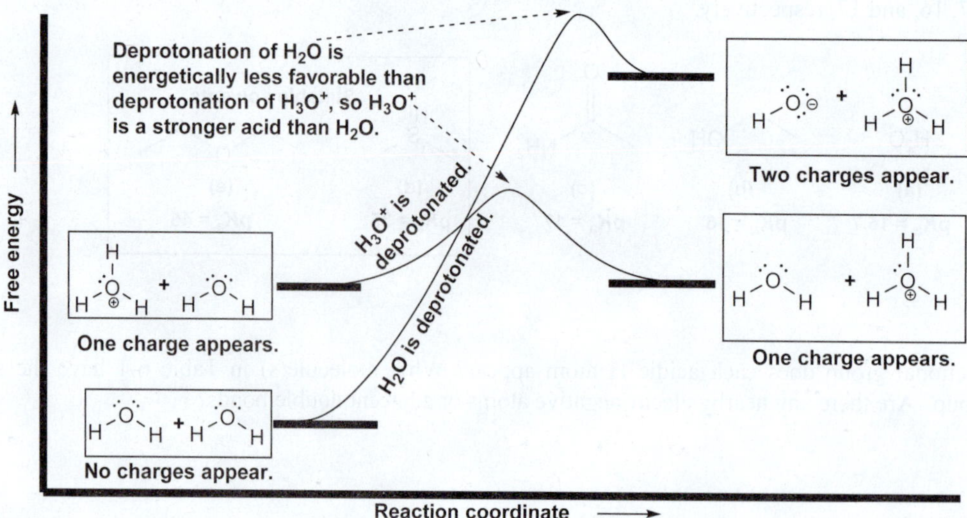

Problem 6.14

Think

What are the products of the reaction of each acid with water? What are the relative stabilities of the reactants? Which is more stable, Br^- or I^-? On the basis of their relative stabilities, which anion's conjugate acid is deprotonated more favorably? How does that correspond to the relative acid strength?

Solve

HI is the stronger acid. The energy diagram that compares the reactions in which they behave as acids is shown below. I^- is lower in energy than Br^-, because I is the larger atom. The larger ion has the less concentrated charge and is more stable. As we can see, deprotonation of HI is the more energetically favorable reaction.

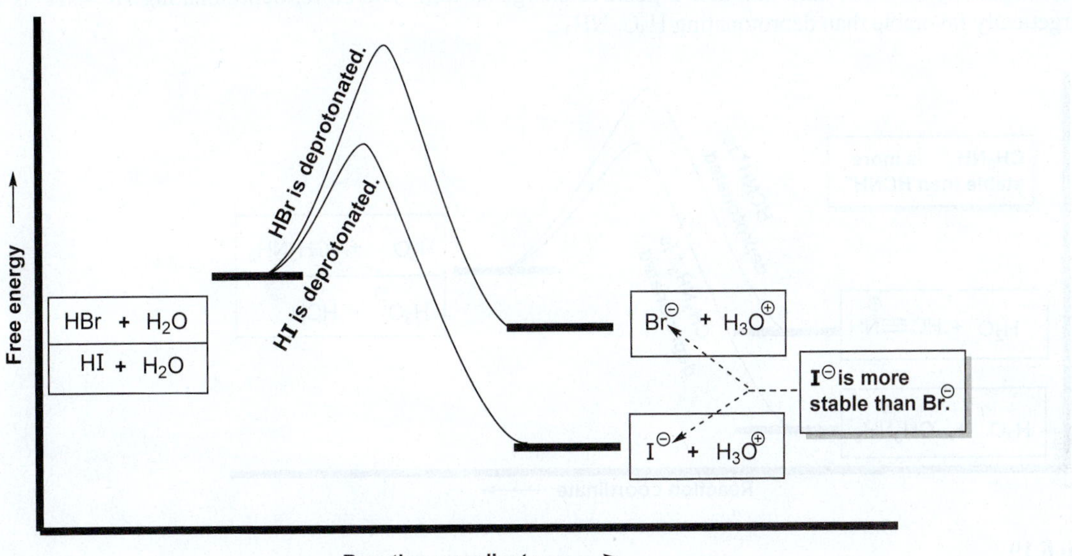

Problem 6.15

Think

What are the products of the reaction of each acid with water? What are the relative stabilities of the reactants? Which is more stable, H_2P^- or CH_3^-? On the basis of their relative stabilities, which anion's conjugate acid is deprotonated more favorably? How does that correspond to the relative acid strength?

Solve

The energy diagram comparing the deprotonation of CH_4 and PH_3 is below. H_2P^- is more stable than H_3C^-. We know that H_2P^- is more stable than H_2N^-, because P, being below N in the periodic table, is larger and can accommodate the negative charge better. And we know that H_2N^- is more stable than H_3C^-, because N is more electronegative than C. Because deprotonation of PH_3 is more energetically favorable, PH_3 is the stronger acid.

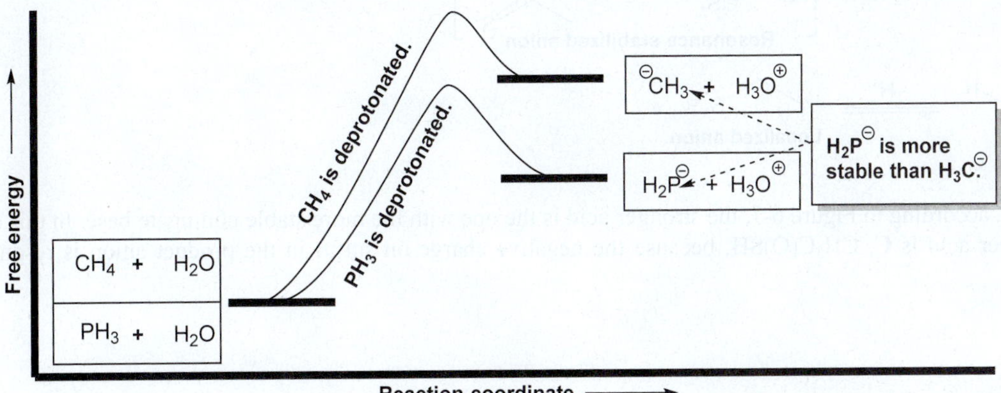

Problem 6.17

Think

What are the complete reactions? What are the relative stabilities of the two sets of reactants? Of the products? Do the species have the same charges? Are the charges on the same atoms? How does the hybridization of each N atom affect the stability of the charge? What is the effect of hybridization on acid strength?

Solve

$HC{\equiv}NH^+$ is the stronger acid. In the figure below, $HC{\equiv}NH^+$ is less stable than $H_3C{-}NH_3^+$ because the positive charge cannot be accommodated as easily on an *sp*-hybridized N atom. The *sp*-hybridized N atom has a greater effective electronegativity than the sp^3-hybridized N atom, and an atom with a greater effective electronegativity does not accommodate a positive charge as well. Therefore, deprotonating $HC{\equiv}NH^+$ is more energetically favorable than deprotonating $H_3C{-}NH_3^+$.

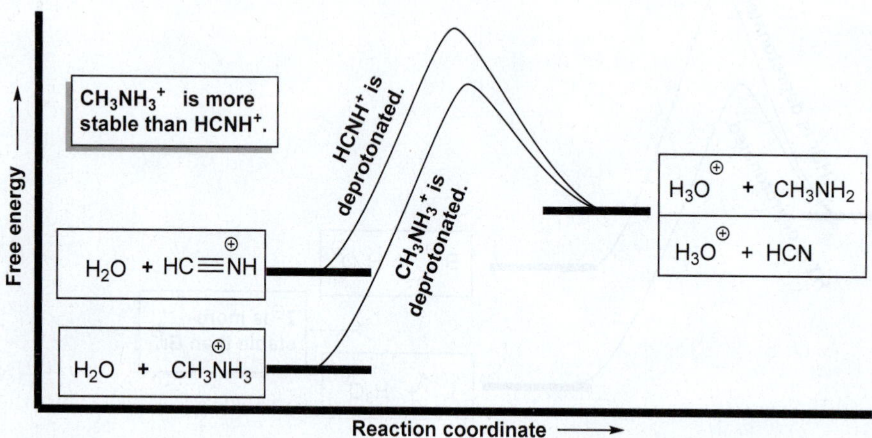

Problem 6.19

Think

Write reactions that depict each acid being deprotonated by a base. Do you expect a significant difference in energy between the reactants of one and the reactants of the other? Between the products of one and the products of the other? Are the charge-bearing atoms different in these two reactions? In the conjugate bases, do charge-bearing atoms have different effective electronegativities? Do the ions differ in resonance delocalization of the charge?

Solve

Being uncharged acids, the reactions have the same forms as Equations 6-19a and 6-19b.

C $\xrightleftharpoons{-H^+}$ Resonance stabilized anion

D $\xrightleftharpoons{-H^+}$ Localized anion

Therefore, according to Figure 6-7, the stronger acid is the one with the more stable conjugate base. In this case, the stronger acid is **C**, $CH_3C(O)SH$, because the negative charge on sulfur in the product anion is resonance stabilized.

Problem 6.21

Think

Being uncharged acids, their reactions have the same forms as Equations 6-19a and 6-19b. Therefore, according to Figure 6-7, the stronger acid is the one with the more stable conjugate base. Draw the conjugate base for each acid. Which conjugate base experiences more resonance stabilization of the charge that develops?

Solve

HNO_3 is a stronger acid than CH_3CO_2H. Both product anions are resonance stabilized, with the negative charge being shared over different O atoms. But NO_3^- has one additional resonance structure than $CH_3CO_2^-$, making NO_3^- more stable. So deprotonating HNO_3 is more energetically favorable than deprotonating CH_3CO_2H.

$-H^+$

$-H^+$

Problem 6.22

Think

Draw the complete reactions. Should the acidities of these acids be governed by the relative stabilities of the acids themselves or of their respective conjugate bases? Do the same charges appear? Are they on the same types of atoms? Is there a difference in charge delocalization by resonance? What is the inductive effect? Are there electronegative atoms present in one molecule but not the other?

Solve

Being charged acids, the complete reactions are analogous to those in Equations 6-20a and 6-20b, so their relative acidities are governed by the relative stabilities of the acids themselves, as shown in Figure 6-8. In this case, the stronger acid is **B** because inductive effects make **B** less stable than **A**. The Cl atom is electron withdrawing compared to H and removes negative charge from the positively charged O atom. This intensifies the positive charge on the O atom and destabilizes the species. The free energy diagram for these two acids is shown below.

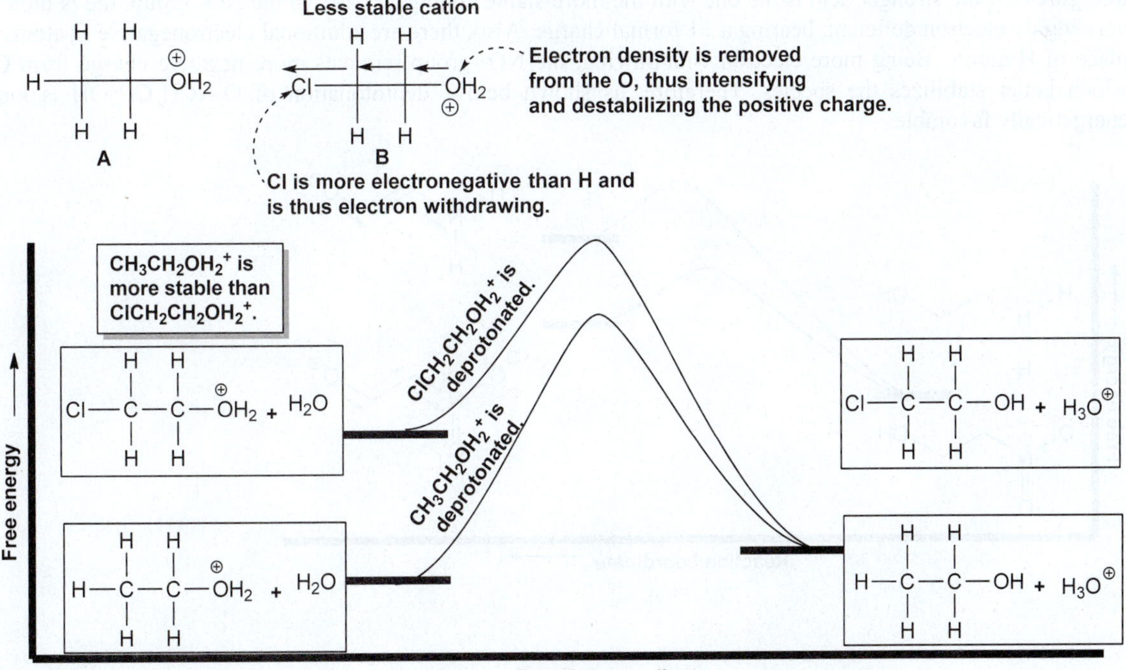

Problem 6.23

Think

Write reactions that depict each acid being deprotonated by a base. Do you expect a significant difference in energy between the reactants of one and the reactants of the other? Between the products of one and the products of the other? Are the charge-bearing atoms different in these two acids? Do the ions differ in resonance delocalization of the charge? Do they have different effective electronegativities?

Solve

The stronger acid is H_2S owing to inductive effects. Being uncharged acids, the reactions have the same forms as Equations 6-19a and 6-19b. Therefore, according to Figure 6-7, the stronger acid is the one with the more stable conjugate base. The CH_3CH_2 group in CH_3CH_2SH is electron donating compared to H and, therefore, destabilizes the nearby negative charge in the product anion. Therefore, deprotonating H_2S is more energetically favorable than deprotonating CH_3CH_2SH, as shown below.

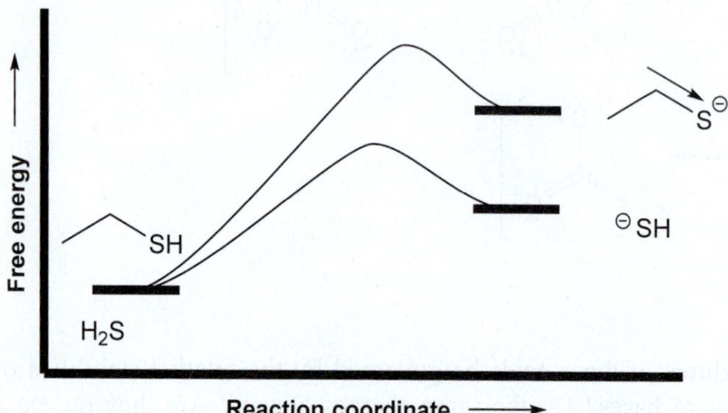

Problem 6.25

Think

Draw out the Lewis structure for each acid. Does the stability of the acid or the conjugate base dictate the pK_a? Do electron-donating or electron-withdrawing effects stabilize those species? Are NO_2 and NH_2 substituents electron donating or electron withdrawing? Which substituent invokes stronger inductive effects?

Solve

The stronger acid is $O_2NCH_2CH_2OH$ because the NO_2 group is more electron withdrawing than the NH_2 group. Being uncharged acids, the reactions have the same forms as Equations 6-19a and 6-19b. Therefore, according to Figure 6-7, the stronger acid is the one with the more stable conjugate base. In the NO_2 group, the N atom is very highly electron deficient, bearing a +1 formal charge. Also, there are additional electronegative O atoms in place of H atoms. Being more electron withdrawing, the NO_2 group removes more negative charge from O^-, which better stabilizes the species. Therefore, as shown below, deprotonation of $O_2NCH_2CH_2OH$ is more energetically favorable.

Free energy

Reaction coordinate

Problem 6.26

Think

Draw the conjugate base for each carboxylic acid. Does the stability of the acid or the conjugate base dictate the pK_a? Do electron-donating or electron-withdrawing effects stabilize those species? Is the C=O substituent electron donating or electron withdrawing? How does the location of the C=O affect the stability of the conjugate base?

Solve

In both molecules, the most acidic functional group is the CO_2H group. Being uncharged acids, the reactions have the same forms as Equations 6-19a and 6-19b. Therefore, according to Figure 6-7, the stronger acid is the one with the more stable conjugate base. The nearby O atom that is part of the C=O bond is electron withdrawing, helps to stabilize the resulting negative charge in the product anion, and increases the acid strength. The stronger acid is **B**, because the C=O group at the top of the ring is closer to the carboxylate group.

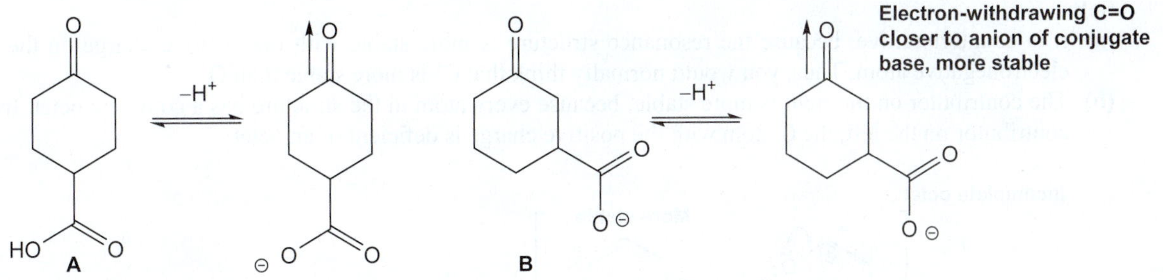

A —H^+→ ; B —H^+→

Electron-withdrawing C=O closer to anion of conjugate base, more stable

Problem 6.27

Think

Which electrons/bonds move between the two resonance structures? How can you show this movement using curved arrows? Do all nonhydrogen atoms have an octet? Do additional atoms with nonzero formal charges lead to a greater or lesser contribution by a resonance structure?

Solve

One lone pair on the Cl atom becomes a π bond, and the π bond between the C≡C becomes a lone pair on the C atom. The contributor on the right is less stable because it has two formal charges, whereas that on the left has no formal charges. Therefore, the structure on the left has the greater contribution.

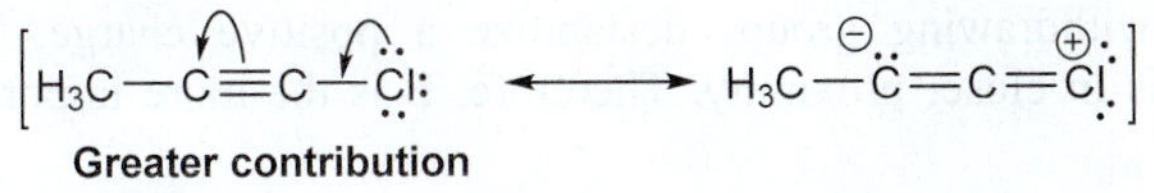

[$H_3C-C\equiv C-\ddot{Cl}$: ⟷ $H_3C-\ddot{C}^{\ominus}=C=\ddot{Cl}^{\oplus}$]

Greater contribution

Problem 6.28

Think

Do all nonhydrogen atoms have an octet? Does one structure have fewer atoms with nonzero formal charges? Are there more covalent bonds in one structure? What is the stability of the positive charge on the O atom versus the N atom?

Solve

The more stable contributor is **B** (see figure on the next page). The positive charge is on the N atom in one resonance structure and on the O atom in the other. The structure is more stable with the positive charge on N, because N is the less electronegative atom. All other stability factors are the same (octets, formal charges, and number of covalent bonds).

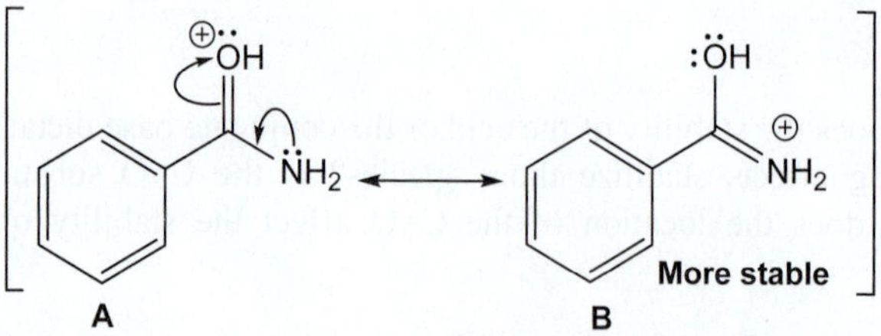

A

B

More stable

Greater contribution

Problem 6.29

Think

Do all atoms have an octet? Does one structure have fewer atoms with nonzero formal charges? Are there more covalent bonds in one structure? What is the stability of the positive charge on the O atom versus the C atom?

Solve

(a) It is counterintuitive, because the resonance structure is more stable with the positive charge on the *less* electronegative atom. Thus, you would normally think that C^+ is more stable than O^+.

(b) The contributor on the right is more stable, because every atom in the structure has a complete octet. In the contributor on the left, the C atom with the positive charge is deficient of an octet.

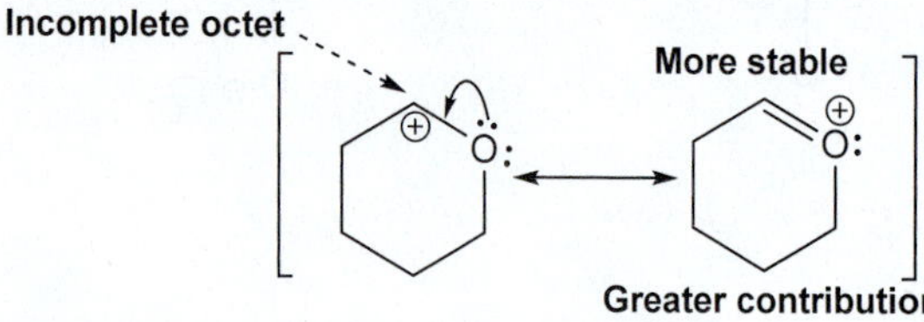

Incomplete octet

More stable

Greater contribution

Problem 6.30

Think

Do all atoms have an octet? Does one structure have fewer atoms with nonzero formal charges? Are there more covalent bonds in one structure? Is CF_3 an electron-withdrawing or electron-donating group? Is the positive charge more stable on the C atom next to the CF_3, as in resonance structure **A**, or farther away from the CF_3, as in structure **B**?

Solve

CF_3 is an electron-withdrawing group. Electron-withdrawing groups destabilize a positive charge. The carbocation C^+ is more destabilized when the CF_3 is in closer proximity. Therefore, **B** is the more important resonance contributor.

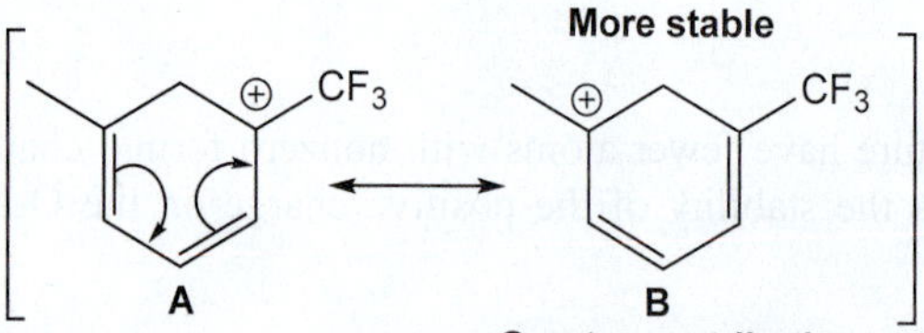

More stable

A

B

Greater contribution

Problem 6.31

Think

What is the structure of alanine in its fully protonated form (i.e., at the most acidic pH)? What are the pK_a values of alanine? What is the relationship between the pH and the pK_a values? Which proton will be removed first, and at what pH? Which will be removed second?

Solve

Alanine's structures at various pH values are shown below. The pK_a values from Table 6-2 are 2.35 and 9.87, so when the solution pH is significantly above those values, the corresponding proton present will almost entirely be dissociated.

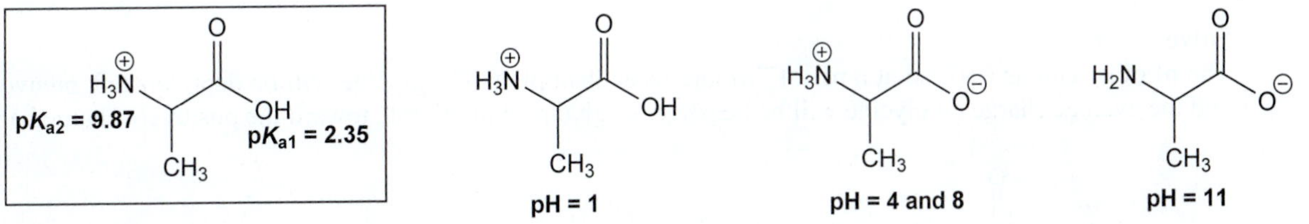

Problem 6.32

Think

What is the structure of glutamic acid in its fully protonated forms (i.e., at the most acidic pH)? What are the pK_a values of each H? What is the relationship between the pH and the pK_a values? Which proton will be removed first, and at what pH? Which will be removed second? Third?

Solve

The fully protonated form of glutamic acid is shown below on the left. Deprotonations for glutamic acid occur at the pH = pK_a values 2.10, 4.07, and 9.47. So the species below are the most abundant at the given pH values.

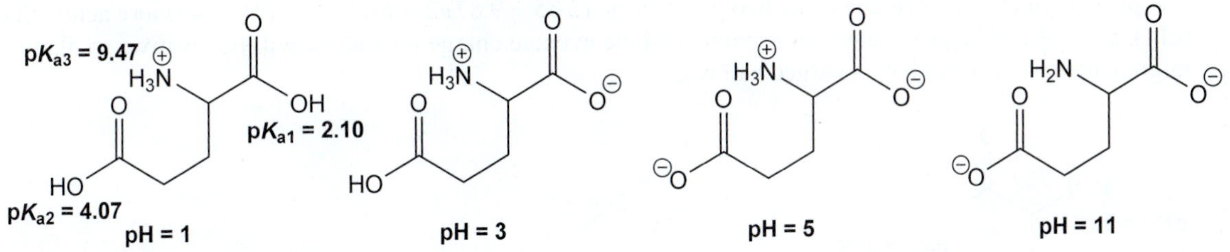

Problem 6.33

Think

What is the structure of arginine in its fully protonated forms (i.e., most acidic pH)? What are the pK_a values of each H? What is the relationship between the pH and the pK_a values? Which proton will be removed first, and at what pH? Which will be removed second? Third?

Solve

The fully protonated form of arginine is shown below on the left. Deprotonations for arginine occur at the pH = pK_a values 2.01, 9.04, and 12.48. So the species below are the most abundant at the given pH values.

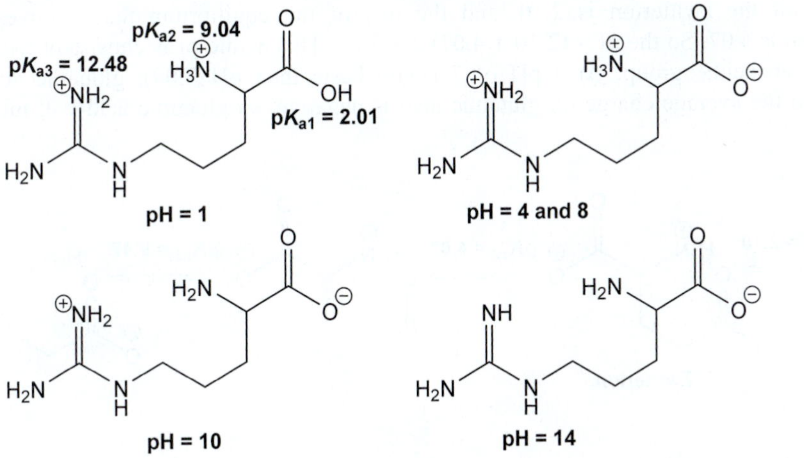

Problem 6.34

Think

What are the pK_a values for glycine? What is the isoelectric point (pI) of glycine? Is the pH greater than, less than, or equal to the pI? Is this species positive, negative, or uncharged on average? Is the cathode positive or negative? What is the charge at the anode?

Solve

The pI of glycine is 6.07, so at a pH of 7 (more basic than pH 6.07), glycine will be deprotonated, on average, and the average charge on glycine will be negative, so glycine will migrate toward the positively charged anode.

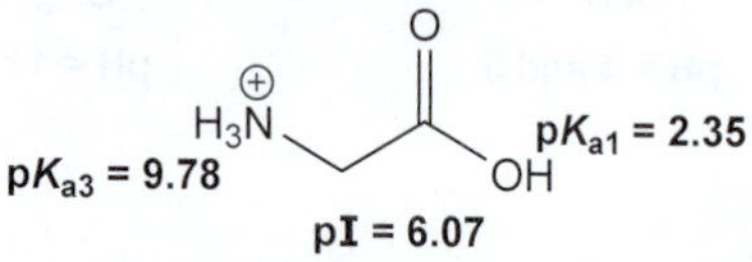

pK_{a3} = 9.78 pK_{a1} = 2.35 pI = 6.07

Problem 6.35

Think

What are the pK_a values for alanine? How do you calculate the pI of alanine from those values? Is the pH greater than, less than, or equal to the pI? Is this species positive, negative, or uncharged on average? Is the cathode positive or negative? What is the charge at the anode?

Solve

The pI of alanine is the average of its two pK_a values: (2.35 + 9.87)/2 = 6.11. At a pH of 4 (more acidic than pH 6.11), alanine will be protonated, on average, and the average charge on alanine will be positive, so alanine will migrate toward the negatively charged cathode.

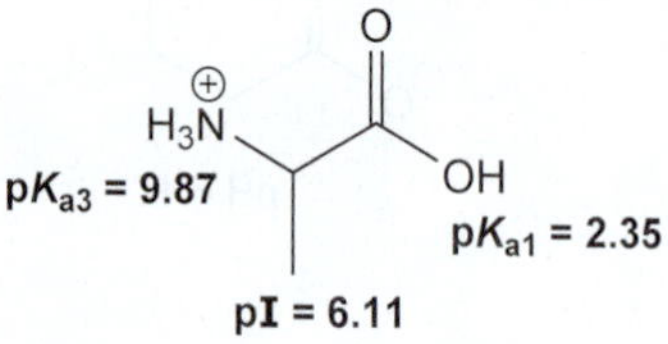

pK_{a3} = 9.87 pK_{a1} = 2.35 pI = 6.11

Problem 6.36

Think

What are the pK_a values for glutamic acid? Which two pK_a values involve the zwitterion? How do you calculate pI from those pK_a values? Is the pH greater than, less than, or equal to the pI? Is this species positive, negative, or uncharged on average? Is the cathode positive or negative? What is the charge at the anode?

Solve

The species that exist in solution for glutamic acid are shown below. The pK_a of the proton transfer equilibrium that leads to the formation of the zwitterion is 2.10, and the pK_a of the equilibrium that involves the deprotonation of the zwitterion is 4.07. So the pI = (2.10 + 4.07)/2 = 3.09. This acidic pI is consistent with the fact that the side chain has an acidic group. At a pH of 7 (more basic than pH 3.09), glutamic acid is deprotonated, on average, and the average charge on glutamic acid is negative, so glutamic acid will migrate toward the positive anode.

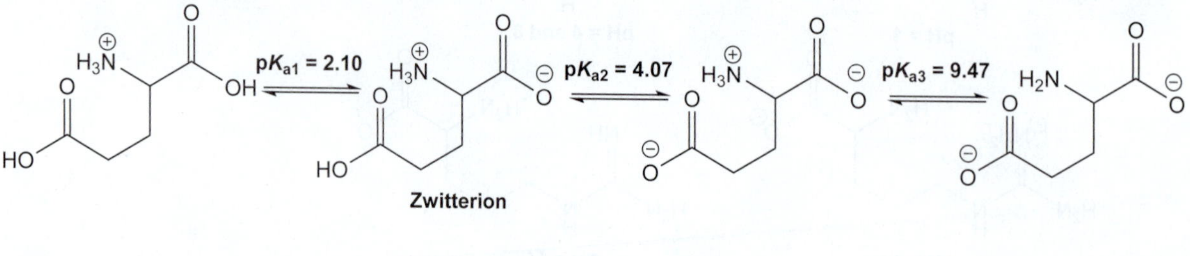

pK_{a1} = 2.10 pK_{a2} = 4.07 pK_{a3} = 9.47

Zwitterion

Problem 6.37

Think

Is the pH greater than, less than, or equal to the pI? Is this species positive, negative, or uncharged on average? Is the cathode positive or negative? What is the charge at the anode? At what pH is lysine uncharged?

Solve

At a pH of 12, the pH is above the pI, so lysine will be deprotonated, on average, and the average charge on the species will be negative, so the species will migrate toward the positively charged anode. At a pH value of 1 or 7, the opposite is true. The species will not migrate at a pH of 9.74, the value of its pI.

End of Chapter Problems

Sections 6.1–6.3 The Proton Transfer Reaction, Equilibrium, and Thermodynamics

Problem 6.38

Think

What do the curved arrows show? Which bonds are broken and formed in each proton transfer reaction?

Solve

Curved arrows show movement of two electrons. In each reaction, the base accepts a proton to form the conjugate acid, and the acid donates a proton to form the conjugate base. Products for reactions **(a)**–**(d)** are shown below.

(a)

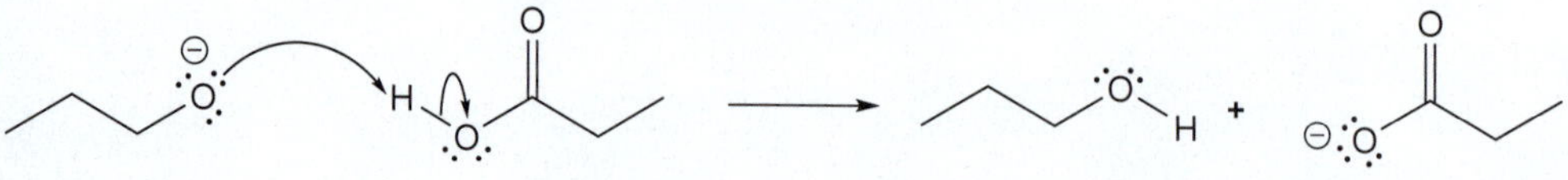

(b)

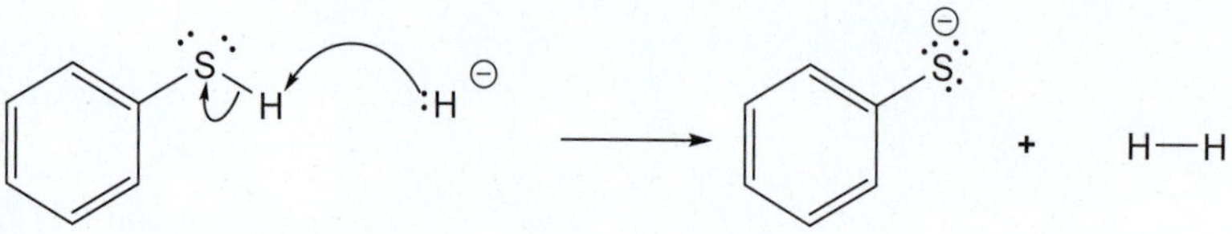

(c)

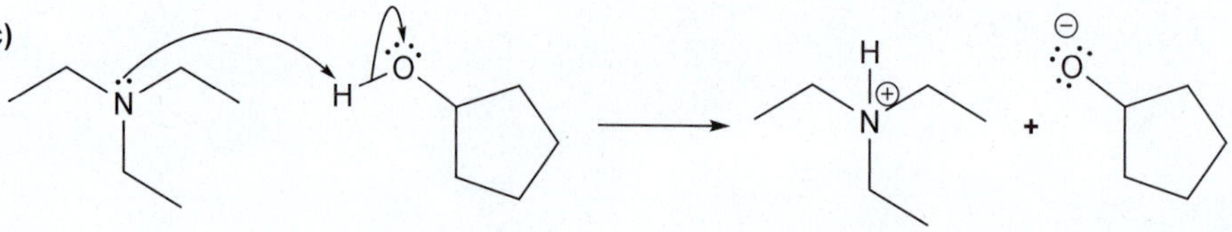

(d)

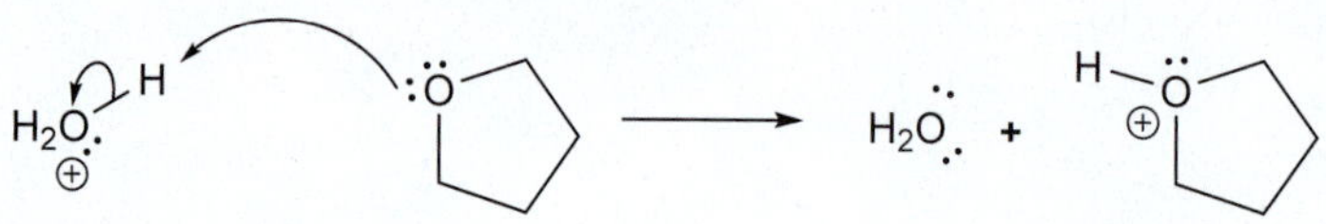

Problem 6.39

Think

Do any lone pairs of electrons in the reactants become bonding pairs in the products? Do any bonding pairs of electrons in the reactants become lone pairs in the products? How do you show these conversions using curved arrows? Which bonds are broken and formed in each proton transfer reaction?

Solve

Curved arrows and missing nonbonding electrons are shown below and on the next page. Each proton transfer reaction requires two arrows. One arrow shows the bond forming from the lone pair of the base to the H atom of the acid, and the other arrow shows the bond breaking between the H atom and the acid, with the bonding pair of electrons ending up as a lone pair.

(a)

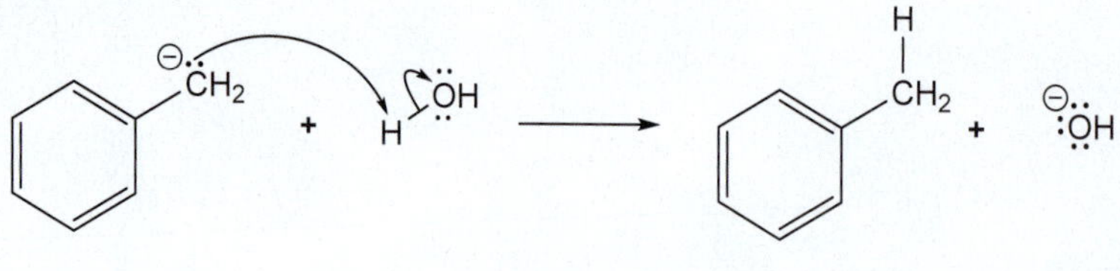

(b)

(c)

(d)

Problem 6.40

Think

What is the concentration of the acid initially? What is the K_a of the acid? Does the K_a expression involve initial concentrations or equilibrium concentrations? How can you relate the two concentrations using a variable? How do the initial and equilibrium concentrations relate to the percent ionization?

Solve

The K_a expression involves equilibrium concentrations, but the initial and equilibrium concentrations are related by a difference x. The percent ionization is the amount of acid that has dissociated, x, divided by the initial amount of the acid.

	$[C_6H_5OH]$		$[C_6H_5O^-]$	$[H_3O^+]$
Initial	0.100 M	$\rightleftharpoons$	0 M	0 M
Change	$-x$		$+x$	$+x$
Equilibrium	$0.100 - x$		x	x

So at equilibrium, $x^2/(0.100 - x) = K_a = 10^{-pK_a} = 10^{-10.0}$.
Solving, $x = 3.16 \times 10^{-6}$ M $= [C_6H_5O^-]_{eq}$.
% dissociation = $(3.16 \times 10^{-6}$ M$)/(0.100$ M$) \times 100\% = 0.00316\%$

Problem 6.41

Think

What bonds are broken and formed in an acid/base (proton transfer) reaction? Which side of the reaction is favored at equilibrium: the side that has the stronger acid or the side opposite the stronger acid? How do pK_a differences relate to the numerical factor by which that side is favored?

Solve

(a) The products of each reaction are shown on the next page, as are the pK_a for each acid involved in the respective equilibria.

(b) Products are favored for the reactions in **(iii)**, **(iv)**, **(v)**, **(vi)**, and **(vii)**, because the acid on the reactant side is stronger (lower pK_a) than the one on the product side. The opposite is true for **(i)** and **(ii)**.

(c) The numerical factor by which a particular side is favored is 10 raised to the pK_a difference between the acids on either side of the equilibrium.

(i) NH_2 + $^{\ominus}OH$ ⇌ $NH^{\ominus}$ + H_2O Reactant side favored by $10^{1.3} = 5.0 \times 10^2$

pK_a~17 $pK_a = 15.7$

(ii) OH + $Cl^{\ominus}$ ⇌ $O^{\ominus}$ + H—Cl Reactant side favored by $10^{23.5} = 3.2 \times 10^{24}$

pK_a ~ 16.5 $pK_a = -7$

(iii) $N^{\ominus}$ + ⇌ H N + $\ominus$ Product side favored by 10^{22}

pK_a ~16 $pK_a = 38$

(iv) $H^{\ominus}$ + ⇌ H_2 + $\ominus$ Product side favored by 10^{10}

pK_a ~ 25 $pK_a = 35$

(v) O $^{\ominus}O$ + $H_3O^{\oplus}$ ⇌ O HO + H_2O Product side favored by $10^{6.7} = 5 \times 10^6$

$pK_a = -1.7$ pK_a ~5

(vi) $\ominus$: + HO ⇌ H + $^{\ominus}O$ Product side favored by 10^{27}

pK_a ~ 16 $pK_a = 43$

(vii) O OH + $H^{\ominus}$ ⇌ O $^{\ominus}O$ + H_2 Product side favored by $10^{30.25} = 1.8 \times 10^{30}$

$pK_a = 4.75$ $pK_a = 35$

Problem 6.42

Think

What species is the acid and what species is the base? Which side of the reaction is favored at equilibrium: the side that has the stronger acid or the side opposite the stronger acid? How do pK_a differences relate to the numerical factor by which that side is favored?

Solve

The pK_a values in Table 6-1 are 35 for H_2 and 16 for CH_3CH_2OH, so the latter is the stronger acid (lower pK_a). The difference in pK_a values is 19, and therefore, the reaction favors the products by a factor of 10^{19}.

$H:^{\ominus}$ + H–Ö– ⟶ H—H + $^{\ominus}$:Ö–

Hydride anion Ethanol Ethoxide

Problem 6.43

Think

When the pH equals the pK_a, what is the percent ionization? What is the ratio of [A^-]/[HA] when the percent ionization is 90% and 10%?

Solve

An acid dissociates 50% (i.e., [A^-] = [HA]) when the pH equals the pK_a. That is, pH = 0.77. The acid dissociates 90% (i.e., [A^-]/[HA] = 9) when the pH is more basic than pK_a by one unit (i.e., pH = 1.77). The acid dissociates 10% (i.e., [A^-]/[HA] = 0.111) when the pH is more acidic than pK_a by one unit (i.e., pH = −0.23).

Problem 6.44

Think

What does the pK_a indicate about the pH at which the proton is lost? How many pH units above the pK_a does the solution need to be for the acid to be 99% deprotonated? How many pH units below the pK_a does the solution need to be for the acid to be 99% protonated? Use Figure 6-1 as a guide.

Solve

We expect the dominant form to be the protonated form (>99%) when pH is more acidic than pK_a by at least two units, or pH < 2.6. We expect the deprotonated form to be dominant (>99%) when pH is more basic than pK_a by more than two units, or pH > 6.6. We expect equal amounts of the two forms when pH = pK_a = 4.6.

Problem 6.45

Think

How does the leveling effect apply when ethanol is the solvent? What is the pK_a of ethanol? What is the pK_a of the conjugate acid of each species listed for comparison? Is it desirable for the solvent to react with the solute or to remain unreacted? Are you looking for a stronger or weaker acid compared to ethanol?

Solve

With ethanol as the solvent, the strongest base that can exist in solution to any appreciable extent is the conjugate base of ethanol, $CH_3CH_2O^-$. Any base that is stronger than $CH_3CH_2O^-$ will deprotonate ethanol readily. Our desire, however, is for the bases to remain unreacted when dissolved in ethanol, so only those bases that will not deprotonate ethanol (pK_a = 16) are acceptable. Their conjugate acids, therefore, must be stronger than ethanol, or <16. Only **(b)**, **(c)**, **(d)**, and **(e)** are acceptable.

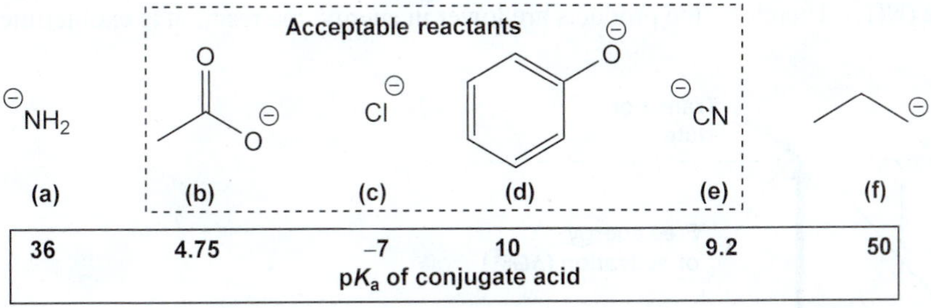

Problem 6.46

Think

How does the leveling effect apply when ethanamine is the solvent? What is the pK_a of ethanamine? What is the pK_a of the conjugate acid of ethanamine, $CH_3CH_2NH_3^+$? For comparison, what is the pK_a of each acid listed and the pK_a of each base's conjugate acid? Is it desirable for the solvent to react with the solute or to remain unreacted? Are you looking for a stronger or weaker acid compared to ethanamine? A stronger or weaker acid compared to the conjugate acid of ethanamine?

Solve

With ethanamine as the solvent, the strongest base that can exist is the conjugate base of ethanamine, $CH_3CH_2NH^-$, and the strongest acid that can exist is the conjugate acid of ethanamine, $CH_3CH_2NH_3^+$. Only those bases that will not deprotonate ethanamine ($pK_a \approx 38$) are acceptable, and only those acids that will not protonate ethanamine to produce $CH_3CH_2NH_3^+$ ($pK_a = 10.6$) are acceptable. The conjugate acids of the bases, therefore, must be stronger than ethanamine, or <38. Bases **(a)**, **(d)**, and **(f)** are acceptable bases (dashed boxes); **(b)** and **(e)** are acids, and to remain unreacted, they must be less acidic than $CH_3CH_2NH_3^+$, or >10.6. Both are more acidic, however, so they are unacceptable. Base **(c)** is unacceptable because its strength is too strong and will deprotonate ethanamine.

	(a)	(b)	(c)	(d)	(e)	(f)
pK_a of acid		−7			9.4	
Species	Cl^-	HCl	CH_3^-	NH_2^-	NH_4^+	OH^-
pK_a of conjugate acid	−7		48	36		15.7

Problem 6.47

Think

Which species is the acid and which species is the base? What are the products of the reaction? On the basis of pK_a values, which acid is stronger: the one on the reactant side or the conjugate acid on the product side? What does the acid strength indicate about the stability of the species?

Solve

(a) The curved arrow notation and products are given below.

H + $:\ddot{N}H_2^-$ → + $\ddot{N}H_3$

$pK_a = 25$ $\qquad$ $pK_a = 38$

(b) The reaction heavily favors products because the acid on the reactant side (RC≡CH) is much stronger than that on the product side (NH_3). Therefore, the products are lower in energy; the reaction is exothermic.

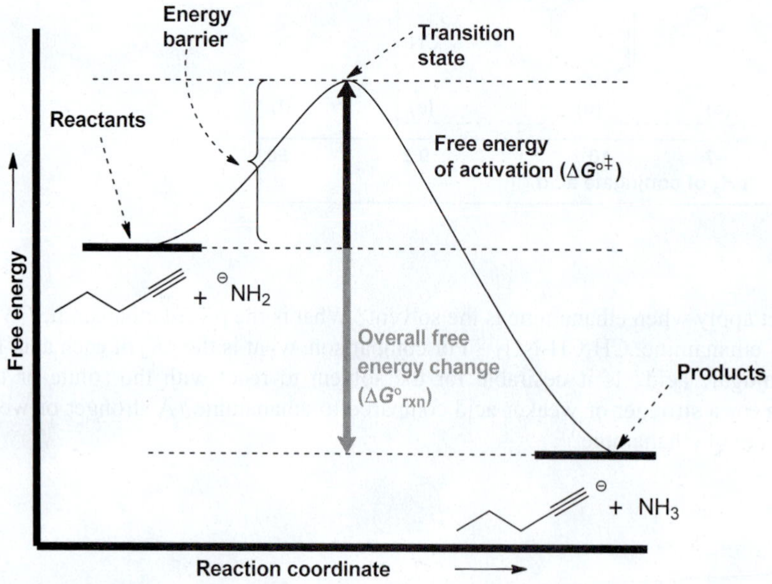

Problem 6.48

Think

Which species is the acid and which species is the base? What are the products of the reaction? On the basis of pK_a values, which acid is stronger: the one on the reactant side or the conjugate acid on the product side? What does the acid strength indicate about the stability of the species?

Solve

(a) The curved arrow notation and products are given below.

pK_a = 16 pK_a = 20

(b) The reaction is significantly product favored because the acid on the reactant side (CH_3CH_2OH) is stronger than that on the product side (ketone). Therefore, the products are lower in energy; the reaction is exothermic.

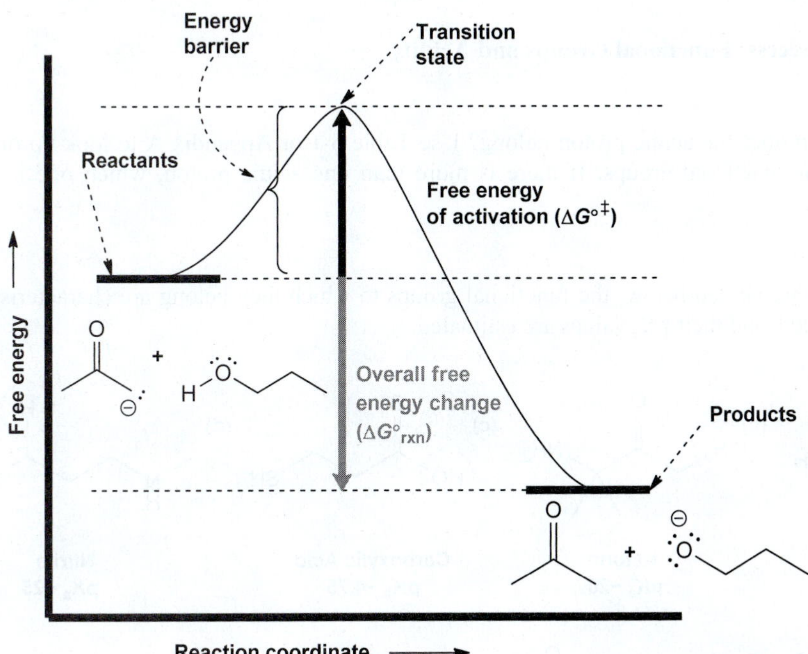

Problem 6.49

Think

Which product is lower in energy? Which product is more stable? What does the stability and energy content of the product indicate about the extent of formation of the product at equilibrium?

Solve

In both reactions, Y is in greater abundance at equilibrium because it is more stable (lower in energy). For the energy diagram on the left, the formation of Y has a greater energy barrier, but the energy barrier is related to reaction rates, not the abundance of a species at equilibrium.

Free energy
W
X
Y
Lower in energy, greater abundance
Reaction coordinate

Free energy
W
X
Y
Lower in energy, greater abundance
Reaction coordinate

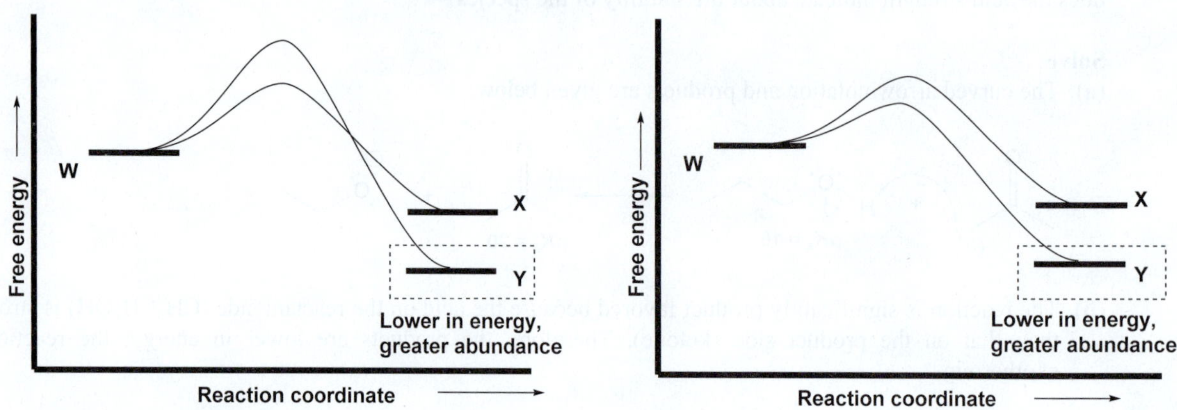

Section 6.4 Strategies for Success: Functional Groups and Acidity

Problem 6.50

Think

To what functional group does the acidic proton belong? Use Table 6-1 or Appendix A to look up or estimate the pK_a based on similar functional groups. If there is more than one acidic proton, which one is the most acidic?

Solve

The most acidic protons are circled below; the functional groups to which they belong are characteristic of the compound classes provided, and their pK_a values are estimated.

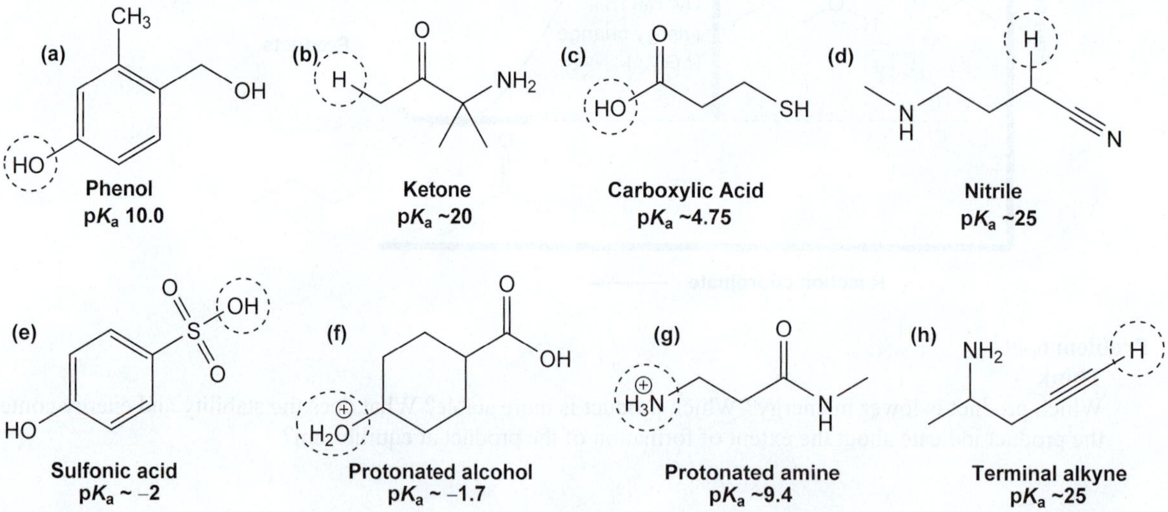

(a)
CH3
OH
HO
Phenol
pKa 10.0
(b)
H
O
NH2
Ketone
pKa ~20
(c)
O
HO
SH
Carboxylic Acid
pKa ~4.75
(d)
H
N
H
N
Nitrile
pKa ~25
(e)
O
S
OH
O
HO
Sulfonic acid
pKa ~ −2
(f)
O
OH
H2O
Protonated alcohol
pKa ~ −1.7
(g)
O
H3N
N
H
Protonated amine
pKa ~9.4
(h)
NH2
H
Terminal alkyne
pKa ~25

Problem 6.51

Think

Which sites can accept a proton? What would the conjugate acid be? What is the relationship between the pK_a of the conjugate acid and base strength? How can you estimate the pK_a value of each conjugate acid?

Solve

To determine relative base strength, we can use pK_a values of their conjugate acids. The pK_a relationship is such that the stronger base is associated with the weaker conjugate acid. The most basic site of each species is circled below; the functional groups to which their conjugate acids belong are characteristic of the compound classes provided, and those pK_a values are estimated.

(a) ROH (pK_a ~16) is a weaker acid than RNH_3^+ (pK_a ~10.6).

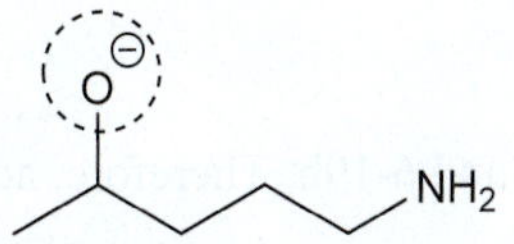

(b) RNH_3^+ (pK_a ~10.6) is a weaker acid than ROH_2^+ (pK_a ~ –1.7).

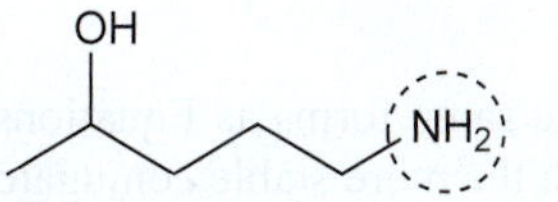

(c) RSH (pK_a ~7.2) is a weaker acid than ROH_2^+ (pK_a ~ –1.7).

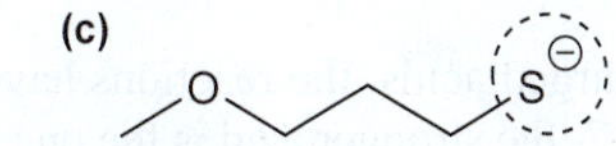

(d) A ketone (pK_a ~20) is a weaker acid than a carboxylic acid (pK_a ~4.75).

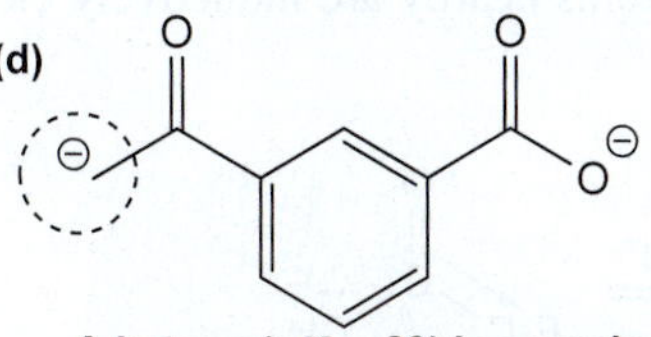

(e) RNH_2 (pK_a ~ 38) is a weaker acid than ROH (pK_a ~16).

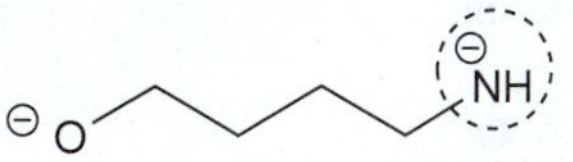

Problem 6.52

Think

Draw the complete deprotonation reactions. Should the differences in acidity be due primarily to the stabilities of the acids themselves or to those of the conjugate bases? Which alcohol shares similar resonance and inductive properties?

Solve

Being uncharged acids, the reactions have the same forms as Equations 6-19a and 6-19b. Therefore, according to Figure 6-7, the stronger acid is the one with the more stable conjugate base. The acidity of molecule **A** should be the most similar to cyclohexanol. In both cases, the negative charge that develops in the conjugate base on the O atom is part of an OH group attached to an sp^3-hybridized C atom, and that C atom is adjacent to two alkyl groups. In **B**, the analogous C atom is bonded to only one alkyl group, and there is a nearby electronegative atom (part of the C=O) that will increase the acidity. In **C**, the nearby Cl will increase the acidity as well. The molecule whose acidity is most different from cyclohexanol is **D**, because the OH group is part of a different functional group altogether—a CO_2H group, which is characteristic of carboxylic acids.

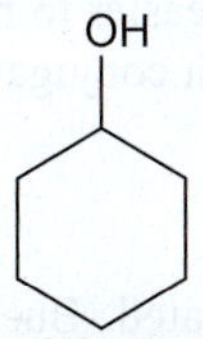

Cyclohexanol 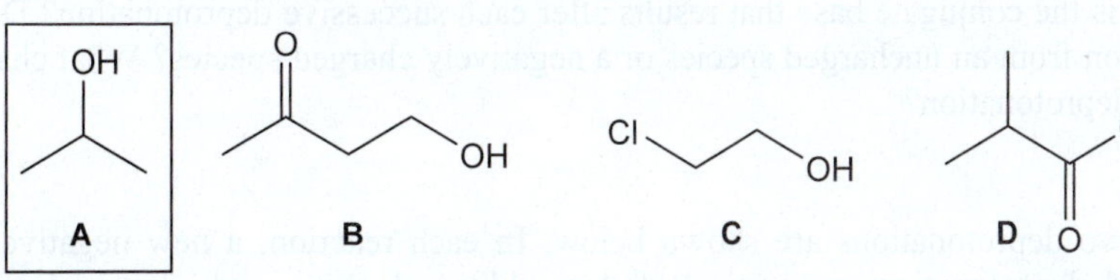 | **A** | **B** | **C** | **D**

Sections 6.5–6.6 Relative Acidities and Factors of Charge Stability

Problem 6.53

Think

Draw the complete deprotonation reactions. Should the differences in acidity be due primarily to the stabilities of the acids themselves or to those of the conjugate bases? Which anion, Cl^-, Br^-, or I^-, is most stable? Where is each located on the periodic table relative to the others? Which anion is the largest?

Solve

All reactions are of the form $HX \rightarrow H^+ + X^-$, where X changes. Being uncharged acids, the reactions have the same forms as Equations 6-19a and 6-19b. Therefore, according to Figure 6-7, the stronger acid is the one with the more stable conjugate base. The major difference in stability is with X^- on the product side. The more stable X^- is, the stronger the acid HX is. I^- is the largest anion (farthest down in the periodic table) and, therefore, is the most stable. Thus, acidity increases going down the periodic table column, **HCl < HBr < HI**.

Problem 6.54

Think

Identify the most acidic proton in each species and draw the complete deprotonation reactions. For uncharged acids, is acidity governed primarily by the stability of the acid itself or by that of its conjugate base? Does one structure have fewer atoms with nonzero formal charges? What is the stability of the ion (size and electronegativity of the atom having the charge, resonance, inductive effect)?

Solve

Being uncharged acids, the reactions have the same forms as Equations 6-19a and 6-19b. Therefore, according to Figure 6-7, the stronger acid is the one with the more stable conjugate base.

(a) The molecule on the right is more acidic. In both cases, the acidic H is on a CH_3 group adjacent to a C=O bond. In both conjugate bases, there is another resonance structure that allows the negative charge to be shared on the O of the C=O bond. But the highly electronegative F atoms nearby are inductively electron withdrawing and stabilize anions.

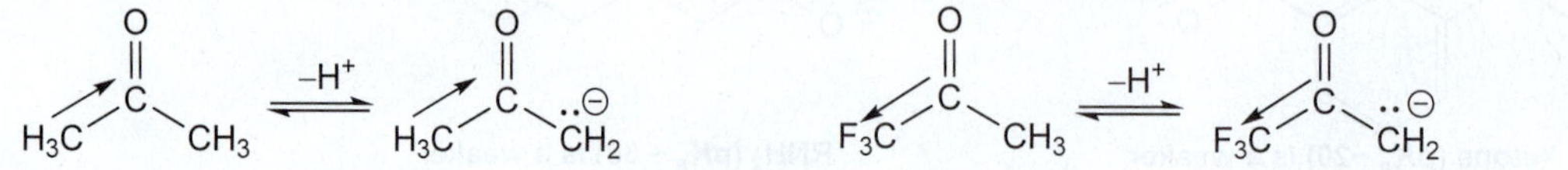

More stable conjugate base

(b) The molecule on the left is more acidic. In both cases, the acidic H is on an sp^3-hybridized C atom. In the conjugate base of the molecule on the left, that C atom is adjacent to two C=C double bonds (part of the ring), whereas in the molecule on the right, it is adjacent to only one. As we learned in the chapter, each adjacent multiple bond increases the strength of the acid as a result of the resonance delocalization of the charge that develops on deprotonation.

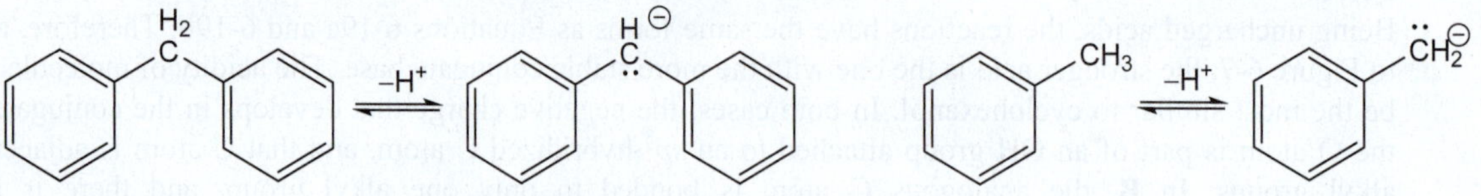

More stable conjugate base

Problem 6.55

Think

What is the conjugate base that results after each successive deprotonation? Do you think it is easier to remove a proton from an uncharged species or a negatively charged species? What charges exist in each conjugate base after deprotonation?

Solve

The two deprotonations are shown below. In each reaction, a new negative charge is generated. But in the second deprotonation, generation of that additional charge introduces charge repulsion among the two −1 formal charges that doesn't exist in HSO_4^-. This additional destabilization in SO_4^{2-} makes the second deprotonation less energetically favorable than the first.

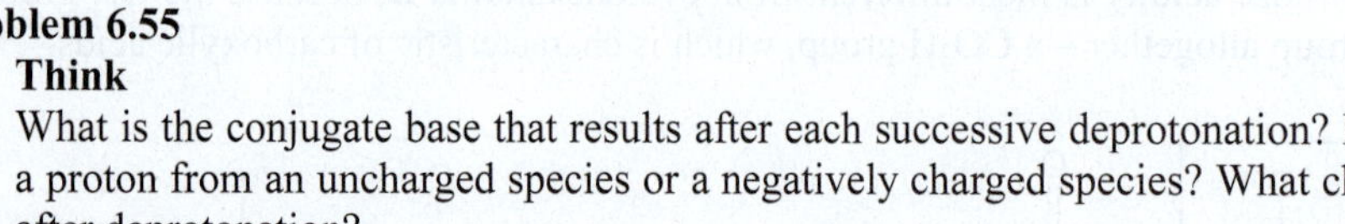

$pK_a = -9$

$pK_a = 2$
One negative charge

Two negative charges

Problem 6.56

Think

Which acid is stronger? How does the strength of the electron-withdrawing group affect the strength of the acid? Consider the stability of the conjugate base.

Solve

Because the pK_a is lower (stronger acid) with the NO_2 group, the NO_2 group must better stabilize the negative charge that develops in the conjugate base. This can happen only if the NO_2 group is a stronger electron-withdrawing group, more effectively reducing the concentration of negative charge that is produced. As we can see in the structures below, the N atom has a +1 formal charge, making it very electron deficient. The C=O carbon has just a partial positive charge, so it is not as electron deficient. Stronger electron-withdrawing groups, in general, increase the stability of the anion conjugate base, and thus the acid is stronger.

pK_a = 1.68 pK_a = 2.83

Problem 6.57

Think

Which color (red or blue) indicates more electron density? Which O^-, according to the electrostatic potential maps, bears more electron density? What does the greater electron density indicate about the strength of the electron-withdrawing group?

Solve

The red region (representing a buildup of negative charge or electron density) is slightly larger in $CF_3CH_2O^-$ than in $NCCH_2O^-$. This suggests that the CN group more effectively delocalizes the negative charge over the rest of the molecule, so the CN group is more electron withdrawing.

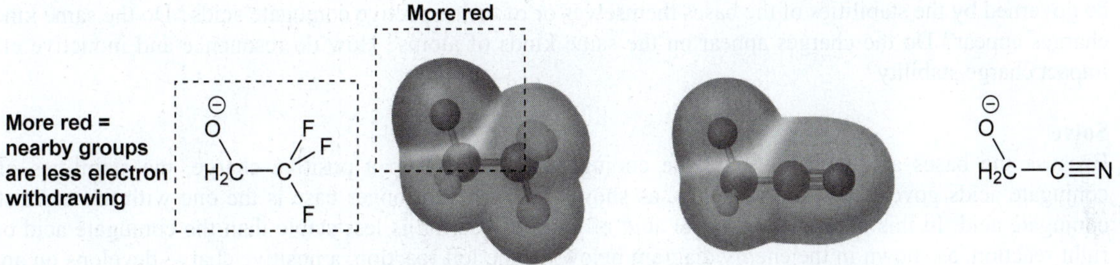

Problem 6.58

Think

Draw out the Lewis structure for each. Draw out the reaction where each base picks up a proton. Does the basic atom begin with a formal charge? Does it pick up a formal charge?

Solve

HNC is the stronger base because it has a −1 formal charge on the C atom (see below). When it picks up a proton, that negative charge becomes 0. In HCN, a 0 charge on the N atom becomes +1 when it picks up a proton. It is more energetically favorable to neutralize a charge than it is to produce a charge.

$H{-}C{\equiv}N: \xrightleftharpoons{+H^+} H{-}C{\equiv}\overset{\oplus}{N}H$ $\quad H{-}\overset{\oplus}{N}{\equiv}\overset{\ominus}{C}: \xrightleftharpoons{+H^+} H{-}\overset{\oplus}{N}{\equiv}CH$

Stronger base

Problem 6.59

Think

Should the relative acidities be governed by the stabilities of the acids themselves or of their respective conjugate bases? Are there differences in the charges that appear? In the types of atoms on which the charges appear? In resonance effects? What is the difference in the inductive effect of Cl, Br, and I? Which one is the strongest electron-withdrawing group? What is the effect of an electron-withdrawing group on the stability of the conjugate base anion?

Solve

Being uncharged acids, the reactions have the same forms as Equations 6-19a and 6-19b. Therefore, according to Figure 6-7, the stronger acid is the one with the more stable conjugate base. In each case, OH becomes O^-, so the same charge is produced and on the same atom, O. The negative charge is delocalized by the same number of resonance structures involving the benzene ring. The main difference is inductive effects. Each halogen is electron withdrawing and thus stabilizes the negative charge generated in the conjugate base and increases the strength of the acid. Because electronegativity increases in the order I < Br < Cl, **A** should have the most stable conjugate base and, thus, should be the most acidic. Its pK_a is 9.0. **C** should have the least stable conjugate base and should be the weakest acid. Its pK_a is 9.2. The *bromo* compound, **B**, has a pK_a of 9.1.

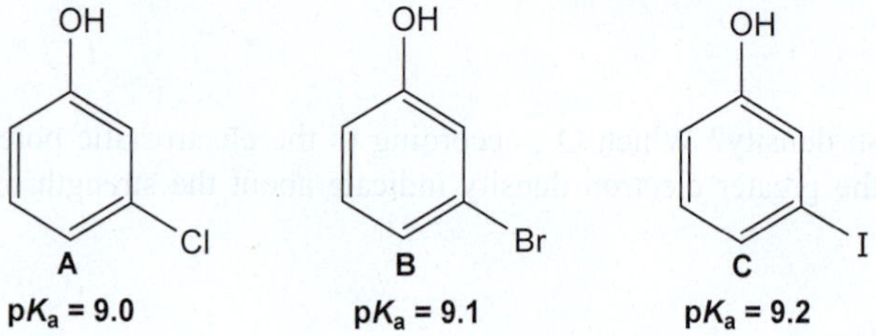

OH, Cl — **A** — $pK_a = 9.0$

OH, Br — **B** — $pK_a = 9.1$

OH, I — **C** — $pK_a = 9.2$

Problem 6.60

Think

What is the definition of a base? Draw the conjugate acid for each. For an uncharged base, should base strength be governed by the stabilities of the bases themselves or of the respective conjugate acids? Do the same kinds of charges appear? Do the charges appear on the same kinds of atoms? How do resonance and inductive effects impact charge stability?

Solve

Because the bases are uncharged and the conjugate acids pick up a positive charge, the stabilities of the conjugate acids govern the base strengths, as shown below; the stronger base is the one with the more stable conjugate acid. In this case, the conjugate acid of the left reaction is less stable than the conjugate acid of the right reaction, as shown in the energy diagram below. In the left reaction, a positive charge develops on an sp^2-hybridized N atom, whereas in the right reaction, it is on an sp^3-hybridized N atom. So, the N atom in the first reaction has a greater effective electronegativity and cannot accommodate the positive charge as well. Therefore, the second reaction is more energetically favorable, making the base in the second reaction stronger.

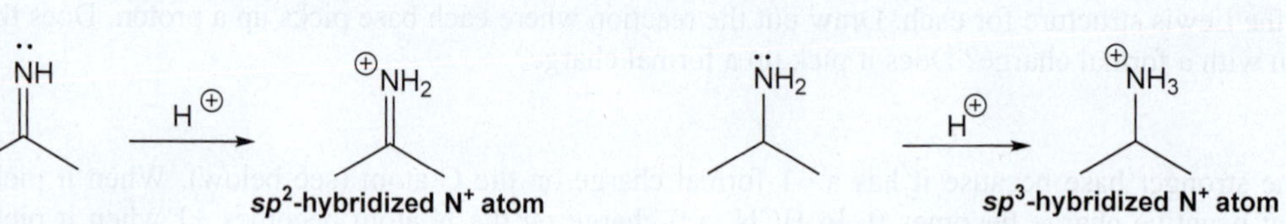

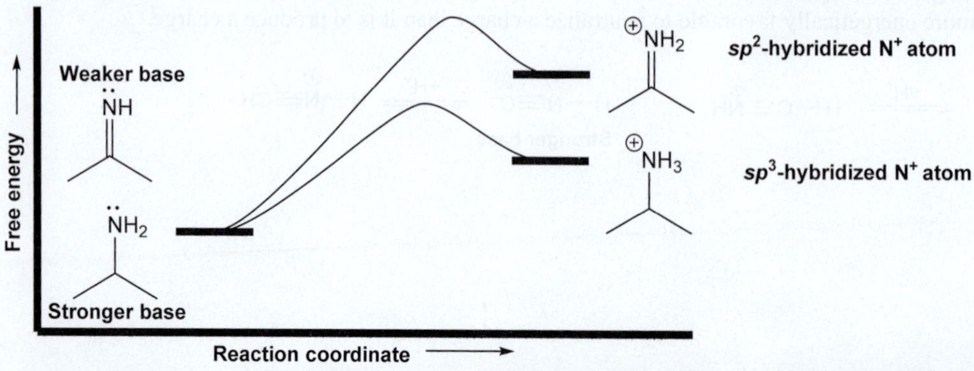

NH $\xrightarrow{H^+}$ NH_2^+ — **sp^2-hybridized N^+ atom**

NH_2 $\xrightarrow{H^+}$ NH_3^+ — **sp^3-hybridized N^+ atom**

Free energy

Weaker base NH

NH_2 **Stronger base**

NH_2^+ **sp^2-hybridized N^+ atom**

NH_3^+ **sp^3-hybridized N^+ atom**

Reaction coordinate

Section 6.7 Strategies for Success: Ranking Acid and Base Strengths—The Relative Importance of Effects on Charge

Problem 6.61

Think

How does acid strength relate to pK_a? Does the stronger acid have the F atom or the Cl atom? How does the acid strength change when the F is changed to a Cl? How does the acid strength change when the F and Cl atoms are moved farther away?

Solve

Distance from the reaction center is more important. Going from F to Cl, the pK_a changes by fewer than 0.3 units. Going from two atoms away from the reaction center to three atoms away, the pK_a changes by more than one unit.

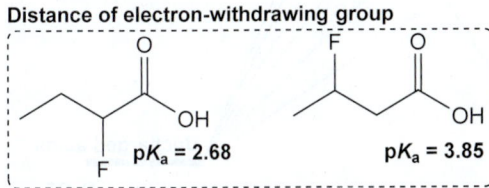

F → Cl

pK_a = 2.68   pK_a = 2.95

F → Cl

pK_a = 3.85   pK_a = 3.97

Distance of electron-withdrawing group

pK_a = 2.68   pK_a = 3.85

Distance of electron-withdrawing group

pK_a = 2.95   pK_a = 3.97

Problem 6.62

Think

Draw each complete protonation reaction. Does charge stability affect the reactant side? The product side? What factors affect the stability of a charged species? What is the nature of the atom with the negative charge? Which anions have resonance delocalization of the charge? Are inductive effects present? Which protonation reaction is most energetically favorable? Least?

Solve

As shown below, bases **A–F** are negatively charged and become uncharged upon protonation, whereas **G** is uncharged and becomes positively charged.

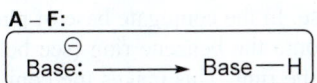

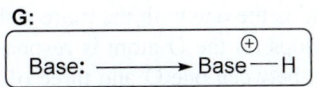

In the free energy diagram on the next page, therefore, base **G** appears lower in energy than bases **A–F**. Bases **B** and **D** are lower in energy than **C**, **E**, **A**, and **F**, because of the atom on which the negative charge appears. A negative charge on O is more stable than a negative charge on C. Base **D** is more stable than base **B** because the F atoms inductively stabilize the nearby negative charge. Of bases **C**, **E**, **A**, and **F**, **F** is the least stable. Although the negative charge on bases **C**, **E**, **A**, and **F** are all resonance delocalized, the negative charge on **F** is delocalized onto N, whereas that on **C**, **E**, and **A** is delocalized onto the O atom. N is less electronegative than O, so it cannot handle a negative charge as well. Base **C** is more stable than **A** or **E** owing to resonance. In base **C**, the negative charge is resonance delocalized onto two O atoms, whereas in **A** and **E**, the negative charge is delocalized onto only one O atom. Base **A** is less stable than base **E** because the CH_3 groups inductively destabilize the negative charge. On the product side, conjugate acid **G** is positively charged, making it less stable than conjugate acids **A–F**. With the relative order of the bases and the conjugate acids established, we can rank the base strengths according to how energetically favorable their protonation reactions are: **G** < **D** < **B** < **C** < **E** < **A** < **F**.

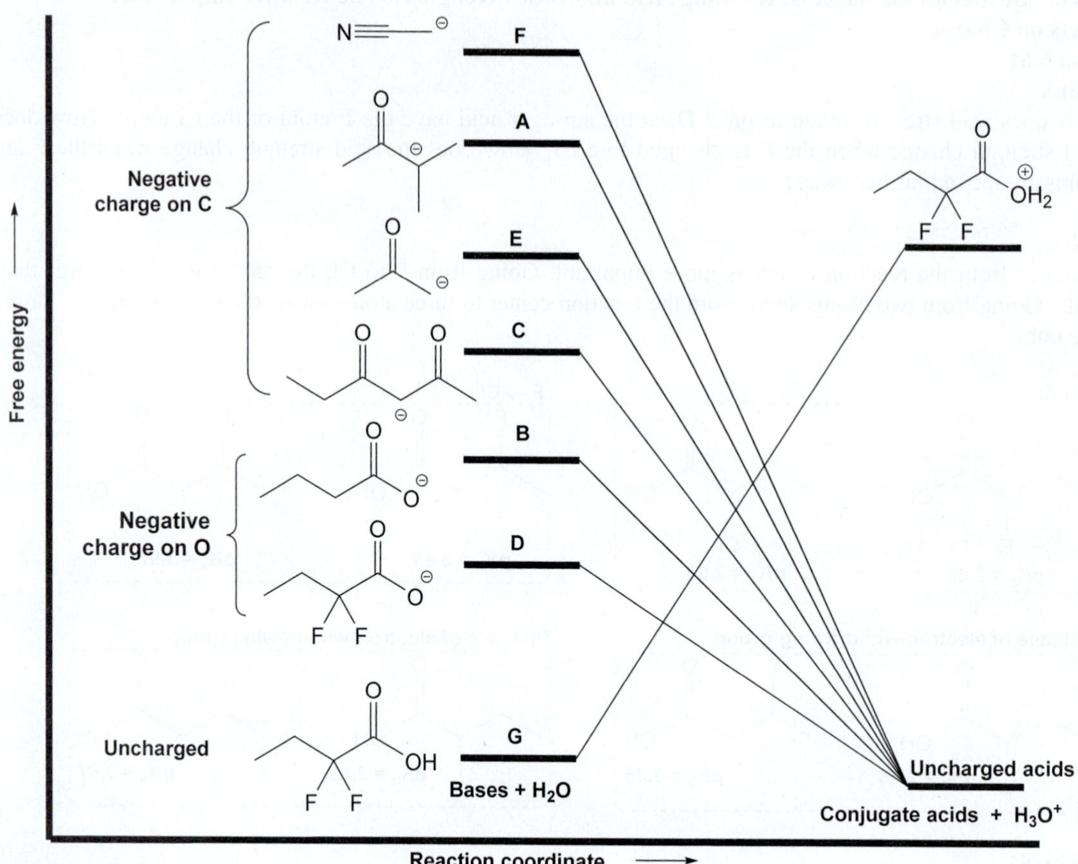

Problem 6.63

Think

Draw the complete deprotonation reactions. Should the differences in acidity be due primarily to the stabilities of the acids themselves or to those of the conjugate bases? Draw the conjugate base of each. Does the negative charge appear on the same atom? Does resonance delocalize the negative charge equally? What about inductive effects?

Solve

Being uncharged acids, the reactions have the same forms as Equations 6-19a and 6-19b. Therefore, according to Figure 6-7, the stronger acid is the one with the more stable conjugate base. In the conjugate base of phenol, the negative charge that develops on the O atom is resonance delocalized onto the benzene ring (see below). The negative charge is shared between one O and three of the C atoms on the ring. This makes the conjugate base of phenol substantially more stable than the conjugate base of methanol, CH_3O^-, for which the negative charge is localized just on O.

Etc.

In benzoic acid (see structures on the next page), the benzene ring does not participate in resonance with the negative charge that develops. The conjugate bases of benzoic acid and acetic acid therefore have similar charge stabilities because they exhibit the same types of resonance structures, so the acids have similar strengths.

No resonance around the ring

Benzoic acid and acetic acid are stronger acids compared to phenol because the negative charge is shared among two O atoms that are much more electronegative compared to C and, therefore, can handle the negative charge better.

Problem 6.64

Think

Draw the complete protonation reactions. Which base has more charge stability? Which acid has more charge stability? Consider the presence of a charge, the type of atom on which the charge appears, resonance, and inductive effects. What is the strength of the resulting conjugate acid, and how does that relate to base strength?

Solve

Stronger Base (boxed) — **Reasoning**

(a)

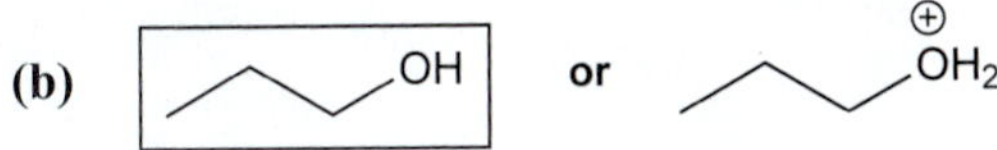

The charged base on the right is less stable than the one on the left owing to the presence of the charge. The opposite is true of their conjugate acids. Therefore, it is more energetically favorable for the species on the right to accept a proton, making it a stronger base.

(b)

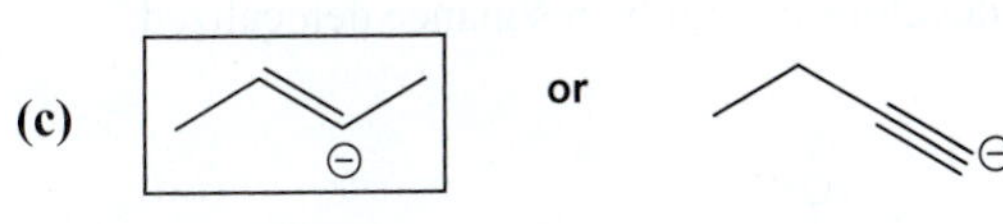

After accepting a proton, the base on the left will have an atom with a +1 charge, whereas the one on the right will have an atom with a +2 charge.

(c)

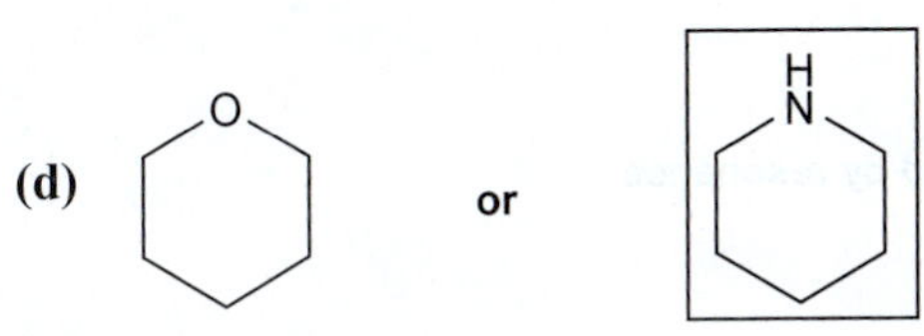

Both bases are negatively charged and become uncharged upon protonation, so the difference in base strength is determined primarily by the stability of the bases themselves. The first base has a negative charge on an sp^2-hybridized C atom, and the second base has a negative charge on an *sp*-hybridized C atom. With decreased *s* character, the C can't handle the negative charge as well, so it is higher in energy, less stable, and thus a stronger base.

(d)

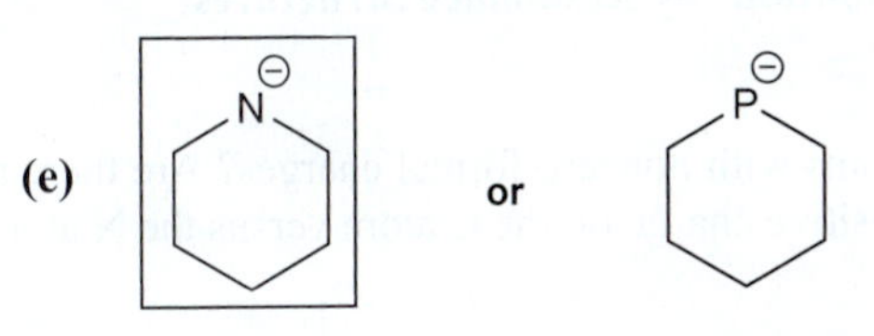

Of the two conjugate acids, $R_2NH_2^+$ is more stable than R_2OH^+, because the positive charge is on a less electronegative atom. Because it produces a more stable conjugate acid, R_2NH is the stronger base.

(e)

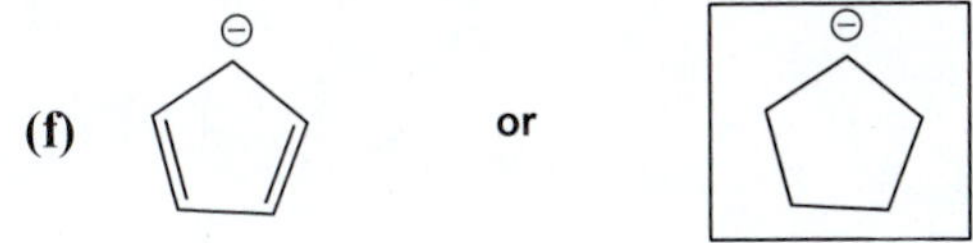

A negative charge is more stable on P than on N because P is lower in the periodic table and, thus, a larger atom.

(f) or

The negative charge on the first anion is stabilized by resonance but is not on the second anion.

Stronger Base (boxed) | **Reasoning**

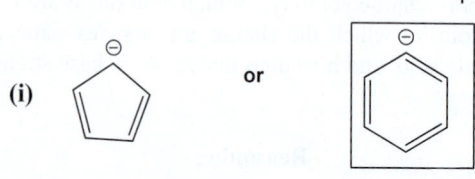

(g) or — The first anion is stabilized by resonance and the second anion is stabilized by the inductive effect, because F is electron withdrawing. The inductive effect is not as important as resonance stabilization.

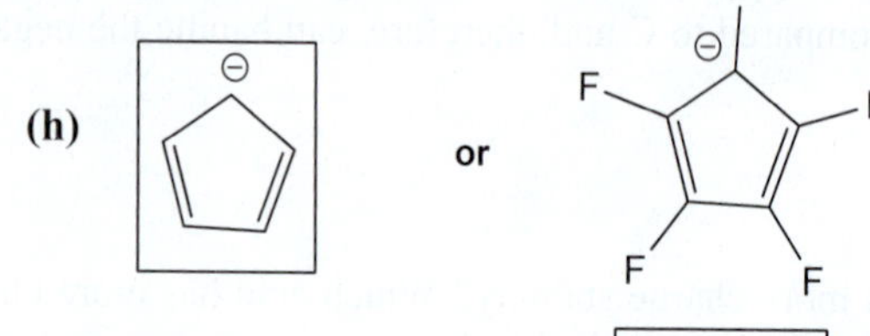

(h) or — The F atoms are electron withdrawing, which stabilizes the anion.

(i) or — The negative charge in the first anion is delocalized over the five C atoms via resonance. The second anion has resonance, but the charge remains localized.

Problem 6.65

Think

Draw the complete protonation reactions. Should the differences in base strength be due primarily to the stabilities of the bases themselves or to those of the conjugate acids? What is the conjugate acid for each? Are the charges the same? Do the charges appear on the same types of atoms? How do resonance and inductive effects impact charge stability?

Solve

Because the basic sites are uncharged and the conjugate acids pick up a positive charge, the stabilities of the conjugate acids govern the base strengths. When the O atom of the C=O group is protonated, a lone pair from the other O atom can participate in resonance to delocalize the positive charge that is generated (see below). When the OR group is protonated, the positive charge that develops cannot be resonance delocalized.

Ester

Not stabilized by resonance

Section 6.8 Strategies for Success: Determining Relative Contributions by Resonance Structures

Problem 6.66

Think

Do all atoms have an octet? Does one structure have fewer atoms with nonzero formal charges? Are there more covalent bonds in one structure? What is the stability of the positive charge on the C atom versus the N atom?

Solve

D has the greatest contribution to the resonance hybrid because it is the only resonance structure in which all atoms have a complete octet. Even though a C atom is less electronegative than a N atom, suggesting that **D** might have a lesser contribution, having complete octets is generally more important than charge stability.

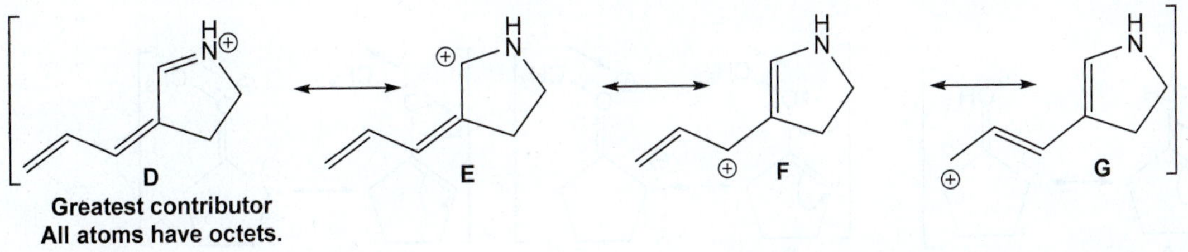

D E F G

Greatest contributor
All atoms have octets.

Problem 6.67

Think

Are there any π bonds adjacent to an atom with an incomplete octet? Do all resonance structures have the same number of atoms lacking an octet? Is the charge located on the same type of atom in each resonance structure? How do inductive effects impact the stability of the different C^+ species?

Solve

The species has three resonance structures, shown below. All resonance structures have a single C^+ atom lacking an octet. The rightmost structure is the most stable owing to electron-donating effects. The rightmost structure has two electron-donating alkyl groups directly attached to C^+, whereas the other structures do not.

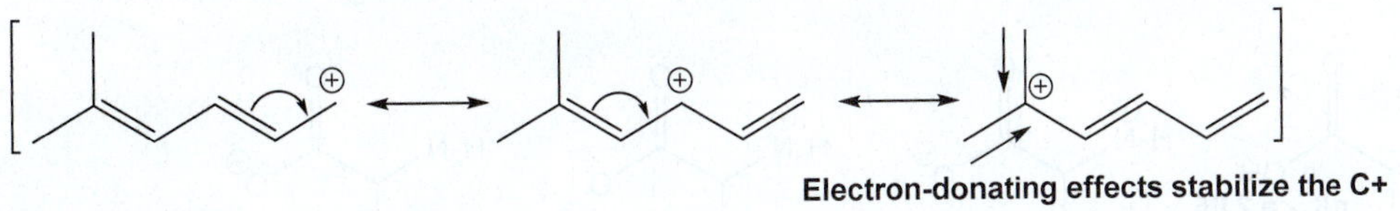

Electron-donating effects stabilize the C+

Problem 6.68

Think

Are there any lone pairs adjacent to a double bond? Do any of the double bonds convert to lone pairs? Is CF_3 electron donating or withdrawing? Does an electron-withdrawing group stabilize or destabilize a negative charge? Is the negative charge more or less stable on an O atom compared to a C atom?

Solve

There are three resonance structures possible for the ion in this problem, drawn below and labeled **i–iii**. CF_3 is an electron-withdrawing group, and electron-withdrawing groups stabilize nearby negative charges. The more stable structures have the negative group closer to the electron-withdrawing CF_3 group. A negative charge on an O atom is more stable than a negative charge on a C atom, because oxygen is more electronegative than carbon. Therefore, the strongest contributor is **iii**, followed by **ii**, then **i**.

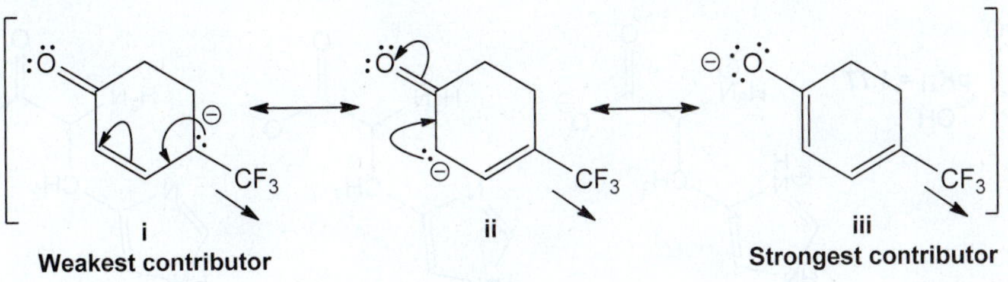

i ii iii

Weakest contributor **Strongest contributor**

Problem 6.69

Think

Draw the resonance structure for each that has the separated positive and negative charges. Is CH_3 electron donating or withdrawing? What about CF_3?

Solve

The resonance structures with the separated positive and negative charges for **A–C** are drawn below. In all cases, the negative charge is on C and the positive charge is on O. CH_3 is an electron-donating group, which stabilizes O^+, whereas CF_3 is an electron-withdrawing group, which destabilizes O^+. Therefore, **B** is the strongest contributor, and **C** is the weakest.

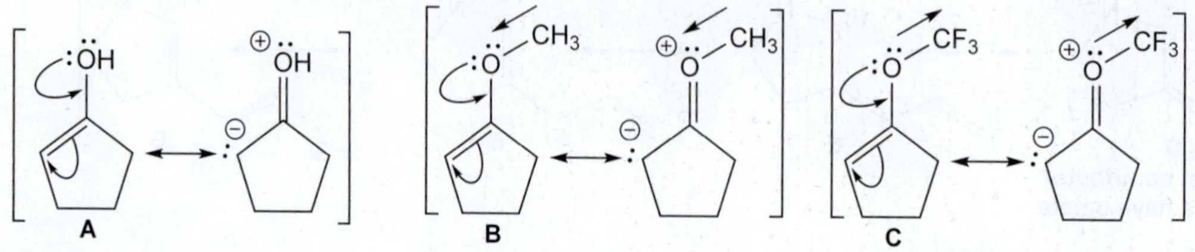

A **B** **C**

Sections 6.9–6.10 The Organic Chemistry of Biomolecules

Problem 6.70

Think

What is the structure of cysteine in its fully protonated form (i.e., at the most acidic pH)? What are the pK_a values of each H? What is the relationship between the pH and the pK_a values? Which proton will be removed first, and at what pH? Which will be removed second? Third?

Solve

The fully protonated form of cysteine is shown below on the left. Deprotonations for cysteine occur at the pH = pK_a values 2.05, 8.00, and 10.25. So the species below are the most abundant at the given pH values.

pK_{a3} = 10.25 pK_{a1} = 2.05 pK_{a2} = 8.00

pH = 1 pH = 4 and 6 pH = 9 pH = 11

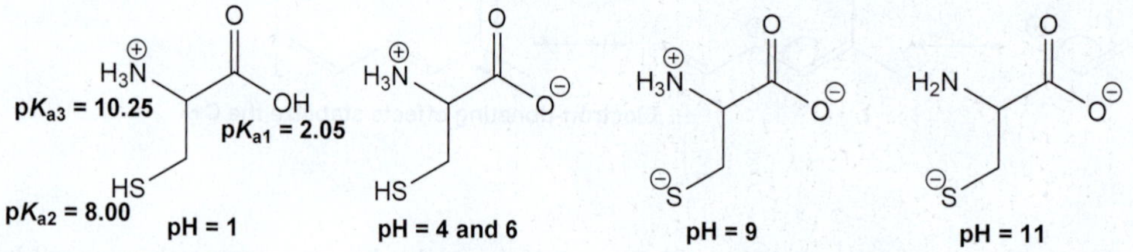

Problem 6.71

Think

What is the structure of histidine in its fully protonated form (i.e., most acidic pH)? What are the pK_a values of each H? What is the relationship between the pH and the pK_a values? Which proton will be removed first, and at what pH? Which will be removed second? Third?

Solve

The fully protonated form of histidine is shown below on the left. Deprotonations for histidine occur at the pH = pK_a values 1.77, 6.10, and 9.18. Therefore, the species below are the most abundant at the given pH values.

pK_{a3} = 9.18 pK_{a1} = 1.77 pK_{a2} = 6.10

pH = 1 pH = 3 and 5 pH = 7 pH = 11

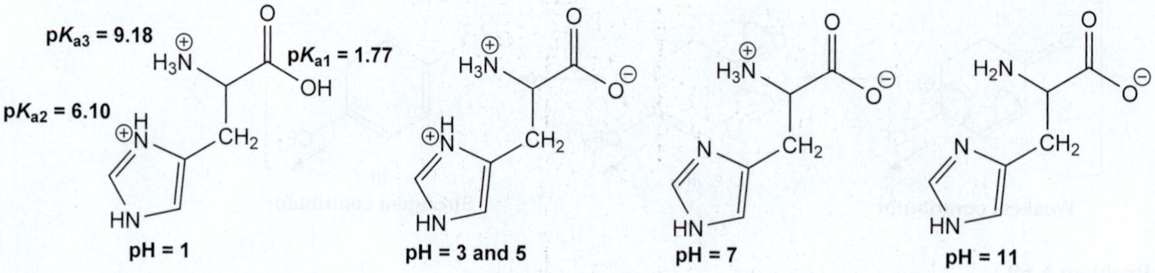

Problem 6.72

Think

What are the pK_a values for proline? What is the isoelectric point (pI) of proline? Is the pH greater than, less than, or equal to the pI? Is this species positive, negative, or uncharged on average at a pH of 6? Is the cathode positive or negative? What is the charge at the anode?

Solve

The pI of proline is 6.30, so at a pH of 6 (more acidic than pI), proline will be protonated, on average, and the average charge on proline will be positive, so proline will migrate toward the negatively charged cathode.

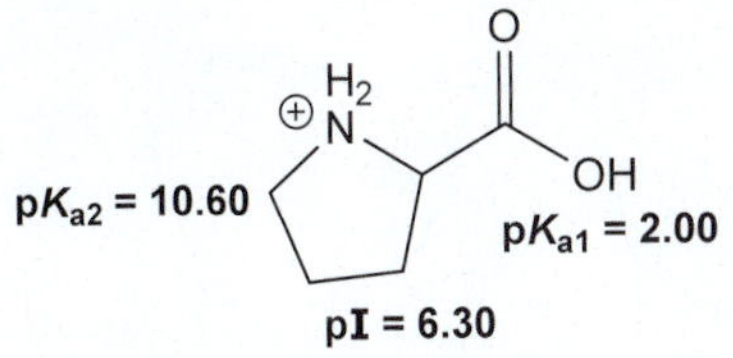

pK_{a2} = 10.60

pK_{a1} = 2.00

pI = 6.30

Problem 6.73

Think

What are the pK_a values for serine? What is the isoelectric point (pI) of serine? Is the pH greater than, less than, or equal to the pI at a pH of 6? Is this species positive, negative, or uncharged on average? Is the cathode positive or negative? What is the charge at the anode?

Solve

The pI of serine is 5.68, so at a pH of 6 (more basic than pI), serine will be deprotonated, on average, and the average charge on serine will be negative, so serine will migrate toward the positively charged anode.

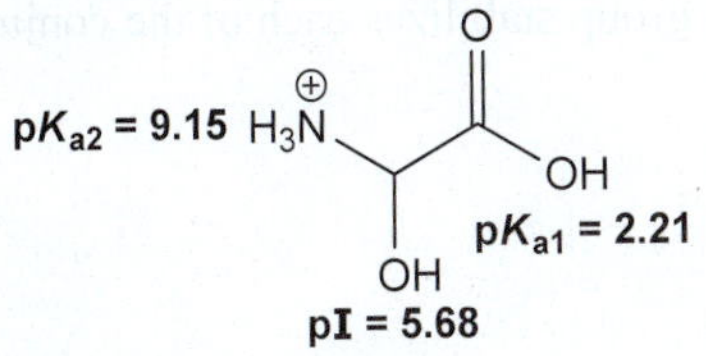

pK_{a2} = 9.15

pK_{a1} = 2.21

pI = 5.68

Problem 6.74

Think

What are the pK_a values for tyrosine? Which two pK_a values involve the zwitterion? How do you calculate pI from those pK_a values? Is the pH greater than, less than, or equal to the pI? Is this species positive, negative, or uncharged on average at a pH of 7? Is the cathode positive or negative? What is the charge at the anode?

Solve

The species that exist in solution and the pK_a values for tyrosine are shown below. The zwitterion is the second molecule from the left. The pK_a values of the equilibria involving the zwitterion are 2.20 and 9.11, so the pI = (2.20 + 9.11)/2 = 5.66. This value is consistent with the fact that the side chain is neutral. At a pH of 7 (more basic than pI), tyrosine is deprotonated, on average, and the average charge on tyrosine is negative, so tyrosine will migrate toward the positive anode.

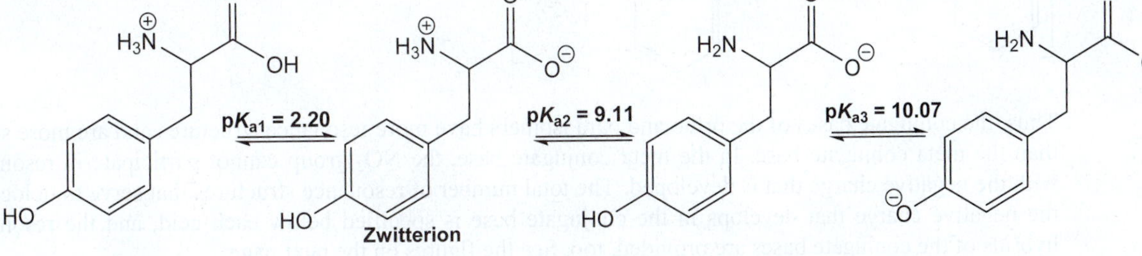

pK_{a1} = 2.20 pK_{a2} = 9.11 pK_{a3} = 10.07

Zwitterion

Integrated Problems

Problem 6.75

Think

Which base is stronger: HO^- or $(CH_3)_3CO^-$? Is an R group electron donating or withdrawing? Is the conjugate base stabilized or destabilized by an electron-donating group?

Solve

$NaOC(CH_3)_3$ (conjugate acid pK_a = 19) is a stronger base than NaOH (conjugate acid pK_a = 15.7). A stronger base is required to deprotonate the alkyl-substituted derivative, because it is less acidic than the unsubstituted molecule. It is less acidic because the alkyl group is electron donating and destabilizes the negative charge in the resulting conjugate base, as shown below.

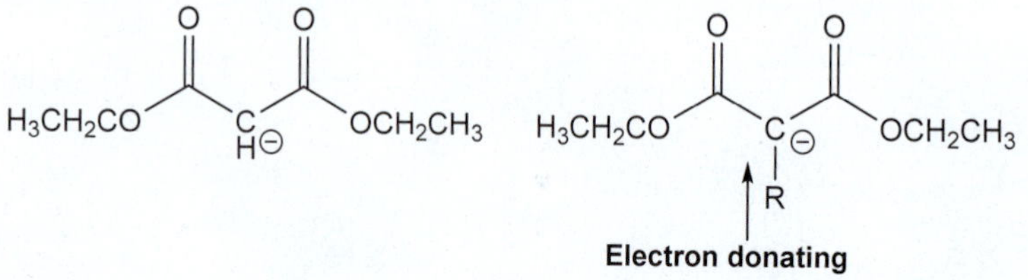

H3CH2CO … OCH2CH3 H3CH2CO … OCH2CH3 R

Electron donating

Problem 6.76

Think

Is the NO_2 an electron-donating group or an electron-withdrawing group (EWG)? What is the effect of the NO_2 group on the stability of the conjugate base anion? How does the position of the NO_2 group affect the stability of the conjugate base? Consider the number of possible resonance structures.

Solve

(a) All of the nitrophenols are more acidic than phenol, because the NO_2 group stabilizes each of the conjugate bases through inductive effects, as shown below.

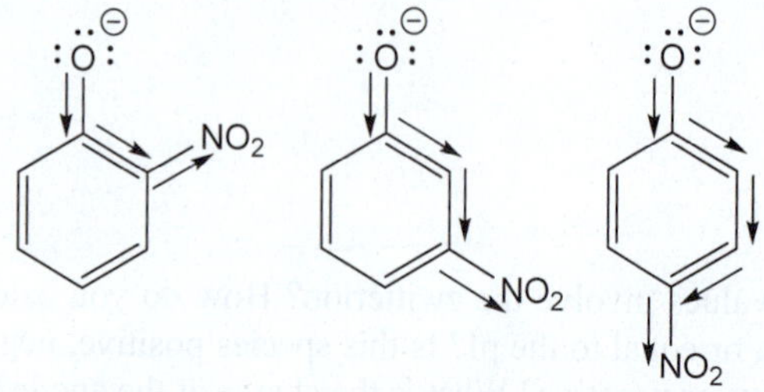

NO2 NO2 NO2

(b) The ortho and para nitrophenols are more acidic than the meta, because in the conjugate bases of the ortho and para compounds, the NO_2 group can participate in resonance with the negative charge that is generated in the conjugate base. This is exemplified by the ortho conjugate base below.

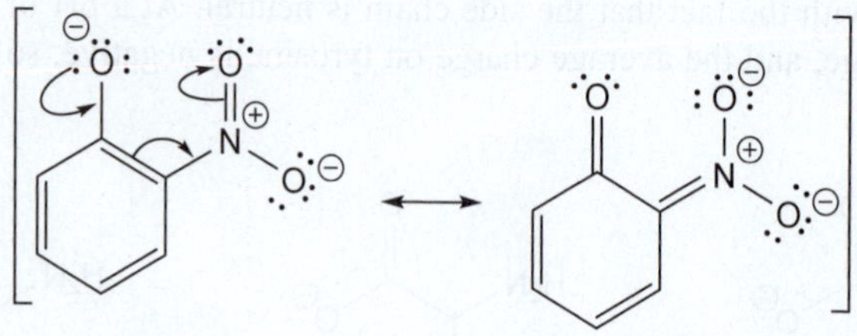

Thus, the conjugate bases of the ortho and para isomers have more resonance structures and are more stable than the meta conjugate base. In the meta conjugate base, the NO_2 group cannot participate in resonance with the negative charge that is developed. The total number of resonance structures that serve to delocalize the negative charge that develops in the conjugate base is specified below each acid, and the resonance hybrids of the conjugate bases are provided, too. See the figures on the next page.

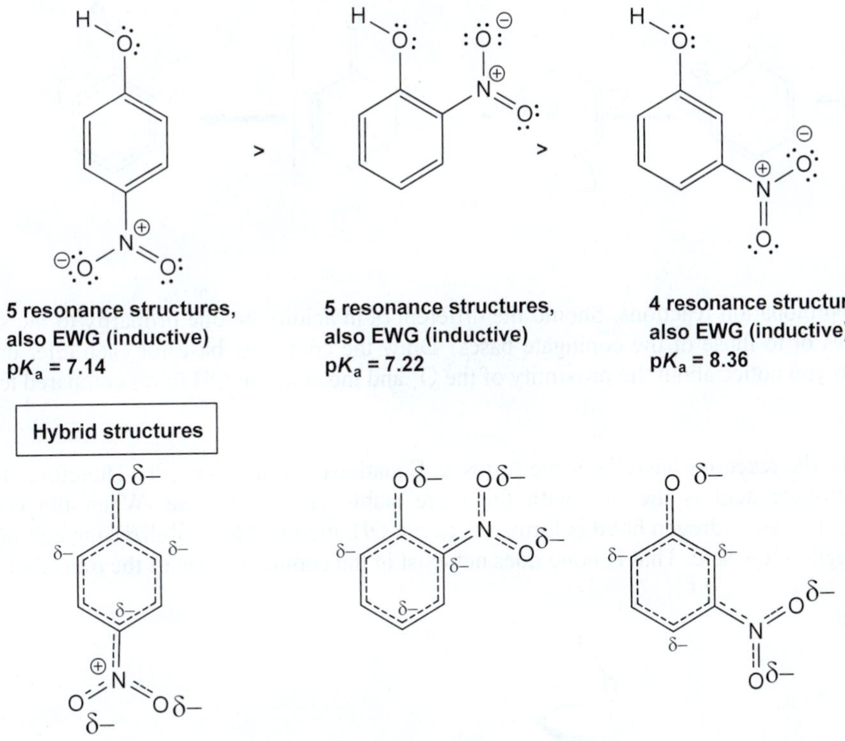

5 resonance structures, also EWG (inductive)
pK_a = 7.14

>

5 resonance structures, also EWG (inductive)
pK_a = 7.22

>

4 resonance structures, also EWG (inductive)
pK_a = 8.36

Hybrid structures

Problem 6.77

Think

Is the CH_3 group electron donating or withdrawing? What is the effect of the CH_3 group on the stability of the conjugate base anion? How does the position of the CH_3 group affect the stability of the conjugate base? Consider the number of possible resonance structures.

Solve

(a) All three methyl-substituted phenols are less acidic than phenol because the methyl group is electron donating and destabilizes the negative charge generated in the conjugate base, shown below.

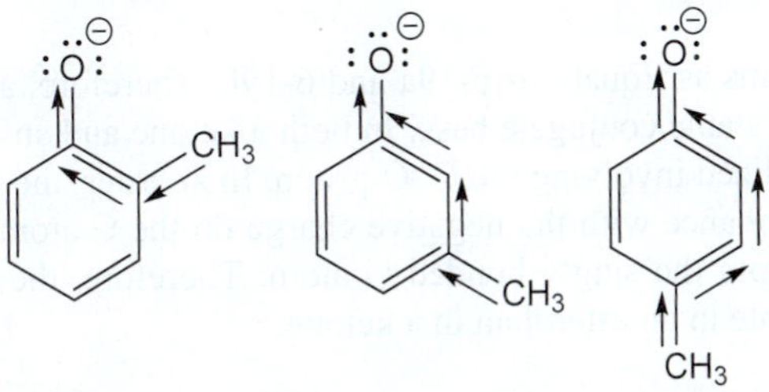

(b) The ortho and para compounds are less acidic than the meta compound because this inductive destabilization is more pronounced in ortho and para than it is in meta. We can see why by looking at the resonance hybrid of phenoxide itself (shown on the next page), which delocalizes the negative charge onto carbons 2, 4, and 6 of the ring. When a methyl group is on one of these carbons that has a share of the negative charge, as in the ortho and para conjugate bases, the negative charge becomes more concentrated, and the anion becomes less stable. In the meta conjugate base, none of these C atoms sharing the negative charge is bonded to the methyl group.

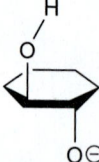

Problem 6.78

Think

Draw the complete deprotonation reactions. Should the differences in acidity be due primarily to the stabilities of the acids themselves or to those of the conjugate bases? Draw the conjugate base for each after one OH is deprotonated. What do you notice about the proximity of the O^- and the adjacent OH in cis compared to trans?

Solve

Being uncharged acids, the reactions have the same forms as Equations 6-19a and 6-19b. Therefore, according to Figure 6-7, the stronger acid is the one with the more stable conjugate base. When the cis diol is deprotonated, a strong internal hydrogen bond is formed between OH and O^-. This stabilizes the conjugate base and increases the strength of the acid. That H bond does not exist in the conjugate base of the trans diol.

Internal hydrogen bond

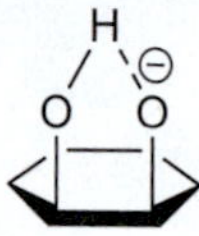

Problem 6.79

Think

Which proton is acidic? Should the differences in acidity be due primarily to the stabilities of the acids themselves or to those of the conjugate bases? In the conjugate base, does resonance serve to delocalize the negative charge of the same atoms or different atoms? Does the OR group affect the ability of resonance to delocalize that negative charge? How does this impact the stability of the conjugate base and the strength of the acid?

Solve

Being uncharged acids, the reactions have the same forms as Equations 6-19a and 6-19b. Therefore, according to Figure 6-7, the stronger acid is the one with the more stable conjugate base. In both a ketone and an ester, the negative charge in the conjugate base is resonance stabilized involving the C=O group. In an ester, the electrons in the C=O bond are less available to participate in resonance with the negative charge on the C atom because they are also involved in resonance with a lone pair from the singly bonded O atom. Therefore, the negative charge that develops is more localized on C and less stable in an ester than in a ketone.

Problem 6.80

Think

Draw the complete deprotonation reactions. Should the differences in acidity be due primarily to the stabilities of the acids themselves or to those of the conjugate bases? Draw the conjugate base of each. Does the negative charge appear on the same atom? Does resonance delocalize the negative charge equally? What can be said about inductive effects?

Solve

Being uncharged acids, the reactions have the same forms as Equations 6-19a and 6-19b. Therefore, according to Figure 6-7, the stronger acid is the one with the more stable conjugate base. The conjugate base of the para compound is more stable because the C=O bond can participate in resonance to provide additional resonance delocalization to the negative charge that develops (see below). This cannot happen in the meta compound. Therefore, the para compound is the stronger acid.

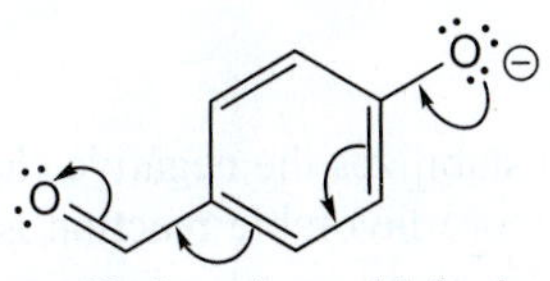

***p*-Hydroxybenzaldehyde**

Problem 6.81

Think

Draw the complete deprotonation reactions. Should the differences in acidity be due primarily to the stabilities of the acids themselves or to those of the conjugate bases? What is the conjugate base for each? Are the charges the same? Do the charges appear on the same types of atoms? How do resonance and inductive effects impact charge stability?

Solve

CH_3NC is the stronger acid. Being uncharged acids, the reactions have the same forms as Equations 6-19a and 6-19b. Therefore, according to Figure 6-7, the stronger acid is the one with the more stable conjugate base. In both cases, a negative charge is formed on the C atom on deprotonation. In CH_3NC, a +1 formal charge exists on the N atom, which very strongly inductively stabilizes the negative charge that is formed. In CH_3CN, the neighboring C atom has a 0 formal charge.

$H_3C{-}C{\equiv}N: \xrightarrow{-H^+} H_2\overset{\ominus}{C}{-}C{\equiv}N:$ $\qquad$ $H_3C{-}\overset{\oplus}{N}{\equiv}\overset{\ominus}{C}: \xrightarrow{-H^+} H_2\overset{\ominus}{C}{-}\overset{\oplus}{N}{\equiv}\overset{\ominus}{C}:$

Stronger acid

Problem 6.82

Think

Identify the acidic proton in formaldehyde and the acidic proton in each of the structures listed for comparison. Draw the complete deprotonation reactions. Should the differences in acidity be due primarily to the stabilities of the acids themselves or to those of the conjugate bases? What is the conjugate base for each? Are the charges the same? Do the charges appear on the same types of atoms, including hybridization? How do resonance and inductive effects impact charge stability?

Solve

(a) The pK_a of formaldehyde should be closest to $H_2C{=}CH_2$, **D**, whose pK_a is 44. Both molecules are uncharged, so their relative acidities should be governed primarily by the stabilities of their conjugate bases. In both conjugate bases, the negative charge develops on the same type of atom (an sp^2-hybridized C atom) with no resonance.

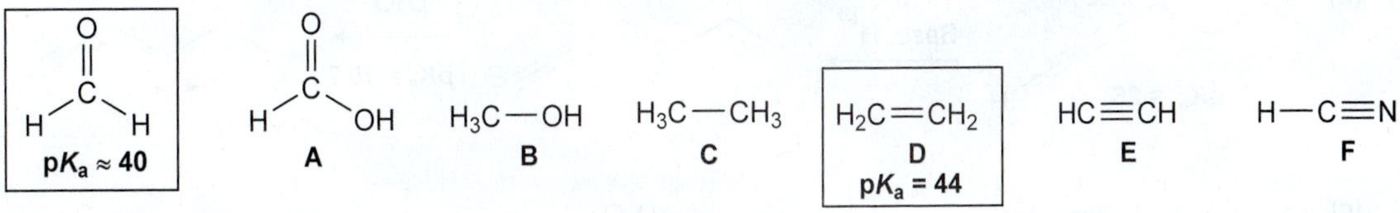

$H_2C{=}O$ (p$K_a \approx 40$) $\quad$ HCOOH **A** $\quad$ $H_3C{-}OH$ **B** $\quad$ $H_3C{-}CH_3$ **C** $\quad$ $H_2C{=}CH_2$ **D** (pK_a = 44) $\quad$ $HC{\equiv}CH$ **E** $\quad$ $H{-}C{\equiv}N$ **F**

(b) The only difference is inductive effects from the nearby O atom, which stabilizes the negative charge in the conjugate base and, hence, increases the acid strength. But recall that inductive effects typically have a fairly small effect on pK_a of just a few pK_a units. We should, therefore, estimate the pK_a to be ~40.

Problem 6.83

Think

In each species, what are the possible acidic protons? What are the possible basic sites? Draw out the products for each reaction. On which atom is a negative charge more stable? On which atom is a positive charge more stable?

Solve

(a) The products for the reaction are listed below.

(b) Oxygen is more electronegative compared to nitrogen. Therefore, the O atom stabilizes the negative charge better, and the N atom accommodates the positive charge better. Thus, the more favorable reaction is the first one, in which the alcohol acts as the acid and the amine acts as the base.

Acid + Base

Base + Acid

Problem 6.84

Think

What was the conjugate base that must have picked up a D^+ from D_2O? From the conjugate base, what was the structure of the original acid? What is the pK_a of D_2O and each original acid? What base can deprotonate the original acids?

Solve

In all cases, a D^+ must replace a H^+. This can be done in two successive proton transfer reactions. The first reaction removes the appropriate H^+, and the second reaction deposits a D^+. For each of these successive reactions to be efficient, K_{eq} should favor products. The proper choice of base in the first step, therefore, is that in which the pK_a of its conjugate base is significantly greater than the pK_a of the reactant acid. H^- works in all three cases, because the pK_a of H_2 is 35. In the second proton transfer, the acid is D_2O, whose pK_a is similar to that of H_2O, or 15.7. So, in all three cases, the second proton transfer does, indeed, favor products.

(a) pK_a = 20; Base, H^-; D_2O, pK_a = 15.7

(b) pK_a = 25; Base, H^-; D_2O, pK_a = 15.7

(c) OH, pK_a = 19; Base, H^-; D_2O, pK_a = 15.7; OD

Problem 6.85

Think

Draw the complete deprotonation reactions. Should the differences in acidity be due primarily to the stabilities of the acids themselves or to those of the conjugate bases? Review Section 2.9 and consider the ability of each solvent to solvate cations and anions. Draw the structures of water and dimethyl sulfoxide (DMSO). Which solvent better stabilizes the conjugate base of acetic acid?

Solve

Being an uncharged acid, the reaction has the same form as Equations 6-19a and 6-19b. Therefore, according to Figure 6-7, the acid becomes stronger as the conjugate base becomes more stable. The acetate anion conjugate base that forms is solvated—and therefore stabilized—better by water than by DMSO. Water solvates anions better than DMSO because water is a polar protic solvent, whereas DMSO is an aprotic solvent. With the conjugate base better stabilized in water, the acid becomes stronger. This explains why the pK_a is 4.75 in water and 12.6 in DMSO.

Problem 6.86

Think

Draw the complete deprotonation reactions. Should the acidity be due primarily to the stability of the acid itself or to that of the conjugate base? Review Section 2.9 and consider the ability of each solvent to solvate cations and anions. Draw the structures of water and DMSO. Why is this effect not felt in NH_4^+?

Solve

Being a charged acid, the complete reaction is analogous to those in Equations 6-20a and 6-20b, so the relative acidity is governed by the relative stability of the acid itself, as shown in Figure 6-8. Water is a protic solvent, whereas DMSO is an aprotic solvent. Even though protic solvents solvate anions much better than polar aprotic solvents, they both solvate cations well. Because cations are present in this case, changing the solvent doesn't have a dramatic effect on charge stability and, therefore, doesn't have a dramatic effect on acidity.

$^{\oplus}NH_4$ **pK_a = 9.4 (in water)**

$^{\oplus}NH_4$ **pK_a = 10.5 (in DMSO)**

Dimethyl sulfoxide
DMSO

Problem 6.87

Think

Is the strength of CH_3NH_2 as a base governed primarily by the stability of the base itself or by the stability of the conjugate acid? Which solvent has the greatest effect on the stability of that species?

Solve

Being an uncharged base, its strength is governed primarily by the stability of its conjugate acid, $CH_3NH_3^+$, owing to the presence of the positive charge. The conjugate base is better stabilized as the intermolecular interactions with the solvent become stronger. All the solvents except **E** (CCl_4) can undergo hydrogen bonding and/or ion–dipole interactions with the conjugate base. Therefore, $CH_3NH_3^+$ will be the least stable in **E**, making CH_3NH_2 the weakest base in that solvent.

Problem 6.88

Think

Draw the complete reactions. Should the acidities of these acids be governed by the relative stabilities of the acids themselves or by those of their respective conjugate bases? Are alkyl groups electron withdrawing or donating? What is the effect of the alkyl group on the stability of the N^+? How would the additional alkyl groups affect the solvation by water?

Solve

(a) Being charged acids, the complete reactions are analogous to those in Equations 6-20a and 6-20b, so their relative acidities are governed by the relative stabilities of the acids themselves, as shown in Figure 6-8. Alkyl groups are electron donating and should stabilize the N cation, making the acid weaker with each additional alkyl group. This is what we observe with the addition of the first two alkyl groups, where the pK_a increases slightly.

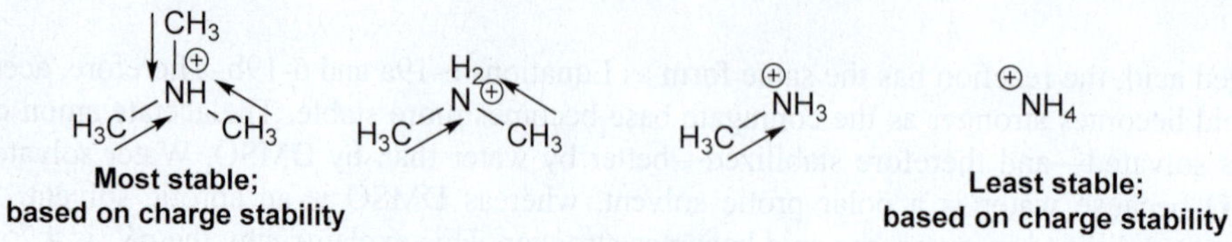

$(CH_3)_3NH^+$ $(CH_3)_2NH_2^+$ $CH_3NH_3^+$ NH_4^+

Most stable; based on charge stability

Least stable; based on charge stability

(b) The addition of the third alkyl group should increase the stability of the cation, weakening the R_3NH^+ acid. However, the pK_a decreases, signifying a stronger acid. This has to do with differences in solvation by water. The substantial bulkiness surrounding the N^+ prevents solvation by water, which decreases stability and increases acidity.

CHAPTER 7 | An Overview of the Most Common Elementary Steps

Your Turn Exercises

Your Turn 7.1

Think

What kinds of charges characterize an electron-rich atom? Are any formal charges present? Any strong partial charges? In which direction do electrons flow—from electron rich to electron poor, or the opposite? How is a curved arrow used to denote the flow of electrons?

Solve

When the S atom bears a full negative charge, it is quite electron rich and not particularly stable owing to the mutual repulsion of the extra electron density. The H in H–Br bears a partial positive charge, δ^+, because Br is more electronegative than H. The electrons in HS^- are attracted to the proton in H–Br. The bond between H–Br simultaneously breaks as the HS–H bond forms.

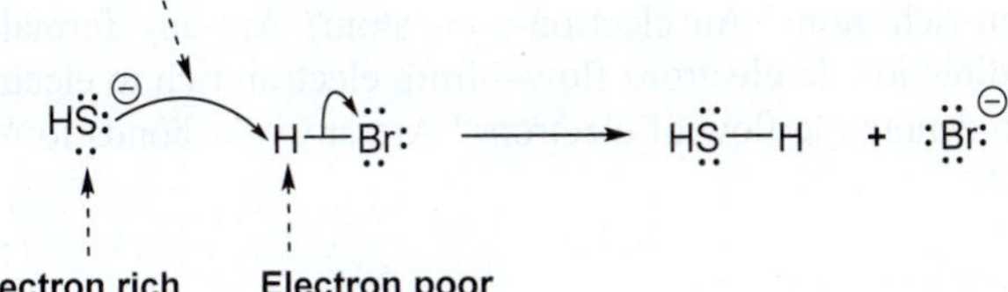

Your Turn 7.2

Think

According to the curved arrow notation, what bond is formed? What atom receives a share of an additional pair of electrons? Do any bonds break from that atom? Are additional arrows needed?

Solve

The HO^- is electron rich and donates an electron pair to the $H^{\delta+}$ in H–Cl. A new HO–HCl bond is formed. No bonds are shown to break, however. This is faulty, because H cannot form two covalent bonds. Shown below is the product from the faulty arrows.

Faulty arrows

HO: H—Cl: ⟶ HO—H—Cl:

To yield an acceptable product, the H–Cl bond would simultaneously need to break, and the two electrons would become an additional lone pair on the Cl to yield the products shown: water, H_2O, and the chloride anion, Cl^-.

Corrected arrows

HO: H—Cl: ⟶ HO—H + :Cl:

Your Turn 7.3

Think

Draw out the bonds connecting each atom, and fill in lone pairs of electrons to fulfill each atom's octet. How do you determine formal charge? (Consult Section 1.9, if necessary.) For the uncharged nucleophiles, consider how electronegativity differences in atoms lead to a polar bond.

Solve

The complete Lewis structures for all of the negatively charged and uncharged nucleophiles are given below. The uncharged nucleophiles have a −δ symbol on the atom bearing the partial negative charge.

Negatively charged nucleophiles

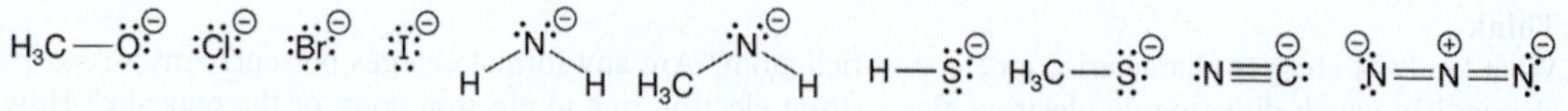

Uncharged nucleophiles

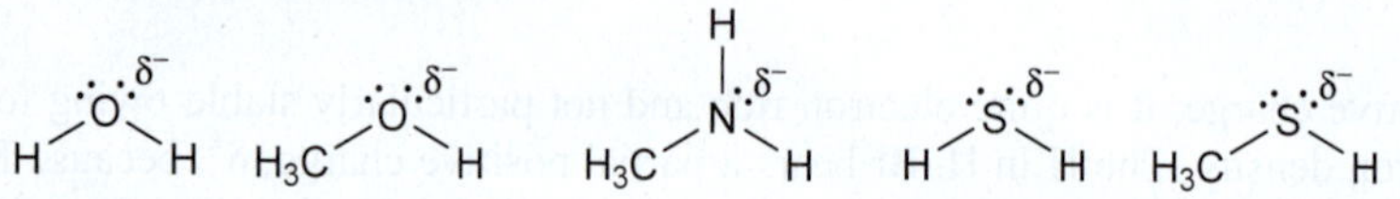

Your Turn 7.4

Think

What kinds of charges characterize an electron-rich atom? An electron-poor atom? Are any formal charges present? Any strong partial charges? In which direction do electrons flow—from electron rich to electron poor, or the opposite? How is a curved arrow used to denote the flow of electrons? Are any new bonds formed from lone pairs of electrons?

Solve

When the Cl atom bears a full negative charge, it is quite electron rich and not particularly stable owing to the mutual repulsion of the extra electron density. The C atom in $H_3C–Br$ bears a partial positive charge, δ^+, because Br is more electronegative than C. The electrons in Cl^- are attracted to the C atom in $H_3C–Br$, and a new $Cl–CH_3$ bond forms. The bond between $H_3C–Br$ simultaneously breaks as the new bond forms.

Your Turn 7.5

Think

What are the electron-rich and electron-poor sites in each compound? How does lacking an octet affect the electron richness of an atom? Do you think it can more easily accept or donate electrons? Are Lewis bases electron-pair donors or acceptors? Lewis acids? Are any new bonds formed from lone pairs of electrons?

Solve

The Cl in the $CH_3C(O)Cl$ compound is electron rich because Cl has lone pairs of electrons and, being more electronegative compared to the C atom, bears a partial negative charge. The Al in $AlCl_3$ lacks an octet and is, therefore, electron poor and accepts an electron pair from the Cl atom. The electron movement is shown below. Because electrons flow from the electron-rich Cl, $CH_3C(O)Cl$ is the Lewis base, and because electrons flow to the Al, $AlCl_3$ is the Lewis acid.

Coordinationstep

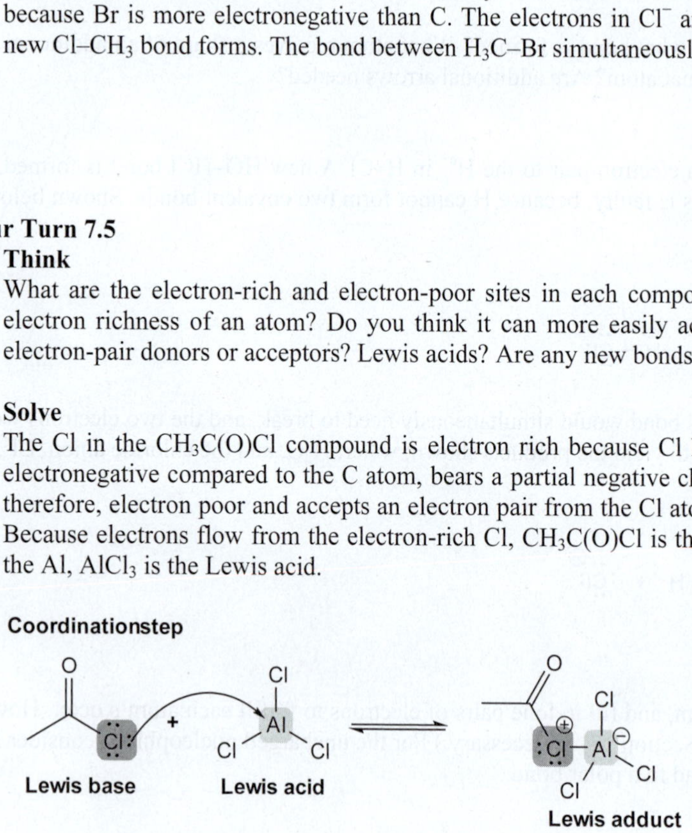

Lewis base   Lewis acid   Lewis adduct

Your Turn 7.6

Think

In looking at an atom, how can you tell if the atom lacks an octet? How do you count bonding electrons? Nonbonding electrons? When a (+) formal charge is drawn on a carbon atom, how many bonds and lone pairs does the carbon atom have?

Solve

Each bond contributes two electrons to an atom's octet, and each lone pair contributes two electrons. A positively charged C atom has three bonds and no lone pairs, for a total of six electrons, so each C^+ lacks an octet. The same is true for an uncharged Al atom. The atoms lacking octets are shown below.

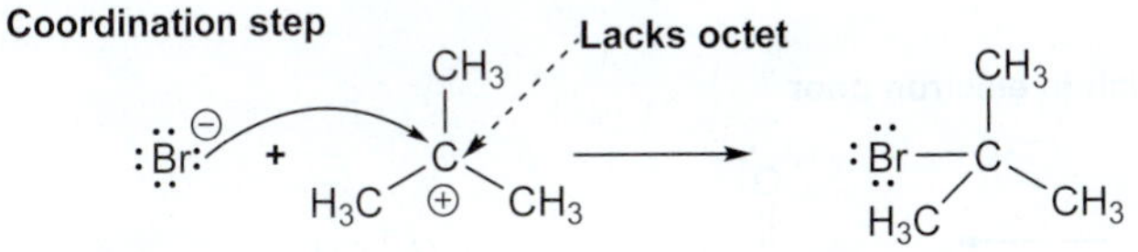

Coordination step

Lacks octet

(7-5)

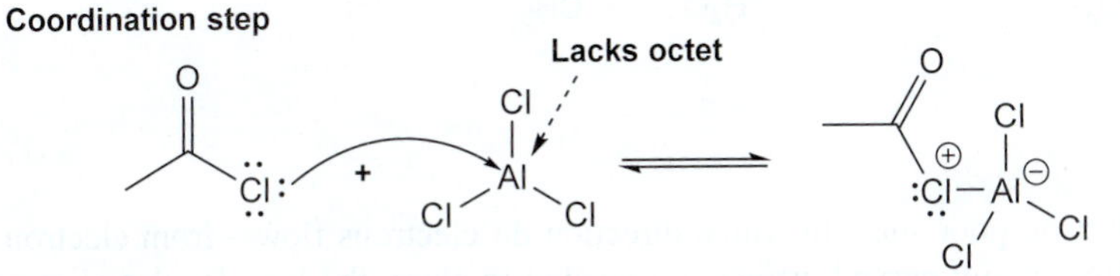

Coordination step

Lacks octet

(7-6)

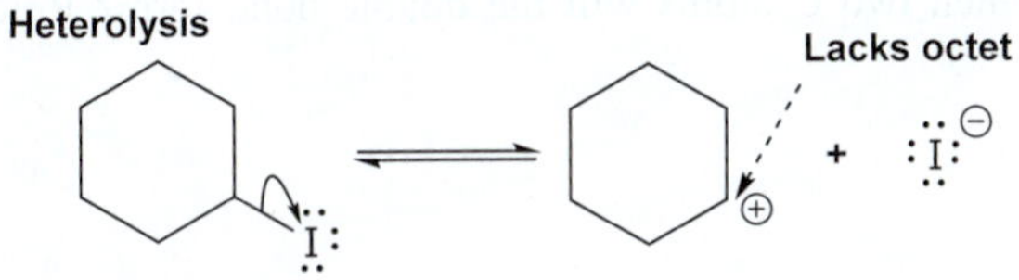

Heterolysis

Lacks octet

(7-7)

Heterolysis

Lacks octet

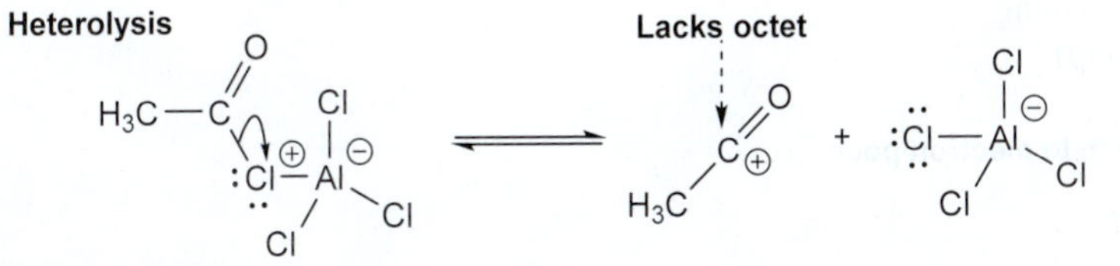

(7-8)

Your Turn 7.7

Think

What kinds of charges characterize an electron-rich atom? Are any formal charges present? Any strong partial charges? Which bonds are broken, and which bonds are formed? How is a curved arrow used to denote the flow of electrons? In which direction do electrons flow—from electron rich to electron poor, or the opposite?

Solve

A C atom bearing a full negative charge is quite electron rich and not particularly stable owing to the mutual repulsion of the extra electron density. The C atom in C=O bears a large, partial positive δ^+ owing to the electronegativity difference of C and O. The electrons flow from the $H_3C:^-$ to the C atom of the C=O bond. A second curved arrow (to illustrate the breaking of the π bond between the C and O atoms) is necessary to avoid exceeding an octet on the less electronegative C atom.

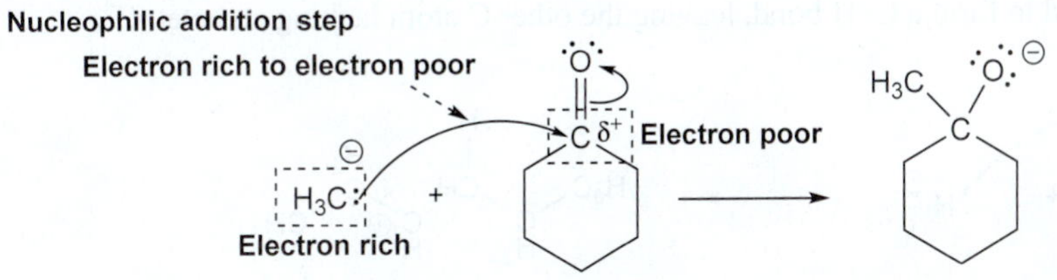

Nucleophilic addition step

Electron rich to electron poor

Electron poor

Electron rich

Your Turn 7.8

Think

What kinds of charges characterize an electron-rich atom? Are any formal charges present? Any strong partial charges? Which bonds are broken, and which bonds are formed? How is a curved arrow used to denote the flow of electrons? In which direction do electrons flow—from electron rich to electron poor, or the opposite?

Solve

The O^- is the electron-rich atom, and the C atom bonded to two O and two C atoms is electron poor, because it bears a partial positive charge, δ^+. A lone pair of electrons on O^- folds down to make a C=O bond. A second curved arrow is necessary to avoid exceeding an octet on C, and the C–OCH_3 bond is broken, leaving CH_3O^- as the charged product.

Nucleophile elimination step

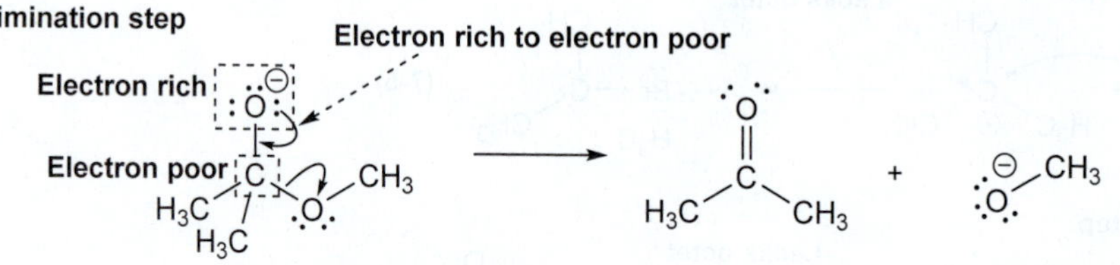

Electron rich to electron poor

Electron rich

Electron poor

Your Turn 7.9

Think

Identify the electron-rich and the electron-poor sites. In which direction do electrons flow—from electron rich to electron poor, or the opposite? How many curved arrows are needed to show the bimolecular elimination (E2) step? What can act as the leaving group? On which two C atoms will the double bond form? Refer to Equation 7-17 as a guide.

Solve

The E2 reaction requires three arrows:
- Base–H bond formation
- H–C bond breaking and C=C bond forming
- C–L bond breaking (L = leaving group)

E2 step

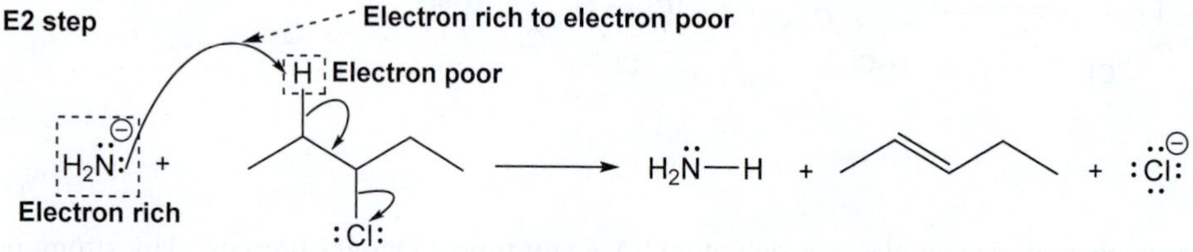

Electron rich to electron poor

Electron poor

Electron rich

Your Turn 7.10

Think

Are π bond electrons in a C=C bond electron rich or electron poor? Do you think H^+ is electron rich or electron poor? Which bonds are broken, and which bonds are formed? In which direction do electrons flow—from electron rich to electron poor, or the opposite?

Solve

The curved arrow originates from the electron-rich double bond and points to the electron-poor H^+. This is an example of electrophilic addition: The nonpolar C=C π bond gains a strongly electron-deficient species, H^+. The π bond in C=C breaks and is used to form a C–H bond, leaving the other C atom lacking an octet, C^+.

Electrophilic addition step

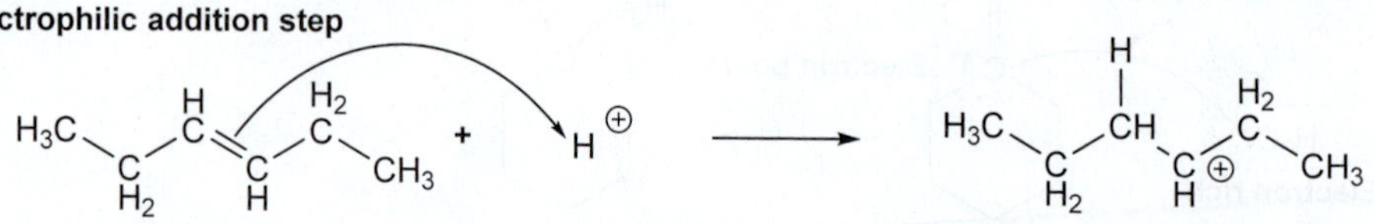

Your Turn 7.11

Think

Which bond has to break to re-form the electrophile? Where is a new bond formed? How do you use curved arrow notation to show the breaking of a bond? The formation of a bond?

Solve

To show the C–H bond breaking, a curved arrow originates from the center of the C–H bond. To show the pair of electrons ending up in the C=C double bond, the curved arrow points to the center of the C–C bond. This is an example of electrophile elimination, in which the H^+ electrophile is eliminated and generates a stable, uncharged organic species.

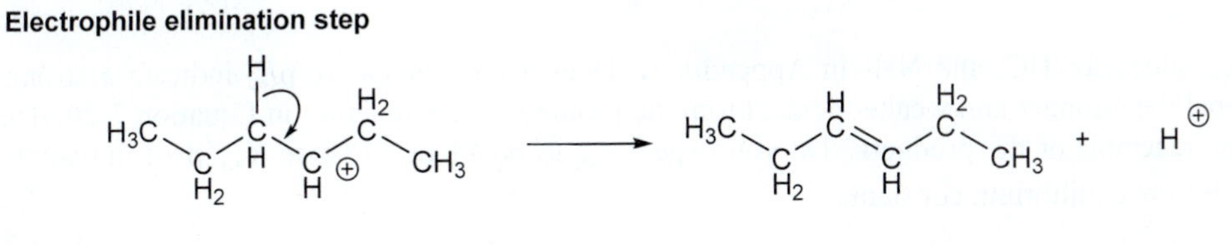

Electrophile elimination step

Your Turn 7.12

Think

What kinds of charges characterize an electron-rich atom? An electron-poor atom? Are any formal charges present? Are there any nonpolar double or triple bonds? Is the electrophile electron rich or electron poor? Do the curved arrow directions show movement from electron rich to electron poor, or vice versa?

Solve

In electrophilic addition, the H^+ electrophile is the electron-poor site, characterized by the positive charge. The π electrons in the C=C bond are the electron-rich site, because four electrons are confined between the two atoms.

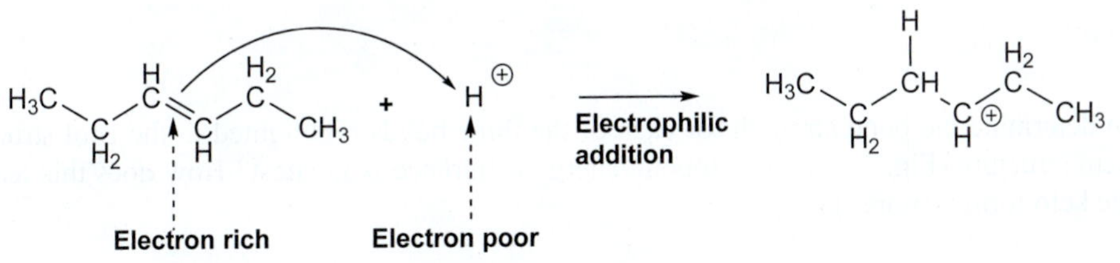

Electron rich Electron poor Electrophilic addition

In electrophile elimination, the carbocation C^+ is the electron-poor site, and the C–H bond is the electron-rich site.

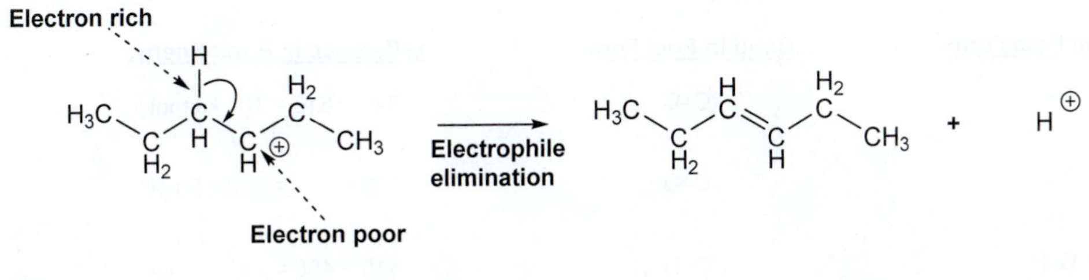

Electron rich Electron poor Electrophile elimination

Your Turn 7.13

Think

On what type of C atom—1°, 2°, or 3°—is the carbocation located in each structure? It might be helpful to draw out implied H atoms. Which H atom shifted? How do you represent the breaking of one C–H bond with the formation of another C–H bond using curved arrows? In which direction do electrons flow—from electron rich to electron poor, or the opposite?

Solve

The single curved arrow shows a C–H bond breaking from the 3° C atom and simultaneously forming a bond to the 2° C atom. The curved arrow points from the electron-rich single bond to the electron-poor C^+ atom.

Your Turn 7.14

Think

Look up the pK_a values for HCl and NH_3 in Appendix A. Does a more negative pK_a indicate a stronger or weaker acid? Label the stronger and weaker acids. From the proton transfer reaction in Equation 7-29, does the reaction favor the reactants or the products? Do you expect K_{eq} to be >1 or <1? Use K_{eq}(proton transfer) = $10^{\Delta pK_a}$ to solve for the equilibrium constant.

Solve

HCl's $pK_a = -7$, and NH_3's $pK_a = 36$. HCl is a stronger acid than NH_3 (it has a more negative pK_a), and the reaction favors the product side (away from the stronger acid). This is confirmed by the large, positive K_{eq} $[10^{36-(-7)} = 10^{43}]$.

$$H_2N:^{\ominus} + H{-}Cl: \longrightarrow H_2N{-}H + :Cl:^{\ominus} \qquad K_{eq} = 1 \times 10^{43} \qquad (7\text{-}31)$$

$pK_a = -7$ **Stronger acid**; $pK_a = 36$ **Weaker acid**

Your Turn 7.15

Think

Consult Figure 7-4 to determine the bond strength for each of the three bonds highlighted in the enol structure (Fig. 7-4a) and the keto structure (Fig. 7-4b). Which bond energy difference is greatest? How does this lead to the conclusion that the keto form is more stable?

Solve

The completed table is below. From the table, it appears that, because the difference is greatest between the C=O and C=C bond energies, the C=O bond (being the stronger of the two) has the most influence on the outcome of the reaction.

Bond in Keto Form	Bond in Enol Form	Difference in Bond Energy
C=O	C=C	720 − 619 = 101 kJ/mol
C–C	C–O	339 − 351 = −12 kJ/mol
C–H	O–H	418 − 460 = −42 kJ/mol

In Chapter Problems

Problem 7.2

Think

What kinds of charges characterize an electron-poor atom? Are any formal charges present? Any strong partial charges? When $(CH_3)_3N$ and H_2O are combined, what curved arrow can we draw to depict the flow of electrons from an electron-rich site to an electron-poor site?

Solve

Neither molecule bears any full charges. O is much more electronegative than H; thus, the H atom is an electron-poor site, and the O atom is an electron-rich site. N is more electronegative than C, so the C atom is an electron-poor site, and the N atom is an electron-rich site. Only one proton transfer reaction makes sense in terms of "electron rich to electron poor," as shown below.

Electron rich to electron poor

Rich

Poor

Rich

Poor

Proton transfer

Problem 7.4

Think

What are the electron-rich and electron-poor sites in each compound? Can we make simplifying assumptions? Are any spectator ions present?

Solve

NaSH is an ionic compound that dissolves in solution as Na^+ and HS^-. Na^+ is treated as a spectator ion, and HS^- is treated as being electron rich at the S. CH_3CO_2H has an electron-poor H atom because of the high electronegativity of the O atoms. A curved arrow is drawn from the electron-rich S atom to the electron-poor H atom to indicate bond formation in a proton transfer. A second curved arrow is drawn to break the O–H bond to avoid two bonds to the H atom.

Electron rich to electron poor

Proton transfer

Problem 7.6

Think

What are the electron-rich and electron-poor sites in each compound? Can we make any simplifying assumptions? Are any spectator ions present?

Solve

Although the C–Li bond is polar covalent, it can be simplified by treating Li^+ as a spectator ion, in which case, it is omitted from the scheme, leaving H_3C^-. Therefore, H_3C^- is highly electron rich, and the H of H–OCH_3 is electron poor because of the highly electronegative O atom. A C–H bond forms, and an O–H bond breaks.

Proton transfer

Problem 7.7

Think

What are the electron-rich and electron-poor sites in each compound? Can we make any simplifying assumptions? Are any spectator ions present?

Solve

$LiAlH_4$ may be treated as H^-, which is electron rich. The H in H_2O is electron poor because of the highly electronegative O atom. A H–H bond forms and a H–O bond breaks.

H:⁻ + H–OH → (Proton transfer) H–H + ⁻:OH

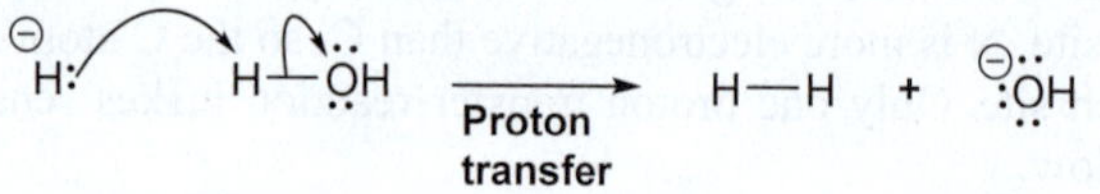

$NaBH_4$ may be treated as H^-, which is electron rich. The H in H–O–C_6H_5 is electron poor because of the highly electronegative O atom. A H–H bond forms, and a H–O bond breaks.

H:⁻ + H–O–C_6H_5 → (Proton transfer) H–H + ⁻:O–C_6H_5

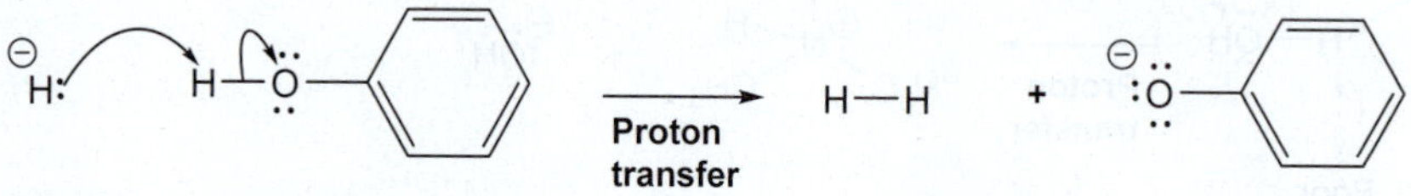

Problem 7.9

Think

Which atoms carry a partial or full negative charge? Do those atoms have pairs of electrons that can be used to form bonds to other atoms?

Solve

The Lewis structures, omitting the metal atoms, appear as follows.

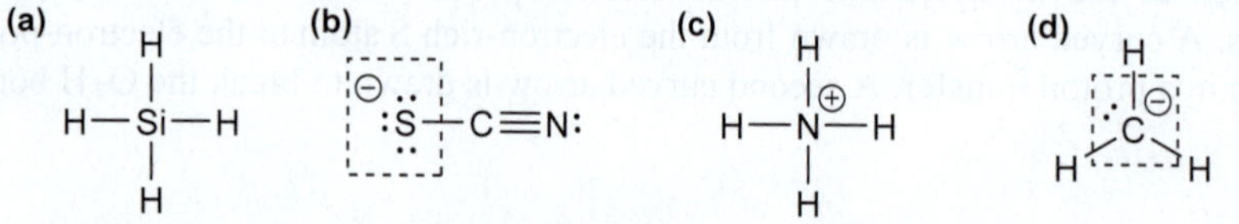

(a) SiH_4 (b) ⁻:S–C≡N: (c) NH_4^+ (d) ⁻:CH_3

Only **(b)** and **(d)** can behave as nucleophiles, because each has an atom with a lone pair of electrons and some excess negative charge (in these cases, a formal −1 charge). The nucleophilic atoms are boxed. In **(a)** and **(c)**, no atom is especially electron rich, and no atom possesses a lone pair of electrons.

Problem 7.11

Think

Which species is the nucleophile? Which is the substrate? What do we do with the metal atom? Which species is electron rich? Which is electron poor?

Solve

The Cl atom is electronegative and pulls electron density away from the C atom. Thus, the C atom is electron poor and bears a δ^+ charge, and the Cl atom bears a δ^- and is a good leaving group. Therefore, the species with Cl is the substrate. The curved arrow notation is shown below. The NC^- nucleophile forms a bond to the C atom at the same time as the C–Cl bond breaks and the Cl^- leaves.

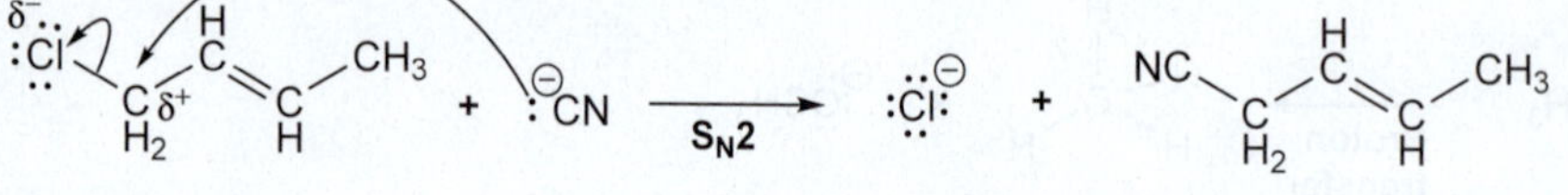

$Cl^{\delta-}$–$CH_2^{\delta+}$–CH=CH–CH_3 + ⁻:CN → (S_N2) :Cl:⁻ + NC–CH_2–CH=CH–CH_3

Problem 7.12

Think

Which sites are electron rich? Which are electron poor? How many curved arrows are needed to show the coordination step? Are curved arrows drawn from electron rich to electron poor, or vice versa? Which species is donating a pair of electrons, and which species is accepting a pair? How are the heterolysis step and the coordination step related? How many arrows are needed to show the heterolysis step? Which bond is involved?

Solve

(a) The reaction is shown below. A single curved arrow is used to show the bond formation and is drawn from the electron-rich Cl^- toward the electron-poor Fe. Cl^- is the Lewis base, because it is donating a pair of electrons to form the bond. $FeCl_3$ is the Lewis acid, because it accepts that pair of electrons.

Coordination

(b) The heterolysis step is shown below and is the opposite of the coordination step for this example. The Cl–Fe bond is the bond forming in **(a)** and breaking in **(b)**. To show the bond breaking, a single curved arrow is drawn from the center of the Fe–Cl bond and points toward Cl.

Heterolysis

Problem 7.13

Think

What simplifying assumptions can we make regarding the metals? Which sites are electron rich? Which are electron poor? Are curved arrows drawn from electron rich to electron poor, or vice versa? How many curved arrows are needed to show the nucleophilic addition step?

Solve

Two arrows are necessary to show the nucleophilic addition step:

- Nu–C bond formation (addition of the nucleophile to the electron-poor site)
- C=X π bond breaking (avoids exceeding an octet at the C atom)

(a) PhMgBr can be thought of as a source of C:$^-$; thus, the $MgBr^+$ combination is left out. The phenyl carbanion is the nucleophile, and the carbon of the C=O bond is the electrophile (electron-poor species), because it is bonded to a highly electronegative O atom.

Nucleophilic addition

(b) The Na^+ in $NaOCH_3$ is a spectator ion and, therefore, is omitted. CH_3O^- is electron rich and, thus, is the nucleophile. The carbon of the C≡N is the electrophile (electron-poor species), because it is bonded to an electronegative N atom. The mechanism is drawn on the next page.

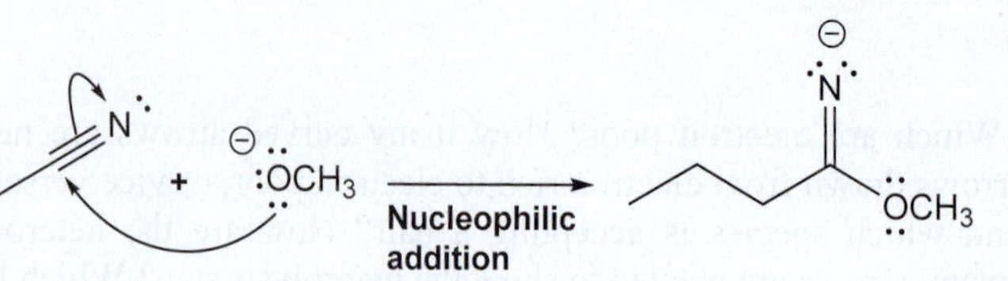

Nucleophilic addition

Problem 7.14

Think

Which species are electron rich? Which are electron poor? Are curved arrows drawn from electron rich to electron poor, or vice versa? How many curved arrows are needed to show the nucleophile elimination step?

Solve

Two arrows are necessary to show the nucleophile elimination step:

- C=X π bond formation
- C–L bond breaking (avoids exceeding an octet at the C atom)

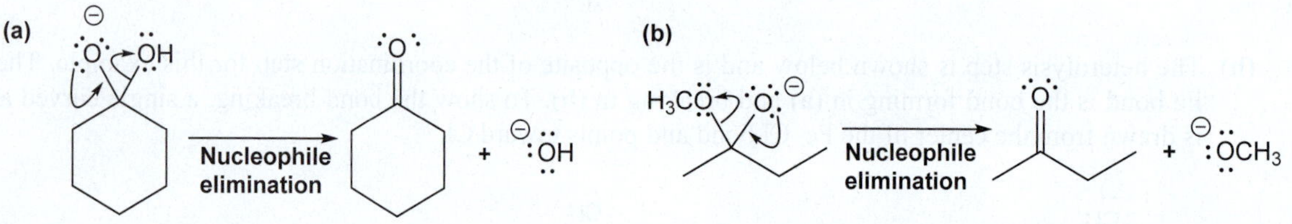

(a) Nucleophile elimination

(b) Nucleophile elimination

Problem 7.15

Think

What simplifying assumptions can we make regarding the metals? Identify the electron-rich and the electron-poor sites. How many curved arrows are needed to show the bimolecular elimination (E2) step? What can act as the leaving group? On which two C atoms will the double bond form? Refer to Equation 7-17 as a guide.

Solve

The E2 reaction requires three arrows:

- Base–H bond formation
- H–C bond breaking and C=C bond forming
- C–L bond breaking (L = leaving group)

(a) Br^- is the leaving group; Na^+ is a spectator ion, so HO^- is the base and attacks the H atom on the carbon next to the carbon bearing the Br atom.

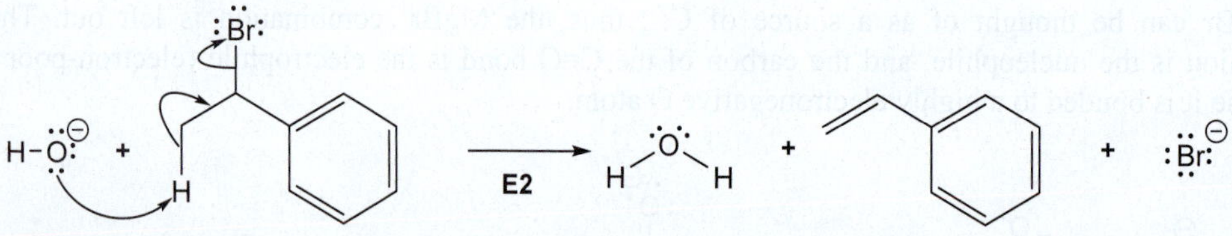

E2

(b) Cl^- is the leaving group; CH_3–Li is treated as H_3C^-, which is the base, and attacks the H atom on the carbon next to the carbon bearing the Cl atom. A triple bond is formed.

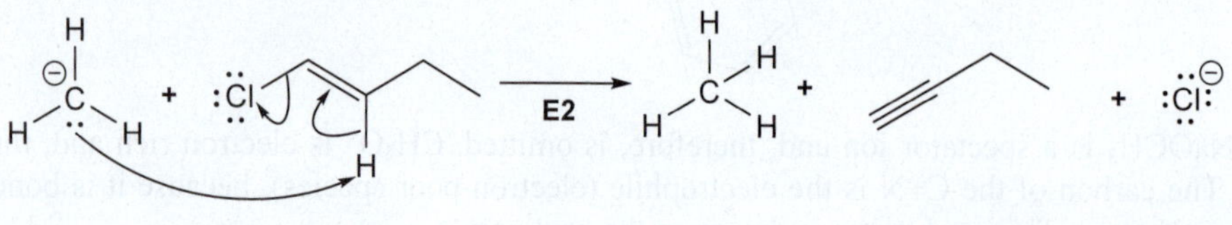

E2

Problem 7.16

Think

Which sites are electron rich? Which are electron poor? Are curved arrows drawn from electron rich to electron poor, or vice versa? How many curved arrows are needed to show the electrophilic addition step?

Solve

One or two arrows are necessary to show the electrophilic addition step:

- Nonpolar π electrons and electrophile (E^+) bond formation
- E–Y bond breaking (only necessary to avoid exceeding an octet or duet at E)

(a) The electron-rich, nonpolar C=C π bond electrons attack the electron-poor carbon of the carbocation. Only one arrow is necessary, as the electrophile lacks an octet. The carbocation is now on the phenyl ring.

Electrophilic addition

(b) The electron-rich, nonpolar C≡C π bond electrons attack the electron-poor H atom of H–Br. The H–Br bond must break to avoid exceeding the duet for H, so two arrows are necessary.

H–Br: Electrophilic addition H + :Br:⊖

Problem 7.17

Think

Which sites are electron rich? Which are electron poor? Are curved arrows drawn from electron rich to electron poor, or vice versa? How many curved arrows are needed to show the electrophile elimination step?

Solve

Two arrows are necessary to show the electrophile elimination step when the electrophile leaves as H^+:

- Breaking the C–E bond and forming the C=C π bond
- Forming a H–base bond

(a) H :Base Electrophile elimination + H—base

(b) H :Base Electrophile elimination + H—base

Problem 7.18

Think

Which bond is broken and formed in a 1,2-hydride shift? A 1,2-methyl shift? How do you show the electron movement via curved arrows? Are curved arrows drawn from electron rich to electron poor, or vice versa?

Solve

In a 1,2-hydride shift, the bond that breaks is the C–H bond adjacent to C^+, and the new bond that forms is the C^+ and hydride C–H bond. In a 1,2-methyl shift, the bond that breaks is the C–CH_3 bond adjacent to C^+, and the new bond that forms is the C^+ and CH_3 C–CH_3 bond.

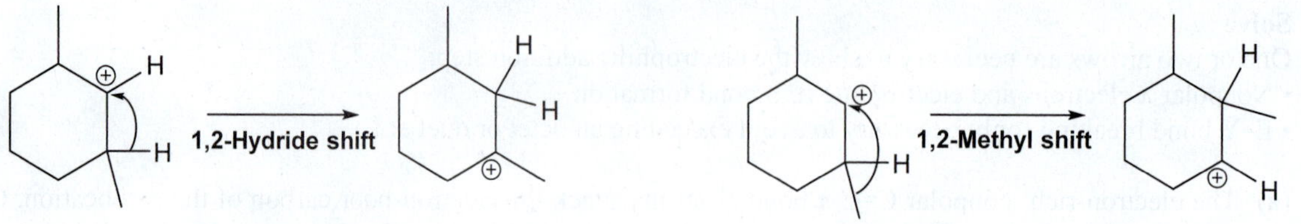

1,2-Hydride shift

1,2-Methyl shift

Problem 7.20

Think

On the two sides of the reaction, is there a difference in charge stability? Is there a difference in total bond energy?

Solve

The favored side of each reaction is boxed.

(a) The + charge on the left is secondary, and the + charge on the right is secondary benzylic. Benzylic carbocations are more stable owing to resonance delocalization of the charge around the ring.

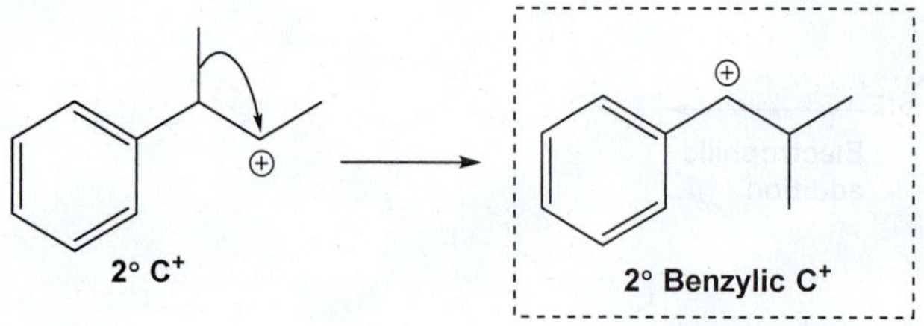

2° C^+

2° Benzylic C^+

(b) The + charge on the left is secondary, and the + charge on the right is tertiary. Tertiary carbocations are more stable owing to electron donation by more alkyl groups.

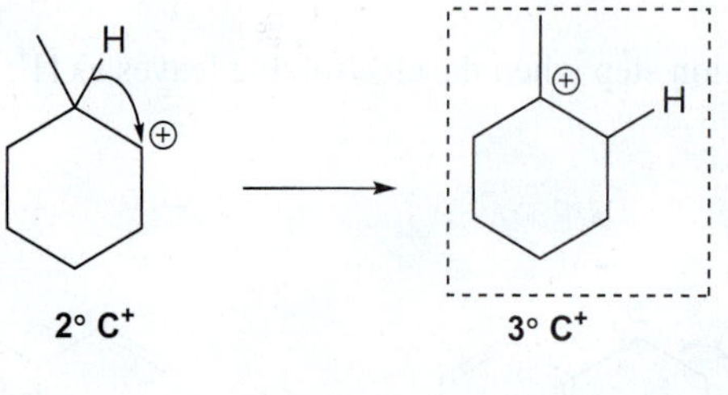

2° C^+

3° C^+

Problem 7.21

Think

What is the keto form of the enol that immediately forms after decarboxylation? Which form is typically more stable?

Solve

The enol form and keto form are shown below. Most keto forms are more stable than their enol forms.

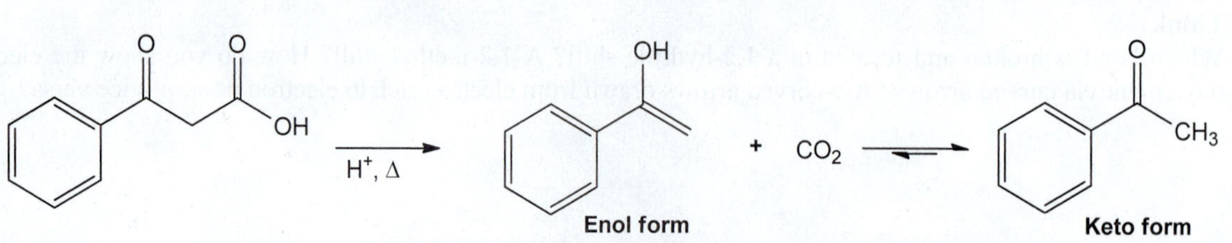

H^+, Δ

+ CO_2

Enol form

Keto form

End of Chapter Problems
Sections 7.1–7.7 Curved Arrow Notation and Elementary Steps
Problem 7.22

Think

What simplifying assumptions can we make regarding the metals? Are the species listed electron rich or electron poor? Toward which kinds of sites are electron-rich species attracted? Toward which kinds of sites are electron-poor species attracted? Which site on phenol is most likely to react with each of the following species?

Solve

The compounds given in the problem can be treated as the electron-rich species shown below.

(a) CH_3O^- (b) $^-CH_3$ (c) $C_6H_5^-$ (d) H^-

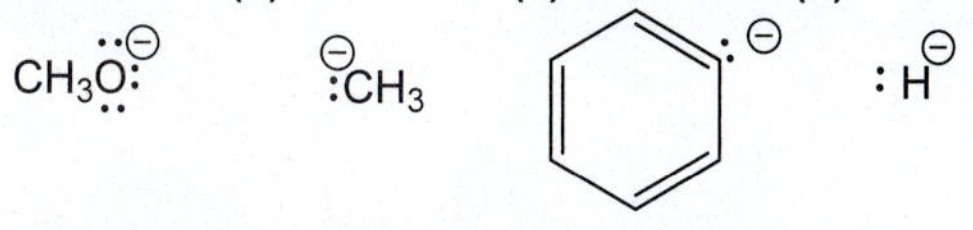

They can react with an electron-poor site on C_6H_5OH. As we can see below, there are two such sites.

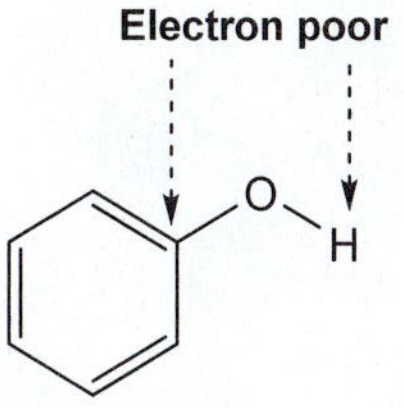

Electron poor

However, the proton is more easily accessed than the C is, and the sp^2 hybridization of C makes the C–O bond stronger than usual. Therefore, the predominant reaction is from attack on the proton, leading to a proton transfer reaction.

(a) $C_6H_5OH + CH_3O^- \xrightarrow{\text{Proton transfer}} C_6H_5O^- + H{-}OCH_3$

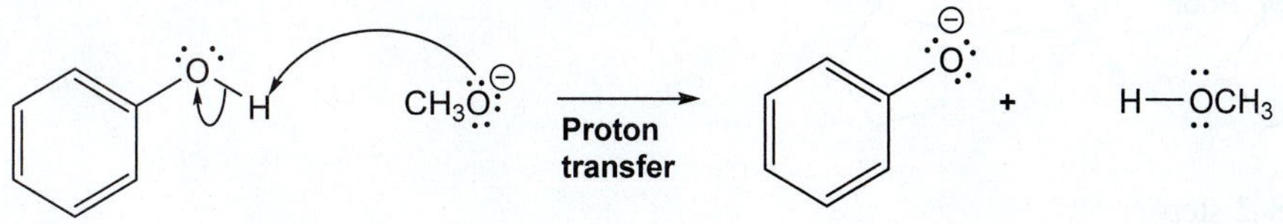

(b) $C_6H_5OH + {}^-CH_3 \xrightarrow{\text{Proton transfer}} C_6H_5O^- + H{-}CH_3$

(c) $C_6H_5OH + C_6H_5^- \xrightarrow{\text{Proton transfer}} C_6H_5O^- + C_6H_5{-}H$

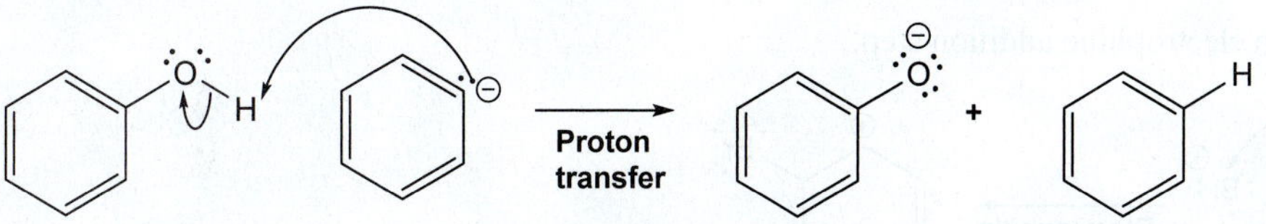

(d) $C_6H_5OH + H^- \xrightarrow{\text{Proton transfer}} C_6H_5O^- + H{-}H$

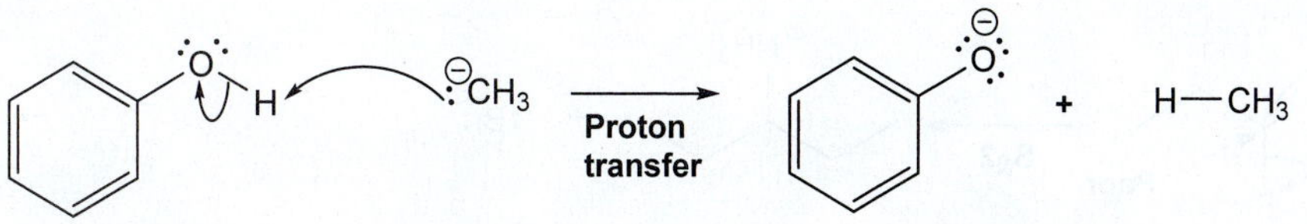

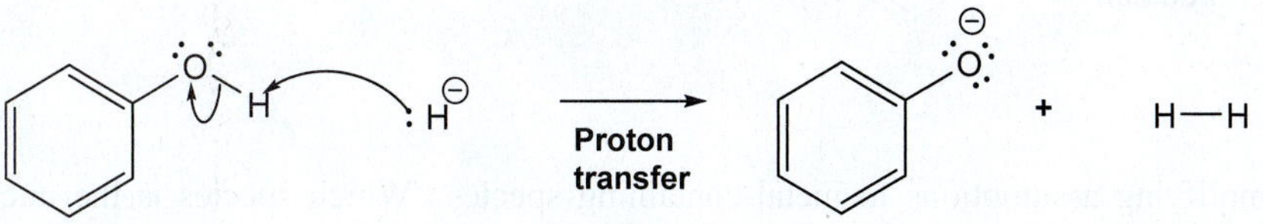

Problem 7.23

Think

Use curved arrow notation to follow the bond making and breaking. Do the curved arrows show the movement of electrons? Do the arrows flow from electron rich to electron poor? Which of the 10 elementary steps (if any) is represented? Do any of the products violate the duet and octet rule?

Solve

Products are drawn for all reactions. The unacceptable reactions are marked with an X.

(a) No. The curved arrow is drawn from electron poor to electron rich. The curved arrow shows the movement of atoms, not of electrons.

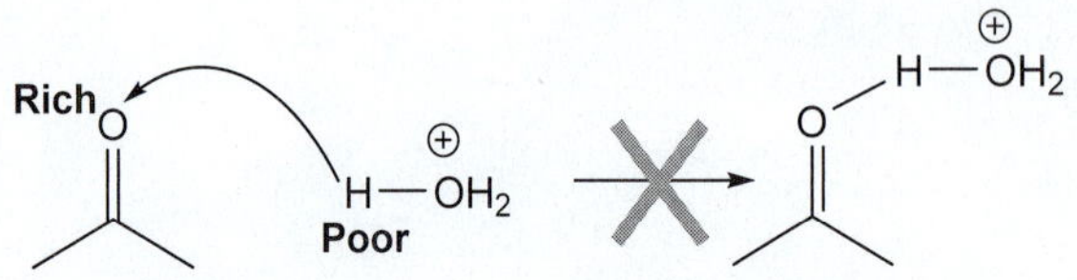

Rich Poor

(b) No. In the products, there would be four electrons on the H atom—a bond and a lone pair.

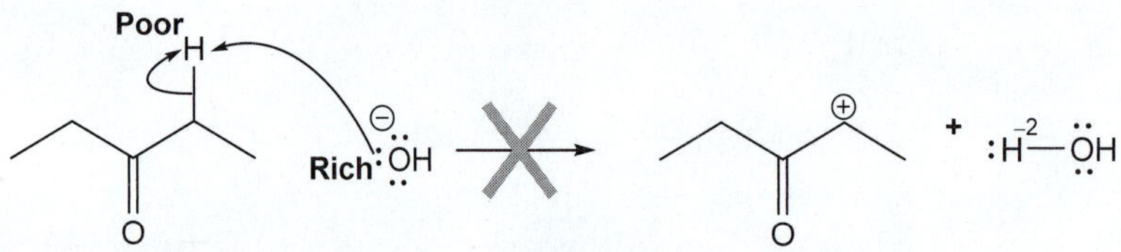

Poor Rich

(c) No. A bond is formed to a C atom with four bonds, so a bond must be broken simultaneously. Otherwise, the C atom would end up with five bonds.

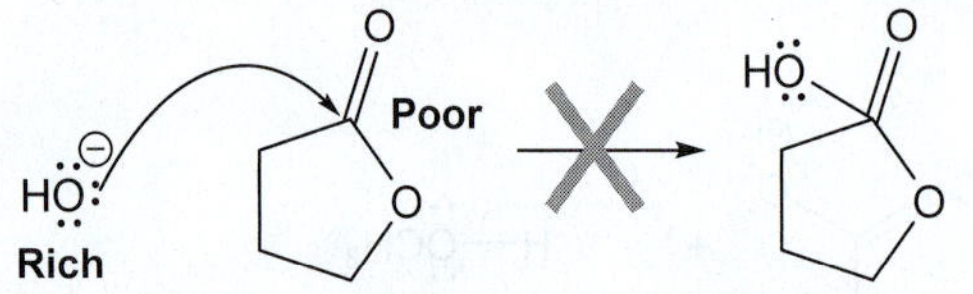

Poor Rich

(d) Yes. This is an S_N2 step.

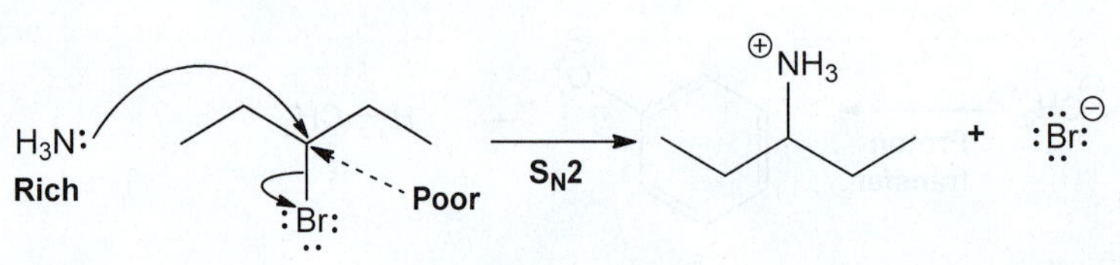

Rich Poor S_N2

(e) Yes. This is an electrophilic addition step.

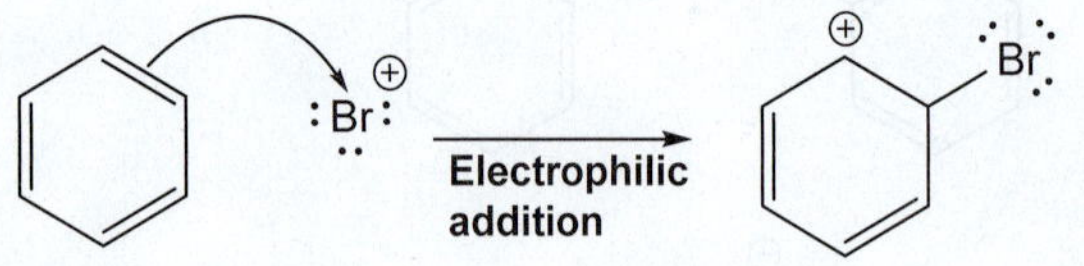

Electrophilic addition

Problem 7.24

Think

Can you make any simplifying assumptions to metal-containing species? Which species is the nucleophile? Which is the substrate? Which species is electron rich? Which is electron poor? How many curved arrows are necessary to show the S_N2 step? Are curved arrows drawn from electron rich to electron poor, or vice versa?

Solve

The reactions are shown below. The Na^+ and K^+ ions are spectator ions and, therefore, are omitted from the reaction mechanism. Two curved arrows are necessary in an S_N2 step—one to show the Nu–C bond formation and another to show the C–L bond breaking.

Problem 7.25

Think

Which sites are electron rich? Which are electron poor? Are curved arrows drawn from electron rich to electron poor, or vice versa? Which species donates an electron pair? Which species accepts an electron pair? How many curved arrows are needed to show the coordination step? Which bond forms in the coordination step? Which bond breaks in the heterolysis (reverse of coordination) step?

Solve

The Lewis acid is the electron-poor species and accepts the electron pair from the Lewis base, which is the electron-rich species (labeled below). The bond that forms in the Lewis acid/base coordination step is the same bond that breaks in the heterolysis step.

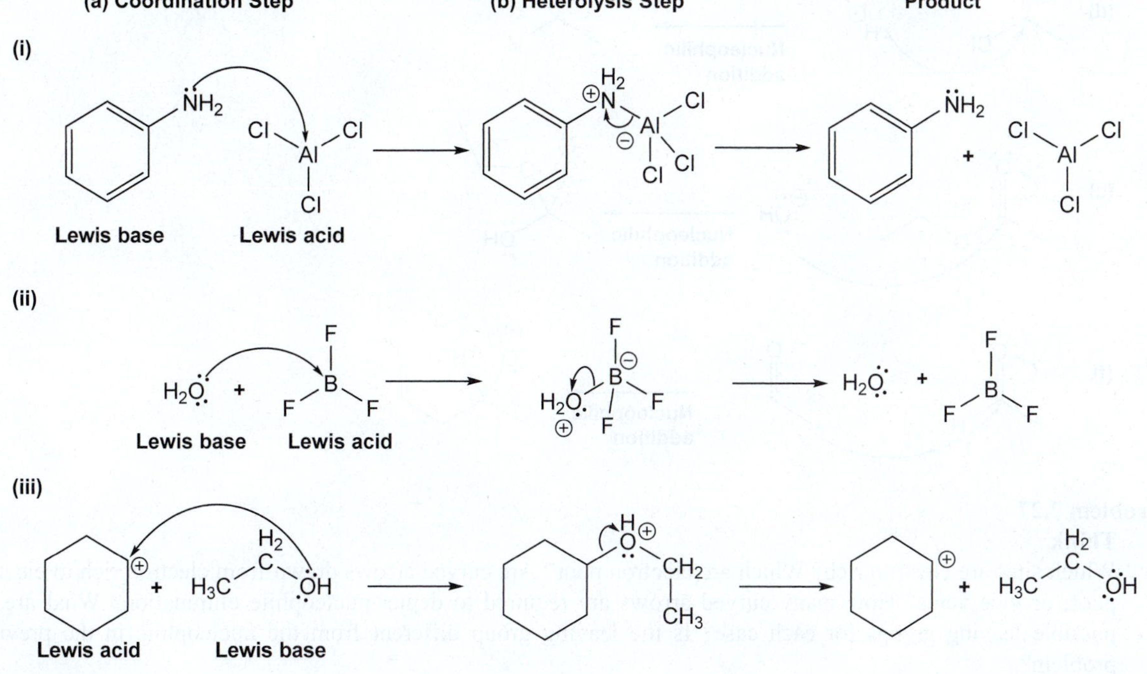

Problem 7.26

Think

Can you make any simplifying assumptions to metal-containing species? Which species is the nucleophile? Which one contains a polar π bond? Which species is electron rich? Which is electron poor? How many curved arrows are necessary to show the nucleophilic addition step? Are curved arrows drawn from electron rich to electron poor, or vice versa?

Solve

The products of each nucleophilic addition are shown below. In **(a)**, CH_3OK is treated as CH_3O^-. In **(b)**, CH_3Li is treated as CH_3^-. In **(c)**, C_6H_5MgBr is treated as $C_6H_5^-$. In **(d)**, $NaBH_4$ is treated as H^-. In **(e)**, NaOH is treated as HO^-. In each case, two curved arrows are necessary. One curved arrow originates from the electron-rich nucleophile and points to the electron-poor C atom of C=O or C≡N. The second curved arrow originates from the middle of the C=O or C≡N bond and points to the heteroatom.

(a) CH_3O^- Nucleophilic addition

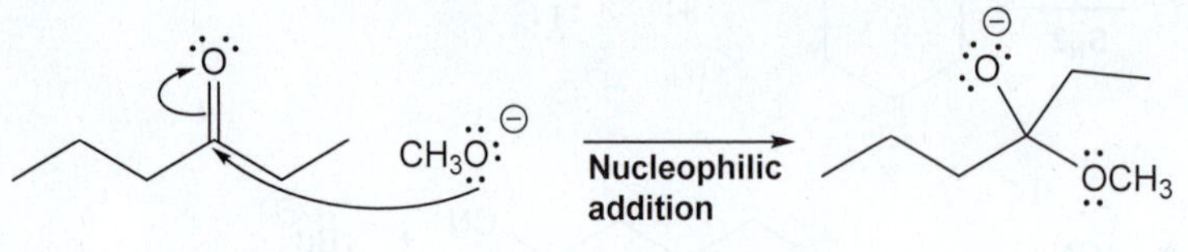

(b) $:CH_3^-$ Nucleophilic addition

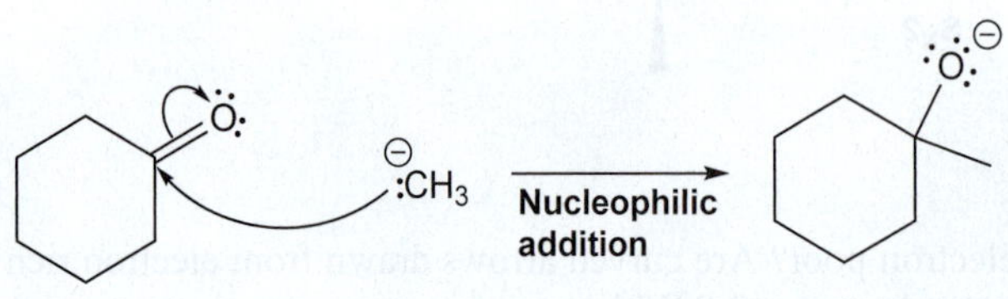

(c) Nucleophilic addition

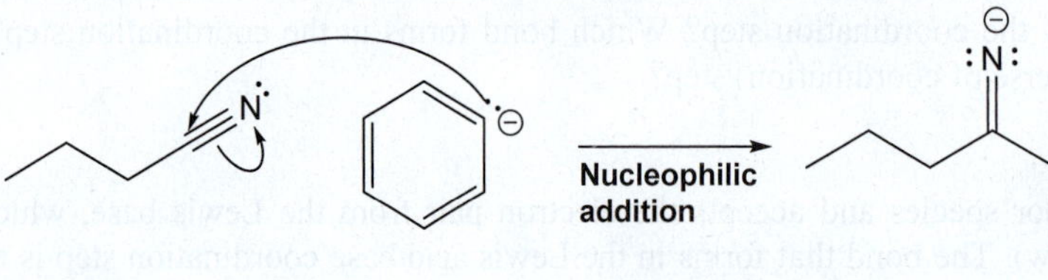

(d) Cl :H⁻ Nucleophilic addition Cl H

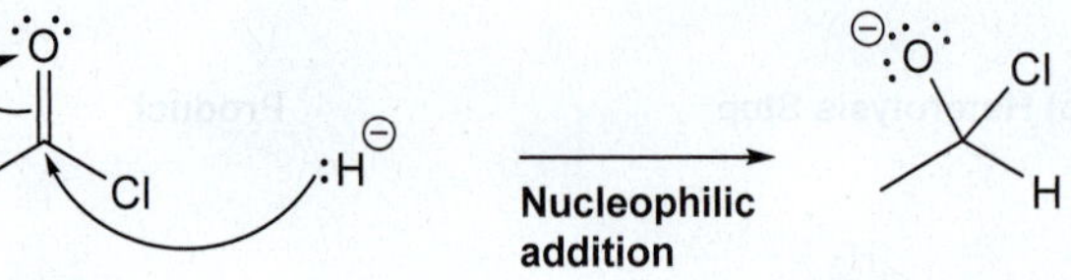

(e) :OH⁻ Nucleophilic addition OH

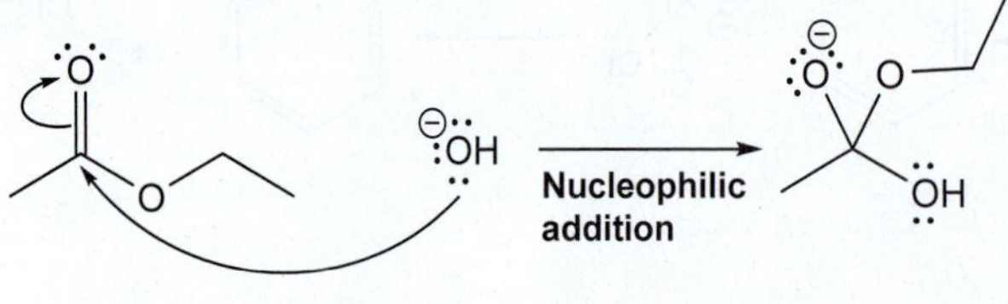

(f) H Nucleophilic addition H

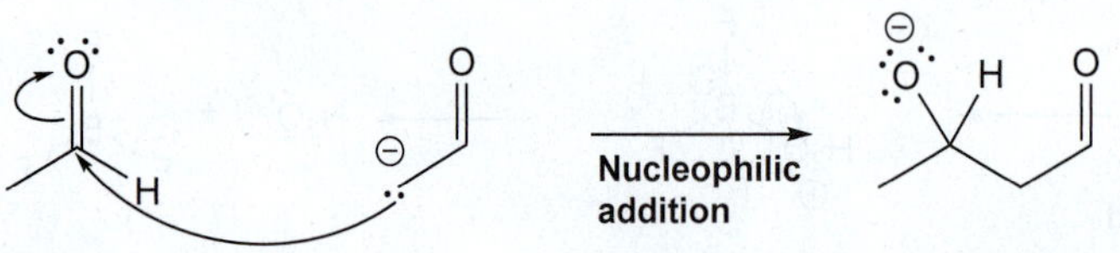

Problem 7.27

Think

Which sites are electron rich? Which are electron poor? Are curved arrows drawn from electron rich to electron poor, or vice versa? How many curved arrows are required to depict nucleophile elimination? What are the possible leaving groups for each case? Is the leaving group different from the nucleophile in the previous problem?

Solve

The nucleophile elimination steps shown on the next two pages produce a compound that is different from the reactants. These steps involve expelling a leaving group that is different from the nucleophile that attacks in the previous problem. *Note:* As we will discuss further in Chapter 21, nucleophile elimination is unfeasible when H^- or R^- is eliminated. For these reactions, an X is drawn on the reaction arrow.

(a)

OCH$_3$

H_3CO

Nucleophile elimination

OCH$_3$

OCH$_3$

Nucleophile elimination

(b)

CH$_3$

CH$_3$

Nucleophile elimination

(c)

Nucleophile elimination

(d)

H

Cl

Nucleophile elimination

H

Cl

H

Nucleophile elimination

Cl

H

CH_3

(e)

OH

Nucleophile elimination

HO

CH_3

OH

Nucleophile elimination

OH

(f)

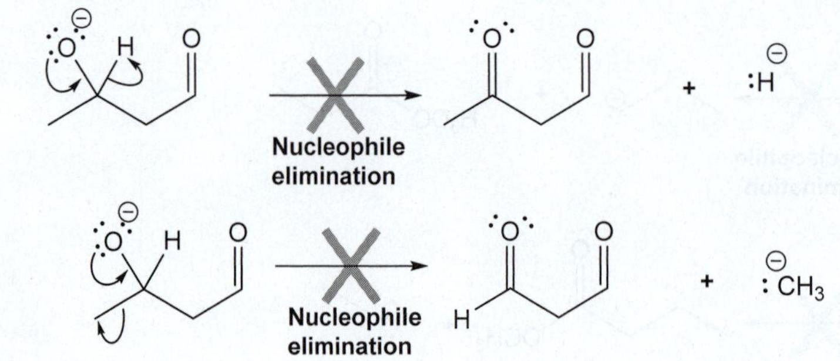

Nucleophile elimination

+ :H

Nucleophile elimination

+ : CH_3

Problem 7.28

Think

Which sites are electron rich? Which are electron poor? Are curved arrows drawn from electron rich to electron poor, or vice versa? How many curved arrows are used to show nucleophile elimination? Do the products exhibit different charge stabilities? How does charge stability affect the major product outcome?

Solve

The possible products are shown below. In each case, a curved arrow originates from O^- and points to the center of the C–O bond, because O^- is electron rich and C is electron poor. The second curved arrow is necessary to avoid exceeding the octet on carbon. The products of the second reaction are favored, because Cl is substantially larger than C or O, so the negative charge is more stable on Cl than it is on either C or O. Moreover, Cl^- is a weak base, because it is the conjugate base of the strong acid H–Cl. C^- and O^- are strong bases.

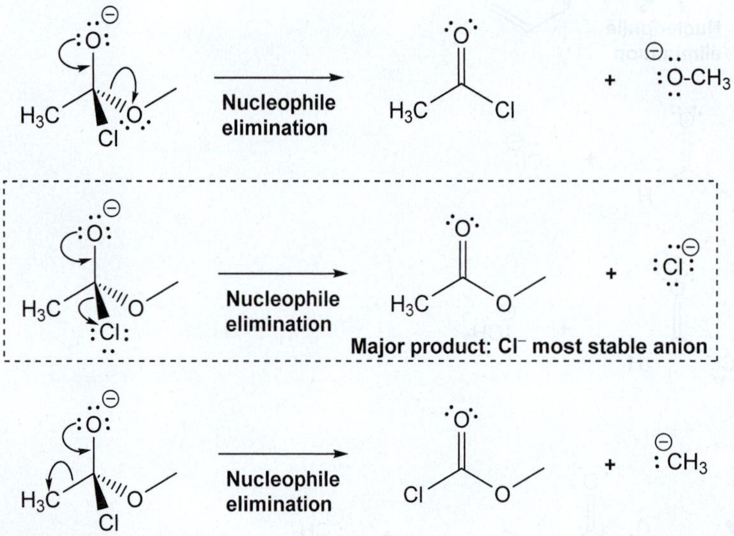

Nucleophile elimination

+ :O-CH_3

Nucleophile elimination

+ :Cl:

Major product: Cl^- most stable anion

Nucleophile elimination

+ : CH_3

Problem 7.29

Think

What are the requirements for bimolecular elimination (E2)? Is there a hydrogen on the carbon adjacent to the carbon with the leaving group? How many curved arrows are needed to show the E2 step? On which two C atoms will the double bond form? Refer to Equation 7-17 as a guide.

Solve

An E2 step can take place when a strong base is in the presence of a substrate in which a leaving group (L) and a hydrogen atom are on adjacent carbon atoms. Substrates **(a)** and **(d)** cannot undergo E2, because there is no H adjacent to the carbon with the leaving group. Substrates **(b)**, **(c)**, and **(e)** can undergo E2, and the reactions are shown on the next page.

(b) Cl^- is the leaving group; H_2N^- is the base and attacks the H atom on the carbon next to the carbon bearing the Cl atom. A double bond is formed.

H_2N^- + [alkyl chloride] $\xrightarrow{E2}$ H_3N + [alkene] + Cl^-

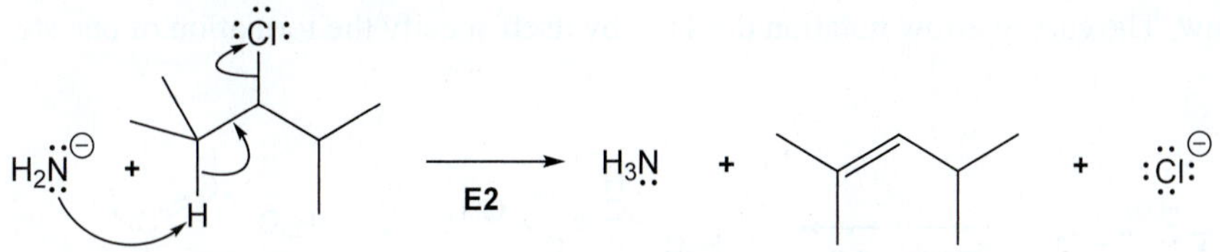

(c) I^- is the leaving group; H_2N^- is the base and attacks the H atom on the carbon next to the carbon bearing the I atom. A double bond is formed.

H_2N^- + [alkyl iodide] $\xrightarrow{E2}$ H_3N + [alkene] + I^-

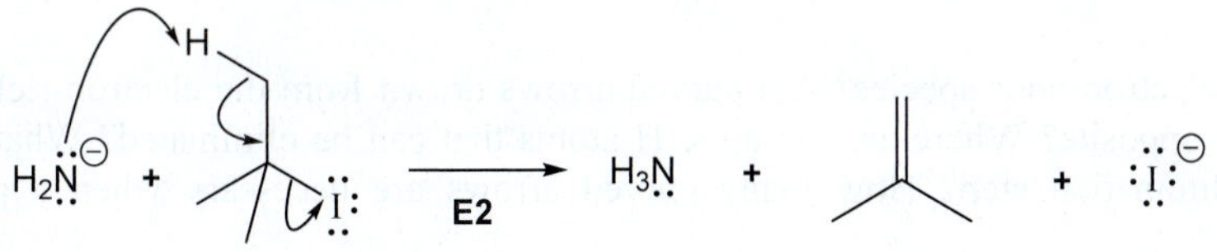

(e) Br^- is the leaving group; H_2N^- is the base and attacks the H atom on the carbon next to the carbon bearing the Br atom. A triple bond is formed.

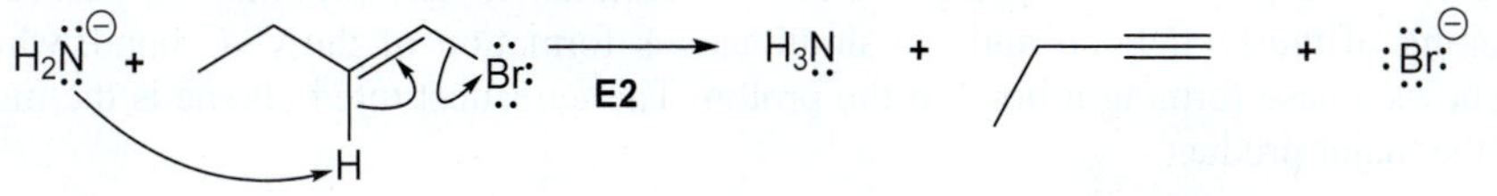

H_2N^- + [vinyl bromide] $\xrightarrow{E2}$ H_3N + [alkyne] + Br^-

Problem 7.30

Think

On which carbons can the new C–H bond form? If the proton adds to one C atom, what is generated on the other C atom? On which C atom is a positive charge better stabilized? Why?

Solve

The H^+ can be deposited onto either of the alkene C atoms. A positive charge results on the other C atom. The two carbocations that can be formed are shown below. The second product is more stable, because it can participate in resonance with the lone pairs on the O atom (shown).

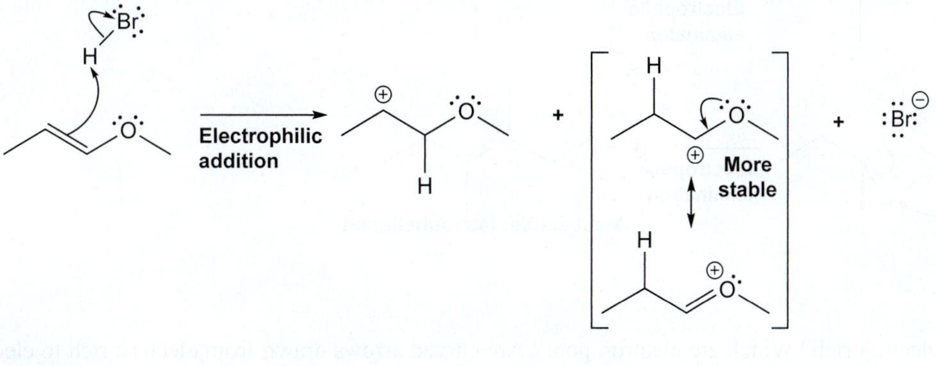

Electrophilic addition

More stable

Problem 7.31

Think

What are the electron-rich and electron-poor species? Are curved arrows drawn from the electron-rich species to the electron-poor one, or the opposite? If the indicated H atom leaves as H^+, where do the electrons go? What type of bond results? What is the definition of a diastereomer?

Solve

Water is electron rich, and the carbocation is electron poor, so the curved arrows are drawn from water toward the carbocation. Because free rotation can occur about the central C–C bond, both cis and trans isomers are formed, as shown below. The curved arrow notation does not by itself specify the formation of one stereoisomer or the other.

Electrophile elimination

trans

cis

Problem 7.32

Think

What are the electron-rich and electron-poor species? Are curved arrows drawn from the electron-rich species to the electron-poor one, or the opposite? Where are the three H atoms that can be eliminated? What product results from an electrophile elimination step? How many curved arrows are necessary when a proton is eliminated?

Solve

The three reactions and their products are shown below. Each reaction is the result of eliminating a different proton. Notice that two curved arrows are necessary to show the elimination of a proton: One curved arrow is used to show the breaking of the H–C bond and the simultaneous formation of the C–C bond, whereas the second curved arrow shows a base forming a bond to the proton. The tetrasubstituted alkene is the most stable and, therefore, will be the major product.

Electrophile elimination

Electrophile elimination

Electrophile elimination

Most stable: tetrasubstituted

Problem 7.33

Think

Which sites are electron rich? Which are electron poor? Are curved arrows drawn from electron rich to electron poor, or vice versa? How many curved arrows are used to show electrophilic addition? If the electrophile adds to one C atom of a C=C, then what appears on the other C?

Solve

The reactions and their products are shown on the next page. The C=C is electron rich, and NO_2^+ is electron poor. A single curved arrow is drawn from the center of the C=C bond and points to NO_2^+. When NO_2^+ adds to one C atom of the C=C bond, a + charge is generated on the other C atom.

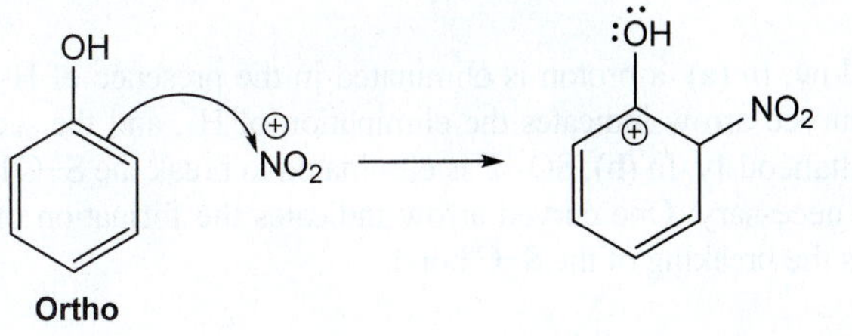

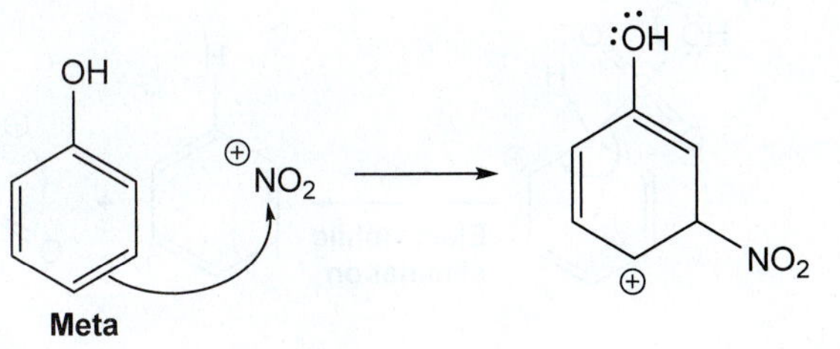

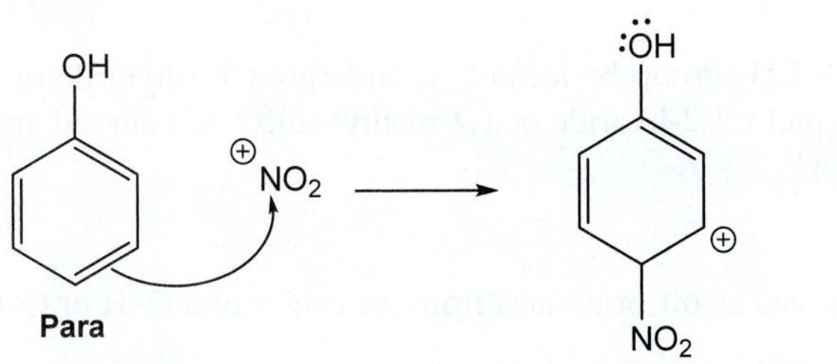

Ortho

Meta

Para

Problem 7.34

Think

Is the alkyne electron rich or electron poor? From where do the electrons originate to form the new bond with the H^+ from CF_3SO_2OH? When one C atom of the alkyne forms the new C–H bond, what is generated on the other C atom? Which product exhibits greater charge stability?

Solve

The curved arrow notation and products are shown below. This electrophilic addition step requires two curved arrows: One originates from the center of the C≡C bond and points to the electron-poor H, whereas the second indicates the breaking of the H–O bond, originating from the center of the H–O bond and pointing to the O. When one C atom of C≡C picks up a proton, the other C atom gains a positive charge. The first carbocation is more stable, because the positively charged carbon is attached to one additional alkyl group, which, as indicated, is electron donating and thus provides stability to the carbocation.

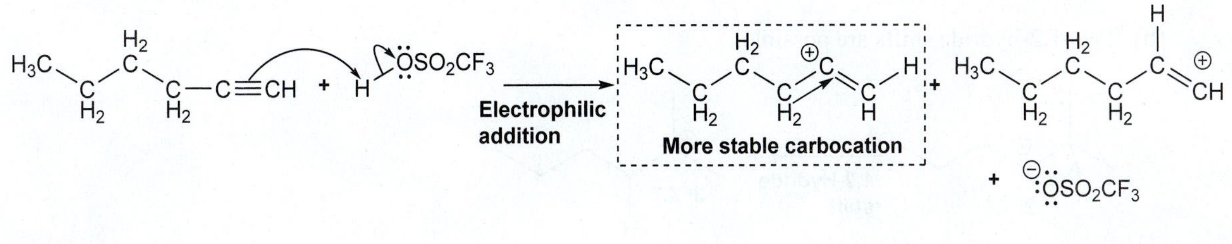

Electrophilic addition

More stable carbocation

Problem 7.35

Think

What is the elementary step to eliminate H^+ or SO_3H^+? What type of bond forms on the ring? Which sites are electron rich? Which are electron poor? Are curved arrows drawn from electron rich to electron poor, or vice versa?

Solve

The electrophile elimination mechanisms are shown below. In **(a)**, a proton is eliminated in the presence of H_2O, a weak base, so two curved arrows are necessary: One curved arrow indicates the elimination of H^+, and the second curved arrow indicates water picking up the proton simultaneously. In **(b)**, SO_3H^+ is eliminated to break the S–C bond and form a C=C double bond. Two curved arrows are necessary: One curved arrow indicates the formation of the S=O double bond, and the second curved arrow indicates the breaking of the S–C bond.

(a) HO, O, S, O, H, ⊕, $H_2\ddot{O}$: → Proton transfer; OH, O=S=O; + H—$\ddot{O}H_2$ ⊕

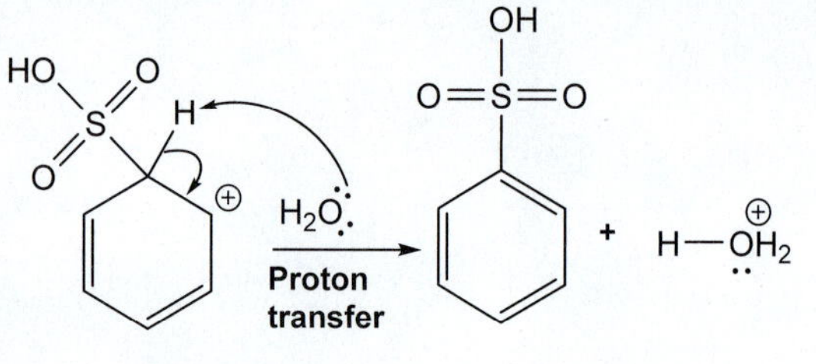

(b) H$\ddot{O}$, O, S, O, H, ⊕ → Electrophile elimination; H; + ⊕OH, S, O, O

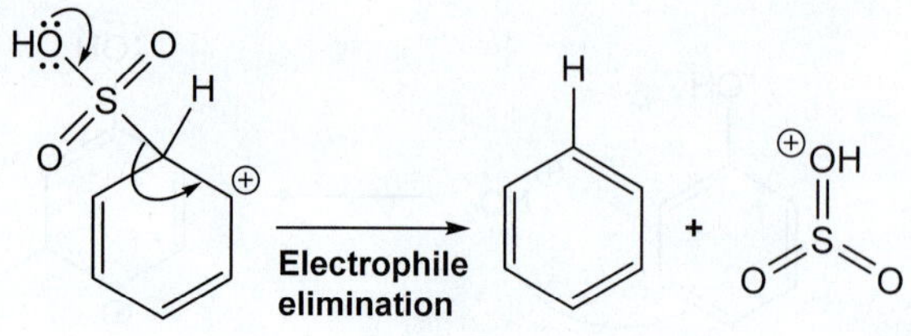

Problem 7.36

Think

Relative to the position of the C^+, where must a H or CH_3 group be located to undergo a 1,2-hydride or 1,2-methyl shift? How many curved arrows are used to depict a 1,2-hydride or 1,2-methyl shift? Are curved arrows drawn from electron rich to electron poor, or vice versa?

Solve

To depict a 1,2-hydride or 1,2-methyl shift, a single curved arrow originates from the center of a C–H or C–CH_3 bond that is adjacent to a C^+, and it points to the C^+.

All possible 1,2-hydride shifts

(a) Two 1,2-hydride shifts are possible.

2° C^+ ⊕ H → 1,2-Hydride shift → H ⊕ 2° C^+

2° C^+ ⊕ H → 1,2-Hydride shift → H ⊕ 2° C^+

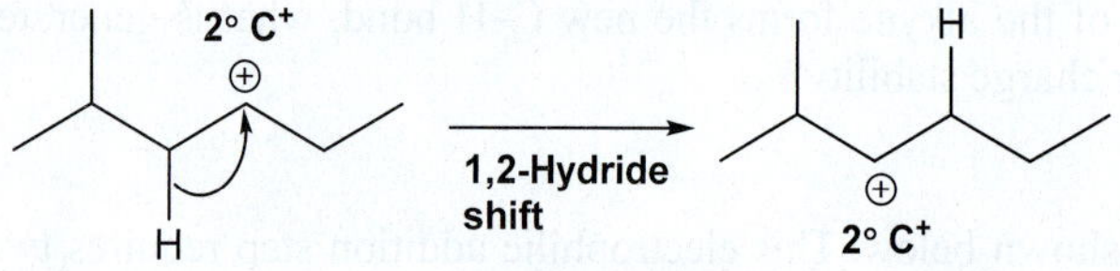

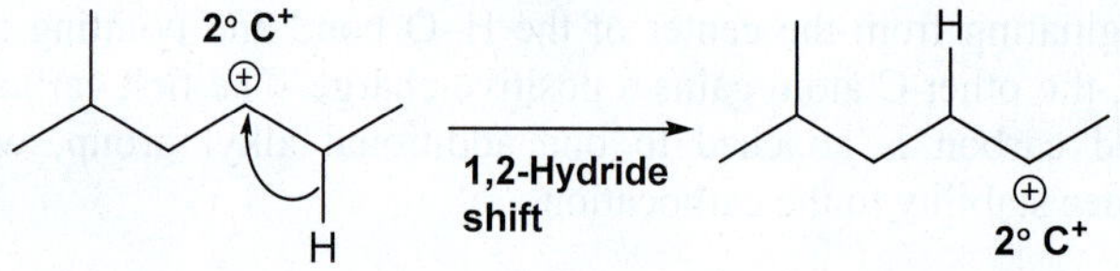

(b) Two 1,2-hydride shifts are possible.

⊕ H 2° C^+ → 1,2-Hydride shift → ⊕ 3° C^+ H

H ⊕ 2° C^+ → 1,2-Hydride shift → 2° C^+ ⊕ H

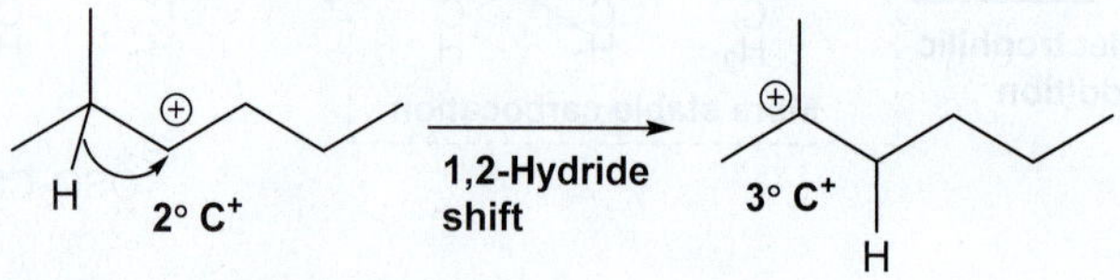

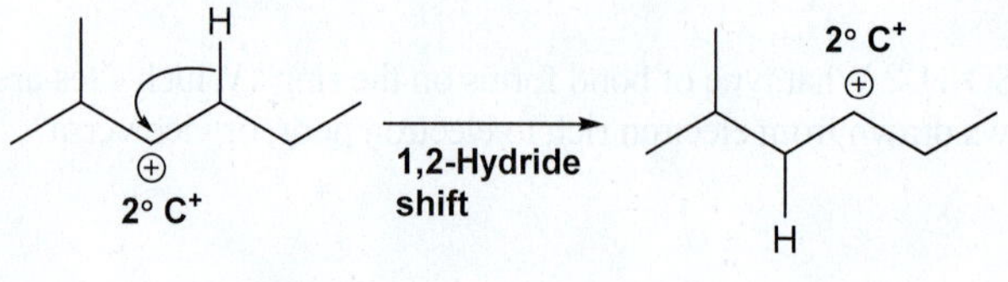

(c) Two 1,2-hydride shifts are possible.

3° C$^+$ → 1,2-Hydride shift → 3° C$^+$

3° C$^+$ → 1,2-Hydride shift → 2° C$^+$

(d) One 1,2-hydride shift is possible.

2° C$^+$ → 1,2-Hydride shift → 2° C$^+$

(e) Two 1,2-hydride shifts are possible.

1,2-Hydride shift

Isolated 3° C$^+$ Benzylic 2° C$^+$: resonance stabilized

3° C$^+$ → 1,2-Hydride shift → 2° C$^+$

(f) Three 1,2-hydride shifts are possible.

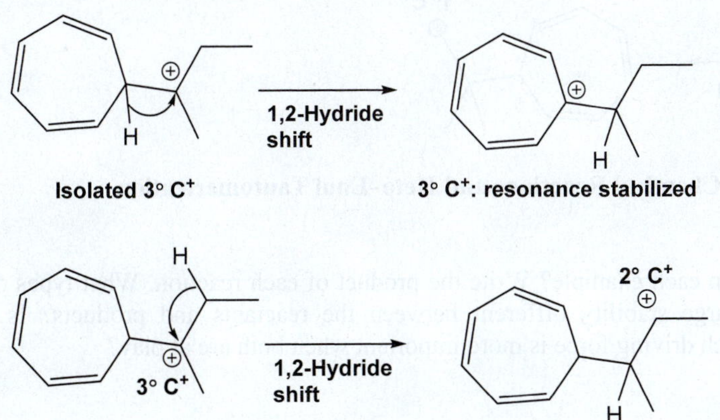

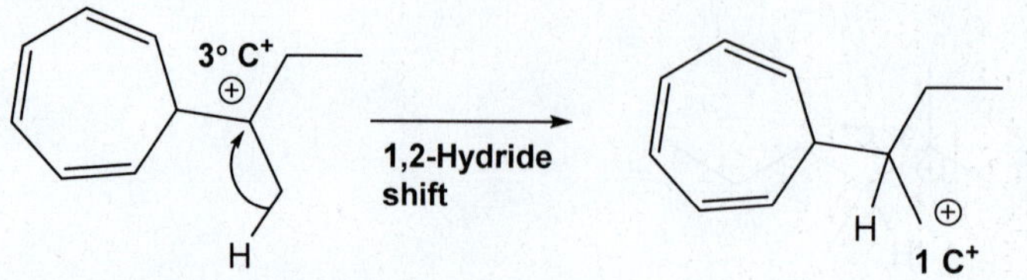

3° C+
1,2-Hydride shift
H
1 C+

All possible 1,2-methyl shifts

(a) One 1,2-methyl shift is possible.

2° C+
1,2-Hydride shift
1° C+

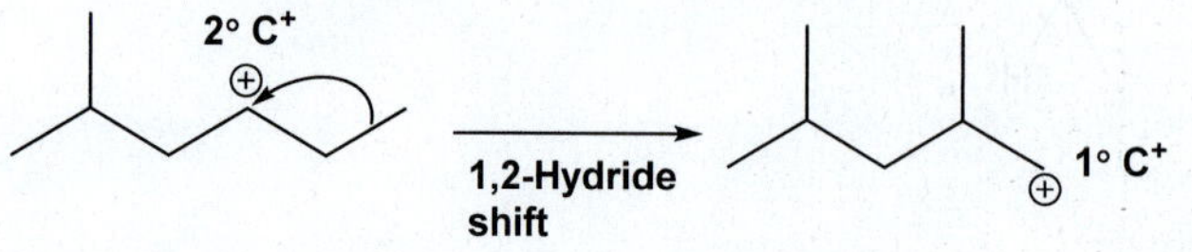

(b) One 1,2-methyl shift is possible.

2° C+
1,2-Methyl shift
2° C+

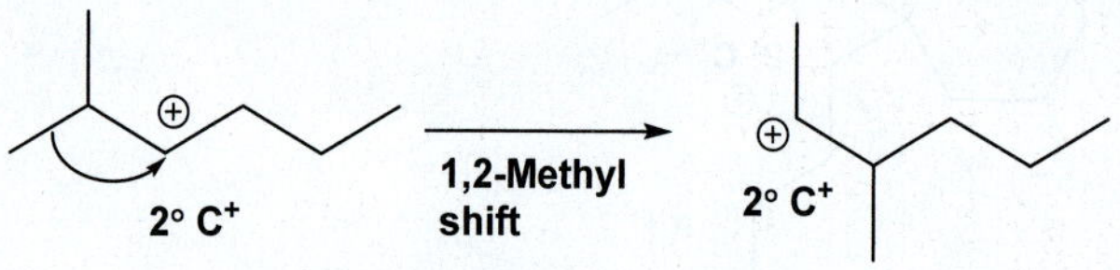

(c) One 1,2-methyl shift is possible.

3° C+
1,2-Methyl shift
2° C+

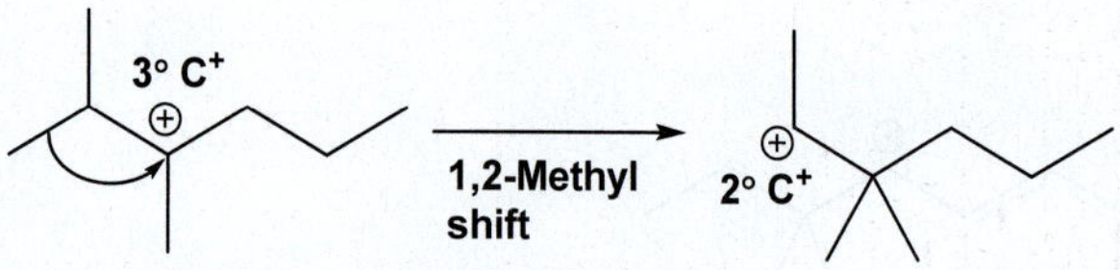

(d) No 1,2-methyl shift is possible.

(e) One 1,2-methyl shift is possible.

3° C+
1,2-Methyl shift
1° C+

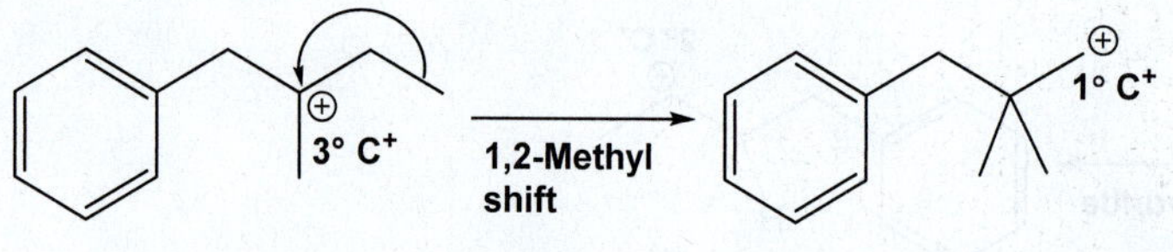

(f) One 1,2-methyl shift is possible.

3° C+
1,2-Methyl shift
1° C+

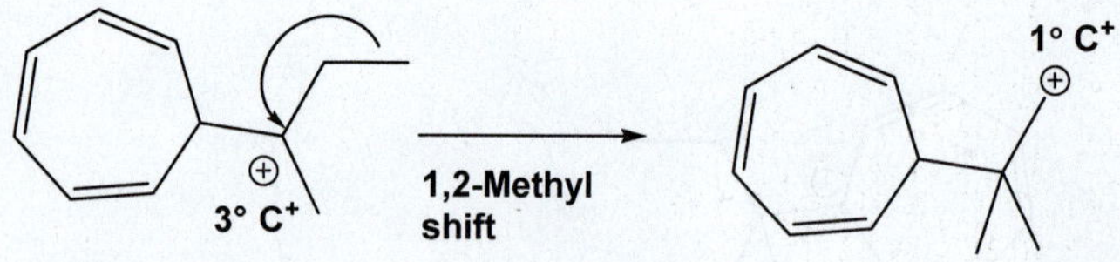

Sections 7.8–7.9 The Driving Force for Chemical Reactions and Keto–Enol Tautomerization

Problem 7.37

Think

What is the elementary step shown in each example? Write the product of each reaction. What types of bonds broke? What types formed? Is charge stability different between the reactants and products? Is there a difference in total bond energy? Which driving force is more important when both are at play?

Solve

(a) The attempted elementary step is an S_N2. The negative charge is more stable on the Cl^- than it is on the HO^-, and the reaction is not favored in the forward direction. Therefore, the reactant side is favored.

Favored

S_N2

(b) The attempted elementary step is an E2. The products form two additional charges and replace a σ bond for a π bond. Both of these driving forces favor the reactant side.

Favored

E2

(c) The elementary step is a nucleophilic addition. The negative charge is more stable on the O than it is on the C, so the reaction is favored in the forward direction.

Nucleophilic addition

Favored

(d) The elementary step is an electrophilic addition. The methyl C^+ on the reactants is less stable than it is on the tertiary C^+ formed in the products. Therefore, the product side is favored.

Methyl C^+

Electrophilic addition

3° C^+

Favored

(e) The elementary step is an electrophile elimination. The carbocation is lacking its octet in the reactants, whereas all atoms in the products have their octets. Therefore, the product side is favored.

Electrophile elimination

Favored

(f) The elementary step is coordination. The secondary C^+ on the reactants lacks its octet, so it is less stable than the RNH_3^+ ion formed in the products, in which all atoms have their octets. Therefore, the product side is favored.

Coordination

Favored

Problem 7.38

Think

If a charged species results as a product, compare the charge stability of the ions in each reaction. What types of bonds broke? What types formed? Compare the bond energies in the products and reactants for each reaction. Which driving force is in play in each example? When both are in play, which driving force is more important?

Solve

(a) There is greater charge stability in the products of the second reaction. The primary carbocation that is produced in the first reaction is too unstable, because only one alkyl group inductively stabilizes the positive charge. In the second reaction, a tertiary carbocation is produced in which three alkyl groups stabilize the positive charge.

CH_3CH_2OH 1° C^+ CH_3CH_2OH 3° C^+

(b) Charge stability and bond energies both favor the second reaction substantially more than the first. In the first reaction, the C atom that has the leaving group is *sp*²-hybridized, so the C–Cl bond is too strong and the positive charge that is formed is too unstable (the C atom has a fairly high effective electronegativity). In the second reaction, the bond that breaks is weaker, because it involves an *sp*³-hybridized C atom, and the product is a benzylic carbocation in which the positive charge is delocalized by resonance.

Localized charge Resonance delocalized charge
*sp*² C CH_3CH_2OH *sp*³ C CH_3CH_2OH

(c) DMSO is an aprotic solvent and does not stabilize the resulting ions very well via solvation. CH_3CH_2OH (or any alcohol) is a protic solvent and strongly stabilizes cations and anions via solvation.

DMSO Aprotic solvent 3° C^+ CH_3CH_2OH Protic solvent 3° C^+

(d) The second reaction is favored much more by charge stability. H is a very small atom and is not very electronegative. The negative charge is more stable on I^-, because the I atom is much larger. I^- is the conjugate base of the strong acid HI.

CH_3CH_2OH CH_3CH_2OH

Problem 7.39

Think

What impact does an O atom attached to C^+ have on charge stability? On resonance? On inductive effects? What impact does a CH_2 group attached to C^+ have on charge stability?

Solve

The O atom is electron withdrawing, which destabilizes a carbocation C^+ via inductive effects. However, O also has lone pairs that can participate in resonance with C^+. In the resonance structure, moreover, all atoms have octets, which provides significant stabilization. Resonance and inductive effects work in opposite directions in this case, but, as we learned in Chapter 6, resonance is generally more important than inductive effects, so the O atom provides substantial stability. This kind of resonance stabilization is not available in the product of the second reaction.

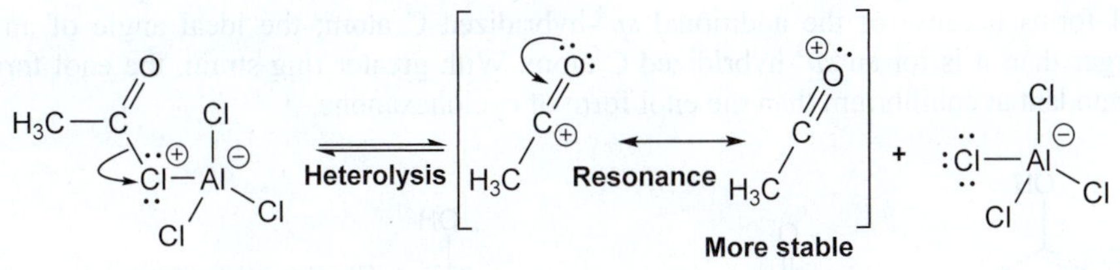

Heterolysis

Resonance

More stable

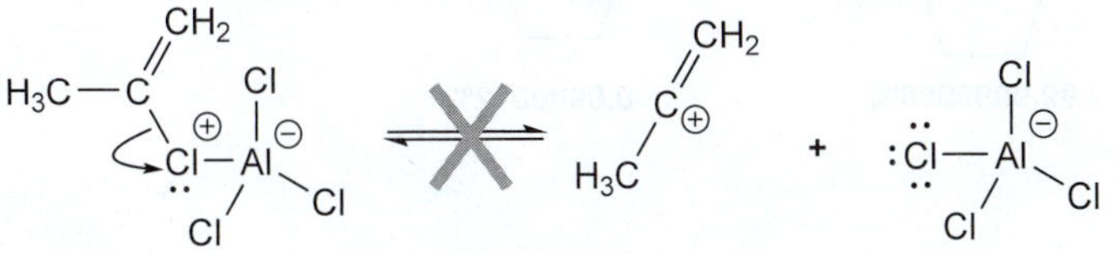

Problem 7.40

Think

What type of carbocation is initially present? What type of carbocation is produced from a 1,2-hydride or 1,2-alkyl shift? What are the effects of resonance on charge stability? What are the effects of electron donation by the alkyl groups on charge stability? Does the carbocation rearrangement lead to a more stable carbocation?

Solve

Only **(b)**, **(e)**, and **(f)** would undergo rearrangements that produce a significantly more stable carbocation. In **(b)**, a 1,2-hydride shift converts a secondary carbocation into a tertiary one. In both **(e)** and **(f)**, the resulting carbocations are resonance stabilized, whereas the initial carbocations have the positive charge localized on a single C atom. In none of the other species does a 1,2-hydride or 1,2-methyl shift produce a more stable carbocation.

Problem 7.41

Think

The keto form of each species was given. What are their corresponding enol forms? Do the alkene portions of the enols have the same degree of alkyl substitution?

Solve

Greater alkyl substitution of the alkene group (C=C) in the enol form provides more stability to the molecule. Because the enol of the first molecule is better stabilized, more of that enol will be present at equilibrium than the enol of the second molecule, so less of the keto form will be present.

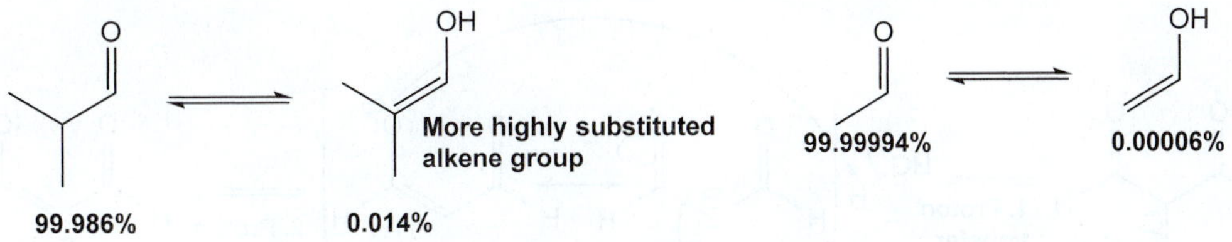

More highly substituted alkene group

99.986% 0.014% 99.99994% 0.00006%

Problem 7.42

Think

The keto form of each species was given. What are their corresponding enol forms? What impact does ring size have on the stability of the molecule? Will ring size affect the keto and enol forms equally?

Solve

The keto and enol forms of each molecule are shown below. The interior angle of a five-membered ring is smaller than it is for a six-membered ring, so the five-membered ring is more strained. That ring strain is more pronounced for the enol forms because of the additional sp^2-hybridized C atom; the ideal angle of an sp^2-hybridized C atom is larger than it is for an sp^3-hybridized C atom. With greater ring strain, the enol form of cyclopentanone is less abundant at equilibrium than the enol form of cyclohexanone.

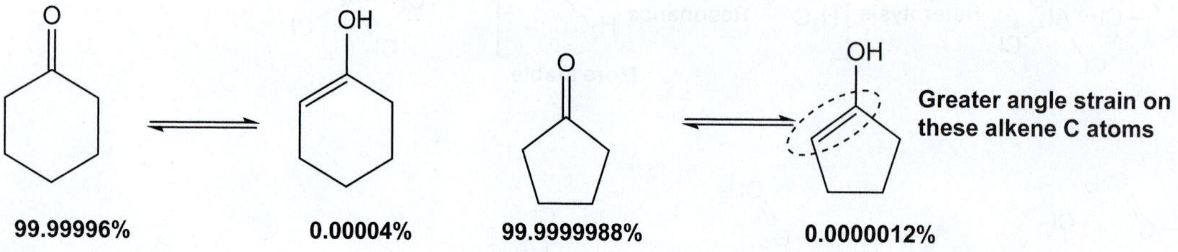

O OH O OH
Greater angle strain on these alkene C atoms
99.99996% 0.00004% 99.9999988% 0.0000012%

Problem 7.43

Think

The keto form of the species was given. What is its corresponding enol form? Draw the tautomeric product. What type of intermolecular interaction is possible with an OH group? What does the presence of an OH group and an O atom permit when they are nearby in the same molecule? Is that interaction possible in the keto form?

Solve

In the enol form (below), an internal hydrogen bond can form, making the enol more stable than usual. That internal hydrogen bond is not possible in the keto form, which has two H-bond acceptors only.

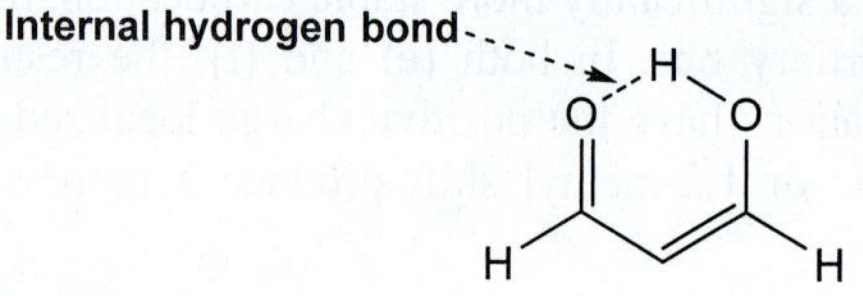

Internal hydrogen bond

Problem 7.44

Think

Refer to Equations 7-34 to 7-37 to review the keto–enol tautomerization mechanisms in both directions and in acidic and basic media.

Solve

Conversion of propanedial in its keto form to its enol form under basic conditions proceeds as follows: In Step 1, the strong base removes a proton from the α carbon, producing an enolate anion; in Step 2, in the resonance structure on the right, the O atom is electron rich and picks up a proton from water, which acts as the acid.

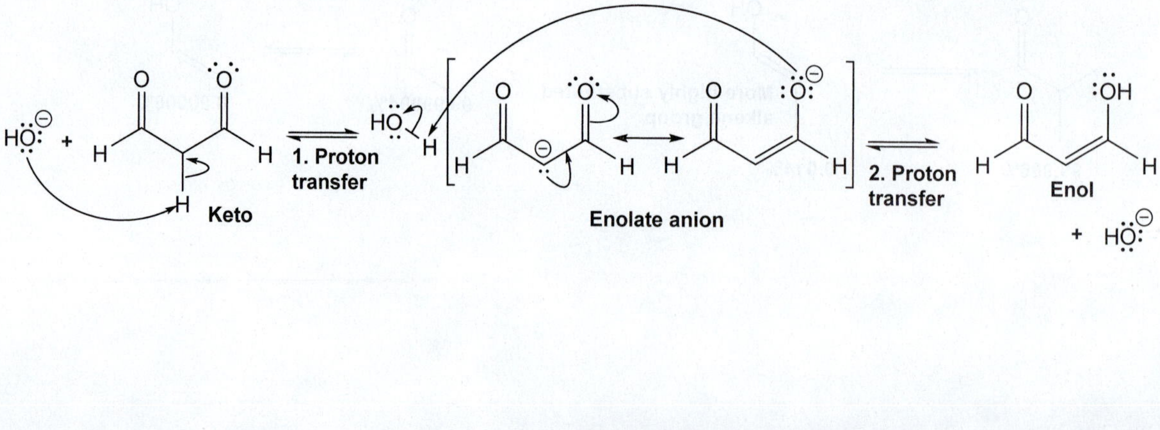

1. Proton transfer
Keto
Enolate anion
2. Proton transfer
Enol

Conversion of propanedial in its keto form to its enol form under acidic conditions proceeds as follows: In Step 1, the strong acid that is present donates a proton to the O atom of the C=O group; in Step 2, water (a weak base) removes the proton from the α carbon to produce the uncharged enol product.

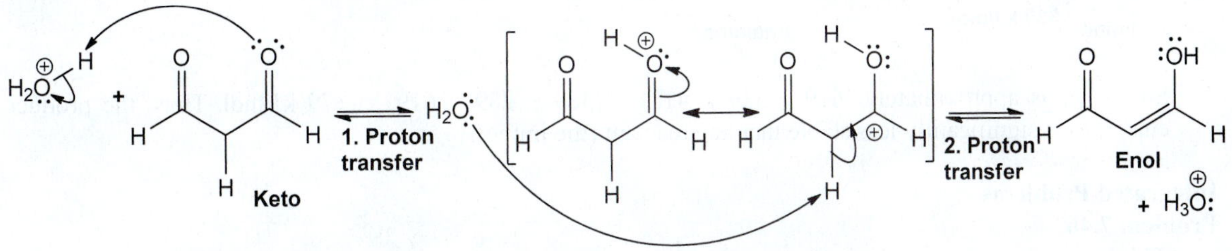

The tautomerization reactions shown above are in equilibrium and can also take place in the reverse direction. The mechanisms below show how the enol form produces the keto form under basic and acidic conditions, respectively. The steps for the mechanisms below are the same as the steps for the keto to enol conversions, but in reverse order.

Conversion of propanedial in its enol form to its keto form under basic conditions proceeds as follows: In Step 1, the strong base removes a proton from the HO group on the enol, producing an enolate anion; in Step 2, in the resonance structure on the right, the C atom is electron rich and picks up a proton from water, which acts as the acid.

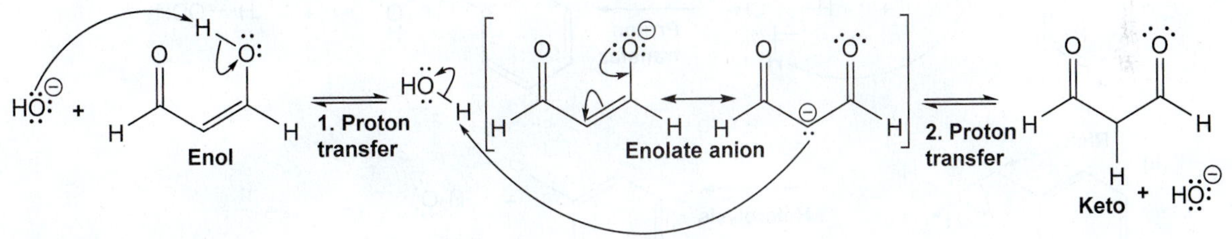

Conversion of propanedial in its enol form to its keto form under acidic conditions proceeds as follows: In Step 1, the strong acid that is present donates a proton to the α C atom of the C=C group; in Step 2, water (a weak base) removes the proton from the HO group to produce the uncharged keto product.

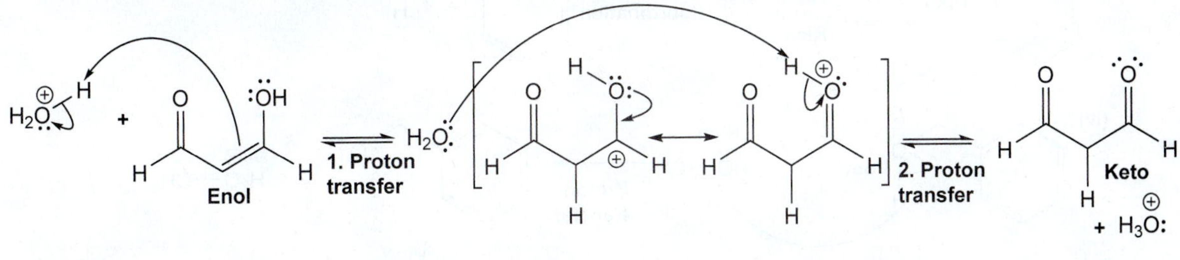

Problem 7.45

Think

Which bonds appear in the imine that are not in the enamine? Which bonds appear in the enamine that are not in the imine? What is the difference in bond energy between the imine and enamine forms? Which form is favored?

Solve

The bonds that are different between the reactants and products are shown on the next page. The enthalpy of the reaction is estimated by taking the sum of the energies of the bonds that are lost from the reactants minus the sum of the energies of the bonds that are gained on the product side.

NH 619 kJ/mol
H 418 kJ/mol
339 kJ/mol
Imine

H 389 kJ/mol
HN 289 kJ/mol
619 kJ/mol
Enamine

So ΔH°_{rxn} is approximately (619 + 339 + 418) − (389 + 289 + 619) = +79 kJ/mol. Thus, the product (the enamine) is significantly less stable than the reactant (the imine).

Integrated Problems

Problem 7.46

Think

Which sites have partial or full negative charges? Which have partial or full positive charges? Which atoms have a lone pair of electrons? Which atoms lack an octet? What type of bond forms or breaks? How many curved arrows are needed to show the elementary step? Name the elementary step.

Solve

The pertinent electron-rich and electron-poor sites are labeled below as "rich" and "poor," respectively. The type of elementary step is identified under each reaction arrow.

(i) Rich, Poor — Proton transfer

(ii) Rich, Poor — Heterolysis — H_2O

(iii) Poor, Rich — $HO-CH_3$ — Coordination

(iv) Poor, Rich — $HO-CH_3$ — Proton transfer — H_2O-CH_3

Problem 7.47

Think

Which sites have partial or full negative charges? Which have partial or full positive charges? Which atoms have a lone pair of electrons? Which atoms lack an octet? What type of bond forms or breaks? How many curved arrows are needed to show the elementary step? Name the elementary step.

Solve

The pertinent electron-rich and electron-poor sites are labeled below as "rich" and "poor," respectively. The type of elementary step is identified under each reaction arrow.

(i) Rich :OH (cyclohexanol) + Poor $H-\overset{\oplus}{O}CH_2CH_3$ (H) → Proton transfer → $:\overset{\oplus}{O}H_2$ (cyclohexyl) + $H-\ddot{O}CH_2CH_3$

(ii) Poor $\oplus$ OH_2 Rich (cyclohexyl) → Heterolysis → cyclohexyl cation ($\oplus$) + $H_2\ddot{O}$

(iii) Poor $\oplus$ (cyclohexyl cation) H + $H-\ddot{O}CH_2CH_3$ Rich → Proton transfer → cyclohexene + $H-\overset{\oplus}{O}CH_2CH_3$ (H)

Problem 7.48

Think

Which sites have partial or full negative charges? Which have partial or full positive charges? Which atoms have a lone pair of electrons? Which atoms lack an octet? What type of bond forms or breaks? How many curved arrows are needed to show the elementary step? Name the elementary step.

Solve

The pertinent electron-rich and electron-poor sites are labeled below as "rich" and "poor," respectively. The type of elementary step is identified under each reaction arrow.

(i) Rich (epoxide O, H_3C, phenyl) + H–Br: (Poor) → Proton transfer → protonated epoxide ($\oplus$) + $:\ddot{Br}:^{\ominus}$

(ii) Poor (protonated epoxide, $\oplus$, H_3C) + Rich $:\ddot{Br}:^{\ominus}$ → S_N2 → $H\ddot{O}$, H_3C, Br (bromohydrin with phenyl)

Problem 7.49

Think

Which sites have partial or full negative charges? Which have partial or full positive charges? Which atoms have a lone pair of electrons? Which atoms lack an octet? What type of bond forms or breaks? How many curved arrows are needed to show the elementary step? Name the elementary step.

Solve

The pertinent electron-rich and electron-poor sites are labeled below as "rich" and "poor," respectively. The type of elementary step is identified under each reaction arrow.

(i) Poor; Rich; S_N2

(ii) Rich; Poor; Proton transfer

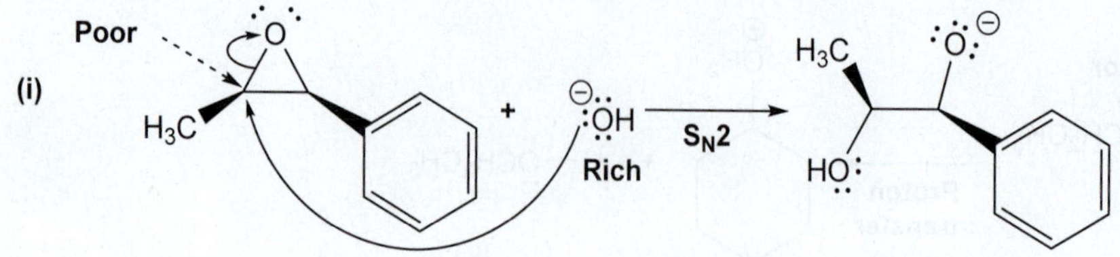

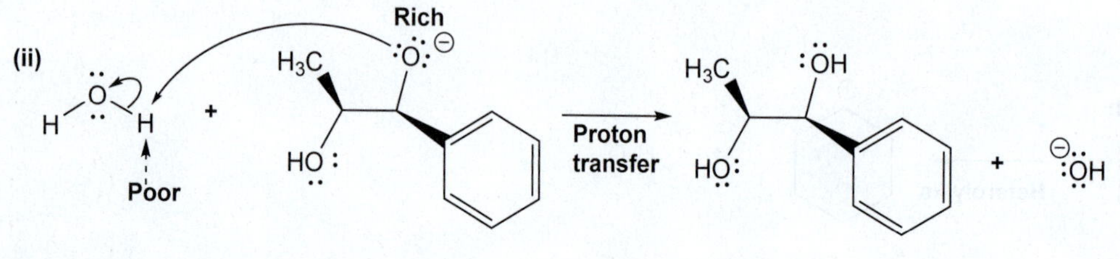

Problem 7.50

Think

Which sites have partial or full negative charges? Which have partial or full positive charges? Which atoms have a lone pair of electrons? Which atoms lack an octet? What type of bond forms or breaks? How many curved arrows are needed to show the elementary step? Name the elementary step.

Solve

The pertinent electron-rich and electron-poor sites are labeled below as "rich" and "poor," respectively. The type of elementary step is identified under each reaction arrow.

(i) Rich; Poor; Proton transfer

(ii) Poor; Rich; S_N2

(iii) Poor; Rich; Proton transfer

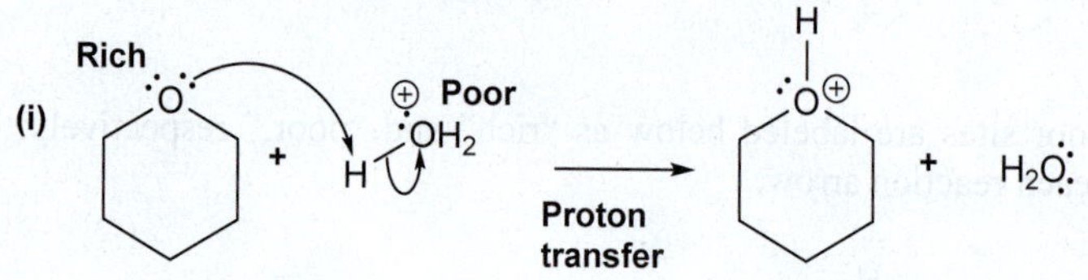

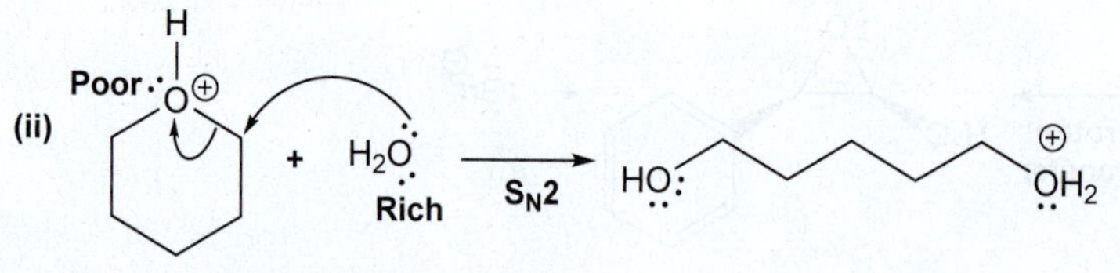

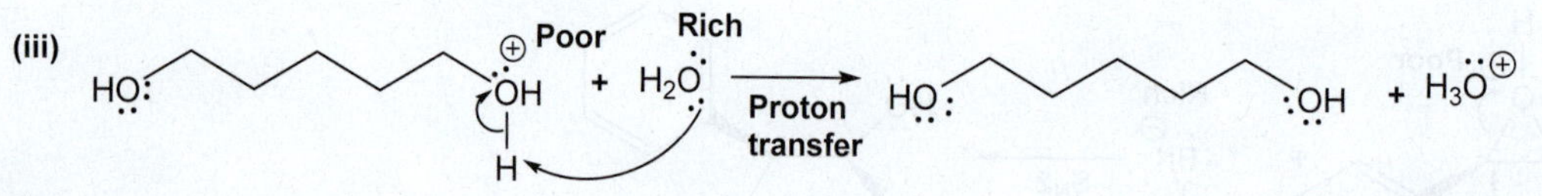

Problem 7.51

Think

Which sites have partial or full negative charges? Which have partial or full positive charges? Which atoms have a lone pair of electrons? What type of bond forms or breaks? How many curved arrows are needed to show the elementary step? Name the elementary step.

Solve

The pertinent electron-rich and electron-poor sites are labeled below as "rich" and "poor," respectively. The type of elementary step is identified under each reaction arrow.

(i) Rich; Poor; S_N2

(ii) Poor; Rich; Nucleophilic addition

(iii) Rich; Poor; Nucleophile elimination

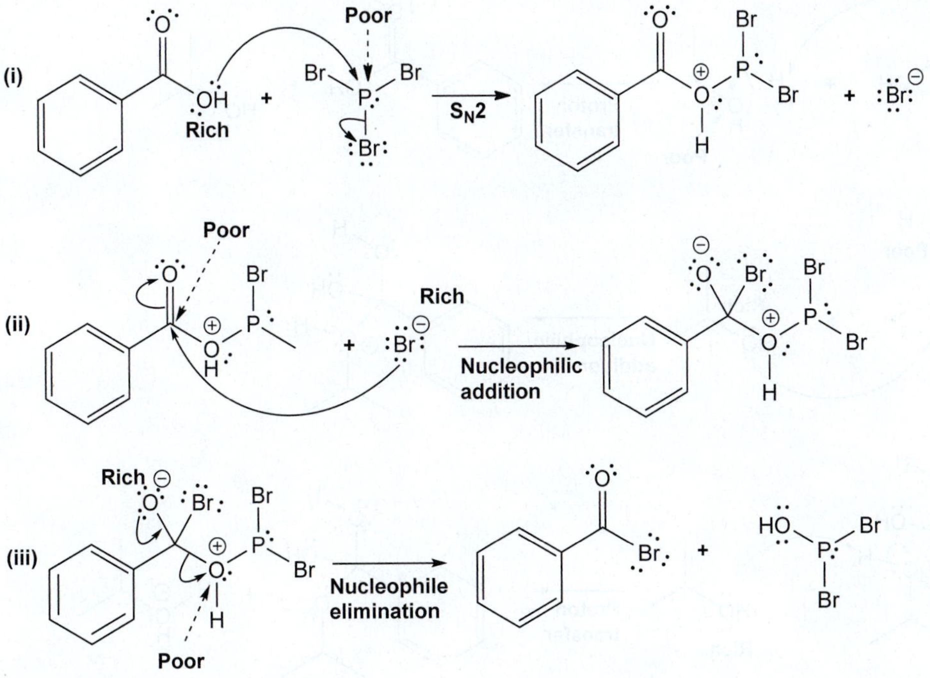

Problem 7.52

Think

Which sites have partial or full negative charges? Which have partial or full positive charges? Which atoms have a lone pair of electrons? Which atoms lack an octet? What type of bond forms or breaks? How many curved arrows are needed to show the elementary step? Name the elementary step.

Solve

The pertinent electron-rich and electron-poor sites are labeled below as "rich" and "poor," respectively. The type of elementary step is identified under each reaction arrow.

(i) $(C_6H_5)_3P$: (Rich) + alkyl bromide (Poor) → S_N2 → $P(C_6H_5)_3^+$ product + Br^-

(ii) Poor; Rich $(CH_3)_3C^-$; Proton transfer → ylide $P(C_6H_5)_3$ + $(CH_3)_3CH$

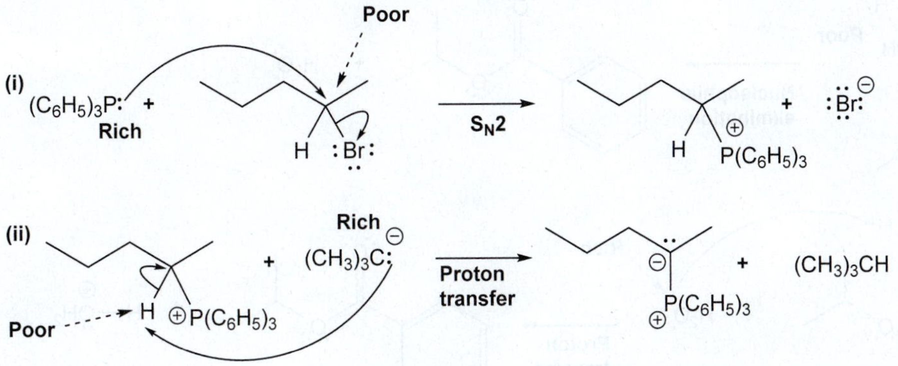

Problem 7.53

Think

Which sites have partial or full negative charges? Which have partial or full positive charges? Which atoms have a lone pair of electrons? Which atoms lack an octet? What type of bond forms or breaks? How many curved arrows are needed to show the elementary step? Name the elementary step.

Solve

The pertinent electron-rich and electron-poor sites are labeled below as "rich" and "poor," respectively. The type of elementary step is identified under each reaction arrow.

(i) Rich, Poor — Proton transfer

(ii) Poor, Rich — Nucleophilic addition

(iii) Poor, Rich — Proton transfer

(iv) Rich, Poor — Proton transfer

(v) Rich, Poor — Nucleophile elimination

(vi) Poor, Rich — Proton transfer

Problem 7.54

Think

Which sites have partial or full negative charges? Which have partial or full positive charges? Which atoms have a lone pair of electrons? Which atoms lack an octet? What type of bond forms or breaks? How many curved arrows are needed to show the elementary step? Name the elementary step.

Solve

The pertinent electron-rich and electron-poor sites are labeled below as "rich" and "poor," respectively. The type of elementary step is identified under each reaction arrow.

(i) Rich, Poor, H_3O^+ → Proton transfer → + H_2O

(ii) Poor, Rich, H_2O → Nucleophilic addition

(iii) Poor, H_2O Rich → Proton transfer → + H_3O^+

(iv) Rich, Poor, H_3O^+ → Proton transfer → + H_2O

(v) Rich, Poor → Nucleophile elimination

(vi) Poor, Rich → Proton transfer

Problem 7.55

Think

Which sites have partial or full negative charges? Which have partial or full positive charges? Which atoms have a lone pair of electrons? Which atoms lack an octet? What type of bond forms or breaks? How many curved arrows are needed to show the elementary step? Name the elementary step.

Solve

The pertinent electron-rich and electron-poor sites are labeled below as "rich" and "poor," respectively. The type of elementary step is identified under each reaction arrow.

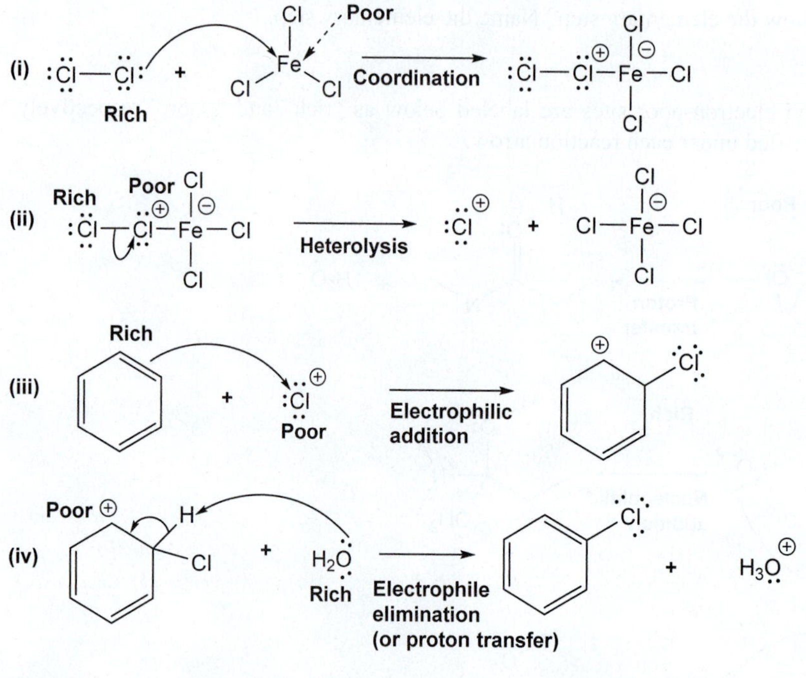

(i) :Cl–Cl: (Rich) + $FeCl_3$ (Poor) —Coordination→ :Cl–Cl⊕–Fe⊖Cl_3

(ii) Rich :Cl–Cl⊕ (Poor)–Fe⊖Cl_3 —Heterolysis→ :Cl⊕ + $Cl-Fe^{\ominus}Cl_3$

(iii) Benzene (Rich) + :Cl⊕ (Poor) —Electrophilic addition→ chlorinated carbocation

(iv) Poor ⊕ carbocation (H, Cl) + H_2O (Rich) —Electrophile elimination (or proton transfer)→ chlorobenzene + H_3O^{+}

Problem 7.56

Think

Which sites are electron rich? Which are electron poor? For the proton transfer reaction, which species is the acid, and which is the base? What is the product of this elementary step? How does the nucleophile Cl^- react with the product of the proton transfer step?

Solve

The curved arrow notation and products for each elementary step are shown below the sequence of two elementary steps: In Step 1, the strong acid HCl donates a proton to the O atom of the ROH, producing a good leaving group; in Step 2, the Cl^- produced from Step 1 acts as a nucleophile to attack the electron-poor C. A new C–Cl bond forms at the same time the C–OH_2^+ bond breaks.

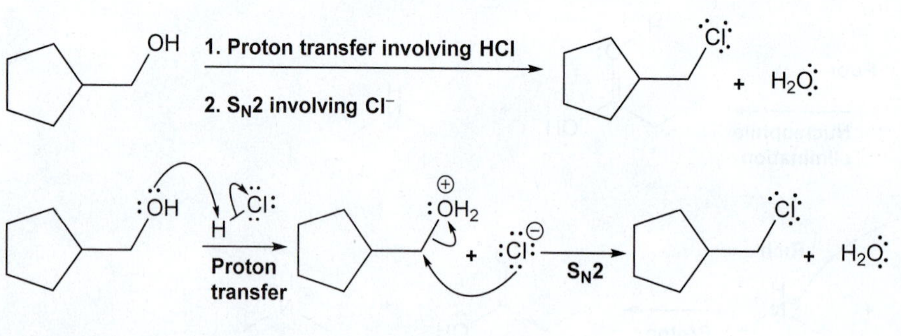

Cyclopentylmethanol (OH) —1. Proton transfer involving HCl; 2. S_N2 involving Cl^-→ (chloromethyl)cyclopentane (Cl) + H_2O

:OH + H–Cl: —Proton transfer→ :OH_2^{+} + :Cl:⊖ —S_N2→ Cl product + H_2O

Problem 7.57

Think

Which sites are electron rich? Which are electron poor? For the electrophilic addition step, where is the new C–H bond formed, and on which carbon is the C^+ formed? In the product of the first step, is there a methyl group on a C atom that is adjacent to the C^+? In the product of Step 2, is there a H on a C atom that is adjacent to the C^+? How many curved arrows should be added for each step? Should those curved arrows flow from electron rich to electron poor, or the opposite?

Solve

The curved arrow notation and products for each elementary step are shown below the sequence of three elementary steps: In Step 1, a single curved arrow is drawn from the nonpolar π bond of the C=C bond to the electron-deficient H^+, forming a new C–H bond, and a C^+ results on the other carbon; in Step 2, a single curved arrow is added to show that the CH_3 on the adjacent carbon shifts to transform the 2° C^+ into a more stable 3° C^+; in Step 3, a single curved arrow is drawn to show the elimination of a H^+ to form a new C=C bond.

1. Electrophilic addition of H^+
2. 1,2-Methyl shift
3. Electrophile elimination of H^+

Electrophilic addition → **1,2-Methyl shift** → **Electrophile elimination**

Problem 7.58

Think

Can you make any simplifying assumptions with the metal-containing species? Which sites are electron rich? Which are electron poor? What is the nucleophile in the nucleophilic addition step? What is the leaving group in the nucleophile elimination step? What is the nucleophile in the second nucleophilic addition step? In the fourth step, can you identify an appropriate acid and base? How many curved arrows should be added for each step? Should those curved arrows flow from electron rich to electron poor, or the opposite?

Solve

The curved arrow notation and products for each elementary step are shown below the sequence of four elementary steps: In Step 1, two curved arrows are added to show the electron-rich $H_3C:^-$ (simplified from CH_3MgBr) attacking the electron-poor C of the polar C=O bond, forming a new C–CH_3 bond, and to show the breaking of the π bond to form an O^-; in Step 2, two curved arrows are drawn to show the electrons on the O^- re-forming the C=O bond and elimination of CH_3O^-; Step 3 is another nucleophilic addition step similar to that in Step 1; in Step 4, two curved arrows are drawn, one from the electron-rich O^- of the base and one to show the breaking of the O–H bond of the acid.

1. Nucleophilic addition involving CH_3MgBr
2. Nucleophile elimination
3. Nucleophilic addition involving CH_3MgBr
4. Proton transfer involving H_3O^+

Nucleophilic addition → **Nucleophile elimination** (+ $:\ddot{O}CH_3^-$) → **Nucleophilic addition** → **Proton transfer**

Problem 7.59

Think

Which sites are electron rich? Which are electron poor? What is the base that will receive the proton in the first proton transfer step? What is the leaving group in the heterolysis step? What is the atom lacking an octet in the coordination step? In the final proton transfer step, can you identify the acid and base? How many curved arrows should be added for each step? Should those curved arrows flow from electron rich to electron poor, or the opposite?

Solve

The curved arrow notation and products for each elementary step are shown below the sequence of four elementary steps: In Step 1, two curved arrows are drawn to show the oxygen of the ether being protonated by H_3O^+, producing a good leaving group; in Step 2, a single curved arrow is drawn to show the C–O bond of the protonated ether being broken, in which the alcohol is the leaving group and a C^+ results; in Step 3, a single curved arrow is added to show water forming a new C–O bond to the C^+; in Step 4, two curved arrows are drawn to show the final proton transfer step, which is necessary to produce an uncharged product.

1. Proton transfer involving H_3O^+
2. Heterolysis
3. Coordination involving H_2O
4. Proton transfer involving H_2O

OH + HO

$\oplus OH_2$ H Proton transfer → Heterolysis → $\oplus$ H_2O: Coordination + HO

$\oplus$ H H_2O: Proton transfer → OH

Problem 7.60

Think

Can you make any simplifying assumptions with the metal-containing species? Which sites are electron rich? Which are electron poor? What are the leaving group and base in the first E2 reaction? What new bond forms? What are the leaving group and base in the second E2 reaction? In each proton transfer step, can you identify the appropriate acid and base? How many curved arrows should be added for each step? Should those curved arrows flow from electron rich to electron poor, or the opposite?

Solve

On the next page, the curved arrow notation and products for each elementary step are shown below the sequence of four elementary steps: In Step 1, three curved arrows are drawn to show the electron-rich $H_2N:^-$ removing a proton from the carbon adjacent to the carbon with the Cl leaving group, forming a new C=C bond; in Step 2, three curved arrows are drawn again to show the electron-rich $H_2N:^-$ removing a proton from the carbon adjacent to the carbon with the Cl leaving group, this time producing a new C≡C bond; in Step 3, two curved arrows are drawn to show the acidic H of the C≡CH bond being deprotonated by the strong base $H_2N:^-$ to yield $C{\equiv}C^-$; in Step 4, two curved arrows are drawn to show the protonation of the $C{\equiv}C^-$ bond by the acid H_3O^+ to yield an uncharged alkyne product.

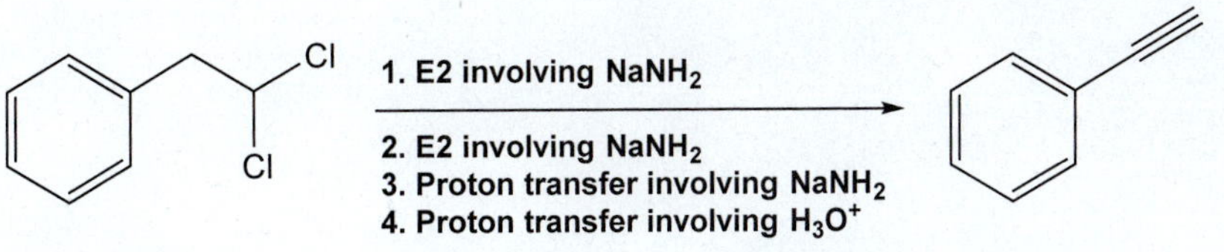

Cl

Cl

1. E2 involving $NaNH_2$

2. E2 involving $NaNH_2$
3. Proton transfer involving $NaNH_2$
4. Proton transfer involving H_3O^+

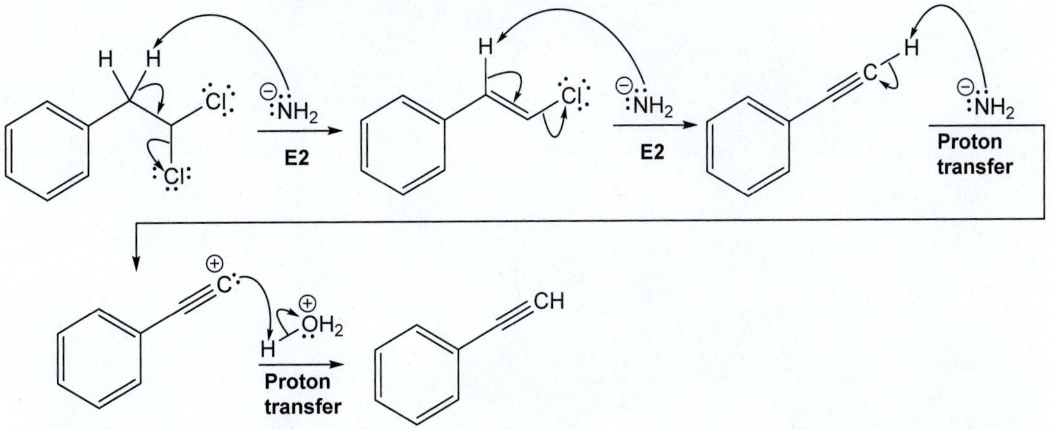

H H

:Cl:

$:\ddot{N}H_2^-$

E2

:Cl:

H

:Cl:

$:\ddot{N}H_2^-$

E2

C

H

$:\ddot{N}H_2^-$

Proton transfer

C^+

$\ddot{O}H_2^+$

H

Proton transfer

CH

INTERCHAPTER D | Molecular Orbital Theory, Hyperconjugation, and Chemical Reactions

Your Turn Exercises

Your Turn D.1

Think

Refer to Figure D-1b as a reference. What is the filled σ bond orbital of $(CH_3)_3C$ that can interact with the empty *p* AO of CH_2^+?

Solve

In $(CH_3)_3CCH_2^+$, the empty *p* AO overlaps with the σ orbital of the adjacent C–CH_3 bond, so the two orbitals can interact.

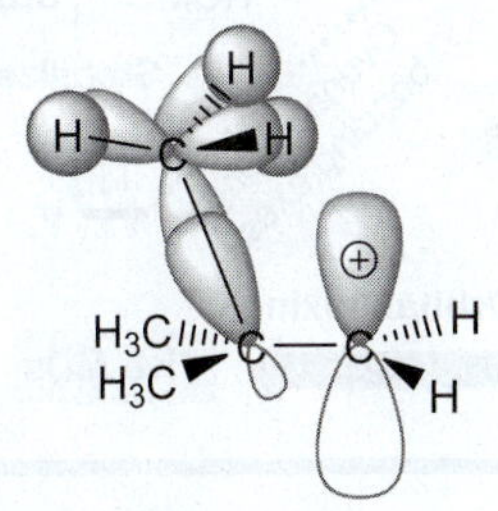

Your Turn D.2

Think

What is hyperconjugation? Refer to Figure D-2b and D-2c as a reference. What is the filled σ bonding orbital of $(CH_3)_3C$ that can interact with the empty π* MO of the double bond?

Solve

(a) In $H_2C{=}CH{-}C(CH_3)_3$, the empty π* MO interacts with the σ bonding orbital of the C–CH_3 bond in $C(CH_3)_3$.

(b) The orbital interaction in $H_2C{=}CH{-}C(CH_3)_3$, called hyperconjugation, stabilizes the electrons from the σ MO.

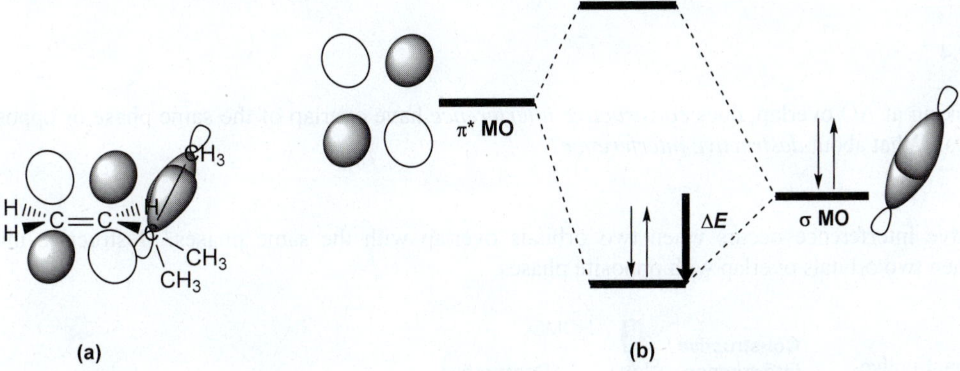

Your Turn D.3

Think

What is the hybridization of the H, O, and Cl atoms, and what orbitals are contributed from each atom? What type of overlap occurs between H and O? Between H and Cl? What type of bonding correlates to this type of overlap? How many total orbitals should be produced from the atomic orbital (AO) mixing?

Solve

H is not hybridized and contributes a 1*s* orbital, whereas O and Cl are both sp^3 hybridized. End-on overlap occurs in both H–O⁻ and H–Cl and, therefore, results in a sigma (σ) bond. When the two AOs overlap, two molecular orbitals (MOs) are produced: a σ bonding orbital and a σ* antibonding (σ*) orbital. The three sets of lone pairs on Cl and O are in sp^3 hybrid, nonbonding orbitals. The highest occupied MO (HOMO) from HO^- is, therefore, a nonbonding orbital, and the lowest unoccupied MO (LUMO) from HCl is a σ* MO, in agreement with Figure D-5.

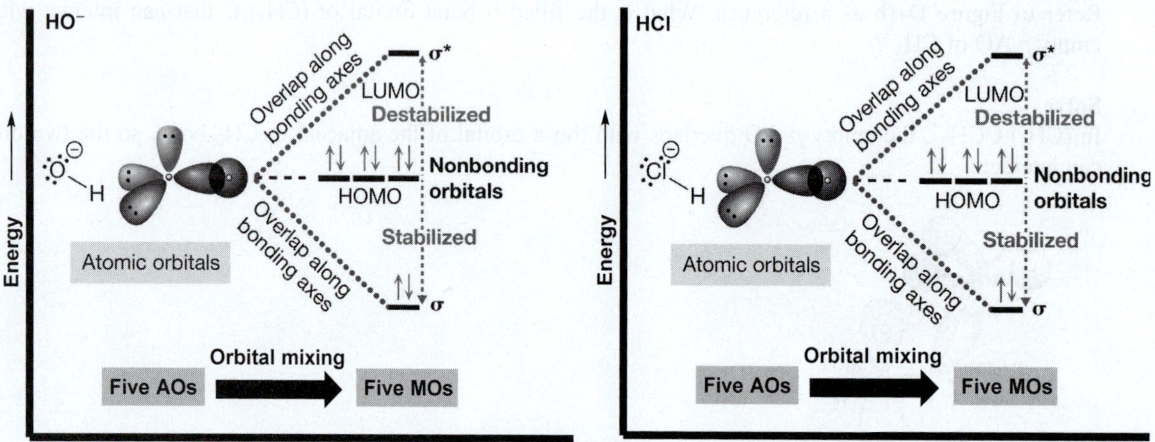

The nonbonding orbital of HO^- comes from a hybrid AO. The LUMO of HCl is the σ* MO, the result of destructive interference, leaving a node in the internuclear region.

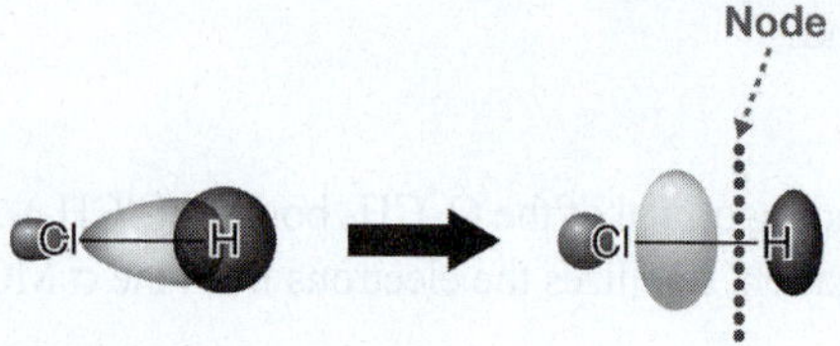

Your Turn D.4

Think

When looking at AO overlap, does *constructive interference* have overlap of the same phase or opposite phase of the AOs? What about *destructive interference*?

Solve

Constructive interference occurs when two orbitals overlap with the same phases. Destructive interference occurs when two orbitals overlap with opposite phases.

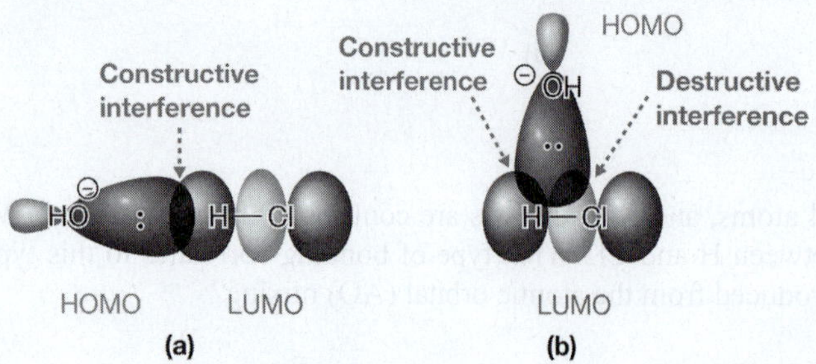

Your Turn D.5

Think

What is the hybridization of the H, C, and Cl atoms, and what do the orbitals look like for each atom? What type of overlap occurs between C and Cl? What type of bonding correlates to this type of overlap? How many total orbitals should be produced from the AO mixing?

Solve

C and Cl are both sp^3 hybridized, and each H atom contributes a 1*s* orbital. The 11 AOs produce 11 new MOs. End-on overlap occurs in the C–Cl and C–H bonding, and each overlap produces two MOs, σ and σ*, which account for eight MOs. The remaining three orbitals are from the noninteracting sp^3 AOs from Cl, producing nonbonding orbitals. The LUMO is, therefore, a σ* MO, in agreement with Figure D-6.

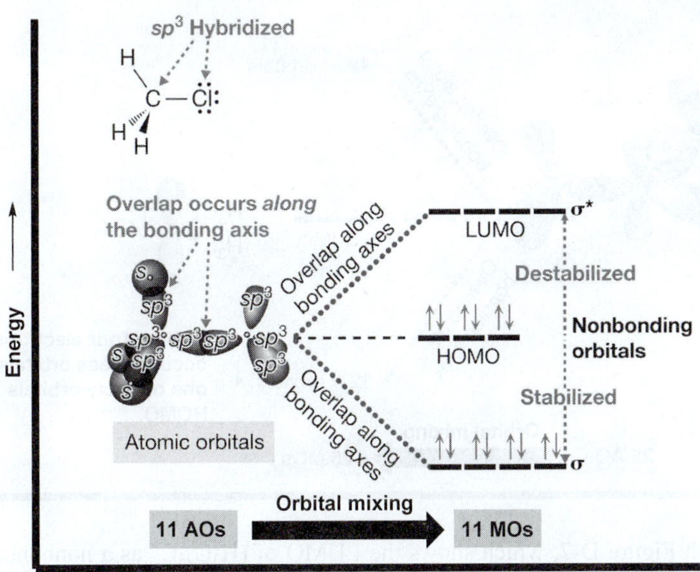

The C–Cl σ* can be thought of as resulting from the destructive interference between two sp^3 AOs, producing a node in the internuclear region:

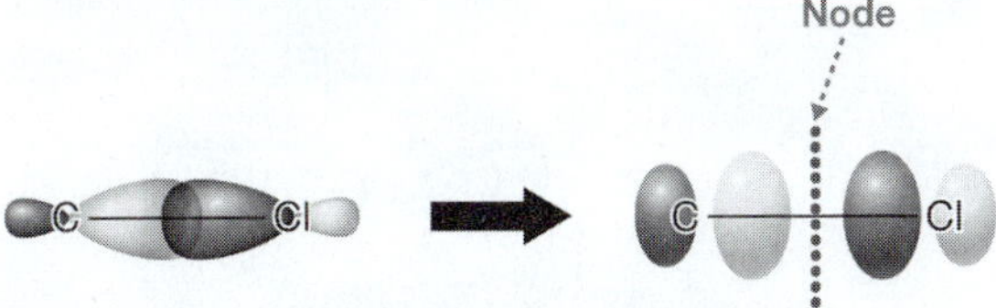

Your Turn D.6

Think

When looking at AO overlap, does *constructive interference* have overlap of the same phase or opposite phase of the AOs? What about *destructive interference*?

Solve

Constructive interference occurs when two orbitals overlap with the same phases. Destructive interference occurs when two orbitals overlap with opposite phases.

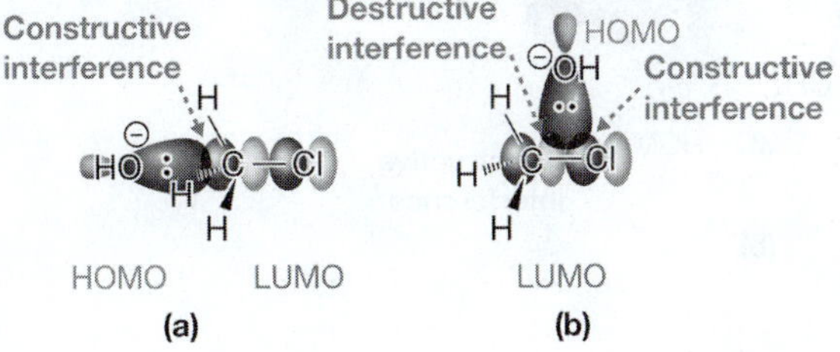

Your Turn D.7

Think

What is the hybridization of each of the C atoms? What do the orbitals look like for each atom? What type of overlap occurs between the C–C and C–H bonds? What type of bonding correlates to this type of overlap? How many total orbitals should be produced from the AO mixing? Are there any unhybridized orbitals on the C atom?

Solve

The central C is sp^2 hybridized and has an unhybridized pure p orbital. The CH_3 carbon atoms are all sp^3 hybridized, and each H atom contributes a 1s orbital. End-on overlap occurs in the C–H and C–C bonding and, therefore, results in 12 σ bonds. When the 24 AOs overlap, 24 MOs are produced: 12 σ and 12 σ*.

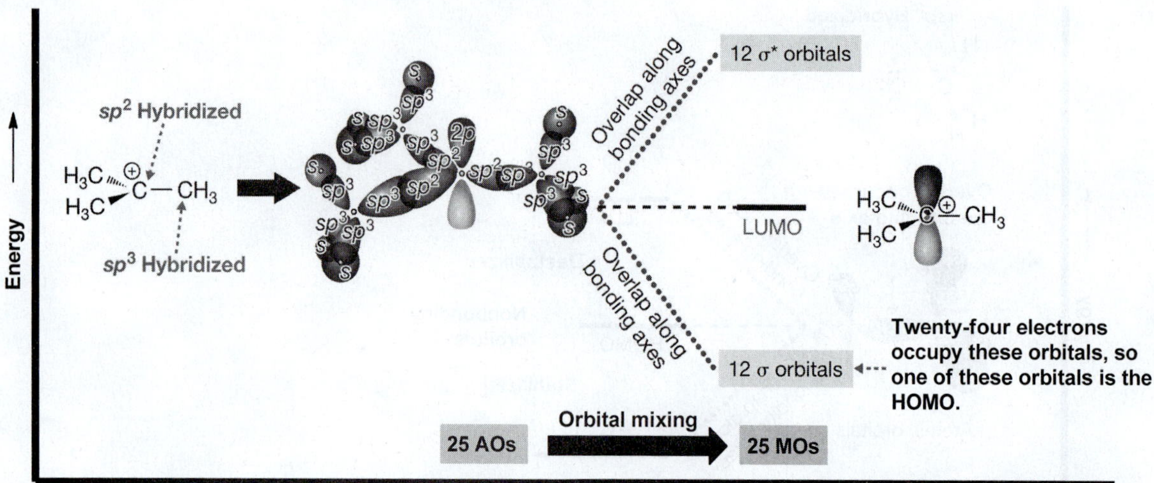

Yes, these figures are in agreement with Figure D-7, which shows the LUMO of $(H_3C)_3C^+$ as a nonbonding p orbital, as shown below.

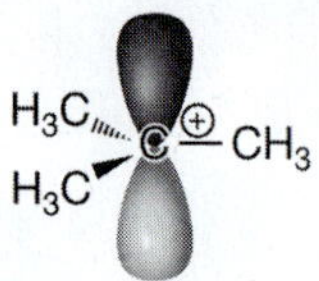

Your Turn D.8

Think

When looking at AO overlap, does *constructive interference* have overlap of the same phase or opposite phase of the AOs? What about *destructive interference*?

Solve

Constructive interference occurs when two orbitals overlap with the same phases. Destructive interference occurs when two orbitals overlap with opposite phases. Figure D-7a and D-7b shows constructive interference.

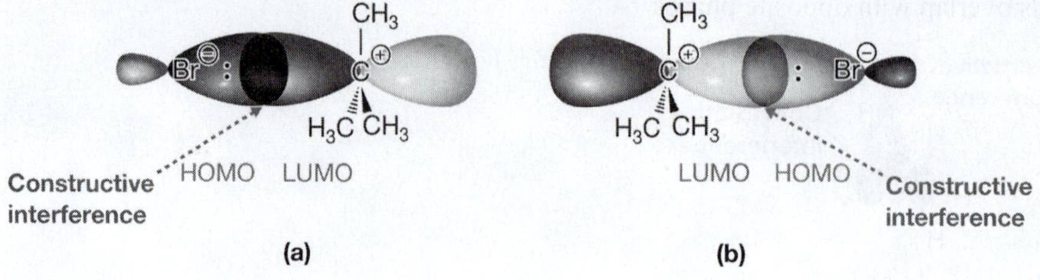

Your Turn D.9

Think

What is the hybridization of each of the H atoms, the C atoms, the Cl atom, and the O atom? What do the contributed orbitals look like for each atom? What type of overlap occurs between the H–C, C–CH_3, C–Cl, and C–O bonds? What type of bonding correlates to this type of overlap? How many total orbitals should be produced from the AO mixing? Are there any unhybridized orbitals on the C and O atoms?

Solve

The central C and O atoms are both sp^2 hybridized, and each has an unhybridized pure p orbital. The C atom of CH_3 and the Cl atom are both sp^3 hybridized, and each H atom contributes a 1s orbital. End-on overlap occurs in the C–H and C–C, C–Cl, and C–O bonding and, therefore, results in six σ bonds. Side-by-side overlap occurs between the two pure p orbitals of C and O. This results in a π bond. When the 14 AOs overlap, 14 MOs are produced: six σ, six σ*, one π bonding, and one π* antibonding.

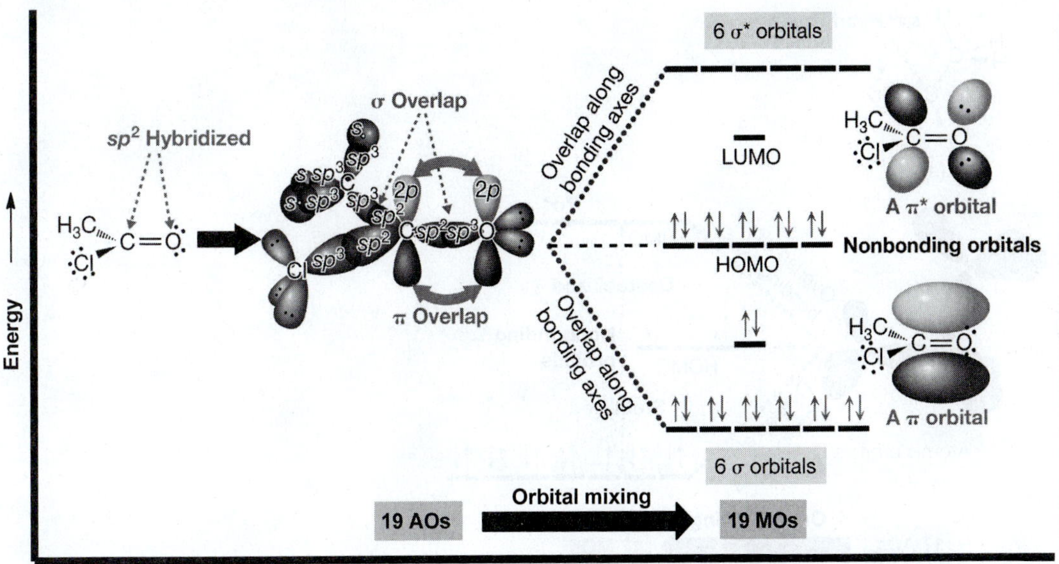

The LUMO is the π* MO, in agreement with the LUMO shown in Figure D-8a:

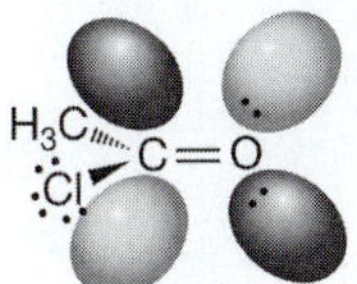

Your Turn D.10

Think

When looking at AO overlap, does *constructive interference* have overlap of the same phase or opposite phase of the AOs? What about *destructive interference*?

Solve

Constructive interference occurs when two orbitals overlap with the same phases. Destructive interference occurs when two orbitals overlap with opposite phases. Constructive interference predominantly takes place in Figure D-8a and D-8b.

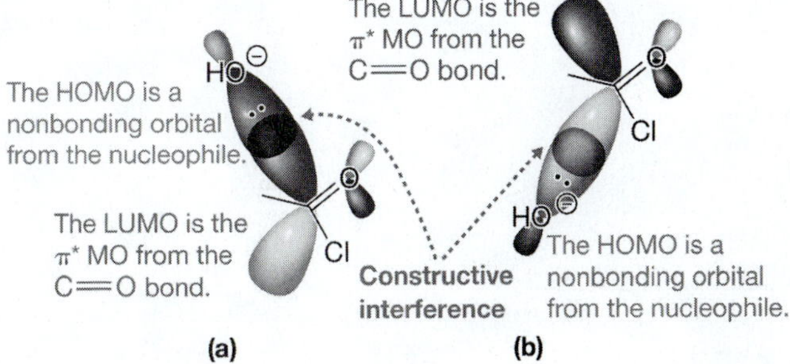

Your Turn D.11

Think

What is the hybridization of each of the C atoms, the H atoms, and the Br atom? What do the contributed orbitals look like for each atom? What type of overlap occurs between the C–CH_3, H–C, and C–Br bonds? What type of bonding correlates to this type of overlap? How many total orbitals should be produced from the AO mixing?

Solve

The two C atoms and the Br atom are all sp^3 hybridized, and each H atom contributes a 1s orbital. End-on overlap occurs in the C–H, C–C, and C–Br bonding and, therefore, results in seven σ bonds. When the 14 AOs overlap, 14 MOs are produced: seven σ bonding orbitals and seven σ* antibonding orbitals. This leaves three nonbonding orbitals. See the energy diagram below.

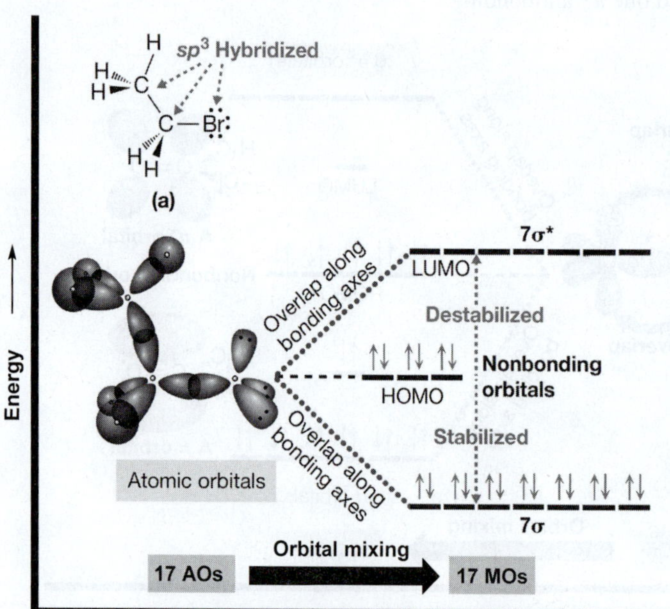

The LUMO of the molecule is a σ* MO, in agreement with Figure D-10a:

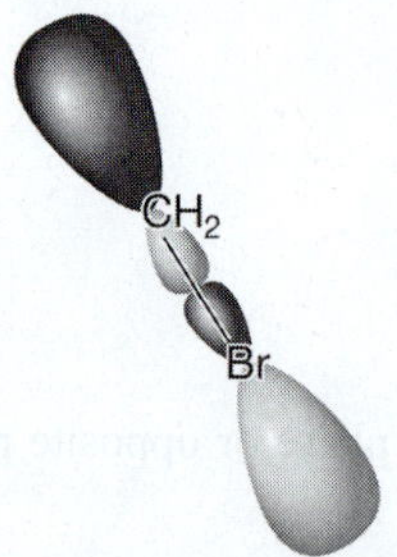

Although the HOMO of the entire molecule is a nonbonding orbital, the HOMO involving the CH_3 group is a σ MO, in agreement with Figure D-10a:

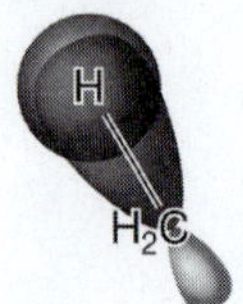

Your Turn D.12

Think

When looking at AO overlap, does *constructive interference* have overlap of the same phase or opposite phase of the AOs? What about *destructive interference*?

Solve

Constructive interference occurs when two orbitals overlap with the same phases. Destructive interference occurs when two orbitals overlap with opposite phases. Predominantly constructive interference takes place in Figure D-10a and D-10b.

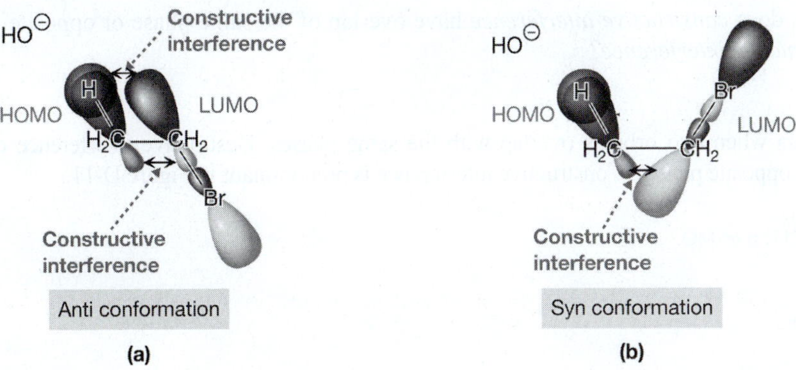

Your Turn D.13

Think

What is the hybridization of each of the H atoms and the C atoms? What do the contributed orbitals look like for each atom? What type of overlap occurs between the H–C and C–C bonds? What type of bonding correlates to this type of overlap? How many total orbitals should be produced from the AO mixing? Are there any unhybridized orbitals on the C atoms?

Solve

The two central C atoms are both sp^2 hybridized, and each has an unhybridized pure p orbital. The CH_3 carbon atoms are sp^3 hybridized, and each H atom contributes a $1s$ orbital. End-on overlap occurs in the C–H and C–C bonding and, therefore, results in 11 σ bonds. Side-on overlap occurs between the two pure p orbitals of the two sp^2-hybridized C atoms. This results in a π bond. When the 24 AOs overlap, 24 MOs are produced: 11 σ bonding orbitals, 11 σ* antibonding orbitals, one π bonding orbital, and one π* antibonding orbital. When the 24 total valence electrons fill the MOs, the π MO is the HOMO and the π* MO is the LUMO.

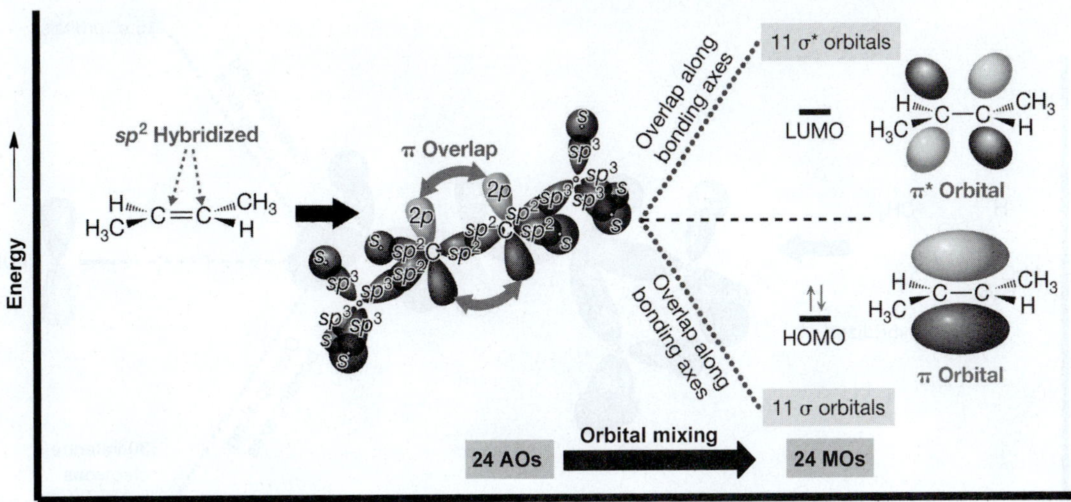

Yes, these figures are in agreement with Figure D-11, which shows the HOMO of $(CH_3)_2C{=}C(CH_3)_2$.

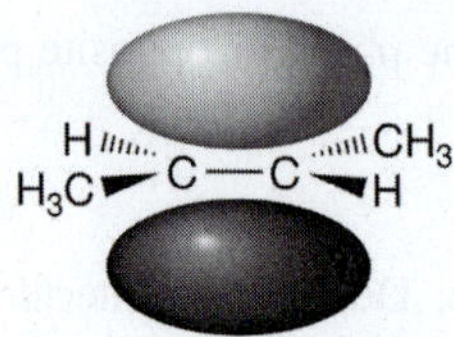

Your Turn D.14

Think

When looking at AO overlap, does *constructive interference* have overlap of the same phase or opposite phase of the AOs? What about *destructive interference*?

Solve

Constructive interference occurs when two orbitals overlap with the same phases. Destructive interference occurs when two orbitals overlap with opposite phases. Constructive interference is predominant in Figure D-11.

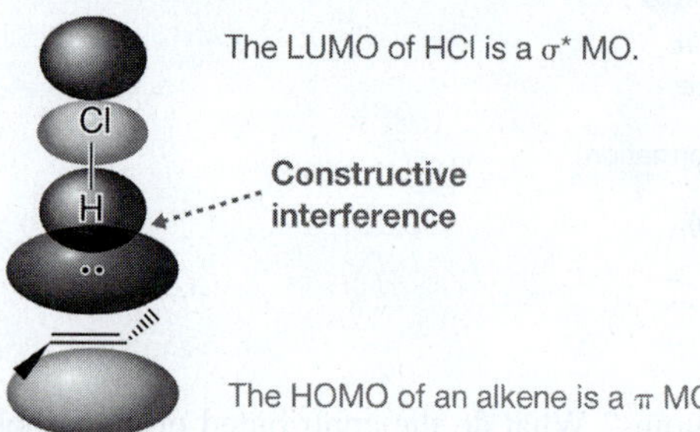

Your Turn D.15

Think

What is the hybridization of each of the C atoms? What do the contributed orbitals look like for each atom? What type of overlap occurs between the C–C and C–H bonds? What type of bonding correlates to this type of overlap? How many total orbitals should be produced from the AO mixing? Are there any unhybridized orbitals on the C atom?

Solve

The carbocation C atom is sp^2 hybridized and has an unhybridized pure p orbital. All the other C atoms are sp^3 hybridized, and each H atom contributes a 1s orbital. End-on overlap occurs in the C–H and C–C bonding and, therefore, results in 15 σ bonds. The unhybridized p AO remains as a nonbonding orbital. The 30 valence electrons fill the 15 σ MOs, so the HOMO is a σ MO and the p AO is the LUMO.

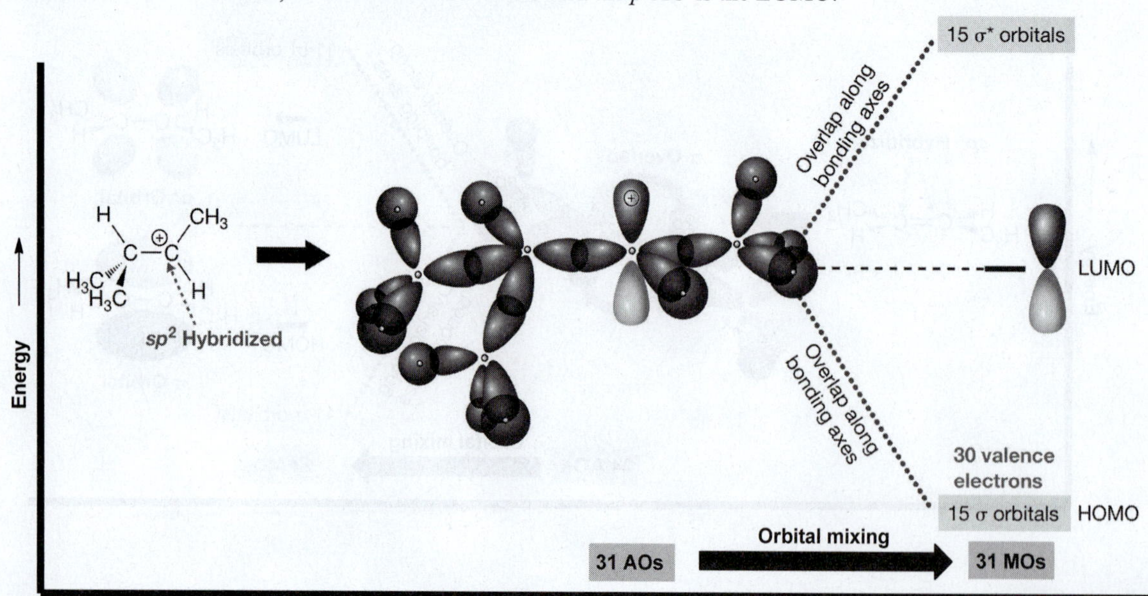

Yes, these figures are in agreement with Figure D-12, which shows the HOMO and LUMO of $(H_3C)_2CHC^+$ as follows:

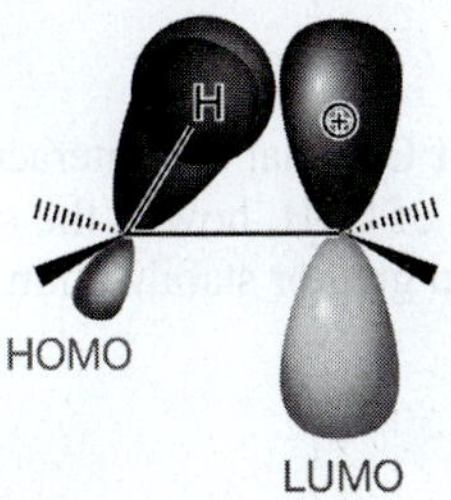
H
⊕
HOMO
LUMO

Your Turn D.16

Think

When looking at AO overlap, does *constructive interference* have overlap of the same phase or opposite phase of the AOs? What about *destructive interference*?

Solve

Constructive interference occurs when two orbitals overlap with the same phases. Destructive interference occurs when two orbitals overlap with opposite phases. Constructive interference is predominant in Figure D-12.

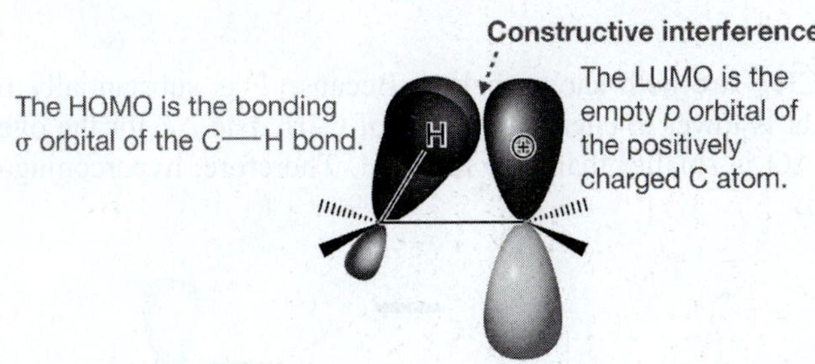
Constructive interference
The HOMO is the bonding σ orbital of the C—H bond.
H
⊕
The LUMO is the empty p orbital of the positively charged C atom.

End of Chapter Problems
Section D.1 Relative Stabilities of Carbocations and Alkenes: Hyperconjugation and Negative Hyperconjugation
Problem D.1

Think

Refer to Figure D-1b and D-1c as a reference. What is the filled σ bonding orbital of CF_3 that can interact with the empty *p* AO of CH_2^+? If the C–F σ bonding orbital is lower in energy than that of C–H, how is the orbital interaction different for $CF_3CH_2^+$ compared to $CH_3CH_2^+$? Does the interaction lead to greater stabilization if the interacting orbitals are closer or further apart in energy?

Solve

In $CF_3CH_2^+$, the empty *p* AO overlaps with the σ bonding orbital of the adjacent C–F bond, so the two orbitals can interact.

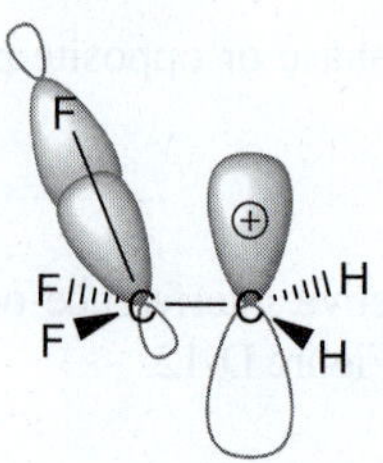

The orbital interactions for $CF_3CH_2^+$ and $CH_3CH_2^+$ are both shown below. Because F is substantially more electronegative than H, the C–F σ bonding orbital is lower in energy than that of C–H. The ΔE for the overlap of the C–F σ bonding orbital and the empty *p* AO is smaller than that for C–H. Therefore, hyperconjugation leads to greater stabilization for the $CH_3CH_2^+$ ion.

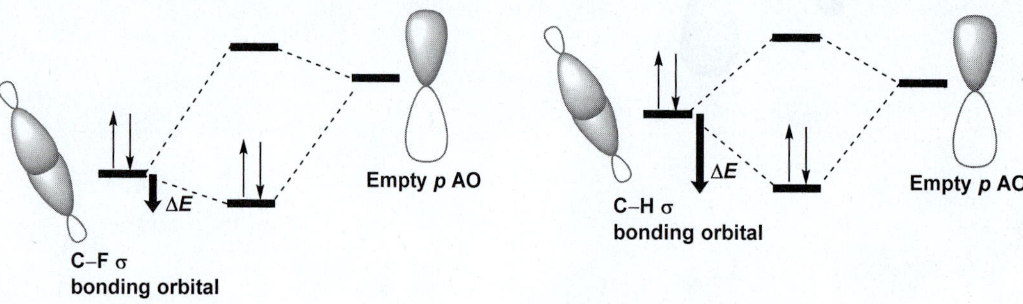

Problem D.2

Think

What is the HOMO in this species? Which adjacent σ* antibonding orbital is the lowest in energy?

Solve

The HOMO is a nonbonding orbital for the lone pair on N, and the adjacent σ* MO that is lowest in energy is the one for C–F. This orbital interaction stabilizes the electrons from the nonbonding MO.

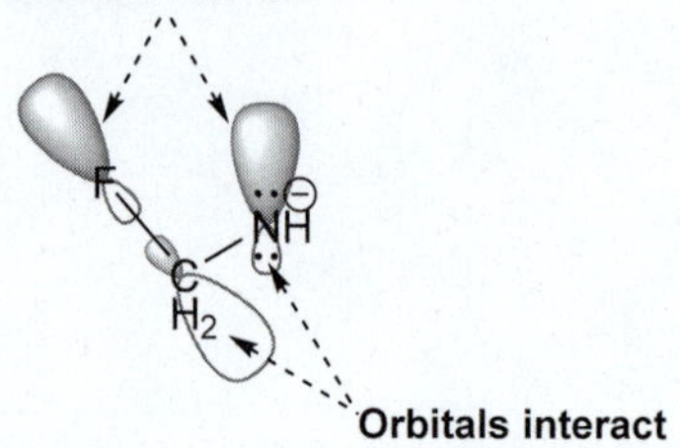

Problem D.3

Think

What is hyperconjugation? What does the empty π* MO in propyne look like? How does this filled σ bond interact with the π* antibonding orbital?

Solve

Hyperconjugation in propyne results from the interaction of an empty π* MO and the adjacent C–H σ bonding orbital (shown below). Note that there is another filled π* MO perpendicular to the one shown below and other C–H σ bonding orbitals. For simplicity, these orbitals are omitted from the drawing.

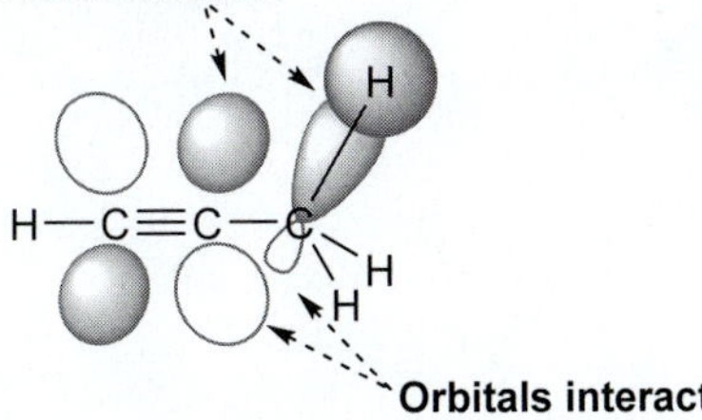

Section D.2 MO Theory and Chemical Reactions

Problem D.4

Think

If the HO⁻ group were to approach the H–Cl bond from the same end as the Cl atom, would there be regions of constructive interference between the HOMO and the LUMO? Would there be regions of destructive interference? Which sites are electron rich, and which are electron poor? If two electron-rich sites approach each other, is this a favored interaction?

Solve

If the O atom were to approach the Cl atom opposite the H atom, the HOMO and the LUMO would have a net interaction. This is shown below, where there is just constructive interference. Therefore, the reaction would be allowed. However, it will not occur readily because of electrostatic repulsion—it would require two electron-rich sites to form a bond.

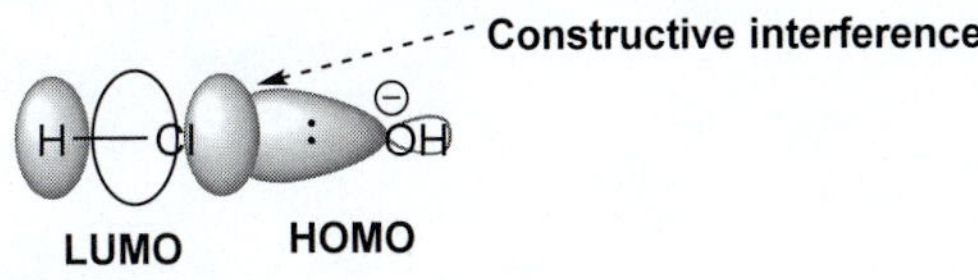

Problem D.5

Think

If the HO⁻ group were to approach the CH_3–Cl bond from the same end as the Cl atom, would there be regions of constructive interference between the HOMO and the LUMO? Would there be regions of destructive interference? Which sites are electron rich, and which are electron poor? If two electron-rich sites approach each other, is this a favored interaction?

Solve

If the O atom were to approach the Cl atom opposite the CH_3 group, the HOMO and the LUMO would have a net interaction. This is shown below, where there is just constructive interference. Therefore, the reaction would be allowed. However, it will not occur readily because of electrostatic repulsion—it would require two electron-rich sites to form a bond.

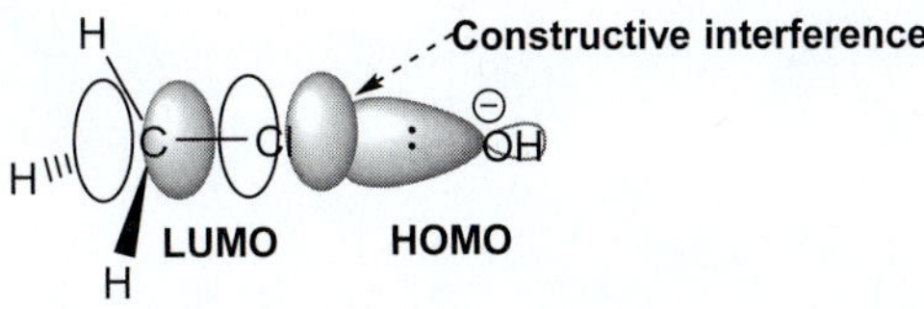

Problem D.6

Think

If the Br^- atom were to approach the $(CH_3)_3C^+$ group from within the plane of the central C atom, would there be regions of constructive interference between the HOMO and the LUMO? Would there be regions of destructive interference? Would there be a net interaction?

Solve

Constructive interference occurs when two orbitals overlap with the same phases. Destructive interference occurs when two orbitals overlap with opposite phases. There are regions of destructive and constructive interference, which cancel each other out. Therefore, there is no net interaction, and the coordination step from this type of approach would not be allowed.

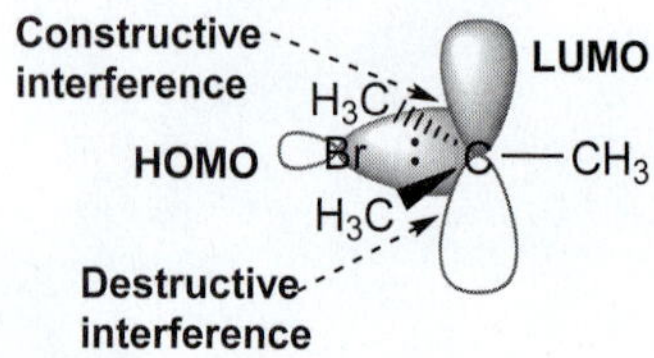

Problem D.7

Think

If the HO^- group were to approach the carbonyl C atom from directly along the C=O bond, would there be regions of constructive interference between the HOMO and the LUMO? Would there be regions of destructive interference? Would there be a net interaction?

Solve

Constructive interference occurs when two orbitals overlap with the same phases. Destructive interference occurs when two orbitals overlap with opposite phases. There are regions of destructive and constructive interference, as shown below, which cancel each other out. Therefore, there is no net interaction, and the nucleophilic addition step from this type of approach would not be allowed.

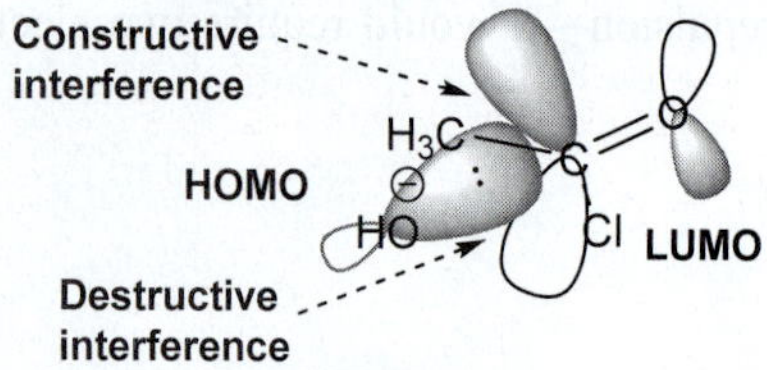

Problem D.8

Think

If the $(CH_3)_3C^+$ group were to approach the $CH_3CH{=}CHCH_3$ bond from above the C=C bond, would there be regions of constructive interference between the HOMO and the LUMO? Would there be regions of destructive interference? Would there be a net interaction?

Solve

Constructive interference occurs when two orbitals overlap with the same phases. Destructive interference occurs when two orbitals overlap with opposite phases. When $(CH_3)_3C^+$ approaches the C=C bond from above, there is a region of constructive interference and essentially no destructive interference. Therefore, there is a net interaction, and the reaction is allowed.

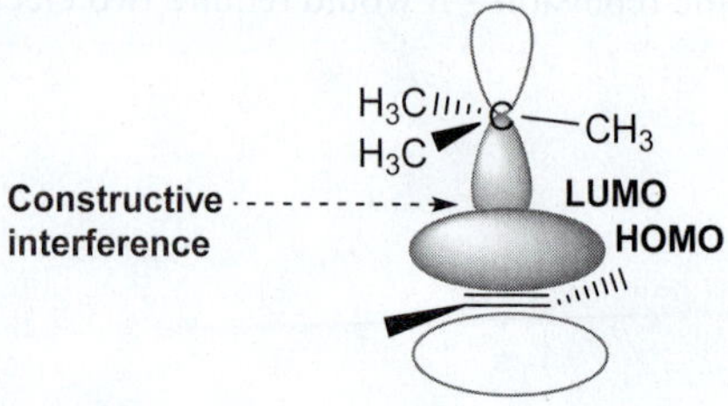

Problem D.9

Think

If the CH_3 group were to shift to the C^+ atom, would there be regions of constructive interference between the HOMO and the LUMO in the transition state? Would there be regions of destructive interference? Would there be a net interaction?

Solve

The HOMO is the σ MO of the $H_3C–C$ bond, and the LUMO is the empty *p* orbital of the carbocation. The HOMO and the LUMO are in phase; they will have a net interaction, and the shift is allowed. Moreover, the shift will occur readily because a more stable tertiary carbocation is formed.

1,2-Methyl shift

HOMO

LUMO

The HOMO and LUMO have the appropriate symmetries to interact, so the reaction is allowed.

Problem D.10

Think

If the nucleophile were to approach the C=O bond from the same side as the O atom, would there be regions of constructive interference between the HOMO and the LUMO? Would there be regions of destructive interference? Would there be a net interaction? Which sites are electron rich, and which are electron poor? If two electron-rich sites approach each other, is this a favored interaction?

Solve

If the nucleophile were to approach the C=O bond from the side of the O atom, the HOMO and the LUMO would have a net interaction. This is shown below, where there is just constructive interference. Therefore, the reaction would be allowed. However, it will not occur because of electrostatic repulsion—it would require two electron-rich sites to form a bond.

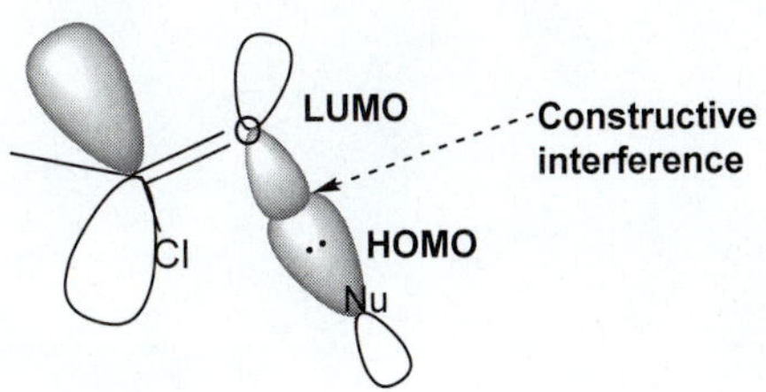

Problem D.11

Think

What is the HOMO and the LUMO for this type of interaction? Are there regions of constructive interference? Destructive interference? Will there be a net interaction?

Solve

For a carbanion, the HOMO is a nonbonding MO containing the lone pair of electrons on C⁻. The LUMO involving an adjacent C–H bond is a σ* MO. Constructive interference occurs when two orbitals overlap with the same phases. Destructive interference occurs when two orbitals overlap with opposite phases. There are regions of destructive and constructive interference, which cancel each other out. Therefore, there is no significant net interaction, and this type of rearrangement is not allowed.

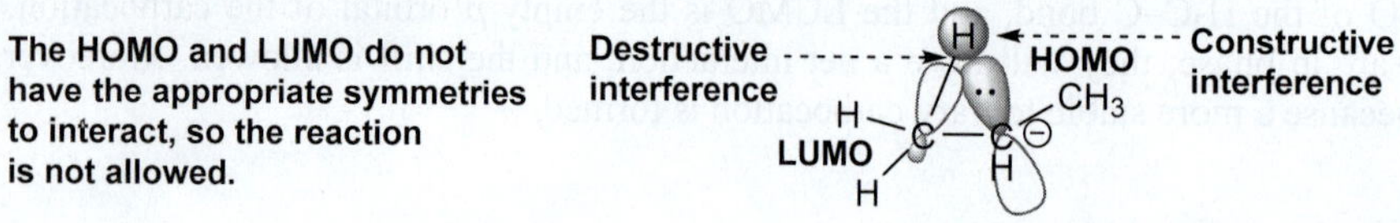

The HOMO and LUMO do not have the appropriate symmetries to interact, so the reaction is not allowed.

Destructive interference **HOMO** **Constructive interference** **LUMO**

INTERCHAPTER E | Naming Compounds with a Functional Group That Calls for a Suffix 1: Alcohols, Amines, Ketones, and Aldehydes

In Chapter Problems

Problem E.1

Think

Of what compound class is the OH group characteristic? How many carbon atoms are in the main chain or ring that contains the OH group? If the OH group were absent, what would be the name of the molecule? What is the suffix for the OH group? Is the locator number necessary?

Solve

The OH group is characteristic of alcohols. If the OH group were absent, the molecule names would be butane for **(a)** and **(b)** and cyclopropane for **(c)**. Drop the final *e* and add the *ol* suffix, so **(a)** and **(b)** become *butanol* and **(c)** becomes *cyclopropanol.* Molecules **(a)** and **(b)** require a locator to establish the location of the OH group, whereas **(c)** does not, because the three carbon atoms are equivalent. The IUPAC names for molecules **(a)–(c)** are given below. The main chain or ring is numbered.

(a) Butan-2-ol **(b) Butan-1-ol** **(c) Cyclopropanol**

Problem E.2

Think

How many carbon atoms are in the main chain or ring? To what functional group does the suffix *ol* correlate? On which C atom is that functional group located?

Solve

The *ol* suffix correlates to the OH functional group. The structures are drawn below. The main chain or ring is numbered. Molecule **(b)** does not require a locator because all four of the carbon atoms are equivalent. Molecules **(a)** and **(c)** both require locators to specify the location of the OH group.

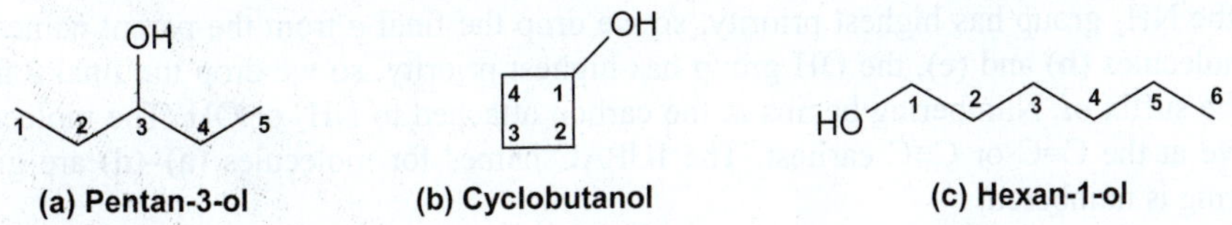

(a) Pentan-3-ol **(b) Cyclobutanol** **(c) Hexan-1-ol**

Problem E.3

Think

Which is a higher-priority functional group, an NH_2 or an R–Cl or R–O–R? How many carbon atoms are in the main chain or ring that contains the NH_2 group? If the NH_2 group were absent, what would be the name of the molecule? What is the suffix for the NH_2 group? How is the molecule numbered to give the NH_2 the lowest number? Is the locator number necessary?

Solve

The NH_2 group is the highest-priority functional group. If the NH_2 group were absent, the molecule names would be butane for **(a)**, ethane for **(b)**, and cyclohexane for **(c)**. Drop the final *e* and add the *amine* suffix, so **(a)** becomes *butanamine*, **(b)** becomes *ethanamine*, and **(c)** becomes *cyclohexanamine*. The molecules are numbered to give the lowest number to the *amine* group. All of the molecules require a locator to establish the location of the NH_2 group. The IUPAC names for molecules **(a)–(c)** are given on the next page. The main chain or ring is numbered.

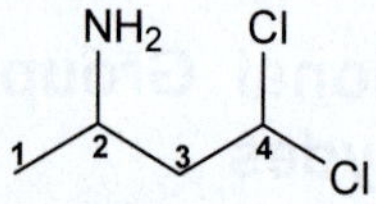

(a) 4,4-Dichlorobutan-2-amine 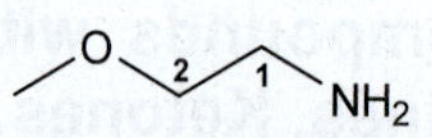 (b) 2-Methoxyethan-1-amine 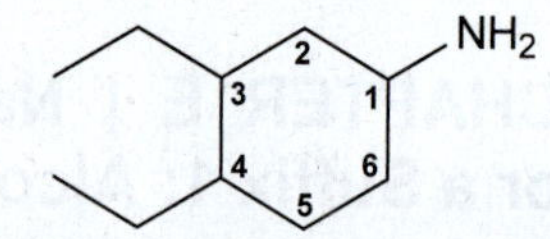 (c) 3,4-Diethylcyclohexan-1-amine

Problem E.4

Think

How many carbon atoms are in the main chain or ring? To what functional group does the suffix *amine* correlate? On which C atom is the NH_2 group located? What are the other substituents present, and where are they located?

Solve

The *amine* suffix correlates to the NH_2 functional group. Molecules **(a)** and **(b)** do not require a locator because all six carbon atoms in **(a)** are equivalent and both carbon atoms in **(b)** are equivalent. Molecules **(c)** and **(d)** require locators to describe the location of the NH_2 group. Structures for IUPAC names **(a)–(d)** are drawn below. The main chain or ring is numbered.

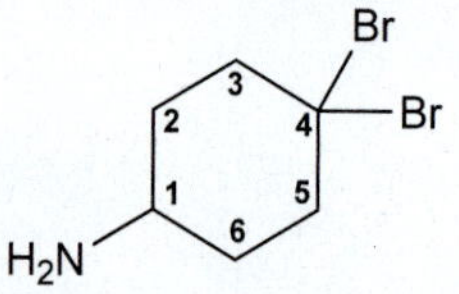

(a) 4,4-Dibromocyclohexanamine 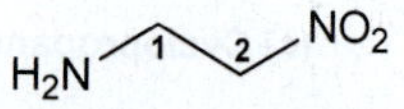 (b) 2-Nitroethanamine 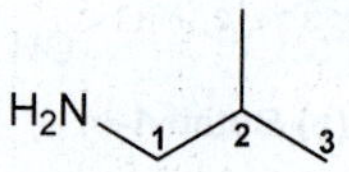 (c) 2-Methylpropan-1-amine 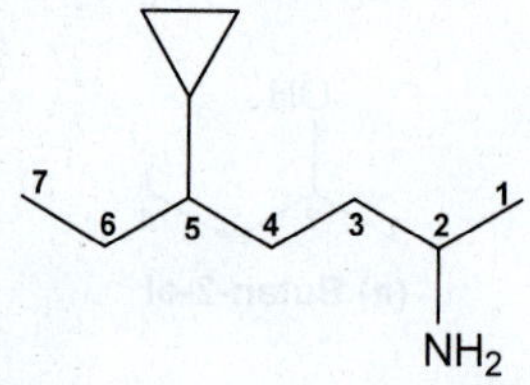 (d) 5-Cyclopropylheptan-2-amine

Problem E.5

Think

Identify the functional groups present in each molecule. Which functional group has the highest priority? How must the molecule be numbered to give the highest-priority functional group the lowest number? After the locator number for NH_2 or OH is established, how can the main chain or ring be numbered to arrive at the C=C or C≡C bond earliest? Review the nomenclature rules from Interchapters A, B, and C.

Solve

In molecules **(a)** and **(d)**, the NH_2 group has highest priority, so we drop the final *e* from the parent name and add the suffix *amine*. In molecules **(b)** and **(c)**, the OH group has highest priority, so we drop the final *e* from the parent name and add the suffix *ol*. Numbering begins at the carbon attached to NH_2 or OH. The molecules are then numbered to arrive at the C=C or C≡C earliest. The IUPAC names for molecules **(a)–(d)** are given below. The main chain or ring is numbered.

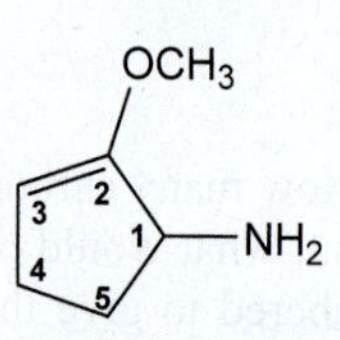

(a) 2-Methoxycyclopent-2-en-1-amine 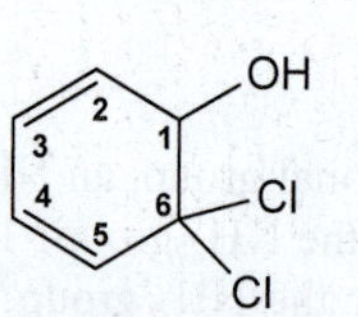 (b) 6,6-Dichlorocyclohexa-2,4-dien-1-ol

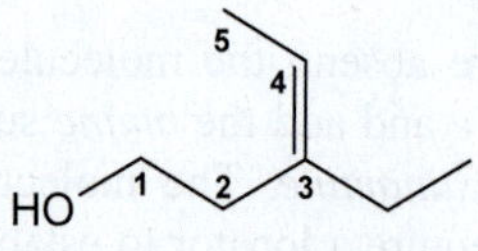

(c) 3-Ethylpent-3-en-1-ol 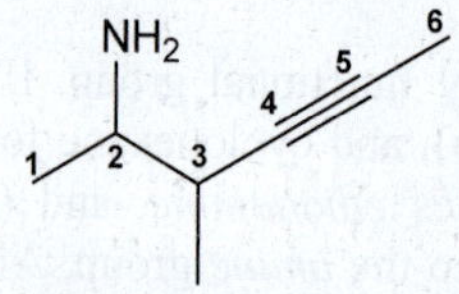 (d) 3-Methylhex-4-yn-2-amine

Problem E.6

Think

How many carbon atoms are in the main chain or ring? To what functional group does the suffix *amine* or *ol* correlate? On which C atom is the NH_2 or OH group located? On which two carbon atoms are each C=C and C≡C located? Review the nomenclature rules from Interchapters A and B.

Solve

The *ol* suffix correlates to the OH functional group, and the *amine* suffix correlates to the NH_2 functional group. In each molecule, the C=C or C≡C is part of the main chain or ring. Structures for IUPAC names **(a)–(d)** are drawn below. The main chain or ring is numbered.

(a) Pent-4-en-1-amine **(b) 3-Cyclopropylcyclopent-3-en-1-amine**

(c) 6-Chlorohexa-1,4-diyn-3-amine **(d) 2-Methylcycloocta-3,6-dien-1-ol**

Problem E.7

Think

If a molecule contains both an OH and a NH_2 group, which one has higher priority? What is the suffix? What prefix is used when the lower-priority group is treated as a substituent? How should the main chain or ring be numbered so that the highest-priority functional groups are encountered the earliest?

Solve

Molecules **(a)–(c)** each contain both an OH and a NH_2 group, both of which call for a suffix to be added. A OH group has a higher priority than a NH_2 group, so the suffix is *ol*. The chain or ring is numbered so that the OH group receives the lowest number. The NH_2 group is named using the *amino* prefix. The IUPAC names for molecules **(a)–(c)** are given below. The main chain or ring is numbered.

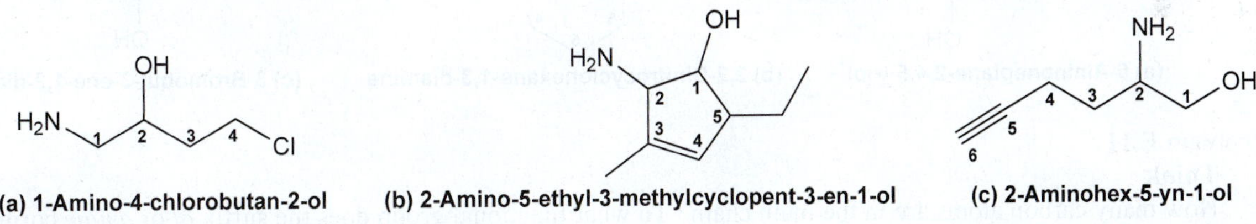

(a) 1-Amino-4-chlorobutan-2-ol **(b) 2-Amino-5-ethyl-3-methylcyclopent-3-en-1-ol** **(c) 2-Aminohex-5-yn-1-ol**

Problem E.8

Think

How many carbon atoms are in the main chain or ring? To what functional group does the suffix *ol* correlate? What is the amino substituent? On which C atom is the NH_2 or OH group located? Review the nomenclature rules from Interchapters A and B.

Solve

The *ol* suffix correlates to the OH functional group, and the *amino* prefix correlates to the NH_2 functional group. Structures for IUPAC names **(a)** and **(b)** are drawn on the next page. The main chain or ring is numbered.

(a) 4,5-Diaminoheptan-2-ol

(b) 2,3,4-Triaminocycloheptanol

Problem E.10

Think

If a molecule contains both an OH and a NH_2 group, which one has higher priority? What is the suffix? What prefix is used when the lower-priority group is treated as a substituent? How can one account for the number of functional groups that are the highest priority? Should the final *e* be dropped before the suffix is added? How should the main chain or ring be numbered so that the highest-priority functional groups are encountered the earliest? Are there any C=C or C≡C bonds within that main chain?

Solve

The IUPAC names for molecules **(a)–(c)** are given below. The main chain or ring is numbered.

(a) This molecule contains three OH and one NH_2 group, all of which call for a suffix to be added. An OH group has a higher priority than a NH_2 group, so the suffix is *ol*. The chain is numbered so that the OH groups receive the lowest numbers, in this case, C2, C4, and C5. Because there are three OH groups, the prefix *tri* is added before the numbers, and the final *e* in *heptane* is retained before adding the suffix. The NH_2 group is named using the *amino* prefix and is located at C6.

(b) This molecule contains two NH_2 groups and two NO_2 groups. The two NH_2 groups have higher priority over the NO_2 groups and require the *amine* suffix. The ring is numbered so that the NH_2 groups receive the lowest numbers, in this case, C1 and C3. Because there are two NH_2 groups, the prefix *di* is added before the numbers, and the final *e* in *cyclohexane* is retained before adding the suffix.

(c) This molecule contains two OH groups, one Br, and one C=C. The two OH groups have higher priority than the other groups and require the *ol* suffix. The chain is numbered so that the OH groups receive the lowest numbers, in this case, C1 and C2. Because there are two OH groups, the prefix *di* is added before the numbers, and the final *e* in *butene* is retained before adding the suffix.

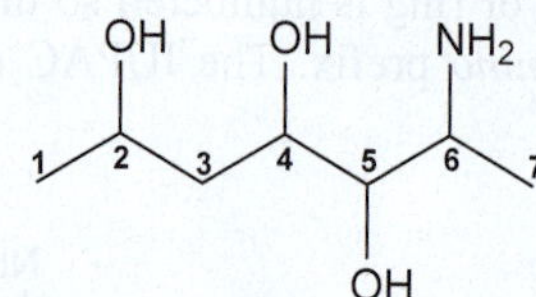

(a) 6-Aminoheptane-2,4,5-triol

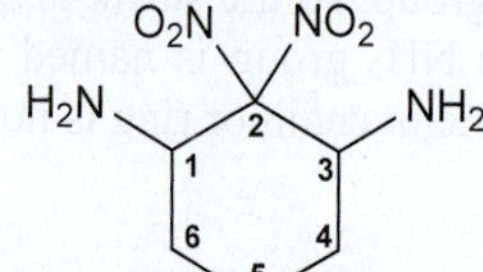

(b) 2,2-Dinitrocyclohexane-1,3-diamine

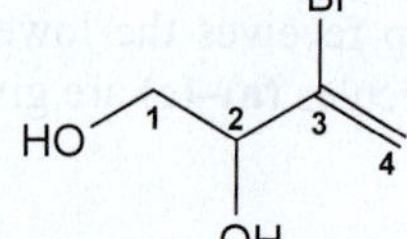

(c) 3-Bromobut-3-ene-1,2-diol

Problem E.11

Think

How many carbon atoms are in the main chain? To what functional group does the suffix *ol* or *amine* correlate? How many of each functional group are present? On which C atoms is the NH_2 or OH group located? Review the nomenclature rules from Interchapters A and B.

Solve

For **(a)** and **(c)**, the *ol* suffix correlates to the OH functional group, and for **(b)**, the *amine* suffix correlates to the NH_2 functional group. Structures for IUPAC names **(a)–(c)** are drawn below. The main chain is numbered.

(a) Propane-1,2-diol

(b) 1-Methoxypentane-2,3-diamine

(c) But-2-ene-1,4-diol

Problem E.12

Think

What is the highest-priority functional group? What suffix is used for the main chain or ring? What is the longest carbon chain or ring attached to the N in each molecule? How are the other alkyl groups treated? What type of locator is given for each of the alkyl substituents attached to N?

Solve

Each molecule is an amine, so each name receives the suffix *amine.* The IUPAC names for molecules **(a)–(c)** are given below. The main chain or ring is numbered.

(a) The longest carbon group that is attached to N is the six-carbon ring with a C=C. The carbon attached to the N is assigned C1, and numbering increases around the ring to have the C=C encountered earliest, so it is a *cyclohex-2-en-1-amine.* There are two methyl substituents attached to N, so methyl appears as a prefix, and each methyl group is given the locator *N.*

(b) There are three propyl groups attached to the N. Only one propyl group can be the main chain; the other two are considered substituents. The carbon to which N is attached is assigned C1. The two propyl substituents are each given the locator *N.*

(c) The longest carbon chain that is attached to N has five carbons and a C≡C. The carbon attached to the N has the highest priority, and the chain is numbered from right to left, so it is a *pent-4-yn-1-amine.* There are two ethyl substituents in the molecule, one attached to N and one attached to C2, so diethyl appears in the prefix, along with the locators *N*,2.

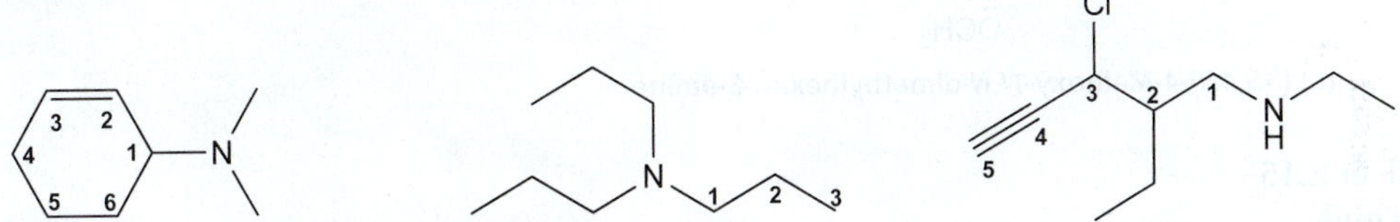

(a) *N,N*-Dimethylcyclohex-2-en-1-amine (b) *N,N*-Dipropylpropan-1-amine (c) 3-Chloro-*N*,2-diethylpent-4-yn-1-amine

Problem E.13

Think

How many carbon atoms are in the main chain? To what functional group does the suffix *amine* correlate? Are any *alkyl* groups attached to the N?

Solve

Structures for IUPAC names **(a)** and **(b)** are drawn below. The main chain or ring is numbered.

(a) The main chain has seven carbon atoms, and the N is located at C4 and two C=C groups are located at C1=C2 and C6=C7. A methyl group is attached to the N.

(b) The main ring is a *cyclobutane*. There are three *cyclopropyl* substituents: Two are located on the N, and the remaining one is on C3 of the *cyclobutane* ring.

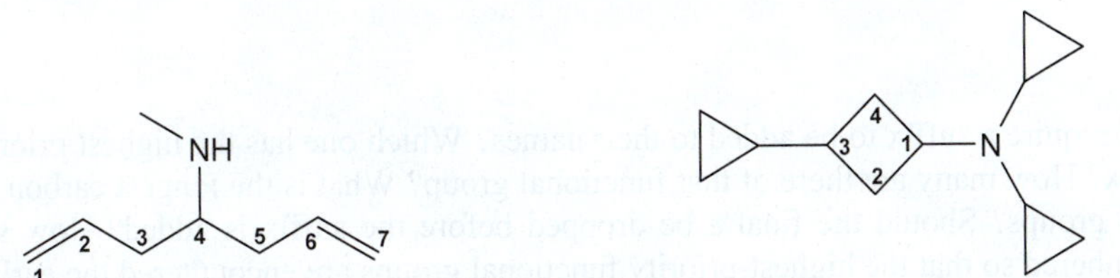

(a) *N*-Methylhepta-1,6-dien-4-amine (b) *N,N*,3-Tricyclopropylcyclobutanamine

Problem E.14

Think

Review the rules for naming compounds with stereochemical designations in Interchapter C and the rules for naming molecules with OH and NH_2 groups thus far in this chapter.

Solve

Stereochemical designations and their locators are enclosed in parentheses at the very beginning of the name. The IUPAC names for molecules **(a)–(c)** are given below. The main chain or ring is numbered.

(a) The OH carbon at C3 has the *S* configuration, and the C=C bond at C4 has the *Z* configuration.

(b) The OH carbon at C1 and the NH_2 at C3 both have the *R* configuration.

(c) The C3 carbon attached to N has the *S* configuration, and the C4 carbon attached to OCH_3 has the *R* configuration.

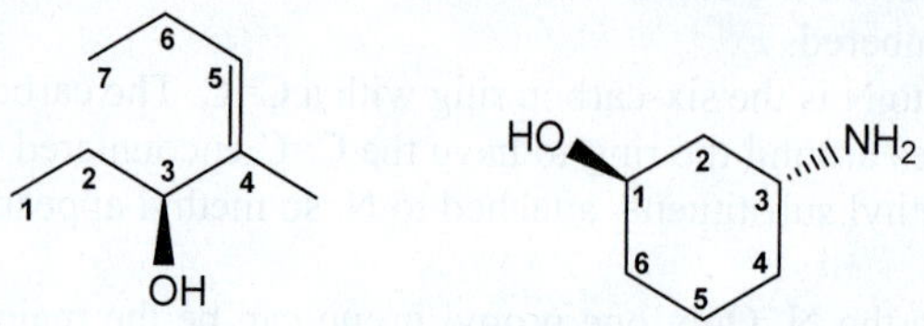

(a) (*S*,*Z*)-4-Methylhept-4-en-3-ol **(b) (1*R*,3*R*)-3-Aminocyclohexan-1-ol**

(c) (3*S*,4*R*)-4-Methoxy-*N*,*N*-dimethylhexan-3-amine

Problem E.15

Think

Draw the molecule that corresponds to each name without drawing the configuration. Then consider the rules for *R*/*S* and *E*/*Z* configurations.

Solve

Structures for IUPAC names **(a)** and **(b)** are drawn below. The main chain or ring is numbered.

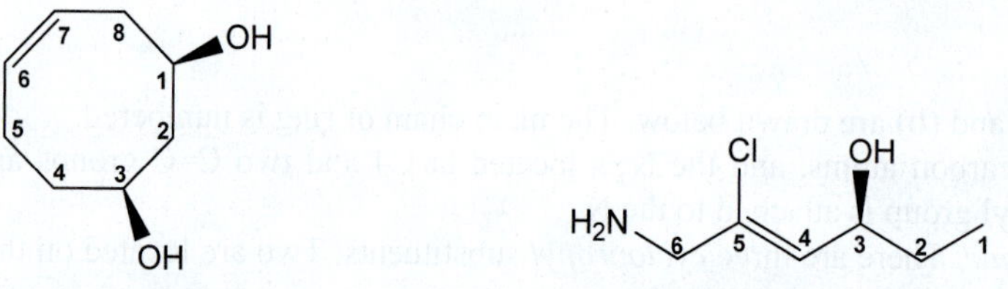

(a) (1*R*,3*S*)-Cyclooct-6-ene-1,3-diol **(b) (3*S*,*Z*)-6-Amino-5-chlorohex-4-en-3-ol**

Problem E.16

Think

Do any functional groups require a suffix to be added to their names? Which one has the highest priority (Table E-1), and what is its suffix? How many are there of that functional group? What is the longest carbon chain that contains those functional groups? Should the final *e* be dropped before the suffix is added? How should the main chain or ring be numbered so that the highest-priority functional groups are encountered the earliest? How is a cyclic aldehyde named?

Solve

The IUPAC names for molecules **(a)–(d)** are given on the next page. The main chain or ring is numbered.

(a) This molecule has both a ketone C=O and an aldehyde C=O, both of which call for a suffix. The priority goes to the aldehyde, so the C=O of the ketone is treated as an *oxo* substituent. A locator number is required for the ketone C=O but not for the aldehyde C=O.

(b) This molecule has an aldehyde C=O and an OH group, both of which call for a suffix. The priority goes to the aldehyde (C1 must be the ring carbon attached to the aldehyde C=O, so no locator is necessary), so the OH group is treated as a *hydroxyl* substituent at C2. The name of a cyclic aldehyde takes the form *cycloalkanecarbaldehyde*. The cycloalkane ring is a cyclohexane, so the molecule is a *cyclohexanecarbaldehyde*.

(c) This molecule contains two ketone C=O groups and one NH_2 group. The priority goes to the two ketone C=O groups, so the NH_2 group is treated as an amino substituent. *Di* is added to account for the two ketone C=O groups, and the *e* in *ane* is retained.

(d) Only one functional group in the molecule requires a suffix, the ketone C=O. The C=C makes the ring a *cyclohexene*. The ketone C=O takes priority and is therefore C1.

(a) 4-Oxoheptanal

(b) 2-Hydroxycyclohexanecarbaldehyde

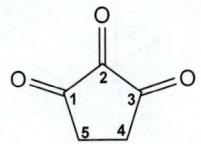

(c) 1-Amino-3-propylpentane-2,4-dione

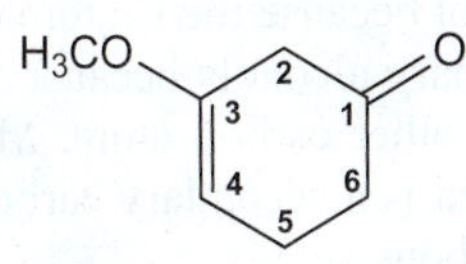

(d) 3-Methoxycyclohex-3-en-1-one

Problem E.17

Think

How many carbon atoms are in the main chain or ring? To what functional group does each suffix correlate? How many of each are present? How do you draw a cyclic aldehyde?

Solve

Structures and IUPAC names for **(a)–(d)** are drawn below. The main chain or ring is numbered.

(a) There are five carbon atoms in the main ring and three ketone C=O groups.

(b) The main chain has three carbon atoms, and the aldehyde C=O is at C1. The *oxo* prefix is a ketone C=O.

(c) There are seven carbons in the main ring and one additional carbon for the *aldehyde* C=O attached to the ring.

(d) There are seven carbons in the main chain, which contain two aldehyde C=O groups. The chain is numbered left to right to have the C≡C earliest.

(a) Cyclopentane-1,2,3-trione

(b) 3,3,3-Trichloro-2-oxopropanal

(c) 2,4,4-Trinitrocycloheptanecarbaldehyde

(d) 4-Hydroxyhept-2-ynedial

Problem E.18

Think

For each trivial name, how many alkyl groups are bonded to C–OH? What are they? Review the trivial names for alkyl groups in Interchapter A.7. How many carbon atoms are in the main chain? On which C atom is the OH group located?

Solve

Structures and IUPAC names for trivial names **(a)–(d)** are drawn below. The main chain or ring is numbered.

(a) Trichloromethyl alcohol
Trichloromethanol

(b) Isobutyl alcohol
2-Methylpropan-1-ol

(c) Pentyl alcohol
Pentan-1-ol

(d) *sec*-Butyl alcohol
Butan-2-ol

Problem E.19

Think

What is the degree of alkyl substitution of the carbon to which the OH group is attached in molecules **(a)–(d)** in Problem E.18?

Solve

Molecule **(a)** is a *methyl* alcohol because the C atom to which OH is attached is not bonded to any alkyl groups. Molecules **(b)** and **(c)** are primary alcohols because the carbon atom to which the OH is attached is a primary carbon—attached to only one other carbon atom. Molecule **(d)** is a secondary alcohol because the carbon to which the OH group is attached is a secondary carbon—attached to two carbon atoms. The shading highlights the C–OH and the attached carbons.

(a) Methyl **(b) Primary** **(c) Primary** **(d) Secondary**

Problem E.20

Think

For each trivial name, how many alkyl groups are attached to N? Review the trivial names for alkyl groups in Interchapter A.7 and for alkenes, alkynes, and substituted benzenes in Interchapter B.4. How many carbon atoms are in the main chain? On which C atom is the amine N located?

Solve

Structures and IUPAC names for trivial names **(a)–(e)** are drawn below. The main chain or ring is numbered.

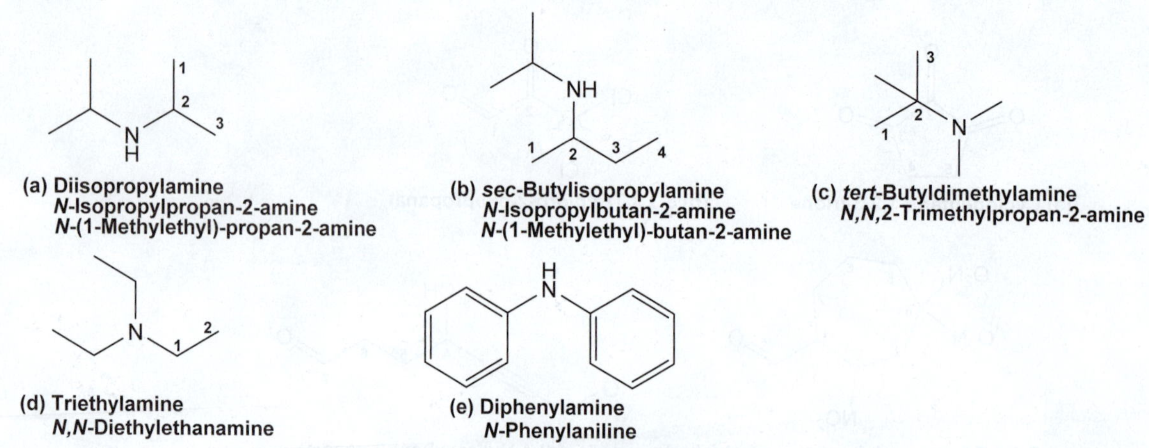

(a) Diisopropylamine
***N*-Isopropylpropan-2-amine**
***N*-(1-Methylethyl)-propan-2-amine**

(b) *sec*-Butylisopropylamine
***N*-Isopropylbutan-2-amine**
***N*-(1-Methylethyl)-butan-2-amine**

(c) *tert*-Butyldimethylamine
***N,N*,2-Trimethylpropan-2-amine**

(d) Triethylamine
***N,N*-Diethylethanamine**

(e) Diphenylamine
***N*-Phenylaniline**

Problem E.21

Think

How many alkyl groups are directly attached to the N atom? How does the number of alkyl groups directly attached to the N atom classify the type of amine?

Solve

None of the molecules are primary amines, in which one alkyl group is attached to the N atom. Molecules **(a)**, **(b)**, and **(e)** are secondary amines, because two alkyl groups are attached to the N atom. Molecules **(c)** and **(d)** are tertiary amines, because three alkyl groups are attached to the N. The bonds connecting N to an alkyl group are shaded.

(a) Secondary **(b) Secondary** **(c) Tertiary** **(d) Tertiary** **(e) Secondary**

Problem E.22

Think

For each trivial name, how many C atoms are present in the main chain? Review the trivial names for alkyl groups in Interchapter A.7 and for alkenes, alkynes, and substituted benzenes in Interchapter B.4. How many carbon atoms are in the main chain? Which C atom is part of the C=O group? Where are the substituents located?

Solve

Structures and IUPAC names for trivial names **(a)–(c)** are drawn below. The main chain or ring is numbered.

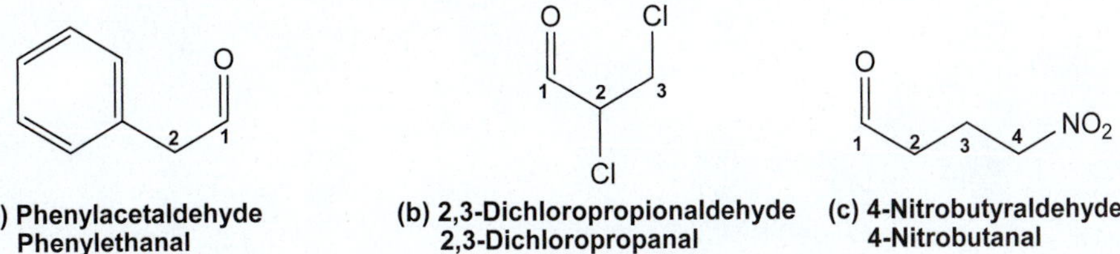

(a) Phenylacetaldehyde
Phenylethanal

(b) 2,3-Dichloropropionaldehyde
2,3-Dichloropropanal

(c) 4-Nitrobutyraldehyde
4-Nitrobutanal

Problem E.23

Think

For each trivial name, what alkyl groups are bonded to the ketone C=O carbon? Review the trivial names for alkyl groups in Interchapter A.7 and for alkenes, alkynes, and substituted benzenes in Interchapter B.4. How many carbon atoms are in the main chain? Which C atom is part of the C=O group? Where are the substituents located?

Solve

Structures and IUPAC names for trivial names **(a)–(e)** are drawn below and the next page. The main chain or ring is numbered.

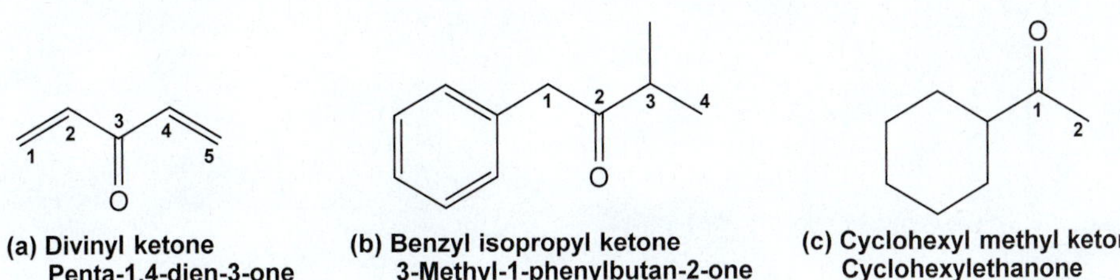

(a) Divinyl ketone
Penta-1,4-dien-3-one

(b) Benzyl isopropyl ketone
3-Methyl-1-phenylbutan-2-one

(c) Cyclohexyl methyl ketone
Cyclohexylethanone

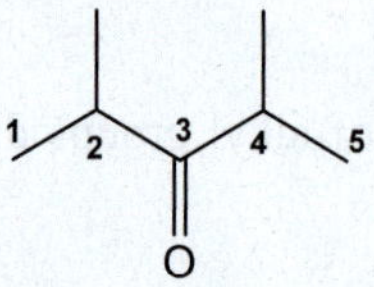

1 2 3 4 5

O

(d) Diisopropyl ketone
2,4-Dimethylpentan-3-one

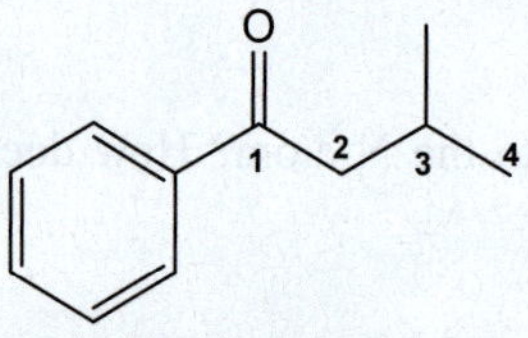

O

1 2 3 4

(e) Isobutyl phenyl ketone
3-Methyl-1-phenylbutan-1-one

End of Chapter Problems
Section E.1 Alcohols and Amines
Problem E.24

Think

Review the rules in Section E.1 for naming alcohols. What suffix is used for an alcohol? How many carbon atoms are in the main chain that contains the OH group? What substituents are present, and where are they located?

Solve

The IUPAC names for molecules **(a)**–**(d)** are given below. The main chain or ring is numbered.

(a) Butan-1-ol **(b) 2-Methylbutan-1-ol** **(c) 2,3-Dinitrocyclopentan-1-ol** **(d) 2-Ethylpentan-1-ol**

Problem E.25

Think

Review the rules in Section E.1 for naming amines. What suffix is used for an amine? How many carbon atoms are in the main chain that contains the NH_2 group?

Solve

The IUPAC names for molecules **(a)**–**(d)** are given below. The main chain or ring is numbered.

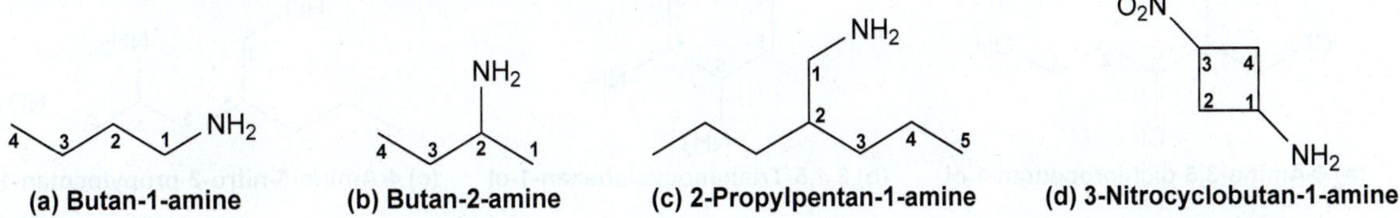

(a) Butan-1-amine **(b) Butan-2-amine** **(c) 2-Propylpentan-1-amine** **(d) 3-Nitrocyclobutan-1-amine**

Problem E.26

Think

How many carbon atoms are in the main chain or ring? To what functional group does the suffix amine or *ol* correlate? On which C atom is the NH_2 or OH group located?

Solve

Structures for IUPAC names **(a)**–**(d)** are drawn below. The main chain or ring is numbered.

(a) 3,3-Dipropoxypentan-1-amine **(b) 2,3,4-Trichlorocyclohexanol**

(c) 3-Cyclopropylpentan-1-ol **(d) 3-(1-Methylethyl)cycloheptanamine**

Problem E.27

Think

Identify the functional groups present in each molecule. Which functional group has the highest priority? How must the molecule be numbered to give the highest-priority functional group the lowest number? After the C bonded to NH_2 or OH is numbered, how can the other carbons of the main chain or ring be numbered to arrive at the C=C or C≡C bond earliest?

Solve

The IUPAC names for molecules **(a)–(c)** are given below. The main chain or ring is numbered.

(a) 5-Chlorohex-2-yn-1-amine **(b) 4-Methoxycyclohexa-2,4-dien-1-ol** **(c) 2-Methylbut-3-en-1-amine**

Problem E.28

Think

If a molecule contains both an OH and a NH_2 group, which one has higher priority? What is the suffix? What prefix is used when the lower-priority group is treated as a substituent? How should the main chain or ring be numbered so that the highest-priority functional groups are encountered the earliest?

Solve

The IUPAC names for molecules **(a)–(c)** are given below. The main chain or ring is numbered.

(a) 4-Amino-3,5-dichloropentan-1-ol

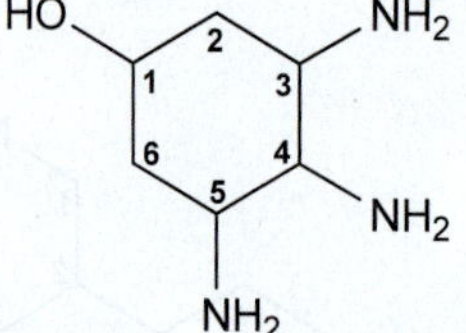

(b) 3,4,5-Triaminocyclohexan-1-ol

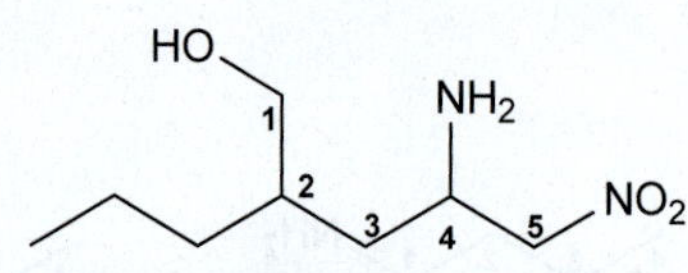

(c) 4-Amino-5-nitro-2-propylpentan-1-ol

Problem E.29

Think

How many carbon atoms are in the main chain or ring? To what functional group does the suffix *ol* correlate? To what functional group does the prefix *amino* correlate? On which C atom is the NH_2 or OH group located?

Solve

Structures for IUPAC names **(a)–(d)** are drawn below. The main chain or ring is numbered.

(a) 5-Amino-2,3,4-trimethylpentan-1-ol **(b) 3-Amino-4,5-diethoxyoctan-1-ol**

(c) 3,4-Diamino-5-bromocyclohexanol **(d) 4-Amino-3,3-diethylhexan-1-ol**

Problem E.30

Think

If a molecule contains both a OH and a NH_2 group, which one has higher priority? What is the suffix? What prefix is used when the lower-priority group is treated as a substituent? How should the main chain or ring be numbered so that the highest-priority functional groups are encountered the earliest? Should you drop the final *e* of the name when there are two or more of the highest-priority functional groups?

Solve

The IUPAC names for molecules **(a)–(d)** are given below. The main chain or ring is numbered.

(a) 3,5-Dichloropentane-1,4-diol **(b) 5-Bromo-2-propylpentane-1,3,4-triamine** **(c) 2,6-Diaminocyclohexane-1,4-diol**

Problem E.31

Think

What is the highest-priority functional group, and to what suffix does it correspond? What is the longest carbon chain or ring attached to the N in each molecule? How are the other alkyl groups treated? What type of locator is given for each of the alkyl substituents? What suffix is used for the main chain or ring?

Solve

The IUPAC names for molecules **(a)–(d)** are given below. The main chain or ring is numbered.

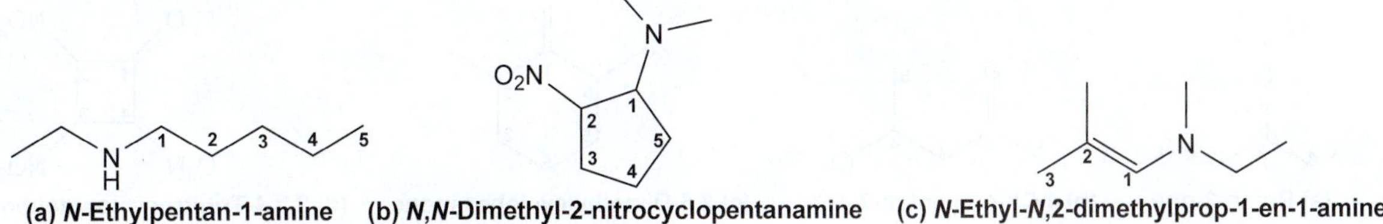

(a) *N*-Ethylpentan-1-amine **(b) *N*,*N*-Dimethyl-2-nitrocyclopentanamine** **(c) *N*-Ethyl-*N*,2-dimethylprop-1-en-1-amine**

Problem E.32

Think

Review the rules for naming compounds with stereochemical designations in Interchapter C and the rules for naming molecules with OH and NH_2 groups in this chapter.

Solve

The IUPAC names for molecules **(a)–(d)** are given below. The main chain or ring is numbered.

(a) (3*S*,4*R*)-3,5-Dichloropentane-1,4-diol **(b) (2*R*,3*S*,4*S*)-5-Bromo-2-propylpentane-1,3,4-triamine**

(c) (1*R*,2*S*,5*S*)-2-Amino-5-nitrocyclohexanol

Problem E.33

Think

Draw the molecule that corresponds to each name, temporarily disregarding the stereochemical configurations. Then consider the rules for *R*/*S* and *E*/*Z* configurations.

Solve

Structures for IUPAC names **(a)–(c)** are drawn below. The main chain or ring is numbered.

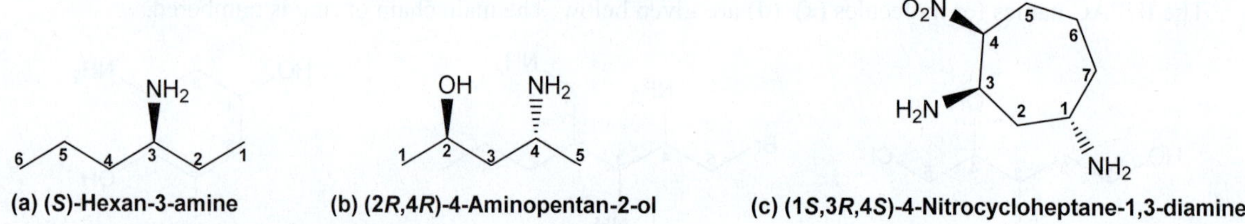

Section E.2 Ketones and Aldehydes

Problem E.34

Think

What suffix corresponds to a ketone (Table E-1)? What is the longest carbon chain or largest ring that contains the ketone C=O group? Should the final *e* be dropped before the suffix is added? How should the main chain or ring be numbered so that the ketone C=O group is encountered the earliest? How is a cyclic aldehyde named?

Solve

The IUPAC names for molecules **(a)–(d)** are given below. The main chain or ring is numbered.

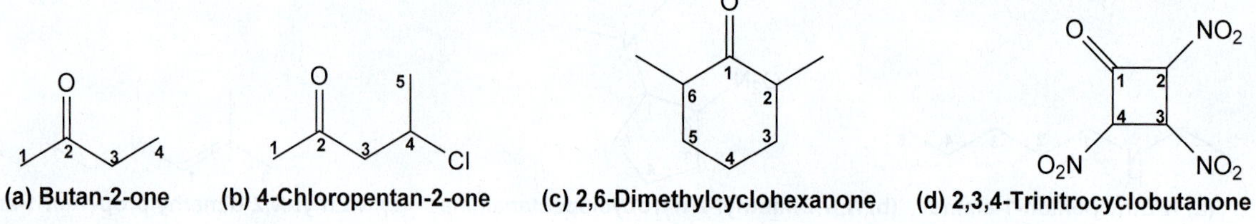

Problem E.35

Think

What suffix corresponds to an aldehyde (Table E-1)? What is the longest carbon chain that contains the aldehyde C=O group? Should the final *e* be dropped before the suffix is added? How should the main chain or ring be numbered so that the aldehyde C=O group is encountered the earliest?

Solve

The IUPAC names for molecules **(a)–(e)** are given below. The main chain or ring is numbered.

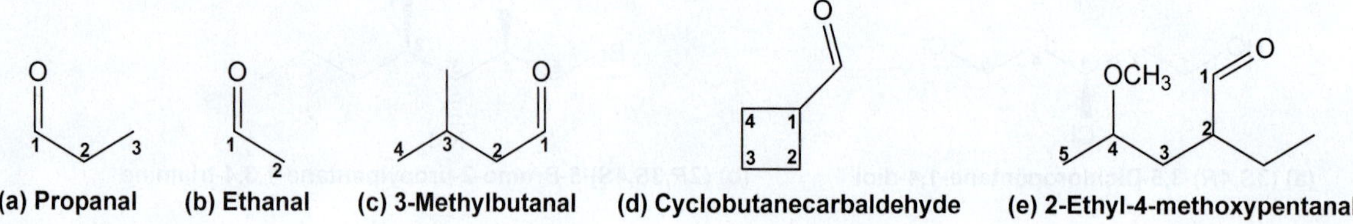

Problem E.36

Think

How many carbon atoms are in the main chain or ring? To what functional group does each suffix correlate? How many of each are present? How do you draw the cyclic aldehyde? What substituents are present, and where are they located?

Solve

Structures for IUPAC names **(a)–(d)** are drawn below. The main chain or ring is numbered.

(a) 2,3-Dimethylcyclopentanone **(b) 4,4-Difluoroheptanal**

(c) 1,1,1-Trichloropentan-3-one **(d) (1*S*, 3*S*)-3-Ethoxycyclohexanecarbaldehyde**

Problem E.37

Think

Do any functional groups require a suffix to be added to their names? Which one has the highest priority (Table E-1), and what is its suffix? How many are there of that functional group? What is the longest carbon chain that contains those functional groups? Should the final *e* be dropped before the suffix is added? How should the main chain or ring be numbered so that the highest-priority functional groups are encountered the earliest? How is a cyclic aldehyde named?

Solve

The IUPAC names for molecules **(a)–(e)** are given below. The main chain or ring is numbered.

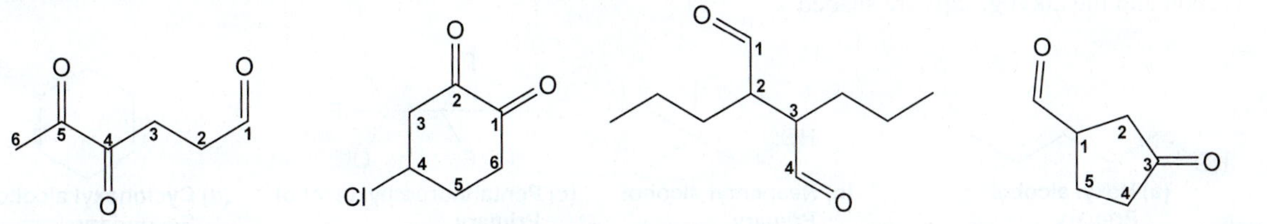

(a) 4,5-Dioxohexanal **(b) 4-Chlorocyclohexane-1,2-dione** **(c) 2,3-Dipropylbutanedial** **(d) 3-Oxocyclopentanecarbaldehyde**

Problem E.38

Think

Do any functional groups require a suffix to be added to their names? Which one has the highest priority (Table E-1), and what is its suffix? How many are there of that functional group? What is the longest carbon chain that contains those functional groups? Should the final *e* be dropped before the suffix is added? How should the main chain or ring be numbered so that the highest-priority functional groups are encountered the earliest?

Solve

The IUPAC names for molecules **(a)–(d)** are given below. The main chain or ring is numbered.

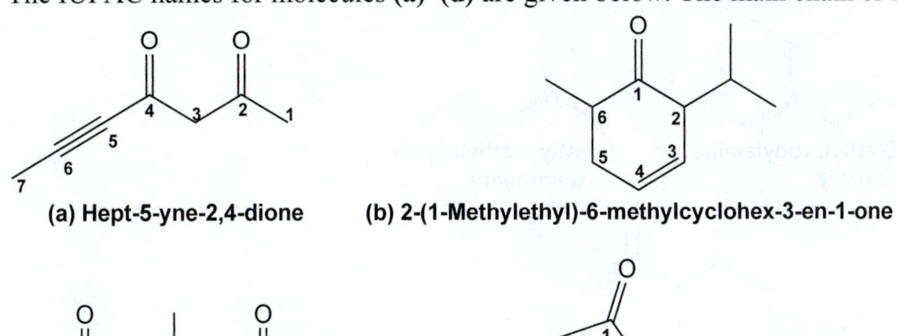

(a) Hept-5-yne-2,4-dione **(b) 2-(1-Methylethyl)-6-methylcyclohex-3-en-1-one**

(c) (*E*)-3-Methylpent-2-enedial **(d) Cyclopenta-2,4-dien-1-one**

Problem E.39

Think

How many carbon atoms are in the main chain or ring? To what functional group does each suffix correlate? How many of each are present? How do you draw the cyclic aldehyde? What substituents are present, and where are they located? Review the rules for naming compounds with stereochemical designations in Interchapter C.

Solve

Structures for IUPAC names **(a)–(d)** are drawn below. The main chain or ring is numbered.

(a) (*Z*)-Hept-3-enedial **(b) 2,2-Diethoxy-4-oxopentanal** **(c) Cyclohept-4-ene-1,3-dione** **(d) Pent-4-ynal**

Section E.3 Trivial Names

Problem E.40

Think

For each trivial name, what are the alkyl groups bonded to the C–OH carbon? How many of those alkyl groups are there? Review the trivial names for alkyl groups in Interchapter A.7.

Solve

Structures and types of alcohols for trivial names **(a)–(d)** are drawn below. The bonds between the C–OH carbon and the alkyl groups are shaded.

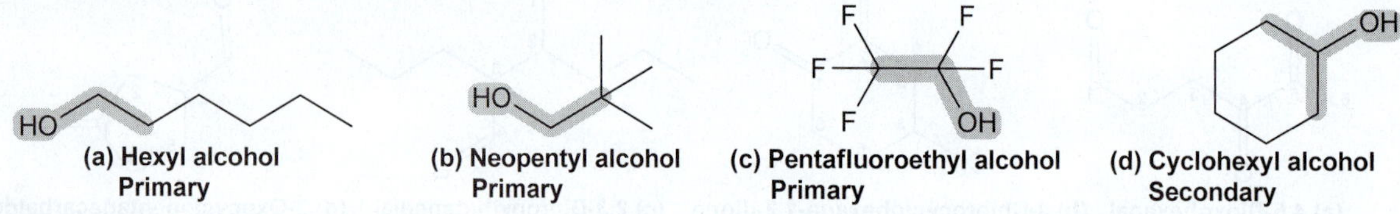

(a) Hexyl alcohol Primary **(b) Neopentyl alcohol** Primary **(c) Pentafluoroethyl alcohol** Primary **(d) Cyclohexyl alcohol** Secondary

Problem E.41

Think

For each trivial name, alkyl groups are bonded to the amine N; how many of them are there? How does the number of alkyl groups directly attached to the N atom classify the type of amine? Review the trivial names for alkyl groups in Interchapter A.7 and for alkenes, alkynes, and substituted benzenes in Interchapter B.4.

Solve

Structures and types of amines for trivial names **(a)–(e)** are drawn below. The bonds between the amine N and the alkyl groups are shaded.

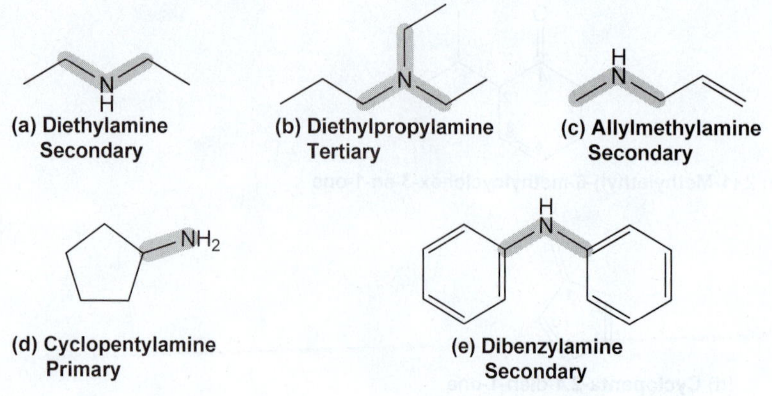

(a) Diethylamine Secondary **(b) Diethylpropylamine** Tertiary **(c) Allylmethylamine** Secondary **(d) Cyclopentylamine** Primary **(e) Dibenzylamine** Secondary

Problem E.42

Think

For each trivial name, what are the alkyl groups bonded to the carbonyl carbon? Review the trivial names for alkyl groups in Interchapter A.7 and for alkenes, alkynes, and substituted benzenes in Interchapter B.4. What is the structure of unsubstituted acetone?

Solve

Structures for trivial names **(a)–(e)** are drawn below.

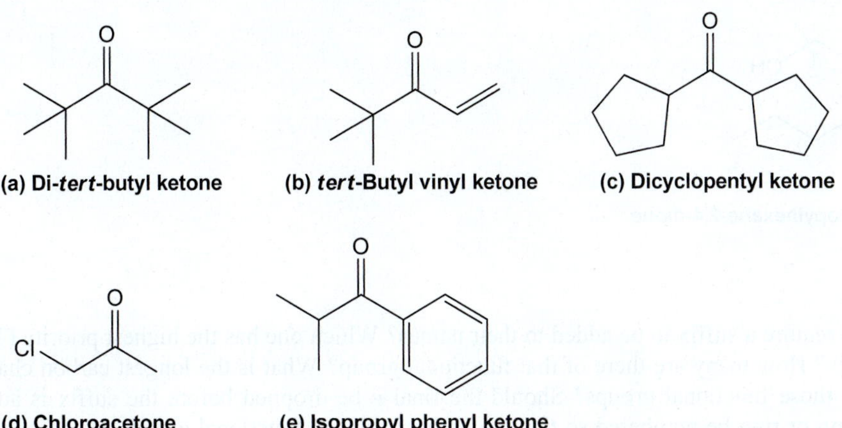

(a) Di-*tert*-butyl ketone **(b) *tert*-Butyl vinyl ketone** **(c) Dicyclopentyl ketone**

(d) Chloroacetone **(e) Isopropyl phenyl ketone**

Problem E.43

Think

To what functional group does the suffix aldehyde correspond? Where should that functional group be located in a chain or ring? For each trivial name, how many C atoms are present in the main chain? What rings are present? Review the trivial names for alkyl groups in Interchapter A.7 and for alkenes, alkynes, and substituted benzenes in Interchapter B.4.

Solve

Structures for trivial names **(a)–(c)** are drawn below.

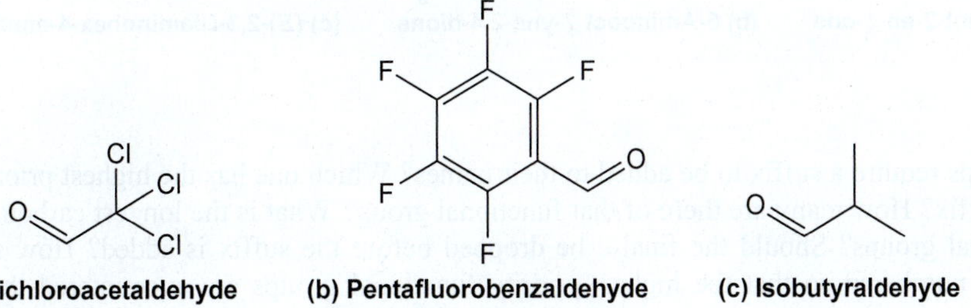

(a) Trichloroacetaldehyde **(b) Pentafluorobenzaldehyde** **(c) Isobutyraldehyde**

Integrated Problems

Problem E.44

Think

Do any functional groups require a suffix to be added to their names? Which one has the highest priority (Table E-1), and what is its suffix? How many are there of that functional group? What is the longest carbon chain or largest ring that contains those functional groups? Should the final *e* be dropped before the suffix is added? How should the main chain or ring be numbered so that the highest-priority functional groups are encountered the earliest?

Solve

The IUPAC names for molecules **(a)–(c)** are given below. The main chain or ring is numbered.

(a) 4-Hydroxy-3-nitro-6-oxoheptanal **(b) 4-Amino-2,5-dihydroxycyclohexan-1-one**

(c) 6-Amino-1-hydroxy-3-propylhexane-2,4-dione

Problem E.45

Think

Do any functional groups require a suffix to be added to their names? Which one has the highest priority (Table E-1), and what is its suffix? How many are there of that functional group? What is the longest carbon chain or largest ring that contains those functional groups? Should the final *e* be dropped before the suffix is added? How should the main chain or ring be numbered so that the highest-priority functional groups are encountered the earliest?

Solve

The IUPAC names for molecules **(a)–(c)** are given below. The main chain or ring is numbered.

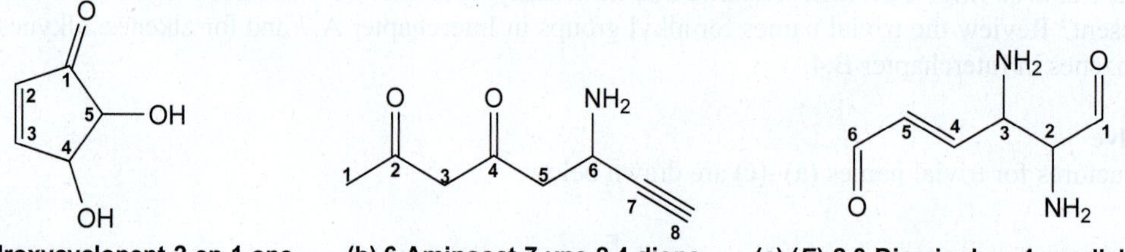

(a) 4,5-Dihydroxycyclopent-2-en-1-one **(b) 6-Aminooct-7-yne-2,4-dione** **(c) (*E*)-2,3-Diaminohex-4-enedial**

Problem E.46

Think

Do any functional groups require a suffix to be added to their names? Which one has the highest priority (Table E-1), and what is its suffix? How many are there of that functional group? What is the longest carbon chain that contains those functional groups? Should the final *e* be dropped before the suffix is added? How should the main chain or ring be numbered so that the highest-priority functional groups are encountered the earliest? Review the rules for assigning stereochemical configurations in Interchapter C.

Solve

The IUPAC names for molecules **(a)–(d)** are given below and on the next page. The main chain or ring is numbered.

(a) (3*R*,4*R*,5*R*)-5-Amino-3-hydroxy-4-nitrohexan-2-one **(b) (1*S*,3*R*,6*R*)-3-Hydroxy-6-methyl-2-oxocyclohex-4-enecarbaldehyde**

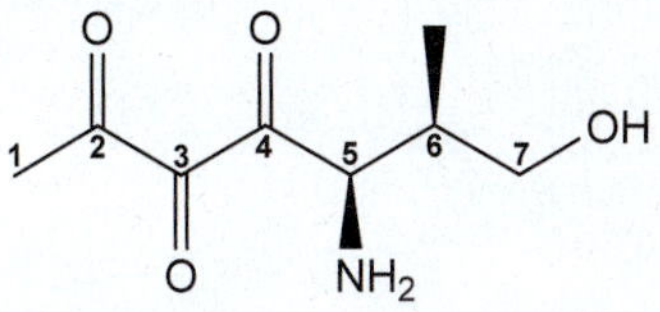

(c) (5*R*,6*R*)-5-Amino-7-hydroxy-6-methylheptane-2,3,4-trione

Problem E.47

Think

Do any functional groups require a suffix to be added to their names? Which one has the highest priority (Table E-1), and what is its suffix? How many are there of that functional group? What is the longest carbon chain or largest ring that contains those functional groups? Should the final *e* be dropped before the suffix is added? How should the main chain or ring be numbered so that the highest-priority functional groups are encountered the earliest?

Solve

The IUPAC names for molecules **(a)** and **(b)** are given below. The main chain or ring is numbered.

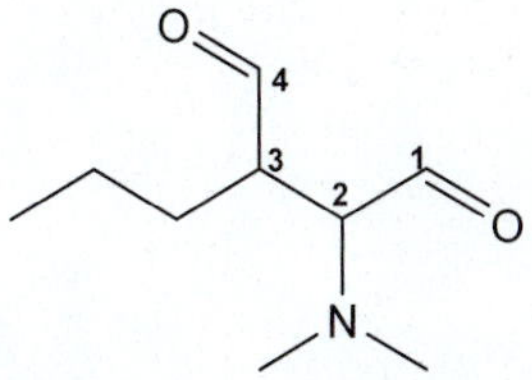

(a) 2-(Dimethylamino)-3-propylbutanedial

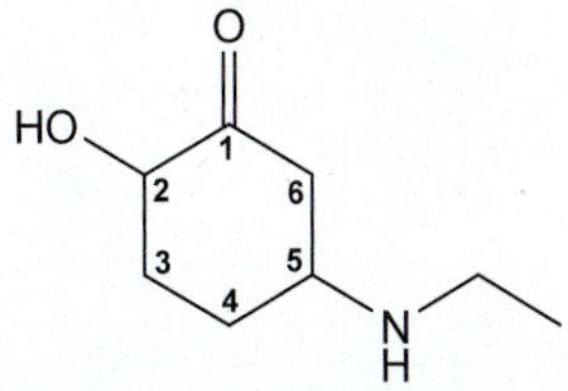

(b) 5-(Ethylamino)-2-hydroxycyclohexanone

Problem E.48

Think

How many carbon atoms are in the main chain or ring? To what functional group does each suffix correlate? How many of each are present? Draw the molecule that corresponds to each name without drawing the configuration. Consider the rules for *R*/*S* and *E*/*Z* configurations from Interchapter C.

Solve

Structures for IUPAC names **(a)–(d)** are drawn below. The main chain or ring is numbered.

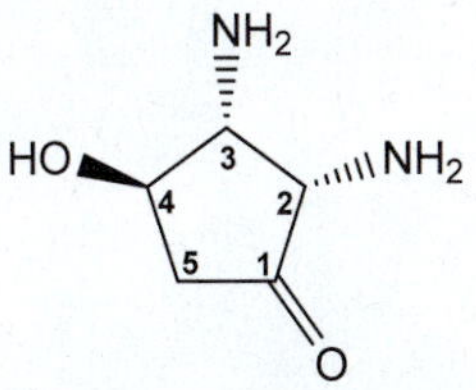

(a) (2*S*,3*R*,4*R*)-2,3-Diamino-4-hydroxycyclopentanone

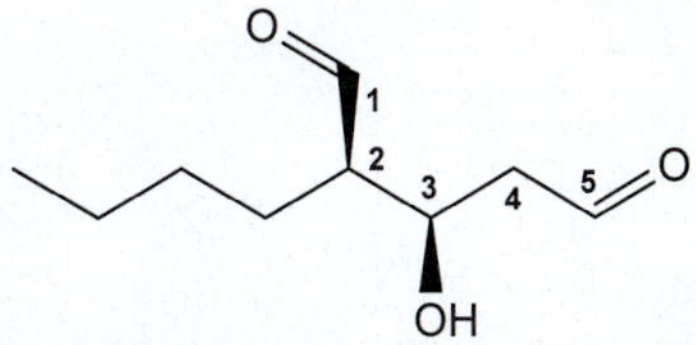

(b) (2*R*,3*R*)-2-Butyl-3-hydroxypentanedial

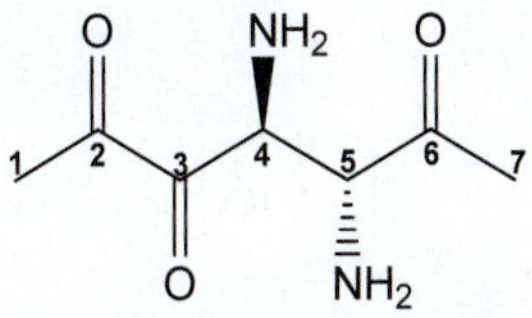

(c) (4*S*,5*R*)-4,5-Diaminoheptane-2,3,6-trione

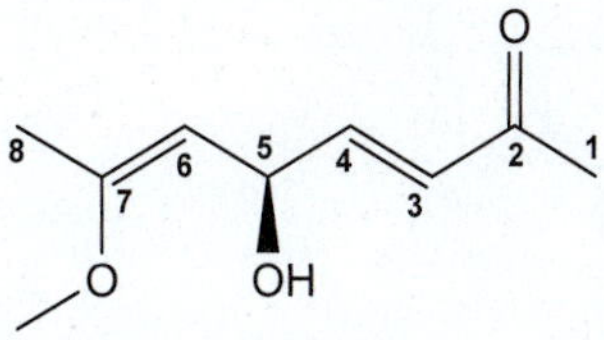

(d) (3*E*,5*R*,6*Z*)-5-Hydroxy-7-methoxyocta-3,6-dien-2-one

CHAPTER 8 | An Introduction to Multistep Mechanisms: S_N1 and E1 Reactions and Their Comparisons to S_N2 and E2 Reactions

Your Turn Exercises

Your Turn 8.1

Think

Consult Chapters 6 and 7 to review the 10 elementary steps. What kinds of bonds are breaking or forming? Are any protons or π bonds involved?

Solve

In Equation 8-2a, the C–L bond breaks. This is a heterolysis step that forms a carbocation and the L^- anion. In Equation 8-2b, the Nu–C bond forms. This is a coordination step in which electrons flow directly from the electron-rich $Nu:^-$ to the electron-poor C^+.

S_N1 mechanism

Leaving group leaves → Heterolysis (8-2a)

Nucleophile Substrate

Nucleophile attacks → Coordination (8-2b)

Your Turn 8.2

Think

Sum up the steps in the reaction first. Which species appears as a product in one step and as a reactant in another and, therefore, gets canceled? Which species appear in the overall reaction? Which species appear in the mechanism but not in the overall reaction?

Solve

In the first step (heterolysis), the C–Br bond is broken, producing an allylic carbocation and Br^-. In the second step (coordination), the electron-rich I^- donates two electrons to the electron-poor allylic C^+, producing the C–I bond. The allylic C^+ is a product in the first step and a reactant in the second, so it is canceled when you sum the steps. Because the allylic C^+ appears only in the reaction mechanism and not in the overall reaction, it is an intermediate.

Overall reactant Intermediate Overall product

Intermediate Overall reactant Overall product

Your Turn 8.3

Think

Do transition states appear at energy maxima or minima? How many transition states appear for the two-step reaction shown? Do intermediates appear at energy minima or maxima? How many intermediate stages appear for the two-step reaction shown?

Solve

Transition states appear at energy maxima and show the bond breaking and bond making in progress. Intermediates appear at energy minima. There are two transition states and one intermediate for a two-step mechanism.

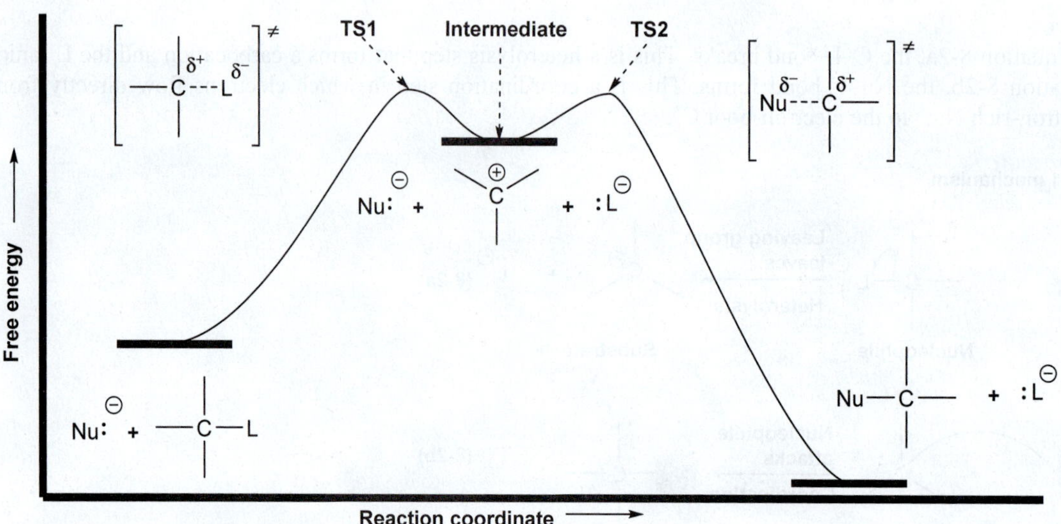

Your Turn 8.4

Think

Do transition states appear at energy maxima or minima? Do intermediates appear at energy minima or maxima? Should overall reactants and products appear in a free energy diagram at the endpoints of the curve or in between? How is the number of transition states and/or number of intermediates related to the number of steps in a mechanism?

Solve

For a mechanism that contains n number of steps, there must be n number of transition states and $n - 1$ number of stages representing intermediates. Transition states occur at energy maxima, and there are three transition states for this mechanism. Intermediates occur at energy minima, and there are two intermediate stages for this mechanism. Therefore, there are three elementary steps in this mechanism.

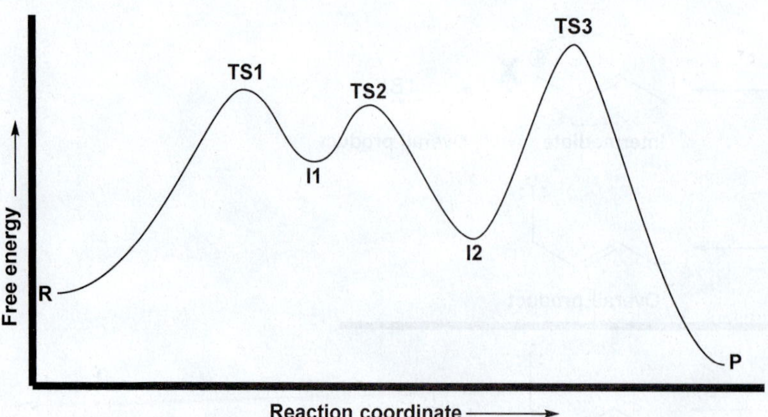

Your Turn 8.5

Think

Consult Chapters 6 and 7 to review the 10 elementary steps. What kinds of bonds are breaking or forming? Are any protons or π bonds involved?

Solve

In Equation 8-5a, the C–L bond is breaking. This is a heterolysis step that forms a carbocation and the L^- anion. In Equation 8-5b, the H–C bond is breaking and the C=C π bond is forming. This is an electrophile elimination step in which electrons flow from the electron-rich generic base $B:^-$ to the H atom of the electron-poor carbocation to form the B–H bond. Simultaneously, the H–C bond breaks, and the electrons fold over to form a C=C π bond.

E1 mechanism

Leaving group leaves / Heterolysis (8-5a)

Overall reactant — Intermediate cancels out — Overall product

Base attacks / Electrophile elimination (8-5b)

Overall reactant — Intermediate cancels out — Overall product — Overall product

Your Turn 8.6

Think

Sum up the steps in the mechanism first. Which species appears as a product in one step and as a reactant in another and, therefore, gets canceled? Which species appear in the overall reaction? Which species appear in the mechanism but not in the overall reaction?

Solve

The carbocation C^+ appears as a product in the first step and as a reactant in the second, so it is canceled when the two steps are summed to yield the overall reaction. The carbocation appears only in the reaction mechanism and not in the overall reaction; therefore, it is an intermediate. See the solution of Your Turn 8.5 for the labeled figure.

Your Turn 8.7

Think

Do transition states appear at energy maxima or minima? How many transition states appear for the two-step reaction shown? Do intermediates appear at energy minima or maxima? How many intermediate stages appear for the two-step reaction shown?

Solve

Transition states appear at energy maxima and show the bond breaking and bond making in progress. Intermediates appear at energy minima. There are two transition states (drawn on the next page) and one intermediate for a two-step mechanism.

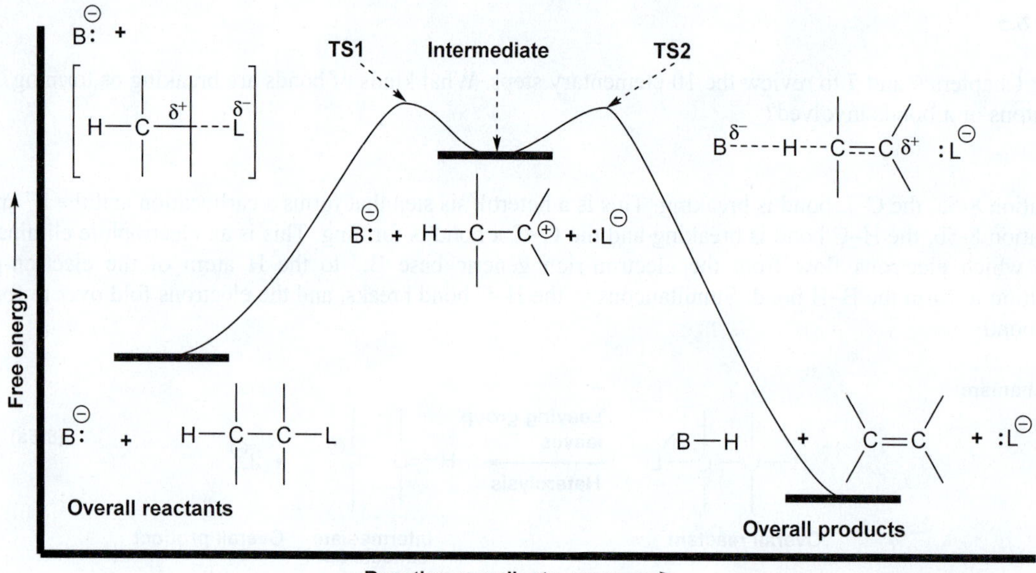

Your Turn 8.8

Think

What is the empirical rate law for an S_N2 reaction? Refer to Equation 8-7. What is the role of $NaSCH_3$ in the reaction? Does this species appear in the empirical rate law? If so, how does doubling the concentration affect the overall rate of the reaction?

Solve

The empirical rate law for an S_N2 reaction is Rate (S_N2) = $k_{SN2}[Nu^-][R–L]$. In the S_N2 reaction of $BrCH_2CH_2CH_2CH_3$ with $NaSCH_3$, $BrCH_2CH_2CH_2CH_3$ is the substrate [R–L] and $NaSCH_3$ is the nucleophile $[Nu^-]$. Therefore, the rate law for this S_N2 reaction is expressed as Rate (S_N2) = $k_{SN2}[NaSCH_3][BrCH_2CH_2CH_2CH_3]$. Doubling the concentration of $NaSCH_3$ doubles the rate of the reaction. The rate of an S_N2 reaction is directly proportional to the concentration of the nucleophile.

Your Turn 8.9

Think

Consult Figure 8-5. Which line corresponds to 100 °C? At a 15 kJ/mol energy barrier (on the *x* axis), what is the corresponding value on the *y* axis (molecules able to surmount the energy barrier, %)? What is the *y*-axis value when the *x*-axis value is 25 kJ/mol? Do you think that the larger energy barrier will lead to a larger or smaller percentage of molecules that can surmount the energy barrier?

Solve

The higher temperature corresponds to the top (red) curve. At a 15 kJ/mol energy barrier, the percentage of molecules that can surmount the energy barrier is 2%. At a 25 kJ/mol energy barrier, the percentage of molecules that can surmount the energy barrier is ~0.1%. When the energy barrier increases by 10 kJ/mol, the percentage of molecules that are able to surmount the energy barrier decreases by a factor of ~20.

Your Turn 8.10

Think

Consult Figure 8-5. Which line corresponds to 100 °C? Which corresponds to 0 °C? At a 20 kJ/mol energy barrier (on the *x* axis), what is the corresponding value on the *y* axis for each curve? Do you think that increasing the temperature from 0 °C to 100 °C will lead to a larger or smaller percentage of molecules that can surmount the energy barrier?

Solve

The bottom (blue) curve corresponds to 0 °C. At a 20 kJ/mol energy barrier (*x* axis), <0.1% of the molecules are able to surmount the energy barrier. The top (red) curve corresponds to 100 °C. At a 20 kJ/mol energy barrier, ~0.4% of the molecules are able to surmount the energy barrier. Increasing the temperature leads to a larger percentage of the molecules with enough kinetic energy to surmount the energy barrier (~0.4% vs. <0.1%).

Your Turn 8.11

Think

Refer back to Chapter 5 to recall what makes a carbon atom a chiral center. In the first step of a unimolecular substitution (S_N1) reaction, what happens to the number of groups bonded to the carbon with the leaving group? What happens to the geometry of that carbon? When the nucleophile attacks the C^+ from in front and behind, what happens to the number of groups bonded to that carbon? What happens to its geometry?

Solve

A chiral center is bonded to four different groups. The overall reactant has one chiral center and, therefore, is chiral. In the first step of an S_N1 reaction, the leaving group departs and leaves behind a carbocation C^+. The carbocation C^+ is no longer a chiral center and is trigonal planar, so the carbocation becomes achiral. The nucleophile I^-, therefore, can attack from either side of the C^+ plane. The C^+, therefore, becomes a chiral center again and leads to formation of both the *R* and *S* enantiomers in this example. Both products are chiral.

S_N1

I^- can attack from in front of or from behind the plane of C^+.

(S) Chiral → $:\ddot{Cl}:^-$ + Achiral (Planar sp^2-hybridized C)

Attack behind → (R) Chiral (8-18a)

Attack in front → (S) Chiral (8-18b)

Your Turn 8.12

Think

From how many sides of the C^+ plane can the I^- approach the C^+? Is a new chiral center formed from the new C–I bond? Is the intermediate C^+ chiral or achiral? How does the chiral nature of the intermediate affect the ratio of stereoisomers produced?

Solve

Coordination with I^- and carbocation **A** produces an *equal* mixture of stereoisomers; because the carbocation intermediate is achiral and is assumed to be free from the leaving group, there is an equal likelihood of nucleophilic attack from the two sides of the C atom.

Can attack in front or behind

A → (*S*)-3-Iodo-3-methylhexane + (*R*)-3-Iodo-3-methylhexane

An equal mixture of enantiomers would be produced, assuming the carbocation is free from the leaving group

Coordination with I⁻ and carbocation **B** produces an *unequal* mixture of stereoisomers; because the carbocation intermediate is chiral, there is no plane of symmetry. The stereochemical configuration of the wedge C–CH_3 remains unchanged. I⁻ can approach from either side of the C atom's plane, and both the *R* and *S* configurations can be produced. The result is a mixture of two diastereomers, and they are produced in *unequal* amounts, because the carbocation intermediate is chiral.

Can attack in front or behind

B

(1*S*,2*R*)-1-Iodo-1,2-dimethylcyclopentane

+

(1*R*,2*R*)-1-Iodo-1,2-dimethylcyclopentane

Unequal mixture of diastereomers

Coordination with I⁻ and carbocation **C** produces a single product, because the C^+ intermediate and the product of the reaction with I⁻ are achiral.

C

Iodocyclopentane

Single achiral product

Your Turn 8.13

Think

In how many elementary steps does an S_N2 reaction occur? Is an S_N2 reaction stereospecific? From how many sides can the nucleophile attack the substrate? From which side does the nucleophile attack occur?

Solve

The S_N2 reaction takes place in one single elementary step; therefore, the reaction is stereospecific when a new chiral center is produced. The nucleophile attacks from the side opposite the leaving group. This is called *backside attack*. In this example, the C–Br bond is a wedge, and the Br^- leaving group departs from the front. The HO^- nucleophile attacks from behind the plane of the paper, and therefore, the new C–OH bond is the dashed product **B**.

Backside attack

S_N2

B

Your Turn 8.14

Think

Use a model kit to construct the alkyl halide substrate in Equation 8-24. Rotate the model in order to have the bond connecting the two tetrahedral carbon atoms perpendicular (similar to a Newman projection orientation). Rotate about this C–C bond in order to have the H and Br atoms that are eliminated anti to each other.

Solve

The alkylhalide in Equation 8-24 is shown (next page) on the left side of the figure. The molecule is rotated 90° clockwise in order to have the C–C bond of the bond connecting the two chiral centers perpendicular to the plane of the paper. The eliminated H and Br are anti to each other (dash boxes). This is the correct conformation necessary to undergo E2. Notice that the two Ph groups (shaded boxes) are anti to each other. This supports the product shown in Equation 8-24, where the Ph groups are trans to each other.

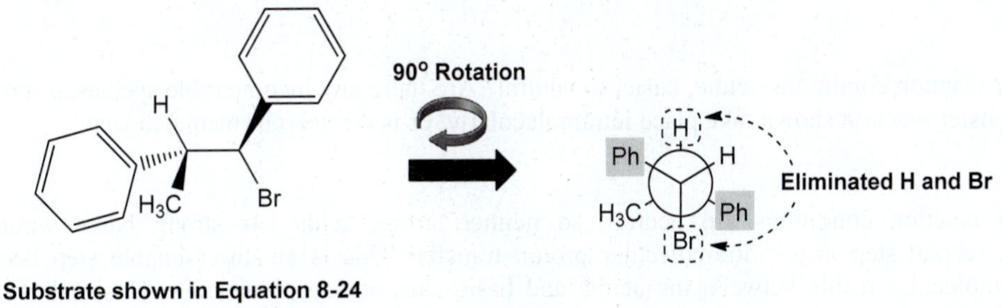

Your Turn 8.15

Think

The reaction takes place under basic conditions. What kinds of species are not compatible with basic conditions: weak acids, strong acids, weak bases, or strong bases? How do you determine whether an acid is strong or weak? How do you determine whether a base is strong or weak?

Solve

If this reaction occurs as shown, the first step yields R_2OH^+, which is an acid that is about as strong as H_3O^+, so it is a strong acid. This strong acid is incompatible in the basic HO^- conditions. Strong acids are incompatible with basic conditions because proton transfer reactions are fast and the acid–base reaction will take place before the organic reaction. This leads to an unreasonable mechanism.

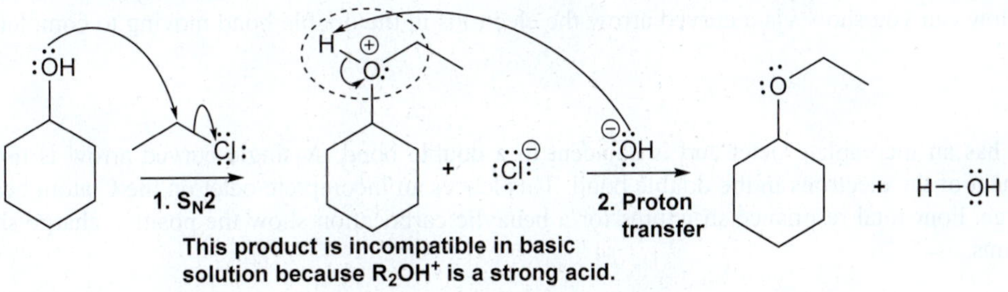

Your Turn 8.16

Think

The reaction takes place under acidic conditions. What kinds of species are not compatible with acidic conditions: weak acids, strong acids, weak bases, or strong bases? How do you determine whether an acid is strong or weak? How do you determine whether a base is strong or weak?

Solve

If this reaction occurs as shown, the first step yields RO^-, which is a base that is slightly stronger than HO^-, so it is a strong base. This strong base is incompatible with the acidic H_3O^+ conditions. Strong bases are incompatible with acidic conditions because proton transfer reactions are fast and the acid–base reaction will take place before the other elementary steps. This leads to an unreasonable mechanism.

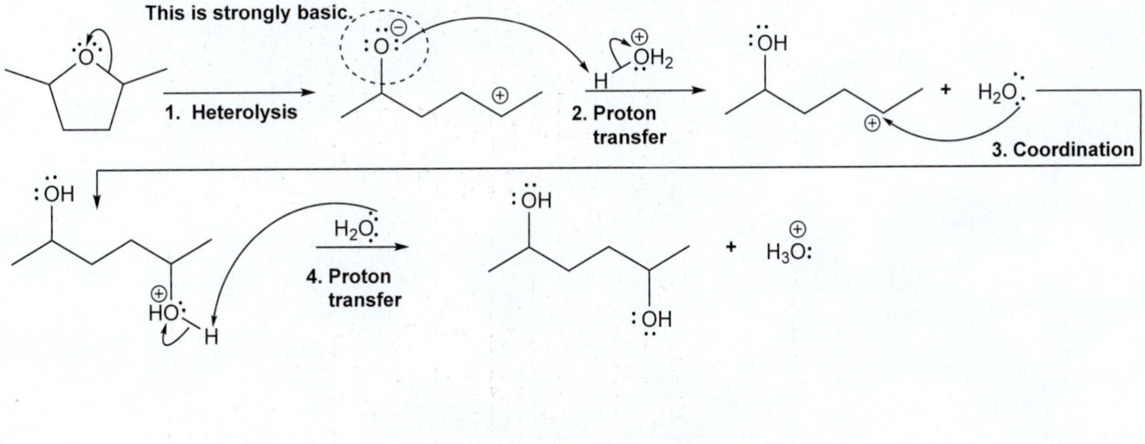

Your Turn 8.17

Think

Are the reagents or reaction conditions acidic, basic, or neutral? Are there any incompatible species as a result? Does the proton transfer reaction shown take place intramolecularly, or is it a solvent-mediated step?

Solve

The reagents and reaction conditions are neutral, so neither strong acids nor strong bases would be incompatible. The second step is an intramolecular proton transfer. This is an unreasonable step because, typically, solvent molecules reside between the acidic and basic sites at any given time. This makes *direct* transfer of protons within a molecule difficult and not likely to occur.

This is an intramolecular proton transfer and is unreasonable.

S_N2

Proton transfer

Your Turn 8.18

Think

Do any of the atoms have an incomplete octet? If so, is the atom with the incomplete octet adjacent to a double or triple bond? How can you show via a curved arrow the electrons in the double bond moving to complete the incomplete octet?

Solve

The carbocation has an incomplete octet and is adjacent to a double bond. A single curved arrow is used to show the movement of the electrons in the double bond. This leaves an incomplete octet on the C atom bearing the positive charge. Four total resonance structures for a benzylic carbocation show the positive charge shared over multiple atoms.

Benzylic C^+

Your Turn 8.19

Think

For Equation 8-40, is a primary or tertiary carbocation more stable? For Equation 8-41, is a tertiary carbocation or tertiary benzylic carbocation more stable? How is stability reflected in a free energy diagram?

Solve

Equation 8-40: A tertiary carbocation is more stable than a primary carbocation, so the product carbocation is higher in energy.

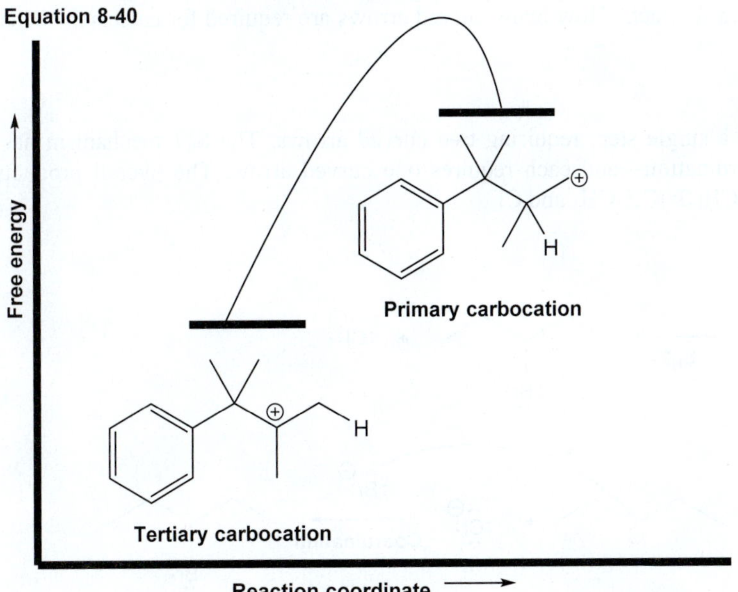

Equation 8-41: A regular tertiary carbocation is less stable than a tertiary benzylic carbocation, so the product carbocation is lower in energy.

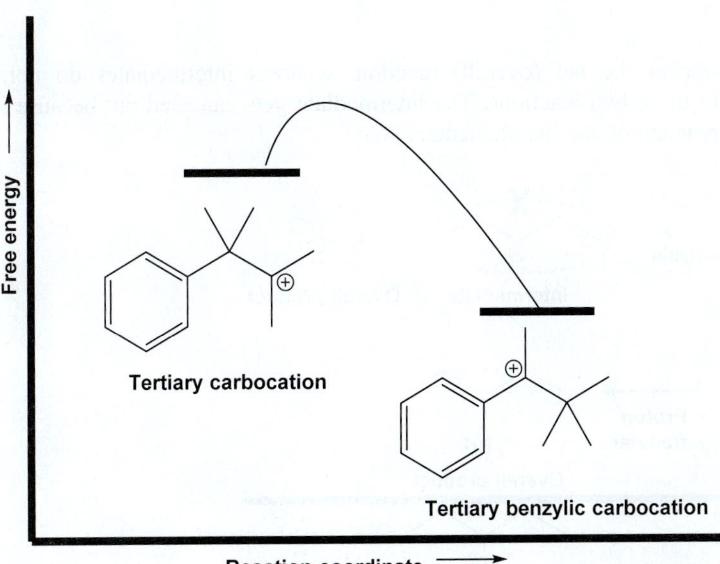

In Chapter Problems

Problem 8.1

Think

How many steps take place in a bimolecular nucleophilic substitution (S_N2) reaction? In an S_N1? Which bonds break and form in the overall reaction for each? How many curved arrows are required for each step? What are the products of the reaction?

Solve

The S_N2 mechanism takes place in a single step, requiring two curved arrows. The S_N1 mechanism has two steps—heterolysis followed by coordination—and each requires one curved arrow. The overall products for each reaction are the same: $CH_3CH_2CH(Br)CH_2CH_3$ and Cl^-.

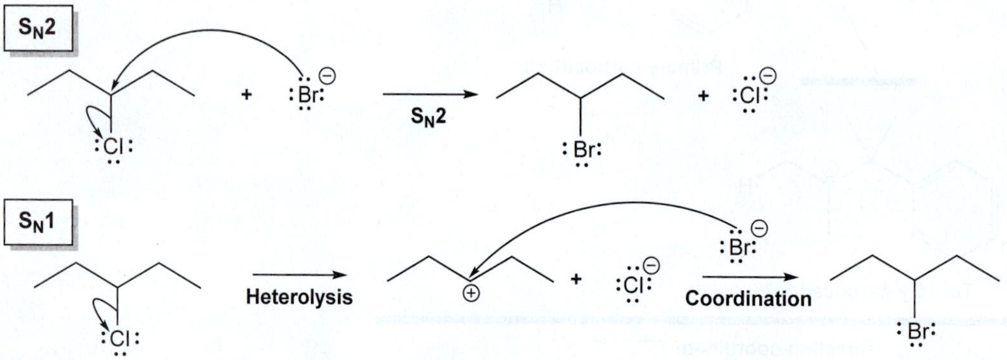

S_N2

:Cl: + :Br: → (S_N2) :Br: + :Cl:

S_N1

:Cl: → Heterolysis → ⊕ + :Cl: → :Br: Coordination → :Br:

Problem 8.2

Think

What is the overall reaction? Which species appears in one of the reaction steps but does not appear in the overall reaction? How can you tell if a species gets canceled out and does not appear in the overall reaction?

Solve

Overall reactants and products appear in the net (overall) reaction, whereas intermediates do not. The carbocation is the only intermediate in these two reactions. The intermediate gets canceled out because it is a product of one elementary step and a reactant of another elementary step.

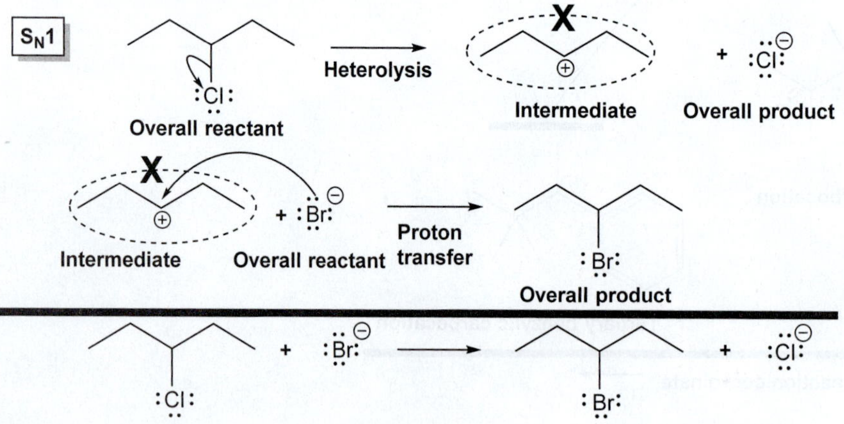

S_N1

:Cl: Overall reactant → Heterolysis → X Intermediate + :Cl: Overall product

X Intermediate + :Br: Overall reactant → Proton transfer → :Br: Overall product

:Cl: + :Br: → :Br: + :Cl:

Problem 8.3

Think

In how many steps does the S_N1 reaction in Your Turn 8.2 occur? How many intermediates are there? How many transition states? What is the intermediate? Do intermediates occur at a local minimum or maximum? Is the energy of the intermediate higher or lower than the energy of the reactants/products? Where does a transition state lie on a free energy diagram?

Solve

The S_N1 reaction occurs in two steps; it has two transition states (local maxima) and one intermediate (local minimum), and the intermediate is higher in energy than the reactants or products. The energy diagram is shown below.

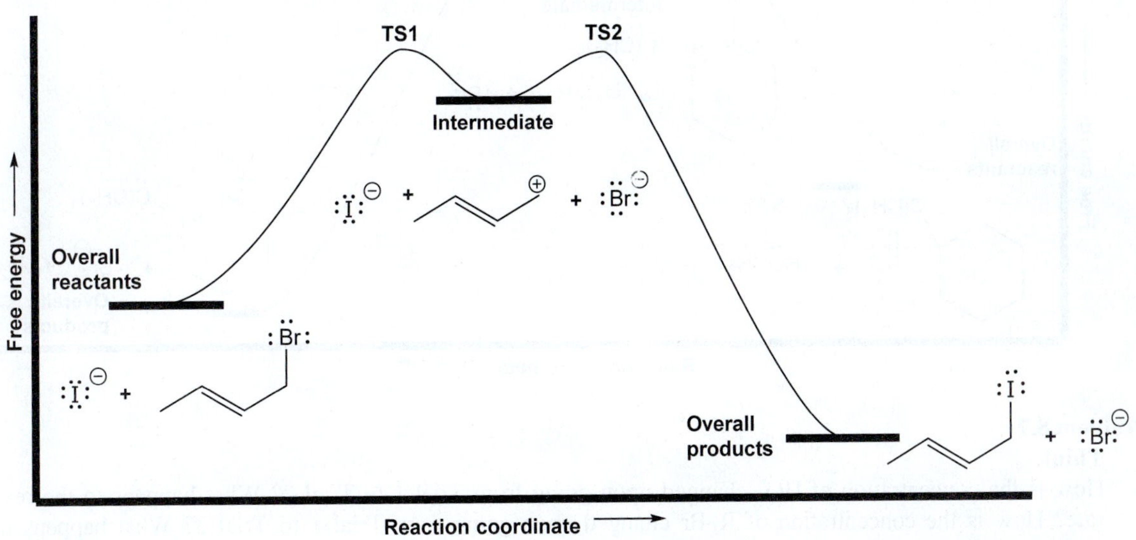

Problem 8.4

Think

In how many steps does the E1 reaction occur? What steps make up an E1 mechanism? What is the leaving group? What species appears in the mechanism but not in the overall reaction? What acts as the base?

Solve

The E1 reaction occurs in two steps:

1. The Br leaving group leaves and forms a tertiary carbocation intermediate on the ring.

2. The $HOCH_3$ acts as a base and deprotonates the adjacent hydrogen, forming a C=C bond on the ring:

$C(CH_3)_3$ Br: — Heterolysis → $C(CH_3)_3$ + :Br:⊖ — $HOCH_3$ Overall reactant, Electrophile elimination → $C(CH_3)_3$ + $H_2OCH_3^+$

Overall reactant | Intermediate | Overall product | Overall product | Overall product

Problem 8.5

Think

In how many steps does the E1 reaction in Problem 8.4 occur? How many intermediate stages should there be? How many transition states? What is the intermediate? Do intermediates occur at a local minimum or maximum? Is the energy of the intermediate higher or lower than the energy of the reactants/products? Where does the transition state lie on the free energy diagram? What functional group results in the product?

Solve

The E1 reaction occurs in two steps; it has two transition states (local maxima) and one intermediate (local minimum), and the intermediate is higher in energy than the reactants or products. An alkene is the product that results. See the energy diagram on the next page.

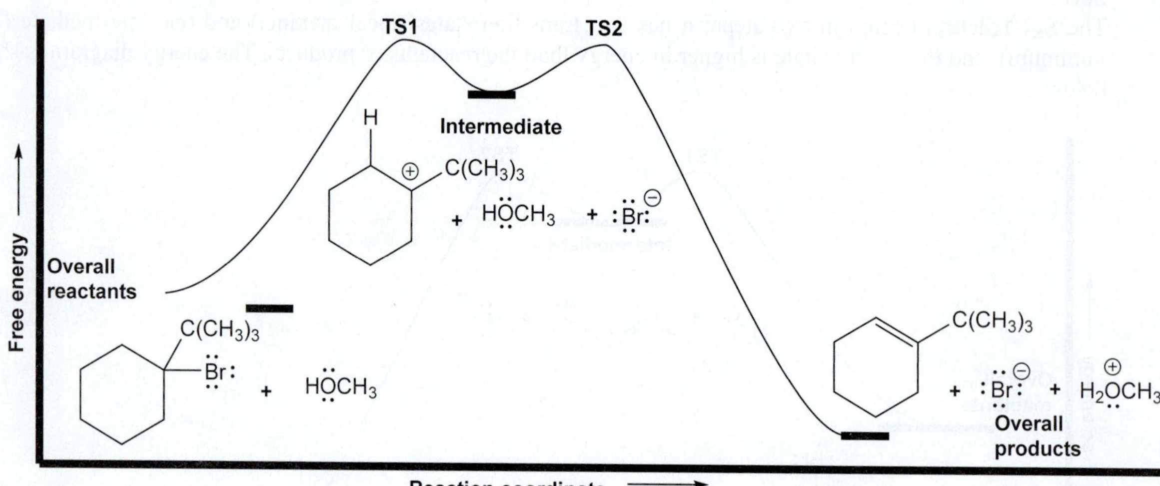

Problem 8.7

Think

How is the concentration of HO^- changed upon going from Trial 1 to Trial 2? What happens to the reaction rate? How is the concentration of R–Br changed upon going from Trial 1 to Trial 3? What happens to the reaction rate? What is the corresponding empirical rate law? Is it consistent with the rate law for an S_N1 or S_N2 reaction?

Solve

Going from Trial 1 to Trial 2, the concentration of the HO^- halves and the concentration of R–Br remains the same. This causes no change in the reaction rate, suggesting that the rate is *not* dependent on HO^-. Going from Trial 1 to Trial 3, the concentration of HO^- remains constant and the concentration of R–Br doubles. This causes the rate to double—from 5.5×10^{-7} M/s to 1.0×10^{-6} M/s. Thus, the rate is directly proportional to the concentration of R–Br. These results indicate an empirical rate law of the form Rate = k[R–Br], which is consistent with an S_N1 reaction.

Trial number	[R—Br]	[HO^-]	Rate (M/s)
1	0.10 M	0.10 M	5.5×10^{-7}
2	0.10 M	0.05 M	5.5×10^{-7}
3	0.20 M	0.10 M	1.0×10^{-6}

Constant
Doubled
Halved
Constant
No change in reaction rate
Reaction rate doubles

Problem 8.9

Think

What would the rate-determining step be under this assumption? What is the theoretical rate law for this step?

Solve

If the second step of an E1 reaction were much slower than the first step, then the second step would be the rate-determining step. In that step, the base and the carbocation are reactants, so the rate of that step would be Rate = $k[B^-][R^+]$, as shown in Equation 8-15b. Because this is the rate law for the rate-determining step, the rate law would describe the overall reaction as well and would suggest that the rate of the overall reaction should be directly proportional to the concentration of the base. This disagrees with the actual rate dependence.

Problem 8.11

Think

What can act as the leaving group? What can act as the nucleophile? What are the steps that compose an S_N1 mechanism? Does the reaction take place at a chiral center?

Solve

The Br substituent can act as a leaving group, coming off as Br^-, which is a relatively stable anion. The HS^- ion can act as a nucleophile. The S_N1 mechanism takes place in two steps. First, the leaving group leaves via heterolysis, yielding a planar carbocation, then HS^- attacks C^+ via a coordination step, yielding the overall products.

Notice that there are two chiral centers in the alkyl halide reactant—one is affected during the reaction and the other one is left alone. The stereochemical configuration at C5 is retained. The stereochemical configuration at C2, however, which is where the reaction takes place, is scrambled. In the carbocation intermediate, the C^+ that is attacked by the nucleophile is planar, so both *R* and *S* configurations of the new chiral center are produced. The product is a mixture of diastereomers, and they are produced in *unequal* amounts because the carbocation intermediate is chiral.

Chiral center
Chiral center
Planar C^+
Chiral intermediate
Can attack in front or behind
Chiral center
Chiral center
Chiral center
Chiral center
Diastereomers of each other

Problem 8.13

Think

What can act as the leaving group? What can act as the nucleophile? How many steps make up an S_N2 mechanism? How does the nucleophile approach the substrate during attack?

Solve

In each S_N2 reaction, the nucleophile attacks from the side opposite the leaving group.

(a) S_N2

(b) S_N2

(c)

Problem 8.15

Think

What can act as the leaving group? What can act as the base? What proton can be removed in the second step of an E1 mechanism? Do *E* and *Z* configurations exist for this product? If so, can both be formed?

Solve

In both reactions, Cl is the leaving group and H_2O acts as the base to remove the proton adjacent to the C^+ that forms in Step 1. The products are the alkene, H_3O^+, and Cl^-.

(a) Notice that both *E* and *Z* isomers exist for the product of this first reaction. Because this is an E1 mechanism and the single bonds to C^+ can rotate, both stereoisomers are produced.

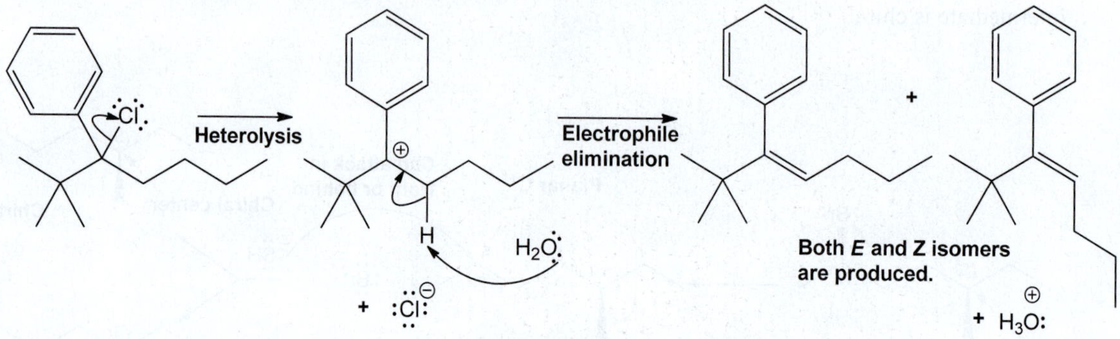

Heterolysis

Electrophile elimination

Both *E* and *Z* isomers are produced.

(b) In the second reaction, no *E*/*Z* isomers exist about the double bond, so stereochemistry is not a concern for this reaction.

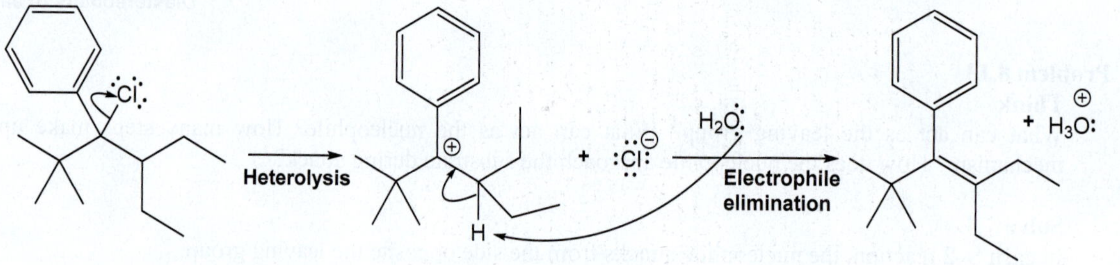

Heterolysis

Electrophile elimination

Problem 8.16

Think

Which group on each side of the C=C is larger? Which isomer has less steric strain? How does steric strain relate to stability? How does the stability of the product relate to the major product?

Solve

In Problem 8.15, only **(a)** has a mixture of diastereomers possible. On the left end of the C=C bond, the *t*-butyl group is bulkier than the Ph group; on the right end of the C=C bond, the propyl group is bulkier than the H atom. The more stable isomer has the larger groups on opposite sides of the C=C bond. Thus, in this example, the *Z* isomer is more stable. The more stable isomer is the major product.

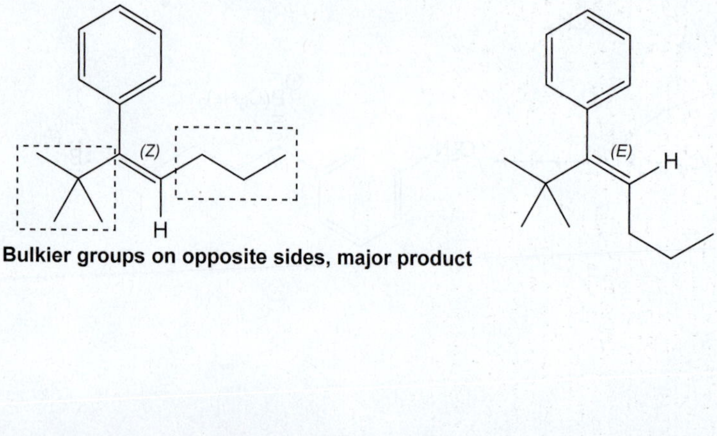

(*Z*) (*E*)

Bulkier groups on opposite sides, major product

Problem 8.17

Think

What is the diastereomer of the alkene product in Equation 8-24? What is the preferred conformation of the substrate for E2? Which groups have to be in this conformation? Which group is the leaving group, and which H atom is adjacent to the leaving group?

Solve

E2 occurs in one step, and the Br leaving group and adjacent H atom have to be anticoplanar (shown below). In comparison to the reaction in Equation 8-24, the other diastereomer (both phenyl groups on the same side of the C=C bond) can be produced from the substrate in which the CH_3 and C_6H_5 groups on the left-hand carbon have been interchanged. Thus, these two groups are interchanged in the product as well.

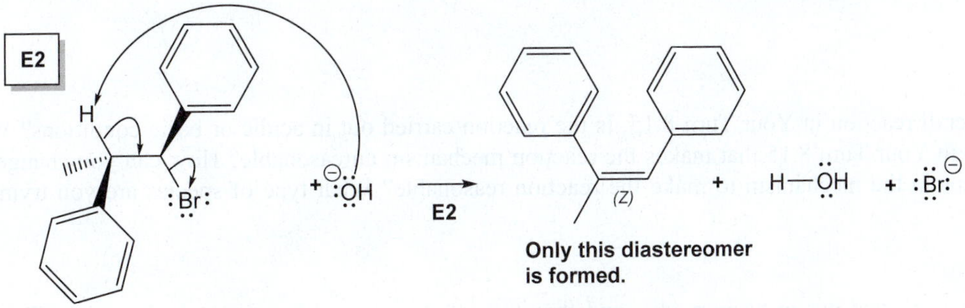

E2

Only this diastereomer is formed.

Problem 8.19

Think

What conformations of the substrate facilitate an E2 reaction? In those conformations, what dictates which atoms/groups are on the same side of the double bond in the products?

Solve

An E2 reaction is facilitated by an anticoplanar conformation involving H (or D), Br (the leaving group), and the two C atoms to which they are bonded. There are two such conformations, as shown on the left in the reaction below: One involves H, and the other involves D. In **(a)**, the H and Br atoms are anticoplanar, so they can be eliminated in an E2 step. In that conformation, the CH_3 groups point in opposite directions, so they end up on opposite sides of the double bond in the product. In **(b)**, the D and Br atoms are anticoplanar, so they will be eliminated in an E2 step. In that conformation, the CH_3 groups both point toward you, so they end up on the same side of the double bond in the product.

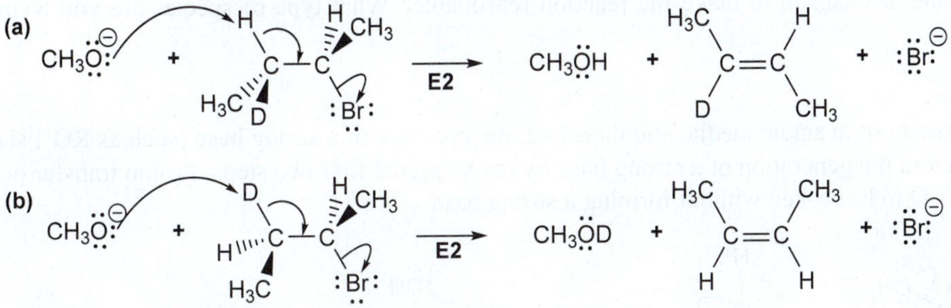

Problem 8.20

Think

On which C atom is the leaving group located? On the two neighboring C atoms, are there H atoms present in the correct orientation to be eliminated? What is the correct orientation of the leaving group and the eliminated H for E2?

Solve

Molecule **A** should undergo E2 elimination more readily because the Cl substituent in molecule **B** is not anti to any H atoms on an adjacent carbon. The only H atom on an adjacent atom is on the *same* side of the ring. By contrast, in molecule **A**, the H and Cl atoms indicated are on opposite sides of the ring and, therefore, can attain the anticoplanar arrangement.

H_3C CH_3 Cl Anti coplanar orientation H A CH_3 Faster E2

H_3C CH_3 Cl Cannot be anti to each other CH_3 B H

Problem 8.21

Think

Refer to the overall reaction in Your Turn 8.15. Is the reaction carried out in acidic or basic conditions? What species appears in Your Turn 8.15 that makes the reaction mechanism unreasonable? How can you change the order of the steps in the mechanism to make the reaction reasonable? What type of species are you trying to avoid forming?

Solve

The mechanism is carried out in basic media, and therefore, the presence of a strong acid (such as R_2OH^+) should be avoided. We can accomplish the same net reaction, but avoid the generation of a strong acid, by first deprotonating cyclohexanol. Then, in the second step, the S_N2 reaction can take place.

:OH Proton transfer S_N2 Cl

Problem 8.22

Think

Refer to the overall reaction in Your Turn 8.16. Is the reaction carried out in acidic or basic conditions? What species appears in Your Turn 8.16 that makes the reaction mechanism unreasonable? How can you change the order of the steps in the mechanism to make the reaction reasonable? What type of species are you trying to avoid forming?

Solve

The mechanism is carried out in acidic media, and therefore, the presence of a strong base (such as RO^-) should be avoided. We can avoid the generation of a strong base by reversing the first two steps. Proton transfer occurs first; this allows the C–O to be broken without forming a strong base.

OH_2 H Proton transfer H Heterolysis :OH + H_2O: Coordination

:OH H_2O: HO H Proton transfer :OH :OH + H_3O:

Problem 8.24

Think

Refer to the overall reaction in Solved Problem 8.23. How many steps generally make up an E2 mechanism? What incompatible species would appear in such a mechanism for this reaction? How can you incorporate a proton transfer reaction in a reasonable way to avoid the formation of such a species?

Solve

An E2 mechanism generally takes place in a single step. In this case, if the E2 step were to take place first, CH_3O^- would be produced, which is strongly basic.

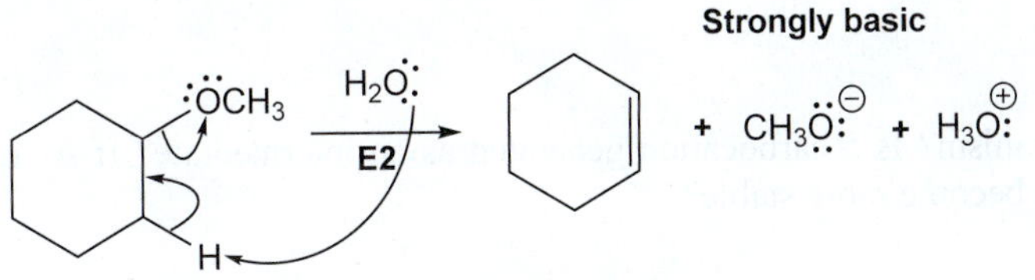

The reaction, however, takes place under acidic conditions, making CH_3O^- incompatible. A proton transfer step needs to occur first to avoid the generation of a strong base when the leaving group leaves. By protonating the oxygen in the ether first, you allow it to leave as an uncharged alcohol and avoid the formation of a basic alkoxide.

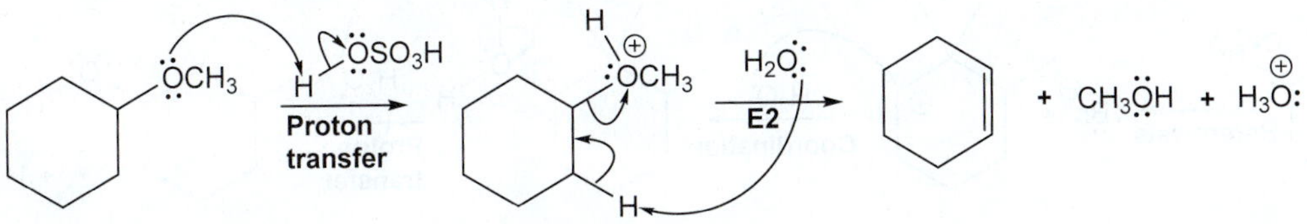

Problem 8.25

Think

Refer to the overall reaction in Your Turn 8.17. Is proton transfer occurring via intramolecular proton transfer or solvent-mediated proton transfer? Which type of proton transfer is reasonable? Why?

Solve

An intramolecular proton transfer is proposed and is unreasonable because the solvent molecules are abundantly present and exist in between the two sites involved in the proton transfer. Therefore, a solvent-mediated proton transfer is more reasonable, as shown below.

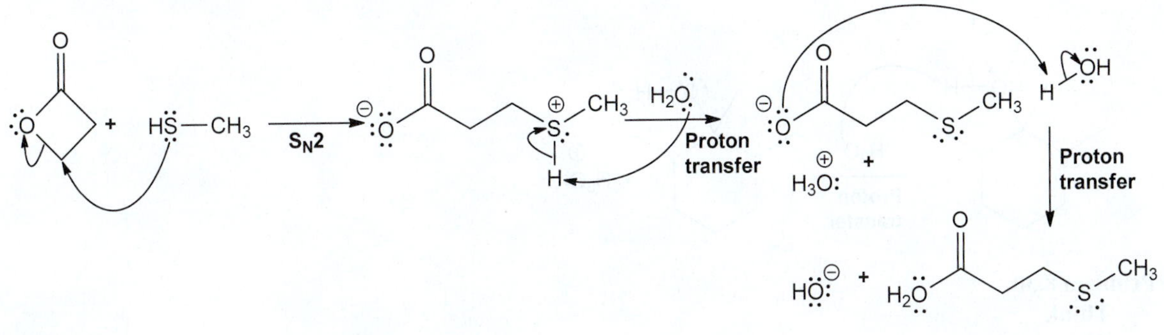

Problem 8.27

Think

What is the molecularity of the mechanism proposed? Is this likely to occur? Can the mechanism take place in more steps to avoid this problem?

Solve

The step is unreasonable because it is a termolecular step—that is, there are three reactants in this step. It is more reasonable to split it up into two steps, while avoiding generating a strongly basic species, given that the reaction takes place under acidic conditions. This requires the OH to be protonated in the first step.

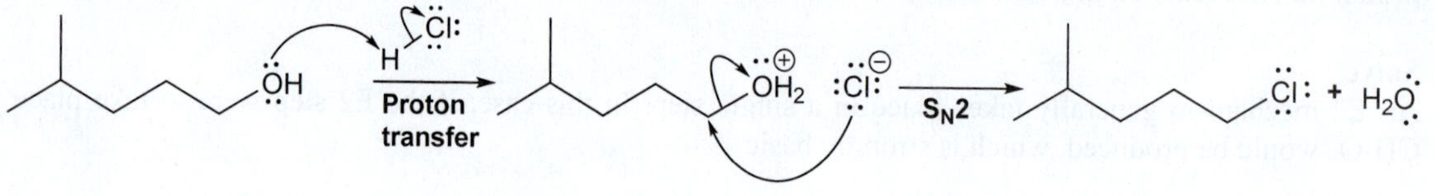

Proton transfer

S_N2

Problem 8.29

Think

What are the normal steps of an S_N1 mechanism? Is a carbocation generated as an intermediate? If so, can it undergo a 1,2-hydride or 1,2-methyl shift to become more stable?

Solve

The normal S_N1 reaction with an uncharged nucleophile takes place in three steps. The leaving group leaves in the first step, the nucleophile coordinates to the carbocation in the second step, and proton transfer takes place in the third step to form an uncharged organic product.

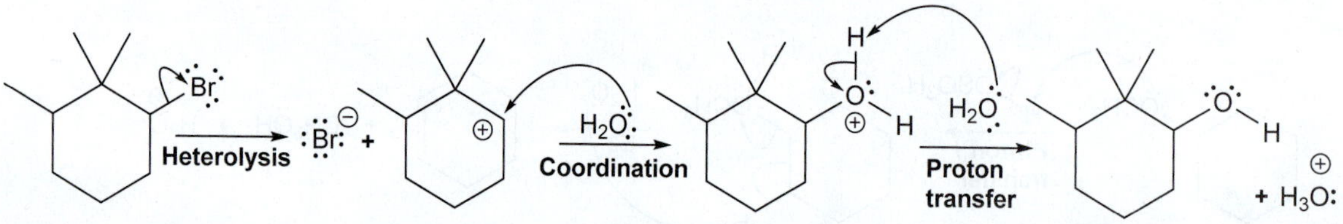

Heterolysis

Coordination

Proton transfer

Notice, however, that a secondary carbocation intermediate can rearrange to become more stable, via a 1,2-methyl shift, as shown below. After this rearrangement, coordination and proton transfer follow to form the tertiary alcohol product.

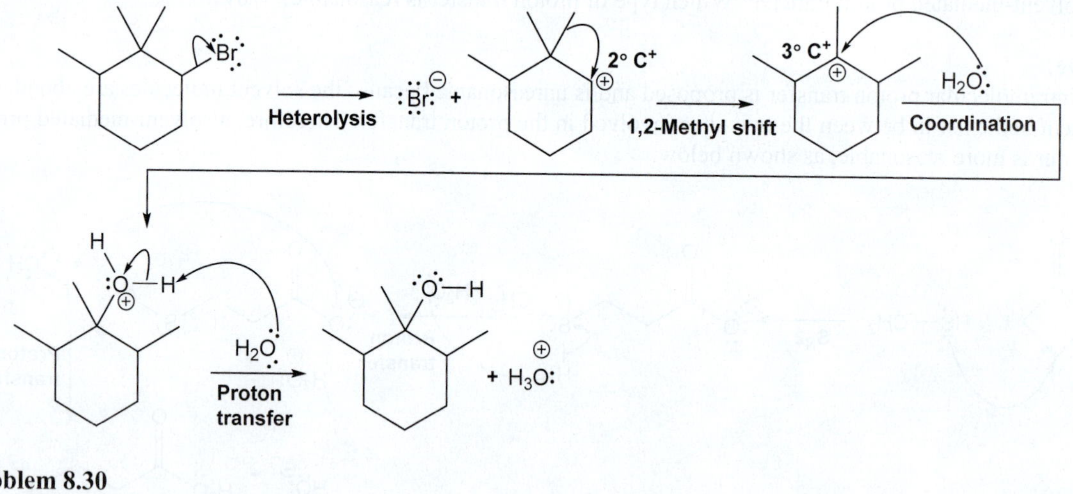

Heterolysis

2° C⁺

1,2-Methyl shift

3° C⁺

Coordination

Proton transfer

Problem 8.30

Think

What are the normal steps of an E1 mechanism? Is a carbocation generated as an intermediate? If so, what type of carbocation is formed? Are additional resonance structures possible? Which resonance structures can you use in a mechanism? Where are the double bonds in the starting material in relation to the double bonds in the product?

Solve

The normal E1 reaction with an uncharged nucleophile takes place in two steps. The leaving group leaves in the first step, and the weak base removes the adjacent proton in the electrophile elimination step. This proton on the adjacent carbon cannot be removed in this case, however, for two reasons—the weak base cannot remove a proton from an sp^2-hybridized carbon, and two adjacent double bonds would have to form in the ring.

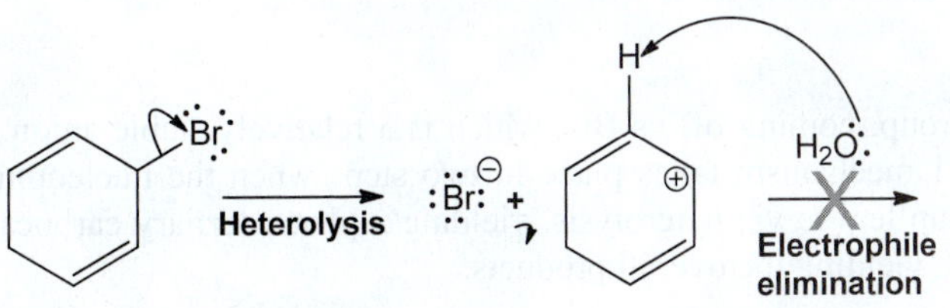

Heterolysis

Electrophile elimination

Notice, however, that the carbocation that is produced has two resonance structures. The second resonance structure has an sp^3-hybridized C–H bond adjacent to the C^+, which can be removed during an electrophile elimination step to produce the product (benzene) shown in this problem.

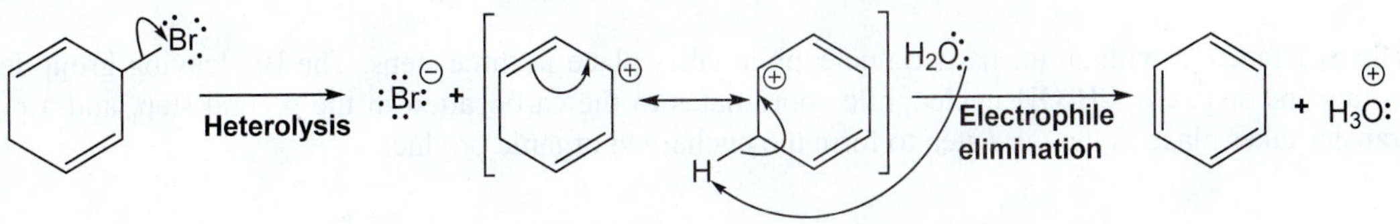

Heterolysis

Electrophile elimination

End of Chapter Problems

Sections 8.1 and 8.2 S_N1 and E1 Mechanisms; Free Energy Diagrams

Problem 8.31

Think

What can act as the leaving group? What can act as the nucleophile? What are the steps that compose an S_N1 mechanism? Is a carbocation generated as an intermediate?

Solve

(a) The Br substituent can act as a leaving group, coming off as Br^-, which is a relatively stable anion. The Cl^- ion can act as a nucleophile. The S_N1 mechanism takes place in two steps when the nucleophile is negatively charged. First, the leaving group leaves via heterolysis, yielding a planar tertiary carbocation, then Cl^- attacks C^+ via a coordination step, yielding the overall products.

Heterolysis Coordination

(b) The S_N1 reaction with an uncharged nucleophile takes place in three steps. The Br^- leaving group leaves in the first step, the CH_3OH nucleophile coordinates to the carbocation in the second step, and a proton transfer takes place in the third step to form the uncharged organic product.

Heterolysis CH_3OH Coordination CH_3OH Proton transfer + $CH_3OH_2^+$

Problem 8.32

Think

In how many steps do the S_N1 reactions in Problem 8.31 **(a)** and **(b)** occur? How many intermediate stages are there? How many transition states? What is the intermediate? Do intermediates occur at a local minimum or maximum? Is the energy of the intermediate higher or lower than the energy of the reactants/products? Where does a transition state lie on a free energy diagram?

Solve

(a) The S_N1 mechanism in Problem 8.31 **(a)** takes place in two steps when the nucleophile is negatively charged; therefore it has two transition states (local maxima) and one intermediate (local minimum), and the intermediate is higher in energy than the reactants or products. The energy diagram is shown below.

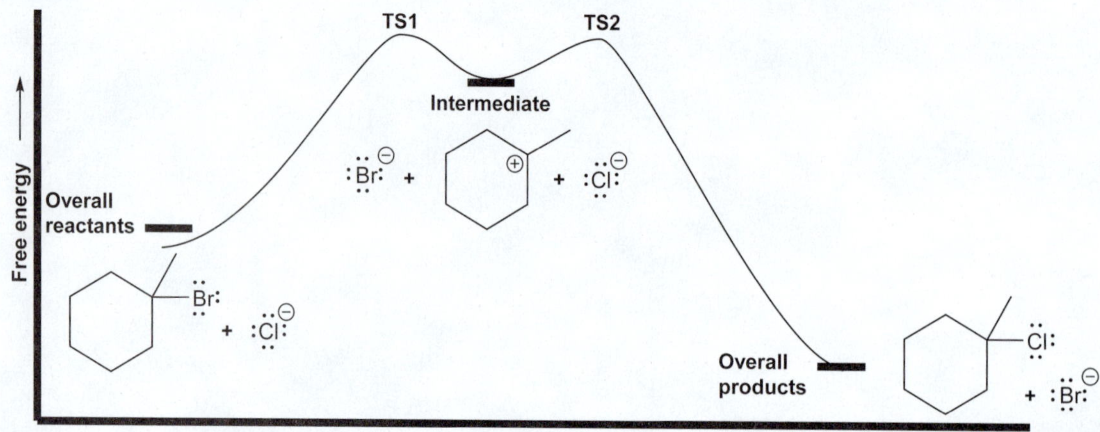

(b) The S_N1 mechanism in Problem 8.31 **(b)** takes place in three steps when the nucleophile is uncharged and therefore has three transition states (local maxima) and two intermediates (local minimum), and the intermediate is higher in energy than the reactants or products. The energy diagram is shown below.

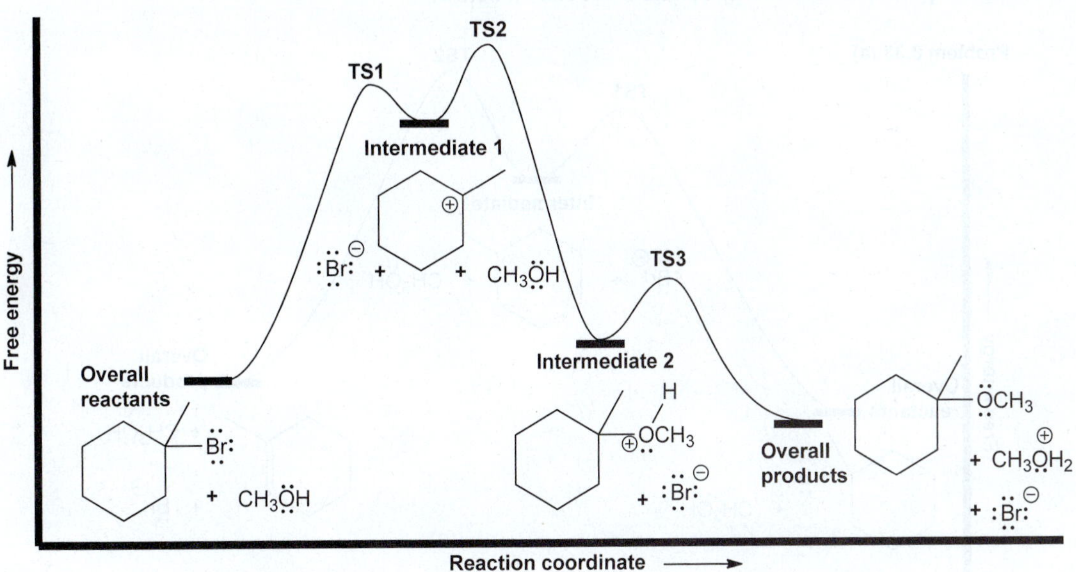

Problem 8.33

Think

In how many steps does the E1 reaction occur? What steps make up an E1 mechanism? What is the leaving group? What is the intermediate? What acts as the base?

Solve

(a) This E1 reaction occurs in two steps. First, the Br leaving group leaves in a heterolysis step and forms a tertiary carbocation intermediate on the ring. Second, the $HOCH_3$ acts as a base and deprotonates the adjacent hydrogen, forming a C=C bond on the ring in an electrophile elimination step.

Heterolysis; CH_3OH Electrophile elimination; $+ H_2OCH_3^{+}$

(b) This E1 reaction occurs in two steps. First, the I leaving group leaves in a heterolysis step and forms a secondary carbocation. Second, the H_2O acts as a base and deprotonates the adjacent hydrogen, forming a C=C bond on the ring in an electrophile elimination step.

Heterolysis; H_2O Electrophile elimination; $+ H_3O^{+}$

Problem 8.34

Think

In how many steps do the E1 reactions in Problem 8.33 **(a)** and **(b)** occur? How many intermediate stages are there? How many transition states? What is the intermediate? Do intermediates occur at a local minimum or maximum? Is the energy of the intermediate higher or lower than the energy of the reactants/products? Where does a transition state lie on a free energy diagram?

Solve

The E1 mechanisms in Problem 8.33 **(a)** and **(b)** take place in two steps; therefore, they have two transition states (local maxima) and one intermediate (local minimum), and the intermediate is higher in energy than the reactants or products. The energy diagrams are shown below.

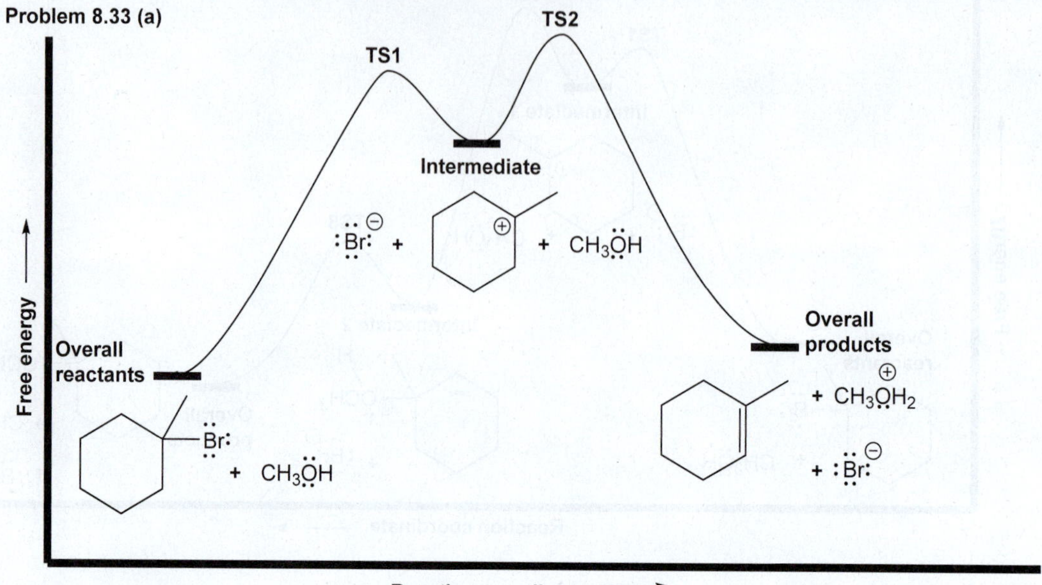

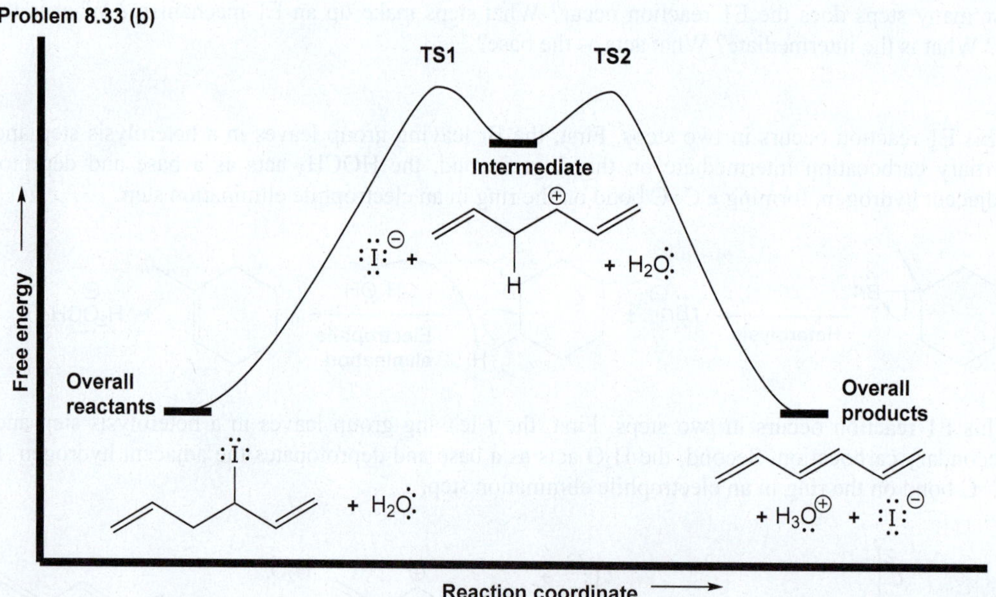

Problem 8.35 (SYN)

Think

On which carbon was the leaving group? Does the product have an OH or OCH_3 group? Which group requires the H_2O nucleophile, and which group requires the CH_3OH nucleophile?

Solve

(a) In the figure below, the carbon where there must have been a halide leaving group is labeled with a shaded box. The product has an OH group and therefore requires H_2O (water) as the nucleophile.

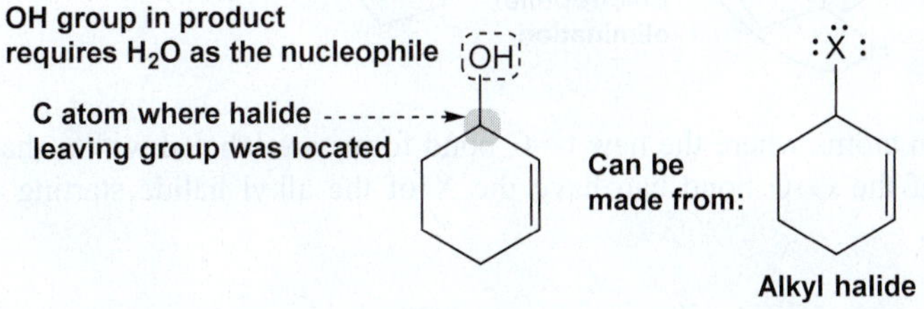

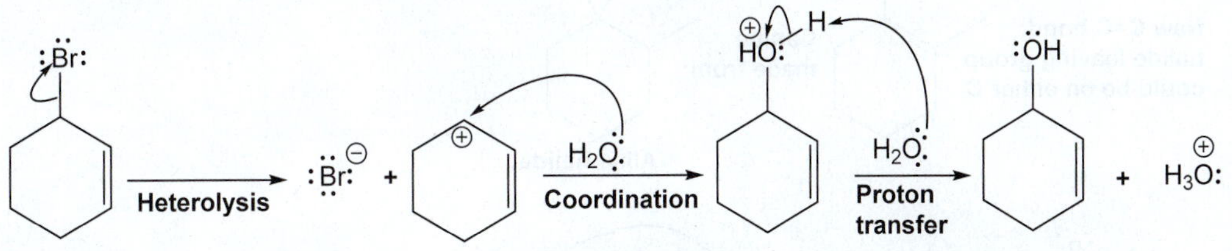

OH group in product requires H_2O as the nucleophile

C atom where halide leaving group was located

Can be made from:

+ H_2O

Alkyl halide

Heterolysis

Coordination

Proton transfer

(b) In the figure below, the carbon where there must have been a halide leaving group is labeled with a shaded box. The product has an OCH_3 group and therefore requires CH_3OH (methanol) as the nucleophile.

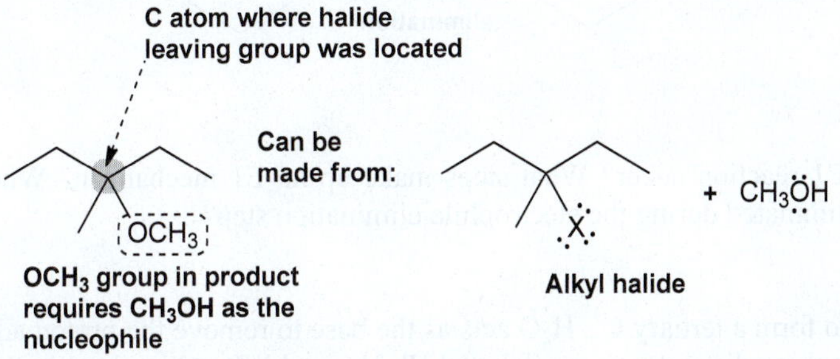

C atom where halide leaving group was located

Can be made from:

+ CH_3OH

OCH_3 group in product requires CH_3OH as the nucleophile

Alkyl halide

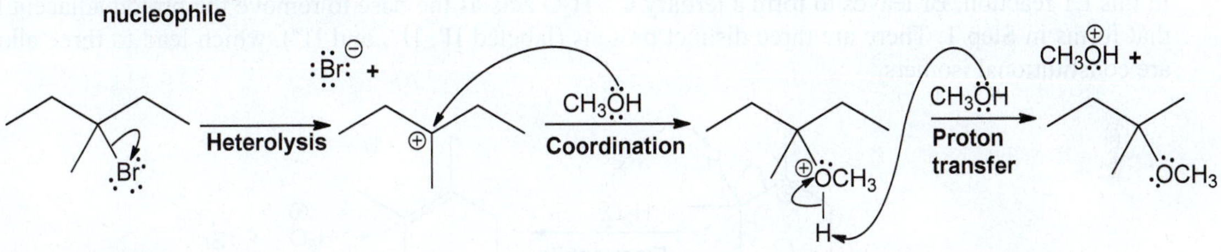

Heterolysis

Coordination

Proton transfer

Problem 8.36 (SYN)

Think

On which two carbons is the new C=C bond formed? To which carbon could the leaving group have initially been attached? In how many steps does the E1 reaction occur? What steps make up an E1 mechanism?

Solve

(a) In the figure below, the two carbon atoms where the new C=C bond forms are labeled with a shaded box. Either one of the carbon atoms of the C=C bond could have the X of the alkyl halide starting material. The better choice, however, is the one shown below, because that alkyl halide yields only the alkene shown in this problem. The mechanism is shown on the next page.

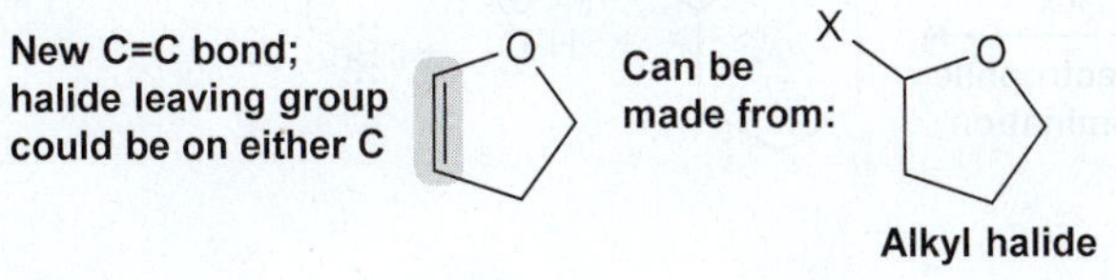

New C=C bond; halide leaving group could be on either C

Can be made from:

Alkyl halide

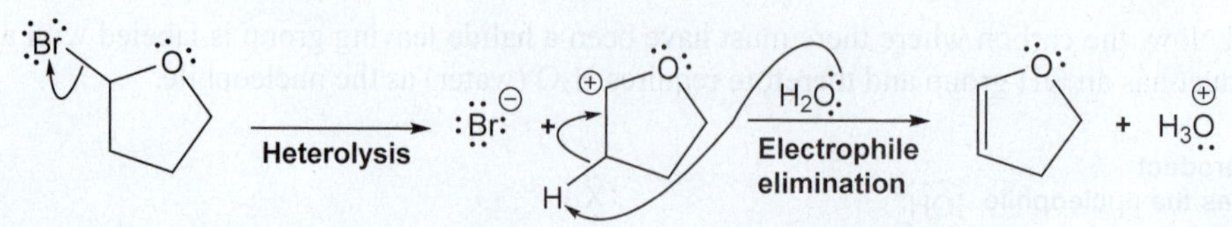

:Br: Heterolysis Electrophile elimination + H_3O^+

(b) In the figure below, the two carbon atoms where the new C=C bond forms are labeled with a shaded box. Either one of the carbon atoms of the C=C bond can have the X of the alkyl halide starting material, because the molecule is symmetric.

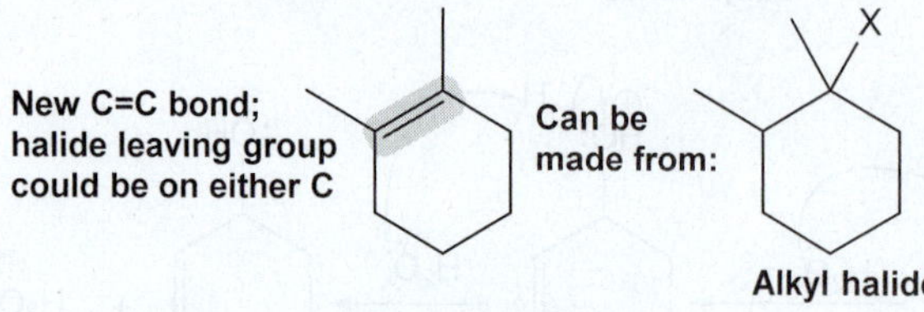

New C=C bond; halide leaving group could be on either C — **Can be made from:** — **Alkyl halide**

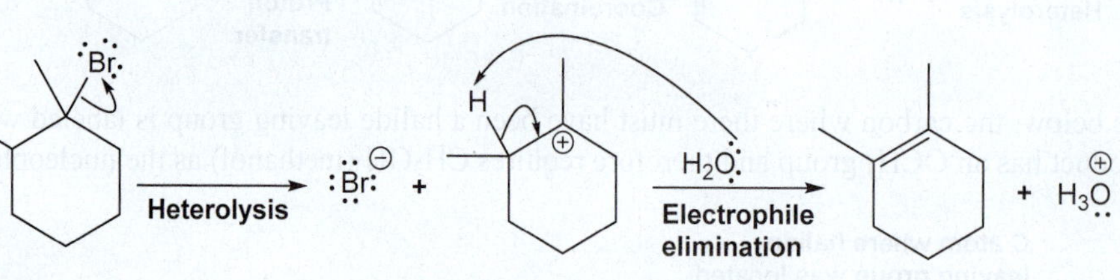

Heterolysis — **Electrophile elimination**

Problem 8.37

Think

In how many steps does the E1 reaction occur? What steps make up an E1 mechanism? Where does the C^+ form? Which protons can be eliminated during the electrophile elimination step?

Solve

In this E1 reaction, Br leaves to form a tertiary C^+. H_2O acts as the base to remove the proton adjacent to the C^+ that forms in Step 1. There are three distinct protons (labeled H′, H″, and H‴), which lead to three alkenes that are constitutional isomers.

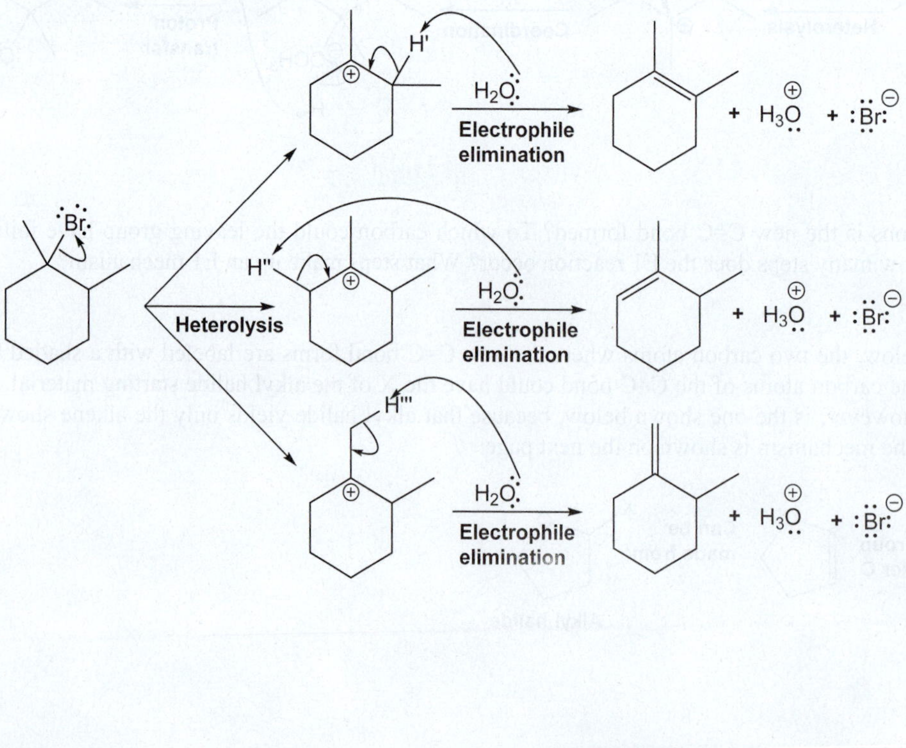

Heterolysis — **Electrophile elimination**

Section 8.4 The Kinetics of S_N2, S_N1, E2, and E1 Reactions

Problem 8.38

Think

Which mechanism, S_N1 or S_N2, does not depend on the nucleophile? If the mechanism does not depend on the nucleophile, would changing the concentration of the nucleophile change the rate?

Solve

(a) The mechanism is S_N1, because the reaction kinetics do not appear to depend on the nature of the nucleophile.

(b) The mechanisms are shown below. Notice that a proton transfer step is required after attack of methanol so that the final product is uncharged.

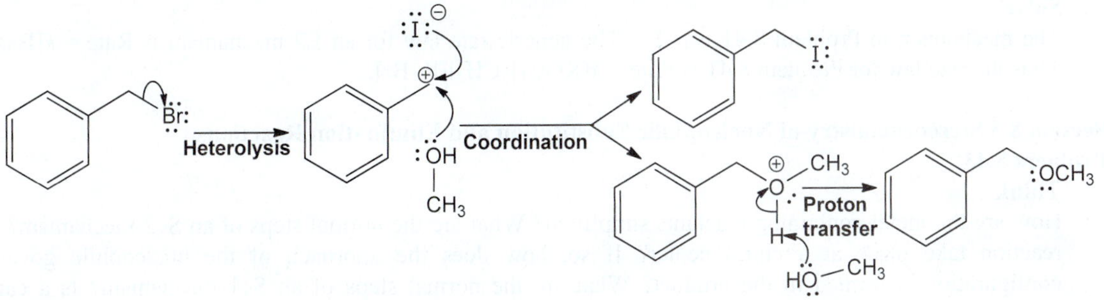

Heterolysis

Coordination

Proton transfer

(c) Because the reaction proceeds by an S_N1 mechanism, the reaction rate is independent of the concentration of the nucleophile. So if the concentration of KI were doubled, the rate of the substitution would stay the same.

Problem 8.39

Think

What happens to the concentration of R–OCH_3 upon going from Trial 1 to Trial 2, and what is the effect on reaction rate? What happens to the concentration of H_2O upon going from Trial 2 to Trial 3, and what is the effect on reaction rate? On which species's concentration, R–OCH_3 and/or H_2O, does the reaction rate depend? What does this suggest about the reaction mechanism?

Solve

Between Trial 1 and Trial 2, the concentration of R–OCH_3 doubles (the concentration of H_2O is constant) and the reaction rate also doubles. This suggests that the reaction rate depends directly on the concentration of R–OCH_3. Between Trial 2 and Trial 3, the concentration of H_2O is halved (R–OCH_3 is constant), and the reaction rate remains constant. This suggests that the reaction rate does not depend on the concentration of H_2O. The mechanism is likely E1.

Problem 8.40

Think

Was the mechanism in Problem 8.39 E1 or E2? What are the generic rate laws for each? How can the rate law be written for the substrate and base in this specific problem?

Solve

The mechanism in Problem 8.39 was E1. The generic rate law for an E1 mechanism is Rate = k[R–L]. Thus the rate law for Problem 8.39 is Rate = k[R–OCH_3].

Problem 8.41

Think

What happens to the concentration of R–Br upon going from Trial 1 to Trial 3, and what is the effect on reaction rate? What happens to the concentration of $KOCH_2CH_3$ upon going from Trial 2 to Trial 3, and what is the effect on reaction rate? On which species's concentration, R–Br and/or $KOCH_2CH_3$, does the reaction rate depend? What does this suggest about the reaction mechanism?

Solve

Between Trial 1 and Trial 3, the concentration of R–Br is halved (the concentration of $KOCH_2CH_3$ is constant), and the reaction rate is also halved. This suggests that the reaction mechanism depends directly on the concentration of R–Br. Between Trial 2 and Trial 3, the concentration of $KOCH_2CH_3$ is doubled (the concentration of R–Br is constant), and the reaction rate doubles. This suggests that the reaction mechanism also directly depends on the concentration of $KOCH_2CH_3$. The mechanism is likely E2.

Problem 8.42

Think

Was the mechanism in Problem 8.41 E1 or E2? What are the generic rate laws for each? How can the rate law be written for the substrate and base in this specific problem?

Solve

The mechanism in Problem 8.41 was E2. The generic rate law for an E2 mechanism is Rate = k[Base][R–L]. Thus the rate law for Problem 8.41 is Rate = $k[KOCH_2CH_3]$[R–Br].

Section 8.5 Stereochemistry of Nucleophilic Substitution and Elimination Reactions

Problem 8.43

Think

How are the metal-containing reactants simplified? What are the normal steps of an S_N2 mechanism? Does the reaction take place at a chiral center? If so, how does the approach of the nucleophile govern which configuration is formed in the product? What are the normal steps of an S_N1 mechanism? Is a carbocation generated as an intermediate? If so, can it undergo a 1,2-hydride or 1,2-methyl shift to become more stable? Does an S_N1 mechanism produce a single stereochemical configuration or a mixture of both?

Solve

Na^+ and K^+ are treated as spectator ions. Each S_N2 mechanism consists of a single step, whereas each S_N1 mechanism consists of two steps. In an S_N2 reaction, the nucleophile attacks the substrate from the side opposite the leaving group to produce only a single stereoisomer; in an S_N1 mechanism, a mixture of stereoisomers can form. Notice that stereochemistry is important in **(ii)**, **(iii)**, **(iv)**, and **(v)**. Notice also in **(iv)** that a carbocation rearrangement will take place in an S_N1 mechanism but not in an S_N2 mechanism. The mechanisms for **(i)–(v)** are shown below and on the two next pages.

(i)

S_N2

Br + ⊖OH → (S_N2) OH + Br⊖

S_N1

Br → (Heterolysis) ⊕ + Br⊖ → (Coordination) OH

(ii)

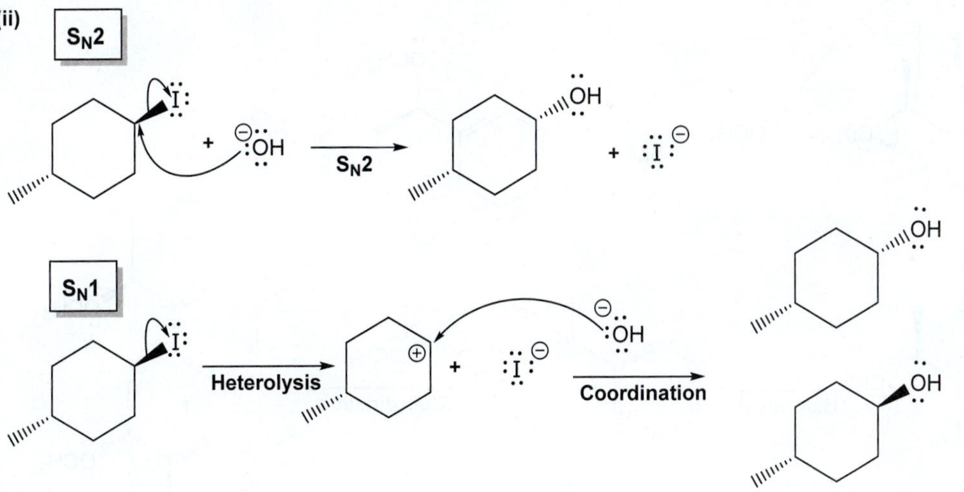

S_N2

S_N2

S_N1

Heterolysis

Coordination

(iii)

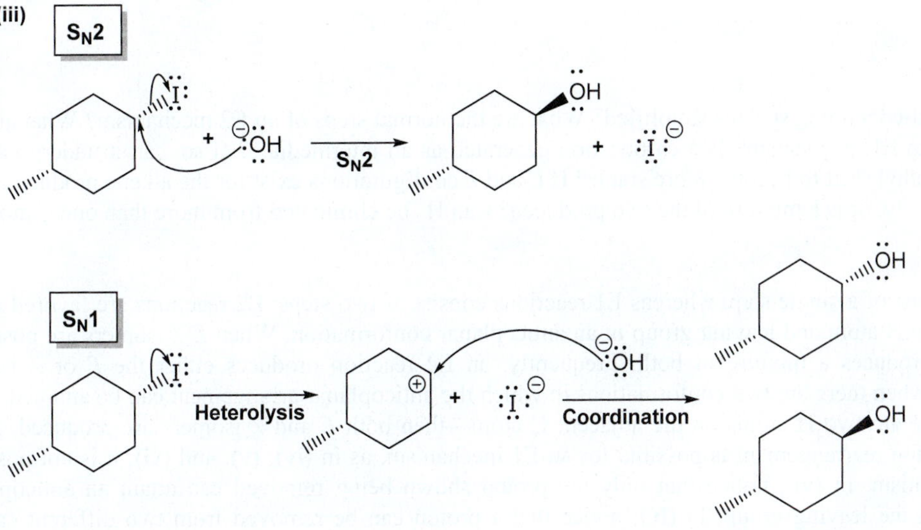

S_N2

S_N2

S_N1

Heterolysis

Coordination

(iv)

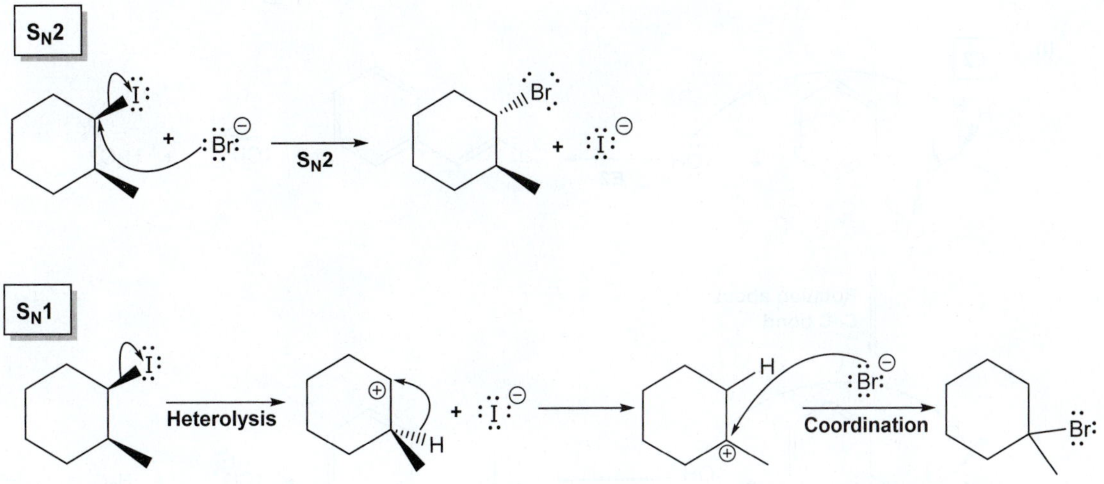

S_N2

S_N2

S_N1

Heterolysis

Coordination

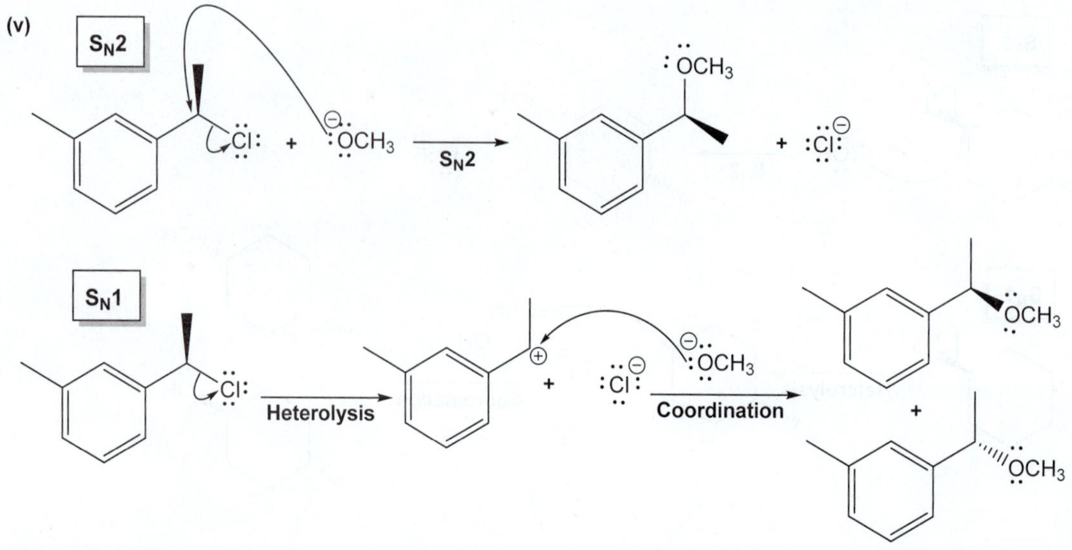

(v) S_N2

+ :OCH3 $\xrightarrow{S_N2}$ + :Cl:

S_N1

Heterolysis + :Cl: + :OCH3 Coordination +

Problem 8.44

Think

How are the metal-containing species simplified? What are the normal steps of an E2 mechanism? What are the normal steps of an E1 mechanism? Is a carbocation generated as an intermediate? If so, can it undergo a 1,2-hydride or 1,2-methyl shift to become more stable? If *E* and *Z* configurations exist for the alkene product, is one produced exclusively, or is a mixture of the two produced? Can H^+ be eliminated from more than one C atom?

Solve

E2 reactions consist of a single step, whereas E1 reactions consist of two steps. E2 reactions are favored when the substrate has a H atom and leaving group in an anticoplanar conformation. When *E*/*Z* isomers are possible, an E1 reaction produces a mixture of both. Frequently, an E2 reaction produces either the *E* or *Z* isomer exclusively. But when there are two conformations in which the anticoplanar arrangement can be attained, as in **(i)**—because there are two H atoms on the adjacent C atom—then both *E* and *Z* isomers are produced. Also, whereas carbocation rearrangement is possible for an E1 mechanism, as in **(iv)**, **(v)**, and **(vi)**, it is not possible for an E2 mechanism. In **(vi)**, notice that only the proton shown being removed can attain an anticoplanar arrangement with the leaving group. In **(iv)**, notice that a proton can be removed from two different carbon atoms. The mechanisms for **(i)**–**(vi)** are shown below and on the next two pages.

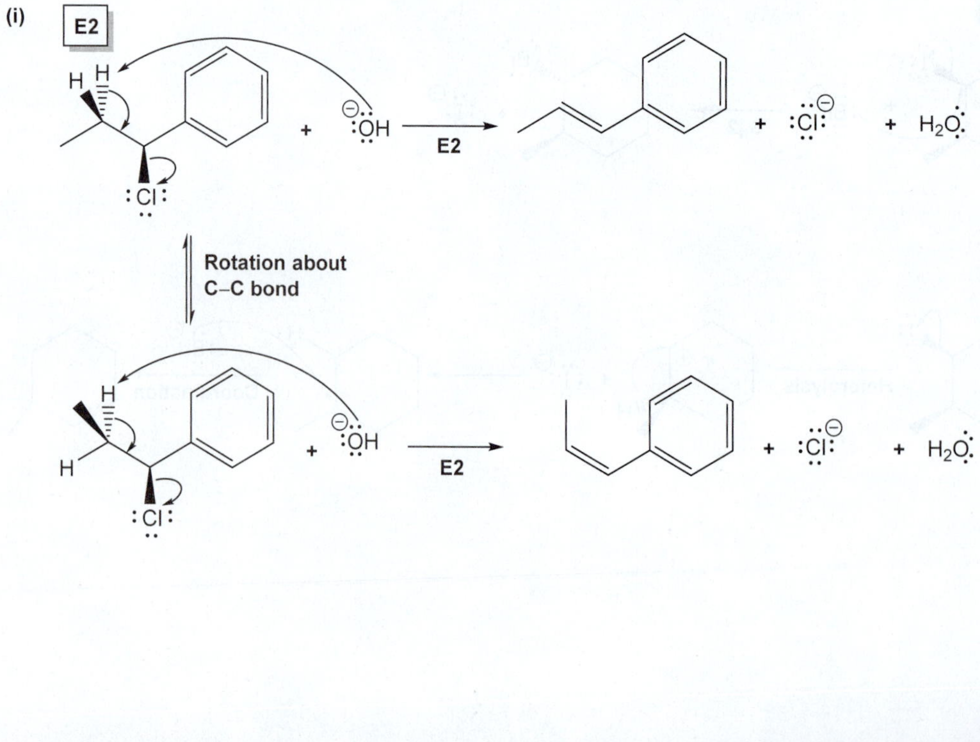

(i) E2

+ :OH $\xrightarrow{E2}$ + :Cl: + H_2O:

Rotation about C–C bond

+ :OH $\xrightarrow{E2}$ + :Cl: + H_2O:

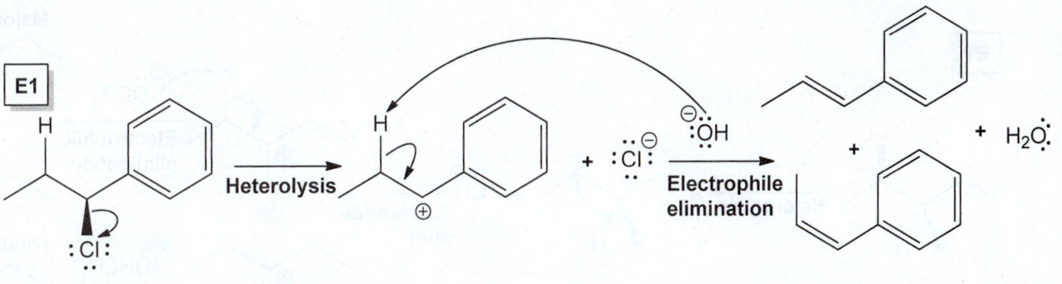

E1

Heterolysis

Electrophile elimination

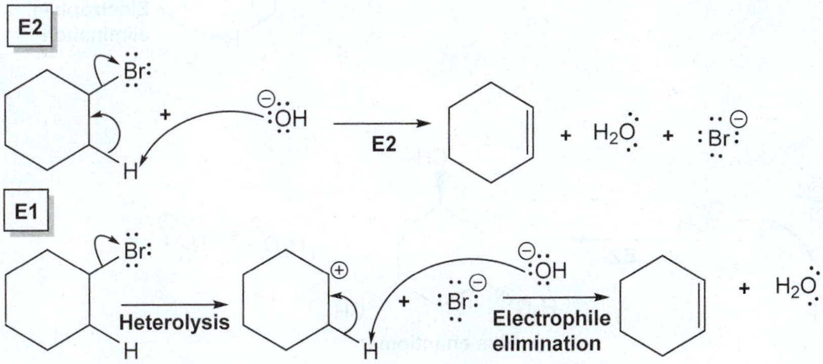

(ii) E2

E2

E1

Heterolysis

Electrophile elimination

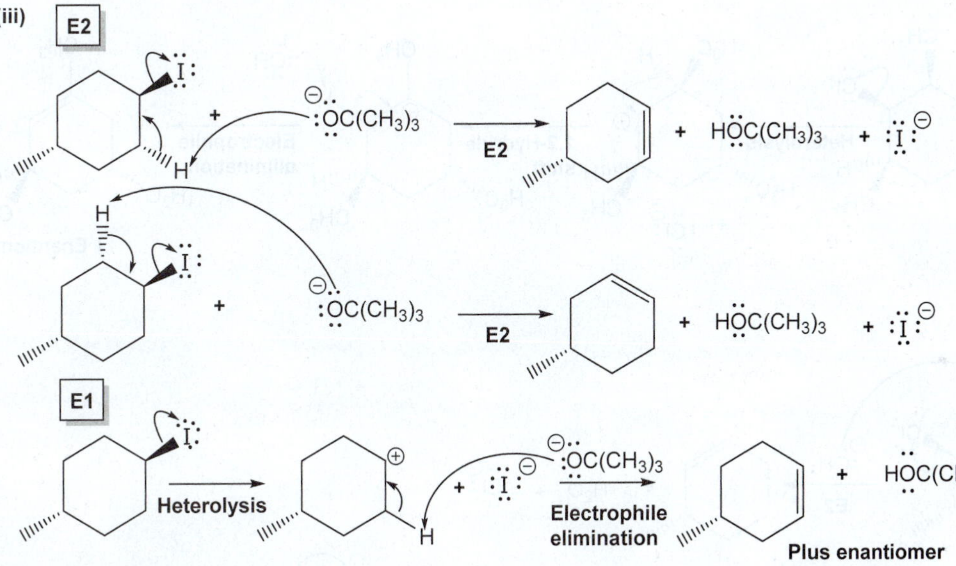

(iii) E2

E2

E2

E1

Heterolysis

Electrophile elimination

Plus enantiomer

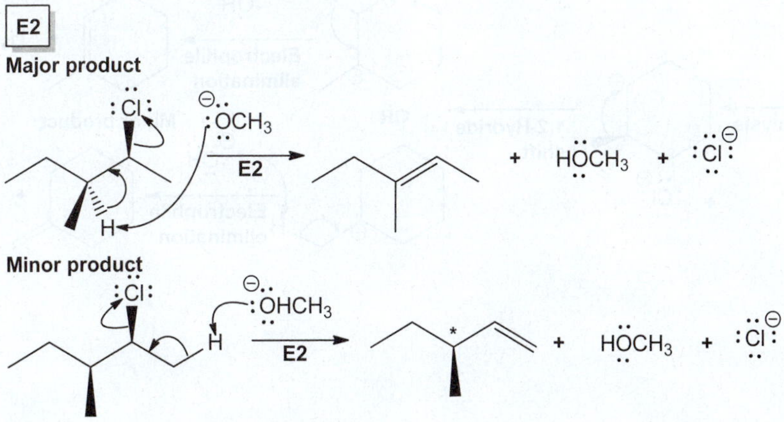

(iv) E2

Major product

E2

Minor product

E2

(v) E2

E1

(vi) E2

E1

Problem 8.45

Think

What is the most stable chair conformation of each isomer? What is the favored orientation for the Br atom and the adjacent H atom in an E2 reaction? Which chair conformation allows the Br and H atoms to be in the correct orientation and still maintain the stable chair conformation?

Solve

The most stable chair conformations of the two compounds are shown below. In both cases, the *t*-butyl group is in the equatorial position, because it is so bulky. In the first molecule, the Br atom is essentially locked in the axial position, whereas in the second molecule, the Br atom is essentially locked in the equatorial position. Notice that in the first molecule, Br is anti to an adjacent H atom, which favors E2 reactions. By contrast, in the second molecule, the H atoms on the opposite side of the ring from Br are equatorial, so the H and Br atoms are gauche to each other. That is, there is not a H atom that is anti to the Br atom on an adjacent carbon. This slows the reaction.

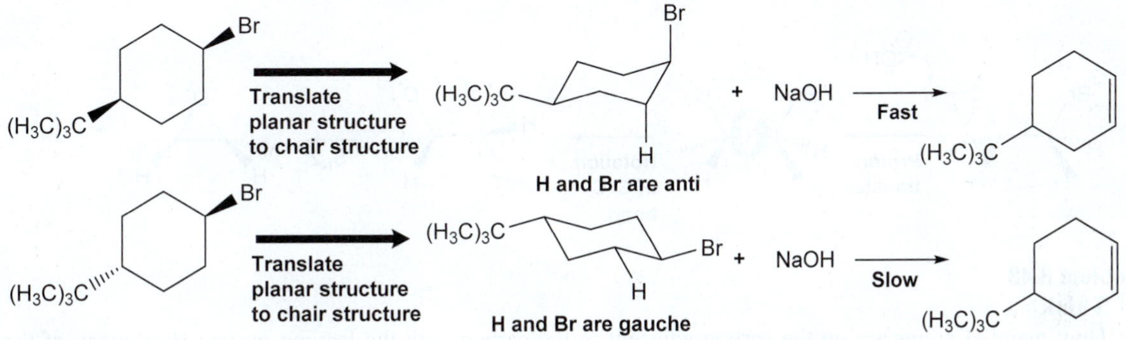

Br
$(H_3C)_3C$
Translate planar structure to chair structure
$(H_3C)_3C$
Br
H
H and Br are anti
+ NaOH
Fast
$(H_3C)_3C$

Br
$(H_3C)_3C$
Translate planar structure to chair structure
$(H_3C)_3C$
Br
H
H and Br are gauche
+ NaOH
Slow
$(H_3C)_3C$

Problem 8.46

Think

Can the reaction take place in a single step, or does it need to take place in multiple steps? For the alcohol carbon to become a mixture of configurations, can it stay tetrahedral throughout the reaction? If not, which of its bonds should break? How can you incorporate proton transfer steps to ensure that all species are compatible with the acidic conditions of the reaction?

Solve

To produce a mixture of configurations at the alcohol carbon, the mechanism must proceed through a step in which that carbon atom is planar—that is, in which it is not a chiral center. This can be explained if that carbon atom is a carbocation in an intermediate, such as in the mechanism below. For the leaving group to leave, it must first be protonated to avoid the production of HO^-, a strong base, in acidic conditions. Notice that a mixture of configurations at the C atom is produced when the nucleophile attacks the carbocation in the third step.

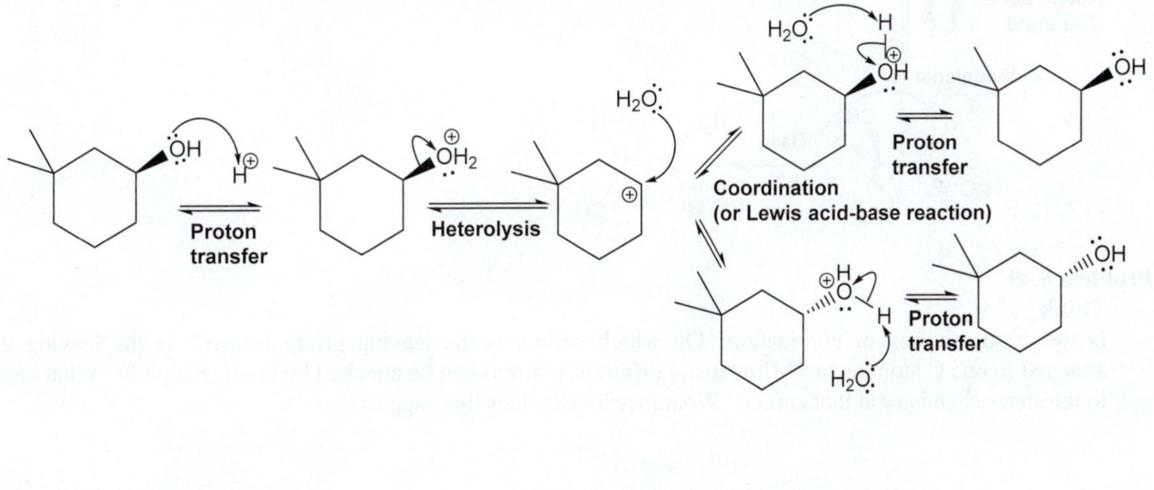

OH
$H^{\oplus}$
Proton transfer
$OH_2^{\oplus}$
Heterolysis
H_2O
Coordination (or Lewis acid-base reaction)
H_2O
H
OH
Proton transfer
OH
H
O
H
Proton transfer
H_2O
OH

Problem 8.47

Think

On which C atom is the leaving group located? Is that C atom a chiral center in the product? If so, is one stereochemical configuration produced, or is a mixture produced? What mechanism does this suggest? Do proton transfer steps need to be incorporated to avoid species that are incompatible with the conditions of the reaction?

Solve

The C atom that was initially bonded to the Br leaving group is a chiral center in the product, and only the configuration shown is produced. This suggests S_N2. Furthermore, a careful examination of the product shows that the C atom on the left has been stereochemically inverted. This is normal for an S_N2 mechanism. Once you are aware of this, you know that backside attack must occur, and the molecule must adopt the conformation shown in the third molecule below for this to happen. Notice also that it would be unreasonable for this S_N2 step to take place first, because it would have produced an R_2OH^+ species, which is a strong acid and is incompatible with the basic conditions of the reaction. Instead, the R_2OH is first deprotonated.

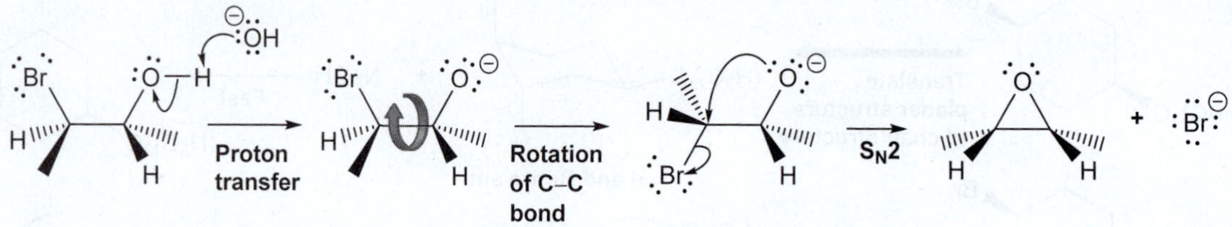

Proton transfer

Rotation of C–C bond

S_N2

Problem 8.48

Think

How many H atoms are on the carbon adjacent to the carbon with the leaving group? How many of those H atoms are able to adopt the anticoplanar conformation? How many products result from an E2? How many result from an E1?

Solve

No, it is not possible to tell whether this reaction took place by E1 or E2. Either H atom on the C atom on the left can become anticoplanar with the Br atom, so either diastereomer of the alkene can be formed in an E2 reaction. A similar result is expected from an E1 reaction, which generally produces both *E* and *Z* isomers.

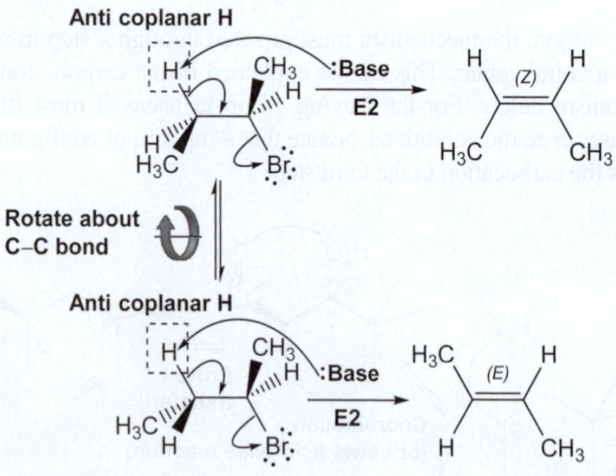

Anti coplanar H

E2

(Z)

Rotate about C–C bond

Anti coplanar H

E2

(E)

Problem 8.49

Think

Is this a substitution or elimination? On which carbon is the leaving group located? Is the leaving group attached to one C atom or two? How many different C atoms can be attacked by the nucleophile? What happens to the stereochemistry at that carbon? What mechanism does this suggest?

Solve

The reaction proceeds by an S_N2 mechanism, followed by a proton transfer. The leaving group can be viewed as Ring–O$^-$. As we can see below, the nucleophile in the S_N2 step must attack the C atom from the side opposite the leaving group, forcing the other substituents on that C atom to flip over to the other side. After protonation, we can see that the product is exactly the stereoisomer that is shown in the problem.

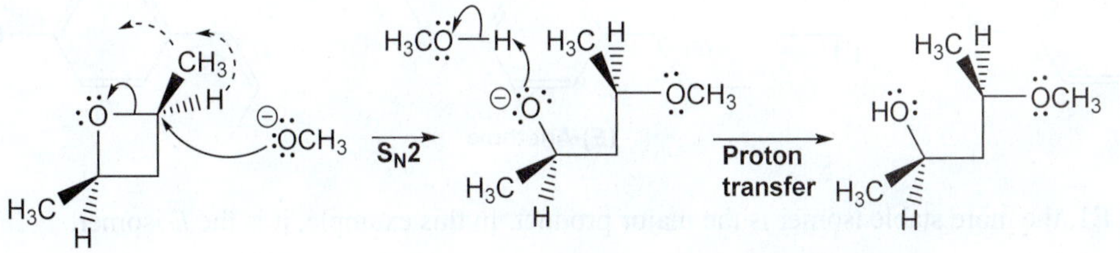

The enantiomer is produced by attack of the nucleophile at the other C atom on the ring, as shown below.

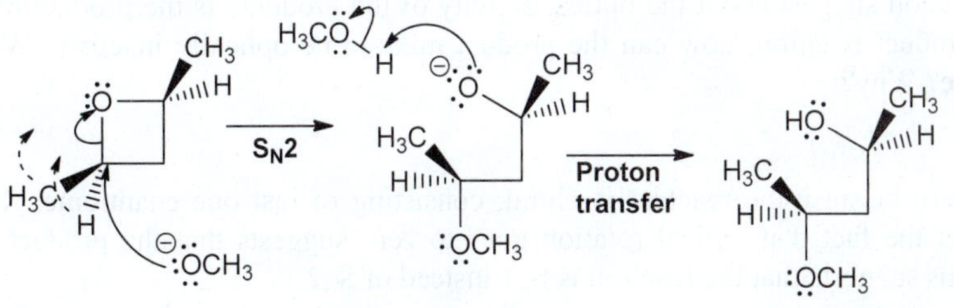

An S_N1 step did not take place, because other diastereomers were not produced. As shown below, a C atom from which the leaving group leaves would become planar, and when the chiral center is regenerated upon attack of the nucleophile, both stereochemical configurations would be produced. An example is shown below. Notice that the product is not one of those given in the problem.

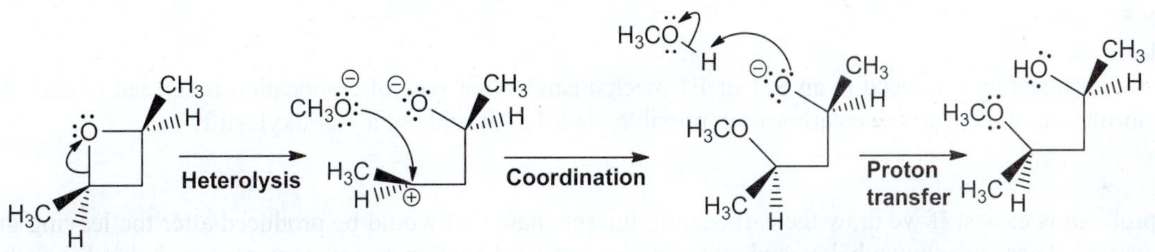

Problem 8.50

Think

When there are *E* and *Z* isomers possible in the product, does an E1 reaction typically produce one isomer exclusively or a mixture? How many H atoms are available for elimination on the adjacent C atom? Can each of them undergo an E2 elimination? What does this mean for the number of stereoisomers possible as products of E2?

Solve

(a) E2 will produce a mixture of *E* and *Z* stereoisomers because there are two H atoms on the C atom adjacent to the C–Br, and each can attain the anticoplanar arrangement that favors E2. This allows for both *E* and *Z* stereoisomers to be produced.

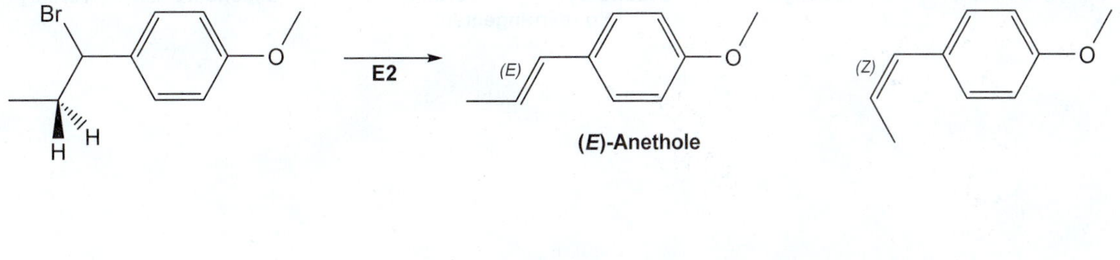

(*E*)-Anethole

(b) E1 will produce a mixture of *E* and *Z* stereoisomers. In Step 1 of the E1 mechanism, the leaving group leaves via heterolysis and forms a carbocation. The electrophile elimination in Step 2 forms the more stable stereoisomer, but also some of the less stable alkene isomer.

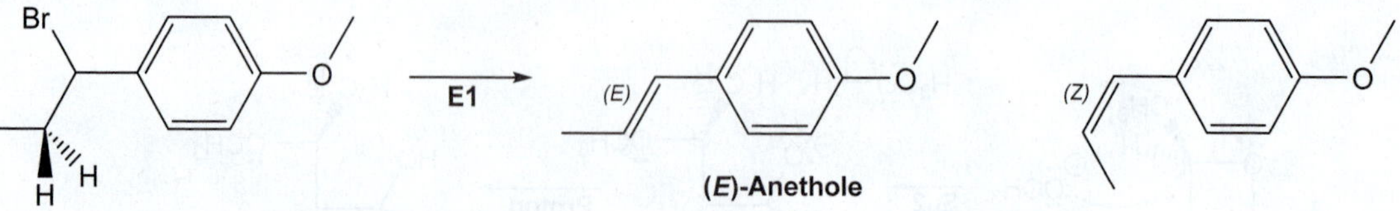

E1 (*E*) (*Z*) (*E*)-Anethole

(c) In both E2 and E1, the more stable isomer is the major product. In this example, it is the *E* isomer.

Problem 8.51

Think

What does a measured angle of rotation suggest about the optical activity of the product? Is the product of the reaction chiral or achiral? If the product is chiral, how can the product mixture be optically inactive? Which mechanism forms a racemic mixture? Why?

Solve

Optical rotation starts out as nonzero because the reactant is chiral, consisting of just one enantiomer. Each product molecule is chiral, too, but the fact that optical rotation goes to zero suggests that the product is a racemic mixture of enantiomers. This suggests that the reaction is S_N1 instead of S_N2.

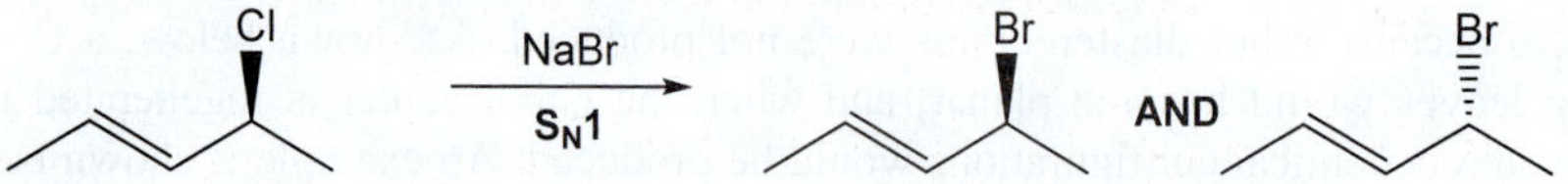

NaBr, S_N1 → AND

Sections 8.6 and 8.7 The Reasonableness of Mechanisms

Problem 8.52

Think

How is a carbocation produced in an S_N1 or E1 mechanism? What type of carbocation is formed in each case? Is the formation of a more stable carbocation possible via a 1,2-hydride or a 1,2-alkyl shift?

Solve

This problem is easiest if we draw the carbocation intermediate that would be produced after the leaving group has departed. These are shown below and on the next page. Carbocation rearrangements are shown for each one that can attain greater stability via a 1,2-hydride shift or a 1,2-methyl shift. In **(a)**, a secondary carbocation rearranges to a tertiary one. The same is true in **(d)** and **(i)**. In **(e)**, a localized carbocation rearranges to a resonance-delocalized one. In **(f)**, a primary carbocation rearranges to a secondary one. In **(g)**, a carbocation that is resonance stabilized by one phenyl ring can rearrange to one that is stabilized by two phenyl rings. Rearrangement does not occur in **(h)**, as ethyl shifts are not common.

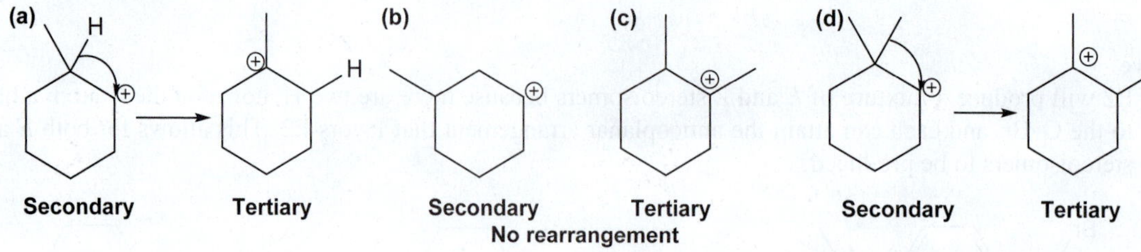

(a) Secondary → Tertiary

(b) Secondary

(c) Tertiary

No rearrangement

(d) Secondary → Tertiary

(e)

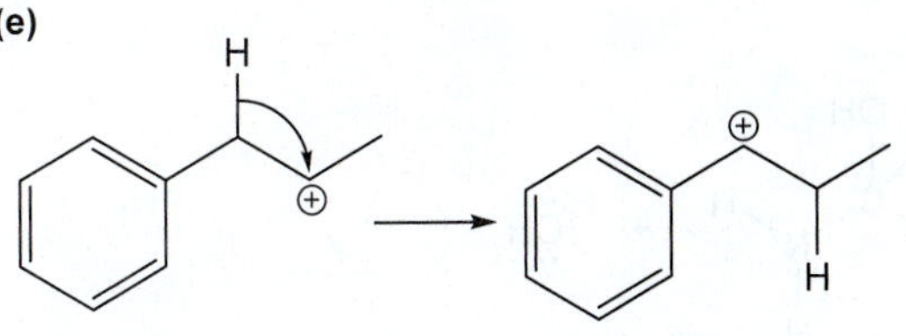

Localized → Resonance delocalized

(f)

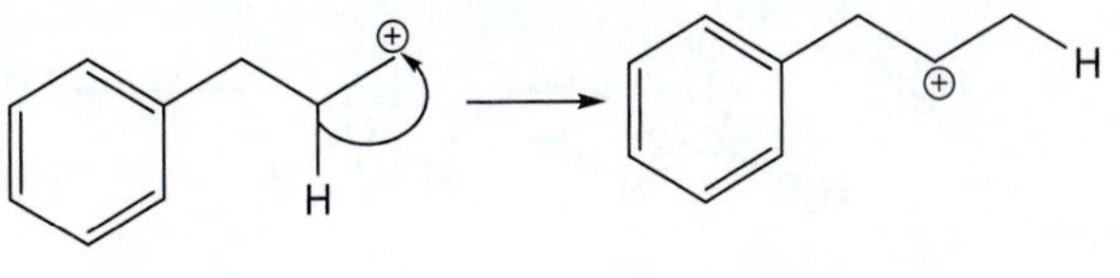

Primary → Secondary

(g)

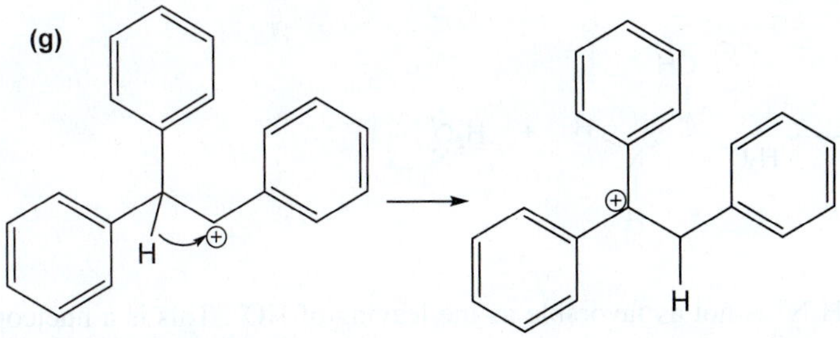

Resonance from one ring → Resonance from two rings

(h)

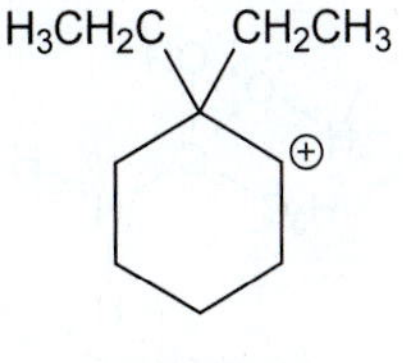

H_3CH_2C CH_2CH_3

Secondary
No rearrangement

(i)

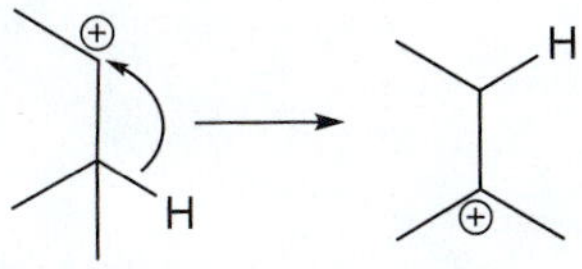

Secondary → Tertiary

Problem 8.53

Think

Consider all possible elementary steps from Chapter 7. Are the curved arrows correct? Review the rules in Section 8.6 for reasonable or unreasonable mechanisms. What are the conditions of the reaction—acidic, basic, or neutral? Would any strong acids or bases be incompatible with the reaction conditions? Are carbocation rearrangements possible? Are there any termolecular steps? What is reasonable for proton transfer steps?

Solve

(i) Reasonable. This is a nucleophilic addition.

$H_3C-C\equiv N$: + :ÖH⁻ → Nucleophilic addition → $H_3C-C(OH)=N^-$

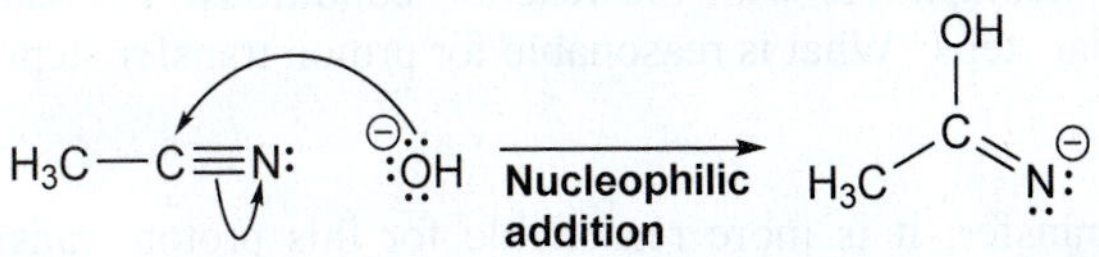

(ii) Not reasonable. H_3O^+ is a strong acid and should not participate in a mechanism that takes place in basic solution (indicated by the presence of HO^- below the reaction arrow).

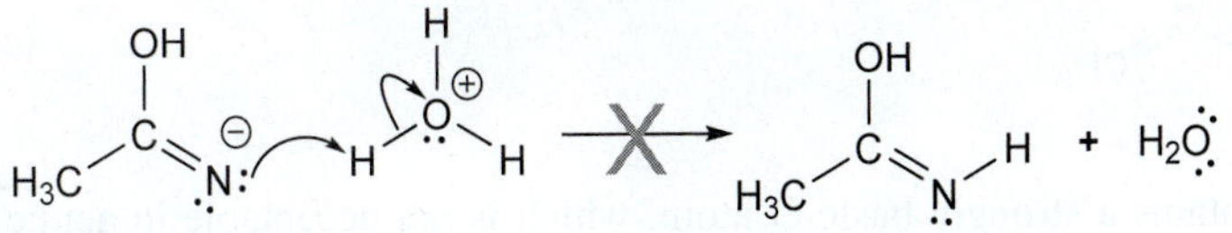

$H_3C-C(OH)=N^-$ + H_3O^+ ✗→ $H_3C-C(OH)=N-H$ + H_2O

(iii) Not reasonable. This is a termolecular step.

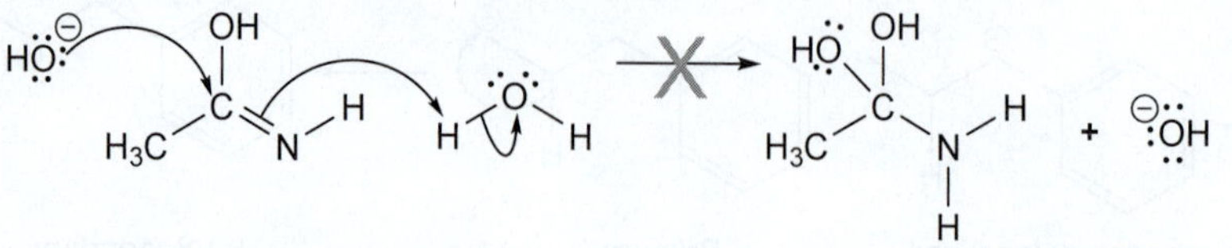

(iv) Not reasonable. The curved arrow shows the movement of an atom instead of electrons.

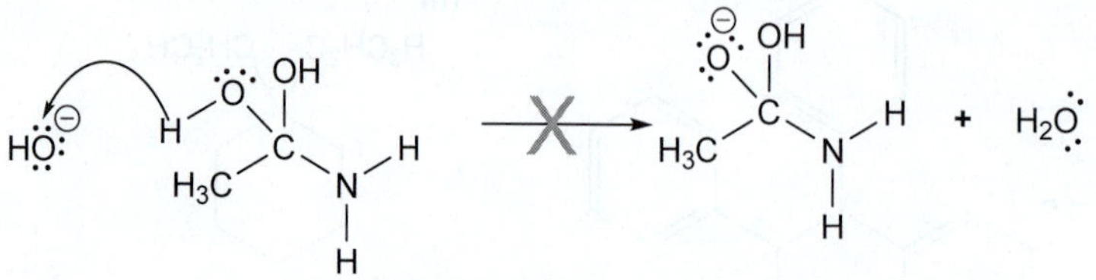

(v) Reasonable, although the leaving of H_2N^- is not as favorable as the leaving of HO^-. This is a nucleophile elimination step.

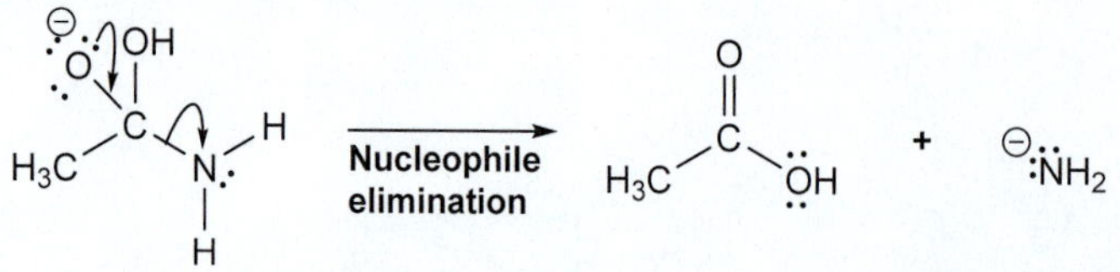

Nucleophile elimination

(vi) Reasonable. A very strong base removes a proton from an acid.

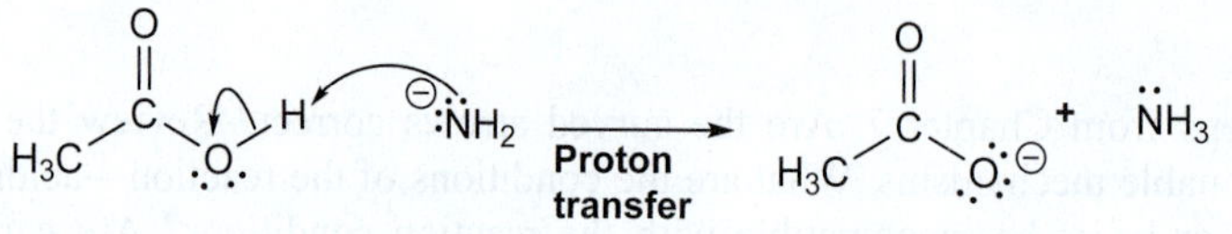

Proton transfer

Problem 8.54

Think

Consider all possible elementary steps from Chapter 7. Are the curved arrows correct? Review the rules in Section 8.6 for reasonable or unreasonable mechanisms. What are the conditions of the reaction—acidic, basic, or neutral? Would any strong acids or bases be incompatible with the reaction conditions? Are carbocation rearrangements possible? Are there any termolecular steps? What is reasonable for proton transfer steps?

Solve

(i) Not reasonable. This is an internal proton transfer; it is more reasonable for this proton transfer to be solvent mediated.

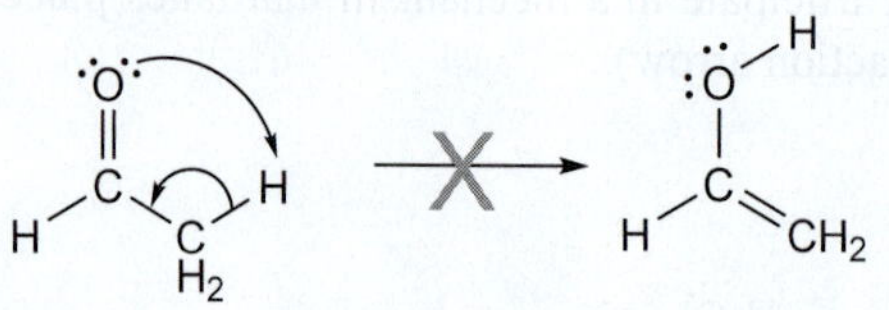

(ii) Not reasonable. The product contains a strongly basic O atom, which is not acceptable in acidic solution (indicated by the H^+ over the reaction arrow).

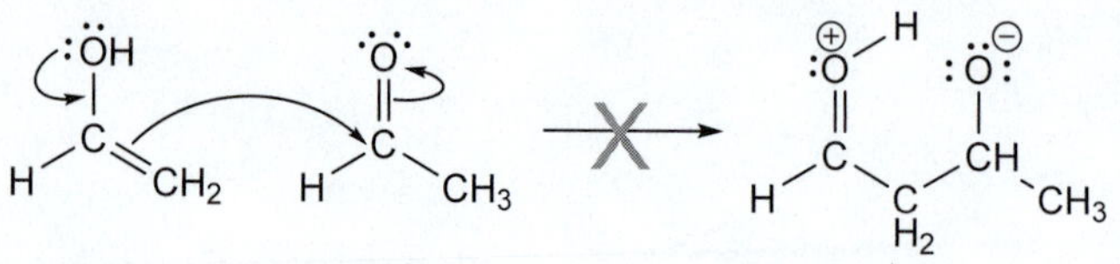

(iii) Not reasonable. This is a termolecular step.

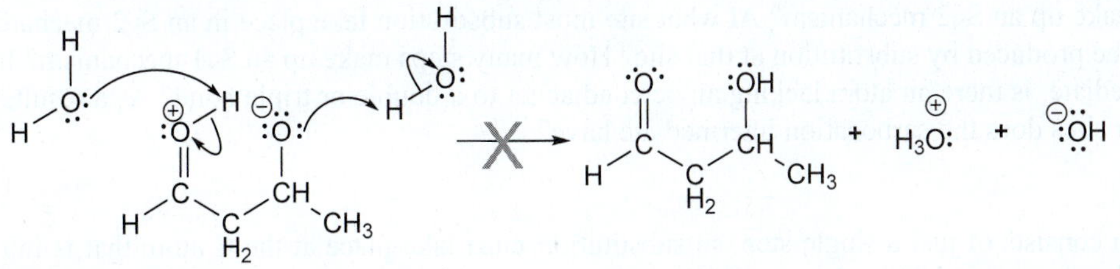

Problem 8.55

Think

How many steps make up an S_N2 mechanism? At what site must substitution take place in an S_N2 mechanism? Can both products be produced by substitution at that site? How many steps make up an S_N1 mechanism? In the carbocation intermediate, is there an atom lacking an octet adjacent to a double or triple bond? As a result, how many electron-poor sites does the carbocation intermediate have?

Solve

(a) If an S_N2 reaction occurred, then only the first of the two products would have been formed, as substitution takes place directly (and only) at the C atom bonded to the leaving group. However, because there is a second substitution product, the mechanism must be an S_N1.

(b) An S_N1 mechanism proceeds through a carbocation intermediate, shown below. As indicated, that carbocation has the positive charge resonance delocalized over two different C atoms. The different products are formed by attack of the nucleophile at each of those two C atoms.

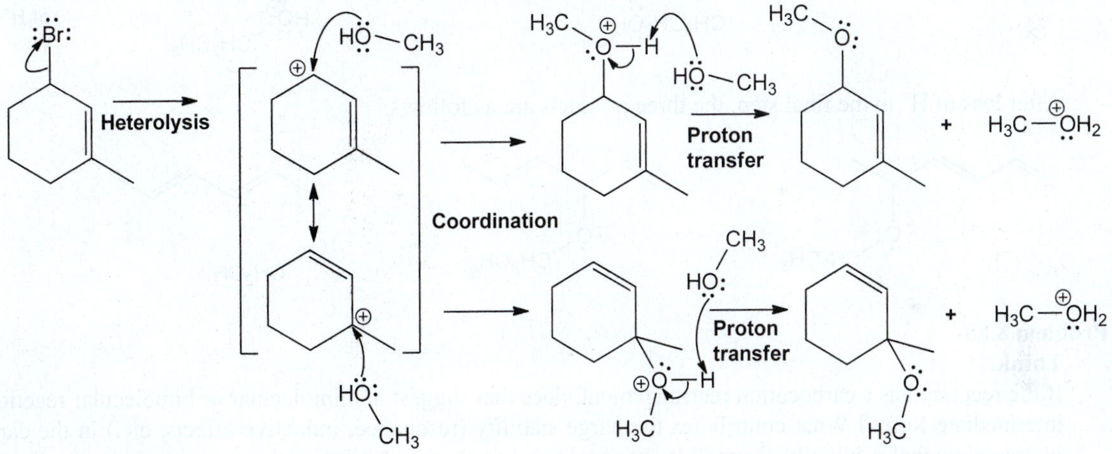

Problem 8.56

Think

On which C atom is the leaving group located? Did the nucleophile attach to the same C atom? Has rearrangement occurred? What mechanism does this suggest?

Solve

A rearrangement of the carbon skeleton has occurred, indicating that a carbocation rearrangement has taken place. Formation of a carbocation can occur only with an S_N1, not with an S_N2.

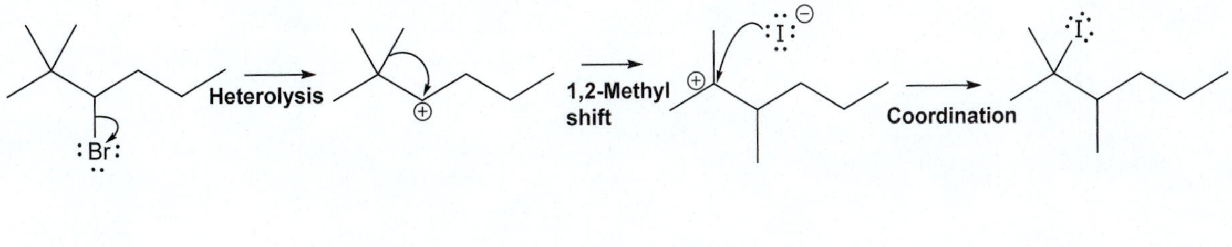

Problem 8.57

Think

How many steps make up an S_N2 mechanism? At what site must substitution take place in an S_N2 mechanism? Can both products be produced by substitution at that site? How many steps make up an S_N1 mechanism? In the carbocation intermediate, is there an atom lacking an octet adjacent to a double or triple bond? As a result, how many electron-poor sites does the carbocation intermediate have?

Solve

An S_N2 mechanism consists of just a single step, so substitution must take place at the C atom that is initially bonded to the Br leaving group. Therefore, an S_N1 mechanism involving a resonance-stabilized carbocation must have taken place. As we can see below, the positive charge of the carbocation intermediate is delocalized by resonance over three different C atoms, so nucleophilic attack can occur at all three of those C atoms.

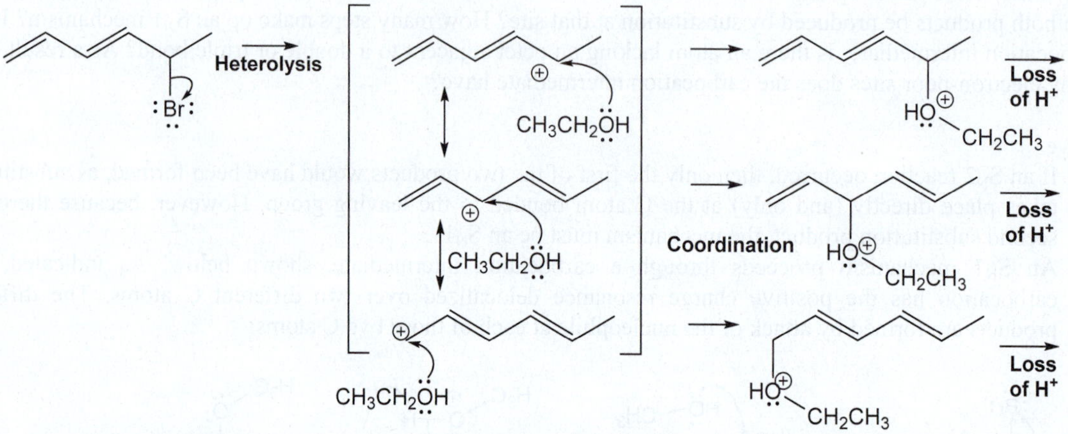

After loss of H^+ in the final step, the three products are as follows:

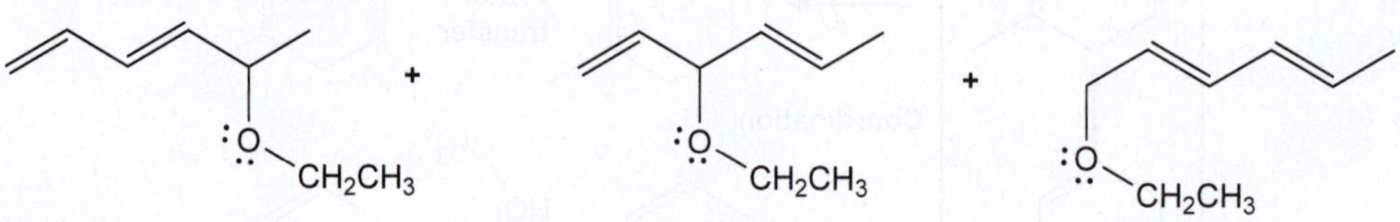

Problem 8.58

Think

If the reaction has a carbocation rearrangement, does that suggest a unimolecular or bimolecular reaction? What intermediate forms? What contributes to charge stability (resonance, inductive effects, etc.) in the carbocation intermediate that is initially formed? What contributes to charge stability in the rearranged intermediate?

Solve

This is a substitution reaction where, overall, Br has been replaced by OCH_3. It is, furthermore, unimolecular, making it S_N1, because a carbocation must have formed prior to the rearrangement. A carbocation rearrangement occurs because after a H atom has migrated, the resulting cation can be stabilized by resonance, involving a lone pair of electrons from the adjacent O atom. That resonance does not exist in the carbocation intermediate that initially formed. See the mechanism on the next page.

Heterolysis

1,2-Hydride shift

Coordination

Proton transfer

Problem 8.59

Think

Would the mechanism be a substitution or an elimination? What is the nucleophile? If the leaving group departs in the first step, will the resulting species be compatible with the conditions under which the reaction is carried out? If not, how can you incorporate proton transfer steps to ensure that the species remain compatible?

Solve

This must be a substitution reaction, where, overall, OH has been replaced by OCH_2CH_3. The mechanism involves proton transfer, an S_N2 step, then another proton transfer. Protonation of the OH group is necessary first because the reaction takes place in acidic conditions. If nucleophilic substitution were to take place first, instead, the leaving group would come off as HO^-, which is a strong base and is not compatible with the acidic conditions.

Proton transfer

S_N2

Proton transfer

Very low boiling point
Leaves reaction mixture
Shifts equilibrium to right

Problem 8.60

Think

Is this a substitution or an elimination reaction? On which carbon does the leaving group leave? Which C atom must have been attacked by a nucleophile to produce the product shown? Is that the same C atom from which the leaving group departed, or is it a different one? Does this suggest that the reaction is unimolecular or bimolecular? Are any proton transfer reactions involved?

Solve

This is an internal S_N1 reaction. It is a substitution reaction because, overall, Br has been replaced by O–Ring. Br is the leaving group on the terminal C atom, but the new O–Ring bond formed at a different C atom, so this must be unimolecular, making it S_N1. The intermediate carbocation produced in the mechanism is resonance stabilized, and the positive charge is shared over two different C atoms. Thus, the OH group can attack either C atom. It preferentially attacks the one shown, because it produces a relatively stable, five-membered ring. Attack of the other C atom would yield a seven-membered ring. A final proton transfer is necessary to produce the uncharged product.

Heterolysis

Coordination

Proton transfer

Integrated Problems

Problem 8.61

Think

In how many steps does an E1 reaction occur? Do proton transfer steps need to be incorporated to avoid incompatible species? Is rearrangement likely? How many transition states and intermediates are present?

Solve

(a) The leaving group (i.e., OH) cannot leave as HO^- because this is a strong base that would be produced in acidic conditions (indicated by H_3O^+ below the reaction arrow). Instead, the OH group is first protonated to create OH_2^+, which would leave in the form of H_2O. Finally, deprotonation of the carbocation intermediate forms the alkene product. The Roman numerals are used to identify the intermediate structures on the energy diagram for **(b)**.

Proton transfer → I → Heterolysis → II → Electrophile elimination → + H_3O^+

(b) Because there are three steps, there are three transition states and two intermediates. The overall reaction is endothermic, because the C=C π and O–H σ bonds that form are collectively weaker than the H–C and C–O σ bonds that break.

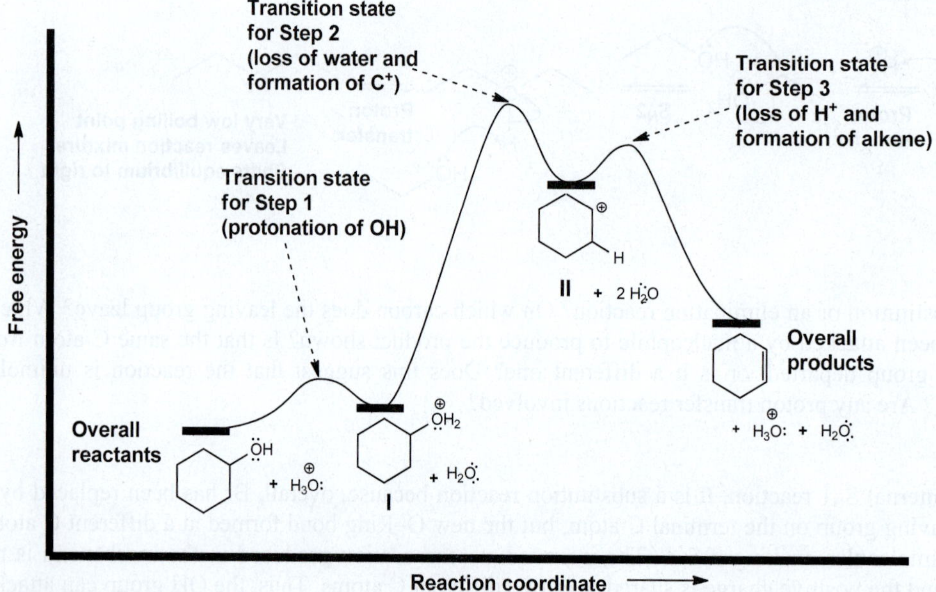

Problem 8.62

Think

Can the reaction take place in a single step, or does it need to take place in multiple steps? For the α carbon to become a mixture of configurations, can it stay tetrahedral throughout the reaction? If not, which of its bonds should break? What species present can cause that bond to break? How can you incorporate proton transfer steps to ensure that all species are compatible with the basic conditions of the reaction?

Solve

Because there are equal amounts of enantiomers produced (a racemic mixture), the chiral center that we see must be formed from an achiral precursor. That achiral intermediate can be formed by deprotonating the α carbon, as shown below. In doing so, the α carbon becomes planar because of the contribution by its resonance structure and, thus, is not a chiral center. In the second step, the α carbon is protonated, and when it is, both *R* and *S* configurations are produced (as a result of protonating the α carbon from either side of carbon's plane).

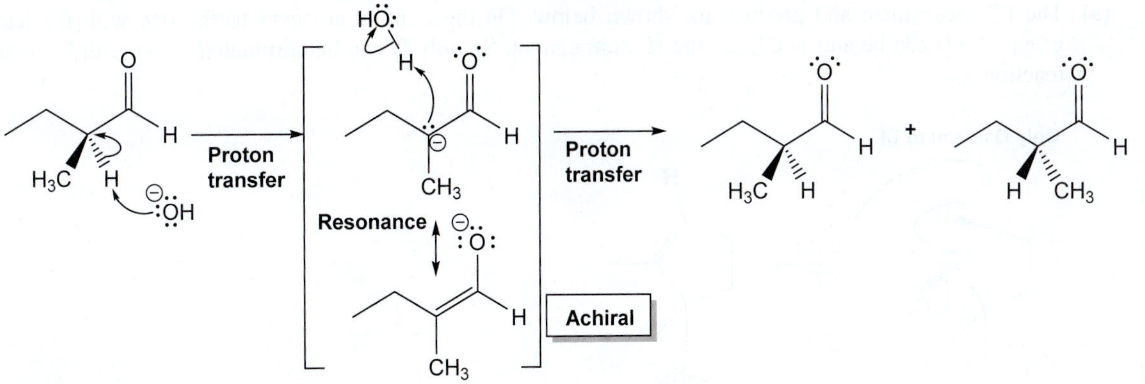

Proton transfer

Resonance

Achiral

Proton transfer

Problem 8.63

Think

Can the reaction take place in a single step, or does it need to take place in multiple steps? For the α carbon to become a mixture of configurations, can it stay tetrahedral throughout the reaction? If not, which of its bonds should break? What species present can cause that bond to break? How can you incorporate proton transfer steps to ensure that all species are compatible with the acidic conditions of the reaction?

Solve

The process is very similar to the previous problem. To form a racemic mixture of enantiomers, the chiral center must be formed from an achiral precursor. That precursor is formed from two successive proton transfer reactions, shown below. Deprotonation cannot take place first, because that would generate a negative charge on the C atom (or elsewhere on the molecule), which corresponds to a strong base. This cannot happen, because the reaction takes place in acidic conditions (indicated by the H_3O^+ below the reaction arrow). So the O atom is protonated first, followed by deprotonation at the α carbon. Once the achiral intermediate is formed, the first two steps are reversed—first is protonation of the C atom, followed by deprotonation at the O atom. When protonation at the C atom takes place, the chiral center is regenerated, and a mixture of configurations is produced.

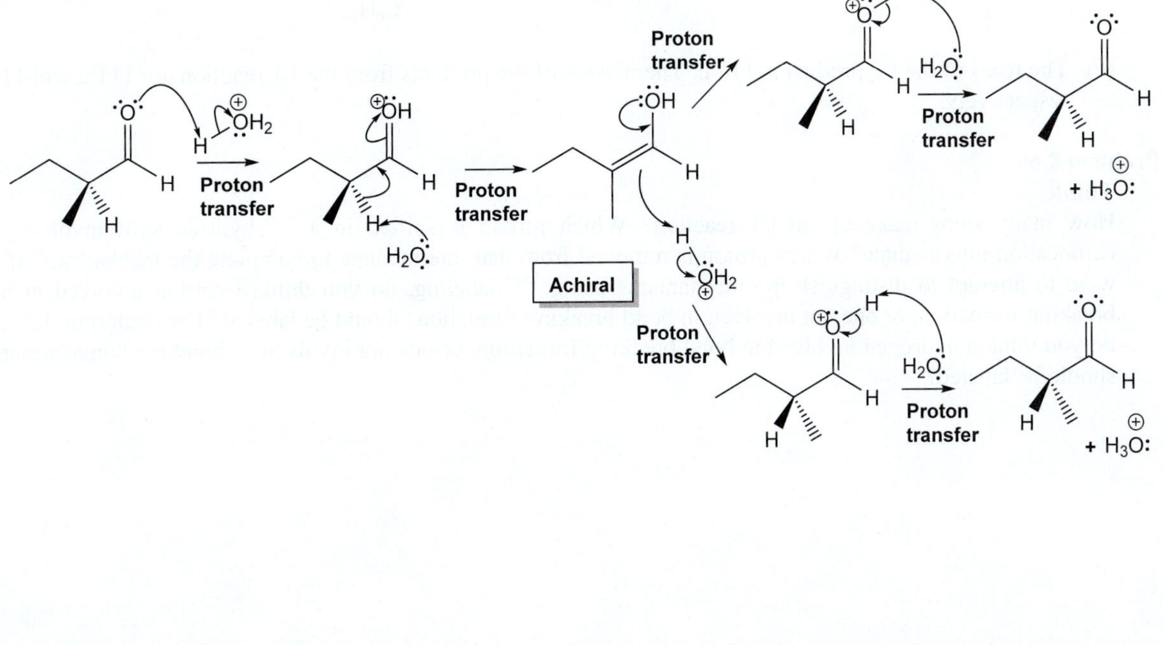

Proton transfer

Proton transfer

Achiral

Proton transfer

Proton transfer

Proton transfer

Proton transfer

$+ H_3O^+$

$+ H_3O^+$

Problem 8.64

Think

On which carbon is the leaving group located? Is rearrangement likely to occur in an E1 mechanism? In an E2? Which neighboring C atoms have H or D atoms available for elimination? Is an anticoplanar arrangement of the appropriate atoms possible for an E2 reaction? How many products are possible for E2? For E1?

Solve

(a) The E2 mechanism and product are shown below. On the C atom adjacent to the one with the leaving group, the D can be anti to Cl, but the H atom cannot. So only D can be eliminated along with Cl in an E2 reaction.

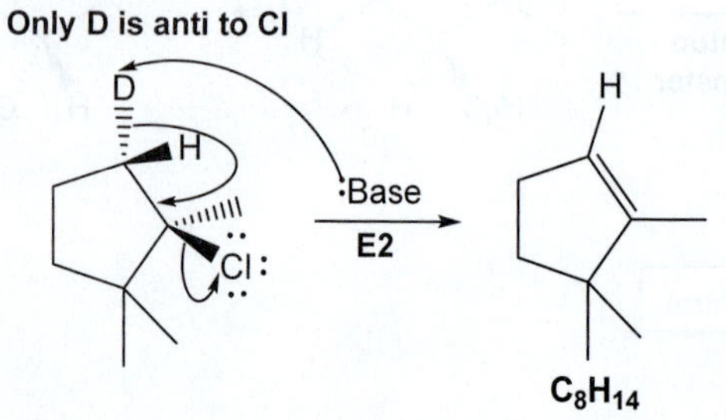

Only D is anti to Cl

:Base E2

C_8H_{14}

(b) The E1 mechanism and products are shown below. Both the D and the H atoms can be eliminated from the carbocation intermediate, so two products are formed. The carbocation intermediate is already tertiary, and rearrangement is unlikely.

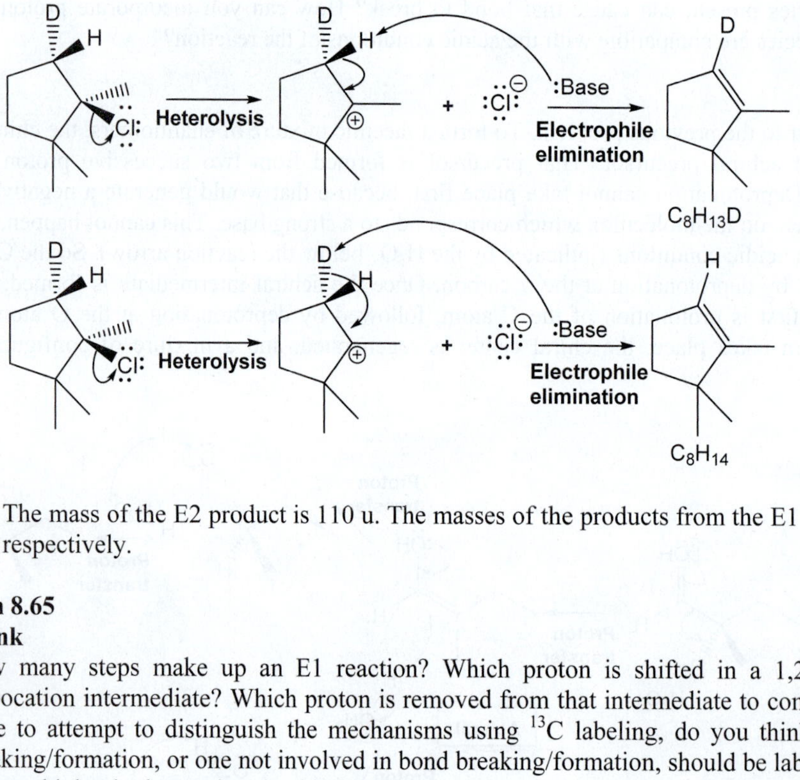

Heterolysis → + :Cl:⁻ :Base Electrophile elimination → $C_8H_{13}D$

Heterolysis → + :Cl:⁻ :Base Electrophile elimination → C_8H_{14}

(c) The mass of the E2 product is 110 u. The masses of the products from the E1 reaction are 111 u and 110 u, respectively.

Problem 8.65

Think

How many steps make up an E1 reaction? Which proton is shifted in a 1,2-hydride shift involving the carbocation intermediate? Which proton is removed from that intermediate to complete the mechanism? If you were to attempt to distinguish the mechanisms using ^{13}C labeling, do you think a carbon involved in bond breaking/formation, or one not involved in bond breaking/formation, should be labeled? For deuterium labeling, do you think a hydrogen involved in bond breaking/formation, or one not involved in bond breaking/formation, should be labeled?

Solve

(a) The first step of an E1 is the leaving of the leaving group, as shown below. The secondary carbocation can then undergo a fast 1,2-hydride shift to yield a more stable tertiary carbocation. In the final step, elimination of H^+ yields the alkene product.

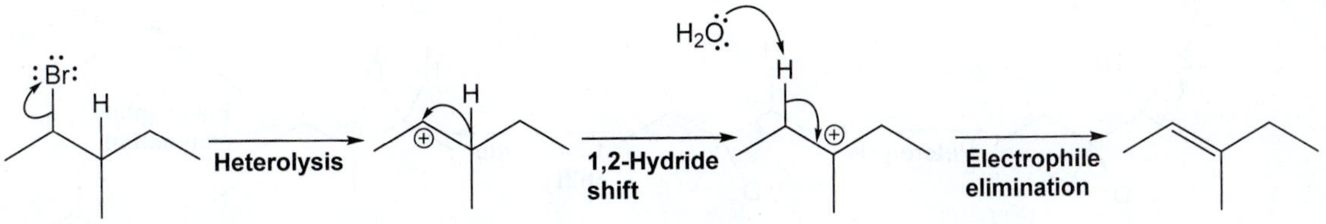

Heterolysis
1,2-Hydride shift
Electrophile elimination

(b) The first step of an E1 is the leaving of the leaving group, as shown below. Without a carbocation rearrangement, the second step would then be elimination of H^+ to yield the alkene product.

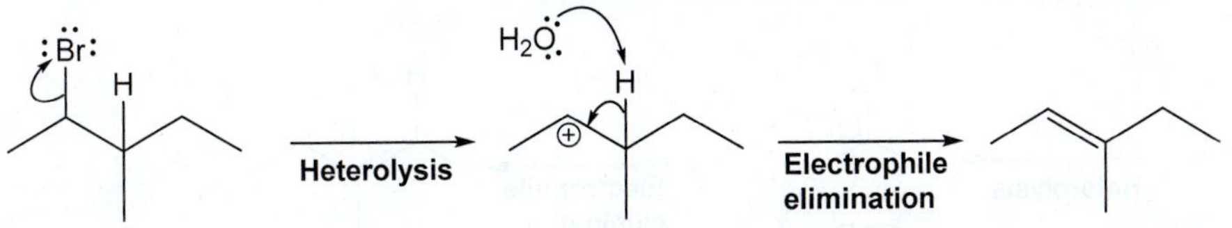

Heterolysis
Electrophile elimination

The same alkene product is produced in **(a)** and **(b)** because the C atom on which the H atom is eliminated and the C^+ occur on the same two C atoms that make up the C=C bond in both mechanisms.

(c) If the C atom bonded to the Br atom is the ^{13}C isotope, then a carbocation rearrangement would yield two different isomeric alkene products, as shown below.

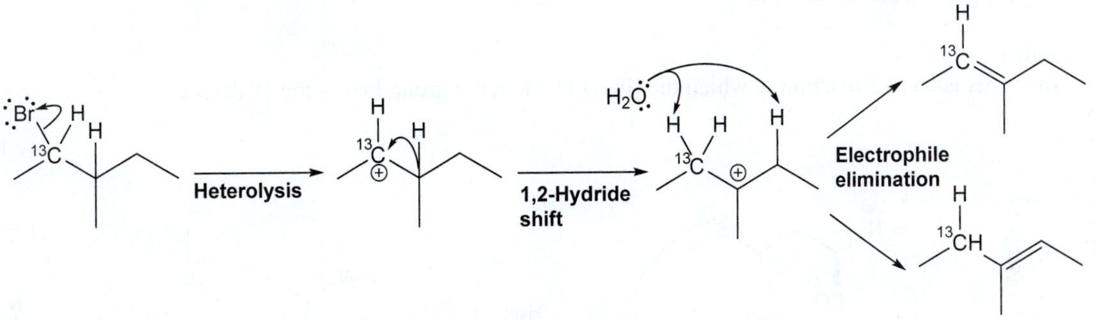

Heterolysis
1,2-Hydride shift
Electrophile elimination

If no rearrangement occurs, then only the product below results.

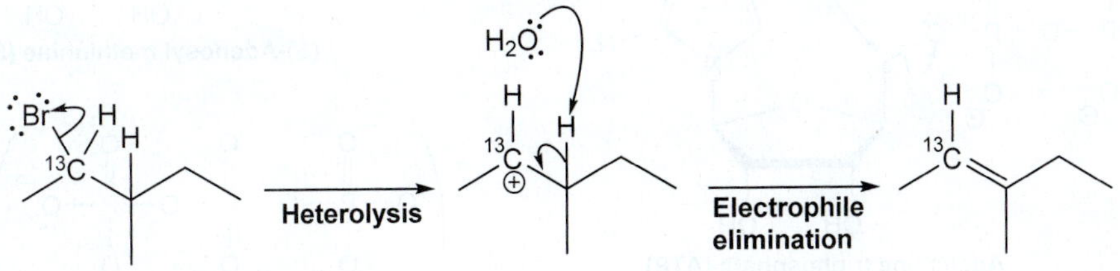

Heterolysis
Electrophile elimination

(d) If the H atom that is proposed to undergo migration is replaced by D, we can tell if the migration occurs by whether we observe any D in the alkene product. If we observe D in a portion of the alkene product, then migration must have occurred.

Heterolysis · 1,2-Hydride shift · Electrophile elimination

If no D remains, then migration must not have occurred.

Heterolysis · Electrophile elimination

Problem 8.66

Think

Which bond formed during the reaction? Which bond broke? How do you show this bond formation and bond breaking using curved arrows?

Solve

(a) This is an S_N2 reaction in which the $(PO_3)_3O^{5-}$ leaving group leaves the 1° carbon.

Methionine + Adenosine triphosphate (ATP) → (S_N2) (*S*)-Adenosyl methionine (SAM) + Leaving group

(b) The O atoms on the methionine are part of a carboxylate group, and they share a negative charge through resonance. Notice that the triphosphate leaving group has several negative charges. Therefore, charge repulsion prevents the CO_2^- portion from acting as the nucleophile. The S atom has no formal charge, so there is less charge repulsion during nucleophilic attack, allowing it to bind to the electrophilic carbon in ATP.

Problem 8.67

Think

For guanidoacetate to gain a N–C bond, is the reaction a substitution or elimination? To make that bond, which species needs to be the electron-rich nucleophile, guanidoacetate or SAM? Which species needs to be the electron-poor substrate? What must be the leaving group? Are any proton transfer reactions involved?

Solve

The reaction must be a substitution because a new single bond is formed and no new double or triple bonds are formed. The CH_3 bound to the S^+ on SAM is the electrophilic (i.e., electron-poor) carbon, and the S^+ on SAM is part of the leaving group. The N atom of the NH group is the electron-rich, nucleophilic site. The S_N2 reaction is shown below.

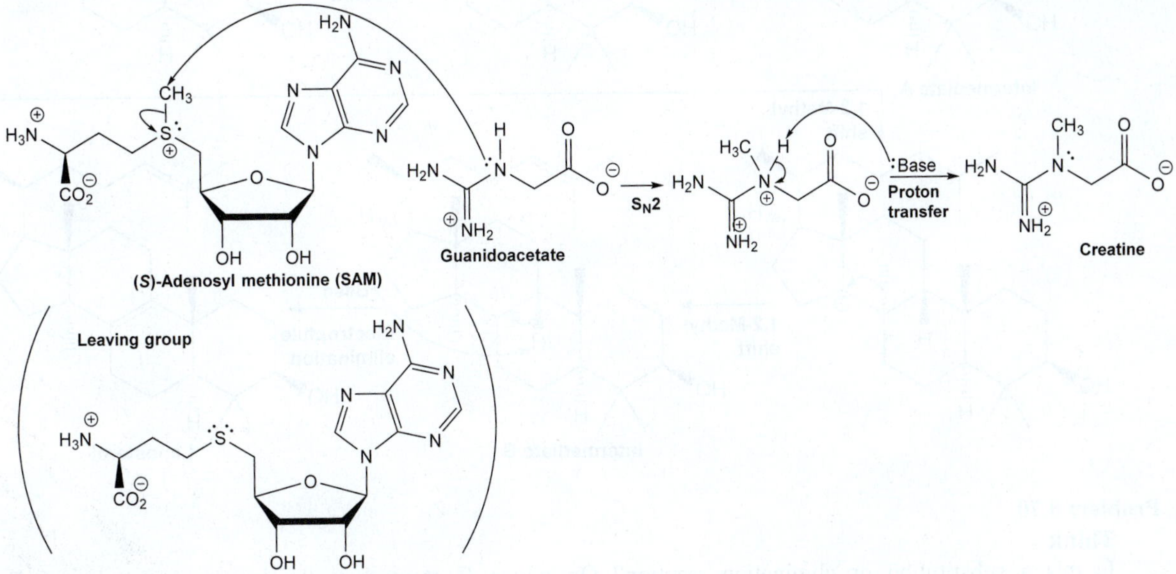

(S)-Adenosyl methionine (SAM)

Guanidoacetate

S_N2

:Base

Proton transfer

Creatine

Leaving group

Problem 8.68

Think

What happens to the stereochemistry at the electrophilic carbon in an S_N2 reaction? Is the CH_3 (R) group a dash or wedge in the product? What must it be, therefore, in the reactant?

Solve

The CH_3 group on L-alanine is dash, and S_N2 reactions invert the stereochemistry at the electrophilic carbon. Therefore, in the starting α-bromo acid, the CH_3 needs to be wedge.

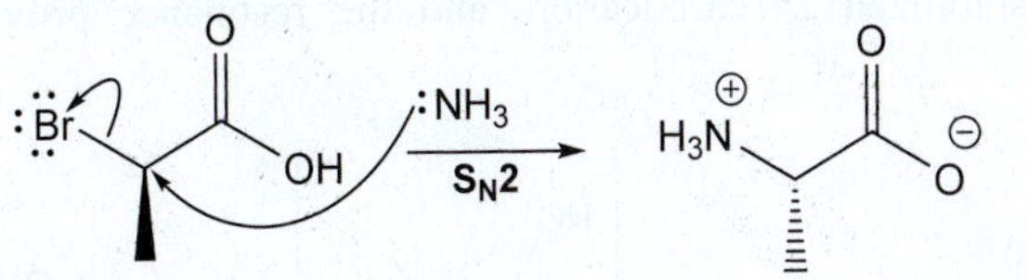

:NH_3

S_N2

Problem 8.69

Think

Identify the H atoms and CH_3 atoms involved in the 1,2-hydride and 1,2-methyl shifts. Where is the positive charge on the carbon located in each shift? How many curved arrows are used for each of these steps? Are the curved arrows drawn from electron rich to electron poor, or vice versa? For the elimination of H^+ involving the base, identify the new C=C bond that is formed.

Solve

The 1,2-hydride shifts, 1,2-methyl shifts, and electrophile elimination step are shown below. Each 1,2-hydride and 1,2-methyl shift requires one curved arrow, originating from the electron-rich bond and pointing to the electron-poor C^+. The final elimination of H^+ requires two curved arrows.

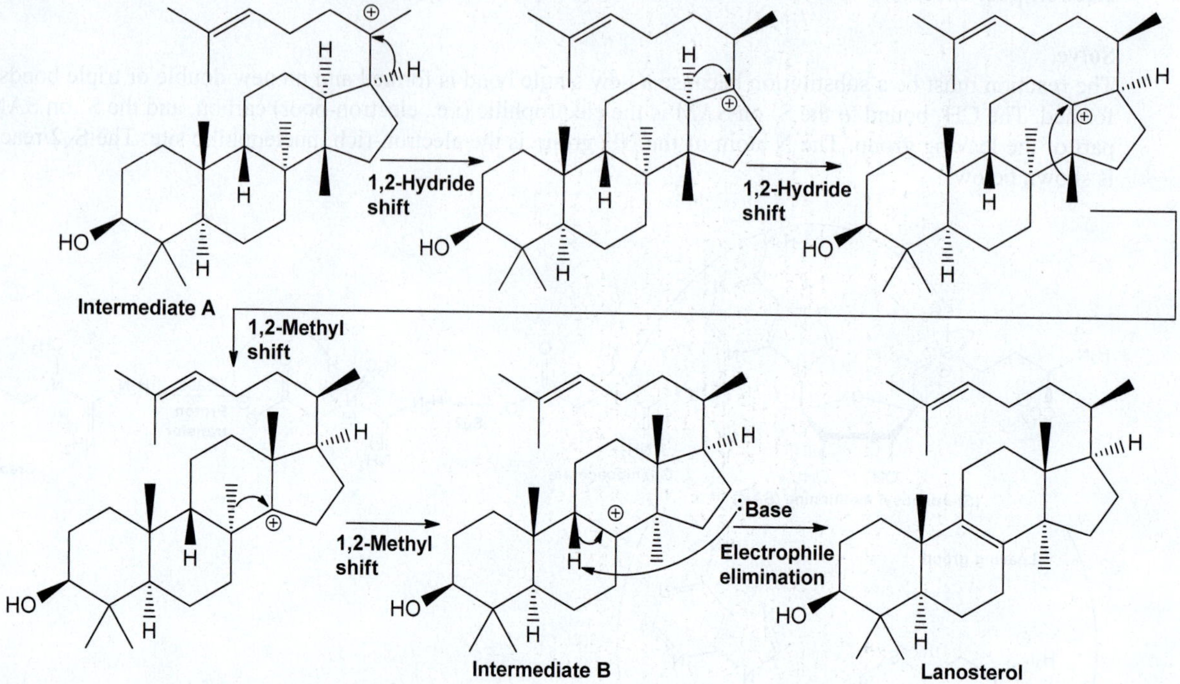

Intermediate A — 1,2-Hydride shift → 1,2-Hydride shift → 1,2-Methyl shift → 1,2-Methyl shift → Intermediate B — :Base, Electrophile elimination → Lanosterol

Problem 8.70

Think

Is this a substitution or elimination reaction? On which C atom does the leaving group leave? For a rearrangement to occur, is the mechanism unimolecular or bimolecular? How can you incorporate proton transfer reactions to avoid the appearance of strong bases in the mechanism, which are incompatible with the acidic conditions of the reaction?

Solve

(a) The mechanism involves an E1 reaction. It is an elimination because an OH group is missing from the products and a new double bond between C and O has been formed. It must be unimolecular because, for a carbocation rearrangement to take place, a carbocation must have formed. Notice that a proton transfer is necessary in the first step because the leaving group does not come off as HO^- under acidic conditions.

(b) The 3° carbocation rearranges to a resonance-stabilized 2° carbocation, and the resonance provides substantial charge stability.

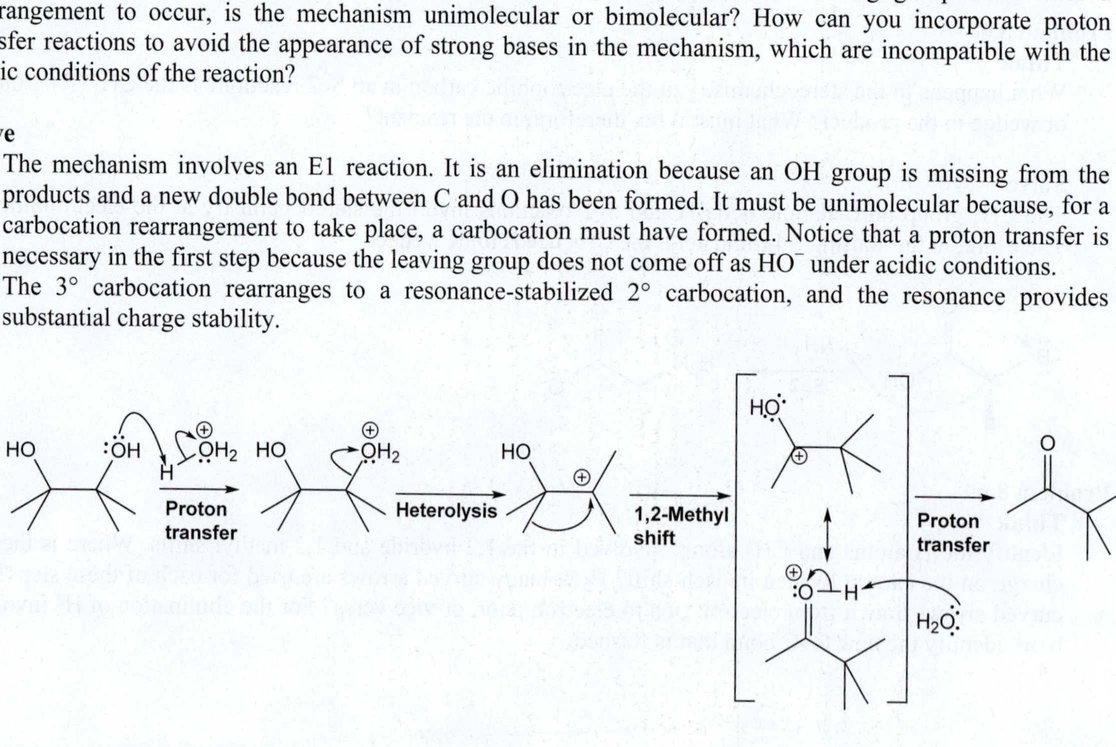

Proton transfer → Heterolysis → 1,2-Methyl shift → Proton transfer (H_2O)

Problem 8.71

Think

Is this a substitution or an elimination reaction? From which C atom does the leaving group leave? Are any proton transfer reactions involved to ensure that the species are compatible with the conditions of the reaction? Does rearrangement occur?

Solve

This reaction involves four substitutions, each of which involves Br being replaced by an O-containing group. Proton transfer steps are necessary in Steps 2 and 5 to ensure that the species remain compatible with the basic solution. Without those proton transfer steps, the product would be a strongly acidic R_2OH^+.

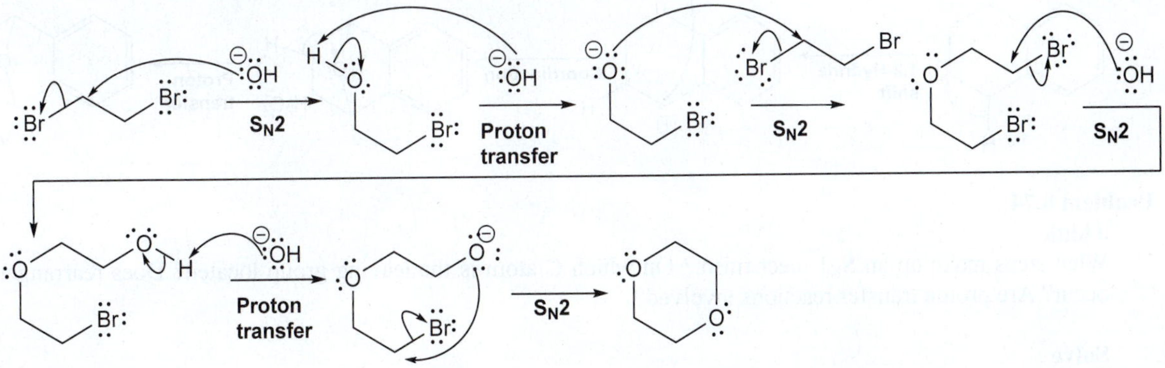

S_N2 Proton transfer S_N2 S_N2

Proton transfer S_N2

Problem 8.72

Think

Which mechanism, S_N2 or S_N1, is dependent on the concentration of the attacking species? How do you account for the optically inactive sample that is produced?

Solve

The fact that the rate of the reaction is directly proportional to the concentration of the attacking species, Br^-, suggests S_N2. The reason that the optical rotation goes to zero has to do with the fact that the S_N2 reaction produces the enantiomer of the reactant, resulting in a racemic mixture of enantiomers. This is shown below.

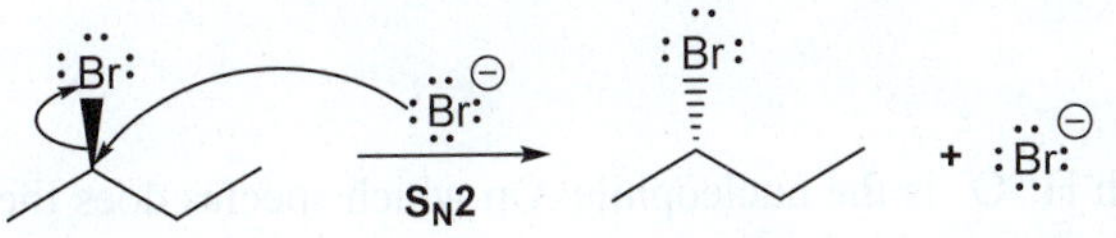

S_N2

Problem 8.73

Think

Does this reaction involve substitution or elimination? How many such substitutions or eliminations must take place? How many leaving groups are present? Are proton transfers involved? What is the attacking species? Are any rearrangements involved?

Solve

The reaction involves substitution, not elimination, because two C–Br bonds are replaced by C–O bonds. Note also that a carbocation rearrangement is necessary, because substitution takes place at a benzylic carbon that does *not* have a leaving group. This suggests S_N1. An important detail to notice is that in the product, an O atom will be bonded to two C atoms adjacent to the aromatic ring. These are benzylic carbons, which can stabilize a positive charge quite well via resonance. And the S_N1 conditions favor the formation of such carbocations. The choice of which Br atom leaves first is *not* arbitrary, because loss of the Br atom that is already next to the aromatic ring leads directly to a resonance-stabilized carbocation. Heterolysis of the other C–Br bond does not have this kind of stabilization initially. See the mechanism on the next page.

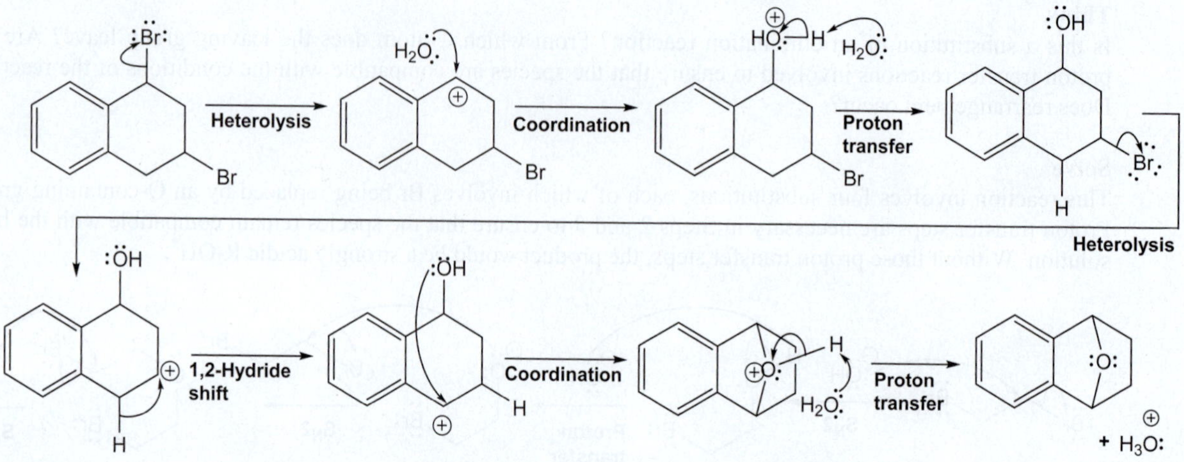

Problem 8.74

Think

What steps make up an S_N1 mechanism? On which C atom is the leaving group located? Does rearrangement occur? Are proton transfer reactions involved?

Solve

A carbocation rearrangement must take place, because the carbon backbone in the product is different from that in the reactants. Therefore, a carbocation must be formed. The product will form as a mixture of stereoisomers, because in Steps 2 and 3, new chiral centers are formed from the addition of a nucleophile to a planar C^+.

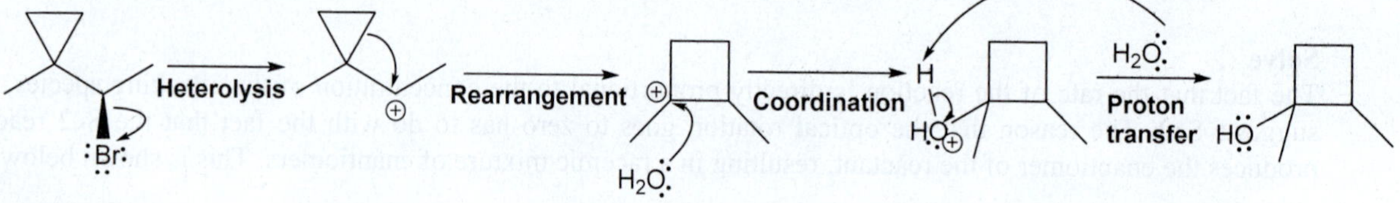

Problem 8.75

Think

Draw out the mechanism for the S_N2 reaction in which $H^{18}O^-$ is the nucleophile. On which species does the ^{18}O label appear if the reaction proceeds via S_N2? Does this agree with the products actually observed?

Solve

These results suggest that the mechanism is *not* S_N2. The S_N2 mechanism that would take place is shown below, and as we can see, the labeled O atom would end up in CH_3OH. The actual mechanism that takes place to account for the experimental results is discussed in Chapter 17.

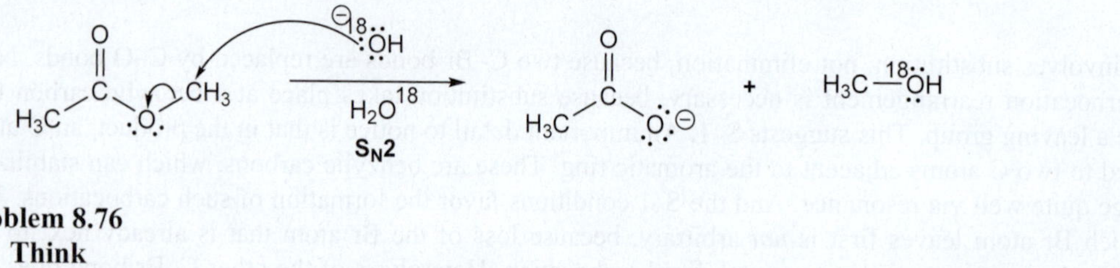

Problem 8.76

Think

Draw out the mechanism for the S_N2 reaction in which $H^{18}O^-$ is the nucleophile. On which species does the ^{18}O label appear if the reaction proceeds via S_N2? Does this agree with the products actually observed?

Solve

These results suggest that the mechanism is S_N2, as shown below. The explanation is provided in the previous problem.

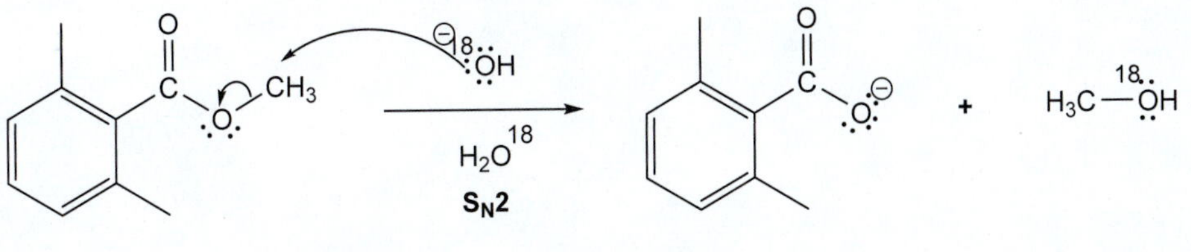

CHAPTER 9 | Nucleophilic Substitution and Elimination Reactions 1: Competition among S_N2, S_N1, E2, and E1 Reactions

Your Turn Exercises

Your Turn 9.1

Think

When an attacking species acts as a base, to what type of electron-poor atom does it bond? When an attacking species acts as a nucleophile, to what type of electron-poor atom does it bond?

Solve

A base uses its lone pair of electrons to form a bond to a hydrogen atom, which is what takes place in elementary step **A**. A nucleophile uses its lone pair of electrons to form a bond to an electron-poor nonhydrogen atom, which is what takes place in elementary step **B**. In this case, a C atom is attacked by NH_3.

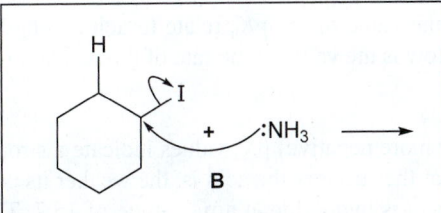

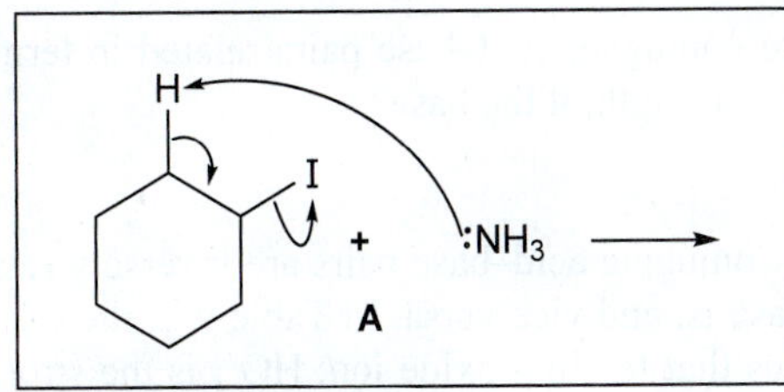

Your Turn 9.2

Think

From where does the arrow begin and end on a free energy diagram to indicate the energy barrier of a reaction? In which figure is the arrow longer? How does the length of the arrow correlate to the value of the energy barrier? How does the length of the arrow correlate to the rate of the reaction?

Solve

In a free energy diagram, the arrow that indicates the energy barrier starts at the energy level of the reactants (or an intermediate, if the reaction proceeds via more than one step) and ends at the energy level of the transition state. The arrow for the energy barrier for the reaction in Figure 9-1a and 9-1b is shown below. The exergonic reaction in Figure 9-1a has a shorter arrow compared to the endergonic reaction in Figure 9-1b. This indicates that the energy barrier for the reaction in Figure 9-1a is smaller than the reaction in Figure 9-1b, and thus the reaction is faster.

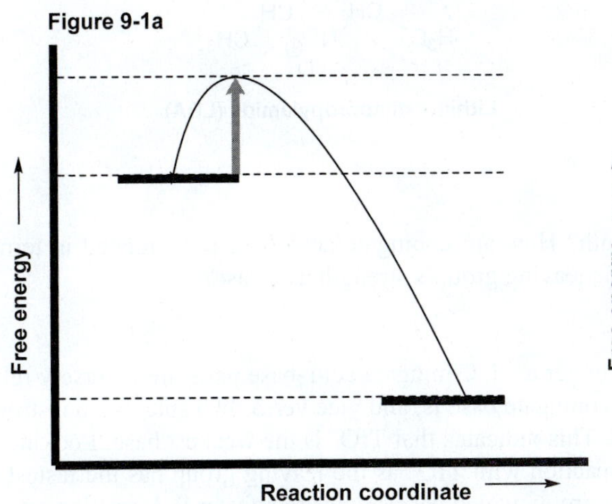

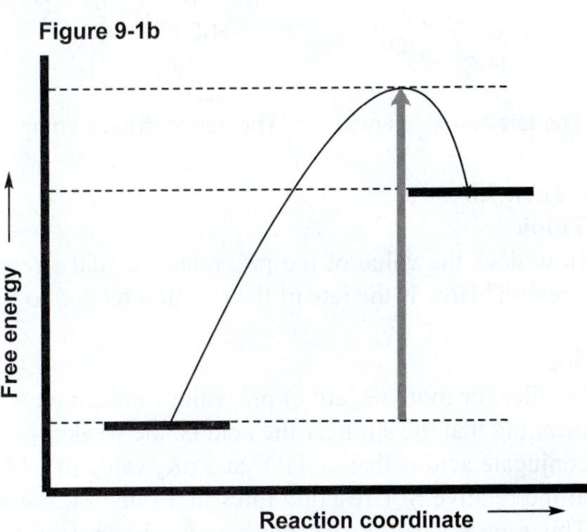

Your Turn 9.3

Think

Looking at the curved arrows, identify which bonds are broken and formed. Is a nucleophilic atom electron rich or electron poor? What kinds of charges are associated with a nucleophile?

Solve

In the reaction below, HO^- acts as a base and attacks the H. The C–H bond breaks and the electrons fold onto the C atom, forming a carbanion. The $C:^-$ is electron rich due to its lone pair and −1 formal charge and, therefore, is a nucleophile.

:N≡C–H + $:\ddot{O}H^-$ → (Proton transfer) :N≡C:$^-$ + H_2O

Hydrocyanic acid

Nucleophilic atom

Your Turn 9.4

Think

How does the value of the pK_a relate to acid strength? How are conjugate acid–base pairs related in terms of strength? How is the value of the rate of the reaction related to the strength of the base?

Solve

Smaller (or more negative) pK_a values indicate a stronger acid. Conjugate acid–base pairs are inversely related, meaning that the stronger the acid is, the weaker its conjugate base is, and vice versa. In Table 9-2, the weakest conjugate acid is hydroxide at a pK_a value of 15.7. This indicates that the hydroxide ion, HO^-, is the strongest base. Looking now at the relative E2 reaction rates in Table 9-2, the hydroxide ion has the fastest rate. This suggests that the stronger base leads to a faster E2 reaction rate.

Your Turn 9.5

Think

Which groups are bulky? Sometimes it helps to draw out the C–H bonds to get a better understanding of the size of a CH_3 group versus a $C(CH_3)_3$ group. How is bulkiness related to steric hindrance?

Solve

Tert-butyl and isopropyl groups are bulky and contribute to the steric hindrance of the bases in this problem.

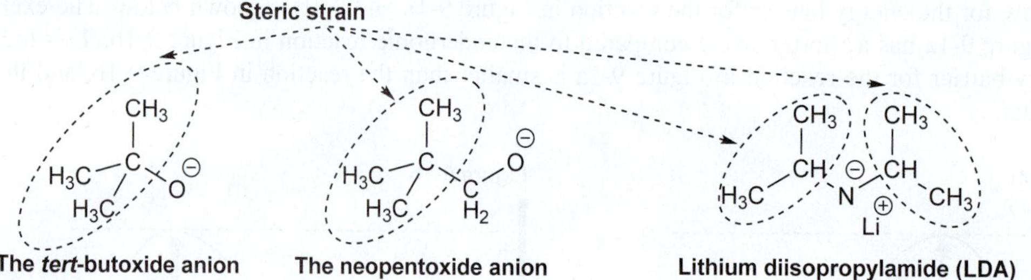

Your Turn 9.6

Think

How does the value of the pK_a relate to acid strength? How are conjugate acid–base pairs related in terms of strength? How is the rate of the reaction related to the leaving group's strength as a base?

Solve

Smaller (or more negative) pK_a values indicate a stronger acid. Conjugate acid–base pairs are inversely related, meaning that the stronger the acid is, the weaker its conjugate base is, and vice versa. In Table 9-4, the strongest conjugate acid is that of TfO^- at a pK_a value of −13. This indicates that TfO^- is the weakest base. Looking now at the relative S_N1 reaction rates in Table 9-4, the reaction with TfO^- as the leaving group has the fastest rate. This suggests that the weaker base is a better leaving group, which corresponds to a faster S_N1 reaction rate.

Your Turn 9.7

Think

Are any lone pairs adjacent to a double or triple bond? Can any of the double bonds convert into lone pairs? How can you show via a curved arrow the electrons in the lone pair moving to form a double bond? How can you show via a curved arrow the electrons in a double bond moving to form a lone pair?

Solve

Recall that atoms do not move on going from one resonance structure to another. Only nonbonding electrons or electrons from multiple bonds move. In the first resonance structure below, the first curved arrow is used to convert the lone pair on the O^- into a covalent double bond, S=O. The second arrow is used to convert a pair of electrons from the double bond, S=O, into a lone pair on the O, resulting in a formal charge of −1. This same pattern is used to convert the second resonance structure into the third.

H_3C—S—O: ⟷ H_3C—S=O: ⟷ H_3C—S=O:

Methyl sulfonate (Mesylate, MsO^-)

Your Turn 9.8

Think

What criteria do you use to judge the ability of a leaving group to depart? What is the importance of charge stability? What types of formal charges are most favored (positive, negative, or uncharged) in good leaving groups?

Solve

The more stable a leaving group is in the form in which it has left, the better its leaving group ability. Good leaving groups are ones that depart as uncharged molecules or as anions in which the −1 charge is stabilized on a large atom or by resonance. Leaving groups tend to be better when their departed forms are weaker bases. In the example below, HO^- or H_2O is the possible leaving group, respectively. HO^- is a strong base and, therefore, is a very poor leaving group. H_2O is an excellent leaving group because it is uncharged and a weak base.

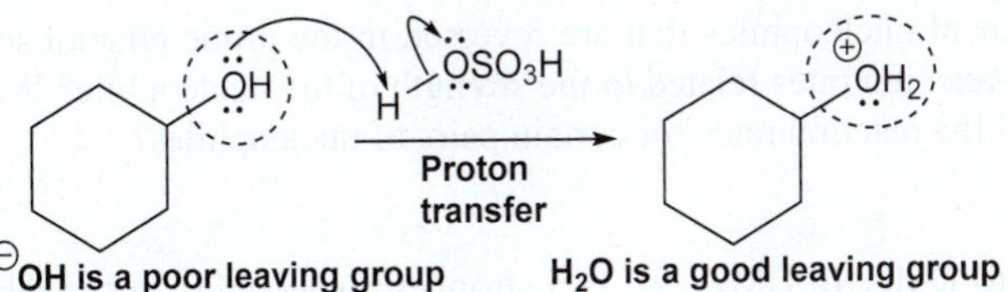

Your Turn 9.9

Think

Consult Figure 3-32 and write down the bond energy for each C–H bond. What is the percent *s* character in an sp^3, sp^2, and *sp* C–H bond? Why do you think that an increasing percent *s* character increases the bond strength?

Solve

Increasing percent *s* character increases the bond strength because pure 2*s* orbitals hold electrons closer to the nucleus than pure 2*p* orbitals do. This leads to more compact orbitals as the percent *s* character increases.

sp^3 C—H	sp^2 C—H	sp C—H
421 kJ/mol	**464 kJ/mol**	**558 kJ/mol**

Your Turn 9.10

Think

Do any of the atoms have an unfilled valence shell? If so, is the atom with the unfilled valence shell adjacent to a double or triple bond? With this kind of structural feature, how many curved arrows are necessary to arrive at the next resonance structure?

Solve

The C with a +1 formal charge has an unfilled valence shell and is adjacent to a double bond. One curved arrow is used to shift a pair of π electrons from the double bond to form a new π bond to the initial C^+. This leaves an unfilled valence shell on a C atom two atoms away from the original carbocation. The resonance can continue around the entire ring because all the C atoms are sp^2 hybridized. The resonance hybrid is shown below.

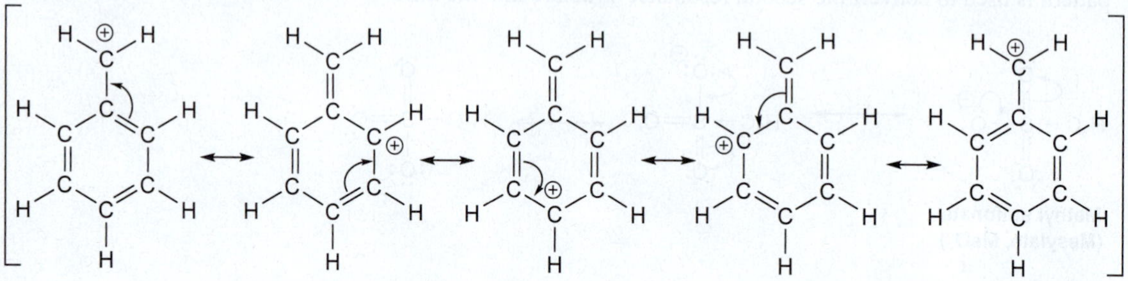

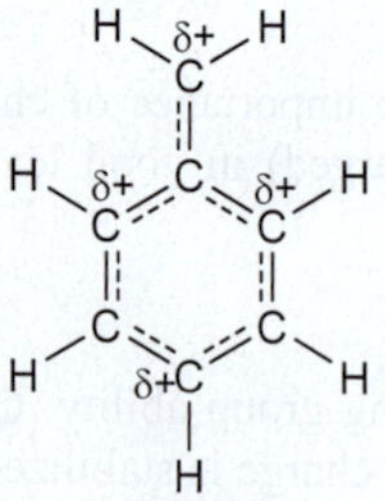

Hybrid

Your Turn 9.11

Think

Compare *relative* reaction rates and select a pair of nucleophiles that are reversed in the protic ethanol solvent compared to the aprotic DMF solvent. How are reaction rates related to the strength of the nucleophile? Why do you think a protic solvent results in a reversal of the reaction rates for certain pairs of nucleophiles?

Solve

Br^- and N_3^- are reversed in ethanol compared to DMF. In DMF, N_3^- is a stronger nucleophile, but in ethanol, Br^- is the stronger nucleophile. This suggests that N_3^- is more strongly solvated by ethanol than Br^- is. Similarly, $CH_3CO_2^-$ is a stronger nucleophile in DMF than either Br^- or Cl^-, but in ethanol, the opposite is true, suggesting that the ethanol solvates $CH_3CO_2^-$ more strongly than it solvates Br^- or Cl^-.

Your Turn 9.12

Think

What is entropy? Count the number of product species. How does the number of product species relate to the entropy?

Solve

Entropy is a measure of energy dispersal in a system and is generally thought of as a measure of "disorder." Greater entropy exists in the elimination products because of the greater number of species—three species in elimination and two species in substitution. See the figure on the next page.

OTs + $HO^{\ominus}$ → OH + $TsO^{\ominus}$

Substitution products (2 species)

→ + $TsO^{\ominus}$ + H_2O

Elimination products (3 species)
Greater entropy

Your Turn 9.13

Think

Is there a suitable leaving group? Does the type of carbon bonded to the leaving group rule out any of the four reactions? What is the attacking species (strong/weak nucleophile/base)? What is the solvent (protic/aprotic)? Fill in the chart based on these factors.

Solve

OTs is a good leaving group suitable for both substitution and elimination. The leaving group is bonded to a secondary carbon, so none of the four reactions can be ruled out. The attacking species, $(CH_3)_3CONa$, is negatively charged and is a strong base, which would normally favor both S_N2 and E2, but because it is also bulky, it favors E2 over S_N2. It is not indicated that the concentration is low and, therefore, bimolecular mechanisms are favored. However, in this case, the attacking species has a large amount of steric hindrance, and E2 is favored. The leaving group ability of OTs is excellent, which promotes all four reactions, but unimolecular mechanisms (S_N1 and E1) are more sensitive to the leaving group ability. The solvent DMF is an aprotic solvent that favors S_N2 and E2.

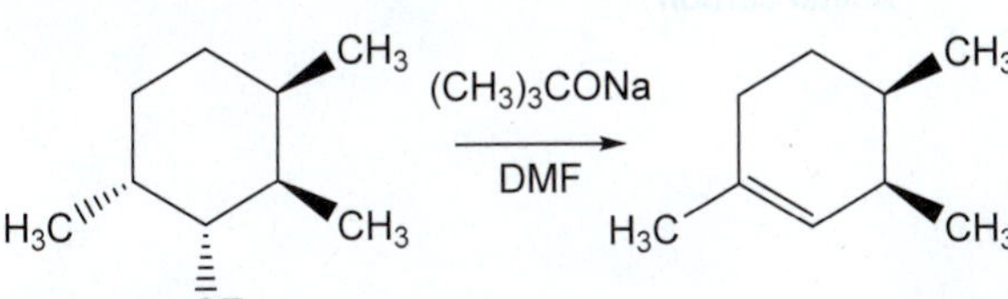

Factor	S_N1	S_N2	E1	E2
Strength				✓
Concentration				✓
Leaving group	✓		✓	
Solvent		✓		✓
Total	1	1	1	**3**

Your Turn 9.14

Think

How many carbon atoms surround each C=C in the alkene products for Equations 9-42 and 9-43? What does this indicate about the degree of alkyl substitution for each?

Solve

In Equation 9-42, hex-2-ene has two alkyl groups surrounding the C=C and is disubstituted; hex-1-ene has one alkyl group surrounding the C=C and is monosubstituted.

$CH_3CH_2CH_2$—C(H)=C(H)—CH_3 $CH_3CH_2CH_2CH_2$—C(H)=CH_2

Hex-2-ene
Disubstituted

Hex-1-ene
Monosubstituted

In Equation 9-43, 2-methylbut-1-ene has two alkyl groups surrounding the C=C and is disubstituted; 2-methylbut-2-ene has three alkyl groups surrounding the C=C and is trisubstituted.

2-Methylbut-1-ene
Disubstituted

2-Methylbut-2-ene
Trisubstituted

Your Turn 9.15

Think

Refer to Figure 9-20 to review the nature of an acetal group. What groups should be bonded to a C atom to characterize it as an acetal carbon?

Solve

An acetal carbon is attached to two OR groups and either an R or H group at the other two bonds. In cellulose, amylose, and amylopectin, C1 in all rings is an acetal carbon. The acetal carbons are labeled and shaded for each structure below.

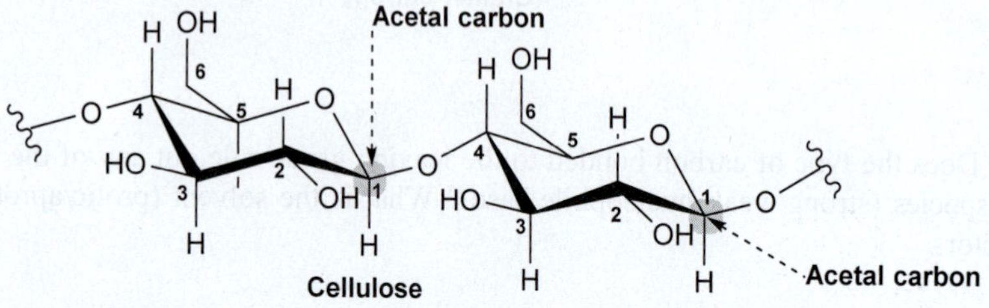

Acetal carbon

Acetal carbon

Cellulose

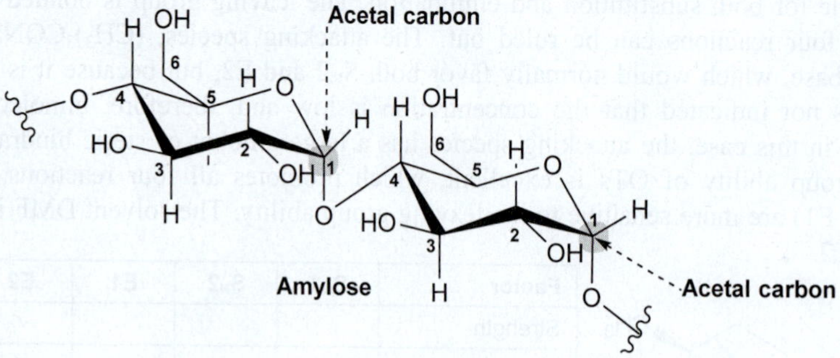

Acetal carbon

Acetal carbon

Amylose

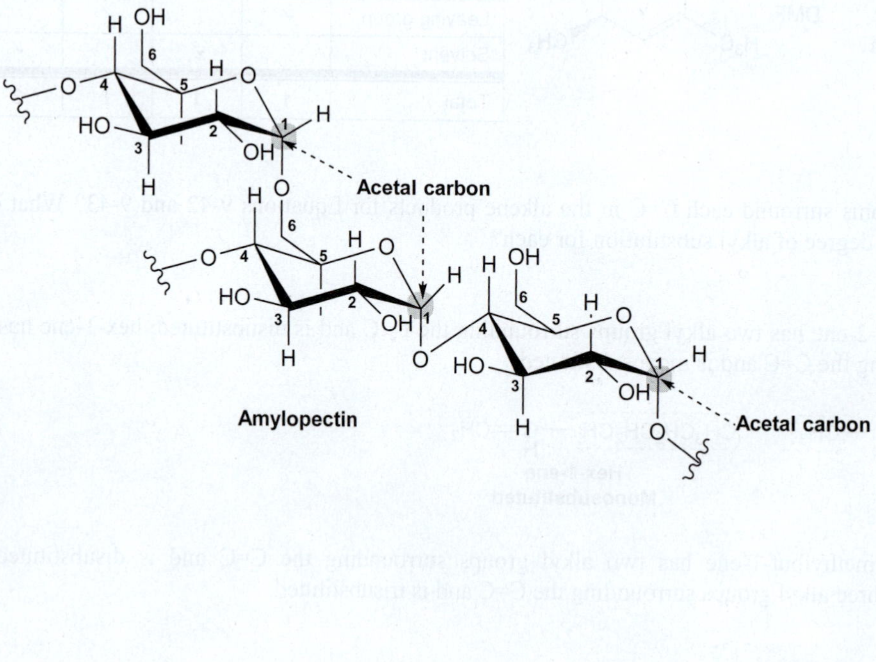

Acetal carbon

Acetal carbon

Amylopectin

In Chapter Problems

Problem 9.1

Think

Review the elementary steps for each mechanism, S_N1, S_N2, E1, and E2. In how many steps does each occur? What is the attacking species? What is the substrate? What is different about the site of attack in substitution versus elimination? If an uncharged nucleophile is involved, are proton transfer steps necessary?

Solve

In the substitution mechanisms, the attacking species (NH_3) acts as a nucleophile and attacks the electrophilic C atom. After a subsequent deprotonation, an amine results as the product. In the elimination mechanisms, the attacking species (NH_3) acts as a base and attacks the H atom adjacent to the electrophilic C atom. An alkene product results.

(a) S_N2: one step followed by a proton transfer, amine product.

S_N2

Proton transfer

(b) S_N1: two steps (heterolysis, coordination) followed by proton transfer, amine product.

Heterolysis

Coordination

Proton transfer

(c) E2: one step, alkene product.

E2

(d) E1: two steps (heterolysis, electrophile elimination), alkene product.

Heterolysis

Electrophile elimination

Problem 9.2

Think

Which nucleophile has a faster S_N2 reaction rate? What does that mean about the energy barrier? Which transition state will have a lower activation energy? Should that lower activation energy correspond to a difference in energy of the reactants or of the products?

Solve

Because the reaction involving NC^- is much faster (250 times faster), the free energy of activation must be smaller. Because the same charged species appears in both sets of products, that difference in activation energy should correspond to a difference in reactant energies. According to the Hammond postulate, for the attack by NC^- to have the lower energy barrier, NC^- is higher in energy than Br^-.

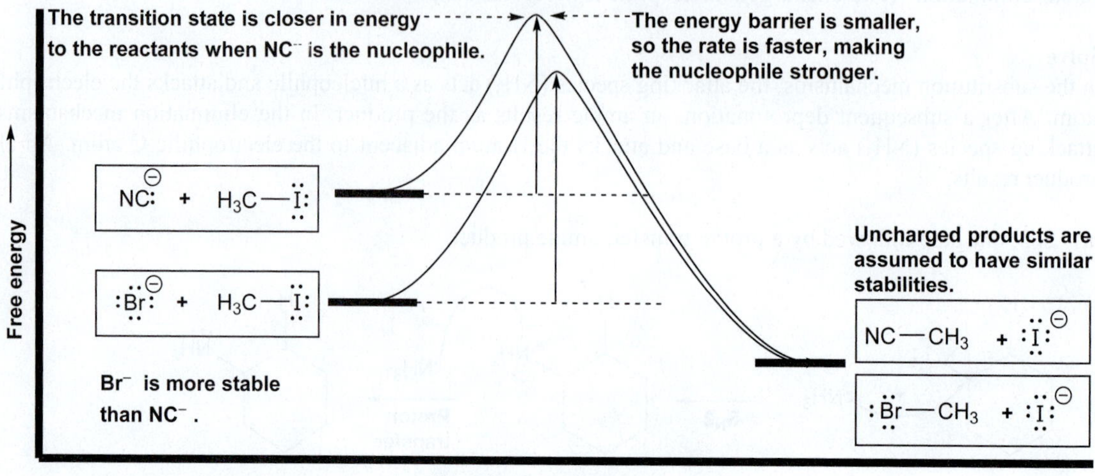

Problem 9.4

Think

Are there differences in charge stability between the two reactions on the reactant side? On the product side? Which ion is more stable? Which ion is higher in energy? How do the relative energies of the reactants or products relate to the energy barriers? How does the energy barrier relate to the S_N2 reaction rate?

Solve

(i) CH_3O^- or $CH_3O_2^-$

(a) There is a significant difference in charge stability on the reactant side but not on the product side. Because the negative charge in $CH_3CO_2^-$ is resonance delocalized, it is more stable than CH_3O^-. Therefore, the pair of free energy diagrams appears as follows. As we can see, the energy barrier involving CH_3O^- is smaller, consistent with what would be suggested by the Hammond postulate, so the reaction is faster. CH_3O^- is a stronger nucleophile.

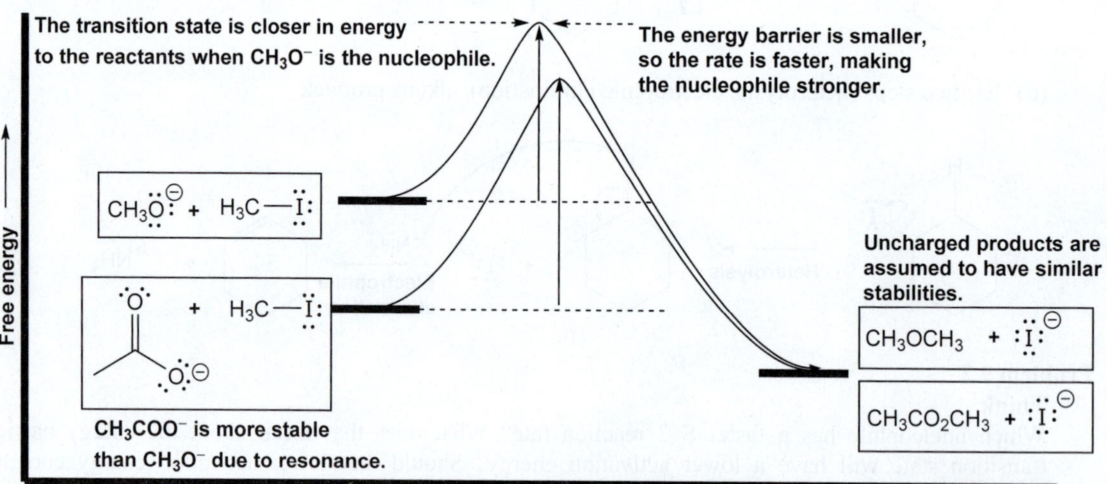

(b) CH_3O^- is a stronger nucleophile compared to $CH_3CO_2^-$.

(ii) H_3N or H_3P

(a) There is a significant difference in charge stability on the product side but not on the reactant side. Because a positive charge is more stable on P than on N (P is the bigger atom), the product of nucleophilic substitution involving PH_3 is more stable, and the Hammond postulate suggests that its reaction should have a smaller energy barrier. Thus it proceeds faster.

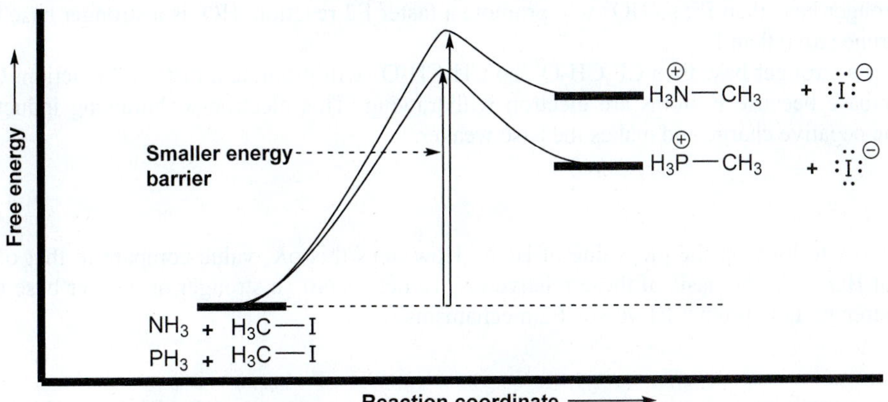

(b) H_3P is a stronger nucleophile compared to H_3N.

Problem 9.5

Think

What kind of charge does this nucleophile have? Is the charge resonance delocalized? What kind of charge is generally associated with a strong nucleophile? Are S_N2 reactions favored by strong or weak nucleophiles? What about S_N1 reactions?

Solve

H_2P^- is a strong nucleophile, because it has a negative charge localized on a single atom (resonance is not possible). Therefore, it will favor the S_N2 mechanism. Strong nucleophiles promote fast S_N2 reactions by forcing the leaving group out. Being a strong nucleophile, H_2P^- will not wait until the leaving group leaves (slow step in S_N1).

Strong nucleophile

Problem 9.6

Think

Does the alkyne have a significantly nucleophilic atom? How does the $H:^-$ react with the terminal alkyne? What is the identity of your nucleophile after $H:^-$ reacts? Which species acts as the electrophile in the S_N2 reaction? How many steps are in the S_N2 reaction?

Solve

The purpose of the NaH step is to deprotonate the terminal alkyne and produce the alkynide ion. In a second step, the S_N2 reaction (one step) takes place, in which the nucleophilic C^- attacks the electrophilic C atom bonded to Br. The six-carbon terminal alkyne is transformed into oct-3-yne.

Hex-1-yne —($:H^{\ominus}$, **Proton transfer**)→ —(Br, **S_N2**)→ **Oct-3-yne**

Problem 9.8

Think

How does the strength of the base relate to the rate of the E2 reaction? Which base is stronger? How can you predict base strength (see Chapter 6)?

Solve

(a) HO^- is a stronger base than F^-, so HO^- will promote a faster E2 reaction. HO^- is a stronger base because O is less electronegative than F^-.

(b) $CH_3CH_2O^-$ is a stronger base than $CF_3CH_2O^-$, so $CH_3CH_2O^-$ will promote a faster E2 reaction. $CF_3CH_2O^-$ is a weaker base because F atoms are electron withdrawing. This electron-withdrawing inductive effect stabilizes the negative charge and makes the base weaker.

Problem 9.10

Think

Consult Appendix A to look up the pK_a value of HCN. How does this pK_a value compare to that of H_2O, the conjugate acid of HO^-? On the basis of these relative pK_a values, is NC^- a stronger or weaker base than HO^-? How does base strength help predict E1 versus E2 mechanisms?

Solve

NC^- is considered a weak base because it is a weaker base than our benchmark, HO^-. We know this because the pK_a of each group's conjugate acid is 9.2 and 15.7, respectively. Weak bases tend to favor the E1 mechanism.

Problem 9.11

Think

Can charge stability distinguish among these three nucleophiles? If not, what other factor is present? What is the effect of nearby methyl groups on the strength of the nucleophile?

Solve

The nucleophilic sites are all negatively charged O atoms. Thus, the charge on the atom is not a factor in this example. The difference is in the proximity of steric hindrance surrounding those sites, represented by the *t*-butyl groups. The nucleophile in **B** has the *t*-butyl group closest to the O atom, so **B** will give rise to the slowest reaction. The nucleophile in **C** has the *t*-butyl group farthest away from O, so it will proceed the fastest. The reaction rate thus increases in the order **B < A < C**.

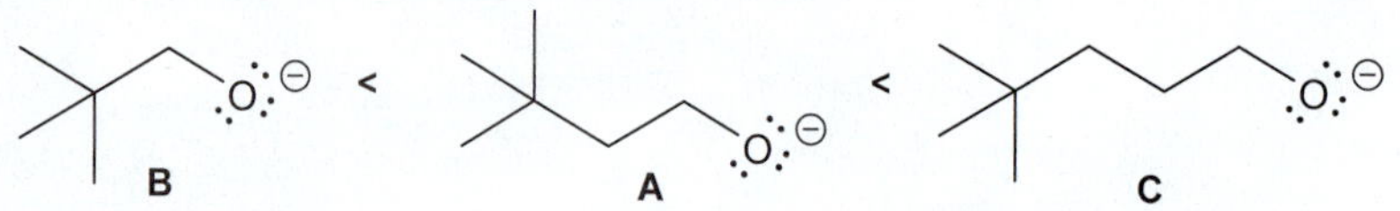

B < A < C

Slowest, *t*-butyl group closest to nucleophilic site

Fastest, *t*-butyl group farthest away from nucleophilic site

Problem 9.13

Think

Is the anion in question a strong nucleophile or a weak nucleophile? Is the anion in question a strong base or a weak base? What relative concentrations favor each reaction? What are the exceptions?

Solve

(a) S_N2, because HO^- is a strong nucleophile in high concentration; E2, because HO^- is a strong base in high concentration

(b) S_N1 and E1, even though the attacking species is a strong nucleophile and a strong base, because the low concentrations of the nucleophile/base greatly slow S_N2 and E2

(c) S_N2, because Br^- is a strong nucleophile in high concentration; E1, because Br^- is a weak base

(d) S_N1, even though Br^- is a strong nucleophile, because the low concentration of the nucleophile greatly slows S_N2; E1, because Br^- is a weak base

(e) S_N2, because $(CH_3)_3CO^-$ is a strong nucleophile in high concentration; E2, because it is a strong base in high concentration—E2 will beat out S_N2, however, because the base is very bulky

(f) S_N1 and E1, for the same reasons as in **(b)**

Problem 9.14

Think

What is the pK_a of each base's conjugate acid? What does that mean for the relative strength of the base? How is the strength of the base related to the leaving group ability? What if you could not look up the pK_a; how else could you evaluate the stability of the anion?

Solve

HCO_2^- is a better leaving group because it is a weaker base. The pK_a of HCO_2H is 3.75, whereas that of C_6H_5OH is 10.0. It is evident that HCO_2^- is a more stable anion because the negative charge can be delocalized over two O atoms. A more stable anion leads to a weaker, less reactive base. $C_6H_5O^-$ does have a delocalized charge, but the charge is delocalized over C atoms, so it is less stable.

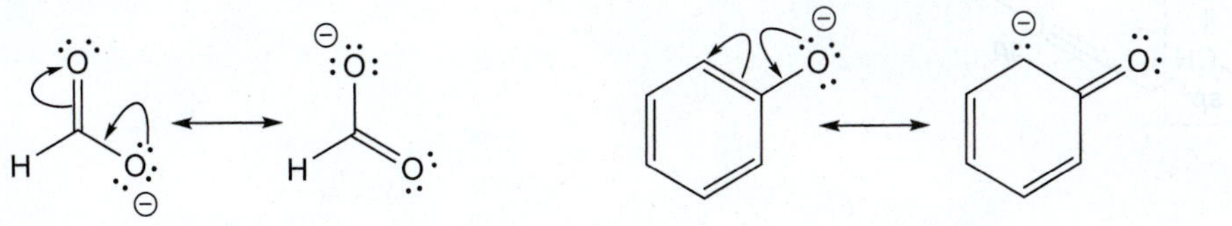

Negative charge delocalization onto oxygen
More stable anion

Negative charge delocalization onto carbon
Less stable anion

Problem 9.16

Think

What are the relative strengths of the bases in **D–F**, and how are the E1 and E2 reactions affected by base strength? What are the relative leaving group abilities of the leaving groups that appear in **D–F**, and how are the E1 and E2 reactions affected by leaving group ability?

Solve

The rate of an E1 reaction depends only on the rate of the departure of the leaving group. The difference among the substrates is only in the identity of the leaving group. Because Br^- is not as good a leaving group as TsO^-, **F** will proceed the slowest via an E1 mechanism. By contrast, an E2 reaction rate depends both on the strength of the base and on the ability of the leaving group to leave. **E** will proceed faster than **D** by E2 because HO^- is a stronger base than NH_3. And **E** will proceed faster than **F** by an E2 because TsO^- is a better leaving group than Br^-.

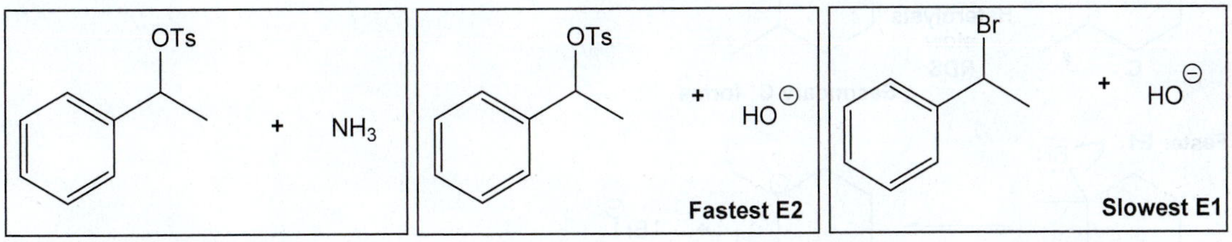

OTs + NH_3 — **D**

OTs + HO^- — **Fastest E2** — **E**

Br + HO^- — **Slowest E1** — **F**

Problem 9.17

Think

How does the basicity of C_6H_5OH compare to that of other leaving groups with which you are familiar? Is a comparable leaving group excellent, moderate, or poor? Which mechanism is more sensitive to leaving group ability, bimolecular or unimolecular?

Solve

C_6H_5OH should have a leaving group ability similar to that of H_2O, which is an excellent leaving group. S_N1 and E1 reactions are more sensitive to leaving group ability than S_N2 and E_2 reactions, so C_6H_5OH acting as a leaving group should favor S_N1 and E1 over S_N2 and E2.

Problem 9.18

Think

Are the leaving groups the same or different? What is the hybridization of the C atom attached to each Br atom? How does the hybridization of that carbon affect the ability of the leaving group to leave?

Solve

Nucleophilic substitution will occur most readily at the sp^3-hybridized C shown below. The leaving group on each C atom is Br^-, so the leaving group identity does not come into play. The other Br leaving groups are on sp^2- and sp-hybridized C atoms, respectively, which tend not to undergo nucleophilic substitution or elimination.

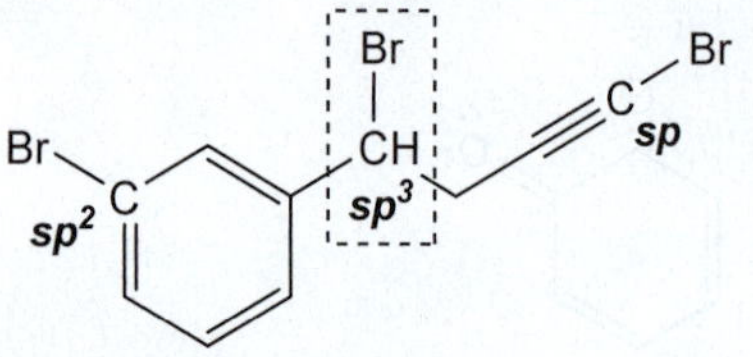

Problem 9.20

Think

What is the rate-determining step in an E1 reaction? Are there any differences in the leaving group? What reactive intermediate forms from the slow step? What stabilizes/destabilizes this reactive intermediate? What happens to the rate when the intermediate is stabilized?

Solve

In an E1 reaction, the rate-determining step is the departure of the leaving group, heterolysis, to form the carbocation. The leaving group, Br^-, is the same in both substrates, so the difference is in the nature of the C atom to which the leaving group is bonded. In the first molecule, the C is secondary, whereas in the second molecule, the C is tertiary. As shown below, the molecule with the tertiary carbon produces the more stable carbocation in the rate-determining step of an E1 reaction, so it will be involved in the faster E1 reaction.

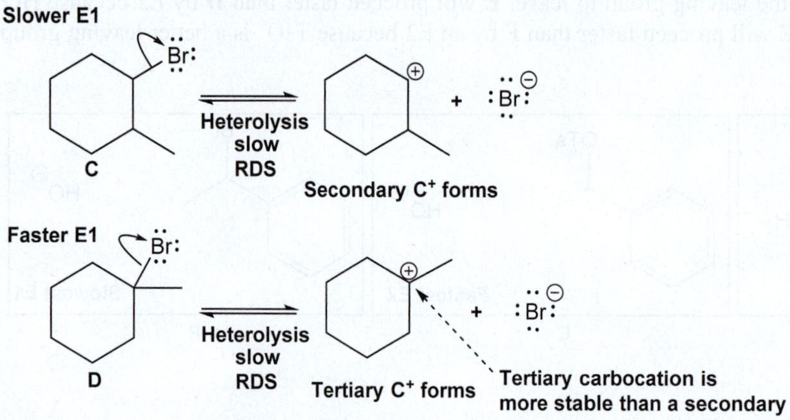

Problem 9.21

Think

Which atom is attacked by the base in an E2? What is the resulting functional group? What is the minimum number of carbon atoms for that functional group to form?

Solve

E2 reactions require a H and a leaving group attached to adjacent atoms. Usually they are on adjacent C atoms, producing a new C=C bond. With a substrate of the form CH_3–L, there is no such adjacent C atom, so no double bond can be formed. See the general examples on the next page.

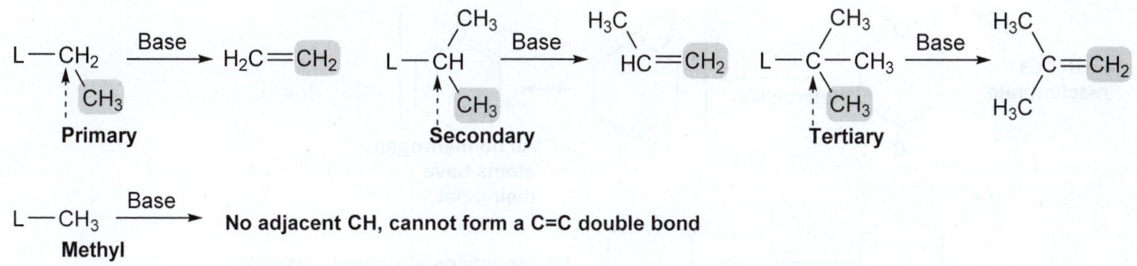

L—CH2(CH3) —Base→ H2C=CH2 (Primary); L—CH(CH3)(CH3) —Base→ H3C–HC=CH2 (Secondary); L—C(CH3)(CH3)(CH3) —Base→ (H3C)2C=CH2 (Tertiary)

L—CH3 —Base→ No adjacent CH, cannot form a C=C double bond (Methyl)

Problem 9.22

Think

Do the substrates differ in their leaving groups? To what type of carbons (primary, secondary, tertiary) is the leaving group attached? What is the rate-determining step for each reaction? For S_N2 reactions, how does steric hindrance involving the substrate R–L affect the rate of the reaction? For S_N1 and E1, how does the stability of the C^+ formed affect the rate of the reaction?

Solve

X reacts fastest by an S_N2 mechanism. The leaving group, Cl^-, is the same in all three cases, so it does not factor in, but the primary carbon is the least hindered. **Z** reacts fastest by an S_N1 and E1 mechanism (same rate-determining step), because it proceeds through the more stable tertiary carbocation, as shown below. **X** and **Y** form less stable primary and secondary carbocations and, therefore, are slower via a unimolecular mechanism.

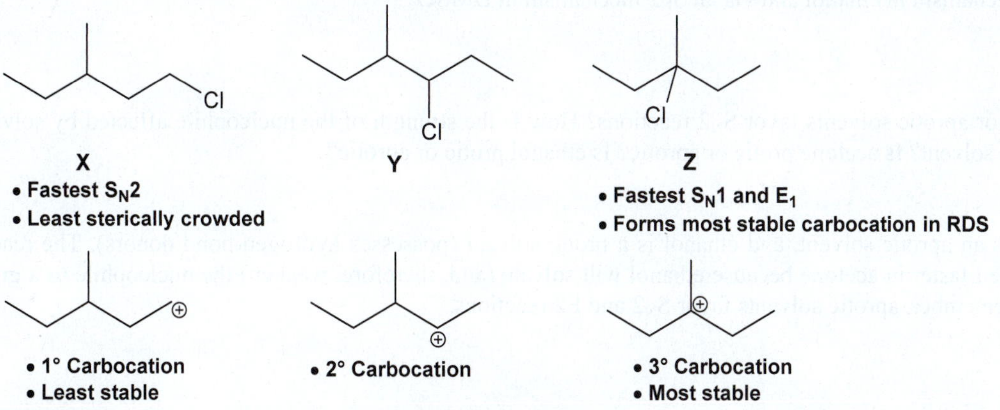

X
- Fastest S_N2
- Least sterically crowded

Y

Z
- Fastest S_N1 and E_1
- Forms most stable carbocation in RDS

- 1° Carbocation
- Least stable

- 2° Carbocation

- 3° Carbocation
- Most stable

Problem 9.24

Think

Are the leaving groups the same or different in this example? What is the carbocation intermediate involved in each E1 mechanism? Which one is more stable? Why? How does the stability of the carbocation affect the E1 reaction rate?

Solve

In both **C** and **D**, the carbocation is generated by the loss of the same leaving group, Br^-. **C** reacts faster than **D**. The carbocation formed from **C** is resonance stabilized by the adjacent O atom, as shown on the next page. This forms a strong resonance contributor with all nonhydrogen atoms having their octet. The carbocation formed from molecule **D** is not resonance stabilized (isolated on C only) and thus is less stable. Formation of the carbocation is the rate-determining step in an E1 reaction, and thus, the formation of a more stable carbocation leads to a faster reaction rate.

Faster E1 reaction rate

Heterolysis

C

All nonhydrogen atoms have their octet.

Heterolysis

D

No resonance structures possible

Problem 9.25

Think

Which solvent is protic? Which solvent is aprotic? Which mechanism is favored by protic solvents? Which mechanism is favored by aprotic solvents? Which mechanism forms ionic intermediates? Which mechanism's rate depends on the strength of the nucleophile?

Solve

Ethanol is a protic solvent, and DMSO is an aprotic solvent. Protic solvents better promote the formation of ions, which is necessary in an S_N1 (or E1) reaction. Protic solvents weaken the strength of nucleophiles, and the rate of an S_N2 reaction depends on the strength of the nucleophile. Therefore, the reaction will occur faster via an S_N1 mechanism in ethanol and via an S_N2 mechanism in DMSO.

Problem 9.26

Think

Do protic or aprotic solvents favor S_N2 reactions? How is the strength of the nucleophile affected by solvation in a protic solvent? Is acetone protic or aprotic? Is ethanol protic or aprotic?

Solve

Acetone is an aprotic solvent, and ethanol is a protic solvent (possesses hydrogen-bond donors). The reaction will proceed faster in acetone because ethanol will solvate (and, therefore, weaken) the nucleophile to a greater degree. Remember, aprotic solvents favor S_N2 and E2 reactions.

Problem 9.27

Think

Which solvent is protic? Which solvent is aprotic? Which mechanisms are favored by protic solvents? Which mechanisms are favored by aprotic solvents? Which mechanism forms ionic intermediates? Which mechanism's rate depends on the strength of the nucleophile?

Solve

Solvent **Y** is protic (it has a N–H covalent bond) and will favor the E1 mechanism more than solvent **Z**, which is aprotic (it has no N–H or O–H bonds), will. Protic solvents solvate ions (especially anions) well, which helps promote the rate-determining step of the E1.

Problem 9.29

Think

Which is a stronger nucleophile *intrinsically* (i.e., without considerations of solvation)? Is ethanol a protic or an aprotic solvent? Is the solvation of F^- in ethanol much different from the solvation of I^-?

Solve

Absent of any solvent, we would predict that F^- would be a stronger nucleophile than I^-, because I is a larger atom than F and thus can better stabilize the negative charge. In a protic solvent like ethanol, however, F^- will be solvated much more strongly than I^- will, because F^- is much smaller and thus has a more highly concentrated negative charge. That solvation of F^- weakens F^- much more than it does I^-. This makes F^- a weaker nucleophile than I^- in ethanol.

Problem 9.30

Think

Which is a stronger nucleophile *intrinsically* (i.e., without considerations of solvation)? Is ethanol a protic or an aprotic solvent? Is the solvation of H_2S in ethanol much different from the solvation of NH_3?

Solve

The nucleophiles are uncharged, so solvation is not a huge factor. Thus, we should expect the relative nucleophilicities to be the same as what they would be absent of any solvent. That is, H_2S is a stronger nucleophile than H_3N, because S, being a larger atom, can accommodate a positive charge better than N can. That positive charge develops in the substitution product, as shown in Solved Problem 9.3.

Problem 9.31

Think

Hydroxide is a strong nucleophile (S_N2) as well as a strong base (E2). This leads to a mixture of the two products at room temperature. Which bimolecular reaction mechanism has more entropy? Does an increase in heat favor the reaction with greater entropy or lesser entropy? Consider the equation $\Delta G°_{rxn} = \Delta H°_{rxn} - T\Delta S°_{rxn}$.

Solve

$\Delta S°_{rxn}$ is more positive for an elimination reaction (three products vs. two products). Raising the temperature will favor elimination products (owing to greater entropy, more positive $\Delta S°_{rxn}$), which leads to the second reaction product. Lowering the temperature will do the reverse, favoring the first product (substitution, S_N2).

Lesser entropy, favored at lowered temperatures

Greater entropy, favored at raised temperatures

Problem 9.33

Think

Is there a suitable leaving group? Does the type of C atom bonded to the leaving group rule out any of the four reactions? What is the attacking species? What is the solvent? Which reaction does each of the factors favor?

Solve

The leaving group is on a benzyl carbon, so elimination (E1 or E2) is not feasible because the benzyl carbon cannot be involved in a new C=C bond. Even though the leaving group is on a primary carbon, we must consider S_N1 along with S_N2, because a benzyl carbocation is relatively stable. CH_3O^- is a strong nucleophile, which favors S_N2. The concentration of the nucleophile is assumed to be high, which also favors S_N2. The leaving group is very good, which favors S_N1. The solvent is aprotic, which favors S_N2. So three factors favor S_N2, whereas only one factor favors S_N1. Thus, the S_N2 reaction below yields the major product. Note the stereochemistry, which is governed by the attack of the nucleophile from the side opposite the leaving group.

Factor	S_N1	S_N2	E1	E2
Strength		✓		
Concentration		✓		
Leaving group	✓			
Solvent		✓		
Total	1	3		

Problem 9.35

Think

In an E2 reaction, which CH protons can be attacked by the base? Draw out the C–H bond for all adjacent CH groups. Where does the double bond in the product appear relative to the leaving group and an adjacent proton? What alkenes form? Which one is most stable?

Solve

(a) The possible E2 products, not considering *E*/*Z* isomerism, are shown below. These are the products of eliminating one of the highlighted H atoms along with the leaving group. The major product is the most highly substituted alkene (most stable), which is tetrasubstituted in this case.

H'
H_2C Br
H
C
C
H" H'''
Adjacent H atoms are highlighted and labeled for emphasis.
From H'
Disubstituted
From H"
Tetrasubstituted most stable, *major*
From H'''
Trisubstituted

(b) There are two possible E2 products, as shown below. The disubstituted alkene is the major product.

H'
CH
H_2
C
H"
Cl
Adjacent H atoms are highlighted and labeled for emphasis.
From H'
Disubstituted most stable, *major*
From H"
Monosubstituted

(c) The two possible E2 products, not considering *E*/*Z* isomerism, are shown below. Both alkenes are disubstituted. The alkene on the top is the major product because it has an additional resonance structure that provides stabilization.

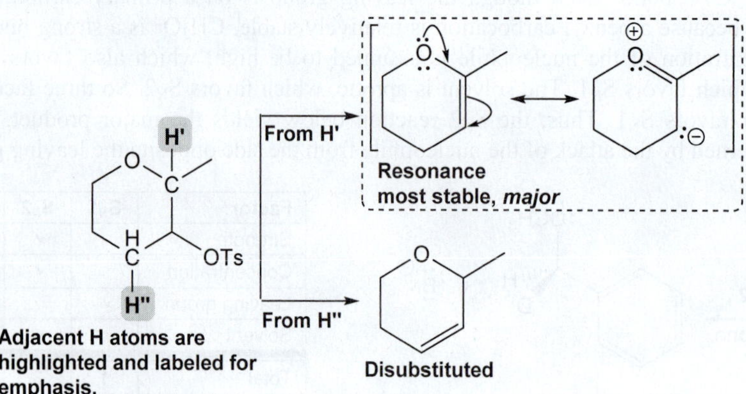

Problem 9.37

Think

Under what conditions does the reaction take place—acidic, basic, or neutral? For an intramolecular nucleophilic substitution reaction under these conditions, what can act as the nucleophile? What can act as the leaving group? For each intramolecular nucleophilic substitution reaction, what is the size of the ring that is formed?

Solve

The reaction takes place under basic conditions, so the alcohol will be deprotonated rapidly. After the alcohol is deprotonated, two different intramolecular S_N2 reactions are possible, as shown below. The one that leads to the formation of a six-membered ring gives the major product.

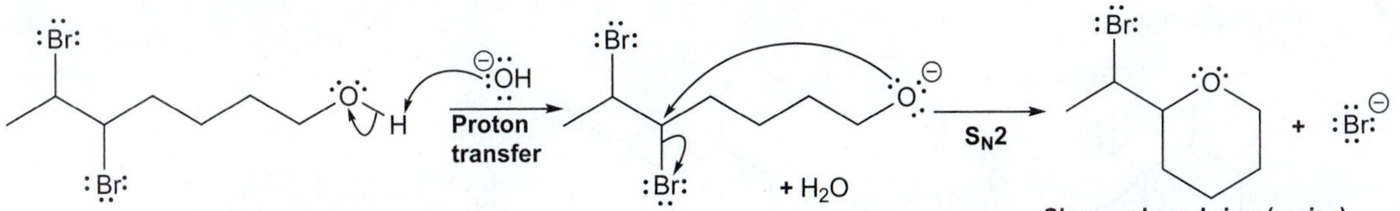

Proton transfer

S_N2

$+ H_2O$

Six-membered ring (major)

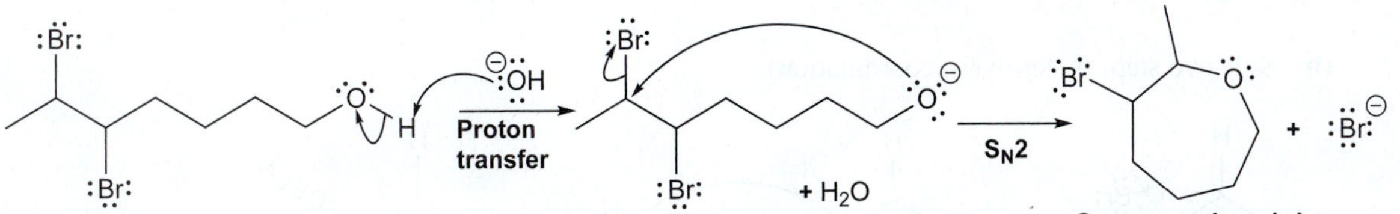

Proton transfer

S_N2

$+ H_2O$

Seven-membered ring

Problem 9.38

Think

What is the charge stability of Br^- on the product side compared to $CH_3CH_2O^-$ on the reactant side? How does charge stability of the reactants and products affect the reversibility of the reaction?

Solve

It is irreversible because charge stability heavily favors the product side of the reaction. This is because Br^- is much more stable than $CH_3CH_2O^-$, because Br is a significantly larger atom than O and can better stabilize the negative charge.

End of Chapter Problems

Section 9.1 The Competition among S_N2, S_N1, E2, and E1 Reactions

Problem 9.39

Think

Review the elementary steps for each mechanism, S_N1, S_N2, E1, and E2. In how many steps does each occur? What is the attacking species? What is the substrate? What is different about the site of attack in substitution versus elimination?

Solve

In the substitution mechanisms, the attacking species (NC:$^-$) acts as a nucleophile and attacks the electrophilic C atom. In the elimination mechanisms, the attacking species (NC:$^-$) acts as a base and attacks the H atom adjacent to the electrophilic C atom. An alkene product results.

(a) S_N2: one step

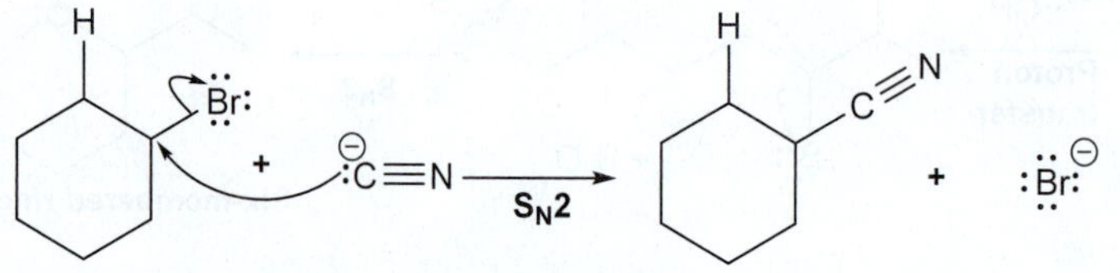

(b) S_N1: two steps (heterolysis, coordination)

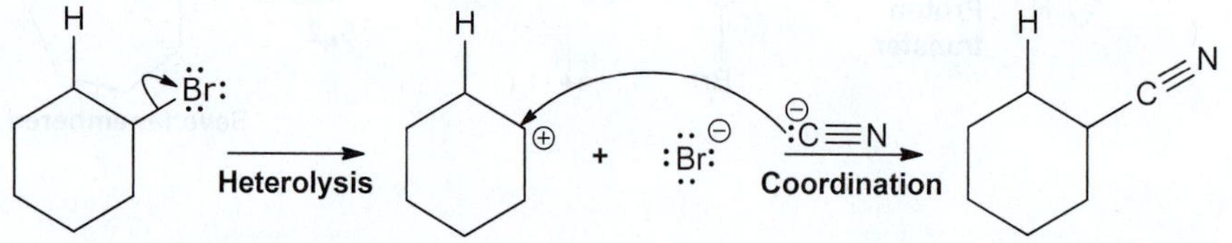

(c) E2: one step, alkene product

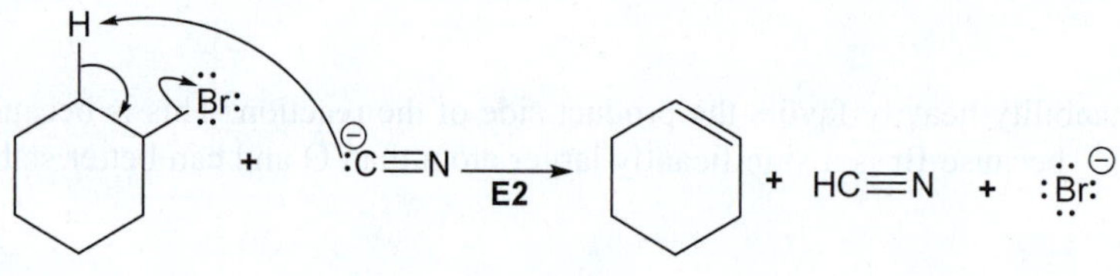

(d) E1: two steps (heterolysis, electrophile elimination), alkene product

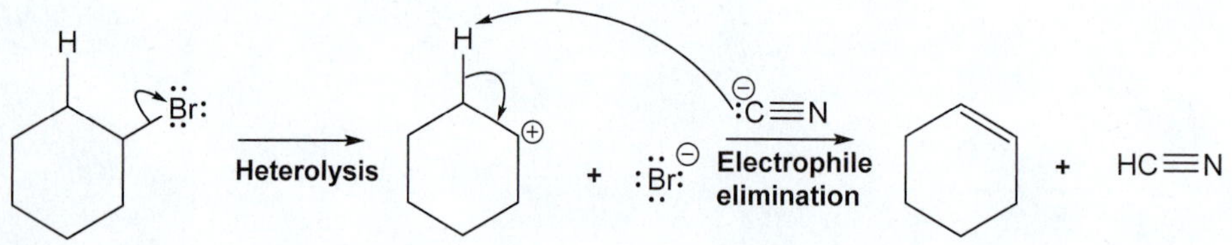

Problem 9.40 (SYN)

Think

Did the products arise from substitution or elimination? Which mechanism gives rise to a stereospecific reaction route? In that reaction, does the stereochemical configuration of the C atom initially bonded to the leaving group remain the same, or does it undergo inversion? On which carbon must the leaving group be located?

Solve

The desired starting compounds are shown below. The reactions should occur by the S_N2 mechanism to ensure that inversion occurs and only one stereoisomer is produced.

Starting alkyl bromide | Given product

(a) Br–cyclohexane–$C(CH_3)_3$ $\xrightarrow{CH_3SNa}$ CH_3S–cyclohexane–$C(CH_3)_3$

(b) Br–cyclohexane–$C(CH_3)_3$ $\xrightarrow{CH_3SNa}$ CH_3S–cyclohexane–$C(CH_3)_3$

Problem 9.41

Think

Are the products a result of substitution or elimination? Did the product arise from a bimolecular or unimolecular mechanism? Did you get inversion of stereochemistry or a mixture of configurations? Does the reaction form a reactive carbocation intermediate?

Solve

The S_N1 mechanism was dominant—it is a nucleophilic substitution reaction that yields a mixture of configurations at the electrophilic carbon. The methyl group at C4 remains wedge in both products, because no bonds were broken or formed at that carbon. Formation of the trigonal planar carbocation intermediate allows the ethanol nucleophile to attack the electrophilic site from the front or back. This leads to a mixture of configurations at that carbon.

Elementary Step 1: A carbocation forms.

Elementary Step 2: Ethanol attacks *either side* of the carbocation's plane, forming protonated ethers with a mixture of configurations.

Elementary Step 3: Each protonated ether loses a proton to a weak base present, such as ethanol, forming the products given.

Heterolysis

Formation of C^+ Intermediate

Coordination

Attack from in front of plane

Attack from behind plane

Proton transfer

$+ CH_3CH_2OH_2^+$
$+ Br^-$

Mixture of configurations

Proton transfer

$+ CH_3CH_2OH_2^+$
$+ Br^-$

Problem 9.42

Think

Are the products a result of substitution or elimination? Did the products arise from a bimolecular or unimolecular mechanism? Does the reaction form a reactive carbocation intermediate? Does the double bond that has formed involve the C atom that was initially attached to the leaving group?

Solve

An E1 mechanism dominated because a carbocation rearrangement must have occurred. We know this because among the products is an alkene in which the double bond does *not* involve the C atom that is initially bonded to the leaving group. For a carbocation rearrangement to happen, a carbocation must be formed. This describes an E1 reaction, not an E2.

Elementary Step 1: The alcohol is protonated.

Elementary Step 2: Water is lost to form a carbocation.

Elementary Step 3: The carbocation rearranges to a more stable one.

Elementary Step 4: A proton is lost from either of two carbons adjacent to C^+ to form the final products.

Proton transfer

Heterolysis

2° C^+

1,2-Hydride shift

3° C^+

Base:

Electrophile elimination

or

:Base

Electrophile elimination

Major product, tetrasubstituted

Problem 9.43

Think

Did substitution or elimination occur? Did the reaction form a single stereoisomer or a mixture of products? What mechanisms do the attacking species, concentration, and solvent favor?

Solve

(a) Because the product's formula is $C_6H_{11}ClO$, substitution must have taken place to replace Cl with Br. The conditions with a high concentration of a strong nucleophile and an aprotic solvent favor S_N2. So we need to choose the appropriate stereochemical configuration in the substrate to account for the inversion of stereochemistry that takes place in an S_N2 reaction.

NaCl, Acetone, S_N2

Chemical formula: $C_6H_{11}BrO$

(b) The product given is the result of elimination, recognized by the fact that the attacking species does not end up in the product. The reaction conditions favor E2, given that the attacking species is a strong, bulky base in high concentration and the solvent is aprotic. Because there are *E*/*Z* isomers possible for the product, we have to choose the stereochemistry appropriately for the substrate. That stereochemistry is chosen so that the H atom and the leaving group (Cl in this case) that are eliminated are in an anticoplanar conformation prior to elimination. The Cl leaving group is on the benzylic carbon. If the Cl leaving group was on the other C of the C=C, a mixture of products that would include the product shown would be produced.

same as

$(CH_3)_3CONa$, DMSO

Chemical formula: $C_{12}H_{17}Cl$

Sections 9.3–9.5 Strength and Concentration of the Attacking Species; Leaving Group Ability

Problem 9.44

Think

What is the basic site in each compound? How should you treat the metal cation species? How does base strength affect the rate of the E2 reaction? Which base is the strongest? Which is the weakest?

Solve

Each Na^+ and K^+ species should be treated as a spectator ion, leaving a negatively charged base. The E2 reaction rate, in general, increases with increasing base strength. The pK_a values of the conjugate acids are shown below. Therefore, the increasing rate of E2 is **B** < **A** < **E** < **C** < **D** < **F**.

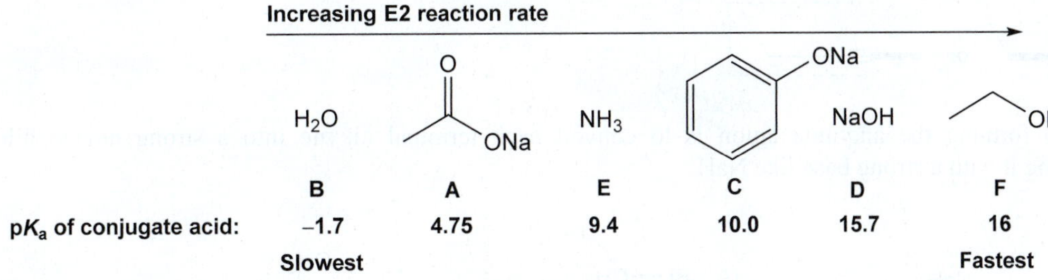

Increasing E2 reaction rate

	H_2O	ONa (acetate)	NH_3	ONa (phenoxide)	NaOH	OK (ethoxide)
	B	**A**	**E**	**C**	**D**	**F**
pK_a of conjugate acid:	**–1.7**	**4.75**	**9.4**	**10.0**	**15.7**	**16**
	Slowest					**Fastest**

Problem 9.45

Think

Is DMSO protic or aprotic? Is solvation a major factor in these examples? How does the rate of an S_N2 reaction depend on the strength of the nucleophile? Is a nucleophile stronger when it is negatively charged or uncharged? How does the size of the atom on which the charge appears affect stability? How does the electronegativity of that atom affect charge stability? Is the charge able to be stabilized by resonance? In the examples where both nucleophiles are uncharged, what factors affect the stability of the positively charged product?

Solve

DMSO is an aprotic solvent, so solvation should not be a major factor in the relative nucleophile strengths. The relative nucleophile strengths should, therefore, be the same as with no solvent. The rate of an S_N2 reaction depends directly on the strength of the nucleophile. Stronger nucleophiles promote a faster S_N2 reaction. In general, negatively charged nucleophiles are stronger than their uncharged counterparts, and charges are more stable on a larger atom. Therefore, RPH^- is a stronger nucleophile than RS^-, and RS^- is stronger than RSH. A resonance-stabilized anion is a weaker nucleophile than a localized anion. For an uncharged nucleophile, you need to look at the difference in charge stability on the product side but not on the reactant side. Because a positive charge is more stable on S than on O (S is the bigger atom), the product of nucleophilic substitution involving RSH is more stable, and the Hammond postulate suggests that its reaction should have a smaller energy barrier. Therefore, RSH is a stronger nucleophile than ROH. The increasing rate of S_N2 is **B** < **D** < **A** < **E** < **C**.

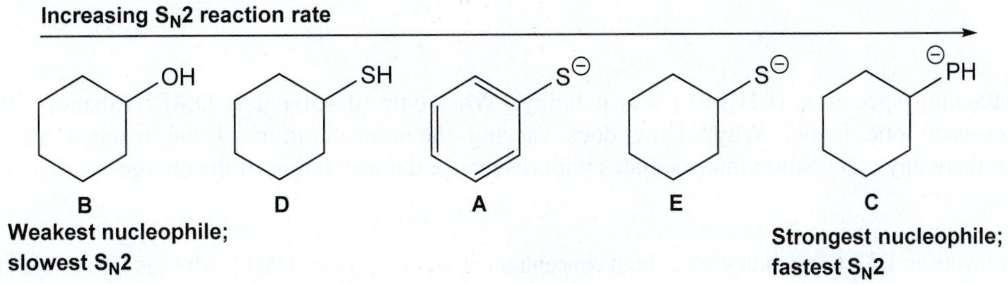

Increasing S_N2 reaction rate

Cyclohexyl–OH	Cyclohexyl–SH	Phenyl–$S^{\ominus}$	Cyclohexyl–$S^{\ominus}$	Cyclohexyl–$PH^{\ominus}$
B	**D**	**A**	**E**	**C**
Weakest nucleophile; slowest S_N2				**Strongest nucleophile; fastest S_N2**

Problem 9.46 (SYN)

Think

If an alkynide anion (RC≡C:⁻) is needed to synthesize pent-2-yne, which carbon–carbon bond is formed? Are there two possible bonds that can be formed? How many carbon atoms must the alkyl bromide then possess?

Solve

Pent-2-yne is not symmetrical, and there are two alkynide anions that can be used as a nucleophile. The C atoms necessary for the alkynide anion are boxed below, and the new C–C bond formed is bold. In the first example, the alkynide anion is $CH_3C{\equiv}C:^-$, and the alkyl bromide, therefore, is CH_3CH_2Br. In the second example, the alkynide anion is $CH_3CH_2C{\equiv}C:^-$, and the alkyl bromide is, therefore, CH_3Br.

or

The key to forming the alkynide anion is to convert each terminal alkyne into a strong nucleophile by deprotonating it with a strong base like NaH.

$$H_3C{-}C{\equiv}CH \xrightarrow{NaH} H_3C{-}C{\equiv}C:^{\ominus} \xrightarrow{Br-CH_2CH_3} H_3C{-}C{\equiv}C{-}CH_2CH_3$$

$$HC{\equiv}C{-}CH_2CH_3 \xrightarrow{NaH} {}^{\ominus}:C{\equiv}C{-}CH_2CH_3 \xrightarrow{H_3C-Br} H_3C{-}C{\equiv}C{-}CH_2CH_3$$

Problem 9.47

Think

What reaction occurs when acetic acid is in the presence of a strong base? What then forms? Consider the C–L in the substrate. Is there a suitable leaving group? To what type of carbon is the leaving group attached? Which mechanisms does this favor or rule out? What is the attacking species?

Solve

Acetic acid in the presence of a strong base, HO⁻, undergoes a fast deprotonation. The carboxylate anion is then formed, which produces a strongly nucleophilic carboxylate anion, RCO_2^-. Elimination reactions are ruled out, as there is no other C or H atom to form a C=C. The reaction conditions favor an S_N2 reaction, as shown below.

Proton transfer — S_N2 — + :Br:⁻

Problem 9.48

Think

What type of attacking species is $(CH_3)_3CO^-$? Is it bulky? What type of solvent is DMF? Ethanol? Which mechanism does each one favor? Why? How does varying the concentration of the reactant alter the mechanism? Are there any carbocation intermediates with resonance delocalization of the charge?

Solve

The first reaction favors an E2 (strong, bulky base; high concentration; aprotic solvent DMF), whereas the second favors an E1 (strong base; low concentration; excellent leaving group; protic solvent; heat). The carbocation that is formed in the E1 mechanism is allylic, so the positive charge is delocalized by resonance over two carbon atoms. This allows the D to be eliminated in the second step. See the mechanisms on the next page.

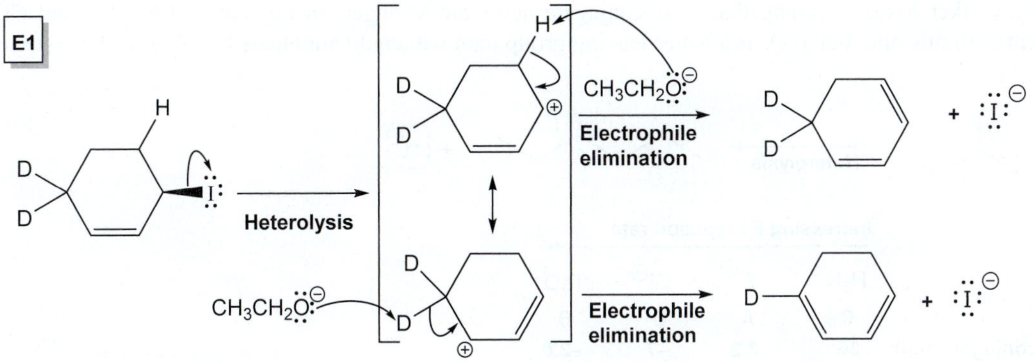

Problem 9.49

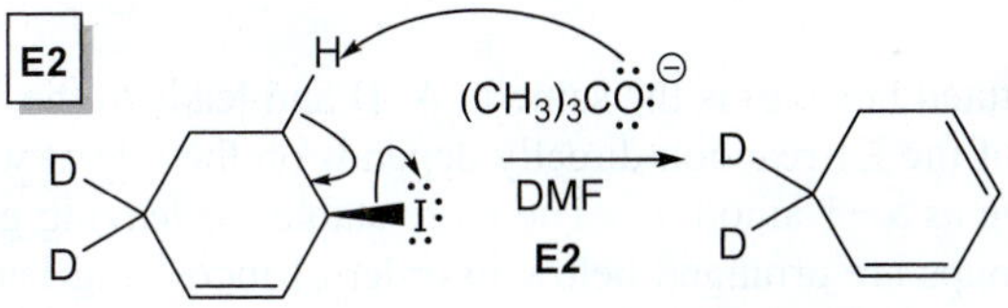

Think

What factors affect the rate of an S_N2 reaction? What are the similarities and differences of each R–L substrate? What is the leaving group in each compound? How does the stability of the leaving group affect the rate of the S_N2 reaction?

Solve

The type of organic substrate in this problem (a substituted cyclopentane) is the same in **A**–**E**, and no nucleophile is given. Therefore, the rate of the S_N2 reaction directly depends on the rate in which the leaving group departs. All of the leaving groups leave as a −1 anion, L^-. The more stable the leaving group is, the better is its leaving group ability. Better leaving groups are weaker bases. pK_a values of the conjugate acids are shown below, and the leaving groups are arranged in order of decreasing basicity, increasing stability, and thus increasing rate of S_N2: **C** < **E** < **D** < **A** < **B**.

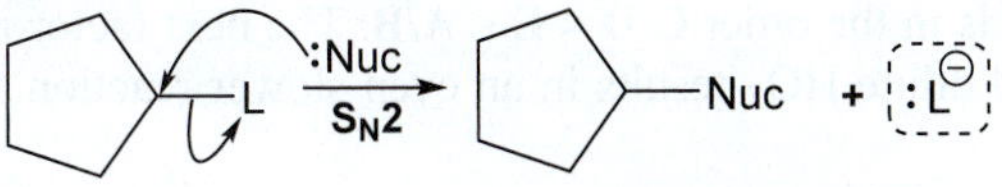

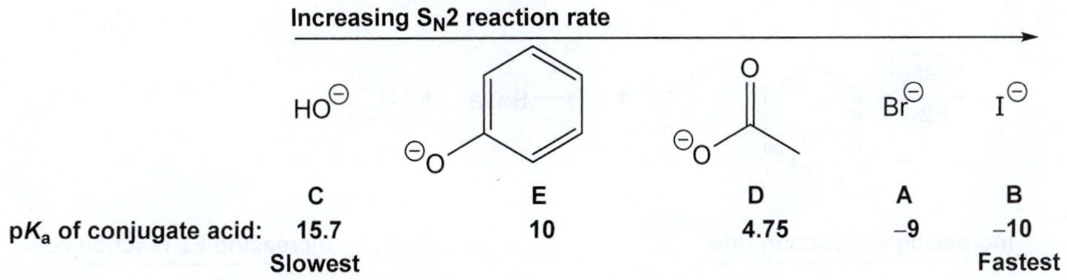

Increasing S_N2 reaction rate

	C	E	D	A	B
pK_a of conjugate acid:	15.7	10	4.75	-9	−10
	Slowest				Fastest

Problem 9.50

Think

What factors affect the rate of an E1 reaction? What are the similarities and differences of each R–L substrate? What is the leaving group in each compound? How does the stability of the leaving group affect the rate of the E1 reaction?

Solve

The type of organic substrate in this problem (a 3-substituted hexane) is the same in **A–D** and leads to the same secondary carbocation intermediate. Therefore, the rate of the E1 reaction directly depends on the rate at which the leaving group departs. All of the leaving groups leave as a −1 anion, L^-. The more stable the leaving group is, the better is its leaving group ability. The leaving groups are arranged below in order of increasing leaving group ability (see Tables 9-3 and 9-4) and thus increasing rate of E1: **C** < **A** < **D** < **B**. Better leaving groups are generally weaker bases, meaning that their conjugate acids are stronger. In this case, H_2N^-, F^-, and Cl^- are consistent with this rule, but TsO^- is a better leaving group than we would anticipate based on its base strength.

Heterolysis

Increasing E1 reaction rate →

	H_2N^-	F^-	Cl^-	TsO^-
	C	**A**	**D**	**B**
pK_a of conjugate acid:	**36**	**3.2**	**−7**	**−2.8**
	Slowest			**Fastest**

Problem 9.51

Think

What factors affect the rate of an E2 reaction? What are the similarities and differences of each R–L substrate? What is the leaving group in each compound? How does the stability of the leaving group affect the rate of the E2 reaction? What is the base in each? Which base is strongest?

Solve

The type of organic substrate in this problem (secondary, benzylic) is the same in **A–E**, the leaving group is attached to the benzylic carbon in each reaction, and all yield styrene as the product. Therefore, the rate of the E2 reactions depends on the rate at which the leaving group departs and on the strength of the base. All of the leaving groups leave as a −1 anion, L^-. The more stable the leaving group is, the better is its leaving group ability. In general, better leaving groups are weaker bases. We can apply this idea to establish that Cl^- (pK_a of HCl = −7) is a better leaving group than PhO^- (pK_a = 10). Then, using Tables 9-3 and 9-4, we know that TsO^- is a better leaving group than Cl^-. Thus, increasing rate of E2 is in the order **C/D** < **E** < **A/B**. The next factor is the strength of the base. HO^- is a weaker base than H_2N^-, and dilute HO^- results in an even slower reaction. This makes the increase in E2 reaction rate **D** < **C** < **E** < **A** < **B**.

:Base → E2; + H—Base + $:L^-$

Increasing E2 reaction rate →

Leaving group	PhO^-	Cl^-	TsO^-
	C/D	**E**	**A/B**
pK_a of conjugate acid:	**10**	**−7**	**−2.8**
	Slowest		**Fastest**

Increasing E2 reaction rate →

Base	HO^-	H_2N^-
	A/C/D*/E	**B**
pK_a of conjugate acid:	**15.7**	**36**
	Slowest	**Fastest**

***Dilute HO^- is slower.**

Section 9.6 Type of Carbon

Problem 9.52

Think

Are the leaving groups the same or different? What is the rate-determining step in an E1 reaction? Consider the stability of the carbocation intermediate from each compound. How does the stability of the carbocation affect the E1 reaction rate?

Solve

Reaction rate increases with increasing stability of the leaving group and of the carbocation that would be formed upon the departure of the leaving group. In this example, the leaving group (Br^-) is the same in each compound and, therefore, does not affect the reaction rate. Reaction occurs most slowly at the aromatic sp^2-hybridized carbon (**D**), because of effective electronegativity of C and bond strength. Reaction occurs fastest at the resonance-stabilized benzylic tertiary carbon (**C**).

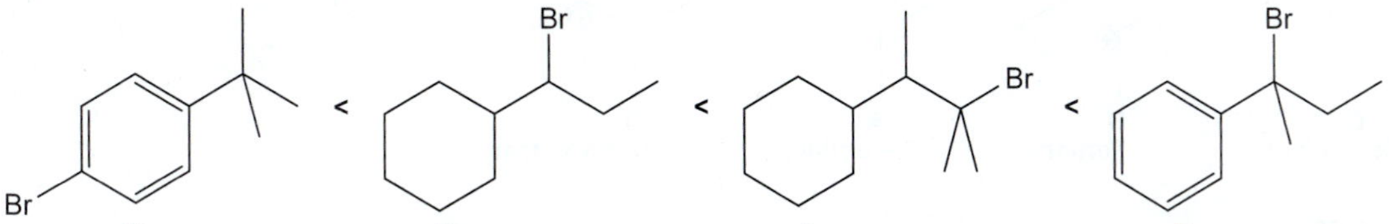

Br Br Br Br

D < B < A < C

Slowest Fastest

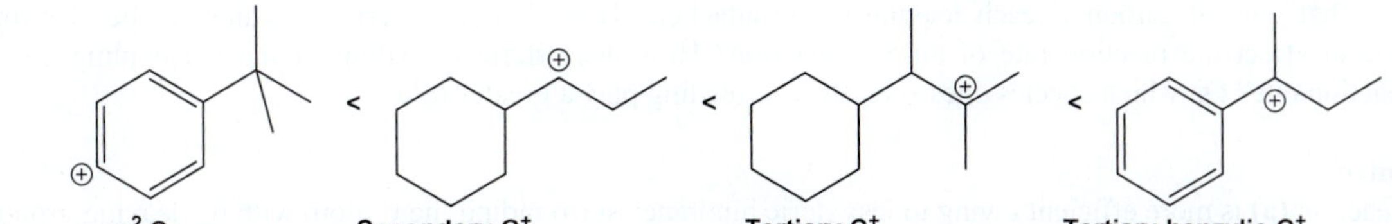

*sp*² C < Secondary C⁺ < Tertiary C⁺ < Tertiary benzylic C⁺

Least stable C⁺, slowest Most stable C⁺, fastest

Problem 9.53

Think

Is the leaving group in each substrate the same or different? To what type of carbon in the substrate is each leaving group attached? How does the hybridization and steric crowding affect the reaction rate of an S_N2 reaction? In how many steps does an S_N2 reaction occur? Are any intermediates formed?

Solve

The reaction occurs most slowly at the sp^2-hybridized carbon because of bond strength and the additional electrostatic repulsion encountered during attack (**B**). For the others, the S_N2 rate increases with less steric hindrance from surrounding alkyl groups, **C** < **D** < **A**.

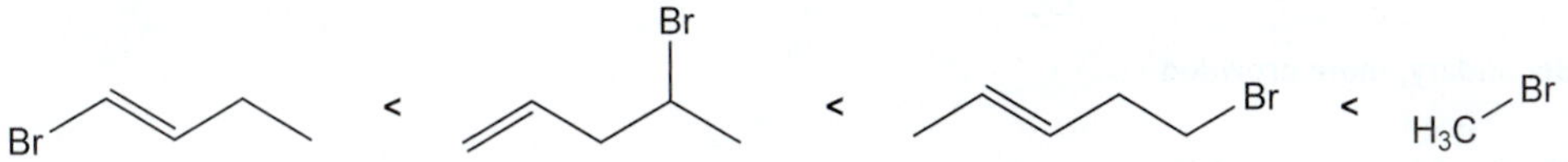

Br Br Br Br

B < C < D < H_3C A

sp^2 C-Br, Slowest — Secondary C-Br — Primary C-Br — Methyl C-Br, Fastest, least crowded

Problem 9.54

Think

Are the leaving groups the same or different in this example? What is the carbocation intermediate involved in each S_N1 mechanism? Which one is more stable? Why? How does the stability of the carbocation affect the S_N1 reaction rate?

Solve

The S_N1 reaction rate is governed by the rate at which the leaving group leaves. Because all of the leaving groups are the same (Br^-), the main differences are in the stabilities of the carbocations that are formed. Carbocations that are formed from **A** and **D** are resonance stabilized, so the reactions involving those cations are faster than those involving **B** and **C**. Because **D** is also a tertiary carbocation, it is more stable than **A**, which is a secondary carbocation. Likewise, **B** is a tertiary carbocation, so it is more stable than the carbocation from **C**, a secondary carbocation. Overall, then, stability of the carbocations increases in the order **C** < **B** < **A** < **D**, which is the same as the order of the reaction rates.

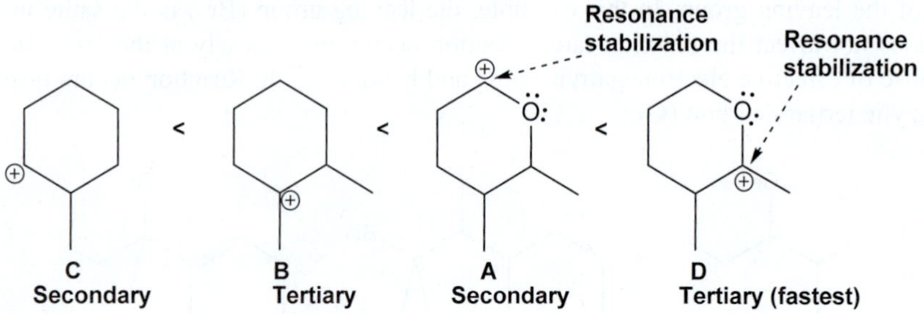

Resonance stabilization
Resonance stabilization

C Secondary < B Tertiary < A Secondary < D Tertiary (fastest)

Problem 9.55

Think

To what type of carbon is each leaving group attached? How does the steric crowding of the electrophilic carbon affect the reaction rate of an S_N2 reaction? How does steric crowding of the nucleophile affect the reaction rate? On which species does such steric crowding play a greater role?

Solve

Reaction **(a)** is more efficient owing to less steric hindrance surrounding the C atom with the leaving group. The C–L in reaction **(a)** is methyl and in reaction **(b)** is secondary. With secondary carbons, S_N2 reactions proceed somewhat slowly owing to the reduced accessibility of the carbon. There is also steric crowding surrounding the nucleophilic O^- in **(a)**, but those sterics have less of an impact on reaction rate, because the bulkiness is farther away from the reaction site.

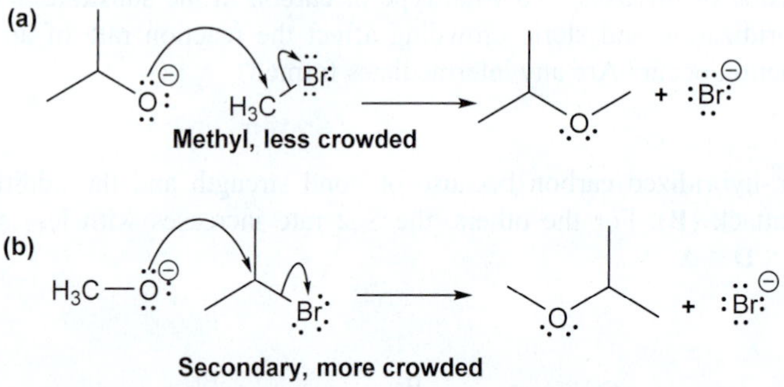

(a)
H_3C—Br → + :Br:⁻
Methyl, less crowded

(b)
H_3C—O⁻ + Br → + :Br:⁻
Secondary, more crowded

Sections 9.7, 9.8 Effects of Solvent and Heat

Problem 9.56

Think

Is acetone protic or aprotic? Is solvation a major factor in these examples? Is a nucleophile stronger when it is positively charged, negatively charged, or uncharged? In the examples where a charge is present on each nucleophile, how does the size of the atom on which the charge appears affect stability? How does the electronegativity of that atom affect charge stability? Is the charge able to be stabilized by resonance? In the examples where both nucleophiles are uncharged, what factors affect the stability of the positively charged product?

Solve

The stronger nucleophile of each pair is circled below. Being in an aprotic solvent, the relative nucleophile strengths should be the same as they are with no solvent because solvation is not a major factor, so we can use arguments of intrinsic charge stability, as we learned in Section 9.3a.

In **(b)**, the species that has a negative charge is the stronger nucleophile compared to the species with the positive charge.

(b) $H_3C—O^{\ominus}$ or $H_3C—\overset{\oplus}{O}H_2$

In **(d)** and **(g)**, the atoms with the negative charge are in the same row of the periodic table, so the stronger nucleophile is the one with the negative charge on the less electronegative atom.

(d) O or O NH (g) F or O

In **(c)**, both nucleophiles are uncharged, so the greater charge stability in the products has the positive charge appear on the less electronegative atom.

(c) $H_3C—OH$ or $H_3C—NH_2$

In **(a)**, the negatively charged nucleophile is less stable.

(a) $H_3C—OH$ or $H_3C—O^{\ominus}$

In **(f)**, **(h)**, **(i)**, and **(j)**, the one that has less charge stability is the one in which the negative charge appears on the smaller atom.

(f) S or O (h) O NH or PH (i) $CH_3S^{\ominus}$ or $CH_3Se^{\ominus}$ (j) $CH_3Se^{\ominus}$ or $Br^{\ominus}$

In **(e)**, the less stable charge is the one that has the charge localized instead of resonance delocalized.

(e) S or S

Problem 9.57

Think

Is ethanol protic or aprotic? Is solvation a major factor in these examples? Is a nucleophile stronger when it is positively charged, negatively charged, or uncharged? In the examples where a charge is present on each nucleophile, how does the size of the atom on which the charge appears affect stability? How does the electronegativity of that atom affect charge stability? Is the charge able to be stabilized by resonance? In the examples where both nucleophiles are uncharged, what factors affect the stability of the positively charged product?

Solve

The results are determined the same way they were in Problem 9.56, except when the nucleophile has a negative charge on atoms of significantly different size, that is, in different rows of the periodic table. Under these circumstances, the protic solvent (ethanol) solvates the nucleophile much more strongly and, hence, weakens the nucleophile strength much more, with the negative charge on the *smaller* atom. So only in **(d)** and **(e)** are the relative nucleophilicities reversed from what we would observe in an aprotic solvent.

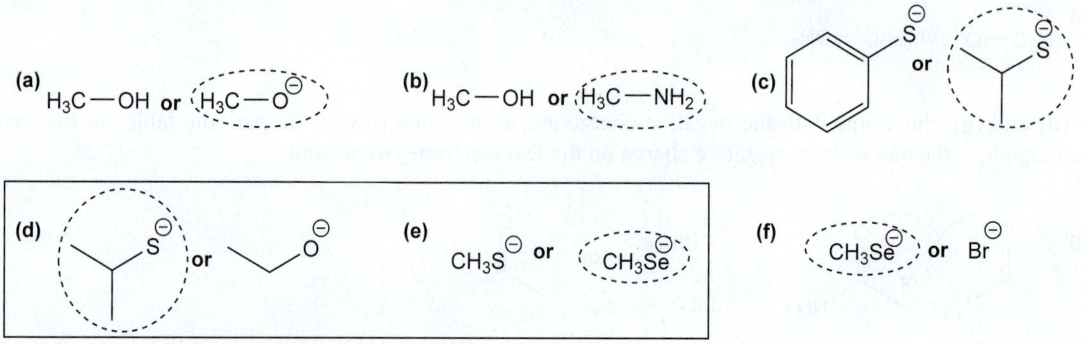

(a) $H_3C{-}OH$ or $H_3C{-}O^{\ominus}$ (b) $H_3C{-}OH$ or $H_3C{-}NH_2$ (c) PhS⁻ or isopropyl-S⁻

(d) isopropyl-S⁻ or ethyl-O⁻ (e) $CH_3S^{\ominus}$ or $CH_3Se^{\ominus}$ (f) $CH_3Se^{\ominus}$ or $Br^{\ominus}$

In a protic solvent, the smaller negatively charged atom is a weaker nucleophile.

Problem 9.58 (SYN)

Think

Which mechanism (S_N1, S_N2, E1, or E2) is occurring in this reaction? Is that mechanism favored by protic or aprotic solvents? Which solvent is protic? Which is aprotic?

Solve

This is a stereospecific nucleophilic substitution reaction in which there is inversion of the configuration at the stereocenter—this is S_N2. The polar aprotic solvent dimethyl sulfoxide (DMSO) will favor this reaction more than the polar protic solvent ethanol will.

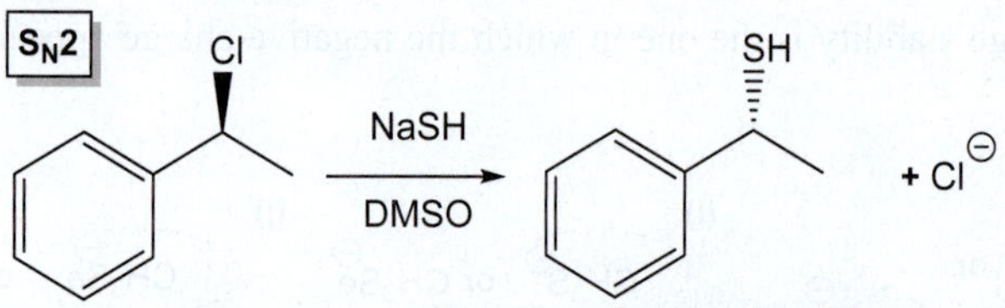

S_N2

Cl — NaSH / DMSO → SH + $Cl^{\ominus}$

Inversion of stereochemistry

Problem 9.59 (SYN)

Think

Which mechanism (S_N1, S_N2, E1, or E2) is occurring in this reaction? Is that mechanism favored by protic or aprotic solvents? Which solvent is protic? Which is aprotic?

Solve

This is an elimination reaction in which there is a 1,2-methyl shift—this is an E1. The polar protic solvent *tert*-butyl alcohol favors an E1 reaction more than the polar aprotic solvent acetone will.

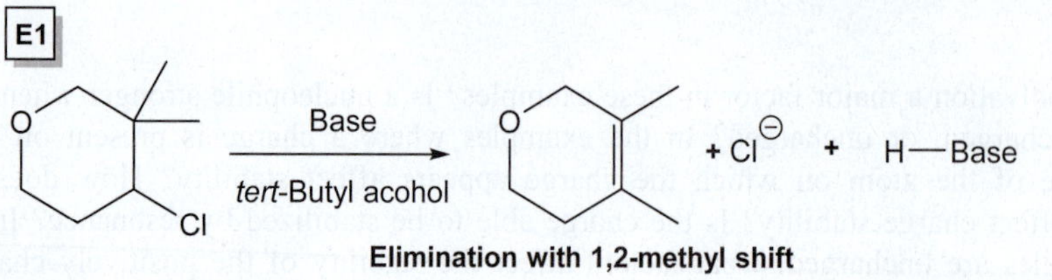

E1

O, Cl — Base / *tert*-Butyl acohol → O + $Cl^{\ominus}$ + H—Base

Elimination with 1,2-methyl shift

Problem 9.60

Think

Which mechanism (S_N1, S_N2, E1, or E2) is occurring in this reaction? Do the reaction conditions favor more than one mechanism? How does adding heat promote one mechanism over the other?

Solve

(a) The reaction shown is an E2, but the reaction conditions (primary C–L and strong nucleophile/base) also favor S_N2. Increasing the temperature of the reaction tends to promote elimination more than substitution and therefore will increase the yield of the E2 product shown.

NaOH
Heat
+ $Cl^{\ominus}$ + H_2O
E2
↑ Heat = ↑ Yield

(b) The reaction shown is an S_N2. Increasing the temperature of the reaction tends to promote elimination more than substitution and therefore will decrease the yield of the S_N2 product shown.

Br
NaO
Heat
+ $Br^{\ominus}$
S_N2
↑ Heat = ↓Yield

Sections 9.9–9.11 Predicting the Outcome of Nucleophilic Substitution and Elimination Reactions; Zaitzev's Rule; Intramolecular versus Intermolecular Reactions

Problem 9.61 (SYN)

Think

Did the product arise from a bimolecular or unimolecular mechanism? Did you get inversion of stereochemistry or a racemic mixture? How does knowing the mechanism (S_N1 vs. S_N2) affect the choice of nucleophile (strong or weak) and choice of solvent (protic or aprotic)?

Solve

(a) This is a stereospecific nucleophilic substitution reaction in which there is inversion of the configuration at the stereocenter—this is S_N2. A polar aprotic solvent like DMSO or DMF will favor such a reaction. Also, a strong nucleophile like $NaOCH_3$ will favor the reaction.

Cl
CH_3ONa
DMSO
OCH_3
+ $Cl^{\ominus}$
Inversion of stereochemistry
S_N2

(b) This nucleophilic substitution generates a mixture of configurations (equal *R* and *S*, racemic), so it is S_N1. A protic solvent will favor such a reaction, as will a weak nucleophile. This can simply be a solvolysis reaction that takes place in CH_3OH, which can act as both the weak nucleophile and the protic solvent.

Cl
CH_3OH
Solvolysis
OCH_3
(*S*)
and
OCH_3
(*R*)
+ $Cl^{\ominus}$
Racemic
S_N1

Problem 9.62

Think

Is there a good leaving group present without the acid? In the presence of an acid, does a good leaving group form? Did the product result via substitution or elimination? Was the mechanism unimolecular or bimolecular? How many times did this reaction occur?

Solve

The mechanism consists of two S_N2 steps and some proton transfer steps. The reactions must be substitution, and because the leaving group is on a primary carbon, the S_N1 reaction is precluded, leaving only S_N2.

Elementary Step 1: The ether is protonated.

Elementary Step 2: Ethanol opens the ring (first S_N2).

Elementary Steps 3 and 4: Solvent-mediated proton transfers occur.

Elementary Step 5: Ethanol displaces water (second S_N2).

Elementary Step 6: A final proton transfer completes the reaction.

See the mechanism below.

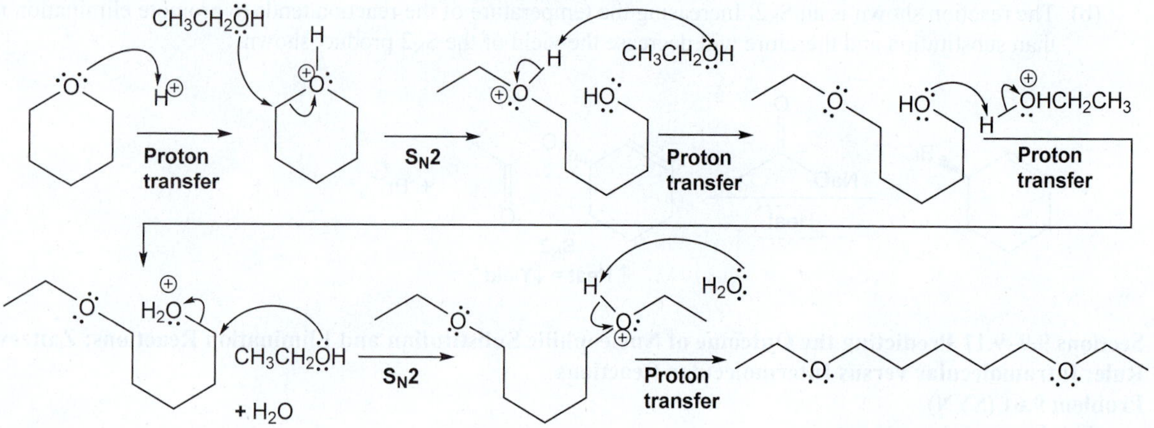

Problem 9.63

Think

Consider the C–L first. Is there a suitable leaving group? To what type of carbon is the leaving group attached? Which mechanisms does this favor or rule out? What is the attacking species: strong base, weak base, strong nucleophile, weak nucleophile, sterically bulky? What is the solvent? Which reaction does each of the factors favor?

Solve

(a) Reaction conditions favor S_N2 and E2 (leaving group is on a secondary C, so we must consider all four mechanisms; attacking species is a strong nucleophile and a strong base; high concentration of the attacking species; aprotic solvent; moderate leaving group). There is a single S_N2 product but four different E2 products we should expect in significant abundance. The different locations of the double bonds are an outcome of the H atoms on C2 and C4 that can be removed, and both alkenes are disubstituted. The cis/trans isomers are possible because on both C2 and C4 are two different H atoms that can attain an anticoplanar conformation with the leaving group. The S_N2 and E2 reaction mechanisms are shown on the next page.

Factor	S_N1	S_N2	E1	E2
Strength		✓		✓
Concentration		✓		✓
Leaving group	✓	✓	✓	✓
Solvent		✓		✓
Total	1	4	1	4

(b) S_N2 reactions are ruled out because the leaving group is on a tertiary C. A strong nucleophile/base favors S_N2 and E2. High concentrations of the attacking species favor S_N2 and E2. The Cl leaving group favors S_N1, S_N2, E1, and E2. Protic solvent favors S_N1 and E1. So overall, the major products should be E2. *E* and *Z* isomers are produced, because both H atoms on the C adjacent to the C–Cl can attain the anticoplanar conformation.

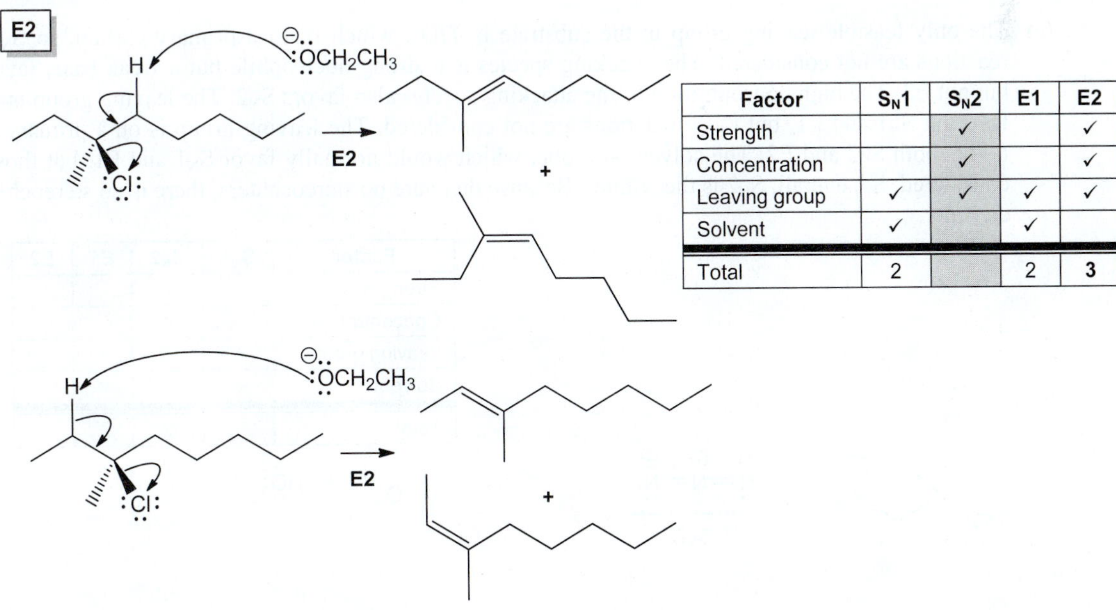

Factor	S_N1	S_N2	E1	E2
Strength		✓		✓
Concentration		✓		✓
Leaving group	✓	✓	✓	✓
Solvent	✓		✓	
Total	2		2	**3**

(c) The substrate has the leaving group on a secondary carbon, so all four mechanisms must be considered. H_3PO_4 or H_2O would be the nucleophile/base, which is weak, so S_N1 and E1 are favored. Concentration of the attacking species is not considered, because it is weak. The leaving group, water, has a very good leaving group ability, favoring S_N1 and E1. The leaving group is on a secondary allylic C, which can stabilize a carbocation by resonance, thus favoring S_N1 and E1. And the solvent is protic, favoring S_N1 and E1. So overall, S_N1 and E1 are favored, but heat tips the balance in favor of E1.

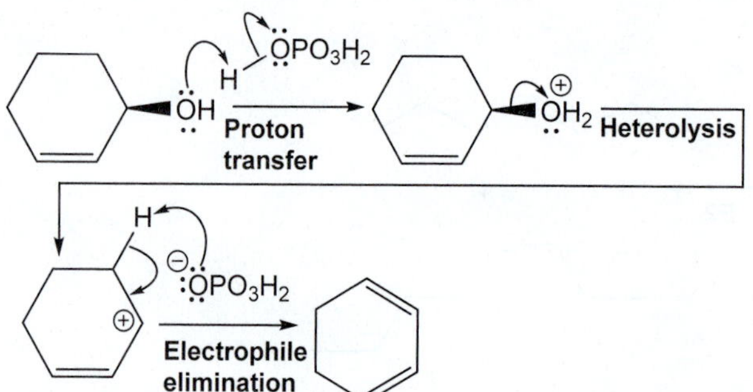

Factor	S_N1	S_N2	E1	E2
Strength	✓		✓	
Concentration				
Leaving group	✓		✓	
Solvent	✓		✓	
Total	**3**		**3**	

(d) The substrate has the leaving group on a secondary C, so all four mechanisms must be considered. The attacking species is $(CH_3)_3CO^-$, which favors E2. The concentration of the attacking species is high, which favors E2. The leaving group is Cl^-, which favors all four mechanisms. The leaving group is on a secondary C, which favors all four mechanisms. And the solvent is aprotic, which favors S_N2 and E2. Overall, E2 is favored. The proton that is removed is one that can achieve an anticoplanar conformation with the leaving group. Normally, the more highly substituted alkene product would be favored, but the bulkiness of the base directs deprotonation to the more accessible side.

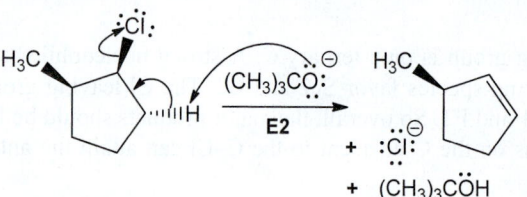

Factor	S_N1	S_N2	E1	E2
Strength				✓
Concentration				✓
Leaving group	✓	✓	✓	✓
Solvent		✓		✓
Total	1	2	1	**4**

(e) The only feasible leaving group in the substrate is TfO^-, which is on a primary carbon, so S_N1 and E1 reactions are not considered. The attacking species is a strong nucleophile but a weak base, favoring S_N2 but not E2. The high concentration of the attacking species also favors S_N2. The leaving group is excellent, favoring S_N1 and E1, but those reactions are not considered. The leaving group is on a primary C, which favors both S_N2 and E2. The solvent is protic, which would normally favor S_N1 and E1, but those are not considered. So, overall, S_N2 is the winner. Because there are no stereocenters, there is no stereochemistry to consider.

Factor	S_N1	S_N2	E1	E2
Strength		✓		
Concentration		✓		
Leaving group	✓		✓	
Solvent	✓		✓	
Total		**2**		

(f) There is no H atom on the C atom adjacent to the C with the leaving group, so we can rule out E1 and E2. The leaving group is on a primary carbon, which would normally prompt us not to consider the S_N1 or E1 mechanisms, but that primary C atom is at the benzylic position, which can accommodate a positive charge quite well. So we must consider S_N1 also. The attacking species is ethanol, which is a weak nucleophile, favoring S_N1 and E1. We ignore the concentration factor, because the nucleophile is weak. The leaving group is MsO^-, which is excellent, favoring S_N1 and E1. The leaving group is on a benzylic C atom, which favors S_N1 and E1. And the solvent is ethanol, which is protic and favors S_N1 and E1. So overall, this is an S_N1 reaction.

Heterolysis
+ MsO^-
CH_3CH_2OH
Coordination
CH_2CH_3
Proton transfer
CH_3CH_2OH
CH_2CH_3

Factor	S_N1	S_N2	E1	E2
Strength	✓			
Concentration				
Leaving group	✓			
Solvent	✓			
Total	3			

Problem 9.64

Think

Consider the C–L first. Is there a suitable leaving group? To what type of carbon is the leaving group attached? Which mechanisms does this favor or rule out? What is the attacking species: strong base, weak base, strong nucleophile, weak nucleophile, sterically bulky? What is the solvent? Which reaction does each of the factors favor?

Solve

(a) S_N1 is favored. TsO^- is a suitable leaving group and is on a secondary carbon, so we consider all four mechanisms. CH_3CH_2OH is a weak nucleophile and weak base, favoring S_N1 and E1. Concentration does not matter. TsO^- is an excellent leaving group, which favors S_N1 and E1. The carbon is secondary, which favors all four reactions. Ethanol is a protic solvent, which favors S_N1 and E1. Because the reaction is not heated, we can expect substitution to be favored over elimination. Notice that a 1,2-methyl shift takes place to produce a more stable carbocation. Note also that a final proton transfer is necessary to produce the uncharged substitution product.

OTs
Heterolysis
+ TsO^-
1,2-Methyl shift
$HOCH_2CH_3$
Coordination
H_3CH_2C
H
$HOCH_2CH_3$
Proton transfer
H_3CH_2C

Factor	S_N1	S_N2	E1	E2
Strength	✓		✓	
Concentration				
Leaving group	✓		✓	
Solvent	✓		✓	
Total	3		3	

(b) S_N2 is favored. TsO^- is a suitable leaving group and is on a secondary carbon, so we consider all four mechanisms. NC^- is a strong nucleophile but a weak base, favoring S_N2 and E1. A high concentration of NC^- favors S_N2. TsO^- is an excellent leaving group, favoring S_N1 and E1. The carbon is secondary, favoring all four reactions. DMF is an aprotic solvent, favoring S_N2 and E2. Notice that the S_N2 reaction leads to inversion of stereochemistry.

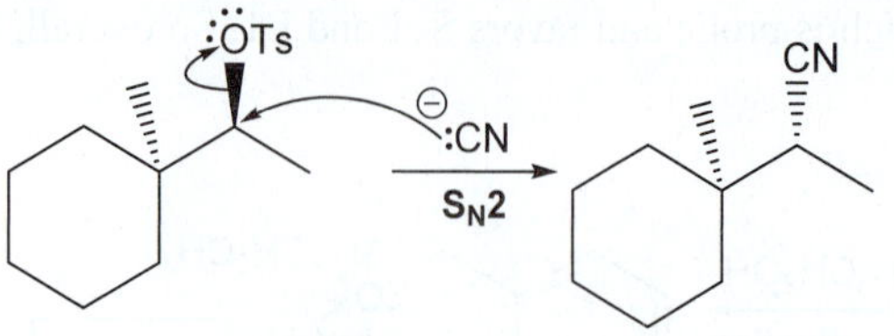

Factor	S_N1	S_N2	E1	E2
Strength		✓	✓	
Concentration		✓		
Leaving group	✓		✓	
Solvent		✓		✓
Total	1	3	2	1

(c) E2 is favored. TsO^- is a suitable leaving group and is on a secondary carbon, so we consider all four mechanisms. The strong bulky base favors E2, as does its high concentration. TsO^- is an excellent leaving group, which favors S_N1 and E1. The carbon is secondary, which favors all four reactions. DMSO is an aprotic solvent, favoring S_N2 and E2. Notice that the terminal CH_3 has the only H atom that can be eliminated.

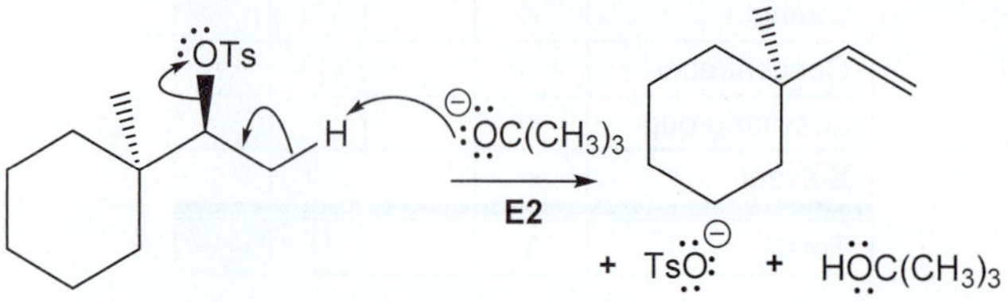

Factor	S_N1	S_N2	E1	E2
Strength				✓
Concentration				✓
Leaving Group	✓		✓	
Solvent		✓		✓
Total	1	1	1	3

(d) E1 is favored. TsO^- is a suitable leaving group and is on a secondary carbon, so we consider all four mechanisms. CO_3^{2-} is a weak base and weak nucleophile, favoring S_N1 and E1. Concentration is ignored. TsO^- is an excellent leaving group, favoring S_N1 and E1. The carbon is secondary, favoring all four reactions. Ethanol is a protic solvent, favoring S_N1 and E1. The reaction is heated, which favors elimination over substitution. A 1,2-methyl shift takes place, producing a more stable carbocation. Also, the H atom that is eliminated is the one that gives the most highly substituted alkene.

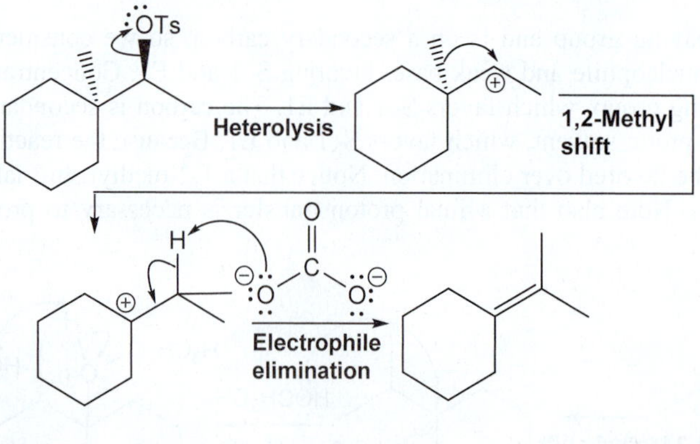

Factor	S_N1	S_N2	E1	E2
Strength	✓		✓	
Concentration				
Leaving Group	✓		✓	
Solvent	✓		✓	
Total	3		3	

(e) E1 is favored. Cl^- is a suitable leaving group and is on a tertiary carbon, so we must rule out S_N2. CH_3CH_2OH is a weak nucleophile and a weak base, favoring S_N1 and E1, and concentration is ignored. Cl^- is a moderate leaving group, favoring all four reactions. The carbon is benzylic, favoring S_N1 and E1. Ethanol is a protic solvent, favoring S_N1 and E1. The reaction is heated to favor elimination over substitution. See the mechanism on the next page.

Factor	S_N1	S_N2	E1	E2
Strength	✓		✓	
Concentration				
Leaving group	✓		✓	✓
Solvent	✓		✓	
Total	3		3	1

(f) E2 is favored. Cl^- is a suitable leaving group and is on a secondary carbon, so we consider all four mechanisms. $CH_3CH_2O^-$ is a strong nucleophile and strong base, favoring S_N2 and E2. Concentration is high, favoring S_N2 and E2. The leaving group is moderate, favoring all four reactions. The carbon attached to the leaving group is benzylic, favoring S_N1 and E1. DMSO is an aprotic solvent, favoring S_N2 and E2. Heat favors elimination over substitution. Notice that only one H atom can be eliminated, and because E2 is the mechanism, it favors the anticoplanar conformation of the H and the leaving group. This gives rise to one configuration about the C=C in the alkene product, exclusively.

Factor	S_N1	S_N2	E1	E2
Strength		✓		✓
Concentration		✓		✓
Leaving group	✓	✓	✓	✓
Solvent		✓		✓
Total	1	4	1	4

Problem 9.65

Think

Consider the C–L first. Is there a suitable leaving group? To what type of carbon is the leaving group attached? Which mechanisms does this favor or rule out? What is the attacking species: strong base, weak base, strong nucleophile, weak nucleophile, sterically bulky? What is the solvent? Which reaction does each of the factors favor?

Solve

(a) Cl^- is a suitable leaving group. There are two sites for substitution or elimination, because there are two separate Cl leaving groups, both on sp^3-hybridized C atoms. The conditions favor S_N2, which governs the site of reaction. Br^- is a strong nucleophile but a weak base, which favors S_N2 and E1. The high concentration also favors S_N2. Also, DMSO is an aprotic solvent, which favors S_N2 and E2. The leaving group is moderate, which favors all four reactions. Because S_N2 is favored, the nucleophile will attack the least sterically hindered site, which is at the secondary carbon. The tertiary C–Cl does not undergo S_N2 and, therefore, is unchanged. Notice that stereochemistry is important here and that the S_N2 reaction leads to inversion at the electrophilic carbon.

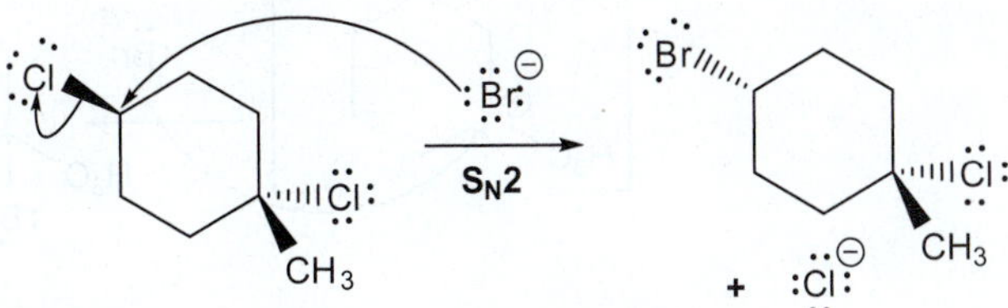

Factor	S_N1	S_N2	E1	E2
Strength		✓		
Concentration		✓		✓
Leaving group	✓	✓	✓	✓
Solvent		✓		✓
Total	1	4	1	3

(b) This reaction favors E2. The attacking species is a strong, bulky base in high concentration. Also, the solvent is aprotic. Only the Cl and F are feasible leaving groups, because the Br atom is bonded to an sp^2-hybridized carbon. Of those, the E2 reaction will favor the Cl leaving group, because it has better leaving group ability. Notice that two H atoms can be eliminated from the adjacent C atom, so both cis and trans isomers can be produced from an anticoplanar conformation of the substrate.

Factor	S_N1	S_N2	E1	E2
Strength				✓
Concentration				✓
Leaving group	✓	✓	✓	✓
Solvent		✓		✓
Total	1	2	1	4

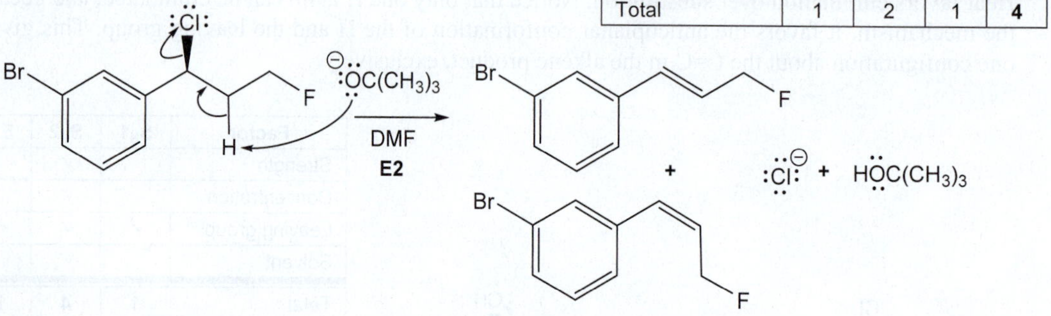

:Cl: Br F H $^{\ominus}$:OC(CH$_3$)$_3$ DMF E2 Br F + :Cl:$^{\ominus}$ + HÖC(CH$_3$)$_3$ Br F

Problem 9.66

Think

Consider the C–L first. Is there a suitable leaving group? In the presence of HBr, is HO able to be made into a suitable leaving group? To what type of carbon is the leaving group attached? What type of carbocation is formed? Is resonance possible?

Solve

To produce products of the given formula, substitution must take place. The reaction conditions favor S_N1 over S_N2. Under the acidic conditions, water is the leaving group, which favors S_N1 over S_N2. The leaving group is on an allylic carbon, which makes a resonance-delocalized carbocation, thus favoring S_N1 over S_N2. And the solvent is an alcohol (the reactant) or water (HBr's solvent), both of which are protic and favor S_N1 over S_N2. After the leaving group leaves, the resonance-delocalized carbocation can be attacked by Br^- at two different locations, leading to the two different products.

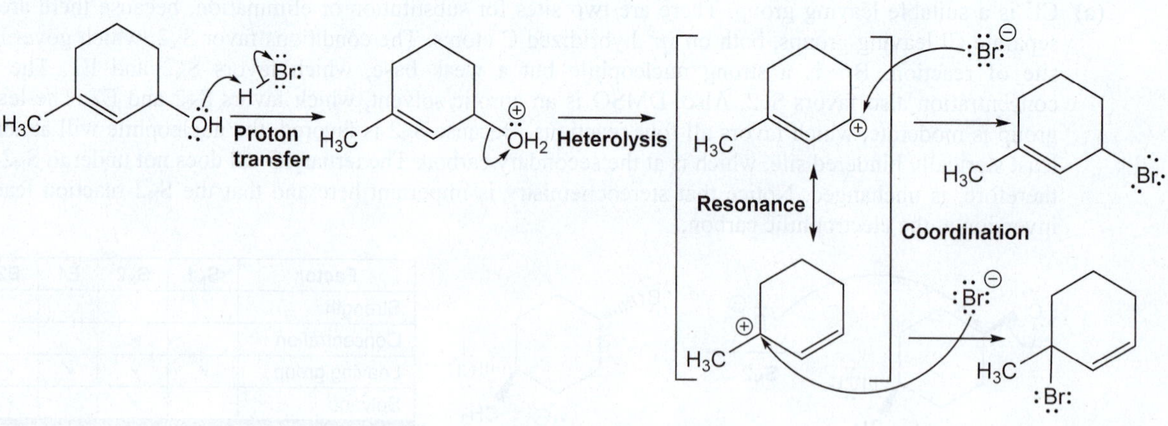

H_3C ÖH H Br: Proton transfer H_3C $\overset{\oplus}{O}H_2$ Heterolysis H_3C ⊕ :Br:$^{\ominus}$ Resonance H_3C ⊕ :Br:$^{\ominus}$ Coordination H_3C Br: H_3C :Br:

Problem 9.67

Think

What was the major mechanism in Problem 9.66? Which mechanism shares the same rate-determining step? How does heating up the reaction change which mechanism is favored?

Solve

S_N1 and E1 reactions are favored by the same factors. Heating the reaction mixture favors elimination over substitution, which is why the products given are elimination products. For the same reasons that S_N1 is favored in the previous problem, E1 is favored under these conditions. Once again, a resonance-delocalized carbocation allows for reaction at two different sites. As shown below, the products that are given in the problem are the result of elimination from one of the carbocation's resonance structures. *Note:* H′ and H″ are used to distinguish the mechanisms that form the two isomeric alkenes, as are the dashed and solid curved arrows.

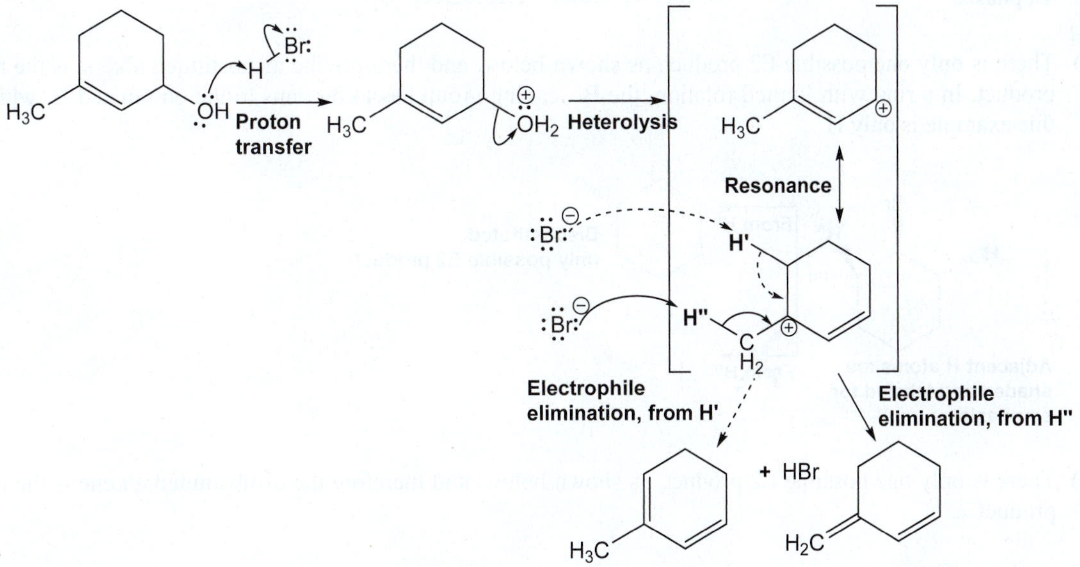

Problem 9.68

Think

In an E2 reaction, which CH protons can be attacked by the base? Draw out the C–H bond for all adjacent CH groups. Where does the double bond in the product appear relative to the leaving group and an adjacent proton? What alkenes form? Which one is most stable?

Solve

(a) The possible E2 products, not considering *E*/*Z* isomerism, are shown below. These are the products of eliminating one of the shaded H atoms along with the leaving group. The major product is the most highly substituted alkene (most stable), which is trisubstituted in this case.

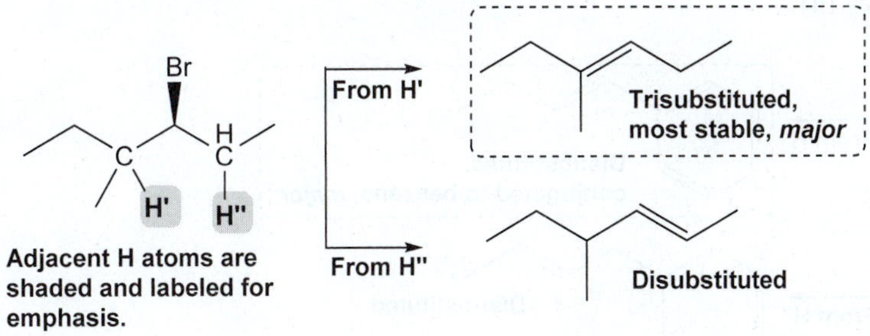

(b) The possible E2 products are shown below. These are the products of eliminating one of the shaded H atoms along with the leaving group. In a ring with limited rotation, the Br leaving group has to be trans to the eliminated H. The major product is the most highly substituted alkene (most stable), which is trisubstituted in this case.

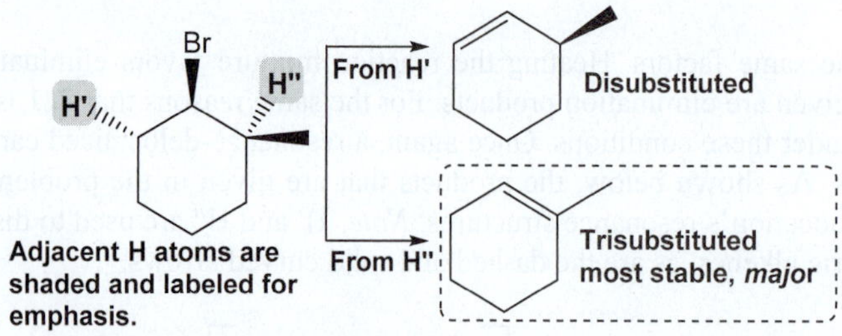

Adjacent H atoms are shaded and labeled for emphasis.

From H' → **Disubstituted**

From H" → **Trisubstituted most stable, *major***

(c) There is only one possible E2 product, as shown below, and therefore the disubstituted alkene is the major product. In a ring with limited rotation, the Br leaving group has to be trans to the eliminated H, which in this example is only H′.

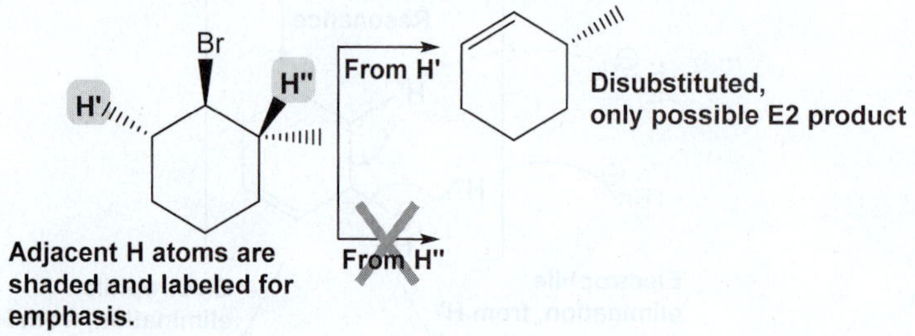

Adjacent H atoms are shaded and labeled for emphasis.

From H' → **Disubstituted, only possible E2 product**

From H" (crossed out)

(d) There is only one possible E2 product, as shown below, and therefore the disubstituted alkene is the major product.

Adjacent H atoms are shaded and labeled for emphasis.

From H' → **Disubstituted, only possible E2 product**

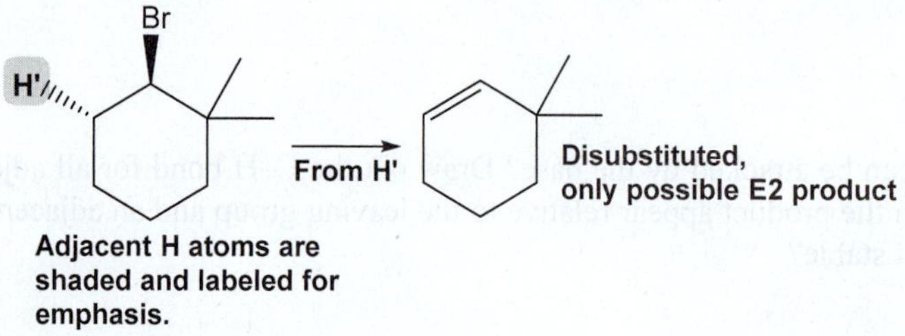

(e) The possible E2 products are shown below. These are the products of eliminating one of the shaded H atoms along with the leaving group. Both alkenes are disubstituted. However, the major product is the alkene that forms the new C=C conjugated to the benzene ring, which provides some resonance stabilization (see also Chapter 14).

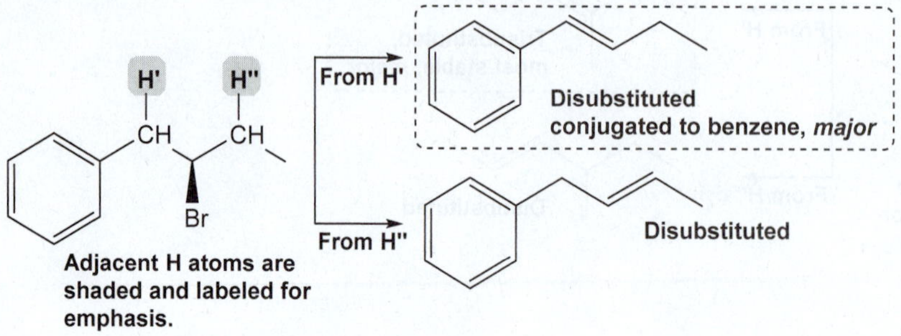

Adjacent H atoms are shaded and labeled for emphasis.

From H' → **Disubstituted conjugated to benzene, *major***

From H" → **Disubstituted**

Problem 9.69

Think

In an E1 reaction, which CH protons can be attacked by the base? Draw out the C–H bond for all adjacent CH groups. Where does the double bond in the product appear relative to the leaving group and an adjacent proton? What alkenes form? Which one is most stable?

Solve

In an E1 mechanism, a carbocation is formed in the first step after heterolysis. Rearrangements must be considered. Only the major product is shown in each example below.

(a) The major E1 product is shown below. Even after a 1,2-hydride shift, the major product is the trisubstituted alkene, same as the major E2 product from Problem 9.68 **(a)**.

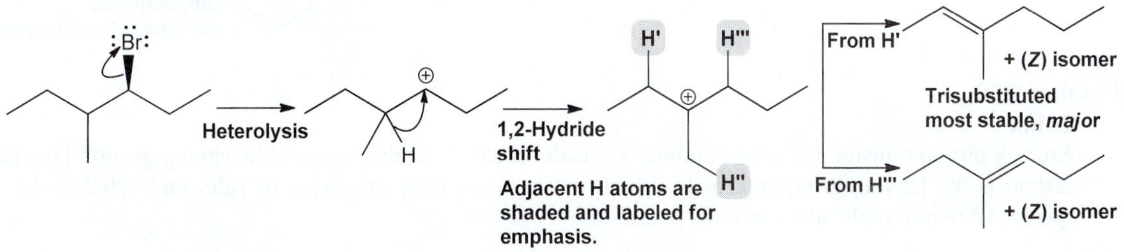

Heterolysis; 1,2-Hydride shift; H'; H'''; H''; Adjacent H atoms are shaded and labeled for emphasis.; From H': + (*Z*) isomer, Trisubstituted most stable, *major*; From H''': + (*Z*) isomer

(b) The major E1 product is shown below. Even after a 1,2-hydride shift, the major product is the trisubstituted alkene, same as the major E2 product from Problem 9.68 **(b)**.

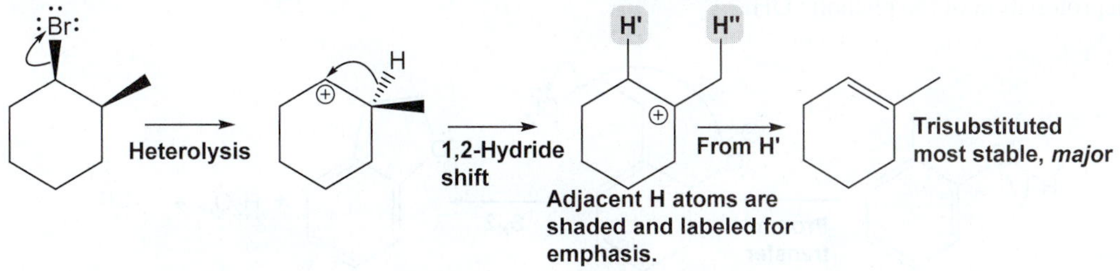

Heterolysis; 1,2-Hydride shift; H'; H''; From H'; Adjacent H atoms are shaded and labeled for emphasis.; Trisubstituted most stable, *major*

(c) The major E1 product is shown below. After a 1,2-hydride shift, the major product is the trisubstituted alkene.

Heterolysis; 1,2-Hydride shift; H'; H''; From H'; Adjacent H atoms are shaded and labeled for emphasis.; Trisubstituted most stable, *major*

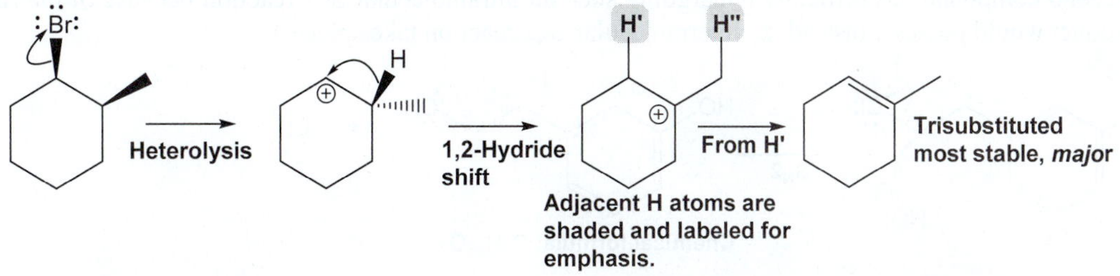

(d) The major E1 product is shown below. After a 1,2-methyl shift, the major product is the tetrasubstituted alkene.

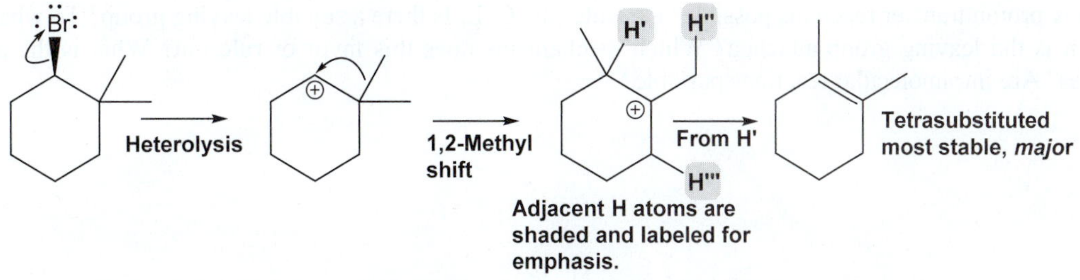

Heterolysis; 1,2-Methyl shift; H'; H''; H'''; From H'; Adjacent H atoms are shaded and labeled for emphasis.; Tetrasubstituted most stable, *major*

(e) The major E1 product is shown below. Even after a 1,2-hydride shift, the major product is the disubstituted conjugated alkene, same as the major E2 product from Problem 9.68 **(e)**.

Heterolysis

1,2-Hydride shift

From H'

Disubstituted conjugated to benzene, *major*

Problem 9.70

Think

Are any proton transfer reactions possible? Consider the C–L. Is there a suitable leaving group? To what type of carbon is the leaving group attached? Which mechanisms does this favor or rule out? What is the attacking species? Are intramolecular reactions possible?

Solve

Both reactions favor the S_N2 mechanism, because the nucleophile is strong and in high concentrations and the leaving group is on a primary carbon. The first compound is produced by an intramolecular S_N2 reaction after a fast deprotonation of the phenolic OH.

Proton transfer

S_N2

+ H_2O + Cl^-

Chemical formula: C_8H_8O

The second compound has difficulty undergoing such an intramolecular S_N2 reaction because of the ring strain the product would possess. Instead, an intermolecular S_N2 reaction takes place.

S_N2

+ Cl^-

Chemical formula: $C_8H_{10}O_2$

Problem 9.71

Think

Are any proton transfer reactions possible? Consider the C–L. Is there a suitable leaving group? To what type of carbon is the leaving group attached? Which mechanisms does this favor or rule out? What is the attacking species? Are intramolecular reactions possible?

Solve

After a fast deprotonation of the carboxyl group, the resulting carboxylate becomes nucleophilic. This reaction favors the S_N2 mechanism, because the nucleophile is strong and in high concentrations, the leaving group is on a secondary carbon, and the solvent is aprotic. Because of the formation of a six-membered ring, an intramolecular S_N2 reaction occurs to yield the product with the formula $C_9H_{14}O_2$.

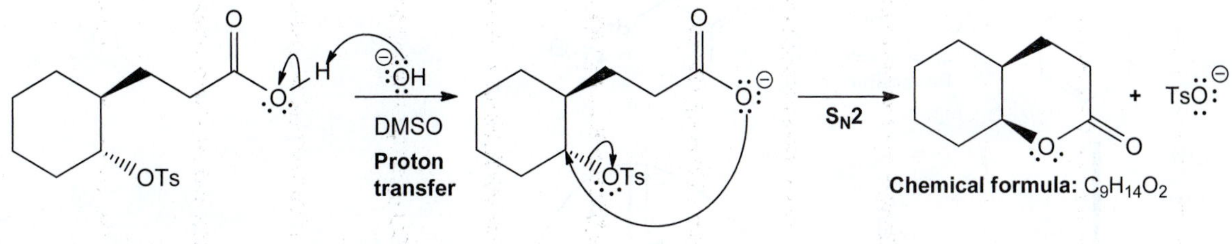

DMSO
Proton transfer
S_N2
Chemical formula: $C_9H_{14}O_2$

Section 9.12 Kinetic Control, Thermodynamic Control, and Reversibility

Problem 9.72

Think

What determines whether a reaction is reversible or irreversible? Complete each reaction by drawing the products. Draw a free energy diagram. Does charge stability favor the reactant or product side? A little or a lot?

Solve

(a) Br^- has greater charge stability compared to CH_3S^-, because Br is a larger atom and the charge is less concentrated. This makes $\Delta G°_{rxn}$ substantially negative. Therefore, $\Delta G°^{\ddagger}_{reverse}$ is very large, so the reverse reaction is excessively slow. Thus, the reaction is *irreversible*.

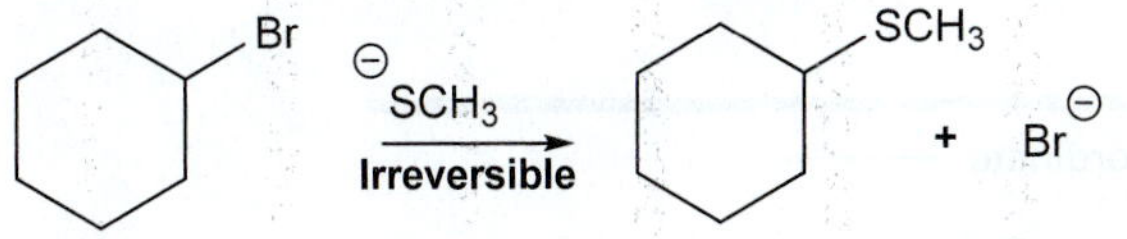

Irreversible

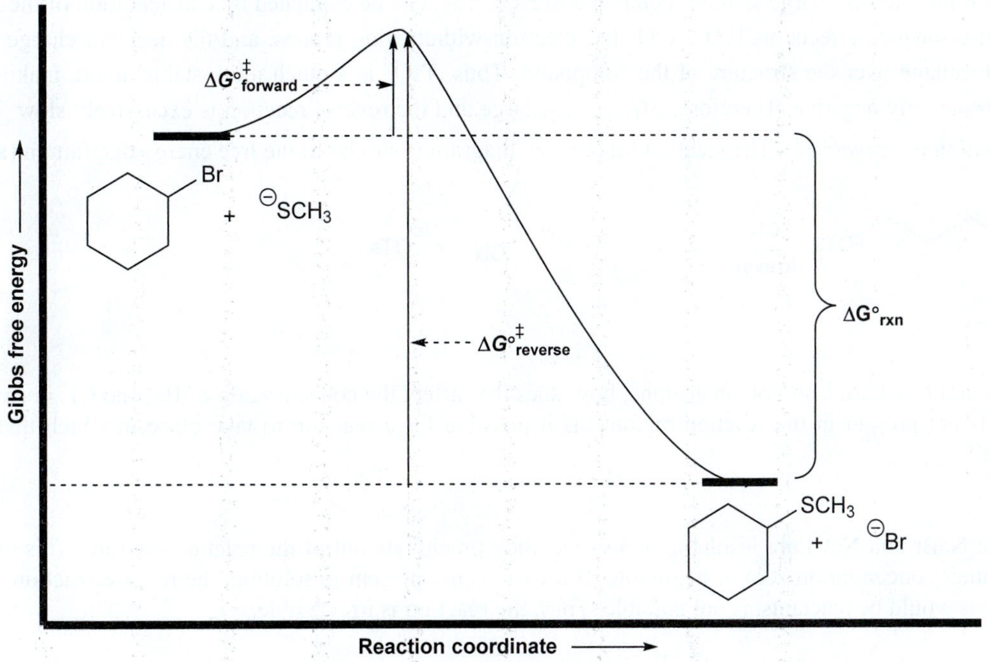

$\Delta G°^{\ddagger}_{forward}$
$\Delta G°^{\ddagger}_{reverse}$
$\Delta G°_{rxn}$
Gibbs free energy
Reaction coordinate

(b) $\Delta G°_{rxn}$ is not substantially negative. To the contrary, Cl^- is slightly less stable than Br^- owing to the smaller atomic size of Cl. This makes the reaction faster in the reverse direction than in the forward direction under standard conditions ($\Delta G°^{\neq}_{forward} > \Delta G°^{\neq}_{reverse}$). If the reaction in the forward direction occurs at a sufficient rate, then so must the reverse reaction, so the reaction is *reversible.*

Cl ⊖Br Br + Cl⊖

Reversible

ΔG°‡forward

ΔG°‡reverse

Gibbs free energy

ΔG°rxn

Cl + ⊖Br

Br + ⊖Cl

Reaction coordinate

(c) TsO^- has greater charge stability compared to NC^-. This can be evaluated by consideration of the inductive and resonance effects on TsO^-. TsO^- has electron-withdrawing groups, and the negative charge is highly delocalized over the structure of the compound. Thus, TsO^- is a much more stable anion, making $\Delta G°_{rxn}$ substantially negative. Therefore, $\Delta G°‡_{reverse}$ is large and the reverse reaction is excessively slow. Thus, the reaction is *irreversible*. The reaction free energy diagram is similar to the free energy diagram in **(a)**.

OTs ⊖CN CN + ⊖OTs

Irreversible

Problem 9.73

Think

If NaBr and NaCl are insoluble in acetone, how does that affect the concentration of Br^- and Cl^- in solution? If an ion is not present in the reaction mixture, is it possible for a reaction to take place in which that ion is a reactant?

Solve

Because NaBr and NaCl are insoluble in acetone, they precipitate out of the reaction mixture. This essentially makes their concentration zero or negligible. If the ion is not present in solution, the reverse reaction, in which those ions would be reactants, is not possible. Thus, the reaction is irreversible.

R—Cl or R—Br

NaI acetone Irreversible

R—I

⊖Cl or ⊖Br

Insoluble in acetone and precipitates out of the solution

Problem 9.74

Think

Write out the complete reactions. How do you evaluate charge stability? What factors stabilize/destabilize an anion? How does base/acid strength factor into charge stability? Consider octet, formal charge location, and resonance.

Solve

(a) In this proton transfer reaction, the reactants and products both have the negative charge on O. No resonance delocalization of the charge is possible, but there are some minor effects on charge stability from the alkyl group. Thus, the reactants and products have similar charge stability, making the reaction reversible. This is reflected by the equilibrium constant we can compute from pK_a values. The reactant is hydroxide (H_2O; $pK_a = 15.7$), and the product is cyclohexanolate (2° OH; $pK_a = {\sim}16.5$). This leads to a K_{eq} = ~0.16.

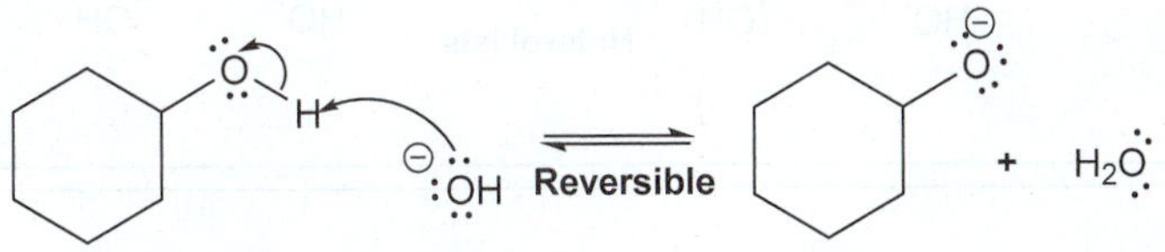

Reversible

(b) An sp^3-hybridized carbanion is the nucleophile in this nucleophilic addition reaction. The negative charge is much more stable on O in the products, owing to the greater electronegativity of O. Thus, the nucleophilic addition products are heavily favored, making this reaction *irreversible.*

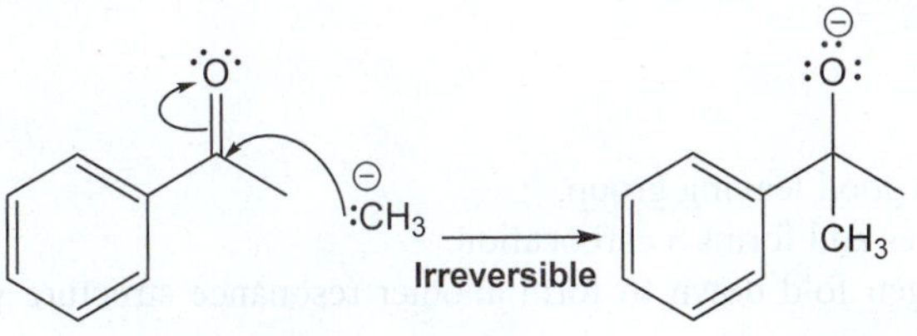

Irreversible

(c) This elementary step is a heterolysis. The structure on the left has all atoms with an octet, and the product on the right has the C atom with an incomplete octet. This leads to a positive $\Delta G°_{rxn}$, and the reaction is thus said to be *reversible.*

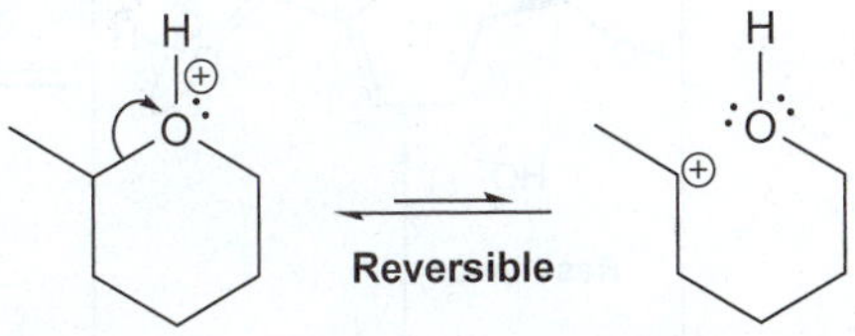

Reversible

Section 9.13 The Organic Chemistry of Biomolecules

Problem 9.75

Think

This is an acid-catalyzed E1 reaction mechanism (Section 8.6). What step must occur before the heterolysis step? Is OH a good leaving group? What species are incompatible in acidic conditions?

Solve

D-Fructose to A:

Elementary Step 1: Protonate OH to form water, a good leaving group.
Elementary Step 2: Heterolysis occurs; water leaves and forms a carbocation.
Elementary Step 3: The adjacent H atom is removed by the HSO_4^- conjugate base to re-form the acid catalyst.

Proton transfer

$-H_2O$ Heterolysis

D-Fructose

Electrophile elimination

+ H_2SO_4

A

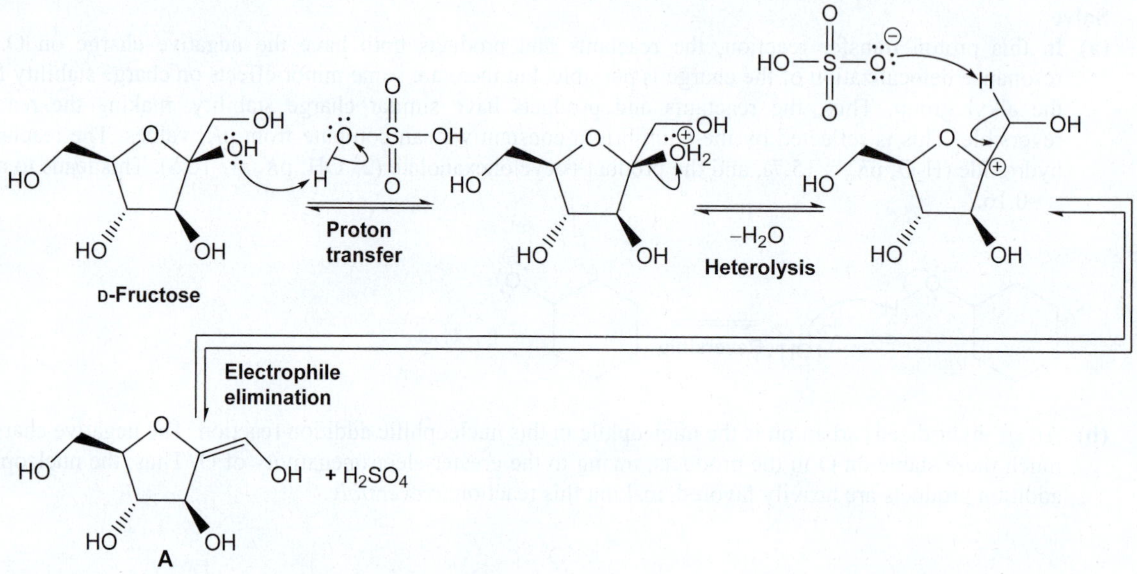

A to B:

Elementary Step 1: Protonate OH to form water, a good leaving group.
Elementary Step 2: Heterolysis occurs; water leaves and forms a carbocation.
Resonance: Nonbonding electrons from the oxygen fold down to form another resonance structure where all atoms have an octet.
Elementary Step 3: The adjacent H atom is removed by the HSO_4^- conjugate base to re-form the acid catalyst.

Resonance

A

Proton transfer

$-H_2O$ Heterolysis

Electrophile elimination

+ H_2SO_4

B

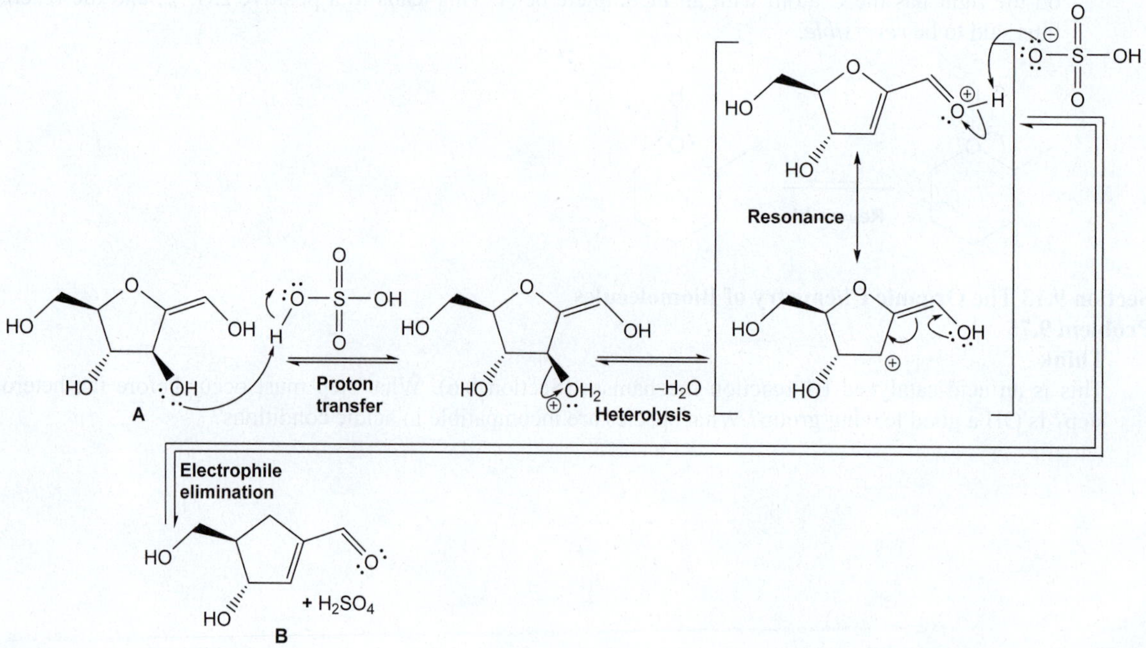

B to HMF:

Elementary Step 1: Protonate OH to form water, a good leaving group.
Elementary Step 2: Heterolysis occurs; water leaves and forms a carbocation.
Elementary Step 3: The adjacent H atom is removed by the HSO_4^- conjugate base to re-form the acid catalyst.

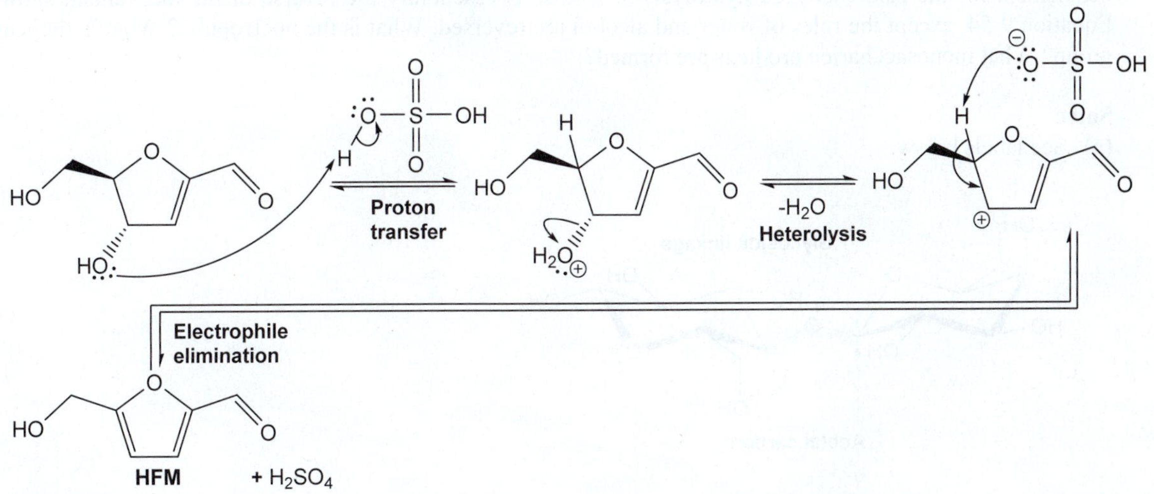

Proton transfer

$-H_2O$
Heterolysis

Electrophile elimination

HFM + H_2SO_4

Problem 9.76

Think

This is an S_N1 mechanism with a weak protic nucleophile in acidic conditions. What step must occur before the heterolysis step? Is OH a good leaving group? What species are incompatible in acidic conditions? In considering stereochemistry of the product, what does an S_N1 mechanism do to the configuration of the carbon stereocenter initially attached to the leaving group?

Solve

The mechanism is drawn and written out below.
Elementary Step 1: Protonate OH to form water, a good leaving group
Elementary Step 2: Heterolysis; water leaves and forms a carbocation
Elementary Step 3: Coordination; CH_3CH_2OH nucleophile can attack from above or below the planar C^+
Elementary Step 4: Proton transfer; deprotonation removes the positive charge from oxygen

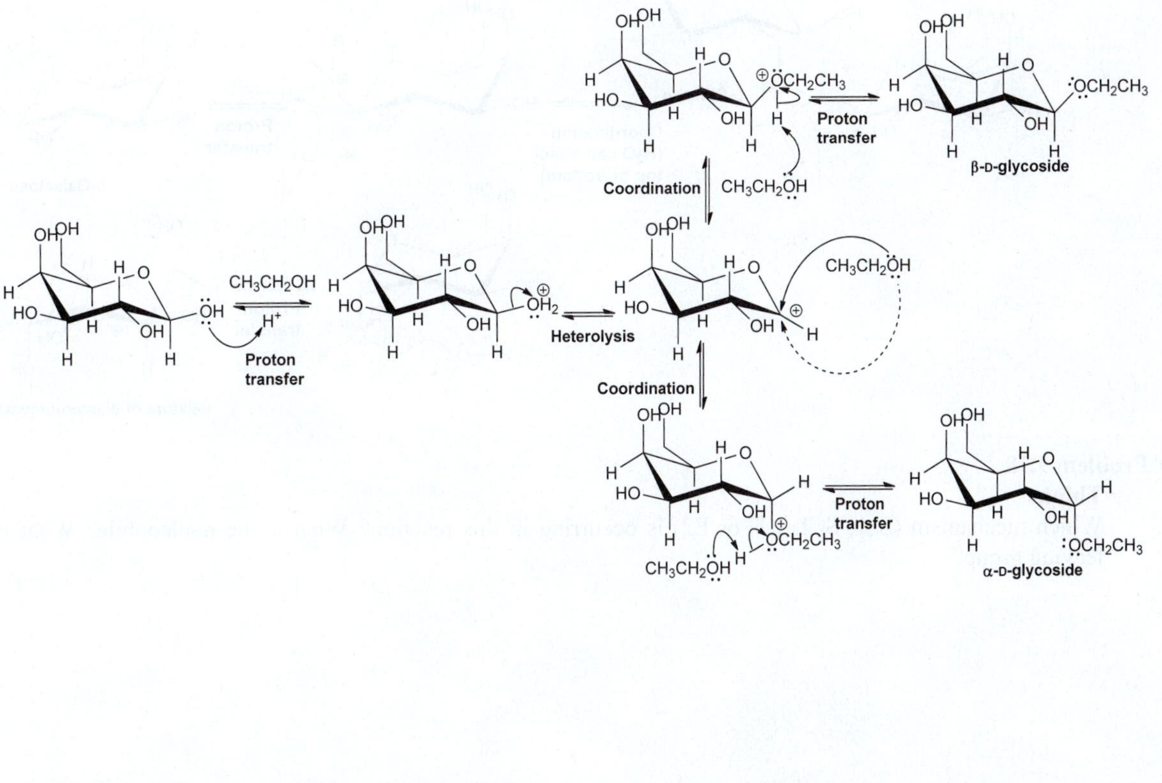

CH_3CH_2OH H^+
Proton transfer

Heterolysis

Coordination CH_3CH_2OH

Proton transfer

β-D-glycoside

Coordination

Proton transfer

α-D-glycoside

Problem 9.77

Think

Can you identify the acetal group, $C(OR)_2$? Does one of those R groups belong to a sugar that is different from the sugar to which C belongs? Are the groups at the glycoside linkage axial (α) or equatorial (β)? The mechanism for the acid-catalyzed hydrolysis of lactose is essentially the reverse of the mechanism shown in Equation 9-54, except the roles of water and alcohol are reversed. What is the nucleophile? What is the leaving group? What monosaccharide products are formed?

Solve

(a) See labels below.

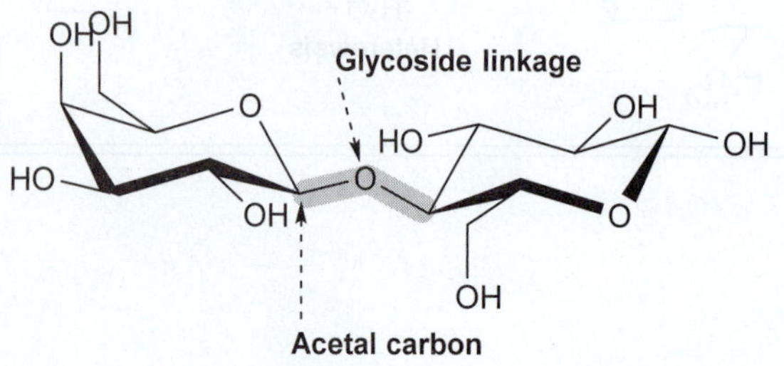

Glycoside linkage

Acetal carbon

(b) β-1,4′-glycosidic linkage. It is β because the OR group on the acetal C is equatorial and is on the same side of the ring as the CH_2OH group. It is 1,4′, because in the ring on the right, the glycosidic linkage is bonded to C4.

(c) Acid-catalyzed mechanism of acetal hydrolysis; hydrolysis of lactose into D-galactose and D-glucose.

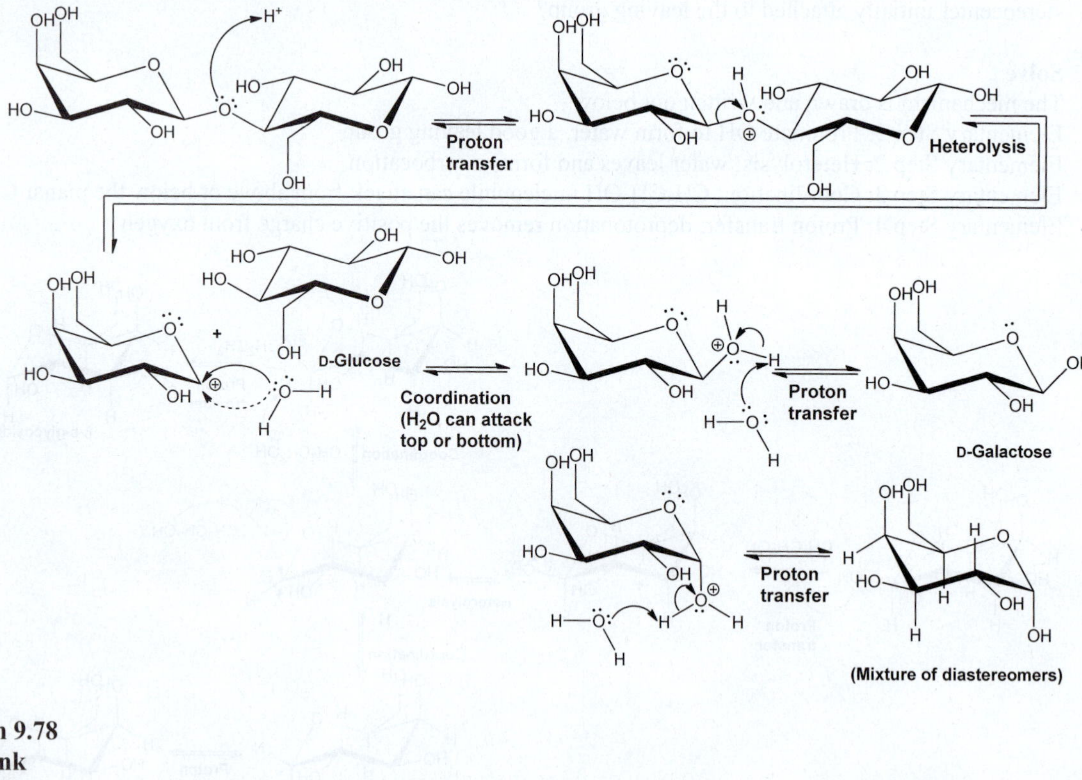

Proton transfer

Heterolysis

D-Glucose

Coordination (H_2O can attack top or bottom)

Proton transfer

D-Galactose

Proton transfer

(Mixture of diastereomers)

Problem 9.78

Think

Which mechanism (S_N1, S_N2, E1, or E2) is occurring in this reaction? What is the nucleophile? What is the leaving group?

Solve

This is an S_N1 mechanism, aided by the fact that the carbocation that is produced has significant resonance stabilization involving the oxygen. Alkylation of the ring nitrogen has converted the nitrogen base into an excellent leaving group, which gets the S_N1 mechanism started. And water, which acts as both the attacking species and the solvent, is a poor weak nucleophile and a protic solvent.

Elementary Step 1: The nitrogen base leaves, forming a resonance-stabilized carbocation (most stable contributor shown).

Elementary Step 2: A water molecule attacks the carbocation.

Elementary Step 3: A proton transfer completes the reaction.

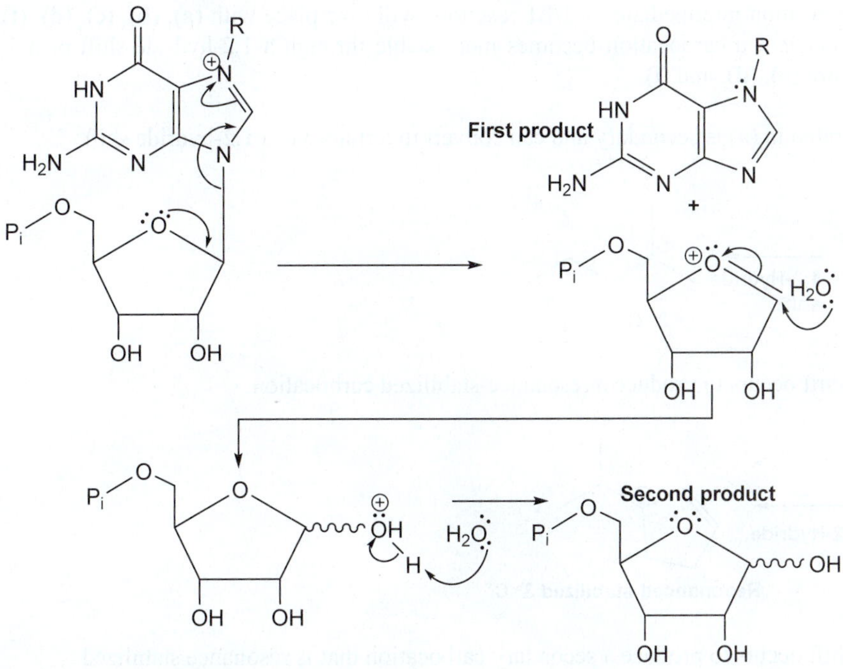

Integrated Problems

Problem 9.79

Think

Do the reaction conditions favor S_N1 or S_N2? How many leaving groups are present on the substrate? How many nucleophilic sites are present on the nucleophile? Will an intramolecular reaction occur?

Solve

Each S atom displaces bromide in an S_N2 mechanism, which we can tell by the inversion of configuration at each of the C–Br bonds. This is an intramolecular reaction that forms a six-membered ring.

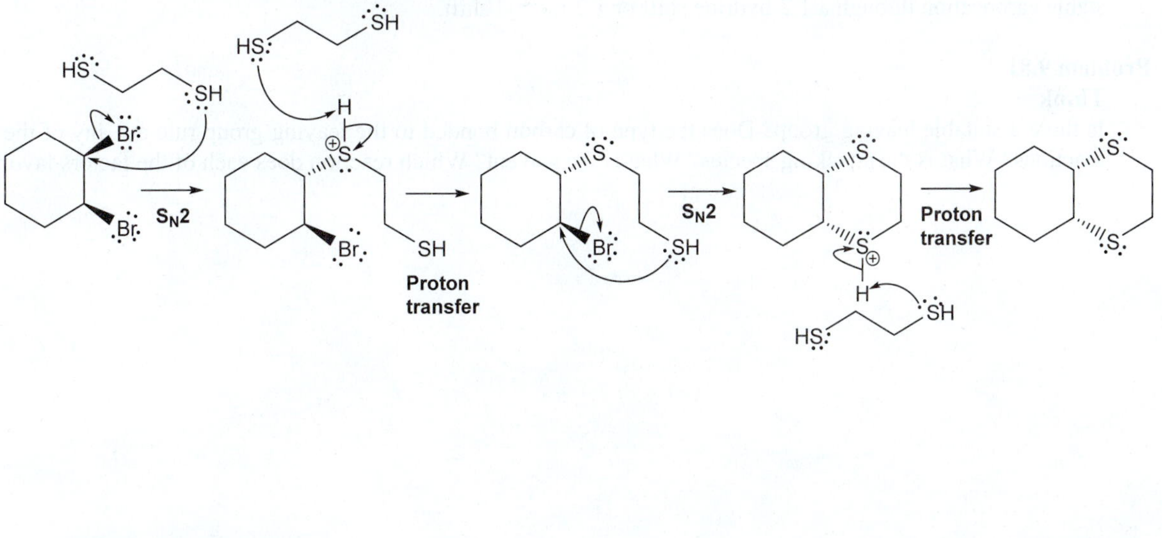

Problem 9.80

Think

In which mechanisms are carbocation rearrangements possible? To favor these kinds of reactions, should the attacking species be strong or weak? Should the leaving group be excellent, moderate, or poor? Should the type of carbon attached to the leaving group be primary, secondary, or tertiary? Should the solvent be protic or aprotic? If a carbocation forms, when is it capable of rearrangement? What type of carbocation is formed? Is a more stable carbocation possible?

Solve

Carbocation rearrangement can occur only in the S_N1 and E1 mechanisms, because those are the ones that proceed through a carbocation intermediate. S_N1/E1 reactions will take place with **(a)**, **(b)**, **(c)**, **(d)**, **(f)**, and **(g)**. Rearrangements will occur if a carbocation becomes more stable through a 1,2-hydride shift or a 1,2-methyl shift. These reactions are **(a)**, **(d)**, and **(f)**.

The carbocation generated in **(a)** is secondary and can convert to tertiary via a 1,2-hydride shift.

(a)
H
⊕
2° C⁺
1,2-Hydride shift
⊕
3° C⁺

In **(d)**, a 1,2-hydride shift occurs to produce a resonance-stabilized carbocation.

(d)
H
⊕
3° C⁺
1,2-Hydride shift
⊕
Resonanced stabilized 3° C⁺

In **(f)**, a 1,2-hydride shift occurs to produce a secondary carbocation that is resonance stabilized.

(f)
H
⊕
2° C⁺
1,2-Hydride shift
⊕
Benzylic C⁺

Rearrangements will not occur in **(b)**, **(c)**, **(e)**, and **(g)**. Formation of a carbocation is not expected in reaction **(e)**, because the reaction conditions favor S_N2; in the other reactions, rearrangement will not produce a more stable carbocation through a 1,2-hydride shift or 1,2-methyl shift.

Problem 9.81

Think

Is there a suitable leaving group? Does the type of carbon bonded to the leaving group rule out any of the four reactions? What is the attacking species? What is the solvent? Which reaction does each of the factors favor?

Solve

(a) The predominant mechanism is E1. The attacking species is an alcohol, which is a weak base and weak nucleophile, favoring S_N1 and E1. The concentration does not matter because the attacking species is weak. The leaving group is water (under acidic conditions), which is an excellent leaving group, and favors S_N1 and E1. The carbon is secondary, which favors all four mechanisms. The solvent is an alcohol, which favors S_N1 and E1. Heat favors elimination over substitution. The complete mechanism is shown below. Notice that the proton that is eliminated is the one that gives the most highly substituted alkene product.

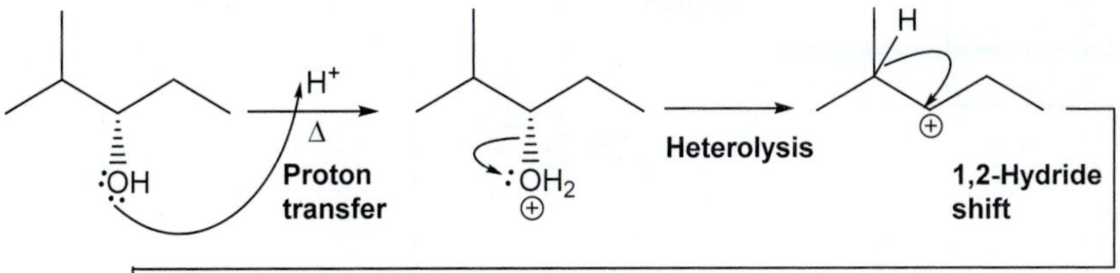

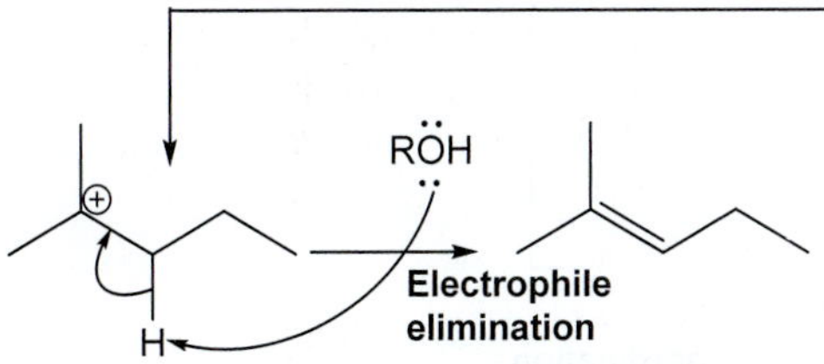

Factor	S_N1	S_N2	E1	E2
Strength	✓		✓	
Concentration				
Leaving group	✓		✓	
Solvent	✓		✓	
Total	3		3	

(b) E1 is favored. The factors are the same as in **(a)**, but the S_N2 column is omitted because the leaving group is on a tertiary carbon. The H atom that is eliminated in the final step yields the most highly substituted alkene product.

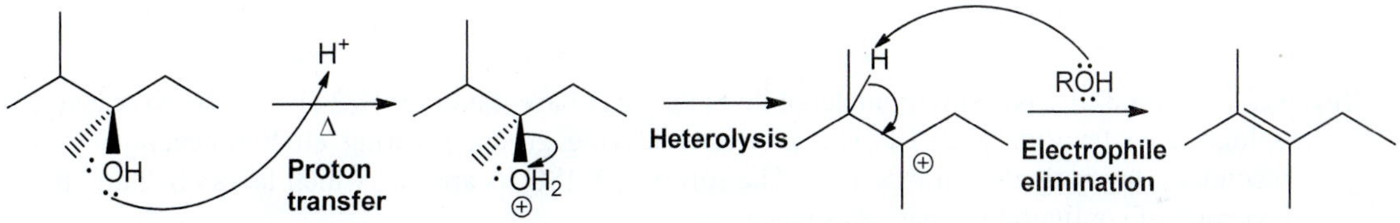

(c) The factors are identical to **(a)**, so E1 is favored, but no carbocation rearrangement is feasible. Notice, however, that both cis and trans isomers are formed.

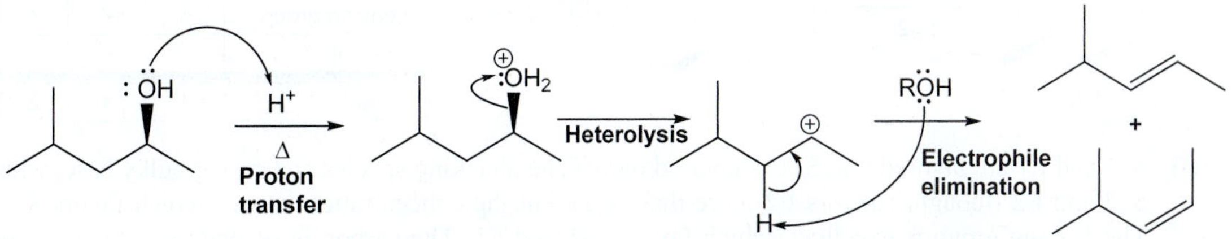

(d) S_N1 is favored. The attacking species is weak as a nucleophile and base, favoring S_N1 and E1. Concentration does not matter, because the attacking species is weak. The leaving group is very good, favoring S_N1 and E1. The carbon is tertiary, shutting down S_N2. The solvent is an alcohol, which is protic and favors S_N1 and E1. The reaction is not heated, so substitution is favored over elimination, though we can expect some elimination product as well. Notice that the carbocation is resonance delocalized, so the nucleophile can attack either carbon bearing the positive charge, producing isomeric products. See the table and mechanism on the next page.

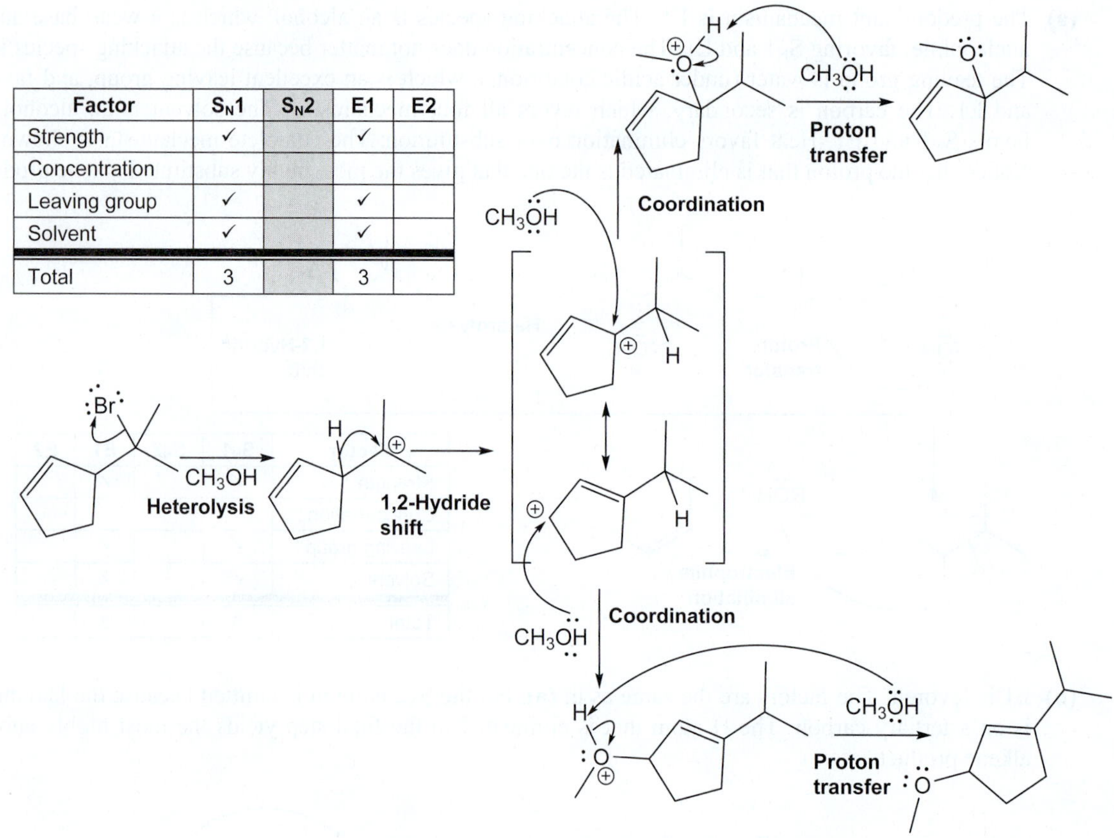

Factor	S_N1	S_N2	E1	E2
Strength	✓		✓	
Concentration				
Leaving group	✓		✓	
Solvent	✓		✓	
Total	3		3	

(e) S_N2 is favored. Br^- is a strong nucleophile but a weak base, favoring S_N2 and E1. The concentration of Br^- is high, also favoring S_N2. Cl^- is a moderate leaving group, favoring all four reactions. The carbon is secondary, favoring all four reactions. The solvent, DMSO, is aprotic, which favors S_N2 and E2. Notice the inversion of configuration that takes place.

Factor	S_N1	S_N2	E1	E2
Strength		✓	✓	
Concentration		✓		
Leaving group	✓	✓	✓	✓
Solvent		✓		✓
Total	1	**4**	2	2

(f) S_N1 and E1 are favored, but S_N1 is favored more. The attacking species is a strong bulky base, which favors S_N2 and E2 (though it favors E2 more than S_N2), but the concentration is low, which favors S_N1 and E1. The leaving group is excellent, which favors S_N1 and E1. The carbon is secondary, which favors all four reactions. The solvent is protic, which favors S_N1 and E1. The reaction is not heated, which tips the balance in the favor of substitution over elimination. Notice that both *R* and *S* configurations are produced for the substitution products, and both cis and trans isomers are produced for the elimination products. See the table and mechanism on the next page.

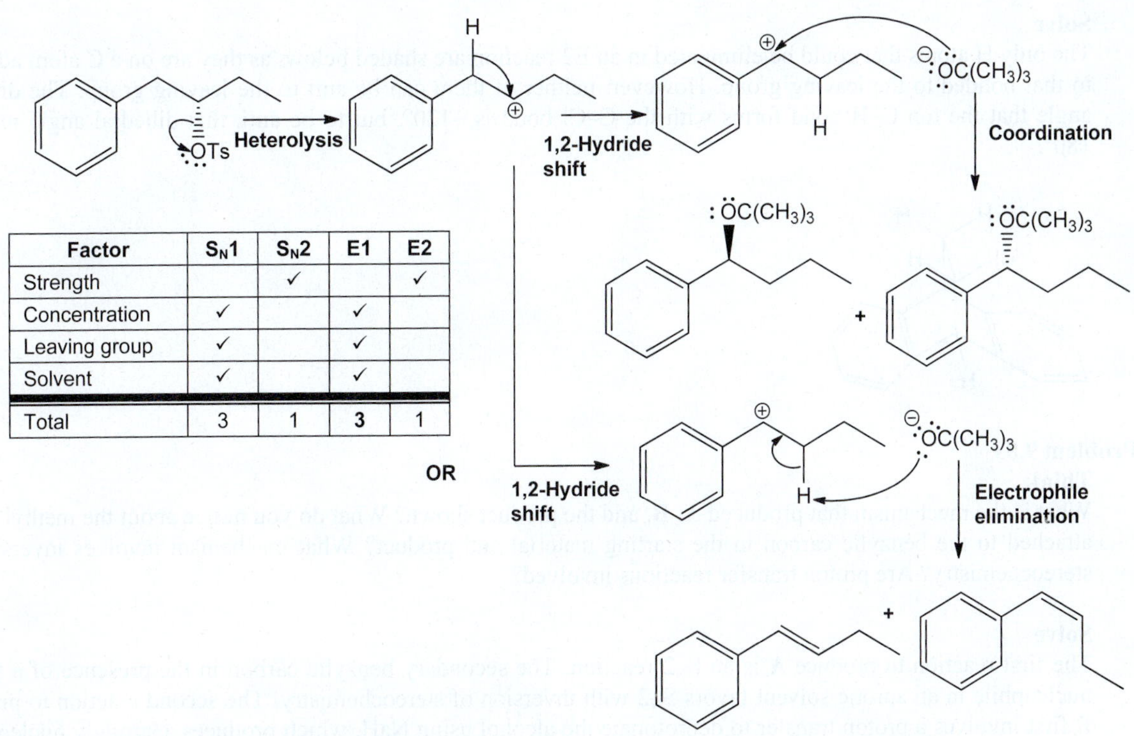

Factor	S_N1	S_N2	E1	E2
Strength		✓		✓
Concentration	✓		✓	
Leaving group	✓		✓	
Solvent	✓		✓	
Total	3	1	3	1

(g) S_N1 is favored. The attacking species is a weak nucleophile and weak base, favoring S_N1 and E1. The concentration does not matter. The leaving group is excellent, which favors S_N1 and E1. The carbon is benzylic, which favors S_N1 and E1. The solvent is protic, which favors S_N1 and E1. The reaction is not heated, so we can expect this to favor S_N1, but E1 products will be mixed in. Notice that the substitution products form a racemic mixture because the ethanol nucleophile can attack the carbocation from either side of the plane of the C^+.

OTs

Heterolysis

$HOCH_2CH_3$

Coordination

H_3CH_2C ⊕ H

Racemic

$HOCH_2CH_3$

Proton transfer

H_3CH_2C O:

Racemic

Factor	S_N1	S_N2	E1	E2
Strength	✓		✓	
Concentration				
Leaving group	✓		✓	
Solvent	✓		✓	
Total	3		3	

Problem 9.82

Think

What must be true about the orientation of the leaving group and the adjacent H atom to be eliminated in an E2? Are there H atoms available in this orientation to the Cl leaving group?

Solve

The only H atoms that could be eliminated in an E2 reaction are shaded below, as they are on a C atom adjacent to that bonded to the leaving group. However, neither of them can be anti to the leaving group. The dihedral angle that the top C–H bond forms with the C–Cl bond is ~120°, but to be anti, that dihedral angle must be 180°.

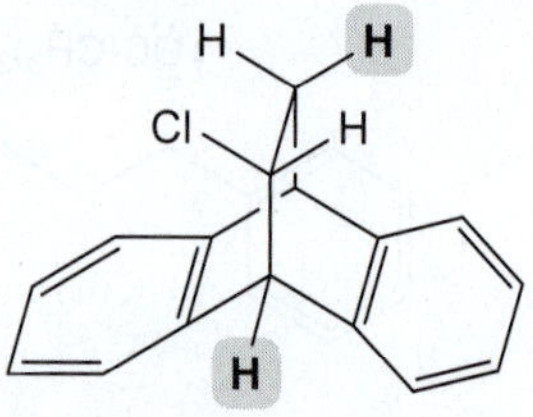

Problem 9.83

Think

What is the mechanism that produced **A**, **B**, and the product shown? What do you notice about the methyl group attached to the benzylic carbon in the starting material and product? What mechanism involves inversion of stereochemistry? Are proton transfer reactions involved?

Solve

The first reaction to produce **A** is an S_N2 reaction. The secondary benzylic carbon in the presence of a strong nucleophile in an aprotic solvent favors S_N2 with inversion of stereochemistry. The second reaction to produce **B** first involves a proton transfer to deprotonate the alcohol using NaH, which produces a strongly nucleophilic alkoxide anion (RO^-), and then an S_N2 follows. The third reaction to produce the final product first involves a proton transfer to deprotonate the alkyne using *n*-butyl lithium, which produces a strongly nucleophilic alkynide anion ($RC{\equiv}C{:}^-$), and then an S_N2 follows. The reaction scheme is given below.

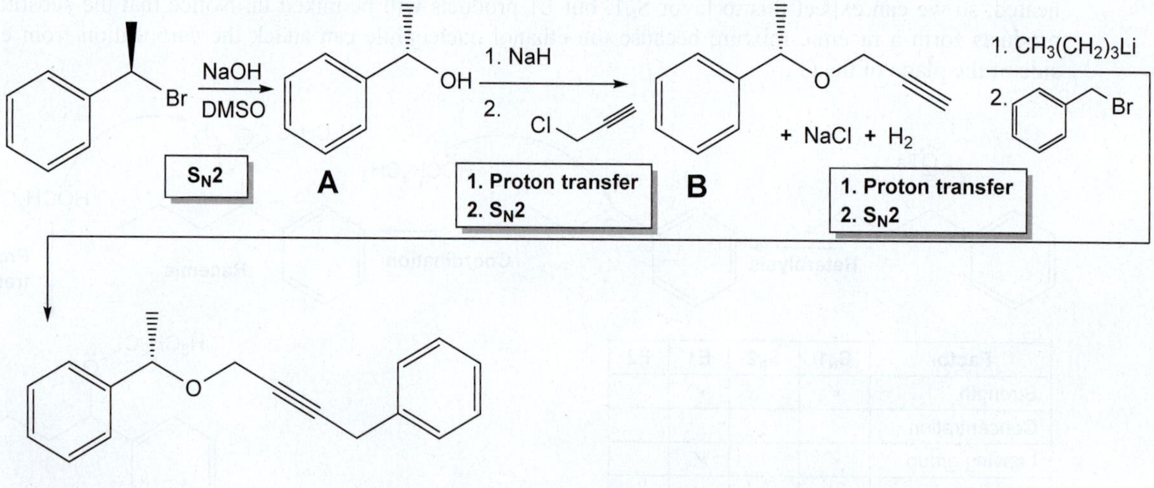

Problem 9.84

Think

Did substitution or elimination occur? Did the reaction produce a single stereoisomer or a mixture of stereoisomers? Does the attacking species match the predicted reaction mechanism? Did proton transfer reactions occur? Is there more than one reactive site?

Solve

(a) The intended leaving group is the OH after protonation, but protonation of the ether oxygen will also produce a CH_3OH leaving group. Thus, reaction can take place at the ether site as well, producing an unwanted product.

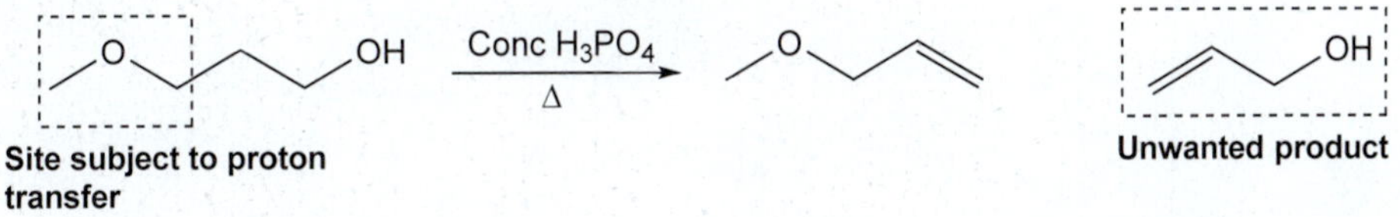

Conc H_3PO_4, Δ

Site subject to proton transfer

Unwanted product

(b) The reaction conditions favor S_N2 (strong nucleophile, high concentration, aprotic solvent), which causes inversion of stereochemistry. The product that is given shows retention of configuration instead.

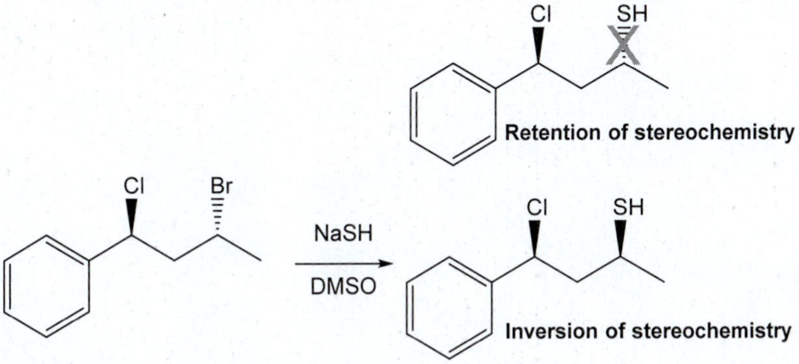

Retention of stereochemistry

NaSH, DMSO

Inversion of stereochemistry

(c) The NaH base deprotonates the terminal carbon to make a negatively charged carbon nucleophile. That nucleophile is intended to react in a substitution involving the chloropropanoic acid as the substrate. However, that negatively charged carbon in the nucleophile is strongly basic and will deprotonate the carboxylic acid instead (remember, proton transfer reactions are fast). The alkyne is then re-formed (see below).

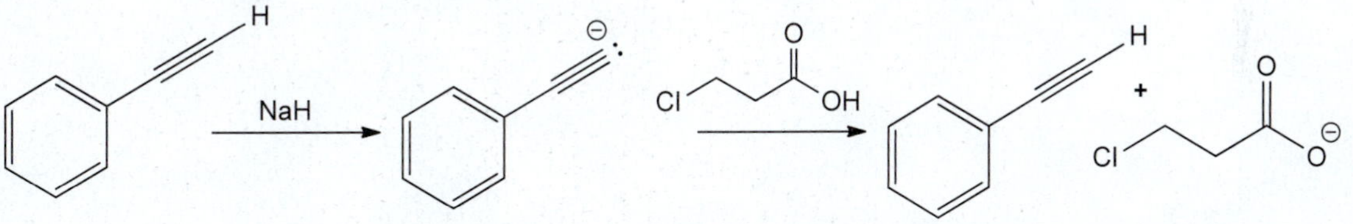

NaH

(d) The intended nucleophile, an alkoxide anion, is also a strong base and can deprotonate the solvent, ethanol, to make $CH_3CH_2O^-$ (remember, proton transfer reactions are fast). Thus, there will be a mixture of two different strong nucleophiles present. This leads to a mixture of two different products.

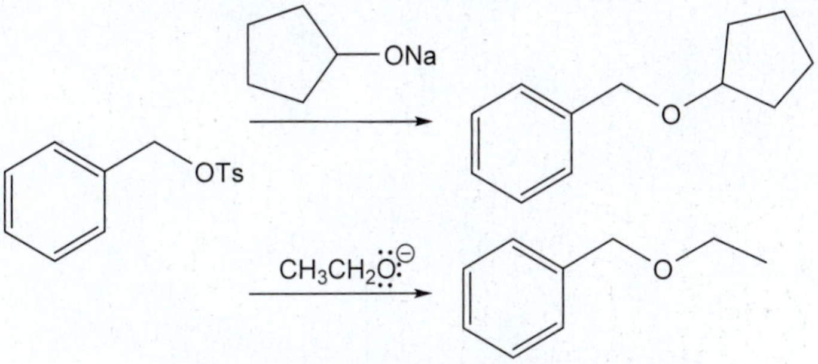

$CH_3CH_2\ddot{O}:^{\ominus}$

INTERCHAPTER F | Naming Compounds with a Functional Group That Calls for a Suffix 2: Carboxylic Acids and Their Derivatives

In Chapter Problems

Problem F.2

Think

What is the highest-priority functional group present and the corresponding suffix that must be added? How many carbon atoms are in the longest carbon chain containing that functional group? Where should the numbering of that chain begin? What other functional groups are present, and how are those named as substituents?

Solve

The IUPAC names for molecules **(a)–(c)** are given below. The main chain is numbered.

(a) The highest-priority functional group in the molecule is a C≡N, so the suffix is *nitrile* and numbering begins at the C≡N's carbon. The other functional group present is a benzene ring, which is treated as a substituent and receives the prefix *phenyl.* There are three carbon atoms in the longest carbon chain containing the C≡N, making the molecule a *propanenitrile*. The *phenyl* is attached to C3.

(b) The highest-priority functional group in the molecule is a CO_2H, so the suffix is *oic acid* and numbering begins at the CO_2H's carbon. The other functional groups present are one C≡N, one C=C, and two C≡Cs. Only the C≡N is treated as a substituent, and it receives the prefix *cyano*. The prefix *cyano* already accounts for the C≡N's C, so there are seven carbon atoms in the longest carbon chain containing the CO_2H group, making the molecule a *heptanoic acid.* The C=C and C≡C are named as part of the main chain, *en* and *diyn*, respectively. The C=C has an *E* configuration.

(c) The highest-priority functional group in the molecule is O=C–N, characteristic of an *amide*, and there are two of them, so the suffix is *diamide* and numbering begins at either carbon of the O=C–N groups. The other substituent present is the branched alkyl group, *1-methylpropyl*, $–CH(CH_3)CH_2CH_3$, at C2.

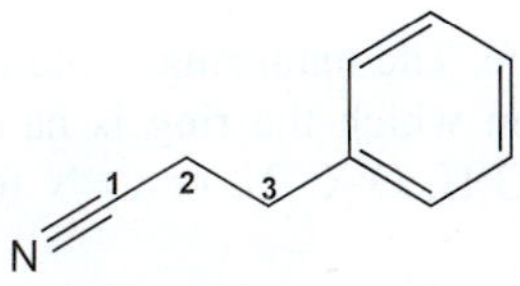

(a) 3-Phenylpropanenitrile 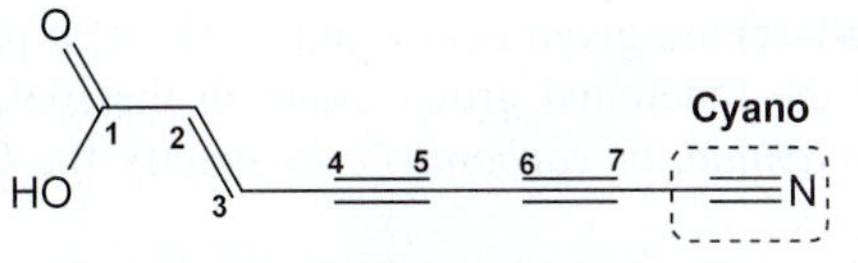 (b) (*E*)-7-Cyanohepta-2-en-4,6-diynoic acid 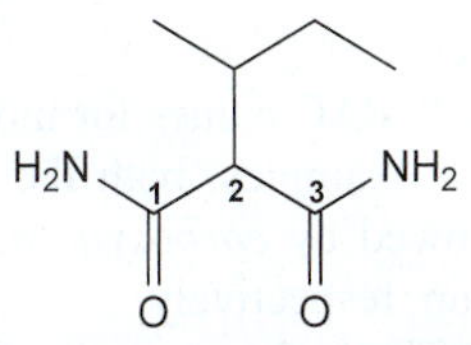 (c) 2-(1-Methylpropyl)propane-1,3-diamide

Problem F.3

Think

How many carbon atoms are in the main chain? To what functional group does the suffix *oic acid*, *amide*, or *nitrile* correlate? Review the nomenclature rules from Interchapters A, B, and C and Table F-1 to identify the substituents and stereochemical configurations. On which C atoms are the substituents located? How are the C atoms in the various functional groups counted—as part of the main chain or ring or separately from it? On which two carbon atoms is each C=C located? Must stereochemical configurations be specified?

Solve

Structures for IUPAC names **(a)–(d)** are drawn below. The main chain or ring is numbered. In each example, CO_2H, O=C–N, or C≡N is the functional group that corresponds to the suffix, so that functional group's C is established as C1. When these functional groups are treated as substituents, on the other hand, that C is not counted as part of the main chain because the prefix already accounts for that C.

(a) *Butanoic acid* is a four-carbon chain with a CO_2H group at one end. The *4-chlorocarbonyl* prefix is O=C–Cl and is located at C4.

(b) *Pentanamide* is a five-carbon chain with a O=C–N group at one end. The prefixes *4-hydroxy* and *5-phenyl* correspond to OH and C_6H_5 groups located at C4 and C5, respectively. *N-Methyl* indicates that there is a CH_3 group attached to the N of the O=C–N group. There is an asymmetric carbon at C4 with an *R* configuration.

(c) *Hex-2-enedinitrile* is a six-carbon chain with a C≡N group at both ends and a C=C between C2 and C3. The prefix *2,3-dimethyl* corresponds to two CH_3 groups at C2 and C3. The *Z* configuration has the two higher-priority groups on the same side of the double bond.

(d) *Butanedioic acid* is a four-carbon chain with two CO_2H groups at the ends. The prefix *2-carboxy* is a CO_2H substituent located at C2.

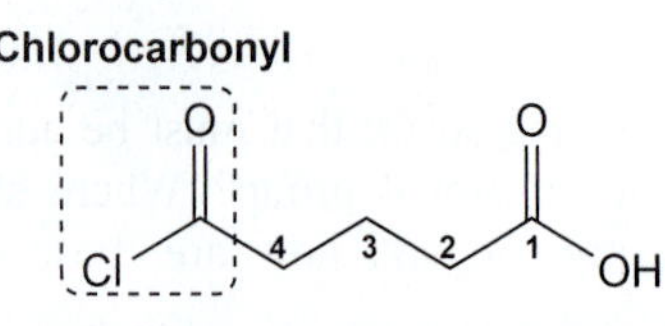

(a) 4-Chlorocarbonylbutanoic acid

(b) (*R*)-4-Hydroxy-*N*-methyl-5-phenylpentanamide

(c) (*Z*)-2,3-Dimethylhex-2-enedinitrile

(d) 2-Carboxy-1,4-butanedioic acid

Problem F.5

Think

What is the highest-priority functional group present? Can it be numbered as part of the root? How do you name a carboxylic acid, an amide, or a nitrile when the group is attached to a ring? What establishes the root—the ring or the carboxylic acid derivative? What other functional groups are present, and how should they be incorporated into the molecule's name? Must stereochemical configurations be specified?

Solve

The IUPAC names for molecules **(a)–(c)** are given below and on the next page. The main ring is numbered. In these examples, both the ring and the functional group establish the root, in which the ring is named first, followed by *carboxylic acid*, *carboxamide*, or *carbonitrile* to specify the CO_2H, O=C–N, or C≡N functional group, respectively.

(a) The highest-priority functional group present is an O=C–N, characteristic of an amide, and, because it is directly attached to a cyclohexane ring, the molecule is a *cyclohexanecarboxamide*. The carbon atoms of the ring are numbered, with C1 being the carbon that is attached to the O=C–N. Two CH_3 groups are attached to the N of the O=C–N group and are therefore named *N,N-dimethyl*.

(b) The highest-priority functional group present is a CO_2H, characteristic of a carboxylic acid, and, because it is directly attached to a cyclopentene ring, the molecule is a *cyclopentenecarboxylic acid*. The carbon atoms of the ring are numbered, with C1 being the carbon that is attached to the CO_2H group. The C=C begins at C2. There is an asymmetric carbon at C1 that has the *S* configuration.

(c) The highest-priority functional group present is a C≡N, characteristic of a nitrile, and, because it is directly attached to a cyclohexane ring, the molecule is a *cyclohexanecarbonitrile*. The carbon atoms of the ring are numbered, with C1 being the carbon that is attached to the C≡N. There is a ketone C=O at C4, which is given the prefix *4-oxo*.

(a) *N,N*-Dimethylcyclohexanecarboxamide

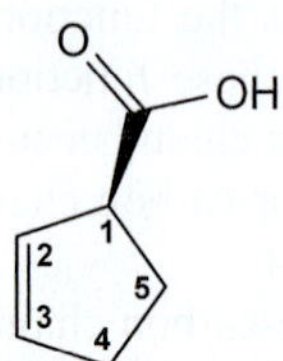

(b) (*S*)-Cyclopent-2-ene-1-carboxylic acid

(c) 4-Oxocyclohexane-1-carbonitrile

Problem F.6

Think

How many carbon atoms are in the main ring? What does the suffix *carboxylic acid*, *carboxamide*, or *carbonitrile* signify? Review the nomenclature rules from Interchapters A, B, and C and Table F-1 to identify the substituents and stereochemical configurations. On which C atoms are the substituents located? Is the C that is part of the highest-priority functional group assigned a locator number? On which carbon atom is each C=C located? Must stereochemical configurations be specified?

Solve

Structures for IUPAC names **(a)–(e)** are drawn below. The main ring is numbered, note the C that is part of the highest-priority functional group. In each example, C1 is the carbon of the ring attached to the CO_2H, O=C–N, or C≡N group.

(a) *Cyclopentanecarboxylic acid* is a five-carbon ring that has a CO_2H group attached to C1. There are two CH_3 groups at C2.

(b) *Cyclohepta-3,5-dien-1-carboxamide* is a seven-carbon ring that has an O=C–N group attached to C1 and two C=C that start at C3 and C5, respectively.

(c) *Cyclooctanecarboxylic acid* is an eight-carbon ring with a CO_2H group attached to C1. The prefix *4-cyano* is a C≡N group at C4. There are asymmetric carbons at C1 and C4 with *R* and *S* configurations, respectively.

(d) *Cyclobutanecarboxamide* is a four-carbon ring with an O=C–N attached to C1. There is a $CH(CH_3)_3$ group attached to the N of the O=C–N group and an OH at C2 of the ring. There are asymmetric carbons at C1 and C2, both with *S* configurations.

(e) *Cyclohexa-2,4-diene-1-carbonitrile* is a six-carbon ring with a C≡N group attached to C1 and double bonds beginning at C2 and C4. There is an asymmetric carbon at C1 with an *R* configuration.

(a) 2,2-Dimethylcyclopentanecarboxylic acid

(b) Cyclohepta-3,5-diene-1-carboxamide

(c) (1*R*,4*S*)-4-Cyanocyclooctane-1-carboxylic acid

(d) (1*S*,2*S*)-2-Hydroxy-*N*-(1-methylethyl)cyclobutane-1-carboxamide

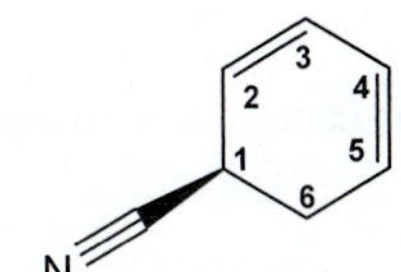

(e) (*R*)-Cyclohexa-2,4-diene-1-carbonitrile

Problem F.8

Think

To name the ester as an *alkyl alkanoate*, which group is the RCO_2 alkanoate? Which group is the alkyl group? How many C atoms are in the longest carbon chain of each group?

Solve

The IUPAC names for molecules **(a)–(c)** are given below. The main chain is numbered. The alkanoate group contains the O=C–O, whereas the alkyl group is singly bonded to the O atom. The ester's name is put together as two separate words, with the alkyl group first and the alkanoate group second—*alkyl alkanoate*. The alkanoate is numbered and the alkyl group is boxed in the figure below.

(a) Alkanoate group = 3C = *propanoate*; alkyl group = 1C = *methyl*

(b) Alkanoate group = 1C = *methanoate*; alkyl group = 4C = *butyl*

(c) Alkanoate group = 6C = *hexanoate*; alkyl group = 2C = *ethyl*

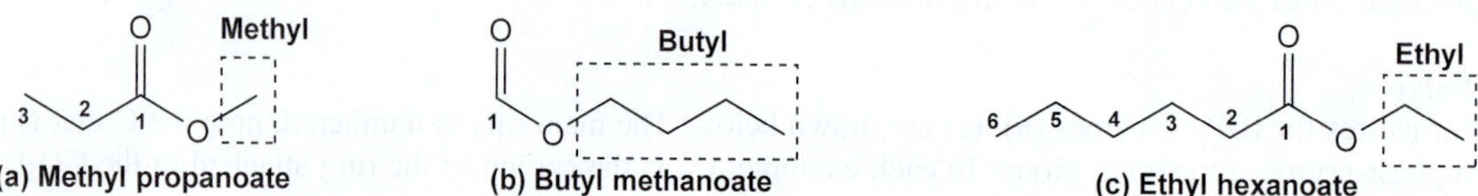

Problem F.9

Think

In the *alkyl alkanoate* format for naming esters, which part corresponds to the RCO_2 group? Which part corresponds to the group singly bonded to the O? How many C atoms are in the longest carbon chain of each group?

Solve

Structures for IUPAC names **(a)–(e)** are drawn below. The alkanoate group contains the O=C–O, and the alkyl group is singly bonded to the O atom. The alkanoate is numbered and the alkyl group is boxed in the figure below.

(a) *Pentyl* = 5C in alkyl group; *pentanoate* = 5C in alkanoate group

(b) *Propyl* = 3C in alkyl group; *butanoate* = 4C in alkanoate group

(c) *Ethyl* = 2C in alkyl group; *methanoate* = 1C in alkanoate group

(d) *Ethyl* = 2C in alkyl group; *penten-2-enoate* = 5C in alkenoate group

(e) *Butyl* = 4C in alkyl group; *propanoate* = 3C in alkanoate group

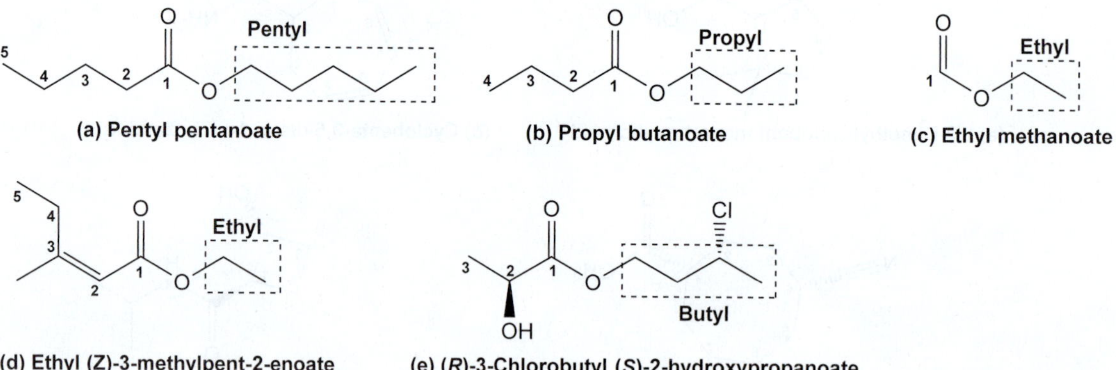

Problem F.10

Think

Is the acid anhydride symmetric or unsymmetric? What is the format for each? How many C atoms are in the longest carbon chain of each group?

Solve

The IUPAC names for molecules **(a)–(c)** are given below. The main chain or ring is numbered. The format for a symmetric acid anhydride is *alkanoic anhydride*, and the format for an unsymmetric acid anhydride is *alkanoic alkanoic anhydride*, where the two alkanoic groups are in alphabetical order. Molecules **(a)** and **(b)** are symmetric, and molecule **(c)** is unsymmetric.

(a) Alkanoic group = 2C = *ethanoic*

(b) Alkanoic group = benzyl = *benzoic*

(c) Alkanoic group = 3C + CH_3 at C2 = 2-*methylpropanoic*; alkanoic group = 3C = *propanoic*

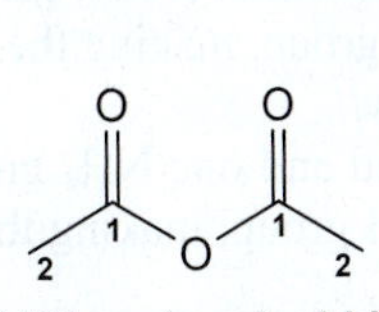

(a) Ethanoic anhydride

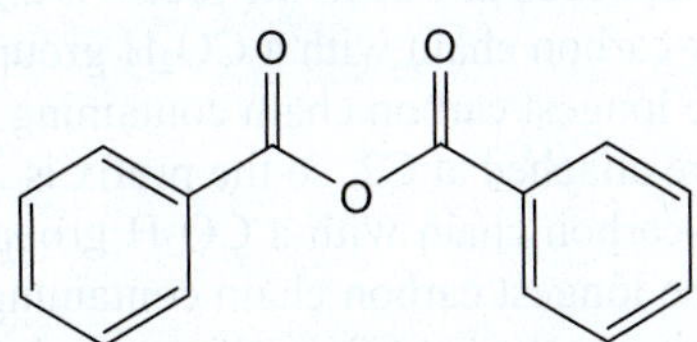

(b) Benzoic anhydride

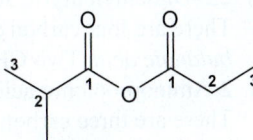

(c) 2-Methylpropanoic propanoic anhydride

Problem F.11

Think

In the *alkanoic anhydride* or *alkanoic alkanoic anhydride* format for naming acid anhydrides, which format is symmetric? Which format is unsymmetric? How many C atoms are in the longest carbon chain of each group?

Solve

Structures for IUPAC names **(a)–(d)** are drawn below. Each alkanoic group is numbered. The format for a symmetric acid anhydride is *alkanoic anhydride*, and the format for an unsymmetric acid anhydride is *alkanoic alkanoic anhydride*. Molecules **(a)** and **(c)** are symmetric, and molecules **(b)** and **(d)** are unsymmetric.

(a) *Butanoic* = 4C in alkanoic group

(b) *Butanoic* = 4C in alkanoic group; *propanoic* = 3C in alkanoic group

(c) 2-*Methylbutanoic* = 4C in alkanoic group with a CH_3 at C2

(d) 2-*Methylbutanoic* = 4C in alkanoic group with a CH_3 at C2; *benzoic* = benzyl group

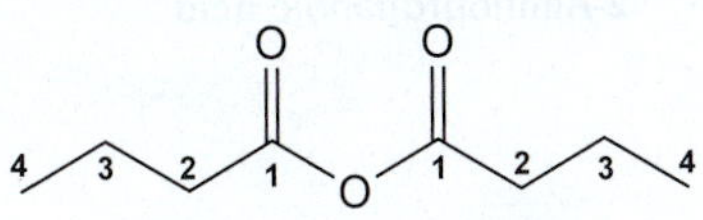

(a) Butanoic anhydride

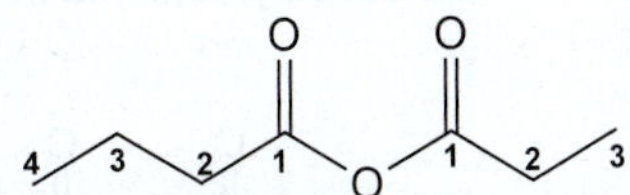

(b) Butanoic propanoic anhydride

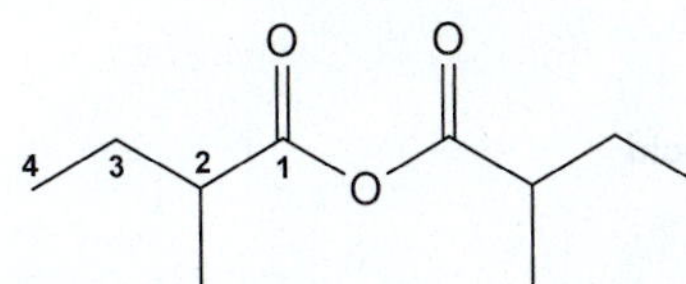

(c) 2-Methylbutanoic anhydride

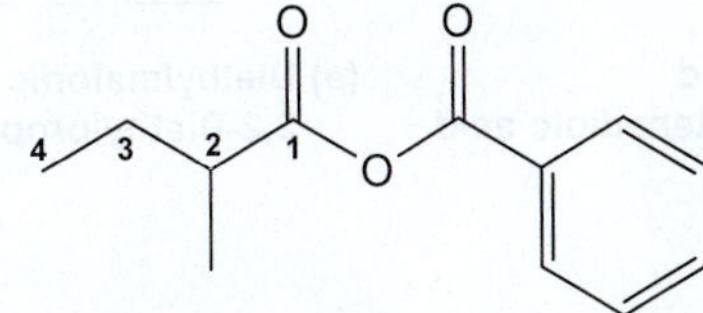

(d) Benzoic 2-methylbutanoic anhydride

Problem F.12

Think

Refer to the figures in Section F.3a to draw the structure of the given trivial name. What is the highest-priority functional group present and the corresponding suffix that must be added? How many carbon atoms are in the longest carbon chain? Where should the numbering of that chain begin? What other functional groups are present, and how are those named as substituents? Must stereochemical configurations be specified?

Solve

Structures and IUPAC names for trivial names **(a)–(e)** are drawn below. The main chain is numbered. In molecules **(a)–(c)**, the highest-priority functional group in the molecule is a CO_2H, so the suffix is *oic acid* and numbering begins at the CO_2H's carbon. In molecules **(d)** and **(e)**, the highest-priority functional group in the molecule is CO_2H, and there are two of them, so the suffix is *dioic acid* and numbering begins at either carbon of the CO_2H groups.

(a) Trichloroacetic acid is a two-carbon chain with a CO_2H group at one end and three Cl groups on C2. There are two carbon atoms in the longest carbon chain containing the CO_2H group, making the molecule an *ethanoic acid.* Three Cl atoms are attached at C2, so the prefix is *2,2,2-trichloro.*

(b) 2,2-Dimethylbutyric acid is a four-carbon chain with a CO_2H group at one end and two CH_3 groups on C2. There are four carbon atoms in the longest carbon chain containing the CO_2H group, making the molecule a *butanoic acid.* Two CH_3 groups are attached at C2, so the prefix is *2,2-dimethyl.*

(c) 2-Aminopropionic acid is a three-carbon chain with a CO_2H group at one end and one NH_2 group on C2. There are three carbon atoms in the longest carbon chain containing the CO_2H group, making the molecule a *propanoic acid.* One NH_2 group is attached at C2, so the prefix is *2-amino.*

(d) Dimethylmaleic acid is a four-carbon chain with two CO_2H groups (one at each end), one C=C between C2 and C3, and two CH_3 groups (attached to C2 and C3, respectively). There are four carbon atoms in the longest carbon chain containing the CO_2H groups, making the molecule a *butenedioic acid.* There are two CH_3 groups, one at C2 and the other at C3, so the prefix is *2,3-dimethyl.* The C=C is named as part of the main chain, *en*. The C=C has an *Z* configuration.

(e) Diethylmalonic acid is a three-carbon chain with two CO_2H groups (one at each end) and two CH_2CH_3 groups at C2. There are three carbon atoms in the longest carbon chain containing the CO_2H groups, making the molecule a *propanedioic acid.* There are two CH_2CH_3 groups at C2, so the prefix is *2,2-diethyl.*

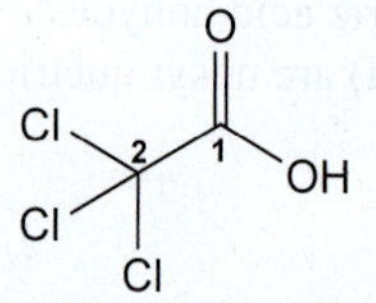

(a) Trichloroacetic acid
2,2,2-Trichloroethanoic acid

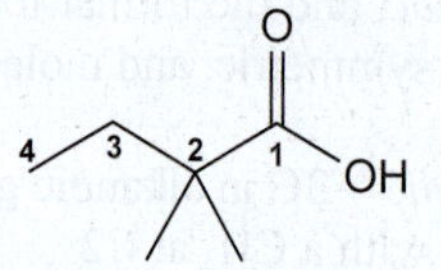

(b) 2,2-Dimethylbutyric acid
2,2-Dimethylbutanoic acid

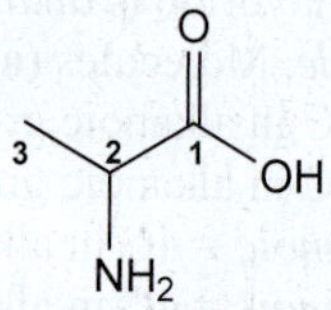

(c) 2-Aminopropionic acid
2-Aminopropanoic acid

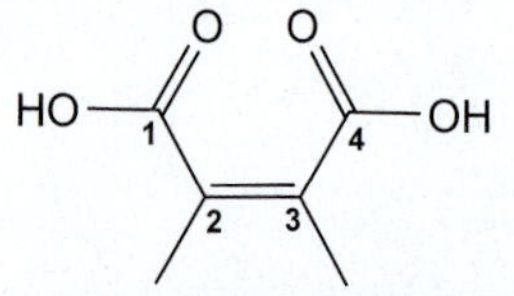

(d) Dimethylmaleic acid
(*Z*)-2,3-Dimethylbutenedioic acid

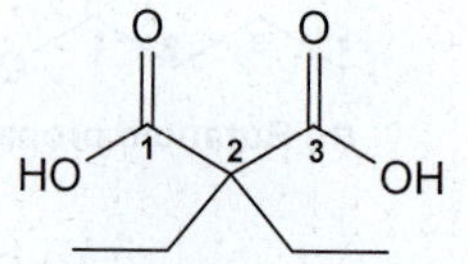

(e) Diethylmalonic acid
2,2-Diethylpropanedioic acid

Problem F.13

Think

Refer to the figures in Section F.3b to draw the structure of the given trivial name. What is the highest-priority functional group present and the corresponding suffix that must be added? How many carbon atoms are in the longest carbon chain? Where should the numbering of that chain begin? What other functional groups are present, and how are those named as substituents? Must stereochemical configurations be specified?

Solve

Structures and IUPAC names for trivial names **(a)**–**(e)** are drawn below. The main chain or ring is numbered. In each example, O=C–N, C≡N, or O=C–O is the functional group that corresponds to the suffix, so that functional group's C is established as C1.

(a) *N,N*-Dimethylacetylamide is a two-carbon chain with a O=C–N group characteristic of an amide at one end and two CH_3 groups attached to the N. There are two carbon atoms in the longest chain containing the O=C–N group, making the molecule an *ethanamide*. Two CH_3 groups are attached at the N, so the prefix is *N,N-dimethyl*.

(b) Methoxyacetonitrile is a two-carbon chain with a C≡N group characteristic of a nitrile at one end and an OCH_3 group at C2. There are two carbon atoms in the longest chain containing the C≡N group, making the molecule an *ethanenitrile*. There is an OCH_3 group at C2, so the prefix is *2-methoxy*.

(c) Ethyl trichloroacetate is an ester with two carbons in the alkyl group and two carbons in the alkanoate group. The alkanoate group has three Cl atoms on C2. There are two carbon atoms in the longest chain containing the O=C–O, making the molecule an *ethanoate*. On C2 of the ethanoate are three Cl atoms, so the prefix is *2,2,2-trichloro*. The alkyl group contains two carbon atoms, so the name is *ethyl*.

(d) Isopropyl formate is an ester with one carbon in the alkanoate group and a branched alkyl group, $CH(CH_3)_2$, attached to the O atom. The O=C–O group contains only one carbon atom, making the molecule a *methanoate*. The alkyl group is a branched alkyl group, $CH(CH_3)_2$, so the prefix is *1-methylethyl*.

(e) *N,N*-Diphenylbenzamide is an amide in which C_6H_5C is attached to the carbon of the O=C–N group and another two C_6H_5 groups are attached to the N. The longest chain containing the O=C–N group is C_6H_5C, making the molecule a *benzamide*. Two C_6H_5 groups are attached at the N, so the prefix is *N,N-diphenyl*. This is an example where the trivial name has been adopted by IUPAC.

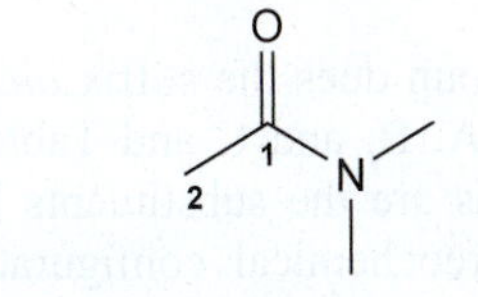

(a) *N,N*-Dimethylacetamide
***N,N*-Dimethylethanamide**

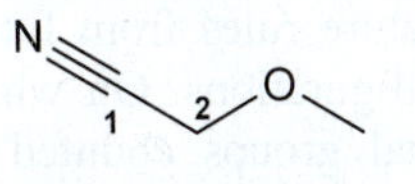

(b) Methoxyacetonitrile
2-Methoxyethanenitrile

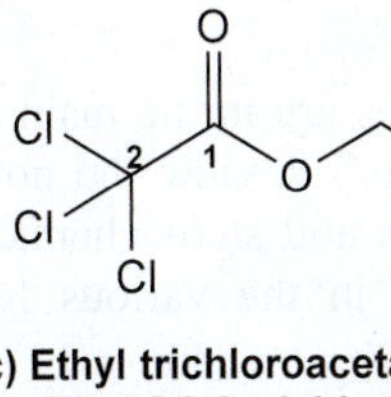

(c) Ethyl trichloroacetate
Ethyl 2,2,2-trichloroethanoate

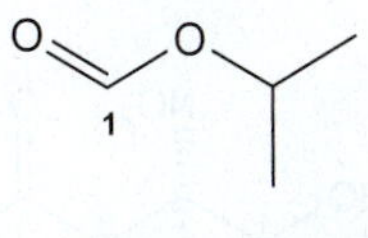

(d) Isopropyl formate
1-Methylethyl methanoate

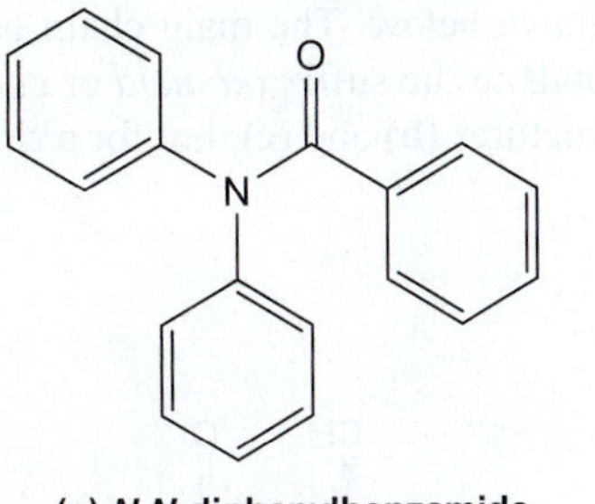

(e) *N,N*-diphenylbenzamide

End of Chapter Problems

Section F.1 Naming Carboxylic Acids, Acid Chlorides, Amides, and Nitriles

Problem F.14

Think

What is the highest-priority functional group present and the corresponding suffix that must be added? How many carbon atoms are in the longest carbon chain containing that functional group? Where should the numbering of that chain begin? What other functional groups are present, and how are those named as substituents? Are there stereochemical configurations that must be specified?

Solve

The IUPAC names for molecules **(a)–(c)** are given below. The main chain is numbered. In molecule **(a)**, the highest-priority functional group in the molecule is a CO_2H, so the suffix is *oic acid* and numbering begins at the CO_2H's carbon. In molecules **(b)** and **(c)**, the highest-priority functional group in the molecule is CO_2H, and there are two of them, so the suffix is *dioic acid* and numbering begins at either carbon of the CO_2H groups for **(b)** and at the top C for **(c)** to have the propyl group come earliest.

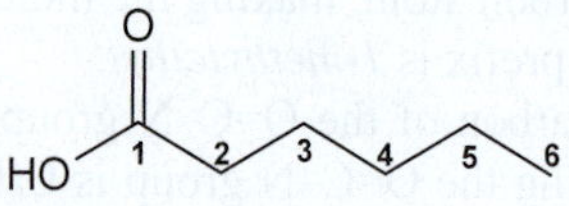

(a) Hexanoic acid 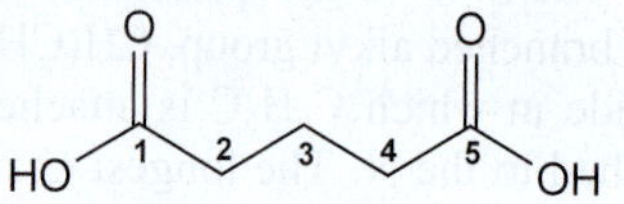 **(b) Pentanedioic acid** 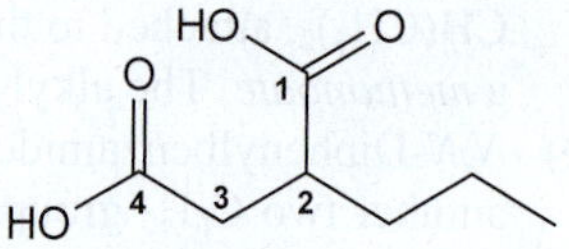 **(c) 2-Propylbutanedioic acid**

Problem F.15

Think

How many carbon atoms are in the main chain or ring? To what functional group does the suffix *oic acid* or *carboxylic acid* correlate? Review the nomenclature rules from Interchapters A, B, and C and Table F-1 to identify the substituents and stereochemical configurations. On which C atoms are the substituents located? How are the C atoms in the various functional groups counted? Must stereochemical configurations be specified?

Solve

Structures for IUPAC names **(a)–(c)** are drawn below. The main chain or ring is numbered. In each example, CO_2H is the functional group that corresponds to the suffix *oic acid* or *carboxylic acid.* The CO_2H's functional group's C is established as C1 for linear structures **(b)** and **(c)**, but for a ring attached to CO_2H, C1 is the carbon directly attached to the CO_2H.

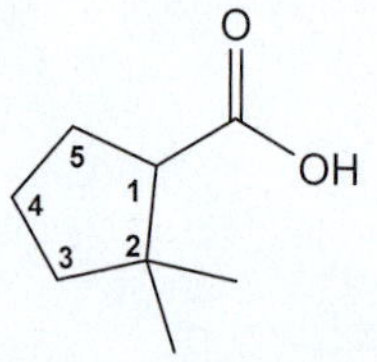

(a) 2,2-Dimethylcyclopentane-1-carboxylic acid 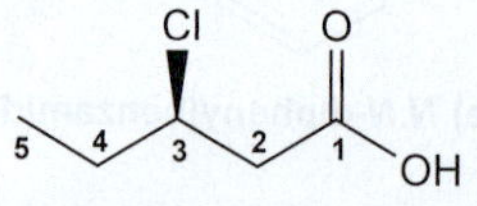 **(b) (*R*)-3-Chloropentanoic acid** 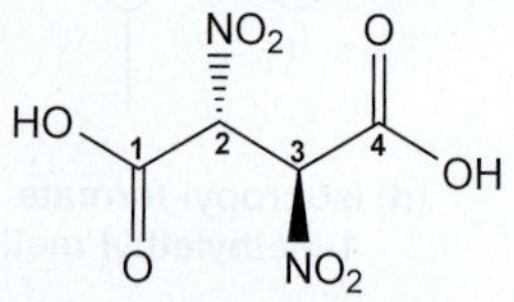 **(c) (2*R*,3*S*)-2,3-Dinitrobutanedioic acid**

Problem F.16

Think

What is the highest-priority functional group present and the corresponding suffix that must be added? How many carbon atoms are in the longest carbon chain containing that functional group? Where should the numbering of that chain begin? What other functional groups are present, and how are those named as substituents? Must stereochemical configurations be specified?

Solve

The IUPAC names for molecules **(a)–(c)** are given below. The main chain is numbered. In molecules **(a)** and **(c)**, the highest-priority functional group in the molecule is a O=C–Cl, characteristic of a carboxylic acid chloride, so the suffix is *oyl chloride* and numbering begins at the O=C–Cl's carbon. In molecule **(b)**, the highest-priority functional group in the molecule is O=C–Cl, and there are two of them, so the suffix is *dioyl chloride* and numbering begins at either carbon of the O=C–Cl groups.

(a) Butanoyl chloride **(b) Propanedioyl chloride** **(c) (*S*)-3-Methyl-4-phenylbutanoyl chloride**

Problem F.17

Think

How many carbon atoms are in the main chain? To what functional group does the suffix *oyl chloride* or *dioyl chloride* correlate? Review the nomenclature rules from Interchapters A, B, and C and Table F-1 to identify the substituents and stereochemical configurations. On which C atoms are the substituents located? How are the C atoms in the various functional groups counted? Must stereochemical configurations be specified?

Solve

Structures for IUPAC names **(a)–(c)** are drawn below. The main chain is numbered. In each example, O=C–Cl is the functional group characteristic of a carboxylic acid chloride, which corresponds to the suffix *oyl chloride* or *dioyl chloride*. The O=C–Cl's functional group's C is established as C1.

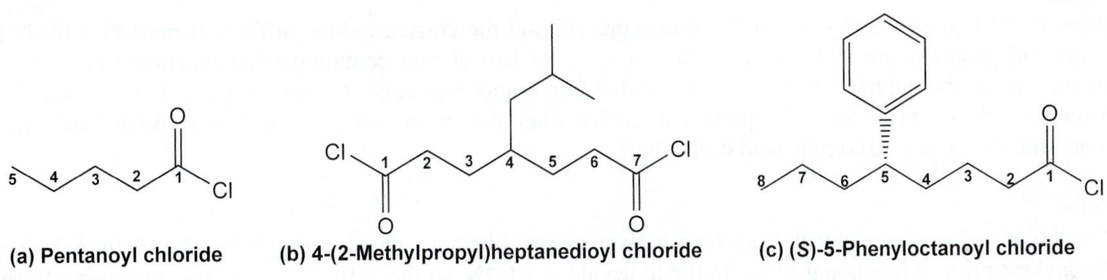

(a) Pentanoyl chloride **(b) 4-(2-Methylpropyl)heptanedioyl chloride** **(c) (*S*)-5-Phenyloctanoyl chloride**

Problem F.18

Think

What is the highest-priority functional group present and the corresponding suffix that must be added? How many carbon atoms are in the longest carbon chain containing that functional group? Where should the numbering of that chain begin? What other functional groups are present, and how are those named as substituents? How do you name an amide when the group is attached to a ring? What establishes the root—the ring or the carboxylic acid derivative?

Solve

The IUPAC names for molecules **(a)–(c)** are given below. The main chain or ring is numbered. In molecule **(a)**, the highest-priority functional group in the molecule is a O=C–N, characteristic of amides, so the suffix is *amide* and numbering begins at the O=C–N's carbon. In molecule **(b)**, both the ring and the functional group O=C–N together establish the root, in which the ring is named first, followed by *carboxamide*. In molecule **(c)**, the highest-priority functional group in the molecule is O=C–N, characteristic of amides, and there are two of them, so the suffix is *diamide* and numbering begins at either carbon of the O=C–N groups.

(a) Pentanamide **(b) *N*,*N*-Dimethylcyclohexanecarboxamide** **(c) Butanediamide**

Problem F.19

Think

How many carbon atoms are in the main chain? To what functional group does the suffix *amide*, *diamide*, or *carboxamide* correlate? Review the nomenclature rules from Interchapters A, B, and C and Table F-1 to identify the substituents and stereochemical configurations. On which C atoms are the substituents located? How are the C atoms in the various functional groups counted—as part of the root or separately from it? Must stereochemical configurations be specified?

Solve

Structures for IUPAC names **(a)**–**(c)** are drawn below. The main chain is numbered. In each example, O=C–N is the functional group that corresponds to the suffix *amide*, *diamide*, or *carboxamide*. The O=C–N's functional group's C is established as C1 for linear structures **(a)** and **(b)** (leftmost C in order to have the substituents come earlier), but for a ring attached to O=C–N, C1 is the carbon directly attached to the O=C–N.

(a) 5-Phenylpentanamide **(b) (2*S*,3*S*)-2,3-Dimethoxyhexanediamide** **(c) *N*-Phenylcyclobutanecarboxamide**

Problem F.20

Think

What is the highest-priority functional group present and the corresponding suffix that must be added? How many carbon atoms are in the longest carbon chain or largest ring containing that functional group? Where should the numbering of that chain or ring begin? What other functional groups are present, and how are those named as substituents? How do you name an amide when the group is attached to a ring? What establishes the root—the ring or the carboxylic acid derivative?

Solve

The IUPAC names for molecules **(a)**–**(c)** are given below. The main chain or ring is numbered. In molecule **(a)**, the highest-priority functional group in the molecule is a C≡N, so the suffix is *nitrile* and numbering begins at the C≡N's carbon. In molecule **(b)**, both the ring and the functional group C≡N together establish the root, in which the ring is named first, followed by *carbonitrile*. In molecule **(c)**, the highest-priority functional group in the molecule is a C≡N, and there are two of them, so the suffix is *dinitrile* and numbering begins at the top C≡N's carbon to encounter the pentyl group the earliest.

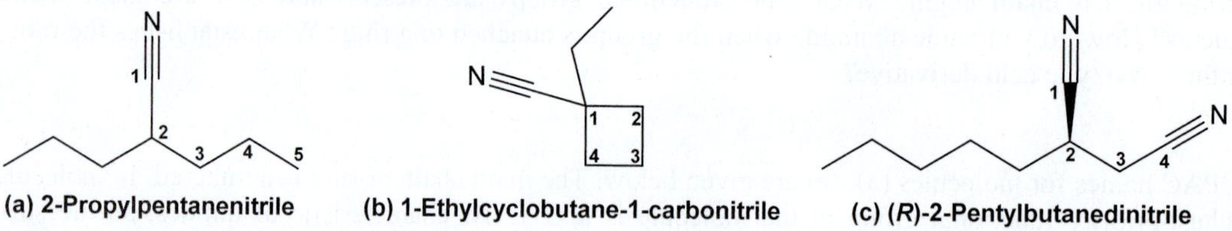

(a) 2-Propylpentanenitrile **(b) 1-Ethylcyclobutane-1-carbonitrile** **(c) (*R*)-2-Pentylbutanedinitrile**

Problem F.21

Think

How many carbon atoms are in the main chain or ring? To what functional group does the suffix *nitrile*, *dinitrile*, or *carbonitrile* correlate? Review the nomenclature rules from Interchapters A, B, and C and Table F-1 to identify the substituents and stereochemical configurations. On which C atoms are the substituents located? How are the C atoms in the various functional groups counted—as part of the root or separately from it? Must stereochemical configurations be specified?

Solve

Structures for IUPAC names **(a)–(c)** are drawn below. The main chain is numbered. In each example, C≡N is the functional group that corresponds to the suffix *nitrile*, *dinitrile*, or *carbonitrile*. The C≡N's functional group's C is established as C1 for linear structures **(a)** and **(b)**, but for a ring attached to C≡N, C1 is the carbon directly attached to the C≡N.

(a) Hexanedinitrile (b) (S)-4-Nitroheptanenitrile (c) 4,4-Diethylcyclohexanecarbonitrile

Section F.2 Naming Esters and Acid Anhydrides

Problem F.22

Think

To name the ester as *alkyl alkanoate*, which group is the RCO_2 alkanoate? Which group is the alkyl group? How many C atoms are in the longest carbon chain of each group?

Solve

The IUPAC names for molecules **(a)–(c)** are given below. The main chain is numbered. The alkanoate group contains the O=C–O, whereas the alkyl group is singly bonded to the O atom. The ester's name is put together with the alkyl group first and the alkanoate group second—*alkyl alkanoate*. The alkanoate is numbered and the alkyl group is boxed in the figure below.

(a) Alkanoate group = 5C = pentanoate; alkyl group = 3C = propyl
(b) Alkanoate group = 2C = ethanoate; alkyl group = C_6H_5 = phenyl
(c) Alkanoate group = 3C = propanoate; alkyl group = $CH(CH_3)_2$ = 1-methylethyl

(a) Propyl pentanoate (b) Phenyl 2-chloroethanoate (c) 1-Methylethyl 3-phenylpropanoate

Problem F.23

Think

In the *alkyl alkanoate* format for naming esters, which part corresponds to the RCO_2 group? Which part corresponds to the group singly bonded to the O? How many C atoms are in the longest carbon chain of each group?

Solve

Structures for IUPAC names **(a)–(c)** are drawn below. The main chain or ring is numbered. The alkanoate group contains the O=C–O, and the alkyl group is singly bonded to the O atom. The alkanoate is numbered and the alkyl group is boxed in the figure below.

(a) *Cyclohexyl* = 6C ring in alkyl group; *butanoate* = 4C in alkanoate group
(b) 1,1-*Dimethylethyl* = branched $CH_2CH(CH_3)_2$ alkyl group; *hexanoate* = 6C in alkanoate group
(c) *Phenyl* = C_6H_5 aryl group; *heptanoate* = 7C in alkanoate group

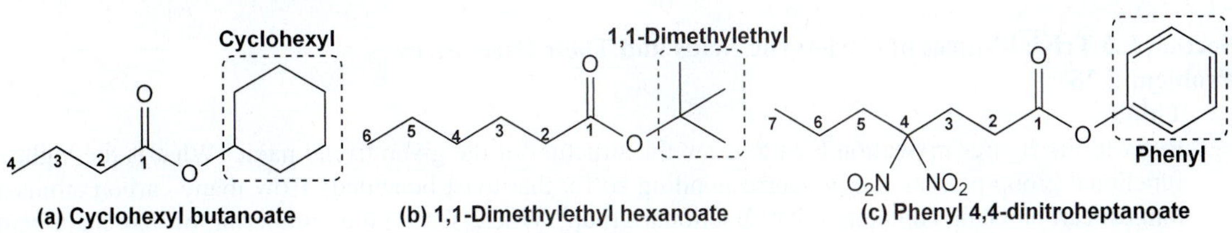

(a) Cyclohexyl butanoate (b) 1,1-Dimethylethyl hexanoate (c) Phenyl 4,4-dinitroheptanoate

Problem F.24

Think

Is the acid anhydride symmetric or unsymmetric? What is the format for the IUPAC name of an acid anhydride? How many C atoms are in the longest carbon chain of each group? Are substituents present along the main chain?

Solve

The IUPAC names for molecules **(a)**–**(c)** are given below. The main chain or ring is numbered. The format for a symmetric acid anhydride is *alkanoic anhydride*, and the format for an unsymmetric acid anhydride is *alkanoic alkanoic anhydride*, where the two alkanoic groups are in alphabetical order. Molecules **(a)** and **(b)** are symmetric, and **(c)** is unsymmetric.

(a) Alkanoic group = 5C = *pentanoic*
(b) Alkanoic group = 3C with two methyl groups on C2 = 2,2-*dimethylpropanoic*
(c) Alkanoic group = benzyl = *benzoic*; alkanoic group = 4C = *butanoic*

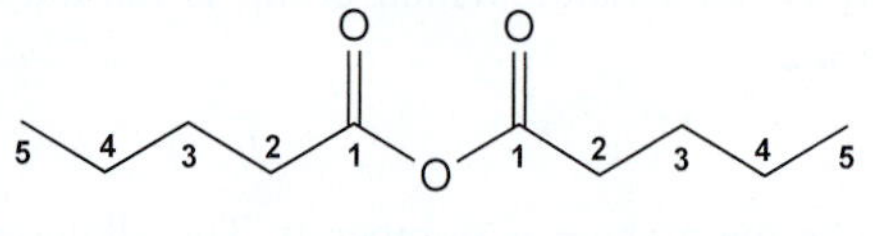

(a) Pentanoic anhydride 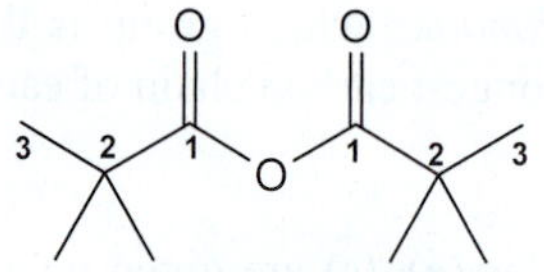 **(b) 2,2-Dimethylpropanoic anhydride** 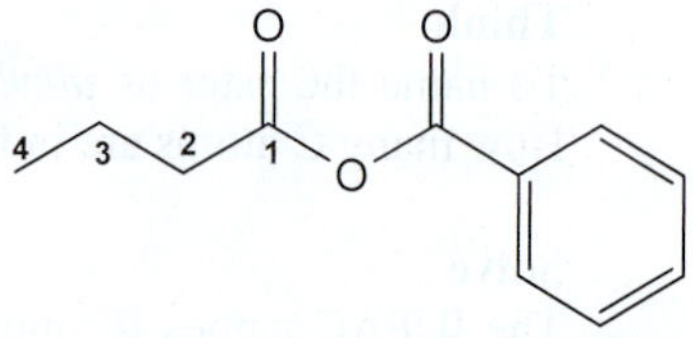 **(c) Benzoic butanoic anhydride**

Problem F.25

Think

In the *alkanoic anhydride* or *alkanoic alkanoic anhydride* format for naming acid anhydrides, which format corresponds to a symmetric anhydride? Which format corresponds to an unsymmetric anhydride? How many C atoms are in the longest carbon chain of each group? Are there any substituents to account for along the main chain?

Solve

Structures for IUPAC names **(a)**–**(c)** are drawn below. Each alkanoic group is numbered. The format for a symmetric acid anhydride is *alkanoic anhydride*, and the format for an unsymmetric acid anhydride is *alkanoic alkanoic anhydride*. Molecule **(a)** is symmetric, and molecules **(b)** and **(c)** are unsymmetric.

(a) *Pentanoic* = 5C in alkanoic group
(b) *Hexanoic* = 6C in alkanoic group; *propanoic* = 3C in alkanoic group
(c) *Ethanoic* = 2 C in alkanoic group; 3-*methylpentanoic* = 5C in alkanoic group with a CH_3 at C3

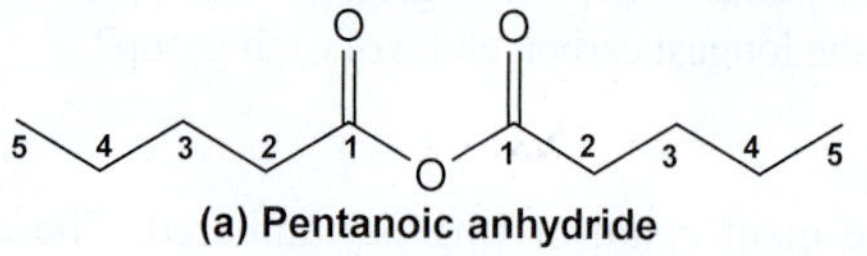

(a) Pentanoic anhydride 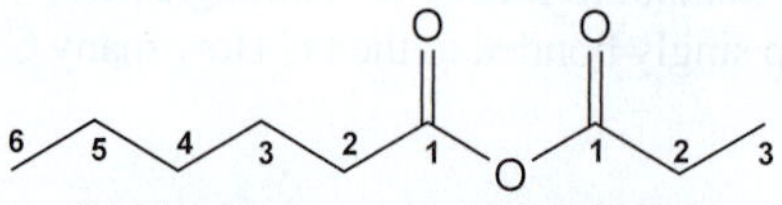 **(b) Hexanoic propanoic anhydride**

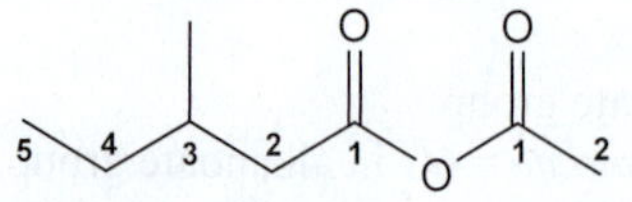

(c) Ethanoic 3-methylpentanoic anhydride

Section F.3 Trivial Names of Carboxylic Acids and Their Derivatives

Problem F.26

Think

Refer to the figures in Section F.3a to draw the structure of the given trivial name. What is the highest-priority functional group present and the corresponding suffix that must be added? How many carbon atoms are in the longest carbon chain containing that functional group? Where should the numbering of that chain begin? What other functional groups are present, and how are those named as substituents? Must stereochemical configurations be specified?

Solve

Structures and IUPAC names for trivial names **(a)**–**(d)** are drawn below. The main chain is numbered. In molecules **(a)**, **(b)**, and **(d)**, the highest-priority functional group in the molecule is a CO_2H, so the suffix is *oic acid*, and numbering begins at the CO_2H's carbon. In molecule **(c)**, the highest-priority functional group in the molecule is CO_2H, and there are two of them, so the suffix is *dioic acid*, and numbering begins at either carbon of the CO_2H groups.

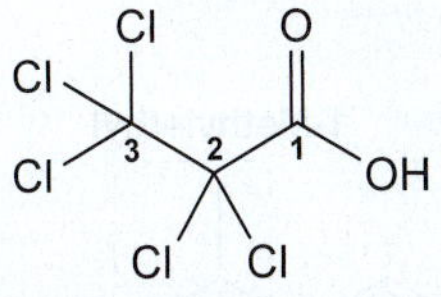

(a) Pentachloropropionic acid
2,2,3,3,3-Pentachloropropanoic acid

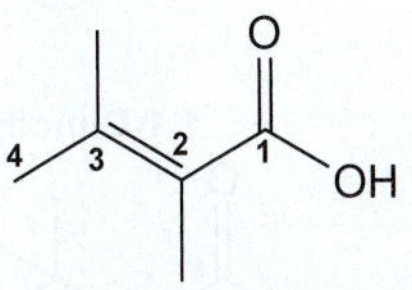

(b) Trimethylacrylic acid
2,3-Dimethylbut-2-enoic acid

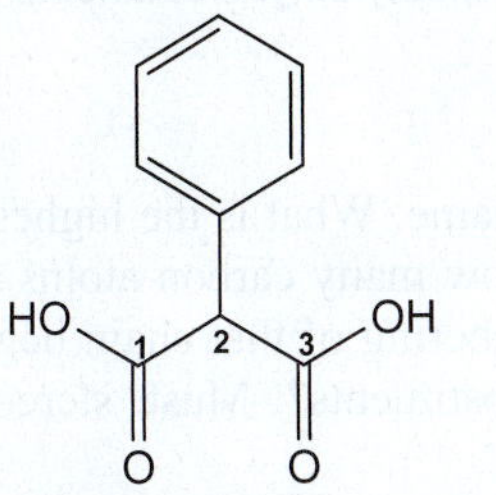

(c) Phenylmalonic acid
2-Phenylpropanedioic acid

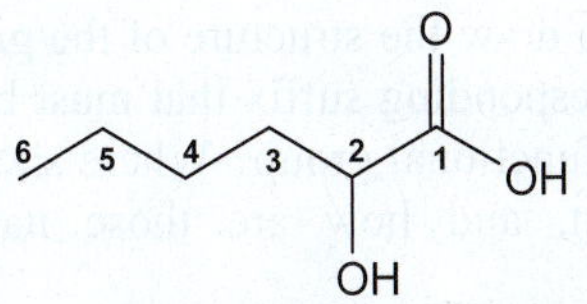

(d) α-Hydroxycaproic acid
2-Hydroxyhexanoic acid

Problem F.27

Think

Refer to the figures in Section F.3b to draw the structure of the given trivial name. What is the highest-priority functional group present and the corresponding suffix that must be added? How many carbon atoms are in the longest carbon chain containing that functional group? Where should the numbering of that chain begin? What other functional groups are present, and how are those named as substituents? Must stereochemical configurations be specified?

Solve

Structures and IUPAC names for trivial names **(a)**–**(c)** are drawn below. The main chain is numbered. The highest-priority functional group in each is O=C–O–C=O, characteristic of an acid anhydride. Each alkanoic group is numbered. The format for a symmetric acid anhydride is *alkanoic anhydride*, and the format for an unsymmetric acid anhydride is *alkanoic alkanoic anhydride*. Molecules **(a)** and **(b)** are symmetric, and molecule **(c)** is unsymmetric.

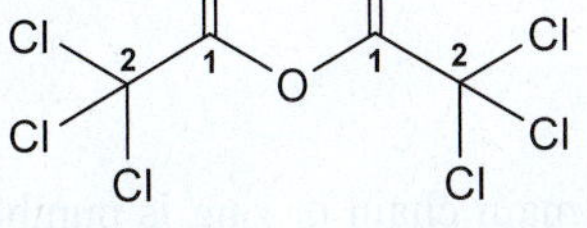

(a) Trichloroacetic anhydride
2,2,2-Trichloroethanoic anhydride

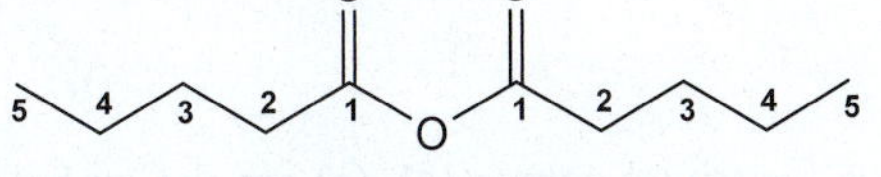

(b) Valeric anhydride
Pentanoic anhydride

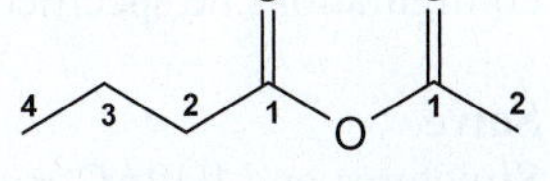

(c) Acetic butyric anhydride
Butanoic ethanoic anhydride

Problem F.28

Think

Refer to the figures in Section F.3b to draw the structure of the given trivial name. What is the highest-priority functional group present and the corresponding suffix that must be added? How many carbon atoms are in the longest carbon chain containing that functional group? Where should the numbering of that chain begin? What other functional groups are present, and how are those named as substituents? Must stereochemical configurations be specified?

Solve

Structures and IUPAC names for trivial names **(a)–(c)** are drawn below. The main chain or ring is numbered. The alkanoate group contains the O=C–O group, and the alkyl group is singly bonded to the O atom. The alkanoate is numbered and the alkyl group is boxed in the figure below.

(a) Isobutyl = *2-methylpropyl* = $CH_2CH(CH_3)_2$ alkyl group; *benzoate* = C_6H_5C in alkanoate group

(b) *tert*-Butyl = 1,1-*dimethylethyl* = $C(CH_3)_3$ alkyl group; acetate = *ethanoate* = 2C in alkanoate group

(c) Isopropyl = 1-*methylethyl* = $CH(CH_3)_2$ alkyl group; formate = *methanoate* = 1C in alkanoate group

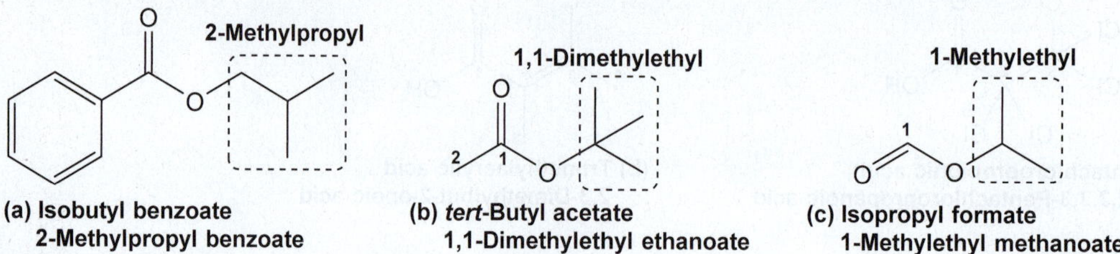

Problem F.29

Think

Refer to the figures in Section F.3b to draw the structure of the given trivial name. What is the highest-priority functional group present and the corresponding suffix that must be added? How many carbon atoms are in the longest carbon chain containing that functional group? Where should the numbering of that chain begin? What other functional groups are present, and how are those named as substituents? Must stereochemical configurations be specified?

Solve

Structures and IUPAC names for trivial names **(a)–(d)** are drawn below. The main chain or ring is numbered. The highest-priority functional group in the molecule is O=C–N, so the suffix is *amide* or *diamide*.

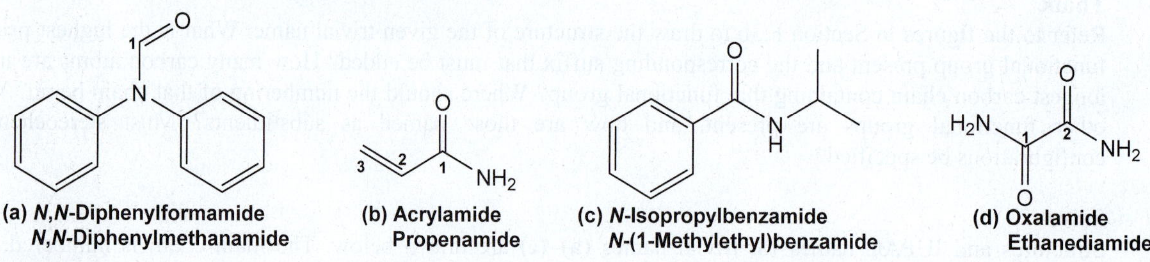

Problem F.30

Think

Refer to the figures in Section F.3a or F.3b to draw the structure of the given trivial name. What is the highest-priority functional group present and the corresponding suffix that must be added? How many carbon atoms are in the longest carbon chain containing that functional group? Where should the numbering of that chain begin? What other functional groups are present, and how are those named as substituents? Must stereochemical configurations be specified?

Solve

Structures and IUPAC names for trivial names **(a)–(e)** are drawn below. The main chain or ring is numbered. The highest-priority functional group in the molecule is a C≡N, so the suffix is *nitrile* or *dicarbonitrile*.

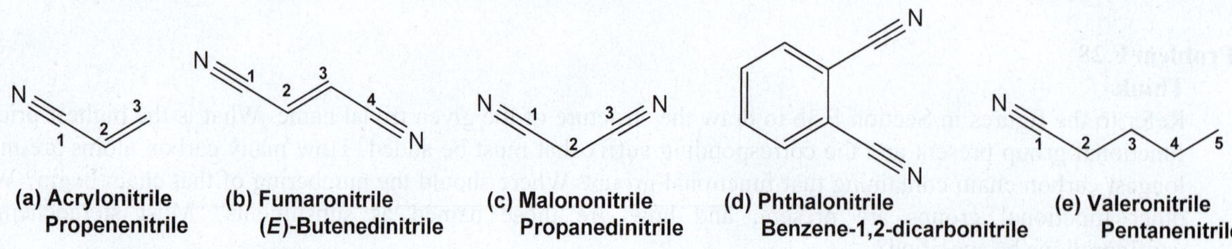

Integrated Problems

Problem F.31

Think

What is the highest-priority functional group present and the corresponding suffix that must be added? How many carbon atoms are in the longest carbon chain containing that functional group? Where should the numbering of that chain begin? What other functional groups are present and how should they be incorporated into the molecule's name? Must stereochemical configurations be specified?

Solve

The IUPAC names for molecules **(a)–(d)** are given below. The main chain or ring is numbered.

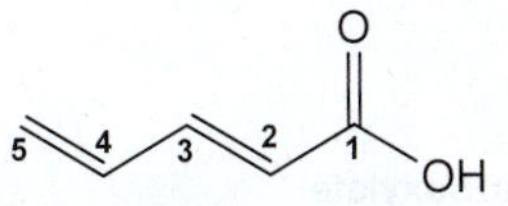

(a) (*E*)-Penta-2,4-dienoic acid

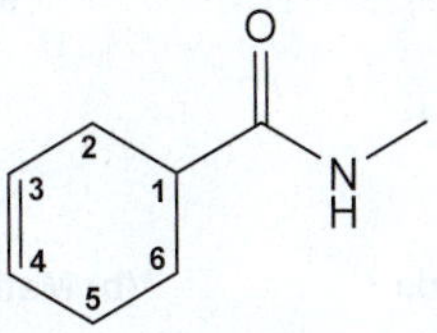

(b) *N*-Methylcyclohex-3-ene-1-carboxamide

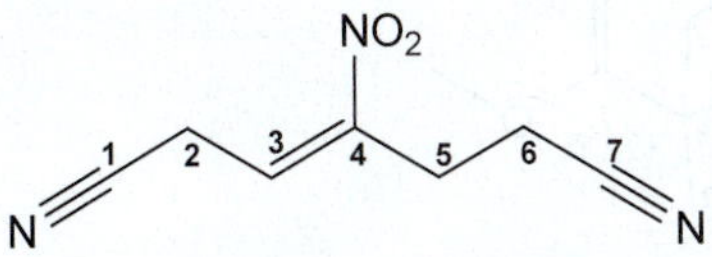

(c) (*Z*)-4-Nitrohept-3-enedinitrile

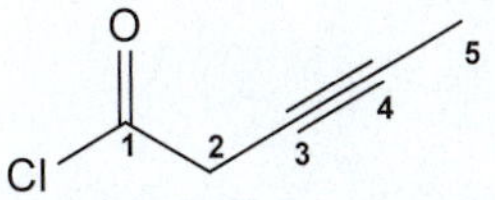

(d) Pent-3-ynoyl chloride

Problem F.32

Think

What is the highest-priority functional group present and the corresponding suffix that must be added? How many carbon atoms are in the longest carbon chain containing that functional group? Where should the numbering of that chain begin? What other functional groups are present, and how should they be incorporated into the molecule's name? Must stereochemical configurations be specified?

Solve

The IUPAC names for molecules **(a)–(d)** are given below. The main chain or ring is numbered.

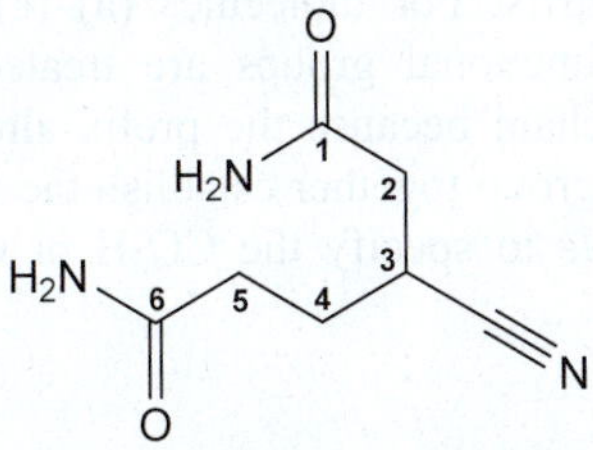

(a) 3-Cyanohexanediamide

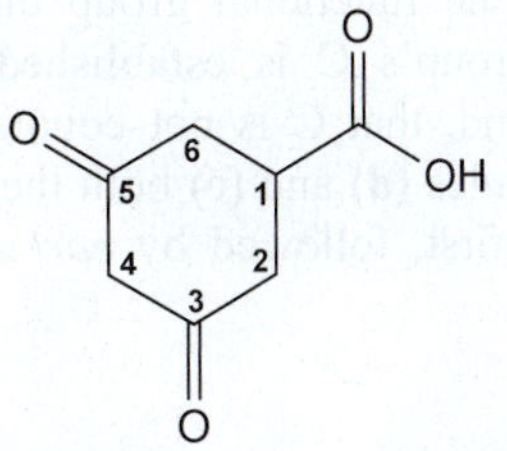

(b) 3,5-Dioxocyclohexane-1-carboxylic acid

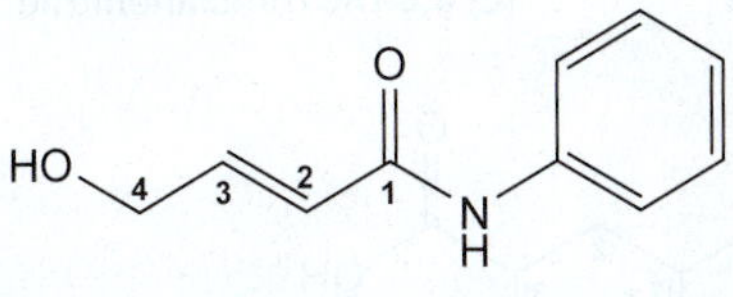

(c) (*E*)-4-Hydroxy-*N*-phenylbut-2-enamide

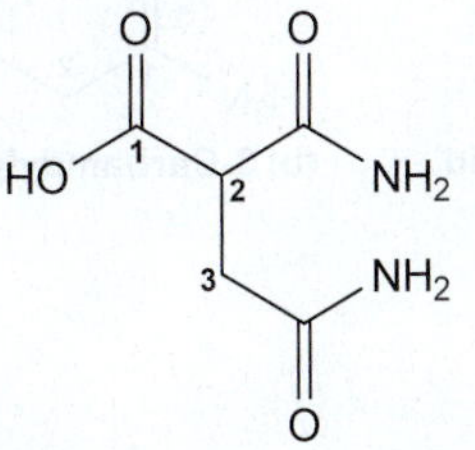

(d) 2,3-Dicarbamoylpropanoic acid

Problem F.33

Think

What is the highest-priority functional group present and the corresponding suffix that must be added? How many carbon atoms are in the longest carbon chain containing that functional group? Where should the numbering of that chain begin? What other functional groups are present, and how should they be incorporated into the molecule's name? Must stereochemical configurations be specified?

Solve

The IUPAC names for molecules **(a)**–**(d)** are given below. The main chain or ring is numbered.

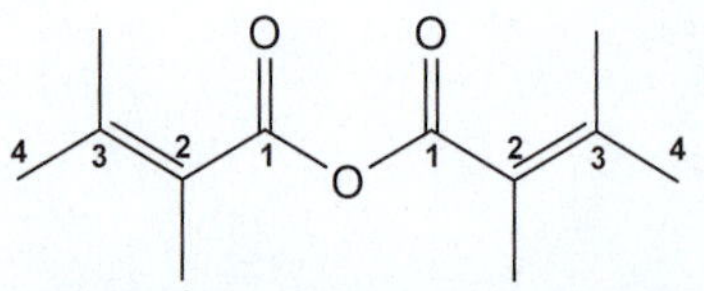

(a) 2,3-Dimethylbut-2-enoic anhydride

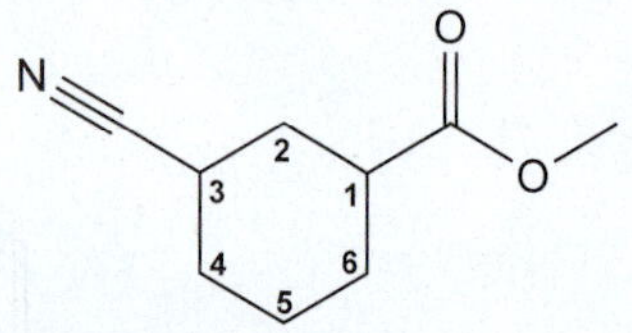

(b) Methyl 3-cyanocyclohexane-1-carboxylate

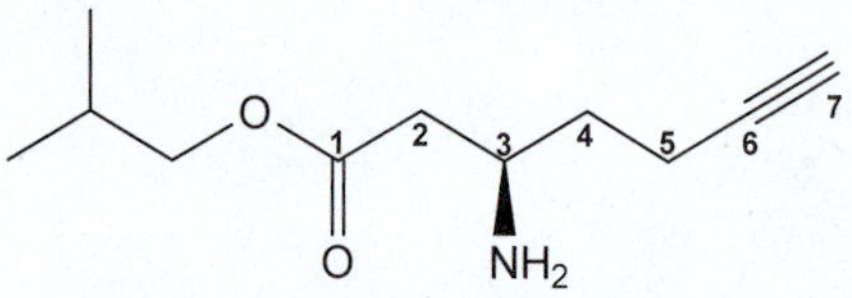

(c) 2-Methylpropyl (*R*)-3-aminohept-6-ynoate

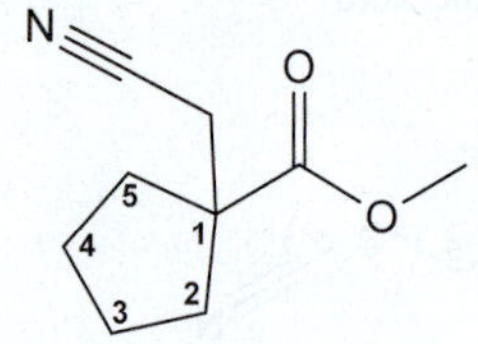

(d) Methyl 1-(cyanomethyl)cyclopentane-1-carboxylate

Problem F.34

Think

How many carbon atoms are in the main chain? To what functional group does the suffix *oic acid*, *amide*, or *nitrile* correlate? Review the nomenclature rules from Interchapters A, B, and C and Table F-1 to identify the substituents and stereochemical configurations. On which C atoms are the substituents located? How are the C atoms in the various functional groups counted—as part of the root or separately from it? On which two carbon atoms is each C=C located? Must stereochemical configurations be specified?

Solve

Structures for IUPAC names **(a)**–**(e)** are drawn below. The main chain or ring is numbered. In each example, CO_2H, O=C–N, or C≡N is the functional group that corresponds to the suffix. For molecules **(a)**–**(c)** the highest-priority functional group's C is established as C1. When these functional groups are treated as substituents, on the other hand, that C is not counted as part of the main chain because the prefix already accounts for that C. In molecules **(d)** and **(e)** both the ring and the functional group together establish the root, in which the ring is named first, followed by *carboxylic acid* or *carbonitrile* to specify the CO_2H or C≡N functional group.

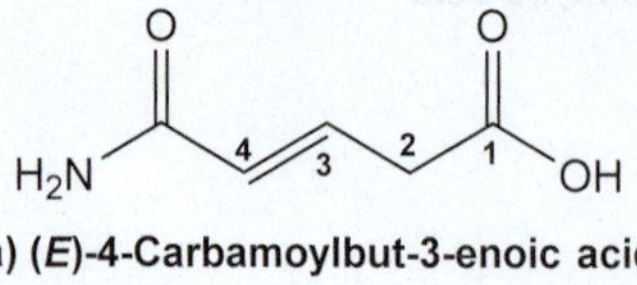

(a) (*E*)-4-Carbamoylbut-3-enoic acid

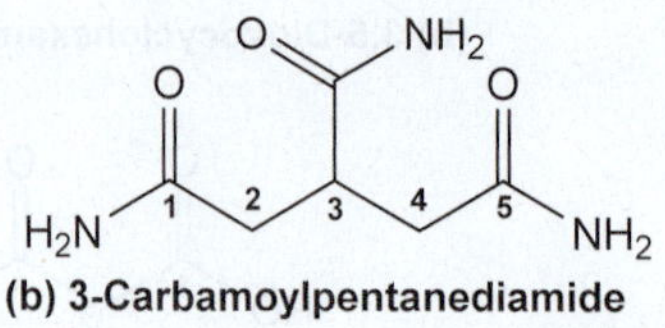

(b) 3-Carbamoylpentanediamide

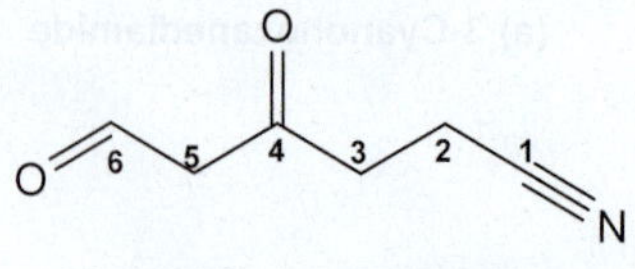

(c) 4,6-Dioxohexanenitrile

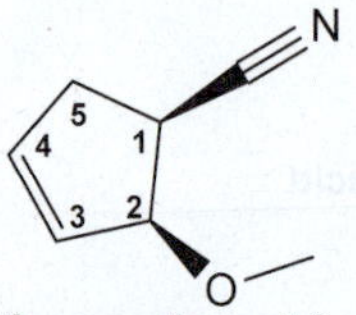

(d) (1*S*,2*S*)-2-Methoxycyclopent-3-ene-1-carbonitrile

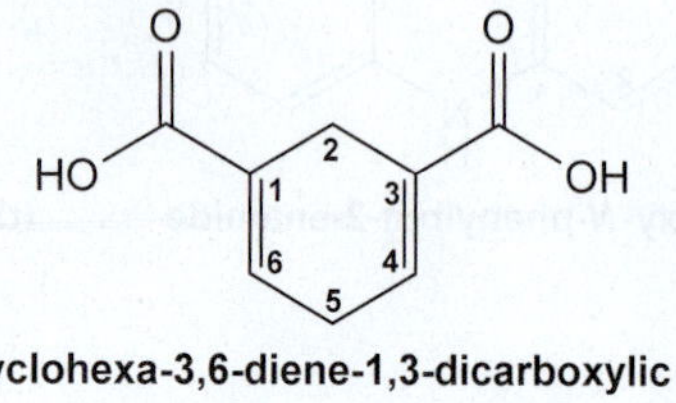

(e) Cyclohexa-3,6-diene-1,3-dicarboxylic acid

Problem F.35

Think

How many carbon atoms are in the main chain? To what functional group does the suffix *oate* or *oic* correlate? Review the nomenclature rules from Interchapters A, B, and C and Table F-1 to identify the substituents and stereochemical configurations. On which C atoms are the substituents located? How are the C atoms in the various functional groups counted—as part of the root or separately from it? On which two carbon atoms is each C=C located? Must stereochemical configurations be specified?

Solve

Structures for IUPAC names **(a)–(e)** are drawn below. The main chain is numbered. Molecules **(a)**, **(c)**, and **(e)** are esters in which the alkanoate group contains the O=C–O and the alkyl group is singly bonded to the O atom. Molecules **(b)** and **(d)** are anhydrides. Each alkanoic group is numbered. The format for a symmetric acid anhydride is *alkanoic anhydride*, and the format for an unsymmetric acid anhydride is *alkanoic alkanoic anhydride*. Molecules **(b)** and **(d)** are both symmetric.

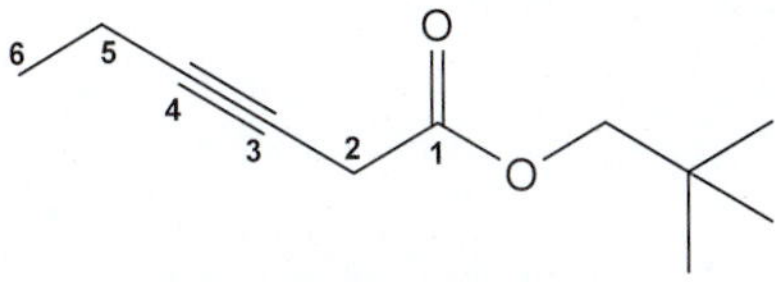

(a) 2,2-Dimethylpropyl hex-3-ynoate

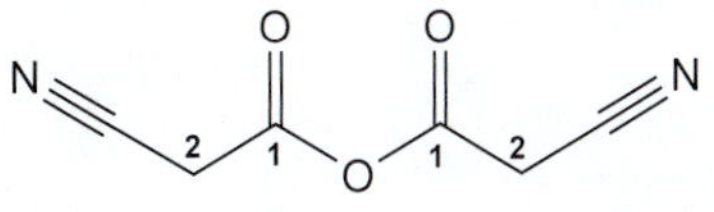

(b) Cyanoethanoic anhydride

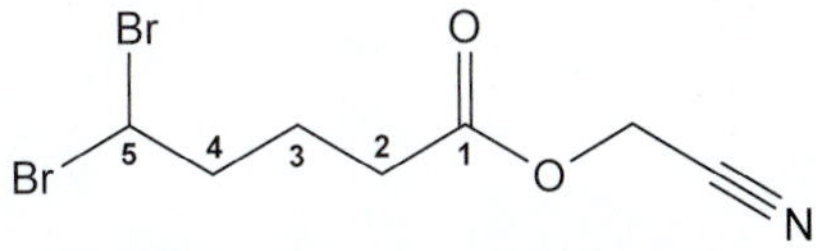

(c) Cyanomethyl 5,5-dibromopentanoate

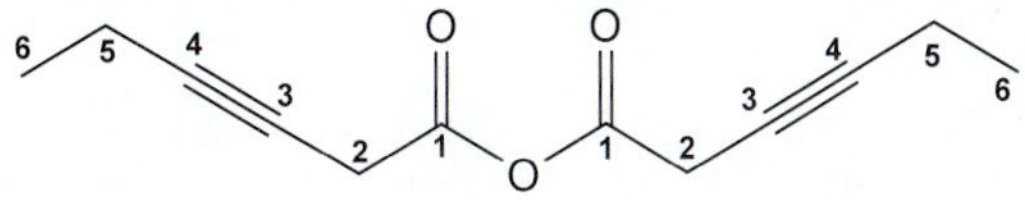

(d) Hex-3-ynoic anhydride

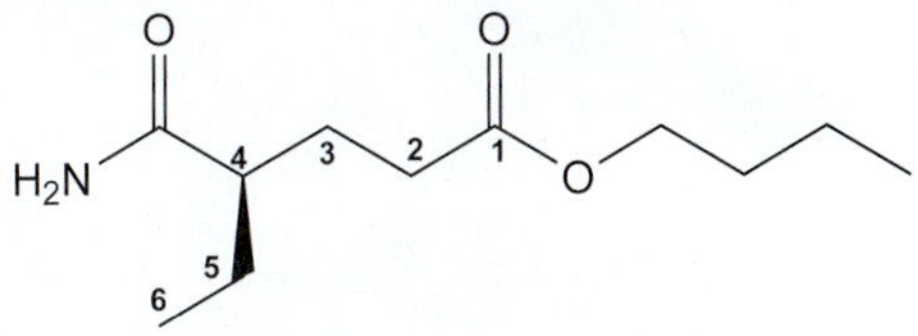

(e) Butyl (*R*)-4-carbamoylhexanoate

CHAPTER 10 | Nucleophilic Substitution and Elimination Reactions 2: Reactions That Are Useful for Synthesis

Your Turn Exercises

Your Turn 10.1

Think

What are the electron-rich and electron-poor species in the reactants? What species will act as the nucleophile? The leaving group? Using the mechanism in Equation 10-8 as a guide, draw curved arrows to show the first step in the reaction. Which elementary step is shown in the first step? Draw the products. Follow the same procedure for Step 2. How is the stereochemistry of the C◀O affected in each step of the reaction?

Solve

The O atom in ROH is electron rich and acts as the nucleophile, and the P atom in PCl_3 is electron poor and acts as the electrophile. The first step is an S_N2 reaction to form the O–P bond and break the P–Cl bond. The C◀O bond is not affected in the first step. In the second step, Cl^- acts as the nucleophile and attacks the electrophilic C◀O via an S_N2 backside attack. This inverts the configuration. The R◀OH is converted to R''''''Cl via two back-to-back S_N2 reactions.

Backside attack inverts the configuration.
Electron rich
Electron poor
1. S_N2
Electron poor
Electron rich
2. S_N2
+ $HOPCl_2$

Your Turn 10.2

Think

Is Cl^- a weak, moderate, or strong nucleophile? Consult Section 9.7b. How does the relative rate of the reaction relate to relative nucleophilicities?

Solve

Cl^- is an example of a moderate nucleophile. The relative nucleophilicities of NH_3 and Cl^- are 320,000 and 23,000, respectively. So NH_3 is, in fact, a stronger nucleophile in water than Cl^- is, making NH_3 a moderate nucleophile, too.

Your Turn 10.3

Think

Identify the bonds that are broken and formed in each step. How do you draw curved arrows to show bond formation? Bond breaking? Which steps involve a Nuc–C bond formation, and which steps involve electron-poor H atoms?

Solve

The first step is an S_N2 reaction in which the secondary amine R_2NH acts as a nucleophile and attacks the electrophilic C, causing Br to leave as Br^-. In the second step, the positively charged tertiary ammonium ion R_3NH^+ is deprotonated by the excess NH_3, and a tertiary amine results: R_3N. The tertiary amine acts as a nucleophile to attack the electrophilic C, and Br leaves, forming Br^- and a quaternary ammonium ion, R_4N^+.

1. S_N2
+ :Br:⊖
NH_3
2. Proton transfer
3. S_N2
+ :Br:⊖

Your Turn 10.4

Think

What are the charged products in Equations 10-11 and 10-13? Are alkyl groups electron donating or withdrawing? Do alkyl groups stabilize or destabilize the positive charge on N? Which product is more stable, the primary or secondary ammonium ion?

Solve

Alkyl groups are electron donating and stabilize positive charges. The negatively charged product is the same in both Equations 10-11 and 10-13: Br^-. The positively charged product is a primary amine ($CH_3CH_2NH_3^+$) in Equation 10-11 and is a secondary amine [$(CH_3CH_2)_2NH_2^+$] in Equation 10-13. The additional alkyl group in the secondary amine is electron donating and thus helps to stabilize the positive charge; the product is lower in energy, and the energy barrier leading to it is smaller.

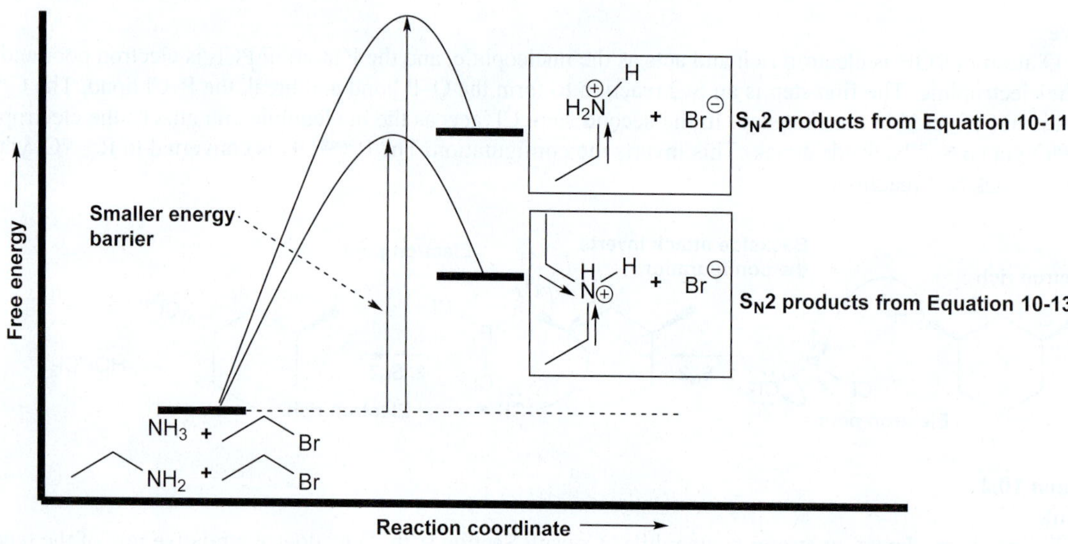

Your Turn 10.5

Think

Identify the bonds that are broken and formed in each step. How do you draw curved arrows to show bond formation? Bond breaking? Which steps involve a Nuc–C bond formation, and which steps involve electron-poor H atoms?

Solve

Hydride, $H:^-$, acts as a base in the first step to deprotonate the α C–H to generate the enolate anion. This is a proton transfer step. The enolate anion acts as a nucleophile in Step 2 and attacks the electrophilic C, causing Br to leave as Br^-. This is an S_N2 step that forms a new C–C bond.

Enolate anion

1. Proton transfer

Resonance

+ H_2

2. S_N2

Your Turn 10.6

Think

Using Equation 10-27 as a guide, what type of reaction occurs in the first step? Are there any protons on the α carbon? What bonds broke? What bonds formed? How is the enolate anion formed? Is resonance possible? What type of reaction occurs in the second step? What bonds broke? What bonds formed?

Solve

In the first step, HO^- acts as a base and deprotonates the α C to form the nucleophilic enolate anion. In the second step, the nucleophilic enolate anion attacks the molecular I–I in an S_N2 reaction, in which I^- departs as the leaving group. Even though I_2 appears not to be an electron-poor substrate, it is polarizable, and an induced dipole is generated in the presence of the nucleophilic enolate anion. See the mechanism below.

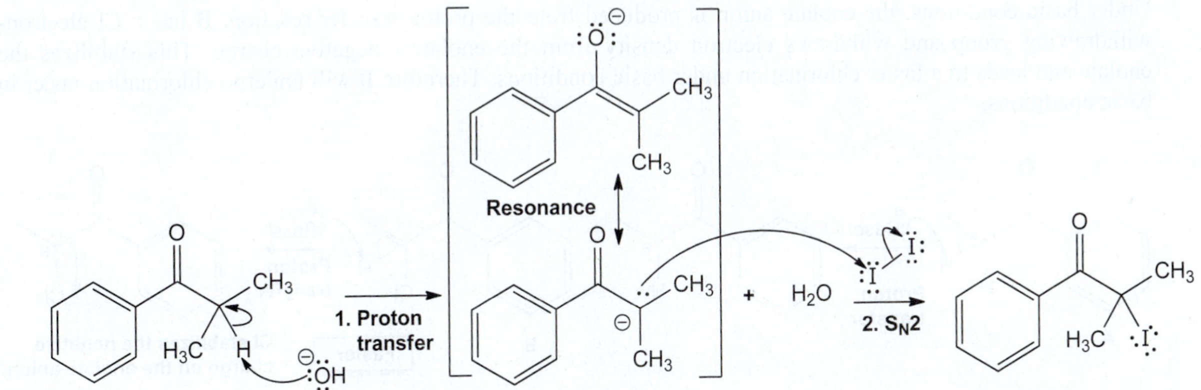

Resonance

1. Proton transfer

2. S_N2

Your Turn 10.7

Think

Identify which bonds broke and formed in each reaction step. When are protons involved (proton transfer)? When does a nucleophile attack an electron-poor nonhydrogen atom (S_N2)?

Solve

The α carbon in this example has two H atoms. In the first step, HO:$^-$ acts as a base and deprotonates the α carbon to form the nucleophilic enolate anion. In the second step, the nucleophilic enolate anion attacks the molecular Br–Br in an S_N2 reaction, in which Br^- departs as the leaving group. In the third step, HO:$^-$ acts as a base and deprotonates the α carbon to form the nucleophilic enolate anion. In the fourth step, the nucleophilic enolate anion attacks the molecular Br–Br in an S_N2 reaction.

Resonance

The halogen atom increases acidity.

1. Proton transfer

2. S_N2

3. Proton transfer

4. S_N2

Br stabilizes the negative charge on the enolate anion.

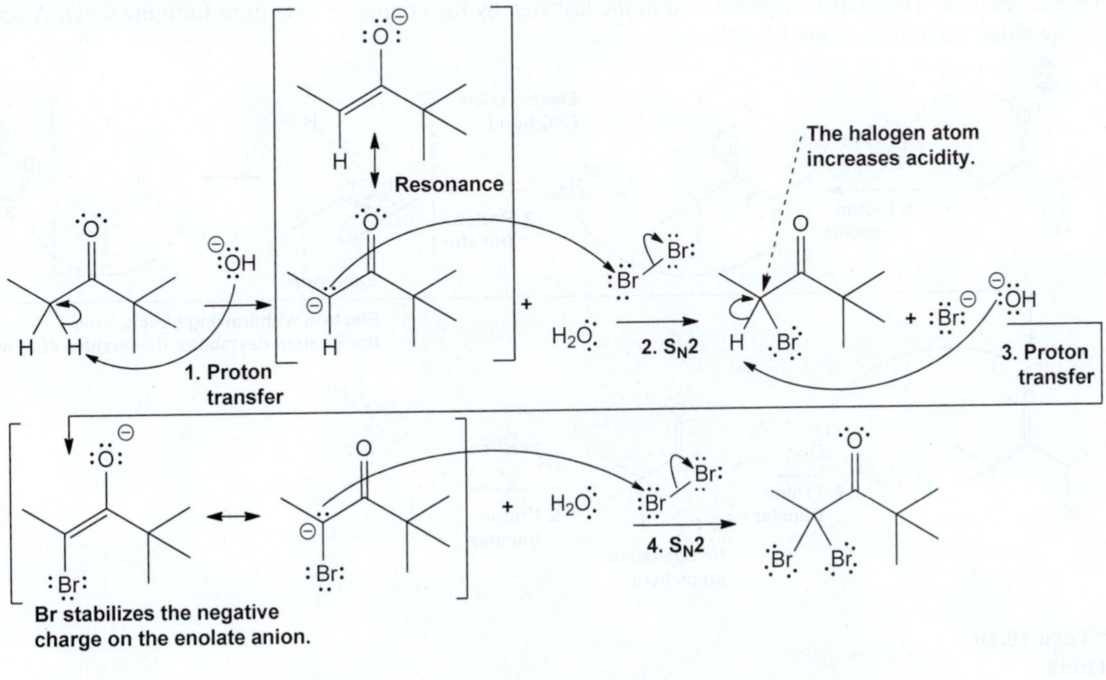

Your Turn 10.8

Think

Under basic conditions, what reaction, **A** or **B**, takes place at an α carbon? What product forms? Is the Cl electron withdrawing or donating? Will that inductive effect stabilize or destabilize that product?

Solve

Under basic conditions, the enolate anion is produced from the proton transfer reaction. **B** has a Cl electron-withdrawing group and withdraws electron density from the enolate's negative charge. This stabilizes the enolate and leads to a faster chlorination under basic conditions. Therefore **B** will undergo chlorination faster in basic conditions.

Your Turn 10.9

Think

In each step, identify the bonds that broke and formed. Use curved arrows to show electron movement. Using Equation 10-31 as a guide, in the presence of acid H–OAc, where is the basic site on the organic molecule? What atom gets protonated? Are there any protons on the α carbon? What reaction takes place at the α carbon? What type of reaction occurs in the third step? In the fourth step?

Solve

The ketone O is protonated in the presence of an acid (proton transfer). The conjugate base deprotonates the α carbon (proton transfer). The enol form acts as the nucleophile because the negatively charged enolate anion cannot exist in any substantial concentration in acidic conditions owing to its basic nature. The enol is electron rich at the α carbon and can polarize the Br–Br. This sets up the electron-rich to electron-poor driving force of the S_N2 reaction. The O–H is deprotonated in the last step by the conjugate base to re-form the C=O. A second halogenation reaction does not take place.

Your Turn 10.10

Think

Under acidic conditions, what reaction takes place at the C=O bond? What product forms? Is Br electron withdrawing or donating? Will that inductive effect stabilize or destabilize the product?

Solve

Ketone **A** will react faster because of the electron-withdrawing Br in ketone **B**. Br destabilizes the positive charge that develops when the carbonyl O is protonated.

A
Faster

B

Br destabilizes the positive charge.

Your Turn 10.11

Think

In each step, identify the bonds that broke and formed. How are curved arrows used to indicate bond breaking and bond forming? Using Equation 10-33 as a guide, in the presence of the electron-rich C atom of $CH_2=N=N$, find the acidic site on the organic molecule. What atom gets protonated? What molecule becomes nucleophilic after the first step?

Solve

The C atom of diazomethane, $CH_2=N=N$, is basic and is protonated by the acidic proton of the carboxylic acid. The carboxylate anion, $CH_3CO_2^-$, that is produced is electron rich and acts as a nucleophile in an S_N2 reaction. N_2 is the leaving group (stable and gaseous).

1. Proton transfer

2. S_N2

$+ N_2(g)$

Your Turn 10.12

Think

Identify the bonds that broke and formed. How do you use curved arrows to show bond breaking and bond formation? Using Equation 10-39 as a guide, determine what type of reaction occurs between ROH and HO^-. In the resulting species, which site is nucleophilic? Where is the leaving group?

Solve

In the first step, the ROH is deprotonated by basic HO^- to form an alkoxide nucleophile. The leaving group Br is on the same molecule that contains the nucleophilic O^-. An intramolecular S_N2 reaction results in a six-membered ring ether.

1. Proton transfer

$+ H—\ddot{O}H$

2. S_N2

$+ :Br:^-$

Your Turn 10.13

Think

Identify the bonds that broke and formed. How do you use curved arrows to show bond breaking and bond formation? Using Equation 10-39 as a guide, determine what type of reaction occurs between ROH and HO^-. In the resulting species, which site is nucleophilic? Where is the leaving group?

Solve

In the first step, the ROH is deprotonated by basic HO^- to form an alkoxide nucleophile and a molecule of H_2O. The leaving group Cl is on the same molecule that contains the nucleophilic O^-. An intramolecular S_N2 reaction results in a three-membered ring ether, an epoxide.

1. Proton transfer
2. S_N2
+ H_2O
+ Cl^-

Your Turn 10.14

Think

In each step, identify the bonds that broke and formed. How are curved arrows used to show bond breaking and bond forming? What type of reaction results from ROH and HCl? What is the leaving group that results? Why does a C^+ not form?

Solve

The acidic conditions cause the ROH group to be protonated, producing ROH_2^+ and generating a good leaving group, H_2O. The leaving group is on a primary C, and thus the nucleophile ROH attacks the electrophilic C in the step where the leaving group departs, the S_N2 reaction, which avoids the formation of an unstable primary carbocation. A final proton transfer step ensues to form the ether product.

OSO_3H / OH / 1. Proton transfer / + $^-OSO_3H$ / OH_2^+ / 2. S_N2 / OH / H_2O / 3. Proton transfer / + H_3O^+

Your Turn 10.15

Think

What species is electron rich? What site is electron poor? In what kind of step is the C–Cl bond formed? In what kind of step is the RO–H bond formed? Use curved arrows to show electron movement.

Solve

Cl^- is a good nucleophile and attacks the electron-poor C of the epoxide. The S_N2 reaction causes the highly strained epoxide ring to open. This compensates for the poor leaving group ability of the RO^-. The alkoxide RO^- is a strong base and abstracts a proton from the methanol solvent.

1° C / Ring opening relieves ring strain. / 1. S_N2 / Poorly stabilized leaving group / $HOCH_3$ / 2. Proton transfer / Cl / OH / + $^-OCH_3$

Your Turn 10.16

Think

Use Equation 10-51 as a reference. In Equation 10-59, which species is electron rich? Which is electron poor? What acts as the nucleophile in the S_N2 elementary step? What new bond is formed? What bond breaks? What is the purpose of the aqueous HCl in the acid workup step? Which species is protonated in this step?

Solve

The alkyllithium and Grignard reagents in Step 1 are extremely strong bases and are present only in the absence of protic species. The $R:^-$ is the electron-rich species that attacks the C atom bonded to the O atom in the ring. The ring opens in the S_N2 step. This leaves a negative charge on the O atom. In the acid workup step, this oxygen atom is protonated by H_3O^+ (produced from HCl and H_2O) to form the alcohol product. A new C–C bond is formed, and a C–O bond is broken.

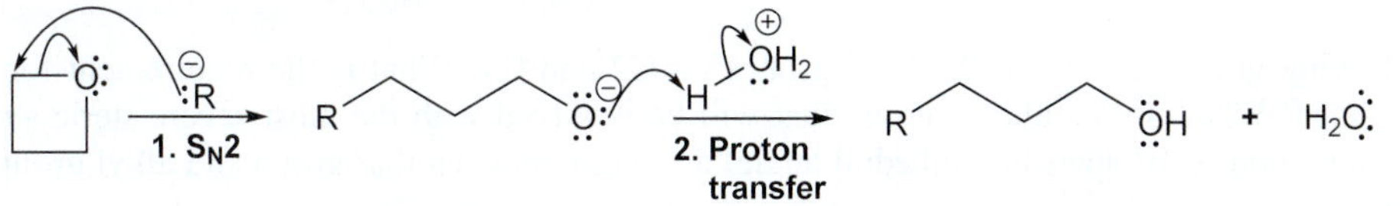

1. S_N2 — 2. Proton transfer

Your Turn 10.17

Think

Consult Table 6-1 to find the pK_a values. Does a lower pK_a value indicate a stronger or weaker acid? Is a proton transfer reaction favored on the same side as or opposite from the stronger acid? How do you use pK_a values of the two acids to determine the numerical factor by which one side of a reaction is favored? How is that numerical factor related to reversibility?

Solve

The pK_a of H_2 is 35, and the pK_a of RC≡CH is about the same as that of HC≡CH, which is 25. Therefore, HC≡CH is the stronger acid. The side opposite the terminal alkyne—the product side—is favored. The factor by which that side is favored is 10 raised to the pK_a difference between the two acids, or 10^{10}. This means that the reverse reaction is difficult, so the forward reaction is irreversible.

Your Turn 10.18

Think

Is the produced alkyne terminal or internal? How does this affect the mechanism? Which new C–C bond is formed in the alkyne product? Which carbon was the nucleophile, and which carbon was the electrophile? In Equations 10-63 and 10-64, which species acts as a base? Which H atom is removed in the E2 step?

Solve

The alkyne product in both Equations 10-63 (below) and 10-64 (next page) is an internal alkyne and does not contain an acidic proton. Therefore, the mechanisms for those reactions would not include the proton transfer steps that appear in Equation 10-66. The mechanism involves an E2, where the Cl is the leaving group and a new π bond is produced, forming the C≡C bond from the C=C bond.

Mechanism for Equation 10-63

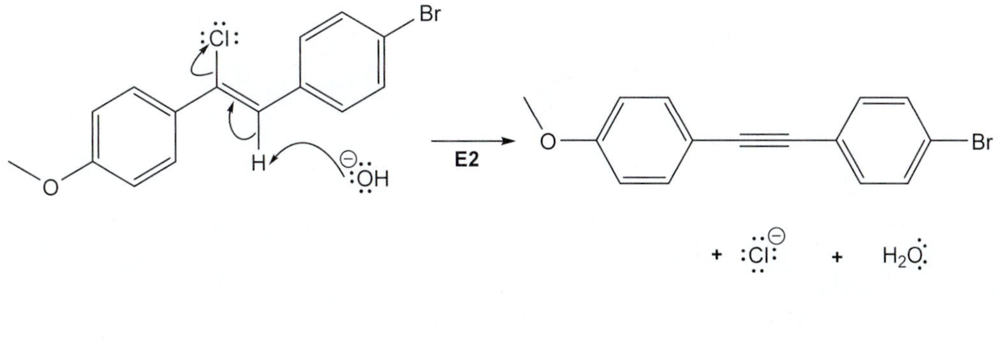

E2 — $+ \; :\ddot{Cl}:^- \; + \; H_2O$

Mechanism for Equation 10-64

$$O_2N\text{-}C_6H_4\text{-}C(Cl)\text{=}CH\text{-}C\equiv C\text{-}C_6H_5 + {:}\ddot{H}^{\ominus} \xrightarrow{E2} O_2N\text{-}C_6H_4\text{-}C\equiv C\text{-}C\equiv C\text{-}C_6H_5 + {:}\ddot{Cl}{:}^{\ominus} + H_2$$

Your Turn 10.19

Think

On which C is the leaving group? Draw out the C–H groups on C2 and C4. What is the requirement for two groups to be anticoplanar? What is the bulkiest group that will be involved with the most severe steric strain? How many gauche interactions (60° apart in a dihedral angle) are there between that group and alkyl groups in each Newman projection?

Solve

The $-N^+(CH_3)_3$ is the leaving group on C3. Therefore, there are adjacent H atoms on C2 or C4 that could be attacked by a base in an elimination reaction. In an E2, the adjacent H atom must be anticoplanar. This can be seen in a Newman projection where the leaving group and the H atom are 180° apart. The $N^+(CH_3)_3$ group is the bulkiest group and will be involved in the most severe steric strain. Comparison of the two Newman projections (**A:** C2–C3 bond or **B:** C3–C4 bond) shows two gauche interactions involving that bulky group with alkyl groups in **A**, as well as another gauche interaction between two alkyl groups. In **B**, there is just one gauche interaction between the $N^+(CH_3)_3$ group and an alkyl group, so there is less steric strain in **B**.

In Chapter Problems

Problem 10.1

Think

What is the first step in an S_N1 mechanism? In the intermediate that forms, what is the geometry of the C that was initially bonded to the leaving group? From how many directions can the nucleophile attack the electrophilic carbocation?

Solve

The mechanism for Equation 10-3 is shown below. When the nucleophile attacks the positively charged C atom, both *R* and *S* configurations can be formed, because that C atom is planar.

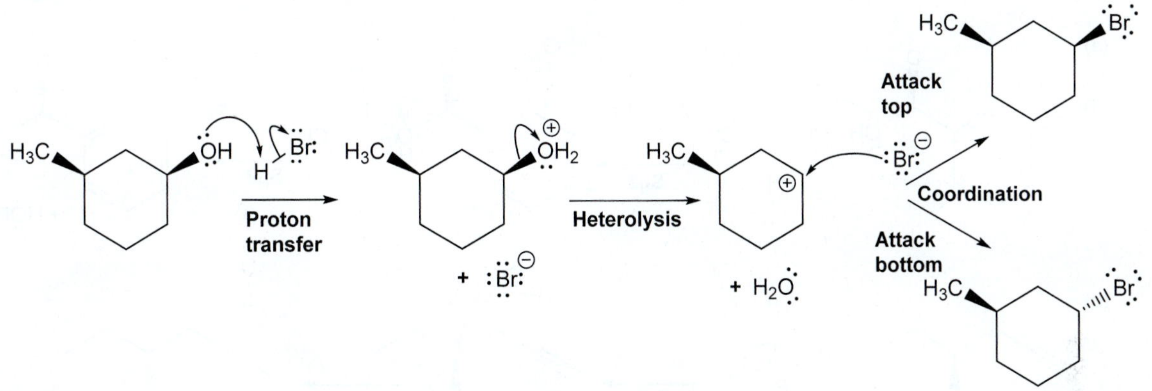

Proton transfer Heterolysis Attack top Coordination Attack bottom

A similar story arises with the mechanism for the reaction in Equation 10-4.

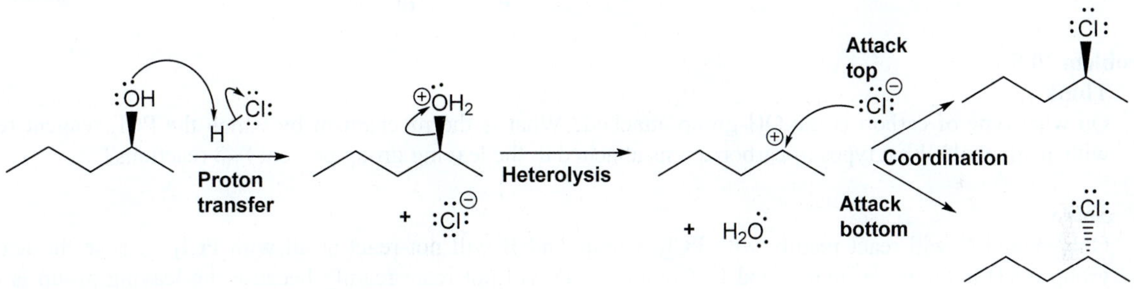

Proton transfer Heterolysis Attack top Coordination Attack bottom

Problem 10.2

Think

What is the first step in an S_N1 mechanism? What type of carbocation forms? Can a more stable intermediate be formed via a 1,2-hydride shift or a 1,2-methyl shift?

Solve

After the HO^- leaving group is turned into a good leaving group, H_2O, the carbocation that is formed can rearrange via a 1,2-hydride shift from a secondary carbocation into a tertiary carbocation. This tertiary carbocation is more stable and thus leads to the major product. The mechanism is given below.

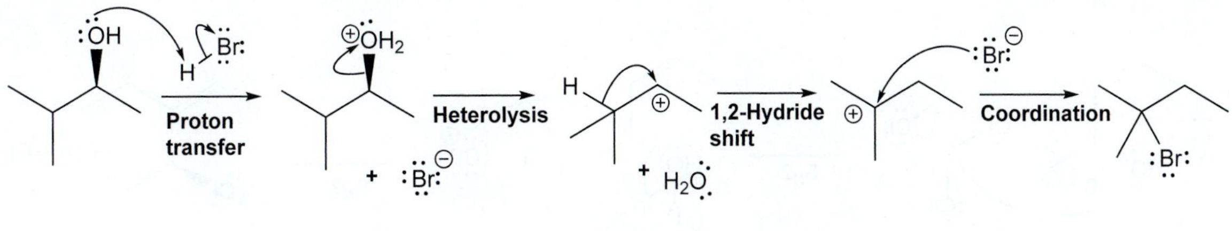

Proton transfer Heterolysis 1,2-Hydride shift Coordination

Problem 10.3

Think

Is HO^- a good leaving group? How does PCl_3 or PBr_3 turn HO^- into a good leaving group? What two elementary steps are involved in the reaction? What happens to the configuration at the electrophilic carbon?

Solve

Recall that D is deuterium, an isotope of hydrogen. It has chemical behavior identical to H but serves as a label in reactions. HO^- is not a good leaving group, and PCl_3 and PBr_3 serve to turn a HO^- leaving group into a $HOPX_2$ leaving group, which is an uncharged leaving group. The reaction occurs in two back-to-back S_N2 steps. This inverts the configuration at the electrophilic carbon. Notice in **(a)** that the C–Br is unaffected because sp^2-hybridized C atoms are resistant to S_N2 reactions.

(a)

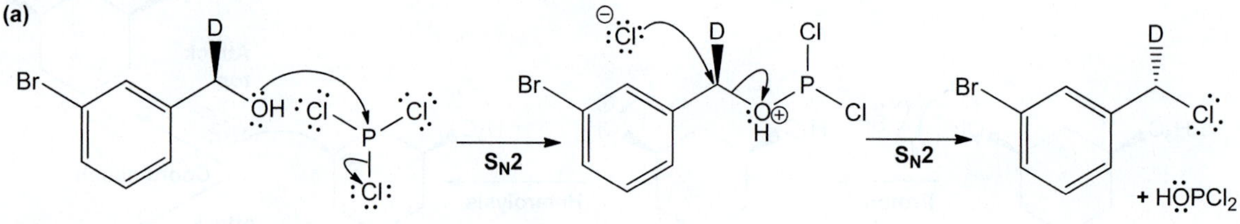

(b)

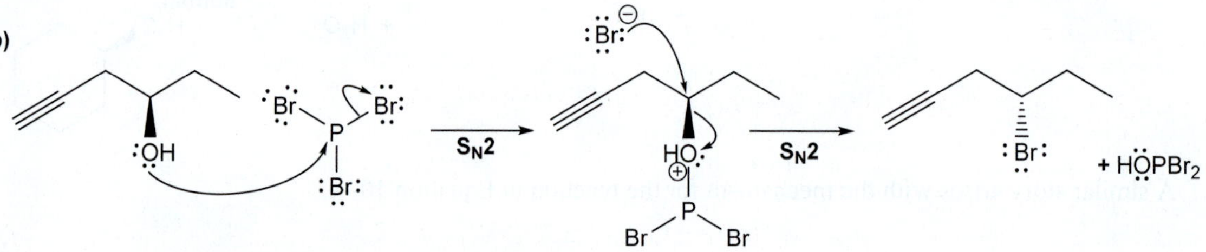

Problem 10.5

Think

On what type of carbon is the OH group attached? What is the mechanism by which the PCl_3 reagent reacts with an alcohol? What types of carbon atoms attached to the leaving group prevent S_N2 reactions?

Solve

Only **A** and **C** will react readily with PCl_3. Compound **B** will not react at all with PCl_3 because the leaving group (OH) is on an sp^2-hybridized C. Compound **D** will not react readily because the leaving group is on a tertiary carbon; S_N2 reactions are unfeasible, and S_N1 generally gives poor yield for these reactions. For the reactions that do proceed, keep in mind that the second S_N2 step causes inversion of the configuration.

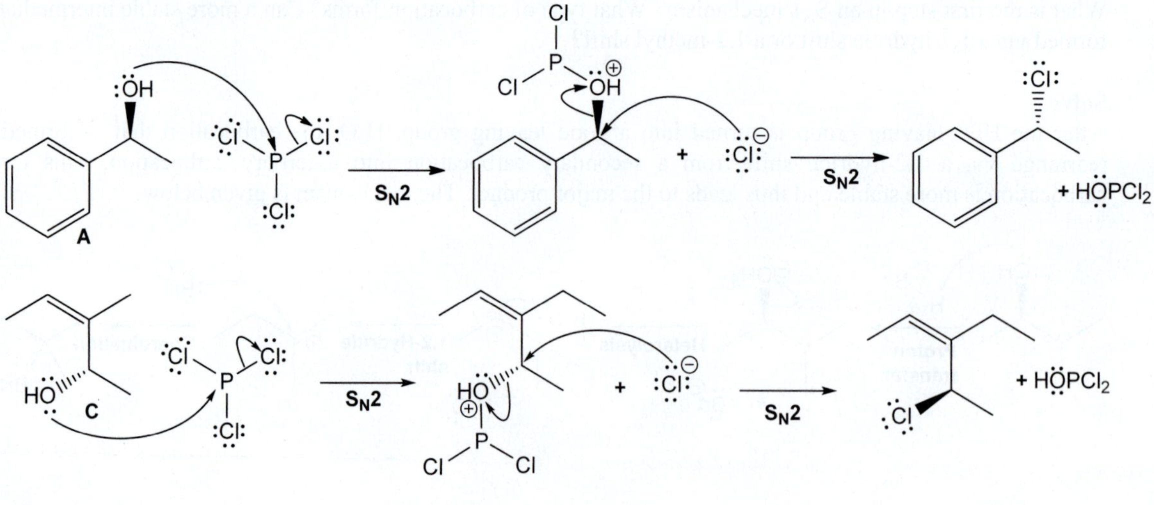

Problem 10.7

Think

Which species is electron rich? Which is electron poor? What kind of reaction will take place between the two? Can the reaction take place a second time?

Solve

The amine N is electron rich (nucleophile), and the C attached to the Br is electron poor (electrophile). The resulting reaction is an S_N2 reaction, yielding a protonated tertiary amine. The product can be deprotonated by another molecule of the amine to yield an uncharged tertiary amine. Because there is excess CH_3Br, a further alkylation generates a quaternary ammonium ion as the major product. No further reaction can take place, because there are no acidic H atoms on the ammonium ion.

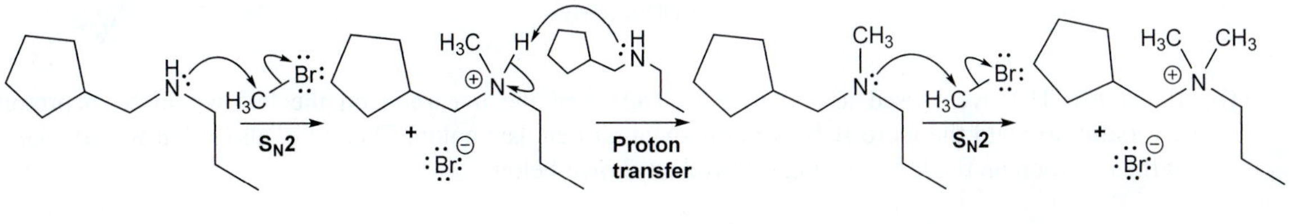

S_N2 — Proton transfer — S_N2

Problem 10.8

Think

How can you simplify $NaNH_2$? What is the role of the resulting species? What is the acidic proton that is removed? What mechanism takes place in Step 2 with the R–Br reagent? What new bond is formed?

Solve

$NaNH_2$ is simplified to $^{-}NH_2$, a strong base, which deprotonates the α carbon to make a relatively strong nucleophile as an enolate anion. Then substitution via S_N2 reaction takes place, involving the alkyl halide.

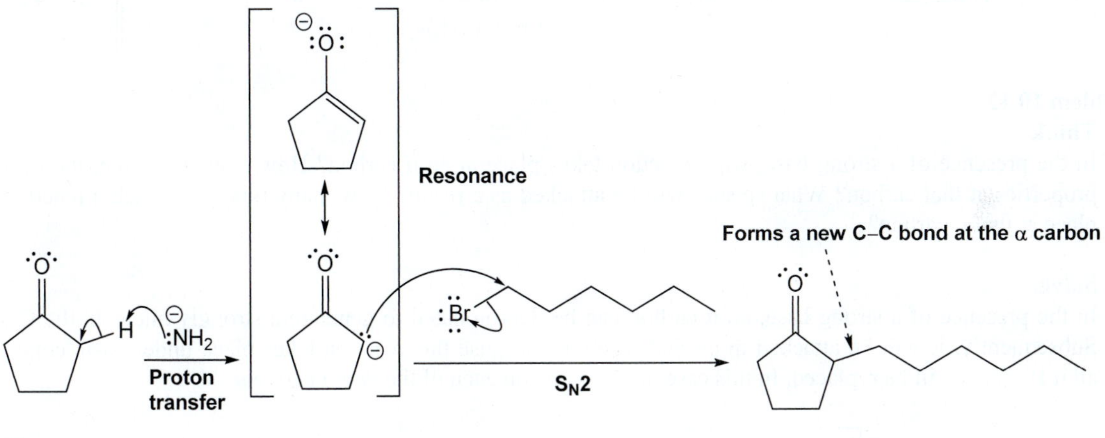

Resonance

Forms a new C–C bond at the α carbon

Proton transfer — S_N2

Problem 10.10

Think

How can you simplify LDA and $KOC(CH_3)_3$? What is the role of the resulting species? Are the two α carbons distinct? How does LDA differentiate between α carbons? How does $KOC(CH_3)_3$ differentiate between those carbons?

Solve

LDA is simplified to $(C_3H_7)_2N^{-}$, and $KOC(CH_3)_3$ is simplified to $^{-}OC(CH_3)_3$. Both anions are strong bases and will deprotonate an α carbon. The two α carbons in the molecule are indeed distinct. The one on the left is a tertiary carbon, part of an isopropyl group, and the one on the right is a secondary carbon, part of a CH_2 group.

(a) LDA is a very strong, bulky base, so it will deprotonate the less sterically hindered α carbon under kinetic control. That α carbon is the one on the right in the given ketone, because it has fewer alkyl groups attached. The enolate anion that is formed then attacks the allyl bromide in an S_N2 reaction, yielding an alkylated ketone. See the mechanism on the next page.

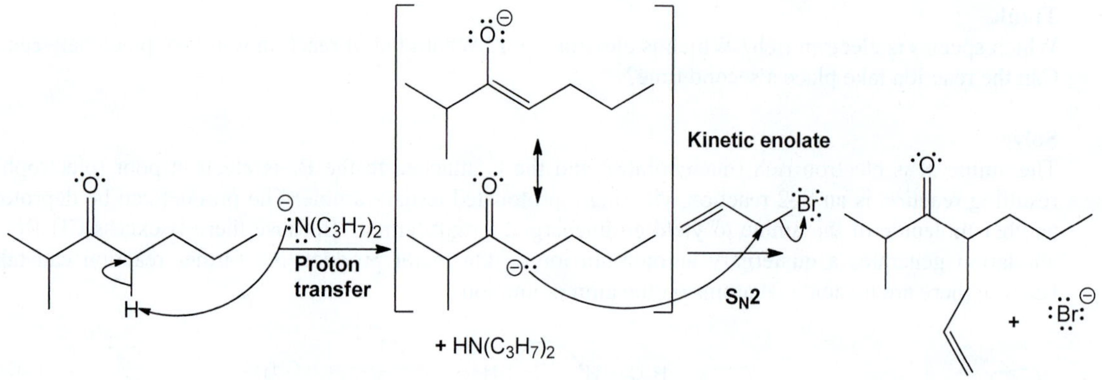

(b) If $NaOC(CH_3)_3$ were used as the base instead, then the α carbon on the left would be deprotonated reversibly to yield the more stable (thermodynamic) enolate anion. Therefore, alkylation would take place at the α carbon on the left, yielding the product shown below.

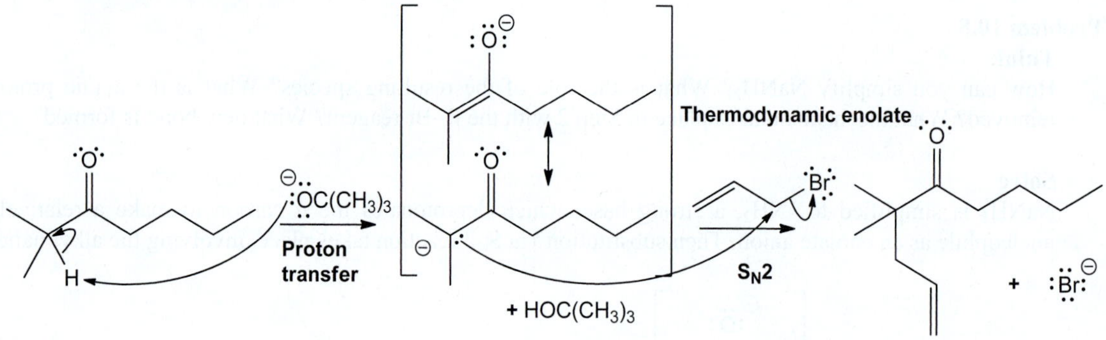

Problem 10.12

Think

In the presence of a strong base, what reaction takes place at an α carbon? How does this affect the chemical properties at that carbon? What species will be attacked as a result? How many times will such a reaction take place at that α carbon?

Solve

In the presence of a strong base, an α carbon can be deprotonated to generate a strongly nucleophilic C atom. Subsequently, I_2 will be attacked in an S_N2 reaction. Because this reaction takes place under basic conditions, all α H atoms will be replaced. In this case, there is one on each of the two α carbons.

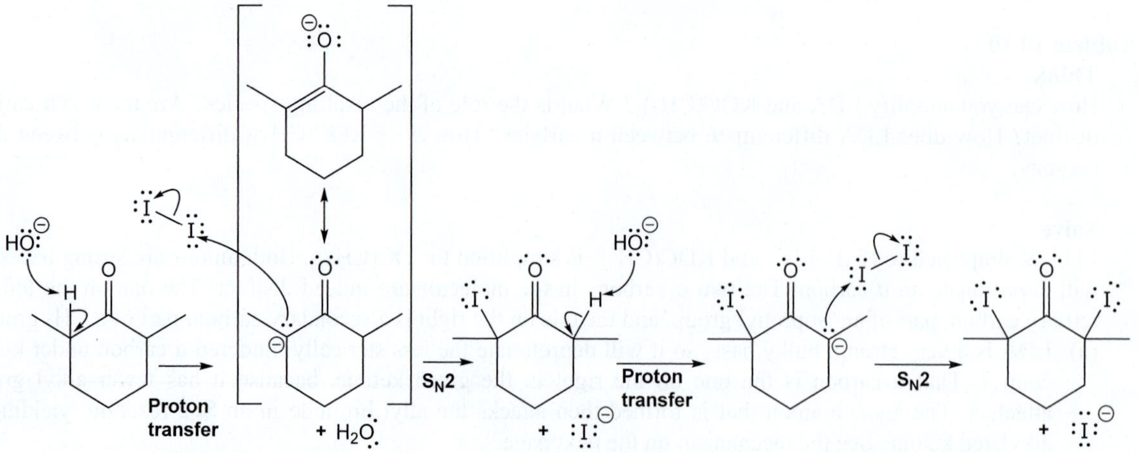

Problem 10.13

Think

Is the reaction taking place under acidic or basic conditions? How is the mechanism different in acidic conditions than in basic? What atom on the ketone acts as the base? How many proton transfer reactions take place before the S_N2 reaction? What atom of the ketone becomes nucleophilic? Will polyhalogenation take place, or will halogenation occur just once?

Solve

The reaction takes place under acidic conditions and follows the same mechanism as shown in Equation 10-31. Two proton transfer reactions occur: The O atom on the carbonyl is protonated by the acid, and then the α carbon is deprotonated to form an enol. The reaction stops with monoiodination, because the electronegativity of the iodine atom destabilizes any positive ion that would lead to a new enol.

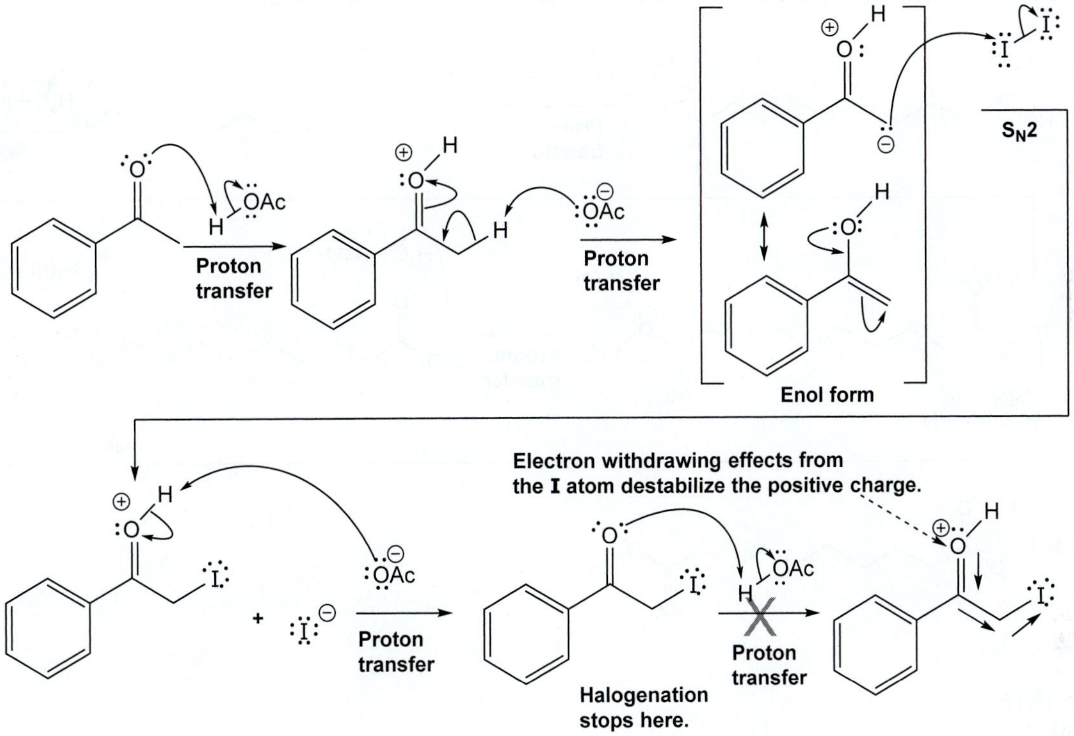

Proton transfer

Proton transfer

Enol form

S_N2

Electron withdrawing effects from the **I** atom destabilize the positive charge.

Proton transfer

Proton transfer

Halogenation stops here.

Problem 10.14

Think

What is the Lewis structure for CH_2N_2? Is the carbon electron rich or electron poor? What is basic on CH_2N_2? What is acidic on the carboxylic acid? What proton transfer reaction takes place? After the proton transfer reaction, what S_N2 reaction can occur?

Solve

In diazomethane, the C atom bears a −1 formal charge and, therefore, is electron rich and basic. In Step 1, the C atom of diazomethane is protonated by the acidic proton of the carboxylic acid. In Step 2, the carboxylate anion is electron rich and acts as the nucleophile in the ensuing S_N2 reaction. The protonated form of $CH_3N_2^+$ is electron poor and acts as the substrate. $N_2(g)$ is a great leaving group. See the mechanisms for **(a)** and **(b)** on the next page.

(a) After diazomethane deprotonates the carboxylic acid, an S_N2 step causes N_2 to be displaced, producing the methyl ester.

Proton transfer

S_N2

CH_3 + $N_2(g)$

(b) This mechanism takes place twice, once for each carboxylic acid functional group.

Proton transfer

S_N2

+ $N_2(g)$

Proton transfer

+ $N_2(g)$

S_N2

Problem 10.16

Think

What does a Williamson ether synthesis require? What type of R group appears on either side of the R–O–R in each case? What combinations of alkyl halide and alkoxide can produce the ether given?

Solve

A Williamson ether synthesis requires an alkyl halide (R–X) as the substrate and an alkoxide anion ($R'–O^-$) as the nucleophile. To produce an unsymmetric ether like the one desired in **(a)**–**(c)**, there are two theoretical combinations, which differ by the choice of R and R′. One or more of the possible combinations might be unfeasible owing to the nature of either the substrate or the alkoxide group.

(a) The two R groups are a benzene ring and an ethyl group. The two combinations are given below. Synthesis 1 is the only feasible option because the substrate is a primary C. In Synthesis 2, the R is the benzene ring, and the reaction is not feasible, because the substrate would have a leaving group on an sp^2-hybridized C.

Synthesis 1 **Synthesis 2**

(b) The two R groups are a benzene ring and a *tert*-butyl group. The two combinations are given below. Neither synthesis is feasible, because the substrate has the leaving group attached to a C atom that is either sp^2 hybridized or tertiary. Neither type of substrate can undergo an S_N2 reaction.

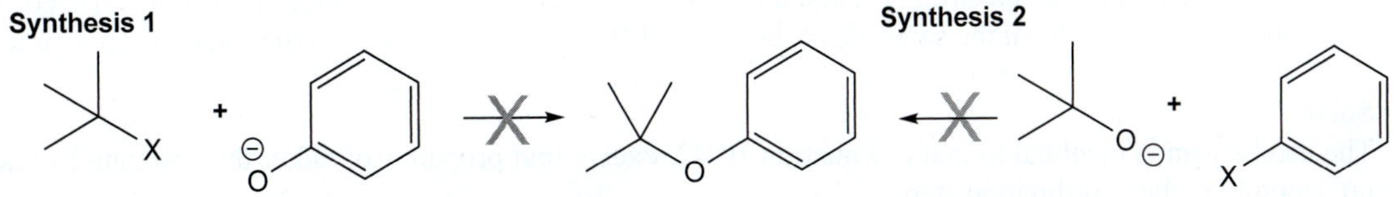

Synthesis 1 Synthesis 2

(c) The two R groups are a cyclohexane ring and an allyl group. The two combinations are given below. Either synthesis is feasible. However, Synthesis 1 is preferable, because an S_N2 reaction takes place more readily when the leaving group is on a primary carbon than when it is on a secondary carbon.

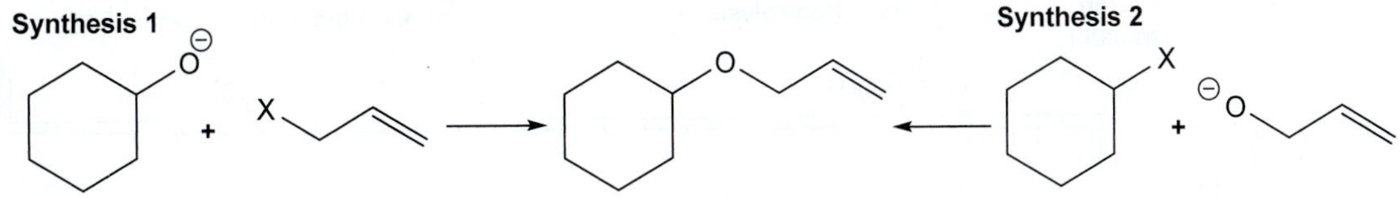

Synthesis 1 Synthesis 2

Problem 10.17

Think

In the presence of a strong base, what type of reaction occurs with an alcohol? What then forms? What are the resulting electron-rich and electron-poor species? Are these species located on the same molecule? Will an intramolecular reaction result?

Solve

In the presence of a strong base, the alcohol is deprotonated to form an alkoxide, which is an electron-rich nucleophile. The carbon of the C–Br bond is electron poor and serves as the electrophile. An intramolecular S_N2 reaction results because the formation of a five-membered ring is possible, and this produces a cyclic ether.

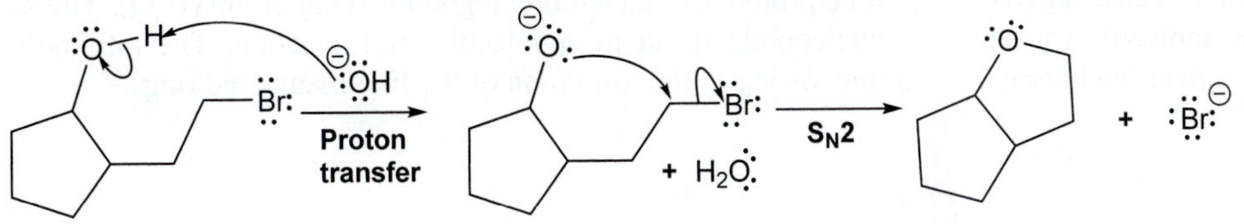

Proton transfer + H_2O S_N2 + :Br:

Problem 10.18

Think

In the presence of an acid, what type of reaction occurs with an alcohol? What then forms? Is there a good leaving group? Is there an electron-rich and an electron-poor species? What functional group is produced?

Solve

Formation of an ether is expected. Because the leaving group is on a benzylic carbon and the solvent is protic, an S_N1 reaction takes place. (However, formation of a primary benzylic carbocation makes other reactions possible, namely, electrophilic aromatic substitution and polymerization, which are covered in future chapters.)

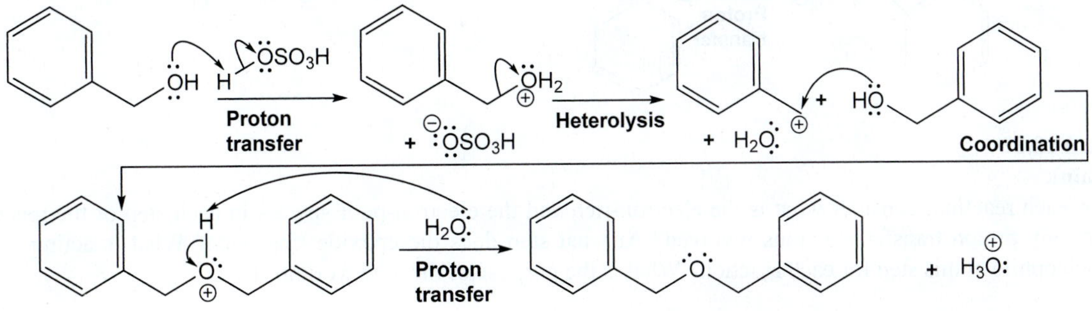

OSO_3H Proton transfer + OSO_3H Heterolysis + H_2O + HO Coordination

H_2O Proton transfer + H_3O:

Problem 10.19

Think

In the presence of an acid, what type of reaction occurs with an alcohol? What then forms? Is there a good leaving group? Is there an electron-rich and an electron-poor species? To produce an unsymmetric ether, should those species originate from the same alcohol or from different alcohols? What functional group is produced?

Solve

The mechanism is identical to that in Equation 10-42, except that propan-1-ol rather than propan-2-ol acts as the nucleophile in the coordination step.

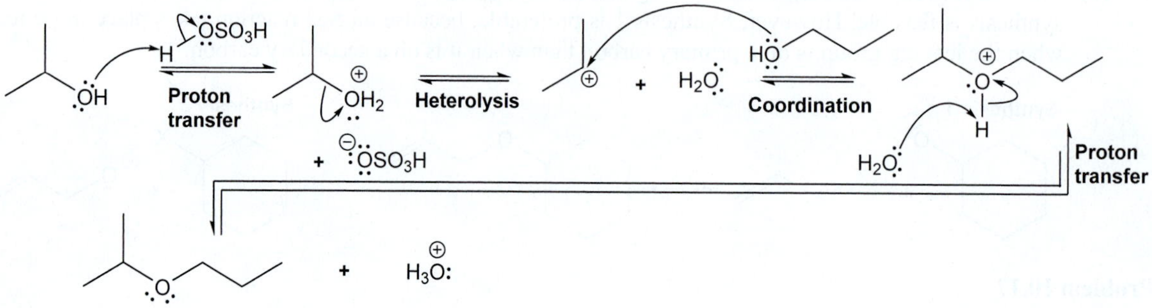

Proton transfer

Heterolysis

Coordination

Proton transfer

Problem 10.21

Think

In the presence of an acid, what type of reaction occurs with an alcohol? In a substitution reaction, what could then act as the nucleophile? What could act as the substrate? Is an S_N1 or S_N2 mechanism favored? Can an intramolecular reaction take place? If so, is it more favorable or less favorable than the corresponding intermolecular reaction?

Solve

The acidic conditions cause an OH group to be protonated, generating a good leaving group (H_2O). The second OH group in the molecule can act as a nucleophile in an intramolecular S_N1 reaction. The intramolecular reaction is favored over the intermolecular one, owing to the formation of the five-membered ring.

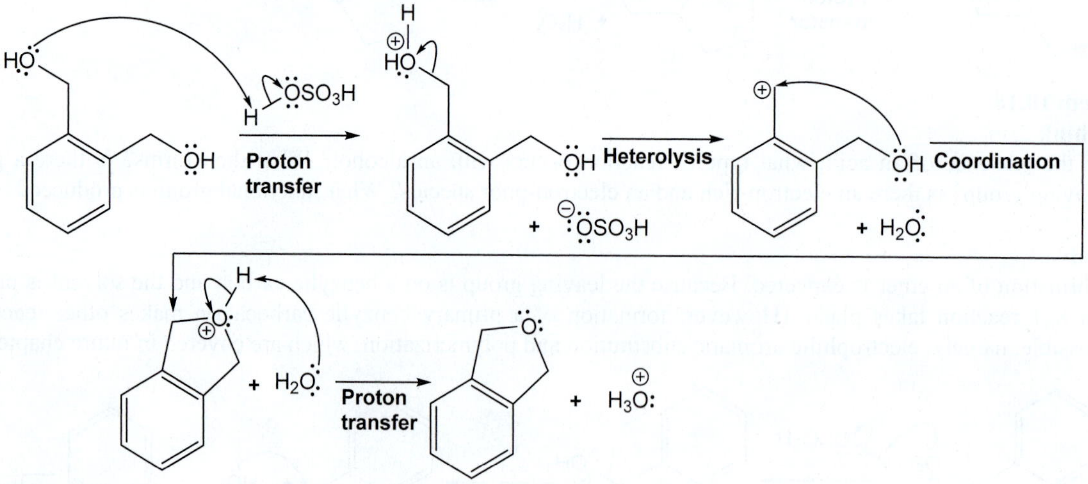

Proton transfer

Heterolysis

Coordination

Proton transfer

Problem 10.22

Think

For each reaction, consider what is the electron-rich and the electron-poor species in each step of the reaction. Are any proton transfer reactions involved? At what step does the epoxide ring open? What is acting as the nucleophile in this step for each reaction? What is the purpose of the acid workup step?

Solve

The mechanisms for each reaction are shown below.

Equation 10-52: The NC^- acts as the nucleophile to open the epoxide ring in an S_N2 reaction. A new C–C bond is formed. The second step is a proton transfer reaction to form an uncharged alcohol product.

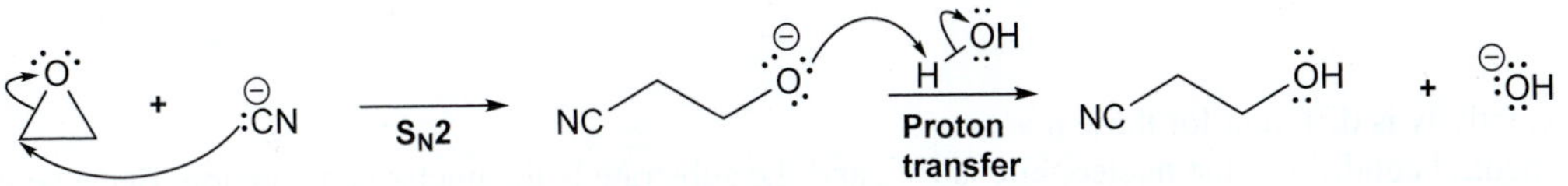

Equation 10-53: The NaH serves as a source of hydride base $H:^-$ to deprotonate the alkyne C–H and form the alkynide anion. The $RC{\equiv}C^-$ acts as the nucleophile to open the epoxide ring in an S_N2 reaction. A new C–C bond is formed. The second step is a proton transfer reaction to form an uncharged alcohol product.

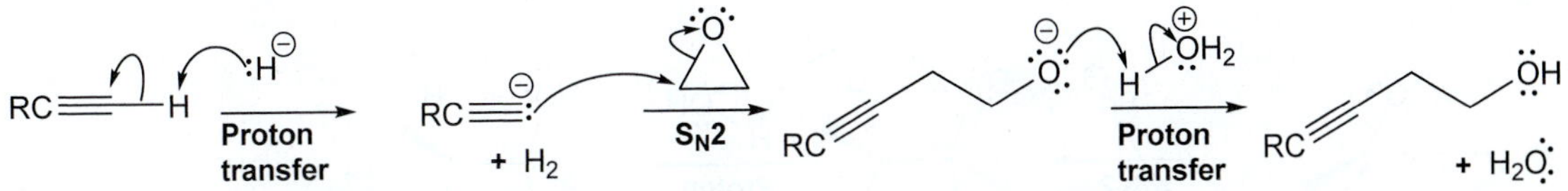

Equation 10-54: $LiAlH_4$ or $NaBH_4$ serves as a source of hydride nucleophile $H:^-$ to open the epoxide ring in an S_N2 reaction. A new C–H bond is formed. The second step is a proton transfer reaction to form an uncharged alcohol product.

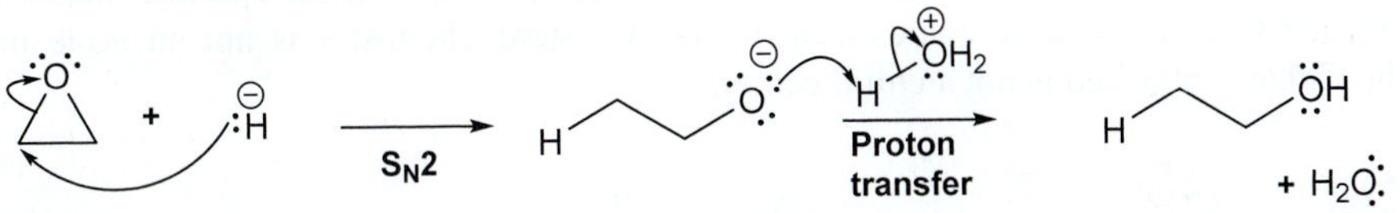

Problem 10.24

Think

What is the nucleophile? Which C atom of the epoxide ring will it attack? What is the stereochemistry of such a reaction?

Solve

The nucleophile is the cyanide anion, NC^-. It will undergo an S_N2 reaction with the epoxide ring, attacking the less sterically hindered C atom, which is the one on the left. The nucleophile attacks from the bottom of the epoxide (i.e., opposite the epoxide O atom), forcing the H and CH_3 substituents upward. The more substituted carbon with the CH_3 and the CH_2CH_3 groups does not undergo bond breaking or formation, so its stereochemical configuration remains the same.

(a)

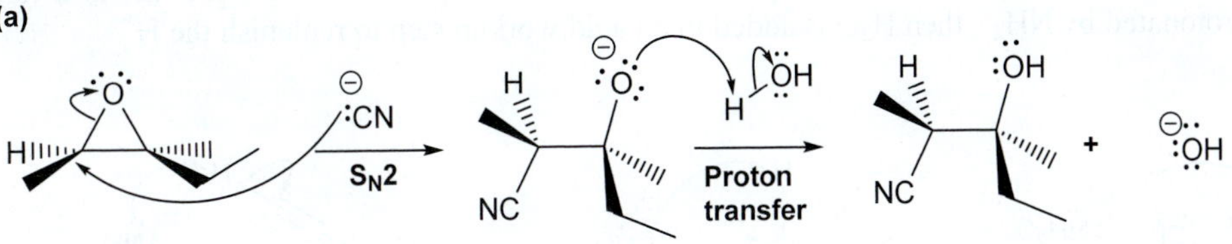

(b)

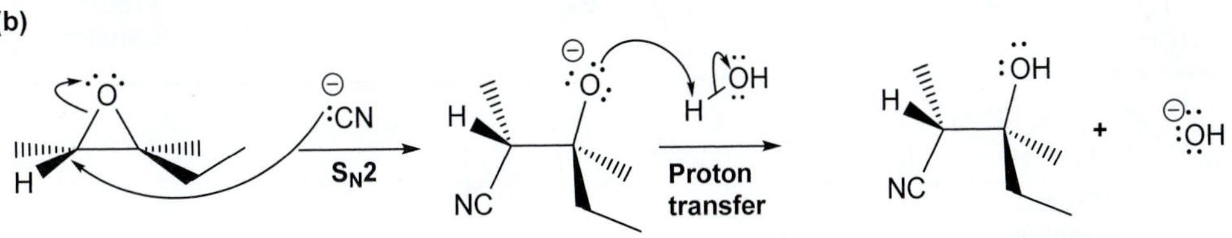

Problem 10.26

Think

What types of species are allowed under acidic conditions, and what types are not allowed? Under neutral conditions? Under each set of conditions, what is the substrate being attacked? What is the nucleophile? From what direction should the nucleophile attack the epoxide ring?

Solve

The regioselectivity is different for these reactions.

(a) Under neutral conditions, the nucleophile is Br^-, and the substrate is the uncharged epoxide. Because Br^- is a stronger nucleophile than water and is high in concentration, we can predict that bromide will open the strained ring to form a bromo alcohol. The mechanism follows the one in Equation 10-56. Because the epoxide being attacked is uncharged, steric hindrance guides the nucleophile to the less substituted carbon of the ring. The nucleophile attacks from the bottom of the epoxide (i.e., opposite the epoxide O atom), forcing the H and CH_3 substituents upward.

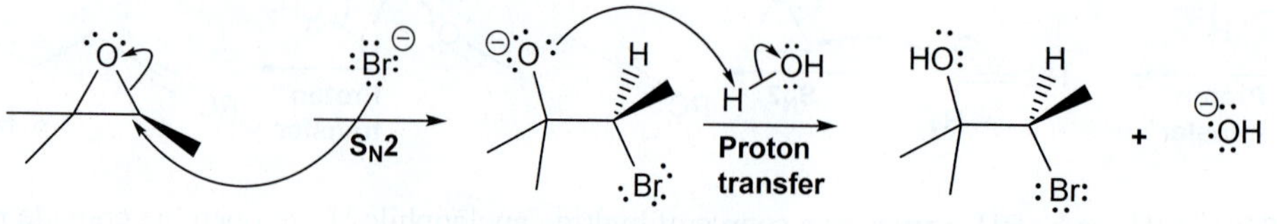

S_N2 — Proton transfer

(b) Under acidic conditions, however, strong bases are not allowed, so we may *not* include species with a localized negative charge on O. Instead, the mechanism in Equation 10-58 takes place, as shown below. Unlike in **(a)**, the nucleophile attacks the more highly alkyl substituted C atom of the epoxide, because the positive charge from the O atom is more delocalized on that C. Stereochemistry is not an issue in this reaction, because the C that is attacked is not a chiral center.

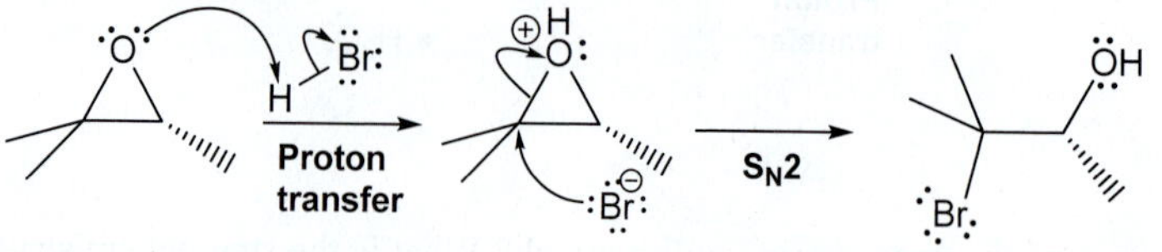

Proton transfer — S_N2

Problem 10.27

Think

How many leaving groups are present? In the presence of a strong base, what reaction mechanism will take place that involves an alkyl halide? How many times will this reaction take place? What functional group results? Are proton transfer reactions involved?

Solve

Each of these reactions is a back-to-back E2 set of reactions.

The reaction in Equation 10-67 requires three equivalents of base, because the initial product is a terminal alkyne that is deprotonated by NH_2^-, then H_2O is added in an acid workup step to replenish the H^+.

Equation 10-67

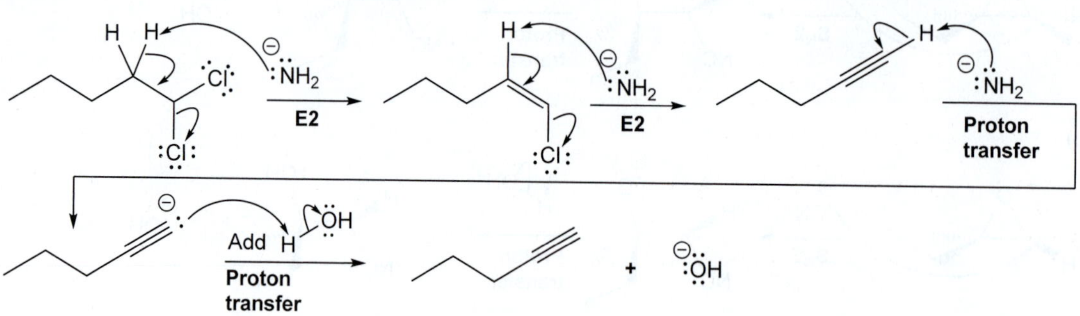

E2 — E2 — Proton transfer — Add H_2O Proton transfer

For Equation 10-68, the Br on the right is shown to be eliminated first, but it is also acceptable if the other Br is eliminated first.

Equation 10-68

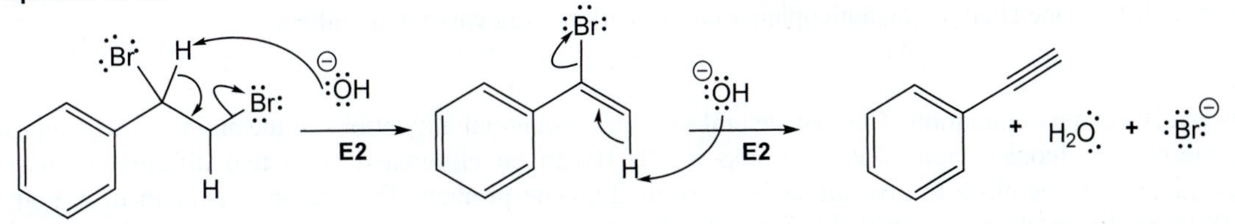

E2 E2 + H_2O + Br^-

Problem 10.28

Think

How many leaving groups are present? In the presence of a strong base, what reaction mechanism will take place that involves an alkyl halide? How many times will this reaction take place? What functional group results? Are proton transfer reactions involved?

Solve

In each problem, three equivalents of H_2N^- are required, because after the first two are used for E2 steps, the terminal alkyne produced is quickly deprotonated by the strong base. Acid workup is necessary to replenish the proton.

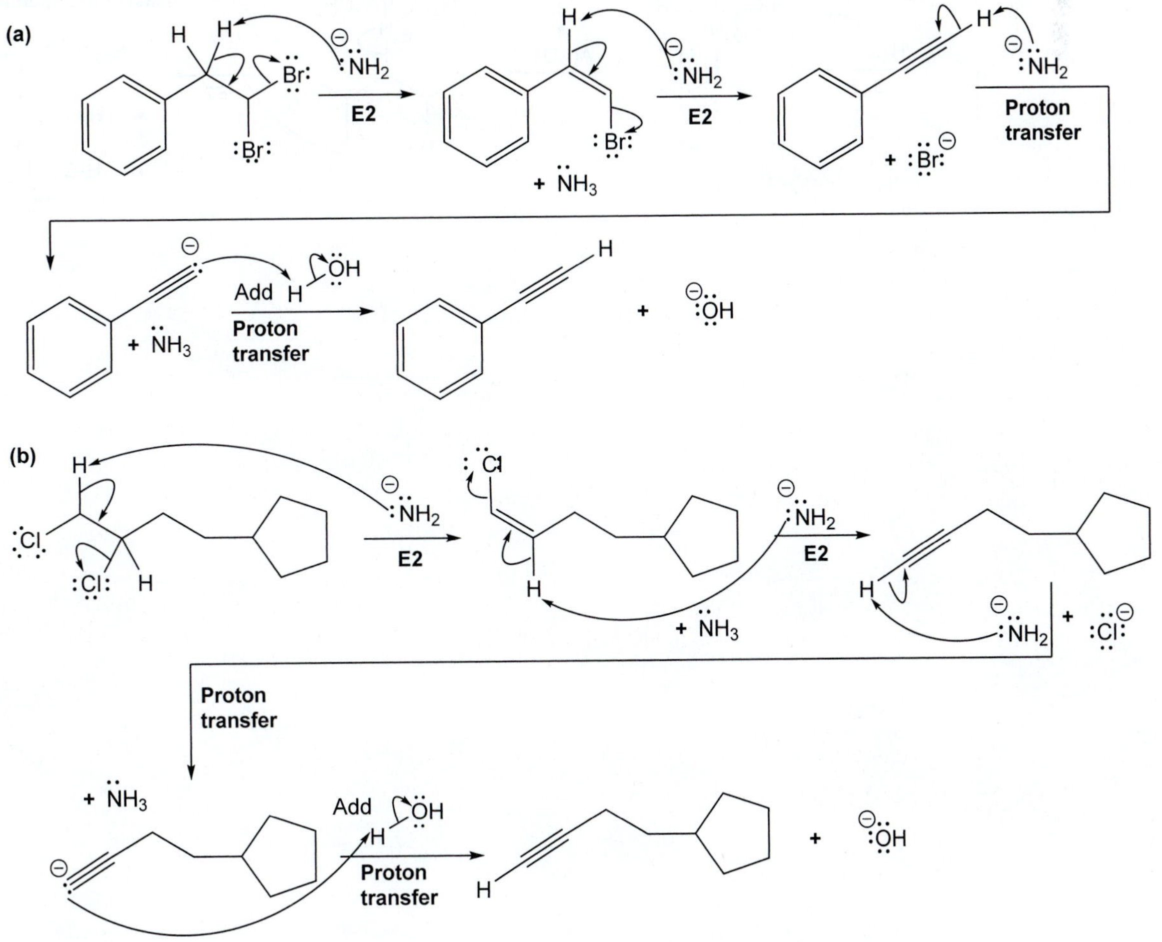

(a)

E2 + NH_3 E2 + Br^- Proton transfer

+ NH_3 Add H–OH Proton transfer + ^-OH

(b)

E2 + NH_3 E2 + Cl^-

Proton transfer

+ NH_3 Add H–OH Proton transfer + ^-OH

Problem 10.29

Think

Is there a good leaving group? In the presence of excess CH_3Br, what results on the N atom? Is there now a good leaving group? A strong base in the presence of heat results in what major reaction? Which H atom is eliminated? Can one H attain the anticoplanar conformation more easily than others?

Solve

This is a Hofmann elimination. The first several steps are sequential alkylations of the amine N, which produces a quaternary ammonium ion. The final step is E2. H can be eliminated from two different C atoms but predominantly takes place to give the least substituted alkene product. The reason is explained in Your Turn 10.18. *Note:* By-products are omitted in Steps 2–4 for clarity.

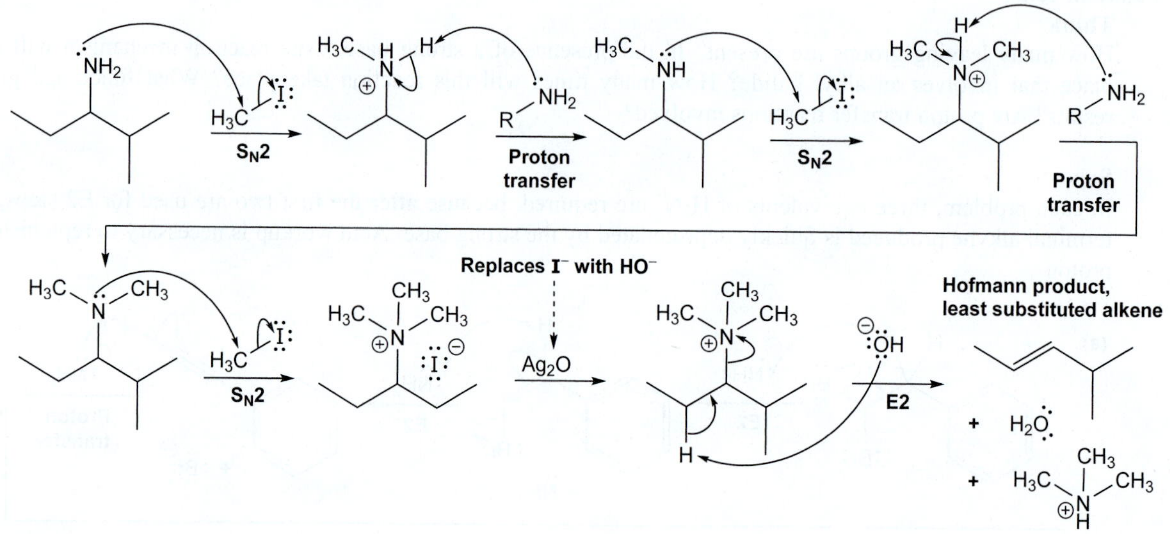

S_N2 · Proton transfer · S_N2 · Proton transfer

S_N2 · Replaces I^- with HO^- · Ag_2O · E2

Hofmann product, least substituted alkene

+ H_2O

+ $(CH_3)_3NH^+$

End of Chapter Problems

Section 10.1 Converting Alcohols to Alkyl Halides Using PBr_3 and PCl_3

Problem 10.30

Think

Is a good leaving group present? On what type of carbon is the leaving group attached? What are the key electron-rich and electron-poor species? Is substitution or elimination favored? Is a unimolecular or bimolecular mechanism favored? Are proton transfer reactions involved?

Solve

HO^- is not a good leaving group, but an S_N2 reaction with PBr_3 will turn HO^- into a $HOPBr_2$ leaving group (a good leaving group). The Br^- then attacks the electrophilic carbon via an S_N2 reaction and inverts the stereochemistry. NC^- is a good nucleophile, and another S_N2 reaction results, which also inverts the stereochemistry at the electrophilic carbon.

S_N2 S_N2 S_N2 + $HOPBr_2$ + Br^-

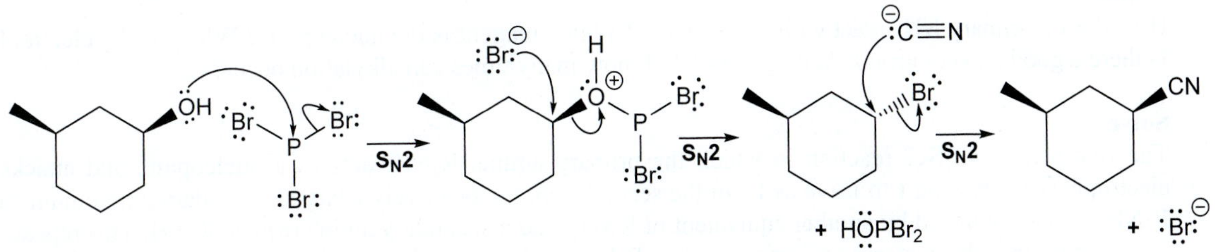

Problem 10.31 (SYN)

Think

Is HO^- a good leaving group? How does PCl_3 or PBr_3 turn HO^- into a good leaving group? What two elementary steps are involved in the reaction? What happens to the stereochemical configuration at the electrophilic carbon?

Solve

Hydroxide is not a good leaving group, and PCl_3 turns the HO^- leaving group into a $HOPCl_2$ leaving group, which leaves as an uncharged compound. The reaction occurs via two back-to-back S_N2 reactions. The first S_N2 reaction breaks the P–Cl bond and forms the O–P bond. The second S_N2 reaction breaks the C–O bond, forms the C–Cl bond, and inverts the stereochemistry of the carbon.

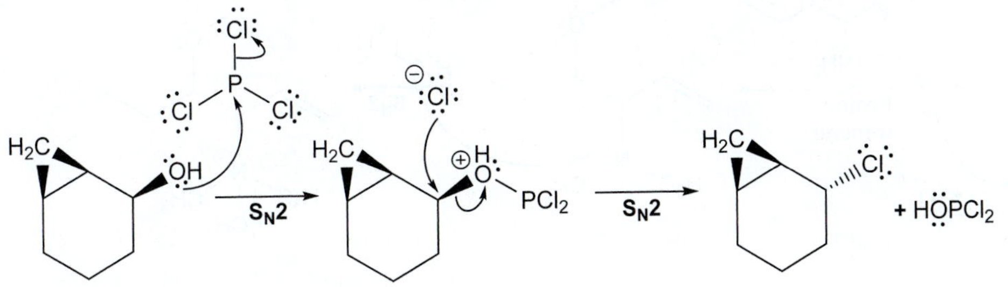

S_N2 S_N2 + $HOPCl_2$

Problem 10.32 (SYN)

Think

How do you transform an R–OH into an R–Cl? Are there any other competing reactions? How can you design your synthesis to avoid reaction with the two other OH groups?

Solve

An R–OH can be transformed into an R–Cl by treatment with either HCl or PCl_3. There are three OH groups in the starting material, and to carry out a successful synthesis, the 3° C–OH and Ph–OH must not react with the reagent. Treatment with HCl, therefore, will not work. PCl_3 reacts with an R–OH to form an R–Cl through two back-to-back S_N2 elementary steps. Only the primary C–OH will react readily with PCl_3. The mechanism is shown on the next page.

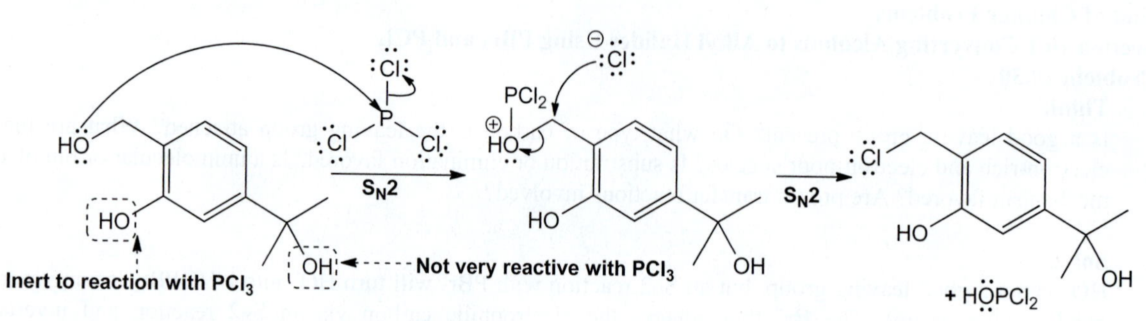

Inert to reaction with PCl_3

Not very reactive with PCl_3

S_N2

S_N2

$+ HOPCl_2$

Section 10.2 Alkylation of Ammonia and Amines

Problem 10.33

Think

How does a primary NH_2 react with excess CH_3I? Which reagent is the nucleophile? Which is the electrophile? Is there a good leaving group? With excess CH_3I, how many times can alkylation occur?

Solve

The first step is an S_N2 reaction in which the primary amine, R_2NH, acts as a nucleophile and attacks the electrophilic C, causing I to leave as I^-. In the second step, the positively charged secondary ammonium cation $R_2NH_2^+$ is deprotonated by another equivalent of RNH_2, and a secondary amine results: R_2NH. This repeats two more times, until the quaternary ammonium ion, R_4N^+, results as the final product.

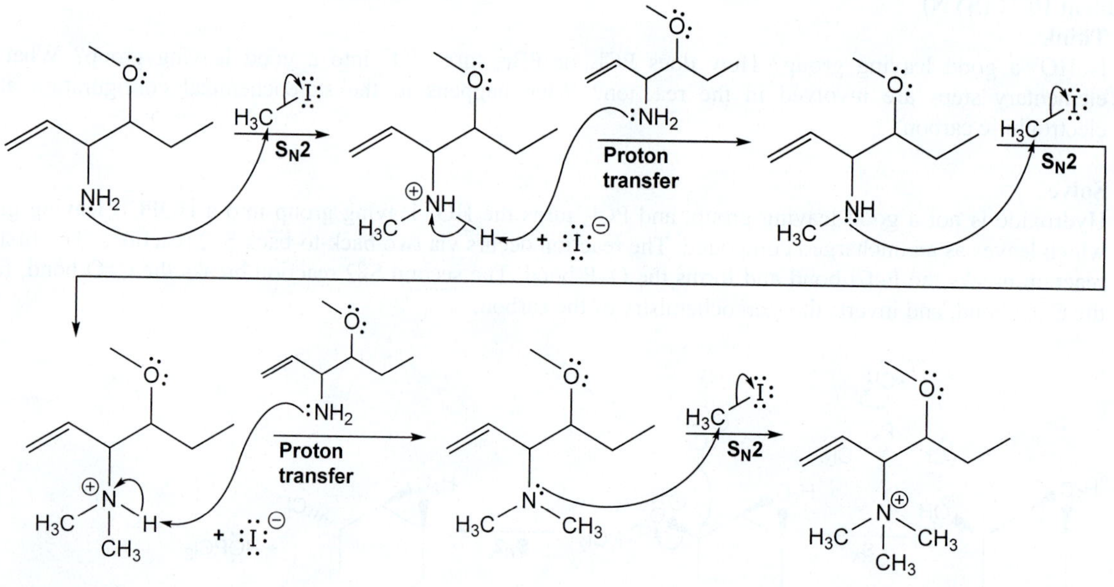

S_N2

Proton transfer

S_N2

Proton transfer

S_N2

Problem 10.34

Think

What is the formula of the product? What is the formula of the starting material? What is the difference? What is the functional group in the product? By what method can this functional group form from reaction with CH_3I? If there is excess CH_3I, how many times does the reaction occur?

Solve

The formula of the product is $C_{12}H_{21}ClN^+$, and the formula of the starting material is $C_{10}H_{16}ClN$. The difference is C_2H_5. The product is a quaternary amine, R_4N^+. Excess CH_3I reacts with amines in alkylation reactions. With excess CH_3I, this reaction can occur several times. Looking at the product, there are two CH_3 groups attached to the N, which came from two alkylation reactions with CH_3I. The R_3NH^+ is deprotonated. The reagent is shown on the next page.

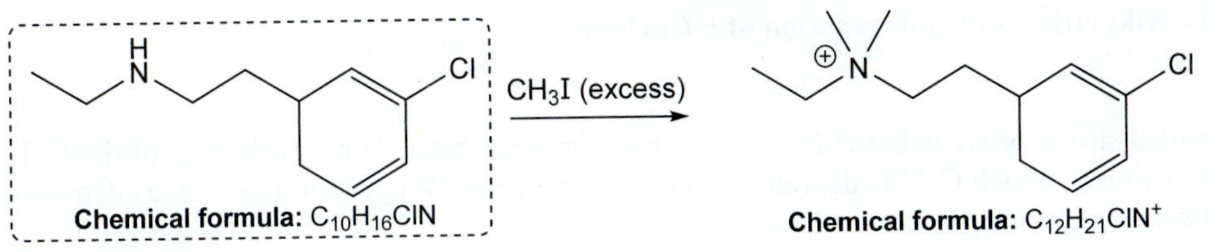

Chemical formula: $C_{10}H_{16}ClN$

CH_3I (excess)

Chemical formula: $C_{12}H_{21}ClN^+$

Problem 10.35 (SYN)

Think

What reagent and conditions will permit alkylation at the N atom to occur? How many times must this alkylation reaction occur? What must be true about the alkyl halide reagent to form a ring at the N?

Solve

Alkylation of the amine N takes place twice. To form a ring, the dibromide below is required.

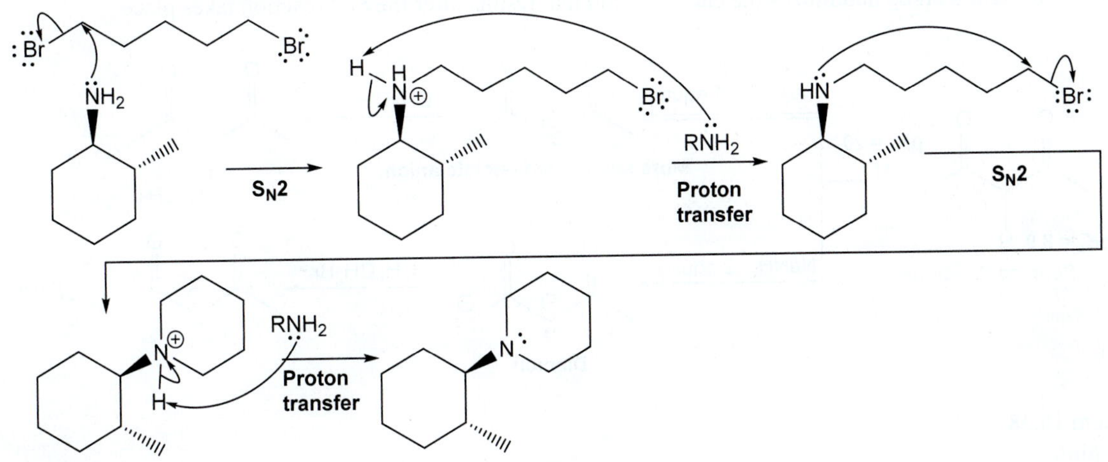

S_N2

RNH_2

Proton transfer

S_N2

RNH_2

Proton transfer

Problem 10.36

Think

What is the formula of the product? What is the formula of the starting material? What is the difference? What is the mechanism by which NH_3 reacts with an R–Cl? How many times can this alkylation reaction occur? If there are two Cl substituents on the substrate, what type of product forms?

Solve

The formula of the product is $C_7H_{15}N$, and the formula of the starting material is $C_7H_{14}Cl_2$. The difference is the addition of NH and the removal of two Cl atoms. The product is therefore a cyclic amine. Alkylation of NH_3 takes place twice to form a ring from the alkyl dichloride. The major organic product and mechanism are shown below.

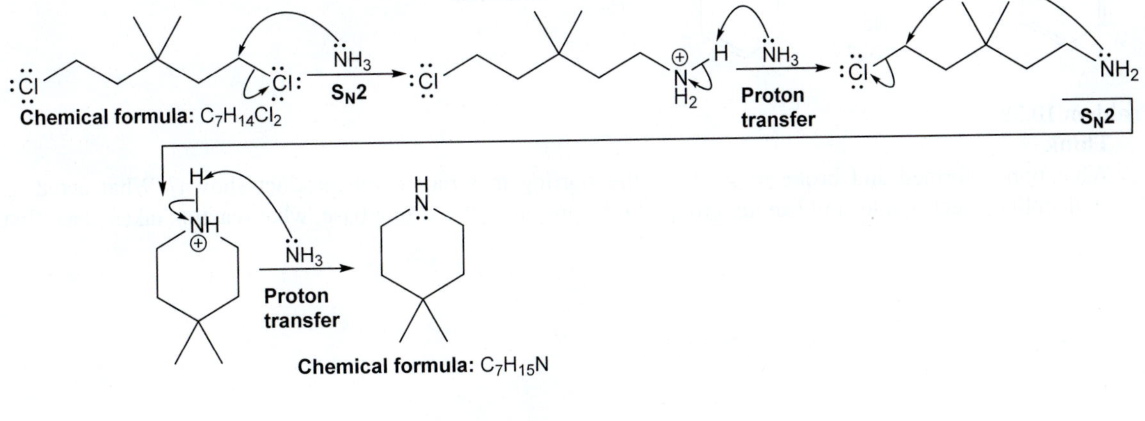

Chemical formula: $C_7H_{14}Cl_2$

NH_3

S_N2

NH_3

Proton transfer

S_N2

NH_3

Proton transfer

Chemical formula: $C_7H_{15}N$

Sections 10.3 and 10.4 Alkylation and Halogenation of α Carbons

Problem 10.37

Think

Is sodium carbonate a strong or weak base? Is H_2N^- a strong or weak base? What is the pK_a of the C–H of a monoketone and diketone? Which C–H is deprotonated in each reaction? Why? Will these bases deprotonate a monoketone or diketone reversibly or irreversibly?

Solve

Sodium carbonate is a weak base (the pK_a of HA is 6.3). The pK_a of a monoketone is about 20, and that of a diketone is 8.9 (Appendix A). Therefore, deprotonation takes place reversibly with sodium carbonate, producing the more highly substituted enoate anion. That proton is on the central carbon, which is why the subsequent alkylation takes place there. By contrast, H_2N^- is a *very* strong base and, with two equivalents, deprotonates both the central carbon and a terminal carbon irreversibly to produce the dianion. The terminal C is more nucleophilic, because it is less sterically hindered, and is the one that participates in the subsequent S_N2 reaction. Acid workup neutralizes the enolate anion that results after the S_N2 reaction takes place.

pK$_a$ = 20
pK$_a$ = 8.9 H
Pentane-2,4-dione
Na_2CO_3 (1 equiv)
More substituted enolate anion
CH_3CH_2Br
CH_2CH_3
$NaNH_2$ (2 equiv)
Dianion
CH_3CH_2Br
CH_2CH_3

Problem 10.38

Think

Under what reaction conditions does alkylation at the α carbon occur? As more alkyl groups are added to the α carbon, what happens to the sterics of that carbon? How does this inhibit multiple alkylation steps?

Solve

Once an alkyl group has been attached to the α carbon, it provides steric hindrance to further alkylation. Alkylations take place by the S_N2 mechanism, which is highly sensitive to steric hindrance, so each subsequent alkylation would take place more slowly.

CH_3
R-Br
NaOH
S_N2
R
$C H_2$
Greater steric hindrance

Problem 10.39

Think

What bonds formed and broke to go from the starting material to the product shown? What acted as the nucleophile, electrophile, and leaving group? In the presence of a strong base, what reaction takes place first?

Solve

Fundamentally, each cyclopropane ring is formed in the same way as an epoxide ring is formed under basic conditions. First, deprotonation produces a strong nucleophile (but in this case, at an α carbon instead of at an OH oxygen), and then the leaving group is displaced in an S_N2 reaction.

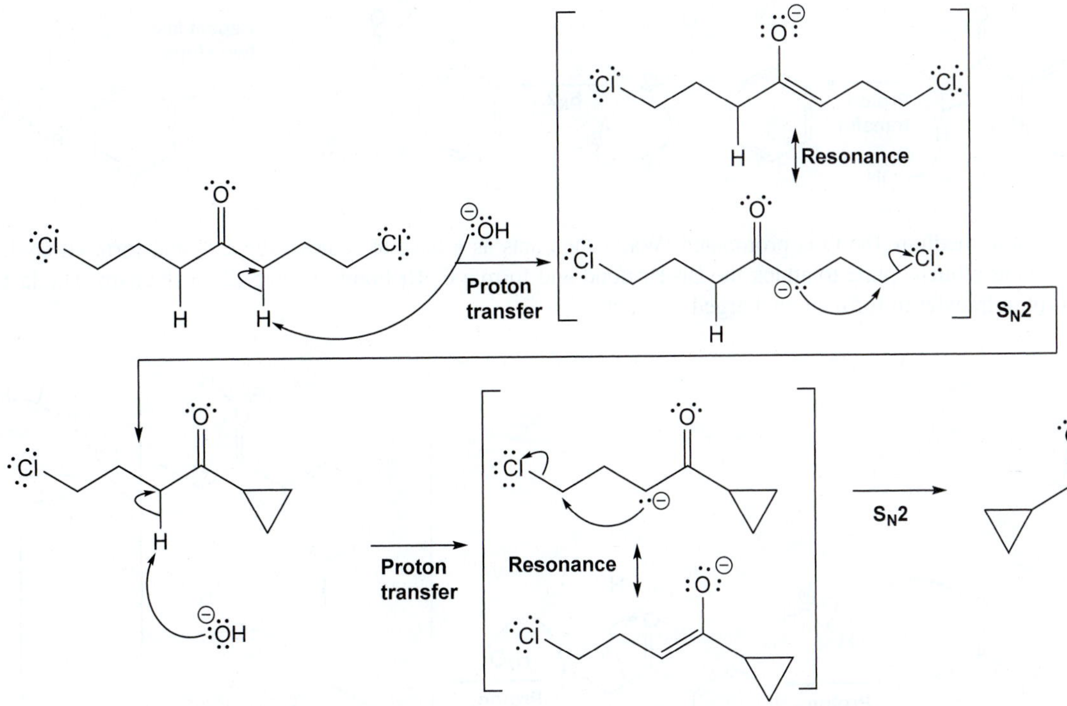

Proton transfer

Resonance

S_N2

Proton transfer

Resonance

S_N2

Problem 10.40

Think

How is this mechanism similar to the alkylation of the α carbon of a ketone? What does the NaH reagent form in the first step? What atom is alkylated?

Solve

This is the same mechanism as alkylation of a ketone. First, the NH proton is removed with the strong base, making the N atom strongly nucleophilic. Then, the nucleophile displaces Cl of the alkyl halide.

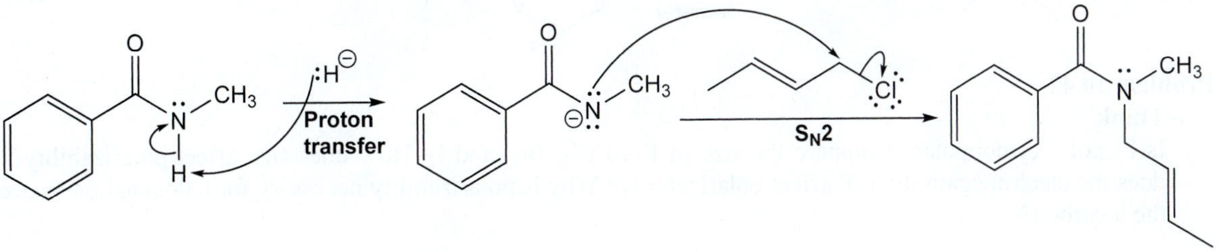

Proton transfer

S_N2

Problem 10.41

Think

Is a good leaving group present? On what type of carbon is the leaving group attached? What are the key electron-rich and electron-poor species? Is substitution or elimination favored? Is a unimolecular or bimolecular mechanism favored? Are proton transfer reactions involved? How does the nature of the media, acidic or basic, change the mechanism?

Solve

In a basic medium, the α H is deprotonated to form the enolate anion. The enolate anion acts as a nucleophile to attack the Br–Br bond and form a C–Br bond via an S_N2 mechanism. The last two steps are repeated to form another C–Br bond.

Proton transfer
S_N2
Repeat last two steps

In an acidic medium, the O is protonated. Water then acts as a base to remove the α H and form an enol. The enol acts as a nucleophile to attack the Br–Br bond and form a C–Br bond via an S_N2 mechanism. The last step is a proton transfer to form an uncharged product.

Proton transfer
Proton transfer
Enol form
S_N2
Proton transfer
Halogenation stops here.

Problem 10.42

Think

Is F_2 polar or nonpolar? Compare the size of F_2 to Cl_2, Br_2, and I_2. How does size affect polarizability? How does the electronegativity of F affect polarizability? Why is polarizability necessary for halogenation to occur at the α carbon?

Solve

The electron-rich to electron-poor driving force for such reactions originates from the induced dipole on the molecular halogen atom, as depicted in Figure 10-4. Because F_2 has relatively few electrons, and also because F is so highly electronegative, F_2 is not very polarizable. Therefore, the induced dipole on F_2 is relatively small, making the driving force for the reaction rather weak.

Section 10.5 Diazomethane Formation of Methyl Esters

Problem 10.43

Think

What is the Lewis structure for CH_2N_2? Is the carbon electron rich or electron poor? What site is basic on CH_2N_2? What site is acidic on the carboxylic acid? What proton transfer reaction takes place? After the proton transfer reaction, what S_N2 reaction can occur?

Solve

In diazomethane, the C atom bears a −1 formal charge and, therefore, is electron rich and basic. In Step 1, the C atom of diazomethane is protonated by the acidic proton of the carboxylic acid. In Step 2, the carboxylate anion is electron rich and acts as the nucleophile in the ensuing S_N2 reaction. The protonated form of $CH_3N_2^+$ is electron poor and acts as the substrate. $N_2(g)$ is a great leaving group. Each of these reactions converts a carboxylic acid into a methyl ester.

(a)

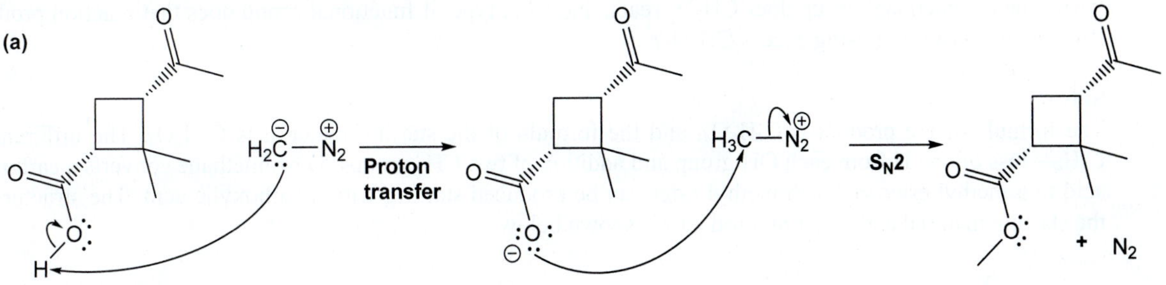

$H_2C^{\ominus}—N_2^{\oplus}$ → Proton transfer → $H_3C—N_2^{\oplus}$ → S_N2 → + N_2

(b)

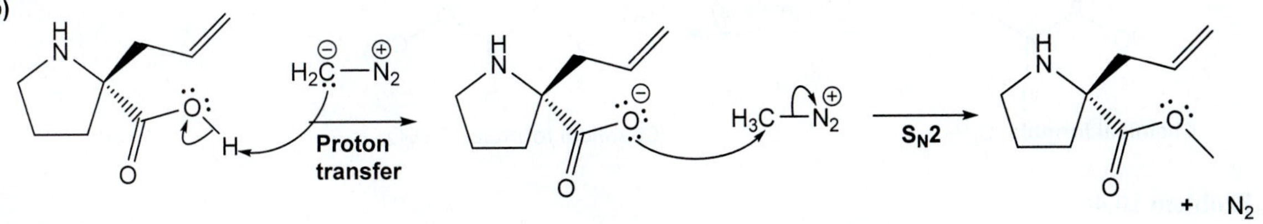

$H_2C^{\ominus}—N_2^{\oplus}$ → Proton transfer → $H_3C—N_2^{\oplus}$ → S_N2 → + N_2

Problem 10.44

Think

What functional group does diazomethane form when reacted with a carboxylic acid? Identify that functional group in each product listed. What bond formed? What carboxylic acid formed the product shown?

Solve

Diazomethane converts a carboxylic acid to a methyl ester. More specifically, the OH of a carboxylic acid is converted to OCH_3 (boxed). So each methyl ester can be produced starting with the corresponding carboxylic acid. Reactions **(a)** and **(b)** are shown below and reaction **(c)** is on the next page.

(a) $\xrightarrow{CH_2N_2}$ (OCH_3 boxed)

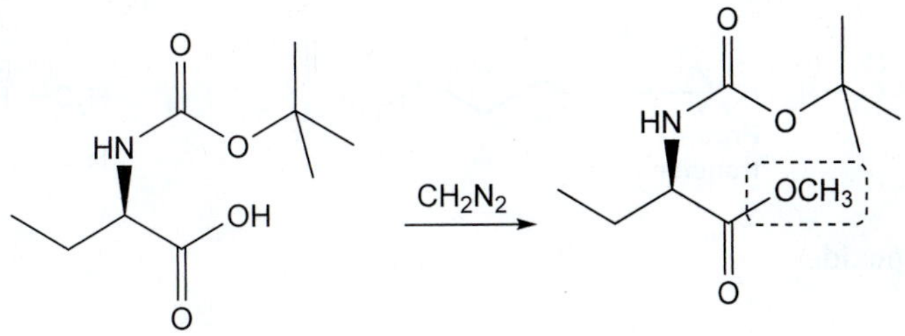

(b) $\xrightarrow{CH_2N_2}$ ($O—CH_3$ boxed)

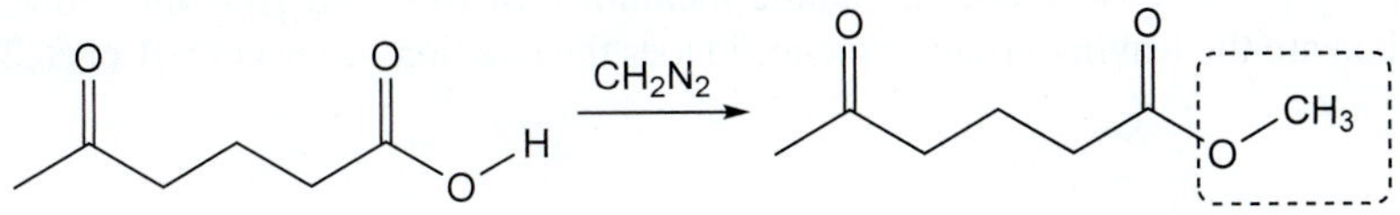

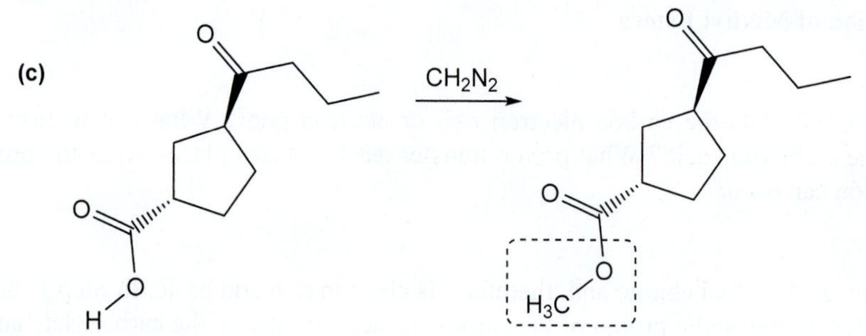

(c) CH_2N_2

Problem 10.45 (SYN)

Think

What is the formula of the product? What is the formula of the starting material? What is the difference? With what type of functional group does CH_2N_2 react, and what type of functional group does that reaction produce? What is the purpose of using excess CH_2N_2?

Solve

The formula of the product is $C_4H_6O_4$, and the formula of the starting material is $C_2H_2O_4$. The difference is C_2H_4—loss of one H from each OH group and addition of two CH_3 groups. Diazomethane converts a carboxylic acid to a methyl ester. So each methyl ester can be produced starting with a carboxylic acid. The structures of the starting material and resulting product are shown below.

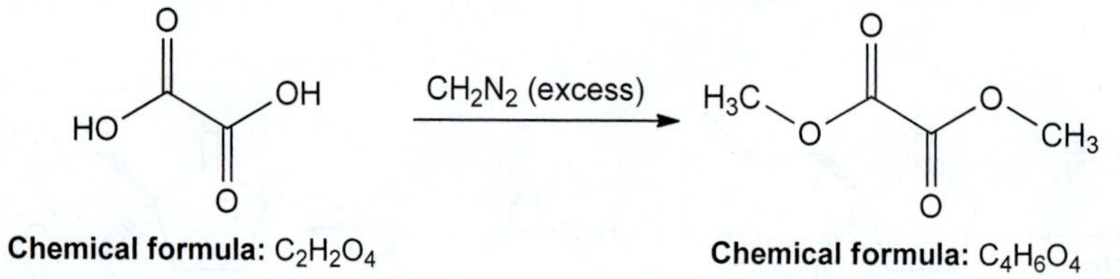

CH_2N_2 (excess)

Chemical formula: $C_2H_2O_4$ **Chemical formula:** $C_4H_6O_4$

Problem 10.46

Think

With what type of functional group does diazomethane react? What is the typical pK_a value of that functional group? How does that compare to the pK_a value of the given compound? What key step of the mechanism would be compromised if the pK_a value were excessively high?

Solve

Diazomethane converts a carboxylic acid to a methyl ester. A key step in the mechanism is the deprotonation of the carboxylic acid, which is possible because a carboxylic acid is moderately acidic (pK_a ~ 5). Given a pK_a of 25 and a formula with one IHD and two O atoms, the given compound is likely an ester (see the structure below). The key proton transfer does not take place with an ester because the proton is not acidic enough.

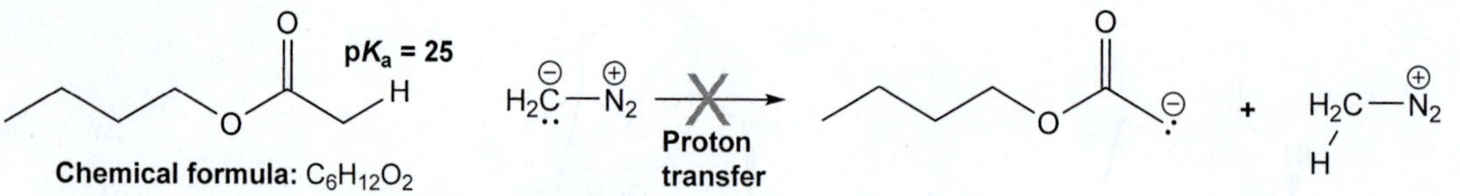

pK_a = 25 **Proton transfer**

Chemical formula: $C_6H_{12}O_2$

Section 10.6 Formation of Ethers and Epoxides

Problem 10.47

Think

Are RO^- and HO^- good leaving groups? How does an acidic medium turn RO^- and HO^- into good leaving groups? On what type of carbon are the leaving groups attached? Does the reaction occur via S_N1 or S_N2?

Solve

All steps are equilibria. Protonation of the O atom makes a good leaving group, ROH. That leaving group is replaced by a CH_3CH_2OH nucleophile in an S_N2 step because the leaving group is on a primary carbon. After two more proton transfers, H_2O is produced as a leaving group, which is then displaced by a second CH_3CH_2OH nucleophile in an S_N2 step. A final proton transfer produces the uncharged product.

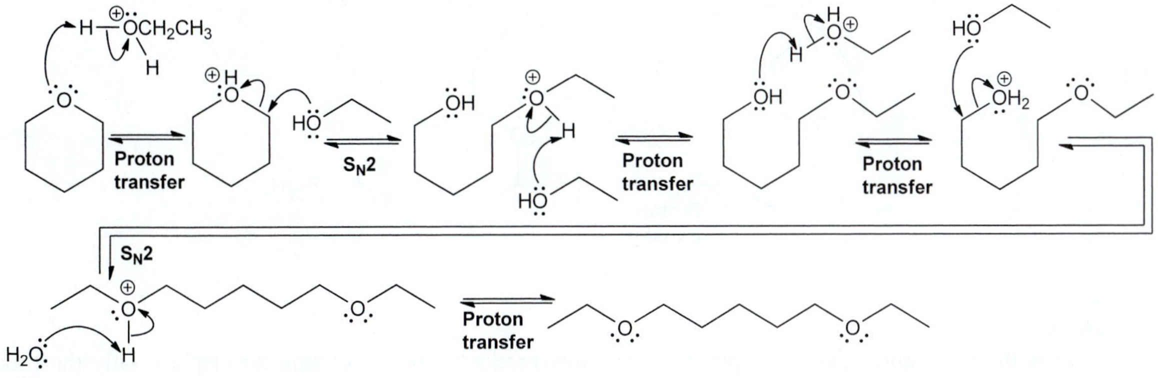

Proton transfer; S_N2; Proton transfer; Proton transfer; S_N2; Proton transfer

Problem 10.48

Think

What is the mechanism by which alcohols form ethers in acidic media? If two alcohols are mixed together in acidic media, is R–O–R′ the only ether possible? What is in solution that can act as a nucleophile? What is in solution that can act as a substrate with a sufficient leaving group?

Solve

In addition to (1-ethylpropoxy)cyclopentane, the desired unsymmetric ether product, two symmetric ethers are produced by self-reaction among the alcohols: cyclopentoxycyclopentane and 3-(1-ethylpropoxy)pentane. The reactions occur via S_N1, and the mechanisms for each are shown below and on the next page.

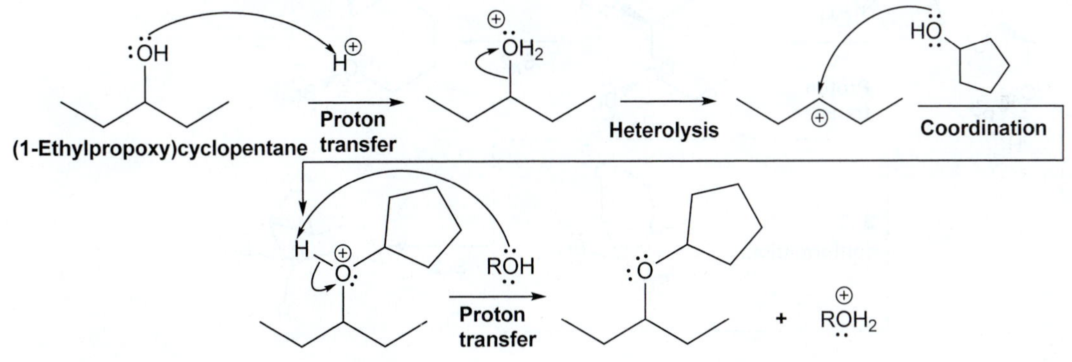

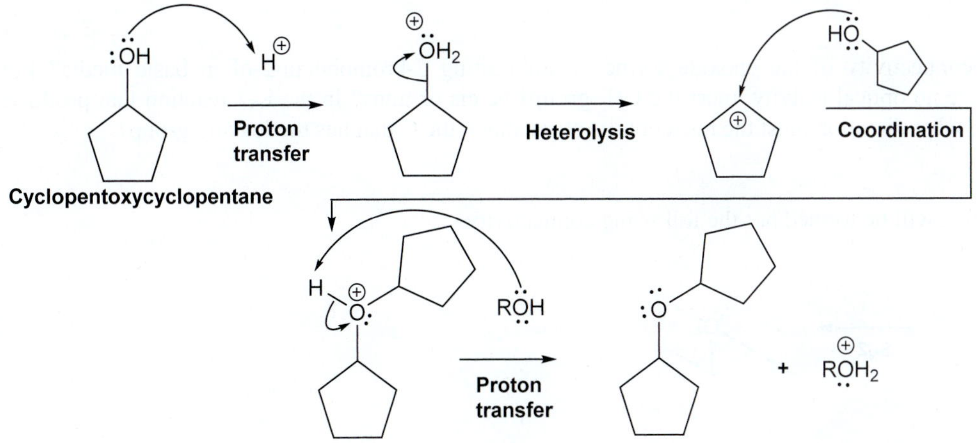

(1-Ethylpropoxy)cyclopentane

Proton transfer; Heterolysis; Coordination; Proton transfer; + ROH_2^+

Cyclopentoxycyclopentane

Proton transfer; Heterolysis; Coordination; Proton transfer; + ROH_2^+

:ÖH
H ⊕
⊕ OH2
HO
3-(1-Ethylpropoxy)pentane
Proton transfer
Heterolysis
Coordination
H ⊕ O
ROH
Proton transfer
+ ROH2

Problem 10.49

Think

What is the mechanism that takes place to form this product? For the O atom to displace only the bromide attached to the labeled ^{13}C, what must be true about the conformation of the O and the Br?

Solve

For O to displace only the Br on the unlabeled carbon, and not the other bromide, an S_N2 reaction must take place, not an S_N1 reaction. For an S_N2 reaction, the molecule must attain a conformation that allows backside attack on the unlabeled carbon. Thus, the Br atom that is lost must be trans to the O atom. The Br on the labeled C must be cis to the O atom to prevent the reaction from occurring there. Either the stereoisomer shown or its enantiomer is correct.

:Br:
O
H
^{13}C
Br
H
⊖ OH
Δ
Proton transfer
Nucleophilic O attacks from the backside.
S_N2
S_N2 conformation:
S_N2

Problem 10.50

Think

What is the connectivity of the epoxide formed from heating 3-bromobutan-2-ol in basic media? For the epoxide to have no optical activity, must the CH_3 groups be cis or trans? In the S_N2 reaction that produces the epoxide, from what direction must the nucleophilic atom attack the C that has the leaving group?

Solve

The epoxide that will be formed has the following connectivity:

:O:⊖
:Br:
S_N2
O

To be optically *inactive*, the epoxide must have a plane of symmetry, as shown below. One haloalcohol that can be used to produce the epoxide is shown below. To arrive at this precursor, we have to take into consideration the fact that, in an S_N2 step, the alkoxide nucleophile must attack the C–Br from the side opposite the Br leaving group.

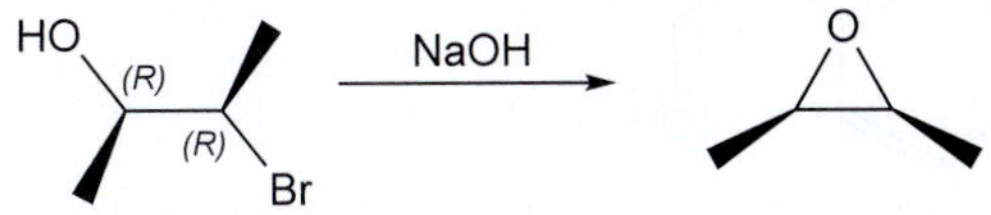

HO (R) (R) Br —NaOH→ O

The enantiomer can also be used.

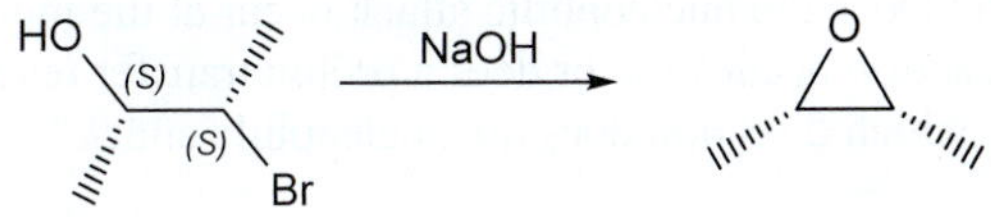

HO (S) (S) Br —NaOH→ O

Problem 10.51

Think

Does the cyclohexane ring exist as a planar structure or a chair? Which group on the ring is the bulkiest? On what position, axial or equatorial, should this group be positioned? How does that affect the OH and Br groups? Why is the S_N2 reaction shut down?

Solve

Formation of an epoxide via an S_N2 reaction is favored by the attack of the O^- from the side opposite the leaving group, Br. In this case, the *t*-butyl group effectively locks the Br and OH groups in equatorial positions because of the *t*-butyl group's bulkiness. In this configuration, the OH and Br groups are gauche to each other, making it impossible for attack to occur.

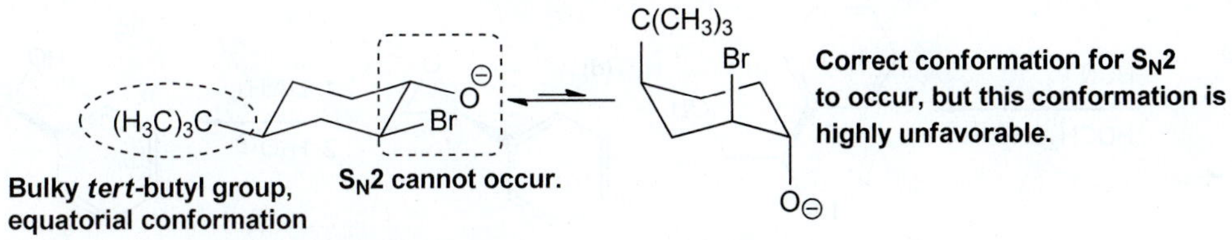

(H3C)3C — O⊖ — Br ⇌ C(CH3)3, Br, O⊖

Bulky *tert*-butyl group, equatorial conformation

S_N2 cannot occur.

Correct conformation for S_N2 to occur, but this conformation is highly unfavorable.

Section 10.7 Epoxides and Oxetanes as Substrates

Problem 10.52

Think

Is a good leaving group present? On what type of carbon is the leaving group attached? What are the key electron-rich and electron-poor species? Is substitution or elimination favored? Is a unimolecular or bimolecular mechanism favored? Are proton transfer reactions involved? How does the nature of the media, acidic or basic, change the mechanism?

Solve

In a neutral reaction medium, the less substituted carbon is attacked by the Cl^- to open the epoxide ring. In the presence of a weakly acidic solvent (water), the alcohol is the final product.

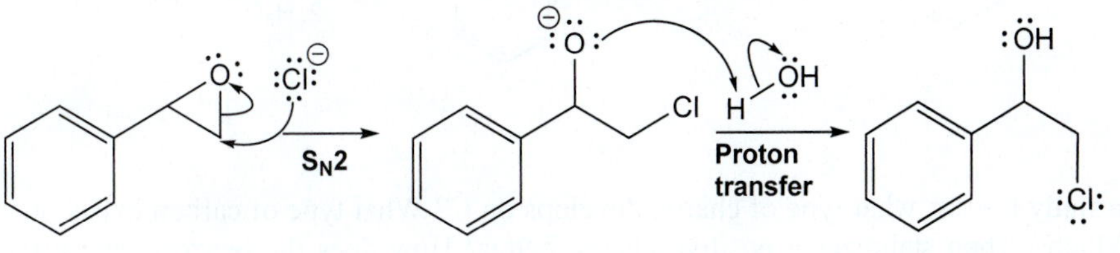

:Cl:⊖ —S_N2→ O⊖, Cl, H–OH —Proton transfer→ :OH, :Cl:

In an acidic reaction medium, the epoxide O is protonated, and the more substituted carbon is attacked by the Cl^-.

Proton transfer

S_N2

Problem 10.53

Think

Does the reaction occur in a basic, neutral, or acidic medium? Does the nucleophilic attack occur at the more or less substituted carbon? Does the nucleophilic substitution reaction occur first, or does a proton transfer reaction occur first? When nucleophilic substitution does occur, from which direction does the nucleophile attack?

Solve

When nucleophilic attack takes place under neutral or basic conditions, the least sterically hindered C is attacked, which is the one on the right. This is the case for **(a)**, **(b)**, **(d)**, **(f)**, **(g)**, and **(h)**. Under acidic conditions, the O atom is protonated first, and the carbon that can best handle a positive charge is the one that is attacked subsequently. This is the benzylic carbon, as shown in **(c)** and **(e)**. In all cases, the nucleophile attacks the epoxide C from the side opposite the C–O bond to open the ring.

(a) NaN_3, H_2O → product with HO, N_3

(b) KCN, $HOCH_2CH_2OH$ → product with HO, CN

(c) HCN, $HOCH_2CH_2OH$ → product with OH, NC

(d) 1. $LiAlH_4$ 2. $H_3O^{\oplus}$ → product with HO, H

(e) CH_3OH, H_2SO_4 → product with OH, H_3CO

(f) CH_3ONa, CH_3OH → product with HO, OCH_3

(g) $C_6H_5CH_2NH_2$, CH_3OH → product with HO, $HN—CH_2C_6H_5$

(h) 1. CH_3Li 2. $H_3O^{\oplus}$ → product with HO, CH_3

Problem 10.54

Think

When the C–O bond partially breaks, what type of charge develops on C? What type of carbon exists on either side of the epoxide? Which carbon stabilizes a positive charge better? How does the benzene ring affect the charge stability?

Solve

The longer C–O bond is the one that involves the carbon that can better accommodate a positive charge, because a partial breaking of the C–O bond places a partial positive charge on C. Both carbons are secondary, but the one on the left is a benzylic carbon as well. So the C–O bond on the left is longer.

Longer bond

Benzylic C, better C^+ charge stability, weaker C–O bond

Problem 10.55

Think

Under acidic conditions, what elementary step takes place first? What nucleophile is produced as a result? Does nucleophilic attack occur at the less sterically hindered C or at the C that can better stabilize a positive charge? How does the allylic C help stabilize the positive charge?

Solve

The epoxide O is protonated first under these acidic conditions. Then, nucleophilic attack takes place at the C that can better stabilize a positive charge. The one on the left is allylic. Similar to how allylic carbocations are stabilized by resonance delocalization of the positive charge, the partial positive charge, δ^+, that builds up is better handled on the allylic C here.

Proton transfer

S_N2

On allyl C, a positive charge is better stabilized.

Problem 10.56

Think

Does the reaction occur in a basic, neutral, or acidic medium? Does the nucleophilic attack occur at the more or less substituted carbon? Does the nucleophilic substitution reaction occur first, or does a proton transfer step occur first?

Solve

The conditions for nucleophilic attack are *not* acidic, so we should expect attack to take place at the less sterically hindered carbon of the aziridine ring. Notice that the proton transfer from S to N takes place intermolecularly instead of intramolecularly.

S_N2

Proton transfer

NHR_2

Proton transfer

Section 10.8 Generating Alkynes via Elimination Reactions

Problem 10.57

Think

How can you simplify $NaNH_2$? How many leaving groups are present? In the presence of a strong base, what reaction mechanism will take place involving a vinylic halide? How many times will this reaction take place? What functional group results? Are proton transfer reactions involved?

Solve

$NaNH_2$ is simplified to $^{-}NH_2$, a strong base, which removes the H and the leaving group on adjacent carbons in an E2 step.

(a) This is an E2 reaction, in which there is only one adjacent H. Therefore, only one E2 product is possible, and no proton transfer reactions are involved.

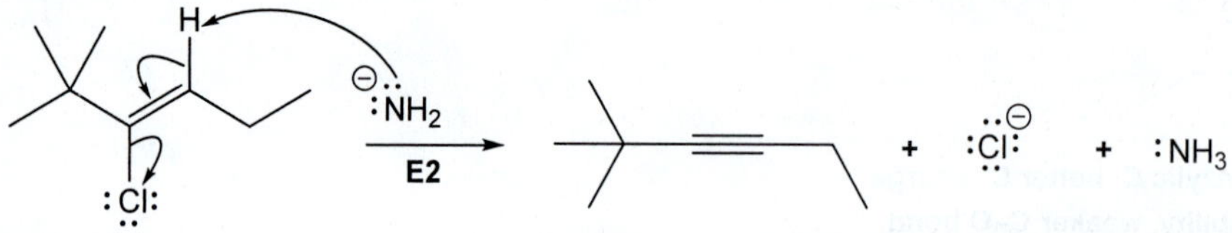

(b) This requires two equivalents of base, because the initial product is a terminal alkyne that is deprotonated by NH_2^-; then, H_2O is added in an acid workup step to replenish that H^+.

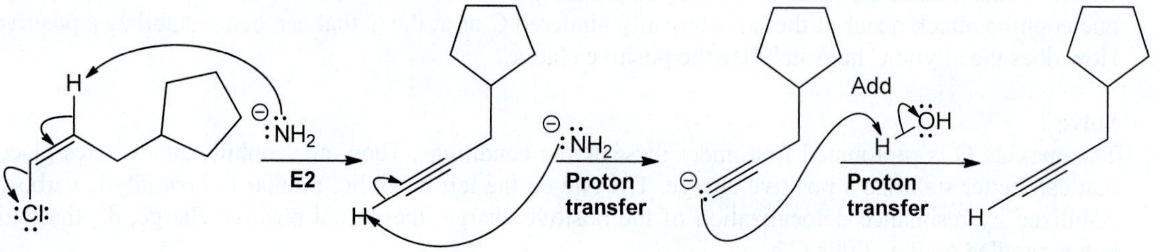

Problem 10.58

Think

What is the mechanism for each reaction? What are the stereochemical requirements for such a reaction?

Solve

The mechanism for each reaction is E2, as shown below. E2 reactions are favored when the H and the leaving group on adjacent atoms are anticoplanar. They are anticoplanar in the *Z* isomer but not in the *E* isomer. Therefore, the reaction is more difficult for the *E* isomer, which is why much higher temperatures are required.

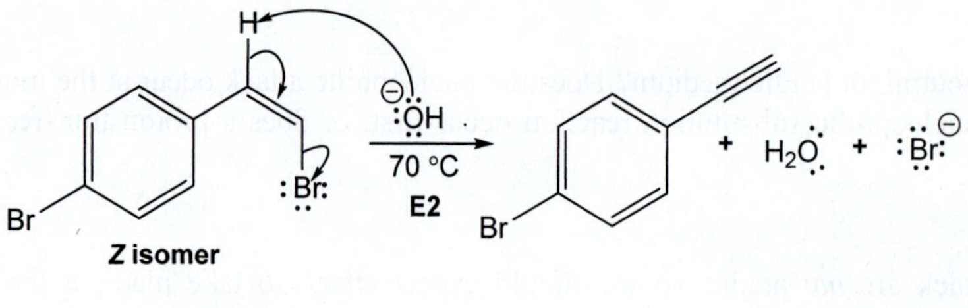

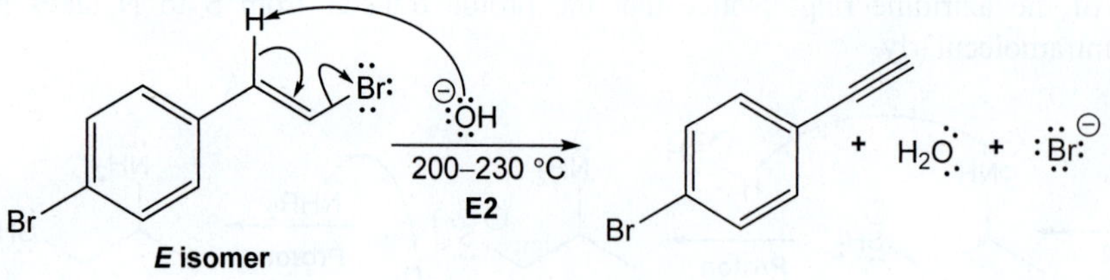

Problem 10.59

Think

How many E2 reactions take place for a vicinal dihalide in the presence of a $NaNH_2$? What is at the end of a terminal alkyne that is not present in an internal alkyne? What reaction takes place between $NaNH_2$ and a terminal alkyne?

Solve

Vicinal dihalide → Internal alkyne

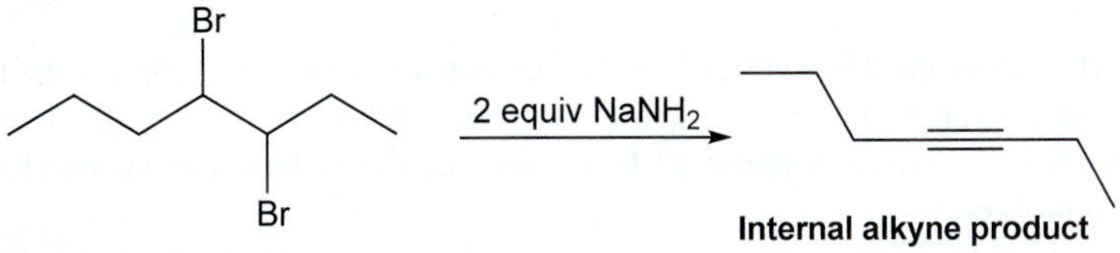

Br

2 equiv $NaNH_2$

Br

Internal alkyne product

One equivalent of $NaNH_2$ is used for each elimination of HBr, as shown below. Because there are two HBr eliminations, two equivalents of $NaNH_2$ are required.

General example

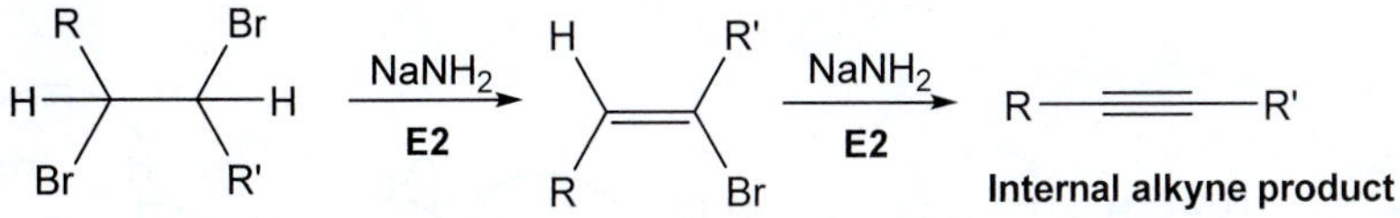

R Br H R'

$NaNH_2$ $NaNH_2$

H— —H **E2** **E2** R——R'

Br R' R Br **Internal alkyne product**

Vicinal dihalide → Terminal alkyne

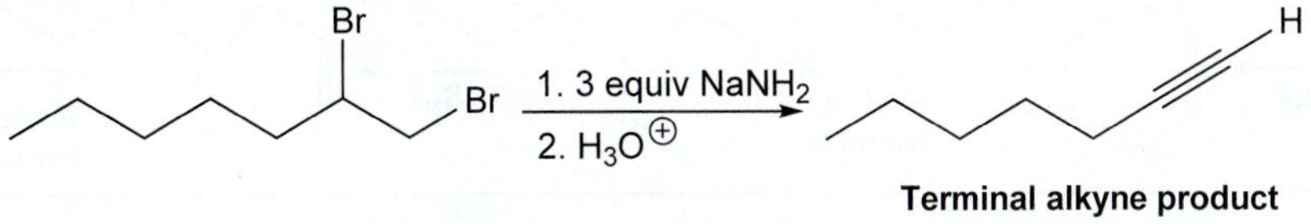

Br H

Br 1. 3 equiv $NaNH_2$

2. $H_3O^{\oplus}$

Terminal alkyne product

If the product is a terminal alkyne, an acidic H is present, whose pK_a is about 25. This is easily deprotonated by $NaNH_2$. Thus, a third equivalent of $NaNH_2$ is spent in such a deprotonation.

General example

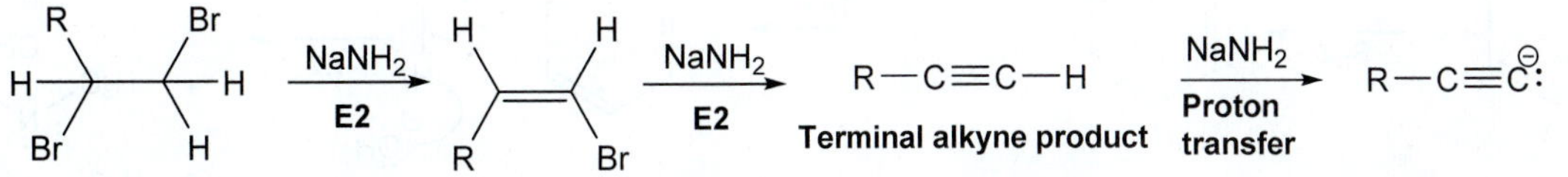

R Br H H

$NaNH_2$ $NaNH_2$ $NaNH_2$

H— —H **E2** **E2** R—C≡C—H **Proton transfer** R—C≡C:⊖

Br H R Br **Terminal alkyne product**

Problem 10.60

Think

What functional group results from this reaction? What is the geometry of the carbon atoms in that functional group? How does ring strain factor into the stability of the products? Will this increase or decrease the rate of product formation?

Solve

These reaction conditions normally convert an alkene into an alkyne via E2. The intended products for each reaction are shown below. In the case of the cyclic alkene reactant, a cyclic alkyne would be produced, and with only six atoms making up the ring, the structure is extremely strained. This will significantly decrease the rate of its formation.

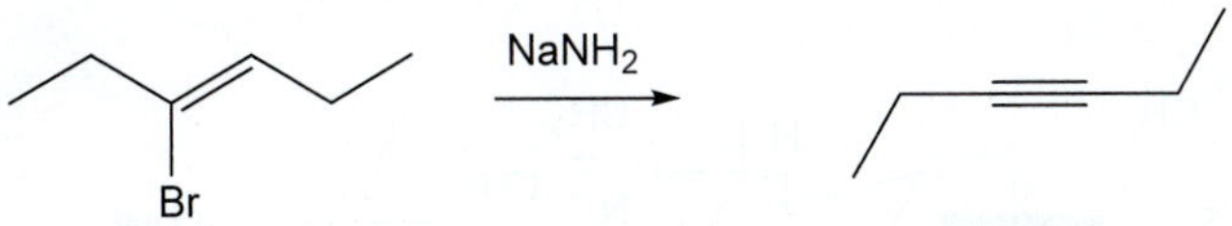

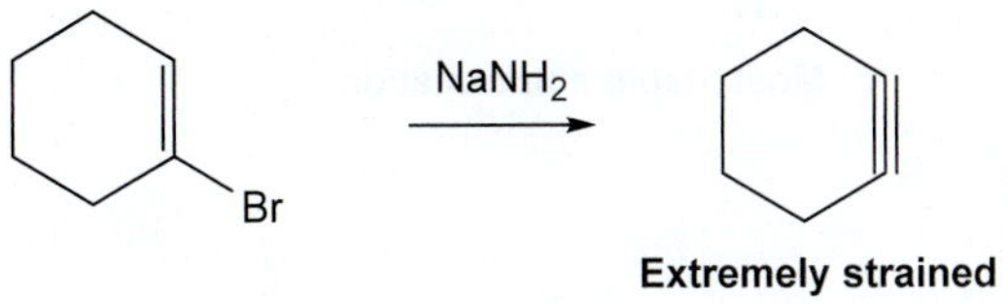

$NaNH_2$

Br

$NaNH_2$

Br

Extremely strained

Section 10.9 Hofmann Elimination

Problem 10.61

Think

Is a good leaving group present initially? How do the reaction conditions convert a poor leaving group into a suitable leaving group? On what type of carbon is the leaving group attached? What are the key electron-rich and electron-poor species? Is substitution or elimination favored? Is a unimolecular or bimolecular mechanism favored? Are proton transfer reactions involved?

Solve

This is a Hofmann elimination. The first several steps are sequential alkylations of the amine N, which produce a quaternary ammonium ion. The quaternary ammonium ion has a suitable leaving group. The final step is E2, which, owing to the excessive bulkiness of the leaving group, takes place to give the least substituted alkene product.

S_N2 | Proton transfer | S_N2 | Proton transfer | S_N2 | Replaces I^- with HO^- Ag_2O | E2 | + $(CH_3)_3N$ + H_2O

Problem 10.62

Think

Does a ring inhibit the nitrogen methylation step? What forms on the N atom in the presence of excess CH_3I? Is this sterically bulky? In a cyclohexane ring, what is the most stable conformation of a bulky group? How does this conformation limit/shut down E2?

Solve

Although the methylation of N proceeds normally (multiple S_N2 steps), in the resulting chair conformation of the ammonium ion, the trimethylammonium substituent is essentially locked in the equatorial position, just as a *t*-butyl group would be. In that conformation, no H atoms on adjacent C atoms can be anticoplanar to the N leaving group, so E2 is disfavored.

1. CH_3I (excess) 2. Ag_2O 3. Δ → E2 → SLOW

Most stable conformation

Problem 10.63

Think

What substrate is present initially? What nucleophile is present? What product forms from the reaction of a primary amine in the presence of excess CH_3I? What is the leaving group that is produced as a result of that reaction? A strong base and heat result in what reaction? How many times does this reaction occur?

Solve

This reaction is a double Hofmann elimination. One NH_2 group is alkylated by CH_3I three times to form a quaternary ammonium ion, then the other NH_2 group is alkylated three times to form a second quaternary ammonium ion. Next, Ag_2O introduces HO^-, which is a strong base, and both $N(CH_3)_3$ groups act as leaving groups. Back-to-back E2 steps produce an alkyne.

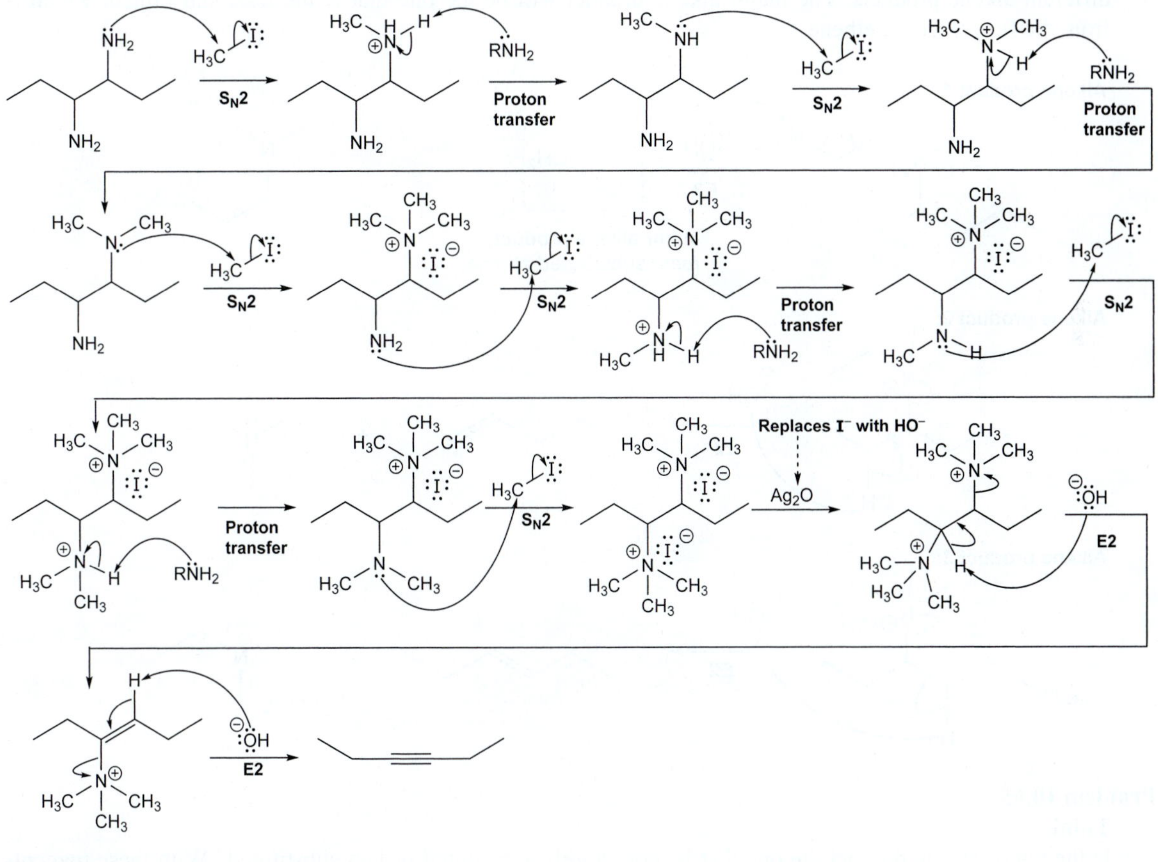

S_N2 Proton transfer S_N2 Proton transfer S_N2 S_N2 Proton transfer S_N2 Proton transfer S_N2 Replaces I⁻ with HO⁻ Ag_2O E2 E2

Problem 10.64

Think

What substrate is present initially? What nucleophile is present? What product forms from the reaction of a primary amine in the presence of excess CH_3I? What is the leaving group produced in that reaction? A strong base and heat result in what reaction? How many different adjacent H atoms are present that can be eliminated along with the leaving group? How many alkene products can result?

Solve

The first part of the Hofmann elimination is alkylation of the amine N to produce a quaternary ammonium ion.

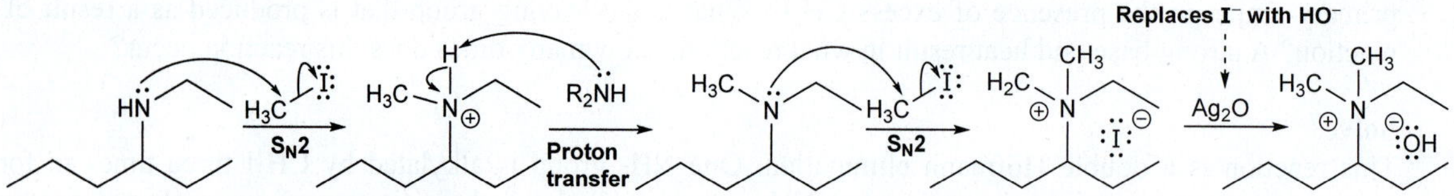

An E2 step can then take place by eliminating the N-containing leaving group along with a H atom on a C atom adjacent to the one attached to N. There are three such C atoms, which, as shown below, can lead to three different alkene products. The major alkene product will be the one that is the least substituted, which is that from the first reaction: ethene.

Alkene product 1:

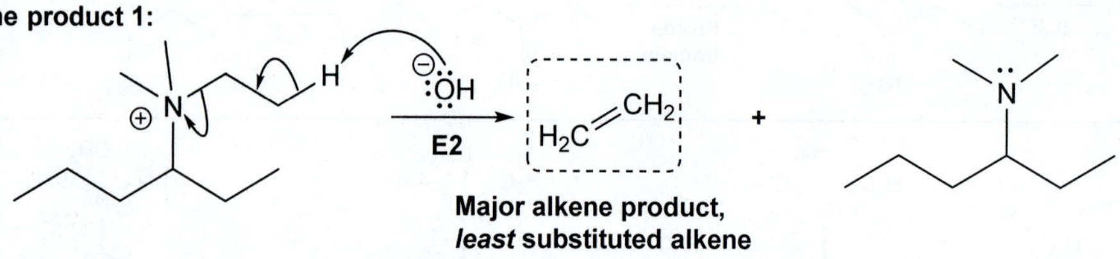

Major alkene product, *least* substituted alkene

Alkene product 2:

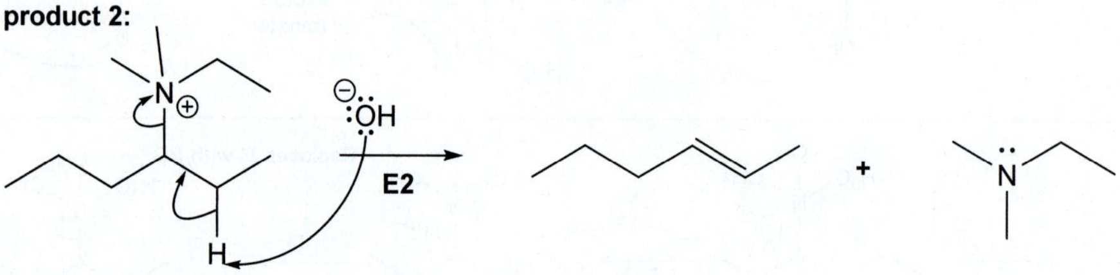

Alkene product 3:

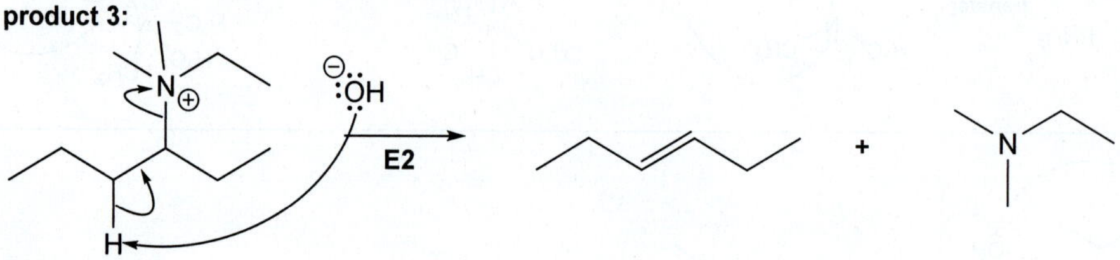

Problem 10.65

Think

Is the major alkene product the one that is more highly substituted or less substituted? With these reagents and conditions, what type of elimination occurs? Why are two alkene products possible? What is the leaving group? In the starting material, how many C atoms adjacent to the one with the leaving group have an attached H?

Solve

These are the conditions for a Hofmann elimination. The two alkene products that are given are the products of E2 elimination, and notice that the major product is the less highly substituted alkene. To give both of those products, H atoms must be eliminated from different C atoms. The precursor is given below with those H atoms drawn out explicitly.

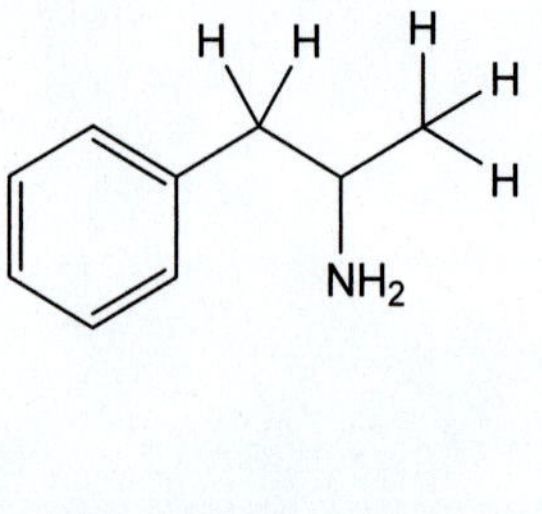

Problem 10.66

Think

With these reagents and conditions, what type of elimination occurs? What is the leaving group that is produced from the reaction of excess CH_3I and an amine? In the product of the first elimination, what functional group present will react with excess CH_3I?

Solve

Each of these provides the conditions necessary for Hofmann elimination. The quaternary ammonium ion intermediate through which each elimination proceeds is given below. Notice that the product of the first Hoffman elimination is an amine that can subsequently participate in another Hoffman elimination.

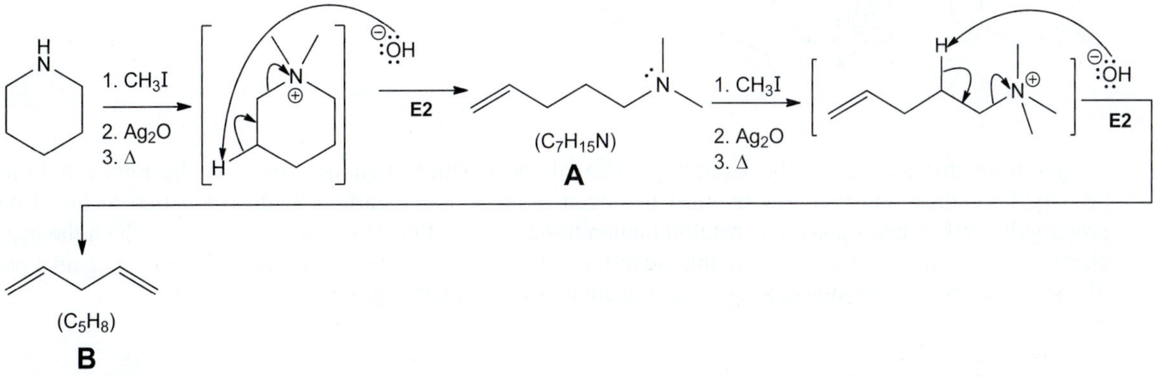

1. CH_3I 2. Ag_2O 3. Δ; E2; ($C_7H_{15}N$) **A**; 1. CH_3I 2. Ag_2O 3. Δ; E2; (C_5H_8) **B**

Integrated Problems

Problem 10.67 (SYN)

Think

To form a stereospecific product, is S_N2 or S_N1 required? In this example, is the stereochemistry inverted or retained? How many reactions need to take place?

Solve

To replace an atom stereospecifically, we need an S_N2 reaction, but S_N2 occurs with inversion. The solution is to do two inversions. First, treat the bromoalcohol with base to form an epoxide, then treat the epoxide with aqueous acid to open the ring.

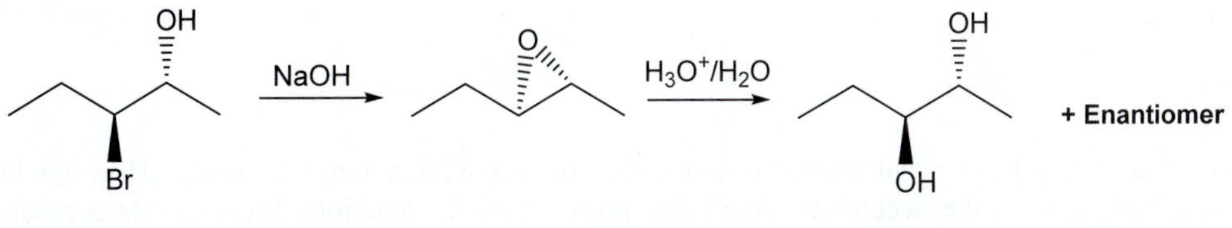

NaOH; H_3O^+/H_2O; **+ Enantiomer**

Problem 10.68

Think

What orientation of the leaving group and the adjacent H favors an E2 reaction? Can this occur in an epoxide? Is rotation limited in rings?

Solve

An E2 mechanism is favored when the proton on the one carbon and the leaving group on an adjacent carbon are anticoplanar, giving a 180° dihedral angle. The presence of the small epoxide ring forces a dihedral angle of 120° between these two groups, preventing the anticoplanar conformation, thus disfavoring elimination.

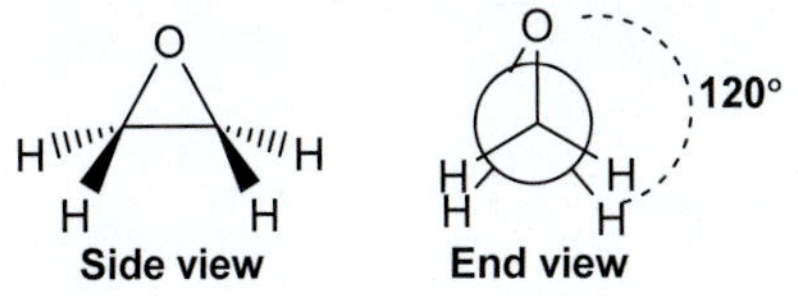

Side view **End view** 120°

Problem 10.69

Think

In the presence of a strong base, what reaction "normally" occurs with a vinylic chloride? How does the product of that reaction compare to the overall product that was given? Show the reversible deprotonation of the H on the propargylic carbon (C≡C–CH) in the product. Are there resonance structures that become important?

Solve

This would be the product resulting from a "normal" E2 reaction, with loss of H on C2 and chlorine on C1.

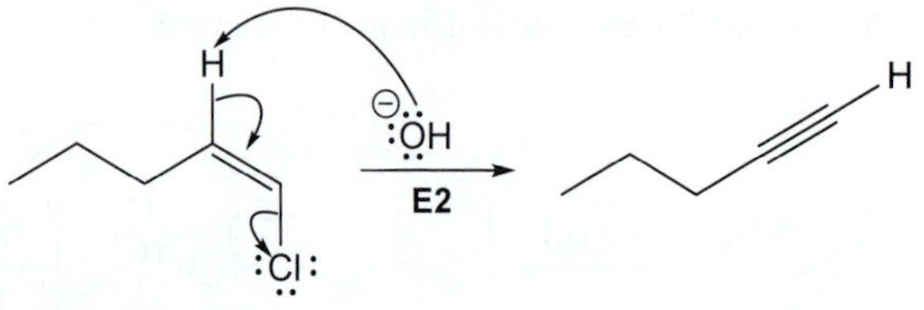

To get from this product to the overall product of the reaction, two protons must be removed from the propargylic carbon, and two protons must be added to the terminal carbon. If deprotonation of the H on the propargylic carbon takes place, the resulting anion has a contributing resonance structure in which the negative charge appears on the terminal C, making the terminal C basic. This allows the terminal carbon to gain a proton. The same two proton transfer steps can occur again to give the overall product.

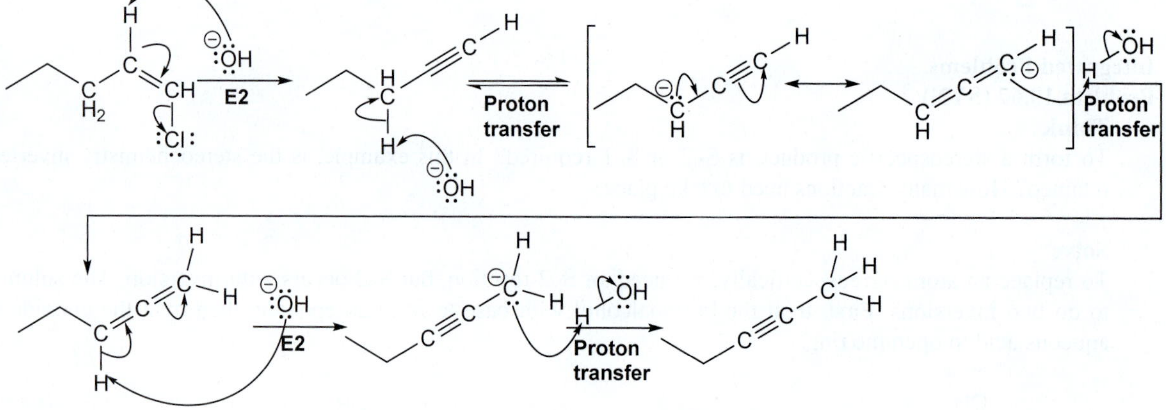

Problem 10.70

Think

In the presence of a strong base, what reaction "normally" occurs with a vinyl bromide? How could the deuterium be removed to form the second product? Are proton transfer reactions involved? Is a resonance structure possible in any of the intermediates?

Solve

The first product results from a "normal" E2 reaction, with loss of H on C2 and bromine on C3.

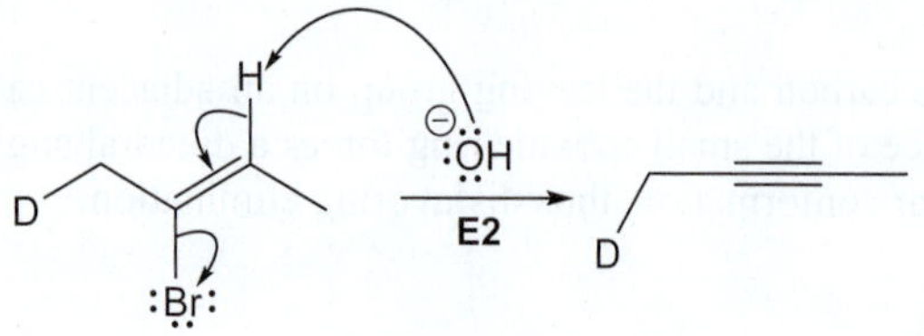

The second product could be the result of an acid–base exchange reaction, where H is traded for D on C1 (the propargylic carbon). This acid–base exchange is facilitated by the resonance stabilization in the propargylic anion and by the heat that is supplied. This is typically a very endothermic proton transfer.

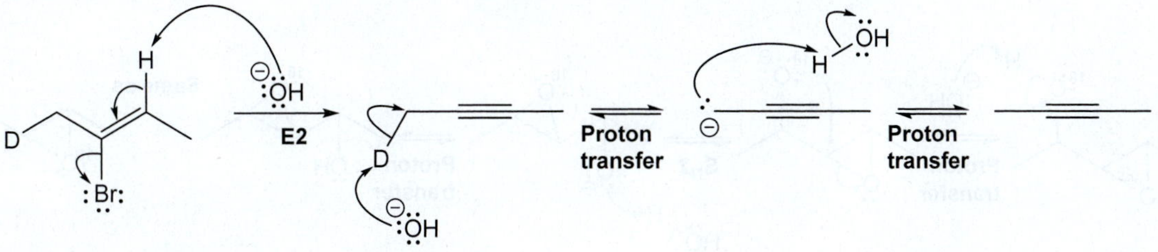

E2
Proton transfer
Proton transfer

Problem 10.71 (SYN)

Think

Refer to Problem 10.48 to review the reaction conditions. What is another possible route to synthesize an ether? What is the mechanism of that reaction? Which R group should be the alkyl halide, and which R group should be the alkoxide? Is more than one route possible?

Solve

The Williamson synthesis of ethers is ideal for this reaction. The reaction requires an alkoxide anion (which can be produced from an alcohol) and an alkyl halide (synthesized from the alcohol using PBr_3), as shown below, and can be done in two different ways.

Method 1

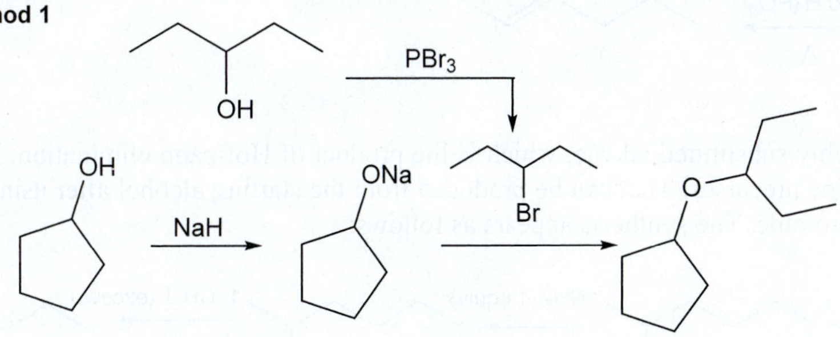

PBr_3
NaH

Method 2

PBr_3
NaH

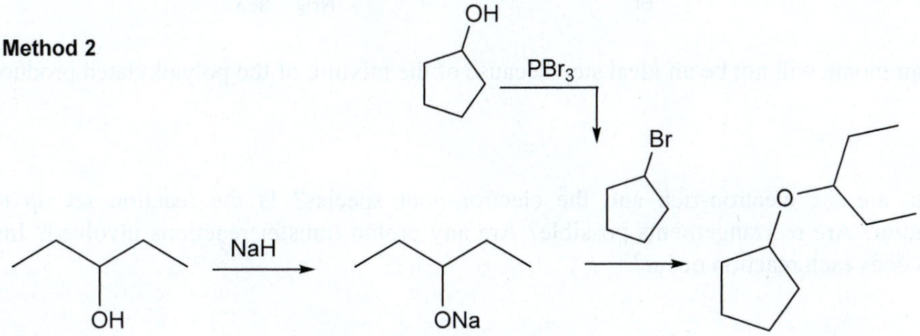

Problem 10.72

Think

In each product, what bonds broke and formed to make the product from the starting material? In the presence of a strong base, how does an alcohol react? What electron-rich and electron-poor products result? How does an epoxide react with an alkoxide anion? How does the epoxide form with the labeled ^{18}O?

Solve

The first "product" is the original compound, so the mechanism simply needs to explain how the second product forms. A proton transfer creates a nucleophilic $^{18}O^{-}$, which then displaces the unlabeled O of the epoxide, trading one strained ring for another. All steps are equilibria, so a mixture of the two compounds results.

Proton transfer · S_N2 · Proton transfer · Same as

Problem 10.73 (SYN)

Think

Does substitution or elimination have to occur to form the final alkene product? The starting material is the same. How is it possible to carry out an elimination reaction to put the alkene in two different locations? Which alkene forms from Zaitsev elimination, and which alkene forms from Hofmann elimination?

Solve

(a) This is a straightforward Zaitsev elimination, in which the product is the most highly substituted alkene. So we can simply heat the alcohol in the presence of concentrated acid.

conc H_3PO_4, Δ

(b) The product is the least highly substituted alkene, which is the product of Hofmann elimination. Hofmann elimination requires an amine precursor. That can be produced from the starting alcohol after using PBr_3 to convert the alcohol to the bromide. The synthesis appears as follows:

PBr_3 → NH_3 (1 equiv) → 1. CH_3I (excess) 2. Ag_2O 3. Δ

The alkylation of ammonia will not be an ideal step because of the mixture of the polyalkylated products.

Problem 10.74

Think

In each reaction, what are the electron-rich and the electron-poor species? Is the reaction set up to do elimination or substitution? Are rearrangements possible? Are any proton transfer reactions involved? In how many elementary steps does each reaction occur?

Solve

In the first reaction, the Grignard reagent, a strong nucleophile, opens the strained ring, producing the alcohol after acid workup. The alcohol is converted to the alkyl bromide using PBr_3.

1. (oxetane) 2. $H_3O^{\oplus}$ → **A** → PBr_3, Diethyl ether → **B**

$NaOC(CH_3)_3$ in the next reaction deprotonates the ketone reversibly at the more highly substituted α carbon, making an enolate nucleophile. That nucleophile attacks the alkyl halide carbon of compound **B**, which alkylates the α carbon. The next step is bromination of the α carbon, and the alkyllithium reagent in the final step replaces Br in a nucleophilic substitution.

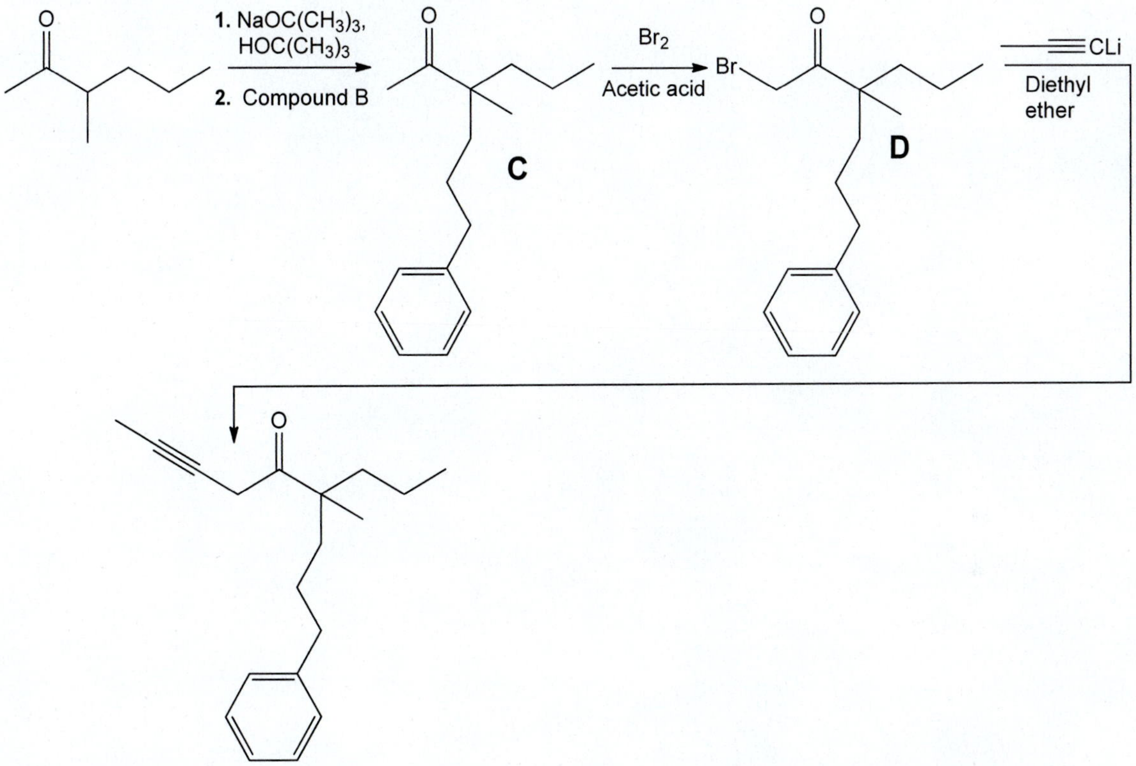

1. $NaOC(CH_3)_3$, $HOC(CH_3)_3$
2. Compound B

C

Br_2

Acetic acid

Br

D

≡CLi

Diethyl ether

CHAPTER 11 | Electrophilic Addition to Nonpolar π Bonds 1: Addition of a Brønsted Acid

Your Turn Exercises

Your Turn 11.1

Think

Consult Equations 11-1 and 11-2. Which bonds are broken and formed? What is the difference in the number and types of atoms between the reactant and product? What happens to the π bond of the C=C?

Solve

In the product, there is an additional H atom and a Br atom that does not appear in the reactant. HBr adds across the C=C double bond of but-2-ene to produce 2-bromobutane. In doing so, new C–Br and C–H bonds form, the π bond of the C=C breaks, and the H–Br bond breaks.

Your Turn 11.2

Think

Which bonds broke and formed to result in the formation of a C^+ and Br^-? What is the electrophile? What is the nucleophile? How can you use curved arrow notation to show the electron movement to form the new C–Br bond in Step 2? If π bond electrons from the nonpolar C=C are involved in the first step, what is the name of the elementary step? Are π bonds electron rich or poor?

Solve

Step 1 is an electrophilic addition in which a pair of electrons from the electron-rich C=C π bond forms a bond to the acid's electron-poor H atom. This leaves a C^+ (electron poor) on the other C of the C=C and a Br^- anion (electron rich). Step 2 is a coordination step whereby Br^- forms a bond to the C^+.

Electron rich

Electron poor

1. Electrophilic addition

Electron poor

Electron rich

2. Coordination

Your Turn 11.3

Think

Identify the bonds that broke and the bonds that formed. Consult Table 1-2 and Table 1-3 to fill in the appropriate energy values for each. Is a reaction energetically favorable if the enthalpy is negative (exothermic) or positive (endothermic)?

Solve

The C=C and H–Cl bonds are lost. The C–H, C–C, and C–Cl bonds are gained. From the bond energy values found in Table 1-1, the reaction is exothermic by 38 kJ/mol, which favors products.

$$\Delta H^\circ_{rxn} = (\underset{C=C}{\underline{619}}\ \text{kJ/mol} + \underset{H-Cl}{\underline{431}}\ \text{kJ/mol}) - (\underset{C-C}{\underline{339}}\ \text{kJ/mol} + \underset{H-C}{\underline{418}}\ \text{kJ/mol} + \underset{C-Cl}{\underline{331}}\ \text{kJ/mol})$$

Sum of energies of bonds lost

Sum of energies of bonds gained

$$= \underline{-38}\ \text{kJ/mol}$$

Your Turn 11.4

Think

Draw the carbocation that results from formation of the C–H bond on either end of the C=C bond. Are alkyl groups electron donating or withdrawing? How does an additional alkyl group affect the stability of a nearby positive charge? Is the more stable carbocation higher or lower energy?

Solve

Each carbocation intermediate is produced by the addition of a H^+ to one of the C=C carbons, resulting in a positive charge on the other C. The primary carbocation is less stable (higher energy), and the secondary carbocation is more stable (lower energy). Alkyl groups are electron donating and stabilize the carbocation. A secondary C^+ has two alkyl groups, whereas a primary C^+ only has one.

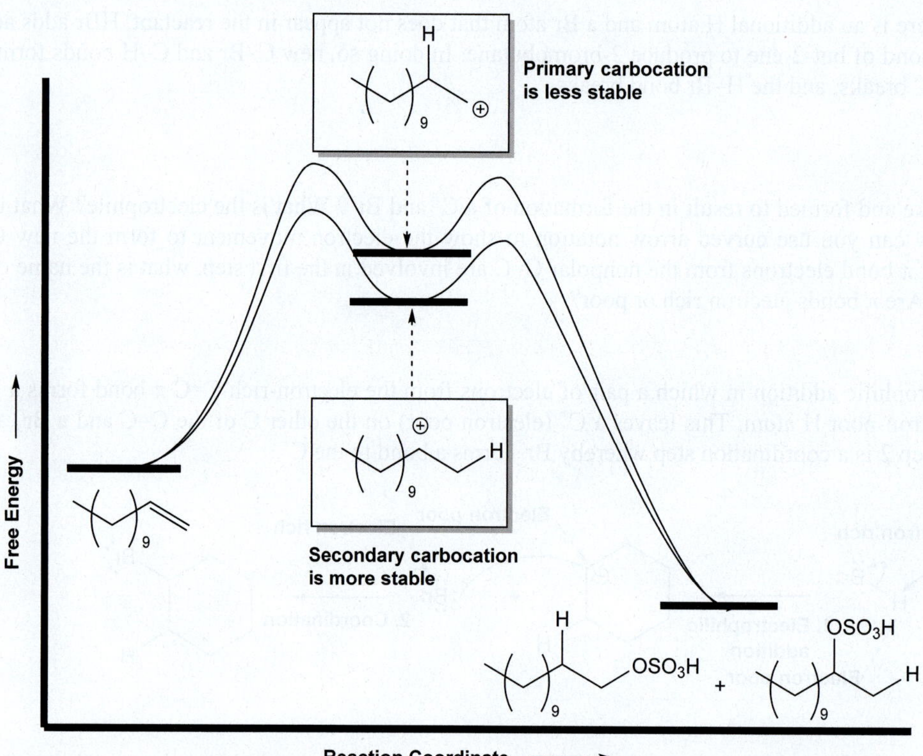

Your Turn 11.5

Think

Do any of the atoms have an unfilled valence shell? If so, is the atom with the unfilled valence shell adjacent to a double bond, a triple bond, or a lone pair? How can you show via a curved arrow the movement of electrons from the π bond to the atom with the unfilled valence? On which atom does the formal charge now reside? How many more times can this movement be repeated?

Solve

In Solved Problem 11.3, carbocation **B** has an unfilled valence and is adjacent to a double bond. The electrons in the π bond can move from the C=C to the C–C to form a new π bond. This movement can continue around the benzene ring, and four total resonance structures are possible.

B

Your Turn 11.6

Think

What are the requirements for a chiral center? How many unique groups must be bonded to the C?

Solve

Four unique groups must be bonded to the C to make it a chiral center. The two possible products from the addition of H_2SO_4 across the double bond are shown. The major product (right) results in the formation of a new chiral center (C*).

H_2SO_4 OSO$_3$H OSO$_3$H New chiral center H H 9 9 9

Your Turn 11.7

Think

What is electron rich (nucleophilic)? Electron poor (electrophilic)? What are the two elementary steps involved in the addition of a Brönsted acid like H–OH across the double bond? What two charged species form in the first step? How do the stabilities of these species compare to the species produced in electrophilic addition reactions encountered previously?

Solve

The C=C bond is electron rich, and the proton on H_2O is electron poor, so we can envision an electrophilic addition step. The HO^- that is produced might then add to the carbocation in a coordination step, resulting in the overall addition of water across the double bond. The electrophilic addition products in Step 1 are a carbocation C^+ and hydroxide HO^-. Hydroxide is significantly less stable than anions produced in other electrophilic addition reactions, such as Cl^- or Br^-, because O, being a significantly smaller atom, cannot accommodate the negative charge as well.

OH H H OH Unstable OH H

1. Electrophilic addition

2. Coordination

Your Turn 11.8

Think

Is the C^+ in the vinylic cation adjacent to a double bond? If so, how does the presence of resonance stabilize the vinylic cation?

Solve

The proton adds to the terminal C to produce a secondary vinylic and benzylic carbocation that is stabilized by resonance involving the phenyl ring. The resonance structures are shown below.

Your Turn 11.9

Think

In how many steps does the mechanism take place? How many intermediates and transition states result from this mechanism? Which intermediates are carbocations? How does the stability of a carbocation intermediate compare with that of an intermediate with no formal charge? In comparing the carbocation intermediates from Steps 1 and 3, which one is more stable? Why? How is stability reflected in the energy diagram?

Solve

The mechanism in Equation 11-17 takes place in four elementary steps. This results in three intermediates and four transition states. The carbocation intermediate in Step 1 is produced by the addition of a H^+ to the terminal carbon of the C≡C. This results in a positive charge on the secondary benzylic carbon. The second intermediate is a vinylic chloride in which Cl is attached directly to an alkene C. There is no formal charge, and this intermediate is more stable than either carbocation intermediate. The carbocation intermediate in Step 3 is produced by the addition of a H^+ to the terminal carbon of the C=C. This results in a positive charge on the secondary benzylic carbon. This carbocation is stabilized via resonance from the benzene ring as well as resonance from the attached Cl.

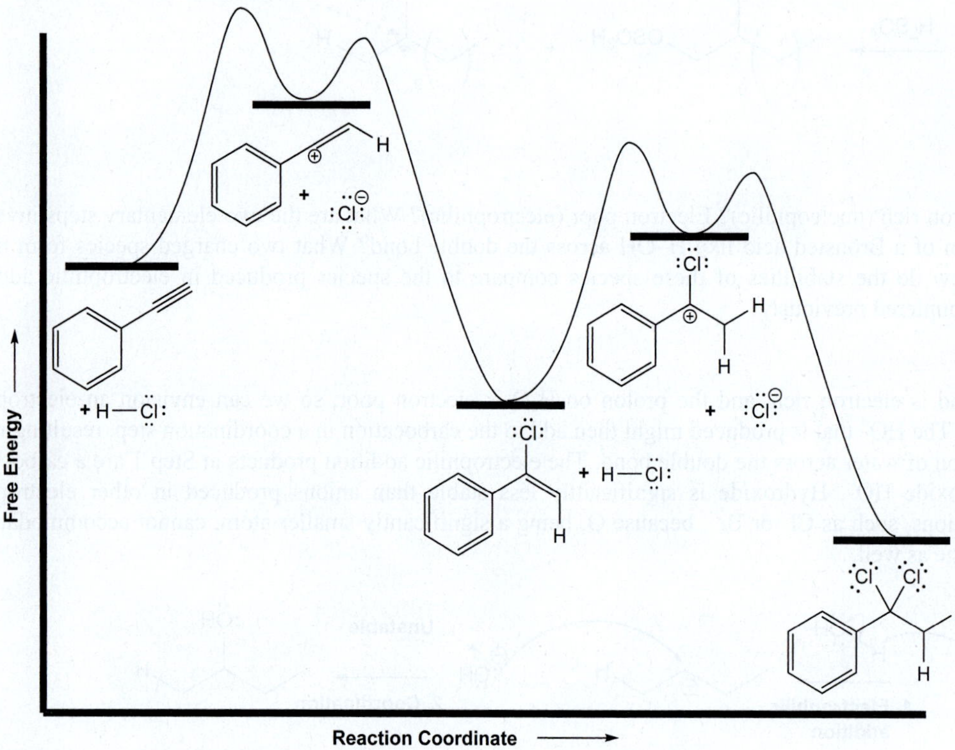

Free Energy

Reaction Coordinate

Your Turn 11.10

Think

Identify which C atom of a terminal alkyne C≡C has to form the bond to O to produce an aldehyde. For that to happen, which C (terminal or internal) in the first step of electrophilic addition must gain the positive charge? Which C atom must gain the C–H bond (terminal or internal)? Compare the stability of this C^+ to that in Equation 11-21.

Solve

To produce an aldehyde, the O atom from water would have to form a bond to the terminal C, meaning that the terminal C would have to have the positive charge in a previous step, as shown below. This is a less stable carbocation than the one produced in Equation 11-21, because the charge in this carbocation intermediate is not resonance stabilized. The carbocation in Equation 11-21, on the other hand, is resonance stabilized.

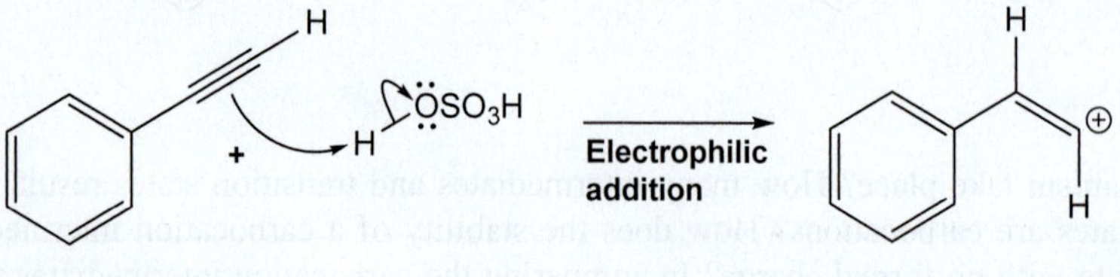

Electrophilic addition

Your Turn 11.11

Think

Can an acid exist in solution if it is a stronger acid than the protonated solvent? What is the protonated solvent if CF_3CH_2OH is the solvent? Are the F atoms electron donating or withdrawing? Do they stabilize or destabilize a nearby positive charge?

Solve

According to the leveling effect, the strongest acid that can exist in solution is the protonated solvent. With CF_3CH_2OH as the solvent, the strongest acid that can exist is $CF_3CH_2OH_2^+$. As an acid, $CF_3CH_2OH_2^+$ is stronger than a similar alcohol, ROH_2^+, because the electron-withdrawing F atoms destabilize the nearby positive charge. The acidity of ROH_2^+ is similar to that of H_3O^+, so the strongest acid that can exist in the solvent CF_3CH_2OH is stronger than H_3O^+.

Your Turn 11.12

Think

Which C atom gains the H from HCl? Identify the C atom that is C1 in both 1,2- and 1,4-electrophilic addition. Which C atoms would then be numbered C2, C3, and C4? (Remember these are just relative numbering labels and do not correlate with nomenclature numbering rules.) Which of those C atoms gains the Cl?

Solve

The first is from 1,2-addition, and the second is from 1,4-addition. In both cases, the C that gains the H atom is assigned C1. In 1,2-addition, C2 gains the Cl, and the remaining double bond will be in the same location as in the overall reactant. In 1,4-addition, C4 gains the Cl, and the double bond will be in a different location. The double bond moves owing to the resonance that exists in the intermediate as a result of a C atom with an incomplete octet next to a double bond.

Your Turn 11.13

Think

Do warm temperatures or cold temperatures facilitate a kinetically controlled reaction? A thermodynamically controlled reaction? Which product, that from 1,2-addition or 1,4-addition, is the major product at cold temperatures? At warm temperatures?

Solve

The reaction in Equation 11-27 takes place under thermodynamic control (warm temperatures), so its major product must be the thermodynamic product, that is, the most stable product. The reaction in Equation 11-28 takes place under kinetic control (cold temperatures), so its major product must be the kinetic product, that is, the product that forms the fastest. This is indicated on the figure on the next page.

Major product at warm temperatures

HCl, 20 °C: Buta-1,3-diene → 3-Chlorobut-1-ene (25% of adduct mixture) + 1-Chlorobut-2-ene (75% of adduct mixture) ← Thermodynamic product

Major product at cold temperatures

HCl, -80 °C: Buta-1,3-diene → 3-Chlorobut-1-ene (80% of adduct mixture) ← Kinetic product + 1-Chlorobut-2-ene (20% of adduct mixture)

Your Turn 11.14

Think

Label each product 1,2-addition or 1,4-addition. Is 1,2- or 1,4-addition faster? Which alkene is the most stable (mono-, di-, tri-, or tetrasubstituted)?

Solve

The first is from 1,4-addition, and the second is from 1,2-addition. In 1,2-addition, the remaining double bond will be in the same location as in the overall reactant. In 1,4-addition, it will be in a different location. Because 1,2-addition is faster, it gives the kinetic product. The 1,4-adduct has the more highly substituted alkene product, so it is more stable and is the thermodynamic product.

HCl →

1,4-Addition
Thermodynamic product

1,2-Addition
Kinetic product

In Chapter Problems

Problem 11.1

Think

What is electron rich in the first step? What is electron poor (electrophilic)? What elementary step occurs in each step of this two-step reaction mechanism? What functional group is formed?

Solve

The reaction occurs in two steps. The first elementary step is electrophilic addition, where the π electrons of the C=C define the electron-rich species and the H atom of the H–Y of the acid defines the electron-poor species (electrophile). A carbocation forms in the first step. The second elementary step is coordination, where the conjugate base of the H–Y acid acts as the nucleophile.

(a) H–Cl: Electrophilic addition; + :Cl:⁻ Coordination

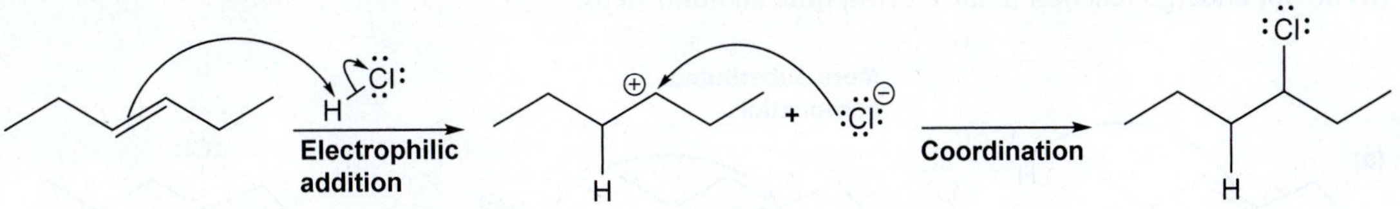

(b) H–I: Electrophilic addition; + :I:⁻ Coordination

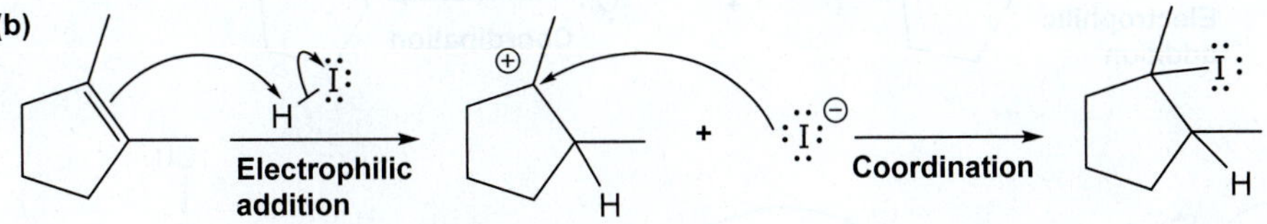

(c) H–OSO$_3$H: Electrophilic addition; + :OSO$_3$H⁻ Coordination

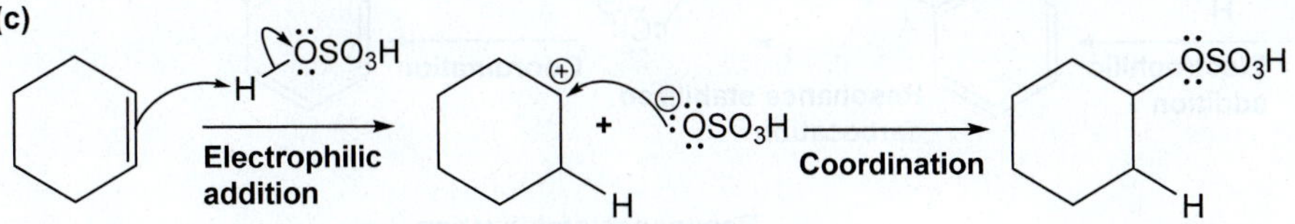

(d) H–Br: Electrophilic addition; + :Br:⁻ Coordination

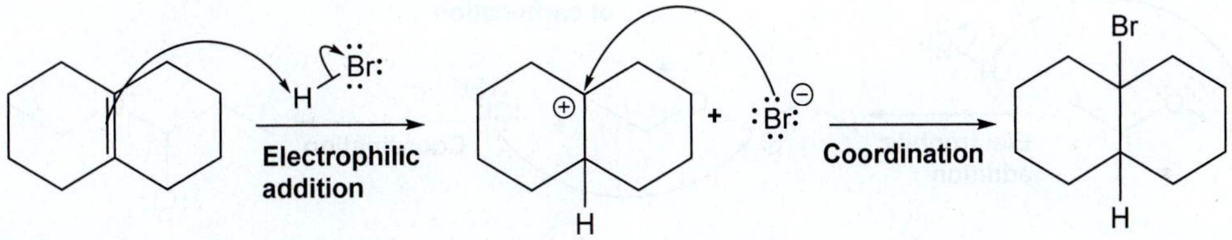

Problem 11.2

Think

What type of reaction starts with an alkene and ends with the functional group shown? On which two C atoms must the C=C be located in the starting material?

Solve

Each compound can be made from an alkene, as shown below. Each alkene is devised by removing the nucleophile that would have added, along with a neighboring H atom, and connecting the two C atoms with an additional π bond.

(a) HCl

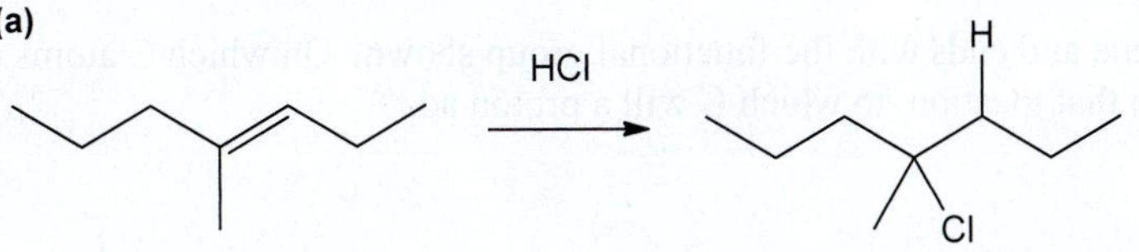

(b) HBr

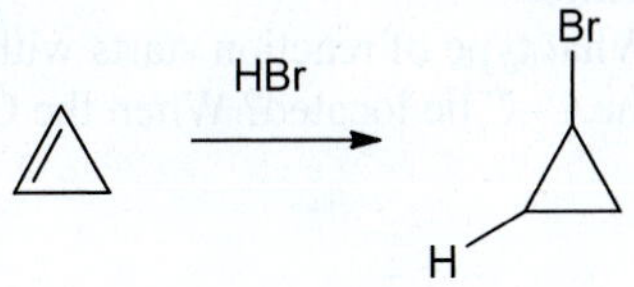

(c) H_2SO_4 → OSO$_3$H

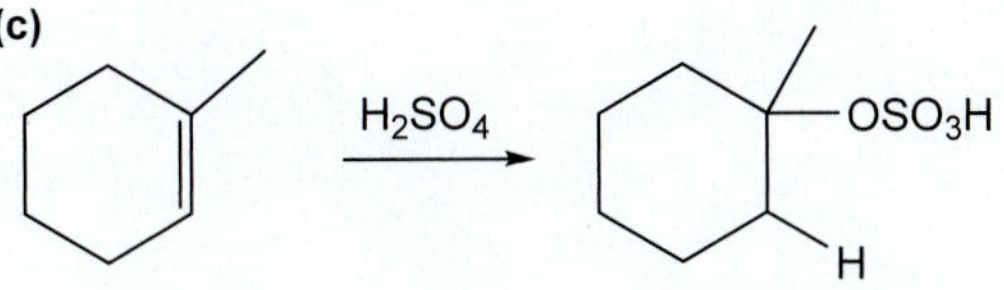

Problem 11.4

Think

Which C=C double bond will undergo electrophilic addition? What are the *possible* products and the corresponding carbocation intermediates from which they derive? Which carbocation intermediate is more stable?

Solve

Mechanisms are shown below. In the first step of each reaction, electrophilic addition of H^+ takes place to produce the more stable carbocation. In **(a)**, the more highly substituted carbocation is more stable. In **(b)** and **(c)**, the carbocation is resonance stabilized. In **(d)**, the NO_2 group would destabilize a neighboring positive charge, so the better choice is having the positive charge farther away. Notice that the phenyl rings in **(b)** and **(d)** do not undergo reaction in the electrophilic addition steps.

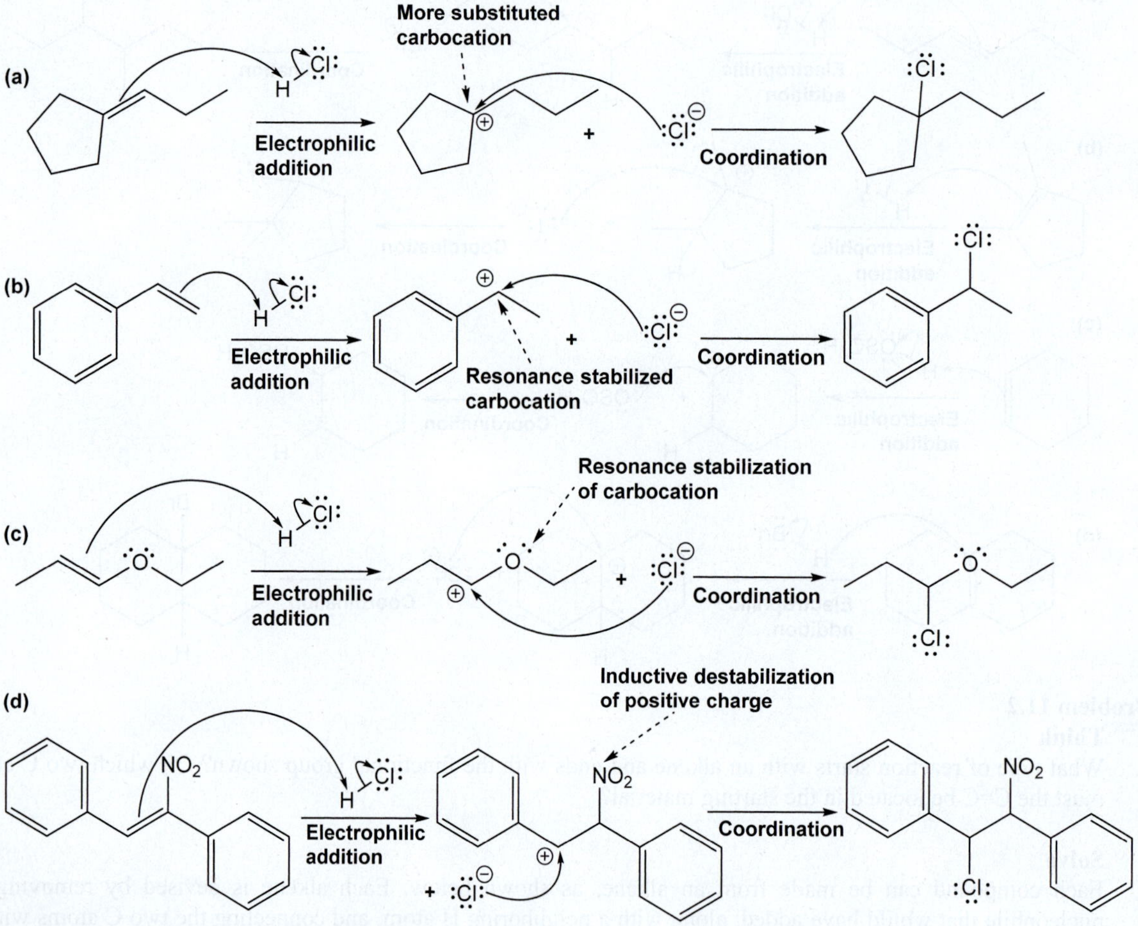

(a) More substituted carbocation; Electrophilic addition; Coordination

(b) Resonance stabilized carbocation; Electrophilic addition; Coordination

(c) Resonance stabilization of carbocation; Electrophilic addition; Coordination

(d) Inductive destabilization of positive charge; Electrophilic addition; Coordination

Problem 11.5

Think

What type of reaction starts with an alkene and ends with the functional group shown? On which C atoms could the C=C be located? When the C=C is in that location, to which C will a proton add?

Solve

The alkene precursors have the halogen and H atoms on adjacent carbons removed and an additional π bond between the C atoms. In each case, the two different alkenes produce the same product because the most stable carbocation that is produced is the same.

(a) or HCl Cl

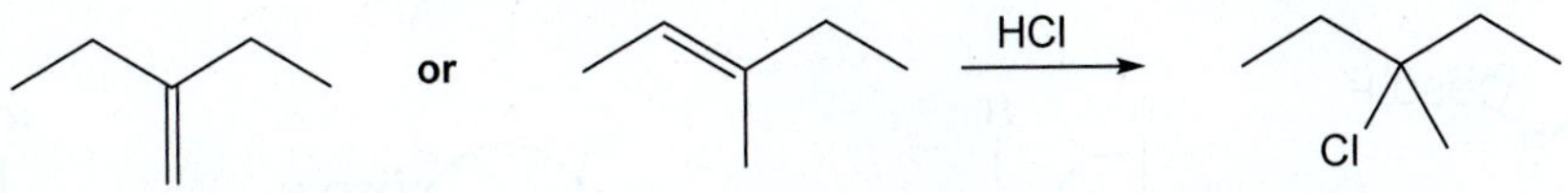

(b) O or O HBr O Br

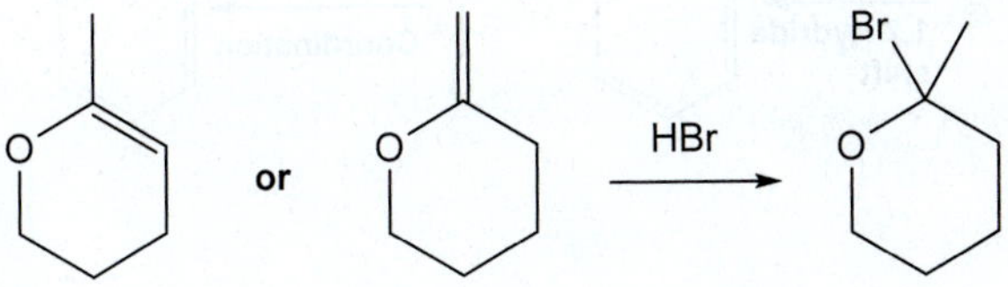

(c) or HCl Cl

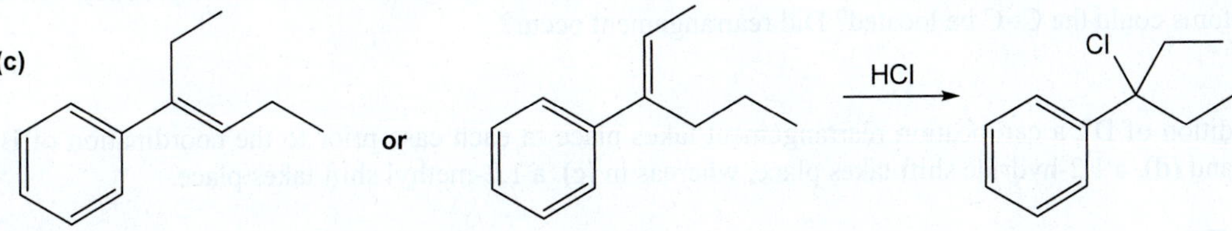

Problem 11.7

Think

What carbocation intermediate is produced on addition of H^+? Can that carbocation intermediate undergo a 1,2-hydride shift or a 1,2-methyl shift to attain greater stability? What species produced from the first step will react with the carbocation intermediate?

Solve

(a) In Step 1 of this electrophilic addition reaction, H^+ adds to either alkene C to produce a secondary carbocation intermediate, as shown below. The secondary carbocation intermediate converts to a more stable tertiary one via a 1,2-methyl shift in Step 2. In Step 3, the I^- coordinates to the carbocation to complete the reaction.

H I: Electrophilic addition ⊕ 1,2-Methyl shift ⊕ :I:⊖ Coordination I:

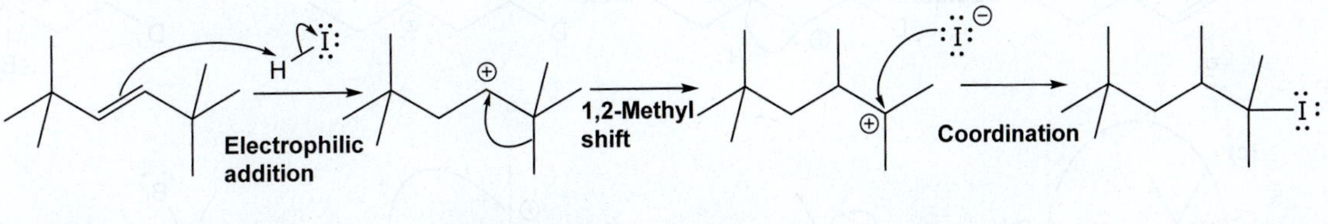

(b) In Step 1 of this electrophilic addition reaction, H^+ adds to either alkene C to produce a secondary carbocation intermediate. The secondary carbocation intermediate converts to a more stable tertiary one via a 1,2-hydride shift in Step 2. In Step 3, the Cl^- coordinates to the carbocation to complete the reaction.

H Cl: Electrophilic addition H ⊕ 1,2-Hydride shift ⊕ H :Cl:⊖ Coordination Cl:

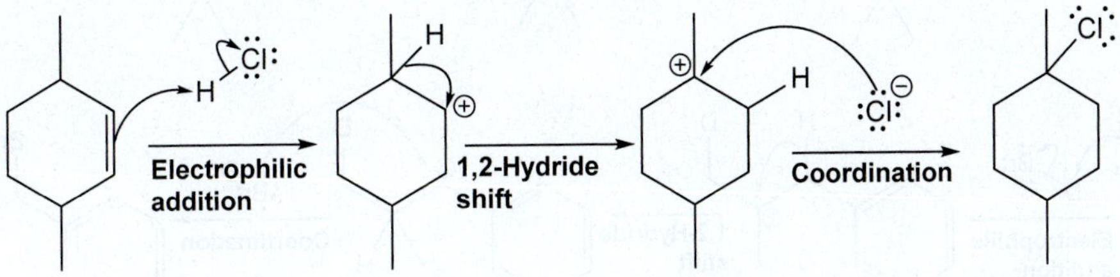

(c) In Step 1 of this electrophilic addition reaction, H^+ adds to the terminal alkene C to produce a secondary carbocation intermediate, as shown below. (Addition of H^+ to the internal alkene C would, instead, produce a less stable primary carbocation intermediate.) The secondary carbocation intermediate converts to a more stable benzylic resonance-stabilized one via a 1,2-hydride shift in Step 2. In Step 3, the HSO_4^- coordinates to the carbocation to complete the reaction.

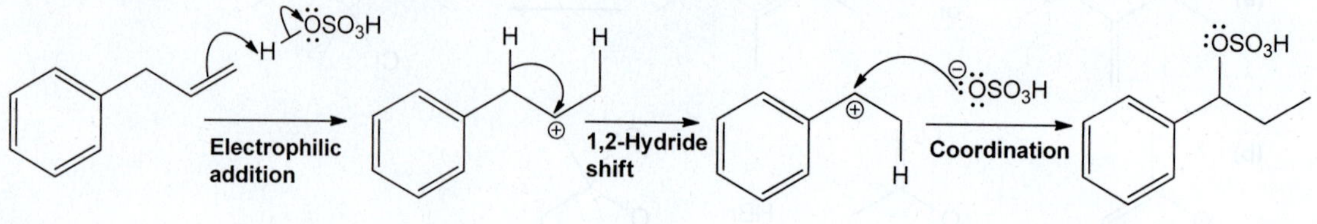

Electrophilic addition → 1,2-Hydride shift → Coordination

Problem 11.8

Think

On which carbon atoms are the Br and D located? Are they on adjacent C atoms in the product? On which carbon atoms could the C=C be located? Did rearrangement occur?

Solve

After addition of D^+, a carbocation rearrangement takes place in each case prior to the coordination of Br^-. In **(a)**, **(b)**, and **(d)**, a 1,2-hydride shift takes place, whereas in **(c)**, a 1,2-methyl shift takes place.

(a) Electrophilic addition → 1,2-Hydride shift → Coordination

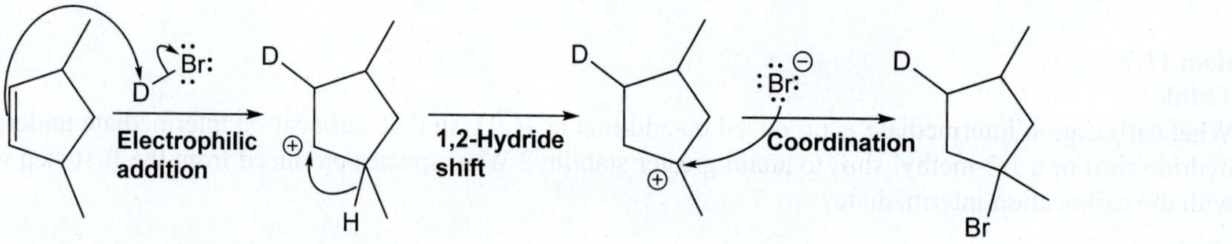

(b) Electrophilic addition → 1,2-Hydride shift → Coordination

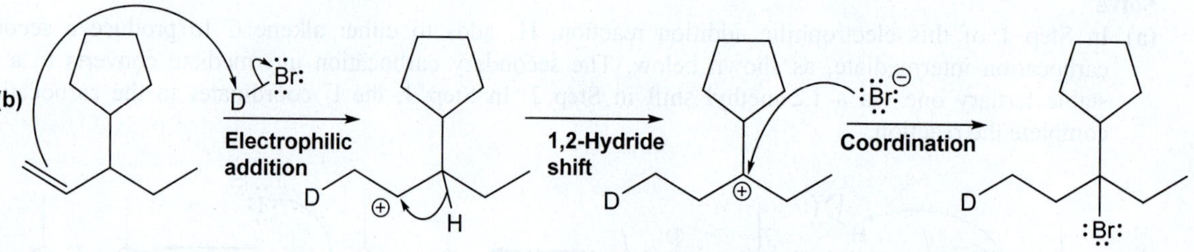

(c) Electrophilic addition → 1,2-Methyl shift → Coordination

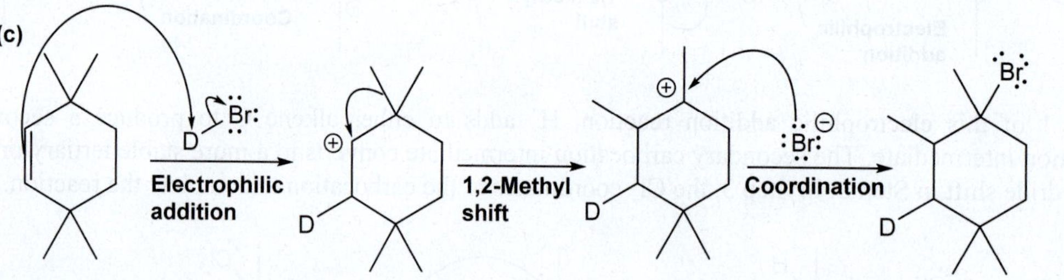

(d) Electrophilic addition → 1,2-Hydride shift → Coordination

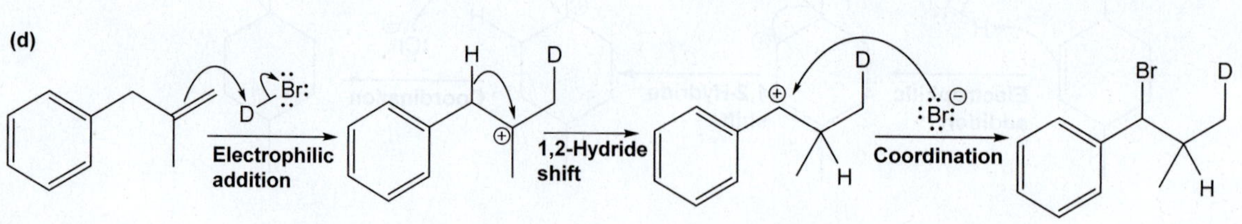

Problem 11.9

Think

To which C=C carbon will the electrophile predominantly attach? From how many directions can the electrophile approach the C=C? What intermediate is formed first? Is there a chiral center in that intermediate? From how many directions can the nucleophile attack the intermediate? Is a chiral center formed in that step?

Solve

After electrophilic addition of HBr to the C=C, there are two possible C^+ intermediates that can form, a 2° and a 3°. In the 2° C^+, one new chiral center formed, so this produces a mixture of *R* and *S* configurations. This is because the Br^- can approach the C^+ from either side of the plane. The wedge CH_3 group does not participate in the reaction and therefore retains the same *R* configuration in all products. From the 3° C^+, two chiral centers are formed in the reaction, one in the first step and the other in the second step. Each chiral center is produced as a mixture of *R* and *S* configurations. This is because the electrophile, H^+, can approach the alkene C from either side of the carbon's plane, and the Br^- can approach the C^+ from either side of the plane. This leads to four additional stereoisomers as overall products. The products formed from the 3° C^+ are the major products because the 3° C^+ is more stable than the 2° C^+.

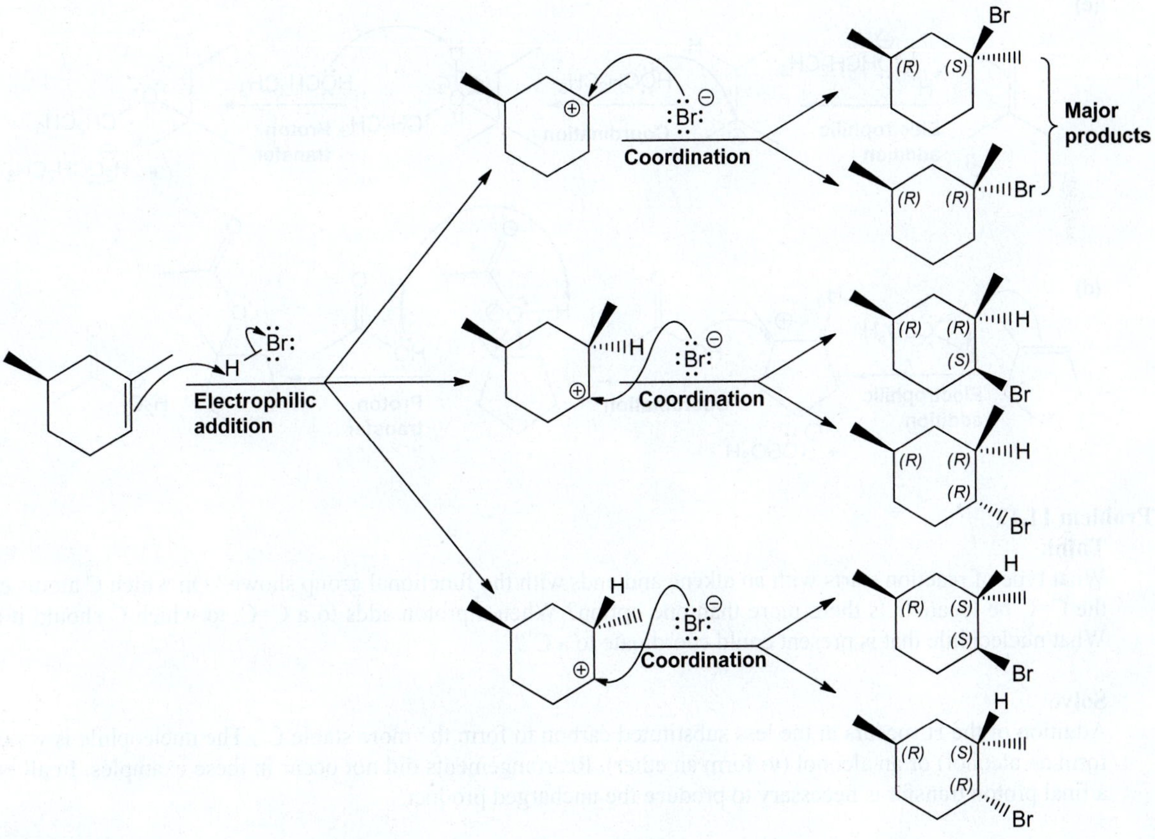

Problem 11.11

Think

What is the electrophile? What is the nucleophile? To which alkene carbon does the electrophile add? Where is the C^+ formed? Are rearrangements possible?

Solve

In each case, the strong acid adds H^+ to the double bond in the first step—electrophilic addition. The product is the most stable carbocation. In the second step, the weak nucleophile undergoes coordination to the carbocation, and in the final step, a weak base deprotonates the positively charged O atom.

(a)

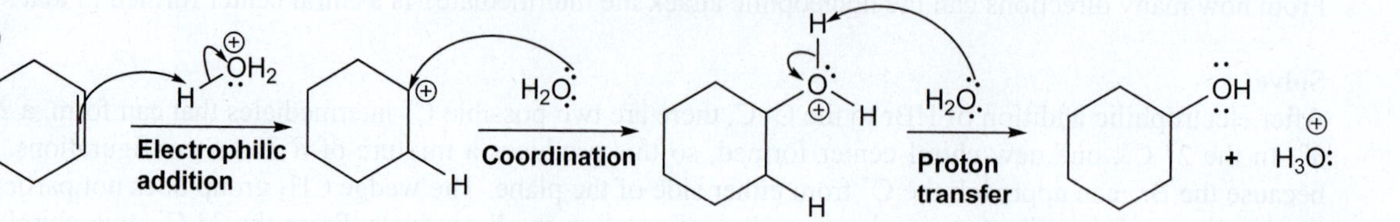

(b)

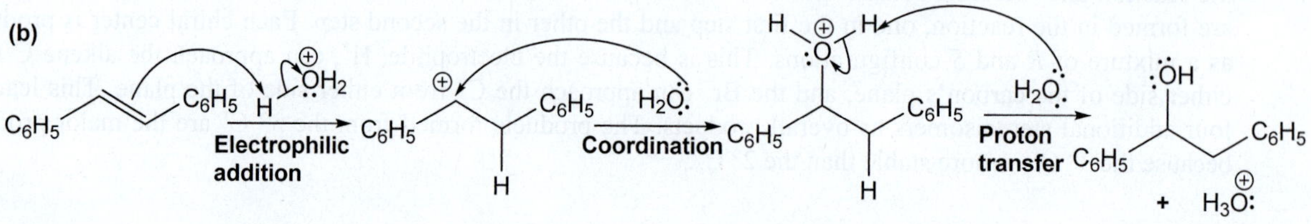

(c)

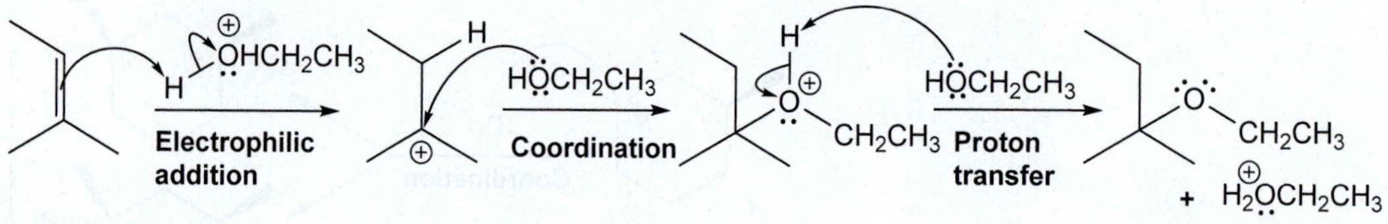

(d)

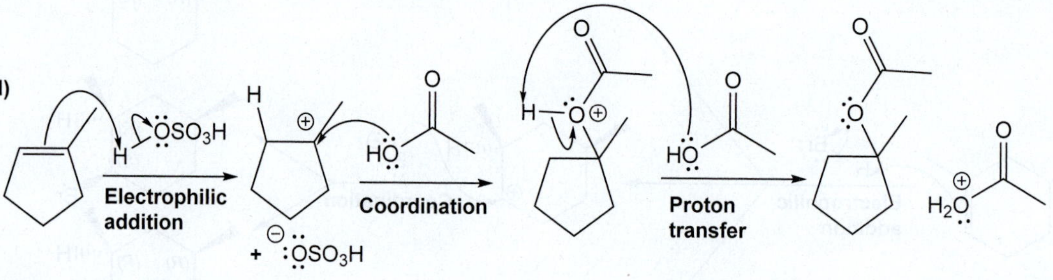

Problem 11.12

Think

What type of reaction starts with an alkene and ends with the functional group shown? On which C atoms could the C=C be located? Is there more than one option? When a proton adds to a C=C, to which C should it add? What nucleophile that is present could coordinate to a C^+?

Solve

Addition of the H^+ occurs at the less substituted carbon to form the more stable C^+. The nucleophile is water (to form an alcohol) or an alcohol (to form an ether). Rearrangements did not occur in these examples. In all cases, a final proton transfer is necessary to produce the uncharged product.

(a)

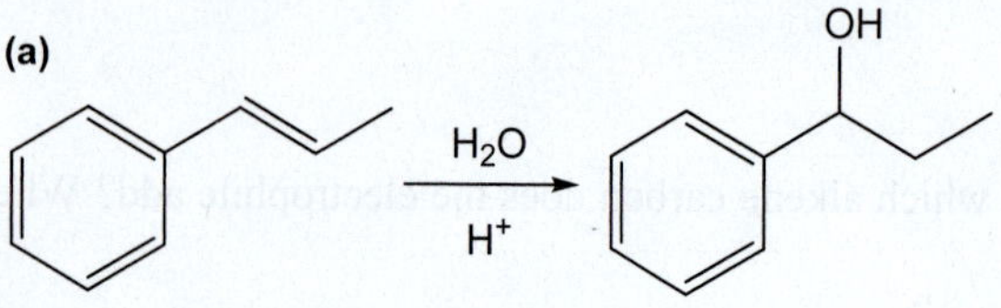

(b)

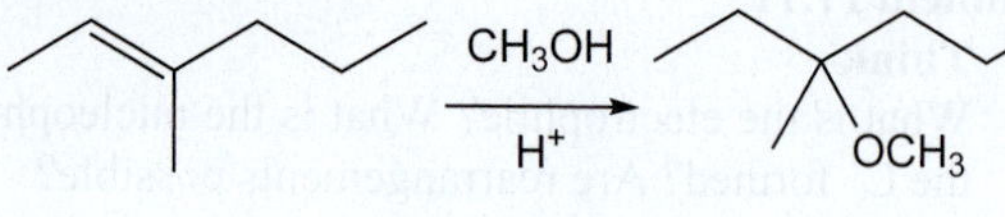

(c)

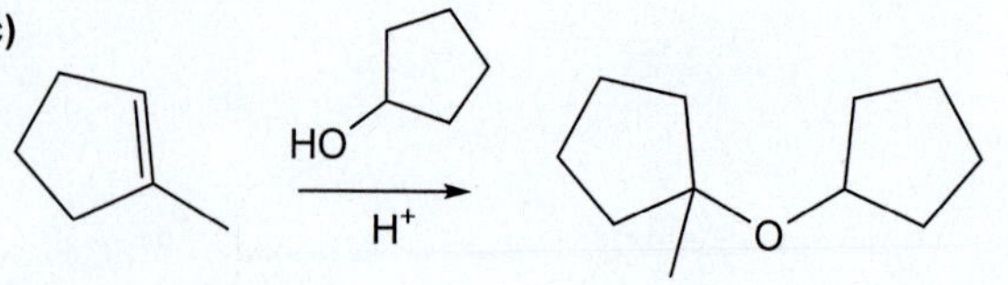

Problem 11.13

Think

On the basis of the electron-rich and electron-poor sites in the reactants, what type of reaction will take place? How many times will that reaction take place? What aspects of regiochemistry should be considered?

Solve

The alkyne is relatively electron rich, and the H atom of HCl or HBr is relatively electron poor. So the conditions favor an electrophilic addition reaction in which HCl or HBr adds across the multiple bond. In **(a)**, the product of that addition to the alkyne is a vinylic halide, which, in the presence of excess HCl, can undergo a second electrophilic addition. In **(b)**, the addition is to the C=C of the vinylic halide. Notice that in each addition to the vinylic halide, the proton adds to the C that is not bonded to the halogen so that the resulting C^+ is resonance stabilized by the halogen, so the product is generated in high yield.

(a)

Electrophilic addition

Coordination

Electrophilic addition

Coordination

(b)

Electrophilic addition

Coordination

Resonance

Problem 11.14

Think

On which C atoms are the halogen atom and H located? On which C atoms could the C≡C be located? How many times did the addition reaction occur?

Solve

In **(a)**, the alkyne must be a terminal alkyne to control the regiochemistry via Markovnikov's rule. HCl adds twice, and the resonance stabilization of the carbocation that is produced in the second addition is responsible for generating the product in high yield. In **(b)**, the alkyne must be symmetric to avoid a mixture of isomers.

(a) Excess HCl

(b) Excess HBr

Problem 11.15

Think

What is electron rich (nucleophilic)? Electron poor (electrophilic)? Did proton transfer reactions occur? What is the product that forms after the acid-catalyzed addition of water across the triple bond? Is this product stable? If not, what type of rearrangement occurs?

Solve

The first few steps compose the acid-catalyzed addition of water across the triple bond. Notice that the proton adds to the terminal C so that the positive charge that develops is stabilized by the attached alkyl group. Then, acid-catalyzed tautomerization converts the enol into the keto form.

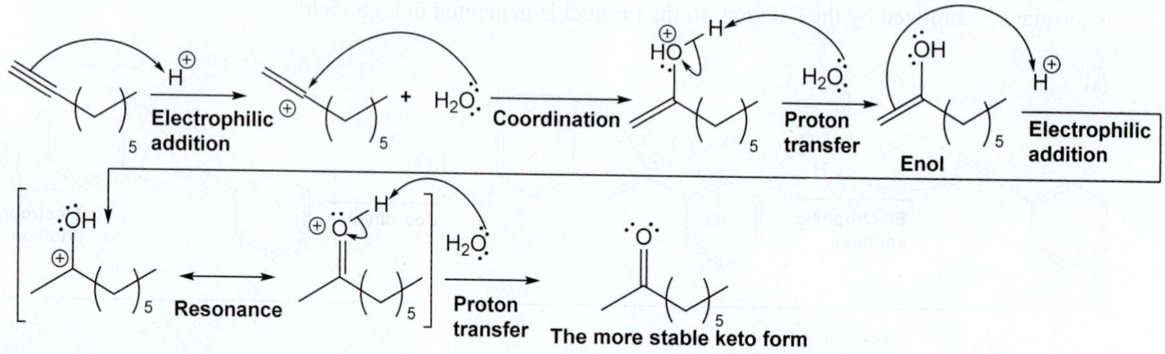

Electrophilic addition
Coordination
Proton transfer
Enol
Electrophilic addition
Resonance
Proton transfer
The more stable keto form

The first two steps likely occur in a single step, according to Equation 11-19.

Problem 11.17

Think

How many distinct carbocation intermediates are possible from protonation of a double bond? Which one is the most stable? What are the species that can be produced on nucleophilic attack of that carbocation intermediate?

Solve

Each of the C=C groups can gain a proton at either of its C atoms, giving rise to two possible carbocation intermediates (**A** and **B**). Intermediate **A** is more stable than intermediate **B** owing to resonance delocalization of the positive charge. With the most stable intermediate identified, the 1,2- and 1,4-addition products are obtained by attack of the nucleophile, Br^-, on the C atoms sharing the positive charge.

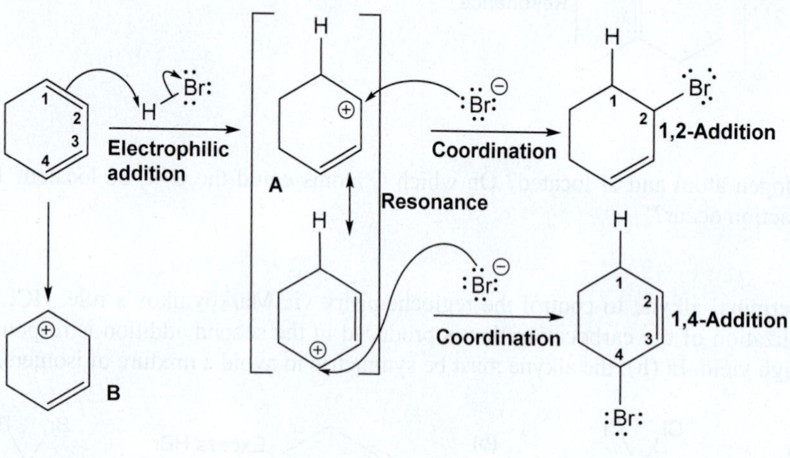

Electrophilic addition
A
Resonance
Coordination
1,2-Addition
Coordination
1,4-Addition
B

Problem 11.18

Think

Which product was formed from 1,2-addition, and which product was formed from 1,4-addition? On which C atom was the H^+ added? Which product results from resonance movement of the C=C?

Solve

The diene below could produce the mixture of products yielded by 1,2-addition (first product) and 1,4-addition (second product). The reaction proceeds through a carbocation intermediate that is the most stable, which is 3° and resonance stabilized.

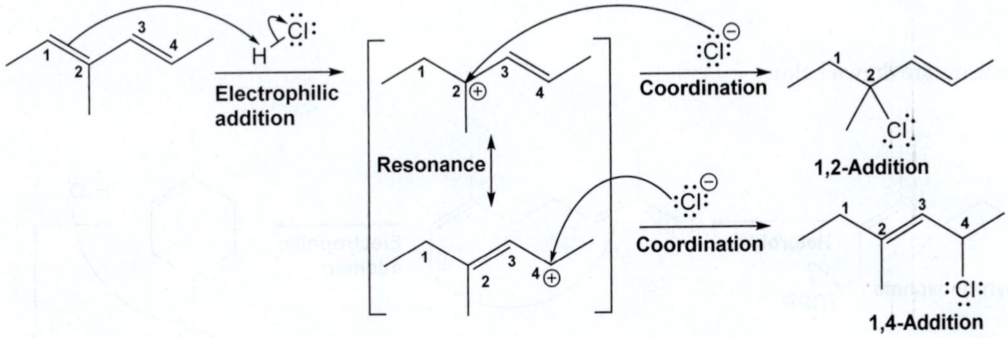

Problem 11.20

Think

What are the *possible* carbocation intermediates? What is the most stable carbocation intermediate? What are the products that can be produced on nucleophilic attack of that carbocation intermediate? Which of those products is produced faster? Which is the more stable product?

Solve

Each of the C=C groups can gain a proton at either of its C atoms, giving rise to four possible carbocation intermediates (**A**, **B**, **C**, and **D**). Intermediates **B** and **D** are more stable than **A** and **C** owing to resonance delocalization of the positive charge. Intermediate **D** is more stable than intermediate **B** (and is the most stable of all four) because its positive charge is shared on a tertiary C atom. Adding Cl^- to that carbocation intermediate yields the 1,2- and 1,4-adducts. As usual, the 1,2-adduct is the kinetic product. It is also the most stable alkene product because it is the most highly alkyl substituted. Therefore, the 1,2-adduct is the thermodynamic product, too.

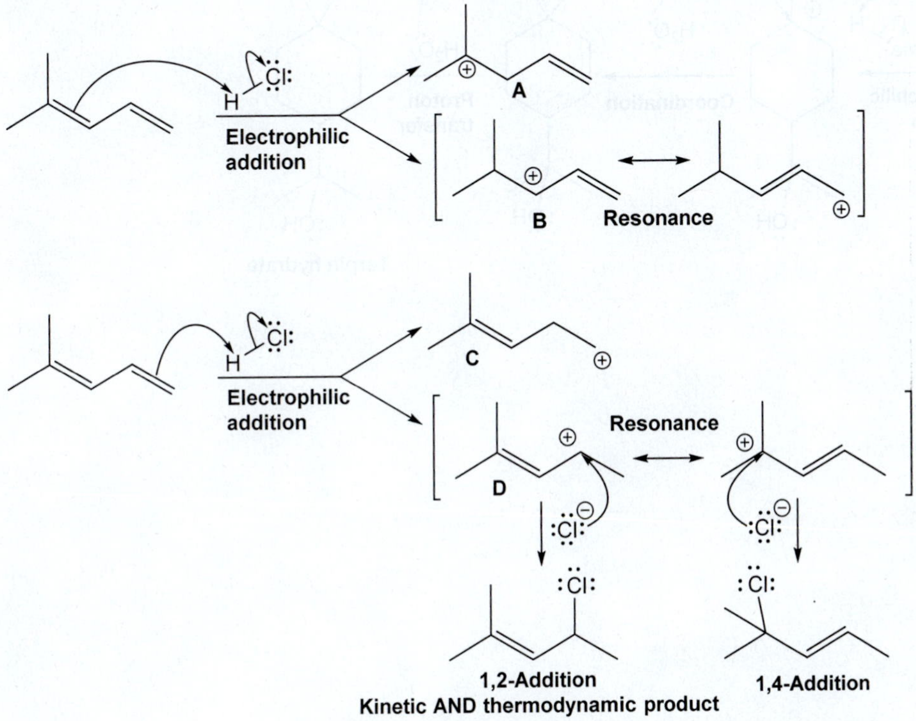

Problem 11.21

Think

What bonds broke and formed in the reaction of neryl phosphate to form α-terpinenol? What C=C bond underwent electrophilic addition? What acted as the nucleophile to form the C–OH bond? Are proton transfer reactions involved? What bonds broke and formed in the reaction of α-terpinenol to form terpin hydrate?

Solve

The mechanisms are drawn below.

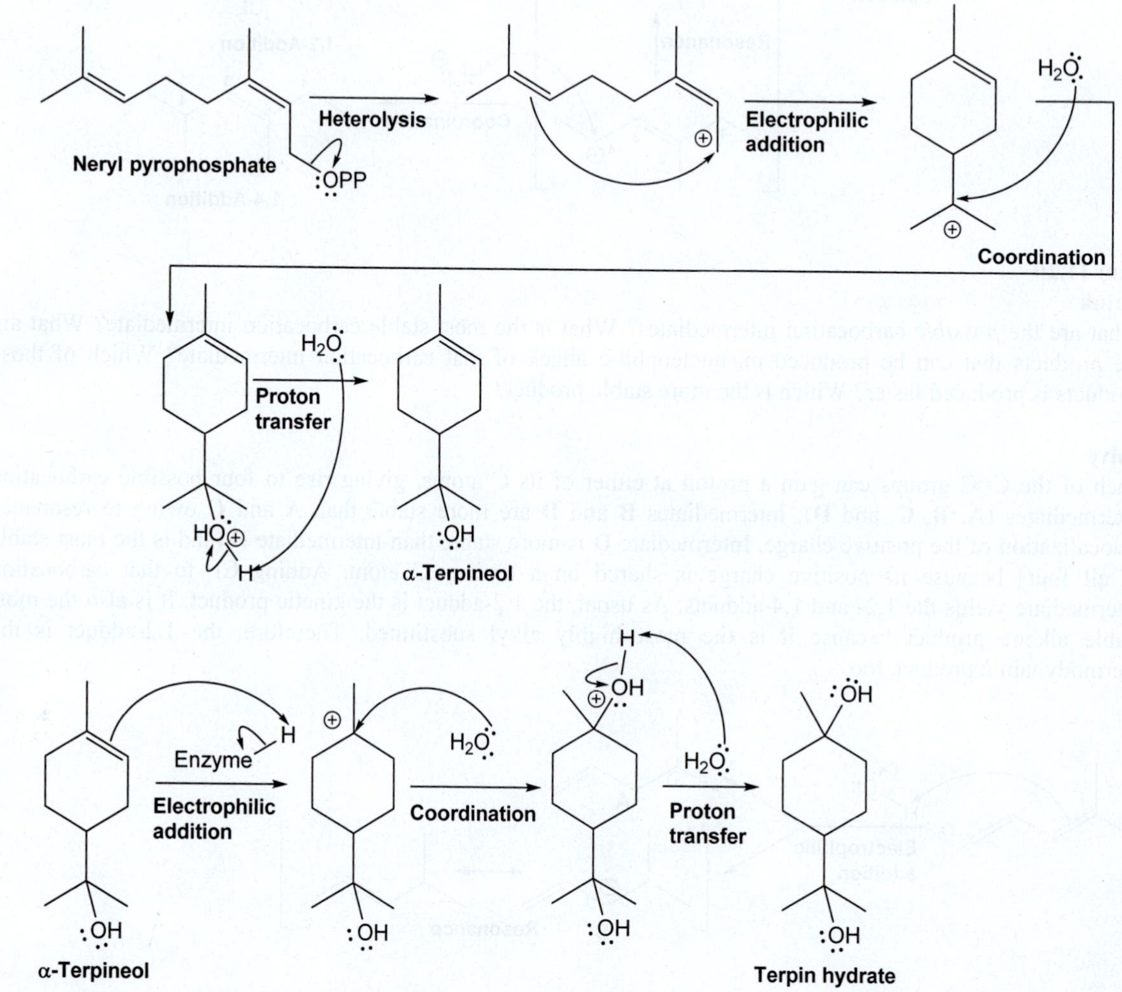

Neryl pyrophosphate
:OPP
Heterolysis
Electrophilic addition
$H_2\ddot{O}$
Coordination
$H_2\ddot{O}$
Proton transfer
HO⊕ H
:OH
α-Terpineol

Enzyme
H
Electrophilic addition
$H_2\ddot{O}$
Coordination
H
:OH
$H_2\ddot{O}$
Proton transfer
:OH
:OH
α-Terpineol
:OH
:OH
:OH
Terpin hydrate

End of Chapter Problems
Section 11.1 The General Electrophilic Addition Mechanism: Strong Brønsted Acids
Problem 11.22

Think

What are the two steps in the addition of a HX acid to an alkene? Of the two steps, which step produces new charges? Which step reduces the number of charges? Are the charges stabilized the same? Which atom that gains the negative charge is largest?

Solve

The reaction rate is determined by how fast the H–X bond breaks in the first (rate-determining) step, generating X^-. The more stable the anion that is generated, the lower the energy barrier, as suggested by the Hammond postulate. Because the order of charge stability is $F^- < Cl^- < Br^- < I^-$, the order of reactivity is **HF < HCl < HBr < HI**.

Problem 11.23

Think

What is the first step in addition of a HX acid to an alkene? What intermediate forms? Which intermediate is most stable? How does the stability of the intermediate affect the rate of the reaction?

Solve

The reaction rate is determined by how fast the carbocation forms in the first (rate-determining) step. The more stable the carbocation is, *the more stable is the transition state leading to it* (according to the Hammond postulate) and the faster the reaction. Because of resonance and the (+) charge on a tertiary carbon, carbocation **B** is by far the most stable: **C < D < A < B**.

Forms resonance-stabilized benzylic 3° C⁺

Forms 1° C⁺ — $H_2C{=}CH_2$ **C** < Forms 2° C⁺ **D** < Forms 3° C⁺ **A** < **B**

$H_2\overset{+}{C}{-}CH_3$ < < <

Problem 11.24

Think

What is electron rich in the first step? What is electron poor (electrophilic)? What elementary step occurs in each step of this two-step reaction mechanism? What functional group is formed?

Solve

The reaction occurs in two steps. The first elementary step is electrophilic addition, where the π electrons of the C=C bond define the electron-rich species and the H atom of the H–Y bond of the acid defines the electron-poor species (electrophile). A carbocation forms in the first step. The second elementary step is coordination, where the conjugate base of the H–Y acid acts as the nucleophile. The mechanisms are shown below and on the next page.

(a) H–Cl: → Electrophilic addition → + :Cl:⁻ → Coordination → :Cl

(b)

Electrophilic addition

Coordination

(S) (R)

(c)

Electrophilic addition

Coordination

(S) (R)

Problem 11.25

Think

Refer to Section 4.13 for strategies to draw all constitutional isomers given a formula. What is the IHD for both the reactant and product? What are possible structures for each? If the reaction conditions use an excess of HCl but only one Cl appears in the formula, how many addition reactions occurred? What does that suggest for the source of the IHD?

Solve

The IHD for the reactant (C_5H_8) is 2 and the product (C_5H_9Cl) is 1. The difference in formula is HCl. Therefore only one electrophilic addition reaction occurred, and the source of the IHD in the reactant is one ring and one double bond. All possible structures for the initial compound are drawn below.

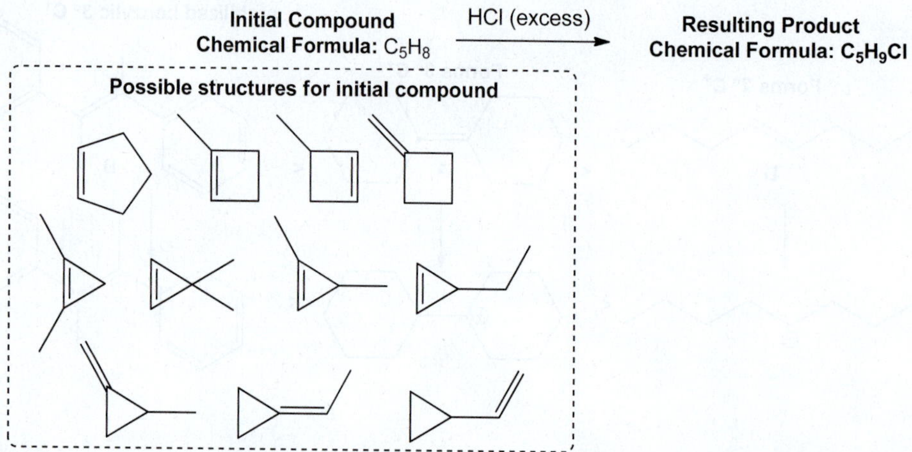

Sections 11.2–11.5 Benzene Rings, Regiochemistry, Carbocation Rearrangements, and Stereochemistry

Problem 11.26

Think

Draw out 2-chloro-3-methyl-2-phenylbutane. What carbocation intermediate forms in the first step when each of the reactants is treated with HCl? If there were two carbocations that could form in the first step, which carbocation would be favored? Are rearrangements possible?

Solve

The best way to answer this question is to draw the mechanism for each reaction. **E**, **F**, and **G** give 2-chloro-3-methyl-2-phenylbutane, because Cl^- attacks the exact same carbocation in each case. In **E** and **G**, the initial carbocation formed undergoes a fast rearrangement to a more stable carbocation prior to attack of Cl^-. **H** and **I** have a longer carbon chain than four carbon atoms and will not form 2-chloro-3-methyl-2-phenylbutane. The mechanisms are shown on the next page.

Electrophilic addition
1,2-Methyl shift
Coordination
E
2-Chloro-3-methyl-2-phenylbutane

Electrophilic addition
Coordination
F
2-Chloro-3-methyl-2-phenylbutane

Electrophilic addition
1,2-Hydride shift
Coordination
G
2-Chloro-3-methyl-2-phenylbutane

Electrophilic addition
Coordination
H
Does not form 2-chloro-3-methyl-2-phenylbutane

Electrophilic addition
Coordination
I
Does not form 2-chloro-3-methyl-2-phenylbutane

Problem 11.27 (SYN)

Think

What is the structure of the compound given? What functional group is present, and what nucleophile formed that functional group? On which carbon atoms are the halogen and H from the addition reaction located? On which carbon atoms could the C=C be located in the precursor?

Solve

The overall reaction and mechanism are both shown below for each of the compounds given. In these examples, the product is an alkyl halide formed from an electrophilic addition reaction of H^+ followed by coordination of the halide anion.

(a) HCl

Cl

:Cl: H Electrophilic addition :Cl: Coordination :Cl:

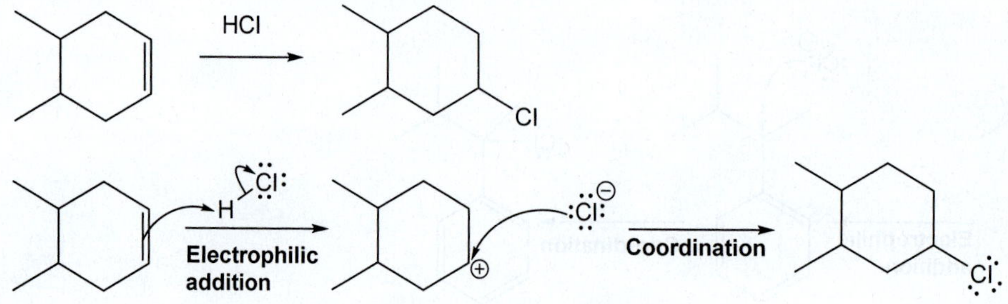

(b) HCl

Cl

:Cl: H Electrophilic addition :Cl: Coordination :Cl:

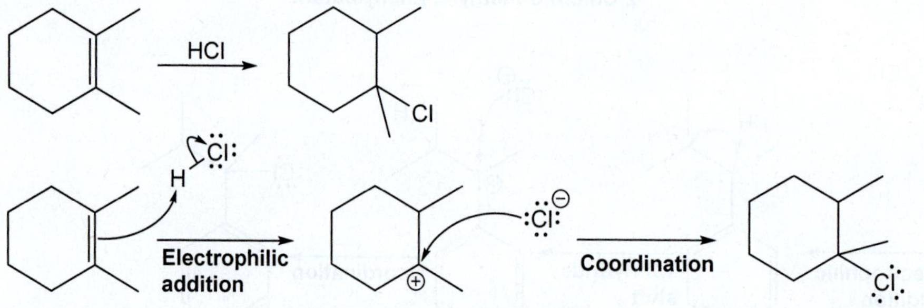

(c) C_6H_5 HBr C_6H_5 C_6H_5

C_6H_5 Br

C_6H_5 :Br: H Electrophilic addition C_6H_5 C_6H_5 :Br: Coordination C_6H_5 C_6H_5 :Br:

C_6H_5

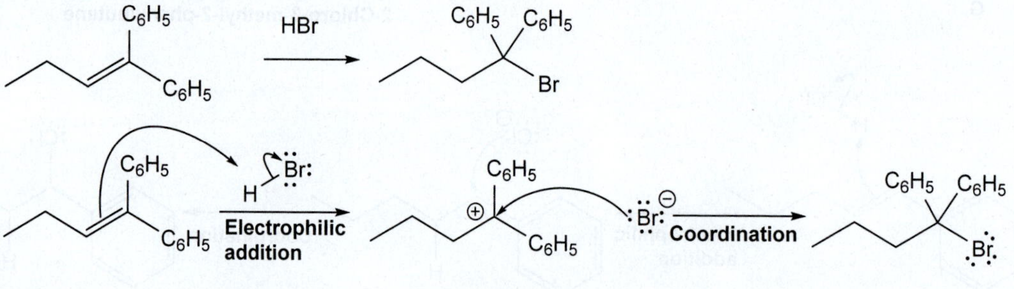

(d) HCl

Cl Cl Cl

:Cl: H Electrophilic addition :Cl: Coordination

Cl :Cl: :Cl: :Cl:

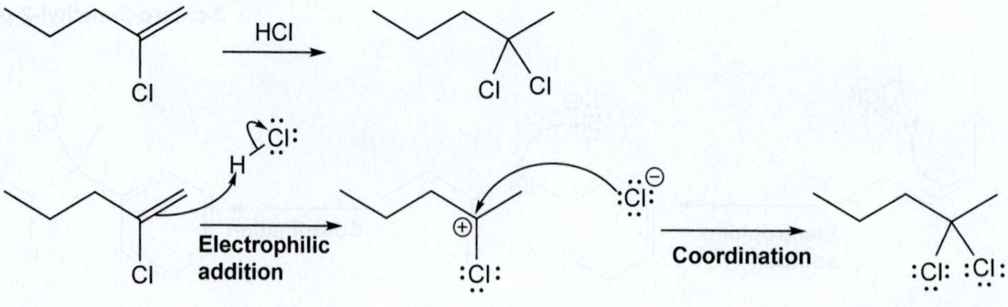

Problem 11.28

Think

How many chiral centers are produced? How many stereoisomers are possible? Can H^+ add to either C atom of the C=C? What intermediate is formed? Can the nucleophile add to either side of the C^+ plane? Which isomer(s) has/have an internal plane of symmetry? How does this affect the optical activity of the compound?

Solve

The alkene is not symmetrical, and addition of H^+ can form two different, secondary carbocation intermediates that are of similar stability. The carbocation intermediate is trigonal planar, and the Cl^- nucleophile can attack the C^+ from in front (wedge) or from behind (dash). This leads to four products. The original wedge C–Cl is not broken and, therefore, has to remain the same configuration in the product. Only *trans*-1,3-dichlorocyclohexane is optically active. All of the other isomers contain an internal plane of symmetry.

Electrophilic addition

Coordination

Coordination

Optically active

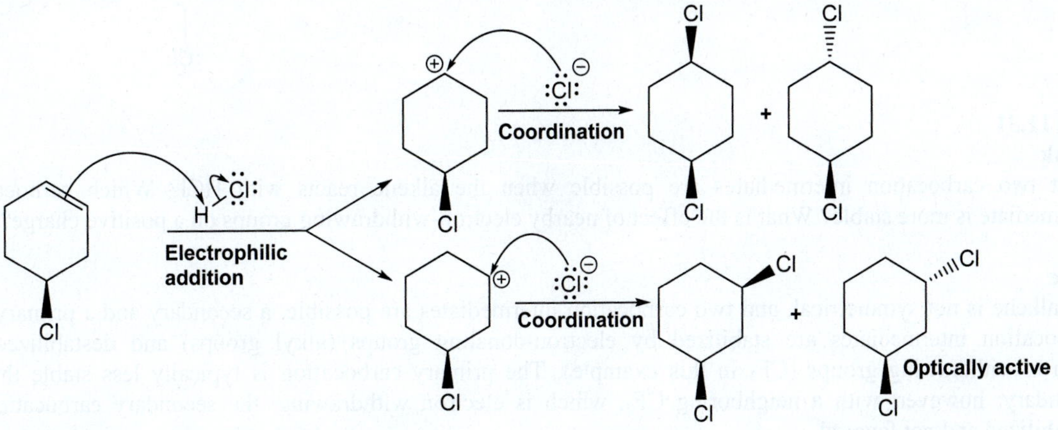

Problem 11.29

Think

On addition of H^+, how many carbocations can form? Which carbocation is more stable? Are resonance structures possible? What acts as the nucleophile in the second step?

Solve

Two possible carbocations can be formed by addition of H^+ to the C=C double bond (**A** or **B**). **B** is the more stable carbocation intermediate because it is resonance stabilized by the lone pair from O. This leads to the major product. Then chloride attacks the carbocation to give the overall products shown below.

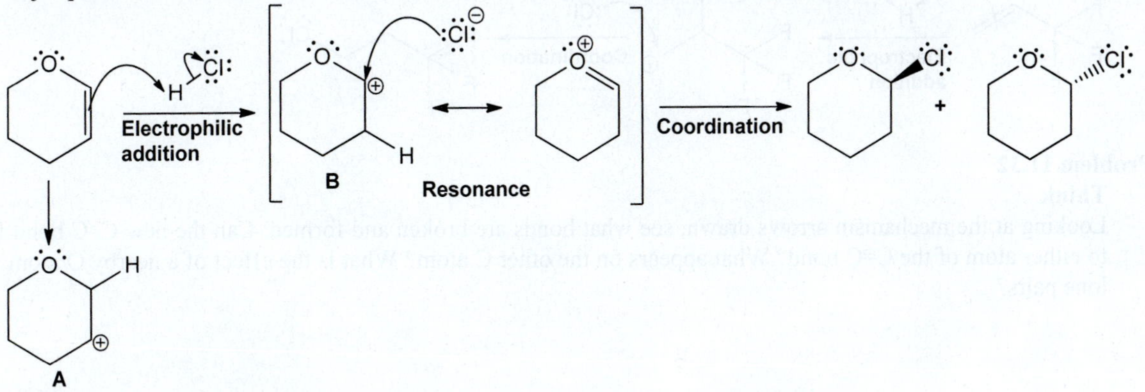

Electrophilic addition

B

Resonance

Coordination

A

Problem 11.30

Think

What carbocation intermediate forms with reaction of penta-1,4-diene and H^+? Is the carbocation able to rearrange? What factor drives this rearrangement? How is the C–Cl bond formed?

Solve

After a π bond undergoes electrophilic addition by the proton in HCl, the secondary C^+ undergoes a 1,2-hydride shift to form a resonance-stabilized secondary C^+. The Cl^- then coordinates to the terminal C^+ to yield the internal alkene, which is more highly substituted and, therefore, more stable than the terminal alkene.

Electrophilic addition

1,2-Hydride shift

Resonance

Coordination

Problem 11.31

Think

What two carbocation intermediates are possible when the alkene reacts with HCl? Which carbocation intermediate is more stable? What is the effect of nearby electron-withdrawing groups on a positive charge?

Solve

The alkene is not symmetrical, and two carbocation intermediates are possible, a secondary and a primary C^+. Carbocation intermediates are stabilized by electron-donating groups (alkyl groups) and destabilized by electron-withdrawing groups (CF_3 in this example). The primary carbocation is typically less stable than a secondary; however, with a neighboring CF_3, which is electron withdrawing, the secondary carbocation is destabilized and not favored.

2° C^+
Less stable

1° C^+
More stable

Electrophilic addition

Coordination

Problem 11.32

Think

Looking at the mechanism arrows drawn, see what bonds are broken and formed. Can the new C–C bond form to either atom of the C=C bond? What appears on the other C atom? What is the effect of a nearby O atom with lone pairs?

Solve

The new C–C bond could form to either atom of the C=C bond. A positive charge appears on the other C atom. The neighboring O atom with lone pairs provides the ability of the C^+ to be resonance stabilized. The mechanism is shown below.

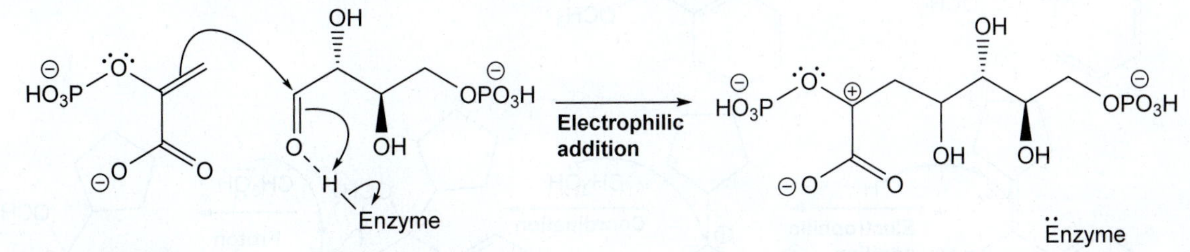

Electrophilic addition

Enzyme

Phosphophenolpyruvate (PEP)

Erythrose-4-phosphate

Section 11.6 Addition of a Weak Acid: Acid Catalysis

Problem 11.33 (SYN)

Think

What is the structure of the compound given? What functional group is present, and what nucleophile formed that functional group? On which carbon atoms are the OH/OR and H from the addition reaction located? On which carbon atoms could the C=C be located in the precursor?

Solve

The overall reaction and mechanism are both shown below and on the next page for each of the compounds given. In these examples, the product is an alcohol or ether formed from an electrophilic addition reaction of H^+ followed by coordination of the water or an alcohol. A final proton transfer reaction takes place to form an uncharged product.

(a) H_2O, H^+

Electrophilic addition — Coordination — Proton transfer

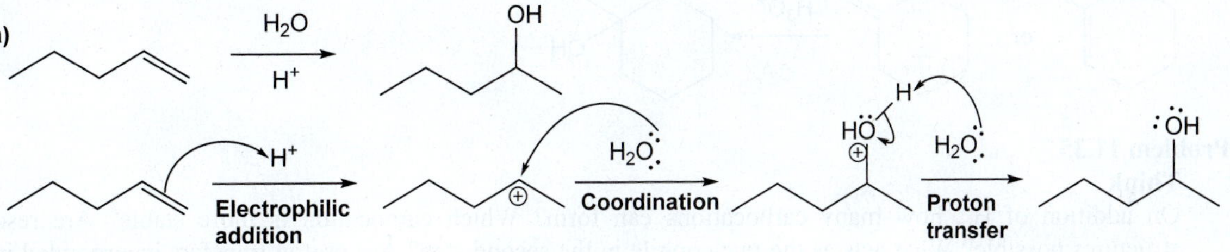

(b) H_2O, H^+

Electrophilic addition — Coordination — Proton transfer

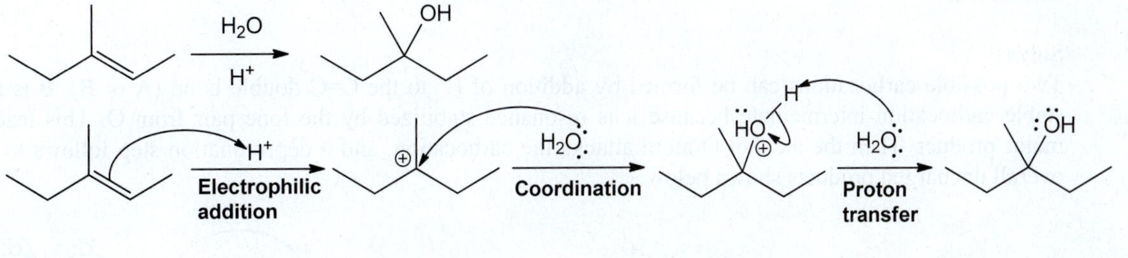

(c) CH_3OH, H^+

Electrophilic addition — Coordination — Proton transfer

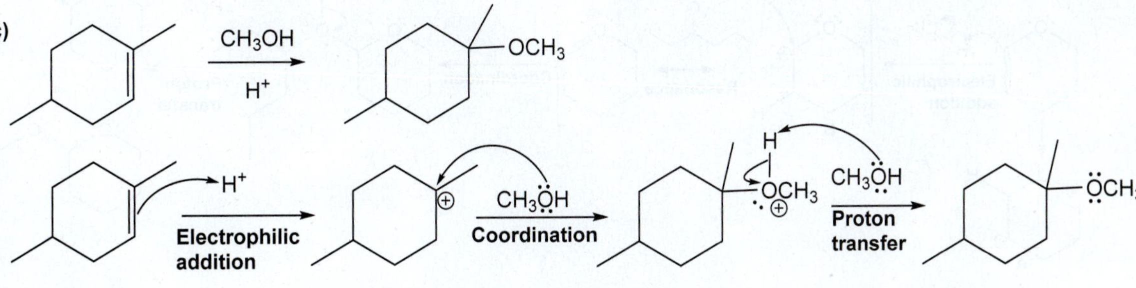

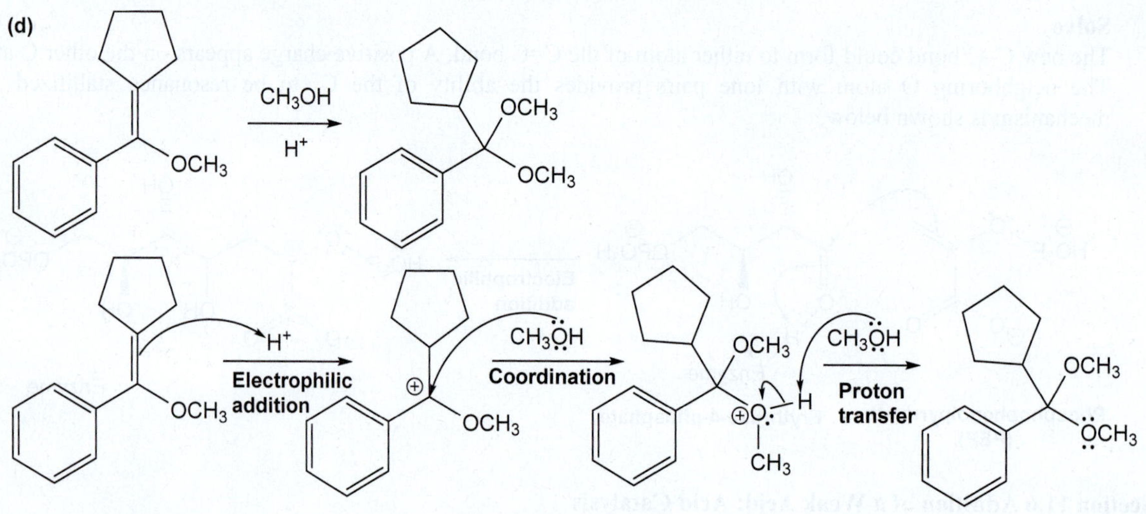

(d)

CH$_3$OH
H$^+$
OCH$_3$
OCH$_3$
OCH$_3$

H$^+$
Electrophilic addition
OCH$_3$
CH$_3$ÖH
Coordination
OCH$_3$
OCH$_3$
CH$_3$
H
CH$_3$ÖH
Proton transfer
OCH$_3$
ÖCH$_3$

Problem 11.34 (SYN)

Think

What is the structure of 1-methylcyclohexanol? On which C atom is the HO located? Does this C atom have to have been part of the C=C? What other C atom(s) could compose the other half of the C=C?

Solve

Addition of water in acid to either of the two alkenes below yields exactly the same carbocation intermediate and, thus, the same product. The C atom to which the OH is attached in the product is the one that was positively charged in the carbocation intermediate. That C$^+$ is produced, in turn, by the addition of H$^+$ on the other C of the initial C=C bond.

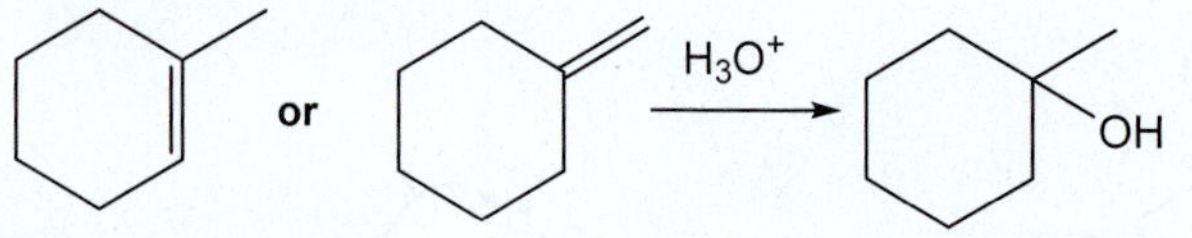

or
H$_3$O$^+$
OH

Problem 11.35

Think

On addition of H$^+$, how many carbocations can form? Which carbocation is more stable? Are resonance structures possible? What acts as the nucleophile in the second step? Are proton transfers incorporated into the mechanism?

Solve

Two possible carbocations can be formed by addition of H$^+$ to the C=C double bond (**A** or **B**). **B** is the more stable carbocation intermediate because it is resonance stabilized by the lone pair from O. This leads to the major product. Then the alcohol O atom attacks the carbocation, and a deprotonation step follows to give the overall uncharged products shown below.

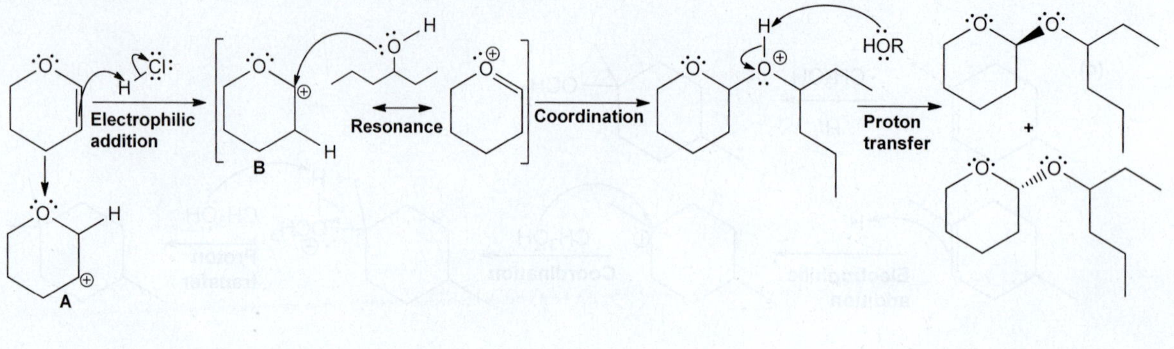

H
Cl:
Electrophilic addition
H
B
H
Resonance
Coordination
H
HÖR
Proton transfer
+
H
A

Problem 11.36

Think

Upon addition of H^+, how many carbocations can form? Which carbocation is more stable? What acts as the nucleophile in the second step? Are proton transfers incorporated into the mechanism? How is the ring formed?

Solve

The strong acid H_2SO_4 adds H^+ to the double bond in the first step—electrophilic addition. The product is the most stable carbocation, which is produced when H^+ adds to the terminal carbon of the C=C bond. In the second step, the nucleophilic OH undergoes coordination to the carbocation to form a ring, and in the final step, a weak base deprotonates the positively charged O atom.

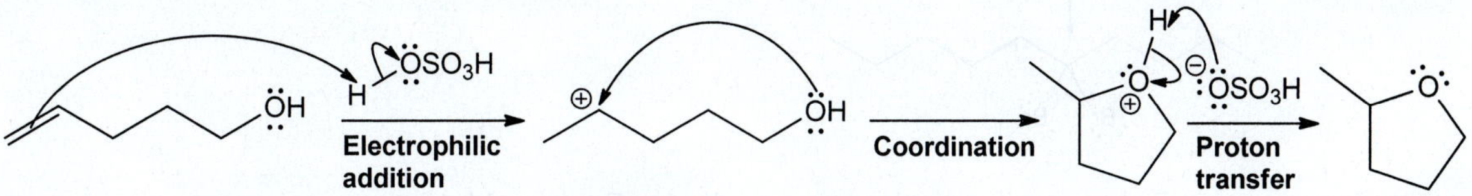

Electrophilic addition → Coordination → Proton transfer

Sections 11.7 and 11.8 Electrophilic Addition to an Alkyne

Problem 11.37

Think

On the basis of the electron-rich and electron-poor sites in the reactants, what type of reaction will take place? How many times will that reaction take place? What aspects of regiochemistry should be considered?

Solve

The alkyne is relatively electron rich, and the H atom of HCl or HBr is relatively electron poor. So the conditions favor an electrophilic addition reaction in which HCl or HBr adds across the multiple bond. In **(a)** and **(b)**, the product of the first addition to the alkyne is a vinylic halide, which, in the presence of excess HCl or HBr, can undergo a second electrophilic addition. The positive charge formed in **(a)** appears on the benzylic carbon. Notice that in each addition to the vinylic halide, the proton adds to the C that is not bonded to the halogen so that the resulting positive charge is resonance stabilized by the halogen and the product is generated in high yield. The mechanisms are shown below and on the next page.

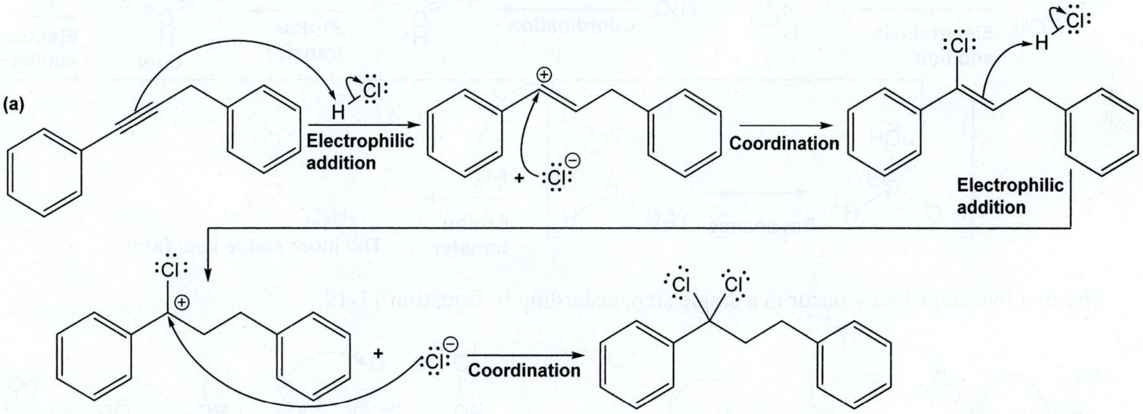

(a)

Electrophilic addition → Coordination → Electrophilic addition → Coordination

(b)

Electrophilic addition

Coordination

Electrophilic addition

Coordination

The first two steps likely occur in a single step, according to Equation 11-19.

Problem 11.38

Think

What is electron rich (nucleophilic)? Electron poor (electrophilic)? Do proton transfer reactions occur? What is the product that forms after the acid-catalyzed addition of water across the triple bond? Is this product stable? If not, what type of rearrangement occurs?

Solve

The first few steps compose the acid-catalyzed addition of water across the triple bond. Notice that the proton adds to the terminal C so that the positive charge that develops is stabilized by any attached alkyl group. Then, acid-catalyzed tautomerization converts the enol into the keto form. In reaction **(b)**, D_3PO_4 + D_2O yields D_3O^+ as the strongest acid in solution.

(a)

Electrophilic addition

Coordination

Proton transfer

Enol

Electrophilic addition

Resonance

Proton transfer

The more stable keto form

The first two steps likely occur in a single step, according to Equation 11-19.

(b)

Electrophilic addition

Coordination

Proton transfer

Enol

Electrophilic addition

Resonance

Proton transfer

The more stable keto form

The first two steps likely occur in a single step, according to Equation 11-19.

(c)

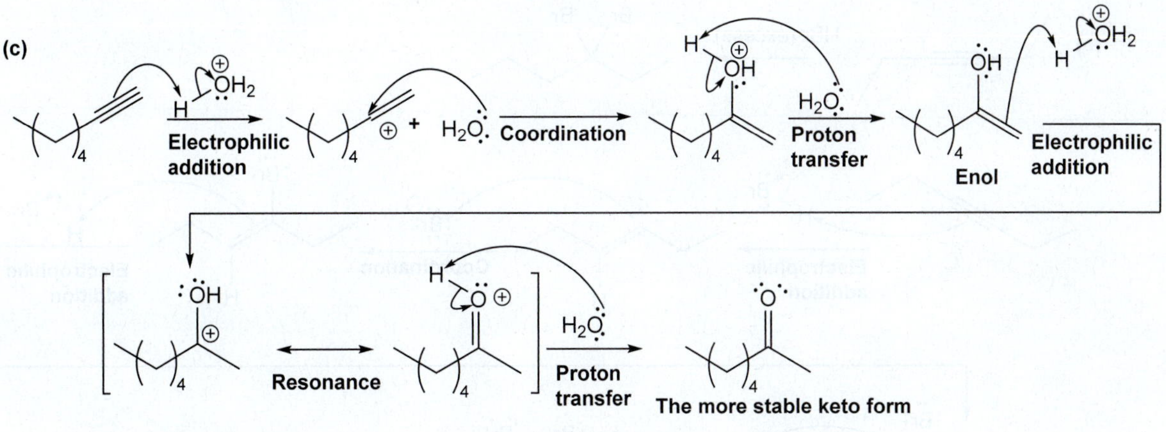

Electrophilic addition

Coordination

Proton transfer

Enol

Electrophilic addition

Resonance

Proton transfer

The more stable keto form

The first two steps likely occur in a single step, according to Equation 11-19.

Problem 11.39 (SYN)

Think

What is the structure of the compound given? What functional group is present, and what nucleophile formed that functional group? On which carbon atoms are the halide and H from the addition reaction located? On which carbon atoms could the C≡C be located in the precursor? How many times did the addition reaction occur?

Solve

The overall reaction and mechanism are both shown below and on the next page for each of the compounds given. In these examples, the product is an alkyl halide (vinylic or geminal) formed from an electrophilic addition reaction of H^+ followed by coordination of the halide anion. In the geminal dihalide products **(a)**, **(b)**, and **(d)**, the addition reaction occurred twice.

(a)

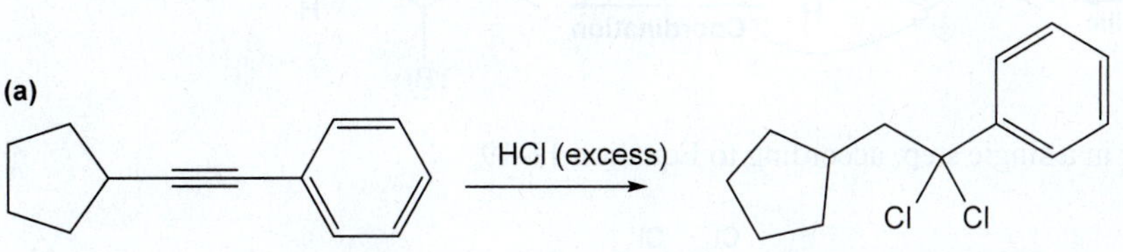

HCl (excess)

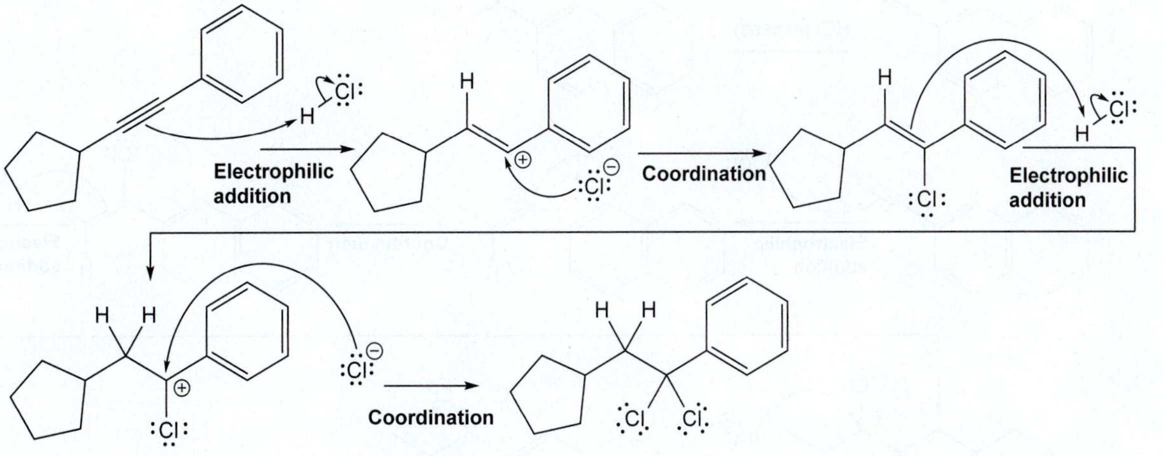

Electrophilic addition

Coordination

Electrophilic addition

Coordination

(b)

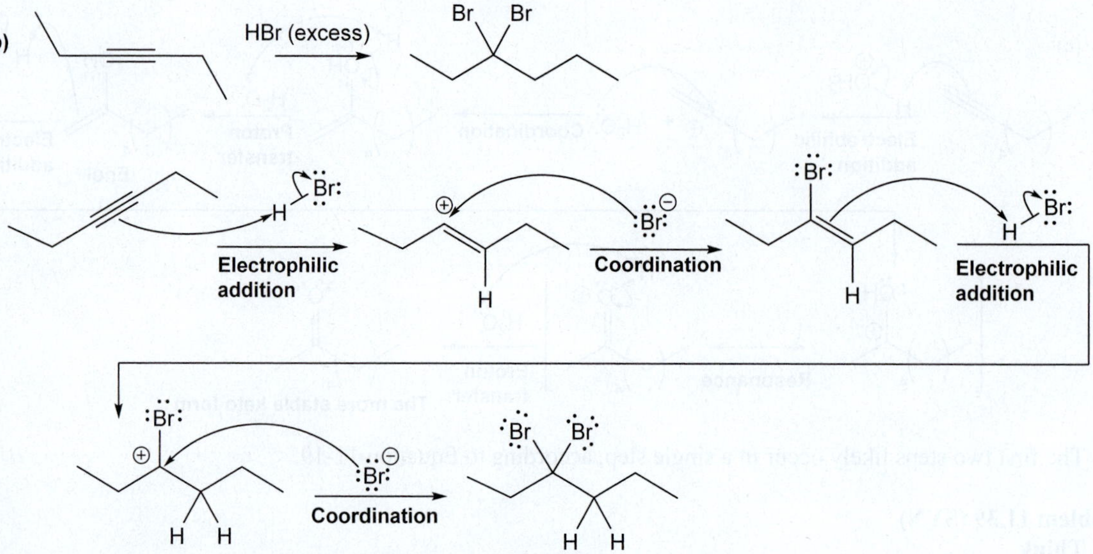

HBr (excess)

Br Br

:Br: H

Electrophilic addition

Coordination

Electrophilic addition

Coordination

The first two steps likely occur in a single step according to Equation 11-19.

(c)

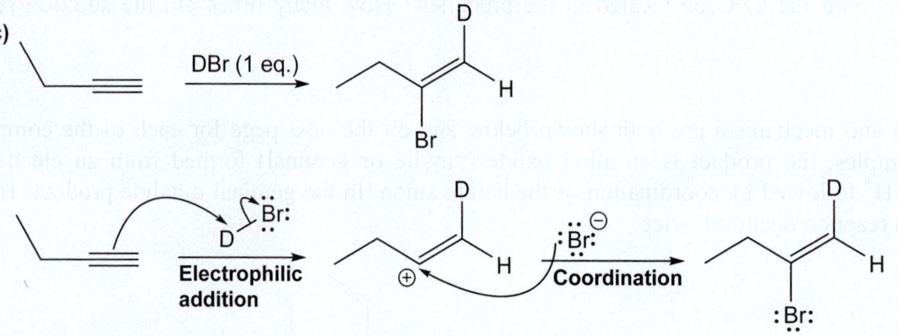

DBr (1 eq.)

D H Br

Electrophilic addition

Coordination

These two steps likely occur in a single step, according to Equation 11-19.

(d)

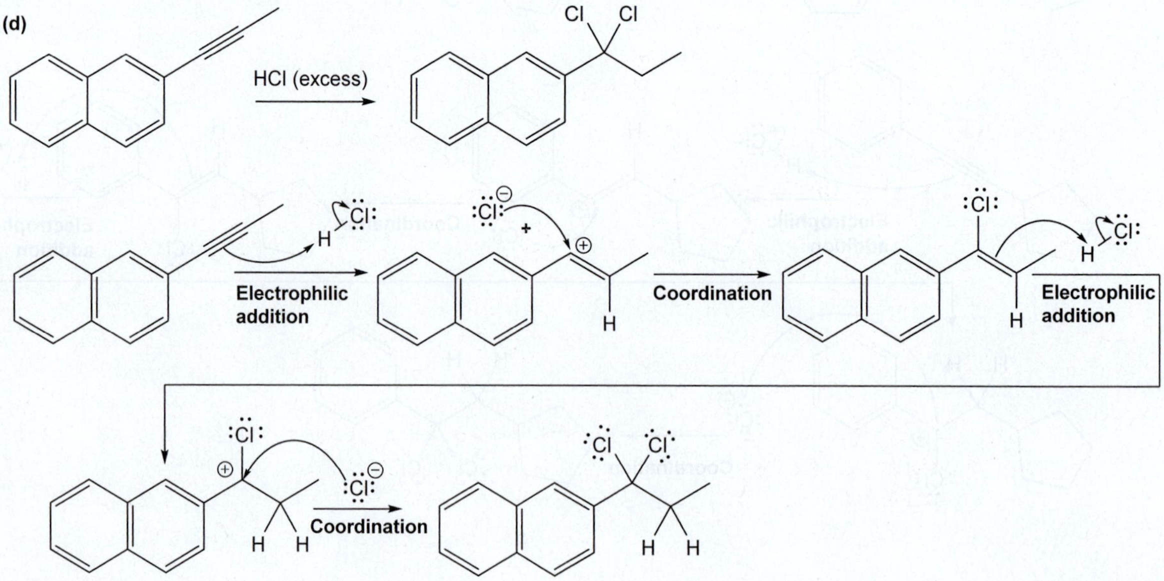

HCl (excess)

Cl Cl

Electrophilic addition

Coordination

Electrophilic addition

Coordination

Problem 11.40 (SYN)

Think

From what type of reaction is a ketone formed given an alkyne starting material? What nucleophile adds to the C^+? On which two C atoms must the C≡C be located in the precursor to produce the positive charge at the appropriate site? What reagent is necessary?

Solve

Each overall reaction and mechanism are shown below. Each is simply an acid-catalyzed addition of water across the triple bond, followed by an acid-catalyzed keto–enol tautomerization.

(a) H_2O, H^+

H^+ Electrophilic addition; H_2O Coordination; H_2O Proton transfer; H^+

Electrophilic addition; Resonance; H_2O Proton transfer

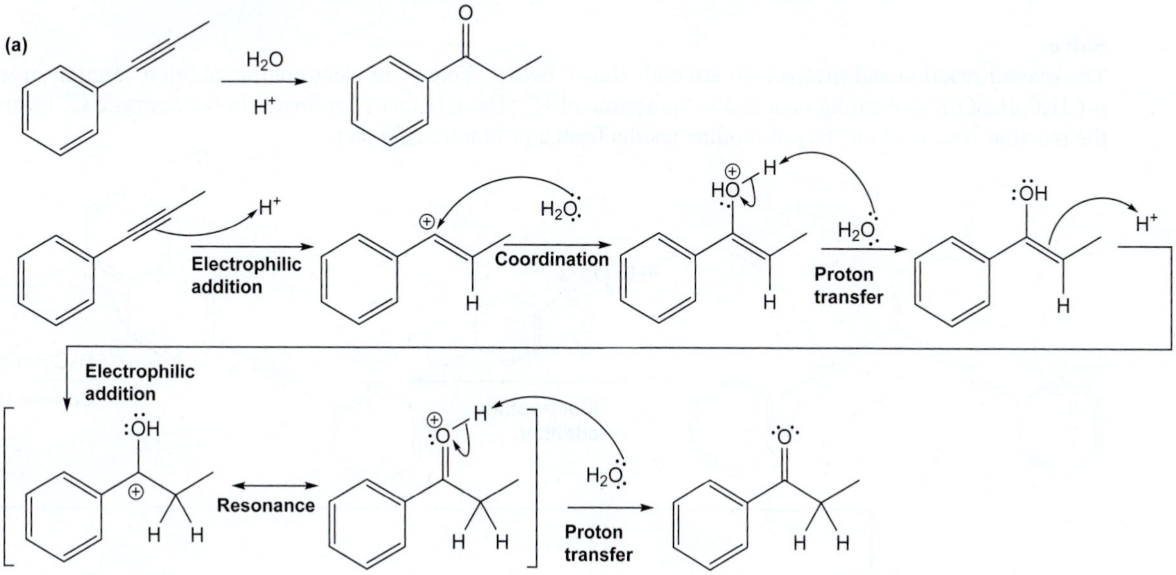

(b) H_2O, H^+

C_6H_5 H^+ Electrophilic addition; H_2O Coordination; H_2O Proton transfer; H^+

Electrophilic addition; Resonance; H_2O Proton transfer

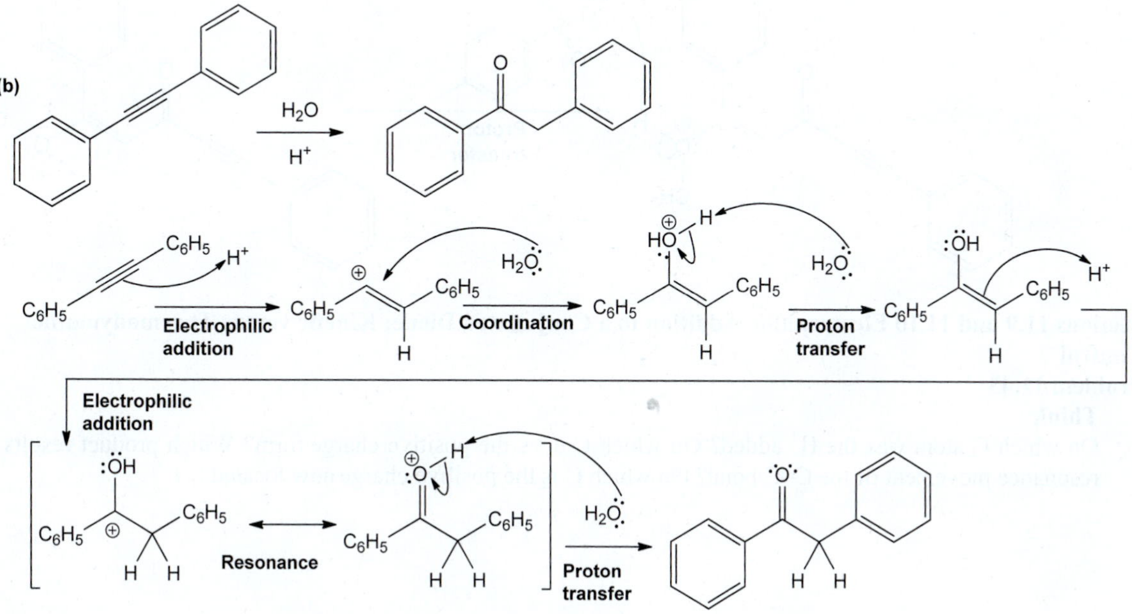

Problem 11.41

Think

What cation must form in the first step of the electrophilic addition reaction? Are two cation options possible in an unsymmetrical alkyne? Which cation is more stable? What nucleophile adds to the C^+? What rearrangement subsequently occurs?

Solve

The most stable carbocation results from addition of a proton to C2 of 1-phenylpropyne, because it places a positive charge on C1, where it can be resonance stabilized. The ketone is C_6H_5–CO–CH_2CH_3. This can be seen in the mechanism in **(a)** of Problem 11.40.

Problem 11.42

Think

What is the structure of the compound given? What functional group is present, and what nucleophile formed that functional group? On which carbon atoms are the CH_3O and H from the addition reaction located? What type of compound is *p*-$CH_3C_6H_4SO_3H$?

Solve

The overall reaction and mechanism are both shown below. This is an electrophilic addition reaction in which *p*-$CH_3C_6H_4SO_3H$ is a strong acid and is the source of H^+. The CH_3OH coordinates to the benzylic C^+ formed in the reaction. The final uncharged product results from a proton transfer step.

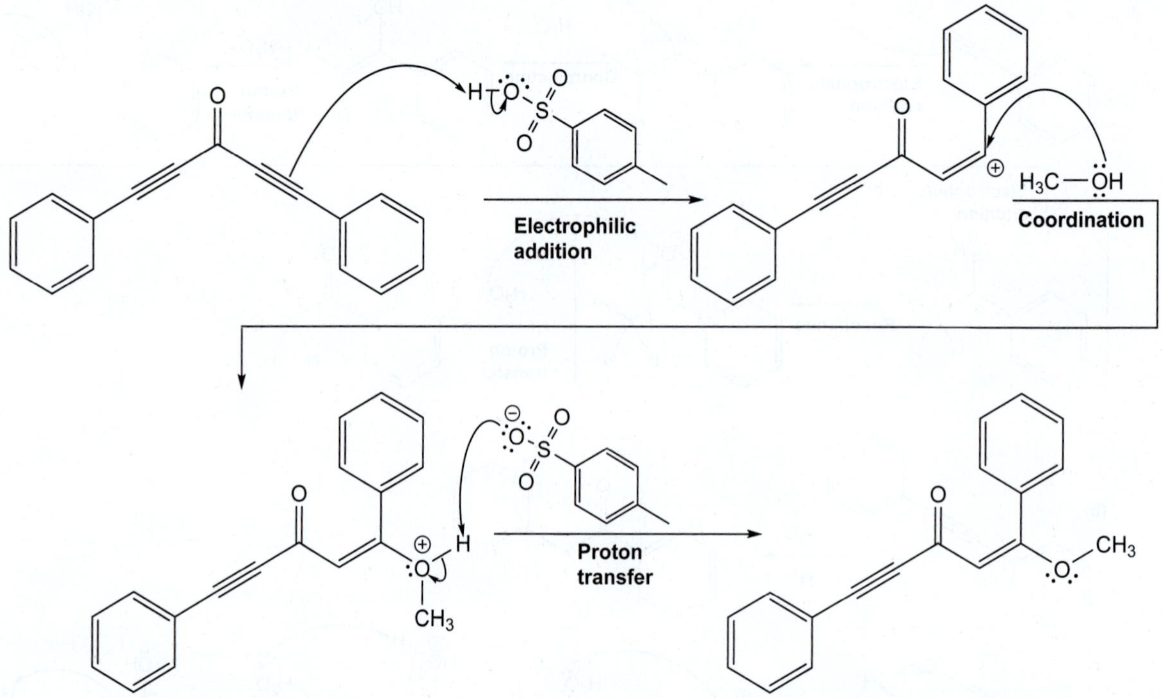

Electrophilic addition

Coordination

Proton transfer

Sections 11.9 and 11.10 Electrophilic Addition to a Conjugated Diene; Kinetic versus Thermodynamic Control

Problem 11.43

Think

On which C atom was the H^+ added? On which C does the positive charge form? Which product results from resonance movement of the C=C bond? On which C is the positive charge now located?

Solve

The first product is formed by 1,2-addition and the second product by 1,4-addition. The mechanism for each addition reaction is shown below.

Problem 11.44

Think

How many C atoms are part of a C=C? On what C does the positive charge form? Which intermediate is most stable? What factors contribute to stability? Are resonance structures possible? In the most stable carbocation intermediate, is there more than one site for the Br^- to attack? Which alkene product is most stable? How does temperature affect the product distribution?

Solve

(a) H^+ can add to any of the four C atoms that are part of the two C=C bonds, producing carbocations **A**–**D**. **D** is the most stable cation because it is resonance stabilized, with a tertiary carbocation resonance contributor. **A** is also resonance stabilized, but both contributors are secondary carbocations.

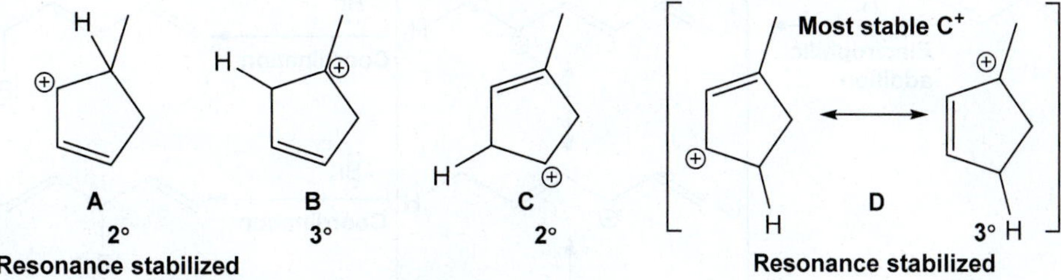

(b) Now that the most stable intermediate is identified, the nucleophile attack occurs on that intermediate. Attack of Br^- forms two (racemic) products. One pair of enantiomers results from 1,2-addition, and another pair of enantiomers results from 1,4-addition.

(c) The 1,2-adduct is the kinetic product and will be the major product at low temperatures.
(d) The 1,2-adduct is also the thermodynamic product and will be the major product at high temperatures. This is because the 1,2-adduct has a more highly alkyl-substituted C=C bond than does the 1,4-adduct, so the 1,2-adduct is the more stable one.

Problem 11.45

Think

How many C atoms are part of a C=C? On what C does the positive charge form? Which intermediate is most stable? What factors contribute to stability? Are resonance structures possible? In the most stable carbocation intermediate, is there more than one site for the Br^- to attack? Which alkene product is most stable? How does temperature affect the product distribution?

Solve

(a) A proton can add to three different C atoms, giving rise to three different carbocation intermediates, **A**–**C**. The most stable ion has the greatest resonance stabilization of the positive charge (**B**). You should draw the resonance contributors to convince yourself.

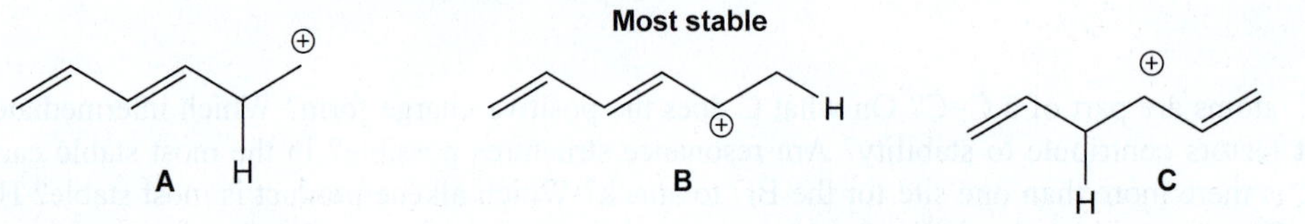

Most stable

A H B H C H

(b) Br^- can attack any of the three C atoms that share the positive charge via resonance delocalization, giving rise to the three products below. The first and second products below will be racemic, because a new chiral center is formed.

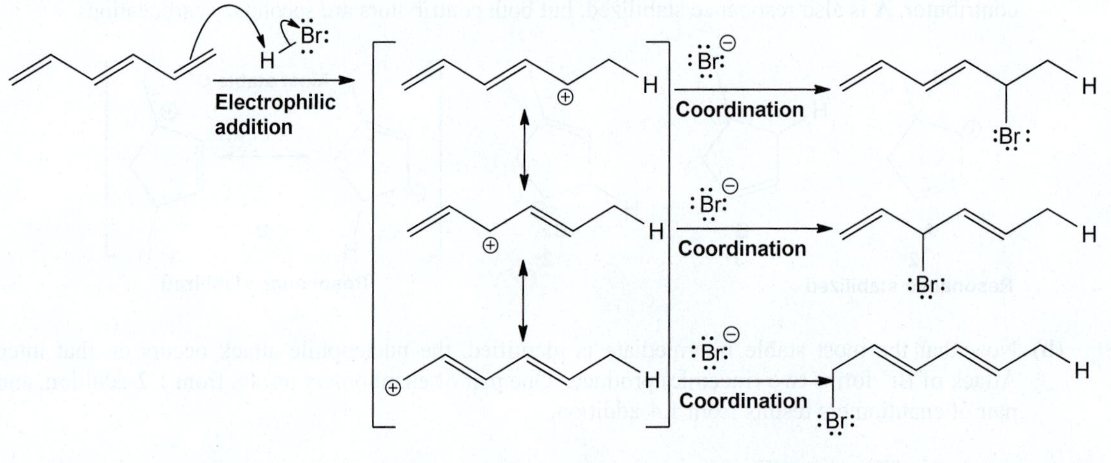

:Br: H Electrophilic addition H :Br: Coordination H :Br: H :Br: Coordination H :Br: H :Br: Coordination H :Br:

(c) The 1,2-adduct (the first product) will be the major product at low temperatures, because it is the kinetic product.
(d) The 1,6-adduct (the third product) will be the major product at high temperatures, because it has the most stable double bonds (they are the most highly substituted, and as we will learn in Chapter 14, they are conjugated) and is the thermodynamic product.

Problem 11.46

Think

To have resonance movement of the C=C bond, what intermediate must form? What elementary step must, therefore, take place? What mechanism allows for this isomerization reaction? Which of the two alkenes is more stable? Should that alkene be favored at low temperatures or high temperatures?

Solve

It is an S_N1 mechanism, taking place in two steps. Bromide leaves, forming a resonance-stabilized cation. Bromide then attacks either positively charged C in the cation but favors attack at the primary C at high temperatures, because it gives rise to the more stable (thermodynamic) product.

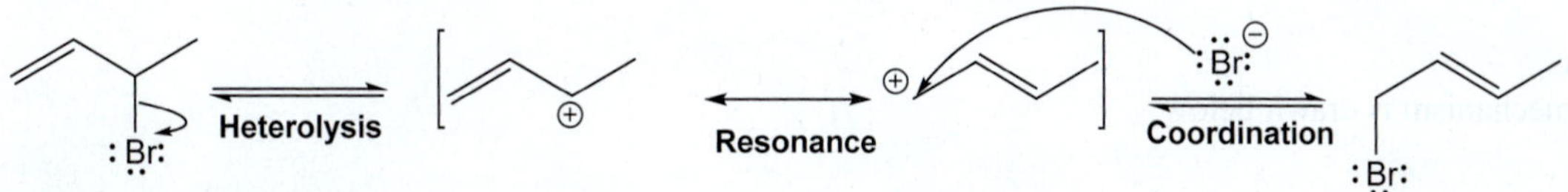

Heterolysis

Resonance

Coordination

Section 11.11 The Organic Chemistry of Biomolecules

Problem 11.47

Think

On which carbon of the C=C does the positive charge form when a proton adds? What acts as the nucleophile in the second step? Are proton transfer reactions involved? In the intermediate that forms the 1,4-cineole product, are any rearrangements possible? If so, why?

Solve

The carbocation forms with the positive charge on the tertiary C when the proton adds to the secondary C. The alcohol acts as the nucleophile and attacks the tertiary C^+ in an intramolecular reaction to produce a new ring. A proton transfer reaction follows to form the uncharged 1,8-cineole product. To form the 1,4-cineole product, the intermediate undergoes a 1,2-hydride shift. The alcohol acts as the nucleophile and attacks the tertiary C^+. A proton transfer reaction follows to form the uncharged 1,4-cineole product.

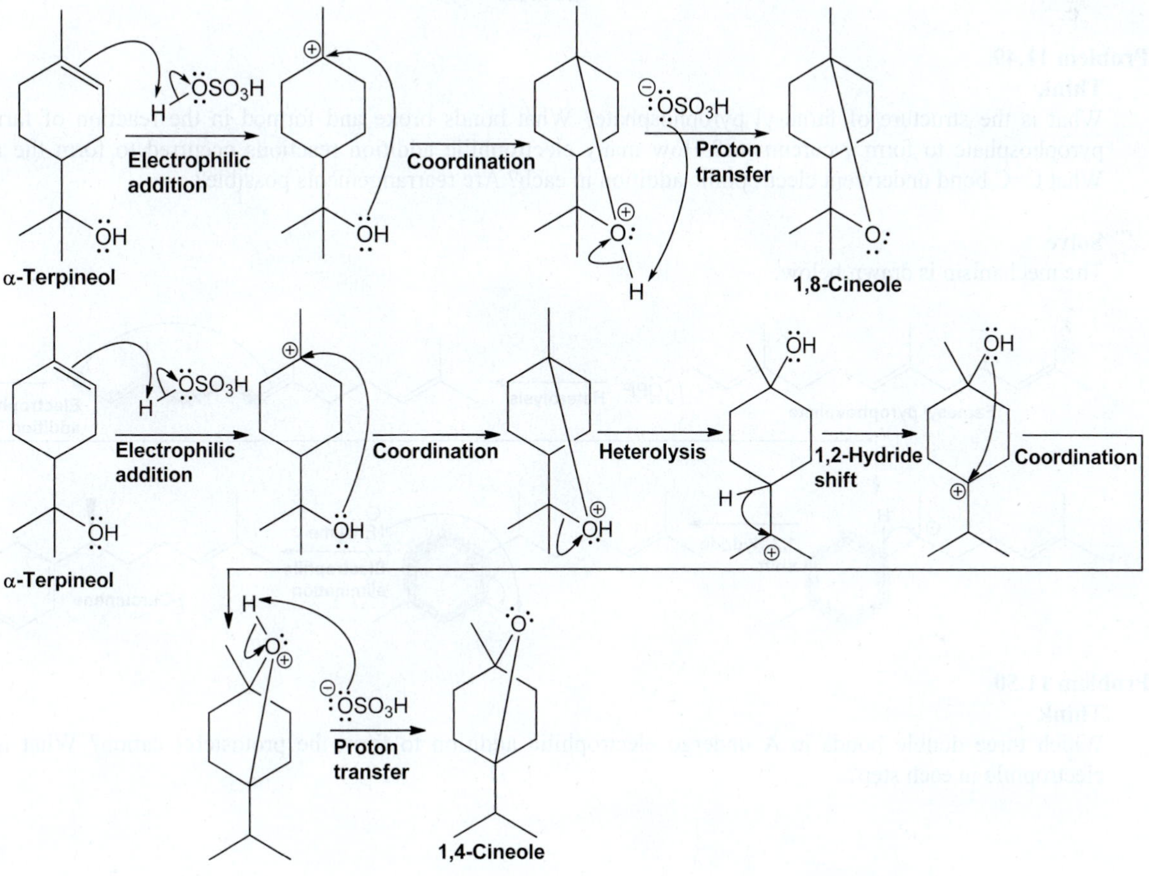

α-Terpineol — Electrophilic addition — Coordination — Proton transfer — 1,8-Cineole

α-Terpineol — Electrophilic addition — Coordination — Heterolysis — 1,2-Hydride shift — Coordination — Proton transfer — 1,4-Cineole

Problem 11.48

Think

What is the structure of geranyl pyrophosphate? What bonds broke and formed in the reaction of geranyl pyrophosphate to form α-pinene? How many rings are present? How many electrophilic addition reactions occurred to form all the rings? What C=C bond underwent electrophilic addition in each?

Solve

The mechanism is drawn below.

Geranyl pyrophosphate — Heterolysis — Electrophilic addition — Electrophilic addition — Enzyme — Electrophile elimination — α-Pinene

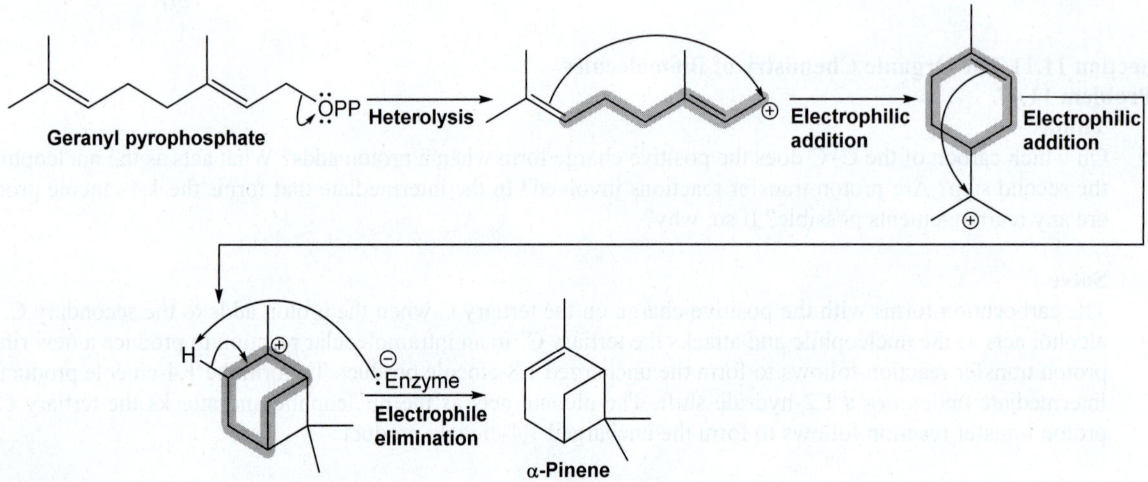

Problem 11.49

Think

What is the structure of farnesyl pyrophosphate? What bonds broke and formed in the reaction of farnesyl pyrophosphate to form γ-curcumene? How many electrophilic addition reactions occurred to form the ring? What C=C bond underwent electrophilic addition in each? Are rearrangements possible?

Solve

The mechanism is drawn below.

Farnesyl pyrophosphate — Heterolysis — Electrophilic addition — 1,2-Hydride shift — Enzyme — Electrophile elimination — γ-Curcumene

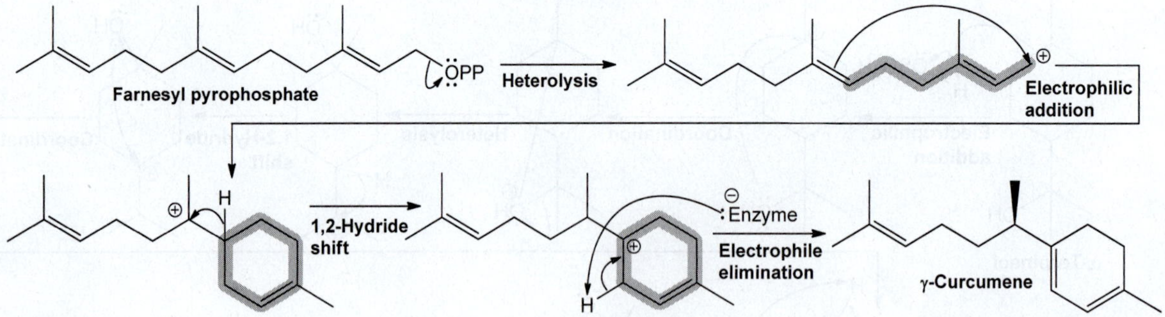

Problem 11.50

Think

Which three double bonds in **A** undergo electrophilic addition to form the protosterol cation? What is the electrophile in each step?

Solve

The three electrophilic addition reactions are shown below with the three C=C bonds highlighted.

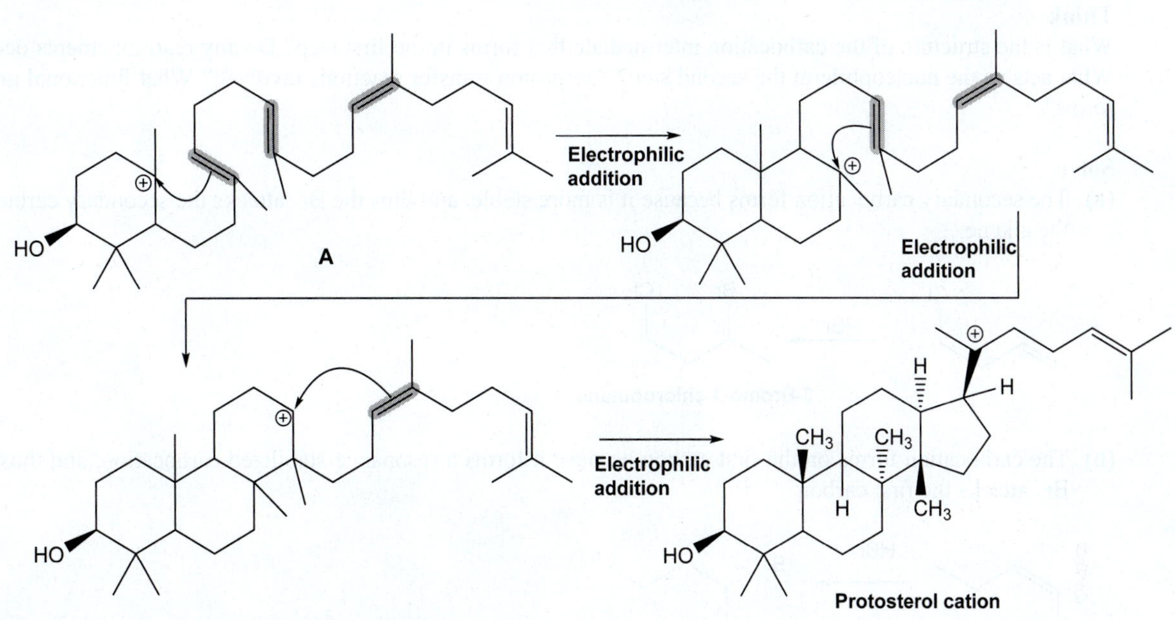

Problem 11.51

Think

Where is the positive charge in **B**? Where is the positive charge in **C**? What type of rearrangement occurred in each step?

Solve

The four C atoms that participate in the three carbocation arrangements are highlighted below. The first rearrangement is a 1,2-hydride shift, and the next two steps are 1,2-methyl shifts. The mechanism is shown below.

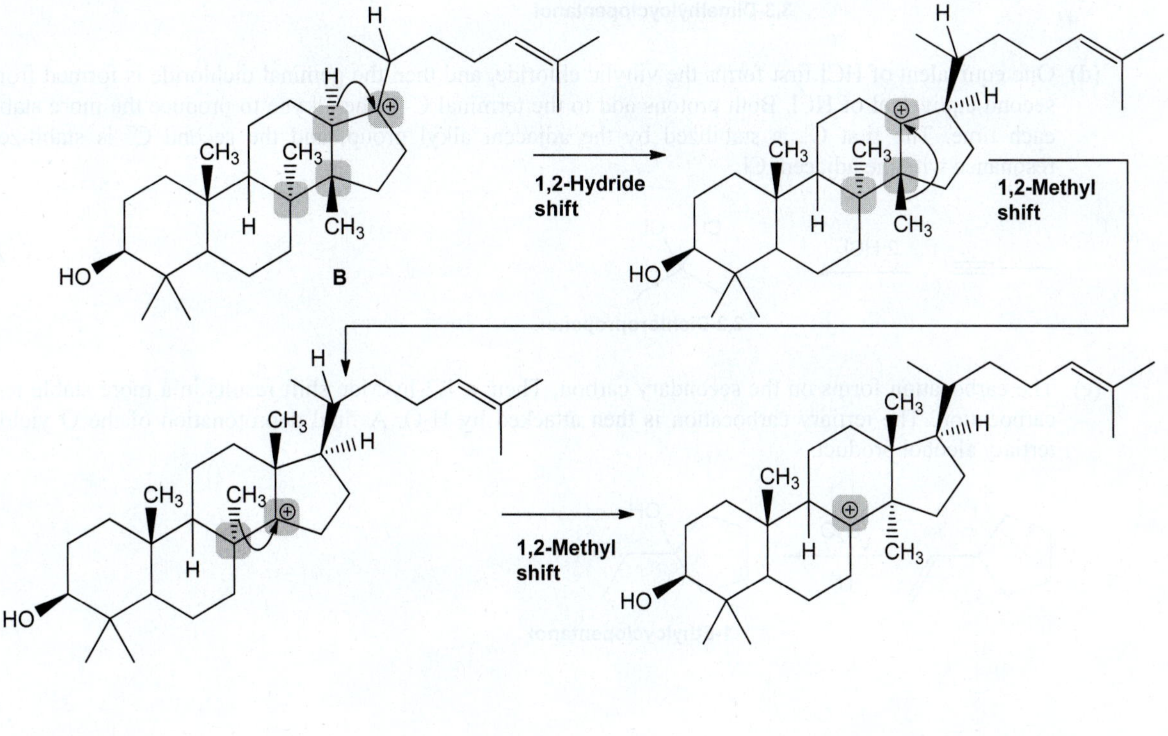

Integrated Problems

Problem 11.52

Think

What is the structure of the carbocation intermediate that forms in the first step? Do any rearrangements occur? What acts as the nucleophile in the second step? Are proton transfer reactions involved? What functional group forms?

Solve

(a) The secondary carbocation forms because it is more stable, and thus the Br^- attacks the secondary carbon of the alkene.

Cl HBr Br Cl

3-Bromo-1-chlorobutane

(b) The carbocation forms on the first carbon because it forms a resonance-stabilized carbocation, and thus the Br^- attacks the first carbon.

HBr Br Cl Cl

1-Bromo-1-chlorobutane

(c) The alkene is symmetrical, and the carbocation can form on either carbon to produce the same product. Water attacks that C^+, and the O is deprotonated to yield the alcohol.

H_2O H^+ HO

3,3-Dimethylcyclopentanol

(d) One equivalent of HCl first forms the vinylic chloride, and then the geminal dichloride is formed from the second equivalent of HCl. Both protons add to the terminal C of the alkyne to produce the more stable C^+ each time. The first C^+ is stabilized by the adjacent alkyl group, and the second C^+ is stabilized by resonance with the adjacent Cl.

2 HCl Cl Cl

2,2-Dichloropropane

(e) The carbocation forms on the secondary carbon, Then, a 1,2-hydride shift results in a more stable tertiary carbocation. The tertiary carbocation is then attacked by H_2O. A final deprotonation of the O yields the tertiary alcohol product.

H_2O H^+ OH

1-Ethylcyclopentanol

Problem 11.53 (SYN)

Think

If the reaction starts and ends with an alkene, what two types of reactions might have occurred? Did the carbon skeleton change? Did a rearrangement occur? What reagents could accomplish the addition reaction? What reagents could accomplish the elimination reaction? Is the least substituted or most substituted alkene formed?

Solve

(a) Addition of water follows Markovnikov's rule (proceeds through the most stable carbocation intermediate), and elimination of water under acidic conditions follows Zaitsev's rule (forms the most stable alkene product).

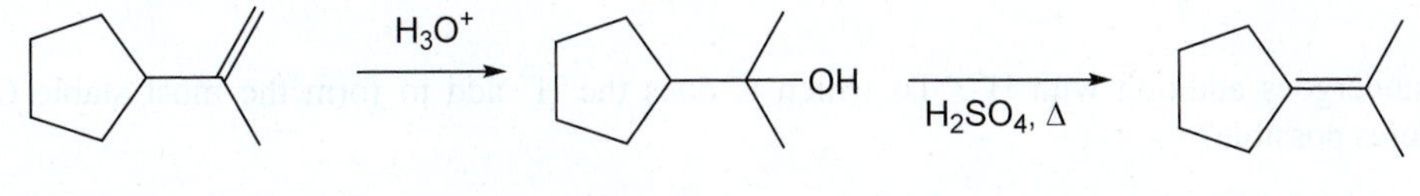

H_3O^+ OH H_2SO_4, Δ

(b) HBr adds according to Markovnikov's rule, and elimination with LDA yields the anti-Zaitsev product because LDA is a very strong and bulky base.

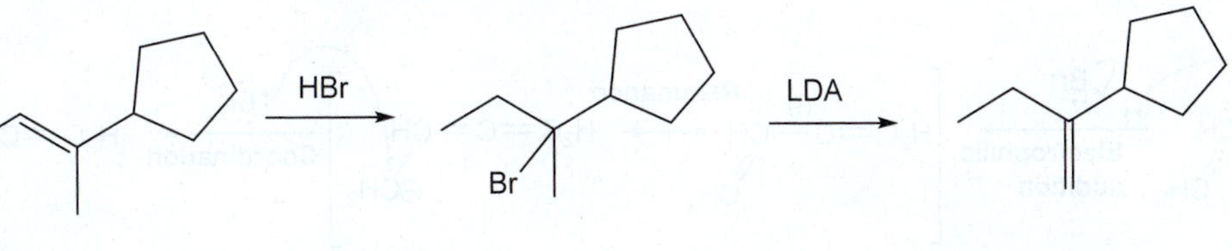

HBr Br LDA

Problem 11.54

Think

What is the electrophile in the electrophilic addition step? In Step 2, what acts as the nucleophile to coordinate to the C^+?

Solve

(a) The π bond undergoes electrophilic addition with the carbocation R^+, then chloride attacks.

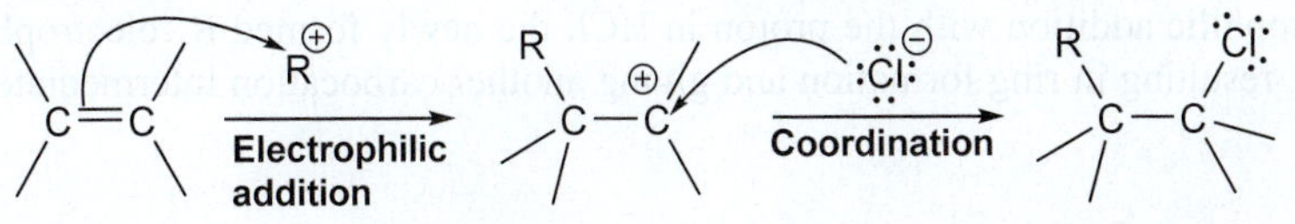

R R :Cl: R :Cl: C=C C–C C–C Electrophilic addition Coordination

(b) The major product is 2-chloro-2,4-dimethylpentane. $AlCl_3$ produces the isopropyl cation, $CH_3CH^+CH_3$, which adds to the terminal carbon of the double bond to yield the more stable tertiary carbocation. Finally, Cl^- coordinates to the new carbocation.

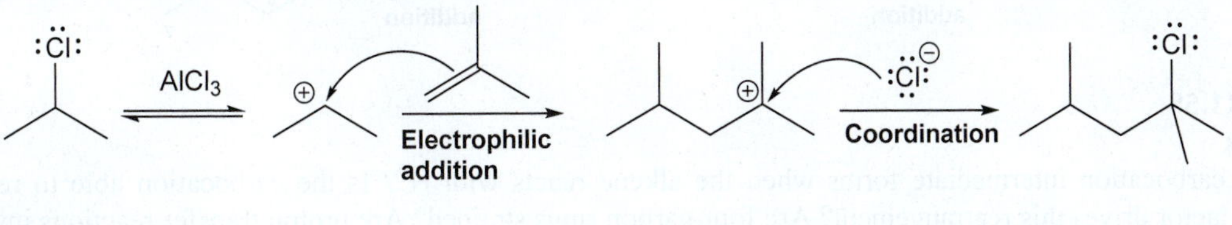

:Cl: $AlCl_3$ Electrophilic addition :Cl: Coordination :Cl:

Problem 11.55

Think

In acid, what is the electrophile that adds to the C=C? To which C atom does it add to produce the most stable carbocation? If the carbocation that is produced reacts with an alkene, what reaction can take place?

Solve

The π bond in styrene undergoes electrophilic addition with a proton, forming a resonance-stabilized carbocation. Successive molecules of styrene add to the cation, lengthening the polymer chain.

H⁺ Electrophilic addition; Electrophilic addition; Electrophilic addition

C_6H_5 C_6H_5 C_6H_5 C_6H_5 C_6H_5 C_6H_5 C_6H_5 C_6H_5 C_6H_5

Problem 11.56

Think

Which π bond undergoes addition with H^+? To which C does the H^+ add to form the most stable C^+? Are resonance structures possible?

Solve

This is a two-step addition of HBr to a multiple bond. Addition of H^+ to the terminal C of the triple bond yields a resonance-delocalized carbocation, which then undergoes coordination with Br^-.

HC≡C—CH=CH₂ + H—Br: → (Electrophilic addition) [H₂C=C⁺—CH=CH₂ ↔ (Resonance) H₂C=C=CH—⁺CH₂] → :Br:⁻ (Coordination) H₂C=C=CH—H₂C—Br:

Problem 11.57

Think

What carbocation intermediate forms with reaction of hepta-1,6-diene and H^+? To which C atom of the C=C bond does H^+ add to produce the most stable C^+? What intramolecular reaction can then take place to form the ring? What is the structure of the carbocation intermediate now? What can then act as the nucleophile?

Solve

After a π bond undergoes electrophilic addition with the proton in HCl, the newly formed R^+ electrophile adds to the second C=C double bond, resulting in ring formation and giving another carbocation intermediate. Attack by Cl^- finishes the mechanism.

Electrophilic addition; Electrophilic addition; Coordination

Problem 11.58

Think

What carbocation intermediate forms when the alkene reacts with H^+? Is the carbocation able to rearrange? What factor drives this rearrangement? Are four-carbon rings strained? Are proton transfer reactions involved?

Solve

After a π bond undergoes electrophilic addition of the proton in HBr, the tertiary C^+ undergoes a rearrangement twice. Each rearrangement is a 1,2-alkyl shift and is driven by the relief of ring strain on going from a four-membered ring to a five-membered ring. The water coordinates to the carbocation, and a final proton transfer reaction is involved to form an uncharged alcohol product. The mechanism is on the next page.

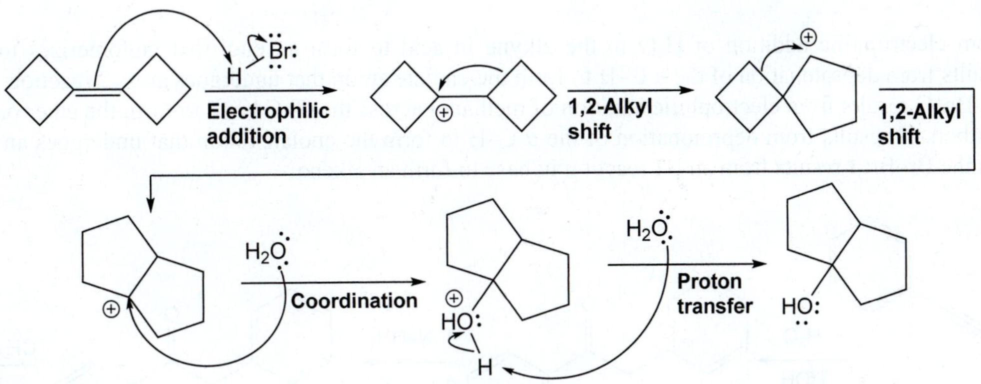

H
Br:
Electrophilic addition
1,2-Alkyl shift
1,2-Alkyl shift
H_2O:
Coordination
HO:
H
H_2O:
Proton transfer
HO:

Problem 11.59

Think

What is the electron-rich and electron-poor species in each reaction? Does addition, substitution, or elimination occur? Are rearrangements possible?

Solve

A results from the electrophilic addition of water to a C=C bond to form an alcohol on the benzylic C atom (most stable C^+ intermediate). **B** results from deprotonation of the O–H bond by hydride in the first step to form the alkoxide nucleophile. The epoxide ring is opened by attack of the RO^- at the less substituted C atom. **C** results from two S_N2 reactions to turn the ROH into an RBr. **D** results from an E2 reaction to form the most substituted alkene.

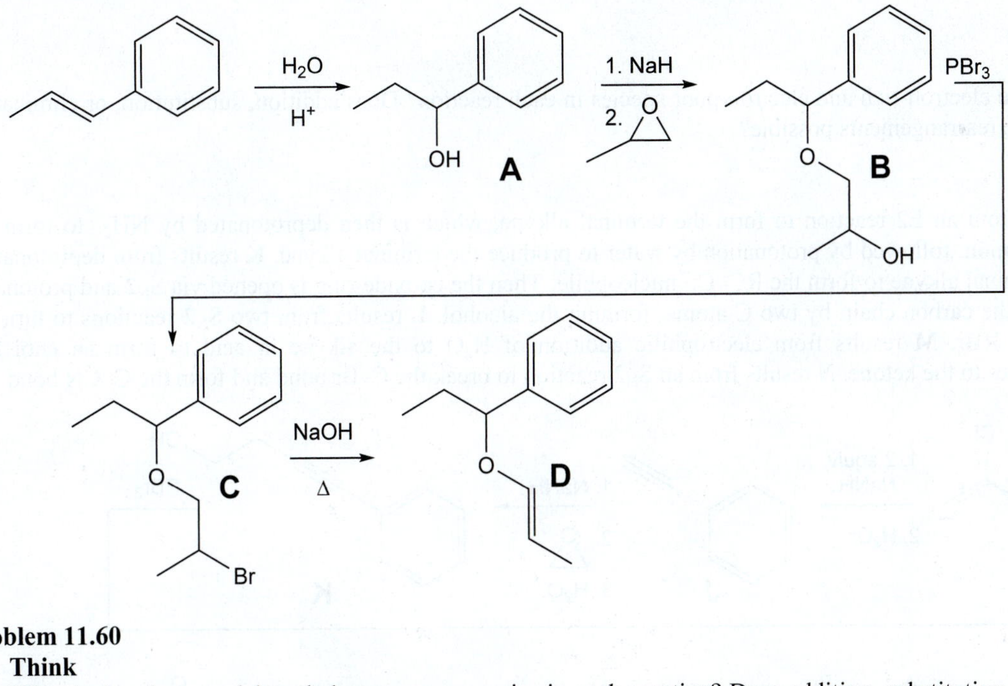

H_2O
H^+
OH
A
1. NaH
2. O
O
B
OH
PBr_3
O
C
Br
NaOH
Δ
O
D

Problem 11.60

Think

What is the electron-rich and electron-poor species in each reaction? Does addition, substitution, or elimination occur? Are rearrangements possible?

Solve

E results from electrophilic addition of H_2O to the alkyne in acid to form an enol that tautomerizes to the ketone. **F** results from deprotonation of the α C–H to form the enolate anion that undergoes an S_N2 reaction with the allylic C–Br. **G** results from electrophilic addition of methanol across the C=C bond to form the ether on the secondary carbon. **H** results from deprotonation of the α C–H to form the enolate anion that undergoes an S_N2 reaction with the Br–Br. **I** results from an E2 reaction in base to form an alkene.

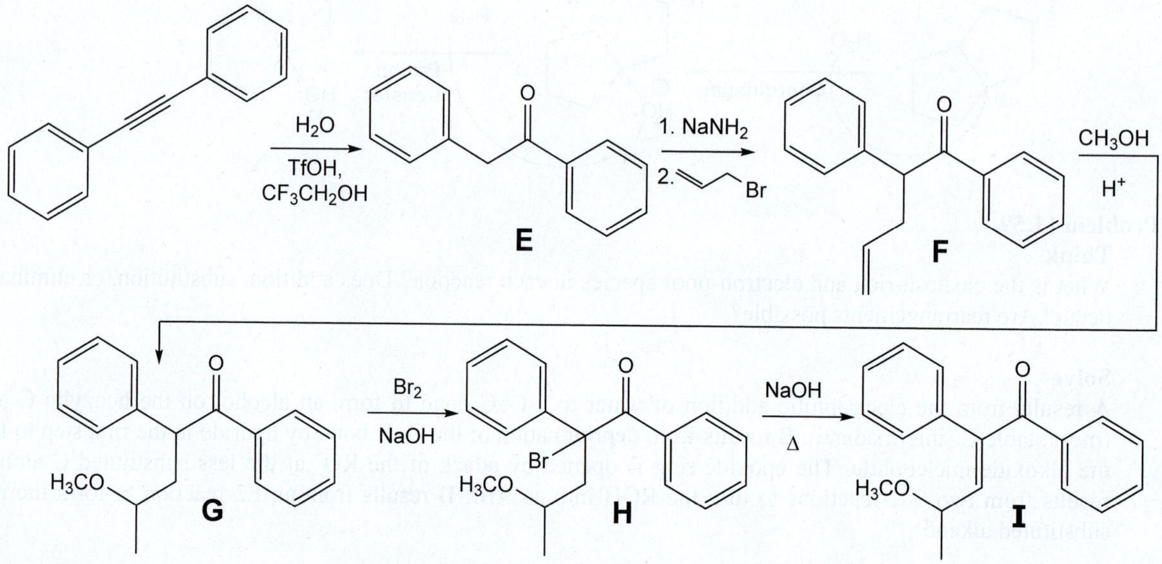

Problem 11.61

Think

What is the electron-rich and electron-poor species in each reaction? Does addition, substitution, or elimination occur? Are rearrangements possible?

Solve

J results from an E2 reaction to form the terminal alkyne, which is then deprotonated by NH_2^- to form the alkynide anion, followed by protonation by water to produce the terminal alkyne. **K** results from deprotonation of the terminal alkyne to form the RC≡C:$^-$ nucleophile. Then the epoxide ring is opened via S_N2 and protonated to extend the carbon chain by two C atoms, forming the alcohol. **L** results from two S_N2 reactions to turn the ROH into RBr. **M** results from electrophilic addition of H_2O to the alkyne in acid to form an enol that tautomerizes to the ketone. **N** results from an S_N2 reaction to break the C–Br bond and form the C–CN bond.

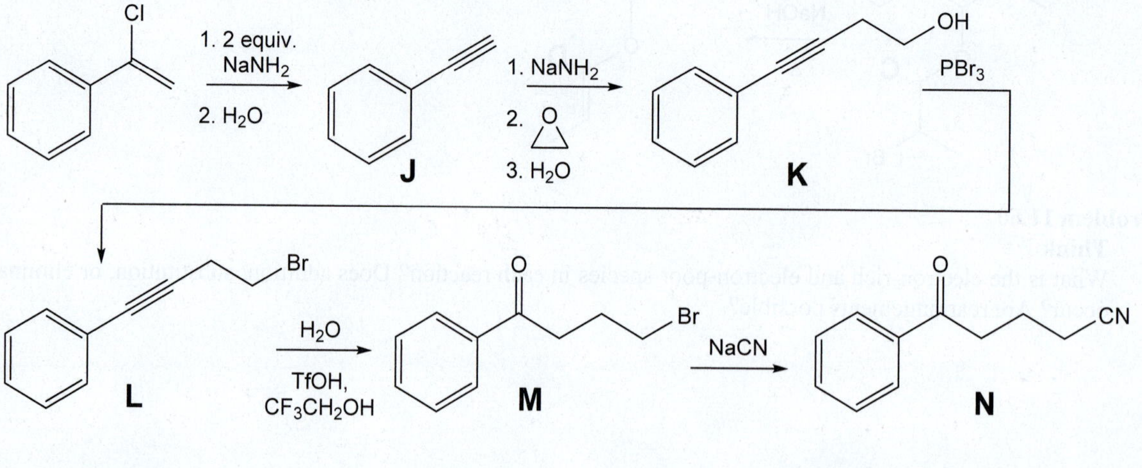

CHAPTER 12 | Electrophilic Addition to Nonpolar π Bonds 2: Reactions Involving Cyclic Transition States

Your Turn Exercises

Your Turn 12.1

Think

How many single bonds does each highlighted C have? How many bonds does a double bond represent?

Solve

In the cyclic product, each highlighted C atom has four single bonds, for four total bonds. In the initial alkene, each highlighted C atom has two single bonds and one double bond. Each double bond represents two bonding pairs of electrons, or two bonds, so each highlighted C in the initial alkene also has four total bonds.

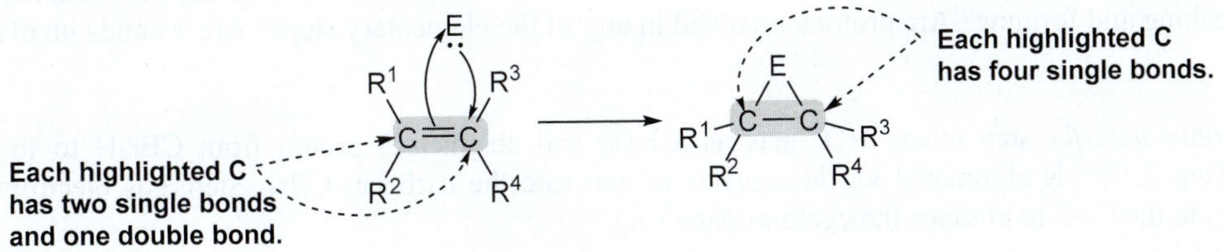

Each highlighted C has two single bonds and one double bond.

Each highlighted C has four single bonds.

Your Turn 12.2

Think

Is there free rotation about the C–C bond in the intermediate? What should be the cis/trans relationship of the CH_3 groups in the alkene compared to that of the product: conserved, inverted, or scrambled?

Solve

The cis/trans relationship for the CH_3 groups attached to the C=C double bond should be conserved. Groups that are on the same side of the C=C double bond (cis) should both be wedge (or both dash) in the product. For groups that are on opposite sides, one bond should be wedge and one should be dash. The cyclic product for the trans alkene is chiral—specifically, enantiomers are produced (the other possible product is shown below). Therefore, the reaction product mixture is racemic.

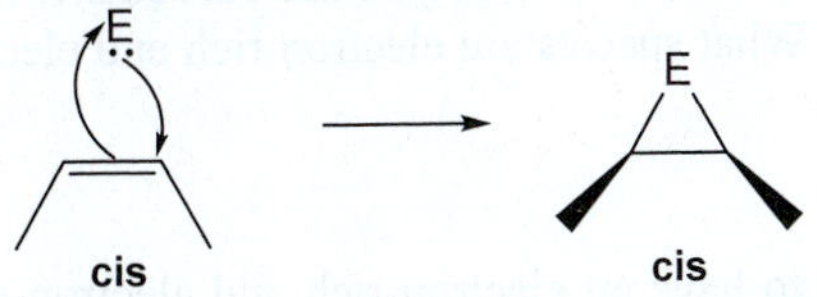

cis cis

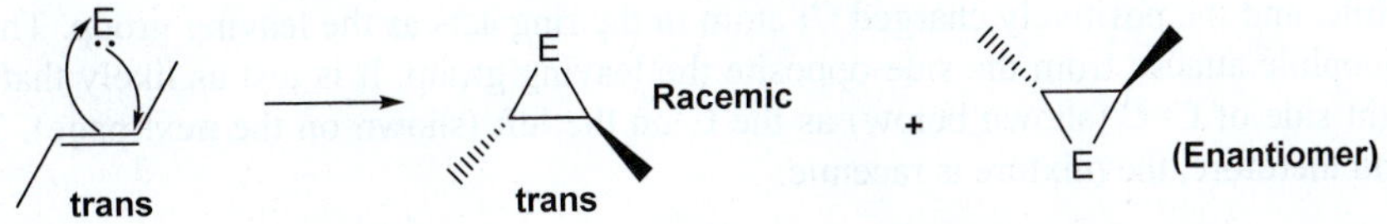

trans trans Racemic (Enantiomer)

Your Turn 12.3

Think

In the formal charge method, how many electrons in a bond are assigned to each atom? How many electrons in a lone pair are assigned to an atom? How many electrons does C have in the carbene molecule? How does this compare to the number of valence electrons in an isolated carbon atom?

Solve

Carbon has four valence electrons. In the formal charge method, half of the electrons in a bond are assigned to each atom. The carbene C has two bonds, thus two electrons. Each nonbonding electron (dot) represents one electron. There are two electrons in the lone pair. Thus, the carbene C is assigned four valence electrons, the same as in an isolated C atom, so the formal charge is 0.

1 electron
2 electrons
1 electron
A carbene

Carbene C
4 electrons
Formal charge = 0

Your Turn 12.4

Think

Using Equation 12-8 as a guide, what bonds broke and formed in each step? How do you use curved arrows to show bonds breaking and forming? Are protons involved in any of the elementary steps? Are π bonds involved?

Solve

Step 1 is a *proton transfer* step where HO^- acts as a base and abstracts a proton from CBr_3H to form a carbanion. In Step 2, Br^- is eliminated via *heterolysis* to generate the carbene, CBr_2. Step 3 is electrophilic addition of CBr_2 to the C=C to produce the cyclopropane ring.

This step is slow because the product carbene is highly unstable.
Dibromocarbene
A new cyclopropane ring is produced.
1. Proton transfer
2. Heterolysis
3. Electrophilic addition

Your Turn 12.5

Think

Using Equation 12-10 as a guide, what bonds broke and formed in each step? How do you use curved arrows to indicate bond breaking and bond formation? Are π bonds involved? What species are electron rich and electron poor in each step?

Solve

In Step 1, the C=C π bond is electron rich and polarizes the Cl–Cl to have an electron-rich and electron-poor site. In a concerted electrophilic addition step, the chloronium ion intermediate and Cl^- ion are formed. In Step 2, the Cl^- acts as a nucleophile, and the positively charged Cl atom in the ring acts as the leaving group. This is an S_N2 step where the nucleophile attacks from the side opposite the leaving group. It is just as likely that Cl^- will attack the C on the right side of C=C (shown below) as the C on the left (shown on the next page). This produces the enantiomer, and therefore, the mixture is racemic.

Attack right C atom:

Chloronium ion intermediate
1. Electrophilic addition
2. S_N2
Same as

Attack left C atom:

Chloronium ion intermediate

1. Electrophilic addition

2. S_N2

Same as

Your Turn 12.6

Think

Using Equation 12-18 as a guide, identify which bonds broke and formed in each step. How do you use curved arrows to show bond formation/breaking? What species are electron rich and electron poor in each step? What nucleophiles are present? Do any steps involve proton transfers?

Solve

Step 1 is an electrophilic addition that produces the chloronium ion intermediate, and Step 2 is an S_N2 step that opens the ring as H_2O acts as the nucleophile and attacks the electrophilic C of the ring. *Note:* Either the left or right side of the ring could be attacked, but this example shows just the right-side C being attacked. In this example, H_2O is the solvent and is present in a much larger concentration than Cl^- and, therefore, wins out over Cl^- as the nucleophile. Step 3 is a proton transfer step to form the chlorohydrin.

Chloronium ion intermediate

1. Electrophilic addition

2. S_N2

3. Proton transfer

Same as

Your Turn 12.7

Think

It is helpful to draw out the C–H bonds that are not shown in the line structure. Recall from Section 11-1 an alkene reaction with a hydrogen halide acid. What is the acid that exists when HCl is dissolved in water? To which C does the proton add? What is the product that results? How can you show the carbocation rearrangement? Why does the carbocation rearrangement take place? If water is the solvent, what acts as the nucleophile in the third step?

Solve

When HCl is dissolved in water, the actual acid that exists is H_3O^+ due to the leveling effect. Step 1 is electrophilic addition of H^+ to C=C to result in the formation of a secondary carbocation. Step 2 is the rearrangement of the secondary carbocation to a tertiary carbocation via a 1,2-hydride shift and is driven by the greater stability of the tertiary carbocation. Step 3 is coordination, where H_2O acts as the nucleophile to attack the tertiary carbocation. Step 4 is proton transfer to form the ROH as the final product.

1. Electrophilic addition

2. 1,2-Hydride shift

3. Coordination

4. Proton transfer

Your Turn 12.8

Think

Consult Equation 12-23 to look at the mechanism for oxymercuration–reduction. Which C atom of the C≡C triple bond can accommodate the partial positive charge better? What determines the regiochemistry? Look at the ketone that is formed, and take the reaction one step back. Where is the OH located? What characterizes an enol?

Solve

A δ+ charge on the first structure would be better accommodated by the benzylic carbon. This is the C atom that the H_2O attacks and is the site of the enol—a species in which OH (characteristic of an alcohol) is attached to a C=C bond (characteristic of an alkene). A δ+ charge on the second structure would be better accommodated by C2 because it is more highly substituted. This is the carbon that the H_2O attacks and is the site of the enol.

OH OH

Equation 12-27 Equation 12-28

Your Turn 12.9

Think

Consult Table 1-2 on page 10 to look up the value of the O–O and C–C bond strengths.

Solve

O–O = 138 kJ/mol; C–C = 339 kJ/mol. The O–O bond is less than half the strength of the C–C bond, and this verifies that the O–O bond is weak.

Your Turn 12.10

Think

Which atom is electron poor in H–BH_2 and H–Cl? Which new electrophile–C bond is formed? What results on the other C of the C=C bond? Which carbon of the C=C bond can better accommodate a positive charge?

Solve

For the reaction of the C=C bond with H–BH_2, the electrophile is the B of H–BH_2, because it lacks an octet and is very electron poor. A C–B bond is formed from the end of the C=C that is less crowded, and a new C–H bond is formed from the end of the C=C that is more sterically crowded. The partial positive charge in the transition state forms on the more substituted carbon. For the reaction of the C=C with H–Cl, the electrophile is the H of H–Cl. A C–H bond forms so that the C^+ that is produced is on the more stable tertiary carbon. In both cases, the electrophilic atom adds to the same carbon of the C=C bond. Also, in both cases, either the partial positive charge in the transition state or the positive charge in the carbocation intermediate forms on the carbon that is more substituted.

Electrophilic atom

H–BH_2

R H

Partial positive charge on more substituted C

R H

Electrophilic atom

:Cl–H

R H

Positive charge on more substituted C

R H

Your Turn 12.11

Think

Consult Equations 12-36 and 12-37 as guides. In how many steps does the reaction occur? Is borane (BH_3) electron rich or electron poor? Is the π bond electron rich or electron poor? Does an intermediate form? What is the influence of the proximity of BH_2–H to the CH_3 group attached to the C=C bond? On which C atom does a partial positive charge develop in the transition state?

Solve

The reaction occurs in one *concerted* step. The B atom of borane is electron poor, and the C=C π bond is electron rich. Steric repulsion directs the larger group, BH_2, to the C atom with the fewer number of alkyl groups. Also, in the transition state, a partial positive charge develops on the C atom that gains the H. In the hydroboration steps below, the first reaction has significant steric repulsion, and a partial positive charge develops on the less alkyl-substituted C, which would, therefore, lead to the minor product. The second hydroboration minimizes the steric repulsion, and a partial positive charge develops on the more alkyl-substituted carbon, which would, therefore, lead to the major product.

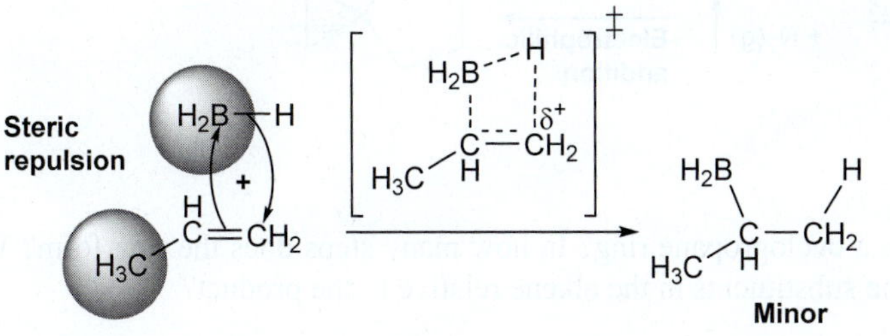

Steric repulsion

Minor

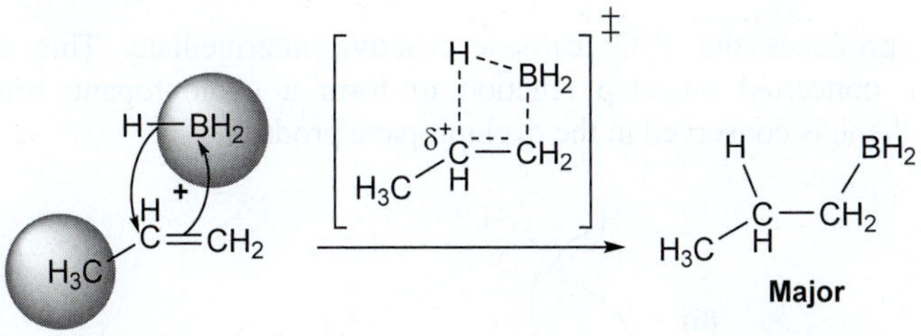

Major

Your Turn 12.12

Think

In the oxidation step, what group replaces B for each B–C bond? How does that replacement take place? Is there retention or inversion of the stereochemical configuration at each of those C atoms?

Solve

In Step 2 of the mechanism, the C–B bond breaks at the same time the C–O bond forms. In this concerted process, the O assumes the same position about the C atom as the original B atom did. The configuration about the C atom does not change. Each C–B bond is replaced by a C–OH bond, and the stereochemistry is retained.

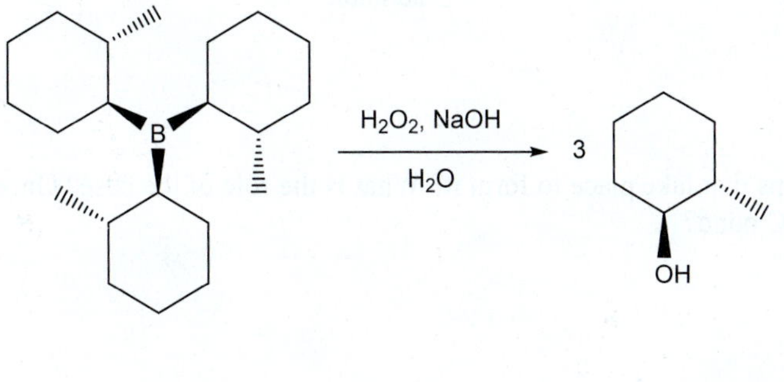

H_2O_2, NaOH; H_2O → 3

In Chapter Problems

Problem 12.1

Think

What is the Lewis structure of CH_2N_2? In the presence of heat, what bond is broken? What reactive intermediate forms? Does the electrophilic atom have a lone pair of electrons? How does this affect the intermediate formed?

Solve

The $H_2C–N_2$ bond breaks in a heterolysis step to form the reactive carbene intermediate. This electrophilic species has a lone pair on the electrophilic atom, and therefore, electrophilic addition proceeds to form a cyclic intermediate to avoid losing an octet. Two new σ C–C bonds are formed simultaneously as the C=C π bond is broken.

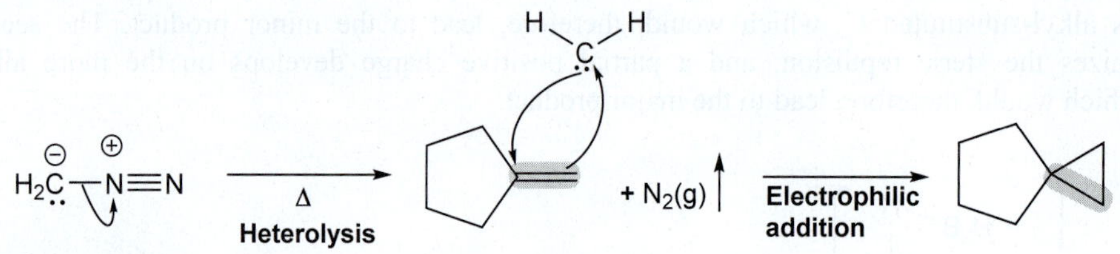

$H_2C–N≡N$ → Δ, Heterolysis → + $N_2(g)$ → Electrophilic addition

Problem 12.2

Think

What reagent reacts with an alkene to form a cyclopropane ring? In how many steps does the ring form? What must be true about the stereochemistry of the substituents in the alkene relative to the product?

Solve

Diazomethane in the presence of heat produces the H_2C: carbene reactive intermediate. This carbene intermediate reacts with an alkene in a concerted one-step reaction to form a cyclopropane ring. The stereochemistry of the substituents in the alkene is conserved in the cyclopropane product.

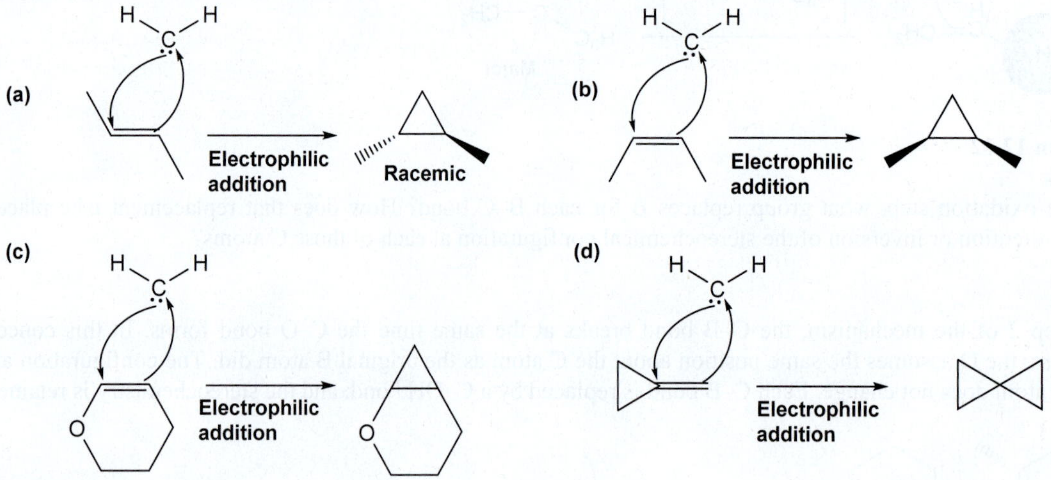

(a) Electrophilic addition → Racemic

(b) Electrophilic addition

(c) Electrophilic addition

(d) Electrophilic addition

Problem 12.3

Think

What carbene is formed, and what are the steps that take place to form it? What is the role of the base? Once the carbene forms, how does it react with the C=C bond?

Solve

The $(CH_3)_3CO^-$ acts as a base to deprotonate the $CHBr_3$ reagent and form the conjugate base. A heterolysis step follows to break the C–Br bond, producing the reactive Br_2C: carbene. The electrophilic addition step is next to form the cyclopropane product. Notice that Br_2C: can approach the C=C bond from either side of the bond's plane and that diastereomers are formed.

Proton transfer

Heterolysis

Slow

Electrophilic addition

Fast

These are diastereomers.

Problem 12.4

Think

How many cyclopropane rings are present? What is the starting carbene-producing reagent when just CH_2 is added to form a cyclopropane? If a CH_2 group is disconnected from a cyclopropane ring, then what atoms are left for the alkene? What is the starting carbene-producing reagent when CCl_2 is added to form a cyclopropane? If a CCl_2 group is disconnected from a cyclopropane ring, then what atoms are left for the alkene?

Solve

Because the product has two cyclopropane rings, we could envision each one as the result of a carbene addition. The first reaction is the result of the addition of CCl_2, produced from $CHCl_3$. The second reaction is the result of the addition of CH_2, produced from CH_2N_2.

$CHCl_3$, $NaOC(CH_3)_3$ — CH_2N_2, light

Problem 12.6

Think

What reaction takes place between Br_2 and an alkene? Do the Br atoms add to the C=C bond in a syn or anti fashion? Are the product molecules chiral? How does chirality relate to optical activity?

Solve

Br_2 undergoes anti addition across the double bond, yielding mixtures of products. The products are shown below and on the next page. Only the product mixture of **(b)** and **(d)** will be optically active, because the products are diastereomers. The product mixtures from **(a)** and **(c)** are optically inactive. The product from **(a)** is a meso compound and the products from **(c)** are equal mixtures of enantiomers (i.e., racemic mixtures).

(a) Br_2, CCl_4 — *meso* Optically inactive — Rotate about C–C bond — Plane pf symmetry

(b) Br_2 / CCl_4

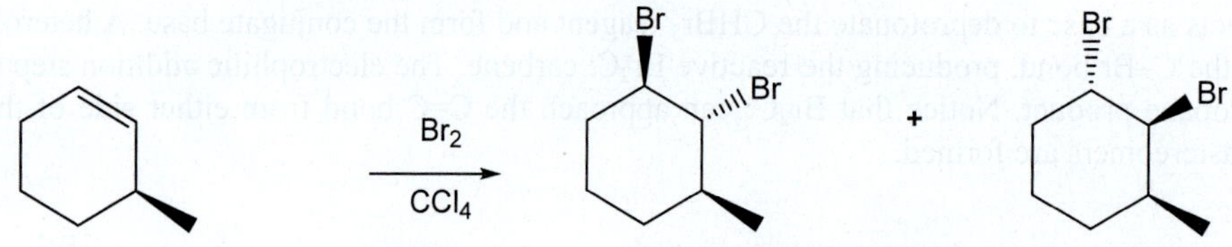

Mixture of diastereomers, optically active

(c) Br_2 / CCl_4

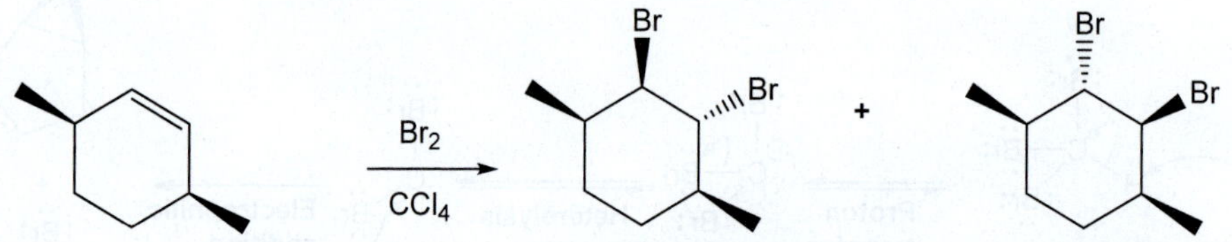

Mixture of enantiomers, optically inactive

(d) Br_2 / CCl_4

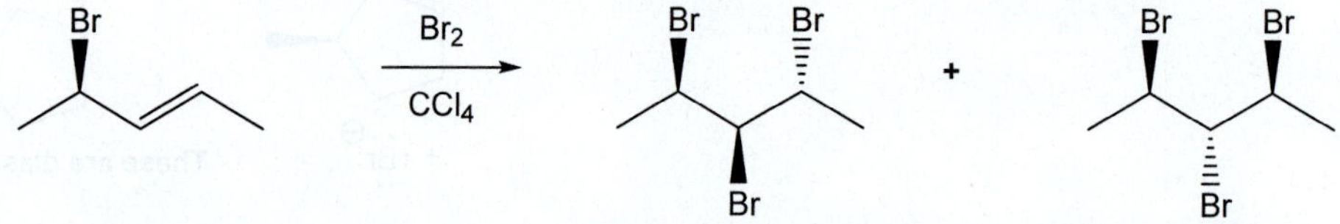

Mixture of diastereomers, optically active

Problem 12.7

Think

What is the structure of (*R*,*R*)-3,4-dichloro-2-methylhexane? Of its enantiomer? Are rearrangements permitted with addition of Cl_2 to an alkene? Does the addition reaction occur syn or anti? On what two C atoms must the C=C bond be located in the alkene precursor? What is the stereochemistry of the substituents on the C=C bond in the alkene precursor?

Solve

The *R*,*R* and *S*,*S* products are shown below on the right. When the C–C bond is rotated so the Cl atoms are anti to each other (which is what is directly produced in the chlorination reactions), the two alkyl groups appear on the same side of the Cl–C–C–Cl plane, and so do the two H atoms. The starting alkene, therefore, has the two alkyl groups cis to each other.

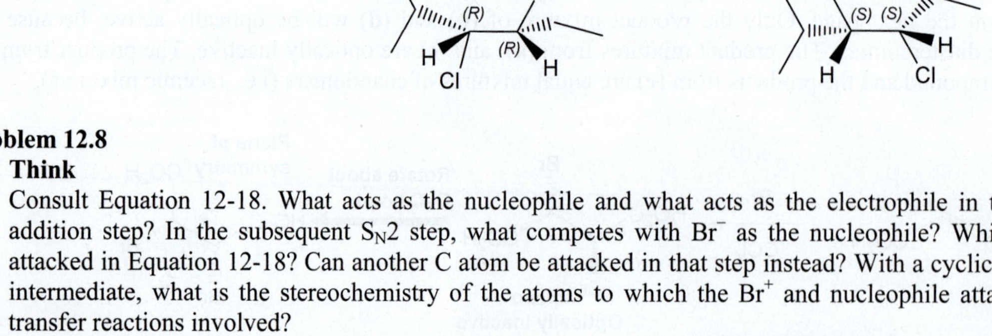

Cl_2 / CCl_4

(3*R*,4*R*)-3,4-Dichloro-2-methylhexane + **(3*S*,4*S*)-3,4-Dichloro-2-methylhexane**

Same as

Same as

Problem 12.8

Think

Consult Equation 12-18. What acts as the nucleophile and what acts as the electrophile in the electrophilic addition step? In the subsequent S_N2 step, what competes with Br^- as the nucleophile? Which C atom was attacked in Equation 12-18? Can another C atom be attacked in that step instead? With a cyclic bromonium ion intermediate, what is the stereochemistry of the atoms to which the Br^+ and nucleophile attach? Are proton transfer reactions involved?

Solve

In the first step, electrophilic addition to the C=C bond produces the bromonium ion intermediate. In H_2O, which is a nucleophilic solvent, the water nucleophile competes with the Br^- to attack the electrophilic C atom in the intermediate. In Equation 12-18, the C atom on the left was attacked in the S_N2, but the one on the right can also be attacked, as shown below. The Br and OH are anti to each other in the final product.

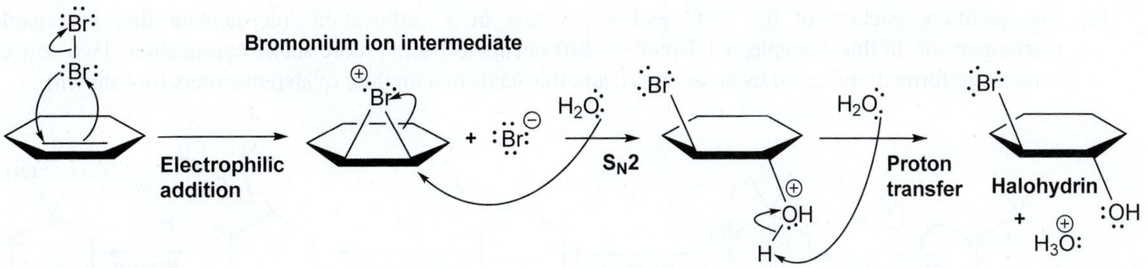

Bromonium ion intermediate

Electrophilic addition

S_N2

Proton transfer

Halohydrin

Problem 12.10

Think

For the reaction in Equation 12-16, what was the role of H_2O? Can ethanol assume the same role? Is the reaction stereoselective?

Solve

This mechanism is identical to the mechanism in Problem 12.8, except that the nucleophilic solvent is ethanol instead of water. A haloether forms rather than a halohydrin. The enantiomer of the product is also generated by attack at the other C atom in the second step, producing a racemic mixture.

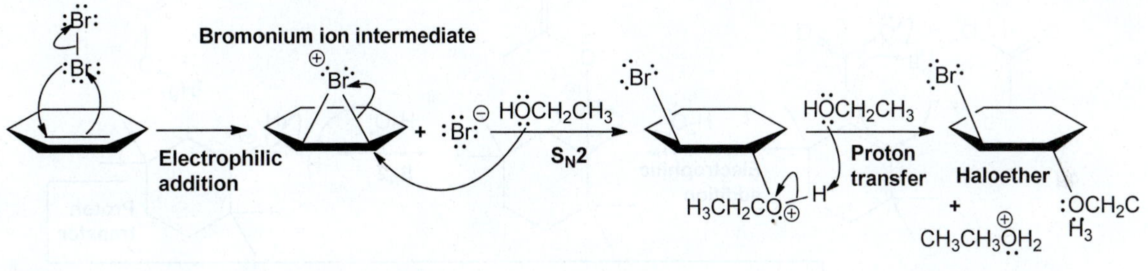

Bromonium ion intermediate

Electrophilic addition

S_N2

Proton transfer

Haloether

Problem 12.11

Think

Are the two C atoms in the C=C bond equivalent or distinct? Which C atom can accommodate the partial positive charge better? Why? Which C atom is the nucleophile more likely to attack? Is the addition reaction syn, anti, or a mixture of both?

Solve

In the chloronium ion intermediate, there is greater concentration of positive charge on the tertiary carbon than on the secondary carbon. Therefore, H_2O attacks the tertiary carbon. The opening of the ring in the S_N2 step ensures that the Cl atoms add anti overall.

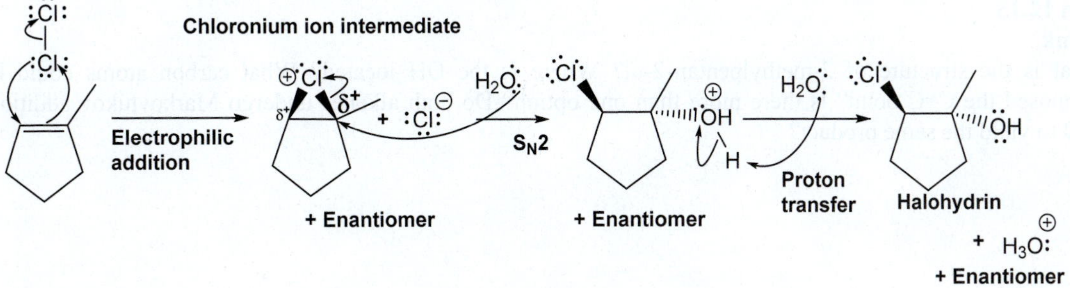

Chloronium ion intermediate

Electrophilic addition

S_N2

+ Enantiomer

+ Enantiomer

Proton transfer

Halohydrin

+ Enantiomer

Problem 12.12

Think

For each reaction, what intermediate forms after the electrophilic addition step? Are rearrangements possible? Is a new stereocenter generated? Is the addition reaction syn, anti, or a mixture of both?

Solve

(a) The addition reaction of the C=C and H^+ results in a carbocation intermediate that is capable of rearrangement. In this example, a 1,2-methyl shift occurs to form a more stable carbocation. Two new chiral centers are formed (indicated by an asterisk), and this leads to a mixture of stereoisomers (not shown).

H_2O Electrophilic addition 1,2-Methyl shift Coordination Proton transfer H_2O OH

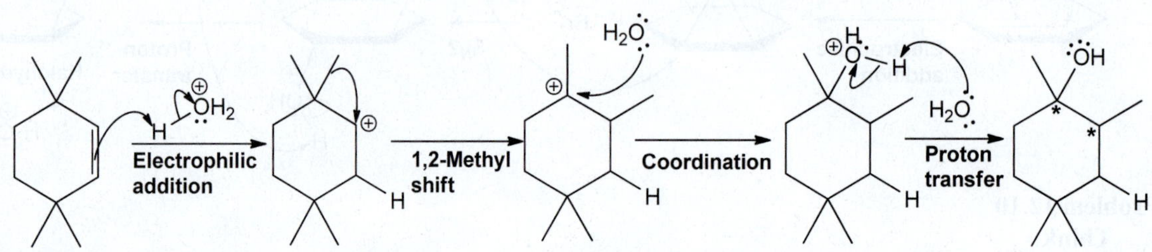

(b) The addition reaction of the C=C and $Hg(OAc)_2$ results in a cyclic mercurinium ion intermediate that is not capable of rearrangement. In Step 2 of the mechanism, H_2O acts as a nucleophile to open the three-membered ring, and in Step 3, the positively charged O atom is deprotonated. Subsequent reduction with sodium borohydride ($NaBH_4$) replaces the Hg-containing substituent with H. One new chiral center is formed (indicated by an asterisk), and this leads to a mixture of stereoisomers (not shown).

H_2O Electrophilic addition S_N2 Proton transfer $NaBH_4$

$NaBH_4$ reduces the C atom, replacing the Hg group with a H atom.

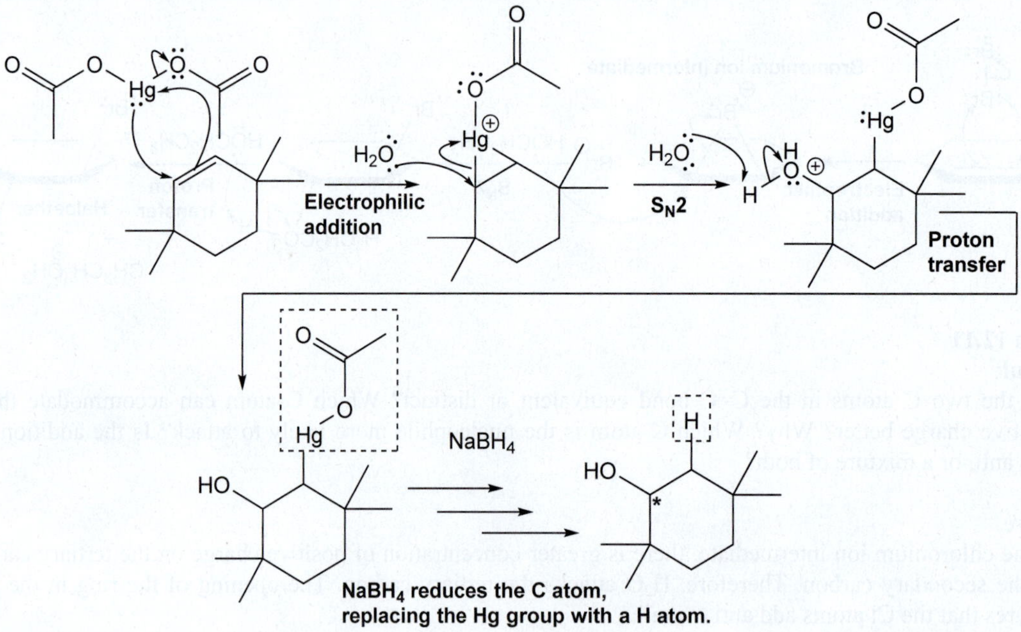

Problem 12.13

Think

What is the structure of 2-methylpentan-2-ol? Where is the OH located? What carbon atoms could have composed the C=C bond? Is there more than one option? Do both alkenes undergo Markovnikov addition of H_2O to yield the same product?

Solve

The OH group is located on C2, which is the most substituted carbon. Therefore, the C=C bond in the starting alkene could have occurred at C1 and C2 or at C2 and C3. In the Markovnikov addition of H_2O to each of these alkenes, the OH adds to C2 to yield the same alcohol product.

or

1. $Hg(OAc)_2$, H_2O/THF

2. $NaBH_4$, ethanol

OH

Problem 12.14

Think

On what C atom did the OH add? Does this represent Markovnikov addition? If this formed a carbocation intermediate, is rearrangement likely to occur to produce a more stable carbocation? If rearrangement is not desired, what reagents are needed?

Solve

This is the result of Markovnikov addition of water across the C=C double bond. We cannot carry out this transformation with just an acid-catalyzed hydration, because the carbocation that would be formed would rearrange to a more stable benzylic carbocation, as shown below

H H^+ Electrophilic addition H 1,2-Hydride shift H

Benzylic carbocation

We, therefore, must perform an addition of water via oxymercuration–reduction in which the water adds to the most substituted carbon but rearrangements do not occur.

1. $Hg(OAc)_2$, H_2O/THF

2. $NaBH_4$, ethanol

OH

Problem 12.16

Think

How does the mechanism for an alkoxymercuration–reduction reaction of an alkyne compare to the one shown for an alkene in Solved Problem 12.15? What considerations should be made about *regiochemistry*? What considerations should be made about rearrangements?

Solve

The mechanism for this alkoxymercuration–reduction is identical to the one shown in Solved Problem 12.15, the only difference being the presence of a C≡C bond instead of a C=C bond. In this case, H will add to the terminal C, and OCH_2CH_3 will add to the adjacent C. There will be no keto–enol tautomerization, as is observed with oxymercuration–reduction, because the product of this reaction is not an enol; it is a vinyl ether. See the reaction on the next page.

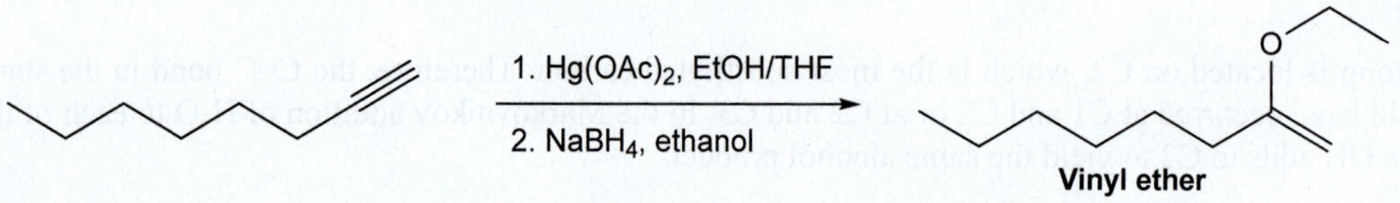

1. $Hg(OAc)_2$, EtOH/THF
2. $NaBH_4$, ethanol

Vinyl ether

Problem 12.17

Think

What is the structure of MCPBA? What functional group is present? How does this functional group react with an alkene? What functional group is produced? In how many steps does the reaction occur? If the reaction is concerted, what happens to the orientation of the substituents on the C=C bond?

Solve

MCPBA is a peroxyacid and forms an epoxide in a one-step (concerted) reaction with an alkene. Rearrangements and rotations about the C=C bond are not permitted. Therefore, the stereochemistry describing the C=C bond remains the same in the product. In this case, the trans configuration about the double bond produces a trans configuration about the epoxide ring. A racemic mixture of enantiomers is produced.

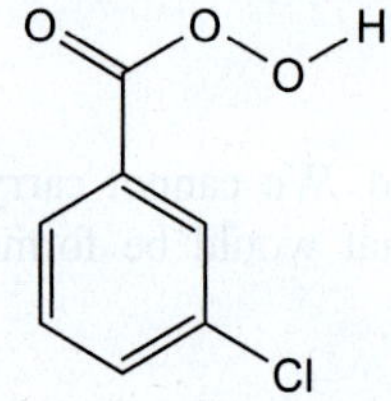

m-Chloroperbenzoic acid
MCPBA

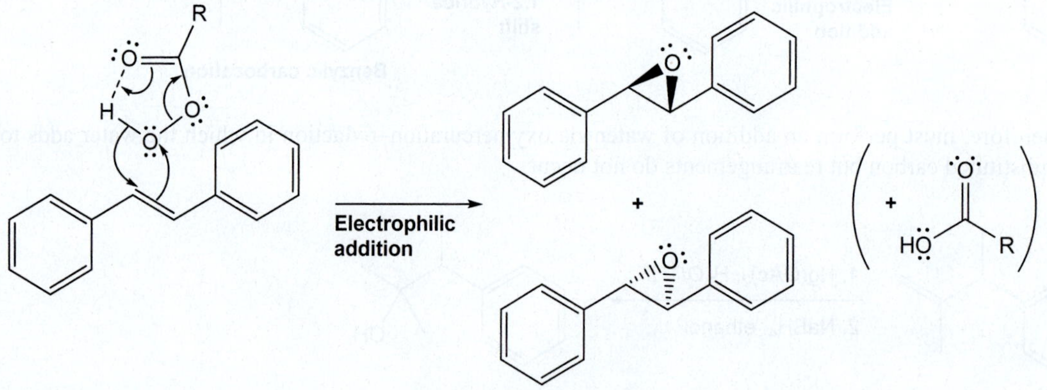

Electrophilic addition

Problem 12.18

Think

What is the orientation of the substituents in the epoxide, cis or trans? What must be the orientation of the same substituents about the C=C bond in the alkene?

Solve

Because the two groups are on the same side of the epoxide ring, the starting alkene must be cis.

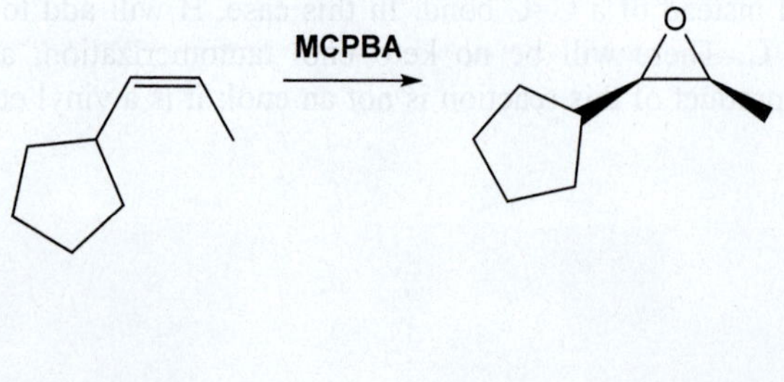

MCPBA

Problem 12.20

Think

In the first reaction, what is the nature of the leaving group? Of the attacking species? Of the solvent? What type of reaction does heat promote? For the second reaction, what type of reaction takes place between an alkene and a peroxyacid? In the third reaction, what acts as the nucleophile? What acts as the substrate?

Solve

In the first reaction, E2 is the major mechanism owing to the high concentration of the strong base and the heat that is added. The E2 reaction forms the most substituted alkene. Because E2 requires the groups to be anticoplanar, the ethyl groups are cis in product **D**. In the second reaction, MCPBA adds to the C=C bond in a one-step (concerted) reaction to form the epoxide **E**, where the ethyl groups are still cis. CH_3MgBr is a source of nucleophilic carbon, $H_3C:^-$. In an S_N2 reaction, the $H_3C:^-$ attacks the *less* substituted C atom of the epoxide, and the ring is opened where the O and the CH_3 become anti to each other. An acid workup step is required to form the uncharged alcohol product.

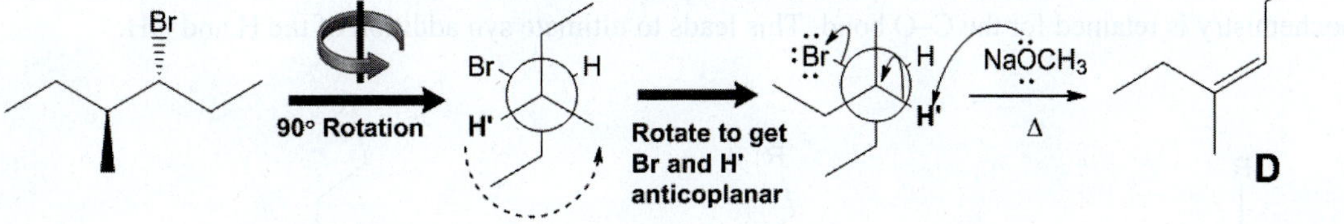

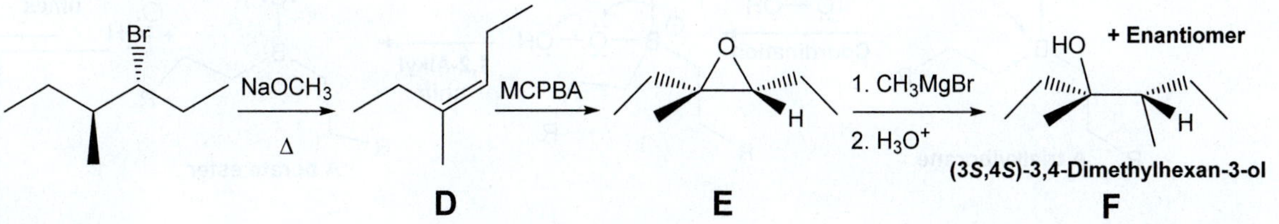

Problem 12.21

Think

In how many steps does the monoalkyl borane compound form? Does the boron attach to the more or less substituted carbon? Why?

Solve

The monoalkyl borane is formed in one step. The BH_2 attaches to the less substituted C atom because of sterics of the transition state and the fact that the partial positive charge in the transition state is more stable on the more substituted C atom. The product of that step has two more B–H bonds, so two more hydroborations occur to produce the trialkylborane.

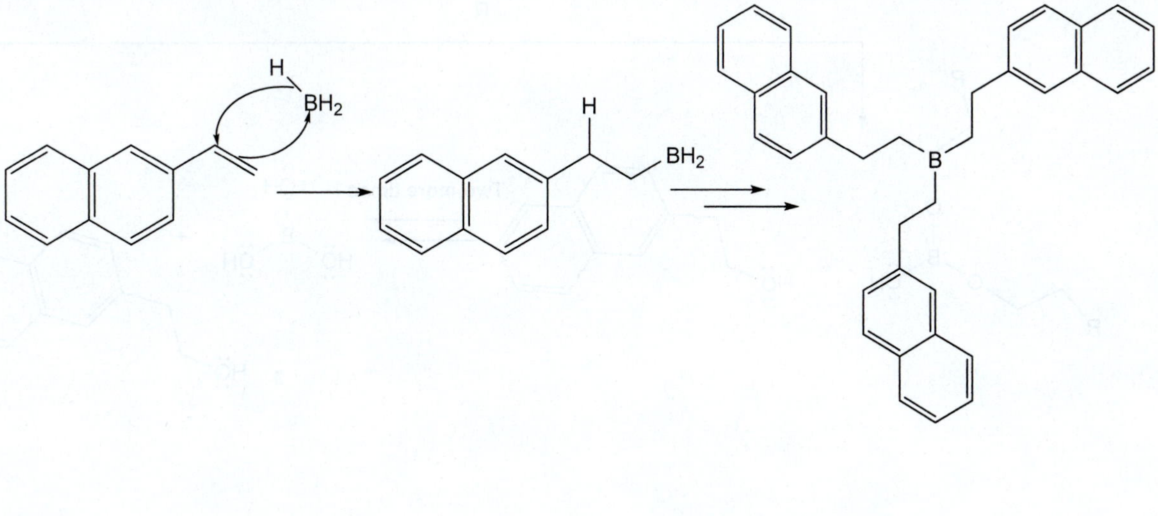

Problem 12.22

Think

What is the reactive species in a basic solution of H_2O_2? How does that react with the electron-poor trialkylborane? How is the C–O bond formed and the B–O bond broken? How many times does this occur? How is the O–H bond formed and the B–O bond broken? Compare this mechanism to the one shown in Equation 12-40.

Solve

This reaction mechanism takes 15 steps overall. The R group represents the naphthyl portion, $C_{10}H_7$. The reactive species is HOO^-, and the first two steps are coordination of the HOO^- to the electron-poor boron atom followed by a 1,2-alkyl shift. These are then repeated two more times to form the trialkylborate ester (Steps 1–6). The HO^- coordinates to the electron-poor boron atom followed by heterolysis of the B–OR bond. A proton transfer reaction occurs on the alkoxide to form the uncharged alcohol product (Steps 7–9). These three steps are then repeated two more times (Steps 10–15) to form three equivalents of the ROH and $B(OH)_3$, which goes on to produce $B(OH)_4^-$. The alcohol OH forms on the less substituted carbon because the O–C bond forms at the same C where the C–B bond broke. In this example, the alcohol is not chiral, but it should be noted that stereochemistry is retained for the C–O bond. This leads to ultimate syn addition of the H and OH.

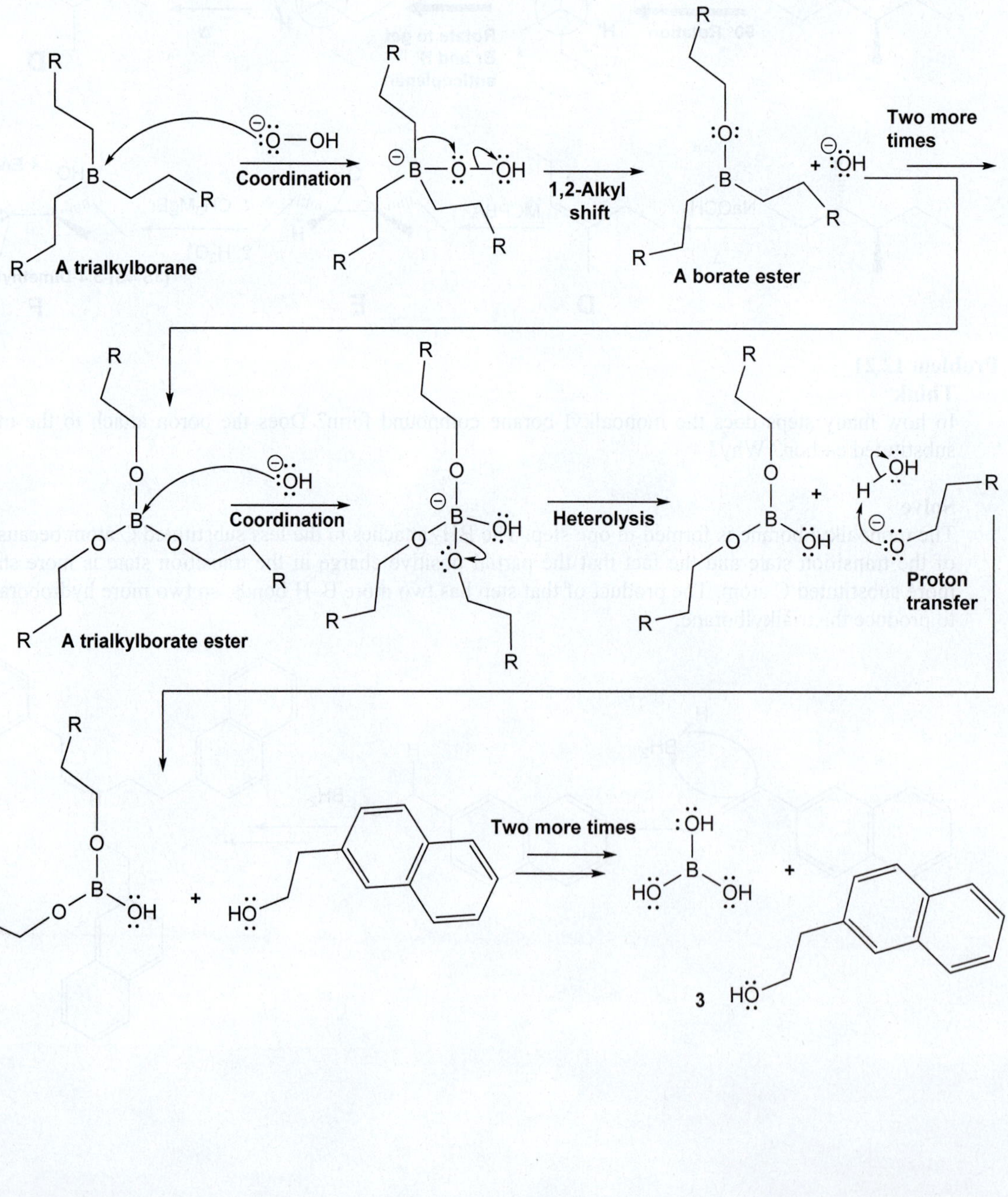

Problem 12.24

Think

In each case, there is a net addition of what molecule across the C=C bond? What is the regiochemistry required in each reaction? Are carbocation rearrangements a concern?

Solve

(a) The net addition is H_2O across the C=C bond, and the OH adds to the less substituted carbon. This is an example of anti-Markovnikov addition and can be carried out with hydroboration–oxidation.

1. BH_3•THF
2. H_2O_2, NaOH, H_2O
OH

(b) The net addition is H_2O across the C=C bond. In this example, a rearrangement occurred because the OH appears on the benzylic C atom, which was not previously part of the double bond. The reaction conditions must be hydration in acidic water.

OH
H_3O^+/H_2O

(c) The net addition is H_2O across the C=C bond, and the OH adds to the more substituted carbon with no rearrangement. This is an example of Markovnikov addition without rearrangement and can be carried out with oxymercuration–reduction.

1. $Hg(OAc)_2$, H_2O/THF
2. $NaBH_4$
OH

Problem 12.25

Think

From what functional group does an aldehyde tautomerize? Where must the OH be located in that functional group? In the presence of disiamylborane, does the more or less substituted C–B bond form? Is the alkyne terminal or internal?

Solve

Treatment of a terminal alkyne with $(C_5H_{11})_2BH$ results in the C–B bond forming at the less-substituted C atom. Treatment of **B** in basic solution results in a terminal enol that tautomerizes to the aldehyde.

$(C_5H_{11})_2BH$
THF
H
$(C_5H_{11})_2B$
$HO^{\ominus}$, H_2O_2
H
HO
O
H
H
A
B
Enol
Aldehyde

End of Chapter Problems
Section 12.2 Electrophilic Addition of Carbenes: Formation of Cyclopropane Rings
Problem 12.26

Think

In each case, there is a net addition of what molecule across the C=C bond? What is the stereochemistry required in each reaction? Are carbocation rearrangements a concern? What does diazomethane CH_2N_2 form in the presence of light (*hv*)?

Solve

Diazomethane in the presence of light (*hv*) is a source of H_2C: carbene.

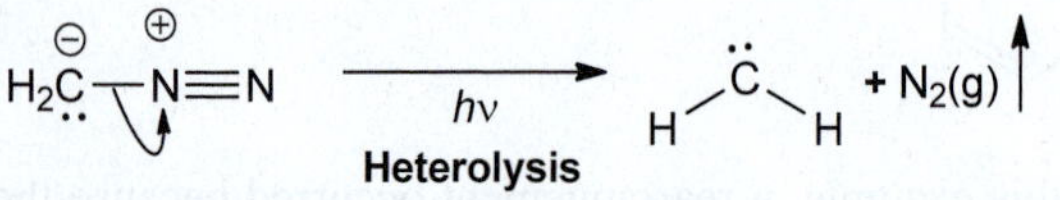

$H_2C^{\ominus}-N^{\oplus}\equiv N \xrightarrow{h\nu} H_2C: + N_2(g)\uparrow$

Heterolysis

Addition of H_2C: to an alkene results in the formation of a cyclopropane ring. In reactions **(a)** and **(b)**, the wedge bond to CH_3 remains in a wedge because no bonds were broken or formed at that carbon, so a mixture of diastereomers forms. In reaction **(d)**, the ethyl groups are trans in the alkene, and a mixture of enantiomers forms.

(a)

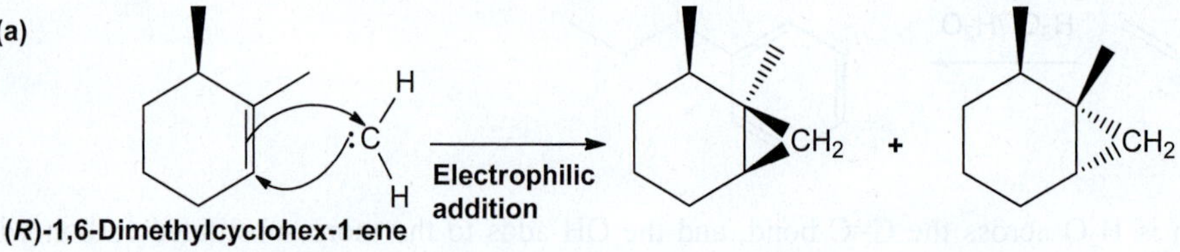

(*R*)-1,6-Dimethylcyclohex-1-ene + :CH$_2$ → (Electrophilic addition) CH$_2$ + CH$_2$

(b)

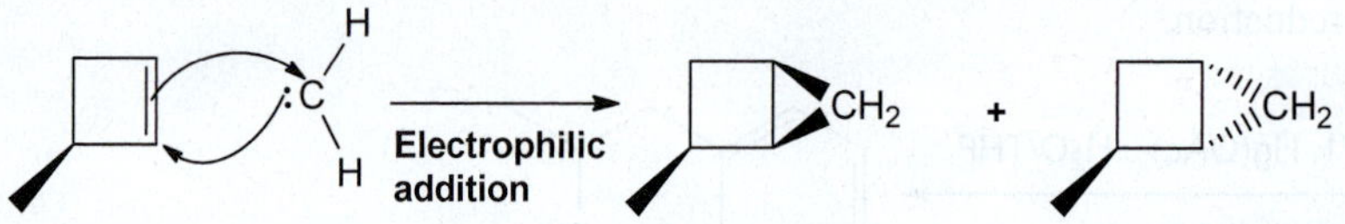

:CH$_2$ → (Electrophilic addition) CH$_2$ + CH$_2$

(c)

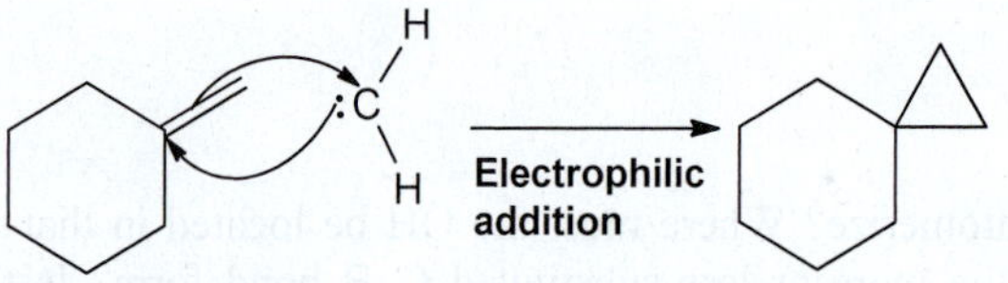

:CH$_2$ → (Electrophilic addition)

(d)

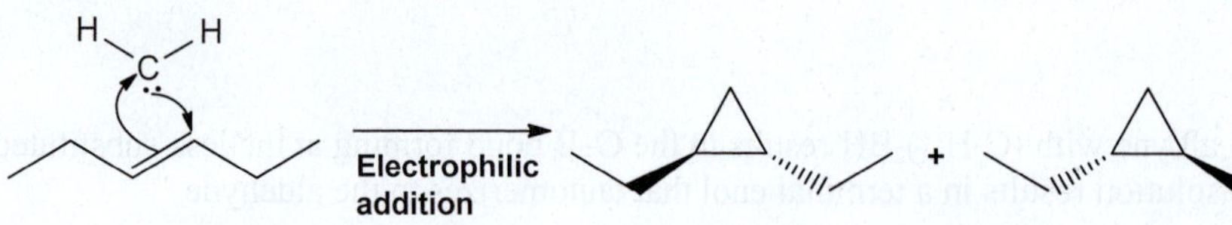

:CH$_2$ → (Electrophilic addition) +

(e)

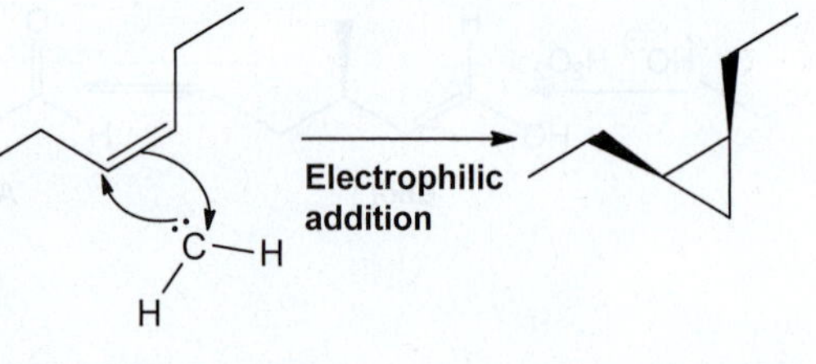

:CH$_2$ → (Electrophilic addition)

Problem 12.27

Think

In each case, there is a net addition of what molecule across the C=C or C≡C bond? What is electron rich, and what is electron poor? What is the stereochemistry required in each reaction? Are carbocation rearrangements a concern? How many times does the addition reaction occur?

Solve

In reactions **(a)–(c)**, the CH_3O^- acts as a base to deprotonate the $CHBr_3$ or $CHCl_3$ reagent and form the conjugate base. A heterolysis step follows to break the C–Br or C–Cl bond, resulting in the reactive Br_2C: or Cl_2C: carbene. An electrophilic addition step is next to form the cyclopropane product.

(a) The trans configuration about the C=C bond results in a trans configuration about the plane of the cyclopropane ring in the product. Electrophilic addition can occur either from above or below the plane of the C=C bond, and therefore, a mixture of enantiomers results.

Proton transfer → Heterolysis → Electrophilic addition → + Enantiomer

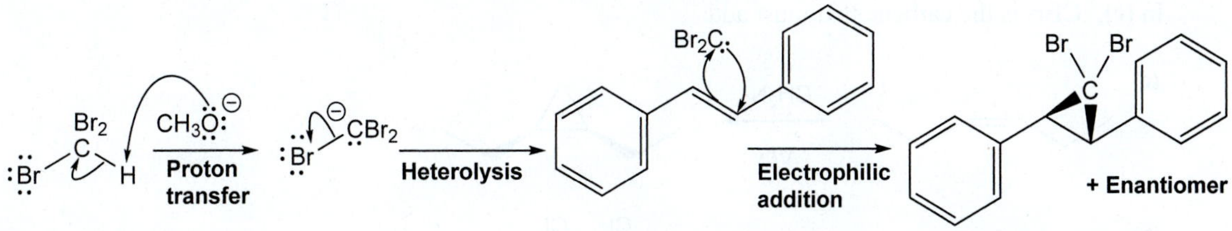

(b) Electrophilic addition can occur either from above or below the plane of the C=C bond, and therefore, a mixture of enantiomers results.

Proton transfer → Heterolysis → Electrophilic addition → + Enantiomer

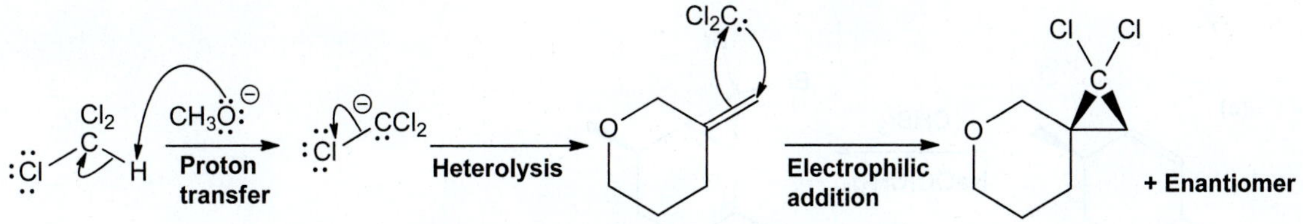

(c) There are no chiral centers in the product, and therefore, only one product is formed.

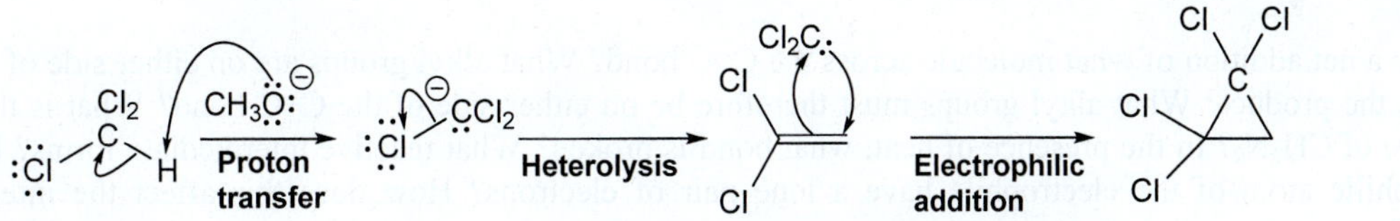

Proton transfer → Heterolysis → Electrophilic addition

Problem 12.28 (SYN)

Think

What reactive intermediate reacts to produce a cyclopropane ring? To what type of functional group does that intermediate add? What is the starting material that reacts to form a carbene? How many C atoms are added per reaction with a carbene? How many cyclopropane rings are present in each compound?

Solve

(a) The carbene addition reaction occurs twice to form two cyclopropane rings from buta-1,3-diene.

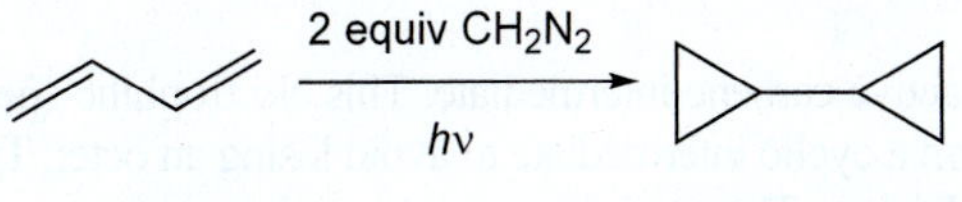

2 equiv CH_2N_2, $h\nu$

(b) The carbene addition reaction occurs twice to form two cyclopropane rings from propa-1,2-diene.

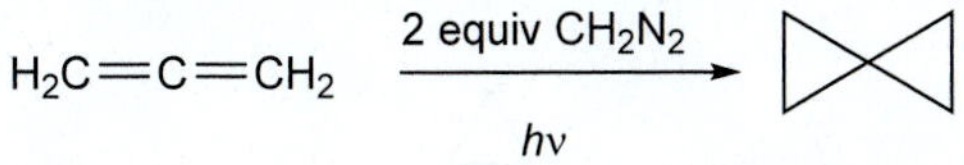

$H_2C{=}C{=}CH_2$ → 2 equiv CH_2N_2, $h\nu$

Problem 12.29 (SYN)

Think

In each case, there is a net addition of what molecule across the C=C bond? What is the stereochemistry required in each reaction? Are carbocation rearrangements a concern?

Solve

Syntheses are shown below. In **(a)**, :CH_2 is the carbene that must add, and the cis configuration in the cyclopropane ring requires a cis configuration from the starting alkene. In **(b)**, :CCl_2 is the carbene that must add, and the trans configuration in the cyclopropane ring requires a trans configuration from the starting alkene. In **(c)**, :CBr_2 is the carbene that must add.

(a) CH_2N_2, Δ/hν

(b) $CHCl_3$, $NaOC(CH_3)_3$ — Cl Cl

(c) $CHBr_3$, $NaOC(CH_3)_3$ — Br Br

Problem 12.30

Think

There is a net addition of what molecule across the C≡C bond? What alkyl groups are on either side of the C=C bond in the product? What alkyl groups must therefore be on either side of the C≡C bond? What is the Lewis structure of CH_2N_2? In the presence of heat, what bond is broken? What reactive intermediate forms? Does the electrophilic atom of the electrophile have a lone pair of electrons? How does this affect the intermediate formed?

Solve

Across the C≡C bond, there is a net addition of CH_2. The alkyl groups on either side of the C=C bond in the cyclopropene product are a CH_3 and a $CH_2CH_2CH_3$. Therefore, the starting alkyne compound is hex-2-yne.

Alkyne starting compound

CH_2N_2, Δ/hν

Hex-2-yne

The $H_2C–N_2$ bond breaks in a heterolysis step to form the reactive carbene intermediate. This electrophilic species has a lone pair, and therefore, electrophilic addition proceeds to form a cyclic intermediate to avoid losing an octet. Two new σ C–C bonds are formed simultaneously as one C≡C π bond is broken. The product is a cyclopropene ring.

Mechanism

$H_2C^{\ominus}–N^{\oplus}≡N$ —Δ, **Heterolysis**→ H–C–H (carbene) + $N_2(g)$↑ —**Electrophilic addition**→

Problem 12.31

Think

What is the charge on the C atom in the C–Li bond? How does this react with the H_2CCl_2 to form the carbene? Compare the acidity of $CHCl_3$ to CH_2Cl_2. What is the base strength of a carbanion compared to HO^-? From how many directions can the :CHCl carbene attack the C=C bond? What is the stereochemistry involving the new bonds that form the cyclopropane ring compared to the wedge bond to the methyl group on C4?

Solve

(a) Butyl lithium is a source of strongly basic carbanions. Bu–Li reacts with H_2CCl_2 via a proton transfer reaction, followed by heterolysis.

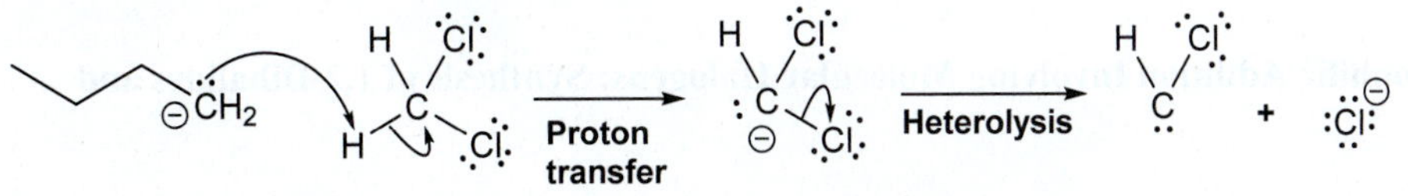

Proton transfer

Heterolysis

(b) With fewer Cl atoms, CH_2Cl_2 is less acidic than $CHCl_3$, so the base must be stronger. A base that has a negatively charged sp^3 C atom is stronger than a base that has a negatively charged sp^3 O atom.

(c) The four products are shown below. The carbene attack can occur from the same side as the wedge methyl (left two molecules) or opposite from the wedge methyl (right two molecules).

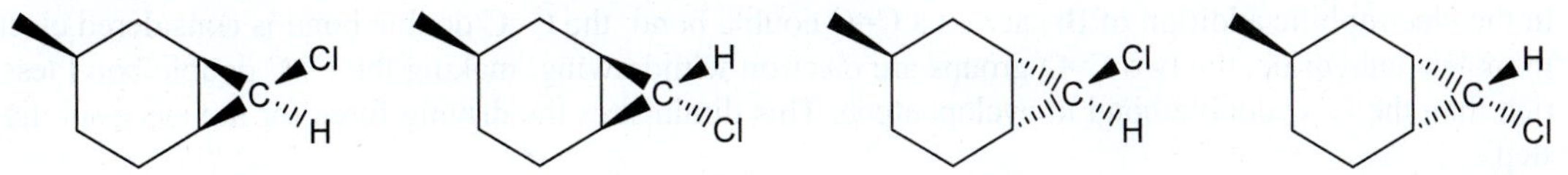

Problem 12.32

Think

In the presence of light, what bond is broken in the compound with the N_2? What reactive intermediate forms? Does the electrophilic atom of the electrophile have a lone pair of electrons? How does the electrophile react with the C≡C bond?

Solve

The C–N_2 bond breaks in a heterolysis step to form the reactive carbene intermediate. This carbene species has a lone pair on the electrophilic C, and therefore, electrophilic addition proceeds to form a cyclic product to avoid losing an octet. Two new σ C–C bonds are formed simultaneously as one C≡C π bond is broken. The product is a cyclopropene ring.

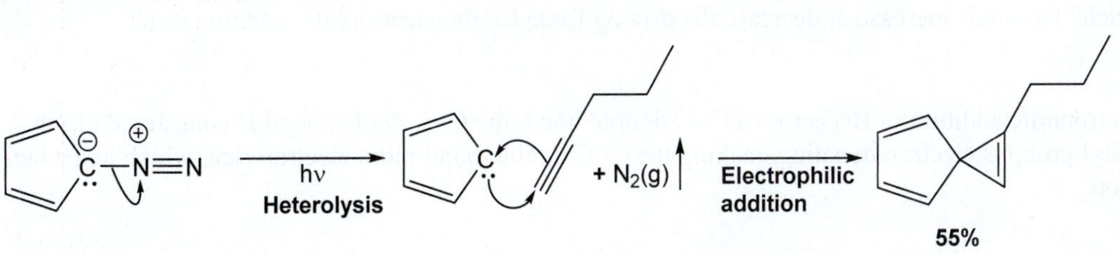

hν

Heterolysis

+ N_2(g)↑

Electrophilic addition

55%

Problem 12.33

Think

In the presence of light, what bond is broken in the $H_2C–N_2$? What reactive species results? How many C=C groups in the starting organic compound can react with excess CH_2N_2? How many times does the reaction occur? What are the stereochemical requirements for addition of H_2C: across a C=C bond?

Solve

Diazomethane in the presence of light produces the H_2C: carbene reactive intermediate. The starting material has two C=C groups, and each reacts in a concerted (one-step) reaction to form a cyclopropane ring in the presence of excess H_2C:. The H_2C: can add to the C=C from above or below the plane of the ring. A mixture of configurations will be produced at each of the two new stereocenters, resulting in a mixture of three stereoisomers (one is meso).

CH_2N_2 (excess), $h\nu$

Section 12.3 Electrophilic Addition Involving Molecular Halogens: Synthesis of 1,2-Dihalides and Halohydrins

Problem 12.34

Think

In 1,2-dibromination reactions, is an alkene relatively electron rich or electron poor? Are C=O groups electron-donating or electron-withdrawing groups? What is the effect of these groups on the extent to which the C=C bond is electron rich? Does this increase or decrease the driving force for the electrophilic addition step?

Solve

In the electrophilic addition of Br_2 across a C=C double bond, the C=C double bond is considered electron rich. In maleic anhydride, the two C=O groups are electron withdrawing, making the C=C double bond less electron rich than the C=C double bond in cyclopentene. This diminishes the driving force for the electrophilic addition step.

Br_2

Less electron rich, slower addition of Br_2

Problem 12.35

Think

In 1,2-dibromination reactions, is an alkene relatively electron rich or electron poor? Are CH_3 alkyl groups electron donating or withdrawing? What is the effect of these groups on the extent to which the C=C bond is electron rich? Does this increase or decrease the driving force for the electrophilic addition step?

Solve

In the electrophilic addition of Br_2 across a C=C double bond, the C=C double bond is considered electron rich. Each methyl group is electron donating, making the C=C double bond more electron rich, which helps facilitate the reaction.

More electron rich, faster addition of Br_2

Increasing reaction rate with Br_2

Problem 12.36

Think

When X_2 (Br_2 or Cl_2) adds to a C=C bond, what intermediate is formed? What nucleophile opens up the ring? Which carbon can better accommodate a partial positive charge? Both are secondary, but is there another factor to consider?

Solve

(a) A bromonium ion intermediate is formed. Water then attacks the carbon adjacent to the ring, because that carbon can stabilize a positive charge better than the other carbon of the initial C=C bond, owing to resonance with the ring.

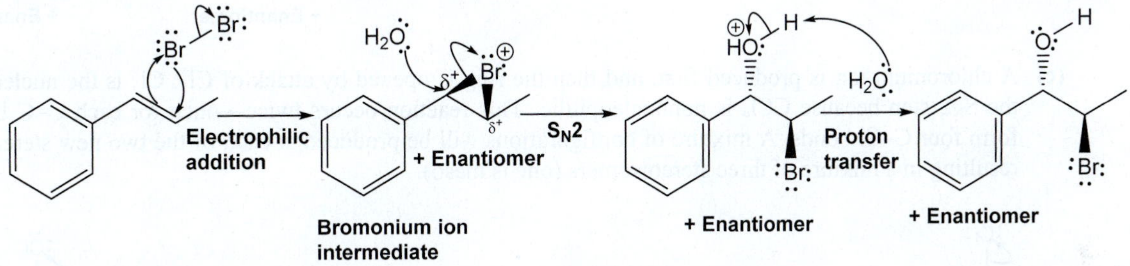

(b) A chloronium ion is produced first. Water then attacks the carbon adjacent to the O atom, because the lone pairs on O can help stabilize the positive charge, owing to resonance with the lone pair of electrons on the O atom.

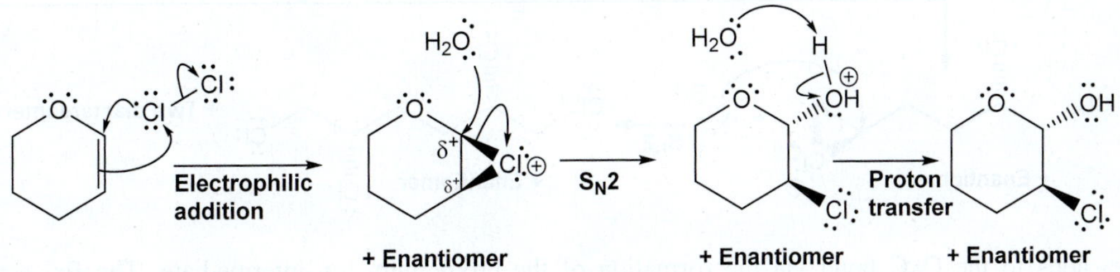

Problem 12.37

Think

What intermediate is formed? In each case, there is a net addition of what molecule across the C=C bond? What is the regiochemistry required in each reaction? The stereochemistry? Are carbocation rearrangements a concern? How many times does the addition reaction occur? How does the solvent affect which species acts as the nucleophile in the S_N2 reaction?

Solve

In each reaction (shown below and on the next page), a bromonium or chloronium ion is produced first, then the ring is opened by S_N2 attack of a nucleophile.

(a) Br_2 in CCl_4, an aprotic solvent, undergoes anti addition to add a Br to each C of the C=C bond. This yields a mixture of enantiomers. Br^- is the nucleophile in the S_N2 step because CCl_4 is nonnucleophilic.

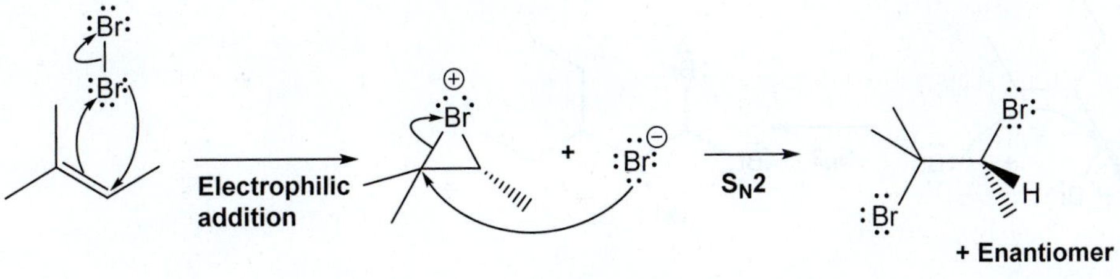

(b) Br_2 adds to the C=C bond to form the bromonium ion intermediate. The H_2O acts as the nucleophile because, being the solvent, it is so abundant. H_2O attacks the more substituted C atom, yielding a mixture of enantiomers. A final proton transfer step is necessary to produce the uncharged halohydrin product.

Electrophilic addition

S_N2

Proton transfer

+ Enantiomer

+ Enantiomer

(c) A chloronium ion is produced first, and then the ring is opened by attack of Cl^-. Cl^- is the nucleophile in the S_N2 step because CCl_4 is nonnucleophilic. This reaction occurs twice—once for each C=C bond—to form four C–Cl bonds. A mixture of configurations will be produced at each of the two new stereocenters, resulting in a mixture of three stereoisomers (one is meso).

Electrophilic addition

S_N2

+ Enantiomer

Electrophilic addition

+ Enantiomer

S_N2

+ Enantiomer

+ Two diastereomers

(d) Br_2 adds to the C≡C bond via the formation of the bromonium ion intermediate. The Br^- acts as the nucleophile and adds to the more substituted C atom. Br^- is the nucleophile in the S_N2 step because CCl_4 is nonnucleophilic. Rearrangements are not possible. The reaction occurs two times to form the achiral tetrabrominated alkane product.

Electrophilic addition

S_N2

Electrophilic addition

S_N2

Problem 12.38

Think

Are both C–Br bonds of the intermediate formed on the same side of the six-membered ring or on opposite sides? What is the second elementary step in the addition reaction of Br_2 to an alkene? What is true about the stereochemistry of the two groups in this type of step? Does a mixture of cis and trans support this?

Solve

(a) In the first step, both bonds to Br must form on the same side of the ring's plane. In the second step, which is an S_N2 step, the nucleophile must come in from the opposite side of the C–Br bond and so must yield a trans dibromide. This is shown below.

(b) The observations from experiment, therefore, discredit the proposed mechanism.

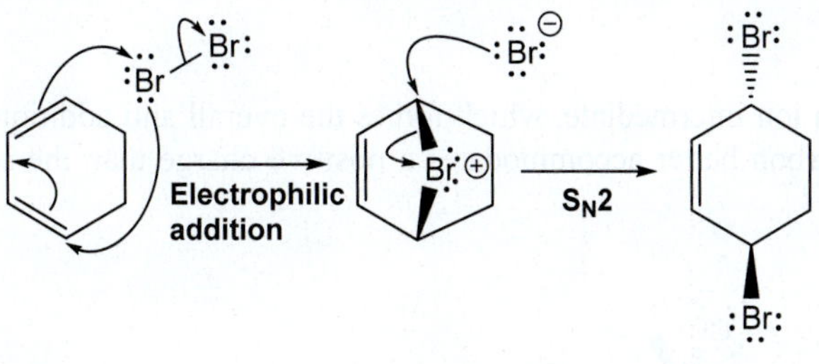

Problem 12.39

Think

How are the Br and $OC(CH_3)_3$ groups arranged relative to each other? How are the ethyl groups arranged in the starting material and the product? What does this suggest about the type of intermediate that must have formed? To what mechanism is this similar?

Solve

The anti addition of the Br and OR groups suggests a bromonium ion intermediate. Therefore, the mechanism is the same as the reaction with Br_2, except the nucleophile generated is RO^- instead of Br^-.

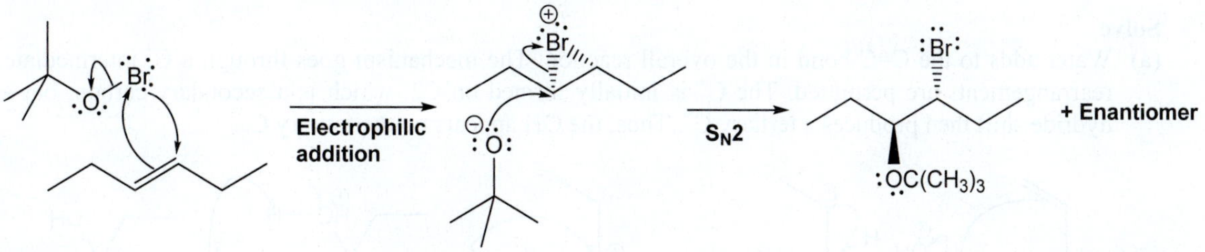

Problem 12.40

Think

What intermediate forms first in the reaction of Br_2 with an alkene? How does that intermediate react in the presence of a nucleophile? Does the ROH compete with the Br^- for nucleophilic attack? Will an intramolecular or intermolecular nucleophilic attack be favored?

Solve

Attack of the alkene by Br_2 leads to the bromonium ion intermediate. In the presence of a nucleophile (ROH), the more substituted C atom is attacked to form the five-membered ether ring. This is the same mechanism by which a halohydrin is formed.

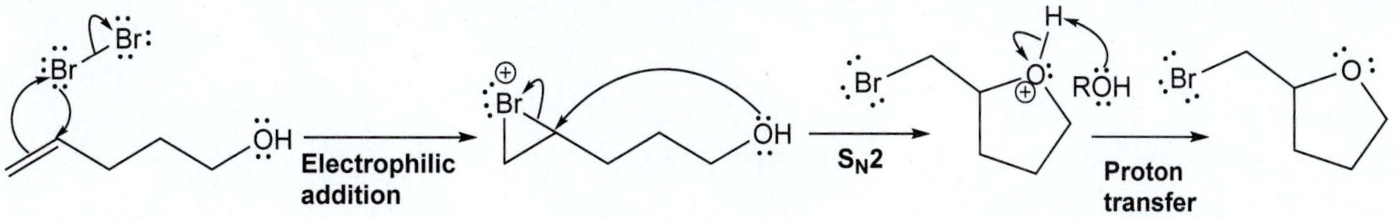

Problem 12.41

Think

Which atom, I or Cl, is more electronegative? In which direction does the dipole point? Which atom will then act as the electrophile to engage in an electrophilic addition reaction with the alkene? What intermediate forms? What acts as the nucleophile in the second step? Is the more or less substituted C atom attacked?

Solve

The I atom is more electrophilic because the more highly electronegative Cl atom leaves it with a partial positive charge.

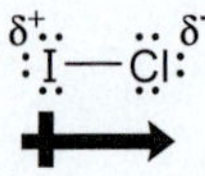

δ^+ I—Cl δ^-

The mechanism proceeds through an iodonium ion intermediate, which forces the overall anti addition of I–Cl. Cl^- attacks the tertiary carbon because that carbon better accommodates a positive charge than the secondary carbon will. The mechanism is shown below.

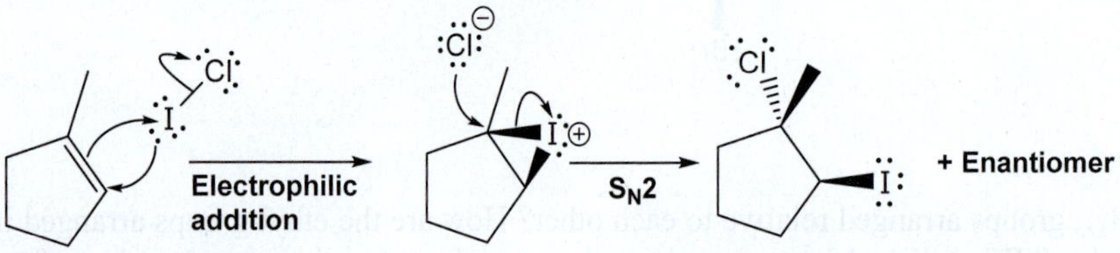

Electrophilic addition

S_N2

+ Enantiomer

Section 12.4 Oxymercuration–Reduction: Addition of Water

Problem 12.42

Think

For each reaction, what intermediate forms after the electrophilic addition step? Are rearrangements possible? Is a new stereocenter generated? Is the addition reaction syn, anti, or a mixture of both?

Solve

(a) Water adds to the C=C bond in the overall reaction. The mechanism goes through a C^+ intermediate, and rearrangements are permitted. The C^+ is initially formed on C2, which is a secondary carbon, but a 1,2-hydride shift then produces a tertiary C^+. Thus, the OH appears on the tertiary C.

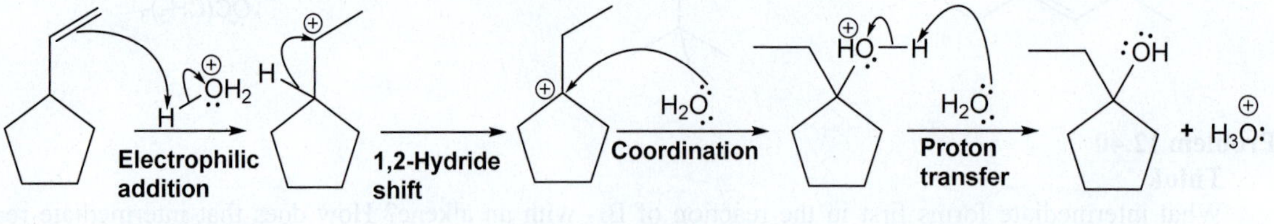

Electrophilic addition

1,2-Hydride shift

Coordination

Proton transfer

+ H_3O^+

(b) Water adds to the C=C bond in the overall reaction. The mechanism goes through a mercurinium ion intermediate, and rearrangements are not permitted. The reaction follows Markovnikov addition. Thus, the water adds to the more substituted secondary C.

Hg
Electrophilic addition
H_2O
S_N2
:Hg
H_2O
Proton transfer
Hg
OH
$NaBH_4$
H
OH
$NaBH_4$ reduces the C atom, replacing the Hg group with a H atom.

(c) Water adds to the C=C bond in the overall reaction. The mechanism goes through a C^+ intermediate, and rearrangements are permitted. The C^+ is initially formed on a secondary carbon, but a 1,2-methyl shift then produces a tertiary benzylic C^+. Thus, the OH appears on the tertiary benzylic C.

OH_2
H
Electrophilic addition
1,2-Methyl shift
H_2O
Coordination
HO—H
H_2O
Proton transfer
OH
+ H_3O:

(d) Water adds to the C=C bond in the overall reaction. The mechanism goes through a mercurinium ion intermediate, and rearrangements are not permitted. The reaction follows Markovnikov addition. Thus, the water adds to the more substituted secondary C. See the mechanism on the next page.

Electrophilic addition

$H_2\ddot{O}$:
S_N2

$H_2\ddot{O}$:
Proton transfer

$NaBH_4$

$NaBH_4$ reduces the C atom, replacing the Hg group with a H atom.

Problem 12.43 (SYN)

Think

Where is the OH located? What carbon atoms could have composed the C=C bond in an initial alkene? Is there more than one option? What is the regiochemistry required in each reaction? Are carbocation rearrangements a concern?

Solve

(a) The OH group is on a secondary C adjacent to a quaternary C that has attached methyl groups. Therefore, the mechanism must not go through a C^+ intermediate, because doing so would lead to a 1,2-methyl shift. The starting C=C is symmetrical, and regiochemistry is therefore not an issue. The net addition is H_2O across the C=C bond, and the OH adds with no rearrangement. This is an example of addition without rearrangement and can be carried out with oxymercuration–reduction.

1. $Hg(OAc)_2$, H_2O/THF
2. $NaBH_4$

(b) The OH group is on a secondary C adjacent to a quaternary C that has an attached methyl group. Therefore, the mechanism must not go through a C^+ intermediate, because doing so would lead to a 1,2-methyl shift. The net addition is H_2O across the C=C bond, and the OH adds to the more substituted carbon with no rearrangement. This is an example of Markovnikov addition without rearrangement and can be carried out with oxymercuration–reduction.

1. $Hg(OAc)_2$, H_2O/THF
2. $NaBH_4$

Problem 12.44

Think

In each case, there is a net addition of what molecule across the C=C bond? What is the regiochemistry required in each reaction? Are carbocation rearrangements a concern? Which mechanism goes through a carbocation intermediate? Which mechanism does not go through a carbocation intermediate?

Solve

(a) The OH group is on a secondary carbon that was part of the original C=C bond. If a C^+ formed during the mechanism, rearrangement to the more stable benzylic C^+ would have occurred. This is an example of Markovnikov addition without rearrangement and can be carried out with oxymercuration–reduction.

1. $Hg(OAc)_2$, H_2O/THF
2. $NaBH_4$

OH

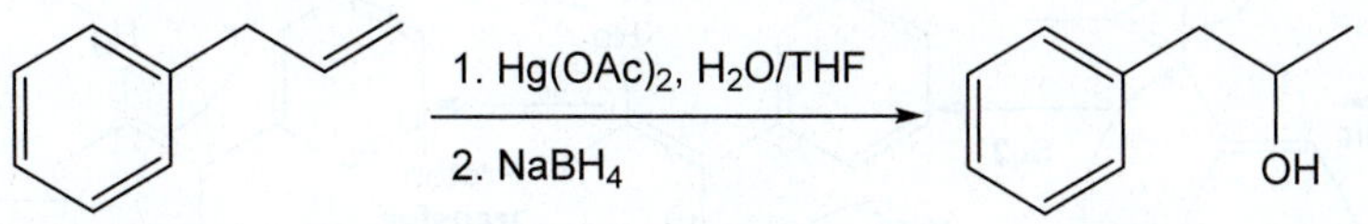

(b) The OH group is on a secondary benzylic carbon that was not part of the original C=C. Therefore, a secondary C^+ formed during the mechanism that rearranged via 1,2-hydride shift to the more stable secondary benzylic C^+. This is an example of Markovnikov addition with rearrangement and can be carried out via acid-catalyzed hydration.

OH

H^+/H_2O

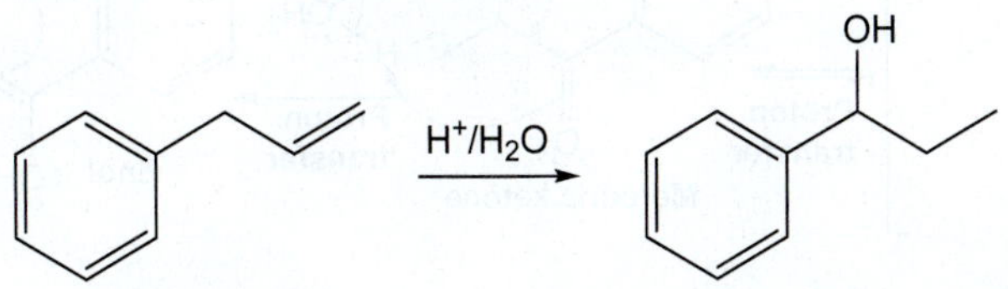

(c) The OH group is on a tertiary carbon that was not part of the original C=C. Therefore, a secondary C^+ formed during the mechanism that rearranged via 1,2-hydride shift to the more stable tertiary C^+. This is an example of Markovnikov addition with rearrangement and can be carried out via acid-catalyzed hydration.

H^+/H_2O

OH

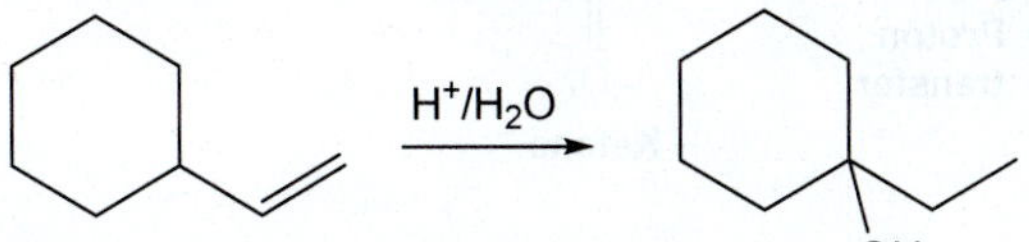

(d) The OH group is on a secondary carbon that was part of the original C=C. If a C^+ formed during the mechanism, rearrangement to the more stable tertiary C^+ would have occurred. This is an example of Markovnikov addition without rearrangement and can be carried out with oxymercuration–reduction.

1. $Hg(OAc)_2$, H_2O/THF
2. $NaBH_4$

OH

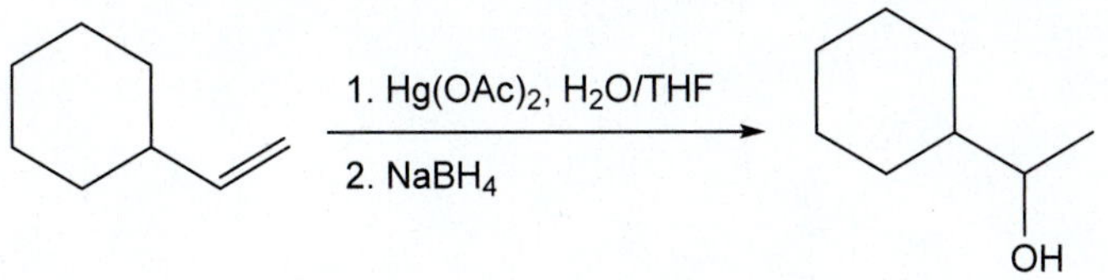

Problem 12.45

Think

What is produced when an alkyne undergoes oxymercuration? Is this product stable? If not, what is produced? Refer to Equation 7-37 to review the keto–enol tautomerization mechanism in acidic media.

Solve

Alkynes can also undergo Markovnikov addition of water via oxymercuration; an unstable mercuric enol is produced initially. Subsequent tautomerization converts the mercuric enol into the more stable mercuric keto form. The mercuric ketone is hydrolyzed by water to produce an enol. The enol then tautomerizes to the ketone. The mechanism for **(a)** is shown below, and the mechanism for **(b)** is shown on the next page.

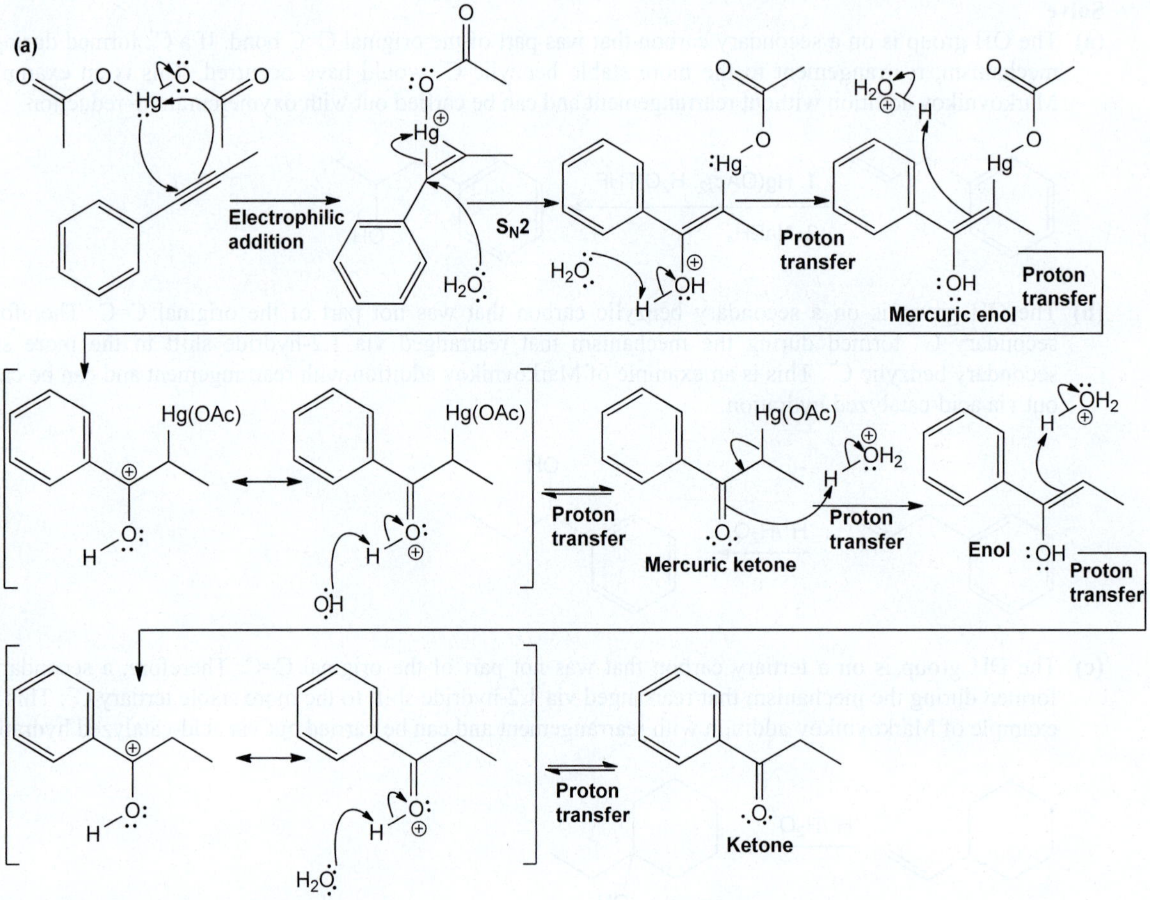

(a)

Electrophilic addition

S_N2

Proton transfer

Proton transfer

Mercuric enol

Hg(OAc)

Proton transfer

Mercuric ketone

Proton transfer

Enol

Proton transfer

Proton transfer

Ketone

(b)

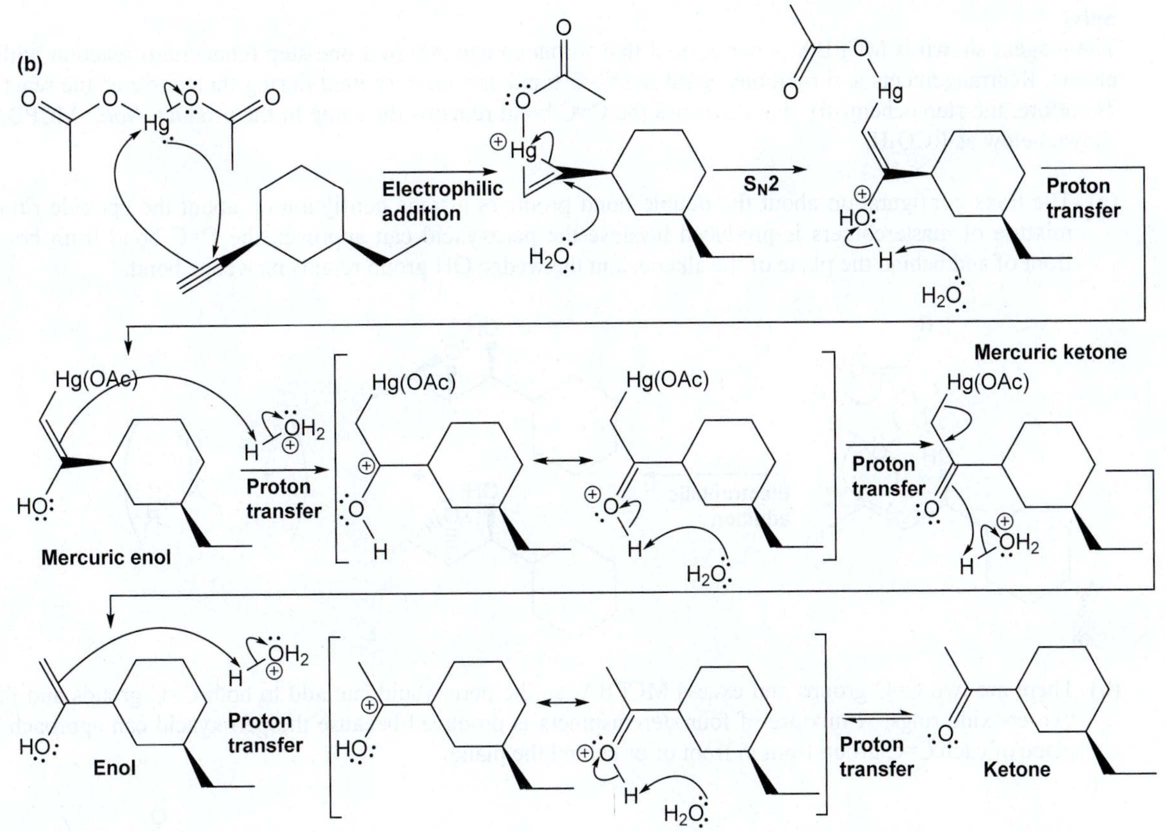

Electrophilic addition

S_N2

Proton transfer

Mercuric enol

Proton transfer

Proton transfer

Mercuric ketone

Enol

Proton transfer

Proton transfer

Ketone

Problem 12.46

Think

What intermediate is formed in an oxymercuration addition reaction with an alkene? Is a CO_2R an electron-donating or electron-withdrawing group? What is the effect of the CO_2R group on the ability of the α carbon atom in the mercurinium ion ring to accommodate a partial positive charge?

Solve

The CO_2R group is an electron-withdrawing group and, therefore, destabilizes nearby positive charges. This hinders the ability of the α carbon atom in the mercurinium ion intermediate to accommodate a partial positive charge. Therefore, the β carbon atom has a larger δ^+ and will be the site that water attacks. The partial mechanism is shown below.

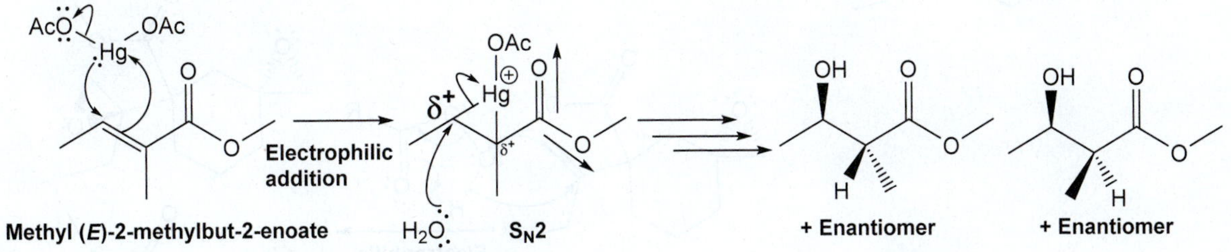

Methyl (*E*)-2-methylbut-2-enoate

Electrophilic addition

S_N2

+ Enantiomer

+ Enantiomer

Section 12.5 Epoxide Formation Using Peroxyacids

Problem 12.47

Think

What functional group is present in the reagent, and how does it react with an alkene? What functional group is produced? In how many steps does the reaction occur? If the reaction is concerted, what happens to the orientation of the substituents on the C=C bond?

Solve

The reagent shown is MCPBA, a peroxyacid that forms an epoxide in a one-step (concerted) reaction with an alkene. Rearrangements and rotations about the C=C bond are not permitted during the course of the reaction. Therefore, the stereochemistry that describes the C=C bond remains the same in the product. *Note:* MCPBA is shown below as RCO_3H.

(a) The trans configuration about the double bond produces a trans configuration about the epoxide ring. A mixture of diastereomers is produced because the peroxyacid can approach the C=C bond from both in front of and behind the plane of the alkene, and the wedge OH group retains its wedge bond.

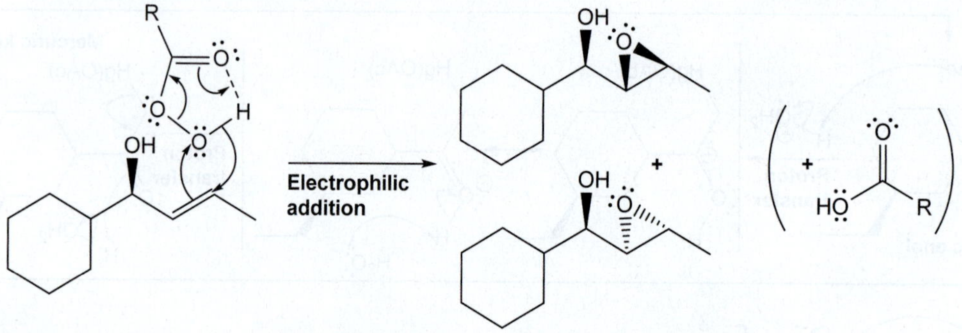

R
OH
H
Electrophilic addition
+

(b) There are two C=C groups and excess MCPBA, so the peroxyacid can add to both C=C groups and form two epoxide rings. A mixture of four stereoisomers is produced because the peroxyacid can approach the plane of each C=C group from in front of or behind the plane.

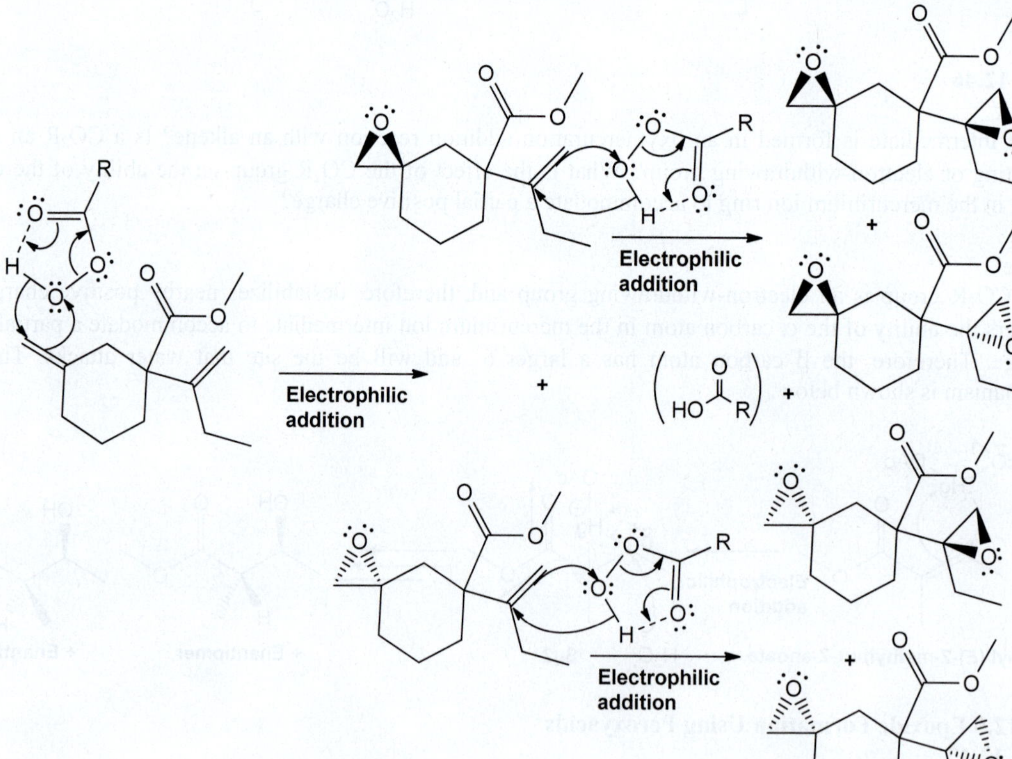

Electrophilic addition
Electrophilic addition
Electrophilic addition
HO R
+

Problem 12.48 (SYN)

Think

In each case, there is a net addition of what group across the C=C bond? What is the orientation of the substituents in the epoxide relative to the plane of the ring, cis or trans? What must be the orientation of the same substituents in the initial alkene?

Solve

The product of each reaction is an epoxide. Therefore, the reagent must be a peroxyacid, abbreviated as RCO_3H.

(a) The two groups are on the same side of the epoxide ring, so the starting alkene must be cis.

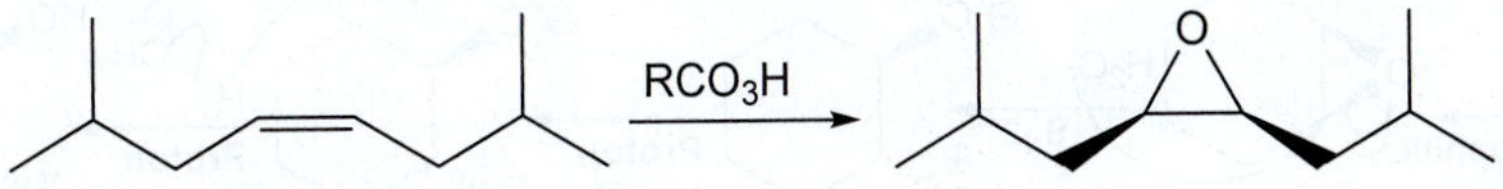

RCO_3H

(b) There are two epoxide rings produced; therefore, excess RCO_3H is necessary, and there must be two C=C groups on the starting alkene compound.

RCO_3H (excess)

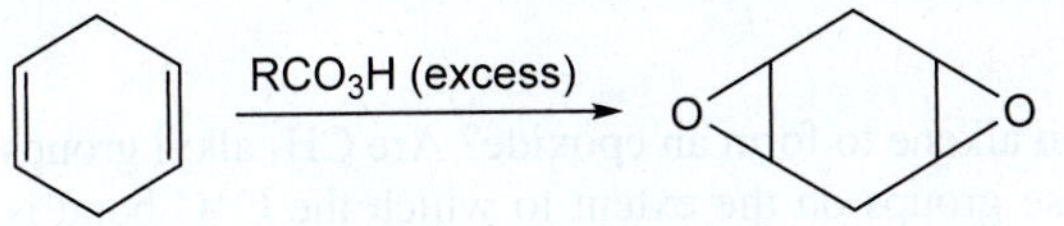

(c) There is just one epoxide ring produced (the other oxygen-containing ring's size is larger than a three-membered ring). Therefore, RCO_3H is necessary, and there must be one C=C group in the starting alkene compound. The benzene ring is inert to epoxidation.

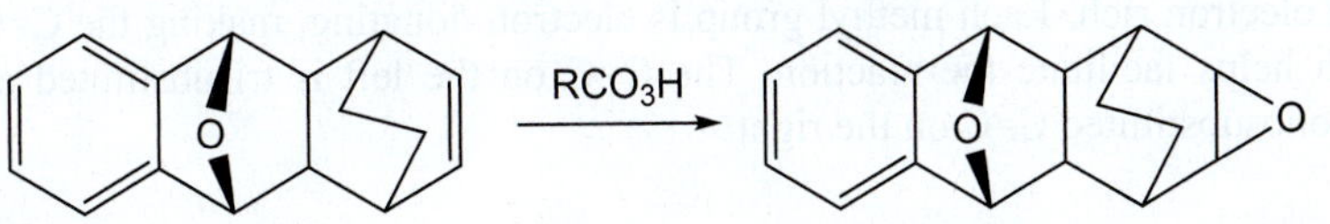

RCO_3H

Problem 12.49

Think

How are the reaction conditions similar to and different from the epoxide formation with MCPBA? What bonds are broken and formed in MCPBA and the starting alkene to form the new O–C bond and OH group in the product? What nucleophile attacks the epoxide ring?

Solve

The mechanism is shown below.

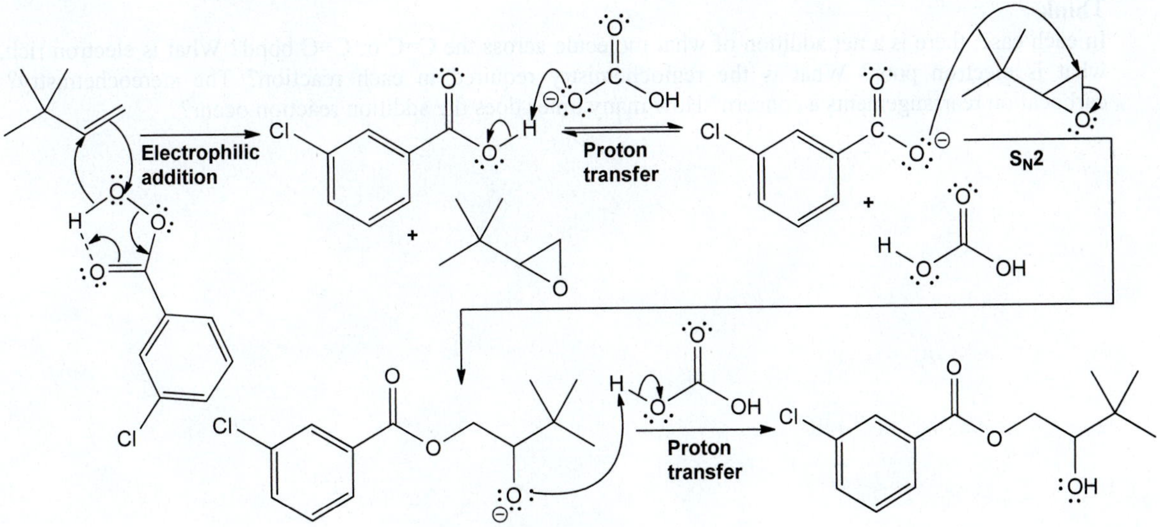

Electrophilic addition

Proton transfer

S_N2

Proton transfer

Problem 12.50

Think

What product is formed with the reaction of a peroxyacid in an alkene? What nucleophile attacks the epoxide? Why are the two OH groups anti?

Solve

MCPBA is a peroxyacid and forms an epoxide through reaction with an alkene. Water acts as a nucleophile to open the epoxide ring. Two proton transfer steps then take place to form the uncharged trans diol product.

Electrophilic addition
S_N2
Proton transfer
Proton transfer
+ Enantiomer
+ H_2O

Problem 12.51

Think

What is the mechanism by which MCPBA adds to an alkene to form an epoxide? Are CH_3 alkyl groups electron donating or withdrawing? What is the effect of these groups on the extent to which the C=C bond is electron rich? Does this increase or decrease the driving force for the electrophilic addition step?

Solve

Alkyl groups are electron donating, and in the electrophilic addition of MCPBA across a C=C double bond, the C=C double bond is considered electron rich. Each methyl group is electron donating, making the C=C double bond more electron rich, which helps facilitate the reaction. The C=C on the left is trisubstituted and more electron rich compared to the monosubstituted C=C on the right.

More electron rich, faster addition of MCPBA

Sections 12.6 and 12.7 Hydroboration–Oxidation of Alkenes and Alkynes

Problem 12.52

Think

In each case, there is a net addition of what molecule across the C=C or C≡C bond? What is electron rich, and what is electron poor? What is the regiochemistry required in each reaction? The stereochemistry? Are carbocation rearrangements a concern? How many times does the addition reaction occur?

Solve

(a) Water adds to the C=C bond in the overall reaction. The first step of the reaction forms an alkyl borane where the C–B bond forms at the less substituted C. This occurs three times to make the trialkylborane. Rearrangements are not permitted. Then oxidation occurs to replace the C–B bond with a C–OH bond. Thus, the OH forms on the primary C.

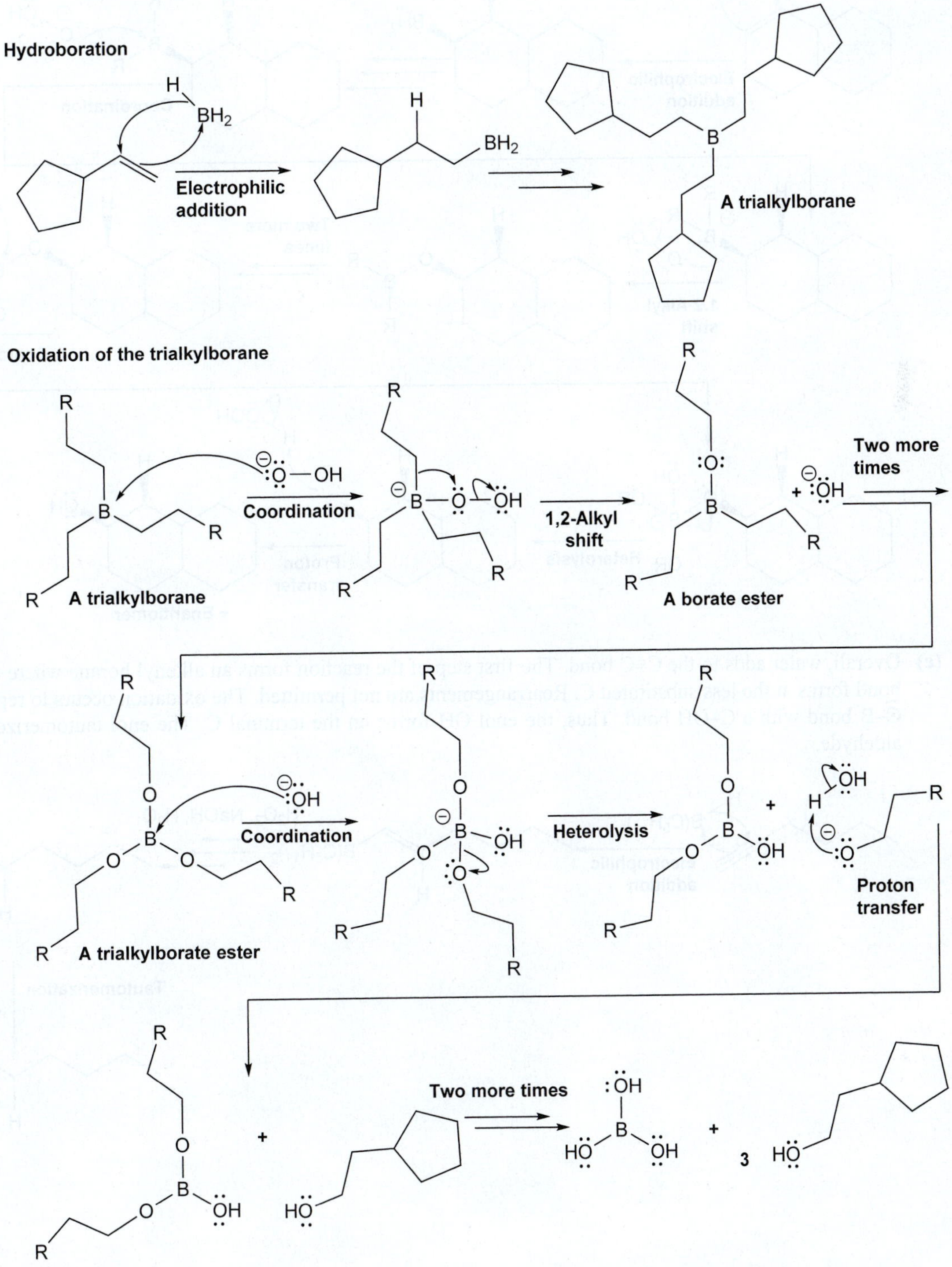

Hydroboration

H BH2 Electrophilic addition H BH2 B A trialkylborane

Oxidation of the trialkylborane

A trialkylborane Coordination 1,2-Alkyl shift A borate ester Two more times

A trialkylborate ester Coordination Heterolysis Proton transfer

Two more times 3

(b) BH_3 adds to the less sterically hindered carbon of the C=C bond, and H adds to the other carbon. The triakylborane forms and then is oxidized to form the alcohol.

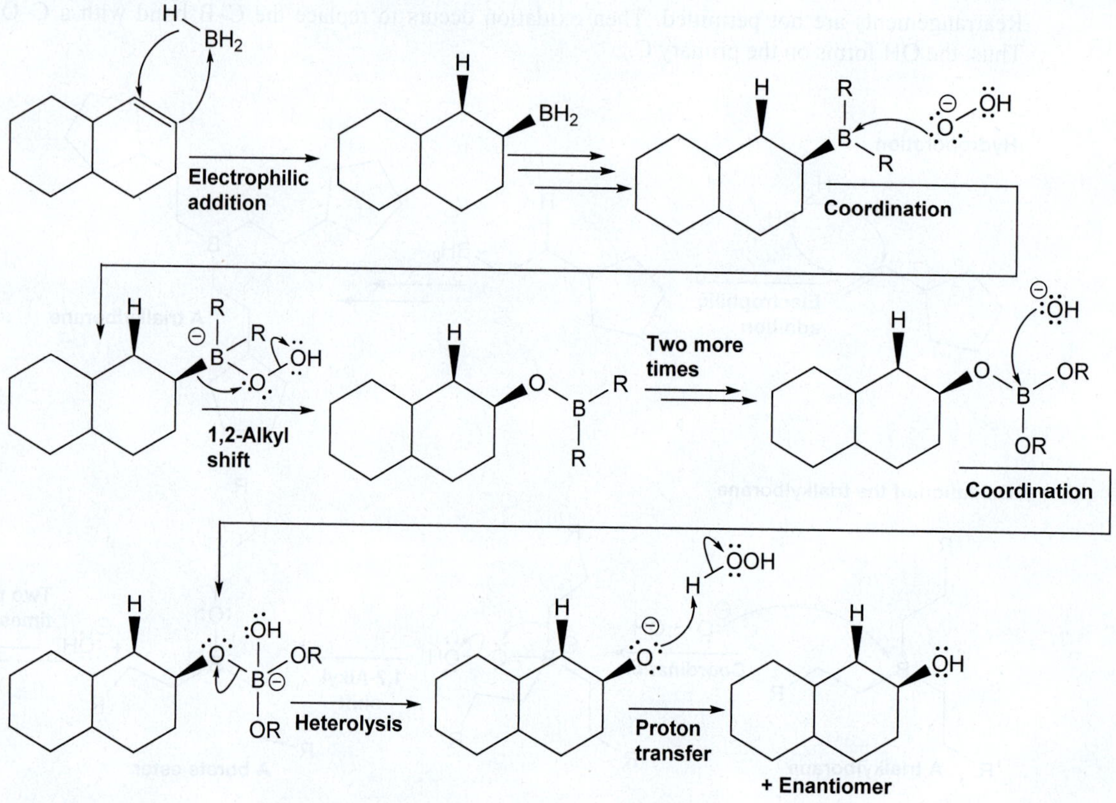

Electrophilic addition

Coordination

1,2-Alkyl shift

Two more times

Coordination

Heterolysis

Proton transfer

+ Enantiomer

(c) Overall, water adds to the C≡C bond. The first step of the reaction forms an alkenyl borane where the C–B bond forms at the less substituted C. Rearrangements are not permitted. The oxidation occurs to replace the C–B bond with a C–OH bond. Thus, the enol OH forms on the terminal C. The enol tautomerizes to the aldehyde.

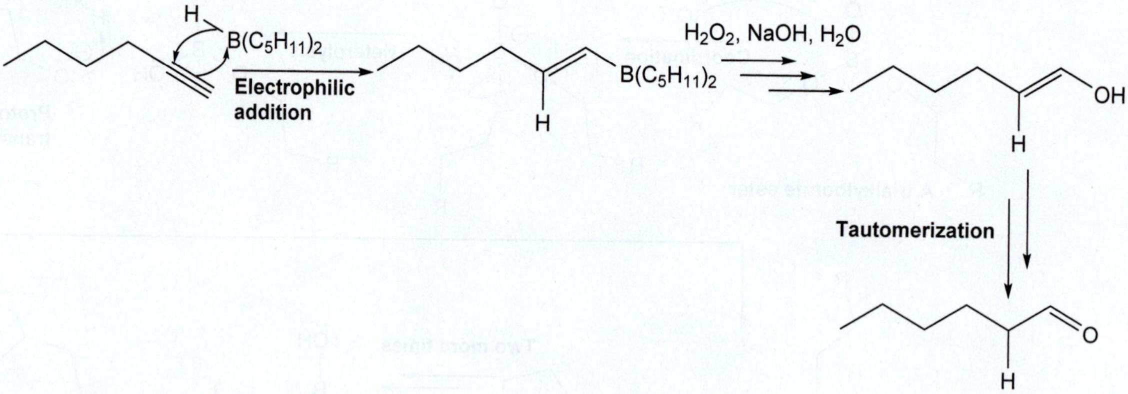

Electrophilic addition

H_2O_2, NaOH, H_2O

Tautomerization

(d) Overall, water adds to the C≡C bond. The first step of the reaction forms an alkyl borane where the C–B bond forms at the less sterically crowded C. Rearrangements are not permitted. The oxidation occurs to replace the C–B bond with a C–OH bond. Thus, the enol OH forms on the terminal C. The enol tautomerizes to the ketone.

H–$B(C_5H_{11})_2$ → Electrophilic addition → $B(C_5H_{11})_2$ → H_2O_2, NaOH, H_2O → OH → Tautomerization → O

Problem 12.53 (SYN)

Think

In each case, there is a net addition of what molecule across the C=C or C≡C bond? What is the regiochemistry required in each reaction? Is stereochemistry a concern? Are carbocation rearrangements a concern?

Solve

Syntheses are shown below.

(a) The net addition is H_2O across the C=C bond, and the OH adds to the less substituted carbon. The H and the OH add syn. This is an example of anti-Markovnikov addition and can be carried out with hydroboration–oxidation.

1. BH_3•THF
2. H_2O_2, NaOH, H_2O

H, OH + Enantiomer

(b) The net addition is H_2O across the C≡C bond, and the OH adds to the less sterically crowded carbon, which is the one on the right. The enol that forms tautomerizes to the ketone. This is an example of anti-Markovnikov addition and can be carried out with hydroboration–oxidation of the alkyne.

1. Disiamylborane, THF
2. H_2O_2, NaOH, H_2O

O

(c) The net addition is H_2O across the C≡C bond, and the OH adds to the less substituted carbon. The enol that forms tautomerizes to the aldehyde. This is an example of anti-Markovnikov addition and can be carried out with hydroboration–oxidation to a C≡C bond.

1. Disiamylborane, THF
2. H_2O_2, NaOH, H_2O

O

(d) The net addition is H_2O across the C=C bond, and the OH adds to the less sterically crowded carbon, which is the one on the right. This is an example of anti-Markovnikov addition and can be carried out with hydroboration–oxidation.

1. BH_3•THF
2. H_2O_2, NaOH, H_2O

OH

Problem 12.54 (SYN)

Think

Using the reagents BD_3•THF and D_2O, what adds to the C=C bond? In hydroboration oxidation, is the OD on the more or less sterically hindered C atom in the final product? Is the addition syn or anti?

Solve

The reagent BD_3•THF adds a D, and D_2O will add an OD. The OD adds to the less substituted C atom syn to the D. The syntheses are shown below.

(a) 1. BD_3•THF 2. H_2O_2, NaOH, D_2O — D, OD

(b) 1. BH_3•THF 2. H_2O_2, NaOH, D_2O — H, OD

(c) 1. BD_3•THF 2. H_2O_2, NaOH, H_2O — D, OH

Problem 12.55

Think

In hydroboration oxidation, does the OH end up on the more or less sterically hindered C atom in the final product? Does this reaction involve a competition between 1,2- and 1,4-addition?

Solve

In hydroboration oxidation, the HO ends up on the less substituted C atom syn to the H. In the step in which the alkylborane is produced, H and BH_2 add simultaneously in a 1,2-positioning only, and therefore, the presence of a conjugated C=C does not lead to a 1,4-product. In the presence of excess BH_3•THF reagent, hydroboration will occur at both C=C groups.

BH_3•THF → BH_2, H → BH_3•THF → H_2B, H, BH_2, H — Hydroboration occurs twice. → NaOH, H_2O_2, H_2O

Oxidation occurs twice.

HO, H, OH, H

Integrated Problems

Problem 12.56

Think

In Section 12.3, what is the intermediate that forms when Cl_2 undergoes addition to an alkene such as but-2-ene? Why does anti addition occur with this intermediate? What type of intermediate must form to have both syn and anti addition? Why does 1-phenylprop-1-ene lead to a different intermediate?

Solve

(a) Anti addition of Cl_2 results if the reaction proceeds through a chlorinium ion intermediate. To have both syn and anti addition, a carbocation (or carbocation-like) intermediate must form. The mechanism is shown below.

(b) 1-Phenylprop-1-ene leads to a different intermediate because of the stability of the benzylic C^+ that forms in the first electrophilic addition step. The benzylic C^+ then undergoes coordination to form the four stereoisomers shown below.

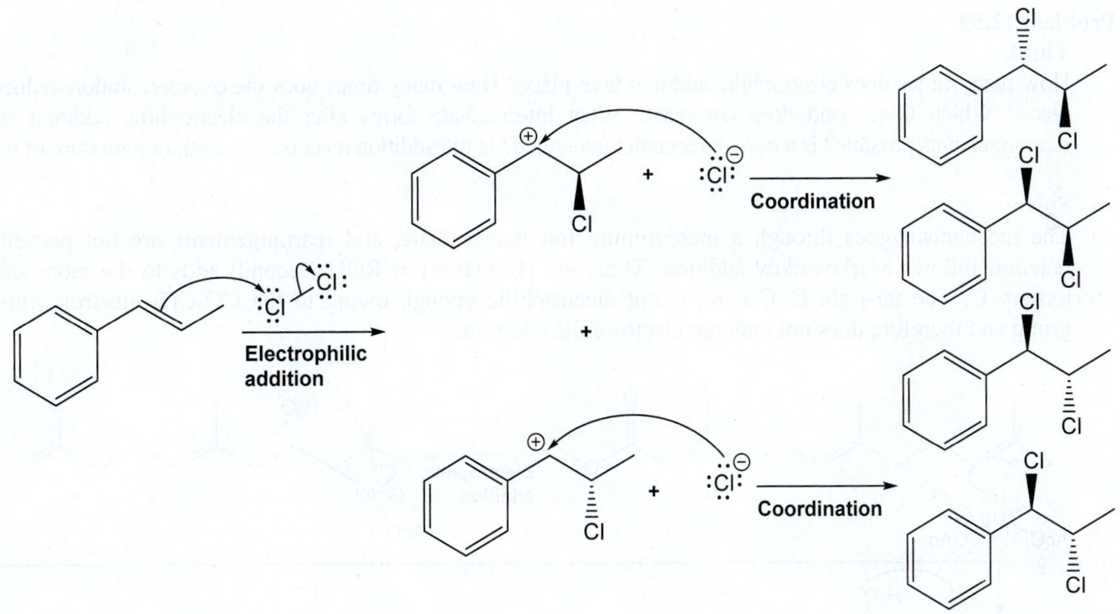

Electrophilic addition

Coordination

Coordination

Problem 12.57

Think

What bonds are broken and formed in I–CH_2–Zn–I to form a cyclopropane ring and I–Zn–I? How is this similar to the peroxyacid mechanism?

Solve

The mechanism is shown below. Broken bonds are Zn–C, C=C, and C–I. Bonds formed are two C–C and Zn–I.

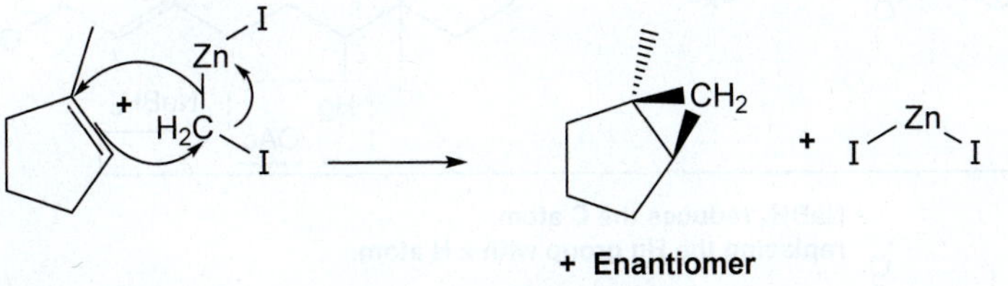

+ Enantiomer

Problem 12.58

Think

How many C atoms are in the starting material compared to the product? What species forms from the reaction of diazomethane and light? How would this reagent react with the C=C bond on benzene? What bonds would need to form and break in that intermediate to form the product?

Solve

The $H_2C–N_2$ bond breaks in a heterolysis step to form the reactive carbene reagent. This electrophilic reagent has a lone pair on the electrophilic atom, and electrophilic addition, therefore, proceeds to form an intermediate that avoids losing an octet. Two new σ C–C bonds are formed simultaneously as the C=C π bond is broken. This forms a cyclopropane ring on the same side of the six-membered ring. This ring undergoes rearrangement to relieve the strain of the three-carbon ring.

$H_2C–N≡N$ → hν, Heterolysis → $+ N_2(g)$ → Electrophilic addition → Rearrangement

Problem 12.59

Think

How many times does electrophilic addition take place? How many times does the oxymercuration–reduction take place? Which C=C bond does not react? What intermediate forms after the electrophilic addition step? Are rearrangements possible? Is a new stereocenter generated? Is the addition reaction syn, anti, or a mixture of both?

Solve

The mechanism goes through a mercurinium ion intermediate, and rearrangements are not permitted. The reaction follows Markovnikov addition. Thus, the H_2O (first) or ROH (second) adds to the more substituted tertiary C. The far-right C=C bond is not electrophilic enough owing to the CO_2CH_3 electron withdrawing group and therefore does not undergo electrophilic addition.

Hg(OAc)$_2$ → Electrophilic addition → H_2O, S_N2 → Proton transfer → HgOAc, Hg(OAc)$_2$ → Electrophilic addition → S_N2 → $NaBH_4$

NaBH$_4$ reduces the C atom, replacing the Hg group with a H atom.

Problem 12.60

Think

What are the various ways BH_3 can approach the C=C bond? How does steric hindrance affect each of those approaches?

Solve

(a) The isomeric products are shown below.

(b) Steric repulsion is minimized when BH_3 approaches the side of the C=C bond with less steric bulk. In the case of norbornene, this is the side opposite of the CH_2 group. In the hydroboration steps below, the first reaction has significant steric repulsion and leads to the minor product. The second hydroboration minimizes the steric repulsion and leads to the major product.

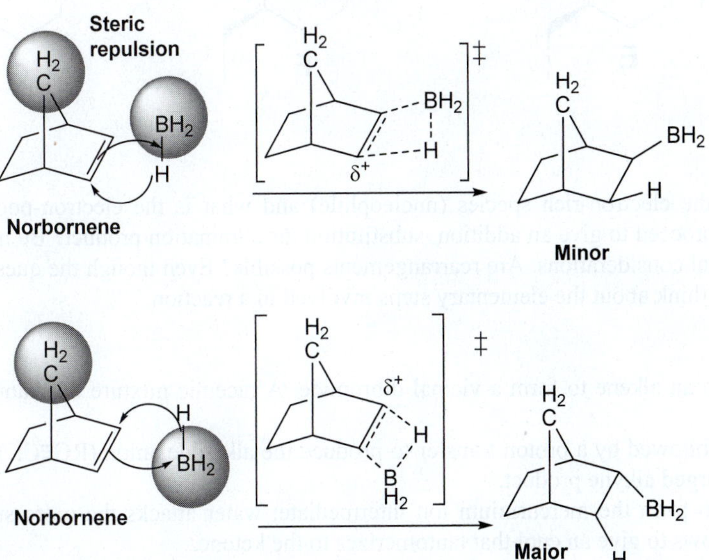

Problem 12.61

Think

In each step, consider what is the electron-rich species (nucleophile) and what is the electron-poor species (electrophile). Will the reaction proceed to give an addition, substitution, or elimination product? Be mindful of regiochemistry and stereochemical considerations. Are rearrangements possible? Even though the question does not ask for a mechanism, always think about the elementary steps involved in a reaction.

Solve

A results from addition of a peroxyacid to an alkene to form an epoxide.
B results from an epoxide ring opening in basic conditions (R group anti to OH) followed by workup in acid to form an uncharged alcohol product. A racemic mixture of enantiomers is produced.
C results from an E1 reaction of an alcohol heated in acidic water.
D results from addition of a peroxyacid to an alkene to form an epoxide. A racemic mixture of enantiomers is produced.
E results from an epoxide ring opening in neutral conditions (less substituted C atom attacked and anti to OH) followed by protonation to form an uncharged alcohol product. A racemic mixture of enantiomers is produced.
F results from the E2 elimination of H and Br to form the alkene product.
G results from the addition of a carbene H_2C: to form the cyclopropane ring in the product. The CH_2 group can add to the same side as OH or to the opposite side, producing diastereomers. Each diastereomer is produced as a racemic mixture of its enantiomers.
The missing compounds in the synthesis scheme are shown on the next page.

MCPBA A 1. CH_3CH_2Li 2. H_3O^+ B H_3O^+ Δ C

MCPBA D NaBr CH_3CH_2OH E NaOH Δ F CH_2N_2 hv G

Problem 12.62

Think

In each step, consider what is the electron-rich species (nucleophile) and what is the electron-poor species (electrophile). Will the reaction proceed to give an addition, substitution, or elimination product? Be mindful of regiochemistry and stereochemical considerations. Are rearrangements possible? Even though the question does not ask for a mechanism, always think about the elementary steps involved in a reaction.

Solve

H results from addition of Br_2 to an alkene to form a vicinal dibromide. A racemic mixture of enantiomers is produced.

I results from two E2 reactions followed by a proton transfer to produce the alkynide anion ($RC{\equiv}C{:}^-$). An acid workup follows to form an uncharged alkyne product.

J results from oxymercuration to form the mercurinium ion intermediate; water attacks the more substituted carbon. A reduction reaction follows to give an enol that tautomerizes to the ketone.

K results from bromination at the α carbon. Under acidic conditions, only a single bromination occurs. A racemic mixture of enantiomers is produced.

L results from an E2 reaction of a strong base in a heated solution.

M results from Br_2 addition of an alkene to yield the bromonium ion intermediate where water attacks the more substituted C atom. A racemic mixture of enantiomers is produced.

Br_2 CCl_4 H 1. $NaNH_2$ 2. H_2O I 1. $Hg(OAc)_2$, H_2O 2. $NaBH_4$ J

Br_2 Acetic acid K NaOH Δ L Br_2 H_2O M + Enantiomer

Problem 12.63 (SYN)

Think

Overall, what gets added to the alkyne? For the two Br atoms to be on the same carbon, what mechanism must the reaction go through? What type of intermediate must form? For the two Br atoms to be on different carbon atoms, what intermediate structure must the reaction go through?

Solve

In both cases, two H atoms and two Br atoms are added to the alkyne.

(a) For the Br atoms to add to the same C atom, the reaction is carried out with excess HBr to form the carbocation on the internal carbon, followed by coordination of Br^-. This occurs two times.

Excess HBr

Br Br

(b) For the Br atoms to add to two different C atoms, the alkyne is first converted to the alkene. This transformation occurs in three synthetic reactions—hydroboration oxidation of the alkyne to form the aldehyde, reduction to form the alcohol, and dehydration to form the alkene. The addition of Br_2 to the alkene forms the vicinal dibromide.

1. $(C_5H_{11})_2BH•THF$

2. $^{\ominus}OH$, H_2O_2

O

$NaBH_4$

EtOH

OH

H_3PO_4

Δ

Br_2

Br

Br

+ Enantiomer

Problem 12.64 (SYN)

Think

What reactions will add Br and OH to adjacent C atoms of an alkene? Is the Br in the target on the more or less substituted C atom? Is the OH on the more or less substituted C atom? Which conditions will favor putting a Br on the less substituted C atom and the OH on the more substituted C atom?

Solve

Route 1: Add MCPBA (or another peroxyacid) to form an epoxide. The epoxide ring is opened in neutral conditions that favor attack at the less substituted C atom (Br^- in a nonnucleophilic solvent). An acid workup is necessary to form an uncharged alcohol product.

MCPBA

O

1. NaBr/DMSO

2. H_3O^+

HO

Br

Route 2: Add Br_2 to form the brominium ion intermediate. In the presence of water, the H_2O attacks the more substituted C atom.

Br_2

H_2O

HO

Br

CHAPTER 13 | Organic Synthesis 1: Beginning Concepts

Your Turn Exercises

Your Turn 13.1

Think

What is the abbreviation commonly used for tosyl? Draw out the organic reactant and product using nomenclature rules previously discussed. Are curved arrows drawn for synthetic steps? Is it necessary to include the inorganic by-product? In drawing out the mechanism, think about the type of carbon to which the leaving group is bonded (1°, 2°, 3°). What type of attacking species is Cl^-? Identify the most likely mechanism.

Solve

(a) The common abbreviation for tosyl is OTs. The synthetic step does not need to show the NaOTs inorganic by-product, nor should the synthetic step include curved arrows.

+ NaCl →

2-Phenyl-2-tosyloxypropane **2-Chloro-2-phenylpropane**

(b) The OTs is located on a 3° C, and Cl^- is a weak base but a good nucleophile. The reaction, therefore, follows an S_N1 mechanism. The first step is heterolysis to form the benzylic 3° C^+ and TsO^-. The Cl^- then attacks the C^+ via a coordination step.

$-TsO^{\ominus}$

1. Heterolysis

2. Coordination

Your Turn 13.2

Think

From the description, identify the reagent and reaction conditions. Relative to the arrow (above or below), where is each commonly written in a synthetic step?

Solve

Water, H_2O, is the solvent and the temperature, 70 °C, is a reaction condition. Both are written below the arrow.

H_2O, 70 °C

Your Turn 13.3

Think

What do the numbers in front of the reagents indicate? How many separate, overall reactions are occurring to accomplish this synthesis?

Solve

The numbers indicate that one reaction is carried out to completion before the next one is initiated. In this case, two separate, overall reactions take place. First, butanoic acid is treated with sodium hydroxide (to yield sodium butanoate). Once that reaction has come to completion, bromoethane is then added (to yield ethylbutanoate). This is a two-reaction synthetic scheme.

Your Turn 13.4

Think

Draw the organic reactant and product using nomenclature rules previously discussed, and draw the two inorganic reagents. In how many separate, overall reactions is this synthesis occurring? What are the reagents, solvents, and conditions? How do you specify each complete reaction using the number convention?

Solve

There are two separate, overall reactions occurring, which can be distinguished using numbers above and below the reaction arrow. First, (*Z*)-hex-3-ene (drawn on the left) is treated with $Hg(OAc)_2$, H_2O (mercury(II) acetate in water), and this can be indicated with a "1." above the reaction arrow. Once that reaction has come to completion, $NaBH_4$ (sodium borohydride) is added in a second step, and this can be indicated with a "2." below the reaction arrow. The final product, hexan-3-ol, is written on the right of the arrow.

(*Z*)-Hex-3-ene — 1. $Hg(OAc)_2/H_2O$; 2. $NaBH_4$ → Hexan-3-ol (OH)

Your Turn 13.5

Think

Review the reaction tables in Chapters 9–12 and note any difficulties you have.

Solve

Review the reaction tables in Chapters 9–12 and note any difficulties you have.

Your Turn 13.6

Think

Identify the functional group(s) present in each step. In each reaction, consider if the functional group is different in the products than it is in the reactants. In each step, look at the carbon skeleton. Were any C–C σ bonds broken and/or formed?

Solve

The synthetic steps are labeled below as being either functional group conversions or reactions that alter the carbon skeleton. In the first reaction, an epoxide is transformed into an alcohol and ether, but no C–C σ bonds are formed or broken. In the second reaction, the alcohol is converted into an alkylbromide, but no C–C σ bonds are formed or broken. In the third reaction, a new C–C bond is formed and the carbon skeleton is increased by two carbons, C≡CH. In the fourth reaction, the alkyne is transformed into an aldehyde, but no C–C σ bonds are formed or broken. In the fifth reaction, the α C is alkylated by CH_2CH_3, which forms a new C–C σ bond.

Epoxide —Functional group conversion→ Alcohol and ether (OH, OCH_3) —Functional group conversion→ Alkyl halide (and ether) (Br, OCH_3) —Alters carbon skeleton→ Addition of C≡CH (and ether) (CH≡C, OCH_3) —Functional group conversion→ Aldehyde (and ether) (O, OCH_3) —Alters carbon skeleton→ Addition of CH_2CH_3 (and ether) (O, OCH_3)

Your Turn 13.7

Think

Identify the functional group(s) present in synthetic steps **B** and **C**. In each reaction, consider if the functional group is different in the products than it is in the reactants. In each step, look at the carbon skeleton. Were any C–C σ bonds broken and/or formed?

Solve

In synthetic step **B**, the epoxide is transformed into an alcohol, and the carbon chain is extended by one carbon through the new C–CH_3 bond formation. This is a reaction that alters the carbon skeleton and, therefore, is found in Appendix C, "Reactions That Alter the Carbon Skeleton," entry 4. In synthetic step **C**, the reaction converts the alcohol (HO) into an alkylbromide (Br). This reaction is found in Appendix D, Table AppD-2, "Reactions That Produce Alkyl and Aryl Halides," entry 16.

1. CH_3MgBr
2. H_2O, H_2SO_4
B
Appendix C
Entry 4

OH, CH_3

PBr_3
C
Appendix D, Table AppD-2
Entry 16

Br, CH_3

Your Turn 13.8

Think

The squiggly line indicates that the C–C bond between the benzylic C and the alkyne C should be disconnected. Once this occurs, one C atom in the precursors should be electron rich, and the other C atom should be electron poor. What are the necessary functional groups in the precursors that accomplish this? (*Hint:* Consult Table 13-1 and Equation 13-11.)

Solve

One precursor should be an alkyl halide (electron-poor C), and the other precursor should be a terminal alkyne (electron-rich C). The alkyl halide is benzyl bromide, $PhCH_2Br$.

Contains 16 C atoms

H_2C–C≡C–CH_2

Undo the C–C bond-forming reaction entry 1 of Table 13–1.

Contains seven C atoms

H_2C—Br

+

Contains nine C atoms

HC≡C—CH_2

Your Turn 13.9

Think

Consult Table 13-2. What are the alkoxide and alkyl bromide precursors if the other C–O bond is disconnected? Does this affect the reaction in Step 2 of the synthesis in Equation 13-13?

Solve

If the other C–O bond is disconnected, CH_3Br and HC≡CCH_2ONa are the precursors. The synthesis in Step 2 is not altered.

Undo ether synthesis.

O

ONa

+ H_3C—Br

The synthesis would proceed in the forward direction, as shown below:

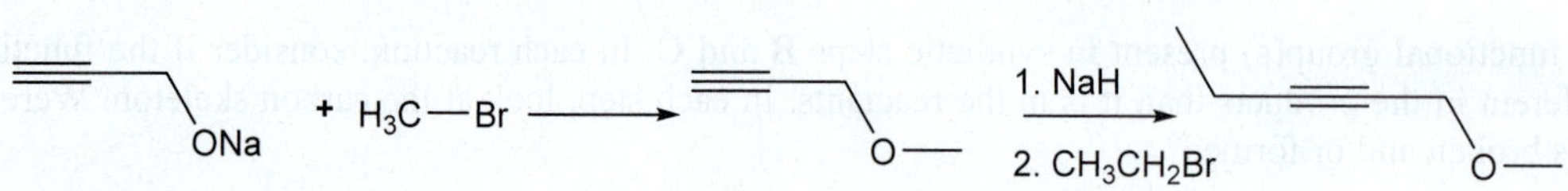

ONa + H_3C—Br → ... 1. NaH 2. CH_3CH_2Br

Your Turn 13.10

Think

What undesirable reaction would be a potential concern when an alcohol solvent (ROH) is used with $CH_3CH_2O^-$? Is there an alcohol solvent that, when added to $CH_3CH_2O^-$, does not produce a new nucleophile even after that reaction takes place?

Solve

Ethanol, CH_3CH_2OH, would be an appropriate solvent. The undesirable reaction is a proton transfer between the alcohol solvent and $CH_3CH_2O^-$. With ethanol as the solvent, a proton transfer would produce species that are identical to the reactant species, as shown below.

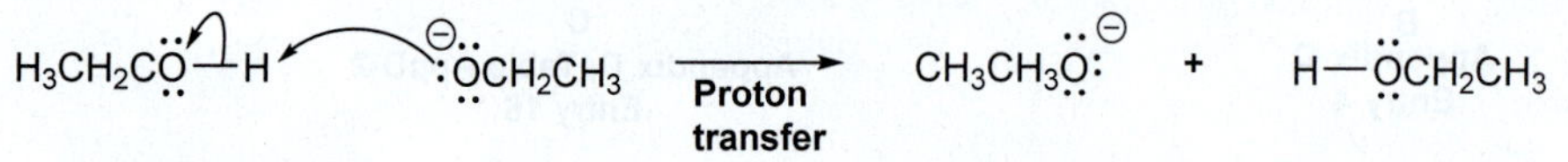

H_3CH_2CO—H + $^{\ominus}$:$\ddot{O}CH_2CH_3$ → (Proton transfer) $CH_3CH_3\ddot{O}$:$^{\ominus}$ + H—$\ddot{O}CH_2CH_3$

Your Turn 13.11

Think

If you perform a 90° counterclockwise rotation of the molecule so that you are looking down the C2–C3 bond, which group is in the front? Which group is in the back? In which plane is the C–C bond? How do you represent the three other bonds on the front C atom? The back C atom?

Solve

If the target alkene molecule is rotated 90° counterclockwise (as viewed from the top) so that you are looking down the C2=C3 bond, C2 is in the front and the C3 group is in the back. In the Newman projection, the front carbon, C2, has the CH_3 at 0° and H at 180°, and C3 has the CH_3 at 0° and CH_2CH_3 at 180°.

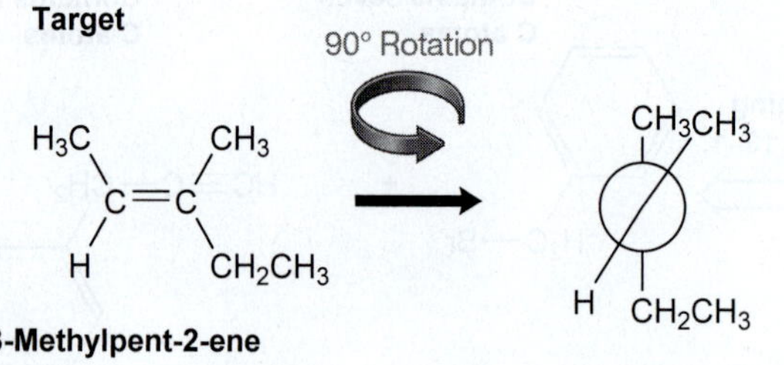

Target

90° Rotation

H_3C, CH_3, C=C, H, CH_2CH_3

CH_3CH_3, H, CH_2CH_3

(*E*)-3-Methylpent-2-ene

If the precursor molecule is rotated 90° counterclockwise (as viewed from the top) so you are looking down the C2–C3 bond, C2 (with the Cl group) is in the front and the C3 group is in the back. In the Newman projection, the front carbon, C2, has the CH_3 at 60°, the Cl at 180°, and the H at 300°, and C3 has the H at 0°, the CH_2CH_3 at 120°, and the CH_3 at 240°.

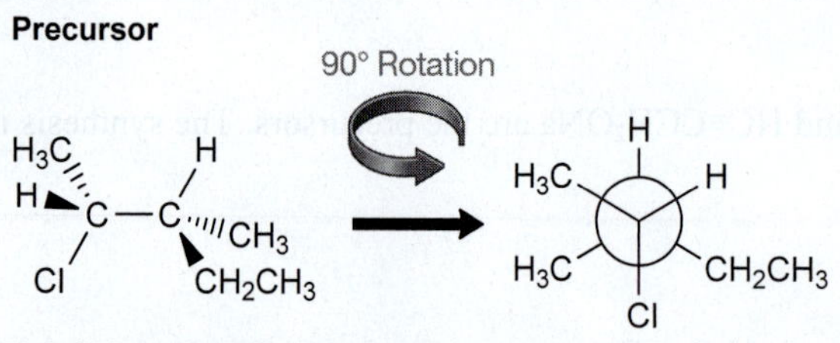

Precursor

90° Rotation

H_3C, H, C—C, CH_3, Cl, CH_2CH_3

H, H_3C, H, H_3C, CH_2CH_3, Cl

(2*R*, 3*R*)-3-Chloro-3-methylpentane

Your Turn 13.12

Think

What is the yield for each step? For a linear synthesis, what is the relationship between the overall product yield and the product yield for each step?

Solve

For a linear synthesis, the overall product yield is the product of each step in the sequence. If each step's yield is 80%, the yield of a six-step synthesis would be $(0.80)^6 = 0.26$, or 26%.

In Chapter Problems

Problem 13.1

Think

Refer to Your Turn 13.6 to determine if the reaction is a functional group conversion or a reaction that alters the carbon skeleton. In each reaction, determine which bonds formed. Did an electrophile or nucleophile react with the organic reactant in each step? Is regiochemistry a concern? Consult the reaction tables indicated in Your Turn 13.5, if necessary.

Solve

The first reaction has CH_3O^- attack the epoxide ring at the less-substituted C atom. Therefore, the reaction occurs in basic conditions. An acid workup step follows to form the uncharged alcohol product. The second reaction is substitution of the HO for Br. Because hydroxide is not a good leaving group, PBr_3 changes it into a good leaving group via back-to-back S_N2 reactions. The third reaction is nucleophilic substitution of the Br for the alkynide anion. The fourth reaction is hydroboration–oxidation to form the terminal enol that tautomerizes to the aldehyde. The fifth and final reaction is alkylation of the α C atom.

1. $NaOCH_3$, CH_3OH; 2. H_3O^+ → OH, OCH_3; PBr_3 → Br, OCH_3; $NaC{\equiv}CH$ → CH≡C, OCH_3; 1. $(C_5H_{11})_2BH$; 2. NaOH, H_2O_2 → O, OCH_3; 1. $NaNH_2$; 2. CH_3CH_2Br → O, OCH_3

Problem 13.2

Think

Which C–C bond in the target could have been formed from the reaction in entry 2 of Table 13-1? Is there more than one choice? From what type of functional group did the OH originate? Was the more- or less-substituted C atom attacked?

Solve

The reaction in entry 2 in Table 13-1 would have formed the bond indicated by the wavy line below, and the reaction involves deprotonation of the alkyne to form the alkynide anion nucleophile. This nucleophile opens the epoxide ring by attack at the less-substituted C. An acid workup step is necessary to form the uncharged alcohol product.

Retrosynthesis

OH ⇒ **Undo an epoxide ring opening.** ⇒ + O

Synthesis in forward direction

NaH → ; 1. O (epoxide); 2. H_3O^+ → OH

Problem 13.4

Think

If you want to make pent-2-yne and have to start with compounds containing two or fewer C atoms, how many reactions that alter the carbon skeleton must you include? How can you form a new C–C bond using an alkynide anion? To what type of atom must a carbon be attached to be an electrophilic carbon?

Solve

Because the target contains five carbons, we will need to form two new carbon–carbon bonds if the starting material is to contain two or fewer carbons. So we must apply transforms that will disconnect two carbon–carbon bonds. An electrophilic carbon is one that is electron poor and, therefore, can be bonded to an electronegative atom that is also a good leaving group. *Note:* Instead of the order shown below, the CH_3CH_2Br could be added first and the CH_3Br second.

Retrosynthesis — **Undo a C–C bond formation.** ⟹ Br + ≡ — **Undo a C–C bond formation.** ⟹ HC≡CH + $BrCH_3$

Synthesis in forward direction

HC≡CH —(1. NaH; 2. CH_3Br)→ —(1. NaH; 2. CH_3CH_2Br)→

Problem 13.5

Think

How do you alkylate the α C atom of a C=O compound? How many times does this alkylation need to take place? What is the identity of your C=O compound after undoing these alkylations?

Solve

Alkylation of the α C atom of a C=O compound first involves a proton transfer to form an enolate anion, and then the enolate anion acts as a nucleophile with an R–L substrate via S_N2. The reaction sequence is performed twice.

H_3C–CHO —(1. $NaNH_2$; 2. $PhCH_2Br$)→ $PhCH_2$–CHO —(1. $NaNH_2$; 2. $PhCH_2Br$)→ $(PhCH_2)_2CH$–CHO

Problem 13.6

Think

What reagent and reaction conditions are necessary to reverse the transform indicated to carry out the reaction in the forward direction? What other functional groups are present? Does that functional group react with the reagent you used? If so, is this desired?

Solve

The first three can proceed as planned.

(a) NaCN can be used as a nucleophile for an S_N2 reaction. A concern might be the alcohol group, but NC^- is not strong enough of a base to deprotonate an alcohol, and HO^- is not sufficient enough of a leaving group to be displaced in an S_N2 reaction. Therefore, the synthesis will proceed as planned.

(b) This presents a situation similar to **(a)**. Therefore, the synthesis will proceed as planned.

(c) The conversion can take place using PBr_3. A concern might be the phenolic OH, but nucleophilic substitution does not take place on an sp^2-hybridized carbon. Therefore, the synthesis will proceed as planned.

(d) This is a synthetic trap. We could attempt to convert the OH group to Br using PBr_3, but both OH groups could be converted, as both are attached to sp^3-hybridized carbons. It is likely, however, that the intended product would be the major product, because the functional group is on a primary carbon. The other is on a secondary carbon, which impedes S_N2 reactions.

(e) This is a synthetic trap. We could attempt to convert the alkene C=C into the alkyl halide using HCl. This could put the Cl at the more-substituted C atom. However, the ether O atom is likely protonated in a strong acid and could lead to an S_N2 reaction in which the Cl opens the protonated ether ring.

(f) This is a synthetic trap. We could attempt to do oxymercuration–reduction of the alkyne to form the internal enol that tautomerizes to the ketone. The problem is that an alkene is also reactive under oxymercuration–reduction reaction conditions, and a secondary alcohol would form.

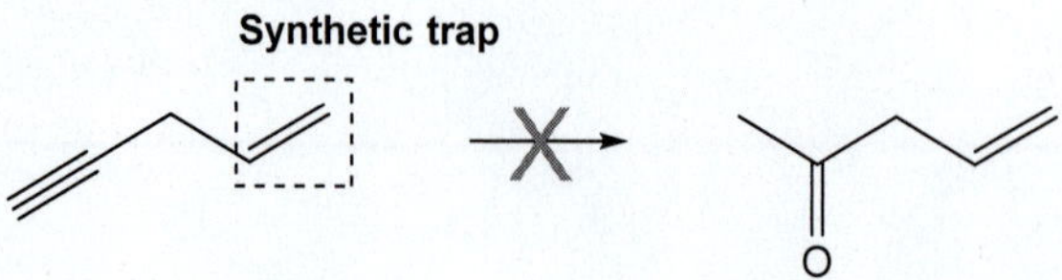

Problem 13.8

Think

What precursor can be used to produce the alkyl bromide target? Is regiochemistry a concern? If so, can the alkyl bromide be produced directly with the desired regiochemistry? If not, what other reaction can be carried out to give the desired regiochemistry?

Solve

We can begin by undoing a HBr addition to arrive at an alkene precursor. If the elimination reaction is also undone, we can arrive at the intended starting material. In the forward direction, elimination proceeds through an elimination dehydration mechanism to form the disubstituted alkene. The subsequent electrophilic addition of HBr will produce the target with the desired regiochemistry.

Retrosynthesis

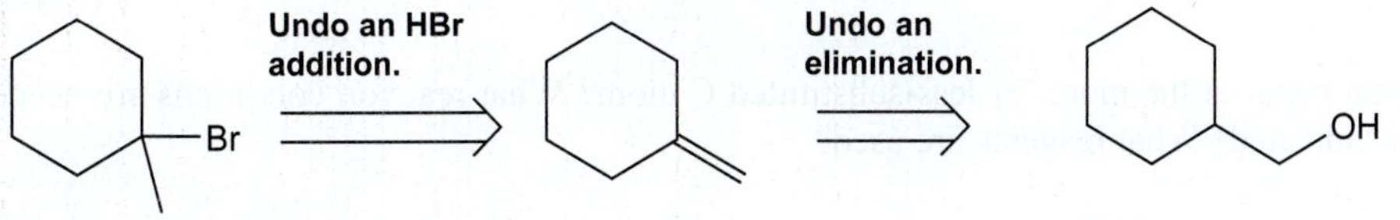

Undo an HBr addition.

Br

Undo an elimination.

OH

Synthesis in forward direction

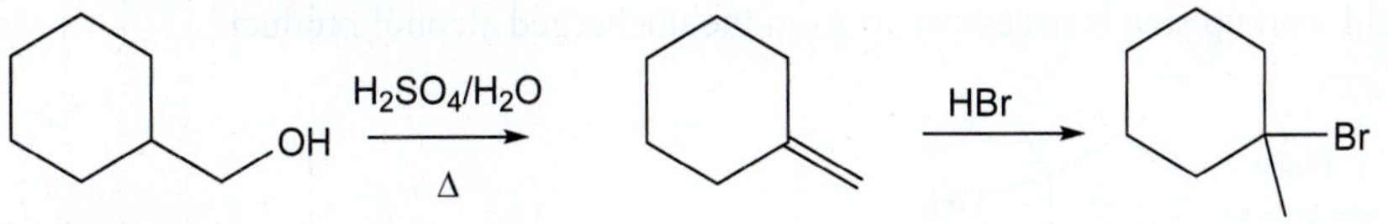

OH

H_2SO_4/H_2O

Δ

HBr

Br

Problem 13.9

Think

Should alkylation occur at the more- or less-substituted C atom? What is the alkyl group that adds? What reaction conditions are needed to form the regioisomer indicated?

Solve

In both reactions, the benzyl group is the alkyl group added, so a benzyl halide can be used as the substrate in an S_N2 reaction with an enolate nucleophile.

(a) Alkylation occurs at the less-substituted C atom. LDA is used because a very strong bulky base is required to favor the kinetic enolate intermediate.

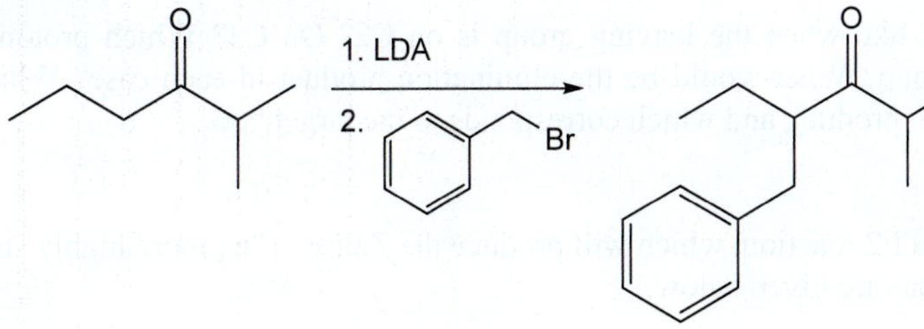

O

1. LDA

2.

Br

O

(b) Alkylation occurs at the more-substituted C atom. *tert*-Butoxide is used because a moderately strong base is required to favor the thermodynamic enolate intermediate.

1. $KOC(CH_3)_3$
2. Br

Problem 13.10

Think

Does bromination occur at the more- or less-substituted C atom? What reaction conditions are needed to form the regioisomer indicated? What reagents are used?

Solve

(a) Bromination occurs at the less-substituted C atom, so the reaction must take place under neutral or basic conditions. An acid workup step is necessary to form the uncharged alcohol product.

O
1. NaBr
2. H_3O^+
OH
Br

(b) Bromination occurs at the more-substituted C atom, so the reaction must take place under acidic conditions.

O
HBr
Br
OH

Problem 13.12

Think

What does the starting material look like when the leaving group is on C2? On C3? Which protons can be eliminated along with the leaving group? What would be the elimination product in each case? Which is the Zaitsev product, which is the Hofmann product, and which corresponds to the target?

Solve

Leaving group on C2: This is a normal E2 reaction, which will produce the Zaitsev (i.e., more highly substituted alkene) product. The reaction conditions are given below.

This H will be deprotonated to produce the more substituted alkene.
H
Br
CH_3CH_2ONa

Leaving group on C3: This is a Hofmann elimination, which will produce the Hofmann product, or the anti-Zaitsev (i.e., less highly substituted alkene) product. The reaction conditions are given below.

NH_2 1. excess CH_3I 2. Ag_2O 3. Δ

H — This H will be deprotonated to produce the less substituted alkene.

Problem 13.13

Think

What is added to the alkene in each reaction? Did 1,2-addition or 1,4-addition occur? How does temperature affect the regiochemistry of the product? In **(c)** and **(d)**, how many addition reactions occurred? What was the second reagent needed to accomplish that transformation?

Solve

(a) 1,4-Addition occurred. Warm temperatures are required. HBr adds to the alkene. H adds at C1, and Br adds at C4. An S_N2 reaction follows, in which Br is substituted by CN.

HBr, High *T* → Br; NaCN → NC

(b) 1,2-Addition occurred. Cold temperatures are required. HBr adds to the alkene. H adds at C1, and Br adds at C2. An S_N2 reaction follows, in which Br is substituted by CN.

HBr, Low *T* → Br; NaCN → CN

(c) 1,4-Addition occurred. Warm temperatures are required. HCl adds to the alkene. H adds at C1, and Cl adds at C4. A second addition reaction occurs, where a carbene $:CH_2$ adds to the C=C to form the cyclopropane ring.

HCl, High *T* → Cl; H_2CN_2, Δ → Cl

(d) 1,2-Addition occurred. Cold temperatures are required. HCl adds to the alkene. H adds at C1, and Cl adds at C2. A second addition reaction occurs, where Br_2 adds to the C=C to form the 1,2-dibromo product.

HCl, Low *T* → Cl; Br_2 → Br, Br, Cl

Problem 13.15

Think

Should the reaction be S_N1 or S_N2? To ensure that this reaction takes place, should you use a strong nucleophile or a weak nucleophile? What protic solvent could be used that would not react with that nucleophile to form a new nucleophile?

Solve

The reaction should be S_N2 because the leaving group is attached to a primary carbon. The nucleophile is ethoxide, $CH_3CH_2O^-$. Ethanol is used as a solvent because deprotonating the solvent yields the same nucleophile that was added.

Br → ($NaOCH_2CH_3$ / $HOCH_2CH_3$) → O

Problem 13.16

Think

In how many steps does an S_N2 reaction take place? Can the leaving group leave from the same side at which the nucleophile attacks the electrophilic carbon? If not, from what side must the nucleophile attack? What does this do to the stereochemistry at the electrophilic carbon?

Solve

S_N2 takes place in one step, and the leaving group leaves opposite the side from which the nucleophile attacks. This leads to inversion of configuration at the electrophilic carbon. The substrate must have the leaving group pointing toward us if the nucleophile is to form a bond pointing away from us.

Retrosynthesis

SH ⇒ Undo an S_N2. ⇒ OTs

Synthesis in forward direction

OTs → NaSH → SH

Problem 13.17

Think

For E2 to occur, the Cl and H must be in what orientation? Are they in this orientation in the precursor given? If so, which groups are cis to each other in the alkene? Which groups are trans? What does this mean for the orientation of those groups in the dash–wedge structure?

Solve

If the carbon with the Cl atom in the precursor is the same carbon as the left-hand alkene carbon in the target, then it must also be bonded to H, so the atom that must be added to the box at the bottom left is H.

H_3C, CH_3, H, CH_2CH_3 ⇒ Undo an E2. ⇒ Cl, H_3C, H

We then recognize that in the target, the two CH_3 groups are on the same side of the double bond. Thus, in the precursor, if one CH_3 group is pointing behind the plane of the paper, so must the CH_3 group bonded to the other carbon. This means that the CH_2CH_3 group must point in front. Notice that this precursor is the enantiomer of the one in Equation 13-43. See the figure on the next page.

Undo an E2.

H₃C, CH₃, H, CH₂CH₃ ⟹ Cl, CH₃, CH₂CH₃, H₃C, H, H

Problem 13.18

Think

Does hydroboration–oxidation add the HO to the more- or less-substituted C? Do the H and the HO add syn or anti? What is the orientation of those groups in (2*S*, 3*R*)-3-phenylbutan-2-ol? Is another conformation necessary to view the target as the immediate product of hydroboration–oxidation?

Solve

Hydroboration–oxidation adds the HO to the less-substituted C, which indicates that the C=C bond must be attached to the benzene ring. The H and the HO add syn in hydroboration–oxidation, and therefore, another conformation is necessary to view the H and HO as syn. Rotation about the C–C bond accomplishes this, and the methyl groups are, therefore, cis.

Rotate about C–C bond. → OH, H and OH same side; Undo a hydroboration–oxidation. ⟹

(2*S*,3*R*)-3-Phenylbutan-2-ol

Problem 13.20

Think

How many carbon atoms does the target have? How many does the starting compound have? In your retrosynthetic analysis, which C–C bond should you consider undoing? What reaction would produce that bond and leave an OH on an adjacent carbon? What precursors would you arrive at by undoing that reaction? How can those precursors be made from propan-1-ol?

Solve

The target has six C atoms, but the starting compound has only three, so let's begin by applying a transform that undoes the formation of the C–C bond shown below. The reaction we choose to undo is one that would open an epoxide to leave a OH group on the carbon adjacent to the one that forms the new bond. Undoing that reaction yields an epoxide and a terminal alkyne as precursors. The precursors, specifically, would be propyne and propylene oxide.

Undo a C–C bond formation. ⟹ **Precursor 1 Propyne** + **Precursor 2 Propylene oxide**

We don't have a reaction that will produce propyne or propylene oxide directly from propan-1-ol, so these involve multistep syntheses. Propyne can be produced from propene by carrying out a bromination followed by elimination. Propylene oxide can be produced from propene by epoxidation using MCPBA. Finally, propene can be produced by dehydrating propan-1-ol. The retrosynthesis for precursors 1 and 2 is shown on the next page.

Precursor 1

Undo an E2 twice. Undo a bromination. Undo a dehydration.

Precursor 2

Undo an epoxidation. Undo a dehydration.

The forward reaction is given below.

H_3PO_4, Δ; Br_2, CCl_4; 1. $NaNH_2$ (excess) 2. H_2O; MCPBA; 1. $NaNH_2$ 2. (epoxide) 3. H_3O^+

Problem 13.21

Think

For a linear synthesis, how is the overall product yield related to the yield of each step in the reaction? What is the percent yield of each step of the eight-step reaction?

Solve

Percent yield in a linear synthesis like this is simply the product of the yields of each of the steps.

% yield = $(0.75)^8 = 0.100 = 10.0\%$

Problem 13.22

Think

For a linear synthesis, how is the overall product yield related to the yield of each step in the reaction? What is the percent yield of each step of the seven-step reaction?

Solve

Percent yield in a linear synthesis like this is simply the product of the yields of each of the steps.

% yield = $(0.90)(0.81)(0.85)(0.98)(0.82)(0.72)(0.94) = 0.337 = 33.7\%$

$$S \xrightarrow[90\%]{} A \xrightarrow[81\%]{} B \xrightarrow[85\%]{} C \xrightarrow[98\%]{} D \xrightarrow[82\%]{} E \xrightarrow[72\%]{} F \xrightarrow[94\%]{} T$$

Problem 13.23

Think

Refer to Figure 13-1 to match Syntheses I and II to either the linear synthesis or the convergent synthesis. Fill in 80% for each reaction in Schemes I and II given in this problem. How do you calculate the overall percent yield for each? What is the longest branch of the synthesis? In general, which type of synthesis produces better yields?

Solve

(a) Scheme I is a convergent synthesis, and Scheme II is a linear synthesis. Notice in Scheme I that the epoxide is produced by the reactions in the first row below and is then added later in the sequence of reactions in the second row below.

(b) The 80% yield for each synthetic step is filled in below. Scheme I is a convergent synthesis, where a four-carbon molecule is produced in three steps and is connected to another four-carbon molecule in Step 2 of another three-step synthesis. Reaching the final target requires six separate steps, but the longest branch of the synthesis is only five steps, so the overall yield would be $(0.80)^5 = 0.33 = 33\%$. Scheme II is a linear synthesis where two carbons are added at a time. Seven steps are required and, with an 80% yield for each step, the overall yield is $(0.80)^7 = 0.21 = 21\%$. In general, the yield of a multistep synthesis is higher if a target can be produced from a convergent synthesis instead of a linear one.

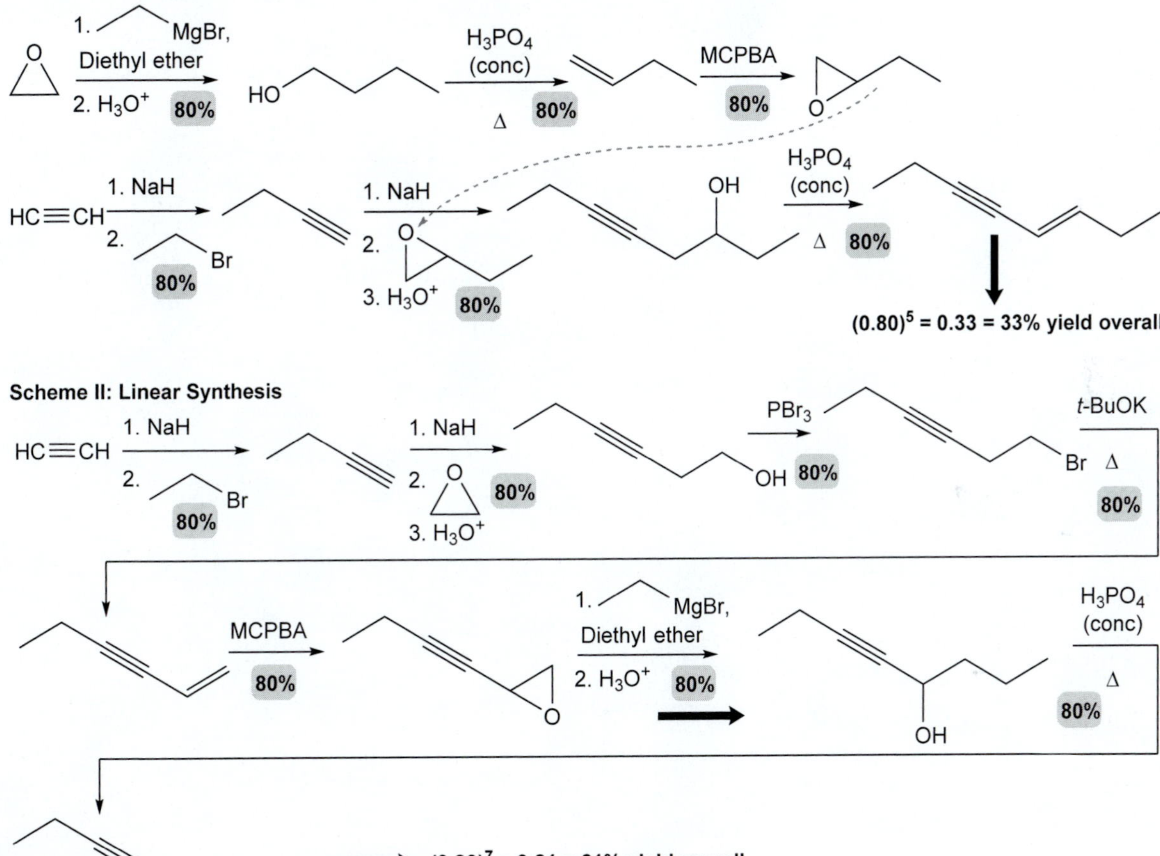

Problem 13.24

Think

Search the Internet to learn about each solvent's toxicity.

Solve

DMF is a highly toxic solvent that is thought to be a carcinogen, linked to liver disease, and is suspected to be a reproductive toxin that can harm a fetus. DMSO, conversely, does not display any of the above risks. Both solvents are polar aprotic, are miscible with water, and have high boiling points. Therefore, DMSO or mixtures of DMSO with other solvents are a suitable replacement for DMF.

Problem 13.26

Think

What is the desired product? What is the mass of its atoms? What is the mass of all the reactant atoms?

Solve

The desired product is 1-bromopentane ($C_5H_{11}Br$), whose molecular weight is 151.1 u. There are three reactant molecules: pentan-1-ol ($C_5H_{12}O$), NaBr, and H_2SO_4. Their molecular weights are 88.2 u, 102.9 u, and 98.1 u, respectively, for a total mass of 289.2 u. The percent atom economy is therefore (151.1 u/289.2 u) × 100% = 52.2%.

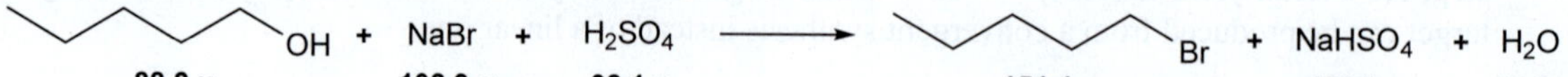

OH + NaBr + H_2SO_4 → Br + $NaHSO_4$ + H_2O

88.2 u 102.9 u 98.1 u 151.1 u 120.1 u 18.0 u

End of Chapter Problems

Sections 13.1–13.3 Writing the Reactions of an Organic Synthesis, Cataloging Reactions, and Retrosynthetic Analysis

Problem 13.27

Think

What is the structure of each organic compound named? What is the structure of each inorganic reagent named? In how many separate steps is each synthesis? How do you denote separate synthetic steps?

Solve

The syntheses are listed below.

(a) 1. LDA; 2. CH_3CH_2Br

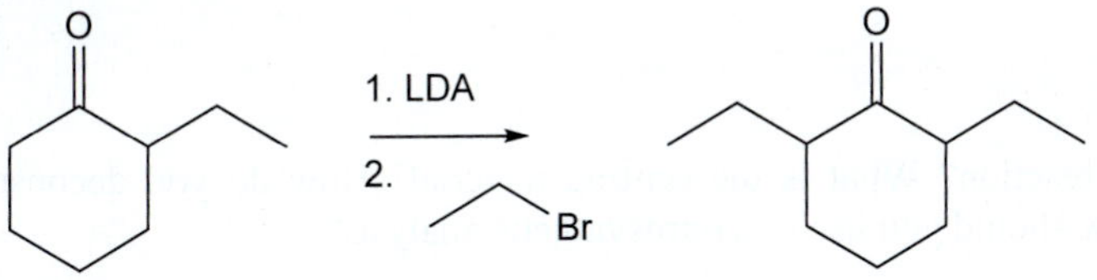

(b) Br_2, CH_3CO_2H; NaCN

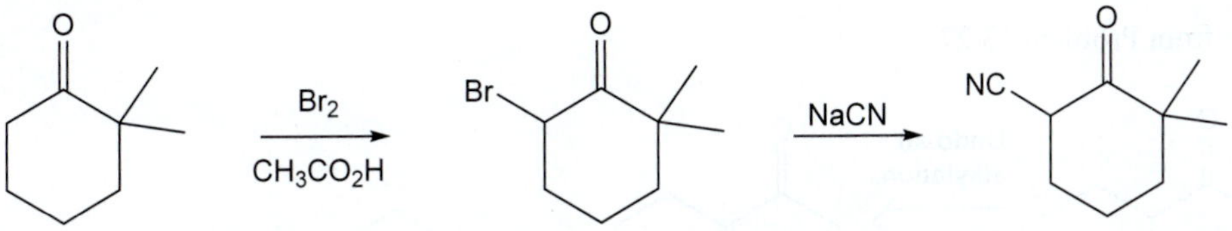

(c) CH_2N_2; 1. NaH; 2. $PhCH_2Br$

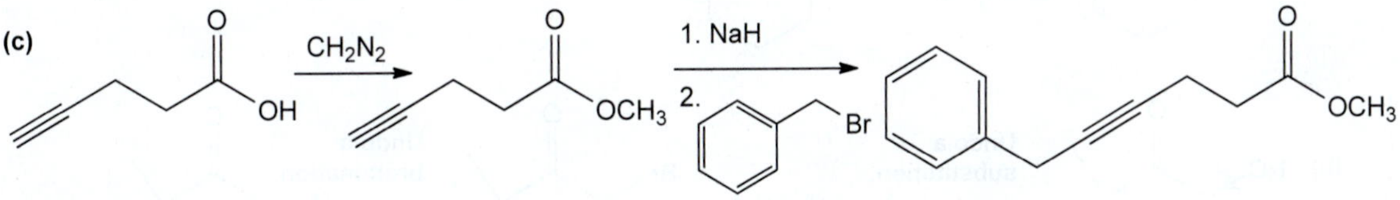

Problem 13.28

Think

What is the name of each organic compound? What is the name of each inorganic reagent? Do any of the reaction conditions need to be specified (e.g., temperature, solvent)? In how many steps does each reaction occur? What products form during intermediate steps?

Solve

(a) To the starting epoxide, add phenylmagnesium bromide in diethyl ether, followed by aqueous acid, to yield 2-phenylcyclopentanol. Treat the 2-phenylcyclopentanol product with concentrated phosphoric acid at 100 °C to yield 1-phenylcyclopentene.

(b) Treat phenylethanone with lithium diisopropylamide, followed by iodomethane, to yield phenylpropanone. Then add molecular bromine in the presence of acetic acid to yield 2-bromo-1-phenylpropanone. Next, add sodium acetate to yield the final product.

(c) Heat 1-(ethoxymethyl)-3-nitrobenzene with aqueous acid. Upon completion of that reaction, add phosphorus tribromide to yield 1-(bromomethyl)-3-nitrobenzene. Finally, add sodium azide to yield the target.

Problem 13.29

Think

How many C atoms are in the product? How many C atoms are in the reactant? Did the number of C atoms change? Are any carbon–carbon σ bonds formed or broken?

Solve

(a) Yes. A carbon–carbon σ bond must break.
(b) No. A carbon–oxygen bond must form, but no carbon–carbon σ bonds are broken or formed.
(c) Yes. A carbon–carbon σ bond must form.
(d) No. A carbon–carbon π bond must break and a carbon–oxygen bond must form, but no carbon–carbon σ bonds must form or break.
(e) Yes. A carbon–carbon σ bond must form.
(f) Yes. Two carbon–carbon σ bonds must form.
(g) No. A carbon–carbon π bond must break and two carbon–oxygen bonds must form, but no carbon–carbon σ bonds must form or break.
(h) Yes. At least two carbon–carbon σ bonds must form.

Problem 13.30

Think

What is the target compound in each reaction? What is the starting material? How do you deconstruct the product in each step? What type of arrow should you use in a retrosynthetic analysis?

Solve

Retrosyntheses from Problem 13.27:

(a) Undo an alkylation. ⟹ + Br

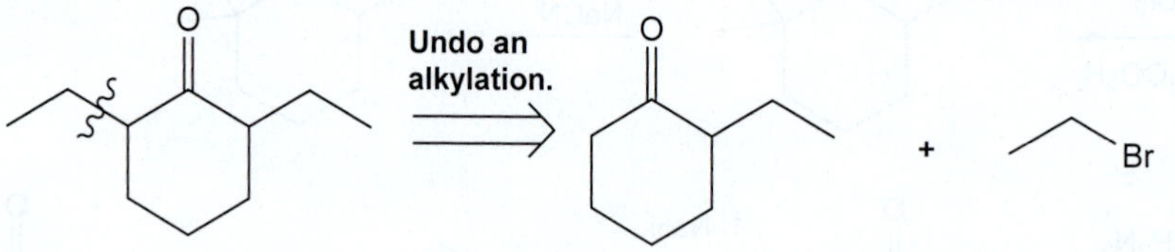

(b) NC — Undo a substitution. ⟹ NaCN + Br — Undo a bromination. ⟹

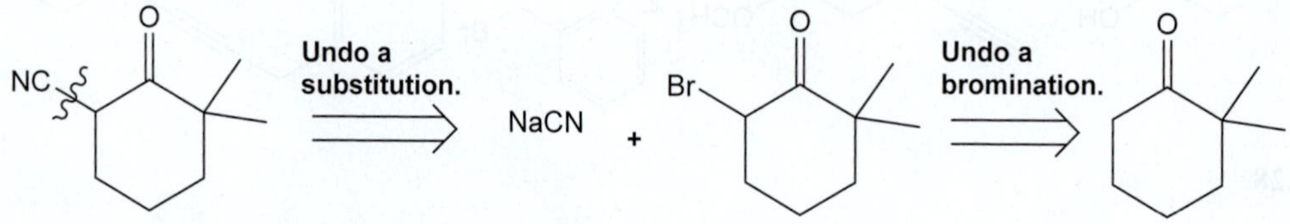

(c) OCH_3 — Undo a substitution. ⟹ Br + OCH_3 — Undo an esterification. ⟹ OH + CH_2N_2

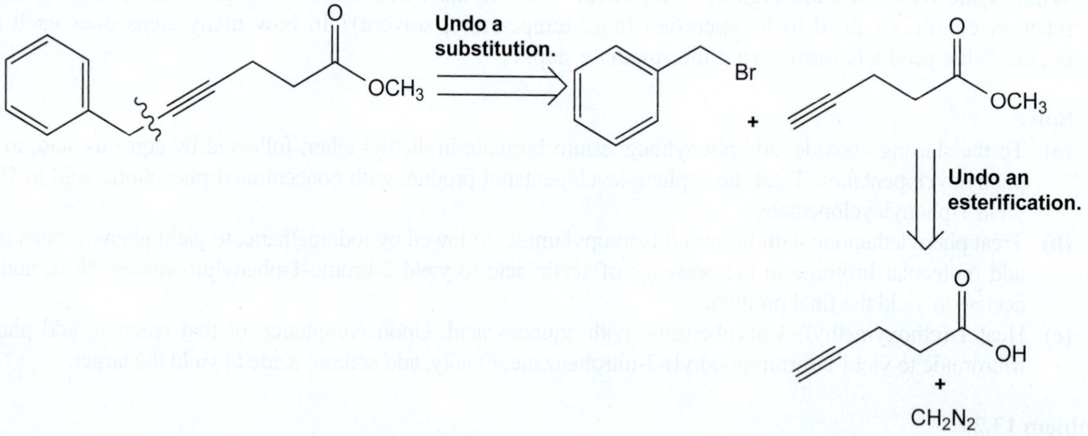

Retrosyntheses from Problem 13.28:

(a) Undo a dehydration. ⟹ OH ⟹ Undo a Grignard reaction. ⟹ MgBr + O

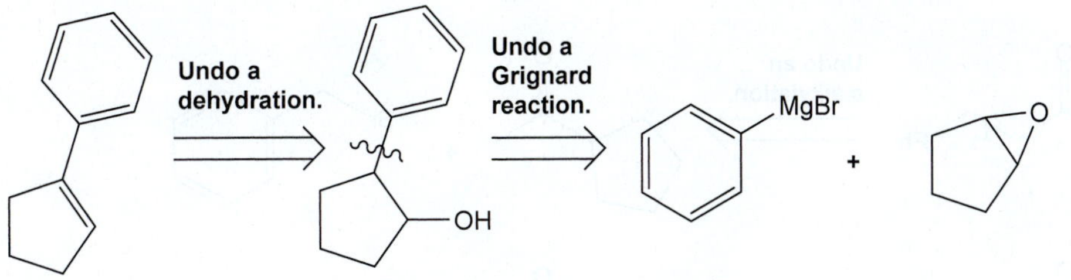

(b) O ⟹ Undo a substitution. ⟹ O, Br ⟹ Undo a bromination. ⟹ O ⟹ Undo an alkylation. ⟹ O

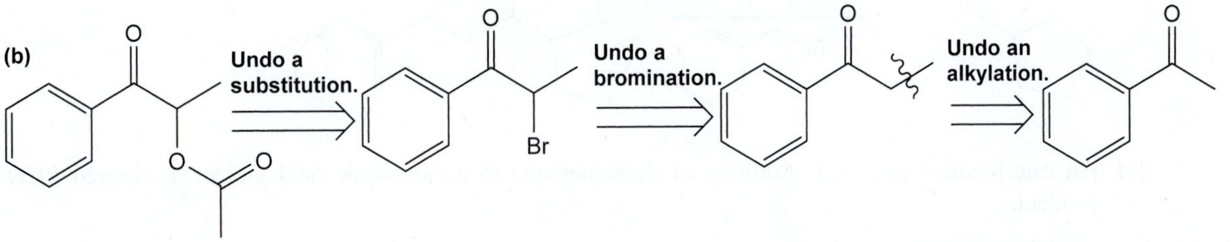

(c) NO_2, N_3 ⟹ Undo a substitution. ⟹ NO_2, Br ⟹ Undo a substitution. ⟹ NO_2, HO ⟹ Undo a hydrolysis. ⟹ NO_2, O, CH_2CH_3

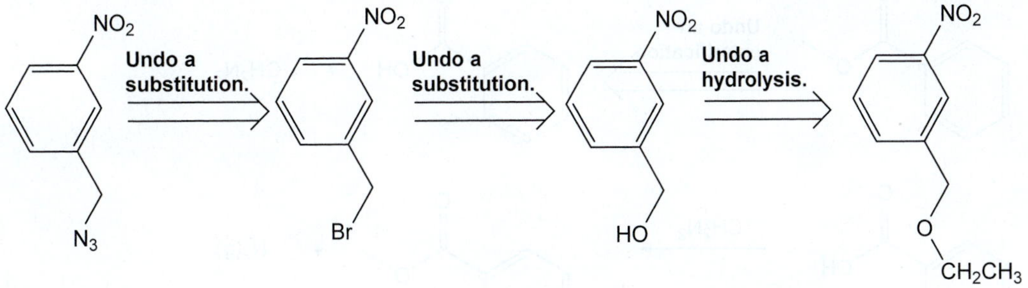

Problem 13.31

Think

What reagent and reaction conditions are necessary to reverse the transform indicated to carry out the reaction in the forward direction? What other functional groups are present?

Solve

The retrosynthesis and forward reaction conditions are given below for **(a)**–**(d)**.

(a) A Hofmann elimination occurred. The leaving group on C2 and the Hofmann product, or anti-Zaitsev (i.e., less highly substituted alkene), is produced. The retrosynthesis and forward reaction conditions are given below.

Undo a Hofmann elimination. ⟹ NH_2

NH_2 → 1. CH_3I (excess) 2. Ag_2O 3. Δ

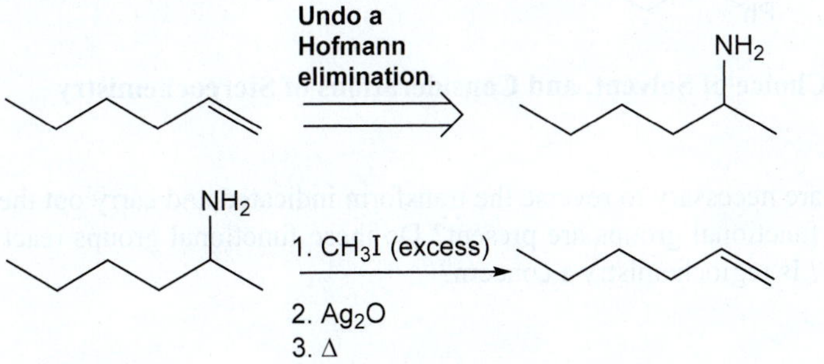

(b) Alkylation occurred at the less-substituted C atom. LDA is used because a very strong bulky base is required to produce the kinetic enolate intermediate.

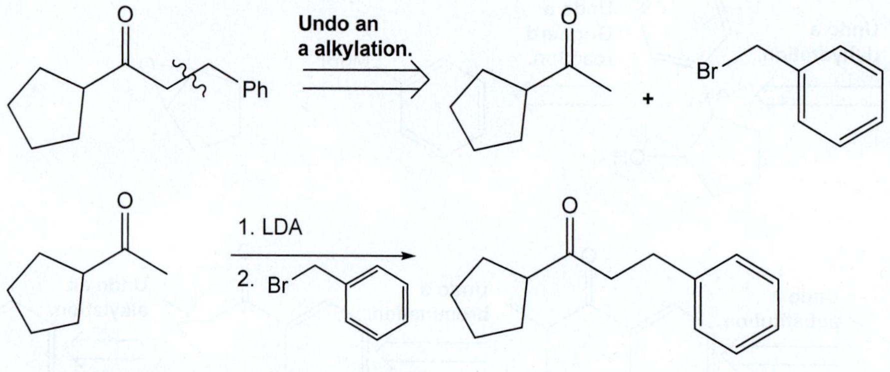

Undo an a alkylation.

1. LDA
2. Br

(c) An esterification occurred. Addition of diazomethane to a carboxylic acid yields the desired RCO_2CH_3 product.

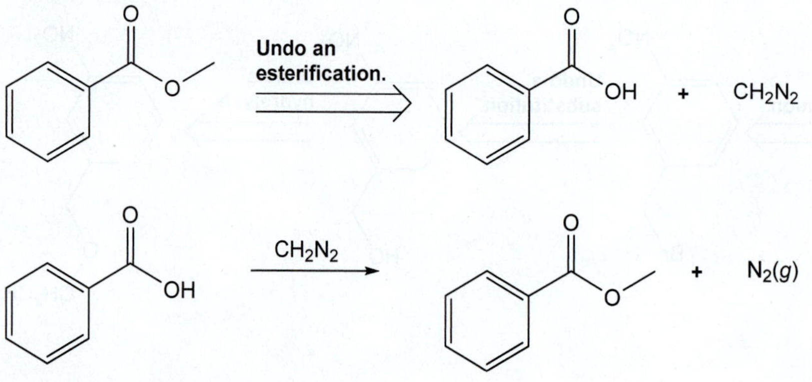

Undo an esterification.

OH + CH_2N_2

CH_2N_2

+ $N_2(g)$

(d) Addition of excess HCl to the C≡C occurred.

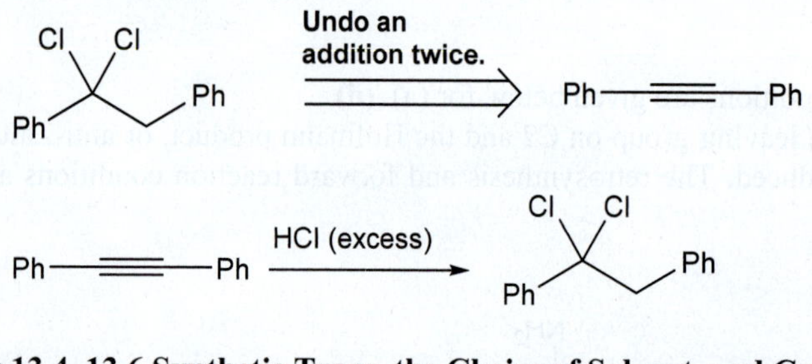

Undo an addition twice.

Ph—≡—Ph

HCl (excess)

Sections 13.4–13.6 Synthetic Traps, the Choice of Solvent, and Considerations of Stereochemistry
Problem 13.32

Think

What reagent and reaction conditions are necessary to reverse the transform indicated and carry out the reaction in the forward direction? What other functional groups are present? Do these functional groups react with the reagent you used? If so, is this desired? Is regiochemistry a concern?

Solve

(a) This is a synthetic trap. Instead of attacking the epoxide, the Grignard reagent could attack the alkyl halide as a base in an E2 reaction or as a nucleophile in an S_N2 reaction.

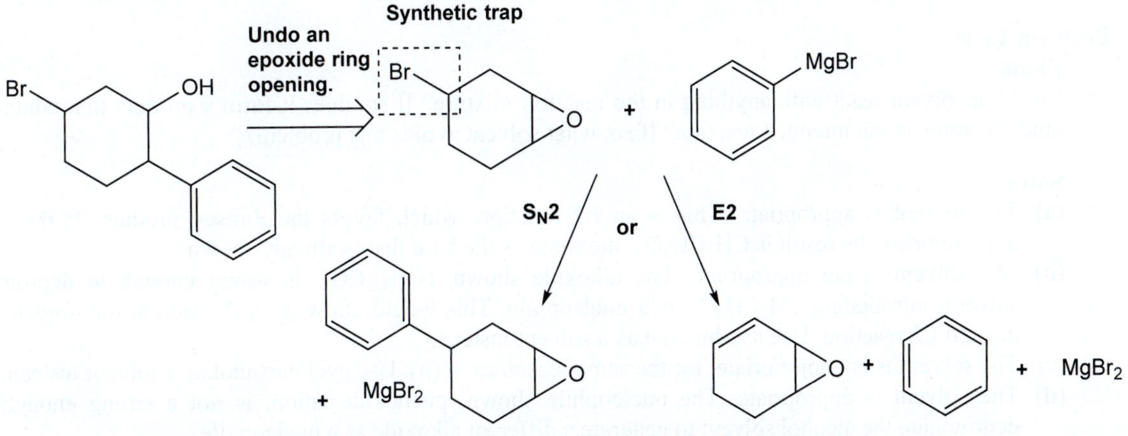

Undo an epoxide ring opening.

Synthetic trap

MgBr

S_N2 or E2

$MgBr_2$

$MgBr_2$

(b) This is a synthetic trap. If the epoxide is treated with NaCN, which would be under neutral or basic conditions, then NC^- would attack the less sterically hindered C atom. To obtain the target, however, the more hindered C atom must be attacked.

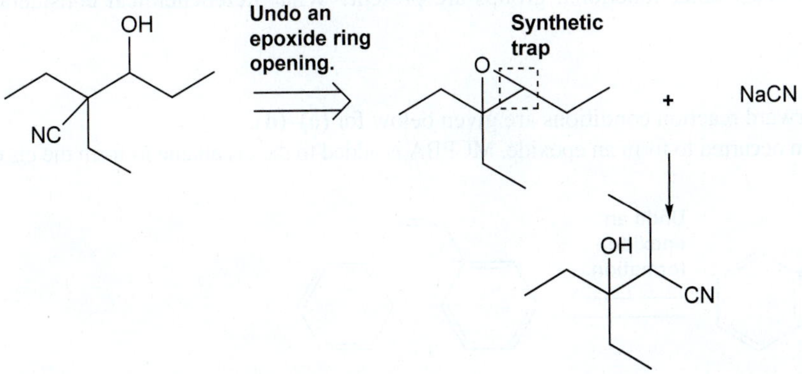

OH

NC

Undo an epoxide ring opening.

Synthetic trap

NaCN

OH

CN

(c) This could proceed as planned. The terminal alkyne would have to be treated with a strong base, such as NaH, to produce the alkynide anion as a nucleophile. When the ethyoxyalkyl halide is added, the alkynide anion will attack only the alkyl halide group. This is because the ether group is resistant to substitution and elimination under basic conditions.

(d) This is a synthetic trap. When the alkoxide anion is added to the bromoalcohol, the alcohol can be deprotonated. This would set up an intramolecular S_N2 reaction, which would produce a six-membered ring, and would thus be more favorable than the desired intermolecular S_N2 reaction.

HO

O

Undo a Williamson synthesis.

HO

Br

ONa

Synthetic trap

O

OH

NaBr

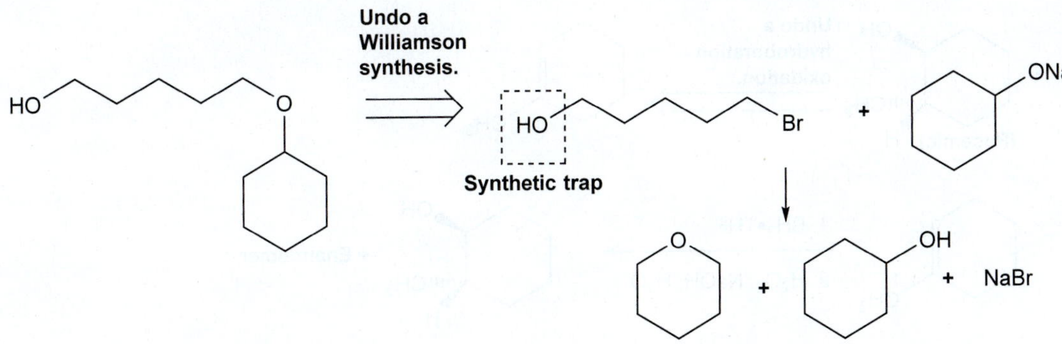

(e) This could proceed as planned by deprotonating the ketone to make an enolate anion nucleophile and then adding the benzyl bromide. The main concern is regiochemistry. The desired alkylation is at the less-substituted carbon, which must proceed through the kinetic enolate. This can be produced by using LDA as the base, not an alkoxide base.

Problem 13.33

Think

Does the solvent react with anything in the reaction mixture? If so, does it form a product that competes with other reagents in the intended reaction? If so, what solvent avoids this problem?

Solve

(a) The solvent is appropriate. This is an E2 reaction, which favors the Zaitsev product. If the solvent is deprotonated, the result is $CH_3CH_2O^-$, the same as the base that is already shown.

(b) The solvent is not appropriate. The alkoxide shown, $(CH_3)_3CO^-$, is strong enough to deprotonate the solvent, introducing $CH_3CH_2O^-$ as a nucleophile. This would allow an S_N2 reaction to compete with the desired E2 reaction. Use *tert*-butanol as a solvent instead.

(c) The solvent is not appropriate, for the same reason as in **(b)**. Use cyclohexanol as a solvent instead.

(d) The solvent is appropriate. The nucleophile shown, phenoxide anion, is not a strong enough base to deprotonate the alcohol solvent to generate a different alkoxide as a nucleophile.

Problem 13.34

Think

What reagent and reaction conditions are necessary to reverse the transform indicated and carry out the reaction in the forward direction? What other functional groups are present? What stereochemical considerations are necessary?

Solve

The retrosynthesis and forward reaction conditions are given below for **(a)**–**(d)**.

(a) A syn addition reaction occurred to form an epoxide. MCPBA is added to the cis alkene to form the cis epoxide.

Undo an epoxide formation.

MCPBA

(b) A hydroboration–oxidation reaction occurred because the HO is on the less-substituted C and the H and the HO are syn.

OH, CH_3, (Racemic) H

Undo a hydroboration–oxidation.

CH_3

1. BH_3•THF

2. H_2O_2, NaOH, H_2O

OH, CH_3, H

+ Enantiomer

(c) A halohydrin reaction occurred because the Br and OH are anti to each other.

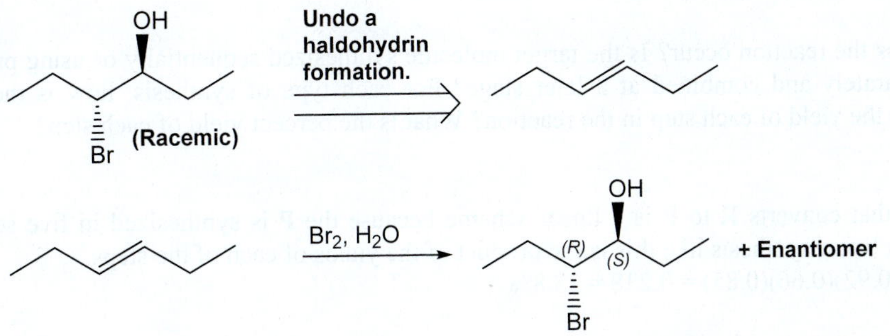
OH
Undo a haldohydrin formation.
Br
(Racemic)
Br2, H2O
OH
(R)
(S)
+ Enantiomer
Br

(d) An S_N2 reaction occurred because the chiral center's configuration is inverted.

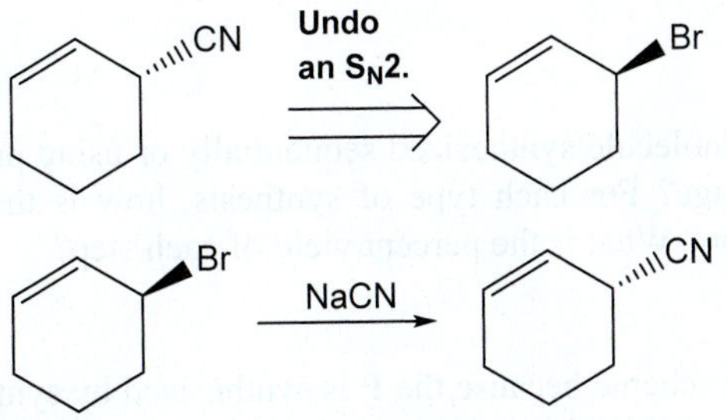
CN
Undo an SN2.
Br
Br
NaCN
CN

Problem 13.35 (SYN)

Think

In how many elementary steps does an E2 reaction occur? Is the Hofmann or Zaitsev product shown? On which C was the leaving group and on which C was the adjacent H? Is a bulky base needed? What is the configuration of the R groups in the alkyl halide starting material?

Solve

E2 occurs in one elementary step. The leaving group and the adjacent H must be anticoplanar in the precursor for the E2 step to be favored. To arrive back at the alkyl halide, we add H and a halogen on the alkene carbons in an anti fashion.

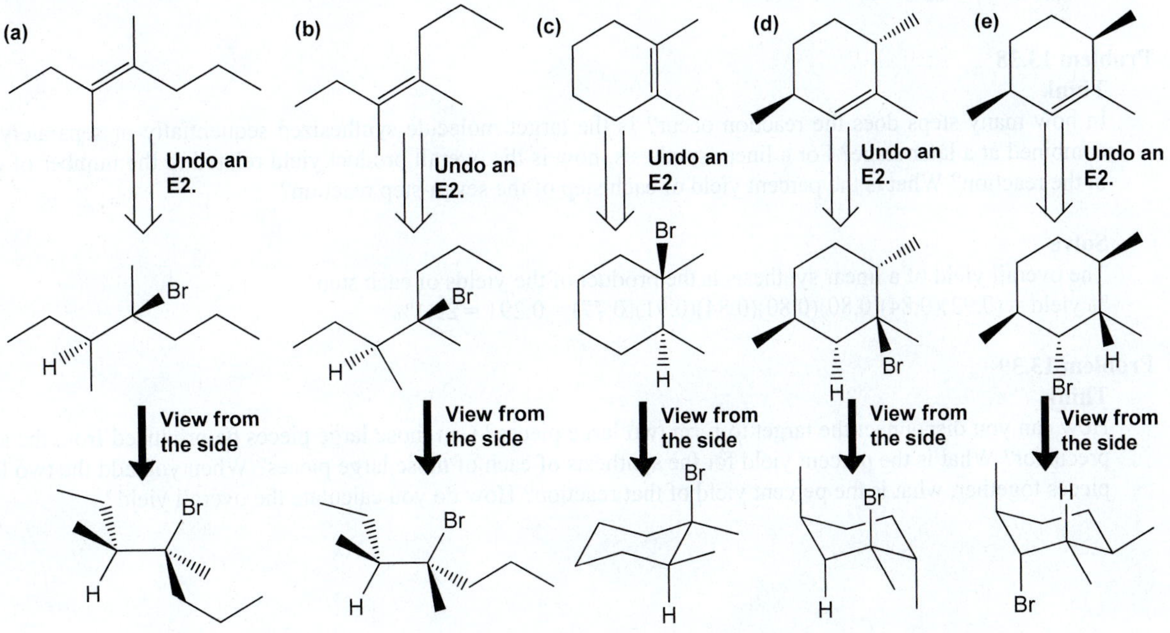
(a)
(b)
(c)
(d)
(e)
Undo an E2.
Undo an E2.
Undo an E2.
Undo an E2.
Undo an E2.
Br
H
Br
H
Br
H
Br
H
H
Br
View from the side
View from the side
View from the side
View from the side
View from the side
Br
H
Br
H
Br
H
Br
H
H
Br

Section 13.8 Percent Yield and Green Chemistry

Problem 13.36

Think

In how many steps does the reaction occur? Is the target molecule synthesized sequentially or using precursors that are produced separately and combined at a later stage? For each type of synthesis, how is the overall product yield related to the yield of each step in the reaction? What is the percent yield of each step?

Solve

The synthesis scheme that converts **R** to **P** is a linear scheme because the **P** is synthesized in five sequential steps. Percent yield in a linear synthesis like this is the product of the yields of each of the steps.

% yield = (0.59)(0.78)(0.92)(0.66)(0.85) = 0.238 = 23.8%

$$\mathrm{R} \xrightarrow[59\%]{\mathrm{V}} \mathrm{A} \xrightarrow[78\%]{\mathrm{W}} \mathrm{B} \xrightarrow[92\%]{\mathrm{X}} \mathrm{C} \xrightarrow[66\%]{\mathrm{Y}} \mathrm{D} \xrightarrow[85\%]{\mathrm{Z}} \mathrm{P}$$

Problem 13.37

Think

In how many steps does the reaction occur? Is the target molecule synthesized sequentially or using precursors that are produced separately and combined at a later stage? For each type of synthesis, how is the overall product yield related to the yield of each step in the reaction? What is the percent yield of each step?

Solve

The synthesis scheme that converts **R** to **P** is a convergent scheme because the **P** is synthesized by synthesizing **B** and **G** separately, each in two steps, and then they are combined in a later synthetic step. Percent yield in a convergent synthesis like this is the product of the yields of the longest branch of the synthesis with the lowest yields. The first precursor, **B**, is produced in a 46.0% yield, and the second precursor, **G**, is produced in a 56.1% yield. Taking **B** to be the limiting reagent, an 81% yield in the final step would give an overall yield of (0.460)(0.81) = 0.373 = 37.3%.

$$\mathrm{R} \xrightarrow[59\%]{\mathrm{V}} \mathrm{A} \xrightarrow[78\%]{\mathrm{W}} \mathrm{B}$$

$$\mathrm{B} + \mathrm{G} \xrightarrow[81\%]{\mathrm{Z}} \mathrm{P}$$

$$\mathrm{E} \xrightarrow[66\%]{\mathrm{X}} \mathrm{F} \xrightarrow[85\%]{\mathrm{Y}} \mathrm{G}$$

Problem 13.38

Think

In how many steps does the reaction occur? Is the target molecule synthesized sequentially or separately and combined at a later stage? For a linear synthesis, how is the overall product yield related to the number of steps in the reaction? What is the percent yield of each step of the seven-step reaction?

Solve

The overall yield of a linear synthesis is the product of the yields of each step.

% yield = (0.92)(0.84)(0.80)(0.80)(0.84)(0.91)(0.77) = 0.291 = 29.1%

Problem 13.39

Think

How can you disconnect the target to form two large pieces? Can those large pieces be produced from the same precursor? What is the percent yield for the synthesis of each of those large pieces? When you add the two large pieces together, what is the percent yield of that reaction? How do you calculate the overall yield?

Solve

To design a convergent synthesis, we can begin by disconnecting the target into two relatively large pieces.

Undo a C–C bond formation.

SCH_3 ⟹ Br + SCH_3

The first large piece is already shown in Problem 13.38 to be produced in three synthetic steps in a yield of (0.92)(0.84)(0.80) = 0.618 = 61.8%. We can perform a retrosynthetic analysis of the second precursor as follows.

SCH_3 ⟹ (**Undo a substitution.**) Br + $NaSCH_3$ ⟹ (**Undo a C–C bond formation.**) OH ⟹ (**Undo an epoxide ring opening.**) + O

In the forward direction, the total synthesis would appear as follows. If we assume the reactions that are similar to the ones in Problem 13.38 also have the same yields, the second precursor could be produced in a 51.7% yield. Taking this to be the limiting reagent, an 80% yield in the final step would give an overall yield of (0.517)(0.80) = 0.4136 = 41.4%. This is much higher than the 29.1% yield calculated in the previous problem.

84%: 1. $NaNH_2$ 2. O 3. HCl → OH; 80%: PBr_3 → Br; 77%: CH_3SH → SCH_3 (51.7%)

92%: 1. $NaNH_2$ 2. CH_3Br →; 84%: 1. $NaNH_2$ 2. O 3. HCl → OH; 80%: PBr_3 → Br (61.8%)

80%: SCH_3 1. $NaNH_2$ 2. Br → SCH_3

Problem 13.40

Think

What is the difference between the two reaction schemes? Search the Internet to learn about each solvent's toxicity.

Solve

The only difference between the reaction schemes is the solvent choice—ethanol versus DMF. Ethanol can have a few potential acute and chronic hazards. However, ethanol is also used in hand sanitizer to help cleanse hands. Ethanol can cause minor irritation and permeation to the skin. However, with proper handling, ethanol is harmless. In general, ethanol is a good solvent, because it is noncorrosive for skin, eyes, and lungs. Recall from Problem 13.24 that DMF is a highly toxic solvent. Ethanol is a greener solvent.

Problem 13.41

Think

What is the difference between the two reaction schemes? Search the Internet to learn about each reagent's toxicity.

Solve

The difference between the reaction schemes is the reagent choice—$Hg(OAc)_2/H_2O$ versus disiamylborane and H_2O_2. Mercury catalysts are more toxic than disiamylborane. Overall, mercury acetate has many hazards: It is fatal if swallowed; if it is inhaled, it may cause drowsiness or dizziness; and it may cause damage to organs through prolonged or repeated exposure. By contrast, disiamylborane has comparatively low toxicity and does not present storage hazards, especially when stored at 0 oC. Hydrogen peroxide in solution is corrosive to the eyes and is a mild skin irritant, and is slightly hazardous in the case of inhalation. Ultimately, synthetic step IV is greener than III.

Problem 13.42

Think

What is the desired product? What is the mass of its atoms? What is the mass of all the reactant atoms?

Solve

The desired C=C product ($C_6H_{10}CH_2$) has a molecular weight of 96.17 u. There is one reactant molecule, $C_6H_{10}(CH_3)N(CH_3)_3OH$, which has a molecular weight of 173.3 u. The percent atom economy is therefore (96.17 u/173.3 u) × 100% = 55.5%.

$\overset{\oplus}{N}(CH_3)_3$ $\overset{\ominus}{O}H$ → + $(CH_3)_3N$ + H_2O

173.3 u **96.17 u** **59.11 u** **18.0 u**

Integrated Problems

Problem 13.43 (SYN)

Think

What is the structure of pent-2-yne? Identify the C–C≡C bonds. Is the alkyne symmetric? If the C–C≡C is formed, how many C atoms should be on the alkyl bromide and alkynide anion precursors? Is there more than one option?

Solve

In a retrosynthetic analysis, the carbon–carbon bond on either side of the triple bond can be disconnected, giving us a terminal alkyne and an alkyl halide as precursors.

Route A

Undo a C–C bond formation. ⟹ + CH_3Br

Route B

Undo a C–C bond formation. ⟹ Br +

The syntheses in the forward direction would appear as follows:

Route A

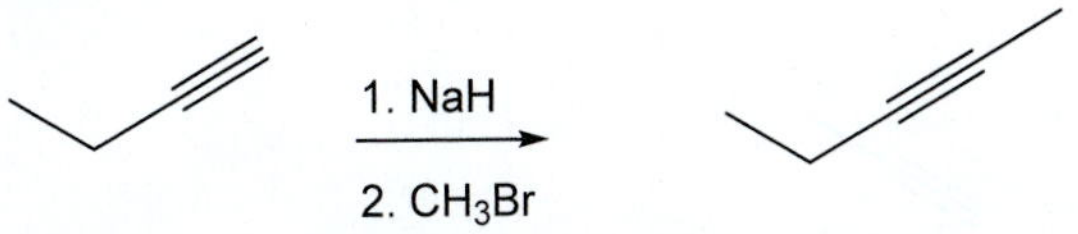

1. NaH
2. CH_3Br

Route B

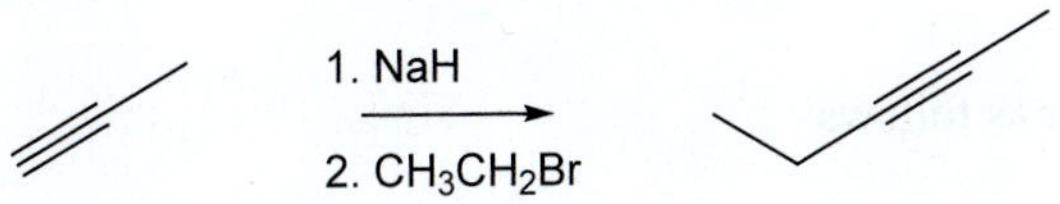

1. NaH
2. CH_3CH_2Br

Problem 13.44 (SYN)

Think

What reaction can be used to form an ether? What reagents would be required? How can an alcohol be converted into an alkylbromide?

Solve

As shown in the retrosynthetic analysis below, the final ether can be produced using a Williamson synthesis, which requires an alkoxide anion and an alkyl halide. The alkoxide would be the phenoxide anion, which can be produced by deprotonating phenol. The alkyl halide can be produced from an alcohol via substitution with PBr_3.

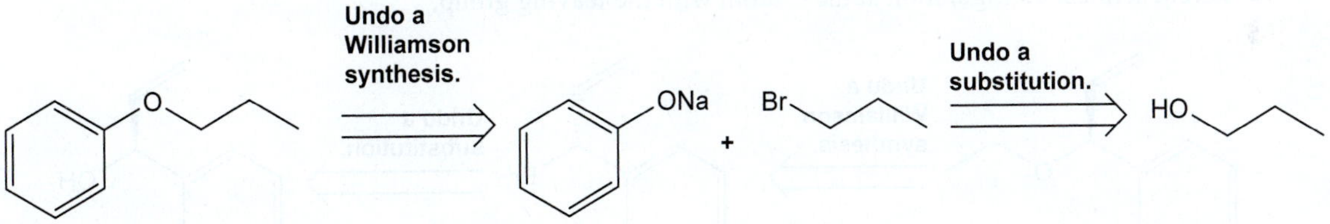

Undo a Williamson synthesis. ONa + Br **Undo a substitution.** HO

In the forward direction, the synthesis would appear as follows.

OH NaH

HO PBr_3 Br

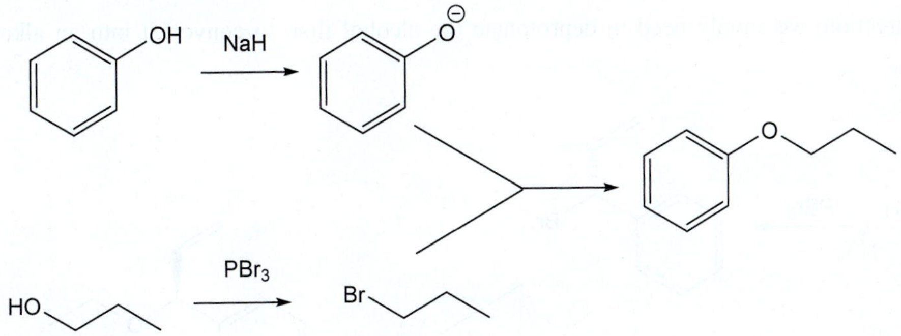

Problem 13.45 (SYN)

Think

How many C atoms appear in your target compound? Did you have to alter the carbon skeleton? What functional group is in your target compound? What functional groups are in your starting compound? Did you have to do an elimination, addition, or substitution? How many steps are needed to accomplish your transformation? Is stereochemistry a concern? Is regiochemistry?

Solve

(a) We can disconnect the carbon–carbon bond as shown.

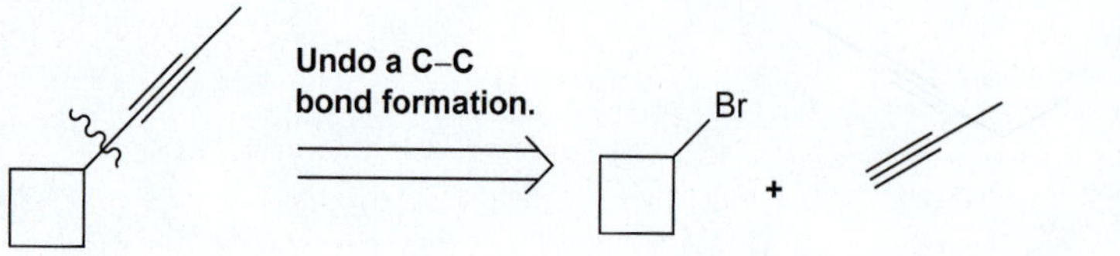

Undo a C–C bond formation.

Br

+

In the forward direction, the synthesis would appear as follows:

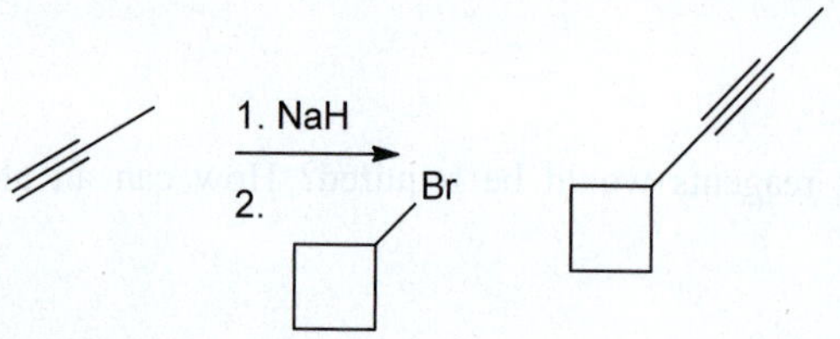

1. NaH

2. Br

(b) An ether can be synthesized from an alcohol and an alkyl halide via a Williamson synthesis. Therefore, we need to undo a Williamson synthesis, as shown. The alkyl halide precursor can be obtained by brominating an alcohol. In each of these reactions, we must take into account the fact that an S_N2 reaction inverts the stereochemical configuration at the C atom with the leaving group.

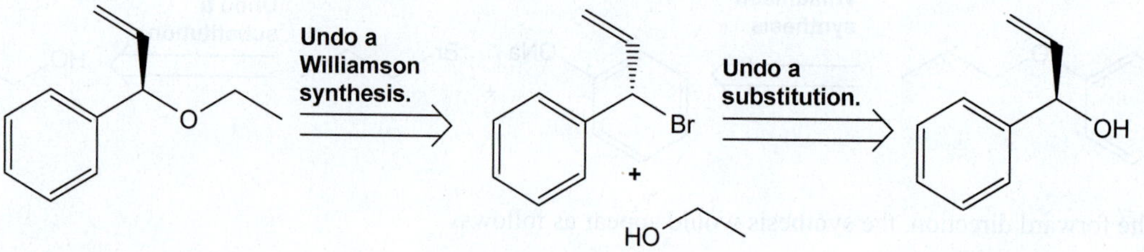

O

Undo a Williamson synthesis.

Br

+

HO

Undo a substitution.

OH

In the forward direction, we simply need to deprotonate the alcohol first, to convert it into an alkoxide nucleophile.

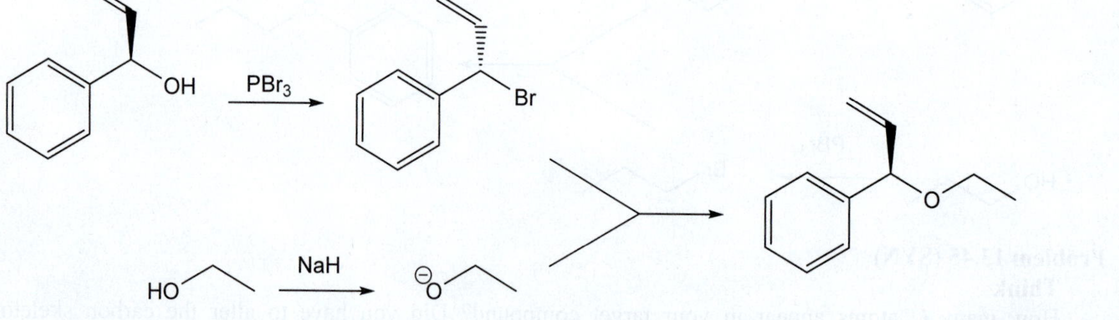

OH

PBr_3

Br

HO

NaH

O

(c) We can disconnect two carbon–carbon single bonds to alkyne carbons to arrive at precursors with five or fewer carbons.

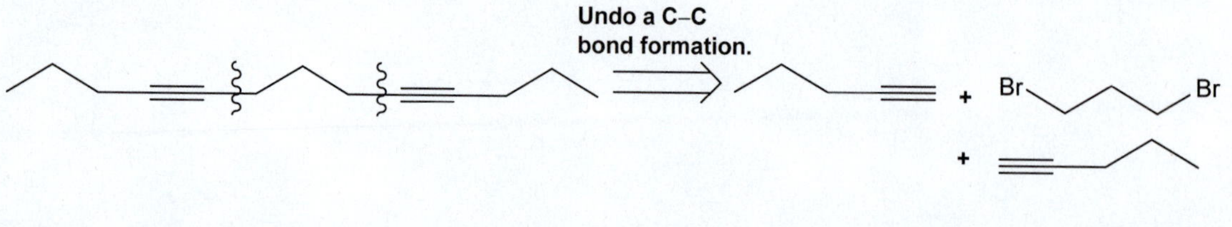

Undo a C–C bond formation.

+ Br Br

+

The synthesis in the forward direction would appear as follows:

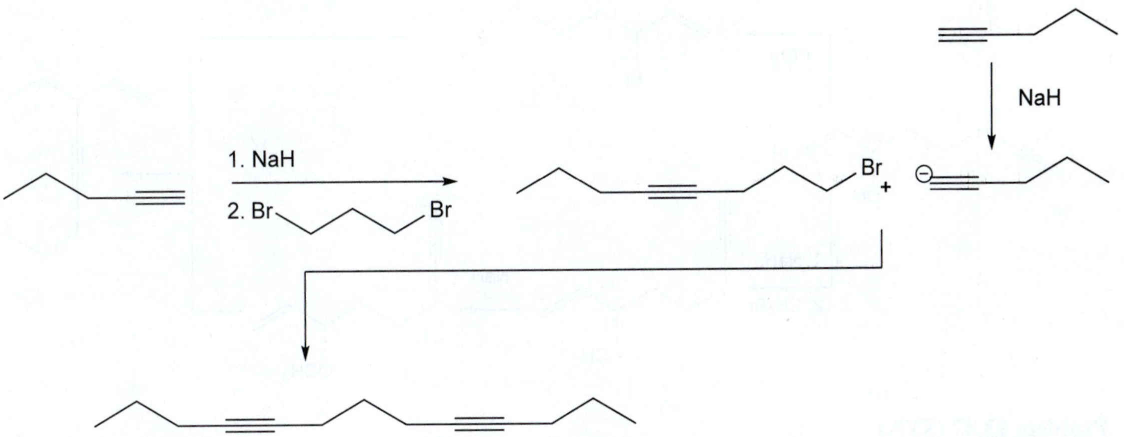

1. NaH
2. Br ⁀⁀ Br

NaH

Br +

Problem 13.46 (SYN)

Think

How many C atoms appear in your target compound? Did you have to alter the carbon skeleton? What functional group is in your target compound? What functional groups are in your starting compound? Did you have to do an elimination, addition, or substitution? Will multiple steps be necessary to accomplish your transformation?

Solve

In performing a retrosynthetic analysis, we could get started by disconnecting the carbon–carbon bond shown below.

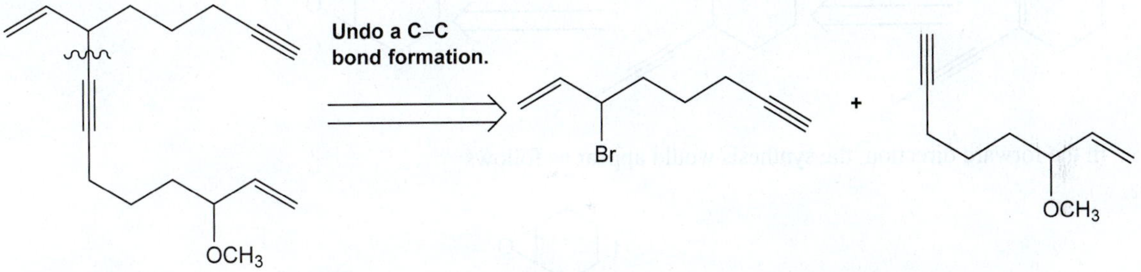

Undo a C–C bond formation.

OCH_3 Br + OCH_3

The precursor on the left could be produced from the starting material using PBr_3 to convert the OH to Br. The precursor on the right could be produced from the starting material using a Williamson ether synthesis.

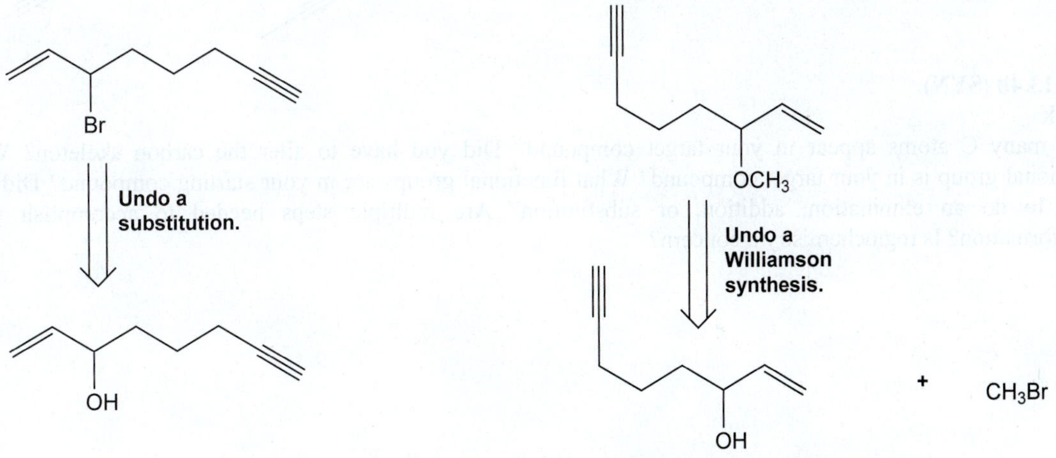

Br

Undo a substitution.

OCH_3

Undo a Williamson synthesis.

OH OH + CH_3Br

In the forward direction, the synthesis would appear as follows:

PBr$_3$

Br

OH

1. NaH
2. CH$_3$Br

OCH$_3$

NaH

OCH$_3$

OCH$_3$

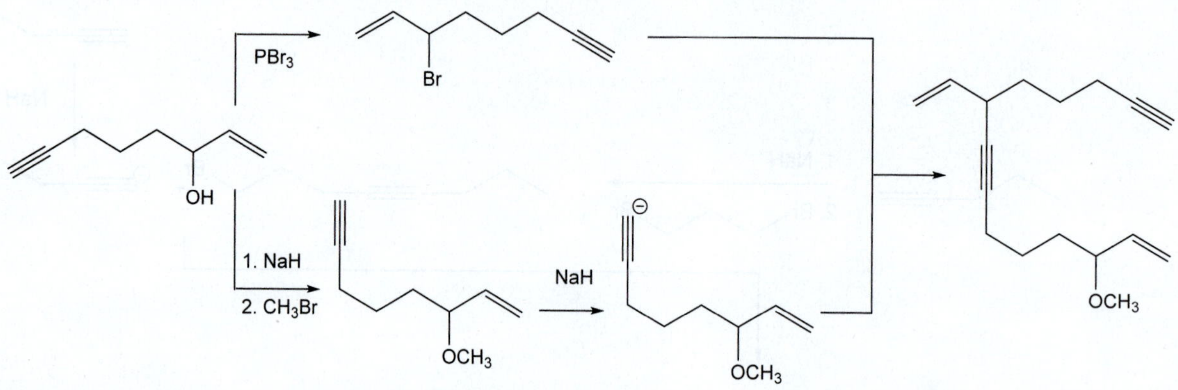

Problem 13.47 (SYN)

Think

How many C atoms appear in your target compound? Did you have to alter the carbon skeleton? What functional group is in your target compound? What functional groups are in your starting compound? Did you have to do an elimination, addition, or substitution? Will multiple steps be needed to accomplish your transformation?

Solve

The retrosynthesis is shown below.

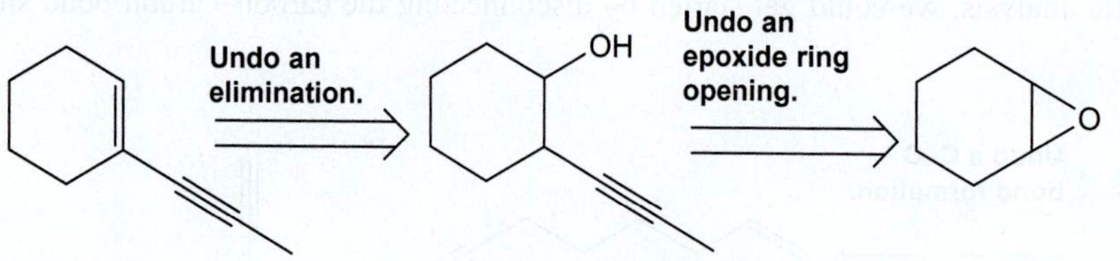

Undo an elimination.

OH

Undo an epoxide ring opening.

O

In the forward direction, the synthesis would appear as follows:

H—C≡ —NaH→ NaC≡ + H$_2$ —1. (epoxide) 2. H_3O^+→ OH —H_2SO_4, Δ→

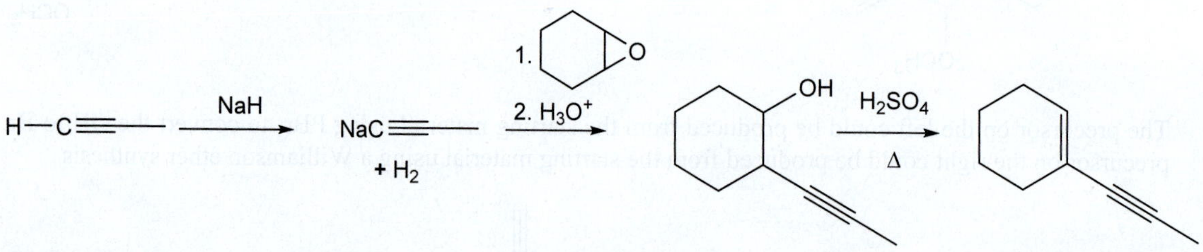

Problem 13.48 (SYN)

Think

How many C atoms appear in your target compound? Did you have to alter the carbon skeleton? What functional group is in your target compound? What functional groups are in your starting compound? Did you have to do an elimination, addition, or substitution? Are multiple steps needed to accomplish your transformation? Is regiochemistry a concern?

Solve

(a) This is the product of elimination. The immediate precursor can then be obtained from bromination of the corresponding methylcyclohexanone.

Undo an elimination. Br Undo a halogenation. Undo an alkylation.

In the forward direction, we must be concerned with regiochemistry in the halogenation step to ensure that bromination takes place at the more highly substituted α C. To prevent polyhalogenation, we must carry out the halogenation under acidic conditions.

CH_3Br / NaOH; Br_2 / Acetic acid; Br; CH_3ONa / Δ

(b) The only difference from part **(a)** is in the regiochemistry of the halogenation step. Halogenation must instead take place at the less-substituted α C, so we must choose a base to produce the kinetic enolate anion. LDA would work.

CH_3Br / NaOH; 1.LDA / 2. Br_2; Br; $(CH_3)_3CONa$ / Δ

Problem 13.49 (SYN)

Think

How many C atoms appear in your target compound? Did you have to alter the carbon skeleton? What functional group is in your target compound? What functional groups are in your starting compound? Did you have to do an elimination, addition, or substitution? Are multiple steps needed to accomplish your transformation? Is stereochemistry a concern?

Solve

In performing a retrosynthetic analysis, we start by disconnecting one of the CN groups. The fact that the CN groups are trans suggests that the reaction at the end must be stereospecific and, thus, an S_N2 occurred. Furthermore, the functional groups on adjacent C atoms suggest that an epoxide ring opening should be considered. The full retrosynthesis is given below.

CN, CN; Undo an S_N2.; Br, CN; Undo a substitution.; OH, CN; Undo an epoxide ring opening.; O; Undo an addition.

The synthesis in the forward direction is given below.

MCPBA; O; 1. NaCN / 2. H_3O^+; OH, CN; PBr_3; Br, CN; NaCN; CN, CN

Problem 13.50 (SYN)

Think

How many C atoms appear in your target compound? Did you have to alter the carbon skeleton? What functional group is in your target compound? What functional groups are in your starting compound? Did you have to do an elimination, addition, or substitution? Are multiple steps needed to accomplish your transformation? Is stereochemistry a concern?

Solve

In performing a retrosynthetic analysis, we start by disconnecting one of the CN groups. The fact that the CN groups are syn suggests that the reaction at the end must be stereospecific and, thus, an S_N2 occurred. Furthermore, the functional groups on adjacent C atoms suggest that an epoxide ring opening should be considered. The full retrosynthesis is given below.

Undo an S_N2 twice. Undo a substitution. Undo an epoxide ring opening. Undo an addition.

CN, CN, Br, Br, OH, Br, O

The forward reaction is given below.

MCPBA, O, HBr, OH, Br, PBr_3, Br, Br, 2 eqiv.NaCN, CN, CN

Problem 13.51 (SYN)

Think

How many C atoms appear in your target compound? Did you have to alter the carbon skeleton? What functional group is in your target compound? What functional groups are in your starting compound? Did you have to do an elimination, addition, or substitution? Are multiple steps needed to accomplish your transformation? What reactions can you use to obtain the desired stereochemistry?

Solve

A new cyclopropane ring appears in the target, so we should consider undoing a carbene addition. In performing the retrosynthetic analysis where stereochemistry is considered, notice that the target compound has the methyl groups trans to each other. This indicates that, when we carried out a transform that undoes a carbene addition, the methyl groups also were trans to each other in the alkene precursor. E2 is a stereospecific reaction; the H and Br in the alkyl halide have to be anticoplanar, and the methyl groups also still need to be anti. To go from the original alcohol to the alkyl halide, PBr_3 is necessary; this occurs via an S_N2 mechanism, so the methyl groups need to be anti in the starting material.

Retrosynthesis

(S), (R), Undo a carbene addition. (Z), Undo an E2. H, (S), H, (R), Br, Undo a substitution. H, (S), (S), OH, H

Synthesis in forward direction

H, (S), (S), OH, H, PBr_3, H, (S), H, (R), Br, NaOEt, Δ, (Z), N_2CH_2, Δ, (R), (S)

CHAPTER 14 | Orbital Interactions 2: Extended π Systems, Conjugation, and Aromaticity

Your Turn Exercises

Your Turn 14.1

Think

From the valence bond (VB) picture shown below on the right, does the π bond occur between C1 and C2 or between C2 and C3? How do you represent a π bond in a Lewis structure? How is the π bond orbital different from an empty *p* orbital in the picture?

Solve

The resonance structure is shown below. The π bond orbital is shown by the overlap between two *p* orbitals and is spread out over C2 and C3. The π bond on the right suggests a double bond between C2 and C3. The empty *p* orbital belonging to C1 suggests that that is where the positive charge is located.

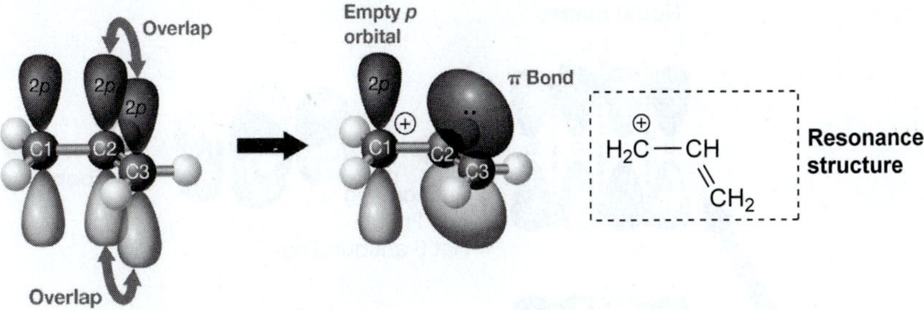

Your Turn 14.2

Think

From the phases of the *p* orbitals, how do you identify a nodal plane? Does constructive or destructive interference result from that type of interaction between those *p* orbitals? In the resulting molecular orbital (MO), what characterizes a nodal plane?

Solve

The nodes are indicated by a vertical dashed line in each MO below. There is one nodal plane in π_2, and there are two nodal planes in π_3. Notice that on either side of each nodal plane, the contributing orbitals have opposite phase. This results in destructive interference. In the resulting MOs, there is complete cancellation at each nodal plane, indicating that an electron in the orbital has zero probability of being found there.

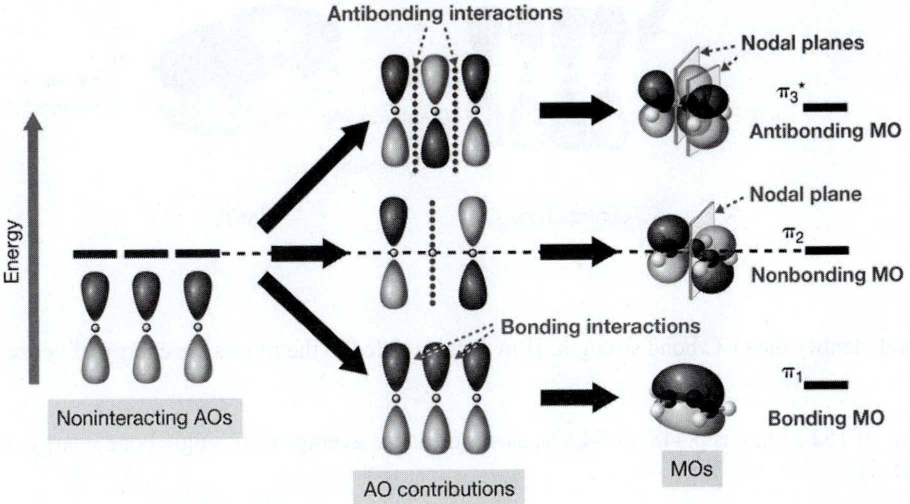

Your Turn 14.3

Think

Do adjacent orbitals with the same phase give rise to a bonding interaction or an antibonding one? What about adjacent orbitals with opposite phases? Once you identify these interactions, consider what is the net interaction relative to the *p* AOs. For example, one antibonding and one bonding interaction have a net interaction that essentially leaves the energy unchanged from the *p* AO (nonbonding), and two antibonding and one bonding interaction have a net of one antibonding (higher in energy relative to the *p* AO).

Solve

Adjacent orbitals with the same phase give rise to a bonding interaction, and those with opposite phase give rise to an antibonding interaction. In this case, all the antibonding interactions among AOs are indicated by the presence of a nodal plane (dashed line). The bonding interactions are indicated below by dashed ovals. Notice that π_1 has three bonding interactions and no antibonding interactions (net: three bonding); π_2 has two bonding interactions and one antibonding interaction (net: one bonding); π_3 has one bonding interaction and two antibonding interactions (net: one antibonding); and π_4 has no bonding interactions and three antibonding interactions (net: three antibonding). Each increase in antibonding interactions leads to a higher-energy MO.

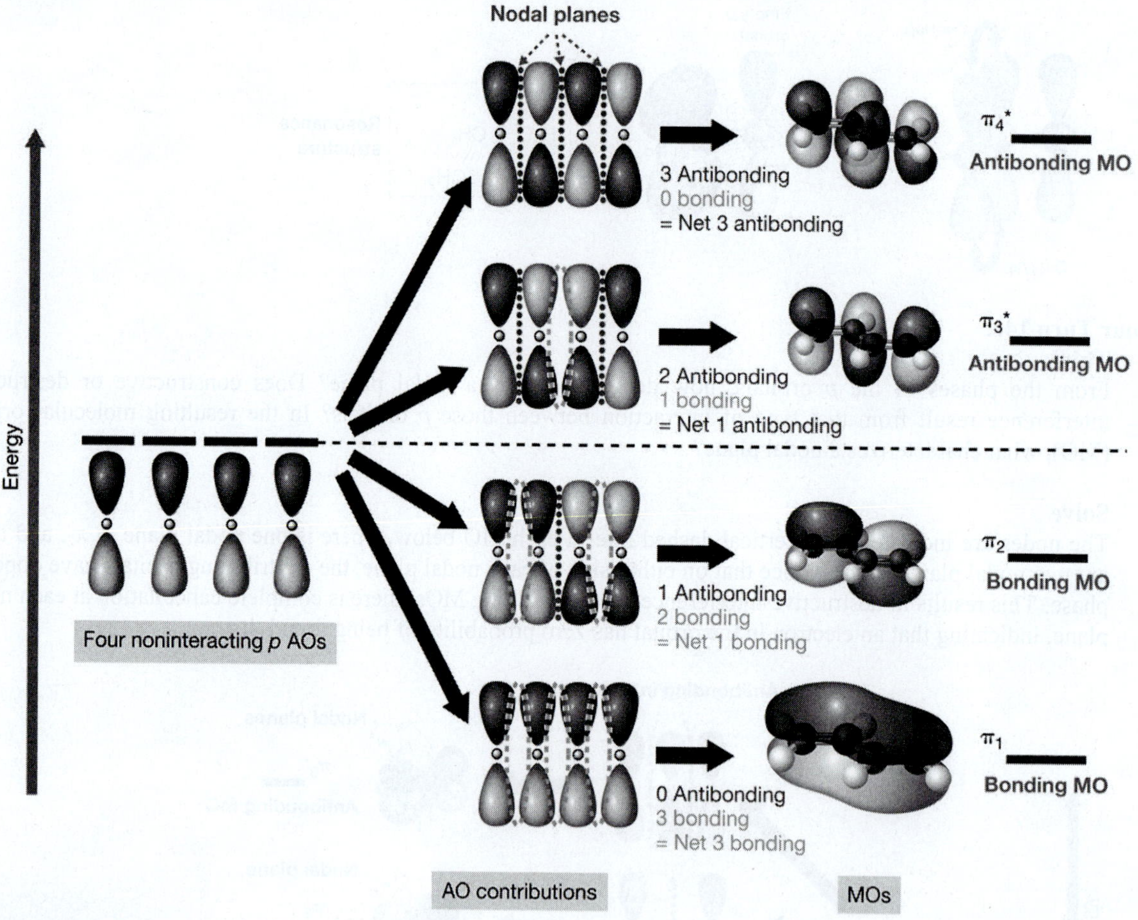

Your Turn 14.4

Think

Consult Table 1-2 and identify the C–C bond strength. How is this related to the resonance energy of benzene?

Solve

The resonance energy of 152 kJ/mol is 0.448, or 44.8%, as strong as the average C–C single bond energy of 339 kJ/mol (from Chapter 1).

Your Turn 14.5

Think

Consult Figure 14-2 and Figure 14-17 for bond lengths. If cyclobutadiene were significantly resonance stabilized, would you expect the single bond lengths in this molecule to be significantly longer or shorter than an isolated C–C?

Solve

If cyclobutadiene were significantly resonance stabilized, we would expect the C–C single bond to have significant double-bond character and be substantially shorter than a typical single bond. The bond length values are 132 pm for a normal C=C bond and 154 pm for a normal C–C bond. The single bonds in cyclobutadiene are 4% longer than average, and the double bonds in cyclobutadiene are the same length as in ethylene. Overall, the numbers are very similar, suggesting that the electrons are *not* significantly resonance delocalized.

Your Turn 14.6

Think

Relative to the *p* AO energies, are bonding MOs lower energy, higher energy, or the same energy? Antibonding MOs? Nonbonding MOs? For orbitals to be degenerate, what is required of their energy levels?

Solve

Bonding orbitals are lower in energy than the *p* AOs (dotted line), whereas antibonding orbitals are higher in energy. There are two pairs of degenerate orbitals, which have the same energy (circled).

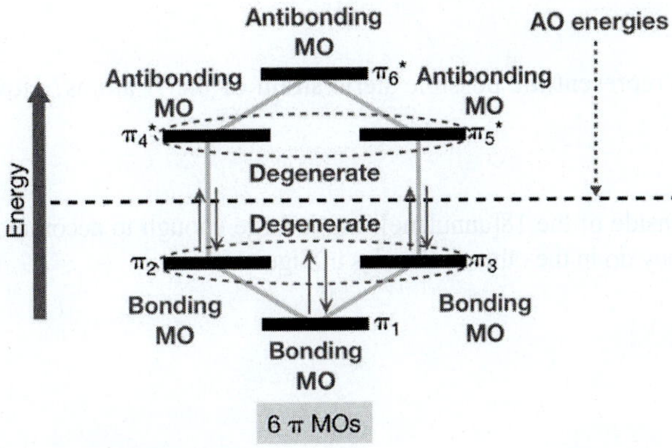

Your Turn 14.7

Think

How many nodal planes are there in each of the first five π MO energy levels? What happens to the number of nodal planes with each increase in MO energy? Is π_6 the highest- or lowest-energy orbital?

Solve

Number of nodes: $\pi_1 = 0$; $\pi_2 = 1$; $\pi_3 = 1$; $\pi_4 = 2$; $\pi_5 = 2$; $\pi_6 = 3$ (highest energy, greatest number of nodal planes). This is because π_6 is higher in energy than π_5 and should have one additional node. Two of the nodal planes are in the same locations as the ones in Figure 14-20d. The third is the one shown in Figure 14-20e that is parallel to the plane of the paper.

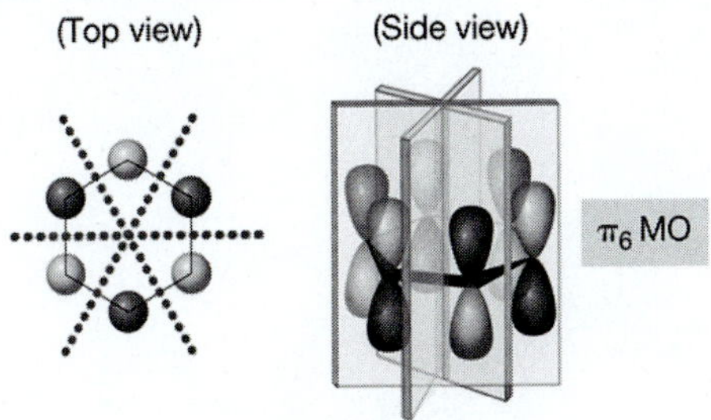

Your Turn 14.8

Think

Relative to the *p* AO energies, are bonding MOs lower energy, higher energy, or the same energy? Antibonding MOs? Nonbonding MOs?

Solve

As shown below, there is one bonding MO (π_1), two nonbonding MOs (π_2 and π_3), and one antibonding MO (π_4). Bonding orbitals are lower in energy than the *p* AOs (dotted line), whereas antibonding MOs are higher in energy than the *p* AOs. Nonbonding MOs appear at the same energy as the contributing AOs.

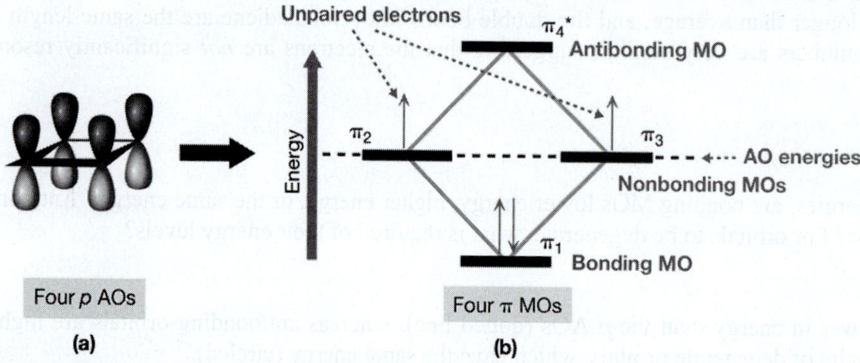

Your Turn 14.9

Think

After you draw in your C–H bonds, how can you represent the possible steric strain of the H atoms? How do you think the ring size affects that steric strain?

Solve

The H atoms are included in the figure below. The inside of the 18[annulene] ring is large enough to accommodate the H atoms so they do not crash into each other, as they do in the other molecules in Figure 14-24.

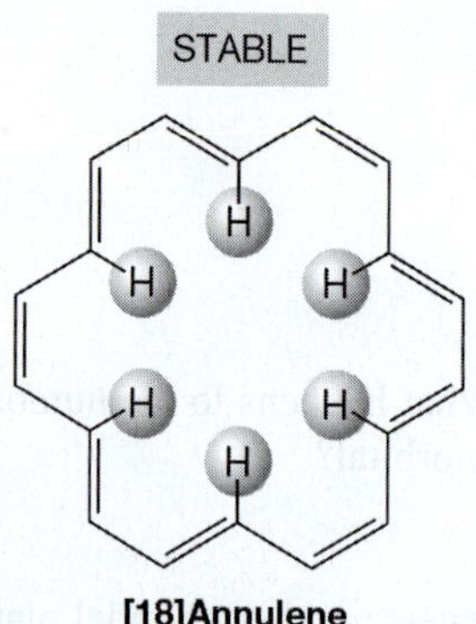

Your Turn 14.10

Think

How can you connect the shaded lobes to draw one closed loop of orbitals? How can you view the fused ring in a single aromatic π system?

Solve

The dotted lines indicate a closed loop of orbitals. Each of these molecules, therefore, is viewed as having a single aromatic π system.

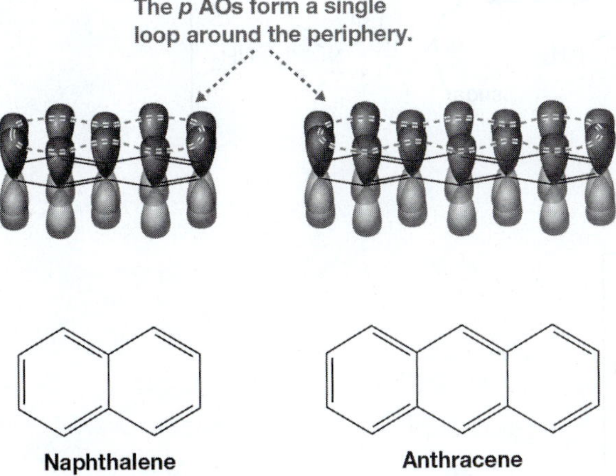

Your Turn 14.11

Think

For naphthalene and anthracene, can you identify a ring of alternating double and single bonds that does not include all π electrons of the system? After you identify such a ring, how can the π electrons be shifted to arrive at a new resonance structure?

Solve

In naphthalene, there is a six-carbon ring of alternating single and double bonds, and its three double bonds can be shifted around the ring to arrive at the new resonance structure shown below. In anthracene, there is a 10-carbon ring of alternating single and double bonds, and its five double bonds can be shifted around the ring to arrive at a new resonance structure. In each new resonance structure, the vertical double bond appears to divide the π electrons into two separate rings, but the *p* orbitals remain in the same locations, so there is still one complete loop of *p* orbitals on the periphery of the molecule.

Naphthalene

Anthracene

Your Turn 14.12

Think

In the structures given, can a Hückel number (an odd number of pairs) of electrons be shifted, involving every atom that composes a ring? If not, can you shift electrons via resonance to arrive at another resonance structure where a Hückel number of electrons can be shifted that way?

Solve

For adenine's resonance structure given in the textbook, five curved arrows can be drawn to move 10 electrons (a Hückel number) completely around the ring, as shown on the next page. For the resonance structures of guanine, thymine, and cytosine given in the textbook, however, it is not easy to see how all of the electrons highlighted in red can be shifted entirely around the ring atoms. In the second resonance structure shown on the next page, for each of these nitrogenous bases, curved arrows can be drawn to shift a Hückel number of electrons completely around the ring atoms: 10 for guanine and six each for thymine and cytosine.

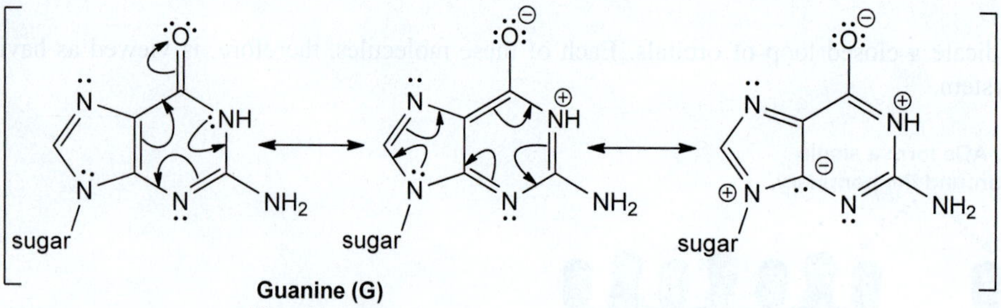

Guanine (G)
10 π electrons
Aromatic

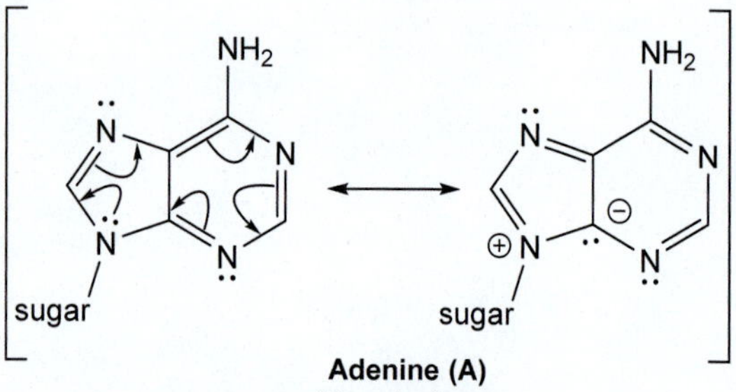

Adenine (A)
10 π electrons
Aromatic

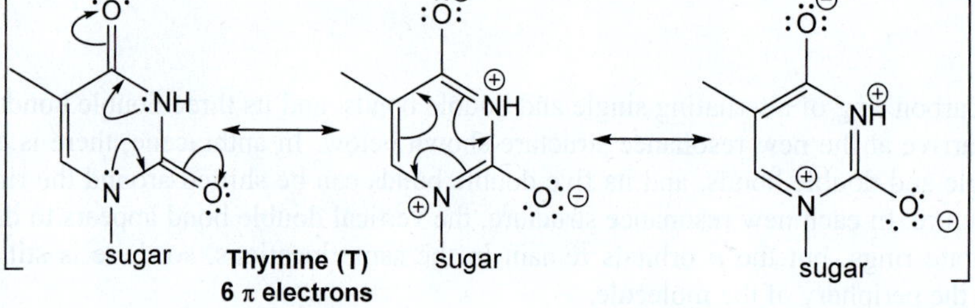

Thymine (T)
6 π electrons
Aromatic

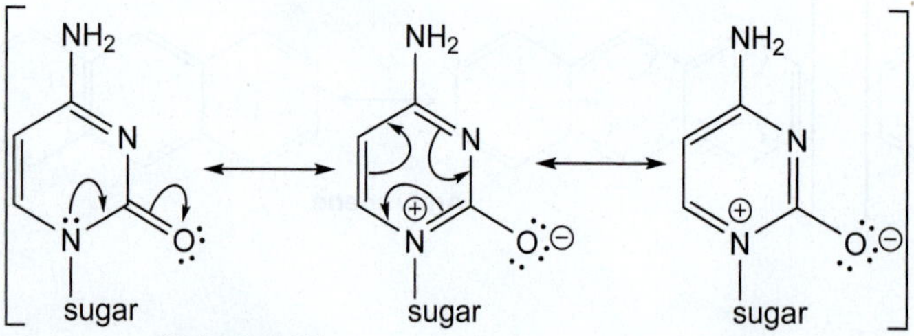

Cytosine (C)
6 π electrons
Aromatic

Your Turn 14.13

Think

Draw both base pairs in their enol forms. Where are the hydrogen bond–donating groups? Where are the hydrogen bond–accepting groups? How many hydrogen-bonding interactions are possible when both groups are in their enol forms? How does this differ from the interactions in Figure 14-31 and 14-33?

Solve

When guanine and cytosine are in their enol forms, only two hydrogen-bonding interactions are possible between the two base pairs. This forms a less stable interaction compared to the three hydrogen-bonding interactions that are possible when both base pairs are in their keto forms.

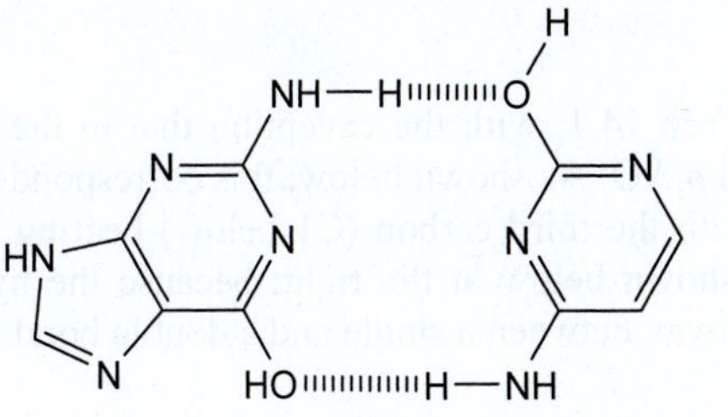

Enol form Guanine

Enol form Cytosine

In Chapter Problems

Problem 14.1

Think

How does the VB picture of the allyl anion compare to the VB picture of the allyl cation? In the VB picture, how do you account for the lone pair? Draw the resonance hybrid. Is the negative charge localized or delocalized?

Solve

The VB picture is identical to that of the allyl cation in Your Turn 14.1, with the exception that in the allyl anion, there is a lone pair of electrons occupying the unhybridized *p* AO. As shown below, this corresponds to a C=C double bond involving two carbons (C2 and C3 below), with the third carbon (C1 below) bearing a −1 formal charge. This does *not* agree with the resonance hybrid shown below at the right, because the hybrid shows the charge delocalized and both C–C bonds equivalent, partway between a single and a double bond.

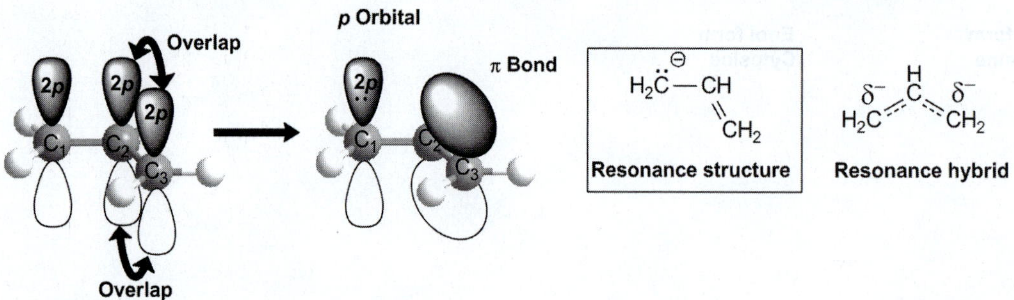

Problem 14.3

Think

How many *p* orbitals are conjugated? How many π MOs should result? How do those π MOs differ from each other? How many total σ and σ* MOs should there be? How many total valence electrons are there, and how should they be arranged in the orbitals?

Solve

As shown below, the MO energy diagram is identical to that for the allyl anion shown in Solved Problem 14.2, with the exception that the allyl radical has one fewer electron to make the species uncharged overall. So there is only one electron in π_2 instead of two. That leaves π_2 as the HOMO and π_3 as the LUMO.

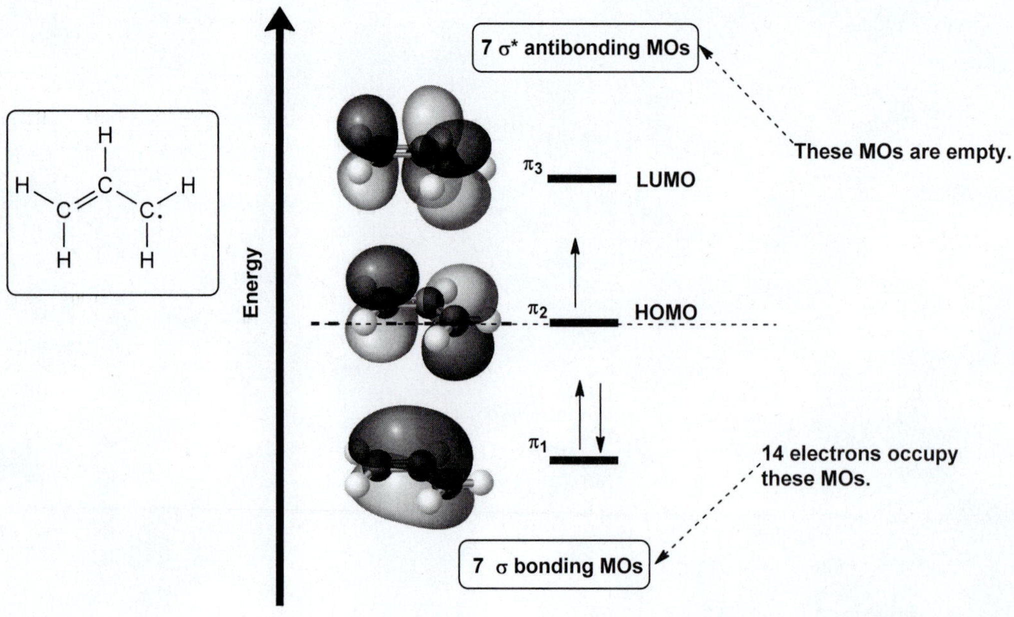

Problem 14.5

Think

How many more electrons are in the allyl radical compared to the allyl cation? Which orbital do those electrons occupy in the VB theory picture? What kind of charge do we associate with an occupied orbital like that? Which orbital does this electron occupy in the MO picture? How is the distribution of charge affected when that multiple-center MO is occupied?

Solve

In the VB theory picture, the one additional electron constitutes a single electron occupying a *p* orbital, which is associated with a formal charge of 0. In the MO theory picture, this electron occupies the π_2 MO, which delocalizes the electron onto C1 and C3. Thus, the unpaired electron is shared equally on C1 and C3, suggesting that each of those C atoms should be uncharged.

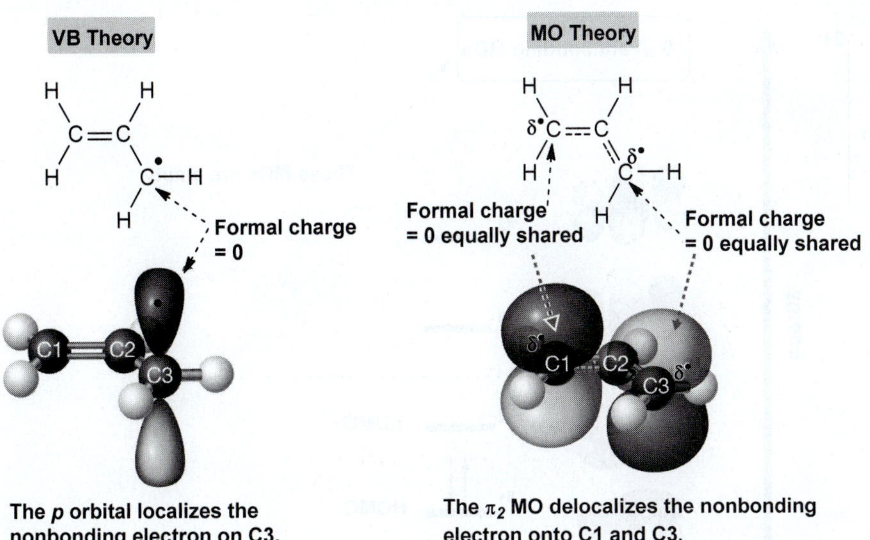

Problem 14.6

Think

Is a π-bonding interaction represented by constructive or destructive interference? Locate each type of interference in the structure drawn. Where are the nodal planes? Do bonding contributions raise or lower the π MO energy relative to the *p* AO? Do antibonding contributions? Count up the number of each type of interaction. What is the net interaction?

Solve

The nodes are drawn below between *p* AOs undergoing destructive interference (opposite phases of *p* orbitals). These represent antibonding interactions, which raise the MO energy relative to the *p* AOs. Constructive interference has the same phase of *p* orbitals. These represent bonding interactions that lower the MO energy relative to the *p* AOs. There are three bonding interactions and two antibonding interactions and, thus, a net of one bonding interaction (one more bonding than antibonding). The orbital, therefore, is a bonding MO.

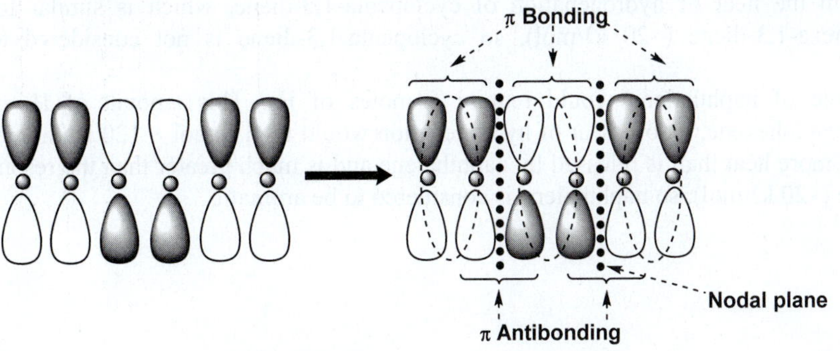

Problem 14.7

Think

If all four carbon atoms are sp^2 hybridized, how many *p* orbitals are conjugated? How many π MOs, therefore, must exist? How many σ and σ^* MOs are present? How many total valence electrons does the butadienyl dication have? How many electrons are in the σ MOs? How many are left? How does this compare to the MO energy diagram for butadiene?

Solve

The MO energy diagram of the butadienyl dication is similar to that for 1,3-butadiene in Figure 14-12. The difference is that the butadienyl dication has two fewer electrons (20 valence electrons total). Therefore, the π_1 MO ends up doubly occupied, but π_2 ends up empty. This makes π_1 the HOMO and π_2 the LUMO, as indicated below.

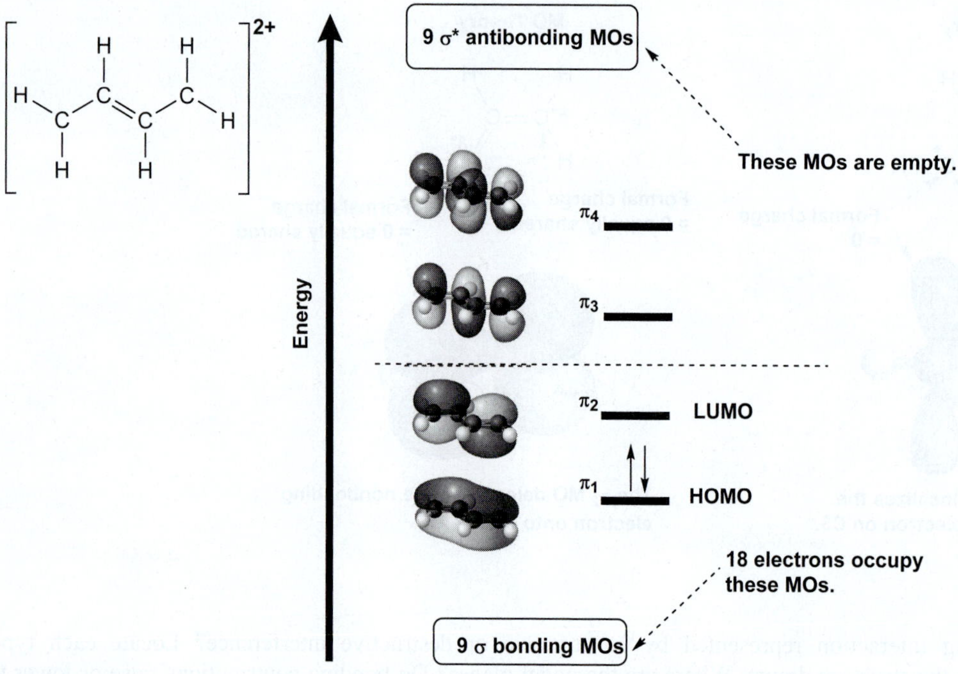

Problem 14.9

Think

How many moles of H_2 are required to completely hydrogenate cyclopenta-1,3-diene and naphthalene? How much heat is released when that number of moles of hydrogen reacts with cyclohexene instead? How does that quantity compare to the heat of hydrogenation for cyclopenta-1,3-diene and naphthalene? How does that quantity compare to the heat of hydrogenation for hexa-1,3-diene, which is nonaromatic?

Solve

(a) Hydrogenation of 1 mole of cyclopenta-1,3-diene would require 2 moles of H_2. This amount of H_2 can also hydrogenate 2 moles of cyclohexene, which would release 2 mol × 120 kJ/mol = 240 kJ/mol. This is only 29 kJ/mol different from the heat of hydrogenation of cyclopenta-1,3-diene, which is similar to the resonance energy for hexa-1,3-diene (~20 kJ/mol), so cyclopenta-1,3-diene is not considered to be aromatic.

(b) Hydrogenation of 1 mole of naphthalene would require 5 moles of H_2. This amount of H_2 could hydrogenate 5 moles of cyclohexene, whose heat of hydrogenation would total 5 mol × 120 kJ/mol = 600 kJ. This is nearly 300 kJ more heat than is released by naphthalene and is much greater than the resonance energy of hexa-1,3-diene (~20 kJ/mol), so naphthalene is considered to be aromatic.

Problem 14.11

Think

Is there a ring of conjugated *p* AOs? Are there any sp^3 C atoms that break up the conjugated π system? Is the structure planar? If any of these questions is answered no, how is the structure categorized? If all of the questions are answered yes, how many electrons are present in the cyclic π system? Does it contain a Hückel or anti-Hückel number of π electrons?

Solve

(a) Nonaromatic; although there are six total π electrons (a Hückel number), the π system is not completely cyclic owing to the presence of four sp^3 C atoms (labeled).

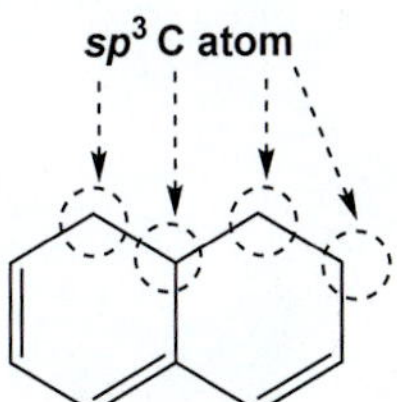

6 π electrons
Nonaromatic

(b) Aromatic; six π electrons (a Hückel number) are completely conjugated in a ring. Even though there are sp^3 C atoms present, these are outside the ring and not part of the conjugated π system.

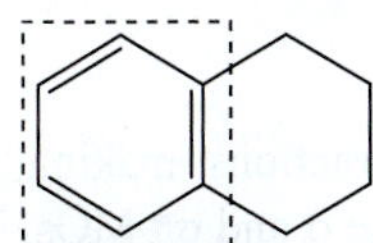

6 π electrons
Aromatic

(c) Nonaromatic; the π system is not conjugated in a complete ring because of the sp^3-hybridized C atom.

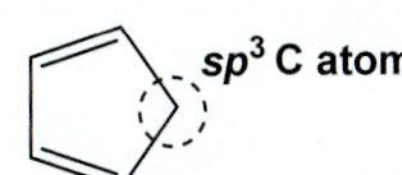

4 π electrons
Nonaromatic

(d) Nonaromatic, for the same reason as in **(c)**.

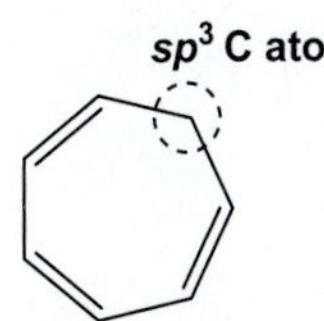

6 π electrons
Nonaromatic

(e) Nonaromatic; two sp^3-hybridized C atoms are preventing conjugation in a complete ring.

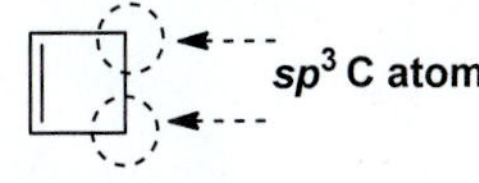

2 π electrons
Nonaromatic

(f) Nonaromatic; the π system is not cyclic.

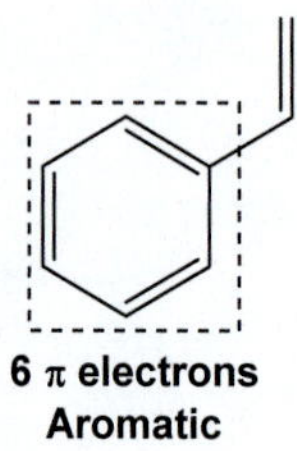

(g) Aromatic; although there are eight total π electrons, they are in two different π systems—one that is part of the benzene ring, which has six π electrons completely conjugated around a ring, and the C=C double bond, which has two π electrons.

6 π electrons
Aromatic

Problem 14.12

Think

How many valence electrons are in benzene, C_6H_6? How many σ bonds are present in benzene? How many σ and σ* MOs do they represent? Where should the σ and σ* MOs appear in the energy diagram relative to the π MOs for benzene (Fig. 14-19)? How many electrons should fill the σ MOs?

Solve

The diagram is shown below. There are six C–C σ interactions and six C–H σ interactions, making 12 total σ MOs and 12 total σ* MOs. The π MOs from Figure 14-19 should appear between the σ and σ* MOs. The six C atoms contribute four valence electrons each; the six H atoms contribute one electron each. The 30 valence electrons are distributed into 12 σ MOs and three π MOs. The π antibonding MOs are empty. The existence of all paired π electrons in the bonding MOs makes benzene very stable.

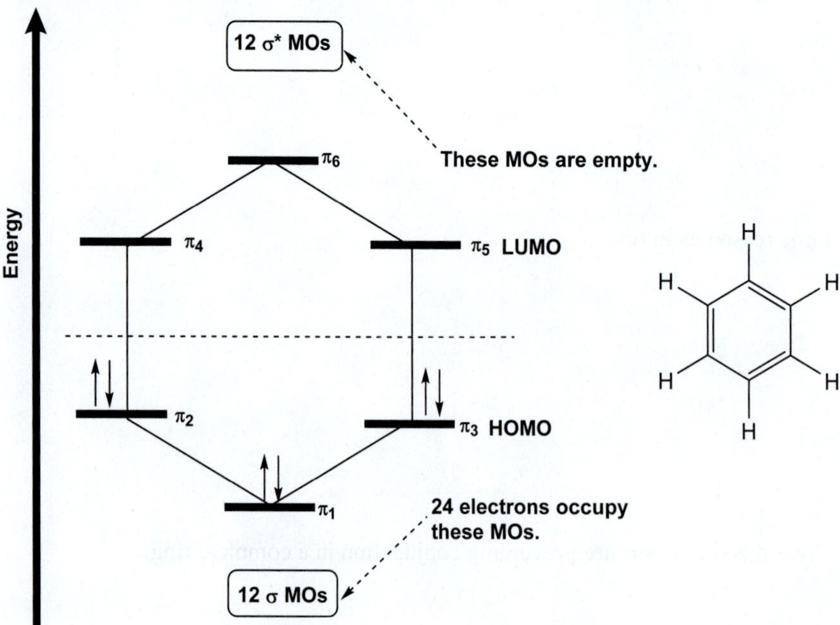

Problem 14.13

Think

How many valence electrons are in cyclobutadiene, C_4H_4? How many σ bonds are present in cyclobutadiene? How many σ and σ* MOs do they represent? Where should the σ and σ* MOs appear in the energy diagram relative to the π MOs for cyclobutadiene (Fig. 14-21)? How many electrons should fill the σ MOs?

Solve

The diagram is shown below. There are four C–C σ interactions and four C–H σ interactions, making eight total σ MOs and eight total σ* MOs. The π MOs from Figure 14-21 should appear between the σ and σ* MOs. The four C atoms contribute four valence electrons each; the four H atoms contribute one electron each. The 20 valence electrons are distributed into eight σ MOs, one π MO, and two π nonbonding MOs. The existence of the unpaired π electrons in the nonbonding MOs makes cyclobutadiene unstable. (Because of this instability, cyclobutadiene distorts from square to rectangular, which lowers the energy of one nonbonding orbital and raises the energy of the other in order to become more stable, but this is not accounted for in the Frost diagram.)

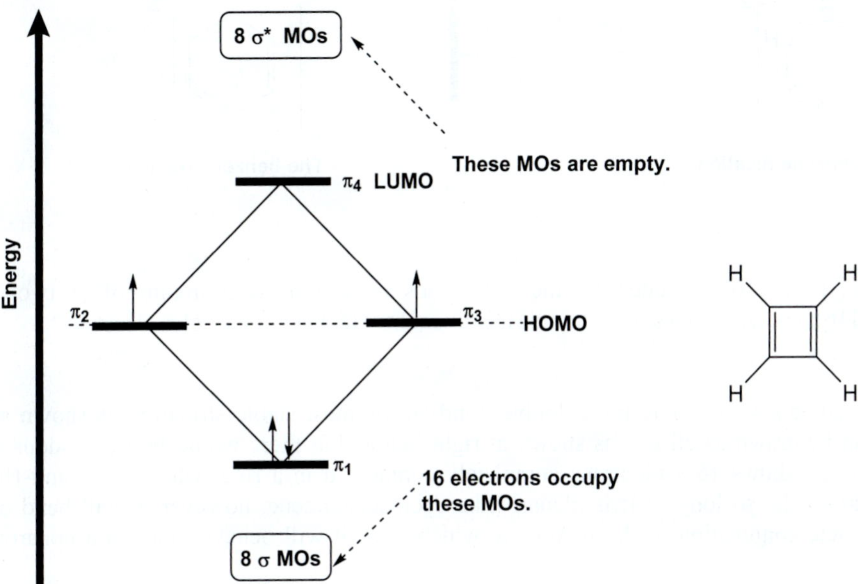

Problem 14.15

Think

Do the benzene dication and benzene dianion have cyclic and fully conjugated π systems? How can the Frost method be applied to these species to derive their π MO energy diagrams? How many π electrons are present? How do you fill in the electrons in the MO diagram? What does the HOMO look like? Is it filled?

Solve

The energy diagrams for the π systems of the two species are shown on the next page. We begin with the energy diagram from Figure 14-19 (Frost method), in which there are six π electrons. The dication, therefore, must have four π electrons. Hund's rule says that one should be in π_2 and the other in π_3. Even though all electrons are in bonding MOs, the instability of the unpaired electrons makes the dication antiaromatic. For the same reason, the dianion should be antiaromatic. It has eight π electrons. The three lowest-energy π MOs are completely filled, but π_4 and π_5 each has one unpaired electron.

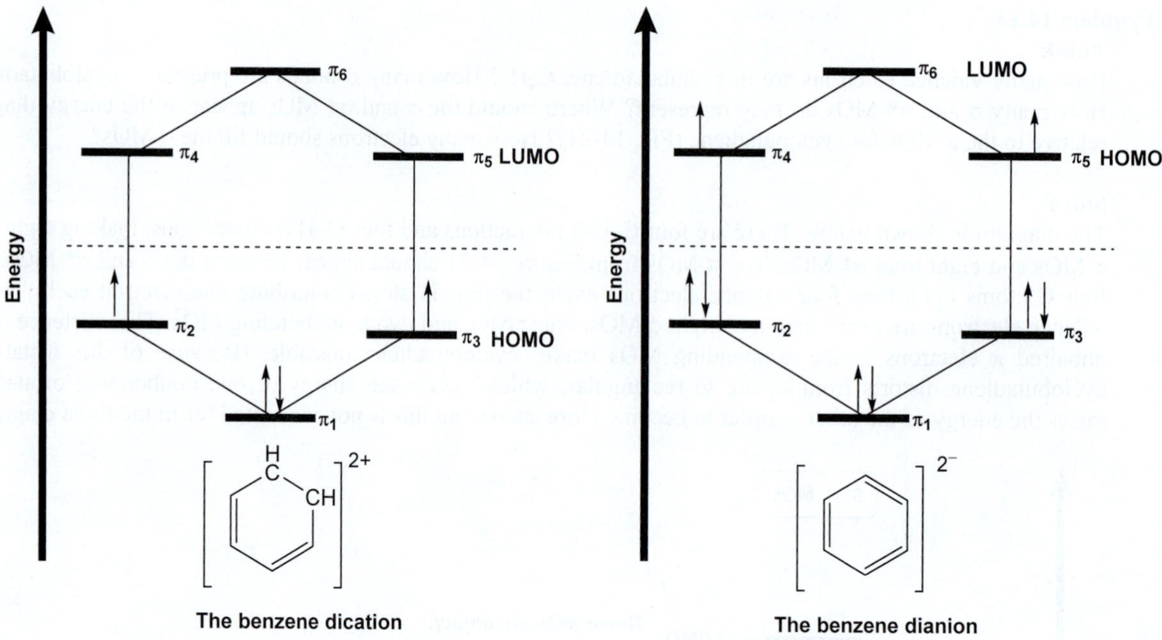

Problem 14.16

Think

How many carbon atoms are in [16]annulene? Are the C=C bonds cis or trans, or a mixture of the two? How many π electrons are in [16]annulene? Does it contain a Hückel or anti-Hückel number of π electrons?

Solve

[16]Annulene will have some cis and some trans double bonds in its most stable structure, as shown at left below. The structure can be drawn as all cis, as shown at right below, but there would be tremendous angle strain. Either way, however, shows 16 π electrons completely conjugated in a ring, which is an anti-Hückel number, making it antiaromatic, so long as it is planar. Like cyclooctatetraene, however, it will bend out of plane to avoid the complete conjugation of the *p* AOs, in which case, it will behave more as a nonaromatic compound.

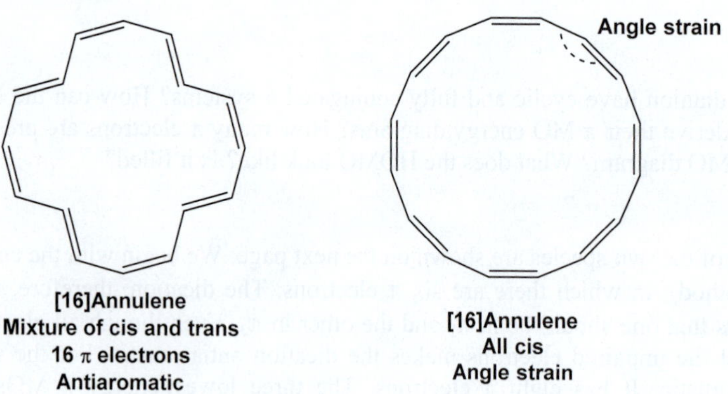

Problem 14.18

Think

Does the molecule contain a π system formed from a ring of fully conjugated *p* orbitals? How many electrons are in that π system?

Solve

(a) Antiaromatic; the outside of the fused ring system has a completely conjugated π system and contains eight π electrons, which is an anti-Hückel number (an even number of pairs). The constraints of the relatively small fused rings prevent the molecule from distorting out of plane to achieve a nonaromatic molecule like cyclooctatetraene.

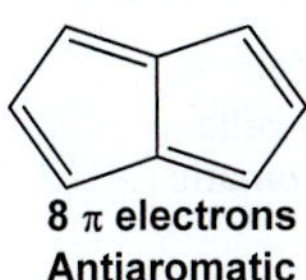

(b) Aromatic; the outside of the fused ring system is a completely conjugated π system that contains 14 electrons, a Hückel number (an odd number of pairs).

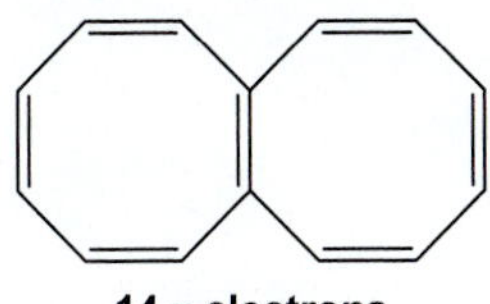

(c) Nonaromatic; the π system is not completely conjugated around the outside of the fused ring system owing to the presence of two sp^3 C atoms.

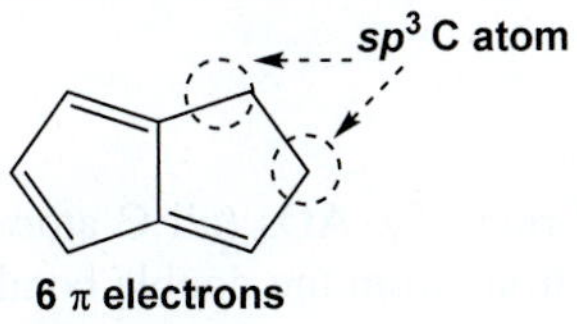

Problem 14.20

Think

How many lone pairs from the heteroatom are part of the π system? How many total π electrons are there? Is that a Hückel number, an anti-Hückel number, or neither? Is the ring planar?

Solve

The answers are given on the next page. For each that is aromatic or antiaromatic, the electron pairs that are part of the π system are indicated by arrows. In **(a)** and **(c)**, the O is sp^2 hybridized, so one lone pair is part of the π system and the other lone pair is not (it is perpendicular). In **(b)**, the lone pair on N is not part of the π system, because it resides in an sp^2 orbital that is perpendicular to the π system. In **(f)**, the lone pair is part of the π system, because the sp^2-hybridized orbitals, used to make the three bonds to N, are in the plane perpendicular to the π system. In **(g)**, there is a set of *p* AOs conjugated around the benzene ring only, composing a system that has just the three π bonds, for six total π electrons (the other double bond and lone pair involving the N are not part of that π system). The remaining ones, **(d)** and **(e)**, are nonaromatic because the π systems do not form a complete ring. Molecule **(c)** would be antiaromatic if planar, but, like cyclooctatetraene, it bends out of plane to avoid being antiaromatic. So it is more accurately described as nonaromatic.

(a) 4 π electrons Antiaromatic

(b) 10 π electrons Aromatic

(c) 8 π electrons Nonaromatic (not planar)

(d) Not fully conjugated Nonaromatic

(e) Acyclic Nonaromatic

(f) 10 π electrons Aromatic

(g) 6 π electrons Aromatic

Problem 14.21

Think

What is the hybridization of the C^+ atom? Draw in any lone pairs not shown. How many lone pairs from the heteroatom are part of the π system? How many total π electrons are there? Is that a Hückel number, an anti-Hückel number, or neither? Is the ring planar?

Solve

Cation **(a)** is aromatic, with two π electrons and a completely conjugated ring of *p* AOs (all C atoms are sp^2 hybridized). Molecule **(b)** is antiaromatic, with four π electrons (two π electrons from the double bond and two π electrons from the pair of electrons shown on N).

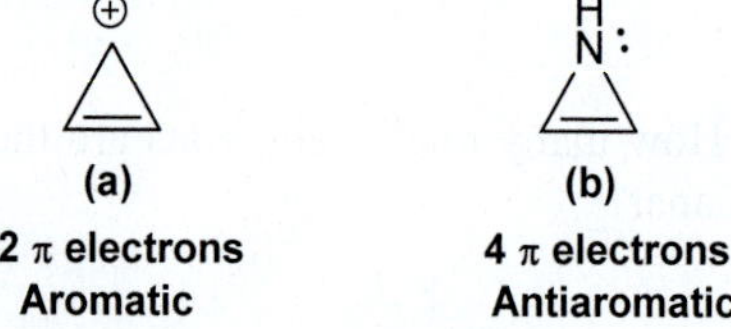

Problem 14.22

Think

How many double bonds are there? How many π electrons are in each C=C? Can all those electrons be shifted simultaneously in resonance?

Solve

The system contains one π system, because a resonance structure can be drawn in which all the π electrons are shifted at once, as shown below. That system contains 22 electrons—two for each curved arrow shown.

β-Carotene
22 π electrons
1 π system

Problem 14.24

Think

Can we draw a resonance structure by shifting electrons fully around the ring? If so, how many electrons must be shifted? Is that a Hückel number, an anti-Hückel number, or neither?

Solve

If the molecule is planar, it is antiaromatic, because the π system that is completely conjugated in a ring contains 12 electrons. This is indicated by the resonance structures below, where each curved arrow represents two π electrons.

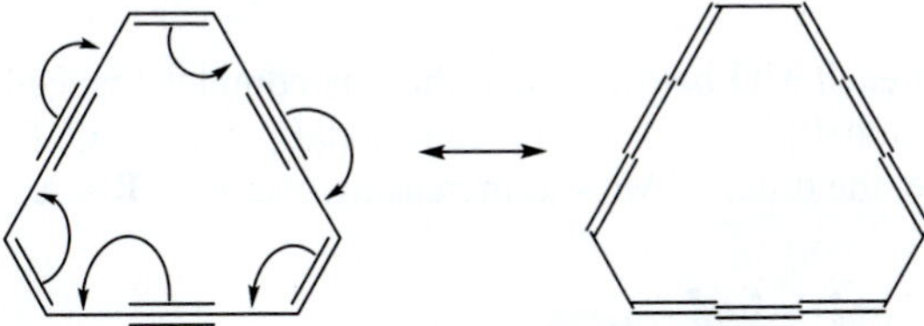

Problem 14.25

Think

How many lone pairs from the heteroatom are part of the π system? How many total π electrons are there? How many electrons can be shifted completely around the ring via resonance? Is that a Hückel number, an anti-Hückel number, or neither? Is the ring planar?

Solve

Molecules **(a)** and **(b)** are both aromatic, because the π systems that are conjugated in a ring each contains six total electrons, which is a Hückel number. In both molecules, the lone pairs on N are not part of the π system, evidenced by the fact that resonance structures can be drawn that shift electrons around the ring, and the lone pairs are unaffected.

(a) 6 π electrons Aromatic

(b) 6 π electrons Aromatic

Molecule **(c)** is also aromatic. The lone pair on the N atom with three single bonds shown is part of the π system, indicated by the fact that a resonance structure can be drawn in which that lone pair and the electrons from the double bonds are shifted around the ring. The lone pairs on the doubly bonded N atom are untouched, so they are not part of the π system. Thus, there are six total π electrons in the ring, which is a Hückel number.

6 π electrons Aromatic

End of Chapter Problems

Sections 14.1 and 14.2 The Shortcomings of VB Theory; Multiple-Center MOs

Problem 14.26

Think

From the phases of the *p* orbitals, how do you identify a nodal plane? Does constructive or destructive interference result from that type of interaction between those *p* orbitals? In the resulting MO, what characterizes a nodal plane? Do bonding contributions raise or lower the π MO energy relative to the contributing *p* AOs? What about antibonding contributions? Count the number of each type of interaction. What is the net interaction?

Solve

(a) The nodes are indicated by a vertical dashed line in each MO below, where there is complete cancellation of the MO. There are three nodal planes in **A**, two nodal planes in **B**, and four nodal planes in **C**. Energy increases with increasing number of nodes; therefore, the order of MOs in increasing energy is **B** < **A** < **C**.

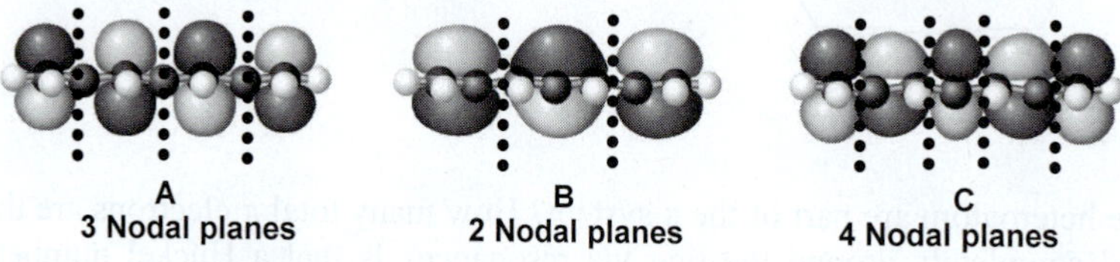

(b) The *p* orbital contributions are shown below. No *p* orbital contribution appears on any atom lying in a nodal plane. Notice that the *p* orbitals on either side of each nodal plane have opposite phases.

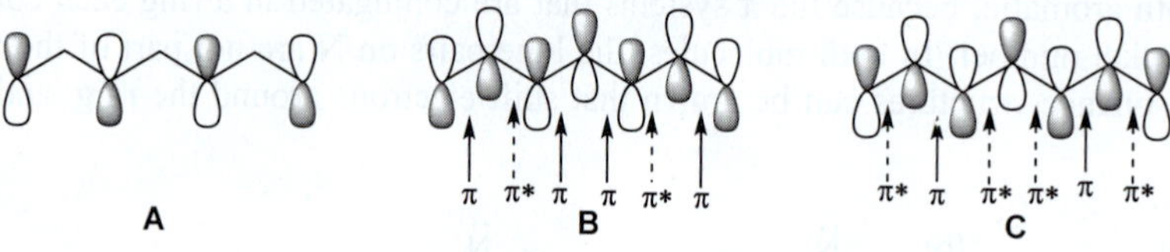

(c) The bonding regions are indicated by the solid arrows above. The antibonding regions are indicated by the dashed arrows above. In MO **A**, no adjacent *p* orbitals interact, so there are no bonding or antibonding interactions. In MO **B** and MO **C**, bonding interactions are identified as regions where the *p*-orbital contribution on either side is of the same phase. In each antibonding interaction, the *p* orbitals contribute opposite phases.

(d) MO **A** is nonbonding because there are no bonding or antibonding interactions at all. MO **B** is a bonding MO because there are four bonding interactions but only two antibonding interactions (net: two bonding interactions). MO **C** is an antibonding MO because there are only two bonding interactions but four antibonding interactions (net: two antibonding interactions).

Problem 14.27

Think

From the phases of the *p* orbitals, how do you identify a nodal plane? Does constructive or destructive interference result from that type of interaction between those *p* orbitals? In the resulting MO, what characterizes a nodal plane? Do bonding contributions raise or lower the π MO energy relative to the contributing *p* AOs? What about antibonding contributions? Count the number of each type of interaction. What is the net interaction?

Solve

(a) The nodes are indicated by the vertical dashed lines below, where there is complete cancellation of the MO. There is one node in the first MO, six nodes in the second, and no nodes in the third. Energy increases with increasing number of nodes; therefore, the order of MOs in increasing energy is **F** < **D** < **E**.

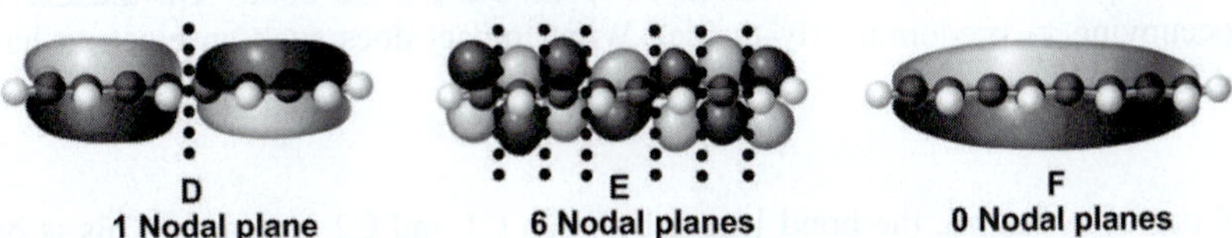

(b) The *p*-orbital contributions are indicated below. Notice that the *p* orbitals on either side of each nodal plane have opposite phases.

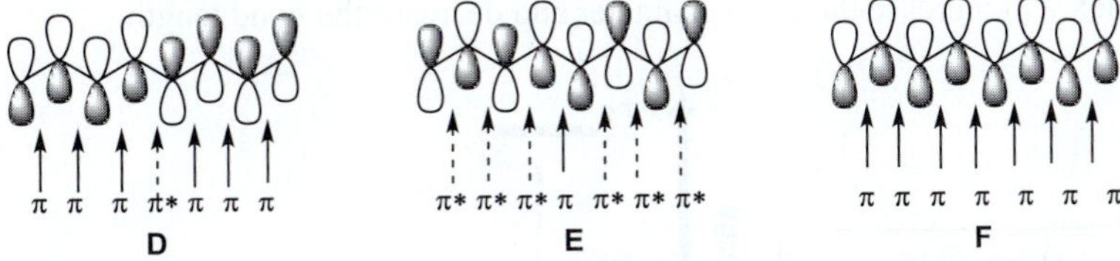

(c) The bonding regions are indicated by solid arrows above. The antibonding regions are indicated by dashed arrows.

(d) MO **D** is bonding because it has six bonding regions and one antibonding region (net: five bonding regions). MO **E** is antibonding because it has one bonding region and six antibonding regions (net: five antibonding regions). MO **F** is a bonding MO because all regions are bonding.

Problem 14.28

Think

What is the hybridization of each C atom? Where are the contributing *p* AOs located? Where do those *p* AOs overlap to form a π bond? Where is there no *p* orbital overlap? In a VB picture, what type of orbital is associated with a positively charged C? A negatively charged C?

Solve

All the C atoms are sp^2 hybridized and, therefore, have an unhybridized *p* AO to contribute. The carbocation and carbanion on the ends do not contribute to any π-bonding interaction. The interaction between the *p* orbitals on C2 and C3 results in a π bond between C2 and C3. A positively charged C is associated with an empty *p* AO on C, whereas a negatively charged C is associated with a filled *p* AO on C. See the valence bond picture for each of the two weak resonance contributors below.

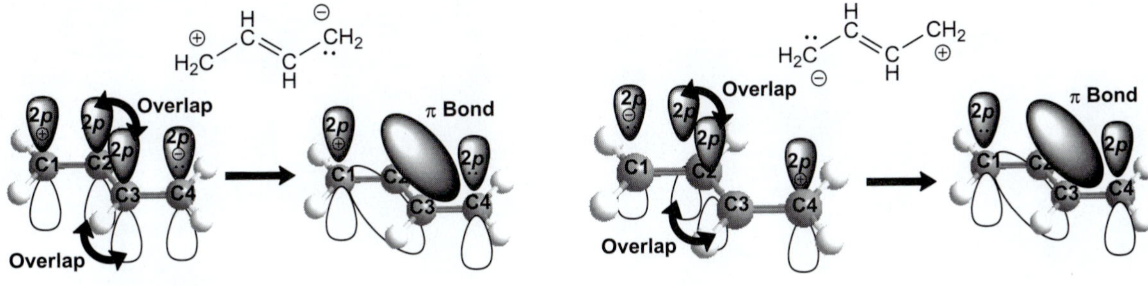

Problem 14.29

Think

Consider Figure 14-11, which shows the π MO energy diagram for buta-1,3-diene. What does the MO energy diagram look like if one electron from π_2 is promoted to π_3? Where does an electron occupying π_2 predominantly reside? What impact does such an electron have on the C1–C2 bond? On the C2–C3 bond? Where does an electron occupying π_3 predominantly reside? What impact does such an electron have on the C1–C2 bond? On the C2–C3 bond?

Solve

If an electron is promoted out of π_2 into π_3, the bond length between C1 and C2 increases. This is because an electron in π_2 resides about half the time in the C1–C2 bonding region and, thus, contributes toward an increased C1–C2 bond character. An electron in π_3, on the other hand, does not reside much in the C1–C2 bonding region, because there is a node there. So, overall, the electron density between C1–C2 decreases with this electron transition. Because the opposite is true for the C2–C3 bonding region, the electron density increases between C2–C3, which will increase the bond character and decrease the bond length.

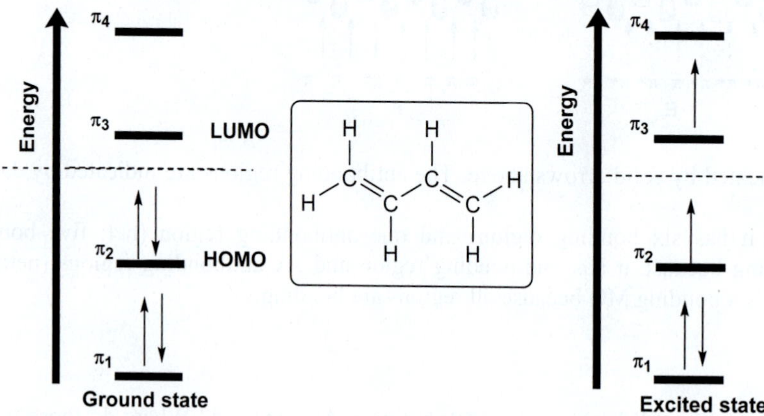

Problem 14.30

Think

What is the hybridization of each C atom? How many *p* AOs are conjugated? How many π MOs will result? How many π electrons are there? How many σ bonds are there in the Lewis structure? How many σ and σ^* MOs does this represent?

Solve

There will be five *p* AOs combining to make five π MOs. Two will be bonding, one will be nonbonding, and two will be antibonding. There are 11 σ bonds in the Lewis structure, corresponding to 11 σ MOs and 11 σ* MOs. The 28 total valence electrons fill the 11 σ MOs and the first three π MOs. The six π electrons are observable from the Lewis structure—two π bonds and one lone pair of electrons conjugated to it. The HOMO is π_3, and the LUMO is π_4.

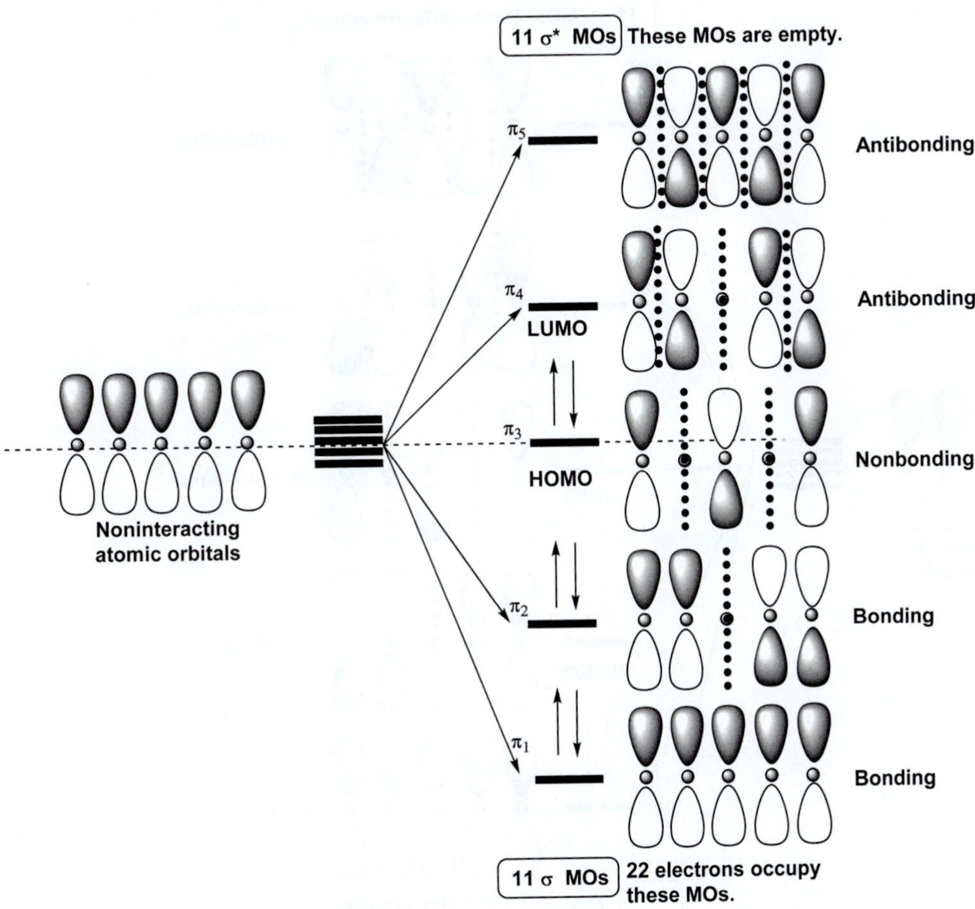

Problem 14.31

Think

What is the hybridization of each C atom? How many *p* AOs are conjugated? How many π MOs will result? How many π electrons are there? How many σ bonds are there in the Lewis structure? How many σ and σ* MOs does this represent?

Solve

There will be five p AOs combining to make five π MOs. Two will be bonding, one will be nonbonding, and two will be antibonding. There are 11 σ bonds in the Lewis structure, corresponding to 11 σ MOs and 11 σ^* MOs. The 26 total valence electrons fill the 11 σ MOs and the first two π MOs. The four π electrons are observable from the Lewis structure—the two π bonds. The HOMO is π_2, and the LUMO is π_3.

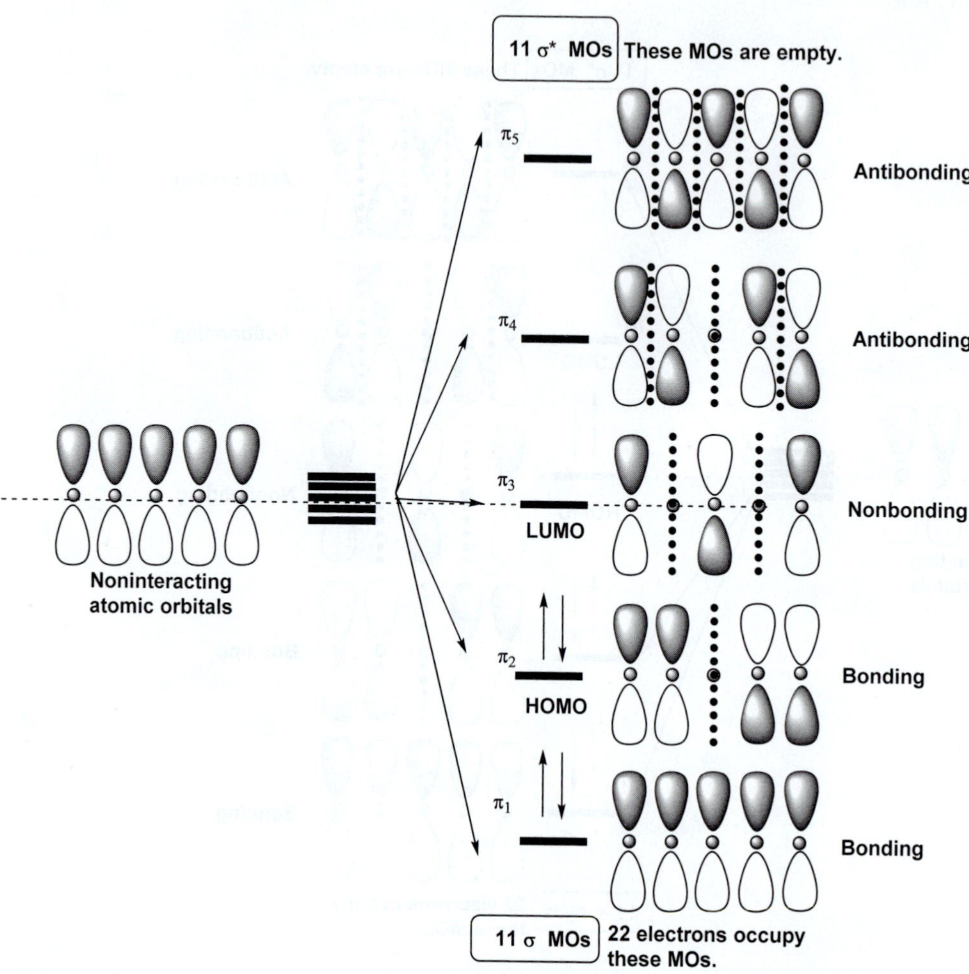

Problem 14.32

Think

What is the hybridization of each C atom? How many p AOs are conjugated? How many π MOs will result? How many π electrons are there? How many σ bonds are there in the Lewis structure? How many σ and σ^* MOs does this represent?

Solve

There will be six p AOs combining to make six π MOs. Three will be bonding, and three will be antibonding. There are 13 σ bonds in the Lewis structure, corresponding to 13 σ MOs and 13 σ^* MOs. The 32 total valence electrons fill the 13 σ MOs and the first three π MOs. The six π electrons are observable from the Lewis structure—three π bonds. The HOMO is π_3, and the LUMO is π_4. See the figure on the next page.

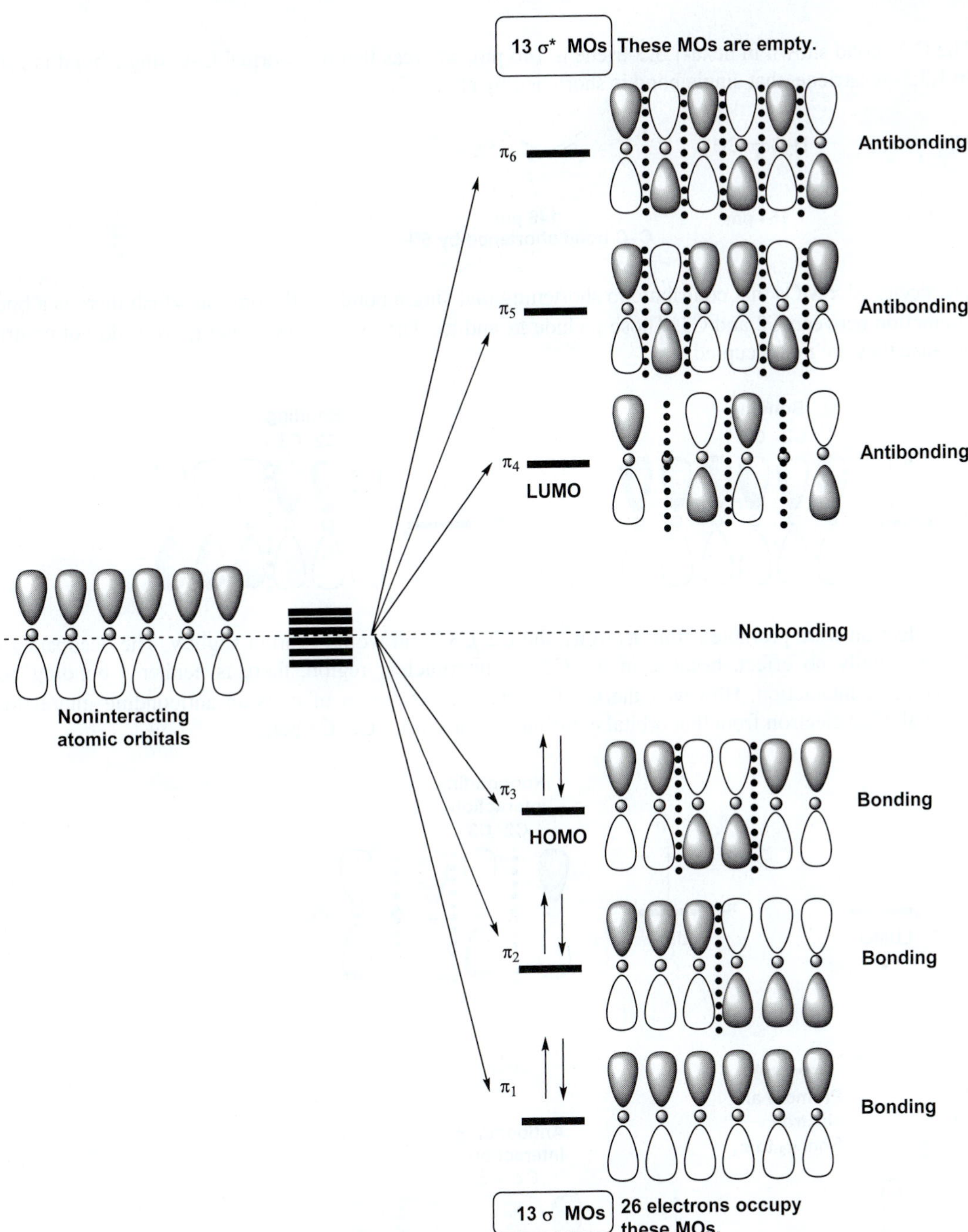

Problem 14.33

Think

See Figure 14-2 for the C–C bond length of ethane. Which π MOs are occupied with electrons (see the solution to Problem 14.32)? In the MO picture from Problem 14.32, what type of interaction occurs between C2 and C3 in each occupied π MO? How does putting an electron in each of these MOs affect the bond length?

Solve

(a) The C–C bond shown in hexa-1,3,5-triene is 146 pm, whereas that of a normal C–C single bond is 154 pm. In 1,3,5-hexatriene, that single bond is shortened by about 5%.

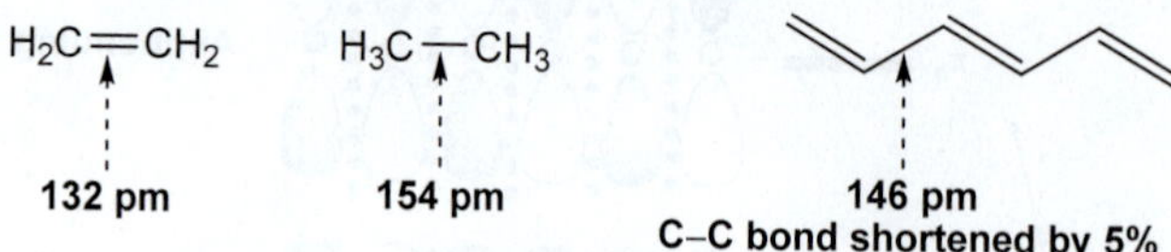

(b) The occupied π MOs that contribute to shortening that single bond are the ones in which there is a bonding interaction between C2 and C3. Those include π_1 and π_2. The three highest-energy MOs do not contribute, because they are not occupied.

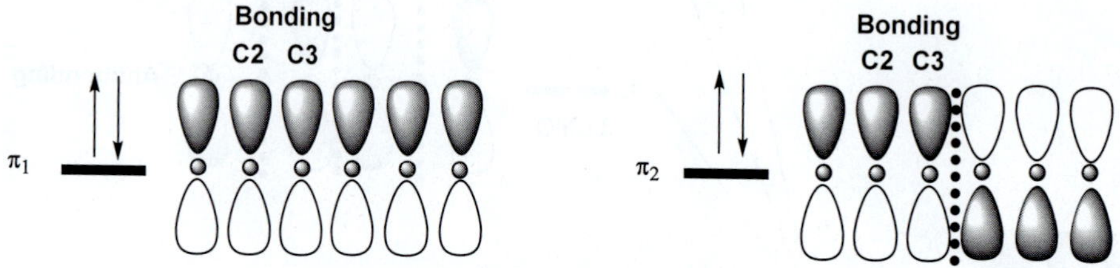

(c) If an electron were promoted from π_3 to π_4, the C2–C3 bond would shorten slightly. The occupancy of π_4 has essentially no effect, because in the C2–C3 internuclear region, there is neither a bonding nor an antibonding interaction. However, the C2–C3 internuclear region of π_3 is an antibonding interaction, so removal of an electron from that orbital contributes to a shorter C2–C3 bond.

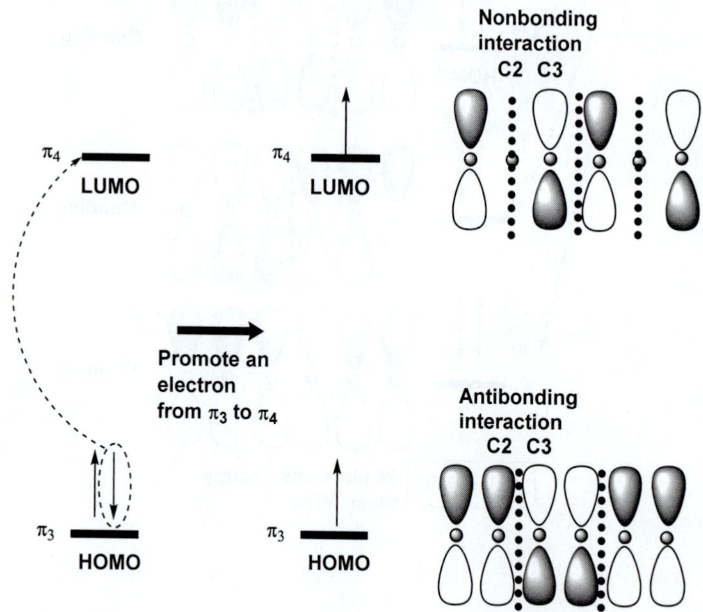

Problem 14.34

Think

How many double bonds are present? In each C=C, how many π electrons are present? Are there any sp^3 C atoms in between double bonds? If so, how does this affect the conjugation of the π system? In a triple bond, what is the orientation of the four *p* orbitals? If the *p* orbitals are not parallel, are they conjugated?

Solve

There are 12 total π electrons—two from each double bond and four from the triple bond. These will occupy four separate π systems. The double bonds between C2 and C3 and between C4 and C5 are conjugated and make up one π system. The π system between C1 and C2 is a separate π system because the *p* orbitals are perpendicular to the first π system. The third π system is made up of the double bond between C7 and C8 and one of the π bonds between C9 and C10, as those are conjugated together. The second π bond of the triple bond is a separate π system, because its *p* orbitals are perpendicular to the first π bond of the triple bond.

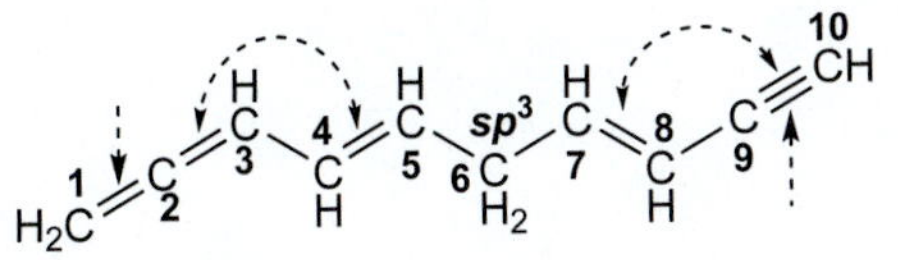

Sections 14.3–14.9 Aromaticity and Molecular Orbitals in Cyclic π Systems

Problem 14.35

Think

Is the ring fully conjugated? Are there any sp^3 C atoms that break up the conjugated π system? Is the structure planar? Counting the total number of π electrons, does it contain a Hückel or an anti-Hückel number of π electrons? Counting only the π electrons on the periphery, does it contain a Hückel or an anti-Hückel number of π electrons?

Solve

It might appear to break Hückel's rule, because there are a total of 16 π electrons, which is an anti-Hückel number. However, these 16 electrons can compose separate π systems—one on the outside of the ring, containing 14 electrons, and one on the inside of the ring, containing two. The 14 electrons on the outside of the ring are a Hückel number, making it aromatic.

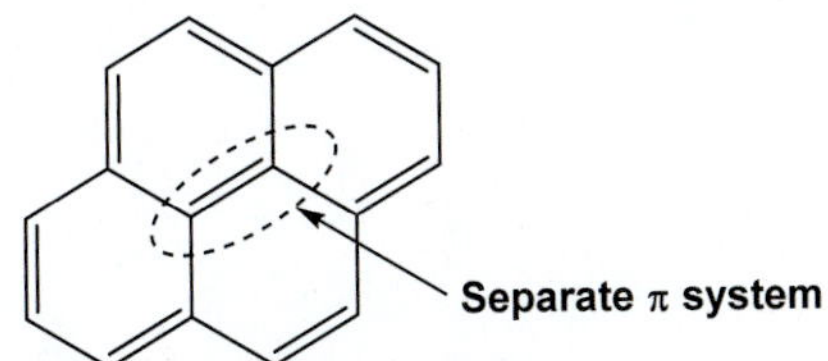

Problem 14.36

Think

Considering the Frost method, what is the polygon shape of [10]annulene? What does the center of that polygon represent? What does each of the vertices represent? How many π electrons are in [10]annulene? Is the HOMO filled or half filled? How is this related to aromaticity?

Solve

The molecule, with five pairs of (or 10 total) π electrons, should be aromatic, as long as the molecule is planar. The MO energy diagram agrees. All MOs are completely filled, and all occupied orbitals are bonding. The π MO energies are derived from the Frost method by orienting the decagon so a vertex points straight downward, the center of the decagon is taken to be the *p* AO energies, and each π MO energy is located at a vertex. See the figure on the next page.

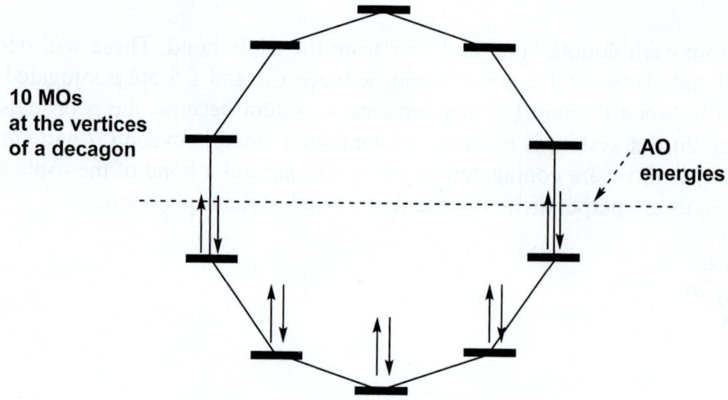

Problem 14.37

Think

Considering the Frost method, what is the polygon shape of [8]annulene? What does the center of that polygon represent? What does each of the vertices represent? How many π electrons are in [8]annulene? Is the HOMO filled or half filled? How is this related to aromaticity?

Solve

With four pairs of (or eight total) π electrons, this molecule should be antiaromatic, as long as the molecule is planar. The MO energy diagram agrees. Not all MOs are filled, leaving very reactive unpaired electrons. And not all occupied orbitals are bonding. The π MO energies are derived from the Frost method by orienting the octagon so a vertex points straight downward, the center of the octagon is taken to be the *p* AO energies, and each π MO energy is located at a vertex.

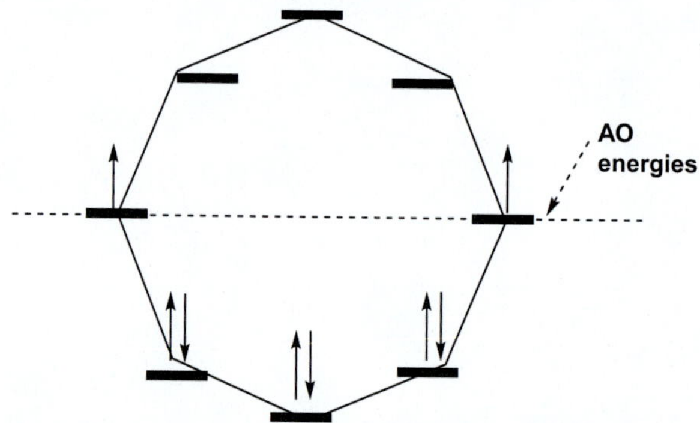

Problem 14.38

Think

Considering the Frost method, what is the polygon shape of each compound? What does the center of that polygon represent? What does each of the vertices represent? How many π electrons are in each compound? Is the HOMO filled or half filled? How is this related to aromaticity?

Solve

The number of π electrons in each system and the aromatic or antiaromatic character are listed below each MO diagram. The π MO energies of each molecule are derived from the Frost method by orienting the polygon so a vertex points straight downward, the center of the polygon is taken to be the *p* AO energies, and each π MO energy is located at a vertex. See the figures on the next page.

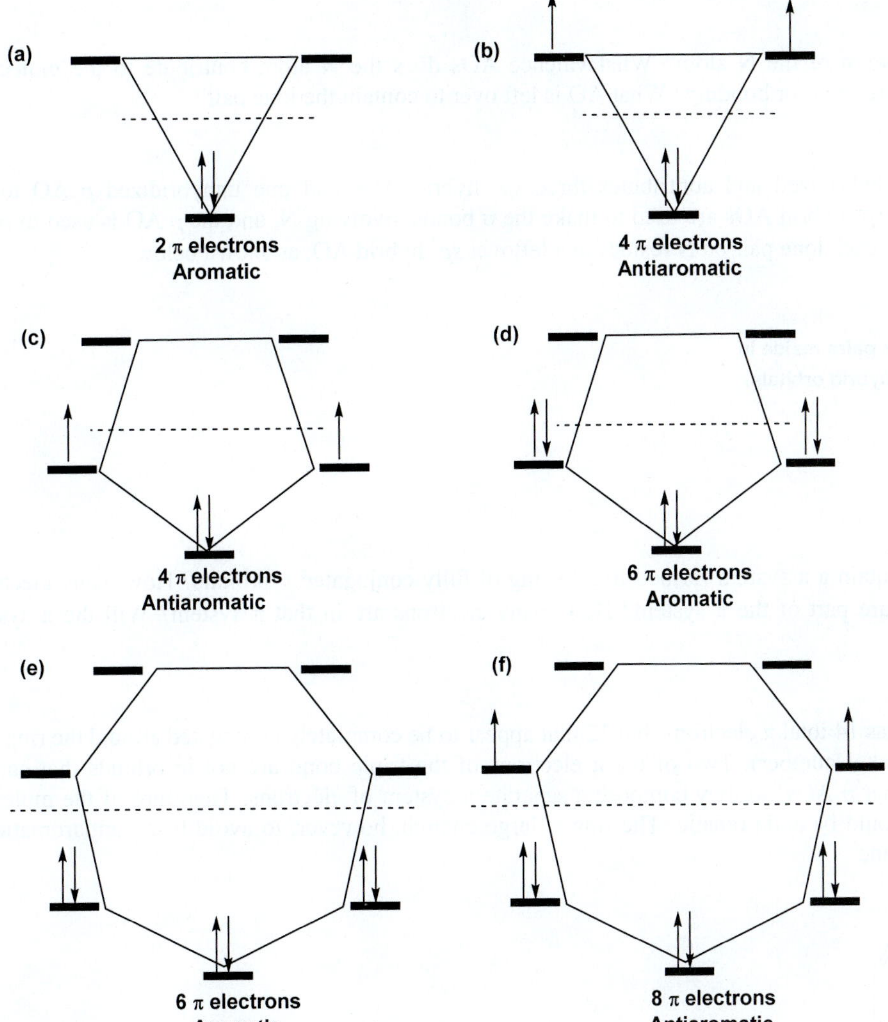

Problem 14.39

Think

What is the hybridization of each atom? Is the ring fully conjugated? Are there any lone pairs on the N or B atoms? Should those lone pairs be part of the same π system as the double bonds, or are they perpendicular to it? How many π electrons are in the conjugated system?

Solve

Boron has no lone pairs and is sp^2 hybridized. The lone pair on each N atom is in an sp^2-hybridized orbital, which is perpendicular to the π system of the ring, and so each lone pair cannot be part of the π system. The molecule, with three pairs of (or six total) π electrons, is aromatic. It is cyclic and conjugated, just like benzene.

6 π electrons
Aromatic

Problem 14.40

Think

What is the hybridization of the N atom? What valence AOs does the N atom contribute to the molecule? Which of those AOs are used for bonding? What AO is left over to contain the lone pair?

Solve

Each N atom is sp^2 hybridized and contributes three sp^2 hybrid AOs and one unhybridized *p* AO to the molecule. Two of the sp^2 hybrid AOs are used to make the σ bonds involving N, and the *p* AO is used to make the π bond. Therefore, each lone pair on N resides in a leftover sp^2 hybrid AO, as shown below.

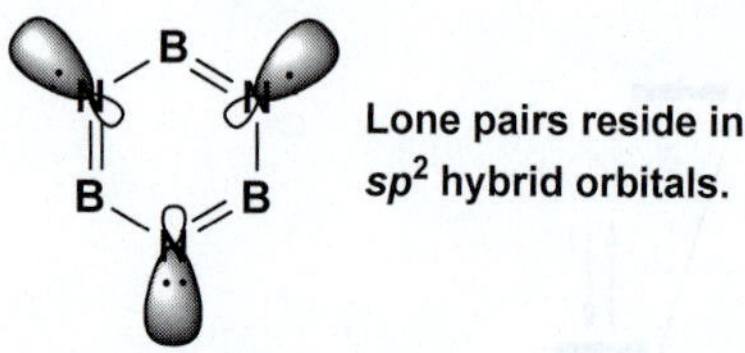

Problem 14.41

Think

Does the molecule contain a π system formed from a ring of fully conjugated *p* orbitals? How many electrons from the heteroatom are part of the π system? How many electrons are in that π system? Will the π system remain planar?

Solve

(a) Nonaromatic: It has 14 total π electrons, but 12 that appear to be completely conjugated around the ring (six pairs: an anti-Hückel number). Two of the π electrons of the triple bond are not in orbitals that can be parallel to the other *p* AOs, so they compose a separate π system of electrons. Therefore, if the molecule were planar, it would be antiaromatic. The ring is large enough, however, to avoid being antiaromatic by bending out of plane.

14 π electrons
Nonaromatic
(not planar)

(b) Nonaromatic: It has 12 π electrons completely conjugated around the periphery of the ring system, which is an anti-Hückel number (six pairs); the center carbon cannot participate in resonance, so there is no way to change the number of π electrons around the outer ring carbons. Therefore, if the molecule were planar, it would be antiaromatic. The ring is large enough, however, to avoid being antiaromatic by bending out of plane.

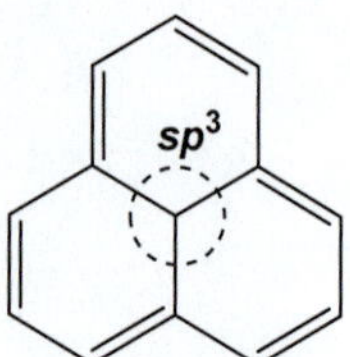

12 π electrons
Nonaromatic
(not planar)

(c) Aromatic: There are 10 total π electrons completely conjugated around the periphery of the molecule, which is a Hückel number (five pairs).

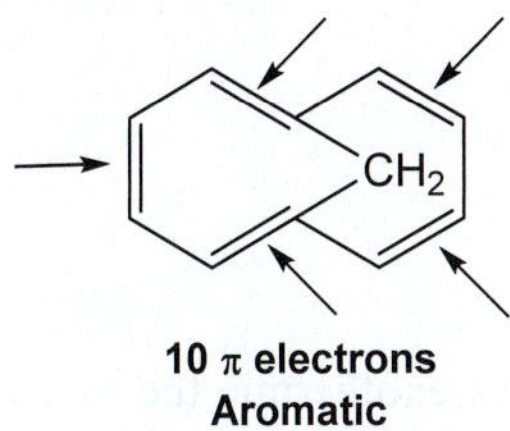

(d) Nonaromatic: Although conjugated with six total π electrons, it is not cyclic.

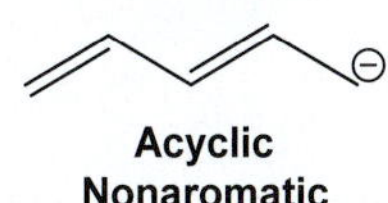

(e) Antiaromatic: One of the lone pairs of electrons on the O atom can participate in resonance, bringing the total number of π electrons to four, which is an anti-Hückel number. And the π electrons are completely conjugated around the ring. The ring is not large enough to bend out of plane to become nonaromatic.

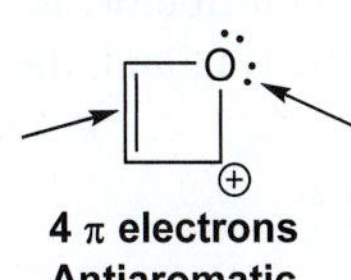

(f) Nonaromatic: There are two total π electrons from the double bond, but the sp^3-hybridized N prevents the π system from being completely conjugated around a ring.

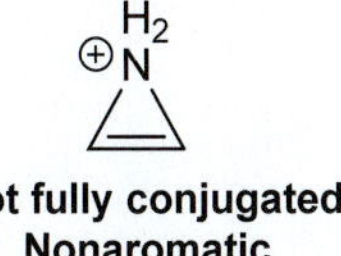

(g) Nonaromatic: Two of the lone pairs of electrons on the O atom can participate in resonance, bringing the total number of π electrons to six; however, the sp^3-hybridized C atom prevents those electrons from being completely conjugated around the ring.

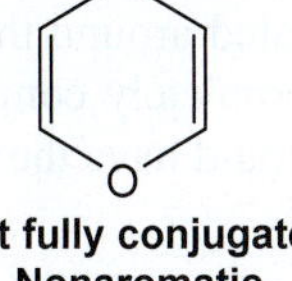

Problem 14.42

Think

What is the formula for each compound? What is the product of the hydrogenation reaction? Which reactant is most stable? Which is least stable? How does aromaticity factor in?

Solve

In all cases, the product of complete hydrogenation is the following bicyclo compound.

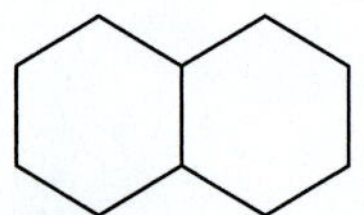

Bicyclo[3.3.0]decane

So the more stable the reactant, the smaller the heat of hydrogenation (the less exothermic the reaction). **A** is aromatic and, therefore, the most stable, and it will have the smallest heat of combustion. The other three compounds are nonaromatic. **B** has three conjugated double bonds, **C** has two conjugated double bonds, and **D** has no conjugated double bonds. The order of increasing heat of hydrogenation is **A** < **B** < **C** < **D**.

Problem 14.43

Think

When the double bond is outside the ring, are its electrons considered part of the conjugated π system? Consider another resonance structure for each species.

Solve

E has a resonance contributor that has a cyclopentidenyl cation. This cation has four π electrons completely conjugated about the ring and, therefore, is antiaromatic. **F** has a resonance contributor that has a cycloheptatrienyl cation. This cation has six π electrons completely conjugated about the ring and, therefore, is aromatic. Consequently, **F** has the more stable π system.

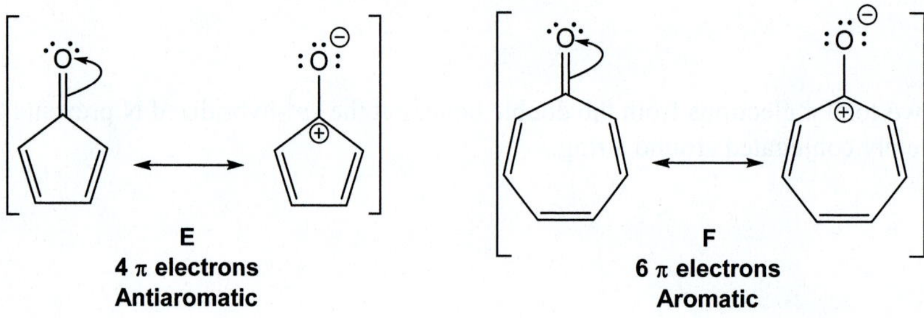

Problem 14.44

Think

How is the measure of the heat of hydrogenation related to the stability of a compound's π system? Which compound(s) is/are aromatic? Does aromaticity increase or decrease a π system's stability?

Solve

Molecule **G** is aromatic because it has 10 π electrons (five pairs) completely conjugated around the outside of the ring structure. Molecule **H** is nonaromatic because the 10 π electrons are not completely conjugated—the sp^3-hybridized C atom prevents complete conjugation. So the molecule on the right should have the less stable π system, which would give it the greater heat of hydrogenation.

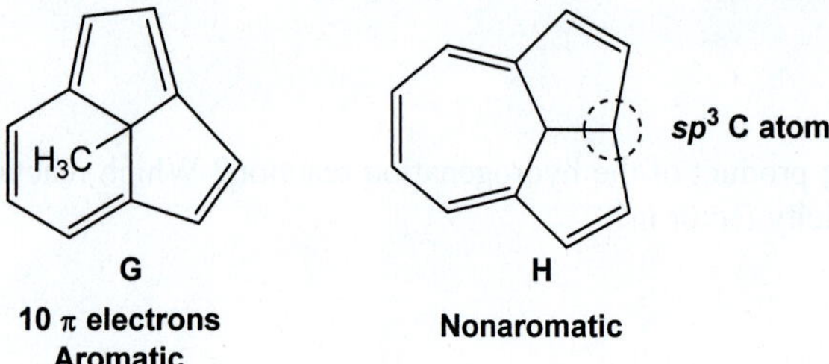

Problem 14.45

Think

How many total π electrons are present in coronene? How many of these π electrons are on the periphery? Is there complete conjugation around the outside of the molecule? Around the inside of the molecule?

Solve

There are 24 total π electrons (12 pairs). According to Hückel's rule, this should be antiaromatic. The observation that it is aromatic, however, is reconciled by the fact that there are two separate aromatic π systems. There is a central benzene-type ring with six π electrons, highlighted in a gray square below. There is also an aromatic outer ring of 18 electrons (nine pairs).

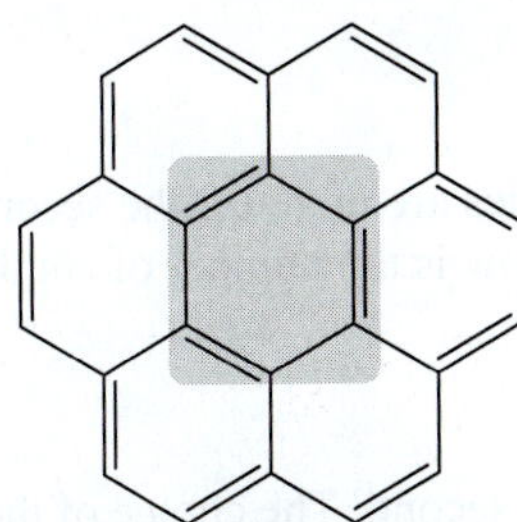

Coronene
24 total π electrons
18 π electrons on periphery (aromatic)
6 π electrons inside (aromatic)

Problem 14.46

Think

How many σ bonds does the N atom possess? Are any lone pairs of electrons present? What is the hybridization of the N atom? Is molecule **I** fully conjugated?

Solve

The N atom has four bonds to it and is, therefore, sp^3 hybridized. Thus, it has no *p* orbital to contribute, and there is no complete conjugation of the π system completely around the ring.

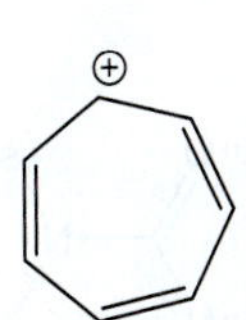

The tropylium ion
6 π electrons
Aromatic

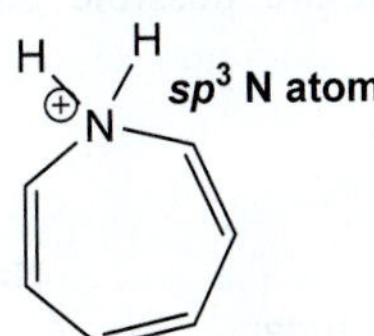

I
Nonaromatic

Problem 14.47

Think

Draw the structure of the $C_8H_8^{2-}$ dianion. How many π electrons are present? What must be true of the shape of aromatic compounds?

Solve

(a) The anion is easy to make because the reactant, cyclooctatetraene, is nonaromatic (owing to nonplanarity) but becomes aromatic when two more electrons are added. That's because the two additional electrons must go in the π system, giving it a total of 10 π electrons (a Hückel number).

(b) The dianion product should be planar to make it aromatic.

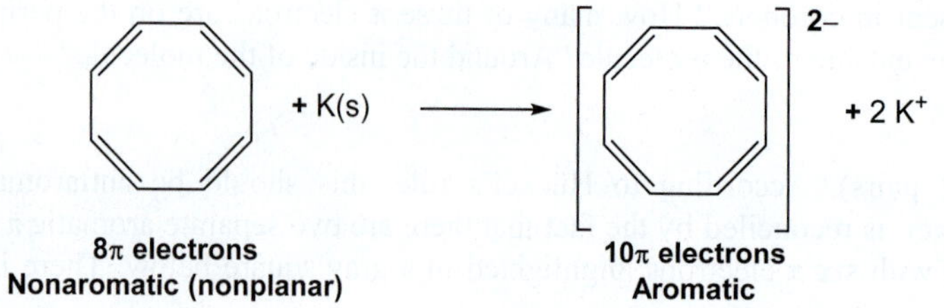

+ K(s) → [ ]$^{2-}$ + 2 K^+

8π electrons
Nonaromatic (nonplanar)

10π electrons
Aromatic

Section 14.11 The Organic Chemistry of Biomolecules: Aromaticity and DNA

Problem 14.48

Think

How many possibilities are there for the first nucleotide? How many possibilities are there for the second? Does the choice for the first nucleotide affect the possible choices for the second? How is the number of combinations possible affected by each additional nucleotide?

Solve

There are four possibilities for the first nucleotide and four possibilities for the second. The choice of the second nucleotide is not impacted by the choice of the first, so the number of possible sequences is multiplied by four with each added nucleotide. A sequence of 200 nucleotides with four choices per position is calculated using the formula $4^{\#\text{ of nucleotides}}$ and would be as follows: $4^{200} = 2.58 \times 10^{120}$. This number is 40 orders of magnitude larger than the number of molecules in the universe—quite a bit larger.

Problem 14.49

Think

Where are the hydrogen bond donors in guanine and adenine? Where are the hydrogen bond acceptors in guanine and adenine? How do these hydrogen bond interactions differ from the interactions of guanine and cytosine in Figure 14-31?

Solve

When guanine pairs with adenine, two hydrogen bonds are possible. This forms a less stable interaction compared to the three hydrogen-bonding interactions that are possible guanine pairs with cytosine. Both interactions are shown below for comparison.

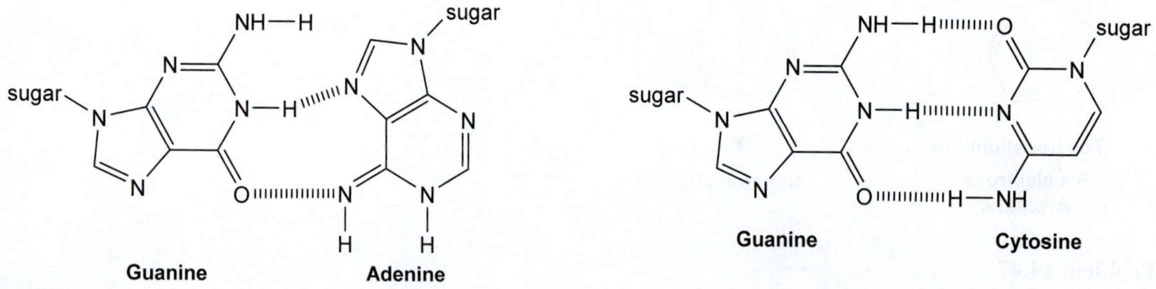

Guanine Adenine

Guanine Cytosine

Problem 14.50

Think

What nitrogenous base is complementary to A? To C? To G? To T?

Solve

In a complementary DNA strand, A pairs with T and C pairs with G.

Initial strand	CGGATACATTTGC
Complementary strand	GCCTATGTAAACG

Problem 14.51

Think

What type of geometry does each nitrogenous base have, which allows for highly efficient stacking of the base pairs on the inside of DNA's double helix? Is there a portion of doxorubicin that has the same type of geometry, which would allow it to intercalate between bases?

Solve

Doxorubicin intercalates into the DNA double helix via the planar portion of the doxorubicin drug (anthracene core, shown below). Being planar, that portion of doxorubicin can easily fit between two planar DNA bases and can be stabilized by p stacking

Aromatic

Doxorubicin

The figure below shows bound doxorubicin (gray = planar portion, black = rest of molecule) and how the double helix does not re-form properly after the strands are separated.

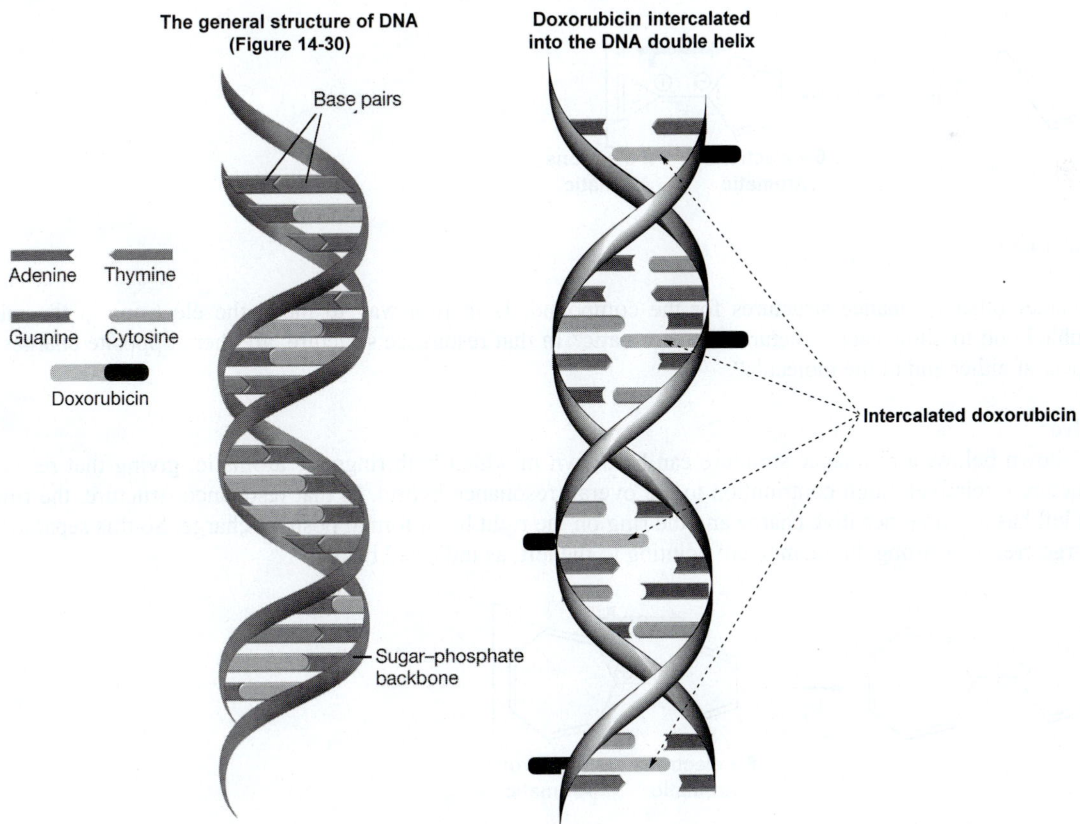

Integrated Problems

Problem 14.52

Think

Are the *p* orbitals in the bicyclic ring system parallel? If the *p* orbitals are not parallel, can they be conjugated?

Solve

The bicyclic ring system does not allow the *p* orbitals on the sp^2-hybridized carbons to be parallel. Thus, there can be no significant overlap to form conjugated π MOs, as shown below.

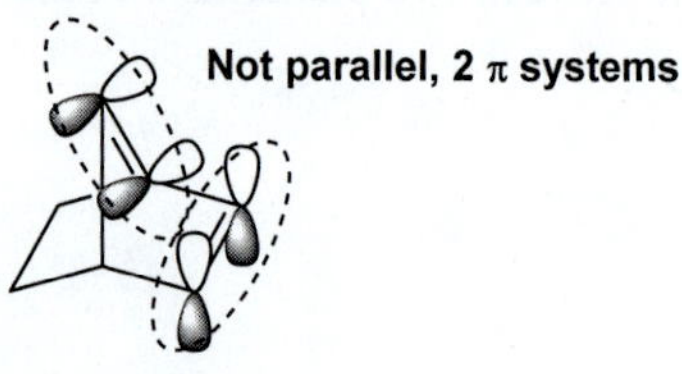

Problem 14.53

Think

Consider other resonance structures for the compound. Is there a way to move the electrons in the middle double bond to allow each structure to be aromatic? In that resonance structure, are there opposite charges that appear at either end of the molecule?

Solve

Another resonance contributor can be drawn for this molecule. Even though it has significant charge separation (destabilizing), both rings become aromatic, with six π electrons and two π electrons, respectively (each a Hückel number). The aromaticity is highly stabilizing, so that resonance structure has a substantial contribution to the resonance hybrid. This separation of charge is consistent with the dipole shown, as the red (on the left side of the molecule) in the figure in the textbook indicates an excess of negative charge and the blue (on the right side of the molecule) indicates an excess of positive charge.

6 π electrons Aromatic | 2 π electrons Aromatic

Problem 14.54

Think

Consider other resonance structures for the compound. Is there a way to move the electrons in the middle double bond to allow each structure to be aromatic? In that resonance structure, are there opposite charges that appear at either end of the molecule?

Solve

As shown below, a resonance structure can be drawn in which both rings are aromatic, giving that resonance structure a relatively high contribution to the overall resonance hybrid. In that resonance structure, the ring on the left has a formal negative charge and the ring on the right has a formal positive charge. So this separation of charge creates a strong dipole moment pointing to the left, as indicated below.

6 π electrons Aromatic | 6 π electrons Aromatic

Problem 14.55

Think

Consider other resonance structures for the compounds. Is there a way to move the electrons to form a resonance structure that is aromatic? In that resonance structure, where are the formal charges located?

Solve

The dipole moment of the species **K** is primarily due to the highly polar C=O bond. This is true of the compound **J** as well, but in addition, there is a relatively strong resonance contributor that adds to that dipole moment. As shown below, that resonance contributor has a cyclopropenyl cation, which has two electrons completely conjugated about the ring and, therefore, is aromatic. This contributor places an additional positive charge on the ring and an additional negative charge on O. Species **K** also has a similar resonance contributor, but that structure is not aromatic, so its contribution is not substantial.

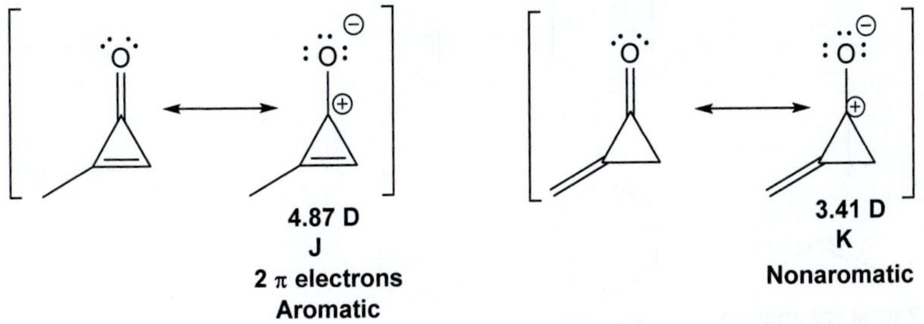

4.87 D
J
2 π electrons
Aromatic

3.41 D
K
Nonaromatic

Problem 14.56

Think

For an organic compound to be water soluble, what intermolecular forces could be present between the compound and water? With the structure shown, is this possible? Is Br a good leaving group? What results after a heterolysis step?

Solve

For the compound with several C atoms to be soluble in water, there needs to be extensive hydrogen-bonding and/or significant ion–dipole interactions. Neither of these interactions is possible with the compounds as given. However, the C–Br bond in the compound on the right dissociates easily because the product cation is aromatic, and the positive charge is delocalized around the ring. When this takes place in water, that cation and Br^- are formed, which are well solvated by water. By contrast, the C–Br bond in the compound on the left does *not* dissociate easily in water, because the product has a localized positive charge on the C atom.

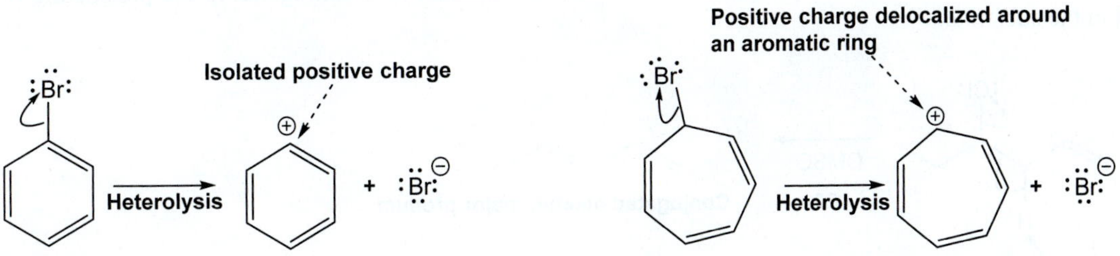

Isolated positive charge

Heterolysis

Positive charge delocalized around an aromatic ring

Heterolysis

Problem 14.57

Think

Does a lower pK_a indicate a stronger or weaker acid? After the acid donates its proton, what is left on the carbon? These acids are all uncharged, so should acid strength be governed by the stabilities of the acids or of their conjugate bases? Are the conjugate bases aromatic, nonaromatic, or antiaromatic?

Solve

All of the acids given are uncharged, so the difference in acid strength is due primarily to differences in conjugate base stability once the H leaves as a proton. All acidic protons are on an sp^3-hybridized C atom. Different stabilities of the anions are due largely to resonance delocalization of the negative charge. The strongest acid is cyclopentadiene (**O**), because the conjugate base is an aromatic anion. Cyclopropene (**P**) is the weakest acid, because the conjugate base is an antiaromatic anion. The stabilities of the other conjugate bases decrease with fewer resonance structures that delocalize the negative charge.

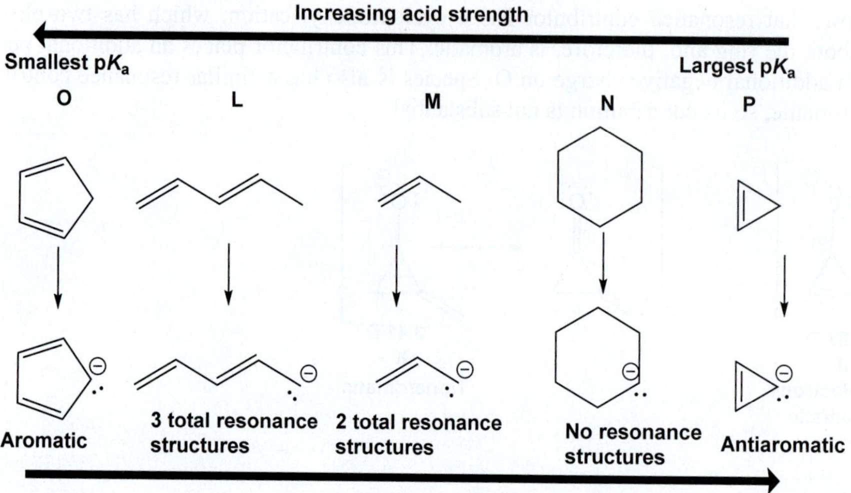

Problem 14.58

Think

Consider an $S_N1/S_N2/E1/E2$ competition. Is there a sufficient leaving group? On what type of carbon is the leaving group located? Is the attacking species a strong or weak nucleophile? A strong or weak base? Is the solvent protic or aprotic? What reactions does heat favor? What is the major mechanism? How many adjacent H atoms are present? How many alkene products could result? Which alkene product is most stable?

Solve

A polar aprotic solvent with a strong nucleophile and strong base will favor the S_N2 or E2 mechanisms, respectively. In a heated solution, elimination is favored, so CH_3O^- acts as a base to remove a proton. The double bond that forms in the major product is more stable because it is conjugated to the preexisting double bond in the substrate.

:Cl:

H

CH_3O^-

DMSO

E2

Conjugated alkene: major product

Problem 14.59

Think

Is there a significant difference in stability of the two charged acids? Is there a significant difference in stability of the uncharged conjugate bases after the acids donate their protons? How do these stability differences affect acid strength?

Solve

The two charged acids are very similar in stability because, in both cases, the positive charge is located on N, neither charge is delocalized by resonance, and there are similar inductive effects for both. **Q** is a stronger acid, however, because its conjugate base is an uncharged, aromatic molecule, which heavily stabilizes the conjugate base. The conjugate base of the molecule on the right is uncharged, too, but not aromatic. The deprotonation reaction is shown below.

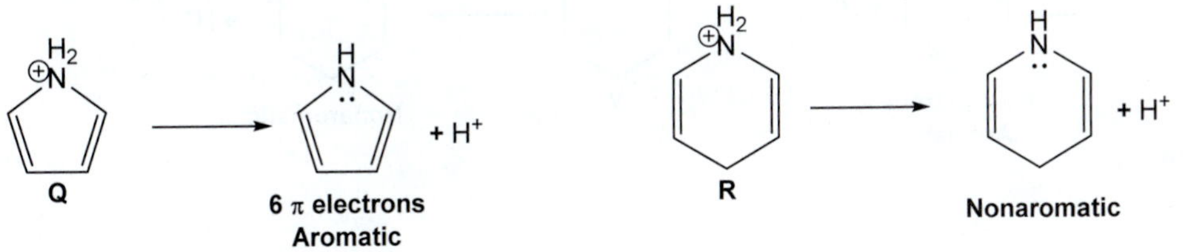

Q → 6 π electrons Aromatic + H^+

R → Nonaromatic + H^+

Problem 14.60

Think

Is another resonance structure possible? Does this resonance structure increase the electron density at one O atom over the other? Is that resonance structure aromatic, antiaromatic, or nonaromatic? How does this correspond to the importance of that resonance structure?

Solve

The carbonyl O is the more basic oxygen. As shown below, there is a relatively strong resonance contributor that places a negative charge on the carbonyl oxygen, making it relatively strongly basic. That resonance structure has a strong contribution because its ring is aromatic, being completely conjugated and having six π electrons.

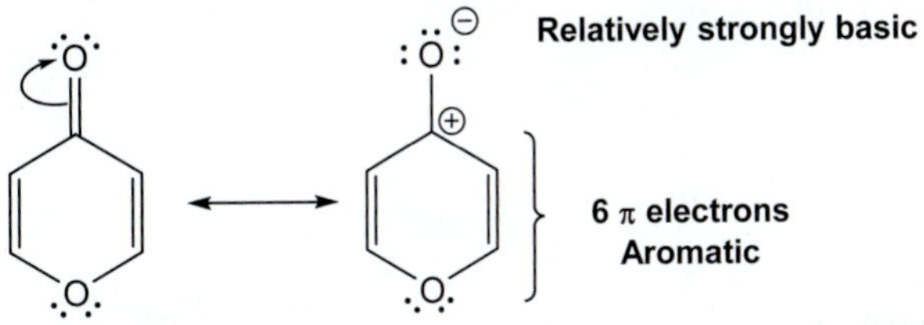

Relatively strongly basic

6 π electrons
Aromatic

Problem 14.61

Think

Is there a significant difference in stability of the two uncharged acids? Is there a significant difference in stability of the charged conjugate bases after the acids donate their protons? How do these stability differences affect acid strength?

Solve

(a) Compound **T** should be more acidic because its resulting enolate anion is more stable. The enolate anion of the ketone on the left, **S**, is antiaromatic owing to the four π electrons occupying the completely conjugated ring.

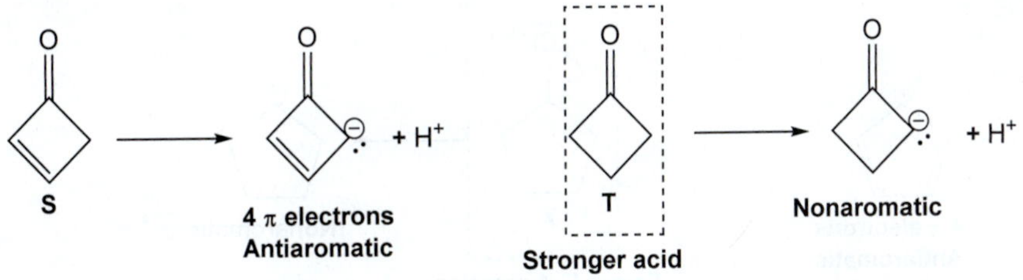

S → 4 π electrons Antiaromatic + H^+

T (Stronger acid) → Nonaromatic + H^+

(b) Compound **U** is more acidic because its resulting enolate anion is aromatic, having six π electrons conjugated around the ring, whereas the enolate anion of the ketone on the right is nonaromatic.

U — Stronger acid → 6 π electrons, Aromatic + H^+

V → Nonaromatic + H^+

Problem 14.62

Think

For a species to act as a nucleophile, what must one of its lone pairs be able to do? If the electron pair is part of the aromatic π system, what does that do to the availability of those electrons? What does that do to the nucleophile strength?

Solve

Pyridine is a stronger nucleophile than pyrrole owing to the orbital in which each lone pair is located. Pyridine's lone pair is located in an sp^2 orbital and, therefore, is not part of the conjugated π system. The lone pair on pyrrole is located in an unhybridized *p* orbital and is part of the conjugated π system. Each compound has a lone pair on N that can be used to form a bond, thus enabling it to act as a nucleophile. However, the lone pair on pyrrole is required to make the compound aromatic, so that lone pair is more stabilized, making it difficult to be usable to make bonds. The lone pair in pyridine, however, is not part of the aromatic π system, leaving it more available for bonding.

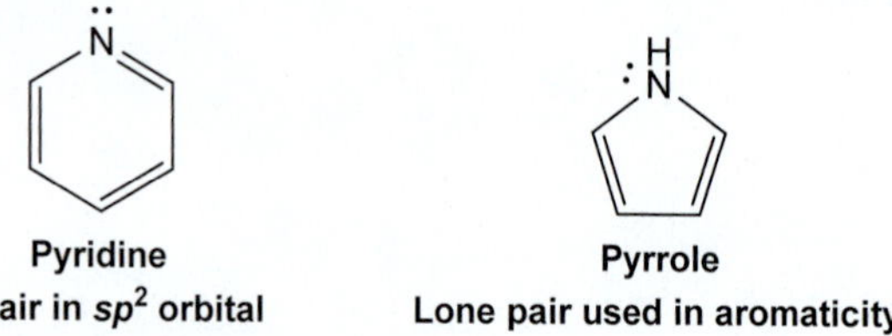

Pyridine
Lone pair in sp^2 orbital

Pyrrole
Lone pair used in aromaticity

Problem 14.63

Think

What is the rate-limiting step in an S_N1 mechanism? What intermediate forms? Is the cation aromatic, antiaromatic, or nonaromatic? How does the stability of the intermediate affect the rate of the reaction?

Solve

(a) Molecule **X** will undergo S_N1 reactions faster. In an S_N1 reaction, the departure of the leaving group is the rate-determining step. These steps are shown for the two substrates below. **W** produces an antiaromatic carbocation, which is less stable than the nonaromatic carbocation produced from the substrate **X**.

W → 4 π electrons, Antiaromatic

X — Faster S_N1 reaction → Nonaromatic

(b) Molecule **Y** will undergo S_N1 reactions faster. As shown below, its carbocation is aromatic and, therefore, more stable than the carbocation that is formed from molecule **Z**.

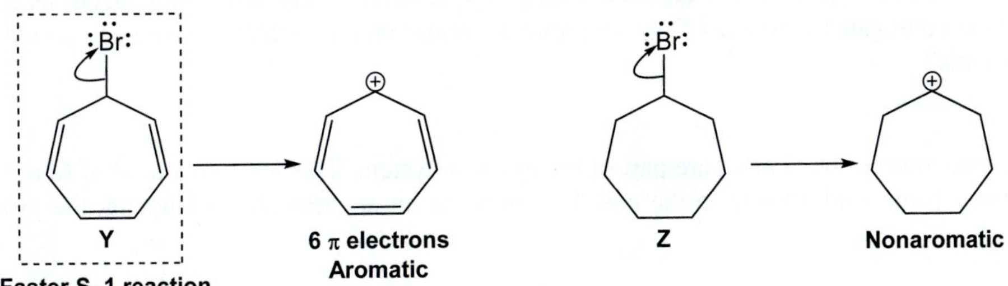

Problem 14.64

Think

For a dehydration reaction to occur on these compounds, what is the leaving group? What is the final product? Is the product compound aromatic, antiaromatic, or nonaromatic?

Solve

Dehydration occurs faster with the molecule **B** because the product is aromatic, whereas the reactant is nonaromatic.

OH

$-\ H_2O$

A

Nonaromatic

OH

$-\ H_2O$

B

Faster E1 dehydration reaction

6 π electrons
Aromatic

Problem 14.65

Think

Once the two carbocation intermediates are drawn, label the type of carbocation intermediate. Is either of these intermediates capable of resonance? If so, how does that affect the stability of the cation?

Solve

(a) The two possible carbocation intermediates are shown below. The one on top is from addition of H^+ to the terminal C, whereas the one on the bottom is from addition of H^+ to the central C.

$H_2C{=}C{=}CH_2$ + HBr

Propadiene

$H_2C{=}\overset{\oplus}{C}{-}CH_3$

Vinylic C^+

$H_2C{=}\underset{H}{C}{-}\overset{\oplus}{C}H_2$

Allylic C^+

Carbocation intermediates

(b) In the vinylic carbocation, the carbon on the far right is sp^3 hybridized and is an electron-donating group (EDG). This stabilizes the carbocation. At first glance, the allylic carbocation might appear to be more stable owing to resonance delocalization of the positive charge. However, even though all three C atoms are sp^2 hybridized, the CH_2 on the far right is not in the same plane as the CH_2 on the left. Therefore, the *p* AOs are not all conjugated, and charge cannot be resonance delocalized (see Section 14.2a for a discussion regarding conjugation). As a result, the vinylic carbocation that is produced on protonation is the more stable carbocation.

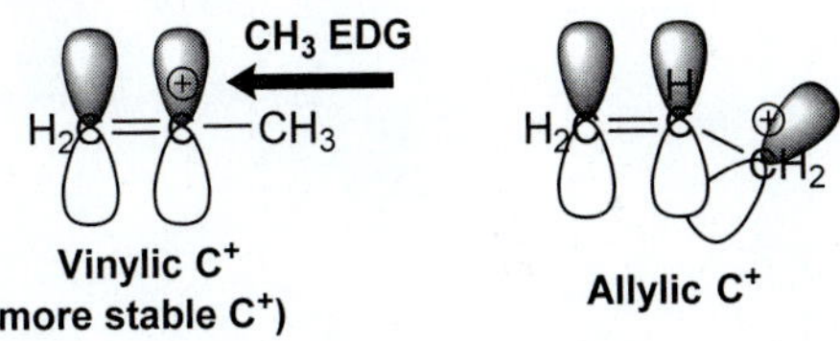

Problem 14.66

Think

In the product from the attempted alkyne addition with peroxyacid (RCO_3H), how many electrons from the O atom are part of the conjugated π system? Does an aromatic, nonaromatic, or antiaromatic compound result? Is this stable or unstable?

Solve

Two of the four electrons on the O atom are part of the cyclic π system. This leads to a total of four π electrons and an antiaromatic compound. Owing to the unstable nature of an antiaromatic compound, the product does not form.

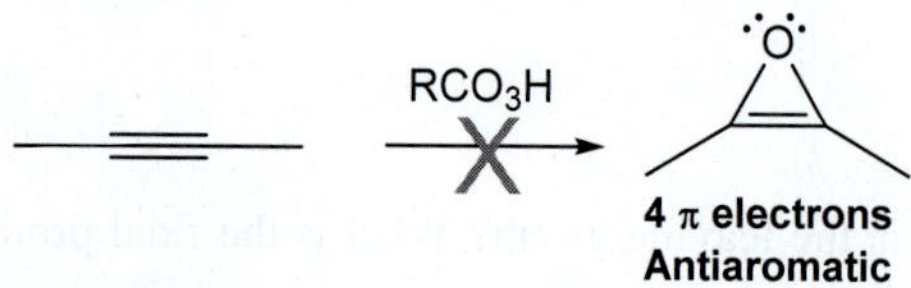

CHAPTER 15 | Structure Determination 1: Ultraviolet–Visible and Infrared Spectroscopies

Your Turn Exercises

Your Turn 15.1

Think

Consult Equation 15-1. With a higher %*T*, does more light or less light passing through the sample arrive at the detector? How does %*T* relate to absorbance?

Solve

With a higher %*T*, more light arrives at the detector, meaning that less light is absorbed by the sample: $\%T = 20\%$ represents more light being absorbed than $\%T = 40\%$. Absorbance, on the other hand, increases with the amount of light absorbed by the sample, so $A = 0.750$ represents more light being absorbed than $A = 0.500$.

Your Turn 15.2

Think

What variable in Equation 15-3 represents molar absorptivity? How is that variable related to absorbance, A?

Solve

Molar absorptivity is represented by ε and is directly proportional to absorbance, A. Therefore, if molar absorptivity increases by a factor of 3, absorbance also increases by a factor of 3.

Your Turn 15.3

Think

In Equation 15-5, which variable represents the wavelength of light? According to the equation, is that variable directly proportional or inversely proportional to the energy of the photon?

Solve

The wavelength of light is represented by λ_{photon}, which is inversely proportional to E_{photon}. So E_{photon} increases as λ_{photon} decreases. Therefore, a 375-nm photon has more energy than a 530-nm photon.

Your Turn 15.4

Think

According to Figure 15-4, which MOs are occupied by electrons? Which MOs are empty? How many electrons can occupy a single MO?

Solve

According to Figure 15-4, the bottom three MOs are occupied with two electrons each. The top three MOs are empty. Up to two electrons can occupy the same MO. Therefore, an electron transition can take place from a lower-energy MO to a higher-energy MO if the lower-energy MO has at least one electron and the higher-energy MO has either zero electrons or one electron. All of the transitions that fit these criteria are shown below.

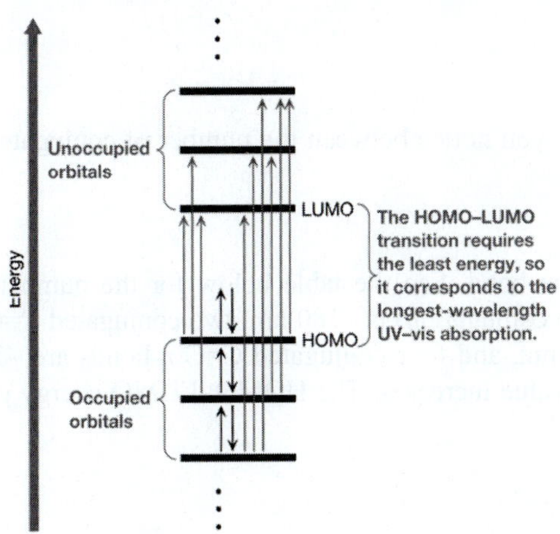

Your Turn 15.5

Think

In the figure, how is the required photon energy represented? How do those representations of photon energies you drew for Your Turn 15.4 compare to the representation for the HOMO–LUMO transition?

Solve

The required photon energies are represented by the lengths of the arrows in Your Turn 15.4. The longer the arrow, the greater the energy difference between the MOs involved in the electron transition and the greater the photon energy that is required. Of all the arrows drawn to represent these transitions, the one representing the HOMO–LUMO transition is the shortest, meaning that the HOMO–LUMO transition corresponds to the smallest-energy absorbed photon.

Your Turn 15.6

Think

Which orbital in Figure 15.5a lost an electron? Which orbital in Figure 15.5b gained an electron? Relative to the *p* AO energies, are bonding MOs lower energy, higher energy, or the same energy? Antibonding MOs? Nonbonding MOs?

Solve

The electron is circled below, and the transition is indicated by the curved arrow. The lowest two π MOs are bonding because they are below the energy of the isolated *p* orbitals (the horizontal dashed line), whereas the highest two π MOs are antibonding because they are above that line.

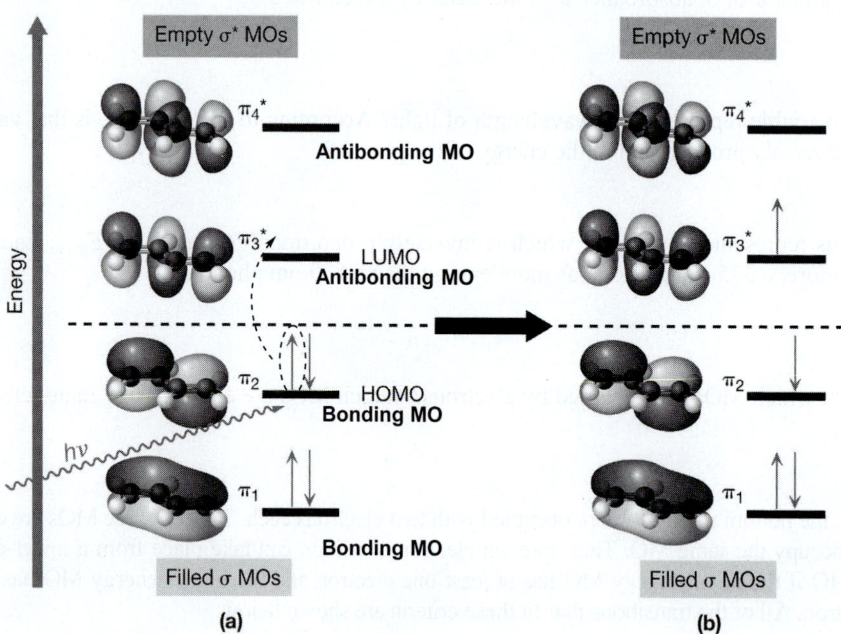

Your Turn 15.7

Think

What is required for π bonds to be conjugated? What trend do you notice between the number of conjugated π bonds and the wavelength?

Solve

Double bonds are conjugated when they are separated by one bond. See the table below for the number of conjugated double bonds in each species. One C=C π bond (no conjugation) is ~180 nm, two conjugated C=C π bonds are ~225 nm, three conjugated C=C π bonds are ~275 nm, and four conjugated C=C π bonds are ~290 nm. As the number of conjugated π bonds increases, the λ_{max} value increases. The HOMO–LUMO energy gap, therefore, decreases. See the table on the next page.

Compound	λ_{max} (nm)	Conjugated π Bonds	Compound	λ_{max} (nm)	Conjugated π Bonds
Alkanes and cycloalkanes	<150	0	*cis*-Penta-1,3-diene	223	2
H H C=C H H Ethene	161	0	*trans*-Penta-1,3-diene	223.5	2
Hex-1-ene	177	0	2-Methylbuta-1,3-diene (isoprene)	224	2
Penta-1,4-diene	178	0	Cyclopentadiene	239	2
Cyclohexene	182	0	Cyclohexa-1,3-diene	256	2
Hex-1-yne	185	0	Hexa-1,3,5-triene	274	3
Buta-1,3-diene	217	2	Octa-1,3,5,7-tetraene	290	4
β-Carotene				455	11

Your Turn 15.8

Think

How does the rate of appearance of the peak at λ_{max} = 244 nm relate to the rate of product formation? If doubling the concentration of the base results in no change in the spectrum, is the base part of the rate law? If the concentration of the base species has no effect on the reaction rate, is the mechanism unimolecular (E1) or bimolecular (E2)?

Solve

The styrene product has a λ_{max} of 244 nm, so as the reaction progresses, the absorbance at that wavelength increases. Therefore, if the concentration of the base, $NaOCH_3$, is doubled and the absorbance at 244 nm that is measured over an initial time period does not change, then the amount of product formed during that time does not change. Thus, the base is not part of the rate law, and the rate of appearance of the product is directly proportional only to the concentration of the organic substrate and not the base. The reaction must therefore proceed by the E1 mechanism, not E2.

Your Turn 15.9

Think

Locate another peak in the spectrum. How do you determine the frequency value for the photon that is absorbed? How is the photon frequency of an absorbed photon related to the type of vibration responsible for absorption of that photon? What are the units?

Solve

The peak at 1320 cm^{-1} is labeled on the figure below. However, several other peaks could just as equally have been selected. The frequency value of the absorbed photon is the value of the peak at the *x* axis, and the units are wavenumbers, cm^{-1}. These are the same frequencies for the type of vibration responsible for photon absorption.

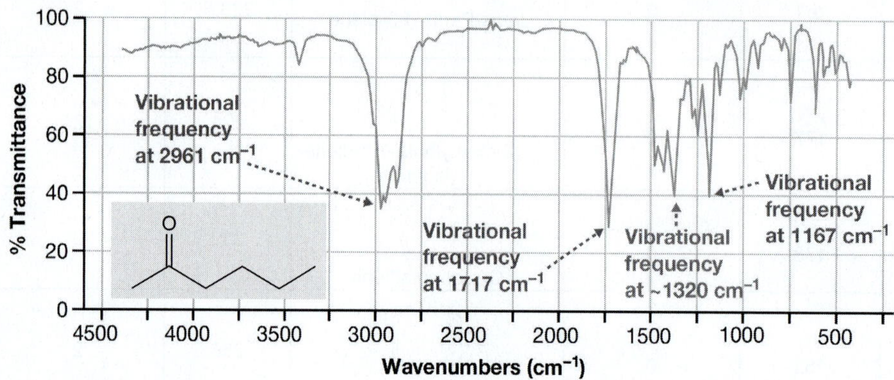

Your Turn 15.10

Think

Around what frequency does the C=O carbonyl peak occur? Is the peak strong or weak? Broad or narrow?

Solve

The C=O peak generally appears at ~1720 cm^{-1} as a strong, sharp peak. The benzaldehyde C=O peak specifically occurs at ~1700 cm^{-1}.

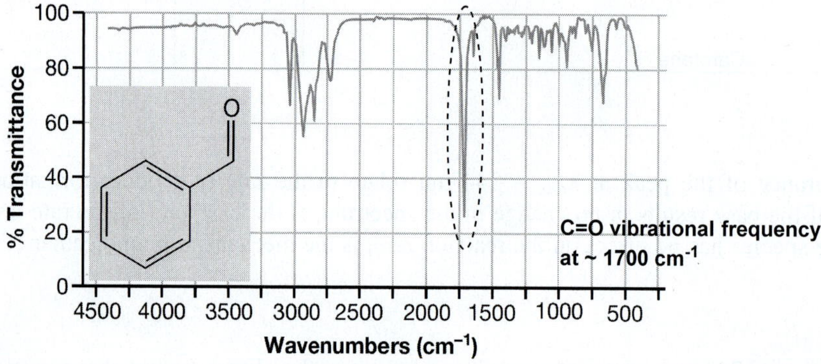

Your Turn 15.11

Think

What wavenumber range does the fingerprint region encompass? Which regions in Figure 15-14 are included in that range?

Solve

The fingerprint region encompasses the frequencies below ~1400 cm^{-1}. In Figure 15-14, the two regions that are in this range are the single-bond stretches and the bending modes (the blue and purple regions).

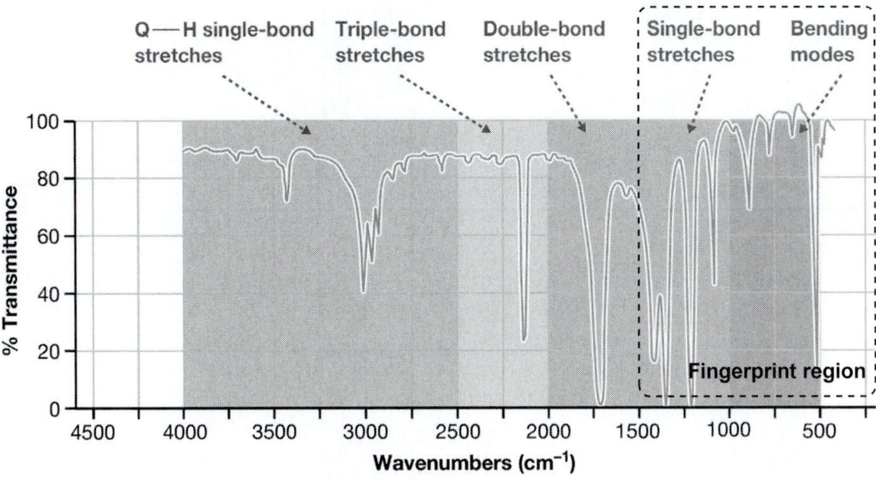

Your Turn 15.12

Think

Consult the IR spectrum in Figure 15-16a for hept-1-ene and Table 15-2 to help you determine where a C=C stretch could appear. Do you see a peak arising from a C=C stretch for hept-3-ene? Why or why not?

Solve

The general region for a C=C stretch is 1620–1680 cm^{-1}. Unlike the IR spectrum for hept-1-ene, hept-3-ene contains no peak in this region. This is due to the larger extent of symmetry about the C=C bond in hept-3-ene. (The annotation at the left of the spectrum corresponds to Your Turn 15.14.)

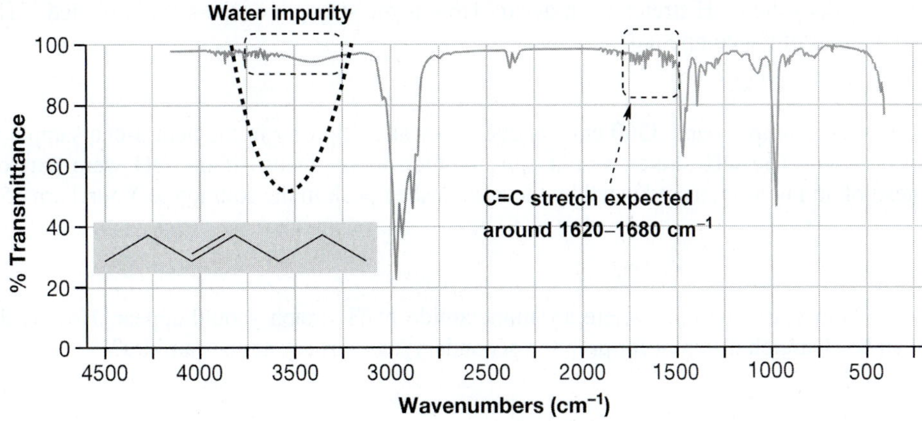

Your Turn 15.13

Think

In what frequency range does the O–H stretch peak occur? Is the peak strong or weak? Broad or narrow?

Solve

The H–O stretch bands are intense, broad, and centered around 3300 cm^{-1} for RO–H (alcohols) and 3000 cm^{-1} for ROO–H (carboxylic acid). The OH stretches are circled below. Notice that in a carboxylic acid, the OH stretch is shifted to a lower frequency and is much broader than in an alcohol; therefore, it overlaps the alkane C–H stretches (2800–3000 cm^{-1}). See the figures on the next page.

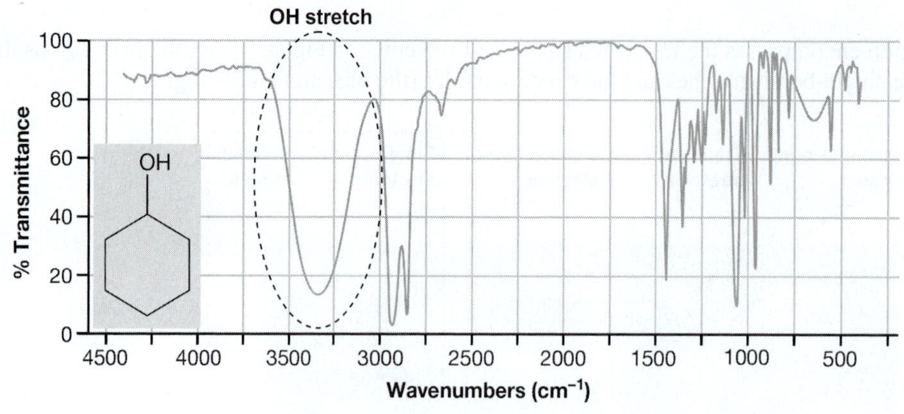

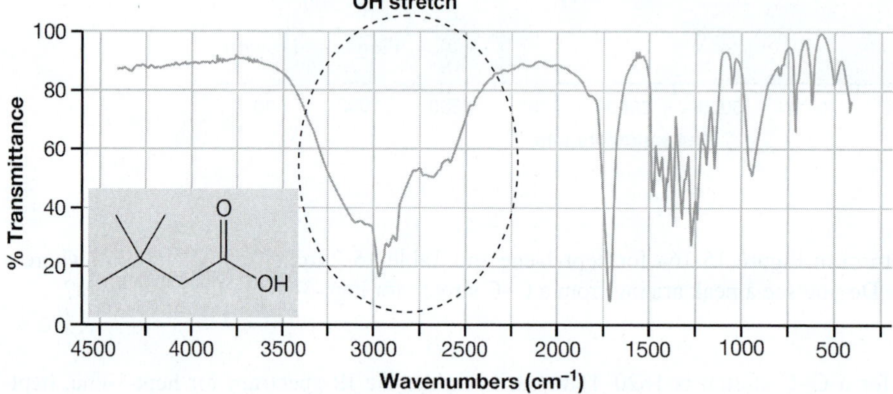

Your Turn 15.14

Think

In what frequency range does the O–H stretch peak occur? How is the strength of the signal affected if H_2O is just an impurity and not part of the compound?

Solve

In Figure 15-16b, the small bump around 3300 cm^{-1} is due to a water impurity in the hept-3-ene sample when the IR spectrum was taken. The OH stretch would appear much more intense if the OH were part of the molecule itself instead of an impurity, as indicated by the bold dashed peak in the solution to Your Turn 15.12.

Your Turn 15.15

Think

Consult Table 15-2 to help you determine where an amine/amide N–H stretch should appear. How is the IR absorption band of an N–H stretch different for primary, secondary, and tertiary amines/amides?

Solve

The N–H stretching modes are the moderately intense, moderately broad peaks near 3300 cm^{-1}. There are two in the first spectrum **(a)** (primary amide), one in the second spectrum **(b)** (secondary amide), and none in the third spectrum **(c)**. See the figures on the next page.

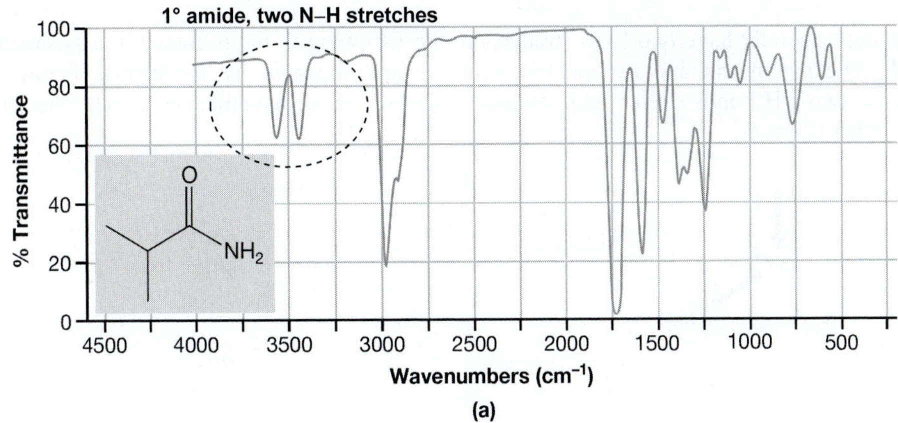

(a)

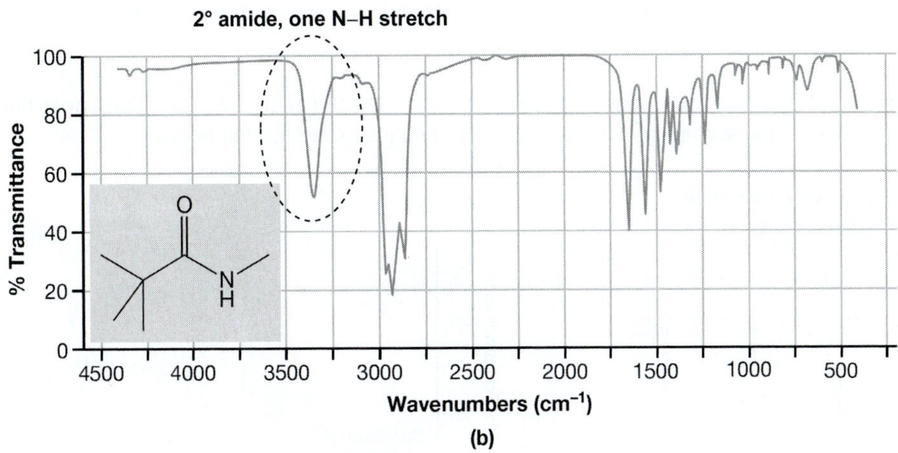

(b)

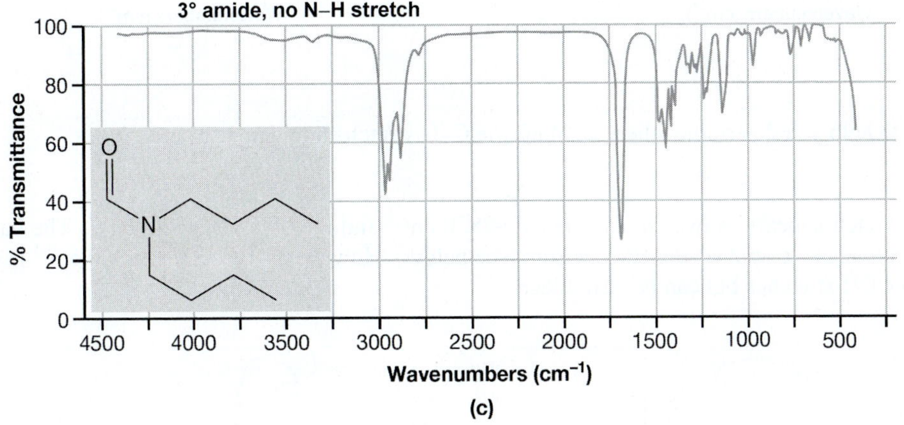

(c)

Your Turn 15.16

Think

Consult Figure 15-11 to help you depict the symmetric N–H and the asymmetric N–H stretch. In a symmetric stretch, should the two bonds stretch at the same time or different times? In an asymmetric stretch?

Solve

Primary amines/amides generally have two N–H stretches in the IR owing to the presence of a symmetric and asymmetric stretch. The symmetric and asymmetric stretches are indicated by the arrows below. In the symmetric stretch, the two NH bonds stretch and compress together. In the asymmetric stretch, one NH bond stretches, while the other shortens.

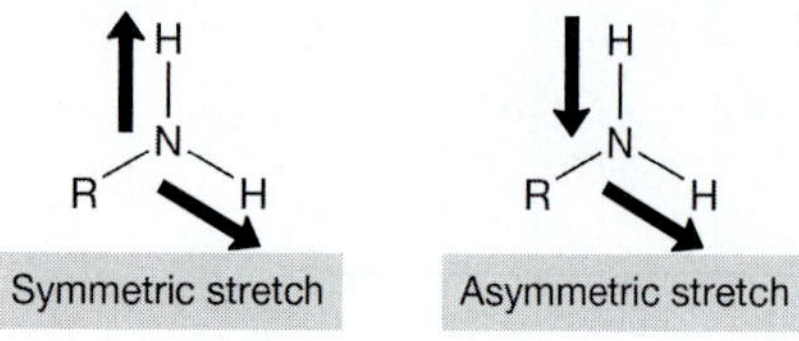

Your Turn 15.17

Think

Consult Figure 15-14 and Table 15-2 to help you determine where a C≡C stretch could appear.

Solve

Triple-bond stretching modes appear between 2000 and 2500 cm^{-1}. In the first spectrum, that triple bond has moderate intensity, whereas in the second spectrum, it is weak. The peaks are circled below.

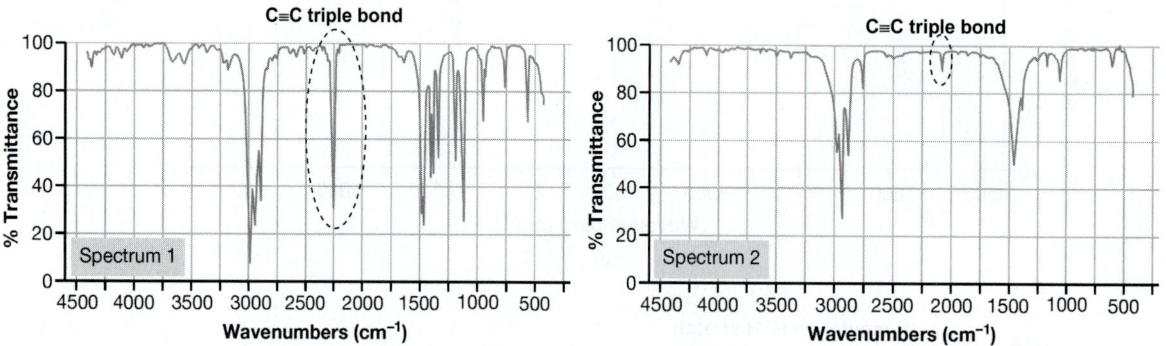

Your Turn 15.18

Think

Consult Table 15-2 to help you determine where an aldehyde C–H stretch should appear.

Solve

The aldehyde C–H stretch appears as two bands—one at ~2820 cm^{-1} and the other at ~2720 cm^{-1}. The band at ~2720 cm^{-1} is generally easy to spot on the IR spectrum (see below). However, the band at ~2820 cm^{-1} is often masked by the alkane CH stretches but can be a shoulder.

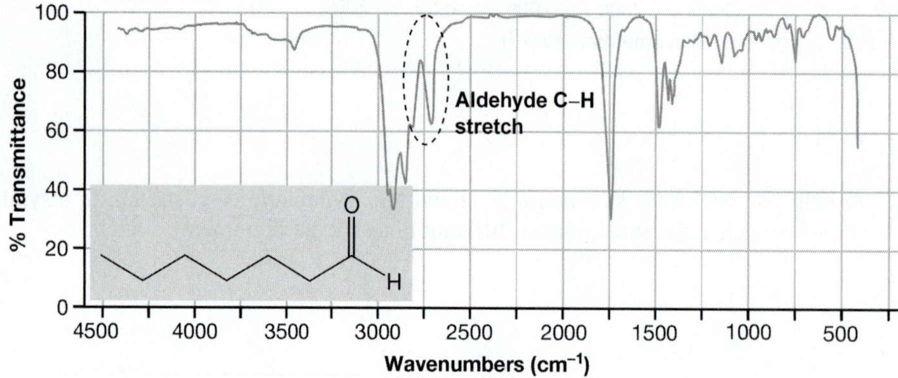

Your Turn 15.19

Think

Consult Table 15-2 to help you determine where alkane *sp*^{3} C–H and alkene *sp*^{2} C–H stretches appear. Do the C–H stretches present in the spectra correspond to the type of C–H bonds in the molecule?

Solve

From Table 15-2, alkane *sp*^{3} C–H stretches occur between 3000 and 2800 cm^{-1} and vary in intensity. Alkene *sp*^{2} C–H stretches occur between 3100 and 3000 cm^{-1} and are generally weak. A dashed line is drawn on each IR spectrum here to assist in visualizing the dividing line between alkane and alkene C–H stretches.

(a) Alkane C–H stretches are intense and occur at 3000–2800 cm^{-1}.

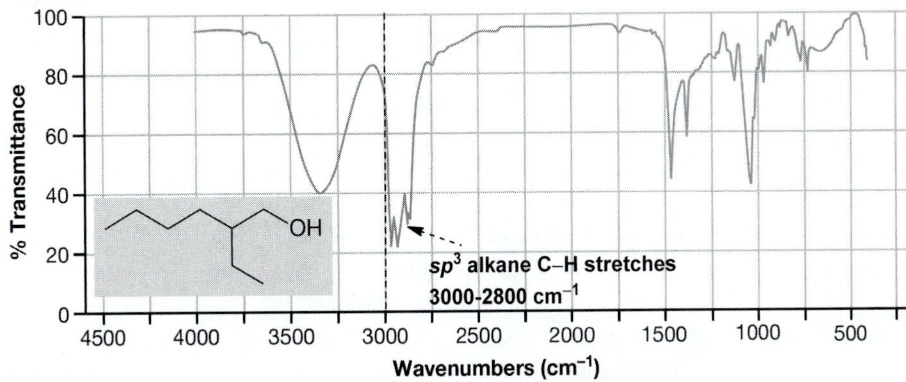

(b) Alkane C–H stretches are moderately intense and occur at 3000–2800 cm^{-1}.

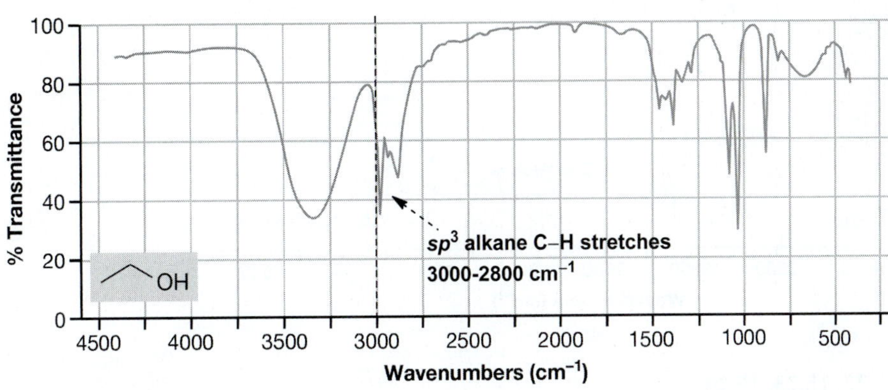

(c) Alkene C–H stretches are weak and occur around 3100 cm^{-1}; no alkane C–H stretches are present.

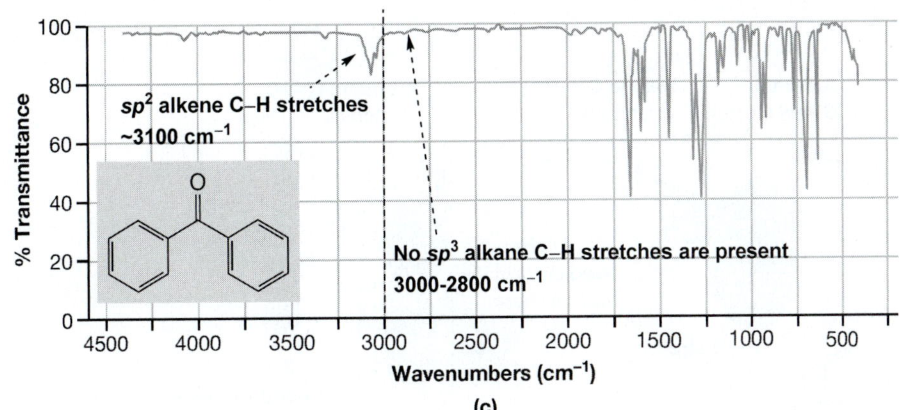

(c)

Your Turn 15.20
Think
In which region of the IR spectrum do bending-mode absorptions appear? What specific bending mode should be different for the two compounds in Figures 15-21a and 15-21b? Use Table 15-3 as a reference.

Solve
Bending-mode absorptions occur below 1000 cm^{-1}. The two compounds in question, *sec*-butylbenzene and 1,4-di-*tert*-butylbenzene, will differ in the bending-mode region with respect to the number and location of substituents on the benzene ring. *Sec*-butylbenzene is monosubstituted and, according to Table 15-3, should exhibit two strong bands around 700 and 750 cm^{-1}. 1,4-Di-*tert*-butylbenzene is para substituted and, according to Table 15-3, should exhibit one strong band at 830 cm^{-1}. These absorptions are labeled on each figure below.

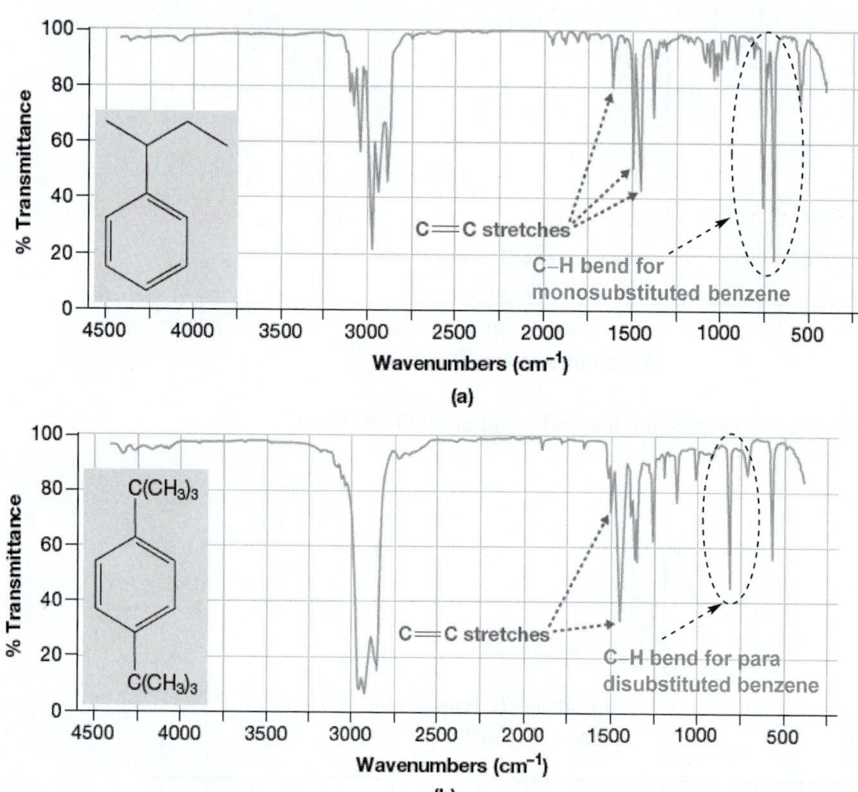

Your Turn 15.21, 15.22, 15.23, 15.24
Think
Consult Tables 15-2 and 15-3 to identify the presence and/or absence of the stretches/bends indicated.

Solve

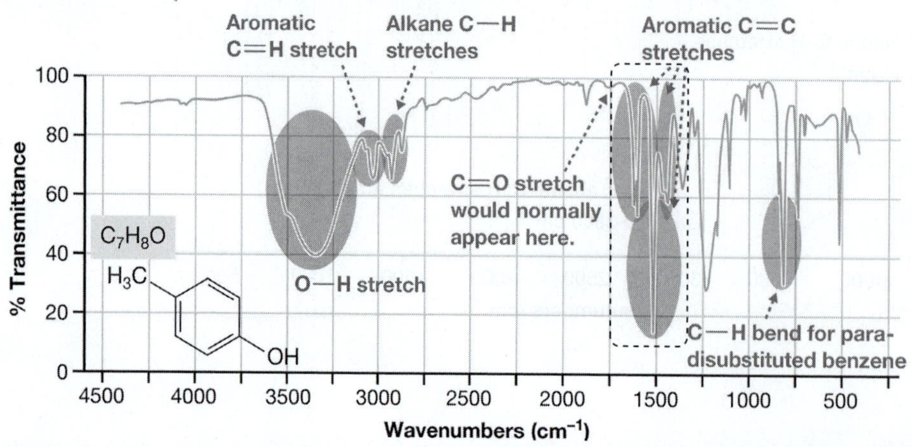

Your Turn 15.25, 15.26, 15.27, 15.28

Think

Consult Tables 15-2 and 15-3 to identify the presence and/or absence of the stretches/bends indicated.

Solve

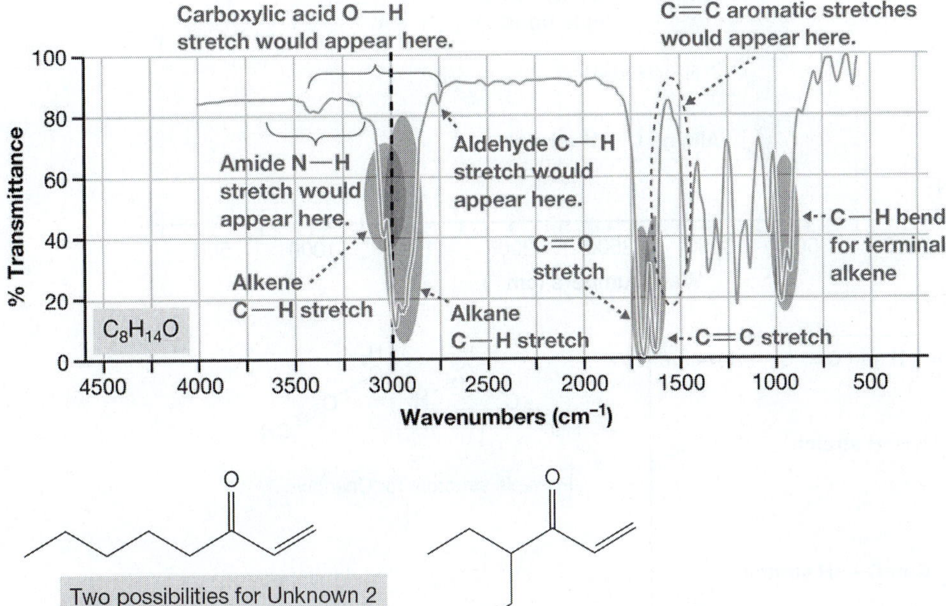

Your Turn 15.29

Think

In a saturated compound, how many bonds do each of the C and N atoms have? Are rings possible? Are multiple bonds possible?

Solve

A saturated molecule with six C atoms and one N atom is shown below. Each C atom should have four bonds and no lone pairs, and the N should have three bonds and one lone pair. The saturated molecule has 15 H atoms.

$H_3C-CH_2-CH_2-CH_2-CH_2-CH_2-NH_2$

Your Turn 15.30

Think

Consult Tables 15-2 and 15-3 to identify the presence and/or absence of the stretches/bends indicated. What is the IHD? How does this help you in proposing a possible structure?

Solve

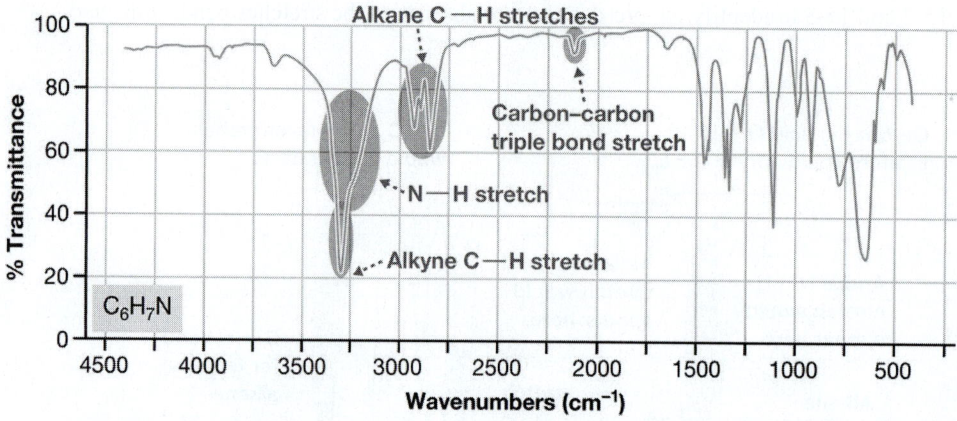

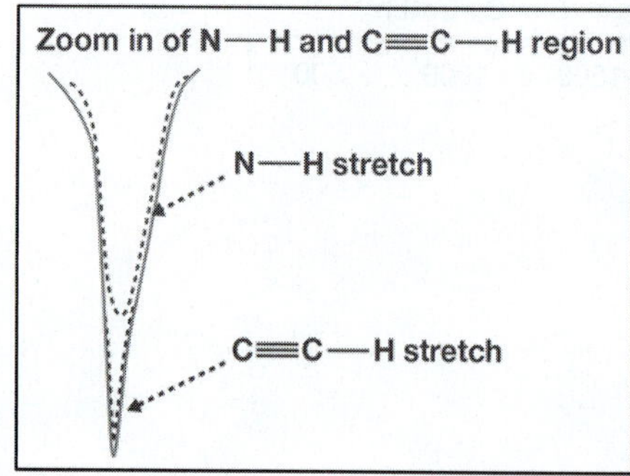

Possible structure for Unknown 3

In Chapter Problems

Problem 15.1

Think

What is the intensity of the light source (I_{source})? What is the intensity of the detected light ($I_{detected}$)? How do these values relate to transmittance? See Equation 15-1. What is the relationship between the absorbed light and the percentage of light transmitted? See Equation 15-2.

Solve

Using the formula from Equation 15-1, $\%T = (I_{detected}/I_{source}) \times 100\% = (2.5 \times 10^{-5}/1.0 \times 10^{-4}\ \text{W}) \times 100\% = 25\%$. Then, using the formula from Equation 15-2, $A = 2 - \log(\%T) = 2 - \log(25) = 0.60$.

Problem 15.2

Think

What is the Beer–Lambert law? How do you solve this equation for molar absorptivity (ε)? What units are given for the absorbance, concentration, and path length?

Solve

The Beer–Lambert law is given in Equation 15-3: $A = \varepsilon lc$, where ε is the molar absorptivity. Solve the equation for molar absorptivity:

$$\varepsilon = \frac{A}{lc} = \frac{0.78}{(1.00\ \text{cm})(6.00\times10^{-6}\ \text{M})} = 130{,}000\,\text{M}^{-1}\text{cm}^{-1}$$

Problem 15.4

Think

Which is the longest-wavelength absorption? What is its λ_{max}? What energy does a photon with that wavelength possess? How does this energy relate to the energy difference between the HOMO and the LUMO?

Solve

The HOMO–LUMO transition is indicated by the longest-wavelength UV–vis absorption. In the spectrum, that absorption appears at 280 nm, whose photon energy is

$$E = \frac{hc}{\lambda} = \frac{(6.626\times10^{-34}\ \text{J}\cdot\text{s})\left(3.00\times10^{-8}\ \dfrac{\text{m}}{\text{s}}\right)}{280\times10^{-9}\ \text{m}} = 7.00\times10^{-19}\ \text{J}$$

Problem 15.5

Think

What does the MO energy diagram look like for $CH_2{=}CH_2$ (see Chapter 3)? How many π electrons are present? What happens to the electron in the HOMO when it absorbs a photon of energy?

Solve

The absorption of a UV–vis photon in ethene is shown below. The ground state configuration is shown on the left. The electron in the bonding π_1 HOMO is promoted to the antibonding π_2 orbital. This molecule underwent a $\pi \rightarrow \pi^*$ transition.

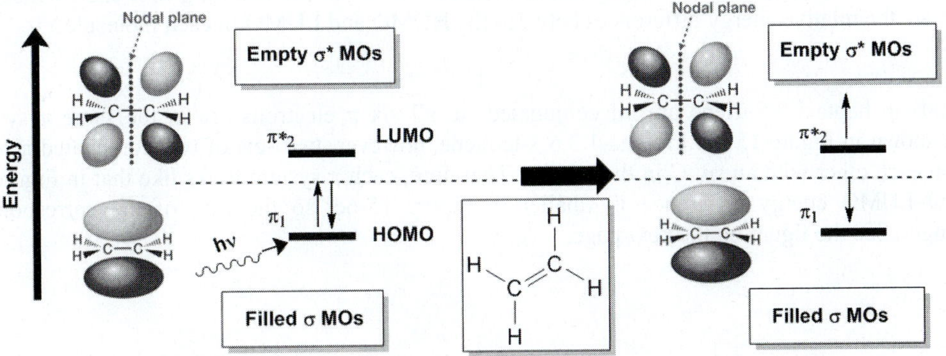

Problem 15.6

Think

How many conjugated π bonds are present? What trend do you observe in λ_{max} with the number of conjugated π bonds? What is the value of λ_{max} in the table that has the closest number of conjugated π bonds compared to deca-1,3,5,7,9-pentaene? Would you expect the λ_{max} value to be higher or lower?

Solve

A trend can be seen:

# Conjugated C=C Bonds	UV λ_{max}	Change in λ_{max}
1	161	–
2	217	56
3	274	57
4	290	16

The compound given, deca-1,3,5,7,9-pentaene, has five conjugated C=C double bonds, so we should expect the wavelength to be longer than 290 nm. But with each additional conjugated double bond, the change is less significant. Because going from three to four conjugated double bonds was about a 16-nm change, going from four to five conjugated double bonds will be about a 10-nm change, giving about 300 nm.

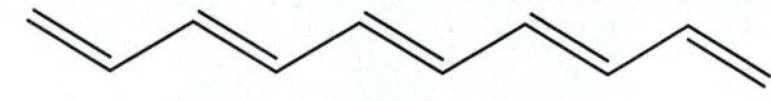

Deca-1,3,5,7,9-pentaene
5 conjugated π bonds
λ_{max} ~ 300 nm

Problem 15.7

Think

How many conjugated π bonds are present? What trend do you observe in λ_{max} with the number of conjugated π bonds in aldehydes? What is the value of λ_{max} of the aldehyde in the table that has the closest number of conjugated π bonds compared to penta-2,4-dienal? Would you expect the λ_{max} value to be higher or lower?

Solve

This compound has two C=C double bonds conjugated to an aldehyde. With no C=C double bonds conjugated, the wavelength is 280 nm. Adding one C=C double bond increases the wavelength by 60 nm, to 340 nm. Adding the second conjugated double bond will increase the wavelength even more, but by <60 nm (e.g., perhaps by 40 nm). So we estimate that the wavelength would be 380 nm.

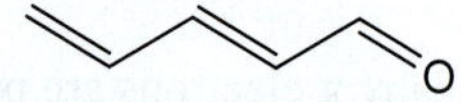

Penta-2,4-dienal
2 conjugated π bonds to C=O
λ_{max} ~ 380 nm

Problem 15.9

Think

In each compound, are the double bonds conjugated or isolated? What do the resulting MO energy diagrams look like? What are the relative energy differences between the HOMO and LUMO in each molecule?

Solve

The double bonds in hepta-1,3,5-triene are all conjugated, so all six π electrons are in the same π system, resembling that shown in Figure 15-6c. In nona-1,3,6,8-tetraene, however, two sets of two conjugated π bonds are isolated from each other with an sp^3 C in the middle. Therefore, each π system looks like that in Figure 15-6b. The HOMO–LUMO energy difference is smaller in Figure 15-6c, so the λ_{max} of the corresponding absorption is longer. See the figure on the next page.

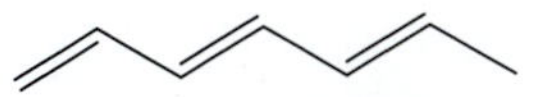

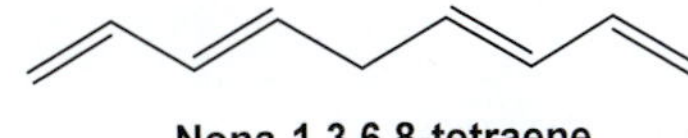

Problem 15.11

Think

For each compound, does the HOMO–LUMO transition correspond to a $\pi \rightarrow \pi^*$ transition or an $n \rightarrow \pi^*$ transition? In general, what are the relative energies of these types of transitions?

Solve

The HOMO–LUMO transition for $H_2C{=}NH_2^+$ is $\pi \rightarrow \pi^*$, as shown in Figure 15-8a. The transition for $H_2C{=}OH^+$, however, is $n \rightarrow \pi^*$, similar to Figure 15-8b, given the presence of both a lone pair on O and the π bond. Accordingly, the HOMO–LUMO transition for $H_2C{=}OH^+$ should require less energy and will thus appear at a longer wavelength.

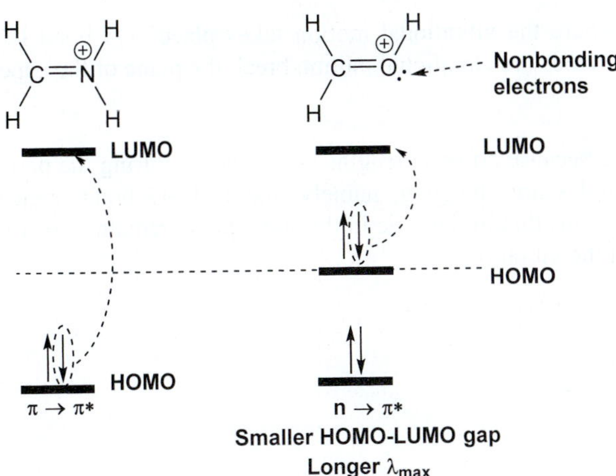

Problem 15.12

Think

Refer to Figure 15-1 to determine the color of light to which 420 nm corresponds. What is the complementary color? Which color do our eyes see? What color, therefore, is the solution of crystal violet?

Solve

From Figure 15-1, 420 nm corresponds to a purple-blue color. The eye registers the complementary color, opposite the purple-blue color on the wheel in Figure 15-9, which is yellow. Therefore, the solution at pH −1 is yellow.

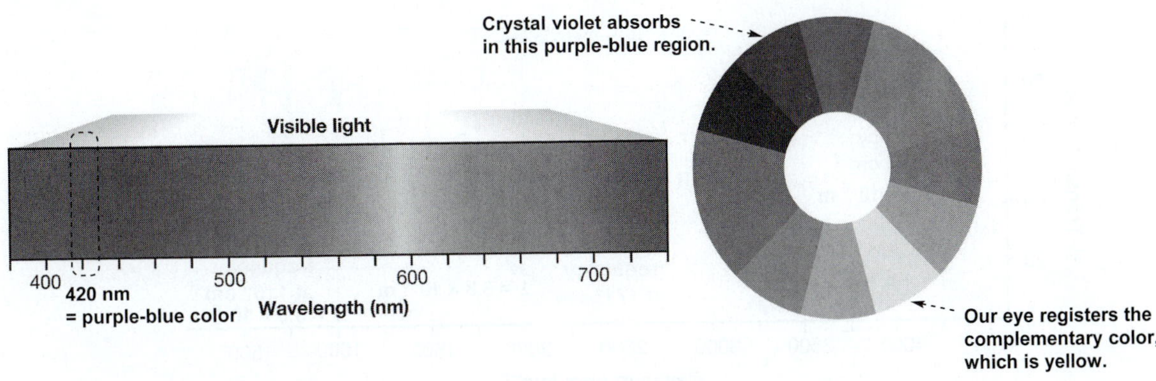

Problem 15.13

Think

What does the disappearance of the peak at 244 nm indicate about the course of the reaction? How does doubling the concentration of HCl affect the amount of reactant left after the allotted time period? If there were no change in the absorbance in the same given time period, what would that suggest about whether HCl is or is not involved in the rate-determining step?

Solve

The reactant styrene absorbs at 244 nm, so the absorbance at 244 nm represents the relative amount of styrene present, and a decrease in this peak's absorbance over a given time period indicates the amount of styrene that has reacted away. Thus, the decrease in absorbance at 244 nm represents the progression of the addition reaction. For the first trial, absorbance dropped from 0.60 to 0.50, a decrease of 0.10. When the concentration of HCl was doubled, absorbance dropped from 0.60 to 0.40, a decrease of 0.20. Over that time period, twice as much styrene reacted, meaning that the rate doubled. Because doubling the concentration of HCl doubled the reaction rate, HCl is involved in the rate-determining step; rate = k[HCl][styrene].

Problem 15.14

Think

What is the geometry about the central C atom where the vibrational motion takes place? Do bond lengths change? Do bond angles or dihedral angles change? Does the vibrational motion break the plane of the paper?

Solve

This is an in-plane bend. It is not a stretching mode because no bond lengths are changing during the period of the vibration. It is bending because the bond angles are changing, namely, the H–C–O bond angles are changing. It is in-plane bending because all of the atoms that initially lie in the same plane remain in that plane (C atom is trigonal planar) throughout the course of the vibration.

In-plane bend

Problem 15.15

Think

What is the relationship between wavenumbers and frequency in Hz? How is frequency in Hz related to wavelength?

Solve

According to Equations 15-6 and 15-7, wavenumbers (cm^{-1}) = $\nu(\text{Hz})/100c = (c/\lambda)/(100c) = 1/(100\ \lambda)$.
So $\lambda = 1/(100 \times \text{wavenumbers})$. For the three absorptions in Figure 15-10, we solve

$\lambda = 1/(100 \times 2961) = 3.4 \times 10^{-6}$ m

$\lambda = 1/(100 \times 1717) = 5.8 \times 10^{-6}$ m

$\lambda = 1/(100 \times 1167) = 8.6 \times 10^{-6}$ m

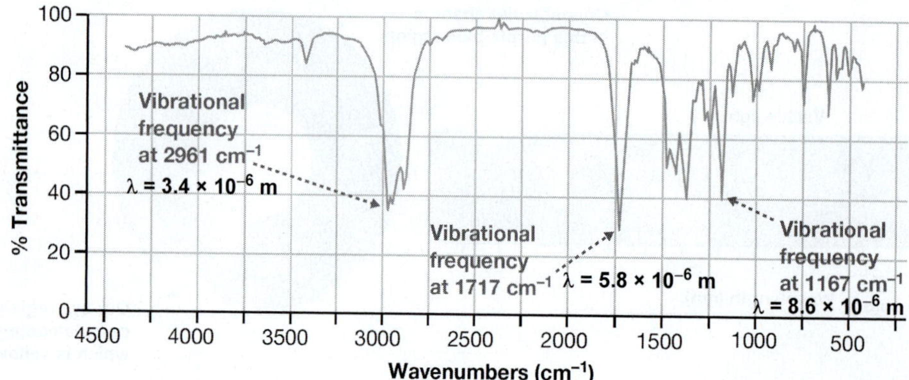

Problem 15.17

Think

What type of bond does each arrow indicate? Is there more than one type of functional group to which each of those bonds could belong?

Solve

(a) Aromatic C=C, 1450–1550 cm^{-1}
(b) Aldehyde C–H, 2820 and 2720 cm^{-1}
(c) Aldehyde C=O, 1720 cm^{-1}
(d) Alkyne C≡C, 2200 cm^{-1}
(e) Alkyne C–H, 3300 cm^{-1}
(f) Secondary amine, N–H, 3400 cm^{-1}
(g) Alcohol C–O, 1100 cm^{-1}
(h) Alcohol O–H, 3400 cm^{-1}
(i) Nitrile C≡N, 2200 cm^{-1}

Problem 15.19

Think

Considering the model of masses connected by a spring, are the masses the same? Is the stiffness of each spring the same? How do these factors govern vibrational frequency? How does vibrational frequency affect IR absorption frequencies?

Solve

(a) Because it is stronger than the double bond, the triple bond acts as a stiffer spring and has the higher absorption frequency, and it vibrates faster. With faster vibration, the C≡N bond absorbs the higher-frequency IR photons.

(b) The compound with the C=N double bond has the CN stretch at a higher frequency. Because a C=N double bond is stronger than a C–N single bond, it acts as a stiffer spring and has the faster vibration. With faster vibration, the C=N bond absorbs the higher-frequency IR photons.

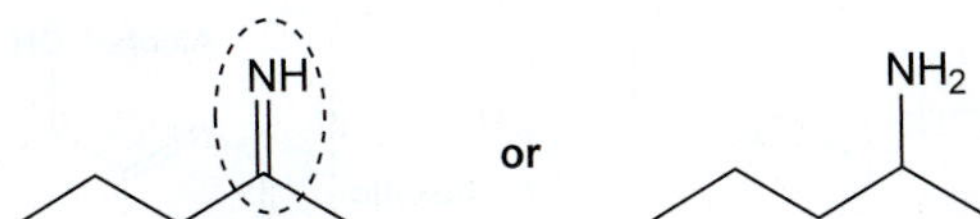

(c) The triple bond with ^{12}C has the higher vibrational frequency because ^{12}C has slightly less mass than ^{13}C, allowing the bond to vibrate faster. With faster vibration, the triple bond with ^{12}C absorbs the higher-frequency IR photons.

Problem 15.20

Think

Where is the C=C bond located in each spectrum? Which C=C peak has a higher intensity? Which alkene is more symmetrical? Which alkene, when stretched, has a larger change in dipole? What is the relationship between peak intensity and dipole change?

Solve

Spectrum A corresponds to 1-methylcyclohexene and Spectrum B corresponds to methylenecyclohexane. The small intensity of the C=C stretch around 1650 cm^{-1} in Spectrum A is indicative of a C=C stretch for which the dipole moment of the molecule does *not* change much during the period of the stretching vibration. In Spectrum B, the intensity of that peak is much greater, signifying a more profound change in dipole moment during the vibration. The C=C stretch of 1-methylcyclohexene should bring about a smaller change in dipole moment during the vibration because the molecule is more symmetric about the double bond than is methylenecyclohexane. 1-Methylcyclohexene is of the form $R_2C{=}CRH$, whereas methylenecyclohexane is of the form $R_2C{=}CH_2$.

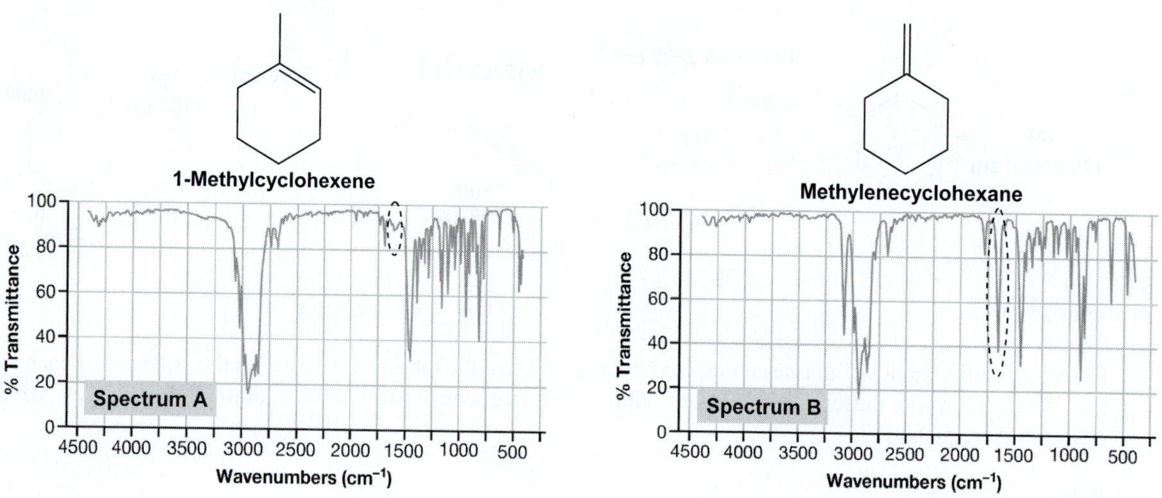

Problem 15.21

Think

What similarities and differences do you see in the structure of each isomer? What functional groups are present? How is the H–O stretch different for an alcohol versus a carboxylic acid?

Solve

The spectrum is consistent with isomer **A** because the OH stretch, overlapping with the alkane C–H stretches, indicates a carboxylic acid. Molecule **A** is a carboxylic acid, whereas molecule **B** is not.

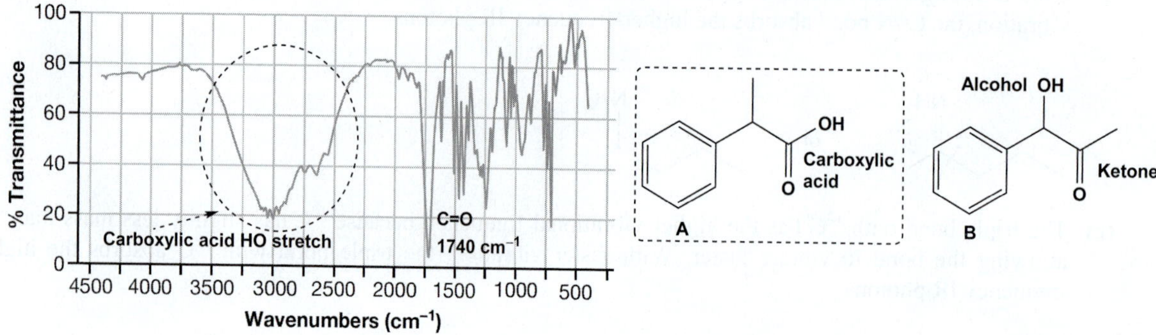

Problem 15.22

Think

What similarities and differences do you see in the structure of each isomer? Where is the C=O peak typically located in an IR spectrum? Which C=O is conjugated to the aromatic ring? How does conjugation affect the frequency of the IR stretch?

Solve

The spectrum corresponds to compound **A**. The absorption at 1686 cm^{-1} corresponds to a C=O stretch that is conjugated to a C=C double bond. An isolated C=O of a ketone normally appears around 1720 cm^{-1}. In the compound on the left, the C=O group is conjugated to the benzene ring, but in the compound on the right, it is not.

A
Conjugated C=O, 1686 cm^{-1}

B
Unconjugated C=O, ~1720 cm^{-1}

Problem 15.23

Think

What are the characteristic peaks for an aromatic ring? What are the characteristic peaks for the C=C of an alkene?

Solve

Spectrum 1 corresponds to the nonaromatic compound. The C=C stretch appears around 1650 cm^{-1}, indicative of a regular alkene C=C stretch. The C=C stretch in Spectrum 2 appears as three peaks between 1450 and 1600 cm^{-1}, typical of aromatic rings.

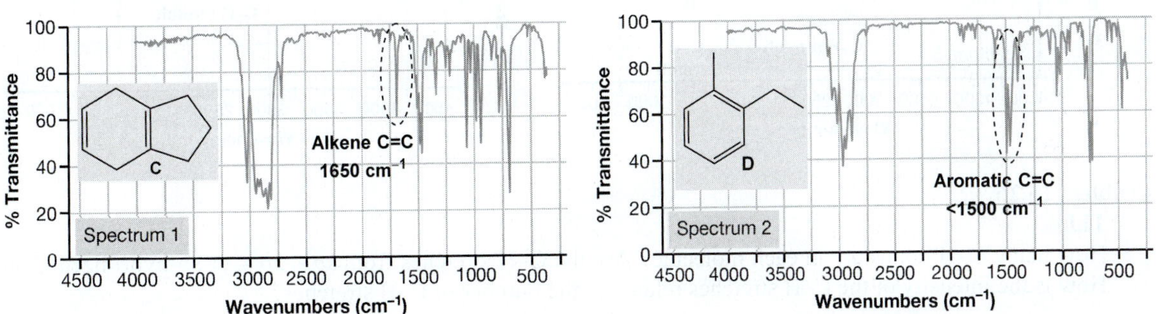

Problem 15.24

Think

In which region of the IR spectrum does a C≡N triple bond stretch occur? Would this stretch be more or less intense than a C≡C? Which bond has a larger change in dipole when stretched? For the spectrum that corresponds to the alkyne, is there an alkyne C–H peak at ~3300 cm^{-1}? What does that indicate about whether the alkyne is terminal or internal?

Solve

(a) Spectrum 3 in Your Turn 15.17 most likely corresponds to a nitrile. This is because the C≡N bond has a relatively large bond dipole, and as it stretches and compresses, the bond dipole oscillates. As mentioned in the chapter, such an oscillation of the bond dipole enhances the probability that a photon will be absorbed, leading to an intense peak. By contrast, the weak absorption in the spectrum on the right is likely a C≡C triple bond, which often has a relatively small bond dipole.

(b) There is no alkyne C–H peak at ~3300 cm^{-1}, therefore the alkyne is internal. Also, the intensity of the C≡C peak would be much greater if the C≡C were terminal because, being of the form RC≡CH, the two ends would be rather unsymmetric.

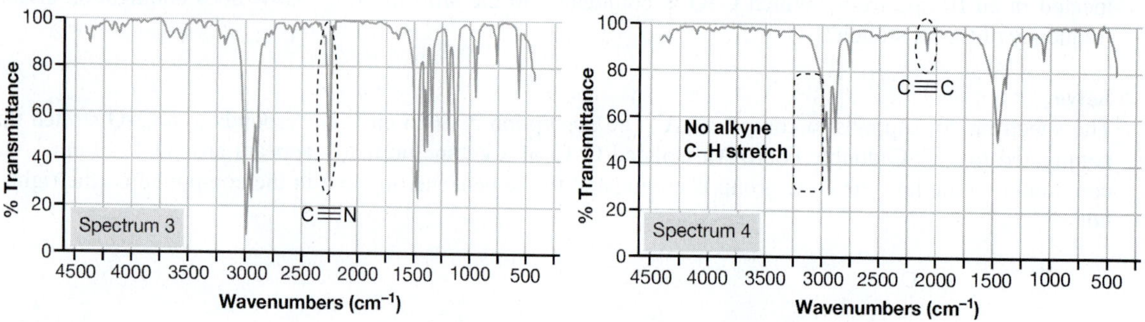

Problem 15.25

Think

Is there a peak suggestive of a C=C stretch? Are there peaks suggestive of *sp*^{2} C–H stretches? Are there strong *sp*^{2} C–H bending stretches?

Solve

Spectrum 5 indicates the presence of a C=C double bond. It has a clear C–H stretch above 3000 cm^{-1}, a C=C stretch at 1650 cm^{-1}, and strong C–H bending bands near 1000 cm^{-1}. Spectrum 6, although it has a peak that is consistent with a C=C stretch, does not have a C–H stretch above 3000 cm^{-1}.

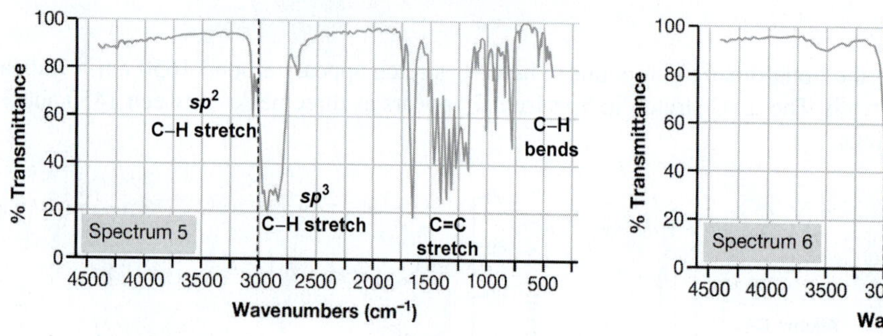

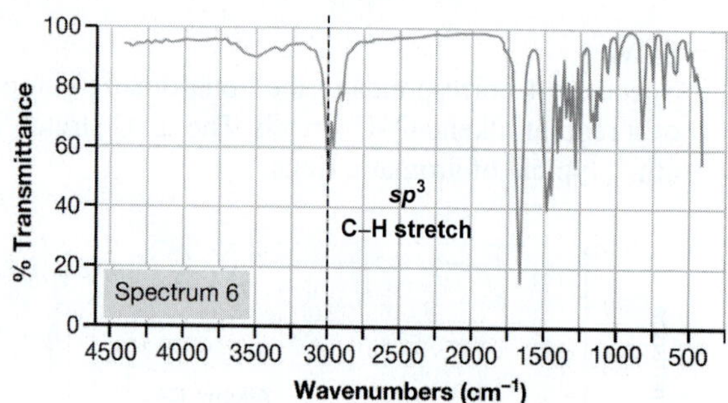

Problem 15.26

Think

How many C–H bonds are in each molecule? Are the C–H stretches from *sp*^{3}-, *sp*^{2}-, or *sp*-hybridized C atoms? How is the intensity of the C–H stretches related to the number of C–H stretches?

Solve

In all cases, there are C=C double bonds and C–O single bonds. The difference is in the ratio of alkane C–H bonds (stretches slightly below 3000 cm^{-1}, marked as a dashed line) to alkene C–H bonds (stretches slightly above 3000 cm^{-1}). In Spectrum 9, there are no alkane C–H stretches, corresponding to no alkane C–H bonds. Therefore, Spectrum 9 corresponds to molecule **E**. In the other two spectra, we see both types of stretches. Spectrum 8 shows a greater proportion of alkane C–H stretches than Spectrum 7, corresponding to a greater proportion of H atoms on *sp*^{3}-hybridized C atoms. In molecule **G**, there are 15 alkane C–H bonds and three alkene C–H bonds. In molecule **F**, that ratio drops to three alkane C–H bonds and three alkene C–H bonds. Therefore, Spectrum 7 corresponds to molecule **F**, and Spectrum 8 corresponds to molecule **G**. See the spectra on the next page.

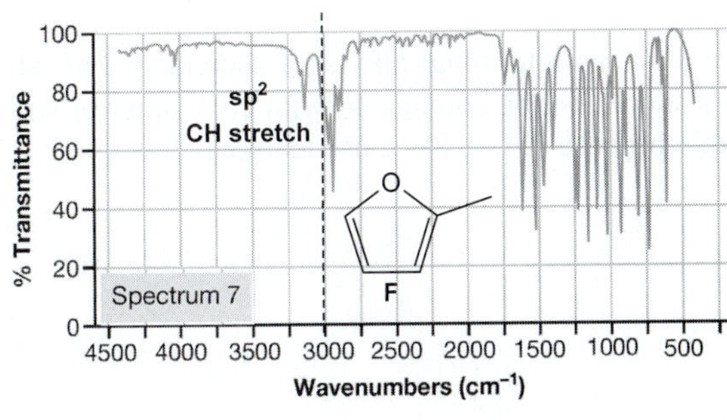

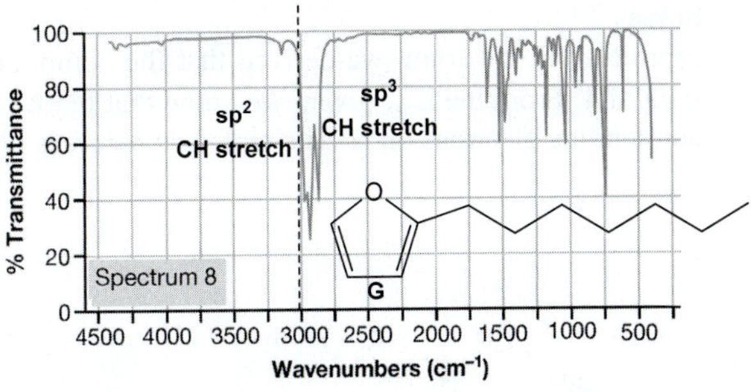

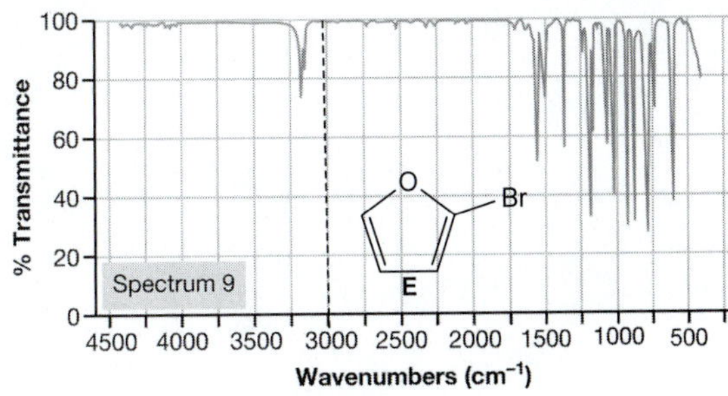

Problem 15.28

Think

In Solved Problem 15.27, what is the basic structural unit that we derived from the IR spectrum? How can the remaining atoms be attached without disturbing that basic structure? What kinds of bonds must connect those atoms?

Solve

The first structure below is the basic structural unit we derived from the spectrum, which accounts for four C atoms, three H atoms, and one O atom.

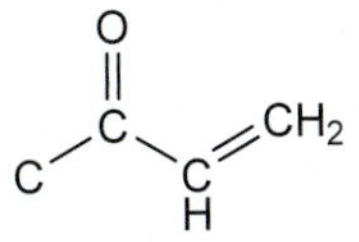

That leaves four C atoms and 11 H atoms yet to add, all of which can be attached only by single bonds—otherwise, we would exceed the compound's IHD of 2. Other possibilities include the following:

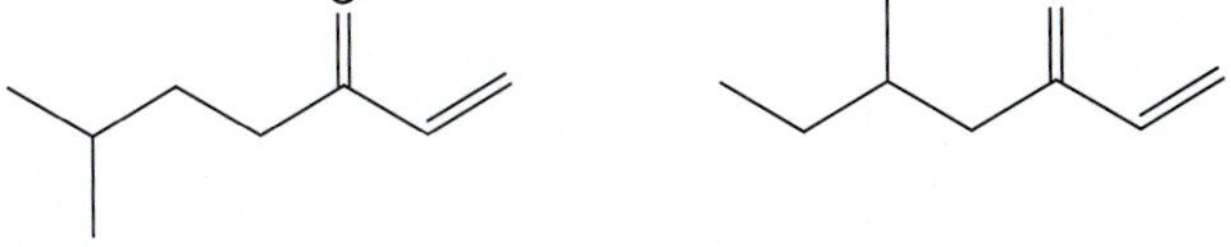

Problem 15.29

Think

What is the basic structural unit that we derived from the IR and UV–vis spectra? How can the remaining atoms be attached without disturbing that basic structure? What kinds of bonds must connect those atoms?

Solve

From the IR spectrum, we derived that the compound must have a terminal alkyne, a secondary NH, and an IHD of 4. From the λ_{max} given, we know that the π bonds should not be conjugated with each other or with the nonbonding electrons on N. The molecule below is possible because it, too, has these features.

CH
|||
C
|
CH — N(H) — CH_3
HC≡C

End of Chapter Problems
Sections 15.1–15.3 UV–Vis Spectroscopy
Problem 15.30

Think

Which structure has more conjugated C=C bonds? How is the number of conjugated C=C bonds related to the longest wavelength of UV–vis absorption?

Solve

The β isomer has the longer-wavelength UV–vis absorption because it has a greater extent of conjugation. In the β isomer, there are 11 conjugated double bonds. In the α isomer, 10 double bonds are conjugated, and one is isolated.

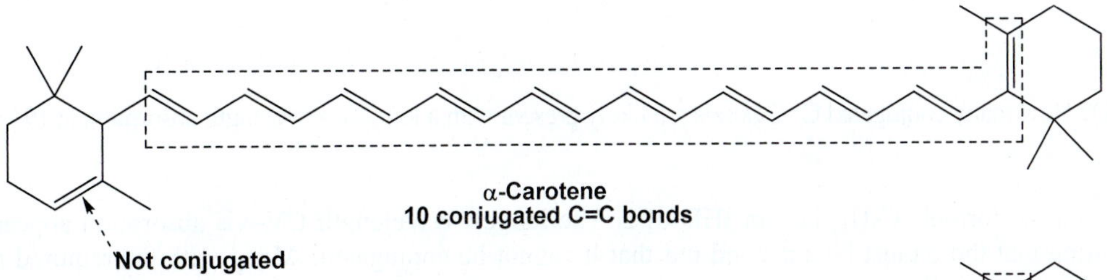

α-Carotene
10 conjugated C=C bonds

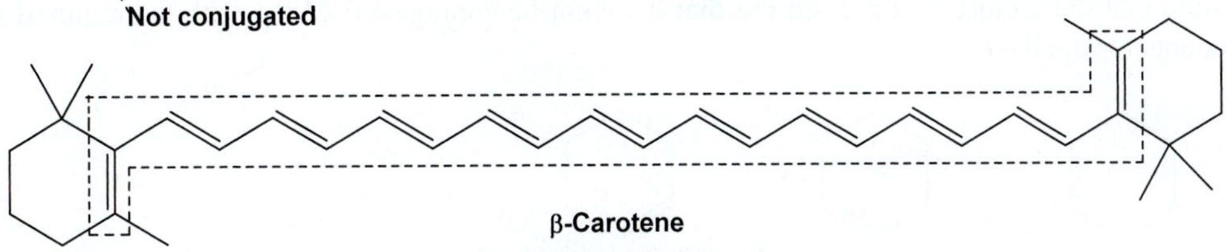

Not conjugated

β-Carotene
11 conjugated C=C bonds
Longer λ_{max}

Problem 15.31

Think

Consult Table 15-1 to obtain the value of the longest-wavelength absorption for but-1,3-diene. Are the bonds in but-1,3-diene conjugated? Are the bonds in buta-1,2-diene conjugated? What does conjugation do to the wavelength of maximum absorption?

Solve

Buta-1,3-diene's longest-wavelength absorption is at 217 nm. Buta-1,2-diene's longest-wavelength absorption appears at 178 nm. The difference comes from the fact that the double bonds in buta-1,2-diene are not conjugated, whereas those in buta-1,3-diene are. The adjacent double bonds in buta-1,2-diene are formed from perpendicular *p* orbitals, and to be conjugated, they must be parallel.

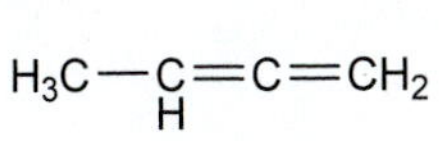

$H_3C-\underset{H}{C}=C=CH_2$

Buta-1,2-diene
λ_{max} = 178 nm
Not conjugated

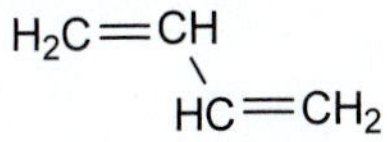

$H_2C=CH-HC=CH_2$

Buta-1,3-diene
λ_{max} = 217 nm
Conjugated

Problem 15.32

Think

What is the IHD? How many conjugated C=C bonds are likely present with a longest-wavelength absorption at 215 nm?

Solve

The IHD is 3 because the saturated compound with five C atoms has 12 H atoms. The compound should have two conjugated π bonds, because buta-1,3-diene, which has two conjugated C=C bonds, absorbs at 217 nm, which is very close to the wavelength of our unknown compound. The two conjugated π bonds account for an IHD of 2. To account for the remaining IHD of 1, the compound must have another π bond or a ring. If a third π bond is added, it must not be conjugated to the other two. See possible structures below.

Problem 15.33

Think

What is the IHD? How many conjugated C=C bonds are likely present with a longest-wavelength absorption at 191 nm?

Solve

A compound with the formula C_7H_{12} has an IHD of 2. The longest-wavelength UV–vis absorption appears at 191 nm, indicating that there must be a π bond but that it cannot be conjugated. Along with the required ring, that π bond accounts for the IHD.

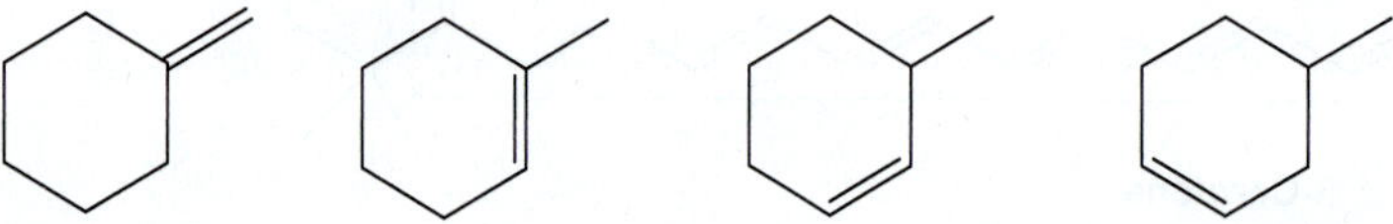

Problem 15.34

Think

How many conjugated C=C are present in each structure? What is likely the longest wavelength of UV–vis absorption? Which of these absorptions is in the visible region? Would this correspond to a color observed by our eyes?

Solve

When the pH is <8 or >14, the π systems on the rings are not conjugated through the central carbon owing to the sp^3 hybridization of that C atom. So the longest-wavelength absorption remains relatively short, lying in the UV region. When absorption takes place in the UV region, our eyes do not register any wavelengths missing (our eyes only detect light in the visible region), so the solution does not appear colored. Between a pH of 9 and 13, however, the central C atom is sp^2 hybridized, so all three π ring systems are conjugated. The much greater extent of conjugation increases the wavelength of the longest-wavelength absorption, placing it in the visible region of the spectrum. When light is absorbed from the visible region, our eyes detect the missing light, which registers as a color.

Phenolphthalein pH <8 — sp^3 C atom — λ_{max} = UV region, Colorless

NaOH → **pH 9–13** — λ_{max} = Visible region, Colored

NaOH → **pH >14** — sp^3 C atom — λ_{max} = UV region, Colorless

Problem 15.35

Think

What is the difference in structure between propyne and acetonitrile? How do the lone pairs on the N atom in acetonitrile affect the longest-wavelength absorption?

Solve

Acetonitrile has the longer wavelength absorption. Both molecules have isolated triple bonds, but acetonitrile also has a lone pair. So whereas propyne's longest-wavelength absorption corresponds to a $\pi \rightarrow \pi^*$ transition, that in acetonitrile corresponds to an $n \rightarrow \pi^*$ transition. All else equal, the $n \rightarrow \pi^*$ transition requires lower energy or longer-wavelength light.

$H_3C—C{\equiv}CH$
Propyne
π-to-π^* transition
Shorter λ_{max}

$H_3C—C{\equiv}N:$
Acetonitrile
n-to-π^* transition
Longer λ_{max}

Problem 15.36

Think

How many conjugated π bonds are present in the starting material? In the product? Are any lone pairs conjugated to π bonds? How does the number of conjugated π bonds affect the longest-wavelength absorption?

Solve

In reactions **(b)** and **(c)**, the longest λ_{max} would shift to a shorter wavelength because the product is less conjugated. In reactions **(a)** and **(e)**, the longest λ_{max} would shift to a longer wavelength because the product is more conjugated. In **(d)**, the longest λ_{max} would shift to a longer wavelength because an $n \rightarrow \pi^*$ transition is introduced as a result of the lone pair of electrons on the N atom of NH_2.

(a) Base → **Shift to longer λ_{max}**

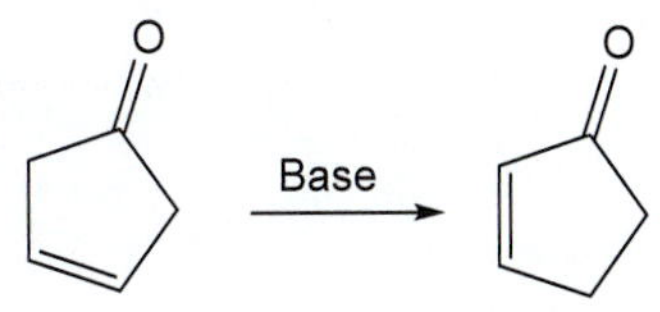

(b) 1. $LiAlH_4$ 2. H_3O^+ → **Shift to shorter λ_{max}**

(c) Li, NH_3/EtOH → **Shift to shorter λ_{max}**

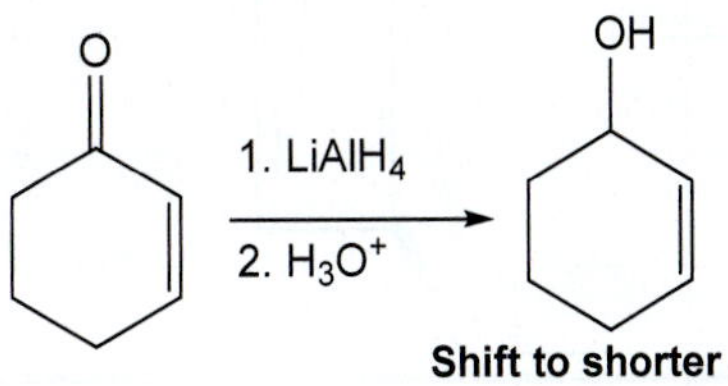

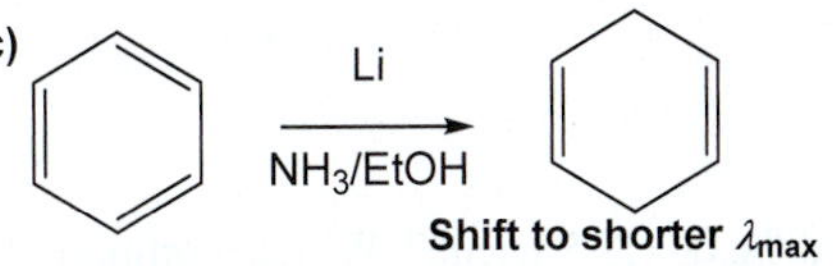

(d) NO_2; Fe, HCl → NH_2 **Shift to longer λ_{max}**

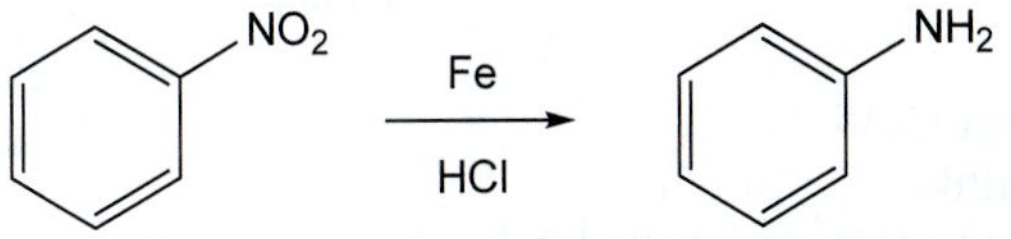

(e) Cl(C=O)CH_3, $AlCl_3$ → **Shift to longer λ_{max}**

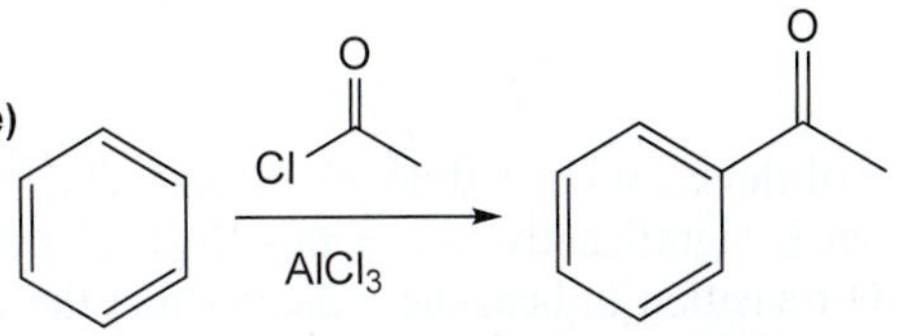

Problem 15.37

Think

What is the structure of propenal? What does the energy diagram of the π MOs look like? What transition in propenal occurs at 340 nm? At 202 nm, is that a shorter or longer wavelength? Higher- or lower-energy photons? How do the lone pairs on the oxygen atom affect the energy required for a transition?

Solve

The molecule and energy diagram of the π system of MOs is shown below.

H_2C=CH–CH=O

π^*_4

π^*_3

Lone pairs

π_2

π_1

The longest-wavelength absorption at 340 nm corresponds to the n→π* transition, in which an electron is promoted from a nonbonding MO to π_3 (lower-energy photon, longer wavelength). The absorption at 202 nm, being lower in wavelength, corresponds to a higher-energy transition, consistent with a regular π→π* transition from the conjugated pair of double bonds. This would specifically be from π_2 to π_3. Notice that in 1,3-butadiene, which does not have any lone pairs and, therefore, does not have a possible n→π* transition, the longest-wavelength absorption is at 217 nm, consistent with the 202 nm π→π* transition in acrolein.

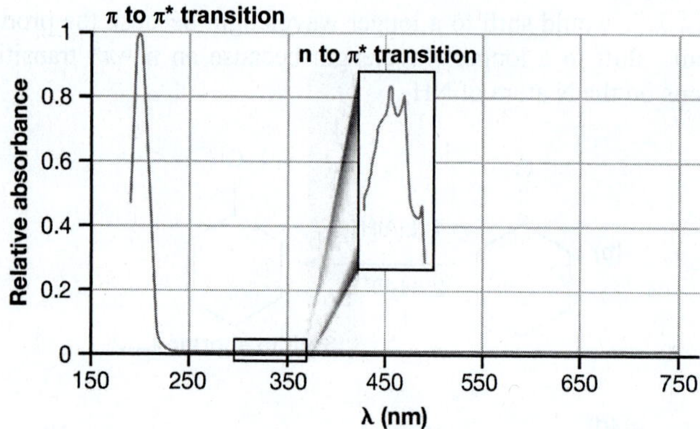

Problem 15.38

Think

How many conjugated π bonds are present in both benzene and hexa-1,3,5-triene? Which compound has a longer-wavelength absorption? Which compound has a lower-energy π→π* transition? How does the aromaticity of benzene affect the HOMO–LUMO transition?

Solve

Benzene and hexa-1,3,5-triene both have a π system consisting of three conjugated double bonds. The longest-wavelength absorption of hexa-1,3,5-triene is at 274 nm, which is significantly longer than that for benzene. The shorter-wavelength light necessary for the HOMO–LUMO transition in benzene indicates that the energy difference between benzene's HOMO and LUMO is greater than the difference for hexa-1,3,5-triene. In turn, this reflects the unusual stability of benzene's π system from its aromaticity. By contrast, hexa-1,3,5-triene is nonaromatic.

Problem 15.39

Think

What is the initial absorbance? At what point in the spectrum has the absorbance decreased by half? To which pH does this correspond? How is this pH related to the pK_a?

Solve

The initial absorbance and the half absorbance are marked on the spectrum below. Because the absorbance at 312 nm corresponds to the concentration of acid present, half the absorbance represents the acid having dissociated into an equal amount of its conjugate base. The pK_a of an acid is equal to the pH of the solution at which half the acid has dissociated into its conjugate base. In this example, the pK_a is 9.3.

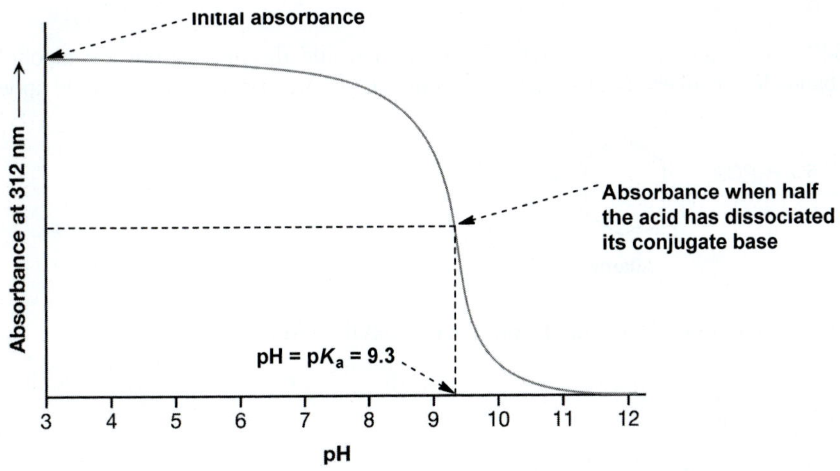

Sections 15.4 and 15.5 IR Spectroscopy

Problem 15.40

Think

Identify the functional group to which each indicated bond belongs. Consult Table 15-2 for a listing of characteristic absorption frequencies in IR spectroscopy.

Solve

The approximate stretching frequencies are given below.

(a) Alkene C=C, 1650 cm^{-1}
(b) Ketone C=O, 1720 cm^{-1}
(c) Alcohol C–O, 1100 cm^{-1}
(d) Alcohol O–H, 3400 cm^{-1}
(e) Alkyne C–H, 3300 cm^{-1}
(f) Alkyne C≡C, 2200 cm^{-1}
(g) Amide C=O, 1650 cm^{-1}
(h) Amide N–H, 3300 cm^{-1}
(i) Aromatic C=C, 1500 cm^{-1}
(j) Aldehyde C–H, 2720 and 2820 cm^{-1}
(k) Aromatic sp^2 C–H, 3050 cm^{-1}
(l) Carboxylic acid C=O, 1740 cm^{-1}
(m) Carboxylic acid O–H, 2800 cm^{-1}

(b) 1720 cm^{-1} O
(a) 1650 cm^{-1}
(c) 1100 cm^{-1}
O–H
(d) 3400 cm^{-1}

(e) 3300 cm^{-1} H
(g) 1650 cm^{-1} O
N
(f) 2200 cm^{-1}
(h) 3300 cm^{-1} H

(k) 3050 cm^{-1} H
O (l) 1740 cm^{-1}
O–H
(i) 1500 cm^{-1}
(m) 2800 cm^{-1}
(j) 2820 and 2720 cm^{-1}
H O

Problem 15.41

Think

Identify the different functional groups present in the reactant and product. Consult Table 15-2 for a listing of characteristic IR absorption frequencies for bonds in those functional groups. What peak will either appear or disappear during the course of the reaction?

Solve

(a) A OH stretch (3400 cm^{-1}) and a C–O stretch (1100 cm^{-1}) would disappear from the alcohol, and the characteristic IR bands for an alkene (CH stretch, 3050 cm^{-1}; C=C stretch, 1650 cm^{-1}) would appear.

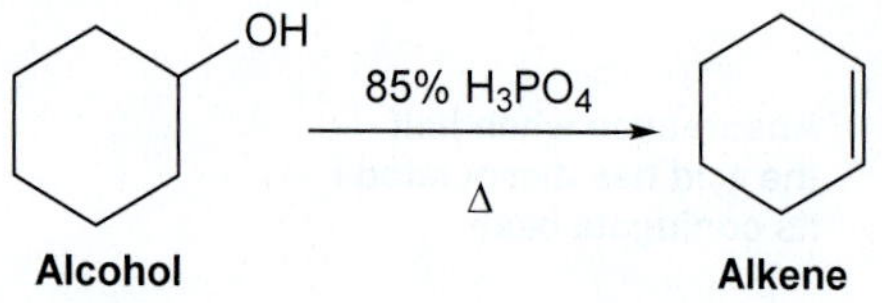

OH

85% H_3PO_4

Δ

Alcohol **Alkene**

(b) The very broad OH stretch (3000–2500 cm^{-1}) from the carboxylic acid would disappear.

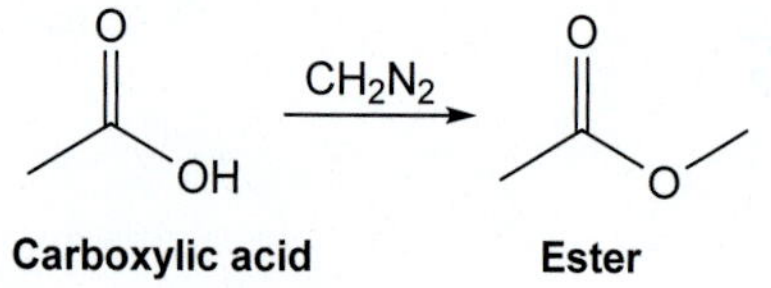

O OH CH_2N_2 O O

Carboxylic acid **Ester**

(c) Two NH stretching bands (3400 cm^{-1}) from the primary amine would become a single band from the secondary amine.

NH_2 Br N H

Primary amine **Secondary amine**

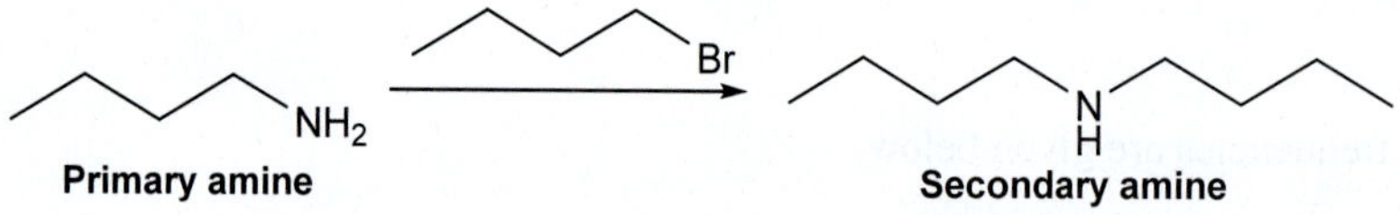

(d) A single NH stretching band (3400 cm^{-1}) would disappear from the secondary amine. Tertiary amines have no characteristic IR absorption bands above 3000 cm^{-1} because they have no N–H bonds.

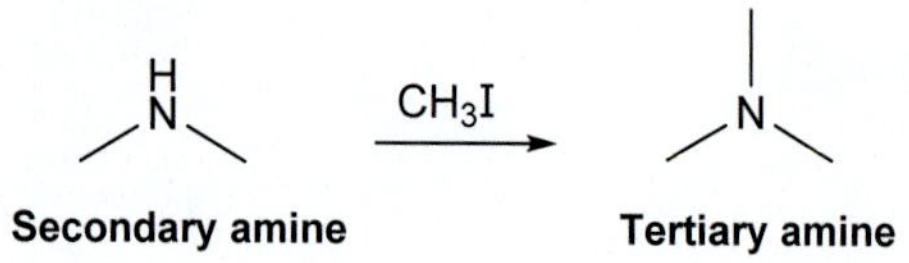

H N CH_3I N

Secondary amine **Tertiary amine**

(e) A OH stretching band (3400 cm^{-1}) would appear for the alcohol.

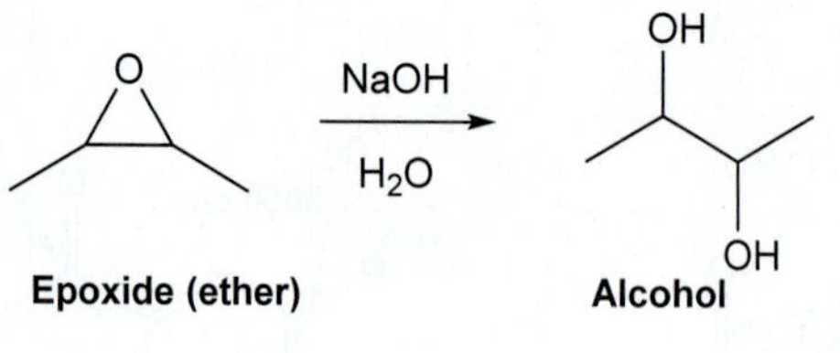

O NaOH H_2O OH OH

Epoxide (ether) **Alcohol**

(f) A C=C stretching band (1600 cm^{-1}) and an alkene C–H stretch (3050 cm^{-1}) from the conjugated ketone would disappear. Also, the C=O stretch would move higher by about 30 cm^{-1}, because the conjugation in the reactant is not present in the product.

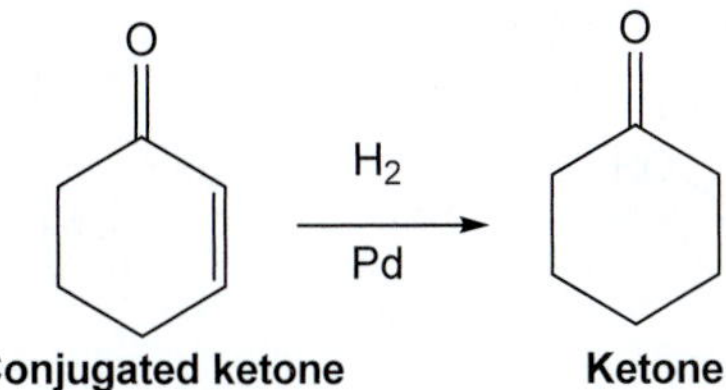

(g) The C=C stretching bands for an aromatic compound (1450–1550 cm^{-1}) would become a simpler C=C stretching band (1650 cm^{-1}) for the alkene. Also, sp^3 C–H stretches would appear at 2800–3000 cm^{-1}.

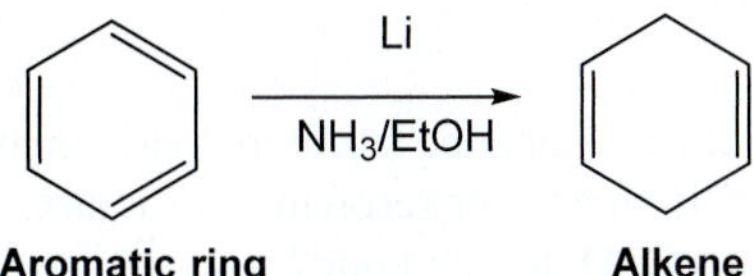

(h) The strong alkyne *sp* C–H stretch would disappear (at 3300 cm^{-1}). Also, the C≡C stretch would become less intense, because the two C atoms of the internal alkyne are more symmetric than they are in the terminal alkyne.

≡CH NaOH, Δ

Terminal alkyne **Internal alkyne**

(i) The conjugated ketone C=O stretch (1680 cm^{-1}) would disappear, and a OH stretch (3400 cm^{-1}) and C–O stretch (~1100 cm^{-1}) from the alcohol would appear.

O OH

1. $LiAlH_4$

2. H_3O^+

Conjugated ketone **Alcohol + alkene**

Problem 15.42

Think

What functional group is present in each compound? How do the bonds in those functional groups differ in their characteristic IR stretches?

Solve

The easiest way to distinguish the two compounds is by the appearance of the OH stretching band. In the compound on the left, the OH group is part of an alcohol, which is relatively broad and appears between 3200 and 3600 cm^{-1}. The OH group in the molecule on the right is part of a carboxylic acid, so the OH stretching band is significantly broader and appears between 2500 and 3000 cm^{-1}.

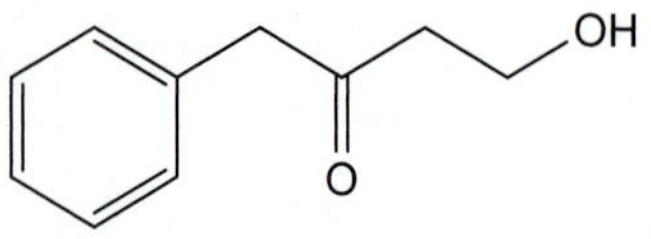

A
Aromatic ring
Ketone
Alcohol

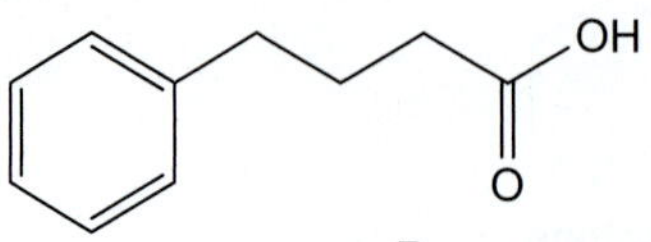

B
Aromatic ring
Carboxylic acid

Problem 15.43

Think

What structural feature is present that indicates there should be another resonance structure? How many curved arrows must be applied to convert to the other resonance structure? In that other resonance structure, what type of bond connects the C and O? Is that bond stronger or weaker than a C=O double bond?

Solve

The amide N atom has a lone pair, which can participate in resonance with the C=O bond, as shown below. The amide resonance structure introduces significant C–O single-bond character to the hybrid, which weakens the C=O bond and lowers the C=O stretching frequency.

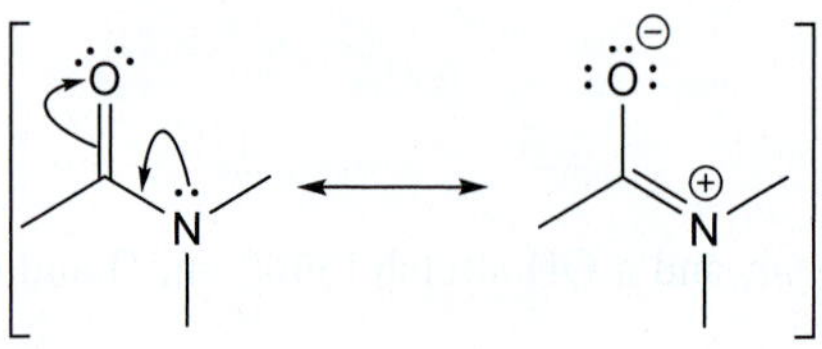

Problem 15.44

Think

Consult the solution to Problem 15.43. How is the resonance structure similar for the amide and the ester? Is the O atom or N atom more electronegative? On which atom is the positive charge more stable? How does this affect the contribution of the resonance structure? How is this reflected in the C=O stretching frequency?

Solve

(a) The resonance structure for methyl acetate is shown below.

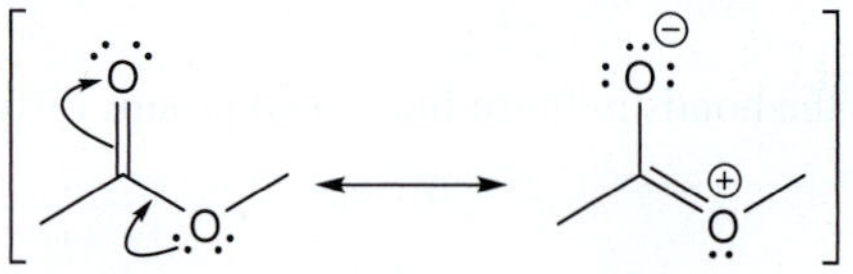

(b) Since the C=O stretching frequency of the ester is higher than that of the amide, it suggests that the resonance structure that introduces C–O single-bond character has a *less* significant contribution. Oxygen is more electronegative than nitrogen, and the positive charge is less stable on the O atom compared to the N atom. This results in less contribution of the resonance structure.

Problem 15.45

Think

Consult the solution to Problems 15.43. How is the resonance structure similar for the acid chloride and the amide? From which orbital in Cl do the π electrons in the resonance structure originate? How effective is the overlap of the orbitals used to make the $C{=}Cl^+$ π bond? Is this more or less effective than in the $C{=}N^+$ π bond?

Solve

The C=O bond in the acid chloride is stronger than the C=O bond in the amide, evidenced by the higher stretching frequency (1806 cm^{-1} compared to 1650 cm^{-1}). This suggests that the lone pair on the Cl atom is less able to contribute to the resonance hybrid than the N atom lone pairs. The lone pair on Cl does not contribute as much because the electrons on the Cl atom originate in the 3*p* orbital, which does not overlap effectively with the 2*p* orbital on C.

π Bond

$C_{2p} + Cl_{3p}$

Problem 15.46

Think

If the C=O bond is conjugated to a C=C, is a resonance structure possible? What kind of bond connects C and O in that resonance structure? How does this lower the stretching frequency of the C=O?

Solve

The resonance structure is shown below.

The resonance structure introduces some C–O single-bond character into the resonance hybrid. This lowers the C=O stretching frequency because a single bond is weaker compared to a double bond.

Problem 15.47

Think

Draw the structures of both cyclohepta-1,3-diene and 3-methylcyclohexa-1,4-diene. Which compound has conjugated C=C bonds? What does conjugation do to the stretching frequency of the C=C bonds in the IR spectrum?

Solve

Cyclohepta-1,3-diene has conjugated C=C bonds and 3-methylcyclohexa-1,4-diene does not. Conjugation lowers the stretching frequency of the C=C in the same way that it lowers the frequency of a C=O bond (see the solution to Problem 15.46). Thus, the stretch at 1618 cm^{-1} belongs to cyclohepta-1,3-diene, and the stretch at 1648 cm^{-1} belongs to 3-methylcyclohexa-1,4-diene.

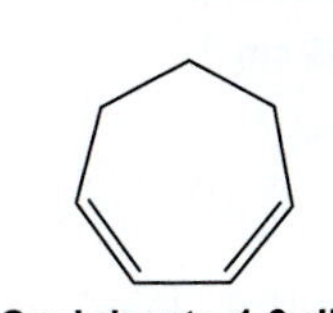

Cyclohepta-1,3-diene
1618 cm^{-1}

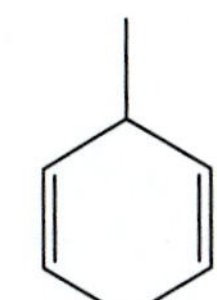

3-Methylcyclohexa-1,4-diene
1648 cm^{-1}

Problem 15.48

Think

Which compound has conjugated C=C bonds? What does conjugation do to the stretching frequency of the C=C bonds in the IR spectrum?

Solve

Aromatic rings have three conjugated C=C bonds and are aromatic. Conjugation lowers the vibrational frequency of a C=C bond in the same way that it does a C=O bond (see the solution to Problem 15.46).

Problem 15.49

Think

Which functional groups can participate in resonance? Is there C=O double-bond character introduced into the hybrid? If so, how does this affect the C–O stretching frequency?

Solve

The ester and carboxylic acid both have resonance structures that introduce C=O double-bond character into the C–O. This increases the stretching frequency of the C–O.

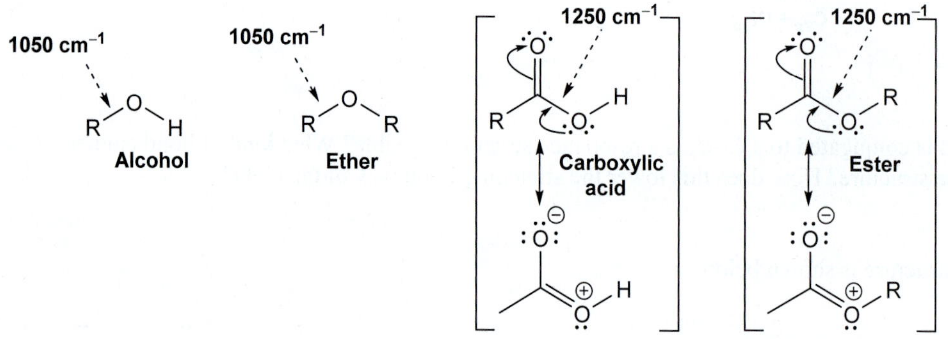

1050 cm^{-1} Alcohol

1050 cm^{-1} Ether

1250 cm^{-1} Carboxylic acid

1250 cm^{-1} Ester

Problem 15.50

Think

Is there a resonance structure involving the C=O bond? In that resonance structure, what type of bond connects the C and O? Does contribution from that resonance contributor increase or decrease the C=O stretch frequency? In which functional group—ketone or aldehyde—is that resonance structure better stabilized?

Solve

Both the ketone and aldehyde have resonance contributors, as shown below. The contribution by that resonance structure is greater for the ketone because of the additional stabilization from the electron-donating alkyl groups attached to C^+. With greater contribution from that resonance structure, the hybrid has more single-bond character in the C=O bond, which weakens the bond and lowers the vibrational frequency.

Butanal 1746 cm^{-1}

More stable Greater contribution

Butan-2-one 1735 cm^{-1}

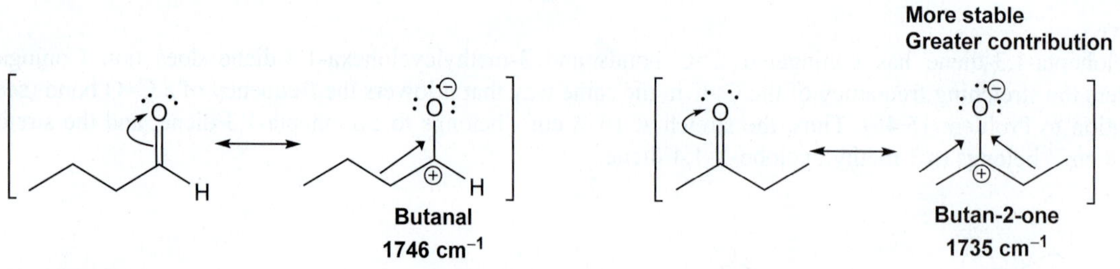

Problem 15.51

Think

In which molecule is resonance more prominent? Are either one of the structures aromatic? If so, how does this affect the resonance structures? Which molecule has more C=O character?

Solve

In both compounds **C** and **D**, the conjugation of the lone pair contributes to strengthening the C–O bond, as suggested by the resonance structure with the C=O bond. In furan (**C**), however, the lone pair is involved in aromaticity, whereas in pyran (**D**), it is not. Aromaticity serves to stabilize the π system more than the regular conjugation, which we associate with a stronger π bond. The stronger π bond gives rise to a higher vibrational frequency. Compound **C** will have a higher-frequency C–O stretch.

Stronger C=O π bond character
Increased IR C–O frequency

Problem 15.52

Think

What affects the intensity of the peak? Are there any significant resonance structures you can draw for these molecules? Does the resonance structure create opposite charges on either side of the C=C bond? How does that separation of opposite charges affect the dipole change during the C=C stretch?

Solve

The C=C stretch in molecule **E** should have a higher intensity. This is because the nearby O atom, which is highly electronegative, places a partial positive charge on the tertiary carbon, as shown by the resonance structures below.

Therefore, as the C=C stretches and compresses, that partial positive charge moves back and forth, creating a significant oscillation in the C=C bond dipole. As we learned in the chapter, a greater oscillation in a dipole moment enhances the probability of photon absorption and, thus, enhances the absorption intensity. Such oscillation in the bond dipole is not present in molecule **F**.

Problem 15.53

Think

What is the value of the C=O stretching frequency in both compounds? Which one has a higher frequency? How is the frequency of the stretch related to the bond strength?

Solve

The C=O bond in cyclobutanone is stronger because its C=O stretch occurs at a higher frequency—1780 cm^{-1} compared to 1720 cm^{-1}.

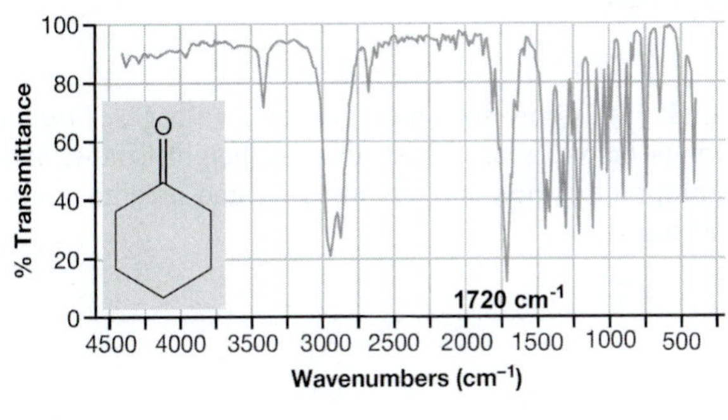

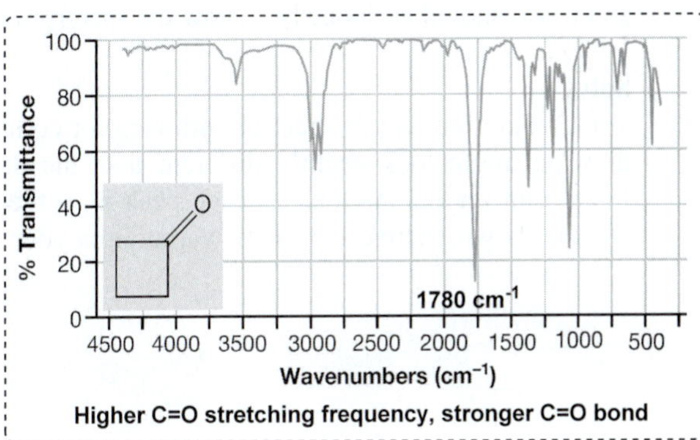

Higher C=O stretching frequency, stronger C=O bond

Problem 15.54

Think

Does the involvement of the N atom's lone pair in resonance give rise to a stronger or weaker C=O bond (consult the solution to Problem 15.43)? Does that result in a higher or lower C=O vibrational frequency? In which molecule does the resonance involving the lone pair on N have a greater impact on a given C=O?

Solve

The C=O stretch will be at a higher frequency for molecule **G**. In an amide, the lone pair on N is involved in resonance with the C=O and, therefore, is responsible for giving the C=O bond more single-bond character, thus weakening the C=O bond and lowering the frequency of the C=O stretch. In molecule **G**, the lone pair on N is involved in resonance with two different C=O groups, and so the frequency-lowering effect is diminished.

G

H
Lower C=O frequency

Problem 15.55

Think

What type of bond exists between the O atom and the Na atom? Is a resonance structure possible? Is the contribution by that resonance structure greater in sodium acetate or in an ester?

Solve

Sodium acetate has an ionic bond between the O atom and the Na atom. The carboxylate anion has another resonance structure possible with equal contribution. This gives the bond half C–O and half C=O character. Thus, the C=O is significantly weaker than a pure C=O and shows up as a lower frequency in the IR. An ester is not ionic, and the resonance structure with two charges has less contribution to the hybrid.

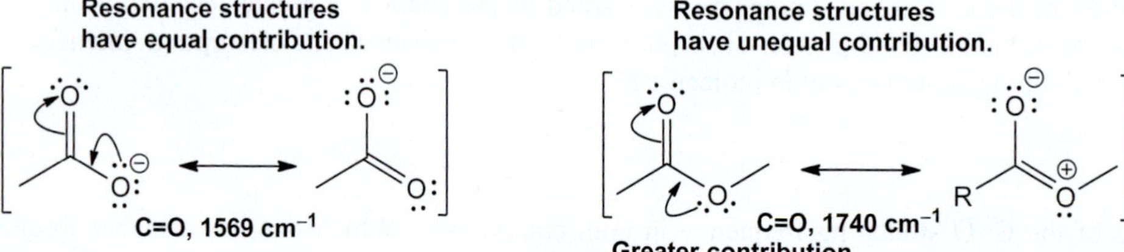
Resonance structures have equal contribution.
C=O, 1569 cm−1
Resonance structures have unequal contribution.
R
C=O, 1740 cm−1
Greater contribution

Problem 15.56

Think

What are the functional groups present in each compound? Refer to Tables 15-2 and 15-3 for common stretching and bending frequencies. At what frequencies should peaks appear in the IR spectrum? How intense and broad should each one be? Draw the spectrum that results.

Solve

Tables and a sample IR spectrum drawing for compounds **(a)–(d)** are shown on the next two pages. *Note:* Your drawing might look slightly different than the drawings below with respect to peak height, shape, and the fingerprint region. As long as each peak is in the appropriate region and has intensity and breadth similar to what is shown, there can be some variation in your drawing.

Molecule **(a)**

Type of bond	Compound class	Frequency range (cm^{-1})	Appearance
O–H	Alcohol	3200–3600	Broad, strong
Conjugated C=O	Ketone	~1690	Strong
C=C	Aromatic	1450–1550	Variable, 2–3 peaks
sp^3 C–H	Alkane	2800–3000	Variable
sp^2 C–H	Alkene	3000–3100	Weak
sp^2 C–H bend	Aromatic (para)	830	Strong

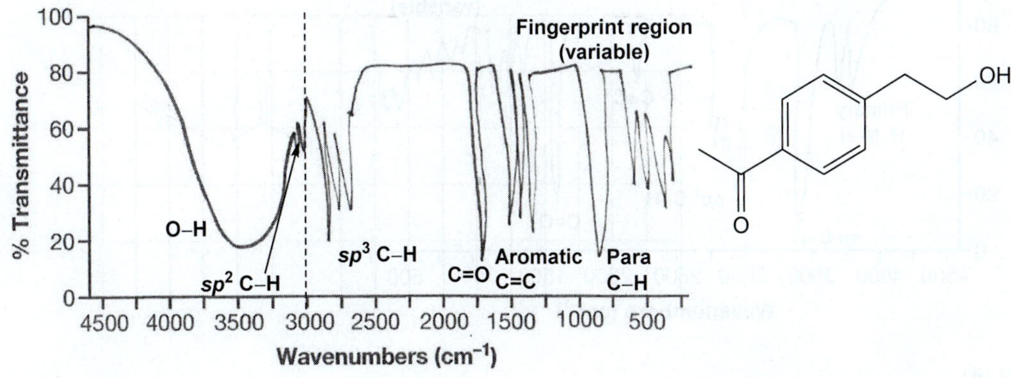

Molecule **(b)**

Type of bond	Compound class	Frequency range (cm^{-1})	Appearance
N–H	Seondary amine	3300–3500	Medium, one peak
C=O	Aldehyde	1680–1750	Strong
sp^3 C–H	Alkane	2800–3000	Variable
sp^2 C–H	Aldehyde	2720 and 2820	Strong

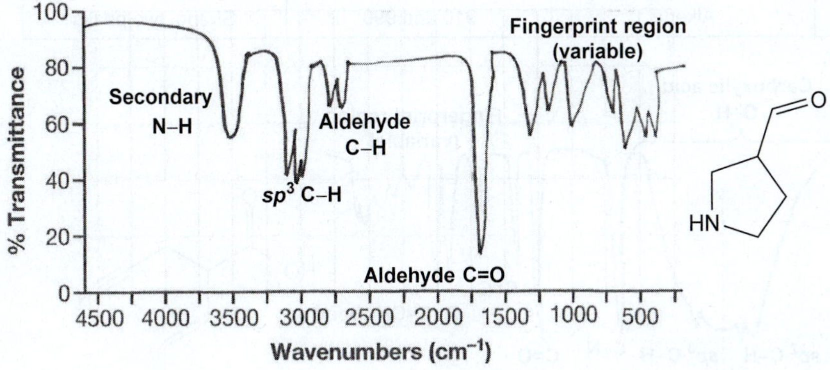

Molecule **(c)**

Type of bond	Compound class	Frequency range (cm^{-1})	Appearance
N–H	Primary Amide	3350–3500	Medium, two peaks
C=O	Amide	1630–1690	Strong
C≡C	Alkyne	2100–2260	Variable
sp^3 C–H	Alkane	2800–3000	Variable
sp C–H	Alkyne	~3300	Strong

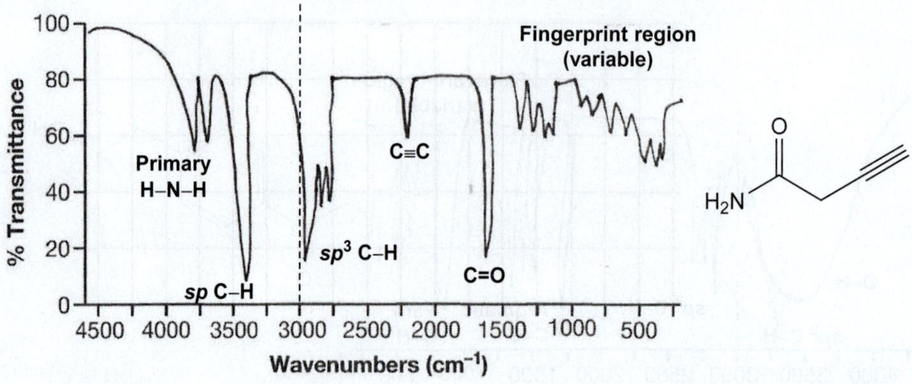

Molecule **(d)**

Type of bond	Compound class	Frequency range (cm^{-1})	Appearance
O–H	Carboxylic acid	2500–3000	Broad, strong
C=O	Carboxylic acid	1710–1780	Strong
C≡N	Nitrile	2210–2260	Medium
C=C	Alkene	1620–1680	Variable
sp^3 C–H	Alkane	2800–3000	Variable
sp^2 C–H	Alkene	3000–3100	Weak
R–C=CH_2 bend	Alkene	910 and 990	Strong, two peaks

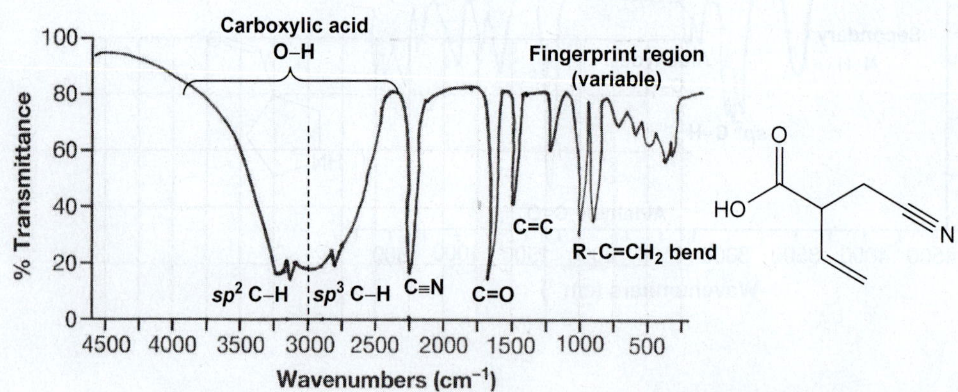

Section 15.6 Structure Elucidation Using IR Spectroscopy; Integrated Problems

Problem 15.57

Think

What similarities do the two isomers possess in terms of functional groups? What differences are present between these two isomers? How many conjugated C=C bonds are present in each isomer? Is this difference evident in the IR spectrum? In the UV–vis spectrum?

Solve

In the IR spectrum, we would expect to see C=C stretching bands and alkane C–H stretching bands for both molecules. Also, neither spectrum would contain an alkene C–H stretch. UV–vis spectroscopy, however, can distinguish between the two compounds because the extent of conjugation is different. In molecule **I**, the three double bonds are conjugated, whereas in molecule **J**, two are conjugated and one is isolated. So the longest-wavelength UV–vis absorption would be longer for **I** than for **J**.

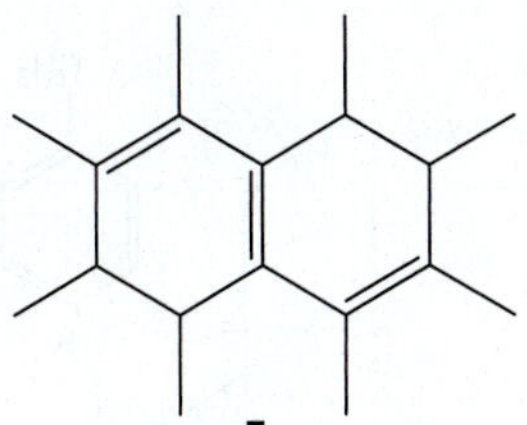

I
C=C
No *sp*^{2} CH
3 conjugated π bonds
Longer λ_{max}

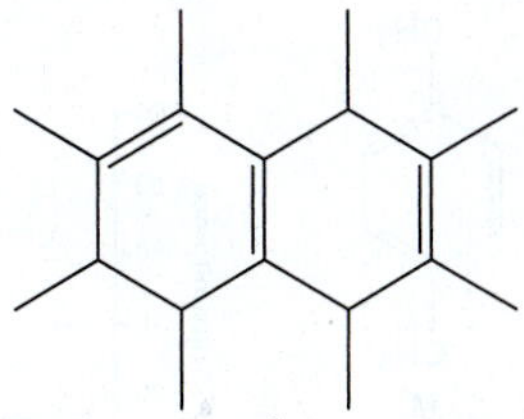

J
C=C
No *sp*^{2} CH
2 conjugated π bonds
Shorter λ_{max}

Problem 15.58

Think

A strong base in the presence of heat results in what reaction mechanism? Which adjacent H atoms are able to be attacked by the base to form the alkene? Which base is bulkier? Which H atom is more likely to be attacked by the bulky base? By the less bulky base? Which product results in a conjugated diene?

Solve

The reaction mechanism is an E2, which will lead to the formation of a diene. Lithium diisopropyl amide (LDA) is a bulky base and irreversibly deprotonates the less substituted C. This leads to a diene that is not conjugated. Sodium hydroxide is a less bulky base and reversibly deprotonates the more substituted C. This leads to a diene that is conjugated. Conjugation increases the wavelength of the longest-wavelength absorption in the UV–vis spectrum, as a result of a smaller HOMO–LUMO energy gap. Therefore, the reaction with LDA has a λ_{max} = 180 nm, and the reaction with NaOH has a λ_{max} = 220 nm.

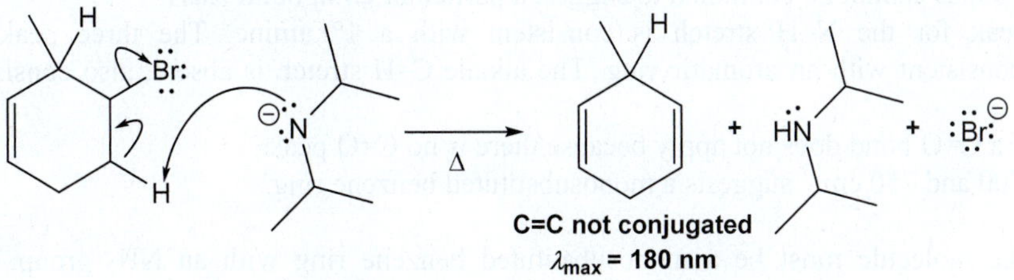

H, Br:, H, N:, Δ, + HN, + :Br:

C=C not conjugated
λ_{max} = 180 nm

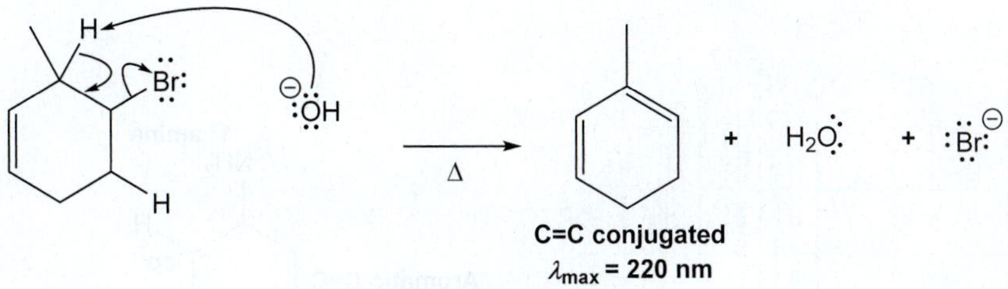

H, Br:, H, :OH, Δ, + H_2O, + :Br:

C=C conjugated
λ_{max} = 220 nm

Problem 15.59

Think

What similarities are present in each structure? What differences are present? Which compound is ortho, which is meta, and which is para? Which region of the IR is examined to determine differences in substitution on the benzene ring? Consult Table 15-3.

Solve

The major difference we should expect in the IR spectra for these compounds is indication of the ortho, meta, or para substitution on the benzene ring. Information about these substitutions appears as absorptions by bending vibrational modes in the fingerprint regions. The ortho isomer should have one band at 880 cm^{-1}, which is consistent with Spectrum 2. The meta isomer should have two bands, at 700 and 780 cm^{-1}, consistent with Spectrum 3. And the para isomer should have one band at 830 cm^{-1}, consistent with Spectrum 1.

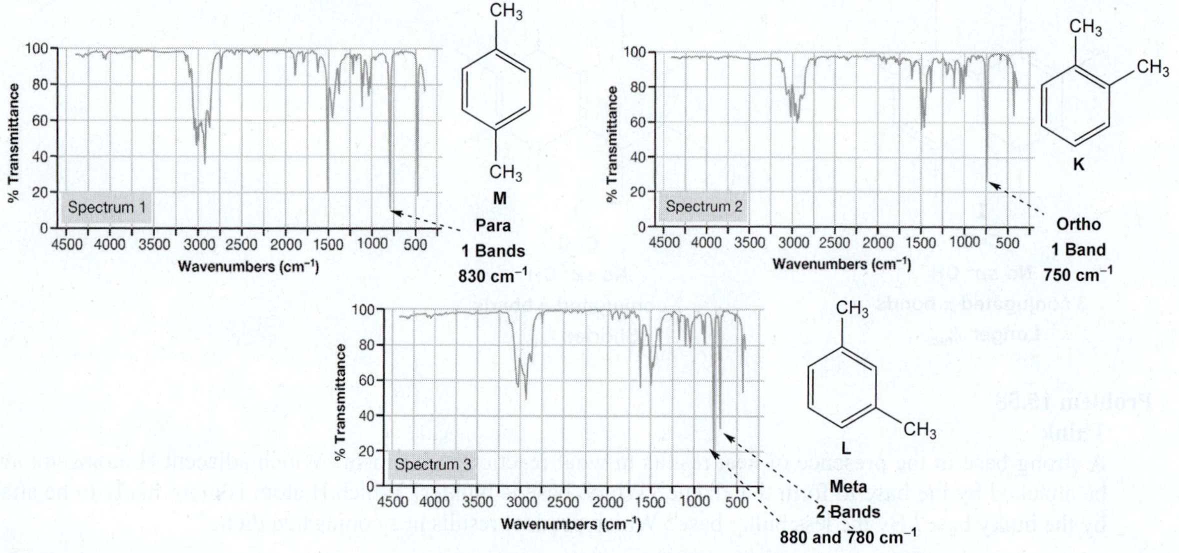

Problem 15.60

Think

What is the IHD? Consider the five questions on page 756. What answers do you get for those questions?

Solve

The IHD of the molecular formula C_6H_7N is 4.

Question 1. N–H stretch ~3400 cm^{-1}; sp^2 alkene/aromatic C–H stretch ~3050 cm^{-1}; C=C stretch ~1450–1600 cm^{-1}

Question 2. These types of bonds cannot be combined to suggest a particular compound class.

Question 3. The double peak for the N–H stretch is consistent with a 1° amine. The three peaks at ~1450–1600 cm^{-1} are consistent with an aromatic ring. The alkane C–H stretch is absent, also consistent with an aromatic ring.

Question 4. Conjugation of a C=O bond does not apply because there is no C=O peak.

Question 5. Two peaks at 700 and 750 cm^{-1} suggests a monosubstituted benzene ring.

Putting it all together: The molecule must be a monosubstituted benzene ring with an NH_2 group. The molecule must therefore be aniline, $C_6H_5NH_2$.

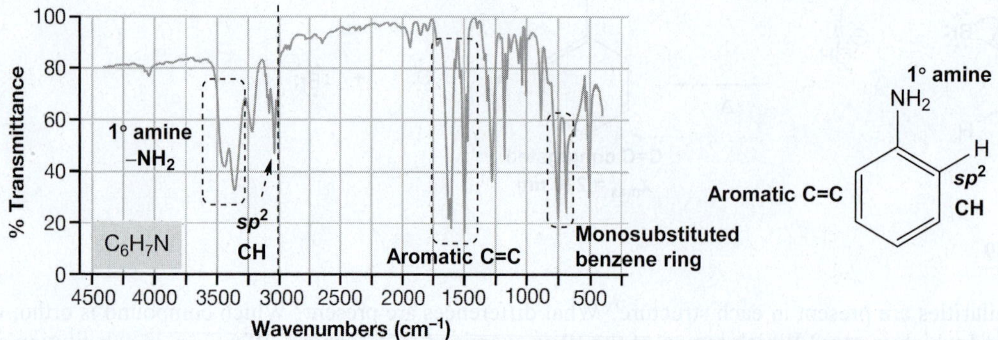

Problem 15.61

Think

What is the IHD? Consider the five questions on page 756. What answers do you get for those questions?

Solve

The IHD of the molecular formula C_8H_8O is 5.

Question 1. C=O stretch ~1720 cm^{-1}; sp^2 alkene/aromatic C–H stretch ~3050 cm^{-1}; sp^2 aldehyde C–H stretch ~2820 and 2720 cm^{-1}; C=C stretch ~1450–1600 cm^{-1}

Question 2. The C=O stretch at ~1720 cm^{-1} and the two aldehyde C–H stretch at ~2820 and 2720 cm^{-1} are consistent with the aldehyde compound class.

Question 3. Three peaks at ~1450–1600 cm^{-1} are consistent with an aromatic ring. The alkane C–H stretch is weak, also consistent with an aromatic ring.

Question 4. The C=O stretch at 1720 cm^{-1} indicates that it is *not* conjugated to a double bond.

Question 5. Two peaks at 700 and 750 cm^{-1} suggest a monosubstituted benzene ring.

Putting it all together: The molecule must be a monosubstituted benzene ring with a nonconjugated –C(O)H group characteristic of an aldehyde. The C_6H_5 and –C(O)H groups account for seven C atoms, six H atoms, and the O atom, leaving one C atom and two H atoms yet to account for. This can be a CH_2 group separating the C_6H_5 and –C(O)H groups, as shown below.

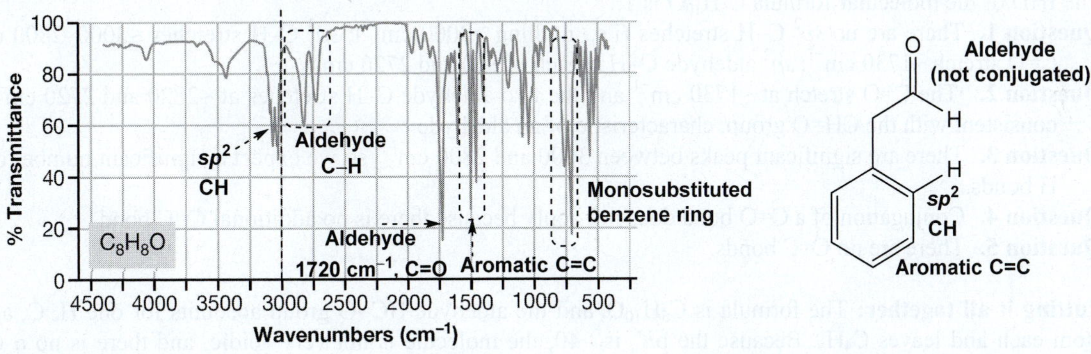

Problem 15.62

Think

What is the IHD? Consider the five questions on page 756. What answers do you get for those questions?

Solve

The IHD of the molecular formula $C_3H_2O_2$ is 3.

Question 1. O–H stretch ~3400–2500 cm^{-1}; C=O stretch ~1700 cm^{-1}; *sp* alkyne C–H stretch ~3300 cm^{-1}; C≡C triple bond stretch ~2100 cm^{-1}

Question 2. The C=O stretch at 1700 cm^{-1} and the OH stretch that extends down to ~2500 cm^{-1} can suggest a carboxyl CO_2H group. The C≡C triple bond stretch at ~2100 cm^{-1} and *sp* alkyne C–H stretch at ~3300 cm^{-1} indicate a terminal alkyne.

Question 3. The very broad OH stretch that extends down to ~2500 cm^{-1} supports a carboxyl group. An alcohol OH group would have been entirely above 3000 cm^{-1}.

Question 4. The C=O of the CO_2H group must be conjugated to the C≡C, because there are only three C atoms.

Question 5. There are no C=C bonds.

Putting it all together: The formula is $C_3H_2O_2$ and accounts for all of the atoms in the two functional groups—the terminal alkyne HC≡C– and the –CO_2H groups characteristic of a carboxylic acid. The CO_2H group has the most acidic proton in the molecule, which should have an estimated pK_a of ~4. See the figure on the next page.

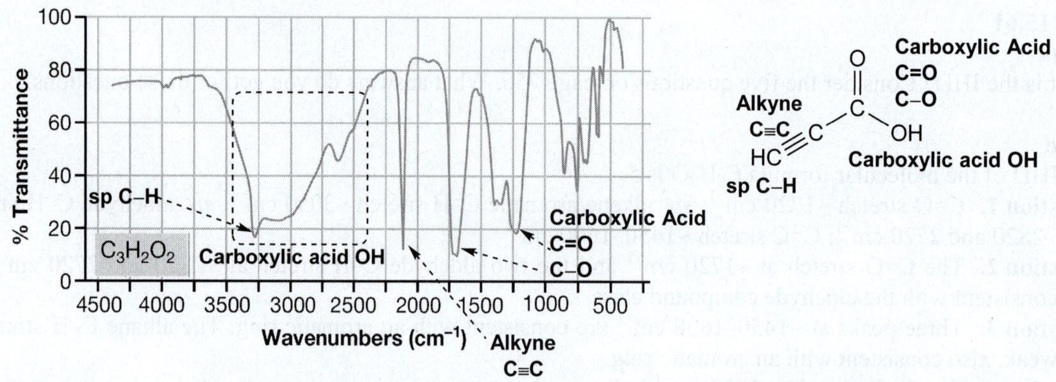

Problem 15.63

Think

What is the IHD? Consider the five questions on page 756. What answers do you get for those questions? If the pK_a is >40, are there any acidic protons?

Solve

The IHD of the molecular formula $C_5H_{10}O$ is 1.

Question 1. There are no sp^2 C–H stretches (i.e., nothing >3000 cm^{-1}); sp^3 C–H stretches ~3000–2800 cm^{-1}; C=O stretch ~1730 cm^{-1}; sp^2 aldehyde C–H stretch ~2820 and 2720 cm^{-1}

Question 2. The C=O stretch at ~1730 cm^{-1} and the two aldehyde C–H stretches at ~2820 and 2720 cm^{-1} are consistent with the CH=O group, characteristic of an aldehyde.

Question 3. There are significant peaks between 3000 and 2800 cm^{-1}, so we expect a significant number of C–H bonds.

Question 4. Conjugation of a C=O bond does not apply because there is no additional C=C bond.

Question 5. There are no C=C bonds.

Putting it all together: The formula is $C_5H_{10}O$, and the aldehyde HC=O group accounts for one H, C, and O atom each and leaves C_4H_9. Because the pK_a is >40, the molecule is not very acidic, and there is no α C–H. Ketone, aldehyde, and esters with α H–C protons have pK_a values in the range 20–25. Therefore, the α C must have no H atoms. The structure is $(CH_3)_3CC(O)H$.

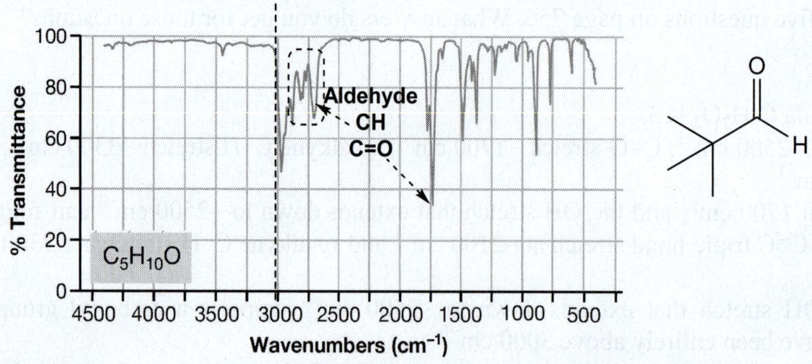

Problem 15.64

Think

Consider the five questions on page 756. What answers do you get for those questions? A strong base in the presence of heat results in what mechanism? How many leaving groups are present? How many times can the E2 reaction occur?

Solve

Question 1. *sp*^{2} alkene/aromatic C–H stretch ~3050 cm^{-1}; C=C stretch ~1450–1600 cm^{-1}; *sp* alkyne C–H stretch ~3300 cm^{-1}; C≡C triple bond stretch ~2200 cm^{-1}

Question 2. The C≡C triple bond stretch at ~2200 cm^{-1} and *sp* alkyne C–H stretch at ~3300 cm^{-1} indicate a terminal alkyne. The three peaks at ~1450–1600 cm^{-1} are consistent with an aromatic ring. The alkane C–H stretch is absent, also consistent with an aromatic ring.

Question 3. There are no *sp*^{3} C–H peaks present, so we should expect no alkane C–H bonds.

Question 4. Conjugation of a C=O bond does not apply because there is no C=O peak.

Question 5. Two peaks at 700 and 750 cm^{-1} suggest a monosubstituted benzene ring.

Putting it all together: The molecule must be a monosubstituted benzene ring with a terminal alkyne C≡CH group. There are two leaving groups present, and the E2 reaction occurs two times to form the terminal alkyne. The formation of the terminal alkyne is supported by the sharp, intense peak at 3300 cm^{-1} (*sp* C–H) and the small peak at 2200 cm^{-1} (C≡C).

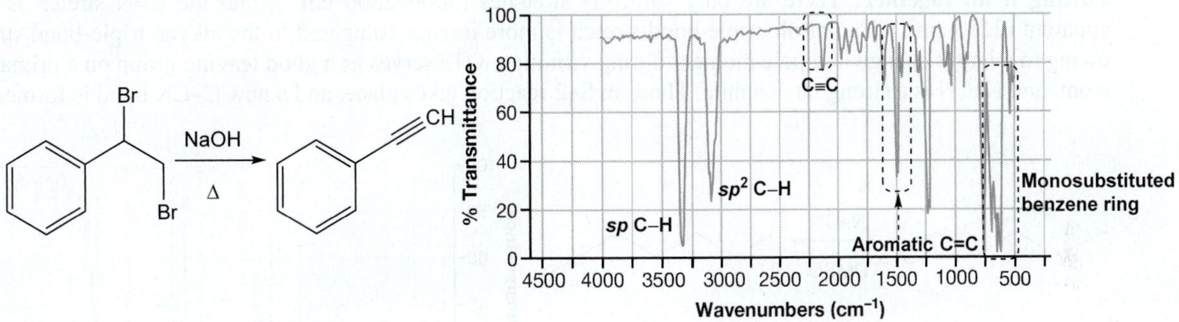

Problem 15.65

Think

Consider the five questions on page 756. What answers do you get for those questions? In the product of the reaction in Problem 15.64, is there an acidic proton to react with NaH? What forms? How does the resulting nucleophilic carbon react with CH_3I? What bond is formed? What functional group results? Is this supported by the IR spectrum?

Solve

Question 1. *sp*^{3} alkane C–H stretch ~2800–3000 cm^{-1}; *sp*^{2} alkene/aromatic C–H stretch ~3050 cm^{-1}; C=C stretch ~1450–1600 cm^{-1}; C≡C triple bond stretch ~2200 cm^{-1}

Question 2. The C≡C triple bond stretch at ~2200 cm^{-1} and no *sp* alkyne C–H stretch at ~3300 cm^{-1} indicates an internal alkyne. The C=C peaks and *sp*^{2} C–H peaks could represent an alkene or aromatic ring.

Question 3. The three peaks at ~1450–1600 cm^{-1} are consistent with an aromatic ring. There are *sp*^{3} C–H peaks present, so we expect alkane C–H bonds.

Question 4. Conjugation of a C=O bond does not apply, because there is no C=O peak.

Question 5. Two peaks at 700 and 750 cm^{-1} suggest a monosubstituted benzene ring.

Putting it all together: The molecule must be a monosubstituted benzene ring with an internal alkyne C≡C group. The *sp* C≡CH is deprotonated by NaH to form the alkynide anion, C≡C:$^-$. This is a nucleophilic carbon that reacts with CH_3I in an S_N2 reaction to form a new C–C bond. There are now *sp*^{3} C–H bonds present. Notice, too, that there is no alkyne C–H stretch near 3300 cm^{-1}.

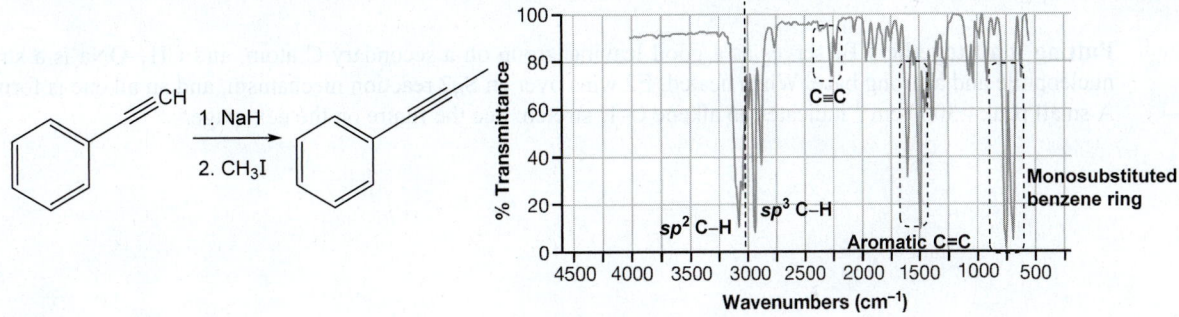

Problem 15.66

Think

Consider the five questions on page 756. What answers do you get for those questions? Is there a good leaving group? What is the attacking species? What bond is formed? What functional group results?

Solve

Question 1. *sp*^{3} alkane C–H stretch ~2800–3000 cm^{-1}; C≡N triple bond stretch ~2200 cm^{-1}

Question 2. These types of bonds cannot be combined to suggest a particular compound class other than a nitrile.

Question 3. The triple bond is consistent with a C≡N. The *sp*^{3} C–H peaks present are intense and broad, so we should expect a considerable number of alkane C–H bonds.

Question 4. Conjugation of a C=O bond does not apply, because there is no C=O peak.

Question 5. There are no C=C bonds.

Putting it all together: There are only *sp*^{3} C–H stretches (3000–2800 cm^{-1}), and the C≡N stretch is also apparent (2250 cm^{-1}). The nitrile triple-bond stretch is more intense compared to the alkyne triple-bond stretch owing to a larger change in dipole moment during vibration. OTs serves as a good leaving group on a primary C atom, and NaCN is a strong nucleophile. Thus, an S_N2 reaction takes place, and a new C–CN bond is formed.

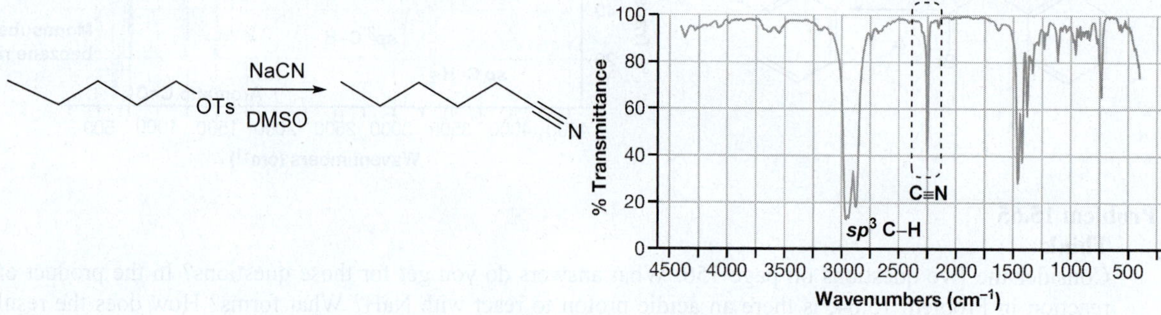

Problem 15.67

Think

Consider the five questions on page 756. What answers do you get for those questions? Is there a good leaving group? What is the attacking species? With a strong base in the presence of heat, what mechanism results? What functional group results? Is this supported by the IR spectrum?

Solve

Question 1. *sp*^{3} alkane C–H stretch ~2800–3000 cm^{-1}; *sp*^{2} alkene C–H stretch ~3050 cm^{-1}; no strong evidence of an alkene C=C stretch

Question 2. These types of bonds cannot be combined to suggest a particular compound class other than an alkene.

Question 3. There are *sp*^{3} and *sp*^{2} C–H peaks present. The alkane absorptions are fairly intense, suggesting a substantial number of alkane C–H bonds.

Question 4. Conjugation of a C=O bond does not apply, because there is no C=O peak.

Question 5. There is no strong evidence of an alkene C=C stretch, but this should not be surprising if there is substantial symmetry about the C=C bond. In the fingerprint region, the C–H bend peak at 970 cm^{-1} suggests that the trans isomer forms.

Putting it all together: Br serves as a good leaving group on a secondary C atom, and CH_3–ONa is a strong nucleophile and a strong base. When heated, E2 wins over an S_N2 reaction mechanism, and an alkene is formed. A small peak >3000 cm^{-1} indicates an alkene C–H stretch. See the figure on the next page.

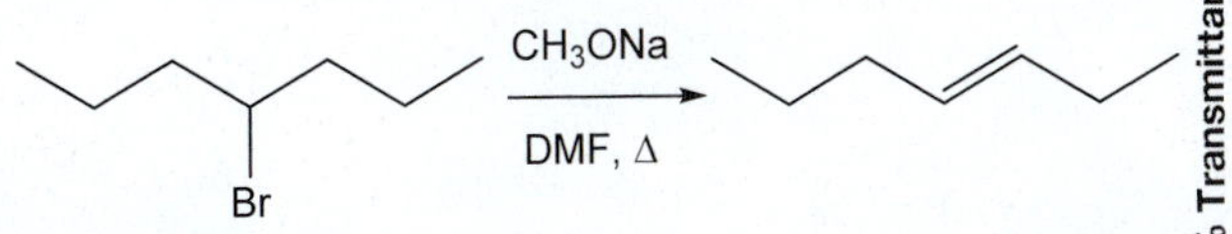

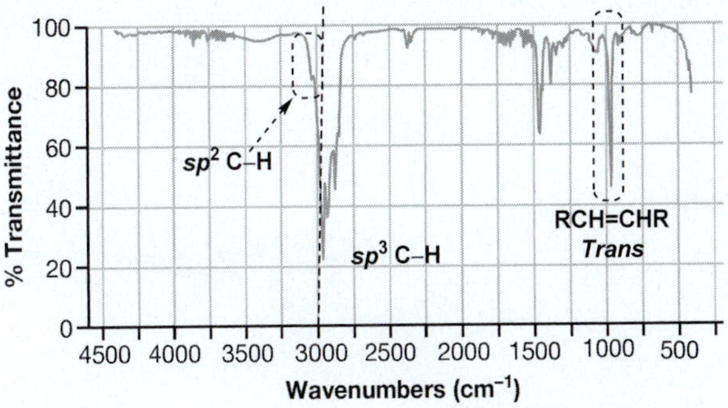

Problem 15.68

Think

What is the C–C–C bond angle in the ring of each compound? How does that bond angle compare to the ideal angle? If the bond angle is smaller, would that require more *p* character or less *p* character for the orbitals that make the C–C–C bonds? Would that leave more *s* character or less *s* character for the orbital used to make the C–O σ bond? How does that agree with the result from Problem 15.53?

Solve

The cyclohexanone ring is relatively unstrained, so the C–C–C bond angle is about 120°. The C–C–C bond angle for cyclobutanone is about 90°. To achieve that bond angle, the orbitals used to make the C–C–C bonds must have excess *p* character, given that *p* orbitals are naturally 90° apart. That means that the orbital used for the C–O single bond must have extra *s* character. With additional *s* character, the C–O bond will be shorter and stronger, which is consistent with cyclobutanone having the stronger C=O bond, as we learned in Problem 15.53.

CHAPTER 16 | Structure Determination 2: Nuclear Magnetic Resonance Spectroscopy and Mass Spectrometry

Your Turn Exercises

Your Turn 16.1

Think

Does a larger ΔE_{spin} occur going left or right on the x axis? At the new location on the x axis, where do you draw the new horizontal line to represent the new α spin state? The new β spin state? How does the vertical distance between them compare to the bracket already shown in the figure?

Solve

A stronger magnetic field appears to the right of the states already shown in the figure. As shown below, this creates a larger energy difference between the α and β spin states.

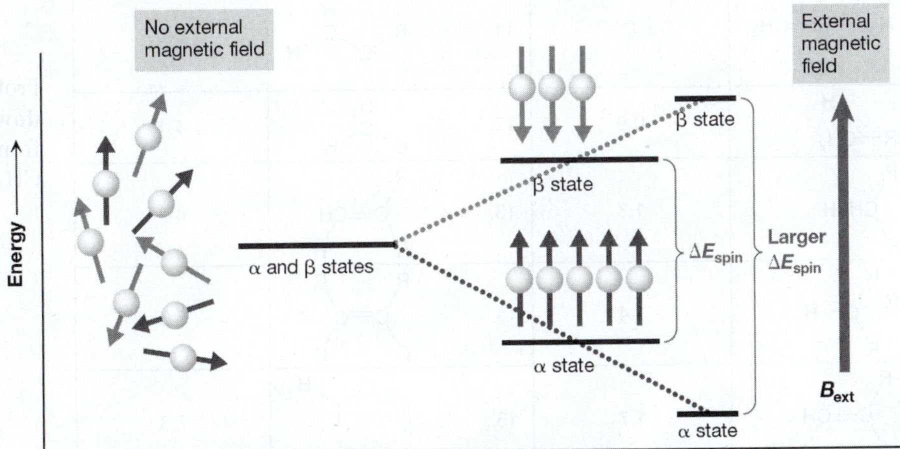

Your Turn 16.2

Think

Can you divide the benzene molecule in two such that one half is the mirror image of the other half? Can you rotate the benzene molecule less than one full rotation so that the H atoms appear to occupy the same locations as they did before rotation?

Solve

(a) One plane of symmetry that benzene has is indicated by the dashed line below. As shown, with respect to that plane of symmetry, H2 and H6 reflect through the plane of symmetry, as do H3 and H5. *Note:* There are other possible planes of symmetry not shown.

(b) As indicated below on the right, after benzene is rotated 60° in the plane of the paper, the molecule appears unchanged, but all six H atoms end up occupying a location that a different H atom occupied before rotation. So the answer to question **(b)** is yes.

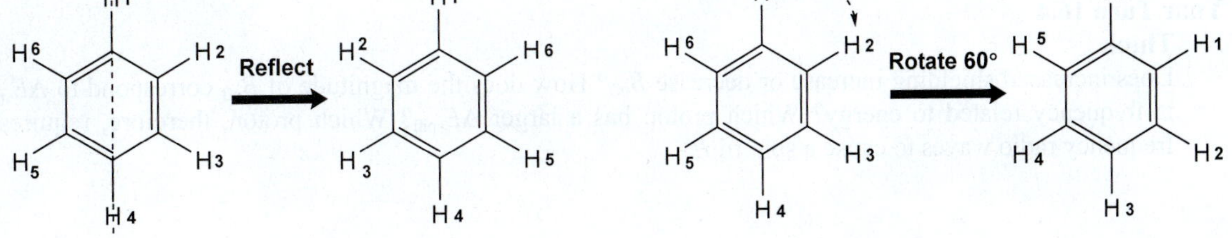

Reflect

Rotate 60°

Your Turn 16.3

Think

How is the x axis on the NMR spectrum numbered? By convention, do the numbers increase or decrease going left to right? From the reference point of CH_3Cl, do downfield signals appear to the right or to the left of the signal at 3.1 ppm?

Solve

By convention, the chemical shift increases from right to left, and the term *upfield* is associated with signals to the right and *downfield* is associated with signals to the left. Thus, 0 ppm is on the far right (most upfield) and 12 ppm is on the far left (most downfield). When comparing signals in Table 16-1 to the CH_3Cl proton signal, the chemical shifts with smaller values are upfield relative to CH_3Cl, and the chemical shifts with larger values are downfield relative to CH_3Cl.

	Type of Proton	Chemical Shift (ppm)		Type of Proton*	Chemical Shift (ppm)
1.	$H_3C-Si(CH_3)_2-CH_2-$**H**	0	11.	$R-O-CH_2-$**H**	3.3
2.	$R-CH_2-$**H**	0.9	12.	$F-CH_2-$**H**	4.1
3.	R_2CH-**H**	1.3	13.	$R_2C{=}CH-$**H**	4.7
4.	R_3C-**H**	1.4	14.	$R_2C{=}CR-$**H**	5.3
5.	$R_2C{=}CH-H_2C-$**H**	1.7	15.	C_6H_5-**H**	7.3
6.	$R-C({=}O)-CH_2-$**H**	2.1	16.	$R-C({=}O)-$**H**	9–10
7.	$C_6H_5-CH_2-$**H**	2.3	17.	$R-NH-$**H**	1.5–4
8.	$R-C{\equiv}C-$**H**	2.4	18.	$R-O-$**H**	2–5
9.	$Br-CH_2-$**H**	2.7	19.	$Ar-O-$**H**	4–7
10.	$Cl-CH_2-$**H**	3.1	20.	$R-C({=}O)-O-$**H**	10–12

Protons **upfield** from CH_3Cl (entries 1–10)

Protons **downfield** from CH_3Cl (entries 11–16)

Protons **downfield** from CH_3Cl (entries 18–20)

Your Turn 16.4

Think

Does increased shielding increase or decrease B_{eff}? How does the magnitude of B_{eff} correspond to ΔE_{spin}? How is frequency related to energy? Which proton has a larger ΔE_{spin}? Which proton, therefore, requires higher-frequency radio waves to cause a spin flip?

Solve

As shown below, the nucleus at the left is shielded to a greater extent. This is reflected by the fact that its spin states are closer together in energy (smaller ΔE_{spin}), meaning that the nucleus experiences less of the external magnetic field. For a photon that is absorbed by the nucleus, energy and frequency are directly proportional, $E = h\nu$. Therefore, proton 2 has a larger ΔE_{spin} and will require higher-frequency radio waves to cause a spin flip.

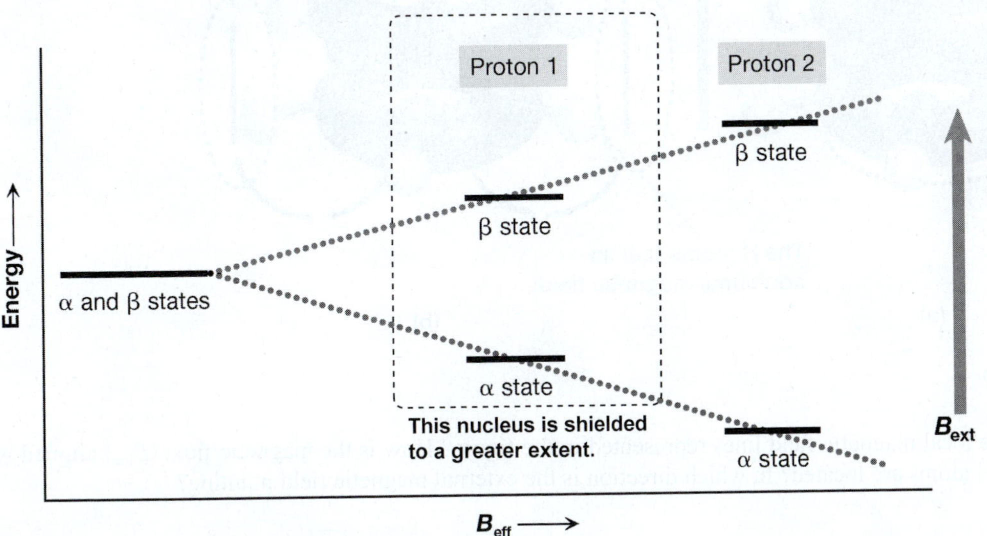

Your Turn 16.5

Think

How does increasing the electronegativity of an atom attached to CH affect the electron density on that H? How does that relate to the strength of the deshielding (more or less deshielded)? What relationship do you observe between electronegativity and chemical shift for XCH types of protons?

Solve

The extent of deshielding depends, in part, on the electronegativity of a nearby atom. Increased electronegativity leads to increased deshielding. The electronegative atom is electron withdrawing and removes electron density from the nearby proton. Increased deshielding causes the signal to shift downfield.

Electronegativity increases / Deshielding increases (↑)

Structure	Chemical Shift	Electronegativity of Halogen
FCH_3	4.1	3.98
$ClCH_3$	3.1	3.16
$BrCH_3$	2.7	2.96

Your Turn 16.6

Think

In which direction is the external magnetic field pointing? How are the local magnetic field lines represented in the figure? Which local magnetic field line is oriented in the same direction? Where is this local magnetic field line relative to the benzene H atoms?

Solve

The local magnetic field lines are the blue arrows in Figure 16-10. The ones that are in the same direction as the external magnetic field are circled on the next page. Notice that these local magnetic field lines appear where benzene's H atoms are located. This leads to additional deshielding of the protons on benzene, and the signal appears more downfield than it otherwise would be.

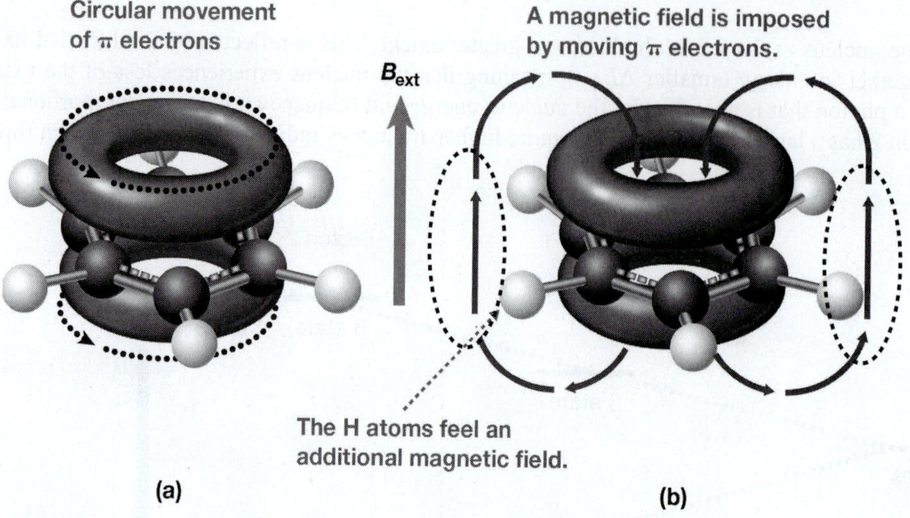
Circular movement of π electrons
A magnetic field is imposed by moving π electrons.
Bext
The H atoms feel an additional magnetic field.
(a)
(b)

Your Turn 16.7

Think

How are the local magnetic field lines represented in the figure? How is the magnetic field (B_{loc}) aligned where the alkyne H atoms are located? In which direction is the external magnetic field pointing?

Solve

The local magnetic field lines are represented by the blue arrows in Figure 16-12b. At each alkyne H, magnetic field lines appear on either side of the H, both pointing downward. Therefore, the alkyne protons lie where B_{loc} opposes B_{ext}. This results in a shielding, not a deshielding, effect.

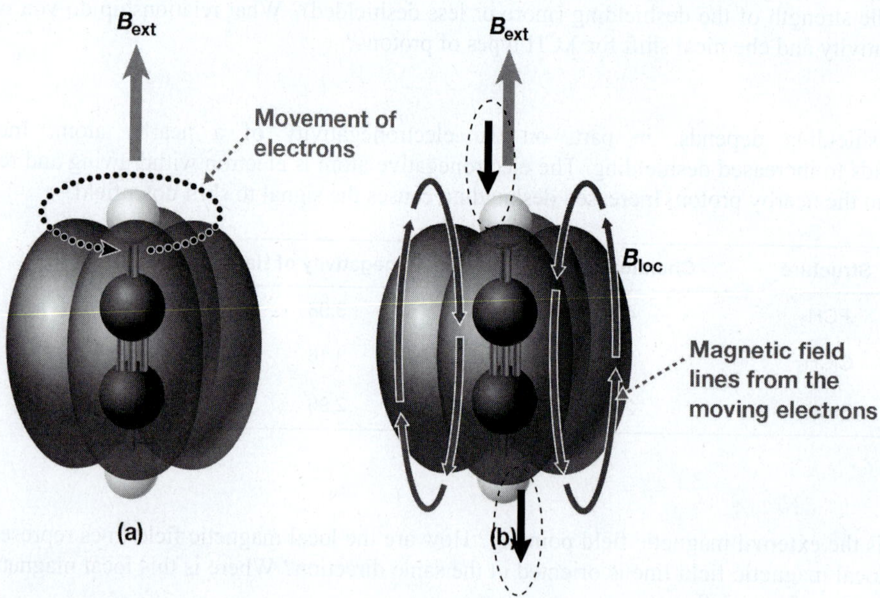
Bext
Movement of electrons
Bext
Bloc
Magnetic field lines from the moving electrons
(a)
(b)

Your Turn 16.8

Think

Does an electronegative atom shield or deshield a nearby CH proton? What is the effect on the chemical shift? What is the effect when the number of these electronegative atoms increases?

Solve

CH_2Cl_2 has two electronegative Cl atoms that remove electron density from the CH protons, causing those CH protons to be deshielded. This results in a downfield chemical shift (5.3 ppm) compared to CH_3Cl (3.1 ppm [see Table 16-1]). $CHCl_3$ has three electronegative Cl atoms that deshield the proton even more and result in a chemical shift farther downfield (7.3 ppm).

Your Turn 16.9

Think

Use a ruler to help measure the height of each signal in mm, and divide each by the smallest height to determine the ratio. How does the ratio of the signal heights inform you about the number of protons to which each corresponds?

Solve

In the figure below, dashed lines are used to illustrate the 1.5:1 ratio. The taller integration is ~19 mm (~0.75 in), and the smaller one is ~13 mm (~0.5 in). Dividing both numbers by 13 gives a 1.46:1 ratio, or about 1.5:1. The molecule has 10 protons total and thus a ratio of 6:4, which correlates to a 1.5:1 ratio.

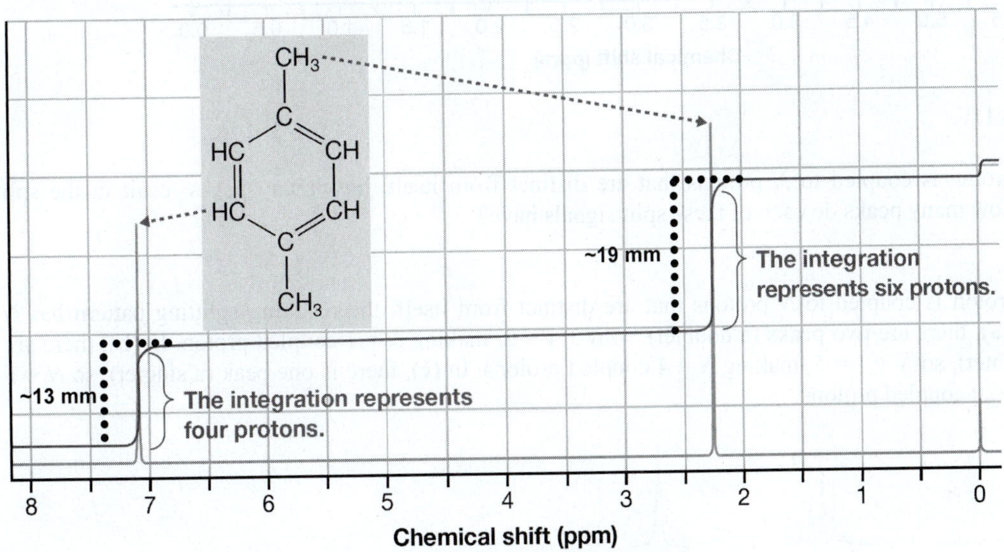

Your Turn 16.10

Think

To identify protons responsible for splitting the signal at ~4.0 ppm, is that signal generated by protons on C1 or C2? Is there a distinct set of protons within three bonds from those protons? Draw a circle around them. To identify protons responsible for splitting the signal at ~5.8 ppm, is that signal generated by protons on C1 or C2? Is there a distinct set of protons within three bonds from those protons? Draw a box around them. How is the splitting of each signal affected by the neighboring protons?

Solve

The signal at ~4.0 ppm is generated by protons on C2, as indicated in Figure 16-15. The splitting of that signal is caused by the proton (circled) attached to C1, which is within three bonds of the protons on C2. There is one proton attached to C1 ($N = 1$), and therefore the signal from the C2 protons is a doublet ($N + 1 = 2$). The signal at ~5.8 ppm is generated by the proton on C1. The splitting of that signal is caused by the protons (boxed) attached to C2. There are two protons attached to C2 ($N = 2$), and therefore, the signal from the C1 proton is a triplet ($N + 1 = 3$). See the figure on the next page.

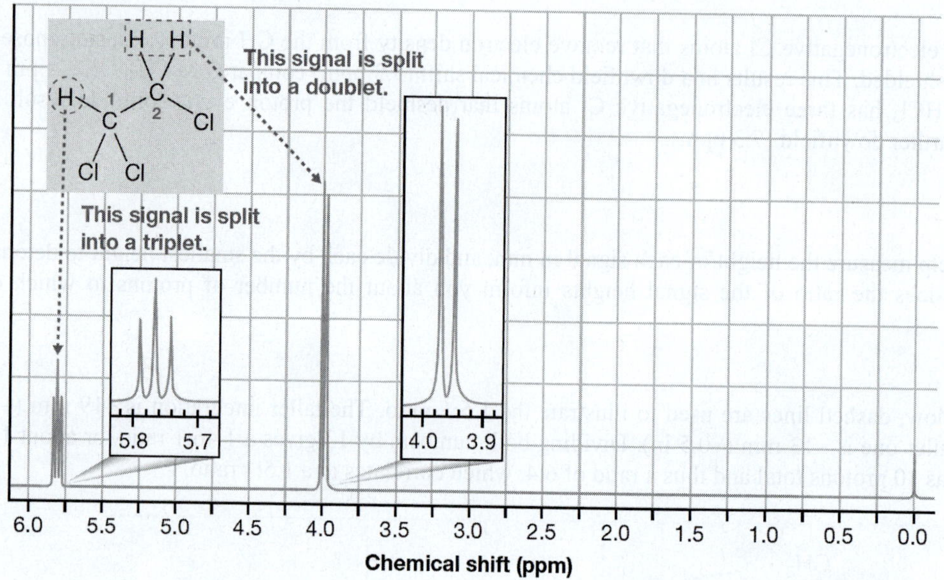

Your Turn 16.11

Think

When a proton is coupled to N protons that are distinct from itself, how many peaks result in the splitting pattern? How many peaks do each of these split signals have?

Solve

When a proton is coupled to N protons that are distinct from itself, the resulting splitting pattern has $N + 1$ peaks. In **(a)**, there are two peaks (a doublet), so $N + 1 = 2$, making $N = 1$ coupled proton. In **(b)**, there are five peaks (quintet), so $N + 1 = 5$, making $N = 4$ coupled protons. In **(c)**, there is one peak (a singlet), so $N + 1 = 1$, making $N = 0$ coupled protons.

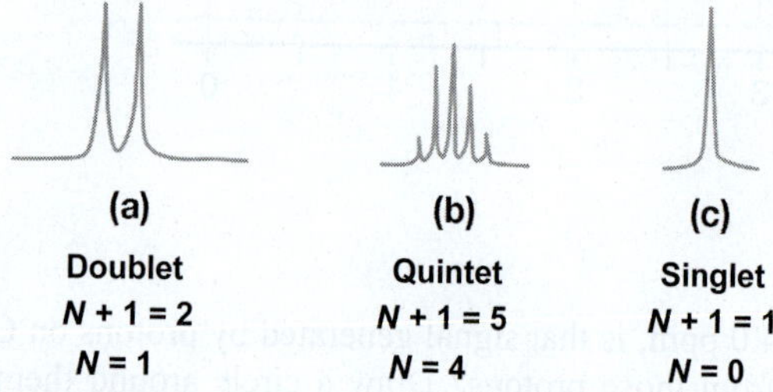

Your Turn 16.12

Think

For protons that are coupled together, should the peaks in their split signals be separated by the same coupling constant or different coupling constants? What are the coupling constants for the signals that correspond to protons **A**, **B**, **C**, and **D**?

Solve

Signals of protons that are coupled together exhibit the same coupling constant. The doublet at 7.36 ppm (protons **A**) and the doublet at 7.05 ppm (protons **B**) both have a coupling constant equal to 8.5 Hz and, therefore, are coupled. The quartet at 2.58 ppm (protons **C**) and the triplet at 1.20 ppm (protons **D**) both have a coupling constant equal to 7.6 Hz and, therefore, are coupled. The signals for protons **B** and **C** do not have the same coupling constant, so those protons are not coupled together. This agrees with the structure, because those protons are separated by four bonds. See the figure on the next page.

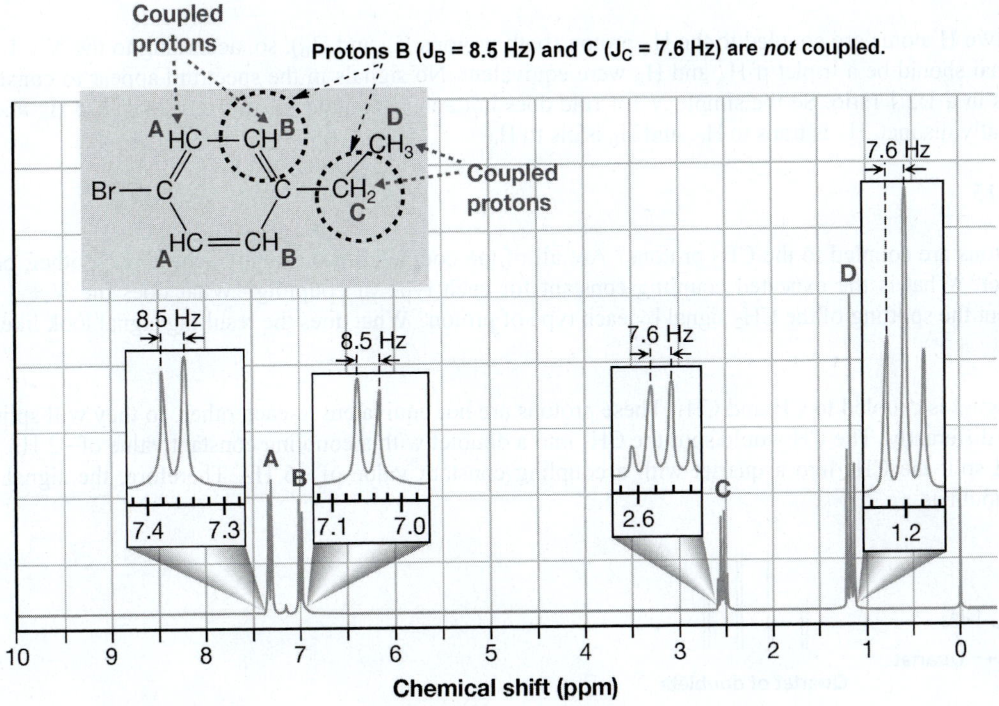

Your Turn 16.13

Think

For protons that are coupled together, should the peaks in their split signals be separated by the same coupling constant or different coupling constants?

Solve

Signals of protons that are coupled together exhibit the same coupling constant. Signal **E**, the quartet at 4.15 ppm, and signal **G**, the triplet at 1.25 ppm, both have a coupling constant equal to 7.1 Hz and, therefore, are coupled. Signal **F**, the quartet at 2.33 ppm, and signal **H**, the triplet at 1.15 ppm, both have a coupling constant equal to 7.5 Hz and, therefore, are coupled.

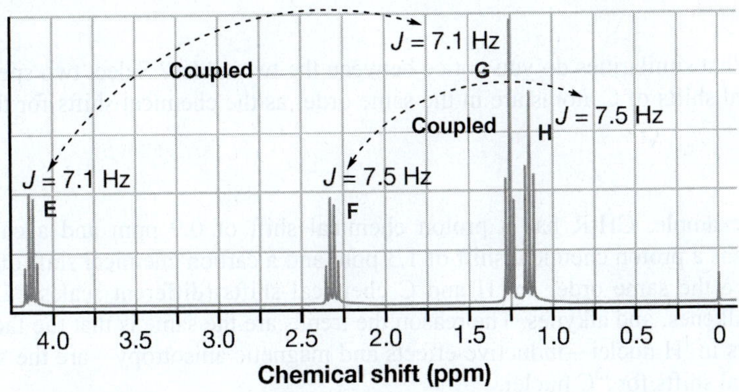

Your Turn 16.14

Think

How many total protons are coupled to H_C? How many signals would the $N + 1$ rule predict?

Solve

A total of two H atoms are coupled to the H_C proton (both protons H_A and H_B), so, according to the $N + 1$ rule, the H_C signal should be a triplet if H_A and H_B were equivalent. No signals in the spectrum appear to consist of three peaks in a 1:2:1 ratio. So the simple $N + 1$ rule does not appear to hold for this case, because H_A and H_B are chemically distinct. H_A is trans to H_C, and H_B is cis to H_C.

Your Turn 16.15

Think

Which protons are coupled to the CH_2 protons? Are all of the coupled protons equivalent to each other, or are they distinct? What is the expected coupling constant for each type of coupling? What does the $N + 1$ rule predict about the splitting of the CH_2 signal by each type of proton? What does the resulting signal look like?

Solve

The CH_2 signal is coupled to CH and CH_3. These protons are not equivalent to each other, so they will split the CH_2 signal differently. The CH would split the CH_2 into a doublet with a coupling constant value of ~2 Hz. The CH_3 would split the CH_2 into a quartet with a coupling constant value of ~6 Hz. Therefore, the signal is a quartet of doublets.

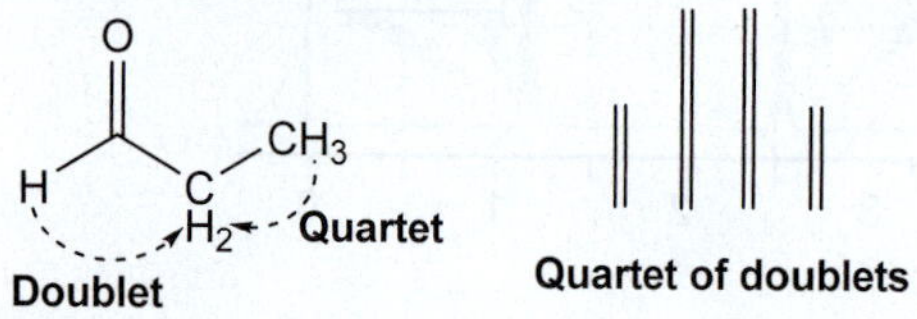

Your Turn 16.16

Think

How many signals do you observe? What is the relationship in a carbon NMR spectrum of the number of signals to the number of chemically distinct carbon atoms? Which signal is from the solvent? Which signal is from TMS?

Solve

There are eight signals in the ^{13}C NMR spectrum, but the one at 77 ppm is from the solvent, and the one at 0 ppm is from TMS. Therefore, the compound gives rise to six signals, so it has six chemically distinct C atoms.

Your Turn 16.17

Think

Consult Table 16-1 and Table 16-4. What similarities do you notice between the two tables? Select two specific examples of molecules whose chemical shifts of C atoms are in the same order as the chemical shifts for the H atoms.

Solve

Several examples are possible. For example, CH_3R has a proton chemical shift of 0.9 ppm and a carbon chemical shift of 10–25 ppm. CH_2R_2 has a proton chemical shift of 1.3 ppm and a carbon chemical shift of 20–45 ppm. Thus, CH_3R and CH_2R_2 have the same order for H and C chemical shifts (different scales). Other examples include $BrCH_3$ and $ClCH_3$, alkenes, and alkynes. The reason the trends are the same is that the factors that dictate the order of chemical shifts in ^1H nuclei—inductive effects and magnetic anisotropy—are the same factors that dictate the order of chemical shifts for ^{13}C nuclei.

Your Turn 16.18

Think

From the DEPT-135 spectrum (Fig. 16-30c), what signals are positive? Negative? What are the signals that appear in the DEPT-90 spectrum (Fig. 16-30b)? What, then, can you deduce about the two unlabeled ^{13}C signals?

Solve

In DEPT-135 spectra, CH_3 and CH appear as normal positive signals, and CH_2 appears as a negative (below the baseline) signal. In DEPT-90 spectra, only CH signals appear. Comparison of both a DEPT-135 spectrum and a DEPT-90 spectrum reveals the signals that correspond to CH, which appear only in the DEPT-90 spectrum and are positive in DEPT-135; the signals that correspond to CH_2, which are negative in the DEPT-135 spectrum; and the signals that correspond to CH_3, which are positive in the DEPT-135 spectrum and missing from the DEPT-90 spectrum.

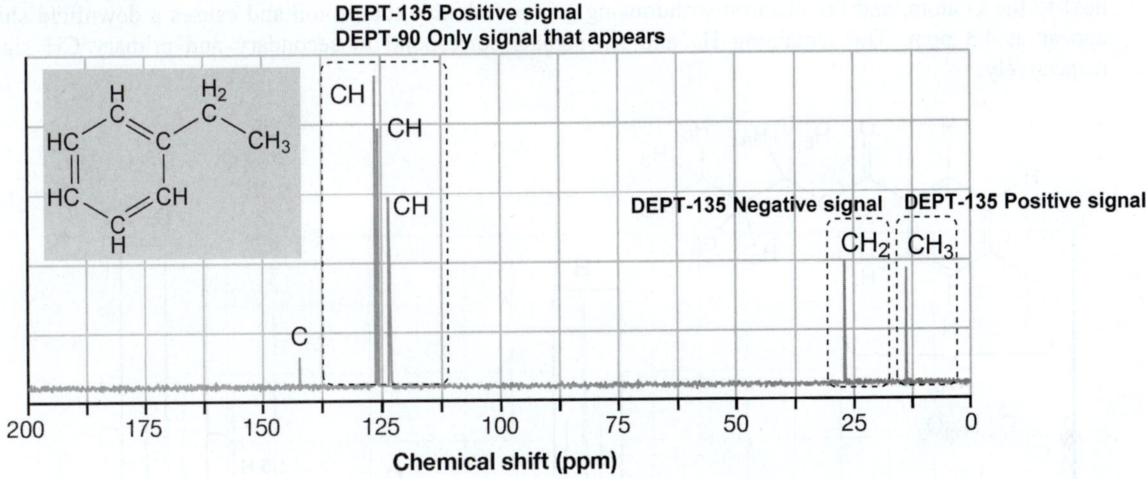

Your Turn 16.19

Think

Review Table 16-1. What is the first entry that has a proton part of a double bond? What is the first entry that has a proton on a carbon attached to an electronegative atom?

Solve

The first entry that has a proton part of a double bond is Entry 13, $R_2C{=}CH{-}\mathbf{H}$, $\delta = 4.7$ ppm. The first entry that has a proton on a carbon attached to an electronegative atom is Entry 9, BrCH2–**H**, $\delta = 2.7$ ppm.

Your Turn 16.20

Think

Review Table 16-1 and locate a proton that has the fragment H–C–O. What is the chemical shift? What is the chemical shift of Entry 3 in Table 16-5? How is the chemical shift different? What causes chemical shifts to move downfield?

Solve

The fragment **H**–C–O is found in Entry 11 in Table 16-1 and has a chemical shift $\delta = 3.3$ ppm. Entry 3 in Table 16-5 with the same fragment **H**–C–O has a chemical shift $\delta = 4.2$ ppm. Proximity of protons to additional electron-withdrawing groups can deshield the proton and cause a downfield shift. In this unknown, the cause could possibly be a benzene ring or an additional oxygen.

Your Turn 16.21

Think

Review Table 16-1 and locate a proton that is on a benzene ring. What is the chemical shift? What is the chemical shift of Entry 4 in Table 16-5? How is the chemical shift different? What causes chemical shifts to move downfield?

Solve

The normal chemical shift of a proton on a benzene ring (Entry 11) is 7.3 ppm. Entry 4 in Table 16-5 has a range of chemical shifts, 7.4–8.1 ppm. This is farther downfield compared to a normal benzene ring. The downfield shift is due to the oxygen atoms in the compound. Electronegative atoms deshield protons and cause downfield shifts.

Your Turn 16.22

Think

Which protons are chemically distinct? It is permissible to assign all benzene protons with one label? Which signals are farthest downfield in the structure? Which protons are next to the O atom, and how does that proximity to the O atom affect the chemical shift of those protons?

Solve

The benzene protons are labeled H_A and correlate to the signals between 7.5 and 8 ppm. The H_B protons are next to the O atom, and the electron-withdrawing group deshields the proton and causes a downfield shift to appear at 4.3 ppm. The remaining H_C and H_D signals correspond to secondary and primary CH signals, respectively.

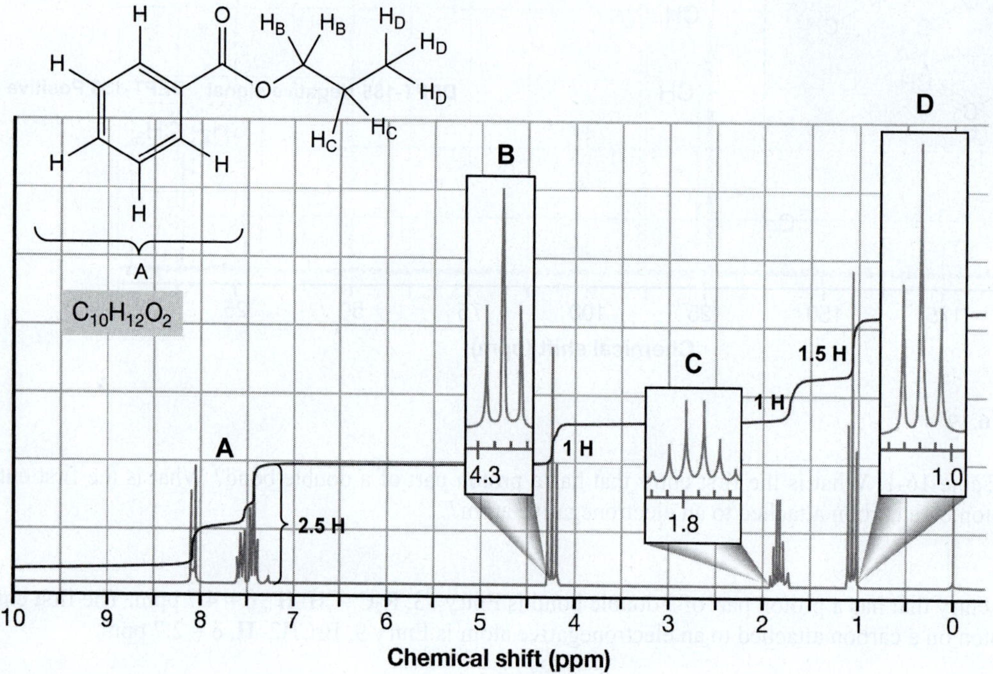

Your Turn 16.23

Think

Review Table 16-4. What is the first entry that has a carbon part of a double bond? What is the first entry that has a carbon attached to an electronegative atom?

Solve

The first entry that has a carbon part of a double bond is Entry 11, **C=C**, δ = 105–150 ppm. The first entry that has a carbon attached to an electronegative atom is Entry 6, Br–CH_2R, δ = 25–35 ppm.

Your Turn 16.24

Think

Review Table 16-4. What are the chemical shift ranges for an alkyne carbon or a carbon singly bonded to an oxygen? What are the chemical shift ranges for a carbon that is part of a double bond (see Your Turn 16.23)?

Solve

The chemical shift range for an alkyne carbon, RC≡CR (Entry 10), is 65–85 ppm. The chemical shift range for a carbon singly bonded to an oxygen, R_3C–OH (Entry 9), is 55–70 ppm. Both of these ranges are reasonable to suggest that the signal found at ~73 ppm could be due to a carbon part of an alkyne or singly bonded to an oxygen. The chemical shift range for a carbon part of a double bond (105–150 ppm) is much farther downfield than 73 ppm. Therefore, it is inconsistent to suggest that the signal at 73 ppm is due to a C=C.

Your Turn 16.25

Think

Refer to Table 15-2. Where does the H–O stretch signal occur in the IR spectrum of an alcohol? Where does the C≡C stretch signal occur in the IR spectrum of an alkyne? Where does the C–H stretch signal occur in the IR spectrum of an alkyne? Are these signals broad or sharp? Weak or strong?

Solve

In general, the H–O signal occurs between 3600 and 3200 cm^{-1} and gives rise to a broad and strong signal. *In general*, the C≡C signal occurs between 2100 and 2260 cm^{-1} and gives rise to signals varying in appearance. *In general*, the C–H signal from an alkyne occurs at ~3300 cm^{-1} and gives rise to a strong signal. In this example, the H–O signal is centered around 3400 cm^{-1}, and no signals are present for a C≡C or an alkyne C–H (although it is possible that an alkyne C–H signal could be buried under the H–O signal).

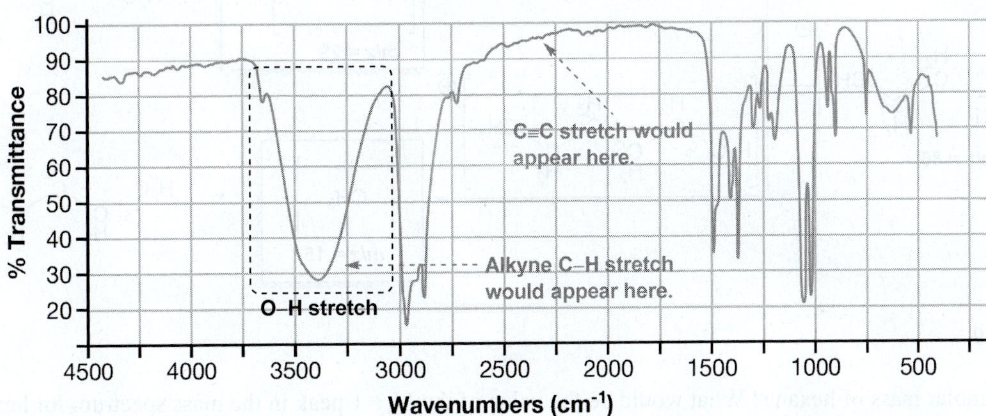

Your Turn 16.26

Think

If an ion is too light, is it deflected too much or not enough? How is this represented in the figure? What about ions that are too massive?

Solve

Ions that are too light are deflected too much (sharper arc), and ions that are too massive are not deflected enough (arc not as sharp). Neither of these types of ions will reach the detector.

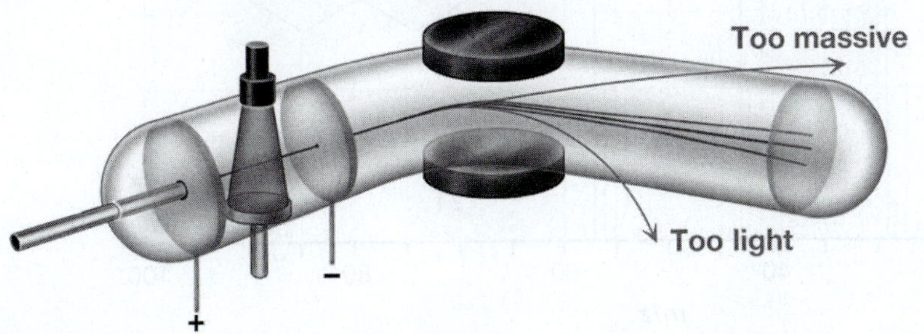

Your Turn 16.27

Think

Does the compound have an odd or even molar mass? According to the nitrogen rule, should this correspond to an odd or even number of N atoms?

Solve

CH_3–CO–$N(CH_3)_2$ = C_4H_9NO = 12(4) + 9(1) + 14 + 16 = 87 u, which is an odd number. There is one N atom, which is also an odd number. This is consistent with the nitrogen rule.

Your Turn 16.28

Think

Where does the fragmentation occur? What part of the original structure is missing that correlates to the *m*/*z* values indicated in the box?

Solve

The fragmentation pathways are shown below.

***m*/*z* = 86**

***m*/*z* = 29**

***m*/*z* = 15**

Your Turn 16.29

Think

What is the molar mass of hexane? What would be the value of the M + 1 peak in the mass spectrum for hexane (Fig. 16-35)? What is the molar mass of ethylbenzene? What would be the value of the M + 1 peak in the mass spectrum for ethylbenzene (Problem 16.33)?

Solve

The molar mass of hexane, $C_6H_{14} = (6 \times 12) + (14 \times 1) = M = 86$, so $M + 1 = 87$, as shown below.

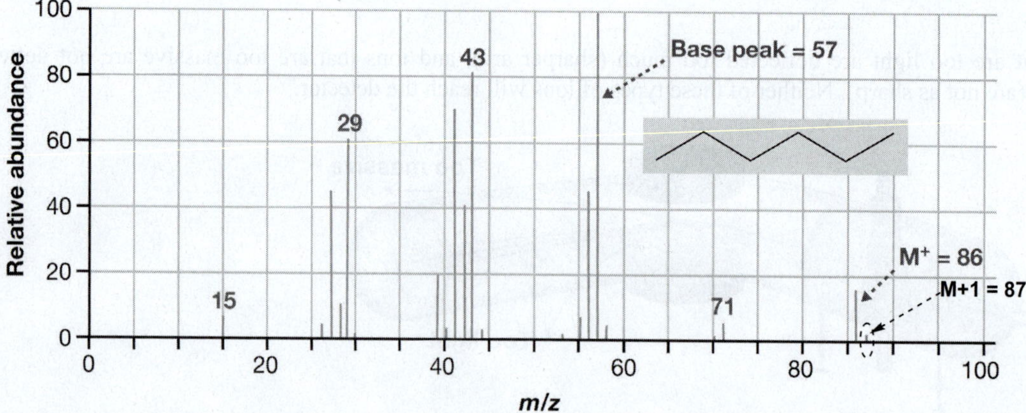

The molar mass of ethylbenzene, C_8H_{10} = (8 × 12) + (10 × 1) = M = 106, so M + 1 = 107, as shown below.

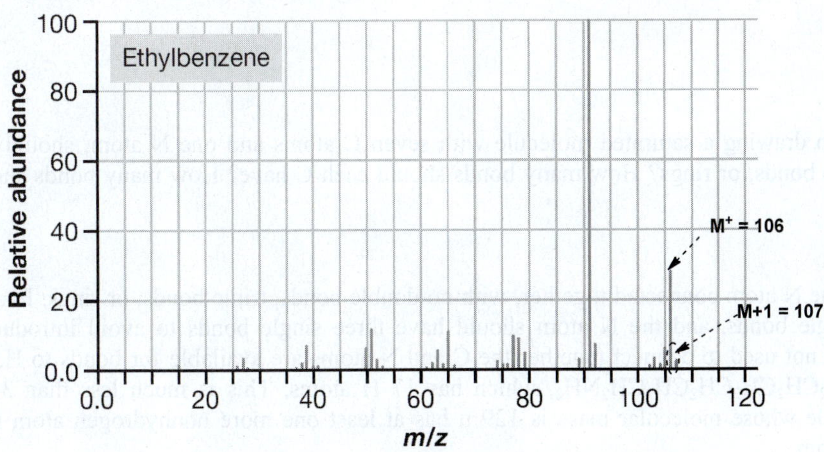

Your Turn 16.30

Think

What is the mass of ^{12}C compared to ^{13}C? How many ^{12}C atoms are present, and what is the mass of all ^{12}C atoms? How many ^{13}C atoms are present, and what is the mass of all ^{13}C atoms? What is the mass of 14 H atoms?

Solve

The mass of ^{12}C is 12 u and of ^{13}C is 13 u. There are five ^{12}C atoms (= 60 u) and one ^{13}C atom (= 13 u). The mass of the 14 H atoms is 14 u. The sum of these is 87. This is the M + 1 peak in the mass spectrum of hexane (C_6H_{14}).

Your Turn 16.31

Think

How can you estimate the peak intensity from Figure 16-35 (hexane) and Figure 16-37 (dodecane)? Locate the M and M + 1 peaks by calculating the molar mass of each.

Solve

Hexane: M^+ = 86 u (intensity ~15.5%) and M + 1 = 87 u (intensity ~1.0%); (1.0/15.5) × (100%/1.1%) = 5.87, which rounds to 6

Dodecane: M^+ = 160 u (intensity ~5.9%) and M + 1 = 161 u (intensity ~0.8%); (0.8/5.9) × (100%/1.1%) = 12.33, which rounds to 12

Your Turn 16.32

Think

What is the mass of ^{79}Br compared to ^{81}Br? Calculate the mass of the C_6H_5 portion. Where is the M + 1 peak for the mass of bromobenzene with ^{79}Br and bromobenzene ^{81}Br?

Solve

The mass of ^{79}Br is 79 u, and the mass of ^{81}Br is 81 u. The mass of the C_6H_5 portion is 77 u, which is added to either 79 u or 81 u, depending on the isotope of Br. The mass of $C_6H_5{}^{79}Br$ is 156 u and corresponds to the M^+ peak. The mass of $C_6H_5{}^{81}Br$ is 158 u and corresponds to the M + 2 peak. Each of those M^+ ions could have a single ^{13}C isotope, which would add another 1 u to the mass, so the peak at *m*/*z* = 157 is the M + 1 peak for $C_6H_5{}^{79}Br$ and *m*/*z* = 159 is the M + 1 peak for $C_6H_5{}^{81}Br$.

Your Turn 16.33

Think

What are the given estimated peak intensities from Figure 16-40? Locate the M^+ and M + 1 peaks by calculating the molar mass of each.

Solve

Unknown 3: M^+ = 129 u (intensity ~30.0%) and M + 1 = 130 u (intensity ~2.4%); (2.4/30) × (100%/1.1%) = 7.3, which rounds to 7

Your Turn 16.34

Think

Review Section 4-12. When drawing a saturated molecule with seven C atoms and one N atom, should you include double bonds, triple bonds, or rings? How many bonds should each C have? How many bonds should the N have?

Solve

Draw seven C atoms and one N atom connected together, with no double bonds, triple bonds, or rings. Each C atom should have four single bonds, and the N atom should have three single bonds to avoid introducing charges. The bonds that are not used to connect together the C and N atoms are available for bonds to H. An example would be $CH_3CH_2CH_2CH_2CH_2CH_2CH_2NH_2$, which has 17 H atoms. This is much less than 31 H atoms. Therefore, a molecule whose molecular mass is 129 u has at least one more nonhydrogen atom than seven C atoms and one N atom.

In Chapter Problems

Problem 16.1

Think

How is the energy difference between the α and β spin states, ΔE_{spin}, related to the applied external magnetic field?

Solve

The energy difference will be greater at 5.0 T than at 0.5 T. According to Equation 16-1, ΔE_{spin} is directly proportional to B_{ext}.

Problem 16.3

Think

In Equation 16-1, what value should be substituted for γ? For h? For B_{ext}?

Solve

As mentioned in the text, γ for a proton is $2.67512 \times 10^8\ T^{-1}\ s^{-1}$; h is 6.626×10^{-34} J·s. In the problem, the value for B_{ext} is given as 2.2 T.

Therefore,

$$\Delta E_{spin} = \gamma h B_{ext}/2\pi$$
$$= (2.67512 \times 10^8\ T^{-1}\ s^{-1})(6.626 \times 10^{-34}\ J{\cdot}s)(2.2\ T)/2\pi$$
$$= 6.2 \times 10^{-26}\ J$$

The units s and T both cancel, leaving energy in units of J.

Problem 16.4

Think

Do the two signals occur at the same chemical shift? If not, what does this mean about the chemical equivalence of the two protons?

Solve

No, the protons in benzene are chemically distinct from those in TMS. We know this because the protons from the two compounds have different chemical shifts—7.3 ppm and 0.0 ppm, respectively.

Problem 16.6

Think

How many signals appear in the proton NMR spectrum? Of those signals, which ones are generated by *trans*-1,2-dichloroethene?

Solve

The compound has just one type of chemically distinct proton. There are two signals total in the spectrum, but the one at 0 ppm belongs to the protons in TMS, which is the reference compound that was added. So the sample gives rise to just one signal, and we know that the number of signals equals the number of chemically distinct types of protons.

H
Cl C=C Cl
H

1 signal in ^{1}H NMR
= chemically equivalent

Problem 16.8

Think

What molecules are generated by substituting each H atom separately by "X"? Which of those molecules are identical? Enantiomers? Diastereomers? Constitutional isomers? How many signals result?

Solve

(a) Molecules **I** and **II** are enantiomers—they are mirror images. According to Step 3 in the chemical distinction test, these H atoms are *not* chemically distinct. Molecules **I/II** and **III** are constitutional isomers. According to Step 3 in the chemical distinction test, these H atoms are chemically distinct. Therefore, there are *two* signals in the ¹H NMR spectrum, one for the CH_a protons and one for the methyl CH_3 protons.

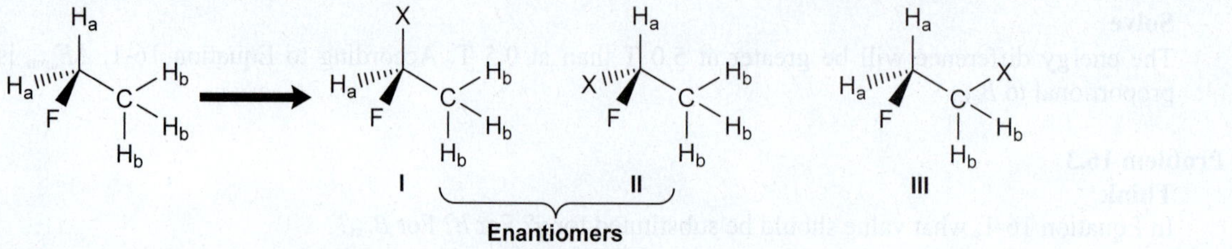

I II III
Enantiomers

(b) Molecules **I, II/III, IV/V,** and **VI** are constitutional isomers. According to Step 3 in the chemical distinction test, these H atoms are chemically distinct. Molecules **II** and **III** are enantiomers—they are mirror images. According to Step 3 in the chemical distinction test, these H atoms are *not* chemically distinct. Molecules **IV** and **V** are enantiomers, too, and are *not* chemically distinct. This gives rise to *four* signals in the ¹H NMR spectrum.

I II III
Enantiomers
IV V VI
Enantiomers

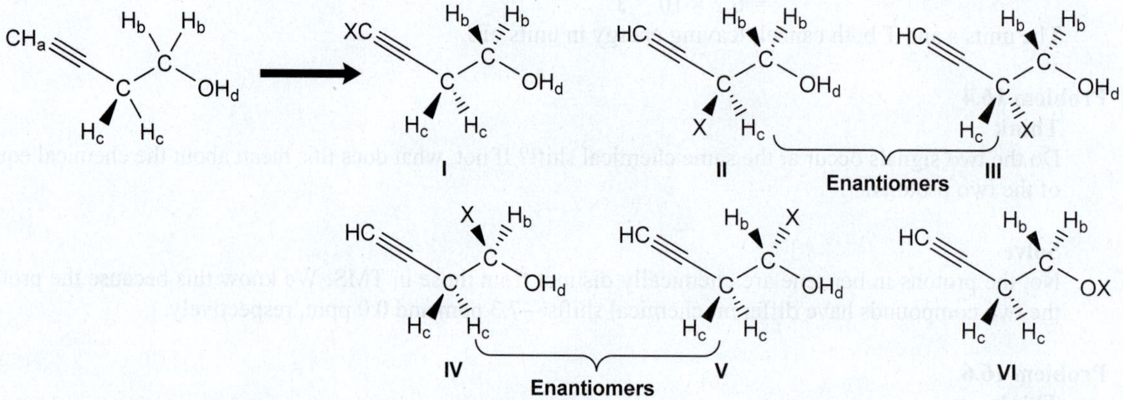

(c) Molecules **I** and **II** are the same molecule. Molecules **III** and **IV** are the same molecule. Molecules **I/II**, **III/IV**, and **V** are constitutional isomers. According to Step 3 in the chemical distinction test, these H atoms are chemically distinct. There are *three* signals in the ¹H NMR spectrum.

I II III IV
Same Same
V

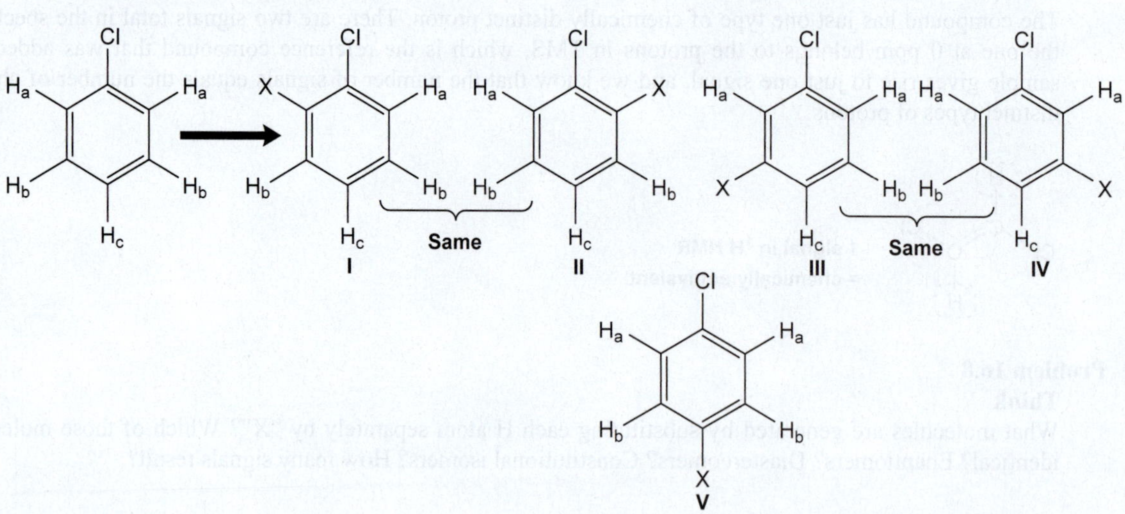

(d) Molecules **II** and **III** are the same molecule. Molecules **I**, **II/III**, and **IV** are constitutional isomers. According to Step 3 in the chemical distinction test, these H atoms are chemically distinct. There are *three* signals in the ^{1}H NMR spectrum.

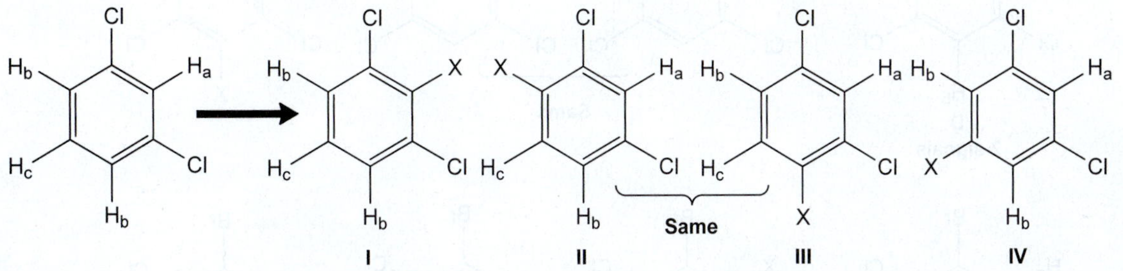

Same

I **II** **III** **IV**

(e) Molecules **I**, **II**, **III**, and **IV** are the same. According to Step 3 in the chemical distinction test, these H atoms are *not* chemically distinct. There is *one* signal in the ^{1}H NMR spectrum.

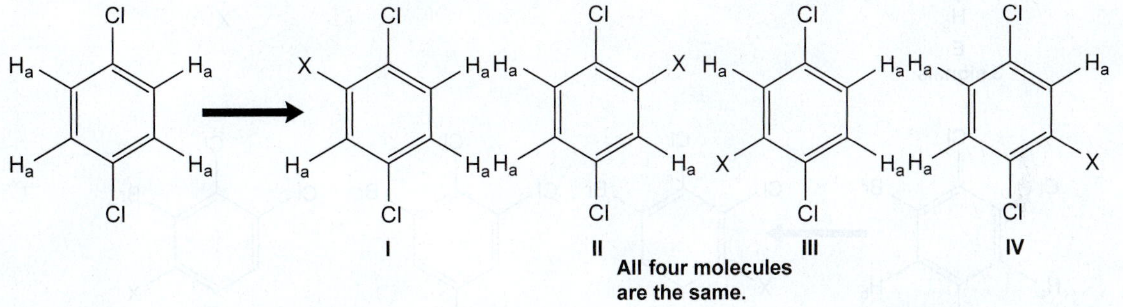

I **II** **III** **IV**

All four molecules are the same.

(f) Molecules **I** and **II** are the same molecule. Molecules **III** and **IV** are the same molecule. Molecules **I/II** and **III/IV** are constitutional isomers. According to Step 3 in the chemical distinction test, these H atoms are chemically distinct. There are *three* signals in the ^{1}H NMR spectrum, two from aromatic C–H and one from the OH.

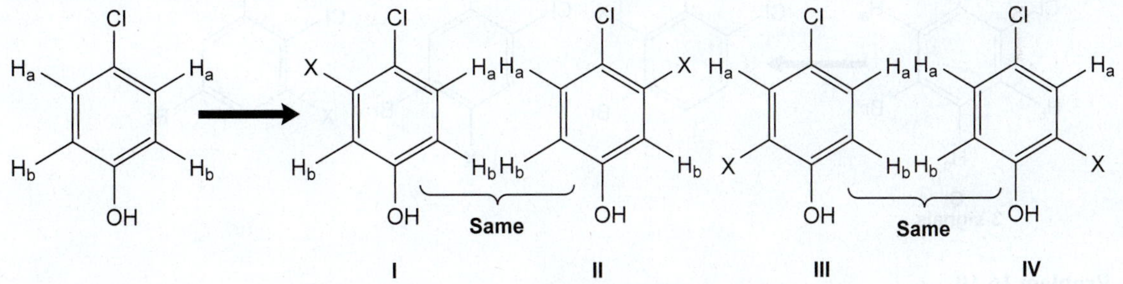

Same Same

I **II** **III** **IV**

Problem 16.9

Think

What molecules are generated by substituting each H atom separately by "X"? Which of those molecules are identical? Enantiomers? Diastereomers? Constitutional isomers? How many signals result?

Solve

Molecule **D** is the only isomer with symmetry that would cause the three protons to give rise to two signals in the ^{1}H NMR spectrum. All other molecules give three signals in the ^{1}H NMR spectrum. This can be seen with the chemical distinction test in the figures on the next page.

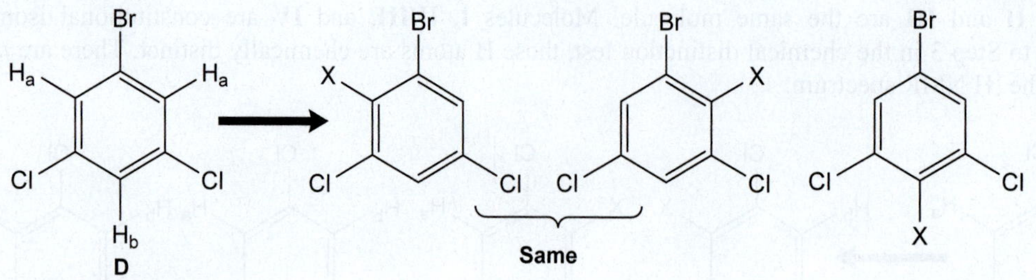

Same

D
2 signals

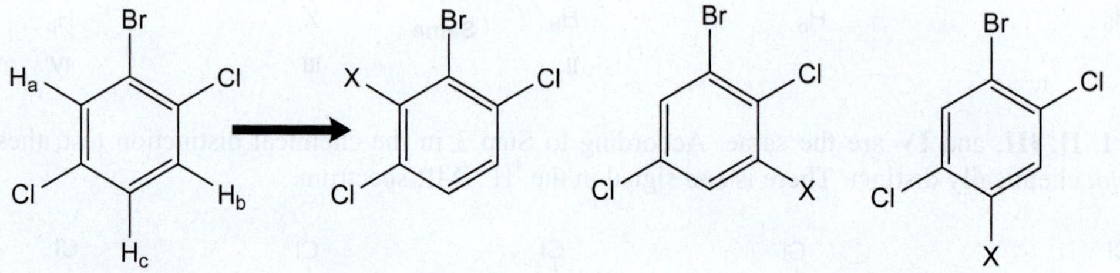

E
3 signals

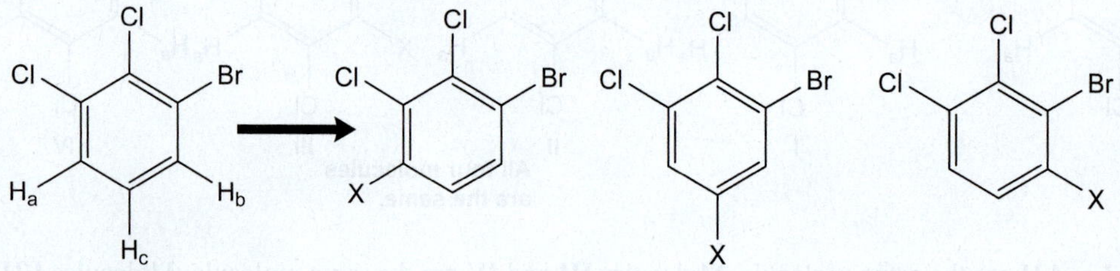

F
3 signals

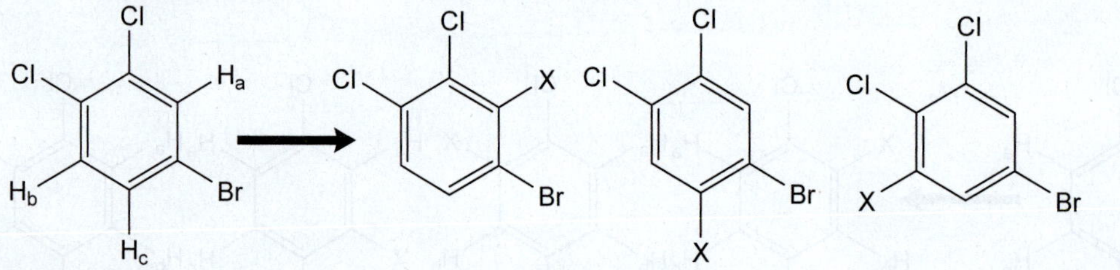

G
3 signals

Problem 16.10

Think

In Equation 16-4, what value should be substituted for γ? For B_{ext}?

Solve

As mentioned in the text, γ for a proton is $2.67512 \times 10^8\ \mathrm{T^{-1}\ s^{-1}}$. In the problem, the value for B_{ext} is given as 11.74 T.

Therefore,

$$\begin{aligned}\nu_{op} &= \gamma B_{ext}/2\pi \\ &= (2.67512 \times 10^8\ \mathrm{T^{-1}\ s^{-1}})(11.74\ \mathrm{T})/2\pi \\ &= 500 \times 10^6\ \mathrm{s^{-1}} = 500\ \mathrm{MHz}\end{aligned}$$

The unit T cancels, leaving the frequency units, $\mathrm{s^{-1}}$ or Hz.

Problem 16.11

Think

What is the operating frequency of the NMR spectrometer using a 7.046-T magnet? In Equation 16-3, what values should be substituted for ($\nu_{sample} - \nu_{TMS}$) and ν_{op}? What is the resulting chemical shift in ppm?

Solve

In the problem, the value for B_{ext} is given as 7.0 T.
Therefore,

$$\nu_{op} = \gamma B_{ext}/2\pi$$
$$= (2.67512 \times 10^8\ T^{-1}\ s^{-1})(7.046\ T)/2\pi$$
$$= 300 \times 10^6\ s^{-1} = 300\ MHz$$

At that magnetic field, the operating frequency is 300 MHz, or 3×10^8 Hz.
Because the signal of the proton of interest is 2200 Hz higher than those in TMS, ($\nu_{sample} - \nu_{TMS}$) = 2200 Hz.
Using Equation 16-3, we can solve for the chemical shift in ppm:

$$= [(\nu_{sample} - \nu_{TMS})/\nu_{op}] \times 10^6$$
$$= [(2200\ Hz/(3 \times 10^8\ Hz)] \times 10^6 = 7.33\ ppm$$

Problem 16.12

Think

Using Equation 16-3, how can you rearrange the equation to solve for the frequency (Hz) of the proton? In Equation 16-3, what values should be substituted for chemical shift (ppm), ν_{TMS}, and ν_{op}?

Solve

2.4 ppm = $[(\nu_{sample} - \nu_{TMS})/\nu_{op}] \times 10^6 = [(\nu_{sample} - \nu_{TMS})/3 \times 10^8\ Hz] \times 10^6$
Rearranged equation = $(\nu_{sample} - \nu_{TMS}) = (2.4\ ppm)(3 \times 10^8\ Hz)/(10^6\ Hz) = 720$ Hz higher

Problem 16.13

Think

Which atom, Br or I, is more electronegative? Which H atoms are more deshielded? Which atoms are, therefore, farther downfield?

Solve

The chemical shift will be greater for the protons in CH_3Br. This is because Br has a higher electronegativity than I, so Br will cause more deshielding, exposing the H atoms to a greater extent of the external magnetic field.

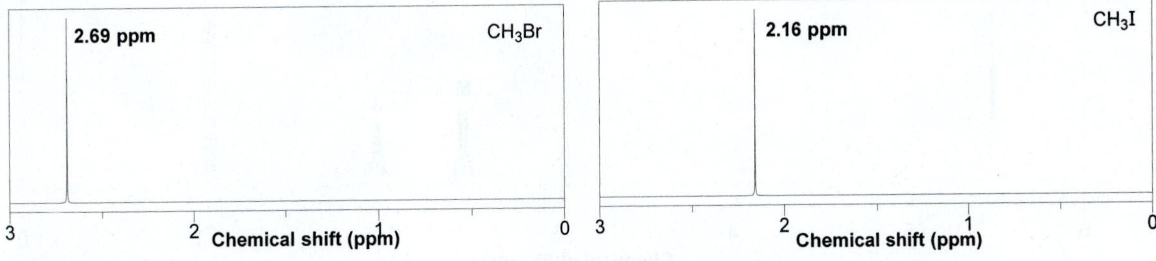

Problem 16.14

Think

Are the effects that cause deshielding additive? If so, which compound has more groups that deshield the H atom?

Solve

Each C=O group deshields the adjacent CH proton via both inductive effects and magnetic anisotropy, which shifts the signal downfield. Protons **C**, therefore, have a higher chemical shift than protons **B**. Proton **A** has an even higher chemical shift, because the CH proton has three adjacent C=O groups.

Lowest chemical shift **Highest chemical shift**

Problem 16.16

Think

What groups present are responsible for deshielding nearby protons? Do those groups deshield protons to the same extent? How severely do those deshielding effects fall off with distance?

Solve

There are two groups present that cause significant deshielding: the C=C group and the CF_3 group. From Table 16-1, we can see that the C=C group is likely to have a much greater effect on deshielding because the chemical shift of H–CR=CR_2 is 5.3 ppm. Therefore, **L** will have the most downfield peak, attached to C=C and closer to CF_3. **K** would be next. **L** and **K** are so close in chemical shift that the signals overlap. **M** is adjacent to the electron-withdrawing CF_3 group and, therefore, would come next. These deshielding effects fall off very quickly with distance, so proton **J** is next, followed by **I**. The chemical shifts of these protons decrease in the order **L** > **K** > **M** > **J** > **I**.

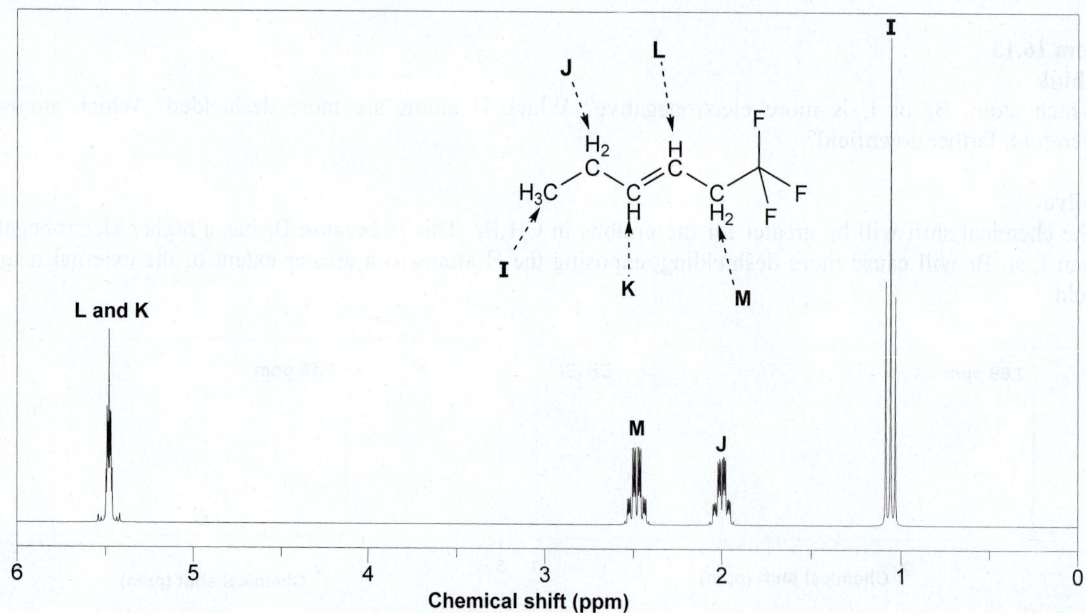

Problem 16.17

Think

How is the area under the absorption peak related to the number of signals that generate that peak? How can you use the integral trace to determine the relative areas under the peaks? Which peak has the largest area under the curve?

Solve

The signal at 1.9 ppm (signal **C**) represents the most H atoms because the stairstep in the integral trace is the largest, as indicated below. The signal at 6.1 ppm (signal **A**) represents the fewest H atoms because the height of the stairstep is the smallest. The ratio of the stair-step heights, and, thus, the number of the corresponding protons, is roughly 3:2:1 for C:B:A.

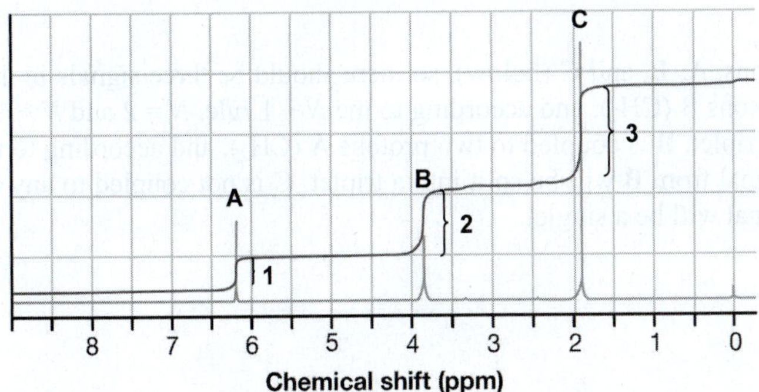

Problem 16.18

Think

What is the ratio of the stair-step heights for the peaks? Do those numbers add up to the number of H atoms given in the problem? If not, by what factor can you multiply those numbers so that the values add up to the total number of H atoms given?

Solve

The values in the 3:2:1 ratios already add up to six, so the signals at 6.1, 3.9, and 1.9 are due to one H, two H, and three H atoms, respectively. See Problem 16.17. On the other hand, if the molecule has a total of 12 H atoms, then each number in the ratio would need to be multiplied by a factor of 2, in which case the signals at 6.1, 3.9, and 1.9 would be due to two H, four H, and six H atoms, respectively.

Problem 16.20

Think

How many chemically distinct protons are in each compound? How many signals should each chemically distinct proton generate? Which of those protons are coupled to other protons that are chemically distinct from themselves?

Solve

(a) There are three chemically distinct protons, **A**, **B**, and **C** (below), so there should be three signals in the ^{1}H NMR spectrum. **A** is coupled to one proton **B** (CH), so $N = 1$, and according to the $N + 1$ rule, $N + 1 = 2$, and the signal from **A** will be split into a doublet. **B** is coupled to three protons **A** (CH_3), so $N = 3$, and according to the $N + 1$ rule, $N + 1 = 4$, and the signal from **B** will be split into a quartet. **C** is not coupled to any other protons, and therefore, $N = 0$, $N + 1 = 1$, and the signal will be a singlet.

Quartet
B
H_3C
A
Doublet
H
C
Cl
O
CH_3
C
Singlet

(b) There are two chemically distinct protons, **A** and **B** (next page), so there should be two signals in the ^{1}H NMR spectrum. Each proton **A** is coupled to two protons **B** (CH_2), and according to the $N + 1$ rule, $N = 2$ and $N + 1 = 3$, so the signal from **A** will be split into a triplet. Each pair of protons **B** is coupled to one proton **A** (CH), and according to the $N + 1$ rule, $N = 1$ and $N + 1 = 2$, so the signal from **B** will be split into a doublet. **B** is also next to another CH_2, but this is a **B** proton; protons that are chemically equivalent do not split each other.

(c) There are three chemically distinct protons, **A**, **B**, and **C** (below), so there should be three signals in the ^{1}H NMR spectrum. **A** is coupled to two protons **B** (CH_2), and according to the $N + 1$ rule, $N = 2$ and $N + 1 = 3$, so the signal from **A** will be split into a triplet. **B** is coupled to two protons **A** (CH_2), and according to the $N + 1$ rule, $N = 2$ and $N + 1 = 3$, so the signal from **B** will be split into a triplet. **C** is not coupled to any other protons, so $N = 0$, $N + 1 = 1$, and the signal will be a singlet.

(d) There is one chemically distinct proton, **A** (below), so there should be one signal in the ^{1}H NMR spectrum. Each pair of protons **A** is next to another CH_2, but that other pair is also **A** protons. Chemically equivalent protons do not split each other, so there will be one singlet signal.

Problem 16.22

Think

What should the first and last numbers in the set of ratios be? How is each remaining number derived from the line above it in Pascal's triangle?

Solve

The first and last numbers should be 1. In Pascal's triangle, the ratios for a septet would appear just below those for a sextet. So, each remaining number in the septet's ratios is computed by adding each adjacent pair of numbers in the sextet's ratios. That is, 1 + 5 = 6, 5 + 10 = 15, 10 + 10 = 20, 10 + 5 = 15, and 5 + 1 = 6. Therefore, the ratios are 1:6:15:20:15:6:1.

Problem 16.23

Think

How many signals will appear in each ^{1}H NMR spectrum? What would the aromatic peaks look like? How many signals would there be from alkene C–H protons? How is each signal split? Where would the CH_3 signal appear in each spectrum? Are the CH_3 protons coupled to other H atoms? Is there an O–H signal in both? What happens to an O–H signal when D_2O is added to the sample?

Solve

In both compounds, we would expect the aromatic H atoms to have similar chemical shifts and similar splitting patterns because both rings are para-disubstituted, and the ring has an attached O and an attached alkene carbon in both cases. The CH_3 signals would be slightly different in chemical shift and would also be different in splitting, because in **A**, the CH_3 carbon is bonded to an alkene CH, whereas in **B**, it is bonded to an O. The alkene H signals would have different splitting patterns, because alkene **B** consists of CH=CH_2 and alkene **A** consists of CH=CH–CH_3. A major difference is that in the first molecule, the OH signal will be a broad singlet, and no such signal exists for the second molecule. If D_2O were added, a portion of molecules would have OD instead of OH, so the OH signal would diminish in size. With enough H/D exchange, the OH signal would disappear entirely.

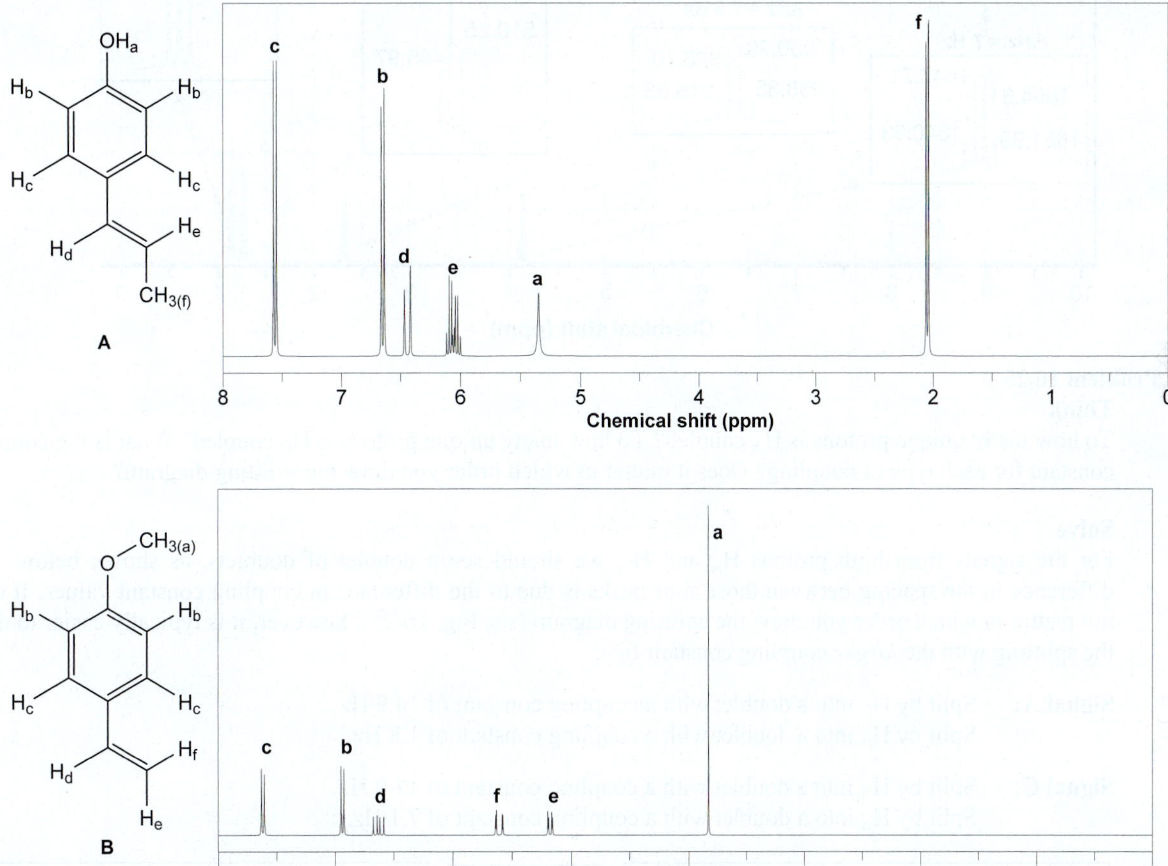

Problem 16.24

Think

If signals are coupled together, what is true about the value of their coupling constants? How can you determine the coupling constant value given the frequency of each absorption in the spectrum?

Solve

The signal at 4.1 ppm exhibits a coupling constant of about 7 Hz, which is computed as the difference in frequency between two adjacent peaks in the signal (e.g., 1647.71 Hz – 1640.63 Hz). The signals at 2.3 ppm, 1.2 ppm, and 1.1 ppm exhibit coupling constants of about 7.5 Hz, 7 Hz, and 7.5 Hz, respectively. So the H atoms giving rise to the signal at 4.1 ppm are coupled to the protons giving rise to the signal at 1.2 ppm.

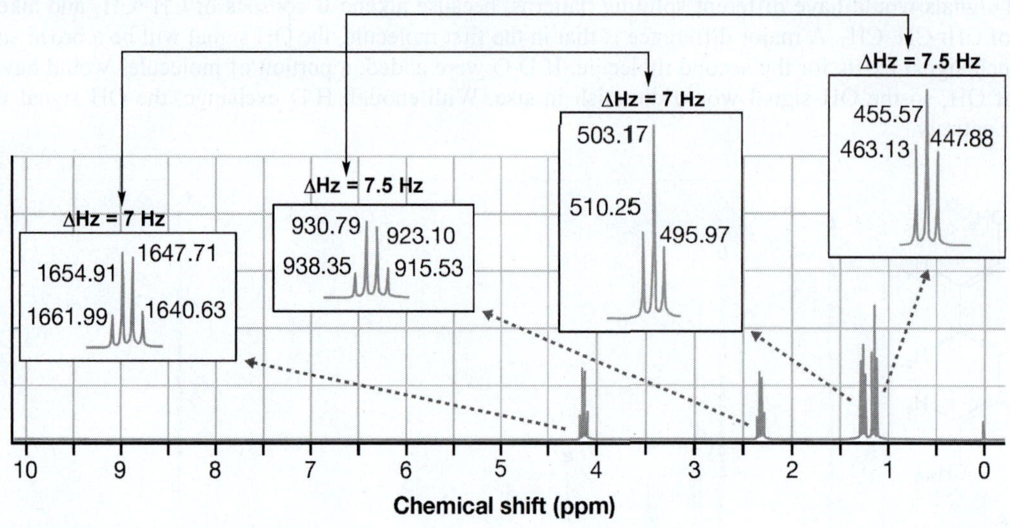

Problem 16.25

Think

To how many unique protons is H_A coupled? To how many unique protons is H_C coupled? What is the coupling constant for each type of coupling? Does it matter in which order you draw the splitting diagram?

Solve

For the signals from both protons H_A and H_C, we should see a doublet of doublets, as shown below. The difference in the spacing between those four peaks is due to the difference in coupling constant values. It does not matter in which order you draw the splitting diagram (see Fig. 16-25); however, it is typically easier to draw the splitting with the *larger* coupling constant first.

Signal A: Split by H_C into a doublet with a coupling constant of 14.9 Hz.
Split by H_B into a doublet with a coupling constant of 1.8 Hz.

Signal C: Split by H_A into a doublet with a coupling constant of 14.9 Hz.
Split by H_B into a doublet with a coupling constant of 7.1 Hz.

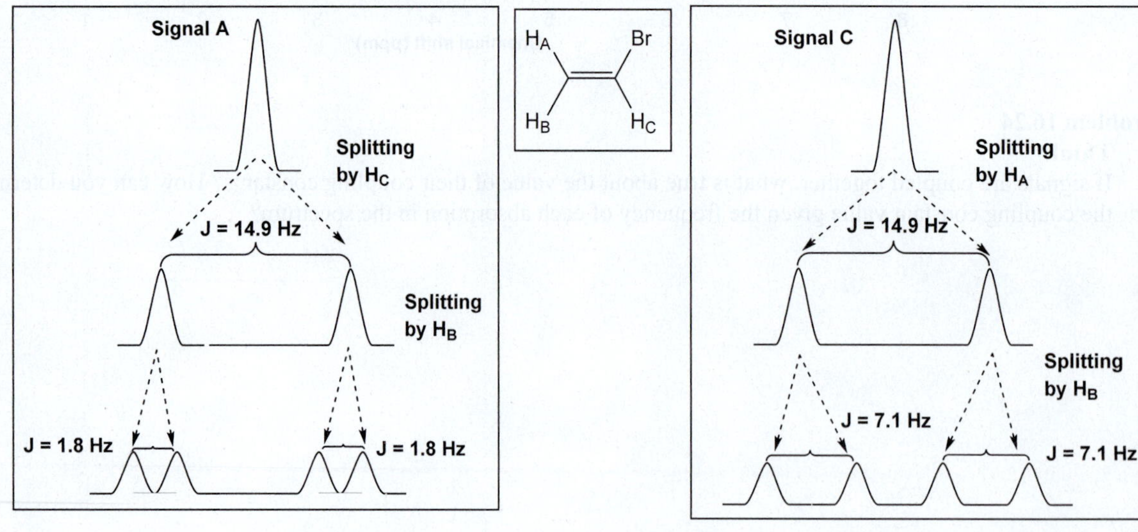

Problem 16.26

Think

To achieve a quartet of triplets, into how many peaks should the signal be split by the first coupling? Into how many peaks should each of the resulting peaks be split by the second coupling? How do the coupling constants for each coupling compare? If the coupling constants are the same, do all of the resulting peaks have different frequencies, or do some of them have the same?

Solve

In a quartet of triplets, the first coupling splits the signal into four peaks, as shown below. The second coupling then splits each of those peaks into three peaks. Because the two J values are the same, the 12-line pattern collapses into a six-line pattern. This is essentially the same as we would obtain using the $N + 1$ rule with $N = 5$ (a sextet), because protons that give rise to a quartet of triplets are coupled to five protons (three of one type and two of another). Note that the greater area under the peaks toward the center of the "sextet" reflects how the area from the quartet is distributed.

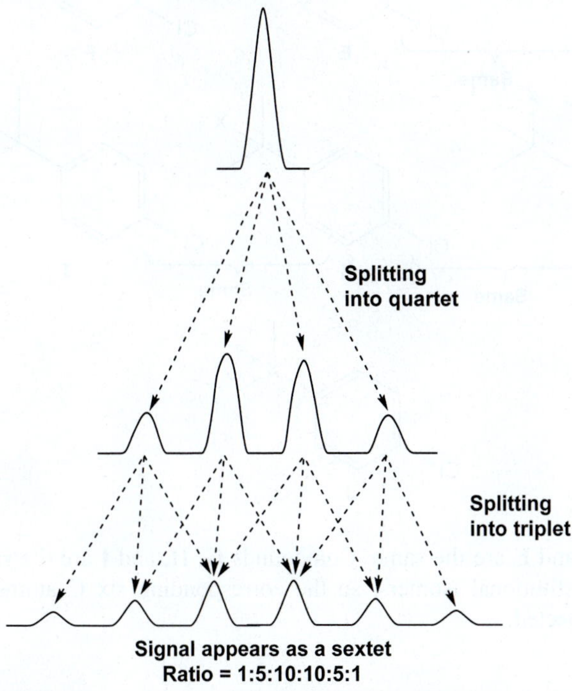

Problem 16.28

Think

When you apply the chemical distinction texs, what molecules are obtained by separately replacing each C atom with X? What are the relationships among the resulting molecules?

Solve

Replacing each of the C atoms with X, we get compounds **A–J**, as follows:

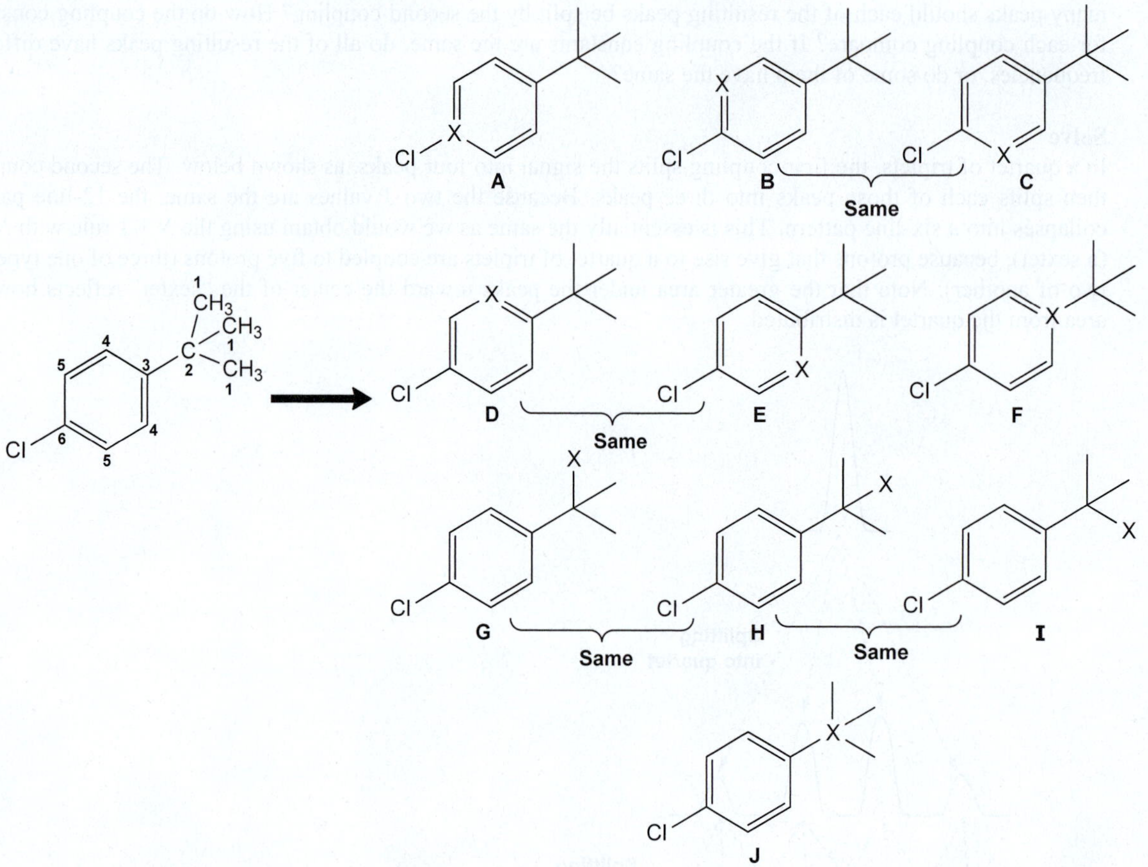

Compounds **B** and **C** are the same. Compounds **D** and **E** are the same. Compounds **G**, **H**, and **I** are the same. Compounds **A**, **B/C**, **D/E**, **F**, **G–I**, and **J** are constitutional isomers, so the corresponding six C atoms are chemically distinct; therefore, six ^{13}C signals are expected.

Problem 16.29

Think

How are the trends in chemical shift similar for ^{1}H and ^{13}C NMR spectroscopy? Which C atoms in the molecule shown experience ring current the most from the aromatic π electrons? Which C atom in the ring also experiences a deshielding effect due to the Cl atom? Refer to Table 16-4.

Solve

As we can see in Table 16-4, aromatic carbons have chemical shifts around 130 ppm, whereas alkyl halide carbons have chemical shifts of 35 to 55 ppm. Magnetic anisotropy has a greater impact on chemical shift than inductive effects do, which is why the aromatic C signals appear at higher chemical shifts. Not all the aromatic carbons have the same chemical shift, however. The one closest to the Cl atom will be most deshielded, due to inductive effects, so its signal will be the most downfield shifted, and it will have the highest chemical shift.

Highest chemical shift

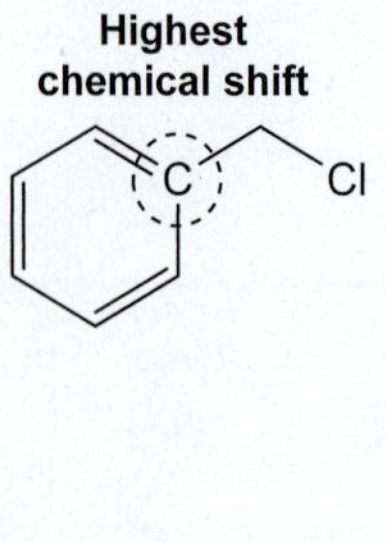

Problem 16.30

Think

What are the structures for $CH_3CH_2CH_2CO_2C_6H_5$ and $CH_3CH_2OCH_2(C{=}O)C_6H_5$? How many signals are expected in the ^1H NMR for each? What is the relative integration? What is the value of the chemical shift predicted for each signal? What is the splitting for each signal? Do either of these predicted ^1H NMR spectra match the spectrum shown in Figure 16-32?

Solve

The structures for $CH_3CH_2CH_2CO_2C_6H_5$ and $CH_3CH_2OCH_2(C{=}O)C_6H_5$ are shown below. Both compounds are a monosubstituted benzene; therefore, the aromatic region of the ^1H NMR will be similar for each. The difference in the structures of the two compounds arises from the difference in connectivity of the CH_3 and two CH_2 groups. Compound $CH_3CH_2CH_2CO_2C_6H_5$ will have a 3H split into a triplet, a 2H split into a multiplet, and a 2H split into a triplet. Compound $CH_3CH_2OCH_2(C{=}O)C_6H_5$ will have a 3H split into a triplet, a 2H split into a quartet, and a 2H as a singlet. The spectrum shown in Figure 16-32 matches these splitting patterns and integrations we would see with $CH_3CH_2CH_2CO_2C_6H_5$. However, the chemical shift of the triplet signal in $CH_3CH_2CH_2CO_2C_6H_5$ (protons **B** below) does not match what is given in the spectrum very well. In $CH_3CH_2CH_2CO_2C_6H_5$, the chemical shift would be around 2.1 ppm (Entry 6 in Table 16-1) or higher, whereas the triplet in the spectrum appears at 4.3 ppm. The chemical shift of 4.3 ppm is more consistent with protons of the form H–C–O (Entry 11 in Table 16-1) that typically appear at 3.3 ppm or higher.

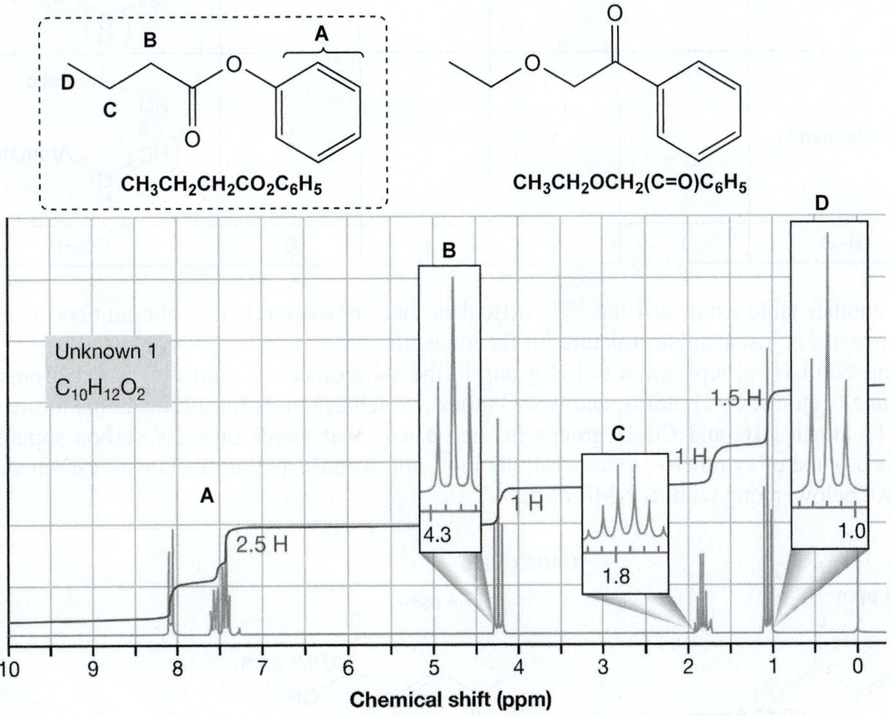

Problem 16.31

Think

What is the IHD? Consider the four questions regarding interpreting ^1H NMR spectra (p. 812) and the four questions regarding interpreting ^{13}C NMR spectra (p. 815). What answers do you get for those questions?

Solve

The IHD for a compound with the formula $C_9H_{10}O_2$ is 5.

Question 1. In the ^{1}H NMR spectrum, there are four signals and so four distinct types of hydrogen. In the ^{13}C NMR spectrum, there are six signals. Because there are nine C atoms but only six carbon signals, some of the C atoms must be equivalent.

Questions 2 and 3. The ^{1}H NMR data are summarized in the table below. Notice that the relative integration is the same as the number of protons because the sum of the relative integrations equals the total number of hydrogen atoms in the molecular formula.

In the "Molecular Fragment" column, the bold H atoms that have gray screens are responsible for the signal, and the H atoms enclosed in parentheses represent the coupled protons.

Entry	δ (ppm)	Nearby Double Bonds or EN Atoms?	Relative Integration	Number of Protons	Splitting Pattern	Number of Neighboring Distinct H	Molecular Fragment
1	2.3	H–C– (Aromatic)	6	6	s	0	H_3C—Aromatic—CH_3
2	7.0	H– (Aromatic)	2	2	d	1	Aromatic–(HC)=**HC**–(C–H)=Aromatic
3	7.2	H– (Aromatic)	1	1	t	2	Aromatic–**HC**=(HC)–**C–H**=Aromatic
4	12.9	H–O	1	1	s	0	CO_2H

We could construct another table using just the ^{13}C NMR data, but, as we see below, the number of carbon signals is sufficient to arrive at just a single structure for the molecule.

Question 4. Attaching two CH_3 groups and a CO_2H group to the six aromatic C atoms and three aromatic H atoms accounts for nine C atoms, 10 H atoms, and two O atoms, which accounts for all the atoms in the given molecular formula. To attach CH_3 and CO_2H groups in such a way that yields only six carbon signals, the molecule should have a plane of symmetry. Otherwise, the molecule would have more than six carbon signals. The molecule is shown below, along with its NMR data.

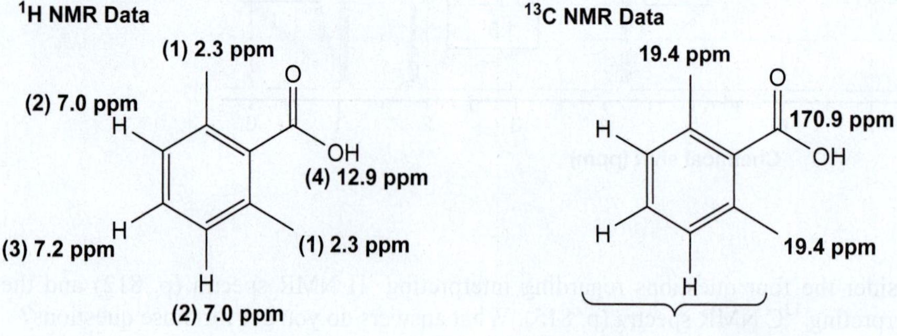

Problem 16.32

Think

What is the IHD? Consider the four questions regarding interpreting ^{1}H NMR spectra (p. 812). What answers do you get for those questions?

Solve

The IHD for a compound with the formula $C_8H_{10}O$ is 4.

Question 1. In the ^1H NMR spectrum, there are four signals and so four distinct types of hydrogen. There are no ^{13}C NMR data given.

Questions 2 and 3. The ^1H NMR data are summarized in the table below. Notice that the relative integration is the same as the number of protons because the sum of the relative integrations equals the total number of hydrogen atoms in the molecular formula. For Entry 3, we know that the signal at 2.4 ppm is an OH proton because that signal disappears when D_2O is added (H/D exchange).

In the "Molecular Fragment" column, the bold H atoms that have gray screens are responsible for the signal, and the H atoms enclosed in parentheses represent the coupled protons.

Entry	δ (ppm)	Nearby Double Bonds or EN Atoms?	Relative Integration	Number of Protons	Splitting Pattern	Number of Neighboring Distinct H	Molecular Fragment
1	1.4	None	3	3	d	1	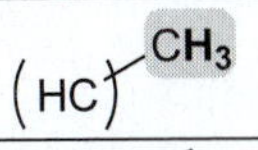 (HC)–**CH₃**
2	2.4	H–O	1	1	s	0	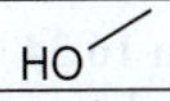 **H**O–
3	4.8	H–C–O	1	1	q	3	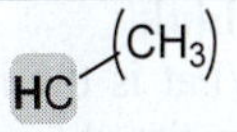 **HC**–(CH₃)
4	7.2–7.4	H– (Aromatic)	5	5	–	–	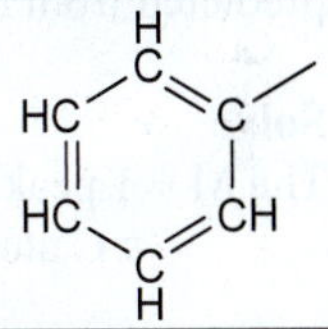 C_6H_5– (ring of HC, CH, C)

Question 4. The eight carbon atoms shown in the above table account for all eight carbon atoms in the given formula. The carbon atoms in Entries 1 and 3 must be the same, so the molecule has a CH–CH_3 fragment. The CH carbon must be bonded to the ring; otherwise, the OH bonded to the ring would have completed the molecule without the $CHCH_3$ group included. Finally, the OH group is bonded to CH to complete the molecule. The completed structure is shown below along with its ^1H NMR data.

^{1}H NMR Data

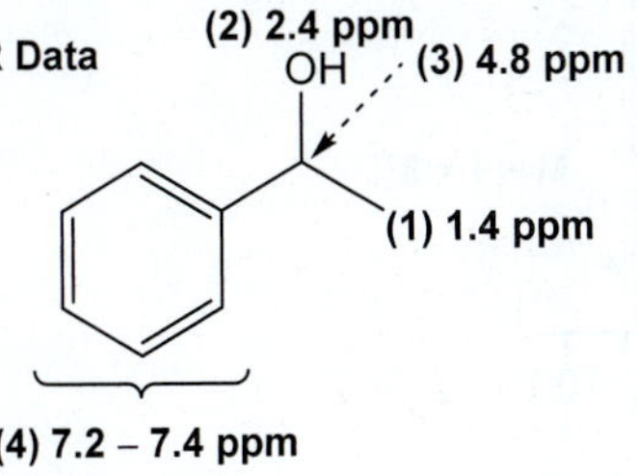

(2) 2.4 ppm OH

(3) 4.8 ppm

(1) 1.4 ppm

(4) 7.2 – 7.4 ppm

Problem 16.33

Think

What is the molecular mass of ethylbenzene? Is there a peak that matches that mass? Which peak is the most intense? What is the difference in mass between those two peaks?

Solve

The mass of ethylbenzene is 106 u, so the M^+ peak has the same *m*/*z* value, which is the second highest *m*/*z* value in the spectrum. The base peak is the tallest peak, at 91 u. The difference in mass between those two peaks is 15 u. As shown below, that difference can come from fragmentation of a CH_3 group from M^+.

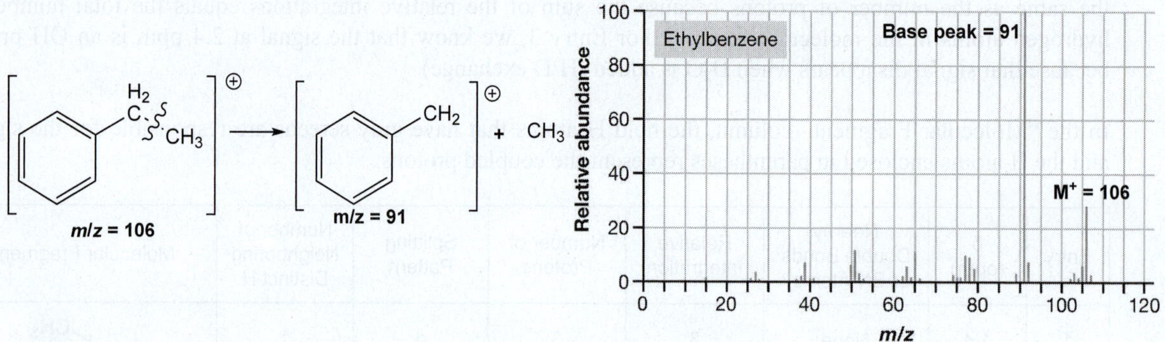

Problem 16.34

Think

What is the intensity of the M + 1 peak? What is the intensity of the M^+ peak? How many carbon atoms are predicted from the formula in Equation 16-7?

Solve

The M + 1 peak appears at *m*/*z* = 95, and its intensity is about 8% of the M^+ peak. So,

$$\# \text{ C atoms} \approx [(\text{intensity of M} + 1)/(\text{intensity of M}^+)] \times (100\%/1.1\%)$$
$$= (8\%/100\%) \times (100\%/1.1\%) = 7.3, \text{ or} \sim 7 \text{ C atoms.}$$

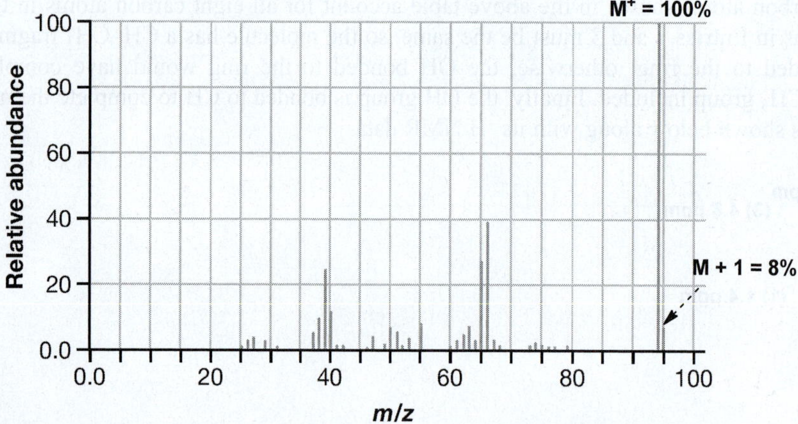

Problem 16.36

Think

What is the formula of the molecule containing the ^{32}S isotope? What is the formula of the molecule containing the ^{34}S isotope? What is the molecular mass of each of those formulas? What is the relative abundance of ^{32}S and ^{34}S (refer to Table 16-5)?

Solve

The formula for the molecule containing the ^{32}S isotope is $C_6H_5{}^{32}SH$, and the one for the molecule containing the ^{34}S isotope is $C_6H_5{}^{34}SH$. The M^+ peak should be at the *m*/*z* value that is the molar mass of the compound with the lighter isotope, which is 110. The ^{34}S isotope would give an M + 2 peak around *m*/*z* = 112. Because the ratio of abundance of ^{34}S to ^{32}S is about 4%, we should expect the M + 2 peak to have an intensity of about 4%.

Problem 16.37

Think

Based on the analysis in the textbook, what are the possible formulas derived from the mass spectrum in Figure 16-40? If there are no peaks >3000 cm^{-1} in the IR, could N–H bonds be present in the molecule? What bond stretch gives the strong absorption at 1670 cm^{-1}?

Solve

The formula is given to us as $C_7H_{15}NO$ or $C_7H_{12}NF$. The absorption at 1670 cm^{-1} is indicative of a C=O amide stretch. This confirms the formula as $C_7H_{15}NO$. If we examine Table 15-2, we see that this carbonyl stretch is at a lower frequency than normal ketones or aldehydes, but it is in the middle of the range in which we normally find amide carbonyl stretches. Because we find no IR absorptions above 3000 cm^{-1}, there should be no N–H bonds. Three molecules that are consistent with this are shown below, and there are others not shown.

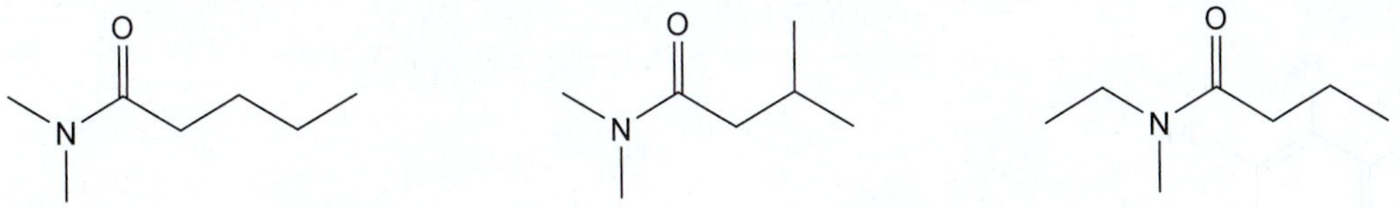

Problem 16.38

Think

Consider the four questions regarding interpreting mass spectra on pages 826 and 827. What answers do you get for those questions? Based on the ^{13}C NMR spectrum, at least how many carbon atoms must there be in the molecule? What type of bond does the IR absorption at 1720 cm^{-1} suggest? What is the total mass of the atoms now accounted for? What is the formula of the compound?

Solve

The IR absorption at 1720 cm^{-1} is due to a C=O stretch. The four ^{13}C NMR signals means there must be at least four carbons in the molecule.

Question 1. There are two mass peaks at the highest-mass end of the spectrum: m/z = 150 and 152. The M^+ peak is at m/z = 150, and the M + 2 peak is at m/z = 152.

Question 2. The intensity of the M + 1 peak is too small to interpret and is not given, so we must skip this question.

Question 3. The molecule must contain no nitrogen atoms, or, if it contains nitrogen, the number of nitrogen atoms must be even. The M + 2 peak's intensity is similar to that of the M^+ peak, so the molecule likely contains Br.

Question 4. After subtracting the mass of ^{79}Br from the M^+ mass of 150 u, we still must account for 71 u. If the molecule contains N, it must contain at least two N atoms. Taking away another 28 u for those N atoms leaves 43 u, which is not enough to accommodate the four C atoms that must be present (48 u). Therefore, the molecule must contain no N atoms. Taking away 48 u for the four C atoms and 16 u for the O atom of the carbonyl group, we are left with 7 u yet to account for, which can be due to seven H atoms. So the compound's formula is C_4H_7OBr. One possible ketone and aldehyde structure consistent with this information are shown below. Other bromo-substituted ketones and aldehydes with four carbons are also possible.

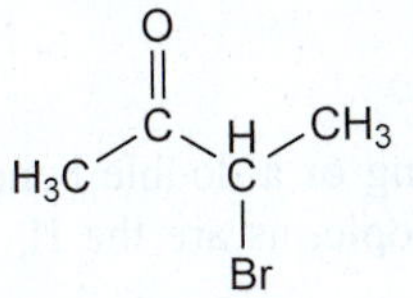

Possible ketone structure 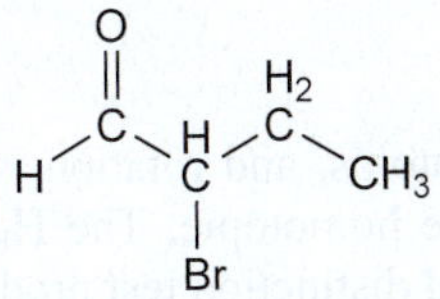 **Possible aldehyde structure**

End of Chapter Problems

Sections 16.3–16.5 Chemical Distinction, the Number of NMR Signals, and the Time Scale of NMR Spectroscopy

Problem 16.39

Think

How many H atoms are present? Does the molecule possess any symmetry? Are there any diastereotopic protons? Are any proton signals in different chemical environments owing to rotation restrictions (double bonds or rings)? If you are unclear about any protons in question, you should apply the chemical distinction test on just those protons.

Solve

(a) **Two signals.** The molecule has two planes of symmetry, and all aromatic H atoms are chemically equivalent, as are the two O–H groups.

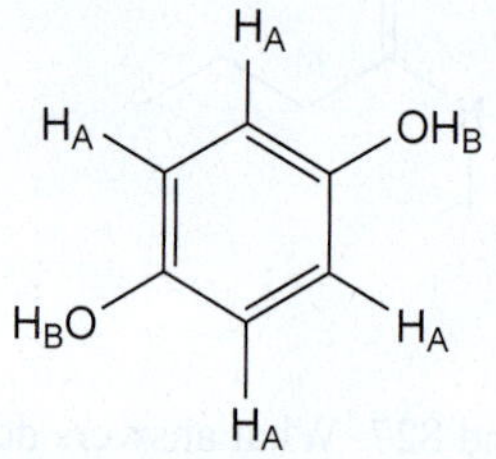

(b) **Five signals.** Protons that are cis to the chlorines are distinct from those that are trans. H atoms that are on the same C in this molecule are diastereotopic; their substitution by X in the chemical distinction test produces diastereomers.

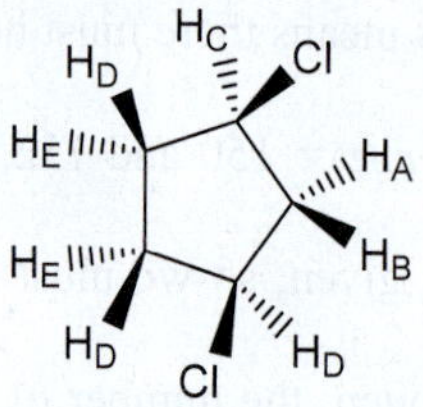

(c) **Eight signals.** All protons are chemically distinct. Protons that are cis to the Cl and Br are distinct from those that are trans. H atoms that are on the same C in this molecule are diastereotopic; their substitution by X in the chemical distinction test produces diastereomers. [*Note:* **(b)** is a meso compound, but **(c)** is not.]

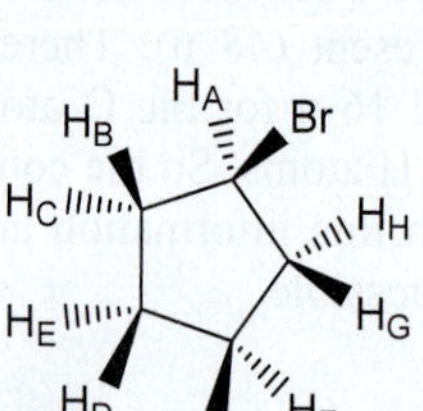

(d) **Four signals.** There are no chiral C atoms, and rotation is not limited by a ring or a double bond, so H atoms bonded to the same C atom are homotopic. The H_B atoms are enantiotopic, as are the H_C atoms: Their replacement by X in the chemical distinction test produces enantiomers.

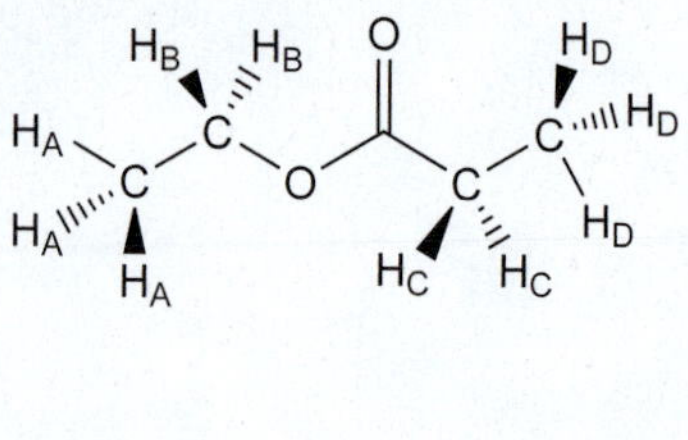

(e) Two signals. The molecule has one plane of symmetry perpendicular to the plane of the page along the C=O bond, so the CH_3 groups are in the same chemical environment, as are the CH_2 groups.

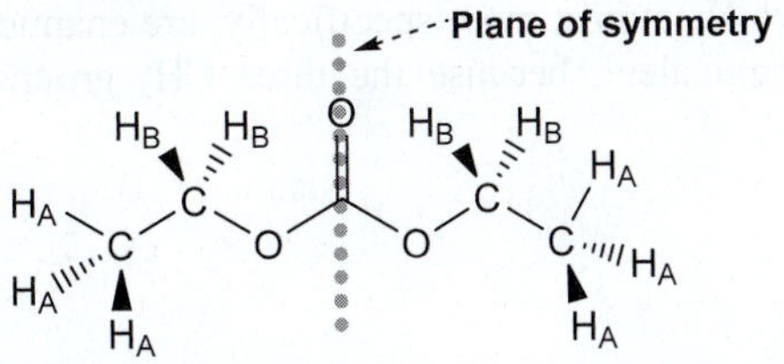

(f) Three signals. The molecule has one plane of symmetry perpendicular to the ring, which contains the Br–C and the C–O bonds, so there are two H signals from the aromatic H atoms and one from the OH.

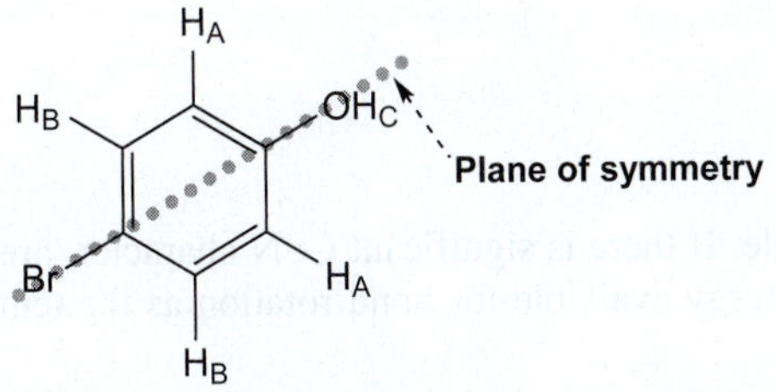

(g) Five signals. All protons are chemically distinct. The molecule has no planes of symmetry other than the plane that contains the ring, and it cannot be rotated less than 360° in such a way as to have all atoms appear unchanged after rotation, so all aromatic H atoms are chemically distinct (four signals). An additional signal arises from the O–H group.

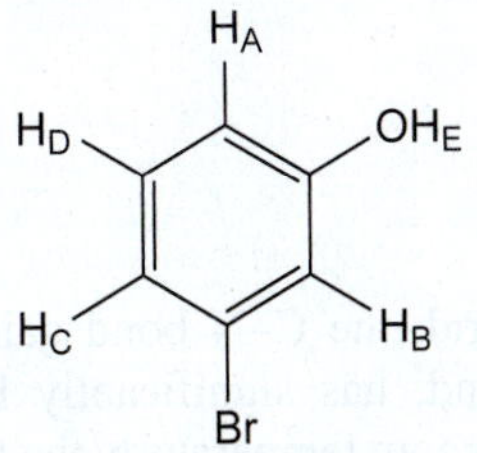

(h) Five signals. All protons are chemically distinct. The molecule has no planes of symmetry other than the one that contains the ring, and it cannot be rotated less than 360° in such a way as to have all atoms appear unchanged after rotation, so all aromatic H atoms are chemically distinct (four signals). An additional signal arises from the O–H group.

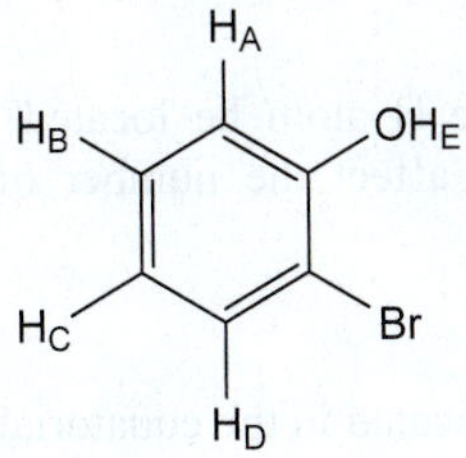

(i) Six signals. The molecule has a plane of symmetry perpendicular to the plane of the ring, which contains the bond that connects the alkyl substituent to the ring. There are three signals from the aromatic H atoms. There are no chiral carbon atoms, and rotation about the C–C single bond takes place freely, so H atoms bonded to the same carbon atom are homotopic. The two H_B atoms, more specifically, are enantiotopic, as are the two H_C protons. The nine CH_3 protons are equivalent, because the three CH_3 groups rapidly interchange locations through single bond rotations.

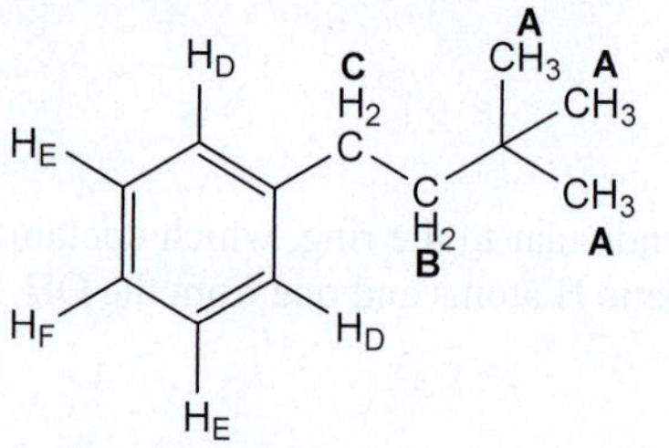

Problem 16.40

Think

Consider the resonance structures of *N,N*-dimethylformamide. If there is significant C=N character, are the two CH_3 signals chemically equivalent? What happens to the energy available for bond rotation as the temperature is increased?

Solve

The lone pair on the N atom of the amide group participates in resonance with the C=O group, as shown below.

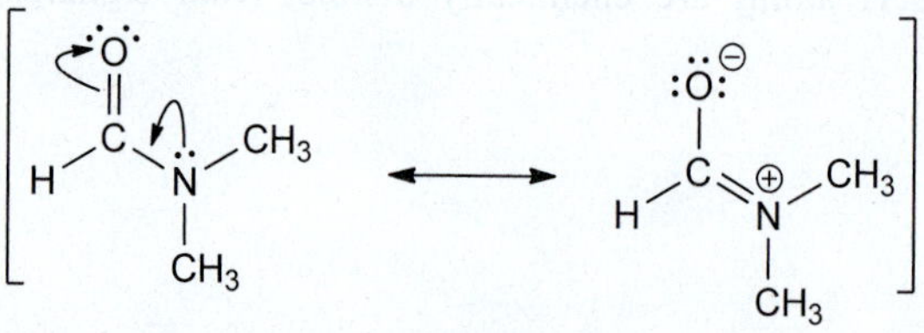

To the extent that the resonance structure on the right contributes to the hybrid, the C–N bond gains more double-bond character. Thus, the C–N bond, although formally a single bond, has significantly hindered rotation about it. If that C–N bond rotation is slower than the NMR time scale at room temperature, the two CH_3 sets of H atoms will give rise to different proton signals, because they are in different chemical environments. One CH_3 group is closer to the O atom than the other CH_3 group is. But when the sample is heated, the C–N bond rotation rate increases, and when the rate of that bond rotation surpasses the NMR time scale, the two CH_3 signals blend into one.

Problem 16.41

Think

What is the structure of cyclohexane-d_{11}? In how many environments can the H atom be located? At low temperature, is this exchange fast or slow? How does the rate of exchange affect the number of signals observed?

Solve

The structure of cyclohexane-d_{11} is shown on the next page. The H atom can be located in the equatorial or axial position. These positions are in two distinct chemical environments and give rise to two signals. At lower temperatures, this exchange between the axial and equatorial positions is slower than the NMR time scale, so two signals appear. At higher temperatures, this exchange is faster than the NMR time scale, and the two signals coalesce into one signal.

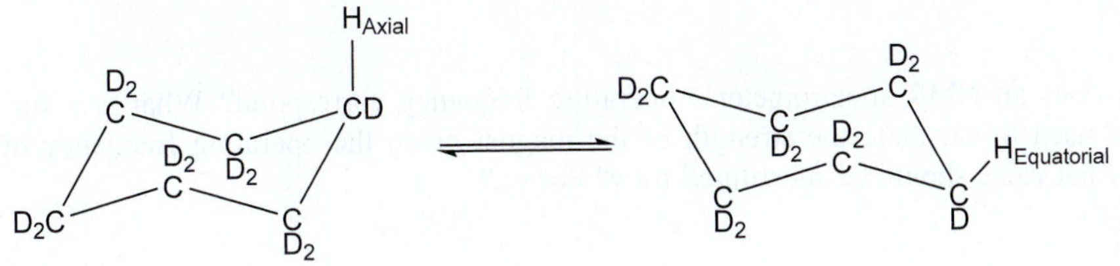

Problem 16.42

Think

Draw the chair structure of *cis*-1,2,3,4,5,6-cyclohexanehexacarboxylic acid. In how many environments can the H atoms bonded to the ring be located? At room temperature, do you think the chair flip that exchanges axial and equatorial positions will be fast or slow? How does the rate of exchange affect the number of signals observed?

Solve

The structure of *cis*-1,2,3,4,5,6-cyclohexanehexacarboxylic acid is shown below. The H atoms on the ring can be located in the equatorial or axial position, which gives two distinct chemical environments for H, which can lead to two signals. A chair flip exchanges the axial and equatorial positions, but the rate of the chair flip is slow because of the six bulky carboxyl groups attached to the cyclohexane ring. With a slow rate of the chair flip, the NMR spectrum exhibits the axial and equatorial H atoms as different signals.

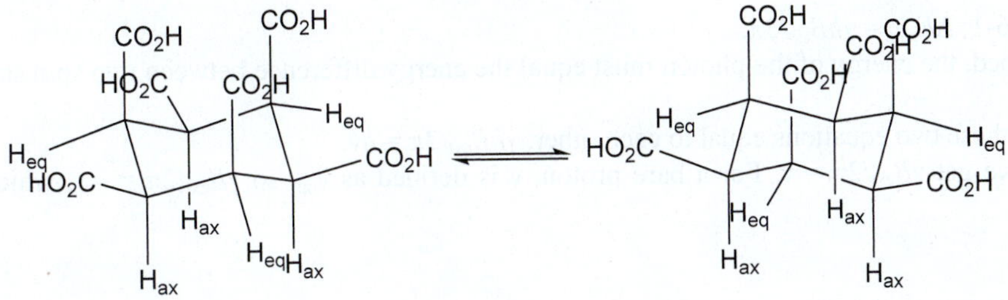

Sections 16.6–16.8 Operating Frequency, Chemical Shift, Shielding, and Deshielding

Problem 16.43

Think

What is the operating frequency of each NMR spectrometer? In Equation 16-3, what value should be substituted for ($\nu_{sample} - \nu_{TMS}$) and ν_{op}?

Solve

The operating frequency for each spectrometer is 300 MHz (300×10^6 Hz) and 90 MHz (90×10^6 Hz). Because the difference in signal for the protons is 150 Hz, we can set ($\nu_{sample} - \nu_{TMS}$) = $\Delta\nu$ = 150 Hz. Using Equation 16-3, we can set up a ratio to solve for the difference for 90 MHz NMR:

$$\frac{150\,\text{Hz}}{300\times10^6\,\text{Hz}} = \frac{\Delta\nu}{90\times10^6\,\text{Hz}} \qquad \Delta\nu = 45\,\text{Hz}$$

This shows that the resolution between the peaks improves as the operating frequency of the NMR increases.

Problem 16.44

Think

In Equation 16-3, what value should be substituted for γ_H? For B_{ext}? For ν_{op} for ^{1}H and ^{13}C? If you have two equations and two unknowns involving B_{ext}, how can you cancel B_{ext}? Solve for γ_C.

Solve

For H and C nuclei, Equation 16-3 gives $\nu_{op}(H) = \gamma_H B_{ext}/2\pi$ and $\nu_{op}(C) = \gamma_C B_{ext}/2\pi$.

If we divide the first equation by the second, we get $\nu_{op}(H)/\nu_{op}(C) = \gamma_H/\gamma_C$, which cancels B_{ext}.

Substituting in for the values we know, 300 MHz/75 MHz = ($2.67512 \times 10^8\ T^{-1}\ s^{-1}$)/$\gamma_C$.

Solving, $\gamma_C = 6.69 \times 10^7\ T^{-1}\ s^{-1}$.

Problem 16.45

Think

To what type of nucleus does an NMR spectrometer's operating frequency correspond? What is γ for that nucleus? What formula is used to calculate the strength of the magnet given the operating frequency of the NMR? In Equation 16-3, what value should be substituted for γ? For ν_{op}?

Solve

The operating frequency of an NMR spectrometer corresponds to the H nucleus, or the proton. As mentioned in the text, γ for a proton is $2.67512 \times 10^8\ T^{-1}\ s^{-1}$. In the problem, the value for ν_{op} is given as 1020×10^6 Hz.

Therefore,

$$\nu_{op} = \gamma B_{ext}/2\pi$$

$$1020 \times 10^6\ \text{Hz} = (2.67512 \times 10^8\ T^{-1}\ s^{-1})(B_{ext})/2\pi$$

$$B_{ext} = 23.96\ T$$

Problem 16.46

Think

Consult Equation 16-1 for the relationship between ΔE_{spin} and B_{ext}. For a photon to be absorbed by a nucleus with spin, how must ΔE_{spin} relate to the energy of the photon? How does the energy of a photon relate to the photon frequency?

Solve

We are given Equation 16-1: $\Delta E_{spin} = \gamma h B_{ext}/2\pi$.

For a photon to be absorbed, the energy of the photon must equal the energy difference between two spin states: $\Delta E_{spin} = E_{photon} = h\nu$.

Setting the right sides of these two equations equal to each other, $\gamma h B_{ext}/2\pi = h\nu$.

The h cancels out, and we get $\gamma B_{ext}/2\pi = \nu$. For a bare proton, ν is defined as ν_{op}, so $\gamma B_{ext}/2\pi = \nu_{op}$, which is Equation 16-3.

Problem 16.47

Think

In Equation 16-2, what values should be substituted for $(\nu_{sample} - \nu_{TMS})$ and ν_{op}?

Solve

Equation 16-2 is Chemical shift (ppm) = $[(\nu_{sample} - \nu_{TMS})/\nu_{op}] \times 10^6$.

The values needed are

$(\nu_{sample} - \nu_{TMS}) = 11{,}250$ Hz $\quad\quad \nu_{op} = 75$ MHz $= 75 \times 10^6$ Hz

Chemical shift (ppm) = $[(11{,}250\ \text{Hz})/(75 \times 10^6\ \text{Hz})] \times 10^6 = 150$ ppm

Problem 16.48

Think

Does a higher signal frequency correspond to a larger or smaller chemical shift? (Consult Equation 16-3.) Which signal is farther downfield? Does a signal farther downfield correspond to a proton that is more shielded or to one that is more deshielded?

Solve

According to Equation 16-2, the signal that is 450 Hz higher than TMS has a smaller chemical shift and is therefore less downfield shifted than the signal 755 Hz higher than TMS. Thus, the signal at 450 Hz represents H atoms that are more shielded, and the signal at 755 Hz represents H atoms that are more deshielded. A greater signal frequency represents a greater ΔE_{spin}, which is a result of a larger B_{eff}. A larger B_{eff} is the result of greater exposure of a nucleus to B_{ext}.

Problem 16.49

Think

Which is more electronegative, C or Si? Which atom shields the CH_3 groups to a larger extent? Does shielding or deshielding lower the chemical shift?

Solve

Si is less electronegative than C, so the CH_3 groups take more electron density from the central Si atom in TMS than they do from the central C atom in $(CH_3)_4C$. Thus, there is more electron density around the CH_3 groups in TMS, making those H atoms (and also the C atoms) more shielded, giving rise to a lower chemical shift.

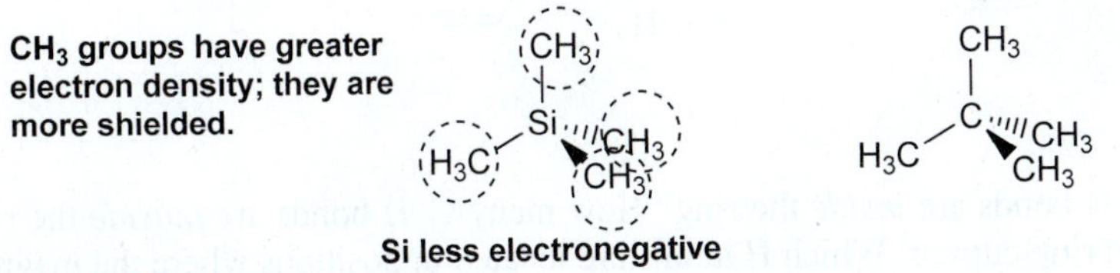

CH_3 groups have greater electron density; they are more shielded.

Si less electronegative

Problem 16.50

Think

What functional group is next to the CH_2 group on C2? To what entry in Table 16-1 does this correlate? How much downfield shifted is this kind of proton relative to an alkane CH_3 proton? If there are two C=O groups next to the CH_2, is the effect additive?

Solve

A proton in a CH_2 next to a carbonyl carbon normally has a chemical shift of 2.1 ppm (Entry 6, Table 16-1), which is an increase of 1.2 ppm from the nominal value of 0.9 ppm for an alkane CH_3 group. Two carbonyl groups will shift the signal about twice as much: 0.9 ppm + 1.2 ppm + 1.2 ppm = 3.3 ppm.

Problem 16.51

Think

How many ^{1}H NMR signals would be produced? How many aromatic H signals would be produced? What is the splitting pattern of each? Can the number of signals and the splitting pattern distinguish these compounds? What is the chemical shift of each signal?

Solve

The only major difference is in the chemical shift of the CH_3 group. In the first compound, the chemical shift will be around 3.3, given the adjacent O atom. In the second compound, it will be about 2.1 ppm, given the adjacent C=O group. The remaining signals will not be very distinguishable based on chemical shift, splitting, or integration.

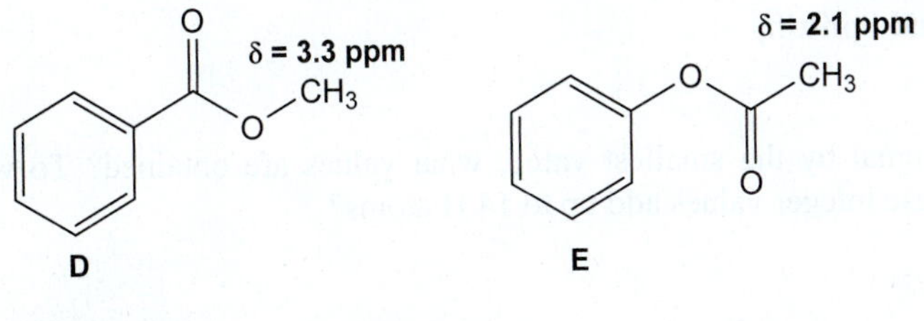

δ = 3.3 ppm **δ = 2.1 ppm**

D **E**

Problem 16.52

Think

Is oxygen inductively electron donating or withdrawing? Does this have a shielding or deshielding effect? Can the O participate in resonance with the aromatic ring? If so, what resonance structures are possible? How does this affect the electron density on the para carbon? Will that serve to shield or deshield the para carbon?

Solve

Oxygen is electronegative and will have an inductive withdrawing effect, which will deshield nearby nuclei. However, oxygen also donates electrons via resonance, indicated by the negative charge on the C atoms in the resonance structures below. Thus, resonance serves to shield the C atoms that share the formal −1 charge. At short range, the inductive effect is stronger, explaining the change from 125 ppm to 153 ppm. At long range, the resonance effect predominates, explaining the (smaller) change from 138 to 130 ppm. See the resonance structures on the next page.

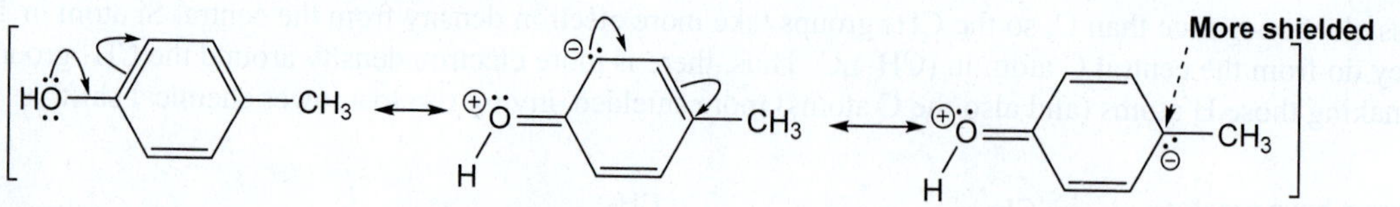

HÖ–C6H4–CH3 ⟷ ⊕O=C6H4(⊖)–CH3 ⟷ ⊕O=C6H4(⊖)–CH3

More shielded

Problem 16.53

Think

Draw each C–H bond. How many C–H bonds are *inside* the ring? How many C–H bonds are *outside* the ring? Consider the magnetic field lines from ring current. Which H atoms are located in positions where the magnetic field lines from ring current are in the same direction as B_{ext}? In the opposite direction?

Solve

An external magnetic field causes the π electrons in the aromatic ring to circulate (solid curved arrows below). This circulation induces a magnetic field (dashed arrows) that is in the same direction as the external magnetic field at the locations outside the ring and induces a magnetic field that is in the opposite direction inside the ring. Similar to benzene, the 12 equivalent protons outside the ring are deshielded and appear downfield at 9.3 ppm. The six protons inside the ring are shielded because the local magnetic field lines oppose B_{ext}, so the corresponding signals appear upfield at −2.9 ppm.

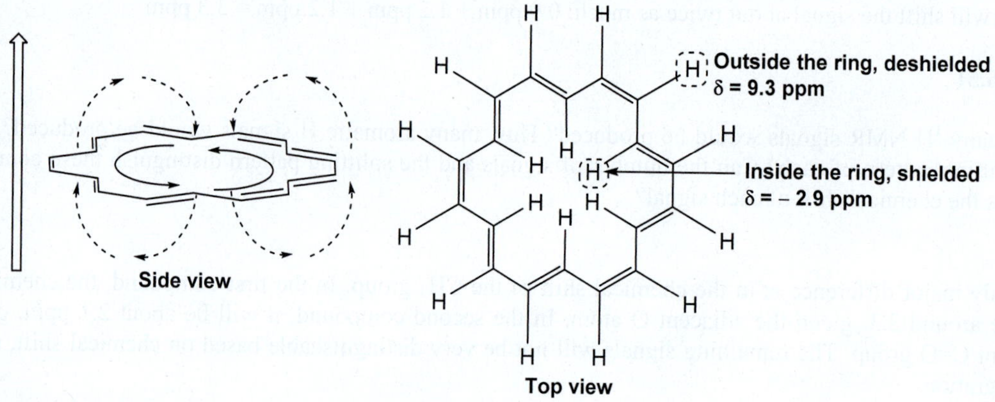

B_{ext}

Side view

Outside the ring, deshielded
δ = 9.3 ppm

Inside the ring, shielded
δ = – 2.9 ppm

Top view

Sections 16.9–16.12 Signal Integration and Signal Splitting

Problem 16.54

Think

If you divide the stair-step height of each signal by the smallest value, what values are obtained? To what integer value does each number round? Do these integer values add up to 14 H atoms?

Solve

Dividing each signal by the smallest value (9) gives the following:

37 mm/9 mm = 4.11, which rounds to 4.

9 mm/9 mm = 1.00.

26 mm/9 mm = 2.89, which rounds to 3.

52 mm/9 mm = 5.78, which rounds to 6.

Sum: 4 H + 1 H + 3 H + 6 H = 14 H. Because this adds up to the total number of H atoms in the molecular formula, the above signals represent four, one, three, and six H atoms, respectively.

Problem 16.55

Think

What are the structures for 1,4-dimethylbenzene and 1,3,5-trimethylbenzene? How many total H atoms are on the benzene ring? How many H atoms are from CH_3 groups? What is the ratio for each?

Solve

The ratio for 1,4-dimethylbenzene is 2:3, and the ratio for 1,3,5-trimethylbenzene is 1:3. See structures below. Therefore, the spectrum of 1,3,5-trimethylbenzene would exhibit signal integrations in a 1:3 ratio, but the spectrum of 1,4-dimethylbenzene would not.

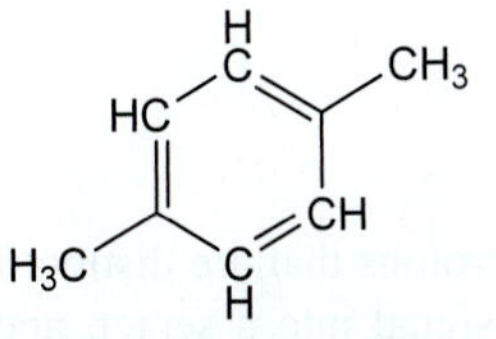

1,4-Dimethylbenzene
CH:CH_3
4:6
2:3

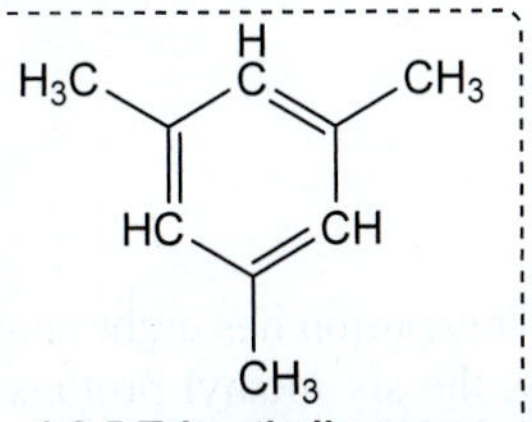

1,3,5-Trimethylbenzene
CH:CH_3
3:9
1:3

Problem 16.56

Think

Are the ratios given in whole numbers? Is there a way to reduce the given ratio? How many hydrogen atoms are in the given formula? If the relative integration of 1 were equal to 1 hydrogen, how many hydrogen atoms must be in the formula?

Solve

The ratio 1:5:5:20:30 is already in whole numbers, and there cannot be fewer than one hydrogen atom. Therefore, the ratio cannot be reduced. The number of hydrogens in the molecule is either the sum of the integrations of all the signals or a multiple of that sum. This would suggest that there are at least 61 (= 1 + 5 + 5 + 20 + 30) hydrogen atoms. However, the formula only has 12 hydrogen atoms. The suggests that the smallest signal is an impurity in the spectrum and can be disregarded. This gives a new ratio of 5:5:20:30. Reduce this ratio to give 1:1:4:6, which sums to the 12 hydrogen atoms in the given formula.

Problem 16.57

Think

If the peaks were generated from one chemically distinct H atom, how many signals would be present? What is the ratio of the intensities of the peaks shown? Can that be the result of the signal being split by one chemically distinct type of proton? Does this match the ratio for a simple quartet? Could the signal be the result of complex splitting? What type of splitting gives rise to peaks of essentially equal height?

Solve

If the peaks came from one chemically distinct type of proton, they would represent one signal. That signal cannot be a simple quartet because, according to Pascal's triangle, the height ratio of the four peaks should be 1:3:3:1. These four peaks, however, have essentially the same height/area. The signal could be the result of complex splitting, where it is split into a doublet twice—once by each of two distinct types of protons. Each splitting into a doublet would produce two peaks of equal height/area. Thus, the signal could be a doublet of doublets.

Problem 16.58

Think

How many neighboring H atoms are three bonds away or closer? Are those H atoms distinct from the ones highlighted? What type of splitting does the $N + 1$ rule predict?

Solve

(a) Singlet. The only H atom within three bonds is another aromatic H, which is equivalent to itself. The O–H bond is more than three bonds away.

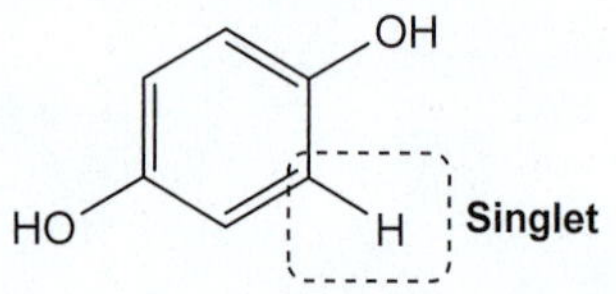

(b) Multiplet (overlapped triplet of septets). This proton has eight neighboring protons that are distinct from itself, giving a multiplet. (More specifically, the six methyl protons split the signal into a septet, and the CH_2 protons split each of those peaks into a triplet, for 21 total peaks. But because the CH_2 protons and the methyl protons have similar coupling constants, the signal will look more like the result of being split by eight protons, resulting in nine peaks—a nonet.)

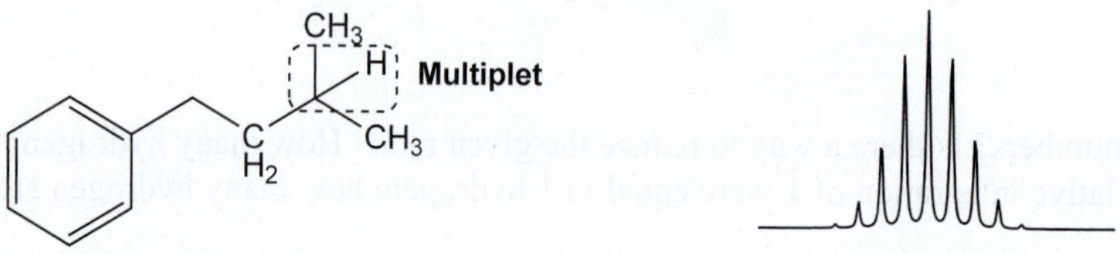

(c) Triplet (left). CH_3 protons (left) are split into a triplet by two adjacent protons.
Quartet (right). The CH_2 protons (right) are split into a quartet by three adjacent protons.

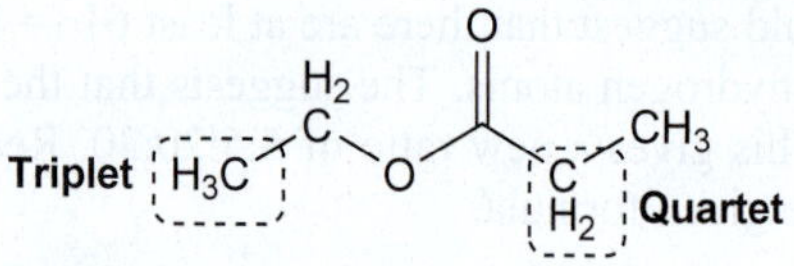

(d) Doublet. The neighboring H on the ring is distinct from the highlighted one. (*Note:* Long-range coupling will make the splitting pattern a bit more complex.)

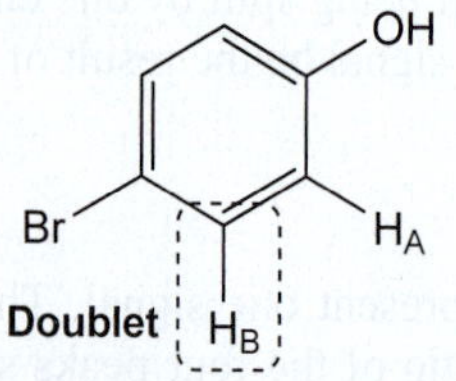

(e) Triplet (left). The CH_2 protons (in the box on the left) are split into a triplet by two adjacent protons.
Singlet (right). The CH_3 protons are a singlet (look closely: there are no protons on the *adjacent* carbon).

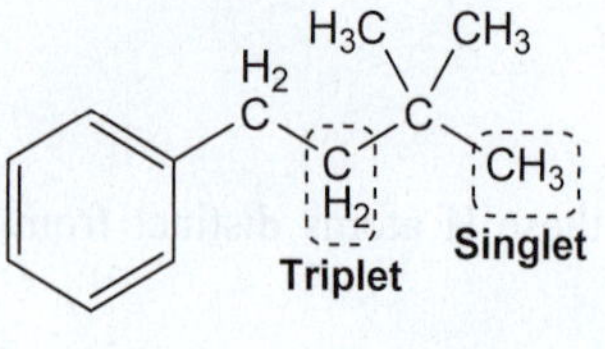

(f) **Triplet of doublets (or doublet of triplets—depends on the J values).** Two H atoms are in the same chemical environment (H_C), which splits the peak into a triplet. There is also another H atom (attached to the same carbon, H_B) that splits the peak into a doublet. H_A and H_B are diastereotopic protons and, therefore, are distinct. If the coupling constants are the same, $J_{AC} = J_{AB}$, then the splitting will result in a quartet. If $J_{AC} > J_{AB}$, the splitting pattern will look like a triplet of doublets, and if $J_{AB} > J_{AC}$, then it will look like a doublet of triplets. The actual peak (shown below) looks like two adjacent triplets and looks more like a doublet of triplets.

Problem 16.59

Think

Using the $N + 1$ rule, how many equivalent H atoms would need to be coupled to the protons giving rise to a sextet? Could this many equivalent H atoms all be attached to the same C atom? To two C atoms? To three? To four?

Solve

Using the $N + 1$ rule, five equivalent H atoms would have to be coupled to the H atoms giving rise to the sextet. It is not possible for all of these neighboring H atoms to be on the same C atom, because a neighboring C atom could have at most three bonds to H. If, on the other hand, five H atoms were distributed over neighboring C atoms, the H atoms could not all be equivalent, because five H atoms cannot be divided equally among two, three, or four neighboring C atoms.

Problem 16.60

Think

To how many unique protons is H_B coupled? Into how many peaks does the first coupling split the signal? What is the coupling constant for that coupling? Into how many peaks does the second coupling split each peak from the first coupling? What is the coupling constant for the second coupling? Does it matter in which order you draw the splitting diagram?

Solve

Signal B is split by H_C into a doublet with a coupling constant of 16 Hz and by H_A into a quartet with a coupling constant of 7 Hz. The signal appears as a multiplet because of the overlap of the doublet of quartets (see below). The difference is in the spacing between those peaks owing to the difference in coupling constant values. It does not matter in which order you draw the splitting diagram (see Fig. 16-25); however, it typically is easier to draw the splitting with the *larger* coupling constant first.

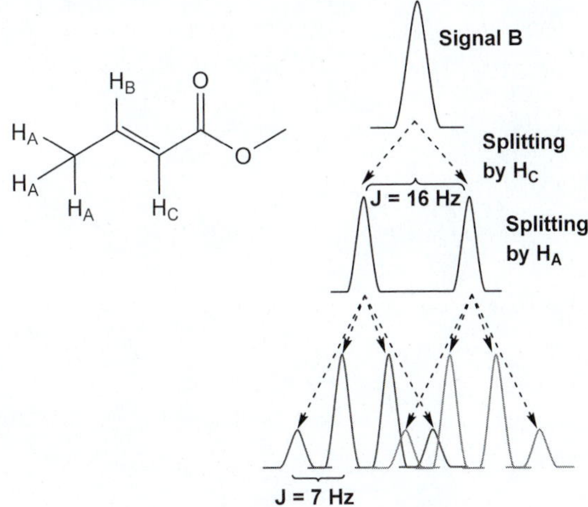

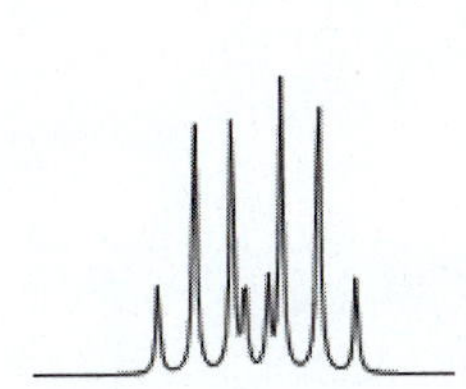

Problem 16.61

Think

What is the splitting pattern of the peaks that are widely spaced? Using the $N + 1$ rule, how many H atoms cause this splitting pattern? What is the splitting pattern of the peaks that are narrowly spaced? Using the $N + 1$ rule, how many H atoms cause this splitting pattern?

Solve

The widely spaced peaks appear as a septet, which is the result of six coupled H atoms. This could be due to two sets of coupled CH_3 protons. The narrowly spaced peaks appear as doublets, which is the result of one coupled H atom. Therefore, the septet of doublets results from coupling to seven protons: six equivalent CH_3 protons and one distinct H atom.

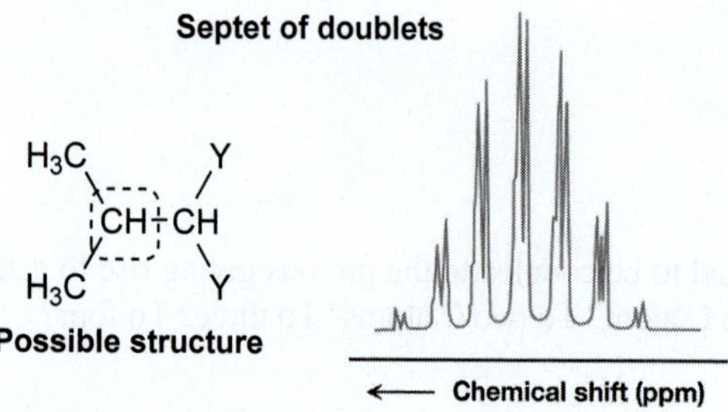

Sections 16.13 and 16.14 ^{13}C NMR Spectroscopy and DEPT ^{13}C NMR Spectroscopy

Problem 16.62

Think

Refer to Problem 16.39 to review the structures of the molecules. How many C atoms are present? Does the molecule possess any symmetry? If you are unclear about any carbons in question, you should apply the chemical distinction test on just those carbons.

Solve

(a) Two signals. One for the carbons bearing the OH groups, and one for the other aromatic ring carbons.

(b) Three signals. Note the difference from protons in the previous problem: cis/trans relationships don't apply to carbons in the ring.

(c) Five signals. All carbons are chemically distinct.

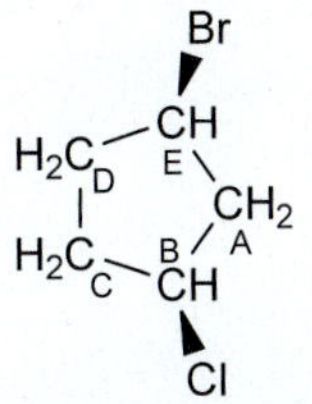

(d) Five signals. Note that the carbonyl carbon gives a signal that has no corresponding proton signal.

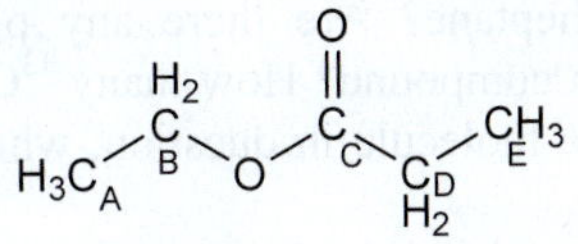

(e) Three signals. Note that the carbonyl carbon gives a signal that has no corresponding proton signal.

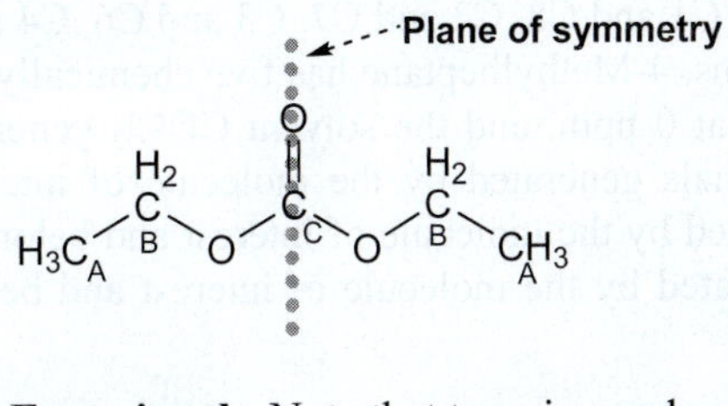

(f) Four signals. Note that two ring carbons give signals that have no corresponding proton signals.

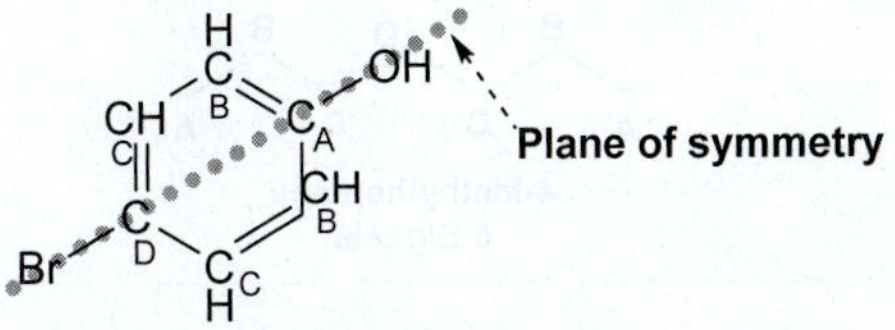

(g) Six signals. All carbons are chemically distinct. Note that some carbons give signals that have no corresponding proton signals.

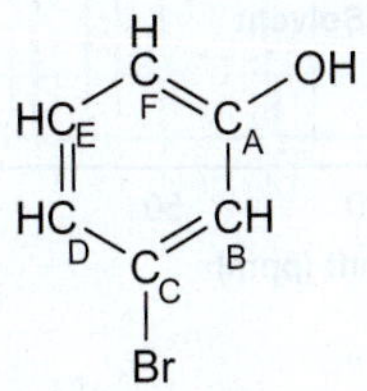

(h) Six signals. All carbons are chemically distinct. Note that some carbons give signals that have no corresponding proton signals.

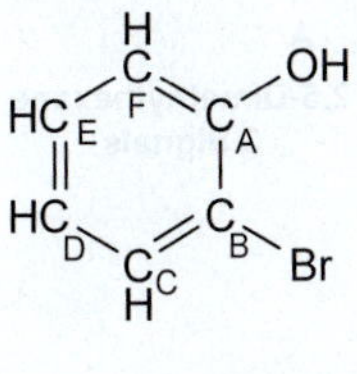

(i) Eight signals. Note that some carbons give signals that have no corresponding proton signals.

Problem 16.63

Think

What are the structures for octane, 2,5-dimethylhexane, and 4-methylheptane? Are there any planes of symmetry in each compound? How many ^{13}C signals are expected for each compound? How many ^{13}C signals are present in each spectrum shown? Which signals were generated by the molecule in question, which were generated by TMS, and which were generated by the solvent?

Solve

The easiest way to do this problem is to focus on the number of chemically distinct C atoms that each isomer must have. Octane has four chemically distinct sets of C atoms (C1 and C8, C2 and C7, C3 and C6, C4 and C5). 2,5-Dimethylhexane has three chemically distinct sets of C atoms. 4-Methylheptane has five chemically distinct sets of C atoms. In each spectrum, TMS generates the signal at 0 ppm, and the solvent $CDCl_3$ generates the signal at 77 ppm. The first spectrum, therefore, has four signals generated by the molecule of interest and belongs to octane; the second spectrum has five signals generated by the molecule of interest and belongs to 4-methylheptane; and the third spectrum has three signals generated by the molecule of interest and belongs to 2,5-dimethylhexane. See below.

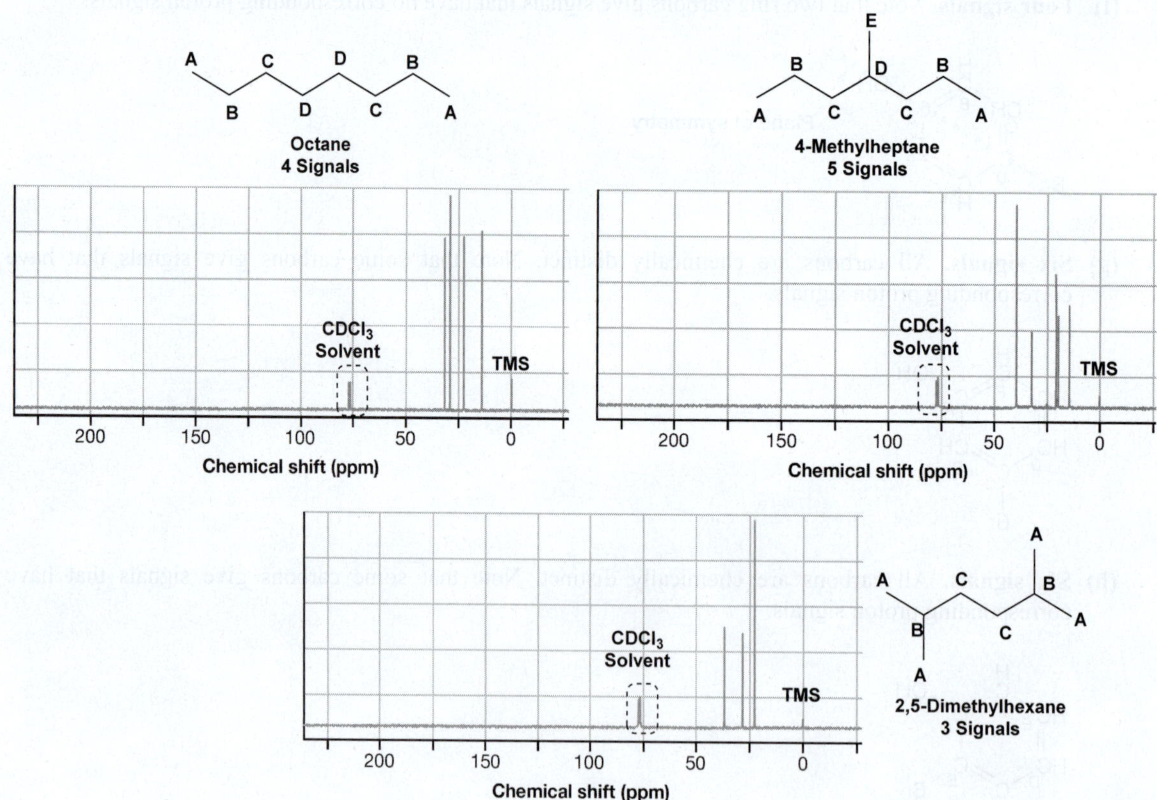

Problem 16.64

Think

What is the structure for chlorocyclohexane and iodocyclohexane? Are there any planes of symmetry in each compound? How many ^{13}C signals are expected for each compound? How many ^{13}C signals are present in each spectrum shown? What is the chemical shift of each signal? What is the difference in electronegativity of Cl compared to I?

Solve

The number of chemically distinct C atoms that each compound has is the same: four signals (labeled **A–D**). Furthermore, each spectrum exhibits four signals generated by the molecule of interest. The difference involves the C atom attached to the electronegative atom, labeled **A** in each example. Cl is a more electronegative atom and causes the signal of the C atom directly attached to the Cl to be farther downfield compared to the C atom attached to the I, 60 ppm versus 40 ppm.

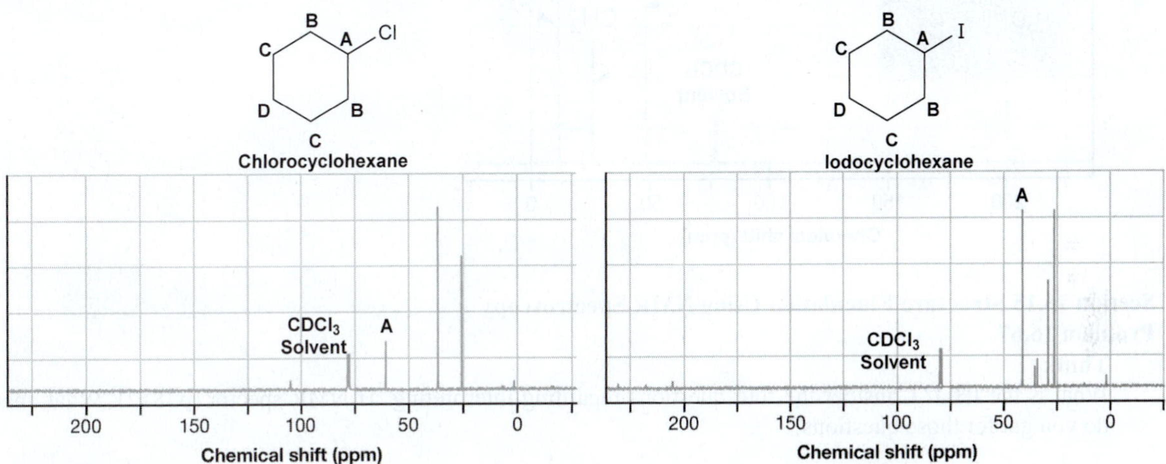

Problem 16.65

Think

Which C atom in the ring experiences a deshielding effect the most due to the Cl atom? What would you expect the chemical shift of that C atom's signal to be? Refer to Table 16-4. How is deshielding affected by proximity to the electronegative atom?

Solve

As we can see in Table 16-4, carbons attached directly to chlorine atoms should have chemical shifts of 35 to 55 ppm. Not all the alkyl carbons have the same chemical shift, however. The one closest to the Cl atom will be most deshielded, owing to inductive effects, so its signal will be the most downfield shifted, and it will have the highest chemical shift. C_B is farther away from the chlorine atom, so the chemical shift of its signal will be lower. C_C is farther away still, so the chemical shift of its signal will be the lowest.

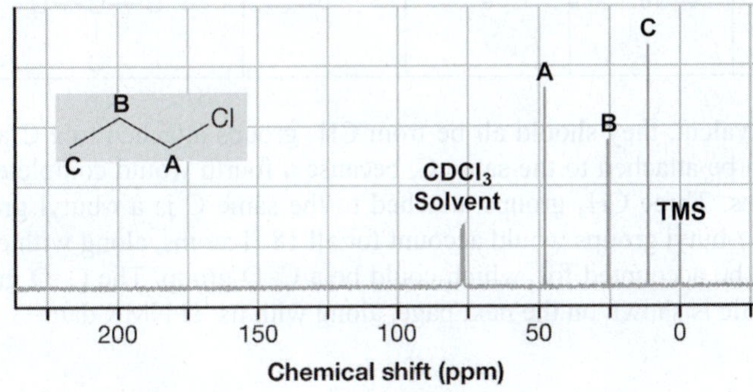

Problem 16.66
Think
Write out all of the C–H bonds for each C atom. What types of signals appear in the DEPT-90 spectrum? In the DEPT-135 spectrum, which signals would appear as positive signals, and which would appear as negative signals?

Solve
In DEPT-90 spectra, only CH signals appear. 1-Chloropropane does not have any CH groups and therefore will exhibit no signals in the DEPT-90 spectra. In DEPT-135 spectra, CH_3 and CH appear as normal positive signals, and CH_2 appears as a negative (below the baseline) signal. The positive and negative signals for 1-chloropropane are shown below.

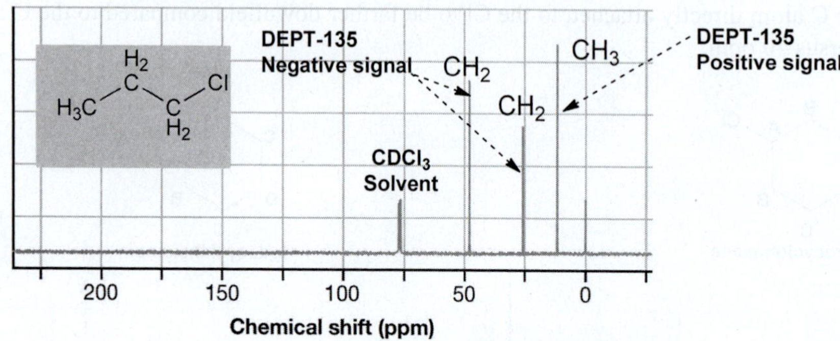

Section 16.15 Structure Elucidation Using NMR Spectroscopy

Problem 16.67
Think
What is the IHD? Consider the four questions regarding interpreting ^{1}H NMR spectra (p. 812). What answers do you get for those questions?

Solve
The IHD for a compound with the formula $C_9H_{18}O$ is 1.
Question 1. In the ^{1}H NMR spectrum, there is one signal, so there must be one chemically distinct type of H.
Questions 2 and 3. The ^{1}H NMR data are summarized in the table below. Notice that the relative integration is 1, because there is only one type of proton. With only one signal, all 18 H atoms in the molecular formula must generate the same signal, so all 18 H atoms must be equivalent. The O must not be bonded to a C that has any H atoms; otherwise, the chemical shift would be around 3.3 ppm or higher. The low chemical shift could signify a CH_3 group that is only slightly deshielded by an O atom a few bonds away. Furthermore, because the signal is a singlet, the neighboring C atom must not be bonded to any H atoms.

Entry	δ (ppm)	Nearby Double Bonds or EN Atoms?	Relative Integration	Number of Protons	Splitting Pattern	Number of Neighboring Distinct H	Molecular Fragment
1	1.25	None	1	18	s	0	CH_3–C(–C)(–C)–C

Question 4. Because all 18 H atoms are equivalent, they should all be from CH_3 groups attached to a C atom with no H atoms. Up to three CH_3 groups can be attached to the same C, because a fourth would complete the molecule, giving it a total of only 12 H atoms. Three CH_3 groups attached to the same C is a *t*-butyl group, $C(CH_3)_3$, and accounts for nine H atoms. Two *t*-butyl groups would account for all 18 H atoms, along with eight C atoms. That leaves one C and one O yet to be accounted for, which could be a C=O group. The C=O group would also count for the IHD of 1. The molecule is shown on the next page, along with its ^{1}H NMR data.

^{1}H NMR Data

(1) 1.25 ppm

Problem 16.68

Think

What is the IHD? Consider the four questions regarding interpreting ^{1}H NMR spectra (p. 812). What answers do you get for those questions?

Solve

The IHD for a compound with the formula $C_6H_{10}O$ is 2.

Question 1. In the ^{1}H NMR spectrum, there are five signals and so five distinct types of hydrogen.

Questions 2 and 3. The ^{1}H NMR data are summarized in the table below. The relative integration data are not given for this molecule. Notice in the spectrum that the signals at 1.9 ppm and 6.1 ppm each largely appear as a doublet (as listed in the table), but there is some long-range coupling, evidenced by the fact that each peak of the double is itself split into a very closely spaced doublet.

In the "Molecular Fragment" column, the bold H atoms that have gray screens are responsible for the signal, and the H atoms enclosed in parentheses represent the coupled protons.

Entry	δ (ppm)	Nearby Double Bonds or EN Atoms?	Relative Integration	Number of Protons	Splitting Pattern	Number of Neighboring Distinct H	Molecular Fragment
1	1.1	None	–	–	t	2	(H_2C)–**CH_3**
2	1.9	None	–	–	d	1	**H_3C**–(CH)
3	2.6	H-C-C=O	–	–	q	3	O=C–**CH_2**–(CH_3)
4	6.1	H-C=C	–	–	d	?	(HC)=**CH**
5	6.9	H-C=C	–	–	Complex	?	(H_3C)–**CH**=(CH)

Question 4. The alkene CH peaks have been shifted from their normal 4.7–5.3 ppm to 6–7 ppm, which can be explained by the presence of the O atom. The CH_3 groups in Entries 1 and 3 are likely the same, as are the CH_2 groups in those same two entries. Therefore, a portion of the compound is $O{=}CCH_2CH_3$. The C=C double bonds in Entries 4 and 5 are likely the same. The CH_3 groups in Entries 2 and 5 are likely the same, so the molecule should have a $CH_3CH{=}CH$ group. The $O{=}CCH_2CH_3$ and $CH_3CH{=}CH$ groups account for six C atoms, 10 H atoms, and one O atom, which are all the atoms in the molecular formula. Those two groups can be attached to give either the *E* or *Z* isomer, as shown below.

^{1}H NMR Data

(5) 6.9 ppm (4) 6.1 ppm (1) 1.1 ppm (2) 1.9 ppm (3) 2.6 ppm

(Z) A

(E) B

Problem 16.69

Think

What is the IHD? Consider the four questions regarding interpreting ^{1}H NMR spectra (p. 812). What answers do you get for those questions?

Solve

The IHD for a compound with the formula $C_7H_{14}O_2$ is 1.

Question 1. In the ^{1}H NMR spectrum, there are five signals and so five distinct types of hydrogen.

Questions 2 and 3. The ^{1}H NMR data are summarized in the table below. Notice that the relative integration is the same as the number of protons, because the sum of the relative integrations equals the total number of hydrogen atoms in the molecular formula.

In the "Molecular Fragment" column, the bold H atoms that have gray screens are responsible for the signal, and the H atoms enclosed in parentheses represent the coupled protons.

Entry	δ (ppm)	Nearby Double Bonds or EN Atoms?	Relative Integration	Number of Protons	Splitting Pattern	Number of Neighboring Distinct H	Molecular Fragment
1	0.9	None	6	6	d	1	**CH_3**–(CH)–**CH_3**
2	1.2	None	3	3	t	2	(H_2C)–**CH_3**
3	1.9	None	1	1	m	?	**CH**
4	2.3	H–C–C=O	2	2	q	3	(H_3C)–**CH_2**–C=O
5	3.75	H–C–O	2	2	d	2	O–**CH_2**–(CH)

Question 4. The CH in Entry 1 is likely the same as the CH appearing in Entries 3 and 5, in which case, we are dealing with the group $OCH_2CH(CH_3)_2$. The CH_3 and CH_2 groups in Entry 2 are likely the same as the ones in Entry 4, which gives the group $CH_3CH_2C{=}O$. These two groups account for seven C atoms, 14 H atoms, and two O atoms, which accounts for all the atoms in the molecular formula. Therefore, we attach those two pieces together, and the structure is isobutyl propionate:

^{1}H NMR Data

(2) 1.2 ppm
(5) 3.75 ppm
(1) 0.9 ppm
(4) 2.3 ppm
(3) 1.9 ppm

Isobutyl propionate

Problem 16.70

Think

What is the IHD? Consider the four questions regarding interpreting ^{1}H NMR spectra (p. 812). What answers do you get for those questions?

Solve

The IHD for a compound with the formula $C_4H_9NO_2$ is 1.

Question 1. In the ^{1}H NMR spectrum, there are three signals and so three distinct types of hydrogen.

Questions 2 and 3. The ^{1}H NMR data are summarized in the table below. Notice that the relative integration is the same as the number of protons, because the sum of the relative integrations equals the total number of hydrogen atoms in the molecular formula.

Entry	δ (ppm)	Nearby Double Bonds or EN Atoms?	Relative Integration	Number of Protons	Splitting Pattern	Number of Neighboring Distinct H	Molecular Fragment
1	2.9	H–C–N	6	6	s	–	CH_3–N–CH_3
2	3.7	H–C–N or H–C–O	2	2	s	–	N—CH_2 or O—CH_2
3	4.8	H–O	1	1	s	–	C—OH

Question 4. If Entry 2 is due to the presence of N, then the N atoms in Entries 1 and 2 must be the same. In that case, one portion of the molecule would be $(CH_3)_2NCH_2$, which would account for three C atoms, eight H atoms, and one N atom. That would leave one C atom, two O atoms, and one H atom yet to account for. To achieve the IHD of 1, this would need to be a CO_2H group, and the molecule would be the first one below. Alternatively, if Entry 2 is due to the presence of O, then the C and O atoms in Entries 2 and 3 could be the same. In that case, a portion of the molecule would be $HOCH_2$. With another portion being $N(CH_3)_2$, that would account for three C atoms, nine H atoms, one O atom, and one N atom. That would leave one C atom and one O atom yet to be accounted for, as well as an IHD of 1, which could all be due to a C=O group. This would lead to the second molecule below. The second molecule is more consistent with the observed chemical shifts, because the carboxyl OH in the first molecule would appear around 10–12 ppm.

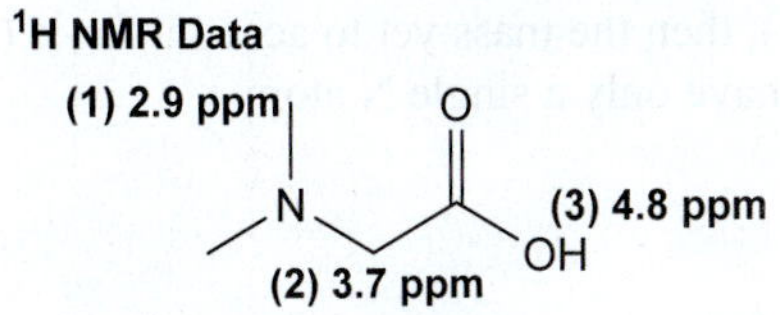

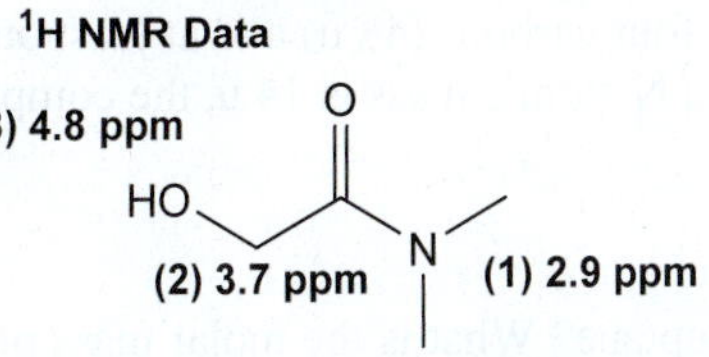

^{1}H NMR Data

(1) 2.9 ppm; (2) 3.7 ppm; (3) 4.8 ppm

^{1}H NMR Data

(3) 4.8 ppm; (2) 3.7 ppm; (1) 2.9 ppm

Problem 16.71

Think

What is the IHD? How many signals appear in the ^{13}C NMR spectrum? Is there any symmetry suggested by the number of signals? From the chemical shifts, what functional groups are likely present?

Solve

The IHD is 4. There are four signals in the ^{13}C NMR spectrum, which means that there is a high degree of symmetry to account for 10 carbon atoms in the molecular formula. The peak at 77 ppm is due to $CDCl_3$. The peaks at ~125 ppm and ~140 ppm are due to C=C carbons characteristic of an alkene or an aromatic ring. Given the high degree of symmetry and the IHD of 4, the molecule probably has a benzene ring, which would account for six C atoms. Para substitution on the benzene ring with the same two groups leads to two aromatic ^{13}C signals. Because no other atoms or functional groups are present aside from sp^3 C atoms, the two groups are CH_2CH_3. Therefore, the structure is 1,4-diethylbenzene (shown on the next page).

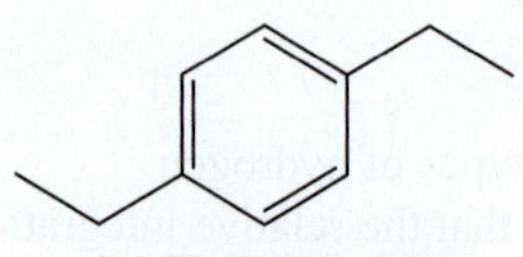

1,4-Diethylbenzene

Sections 16.16–16.19 Mass Spectrometry and Determining a Molecular Formula from a Mass Spectrum

Problem 16.72

Think

How can you use Equation 16-7 to estimate the number of carbons in **X** and **Y**? What are the given peak intensities from this problem?

Solve

Equation 16-7 uses the peak intensities of M + 1 and M^+ and the relative abundances of $^{12}C/^{13}C$ to estimate the number of carbon atoms:

$$\text{Number of C atoms} \approx \frac{\text{Intensity of M+1}}{\text{Intensity of M}^+} \times \frac{100\%}{1.1\%}$$

X $$\text{Number of C atoms} \approx \frac{6.7}{83.2} \times \frac{100\%}{1.1\%} = 7.3 \rightarrow 7 \text{ C atoms}$$

Y $$\text{Number of C atoms} \approx \frac{1.6}{16.9} \times \frac{100\%}{1.1\%} = 8.6 \rightarrow 8 \text{ C atoms}$$

Problem 16.73

Think

What is the mass of the four C atoms? What is the total mass? If there is at least one O atom, what is now the total mass? What is the mass of a N atom? How many N atoms are possible?

Solve

If the compound contains four carbons (48 u) and at least one O (16 u), then the mass yet to account for is (87 − 48 − 16) = 23 u. Because a N atom's mass is 14 u, the compound can have only a single N atom.

Problem 16.74

Think

What is the structure of heptane? What is the molar mass of heptane? What is the formula of the molecular ion, M^+? What bonds can be severed during fragmentation? What is the molar mass of peaks **A–E**? What structure corresponds to each peak? Which peak is the most intense?

Solve

Heptane has an *m/z* value of 100. The fragmentation patterns occur from a C–C bond breaking, as shown below and on the next page. The M^+ occurs at *m/z* = 100, so the M + 1 peak occurs at *m/z* = 101. The base peak is the peak with the highest intensity. This occurs at *m/z* = 43 ($CH_3CH_2CH_2^+$).

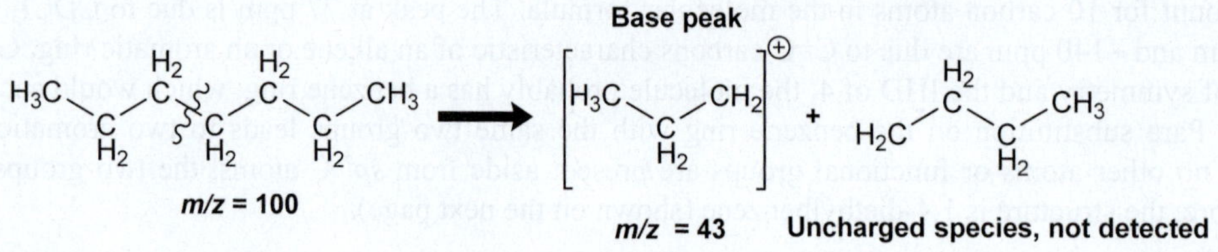

Base peak

m/z = 100

m/z = 43

A

Uncharged species, not detected

m/z = 100 → m/z = 57 **B** + Uncharged species, not detected

m/z = 100 → m/z = 71 **C** + Uncharged species, not detected

m/z = 100 → m/z = 85 **D** + CH_3 Uncharged species, not detected

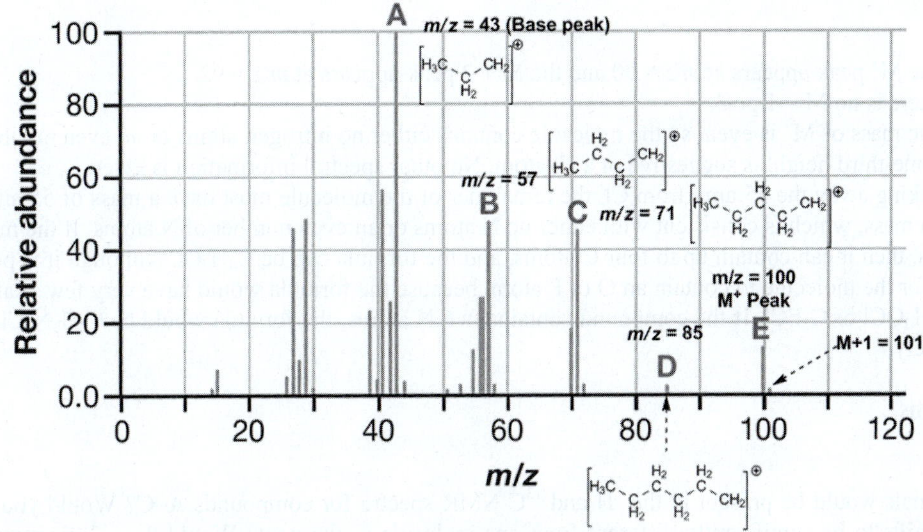

Problem 16.75

Think

Look at the M^+ and M + 2 peak. What is the relative abundance of each? Which isotope matches this relative abundance pattern?

Solve

The compound most likely contains S. The M^+ peak appears to be at m/z = 146. The intensity of the M + 2 peak (m/z = 148) appears to be about 5% of the M^+ peak, which is consistent with it containing S. If the compound contained Br, the ratio would be roughly 1:1; if it were to contain Cl, it would be in roughly a 3:1 ratio.

Problem 16.76

Think

Consider the questions on pages 826 and 827. What answers do you get for those questions?

Solve

(a) Question 1. The M^+ peak appears at $m/z = 72$, and the M + 1 peak appears at $m/z = 73$.

Question 2. The intensity of the M + 1 peak is 1.0% and that of the M^+ peak is 25%. Therefore, the estimate for the number of carbons is (1.0/25.0) × (100%/1.1%) = 3.64, which rounds to 4.

Question 3. The mass of M^+ is even, so the molecule contains either no nitrogen atoms or an even number. There is no M + 2 peak and no other spectral information is given.

Question 4. The four C atoms account for 4(12) = 48 u, leaving 24 u yet to account for. If the molecule contains N, it must contain at least two N atoms, which would require 2(14) = 28 u. This is more than the 24 u that needs to be accounted for, so there must be no N atoms in the molecule. It could contain one O or F atom, however, yielding a formula of C_4H_8O or C_4H_5F.

(b) Question 1. The M^+ peak appears at $m/z = 85$, and the M + 1 peak appears at $m/z = 86$.

Question 2. The intensity of the M + 1 peak is 2.5% and that of the M^+ peak is 40.5%. Therefore, the estimate for the number of carbons is (2.5/40.5) × (100%/1.1%) = 5.6, which rounds to 5.

Question 3. The mass of M^+ is odd, so the molecule contains an odd number of nitrogen atoms. There is no M + 2 peak, and no other spectral information is given.

Question 4. The five C atoms account for 5(12) = 60 u, one N atom 1(14) = 14, leaving 11 u yet to account for. This leaves 11 H atoms, which are the remaining atoms, and the formula is $C_5H_{11}N$.

Problem 16.77

Think

Consider the questions on pages 826 and 827. What answers do you get for those questions?

Solve

Question 1. The M^+ peak appears at $m/z = 90$ and the M + 2 peak appears at $m/z = 92$.

Question 2. There is no M + 1 peak.

Question 3. The mass of M^+ is even, so the molecule contains either no nitrogen atoms or an even number. The M + 2 peak at one-third height is suggestive of a Cl atom. No other spectral information is given.

Question 4. Taking away the 35 amu from Cl, the remainder of the molecule must have a mass of 55 amu. The M^+ has an even mass, which is consistent with either no N atoms or an even number of N atoms. If the molecule has no N atoms, then it can contain up to four C atoms, and the formula can be C_4H_7Cl. Although it is possible, it is not likely for the molecule to contain an O or F atom, because the formula would have very few H atoms or none at all: C_3H_3OCl or C_3FCl. If the compound contains two N atoms, the formula would be $C_2H_3N_2Cl$, which would be unlikely.

Integrated Problems

Problem 16.78

Think

How many signals would be present in the ^1H and ^{13}C NMR spectra for compounds **A–C**? Would you expect the chemical shifts to be significantly different from one molecule to the next? Would the splitting pattern of each ^1H NMR spectrum differ? What would be the integrations for each ^1H NMR signal?

Solve

Proton NMR spectroscopy would be better to distinguish among the three compounds. The carbon spectrum of each would exhibit three signals, given that there are three chemically distinct types of carbons. Moreover, the chemical shifts of those carbons in one molecule will be about the same as the chemical shifts in the other two. A similar story exists with the number of signals and chemical shifts in the proton spectra. However, integration of the ^1H NMR signals will allow us to clearly distinguish among the compounds. In compound **A**, the aldehyde H signal will have an integration of 2H, and the aromatic H signal will have an integration of 4H. In compound **B**, the aldehyde H signal will have an integration of 3H, and the aromatic H signal will have an integration of 3H. In compound **C**, the aldehyde H signal will have an integration of 4H, and the aromatic H signal will have an integration of 2H. See the figure on the next page.

A
H_a:H_b
2:4

B
H_a:H_b
3:3

C
H_a:H_b
4:2

Problem 16.79

Think

How many signals should there be in the ^{1}H and ^{13}C NMR spectra? How many neighboring H atoms are present for each distinct type of proton? What does the $N + 1$ rule predict for each type of coupling? What is the chemical shift of each peak? Are there electron-donating or electron-withdrawing groups nearby? Are there π electrons present? How many H atoms are represented by each peak?

Solve

(a) There are two ^{1}H NMR signals and three ^{13}C NMR signals. The representative spectra are shown below.

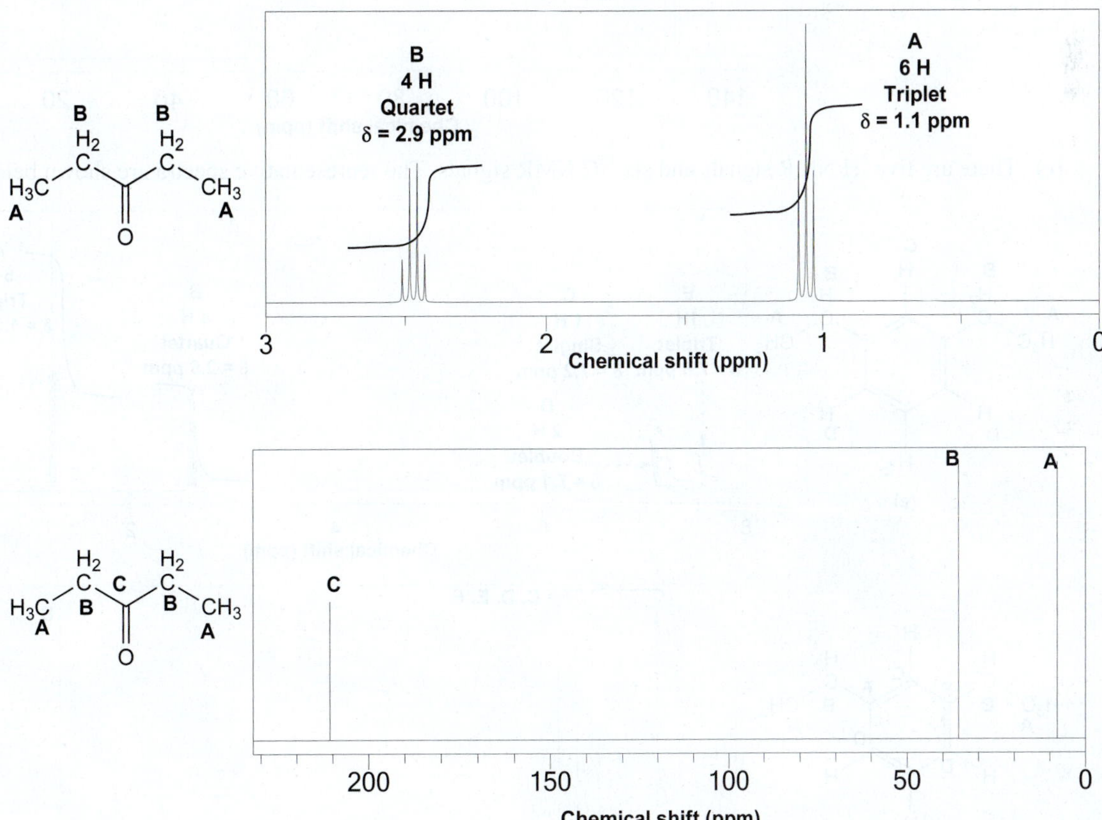

(b) There are three ^{1}H NMR signals and five ^{13}C NMR signals. The representative spectra are shown below.

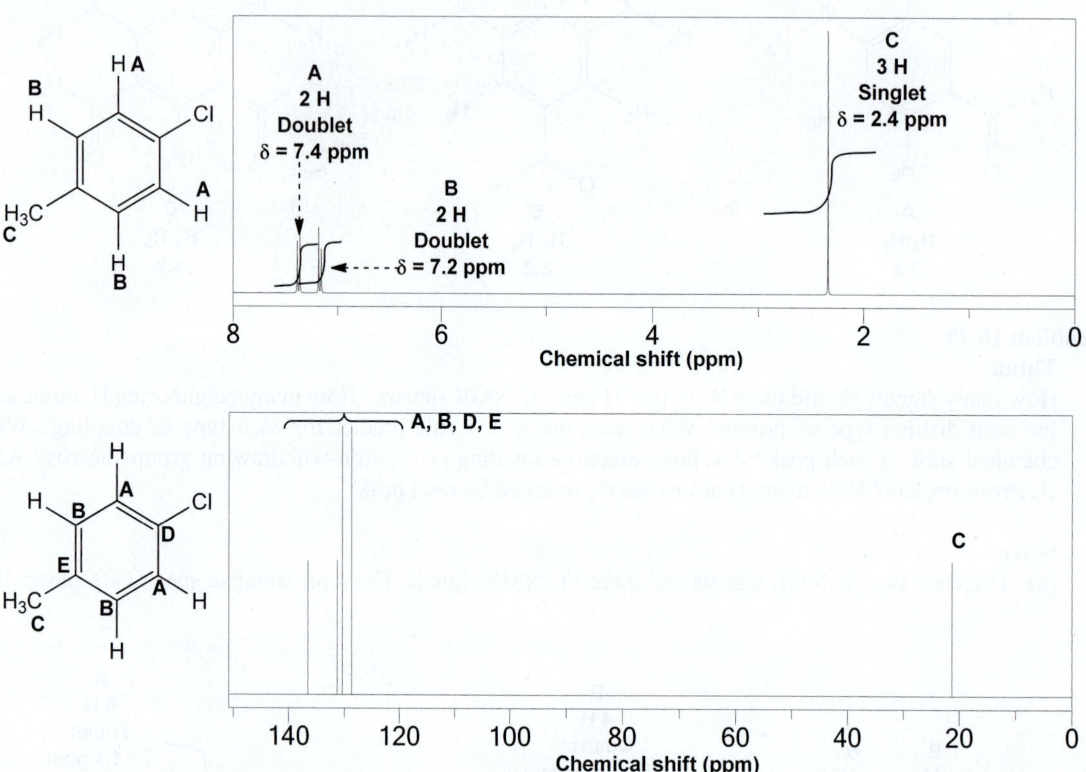

(c) There are five ^{1}H NMR signals and six ^{13}C NMR signals. The representative spectra are shown below.

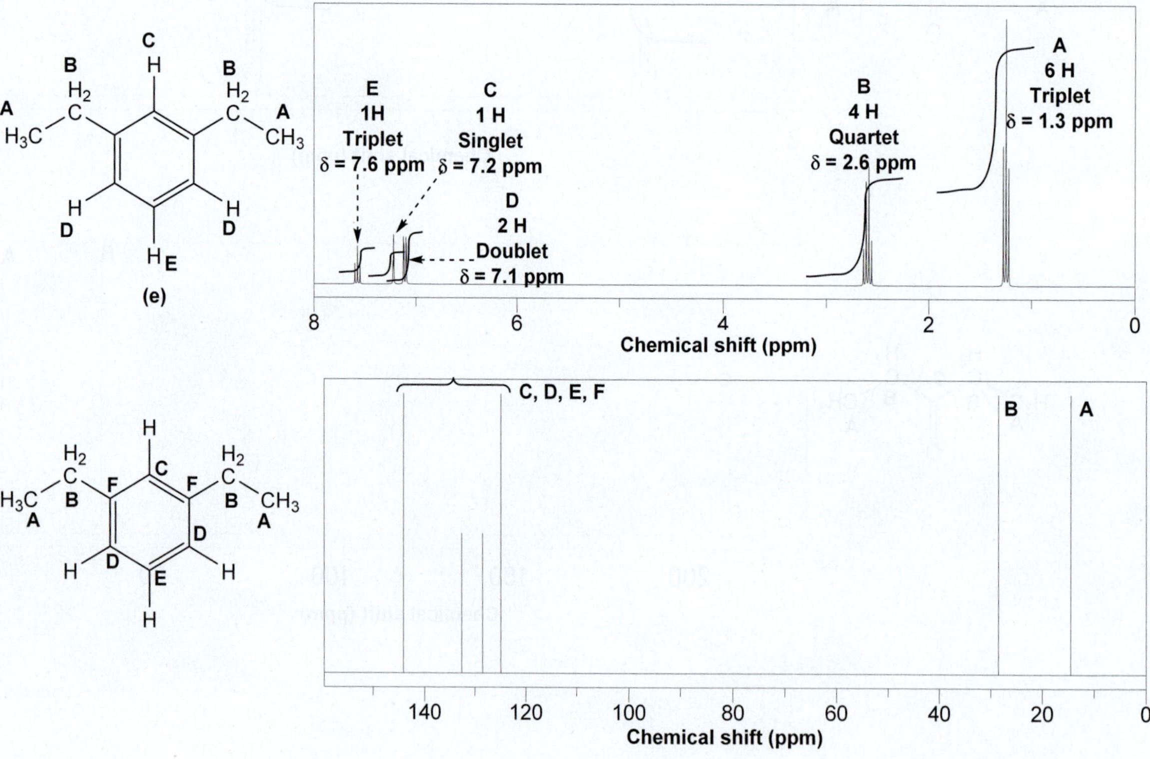

Problem 16.80

Think

What is the IHD? Consider the four questions regarding interpreting ^{1}H NMR spectra (p. 812) and the four questions regarding interpreting ^{13}C NMR spectra (p. 815). What answers do you get for those questions?

Solve

The IHD for a compound with the formula $C_{19}H_{16}O_2$ is 12, so it can contain up to 12 double bonds, up to 12 rings, and up to six triple bonds.

Question 1. In the ^{1}H NMR spectrum, there are two signals. Because one of the signals is an unresolved multiplet, we cannot make an interpretation about the number of distinct H atoms. In the ^{13}C NMR spectrum, there are five signals. To have only five signals for 19 carbons, a high degree of symmetry must be present in the compound. Many carbons must be chemically equivalent.

Questions 2 and 3. The ^{1}H NMR data are summarized in the table below. Notice that the relative integration is the same as the number of protons because the sum of the relative integrations equals the total number of hydrogen atoms in the molecular formula. Insight into the structural feature that gives rise to this symmetry is the fact that several C signals and several H signals appear in the aromatic region, suggesting, perhaps, one or more benzene rings.

Entry	δ (ppm)	Nearby Double Bonds or EN Atoms?	Relative Integration	Number of Protons	Splitting Pattern	Number of Neighboring Distinct H	Molecular Fragment
1	5.5	H–C– (Aromatic)	1	1	s	0	Aromatic—CH
2	6.9–7.44	H– (Aromatic)	15	15	m	?	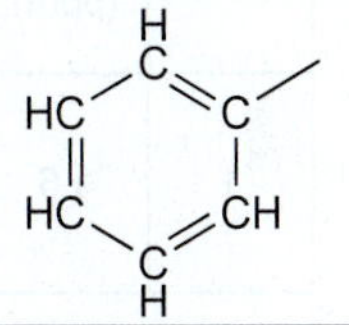 (benzene ring: HC, HC, CH, C–H, C–H, C–)

We could construct another table using just the ^{13}C NMR data, but, as we see below, the number of carbon signals is sufficient to arrive at just a single structure for the molecule—one sp^3 C at 57 ppm, with the remaining signals between 126 and 144 ppm owing to the aromatic C atoms.

Question 4. If each benzene ring accounts for an IHD of 4, we could have three benzene rings. Only one carbon is distinct from these aromatic carbons—the one at 57 ppm in the ^{13}C NMR spectrum. This carbon possesses one chemically distinct proton, which gave the proton NMR signal at 5.5 ppm. Subtracting CH from $C_{19}H_{16}$ gives $C_{18}H_{15}$, which is consistent with three equivalent phenyl groups. The compound is triphenylmethane, $(C_6H_5)_3CH$.

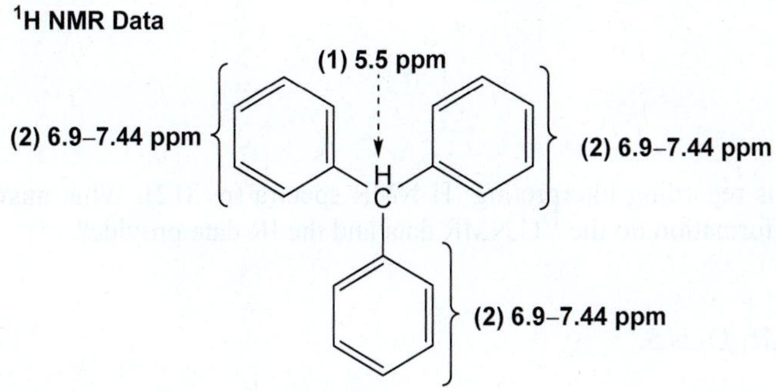

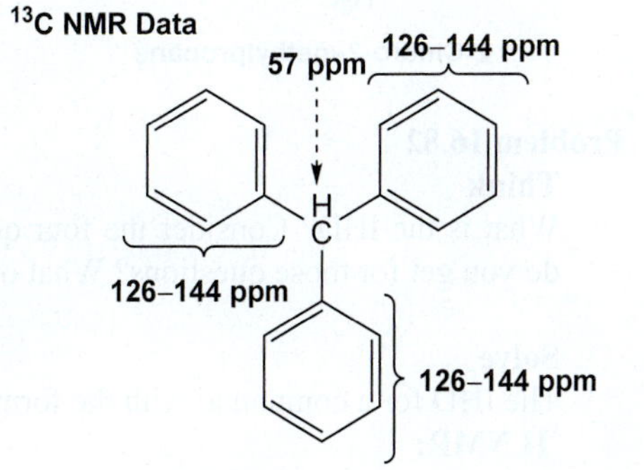

Problem 16.81

Think

Consider the questions on pages 826 and 827 for interpreting mass spectra. What answers do you get for those questions? Consider the four questions regarding interpreting ^{1}H NMR spectra (p. 812). What answers do you get for those questions?

Solve

Mass Spectrum:

Question 1. The M^+ peak appears at $m/z = 92$, and the $M + 2$ peak appears at $m/z = 94$.

Question 2. No $M + 1$ peak data are given.

Question 3. The mass of M^+ is even, so the molecule contains either no nitrogen atoms or an even number. The $M + 2$ peak at one-third height is suggestive of a Cl atom. No other spectral information is given.

Question 4. Taking away the 35 u due to Cl, the remainder of the molecule must have a mass of $92 - 35 = 57$ u. There can be at most four C atoms, because five C atoms would give a mass of 60 u, which would exceed the 57 u limit. If the molecule contains four C atoms, the remaining 9 u would be accounted for by nine H atoms, giving the molecular formula C_4H_9Cl. If, on the other hand, the molecule contains three C atoms, the remaining 21 u could be accounted for by one F atom and two H atoms (C_3H_2ClF) or by one O atom and five H atoms (C_3H_5ClO). It is not likely that the molecule contains fewer than four C atoms, because the additional electronegative atoms that would be necessary would cause higher chemical shifts in the ^{1}H NMR spectrum. Therefore, the formula is likely C_4H_9Cl.

^{1}H NMR:

With a formula of C_4H_9Cl, the IHD is 0.

Question 1. In the ^{1}H NMR spectrum, there is one signal and so only one distinct type of hydrogen.

Questions 2 and 3. The ^{1}H NMR data are summarized in the table below. Notice that the relative integration is 1, because there is only one type of proton. With only one signal, all nine H atoms must be equivalent

Entry	δ (ppm)	Nearby Double Bonds or EN Atoms?	Relative Integration	Number of Protons	Splitting Pattern	Number of Neighboring Distinct H	Molecular Fragment
1	1.6	None	1	9	s	0	—C—CH

Question 4. The low chemical shift could be due to a CH_3 group, slightly deshielded by Cl a few bonds away. If so, then all nine H atoms would make up three equivalent CH_3 groups attached to a C atom that has no H atoms bonded to it, a *t*-butyl group, $C(CH_3)_3$. The molecule would then be *t*-butyl chloride, $Cl–C(CH_3)_3$.

^{1}H NMR Data

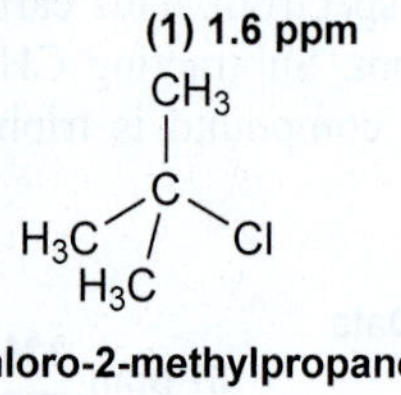

2-Chloro-2-methylpropane

Problem 16.82

Think

What is the IHD? Consider the four questions regarding interpreting ^{1}H NMR spectra (p. 812). What answers do you get for those questions? What other information do the ^{13}C NMR data and the IR data provide?

Solve

The IHD for a compound with the formula $C_9H_{10}O_3$ is 5.

^{1}H NMR:

Question 1. In the ^{1}H NMR spectrum, there are five signals and so five distinct types of hydrogen.

Questions 2 and 3. The ^{1}H NMR data are summarized in the table on the next page. Notice that the relative integration is the same as the number of protons, because the sum of the relative integrations equals the total number of hydrogen atoms in the molecular formula.

In the "Molecular Fragment" column, the bold H atoms that have gray screens are responsible for the signal, and the H atoms enclosed in parentheses represent the coupled protons.

Entry	δ (ppm)	Nearby Double Bonds or EN Atoms?	Relative Integration	Number of Protons	Splitting Pattern	Number of Neighboring Distinct H	Molecular Fragment
1	1.4	None	3	3	t	2	$(H_2C){-}CH_3$
2	4.1	H–C–O	2	2	q	3	$O{-}CH_2{-}(CH_3)$
3	6.9	H– (Aromatic)	2	2	d	1	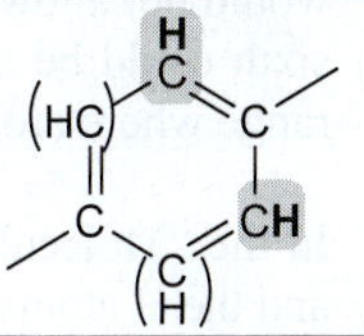 aromatic ring fragment, (HC)–C(H)=C, CH
4	7.9	H– (Aromatic)	2	2	d	1	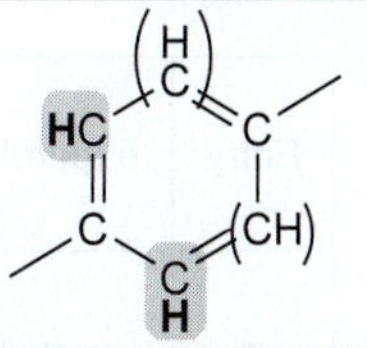 aromatic ring fragment, HC–C(H), (CH)
5	12.2	H–O	1	1	s	0	CO_2H

Question 4. The CH_3 and CH_2 groups in Entry 1 could be the same as those in Entry 2, in which case we have an ethoxy group, OCH_2CH_3. Thus, the three groups OCH_2CH_3, C_6H_4, and CO_2H would account for nine C atoms, 10 H atoms, and three O atoms, which is all of the atoms in the molecular formula. The fact that the aromatic H signals are doublets suggests that the ring is para-disubstituted and the two groups are different, as shown in the molecule below.

^{13}C NMR Spectrum:

There are seven signals in the spectrum, so there are seven distinct C atoms. This agrees with the structures below, especially the para substitution. Ortho or meta substitution would have made all six aromatic carbons distinct, producing nine total signals in the spectrum.

IR Spectrum:

A C=O is evident, with a C=O stretch at around 1700 cm^{-1} in the IR spectrum. There are also C=C stretches around 1450 to 1600 cm^{-1}, which indicate an aromatic ring. Also, the very broad peak that extends from ~2500 cm^{-1} to ~3200 cm^{-1}, which overlaps the C–H stretches, is indicative of a carboxylic acid.

^{1}H NMR and ^{13}C NMR data are shown below for the structure.

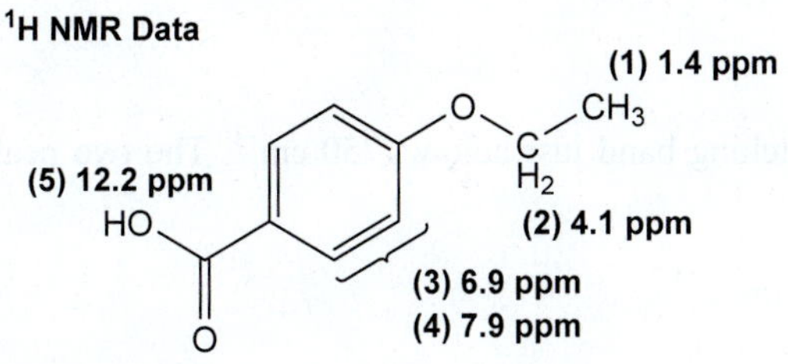

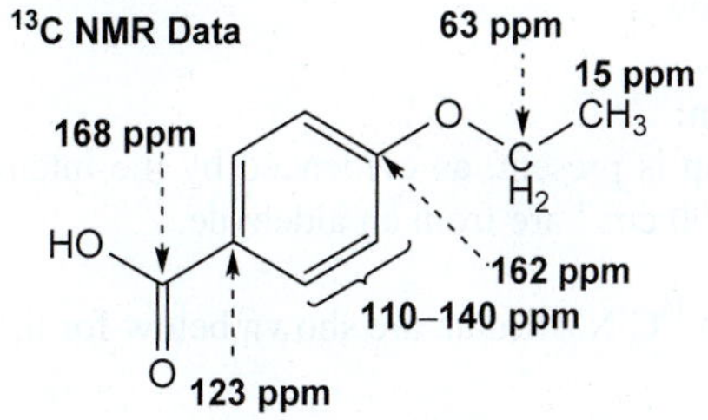

^{1}H NMR Data

(1) 1.4 ppm
(2) 4.1 ppm
(3) 6.9 ppm
(4) 7.9 ppm
(5) 12.2 ppm

^{13}C NMR Data

63 ppm
15 ppm
168 ppm
162 ppm
110–140 ppm
123 ppm

Problem 16.83

Think

What is the IHD? Consider the four questions regarding interpreting ^{1}H NMR spectra (p. 812). What answers do you get for those questions? What other information do the ^{13}C NMR data and the IR data provide?

Solve

The IHD for a compound with the formula C_9H_8O is 6.

^{1}H NMR Spectrum:

Question 1. In the ^{1}H NMR spectrum, there are three signals. The signals between 6.5 and 7 and between 7 and 8 ppm do not appear to be fully resolved, so there are at least three chemically distinct H atoms.

Questions 2 and 3. The ^{1}H NMR data are summarized in the table below. Notice that the relative integration is the same as the number of protons, because the sum of the relative integrations equals the total number of hydrogen atoms in the molecular formula. There appears to be an alkene group, indicated by the signal near 6.6 ppm, Entry 1. The signal at 9.7 ppm, Entry 3, is from an aldehyde H. Entry 3 is likely multiple signals overlapping. The chemical shift is in the range of aromatic protons, but the signal integrates to six hydrogens. If that signal were to represent only aromatic H atoms, then that would complete a molecule of benzene, C_6H_6, and would not allow for other atoms/groups to be attached. Instead, five of these protons could be aromatic, and the sixth could be an alkene H, which is deshielded by the aromatic ring to give it a chemical shift that is in the range where aromatic protons normally appear.

In the "Molecular Fragment" column, the bold H atoms that have gray screens are responsible for the signal, and the H atoms enclosed in parentheses represent the coupled protons.

Entry	δ (ppm)	Nearby Double Bonds or EN Atoms?	Relative Integration	Number of Protons	Splitting Pattern	Number of Neighboring Distinct H	Molecular Fragment
1	6.7	H–C=C	1	1	d of d	?	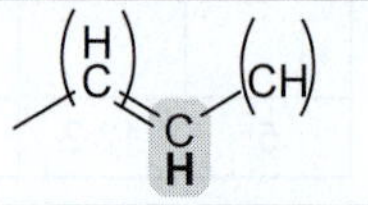 (H)C=CH–(CH)
2	7.3–7.8 Multiple signals	H– (Aromatic) and H–C=C	6	6	Complex	?	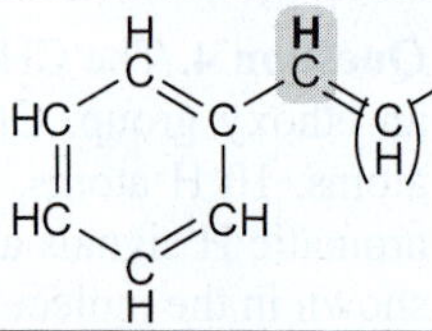 C₆H₅–CH=(C)(H)
3	9.7	H–C=O	1	1	d	2	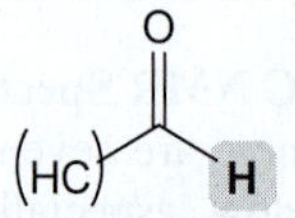 (HC)–CH=O

Question 4. The alkene C=C in Entry 1 could be the same alkene C=C in Entry 2, in which case, we have a portion of the molecule: $C_6H_5CH{=}CHCH$. If the two CHs at the very right of this portion are the same as the two CHs in Entry 3, then we have the molecule $C_6H_5CH{=}CHCH{=}O$, shown below, which has the formula C_9H_8O, the same as the formula we were given. *Note:* Without coupling constant information, the cis isomer is also possible.

^{13}C NMR Spectrum:

The spectrum has seven carbon signals, so there are seven chemically distinct C atoms. This agrees with the structure below.

IR Spectrum:

A C=O group is present, as evidenced by the intense stretching band just below 1750 cm^{-1}. The two peaks at 2700 and 2800 cm^{-1} are from an aldehyde.

^{1}H NMR and ^{13}C NMR data are shown below for the structure.

^{1}H NMR Data

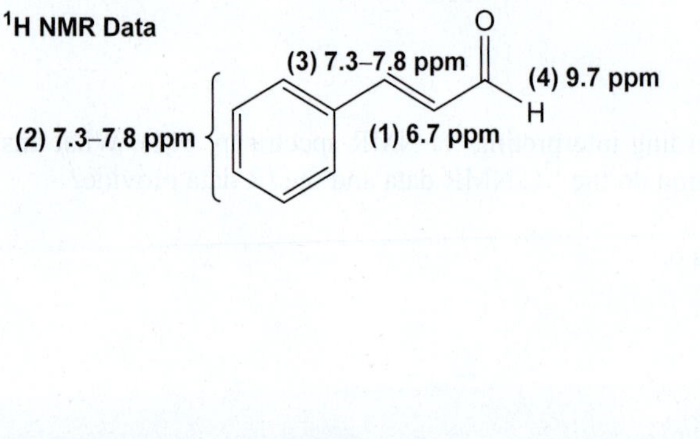

(2) 7.3–7.8 ppm
(3) 7.3–7.8 ppm
(1) 6.7 ppm
(4) 9.7 ppm

^{13}C NMR Data (7 signals)

A B C D E F G

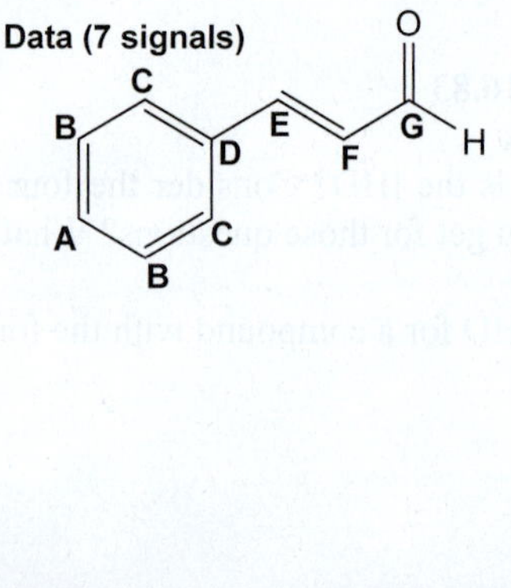

Problem 16.84

Think

What is the IHD? Consider the four questions regarding interpreting ^{1}H NMR spectra (p. 812). What other information do the ^{13}C NMR data and the IR data provide?

Solve

The IHD for a compound with the formula $C_4H_{10}O_2$ is 0.

^{1}H NMR Spectrum:

Question 1. In the ^{1}H NMR spectrum, there are four signals and so four distinct types of hydrogen.

Questions 2 and 3. The ^{1}H NMR data are summarized in the table below. The relative integration is the same as the number of protons.

In the "Molecular Fragment" column, the bold H atoms that have gray screens are responsible for the signal, and the H atoms enclosed in parentheses represent the coupled protons.

Entry	δ (ppm)	Nearby Double Bonds or EN Atoms?	Relative Integration	Number of Protons	Splitting Pattern	Number of Neighboring Distinct H	Molecular Fragment
1	0.85	None	3	3	d	1	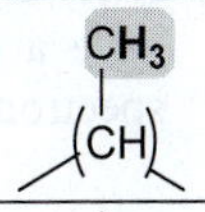 **CH$_3$**–(CH)
2	1.9	None	1	1	broad	?	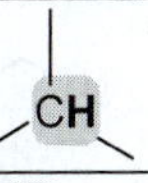 **CH**
3	3.6	H–C–O	4	4	d	1	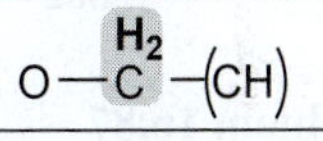 O–**C(H$_2$)**–(CH)
4	4.1	H–O	2	2	t	2	O**H**

Question 4. The CH groups in Entries 1, 2, and 3 are likely the same, in which case, we have a portion of the molecule: CH_3CHCH_2O. Entry 4 suggests that the O at the very right of this portion is OH, so we have CH_3CHCH_2OH. To have four H atoms for the fragment in Entry 2 and to have two H atoms for the fragment in Entry 4, we should consider two equivalent $-CH_2OH$ fragments. The second $-CH_2OH$ fragment would need to attach to the open valence on CH, giving $CH_3CH(CH_2OH)_2$, as shown below.

^{13}C NMR Spectrum:

There are three carbon signals and so three distinct C atoms. This agrees with the structure below.

IR Spectrum:

The IR clearly shows an OH stretch at 3400 cm^{-1}, which corresponds to an alcohol OH. This is also consistent with the structure below.

^{1}H NMR and ^{13}C NMR data are shown below for the structure.

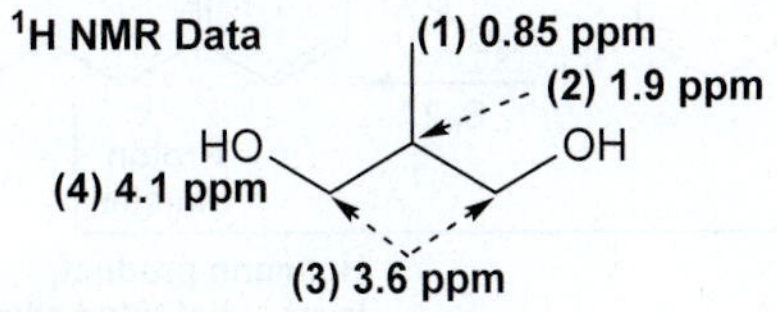

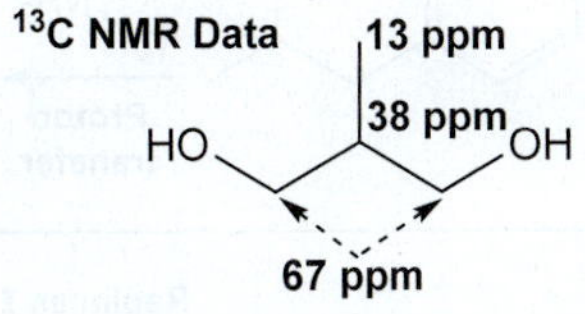

^{1}H NMR Data: HO, OH; (1) 0.85 ppm; (2) 1.9 ppm; (3) 3.6 ppm; (4) 4.1 ppm

^{13}C NMR Data: HO, OH; 13 ppm; 38 ppm; 67 ppm

Problem 16.85

Think

Which compound has a signal that is farther downfield? Which signal is more deshielded? Do electronegative groups cause nearby protons to have more shielded or more deshielded signals? What is the hybridization of each α carbon? What is the ideal bond angle of an atom with this hybridization? What is the actual bond angle of each C–C$_\alpha$–C (think about the ring angle). What atomic orbitals make a 90° angle?

Solve

(a) The α carbon in cyclobutanone has a greater effective electronegativity, because the protons (being farther downfield) are more deshielded.

(b) The α carbon to which the proton is attached should be sp^3 hybridized in each molecule, giving it 25% *s* character. But to have a 90° C–C–C bond angle, the C–C bonds must use orbitals that have more *p* character than normal. That leaves more *s* character than normal for the orbitals used for the C–H bonds. As we learned in Chapter 3, greater *s* character corresponds to higher effective electronegativity.

Problem 16.86

Think

What are the identities of the organic reactant and product? Would these compounds have different or the same ^1H, ^{13}C, and mass spectra? If the HO^- nucleophile were isotopically labeled with $H^{18}O^-$, how would that change the mass of the product?

Solve

(a) Isotopically labeled atoms would be necessary because the products would otherwise be identical to the reactants.

(b) If $H^{18}O^-$ were used, then an isotopically labeled product would be formed. Because the products would have a different molar mass than the reactants, the two compounds could be distinguished using mass spectrometry.

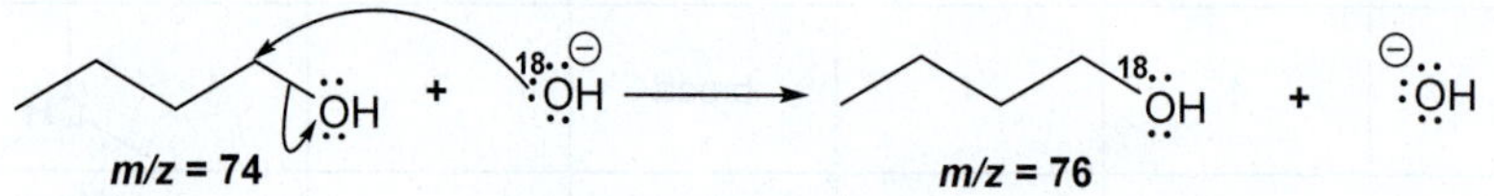

m/z = 74 *m/z* = 76

Problem 16.87

Think

Is a leaving group initially present? If not, how does the reaction with excess methyl iodide lead to the formation of a good leaving group? When the solution is heated in the presence of a strong base, what mechanism results? Does the more or less substituted alkene form? How is this product supported by the ^1H NMR spectrum?

Solve

This is an example of a Hofmann elimination reaction, yielding $C_6H_5CH_2CH{=}CH_2$. The signals from the five aromatic H atoms appear above 7 ppm. The signal at 6 ppm, which integrates to 1H, is the alkene CH, and the one at 5 ppm that integrates to 2H is the alkene CH_2. The doublet is the alkane CH_2, which is split by the alkene CH proton. It is significantly downfield shifted because of the adjacent phenyl ring and alkene group. The complete mechanism is shown below.

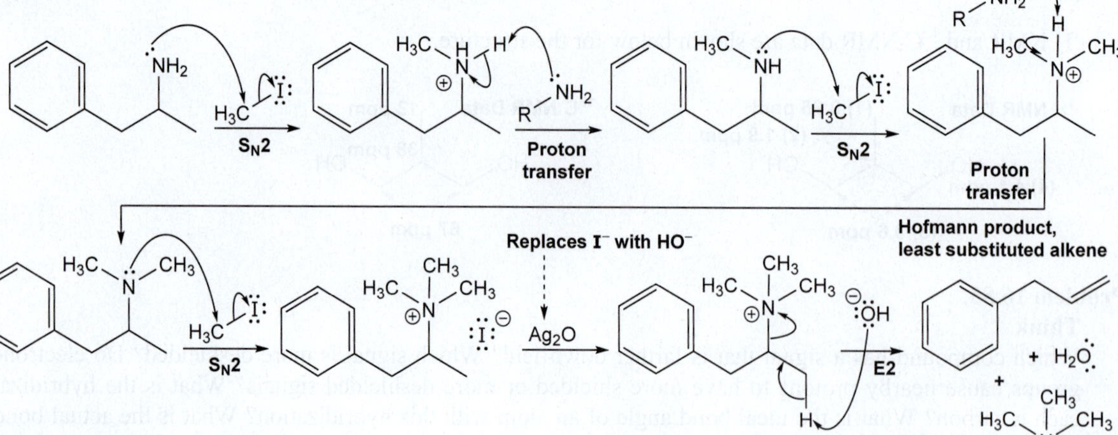

S_N2 Proton transfer S_N2 Proton transfer

S_N2 Replaces I^- with HO^- Ag_2O E2

Hofmann product, least substituted alkene

Problem 16.88

Think

Is the reaction an example of addition, elimination, or substitution? Is there a good leaving group? What are the identities of the two products? On which C atom is the leaving group likely located? Can both products be produced from the same carbocation intermediate? If so, is the mechanism most likely S_N1 or S_N2?

Solve

Because the OH group is replaced by Br, the reaction must be an S_N1 or S_N2 reaction. The product alkenes are shown below.

1-Bromopent-2-ene   3-Bromopent-1-ene

These isomers could be produced from the allylic carbocation intermediate shown below, which has two resonance structures.

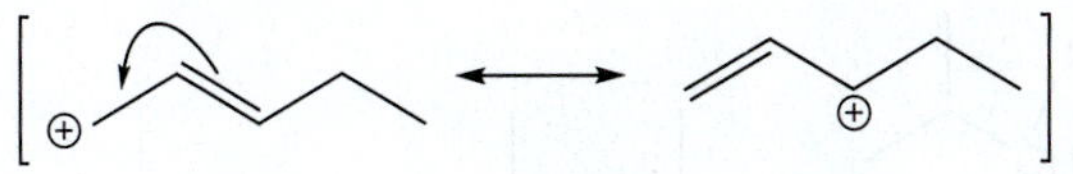

The allylic cation can be produced under S_N1 conditions from the following alcohols.

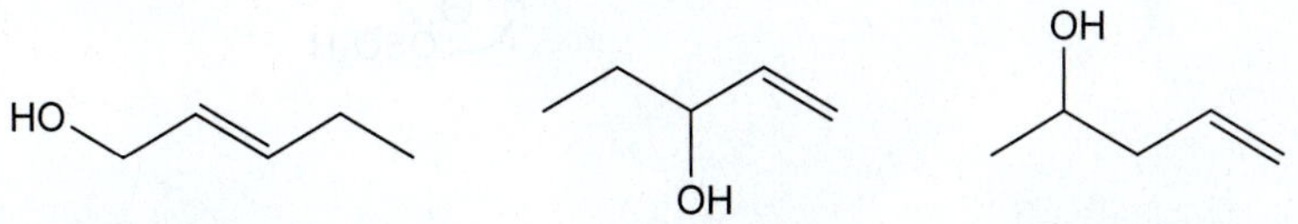

Pent-2-en-1-ol   Pent-1-en-3-ol   Pent-4-en-2-ol

The pent-4-en-2-ol is more likely the starting alcohol, owing to the two alkene proton signals (5.8 ppm and 5.1 ppm) integrating to 1H and 2H, respectively. In the compound pent-2-en-1-ol, the two signals would have been 1H and 1H. The peak at 1.2 ppm is also a doublet, which is due to the CH_3 that is next to a CH. Pent-1-en-3-ol would have the CH_3 split as a triplet. The complete mechanism is shown below.

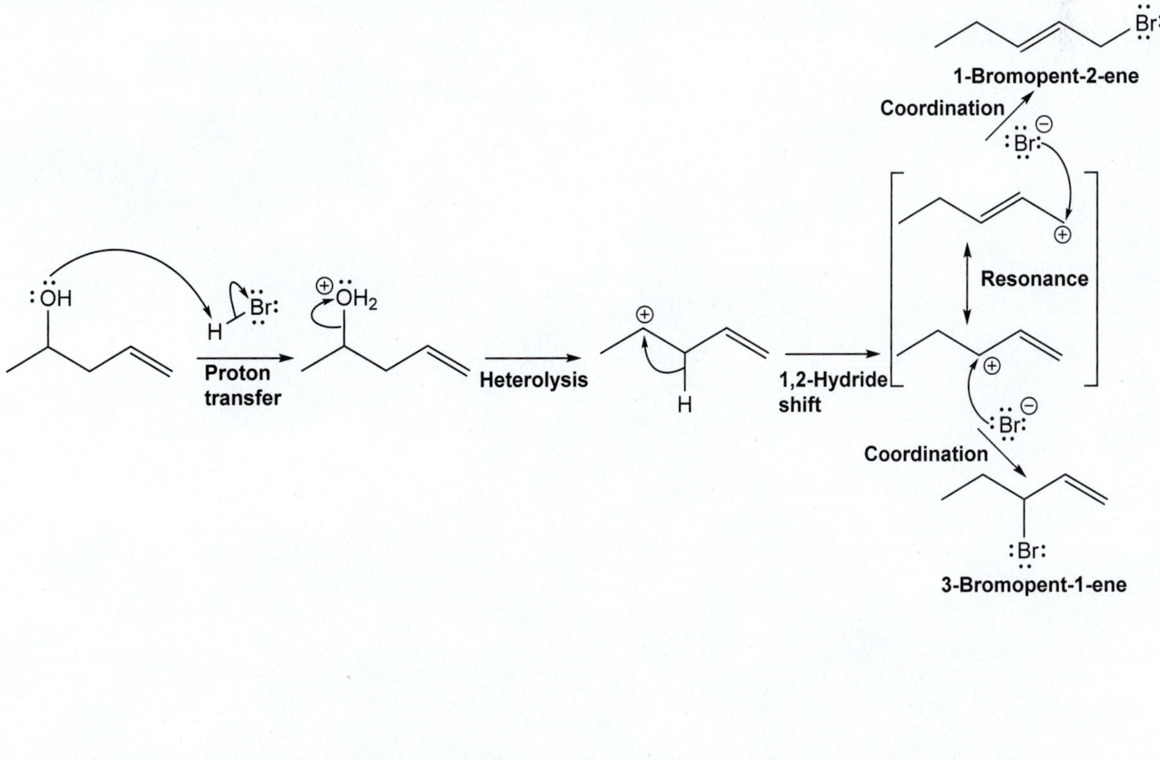

Proton transfer → Heterolysis → 1,2-Hydride shift → Resonance → Coordination → 1-Bromopent-2-ene; Coordination → 3-Bromopent-1-ene

Problem 16.89

Think

What mechanism and functional group result when an alcohol is heated in strong acid? Did substitution or elimination occur? From the IR spectrum, are any sp^2 C–H groups present? Does the ^{1}H NMR spectrum support this? In the ^{1}H NMR spectrum, what causes the downfield signal at 3.6 ppm? In the ^{13}C NMR spectrum, how many signals are present? Does the number of signals suggest a high or low degree of symmetry?

Solve

The ^{13}C–NMR spectrum has only two signals, indicating only two distinct types of carbons in the molecule. One C atom at 68 ppm is clearly attached to an electronegative atom—oxygen—whereas the other is not. The proton NMR spectrum shows only two signals with a relative integral of 1:6 hydrogens. The smaller signal, farther downfield, indicates that a proton is on a C atom attached to O. Furthermore, the signal is split into seven peaks, so it is split by six neighboring protons that are equivalent. We consider the fragment $(CH_3)_2CH$–O. If these are the only types of carbons in the molecule, then the rest of the molecule must have the same carbon structure. Recall from Chapter 10 that heating an alcohol with acid will dehydrate it to form either an alkene or an ether. The former is not indicated by the ^{13}C NMR spectrum, so the compound must be diisopropyl ether, $(CH_3)_2CH$–O–CH–$(CH_3)_2$. The mechanism for its formation is shown below.

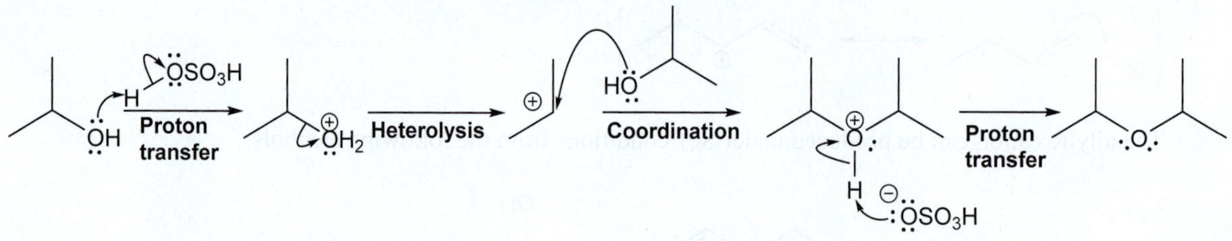

OSO₃H
H
OH
Proton transfer
OH₂
Heterolysis
HO
Coordination
O
H
OSO₃H
Proton transfer
O

CHAPTER 17 | Nucleophilic Addition to Polar π Bonds 1: Addition of Strong Nucleophiles

Your Turn Exercises

Your Turn 17.1

Think

Use Equation 17-1 as a guide. What kinds of charges characterize an electron-rich atom? Are there any formal charges present? Any strong partial charges? Which bonds are broken and which bonds are formed? How is a curved arrow used to denote the flow of electrons? Consult Chapters 6 and 7 to review the 10 most common elementary steps. Are any protons or π bonds involved?

Solve

The C in a C=O bond bears a large partial positive charge, δ^+, owing to the electronegativity difference of C and O. The electrons flow from the electron-rich Nu:$^-$ to the electron-poor C atom of the C=O bond. A second curved arrow (to illustrate the breaking of the π bond between the C and O) is necessary to avoid exceeding an octet on the C. The pair of electrons in the π bond becomes a lone pair on the O and forms an electron-rich basic product. The second reaction is a proton transfer where the electron-rich negative O deprotonates the H–A acid. Overall, the nucleophile is added to the C and a proton is added to the O of the C=O. The ketone is transformed into an alcohol.

Electron rich
Electron poor
Electron rich
Electron poor
1. Nucleophilic addition
2. Proton transfer

Your Turn 17.2

Think

How many different groups must be bonded to a tetrahedral carbon for it to be a chiral center (i.e., an asymmetric carbon)? What is the geometry of the carbonyl C=O carbon? From which side of the carbon's plane can the nucleophile attack?

Solve

The carbon of the C=O forms a new chiral center from a trigonal planar C atom. The carbon becomes a chiral center because its bonds are to four different groups. The nucleophile can attack the C atom from two different sides: from behind the plane and in front of the plane. See below.

Nucleophile can attack from either side of the carbon's plane
Attack from behind the plane
Attack from in front of the plane

Your Turn 17.3

Think

Refer to the general mechanism for nucleophile addition to a polar π bond in Equation 17-1. What are the electron-rich and electron-poor species in the starting reagents? What elementary step takes place first? What is the role of the H_2O? Are proton transfer reactions involved?

Solve

The mechanism is the same for each hydration reaction in Table 17-1. HO^- is a strong nucleophile and will attack the electron-poor carbonyl C, breaking the π bond of the double bond. The resulting O^- is protonated by the weakly acidic water solvent.

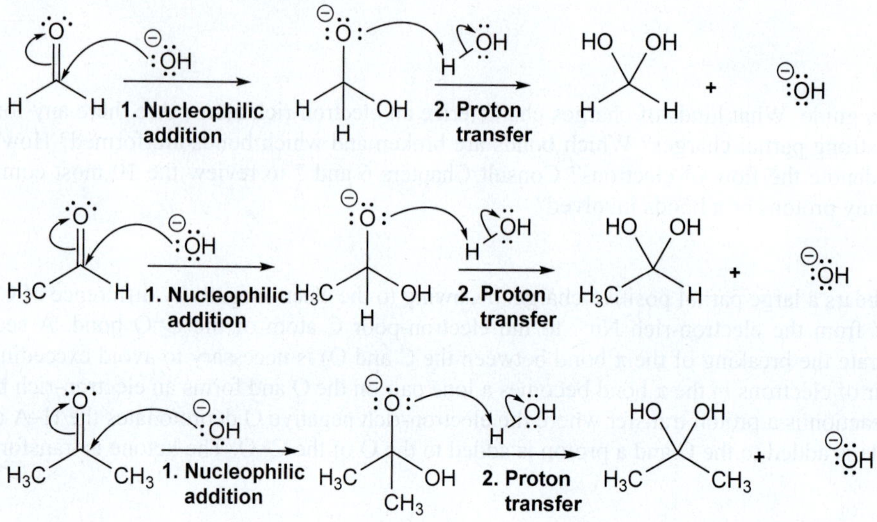

1. Nucleophilic addition
2. Proton transfer

Your Turn 17.4

Think

Are CH_3 groups electron-donating or electron-withdrawing groups? What effect does each CH_3 have on the partial positive charge of the C atom of the C=NH? How does the bulkiness of the CH_3 groups affect the ability of a nucleophile to attack C=NH?

Solve

Alkyl groups are electron donating and decrease the concentration of positive charge at the C atom of the C=NH group. This is noted below by the size of the δ^+ on each C below. With decreased concentration of the positive charge, the C=N group is less reactive toward nucleophiles. Alkyl groups also add more steric hindrance surrounding C and decrease the reactivity at the C=NH. Therefore, the most reactive imine is **B**, followed by **A**, followed by **C**.

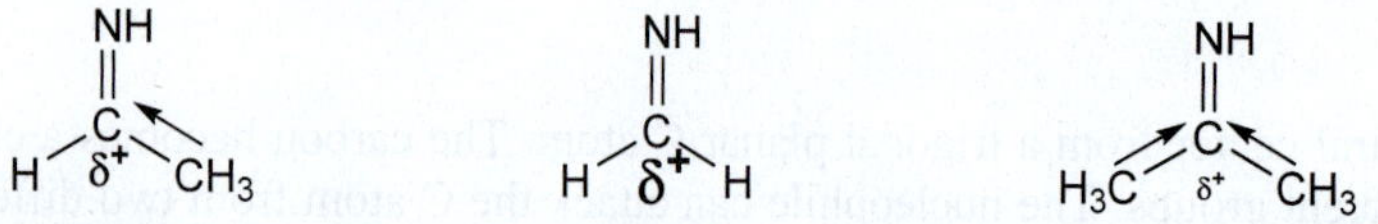

A

B
Highest concentration of positive charge

C
Lowest concentration of positive charge

Your Turn 17.5

Think

To what nucleophile can you simplify $LiAlH_4$ and $NaBH_4$? What is the electrophile? Are proton transfer reactions involved?

Solve

In Equation 17-4, $LiAlH_4$ is simplified to the hydride anion, $:H^-$, as the nucleophile. The first step of the mechanism is nucleophilic addition of the hydride to the C of the carbonyl. A separate acid, NH_4^+, is added to initiate a proton transfer. In Equation 17-5, the $NaBH_4$ is also simplified to $:H^-$ as the nucleophile. The first step of the mechanism is nucleophilic addition of the hydride to the C of the carbonyl. In Step 2, glycerol, the solvent that is already present, protonates the alkoxide anion to produce the final alcohol product. See the figure on the next page.

Equation 17-4

1. Nucleophilic addition

2. Proton transfer

Equation 17-5

1. Nucleophilic addition

2. Proton transfer

Your Turn 17.6

Think

Use Equation 17-9 as a guide. Identify the similarities between BH_4^- and AlH_4^-. How do you show a pair of electrons from the C=O bond becoming an O–Al bond? How do you show a pair of electrons from the Al–H bond becoming a C–H bond?

Solve

BH_4^- and AlH_4^- are similar, and the mechanism for the reduction of butanone is the same for the two reducing agents. In the first step, to show the O–Al bond being formed, a curved arrow is drawn from the center of the C=O bond to the Al atom. At the same time, the H–AlH_3^- bond is broken and the C–H bond is formed, so a curved arrow is drawn from the center of the Al–H bond to the C atom. In the second step, to show the RO–H bond being formed, a curved arrow is drawn from the center of the O–Al bond to the H atom. At the same time, formation of the O–Al bond is shown by drawing a curved arrow from the center of the H–OH bond to the Al atom.

1. Nucleophilic addition

2. Proton transfer

Your Turn 17.7

Think

Use Figure 17-7 as a guide. What is the oxidation state of C2 in butan-2-one? What is the oxidation of C2 in butan-2-ol? Is the oxidation state becoming more or less positive? By how many electrons?

Solve

The oxidation state of C2 in butan-2-one is +2. The oxidation state of C2 in butan-2-ol is 0. This carbon becomes less positive by 2 and is therefore reduced by two electrons.

+2 oxidation state

$NaBH_4$

H_2O

0 oxidation state
Reduced by 2 electrons

Butan-2-one

Butan-2-ol

Your Turn 17.8

Think

Use Equation 17-18 as a guide. What is the partial charge on C in the C–MgBr? The partial charge on H of H_2O? How is the alkyl–Li reagent similar to the alkyl–MgBr reagent?

Solve

The alkyl–metal reagents are similar in that C bears a partial negative charge and, therefore, they can be thought of as R^- *donors*. The H atoms of H_2O bear a partial positive charge. Therefore, the electrons in the C–MgBr bond attack the H in H_2O, illustrated by a curved arrow originating from the middle of the C–MgBr bond and pointing to H. At the same time, the H–O bond breaks, leading to a new C–H bond and HO^- and $MgBr^+$ as products.

MgBr H O H → H + OH + MgBr

Your Turn 17.9

Think

What kinds of charges characterize an electron-rich atom? Are there any formal charges present? Any strong partial charges?

Solve

In the first step, the C in the C=O bears a large partial positive δ^+ owing to the electronegativity difference of C and O. The C of the Wittig reagent is highly nucleophilic and is, therefore, very electron rich. Thus, a curved arrow is drawn from the electron-rich carbon to the electron-poor C of the C=O. In the second step, the alkoxide O is electron rich and the positive P in $^+PR_4$ is electron poor; a curved arrow is drawn from O to P.

Electron poor
Electron rich
1. Nucleophilic addition
Electron rich
Electron poor
A betaine
2. Coordination
Strained 4-membered ring
An oxaphosphetane

Your Turn 17.10

Think

Consult Table 9-10 to verify this statement. Is Br^- considered to be a good nucleophile? What are the relative nucleophilicities of triphenylphosphine and Br^-?

Solve

Br^- is a very good nucleophile owing to its full negative charge. In Table 9-10, the nucleophilicity of Ph_3P is listed as 10,000,000, and that of Br^- is listed as 620,000. Although Ph_3P is uncharged, it is a good nucleophile, because the P atom can accommodate the positive charge in the product rather well, owing both to its large size and its modest electronegativity.

Your Turn 17.11

Think

Identify any polar bonds and consider how bond polarity leads to an electron-deficient site. Draw a possible resonance structure for cyclohex-2-en-1-one that has a C=C conjugated to a polar C=O. How does this resonance structure illustrate the presence of an electron-deficient site?

Solve

The carbonyl C is made electron deficient by the adjacent electronegative O atom. The β C atom is made electron deficient by the resonance structure that places a positive charge on that C atom, as shown below.

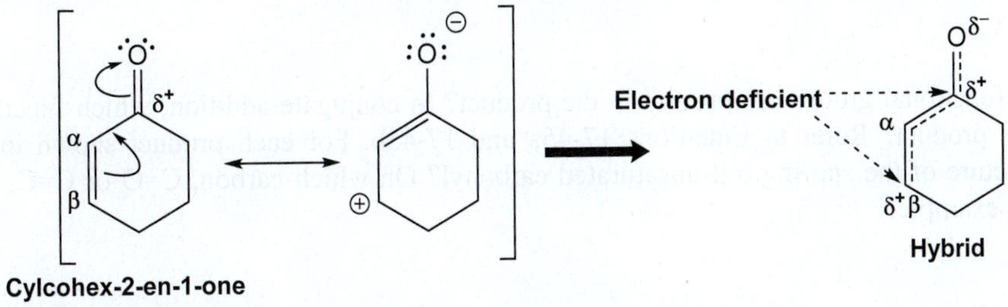

Electron deficient

Hybrid

Cylcohex-2-en-1-one

Your Turn 17.12

Think

Consult Section 7.9. In what type of medium—acidic or basic—does this tautomerization take place? What type of atom is moved in such a tautomerization? Which elementary steps are involved? Can this take place in a single step?

Solve

The enol converts to the keto form under basic conditions. A proton must move from the O to the alkene C in two separate proton transfer steps. A single step is not reasonable, because it would constitute an intramolecular proton transfer. Deprotonation of OH occurs first, not protonation of the C=C bond, to avoid the production of a strong acid under basic conditions.

Basic media

Enol form

Keto form

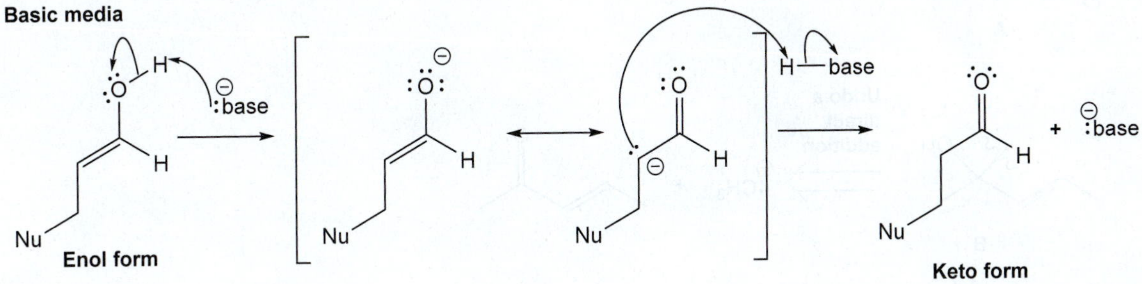

Your Turn 17.13

Think

Which product is the result of the nucleophile bonding to the C atom with a larger partial positive charge? Which product stabilizes the negative charge better? Consider possible resonance structures.

Solve

The first product is a result of conjugate addition, and the second product is a result of direct addition. Direct addition has a lower-energy transition state because the carbonyl C has a larger partial positive charge to which the nucleophile is attracted. This leads to a lower activation energy and is, therefore, the kinetic product. The first product stabilizes the negative charge better, because the negative charge is resonance delocalized, so the first product is the thermodynamic product.

NC: + → +

NC

Conjugate addition
Thermodynamic product

CN

Direct addition
Kinetic product

Your Turn 17.14

Think

In direct addition, which functional groups are present in the product? In conjugate addition, which functional groups are present in the product? Refer to Equations 17-45a and 17-45b. For each product shown in this question, what is the structure of the starting α,β-unsaturated carbonyl? On which carbon, C=O or C=C, does the addition occur in each example?

Solve

In direct addition, the C=C remains, and the product has a C–OH group characteristic of an alcohol. The addition occurs at the carbon of the carbonyl, and this is where the new C–Nu bond forms. In conjugate addition, the C=O remains, characteristic of a ketone or aldehyde. The addition occurs at the β C of the C=C, and this is where the new C–Nu bond forms. In **A**, the product has a Ph group beta to the C=O bond. The given compound can be generated by *conjugate* addition, in which Ph^- attacks at the beta carbon. In **B**, a C=C is adjacent to a C–OH group. The given compound can be generated by *direct* addition, in which CH_3^- attacks the carbon of the carbonyl.

Undo a conjugate addition

Ph

A

:Ph +

Undo a direct addition

OH

B

$:CH_3$ +

In Chapter Problems

Problem 17.2

Think

What nucleophile is generated when $NaSCH_3$ dissolves in ethanol? Which atom will it attack? Which bond is most easily broken in the process? What is the role of the weakly acidic solvent?

Solve

$NaSCH_3$ is ionic, consisting of Na^+ and H_3CS^- ions. In ethanol, therefore, $NaSCH_3$ dissolves as Na^+ and H_3CS^-, with H_3CS^- being a strong nucleophile. H_3CS^- will subsequently attack the electron-poor carbonyl C, breaking the π bond of the double bond. The resulting O^- is protonated by the weakly acidic ethanol solvent.

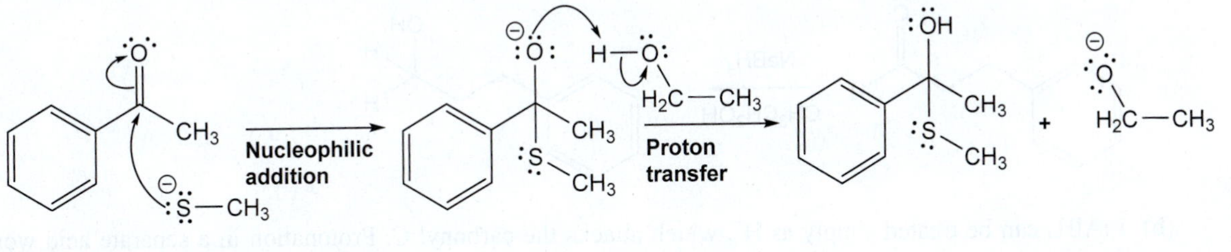

Nucleophilic addition

Proton transfer

Problem 17.3

Think

Is the CH_3 group electron donating or electron withdrawing? Is the CCl_3 group electron donating or withdrawing? Which carbonyl C has a larger concentration of positive charge? Which carbonyl C is more likely to react with the water nucleophile?

Solve

The inductive effect of three electronegative Cl atoms makes the carbonyl C atom very electron poor—more electron poor than the carbonyl C in ethanal. So a nucleophile will be attracted to the carbon in chloral more strongly.

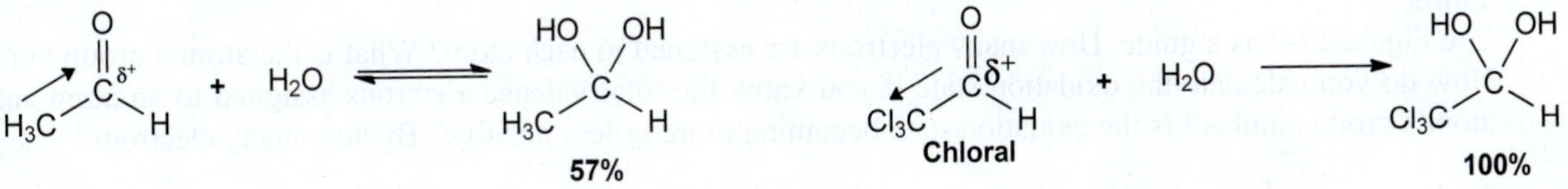

57%

Chloral

100%

Problem 17.4

Think

Compare the steric repulsion of the hydrate formed in each reaction. Which hydrate is more crowded? How does this affect the extent to which the hydrate is produced? Is there a difference in the inductive effect between the methyl and isopropyl groups?

Solve

Molecule **D** (acetone) will be hydrated to a greater extent. The carbonyl C of molecule **E** (2,4-dimethyl-3-pentanone) will be more sterically hindered, making hydration more difficult. The inductive effect, on the other hand, does not differ between the two alkyl groups to a significant extent, because each carbonyl C has two attached alkyl groups.

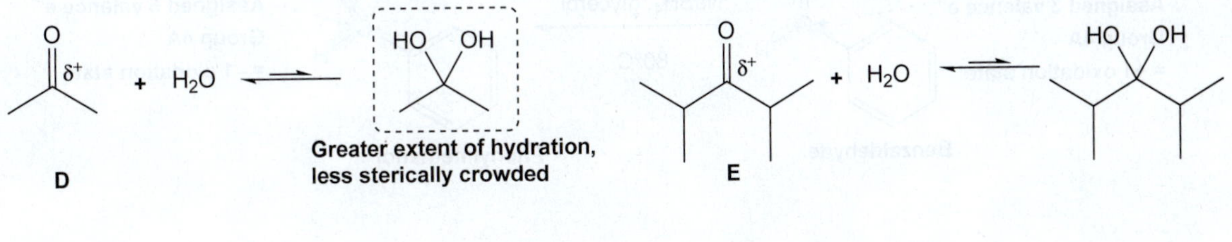

Greater extent of hydration, less sterically crowded

D

E

Problem 17.6

Think

Can $LiAlH_4$ and $NaBH_4$ be treated as a simpler nucleophile? Why can the reaction with $NaBH_4$ be carried out in a protic solvent? Why must the reaction with $LiAlH_4$ be complete before NH_4^+ is added?

Solve

In each reaction, H^- adds, followed by protonation. The result is the reduction of the aldehyde or ketone to an alcohol.

(a) $NaBH_4$ can be treated simply as H^-, which attacks the carbonyl C. Ethanol is a weakly acidic solvent and will protonate the O^- to form an uncharged final alcohol product.

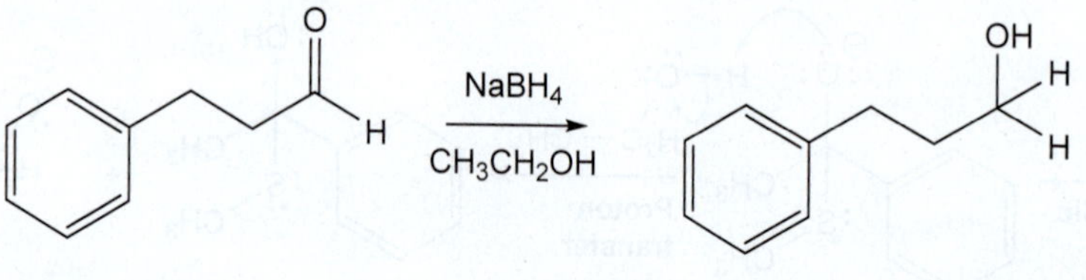

$NaBH_4$ / CH_3CH_2OH

(b) $LiAlH_4$ can be treated simply as H^-, which attacks the carbonyl C. Protonation in a separate acid workup step is required, because $LiAlH_4$ will deprotonate the weakly acidic proton from solvents like water and alcohols.

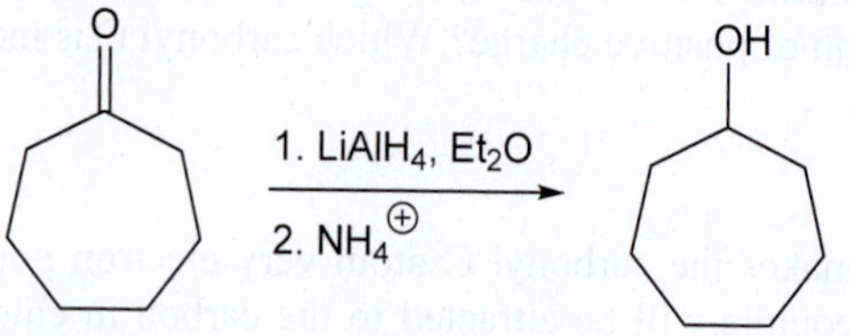

1. $LiAlH_4$, Et_2O
2. $NH_4^{\oplus}$

Problem 17.7

Think

Use Figure 17-7 as a guide. How many electrons are assigned to each atom? What is the atom's group number? How do you calculate the oxidation state if you know the total valence electrons assigned to an atom and the atom's group number? Is the oxidation state becoming more or less positive? By how many electrons?

Solve

The oxidation states of the carbonyl C and O atoms in both the reactant and product in Equation 17-5 are shown below. The carbon atom in the reactant is assigned three electrons (one from the C–C bond, two from the C–H bond, and zero from the C=O bond). The three total valence electrons assigned to C is one less than the group number of 4A, so the oxidation state of C is +1. The carbon atom in the product is assigned five electrons (one from the C–C bond, four from the two C–H bonds, and zero from the C–O bond). The five total valence electrons assigned to C is one more than the group number of 4A, so the oxidation state of C is −1. This carbon becomes less positive by 2 and is therefore reduced by two electrons. The oxygen has a −2 oxidation state in both reactant and product—assigned eight valence electrons and in group 6A.

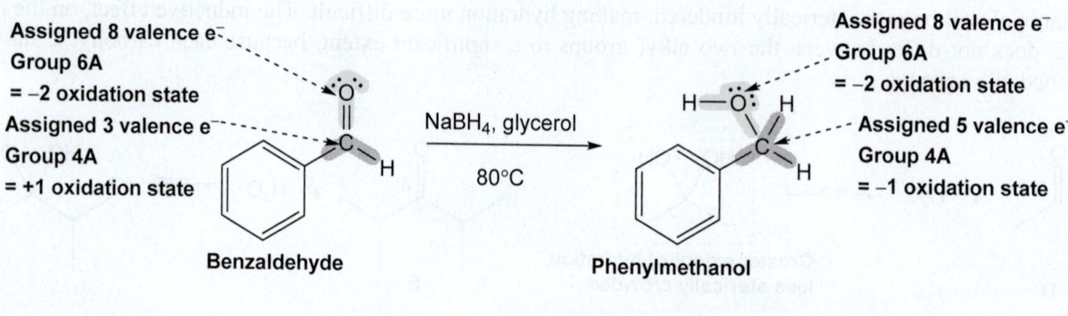

Assigned 8 valence e⁻
Group 6A
= −2 oxidation state
Assigned 3 valence e⁻
Group 4A
= +1 oxidation state

$NaBH_4$, glycerol
80°C

Assigned 8 valence e⁻
Group 6A
= −2 oxidation state
Assigned 5 valence e⁻
Group 4A
= −1 oxidation state

Benzaldehyde **Phenylmethanol**

Problem 17.8

Think

Can you simplify $NaBH_4$ and $LiAlH_4$? What are the electron-rich and electron-poor species in the starting reagents? How is the C=N similar to the C=O in terms of how it might react with a nucleophile? What elementary step takes place first? Are proton transfer reactions involved?

Solve

$NaBH_4$ and $LiAlH_4$ can be simplified to H^-, which adds to the imine carbon in a nucleophilic addition elementary step. The N^- is protonated by the CH_3OH or H_2O. The imine, as a result, is reduced to an amine.

Equation 7-12

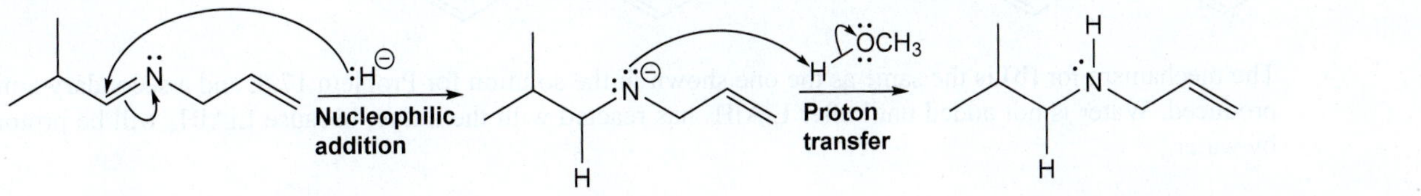

Nucleophilic addition

Proton transfer

Equation 7-13

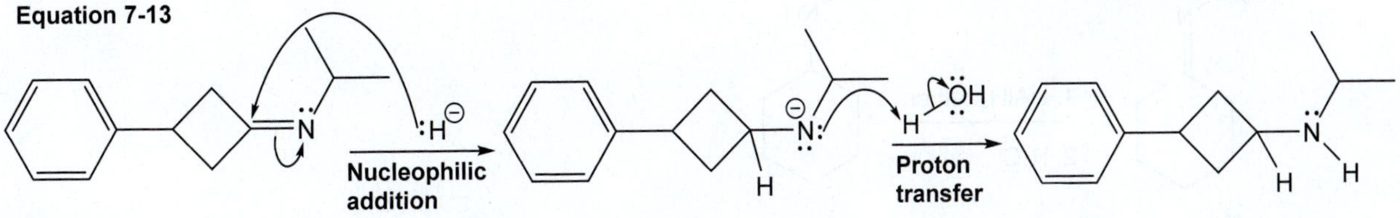

Nucleophilic addition

Proton transfer

Problem 17.9

Think

How can you simplify $LiAlH_4$? What are the electron-rich and electron-poor species in the starting reagents? How is the C≡N similar to the C=O in terms of how it might react with a nucleophile? What elementary step takes place first? Are proton transfer reactions involved? How many times does the nucleophile add?

Solve

$LiAlH_4$ is a source of a very strong H^- nucleophile and can add twice to the triple bond of the nitrile. Subsequently, the N^{2-} is protonated twice, producing a primary amine. The complete mechanism is shown below.

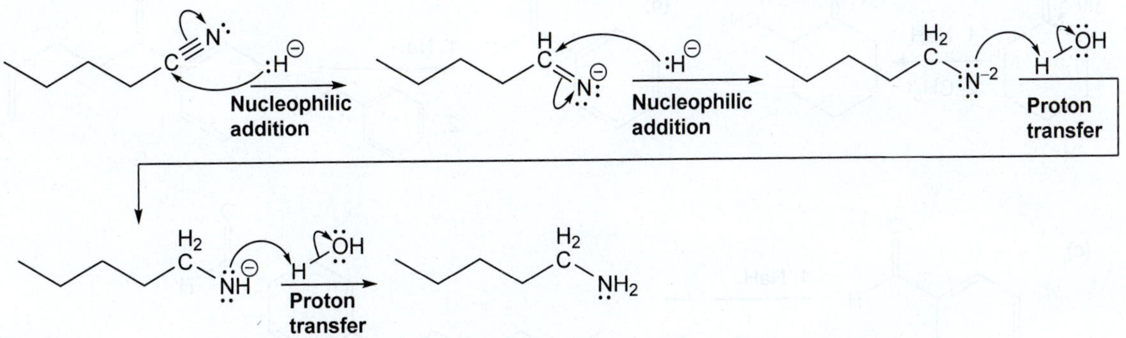

Nucleophilic addition

Nucleophilic addition

Proton transfer

Proton transfer

Problem 17.10

Think

Can $LiAlH_4$ be treated as a simpler nucleophile? How does a nucleophile tend to react with C=N and C≡N groups? How many times does the nucleophilic addition reaction occur? Why must the reaction with $LiAlH_4$ be complete before H_2O is added? What functional group is produced?

Solve

The mechanism for **(a)** is the same as that shown in the solution of Problem 17.9, and a primary amine is produced. The reduction reaction occurs twice, and the proton transfer reaction occurs twice. Water is not added until after $LiAlH_4$ has reacted with the nitrile, because $LiAlH_4$ will be protonated by water.

N C 1. $LiAlH_4$, ether 2. H_2O NH_2 H_2C

The mechanism for **(b)** is the same as the one shown in the solution for Problem 17.8, and a secondary amine is produced. Water is not added until after $LiAlH_4$ has reacted with the imine, because $LiAlH_4$ will be protonated by water.

H_3C N 1. $LiAlH_4$, ether 2. H_2O H_3C NH

Problem 17.12

Think

Will the hydride anion from NaH act as a base or a nucleophile? Which atom will it attack? How will the resulting species behave in the presence of an alkyl halide?

Solve

NaH is a strong base but a poor nucleophile, so it will deprotonate at the α carbon of a ketone or aldehyde. As we learned in Section 10.3, the resulting enolate anion is a strong nucleophile and will displace X from RX in an S_N2 reaction, yielding an alpha-alkylated ketone or aldehyde.

(a) O 1. NaH 2. CH_3I O CH_3

(b) O 1. NaH 2. Cl O

(c) O H O 1. NaH 2. Br O H O

Problem 17.14

Think

To what R^- nucleophile can the Grignard or alkyllithium reagent be simplified? Which atom will it attack? What is the role of NH_4^+, H_3O^+, or CH_3OH?

Solve

(a) The C_6H_5MgBr reagent can be simplified to $C_6H_5^-$, which will undergo nucleophilic addition at the carbonyl C atom. Once this is complete, NH_4^+ is added in an acid workup to protonate the strongly basic O^- generated in the first step.

1. C_6H_5MgBr, Et_2O
2. NH_4Cl, H_2O

(b) The CH_3Li reagent can be simplified to CH_3^-, which will undergo nucleophilic addition at the carbonyl C. Once this is complete, H_3O^+ is added in an acid workup to protonate the strongly basic O^- generated in the first step.

1. CH_3Li, Et_2O
2. H_2O, H_2SO_4

+ Enantiomer

(c) The CH_3Li reagent can be simplified to CH_3^-, which will undergo nucleophilic addition at the nitrile C. Once this is complete, CH_3OH is added in a workup step to protonate the strongly basic N^- generated in the first step. Methanol is used as the proton source, instead of H_3O^+, to avoid the formation of the ketone (see Equation 17-20).

1. CH_3Li, Et_2O
2. CH_3OH

Problem 17.15

Think

What are the electron-rich and electron-poor species in the starting reagents? How can the RMgBr be simplified as a nucleophile? How is the C=O of O=C=O similar to the C=O of ketones and aldehydes in terms of how it might react with a nucleophile? What elementary step takes place first? What is the role of the NH_4Cl? Are proton transfer reactions involved?

Solve

The RMgBr can be treated as R^-, in this case, $H_2C{=}CHCH_2^-$. The C of the CO_2 is very electron poor and highly susceptible to nucleophilic attack. In the presence of a strong nucleophile, the CO_2 undergoes nucleophilic addition followed by protonation by NH_4Cl to form a carboxylic acid as the final product.

Nucleophilic addition

Proton transfer

Problem 17.16

Think

How can RMgBr be simplified as a nucleophile? What is the nucleophile and electrophile in the first step? Which new bonds are formed? Are proton transfer reactions involved?

Solve

The RMgBr Grignard reagent can be simplified to R^-. The C of CO_2 is the electrophile that is attacked by the nucleophilic C of R^-. A new C–C bond is formed. A proton transfer reaction follows to protonate O^- to form an uncharged carboxylic acid product.

(a)

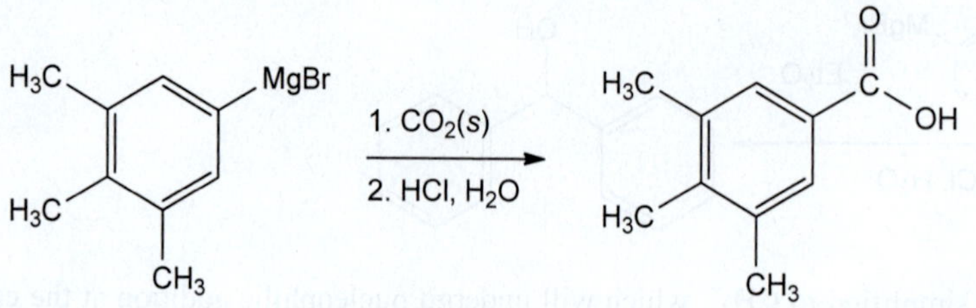

1. $CO_2(s)$
2. HCl, H_2O

(b)

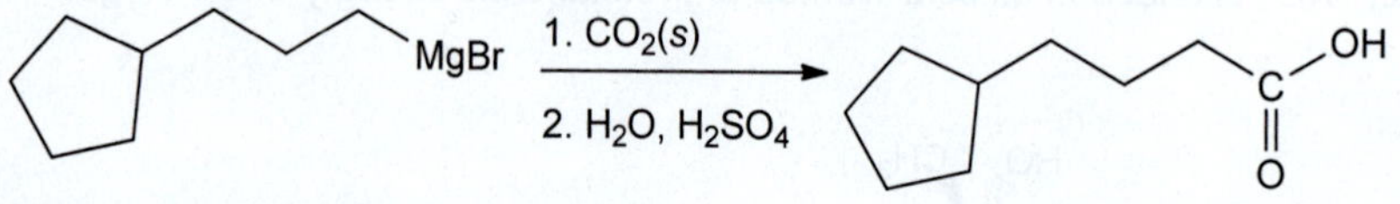

1. $CO_2(s)$
2. H_2O, H_2SO_4

Problem 17.17

Think

What groups are susceptible to deprotonation or nucleophilic attack by the electron-rich C atom of the RMgBr? Does the unfeasible Grignard reagent have such a group? If the reagent reacts with itself, what product forms? Is the Grignard reagent destroyed?

Solve

(a) The OH group is susceptible to deprotonation.

Proton transfer

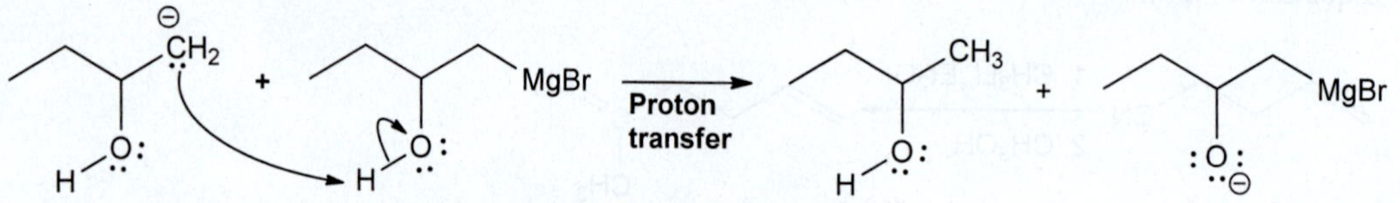

(b) The C=O group is susceptible to nucleophilic attack (nucleophilic addition).

Nucleophilic addition

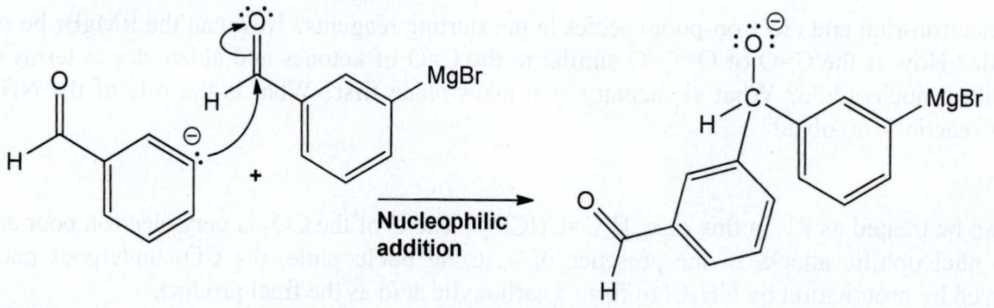

(c) The less crowded C of the epoxide is susceptible to nucleophilic attack (S_N2).

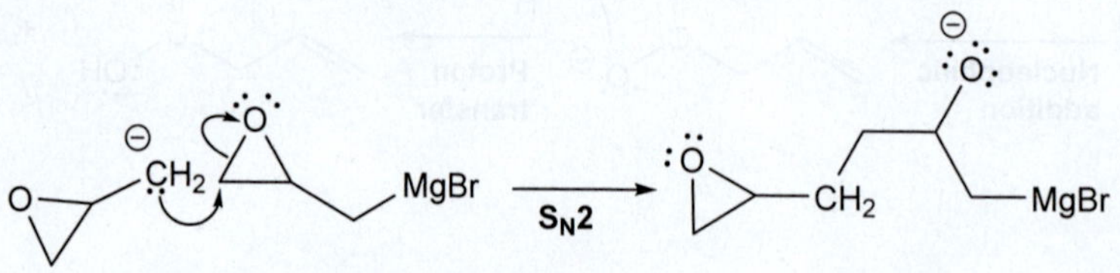

S_N2

Problem 17.18

Think

What are the electron-rich and electron-poor species in the first step? When the nucleophile attacks, what bond is broken? In how many steps does this mechanism occur? What functional group is produced?

Solve

This is an example of a Wittig reaction. The carbonyl carbon of the aldehyde is electron poor, and the Wittig reagent has an electron-rich C atom. This forms a new C–C bond in a nucleophilic addition mechanism, and the π bond breaks to form O⁻. The negative O atom coordinates to the positive P atom. An elimination elementary step follows to form the C=C and the oxaphosphetane. Both the cis and the trans isomers are possible.

Nucleophilic addition

Coordination

Elimination

Problem 17.19

Think

What are the electron-rich and electron-poor species in the first reaction? Is triphenylphosphine a good nucleophile? What is the role of the strongly basic Bu–Li? What elementary step takes place between a Wittig reagent and a C=O group? In how many steps does this mechanism occur? What functional group is produced?

Solve

This is an example of a Wittig reagent synthesis followed by a Wittig reaction. The alkyl halide is attacked by PPh_3 in an S_N2 mechanism followed by deprotonation to form the Wittig reagent. The aldehyde C=O carbon atom is electron poor, and the Wittig reagent has an electron-rich C atom. This forms a new C–C bond in a nucleophilic addition mechanism. The negative O atom coordinates to the positive P atom. An elimination elementary step follows to form the C=C and the triphenylphosphine oxide. Both the *E* and the *Z* isomers are possible.

S_N2

Proton transfer

Nucleophilic addition

Coordination

Elimination

(E)

(Z)

Problem 17.20

Think

Which C atom is charged and strongly nucleophilic? Which C atom was deprotonated? To which C atom was the halide attached?

Solve

The negatively charged C atom of the Wittig reagent is initially bonded to a halogen atom of an alkyl halide. Benzyl bromide is shown, but benzyl chloride or benzyl iodide would also be a reasonable choice.

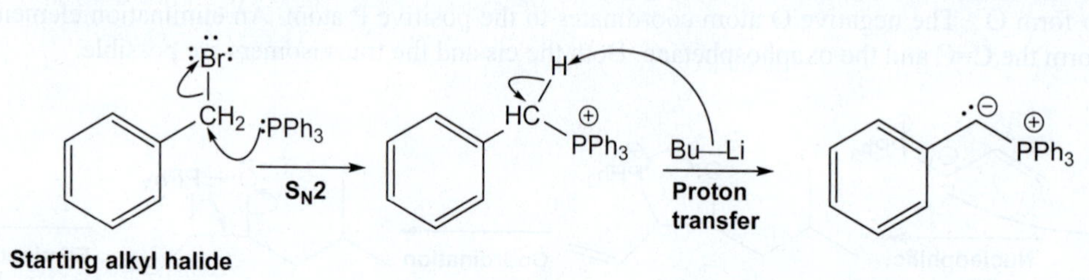

Br: CH2 :PPh3 — SN2 → HC H PPh3 — Bu—Li, Proton transfer → PPh3

Starting alkyl halide

Problem 17.22

Think

What Wittig reagent would you need to carry out this reaction? What alkyl halide could serve as a precursor to that Wittig reagent? How can that alkyl halide be synthesized from benzaldehyde?

Solve

We can begin a retrosynthetic analysis by applying a transform that undoes a Wittig reaction, so we disconnect the C=C bond to give us a carbonyl compound and a Wittig reagent. The Wittig reagent can ultimately be produced from benzaldehyde.

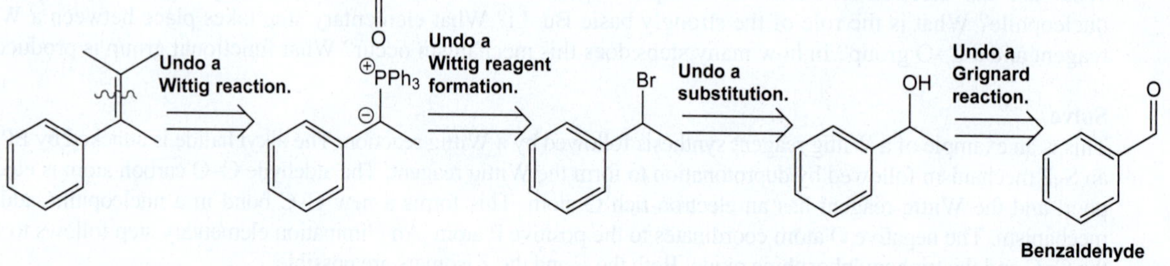

Undo a Wittig reaction. ⇒ O, PPh3 — Undo a Wittig reagent formation. ⇒ Br — Undo a substitution. ⇒ OH — Undo a Grignard reaction. ⇒ O

Benzaldehyde

The synthesis in the forward direction appears as follows:

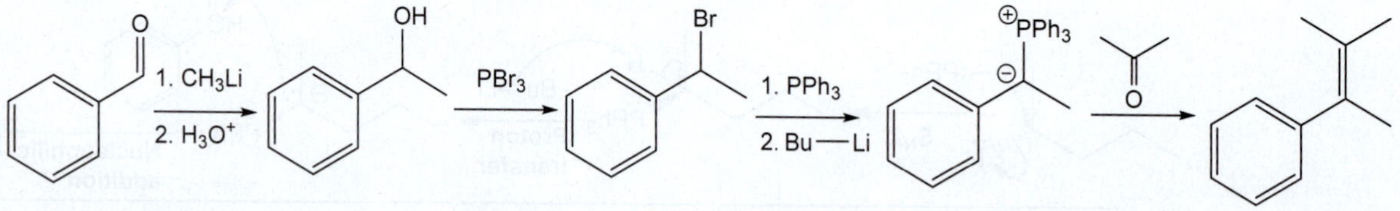

O — 1. CH_3Li, 2. H_3O^+ → OH — PBr_3 → Br — 1. PPh_3, 2. Bu—Li → PPh3 — O →

Problem 17.23

Think

What are the two electron-poor sites on the α,β-unsaturated ketone? What C atom does the nucleophile attack in a 1,2-addition mechanism? In a 1,4-addition mechanism? Are proton transfer steps involved?

Solve

Addition of the nucleophile at the carbonyl C atom is called 1,2-addition or direct addition, and addition of the nucleophile at the β C atom is called 1,4-addition or conjugate addition. This addition occurs via the nucleophilic addition elementary step. The negative O atom is protonated to form an uncharged alcohol product. In 1,4-addition, the enol tautomerizes (back-to-back proton transfer steps) to the aldehyde or ketone.

(a) The HO^- acts as the nucleophile at either the 2 or 4 site in 1,2-addition or 1,4-addition.

Direct addition (1,2-addition)

:ÖH Nucleophilic addition → Proton transfer → HÖ: :ÖH

Conjugate addition (1,4-addition)

Nucleophilic addition → Proton transfer → Tautomerization

(b) The NC^- acts as the nucleophile at either the 2 or 4 site in 1,2-addition or 1,4-addition.

Direct addition (1,2-addition)

:CN Nucleophilic addition → Proton transfer ($H-\ddot{O}SO_3H$) → HÖ: CN

Conjugate addition (1,4-addition)

:CN Nucleophilic addition → NC … Proton transfer ($H-\ddot{O}SO_3H$) → NC … :ÖH → Tautomerization → NC …

Problem 17.24

Think

Is the nucleophile one that leads to reversible or irreversible nucleophilic addition? Does reversible addition favor 1,2-addition or 1,4-addition? Does irreversible addition favor 1,2-addition or 1,4-addition?

Solve

Only the highly reactive R^- and H^- nucleophiles add irreversibly to a carbonyl C atom: **(b)**, **(d)**, and **(e)**. Nucleophiles that add irreversibly to the carbonyl C yield the direct 1,2-addition product. The remaining nucleophiles, **(a)** and **(c)**, are significantly less reactive and add reversibly. Nucleophiles that add reversibly to the carbonyl C yield the conjugate 1,4-addition product. The mechanisms for reactions **(a)–(e)** are given on the next page.

(a) Conjugate addition (1,4-addition)

Nucleophilic addition → Proton transfer → Tautomerization

$H_3C\ddot{S}$; $:\ddot{S}CH_3$; $:\ddot{O}H$

(b) Direct addition (1,2-addition), Wittig reaction

Ylide (Witting reagent)

Nucleophilic addition → Coordination → Elimination

H_3C, C_6H_5, PPh_3, CH_3, Ph_3 P, + $:\ddot{O}-PPh_3$

(c) Conjugate addition (1,4-addition)

Nucleophilic addition → Proton transfer → Tautomerization

CH_2, H_2

(d) Direct addition (1,2-addition)

Nucleophilic addition → Proton transfer

Ph, $H\ddot{O}$

(e) Direct addition (1,2-addition)

Nucleophilic addition → Proton transfer

$:H$, $H\ddot{O}$

Problem 17.26

Think

Which group has the polar π bond? Will the nucleophile add reversibly or irreversibly? How does that affect whether direct addition or conjugate addition takes place?

Solve

The C≡N group has the polar π bond. The nucleophile is NC^-, which, as we know from Table 17-2, adds reversibly. As a result, the thermodynamic product, which is formed via conjugate addition, is favored. The mechanism is essentially the same as that in Equation 17-32.

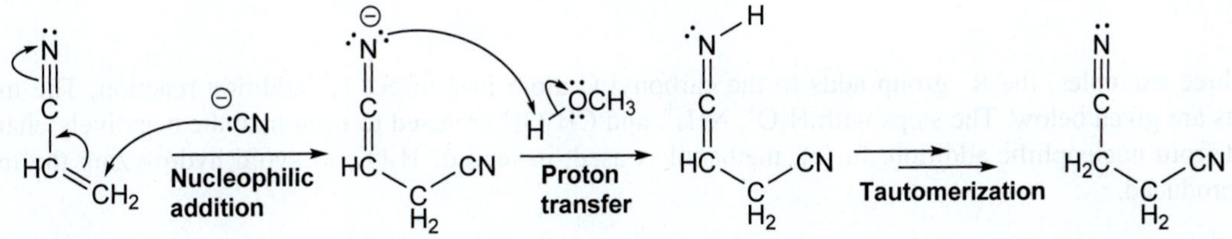

Nucleophilic addition

Proton transfer

Tautomerization

Problem 17.28

Think

To what R^- nucleophile can the lithium dialkylcuprate be simplified? Will it add predominantly via direct addition or conjugate addition? What role does NH_4^+ play after the addition of R^- is complete?

Solve

In **(a)**, **(b)**, and **(c)**, the lithium dialkylcuprate is a source of $^-CH_3$, $^-CH_2CH_3$, or $^-CH(CH_3)_2$, respectively, which will add to the α,β-unsaturated carbonyl group via conjugate addition. The resulting enolate anion is protonated by NH_4^+.

(a)

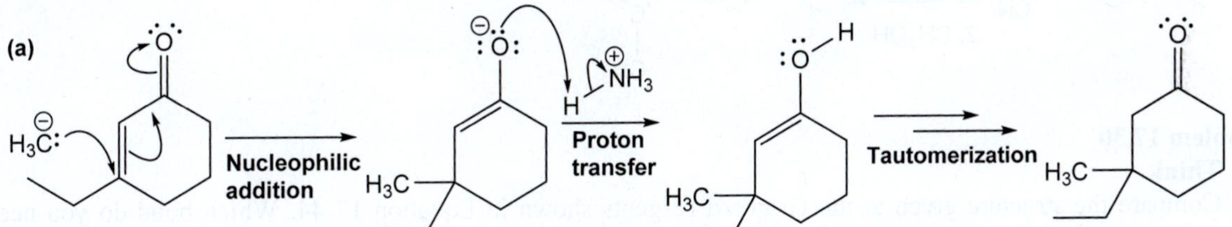

Nucleophilic addition

Proton transfer

Tautomerization

(b)

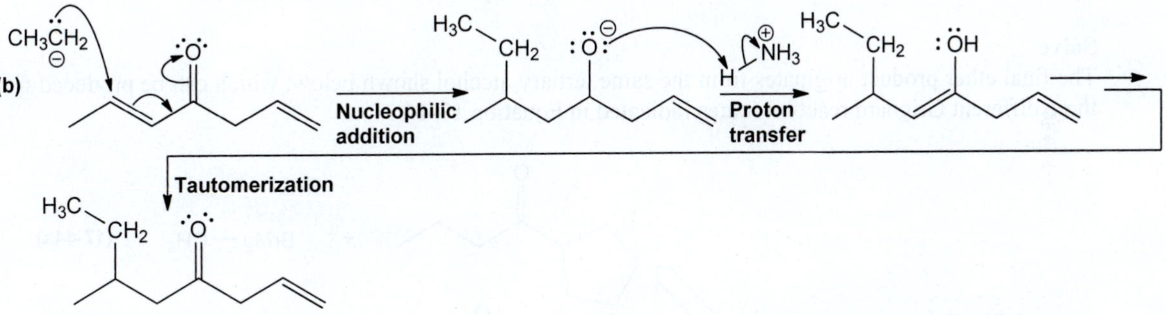

Nucleophilic addition

Proton transfer

Tautomerization

(c)

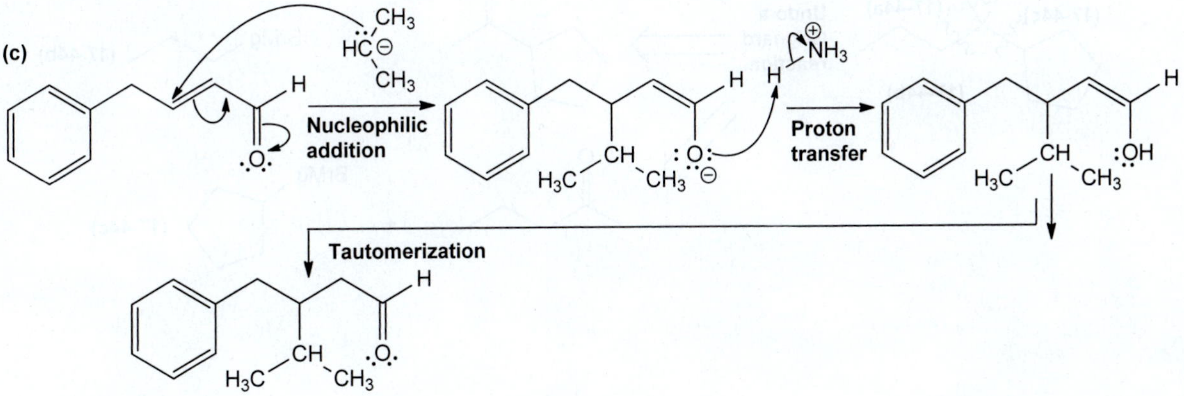

Nucleophilic addition

Proton transfer

Tautomerization

Problem 17.29

Think

To what R⁻ nucleophile can the Grignard and alkyllithium reagent in each example be simplified? Will it add predominantly via direct addition or conjugate addition? What role does H_3O^+, NH_4^+, or CH_3OH play after the addition of R⁻ is complete?

Solve

In all three examples, the R⁻ group adds to the carbonyl C atom in a direct 1,2-addition reaction. The major products are given below. The steps with H_3O^+, NH_4^+, and CH_3OH are used to protonate the negatively charged product from nucleophilic addition. In **(c)**, methanol is used, instead of H_3O^+, to avoid hydrolyzing the imine that is produced.

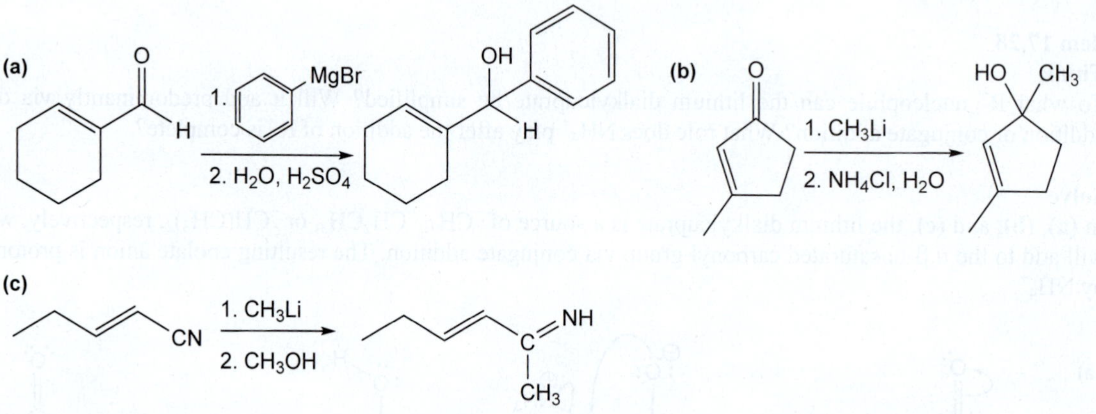
(a)
O
H
1. MgBr
2. H2O, H2SO4
OH
H
(b)
O
1. CH3Li
2. NH4Cl, H2O
HO CH3
(c)
CN
1. CH3Li
2. CH3OH
NH
C
CH3

Problem 17.30

Think

Compare the structure given to the Grignard reagents shown in Equation 17-44. Which bond do you need to "undo" for each reagent? Is the Grignard reagent a source of electrophilic or nucleophilic carbon? Is the product listed an alcohol? If not, how can this functional group be synthesized from an alcohol?

Solve

The final ether product originates from the same tertiary alcohol shown below, which can be produced from the three different Grignard reaction routes indicated in Equation 17-44.

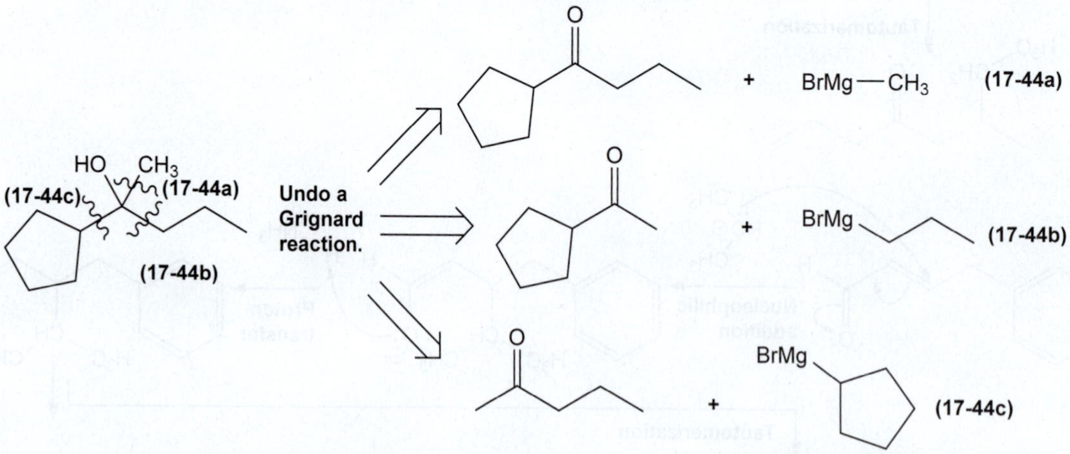
HO CH3
(17-44c)
(17-44a)
(17-44b)
Undo a Grignard reaction.
O
+
BrMg—CH3
(17-44a)
O
+
BrMg
(17-44b)
O
BrMg
+
(17-44c)

An ether is formed from an alcohol first by deprotonating the alcohol with NaH to form a strong anionic nucleophile, followed by an S_N2 reaction. This is a Williamson ether synthesis.

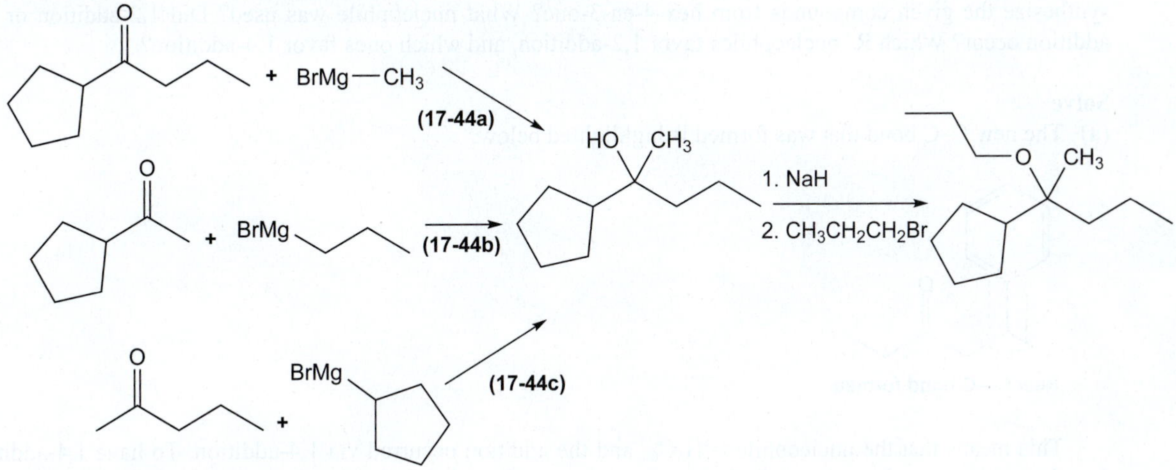

+ BrMg—CH_3 (17-44a) HO CH_3

+ BrMg (17-44b) 1. NaH 2. $CH_3CH_2CH_2Br$ O CH_3

BrMg + (17-44c)

Problem 17.31

Think

Which C atom in the product must have been the carbonyl C? What R groups are attached to that C atom? Can that bond be disconnected in a retrosynthetic analysis to arrive at a carbonyl precursor and an alkyllithium precursor? What, then, is the identity of the LiR reagent?

Solve

The C–OH in the product would have been C=O in the starting material. Disconnecting the C–C bond on either side of the C–OH gives the following:

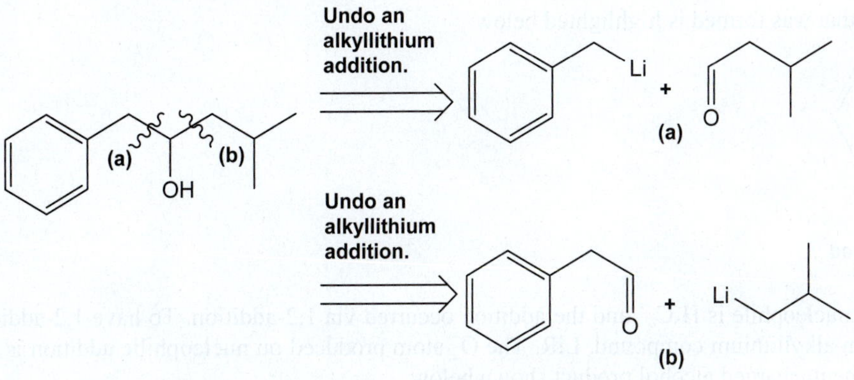

(a) (b) OH

Undo an alkyllithium addition. Li + O (a)

Undo an alkyllithium addition. O + Li (b)

In the forward direction, the synthesis would appear as follows:

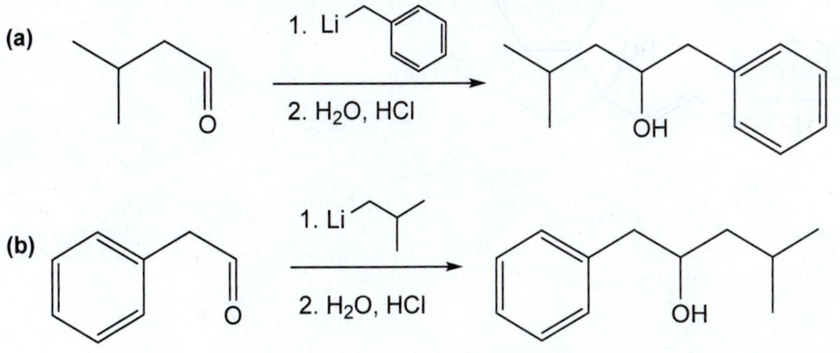

(a) O 1. Li 2. H_2O, HCl OH

(b) O 1. Li 2. H_2O, HCl OH

Problem 17.33

Think

What is the structure of hex-4-en-3-one? How many C atoms does it possess? What bonds were formed to synthesize the given compounds from hex-4-en-3-one? What nucleophile was used? Did 1,2-addition or 1,4-addition occur? Which R^- nucleophiles favor 1,2-addition, and which ones favor 1,4-addition?

Solve

(a) The new C–C bond that was formed is highlighted below.

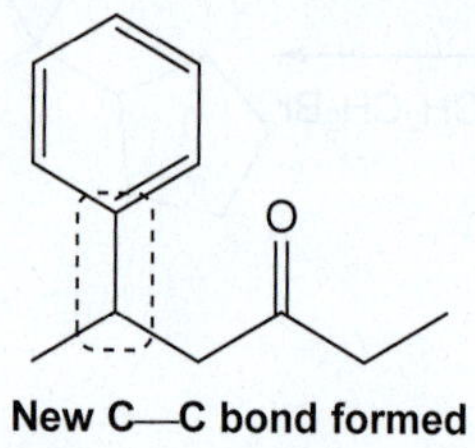

New C—C bond formed

This means that the nucleophile is $H_5C_6^-$ and the addition occurred via 1,4-addition. To have 1,4-addition, the reagent needs to be a dialkylcuprate compound, R_2CuLi. The enolate produced from nucleophilic addition is then protonated to form an enol that tautomerizes back to the ketone, giving the product listed.

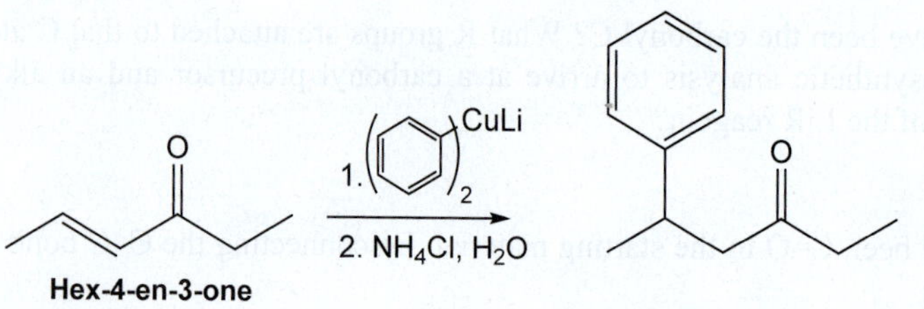

1. $(C_6H_5)_2CuLi$
2. NH_4Cl, H_2O

Hex-4-en-3-one

(b) The new C–C bond that was formed is highlighted below.

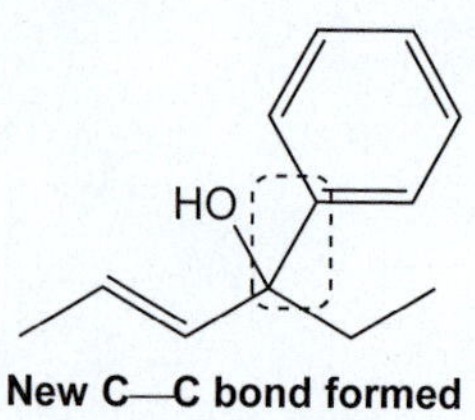

HO

New C—C bond formed

This means that the nucleophile is $H_5C_6^-$ and the addition occurred via 1,2-addition. To have 1,2-addition, the reagent can be an alkyllithium compound, LiR. The O^- atom produced on nucleophilic addition is then protonated to form the uncharged alcohol product shown below.

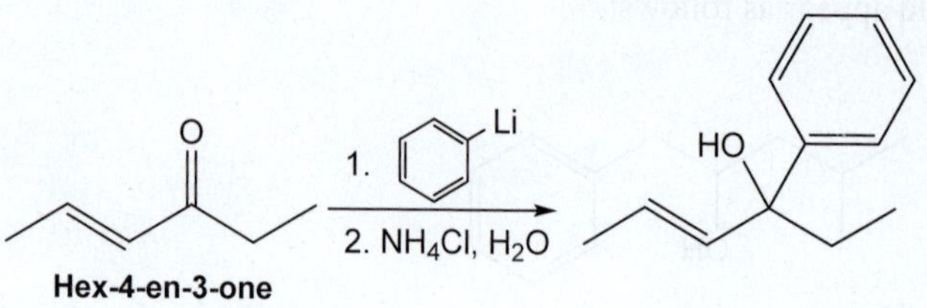

1. C_6H_5Li
2. NH_4Cl, H_2O

HO

Hex-4-en-3-one

End of Chapter Problems

Sections 17.1 and 17.2 Addition of Strong Nucleophiles and Substituent Effects

Problem 17.34

Are the groups attached electron donating or electron withdrawing? Which C atom is more electron poor? Which tetrahedral product would be more sterically crowded?

Solve

(a) The second one, because with fewer alkyl groups attached to the carbonyl C, there is less bulkiness and greater concentration of positive charge.

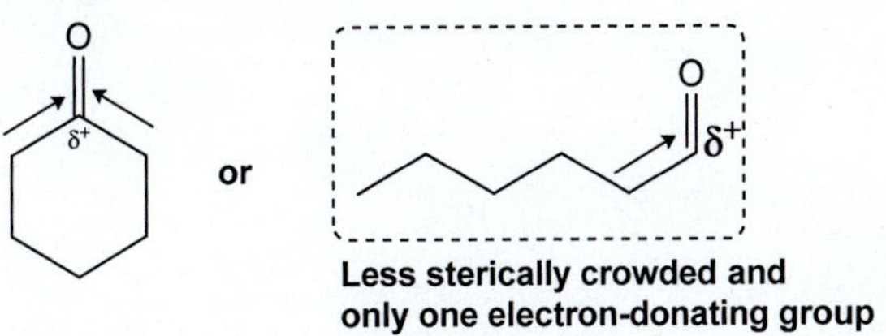

(b) The second one, because the bulkiness at the carbonyl C is about the same for both. However, the CF_3 is electron withdrawing, whereas a CH_3 group is electron donating, so the CF_3 group induces a greater positive charge on the carbonyl C.

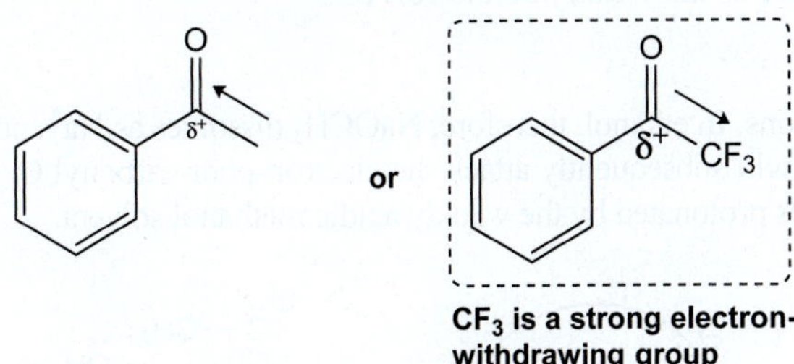

(c) The first one, because with the Cl atom closer to the carbonyl C, the carbonyl C bears a greater concentration of positive charge.

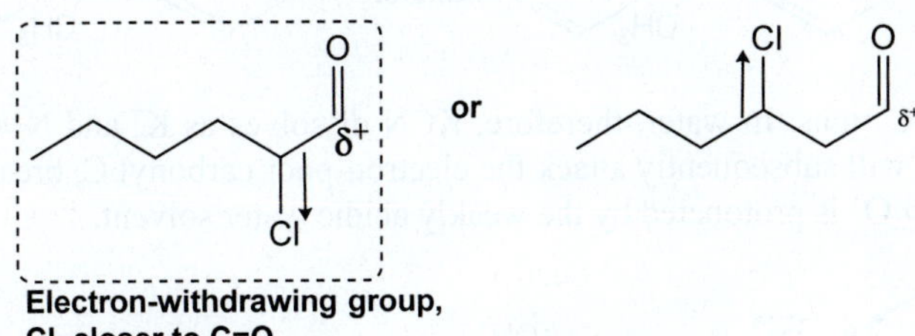

(d) The first one, because the nitrile carbon bears a higher concentration of positive charge owing to the presence of the electron-withdrawing CCl_3 group. The CH_3 group is electron donating.

$\overset{\delta^+}{Cl_3C-C\equiv N}$ or $\overset{\delta^+}{H_3C-C\equiv N}$

CCl_3 is an electron-withdrawing group

Problem 17.35

Think

What is the structure of cyclopropanone? What is the structure of the hydrate of cyclopropanone? Compare the ring strain of the cyclopropanone to the hydrate of cyclopropanone. What are the hybridization and angle of the C atom in the C=O and $C(OH)_2$?

Solve

The hydrate is shown below, alongside cyclopropanone itself. The reason the hydrate is stable is that in the hydrate, there is much less ring strain than in the reactant. The interior angle of the three-membered ring is 60°. The ideal angle at the carbonyl C is 120°, whereas that for the hydrate carbon is 109.5°. The angles, therefore, are better matched in the hydrate.

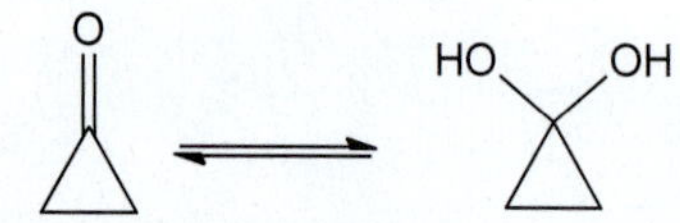

O
HO OH
More ring strain, less stable

Problem 17.36

Think

What nucleophile is generated when each salt dissolves in the solvent? Which atom will it attack? Which bond is most easily broken in the process? What is the role of the weakly acidic solvent?

Solve

(a) $NaOCH_3$ is ionic, consisting of Na^+ and H_3CO^- ions. In ethanol, therefore, $NaOCH_3$ dissolves as Na^+ and H_3CO^-, with H_3CO^- being a strong nucleophile. H_3CO^- will subsequently attack the electron-poor carbonyl C, breaking the π bond of the double bond. The resulting O^- is protonated by the weakly acidic methanol solvent.

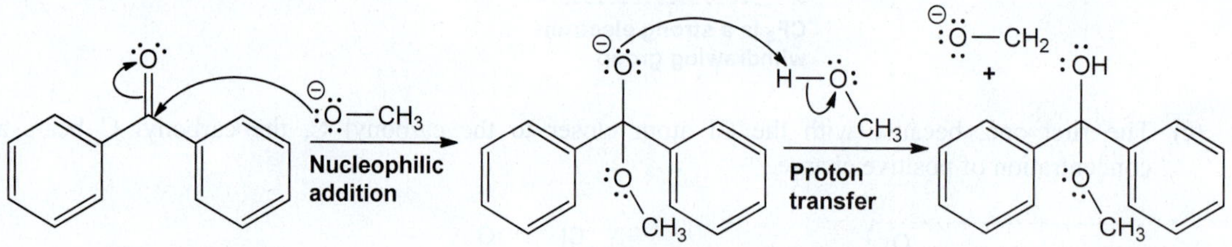

:Ö—CH₃
Nucleophilic addition
H—Ö: CH₃
Proton transfer
:Ö—CH₂ + :ÖH
CH₃

(b) KCN is ionic, consisting of K^+ and $N{\equiv}C^-$ ions. In water, therefore, KCN dissolves as K^+ and $N{\equiv}C^-$, with $N{\equiv}C^-$ being a strong nucleophile. $N{\equiv}C^-$ will subsequently attack the electron-poor carbonyl C, breaking the π bond of the double bond. The resulting O^- is protonated by the weakly acidic water solvent.

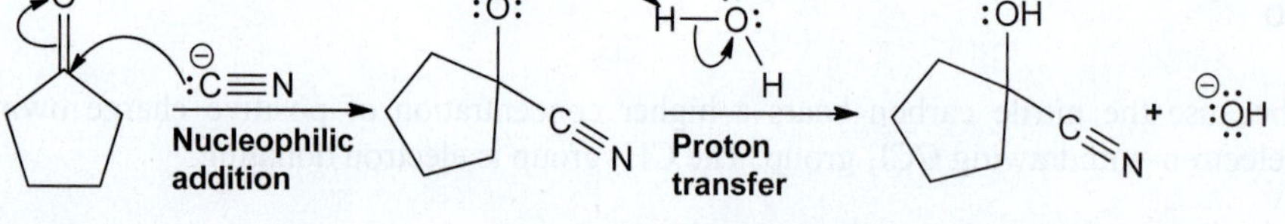

:C≡N
Nucleophilic addition
H—Ö: H
Proton transfer
:ÖH + :ÖH
C≡N

(c) $NaSC_6H_5$ is ionic, consisting of Na^+ and $H_5C_6S^-$ ions. In ethanol, therefore, $NaSC_6H_5$ dissolves as Na^+ and $H_5C_6S^-$, with $H_5C_6S^-$ being a strong nucleophile. $H_5C_6S^-$ will subsequently attack the electron-poor carbonyl C, breaking the π bond of the double bond. The resulting O^- is protonated by the weakly acidic ethanol solvent.

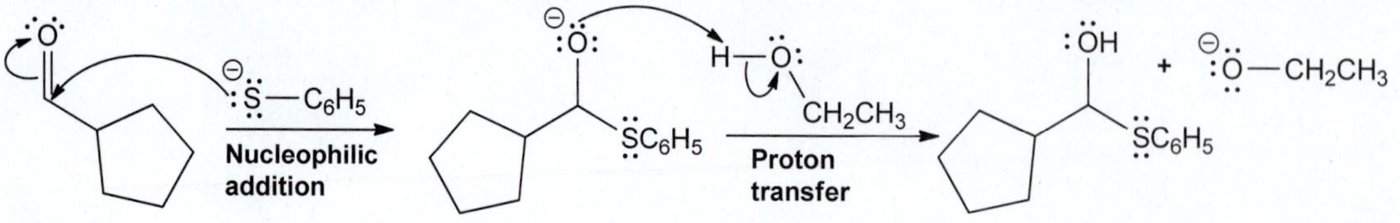

:S̈—C₆H₅
Nucleophilic addition
SC₆H₅
H—Ö: CH₂CH₃
Proton transfer
:ÖH + :Ö—CH₂CH₃
SC₆H₅

Problem 17.37

Think

Is the C atom in the nitrile electron rich or electron poor? What is the identity of the nucleophile in the nucleophilic addition reaction? How many times does the addition reaction occur? Are proton transfer reactions involved?

Solve

The cyano group of the nitrile undergoes nucleophilic addition by NC^-, which is a strong nucleophile, followed by protonation by HCN, a weak acid. This occurs two separate times.

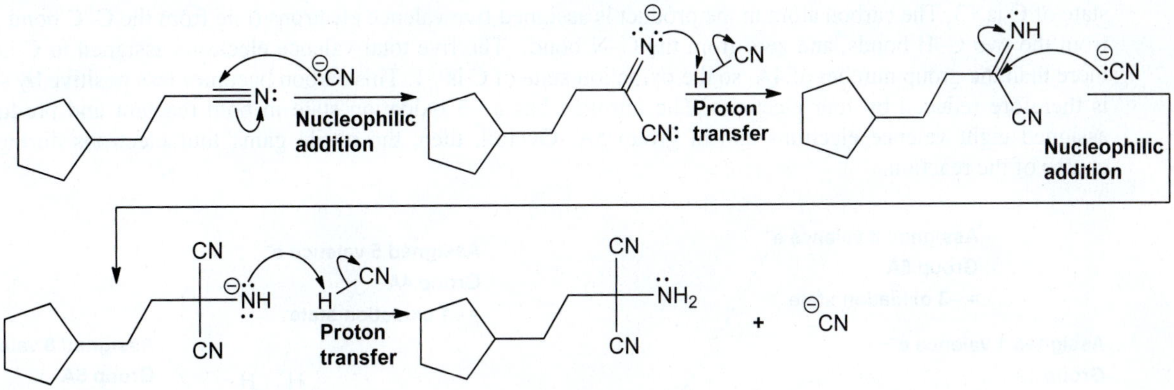

Nucleophilic addition
Proton transfer
Nucleophilic addition
Proton transfer

Sections 17.3 and 17.4 Reactions of Hydride Agents; Oxidation States

Problem 17.38

Think

Use Figure 17-7 as a guide. How many valence electrons are assigned to each atom? What is each atom's group number? How do you calculate the oxidation state using the total electrons assigned and the atom's group number? Is the oxidation state becoming more or less positive? By how many electrons?

Solve

The oxidation states of the C and N atoms of the C=N group in both the reactant and product are shown below. The carbon atom in the reactant is assigned three valence electrons (one from the C–C bond, two from the C–H bond, and zero from the C=N bond). The three total valence electrons assigned to C is one less than the group number of 4A, so the oxidation state of C is +1. The carbon atom in the product is assigned five valence electrons (one from the C–C bond, four from the two C–H bonds, and zero from the C–N bond). The five total valence electrons assigned to C is one more than the group number of 4A, so the oxidation state of C is −1. This carbon becomes less positive by 2 and is therefore reduced by two electrons. The nitrogen has a −3 oxidation state in both reactant and product—assigned eight valence electrons and in group 5A. Overall, then, the imine gains two electrons during the course of the reaction.

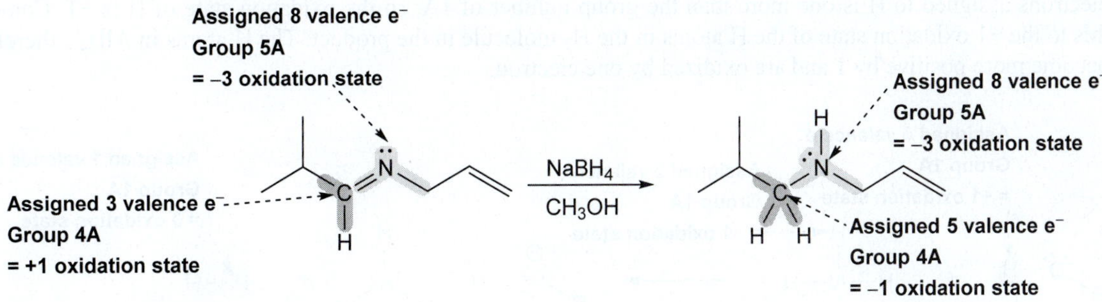

Assigned 8 valence e^-
Group 5A
= −3 oxidation state

Assigned 3 valence e^-
Group 4A
= +1 oxidation state

$NaBH_4$
CH_3OH

Assigned 8 valence e^-
Group 5A
= −3 oxidation state

Assigned 5 valence e^-
Group 4A
= −1 oxidation state

Problem 17.39

Think

Use Figure 17-7 as a guide. How many valence electrons are assigned to each atom? What is each atom's group number? How do you calculate the oxidation state using the total electrons assigned and the group number? Is the oxidation state becoming more or less positive? By how many electrons?

Solve

The oxidation states of the C and N atoms of the C≡N group in both the reactant and product are shown below. The carbon atom in the reactant is assigned one valence electron (one from the C–C bond and zero from the ≡N bond). The one total valence electron assigned to C is one less than the group number of 4A, so the oxidation state of C is +3. The carbon atom in the product is assigned five valence electrons (one from the C–C bond, four from the two C–H bonds, and zero from the C–N bond). The five total valence electrons assigned to C is one more than the group number of 4A, so the oxidation state of C is −1. This carbon becomes less positive by 4 and is therefore reduced by four electrons. The nitrogen has a −3 oxidation state in both reactant and product—assigned eight valence electrons and in group 5A. Overall, then, the nitrile gains four electrons during the course of the reaction.

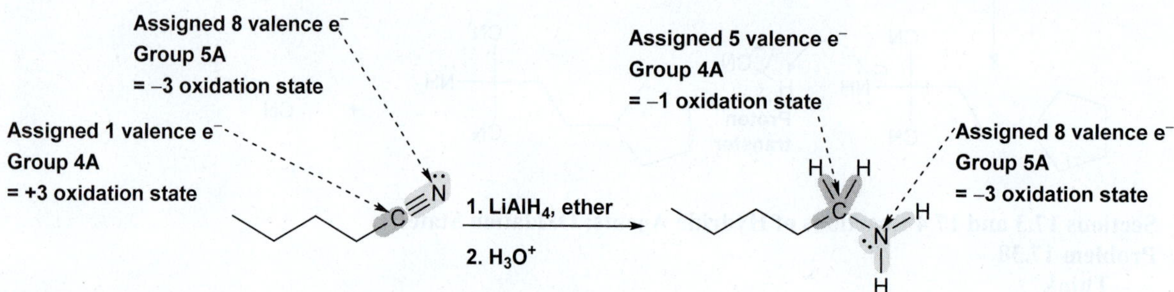

Problem 17.40

Think

How many valence electrons are assigned to each atom? How do you calculate an atom's oxidation state from the number of valence electrons it is assigned and the atom's group number? Do any atoms gain electrons? Do any atoms lose electrons?

Solve

The oxidation states of the atoms that participate in the redox reaction in both the reactant and product are shown below. The H atom in the ROH reactant is assigned zero valence electrons (the O–H). The zero total valence electrons assigned to H is one less than the group number of 1A, so the oxidation state of H is +1. The H atoms in the H_2 molecule in the product are assigned one valence electron each (one from the H–H). The one total valence electron assigned to H atoms in the H_2 molecule is the same as the group number of 1A, so the oxidation state of H is 0. This H is less positive by 1, so the H atom of ROH is reduced by one electron. The H atoms in the AlH_4 reactant are assigned two valence electrons (two from the Al–H bond). The two total valence electrons assigned to H is one more than the group number of 1A, so the oxidation state of H is −1. Compare this to the +1 oxidation state of the H atoms in the H_2 molecule in the product. The H atoms in AlH_4^-, therefore, become more positive by 1 and are oxidized by one electron.

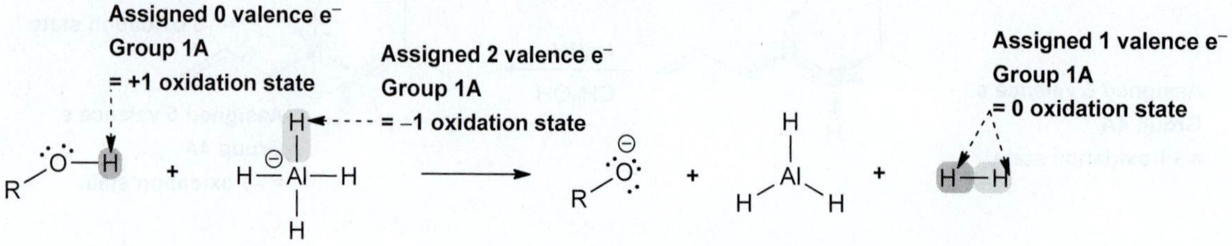

Assigned 0 valence e⁻
Group 1A
= +1 oxidation state

Assigned 2 valence e⁻
Group 1A
= −1 oxidation state

Assigned 1 valence e⁻
Group 1A
= 0 oxidation state

Problem 17.41

Think

What is the nucleophile in each reaction? Can the nucleophile be simplified? Which C atom is electrophilic? Are proton transfer reactions involved?

Solve

The nucleophile in each reaction can be simplified to the hydride anion, :H⁻, from either $NaBH_4$ or $LiAlH_4$. The first step of the mechanism is nucleophilic addition of the hydride to the C of the carbonyl. In **(a)** and **(b)**, the methanol solvent that is already present protonates the alkoxide anion to produce the final alcohol product. In reaction **(c)**, a separate proton transfer reaction, an acid workup, is necessary to form the alcohol product using H_3O^+ (formed from H_2O, H_2SO_4).

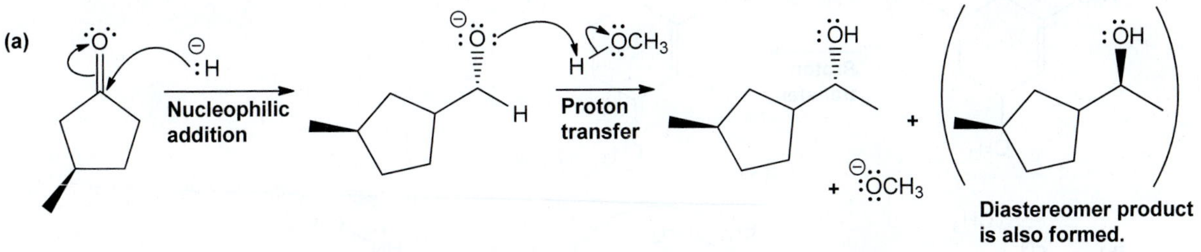

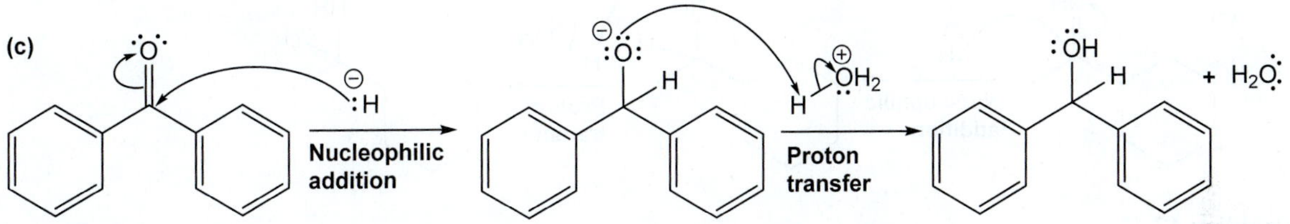

(a) Nucleophilic addition → Proton transfer → + $:\ddot{O}CH_3^-$ + Diastereomer product is also formed.

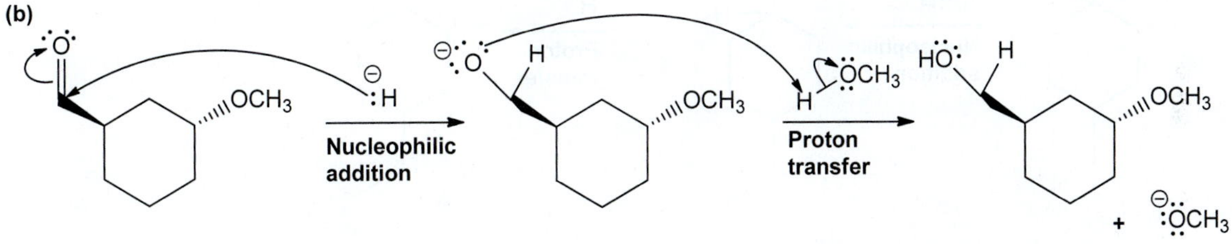

(b) Nucleophilic addition → Proton transfer → + $:\ddot{O}CH_3^-$

(c) Nucleophilic addition → Proton transfer → + H_2O

Problem 17.42

Think

What is the nucleophile in each reaction? Can the nucleophile be simplified? Which C atom is electrophilic? Are proton transfer reactions involved? How many times does the nucleophilic addition take place?

Solve

The nucleophile in each reaction can be simplified to the hydride anion, :H⁻, from either $NaBH_4$ or $LiAlH_4$. The first step of the mechanism is nucleophilic addition of the hydride to the C of the carbonyl. In reactions **(a)** and **(b)**, a separate proton transfer reaction, an acid workup, is necessary to form the uncharged amine product using H_2O. In **(c)**, the water solvent that is already present protonates the N⁻ anion to produce the final amine product. In **(a)**, the nitrile C atom undergoes nucleophilic addition twice, and two proton transfer reactions are involved to form the uncharged amine. In **(b)** and **(c)**, the imine C is electrophilic and undergoes nucleophilic addition once. A proton transfer reaction follows to form the uncharged amine. The mechanisms are given on the next page.

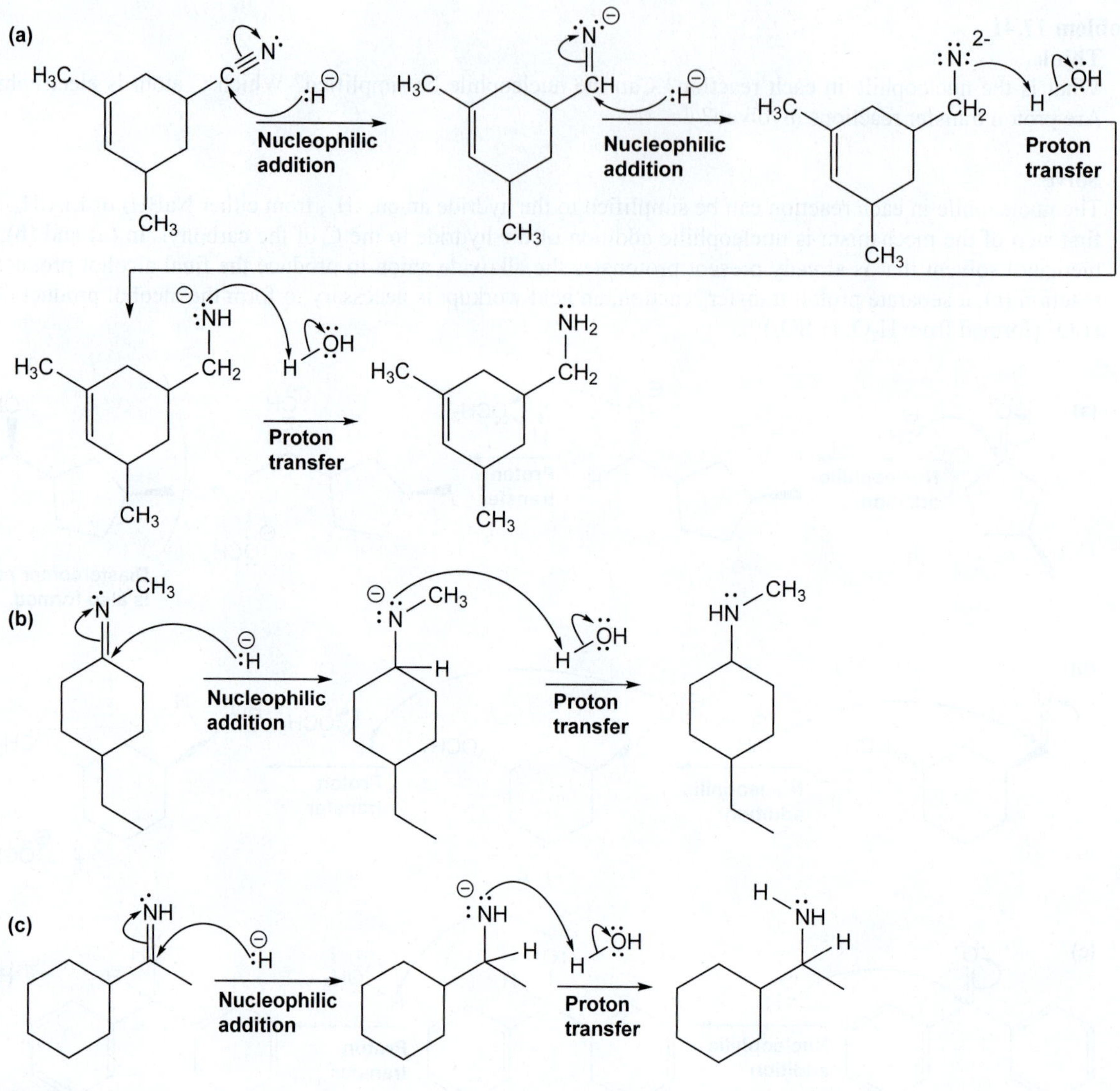

(a)

Nucleophilic addition

Nucleophilic addition

Proton transfer

Proton transfer

(b)

Nucleophilic addition

Proton transfer

(c)

Nucleophilic addition

Proton transfer

Problem 17.43

Think

In each reaction, is the deuterium-containing reagent a source of D^+ (like a proton), a source of $D:^-$ (like a hydride), or not involved in the reaction mechanism? How are these reaction mechanisms similar to the nucleophilic addition reactions you have already completed? Are proton transfer reactions involved?

Solve

(a) The deuterium in D_2O is a source of D^+ (like a proton) and thus adds to the O^- atom to form an OD alcohol.

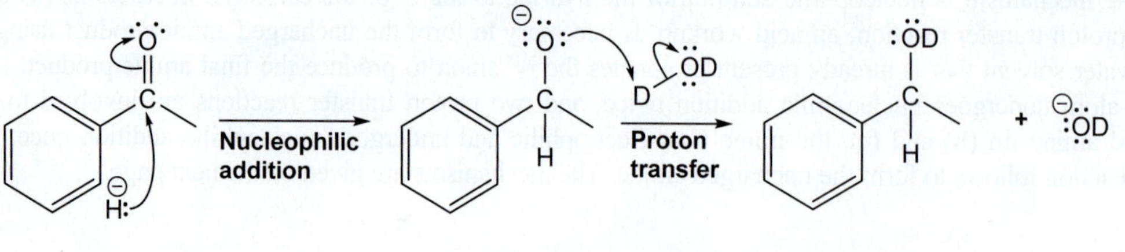

Nucleophilic addition

Proton transfer

(b) The deuterium in $LiAlD_4$ is a source of $D:^-$ (like a hydride) and thus adds to the carbonyl C atom to form a C–D bond.

Nucleophilic addition

Proton transfer

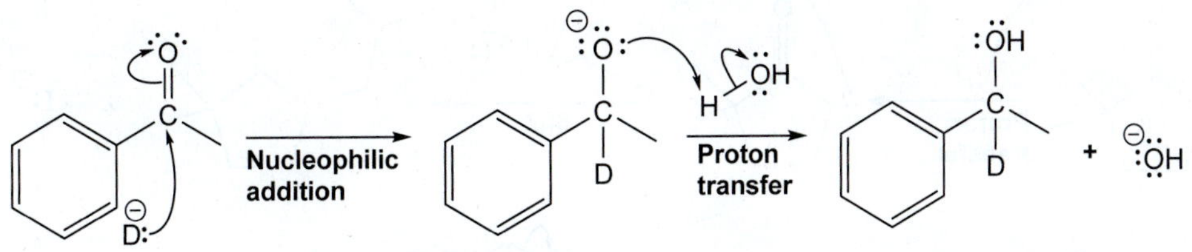

(c) The deuterium in $LiAlD_4$ is a source of $D:^-$ (like a hydride) and thus adds to the carbonyl C atom to form a C–D bond. The deuterium in D_2O is a source of D^+ (like a proton) and thus adds to the O^- atom to form an OD alcohol. The product has two D atoms.

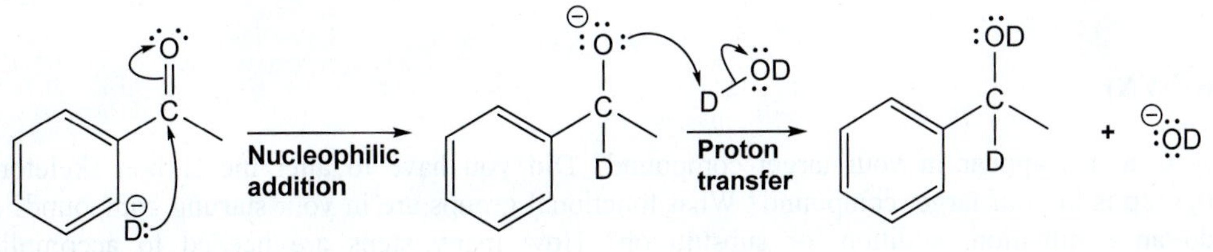

Nucleophilic addition

Proton transfer

Problem 17.44

Think

What is first step of the mechanism? Does NaH behave as a nucleophile or base? Are there any acidic protons? Is the deuterium-containing reagent a source of D^+ (like a proton), a source of $D:^-$ (like a hydride), or not involved in the reaction mechanism? How are these reaction mechanisms similar to other reactions involving NaH you have already encountered?

Solve

NaH is a strong base but a weak nucleophile. The aldehyde has an acidic α carbon that is deprotonated by NaH, which produces a basic enolate anion. The deuterium in D_2O is a source of D^+ (like a proton) and thus adds to the C^- atom to form a C–H bond.

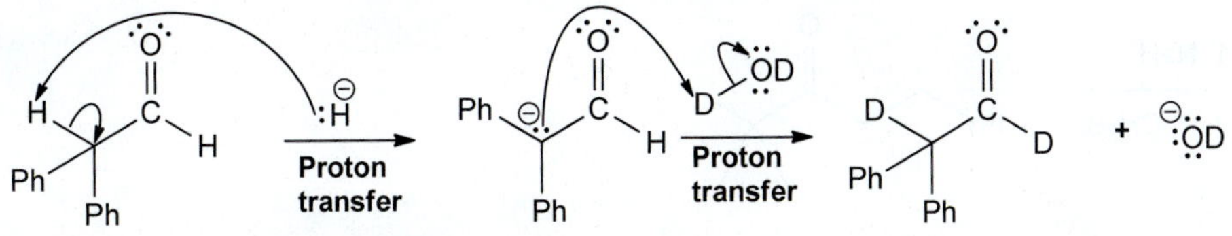

Proton transfer

Proton transfer

Problem 17.45

Think

Will the hydride anion from NaH act as a base or a nucleophile? Which atom will it attack? How will the resulting species behave in the presence of an alkyl halide?

Solve

NaH is a strong base but a poor nucleophile, so it will deprotonate at the α carbon. As we learned in Section 10.3, the resulting enolate anion is a strong nucleophile and will displace X from RX in an S_N2 reaction, yielding an alpha-alkylated ketone. See the figures on the next page.

(a) Proton transfer → S_N2 → + :I:⊖

(b) Proton transfer → S_N2 → + :Br:⊖

Problem 17.46 (SYN)

Think

How many C atoms appear in your target compound? Did you have to alter the carbon skeleton? What functional group is in your target compound? What functional groups are in your starting compound? Did you have to do an elimination, addition, or substitution? How many steps are needed to accomplish your transformation? Is stereochemistry a concern? Regiochemistry?

Solve

(a) In a retrosynthetic analysis, we can disconnect the carbon–carbon bond as shown.

Undo an α alkylation. ⟹ Ph–CH_2–Br +

In the forward direction, the synthesis would appear as follows:

1. NaH
2. $PhCH_2Br$

(b) In a retrosynthetic analysis, we can disconnect the carbon–carbon bond as shown.

Undo an α alkylation. ⟹ allyl Br +

In the forward direction, the synthesis would appear as follows:

1. NaH
2. allyl Br

Sections 17.5 and 17.6 Reactions of Alkyllithium Reagents and Grignard Reagents

Problem 17.47

Think

What is the nucleophile in each reaction? Can the nucleophile be simplified? Which C atom is electrophilic? Are proton transfer reactions involved? How many times does the nucleophilic addition take place?

Solve

The nucleophiles in each reaction are the metal-containing species and can be simplified by treating the metal portions as spectator cations. The nucleophile is therefore simplified to R^-. In each reaction mechanism for **(a)–(c)**, the carbonyl C is electrophilic and undergoes nucleophilic addition by R^-. The O^- atom is then protonated in a separate proton transfer reaction, an acid workup, to form an uncharged alcohol or carboxylic acid. The mechanisms are given below.

(a)
Nucleophilic addition
Proton transfer
+ NH_3

(b)
Nucleophilic addition
Proton transfer
+ NH_3

(c)
Nucleophilic addition
Proton transfer
+ H_2O

Problem 17.48

Think

What is the nucleophile in each reaction? Can the nucleophile be simplified? Which C atom is electrophilic? Are proton transfer reactions involved? How many times does the nucleophilic addition take place?

Solve

The nucleophiles in each reaction are the metal-containing species and can be simplified to R^- by treating the metal portions as spectator cations. See mechanisms for reactions **(a)–(c)** on the next page.

(a) The carbon of the C=N group is electrophilic and undergoes nucleophilic addition. The N$^-$ atom is then protonated via the CH_3CH_2OH in a separate proton transfer reaction, an acid workup, to form an uncharged amine.

Nucleophilic addition

Proton transfer

+ $:\ddot{O}CH_2CH_3$

(b) The carbon of the C=N group is electrophilic and undergoes nucleophilic addition. The N$^-$ atom is then protonated via the CH_3CH_2OH in a separate proton transfer reaction, an acid workup, to form an uncharged amine.

Nucleophilic addition

Proton transfer

+ $:\ddot{O}CH_2CH_3$

(c) The carbon of the C≡N group is electrophilic and undergoes nucleophilic addition. The N$^-$ atom is then protonated in a separate proton transfer reaction, an acid workup, to form an uncharged imine.

Nucleophilic addition

Proton transfer

+ $:\ddot{O}CH_3$

Problem 17.49

Think

In each reaction, is the deuterium-containing reagent a source of D$^+$ (like a proton), a source of D:$^-$ (like a hydride), or not involved in the reaction mechanism? How are these reaction mechanisms similar to the nucleophilic addition reactions you have already completed? Are proton transfer reactions involved?

Solve

(a) CH_3Li can be simplified to H_3C^-. The deuterium in D_2O is a source of D$^+$ (like a proton) and thus adds to the O$^-$ atom to form an OD alcohol.

Nucleophilic addition

Proton transfer

+ $:\ddot{O}D$

(b) The deuterium in D_3C^- is not directly involved in the reaction mechanism. The C–D bonds are not broken, and the nucleophile is the C atom in D_3C^-.

Nucleophilic addition

Proton transfer

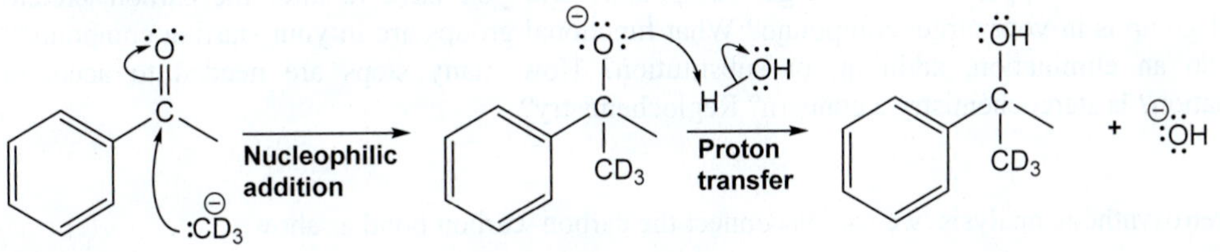

(c) The deuterium in D_3C^- is not directly involved in the reaction mechanism. The C–D bonds are not broken, and the nucleophile is the C atom in D_3C^-. The deuterium in D_2O is a source of D^+ (like a proton) and thus adds to the O^- atom to form an OD alcohol.

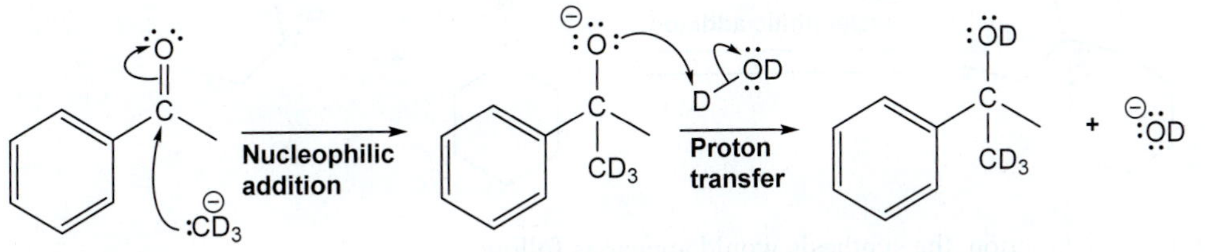

Nucleophilic addition

Proton transfer

Problem 17.50 (SYN)

Think

On which C atom was the C=O located in the starting material? What is the identity of the three R groups off the alcohol C atom in the target? What is the identity of the three Grignard reagents that can be used as precursors if each of those bonds is disconnected in a retrosynthetic analysis? What other two R groups must already be attached to the C=O in the ketone starting material after each of those bonds has been disconnected?

Solve

The alcohol C atom was the site of the C=O ketone in the starting material. The three R groups off the tertiary alcohol are CH_3, C_6H_5, and $CH_3CH{=}CH$, and thus each could have come from an RMgBr Grignard reagent precursor. The three different syntheses are given below.

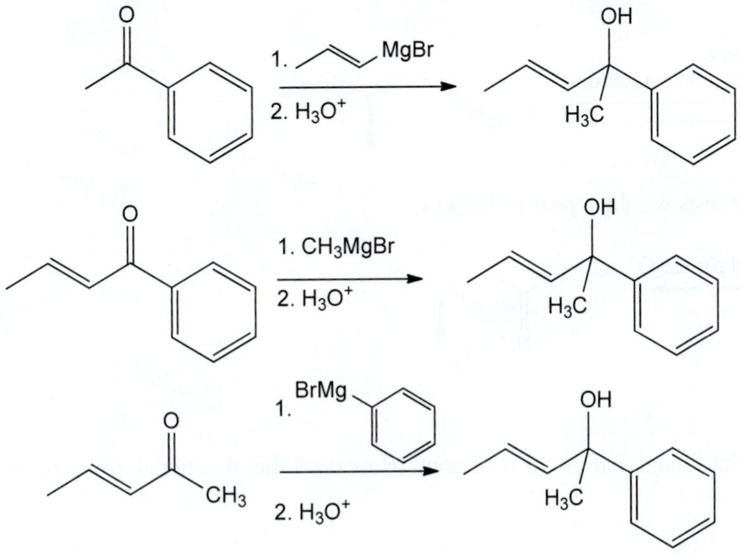

1. MgBr
2. H_3O^+

1. CH_3MgBr
2. H_3O^+

1. BrMg
2. H_3O^+

Problem 17.51 (SYN)

Think

How many C atoms appear in your target compound? Did you have to alter the carbon skeleton? What functional group is in your target compound? What functional groups are in your starting compound? Did you have to do an elimination, addition, or substitution? How many steps are needed to accomplish your transformation? Is stereochemistry a concern? Regiochemistry?

Solve

(a) In a retrosynthetic analysis, we can disconnect the carbon–carbon bond as shown.

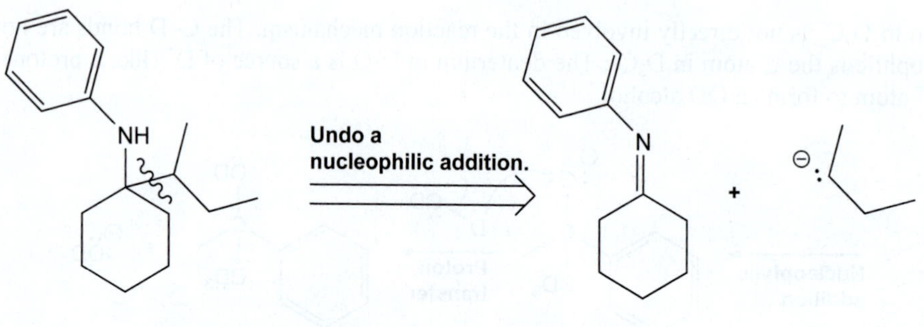

NH

Undo a nucleophilic addition.

N

+

In the forward direction, the synthesis would appear as follows:

N

1. Li

2. CH_3OH

NH

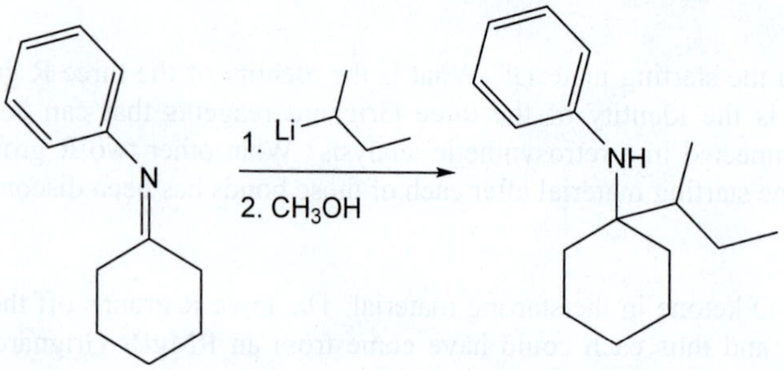

(b) In a retrosynthetic analysis, we can disconnect the carbon–carbon bond as shown.

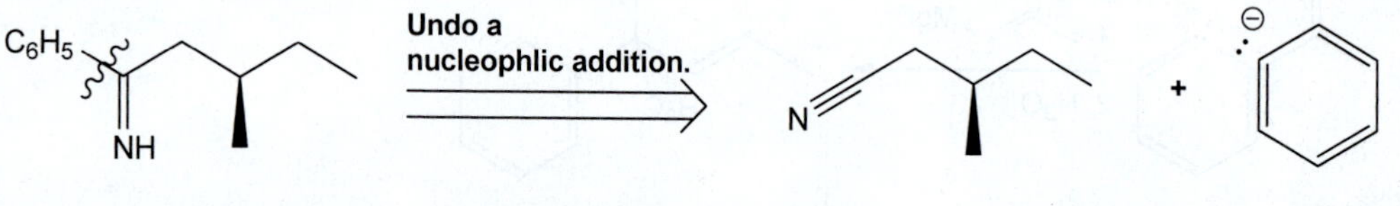

C_6H_5

NH

Undo a nucleophlic addition.

N

+

In the forward direction, the synthesis would appear as follows:

N

1. C_6H_5MgBr, Et_2O

2. H_2O

C_6H_5

NH

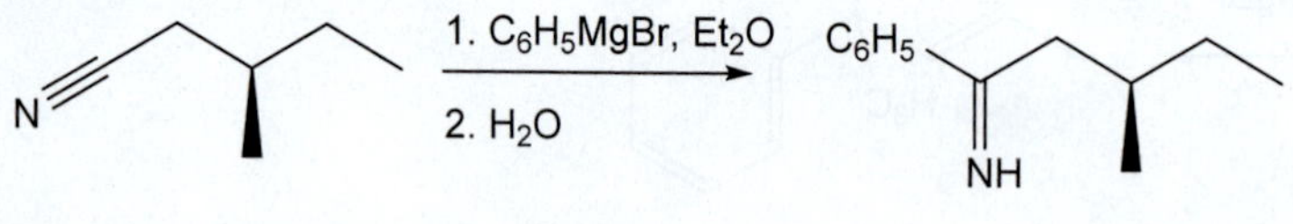

Problem 17.52

Think

What is the structure of acetone? What functional group is present? How does this functional group react with alkyllithium and Grignard reagents?

Solve

Acetone is a ketone, and the carbonyl C of acetone can be attacked by an alkyllithium or Grignard reagent, initiating a nucleophilic addition reaction. Students who dry their glassware with acetone before running a Grignard reaction frequently find this out the hard way.

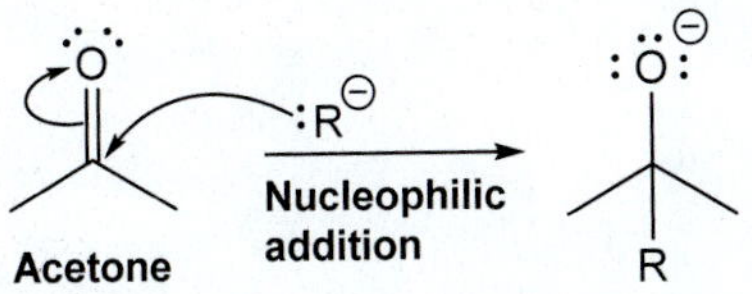

Acetone —:R⊖ / Nucleophilic addition→ alkoxide (:Ö:⊖, R)

Sections 17.7 and 17.8 Wittig Reagents and the Wittig Reaction

Problem 17.53

Think

What is the structure of triphenylphosphine? Is it electron rich or electron poor? How does Ph_3P: react with an R–X compound? Does Bu–Li act as a nucleophile or base?

Solve

Ph_3P: is nucleophilic and attacks the C of the C–X in an S_N2 mechanism. Bu–Li then deprotonates the C to form the Wittig reagent.

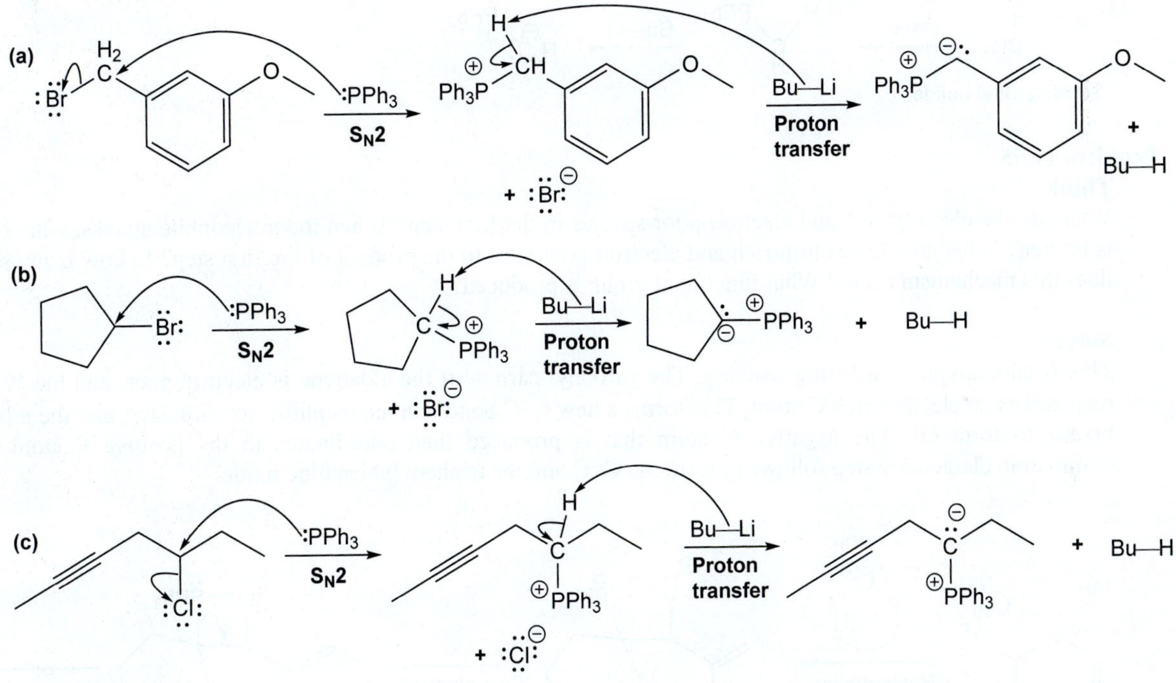

(a) :Br–CH_2–(3-methoxyphenyl) —:PPh_3 / S_N2→ Ph_3P^+–CH(H)–Ar + :Br:⊖ —Bu–Li / Proton transfer→ Ph_3P^+–C⊖H–Ar + Bu–H

(b) Cyclopentyl–Br —:PPh_3 / S_N2→ cyclopentyl(H)–PPh_3^+ + :Br:⊖ —Bu–Li / Proton transfer→ cyclopentylidene C⊖–PPh_3^+ + Bu–H

(c) Alkynyl alkyl chloride (:Cl:) —:PPh_3 / S_N2→ C(H)–PPh_3^+ + :Cl:⊖ —Bu–Li / Proton transfer→ C⊖–PPh_3^+ + Bu–H

Problem 17.54 (SYN)

Think

Which C atom is charged and a strong nucleophile? Which C atom was deprotonated? To which C atom was the halide attached?

Solve

The negatively charged C atom of the Wittig reagent is the one that would have been deprotonated and is also the one that would have initially been bonded to a halogen atom of an alkyl halide. Bromide as the alkyl halide is shown for each example, but chloride or iodide would also be a reasonable choice.

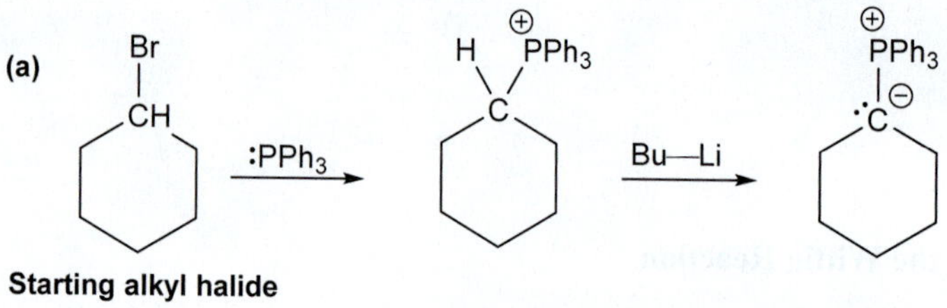

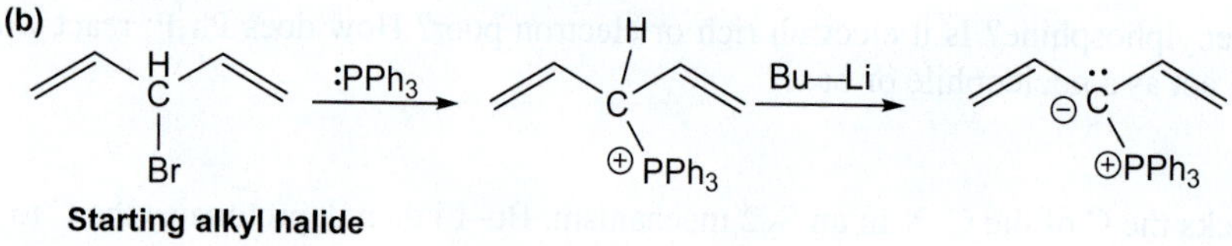

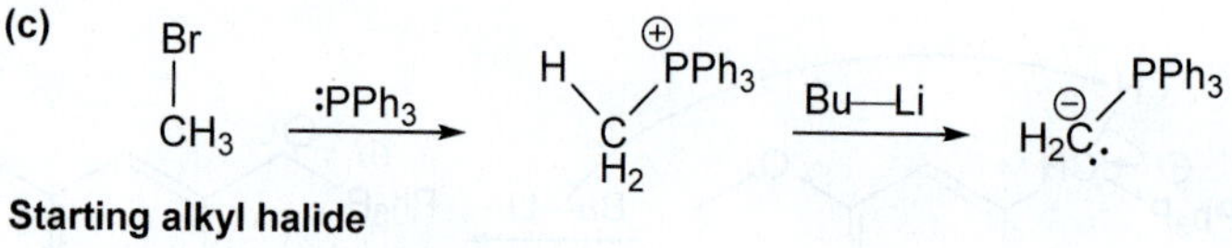

(a) Starting alkyl halide —:PPh3→ —Bu—Li→

(b) Starting alkyl halide —:PPh3→ —Bu—Li→

(c) Starting alkyl halide —:PPh3→ —Bu—Li→

Problem 17.55

Think

What are the electron-rich and electron-poor species in the first step? When the nucleophile attacks, what bond is broken? What are the electron-rich and electron-poor sites in the product of the first step? In how many steps does this mechanism occur? What functional group is produced?

Solve

This is an example of a Wittig reaction. The carbonyl carbon of the aldehyde is electron poor, and the Wittig reagent has an electron-rich C atom. This forms a new C–C bond in a nucleophilic addition step, and the π bond breaks to form O^-. The negative O atom that is produced then coordinates to the positive P atom. An elimination elementary step follows to form the C=C and the triphenylphosphine oxide.

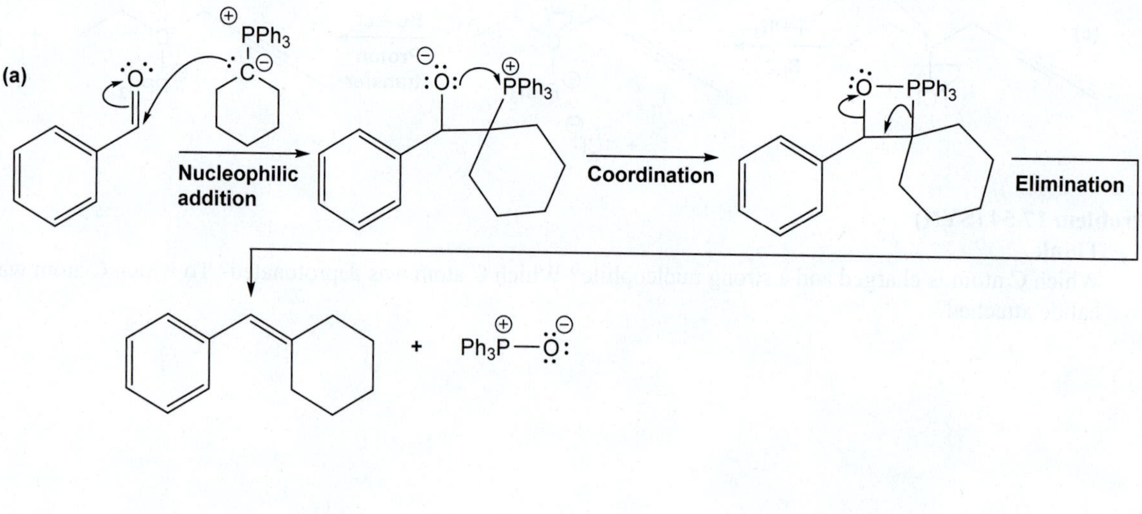

(a) —Nucleophilic addition→ —Coordination→ —Elimination→ + $Ph_3P^+—O^-$

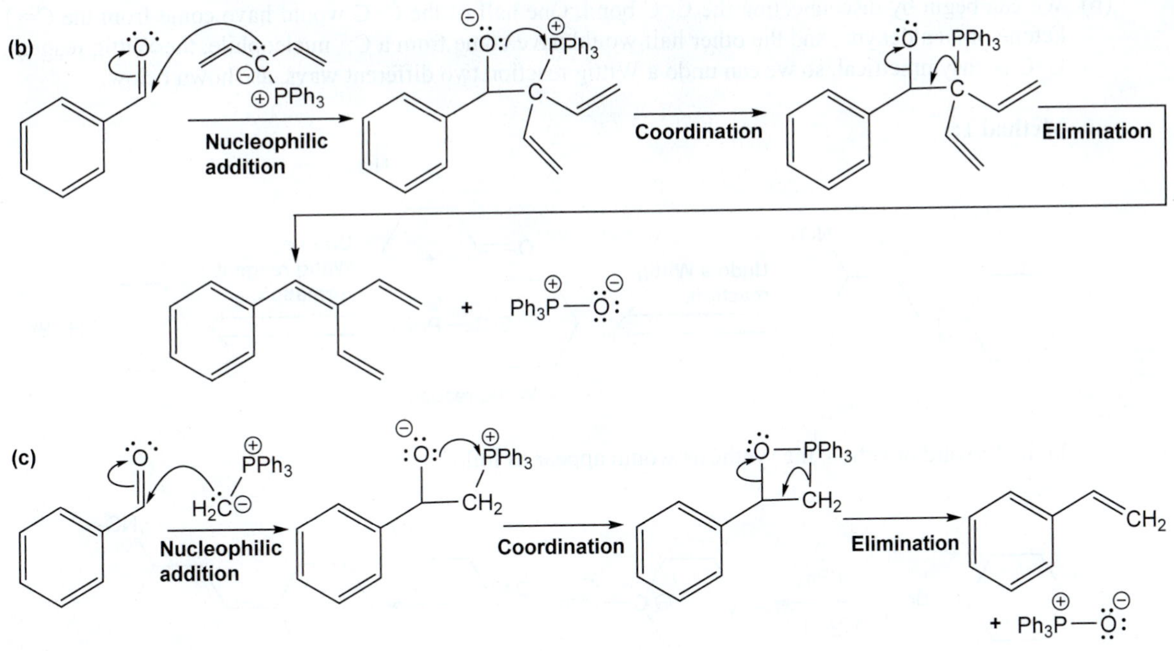

(b)

(c)

Problem 17.56 (SYN)

Think

How does a ketone or aldehyde starting material transform into a C=C in a Wittig reaction? Can you disconnect the C=C bond by undoing a Wittig reaction in a retrosynthetic analysis? Which part of the C=C was part of the aldehyde or ketone? Which part of the C=C was a Wittig reagent?

Solve

In each example, the C=C bond is formed via a Wittig reaction. Undo the C=C to determine the precursors: a C=O compound and a C:$^-$ nucleophile that will be the Wittig reagent. Undo the Wittig reagent formation to determine the alkyl halide. In these examples, an alkyl bromide is used, but an alkyl chloride or iodide would be a suitable choice as well.

(a) We can disconnect the C=C bond as shown below. The C=C is symmetrical, and one half of the C=C is part of the C=O aldehyde precursor while the other half is a C:$^-$ nucleophile precursor, which is the Wittig reagent.

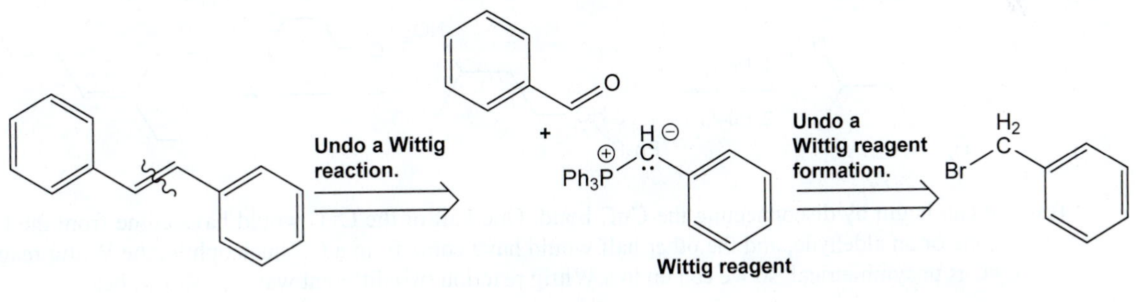

In the forward direction, the synthesis would appear as follows:

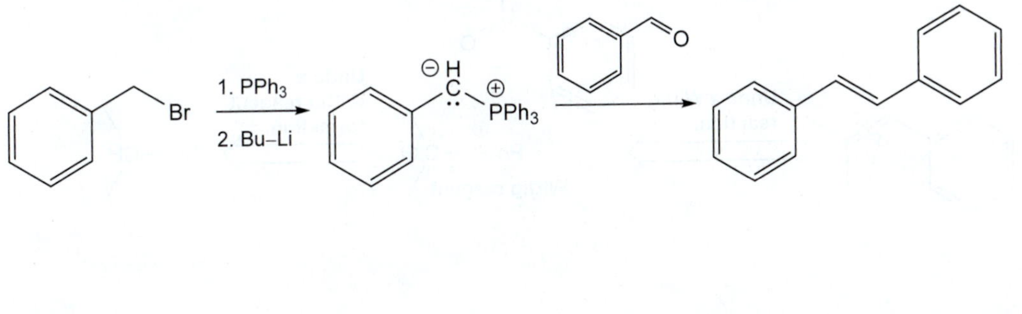

(b) We can begin by disconnecting the C=C bond. One half of the C=C would have come from the C=O of a ketone or an aldehyde, and the other half would have come from a C:⁻ nucleophile, the Wittig reagent. The C=C is unsymmetrical, so we can undo a Wittig reaction two different ways, as shown below.

Method 1:

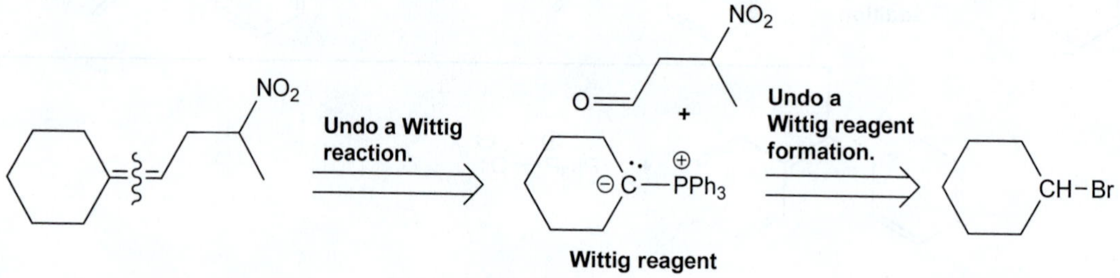

In the forward direction, the synthesis would appear as follows:

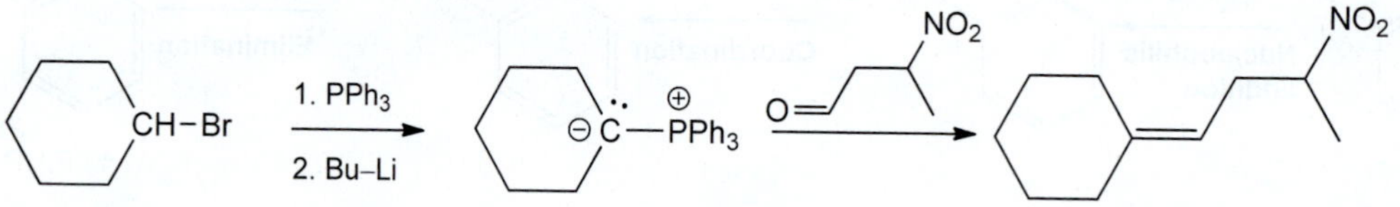

Method 2:

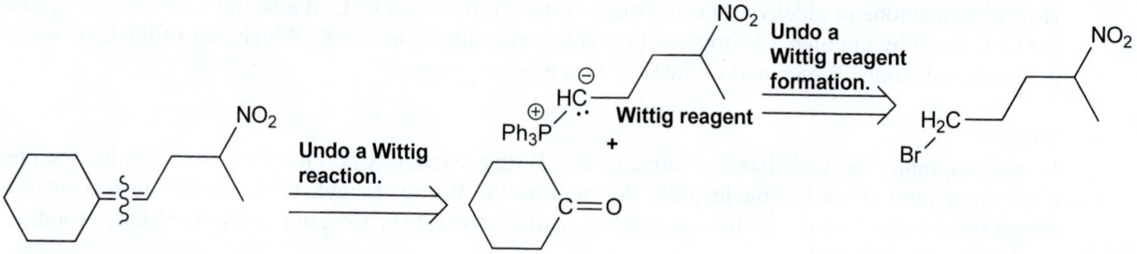

In the forward direction, the synthesis would appear as follows:

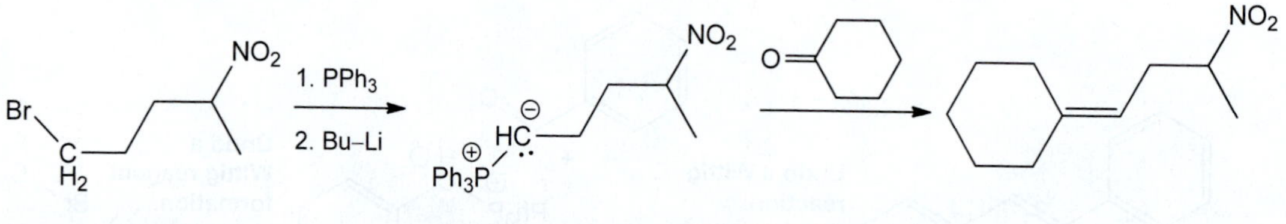

(c) We can begin by disconnecting the C=C bond. One half of the C=C would have come from the C=O of a ketone or an aldehyde, and the other half would have come from a C:⁻ nucleophile, the Wittig reagent. The C=C is unsymmetrical, so we can undo a Wittig reaction two different ways, as shown below.

Method 1:

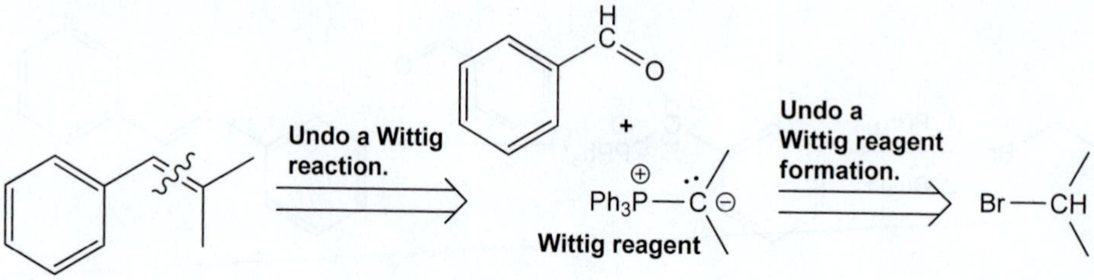

In the forward direction, the synthesis would appear as follows:

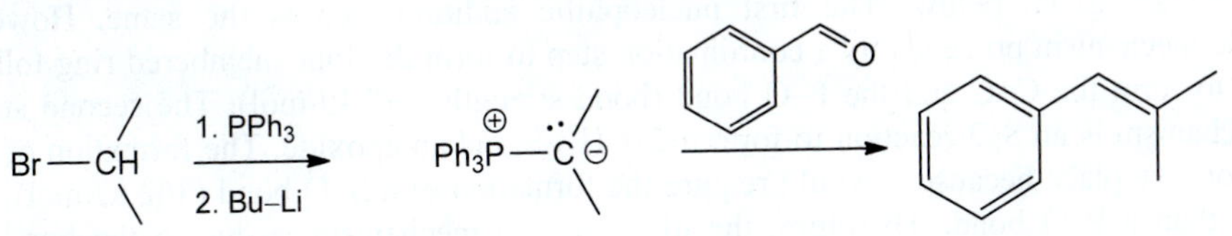

Br—CH 1. PPh_3 2. Bu–Li → Ph_3P—C⊖ →

Method 2:

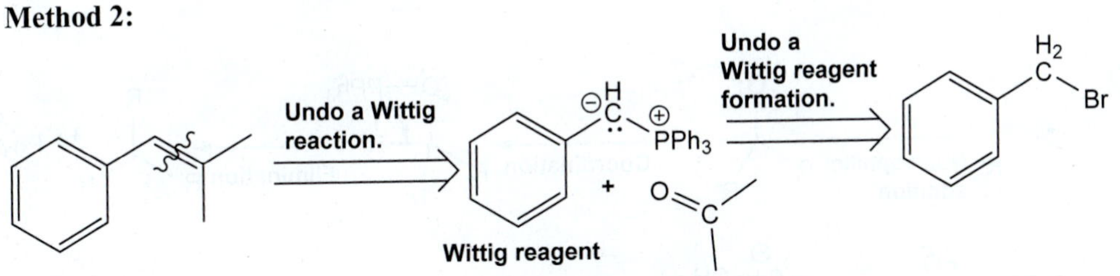

Undo a Wittig reaction. ⇒ Wittig reagent + O=C Undo a Wittig reagent formation. ⇒ H_2C Br

In the forward direction, the synthesis would appear as follows:

H_2C Br 1. PPh_3 2. Bu–Li → PPh_3 →

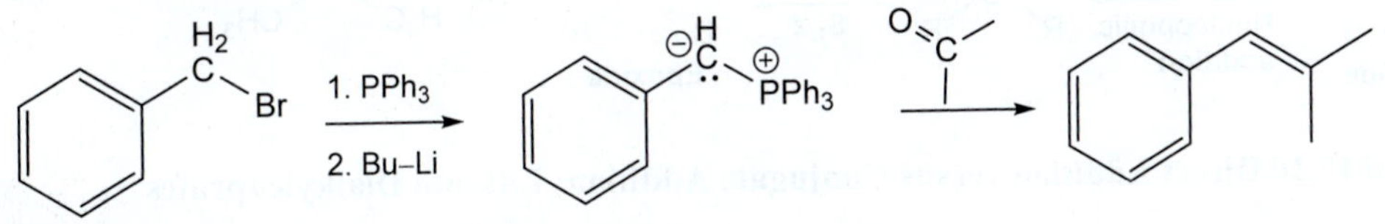

Problem 17.57

Think

What are the electron-rich and electron-poor species in the first step? Is there a good leaving group in the product of that step? How can you add the curved arrows to show the production of the epoxide functional group?

Solve

The sulfonium ylide is synthesized via an S_N2 mechanism followed by a proton transfer. The ketone C=O carbon is electron poor, and the sulfonium ylide C atom is electron rich. A new C–C bond forms. The negative O atom is a good nucleophile, and the $(CH_3)_2S$ group is a good leaving group. The last step is an example of an intramolecular S_N2 reaction. An epoxide functional group results.

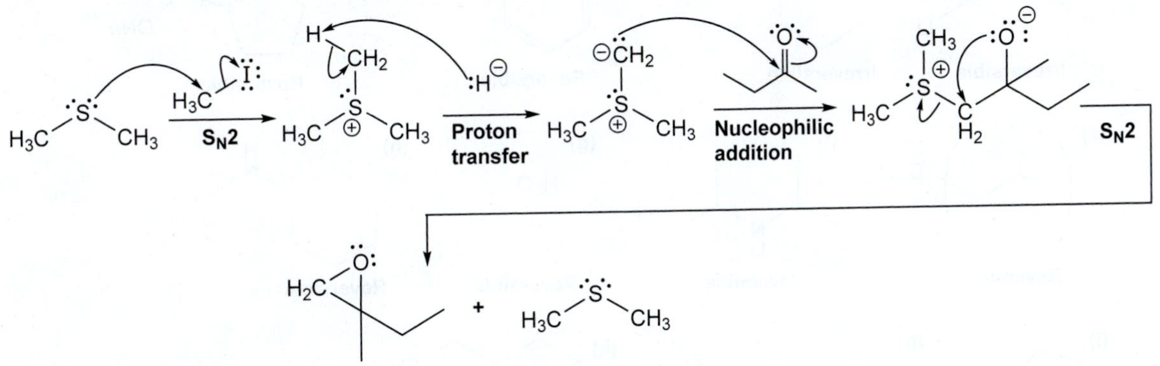

H_3C–S–CH_3 + H_3C–I S_N2 → H_3C–S⊕(CH_3)–CH_3 + :H⊖ Proton transfer → H_3C–S⊕(CH_2⊖)–CH_3 Nucleophilic addition → S_N2 → H_2C–O epoxide + H_3C–S–CH_3

Problem 17.58

Think

What is the mechanism for each reaction? What steps are the same, and what step is different? What is the bond strength of the P–O bond? What is the bond strength of the S–O bond?

Solve

The two mechanisms are given below. The first nucleophilic addition step is the same. However, the phosphoronium ylide mechanism proceeds by a coordination step to form the four-membered ring followed by an elimination step to form the C=C and the P–O bond (bond strength: 537 kJ/mol). The second step in the sulfonium ylide mechanism is an S_N2 reaction to form the $(CH_3)_2S$ and an epoxide. The formation of the C=C double bond does not take place because it would require the formation of a S–O bond (362 kJ/mol), which is 175 kJ/mol weaker than a P–O bond. Therefore, the difference in mechanism is due to the bond strength differences of the P–O and S–O bonds.

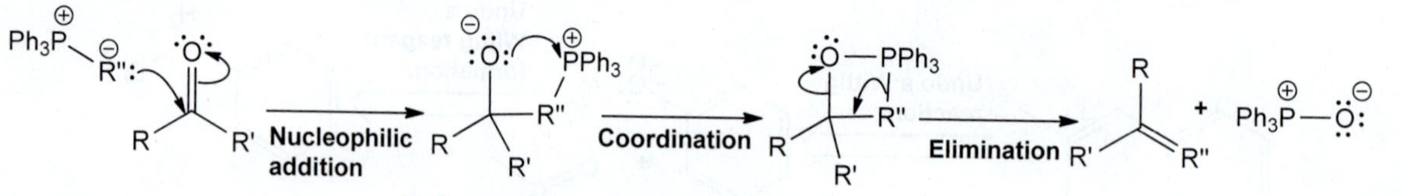

Nucleophilic addition → Coordination → Elimination

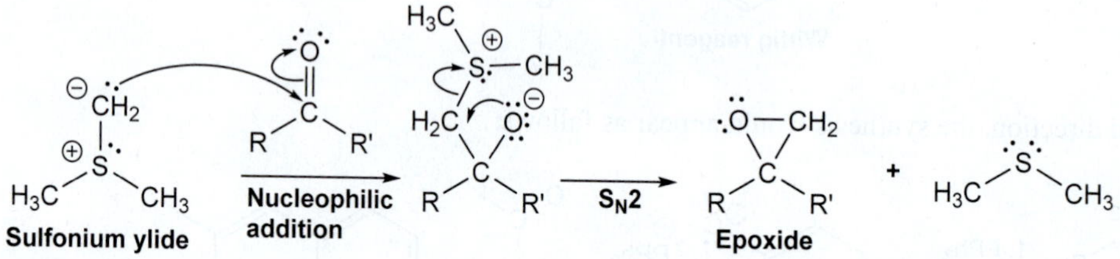

Sulfonium ylide → Nucleophilic addition → S_N2 → Epoxide

Sections 17.9 and 17.10 Direct Addition versus Conjugate Addition; Lithium Dialkylcuprates

Problem 17.59

Think

Which nucleophiles are sources of R^- or H^-? Do those kinds of nucleophiles add reversibly or irreversibly to a carbonyl group? Refer to Table 17-2 for a list of reversible and irreversible nucleophiles.

Solve

Only the highly reactive R^- and H^- nucleophiles add irreversibly to a carbonyl C, which include Grignard reagents, alkyllithium reagents, and Wittig reagents. Therefore, of the reagents given, the only ones that add irreversibly are **(a)** and **(b)**. The remaining nucleophiles are significantly less reactive and add reversibly.

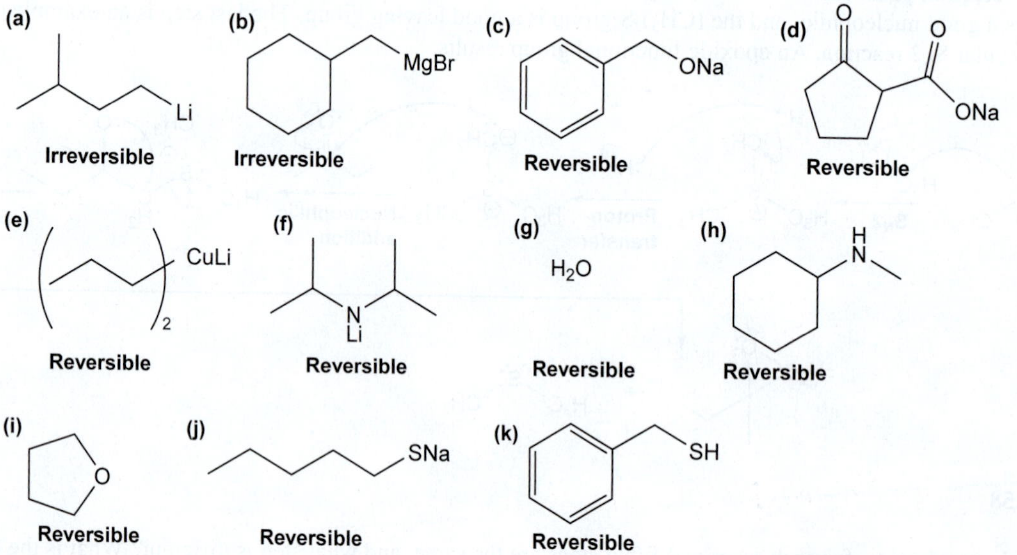

(a) Irreversible

(b) Irreversible

(c) Reversible

(d) Reversible

(e) Reversible

(f) Reversible

(g) H_2O Reversible

(h) Reversible

(i) Reversible

(j) Reversible

(k) Reversible

Problem 17.60

Think

What is the nucleophile in each reaction? Which C atom is electrophilic? Are proton transfer reactions involved? Does the nucleophilic addition occur via 1,2-addition or 1,4-addition?

Solve

In each reaction mechanism for **(a)**–**(e)**, the carbonyl C is electrophilic and can undergo nucleophilic addition. In **(a)**, **(c)**, **(d)**, and **(e)**, the β carbon is also electrophilic, so you must consider both 1,2-addition and 1,4-addition. Reactions **(a)**, **(d)**, and **(e)** undergo 1,4-addition because the nucleophile adds reversibly. Reaction **(c)** undergoes 1,2-addition because the nucleophile adds irreversibly. In all cases, the O^- atom that forms in the nucleophilic addition step is then protonated in a subsequent proton transfer reaction to form an uncharged alcohol. In the conjugate addition mechanism, the enol tautomerizes back to the ketone or aldehyde via back-to-back proton transfer steps. See the mechanisms below and on the next page.

(a)

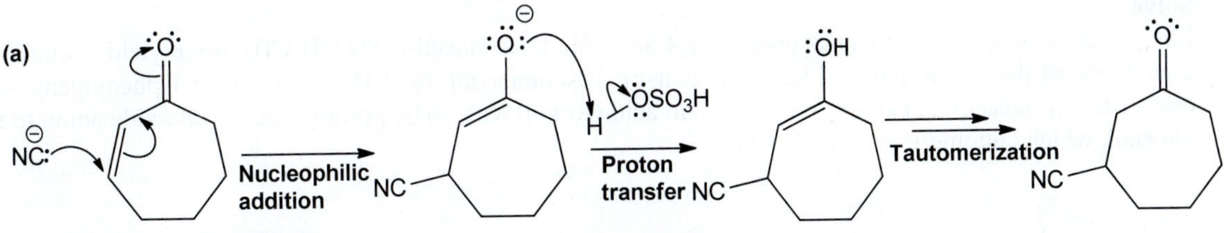

Nucleophilic addition → Proton transfer → Tautomerization

Conjugate (1,4-addition)

(b)

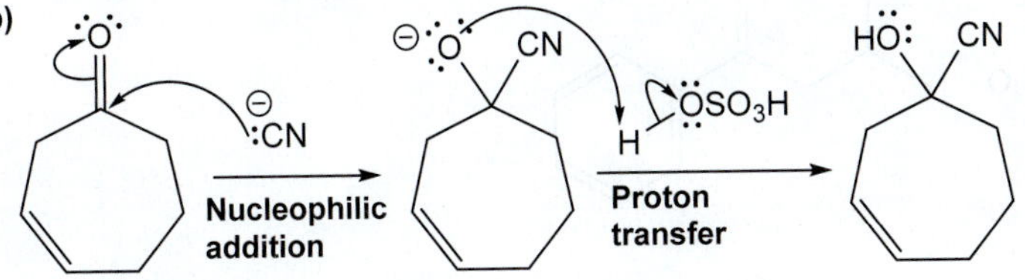

Nucleophilic addition → Proton transfer

Direct (1,2-addition)

(c)

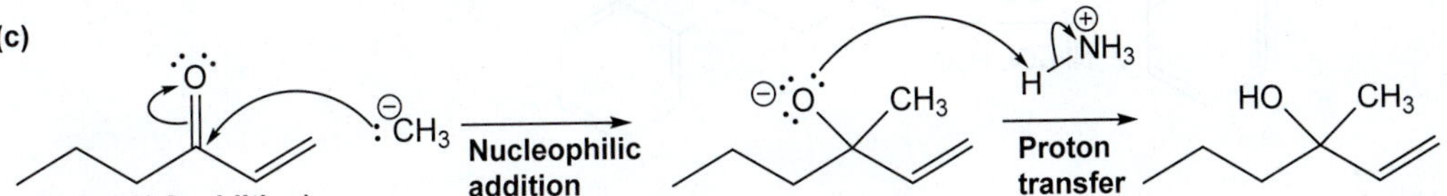

Nucleophilic addition → Proton transfer

Direct (1,2-addition)

(d)

Nucleophilic addition → Proton transfer → Tautomerization

Conjugate (1,4-addition)

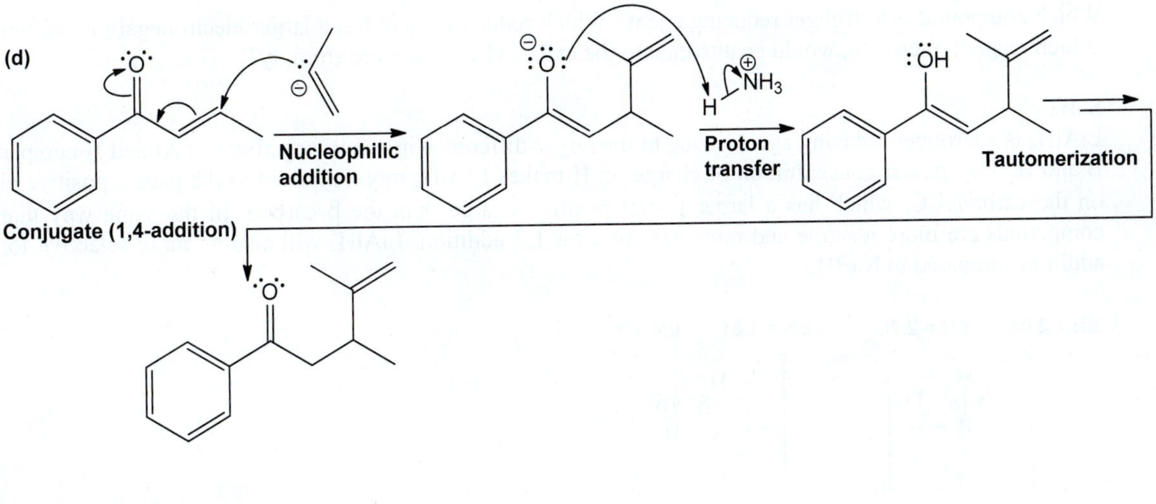

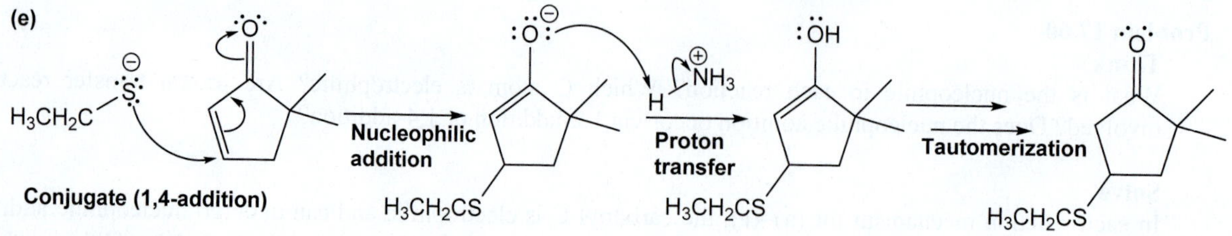

(e)

Conjugate (1,4-addition) — Nucleophilic addition — Proton transfer — Tautomerization

H_3CH_2C–S$^-$; H_3CH_2CS ; NH_3^+ ; :OH

Problem 17.61 (SYN)

Think

Do lithium dialkylcuprate compounds undergo 1,2-addition or 1,4-addition? What would be the structure of the α,β-unsaturated ketone starting material if you were to disconnect the bond between the β carbon and CH_3? What would it be if you were to disconnect the bond to the CH_3CH_2 group instead?

Solve

Lithium dialkylcuprate compounds undergo 1,4-addition. Disconnecting the CH_3CH_2 group yields 1-phenylbut-2-en-1-one as the α,β-unsaturated ketone precursor. Disconnecting the CH_3 group yields 1-phenylpent-2-en-1-one as the α,β-unsaturated ketone precursor. An acid workup with NH_4Cl protonates the alkoxide anion to make the enol, which tautomerizes to the ketone.

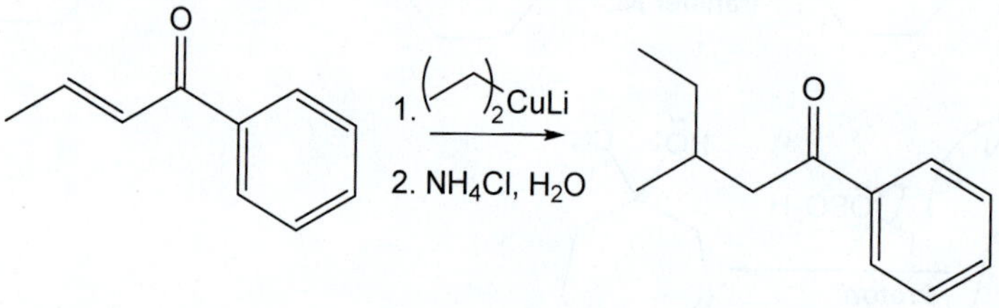

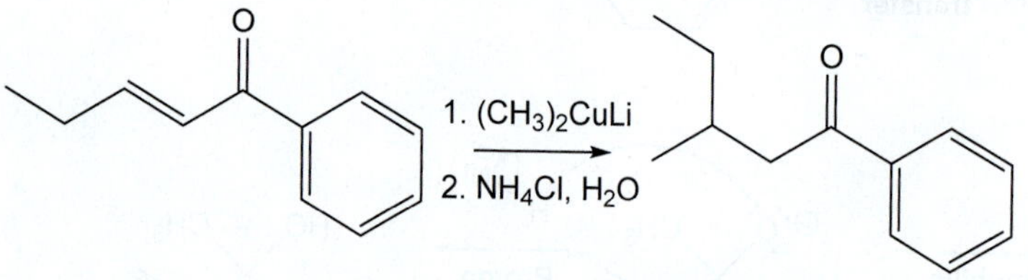

1. $(CH_3CH_2)_2CuLi$
2. NH_4Cl, H_2O

1. $(CH_3)_2CuLi$
2. NH_4Cl, H_2O

Problem 17.62

Think

Which compound is a stronger reducing agent? Which reducing agent has a larger electronegativity difference? Which reagent, therefore, would be attracted to the carbonyl carbon more strongly?

Solve

$LiAlH_4$ is a stronger reducing agent owing to the larger difference in electronegativity of Al and H compared to B and H. The greater concentration of charge on H makes $LiAlH_4$ more attracted to the partial positive charge on the carbonyl C, which has a larger partial positive charge than the β carbon. In the same way that RLi compounds are more reactive and more selective for 1,2-addition, $LiAlH_4$ will also be more selective for 1,2-addition compared to $NaBH_4$.

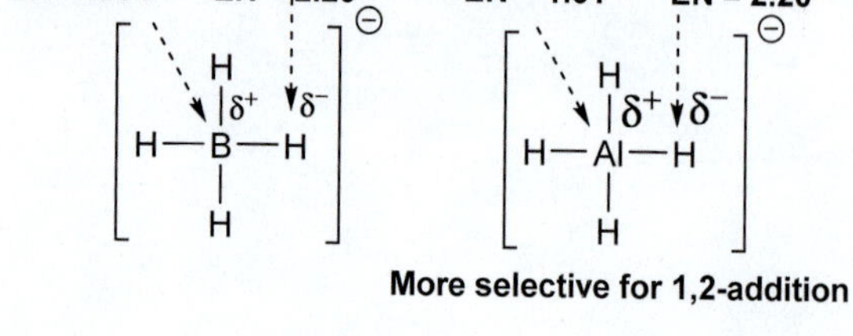

EN = 2.04 EN = 2.20 — $[BH_4]^-$ (δ+ on B, δ− on H)

EN = 1.61 EN = 2.20 — $[AlH_4]^-$ (δ+ on Al, δ− on H)

More selective for 1,2-addition

Problem 17.63

Think

What is the product of a conjugate addition reaction? How does $NaBH_4$ react with this product? How does a lithium dialkylcuprate react with this product? Why is there a difference?

Solve

In the product of conjugate addition, a polar π bond remains, and $NaBH_4$ adds hydride to such bonds via direct addition. A lithium dialkylcuprate, on the other hand, does not react with polar π bonds via direct addition, so no second reaction can take place.

O NaBH4 O NaBH4 OH

O R2CuLi O R

Problem 17.64

Think

What is the product of direct addition? What is the product of conjugate addition? If direct addition takes place first, can the product that is given be produced? If conjugate addition takes place first, can the product that is given be produced?

Solve

$NaBH_4$ is a source of hydride, and CH_3CH_2OH is a source of protons. If direct addition were to take place first, an allylic alcohol would be produced, and no further reaction would take place, so the given product could not be produced. On the other hand, the conjugate addition of $NaBH_4$ to but-3-en-2-one (α,β-unsaturated ketone) results in a ketone, which can be reduced a second time to the alcohol.

H: Nucleophilic addition :O: OCH2CH3 H Proton transfer :OH Tautomerization :O: :H Nucleophilic addition

:O: OCH2CH3 H Proton transfer :OH

Integrated Problems

Problem 17.65

Think

What bonds broke and formed in each reaction? What new atoms are present? Is a C–C bond-forming reaction necessary? Did the reaction require an acid workup step?

Solve

The first reaction forms a new C–C bond, and a butyl group is added. Therefore, the reagent **A** is either butylmagnesium bromide ($CH_3CH_2CH_2CH_2MgBr$) or butyllithium ($CH_3CH_2CH_2CH_2Li$). The uncharged alcohol is the final product, and therefore, an acid workup step is necessary. The second reaction requires acid and heat to promote an E1 reaction. The (*E*) isomer is the major product. See the transformation reactions on the next page.

1. (butyllithium) 2. H_2O **A**; H_2SO_4 (conc), Δ **B**; (E) + (Z)

Problem 17.66 (SYN)

Think

What new atoms are present in the products given? Is a C–C bond-forming reaction necessary? What is the identity of the nucleophile that would be necessary to participate in each nucleophilic addition reaction with phenylethanone to produce the products shown in **(a)**–**(c)**? Are proton transfer reactions involved? What other reagents are necessary?

Solve

(a) The first synthesis is a reduction (hydride nucleophile), followed by replacement of O with a halogen (Chapter 10) using PBr_3 (two S_N2 steps).

O → $NaBH_4$, H_2O → OH → PBr_3 → Br

(b) The product can be made by dehydration of an alcohol (Chapter 9), which can be made via a Grignard or alkyllithium reaction (adds on H_3C^- as the nucleophile).

O → 1. CH_3Li 2. H_3O^+ → HO → H_3PO_4 (conc), Δ →

(c) The first reaction is a reduction (hydride nucleophile), followed by the Williamson synthesis of an ether (Chapter 10).

O → $NaBH_4$, CH_3CH_2OH → OH → 1. NaH 2. CH_3Br → O

Problem 17.67

Think

What is the nucleophile in each reaction? Which C atom is electrophilic? Are proton transfer reactions involved? Does the nucleophilic addition occur via 1,2-addition or 1,4-addition?

Solve

(a) Hydroxide is not a good leaving group, and PBr_3 turns HO^- into $HOPBr_2$, which is a good leaving group. The Br^- attacks the electrophilic C atom in Step 2 to form the alkyl bromide. The ylide Wittig reagent is then formed via an S_N2 reaction followed by a proton transfer. The three-step Wittig reaction follows. The mechanism is shown below.

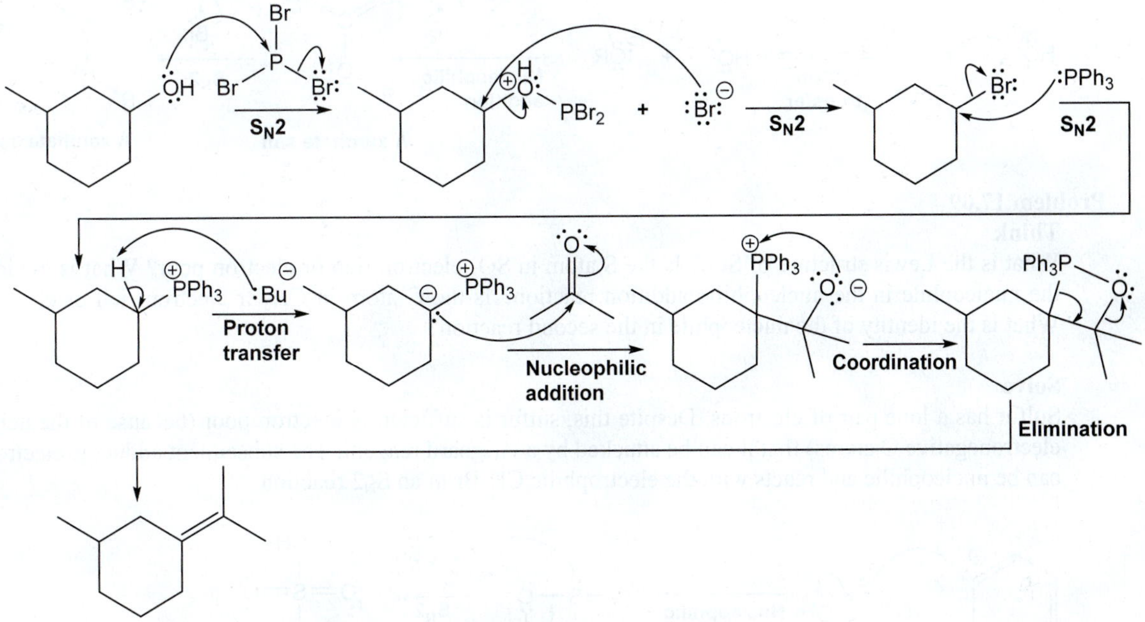

S_N2 PBr_2 + $:Br:^-$ S_N2 $:PPh_3$ S_N2

PPh_3 $:Bu$ Proton transfer PPh_3 Nucleophilic addition PPh_3 Coordination Ph_3P Elimination

(b) The first two steps are a Grignard reaction, in which the Grignard reagent adds to the C=O, followed by proton transfer to form a tertiary alcohol. The second two steps make up a Williamson ether synthesis. First is a deprotonation by NaH to form a negatively charged alkoxide. The resulting alkoxide, a strong nucleophile, then attacks the CH_3CH_2Br via an S_N2 mechanism to form the ether functional group.

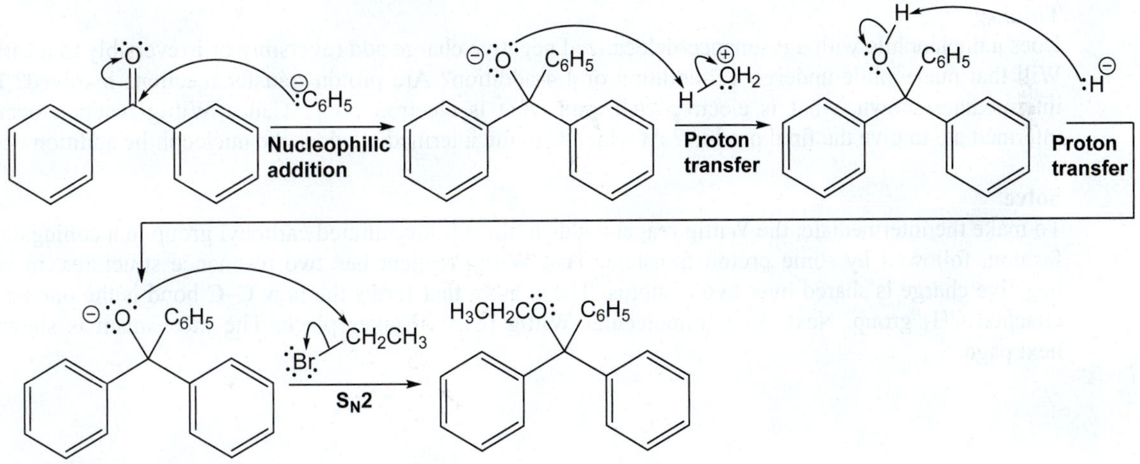

$:C_6H_5$ Nucleophilic addition C_6H_5 OH_2 Proton transfer C_6H_5 $:H$ Proton transfer

C_6H_5 CH_2CH_3 $:Br$ S_N2 $H_3CH_2CO:$ C_6H_5

Problem 17.68

Think

Is the C atom in CS_2 electron rich or electron poor? What reaction occurs between ROH and NaOH? What is the identity of the nucleophile in the nucleophilic addition reaction? How can the xanthate salt react with RBr?

Solve

HO^- is used to convert the weak ROH nucleophile into a strong RO^- nucleophile, as shown in the first step below. RO^- attacks CS_2 in a nucleophilic addition, as shown in the second step. Once the nucleophilic addition is complete, the xanthate salt can do an S_N2 reaction on an alkyl halide (if present) to make the xanthate ester.

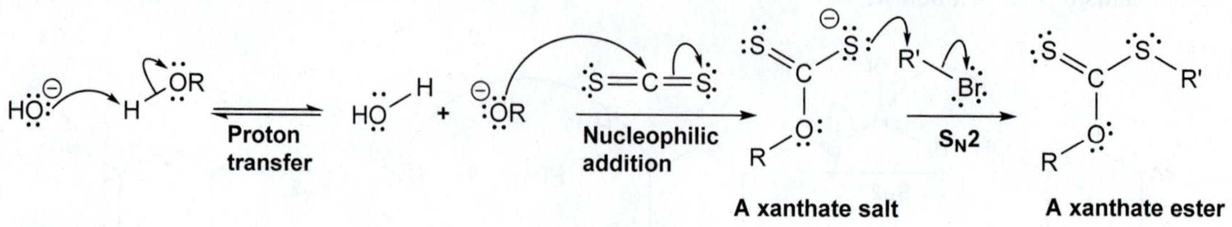

Proton transfer

Nucleophilic addition

S_N2

A xanthate salt

A xanthate ester

Problem 17.69

Think

What is the Lewis structure of SO_2? Is the S atom in SO_2 electron rich or electron poor? What is the identity of the nucleophile in the nucleophilic addition reaction? Is the C atom in CH_3Br electron rich or electron poor? What is the identity of the nucleophile in the second reaction?

Solve

Sulfur has a lone pair of electrons. Despite this, sulfur is sufficiently electron poor (because of the neighboring electronegative O atoms) that it can be attacked by a Grignard reagent. The subsequent adduct is electron rich; it can be nucleophilic and reacts with the electrophilic CH_3Br in an S_N2 reaction.

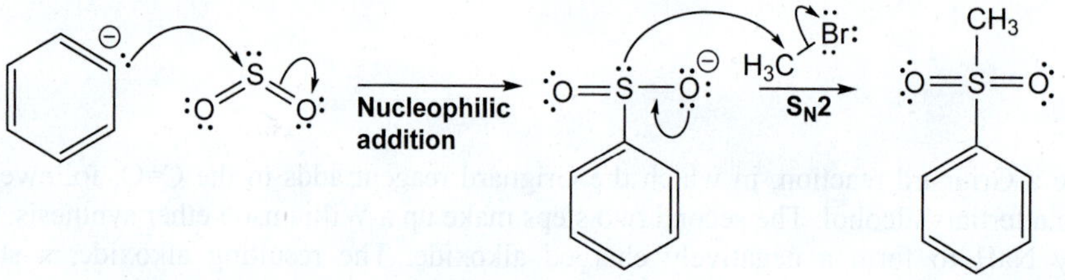

Nucleophilic addition

S_N2

Problem 17.70

Think

Does a nucleophile with a resonance-delocalized negative charge add reversibly or irreversibly to a carbonyl C? Will that nucleophile undergo 1,2-addition or 1,4-addition? Are proton transfer reactions involved? From the intermediate shown, what is electron rich, and what is electron poor? Can a Wittig reaction occur in the intermediate to give the final product? At which C in the intermediate does the nucleophilic addition occur?

Solve

To make the intermediate, the Wittig reagent adds to the α,β-unsaturated carbonyl group in a conjugate addition fashion, followed by some proton transfers. That Wittig reagent has two resonance structures, in which the negative charge is shared over two C atoms. The C atom that forms the new C–C bond is the one that has the attached CH_3 group. Next, an intramolecular Wittig reaction takes place. The mechanism is shown on the next page.

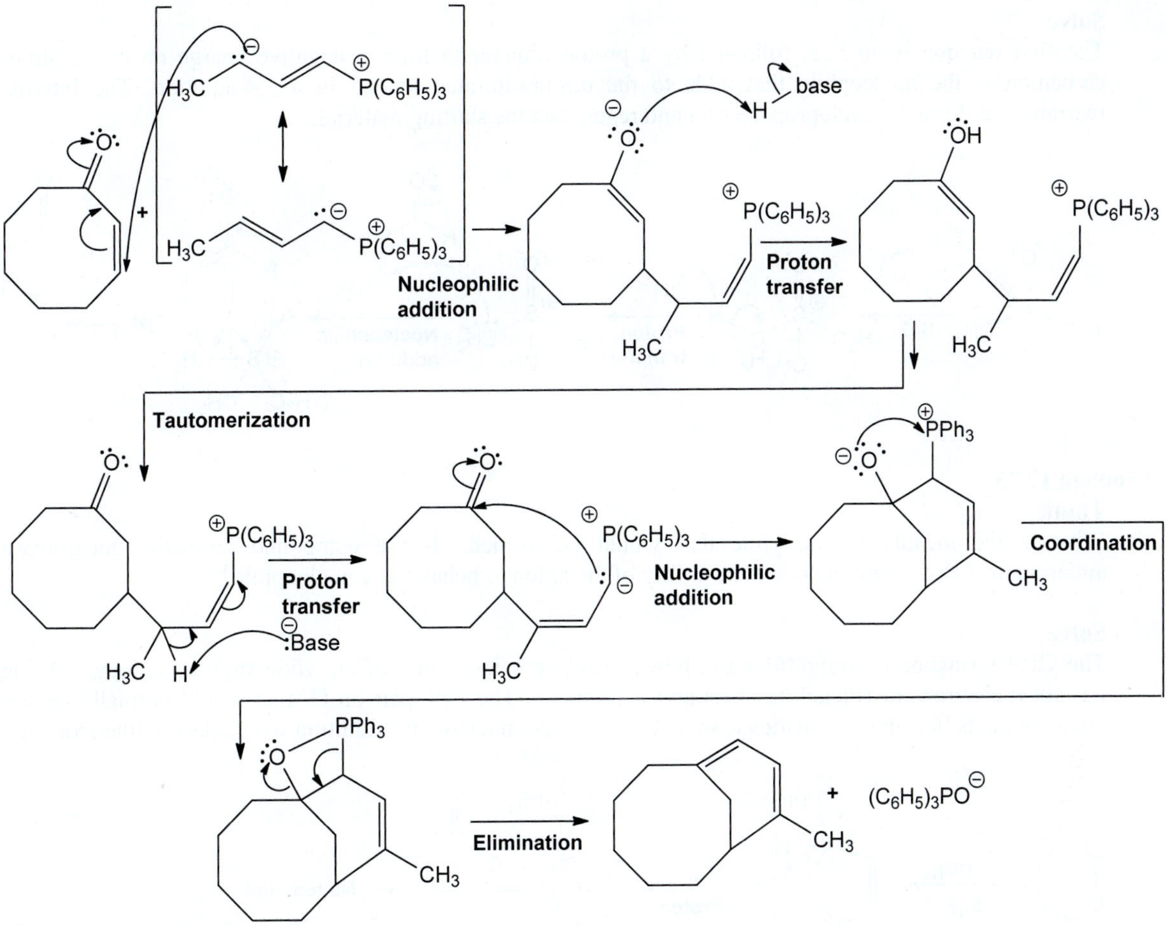

Nucleophilic addition

Proton transfer

Tautomerization

Proton transfer

Nucleophilic addition

Coordination

Elimination

Problem 17.71

Think

In the first reaction, what is the nucleophile, and what is the electrophile? Is this an addition, substitution, or elimination reaction? What is the purpose of NaH? Which proton is most likely to be abstracted? What is the nucleophile that adds to the electrophilic C of the imine?

Solve

The first reaction is an S_N2 followed by a proton transfer to form the sulfonium ylide. The sulfonium ylide reacts with the imine in a two-step mechanism to form the product shown.

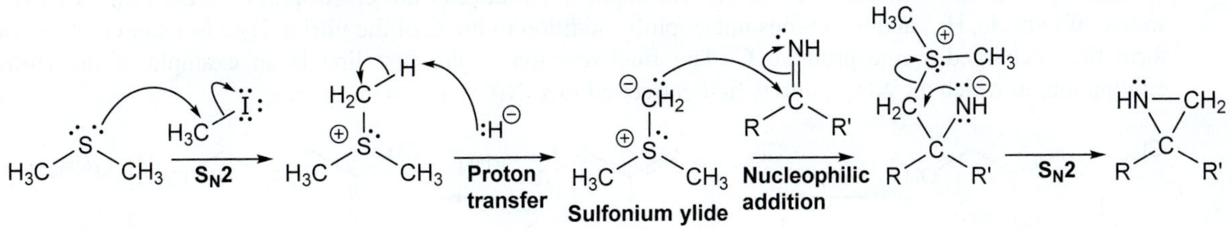

S_N2

Proton transfer

Sulfonium ylide

Nucleophilic addition

S_N2

Problem 17.72

Think

In the first reaction, what is the nucleophile, and what is the electrophile? Is this an addition, substitution, or elimination reaction? What is the purpose of NaH? Which proton is most likely to be abstracted? In the third reaction, what is the identity of the nucleophile? Does the nucleophile react with the α,β-unsaturated ketone via 1,2-addition or 1,4-addition? After nucleophilic addition, can a leaving group be eliminated?

Solve

The first reaction is an S_N2, followed by a proton transfer to form a negative charge on the C atom. The carbanion is the nucleophile that adds to the α,β-unsaturated ketone in a 1,4-addition. The intermediate rearranges to form the cyclopropane ring and regenerate the starting material.

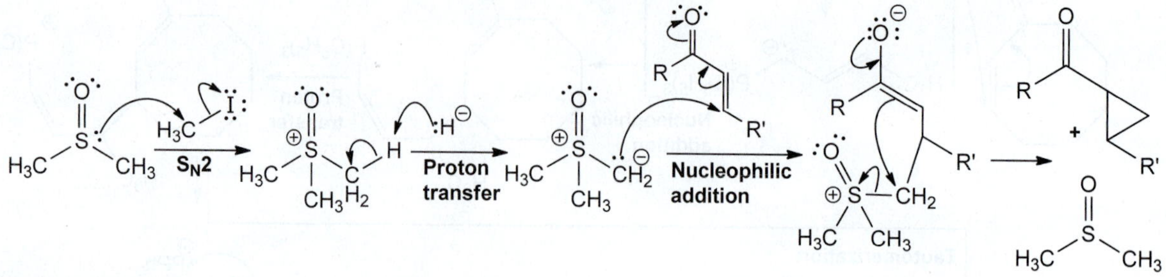

Problem 17.73

Think

What is the identity of the ylide that would be formed? Is the compound aromatic, nonaromatic, or antiaromatic? How does this affect the ability of the anion to behave as a nucleophile?

Solve

The ylide formation is straightforward; however, close examination of the ylide shows a conjugated ring that has six π electrons (a Hückel number) and is aromatic. The lone pair on C^- that would normally be strongly nucleophilic is tied up in aromaticity, so it is unreactive—nucleophilic addition would destroy the aromaticity.

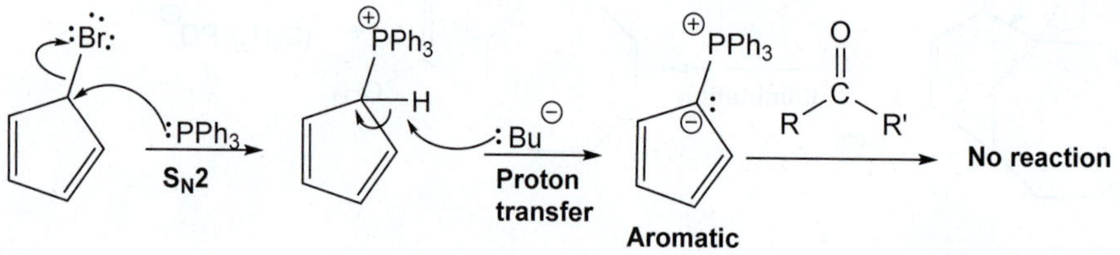

Problem 17.74

Think

What are the electron-rich and electron-poor species in each reaction? Does the reaction mechanism occur via substitution, elimination, or addition? Are proton transfer reactions involved? Be mindful of regiochemistry and stereochemistry, where necessary.

Solve

A is formed from the reaction of PBr_3 with the alcohol (two S_N2 steps), which converts the OH to a Br. **B** is formed from an S_N2 reaction; NC^- is the nucleophile that attacks the electrophilic C–Br. Next, $LiAlH_4$ is a source of hydride, $H:^-$, and undergoes nucleophilic addition to the C of the nitrile. H_2O is a source of protons to form the uncharged amine product, **C**. The final reaction in the first line is an example of the Hofmann elimination, in which the NH_2 group is first converted to a $^+N(CH_3)_3$ leaving group.

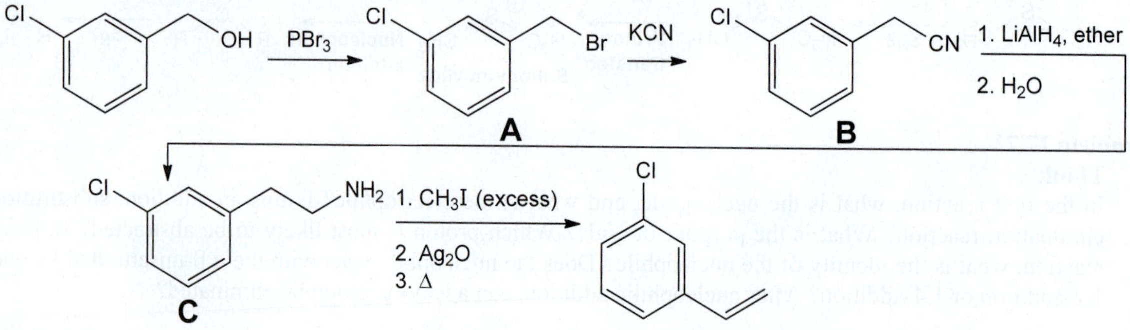

D is formed from a proton transfer reaction to form the strong enolate anion nucleophile that reacts with CH_3CH_2Br in an S_N2 reaction. Next, $NaBH_4$ is a source of hydride that reacts via nucleophilic addition to the aldehyde C. Ethanol is a source of protons to yield **E**, the uncharged alcohol. The same reaction mechanism that formed **D** also forms **F**, but this time, the acidic OH is deprotonated to make a strong alkoxide nucleophile.

1. NaH
2. CH_3CH_2Br

D

$NaBH_4$
CH_3CH_2OH

E

1. NaH
2. CH_3CH_2Br

F

G forms from 1,4-addition of the $CH_2{=}CHCH_2^-$ followed by proton transfer from NH_4Cl, a weak acid. The enol tautomerizes to the ketone. Next, $NaBH_4$ is a source of hydride that reacts via nucleophilic addition to the ketone C. Ethanol is a source of protons to yield **H**, the uncharged alcohol. **I** is formed from the reaction of PBr_3 with the alcohol (two S_N2 steps) to convert OH to Br. The last reaction is an E2 (strong bulky base in the presence of heat). There are two adjacent protons, and thus **J** has two possible isomers. The major product is the first one, because the base is very bulky and will attack the least sterically hindered C–H.

1. CuLi
2. NH_4Cl, H_2O

G

$NaBH_4$
CH_3CH_2OH

H

PBr_3

I

$(CH_3)_3CONa$
Δ

J

Problem 17.75

Think

What are the electron-rich and electron-poor species in each reaction? Does the reaction mechanism occur via substitution, elimination, or addition? Are proton transfer reactions involved? Be mindful of regiochemistry and stereochemistry, where necessary.

Solve

The imine shown was synthesized from the reaction of a nitrile, **K**, with a Grignard reagent, followed by proton transfer using a weak acid. $LiAlH_4$ is a source of hydride, $H:^-$, and undergoes nucleophilic addition to the carbon of the C=N. H_2O is a source of protons to form the uncharged amine product, **L**. Heating **L** with H_3O^+ hydrolyzes the ether to produce a phenol, **M**. NaOH deprotonates the alcohol to form the alkoxide, which then undergoes S_N2 with CH_3CH_2Br to form the ether **N**. See the transformations on the next page.

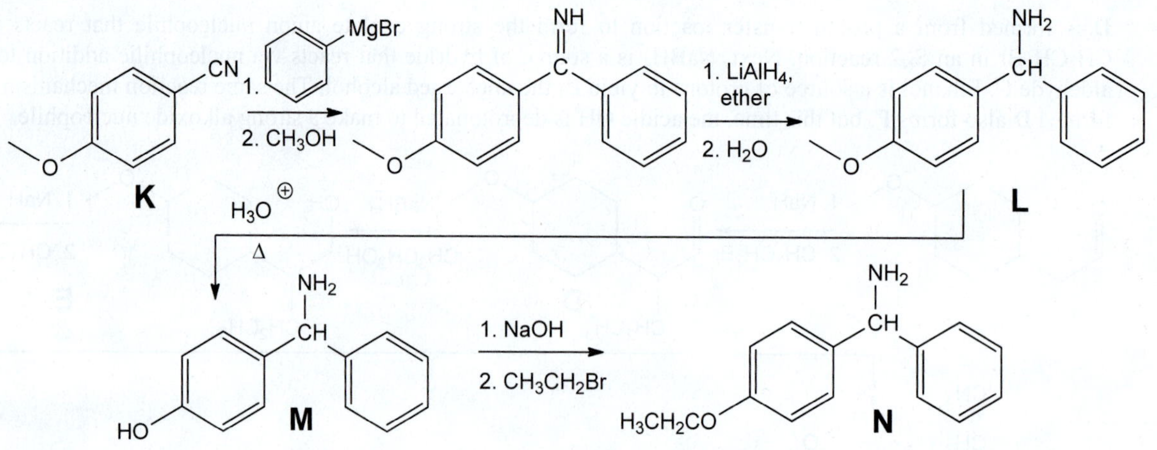

LiAlH$_4$ is a source of hydride, H:$^-$, and undergoes nucleophilic addition to the carbon of the C=O. NH$_4^+$ is a weak acid and protonates the alkoxide to form the uncharged alcohol product, **O**. **P** is formed from the reaction of PBr$_3$ with the alcohol (two S$_N$2 steps), converting OH to Br. **Q** is the phosphonium ylide that forms from the S$_N$2 reaction of the alkybromide and PPh$_3$, followed by deprotonation of the C–H by Bu–Li to form a carbanion. The phosphonium ylide reacts with formaldehyde in a Wittig reaction to form the alkene product, **R**.

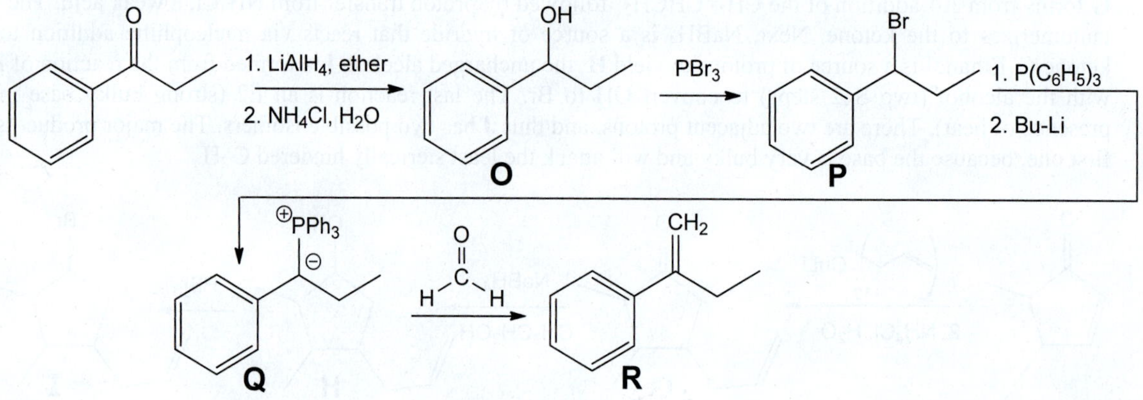

S is formed from the reaction of PBr$_3$ with the alcohol (two S$_N$2 steps), converting OH to Br. **T** is formed from an S$_N$2 reaction; NC$^-$ is the nucleophile that attacks the electrophilic C–Br. Next, the CH$_3^-$ (from the Grignard reagent CH$_3$MgBr) adds via nucleophilic addition to the C of the nitrile. An acid workup follows to form the uncharged imine product, **U**. LiAlH$_4$ is a source of hydride, H:$^-$, and undergoes nucleophilic addition to the carbon of the C=N. H$_2$O is a source of protons to form the uncharged amine product, **V**. The final reaction is an example of the Hofmann elimination, which first converts the NH$_2$ group to a $^+$N(CH$_3$)$_3$ leaving group. Hofmann elimination produces the less alkyl-substituted alkene.

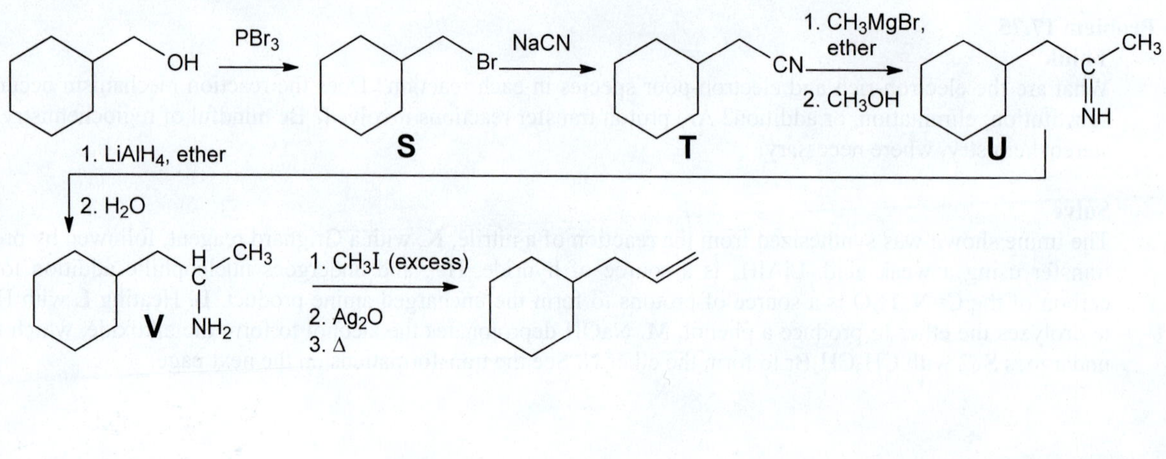

Problem 17.76 (SYN)

Think

What new atoms are present in the products given? How can you react a Grignard reagent to produce a carboxyl group? How can you convert a carboxylic acid into an ester?

Solve

A Grignard addition to CO_2 (followed by the usual acid workup) makes a carboxylic acid. Next, a base can generate the moderately strong carboxylate nucleophile, which can attack CH_3Br in an S_N2 step. Alternatively, diazomethane can be used to make the methyl ester (Chapter 10).

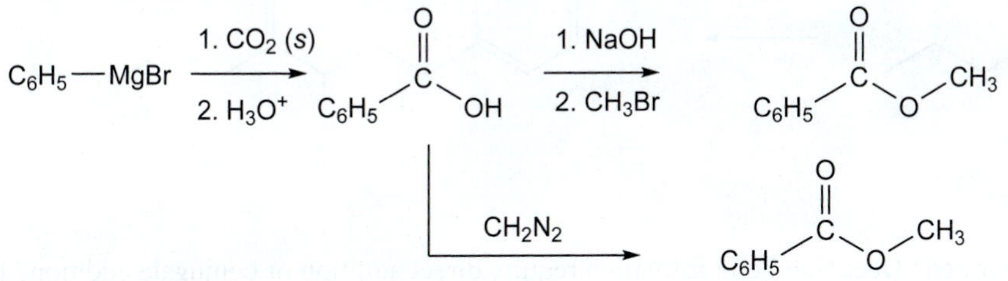

C_6H_5—MgBr → (1. CO_2 (*s*); 2. H_3O^+) $C_6H_5C(=O)OH$ → (1. NaOH; 2. CH_3Br) $C_6H_5C(=O)OCH_3$

$C_6H_5C(=O)OH$ → (CH_2N_2) $C_6H_5C(=O)OCH_3$

Problem 17.77 (SYN)

Think

What new atoms are present in the product given? Which C–C bond needs to be formed? Can the C–OH be produced from a C=O? How can you form a C–C bond at the α carbon of a ketone?

Solve

The five-carbon backbone from the original molecule is shown below. So we need to add a two-carbon piece. Deprotonating the α C atom with a strong base, then performing an alkylation, is sufficient.

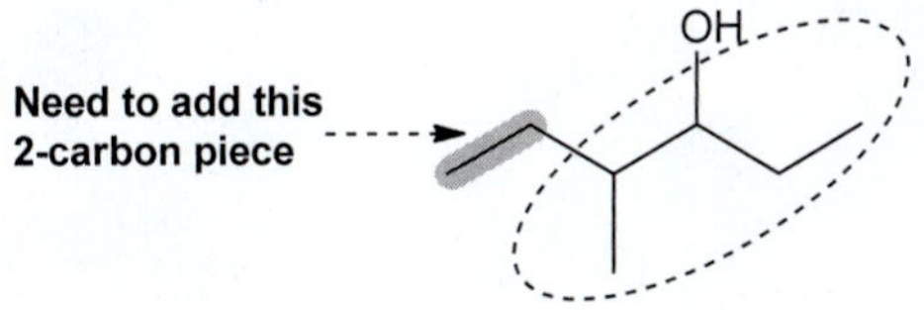

Need to add this 2-carbon piece

OH

We can make the α C nucleophilic by treating the ketone with a strong base like NaH, and we can attach the two-carbon piece through an S_N2 reaction. The C=O bond can be reduced to the alcohol using $NaBH_4$.

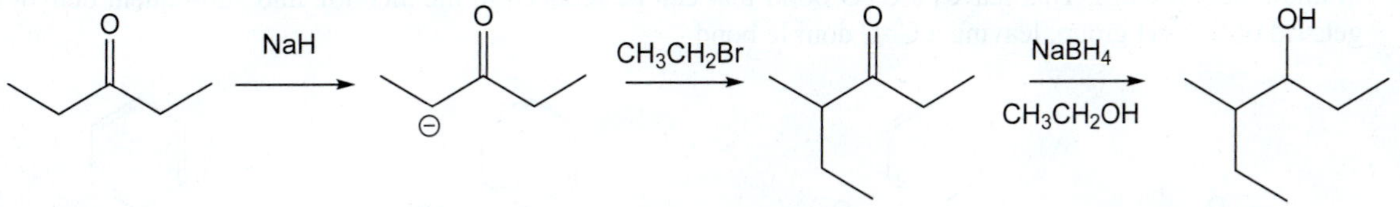

NaH → CH_3CH_2Br → $NaBH_4$, CH_3CH_2OH → OH

Problem 17.78 (SYN)

Think

What new atoms are present in the product given? What C–C bonds need to be formed? Can the C=C be produced from a C=O using a Wittig reaction? How can you add an alkyl group to the β carbon of an α,β-unsaturated ketone? How can you add a methyl group to an α carbon?

Solve

As shown on the next page, this synthesis calls for an α alkylation, a Wittig reaction, and a conjugate addition. The α alkylation can take place first, which would still leave the α,β-unsaturated carbonyl group for subsequent conjugate addition. Then conjugate addition can take place, which still leaves the ketone for a final Wittig reaction.

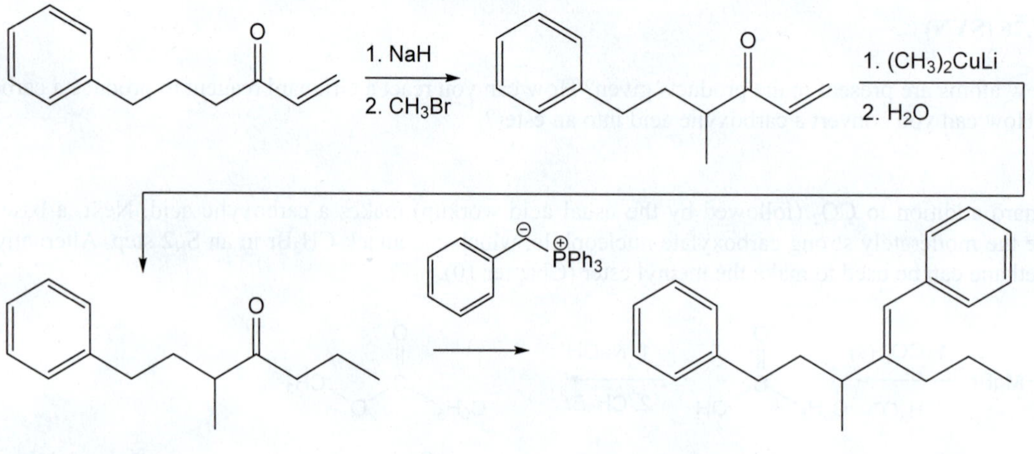

1. NaH
2. CH_3Br

1. $(CH_3)_2CuLi$
2. H_2O

PPh_3

Problem 17.79 (SYN)

Think

What C–C bonds must be formed? Does that bond formation require direct addition or conjugate addition? How can you remove an O atom and leave a C=C bond?

Solve

We can spot the original carbon backbone, which is circled below.

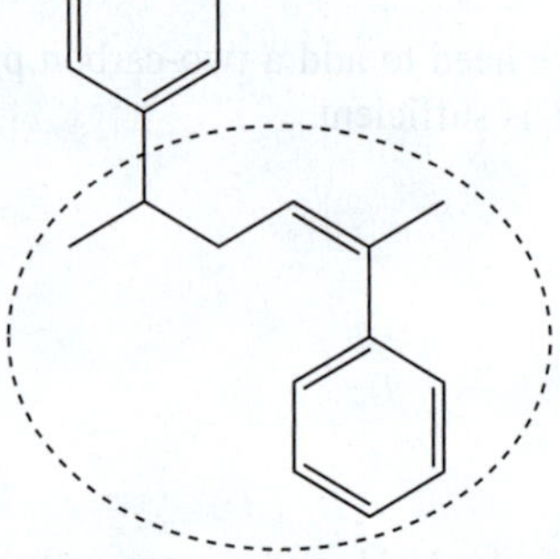

We need to add the benzene ring at what is originally the β C atom, which can be done via conjugate addition—in this case, Ph_2CuLi. That leaves a C=O bond that can be reduced to the alcohol, and subsequent dehydration gets rid of the OH group, leaving a C=C double bond.

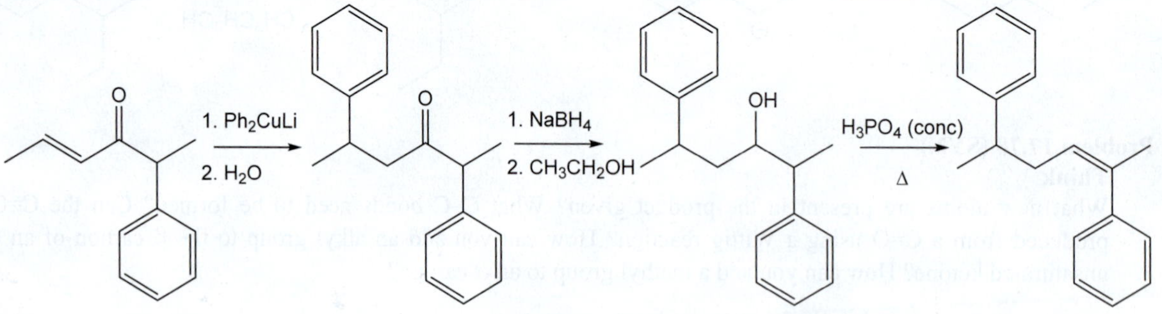

1. Ph_2CuLi
2. H_2O

1. $NaBH_4$
2. CH_3CH_2OH

H_3PO_4 (conc)
Δ

Problem 17.80

Think

What functional group is identified from the broad peak from 3200–3600 cm^{-1}? If the peak at 1650 cm^{-1} is not present in the IR, is the carbonyl group still present in the product? Did the nucleophilic addition occur via 1,2-addition or 1,4-addition?

Solve

(a) The broad peak from 3200–3600 cm^{-1} suggests the presence of an alcohol OH group, and the absence of a peak near 1700 cm^{-1} suggests that the conjugated C=O is no longer present. This leads to the conclusion that the addition reaction occurred via 1,2-addition. 1,4-Addition would have led to an enol that tautomerizes back to the aldehyde, which would have exhibited a C=O peak near 1700 cm^{-1}.

(b) If 1,2-addition occurs, the nucleophile must add irreversibly. If it added reversibly, the 1,4-addition product would have been the major product.

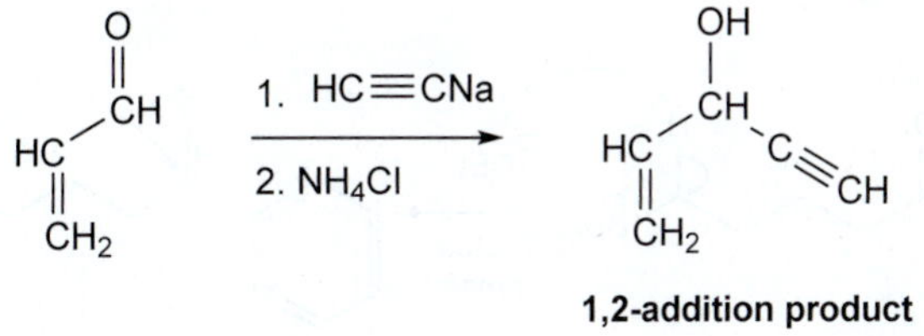

1. HC≡CNa
2. NH_4Cl

1,2-addition product

Problem 17.81

Think

What functional groups are absent if no peaks are present above 3000 cm^{-1} in the IR spectrum and there is no absorption band at 1700 cm^{-1}? How many C atoms are present, based on the number of signals in the ^{13}C NMR spectrum? How many C atoms are present in the starting material? Is there any symmetry in the product? What does this suggest about the structure of the product?

Solve

The absence of a peak above 3000 cm^{-1} in the IR spectrum suggests that there are no H–O bonds and no sp^2 or *sp* C–H atoms. The absence of a peak at 1700 cm^{-1} suggests that there is no carbonyl group. There are five C atoms in the starting material and only three C signals in the ^{13}C NMR of the product. This suggests that there is symmetry in the product, and that is the result of a ring. Therefore, an intramolecular S_N2 reaction occurred after the nucleophilic addition reaction.

Nucleophilic addition

S_N2

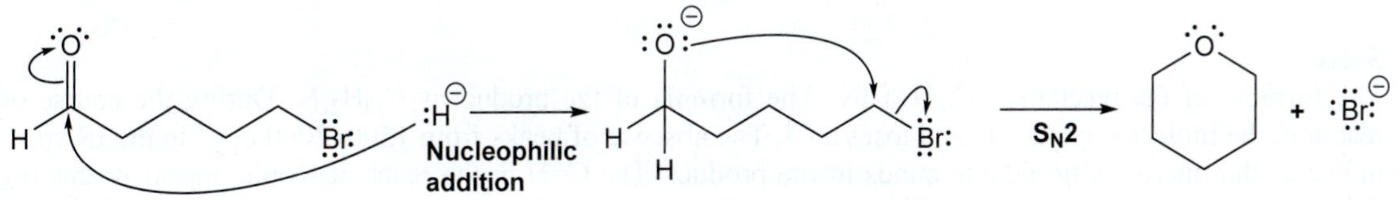

Problem 17.82

Think

Based on the NMR data, how many different CH_3 groups are present? What type of functional group can give rise to the proton signal at 6.1 ppm? Does the nucleophilic addition reaction likely occur via 1,2-addition or 1,4-addition? From the IR spectrum, what causes the strong signals at 1650 and 1600 cm^{-1}?

Solve

Three CH_3 groups are present, evidenced by the three NMR signals that integrate to 3H, and the signal at 6.1 ppm is the sp^2 C–H of a C=C group that is also adjacent to the C=O. The addition reaction occurs via 1,4-addition. The enol tautomerizes back to the ketone.

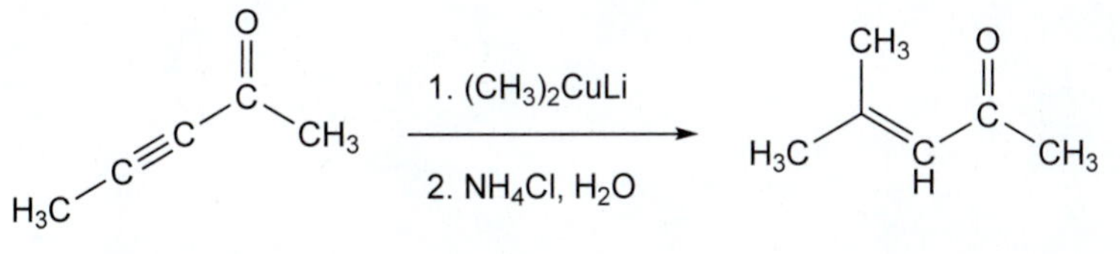

1. $(CH_3)_2CuLi$
2. NH_4Cl, H_2O

Problem 17.83

Think

What is the mechanism by which a Wittig reagent is formed? Can a similar reaction occur here? What is the identity of the nucleophilic C atom in the Wittig reagent? Where is the electrophilic C atom? Are these atoms on the same compound?

Solve

Once the Wittig reagent is produced, an internal Wittig reaction takes place, because the nucleophile and electrophile are on the same compound.

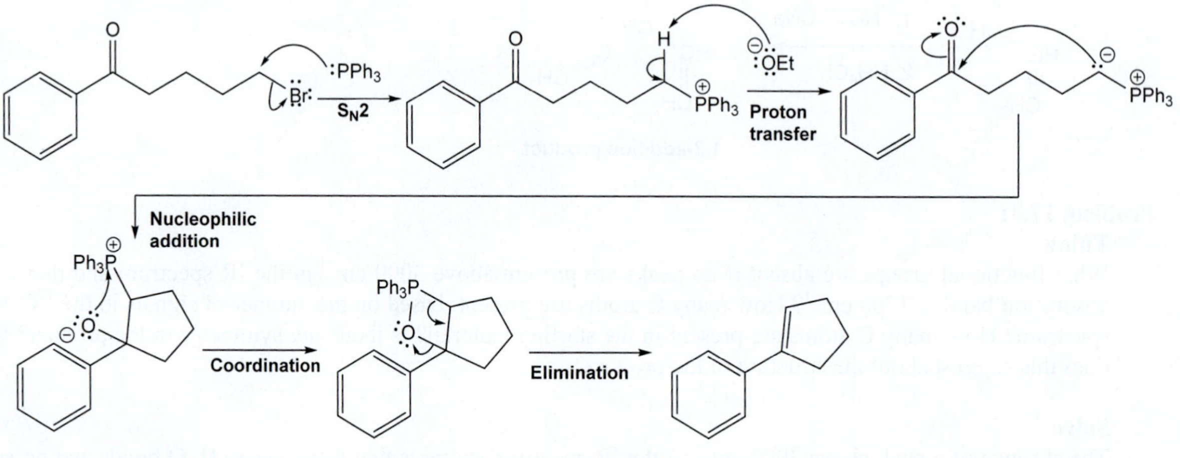

:PPh3
SN2
:OEt
PPh3
Proton transfer
PPh3
Nucleophilic addition
Ph3P
Ph3P
Coordination
Elimination

Problem 17.84

Think

What is the formula of the reactant? How does the formula change in the reaction? What does the absence of peaks from 1500–2000 cm^{-1} in the IR spectrum indicate? Do any peaks appear in this region from the reactant? What must then occur at this bond? What is the nucleophile in the reaction?

Solve

The formula of the reactant is $C_{10}H_{20}ClN$. The formula of the product is $C_{10}H_{21}N$. During the course of the reaction, the molecule gains a H and loses a Cl. The absence of peaks from 1500–2000 cm^{-1} in the IR spectrum indicates that there are no double bonds in the product. The C=N in the reactant would appear in this region; therefore, this bond undergoes a nucleophilic addition reaction. H^- adds to the imine carbon in a nucleophilic addition elementary step. The N^- is nucleophilic and attacks the electrophilic C attached to the Cl in an S_N2 reaction to form a ring. The product and mechanism are shown below.

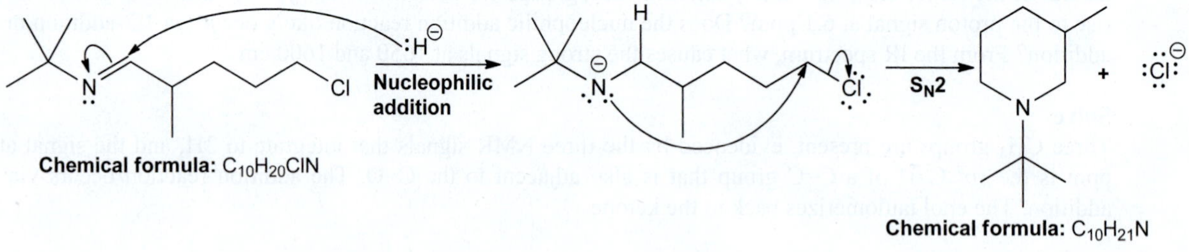

:H
Nucleophilic addition
SN2
Chemical formula: $C_{10}H_{20}ClN$
Chemical formula: $C_{10}H_{21}N$

CHAPTER 18 | Nucleophilic Addition to Polar π Bonds 2: Weak Nucleophiles and Acid and Base Catalysis

Your Turn Exercises

Your Turn 18.1

Think

What is the structure of butan-2-one? What is the general form of a hemiacetal? How many R groups are necessary, and what should they be?

Solve

The structure for butan-2-one is shown below. A hemiacetal is an organic compound in which the carbon is bound to both an OH and an OR group. Therefore only one OR group is necessary. This is different from an acetal, in which the carbon is bound to two OR groups. The hemiacetal in which the OR group is a butoxy ($CH_3CH_2CH_2CH_2O$) is drawn below.

O

HO OCH$_2$CH$_2$CH$_2$CH$_3$

Butan-2-one **Hemiacetal of butan-2-one, OR = OCH$_2$CH$_2$CH$_2$CH$_3$**

Your Turn 18.2

Think

Count the *number* of formal charges present in each step for all reagents involved in the reaction.

Solve

In Equation 18-3, the first stage has no charges and the second stage (after the first reaction step) has two charges. Stage 3 has two charges, too. Stage 4 also has two charges, even though only one is drawn in the mechanism. That is because the alkoxide anion produced in Step 3 of the mechanism would still remain. In Equation 18-4, each of the stages has one charge. In Equation 18-5, each of the stages has one charge.

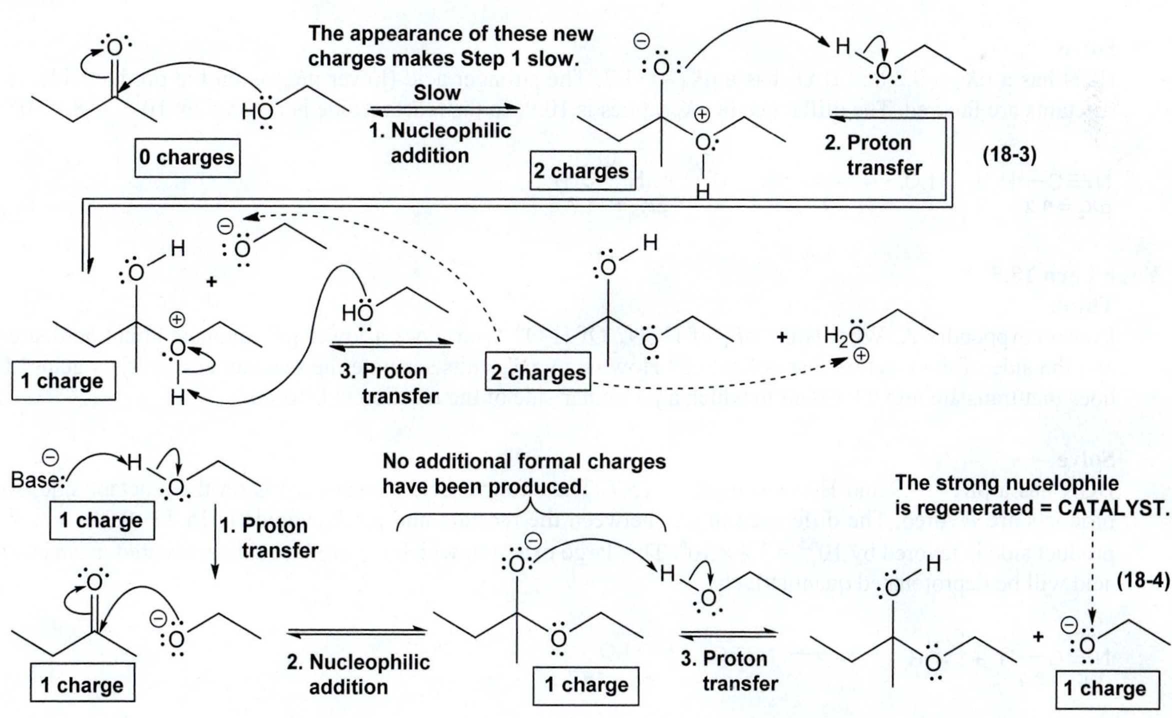

Strong acid

1 charge

1. Proton transfer

No additional formal charges have been produced.

The strong acid is regenerated = CATALYST.

1 charge

2. Nucleophilc addition

1 charge

3. Proton transfer

1 charge

(18-5)

Your Turn 18.3

Think

What type of species (electron rich or electron poor) does a nucleophile attack? In which species is that type of charge greater on the C≡N carbon? How many pK_a units separate the reactant and product acids? How does that translate into the extent that a particular side of the reaction is favored?

Solve

The positively charged species on the right is activated toward nucleophilic attack, because nucleophiles attack electron-poor species. It has a resonance structure that puts a positive charge on the carbon, making the C more susceptible to nucleophilic attack.

Your Turn 18.4

Think

Consult Appendix A. What is the pK_a of HCN? Of H_3O^+? What does a lower pK_a indicate about acid strength and the side of the reaction that is favored?

Solve

HCN has a pK_a = 9.2 and H_3O^+ has a pK_a = −1.7. The stronger acid (lower pK_a) is on the product side, so the reactants are favored. The difference in pK_a values is 10.9, so the reactant side is favored by $10^{10.9} = 8 \times 10^{10}$.

$$N{\equiv}C{-}H + H_2\ddot{O}: \rightleftharpoons N{\equiv}C:^{\ominus} + H{-}\overset{\oplus}{O}H_2$$

pK_a = 9.2 pK_a = −1.7

Your Turn 18.5

Think

Consult Appendix A. What is the pK_a of HCN? Of H_2O? What does a lower pK_a indicate about acid strength and the side of the reaction that is favored? How many pK_a units separate the reactant and product acids? How does that translate into the extent to which a particular side of the reaction is favored?

Solve

HCN has a pK_a = 9.2 and H_2O has a pK_a = 15.7. The stronger acid (lower pK_a) is on the reactant side, so the products are favored. The difference in pK_a between the reactant and product acids is 15.7 − 9.2 = 6.5, so the product side is favored by $10^{6.5} = 3.2 \times 10^6$. This large extent to which the product side is favored means that the acid will be deprotonated quantitatively.

$$N{\equiv}C{-}H + :\ddot{O}H^{\ominus} \rightleftharpoons N{\equiv}C:^{\ominus} + H_2O$$

pK_a = 9.2 pK_a = 15.7

Your Turn 18.6

Think

What is the structure of pentanal? What is the general form of an acetal? How many R groups are necessary, and what should they be?

Solve

The structure of pentanal is shown below. An acetal is an organic compound in which one carbon is bound to two OR groups. The acetal of pentanal in which the OR groups are propoxy ($CH_3CH_2CH_2O$) is drawn below.

O

$H_3CH_2CH_2CO$ $OCH_2CH_2CH_3$

Pentanal

Acetal of pentanal, OR = $OCH_2CH_2CH_3$

Your Turn 18.7

Think

Consult Table 1-6 to identify what defines a hemiacetal and acetal. Locate these two functional groups in the mechanism in Equation 18-13. What elementary steps make up an S_N1 mechanism?

Solve

A hemiacetal is characterized by an $R_2C(OH)(OR)$ or $RCH(OH)(OR)$ grouping. An acetal is characterized by an $R_2C(OR)_2$ or $RCH(OR)_2$ grouping. The hemiacetal forms after Step 3, and the acetal forms after Step 7 (dashed ovals). The two steps that make up the S_N1 mechanism are Step 5 (heterolysis) and Step 6 (coordination) and are noted by the dashed box.

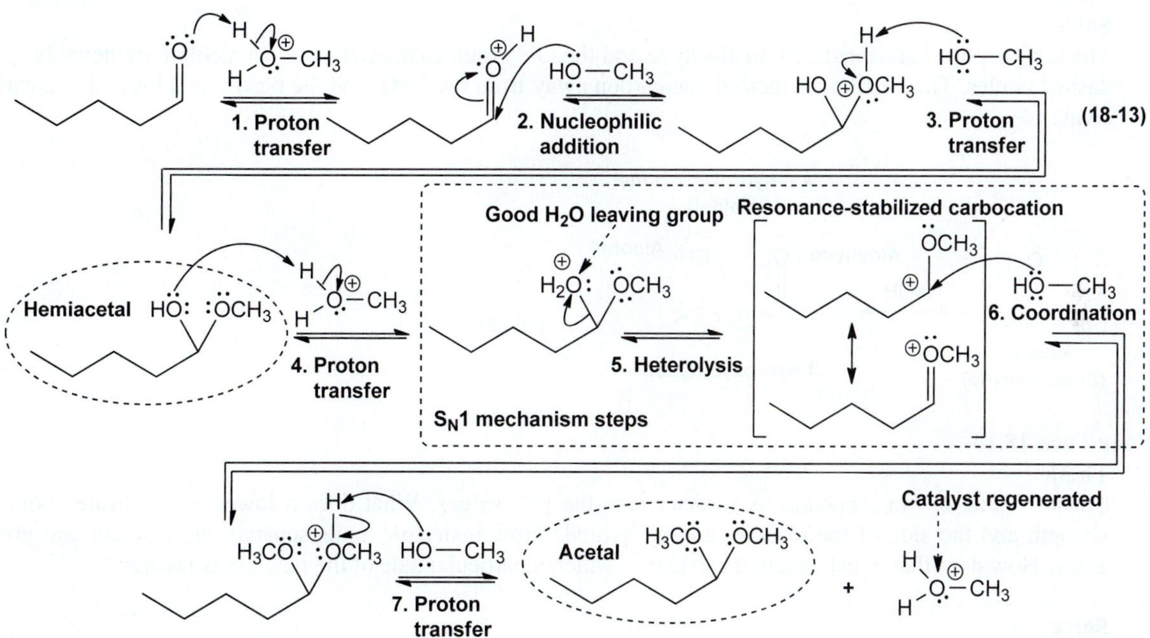

Your Turn 18.8

Think

Consult Section 17.3b to review how to calculate oxidation states. Compare the electronegativity of C versus H and C versus O. Which is more electronegative? How are covalently bonded electrons assigned when the atoms bonded together have different electronegativities? The same electronegativities?

Solve

In a given covalent bond, all electrons are assigned to the more electronegative atom. If the atoms are identical, the electrons are split up evenly. Hydrogen is *less* electronegative compared to C, and O is *more* electronegative compared to C. In cyclopropylethanone, the C of the C=O group is assigned one electron from each C–C bond and no electrons from the C=O bond, for a total of two valence electrons. That C has two less than an uncharged isolated C, giving it an oxidation state of +2. In ethylcyclopropane, the C is assigned two electrons from each C–H bond and one electron from each C–C bond, giving it a total of six valence electrons. This is two more valence electrons than in an uncharged isolated C, giving that C atom an oxidation state of −2. The carbonyl C, therefore, gains four electrons, signifying a reduction.

Wolff–Kishner reduction

Oxidation state = +2

Cyclopropylethanone

1. $H_2N–NH_2$, acetic acid, H_2O/triethylene glycol

2. NaOH, 180–200°C

The carbonyl group is reduced to a CH_2 group.

Oxidation state = −2

72%

Ethylcyclopropane

+ $N_2(g)$ + H_2O **(18-27)**

Your Turn 18.9

Think

Review the functional groups in Table 1-6, if necessary. How many carbons away from the C=O is the α carbon located? The β carbon?

Solve

The C=O group characteristic of an aldehyde and the OH group characteristic of an alcohol are noted below by dashed circles. The α carbon is located one carbon away from the C=O, and the β carbon is located two carbons away.

A β-hydroxycarbonyl (An aldol)

Ethanal (Acetaldehyde)

NaOH

Aldehyde

Alcohol

α β

3-Hydroxybutanal

(18-30)

Your Turn 18.10

Think

Consult Table 6-1 or Appendix A to determine the pK_a values. What does a lower pK_a indicate about acid strength and the side of the reaction that is favored? How many pK_a units separate the reactant and product acids? How does that translate into the extent to which a particular side of the reaction is favored?

Solve

$CH_3CH{=}O$ has a pK_a = 20 and H_2O has a pK_a = 15.7. The stronger acid (lower pK_a) is on the product side, so the reactants are favored. The difference in pK_a between the reactant and product acids is 20 − 15.7 = 4.3, so the reactant side is favored by $10^{4.3} = 2 \times 10^4$.

pK_a = 20

+ H_2O

pK_a = 15.7

Reactant side favored by 2×10^4

Your Turn 18.11

Think

What is a condensation reaction? What is the difference in the atoms that make up the reactants and the organic product? What molecule has that same formula?

Solve

A condensation reaction is one in which two larger molecules bond together (two molecules of propanal) with the elimination of a smaller molecule (in this case, H_2O). The numbers of C, H, and O atoms in the reactant and products are listed in the table below. The difference is two H and one O = H_2O.

Reactant	Product	Difference
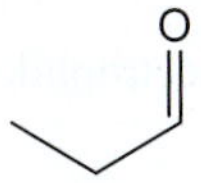 Propanal (2 equiv.)	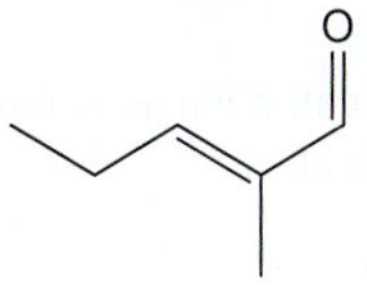 2-Methylpent-2-enal	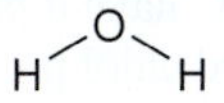 Water
C = 6	C = 6	C: 6 − 6 = 0
H = 12	H = 10	H: 12 − 10 = 2
O = 2	O = 1	O: 2 − 1 = 1

Your Turn 18.12

Think

How many carbons away from the carbonyl C is the α carbon? The β carbon? What functional group is attached to each of those carbons?

Solve

The α and β carbons in the product 4-hydroxy-4-methylpentan-2-one are labeled below. The OH group is attached to the β carbon. Therefore, the product is a β-hydroxycarbonyl. Every aldol addition produces a β-hydroxycarbonyl, and the condensation of two ketones is no exception.

A β-hydroxycarbonyl

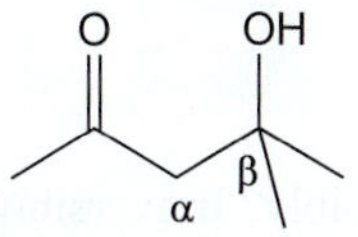

O OH

β

α

4-Hydroxy-4-methylpentan-2-one

Your Turn 18.13

Think

Label the α carbons on both butanal and the three compounds listed. Do those C atoms have attached protons? If a compound has no α protons, is this advantageous when performing a crossed aldol reaction?

Solve

Only compound **A** has no α protons. This is helpful in performing a crossed aldol reaction, because the compound without α protons cannot form an enolate ion.

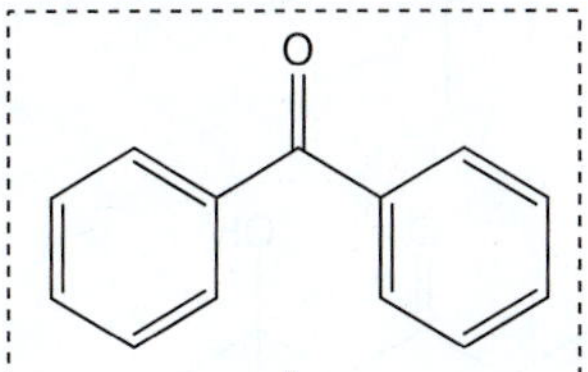

O O O

A 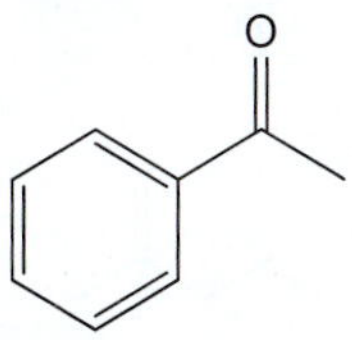 B 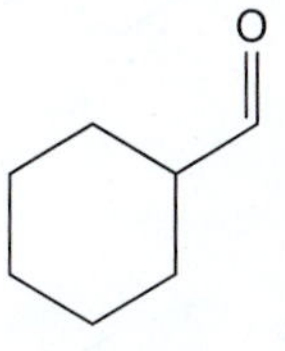 C

Therefore, this compound can act as an electrophile only, producing a single crossed aldol product with butanal, as shown below:

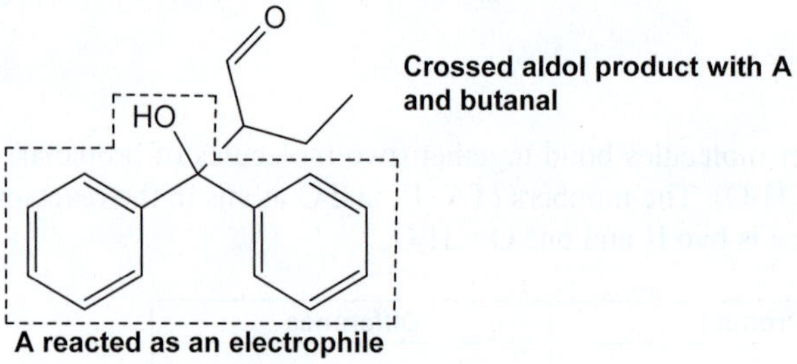

Crossed aldol product with A and butanal

A reacted as an electrophile

Compounds **B** and **C** have α protons, so they can behave as nucleophiles or electrophiles, and can each produce two different crossed aldol products with butanal.

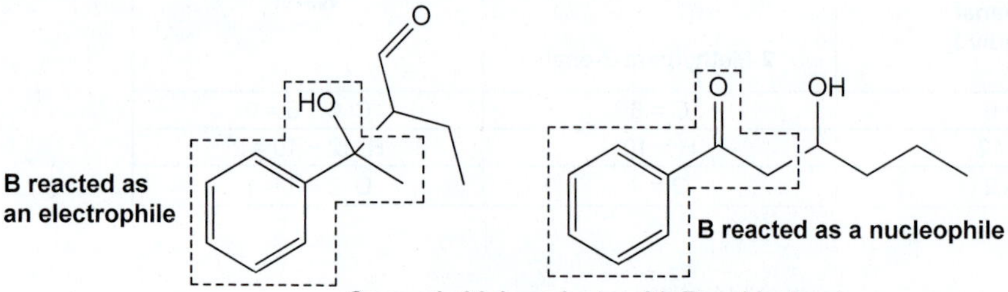

B reacted as an electrophile

B reacted as a nucleophile

Crossed aldol products with B and butanal

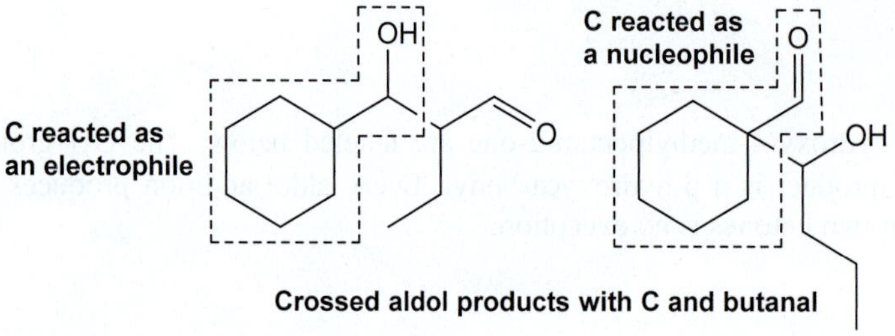

C reacted as an electrophile

C reacted as a nucleophile

Crossed aldol products with C and butanal

Your Turn 18.14

Think

Review Section 10.3. Which base (NaOH or LDA) deprotonates an α carbon reversibly? Irreversibly? Which of those bases favors deprotonating the more highly substituted α carbon? Is this the kinetic or thermodynamic enolate anion?

Solve

NaOH deprotonates an α carbon reversibly to favor the thermodynamic enolate anion, which is formed by deprotonating the more highly substituted α carbon. LDA (a very strong base) deprotonates an α carbon irreversibly to favor the kinetic enolate anion, which is formed by deprotonating the less highly substituted α carbon.

NaOH

LDA

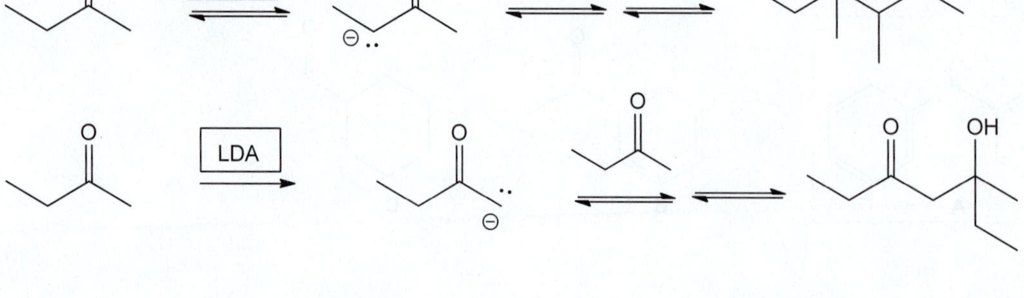

Your Turn 18.15

Think

What functional groups characterize an aldol condensation reaction product? What should the relative locations be for those functional groups? Review Section 18.5.

Solve

The product of an aldol condensation is an α,β-unsaturated carbonyl. This is what we see in the product, where the carbonyl group is conjugated to a C=C.

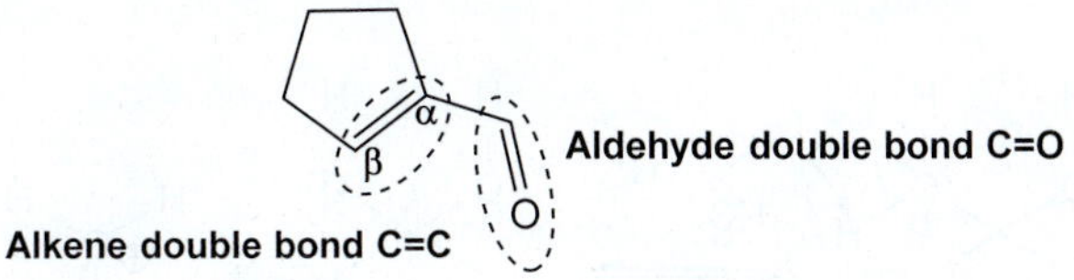

α
β
O
Aldehyde double bond C=O
Alkene double bond C=C

In Chapter Problems

Problem 18.2

Think

What nucleophiles are present under acidic conditions? What is the total number of charges before and after the nucleophilic addition step in the mechanism?

Solve

Under acidic conditions, water acts as the nucleophile *after* the ketone O atom is protonated to increase the electrophilic nature of the carbonyl carbon.

Proton transfer

Nucleophilic addition

Proton transfer

Under neutral conditions, the nucleophilic addition step increases the total number of charges by two, whereas the total number of charges remains the same under acidic conditions; before and after nucleophilic addition, there is a single positive charge. Thus, nucleophilic addition is faster under acidic conditions.

Problem 18.3

Think

What nucleophiles are present under basic conditions? Under acidic conditions? In what order do the nucleophilic addition and proton transfer reactions need to occur to avoid incompatible species in solution? How are the acid and base catalysts regenerated?

Solve

(a) In basic conditions, the HO^- deprotonates the phenol O–H to form the phenoxide anion, which acts as the nucleophile. In basic conditions, the nucleophilic addition to the C=O occurs before the proton transfer reaction to the carbonyl O atom. This order of steps avoids the appearance of strong acids, which are incompatible with basic conditions.

Proton transfer

Nucleophilic addition

Proton transfer

(b) In acidic conditions, phenol acts as the nucleophile *after* the ketone O atom is protonated to increase the electrophilic nature of the carbonyl C atom. This order of steps avoids the appearance of strong bases, which are incompatible with acidic conditions.

Proton transfer

Nucleophilic addition

Proton transfer

Problem 18.4

Think

What nucleophiles are present under these conditions? Are strong acids or bases present? Does the mechanism follow acidic or basic mechanistic conditions? What product is formed?

Solve

In both reaction conditions, the cyanohydrin product is formed.

(a) In these conditions, NC^- is the strongest nucleophile present. These conditions follow the same mechanism as in Equation 18-7 (nucleophilic addition to the C=O followed by proton transfer).

Nucleophilic addition

Proton transfer

A cyanohydrin

(b) In these conditions, HO^- deprotonates HCN quantitatively to produce NC^- as the major nucleophile. Subsequently, the mechanism under basic conditions follows the same format as in Equation 18-7 (nucleophilic addition to the C=O followed by proton transfer).

Proton transfer

Nucleophilic addition

Proton transfer

A cyanohydrin

Problem 18.6

Think

Are the nucleophiles charged or uncharged? Does the nucleophile add to the carbonyl group reversibly or irreversibly (review Table 17-2)? Will this result in direct addition or conjugate addition?

Solve

(a) Even though NC^- is a charged nucleophile, it still adds reversibly to a carbonyl group, so it will favor conjugate addition upon attacking an α,β-unsaturated carbonyl. The HO^- below the reaction arrow indicates basic conditions, so no strongly acidic species should appear in the mechanism. Therefore, we do not show the HCN protonating the carbonyl O atom. Nucleophilic addition occurs first. The enol that is produced then undergoes tautomerization (back-to-back proton transfer steps) to convert to its keto form.

Proton transfer

Nucleophilic addition

Proton transfer

Tautomerization

(b) $PhCH_2SH$, an uncharged nucleophile, adds reversibly to a carbonyl group, so it will favor conjugate addition upon attacking an α,β-unsaturated carbonyl. The reaction is not catalyzed by acidic or basic conditions. The enol that is produced then undergoes tautomerization (back-to-back proton transfer steps) to convert to its keto form.

Nucleophilic addition

Proton transfer

Proton transfer

Tautomerization

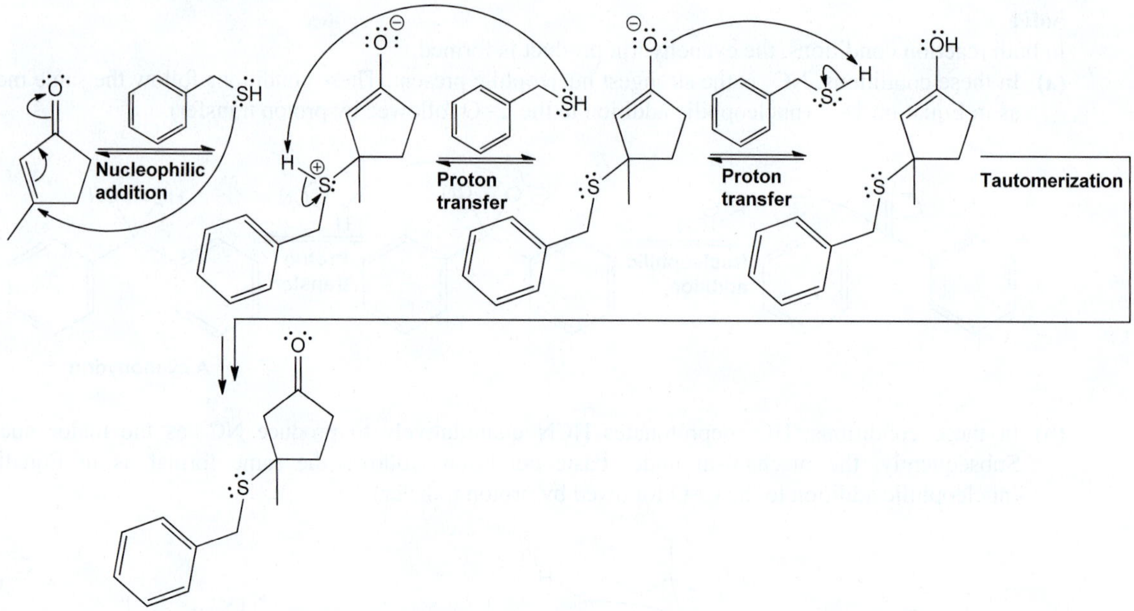

(c) Pyrrolidine, an uncharged nucleophile, adds reversibly to a carbonyl group, so it will favor conjugate addition upon attacking an α,β-unsaturated carbonyl. The reaction is not catalyzed by acidic or basic conditions. The enol that is produced then undergoes tautomerization (back-to-back proton transfer steps) to convert to its keto form.

Nucleophilic addition

Proton transfer

Proton transfer

Tautomerization

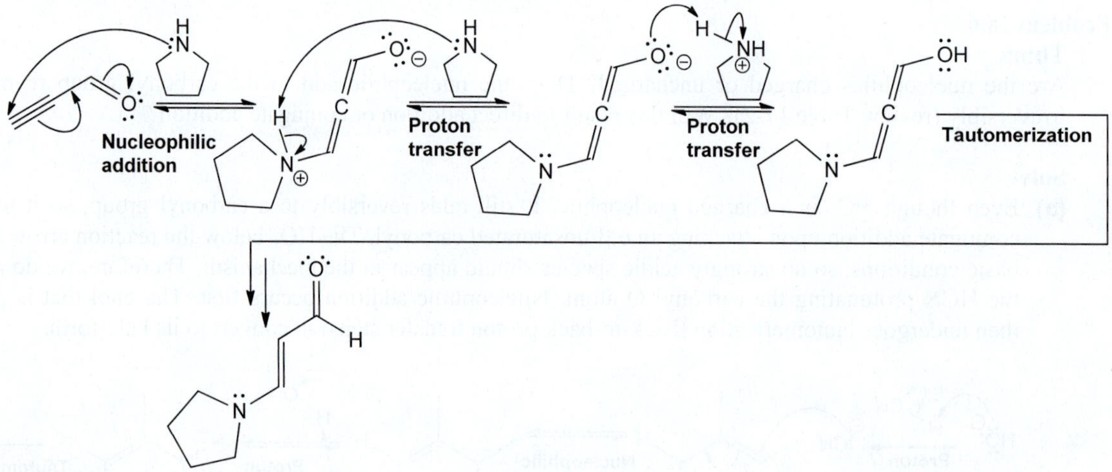

Problem 18.7

Think

This reaction is the reverse of what reaction? Are good leaving groups present? If not, how can a good leaving group be formed? What can act as a nucleophile? In how many steps does this mechanism occur if it is the reverse of the mechanism in Equation 18-13?

Solve

The mechanism for this reaction is the reverse of the mechanism in Equation 18-13. An alkoxide is not a good leaving group. However, in the presence of a strong acid, H_3O^+, the O is protonated and, thus, forms ROH, which is a good leaving group. The mechanism is shown below.

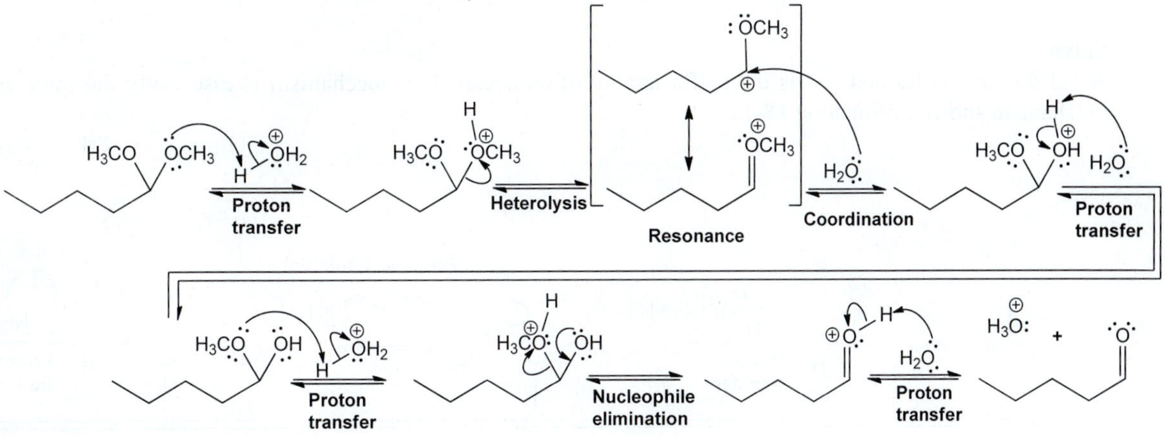

Proton transfer | Heterolysis | Resonance | Coordination | Proton transfer | Proton transfer | Nucleophile elimination | Proton transfer

Problem 18.8

Think

What is the name of the functional group that was produced in this problem? How is this mechanism similar to the mechanism shown in Equation 18-13? How is a good leaving group produced? What acts as the nucleophile?

Solve

An acetal is formed; this mechanism is essentially the same as the mechanism shown in Equation 18-13. The only difference is that the coordination in Step 6 is intramolecular rather than intermolecular.

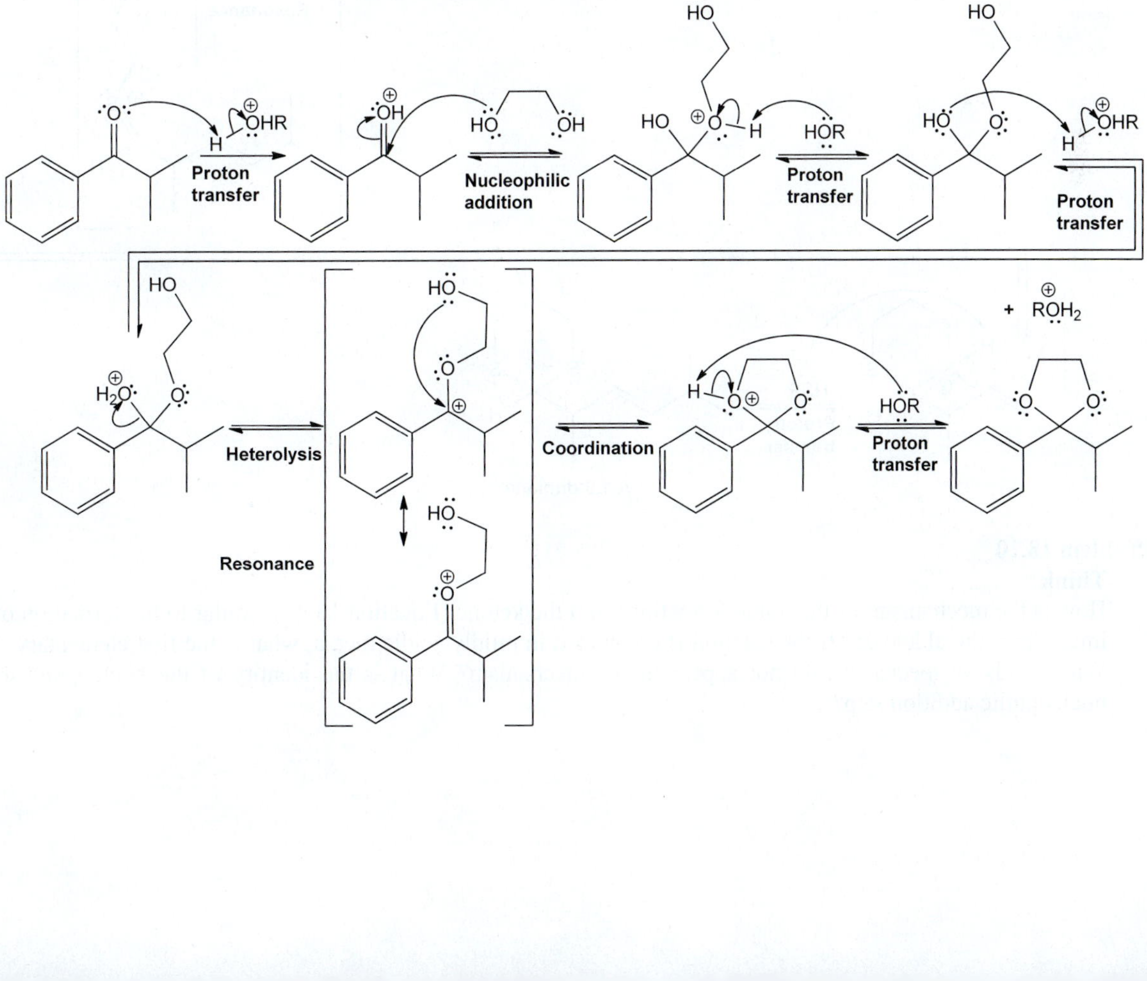

Proton transfer | Nucleophilic addition | Proton transfer | Proton transfer | Heterolysis | Resonance | Coordination | Proton transfer

Problem 18.9

Think

What is the name of the functional group that is produced in this problem? How is this functional group similar to an acetal? How is this mechanism similar to the mechanism shown in Equation 18-13? How is a good leaving group produced under acidic conditions? What is the nucleophile?

Solve

A 1,3-dithiane is formed and is the sulfur analog of an acetal. This mechanism is essentially the same as the mechanism shown in Equation 18-13.

Propane-1,3-dithiol

Proton transfer

Proton transfer

Proton transfer

Heterolysis

Coordination

Resonance

Proton transfer

A 1,3-dithiane

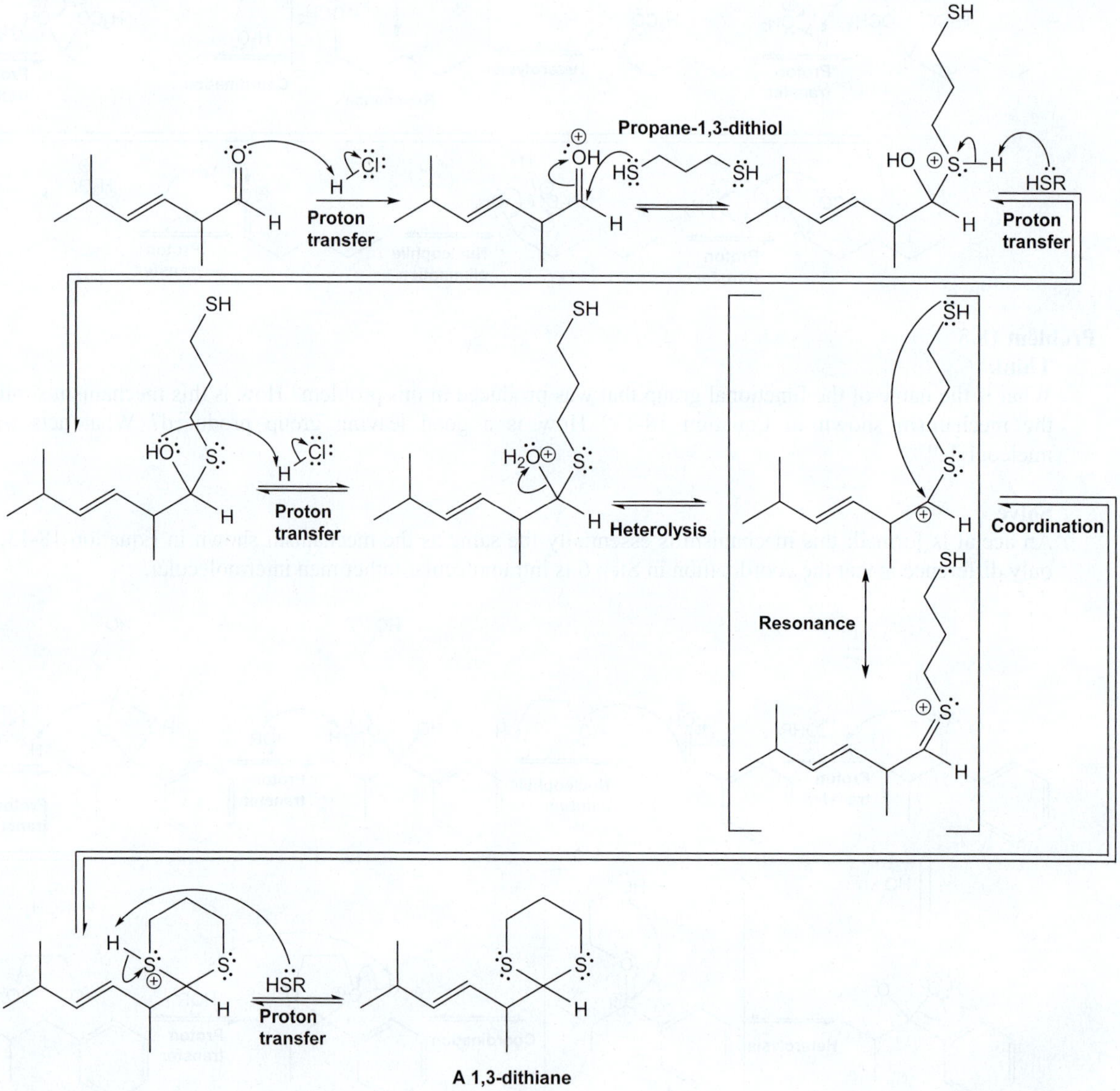

Problem 18.10

Think

How is the mechanism for the imine formation from the ketone (Equation 18-19) similar to the formation of the imine from the aldehyde? If the reaction is conducted in mildly acidic media, what is the first elementary step? What kinds of species should not appear in the mechanism? What is the identity of the nucleophile in the nucleophilic addition step?

Solve

The mechanism for the formation of the imine from the aldehyde is the same mechanism shown in Equation 18-19, the imine formation from the ketone. The mechanism is shown below. Under mildly acidic conditions, the nucleophile is the uncharged amine, and the C=O oxygen should not be protonated prior to nucleophilic attack. The conditions are acidic enough, however, to convert a small amount of the poor HO^- leaving group to a good H_2O leaving group.

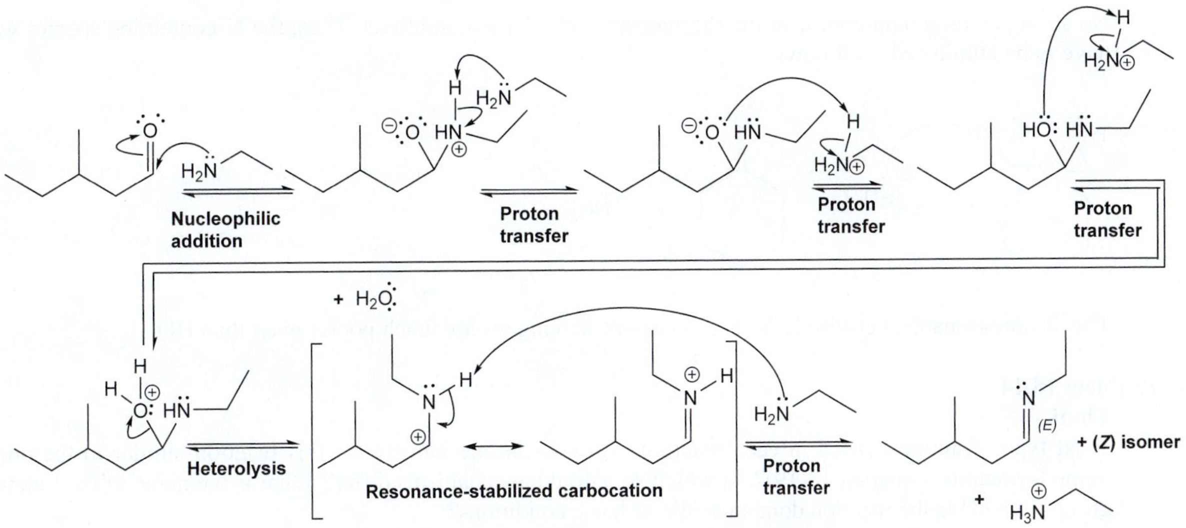

Nucleophilic addition · Proton transfer · Proton transfer · Proton transfer

Heterolysis · Resonance-stabilized carbocation · Proton transfer · (E) + (Z) isomer

Problem 18.11

Think

This reaction is the reverse of what reaction? In how many steps does this mechanism occur if it is the reverse of the mechanism in Equation 18-19? If the reaction is catalyzed by an acid, what is the first elementary step? What kinds of species should not appear in the mechanism?

Solve

This mechanism is the reverse of the mechanism shown in Equation 18-19. A proton transfer reaction is first, followed by nucleophilic addition of water. The mechanism is shown below. No strong bases should appear in this mechanism, because it takes place under acidic conditions.

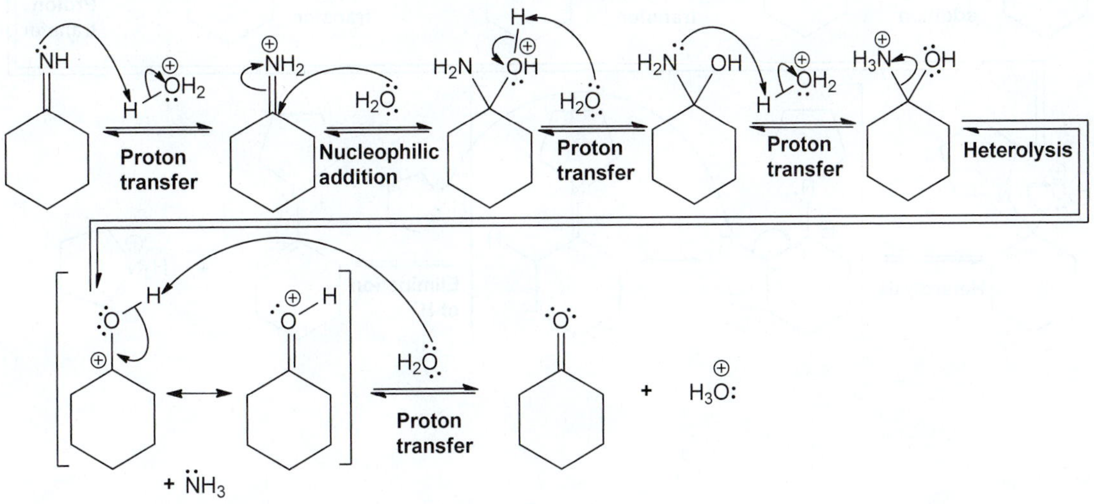

Proton transfer · Nucleophilic addition · Proton transfer · Proton transfer · Heterolysis

Proton transfer · + H_3O: · + NH_3

Problem 18.13

Think

Under basic conditions, what types of species should *not* appear in the mechanism? Are all the steps that must happen for imine formation under these conditions reasonable? What species would have to act as the leaving group? Is this reasonable?

Solve

No strong acids should appear in the mechanism under basic conditions. Thus, the N-containing species would have to be eliminated, as follows:

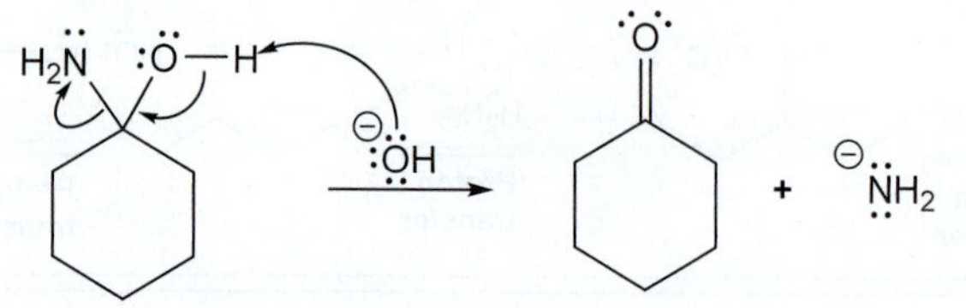

This is unreasonable, because H_2N^- is a very poor leaving group, much poorer even than HO^-.

Problem 18.14

Think

What type of amine is used in each reaction? How is the mechanism for this reaction similar to the one for imine formation (Equation 18-19)? At which step do the mechanisms differ? What is the name of the functional group formed? Is the reaction done in acidic or basic conditions?

Solve

In each case, the amine is a secondary amine. The mechanism is essentially the same as the mechanism in Equation 18-19 up through Step 5. After Step 5, the product has no more H atoms on the N atom. Thus, the final deprotonation occurs at the adjacent C–H to form the enamine product.

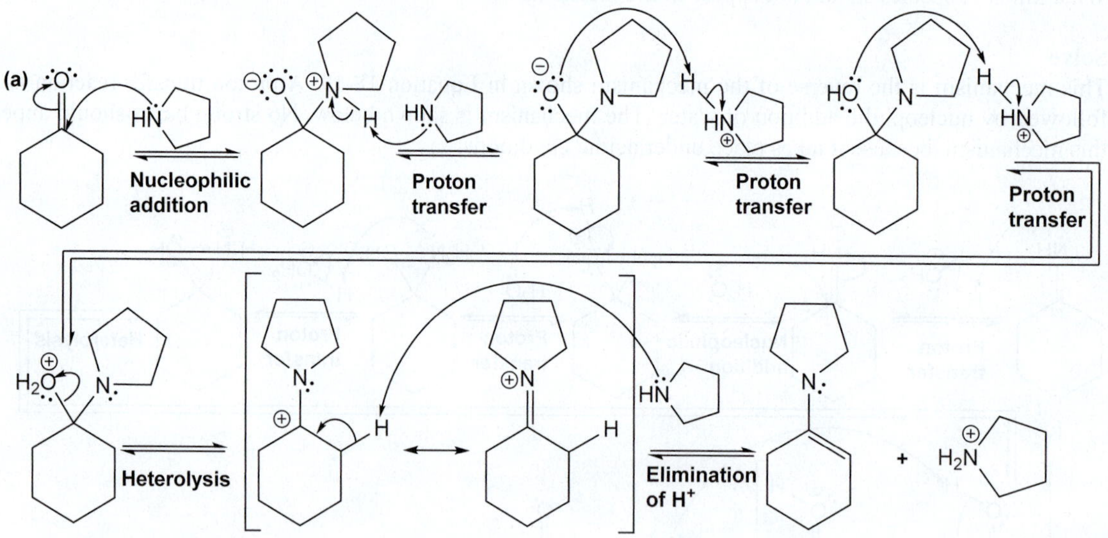

(a)

Nucleophilic addition

Proton transfer

Proton transfer

Proton transfer

Heterolysis

Elimination of H^+

(b)

Nucleophilic addition

Proton transfer

Proton transfer

Proton transfer

Heterolysis

Elimination of H^+

(c)

Nucleophilic addition

Proton transfer

Proton transfer

Proton transfer

Heterolysis

Resonance

Elimination of H^+

Problem 18.16

Think

Which of the steps in Equations 18-19 or 18-22 can the reaction include? Are there any steps that are not possible?

Solve

To form an imine, two H atoms must be attached to N on the initial amine. To form an enamine, a secondary amine must react with a ketone or aldehyde that has at least one α proton. None of these scenarios exist with the reactants given. Therefore, as shown below, the mechanism runs into a dead end because there is no appropriate proton that can be removed to form a stable product.

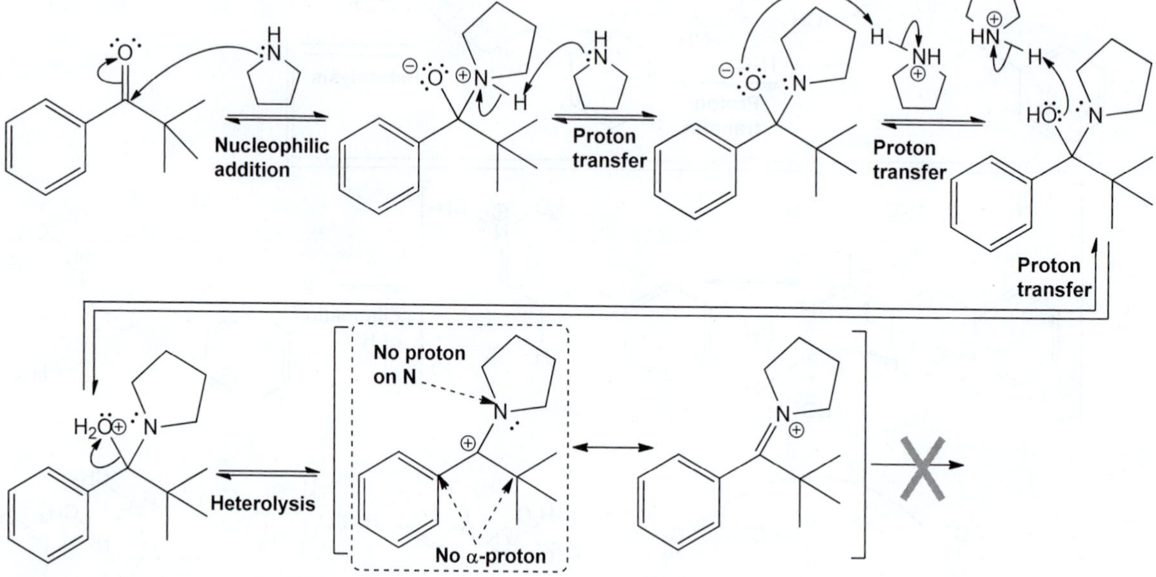

Nucleophilic addition

Proton transfer

Proton transfer

Proton transfer

No proton on N

Heterolysis

No α-proton

Problem 18.17

Think

Does the reaction take place under acidic or basic conditions? In each medium, what species should not appear? Which step is the only step in the mechanism that does not consist of a proton transfer step? In how many steps does the mechanism proceed? What is the functional group produced?

Solve

Both reactions are hydrolysis of the nitrile to produce a primary amide.

(a) Under basic conditions, no strong acids should appear. The first step (nucleophilic addition) is the only step that is not a proton transfer reaction.

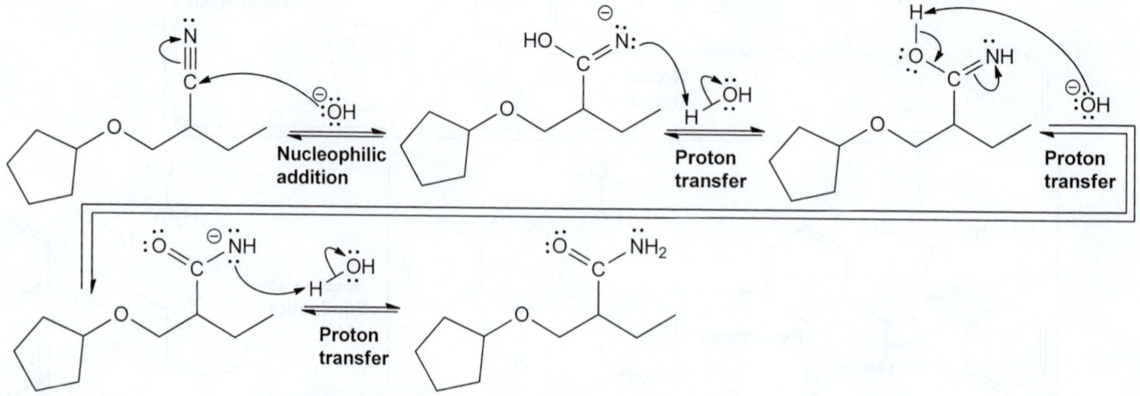

Nucleophilic addition

Proton transfer

Proton transfer

Proton transfer

(b) Under acidic conditions, no strong bases should appear. The second step (nucleophilic addition) is the only step that is not a proton transfer reaction. A proton transfer occurs before nucleophilic addition to avoid the formation of a strong base when the nucleophile adds.

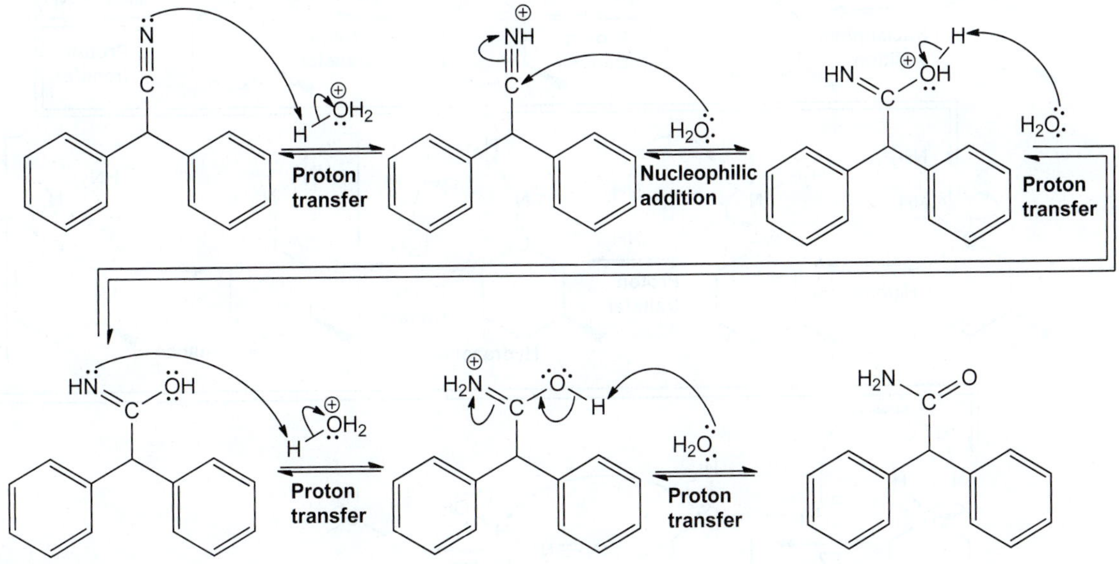

Proton transfer

Nucleophilic addition

Proton transfer

Proton transfer

Proton transfer

Problem 18.18

Think

Does the first reaction take place under acidic or basic conditions? Are the conditions mild or strong? What impact does that have on the nature of the nucleophile? Does the second reaction take place under acidic or basic conditions? Are the conditions mild or strong? What species, therefore, should not appear in the mechanism? What two mechanisms from this chapter combine to show the complete mechanism? What functional group is removed entirely?

Solve

Both of these are Wolff–Kishner reductions, which reduce the C=O bond of a ketone or aldehyde to a methylene (CH_2) group.

(a) The first reaction is similar to imine formation. It takes place under mildly acidic conditions, so the nucleophilic N is not charged, and neither is the carbonyl O. First, nucleophilic addition occurs, followed by elimination of water to form the hydrazone. Next, the basic conditions remove two protons from the hydrazone, creating a great N_2 leaving group. The product is methylcyclohexane, which is no longer chiral. See the mechanism on the next page.

Nucleophilic addition

Proton transfer

Proton transfer

Proton transfer

Heterolysis

Proton transfer

Hydrazone

Proton transfer

Resonance

Proton transfer

E2

$+ :N{\equiv}N:$

Proton transfer

(b) The mechanistic route is the same as **(a)**.

Nucleophilic addition

Proton transfer

Proton transfer

Proton transfer

Heterolysis

Proton transfer

Proton transfer

Resonance

Proton transfer

E2

Proton transfer

$+ :N{\equiv}N:$

Problem 18.19

Think

What functional group reacts with NH_2NH_2/H^+ and $NaOH/H_2O$ to give $-CH_2$? On which C atom in the product are there two H atoms, indicating what could be the original functional group?

Solve

Only one sp^3 C in the target molecule has two H atoms, a characteristic of a Wolff–Kishner reduction product. The functional group that reacts in a Wolff–Kishner reaction is the C=O group of a ketone or aldehyde. The precursor is shown below.

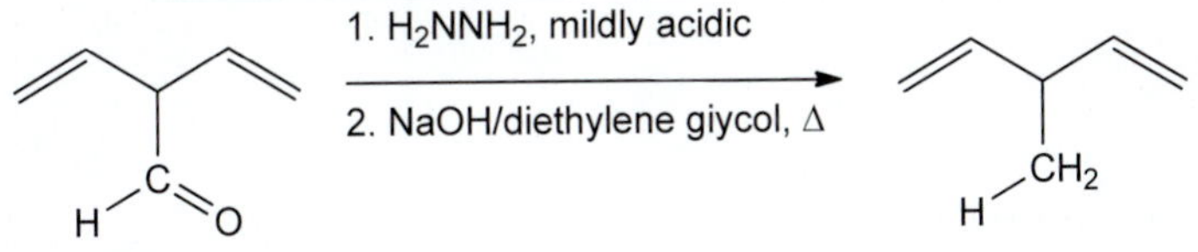

1. H_2NNH_2, mildly acidic
2. NaOH/diethylene giycol, Δ

Problem 18.20

Think

What functional groups appear in the product of every aldol reaction? What should the relative positions be for those functional groups?

Solve

The product of every aldol addition is a β-hydroxycarbonyl compound. As shown below, the only molecule that is a β-hydroxycarbonyl compound is molecule **(b)**.

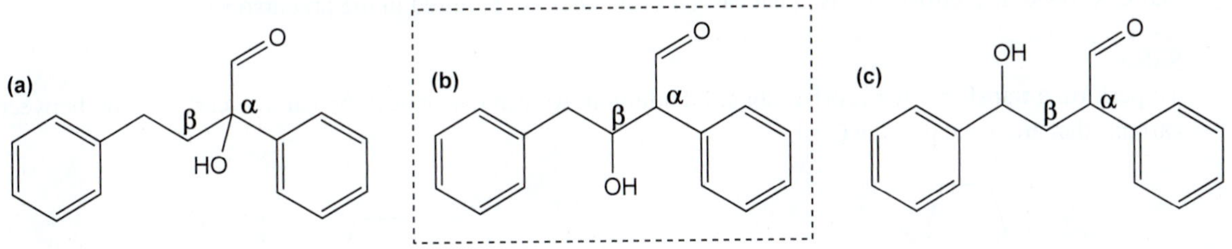

Problem 18.22

Think

How is this reaction similar to the one in Equation 18-30 (and Equation 18-31)? How is it different? What is the role of NaOH? What can act as a nucleophile to add to the C=O group?

Solve

Just as in Equation 18-30, an aldehyde is treated with hydroxide. The only difference is the carbon backbone; in this case, we have a $PhCH_2CH_2CHO$ aldehyde, whereas it was a two-carbon aldehyde in Equation 18-30. The mechanisms, therefore, are essentially identical.

In the first step, HO^- acts as a base to generate an enolate anion. (Note that HO^- could also act as a nucleophile, but as we saw in Chapter 8, proton transfer reactions are very fast, so the dominant reaction is deprotonation of the α carbon.) In the second step, the enolate anion acts as a nucleophile, attacking a second molecule of aldehyde that still has its proton. This step forms a C–C bond between the two species. Finally, the O^- is protonated to yield the β-hydroxycarbonyl compound. See the mechanism on the next page.

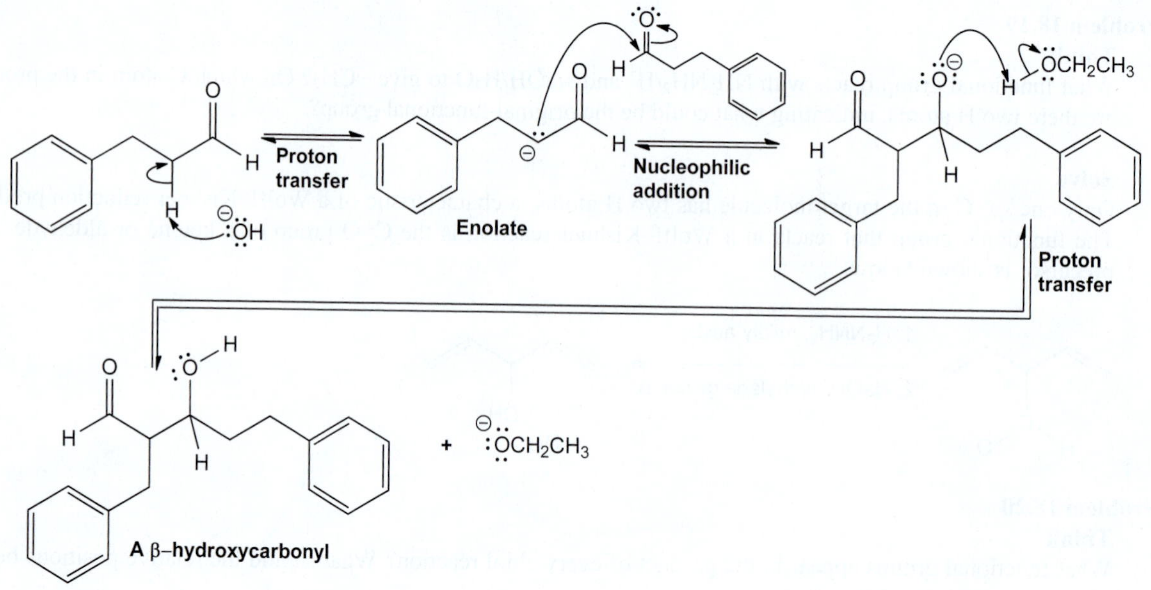

Proton transfer

Nucleophilic addition

Enolate

Proton transfer

$:\overset{\ominus}{\ddot{O}}CH_2CH_3$

A β–hydroxycarbonyl

Problem 18.24

Think

Can you identify the portion of the molecule that characterizes it as a β-hydroxycarbonyl compound? Which C atoms of the β-hydroxycarbonyl compound would have been joined in the aldol addition? Upon disconnecting that C–C bond in a retrosynthesis, what carbon backbones are required in the precursors?

Solve

To perform a transform that reverses an aldol addition, we can disconnect the carbon–carbon bond between the carbons that are α and β to the C=O.

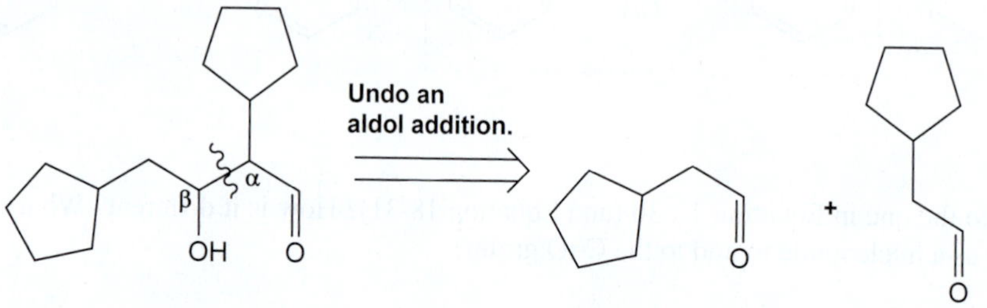

Undo an aldol addition.

The two aldehyde precursors are the same, so we simply need to treat that aldehyde with base.

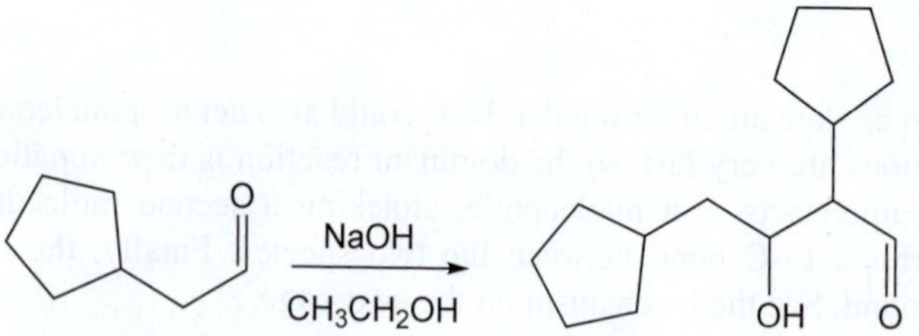

NaOH

CH_3CH_2OH

Problem 18.25

Think

What is the pK_a of the α hydrogen in the aldehyde? What is the pK_a of the α hydrogen in an alkene? What species is acting as the base? What is the pK_a of the conjugate acid? Which side of the proton transfer reaction does the equilibrium favor?

Solve

Although a mechanism analogous to the reaction in Equation 18-33 (E1cb) can be written, the first step would be very unfavorable. In Equation 18-33, it is feasible because the α hydrogen is reasonably acidic, with a pK_a ~ 20. But the analogous proton in the compound given in the problem has a pK_a ~ 40. With the conjugate acid of HO^- (i.e., H_2O) having a pK_a of 15.7, the equilibrium favors the reactants by ~10^{24}, making the reaction unfeasible.

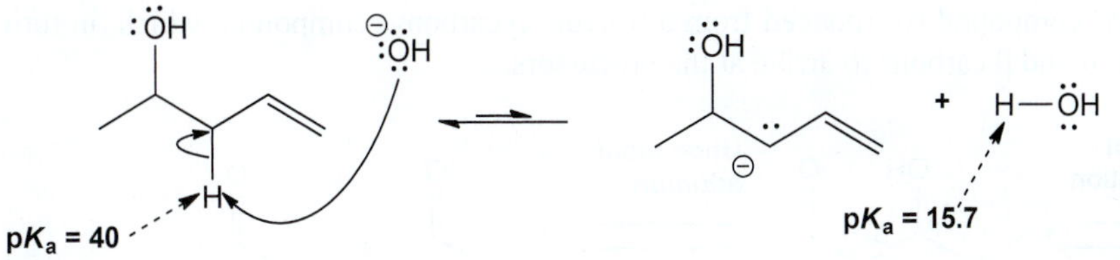

:ÖH ⊖:ÖH :ÖH + H—ÖH

pK_a = 40 pK_a = 15.7

Problem 18.26

Think

Which aldehyde acts as the enolate nucleophile? Which aldehyde is attacked by that enolate nucleophile? Is this an aldol addition or condensation?

Solve

In this example, ethanal is deprotonated at the α position by HO^- and acts as the enolate nucleophile that attacks the propanal carbonyl C. This is an example of an aldol addition that forms a β-hydroxycarbonyl compound. The ethanal atoms are boxed in each step of the mechanism below.

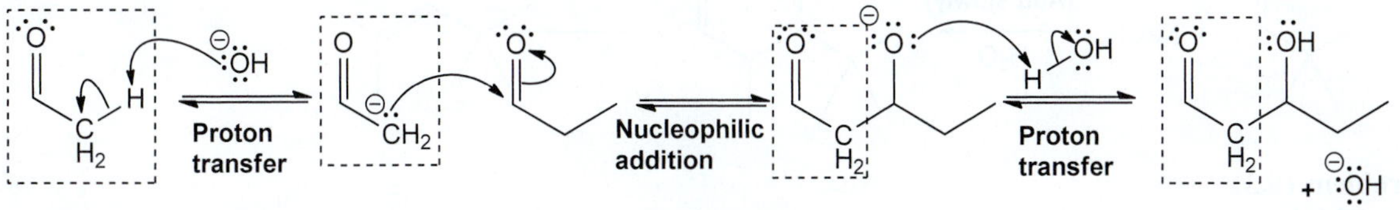

Proton transfer Nucleophilic addition Proton transfer

Problem 18.28

Think

Which aldehyde can form an enolate anion, and which cannot? Which uncharged aldehyde is present in only small concentrations in the reaction mixture? Which uncharged aldehyde is the one that will be attacked?

Solve

Dimethyl propanal has no α hydrogens, so it cannot form an enolate anion. Thus, the only enolate nucleophile that is present is derived from phenylethanal, which does have α hydrogens. Because phenylethanal is added slowly, there is never a substantial concentration of it in its uncharged form, so the major reaction occurs between the phenylethanal enolate anion and dimethylpropanal.

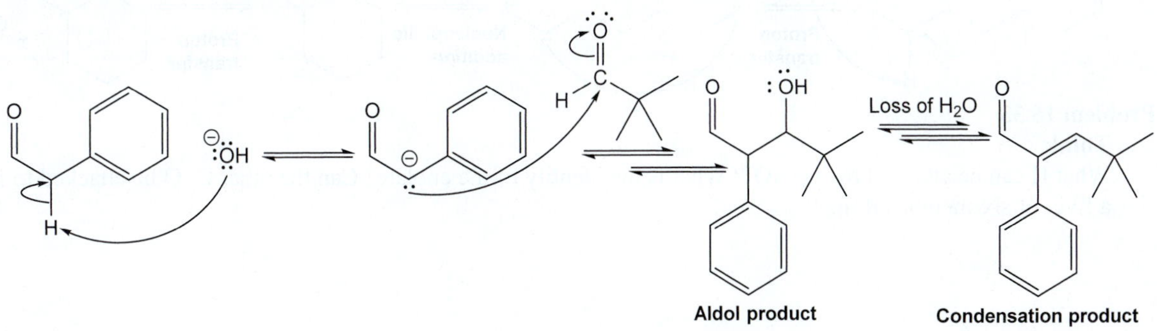

Loss of H_2O

Aldol product Condensation product

Problem 18.30

Think

Can you identify the portion of the molecule that characterizes it as an α,β-unsaturated carbonyl compound? From what β-hydroxycarbonyl compound could it have been generated? Upon applying a transform to that β-hydroxycarbonyl compound to undo an aldol addition, which C–C bond should be disconnected?

Solve

The α,β-unsaturated carbonyl compound is produced from a β-hydroxycarbonyl compound, which, in turn, can be disconnected between the α and β carbons to arrive at the precursors.

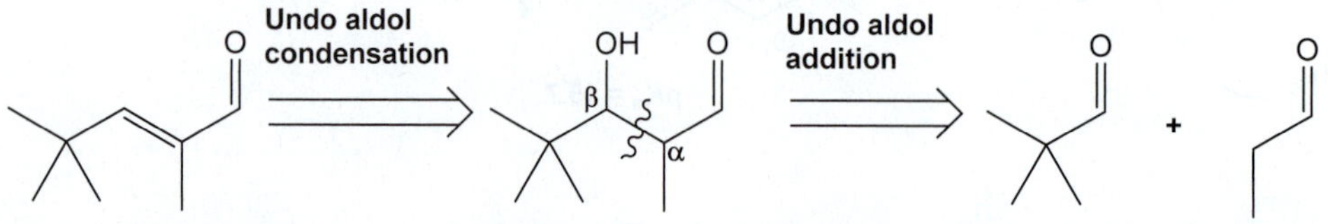

Undo aldol condensation

Undo aldol addition

Because the starting carbonyls are different, we must consider how to execute a crossed aldol reaction. In this case, the first aldehyde has no α hydrogens, so we can treat it with base before adding the second aldehyde. Heating the reaction favors the formation of the condensation product.

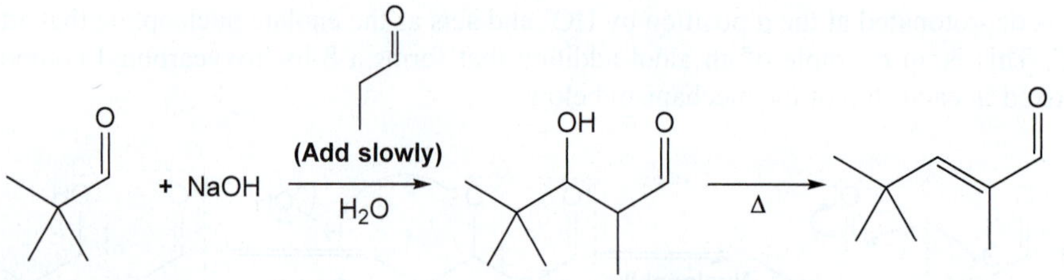

+ NaOH (Add slowly) H_2O Δ

Problem 18.31

Think

What H is attacked by the HO^-? What is the identity of the enolate? What electrophilic atom is attacked by the enolate? How many C atoms are in the ring that forms?

Solve

The mechanism is shown below. The α hydrogen of heptanedial is attacked by the HO^- to form the enolate. The other C of the C=O is attacked by the enolate. The intramolecular aldol reaction is favorable because it produces a six-membered ring.

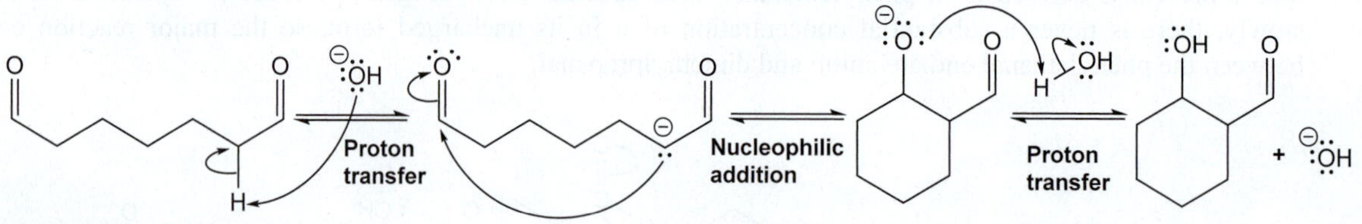

Proton transfer

Nucleophilic addition

Proton transfer

Problem 18.33

Think

What H can be attacked by the HO^-? What is the identity of the enolate? Can the other C=O be attacked to form a five- or six-membered ring?

Solve

Four α C atoms can be deprotonated to produce an enolate anion. The two α C atoms in the middle of the molecule won't lead to an intramolecular aldol reaction because they are too close to the carbonyl groups—they will form a three-membered ring if their enolates attack. Either of the terminal C atoms can be deprotonated to ultimately lead to the formation of a five-membered ring. The molecule is symmetrical, so it does not matter which end is deprotonated in the first step.

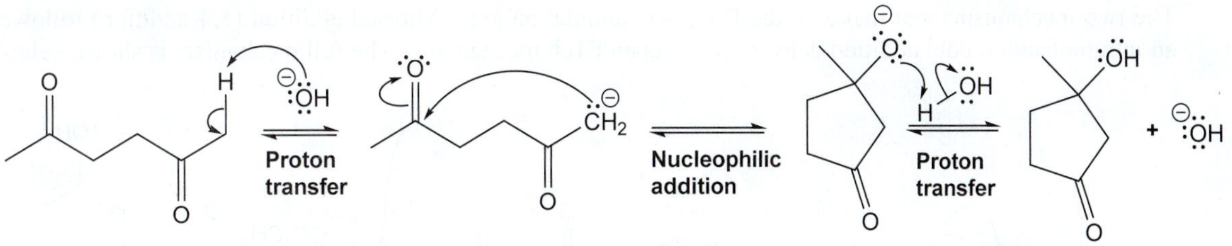

Proton transfer

Nucleophilic addition

Proton transfer

Problem 18.35

Think

Which protons are the most acidic? After deprotonation, which polar π bond undergoes nucleophilic addition? How are proton transfers incorporated into the mechanism?

Solve

The hydrogen that is α to the NO_2 group is the most acidic, so it is removed by the base in the first step. The resulting anion attacks the aldehyde group, and the O^- generated in that step is subsequently protonated.

Proton transfer

Nucleophilic addition

Proton transfer

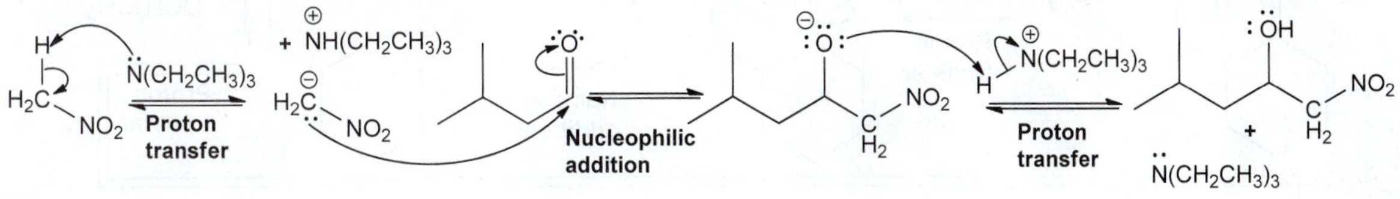

Problem 18.36

Think

Which protons are the most acidic? After deprotonation, which polar π bond undergoes nucleophilic addition? How is a C=C double bond formed in an aldol reaction? Will it be favored in this reaction?

Solve

The H that is α to the CN group is the most acidic, so it is removed by the base in the first step. The resulting enolate anion attacks the C=O group of the aldehyde, and the O^- generated in that step is subsequently protonated. Owing to the conjugation that results from condensation, condensation then takes place via an E1cb mechanism.

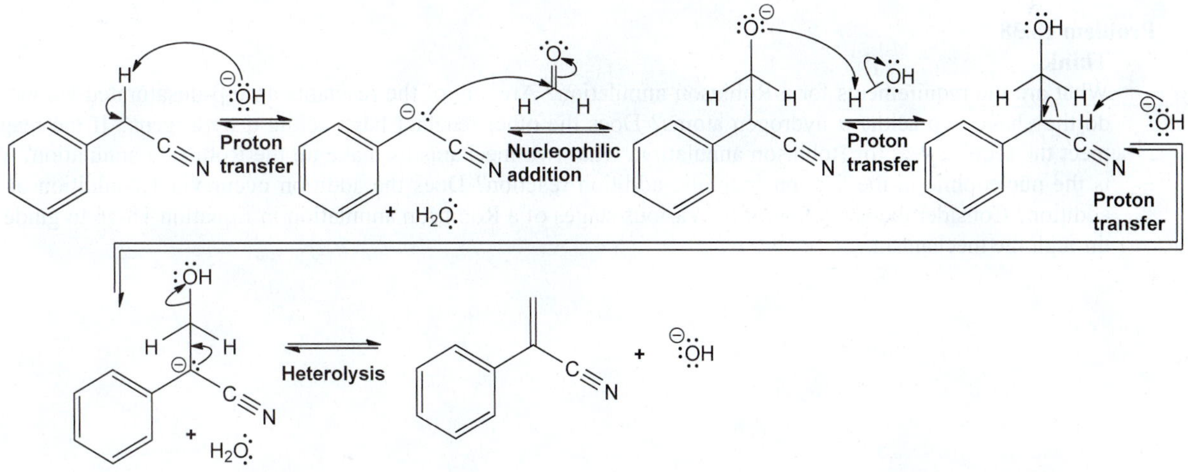

Proton transfer

Nucleophilic addition

Proton transfer

Proton transfer

Heterolysis

Problem 18.37

Think

The Robinson annulation comprises the mechanisms of which two reactions? What is the nucleophile in the first nucleophilic addition reaction? Does the addition occur via 1,2-addition or 1,4-addition? Consider the dissection of the various stages of a Robinson annulation in Equation 18-56 to guide you through the mechanism.

Solve

The two mechanisms that make up the Robinson annulation are a Michael addition (1,4-addition) followed by an intramolecular aldol addition/dehydration (via an E1cb mechanism). The full mechanism is shown below.

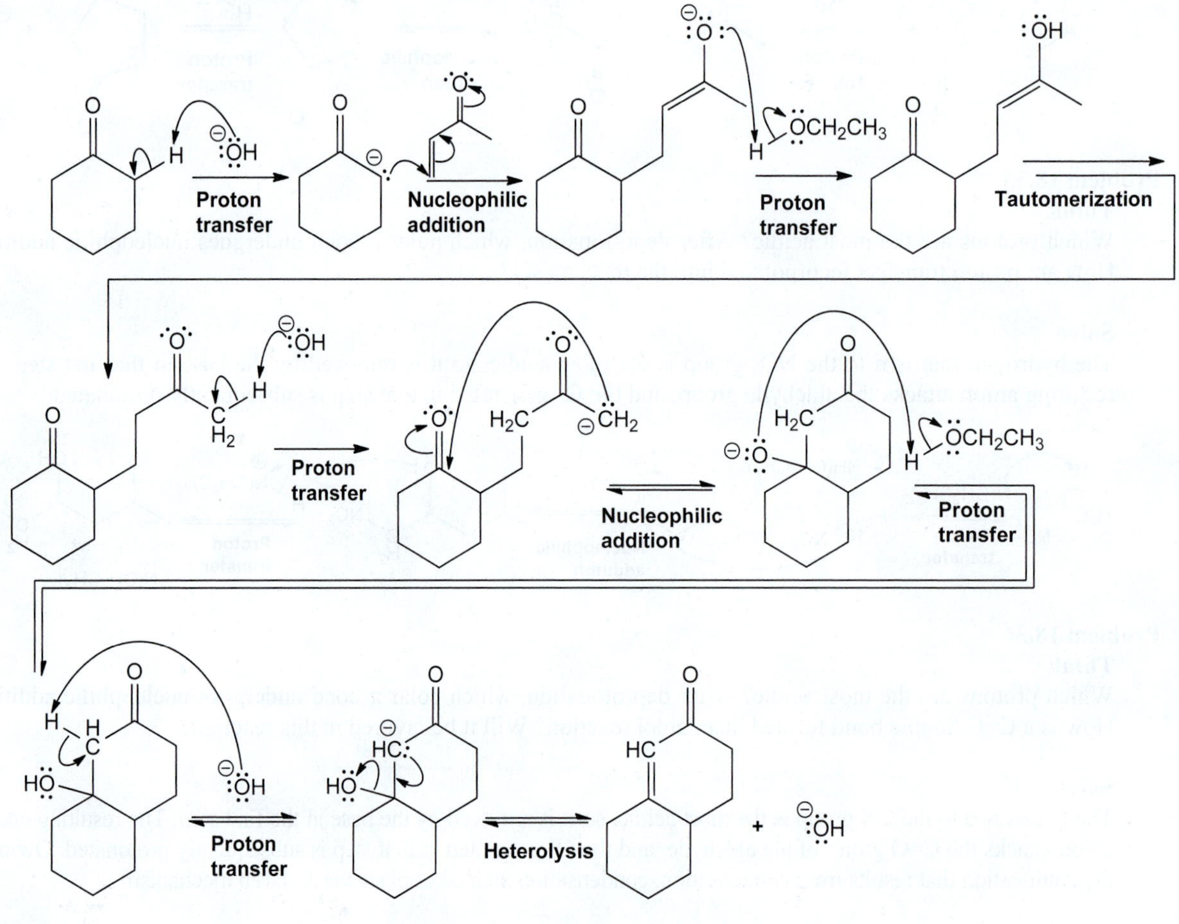

Proton transfer → Nucleophilic addition → Proton transfer → Tautomerization

Proton transfer → Nucleophilic addition ⇌ Proton transfer

Proton transfer ⇌ Heterolysis ⇌ + $:\ddot{O}H^-$

Problem 18.38

Think

What are the requirements for a Robinson annulation? Are any of the reactants an α,β-unsaturated ketone, and do they have two acidic α hydrogen atoms? Does the other reactant have acidic α hydrogens? If the reagents meet the requirement for Robinson annulation, what two mechanisms make up the Robinson annulation? What is the nucleophile in the first nucleophilic addition reaction? Does the addition occur via 1,2-addition or 1,4-addition? Consider the dissection of the various stages of a Robinson annulation in Equation 18-56 to guide you through the mechanism.

Solve

(a) These reagents will not work because benzaldehyde does not have any α protons.

(b) These reagents will not work because the α,β-unsaturated carbonyl does not have two acidic α protons.

(c) These reagents will not work because neither compound is an α,β-unsaturated carbonyl compound.

(d) These reagents fit the criteria for a Robinson annulation. The mechanism is shown below.

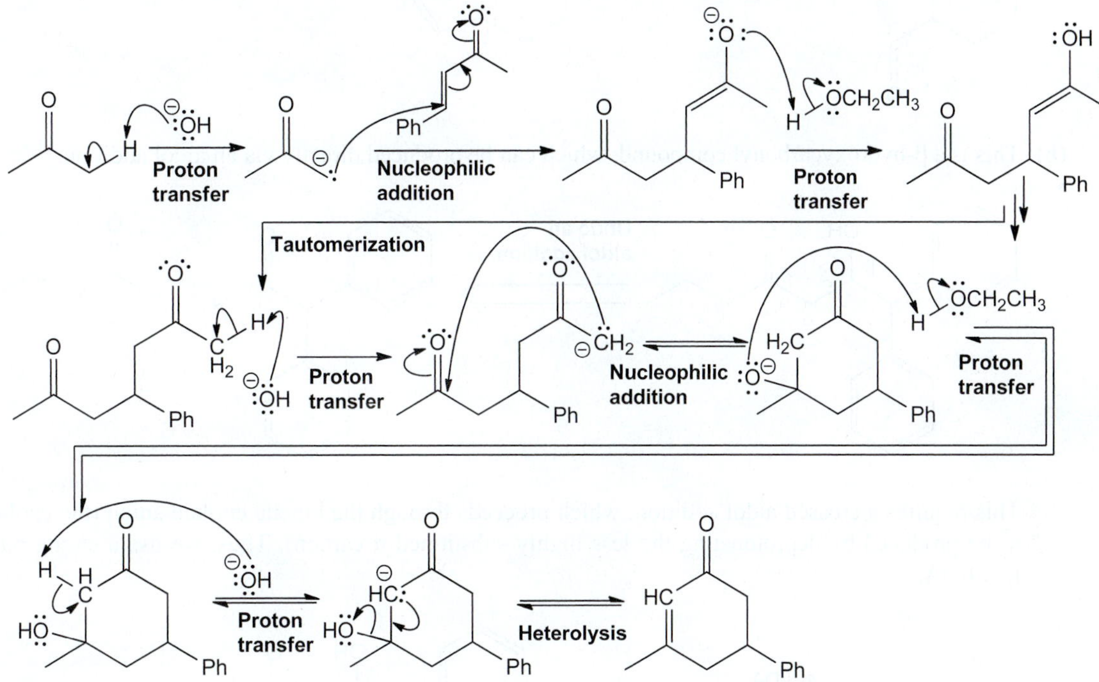

Proton transfer → Nucleophilic addition → Proton transfer → Tautomerization → Proton transfer → Nucleophilic addition ⇌ Proton transfer ⇌ Proton transfer ⇌ Heterolysis

Problem 18.39

Think

Is the product a result of aldol addition or aldol condensation? What are the characteristic features of the products from these reactions? What are the α and β carbon atoms in the product? Which two C atoms formed the new C–C bond? Which carbon atoms would be part of C=O groups in the starting materials?

Solve

(a) This is an α,β-unsaturated carbonyl compound that results from an aldol condensation and can be produced from a β-hydroxycarbonyl compound. The β-hydroxycarbonyl compound is the product of an aldol addition. The retrosynthesis is shown below.

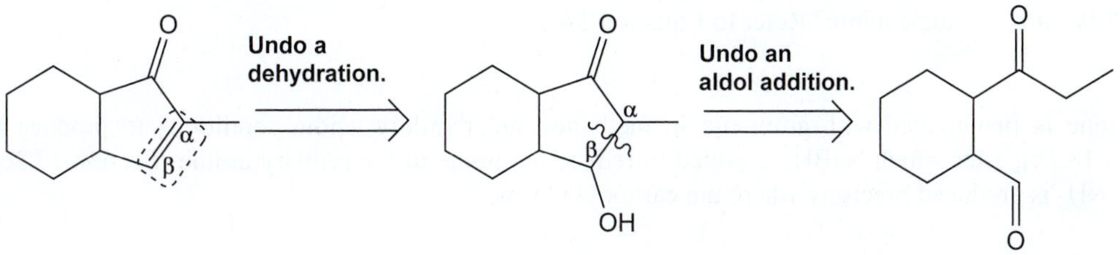

Undo a dehydration. ⇒ Undo an aldol addition. ⇒

In the forward direction, the synthesis would appear as follows:

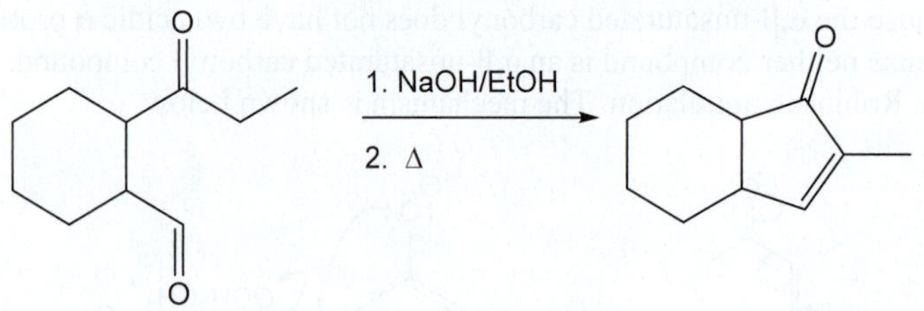

1. NaOH/EtOH
2. Δ

(b) This is a β-hydroxycarbonyl compound, which can be produced directly via an aldol addition.

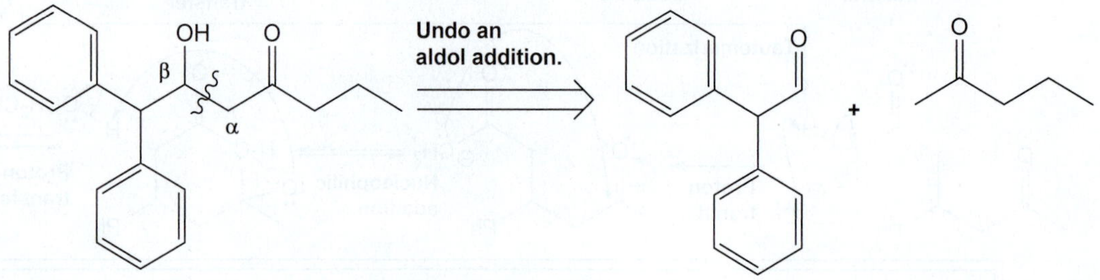

Undo an aldol addition.

This requires a crossed aldol addition, which proceeds through the kinetic enolate anion (the enolate needs to be produced by deprotonating the less highly substituted α carbon). Thus, we use a strong bulky base like LDA.

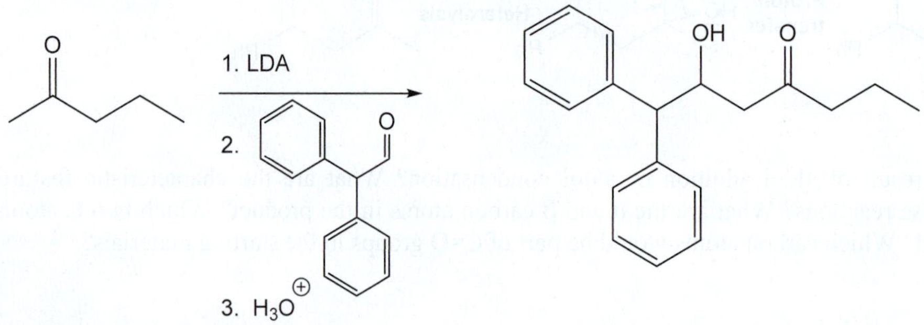

1. LDA
2.
3. $H_3O^{\oplus}$

Problem 18.40

Think

What is produced when NH_3 reacts with the C=O of a ketone or aldehyde? How does this product react with $NaBH_4$? What is the nucleophile? Refer to Equation 18-61.

Solve

The ketone is first treated with ammonia in methanol under mildly acidic conditions to produce an imine (Section 18.2b), after which $NaBH_4$ is added to reduce the imine to the primary amine (Section 17.3c). Notice that the NH_2 is produced precisely where the carbonyl O was.

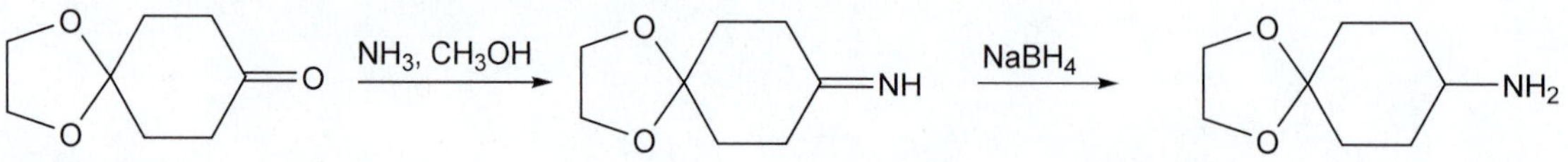

NH_3, CH_3OH → =NH → $NaBH_4$ → –NH_2

Problem 18.41

Think

Where is the N in each product? In the reactant, on which C is the O of the ketone or aldehyde? What reagents can transform a C=O into a C–N? Is more than one synthetic step required?

Solve

The conversions can take place by carrying out a reductive amination on a carbonyl compound, similar to the reaction in Equation 18-61. In amines **(a)**–**(c)**, the N in the product is precisely where the carbonyl O was in the reactant. The syntheses are shown below.

(a) 1. aniline (NH_2), CH_3OH, mildly acidic; 2. $NaBH_4$ → NH

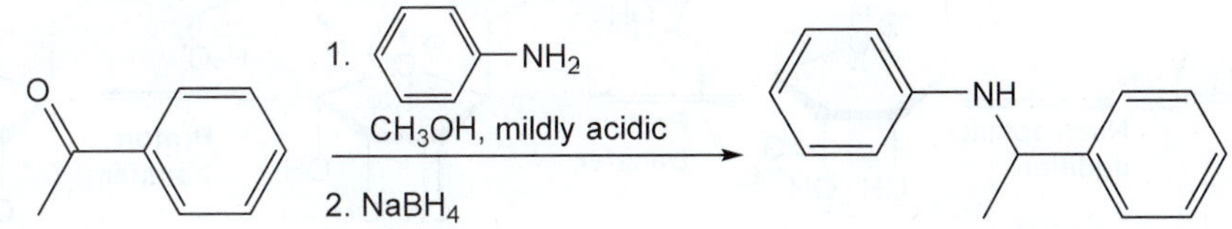

(b) 1. piperidine (NH), CH_3OH, mildly acidic; 2. $NaBH_4$ → N

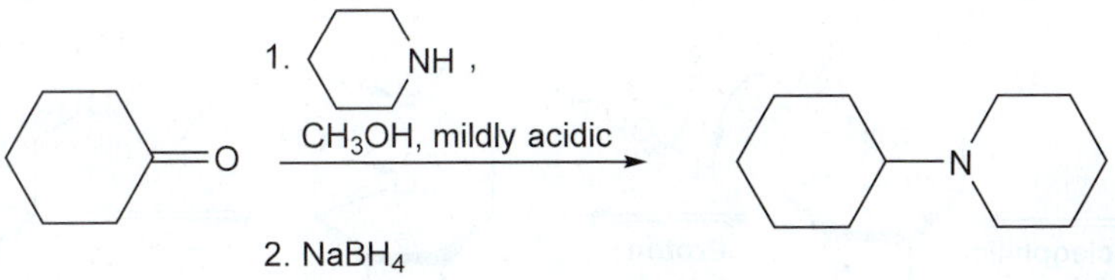

(c) OCH_3 … O; 1. NH_3, CH_3OH, mildly acidic; 2. $NaBH_4$ → OCH_3 … NH_2

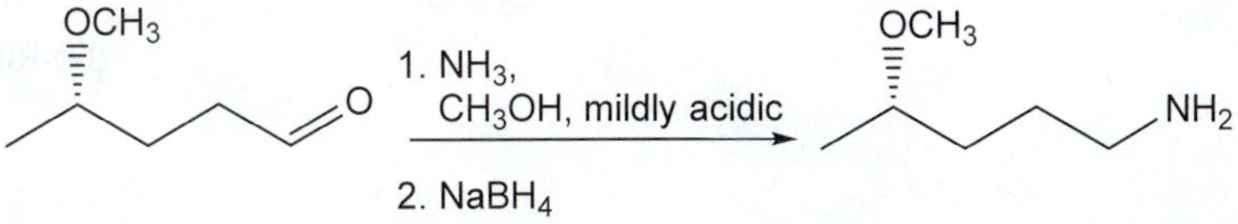

Problem 18.42

Think

Consult Equation 18-65 for the structure of D-ribofuranose. What is the Haworth projection (*Hint:* Review Section 4-9)? Which C is the anomeric carbon? Which anomer has the CH_2OH group on the same side of the ring as the OH group at the anomeric carbon? Consult Equation 18-66b for the general mechanism. What acts as the nucleophile in the nucleophilic addition reaction to close the ring? What C is electrophilic? Are proton transfer reactions involved? What approach is required of the nucleophile to produce each anomer?

Solve

(a) The Haworth projections are shown below. The anomeric C is the one at the far right in each case. The CH_2OH group is on the opposite side from the anomeric OH group in the α-anomer, and they are on the same side in the β-anomer.

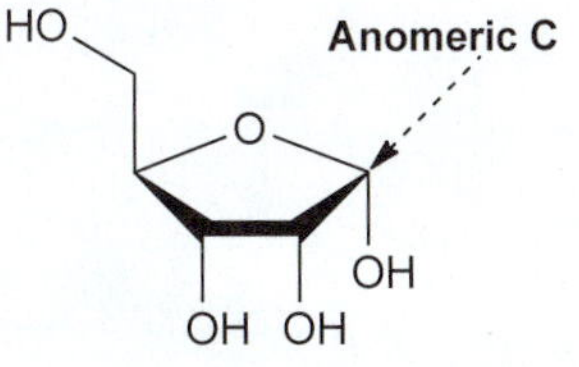

HO, O, Anomeric C, OH, OH OH

α-D-Ribofuranose

HO, O, OH, Anomeric C, OH OH

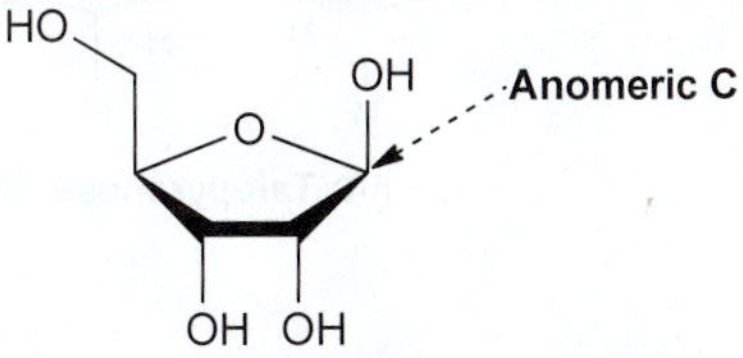

β-D-Ribofuranose

(b) The mechanisms showing the production of each anomer are shown below. To form the α-anomer, nucleophilic attack must come from the top of the C=O plane, and to form the β-anomer, nucleophilic attack must come from the bottom.

Nucleophilic addition → Proton transfer → Proton transfer

α-D-Ribofuranose

Nucleophilic addition → Proton transfer → Proton transfer

β-D-Ribofuranose

Problem 18.43

Think

What is the name of the sugar in the acylic form (see Fig. 5-32)? To help you name the sugar, can you identify it as a specific epimer of another sugar? Is the anomeric OH and CH_2OH on the same side or opposite sides of the ring?

Solve

(a) α-D-Mannopyranose. The CH_2OH group is on the opposite side from the anomeric OH, making it an α-anomer. That this sugar is mannose can be seen from the fact that it is a C2 epimer of glucose.

(b) β-D-Talopyranose. The CH_2OH group is on the same side as the anomeric OH, making it a β-anomer. That this sugar is talose can be seen from the fact that it is a C4 epimer of mannose.

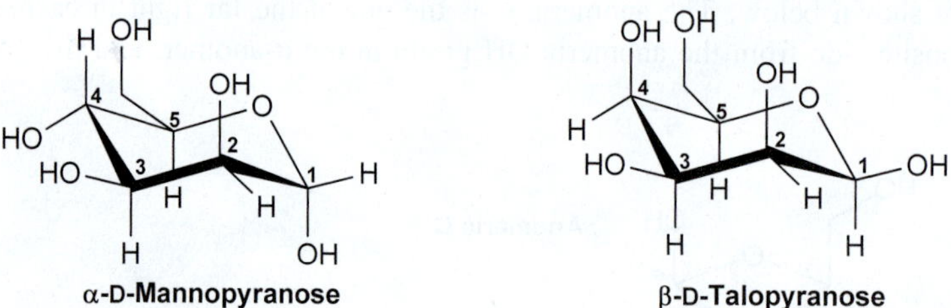

α-D-Mannopyranose β-D-Talopyranose

Problem 18.44

Think

If the specific rotation is +14.5°, which anomer, α-D-mannopyranose ([α] = +29.3°) or β-D-mannopyranose ([α] = −16.3°) is the dominant species in solution? If you set the percentage of α-D-mannopyranose equal to x in Equation 16-69, then what is the percentage of β-D-mannopyranose?

Solve

Let x be the equilibrium percentage of the α-D-mannopyranose and $1 - x$ be the equilibrium percentage of β-D-mannopyranose.

$x(+29.3) + (1 - x)(-16.3) = +14.5$

$29.3x - 16.3 + 16.3x = 14.5$

$45.6x = 30.8$

$x = 0.675$, or 67.5% = α-D-mannopyranose

$1 - x = 0.325$, or 32.5% = β-D-mannopyranose

End of Chapter Problems

Section 18.1 Addition of Weak Nucleophiles: Acid and Base Catalysis

Problem 18.45

Think

Is this mechanism uncatalyzed, base catalyzed, or acid catalyzed? Does the C=O undergo nucleophilic addition or proton transfer first? In the α,β-unsaturated carbonyl in **(b)**, is the nucleophilic addition direct or conjugate?

Solve

(a) This mechanism is uncatalyzed and follows the same steps as Equation 18-3. The nucleophile adds to the C=O carbon in Step 1 to produce an intermediate with two formal charges, and proton transfers in Steps 2 and 3 produce the uncharged product. The water is shown as the acid and base in Steps 2 and 3 because it is in much greater abundance than any other acids or bases present.

Nucleophilic addition — Proton transfer — Proton transfer — + H_3O^+

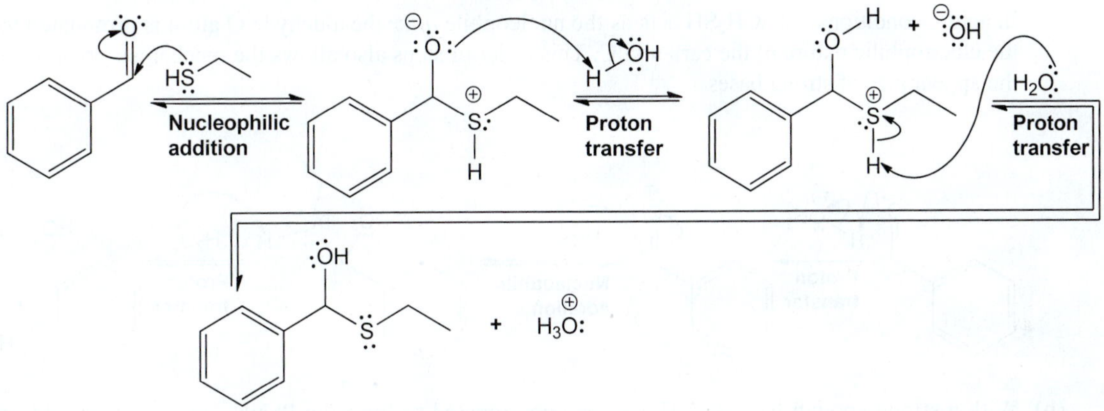

(b) $PhNH_2$, an uncharged nucleophile, adds reversibly to a carbonyl group, so it will favor conjugate addition upon attacking an α,β-unsaturated carbonyl. The reaction is not catalyzed by acidic or basic conditions. The water is shown as the acid and base in Steps 2 and 3 because it is in much greater abundance than any other acids or bases present. The enol that is produced then tautomerizes to the more stable keto form via back-to-back proton transfer steps.

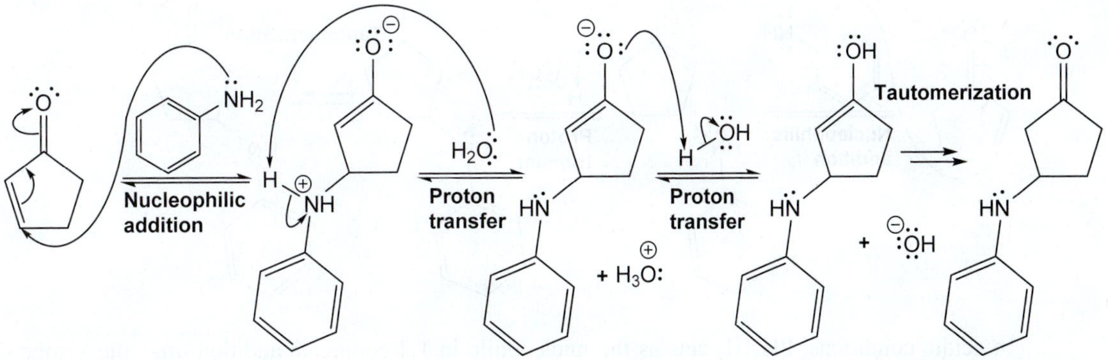

Nucleophilic addition — Proton transfer — + H_3O^+ — Proton transfer — + ^-OH — Tautomerization

Problem 18.46

Think

What nucleophiles are present under basic conditions? Under acidic conditions? In what order do the nucleophilic addition and proton transfer reactions need to occur to avoid incompatible species in solution? How can the acid and base catalysts be regenerated? In the α,β-unsaturated carbonyl in **(b)**, is the nucleophile one that leads to reversible or irreversible nucleophilic addition? Does reversible addition favor 1,2-addition or 1,4-addition? Does irreversible addition favor 1,2-addition or 1,4-addition?

Solve

(a) In basic conditions, the HO^- deprotonates the CH_3CH_2S–H to form the S^- anion, which acts as the nucleophile. In basic conditions, the nucleophilic addition to the C=O occurs before the protonation of the carbonyl O atom to avoid the appearance of a strong acid.

Proton transfer

Nucleophilic addition

Proton transfer

In acidic conditions, CH_3CH_2SH acts as the nucleophile *after* the aldehyde O atom is protonated to increase the electrophilic nature of the carbonyl C. This order of steps also allows the mechanism to proceed without the appearance of strong bases.

Proton transfer

Nucleophilic addition

Proton transfer

(b) With a strong enough base, $PhNH_2$ can be deprotonated to form the $PhNH^-$ anion, which adds reversibly (Table 17-2) to an α,β-unsaturated carbonyl, so it will favor conjugate addition upon attacking. In basic conditions, the nucleophilic addition to the C=O occurs before the protonation of the carbonyl O atom to avoid the appearance of a strong acid. The enol that is formed tautomerizes (back-to-back proton transfer steps) to the more stable keto form.

Nucleophilic addition

Proton transfer

Tautomerization

In acidic conditions, $PHNH_2$ acts as the nucleophile in 1,4-conjugate addition *after* the ketone O atom is protonated to increase the electrophilic nature of the carbonyl C. This order of steps also allows the mechanism to proceed without the appearance of strong bases. The enol that is formed tautomerizes (back-to-back proton transfer steps) to the more stable keto form. See the mechanism on the next page.

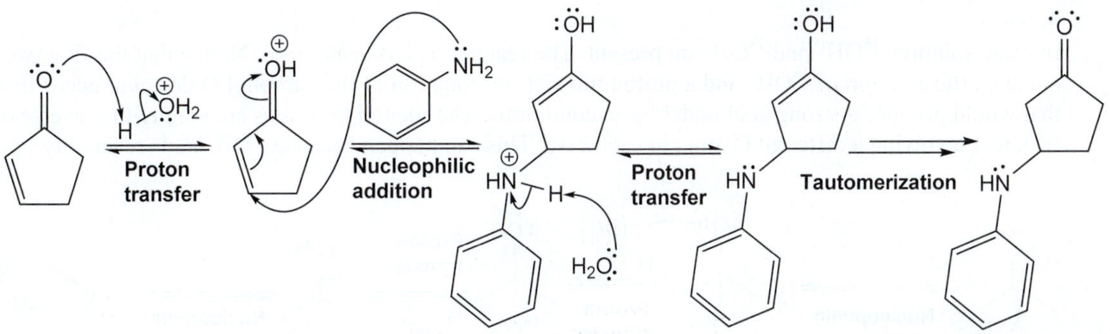

Proton transfer

Nucleophilic addition

Proton transfer

Tautomerization

Problem 18.47

Think

What nucleophiles are present under these conditions? Are strong acids or bases present? On the basis of the conditions, should any types of species not appear in the mechanism? What product is formed? In the α,β-unsaturated carbonyl in **(b)**, does the nucleophile add to the carbonyl group reversibly or irreversibly? Will this result in direct addition or conjugate addition?

Solve

(a) A cyanohydrin forms. In these conditions, CN^- is the strongest nucleophile present. These conditions follow the same mechanism as in Equation 18-7 (nucleophilic addition to the C=O followed by proton transfer).

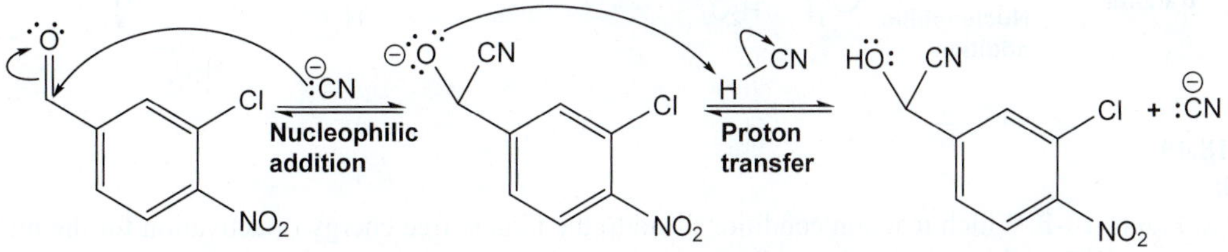

Nucleophilic addition

Proton transfer

(b) Even though NC^- is a charged nucleophile, it still adds reversibly to a carbonyl group, so it will favor conjugate addition upon attacking an α,β-unsaturated carbonyl. The HO^- below the reaction arrow indicates basic conditions, so no strongly acidic species should appear in the mechanism. Therefore, we do not show the HCN protonating the carbonyl O atom. Nucleophilic addition occurs first. The enol that is formed tautomerizes (back-to-back proton transfer steps) to the more stable keto form.

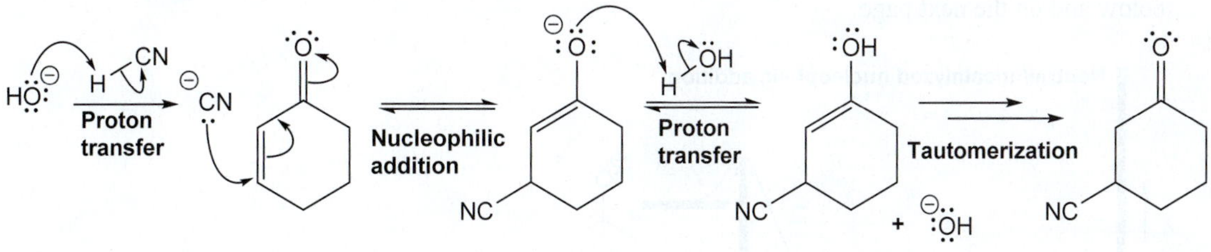

Proton transfer

Nucleophilic addition

Proton transfer

Tautomerization

Problem 18.48

Think

What nucleophiles are present under basic conditions? Under acidic conditions? In what order do the nucleophilic addition and proton transfer steps occur to avoid species that are incompatible with the reaction conditions? How can the acid and base catalyst be regenerated?

Solve

In basic solution, $^{18}OH^-$ and $^{18}OH_2$ are present. The reaction is base catalyzed. Notice that the first two steps are nucleophilic addition of $^{18}OH^-$ and a proton transfer. Protonation of the carbonyl O does *not* occur first because that would produce a strong acid under basic conditions. The second two steps are simply the reverse of the first two, just involving a different O atom in each case. This can happen because $^{18}OH^-$ adds reversibly.

In acidic solution, $H_3{}^{18}O^+$ and $H_2{}^{18}O$ are present. The reaction is acid catalyzed. Again, notice that the last three steps are the reverse of the first three, just involving different O atoms. In this case, the carbonyl group is protonated before nucleophilic addition to avoid producing O^-, which would represent a strong base.

Problem 18.49

Think

Refer to Figure 18-1. Which reaction conditions lead to the largest free energy of activation for the nucleophilic addition step? Why is this step the slowest?

Solve

The dramatic increase in reaction rate under basic and acidic conditions compared to neutral conditions can be understood by comparing the nucleophilic addition step in each mechanism. In the neutral uncatalyzed reaction, the nucleophilic addition step produces two new formal charges, whereas in the base- and acid-catalyzed reactions, the nucleophilic addition step produces no additional formal charges. See the free energy diagrams below and on the next page.

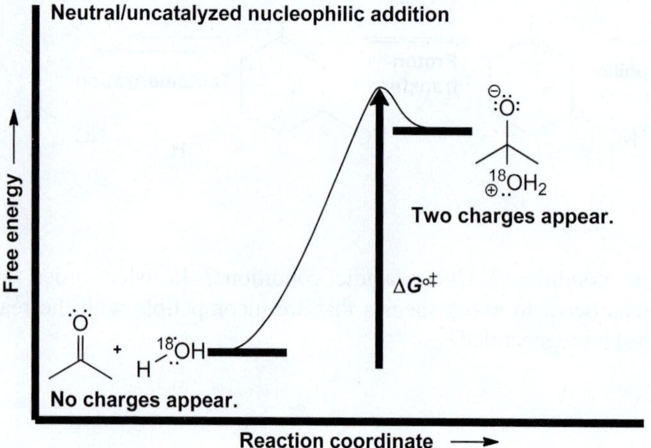

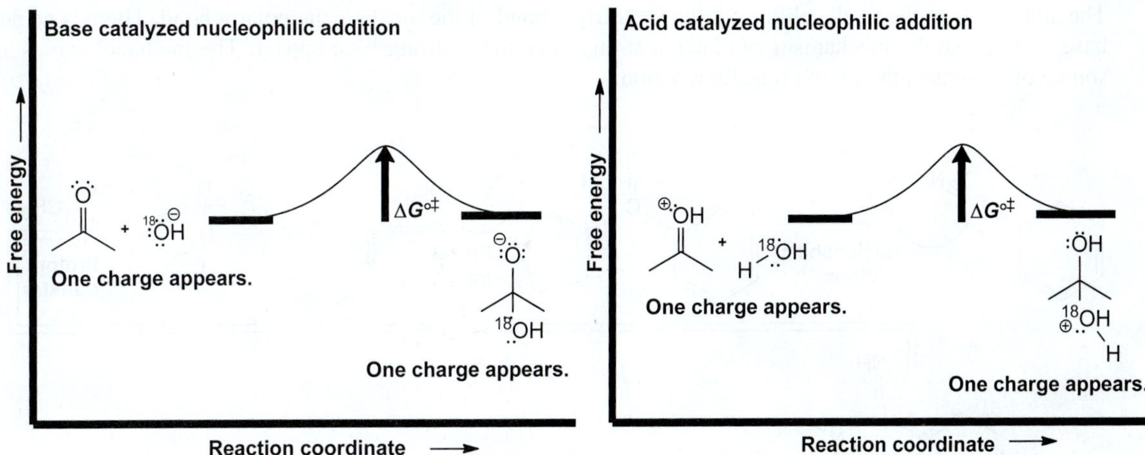

Problem 18.50

Think

What acts as the nucleophile in the nucleophilic addition step? Is the reaction conducted in acidic or basic solution? What species are not permitted in the mechanism?

Solve

The mechanism is a proton transfer followed by a nucleophilic addition, then another proton transfer. The first step has to be protonation to avoid generating any N^- or O^-, which are strongly basic.

Proton transfer
Nucleophilic addition
Proton transfer

Problem 18.51

Think

What acts as the nucleophile in the nucleophilic addition step? What is the polar π bond that is attacked? Is the reaction conducted in acidic or basic solution? What species are not permitted in the mechanism? What is the significance of having the reaction free of water?

Solve

The alcohol R′OH is the nucleophile, and the C≡N triple bond of the nitrile is the polar π bond. The mechanism is a proton transfer followed by a nucleophilic addition, then another proton transfer. The first step has to be proton transfer to avoid generating N^-, which is strongly basic. The reaction is free of water to have the ROH act as the nucleophile and not have water compete for nucleophilic addition.

Proton transfer
Nucleophilic addition
Proton transfer

Problem 18.52

Think

What acts as the nucleophile in the nucleophilic addition step? What is the polar π bond that is attacked? Is the reaction conducted in acidic or basic solution? What species are not permitted in the mechanism?

The amine acts as the nucleophile, and the C≡N triple bond of the nitrile is the polar π bond. There is no acid or base catalyst, so the mechanism can have a strong acid and a strong base appear. The methanol serves as the source of protons in the proton transfer reaction.

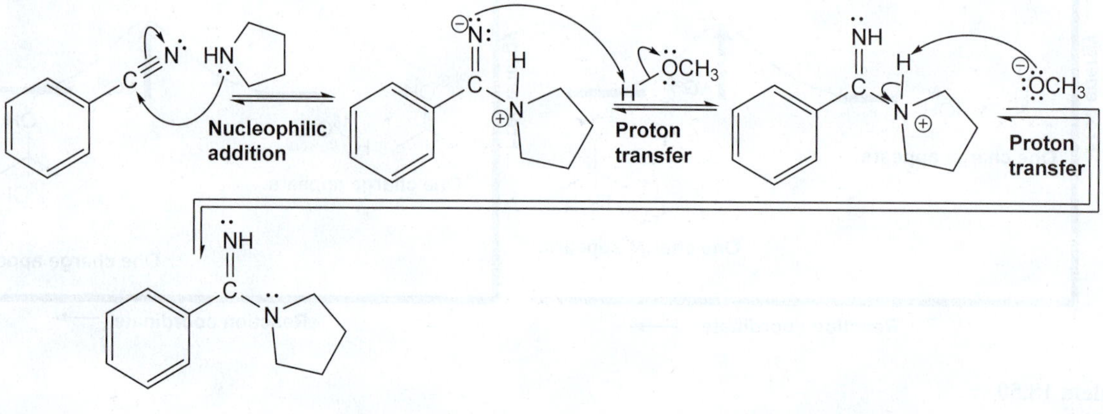

Nucleophilic addition

Proton transfer

Proton transfer

Problem 18.53 (SYN)

Think

Where is the amidine functional group in this problem? What is the structure of benzonitrile? What is the R group in the amine nucleophile that would be required?

Solve

Examining the product, we see that the connectivity of benzonitrile appears twice, as shown below.

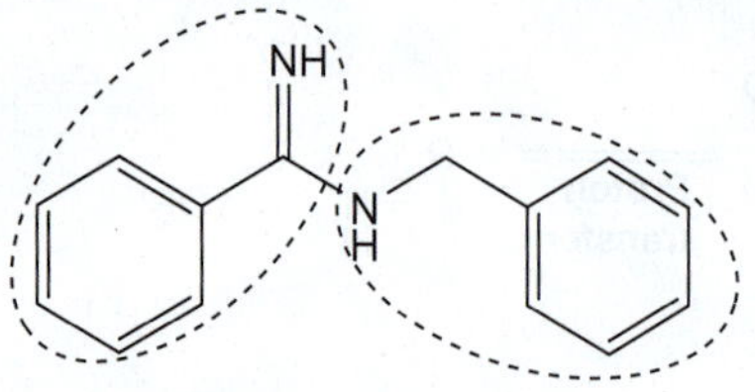

So we can envision the product being the result of a nucleophilic addition of benzylamine to benzonitrile, as shown below.

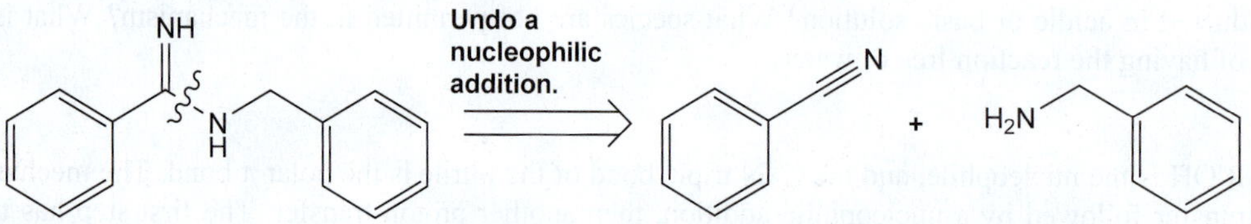

Undo a nucleophilic addition.

Benzylamine can be made by treating benzonitrile with lithium aluminum hydride, followed by acid workup. Then, benzylamine can be treated with benzonitrile, according to the reaction in Problem 18.52. We can subsequently protonate using an alcohol (water can convert the C=N to a C=O, as we saw in the text). (Using acidic conditions is not advisable because a protonated amine is not nucleophilic.) See the reaction in the forward direction on the next page.

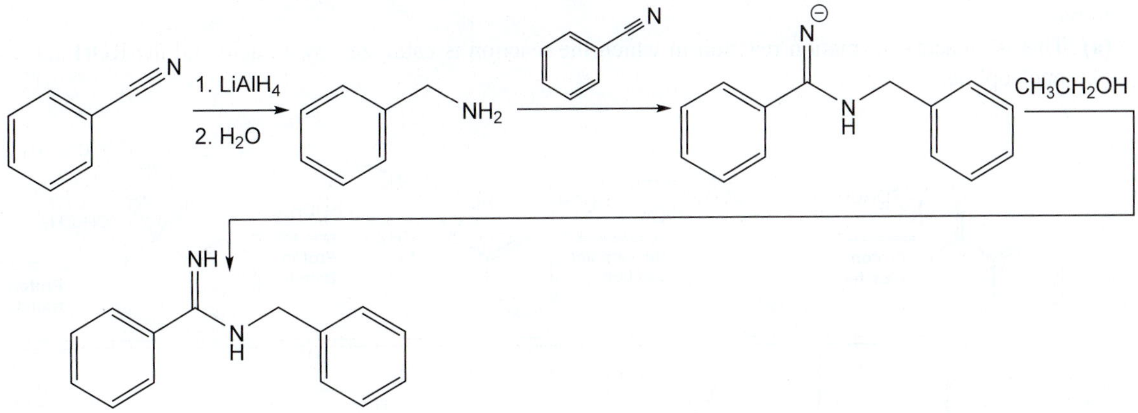

1. $LiAlH_4$
2. H_2O

NH_2

CH_3CH_2OH

NH

N
H

Problem 18.54

Think

What acts as the nucleophile in the nucleophilic addition step? What is the polar π bond that is attacked? How many times does nucleophilic addition need to take place? Is the reaction conducted in acidic or basic solution? What species are not permitted in the mechanism?

Solve

After being protonated, each cyano group undergoes nucleophilic addition by the same N atom. The nitrile has to be protonated first to avoid strongly basic, negatively charged atoms.

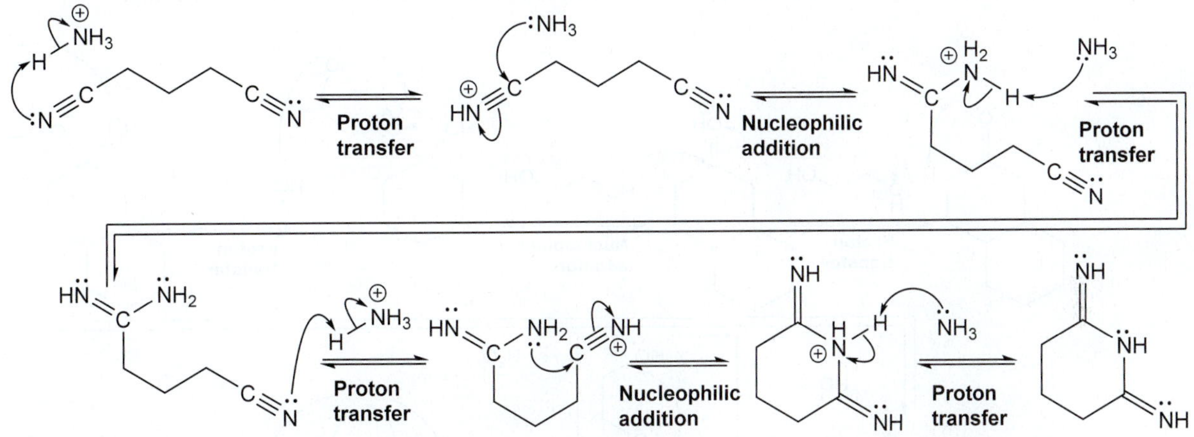

Proton transfer

Nucleophilic addition

Proton transfer

Proton transfer

Nucleophilic addition

Proton transfer

Sections 18.2 and 18.3 Formation and Hydrolysis Reactions Involving Acetals, Imines, Enamines, and Nitriles; the Wolff–Kishner Reduction

Problem 18.55

Think

What nucleophiles are present under acidic conditions? In what order do the nucleophilic addition and proton transfer reactions occur? How can the acid catalyst be regenerated? What are unreasonable species in each reaction medium? What functional group is produced?

Solve

(a) This is an acetal formation reaction in which the reaction is catalyzed by an acid and the ROH acts as the nucleophile.

Proton transfer

Nucleophilic addition

Proton transfer

Proton transfer

Heterolysis

Coordination

Proton transfer

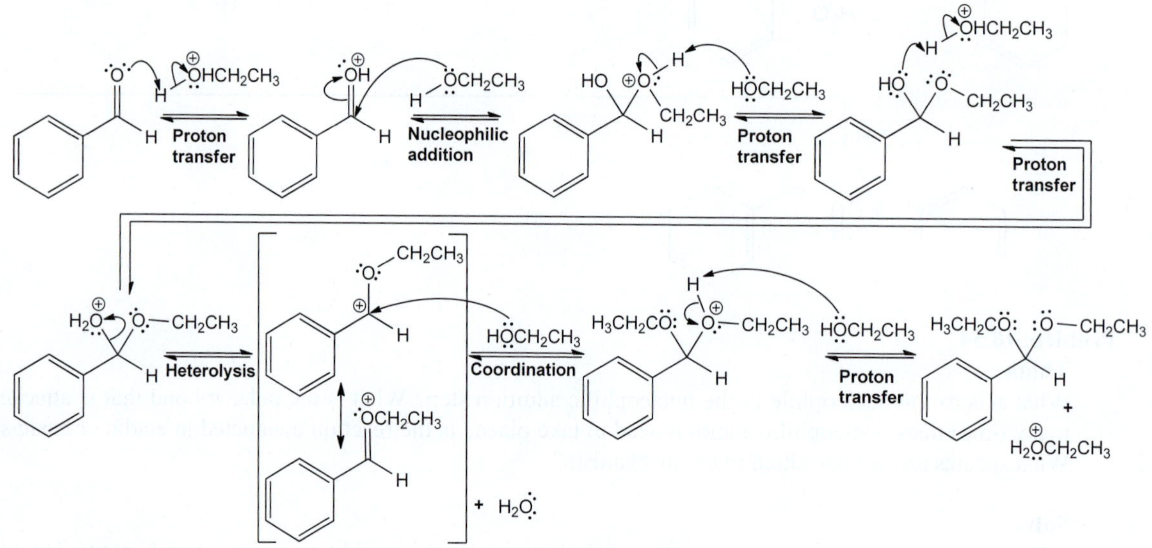

(b) This is an acetal formation reaction in which the reaction is catalyzed by an acid and the diol acts as the nucleophile. The coordination step is intramolecular because it forms a five-membered ring.

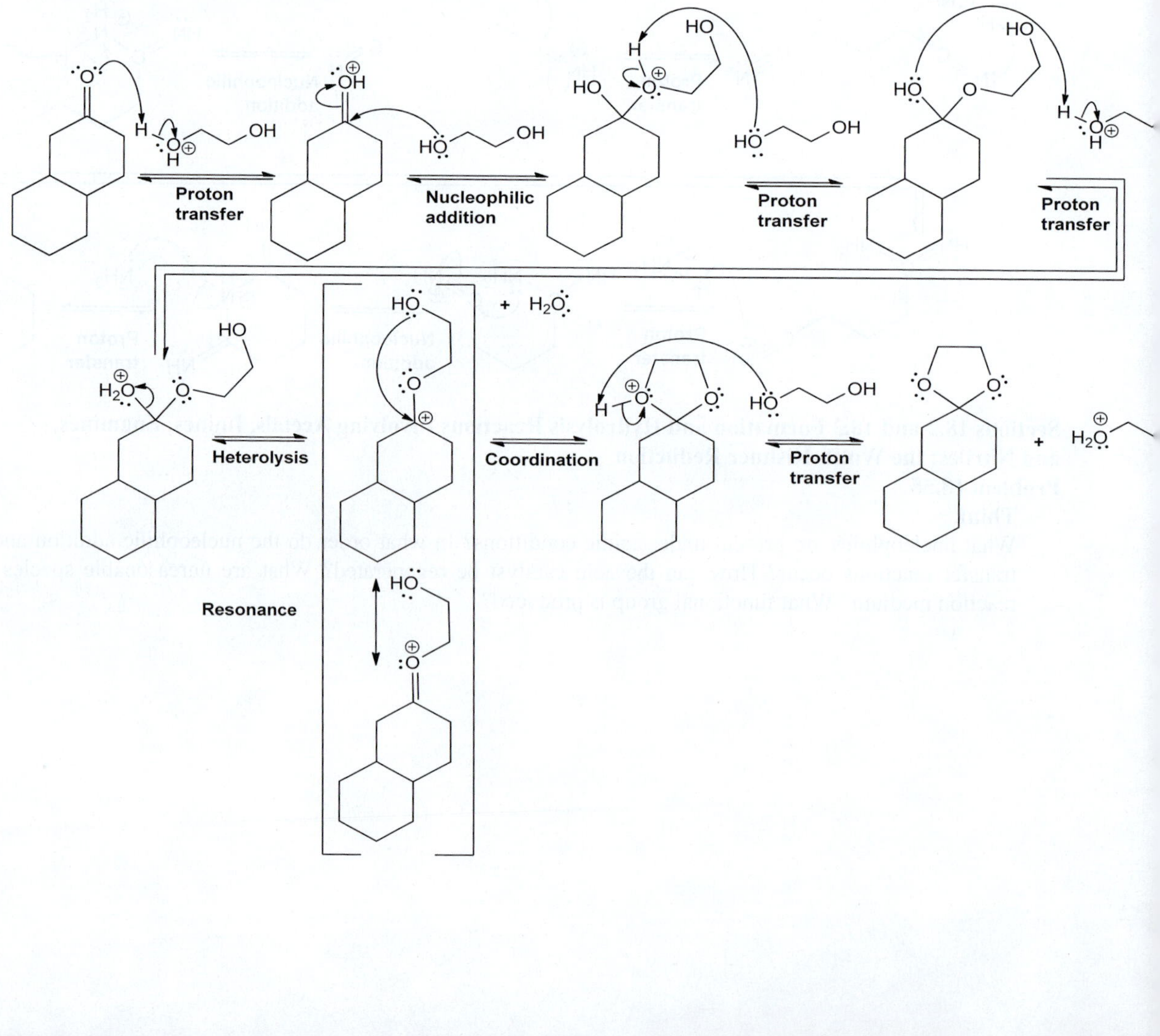

Proton transfer

Nucleophilic addition

Proton transfer

Proton transfer

Heterolysis

Coordination

Proton transfer

Resonance

(c) This is an acetal formation reaction in which the reaction is catalyzed by an acid and the ROH acts as the nucleophile.

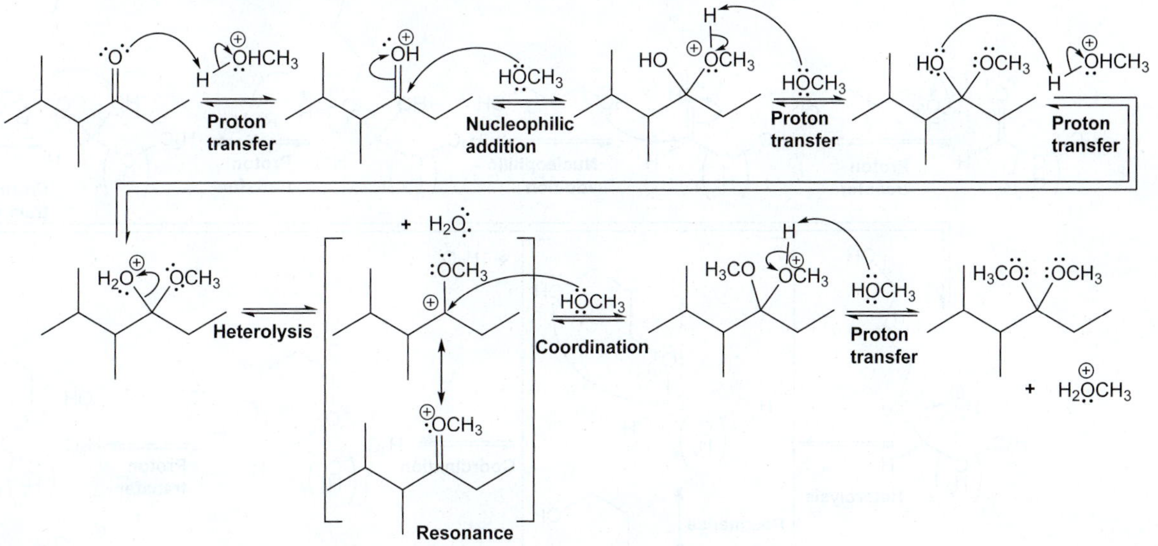

Proton transfer
Nucleophilic addition
Proton transfer
Proton transfer
Heterolysis
Resonance
Coordination
Proton transfer

(d) This mechanism is the same as in **(b)**. The S is more nucleophilic than O because it is larger and can handle the resulting positive charge better, so the S end attacks first. The coordination step is intramolecular because it forms a six-membered ring.

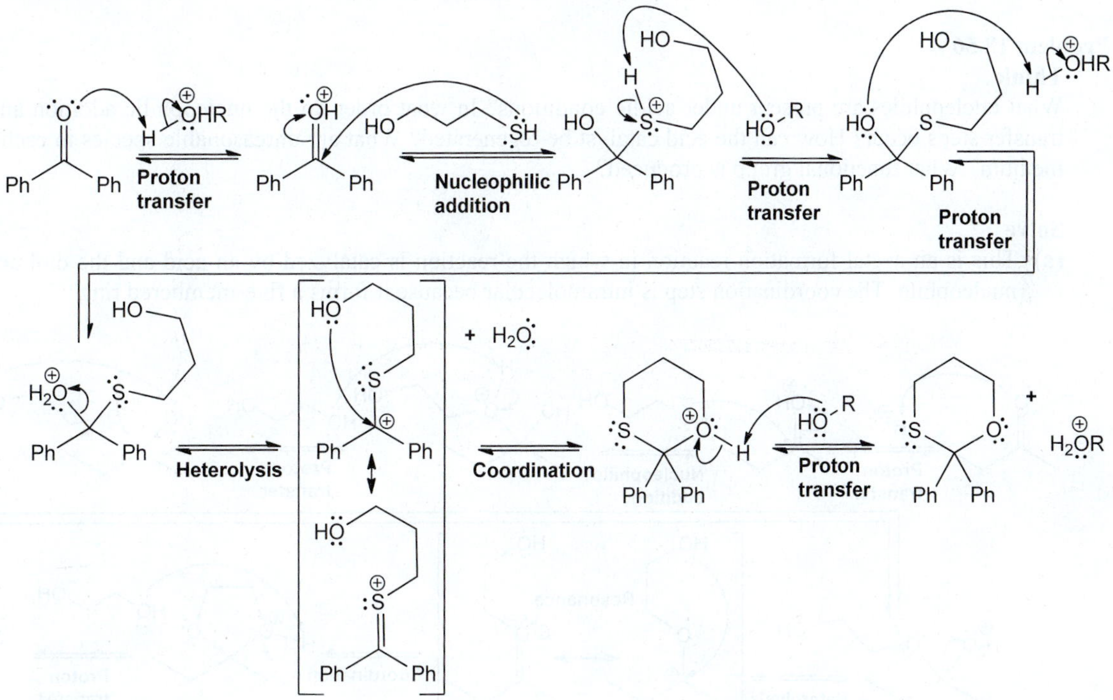

Proton transfer
Nucleophilic addition
Proton transfer
Proton transfer
Heterolysis
Coordination
Proton transfer

(e) Same mechanism as in **(b)**.

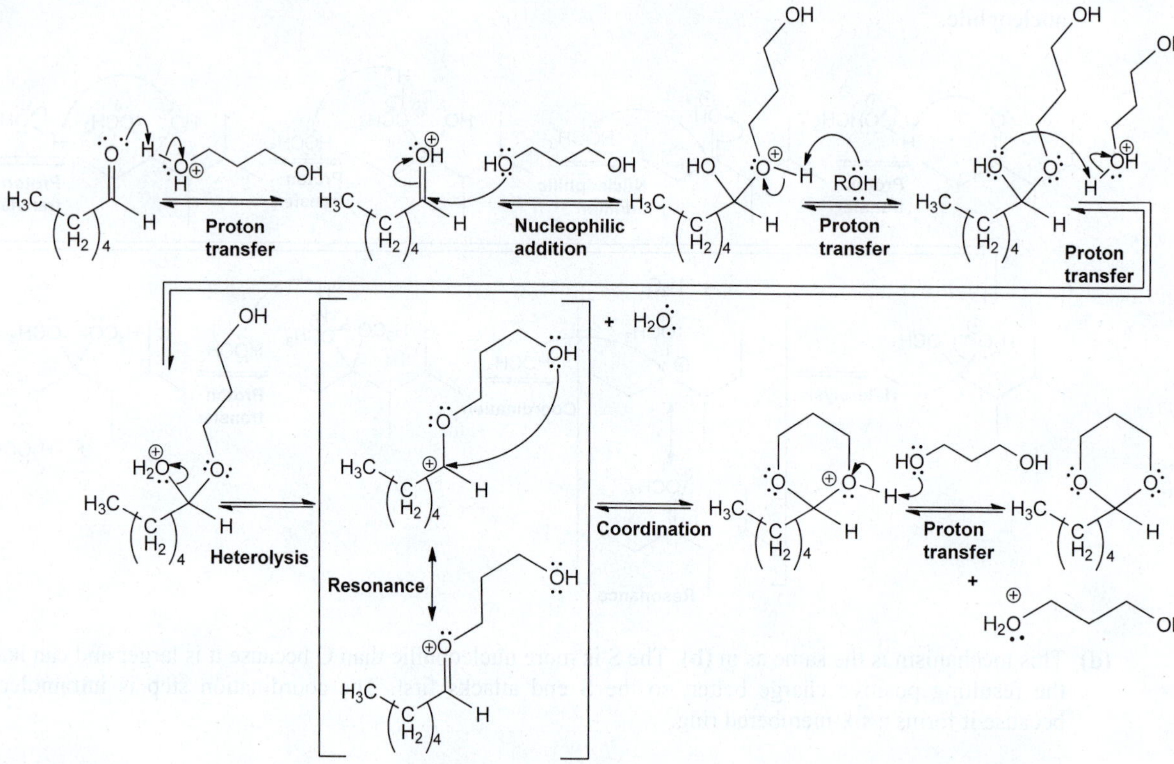

Proton transfer

Nucleophilic addition

Proton transfer

Proton transfer

Heterolysis

Resonance

Coordination

Proton transfer

Problem 18.56

Think

What nucleophiles are present under acidic conditions? In what order do the nucleophilic addition and proton transfer steps occur? How can the acid catalyst be regenerated? What are unreasonable species in each reaction medium? What functional group is produced?

Solve

(a) This is an acetal formation reaction in which the reaction is catalyzed by an acid and the diol acts as the nucleophile. The coordination step is intramolecular because it forms a five-membered ring.

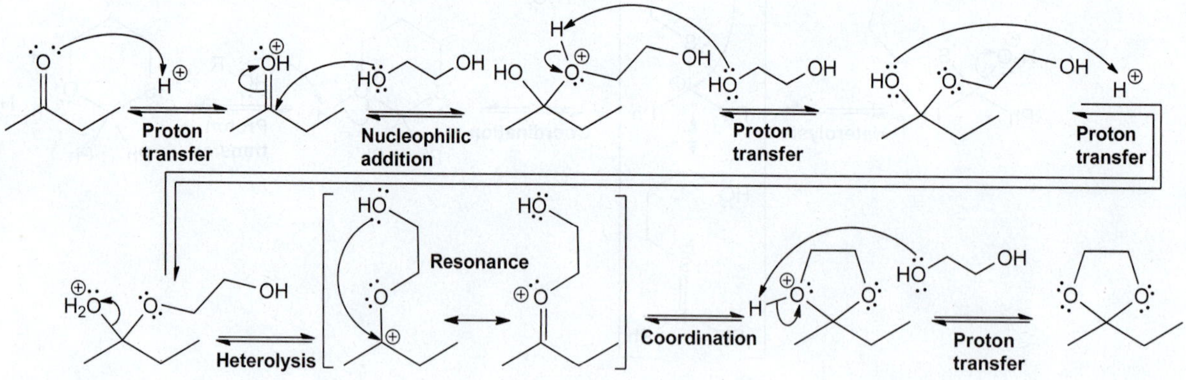

Proton transfer

Nucleophilic addition

Proton transfer

Proton transfer

Heterolysis

Resonance

Coordination

Proton transfer

(b) This is an acetal formation reaction in which the reaction is catalyzed by an acid and the diol acts as the nucleophile. The coordination step is intramolecular because it forms a five-membered ring.

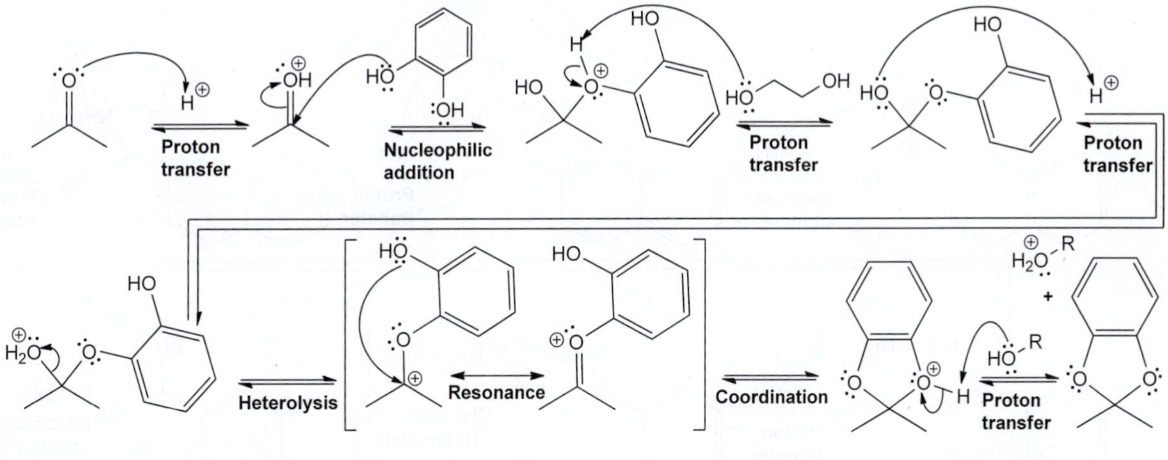

Proton transfer | Nucleophilic addition | Proton transfer | Proton transfer

Heterolysis | Resonance | Coordination | Proton transfer

(c) This is an acetal formation reaction in which the reaction is catalyzed by an acid and the diol acts as the nucleophile. The coordination step is intramolecular because it forms a six-membered ring.

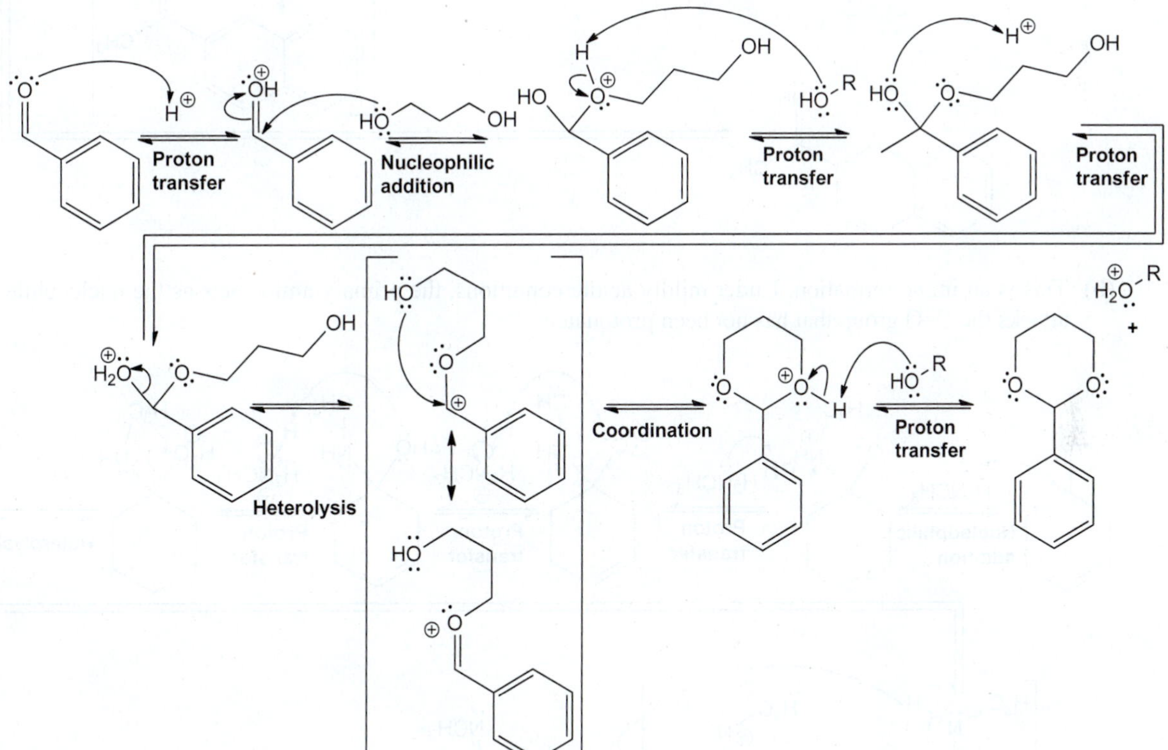

Proton transfer | Nucleophilic addition | Proton transfer | Proton transfer

Heterolysis | Coordination | Proton transfer

Problem 18.57

Think

What nucleophiles are present under mildly acidic conditions? Under strongly acidic conditions? Is the carbonyl O protonated under mildly acidic conditions? Under strongly acidic conditions? In what order do the nucleophilic addition and proton transfer steps occur? How can the acid catalyst be regenerated? What are unreasonable species in each reaction medium? What functional group is produced when a ketone or aldehyde reacts with NH_3 or a primary amine? When a ketone or aldehyde reacts with a secondary amine? When an imine or enamine reacts with water?

Solve

(a) This is an imine formation. Under mildly acidic conditions, NH_3 acts as the nucleophile and attacks the C=O group that has not been protonated.

Nucleophilic addition

Proton transfer

Proton transfer

Proton transfer

Heterolysis

Proton transfer

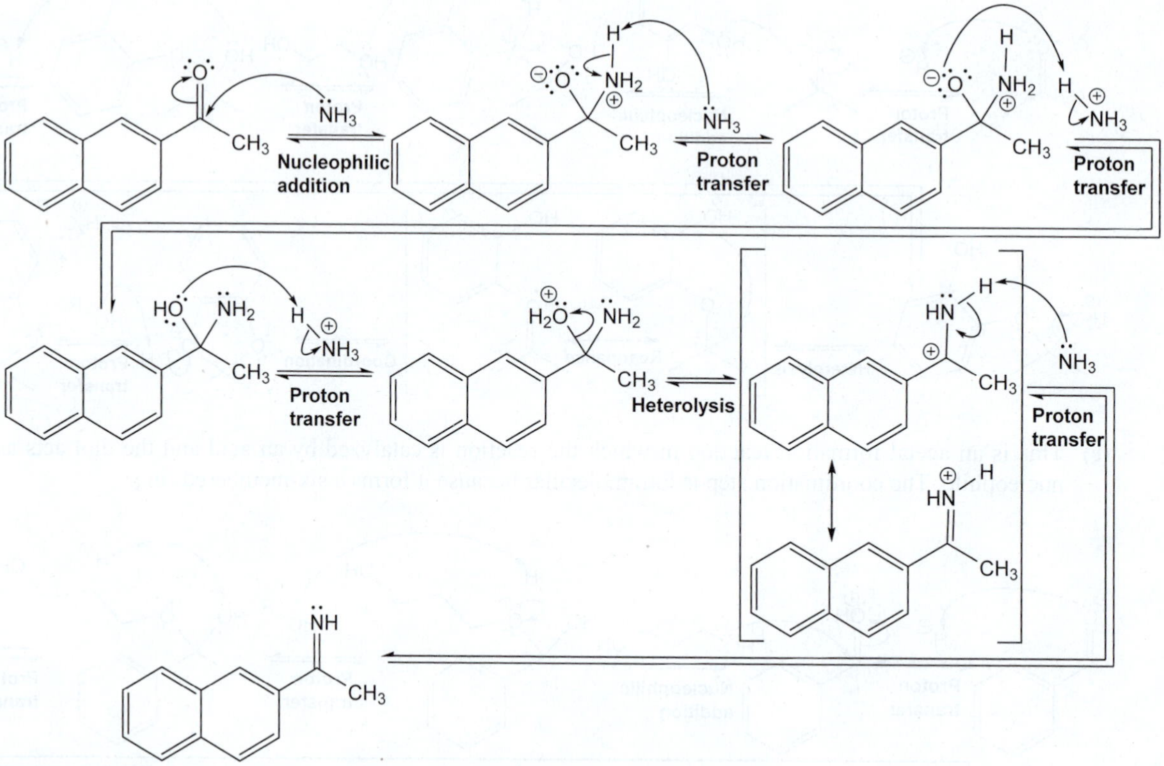

(b) This is an imine formation. Under mildly acidic conditions, the primary amine acts as the nucleophile and attacks the C=O group that has not been protonated.

Nucleophilic addition

Proton transfer

Proton transfer

Proton transfer

Heterolysis

Resonance

Proton transfer

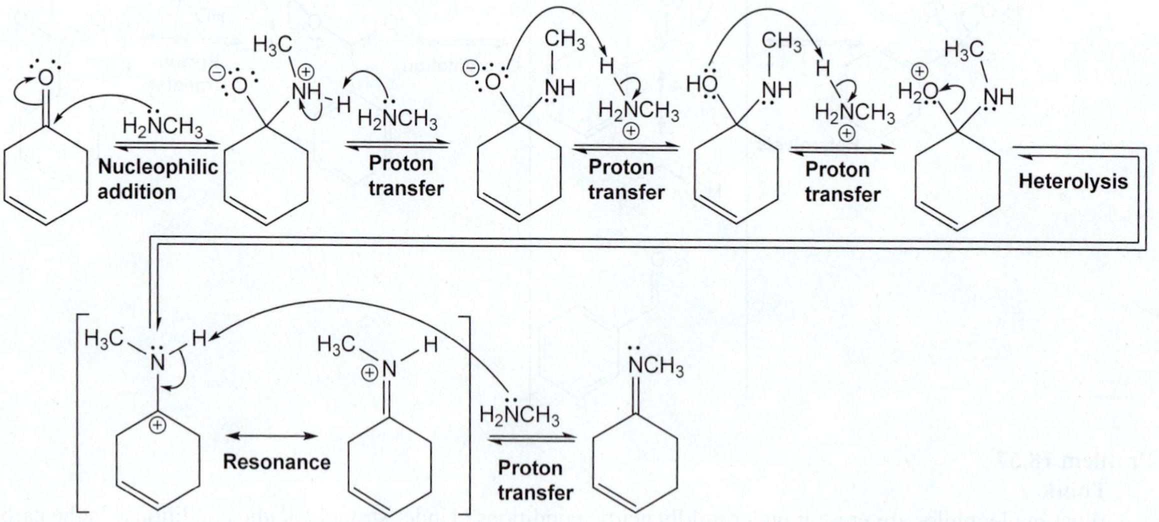

(c) This is an imine hydrolysis to produce a ketone. Under strongly acidic conditions, no strong bases should appear. The C=N is protonated first, and water acts as the nucleophile.

H^+
Proton transfer
H_2O
Nucleophilic addition
H_2O
Proton transfer
H^+
Proton transfer
Heterolysis
H_2O
Proton transfer

(d) This is an enamine formation. Under mildly acidic conditions, the secondary amine acts as the nucleophile and attacks the C=O group that has not been protonated. Because there are not two H atoms on N in the amine, the second proton is lost from C in the final step, instead of from N.

Nucleophilic addition
Proton transfer
Proton transfer
Proton transfer
Heterolysis
Elimination of H^+

(e) This is an enamine hydrolysis. Under strongly acidic conditions, no strong bases should appear. The C=C is protonated first, and water acts as the nucleophile.

Proton transfer

Nucleophilic addition

Proton transfer

Proton transfer

Heterolysis

Proton transfer

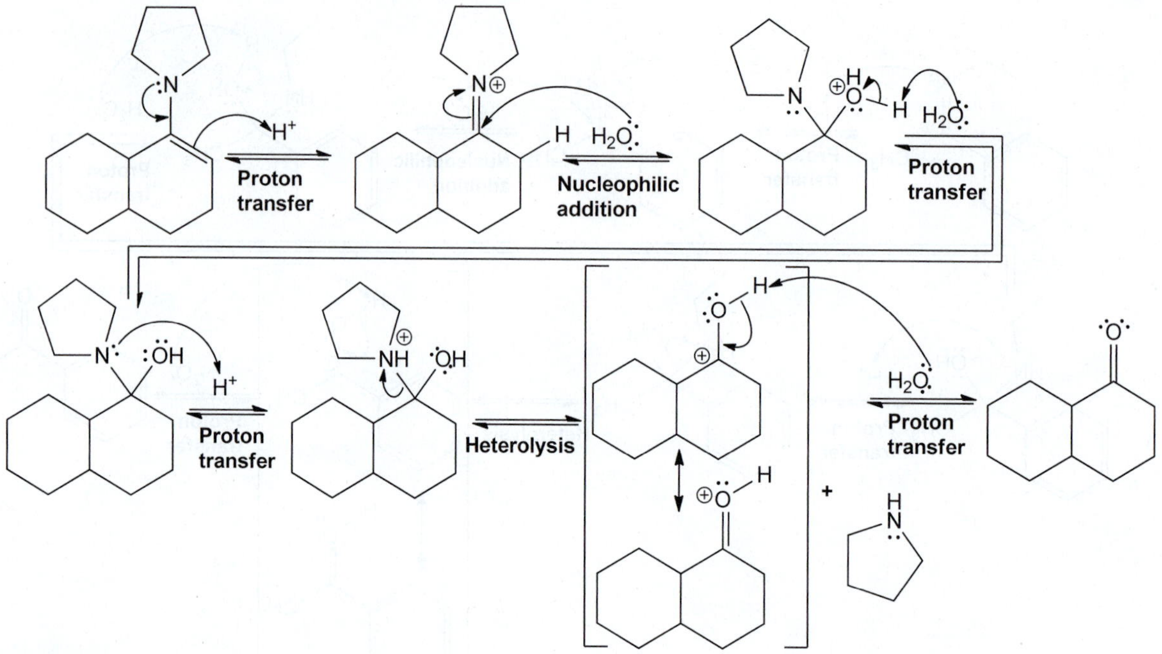

Problem 18.58

Think

What nucleophiles are present under acidic conditions? What are the leaving groups under acidic conditions? In what order do the nucleophilic addition and proton transfer steps occur to avoid unreasonable species in solution? How can the acid catalyst be regenerated? What functional group is produced?

Solve

(a) This is an acetal hydrolysis. Under acidic conditions, no strong bases should appear. The leaving group is an uncharged alcohol, and water acts as the nucleophile.

Proton transfer

Heterolysis

Nucleophilic addition

Proton transfer

Proton transfer

Heterolysis

Proton transfer

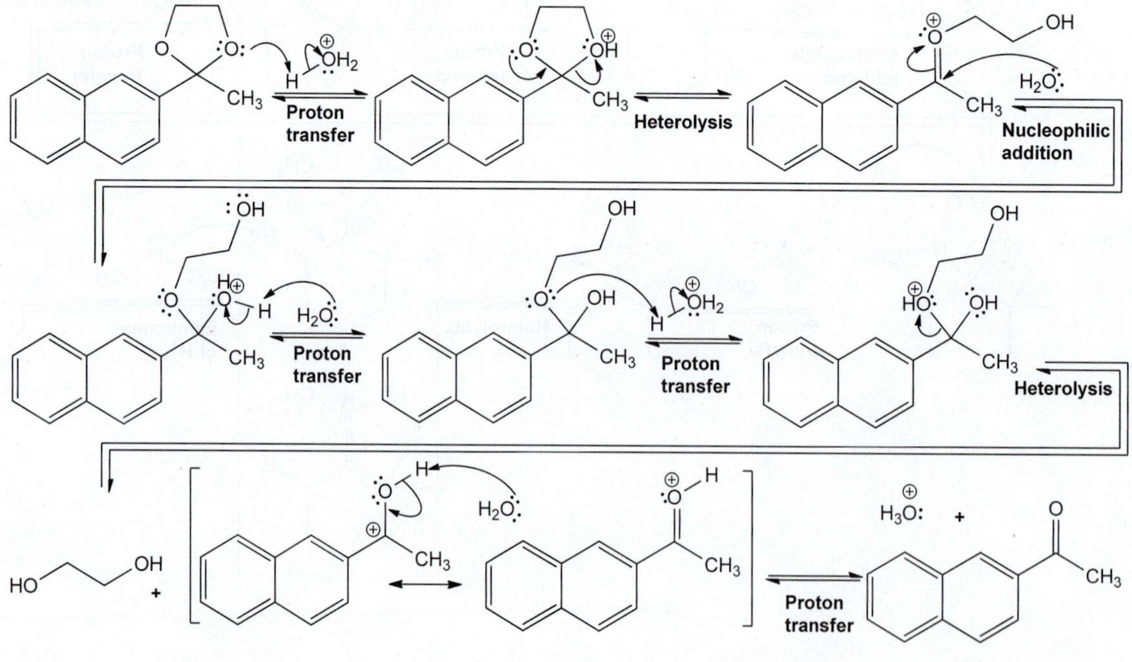

(b) This is an acetal hydrolysis, essentially the same as in **(a)**.

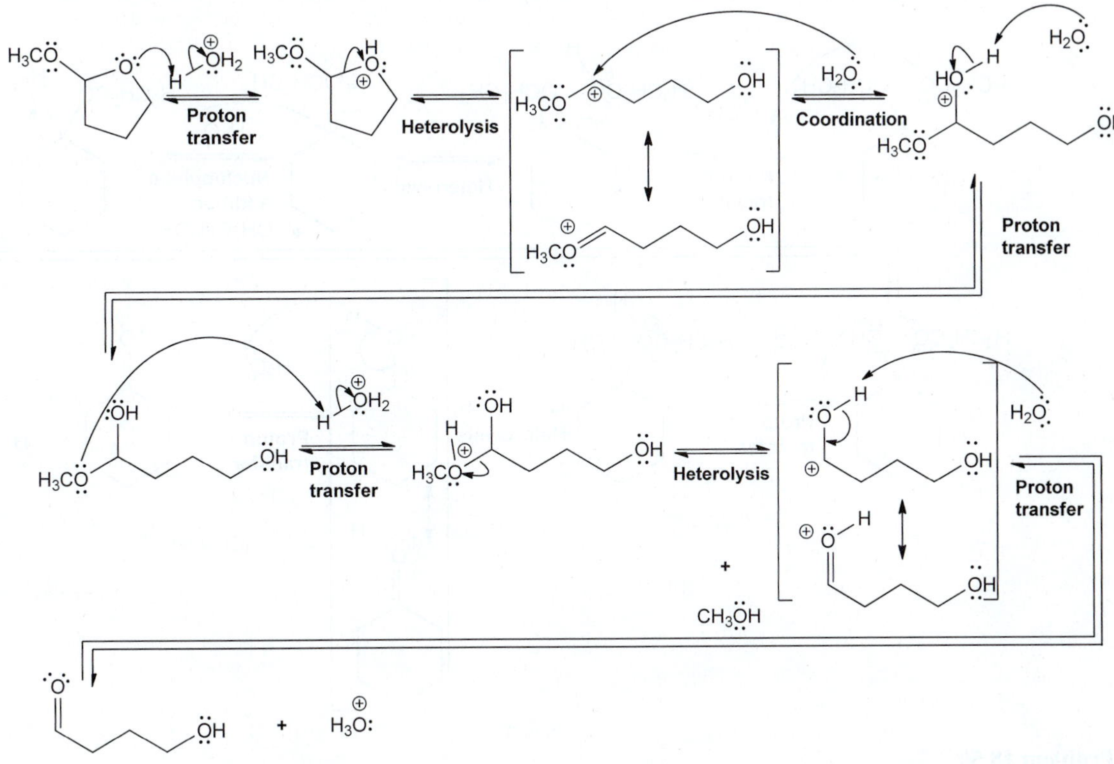

(c) This is hydrolysis of the nitrile to produce an amide. Under acidic conditions, no strong bases should appear. The C≡N is protonated first, and water acts as the nucleophile.

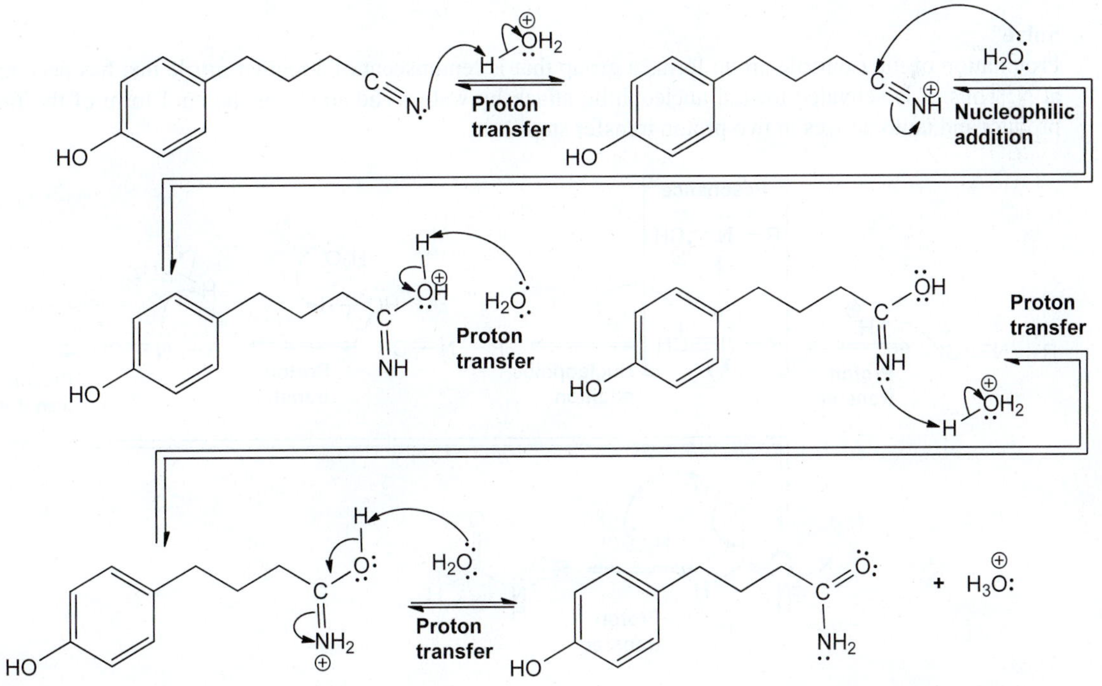

(d) This is a hydrolysis of an acetal. Under acidic conditions, no strong bases should appear. The leaving group is an uncharged alcohol, and water is the nucleophile.

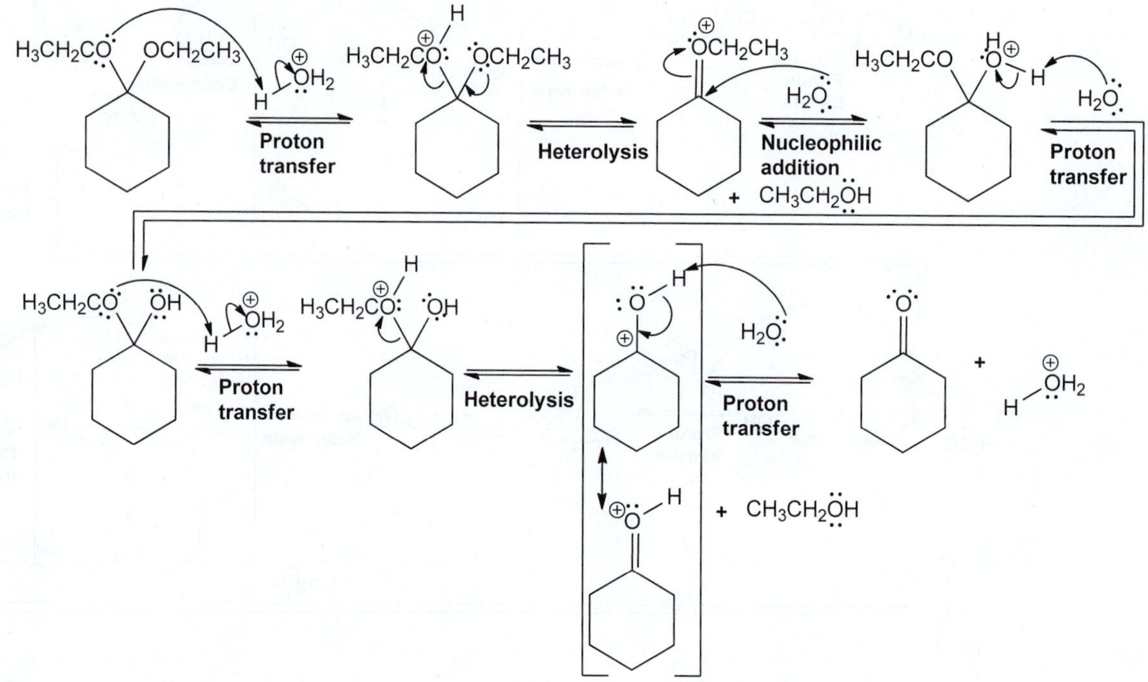

Proton transfer

Heterolysis

Nucleophilic addition

+ CH_3CH_2OH

Proton transfer

Proton transfer

Heterolysis

Proton transfer

+ CH_3CH_2OH

Problem 18.59

Think

How is the isonitrile similar to a carbonyl? What acts as the nucleophile in the nucleophilic addition step? Is the reaction conducted in acidic or basic solution? What species are not permitted in the mechanism?

Solve

Protonation of the isonitrile group forms a group that is reminiscent of a regular nitrile that has been protonated at N. Thus, it is activated toward nucleophilic attack by water. The adduct is the enol form of the final amide product and tautomerizes in two proton transfer steps.

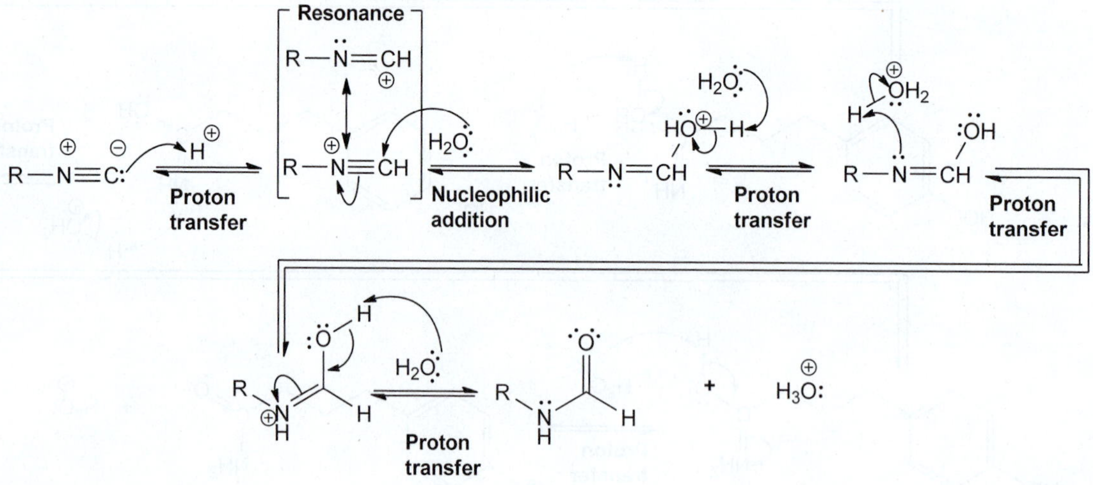

Resonance

Proton transfer

Nucleophilic addition

Proton transfer

Proton transfer

Proton transfer

+ H_3O^+

Problem 18.60

Think

What happens to an amine at low pH? Is the lone pair still present? Is the amine nucleophilic? How is an alcohol different?

Solve

Amines are weak bases, so if the solution pH is too low, then the N of an amine, such as R–NH_2, is protonated to form, for example, R–NH_3^+. A protonated N has no lone pair of electrons and, further, carries a +1 formal charge, so it is a terrible nucleophile. This is not a problem for acetal formation, because the nucleophile is an alcohol, which requires the pH to be much more acidic to become protonated.

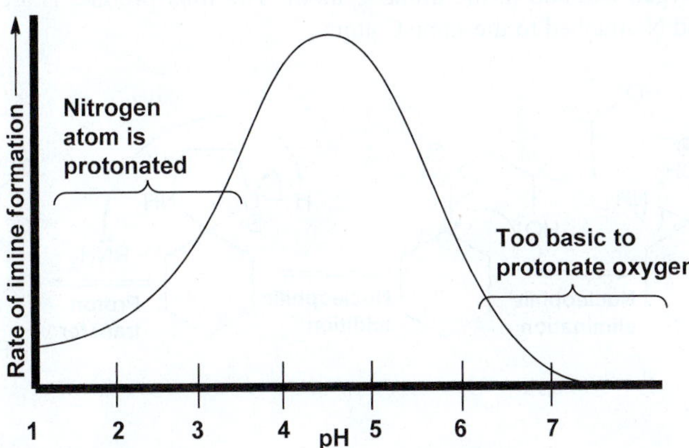

Problem 18.61

Think

What is the Lewis structure of the hydroxylamine? What is the nucleophile in each of the two possible nucleophilic addition reactions? Which atom, N or O, is a stronger nucleophile?

Solve

(a) We will assume a pH of about 5. (See the previous problem for more details as to why.) Acetal formation is shown first.

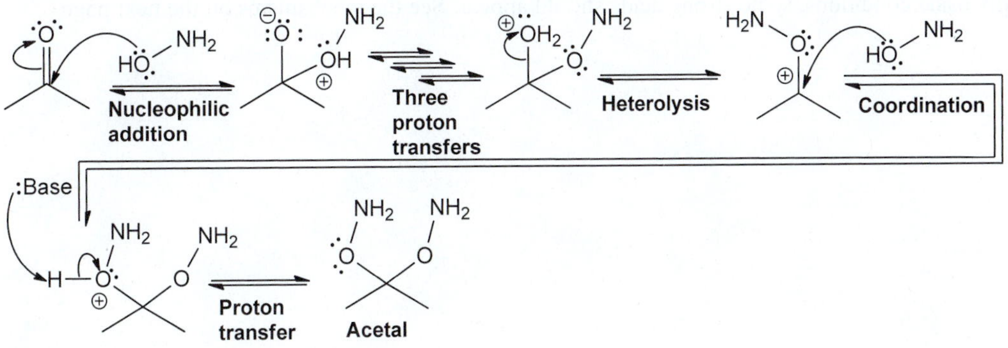

Nucleophilic addition

Three proton transfers

Heterolysis

Coordination

:Base

Proton transfer

Acetal

Next, formation of the oxime is shown.

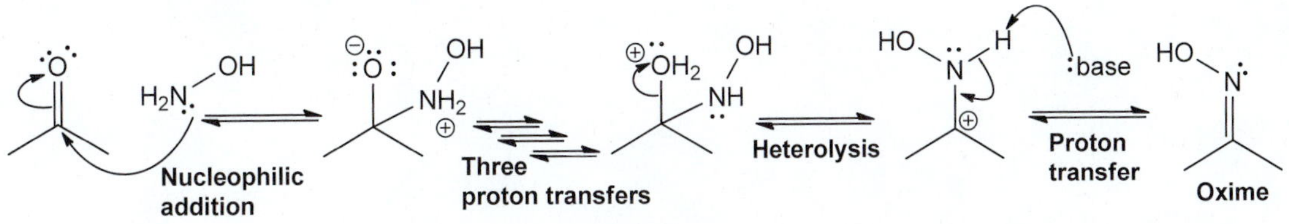

Nucleophilic addition

Three proton transfers

Heterolysis

:base

Proton transfer

Oxime

(b) Formation of the oxime is the favored reaction, because N is more nucleophilic than O.

Problem 18.62

Think

What is TsOH? Under mildly acidic conditions, what acts as the nucleophile in the nucleophilic addition reaction? Would the C=O oxygen be protonated prior to nucleophilic attack? Is oxygen or nitrogen more nucleophilic? Can a new five- or six-membered ring be produced?

Solve

If we assume mildly acidic conditions from just a small amount of added TsOH, then the C=O oxygen would not be protonated prior to nucleophilic attack. Nitrogen is the more nucleophilic atom in aminoethanol, so it adds to the carbonyl C first. The carbonyl O is protonated again to form a good leaving group (water), and C=N forms, characteristic of an imine. Now the oxygen can add to the imino C atom. The final product is a cyclic *hemiaminal*—a functional group that has O and N attached to the same C atom.

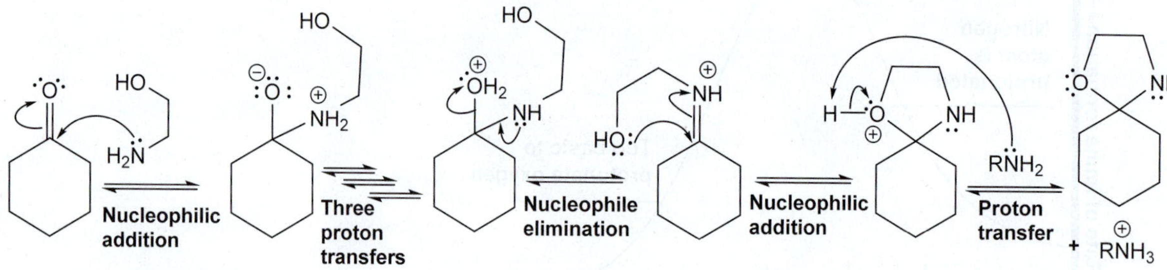

Problem 18.63

Think

What kind of reaction involves hydrazine and a ketone or aldehyde? What nucleophiles are present under mildly acidic conditions? In what order do the nucleophilic addition and proton transfer steps occur? How can the acid catalyst be regenerated? What are unreasonable species in strongly basic conditions? What functional group is produced? How is the base in the second reaction involved?

Solve

Both of these are Wolff–Kishner reductions, which reduce the C=O bond of a ketone or aldehyde to a methylene (CH_2) group. In each case, the first reaction takes place under mildly acidic conditions, so the nucleophilic NH_2 remains uncharged, and so does the C=O oxygen. The second reaction takes place under strongly basic conditions, so no strong acids should appear. See the mechanisms on the next page.

(a)

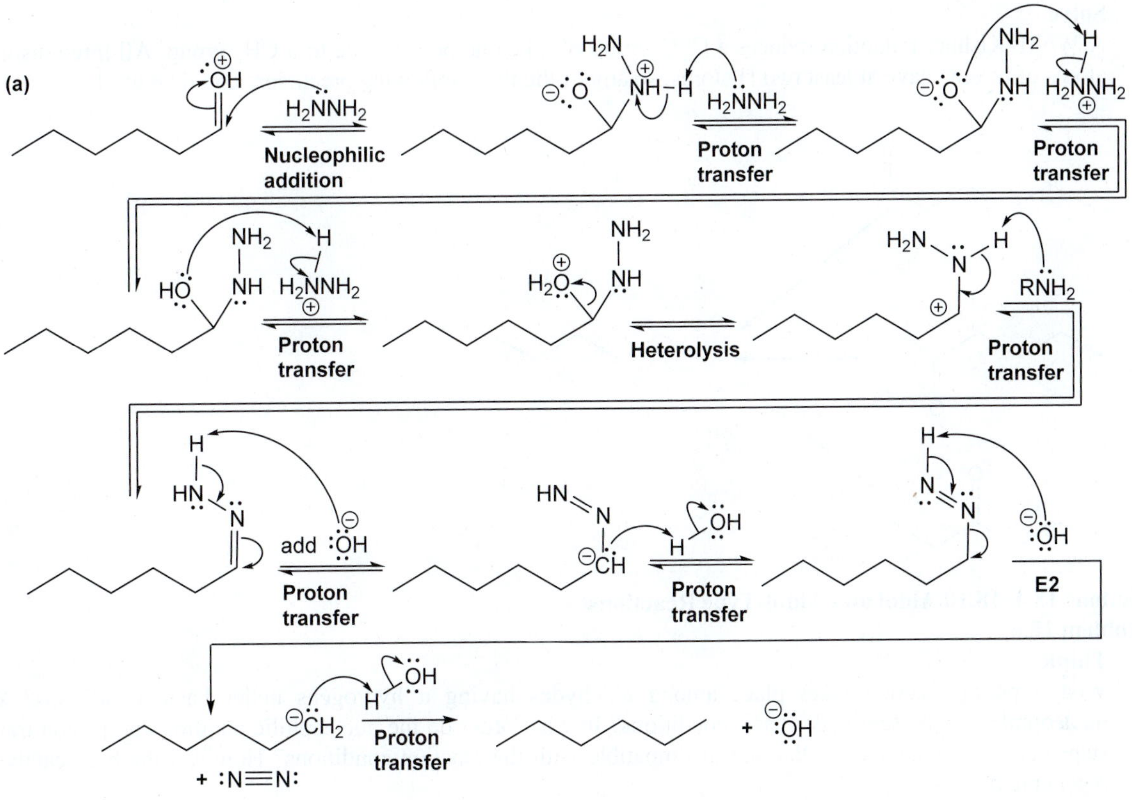

(b)

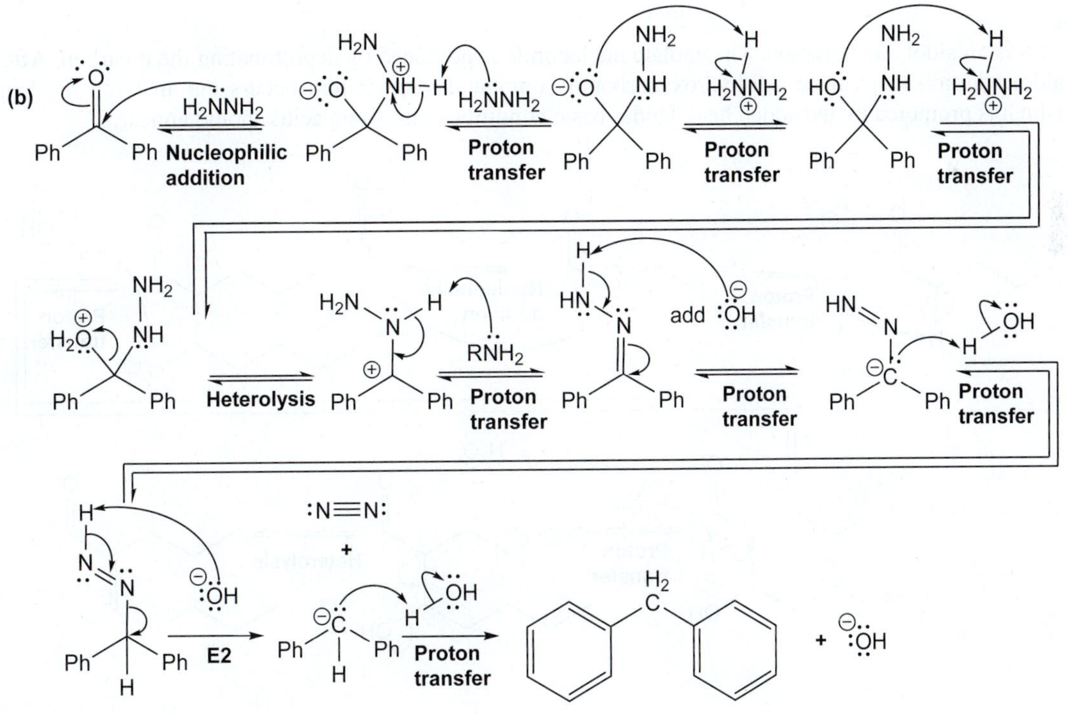

Problem 18.64 (SYN)

Think

What functional group transformation occurs in a Wolff–Kishner reduction? Which C atoms of hexane have at least two H atoms? Can each of those be produced from a C=O using a Wolff–Kishner reduction?

Solve

A Wolff–Kishner reduction reduces a C=O group of a ketone or aldehyde to a CH_2 group. All three distinct C atoms of hexane have at least two H atoms, so any of the three following precursors could be used.

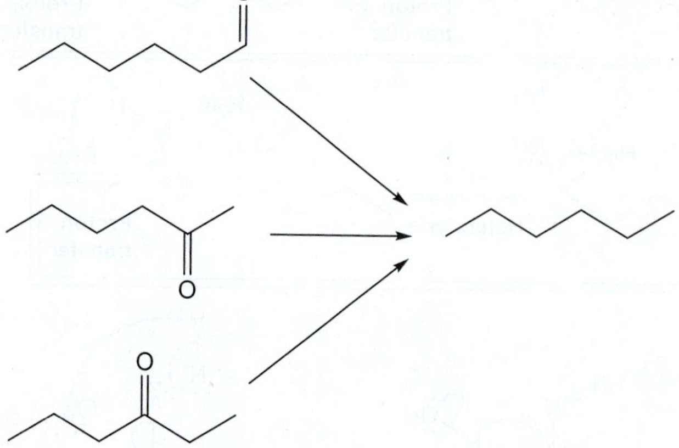

Sections 18.4–18.10 Aldol and Aldol-Type Reactions

Problem 18.65

Think

What type of reaction takes place among aldehydes having α hydrogens under basic conditions? What nucleophiles are present under basic conditions? In what order do the nucleophilic addition and proton transfer steps occur to avoid species that are incompatible with the reaction conditions? How can the base catalyst be regenerated?

Solve

(a) This is an aldol condensation. The enolate nucleophile is produced by deprotonating the α carbon. After the aldol addition to produce a β-hydroxycarbonyl compound, dehydration occurs via an E1cb mechanism, which is promoted by the added heat. Under basic conditions, no strong acids should appear.

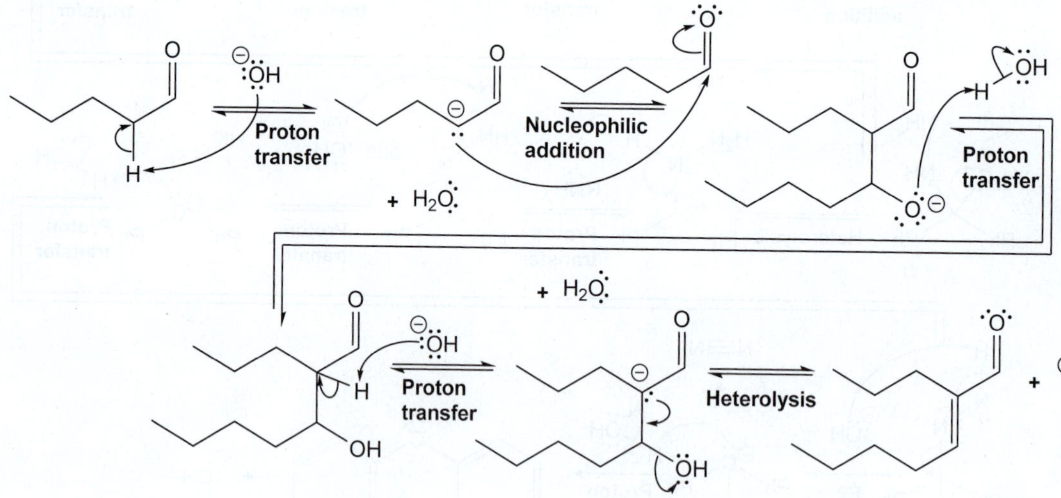

(b) This is an aldol condensation. The enolate nucleophile is produced by deprotonating the α carbon. After the aldol addition to produce a β-hydroxycarbonyl, dehydration occurs via an E1cb mechanism, which is promoted by the added heat. Under basic conditions, no strong acids should appear.

Proton transfer + H_2O — Nucleophilic addition — Proton transfer — O + H_2O — Proton transfer — Heterolysis — + $^{\ominus}OH$

(c) This aldol addition is intramolecular, favored by the formation of the six-membered ring. The enolate nucleophile is produced by deprotonating the α carbon.

Proton transfer — + H_2O — Nucleophilic addition — Proton transfer — + $^{\ominus}OH$

(d) This is an aldol addition. The conditions with the added heat favor an aldol condensation, but an aldol condensation cannot occur, because the β-hydroxycarbonyl product's α C atom does not have any more hydrogen atoms to allow it to undergo the E1cb dehydration mechanism.

Proton transfer — Nucleophilic addition — + H_2O — Proton transfer — + $^{\ominus}OH$

Problem 18.66

Think

What nucleophile is present under acidic conditions? Is it nucleophilic at an α carbon? In what order do the nucleophilic addition and proton transfer steps occur to avoid species that are incompatible with the reaction conditions?

Solve

This aldol condensation reaction takes place under acidic conditions, so no strongly basic species should appear in the mechanism. Therefore, the carbon nucleophile cannot be an enolate anion, which is strongly basic. Instead, the nucleophile has to be the enol, which is nucleophilic at the α carbon. The enol is produced by back-to-back proton transfer steps (tautomerization).

Propanal

Proton transfer

Proton transfer

Nucleophilic addition

Proton transfer

3-Hydroxy-2-methylpentanal

Problem 18.67

Think

Compare the sterics of the two products. Is there a significant difference? Which C=O has a more electron-poor C atom? How does this affect the amount of aldol product in equilibrium?

Solve

The difluoroketone **B** will produce more product at equilibrium. F and H atoms are about the same in size (F is small because it is so highly electronegative), so the two molecules have about the same steric bulk surrounding their respective carbonyl carbons. But the F atoms make the carbonyl carbon more electron deficient, which favors attack of a nucleophile.

H_3C δ^+ CH_3

A

FH_2C δ^+ CH_2F

B

Problem 18.68

Think

Do HO^- and LDA act as bases or nucleophiles in these reactions? What nucleophiles are present under basic conditions? How does the choice of base impact the regiochemistry in the production of an enolate anion? In what order do the nucleophilic addition and proton transfer steps occur to avoid species that are incompatible with the basic conditions? How can the base catalyst be regenerated?

Solve

HO^- and LDA in each of these examples act as bases and deprotonate the α C–H to form the enolate anion.

(a) This is a crossed aldol reaction, where LDA produces the kinetic (less substituted) enolate anion. Condensation occurs, even without heat, owing to the extended conjugation with the aromatic ring in the α,β-unsaturated carbonyl product.

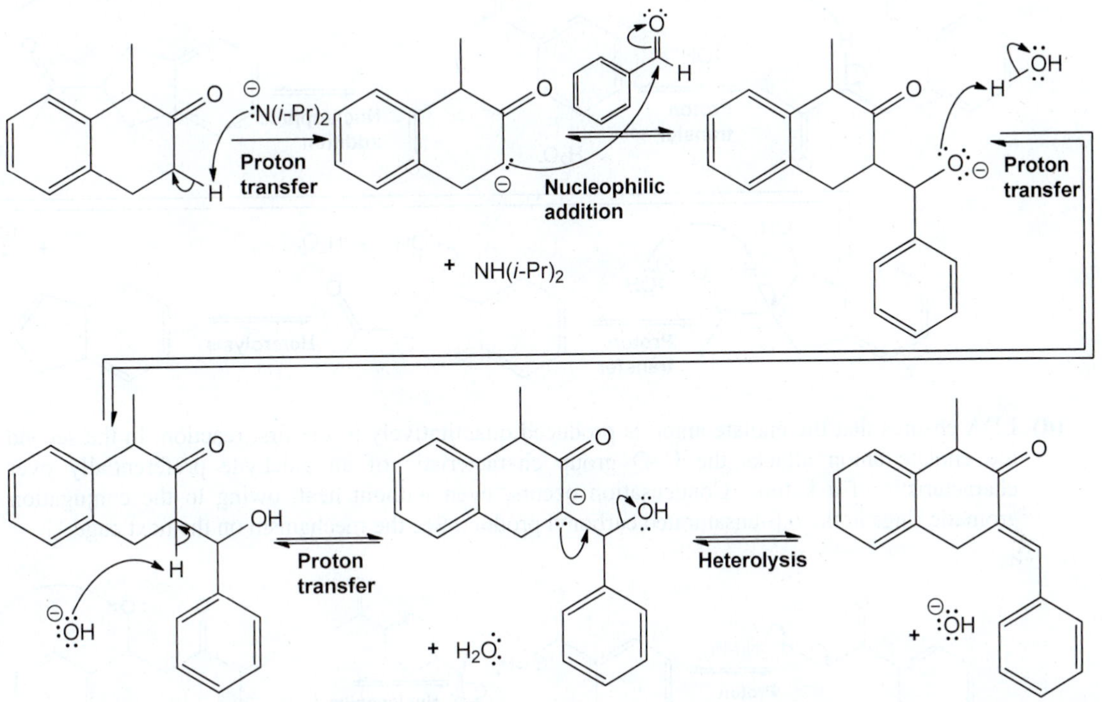

Proton transfer

$:N(i\text{-}Pr)_2$

Nucleophilic addition

$+ NH(i\text{-}Pr)_2$

Proton transfer

Proton transfer

$+ H_2O$

Heterolysis

$+ :OH^-$

(b) This is a crossed aldol reaction, where NaOH produces the thermodynamic (more substituted) enolate anion. An aldol condensation cannot occur because the α C atom in the β-hydroxycarbonyl product does not have any more hydrogen atoms.

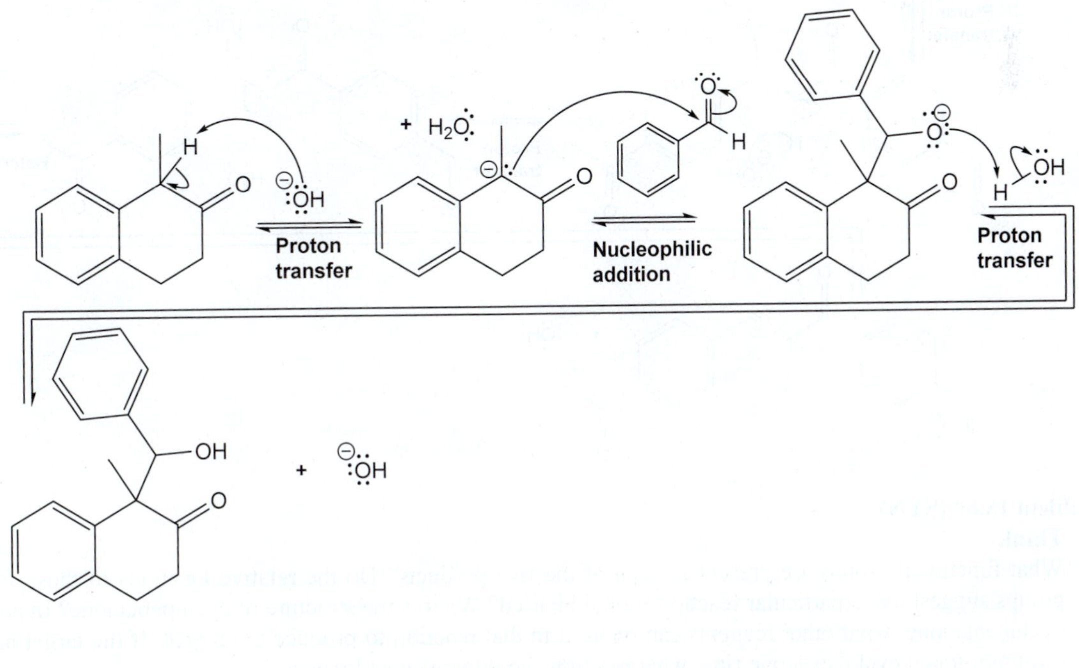

Proton transfer

$+ H_2O$

Nucleophilic addition

Proton transfer

$+ :OH^-$

(c) This is an intramolecular aldol condensation reaction. Two α carbons can be deprotonated. When the one shown below is deprotonated, a five-membered ring is produced. Deprotonating the other carbon would lead to the formation of a less favorable seven-membered ring. The added heat favors condensation.

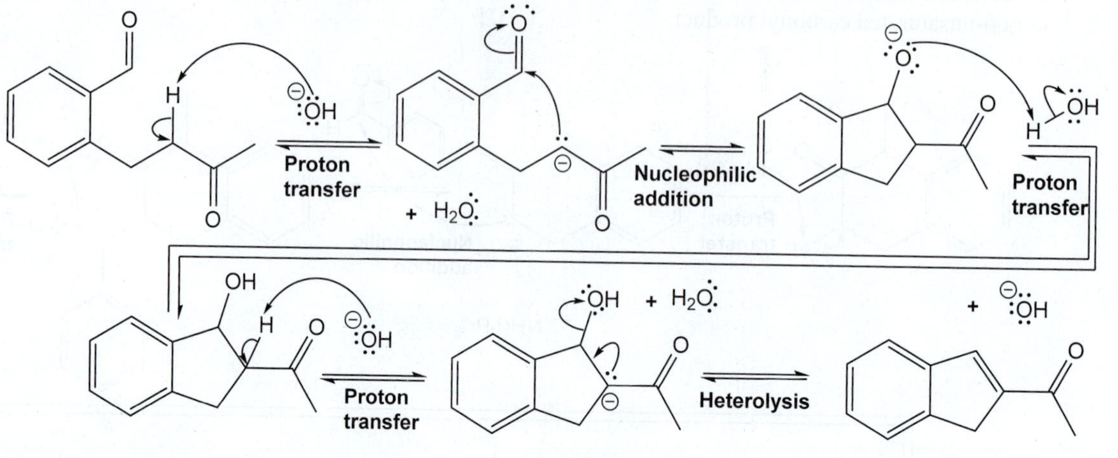

Proton transfer

$+ H_2O$

Nucleophilic addition

Proton transfer

Proton transfer

$+ H_2O$

$+ {}^{-}OH$

Heterolysis

(d) LDA ensures that the enolate anion is produced quantitatively in the first reaction. In the second reaction, the enolate anion attacks the C=O group characteristic of an aldehyde preferentially over the one characteristic of a ketone. Condensation occurs, even without heat, owing to the conjugation with the aromatic rings in the α,β-unsaturated carbonyl product. See the mechanism on the next page.

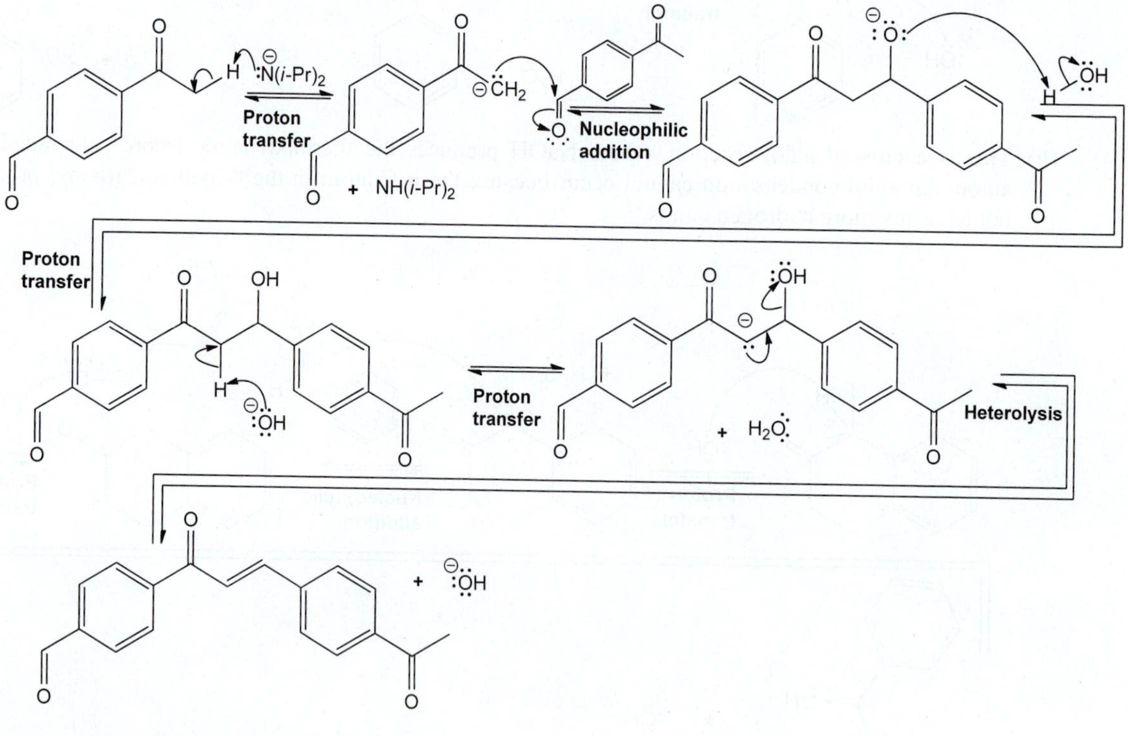

Proton transfer

$+ NH(i\text{-Pr})_2$

Nucleophilic addition

Proton transfer

Proton transfer

$+ H_2O$

Heterolysis

$+ {}^{-}OH$

Problem 18.69 (SYN)

Think

What functional groups are present in each of the five products? Do the relative locations of those functional groups suggest that a particular reaction should be used? What is the structure of cyclopentanone? In addition to cyclopentanone, what other reagents can be used in that reaction to produce the target? If the target has a new α,β-unsaturated cyclohexenone ring, what reaction should you consider using?

Solve

Each of the products is an α,β-unsaturated ketone or nitrile, which can be produced from an aldol or aldol-type condensation reaction with cyclopentanone. The routes to synthesize **(a)–(e)** are given below.

(a) A simple aldol condensation (with heat to promote dehydration) suffices for the first synthesis. Retrosynthetic analysis: Disconnect the bond between the α and β carbons to arrive at the precursors to use in an aldol condensation.

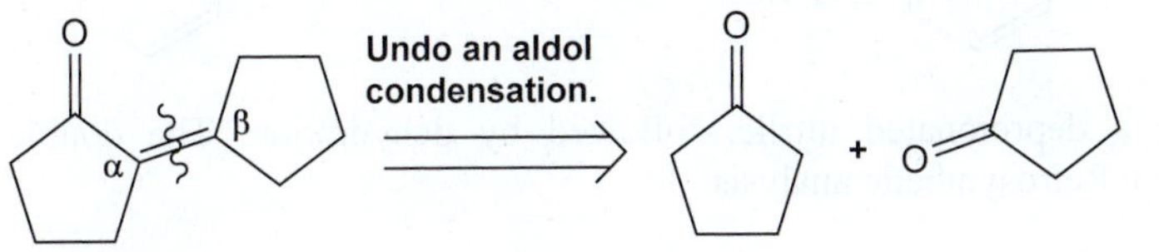

Undo an aldol condensation.

Synthesis:

NaOH, Δ

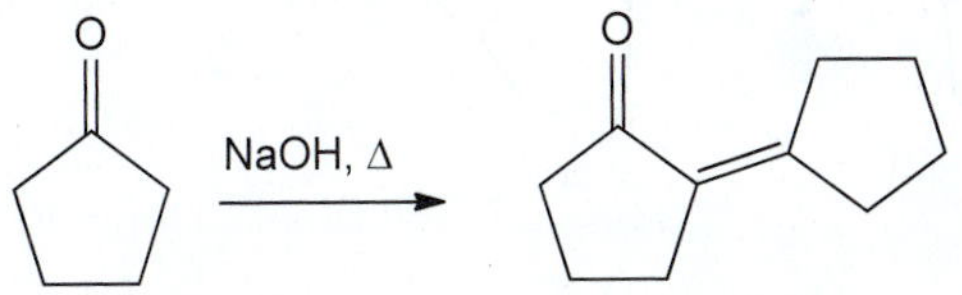

(b) A crossed aldol with dehydration works here. Retrosynthetic analysis:

Undo a crossed aldol condensation.

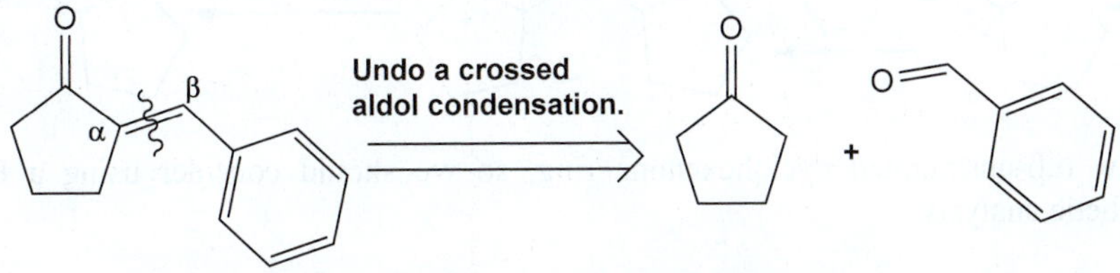

Synthesis: NaOH can be combined with benzaldehyde first, and no reaction will take place because there are no α hydrogens. Cyclopentanone is then added slowly.

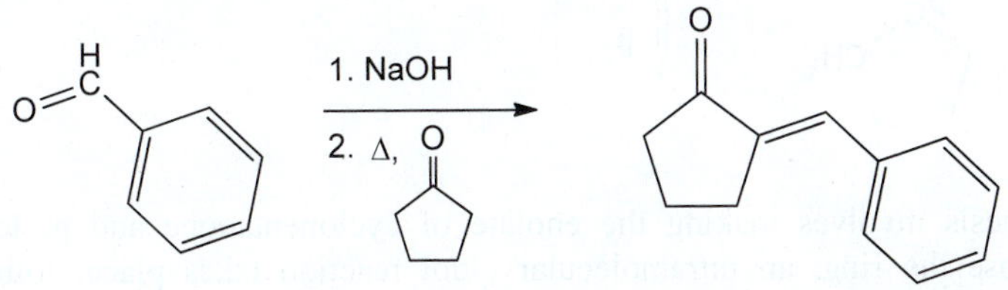

1. NaOH

2. Δ,

(c) A crossed aldol reaction, followed by dehydration. Retrosynthetic analysis:

Undo a crossed aldol condensation.

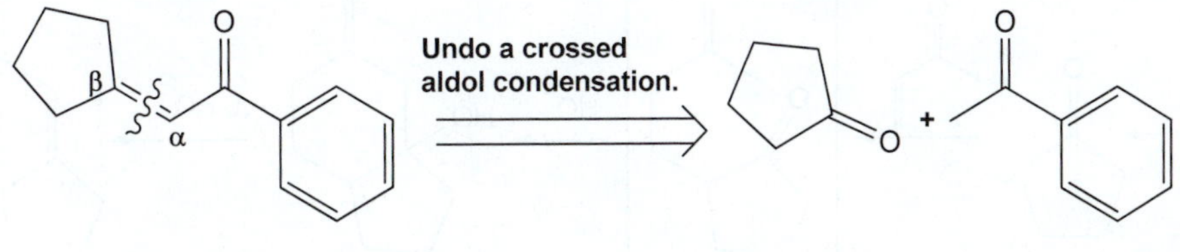

Synthesis: Because both ketones have α carbons, the crossed aldol is achieved by quantitatively deprotonating one ketone first, then adding the second ketone.

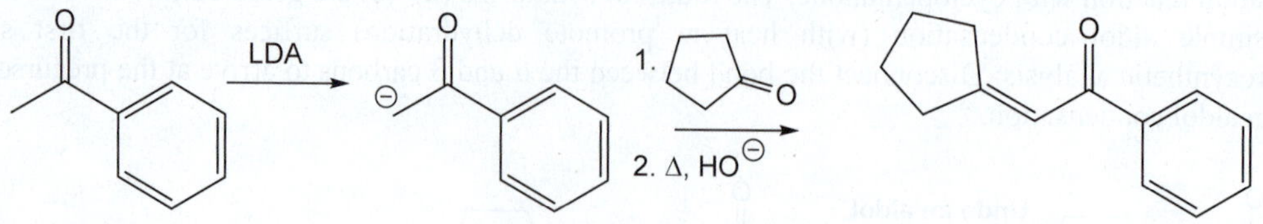

LDA
1.
2. Δ, $HO^{\ominus}$

(d) Aldol-type reaction involving a deprotonated nitrile, followed by dehydration. The double bond is conjugated with the nitrile group. Retrosynthetic analysis:

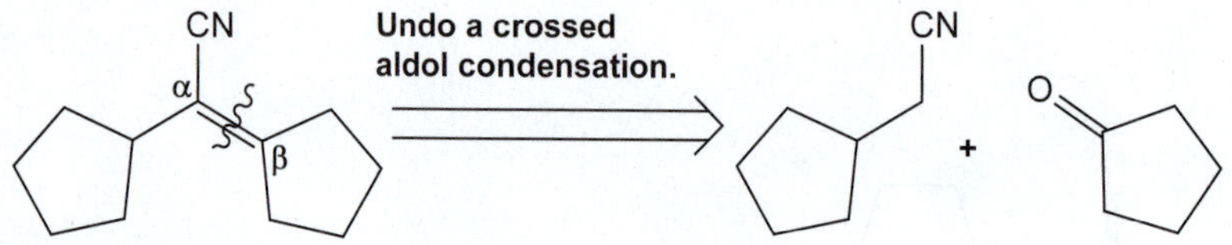

CN α β
Undo a crossed aldol condensation.
CN + O

Synthesis:

CN LDA CN O CN O NaOH, H_2O, Δ CN

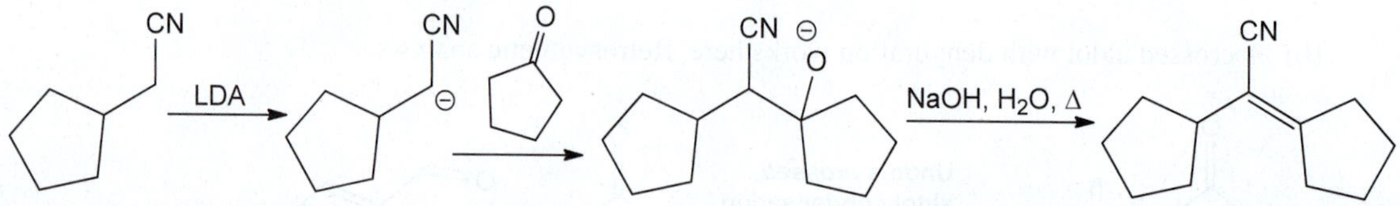

(e) The target has a new α,β-unsaturated cyclohexenone ring, so we should consider using a Robinson annulation. Retrosynthetic analysis:

O α α β β
Undo a Robinson annulation.
O α β C CH_2 + O α β

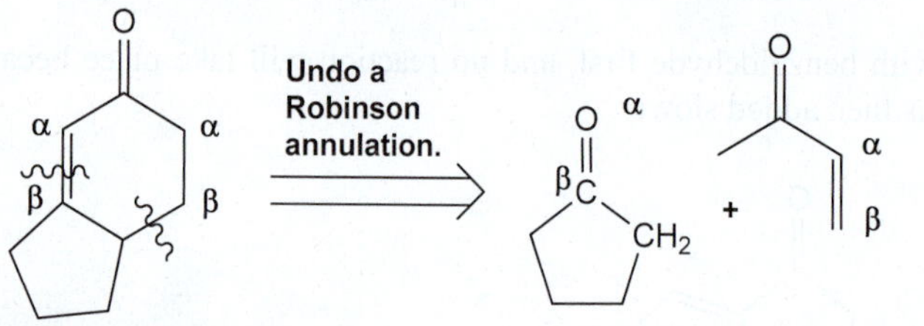

Synthesis: Executing this synthesis involves making the enolate of cyclopentanone and performing a conjugate addition first. To close the ring, an intramolecular aldol reaction takes place, followed by dehydration of the adduct.

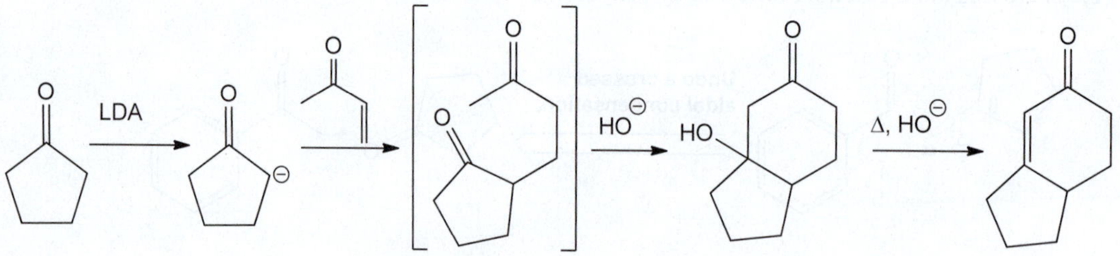

O LDA O O O $HO^{\ominus}$ O HO Δ, $HO^{\ominus}$ O

Problem 18.70

Think

How has the carbon skeleton been altered? What functional group transformation took place? Is the reaction conducted in acidic or basic solution? What species are not permitted in the mechanism? What reaction takes place with NC^{-} and the aldehyde? In the product of that reaction, what does the presence of CN do to the acidity of the proton attached to the initial aldehyde C?

Solve

All the C atoms are accounted for by combining the two reactants, so cyanide must be acting as a catalyst. In the first two steps of the mechanism, NC^- adds to the C=O group, and the resulting O^- is protonated. In that product, the H that was originally the aldehyde proton is much more acidic, owing to the CN group, which stabilizes the deprotonated form via resonance and inductive effects. Therefore, the third step deprotonates the benzylic C to make the carbanion nucleophile.

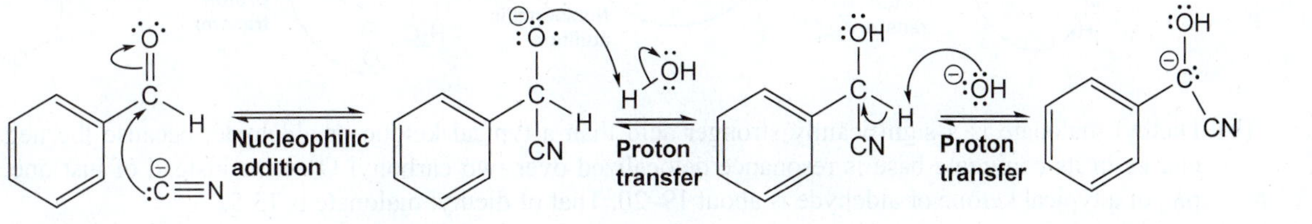

Nucleophilic addition — Proton transfer — Proton transfer

The resonance-stabilized anion attacks another equivalent of benzaldehyde in a nucleophilic addition mechanism. Finally, the catalyst (cyanide) is expelled.

Nucleophilic addition — Proton transfer — Proton transfer — Nucleophile elimination

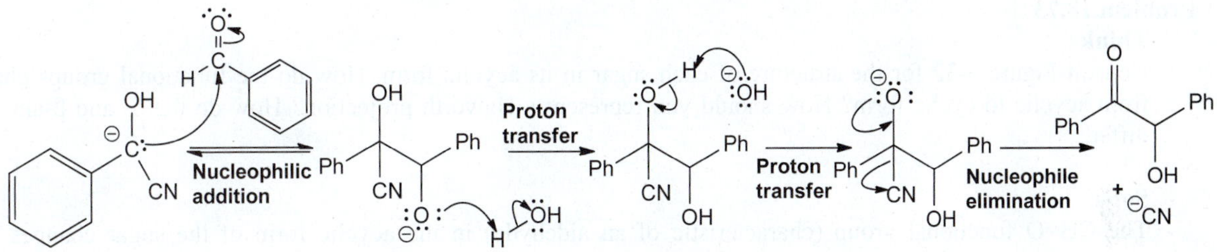

Problem 18.71

Think

How is this reaction similar to the mechanism in Problem 18.70? Which step is different?

Solve

This problem has the same first three steps to form the resonance-stabilized carbanion nucleophile. The carbanion nucleophile then attacks propenenitrile in a conjugate addition. Finally, the catalyst (cyanide) is expelled.

Nucleophilic addition — Proton transfer — Proton transfer — Nucleophile elimination

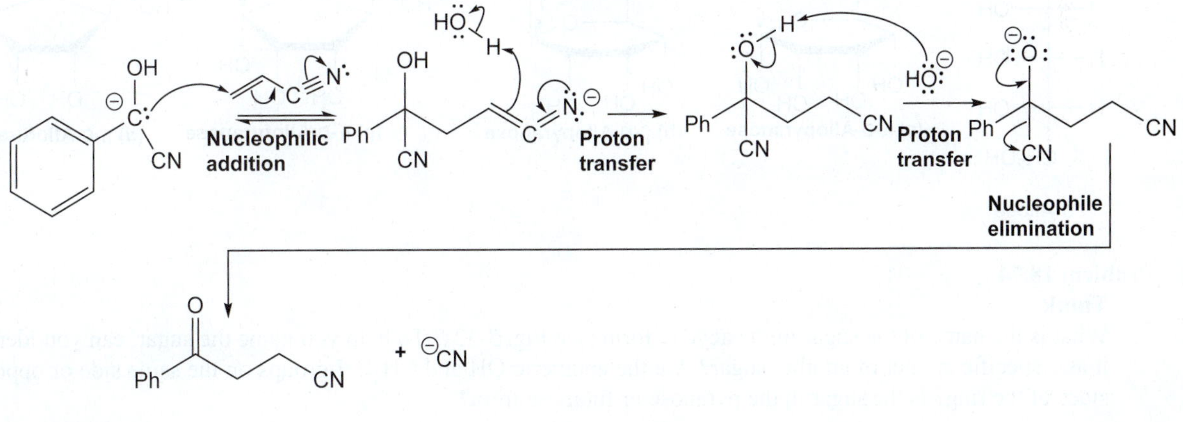

Problem 18.72

Think

What is the mechanism for the crossed aldol reaction? What acts as the nucleophile? Is the reaction conducted in acidic or basic solution? What species are not permitted in the mechanism? How is the pK_a of the α C–H in the β-diester different from an analogous proton in an ester or ketone?

Solve

(a) The mechanism is as follows:

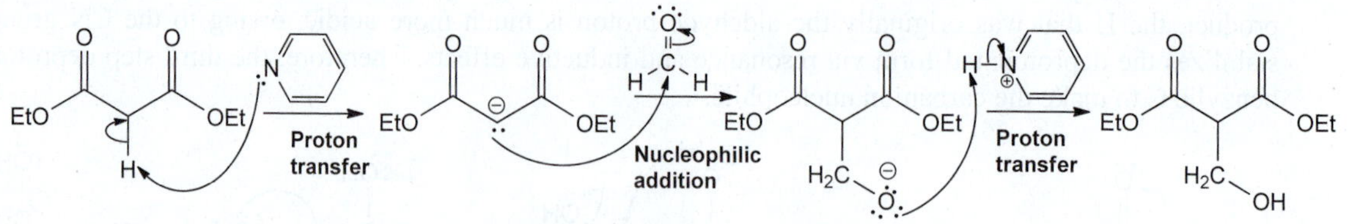

(b) Diethyl malonate is a significantly stronger acid than a typical ketone or aldehyde, because the negative charge in its conjugate base is resonance delocalized over two carbonyl O atoms instead of just one. The pK_a of a typical ketone or aldehyde is about 19–20. That of diethyl malonate is 13.5.

Section 18.13 The Organic Chemistry of Biomolecules: Ring Opening and Closing of Monosaccharides; Mutarotation

Problem 18.73

Think

Consult Figure 5-32 for the structure of each sugar in its acyclic form. How do the functional groups change from acyclic to cyclic form? How should you represent a Haworth projection? How do the α- and β-anomers differ?

Solve

The CH=O functional group (characteristic of an aldehyde) in the acyclic form of the sugar changes to a hemiacetal in the cyclic form. A pyranose has a six-membered ring, and a furanose has a five-membered ring. The difference in an α-anomer and β-anomer is the location of the CH_2OH and OH attached to the anomeric carbon. In the α-anomer, the anomeric OH is on the side of the ring opposite the CH_2OH group, and in a β-anomer, the groups are on the same side. The structure for D-allose and Haworth projections for sugars **(a)**–**(d)** are drawn below.

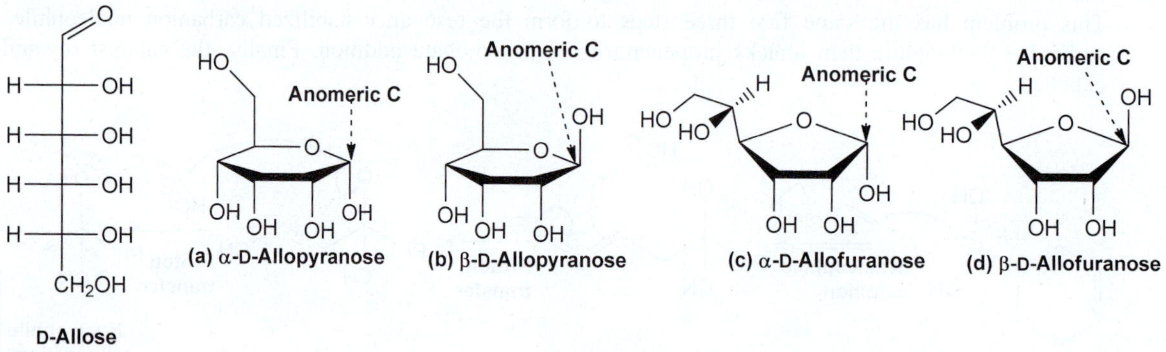

Problem 18.74

Think

What is the name of the sugar in the acyclic form (see Fig. 5-32)? To help you name the sugar, can you identify it as a specific epimer of another sugar? Are the anomeric OH and CH_2OH groups on the same side or opposite sides of the ring? Is the sugar in the pyranose or furanose form?

Solve

(a) α-D-Allopyranose. This sugar is D-allose, in the α-anomer form, and is a six-membered ring, a pyranose.
(b) β-D-Idopyranose. This sugar is D-idose, in the β-anomer form, and is a six-membered ring, a pyranose.
(c) α-D-Talofuranose. This sugar is D-talose, in the α-anomer form, and is a five-membered ring, a furanose.

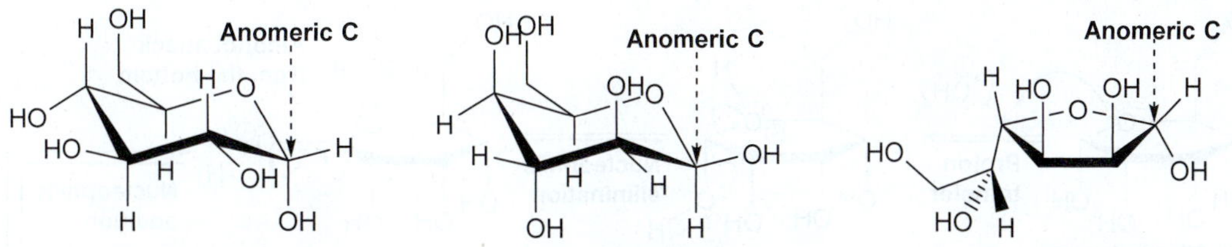

(a) α-D-Glucopyranose (b) β-D-Idopyranose (c) α-D-Talofuranose

Problem 18.75

Think

To what functional group should an anomeric C belong in the cyclic form of the sugar? Can you locate that functional group?

Solve

In the cyclic form of a sugar, the anomeric carbon belongs to a hemiacetal, characterized as a carbon atom that is bonded to both an OH and an OR group. The anomeric C is the one at the far right in each molecule in Problems 18.73 and 18.74 and is labeled in the previous problems.

Problem 18.76

Think

To what functional group should an anomeric C belong in the cyclic form of the sugar? To what functional group would it belong in an acyclic form of the sugar? Can you locate that functional group?

Solve

The anomeric C is the carbon that is part of the C=O group in the open-chain form. The anomeric carbon for D-talose is labeled below.

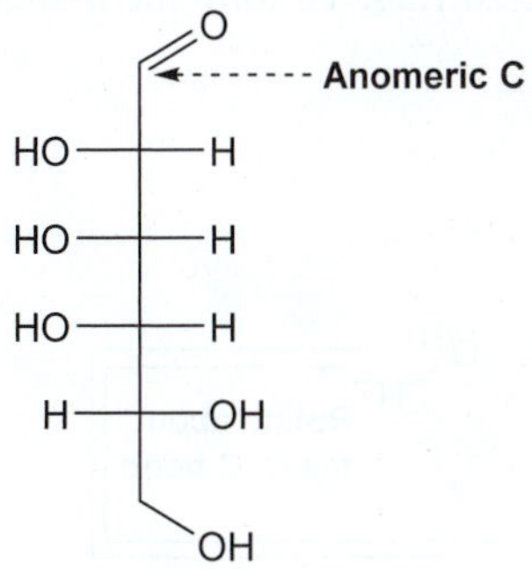

D-Talose

Problem 18.77

Think

Consult the solution to Problem 18.73 for the structures of α-D-allopyranose and β-D-allopyranose. How can the anomeric carbon invert its configuration? Consult Equation 18-66a for the general mechanism of cyclization. What acts as the nucleophile in the nucleophilic addition reaction to close the ring? What C is electrophilic? Are proton transfer reactions involved? What approach is required of the nucleophile to produce each anomer?

Solve

To convert α-D-allopyranose into β-D-allopyranose, the reverse of the cyclization reaction must occur to transform the ring into the open-chain form. Then the open-chain form can cyclize into the β-anomer when the nucleophile attacks from below the C=O plane.

α-D-Allopyranose ⇌ (Proton transfer) ⇌ (Nucleophile elimination) ⇌ (Alcohol attack from the bottom; Nucleophilic addition) ⇌ (Proton transfer) β-D-Allopyranose

Problem 18.78

Think

Consult Problem 18.73 for the structures of α-D-allopyranose and α-D-allofuranose. Consult Equations 18-66a and 18-66b for the general mechanism of cyclization involving the six- and five-membered rings. What acts as the nucleophile in the nucleophilic addition reaction to form a five-membered ring?

Solve

To convert α-D-allopyranose into α-D-allofuranose, the reverse of the cyclicilzation reaction must occur to transform the ring into the open-chain form. Then the open-chain form rotates about the C–C bond shown below to move the OH at C3 closer to the C=OH$^+$ and form the five-membered ring. To form the α-anomer, nucleophilic attack should come from above the C=O plane.

α-D-Allopyranose ⇌ (Proton transfer) ⇌ (Nucleophile elimination) ⇌ (Rotate about the C–C bond) ⇌ (Alcohol attack from the top; Nucleophilic addition) ⇌ (Proton transfer) α-D-Allofuranose

Problem 18.79

Think

If the measured rotation is +80.2°, which anomer, α-D-galactopyranose ([α] = +150.7°) or β-D-mannopyranose ([α] = +52.8°), is the dominant species in solution? If you set the percentage of α-D-galactopyranose equal to x in Equation 18-69, then what is the percentage of β-D-galactopyranose?

Solve

Let x be the equilibrium percentage of the α-D-galactopyranose and $1 - x$ be the equilibrium percentage of β-D-galactopyranose.

$$x(+150.7) + (1 - x)(+52.8) = +80.2$$
$$150.7x + 52.8 - 52.8x = 80.2$$
$$97.9x = 27.4$$
$x = 0.280$, or 28.0% = α-D-galactopyranose
$1 - x = 0.720$, or 72.0% = β-D-galactopyranose

Integrated Problems

Problem 18.80

Think

For **(a)**, does the addition occur via 1,2-addition or 1,4-addition? What is the product of the Wolff–Kishner reduction? For **(b)**, does the HO^- act as a base or nucleophile? What reaction takes place when an aldehyde is treated with base? What is the product of reduction using $NaBH_4$?

Solve

(a) This reaction involves a conjugate 1,4-addition followed by the Wolff–Kishner reduction.

1. $(CH_3)_2CuLi$

2. H_2NNH_2, $HO^{\ominus}$, Δ

O C H3C H H C

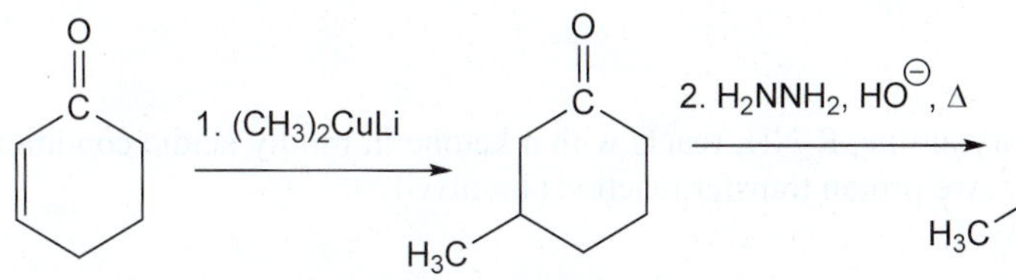

(b) HO^- acts as a base to form the enolate anion. This is, therefore, an aldol addition followed by a reduction.

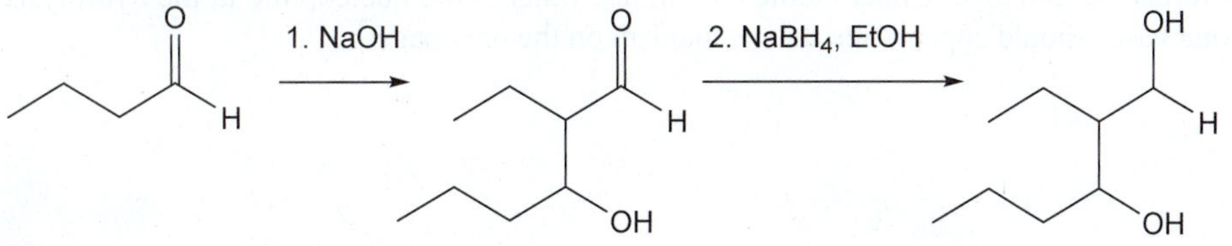

1. NaOH

2. $NaBH_4$, EtOH

O H OH

Problem 18.81

Think

Which C–C bond was formed? What is the identity of the nucleophile in the nucleophilic addition reaction? Did the addition occur via 1,2-addition or 1,4-addition? Are proton transfer reactions involved?

Solve

The nucleophilic carbon of the enamine does a conjugate 1,4-addition to the ketone. Water adds to the C=N double bond, starting a hydrolysis that eventually forms the diketone. Under acidic conditions, water is the nucleophile in the hydrolysis reaction, and no strong bases should appear. See the mechanism on the next page.

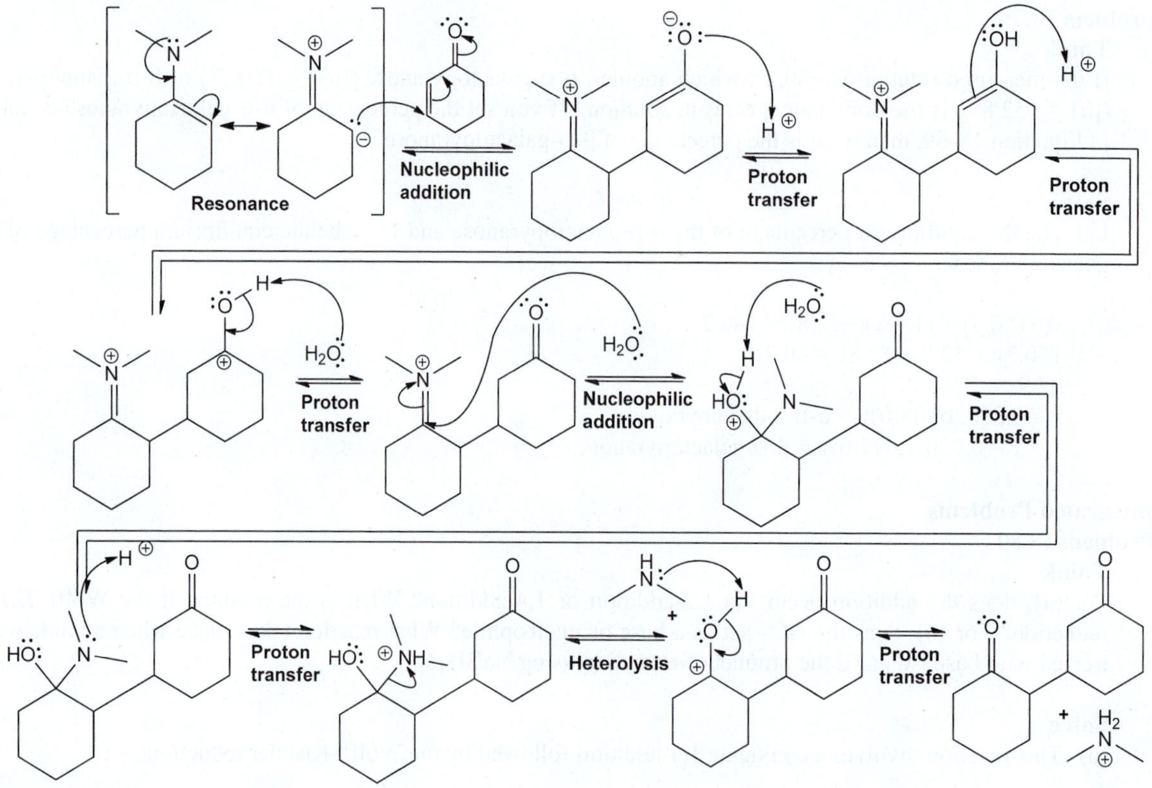

Problem 18.82

Think

What compound forms when a secondary amine, R_2NH, reacts with a ketone in mildly acidic conditions? How is the resulting compound nucleophilic? Are proton transfer reactions involved?

Solve

(a) Nucleophilic addition occurs first, and an enamine is formed. The nucleophilic carbon of the enamine does a conjugate 1,4-addition to the ketone. Water adds to the C=N double bond, starting a hydrolysis that eventually forms the diketone. Under acidic conditions, water is the nucleophile in the hydrolysis reaction, and no strong bases should appear. See the mechanism on the next page.

Nucleophilic addition

Proton transfer

Proton transfer

Proton transfer

Heterolysis

Elimination of H^+

Resonance

Nucleophilic addition

Proton transfer

Tautomerization

Nucleophilic addition

Two proton transfers

Heterolysis

Proton transfer

(b) Nucleophilic addition occurs first, and an enamine is formed. The nucleophilic carbon of the enamine does an S_N2 reaction to the $CH_3CH_2CH_2Br$. Water adds to the C=N double bond, starting a hydrolysis that eventually forms the aldehyde. Under acidic conditions, water is the nucleophile in the hydrolysis reaction, and no strong bases should appear.

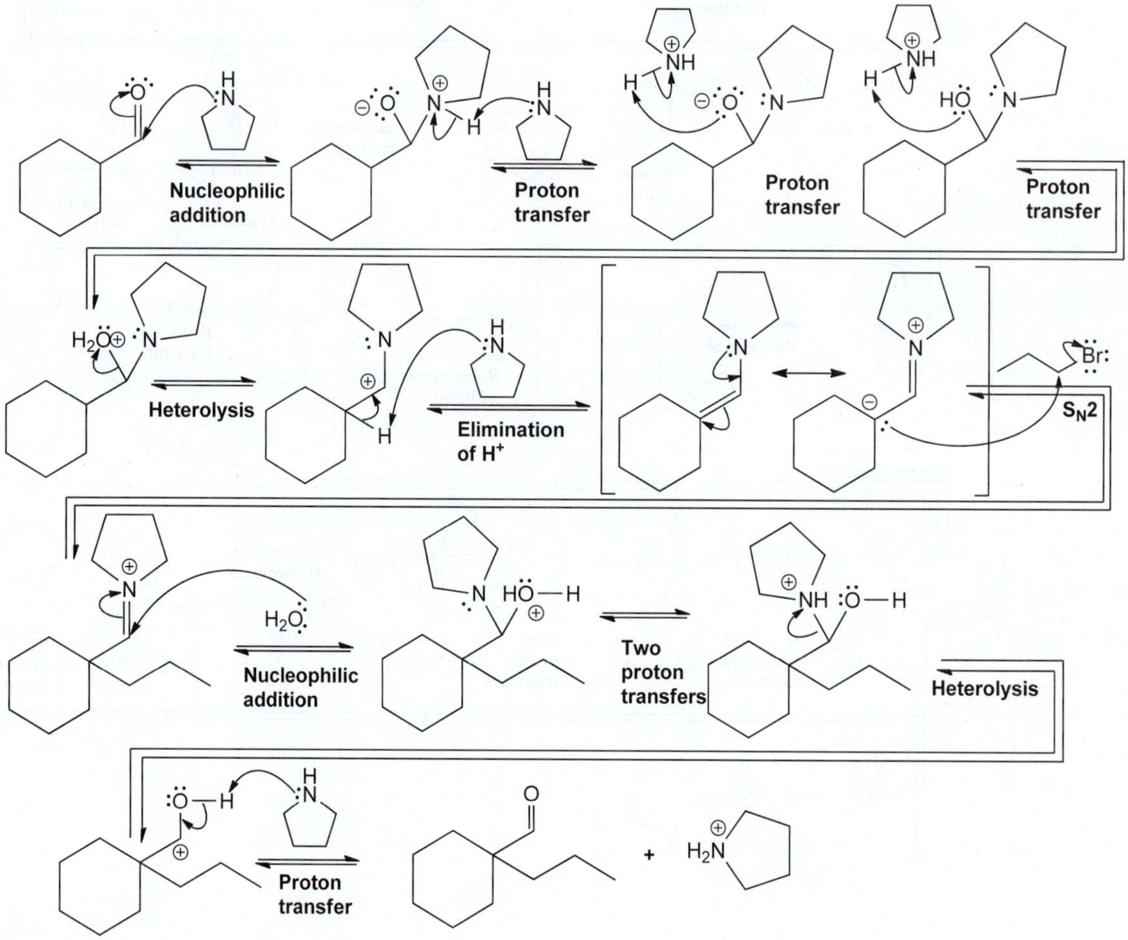

Nucleophilic addition

Proton transfer

Proton transfer

Proton transfer

Heterolysis

Elimination of H^+

S_N2

Nucleophilic addition

Two proton transfers

Heterolysis

Proton transfer

Problem 18.83

Think

Is the nitro group electron donating or withdrawing? How does the nitro group affect the acidity of the α C–H? Is the reaction conducted in acidic or basic medium? Under these conditions, what acts as the nucleophile in the nucleophilic addition reaction? How can a good leaving group be produced under these conditions?

Solve

Protonation of a nitro group makes it even more electron withdrawing than normal. Loss of a proton from the α carbon forms an *N,N*-dihydroxyiminium ion—still quite electron withdrawing. The α carbon is susceptible to nucleophilic addition of water. Two more proton transfers turn N into a good leaving group (*N*-hydroxyhydroxylamine).

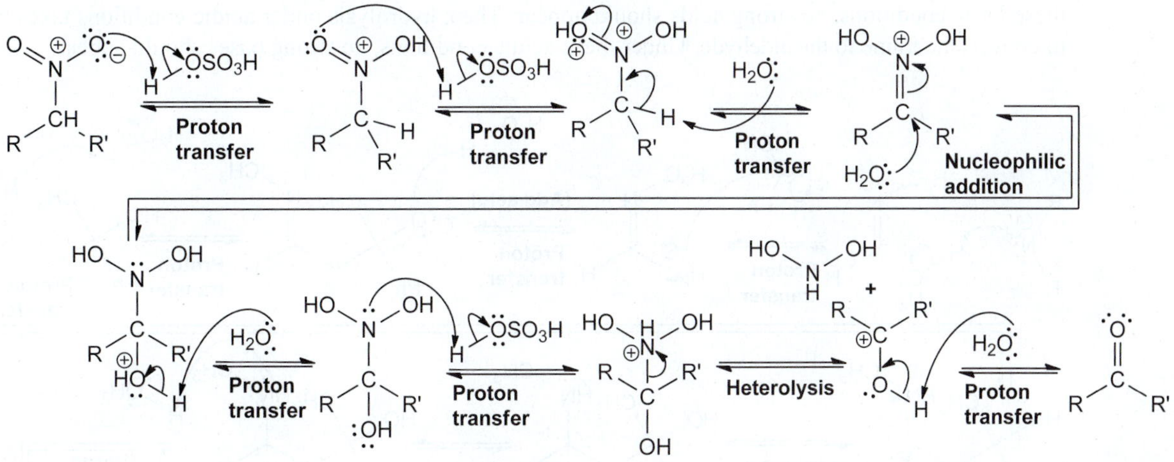

Proton transfer

Proton transfer

Proton transfer

Nucleophilic addition

Proton transfer

Proton transfer

Heterolysis

Proton transfer

Problem 18.84

Think

How is the C≡N group of a nitrile similar to the C=O group of a ketone or aldehyde? What acts as the nucleophile in the nucleophilic addition step? How is the N transformed into a good leaving group? Is the reaction conducted in acidic or basic solution? What species are not permitted in the mechanism?

Solve

Nucleophilic addition of an alcohol to a protonated C≡N triple bond is followed later by nucleophilic addition of water to a protonated C=N double bond. Pay attention to the fact that the reaction takes place under acidic conditions, so no strong bases should appear.

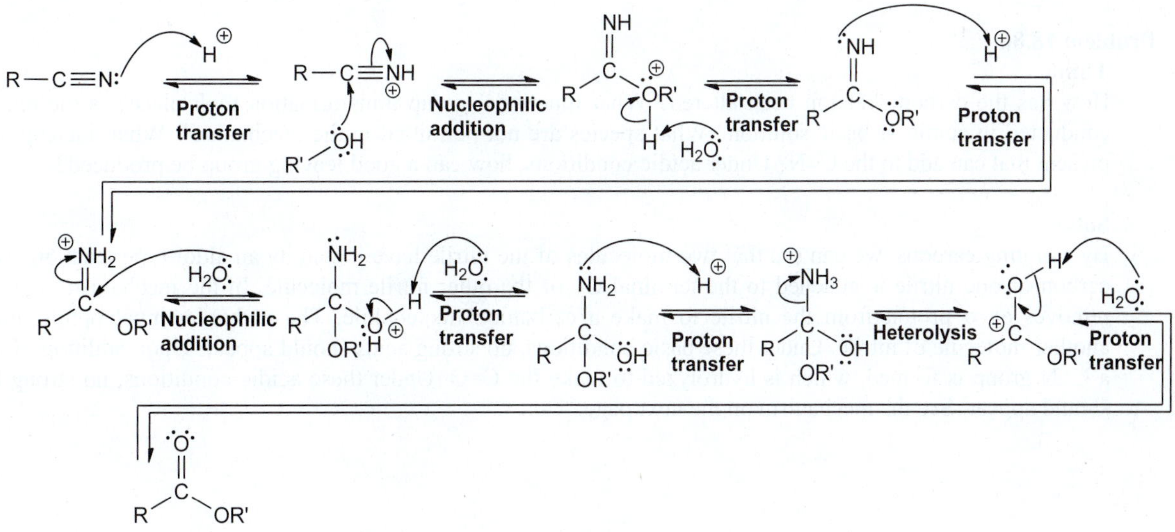

Proton transfer

Nucleophilic addition

Proton transfer

Proton transfer

Nucleophilic addition

Proton transfer

Proton transfer

Heterolysis

Proton transfer

Problem 18.85

Think

Which base is stronger, LDA or HO^-? On the basis of charge stability arguments, which is more acidic, the α C–H of an imine or that of an aldehyde? What is the mechanism for an aldol addition reaction? What species in this reaction are similar to the aldol addition reaction? Is the reaction conducted in acidic or basic solution? What species are not permitted in the mechanism?

Solve

(a) The α C–H hydrogen of an imine is much less acidic than that of a ketone or aldehyde, because a N atom cannot accommodate a negative charge as well as O can. So hydroxide is simply not a strong enough base to generate the enolate anion.

(b) The mechanism that parallels an aldol addition mechanism is shown in the first three steps below. Under these basic conditions, no strong acids should appear. Then, hydrolysis under acidic conditions takes place to convert the imine to the aldehyde. Under these acidic conditions, no strong bases should appear.

Problem 18.86

Think

How has the carbon skeleton been altered? What functional group transformation took place? Is the reaction conducted in acidic or basic solution? What species are not permitted in the mechanism? What nucleophile is present that can add to the C≡N? Under acidic conditions, how can a good leaving group be produced?

Solve

By counting carbons, we can see that two molecules of the nitrile have joined by an aldol-type reaction. The α carbon of one nitrile is attached to the terminal C1 of the other nitrile molecule. In the mechanism, ethoxide removes an α proton from the nitrile to make a carbanion nucleophile. The carbanion nucleophile attacks another molecule of nitrile. Under these basic conditions, no strong acids should appear. Upon addition of acid, a C=N group is formed, which is hydrolyzed to make the C=O. Under these acidic conditions, no strong base should appear. See the mechanism on the next page.

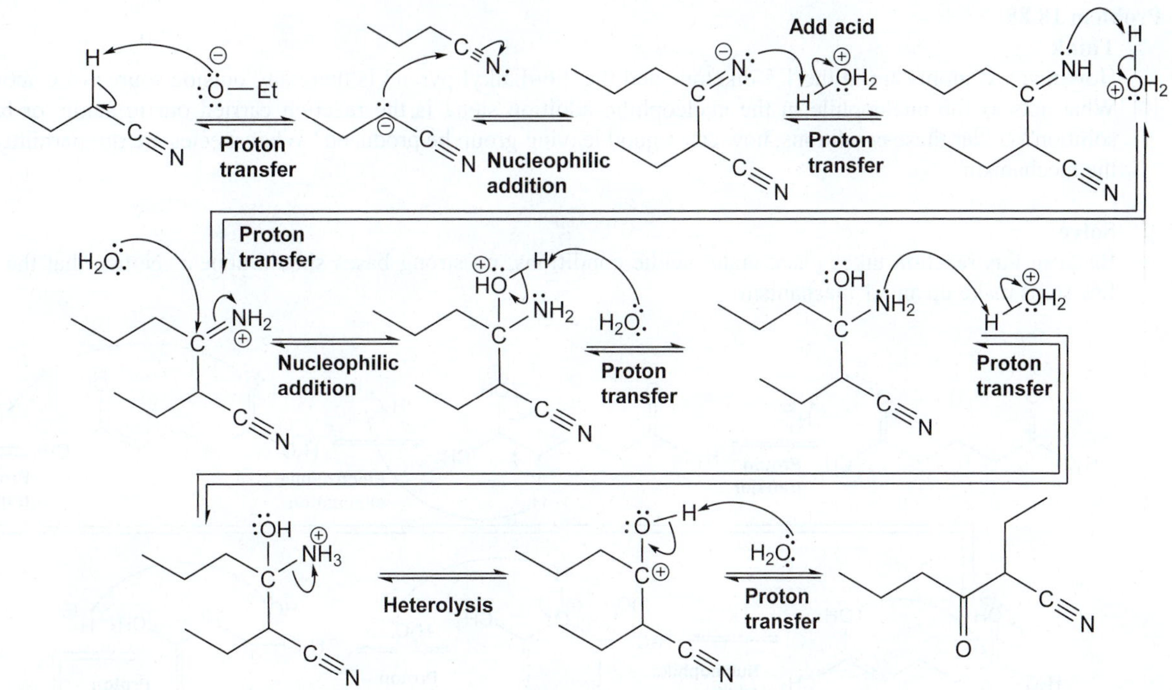

Add acid

Proton transfer

Nucleophilic addition

Proton transfer

Proton transfer

Nucleophilic addition

Proton transfer

Proton transfer

Heterolysis

Proton transfer

Problem 18.87

Think

What C–C bond is formed? What acts as the nucleophile in the nucleophilic addition step that forms that bond? How is the pK_a of the α C–H of a β-dicarboxylic acid different compared to a single carboxylic acid? Is the reaction conducted in acidic or basic solution? What species are not permitted in the mechanism?

Solve

This reaction takes place under mildly acidic conditions, evidenced by the presence of the carboxylic acid groups. The aldehyde reacts with NH_3 in acidic conditions to form the imine according to the mechanism in Equation 18-19, which is in equilibrium with a small amount of its protonated form.

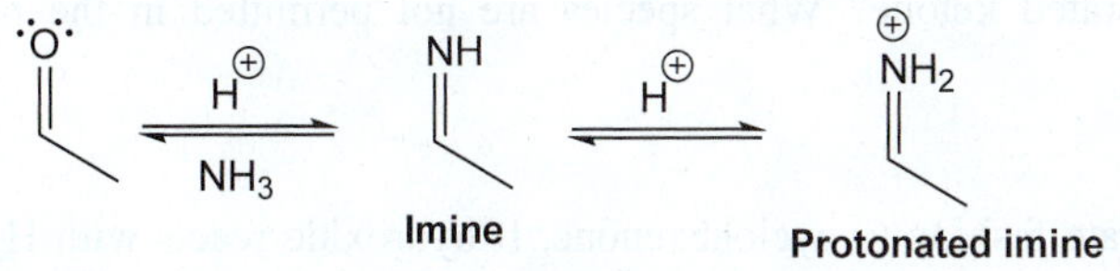

Imine

Protonated imine

The carbon nucleophile will be the enol. The ammonia deprotonates the α C–H of a β-dicarboxylic acid to produce the enol that is nucleophilic at the C. The mechanism is given below.

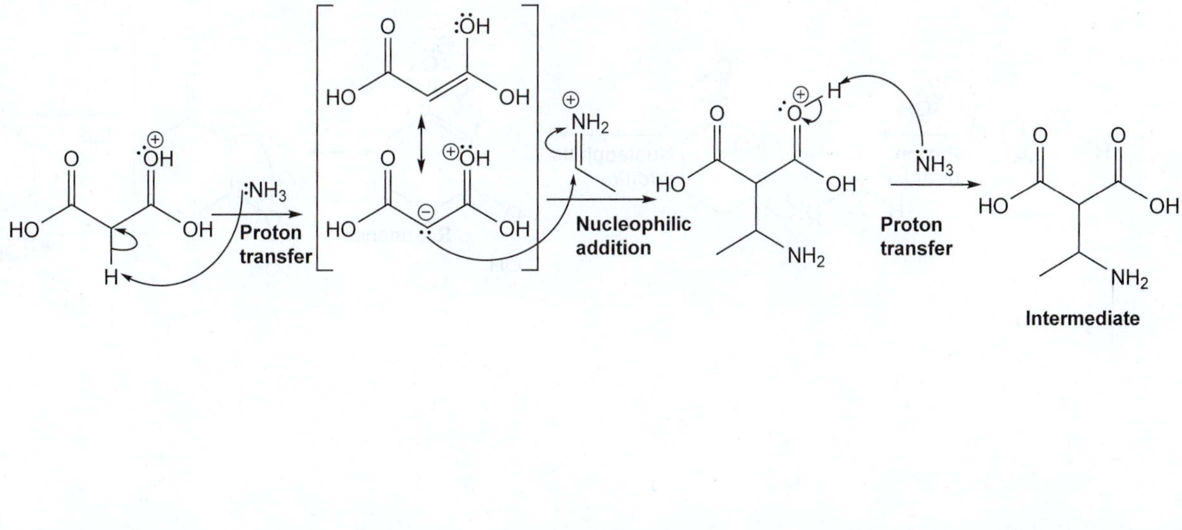

Proton transfer

Nucleophilic addition

Proton transfer

Intermediate

Problem 18.88

Think

How many C atoms are in the 1,5-diketone and the 2,6-dialkyl pyran? Is there any outside source of C atoms? What acts as the nucleophile in the nucleophilic addition step? Is the reaction carried out in acidic or basic solution? Under these conditions, how can a good leaving group be produced? What species are not permitted in the mechanism?

Solve

Because this reaction takes place under acidic conditions, no strong bases should appear. Notice that the last few steps make up an E1 mechanism.

Proton transfer

Electrophile elimination

Proton transfer

Nucleophilic addition

Proton transfer

Proton transfer

Resonance stabilized C^+

Heterolysis

Electrophile elimination

Problem 18.89

Think

What reaction takes place between HO^- and H_2O_2? What nucleophile is produced? Does direct or conjugate nucleophilic addition occur with the α,β-unsaturated ketone? What species are not permitted in the basic solution?

Solve

A nucleophile is required that will add in conjugate fashion to cyclohexenone. If hydroxide reacts with H_2O_2, the hydroperoxide ion, $^-$O–OH, forms. This can add in conjugate fashion. Next, the C^- of the enolate displaces hydroxide to close the epoxide ring in an internal S_N2 step. Normally, hydroxide is a poor leaving group, but this is feasible under these conditions because the O–O bond is very weak. Notice that hydroxide is *regenerated*, making this a *base-catalyzed* reaction.

Proton transfer

Nucleophilic addition

Resonance

S_N2

Problem 18.90

Think

What acts as the nucleophile in the nucleophilic addition step? Is the reaction conducted in acidic or basic solution? What species are not permitted in the mechanism? Is there a good leaving group in the product of nucleophilic addition?

Solve

The first two steps proceed by the normal aldol addition mechanism. The third step is an intramolecular S_N2 step instead of a proton transfer.

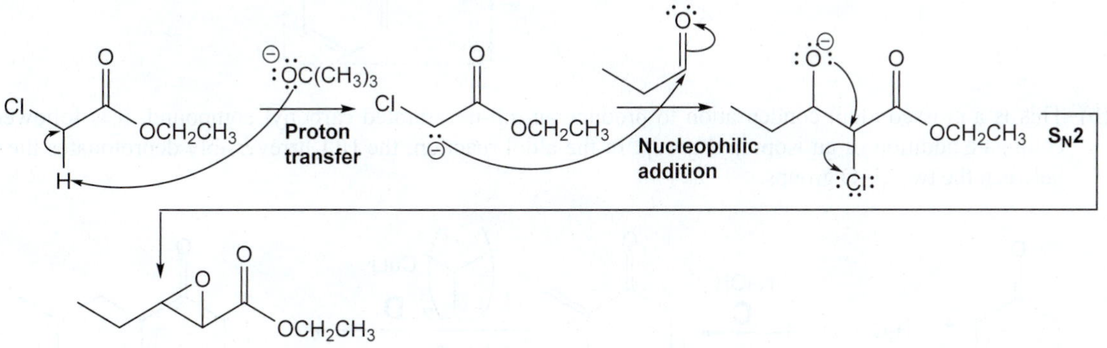

:OC(CH3)3
Proton transfer
Nucleophilic addition
S_N2
OCH2CH3

Problem 18.91

Think

What C–H bond does Bu–Li deprotonate? How is this reactant similar to a Wittig reagent? How is this mechanism similar to the Wittig reaction?

Solve

The P-containing starting material resembles a Wittig reagent. Butyllithium is a strong enough base to deprotonate the C, making a carbon nucleophile that can attack the aldehyde in a nucleophilic addition. In the third step, nucleophilic addition to the P=O bond produces the intermediate. From the intermediate, the products are formed in a single step, similar to the final step of a Wittig reaction.

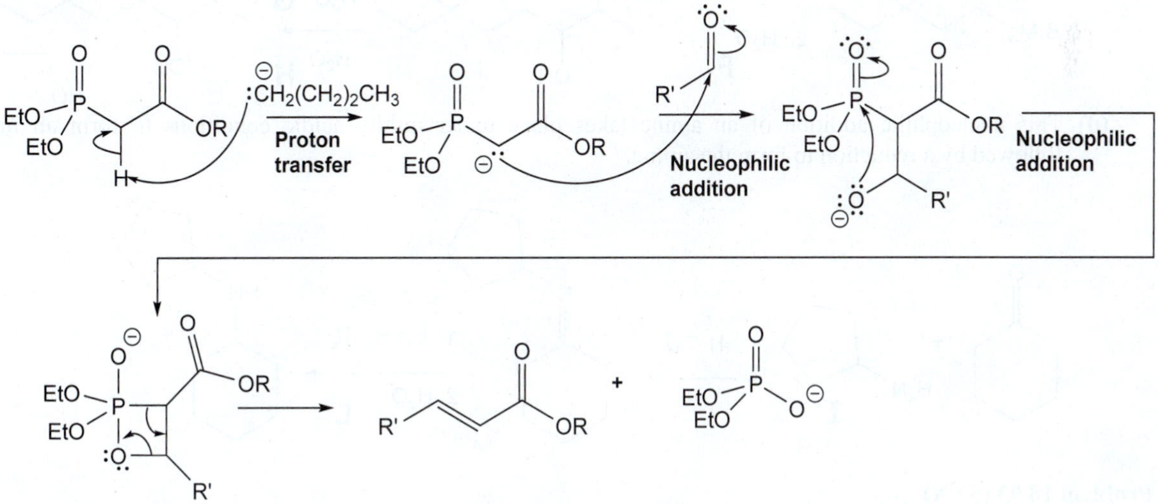

:CH2(CH2)2CH3
Proton transfer
Nucleophilic addition
Nucleophilic addition

Problem 18.92

Think

What are the electron-rich and electron-poor species in each reaction? Is the carbon skeleton altered? Are there any characteristic structural features in the reactants or products? What functional group transformations occur?

Solve

(a) This is a Robinson annulation, characterized by the formation of a six-membered ring with an α,β-unsaturated carbonyl group. The carbon skeleton from **A** is boxed below.

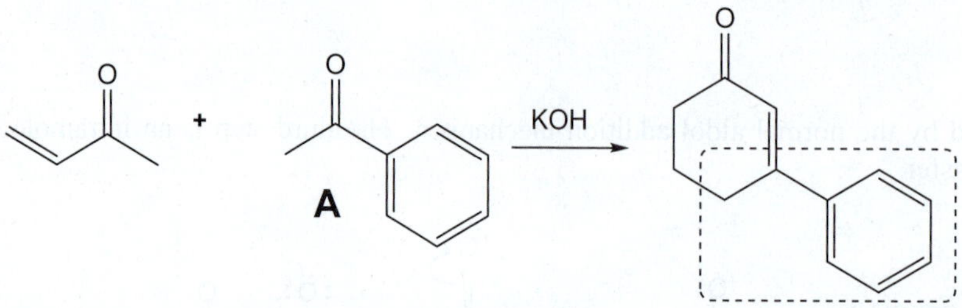

+ KOH

A

(b) This is a crossed aldol condensation to produce an α,β-unsaturated carbonyl compound; it is followed by conjugate addition of an isopropyl group. In the aldol reaction, the HO^- irreversibly deprotonates the CH_2 between the two C=O groups.

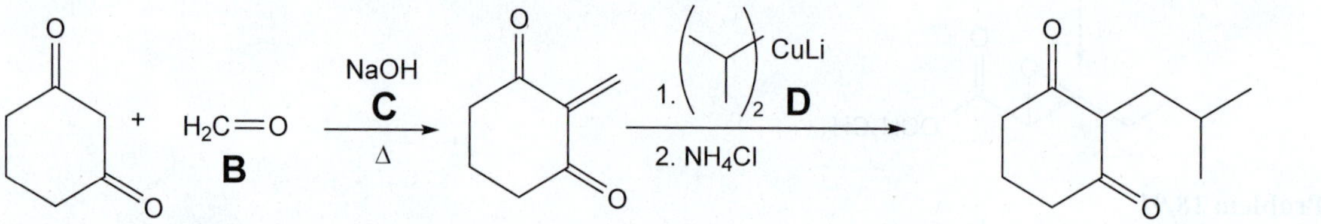

+ $H_2C{=}O$ **B** NaOH **C** Δ 1. CuLi **D** 2. NH_4Cl

(c) A Grignard addition to isobutyraldehyde is followed by aqueous workup. Then the acetal undergoes hydrolysis to produce the ketone. If aqueous H_2SO_4 is used in Step 2, then hydrolysis will take place in that same step.

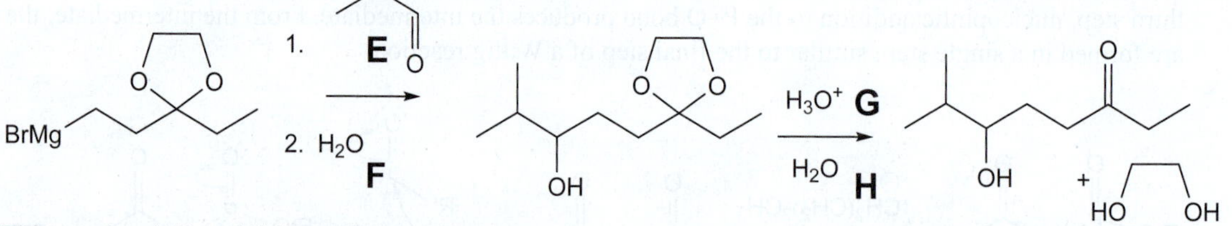

BrMg 1. **E** 2. H_2O **F** OH H_3O^+ **G** H_2O **H** OH + HO OH

(d) This nucleophilic addition of an amine takes place under mildly acidic conditions to form an imine, followed by a reduction to form the amine.

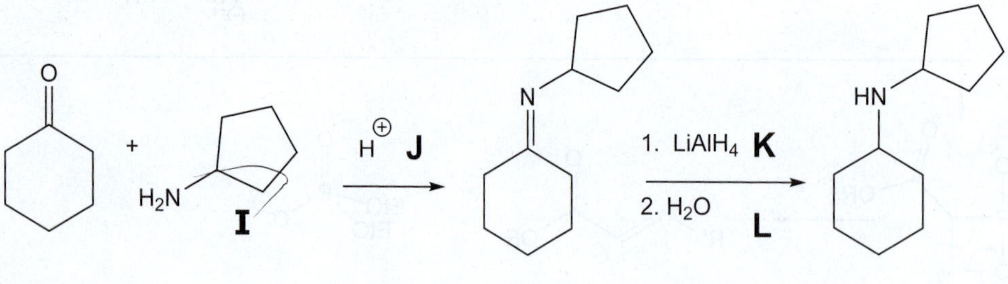

+ H_2N **I** $H^{\oplus}$ **J** N 1. $LiAlH_4$ **K** 2. H_2O **L** HN

Problem 18.93 (SYN)

Think

How many C atoms are in the product? How many C–C bonds must be formed? What other functional groups are present? What is the relative positioning of those functional groups?

Solve

With only two-carbon precursors allowed, and six C atoms in the product, we must join three molecules of two C atoms each. So in the retrosynthetic analysis, we should disconnect the target C atoms two at a time. We can first try to disconnect the C–C bond below.

Br

?

However, we don't know of a reaction that, in the forward direction, generates a C–C bond and leaves us with an alkyl halide on one of the C atoms. On the other hand, a Grignard reaction will make such a bond and leave us with an OH group.

Br **Undo a substitution.** OH **Undo a Grignard reaction.** O + BrMg

Synthesis of the unsaturated aldehyde involves an aldol condensation reaction. The retrosynthesis is given below.

O **Undo an aldol condensation.** O + O

Executing the synthesis involves filling in the necessary reagents for each step.

O NaOH Δ O 1. BrMg 2. H_2O, H^+ OH PBr_3 Br

Problem 18.94 (SYN)

Think

How many C atoms are in the product? How many C–C bonds must be formed? What other functional groups are present? What reactions do you know that will produce a new C=C bond?

Solve

If we disconnect the C=C double bond, we obtain the same carbon backbone as 2-phenylacetaldehyde, so we should consider undoing a Wittig reaction, giving us 2-phenylacetaldehyde and a Wittig reagent as precursors.

Ph Ph **Undo a Wittig reaction.** Ph =O + $Ph_3P^{\oplus}$ $^{\ominus}$ Ph

2-Phenylacetaldehyde

The Wittig reagent is made from an alkyl halide, which can be obtained by reducing benzaldehyde to benzyl alcohol.

Undo ylide formation. Undo a substitution. Undo a reduction.

Synthesis:

NaBH4 PBr3 1. PPh3 2. BuLi

Problem 18.95 (SYN)

Think

How many C atoms are in the product? How many C–C bonds must be formed? What other functional groups are present in the starting material? In the product? What reactions do you know that will remove a functional group entirely?

Solve

Because the starting material is given, the retrosynthesis must work back to that exact compound. The solution becomes clearer if you ignore the numbering in the product and realize that you have to add a methyl group to C5 of the ketone. This can be done by a conjugate addition. Then you have to remove all the functional groups. The C=O group can be converted to CH_2 using a Wolff–Kishner reduction, which is shown, but could also have been reduced using a Clemmensen reduction or a Raney-nickel reduction.

Undo a reduction. Undo conjugate 1,4-addition.

Synthesis:

$(CH_3)_2CuLi$ NH_2NH_2, KOH, Δ

Problem 18.96 (SYN)

Think

How many C atoms are in butanal? How many C atoms are in the product? What functional groups are present in the starting material? In the product? What reactions do you know that will yield an amine from an aldehyde?

Solve

The product has eight C atoms, and the starting material, butanal, has four C atoms. The amine can be formed from the reduction of an imine. The imine is formed from nucleophilic addition of the C=O and an amine. Finally, the amine is formed from a reduction of the N=C, which is formed from a nucleophilic addition of RCH=O and NH_3. See the mechanism on the next page.

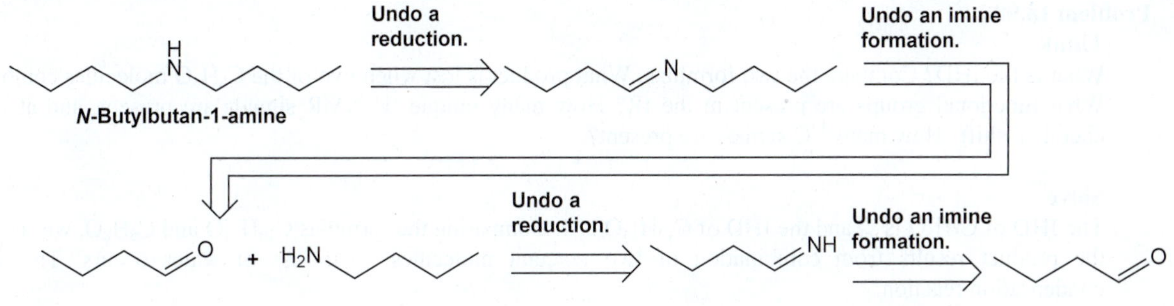

Undo a reduction.

Undo an imine formation.

N-Butylbutan-1-amine

Undo a reduction.

Undo an imine formation.

Synthesis:

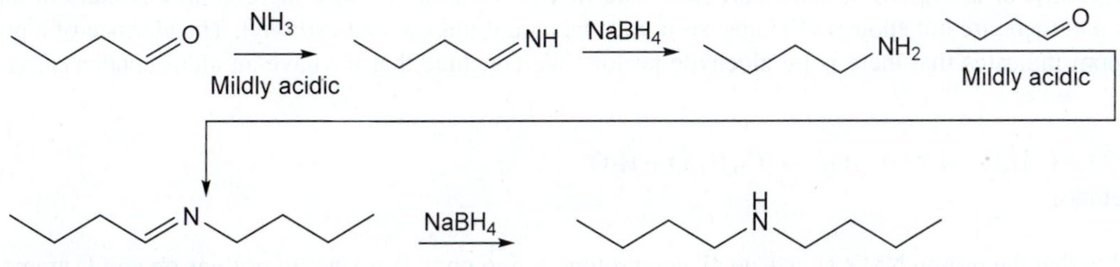

NH_3 Mildly acidic

$NaBH_4$

Mildly acidic

$NaBH_4$

Problem 18.97 (SYN)

Think

How many C atoms are in the product? How many C–C bonds must be formed? What other functional groups are present in the starting material? In the product? What reactions do you know that will yield an amine?

Solve

There are seven C atoms in the product, and the starting material requirement is three or fewer C atoms. Therefore, two C–C bonds must form in the synthesis. The amine is formed from a reduction of the N=C, which is formed from a nucleophilic addition of RCH=O and NH_3. We add a methyl group to C2 of the ketone; this can be done by a conjugate addition. Synthesis of the α,β-unsaturated aldehyde, which has six C atoms, involves an aldol condensation reaction. *Note:* Synthesis routes other than the one shown below are also feasible.

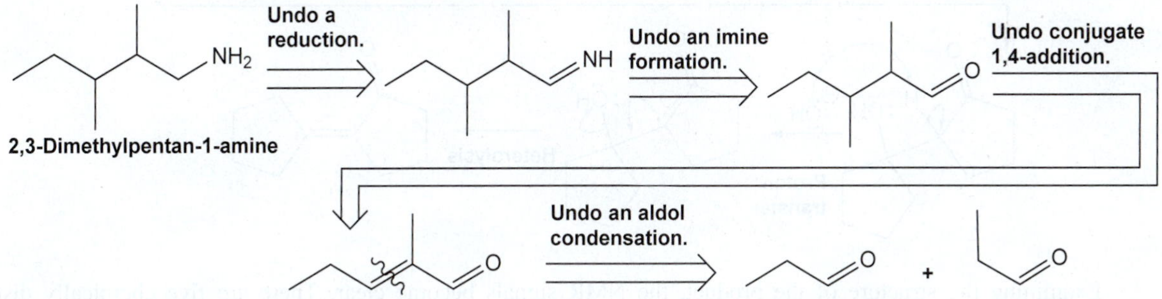

Undo a reduction.

Undo an imine formation.

Undo conjugate 1,4-addition.

2,3-Dimethylpentan-1-amine

Undo an aldol condensation.

Synthesis:

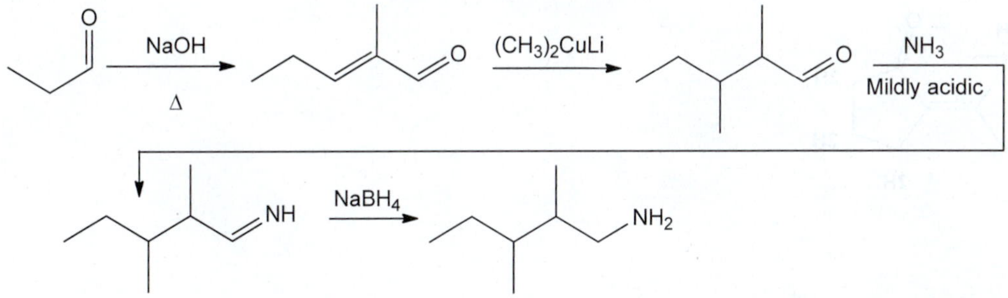

NaOH, Δ

$(CH_3)_2CuLi$

NH_3 Mildly acidic

$NaBH_4$

Problem 18.98

Think

What is the IHD? Compare the two formulas. What product is lost when two of the C_5H_8O molecules combine? What functional groups are present in the IR? How many unique ^1H NMR signals are present, and at what chemical shift? How many ^{13}C signals are present?

Solve

The IHD of C_5H_8O is 2, and the IHD of $C_{10}H_{14}O$ is 4. Comparing the formulas $C_{10}H_{14}O$ and C_5H_8O, we see that the product results from combination of two reactant molecules, then loss of water occurs. This is a condensation reaction.

Because the compound $C_{10}H_{14}O$ is the product of an aldol condensation reaction, it must be either a β-hydroxycarbonyl or an α,β-unsaturated carbonyl. The first is ruled out because there is no OH band in the IR spectrum and there are not enough O atoms, so it must be an α,β-unsaturated carbonyl. The absence of a proton at 8–10 ppm indicates that there is no aldehyde proton. We conclude that we have an aldol condensation of a ketone:

$$C_5H_8O + C_5H_8O \rightarrow C_{10}H_{16}O_2 \rightarrow C_{10}H_{14}O + H_2O$$
(a ketone)

It is notable that the proton NMR shows no alkene protons at 5–6 ppm; there are no protons on any C atoms that have a double bond. Notice that no signal in the proton NMR integrates to three protons—all of them integrate to either two protons or four protons. So the product contains no methyl groups, and the same must be true of the ketone reactant. The only ketone of the formula C_5H_8O with no methyl groups is cyclopentanone. A mechanism becomes evident, as does the most likely product.

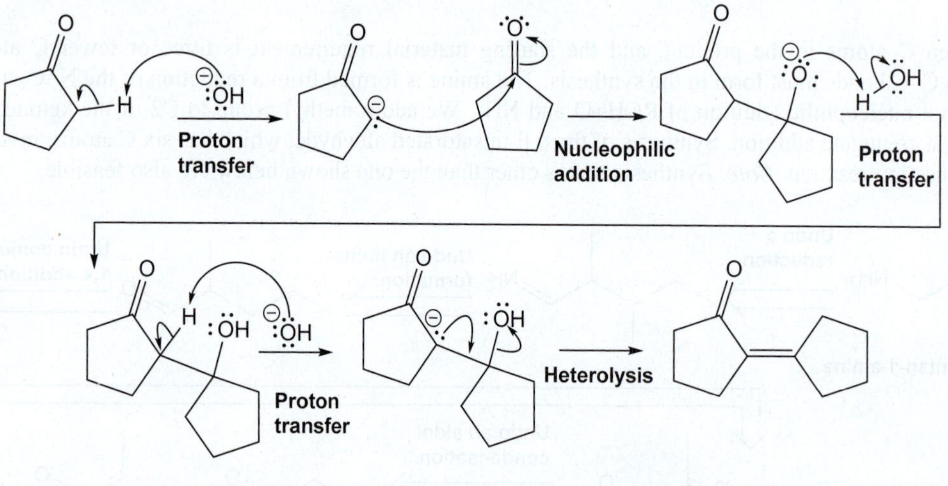

Proton transfer

Nucleophilic addition

Proton transfer

Proton transfer

Heterolysis

Examining the structure of the product, the NMR signals become clear. There are five chemically distinct groups of protons. All the NMR signals in the product are at 1–3 ppm, indicating that the product possesses alkane-type CH_2 protons near a double bond of some type.

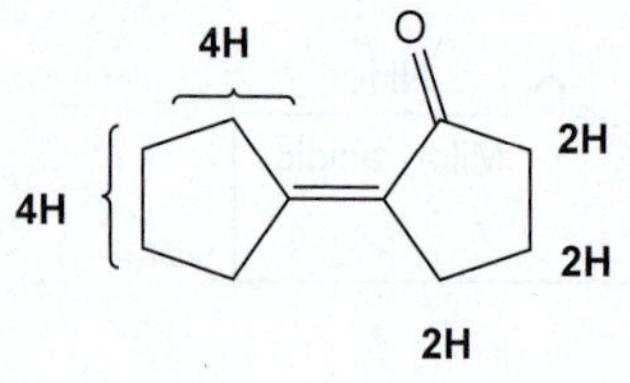

4H

4H

2H

2H

2H

Problem 18.99

Think

What is the IHD? Compare the two formulas. What product is lost when the two reactant molecules combine? What functional groups are indicated by the IR spectrum? How many unique ^{1}H NMR signals are present, and at what chemical shift?

Solve

(a) Comparing the formulas of the reactants to the products gives this equation:

$$C_4H_6O + C_9H_8O \rightarrow C_{13}H_{14}O_2 \rightarrow C_{13}H_{12}O + H_2O$$

The proton NMR assignments are 10 ppm, 1H (aldehyde proton); 7–9 ppm, 5H (phenyl protons); and 5–6 ppm, 6H (alkene protons). The structure of the $C_{13}H_{12}O$ is shown below, after we have used our understanding of aldol condensations to propose a suitable mechanism.

(b) Because the carbon structure of the product seems to result from the joining of the two reactants, this reaction is likely an aldol-type reaction. The only acidic C–H atom in the two reactants is a H atom attached to C4 of but-2-enal, owing to resonance stabilization of the conjugate base. This C can attack the aldehyde C of 3-phenylpropenal.

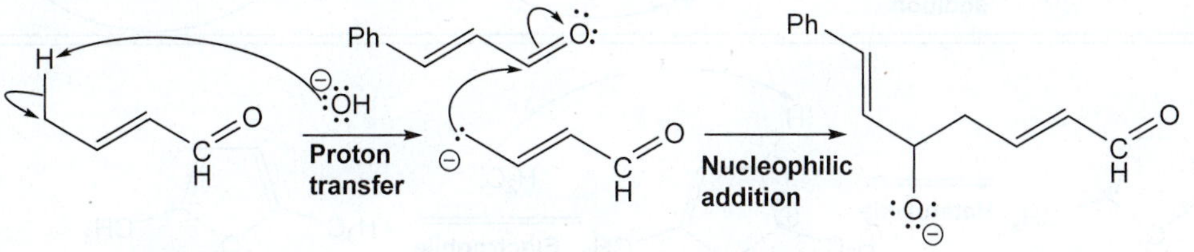

Proton transfer

Nucleophilic addition

The adduct is protonated and loses water to form a product that is completely conjugated. The dehydration occurs via an E1cb mechanism. Notice the six alkene protons in the product.

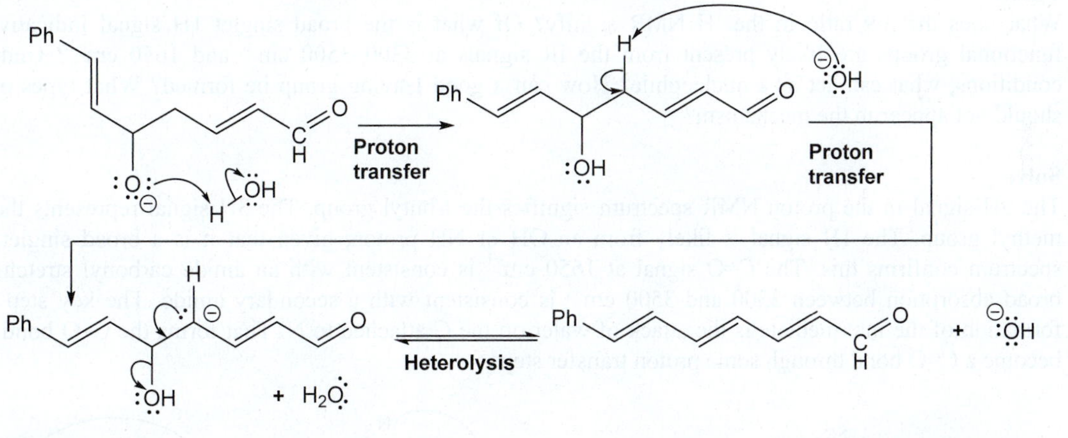

Proton transfer

Proton transfer

Heterolysis

Problem 18.100

Think

What is the formula of the starting material? What is the formula of the product? What molecule is lost during the course of the reaction? How many signals are in the ^{1}H NMR spectrum? What does this suggest about the symmetry of the molecule? What is the relative ratio of the two signals? What do the chemical shifts tell you about the types of protons?

Solve

The starting material's formula is $C_6H_{10}O_2$, so during the course of the reaction, the molecule must lose one O atom and two H atoms. Also, the proton NMR spectrum shows only two signals, so the product must have significant symmetry. The signals are in a 2:6 ratio, and the 2H signal is indicative of alkene hydrogen atoms. The 6H signal is indicative of two CH_3 groups that are slightly deshielded. The necessary symmetry can be obtained, and H_2O can be lost, via a cyclization of the intermediate, followed by an E1 reaction.

Proton transfer
Proton transfer
Nucleophilic addition
Proton transfer
Proton transfer
Heterolysis
Electrophile elimination

Problem 18.101

Think

What does the 3:9 ratio in the ¹H NMR signify? Of what is the broad singlet 1H signal indicative? What functional groups are likely present from the IR signals at 3300–3500 cm^{-1} and 1650 cm^{-1}? Under acidic conditions, what can act as a nucleophile? How can a good leaving group be formed? What types of species should not appear in the mechanism?

Solve

The 9H signal in the proton NMR spectrum signifies the *t*-butyl group. The 3H signal represents the distinct methyl group. The 1H signal is likely from an OH or NH proton, given that it is a broad singlet. The IR spectrum confirms this. The C=O signal at 1650 cm^{-1} is consistent with an amide carbonyl stretch, and the broad absorption between 3300 and 3500 cm^{-1} is consistent with a secondary amide. The key step after the formation of the intermediate is the attack of water on the C attached to N. That forms the C–O bond that will become a C=O bond through some proton transfer steps.

Proton transfer
Heterolysis
Coordination
Nucleophilic addition
Proton transfer
H^+
Proton transfer
Proton transfer

Problem 18.102

Think

What is the only type of proton in the product, as indicated by the ^{1}H NMR? How many types of C atoms does the ^{13}C NMR spectrum indicate are present? What does the odd molecular mass indicate is present? With those heavy atoms, what molecule must have been eliminated during the course of the reaction?

Solve

The proton NMR spectrum informs us that all protons (via their chemical-shift values) are aromatic. The odd molar mass indicates that N should be present. The nine signals in the ^{13}C NMR spectrum indicate nine distinct carbon atoms. With nine C atoms and one N atom, the mass sums to 9(12) + 1(14) = 122 amu. That leaves seven H atoms, so the formula must be C_9H_7N. Compared to the formula of the reactant, C_9H_9NO, we can see that there is a net loss of H_2O during the course of the reaction. This formula can be produced, and all H atoms can be made aromatic, if a second ring is formed, giving 10 π electrons total that are part of the same aromatic system.

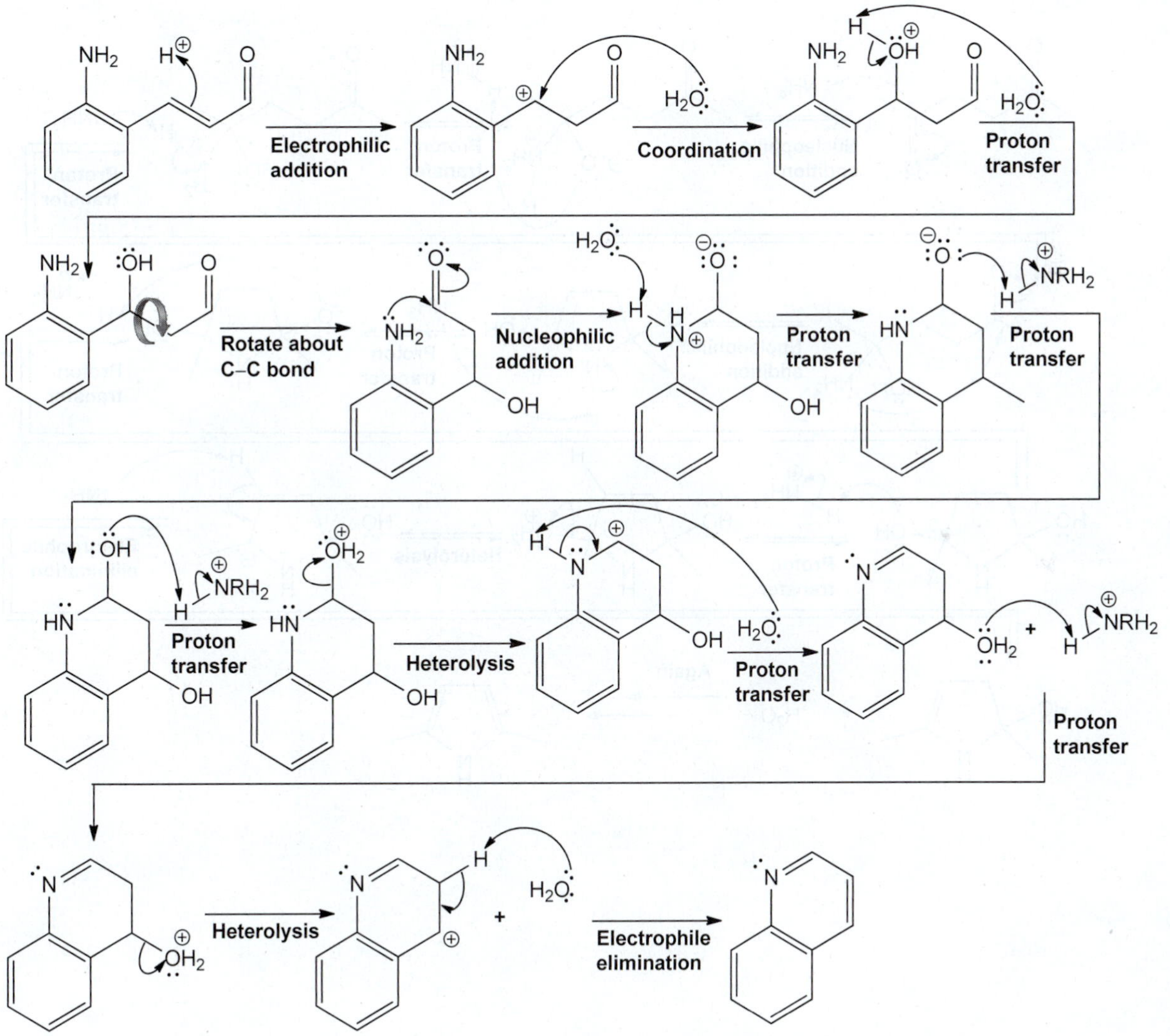

Electrophilic addition → Coordination → Proton transfer → Rotate about C–C bond → Nucleophilic addition → Proton transfer → Proton transfer → Proton transfer → Heterolysis → Proton transfer → Proton transfer → Heterolysis → Electrophile elimination

Problem 18.103

Think

What do the relatively few signals in the ^{1}H NMR signify about the symmetry of the product? What are the integrations of the ^{1}H NMR signals? What do the chemical shifts of those signals tell you? What reaction takes place between NH_3 and the C=O group of a ketone or aldehyde?

Solve

The relatively low number of carbon and proton NMR signals means that the product has a relatively high degree of symmetry. This symmetry can be achieved by a ring. The proton spectrum shows three signals whose integrations are in a 1:2:6 ratio. The 1H signal is broad, indicative of a NH or OH proton. The 2H signal has a chemical shift consistent with alkene protons, and the 6H signal is consistent with two CH_3 groups that are moderately deshielded. The mechanism below begins with nucleophilic addition of NH_3 to a C=O group. The N subsequently attacks the other C=O group to close the ring, which allows one NH proton to remain. The last several steps are consecutive elimination reactions, which form the C=C groups. Note that for those elimination reactions to occur, the solution needs to be at least slightly acidic.

Nucleophilic addition

Proton transfer

Proton transfer

Nucleophilic addition

Proton transfer

Proton transfer

Proton transfer

Heterolysis

Electrophile elimination

Again

CHAPTER 19 | Organic Synthesis 2: Intermediate Topics in Synthesis Design, and Useful Redox and Carbon–Carbon Bond-Forming Reactions

Your Turn Exercises

Your Turn 19.1

Think

Consider the electronegativity of Br compared to that of C and of Li compared to that of C. Consider the electronegativity of Cl compared to that of C and of Mg compared to that of C. Draw a bond dipole and then consider the δ^+ or δ^- charge on C.

Solve

(a) Br is more electronegative compared to C and, therefore, bears the δ^-, leaving a δ^+ on C. Li is less electronegative compared to C and, therefore, bears the δ^+, leaving a δ^- on C.

δ^+ Br   Li(*s*)   Ether δ^- Li

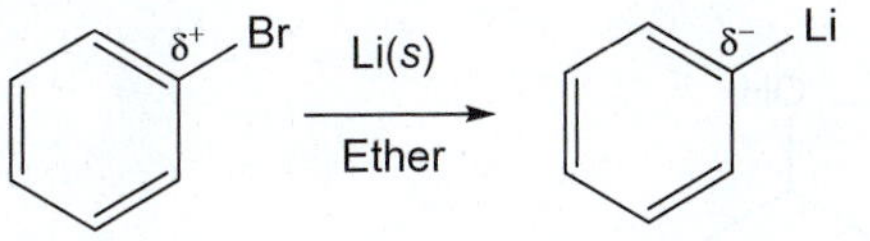

(b) Cl is more electronegative compared to C and, therefore, bears the δ^-, leaving a δ^+ on C. Mg is less electronegative compared to C and, therefore, bears the δ^+, leaving a δ^- on C.

Cl δ^+ Mg(*s*)   Ether   ClMg δ^-

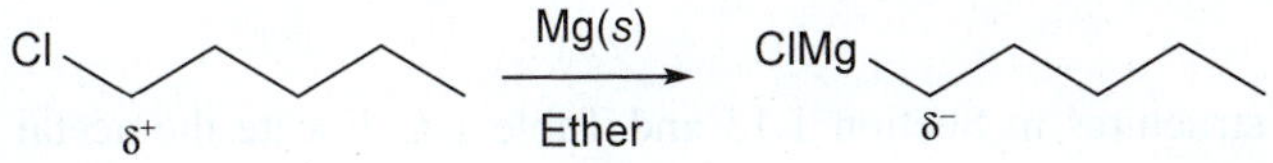

Your Turn 19.2

Think

What does relative positioning mean? Identify the C atom associated with each heteroatom in the product. If one of those C atoms is designated "1" and the C atoms in the chain are numbered sequentially, what number does the C atom in the second heteroatom receive? Review the column labeled "relative positioning" in Table 19-1 to verify your circled C atoms.

Solve

The relative positioning describes the C atoms that are associated with the heteroatoms. The C atoms are circled below, the carbon chain is numbered from one heteroatom to the second, and the relative positioning numbers are indicated.

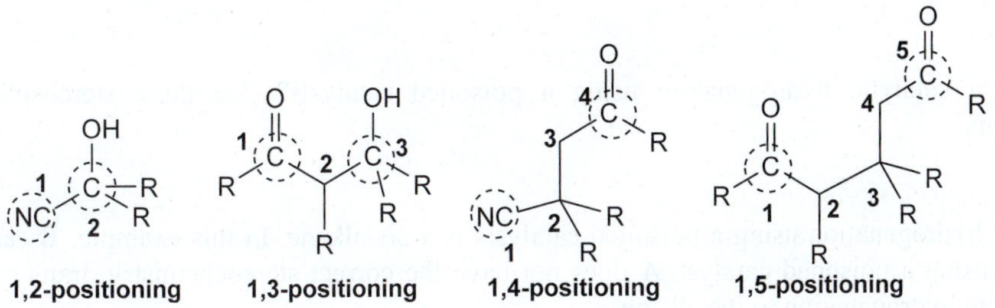

1,2-positioning   **1,3-positioning**   **1,4-positioning**   **1,5-positioning**

Your Turn 19.3

Think

What are the limitations of a Wolff–Kishner reduction? What are the limitations of a Clemmensen reduction? What is the product of the reaction of a nitrile with water in a strong base? In a strong acid? Will the reduction or a hydrolysis take place?

Solve

A nitrile (R–C≡N) can be hydrolyzed to an amide under either acidic or basic conditions. Therefore, a Wolff–Kishner reduction should not be carried out on a ketone or aldehyde containing a C≡N group owing to the strong basic conditions in water. Nor should a reduction via Clemmensen be carried out on such a compound, owing to the strong acidic conditions in water. The C≡N will be hydrolyzed, which would not be the desired reaction.

Your Turn 19.4

Think

What functional group results when the C=O group of a ketone or aldehyde is protected? Where is this functional group in the reactant? What results when this protecting group is removed?

Solve

An aldehyde (or ketone) is protected by converting it to an acetal. The acetal is circled in the figure below. When the protecting acetal group is removed, a C=O group is produced, which is part of an aldehyde in this case.

Your Turn 19.5

Think

Review functional group definitions and structures in Section 1.13 and Table 1-6. Locate the acetal group in Entries 1 and 2.

Solve

An acetal is characterized by a $R_2C(OR)_2$, $RCH(OR)_2$, or $CH_2(OR)_2$ grouping. The acetal group is noted by the dashed ovals below.

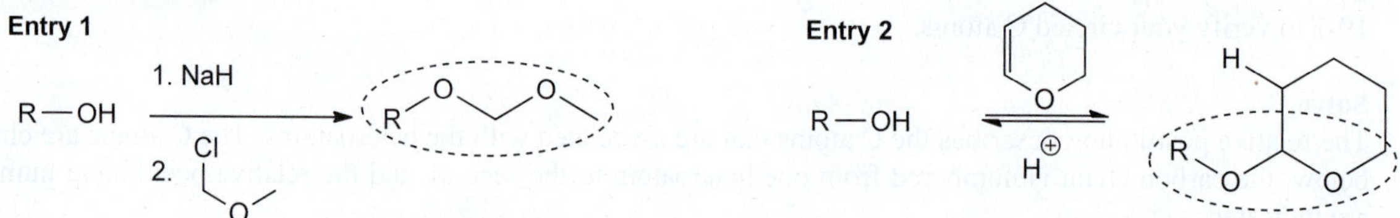

Your Turn 19.6

Think

What is the product of a catalytic hydrogenation using a poisoned catalysis? Are there stereospecific stereochemical requirements?

Solve

The product of a catalytic hydrogenation using a poisoned catalysis is a *cis*-alkene. In this example, **B** can be produced from an alkyne using a poisoned catalyst. **A** does not have the correct stereochemistry, trans rather than cis, and **C** is a complete hydrogenation to the alkane.

Your Turn 19.7

Think

Consult Section 17.3 to review how to calculate oxidation states. Compare the electronegativity of C versus H and of C versus O. Which is more electronegative? How are the covalently bonded electrons assigned when the bonded atoms have different electronegativities? When they have the same electronegativities?

Solve

In a given covalent bond, all electrons are assigned to the more electronegative atom. If the atoms are identical, the electrons are split up evenly. H is *less* electronegative compared to C, and O is *more* electronegative compared to C. In the secondary alcohol, the C atom is assigned two electrons from the C–H bond and one electron from each of the two C–C bonds, for a total of four valence electrons. That is the same number of valence electrons for an isolated uncharged C atom, so that C has an oxidation state of 0. In the ketone, the C is assigned only one electron from each of the two C–C bonds, for a total of two. Because that is two fewer than in an isolated uncharged C atom, that C has an oxidation state of +2. The oxidation state of C increased by two, verifying that the C atom lost two electrons and that the compound was oxidized.

A 2° alcohol; OH; 0; Chromic acid; H_2CrO_4; H_2SO_4; H_2O, acetone; O; +2; A ketone

Cyclooctanol → Cyclooctanone 96%

Your Turn 19.8

Think

What is the functional group of the reactant in both **A** and **B**? What functional groups are produced in each of these reactions? Which oxidizing agent allows the oxidation to stop at the aldehyde? Which oxidizing agent allows the oxidation to produce a carboxylic acid?

Solve

A primary alcohol is the reactant for each of these reactions. **A** yields an aldehyde, which requires the absence of water. PCC will accomplish this oxidation. **B** is oxidized to a carboxylic acid, which is accomplished using chromic acid.

OH; PCC; $\oplus$N–H • $CrO_3Cl^{\ominus}$; CH_2Cl_2; A; O; H

OH; Chromic acid; H_2CrO_4; H_2SO_4; H_2O, acetone; B; O; OH

Your Turn 19.9

Think

What are the organic requirements of R and R′ in R–X and R′–(boron or H) for Suzuki and Heck coupling reactions? What types of carbons are present in the carbon–carbon bond of C_6H_5–C_6H_5?

Solve

The carbon–carbon bond of C_6H_5–C_6H_5 would need to be formed from two aryl carbon atoms, R and R′. The precursors to form each are shown on the next page. In Suzuki coupling, the R in R–X can be vinylic or aryl, whereas R′ can be alkyl, vinylic, or aryl. In Heck coupling, the R in R–X can be vinylic or aryl, whereas R′ can be vinylic but cannot be aryl. Therefore, C_6H_5–C_6H_5 can be formed by a Suzuki coupling reaction but cannot be formed by a Heck coupling reaction.The precursors to form each are shown on the next page.

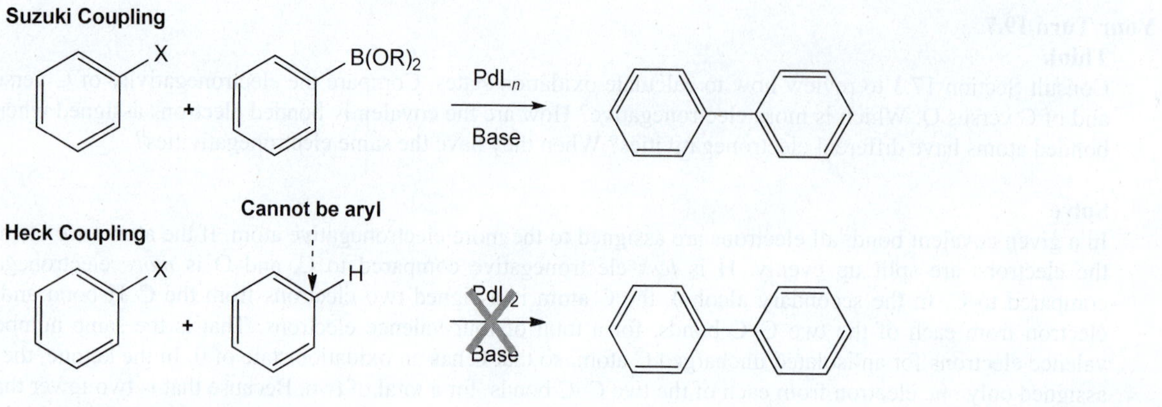

Suzuki Coupling

Heck Coupling

Cannot be aryl

Your Turn 19.10

Think

Refer to Equation 19-50 as a guide. What changes need to be made to the orientation of the CH_2 and CH–R groups relative to M and CHPh?

Solve

To produce $H_2C{=}CH_2$, the CH_2 group must be oriented on the same end as M, and the CH–R group must be oriented on the same end as CHPh. The four steps of the mechanism to produce R–CH=CH–R are the same four steps to produce $H_2C{=}CH_2$. The modified mechanism is shown below.

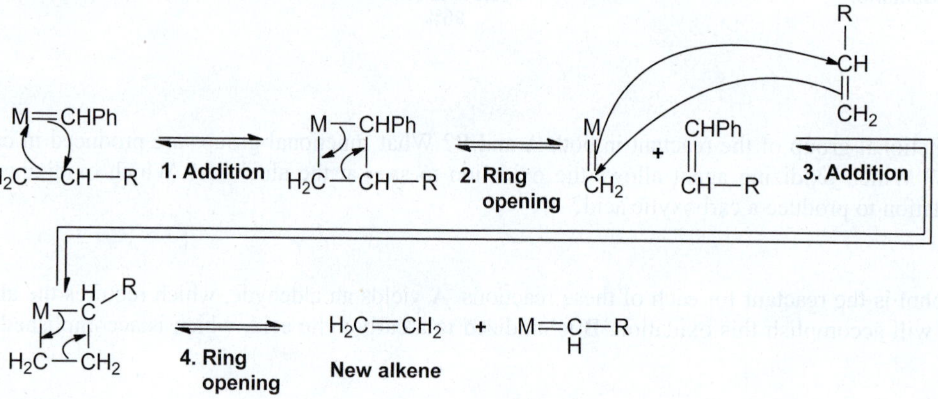

1. Addition
2. Ring opening
3. Addition
4. Ring opening

New alkene

In Chapter Problems

Problem 19.1

Think

Is the C atom in the C–I or C–Br bond electron rich or electron poor? What type of reagent forms when the alkyl, alkenyl, or aryl halide reacts with Mg(*s*) or Li(*s*)? Is the resulting C atom electron rich or electron poor?

Solve

The C atom in the C–I bond is electron poor, and the same is true of the C–Br bond. When an alkyl, alkenyl, or aryl halide reacts with Mg(*s*) or Li(*s*), the resulting C atom is electron rich, because C is more electronegative than the metals. These products are shown in **(a)**–**(c)**. The electron-rich C atom in **(c)** reacts with the CuI to form a lithium dialkylcuprate reagent.

(a) I — Mg(*s*), Ether → MgI

(b) Br — Mg(*s*), Ether → MgBr

(c) I — Li(*s*), THF → Li

(d) Li (Product from (c)) — CuI, THF → Li Cu

Problem 19.2

Think

Which C atom in the product shown was originally bonded to the halide? Did the electronic nature of the C atom change?

Solve

In each case, the C atom bonded to metal was originally bonded to the halide. The C–X bond has an electron-poor C, while the C–M bond (M = metal) has an electron-rich C.

(a) Br — Li(*s*), THF → Li

(b) Cl — Mg(*s*), Ether → MgCl

(c) Br — Mg(*s*), Ether → MgBr

(d) Br — 1. Li(*s*), 2. CuI → Li Cu

Problem 19.4

Think

Do C–C bonds have to be formed in the synthesis? If so, what should the electron-rich species be? What should the electron-poor species be? What carbon–carbon bond-forming reaction produces an alcohol? Can the synthesis take place in a single reaction? If not, what precursor should we choose?

Solve

Because the source of carbon contains two C atoms and the product contains six C atoms, two C–C bonds must be formed. A C–C bond-forming reaction would be difficult to carry out directly between two molecules of ethanal, because such a reaction would entail the formation of a bond between two C atoms bearing the same charge—a partial positive charge. The precursor should bear a C atom with a negative or partial negative charge. It's not immediately obvious what reactions to undo in the second transform. When carrying out the synthesis in the forward direction, we need to figure out how make the ketone. One way to make the ketone begins with an alkyne, which can be produced by elimination.

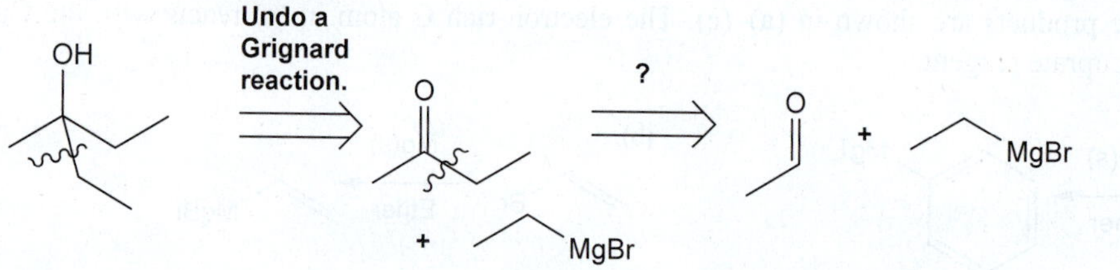

The acetaldehyde starting material can be converted to the alcohol via reduction with $NaBH_4$/ROH. This alcohol can then be converted to an alkyl bromide, which can react to form a Grignard reagent (C atom with a negative charge). Ethanal can then react with an equivalent of CH_3CH_2MgBr, followed by protonation, to form butan-2-ol. This alcohol is dehydrated with H_3PO_4/Δ to yield the alkene, which reacts with Br_2/CCl_4 to give the vicinal dibromide. The vicinal dibromide is converted to the alkyne via a double elimination reaction with excess $NaNH_2$. The alkyne reacts with water and a Hg^{2+} catalyst to produce the ketone. The ketone can then react with another equivalent of CH_3CH_2MgBr, followed by protonation, to form the alcohol product given: 3-methylpentan-3-ol.

The synthesis in the forward direction is as follows:

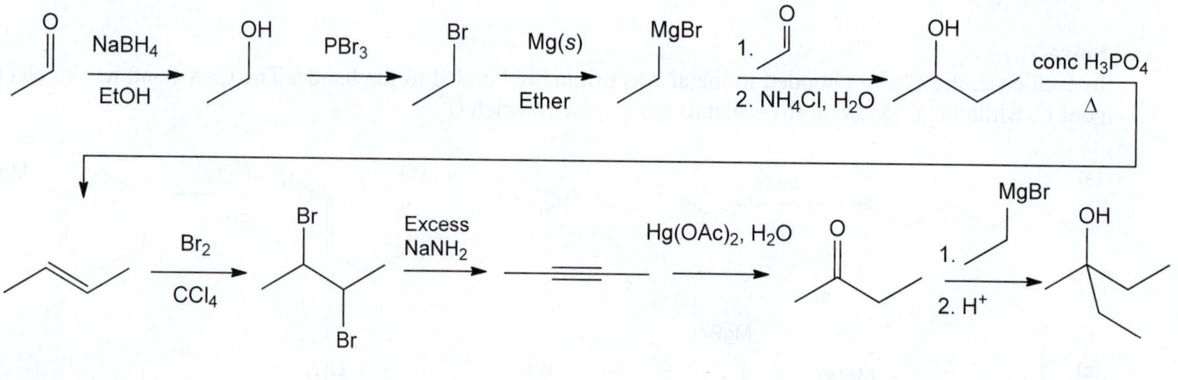

Problem 19.5

Think

Do C–C bonds have to be formed in the synthesis? If so, how many such bonds must form? What should the electron-rich species be? What should the electron-poor species be? Can the synthesis take place in a single reaction? If not, what precursor should we choose?

Solve

Because the other source of carbon other than bromobenzene contains two C atoms and the product contains 10 C atoms, two C–C bonds must be formed. A C–C bond formation reaction would be difficult to carry out directly with bromobenzene and oxirane, because such a reaction would entail the formation of a bond between two C atoms bearing a partial positive charge. Bromobenzene can be converted to the Grignard reagent upon reaction with Mg(*s*). Phenylmagnesium bromide can attack the electron-poor C atom of the epoxide ring and open the compound. The alcohol forms from a proton transfer reaction. The alcohol can be converted to an alkyl bromide via PBr_3, which is not subject to rearrangement. The reaction sequence is then repeated to add the additional two C atoms. See the reaction on the next page.

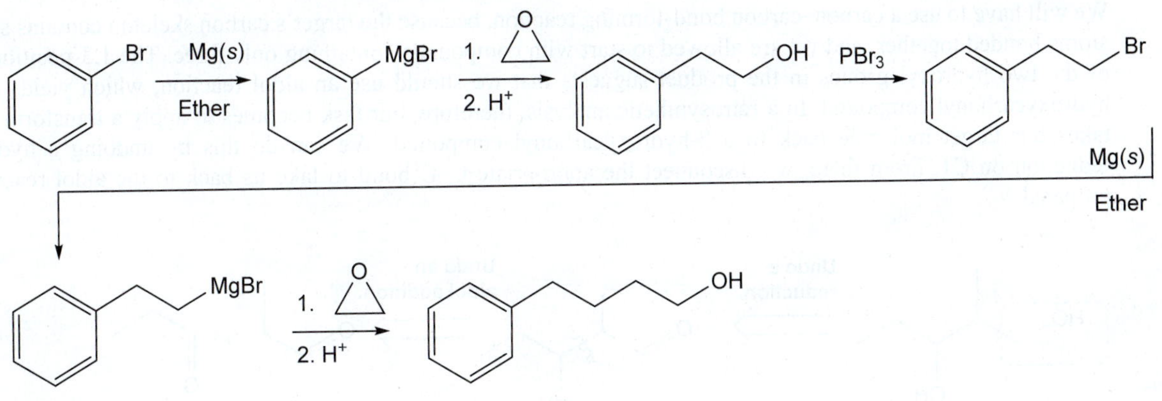

Br $\xrightarrow[\text{Ether}]{Mg(s)}$ MgBr $\xrightarrow[2.\ H^+]{1.\ \text{epoxide}}$ OH $\xrightarrow{PBr_3}$ Br $\xrightarrow[\text{Ether}]{Mg(s)}$ MgBr $\xrightarrow[2.\ H^+]{1.\ \text{epoxide}}$ OH

Problem 19.7

Think

Does the target have a 1,2-positioning of the heteroatoms that could result from the formation of a C–C bond? If so, from what cyanohydrin could this target be synthesized? How could that cyanohydrin be synthesized from the starting compounds?

Solve

The target does indeed have a 1,2-positioning of the NH_2 and OR groups, so we can conceive of obtaining it from a cyanohydrin, as shown in the following retrosynthetic analysis:

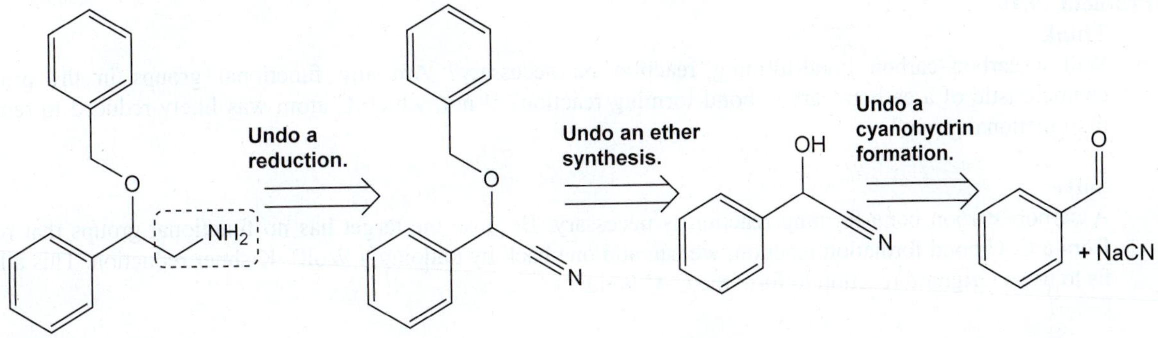

O, NH_2 $\xRightarrow{\text{Undo a reduction.}}$ O, N $\xRightarrow{\text{Undo an ether synthesis.}}$ OH, N $\xRightarrow{\text{Undo a cyanohydrin formation.}}$ O + NaCN

The forward reaction sequence is shown below.

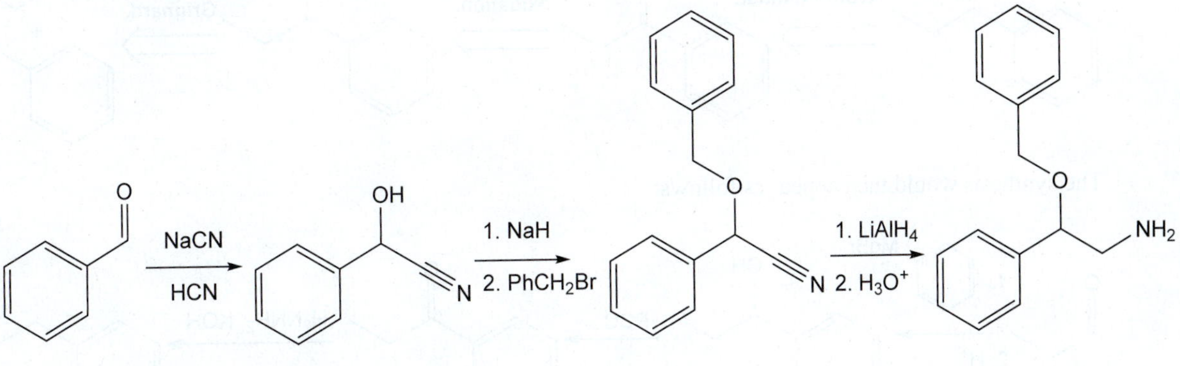

O $\xrightarrow[\text{HCN}]{\text{NaCN}}$ OH, N $\xrightarrow[2.\ PhCH_2Br]{1.\ \text{NaH}}$ O, N $\xrightarrow[2.\ H_3O^+]{1.\ LiAlH_4}$ O, NH_2

Problem 19.9

Think

Will a carbon–carbon bond-forming reaction be necessary? What is the relative positioning of the heteroatoms in the target? Does the appropriate carbon–carbon bond-forming reaction in Table 19-1 leave us with the correct functional groups, or will we need to carry out an additional functional group conversion?

Solve

We will have to use a carbon–carbon bond-forming reaction, because the target's carbon skeleton contains six C atoms bonded together, and we are allowed to start with compounds containing only three. The 1,3-positioning of the two hydroxyl groups in the product suggests that we should use an aldol reaction, which yields a β-hydroxycarbonyl compound. In a retrosynthetic analysis, therefore, our task becomes to apply a transform that takes our target molecule back to a β-hydroxycarbonyl compound. We can do this by undoing a hydride reduction on C1. From there, we disconnect the appropriate C–C bond to take us back to the aldol reactant, propanal.

Undo a reduction. Undo an aldol addition.

HO OH OH O O O +

1,3-positioning **A β-hydroxycarbonyl**

The synthesis in the forward direction is as follows:

1. NaOH 2. NH_4Cl, H_2O — $NaBH_4$ EtOH

O O OH HO OH

Problem 19.10

Think

Will a carbon–carbon bond-forming reaction be necessary? Are any functional groups in the product characteristic of a carbon–carbon bond-forming reaction? If not, which C atom was likely reduced to remove the functional group?

Solve

A carbon–carbon bond-forming reaction is necessary. Because the target has no functional groups that result from a C–C bond formation reaction, we can add one back by undoing a Wolff–Kishner reduction. This allows us to use a Grignard reaction to form the C–C bond.

Undo a Wolff–Kishner. Undo an oxidation. Undo a Grignard.

O OH O + MgBr

The synthesis would then appear as follows:

1. MgBr 2. H^+ — PCC — H_2NNH_2, KOH Δ

O OH O

Problem 19.12

Think

What C–C bond could you disconnect in a transform? If that bond is the result of conjugate addition to an α,β-unsaturated ketone or aldehyde, which CH_2 group could we imagine coming from a C=O group?

Solve

One way to carry out the syntheses is to use a conjugate addition of a lithium dialkylcuprate to produce a ketone that can be reduced by a Wolff–Kishner reduction. Four examples are given below.

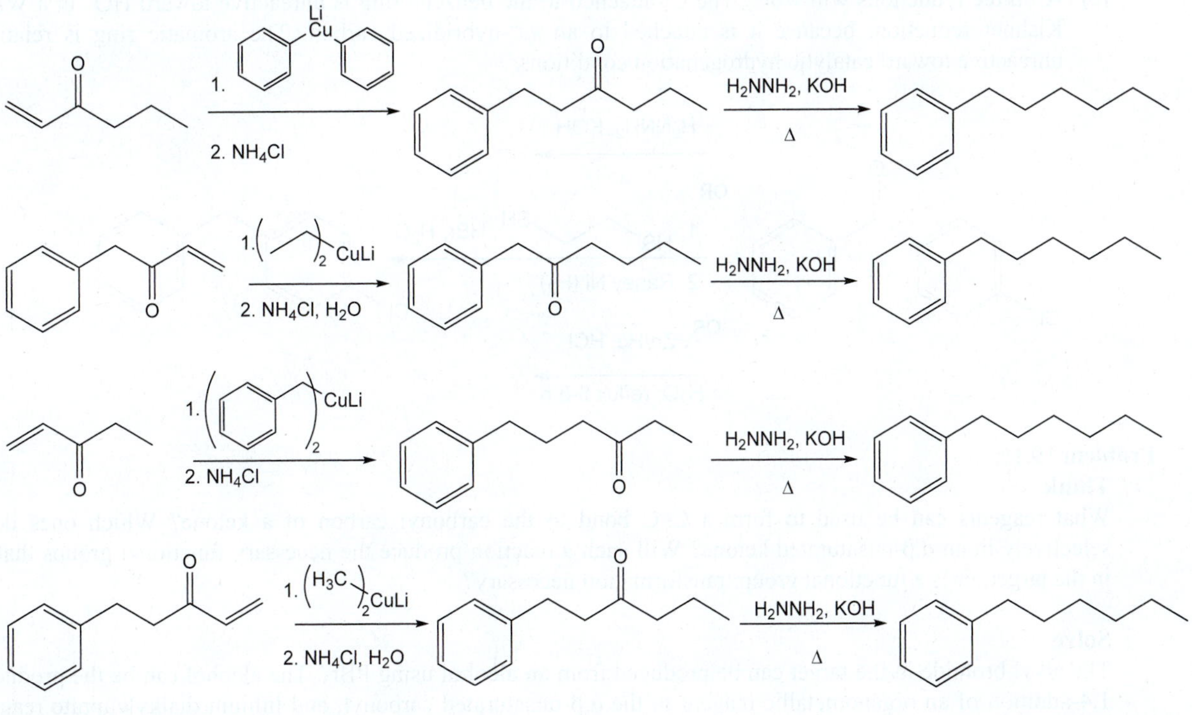

1. Li Cu (diphenyl); 2. NH_4Cl → H_2NNH_2, KOH, Δ

1. (butyl)$_2$CuLi; 2. NH_4Cl, H_2O → H_2NNH_2, KOH, Δ

1. (benzyl)$_2$CuLi; 2. NH_4Cl → H_2NNH_2, KOH, Δ

1. $(H_3C)_2CuLi$; 2. NH_4Cl, H_2O → H_2NNH_2, KOH, Δ

Problem 19.14

Think

Are any functional groups, aside from the carbonyl group, susceptible to reaction in the presence of HO^-? Of HCl? Of thiols? Of $H_2(g)$ and a metal catalyst?

Solve

(a) Only a Wolff–Kishner reduction will work, owing to the C=C double bond that is present. Catalytic hydrogenation will reduce the C=C bond, and as we saw in Chapter 11, the acidic, aqueous conditions of the Clemmensen reduction will lead to reaction with the C=C bond.

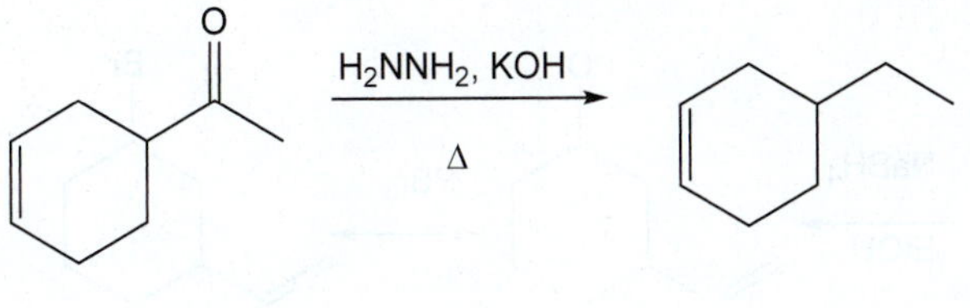

H_2NNH_2, KOH, Δ

(b) The Wolff–Kishner reduction will work. A Clemmensen reduction should be avoided, because hydrolysis of the ether will take place under the acidic, aqueous reaction conditions. Raney-nickel reduction should also be avoided because of the acidic conditions necessary for producing the thioacetal.

H_2NNH_2, KOH

Δ

(c) All three reductions will work. The Cl attached to the benzene ring is unreactive toward HO^- in a Wolff–Kishner reduction, because it is attached to an sp^2-hybridized carbon. The aromatic ring is relatively unreactive toward catalytic hydrogenation conditions.

H_2NNH_2, KOH

Δ

OR

1. HS SH, HBr, H_2O

2. Raney Ni (H_2)

OR

Zn/Hg, HCl

H_2O, reflux 5-8 h

Cl

Cl

Problem 19.16

Think

What reagents can be used to form a C–C bond to the carbonyl carbon of a ketone? Which ones do so selectively in an α,β-unsaturated ketone? Will such a reaction produce the necessary functional groups that are in the target, or is a functional group transformation necessary?

Solve

The alkyl bromide in the target can be produced from an alcohol using PBr_3. The alcohol can be the product of 1,4-addition of an organometallic reagent to the α,β-unsaturated carbonyl, and lithium dialkylcuprate reagents are selective for 1,4-addition rather than 1,2-addition.

Br — **Undo a substitution.** ⟹ OH — **Undo selective reduction.** ⟹ O — **Undo a lithium dialkyl cuprate reaction.** ⟹ O + LiCu (vinyl)$_2$

In the forward direction, the synthesis can be written as follows:

1. (vinyl)$_2$CuLi
2. NH_4Cl, H_2O

$NaBH_4$, EtOH

PBr_3

Problem 19.17

Think

How many C atoms are in acetone? How many C atoms are in hexane-2,5-dione? Which C–C bond is formed? What would be the two starting materials? Are all of the functional groups compatible? If not, how can a protecting group be used?

Solve

Because we are limited to a three-carbon starting compound, a transform must disconnect the C–C bond shown below. We can envision forming that bond in an S_N2 reaction, so one precursor could be an alkyl halide and the other a negatively charged nucleophile.

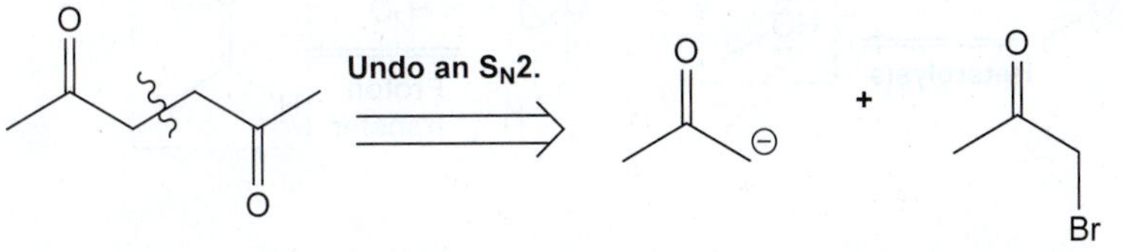

Undo an S_N2.

The desired reaction between the enolate anion and bromoacetone would be S_N2, but it would compete with nucleophilic addition. To prevent nucleophilic addition, we could use a protected form of the C=O instead. The bromoacetone can be made by brominating acetone under acidic conditions.

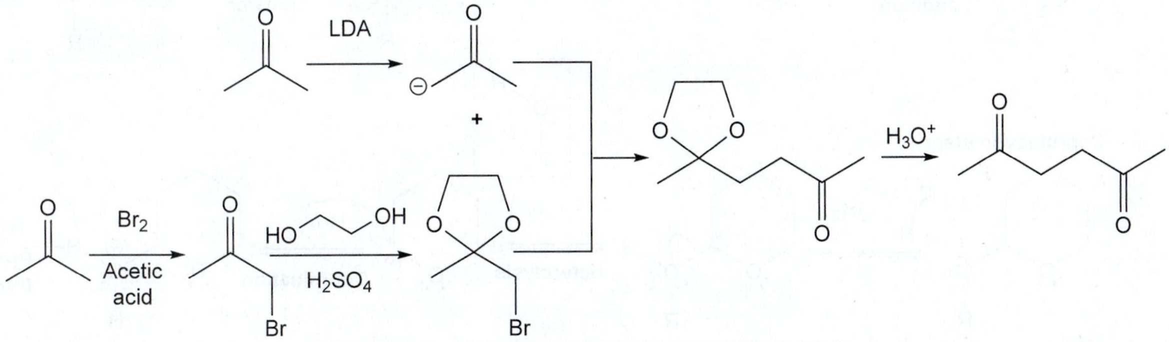

LDA

Br_2 Acetic acid

HO OH H_2SO_4

H_3O^+

Problem 19.18

Think

How is the ROH deprotonated in Entries 1 and 2? In the subsequent step, what is the nucleophile and leaving group? For deprotection in Entries 1 and 2, how do the acidic conditions create a good leaving group? What is the nucleophile?

Solve

The ROH is deprotonated in Entry 1 using NaH, generating a strong RO^- nucleophile that undergoes a substitution reaction with the alkyl halide. In Entry 2, H^+ adds to the C=C to produce the more stable carbocation, after which ROH behaves as a nucleophile to add to the C^+. In the deprotection reactions, O is protonated to make a good leaving group, and water acts as the nucleophile. The protection and deprotection mechanisms are given below and on the next page.

Entry 1

Protection step:

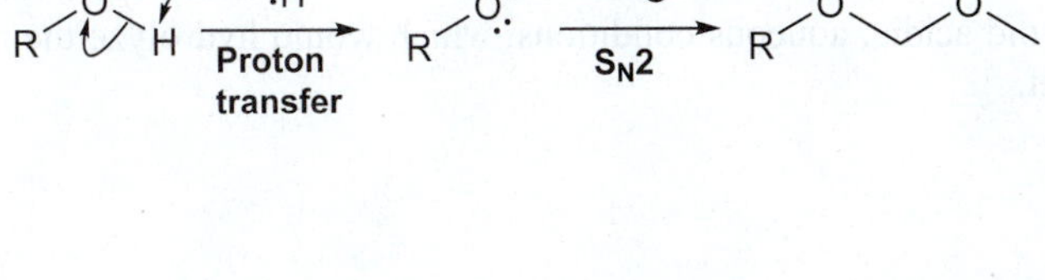

Proton transfer

S_N2

Deprotection step:

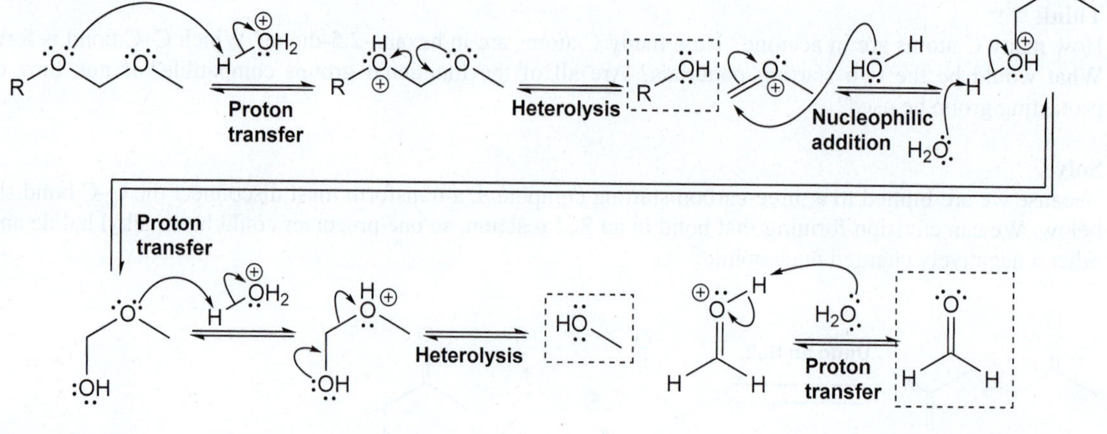

Proton transfer — Heterolysis — Nucleophilic addition — Proton transfer — Heterolysis — Proton transfer

Entry 2

Protection step:

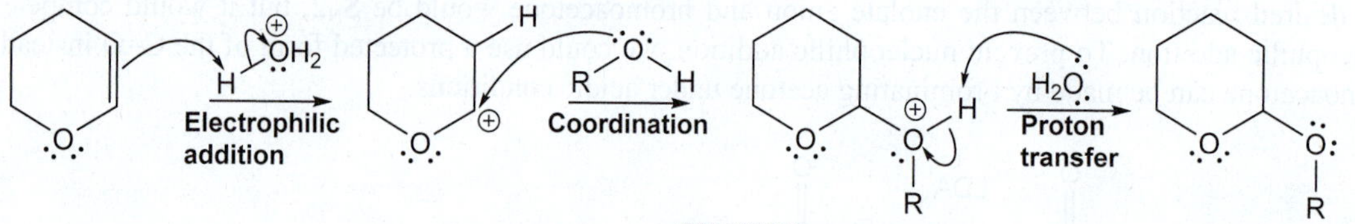

Electrophilic addition — Coordination — Proton transfer

Deprotection step:

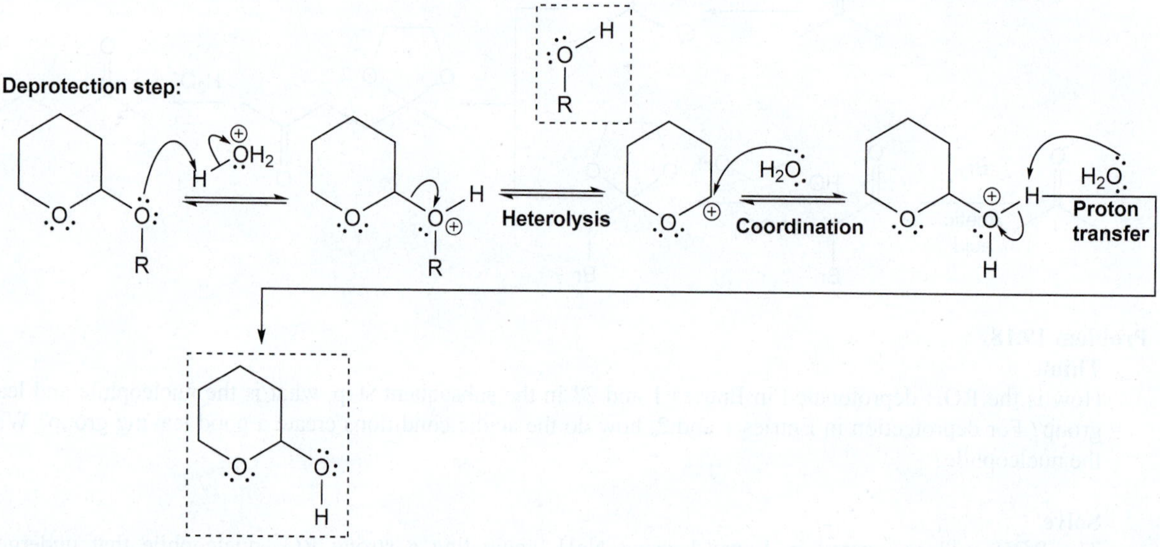

Heterolysis — Coordination — Proton transfer

Problem 19.20

Think

What new bonds are formed? What reagents are required to form this bond? Are these reagents reactive toward any other functional groups present in the compound? If so, how can the functional group be protected?

Solve

The C=O is transformed into an ether. This is accomplished via reduction of the C=O followed by a Williamson ether synthesis on the alcohol. The alcohol originally present needs to be protected, because the Williamson ether synthesis can be carried out on either alcohol. The reaction scheme is shown on the next page. Note that TBDMS is used as the protecting group to avoid acidic, aqueous conditions, which would hydrolyze the newly formed ether that is intended to be in the target.

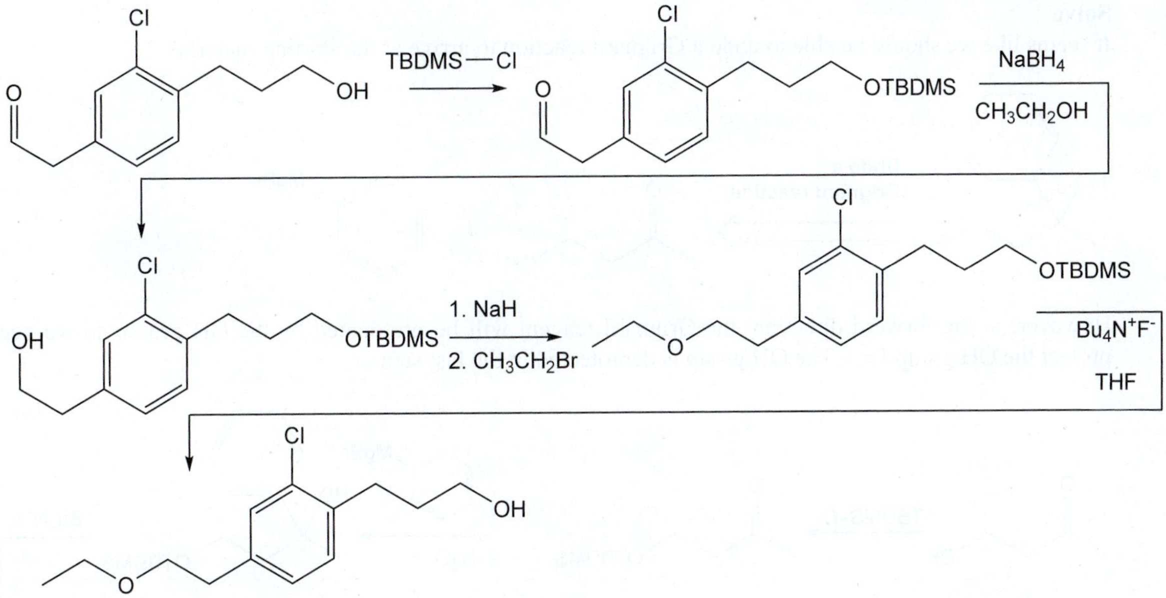

Problem 19.21

Think

What type of reaction occurs in the second reaction sequence? What functional group is formed from an alkylbromide and Mg(*s*)? Will the OH group also react?

Solve

(a) The OH group needs to be protected because in the second reaction sequence, the Grignard reaction that is formed will react with the OH group (see below).

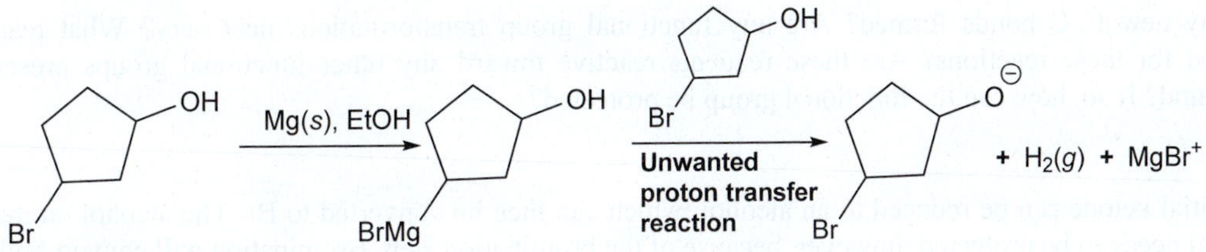

(b) The alcohol could have also been protected prior to performing the Grignard reaction, using dihydropyran (DHP). Once the Grignard reaction is complete, the product can be deprotected. Because the workup step of the Grignard reaction requires acidic aqueous conditions, the acid workup and deprotection occur together. Protection using the MOM ether, on the other hand, should be avoided, because the C–Br bond in the molecule we want to protect would interfere with the required S_N2 reaction in the protection step.

Problem 19.22

Think

What new C–C bonds are formed? What reagents are required to form this bond? Are these reagents reactive toward any other functional groups present in the compound? If so, how can the functional group be protected?

Solve

It seems like we should be able to undo a Grignard reaction to arrive at the starting material.

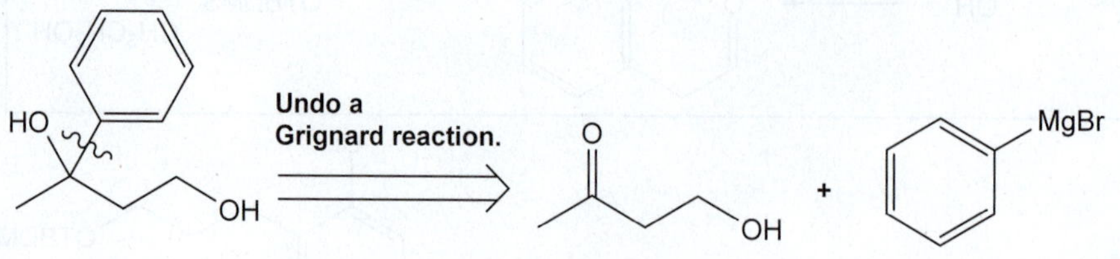

Undo a Grignard reaction.

MgBr

+

However, in the forward direction, the Grignard reagent will be protonated by the OH group, so we need to protect the OH group first. The OH group is deprotected in the last step.

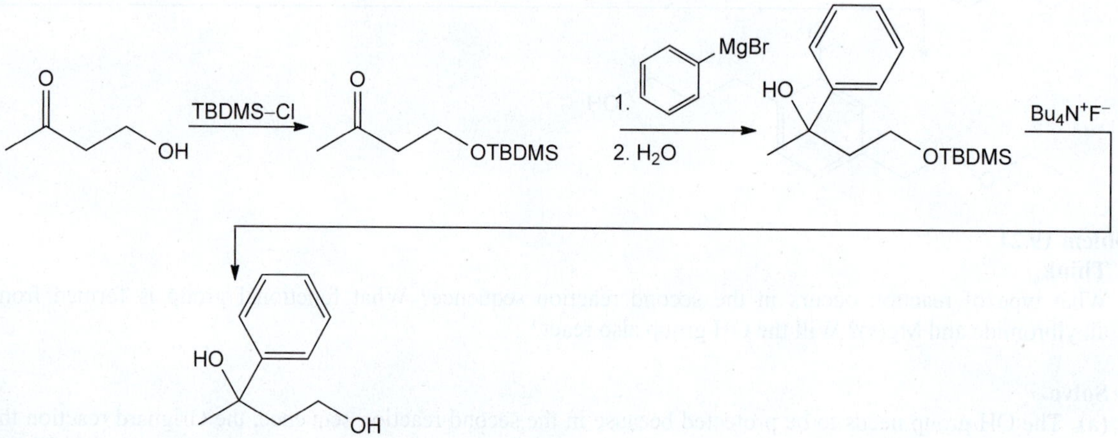

TBDMS–Cl

1. MgBr

2. H_2O

$Bu_4N^+F^-$

Problem 19.23

Think

Are any new C–C bonds formed? Are any functional group transformations necessary? What reagents are required for these reactions? Are these reagents reactive toward any other functional groups present in the compound? If so, how can the functional group be protected?

Solve

The initial ketone can be reduced to an alcohol, which can then be converted to Br. The alcohol on the starting material needs to be protected, however, because of the bromination step. Bromination will convert both alcohol groups to Br if the first one is not protected.

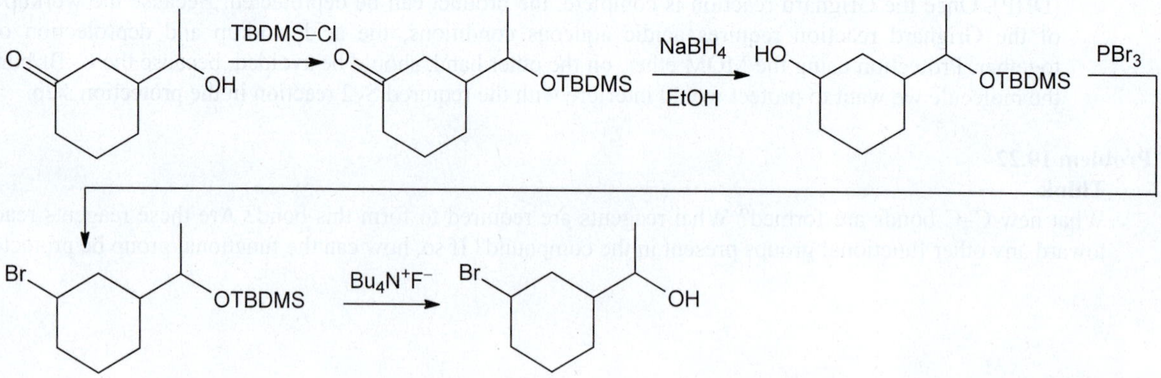

TBDMS–Cl

$NaBH_4$ / EtOH

PBr_3

$Bu_4N^+F^-$

Problem 19.24

Think

What is the structure of 1,2,5-pentanetriol? What new C–C bonds are formed? What reagents are required to form this bond? Are these reagents reactive toward any other functional groups present in the compound? If so, how can the functional group be protected? Are any functional group transformations necessary?

Solve

(a) We want to do a simple S_N2 reaction (a Williamson synthesis).

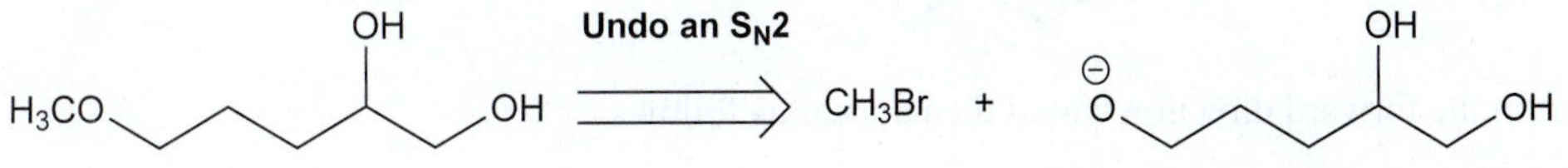

Undo an S_N2

H_3CO OH OH ⟹ CH_3Br + $O^{\ominus}$ OH OH

But to deprotonate only the OH at the left, both at the right have to be protected.

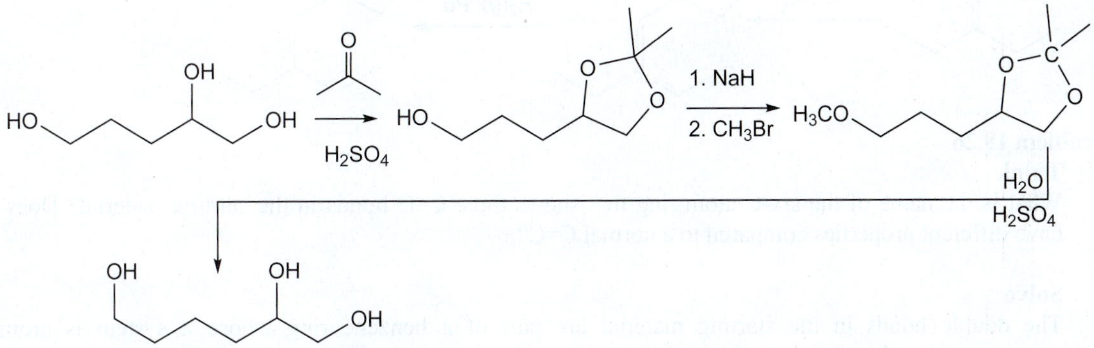

HO OH OH → (acetone, H_2SO_4) → HO (acetonide) → (1. NaH; 2. CH_3Br) → H_3CO (acetonide) → (H_2O, H_2SO_4) → OH OH OH

(b) For this compound, it looks like we want to carry out a Grignard reaction. However, in the forward direction, the two OH groups at the right of the starting compound are not compatible with the Grignard reagent. So those OH groups should be protected.

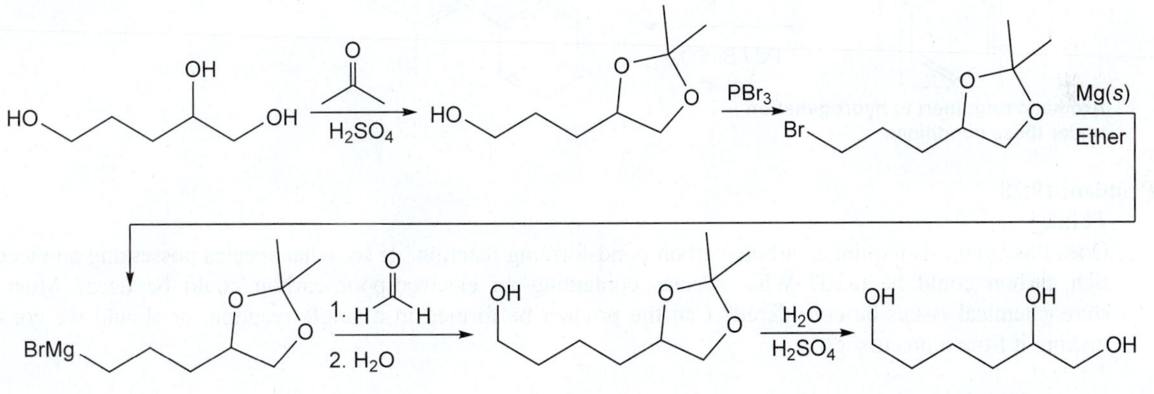

HO OH OH → (acetone, H_2SO_4) → HO (acetonide) → (PBr_3) → Br (acetonide) → (Mg(s), Ether) → BrMg (acetonide) → (1. $H_2C{=}O$; 2. H_2O) → OH (acetonide) → (H_2O, H_2SO_4) → OH OH OH

Problem 19.25

Think

Which C–C bond in the final product was previously the C=C? How do you convert a C=C into a C–C? Which C atom in the C=C was previously the C=O carbon? What is the identity of the R group in the ylide?

Solve

Because the Wittig reaction forms an alkene, we must consider an alkene as a critical synthetic intermediate. The new C=C bond formed in a Wittig reaction could have been the one shown below.

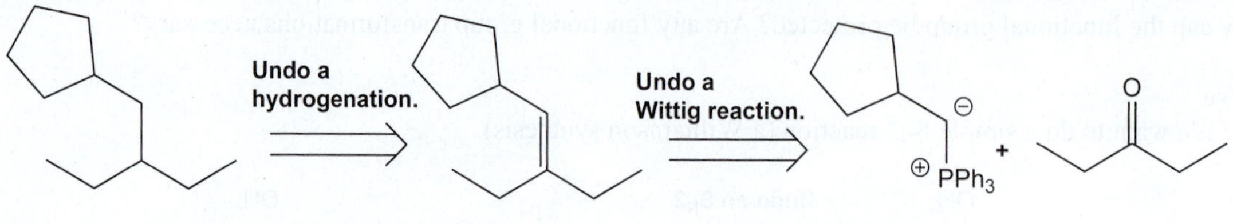

Undo a hydrogenation.

Undo a Wittig reaction.

PPh_3

The synthesis in the forward direction would then appear as follows:

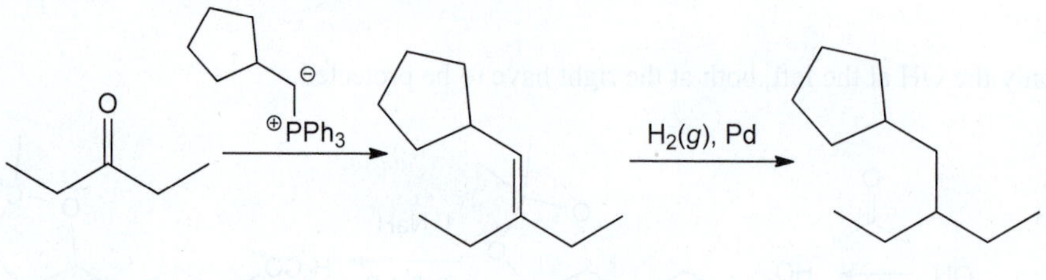

PPh_3

$H_2(g)$, Pd

Problem 19.26

Think

What is the name of the six-C-atom ring that shows three C=C bonds in the starting material? Does this ring have different properties compared to a normal C=C?

Solve

The double bonds in the starting material are part of a benzene ring whose π system is aromatic and exceptionally stable. The conditions required to reduce an aromatic ring are much more extreme than those required to reduce an alkyne.

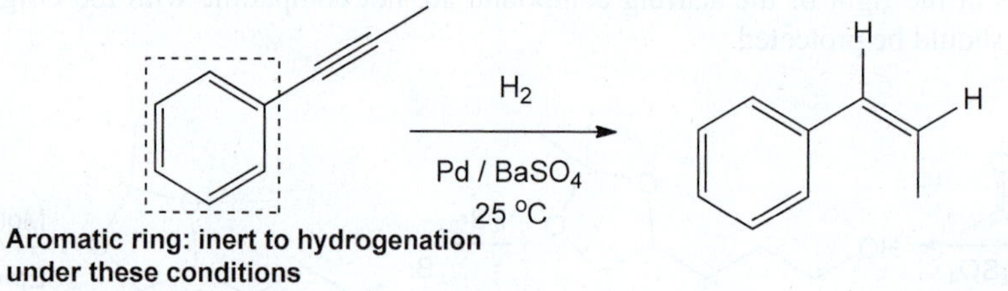

H_2

Pd / $BaSO_4$

25 °C

Aromatic ring: inert to hydrogenation under these conditions

Problem 19.28

Think

Does this synthesis require a carbon–carbon bond-forming reaction? If so, what species possessing an electron-rich carbon could be used? What species containing an electron-poor carbon could be used? Must any stereochemical issues be considered? Can the product be formed in a single reaction, or should we consider making it from a precursor?

Solve

The cis alkene can be the product of reduction using catalytic hydrogenation. The alkyne precursor can be the product of direct addition of $CH_3C{\equiv}C^-$.

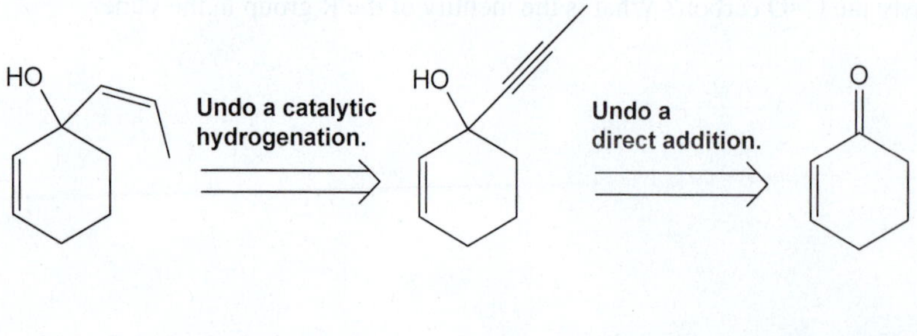

HO

Undo a catalytic hydrogenation.

HO

Undo a direct addition.

O

The synthesis could appear as follows; the reduction involves a poisoned catalyst to stop at the cis alkene:

1. $H_3C-C\equiv CNa$
2. H_2O

H_2
Pd, $BaSO_4$

HO HO

Problem 19.29

Think

Which side of the bicyclic ring is more sterically crowded? How does steric crowding affect which syn product forms?

Solve

The product **A** is favored. For product **B** to form, the top face of the C=C bond must bind to the catalyst surface, but, as indicated, there is excessive steric hindrance provided by the bridged carbon and its methyl groups. To form product **A**, the bottom face of the C=C binds to the catalyst surface, and so do the H atoms.

Lots of steric hindrance

H_2
Pd

α-Pinene A + B

Problem 19.30

Think

Which C=O is more reactive, that of an aldehyde or of a ketone? Which C=O remains unreacted in the final product? How is it possible to selectively react the aldehyde C=O over the ketone C=O?

Solve

The aldehyde C=O needs to be selectively reduced over the ketone C=O to carry out the subsequent bromination. This selective reduction can take place with catalytic hydrogenation, because the aldehyde C=O is more reactive than the ketone C=O. The resulting alcohol can be brominated with PBr_3.

H_2
Pd

PBr_3

Problem 19.32

Think

Which C–C bond can we "disconnect" in a retrosynthetic analysis? By undoing what reaction? Must any oxidations of alcohols be stopped at the aldehyde?

Solve
A Grignard reaction forms a carbon–carbon bond, yielding an alcohol. Our target is a ketone that could be the result of oxidizing an alcohol product which is an immediate product of a Grignard reaction involving C_6H_5MgBr. We can therefore begin the retrosynthesis by disconnecting the C-C bond between the phenyl ring and the C=O

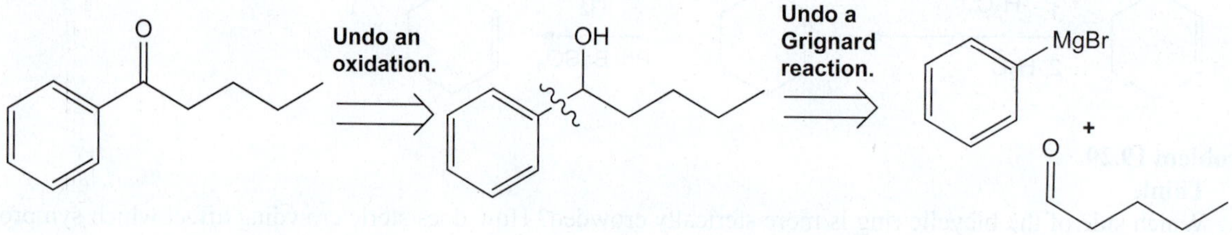

Undo an oxidation.

OH

Undo a Grignard reaction.

MgBr

+

The aldehyde can be formed from an oxidation of an alcohol.

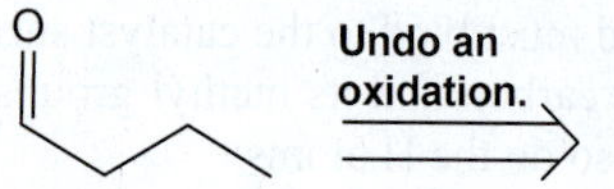

Undo an oxidation.

OH

The synthesis in the forward direction is as shown below.

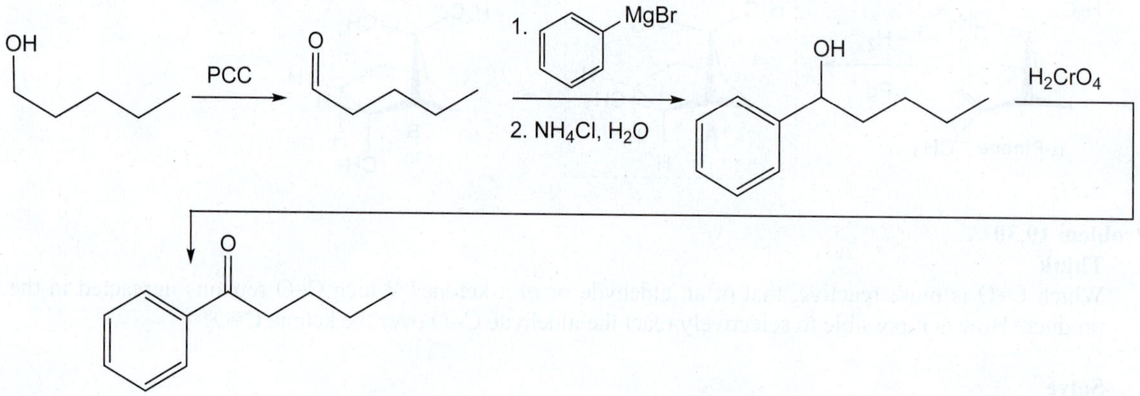

OH

PCC

1. MgBr

2. NH_4Cl, H_2O

OH

H_2CrO_4

Problem 19.33

Think

What functional group results when a secondary alcohol is oxidized by $KMnO_4$? How can this functional group react with a Grignard reagent? What final uncharged product results?

Solve

$KMnO_4$ reacts with a secondary alcohol under the given reaction conditions to form a ketone (**A**). The ketone then reacts with the Grignard reagent to undergo nucleophilic addition, followed by acid workup, to form the tertiary alcohol shown below (**B**).

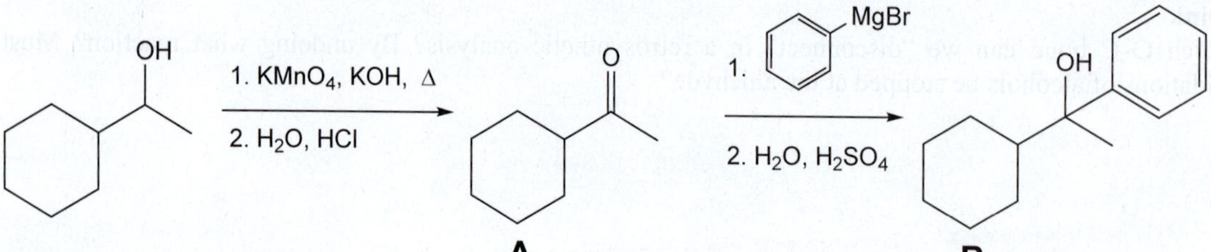

OH

1. $KMnO_4$, KOH, Δ

2. H_2O, HCl

1. MgBr

2. H_2O, H_2SO_4

OH

A

B

Problem 19.34

Think

What new bond forms from reaction of R–X + R'_2–CuLi? Are there stereochemical considerations?

Solve

A new R–R′ bond forms from reaction of R–X with R'_2CuLi with retention of configuration about the C=C bond. The coupling products are shown below. For **(b)**, the stereochemistry at the chiral carbon will be scrambled. Also, notice that there is a leaving group and a hydrogen on adjacent C atoms, so β-elimination would likely occur, producing an alkene instead. The products from reactions **(a)** and **(b)** are given below.

(a) Br + $(CH_3)_2CuLi$ → CH_3 + CH_3Cu + LiBr

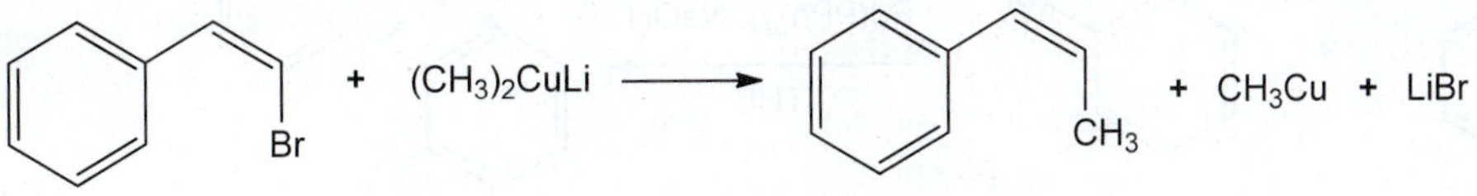

(b) Cl + $(C_6H_5)_2CuLi$ → C_6H_5 + C_6H_5Cu + LiCl

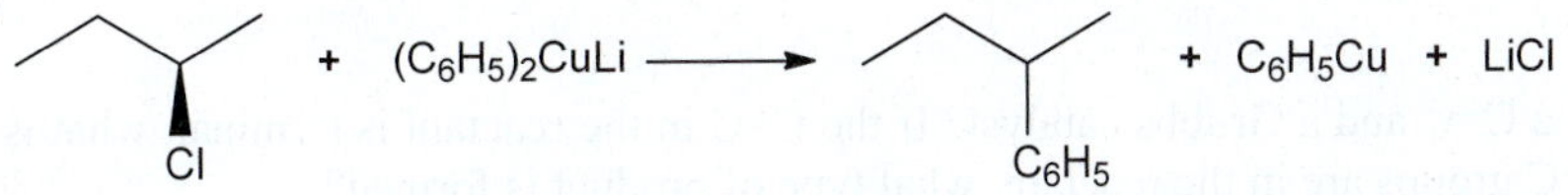

Problem 19.35

Think

What new bond forms from reaction of R–X + R′–B(OR″)$_2$ or H–R′ with a Pd catalyst? Are there stereochemical considerations?

Solve

A new R–R′ bond forms from reaction of R–X + R′–B(OR″)$_2$ (Suzuki reaction) or H–R′ (Heck reaction) with a Pd catalyst. Reactions **(a)** and **(b)** are Heck coupling reactions, where the Ph **(a)** or CN **(b)** substituent on H–R′ is trans to the R substituent from R–X. In reaction **(b)**, the configuration about the C=C in the RI reactant is retained. Reactions **(c)** and **(d)** (shown on the next page) are Suzuki coupling reactions, and the product in **(c)** is formed with retention of configuration.

(a) Br + → ($Pd(OAc)_2$; PPh_3, Et_3N) New C–C bond; Trans configuration

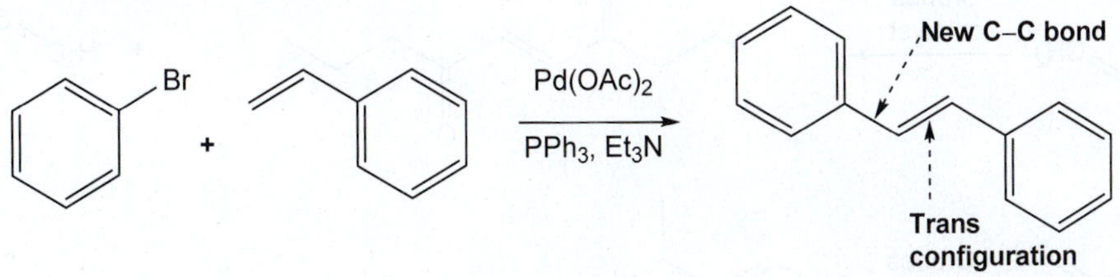

(b) I + CN → ($Pd(OAc)_2$; PPh_3, Et_3N) Retention of configuration; New C–C bond; CN; Trans configuration

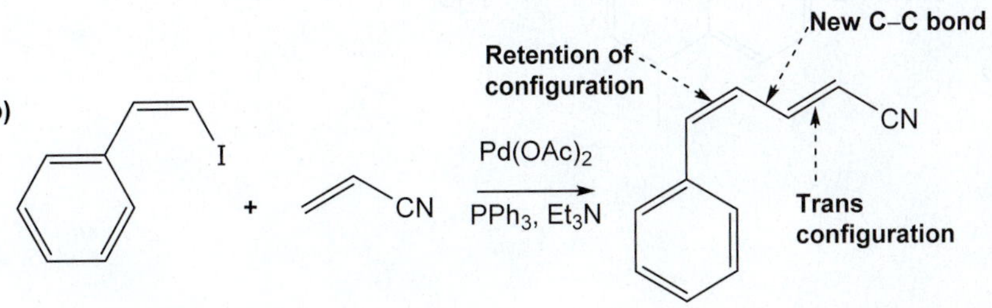

(c) Retention of configuration; New C–C bond; $Pd(PPh_3)_4$, NaOH; THF; Br; O; B; O

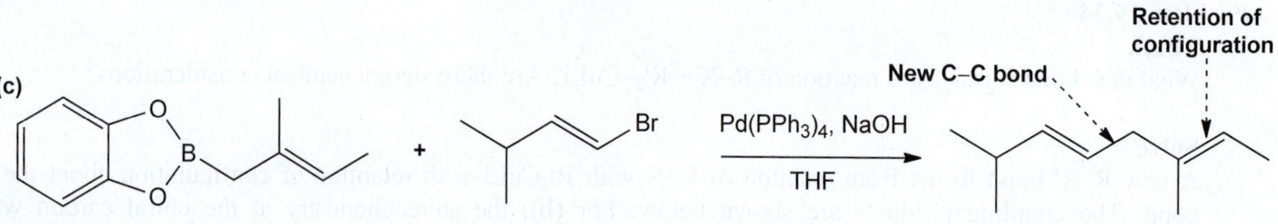

(d) OEt; B; EtO; Cl; $Pd(PPh_3)_4$, NaOH; THF; New C–C bond

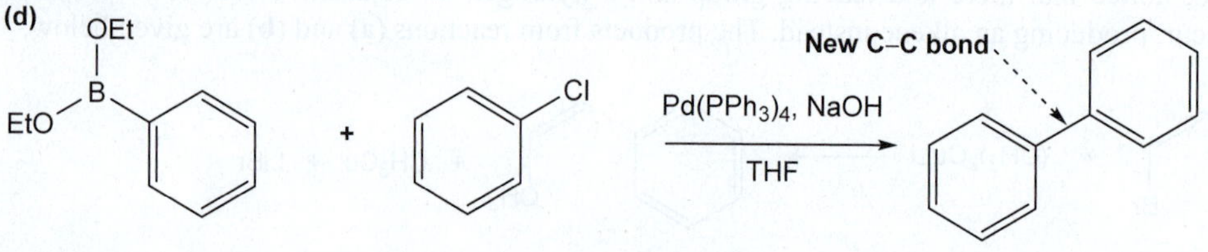

Problem 19.36

Think

What are the products from a C=C and a Grubbs catalyst? If the C=C in the reactant is terminal, what is the by-product? If two terminal C=C groups are in the reactant, what type of product is formed?

Solve

A terminal C=C and a Grubbs catalyst is an example of alkene metathesis where a new C=C forms and $CH_2=CH_2$ is the by-product. A ring forms in reaction **(c)** where the two terminal C=C groups in the reactant are on the same molecule. The major organic products from reactions **(a)–(c)** are shown below.

(a) H_2C; Grubbs catalyst; + $H_2C=CH_2$

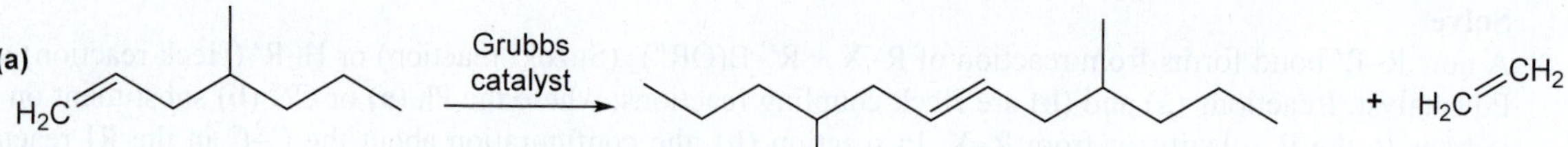

(b) O; CH_2; Grubbs catalyst; O; O; + $H_2C=CH_2$

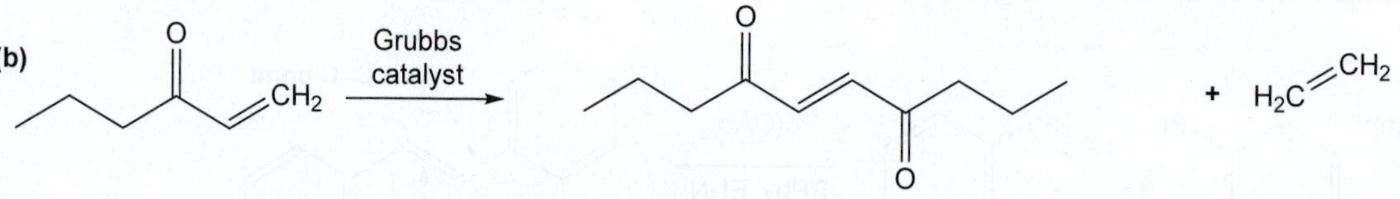

(c) CH_2; CH_2; Grubbs catalyst; + $H_2C=CH_2$

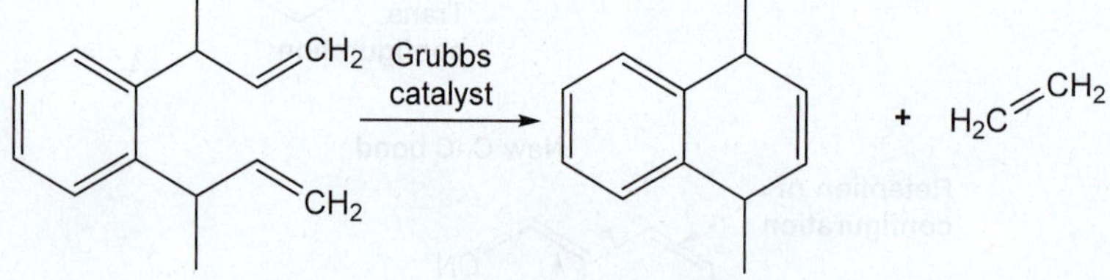

End of Chapter Problems

Section 19.1 Umpolung in Organic Synthesis

Problem 19.37

Think

What is the charge or partial charge of each highlighted C atom? What is the requirement to join two C atoms together to form a new C–C bond? What does umpolung accomplish?

Solve

Umpolung reverses a charge at a particular atom. The partial charges for the highlighted C atoms in the pairs of molecules are shown below. For reactions **(a)** and **(c)**, there are two partial positive charges (δ^+), which are both electrophilic and cannot be joined to form a new C–C. Therefore, umpolung is required to reverse the polarity of one of the reactants.

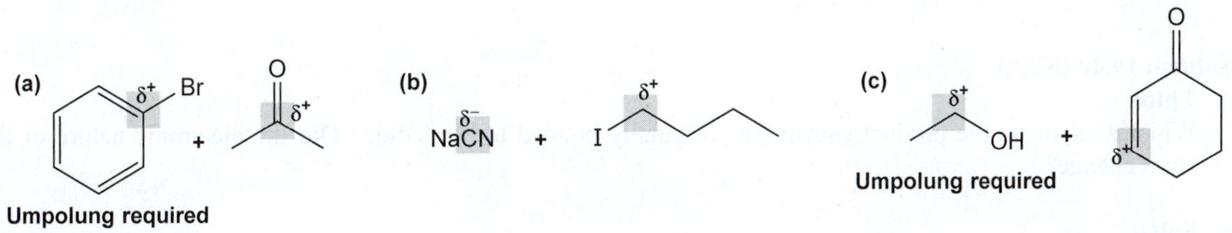

(a) δ^+ Br + O δ^+ — Umpolung required

(b) δ^- NaCN + I δ^+

(c) δ^+ OH + O δ^+ — Umpolung required

Problem 19.38 (SYN)

Think

Which C atom in the product shown was originally bonded to the halide? Did the electronic nature of the C atom change?

Solve

In each case, the C atom bonded to the Mg metal was originally bonded to the halide. The C–X bond has an electron-poor carbon, and the C–Mg bond has an electron-rich carbon.

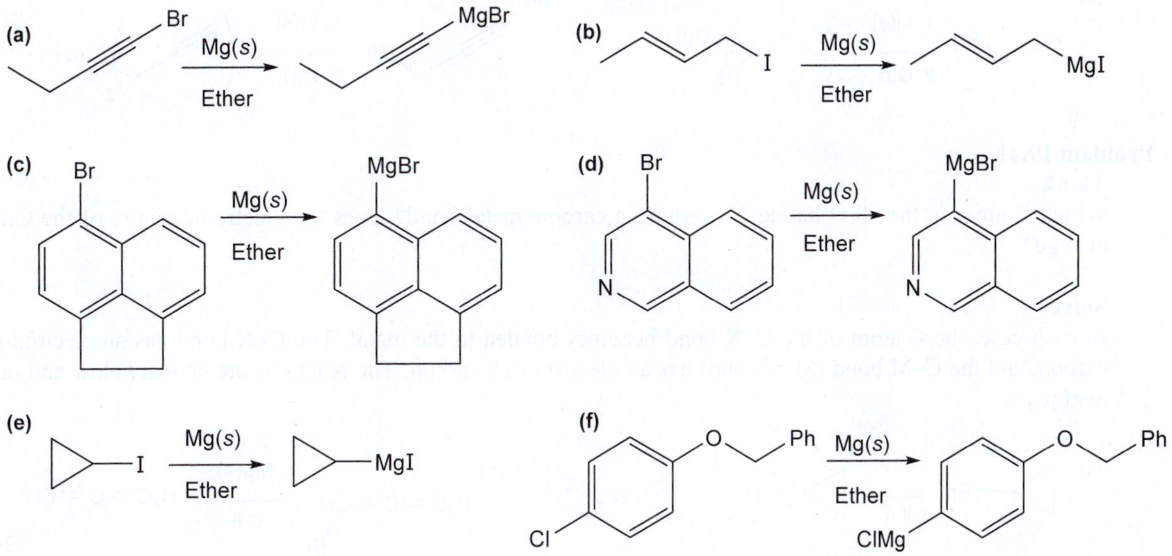

(a) Br → Mg(*s*), Ether → MgBr

(b) I → Mg(*s*), Ether → MgI

(c) Br → Mg(*s*), Ether → MgBr

(d) Br, N → Mg(*s*), Ether → MgBr, N

(e) I → Mg(*s*), Ether → MgI

(f) Cl, O, Ph → Mg(*s*), Ether → ClMg, O, Ph

Problem 19.39 (SYN)

Think

Which C atom in the product shown was originally bonded to the halide? Did the electronic nature of the C atom change?

Solve

In each case, the C atom bonded to the Li metal was originally bonded to the halide. The C–X bond has an electron-poor carbon, and the C–Li bond has an electron-rich carbon.

(a) Li(s), THF
(b) Li(s), THF
(c) Li(s), THF
(d) Li(s), THF

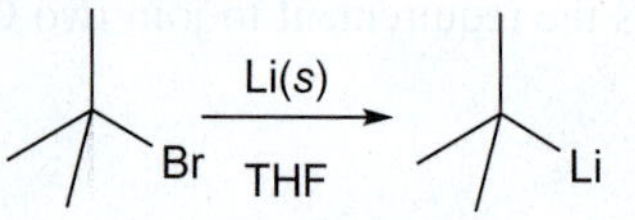

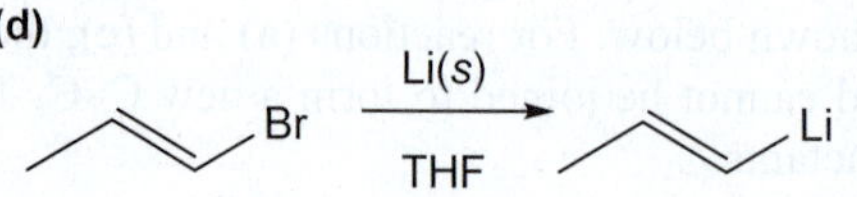

Problem 19.40 (SYN)

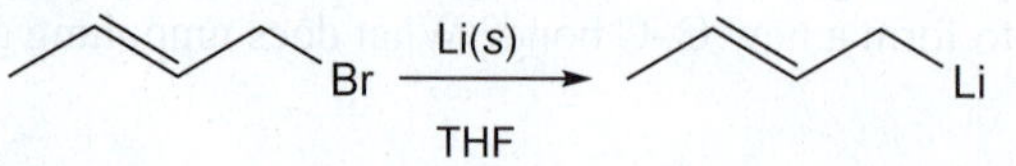

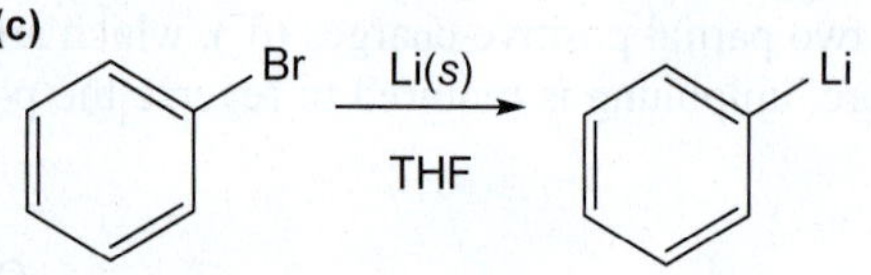

Think

Which C atom in the product shown was originally bonded to the halide? Did the electronic nature of the C atom change?

Solve

In each case, the C atom bonded to Cu was originally bonded to the halide. The C–X bond has an electron-poor carbon, and the C–Cu bond has an electron-rich carbon.

(a) 1. Li(s) 2. CuI → CuLi
(b) 1. Li(s) 2. CuI → CuLi
(c) 1. Li(s) 2. CuI → CuLi
(d) 1. Li(s) 2. CuI → CuLi

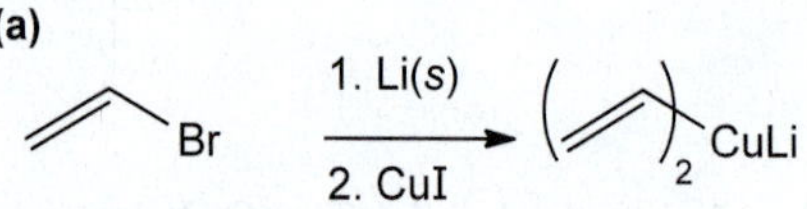

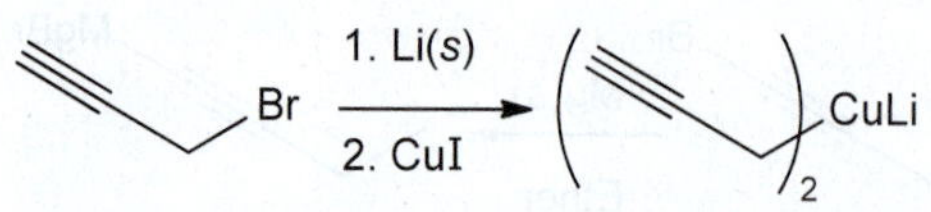

Problem 19.41

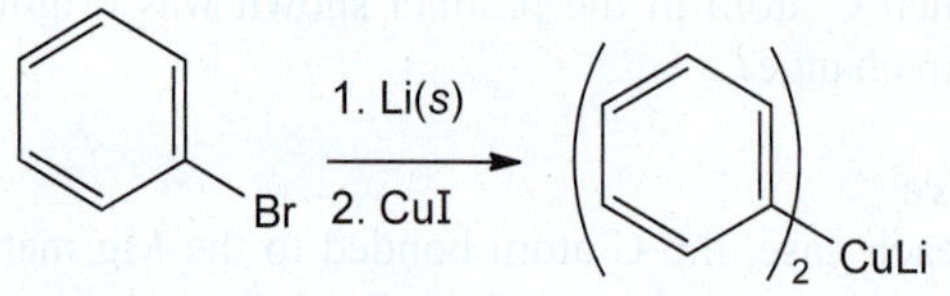

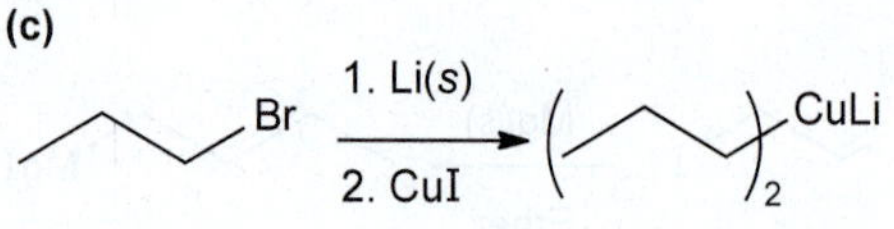

Think

Which C atom in the alkyl halide converts to a carbon–metal bond? Does the electronic nature of the C atom change?

Solve

In each case, the C atom of the C–X bond becomes bonded to the metal. The C–X bond has an electron-poor carbon, and the C–M bond (M = metal) has an electron-rich carbon. The reactions are shown below and on the next page.

(a) Li(s), Ether
(b) $H_2C{=}C{=}CH$–Br → Mg(s), THF → $H_2C{=}C{=}CH$–MgBr
(c) Mg(s), THF → MgBr
(d) Cl → Mg(s), Ether → MgCl

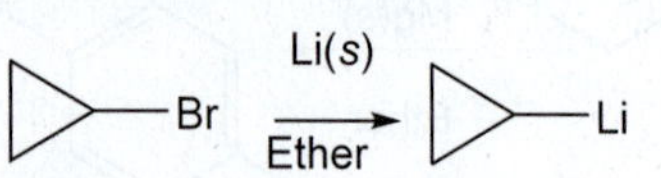

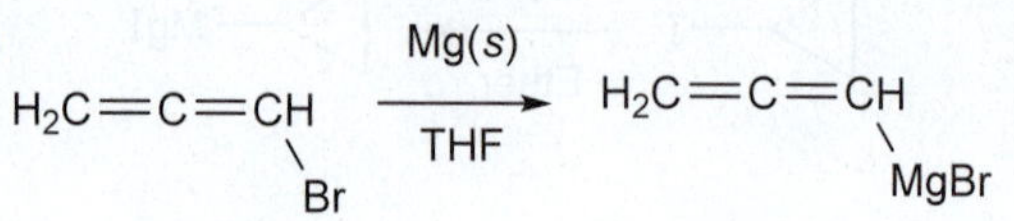

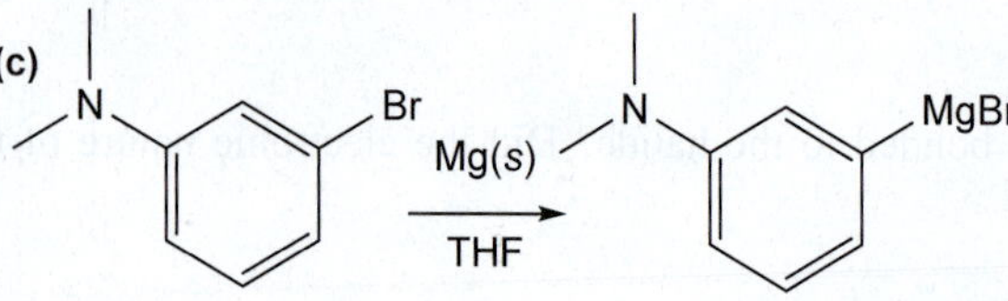

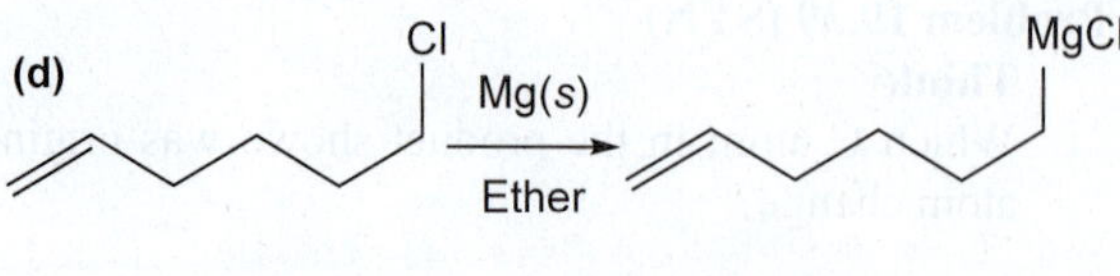

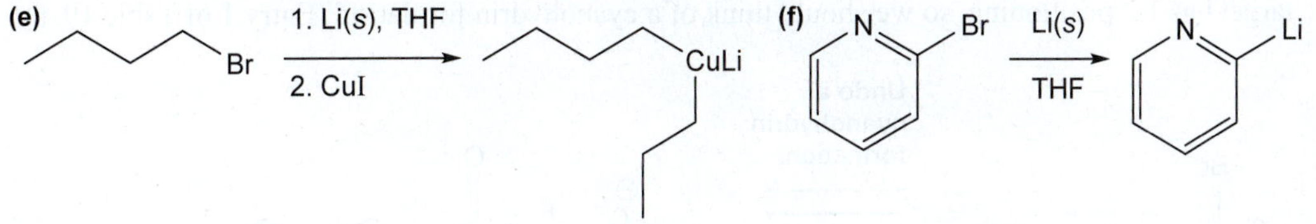
(e)
1. Li(s), THF
2. CuI
Br
CuLi
(f)
N
Br
Li(s)
THF
N
Li

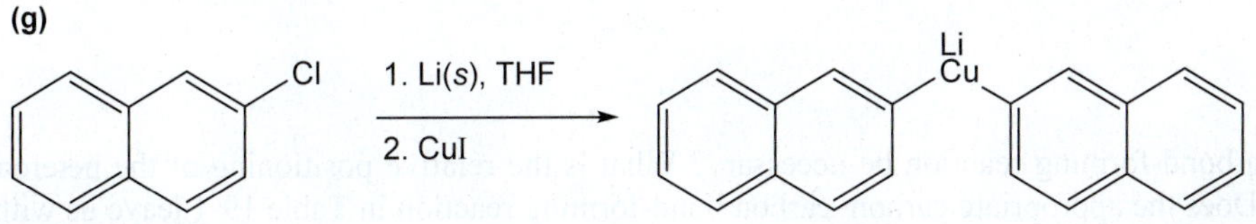
(g)
Cl
1. Li(s), THF
2. CuI
Li
Cu

Sections 19.2 and 19.3 Relative Positioning of Heteroatoms in Carbon–Carbon Bond-Forming Reactions; Reduction of C=O to CH_2

Problem 19.42 (SYN)

Think

Which carbon–carbon bond-forming reaction is necessary? What is the relative positioning of the Br atoms in each target, and does it match one of the carbon–carbon bond-forming reactions in Table 19-1? Does the appropriate carbon–carbon bond-forming reaction in Table 19-1 leave us with the correct functional groups, or will we need to carry out an additional functional group conversion?

Solve

We will have to use a carbon–carbon bond-forming reaction, because each target's carbon skeleton contains seven carbons bonded together, and the requirements are to start with compounds containing six or fewer carbon atoms.

(a) With a 1,3-positioning of the Br groups, we should think of an aldol addition (Entry 2 of Table 19-1).

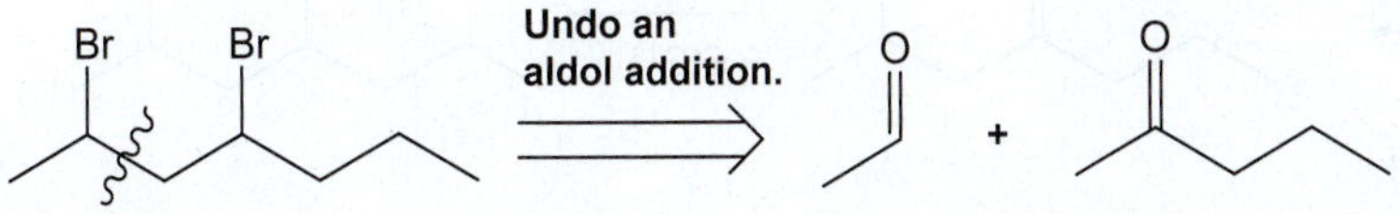
Br
Br
Undo an
aldol addition.
O
+
O

(b) With a 1,4-positioning of the Br groups, we should think of conjugate addition of NC^- to an α,β-unsaturated carbonyl compound (Entry 3 of Table 19-1). An additional carbon–carbon bond-forming reaction would be necessary to achieve the required number of carbon atoms in the target.

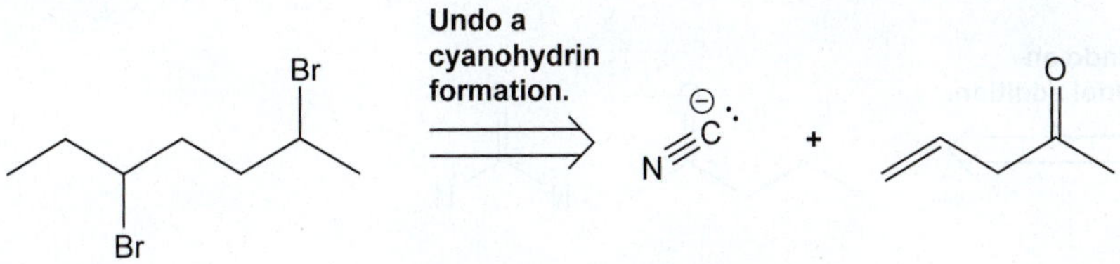
Br
Br
Undo a
cyanohydrin
formation.
N
C
+
O

(c) The target has a 1,5-positioning of the Br groups, so we should consider a conjugate addition with an enolate anion as the nucleophile (Entry 4 of Table 19-1).

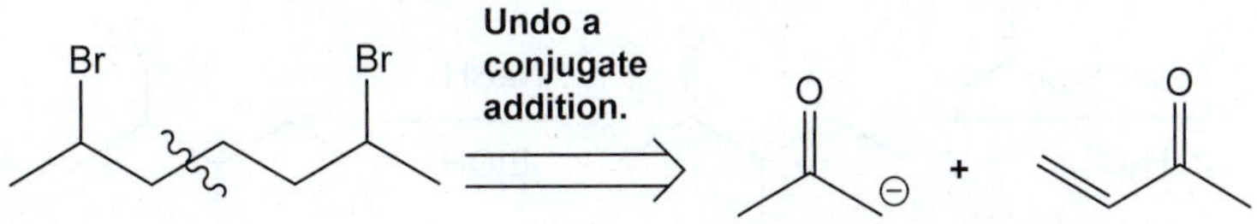
Br
Br
Undo a
conjugate
addition.
O
+
O

(d) The target has 1,2 positioning, so we should think of a cyanohydrin formation (Entry 1 of Table 19-1).

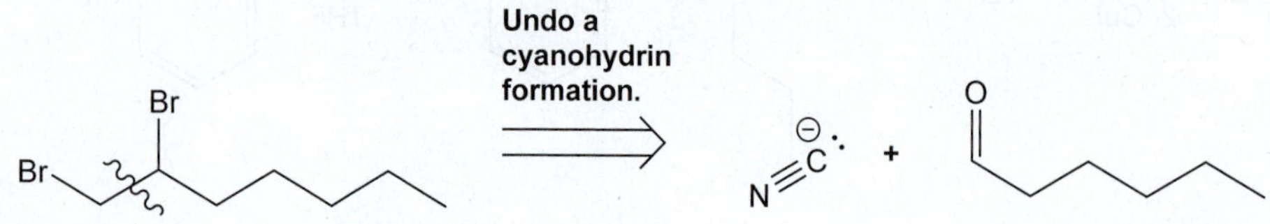

Undo a cyanohydrin formation.

Problem 19.43 (SYN)

Think

Will a carbon–carbon bond-forming reaction be necessary? What is the relative positioning of the heteroatom groups in the target? Does the appropriate carbon–carbon bond-forming reaction in Table 19-1 leave us with the correct functional groups, or will we need to carry out an additional functional group conversion? Are oxidation or reduction reactions needed?

Solve

(a) With a 1,3-positioning of the OH groups, we should think of an aldol addition. *Note:* Synthesis routes other than the one shown below are also feasible.

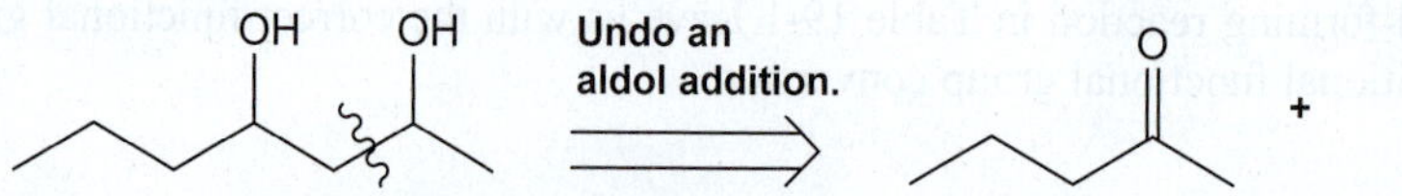

Undo an aldol addition.

In the forward direction, we would need to obtain the kinetic enolate, so LDA is a good choice of base. After the aldol, we reduce the carbonyl.

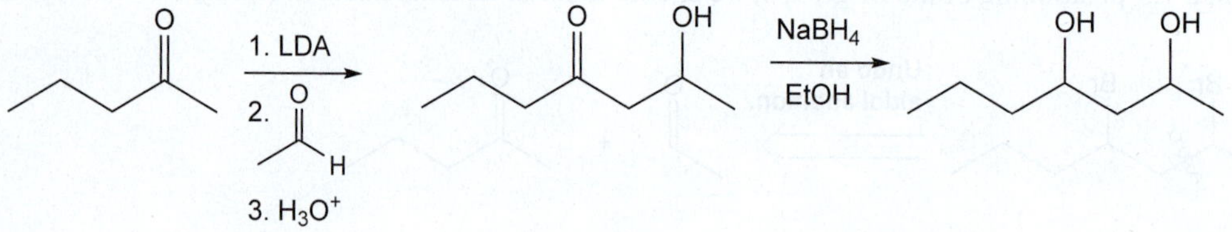

1. LDA; 2. (acetaldehyde); 3. H_3O^+; $NaBH_4$, EtOH

(b) The target also has a 1,3-positioning of the alcohols, so we should undo an aldol, disconnecting the bond shown.

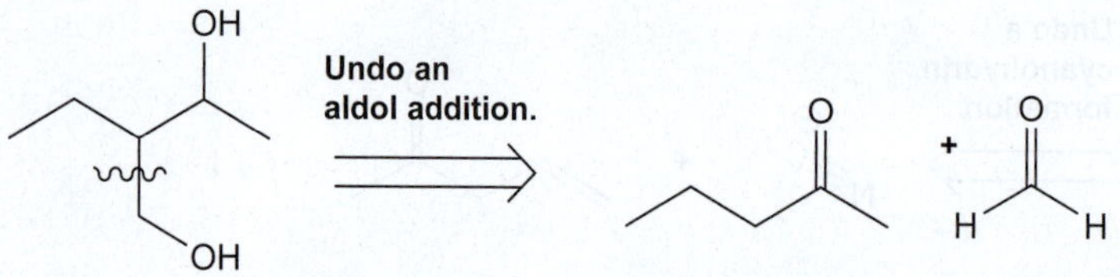

Undo an aldol addition.

In the forward direction, we want to generate the thermodynamic enolate, so $KOC(CH_3)_3$ is a good choice of base.

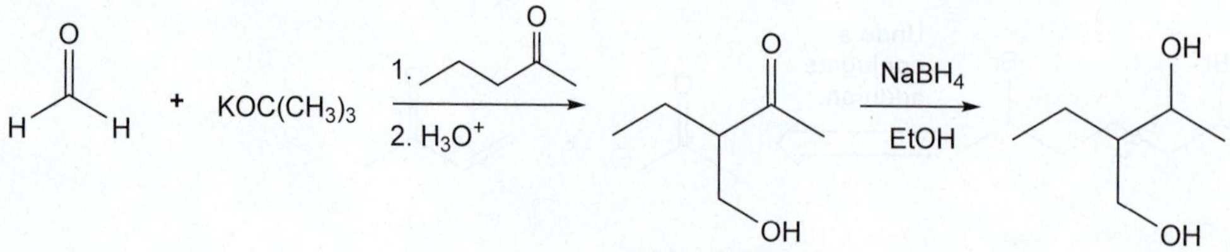

+ $KOC(CH_3)_3$; 1. (2-butanone); 2. H_3O^+; $NaBH_4$, EtOH

(c) The target has a 1,5-positioning of the functional groups, so we should consider a conjugate addition with an enolate anion as the nucleophile.

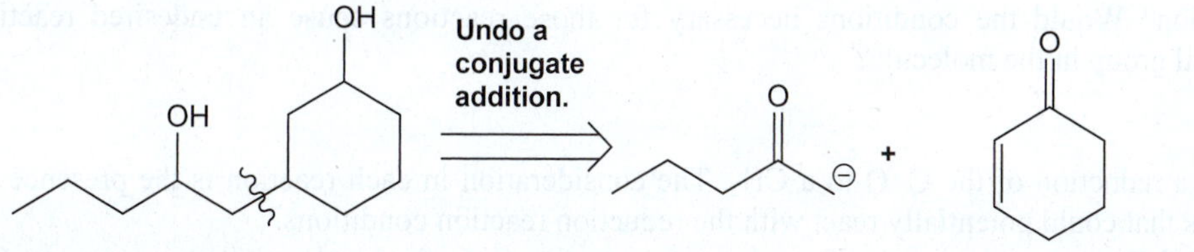
OH
OH
Undo a conjugate addition.
O
O
+

To obtain the kinetic enolate, we use LDA as the base.

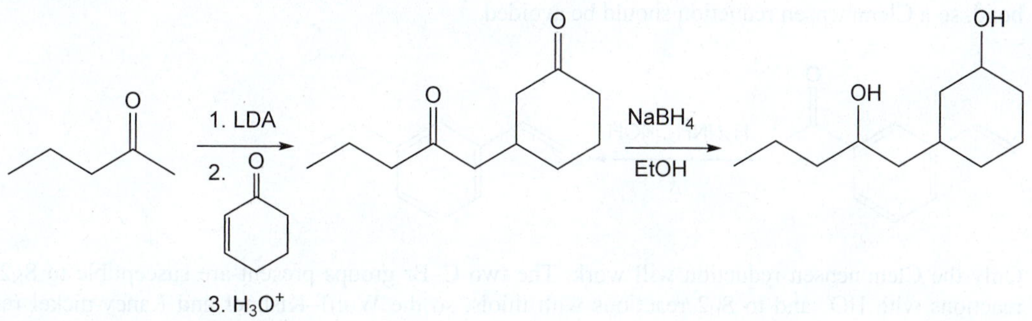
O
1. LDA
2.
3. H_3O^+
O
O
NaBH$_4$
EtOH
OH
OH

Problem 19.44

Think

What bond is reduced by the reagents given? What group forms from the reduction reaction?

Solve

(a) The Clemmensen reaction reduces the ketone C=O group to a methylene group (CH_2).

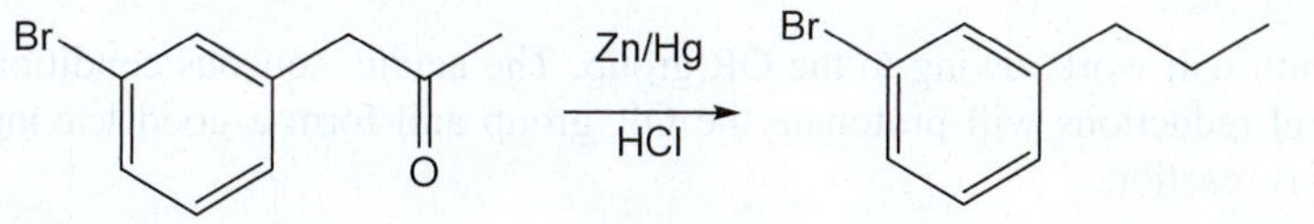
Br
O
Zn/Hg
HCl
Br

(b) The Wolff–Kishner reaction reduces the ketone C=O group to a CH_2 group.

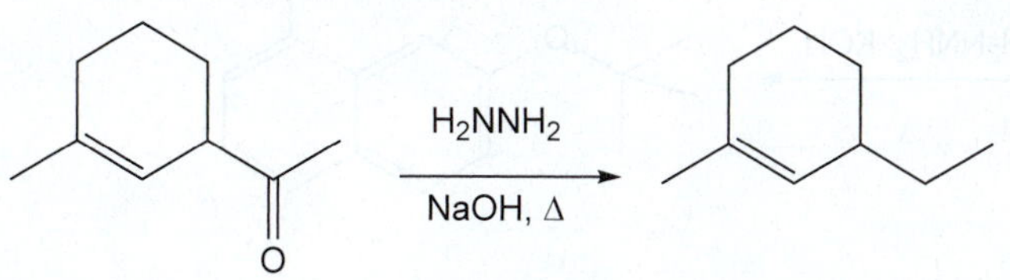
O
H_2NNH_2
NaOH, Δ

(c) The Raney-nickel reaction reduces the ketone C=O group to a CH_2 group.

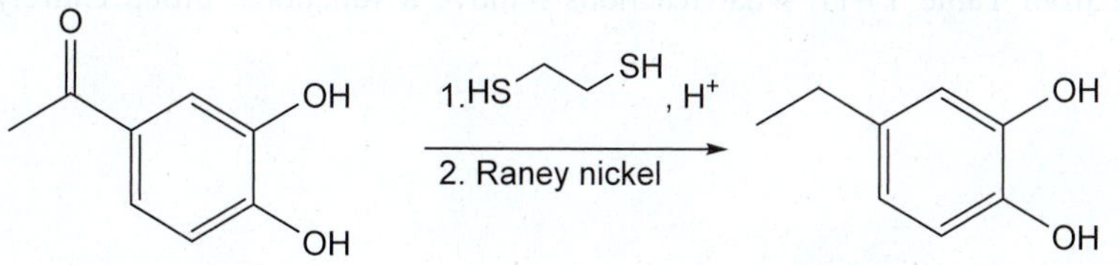
O
OH
OH
1.HS
SH
, H$^+$
2. Raney nickel
OH
OH

Problem 19.45 (SYN)

Think

What are the structural differences between the reactant and product molecules? What reactions will accomplish that transformation? Would the conditions necessary for those reactions cause an undesired reaction with another functional group in the molecule?

Solve

Each reaction is a reduction of the C=O to a CH_2. The consideration in each reaction is the presence of other functional groups that could potentially react with the reduction reaction conditions.

(a) Only a Wolff–Kishner reduction will work, owing to the C=C double bond that is present. Catalytic hydrogenation will reduce the C=C bond, so a Raney-nickel reduction should be avoided. And as we saw in Chapter 11, the acidic, aqueous conditions of the Clemmensen reduction will lead to reaction with the C=C bond, so a Clemmensen reduction should be avoided.

H_2NNH_2, KOH, Δ

(b) Only the Clemmensen reduction will work. The two C–Br groups present are susceptible to S_N2 and E2 reactions with HO^- and to S_N2 reactions with thiols, so the Wolff–Kishner and Raney-nickel reductions should be avoided.

Zn/Hg, HCl; H_2O, reflux 5-8 h

(c) Only a Wolff–Kishner reduction will work, owing to the OR group. The acidic, aqueous conditions of the Clemmensen and Raney-nickel reductions will protonate the OR group and form a good leaving group, causing an undesired hydrolysis reaction.

H_2NNH_2, KOH, Δ

Problem 19.46 (SYN)

Think

Are any new C–C bonds formed? Do any functional groups in the target suggest using a particular carbon–carbon bond-forming reaction from Table 19-1? What reactions remove a functional group entirely from a molecule?

Solve

The new C–C bond is formed at the β carbon from conjugate addition of $(CH_3CH_2CH_2)_2CuLi$ to the α,β-unsaturated ketone reactant. The C=O group is then reduced to CH_2 from any of the three methods in Section 19.3. This reaction is similar to the one shown in Equation 19-10. The Wolff–Kishner reduction is used in the example below, but the Clemmensen and Raney-nickel reductions would work, too.

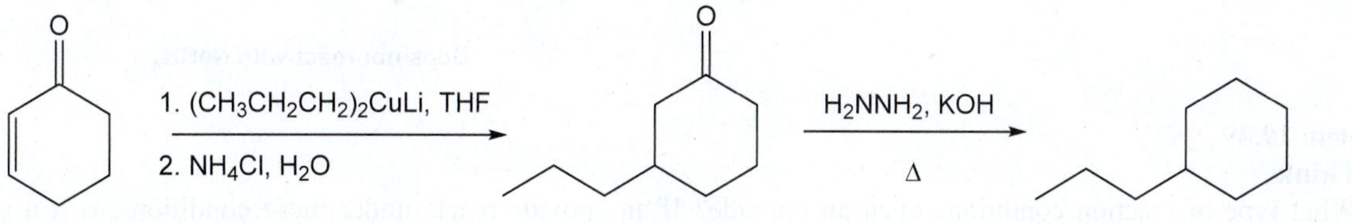

1. $(CH_3CH_2CH_2)_2CuLi$, THF
2. NH_4Cl, H_2O

H_2NNH_2, KOH

Δ

Section 19.4 Selective Reagents and Protecting Groups

Problem 19.47

Think

Is the C=O group intended to react? Would the C=O react in the given reaction conditions? If so, does it need protection?

Solve

The C=O that requires protection in reactions **(a)**–**(d)** is noted below by a dashed oval. If the noted C=O is not protected, the following reactions will occur (not shown):

Reaction **(a)**: The α carbon would be deprotonated and subsequently alkylated by CH_3I to form a new C–C bond.
Reaction **(b)**: The C=O would react with the C–MgBr portion of the molecule in a Grignard reaction.
Reaction **(c)**: The $NaSCH_3$ would attack the C=O in a nucleophilic addition step.
Reaction **(d)**: The aldehyde C=O would be also reduced.

(a) 1. $NaNH_2$ 2. CH_3I

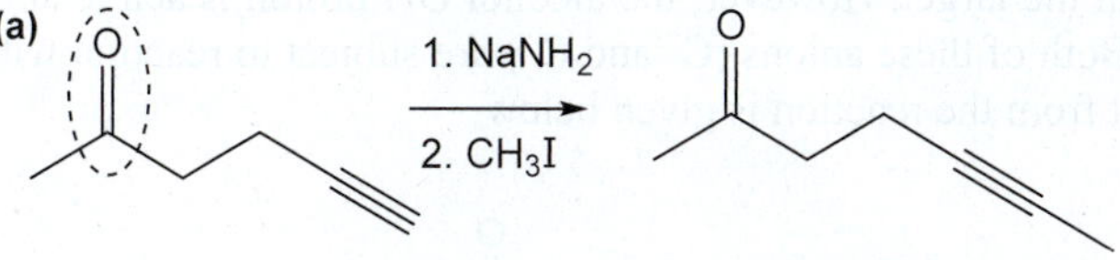

(b) Br → Mg(*s*) → MgBr

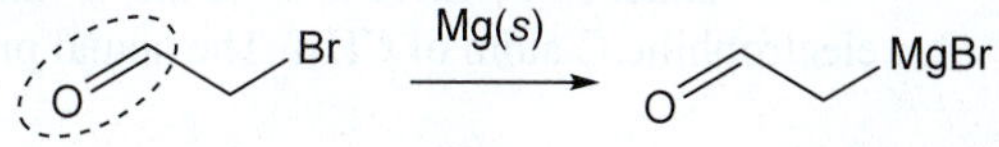

(c) Cl → $NaSCH_3$ → H_3CS

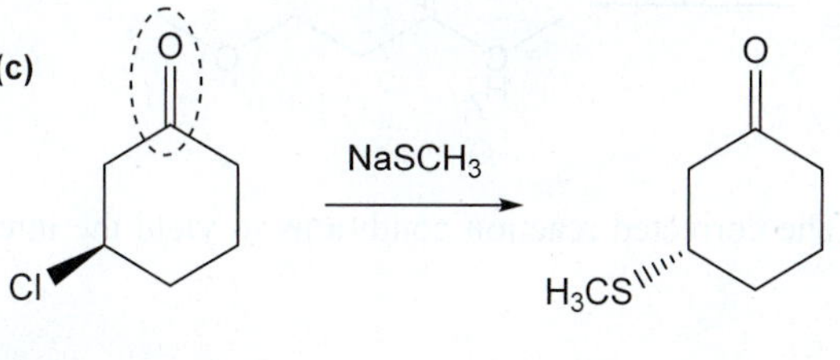

(d) $NaBH_4$, Ethanol → OH

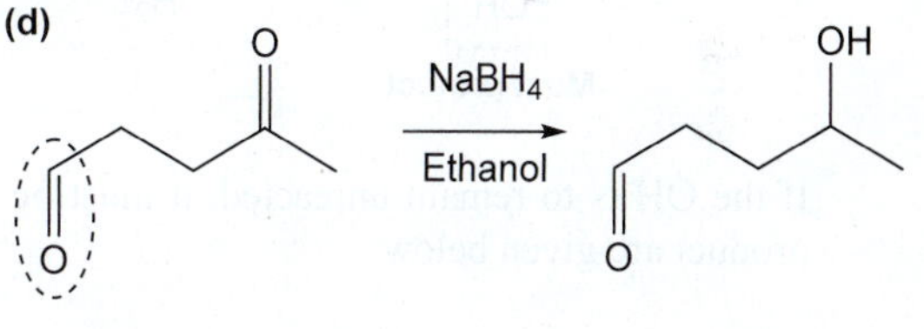

Problem 19.48

Think

Is an OH group intended to react? Would the OH group react in the given reaction conditions? If so, does it need protection?

Solve

The OH that requires protection in reactions **(a)** and **(c)** is noted below by a dashed circle. The OH in **(b)** and **(d)** is unreactive in the given reaction conditions and does not require protection. The reactions are shown below and on the next page.

(a) HO, Br → Mg(*s*) → HO, MgBr

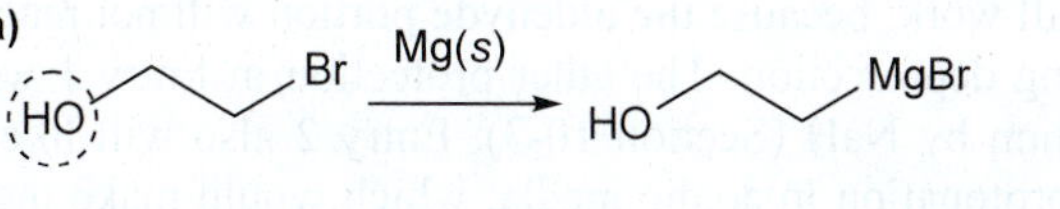

(b) HO, Cl → NaCN → HO, CN

Does not react with NaCN

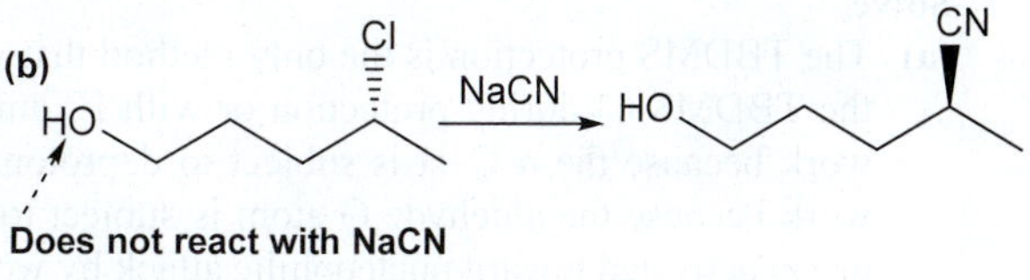

(c)

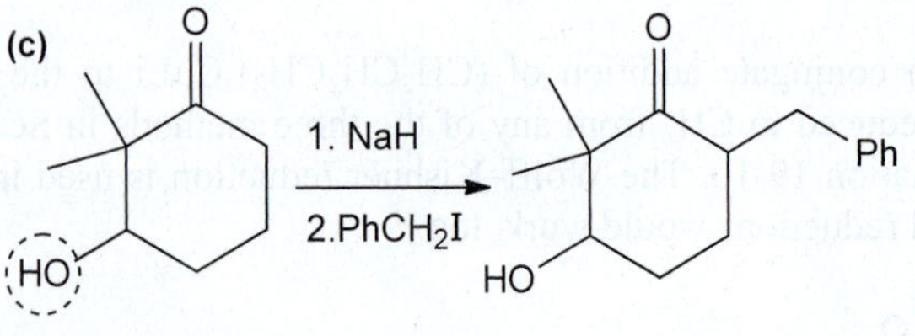

(d)

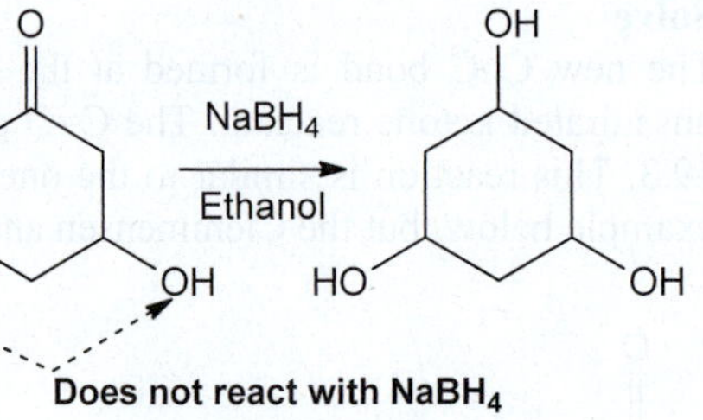

Problem 19.49

Think

What type of reaction conditions open an epoxide? If an epoxide reacts under these conditions, is it a good protecting group? Why would an epoxide be more reactive than the five-member-ring acetal?

Solve

An epoxide has a lot of ring strain and can open easily in an S_N2 reaction with a species like HO^-. So it does a poor job of protecting the C=O group. The five-member-ring acetal is much more stable and is unreactive in the presence of a strong nucleophile.

Problem 19.50

Think

What type of reagent is LDA? What are the acidic H atoms present in the starting compound? Is there more than one acidic H that can react with LDA? How can you protect this group?

Solve

LDA is a strong base and selectively produces the kinetic enolate, which leads to alkylation at the less-substituted α C atom. This is what is proposed in the target. However, the alcohol OH proton is acidic and will also be attacked by LDA to form the O^- anion. Both of these anions (C^- and O^-) are subject to reaction with the electrophilic C atom of CH_3I. The actual product from the reaction is given below.

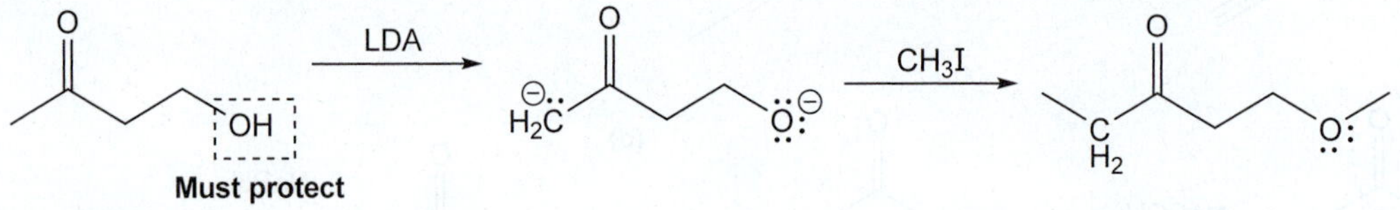

If the OH is to remain unreacted, it must be protected. The corrected reaction conditions to yield the intended product are given below:

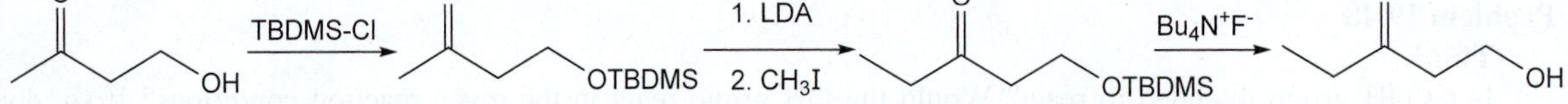

Problem 19.51

Think

In addition to the alcohol OH group that is to be protected, what other functional groups are present? What reaction conditions are required for protection and deprotection in each of the five methods in Table 19-2? Are the other functional groups present reactive under these conditions?

Solve

(a) The TBDMS protection is the only method that will work, because the aldehyde portion will not react with the TBDMS-Cl during protection or with F^- during deprotection. The ether protection in Entry 1 will not work because the α C–H is subject to deprotonation by NaH (Section 10-3). Entry 2 also will likely not work because the aldehyde O atom is subject to protonation in acidic media, which would make the C=O group activated toward nucleophilic attack by weak nucleophiles.

(b) The TBDMS protection will work because the benzene ring and C–Br group will not react with the TBDMS-Cl during protection or with F^- during deprotection. Entry 2 will likely work because the benzene ring and the C–Br group are unreactive in acidic medium. However, these reactions should probably be avoided, because the benzylic bromide could react under S_N1 conditions. The ether protection in Entry 1 will not work because the RO^- that results after reaction with NaH is reactive with the C–Br group.

Problem 19.52 (SYN)

Think

Is the carbon skeleton altered? Does functional group transformation occur? Is there more than one functional group that can react under the necessary conditions for the desired reaction? Are protecting groups necessary? Are strong acids or bases needed? What are the stereochemical considerations?

Solve

In both cases, the OH group on the left should react, leaving the other OH groups alone. Therefore, the cis diol must be protected to target just the one alcohol group on the left. With the protected diol in **(a)**, we must convert the remaining OH group into a good leaving group. PBr_3 reverses the stereochemical configuration. KCN promotes another S_N2 reaction that reverses the configuration, and reduction with $LiAlH_4$ yields the primary amine. In **(b)**, the protected intermediate is oxidized to the ketone before deprotection.

(a)

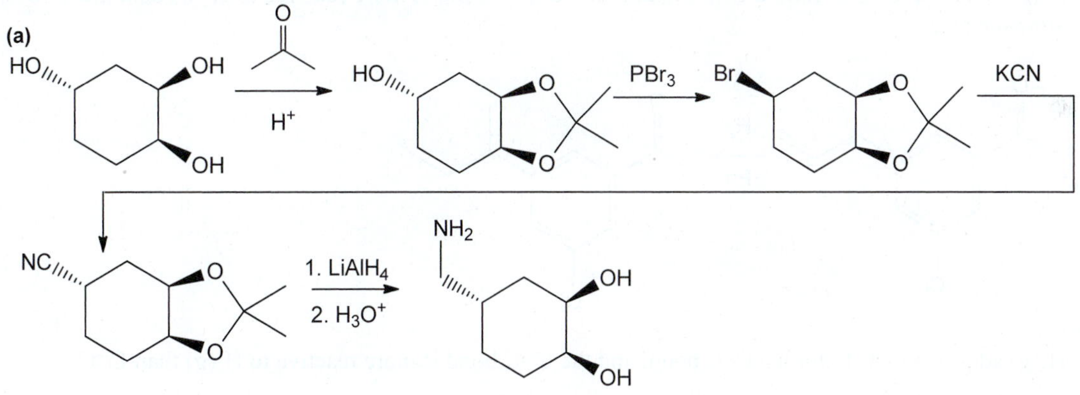

(b)

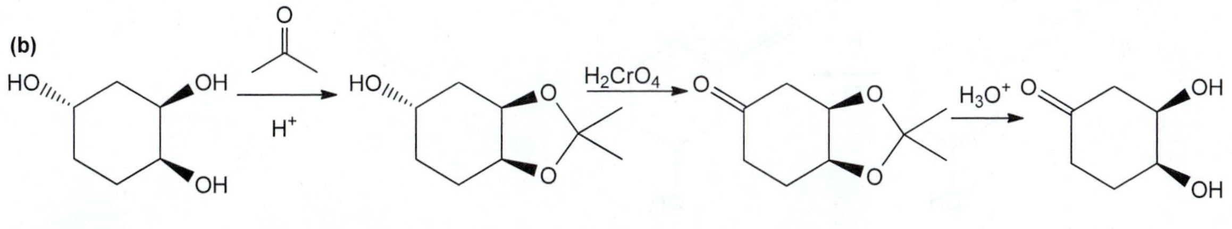

Problem 19.53

Think

What protected functional group does O–TBDMS represent? Is the carbon skeleton altered? Which C atom forms the new C–Ph bond? Does functional group transformation occur? Are protecting groups necessary? Are strong acids or bases needed? What rearrangement occurs when an enol is formed?

Solve

The O–TBDMS group represents a protected OH, so when it is deprotected, an enol will form that will tautomerize to the keto form. However, the initial ketone C=O first needs to be protected with an acetal group before the O–TBDMS group is removed, to avoid an intermediate with two ketone C=O groups that would be similarly reactive. Therefore, after removal of the O–TBDMS and subsequent tautomerization, the new ketone C=O undergoes a Grignard reaction to form the C–Ph bond. Notice that the aqueous acid workup of the Grignard reaction also deprotects the ketone C=O group. See the reaction on the next page.

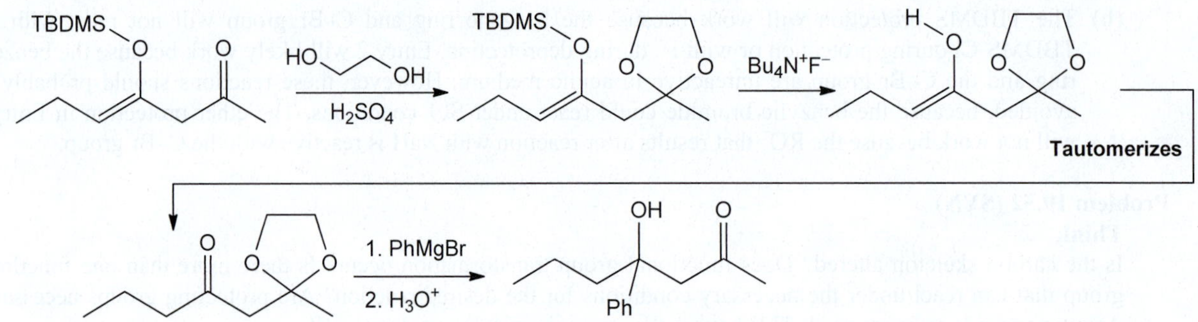

TBDMS O O → HO OH / H_2SO_4 → TBDMS O O O → $Bu_4N^+F^-$ → H O O O → Tautomerizes → O O O → 1. PhMgBr / 2. H_3O^+ → OH O Ph

Sections 19.5 and 19.6 Catalytic Hydrogenation; Oxidation of Alcohols and Aldehydes

Problem 19.54

Think

What adds to the multiple bond? Which bond is subject to reaction? If there is more than one choice, which bond is more selective to the reagents given? Are there any stereochemical considerations?

Solve

(a) $H_2(g)$ adds to a C=C to form a C–C bond, and the C=C bond is more reactive to $H_2(g)$ than the C=O or the aromatic ring.

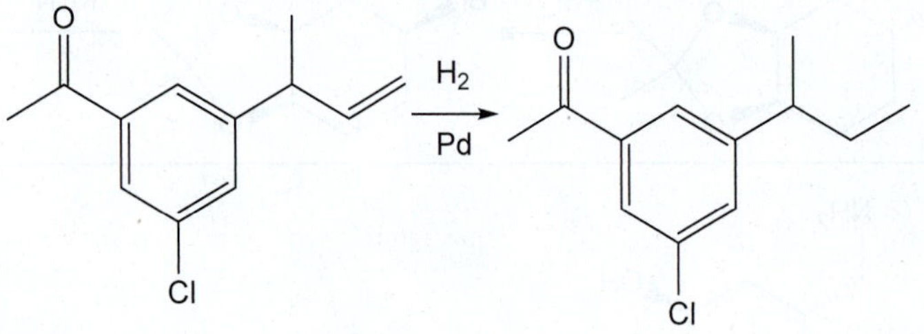

O Cl → H_2 / Pd → O Cl

(b) $H_2(g)$ adds to a C=C to form a C–C bond, and the C=C bond is more reactive to $H_2(g)$ than C≡C.

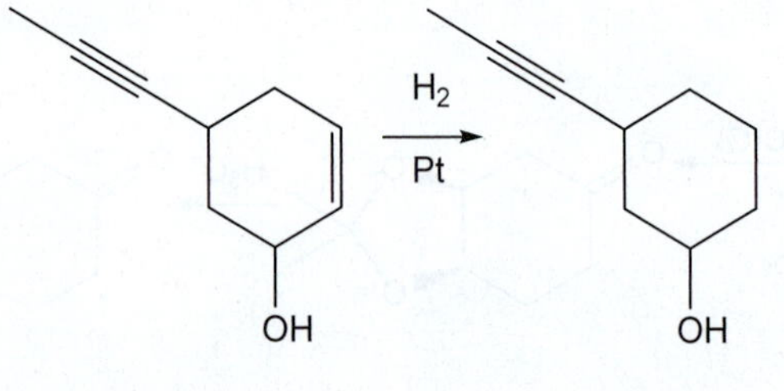

OH → H_2 / Pt → OH

(c) Catalytic hydrogenation is more favored at the *less* sterically hindered C=C bond.

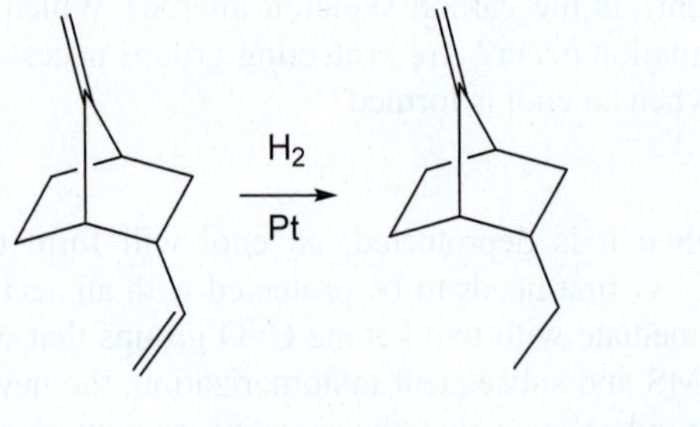

→ H_2 / Pt →

(d) H_2 in the presence of Lindlar catalyst adds to a C≡C to form a cis alkene. The C≡C is selectively reduced over the amide portion.

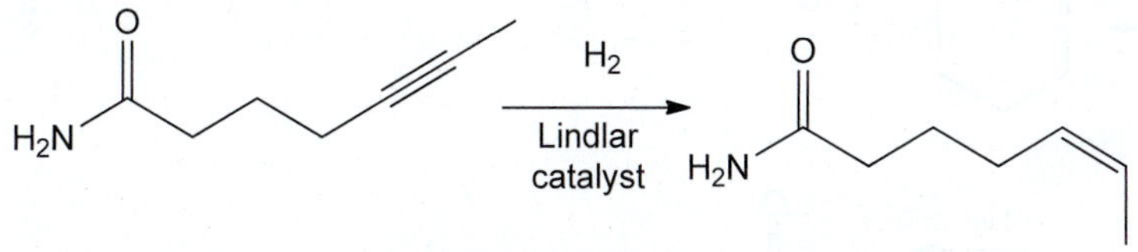

H_2, Lindlar catalyst

(e) The C≡N is less reactive than the HC=O. Therefore, H_2 adds preferentially to the HC=O.

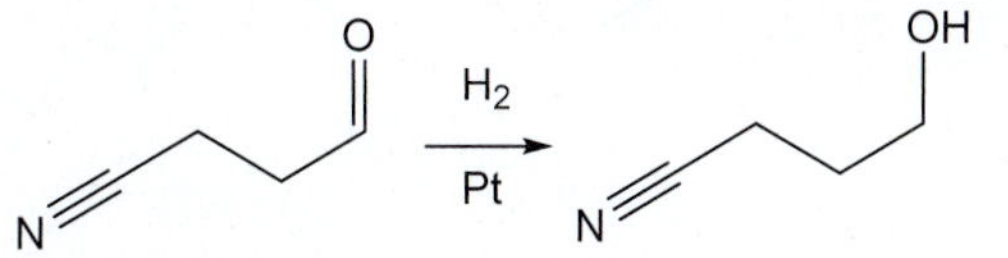

H_2, Pt

(f) Aldehydes and alkenes have similar reactivity toward catalytic hydrogenation, but the aldehyde C=O is less sterically hindered.

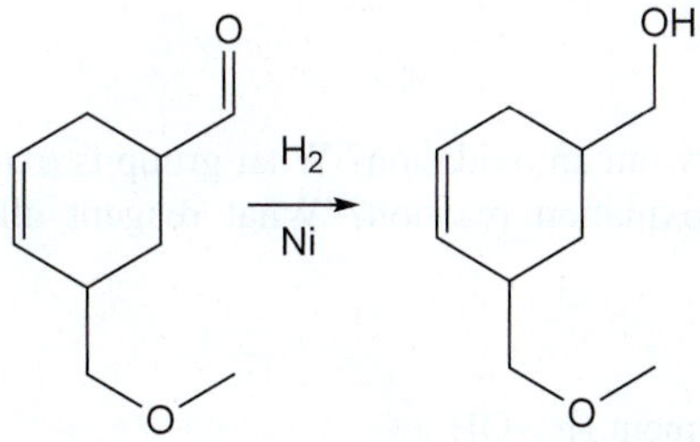

H_2, Ni

(g) With excess H_2, the C≡C is reduced all the way to the C–C. Without more extreme conditions, the amide portion does not react.

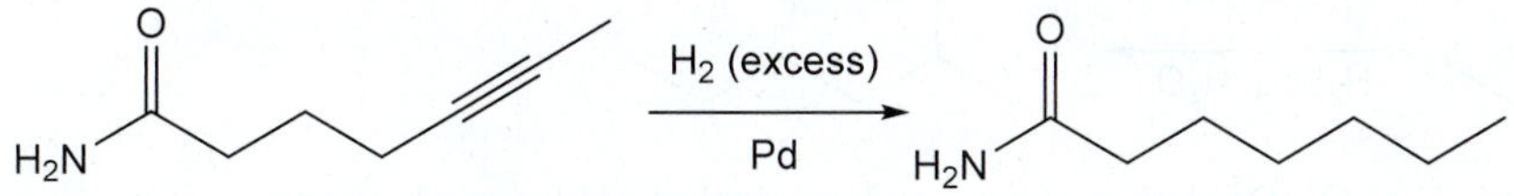

H_2 (excess), Pd

Problem 19.55 (SYN)

Think

What are the structural differences between the reactant and product? What molecule needs to add? What reaction, therefore, must take place? Are there any stereochemical considerations?

Solve

Each reaction is a reduction that requires the addition of H_2 either to a C=O to produce a C–OH, to C=C to produce a C–C, or to C≡C to produce a cis C=C. The required reduction conditions are given for each reaction below. In **(a)**, both the C=C and the C=O are reduced. In **(b)**, the C=C is selectively reduced over the ester portion. In **(c)**, the C=C is selectively reduced over the benzene ring or the amide portion. In **(d)**, two reductions occur with excess H_2. The reactions are shown below and on the next page.

(a) H_2, Pd

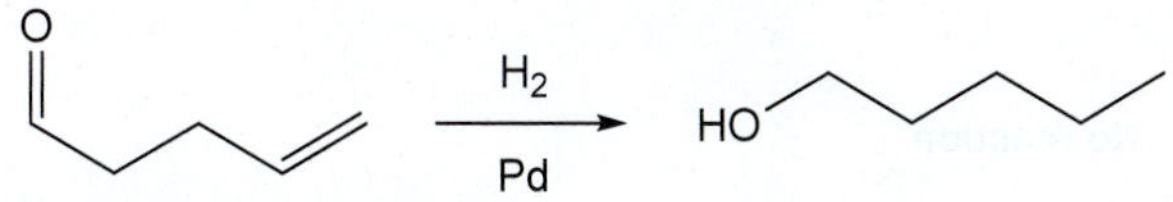

(b)

H_2 / Pd

(c)

H_2N … H_2 / Lindlar catalyst … H_2N

(d)

H_3CO … H_2 (excess) / Pd … H_3CO

Problem 19.56

Think

What arrangement of atoms is necessary for Cr^{6+} or Mn^{7+} to carry out an oxidation? What group is oxidized by the reagents given? What functional group forms from the oxidation reaction? What reagent allows the oxidation of a primary alcohol to stop at the aldehyde?

Solve

In each case, oxidation occurs when the molecule has the arrangement HC–OH.

(a) The oxidizing agent is H_2CrO_4. Oxidation of a primary alcohol produces a carboxylic acid.

HO … $Na_2Cr_2O_7$ / H_2SO_4, H_2O … HO, O

(b) The oxidizing agent is H_2CrO_4. Oxidation of a secondary alcohol produces a ketone.

CrO_3 / H_2SO_4, H_2O

OH … O

(c) There is no oxidation reaction of a tertiary alcohol or a phenol because there is no CH–OH arrangement of atoms.

OH

HO … $Na_2Cr_2O_7$ / H_2SO_4, H_2O … **No reaction**

(d) The aldehyde portion is oxidized to the carboxylic acid using H_2CrO_4. There is no reaction with the ketone portion because it does not contain HC–OH.

H_2CrO_4

F O OH

O

(e) Oxidation of a primary alcohol using $KMnO_4$ in the reaction conditions produces a carboxylic acid. There is no reaction with the ketone portion because there is no arrangement of the form HC–OH.

HO 1. $KMnO_4$, KOH 2. H_2O, HCl HO

O O O

(f) PCC oxidizes the primary alcohol to the aldehyde but leaves the ketone portion alone.

HO PCC O

O O

Problem 19.57 (SYN)

Think

Are any new C–C bonds formed? Are any functional group transformations necessary? Are oxidations or reductions taking place? What oxidizing reagents will carry out the desired oxidations?

Solve

Each reaction is an oxidation of the HC–OH to a C=O (carboxylic acid, aldehyde, or ketone).

(a) Chromic acid or potassium permanganate will oxidize a primary alcohol to a carboxylic acid.

HO 1. $KMnO_4$, KOH 2. H_2O, HCl O HO

(b) PCC oxidizes the primary alcohol to the aldehyde.

HO PCC O

(c) Chromic acid, PCC, or potassium permanganate will oxidize a secondary alcohol to a ketone.

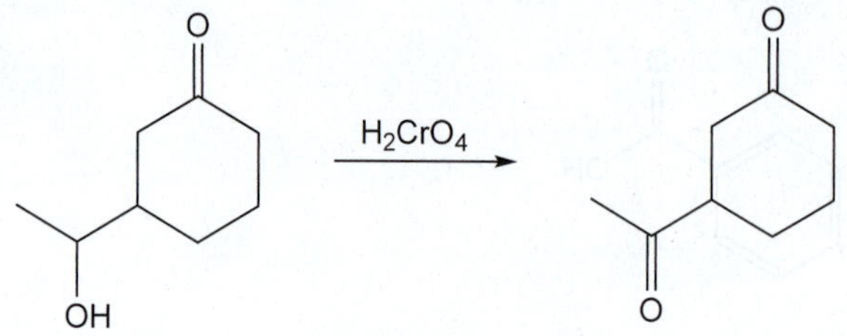

OH

H_2CrO_4

O

O

(d) PCC oxidizes the primary alcohol to the aldehyde.

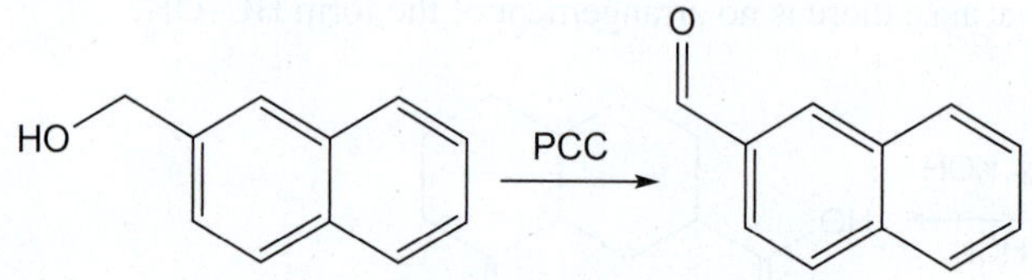

HO

PCC

O

Section 19.7 Coupling and Alkene Metathesis Reactions

Problem 19.58

Think

For a coupling reaction between R'_2CuLi and RX, what groups become bonded together? What are the restrictions for a lithium dialkylcuprate reagent in such a coupling reaction? What are the restrictions for the RX reactant? Are there any stereochemical considerations?

Solve

For a coupling reaction between R'_2CuLi and RX, a new C–C bond forms to join together R and R′. The limitation of this coupling reaction is that the RX cannot be a 3° alkyl halide, so **(b)** has no reaction. Stereochemistry is an issue in **(a)**, and there is retention of configuration.

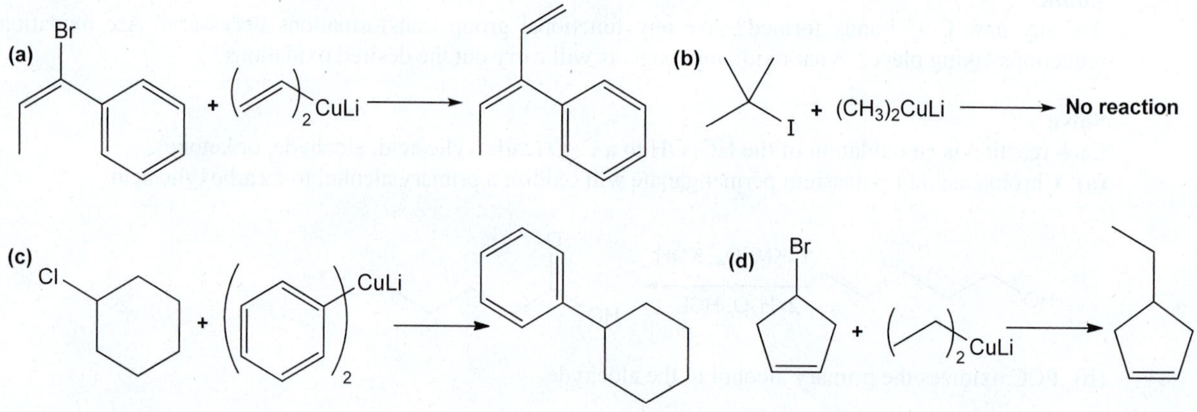

(a) Br + ( )$_2$CuLi →

(b) I + $(CH_3)_2CuLi$ → **No reaction**

(c) Cl + ( )$_2$CuLi →

(d) Br + ( )$_2$CuLi →

Problem 19.59 (SYN)

Think

What is the general form of a lithium dialkylcuprate reagent? What new bond formed in the reaction? What is the identity of the R′ group in the lithium dialkylcuprate reagent?

Solve

The lithium dialkylcuprate reagent has the form R'_2CuLi. A new C–C bond forms between the R′ in the lithium dialkylcuprate reagent and the RX in the reactant. The missing lithium dialkylcuprates necessary to carry out transformations **(a)** and **(b)** are given below.

(a) Br ... $(\!\diagup\!\diagdown)_2$CuLi ... New C–C bond

(b) I ... O ... CuLi ... 2 ... O ... New C–C bond

Problem 19.60

Think

What new bond forms from reaction of R–X + R′–B(OR″)$_2$ or H–R′ with a Pd catalyst? Are there stereochemical considerations?

Solve

A new R–R′ bond forms from reaction of an R–X + R′–B(OR″)$_2$ (Suzuki reaction) or H–R′ (Heck reaction) with a Pd catalyst. Reactions **(a)** and **(b)** are Suzuki coupling reactions, and the product is formed with retention of configuration for the C=C in **(b)**. Reactions **(c)** and **(d)** are Heck coupling reactions where the Ph **(c)** or $CH_3CH_2CH{=}CH$ **(d)** substituent on H–R′ becomes trans to the R substituent from R–X. In reaction **(d)**, the cis configuration about the C=C in the RBr reactant is retained.

(a) O B O + I OCH_3 → Pd(PPh$_3$)$_4$, NaOH / THF → New C–C bond OCH_3

(b) OEt B EtO + I → Pd(PPh$_3$)$_4$, NaOH / THF → New C–C bond; Retention of configuration

(c) I + O O → Pd(OAc)$_2$ / PPh$_3$, Et$_3$N → New C–C bond; O; Trans configuration; O

(d) Br + O → Pd(OAc)$_2$ / PPh$_3$, Et$_3$N → Retention of configuration; New C–C bond; O; Trans configuration

Problem 19.61 (SYN)

Think

In each reaction, what new bond forms on going from reactants to products? Which reaction is a Heck coupling? Which reaction is a Suzuki coupling? What are the structural requirements of R in R–X for Suzuki coupling reactions? What are the structural requirements of R′ in R′–H for Heck coupling reactions?

Solve

In each reaction, a new C–C bond is formed. Reaction **(a)** is a Suzuki coupling reaction where the missing reagent is an R–X compound, namely, (*E*)-2-bromobut-2-ene (Cl and I are other options for the halide). Reaction **(b)** is a Heck reaction where the missing compound is an R′–H compound, namely, styrene.

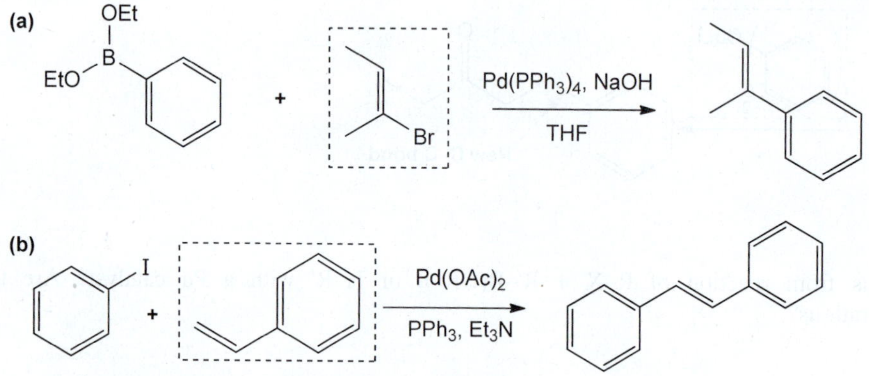

(a) OEt, EtO, B, Br, $Pd(PPh_3)_4$, NaOH, THF

(b) I, $Pd(OAc)_2$, PPh_3, Et_3N

Problem 19.62

Think

What are the products from two C=C bonds and a Grubbs catalyst? If the C=C bonds in the reactant are terminal, what is the by-product? If there are two C=C groups in the same reactant, what type of product is formed?

Solve

Two terminal C=C bonds react in the presence of a Grubbs catalyst to undergo an alkene metathesis, forming a new C=C bond. In this case, $CH_2=CH_2$ is the by-product. A ring forms in reaction **(a)** where the two C=C groups in the reactant are on the same molecule. The carbon atoms of the new ring in reaction **(a)** are highlighted in gray. The major organic products from reactions **(a)** and **(b)** are shown below.

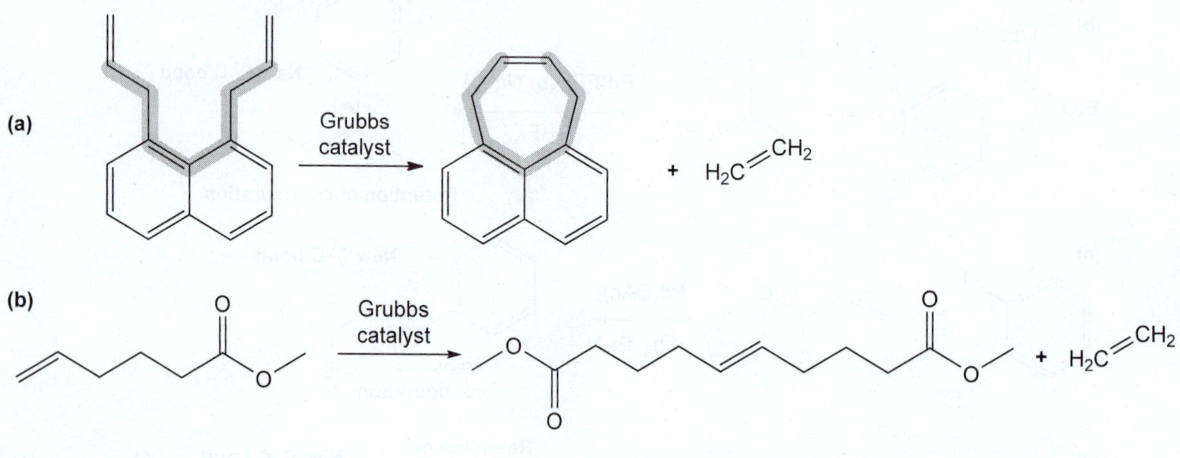

(a) Grubbs catalyst + $H_2C{=}CH_2$

(b) O, O, Grubbs catalyst, O, O, O, O + $H_2C{=}CH_2$

Problem 19.63 (SYN)

Think

What new bond formed in the product from reaction with a Grubbs catalyst? How many C=C bonds need to be in the reactant?

Solve

The diene that reacts with a Grubbs catalyst to yield the products in reactions **(a)** and **(b)** are shown below.

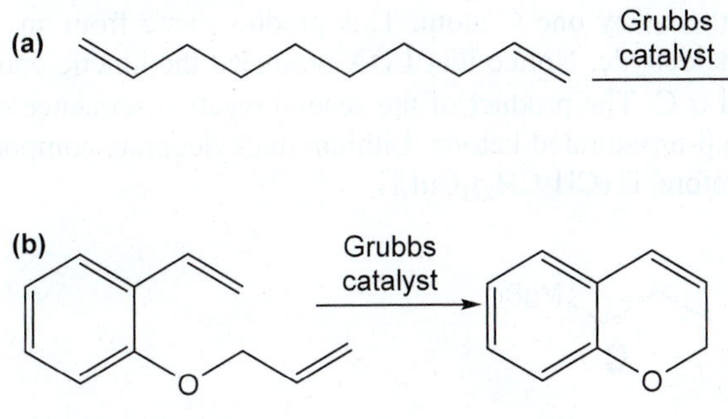

(a) Grubbs catalyst

(b) Grubbs catalyst

Integrated Problems

Problem 19.64

Think

In each reaction, consider which atoms are electron rich and electron poor. Is the reaction a functional group transformation, or does the reaction alter the carbon skeleton? Are proton transfer reactions involved? Are oxidations or reductions taking place?

Solve

A is the alcohol product from a reduction reaction of an aldehyde and $LiAlH_4$. The alcohol is then converted to the alkyl bromide, which reacts with Mg(*s*)/ether to form a Grignard reagent, **B**. A benzyl group is added to alter the carbon skeleton, and the identity of **C** is benzaldehyde (source of electrophilic carbon). Another benzyl group is added to form the benzylic ether in a Williamson ether synthesis, and the identity of **D** is benzyl bromide (source of electrophilic carbon).

1. $LiAlH_4$ 2. NH_4Cl, H_2O → **A** (OH) 1. PBr_3 2. Mg(s), ether → **B** (MgBr) 1. **C** 2. NH_4Cl, H_2O → (OH) 1. NaH 2. **D** (Br) →

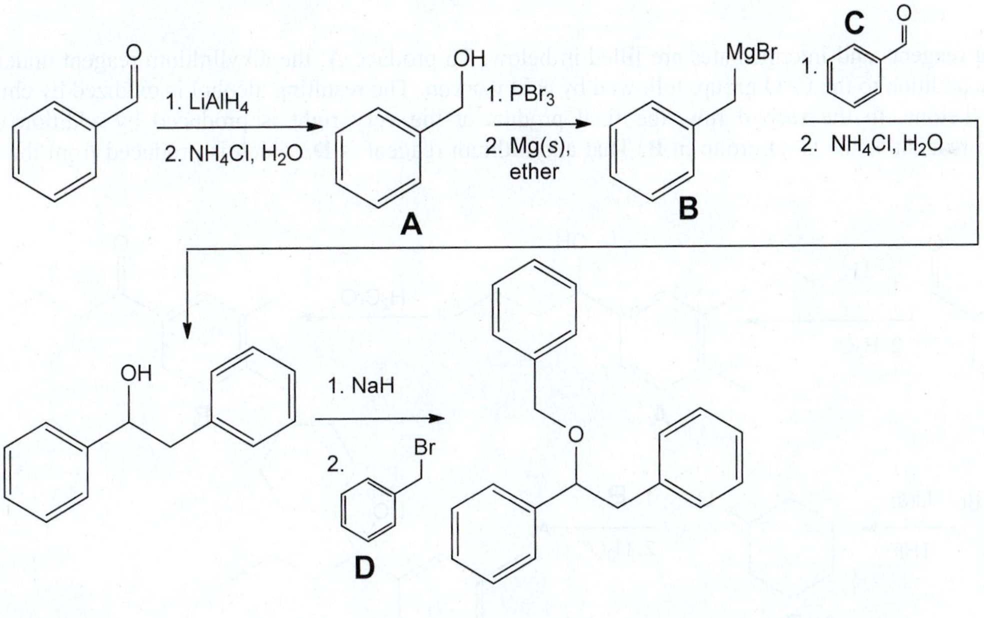

Problem 19.65

Think

In each reaction, consider which atoms are electron rich and electron poor. Is the reaction a functional group transformation, or does the reaction alter the carbon skeleton? Are proton transfer reactions involved? Are there any oxidation or reduction reactions?

Solve

An alkylbromide can react with Mg(*s*)/ether to form a Grignard reagent, **C**. The identity of **A** is PBr_3, which turns the alcohol into alkyl halide, **B**. In the second row, the product of the first reaction sequence is an α,β-unsaturated ketone, where the carbon skeleton was altered by one C atom. This product came from an aldol condensation, and therefore, the identity of **D** is formaldehyde. Notice that LDA produced the kinetic enolate, allowing alkylation to take place at the least hindered α C. The product of the second reaction sequence came from a 1,4-addition of an ethyl (two C atoms) to the α,β-unsaturated ketone. Lithium dialkylcuprate compounds react preferentially via conjugate addition, and **E**, therefore, is $(CH_3CH_2)_2CuLi$.

Problem 19.66

Think

In each reaction, consider which atoms are electron rich and electron poor. Is the reaction a functional group transformation, or does the reaction alter the carbon skeleton? Are proton transfer reactions involved? Are any oxidation or reduction reactions involved?

Solve

The missing reagents and intermediates are filled in below. To produce **A**, the alkyllithium reagent undergoes nucleophilic addition to the C=O group, followed by acid workup. The resulting alcohol is oxidized by chromic acid to the ketone. In the second row, the final product at the very right is produced by addition of an alkyllithium reagent to the C=O group in **B**. That alkyllithium reagent is **D**, which is produced from the alkyl halide **C**.

Problem 19.67

Think

In each reaction, consider which atoms are electron rich and electron poor. Is the reaction a functional group transformation or does the reaction alter the carbon skeleton? Are proton transfer reactions involved? Are there any steps that protect or deprotect functional groups?

Solve

The missing reagents and intermediates are filled in below. **A** is MCPBA to convert the C=C to an epoxide. To open the epoxide, Grignard reagent **B** is used, followed by acid workup. The OH group is then protected using TBDMS-Cl, which is **C**, followed by hydrolysis of the acetal to produce the ketone **D**. Grignard reagent **E** is then added to undergo nucleophilic addition to the C=O group. After acid workup, the OH group is deprotected to produce the final product **F**.

MCPBA (A); B: 1. PhMgBr 2. NH_4Cl; TBDMS-Cl (C); TBDMS; H_3O^+; D; E: 1. (vinyl)MgBr 2. H_3O^+; TBDMS; HO; $Bu_4N^+F^-$; OH; HO; F

Problem 19.68

Think

In each reaction, consider which atoms are electron rich and electron poor. Is the reaction a functional group transformation, or does the reaction alter the carbon skeleton? Are proton transfer reactions involved? Are any oxidation or reduction reactions involved? Is regiochemistry a concern?

Solve

The missing reagents and intermediates are filled in below. Notice that **A** is the result of anti-Markovnikov addition of water and is then oxidized to the aldehyde, **B**. A Wittig reaction then produces **C**, which undergoes Markovnikov addition of water to produce **D**. Then **D** is oxidized to the ketone using $KMnO_4$, and **F** is the result of α alkylation, produced from the kinetic enolate.

1. $BH_3\bullet THF$ 2. NaOH, H_2O_2; OH; A; PCC; O; B; $(C_6H_5)_3\overset{\oplus}{P}-\overset{\ominus}{C}H_2$; C; 1. $Hg(OAc)_2$, H_2O 2. $NaBH_4$; D; OH; $KMnO_4$; E; O; 1. LDA 2. Br(allyl); O; F

Problem 19.69

Think

In each reaction, consider which atoms are electron rich and electron poor. Is the reaction a functional group transformation, or does the reaction alter the carbon skeleton? Are proton transfer reactions involved? Are any oxidation or reduction reactions involved? Is regiochemistry a concern?

Solve

The missing reagents and intermediates are filled in below. **A** is the result of adding Br_2 across the double bond, which then undergoes back-to-back E2 steps, followed by deprotonation and an acid workup, to produce a terminal alkyne **B**. **B** is then deprotonated to produce a nucleophilic alkynide anion, which undergoes S_N2 with alkyl halide **C**. The resulting C≡C bond is reduced to the cis C=C bond to produce **D**, which is epoxidized using MCPBA to produce **E**. The epoxide is opened using a Grignard reagent to produce alcohol **F**, which is oxidized to ketone **G**. There are two products for **F**, because both carbon atoms attached to the epoxide are secondary. This leads to two ketones for **G** as well.

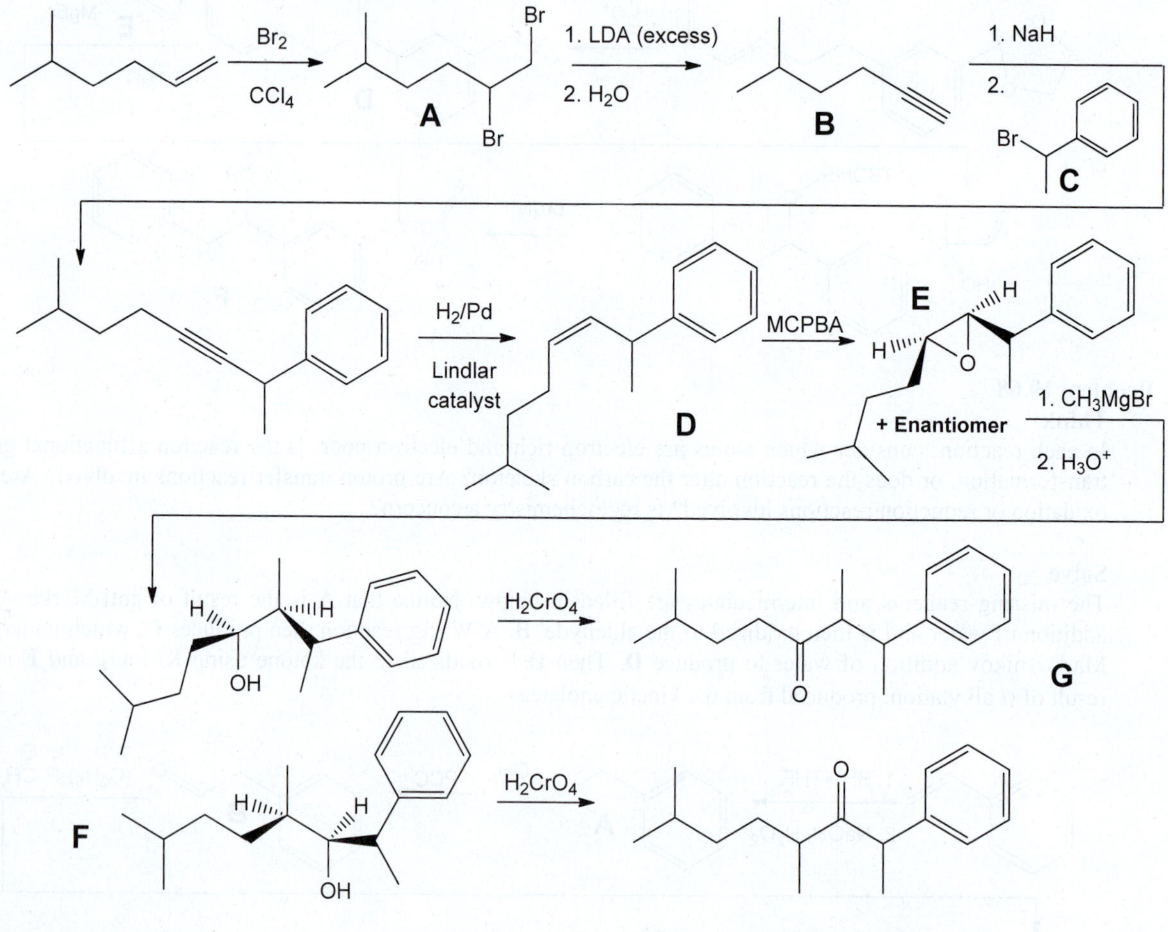

Problem 19.70 (SYN)

Think

Which C–C bond formed from the reaction of acetone with the other organic precursor? Can the synthesis take place in a single reaction with the alcohol? If not, what precursor should we choose? How can this precursor be formed from the alcohol?

Solve

In each of these reactions, the tertiary alcohol can be formed from a Grignard reaction. In **(a)** through **(c)**, a Grignard reagent adds to the C=O of acetone. In **(d)**, acetone becomes the Grignard reagent. The Grignard reagent is formed from the alkylbromide, which can be formed from an alcohol. The syntheses are given on the next two pages.

(a) Cyclohexanol is the alcohol starting material.

Retrosynthesis

Undo a Grignard reaction.

MgBr + O

Reaction in the forward direction

OH PBr_3 → Br; Mg(*s*), Ether → MgBr; 1. (acetone) 2. H_3O^+ → OH

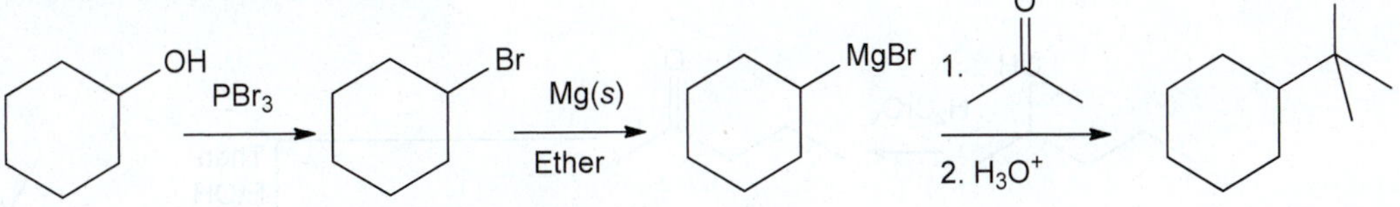

(b) Pentan-1-ol is the alcohol starting material.

Retrosynthesis

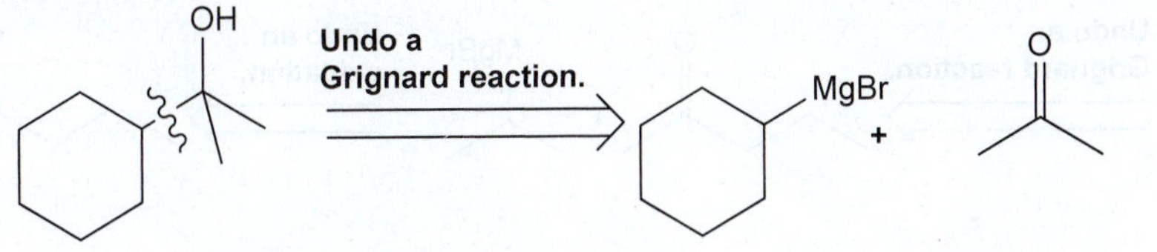

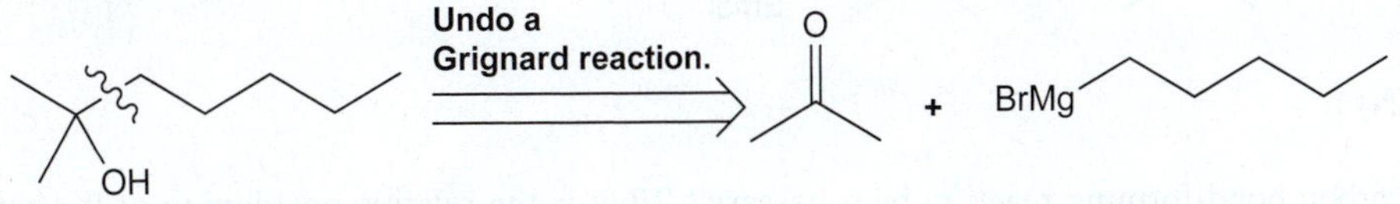

OH; Undo a Grignard reaction. O + BrMg

Reaction in the forward direction

HO; 1. PBr_3 2. Mg(*s*) Ether → BrMg; 1. (acetone) 2. H_3O^+ → OH

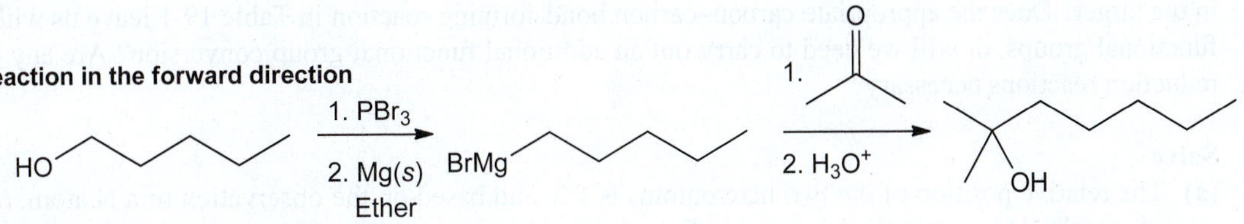

(c) Prop-2-en-1-ol is the alcohol starting material.

Retrosynthesis

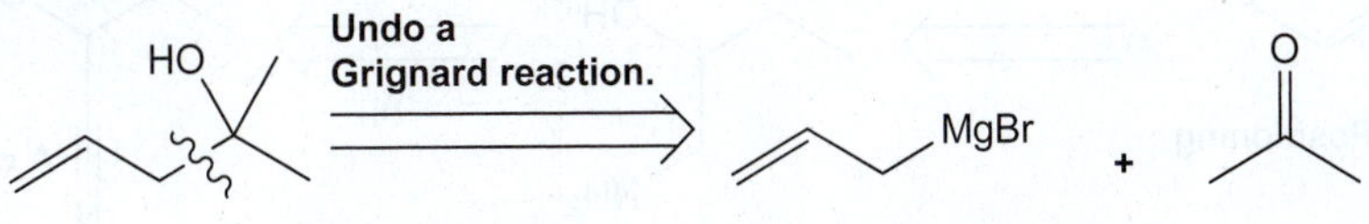

HO; Undo a Grignard reaction. MgBr + O

Reaction in the forward direction

OH; 1. PBr_3 2. Mg(*s*) Ether → MgBr; 1. (acetone) 2. H_3O^+ → HO

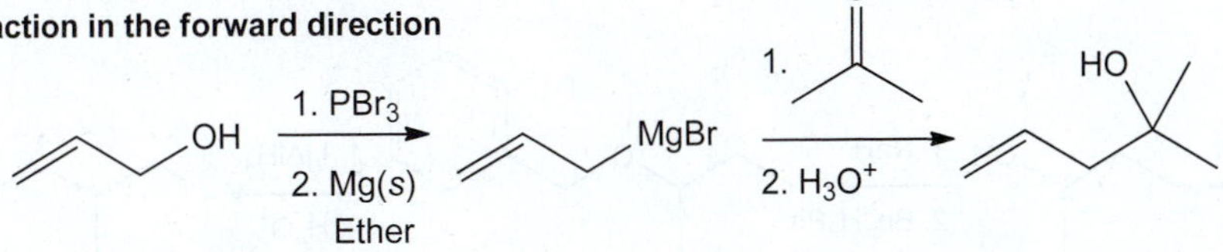

(d) The 2,3-dimethylhexan-3-ol target requires a Grignard reaction using pentan-2-one, which can be synthesized by the oxidation of pentan-2-ol. Acetone is used to form the Grignard reagent.

Retrosynthesis

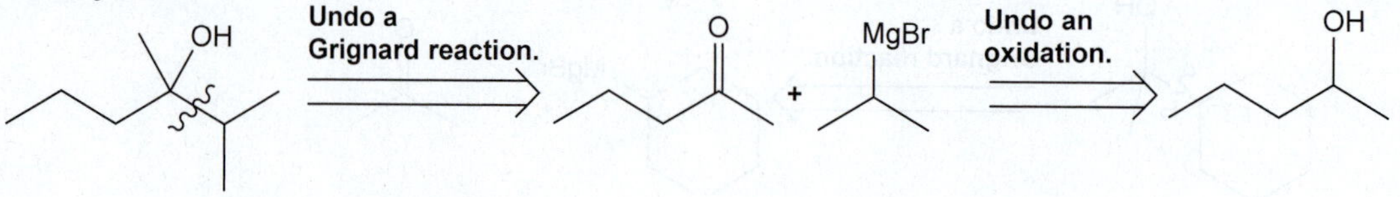

OH — Undo a Grignard reaction. ⟹ O + MgBr — Undo an oxidation. ⟹ OH

Reaction in the forward direction

OH — H_2CrO_4 → O — Then EtOH → OH

O — $NaBH_4$/EtOH → OH — PBr_3 → Br — Mg(s), Ether → MgBr

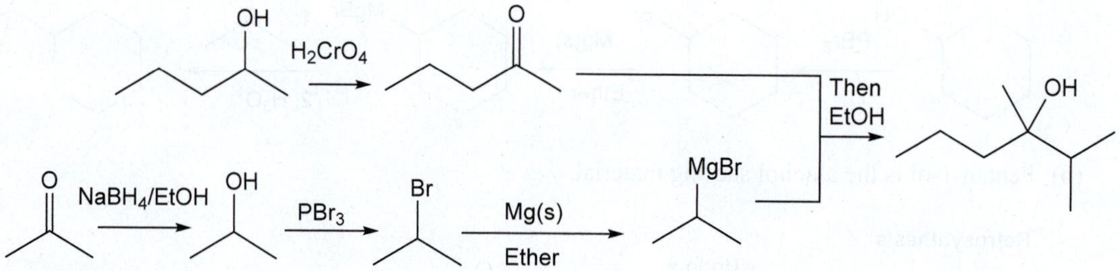

Problem 19.71 (SYN)

Think

Will a carbon–carbon bond-forming reaction be necessary? What is the relative positioning of the heteroatoms in the target? Does the appropriate carbon–carbon bond-forming reaction in Table 19-1 leave us with the correct functional groups, or will we need to carry out an additional functional group conversion? Are any oxidation or reduction reactions necessary?

Solve

(a) The relative position of the two heteroatoms is 1,2, and based on the observation of a N atom, the synthesis can go through a cyanohydrin intermediate.

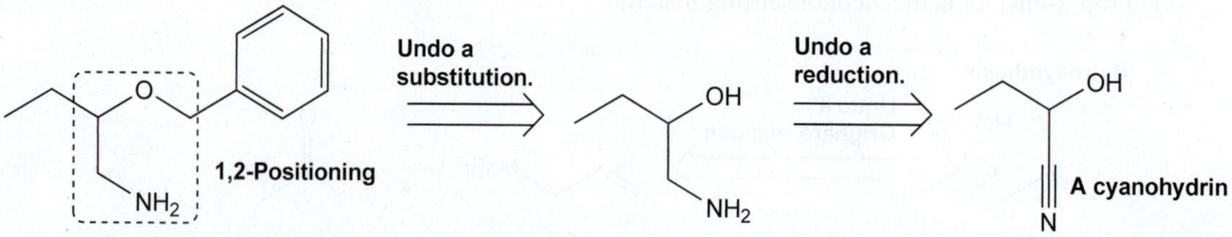

O, NH_2 — **1,2-Positioning** — Undo a substitution. ⟹ OH, NH_2 — Undo a reduction. ⟹ OH, N — **A cyanohydrin**

The synthesis in the forward direction is given below:

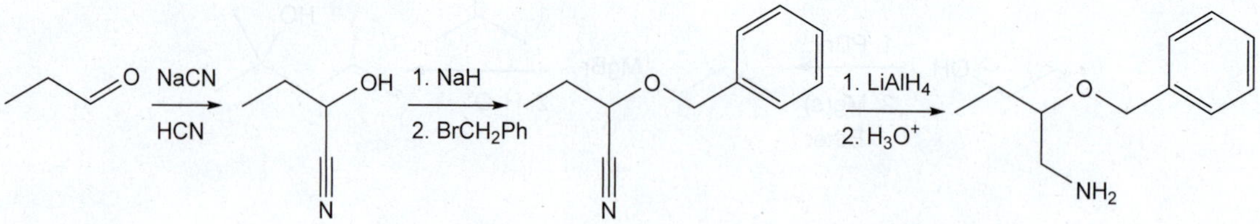

O — NaCN, HCN → OH, N — 1. NaH 2. $BrCH_2Ph$ → O, N — 1. $LiAlH_4$ 2. H_3O^+ → O, NH_2

(b) The relative position of the two heteroatoms is 1,3, and the synthesis likely should include an aldol addition reaction. The aldol OH can then be converted to the thioether.

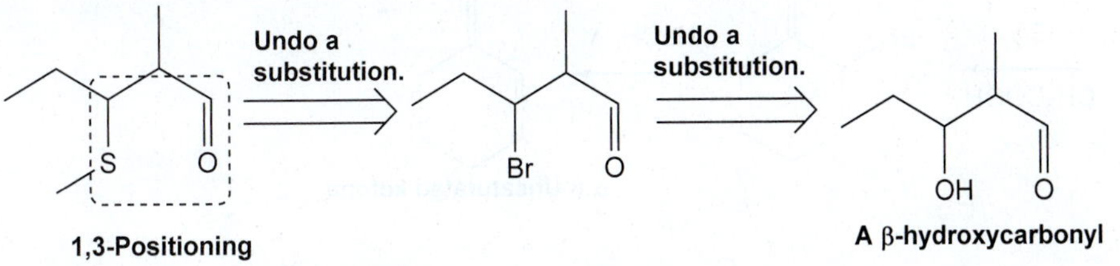

Undo a substitution.
Undo a substitution.
S O
Br O
OH O
1,3-Positioning
A β-hydroxycarbonyl

The synthesis in the forward direction is given below:

O NaOH EtOH
OH O
PBr_3
Br O
$NaSCH_3$
S O

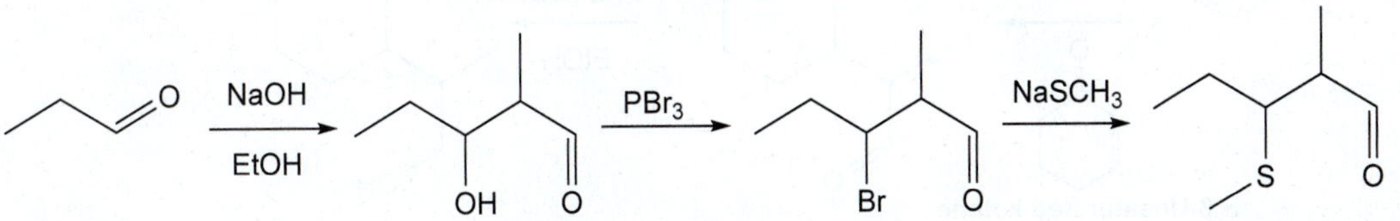

(c) The relative position of the two heteroatoms is 1,3, and the synthesis likely should go through an α,β-unsaturated aldehyde, the product of an aldol condensation.

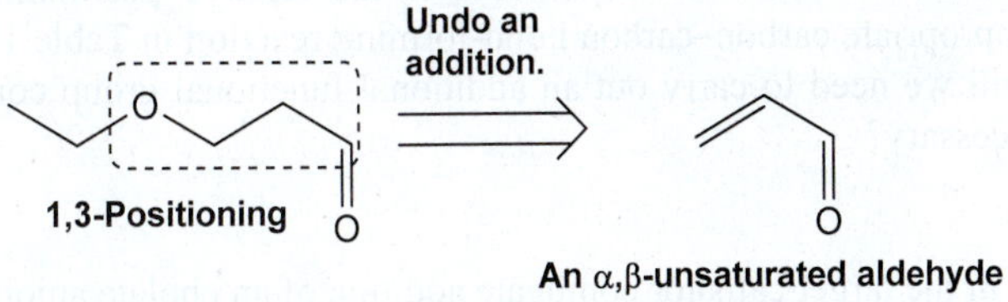

Undo an addition.
O
1,3-Positioning O
O
An α,β-unsaturated aldehyde

The synthesis in the forward direction is given below:

O $NaBH_4$ EtOH
OH
NaH
O Br_2 CH_3COOH
Br O
NaOH, Δ
O
NaO
EtOH
O O

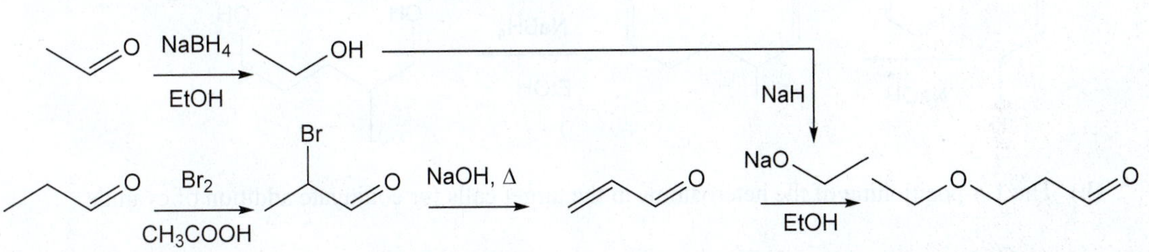

(d) The relative position of the two heteroatoms is 1,5, and the synthesis likely goes through a 1,5-dicarbonyl, which would come from the conjugate addition of an enolate to an α,β-unsaturated ketone.

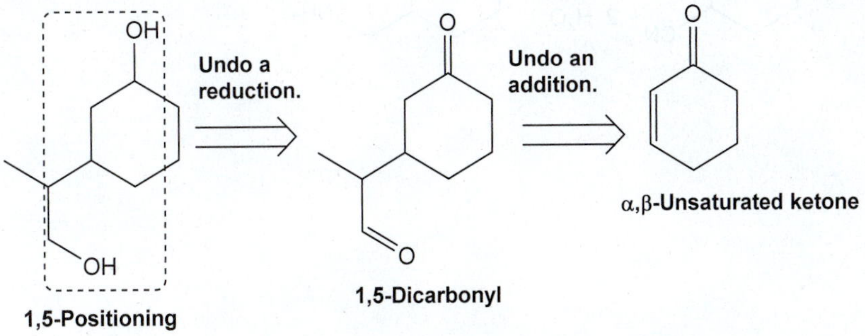

OH
Undo a reduction.
O
Undo an addition.
O
OH
O
1,5-Positioning
1,5-Dicarbonyl
α,β-Unsaturated ketone

The synthesis in the forward direction is given below:

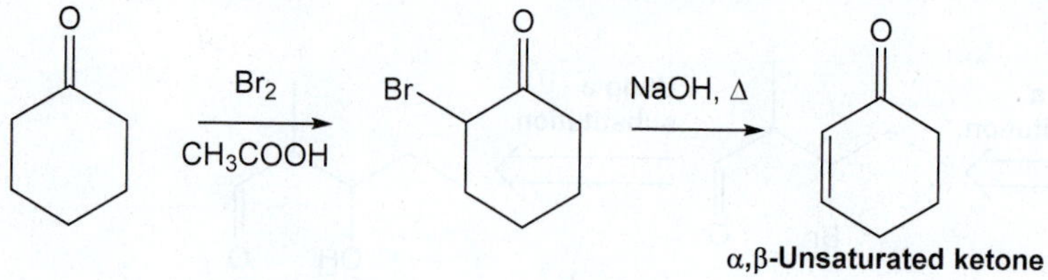

$\xrightarrow[CH_3COOH]{Br_2}$ $\xrightarrow{NaOH,\ \Delta}$

α,β-Unsaturated ketone

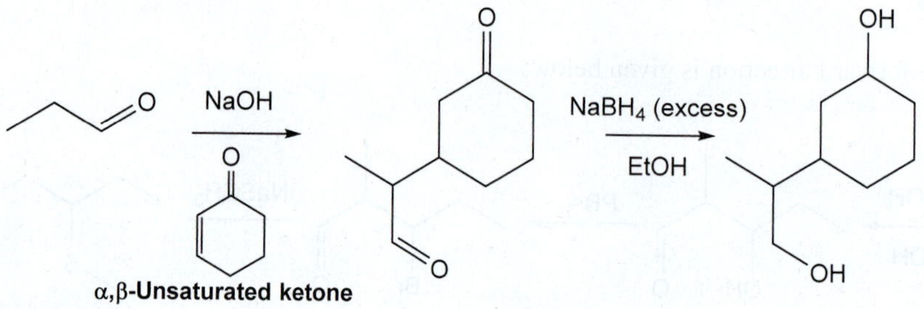

$\xrightarrow{NaOH}$ $\xrightarrow[EtOH]{NaBH_4\ (excess)}$

α,β-Unsaturated ketone

Problem 19.72 (SYN)

Think

How many carbon–carbon bond-forming reactions are necessary? What is the relative positioning of the functional groups in the target? Does the appropriate carbon–carbon bond-forming reaction in Table 19-1 leave us with the correct functional groups, or will we need to carry out an additional functional group conversion? Are any oxidation or reduction reactions necessary?

Solve

(a) The 1,5-positioning of the heteroatoms in the target calls for conjugate addition of an enolate anion.

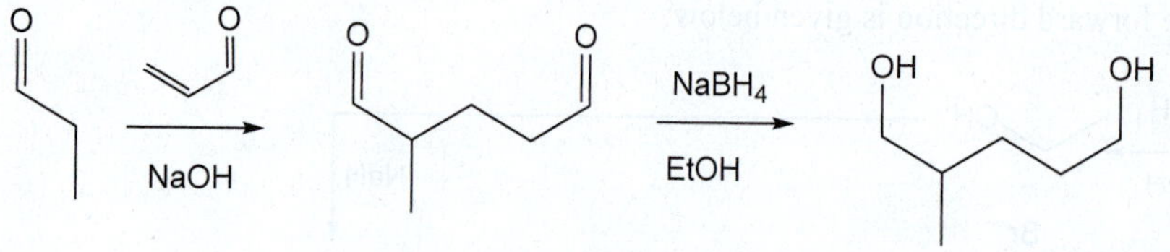

$\xrightarrow{NaOH}$ $\xrightarrow[EtOH]{NaBH_4}$

(b) The 1,4-positioning of the heteroatoms in the target calls for conjugate addition of cyanide.

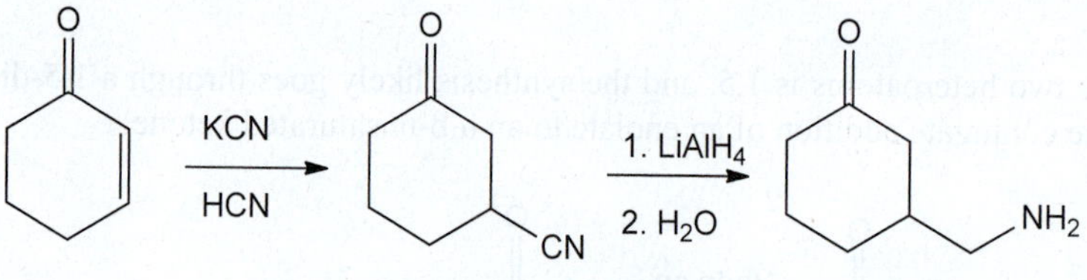

$\xrightarrow[HCN]{KCN}$ $\xrightarrow[2.\ H_2O]{1.\ LiAlH_4}$

(c) The 1,2-positioning of the heteroatoms in the target calls for the formation of a cyanohydrin. The OH of the cyanohydrin can then be protected, and a Grignard reaction can form the necessary C–C bond.

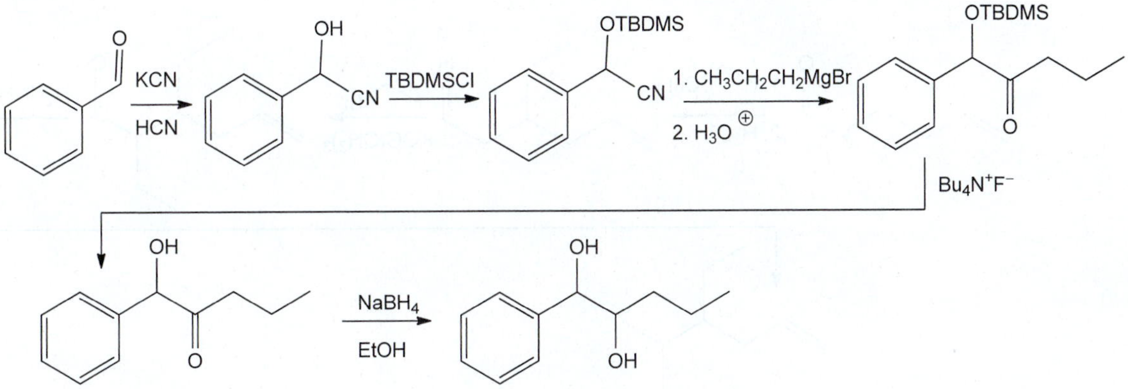

KCN / HCN; TBDMSCl; 1. $CH_3CH_2CH_2MgBr$ 2. $H_3O^{\oplus}$; $Bu_4N^+F^-$; $NaBH_4$, EtOH

(d) The 1,3-positioning of the heteroatoms in the target calls for an aldol addition.

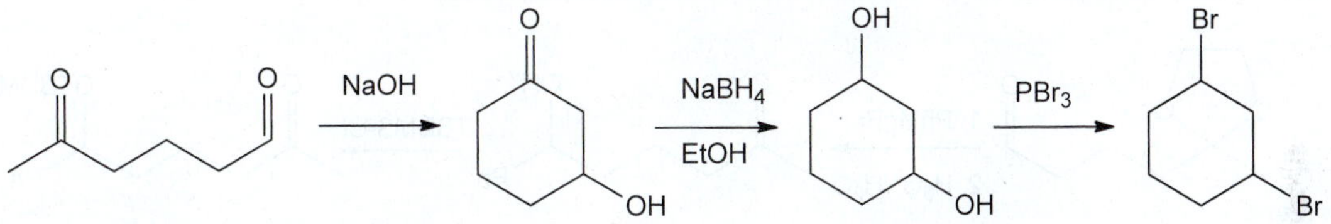

NaOH; $NaBH_4$, EtOH; PBr_3

Problem 19.73 (SYN)

Think

Is the carbon skeleton altered? Does functional group transformation occur? Are protecting groups necessary? Are strong acids or bases needed? What reactions can you use to remove a functional group entirely?

Solve

Syntheses for **(a)**–**(d)** are shown below and on the next page. Other synthesis schemes could also be feasible.

(a) Requires alkylation via the kinetic enolate, which is why LDA is used as the base. To avoid reaction with the ether, the reduction of C=O to CH_2 should avoid the Clemmensen reduction. A Wolff–Kishner or Raney-nickel reduction could be used.

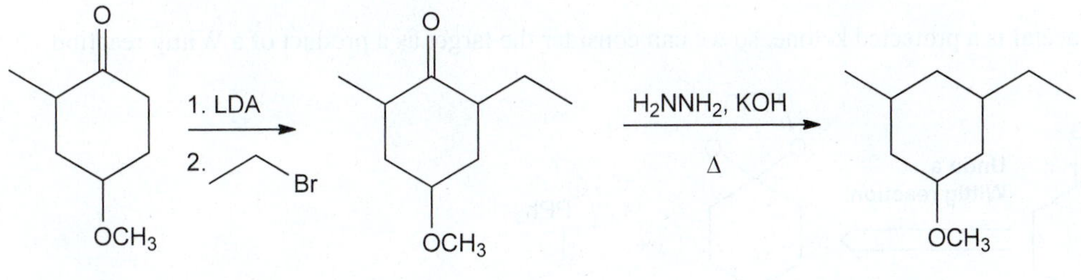

1. LDA 2. Br; H_2NNH_2, KOH, Δ; OCH_3

(b) Requires the thermodynamic enolate, which is why NaOH is used as the base.

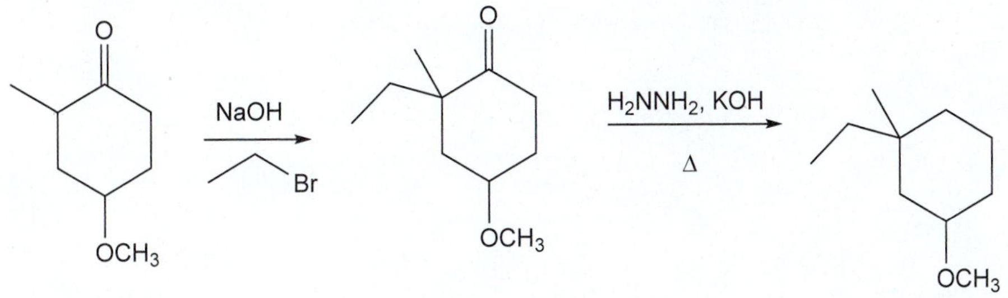

NaOH, Br; H_2NNH_2, KOH, Δ; OCH_3

(c) Conjugate addition is required, so $LiCuR_2$ is used as the alkylating reagent. To keep the alkene unreacted, the Wolff–Kishner reduction conditions are used to avoid the use of strong acids (Clemmensen reduction) or catalytic hydrogenation (Raney nickel).

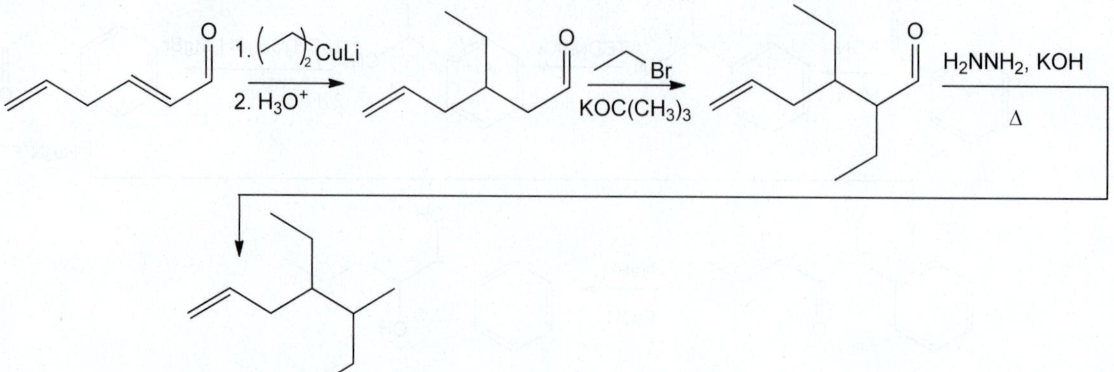

1. $(\text{propyl})_2CuLi$; 2. H_3O^+ → ; Br (ethyl bromide), $KOC(CH_3)_3$ → ; H_2NNH_2, KOH, Δ →

(d) The alcohol is protected to avoid a reaction between it and LDA.

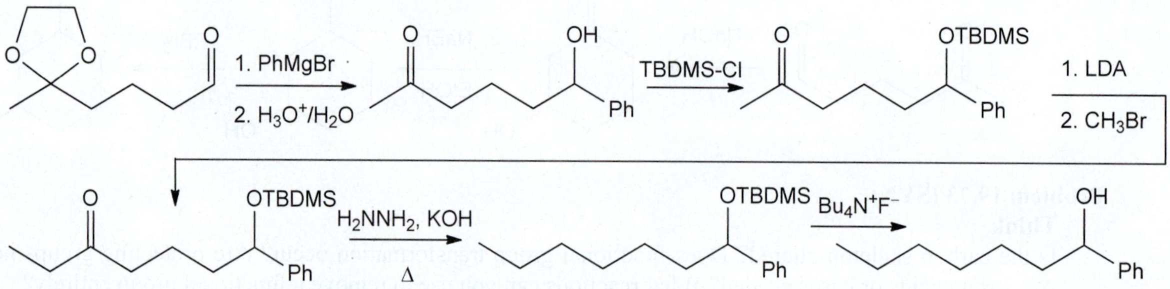

1. PhMgBr; 2. H_3O^+/H_2O → OH, Ph ; TBDMS-Cl → OTBDMS, Ph ; 1. LDA; 2. CH_3Br → OTBDMS, Ph ; H_2NNH_2, KOH, Δ → OTBDMS, Ph ; $Bu_4N^+F^-$ → OH, Ph

Problem 19.74 (SYN)

Think

What functional group is formed in this reaction? Which functional group is protected? How do you remove the protecting group? In 1,4-cyclohexanedione, are both ketones reactive toward an ylide? If so, what product would result?

Solve

(a) The acetal is a protected ketone, so we can consider the target as a product of a Wittig reaction.

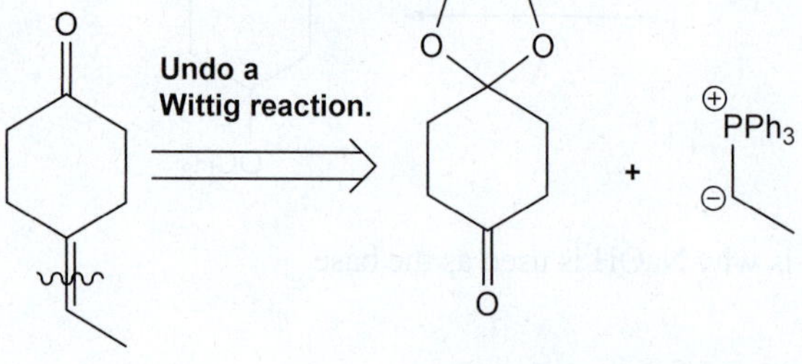

Undo a Wittig reaction. ⟹ + $\overset{\oplus}{P}Ph_3$ ylide

After the Wittig reaction, we can deprotect the ketone. The reaction in the forward direction is as follows:

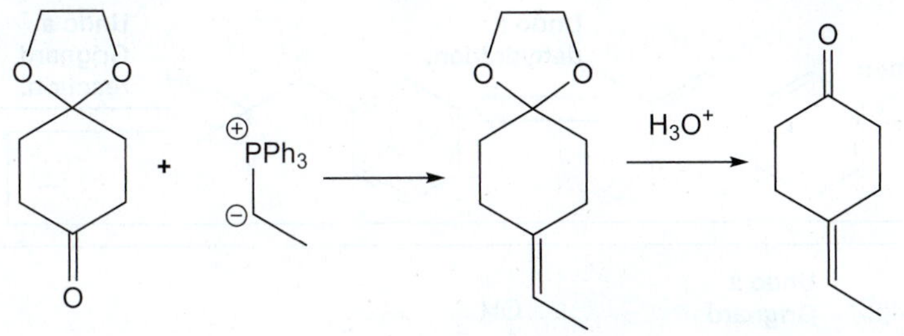

H_3O^+

PPh3

(b) If 1,4-cyclohexanedione were used, the ketone in the product would not remain. Both ketones are susceptible to attack by the ylide to form an alkene. The reaction that would result is given below.

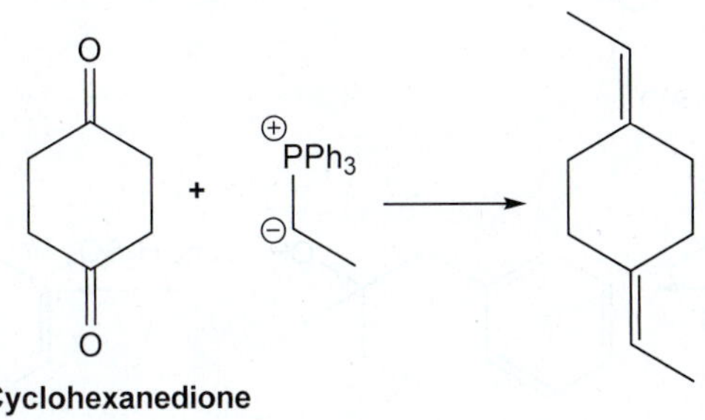

PPh3

1,4-Cyclohexanedione

Problem 19.75 (SYN)

Think

Are any C–C bond-forming reactions necessary? Do you need to include oxidation or reduction reactions? Functional group transformations?

Solve

(a) We can undo an oxidation at the benzylic carbon, as shown below:

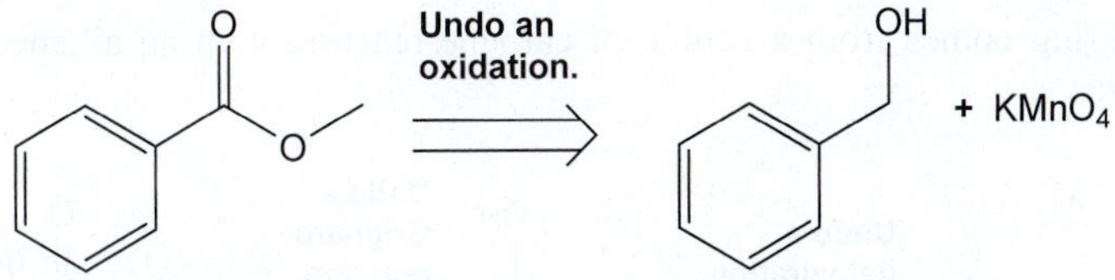

Undo an oxidation.

+ $KMnO_4$

The carboxylic acid is converted to the ester, as shown below:

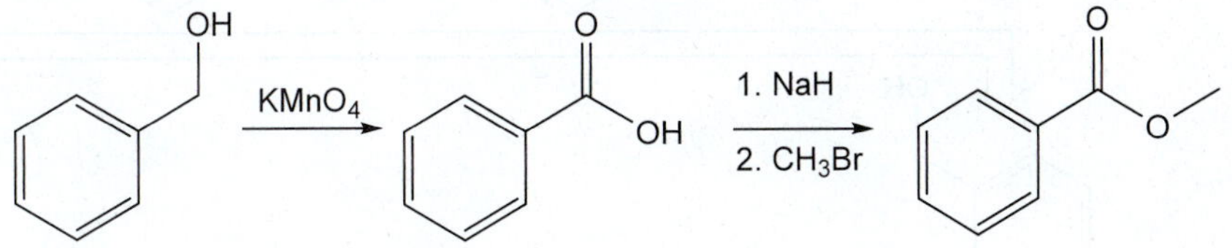

$KMnO_4$

1. NaH
2. CH_3Br

(b) The carbon skeleton is altered by the addition of three C atoms.

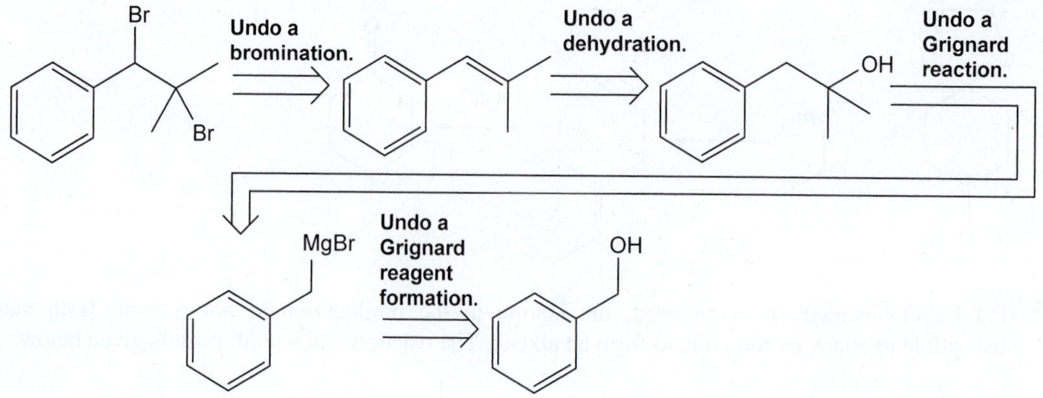

Undo a bromination.

Undo a dehydration.

Undo a Grignard reaction.

Undo a Grignard reagent formation.

The reaction in the forward direction is given below:

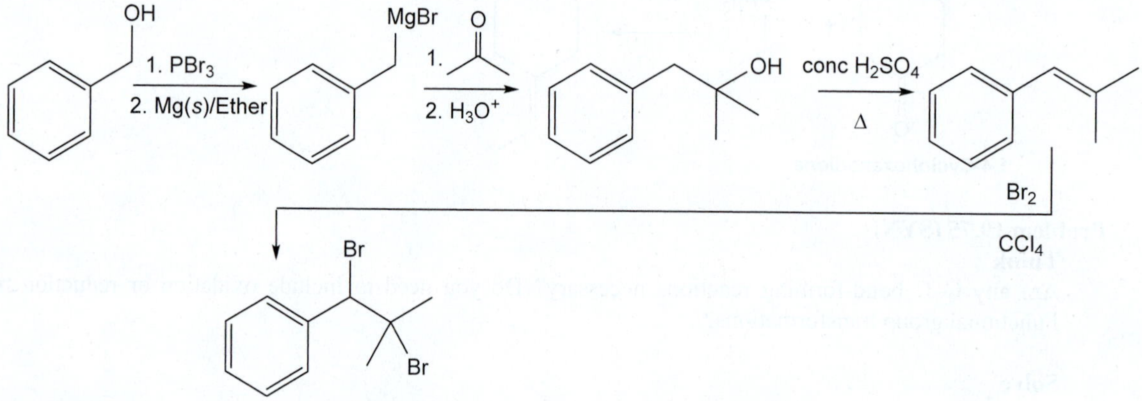

1. PBr_3
2. Mg(*s*)/Ether

1. (acetone)
2. H_3O^+

conc H_2SO_4
Δ

Br_2
CCl_4

(c) The formation of a cyclopropane ring comes from a source of carbene reacting with an alkene. We can undo the reaction, as shown below:

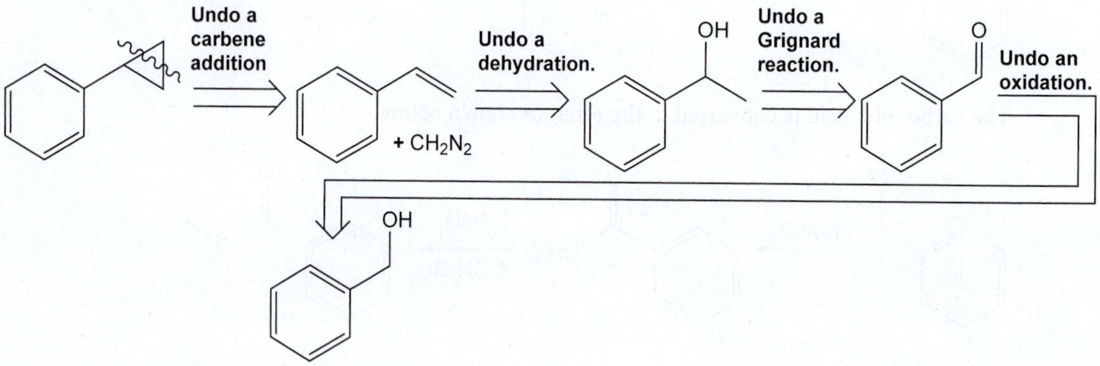

Undo a carbene addition

$+ CH_2N_2$

Undo a dehydration.

Undo a Grignard reaction.

Undo an oxidation.

The reaction in the forward direction is given below:

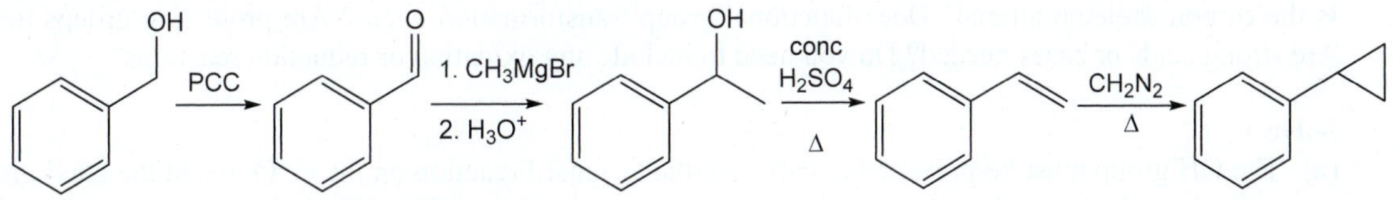

(d) We can undo a Grignard on the target, as shown below:

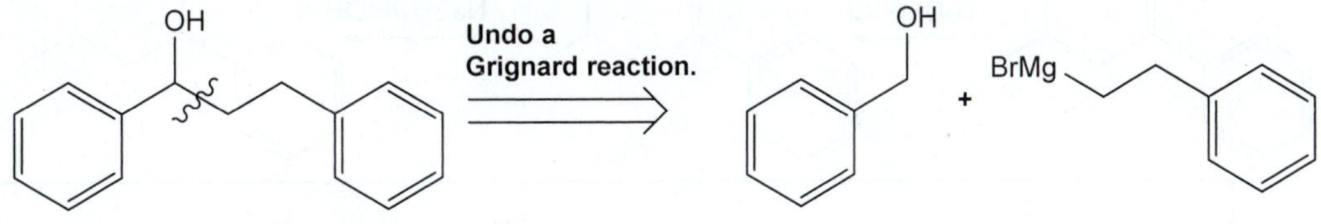

The reaction in the forward direction is shown below:

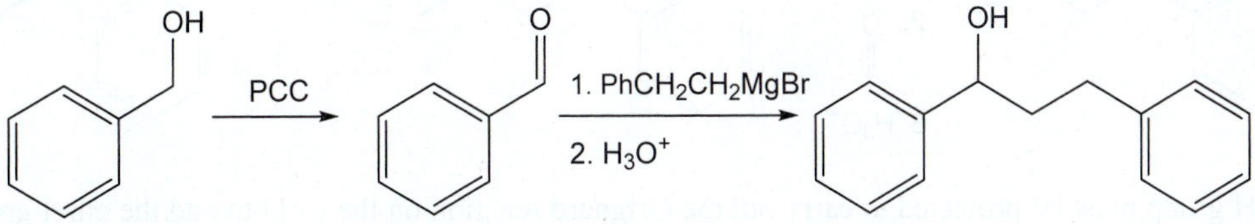

Problem 19.76 (SYN)

Think

Is the carbon skeleton altered? Does functional group transformation occur? Are protecting groups necessary? Are strong acids or bases needed? Will you need to incorporate an oxidation or a reduction reaction?

Solve

Because we are limited only to a starting material with three carbons, we should consider disconnecting the bonds below in a retrosynthetic analysis. The 1,2-diol suggests an epoxidation reaction from an alkene.

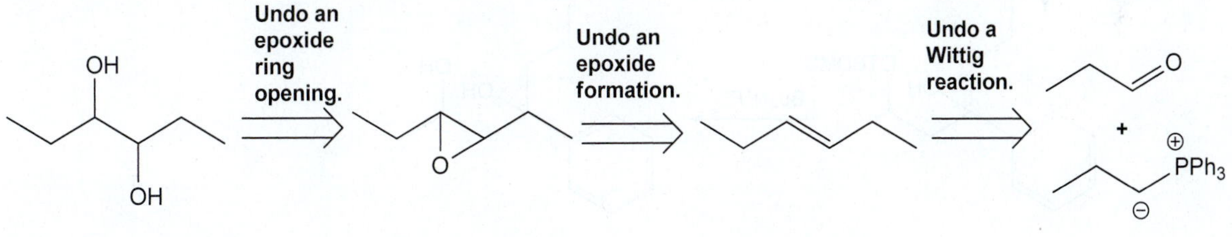

The reaction in the forward direction is given below:

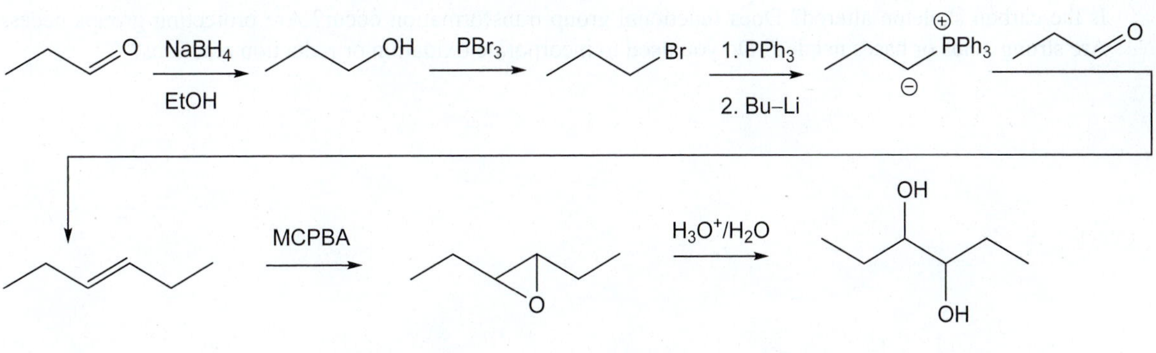

Problem 19.77 (SYN)

Think

Is the carbon skeleton altered? Does functional group transformation occur? Are protecting groups necessary? Are strong acids or bases needed? Do you need to include any oxidation or reduction reactions?

Solve

(a) The OH group must be protected to carry out the Grignard reaction on the C=O to add the ethyl group. The secondary alcohol in the product came from a Grignard reaction. The full synthesis is given below:

TBDMS-Cl; $NaBH_4$/EtOH; PBr_3; 1. Mg(*s*)/Ether 2. (acetaldehyde) 3. H_3O^+; $Bu_4N^+F^-$

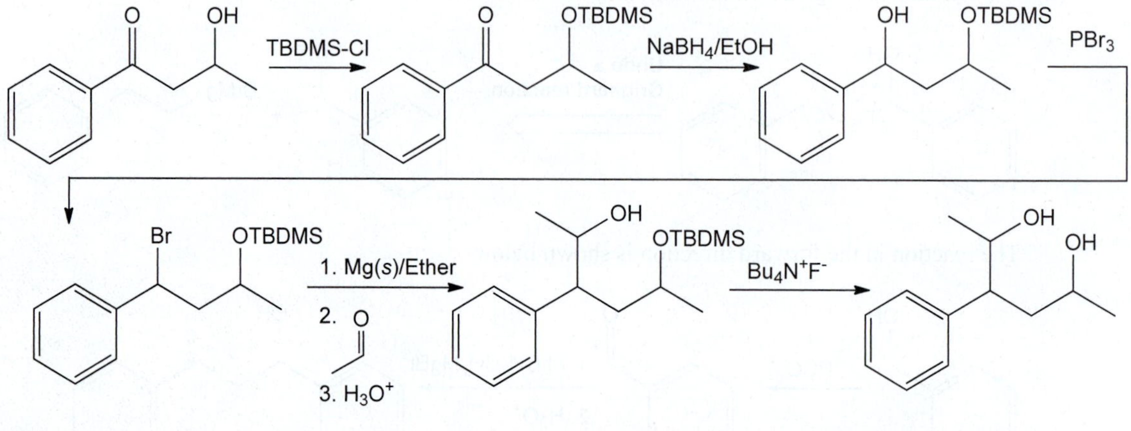

(b) The OH group must be protected to carry out the Grignard reaction on the C=O to add the ethyl group. The OH is on the less-substituted carbon, and this likely came from the hydroboration–oxidation reaction. The full synthesis is given below.

TBDMS-Cl; 1. CH_2=CHMgBr 2. H_3O^+; 1. BH_3•THF 2. H_2O_2/NaOH; $Bu_4N^+F^-$

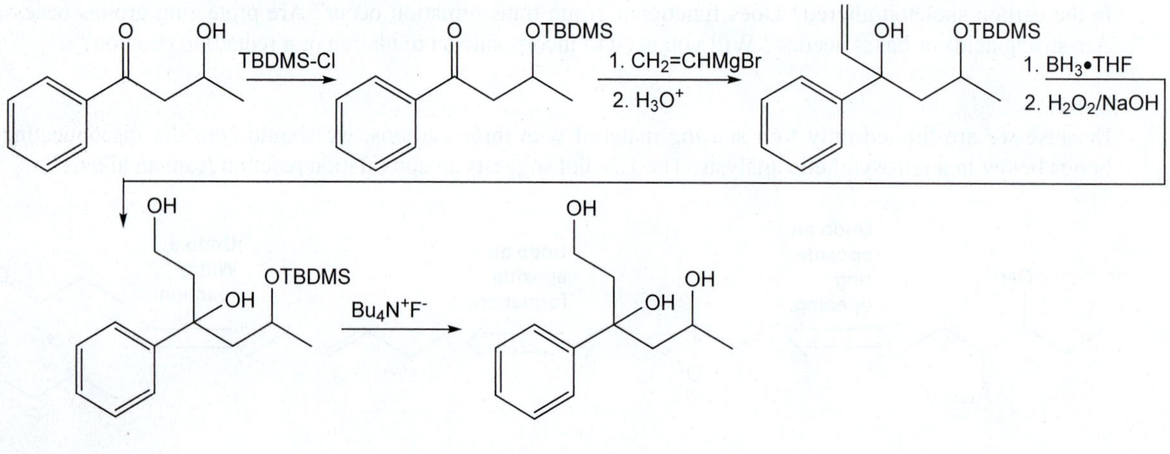

Problem 19.78 (SYN)

Think

Is the carbon skeleton altered? Does functional group transformation occur? Are protecting groups necessary? Are strong acids or bases needed? Do you need to incorporate oxidation or reduction reactions?

Solve

Because we are limited only to a starting material with three carbons, we should consider disconnecting the bonds below in a retrosynthetic analysis.

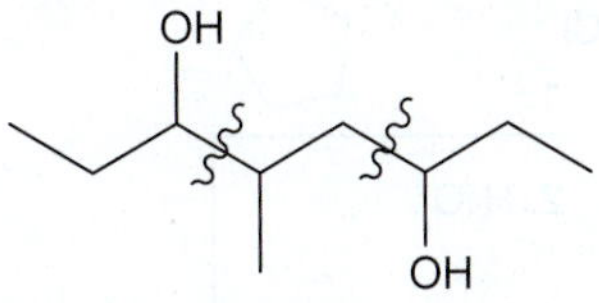

OH

OH

We can disconnect the bond on the left first, undoing a Grignard reaction.

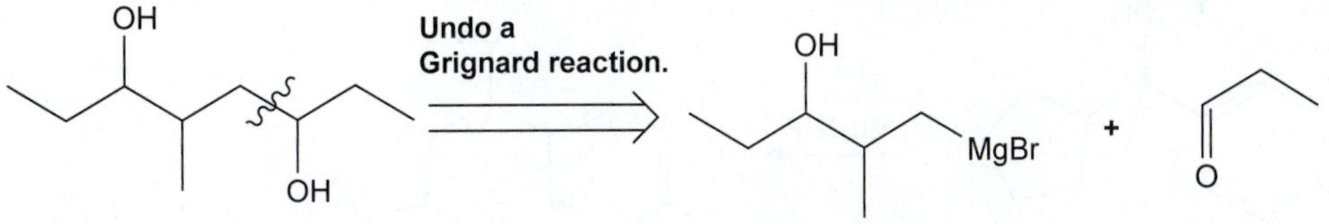

OH **Undo a Grignard reaction.** OH MgBr + O

OH

Not worrying about the incompatibility of the OH group (we'll use a protecting group for it), we see that the Grignard reagent has the heteroatoms in a 1,3-position that could come from an aldol reaction, as shown below:

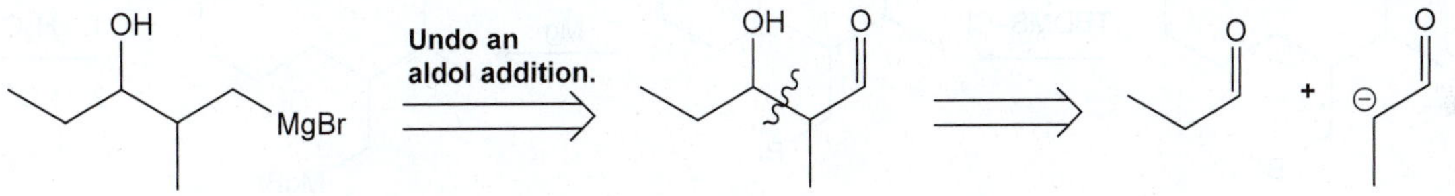

OH **Undo an aldol addition.** OH O O + O

MgBr

In the forward direction, we will need to use protecting groups to prevent incompatibilities:

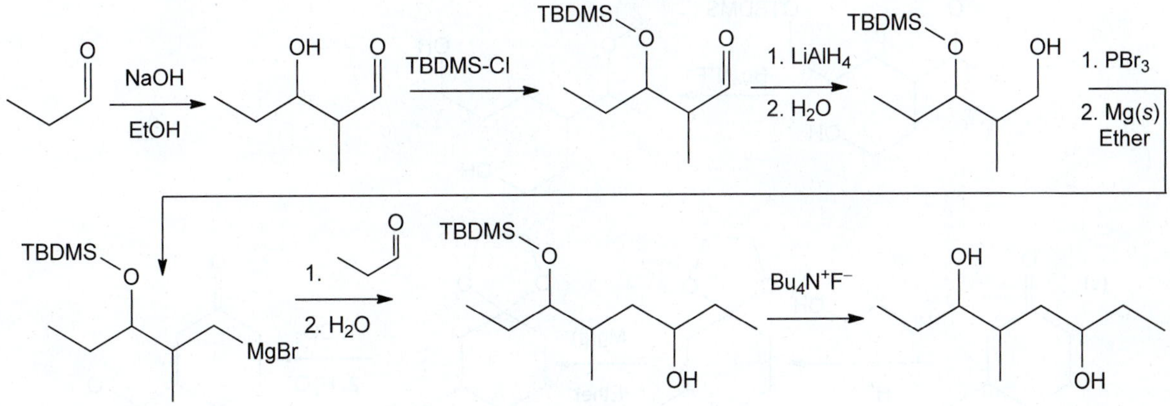

O NaOH EtOH OH O TBDMS-Cl TBDMS O O 1. $LiAlH_4$ 2. H_2O TBDMS O OH 1. PBr_3 2. Mg(*s*) Ether

TBDMS O MgBr 1. O 2. H_2O TBDMS O OH $Bu_4N^+F^-$ OH OH

Problem 19.79 (SYN)

Think

Is the carbon skeleton altered? Does functional group transformation occur? Are protecting groups necessary? Are strong acids or bases needed? Will you need to use oxidation or reduction reactions?

Solve

Syntheses for **(a)–(e)** are shown below and on the next page. Other synthesis schemes could also be feasible.

(a)

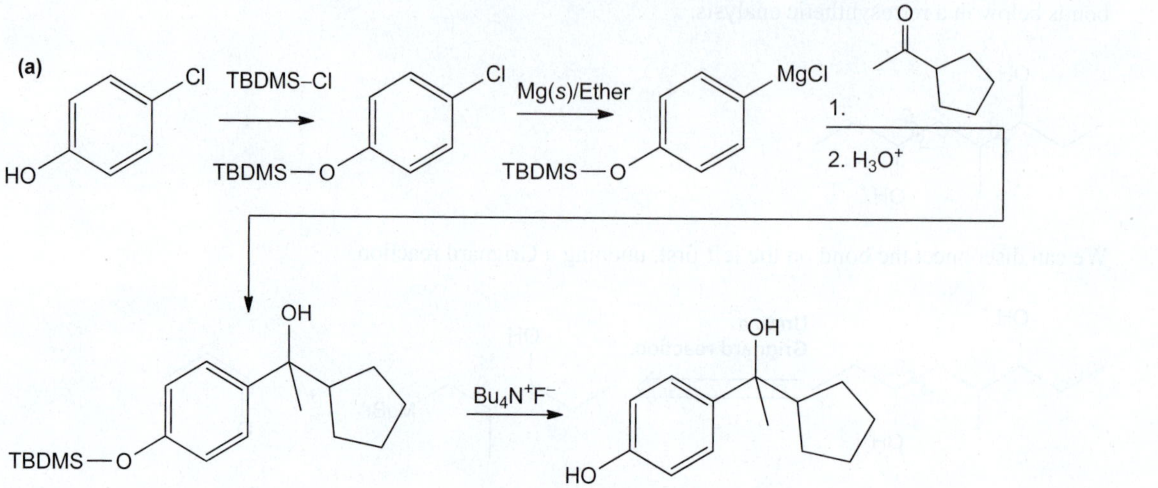

HO–C_6H_4–Cl $\xrightarrow{\text{TBDMS–Cl}}$ TBDMS—O–C_6H_4–Cl $\xrightarrow{\text{Mg}(s)\text{/Ether}}$ TBDMS—O–C_6H_4–MgCl $\xrightarrow{\text{1. cyclopentyl methyl ketone; 2. } H_3O^+}$ TBDMS—O–C_6H_4–C(OH)(CH_3)(cyclopentyl) $\xrightarrow{Bu_4N^+F^-}$ HO–C_6H_4–C(OH)(CH_3)(cyclopentyl)

(b)

OTBDMS, Mg, MgBr, OTBDMS, 1. $H_2C{=}O$, 2. H_3O^+

O, OH, TBDMS–Cl, Br, Br

OTBDMS, $Bu_4N^+F^-$, OH, OH, OH

(c)

HO‿‿OH, H^+; Mg(*s*), Ether; 1. CO_2, 2. H_3O^+

Br, Br, MgBr, OH

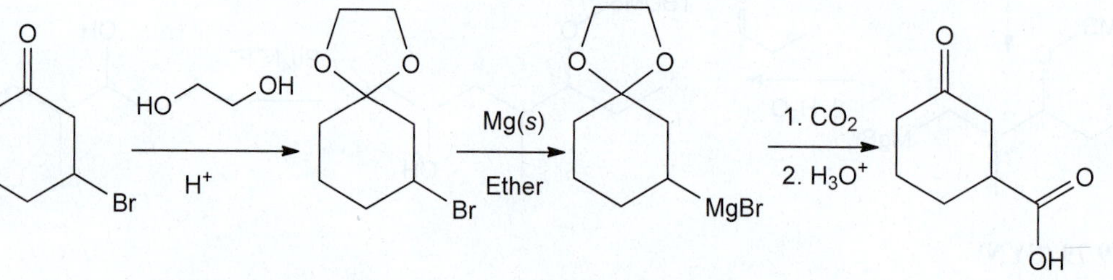

(d)

H_2/Pd; 1. LDA, 2. CH_3I; $Ph_3\overset{\oplus}{P}—\overset{\ominus}{C}H_2$

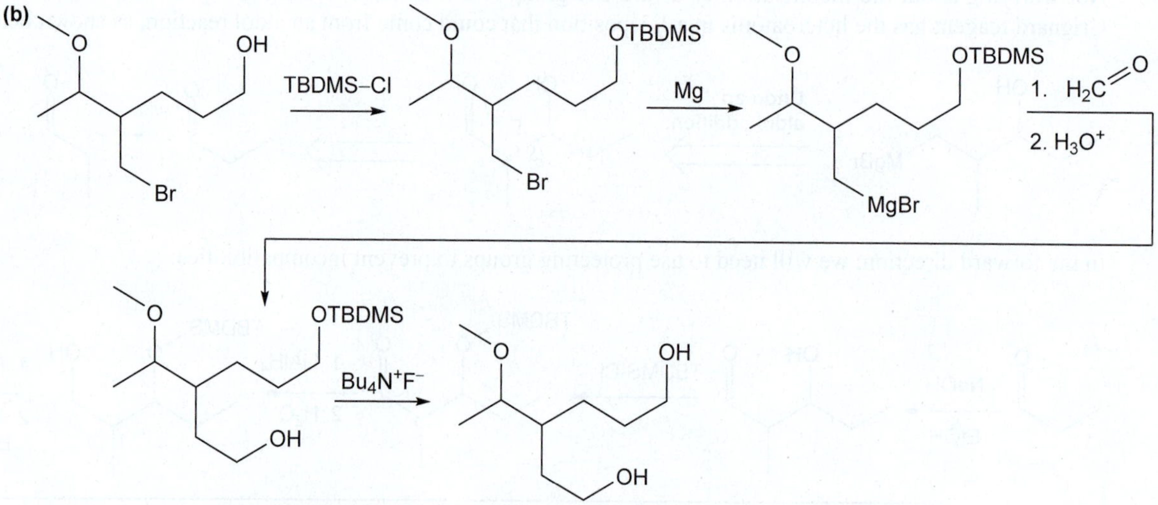

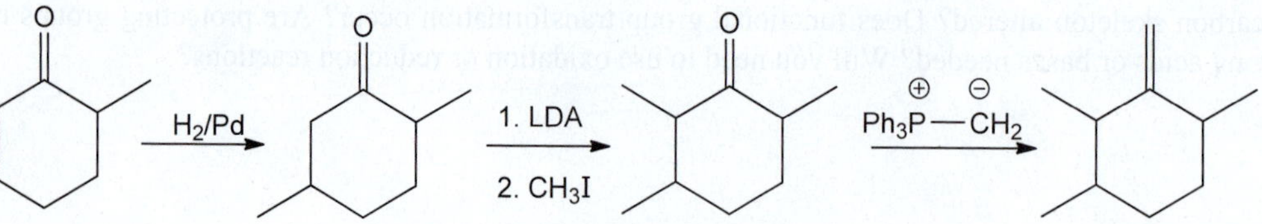

(e)

HO⌒OH, H^+

NaC≡C–$CH_2CH_2CH_3$

H_2, Lindlar catalyst

H_3O^+

CHAPTER 20 | Nucleophilic Addition–Elimination Reactions 1: The General Mechanism Involving Strong Nucleophiles

Your Turn Exercises

Your Turn 20.1

Think

In each step, identify the bonds that are broken and the bonds that are formed. How can you use curved arrows to show the electron movement? Review the 10 elementary steps in Chapter 7. Which elementary step involves attack of a nucleophile at a polar π bond? Which elementary step involves regeneration of the polar π bond?

Solve

Step 1 is nucleophilic addition where the nucleophile attacks the electron-poor C of the C=O. This forces the electrons in the C=O π bond onto the O atom. The product of this step is the *tetrahedral* intermediate. The tetrahedral intermediate in Step 2 undergoes nucleophile elimination, in which one of the lone pairs on the O of the tetrahedral intermediate regenerates the C=O and the leaving group, CH_3O^-, departs.

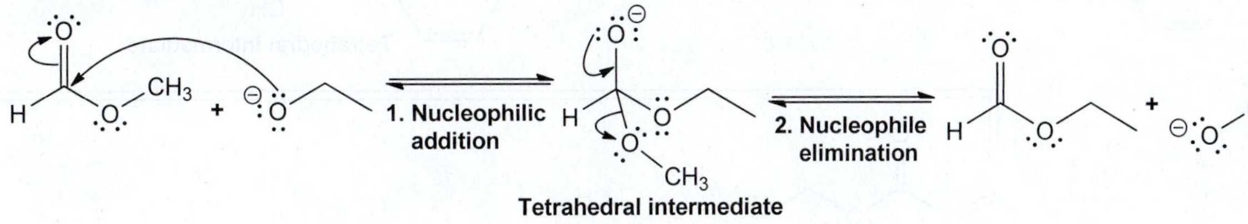

1. Nucleophilic addition

2. Nucleophile elimination

Tetrahedral intermediate

Your Turn 20.2

Think

Use Figure 20-1 as a guide. How do the charges impact the stability of the reactants, products, and intermediates? How do they impact the resonance? How you do represent the energy barrier in each step?

Solve

In each stage of the reaction, one negative charge is localized on O, so charge stability is roughly the same for all three. The C=O bond in both the reactants and the products is resonance stabilized, but the tetrahedral intermediate is not resonance stabilized, so the tetrahedral intermediate is higher in energy. Each energy barrier is identified as the difference in energy between a reactant and the subsequent transition state and is represented by the vertical arrows.

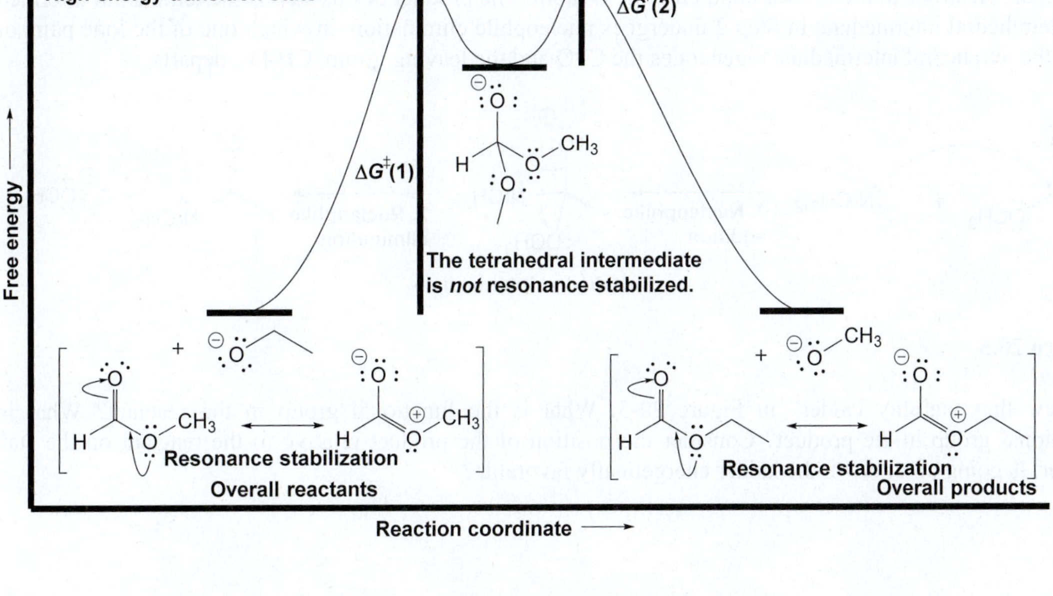

High-energy transition state

$\Delta G^{\ddagger}(1)$

$\Delta G^{\ddagger}(2)$

The tetrahedral intermediate is *not* resonance stabilized.

Free energy

Resonance stabilization

Overall reactants

Resonance stabilization

Overall products

Reaction coordinate

Your Turn 20.3

Think

Use the mechanism in Equation 20-2 as guide. Write the species/steps appearing in Equation 20-2 in the reverse order, and in each step, identify the bonds that are broken and the bonds that are formed. How can you use curved arrows to show the electron movement?

Solve

Step 1 is nucleophilic addition, in which the nucleophile attacks the electron-poor C of the C=O. This forces the electrons in the C=O π bond onto the O atom. The product of this step is the *tetrahedral* intermediate. The tetrahedral intermediate in Step 2 undergoes nucleophile elimination, in which one of the lone pairs on the O of the tetrahedral intermediate regenerates the C=O and the leaving group, CH_3O^-, departs.

CH_3 + 1.Nucleophilic addition NH CH_3 Tetrahedral intermediate 2. Nucleophile elimination NH + CH_3

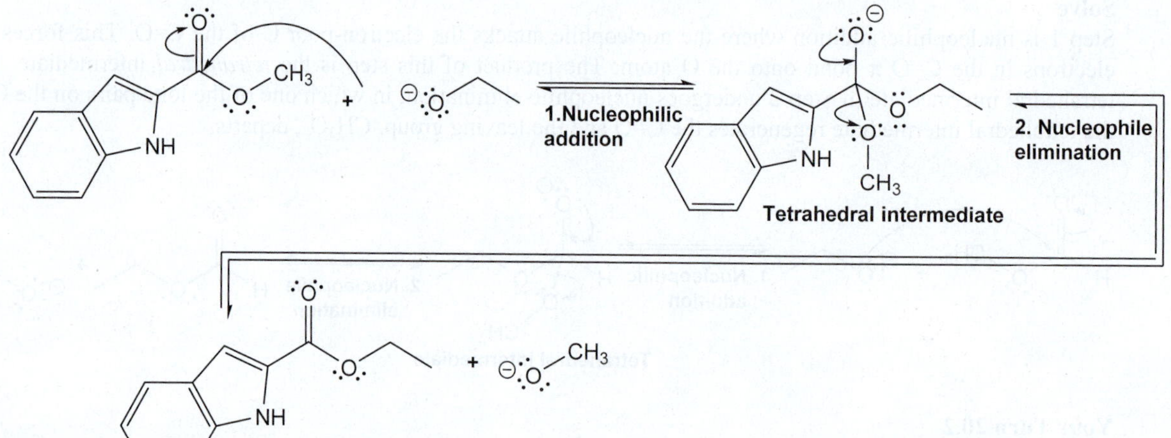

Your Turn 20.4

Think

In each step, identify the bonds that are broken and the bonds that are formed. Which sites are electron rich, and which are electron poor? How can you use curved arrows to show the electron movement? Review the 10 elementary steps in Chapter 7. Which elementary step involves attack of a nucleophile at a polar π bond? Which elementary step involves regeneration of the polar π bond?

Solve

Step 1 is nucleophilic addition, in which the $^-N(CH_3)_2$ nucleophile attacks the electron-poor C of the C=O. This forces the electrons in the C=O π bond onto the O atom. The product of this step is the tetrahedral intermediate. The tetrahedral intermediate in Step 2 undergoes nucleophile elimination, in which one of the lone pairs on the O of the tetrahedral intermediate regenerates the C=O and the leaving group, CH_3O^-, departs.

OCH_3 + $N(CH_3)_2$ 1. Nucleophilic addition $N(CH_3)_2$ OCH_3 Tetrahedral intermediate 2. Nucleophile elimination $N(CH_3)_2$ + OCH_3

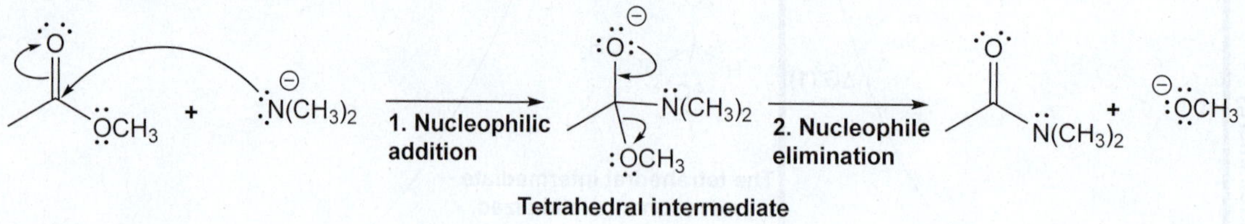

Your Turn 20.5

Think

Review the "stability ladder" in Figure 20-5. What is the functional group in the reactant? What is the functional group in the product? Consider the position of the product relative to the reactant on the stability ladder. Is going up or down the ladder energetically favorable?

Solve

Whether each conversion is favorable is determined by whether the conversion represents going up or down the stability ladder (Fig. 20-5). Going up the ladder is an energetically unfavorable process that typically does not occur readily, whereas going down is favorable and typically occurs readily.

(a)

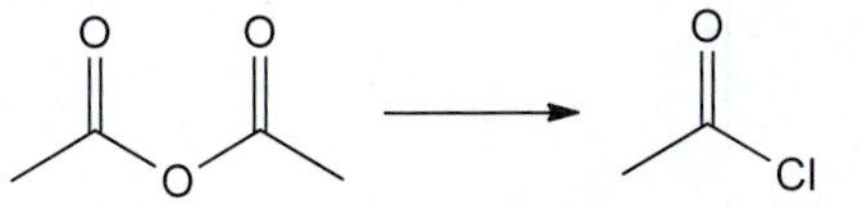

Acid anhydride → Acid chloride
= Up = Unfavorable

(b)

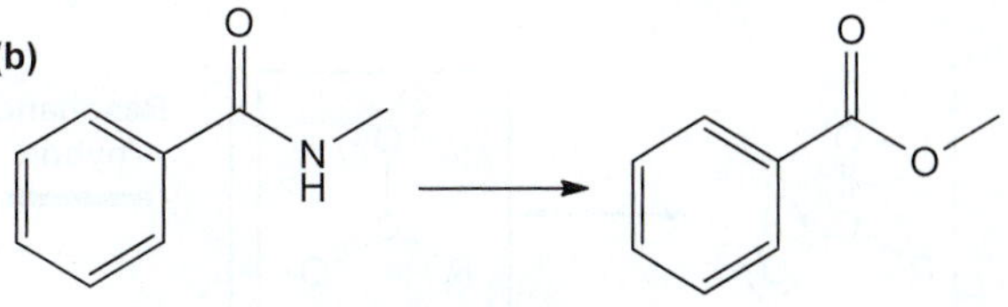

Amide → Ester
= Up = Unfavorable

(c)

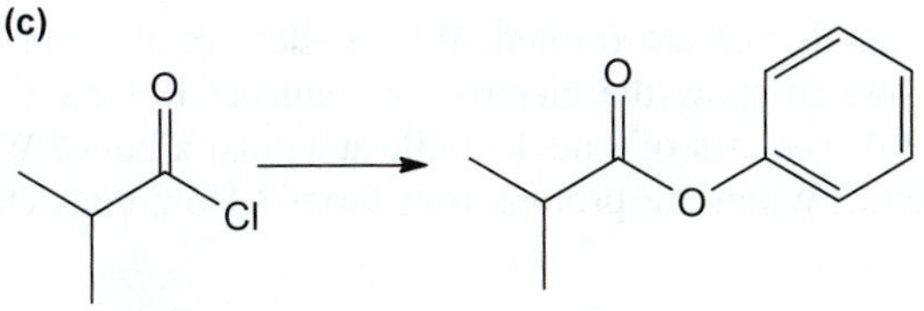

Acid chloride → Ester
= Down = Favorable

(d)

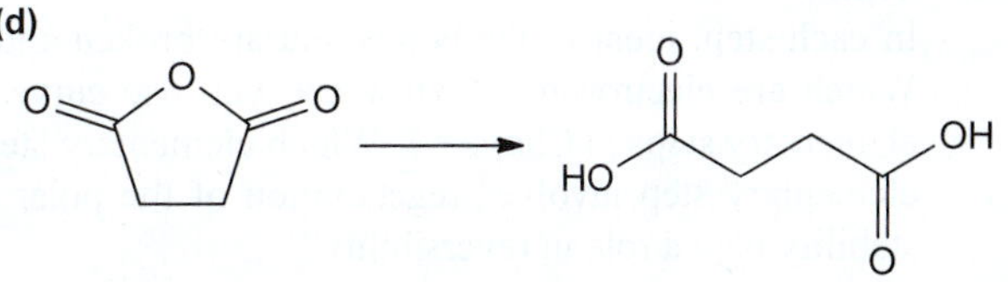

Acid anhydride → Carboxylic acid
= Down = Favorable

Your Turn 20.6

Think

Consult Table 6-1 to determine the pK_a values. What does a lower pK_a indicate about acid strength and the side of the reaction favored? What does the difference in pK_a values tell you about the extent to which that side of the reaction is favored?

Solve

CH_3CO_2H has a pK_a = 4.75 and CH_3CH_2OH has a pK_a = 16. The stronger acid (lower pK_a) is on the reactant side, so the products are favored. The difference in pK_a values is 16 − 4.75 = 11.25, so the products are favored by $10^{11.25} = 1.8 \times 10^{11}$. This shows that the reaction very heavily favors the products, making the reaction irreversible.

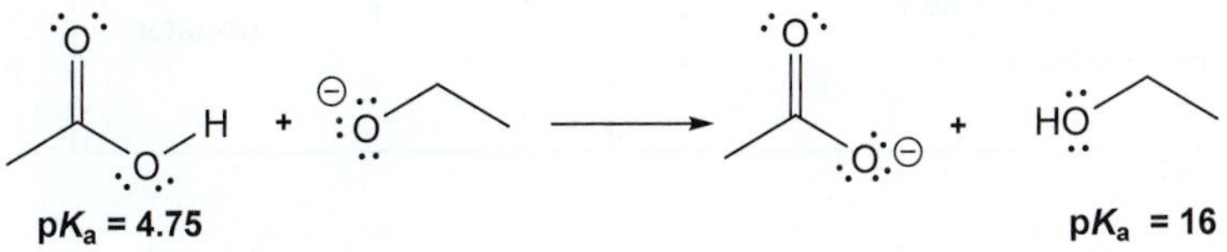

pK_a = 4.75 pK_a = 16

Your Turn 20.7

Think

Review Sections 1.10 and 1.11 to review how to draw resonance structures and resonance hybrids. Are any lone pairs adjacent to a double or triple bond? Do any of the double bonds convert into lone pairs? How can you show via a curved arrow the electrons in the lone pair moving to form a double bond? How can you show via a curved arrow the electrons in a double bond moving to form a lone pair? How can you draw the average of the two resonance structures?

Solve

The lone pairs on the negative O atom are moved to form a C=O π bond. The π bond electrons in the original C=O are moved onto the O to give the top O a −1 formal charge. The hybrid is an average of the two resonance structures. Each O atom sharing the −1 charge receives a δ^-, and each bonding region sharing the π bond receives a partial π bond.

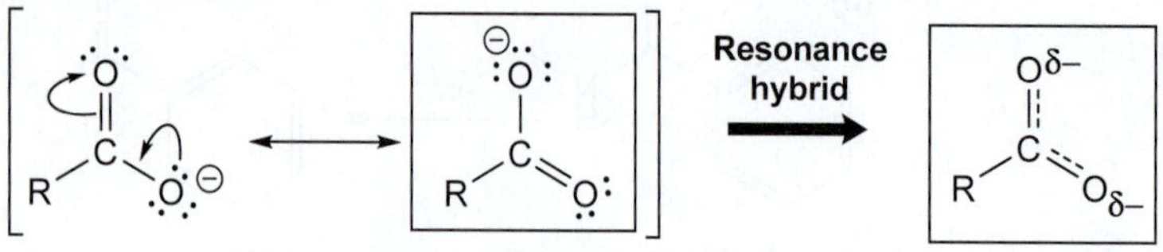

Resonance hybrid

Your Turn 20.8

Think

In each step, identify the bonds that are broken and the bonds that are formed. Which sites are electron rich? Which are electron poor? How can you use curved arrows to show the electron movement? Review the 10 elementary steps in Chapter 7. Which elementary step involves attack of a nucleophile at a polar π bond? Which elementary step involves regeneration of the polar π bond? When are protons transferred? How does charge stability play a role in reversibility?

Solve

Step 1 is nucleophilic addition, where the $^-$OH nucleophile attacks the electron-poor C of the C=O. This forces the electrons in the C=O π bond onto the O atom. The product of this step is the tetrahedral intermediate. The tetrahedral intermediate in Step 2 undergoes nucleophile elimination, where one of the lone pairs on the O of the tetrahedral intermediate regenerates the C=O and the leaving group, $(CH_3)_2N^-$, departs. In Step 3, the carboxylic acid is quickly deprotonated in the basic conditions. Step 4 is an acid workup to regenerate the carboxylic acid. Steps that significantly increase charge stability tend to be irreversible. This does not happen in the first two steps, so the first two steps are reversible. Charge stability is significantly increased in Step 3, going from a localized N^- to a resonance-delocalized O^-, so that step is irreversible.

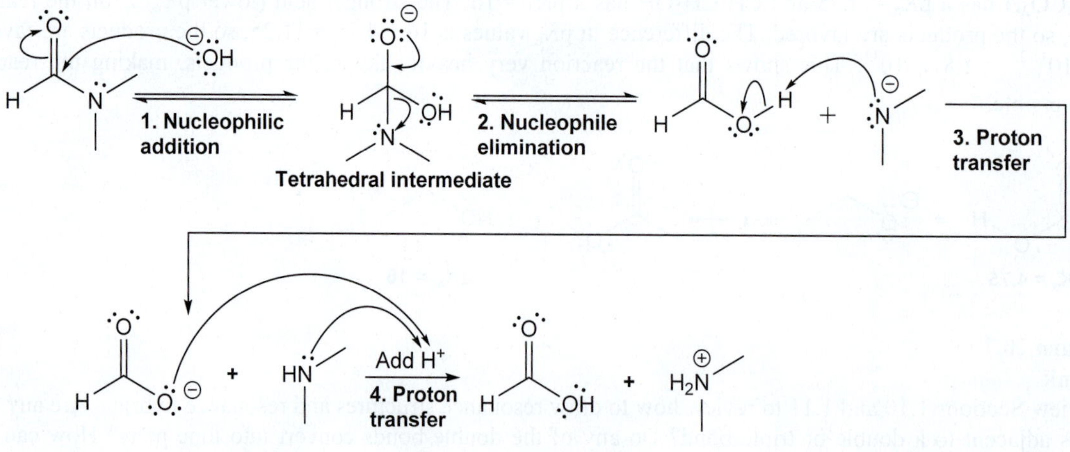

1. Nucleophilic addition

Tetrahedral intermediate

2. Nucleophile elimination

3. Proton transfer

Add H^+

4. Proton transfer

Your Turn 20.9

Think

Use Equation 20-15 as a guide. What is the organic product of the haloform reaction? What is the inorganic product when the halogen molecule changes?

Solve

The organic product of the haloform reaction is a carboxylic acid, and the inorganic product is a haloform (HCX_3). The identity of the haloform produced depends on the identity of the halogen used as a reagent.

(a) 1. Cl_2, NaOH/H_2O 2. HCl → OH + $HCCl_3$

(b) 1. I_2, NaOH/H_2O 2. HCl → OH + HCI_3

Your Turn 20.10

Think

What is the requirement for a C=O compound to undergo a haloform reaction? Why is this requirement present?

Solve

Only methyl ketones or aldehydes can undergo a haloform reaction, which includes compounds **A** and **C**. This requirement is present because CX_3^- is the suitable leaving group in the mechanism, according to Equation 20-16, Step 8. This leaving group can be produced only when three H atoms can be replaced by X on the same α carbon.

Your Turn 20.11

Think

In each step, identify the bonds that are broken and the bonds that are formed. Which sites are electron rich? Which are electron poor? How can you use curved arrows to show the electron movement? Review the 10 elementary steps in Chapter 7. Which elementary step involves attack of a nucleophile at a polar π bond? Which elementary step involves regeneration of the polar π bond? When are protons transferred?

Solve

Step 1 is nucleophilic addition, in which the H:$^-$ nucleophile attacks the electron-poor C of the C=O. This forces the electrons in the C=O π bond onto the O atom. The product of this step is the tetrahedral intermediate. The tetrahedral intermediate in Step 2 undergoes nucleophile elimination, in which one of the lone pairs on the O of the tetrahedral intermediate regenerates the C=O and the leaving group, Cl^-, departs. Step 3 is nucleophilic addition, where the H:$^-$ nucleophile attacks the electron-poor C of the C=O. This forces the electrons in the C=O π bond onto the O atom. Step 4 is an acid workup to protonate the alkoxide and form the primary alcohol product.

From $LiAlH_4$ — 1. Nucleophilic addition — Tetrahedral intermediate — 2. Nucleophile elimination — From $LiAlH_4$ — 3. Nucleophilic addition — Add H^+ — 4. Proton transfer

Your Turn 20.12

Think

In each step, identify the bonds that are broken and the bonds that are formed. Which sites are electron rich? Which are electron poor? How can you use curved arrows to show the electron movement? Review the 10 elementary steps in Chapter 7. Which elementary step involves attack of a nucleophile at a polar π bond? Which elementary step involves regeneration of the polar π bond?

Solve

Step 1 is nucleophilic addition, in which the AlH_4^- (H:⁻) nucleophile attacks the electron-poor C of the C=O at the same time the O–Al bond forms. The product of this step is the tetrahedral intermediate. The tetrahedral intermediate in Step 2 undergoes nucleophile elimination, in which the lone pair on the N generates the C=N iminium ion as the leaving group (OAl^-H_3) departs. Step 3 is nucleophilic addition, where the H:⁻ nucleophile attacks the electron-poor C of the C=N. This forces the electrons in the C=N π bond onto the N atom. The product of this step is the amine.

1. Nucleophilic addition
2. Nucleophile elimination
3. Nucleophilic addition
Tetrahedral intermediate
From $LiAlH_4$

Your Turn 20.13

Think

Use Equation 20-35 as a guide, which is the mechanism for Equation 20-34. How can CH_3Li be simplified? In each step, identify the bonds that are broken and the bonds that are formed. Which sites are electron rich? Which are electron poor? How can you use curved arrows to show the electron movement? Review the 10 elementary steps in Chapter 7. Which elementary step involves attack of a nucleophile at a polar π bond? Which elementary step involves regeneration of the polar π bond? When are proton transfers involved?

Solve

CH_3Li can be simplified to CH_3^-. Step 1 is nucleophilic addition, where the CH_3^- nucleophile attacks the electron-poor C of the C=O. This forces the electrons in the C=O π bond onto the O atom. The product of this step is the tetrahedral intermediate. The tetrahedral intermediate in Step 2 undergoes nucleophile elimination, where one of the lone pairs on the O of the tetrahedral intermediate regenerates the C=O and the leaving group, CH_3O^-, departs. Step 3 is another nucleophilic addition, where the CH_3^- nucleophile attacks the electron-poor C of the C=O. This forces the electrons in the C=O π bond onto the O atom. Step 4 is an acid workup step to protonate the alkoxide O^- and form the tertiary alcohol product.

Methyl hept-6-enoate
1. Nucleophilic addition
Tetrahedral intermediate
2. Nucleophile elimination
Two equivalents of CH_3^- have been added in.
3. Nucleophilic addition
Add H^+
4. Proton transfer
2-Methyloct-7-en-2-ol
99%

In Chapter Problems

Problem 20.1

Think

In each step, identify the bonds that are broken and the bonds that are formed. Which sites are electron rich? Which are electron poor? How can you use curved arrows to show the electron movement? Review the 10 elementary steps in Chapter 7. Which elementary step involves attack of a nucleophile at a polar π bond? Which elementary step involves regeneration of the polar π bond?

Solve

In both of these reactions, the base RO^- attacks the electrophilic carbonyl carbon via nucleophilic addition. The C=O bond is regenerated as the leaving group leaves.

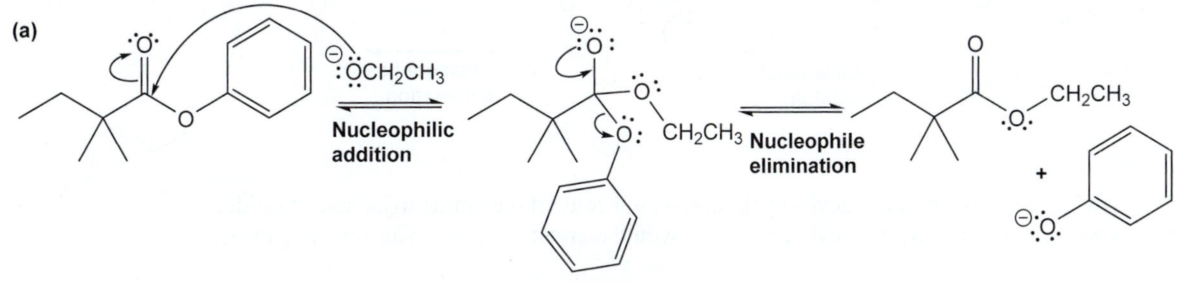

(a) $:\ddot{O}CH_2CH_3$ ⇌ Nucleophilic addition ⇌ Nucleophile elimination; CH_2CH_3; +

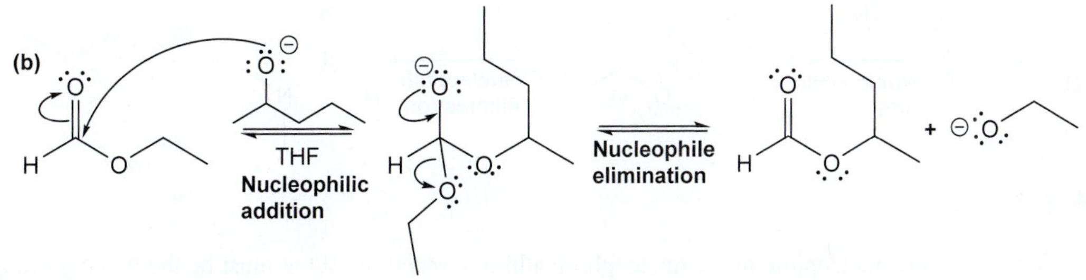

(b) THF Nucleophilic addition ⇌ Nucleophile elimination; H; +

Problem 20.3

Think

What is the nucleophile? What is the leaving group? On what rungs of the stability ladder do you find the reactant and product acid derivatives? Is the potential acyl substitution reaction energetically favorable?

Solve

The acyl substitution reactions in **(a)**, **(b)**, and **(f)** are favored because the product acid derivative is lower on the stability ladder than the starting one. The acyl substitutions in **(d)** and **(e)** are not favorable because they represent going up the stability ladder. Finally, **(c)** converts one carboxylic acid anhydride into another and is reversible.

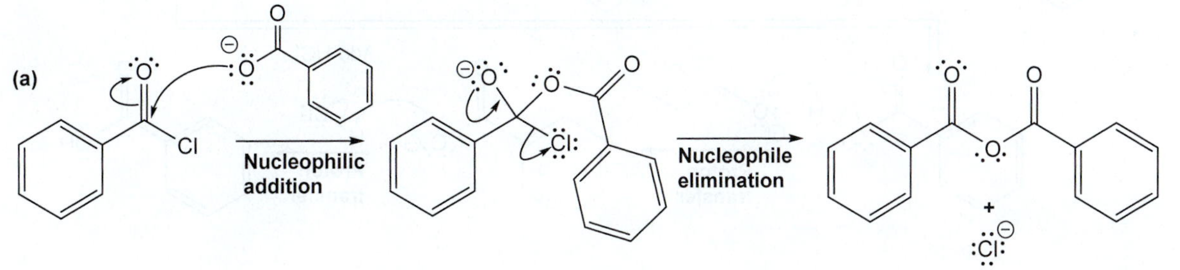

(a) Cl → Nucleophilic addition → Nucleophile elimination; + $:\ddot{Cl}:^-$

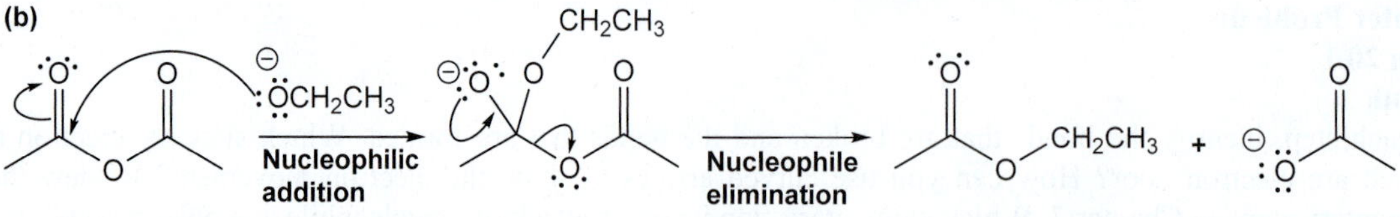

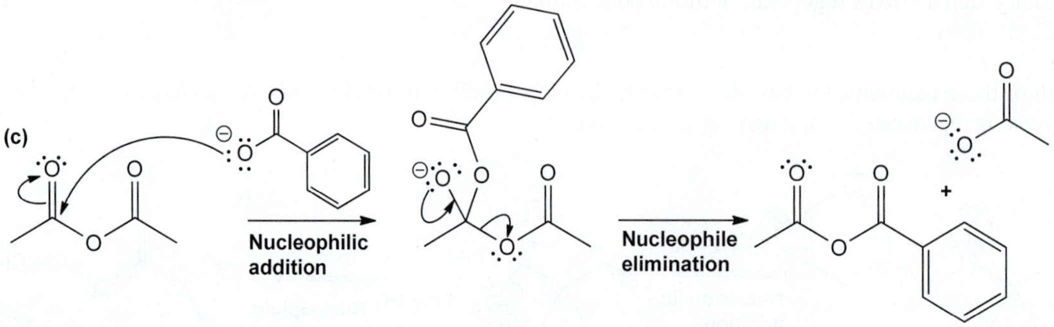

(b)

Nucleophilic addition

Nucleophile elimination

(c)

Nucleophilic addition

Nucleophile elimination

(d) No reaction, because acyl substitution would convert the amide to an acid chloride.
(e) No reaction, because acyl substitution would convert the ester to an acid anhydride.

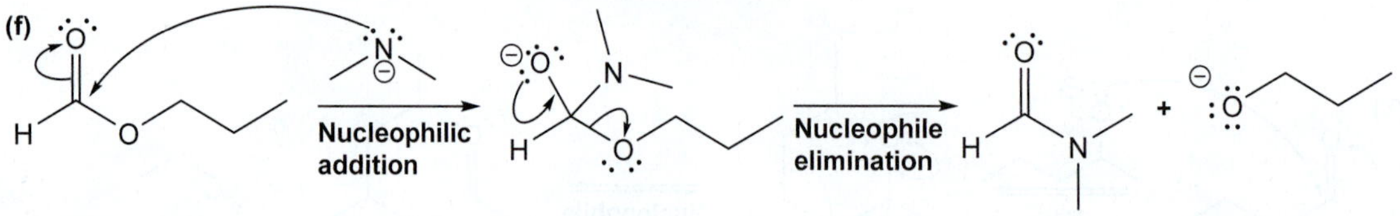

(f)

Nucleophilic addition

Nucleophile elimination

Problem 20.4

Think

What is the identity of the nucleophile in the nucleophilic addition reaction? What must be the leaving group in the nucleophile elimination step for a carboxylic acid to form as the final product? Which step is irreversible? Why is an acid workup necessary?

Solve

These are saponification reactions. After nucleophilic addition–elimination to produce an initial carboxylic acid, the carboxylic acid is deprotonated irreversibly. Acid workup adds the proton back.

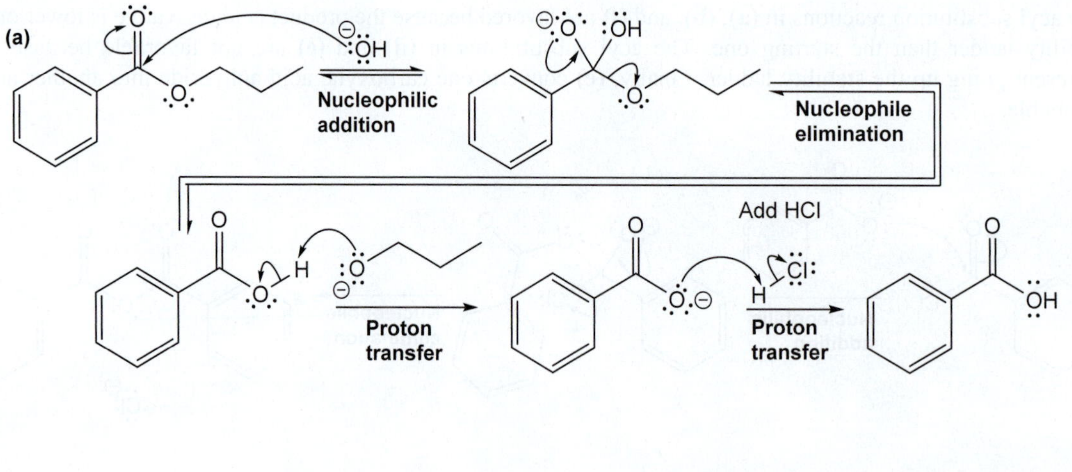

(a)

Nucleophilic addition

Nucleophile elimination

Proton transfer

Add HCl

Proton transfer

(b)

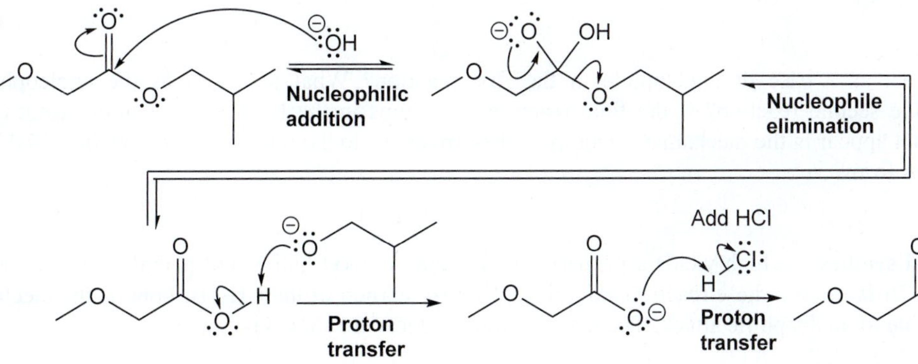

Nucleophilic addition

Nucleophile elimination

Proton transfer

Add HCl

Proton transfer

Problem 20.5

Think

What is the nucleophile? What is the leaving group? Is the potential acyl substitution reaction energetically favorable? Which step is irreversible? Why is an acid workup necessary?

Solve

The HO^- is the nucleophile and attacks the electron-poor C of the amide. After nucleophilic addition–elimination to produce an initial carboxylic acid, the carboxylic acid is deprotonated irreversibly. Acid workup adds the proton back.

(a)

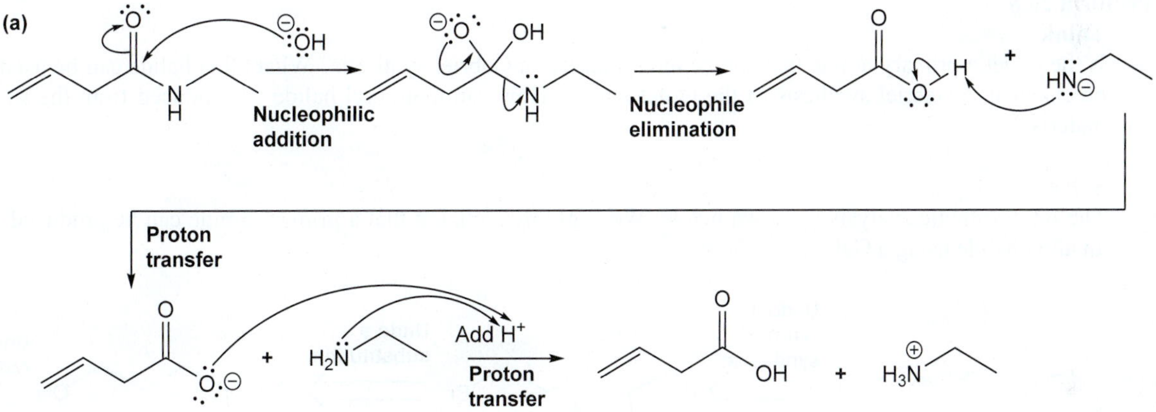

Nucleophilic addition

Nucleophile elimination

Proton transfer

Add H^+

Proton transfer

(b)

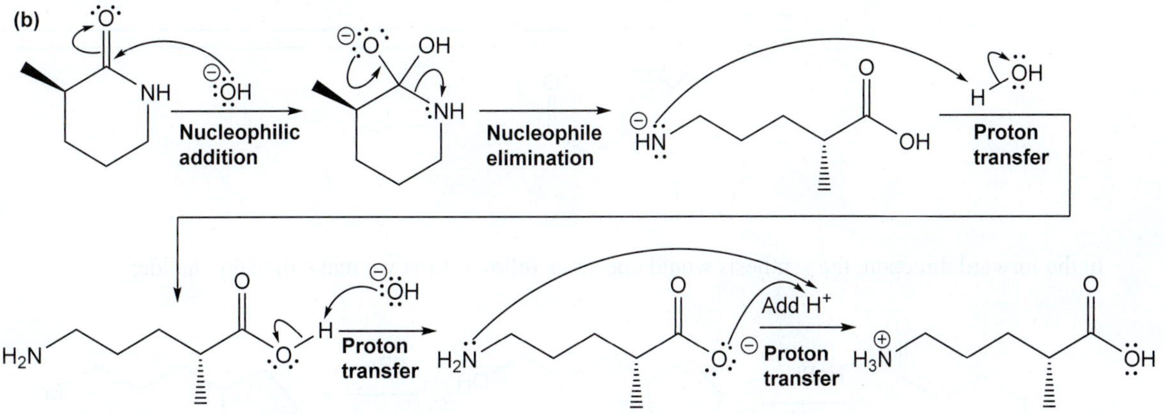

Nucleophilic addition

Nucleophile elimination

Proton transfer

Proton transfer

Add H^+

Proton transfer

Problem 20.6

Think

Does the HO^- act as a base or nucleophile in the first reaction? What then acts as the nucleophile and electrophile in the second reaction? If the final reaction conditions in the final step are basic, what types of species should not appear in the mechanism? Compare these reagents to the ones given in Equation 20-12. What functional groups form?

Solve

This is a Gabriel synthesis, which yields a primary amine from an alkyl halide and phthalate as a by-product. The aromatic C–Br is inert to these reaction conditions. The sp^3 carbon of the CH_2–Br bond is the electrophilic C atom susceptible to nucleophilic attack. The reaction and products are shown below.

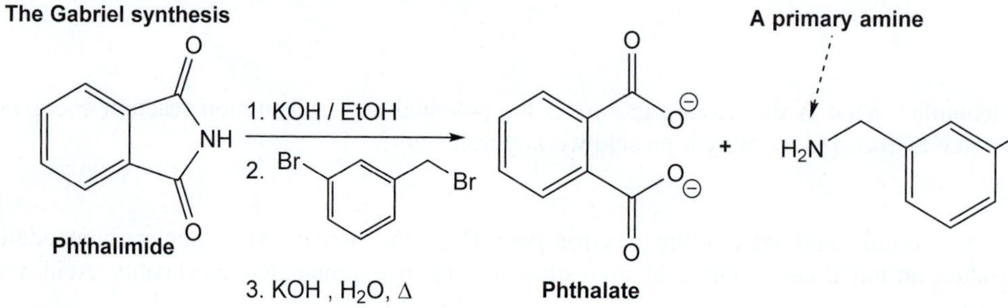

The Gabriel synthesis

Phthalimide — 1. KOH / EtOH; 2. Br–CH₂–C₆H₄–CH₂–Br; 3. KOH, H_2O, Δ → Phthalate + H_2N–CH₂–C₆H₄–CH₂–Br (**A primary amine**)

Problem 20.8

Think

Is the target a primary amine that can be the product of a Gabriel synthesis? What alkyl halide can be used as a precursor in a Gabriel synthesis of the target amine? How can that alkyl halide be produced from the starting material?

Solve

The retrosynthetic analysis is shown below. We begin by realizing that a primary amine can be produced from an alkyl halide using a Gabriel synthesis.

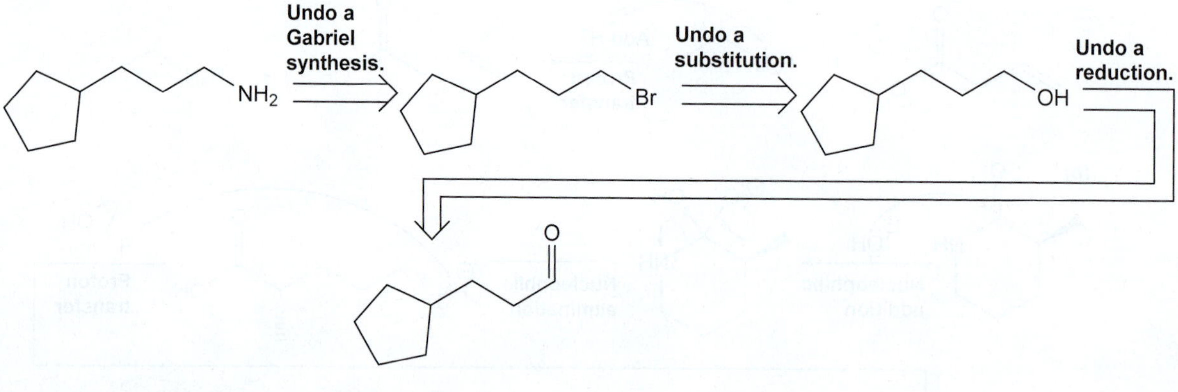

NH_2 ⟹ (**Undo a Gabriel synthesis.**) Br ⟹ (**Undo a substitution.**) OH ⟹ (**Undo a reduction.**) aldehyde (C=O)

In the forward direction, the synthesis would appear as follows. First we make the alkyl halide:

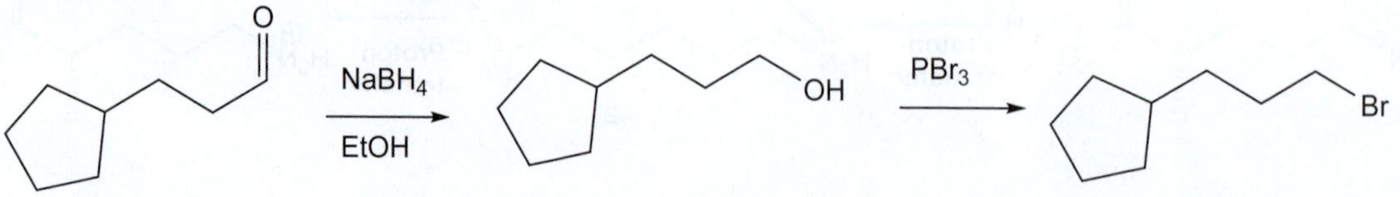

Aldehyde —NaBH₄ / EtOH→ alcohol (OH) —PBr₃→ alkyl bromide (Br)

Then we incorporate the alkyl halide into the Gabriel synthesis:

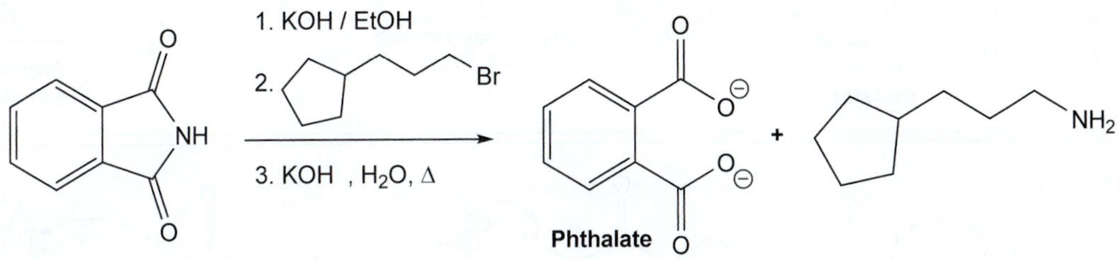

1. KOH / EtOH
2. Br
3. KOH , H_2O, Δ

NH

Phthalate

NH_2

Problem 20.9

Think

Review Section 10.4. How are the α C–H atoms replaced by the halogen? Once the $-CX_3$ group is formed, is it a suitable leaving group? What are the nucleophile and leaving group in the nucleophilic addition–elimination mechanism? What steps are irreversible? What does the acid workup accomplish?

Solve

In both **(a)** and **(b)**, the α C–H bonds are replaced by X (i.e., Br or I) through use of HO^- acting as a base to deprotonate the α C–H and form the carbanion. The C^- then acts as a nucleophile to attack the X–X via an S_N2 mechanism. This same reaction occurs two more times, until the CX_3 is formed at the α carbon. The resulting trihalomethylketone then undergoes nucleophilic addition by attack of HO^- to form the tetrahedral intermediate, which subsequently eliminates CX_3^-. The carboxylic acid that is formed is then irreversibly deprotonated, and acid workup adds that proton back.

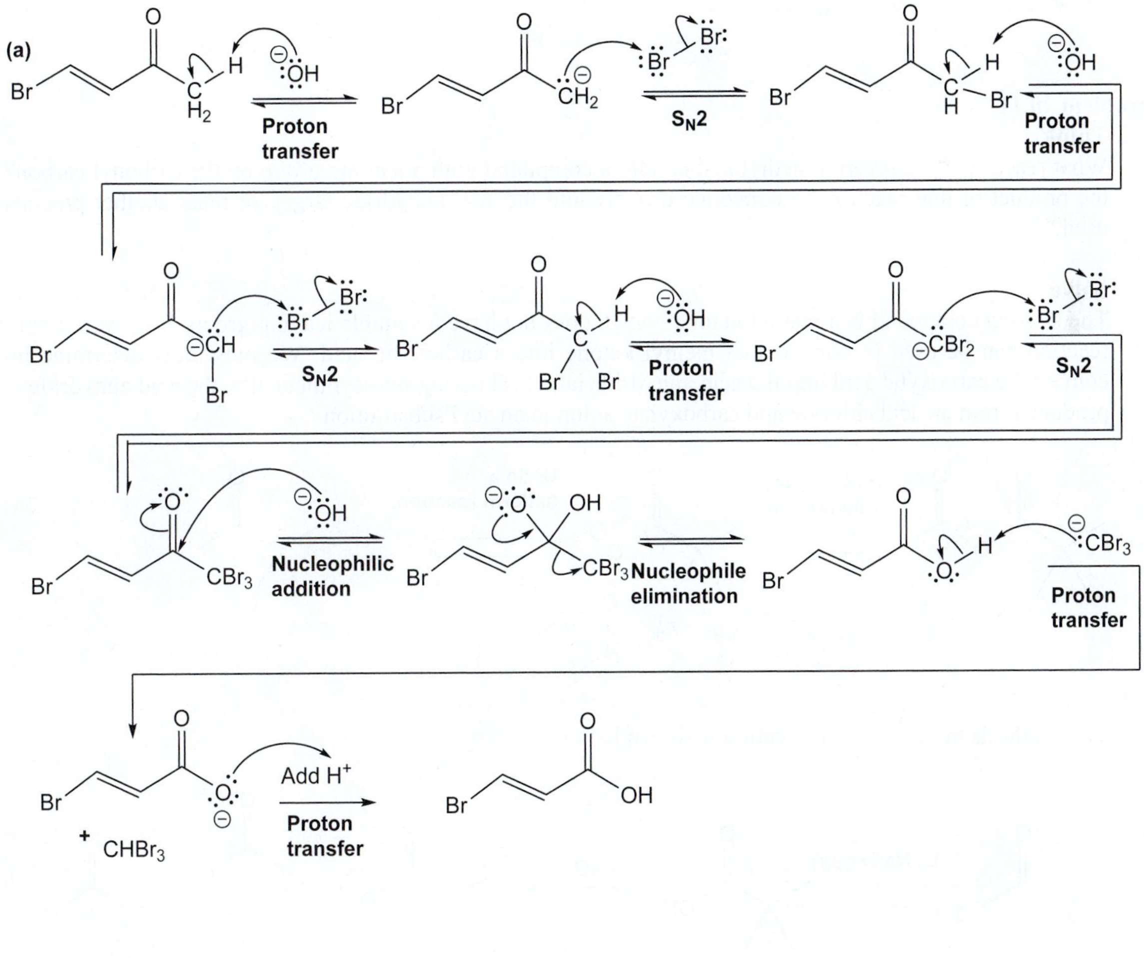

(a)

Proton transfer

S_N2

Proton transfer

S_N2

Proton transfer

S_N2

Nucleophilic addition

Nucleophile elimination

Proton transfer

Add H^+

Proton transfer

+ $CHBr_3$

(b)

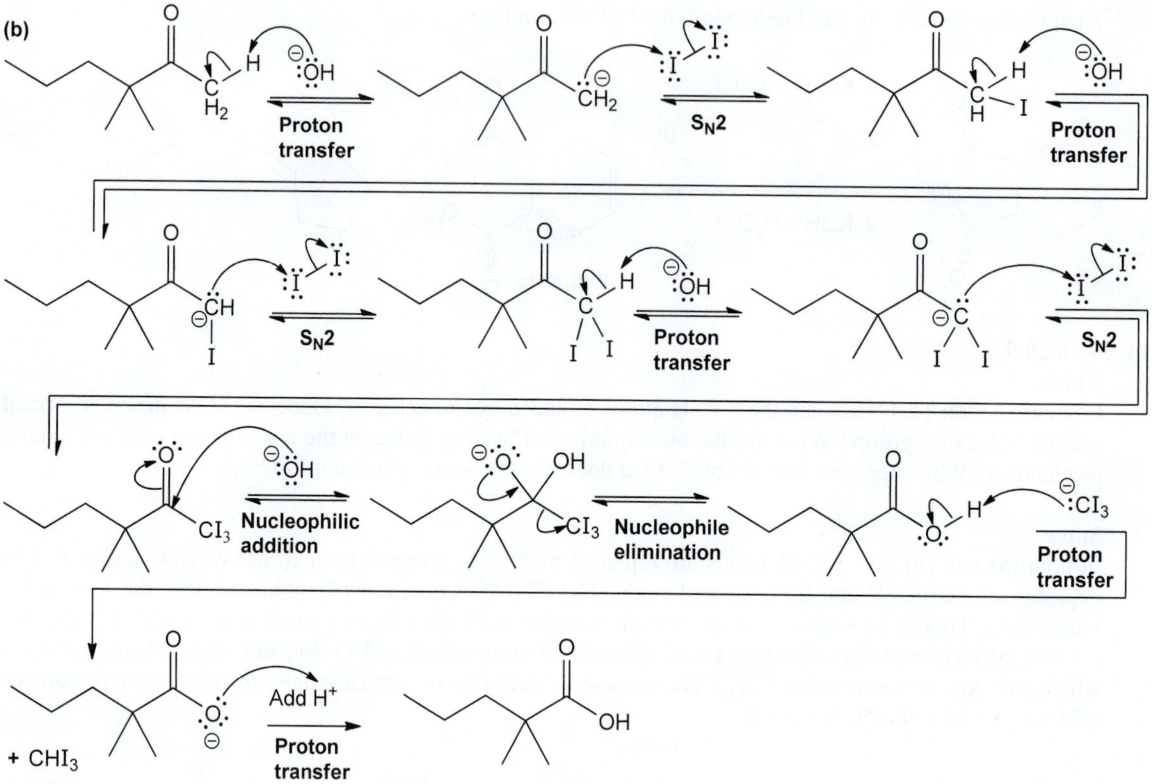

Proton transfer

S_N2

Proton transfer

S_N2

Proton transfer

S_N2

Nucleophilic addition

Nucleophile elimination

Proton transfer

+ CHI_3

Add H^+

Proton transfer

Problem 20.11

Think

What reaction can convert a methyl ketone into a compound with a leaving group on the carbonyl carbon? Can the product of that reaction be converted directly into the acid anhydride target, or must another precursor be used?

Solve

The starting compound is a methyl ketone, which does not have a suitable leaving group. However, a haloform reaction can be used to convert that methyl ketone into a carboxylic acid. We must now determine how to convert the carboxylic acid into the acid anhydride target. Thinking retrosynthetically, the acid anhydride can be produced from an acid chloride and carboxylate anion in an acyl substitution.

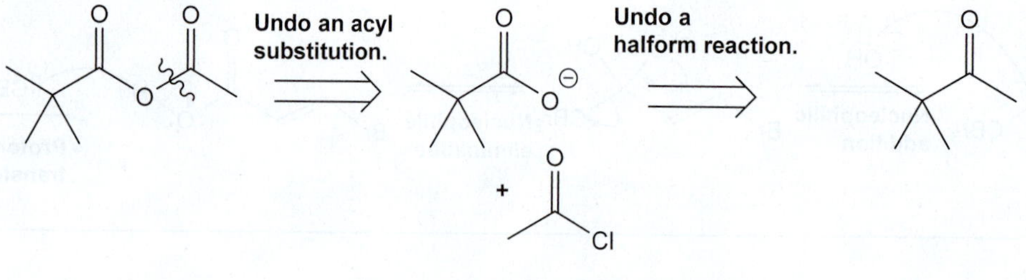

Undo an acyl substitution.

Undo a halform reaction.

The synthesis in the forward direction is shown below:

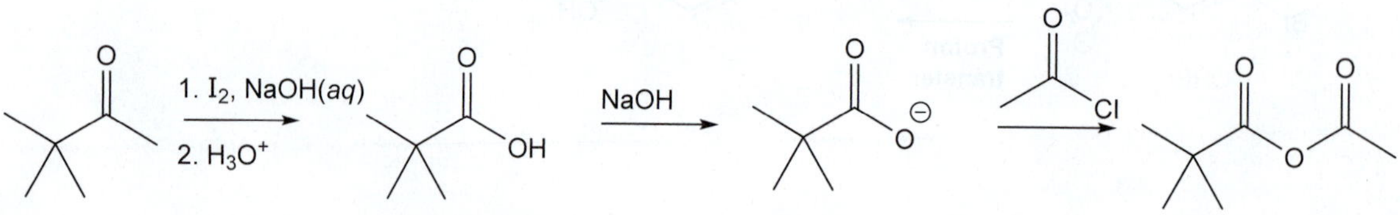

1. I_2, NaOH(*aq*)
2. H_3O^+

NaOH

Problem 20.12

Think

What is the index of hydrogen deficiency (IHD)? On the basis of this number and the molecular formula, what functional group is likely giving the IR signal at 1708 cm^{-1}? If there are 12 H atoms in the formula and the ^1H NMR gives rise to only two signals, is there symmetry in the structure? What functional group reacts with I_2 to give the bright yellow solid?

Solve

Notice that IHD = 1, meaning the presence of either a double bond or a ring. The IR peak is indicative of a C=O double bond, which accounts for the IHD.

With a total of 12 H atoms in a 3:1 ratio, one NMR signal must account for nine equivalent H atoms, whereas the other must account for three equivalent H atoms. This is accomplished by a *t*-butyl group, $C(CH_3)_3$, and a methyl group, CH_3. The CH_3 must be adjacent to the C=O bond to give a methyl ketone, which will react in an iodoform reaction to produce a bright yellow solid precipitate. The methyl ketone also is consistent with the signal at 2.2 ppm, giving the following molecule:

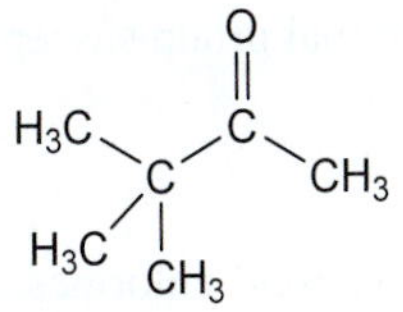

The reaction proceeds as follows. The yellow solid is the HCI_3.

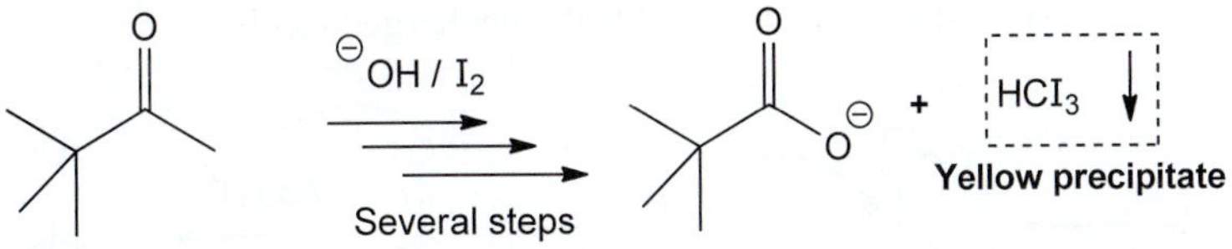

Problem 20.13

Think

Is the reaction a nucleophilic addition–elimination mechanism? What acts as the nucleophile in each case? To what functional group does the hydride reduce the acid chloride? Is this functional group susceptible to further reaction?

Solve

Steps 1 and 2 make up the nucleophilic addition–elimination mechanism, which produces an aldehyde intermediate. Under these reaction conditions, the aldehyde reacts rapidly with another equivalent of hydride (Section 17.3) to produce the alkoxide anion. An acid workup step yields the uncharged alcohol. See the mechanisms below **(a)** and on the next page **(b)**.

(a)

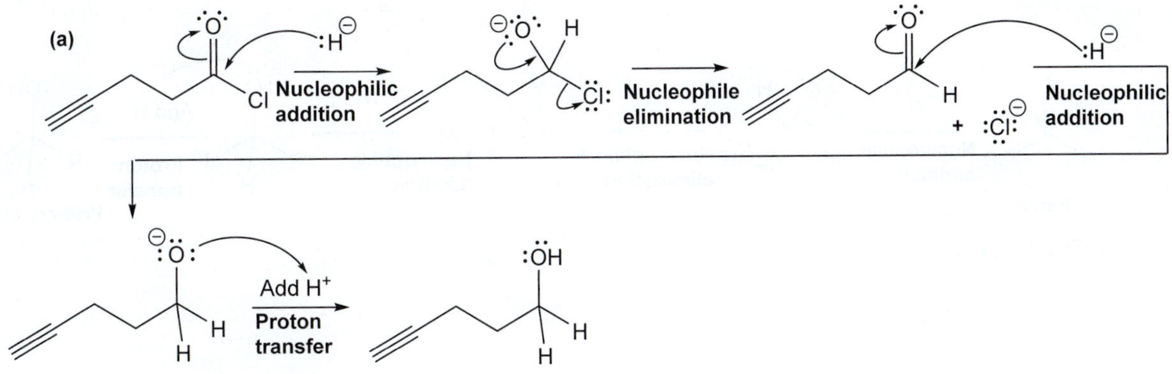

(b) Ph–CH₂CH₂–C(=O)Cl + :H⁻ —Nucleophilic addition→ tetrahedral alkoxide —Nucleophile elimination→ Ph–CH₂CH₂–CHO + :Cl⁻ —:H⁻, Nucleophilic addition→ alkoxide —Add H⁺, Proton transfer→ Ph–CH₂CH₂–CH₂OH

Problem 20.14

Think

Refer to Table 20-1 to see which functional groups are reduced to primary alcohols. Is the reaction a nucleophilic addition–elimination mechanism? What acts as the nucleophile in each case? To what functional group does the hydride initially reduce the carboxylic acid derivative given? Is this functional group susceptible to further reaction?

Solve

Acid chlorides, acid anhydrides, and esters react with $NaBH_4$ to form a primary alcohol. Acid chlorides, acid anhydrides, esters, and carboxylic acids react with $LiAlH_4$ to form a primary alcohol.

Steps 1 and 2 make up the nucleophilic addition–elimination mechanism, which produces an aldehyde intermediate. Under these reaction conditions, the aldehyde reacts rapidly with another equivalent of hydride (Section 17.3) to produce the alkoxide anion. An acid workup step yields the uncharged alcohol.

Nucleophilic addition; Nucleophile elimination; + :Cl⁻; Nucleophilic addition; Add H⁺; Proton transfer

Acid chloride

:H⁻ = $LiAlH_4$ or $NaBH_4$

Primary alcohol

Nucleophilic addition; Nucleophile elimination; +; Nucleophilic addition; Add H⁺; Proton transfer

Acid anhydride

:H⁻ = $LiAlH_4$ or $NaBH_4$

Primary alcohol

Nucleophilic addition; Nucleophile elimination; + :OR⁻; Nucleophilic addition; Add H⁺; Proton transfer

Ester

:H⁻ = $LiAlH_4$ or $NaBH_4$

Primary alcohol

Carboxylic acids have acidic hydrogen atoms. This functional group reacts with a hydride first via proton transfer. The mechanism is shown below. Carboxylic acids produce the primary alcohol when reacted with $LiAlH_4$, but not with $NaBH_4$.

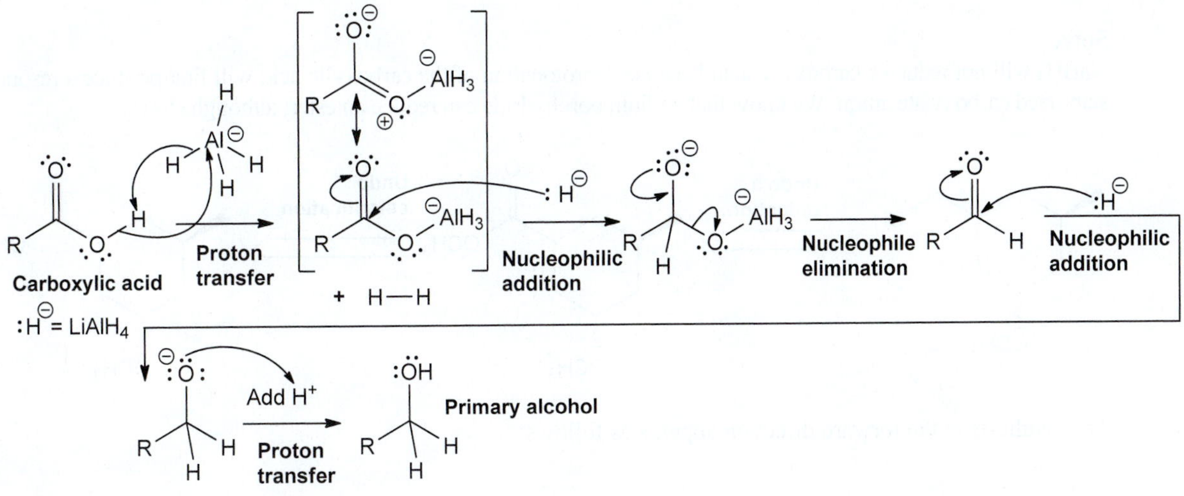

Carboxylic acid → Proton transfer → + H—H → Nucleophilic addition → Nucleophile elimination → Nucleophilic addition → Add H⁺, Proton transfer → Primary alcohol

$:H^{\ominus} = LiAlH_4$

Problem 20.16

Think

Which functional groups are present in each compound? Will they both react with $NaBH_4$? Are the functional groups equally reactive? Which one is more likely to react with $NaBH_4$ in a nucleophilic addition–elimination reaction?

Solve

Sodium borohydride is a selective reducing agent.

(a) $NaBH_4$ will reduce the ketone C=O over the ester. The ketone C=O does not have a suitable leaving group and will undergo only nucleophilic addition followed by proton transfer.

Nucleophilic addition → Add H⁺, Proton transfer

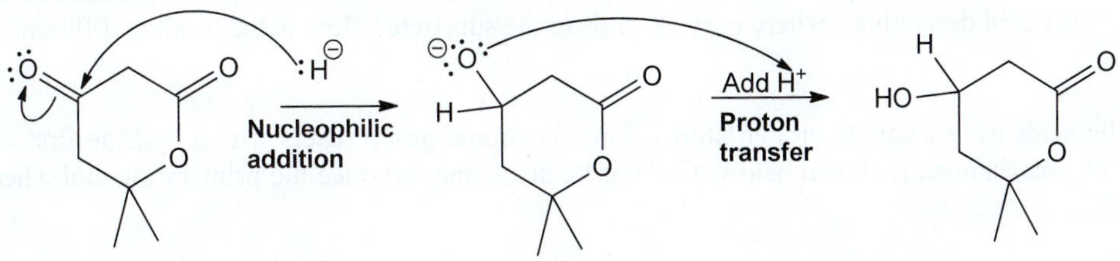

(b) $NaBH_4$ will reduce the aldehyde C=O over the amide. The aldehyde C=O does not have a suitable leaving group and will undergo only nucleophilic addition followed by proton transfer.

Nucleophilic addition → Add H⁺, Proton transfer

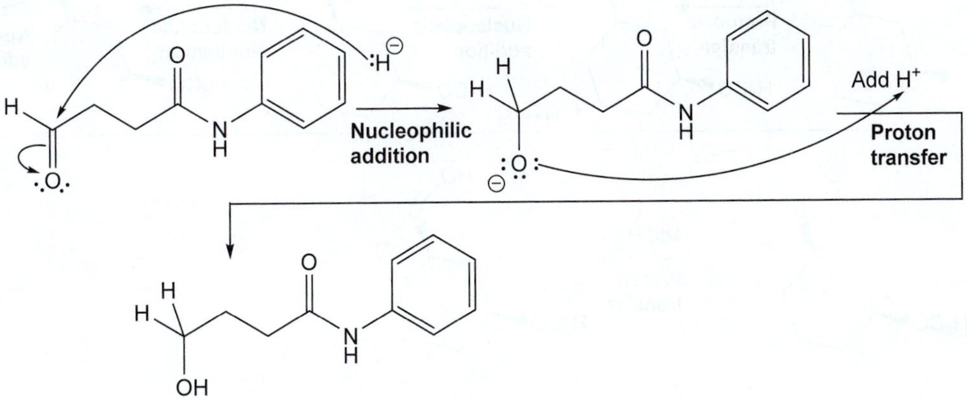

Problem 20.17

Think

Can a carboxylic acid be reduced by $NaBH_4$? If not, can a carboxylic acid be converted to another acid derivative that will be reduced by $NaBH_4$?

Solve

$NaBH_4$ will not reduce a carboxylic acid because deprotonation of the carboxylic acid will first produce a resonance-stabilized carboxylate anion. We know that sodium borohydride can reduce an ester, although slowly.

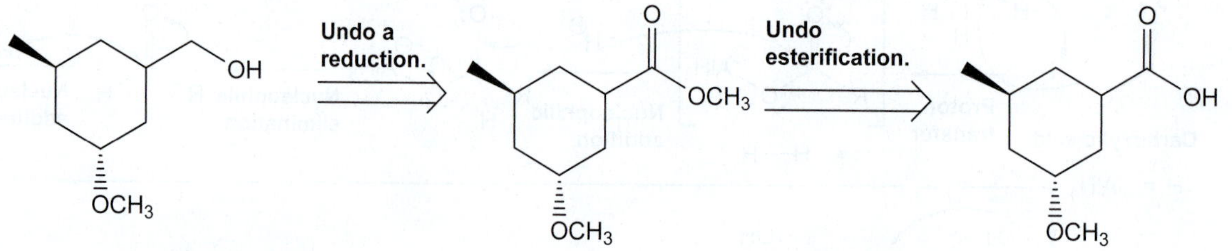

Undo a reduction.

Undo esterification.

The synthesis in the forward direction appears as follows:

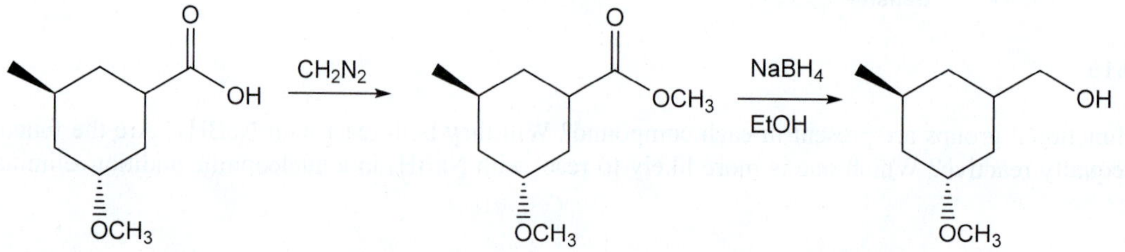

CH_2N_2

$NaBH_4$ / EtOH

Problem 20.18

Think

How is the mechanism for the reduction of a carboxylic acid different from that for other acid derivatives in Table 20-1? How does the acidic proton affect the mechanism? How is the nucleophilic addition of the amide different than other acid derivatives, where only H^- adds to the substrate? How is the product different?

Solve

(a) Carboxylic acids have acidic hydrogen atoms. This functional group reacts with a hydride first via proton transfer. The mechanism is shown below. Carboxylic acids only produce the primary alcohol when reacted with $LiAlH_4$.

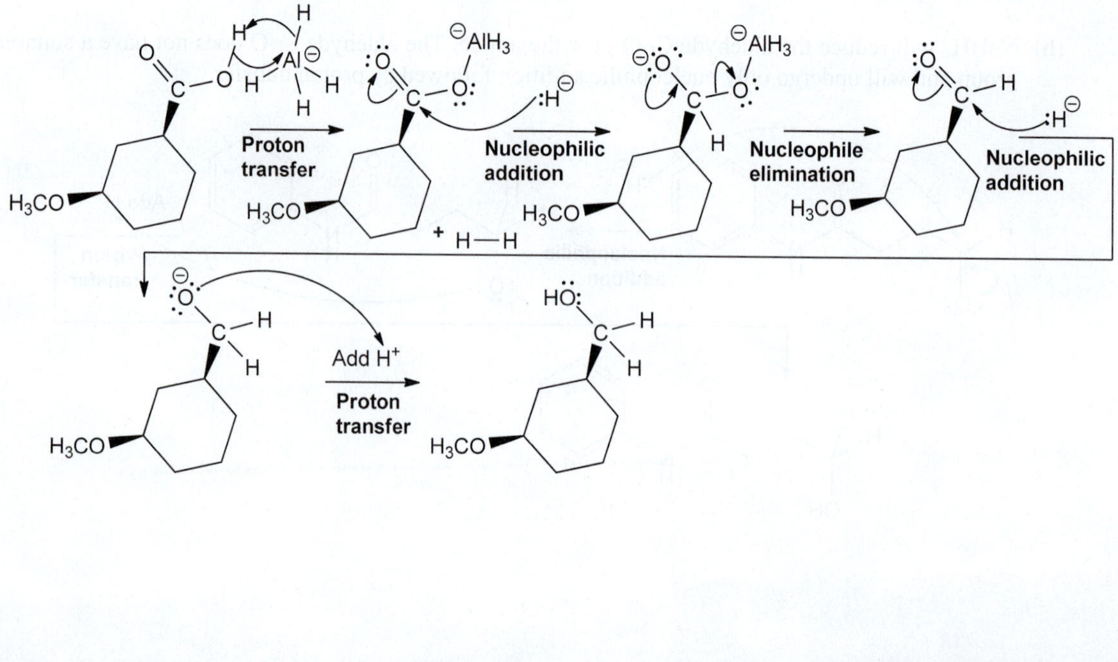

Proton transfer

Nucleophilic addition

Nucleophile elimination

Nucleophilic addition

Add H^+ / Proton transfer

(b) Amides react with $LiAlH_4$ to form amines, not alcohols. The $^{-}OAlH_3$ is the leaving group rather than the R_2N^{-}.

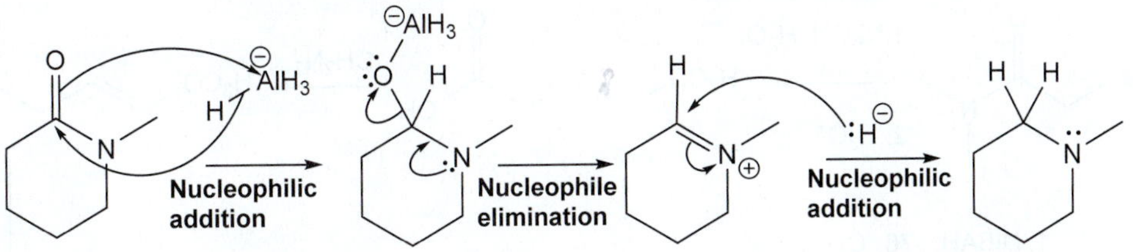

Nucleophilic addition → Nucleophile elimination → Nucleophilic addition

Problem 20.20

Think

Which of the groups in the starting compound must be reduced? Which should remain unchanged? Will the desired reduction also reduce the latter functional group? How can a protecting group be used to prevent the undesired reductions?

Solve

The goal is to reduce the amide to an amine, which can be done by using $LiAlH_4$. However, that would also reduce the ketone C=O. We can protect the ketone C=O first, then carry out the reduction as planned, and finally deprotect the ketone C=O.

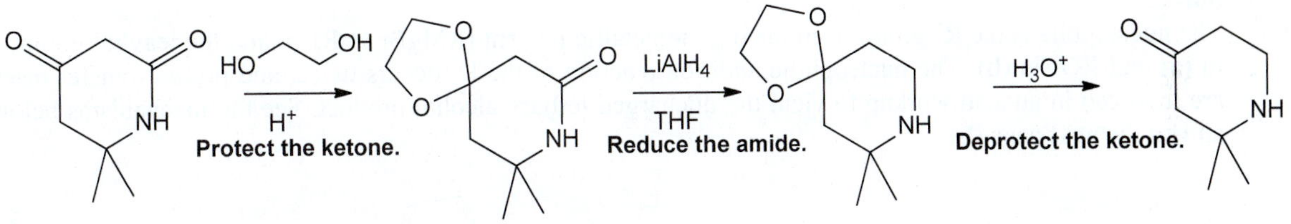

HO–CH₂CH₂–OH, H^+: Protect the ketone. → $LiAlH_4$, THF: Reduce the amide. → H_3O^+: Deprotect the ketone.

Problem 20.22

Think

What reactions do we know that can convert an acid derivative into an aldehyde? Can this be accomplished directly from an amide?

Solve

An aldehyde can be produced from an ester using diisobutylaluminum hydride (DIBAH). Thus, we must convert the initial amide into a carboxylic acid and then into an ester. We have not yet learned a way to convert a carboxylic acid into an acid chloride directly, but we have seen that diazomethane can convert the carboxylic acid into a methyl ester.

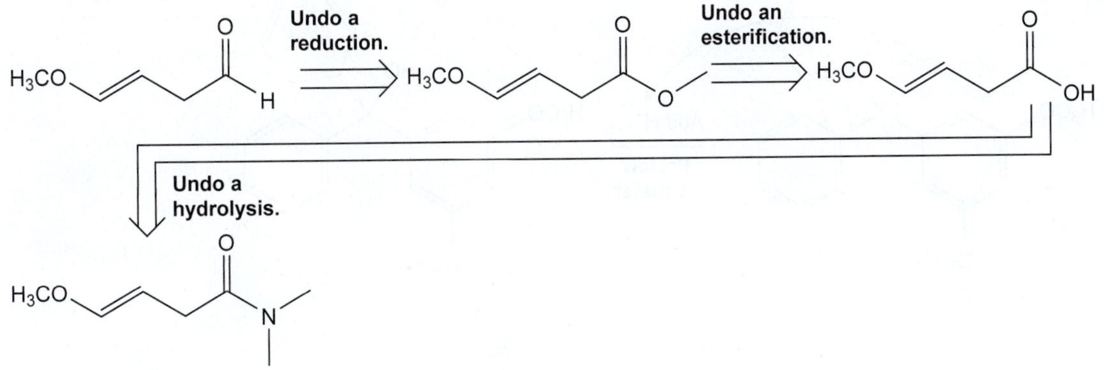

Undo a reduction. ⟹ Undo an esterification. ⟹ Undo a hydrolysis.

The synthesis in the forward direction is given below:

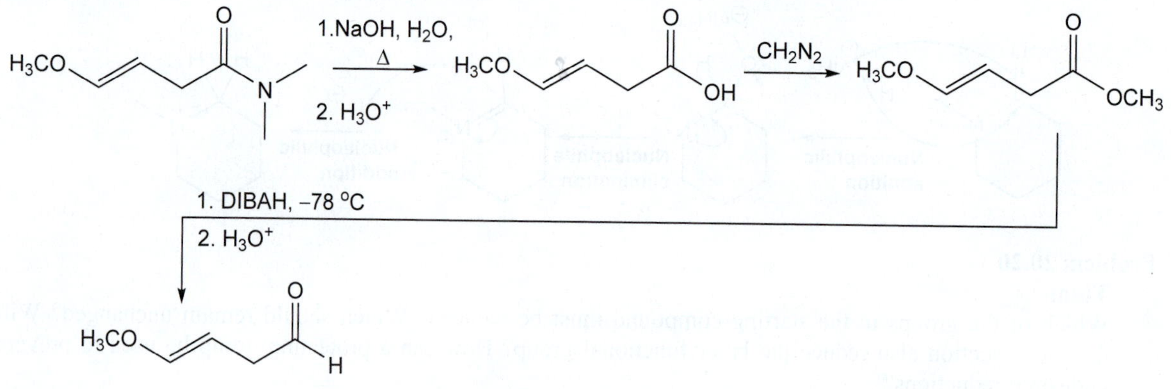

H3CO … 1.NaOH, H_2O, Δ 2. H_3O^+ … H3CO … OH … CH_2N_2 … H3CO … OCH_3

1. DIBAH, –78 °C
2. H_3O^+

H3CO … H

Problem 20.23

Think

What are the nucleophile and leaving group in the nucleophilic addition–elimination mechanism? How many times does the nucleophilic addition reaction occur? Are proton transfer reactions involved?

Solve

The nucleophile is the R^- group from the organometallic reagent (RMgBr or RLi), and the leaving group is Cl^- in **(a)** and RO^- in **(b)**. The nucleophilic addition reaction of the R^- occurs twice, and proton transfer reactions are involved in an acid workup to yield the uncharged tertiary alcohol product. See the mechanisms below **(a)** and on the next page **(b)**.

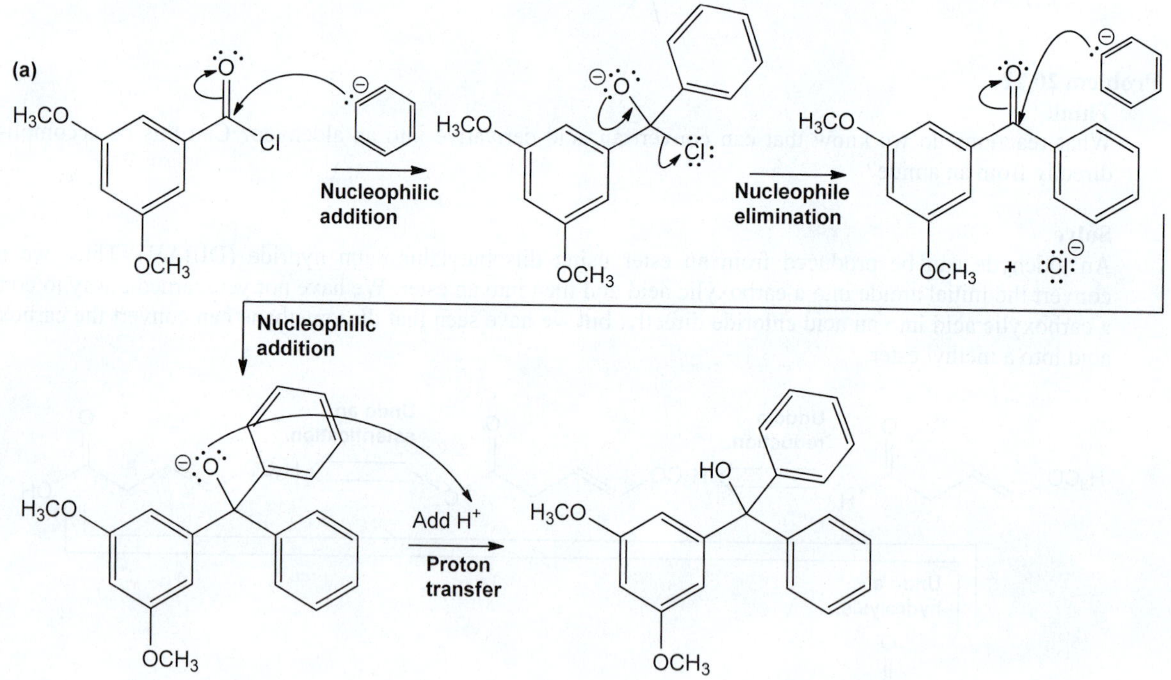

(a)

H3CO … Cl … OCH3 → **Nucleophilic addition** → H3CO … :Cl: … OCH3 → **Nucleophile elimination** → H3CO … OCH3 + :Cl:

Nucleophilic addition

H3CO … OCH3 → **Add H⁺** **Proton transfer** → H3CO … HO … OCH3

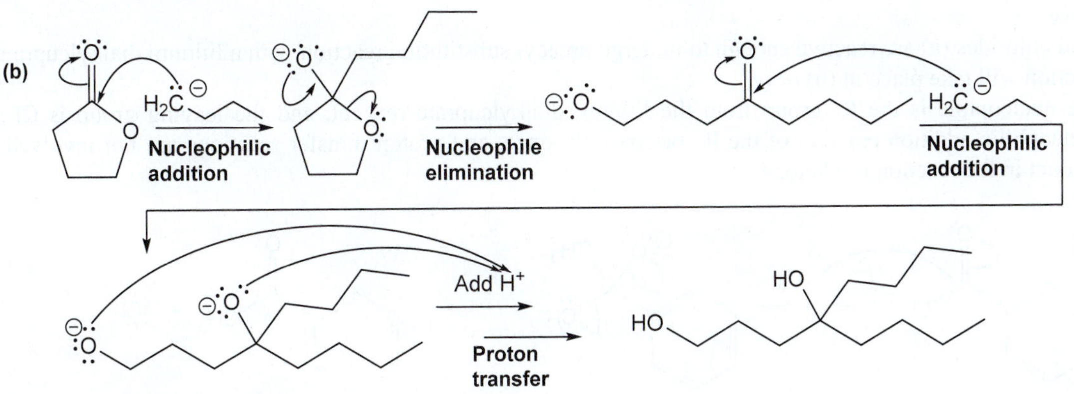

(b)

Problem 20.24

Think

What reactions do you know that can convert a C=O into a tertiary alcohol? What should the nucleophile be? How many times does the nucleophile need to add? Which acid derivatives accomplish this transformation? Can a carboxylic acid accomplish this transformation, or would a side reaction occur? How can you convert the carboxylic acid into a functional group that will react with an organometallic reagent?

Solve

Notice that the tertiary alcohol target has two identical alkyl groups attached to the original carbonyl carbon. This is the expected product of a Grignard reaction involving an ester. So we can first convert the carboxylic acid to an ester. The ketone intermediate in brackets does not last long enough to be isolated; it continues to react with more Grignard reagent.

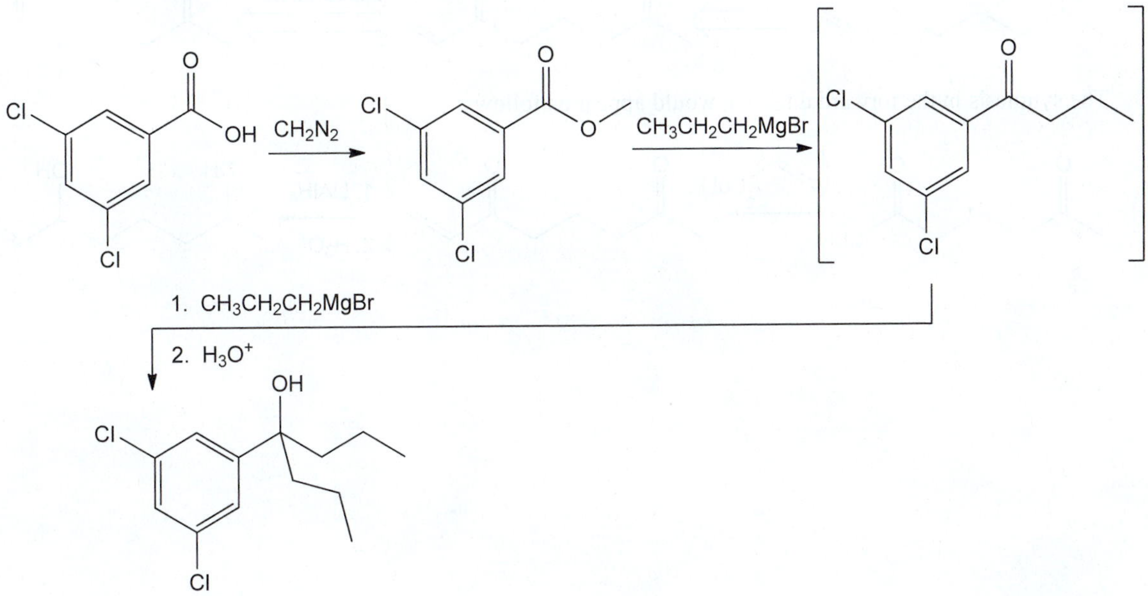

Problem 20.25

Think

Which acid derivative is reactive enough to react with the lithium dialkylcuprate reagent? What is the nucleophile and leaving group in the nucleophilic addition–elimination mechanism? How many times does the nucleophilic addition reaction occur? Are proton transfer reactions involved?

Solve

Acid chlorides **(a)** are reactive enough to undergo an acyl substitution reaction with a lithium dialkylcuprate. No reaction will take place in **(b)** or **(c)**.

The nucleophile is the R^- group from the lithium dialkylcuprate reagent, and the leaving group is Cl^-. The nucleophilic addition reaction of the R^- occurs only once, and proton transfer reactions are not involved. The product in the reaction is a ketone.

(a)

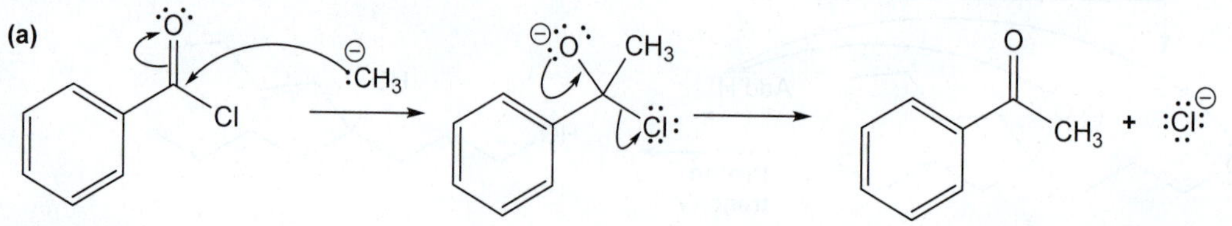

CH_3 CH_3 Cl Cl CH_3 + $:\ddot{Cl}:^-$

Problem 20.27

Think

What reaction converts a C=O into a CH–OH? What precursor can form a ketone from an acid chloride? What reagent must be used to stop at the ketone? Which carbonyl-containing functional group in the starting material must gain an additional carbon–carbon bond? What reagent can be used to selectively react with that functional group?

Solve

The target can be the product of reducing both carbonyls of a diketone. In turn, the diketone can be produced from the starting material using a lithium dialkylcuprate.

OH OH **Undo a reduction.** O O **Undo an acyl substitution.** O O Cl

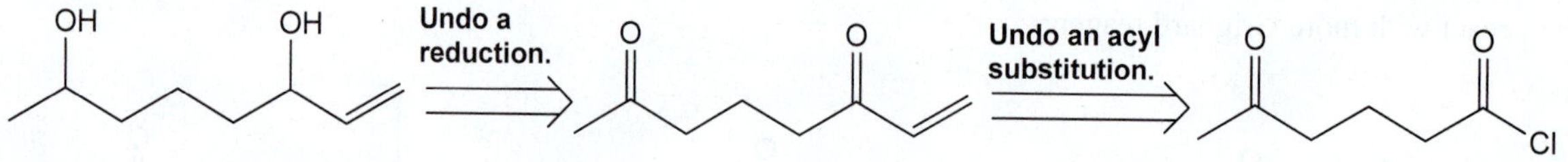

The synthesis in the forward direction would appear as follows:

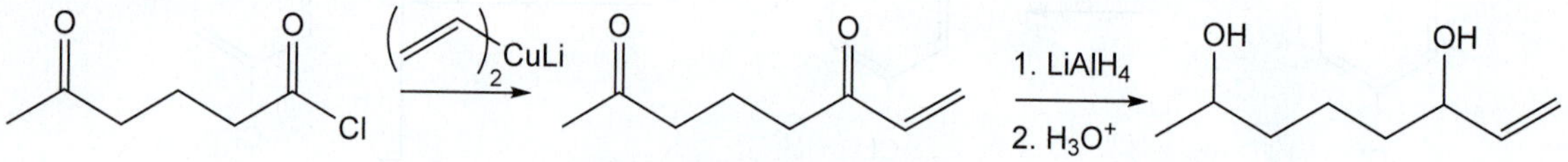

O O Cl $(\text{CH}_2{=}\text{CH})_2$CuLi O O 1. $LiAlH_4$ 2. H_3O^+ OH OH

End of Chapter Problems

Sections 20.1 and 20.3 Transesterification and Saponification

Problem 20.28

Think

What is the identity of the nucleophile in the nucleophilic addition reaction? What is the leaving group in the nucleophile elimination? Will nucleophilic addition–elimination produce a derivative lower on the stability ladder? How many times does the nucleophilic addition reaction occur? Are proton transfer reactions involved?

Solve

The reaction in **(a)** is a saponification, where reversible nucleophilic addition–elimination produces a carboxylic acid that is then irreversibly deprotonated. The reactions in **(b)** and **(c)** are reversible transesterifications. Reaction **(c)** favors the reactant side of the equilibrium because the negative charge is resonance delocalized. The mechanisms are given below:

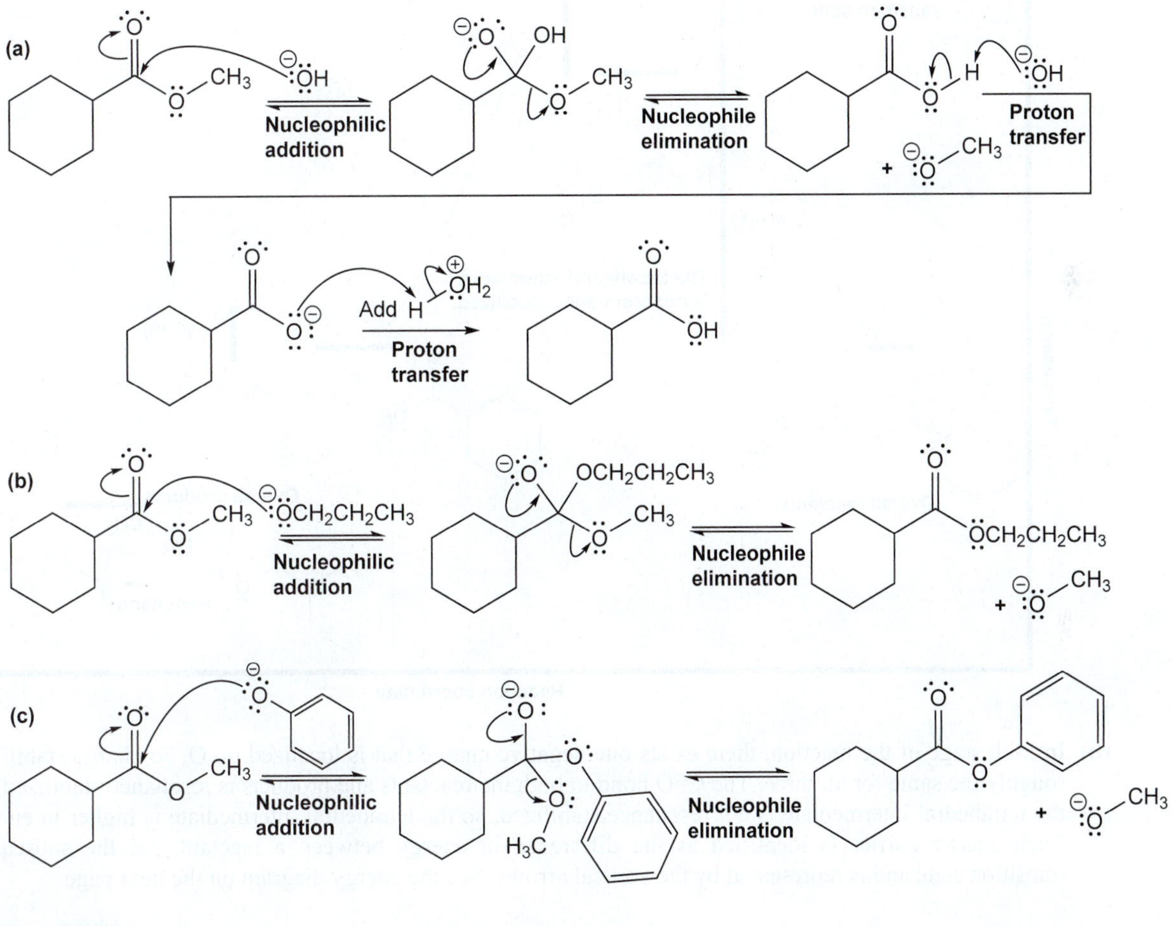

Problem 20.29

Think

What is the identity of the nucleophile in the nucleophilic addition reaction? What is the leaving group in the nucleophile elimination? Are any competing proton transfer reactions involved?

Solve

Products **(b)** and **(c)** from Problem 20.28 will produce methyl cyclohexylmethanoate when treated with $NaOCH_3$. That is because the products in **(b)** and **(c)** are esters, and treatment with CH_3O^- will initiate a transesterification. Product **(a)** will undergo an irreversible deprotonation to yield the carboxylate anion (Equation 20-7).

Problem 20.30

Think

Use Figure 20-1 as a guide. How do the charges impact the stability of the reactants, products, and intermediates? How do they impact the resonance? How you do represent the energy barrier in each step?

Solve

(a) The direct products of nucleophilic addition–elimination, RCO_2H and RO^-, have roughly the same energy as the overall reactants. The tetrahedral intermediate is higher in energy because it has lost resonance stabilization involving the C=O bond. The products of the proton transfer in the third step, however, are substantially lower in energy, owing largely to the resonance delocalization of the negative charge in the carboxylate anion. Each energy barrier is identified as the difference in energy between a reactant and the subsequent transition state and is represented by the vertical arrows.

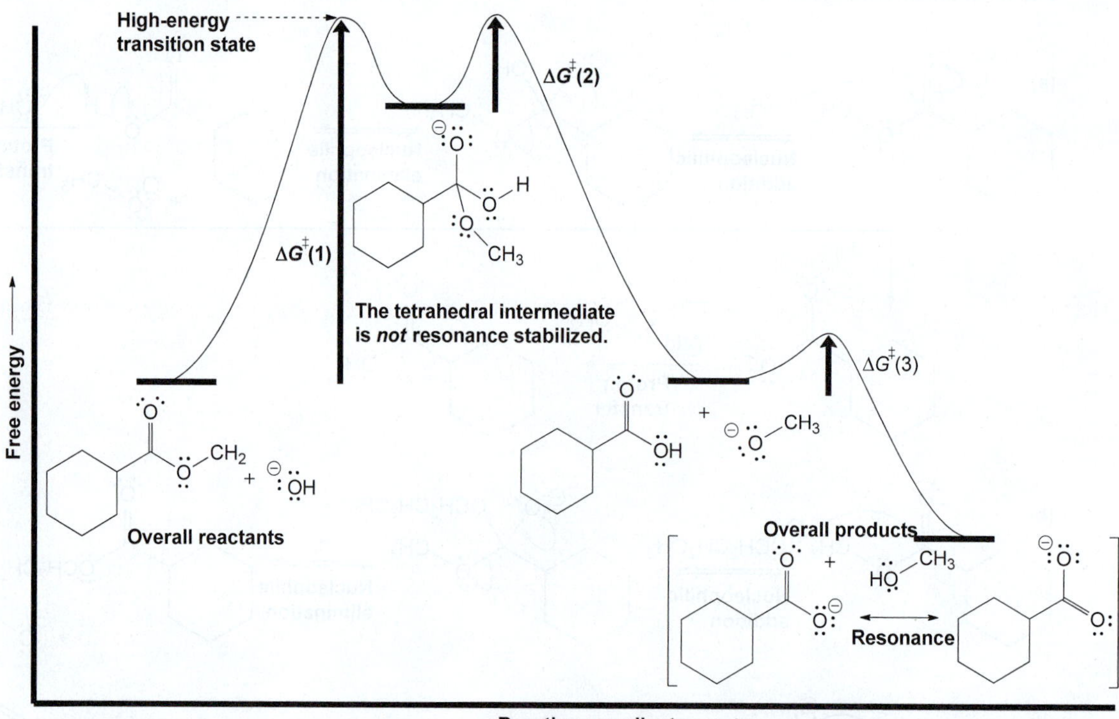

(b) In each stage of the reaction, there exists one negative charge that is localized on O, so charge stability is roughly the same for all three. The C=O bond in both the reactants and products is resonance stabilized, but the tetrahedral intermediate is not resonance stabilized, so the tetrahedral intermediate is higher in energy. Each energy barrier is identified as the difference in energy between a reactant and the subsequent transition state and is represented by the vertical arrows. See the energy diagram on the next page.

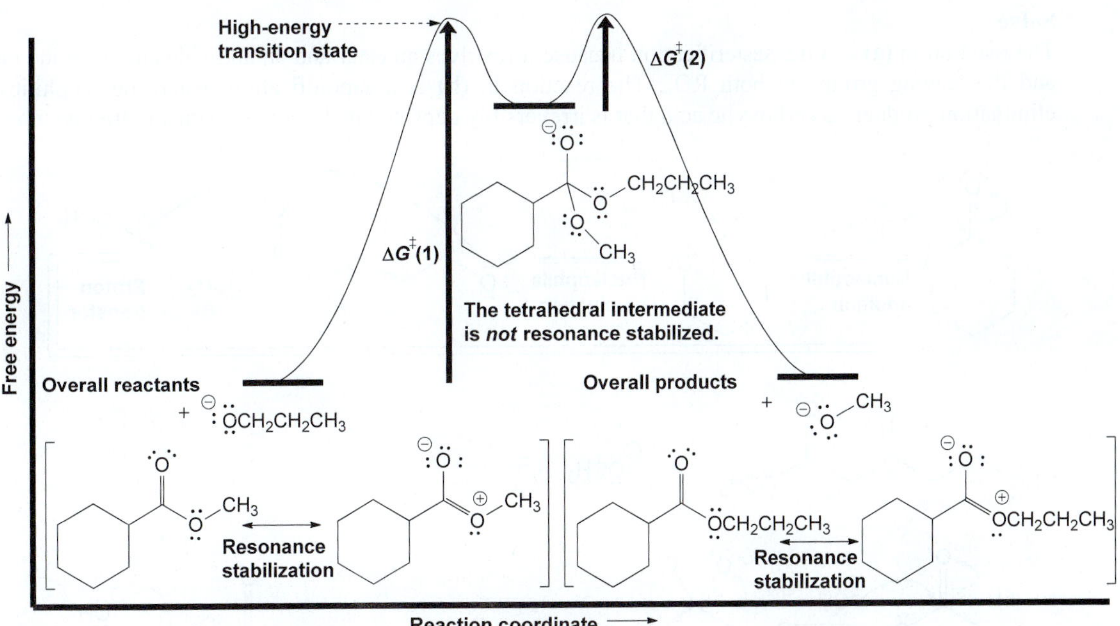

(c) The C=O bond in both the reactants and products is resonance stabilized, but the tetrahedral intermediate is not resonance stabilized, so the tetrahedral intermediate is higher in energy. The negative charge is resonance stabilized on $C_6H_5O^-$ in the reactants, but it is localized on O in CH_3O^- in the products, so the reactants are somewhat more stable than the products. Each energy barrier is identified as the difference in energy between a reactant and the subsequent transition state and is represented by the vertical arrows.

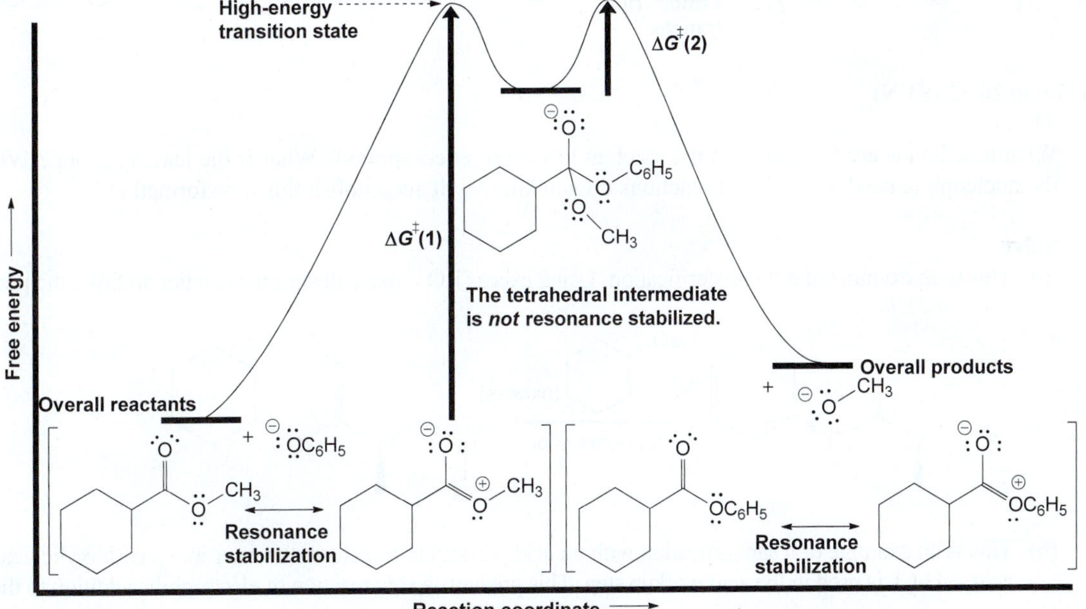

Problem 20.31

Think

What is the identity of the nucleophile in the nucleophilic addition reaction? What is the leaving group in the nucleophile elimination? Will nucleophilic addition–elimination produce a derivative lower on the stability ladder? How many times does the nucleophilic addition reaction occur? Are proton transfer reactions involved?

Solve

The reaction in **(a)** is a transesterification because it involves an ester and an alkoxide anion, so the nucleophile and the leaving group are both RO^-. The reaction in **(b)** is a saponification, where nucleophilic addition–elimination produces a carboxylic acid that is irreversibly deprotonated. The mechanisms are given below:

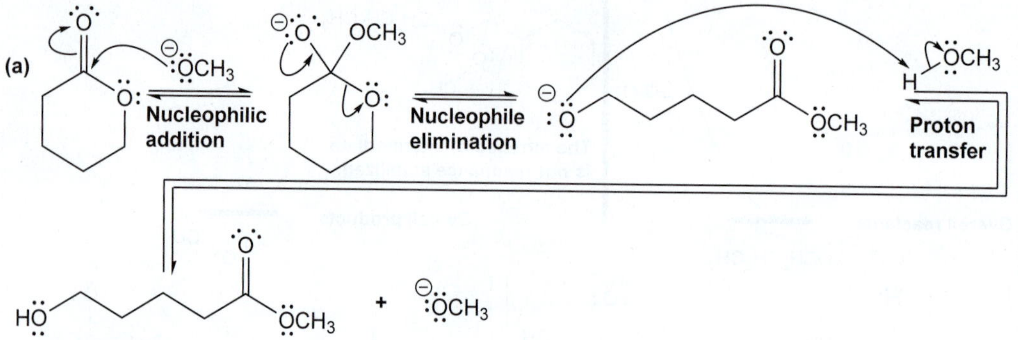

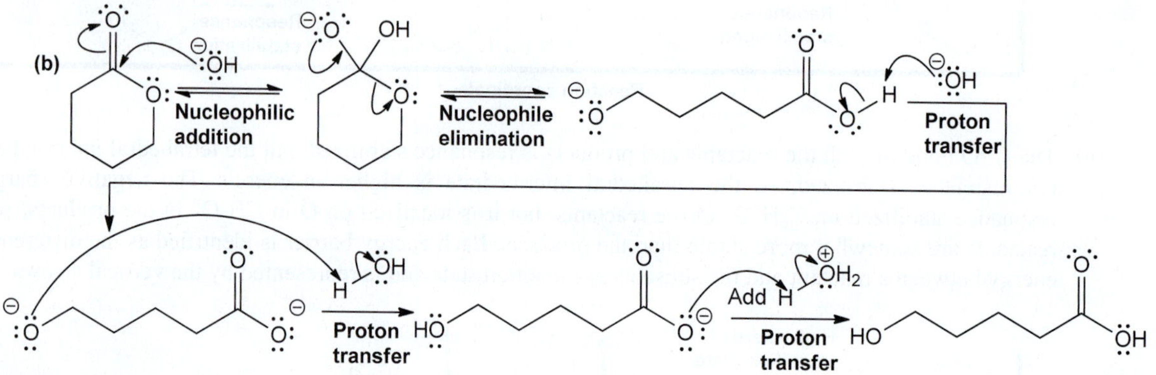

(a) Nucleophilic addition; Nucleophile elimination; Proton transfer

OCH_3 + $^-OCH_3$

(b) Nucleophilic addition; Nucleophile elimination; Proton transfer; Proton transfer; Add H^+; Proton transfer

Problem 20.32 (SYN)

Think

What new bonds are formed from the reactant to the target compound? What is the leaving group? What would the nucleophile need to be? What reactions do you know will accomplish this transformation?

Solve

(a) This is an example of a transesterification. Using excess RO^- base will shift the reaction to favor the products.

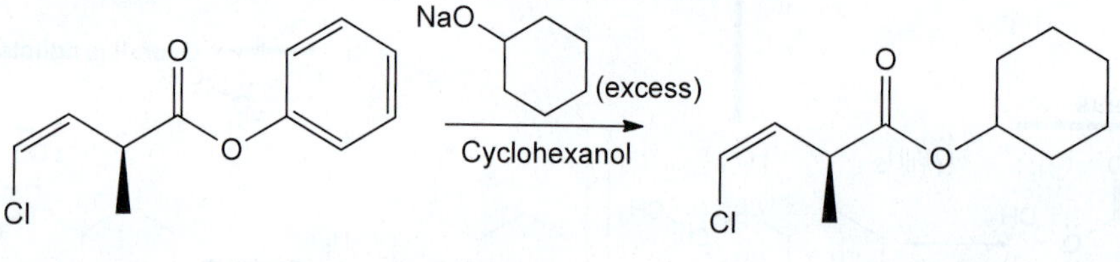

NaO–cyclohexyl (excess), Cyclohexanol

(b) This is an example of a saponification with an acid workup to transform the ester into a carboxylic acid. A weak acid, NH_4Cl, is used in the acid workup step. This prevents a side reaction of electrophilic addition to the C=C.

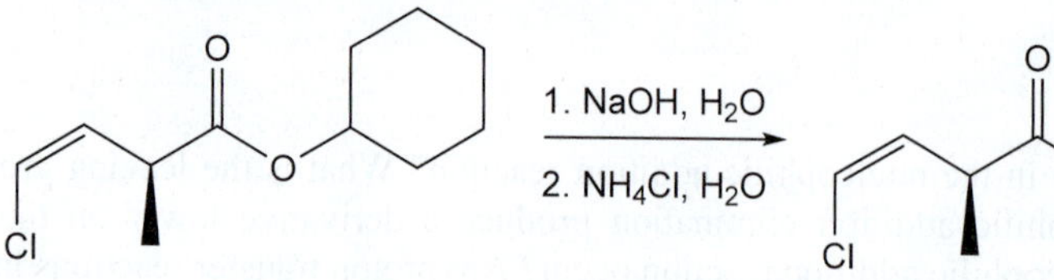

1. NaOH, H_2O
2. NH_4Cl, H_2O

Problem 20.33

Think

Which group, CH_3 or CF_3, is electron withdrawing? Which carbonyl C atom is more electrophilic? How does the electronic nature of the carbonyl carbon affect the rate of nucleophilic addition?

Solve

The fluorinated ester has a carbonyl carbon that is more electron poor, so it will react faster. The greater partial positive charge on the carbonyl C increases the attraction with the nucleophile and lowers the energy barrier for the nucleophilic addition step, which is the rate-determining step.

A B

Problem 20.34

Think

Under basic conditions, what types of species should not appear in the mechanism? What is the nucleophile in the nucleophilic addition–elimination reaction? What is the leaving group? How many times does the nucleophilic addition–elimination reaction occur? Are proton transfer reactions involved?

Solve

The reaction takes place in basic solution, so no strong acids should appear. The H–^{18}OH protonates the carboxylate anion to form the carboxylic acid. The carboxylic acid undergoes nucleophilic addition–elimination with the H^{18}O$^-$ anion. In the nucleophile elimination step, the leaving group is the unlabeled HO$^-$.

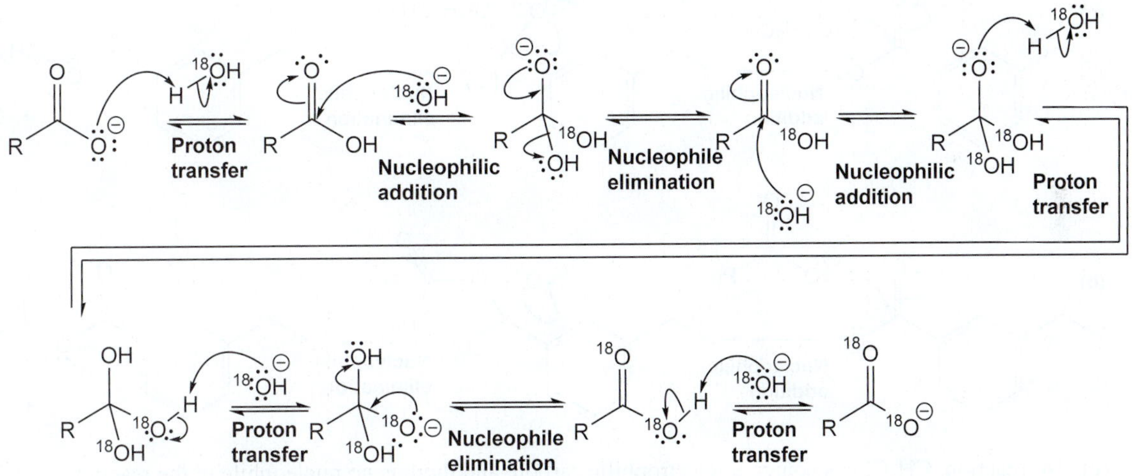

Proton transfer | Nucleophilic addition | Nucleophile elimination | Nucleophilic addition | Proton transfer

Proton transfer | Nucleophile elimination | Proton transfer

Sections 20.2, 20.4, and 20.5 Interconverting Carboxylic Acid Derivatives and the Stability Ladder; The Gabriel Synthesis of Primary Amines; Haloform Reactions

Problem 20.35

Think

What is the identity of the nucleophile in the nucleophilic addition reaction? What is the leaving group in the nucleophile elimination? Will nucleophilic addition–elimination produce a derivative lower on the stability ladder? How many times does the nucleophilic addition reaction occur? Are proton transfer reactions involved?

Solve

The reactions in **(a)** through **(d)** undergo nucleophilic addition–elimination to produce a carboxylic acid or acid derivative that is on a lower rung of the stability ladder than the one in the reactants, so those reactions occur readily. The mechanisms are given below.

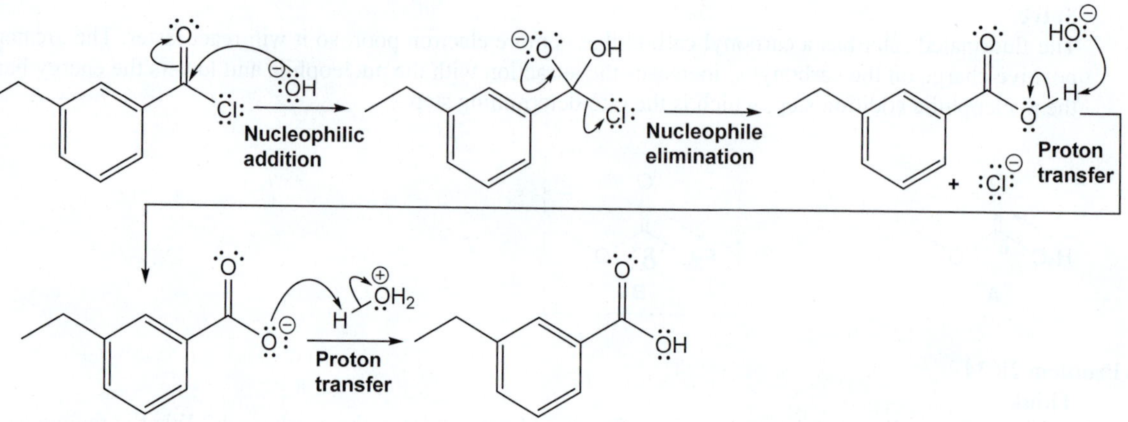

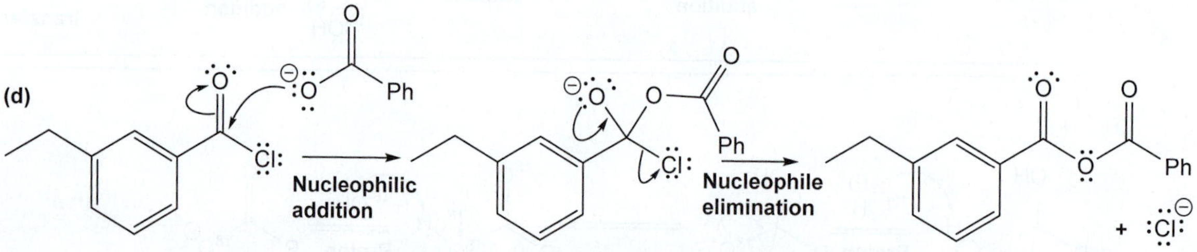

(e) No reaction. CH_3Cl is a source of electrophilic carbon, and there is no nucleophile in the reaction.

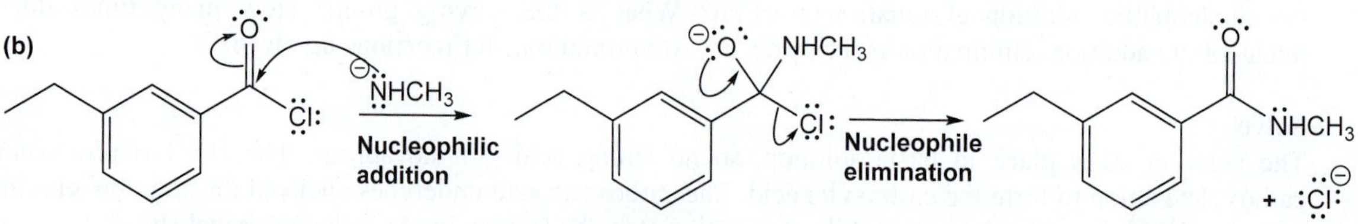

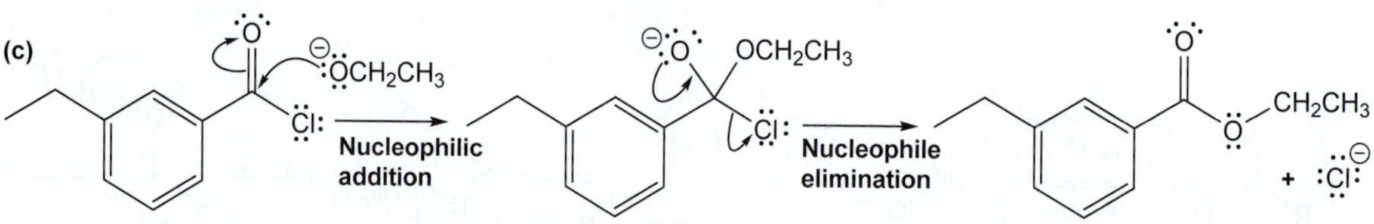

(f) No reaction. An ether is not a strong enough nucleophile. Also, a nucleophilic addition–elimination will not produce a derivative that is lower on the stability ladder.

Problem 20.36

Think

What is the identity of the nucleophile in the nucleophilic addition reaction? What is the leaving group in the nucleophile elimination? Will nucleophilic addition–elimination produce a derivative lower on the stability ladder? How many times does the nucleophilic addition reaction occur? Are proton transfer reactions involved?

Solve

The reactions in **(a)** through **(c)** undergo nucleophilic addition–elimination to produce a carboxylic acid or acid derivative that is on a lower rung of the stability ladder than the one in the reactants, so those reactions occur readily. For the reaction in **(d)**, the reactant and product are both acid anhydrides, so the reaction takes place reversibly. The mechanisms are given below.

(a) Nucleophilic addition → Nucleophile elimination → Proton transfer → Add $^{+}OH_2$, Proton transfer

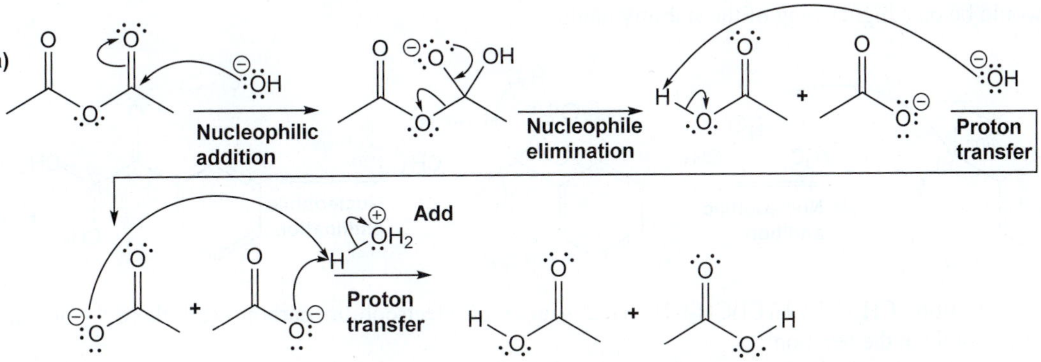

(b) $^{-}NHCH_3$: Nucleophilic addition → Nucleophile elimination

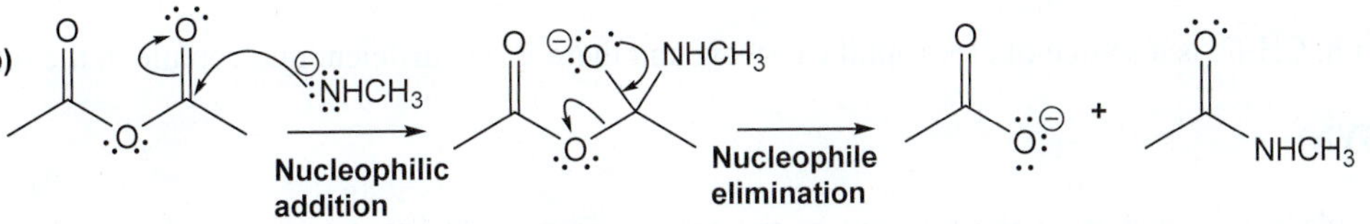

(c) $^{-}OCH_2CH_3$: Nucleophilic addition → Nucleophile elimination

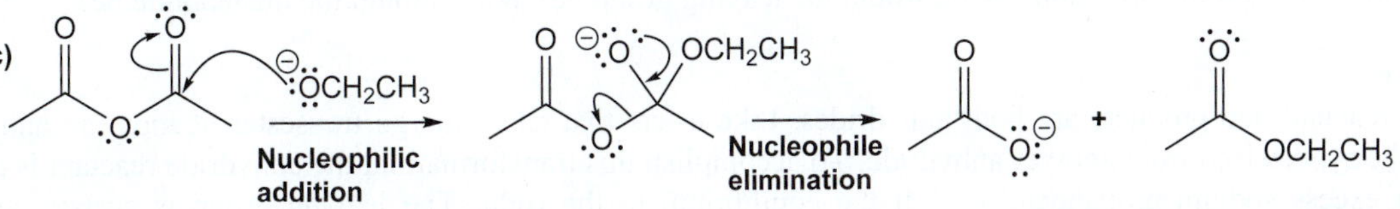

(d) Nucleophilic addition → Nucleophile elimination

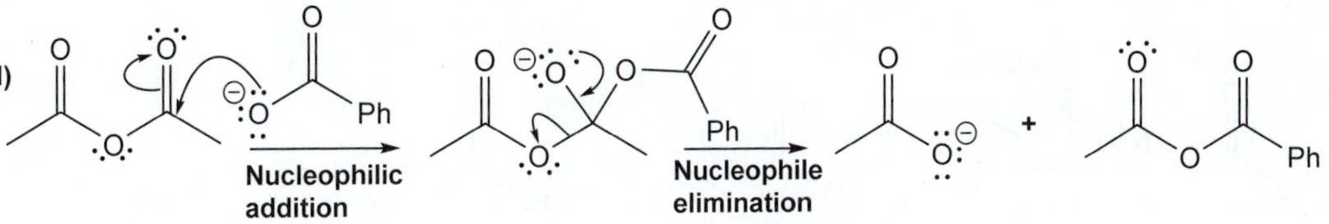

(e) No reaction. This nucleophilic addition–elimination would produce an acid bromide, which is a derivative that would be higher on the stability ladder than the reactant acid anhydride.

(f) No reaction. An ether is not a strong enough nucleophile. Also, this nucleophilic addition–elimination will not produce a derivative that is lower on the stability ladder.

(g) No reaction. $CH_3CH_2CH(Cl)CH_2CH_3$ is a source of electrophilic carbon, and there is no sufficient nucleophile in the reaction.

(h) No reaction. An aldehyde is a source of electrophilic carbon, and there is no sufficient nucleophile in the reaction.

Problem 20.37

Think

What is the identity of the nucleophile in the nucleophilic addition reaction? What is the leaving group in the nucleophile elimination? Will nucleophilic addition–elimination produce a derivative lower on the stability ladder? How many times does the nucleophilic addition reaction occur? Are proton transfer reactions involved?

Solve

The reaction in **(b)** undergoes nucleophilic addition–elimination to produce a carboxylic acid derivative that is on a lower rung of the stability ladder than the one in the reactants, so that reaction occurs readily. The mechanism is given below.

(a) No reaction. This nucleophilic addition–elimination would convert the ester into an acid bromide, which would be on a higher rung of the stability ladder.

(b) Nucleophilic addition; Nucleophile elimination

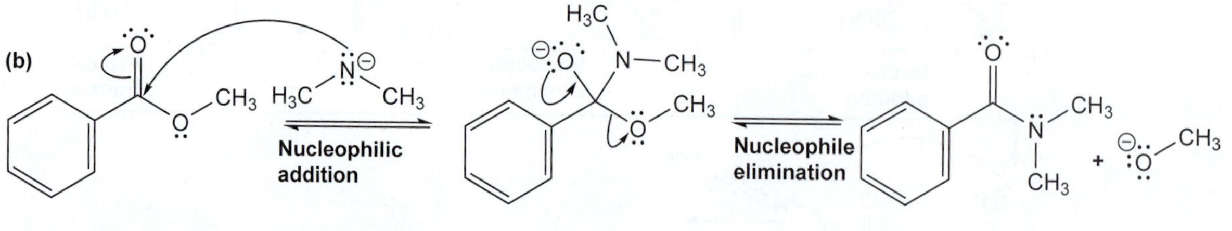

(c) No reaction. $CH_3CH_2CH(Cl)CH_2CH_3$ is a source of electrophilic carbon, and there is no sufficient nucleophile in the reaction.

(d) No reaction. An aldehyde is a source of electrophilic carbon, and there is no sufficient nucleophile in the reaction.

(e) No reaction. CH_3Cl is a source of electrophilic carbon, and there is no sufficient nucleophile in the reaction.

Problem 20.38 (SYN)

Think

What new bonds are formed from the reactant to the target compound? What reactions do you know will accomplish this transformation? What would the leaving group be? What should the nucleophile be?

Solve

The reactant and products are both anhydrides. Like esters that can undergo transesterification, an anhydride can be synthesized from another anhydride. To accomplish this transformation, the anhydride reactant is treated with excess sodium propanoate to shift the equilibrium to the right. The leaving group is acetate, and the nucleophile is propanoate.

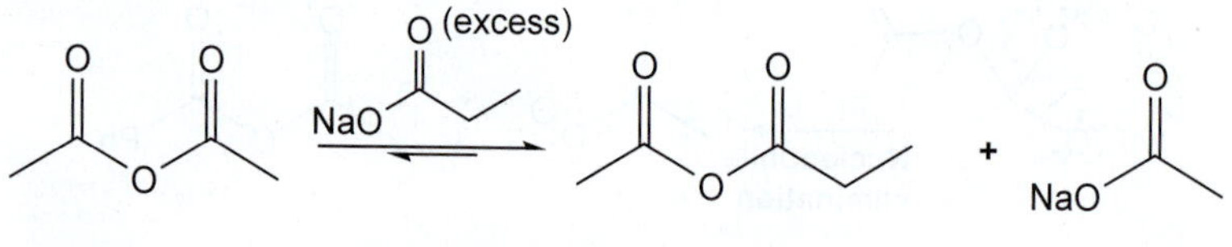

(excess)

Problem 20.39 (SYN)

Think

Review the "stability ladder" in Figure 20-5. What is the functional group in the reactant? What is the functional group in the product? Consider the position of the product relative to the reactant on the stability ladder. Can the product be synthesized by direct acyl substitution, or is more than one step required?

Solve

(a) One acid anhydride can be transformed into another. Both the nucleophile and the leaving group are carboxylate anions. To accomplish this transformation, the anhydride reactant is treated with excess sodium acetate to shift the equilibrium to the right. The leaving group is benzoate, and the nucleophile is acetate.

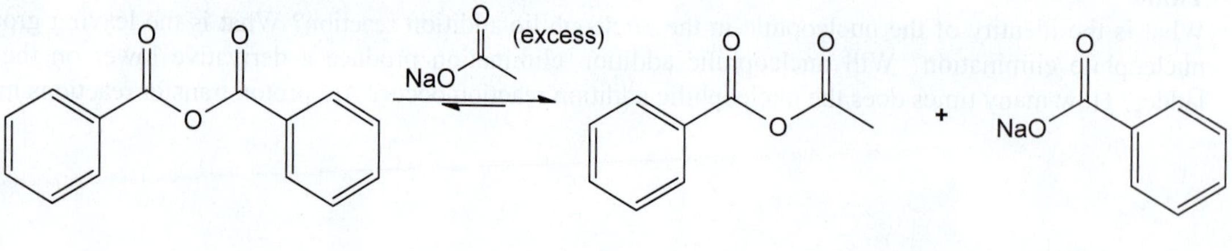

(excess)

(b) An acid chloride can be directly transformed into an ester. The nucleophile is the phenoxide anion, and the leaving group is the chloride anion.

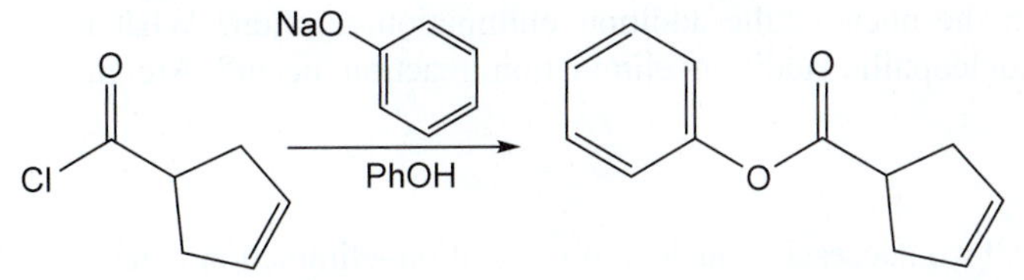

NaO PhOH O Cl O O

(c) An ester can be directly transformed into an amide. The nucleophile should be $(CH_3)_2N^-$, and the leaving group is the ethoxide anion.

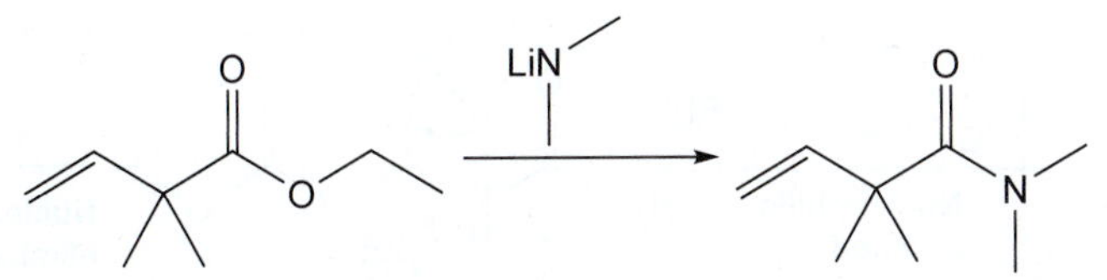

O O LiN O N

(d) An amide can undergo hydrolysis to be transformed into a carboxylic acid. Nucleophilic addition–elimination produces a carboxylic acid that is irreversibly deprotonated under the strongly basic conditions. An acid workup is necessary to add back that proton. Strong acids should be avoided to prevent electrophilic addition to the C=C bond.

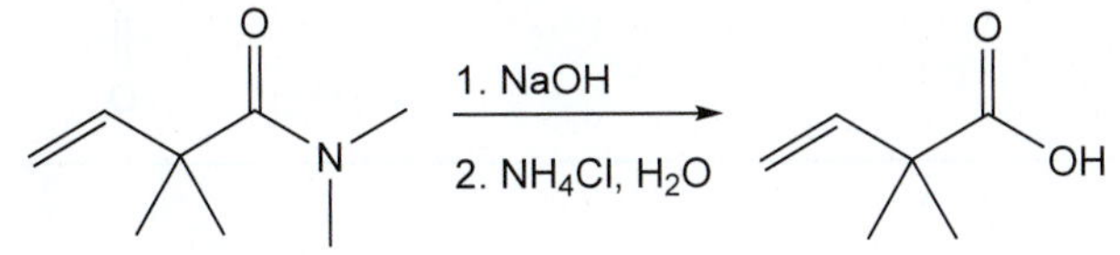

O N 1. NaOH 2. NH_4Cl, H_2O O OH

Problem 20.40

Think

What is the charge on the nitrogen in LiN_3? What is the nucleophile in the nucleophilic addition–elimination reaction? After the first nucleophilic addition–elimination reaction, is the resulting product nucleophilic?

Solve

Li_3N can be treated as the nitride ion (N^{3-}), and this nucleophile reacts with three equivalents of acid chloride to undergo three nucleophilic addition–elimination reactions.

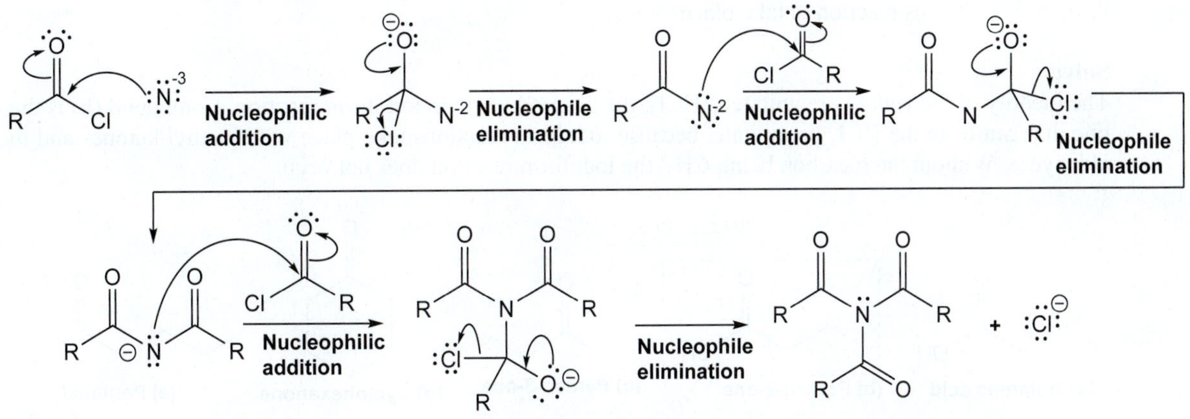

O R Cl :N:-3 **Nucleophilic addition** O R N-2 :Cl: **Nucleophile elimination** O R N-2 O Cl R **Nucleophilic addition** O R N O R Cl: **Nucleophile elimination**

O R N O R O Cl R **Nucleophilic addition** O O R N R :Cl R O **Nucleophile elimination** O O R N R R O + :Cl:

Problem 20.41

Think

What types of species should not appear in the mechanism for a reaction that takes place under basic conditions? What is the nucleophile in the nucleophilic addition–elimination reaction? What is the leaving group? How many times does the nucleophilic addition–elimination reaction occur? Are proton transfer reactions involved?

Solve

The EtO^- deprotonates the amide N–H. Two successive nucleophilic addition–elimination reactions take place in which the RHN^- is the nucleophile and EtO^- is the leaving group. The second one is intramolecular, because it forms a six-membered ring. Note that in basic solution, no strong acids should appear in the mechanism; therefore, the proton transfer reaction occurs before the nucleophilic addition–elimination mechanism does.

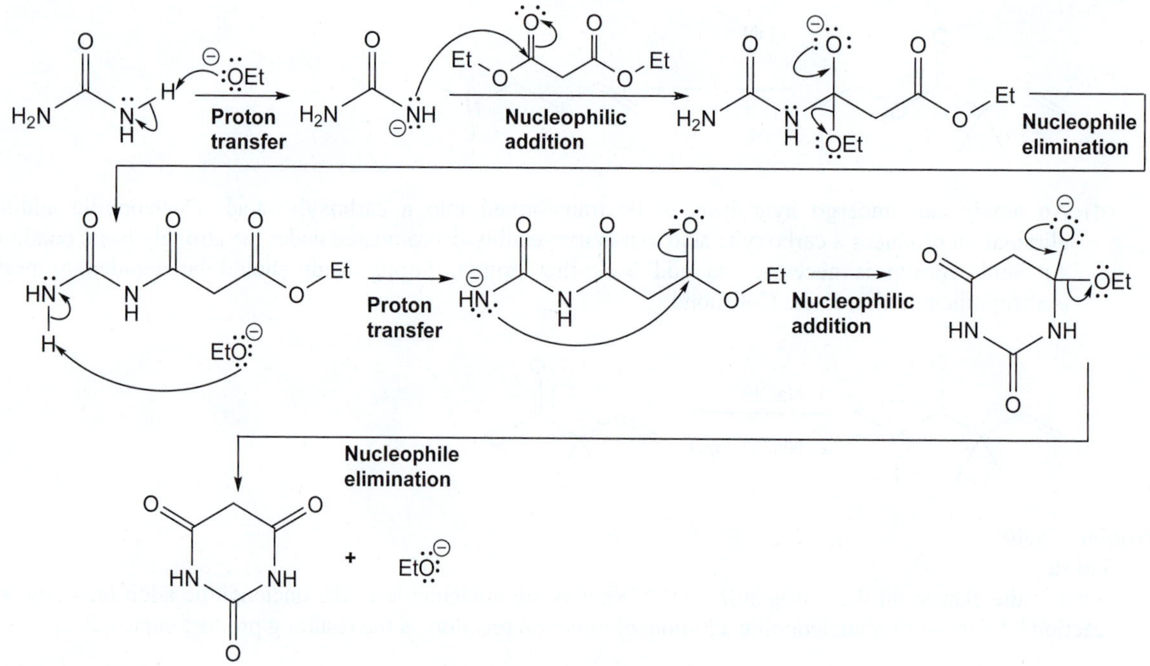

Proton transfer

Nucleophilic addition

Nucleophile elimination

Proton transfer

Nucleophilic addition

Nucleophile elimination

Problem 20.42

Think

What is the identity of the yellow precipitate? What type of reaction does it signify? What is the requirement of the α carbon for this reaction to take place?

Solve

The identity of the yellow precipitate is HCI_3, the by-product of an iodoform reaction. Compound **(b)** is the only one that can form the HCI_3 precipitate, because iodoform reactions take place with methyl ketones and methyl aldehydes. Without the α carbon being CH_3, the iodoform reaction does not occur.

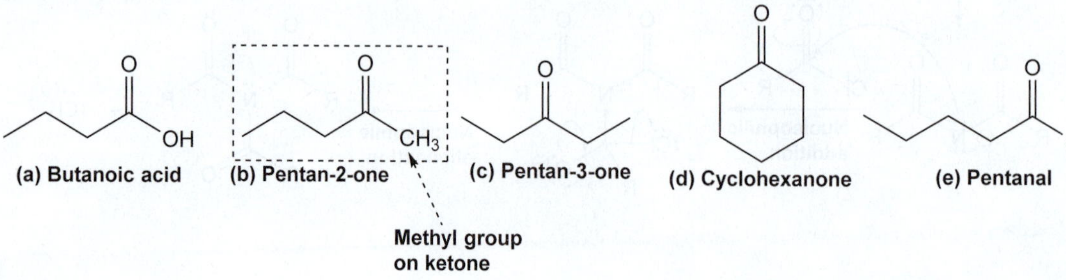

(a) Butanoic acid (b) Pentan-2-one (c) Pentan-3-one (d) Cyclohexanone (e) Pentanal

Methyl group on ketone

Problem 20.43 (SYN)

Think

What functional group is the target in each of these compounds? Is it primary, secondary, or tertiary? What method do you know to form this functional group? What reagents are necessary?

Solve

In a Gabriel synthesis, an alkyl halide is converted to a primary amine. So we must simply choose the appropriate alkyl halide that is analogous to the primary amine target. Notice in **(c)** that the stereochemical configuration is inverted, because the second reaction is an S_N2 that takes place at a carbon stereocenter.

(a)

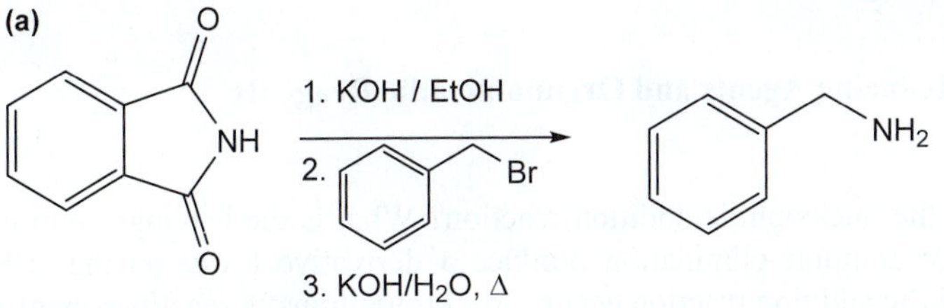

1. KOH / EtOH
2. Br
3. KOH/H_2O, Δ

NH_2

(b)

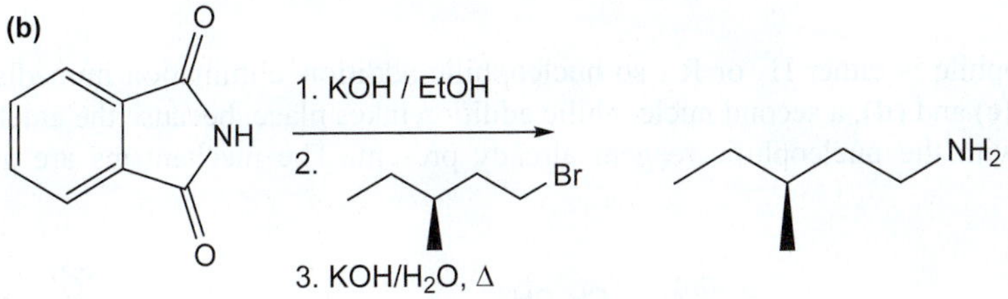

1. KOH / EtOH
2. Br
3. KOH/H_2O, Δ

NH_2

(c)

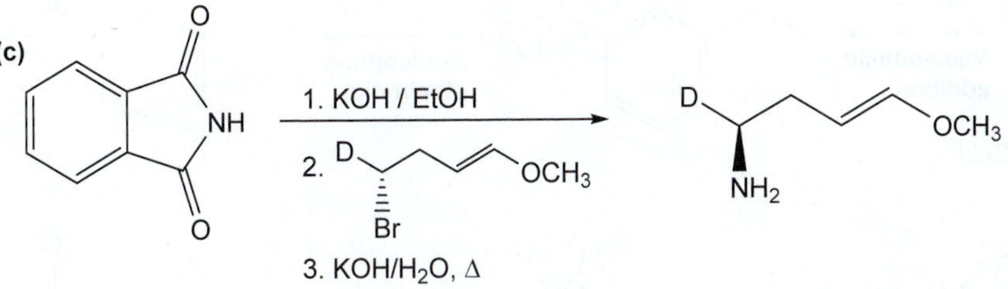

1. KOH / EtOH
2. D, OCH_3, Br
3. KOH/H_2O, Δ

D, OCH_3, NH_2

Problem 20.44

Think

In Steps 6–8 of the mechanism in Equation 20-13, what elementary steps need to occur? Is there a competing proton transfer under strongly basic conditions that would compromise the steps that need to occur?

Solve

In Step 6 of the mechanism, hydroxide needs to act as a nucleophile to add to the C=O group of the amide.

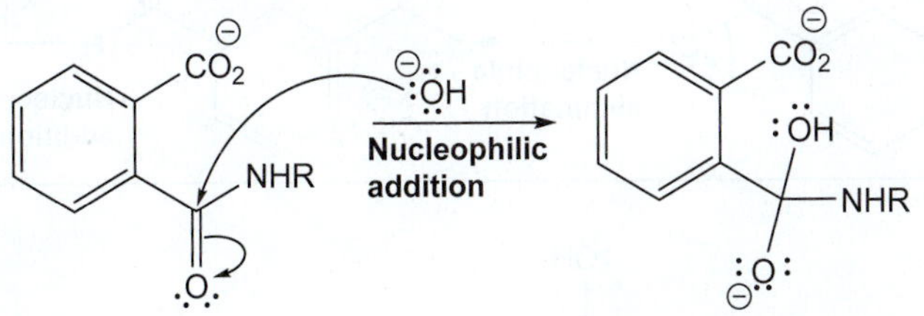

CO_2^-, NHR, ^-OH, **Nucleophilic addition**, CO_2^-, OH, NHR, O^-

However, the amide proton is somewhat acidic and can be deprotonated under strongly basic conditions, as shown below. When this happens, the resulting anion stabilizes the C=O group via resonance and hinders nucleophilic attack by hydroxide.

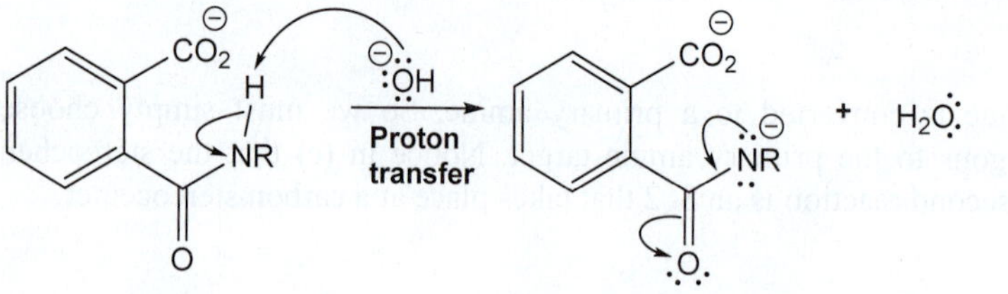

Sections 20.6–20.8 Reactions with Hydride Reducing Agents and Organometallic Reagents

Problem 20.45

Think

What is the identity of the nucleophile in the nucleophilic addition reaction? What is the leaving group in the nucleophile elimination? Will nucleophilic addition–elimination produce a derivative lower on the stability ladder? How many times does the nucleophilic addition reaction occur? Are proton transfer reactions involved?

Solve

In all of these reactions, the nucleophile is either H^- or R^-, so nucleophilic addition–elimination immediately produces an aldehyde or ketone. In **(c)** and **(d)**, a second nucleophilic addition takes place, because the aldehyde and ketone products are reactive with the nucleophilic reagent already present. The mechanisms are given below and on the next page:

(a)

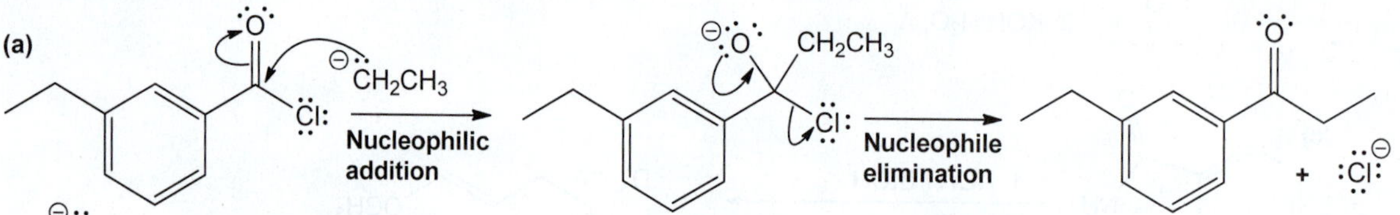

$^{\ominus}CH_2CH_3 = (CH_3CH_2)_2CuLi$

(b)

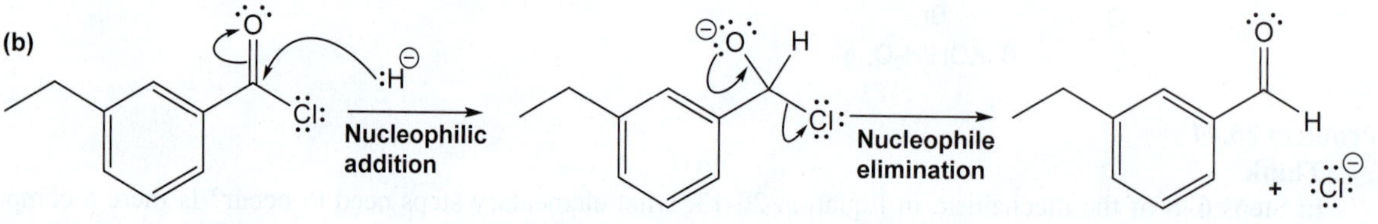

$:H^{\ominus}$ = LiAlH(O-*t*-Bu)$_3$ at –78 °C (stops at aldehyde)

(c)

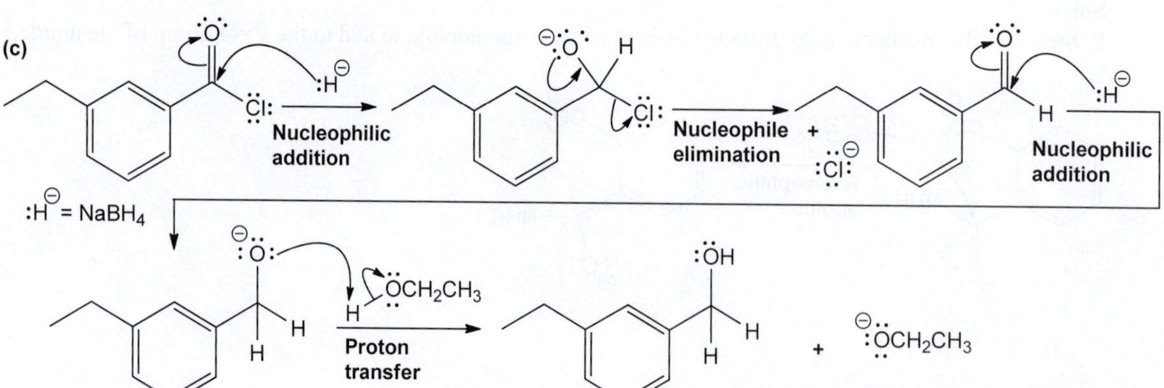

$:H^{\ominus} = NaBH_4$

(d)

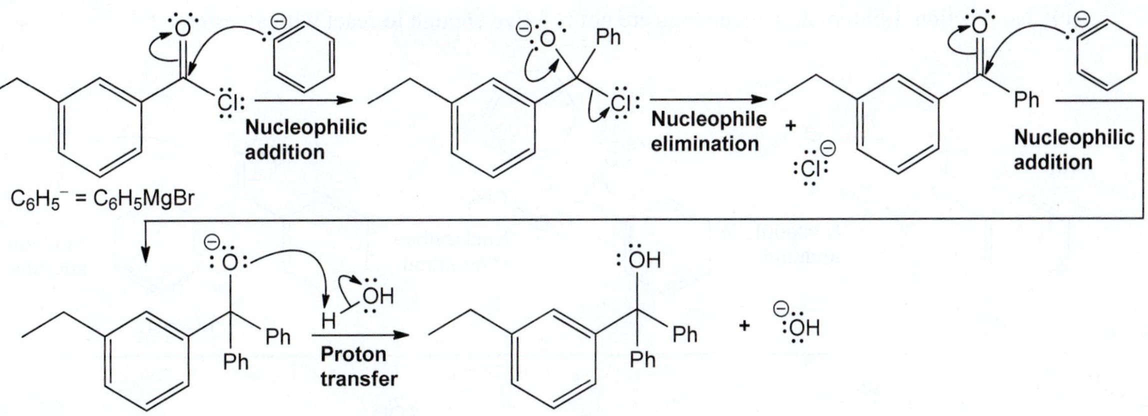

Nucleophilic addition

Nucleophile elimination

Nucleophilic addition

$C_6H_5^- = C_6H_5MgBr$

Proton transfer

Problem 20.46

Think

What is the identity of the nucleophile in the nucleophilic addition reaction? What is the leaving group in the nucleophile elimination? Will nucleophilic addition–elimination produce a derivative lower on the stability ladder? How many times does the nucleophilic addition reaction occur? Are proton transfer reactions involved?

Solve

In all of these reactions, the nucleophile is either H^- or R^-, so nucleophilic addition–elimination would immediately produce an aldehyde or ketone. In **(a)**, **(b)**, and **(d)**, a second nucleophilic addition takes place because the aldehyde and ketone products are reactive with the nucleophilic reagent already present. The mechanisms are given below and on the next page:

(a)

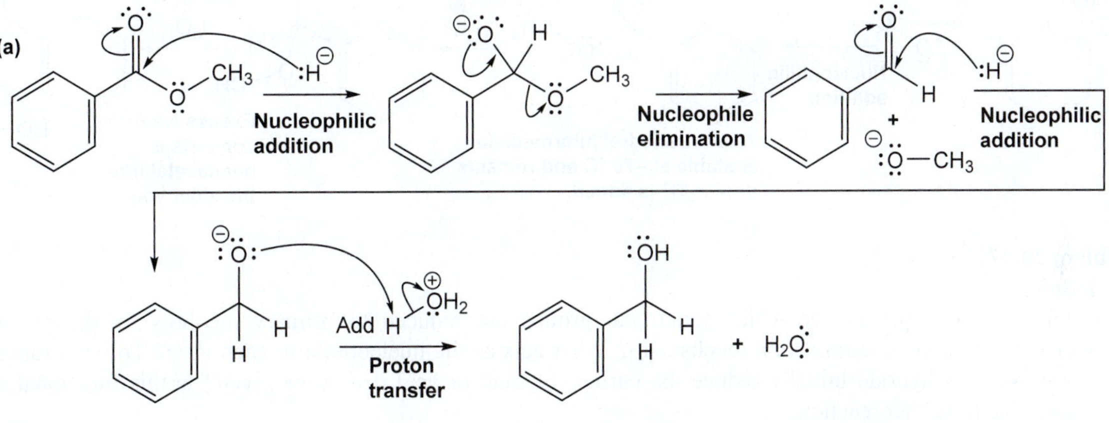

Nucleophilic addition

Nucleophile elimination

Nucleophilic addition

Add H^+ Proton transfer

(b)

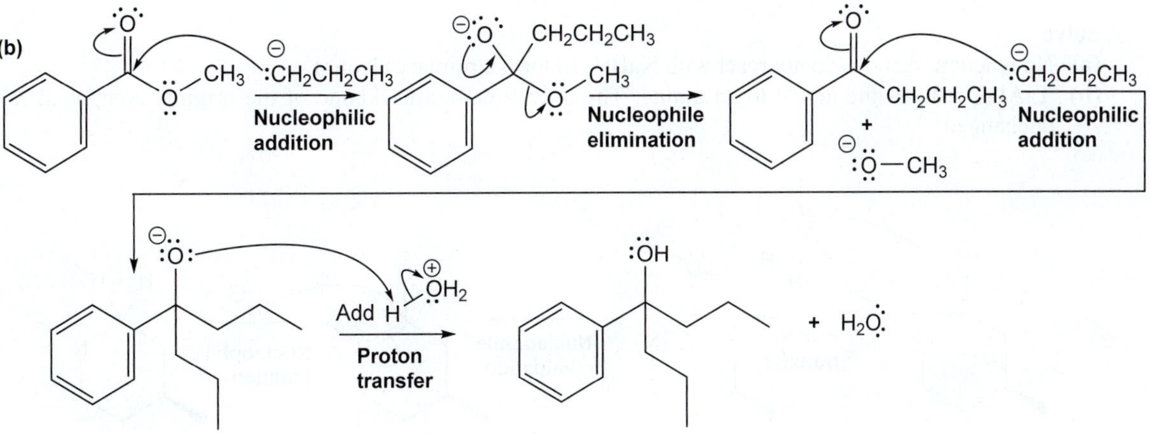

Nucleophilic addition

Nucleophile elimination

Nucleophilic addition

Add H^+ Proton transfer

(c) No reaction. Lithium dialkylcuprates are not reactive enough to react with an ester.

(d)

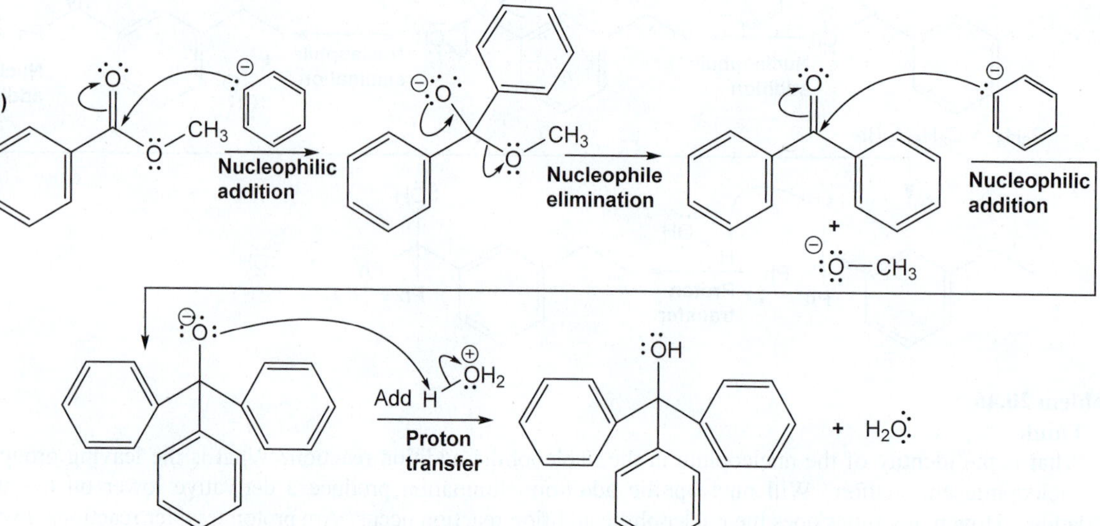

Nucleophilic addition → Nucleophile elimination → + $:\ddot{O}^{\ominus}—CH_3$ → Nucleophilic addition

Add $H—\overset{\oplus}{O}H_2$ → Proton transfer → + $H_2\ddot{O}:$

(e)

Nucleophilic addition → $H_3\overset{\oplus}{O}:$ → $H_3\overset{\oplus}{O}:$ → + HO—CH₃

This tetrahedral intermediate is stable at –78 °C and remains until H_3O^+ is added.

Excess water converts a hemiacetal into the aldehyde.

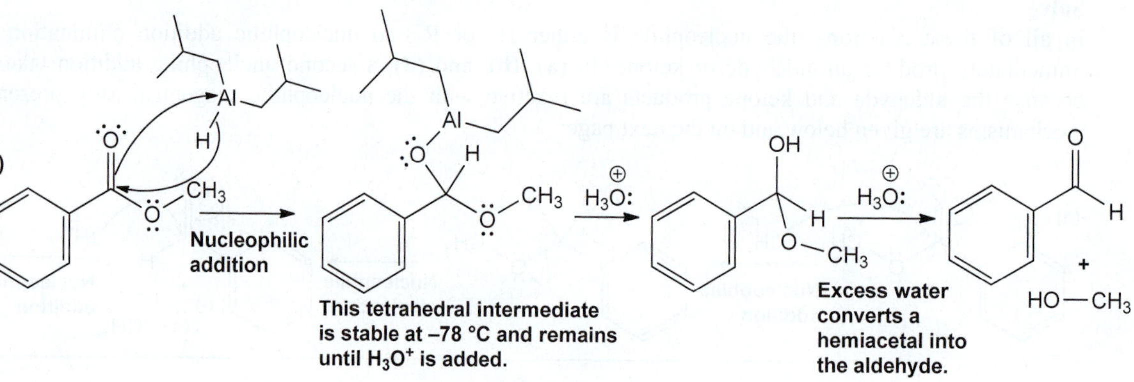

Problem 20.47

Think

Refer to Table 20-1 to see which functional groups are reduced to primary alcohols. Is the reaction a nucleophilic addition–elimination mechanism? What acts as the nucleophile in each case? To what functional group does the hydride initially reduce the carboxylic acid or acid derivative given? Is this functional group susceptible to further reaction?

Solve

(a) No reaction. Amides do not react with $NaBH_4$ to form a primary alcohol.

(b) $LiAlH_4$ reduces the amide to an amine. The overall carbon backbone of the original compound remains unchanged.

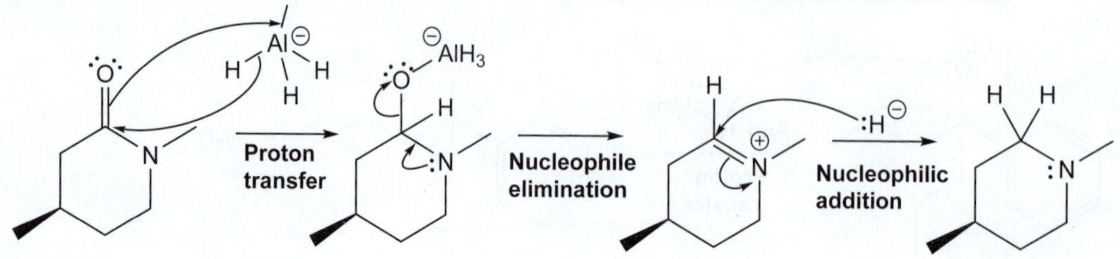

Proton transfer → Nucleophile elimination → :H⁻ Nucleophilic addition

(c) No reduction. Carboxylic acids are deprotonated by $NaBH_4$ to form a carboxylate anion, which does not undergo further reaction.

(d) Carboxylic acids have acidic hydrogen atoms. This functional group reacts with a hydride first via proton transfer. The mechanism is shown below. Carboxylic acids produce the primary alcohol when reacted with $LiAlH_4$.

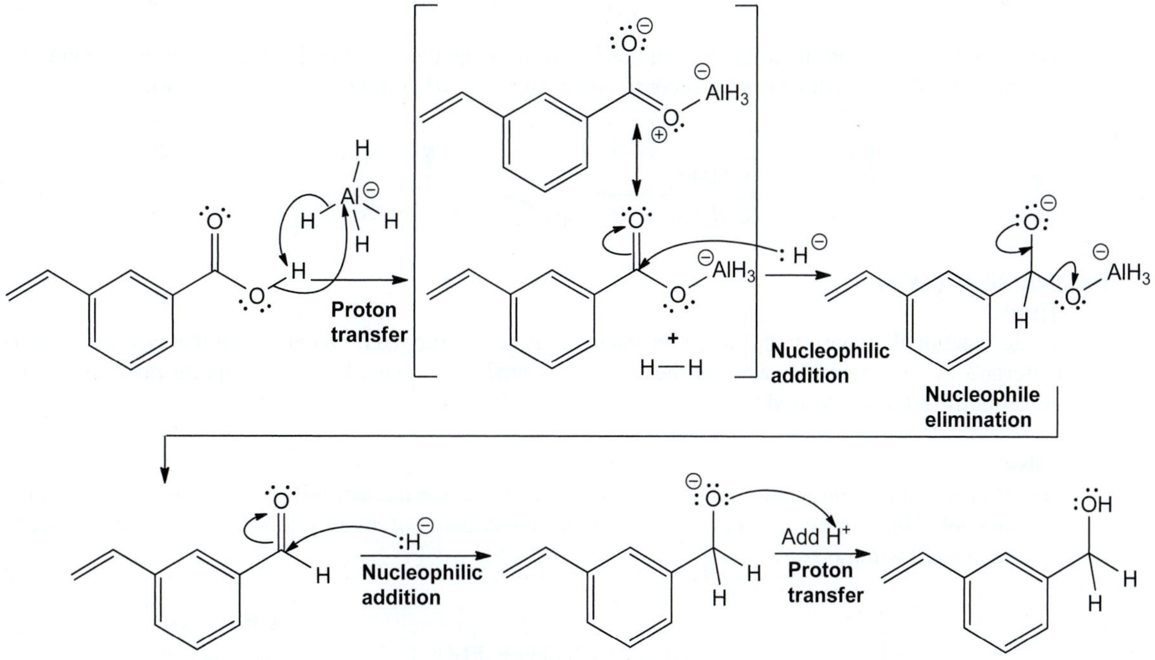

Proton transfer

Nucleophilic addition

Nucleophile elimination

Nucleophilic addition

Add H^+ Proton transfer

Problem 20.48 (SYN)

Think

What functional group is the target in each of these compounds? What is the functional group in the starting material? In each reaction, what is the leaving group? What would the nucleophile need to be? Is a second nucleophilic addition desired?

Solve

(a) This is a nucleophilic addition–elimination with H^- as the nucleophile, transforming an ester to an aldehyde. To avoid a second nucleophilic addition, the reaction requires DIBAH in cold conditions followed by an acid workup. A weak acid, NH_4Cl, is used in the acid workup step. This prevents a side reaction of electrophilic addition to the C=C.

1. DIBAH, –78 °C
2. NH_4Cl, H_2O

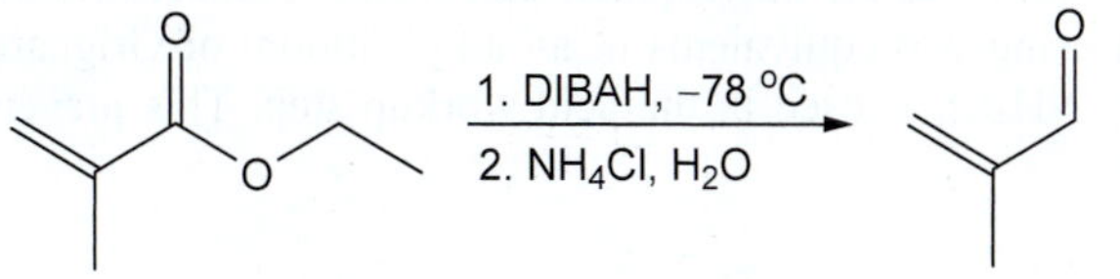

(b) This is a nucleophilic addition–elimination with H^- as the nucleophile, transforming an ester to a primary alcohol, which can be accomplished using $LiAlH_4$ followed by an acid workup. A weak acid, NH_4Cl, is used in the acid workup step. This prevents a side reaction of electrophilic addition to the C=C.

1. $LiAlH_4$
2. NH_4Cl, H_2O

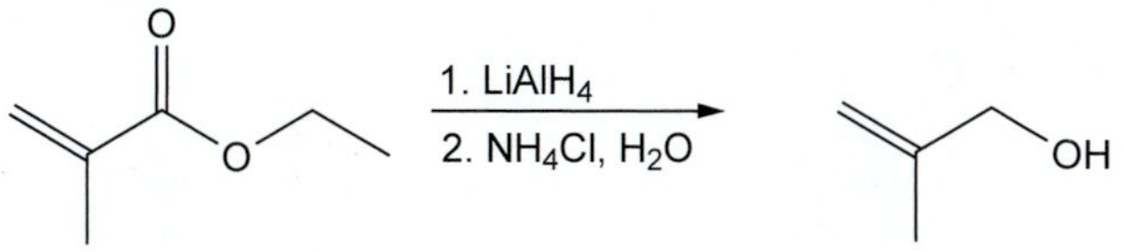

(c) This is a nucleophilic addition–elimination with H^- as the nucleophile, transforming an acid chloride to an aldehyde, which requires [LiAlH(O-*t*-Bu)$_3$] in cold conditions.

LiAlH(O-*t*-Bu)$_3$
−78 °C

(d) This is a nucleophilic addition–elimination with H^- as the nucleophile, transforming an acid chloride to a primary alcohol, which can be accomplished using $LiAlH_4$ followed by an acid workup.

1. $LiAlH_4$
2. NH_4Cl, H_2O

Problem 20.49 (SYN)

Think

What functional group is the target in each of these compounds? What functional group is in the starting material? In each reaction, what is the leaving group? What would the nucleophile need to be? Is a second nucleophilic addition desired?

Solve

(a) This is a nucleophilic addition–elimination with R^- as the nucleophile, transforming an ester into a tertiary alcohol. This can be accomplished using two equivalents of an alkyllithium or Grignard reagent followed by an acid workup.

1. 2 equiv CH_3CH_2MgBr, Et_2O
2. H_2O, HCl

(b) This is a nucleophilic addition–elimination with R^- as the nucleophile, transforming an acid chloride into a ketone. This can be accomplished using a lithium dialkyl cuprate reagent.

THF

(c) This is a nucleophilic addition–elimination with R^- as the nucleophile, transforming an acid chloride into a tertiary alcohol. This can be accomplished using two equivalents of an alkyl lithium or Grignard reagent followed by an acid workup. A weak acid, NH_4Cl, is used in the acid workup step. This prevents a side reaction of electrophilic addition to the C=C.

1. 2 equiv C_6H_5MgBr, Et_2O
2. NH_4Cl, H_2O

(d) This is a nucleophilic addition–elimination with R^- as the nucleophile, transforming an acid anhydride into a tertiary alcohol. This can be accomplished using two equivalents of an alkyl lithium or Grignard reagent followed by an acid workup.

1. 2 equiv CH_3MgBr, Et_2O
2. H_2O, HCl

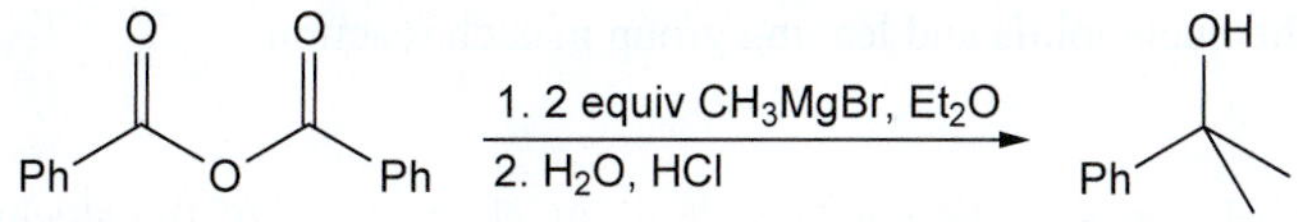

Problem 20.50 (SYN)

Think

What functional group transformation occurs from the reactant to the target? What functional group needs to remain unreacted during the course of the desired transformation? What reagent reacts only with the intended functional group?

Solve

This reaction transforms the acid chloride into a primary alcohol via a reduction reaction. The ester remains unreacted from the reactant to the target product. Reduction using $NaBH_4$ would work because the reduction of an ester using $NaBH_4$ is very slow. $LiAlH_4$, on the other hand, is much more reactive and reduces both acid chlorides and esters quickly.

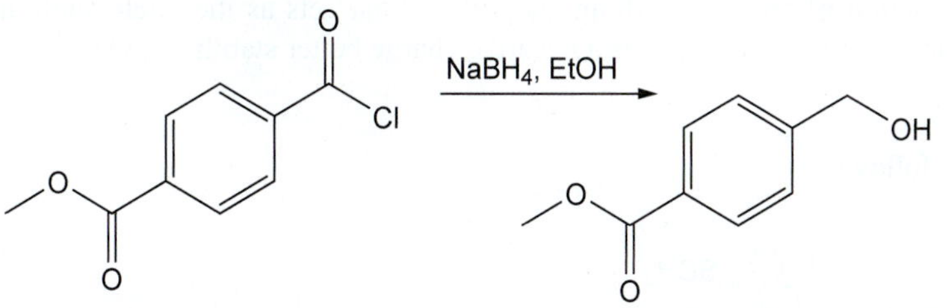

$NaBH_4$, EtOH

Problem 20.51 (SYN)

Think

What functional group is the target in each of these compounds? What method do you know to form this functional group? What reagents are necessary?

Solve

(a) The reaction reduces an amide into an amine. This can be accomplished using $LiAlH_4$.

$LiAlH_4$

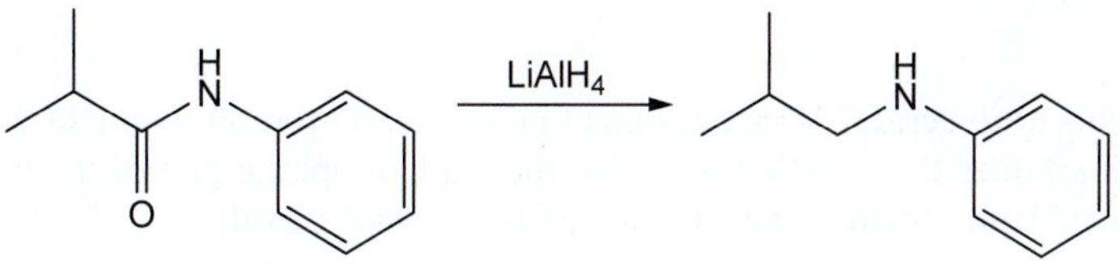

(b) The reaction transforms a carboxylic acid into a primary alcohol. This can be accomplished using $LiAlH_4$ followed by an acid workup. To avoid reaction with the C=C bond, the amount of HCl added should be minimized.

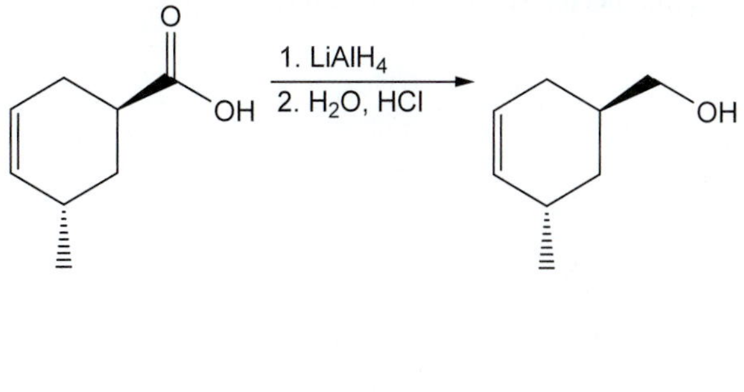

1. $LiAlH_4$
2. H_2O, HCl

Integrated Problems

Problem 20.52

Think

How does the reducing reagent react with the aldehyde? How would the product react with the acid chloride? Identify which species would behave as the nucleophile and leaving group in each reaction.

Solve

$LiAlH_4$ is a reducing agent, and hydride, H:$^-$, acts as the nucleophile to attack the C=O of the aldehyde. The alkoxide product then acts as the nucleophile and yields an ester after a nucleophilic addition–elimination reaction with the acid chloride.

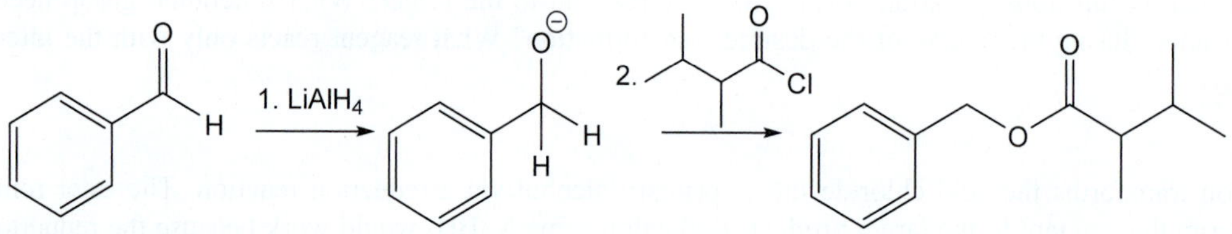

1. $LiAlH_4$

2.

Problem 20.53

Think

How is this reaction similar to the reaction of an ester with an alkoxide? What acts as the nucleophile in the nucleophilic addition–elimination reaction? On which atom is a negative charge better stabilized, O or S?

Solve

(a) The mechanism would appear as follows:

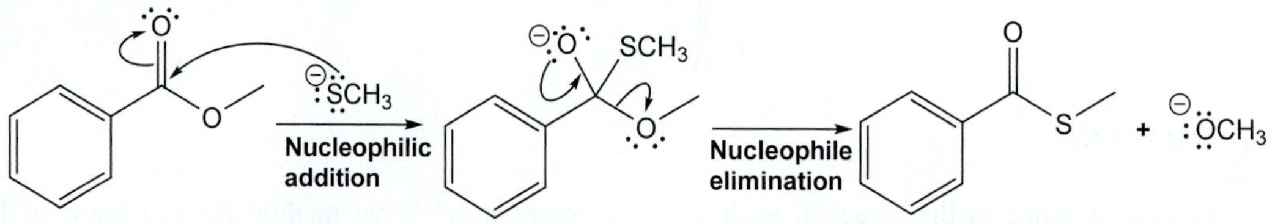

Nucleophilic addition

Nucleophile elimination

(b) The reaction would *not* be energetically favorable, because the product anion is less stable than the reactant anion. In the product, a negative charge appears on O, whereas in the reactants, the negative charge appears on S. The S atom is bigger and can accommodate the negative charge better.

Problem 20.54

Think

What functional group is present in this triglyceride? How does this functional group react with HO$^-$? Are there any irreversible steps? How many times does this reaction occur? Is there a hydrophilic part of the product? Is there a hydrophobic part of the product? How can the product emulsify dirt, grease, or oil?

Solve

The triglyceride has three ester groups, and each undergoes base-catalyzed hydrolysis to form triglycerol and three fatty-acid carboxylate anions. These fatty-acid carboxylates can serve as soap, because there is a very hydrophobic/nonpolar end and a very hydrophilic/polar end. The hydrophobic end dissolves in dirt, grease, or oil, and the hydrophilic end is soluble in water. See the mechanism on the next page.

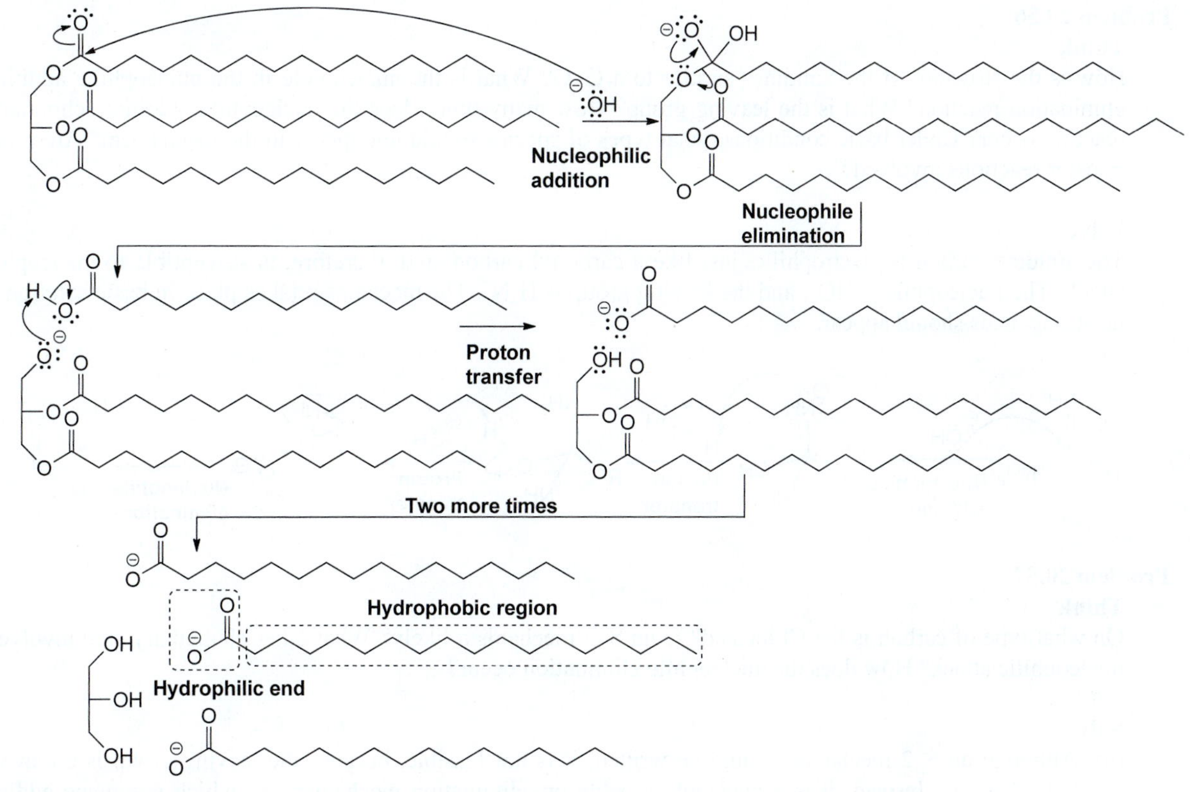

Nucleophilic addition

Nucleophile elimination

Proton transfer

Two more times

Hydrophobic region

Hydrophilic end

Problem 20.55

Think

Consider another resonance structure for the enamine. What is the nucleophile in the nucleophilic addition–elimination reaction? What is the leaving group? In the hydrolysis reaction, what kinds of species should not appear under acidic conditions? What is the nucleophile in the hydrolysis?

Solve

The enamine has a resonance structure in which a negative charge appears on C so that carbon is nucleophilic. It attacks the acid chloride in a nucleophilic addition–elimination reaction in which Cl^- is the leaving group. When H_3O^+ is added, hydrolysis of the C=N bond takes place in a mechanism that is essentially identical to hydrolysis of imines (see Chapter 18). Because the C=N bond is already activated, that hydrolysis is initiated by attack of the C=N carbon by water.

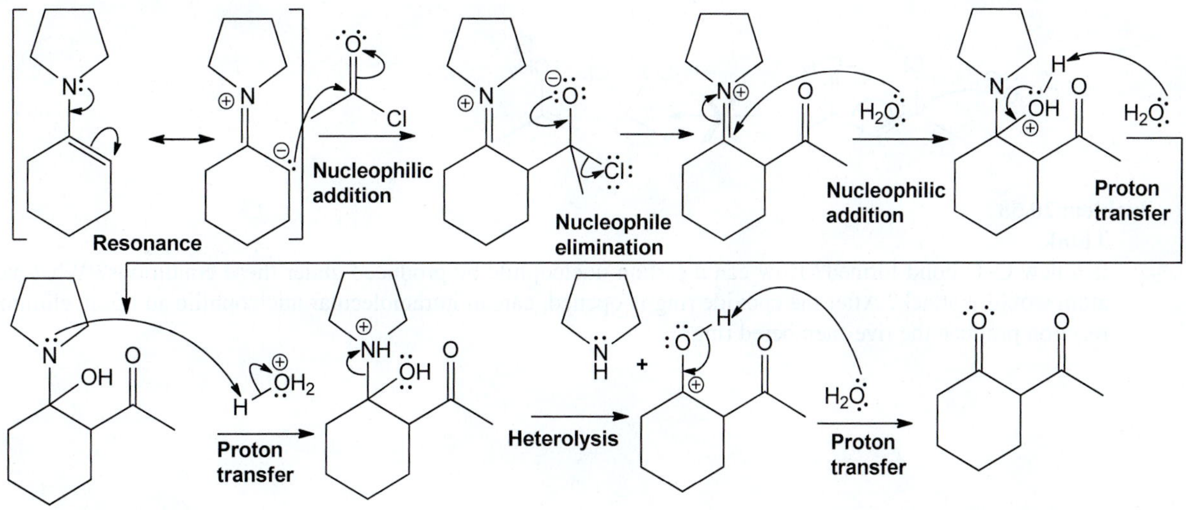

Resonance

Nucleophilic addition

Nucleophile elimination

Nucleophilic addition

Proton transfer

Proton transfer

Heterolysis

Proton transfer

Problem 20.56

Think

How is the structure of the amidines similar to a C=O? What is the nucleophile in the nucleophilic addition–elimination reaction? What is the leaving group? How many times does the nucleophilic addition–elimination reaction occur? Under basic conditions, what types of species should not appear in the mechanism? Are proton transfer reactions involved?

Solve

The amidine carbon is electrophilic, just like a carbonyl carbon, and, therefore, is susceptible to nucleophilic attack. The nucleophile is HO^-, and the leaving group is H_2N^-. The mechanism takes place in basic solution, so no strong acids should appear.

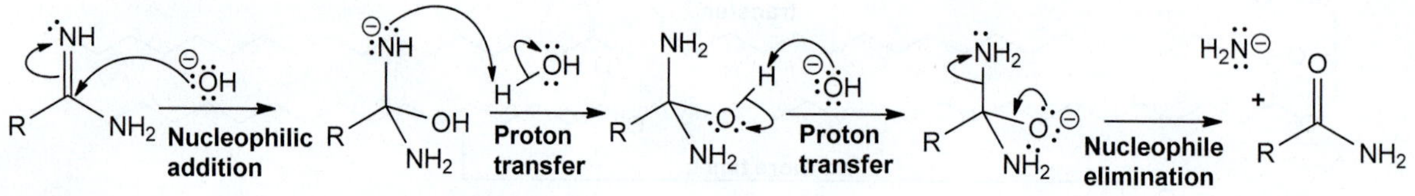

Nucleophilic addition

Proton transfer

Proton transfer

Nucleophile elimination

Problem 20.57

Think

On what type of carbon is the Cl located? Is an S_N2 mechanism likely? What other elementary step involves a nucleophilic attack? How does the nucleophile elimination occur?

Solve

(a) Although an S_N2 mechanism could be written, it is not feasible, because the leaving group is on an sp^2-hybridized C. Instead, it is a nucleophilic addition–elimination mechanism, in which conjugate addition takes place instead of direct addition.

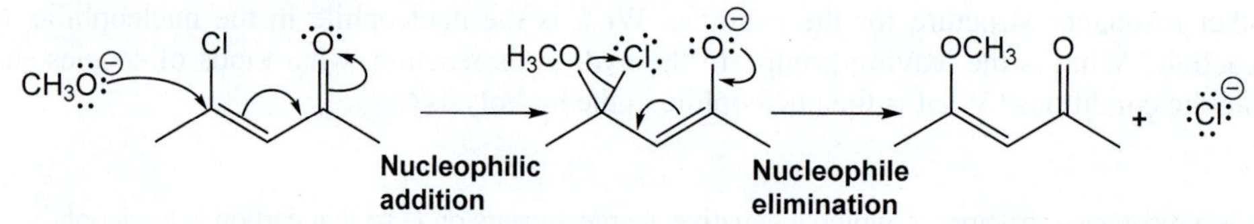

Nucleophilic addition

Nucleophile elimination

(b) The same mechanism is not feasible for the second reaction, because it would require a negative charge to develop on C, which is very unstable.

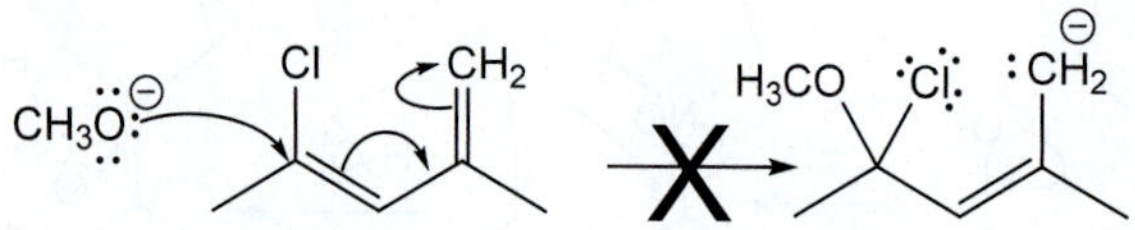

Problem 20.58

Think

Is a new C–C bond formed? How can a carbon nucleophile be produced under these conditions? What carbon atom would it attack? After the epoxide ring is opened, can an intramolecular nucleophilic addition–elimination reaction produce the five-membered ring?

Solve

The central C atom of the diester is mildly acidic, so it is deprotonated to make an enolate anion that is nucleophilic at that C. It attacks the epoxide to ring open the epoxide in an S_N2 mechanism. The alkoxide formed attacks the ester in an intramolecular nucleophilic addition–elimination reaction.

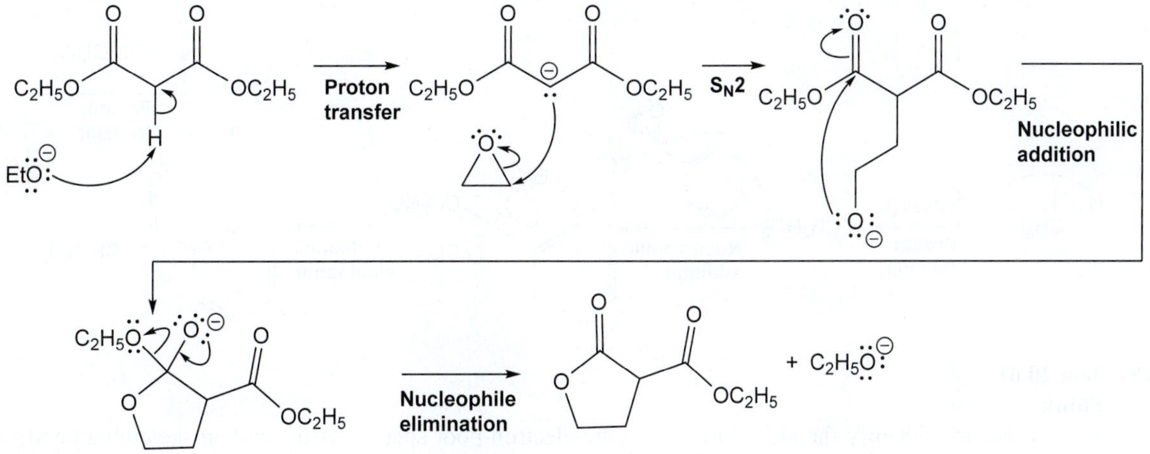

Problem 20.59

Think

Under basic conditions, what types of species should not appear in the mechanism? What is the nucleophile in the nucleophilic addition–elimination reaction? How many times does the nucleophilic addition–elimination reaction occur? Are proton transfer reactions involved?

Solve

LDA deprotonates the NH_2 group to convert it into a strong nucleophile that can attack the C=O group. The mechanism takes place in basic solution, so no strong acids should appear.

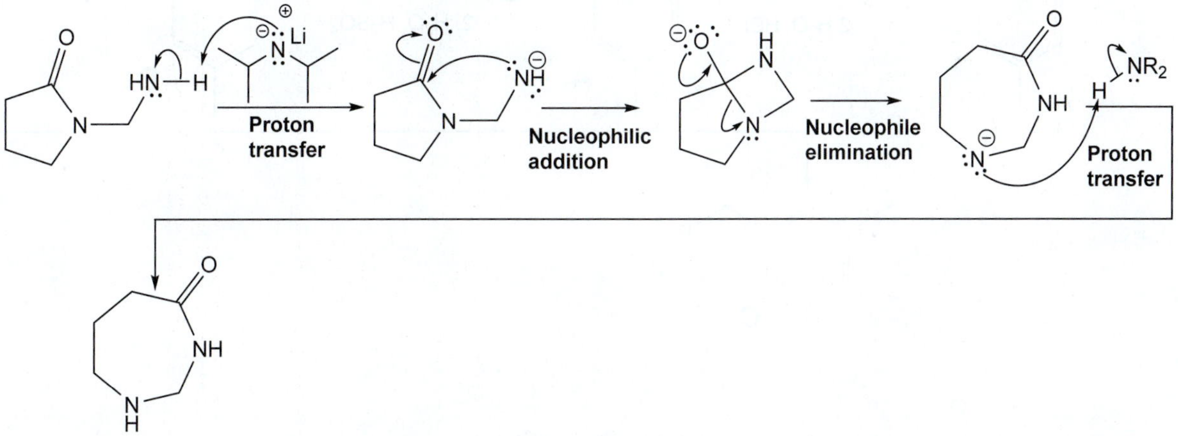

Problem 20.60

Think

Under basic conditions, what types of species should not appear in the mechanism? Is CCl_3 a suitable leaving group? What is the nucleophile in the nucleophilic addition–elimination reaction? How many times does the nucleophilic addition–elimination reaction occur? Are proton transfer reactions involved?

Solve

CH_3NO_2 is moderately acidic, so it can be deprotonated by *t*-butoxide to produce a strong carbon nucleophile, which acts as the nucleophile in the nucleophilic addition–elimination reaction. Cl_3C^- is a suitable leaving group that is protonated in a final proton transfer step to yield Cl_3CH. The mechanism takes place in basic solution, so no strong acids should appear.

Proton transfer

Nucleophilic addition

Nucleophile elimination

Proton transfer

Problem 20.61

Think

In each reaction, identify the electron-rich and the electron-poor species. Is the carbon skeleton altered? What functional group transformations occur? Is the carbon in each reaction the electrophile or the nucleophile? Are proton transfer reactions involved?

Solve

The methyl ketone starting material undergoes a haloform reaction to produce carboxylic acid **A**, which is then reduced with $LiAlH_4$ to produce the primary alcohol **B**. Bromination with PBr_3 converts the OH to a Br to make alkyl halide **C**. That alkyl halide is then used in a Gabriel synthesis to produce primary amine **D**.

1. I_2, NaOH
2 H_2O, HCl

A

1. $LiAlH_4$, THF
2. H_2O, H_2SO_4

B

PBr_3

C

1. NaOH
2. **C**
3. NaOH, Δ

D

Problem 20.62

Think

In each reaction, identify the electron-rich and the electron-poor species. Is the carbon skeleton altered? What functional group transformations occur? Is the carbon in each reaction the electrophile or the nucleophile? Are proton transfer reactions involved?

Solve

A is produced by reducing the starting material to the aldehyde. **A** then reacts with ketone **B** in an aldol condensation reaction. Lithium dimethylcuprate **C** then undergoes conjugate addition, followed by reduction to produce **D**. PBr_3 then converts the OH to a Br to produce **E**. In the Gabriel synthesis involving **E**, the primary amine **F** is produced, which undergoes Hofmann elimination in the final step to produce **G**.

1. DIBAH 2 H_2O, HCl; **A**; **B**; NaOH, Δ; **C** 1. $(CH_3)_2CuLi$ 2. NH_4Cl; $NaBH_4$ EtOH; **D**; PBr_3; **E**; 1. NaOH 2. **E** 3. NaOH, Δ; **F**; 1. CH_3I (excess) 2. Ag_2O, H_2O 3. Δ; **G**

Problem 20.63

Think

In each reaction, identify the electron-rich and the electron-poor species. Is the carbon skeleton altered? What functional group transformations occur? Is the carbon in each reaction the electrophile or the nucleophile? Are proton transfer reactions involved?

Solve

The starting amide undergoes hydrolysis to produce carboxylic acid **A**. Diazomethane converts the carboxylic acid to a methyl ester, which is reduced to the aldehyde **B** by DIBAH. The aldehyde then undergoes a Grignard reaction to make alcohol **C**. The alcohol is then deprotonated to make the alkoxide anion, which is then acylated with acetic anhydride, yielding the target compound **D**.

1. NaOH, Δ 2. H_2O, HCl; **A**; CH_2N_2; DIBAH **B**; 1. vinyl-MgBr 2. NH_4Cl, H_2O; **C**; 1. NaH 2. acetic anhydride; **D**

Problem 20.64

Think

In each reaction, identify the electron-rich and the electron-poor species. Is the carbon skeleton altered? What functional group transformations occur? Is the carbon in each reaction the electrophile or the nucleophile? Are proton transfer reactions involved?

Solve

The starting alkene undergoes hydroboration–oxidation to form the primary alcohol **A**. The alcohol is oxidized to the carboxylic acid **B** by $KMnO_4$, followed by acid workup. The carboxylic acid reacts with CH_2N_2 to form the methyl ester **C**. The ester undergoes a Grignard reaction (addition of two $C_6H_5^-$ nucleophiles) to form the tertiary alcohol **D**. This alcohol undergoes a dehydration reaction to form the trisubstituted alkene. The alkene then undergoes hydroboration–oxidation to form the less-substituted alcohol **F**. The alcohol is deprotonated by NaH to form an O^- nucleophile that reacts with the ester **C** to form the new ester **G**.

1. BH_3•THF
2. NaOH, H_2O_2
OH
A
1. $KMnO_4$, KOH, H_2O
2. HCl
O
OH
B
CH_2N_2
O
O
C
1. C_6H_5MgBr
2. H_2O, HCl
OH
Ph
D Ph
conc
H_3PO_4
Δ
Ph
Ph
E
1. BH_3•THF
2. NaOH, H_2O_2
Ph
Ph
OH
F
1. NaH
2. O O **C**
O
O
Ph
Ph
G

Problem 20.65 (SYN)

Think

To form the product given, what nucleophile is used to undergo a nucleophilic addition–elimination reaction with the acid chloride? Are proton transfer reactions involved?

Solve

The reactions are given below and on the next page:

(a)
O
Cl
O
ONa
O O
O

(b)
O
Cl
NaO
O
O

(c) O
Cl
$LiAlH_4$
O⊖
O
Cl
O
O

(d) 1. LiHN 2. H^+ ; LiAlH$_4$

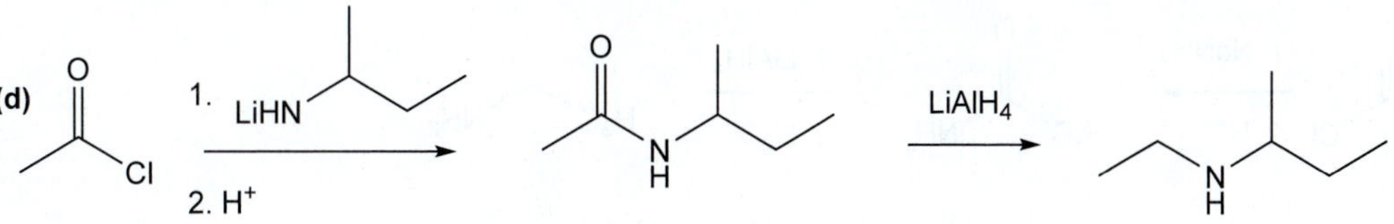

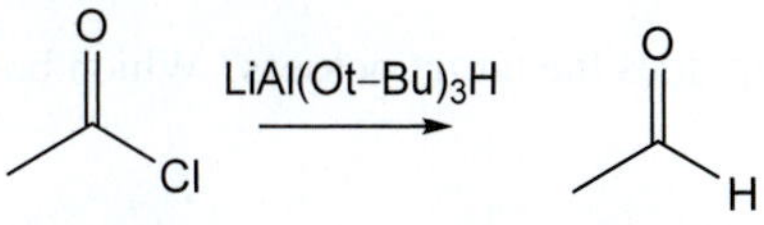

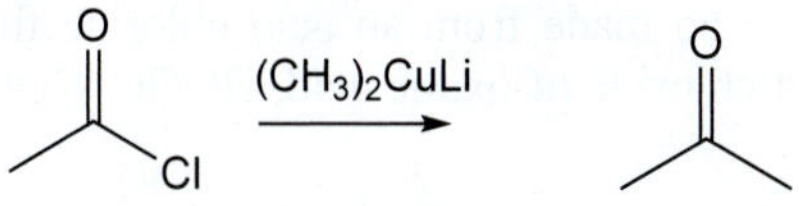

(e) 1. LiAlH$_4$ 2. H_2O ; OH ; H_2SO_4, Δ ; O

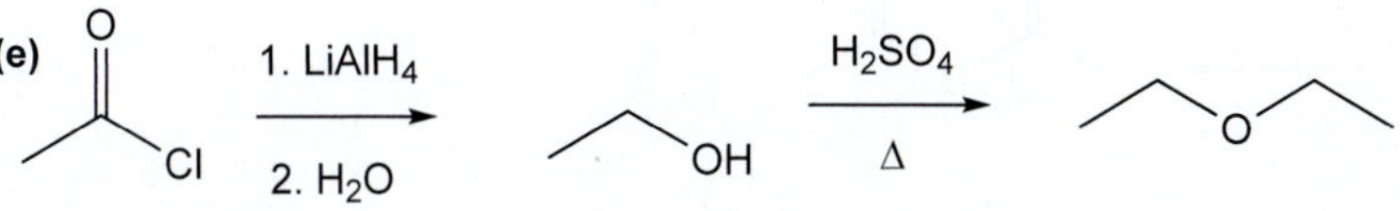

(f) LiAl(Ot–Bu)$_3$H ; H

(g) $(CH_3)_2CuLi$

(h) ONa ; O

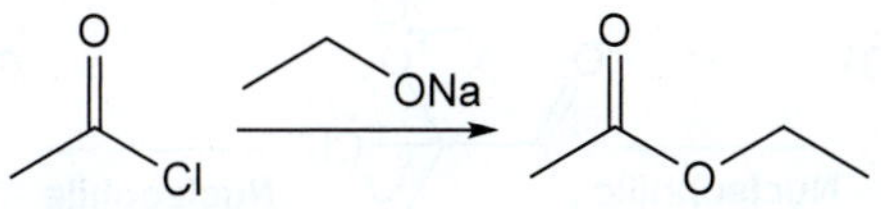

Problem 20.66 (SYN)

Think

To form the product given, what nucleophile is used to undergo a nucleophilic addition–elimination reaction with the acid chloride? Are any carbon–carbon bond-forming reactions necessary? Are proton transfer reactions involved?

Solve

The reactions are given below and on the next page:

(a) Cl ; LiAlH$_4$; Cl ; O

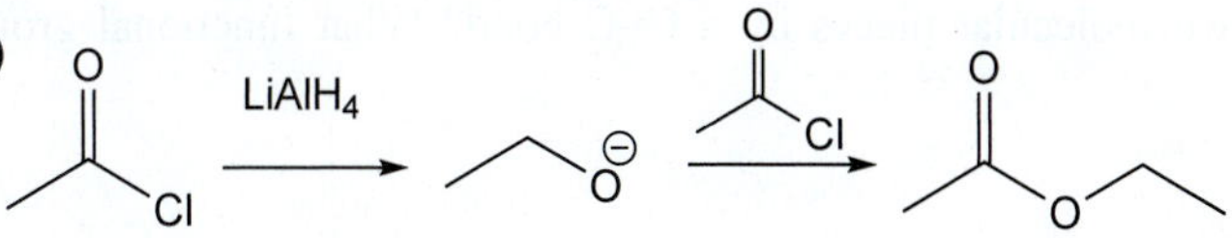

(b) From (a) ; 1. H_3O^+ 2. PBr_3 ; Br ; Li ; Li ; CuI ; CuLi ; Cl ; NaBH$_4$, EtOH ; OH

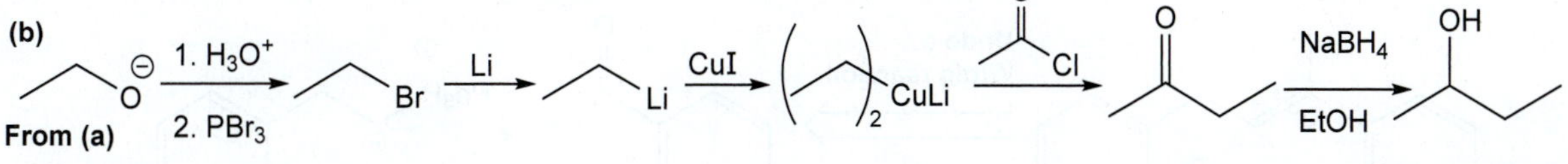

(c) Use the ester from **(a)** and ethyllithium from part **(b)** (or a Grignard reagent) to produce the target.

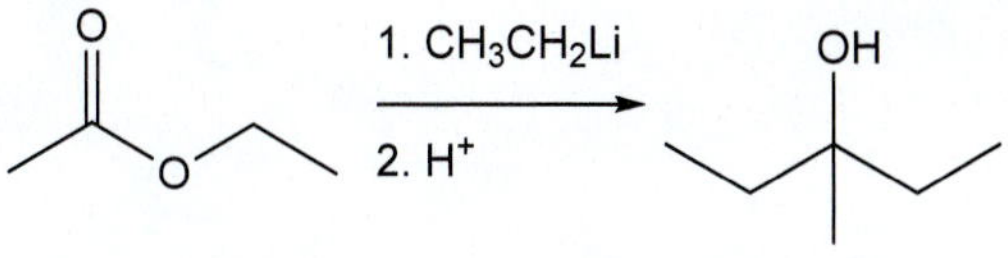

1. CH_3CH_2Li 2. H^+ ; OH

(d) Butan-2-one is made in **(b)**.

(e) H_3C–CO–Cl → 1. $NaNH_2$ 2. H^+ → H_3C–CO–NH_2 → 1. $LiAlH_4$ 2. H_2O → H_3C–CH_2–NH_2

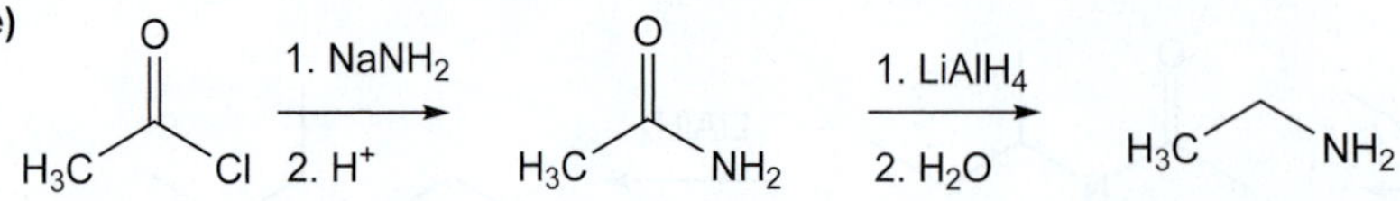

(f) CH_3COCl → NaOH → $CH_3CO_2^-$ → CH_3COCl → CH_3CO–O–$COCH_3$

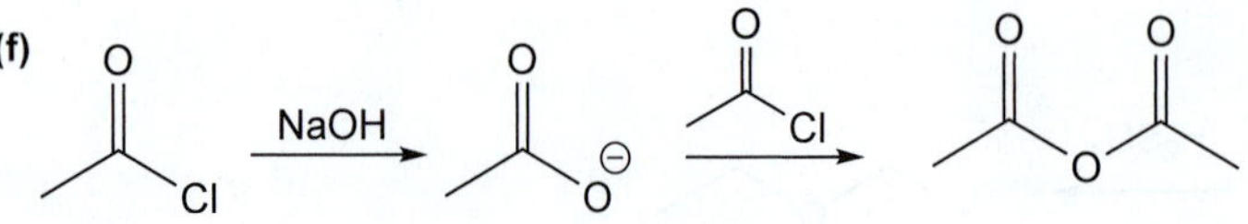

Problem 20.67 (SYN)

Think

How many C atoms does the product possess? What functional group does the target possess? Which bonds are likely formed?

Solve

The product is a diester with four C atoms. Each ester group can be made from an acid chloride functional group and an alkoxide functional group. Oxalyl chloride (the acid chloride of oxalic acid, HOOC–COOH) and ethylene glycol dianion will combine in one reaction.

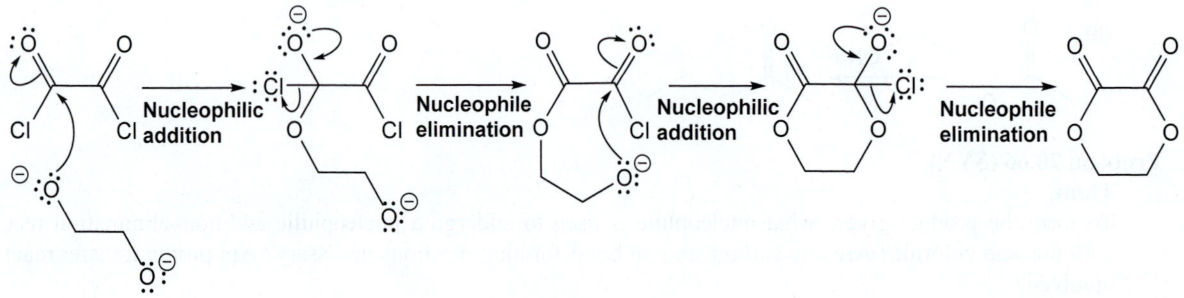

Nucleophilic addition → Nucleophile elimination → Nucleophilic addition → Nucleophile elimination

Problem 20.68

Think

How many C atoms does the product possess? Are carbon–carbon bond-forming reactions necessary? What reactions do you know that will connect two molecular pieces by a C=C bond? What functional groups are required for that reaction?

Solve

The C=C bond can be the last bond formed from a Wittig reaction.

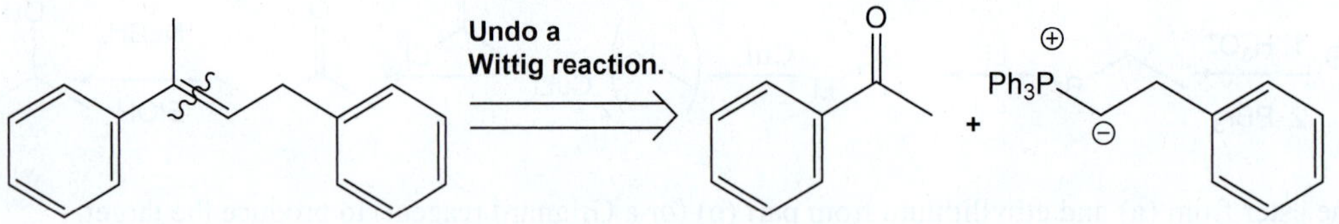

Undo a Wittig reaction. ⟹ acetophenone + Ph_3P^+–CH^-–CH_2–Ph

The two reagents for the Wittig reaction can be synthesized from a carboxylic acid.

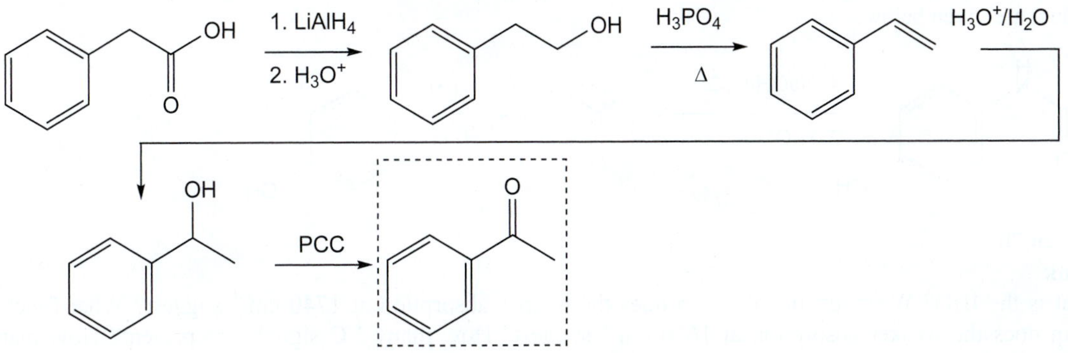

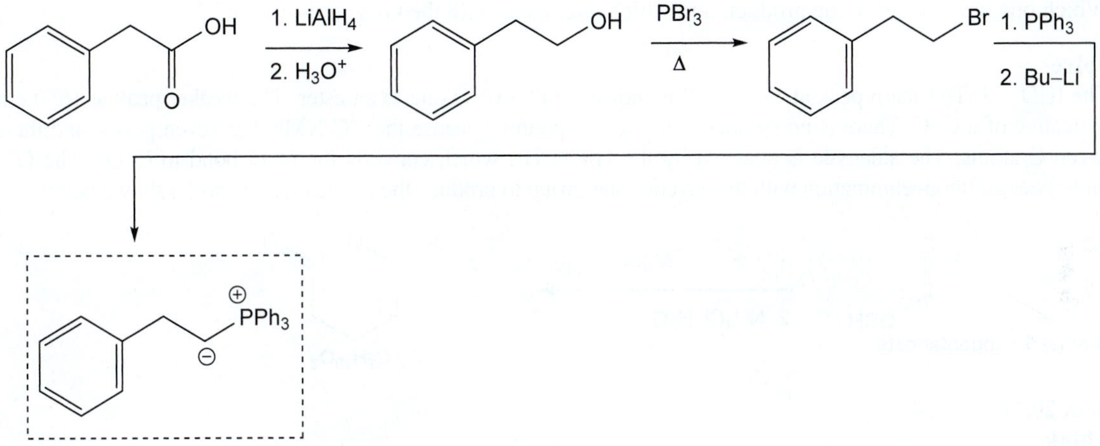

1. $LiAlH_4$
2. H_3O^+

H_3PO_4
Δ

H_3O^+/H_2O

OH

PCC

O

OH

1. $LiAlH_4$
2. H_3O^+

OH

PBr_3
Δ

Br

1. PPh_3
2. Bu–Li

PPh_3

The Wittig reaction in the forward direction is shown below:

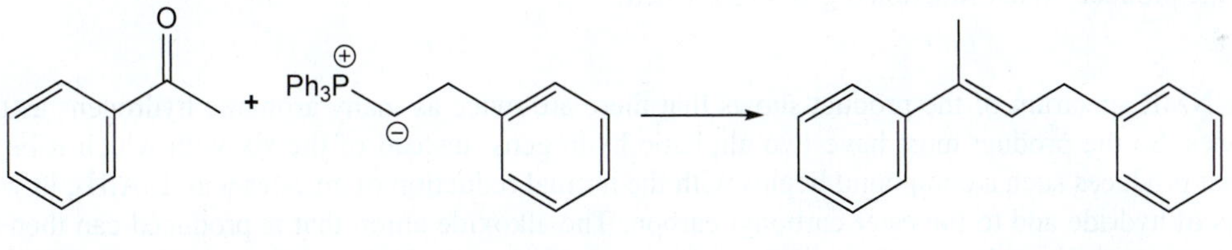

O

+ Ph_3P

Problem 20.69

Think

What is the IHD? What functional group do the two broad IR peaks from 3400–3200 cm^{-1} suggest is present? The multiple IR peaks from 1600–1400 cm^{-1} suggest the presence of what functional group? In the NMR spectrum of the product, what could account for the peaks around 6.5 ppm? For the broad signals around 8.4 and 4.3 ppm? What is the structure of the product? What acid derivative yields this product and acetic acid from the hydrolysis reaction in base?

Solve

The IHD = 5. The IR peaks from 3400–3200 cm^{-1} suggest the presence of a primary amine and/or an alcohol. The several peaks from 1600–1400 cm^{-1} are indicative of an aromatic ring, which accounts for an IHD = 4. Therefore, this hydrolysis product contains an amine and is aromatic. This is further indicated by the peaks around 6.5 ppm in the ^1H NMR spectrum. The other two peaks are broad. One arises from the protons on the amide group, at 4.3 ppm, which integrates to 2 H. The other, at 8.4 ppm, is consistent with an OH group as part of a phenol, as it integrates to 1 H. The C=O group also contributes one to the IHD to bring it to a total of five.

Acetic acid is a carboxylic acid. Therefore, an amide starting material would yield these two products from the hydrolysis reaction in base followed by an acid workup. The structure of the starting material pain reliever and product are given below:

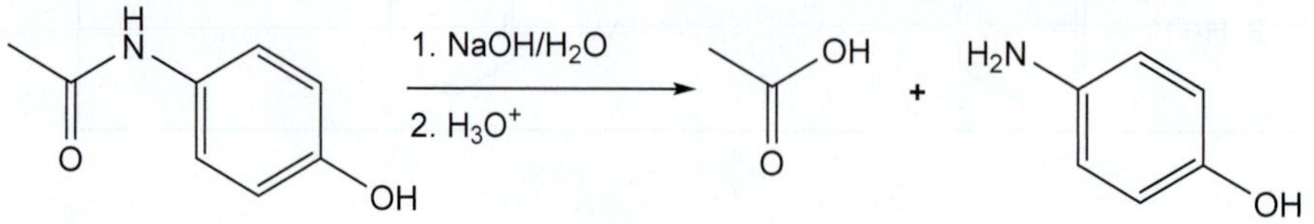

1. NaOH/H_2O
2. H_3O^+

Problem 20.70

Think

What is the IHD? What functional group does the intense absorption at 1740 cm^{-1} suggest? What functional group does the weaker absorption at 1650 cm^{-1} suggest? How many ^{13}C signals are present? How many C atoms are present in the formula? Is there any symmetry? Which C=O in the compound is more reactive? Which one will remain in the product, and which one reacts with the Grignard reagent?

Solve

The IHD = 3. The sharp peak at 1740 cm^{-1} is indicative of a C=O due to an ester. The weaker peak at 1650 cm^{-1} is indicative of a C=C. There is no symmetry in the compound, because the ^{13}C NMR has seven peaks, and there are seven C atoms. The aldehyde is attacked by the $H_2C{=}CH^-$, which converts the C=O bond to C–O$^-$. The O$^-$ then undergoes addition–elimination with the acyclic ester group to produce the cyclic ester (lactone) shown below.

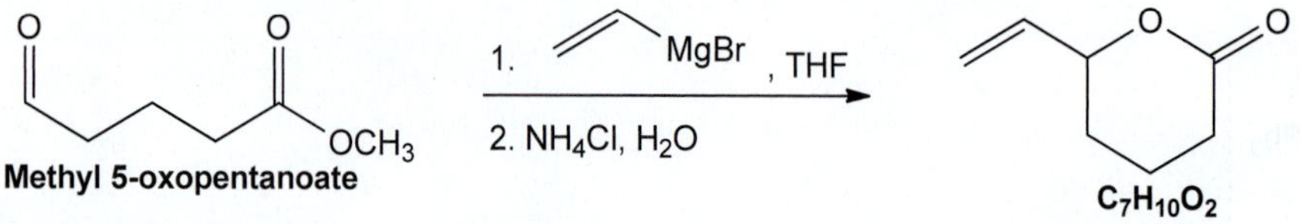

Methyl 5-oxopentanoate

1. MgBr, THF
2. NH_4Cl, H_2O

$C_7H_{10}O_2$

Problem 20.71

Think

What functional group is present in dimethyl phthalate? How does this acid derivative react with $LiAlH_4$? How many signals are in the ^1H NMR? What is the relative ratio of the integration of the peaks? Does the aromatic ring remain in the product? What functional group is formed?

Solve

Notice that the NMR spectrum of the product shows that there are twice as many aromatic hydrogens as there are aliphatic ones. So the product must have two aliphatic hydrogens, instead of the six with which it begins. The reaction that produces such a compound begins with the normal reduction of an ester with $LiAlH_4$, in which two equivalents of hydride add to the ester carbonyl carbon. The alkoxide anion that is produced can then be a nucleophile in a nucleophilic addition–elimination mechanism with the other ester group.

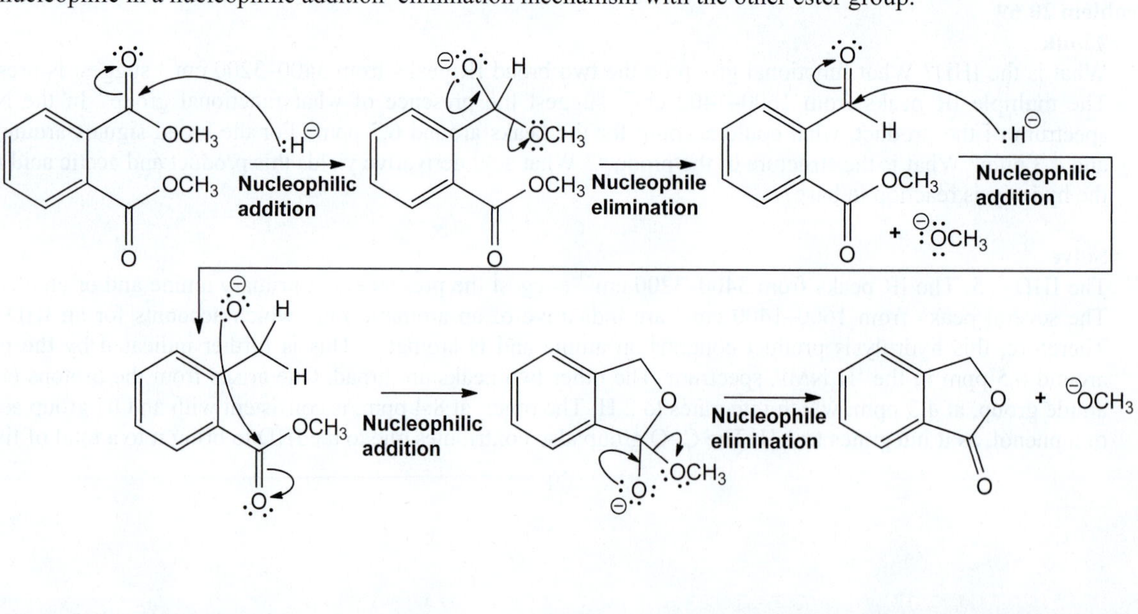

Nucleophilic addition → Nucleophile elimination → Nucleophilic addition → Nucleophilic addition → Nucleophile elimination

CHAPTER 21 | Nucleophilic Addition–Elimination Reactions 2: Weak Nucleophiles

Your Turn Exercises

Your Turn 21.1

Think

Use Equation 21-4 as a guide. What bonds are broken and formed in each step? Which sites are electron rich? Which are electron poor? How can you use curved arrows to show that electron movement? Review the nucleophilic addition–elimination mechanism from Chapter 20. What similarities do you observe? In what steps are protons involved in the reaction?

Solve

Steps 1 and 3 of the mechanism consist of the usual nucleophilic addition and elimination steps. The alcohol $CH_2{=}CH(CH_2)_3{-}OH$ attacks the carbonyl carbon in Step 1 to produce the tetrahedral intermediate. Step 2 involves a proton transfer because this tetrahedral intermediate contains an acidic proton that is rapidly deprotonated by another alcohol molecule. A lone pair of electrons on the O in the tetrahedral intermediate folds down to re-form the C=O and expel the Cl^- leaving group in Step 3.

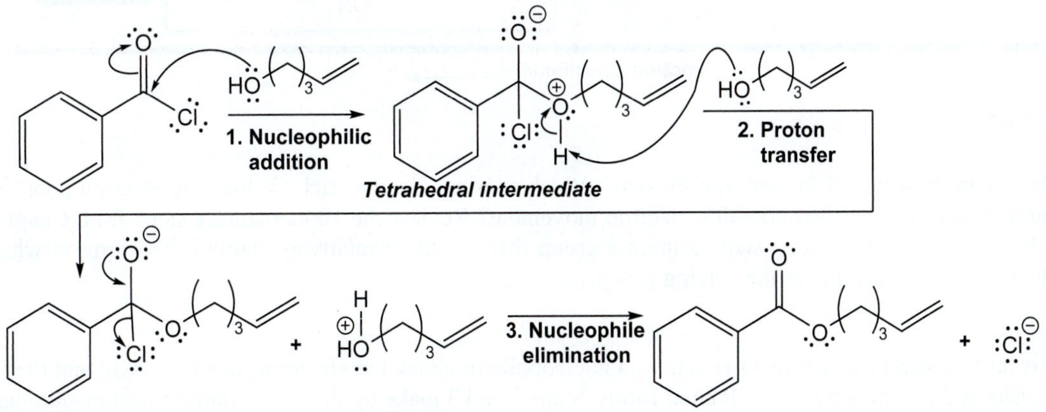

1. Nucleophilic addition

Tetrahedral intermediate

2. Proton transfer

3. Nucleophile elimination

Your Turn 21.2

Think

Using Figure 20-3 and Equation 21-4 as guides, see what intermediate is formed in Step 1 and Step 2. Where do intermediates appear on energy diagrams? Which step has the highest transition state energy, and how does that correspond to its rate?

Solve

The species are added on the free energy diagram on the next page. The tetrahedral intermediate is the product of nucleophilic addition and has a relatively high energy. The first step has the highest energy transition state, which is consistent with the first step (nucleophilic addition) being the slow step.

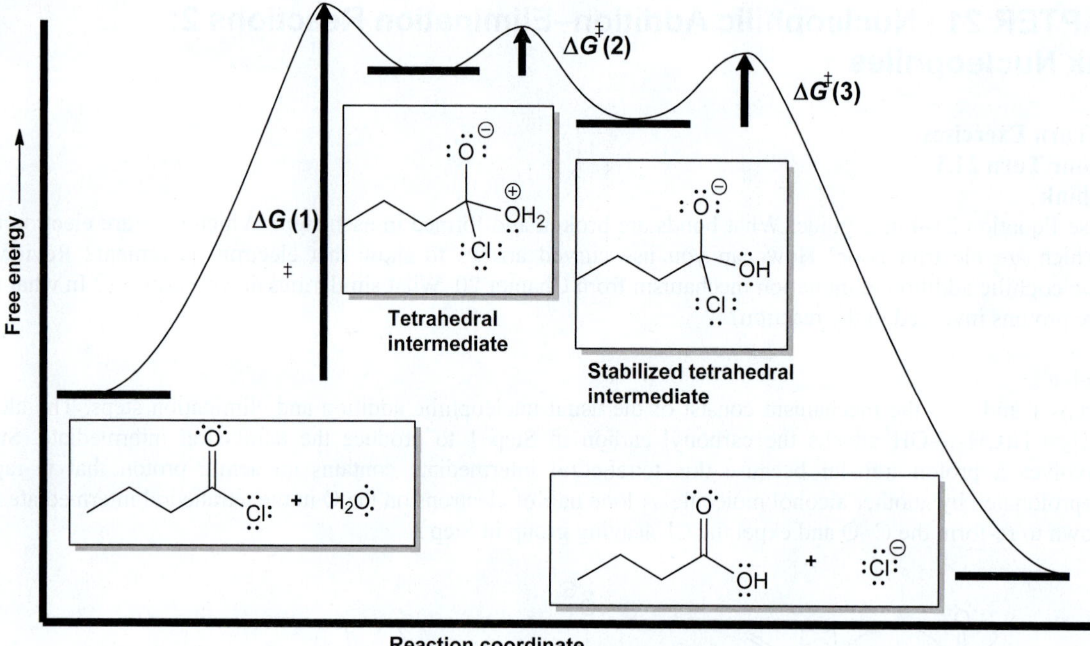

Your Turn 21.3

Think

What bonds are broken and formed in each step? Which sites are electron rich? Which are electron poor? How can you use curved arrows to show that electron movement? Review the 10 elementary steps from Chapters 6 and 7. In the product of the first step, is there a group that would be relatively stable in the form in which it could leave? Label this group as the leaving group.

Solve

Step 1 is an S_N2 step in which RCO_2H acts as a nucleophile to attack the electron-poor P in PCl_3, and the P–Cl bond breaks at the same time the O–P bond forms. Steps 2 and 3 make up the nucleophilic addition–elimination mechanism. The leaving group in the product of Step 1 is circled.

1. S_N2

Leaving group

Resonance

2. Nucleophilic addition

3. Nucleophile elimination

4. Proton transfer

+ $HOPCl_2$

Your Turn 21.4

Think

Refer to the referenced sections in the partial mechanism of Equation 21-18. In each partial mechanism, what is the overall reaction? What elementary steps make up the overall reaction? Which species is electron rich? Which is electron poor? Are proton transfer reactions involved?

Solve

In the first partial mechanism of Equation 21-18, the carboxylic acid is transformed into an acyl bromide using PBr_3. The mechanism is the same as the one shown in Your Turn 21.2 for PCl_3. Step 1 is an S_N2 step in which RCO_2H acts as a nucleophile to attack the electron-poor P in PBr_3, and the P–Br bond breaks at the same time the O–P bond forms. Steps 2 and 3 make up the nucleophilic addition–elimination mechanism. The final step is a proton transfer to yield the uncharged acyl bromide.

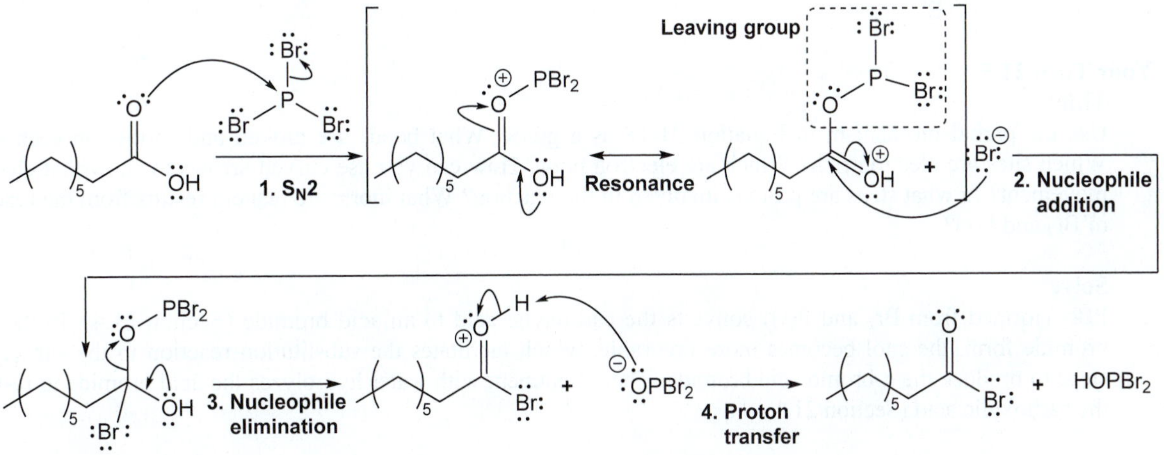

1. S_N2 Resonance Leaving group 2. Nucleophile addition 3. Nucleophile elimination 4. Proton transfer $HOPBr_2$

In the second partial mechanism of Equation 21-18, the acyl bromide is tautomerized into an enol in the presence of the $HOPBr_2$ acid. The strong acid that is present donates a proton to the O atom of the C=O group in Step 1. In Step 2, Br_2PO^- (a weak base) removes the proton from the α carbon to produce the uncharged enol product (Equation 7-35).

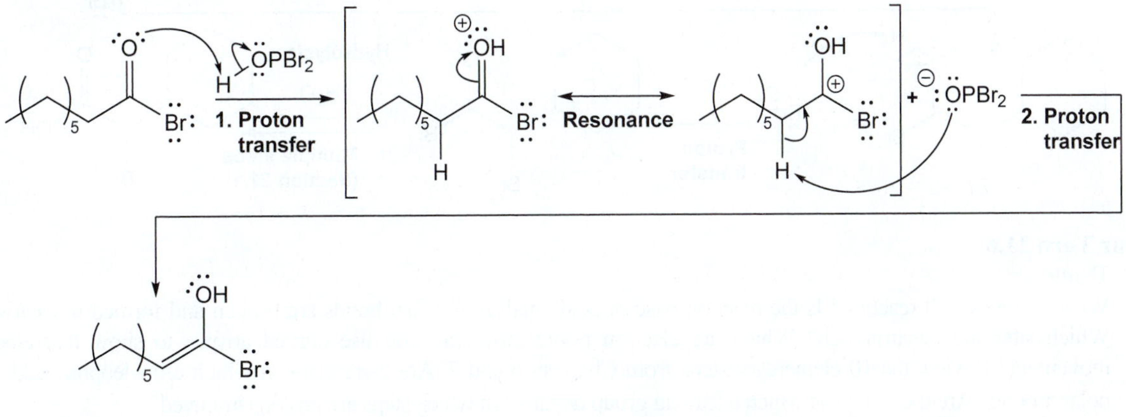

1. Proton transfer Resonance 2. Proton transfer

In the third partial mechanism of Equation 21-18, the acyl bromide is hydrolyzed into a carboxylic acid (Section 21.1). Steps 1 and 3 of the mechanism consist of the usual nucleophilic addition and elimination steps. The water attacks the carbonyl carbon in Step 1 to produce the tetrahedral intermediate. Step 2 involves a proton transfer, because this tetrahedral intermediate contains an acidic proton that is rapidly deprotonated by another alcohol molecule. A lone pair of electrons on the O in the tetrahedral intermediate folds down to re-form the C=O and expel the Br^- leaving group in Step 3. The mechanism is drawn out on the following page.

H_2O 1. Nucleophilic addition

Tetrahedral intermediate

H_2O 2. Proton transfer

3. Nucleophile elimination

Your Turn 21.5

Think

Use the partial mechanism in Equation 21-18 as a guide. What bonds are broken and formed in each step? Which sites are electron rich? Which are electron poor? How can you use curved arrows to show that electron movement? In what steps are protons involved in the reaction? What inorganic reagent results from the reaction of Br_2 and P(*s*)?

Solve

PBr_3 (formed from Br_2 and P(*s*)) converts the carboxylic acid to an acid bromide (Section 21.4). In the acid bromide form, the enol becomes more favorable, which facilitates the substitution reaction in the subsequent steps to produce the α-bromo acid bromide. Then, treatment with water hydrolyzes the acid bromide to re-form the carboxylic acid (Section 21.1).

PBr_3 Multiple steps (Section 21.4) + $HOPBr_2$

Two steps (Section 7.9)

Proton transfer

Hydrolysis H_2O Multiple steps (Section 21.1)

Your Turn 21.6

Think

What is the overall reaction? Is the reaction base or acid catalyzed? What bonds are broken and formed in each step? Which sites are electron rich? Which are electron poor? How can you use curved arrows to show that electron movement? Review the 10 elementary steps from Chapters 6 and 7. Are there steps in which a nucleophile adds to a polar π bond? Are there steps in which a leaving group departs? In which steps are protons involved?

Solve

Step 1 is a proton transfer to the carboxylic acid's carbonyl group to activate the carbonyl. Step 2 is nucleophilic addition of the alcohol to yield the tetrahedral intermediate. Step 3 is a proton transfer to remove the charge from O on the incoming nucleophile. Step 4 is a proton transfer in which the singly bonded O atom of the original carboxylic acid gains a positive charge. This increases the ability of the leaving group to depart. In Step 5, the leaving group departs in a nucleophile elimination step. In Step 6, the carbonyl O is deprotonated, yielding the uncharged product. See the mechanism on the next page.

1.Proton transfer
2. Nucleophilic addition
Tetrahedral intermediate
3.Proton transfer
4. Proton transfer
5. Nucleophile elimination
6. Proton transfer

Your Turn 21.7

Think

Review Sections 20.1 and 20.2 to consider the reactivity of an ester with an alkoxide and a hydroxide. In this example, how can the RO^- and HO^- species act as nucleophiles? As bases? Are any steps irreversible?

Solve

The mechanism for transesterification (Equation 21-39) consists of the usual nucleophilic addition–elimination steps, with an RO^- nucleophile and a separate RO^- leaving group. The mechanism for saponification (Equation 21-40) begins with the same two steps, in which HO^- is the nucleophile and RO^- is the leaving group, followed by an irreversible proton transfer.

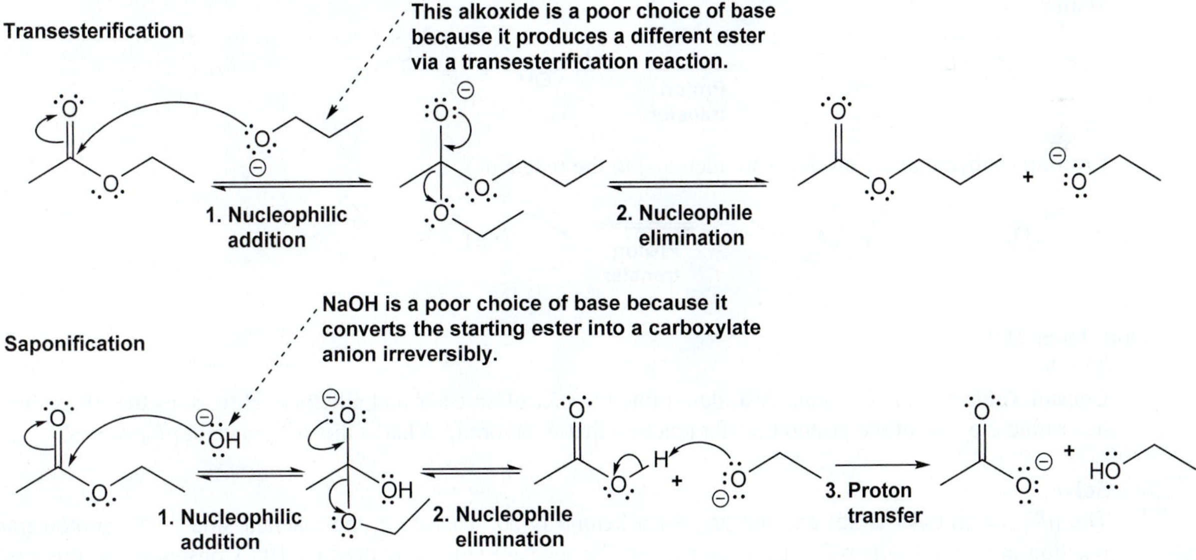
Transesterification
This alkoxide is a poor choice of base because it produces a different ester via a transesterification reaction.
1. Nucleophilic addition
2. Nucleophile elimination
Saponification
NaOH is a poor choice of base because it converts the starting ester into a carboxylate anion irreversibly.
1. Nucleophilic addition
2. Nucleophile elimination
3. Proton transfer

Your Turn 21.8

Think

Review the transesterification mechanism from Section 20.1. How can the $CH_3CH_2O^-$ species act as a nucleophile with a polar π bond? Is there a separate RO^- leaving group? How do these reaction conditions avoid undesired products?

Solve

$CH_3CH_2O^-$ acts as a nucleophile (highlighted in gray below) in Step 1, adding to the C atom of the C=O group. In Step 2, a separate $CH_3CH_2O^-$ species is kicked out by the reformation of the C=O to produce the overall product. Transesterification using ethoxide as the base avoids undesired products because the ester product of the reaction is the same as the ester starting material.

Transesterification

1. Nucleophilic addition

2. Nucleophile elimination

Your Turn 21.9

Think

In each proton transfer reaction, what is the base and what is the acid? How do you use curved arrows to show bond formation and breaking? What products form? If the product is nucleophilic, how does that pose a problem to the desired reaction?

Solve

In each case, $CH_3CH_2O^-$ is the base and picks up a proton from each of the uncharged acids. The products are different from the reactants unless ethanol is the acid. When propan-1-ol is used, the propoxide anion is produced, which can react with the ester in a transesterification reaction. When water is used, hydroxide anion is produced, which can react with the ester in a saponification reaction.

Propan-1-ol

Proton transfer

Water

Proton transfer

Ethanol (notice that the products are identical to the reactants)

Proton transfer

Your Turn 21.10

Think

Consult Table 6-1 or Appendix A to determine the pK_a of an ester and a ketone. How does the pK_a value help determine the side of the proton transfer reaction that is favored? What is the difference in pK_a values?

Solve

The pK_a for an ester is 25, and the pK_a for a ketone is 20. Therefore, the reactant side of the proton transfer reaction is favored. The pK_a difference is 5, so the reactant side is favored by 10^5. Consequently, this reaction will not interfere with the intended Claisen condensation reaction.

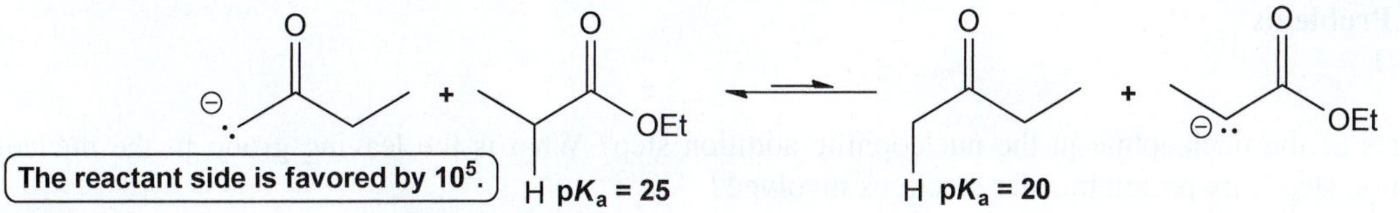

The reactant side is favored by 10^5.

H pK_a = 25

H pK_a = 20

Your Turn 21.11

Think

Consult Appendix A to determine the pK_a of the diester and an alcohol. How do those pK_a values help determine the side of the reaction that is favored? How does the difference in pK_a values correspond to the extent to which that side is favored?

Solve

Malonic ester has a pK_a of 13.5. The pK_a of ethanol is 16. Malonic ester, a reactant, is the stronger acid, so the product side is favored. The difference in pK_a values is 2.5, so the product side is favored by $10^{(16-13.5)} = 10^{2.5} = 3 \times 10^2$.

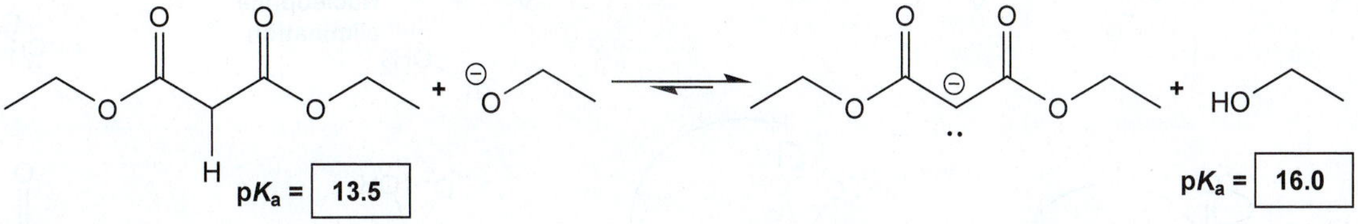

pK_a = 13.5

pK_a = 16.0

Your Turn 21.12

Think

Is an alkyl group electron donating or electron withdrawing? What impact does that have on a nearby negative charge?

Solve

The alkyl group is electron donating, so it increases the concentration of negative charge on the nearby carbon, which destabilizes that anion.

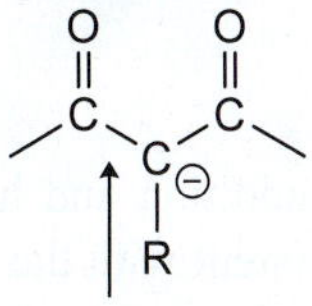

In Chapter Problems

Problem 21.1

Think

What acts as the nucleophile in the nucleophilic addition step? What is the leaving group in the nucleophile elimination step? Are proton transfer reactions involved?

Solve

Reaction of an acid chloride with the nucleophilic alcohol or water produces an ester or a carboxylic acid, respectively. The chloride is the leaving group in the nucleophile elimination reaction.

(a)
Nucleophilic addition
Proton transfer
Nucleophile elimination
+ $PhOH_2^+$
+ Cl^-

(b)
H_2O
Nucleophilic addition
H_2O
Proton transfer
Nucleophile elimination
+ H_3O^+
+ Cl^-

Problem 21.2

Think

Using Figure 20-3 and Equation 21-4 as guides, what is the intermediate formed in Step 1 and Step 2? Where do intermediates appear on energy diagrams? Which step has the largest free energy of activation, and how does that correspond to its rate constant?

Solve

The species are added below. The tetrahedral intermediate is the product of nucleophilic addition and has a relatively high energy. The first step has the largest activation energy, $\Delta G^{\ddagger}(1)$, which is consistent with the first step (nucleophilic addition) being the slow step.

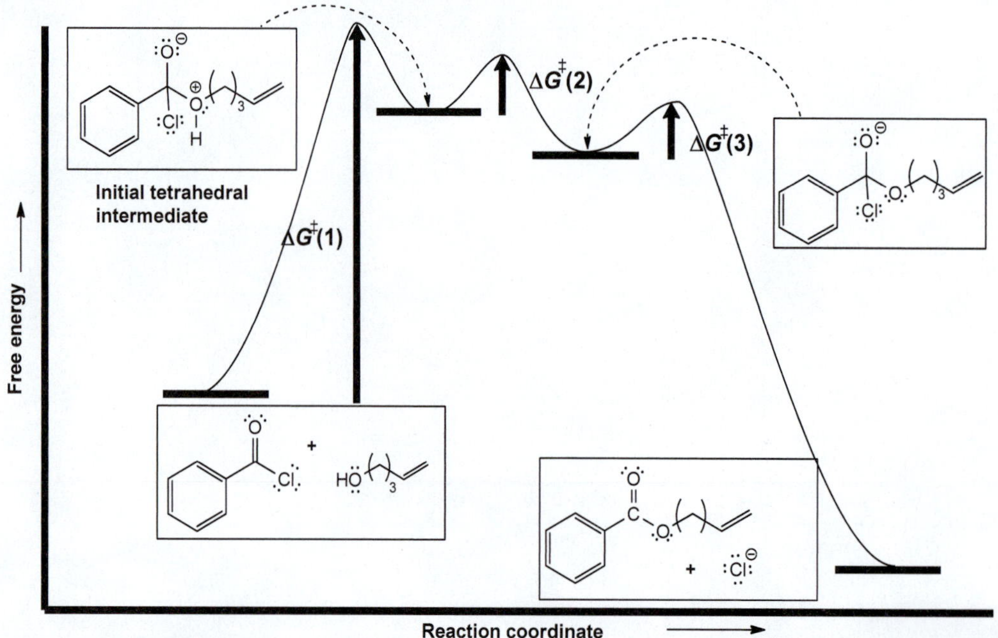

Problem 21.3

Think

What acts as the nucleophile in the nucleophilic addition step? What is the leaving group in the nucleophile elimination step? Are proton transfer reactions involved?

Solve

Reaction of an acid anhydride with a nucleophilic alcohol produces an ester and a carboxylic acid, and reaction of an acid anhydride with the nucleophile water produces two carboxylic acids. In each case, a carboxylate anion is the leaving group in the nucleophile elimination step.

The mechanism for Equation 21-5 is shown below:

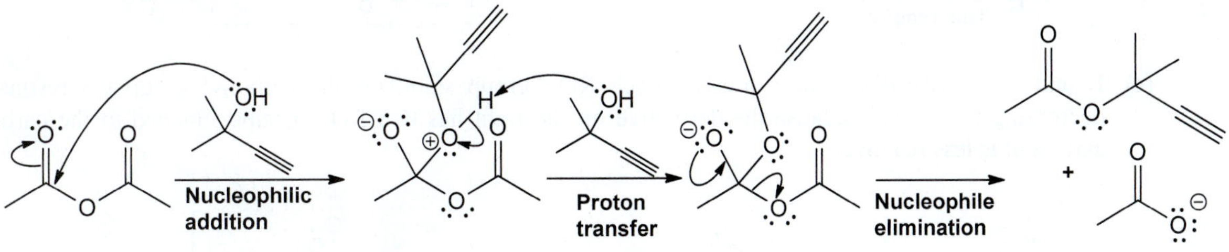

Nucleophilic addition

Proton transfer

Nucleophile elimination

The mechanism for Equation 21-6 is shown below:

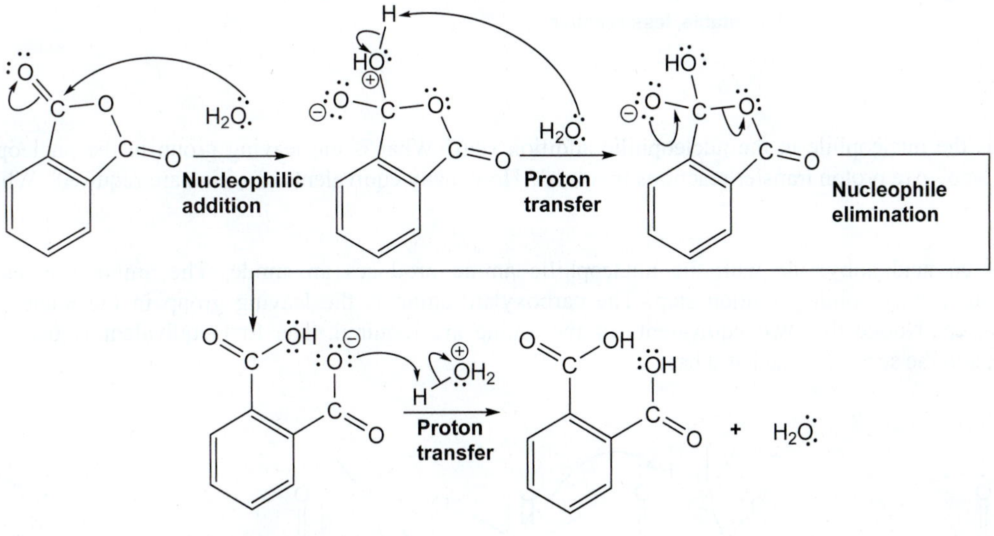

Nucleophilic addition

Proton transfer

Nucleophile elimination

Proton transfer

Problem 21.5

Think

How do resonance and inductive effects play a role in the reactivity of carboxylic acid derivatives? Are the carbonyl groups stabilized differently by resonance? How do inductive effects alter the concentration of partial positive charge on the carbonyl C?

Solve

(a) **A** undergoes hydrolysis more quickly. The difference is whether a CH_3 or CF_3 is attached to the carbonyl carbon. A CF_3 group is inductively electron withdrawing and, thus, increases the concentration of positive charge on the carbonyl carbon, making it more susceptible to reaction with a nucleophile.

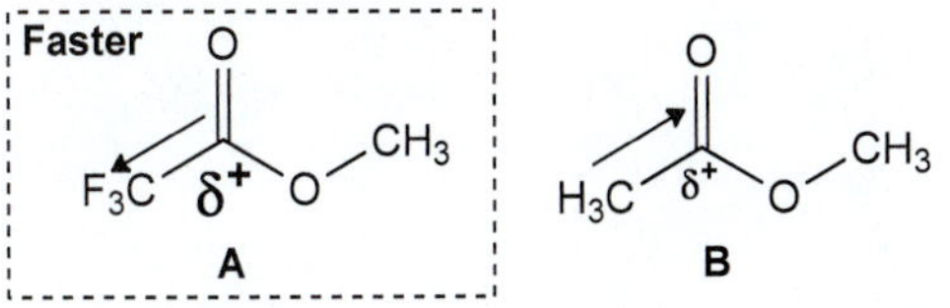

Faster

A

B

(b) **D** undergoes hydrolysis more quickly. The difference is whether a benzene ring or cyclohexane ring is attached to the carbonyl carbon. The benzene ring stabilizes the carbonyl group via resonance, as shown below, which makes the acid derivative less reactive.

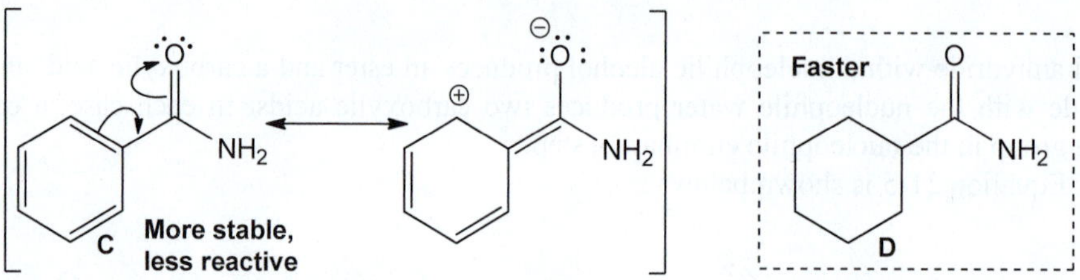

NH_2 NH_2 Faster NH_2
C More stable, less reactive D

(c) **E** undergoes hydrolysis more quickly. Each RO– group stabilizes the carbonyl group via resonance, decreasing reactivity. Because the derivative on the right has two RO– groups attached to the carbonyl carbon, it is less reactive.

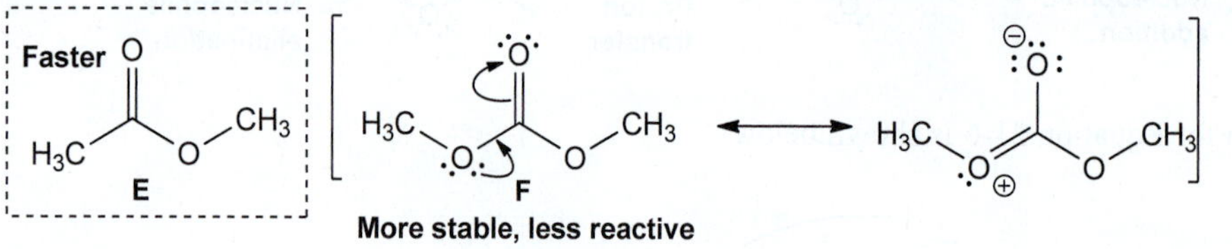

Faster CH_3 H_3C E H_3C CH_3 F H_3C CH_3
More stable, less reactive

Problem 21.6

Think

What acts as the nucleophile in the nucleophilic addition step? What is the leaving group in the nucleophile elimination step? Are proton transfer reactions involved? How many equivalents of amine are required? Why?

Solve

Reaction of an acid anhydride with the nucleophilic amine produces an amide. The amine acts as the nucleophile in a nucleophilic addition step. The carboxylate anion is the leaving group in the nucleophile elimination step. Notice that two equivalents of the amine are required. The first equivalent is used as a nucleophile, and the second is used as a base.

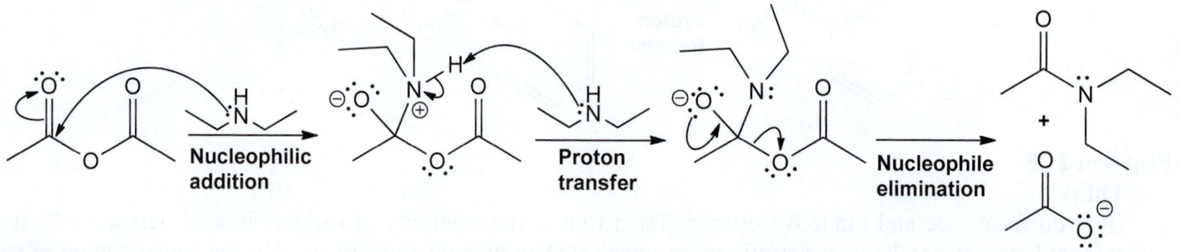

Nucleophilic addition → Proton transfer → Nucleophile elimination

Problem 21.7

Think

What acts as the nucleophile in the nucleophilic addition step? What is the leaving group in the nucleophile elimination step? Are proton transfer reactions involved? How many equivalents of amine or ammonia are required? Why?

Solve

(a) Reaction of an acid chloride with the nucleophilic benzylamine produces an amide. Benzylamine is the nucleophile in the nucleophilic addition step, and the chloride anion is the leaving group in the nucleophile elimination step. Notice that triethylamine is present, which serves as a base.

Nucleophilic addition — Proton transfer — Nucleophile elimination

(b) Reaction of an acid anhydride with nucleophilic ammonia produces an amide and a carboxylic acid. Ammonia is the nucleophile in the nucleophilic addition step, and the carboxylate anion is the leaving group in the nucleophile elimination step. Notice that two equivalents of ammonia are required. The first equivalent is used as a nucleophile, and the second is used as a base.

Nucleophilic addition — Proton transfer — Nucleophile elimination

Problem 21.9

Think

What reactions do we know that transform an acid derivative lower on the stability ladder to one higher on the ladder? Can this transformation be done in a single step? Can an acid chloride be used that is different from the one in Solved Problem 21.8?

Solve

Because $SOCl_2$ can transform any carboxylic acid derivative into another, we can treat acetyl chloride with benzoate to form the target compound. The precursor for benzoate is benzamide.

Undo an acyl substitution. — Undo a hydrolysis.

In the forward direction, the reaction would proceed as follows:

NaOH, H_2O, Δ

Problem 21.10

Think

What is the structure of butanoic acid? What new functional group is added to the α carbon? From what intermediate did this functional group originate?

Solve

All reactions require the Hell–Volhard–Zelinsky (HVZ) reaction to produce the α-bromo acid intermediate. From there, different choices of nucleophiles produce the different targets.

(a)

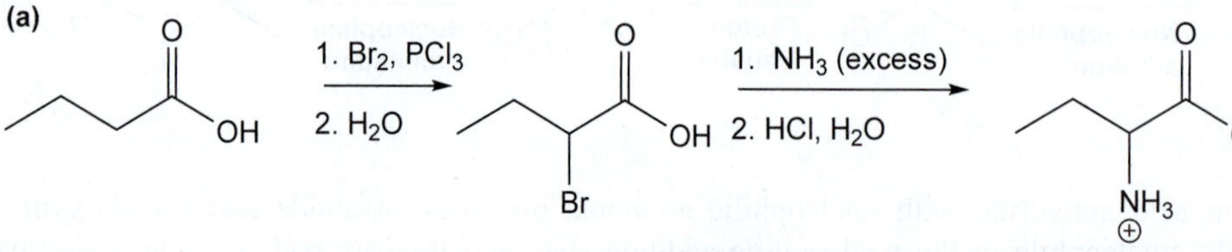

O
OH
1. Br2, PCl3
2. H2O
O
OH
Br
1. NH3 (excess)
2. HCl, H2O
O
OH
NH3

(b)

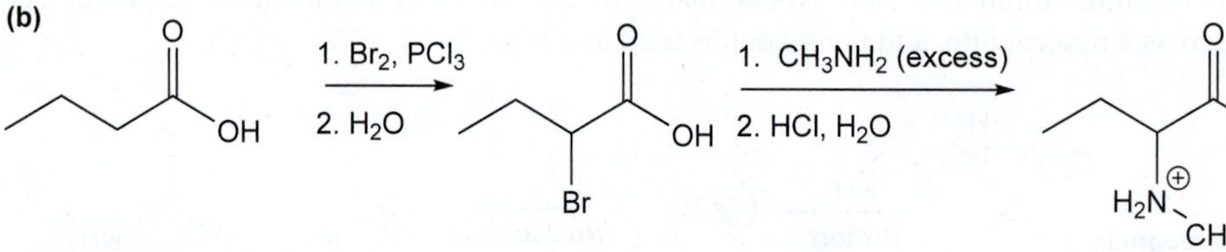

O
OH
1. Br2, PCl3
2. H2O
O
OH
Br
1. CH3NH2 (excess)
2. HCl, H2O
O
OH
H2N
CH3

(c)

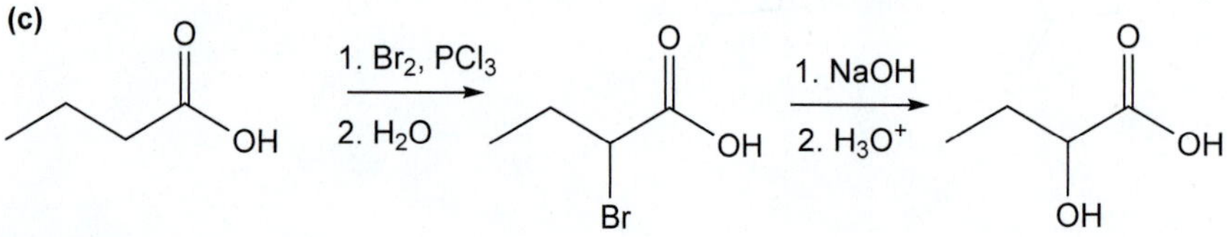

O
OH
1. Br2, PCl3
2. H2O
O
OH
Br
1. NaOH
2. H3O+
O
OH
OH

Problem 21.12

Think

Is there a suitable leaving group on the starting organic compound? If not, how can a good leaving group be formed? What functional group is formed in the target compound? Is this the Z or E isomer? How is this product different from the product in Solved Problem 21.11? What is the stereochemistry associated with each of the above reactions?

Solve

The same strategy can be used here as was used in Solved Problem 21.11, in which the OH must first be converted to a good leaving group, followed by an E2 reaction. However, the target here has the opposite configuration about the C=C bond than does the target in Solved Problem 21.11. Thus, the substrate from which the E2 product is produced must have the opposite configuration at the carbon bonded to the leaving group (see Chapter 8). That substrate can be produced by treating the starting material with PBr_3 instead of TsCl, given that PBr_3 leads to inversion of stereochemistry.

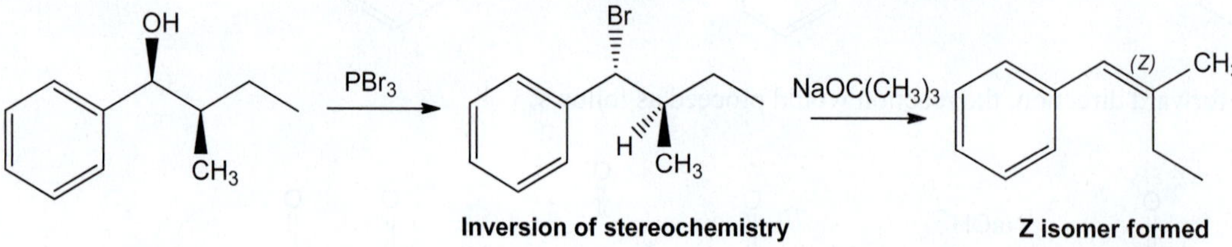

OH
CH3
PBr3
Br
H
CH3
Inversion of stereochemistry
NaOC(CH3)3
(Z)
CH3
Z isomer formed

Problem 21.13

Think

To what functional group is the ester hydrolyzed? What acts as the nucleophile in the nucleophilic addition step? What is the leaving group in the nucleophile elimination step? Is the reaction conducted in basic or acidic conditions? What types of species should not appear in a mechanism in this medium?

Solve

The ester is hydrolyzed to the carboxylic acid in acidic conditions, so no strong bases should appear in the mechanism. Water is the nucleophile, and the leaving group is a molecule of ethanol. Proton transfer steps are incorporated to avoid the appearance of strong bases.

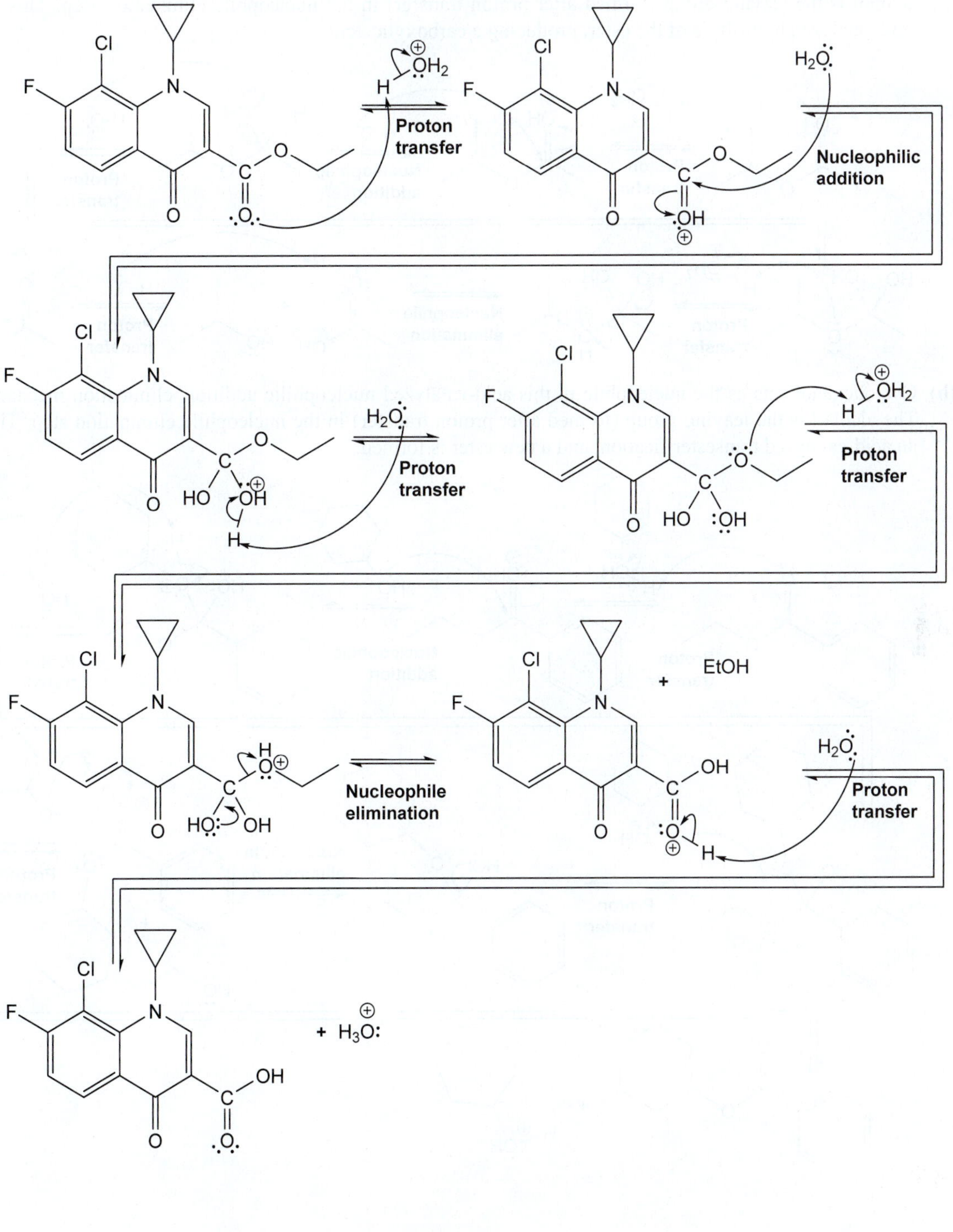

Proton transfer

Nucleophilic addition

Proton transfer

Proton transfer

Nucleophile elimination

EtOH

Proton transfer

+ H_3O^+

Problem 21.14

Think

What functional group is produced? What acts as the nucleophile in the nucleophilic addition step? What is the leaving group in the nucleophile elimination step? Is the reaction conducted in basic or acidic conditions? What species should not appear in a mechanism in this medium?

Solve

All of these reactions are conducted in acidic conditions, so no strong bases should appear in the mechanism. In each reaction, either water or an alcohol is the nucleophile, and either water or an alcohol is the leaving group. Proton transfer steps are incorporated to avoid the appearance of strong bases.

(a) Water acts as the nucleophile in this acid-catalyzed nucleophilic addition–elimination mechanism. The alcohol is the leaving group (formed after proton transfer) in the nucleophile elimination step. This is an acid-catalyzed hydrolysis of the ester, producing a carboxylic acid.

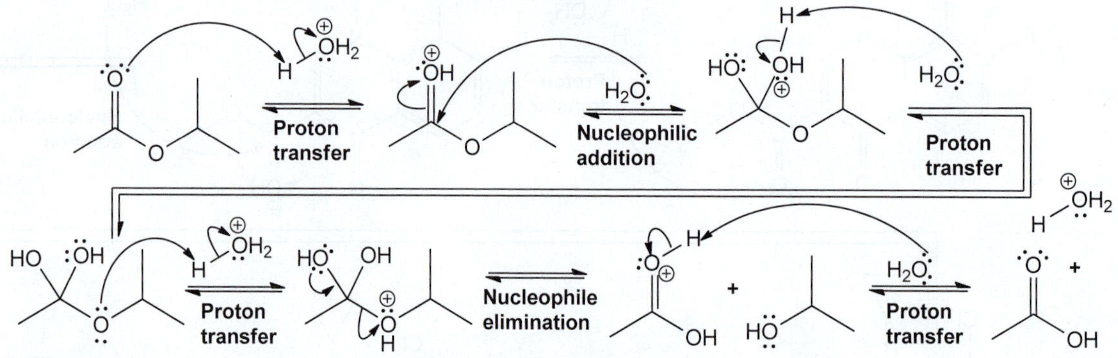

Proton transfer; Nucleophilic addition; Proton transfer; Proton transfer; Nucleophile elimination; Proton transfer

(b) Cyclohexanol acts as the nucleophile in this acid-catalyzed nucleophilic addition–elimination mechanism. The alcohol is the leaving group (formed after proton transfer) in the nucleophile elimination step. This is an acid-catalyzed transesterification, and a new ester is formed.

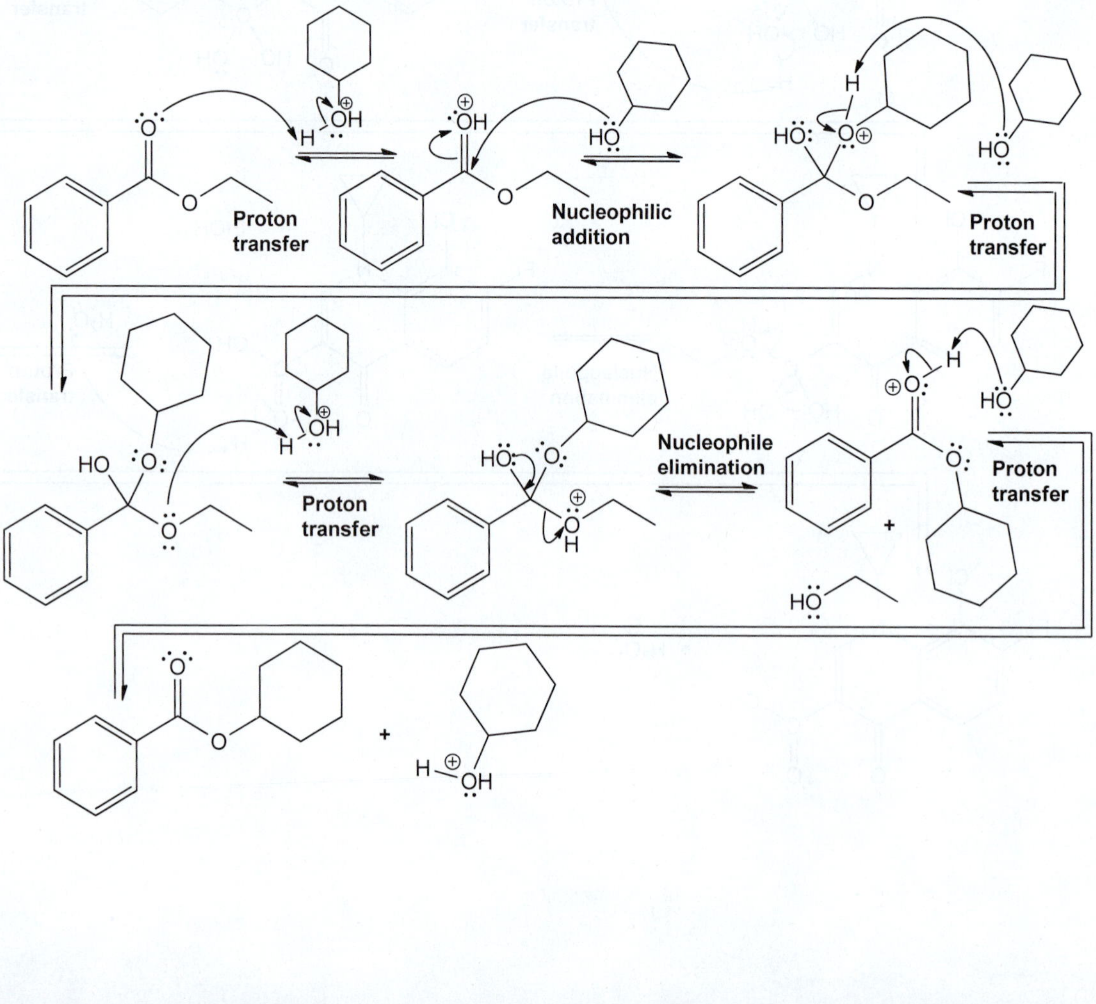

Proton transfer; Nucleophilic addition; Proton transfer; Proton transfer; Nucleophile elimination; Proton transfer

(c) Water acts as the nucleophile in this acid-catalyzed nucleophilic addition–elimination mechanism. The amine is the leaving group (formed after proton transfer) in the nucleophile elimination step. This is an acid-catalyzed amide hydrolysis, producing a carboxylic acid.

Proton transfer

Nucleophilic addition

Proton transfer

Proton transfer

Nucleophile elimination

Proton transfer

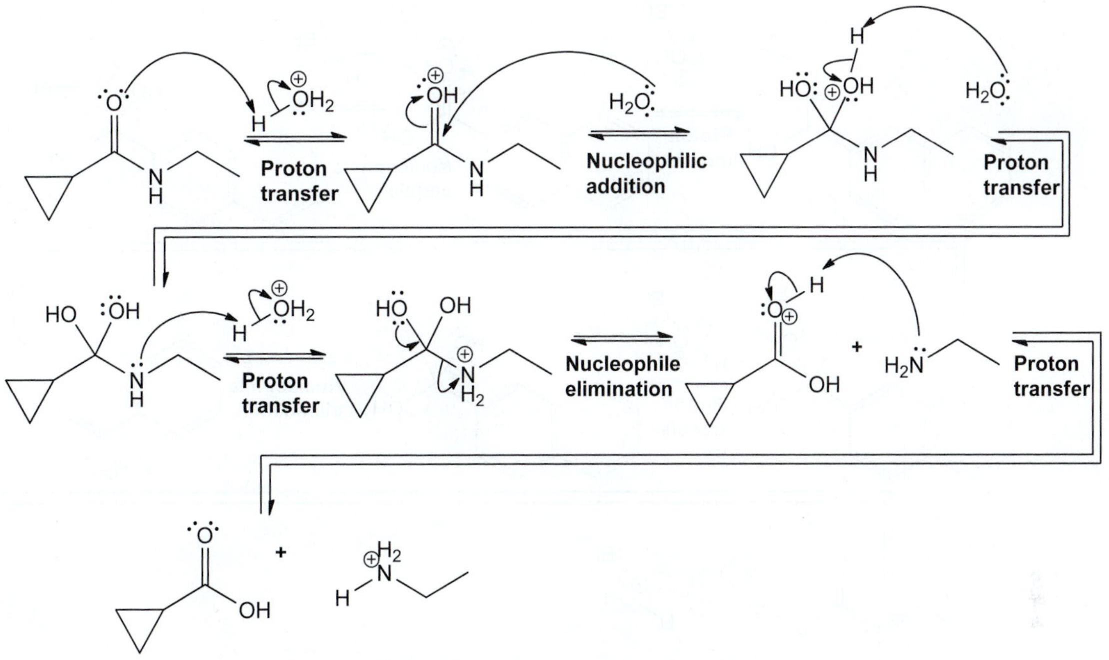

(d) Water acts as the nucleophile in this acid-catalyzed nucleophilic addition–elimination mechanism. The alcohol is the leaving group (formed after proton transfer) in the nucleophile elimination step. This is an acid-catalyzed ester hydrolysis, producing a carboxylic acid.

Proton transfer

Nucleophilic addition

Proton transfer

Proton transfer

Nucleophile elimination

Proton transfer

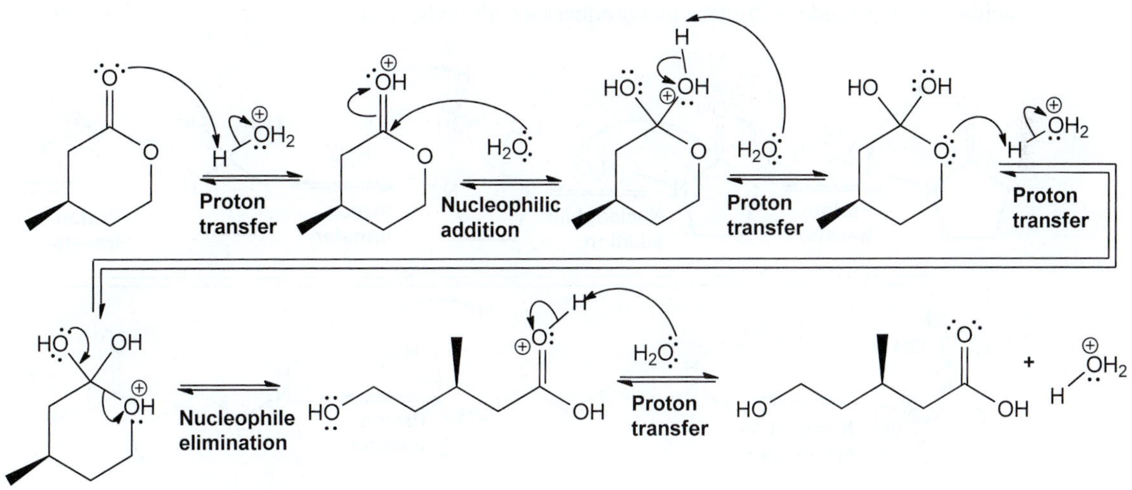

(e) Ethanol acts as the nucleophile in this acid-catalyzed nucleophilic addition–elimination mechanism. Water is the leaving group (formed after proton transfer) in the nucleophile elimination step. This is a Fischer esterification reaction.

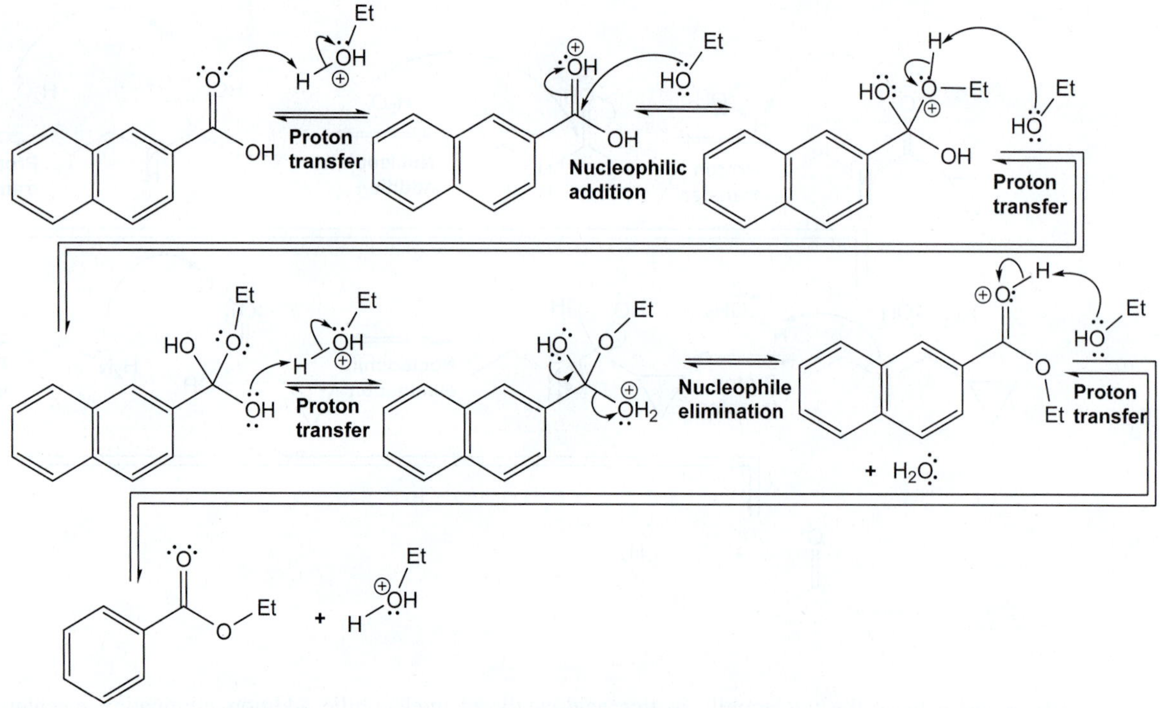

Proton transfer

Nucleophilic addition

Proton transfer

Proton transfer

Nucleophile elimination

Proton transfer

+ H_2O

(f) Water acts as the nucleophile in this acid-catalyzed nucleophilic addition–elimination mechanism. The amine is the leaving group (formed after proton transfer) in the nucleophile elimination step. This is an acid-catalyzed amide hydrolysis that produces a carboxylic acid.

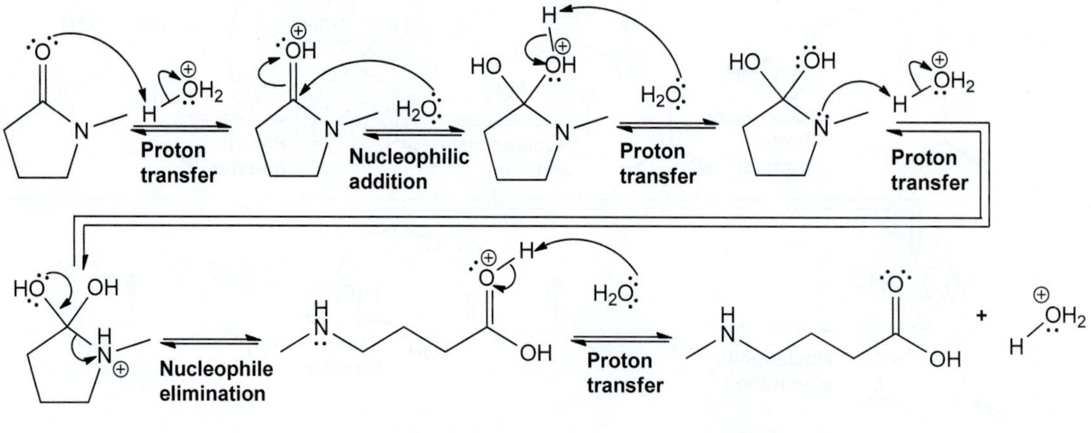

Proton transfer

Nucleophilic addition

Proton transfer

Proton transfer

Nucleophile elimination

Proton transfer

Problem 21.16

Think

What type of reagent is CF_3CO_3H or MCPBA? How does it tend to react with a ketone? Does one side of the ketone favor reaction over the other? If so, which side?

Solve

CF_3CO_3H and MCPBA are peroxyacids and will react with a ketone in a Baeyer–Villiger oxidation. In such a reaction, an O atom from the peroxyacid is inserted between the carbonyl C and one of the groups initially bonded to the carbonyl C.

(a) In this case, the carbonyl C is bonded to a primary alkyl group and to an aryl group. The aryl group has a greater migratory aptitude, so its bond will preferentially break, producing the ester shown below:

Aryl group

CF_3CO_3H

CH_2Cl_2

1° Alkyl group

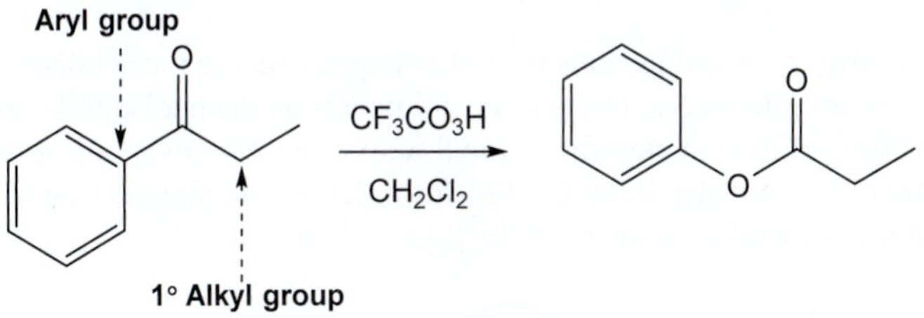

(b) In this case, the carbonyl C is bonded to a primary group and to a tertiary alkyl group. The tertiary alkyl group has a greater migratory aptitude, so its bond will preferentially break, producing the ester shown below:

CF_3CO_3H

CH_2Cl_2

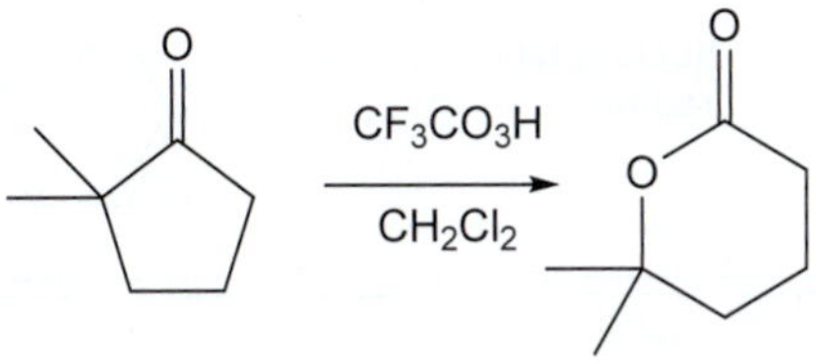

(c) In this case, the carbonyl C is bonded to a benzyl (primary) group and to an aryl group. The aryl group has a greater migratory aptitude, so its bond will preferentially break, producing the ester shown below:

MCPBA

CF_3SO_3H, CH_2Cl_2

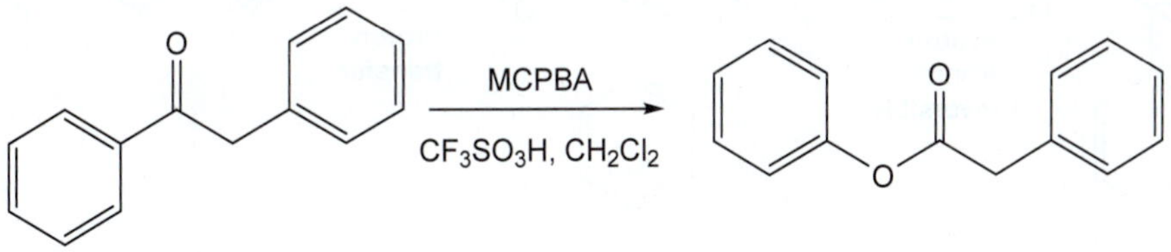

Problem 21.17

Think

Which oxygen lone pair participates in resonance? How does this affect its nucleophile strength?

Solve

The lone pair of electrons on the O adjacent to the C=O is tied up in resonance with the C=O group, as shown here:

This makes the lone pair less available to form a bond to an electrophile, such that the O atom is less nucleophilic.

Problem 21.18

Think

How does the strong base CH_3O^- react with an ester? What product is formed? What acts as the nucleophile in the nucleophilic addition step? What is the leaving group in the nucleophile elimination step? Is the reaction conducted in basic or acidic conditions? What species should not appear in the mechanism for a reaction in this medium? Are any steps irreversible? Why is an acid workup necessary?

Solve

The reaction is conducted under basic conditions, so strong acids should not appear in the mechanism. The base, CH_3O^-, deprotonates the α C–H to form the enolate anion. The enolate anion acts as the nucleophile and attacks another equivalent of ester. The nucleophile elimination step forms the β-keto ester. However, all of these steps are reversible and unfavorable. CH_3O^- then deprotonates the α C–H to form the enolate anion in an irreversible step. The enolate anion is protonated in the acid workup to re-form the β-keto ester.

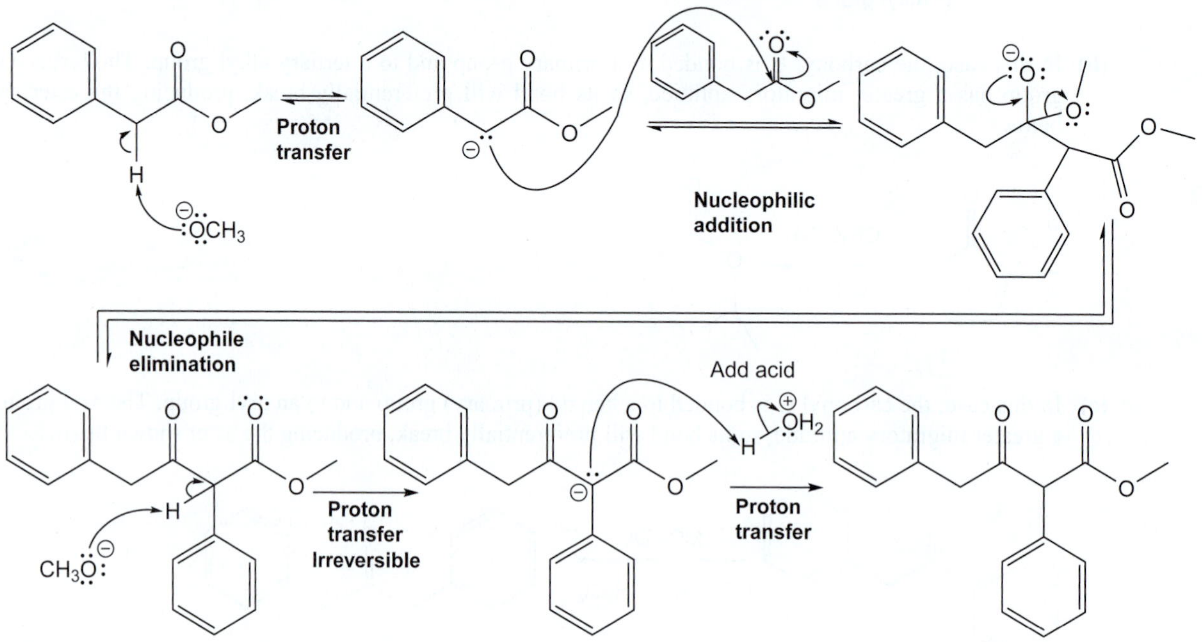

Proton transfer

Nucleophilic addition

Nucleophile elimination

Proton transfer Irreversible

Add acid

Proton transfer

Problem 21.20

Think

Which C–C bond in the β-keto ester product would have been formed in a Claisen condensation? Which C atom would have been bonded to the alkyoxy leaving group?

Solve

The new C–C bond is the one between the α and β carbons, as shown in the following structure. Moreover, the leaving group would have been attached to the β C atom.

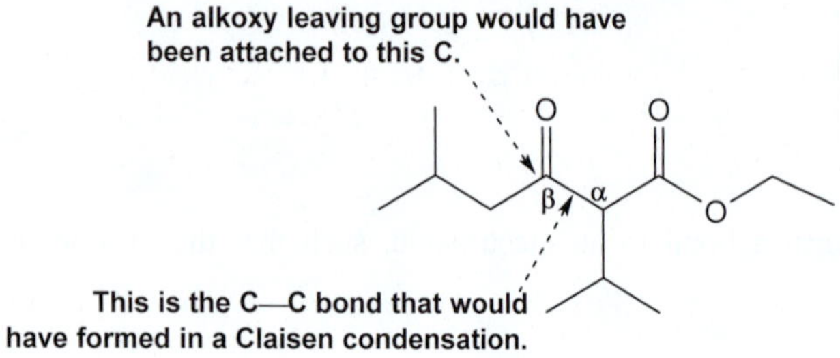

An alkoxy leaving group would have been attached to this C.

This is the C—C bond that would have formed in a Claisen condensation.

Therefore, we can apply a transform that undoes a Claisen condensation by disconnecting that C–C bond and reattaching the leaving group. Doing so yields two identical ester precursors.

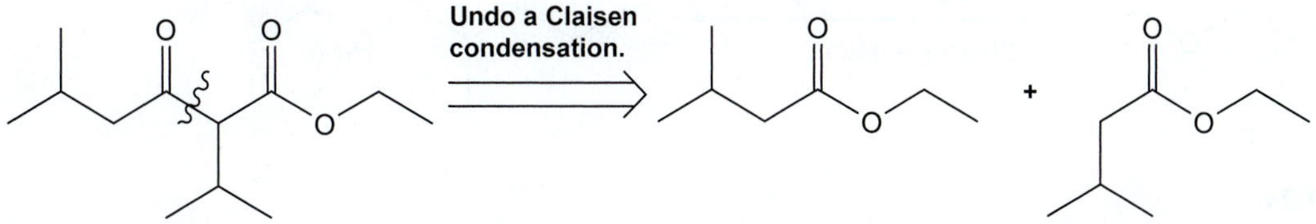

Undo a Claisen condensation.

The reaction in the forward direction would proceed as follows:

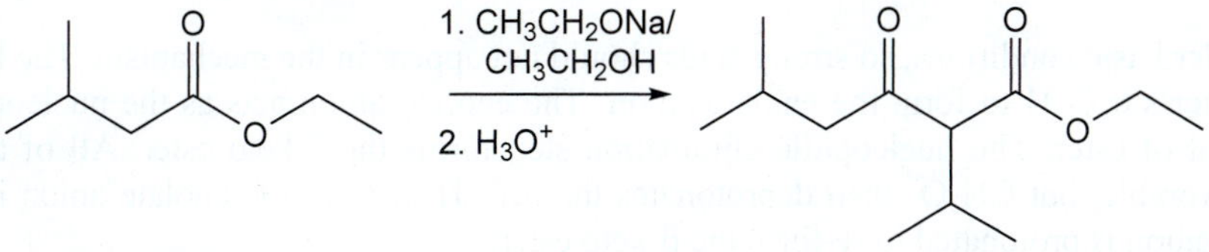

1. CH_3CH_2ONa/ CH_3CH_2OH
2. H_3O^+

Problem 21.22

Think

What is the leaving group on the ester? What choice of base would ensure that a nucleophilic addition–elimination would leave us with the same ester? What solvent could be used so that, if it were deprotonated, the base that would form would be the same as the base that is added?

Solve

The appropriate base should be the same as the leaving group of the ester. The appropriate solvent is the conjugate acid of that base. The Claisen condensation product is shown.

(a) CH_3ONa / CH_3OH

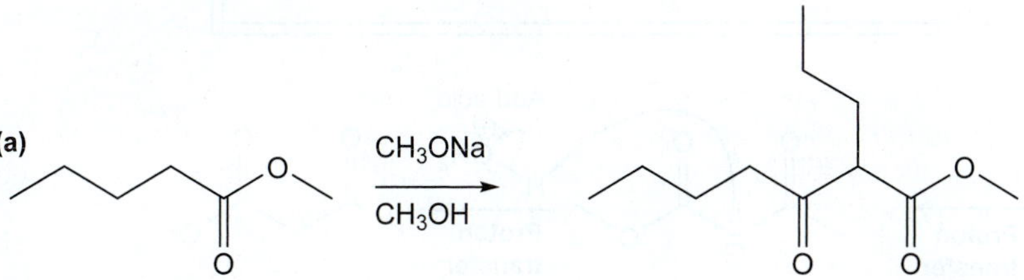

(b) ONa / OH

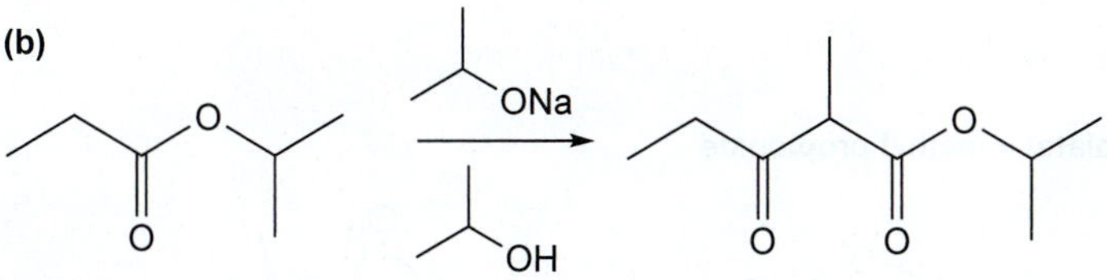

Problem 21.23

Think

Is $(CH_3)_3CO^-$ a good nucleophile? How does the nucleophile strength of the base affect the outcome of the reaction?

Solve

$(CH_3)_3CO^-$ is a relatively poor nucleophile because of the bulkiness provided by the three methyl groups. Therefore, it doesn't attack the ester's carbonyl group to initiate a transesterification. See the figure on the next page.

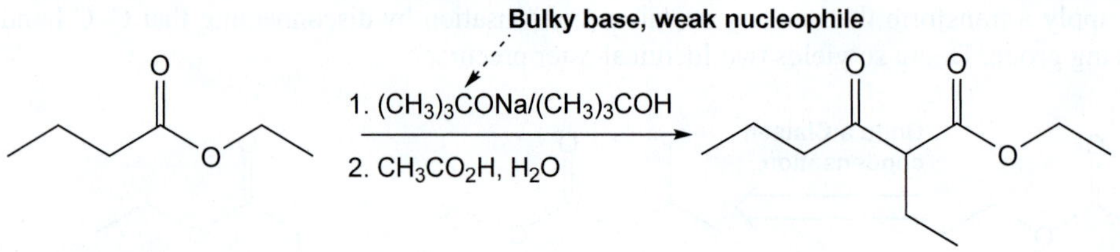

Bulky base, weak nucleophile

1. $(CH_3)_3CONa/(CH_3)_3COH$
2. CH_3CO_2H, H_2O

Problem 21.24

How does the strong base, CH_3O^-, react with each ester? How many nucleophile and electrophile choices are possible? What is the leaving group in the nucleophile elimination step?

Solve

The reaction is conducted under basic conditions, so strong acids should not appear in the mechanism. The base, CH_3O^-, deprotonates each ester's α C–H to form the enolate anion. The enolate anion acts as the nucleophile and attacks another equivalent of ester. The nucleophile elimination step forms the β-keto ester. All of these steps are reversible and unfavorable, but CH_3O^- then deprotonates the α C–H to form the enolate anion in an irreversible step. The enolate anion is protonated to re-form the β-keto ester.

There are two possible ester compounds to form two possible enolate anions (nucleophile) and two possible electrophilic C=O compounds. Therefore, four nucleophilic addition steps are possible, each producing a different overall product. The mechanisms are shown below:

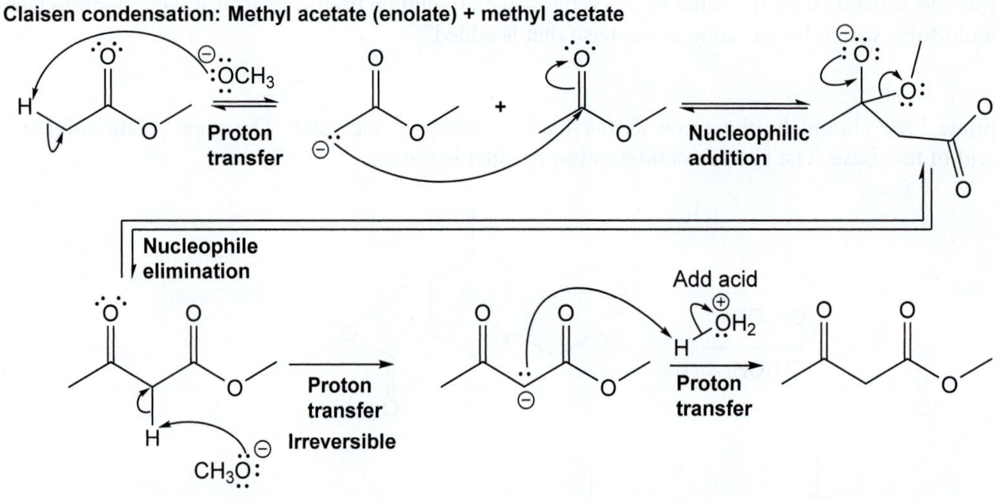

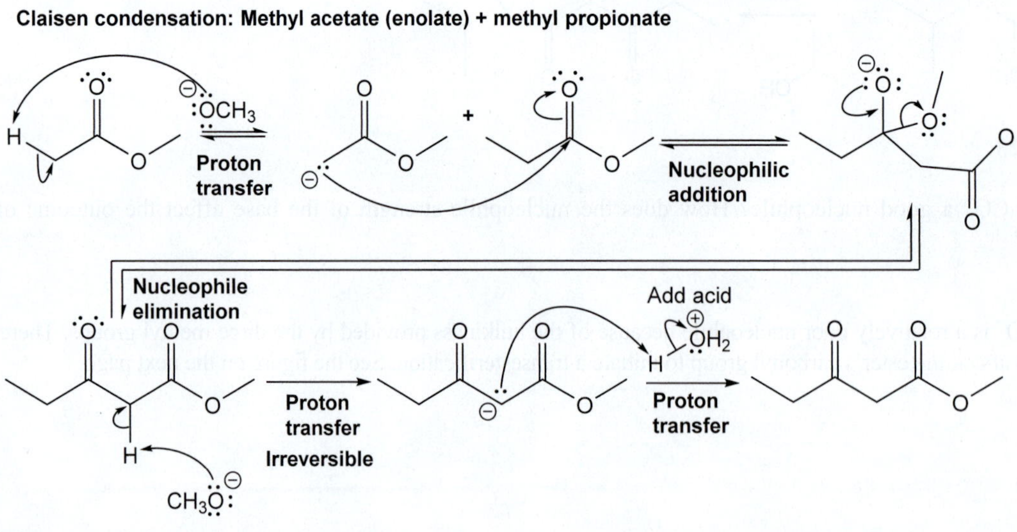

Claisen condensation: Methyl acetate (enolate) + methyl acetate

$:OCH_3$ Proton transfer; Nucleophilic addition; Nucleophile elimination; Proton transfer Irreversible CH_3O^-; Add acid OH_2^+ Proton transfer

Claisen condensation: Methyl acetate (enolate) + methyl propionate

$:OCH_3$ Proton transfer; Nucleophilic addition; Nucleophile elimination; Proton transfer Irreversible CH_3O^-; Add acid OH_2^+ Proton transfer

Claisen condensation: Methyl propionate (enolate) + methyl acetate

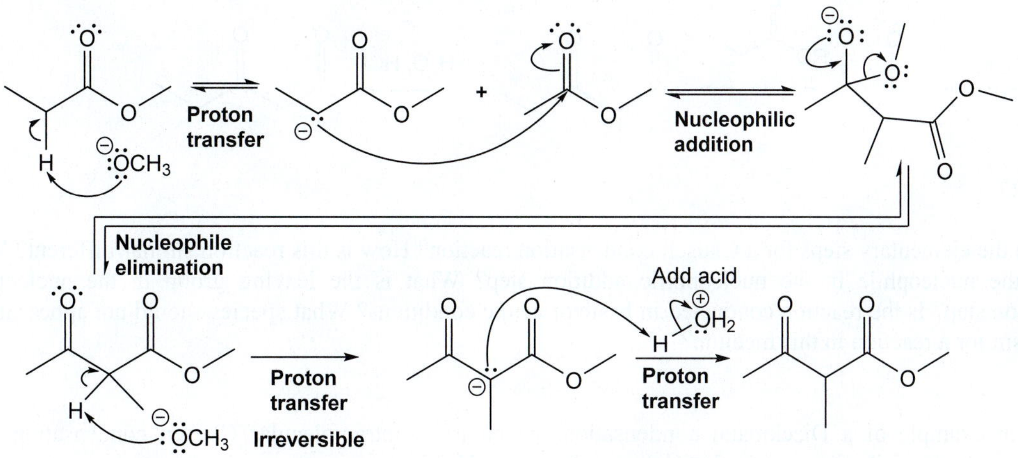

Proton transfer

Nucleophilic addition

Nucleophile elimination

Proton transfer

Irreversible

Add acid

Proton transfer

Claisen condensation: Methyl propionate (enolate) + methyl propionate

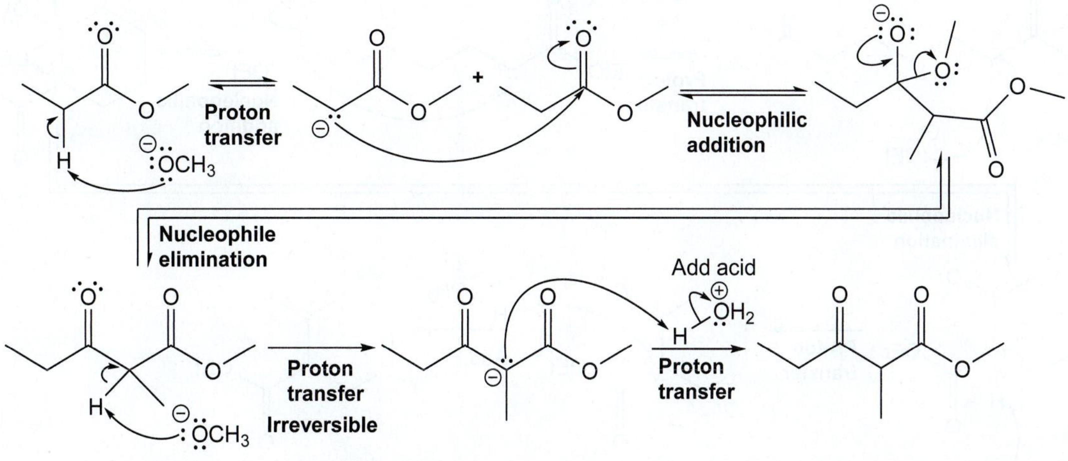

Proton transfer

Nucleophilic addition

Nucleophile elimination

Proton transfer

Irreversible

Add acid

Proton transfer

Problem 21.26

Think

Which C–C bond can be "disconnected" in a transform that undoes a Claisen condensation? Does this transform suggest a self-Claisen condensation reaction or a crossed Claisen? Can we use a base that reversibly deprotonates an α hydrogen, or should we use one that brings about an irreversible deprotonation? How can this be done without using HC(O)OR? Can a C–C bond other than the one in Solved Problem 21.25 be disconnected?

Solve

In applying a transform that reverses a Claisen condensation, one precursor will be an ester and the other will be a ketone. Instead of the ketone and ester precursors in Solved Problem 21.25, we can use the ones below:

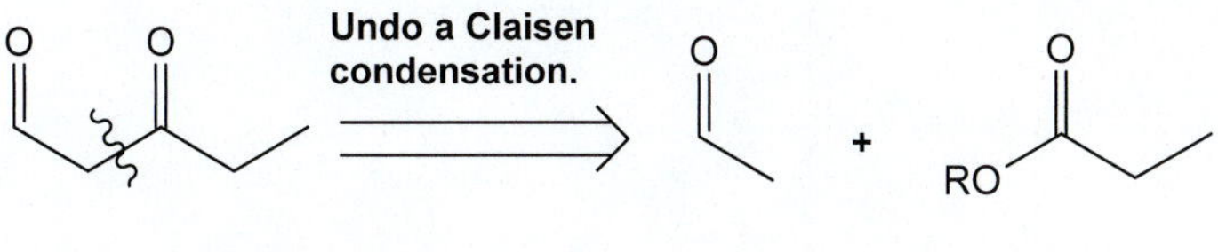

Undo a Claisen condensation.

RO

The synthesis in the forward direction would then appear as follows:

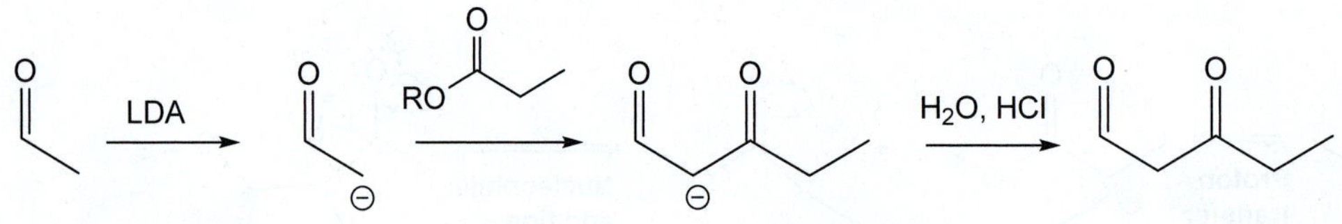

LDA

RO

H_2O, HCl

Problem 21.27

Think

What are the elementary steps for a Claisen condensation reaction? How is this reaction similar/different? What acts as the nucleophile in the nucleophilic addition step? What is the leaving group in the nucleophile elimination step? Is the reaction conducted in basic or acidic conditions? What species should not appear in the mechanism for a reaction in this medium?

Solve

This is an example of a Dieckmann condensation, which is an intramolecular Claisen condensation. The mechanism is essentially the same as that shown in Equation 21-37.

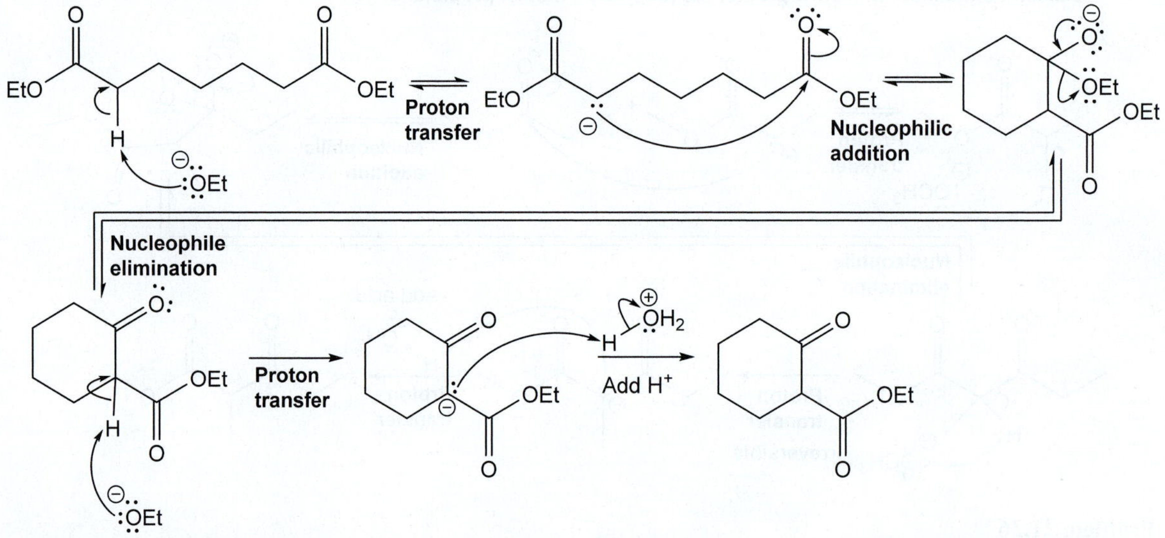

EtO OEt :OEt Proton transfer EtO OEt Nucleophilic addition OEt OEt

Nucleophile elimination

OEt :OEt Proton transfer OEt Add H^+ OH_2 OEt

Problem 21.28

Think

Identify the two functional groups present. Which α C–H is more acidic? Which carbon forms the enolate? Which carbon is the electrophile? Is an intramolecular reaction favorable?

Solve

The α C–H of the ketone portion of the molecule is more acidic compared to that of the ester portion. Thus the α C–H of the ketone portion will be deprotonated and form the enolate anion, and an intramolecular nucleophilic addition–elimination mechanism follows. There are two distinct α C–H atoms of the ketone portion and thus two possible enolates. Deprotonation of the α C–H on the left leads to the formation of the six-membered ring, which is the major product.

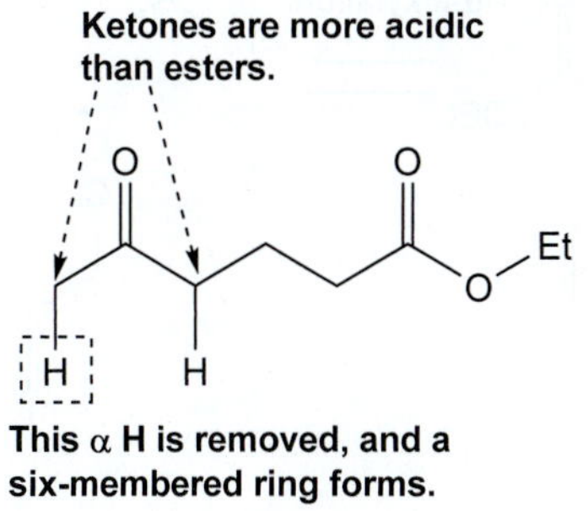

Ketones are more acidic than esters.

This α H is removed, and a six-membered ring forms.

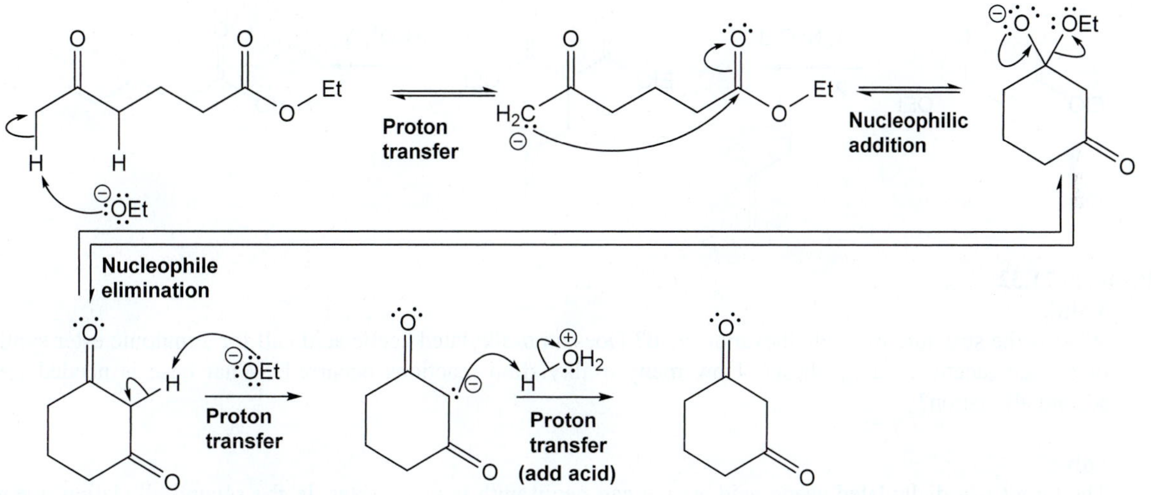

Proton transfer

Nucleophilic addition

Nucleophile elimination

Proton transfer

Proton transfer (add acid)

Problem 21.29

Think

What new C–C bond formed? What functional groups were present in the original acyclic dicarbonyl species? Where should those functional groups be put back?

Solve

To apply a transform that undoes a Dieckmann condensation, we must disconnect the C–C bond indicated. The original compound consisted of an aldehyde C=O and an ester.

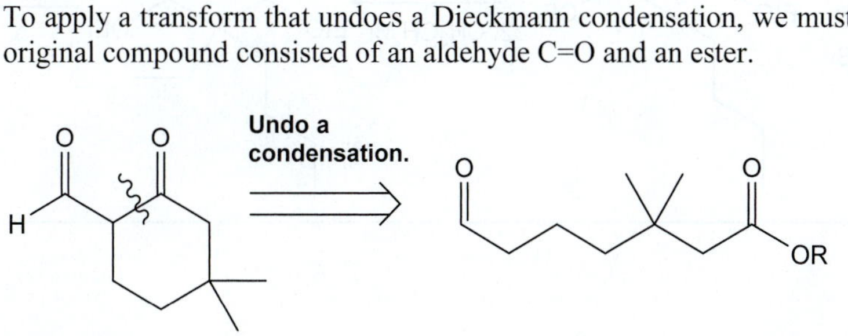

Undo a condensation.

Problem 21.31

Think

What would the precursor look like before decarboxylation? Which precursor, acetoacetic ester or malonic ester, will leave the CO_2H functional group in the target? What R group is added to the α carbon?

Solve

A malonic ester synthesis results in an α-substituted carboxylic acid because it has two ester groups. Both esters are hydrolyzed to the carboxylic acid, but only one undergoes decarboxylation. The retrosynthesis is given below:

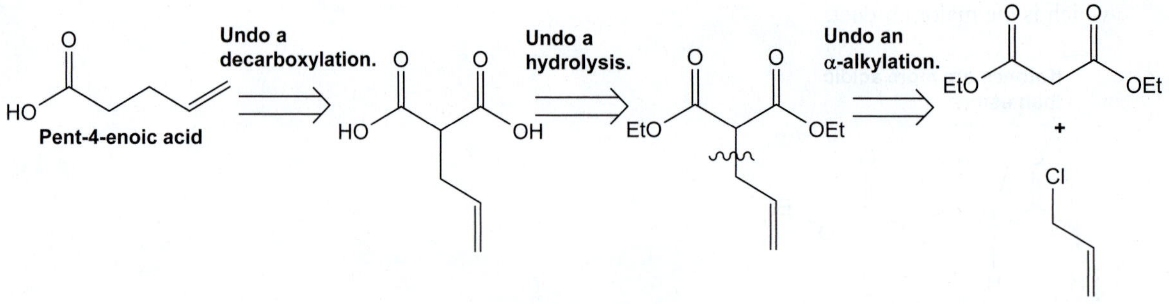

Pent-4-enoic acid ⟹ (Undo a decarboxylation.) ⟹ (Undo a hydrolysis.) ⟹ (Undo an α-alkylation.) EtO–C(O)CH₂C(O)–OEt + Cl–CH₂CH=CH₂

The synthesis in the forward direction would be written as follows:

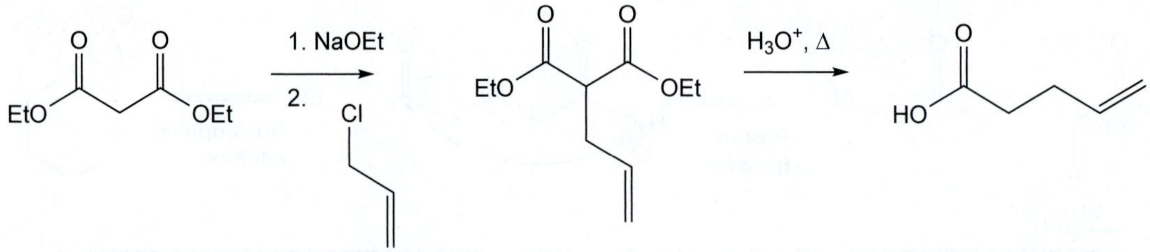

EtO–C(O)CH₂C(O)–OEt —(1. NaOEt; 2. Cl–CH₂CH=CH₂)→ diethyl allylmalonate —(H_3O^+, Δ)→ HO–C(O)CH₂CH₂CH=CH₂

Problem 21.32

Think

What is the structure of 2-ethylhexanoic acid? Does an α-alkylated acetic acid call for a malonic ester synthesis or an acetoacetic ester synthesis? How many α-alkylation reactions occurred? What base is needed for the second alkylation?

Solve

The target is a dialkylated acetic acid, so we can begin with malonic ester. In the second alkylation, we use a stronger alkoxide base than in the first alkylation, because the acidity of the α carbon decreases with the presence of an additional alkyl group. In the synthesis below, the butyl group was added first, but the ethyl group could have been added first instead.

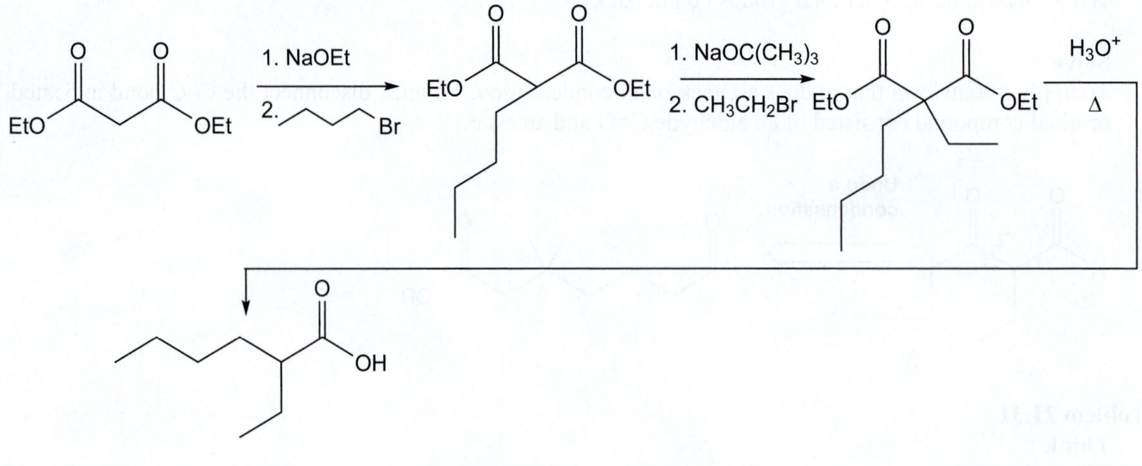

EtO–C(O)CH₂C(O)–OEt —(1. NaOEt; 2. butyl bromide)→ diethyl butylmalonate —(1. NaOC(CH₃)₃; 2. CH₃CH₂Br)→ diethyl butylethylmalonate —(H_3O^+, Δ)→ 2-ethylhexanoic acid (OH)

Problem 21.34

Think

Without considering undesired side reactions, how would you normally carry out this synthesis? Under the conditions for that synthesis, would any functional group undergo an undesired side reaction? If so, how can that functional group be protected?

Solve

The NH_2 and the OH groups are subject to oxidation in the presence of a strong oxidizing agent. The amine, therefore, has to be protected so as only to oxidize the alcohol portion to the carboxylic acid. Refer to Table 21-2, Entry 4 for the protection of an amine.

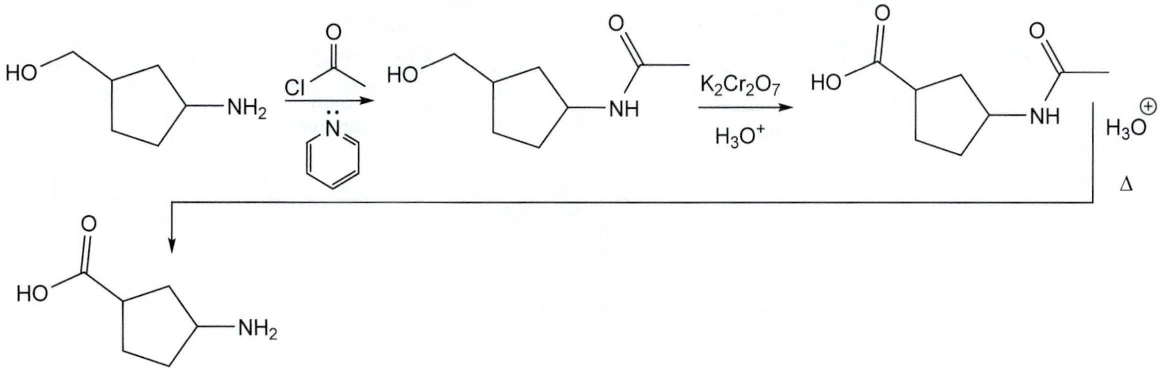

Problem 21.35

Think

Are certain sequences in the peptides the same? Can you assume those identical sequences came from the same portion of the original protein? How can you piece together the sequence?

Solve

Certain sequences in each peptide have overlapping amino acids—specifically, the Asp-Ser sequence in **A** and **C**, the Pro-Asn sequence in **C** and **D**, and the Ile-Val-Met-Pro-Val sequence in **D** and **B**. Each of these portions is highlighted by a gray screen. You can assume that these sequences come from the same portions of the original protein and piece together the overall sequence, shown below:

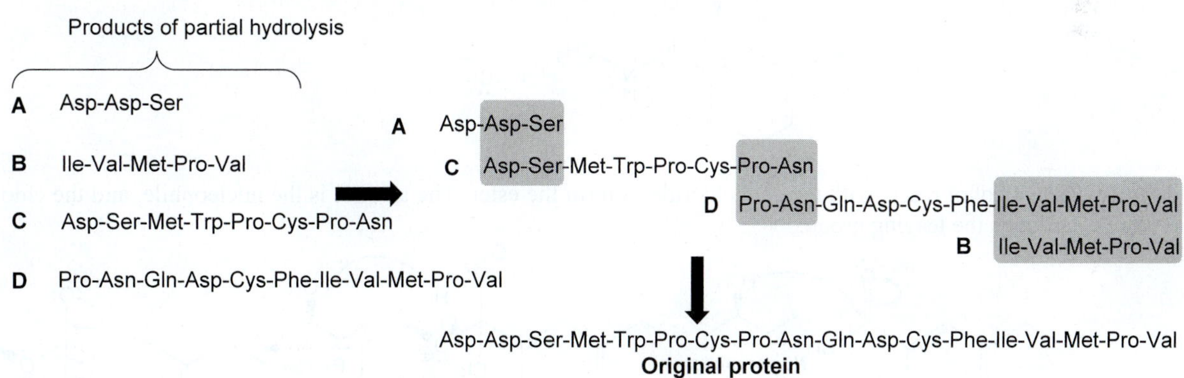

End of Chapter Problems
Sections 21.1–21.3 Alcoholysis, Aminolysis, and Relative Reactivities of Acid Derivatives
Problem 21.36

Think

What acts as the nucleophile in the nucleophilic addition step? What is the leaving group in the nucleophile elimination step? Are proton transfer steps involved?

Solve

(a) Water reacts with the acid chloride to form the carboxylic acid. Water is the nucleophile, and the chloride anion is the leaving group.

H_2O Nucleophilic addition

H_2O Proton transfer

Nucleophile elimination

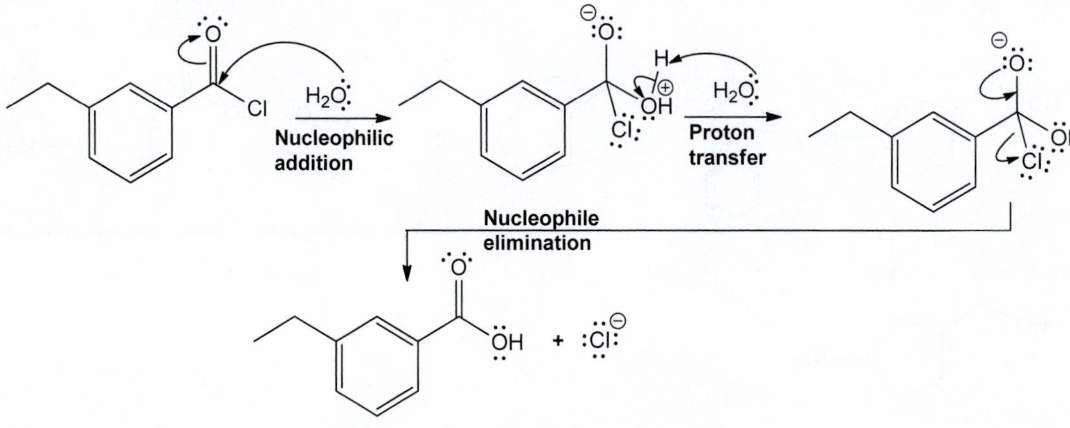

(b) The amine reacts with the acid chloride to form the amide. The amine is the nucleophile, and the chloride anion is the leaving group.

NH_2CH_3 Nucleophilic addition

NH_2CH_3 Proton transfer

Nucleophile elimination

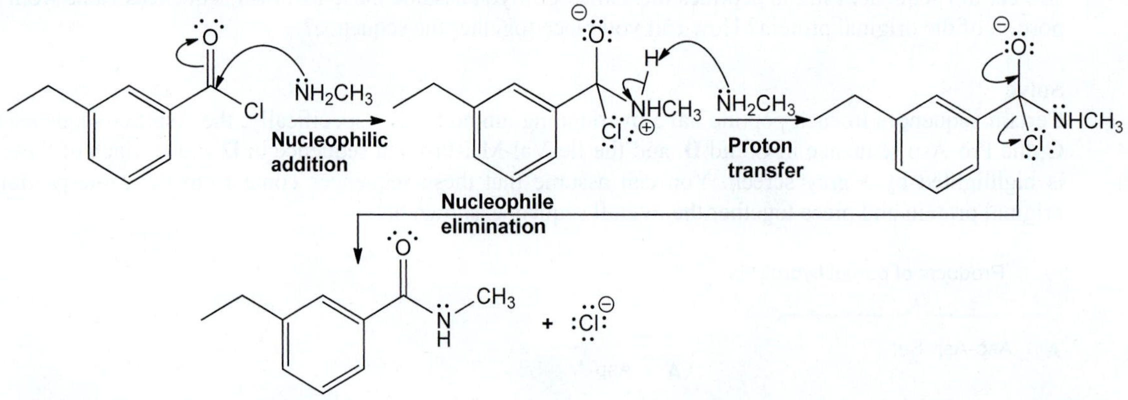

(c) The alcohol reacts with the acid chloride to form the ester. The alcohol is the nucleophile, and the chloride anion is the leaving group.

Nucleophilic addition

Proton transfer

Nucleophile elimination

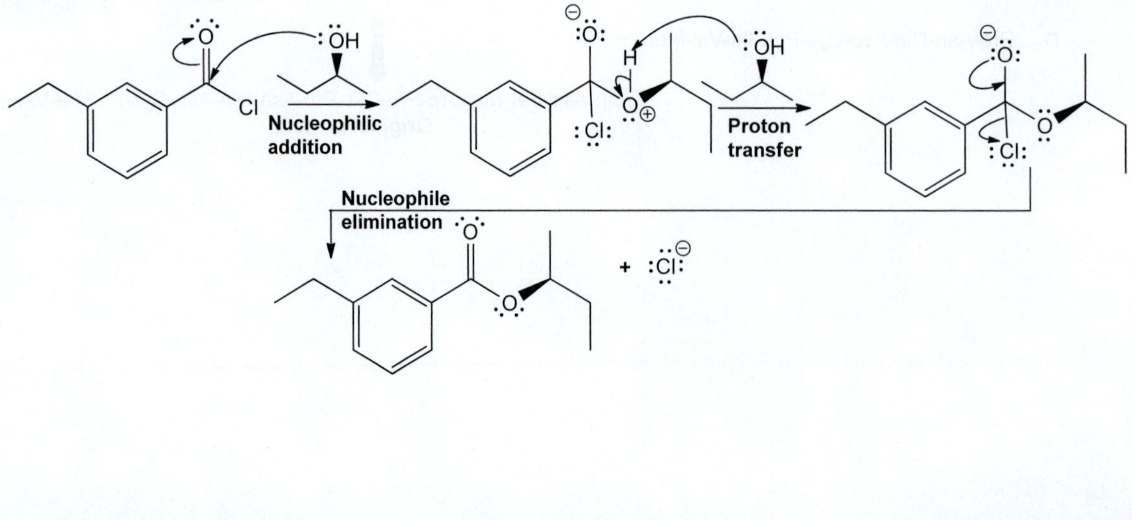

(d) No reaction. Nucleophilic addition–elimination involving the ether as the nucleophile would result in a positive charge on O that cannot be removed by a simple deprotonation, so a stable product cannot be formed.

Problem 21.37

Think

What acts as the nucleophile in the nucleophilic addition step? What is the leaving group in the nucleophile elimination step? Are proton transfer steps involved?

Solve

(a) Water reacts with the acid anhydride to form the carboxylic acid. Water acts as the nucleophile, and the ethanoate anion acts as the leaving group.

H_2O: Nucleophilic addition → H_2O: Proton transfer → Nucleophile elimination

(b) The amine reacts with the acid anhydride to form the amide. The amine acts as the nucleophile, and the ethanoate anion acts as the leaving group. Pyridine acts as the base.

NH_2CH_3 Nucleophilic addition → Proton transfer → Nucleophile elimination

(c) The alcohol reacts with the acid anhydride to form the ester. The alcohol acts as the nucleophile, and the ethanoate anion acts as the leaving group.

Nucleophilic addition → Proton transfer → Nucleophile elimination

(d) No reaction. Nucleophilic addition–elimination involving the ether as the nucleophile would result in a positive charge on O that cannot be removed by a simple deprotonation, so a stable product cannot be formed.

Problem 21.38

Think

What is the structure of acetic anhydride? Which HO group on salicylic acid serves as the nucleophile in the nucleophilic addition step? What is the leaving group in the nucleophile elimination step? How are proton transfer steps involved?

Solve

The phenolic OH, not the carboxyl OH, acts as the nucleophile in the nucleophilic addition step. The phenolic OH is more nucleophilic because the carboxyl C=O group is electron withdrawing via both resonance and inductive effects. The carboxylate is the leaving group in the nucleophile elimination step.

Nucleophilic addition

Salicylic acid

:Base Proton transfer

Nucleophile elimination

Acetylsalicylic acid

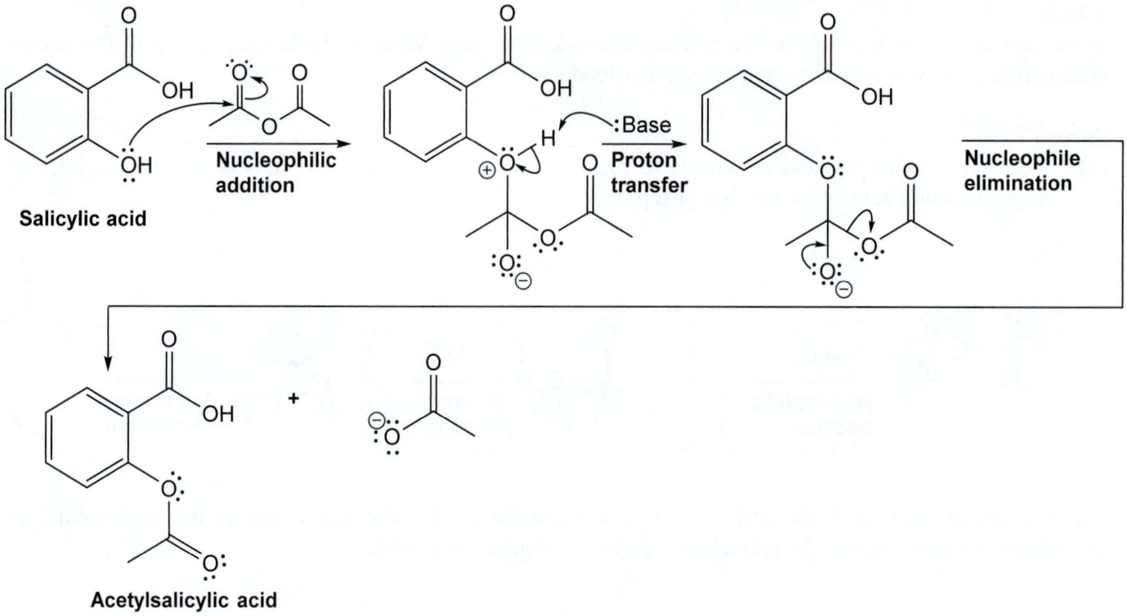

Problem 21.39

Think

What acts as the nucleophile in the nucleophilic addition step? What is the leaving group in the nucleophile elimination step? Are proton transfer reactions involved?

Solve

(a) The reaction leads to the formation of the ester; the alcohol acts as the nucleophile, and the anhydride is the source of electrophilic carbon. The carboxylate anion is the leaving group. The alcohol serves as the base in the proton transfer reaction.

Nucleophilic addition

ROH Proton transfer

Nucleophile elimination

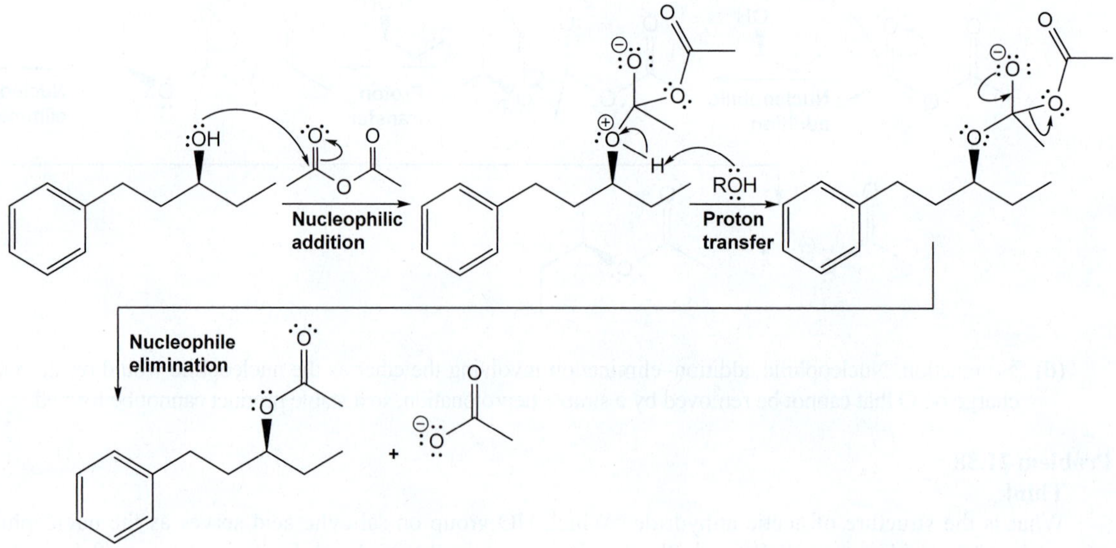

(b) The reaction leads to the formation of the amide; the amine acts as the nucleophile, and the acid chloride is the source of electrophilic carbon. The chloride anion is the leaving group. The amine serves as the base in the proton transfer reaction.

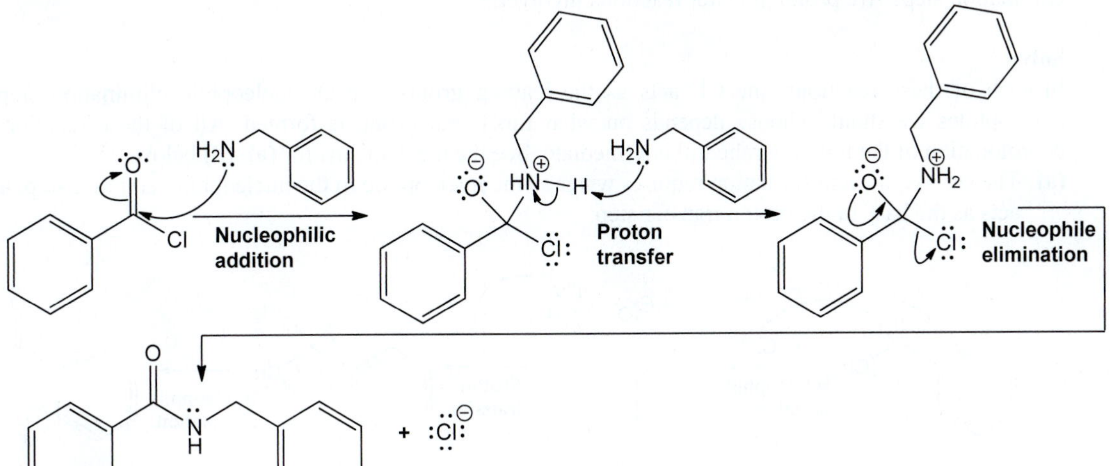

(c) The reaction leads to the formation of the amide; the amine acts as the nucleophile, and the acid chloride is the source of electrophilic carbon. The chloride anion is the leaving group. Pyridine is added to serve as the base in the proton transfer reaction.

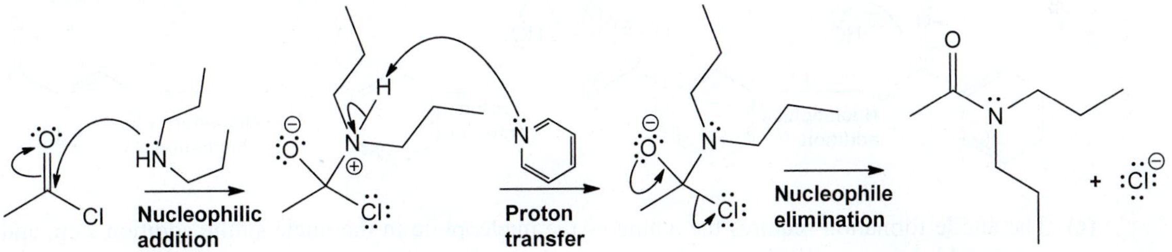

(d) The reaction leads to the formation of the ester; the alcohol acts as the nucleophile, and the anhydride is the source of electrophilic carbon. The carboxylate anion is the leaving group. The alcohol serves as the base in the proton transfer reaction.

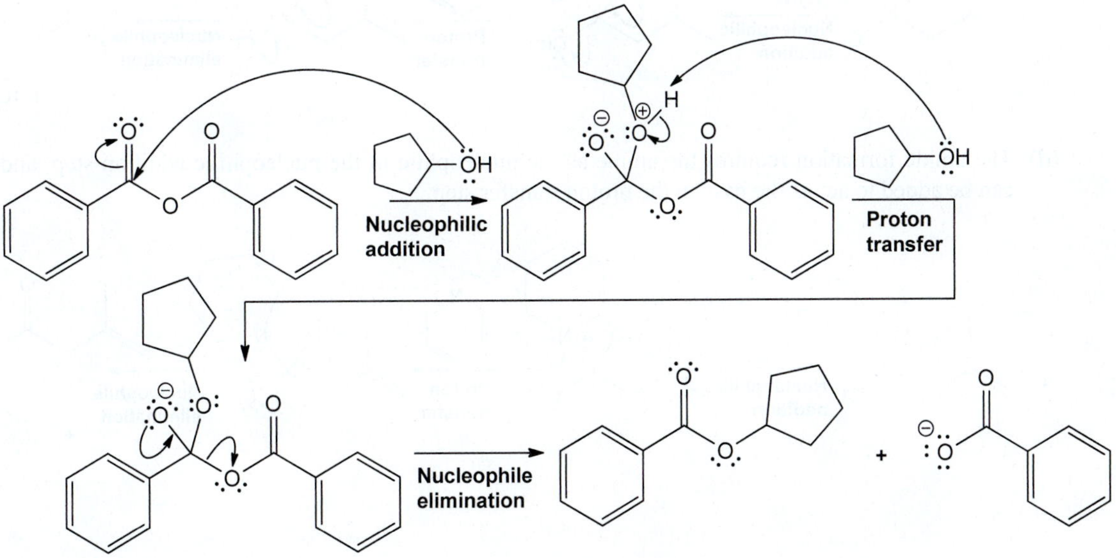

Problem 21.40 (SYN)

Think

What acts as the nucleophile in the nucleophilic addition step? What is the leaving group in the nucleophile elimination step? Are proton transfer reactions involved?

Solve

In each of these reactions, the Cl^- acts as the leaving group in each nucleophile elimination step. Which nucleophiles we should choose depends on what functional group is formed. All of these reactions require deprotonation of the initial tetrahedral intermediate. See the mechanisms for **(a)–(d)** below.

(a) The carboxylic acid formation requires water as the nucleophile in the nucleophilic addition step, and water acts as the base in the proton transfer step.

H_2O: Nucleophilic addition → H_2O: Proton transfer → Nucleophile elimination → + :Cl:⁻

(b) The ester formation requires isopropanol as the nucleophile in the nucleophilic addition step, and the alcohol acts as the base in the proton transfer step.

Nucleophilic addition → Proton transfer → Nucleophile elimination → + :Cl:⁻

(c) The amide formation requires the amine as the nucleophile in the nucleophilic addition step, and pyridine can be added to act as the base in the proton transfer step.

H_2N-CH_3 Nucleophilic addition → Proton transfer → Nucleophile elimination → + :Cl:⁻

(d) The amide formation requires the amine as the nucleophile in the nucleophilic addition step, and pyridine can be added to act as the base in the proton transfer step.

Nucleophilic addition → Proton transfer → Nucleophile elimination → + :Cl:⁻

Problem 21.41

Think

What atom is involved in resonance with the C=O group in an ester? In a thioester? Are those atoms in the same row of the periodic table? How does that affect resonance stabilization? How does the stability of the reactant affect the reaction rate?

Solve

The resonance structures for a thioester and an ester are shown below. In a thioester, S is involved in resonance with the C=O group, whereas in an ester, an O atom is involved in resonance with the C=O group. Because S is in the third shell and the other atoms are in the second shell, resonance is less effective in a thioester, making the thioester less stable and more reactive.

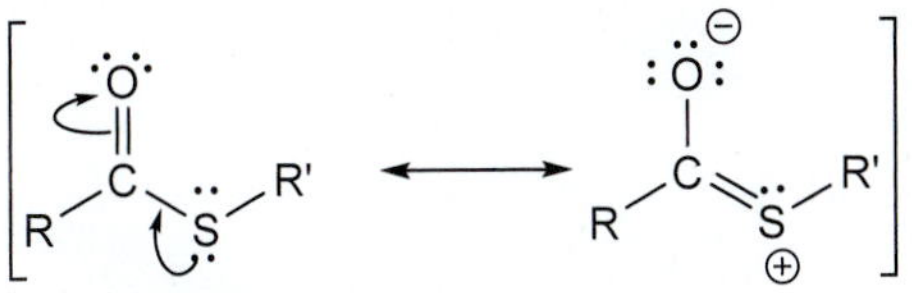

A thioester

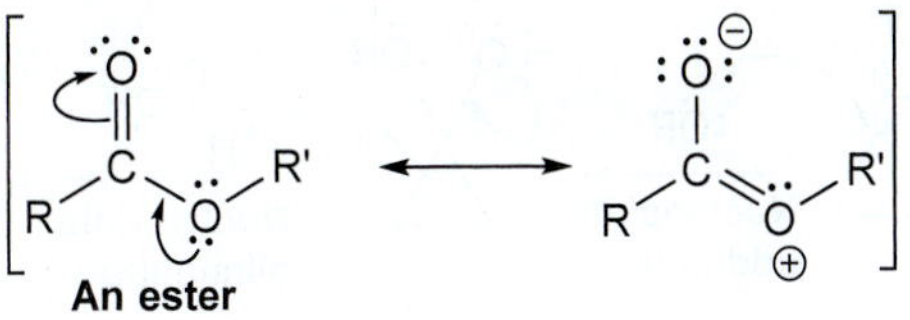

An ester

Problem 21.42

Think

What mechanism occurs in the first reaction? How does ring strain affect the mechanism in the second reaction? Which bonds in the second reaction must have formed and broken?

Solve

Reaction **A** is a normal addition–elimination mechanism.

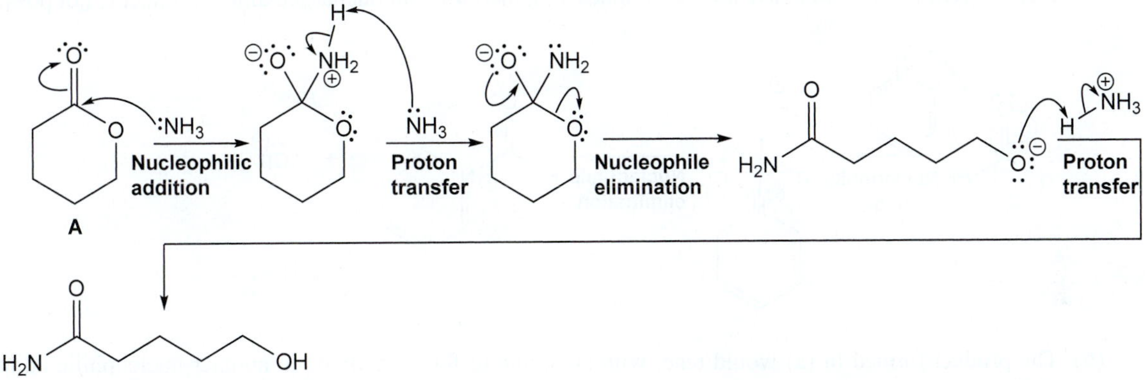

Nucleophilic addition

Proton transfer

Nucleophile elimination

Proton transfer

In reaction **B**, the nucleophile must form a bond to the β carbon and the C–O bond must break. Thus, the strained ring opens in an S_N2 reaction, much like an epoxide ring opens. The opening of the ring is facilitated further by the resonance stabilization of the negative charge in the carboxylate leaving group.

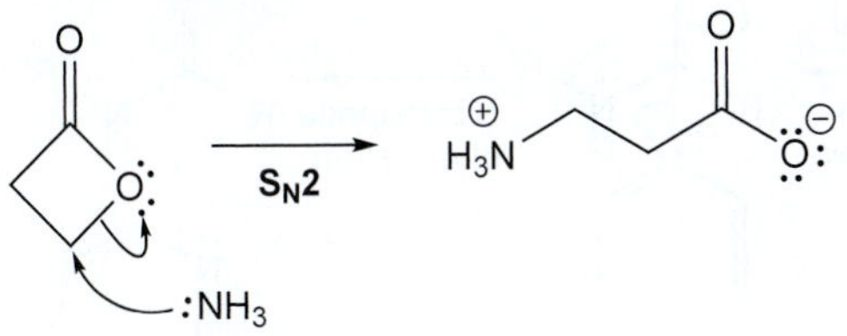

S_N2

Problem 21.43

Think

Is the three-carbon ring starting material stable or unstable? How does the nucleophile elimination step relieve the ring strain? Why would this allow R^- to be a leaving group?

Solve

(a) The mechanism is shown below. R^- can act as a leaving group because the nucleophile elimination step relieves the ring strain.

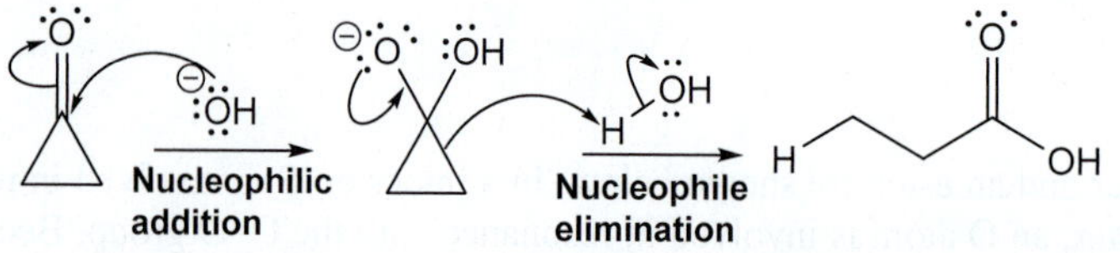

Nucleophilic addition

Nucleophile elimination

(b) To make an ester using this reaction, we can use RO^- as the nucleophile instead of HO^-.

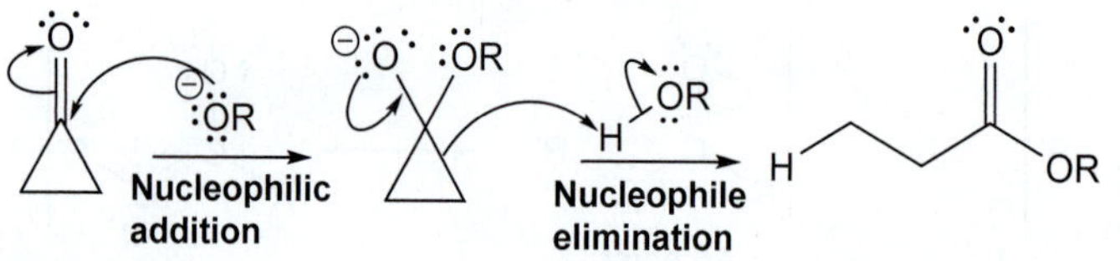

Nucleophilic addition

Nucleophile elimination

Problem 21.44

Think

If pyridine is the nucleophile in the nucleophilic addition–elimination reaction, what is the product that forms? Are proton transfer reactions involved? Does this product have a good leaving group? How would that product react with an amine?

Solve

(a) The nucleophilic addition–elimination reaction using pyridine as the nucleophile is shown below. Notice that the pyridine does not have any N–H bonds and, therefore, an uncharged amide product is not possible.

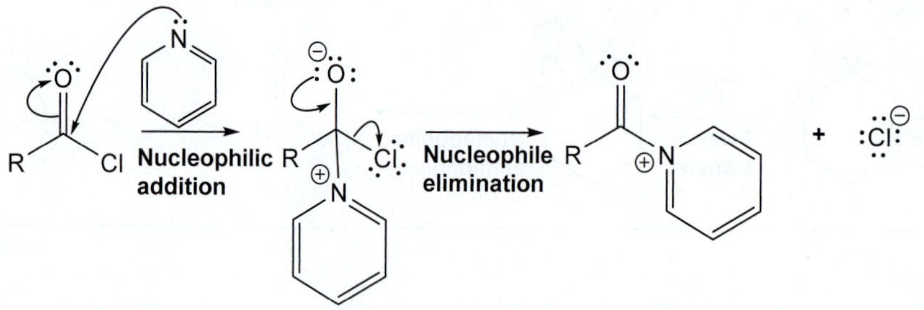

Nucleophilic addition

Nucleophile elimination

(b) The product formed in **(a)** would react with an amine to form an amide in another nucleophilic addition–elimination reaction, because the positively charged N makes pyridine a good leaving group for nucleophile elimination. Therefore, the pyridine does not interfere with the amide formation.

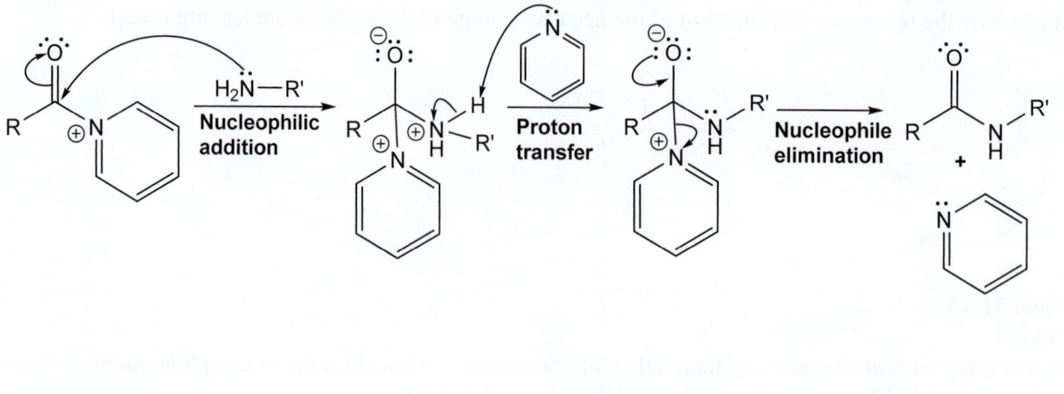

Nucleophilic addition

Proton transfer

Nucleophile elimination

Sections 21.4–21.6 Synthesis of Acid Halides, the Hell–Volhard–Zelinsky Reaction, and Sulfonyl Chlorides

Problem 21.45

Think

What acts as the nucleophile in the nucleophilic addition step? What is the leaving group in the nucleophile elimination step? Are proton transfer reactions involved?

Solve

(a) This reaction consists of back-to-back nucleophilic addition–elimination sequences. The first addition–elimination sequence involves the carboxylic acid and $SOCl_2$. In Step 1, the carbonyl O of the carboxylic acid attacks the polar S=O bond of $SOCl_2$, producing a species in which the positive charge is resonance delocalized. In Step 2, a Cl^- leaving group is eliminated from S, thereby regenerating the S=O bond. In the second nucleophilic addition–elimination reaction, Cl^- is the nucleophile, and SO_2Cl^- is the leaving group.

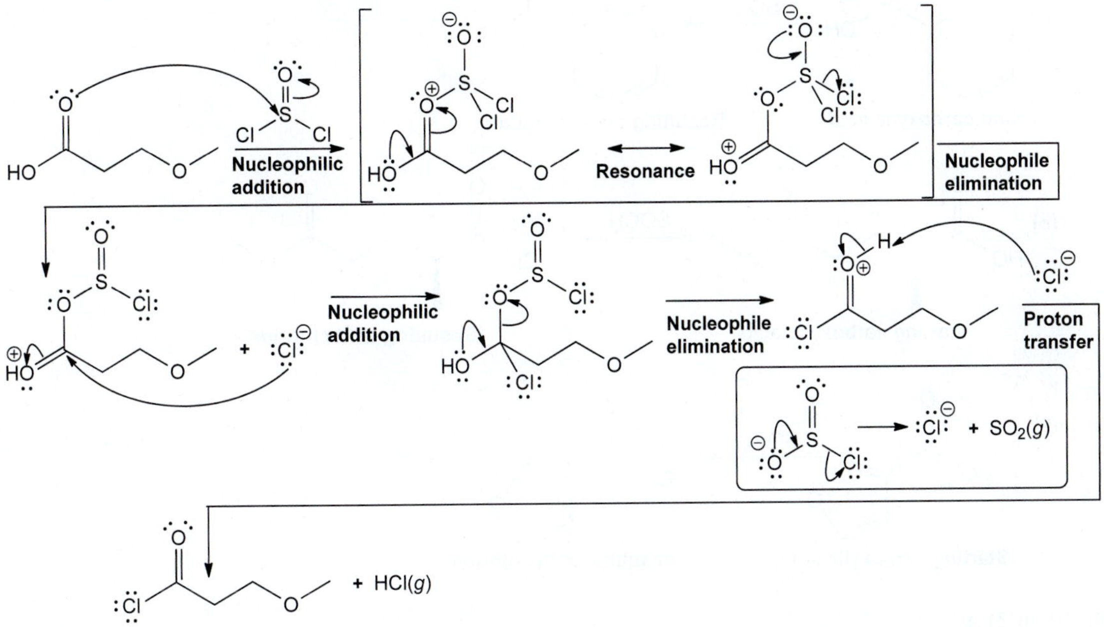

Nucleophilic addition

Resonance

Nucleophile elimination

Nucleophilic addition

Nucleophile elimination

Proton transfer

$\rightarrow$:Cl:$^-$ + $SO_2(g)$

+ HCl(*g*)

(b) Step 1 is an S_N2 step in which RCO_2H acts as a nucleophile to attack the electron-poor P in PCl_3, and the P–Cl bond breaks at the same time the O–P bond forms. Steps 2 and 3 make up the nucleophilic addition–elimination mechanism. The leaving group in the product of Step 1 is circled.

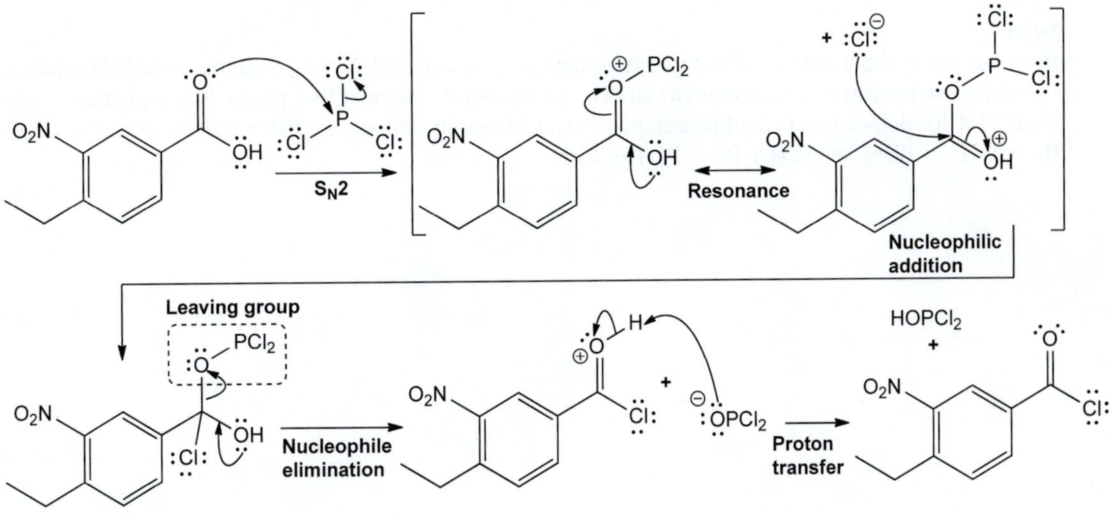

S_N2

Resonance

Nucleophilic addition

Leaving group

Nucleophile elimination

Proton transfer

$HOPCl_2$ +

Problem 21.46 (SYN)

Think

What bonds are broken and formed when a carboxylic acid reacts with $SOCl_2$ to form an acid chloride? Is the carbon skeleton altered?

Solve

In the reaction between a carboxylic acid and $SOCl_2$ to yield an acid chloride, only the bonds around the C=O and C–OH are broken, and C=O and C–Cl are formed. Therefore, the original carbon skeleton remains unchanged. The carboxylic acid that would produce acid chlorides **(a)**–**(c)** when treated with $SOCl_2$ is drawn below.

(a)

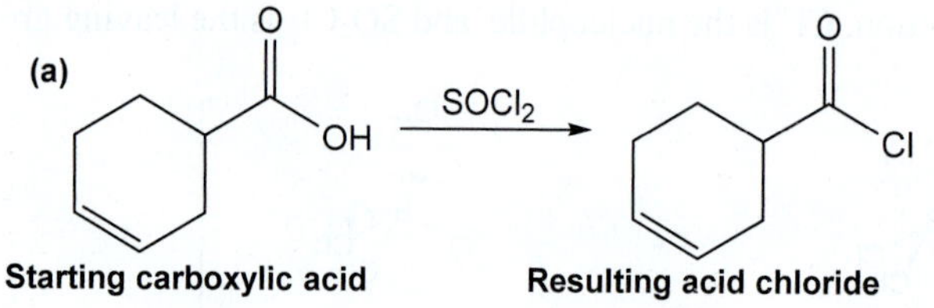

$SOCl_2$

Starting carboxylic acid **Resulting acid chloride**

(b)

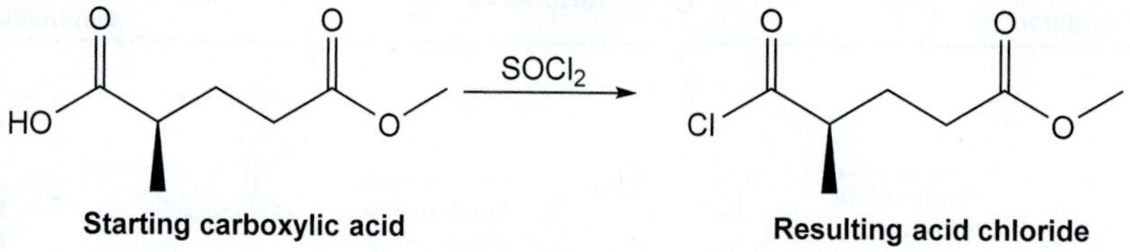

$SOCl_2$

Starting carboxylic acid **Resulting acid chloride**

(c)

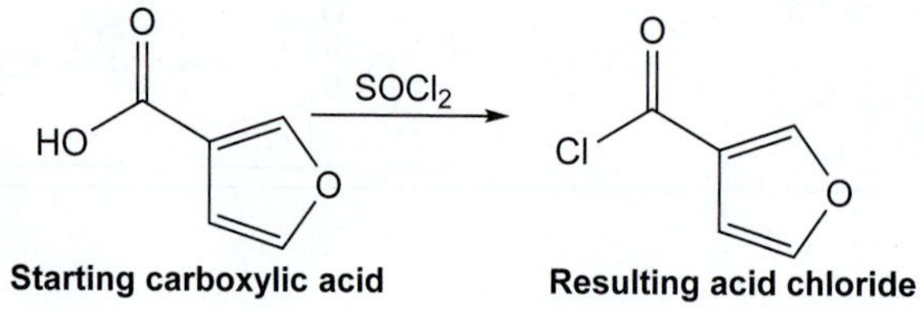

$SOCl_2$

Starting carboxylic acid **Resulting acid chloride**

Problem 21.47

Think

What acts as the nucleophile in the nucleophilic addition step? What is the leaving group in the nucleophile elimination step? Are proton transfer reactions involved? Review Equation 21-16 and Your Turn 21.4 for the details for the complete mechanism.

Solve

Bromination at the α carbon of a carboxylic acid is called a Hell–Volhard–Zelinsky (HVZ) reaction, and the complete mechanisms for reactions **(a)** and **(b)** are shown on the next two pages. See Equation 21-16 and Your Turn 21.4 for details regarding this complete mechanism. In both cases, PBr_3 is generated as a reagent. In **(a)**, $Br_2 + PCl_3 \rightarrow PBr_3$, and in **(b)**, $Br_2 + P(s) \rightarrow PBr_3$.

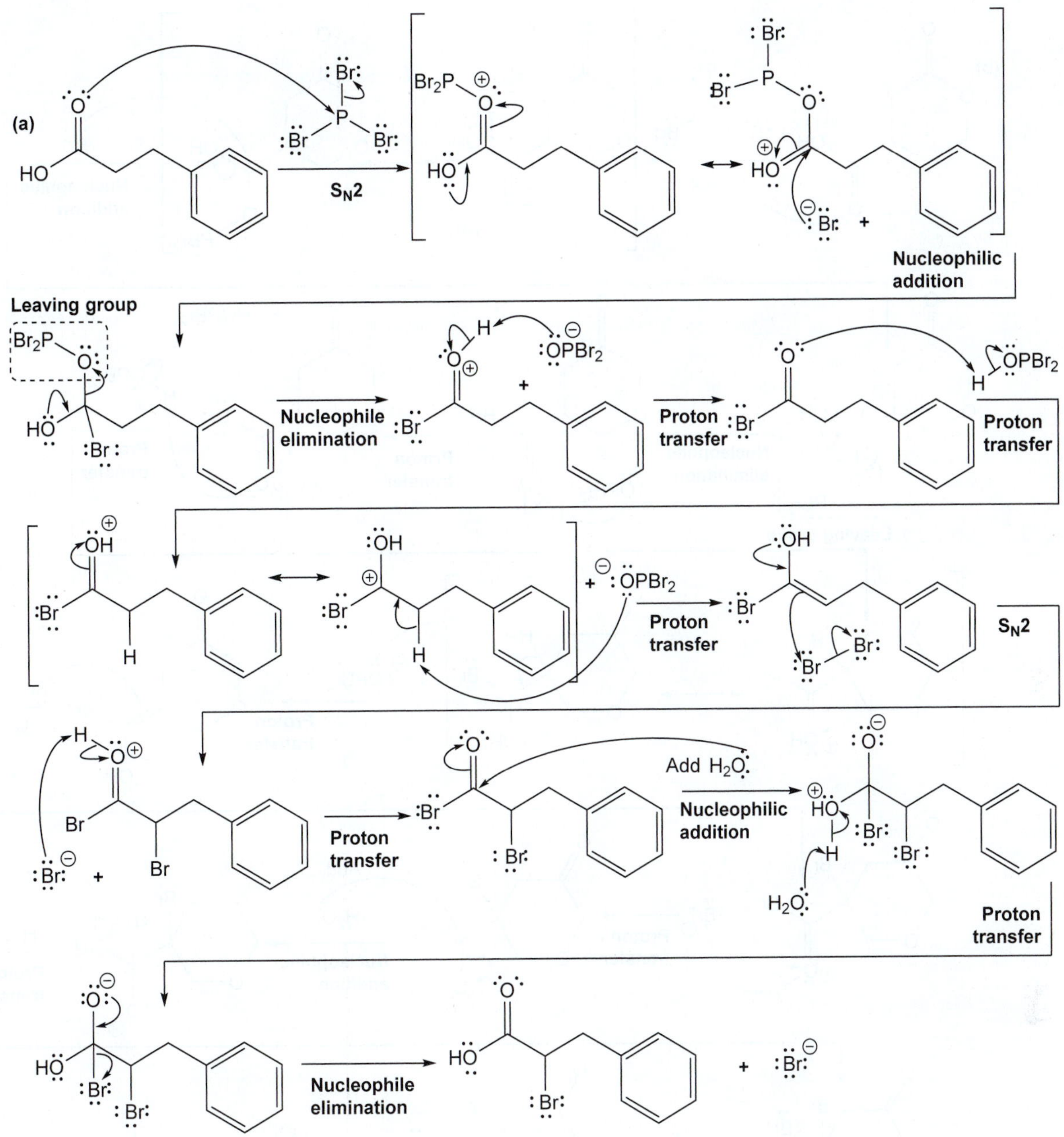

(a)

S_N2

Nucleophilic addition

Leaving group

Nucleophile elimination

Proton transfer

Proton transfer

Proton transfer

S_N2

Proton transfer

Add H_2O

Nucleophilic addition

Proton transfer

Nucleophile elimination

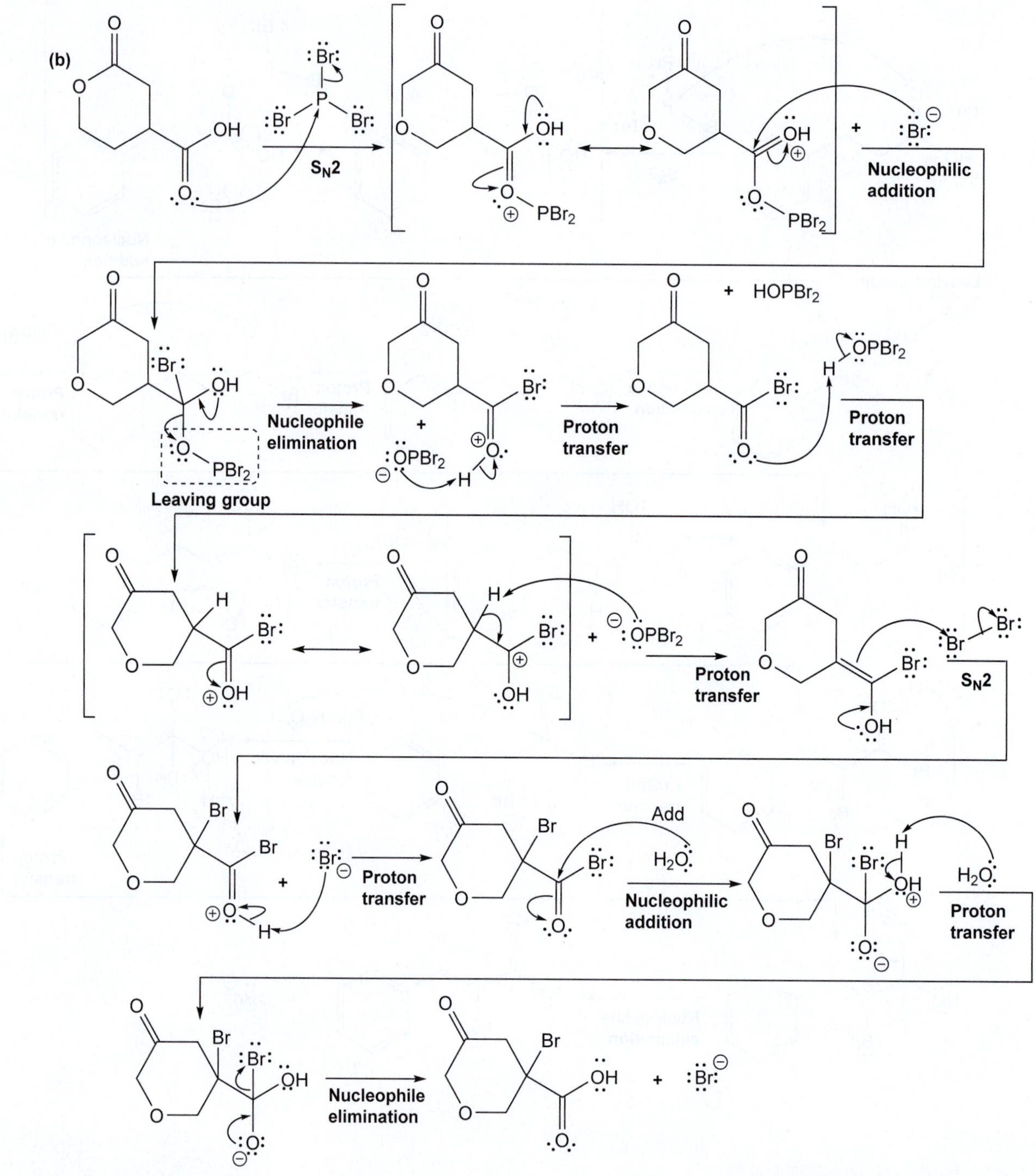

(b)

Problem 21.48 (SYN)

Think

What is the structure of 3-methylpentanoic acid? What new functional group is added to the α carbon? From what intermediate did this functional group originate?

Solve

All reactions require the HVZ reaction to produce the α-bromo acid intermediate. From there, different choices of nucleophiles produce the different targets—$(CH_3)_2NH$ yields the tertiary amine **(a)**, CH_3CH_2OH yields the ether **(b)**, and NaSH yields the thiol **(c)**.

(a) 1. Br_2, PCl_3 2. H_2O → Br, OH → N–H (dimethylamine) → N, OH

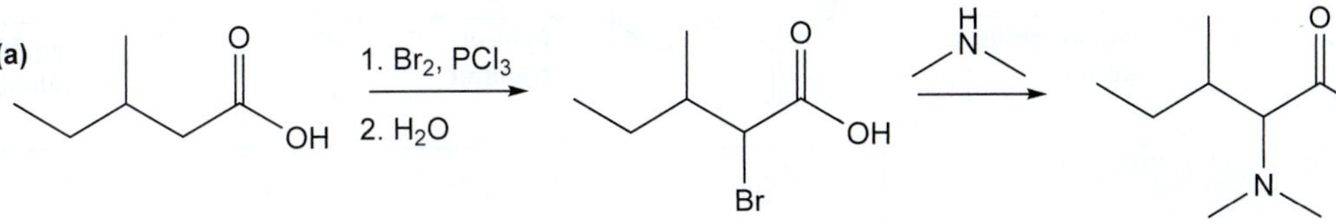

(b) 1. Br_2, PCl_3 2. H_2O → Br, OH → ethanol (OH) → O, OH

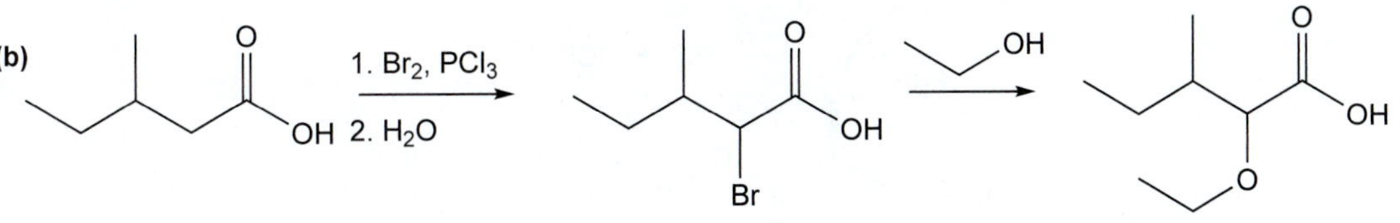

(c) 1. Br_2, PCl_3 2. H_2O → Br, OH → NaSH → SH, OH

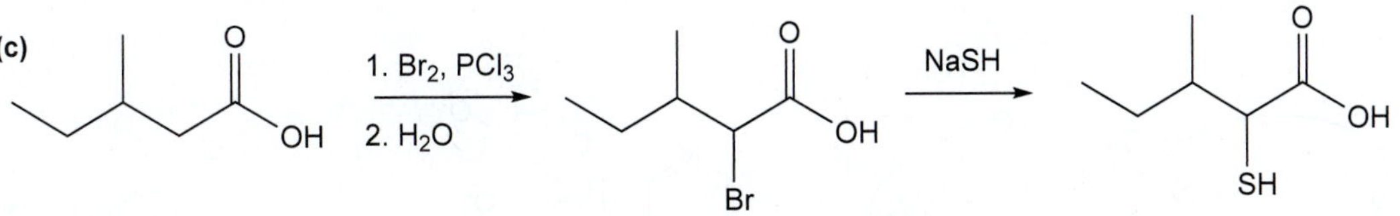

Problem 21.49

Think

What acts as the nucleophile in the nucleophilic addition step? What is the leaving group in the nucleophile elimination step? Are proton transfer reactions involved?

Solve

A sulfonyl chloride ($R–SO_2Cl$) reacts with an alcohol (R′–OH) in the presence of pyridine, a weak base, to produce a sulfonate ester ($R'–O–SO_2R$) via a three-step mechanism—nucleophilic addition, proton transfer, and nucleophile elimination. The mechanisms for reactions **(a)**–**(c)** are shown below and on the next page.

(a)

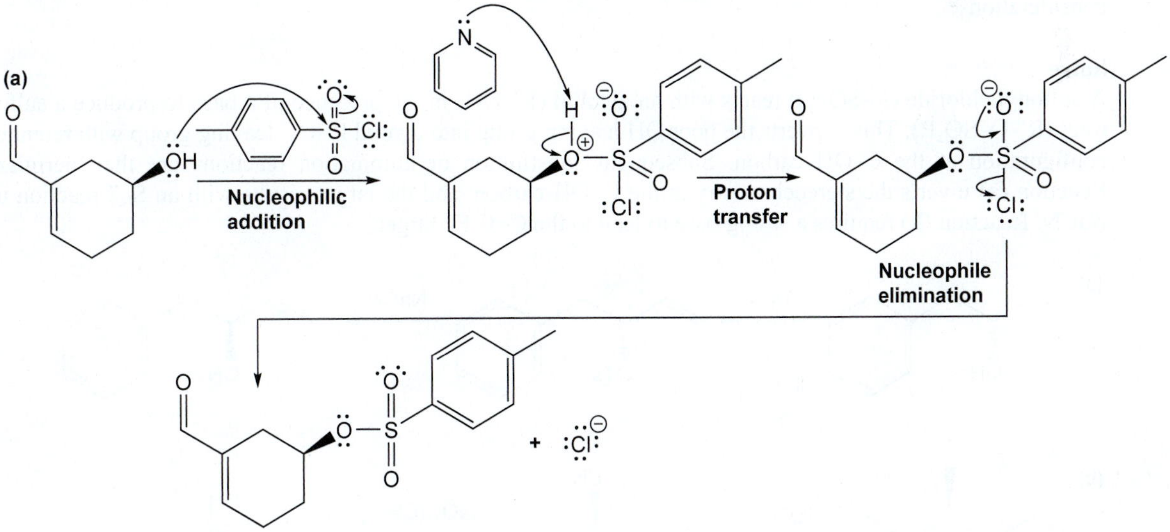

Nucleophilic addition

Proton transfer

Nucleophile elimination

+ $:Cl:^-$

(b)

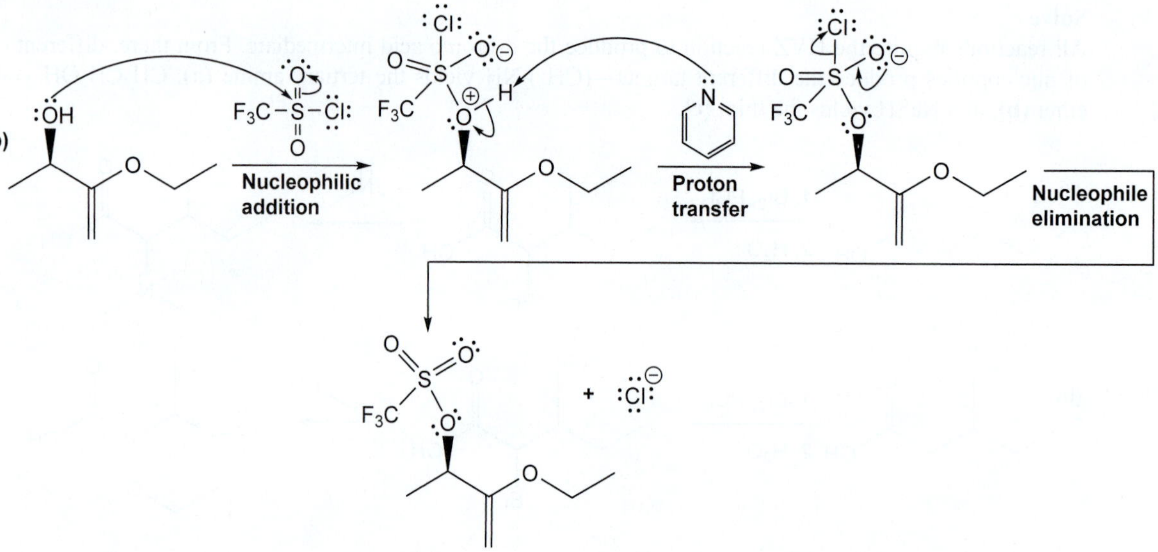

Nucleophilic addition

Proton transfer

Nucleophile elimination

(c)

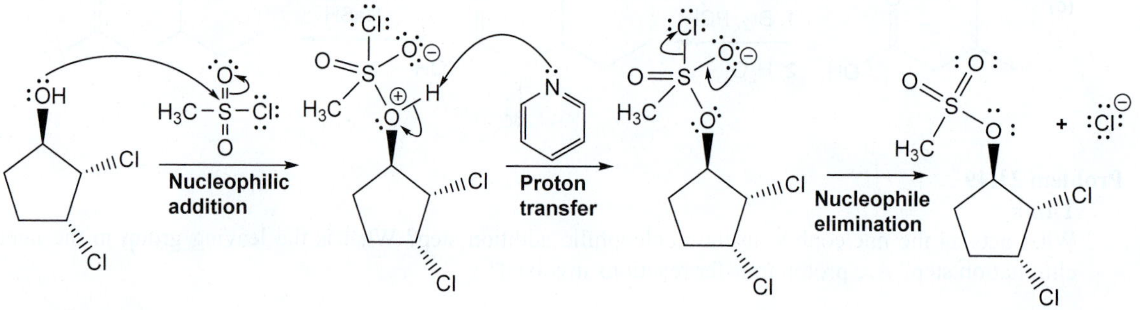

Nucleophilic addition

Proton transfer

Nucleophile elimination

Problem 21.50 (SYN)

Think

Is a OH a good leaving group? How do you transform an OH into a sulfonate ester? What nucleophile is necessary in each reaction to transform a sulfonate ester into the target? What are the stereochemical considerations?

Solve

A sulfonyl chloride (R–SO_2Cl) reacts with an alcohol (R′–OH) in the presence of a base to produce a sulfonate ester (R′–O–SO_2R). This converts the poor OH leaving group into a good RSO_3^- leaving group with retention of configuration at the C–OH carbon. Subsequent substitution or elimination reactions are then permissible. Reaction **(a)** inverts the stereochemistry at the C–OH carbon, and the target results with an S_N2 reaction using NaCN. Reaction **(b)** requires a strong base to lead to the C=C E2 target.

(a)

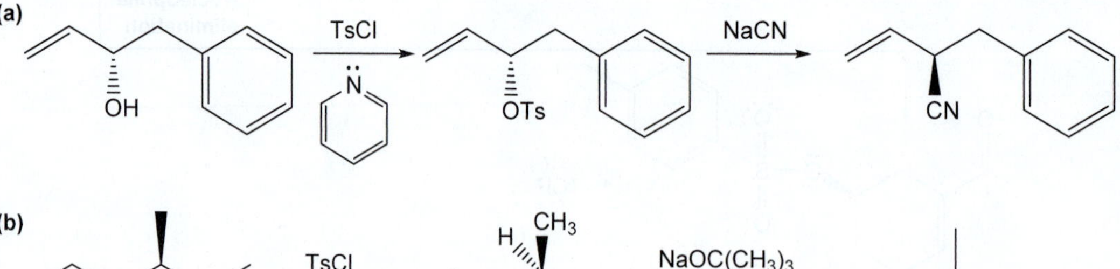

TsCl

NaCN

(b)

TsCl

$NaOC(CH_3)_3$

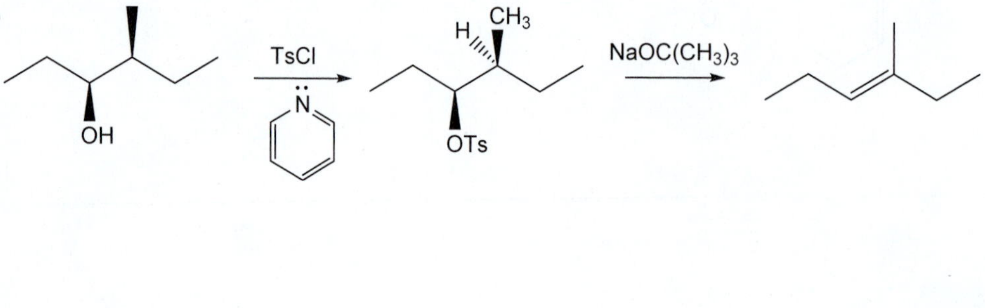

Problem 21.51

Think

What is the identity of the nucleophile in each nucleophilic addition? What are the leaving groups? Are proton transfer reactions involved?

Solve

The mechanism consists of a nucleophilic addition–elimination mechanism with the C=O oxygen as the nucleophile and the sulfur of $SOCl_2$ serving as the electrophilic atom. The Cl^- is eliminated and is the nucleophile in the second nucleophilic addition–elimination reaction.

Nucleophilic addition

Nucleophile elimination

Nucleophilic addition

Nucleophile elimination

Proton transfer

$+ HCl(g)$

$:Cl:^- + SO_2(g)$

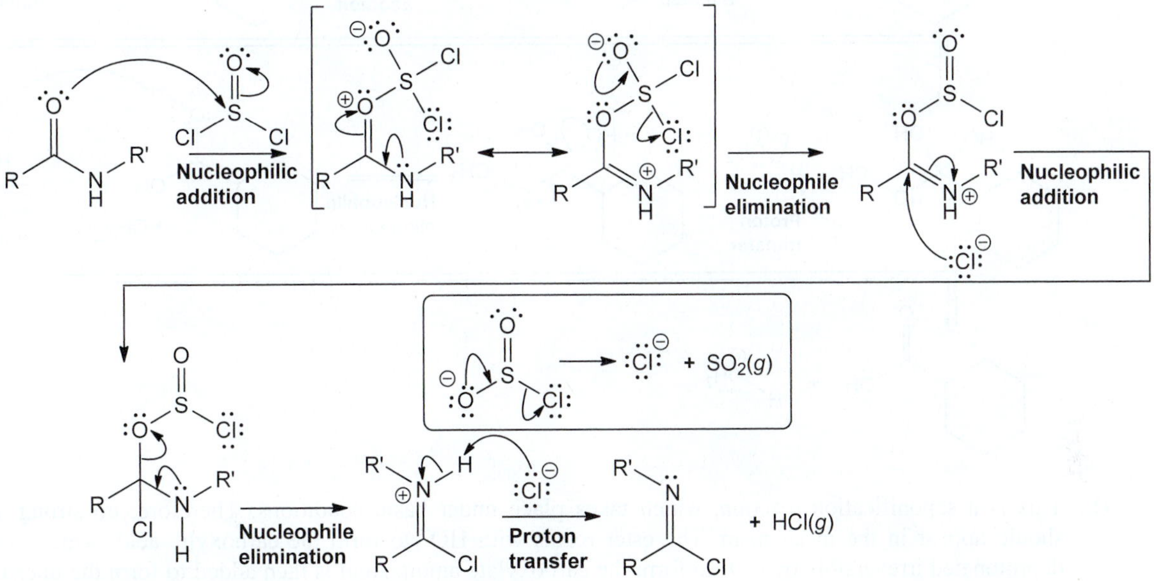

Problem 21.52

Think

What is the nucleophile? Which atom is electrophilic in the nucleophilic addition step? What is the leaving group in the nucleophile elimination step? Are proton transfer reactions involved?

Solve

The sulfur in sulfonyl chloride is electrophilic and is attacked by the amine nitrogen in the nucleophilic addition step. The amine acts as a base in a proton transfer step. The Cl^- is the leaving group in the nucleophile elimination step.

Nucleophilic addition

Proton transfer

Nucleophile elimination

A sulfonamide

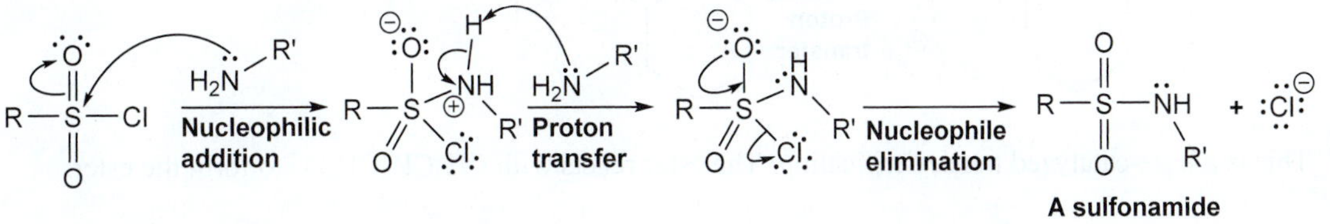

Sections 21.7 and 21.8 Base and Acid Catalysis in Nucleophilic Addition–Elimination Reactions; Baeyer–Villiger Oxidations

Problem 21.53

Think

What acts as the nucleophile in the nucleophilic addition step? What is the leaving group in the nucleophile elimination step? Are proton transfer steps involved? Is the reaction conducted in basic or acidic solution? What types of species should not appear in the mechanism under those conditions? Are there any irreversible proton transfers?

Solve

(a) This is an acid-catalyzed ester hydrolysis. Therefore, no strong bases should appear in the mechanism. The protonated ester reacts with water to form the carboxylic acid. Water acts as the nucleophile, and CH_3OH is the leaving group.

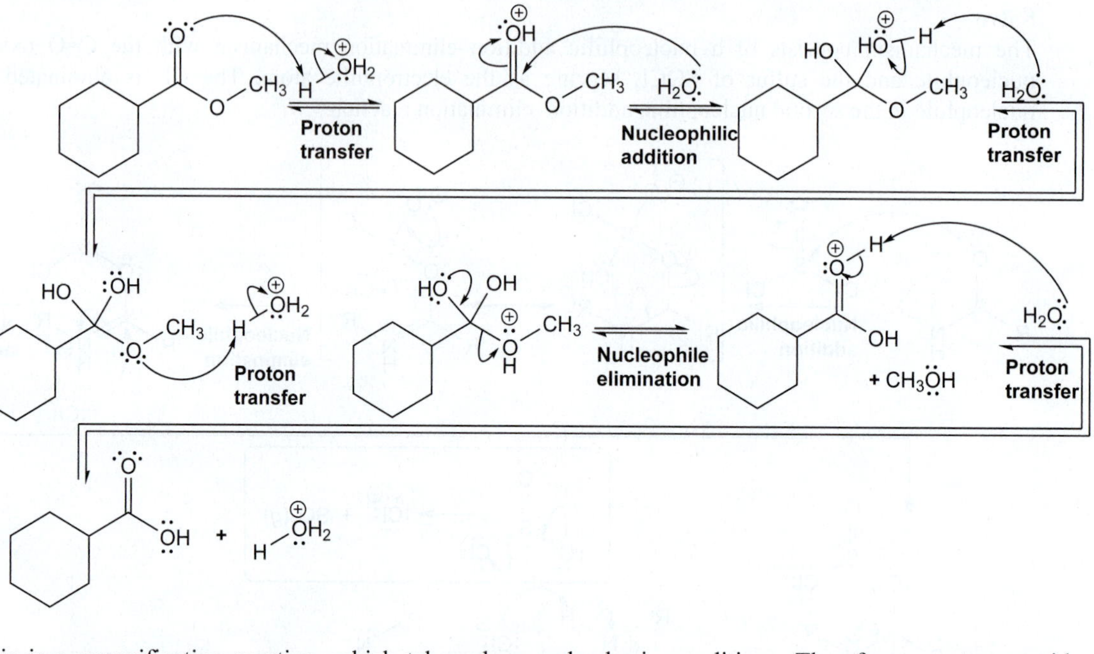

(b) This is a saponification reaction, which takes place under basic conditions. Therefore, no strong acids should appear in the mechanism. The ester reacts with HO^- to form the carboxylic acid, which is then deprotonated irreversibly by HO^- to form the carboxylate anion. Acid is then added to form the uncharged carboxylic acid product.

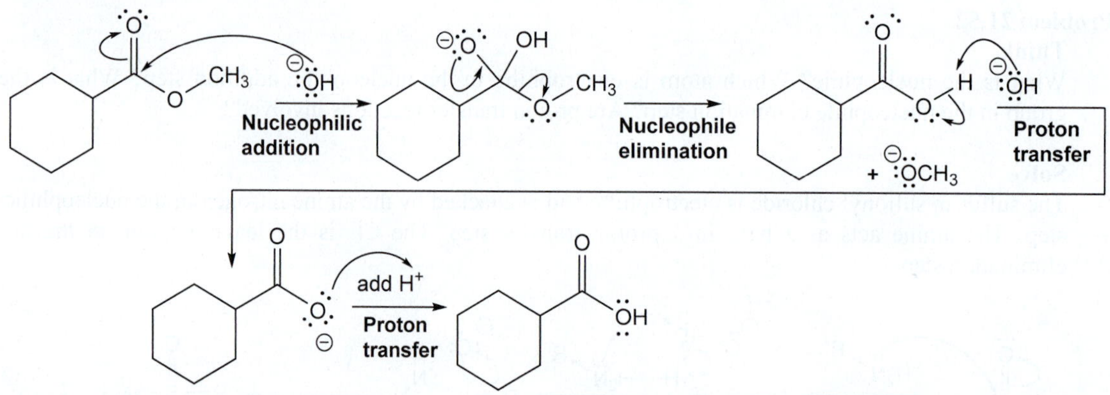

(c) This is a base-catalyzed transesterification. The ester reacts with $CH_3CH_2CH_2O^-$ to form the ester.

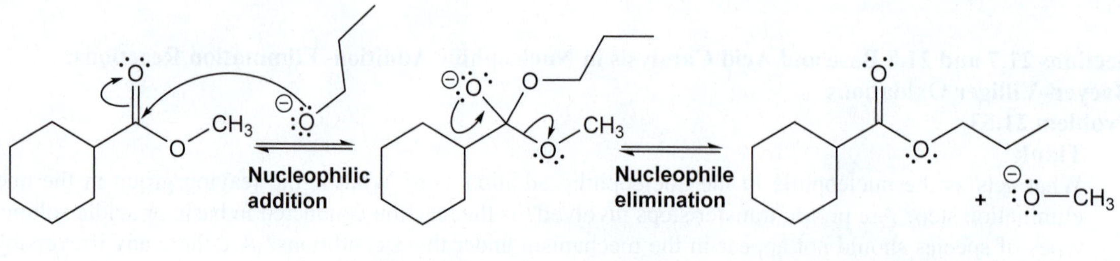

(d) This is an acid-catalyzed transesterification. The protonated ester reacts with the alcohol in a nucleophilic addition–elimination reaction. The leaving group is a molecule of CH_3OH.

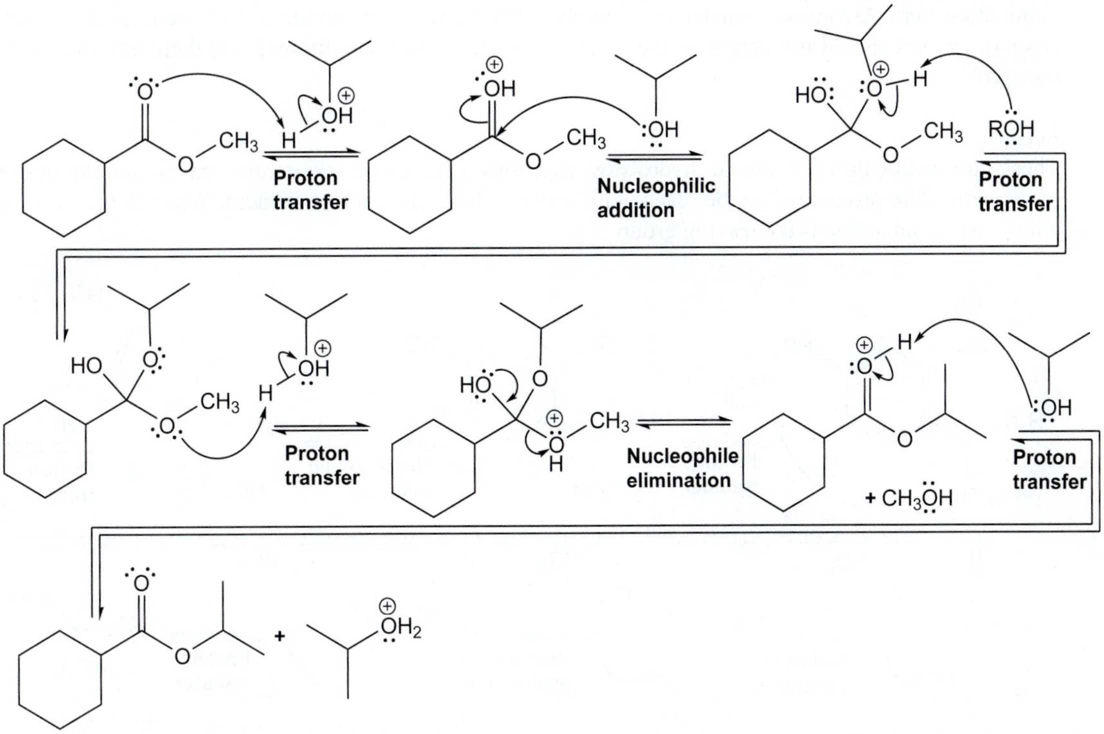

Proton transfer

Nucleophilic addition

Proton transfer

Proton transfer

Nucleophile elimination

+ CH_3OH

Proton transfer

(e) This is an acid-catalyzed aminolysis. The protonated ester reacts with the amine to form the amide. The amine is the nucleophile and the base, and CH_3OH is the leaving group.

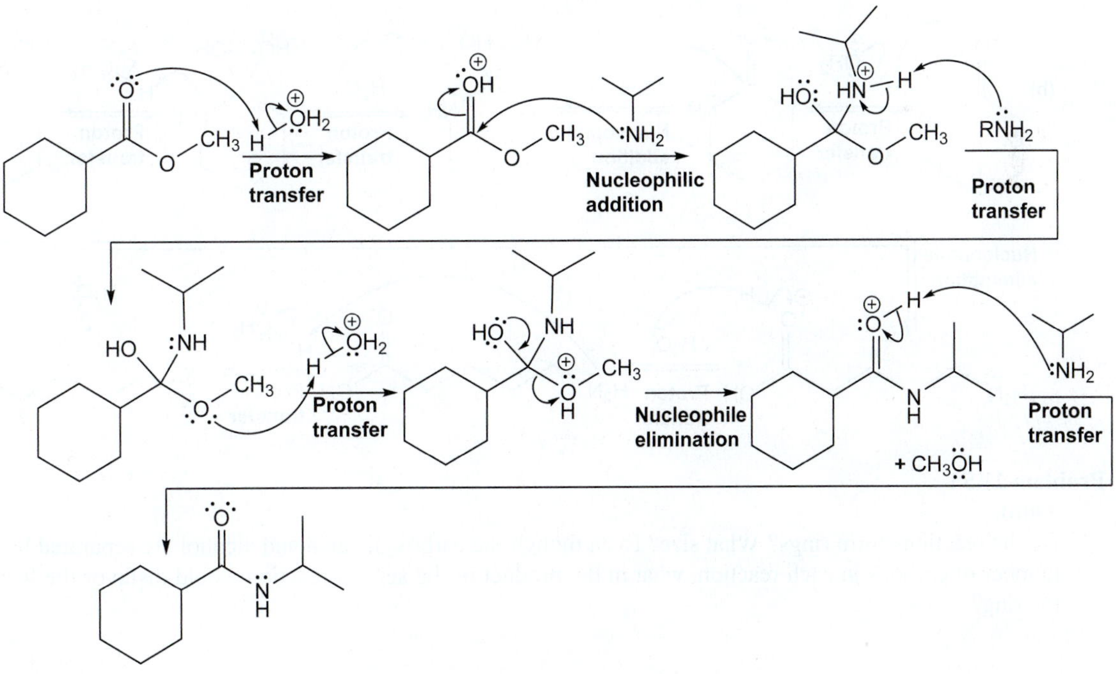

Proton transfer

Nucleophilic addition

Proton transfer

Proton transfer

Nucleophile elimination

+ CH_3OH

Proton transfer

(f) No reaction will take place. Propan-1-ol is a weak nucleophile, and the reaction is not catalyzed by acid or base.

Problem 21.54

Think

What acts as the nucleophile in the nucleophilic addition step? What is the leaving group in the nucleophile elimination step? Are proton transfer steps involved? Is the reaction conducted in basic or acidic solution? What types of species should not appear in the mechanism under those conditions? Are there any irreversible proton transfers?

Solve

These are acid-catalyzed amide hydrolysis reactions. Therefore, no strong bases should appear in the mechanism. The protonated amide reacts with water to form the carboxylic acid. Water is the nucleophile, and either NH_3 or an amine is the leaving group.

(a)

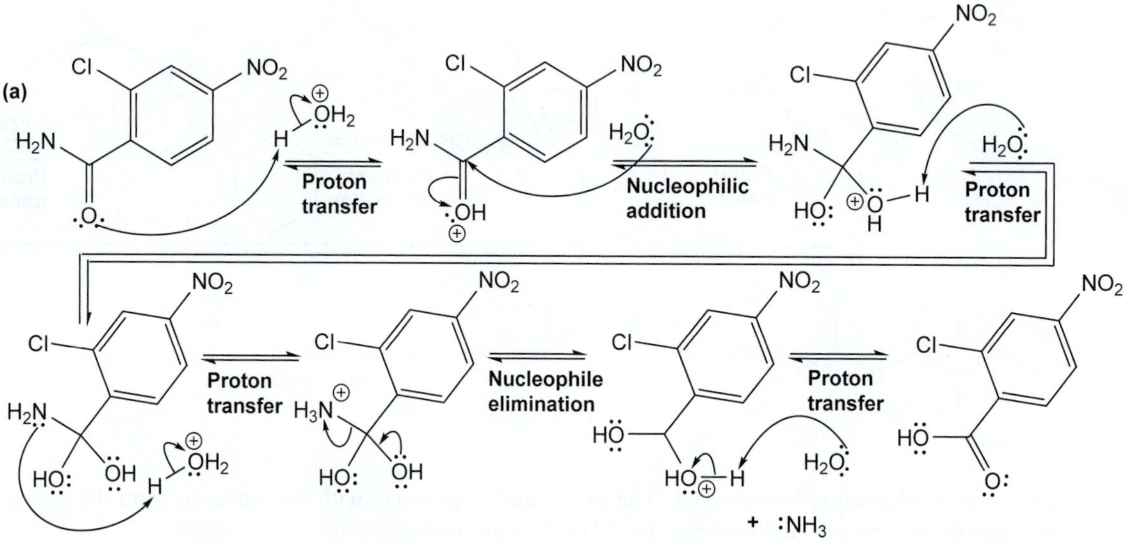

Proton transfer → Nucleophilic addition → Proton transfer → Proton transfer → Nucleophile elimination → Proton transfer

Cl, NO_2, H_2N, H_2O, $+ :NH_3$

(b)

Proton transfer → Nucleophilic addition → Proton transfer → Proton transfer → Nucleophile elimination → Proton transfer → Proton transfer

H_2O +

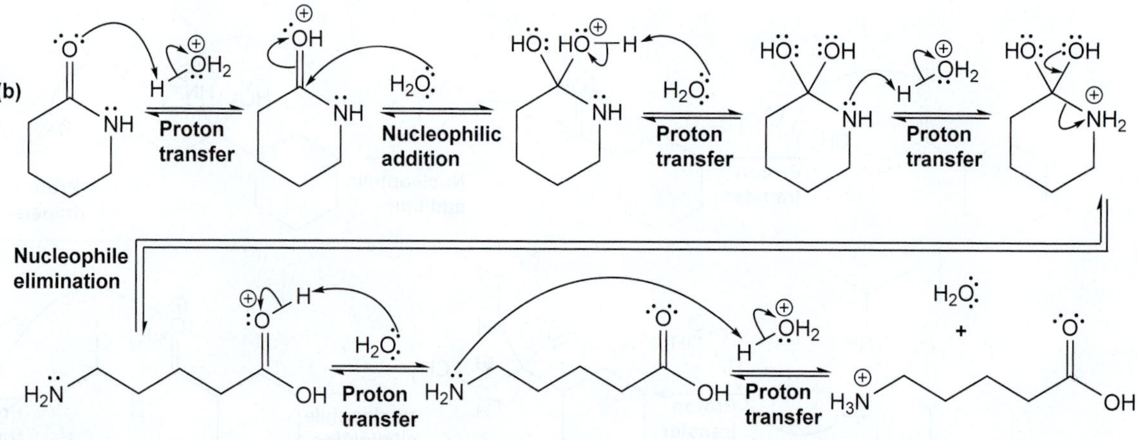

Problem 21.55

Think

Do the reactions form rings? What size? Even though the carboxylic acid and alcohol are separated by the same number of carbons in each reaction, what in the product of the second reaction would disfavor the formation of the ring?

Solve

In the second compound, the alcohol and carboxyl groups are separated by two C–C bonds along the ring. The ring's C–C bonds are more rigid than the C–C bond external to the ring, so the reactant functional groups in **(b)** would be farther away from each other and, therefore, less likely to find each other (greater loss of entropy). Thus, the product would have more strain.

Too far away to react.

H_2SO_4 Forms 5-membered ring

No reaction H_2SO_4 Strained ring, does not form

Problem 21.56

Think

To speed up an aminolysis of an ester in base, how is the nucleophile changed? Can that change to the nucleophile occur if HO^- or RO^- is added as a base? If a stronger base is added, would another proton transfer reaction take place that would prevent the intended nucleophilic addition?

Solve

To speed up a nucleophilic addition–elimination reaction under basic conditions, the job of the base is to convert the weak nucleophile into a strong nucleophile by deprotonation. So under basic conditions, the nucleophile in a hydrolysis reaction, H_2O, would be converted to HO^-. For an aminolysis reaction, the RNH_2 nucleophile would have to be converted to RNH^-. However, HO^- and RO^- are not strong enough as bases to carry out this deprotonation. If a stronger base is used to carry out this deprotonation, there is a separate problem: RNH^- is a very strong base (pK_a of RNH_2 is ~38), strong enough to irreversibly deprotonate the α carbon of an ester to make an enolate anion. Once the enolate anion is made, it is no longer susceptible to nucleophilic attack at the carbonyl carbon.

Problem 21.57

Think

What is the nucleophile in the nucleophilic addition step? What is the leaving group in the nucleophile elimination step? Is the reaction promoted by acid or base? What types of species should not appear in the mechanism? How is the alcohol product different if the ester shown in **(b)** is used? What rearrangement would that alcohol product undergo to gain stability?

Solve

(a) The mechanism is an acid-catalyzed nucleophilic addition–elimination reaction. The carboxylic acid is the nucleophile, and an alcohol is the leaving group. Under acidic conditions, no strong bases should appear in the mechanism.

Proton transfer — Nucleophilic addition — Proton transfer — Proton transfer — Nucleophile elimination — Proton transfer

(b) R″–OH in the products would have the form of an enol, which tautomerizes to the more stable ketone. This effectively removes R″–OH from the products of the nucleophilic addition–elimination and, therefore, drives that equilibrium more toward products.

HO O

Problem 21.58

Think

By examining the products, which bonds broke and formed in the reaction? Which O atom in each reactant species, therefore, must have been protonated by the $H_3{}^{18}O^+$ acid? How does this change the nature of the mechanism? How does the alkyl group bonded to O favor one mechanism over the other?

Solve

In the first case, $H_3{}^{18}O^+$ protonates the carbonyl oxygen, and $H_2{}^{18}O$ attacks the carbonyl carbon in an acid-catalyzed nucleophilic addition–elimination reaction. The initial O–CH_3 bond stays intact, and the ^{18}O label ends up in the carboxylic acid after departure of protonated methanol.

Proton transfer · Nucleophilic addition · Proton transfer · Proton transfer · Nucleophile elimination · Proton transfer

In the second reaction, the *t*-butyl group hinders the nucleophilic attack at the carbonyl carbon, but an S_N1 reaction is favorable owing to the generation of a 3° carbocation; $^{18}OH_2$ attacks the carbocation to form the protonated and labeled alcohol.

Proton transfer · Heterolysis · Coordination · Proton transfer

Problem 21.59

Think

How many nucleophilic addition–elimination reactions occur? With hydrolysis taking place under acidic conditions, what species should not appear in the mechanism? What is the nucleophile? What is the leaving group?

Solve

(a) In acid, the carbonyl oxygen is protonated, and then water is the nucleophile in the nucleophilic addition step. The full mechanism is given below.

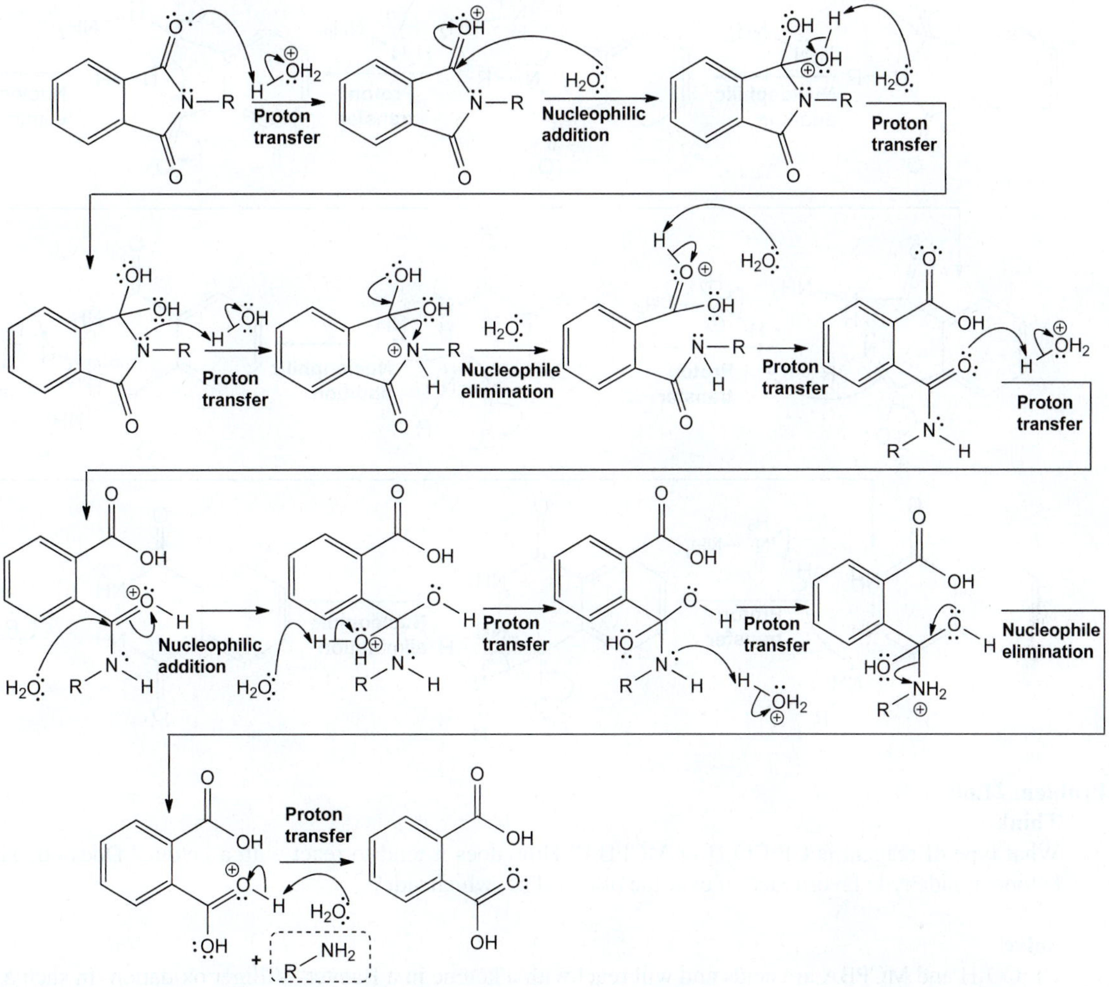

Proton transfer → Nucleophilic addition → Proton transfer → Proton transfer → Nucleophile elimination → Proton transfer → Proton transfer → Nucleophilic addition → Proton transfer → Proton transfer → Nucleophile elimination → Proton transfer

(b) Hydrazine is the nucleophile in the nucleophilic addition step. The full mechanism is given below.

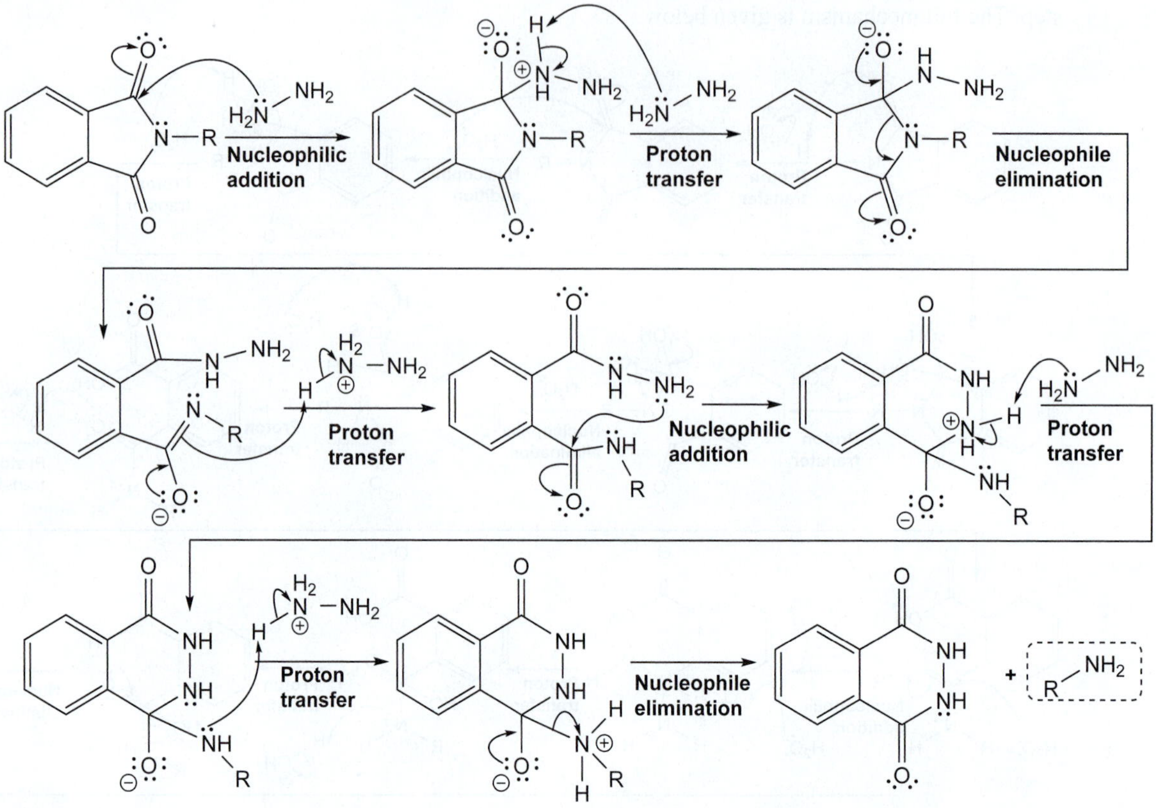

Problem 21.60

Think

What type of reagent is CF_3CO_3H or MCPBA? How does it tend to react with a ketone? Does one side of the ketone or aldehyde favor reaction over the other? If so, which side?

Solve

CF_3CO_3H and MCPBA are acids and will react with a ketone in a Baeyer–Villiger oxidation. In such a reaction, an O atom from the acid is inserted between the carbonyl C and one of the groups initially bonded to the carbonyl C. The C=O is activated by the CF_3CO_3H acid. The weak peroxyacid MCPBA attacks the C=O in a nucleophilic addition step. A proton is removed by the conjugate base, $CF_3CO_3^-$, in Step 3. Step 4 is a nucleophile elimination step in which the R group leaves at the same time as it forms a bond to the O of MCPBA. The weak O–O bond of MCPBA is simultaneously broken in Step 4. The final step is a proton transfer to yield the uncharged ester or carboxylic acid product. Mechanisms for **(a)**–**(c)** are given on the next two pages.

(a) The carbonyl C is bonded to a primary alkyl group and an aryl group. The aryl group has a greater migratory aptitude, so its bond will preferentially break, producing the ester shown below.

Proton transfer

Nucleophilic addition

Proton transfer

Nucleophile elimination

Proton transfer

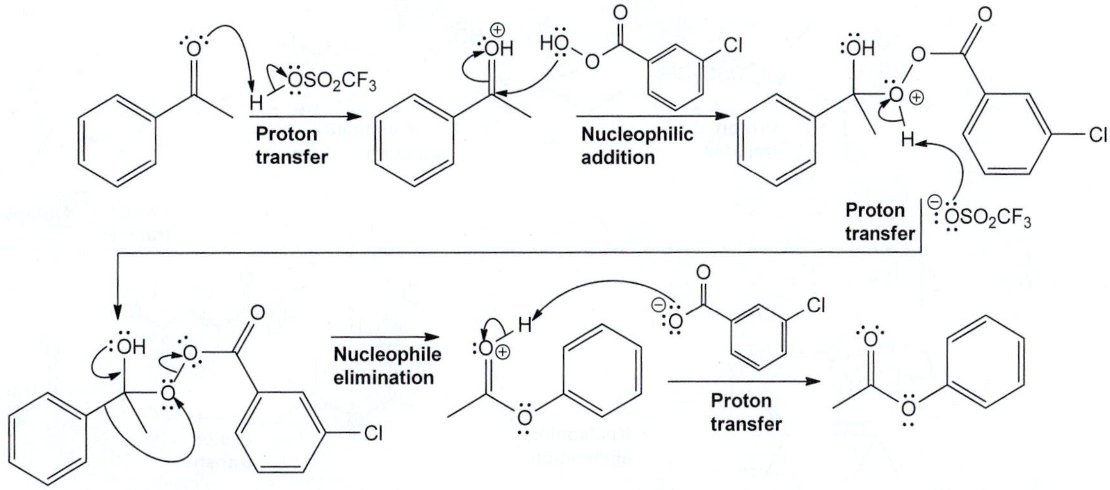

(b) The carbonyl C is bonded to a H atom and an aryl group. The H atom has a greater migratory aptitude, so its bond will preferentially break, producing the ester shown below.

Proton transfer

Nucleophilic addition

Proton transfer

Nucleophile elimination

Proton transfer

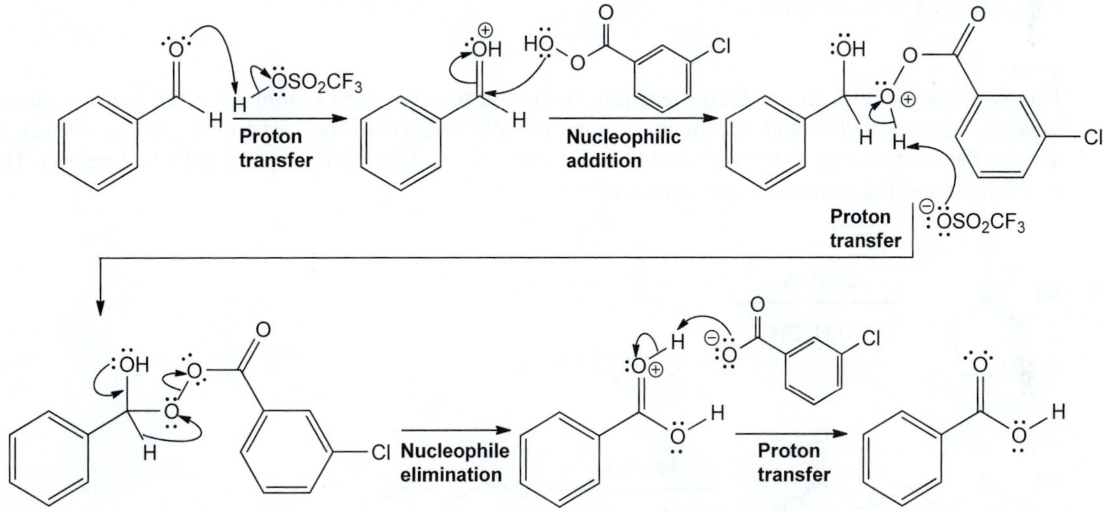

(c) The carbonyl C is bonded to a primary alkyl group and a tertiary alkyl group. The tertiary alkyl group has a greater migratory aptitude, so its bond will preferentially break, producing the ester shown below.

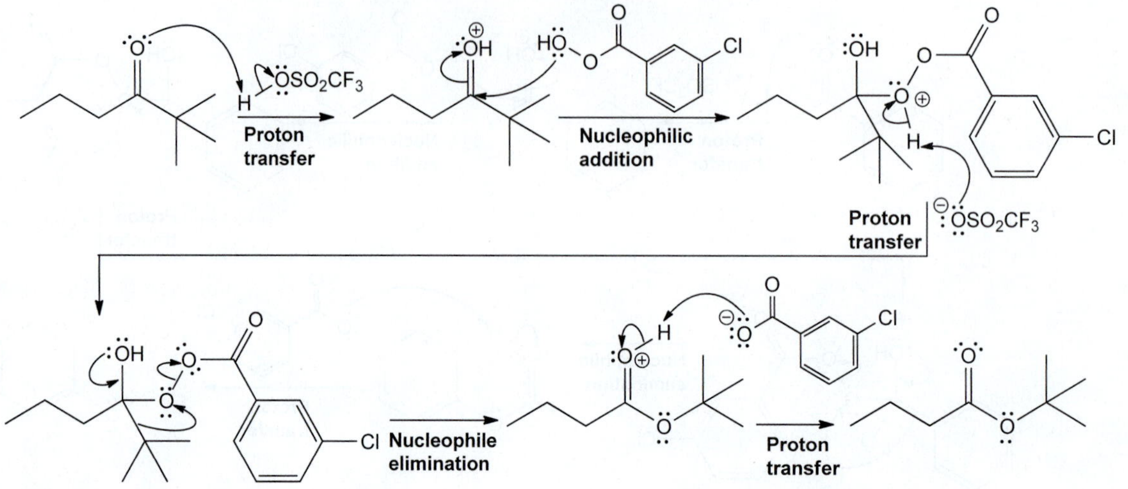

Proton transfer

Nucleophilic addition

Proton transfer

Nucleophile elimination

Proton transfer

Problem 21.61 (SYN)

Think

What bonds are broken and formed when a ketone or aldehyde is transformed into an ester or carboxylic acid using MCPBA in acidic conditions?

Solve

This is a Baeyer–Villiger oxidation reaction. In such a reaction, an O atom from the peroxyacid is inserted between the carbonyl C and one of the groups initially bonded to the carbonyl C. To reverse the reaction, remove the oxygen next to the carbonyl C and re-form the C–C or C–H bond (highlighted below). The ketone or aldehyde starting compounds are shown below.

(a) MCPBA; CF_3SO_3H, CH_2Cl_2

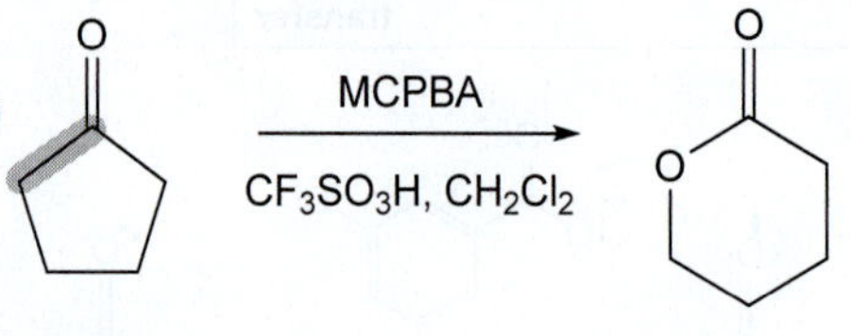

(b) MCPBA; CF_3SO_3H, CH_2Cl_2

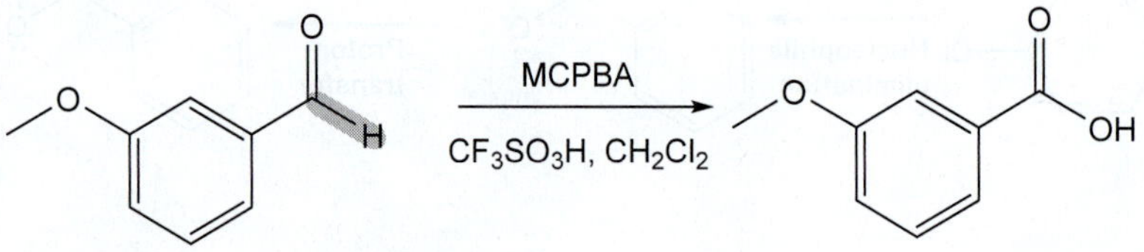

(c) MCPBA; CF_3SO_3H, CH_2Cl_2

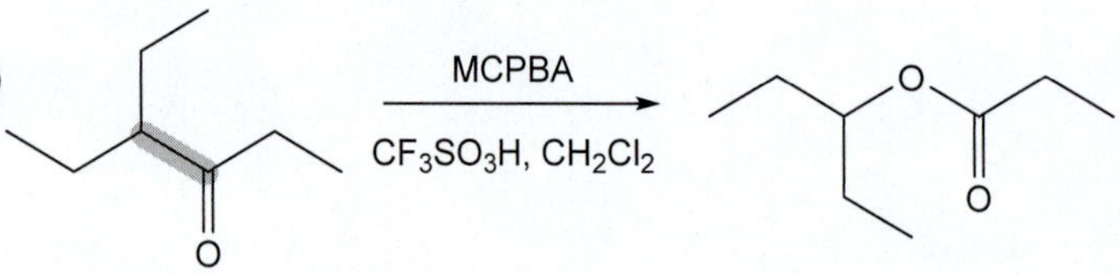

(d) MCPBA; CF_3SO_3H, CH_2Cl_2

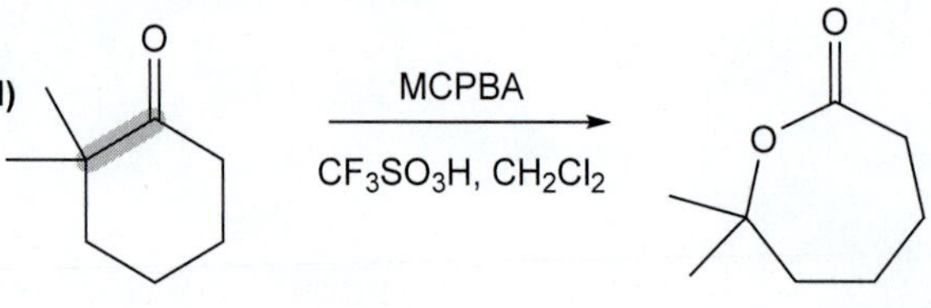

Section 21.9 Claisen Condensations

Problem 21.62

Think

What functional group is present? How does this functional group react with the strong base given in the problem? What nucleophile is formed? With what electrophile will it react? What is the leaving group? Under basic conditions, what types of species should not appear in the mechanism?

Solve

These are all examples of the Claisen condensation reaction. The base deprotonates an α C–H to make an ester enolate anion, which is the nucleophile that will react with another ester molecule in a nucleophilic addition–elimination mechanism. The leaving group in each case is an alkoxide anion. The mechanisms for **(a)–(c)** are given below. Notice that **(a)** is a self-Claisen condensation reaction, whereas **(b)** and **(c)** are crossed Claisen condensations. These crossed Claisen condensations are feasible because one of the esters has no α hydrogen.

(a)

OEt **Proton transfer** OEt **Nucleophilic addition** OEt **Nucleophile elimination** OEt **Proton transfer**

OEt **Proton transfer** Add H⁺ OEt

(b)

EtO **Proton transfer** EtO EtO **Nucleophilic addition** EtO

Nucleophile elimination

EtO **Proton transfer** EtO Add H⁺ **Proton transfer** EtO

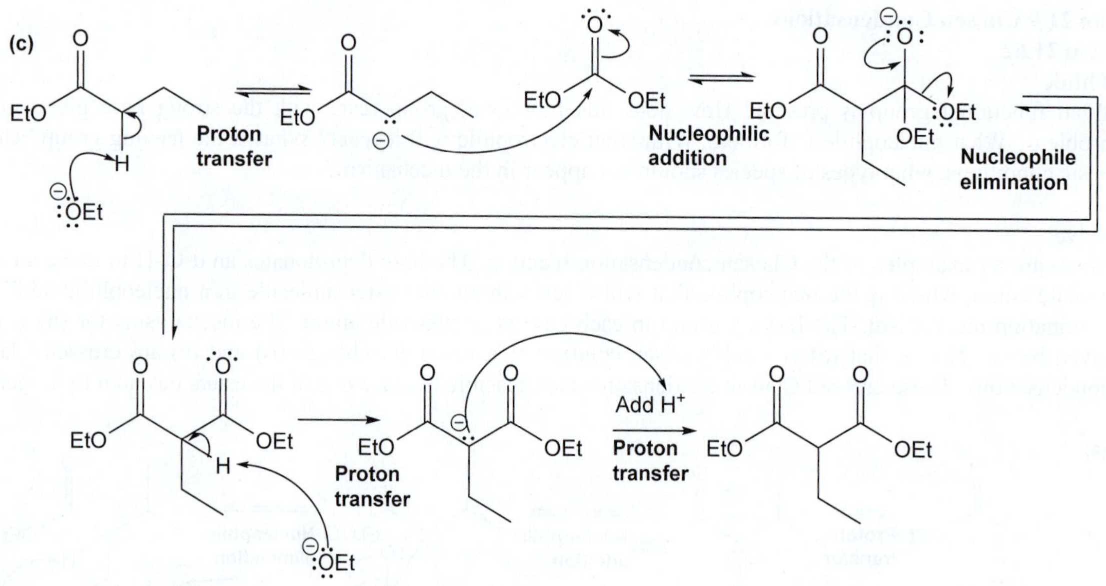

(c)

Problem 21.63

Think

In a Claisen condensation reaction, are the nucleophilic addition and elimination steps reversible or irreversible? Does that equilibrium favor reactants or products? To be a successful Claisen condensation reaction, what must be true of a proton transfer involving the β-keto ester that is initially produced? Is that type of proton transfer possible in the self-Claisen reaction involving methyl propanoate? Involving methyl 2-methylpropanoate?

Solve

In each case, we can envision a self-Claisen condensation reaction. In the second reaction, however, the β-keto ester that is produced does not have an acidic proton on the carbon between the two C=O groups. Therefore, there is no available irreversible proton transfer that can drive the reaction toward products. The nucleophilic addition–elimination equilibrium favors reactants instead. Therefore, whereas a self-Claisen condensation involving methyl propanoate will interfere with a crossed Claisen condensation, a self-Claisen condensation reaction involving methyl 2-methylpropanoate will not interfere with a crossed Claisen condensation.

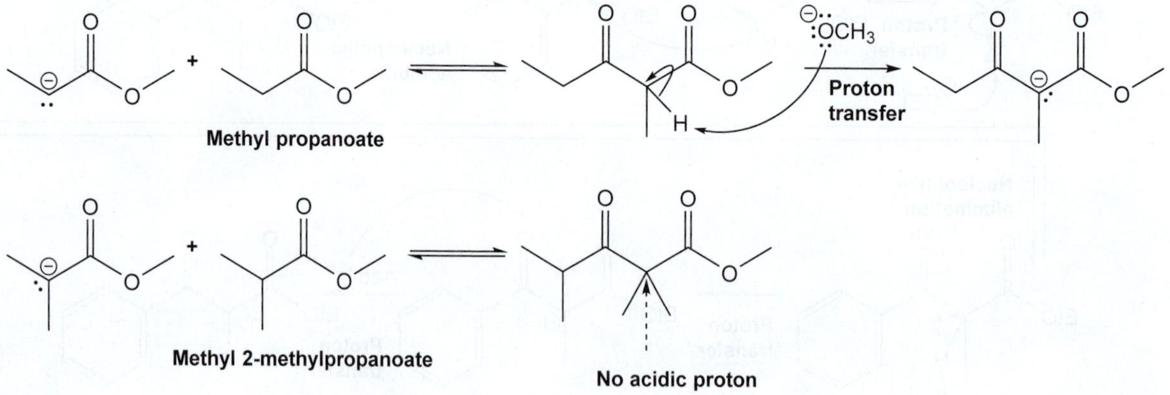

Problem 21.64

Think

What functional group is present? How does this functional group react with the strong base given in the problem? What nucleophile is formed? With what electrophile will it react? Will an intermolecular or intramolecular reaction be favored? Under basic conditions, what types of species should not appear in the mechanism?

Solve

(a) This is an example of a Claisen condensation reaction. Under these basic conditions, no strong acids should appear in the mechanism. The CH_3O^- deprotonates the α C–H to form the enolate, which undergoes nucleophilic addition to the other ester C=O. An intramolecular reaction is favored because of the formation of a five-membered ring. Nucleophile elimination followed by proton transfer forms the enolate anion in an irreversible reaction. An acid workup is necessary to form the uncharged product.

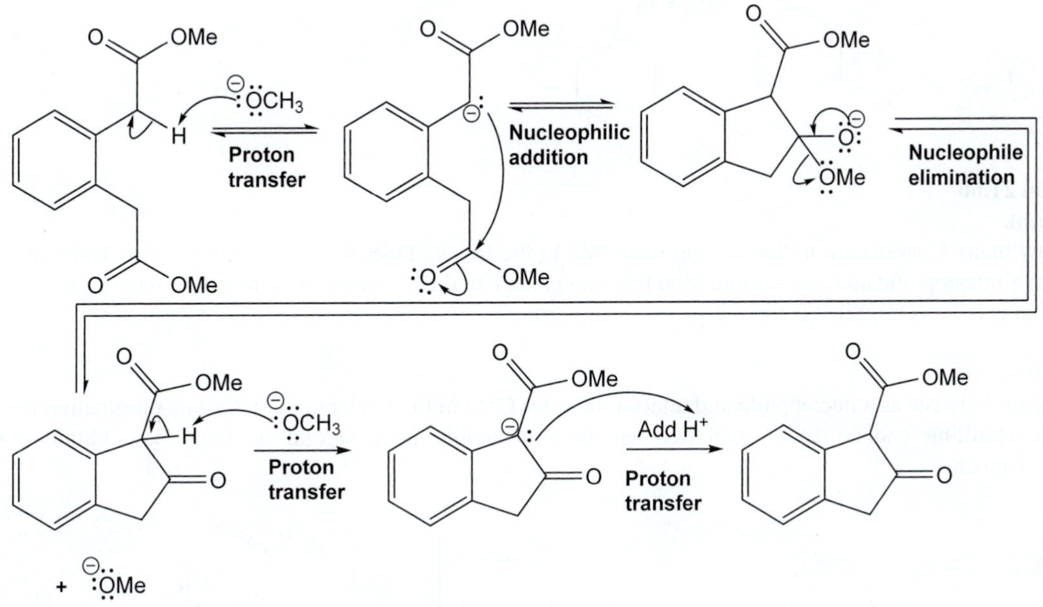

(b) This is a crossed Claisen condensation reaction. LDA deprotonates the α C–H to form the enolate quantitatively, which undergoes nucleophilic addition to the ethyl acetate ester C=O. Nucleophile elimination followed by proton transfer forms the enolate anion in an irreversible reaction. An acid workup is necessary to form the uncharged product.

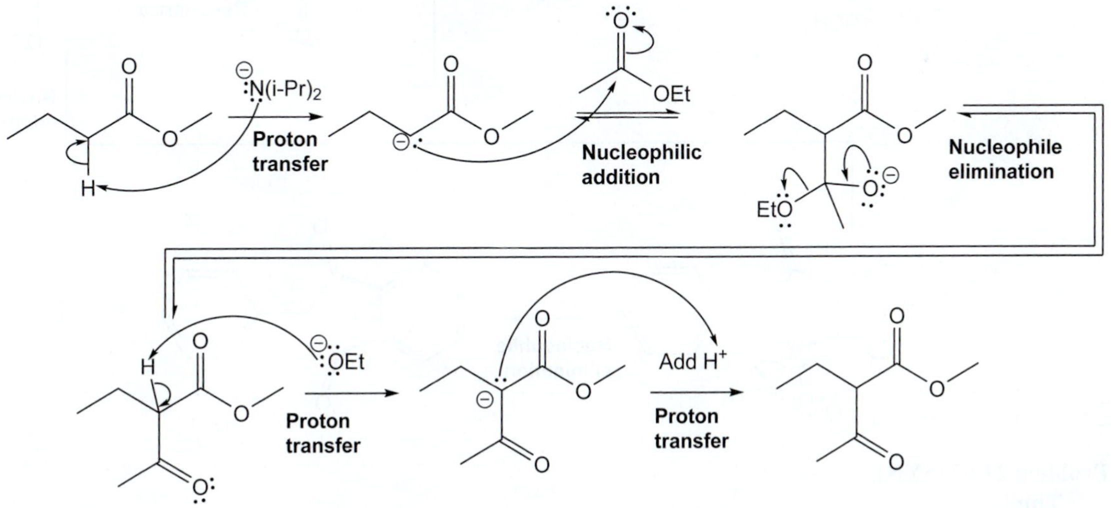

Problem 21.65

Think

In a Claisen condensation reaction, what should the nucleophile be? Can that nucleophile be produced effectively with C_6H_5ONa used as the base?

Solve

C_6H_5ONa is not a strong enough base to remove the α C–H, evidenced by the fact that its conjugate acid is much stronger than an ester. The enolate is not formed and, therefore, the Claisen condensation reaction will not occur.

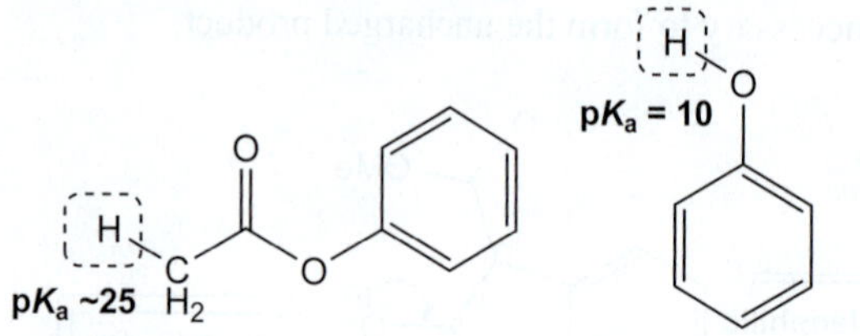

pK$_a$ = 10

pK$_a$ ~25 H_2

Problem 21.66

Think

How many C atoms are in the starting material? In the target? Does CH_3O^- behave as a nucleophile or a base? Can a nucleophilic addition–elimination reaction ensue? In the product of that reaction, what is the nucleophile? What is the electrophile? Is there a good leaving group?

Solve

CH_3O^- behaves as a nucleophile and attacks the ester C=O in the nucleophilic addition–elimination mechanism. The resulting enolate anion then acts as the nucleophile in a second nucleophilic addition–elimination mechanism.

Nucleophilic addition

Nucleophile elimination

Resonance

Nucleophilic addition

Nucleophile elimination

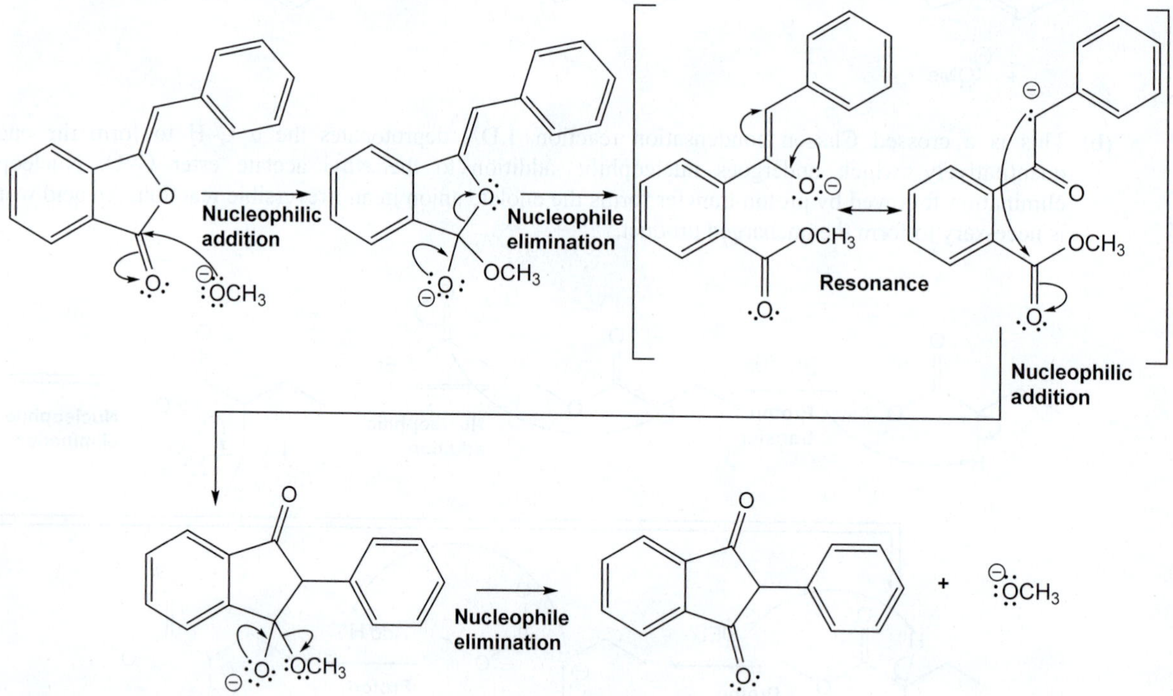

Problem 21.67 (SYN)

Think

Which C–C bond in the β-keto ester product would have been formed in a Claisen or Dieckmann condensation? Which C atom would have been bonded to the alkoxy leaving group?

Solve

In both the Claisen and Dieckmann condensation reactions, the new C–C bond is the one between the α and β carbons. Moreover, the leaving group would have been attached to the β C atom. Therefore, we can apply a transform to undo these condensations by disconnecting that C–C bond and reattaching the leaving group. Doing so yields two ester precursors.

(a) This example yields two identical ester precursors. The base required is $(CH_3)_2CHO^-$.

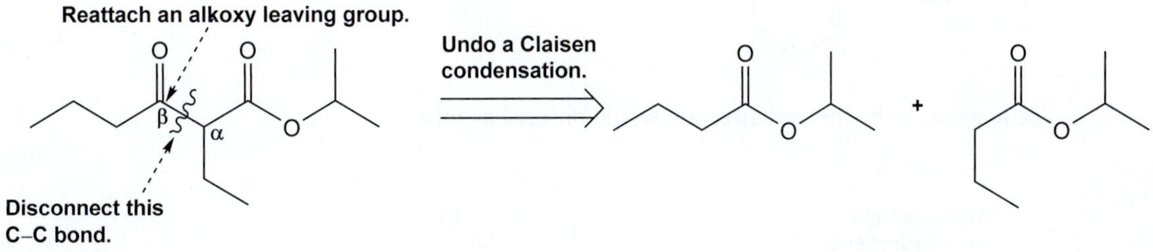

Reattach an alkoxy leaving group.

Undo a Claisen condensation.

Disconnect this C–C bond.

The reaction in the forward direction would proceed as follows:

1. $(CH_3)_2CHONa/(CH_3)_2CHOH$
2. H_3O^+

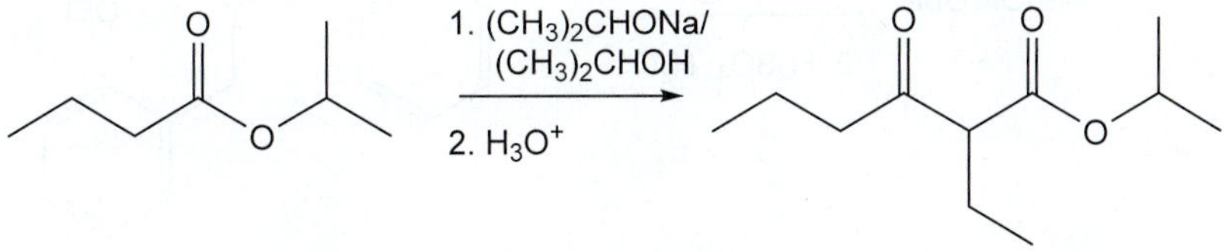

(b) To apply a transform that undoes a Dieckmann condensation, we must disconnect the C–C bond indicated. The original compound consisted of a molecule with two ester groups. The base required is CH_3O^-.

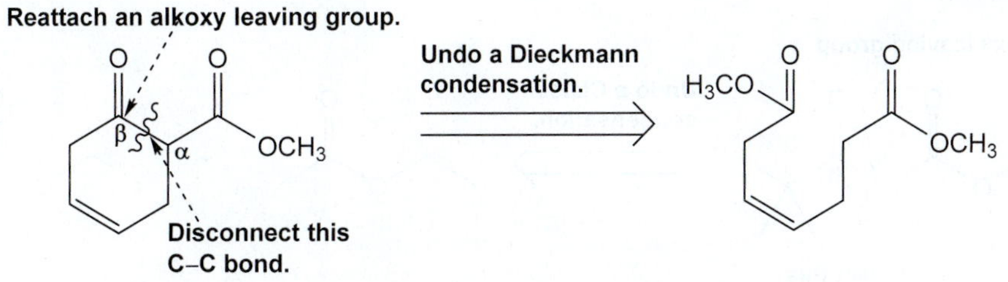

Reattach an alkoxy leaving group.

Undo a Dieckmann condensation.

Disconnect this C–C bond.

The reaction in the forward direction would proceed as follows:

1. CH_3ONa/CH_3OH
2. H_3O^+

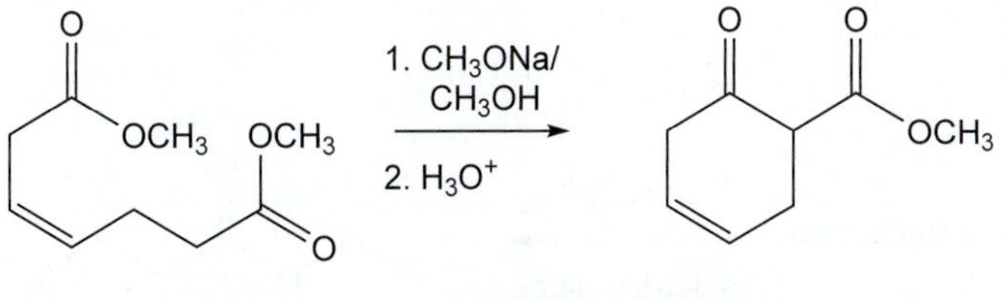

(c) This example yields the two ester precursors shown on the next page. The ester on the left has no α hydrogens and, therefore, the crossed Claisen reaction leads only to the target product. The base used is $(CH_3)_3CO^-$ in DMF. But the base could also be $CH_3CH_2O^-$.

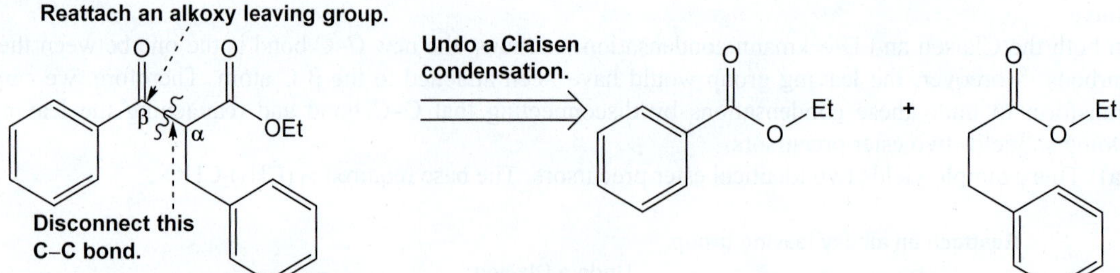

Reattach an alkoxy leaving group.

Undo a Claisen condensation.

Disconnect this C–C bond.

The reaction in the forward direction would proceed as follows:

The ester has no α hydrogens.

1. ; *t*-BuOK/DMF; 2. H_2SO_4, H_2O

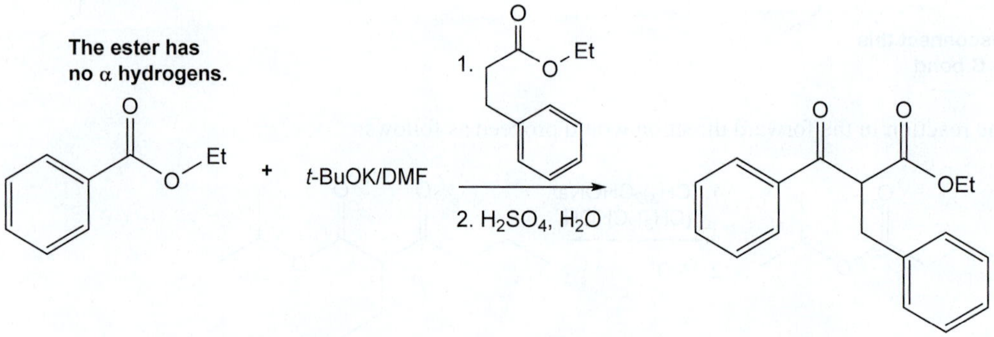

(d) This example yields the two ester precursors shown below. The ester on the right has no α hydrogens and, therefore, the crossed Claisen reaction leads only to the target product. The base used is $(CH_3)_3CO^-$ in DMF. But the base could also be $CH_3CH_2O^-$.

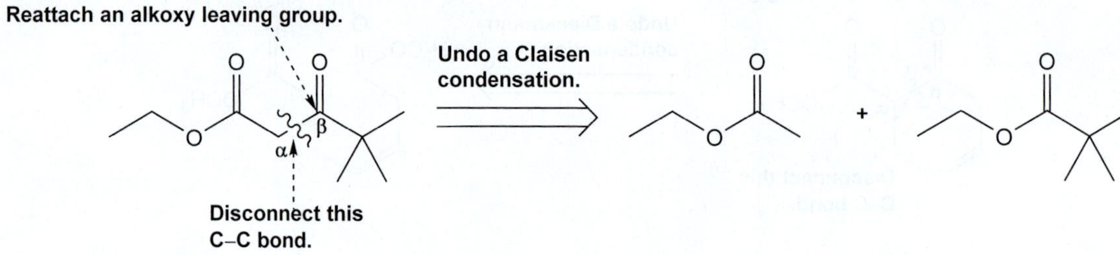

Reattach an alkoxy leaving group.

Undo a Claisen condensation.

Disconnect this C–C bond.

The reaction in the forward direction would proceed as follows:

The ester has no α hydrogens.

1. ; *t*-BuOK/DMF; 2. H_2SO_4, H_2O

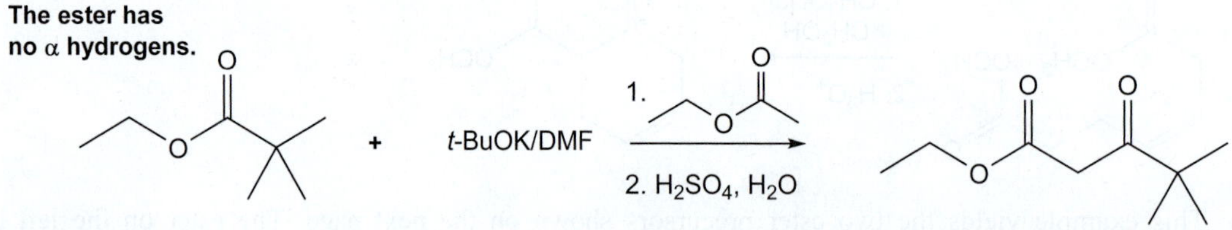

Sections 21.10 and 21.11 The Malonic Ester and Acetoacetic Ester Syntheses; Protecting Carboxylic Acids and Amines

Problem 21.68

Think

In each reaction, what are the electron-rich sites and electron-poor sites? Are there any nucleophiles? What electrophiles will they attack? What functional groups are formed in each reaction? What types of reactions occur—substitution, elimination, addition, oxidation, reduction?

Solve

(a) The α C–H is deprotonated quantitatively by EtO^- to form the enolate, which undergoes an S_N2 reaction with 1-chloro-4-methylpentane to alkylate the α C (product **A**). The β-keto ester undergoes hydrolysis, followed by a decarboxylation reaction to form **B**.

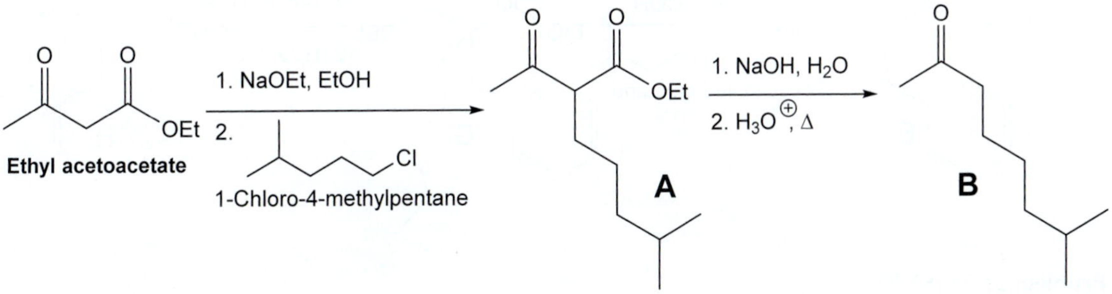

Ethyl acetoacetate → 1. NaOEt, EtOH; 2. 1-Chloro-4-methylpentane → **A** → 1. NaOH, H_2O; 2. $H_3O^{\oplus}$, Δ → **B**

(b) The reaction sequence is the same as that in **(a)**: the β-keto ester product **A** is alkylated at the α C (product **C**), and hydrolysis and decarboxylation follow (product **D**).

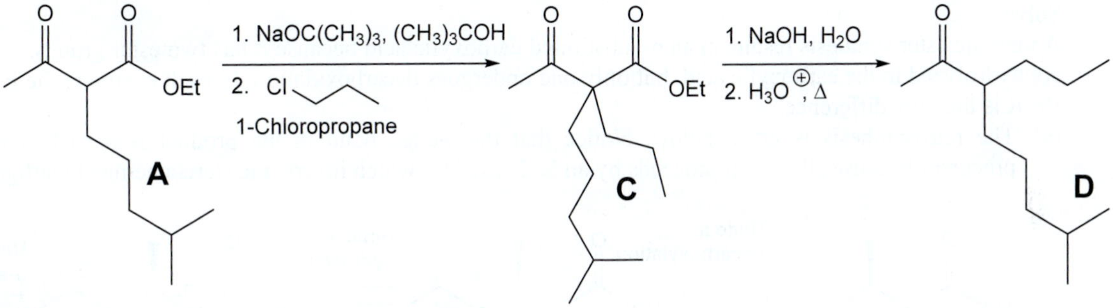

A → 1. $NaOC(CH_3)_3$, $(CH_3)_3COH$; 2. Cl (1-Chloropropane) → **C** → 1. NaOH, H_2O; 2. $H_3O^{\oplus}$, Δ → **D**

Problem 21.69

Think

In each reaction, what are the electron-rich sites and electron-poor sites? Are there any nucleophiles? What electrophiles will they attack? What functional groups are formed in each reaction? What types of reactions occur—substitution, elimination, addition, oxidation, reduction?

Solve

(a) The β-diester diethylmalonate is alkylated at the α C (product **E**). The resulting β-diester undergoes hydrolysis to form two carboxylic acids, and a decarboxylation reaction follows to form **F**.

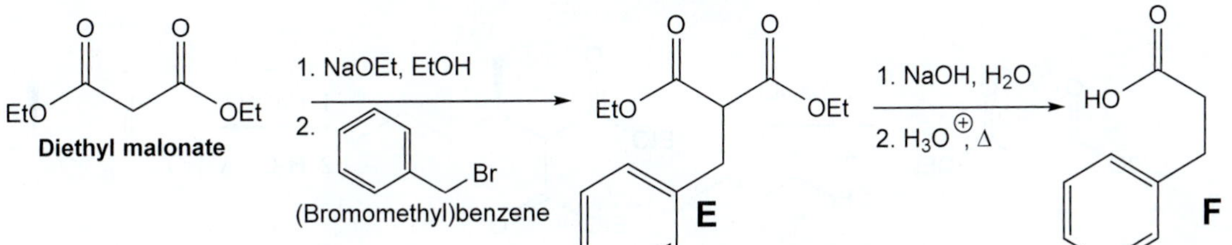

Diethyl malonate → 1. NaOEt, EtOH; 2. (Bromomethyl)benzene → **E** → 1. NaOH, H_2O; 2. $H_3O^{\oplus}$, Δ → **F**

(b) The reaction sequence is the same as that in **(c)**: the β-diester product **E** is alkylated at the α C (product **G**), and hydrolysis/decarboxylation follows (product **H**).

E → 1. $NaOC(CH_3)_3$, $(CH_3)_3COH$; 2. I–(1-Iodopentane) → G → 1. NaOH, H_2O; 2. $H_3O^{\oplus}$, Δ → H

Problem 21.70 (SYN)

Think

What would the precursor look like before decarboxylation? What is the structure of malonic ester? What R group is added to the α carbon? Does the alkylation occur more than once?

Solve

A malonic ester synthesis results in an α-substituted carboxylic acid because it has two ester groups. Both esters are hydrolyzed to the carboxylic acid, but only one undergoes decarboxylation. In each example, the R group of Br–R is the only difference.

(a) The retrosynthesis is given below. Notice that the wedge bond in the product is a dash bond in the precursor because alkylation proceeds by an S_N2 reaction, which inverts the stereochemical configuration

Undo a decarboxylation. ⟹ **Undo a hydrolysis.** ⟹ **Undo an α-alkylation.** ⟹ +

The synthesis in the forward direction would be written as follows:

1. NaOEt; 2. Br– → 1. NaOH; 2. H_3O^+, Δ

(b) The retrosynthesis is given below:

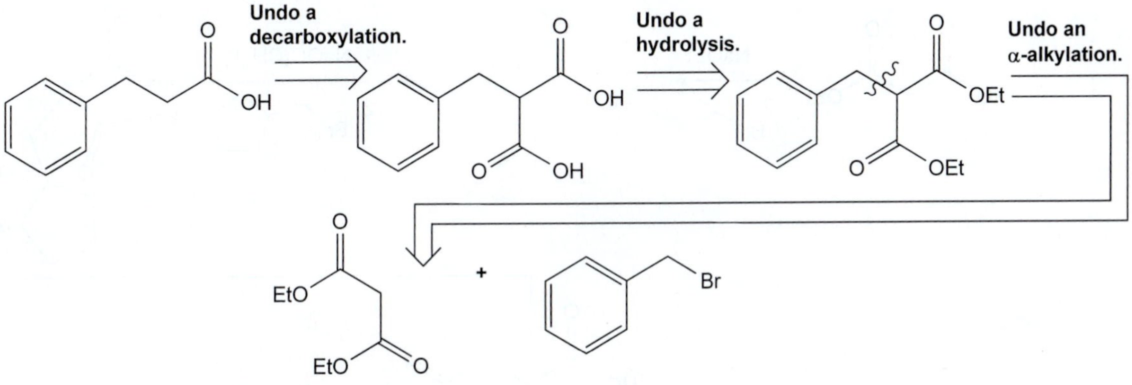
Undo a decarboxylation.
OH
O
OH
O
OH
Undo a hydrolysis.
O
OEt
O
OEt
Undo an α-alkylation.
O
EtO
EtO
O
+
Br

The synthesis in the forward direction would be written as follows:

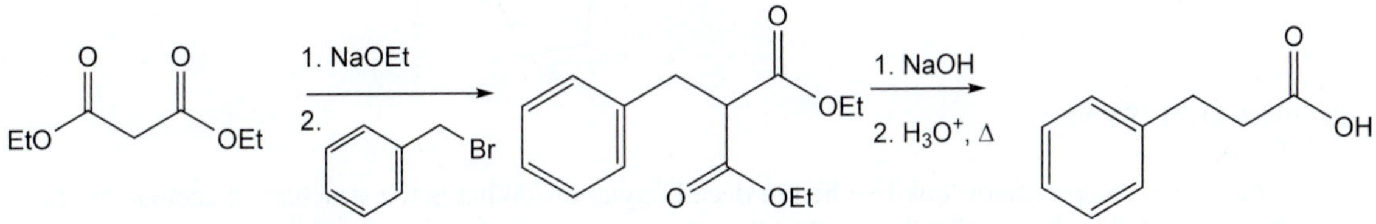
O
O
EtO
OEt
1. NaOEt
2.
Br
O
OEt
O
OEt
1. NaOH
2. H3O+, Δ
O
OH

(c) The retrosynthesis is given below:

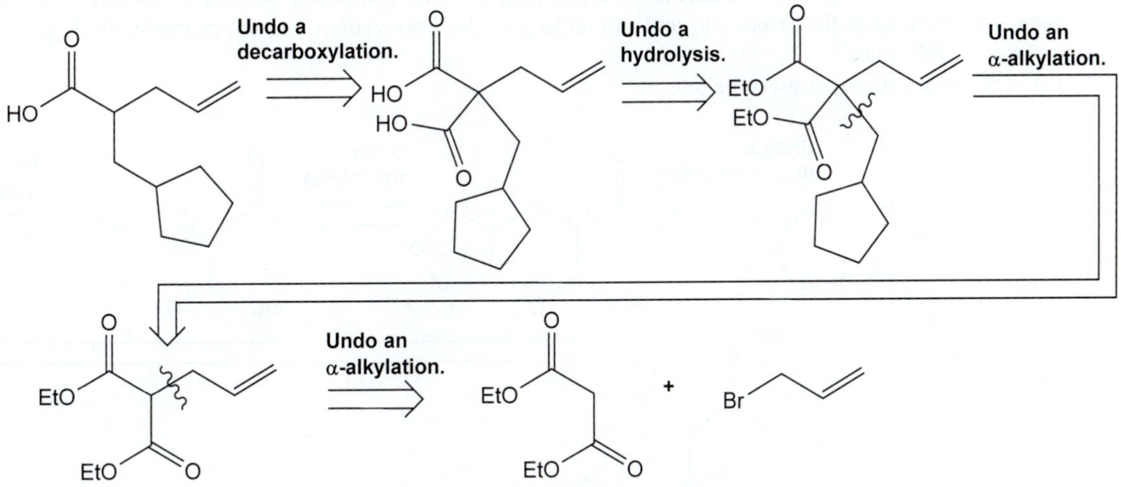
O
HO
Undo a decarboxylation.
O
HO
HO
O
Undo a hydrolysis.
O
EtO
EtO
O
Undo an α-alkylation.
O
EtO
EtO
O
Undo an α-alkylation.
O
EtO
EtO
O
+
Br

The synthesis in the forward direction would be written as follows:

1. NaOEt
2. Br

EtO OEt

1. $NaOC(CH_3)_3$
2. Br

1. NaOH
2. H_3O^+, Δ

HO

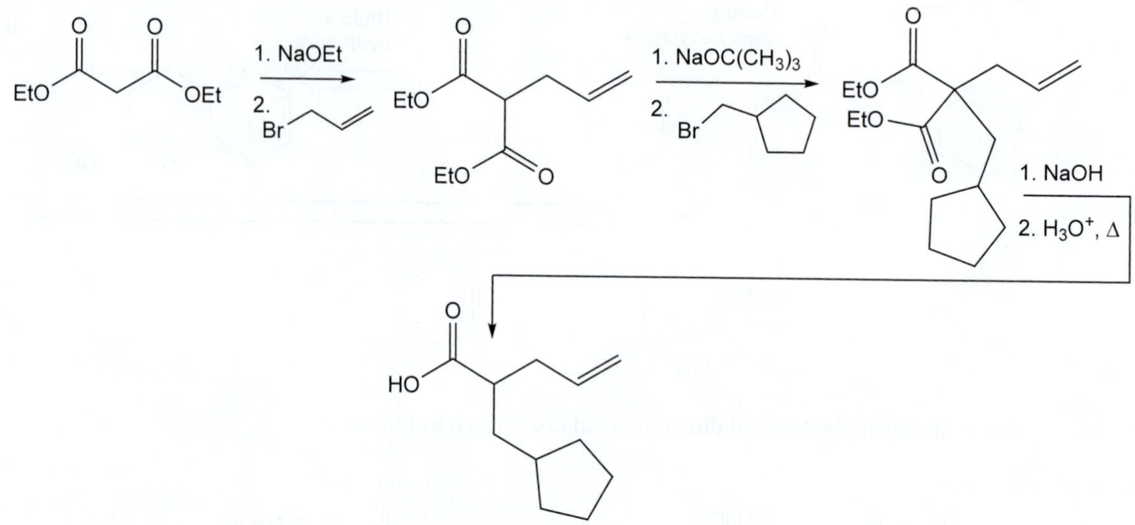

Problem 21.71 (SYN)

Think

What would the precursor look like before decarboxylation? What is the structure of acetoacetic ester? What R group is added to the α carbon? Does the alkylation occur more than once?

Solve

An acetoacetic ester synthesis results in an α-alkylated acetone compound because it has one ester group. The ester is hydrolyzed to the carboxylic acid and undergoes decarboxylation. In each example, the R group of Br–R is the only difference.

(a) The retrosynthesis is given below:

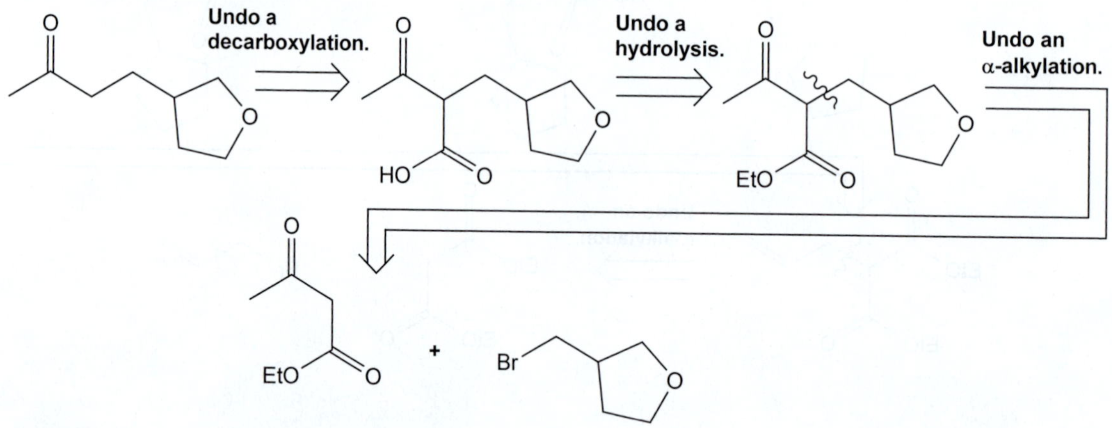

Undo a decarboxylation.

Undo a hydrolysis.

Undo an α-alkylation.

HO EtO EtO Br

The synthesis in the forward direction would be written as follows:

1. NaOEt
2. Br

OEt

1. NaOH
2. H_3O^+, Δ

EtO

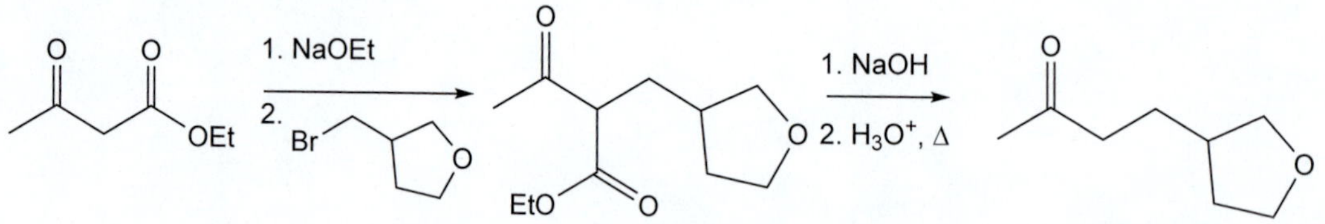

(b) The retrosynthesis is given below:

Undo a decarboxylation.

Undo a hydrolysis.

Undo an α-alkylation.

OH

OEt

Undo an α-alkylation.

+ Br

EtO

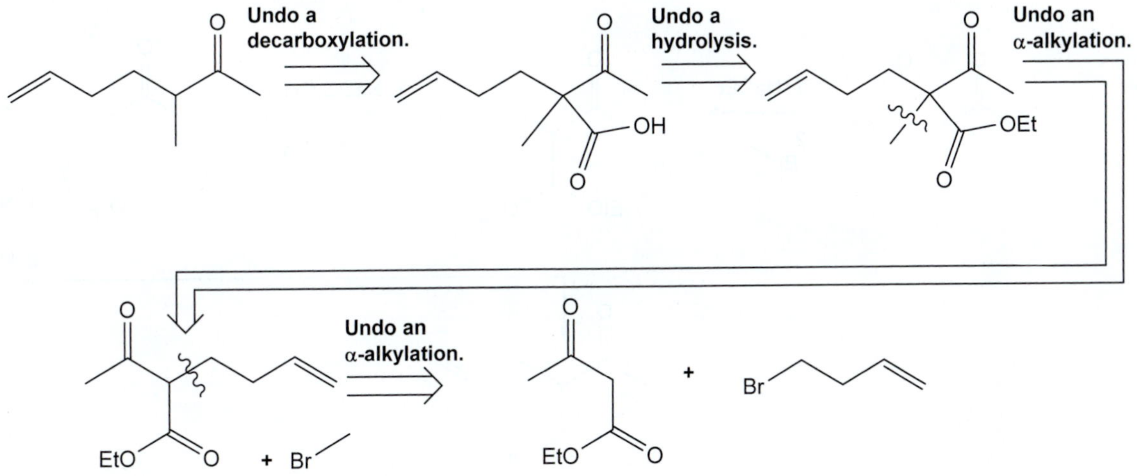

The synthesis in the forward direction would be written as follows (we show the alkylation using 4-bromobut-1-ene first, but it could take place with methyl bromide first instead):

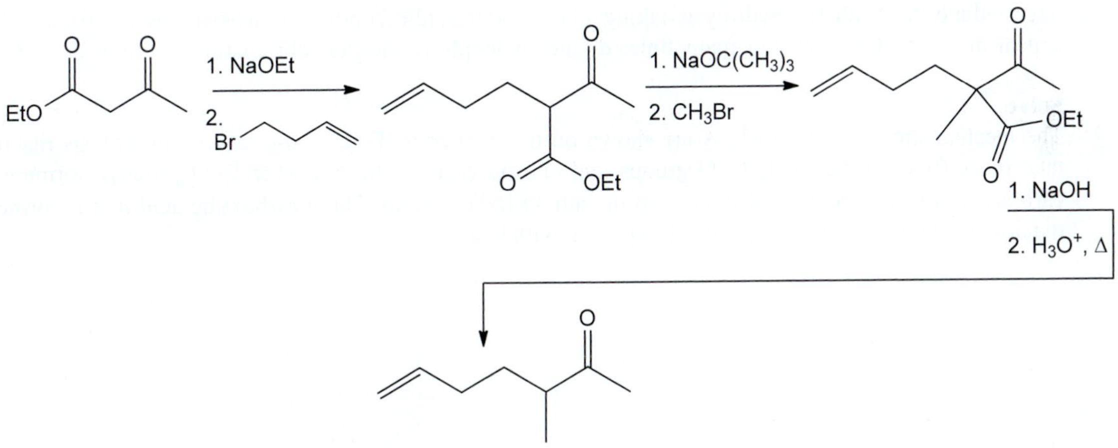

EtO

1. NaOEt
2. Br

OEt

1. $NaOC(CH_3)_3$
2. CH_3Br

OEt

1. NaOH
2. H_3O^+, Δ

(c) The retrosynthesis is given below:

Undo a decarboxylation.

Undo a hydrolysis.

Undo an α-alkylation.

Ph

HO

EtO

Undo an α-alkylation.

+ Br Ph

EtO

+ Br

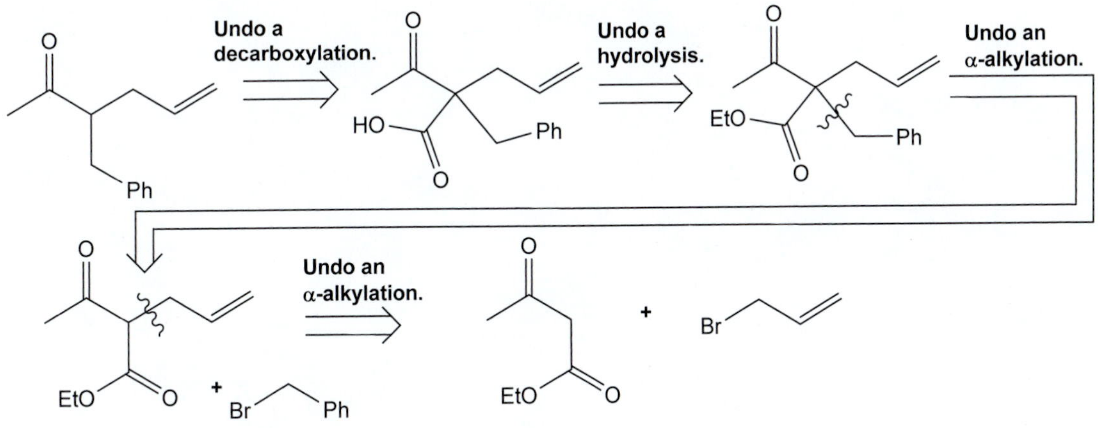

The synthesis in the forward direction would be written as follows (we show the alkylation using 3-bromoprop-1-ene first, but it could take place with benzyl bromide first instead):

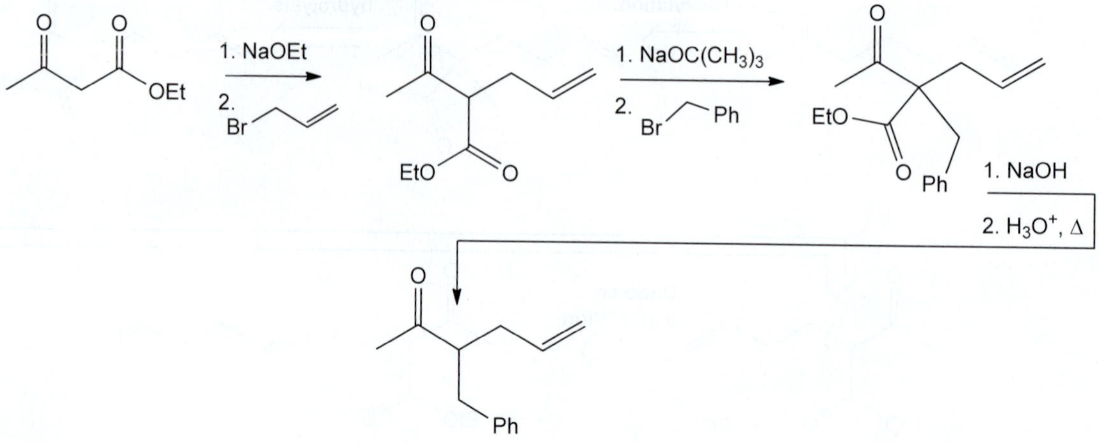

1. NaOEt
2. Br⌒⌒

1. $NaOC(CH_3)_3$
2. Br⌒Ph

1. NaOH
2. H_3O^+, Δ

Problem 21.72

Think

What is the acidic proton that reacts with EtO^-? How is the C–R bond formed? What is the identity of intermediate **A**? With the hydrolysis taking place under acidic conditions, what types of species should not appear in the mechanism? How many times do the nucleophilic addition–elimination reactions occur?

Solve

The mechanistic steps that yield **A** are shown on the next page. Four acid-catalyzed hydrolysis reactions then take place (two on the imide C=O groups and one on each of the two ester C=O groups), forming the four carboxyl (RCO_2H) groups and freeing up the amino (NH_2) group. The dicarboxylic acid that is formed is a β-diacid and, therefore, undergoes decarboxylation with heat.

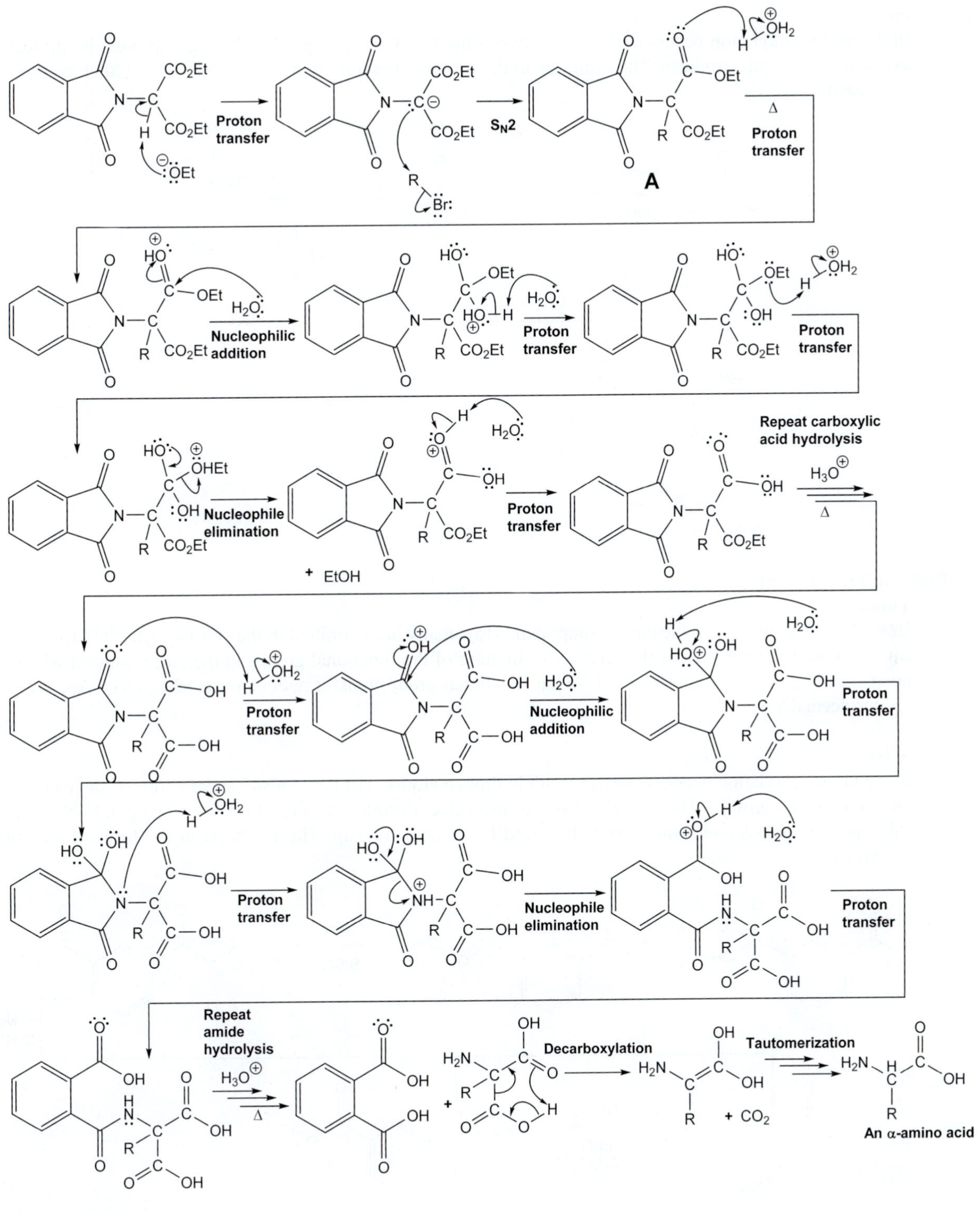

Proton transfer

S_N2

A

Δ

Proton transfer

Nucleophilic addition

Proton transfer

Proton transfer

Nucleophile elimination

+ EtOH

Proton transfer

Repeat carboxylic acid hydrolysis

$H_3O^{\oplus}$

Δ

Proton transfer

Nucleophilic addition

Proton transfer

Proton transfer

Nucleophile elimination

Proton transfer

Repeat amide hydrolysis

$H_3O^{\oplus}$

Δ

Decarboxylation

$+ CO_2$

Tautomerization

An α-amino acid

Problem 21.73 (SYN)

Think

How many C atoms are in the target compound? How many are permitted in the starting material? How was the carbon skeleton altered? Does the relative positioning of the functional groups in the target suggest which C–C bond-forming reaction might be used? What functional group transformations must take place? Are protecting groups needed?

Solve

In the target, alkylation occurs at the α C between the two C=O groups. The NH_2 group must be protected to avoid unwanted side products. The synthesis in the forward direction is shown below. *Note:* Other syntheses are also feasible.

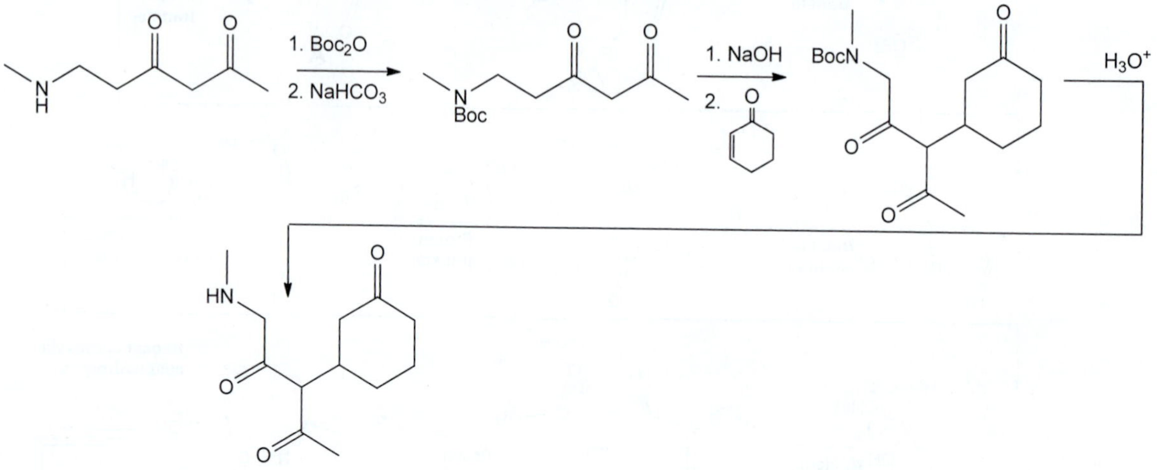

1. Boc_2O
2. $NaHCO_3$

1. NaOH
2.

BocN

H_3O^+

HN

Problem 21.74 (SYN)

Think

How many C atoms are in the target compound? How many are permitted in the starting material? How was the carbon skeleton altered? Does the relative positioning of the functional groups in the target suggest which C–C bond-forming reaction might be used? What functional group transformations must take place? Are protecting groups needed?

Solve

In the target, alkylation occurs at the α C with hydrogen atoms. The target also has transformed the CO_2H group into a CO_2CH_3 group. The CO_2H group is protected during the alkylation step using CH_2N_2 and then subsequently deprotected using NaOH followed by an acid workup. The synthesis in the forward direction is shown below.

OH → CH_2N_2 → O–CH_3 → 1. NaH 2. $BrCH_2Ph$ → Ph ... O–CH_3 → 1. NaOH 2. H^+

Ph ... OH → CH_2N_2, Et_2O, CH_2Cl_2 → Ph ... O

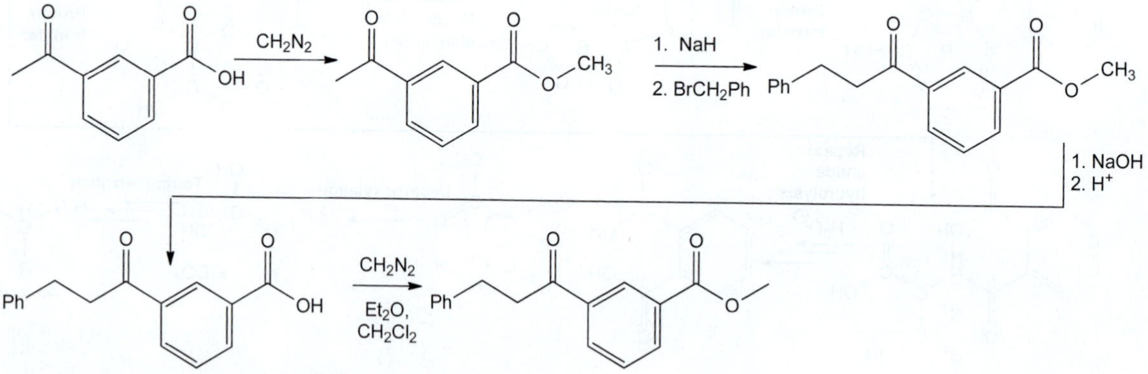

Sections 21.12 and 21.13 The Organic Chemistry of Biomolecules

Problem 21.75

Think

How many amino acids are in a tripeptide? What amino acid could occupy the first position of the tripeptide? The second position? The third?

Solve

A tripeptide has three peptides but is composed of only two amino acids: phenylalanine and glycine. Therefore, the original tripeptide could have either two phenylalanine amino acids and one glycine or two glycine amino acids and one phenylalanine.

Peptide 1–Pep 2–Pep 3 ⟶ Phenylalanine (2) + Glycine or Phenylalanine + Glycine (2)

All possible sequences of the original peptide are drawn below.

Phenylalanine (2) + Glycine

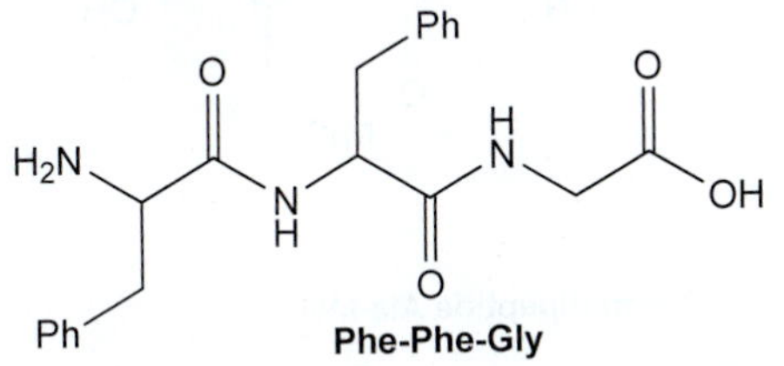

Phe-Phe-Gly

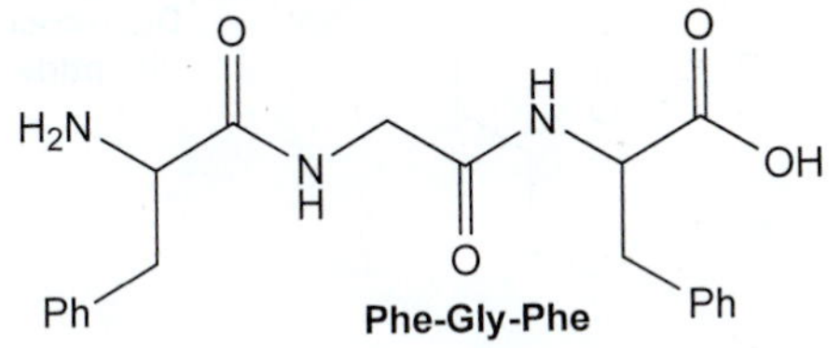

Phe-Gly-Phe

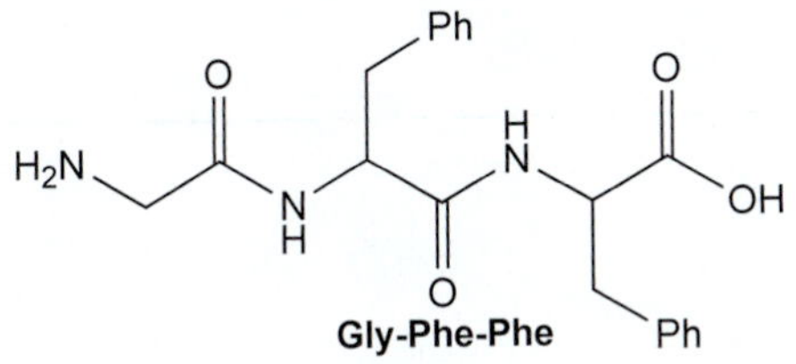

Gly-Phe-Phe

Phenylalanine + Glycine (2)

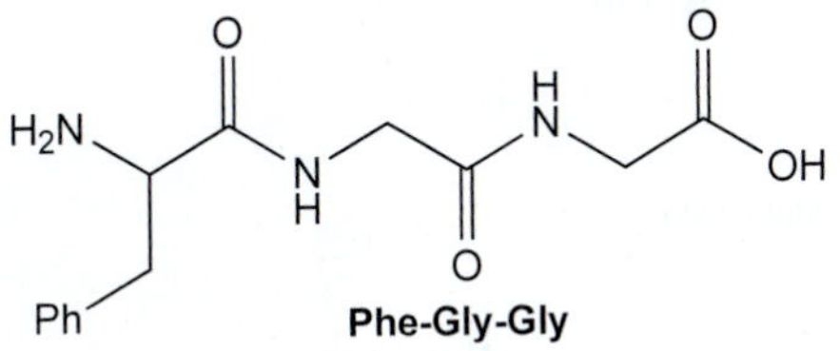

Phe-Gly-Gly

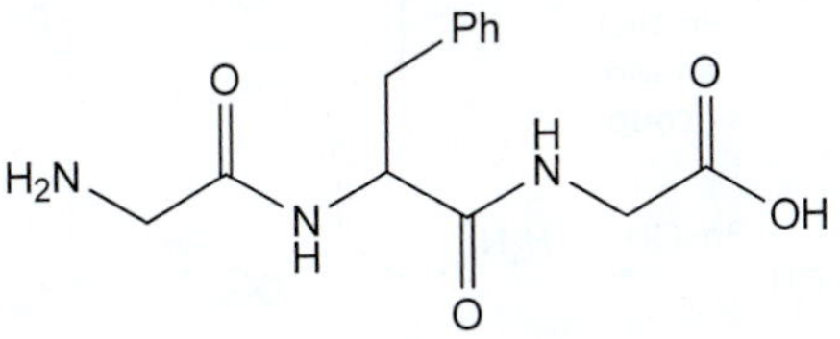

Gly-Phe-Gly

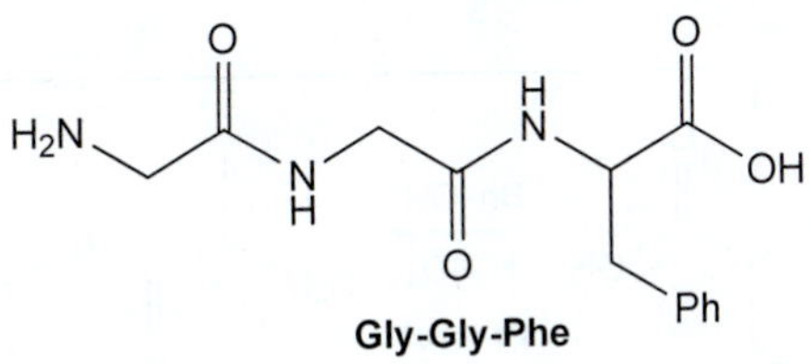

Gly-Gly-Phe

Problem 21.76 (SYN)

Think

What role does DCC play? How are multiple couplings avoided to yield the target peptide sequence? How are protecting groups used? What type of protecting group is used to protect the amino group of an amino acid? The carboxyl group? How are the groups deprotected?

Solve

DCC is the coupling agent that facilitates the reaction of a CO_2H group with a NH_2 group to produce an amide bond (peptide bond) between two amino acids. Each amino acid has both a CO_2H and an NH_2 group and, therefore, multiple coupling arrangements are possible. To produce a specific coupling product, the CO_2H of one amino acid and the NH_2 group of the other amino acid must be protected. Z–Cl is used to protect the NH_2 group of one amino acid, and Bn–OH is used to protect the CO_2H group of another. Once the couplings have taken place, H_2/Pd is used to deprotect. The syntheses for **(a)** Ser-Phe and **(b)** Leu-Ala-Met are shown on the next page.

(a) Ser-Phe

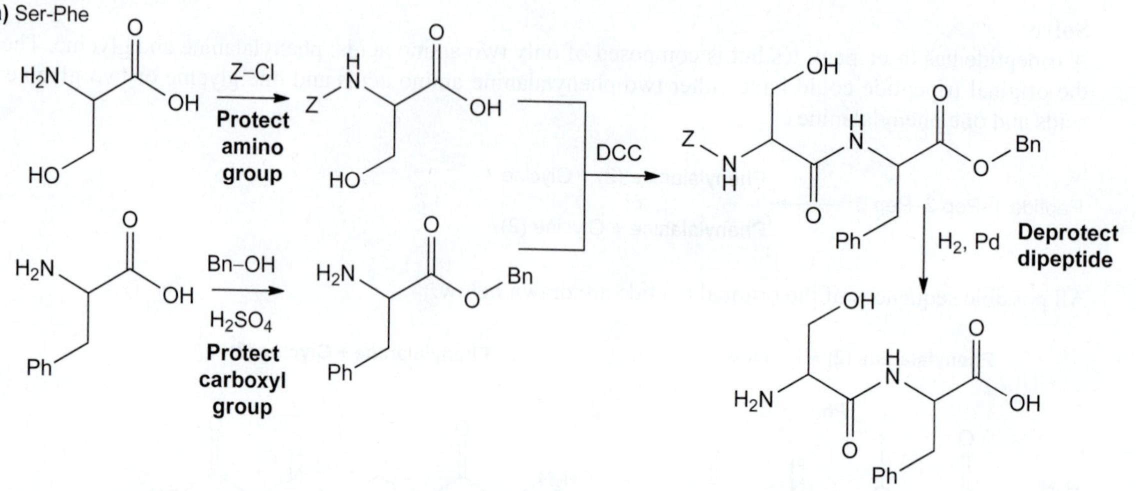

H_2N O OH HO

Z–Cl **Protect amino group**

Z N H O OH HO

O H_2N OH Ph

Bn–OH H_2SO_4 **Protect carboxyl group**

O H_2N O Bn Ph

DCC

OH Z N H H N O O O Bn Ph

H_2, Pd **Deprotect dipeptide**

OH O H N H_2N O OH Ph

(b) Leucine-Alanine-Methionine

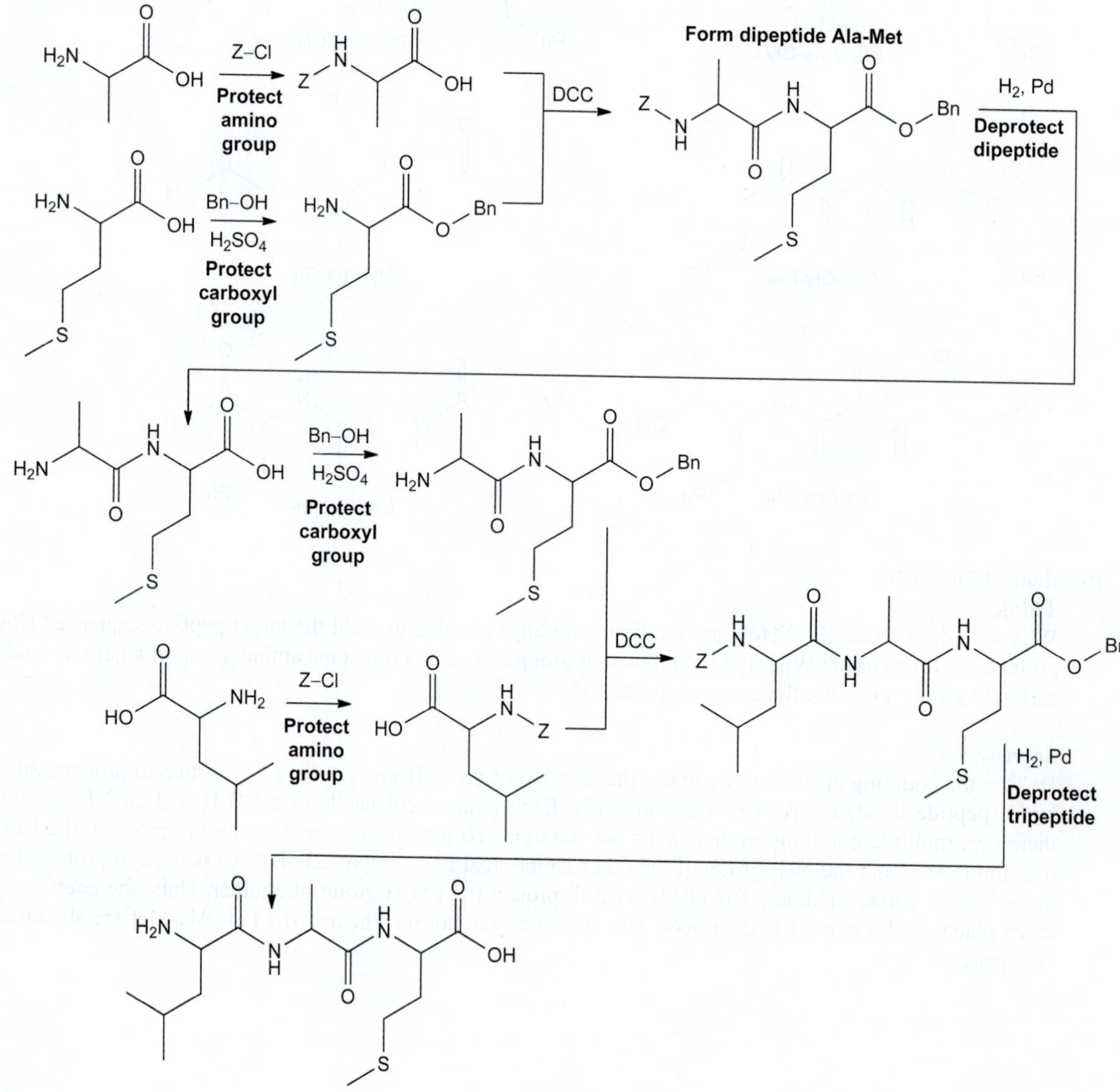

O H_2N OH

Z–Cl **Protect amino group**

Z H N O OH

O H_2N OH S

Bn–OH H_2SO_4 **Protect carboxyl group**

O H_2N O Bn S

DCC

Form dipeptide Ala-Met

Z N H H N O O O Bn S

H_2, Pd **Deprotect dipeptide**

H_2N H N O O OH S

Bn–OH H_2SO_4 **Protect carboxyl group**

H_2N H N O O O Bn S

O HO NH_2

Z–Cl **Protect amino group**

HO O H N Z

DCC

Z H N O H N O H N O O Bn S

H_2, Pd **Deprotect tripeptide**

H_2N H N O O H N OH S

Problem 21.77

Think

Are certain sequences in the peptides the same? Can you assume those identical sequences came from the same portion of the original protein? How can you piece the sequence together?

Solve

Certain sequences in each peptide have overlapping amino acids—specifically, the Phe-Gly-Ala sequence in **A** and **B**, the Met-Ala-Ala-Pro-Trp-Cys sequence in **C** and **D**, and the Ser-Ser sequence in **D** and **B**. You can assume that these sequences come from the same portions of the original protein and piece together the overall sequence, shown below:

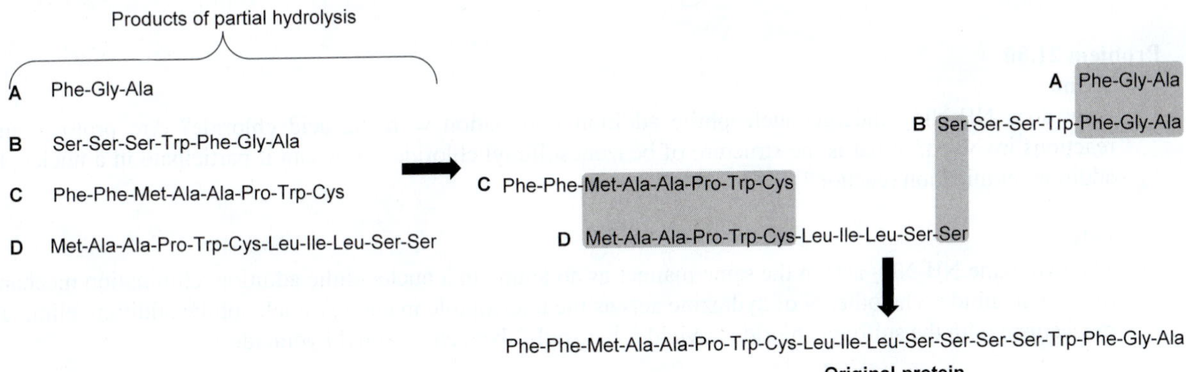

Problem 21.78

Think

What functional groups are present on the side chains of these two amino acids? What happens to these functional groups in hot, concentrated HCl? What is the nucleophile under these conditions? What is the leaving group?

Solve

The side chains of asparagine and glutamine contain amide groups. Like any amide group, the amide group in the respective amino acids is hydrolyzed under acidic, aqueous conditions to carboxylic acids.

Amide

Asparagine $\xrightarrow[\Delta]{H_3O^+}$

Amide

Glutamine $\xrightarrow[\Delta]{H_3O^+}$

Integrated Problems

Problem 21.79

Think

What type of reagent is sodium borohydride, $NaBH_4$? In the first reaction, what is the nucleophile? Is there a leaving group? Are proton transfer reactions involved? How does this product react with the acid chloride?

Solve

Sodium borohydride is the source of nucleophilic hydride and reduces an aldehyde to a primary alcohol. In the subsequent treatment with an acid chloride, the alcohol produced in the first reaction will act as the nucleophile in a nucleophilic addition–elimination mechanism, producing the ester.

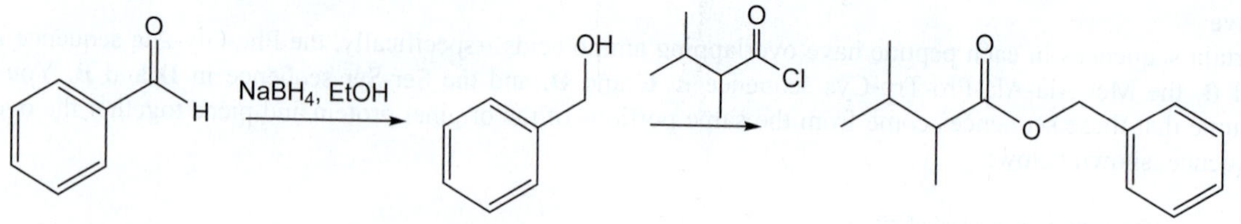

NaBH$_4$, EtOH

Problem 21.80

Think

How does NH$_2$NH$_2$ undergo nucleophilic addition–elimination with the acid chloride? Are proton transfer reactions involved? What is the structure of benzenesulfonyl chloride? How can it participate in a nucleophilic addition–elimination reaction?

Solve

The hydrazine NH$_2$NH$_2$ acts in the same manner as an amine in a nucleophilic addition–elimination mechanism to form an amide. The other N of hydrazine acts as the nucleophile in another nucleophilic addition–elimination mechanism with the sulfonyl chloride to yield 1-benzoyl-2-benzenesulfonyl hydrazide.

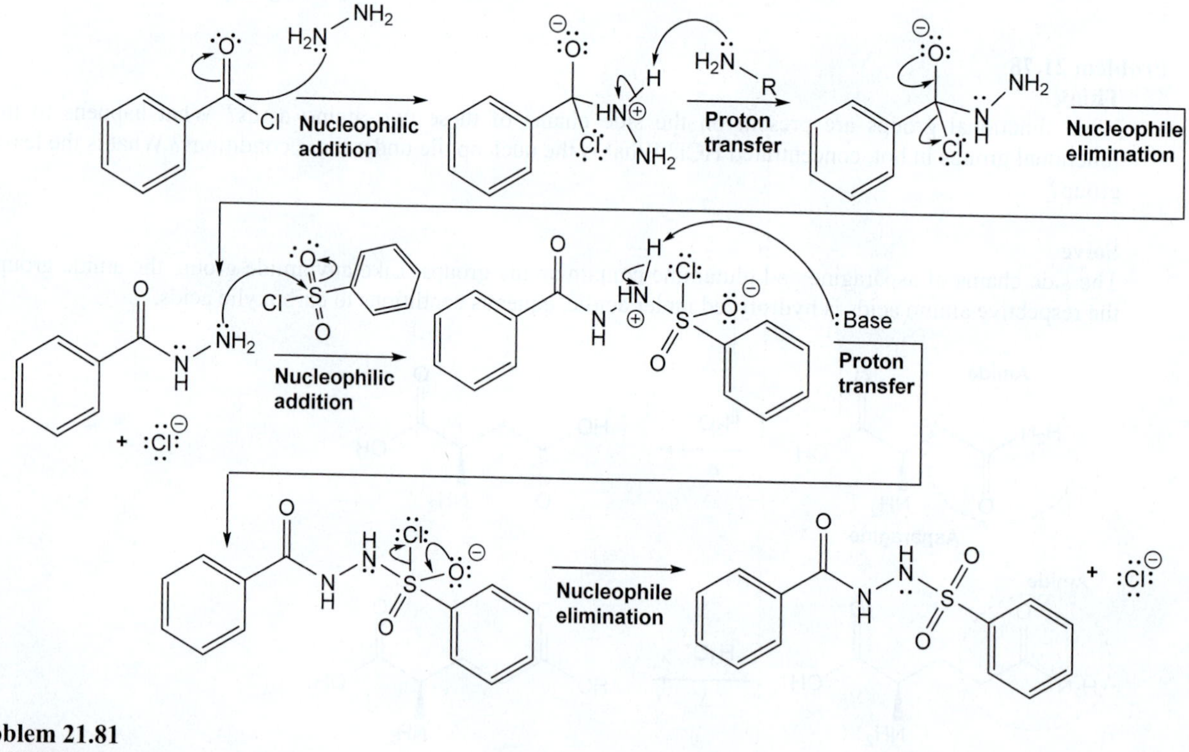

Nucleophilic addition

Proton transfer

Nucleophile elimination

Nucleophilic addition

Proton transfer

Nucleophile elimination

Problem 21.81

Think

What is the nucleophile in the nucleophilic addition step involving 2-pyridinethiol? Is the reaction promoted by acid or base? What types of species should not appear in the mechanism? In the reaction that produces the lactone, how is the stability of the tetrahedral intermediate affected by the N atom? Does the N atom lead to the formation of a better or worse leaving group?

Solve

(a) The mechanism for the formation of the thioester is shown first. The long carbon chain is abbreviated R.

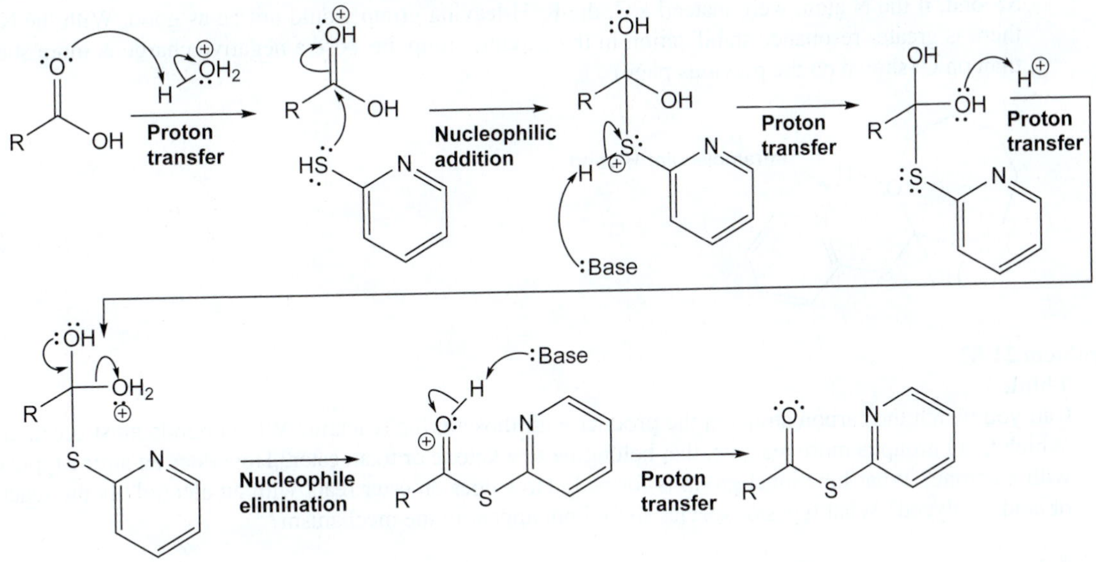

Next comes the formation of the cyclic ester.

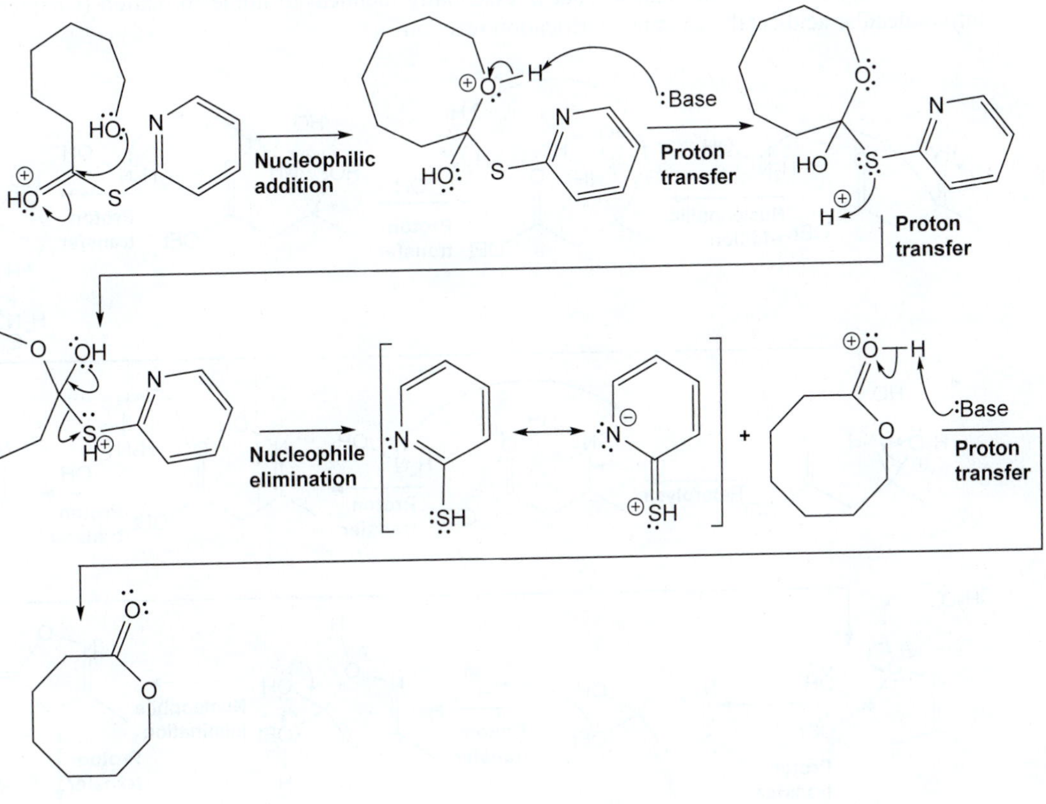

There are two reasons why the reaction is less effective without the N atom in the ring. First, the product of the nucleophilic addition step, shown below, has an intramolecular hydrogen bond involving the N atom. This stabilization of the tetrahedral intermediate wouldn't be possible if the N atom were a C atom instead. Second, if the N atom were instead a C, the RSH leaving group would not be as good. With the N present, there is greater resonance stabilization in the leaving group, because a negative charge is more stable on N than on C (shown on the previous page).

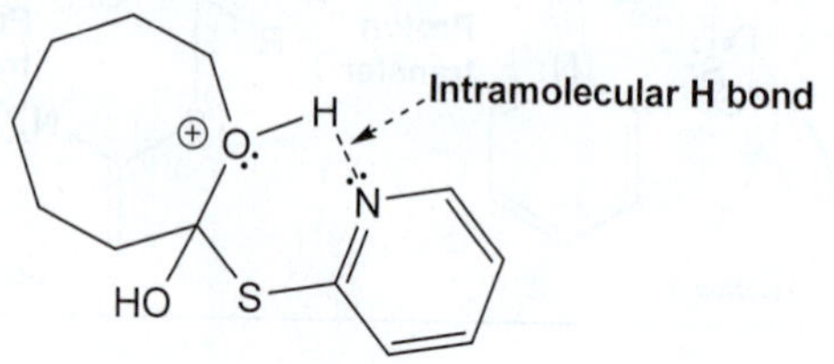

Intramolecular H bond

Problem 21.82

Think

Can you match the carbon atoms in the product with those in the reactant? Which bonds must break and form? Which C=O group is more reactive, that belonging to a ketone or to an ester? How does an amine typically react with a ketone? What functional group is formed? How does an ester react with an alcohol? Is the reaction base or acid catalyzed? What types of species should not appear in the mechanism?

Solve

The C=O belonging to the ketone portion is the more reactive C=O group, and it will react with the amino portion of $H_2N–OH$ in a mechanism that is essentially identical to imine formation (Chapter 18). Next is an intramolecular acid-catalyzed transesterification reaction.

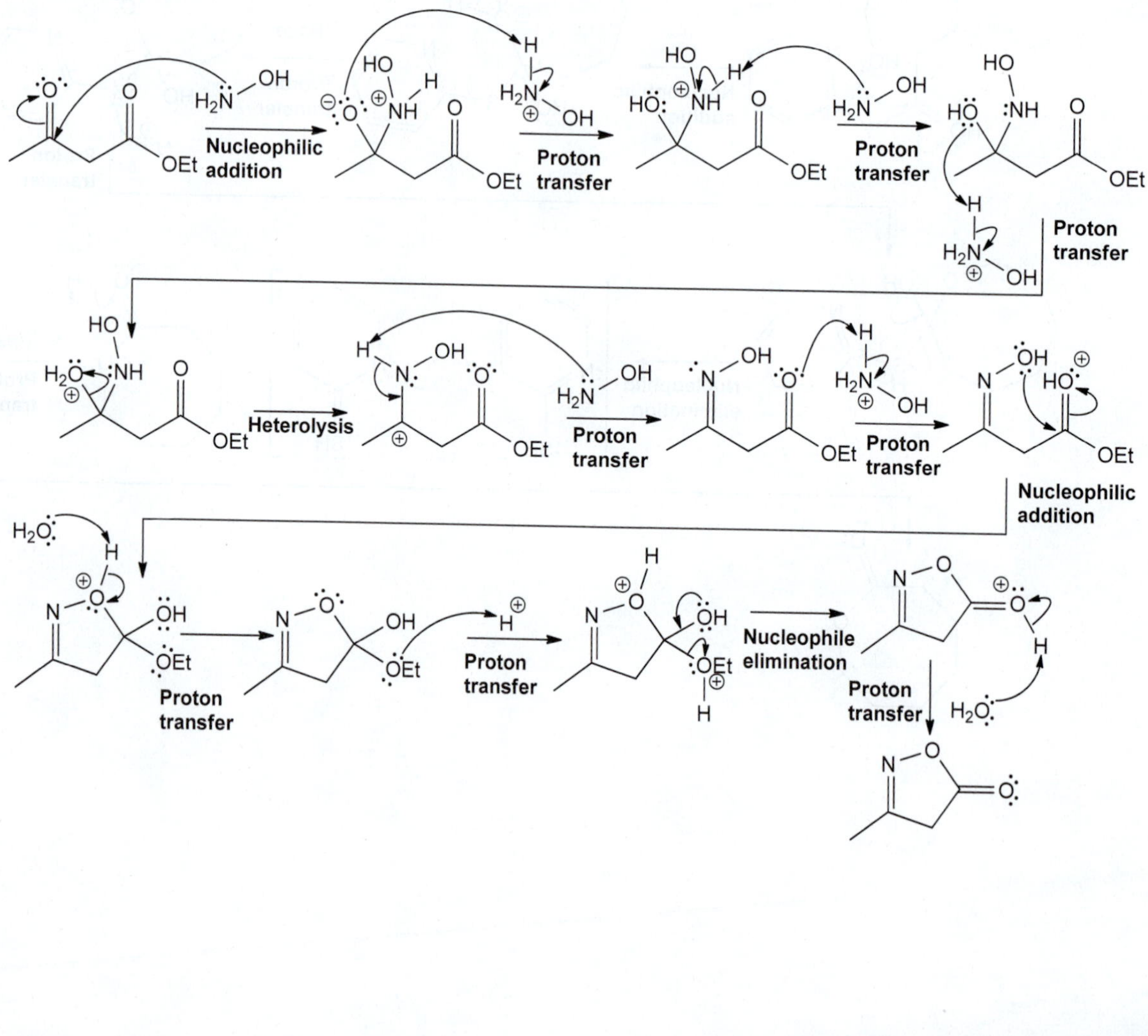

Nucleophilic addition

Proton transfer

Proton transfer

Proton transfer

Heterolysis

Proton transfer

Proton transfer

Nucleophilic addition

Proton transfer

Proton transfer

Nucleophile elimination

Proton transfer

Problem 21.83

Think

How many C atoms are in the reactants and products? Were any C atoms lost? Which C–C bonds formed? Is the reaction conducted under basic or acidic conditions? Which α C–H is most acidic? Does the nucleophilic addition occur via 1,2- or 1,4-addition?

Solve

The α C–H on the β-diester, diethyl malonate, is more acidic than but-3-en-2-one. The enolate is formed from diethyl malonate, which undergoes conjugate nucleophilic addition to but-3-en-2-one. The next step is a proton transfer to the C atom of the enolate anion. Simultaneously, the C=O is re-formed using a lone pair of electrons from the negatively charged O. The next few steps make up an intramolecular Claisen condensation reaction, favored by the formation of a six-membered ring.

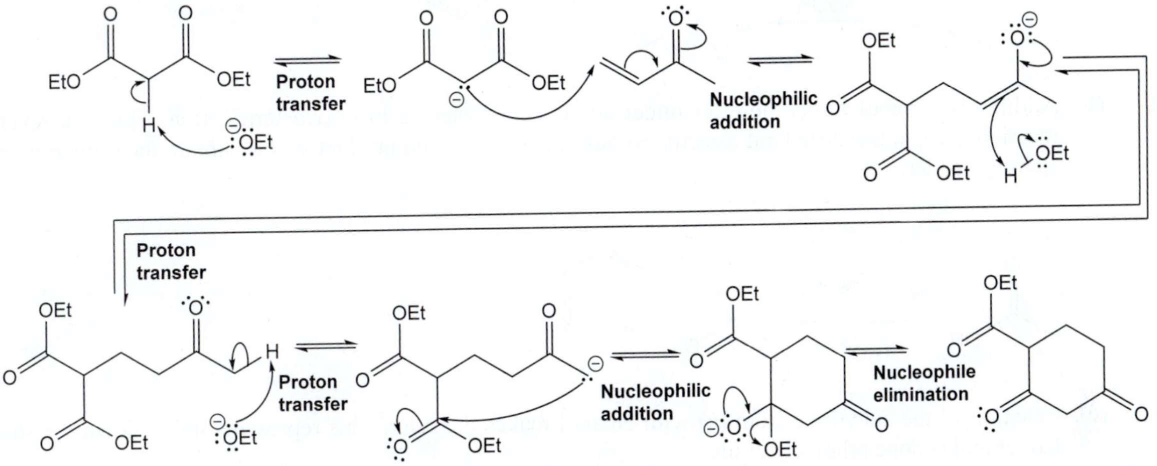

Proton transfer

Nucleophilic addition

Proton transfer

Proton transfer

Nucleophilic addition

Nucleophile elimination

Problem 21.84

Think

What is the Lewis structure for CH_2N_2? What acts as the nucleophile in the nucleophilic addition step? What is the leaving group in the nucleophile elimination step? How does the formation of this leaving group drive the reaction forward?

Solve

The mechanism is much like a nucleophilic addition–elimination mechanism, except that in the elimination step, R^- is never truly formed but rather is shifted via an alkyl shift. This resembles a Baeyer–Villiger oxidation and is driven by the formation of a very stable N_2 molecule that leaves irreversibly as a gas.

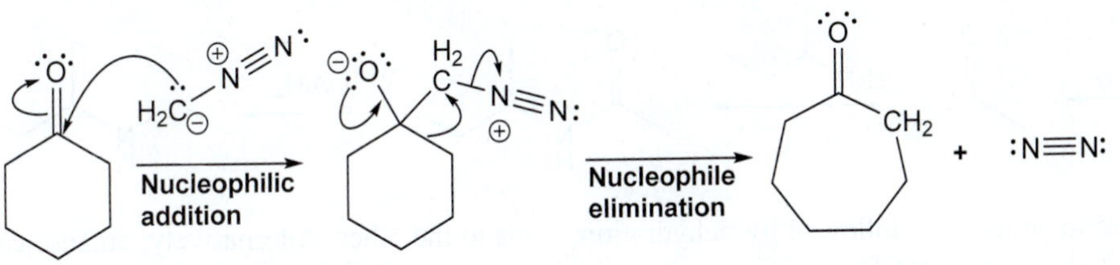

Nucleophilic addition

Nucleophile elimination

Problem 21.85 (SYN)

Think

Which atoms came from acetic acid? What new functional group is formed? What was the identity of the nucleophile in the nucleophilic addition step? Does the acyl substitution require going up or down the stability ladder? What reactions do you know that can accomplish going up the ladder?

Solve

(a) The reaction requires converting a carboxylic acid to an acid anhydride, which represents going up the stability ladder. This is difficult to do directly but can be carried out by first producing an acid chloride. The acid chloride and the salt of acetic acid will make the desired anhydride.

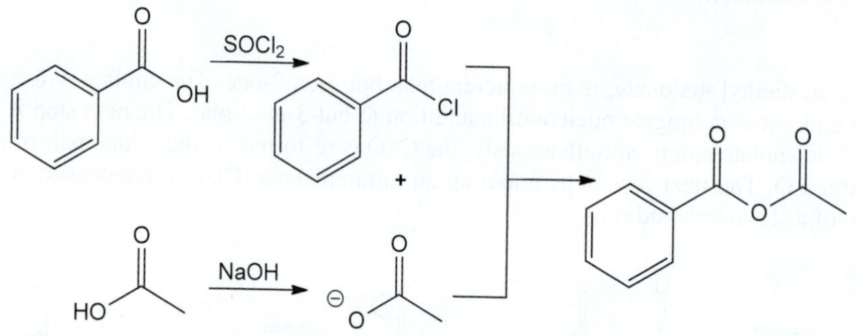

$SOCl_2$

NaOH

(b) Addition of phenol forms the ester under acidic conditions—a Fischer esterification. This is a reversible reaction and can be carried out directly because the reactant and product species are on the same rung of the stability ladder.

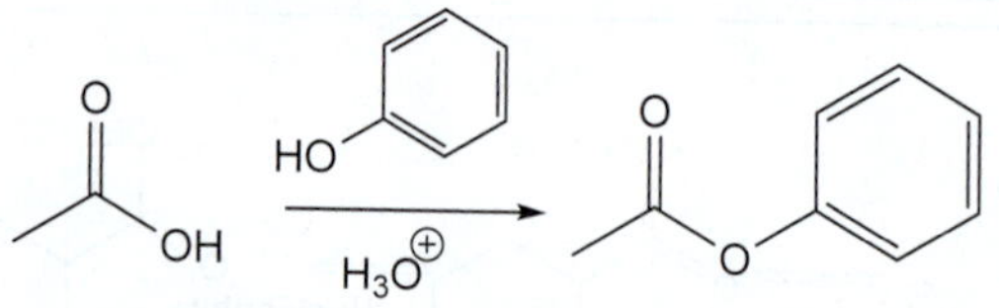

$H_3O^{\oplus}$

(c) Reaction of the anhydride from **(a)** with ethanol makes the ester. This represents going down the stability ladder and is done relatively easily.

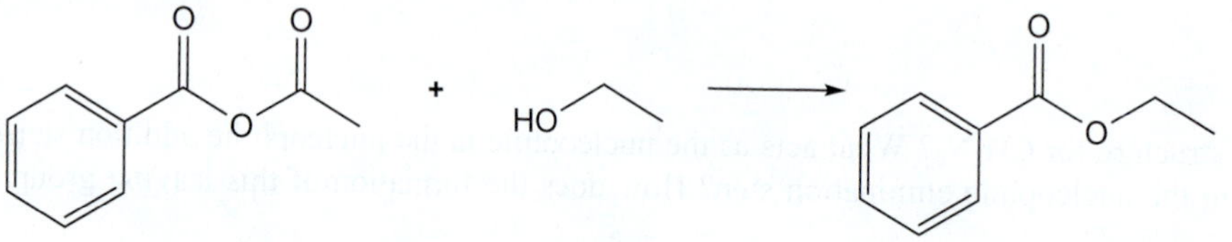

(d) The amine can be produced by reducing the corresponding amide. It is difficult to produce the amide directly from acetic acid because the amine that would be required for the nucleophile would deprotonate the carboxylic acid. The amide can be made relatively easily, however, by first converting acetic acid to acetyl chloride and then treating acetyl chloride with the amine.

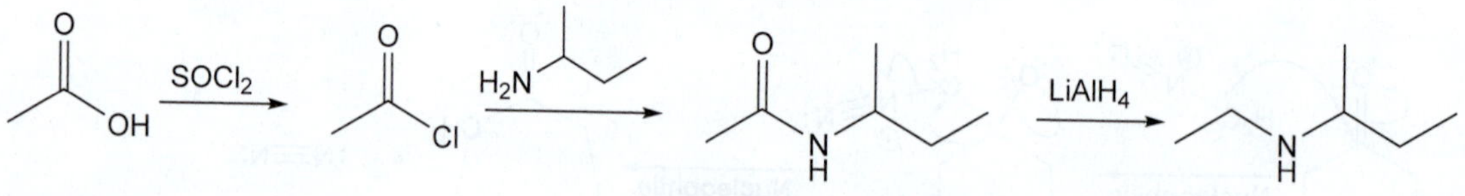

$SOCl_2$ H_2N $LiAlH_4$

(e) Reducing the acid to an alcohol, followed by dehydration, leads to the ether. Alternatively, ethanol can be deprotonated by NaH to produce $CH_3CH_2O^-$, which could then be reacted with CH_3CH_2Br in a Williamson ether synthesis.

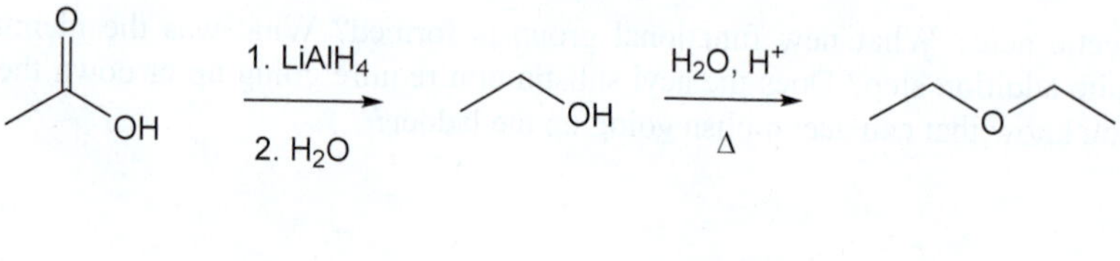

1. $LiAlH_4$
2. H_2O

H_2O, H^+
Δ

(f) The acid chloride plus the specialized reducing agent, $LiAlH(O\text{-}tBu)_3$, forms the aldehyde.

$$CH_3COOH \xrightarrow{SOCl_2} CH_3COCl \xrightarrow{LiAl(O\text{-}tBu)_3H} CH_3CHO$$

(g) A lithium dialkylcuprate forms the ketone from the acid chloride.

$$CH_3COOH \xrightarrow{SOCl_2} CH_3COCl \xrightarrow[2.\ NH_4Cl]{1.\ (CH_3)_2CuLi} CH_3COCH_3$$

Problem 21.86 (SYN)

Think

Which atoms came from the acetic acid? Do any new carbon–carbon bonds need to be formed? What new functional group is formed? Do the necessary acyl substitutions represent going up or down the stability ladder? Are any oxidations or reductions necessary? What was the identity of the nucleophile in the nucleophilic addition step?

Solve

(a) Reduction produces ethanol, which can react with the acid chloride (formed from the reaction of acetic acid and $SOCl_2$).

$$CH_3COOH \xrightarrow[2.\ H_3O^{\oplus}]{1.\ LiAlH_4} CH_3CH_2OH$$
$$CH_3COOH \xrightarrow{SOCl_2} CH_3COCl$$
$$CH_3CH_2OH + CH_3COCl \rightarrow CH_3COOCH_2CH_3$$

(b) We can start with ethanol and ethanoyl chloride from **(a)**.

$$CH_3CH_2OH \xrightarrow{PBr_3} CH_3CH_2Br \xrightarrow{Li} CH_3CH_2Li \xrightarrow{CuI} (CH_3CH_2)_2CuLi$$

$$CH_3COCl \xrightarrow{(CH_3CH_2)_2CuLi} CH_3COCH_2CH_3 \xrightarrow[EtOH]{NaBH_4} CH_3CH(OH)CH_2CH_3$$

(c) Use of the ester from **(a)** and ethyllithium from **(b)** (or a Grignard reagent) is sufficient.

$$CH_3COOCH_2CH_3 \xrightarrow[2.\ H_3O^{\oplus}]{1.\ 2\ CH_3CH_2Li} CH_3C(OH)(CH_2CH_3)_2$$

(d) Butan-2-one is made in **(b)**.

(e) Ethanamine is made using ammonia to form the amide followed by reduction.

$$CH_3COCl \xrightarrow{NH_3} CH_3CONH_2 \xrightarrow[2.\ H_2O]{1.\ LiAlH_4} CH_3CH_2NH_2$$

(f) We can use ethanamine from **(e)** and the ethanoyl chloride from **(a)**.

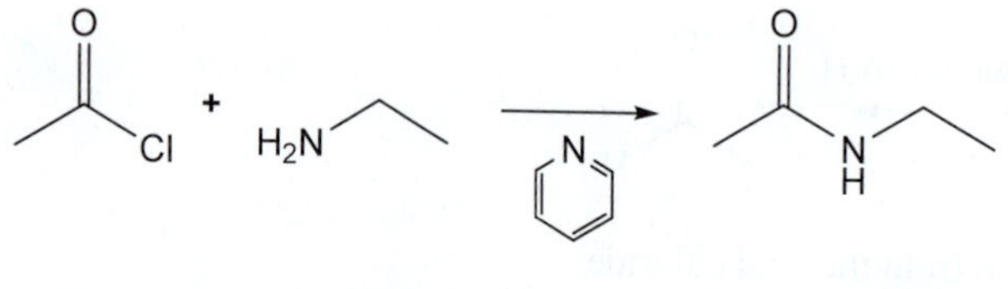

(g) Synthesis of diethylamine requires reduction of the amine produced in **(f)**; then reaction with the acid chloride leads to the desired product.

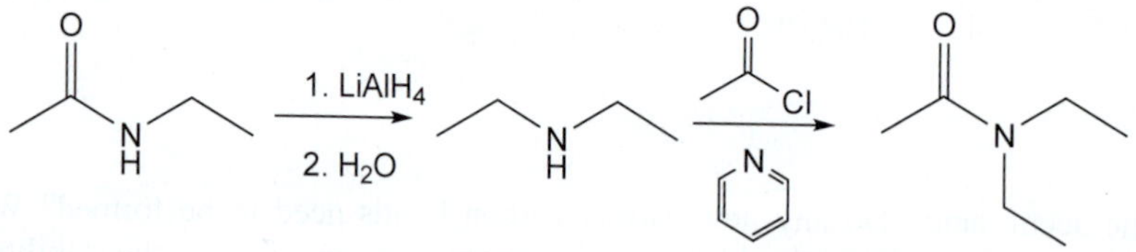

1. $LiAlH_4$
2. H_2O

Problem 21.87

Think

How does $LiAlH_4$ react with a ketone? In the reaction with TsCl, what is the nucleophile, and what is the electrophile? In the reaction with NaCN, is there a good leaving group?

Solve

The ketone is reduced to the alcohol using $LiAlH_4$ with an acid workup. The alcohol OH is not a good leaving group and is transformed into the OTs, which is a good leaving group. An S_N2 reaction follows with NaCN to form the nitrile.

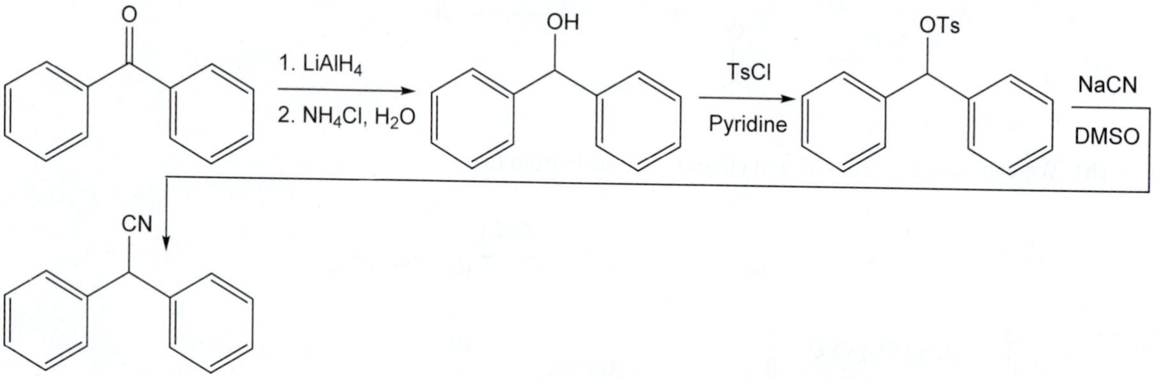

1. $LiAlH_4$
2. NH_4Cl, H_2O

TsCl, Pyridine

NaCN, DMSO

Problem 21.88

Think

In each reaction, what are the electron-rich sites and electron-poor sites? Are there any nucleophiles? What electrophiles will they attack? Are there any good leaving groups? What functional groups are formed in each reaction? What types of reactions occur—substitution, elimination, addition, oxidation, reduction?

Solve

(a) The ester is hydrolyzed to the carboxylic acid **A**. $SOCl_2$ transforms carboxylic acid **A** into the acid chloride **B**. The acid chloride is transformed into the amide **C**, which is reduced by $LiAlH_4$ into the amine **D**.

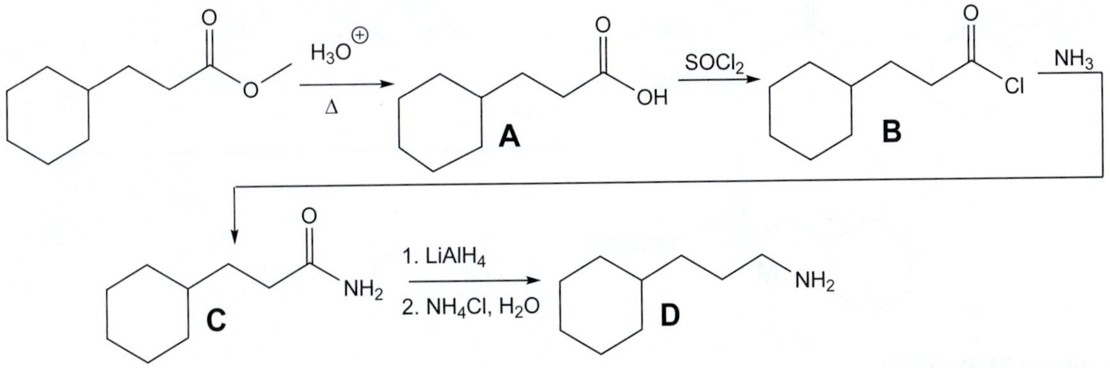

$H_3O^{\oplus}$, Δ; $SOCl_2$; NH_3

A B

1. $LiAlH_4$
2. NH_4Cl, H_2O

C D

(b) The carboxylic acid **A** is brominated at the α C, via a Hell–Volhard–Zelinski reaction, to form **E**. Br is a good leaving group, and an S_N2 reaction with NaCN yields the nitrile **F**. The nitrile is hydrolyzed to the carboxylic acid **G**.

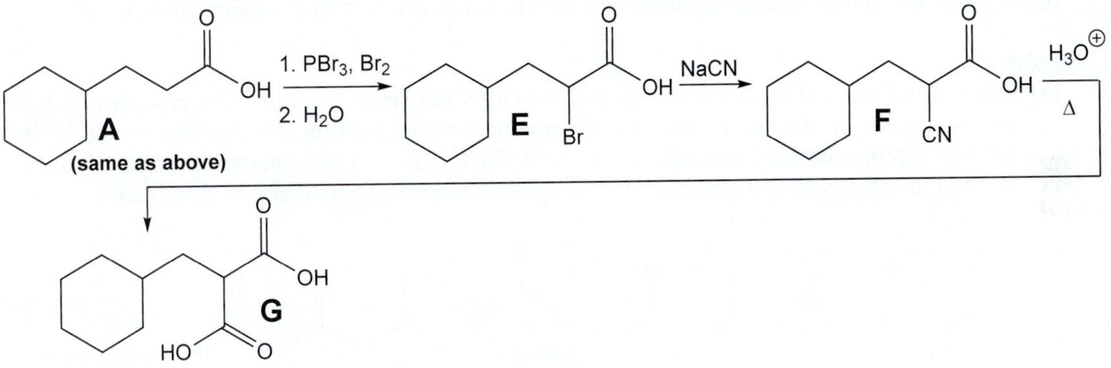

1. PBr_3, Br_2
2. H_2O

NaCN; $H_3O^{\oplus}$, Δ

A (same as above) E F G

(c) DIBAH reduces the ester to the aldehyde **H**, which undergoes a Grignard reaction to form a secondary alcohol **I**. The alcohol is oxidized to the ketone **J**.

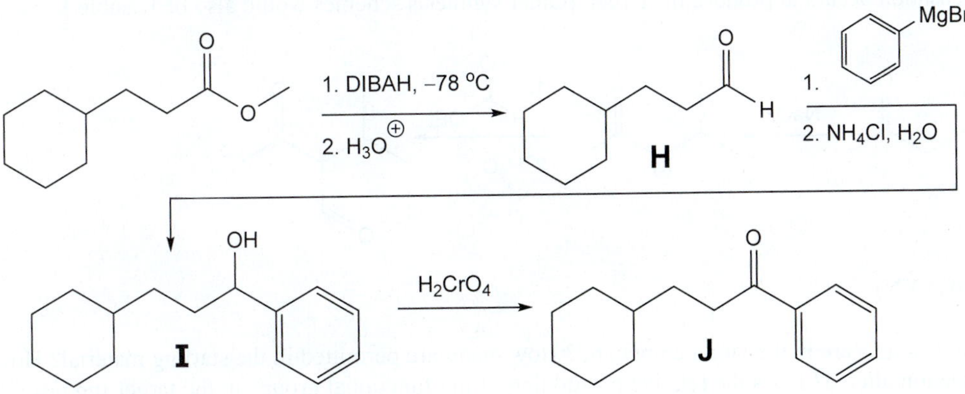

1. DIBAH, −78 °C
2. $H_3O^{\oplus}$

1. PhMgBr
2. NH_4Cl, H_2O

H

H_2CrO_4

I J

(d) $LiAlH_4$ reduces the ester to the alcohol **K**. The HO group is not a good leaving group and is transformed into OTs (product **L**), which can undergo an S_N2 reaction to form the thioether product **M**.

1. $LiAlH_4$ 2. NH_4Cl, H_2O → **K** (OH) → TsCl, Pyridine → **L** (OTs) → CH_3CH_2SH, DMSO → **M** (SCH_2CH_3)

Problem 21.89 (SYN)

Think

How many C atoms are in the target compound? How many are in the starting material? Which α C–H is deprotonated? Does that require a reversible or irreversible deprotonation? What is the relative positioning of the functional groups in the target? What reaction can produce a new C–C bond and result in that relative positioning? Will a functional group transformation be necessary after that reaction takes place?

Solve

(a) The starting material has six C atoms, and the target has seven C atoms. The product is a 1,3-dicarbonyl compound, so we should consider a Claisen condensation reaction. The ketone enolate is formed using LDA to deprotonate the *less*-substituted C atom. A Claisen condensation reaction with a formate ester such as ethyl formate leads to the product. (Other synthesis schemes would also be feasible.)

LDA → 1. HC(O)OEt 2. H_3O^+

(b) The starting material has six C atoms, and the target has seven C atoms. The ketone enolate is formed using NaOEt to deprotonate the *more*-substituted C atom. In the presence of ethyl formate, a Claisen condensation occurs to produce the target. (Other synthesis schemes would also be feasible.)

NaOH → 1. HC(O)OEt 2. H_3O^+

Problem 21.90 (SYN)

Think

How many C atoms are in the target compound? How many are permitted in the starting material? How was the carbon skeleton altered? Does the relative positioning of the functional groups in the target suggest which C–C bond-forming reaction might be used? What functional group transformations must take place? Are protecting groups needed?

Solve

The target has six C atoms; therefore, one reaction must be used to alter the carbon skeleton, owing to the limitation of a three-carbon atom starting material. The alcohol groups are in the 1,3 position, which suggests that an aldol addition reaction occurred. The synthesis in the forward direction is shown below. (Other synthesis schemes would also be feasible.)

PCC; 1. NaOEt, 2. H_3O^+/H_2O; $NaBH_4$, EtOH

2-Methylpentane-1,3-diol

Problem 21.91 (SYN)

Think

How many C atoms are in the target compound? How many are permitted in the starting material? How was the carbon skeleton altered? Does the relative positioning of the functional groups in the target suggest which C–C bond-forming reaction might be used? What functional group transformations must take place? Are protecting groups needed?

Solve

The target has six C atoms; therefore, one reaction altered the carbon skeleton, owing to the limitation of a three-carbon atom starting material. The product is a 1,3-dicarbonyl compound, which suggests that a crossed Claisen condensation reaction could have occurred. The synthesis in the forward direction is shown below. (Other synthesis schemes would also be feasible.)

PCC; $K_2Cr_2O_7$; $SOCl_2$; EtOH; NaOEt

3-Oxo-2-methylpentanal

Problem 21.92 (SYN)

Think

How many C atoms are in the target compound? How many are in the starting material? How was the carbon skeleton altered? What functional group transformations took place? What sequence of reactions can produce an alkyl-substituted acetic acid? Are protecting groups needed?

Solve

The target is an alkyl-substituted acetic acid that can be produced from a malonic ester synthesis. Because three C atoms need to be added to the carbon chain, 1-bromopropane can be used as the alkyl halide. The dicarboxylic acid is first transformed into the diester. The diester is alkylated at the α C, which then undergoes decarboxylation to give the desired target compound.

1,3-Propanedioic acid → EtOH (excess), H^+ → 1. NaOEt, EtOH; 2. 1-bromopropane → 1. NaOH, H_2O; 2. H_3O^+, Δ → **Pentanoic acid**

Problem 21.93 (SYN)

Think

How many C atoms are in the target compound? How many are in the starting material? How was the carbon skeleton altered? What functional group transformations took place? What precursor would produce a cyclic ester (lactone)? Are protecting groups needed?

Solve

The target compound has four C atoms, and the starting material has three C atoms. The α,β-unsaturated aldehyde undergoes 1,4-addition by NaCN to form the nitrile. The nitrile is hydrolyzed to the carboxylic acid. The aldehyde is reduced to the alcohol, which undergoes an acid-catalyzed intramolecular Fischer esterification reaction.

NaCN; H_3O^+; $NaBH_4$ (excess); H_2SO_4, Δ

Problem 21.94

Think

What acts as the nucleophile and the leaving group in the nucleophilic addition–elimination mechanism? How many times does this mechanism occur? What product forms? How many unique H atoms are in the product? Does the chemical environment of these H atoms support the appearance of a signal at 3.8 ppm?

Solve

Two chloride leaving groups can be replaced by methanol. The product is dimethylcarbonate, which has six equivalent H atoms, giving rise to a singlet in the NMR spectrum at 3.8 ppm. The moderate chemical shift reflects the deshielding that occurs from the O atom attached to CH_3.

Nucleophilic addition; Proton transfer; Nucleophile elimination; Nucleophilic addition; Proton transfer; Nucleophile elimination

Problem 21.95

Think

How does diethyl malonate react with a base and an alkyl halide? How many times does this kind of reaction occur? Is an intramolecular reaction favored? What functional groups are evident from the IR spectrum? How many ^{13}C signals are present? Does the number of ^{13}C signals suggest there should be symmetry? What gives rise to the signal at 180 ppm?

Solve

This reaction sequence is a malonic ester synthesis. The second alkylation is intramolecular, producing **B**. Then, hydrolysis and decarboxylation result in the final product, **C**. For product **C**, the IR spectrum is consistent with a carboxylic acid (i.e., a broad OH stretch at 3100 cm^{-1} and a C=O stretch at ~1740 cm^{-1}). There are five C signals in the NMR spectrum, consistent with the five distinct C atoms in the product. The ^{13}C signal at 180 ppm is due to the carbonyl carbon. The reaction sequence is shown below:

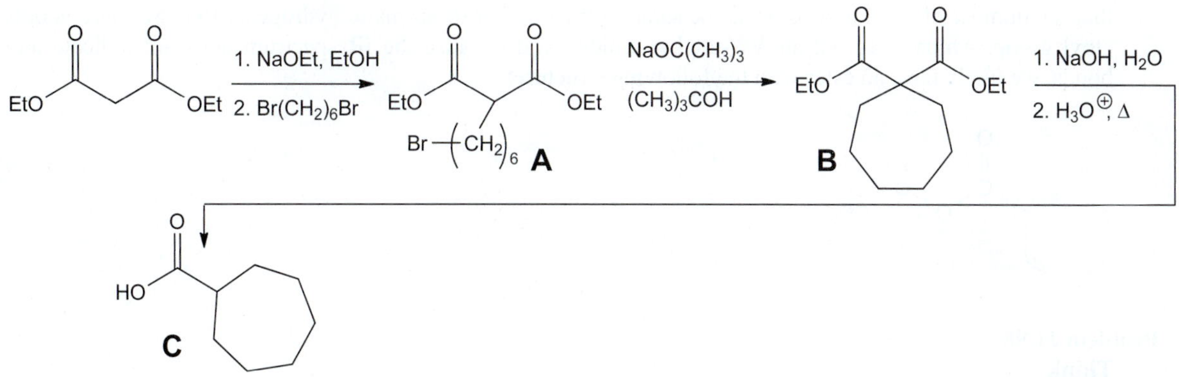

EtO
OEt
1. NaOEt, EtOH
2. Br(CH2)6Br
EtO
OEt
Br
CH2
6
A
NaOC(CH3)3
(CH3)3COH
EtO
OEt
B
1. NaOH, H2O
2. H3O⊕, Δ
HO
C

Problem 21.96

Think

What is the structure of methyl 2-methylpropanoate? What reaction is characteristic of treating an ester with a base? What is the product of the Claisen condensation reaction of an ester? How many signals are in each NMR spectrum? What structural information is evident from the NMR spectra? What functional groups are evident from the IR? Why did the reaction not go as planned?

Solve

(a) The student expected this reaction, which is a Claisen condensation:

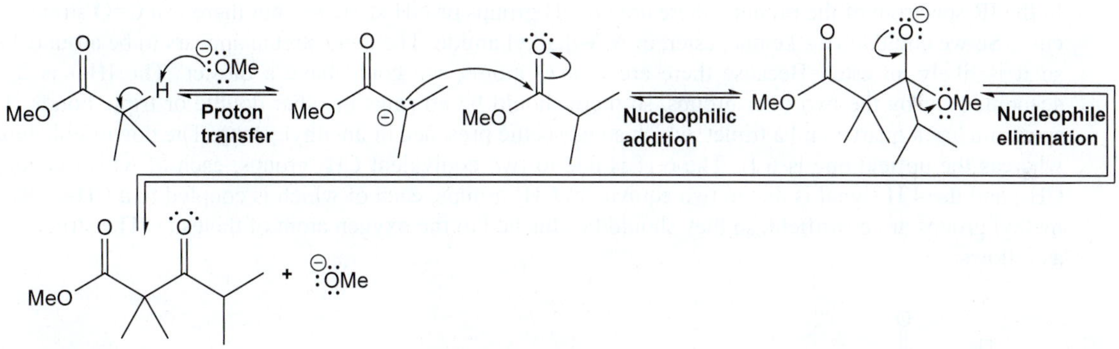

MeO
OMe
H
Proton
transfer
MeO
MeO
Nucleophilic
addition
MeO
OMe
Nucleophile
elimination
MeO
+
OMe

(b) However, the spectra indicate that the student recovered the starting ester. The starting ester would give an M^+ peak at m/z = 102 in the mass spectrum and would give rise to the three signals of 1H, 3H, and 6H in the ^1H NMR spectrum.

(c) Claisen condensation reactions are reversible and favor the reactants unless the product β-keto ester has an acidic proton in between the two carbonyl groups. This one doesn't, so the student could not isolate the intended Claisen condensation product.

Problem 21.97

Think

What functional group is insoluble in water but soluble in basic solution? What functional groups are evident from the IR? How many aromatic H signals are there? What are the relative ratios? What causes the signal at 4.3 ppm to have a downfield shift? What does the quartet:triplet in a 2:3 integration ratio signify?

Solve

An insoluble compound that is soluble under basic conditions suggests a carboxylic acid. That carboxylic acid could be made by hydrolysis of either an ester or an amide, which would also be insoluble in water. Indeed, the absorbance around 1700 cm^{-1} in the IR spectrum suggests a C=O bond. The signals at ~7–8 ppm in the ^{1}H NMR spectrum suggest a benzene ring. In the ^{13}C NMR spectrum, there are three aromatic signals, so we probably have a monosubstituted benzene ring. In the ^{1}H NMR spectrum, the triplet and quartet are consistent with a CH_3CH_2 group attached to an electronegative atom like N or O. The integration of those signals suggests that the number of ethyl hydrogens is the same as the number of aromatic hydrogens: five. So there is only one ethyl group, which rules out an *N,N*-diethyl amide. And because the IR spectrum does not indicate any NH bonds, we likely have an ester with the following structure:

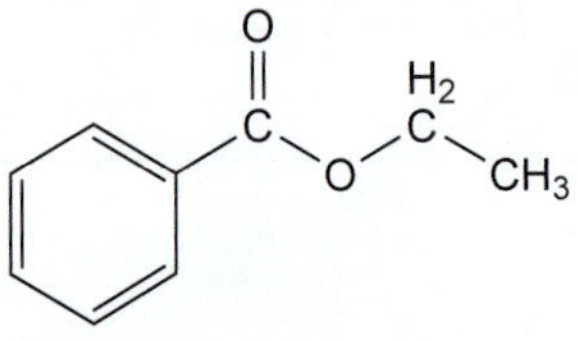

O
H2
C
C
O
CH3

Problem 21.98

Think

What functional groups are evident from the IR spectrum? What is the IHD? On the basis of the given molecular formula and the reactant ethanol, how many C atoms are in the reactant? How many ^{1}H NMR signals are present? What are the splitting patterns, integration ratios, and chemical shifts?

Solve

We are faced with the following question:

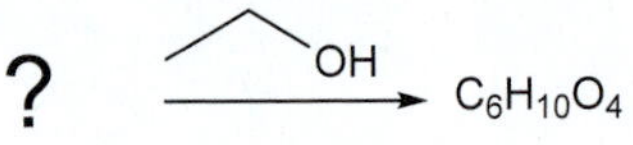

? —OH→ $C_6H_{10}O_4$

In the IR spectrum of the product, there are no OH groups or NH stretches, but there is a C=O stretch near 1700 cm^{-1}. So we could have a ketone, ester, or *N,N*-dialkyl amide. The C=O stretch appears to be around 1740 cm^{-1}, so it is likely an ester. Because there are four O atoms, we could have a diester. The IHD is 2, which is accounted for by the two ester groups, so there should be no rings or other double or triple bonds. The NMR spectrum has a quartet and a triplet, which suggests the presence of an ethyl group. The downfield signal is 4 H, whereas the upfield one is 6 H. The 6 H is due to two equivalent CH_3 groups, each of which is coupled to a CH_2, and the 4 H signal is due to two equivalent CH_2 groups, each of which is coupled to a CH_3. Also, the CH_2 methyl groups are downfield, so they should be attached to the oxygen atom of the ester. The structure could be as follows:

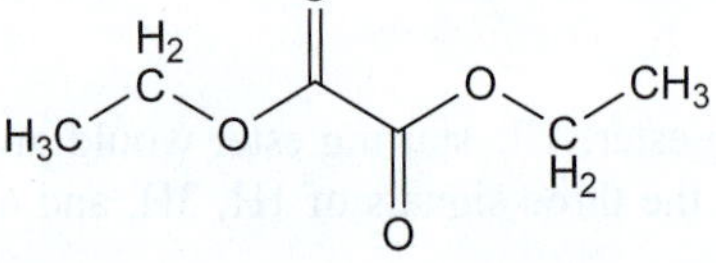

H2 O
C O CH3
H3C O C
H2
O

And this compound could be formed from reaction of acetyl chloride with ethanol, as follows:

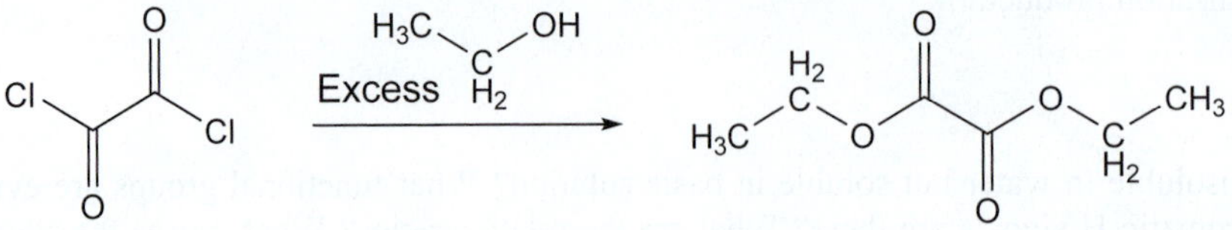

O H3C OH
C
Cl Excess H2 H2 O
CI C O CH3
H3C O C
H2
O O

CHAPTER 22 | Aromatic Substitution 1: Electrophilic Aromatic Substitution on Benzene; Useful Accompanying Reactions

Your Turn Exercises

Your Turn 22.1

Think

What would the product be for the HCl addition to a C=C of benzene? What would the product of Cl_2 addition be? Review the requirements for aromaticity in Section 14.4. Is there still a cyclic, fully conjugated π system containing a Hückel number of electrons? Are these products still aromatic?

Solve

The products of HCl and Cl_2 addition are shown below. In both cases, aromaticity has been lost because the π system is no longer fully conjugated around the ring and no longer contains an odd number of pairs of electrons.

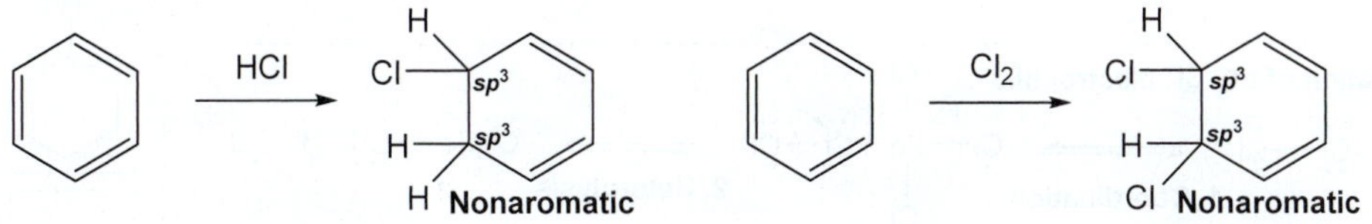

HCl
H
Cl
sp^3
sp^3
H
H Nonaromatic

Cl_2
H
Cl
sp^3
sp^3
H
Cl Nonaromatic

Your Turn 22.2

Think

Review atomic hybridization in Section 3.4. How many electron groups does each C atom have? How does that number of electron groups correspond to electron geometry and to hybridization?

Solve

Each doubly bonded C has three electron groups, and so does the positively charged C. These C atoms have a trigonal planar electron geometry, which corresponds to an sp^2 hybridization. The C atom attached to E has four electron groups, which corresponds to a tetrahedral geometry and an sp^3 hybridization. The arenium ion intermediate has five sp^2-hybridized C atoms and one sp^3-hybridized C atom.

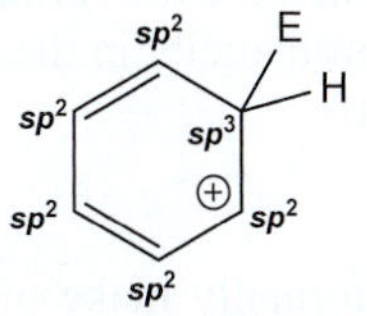

sp^2 E
H
sp^2 sp^3
sp^2 sp^2
sp^2

Your Turn 22.3

Think

Are any atoms with an incomplete octet attached to a double or triple bond? Does the sp^3-hybridized C participate in resonance? In the individual resonance structures, on which carbon atoms does the positive charge appear? How is the positive charge on the individual resonance structures represented in the hybrid?

Solve

The carbocation C^+ is sp^2 hybridized and has an incomplete octet, and it is attached to a C=C bond. Resonance is possible among all five sp^2-hybridized C atoms. The sp^3-hybridized C cuts off resonance. On the individual resonance structures, the positive charge appears on three different carbon atoms. Thus, in the resonance hybrid, the partial positive (δ^+) is shared over three carbon atoms.

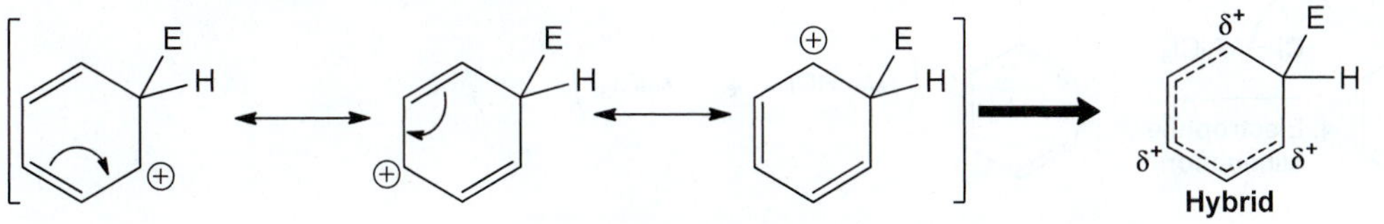

E H ⟷ E H ⟷ E H → δ^+ E H δ^+ δ^+

Hybrid

Your Turn 22.4

Think

Which bonds were broken and formed in each step? Which sites are electron rich and which are electron poor? How can you use curved arrows to show the flow of electrons in each step? Review the 10 elementary steps from Chapters 6 and 7. Which steps involve the formation of the electrophile? What is the electrophile? Which steps temporarily break and re-form aromaticity in the ring, resulting in substitution of H^+ by an electrophile?

Solve

Steps 1 and 2 involve the formation of the Cl^+ electrophile. In Step 1, $FeCl_3$ acts as a Lewis acid (electron-pair acceptor) and complexes with Cl_2 in a coordination step. In Step 2, heterolysis of Cl–Cl takes places slowly to produce Cl^+ and $FeCl_3^-$. Steps 3 and 4 formally make up the electrophilic aromatic substitution mechanism with Cl^+ as the electrophile. Step 3 is electrophilic addition, and Step 4 is electrophile elimination. $FeCl_3$ is regenerated at the end of the reaction and is, therefore, a catalyst. It should be noted that the coordination and heterolysis steps probably occur simultaneously instead of separately, but the electrophile Cl^+ is shown explicitly for emphasis.

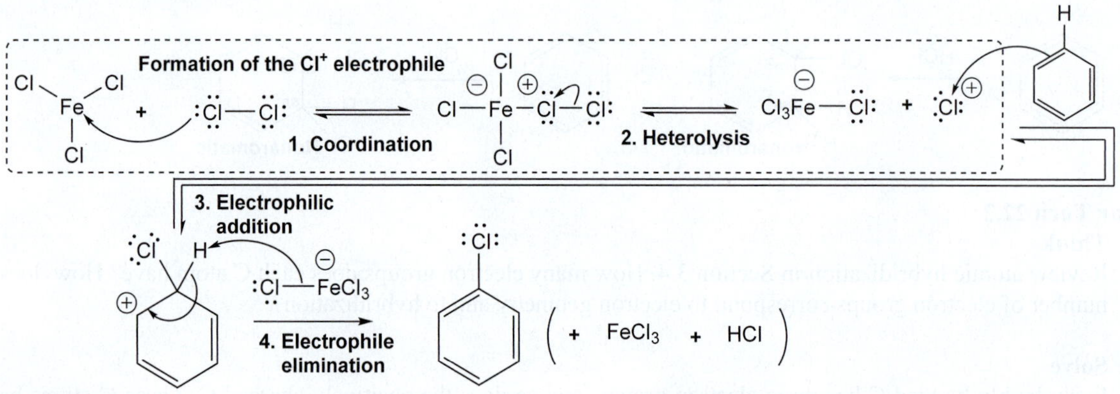

Your Turn 22.5

Think

Which bonds were broken and formed in each step? Which sites are electron rich and which are electron poor? How can you use curved arrows to show the flow of electrons in each step? Which steps involve the formation of the electrophile? What is the electrophile? Which steps temporarily break and re-form aromaticity in the ring, resulting in substitution of H^+ by an electrophile? Review the mechanism in Equation 22-4.

Solve

Steps 1 and 2 involve the formation of the cyclohexyl C^+ electrophile. Steps 3 and 4 formally make up the electrophilic aromatic substitution mechanism with the cyclohexyl C^+ as the electrophile.

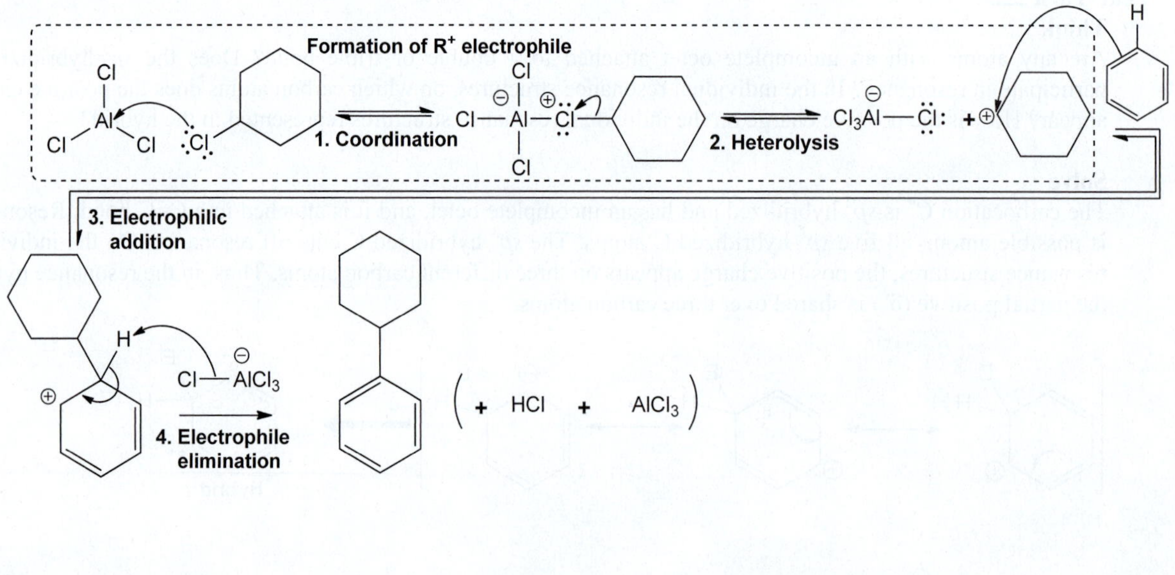

Your Turn 22.6

Think

Consult Figure 3-32 on page 145 for the C–H bond strengths of an sp^3 C–H bond compared to an sp^2 C–H bond.

Solve

The C–H bond energy for an sp^3-hybridized C is 421 kJ/mol, and that for an sp^2-hybridized C is 464 kJ/mol.

Your Turn 22.7

Think

Which bonds were broken and formed in each step? Which sites are electron rich and which are electron poor? How can you use curved arrows to show the flow of electrons in each step? Which steps involve the formation of the electrophile? What is the electrophile? Which steps temporarily break and re-form aromaticity in the ring, resulting in substitution of H^+ by an electrophile? Review the mechanism in Equation 22-4.

Solve

Steps 1 and 2 involve the formation of the acylium ion electrophile. Steps 3 and 4 formally make up the electrophilic aromatic substitution mechanism with the acylium ion as the electrophile.

Your Turn 22.8

Think

Review Section 1.10 to draw a resonance hybrid. How do you represent a bond intermediate between a double and a triple bond? Which atoms share the positive charge?

Solve

A resonance hybrid is an average of all resonance structures. The carbon–oxygen bond is between a double and a triple bond, represented as two solid bonds and a dashed bond. The positive charge is delocalized over both C and O, so each receives a δ^+ in the hybrid.

Your Turn 22.9

Think

Consult Equation 22-24 in writing a complete mechanism for the nitration of benzene in the presence of sulfuric acid, H_2SO_4. How is the mechanism similar/different compared to Equation 22-24? What is the purpose of H_2SO_4?

Solve

Sulfuric acid is a stronger acid compared to HNO_3 and serves to protonate nitric acid in Step 1. This generates a substantially higher concentration of the nitronium electrophile, NO_2^+. Other than that, the mechanism is identical to the mechanism in Equation 22-24.

Formation of the nitronium ion NO_2^+

1. Proton transfer

2. Heterolysis

3. Electrophilic addition

4. Electrophile elimination

$(+ H_3O^+ + HSO_4^-)$

Your Turn 22.10

Think

Consult Equation 22-26. How can you arrange the reactant, products, and intermediates in Equation 22-26 in reverse order? Which sites are electron rich and which are electron poor? How can you use curved arrows to show the flow of electrons in each step?

Solve

Desulfonation is the reverse of Steps 1–4 in Equation 22-26.

1. Electrophilic addition

2. Electrophile elimination

3. Coordination

4. Proton transfer

Your Turn 22.11

Think

Refer to Section 22-4 and review the limitations of Friedel–Crafts alkylations. Is this carbocation intermediate susceptible to a rearrangement? If so, what is the actual product of the Friedel–Crafts alkylation? Is oxidation to a carboxylic acid possible for this product?

Solve

The carbocation intermediate in this problem is susceptible to a 1,2-hydride shift to convert the secondary carbocation into a more stable tertiary carbocation. The actual product of the Friedel–Crafts alkylation is shown below. The resulting product yields a quaternary benzylic carbon; the permanganate oxidation fails, however, when the benzylic C is *quaternary*.

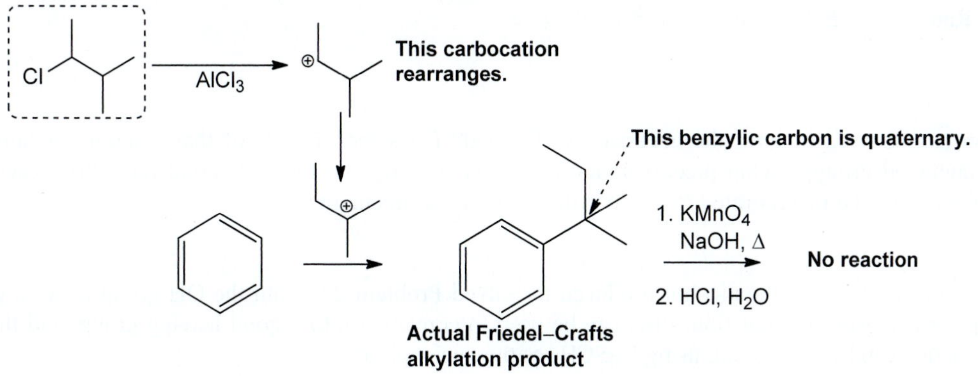

Cl

$AlCl_3$

This carbocation rearranges.

This benzylic carbon is quaternary.

1. $KMnO_4$ NaOH, Δ

2. HCl, H_2O

No reaction

Actual Friedel–Crafts alkylation product

Your Turn 22.12

Think

What is the structure of the alkyldiazonium ion and the benzenediazonium ion? Does one of those structures have an atom with a positive charge attached to a pair of atoms connected by a double bond? Which is more stable?

Solve

The alkyldiazonium ion that would be produced from $(CH_3)_2CHNH_2$ when treated with $NaNO_2$ under acidic conditions is shown below. This alkyldiazonium ion is less stable than the benzenediazonium ion because the positive charge on the nitrogen is localized. The positive charge on the nitrogen in benzenediazonium is delocalized via resonance and is therefore more stable.

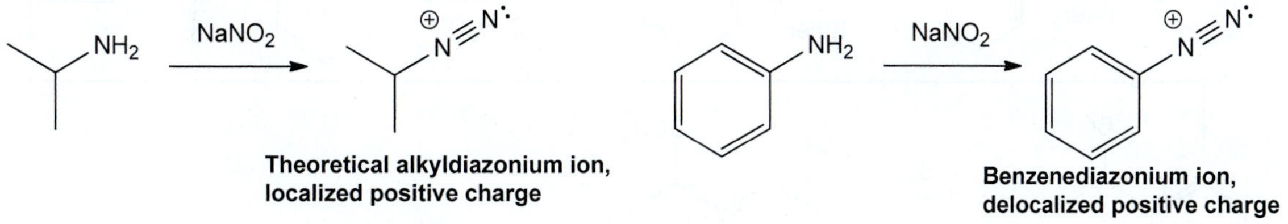

NH_2 $NaNO_2$

Theoretical alkyldiazonium ion, localized positive charge

NH_2 $NaNO_2$

Benzenediazonium ion, delocalized positive charge

In Chapter Problems

Problem 22.1

Think

What is the rate law for the reaction? What is the order of each reactant? How does the reaction rate change when the concentration of a reactant is doubled (electrophile) or tripled (benzene)?

Solve

The rate law for the reaction is Rate = $k[E^+]$[Benzene].

The reaction would be six times faster. Because the reaction is first order with respect to benzene, tripling the concentration of benzene triples the reaction rate. Also, the reaction is first order with respect to the electrophile, so doubling the concentration of the electrophile doubles the reaction rate.

$$\text{Rate} = k[2 \times E^+][3 \times \text{Benzene}] \rightarrow 6\times \text{ the rate}$$

Problem 22.3

Think

What reaction can be used to form the necessary C–C bond? Does the product of that reaction contain the appropriate functional groups? What precursors are necessary to carry out the C–C bond formation reaction? How must halogenation be incorporated to produce the appropriate precursor?

Solve

The product can be made from the alcohol produced in Solved Problem 22.2, but the OH group is not a good leaving group for a substitution reaction. PBr_3 can be used to convert it into a good leaving group, and then a substitution reaction can be carried out, using $NaSCH_3$ as the nucleophile.

SCH_3 **Undo an S_N2.** ⟹ Br **Undo a substitution.** ⟹ OH **Alcohol product from Solved Problem 22.2**

In the forward direction, the synthesis would be written as follows:

Benzene → (Br_2, $FeBr_3$) → Br-benzene → (Mg(*s*), Ether) → MgBr → (1. propanal; 2. H_3O^+) → OH → PBr_3 → Br → $NaSCH_3$ → SCH_3

Problem 22.4

Think

What is the electrophile? What two elementary steps make up the electrophilic aromatic substitution reaction? What acts as the base to remove the H^+ and regenerate the aromatic ring?

Solve

The I^+ cation is the electrophile; once formed, it can be attacked by the aryl ring. A weak base in solution, such as H_2O_2, could act as the base to remove the proton from the arenium ion intermediate.

Electrophilic addition

Electrophile elimination

Problem 22.5

Think

Does the $(CH_3)_3COH$ have a good leaving group? How does a reaction with H_3PO_4 generate a good leaving group? What strong electrophile is produced? What two elementary steps make up the electrophilic aromatic substitution reaction? What acts as the base to remove the H^+ and regenerate the aromatic ring?

Solve

$(CH_3)_3COH$ does not have a good leaving group. Reaction in an acid (H_3PO_4) protonates the alcohol and turns it into $(CH_3)_3CO^+H_2$. Water now acts as the leaving group and thus generates the carbocation via heterolysis. Once the strongly electrophilic carbocation is formed, the electrophilic aromatic substitution mechanism is identical to the corresponding steps in Equation 22-12. The ROH acts as the base to remove the proton and restore the aromaticity.

Proton transfer

Heterolysis

Electrophilic addition

Electrophile elimination

Problem 22.6

Think

What reaction occurs between a strong acid and the C=C bond of an alkene? What strong electrophile results? What two elementary steps make up the electrophilic aromatic substitution reaction? What acts as the base to remove the H^+ and regenerate the aromatic ring?

Solve

Attack of cyclohexene on a proton forms the carbocation electrophile (Chapter 11). Then the electrophilic aromatic substitution mechanism occurs via the usual electrophilic addition–elimination route.

Electrophilic addition

Electrophilic addition

Electrophile elimination

Problem 22.8

Think

What is the electrophile that must substitute for H^+? How can that electrophile be produced from an alkyl halide, an alkene, or an alcohol?

Solve

The electrophile that must substitute for H^+ as a carbocation is as follows:

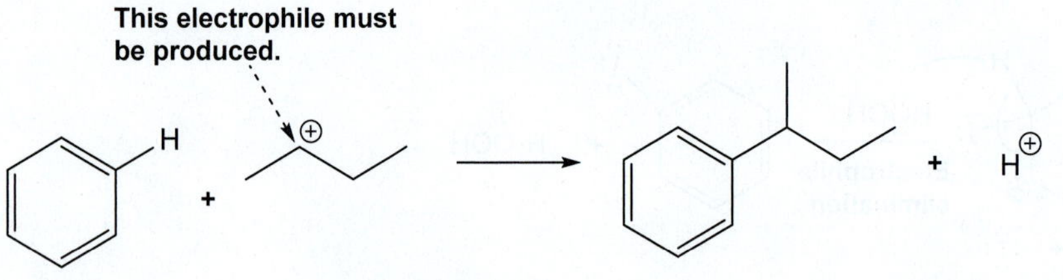

This electrophile must be produced.

The carbocation can be produced from an alkyl halide using $AlCl_3$, a strong Lewis acid catalyst. Alternatively, it can be produced from an alkene or alcohol under acidic conditions.

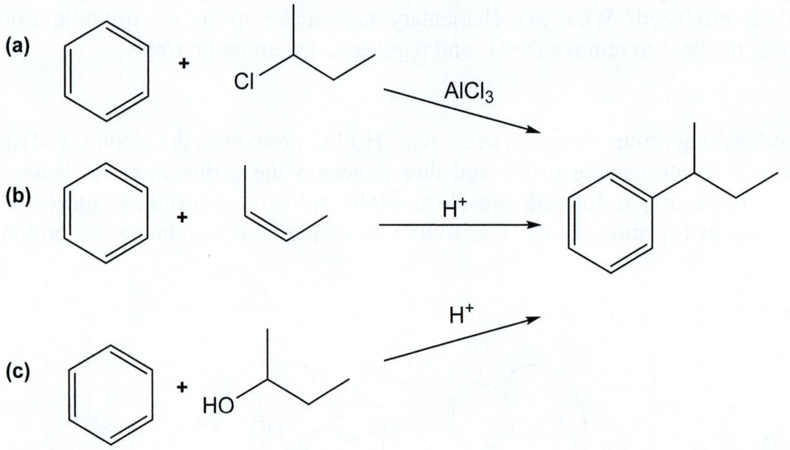

(a) + Cl — $AlCl_3$

(b) + — H^+

(c) + HO — H^+

Problem 22.10

Think

Can the target be produced directly from a simple Friedel–Crafts alkylation? What carbocation intermediate would be produced? Would that carbocation intermediate rearrange? Can another reaction be used to form the appropriate C–C bond?

Solve

The product would need to form the C–C bond from the aromatic ring and a 1° carbon. This cannot occur with a Friedel–Crafts alkylation due to a carbocation rearrangement that would take place prior to electrophilic addition.

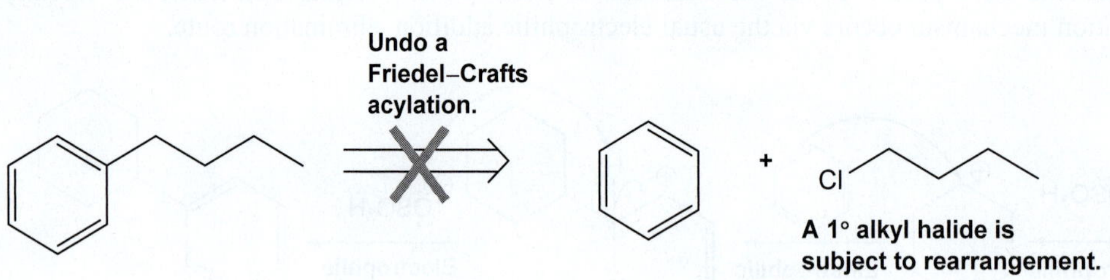

Undo a Friedel–Crafts acylation.

Cl

A 1° alkyl halide is subject to rearrangement.

We can use a Grignard reaction to form the C–C. This is followed by dehydration of the resulting alcohol, then by hydrogenation, to remove the functional groups on the chain.

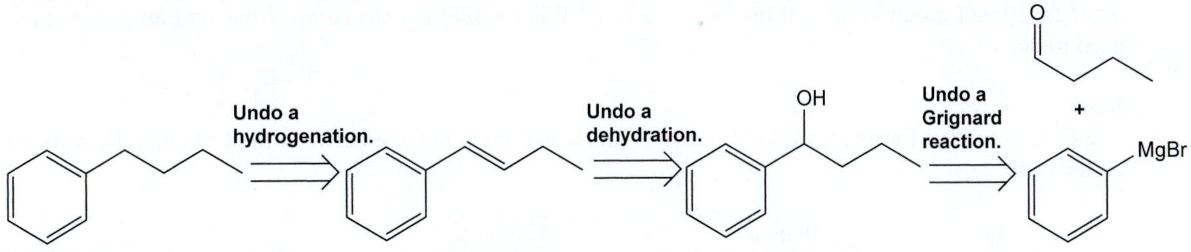

Undo a hydrogenation. Undo a dehydration. OH Undo a Grignard reaction. O + MgBr

In the forward direction, the synthesis would be written as follows:

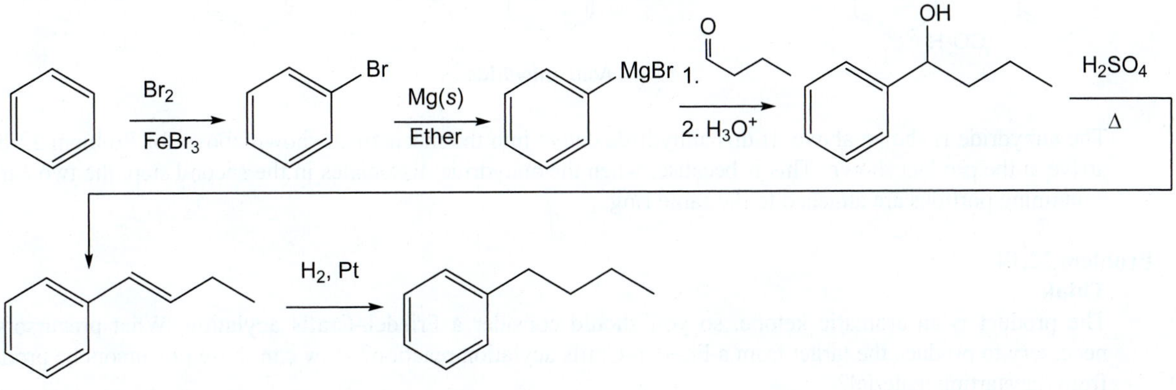

Br_2 / $FeBr_3$ → Br; Mg(*s*) / Ether → MgBr; 1. O 2. H_3O^+ → OH; H_2SO_4 / Δ → ; H_2, Pt →

Problem 22.11

Think

Which atom in the anhydride will undergo coordination with the strong Lewis acid? What can act as the leaving group in that complex? What is the strong electrophile that results? What two elementary steps make up the electrophilic aromatic substitution reaction? What acts as the base to remove the H^+ and regenerate the aromatic ring?

Solve

The mechanism is identical to that with the acid chloride (see Equation 22-22, with the $[CH_3CO_2\text{–}AlCl_3]^-$ as the leaving group instead of $[Cl\text{–}AlCl_3]^-$).

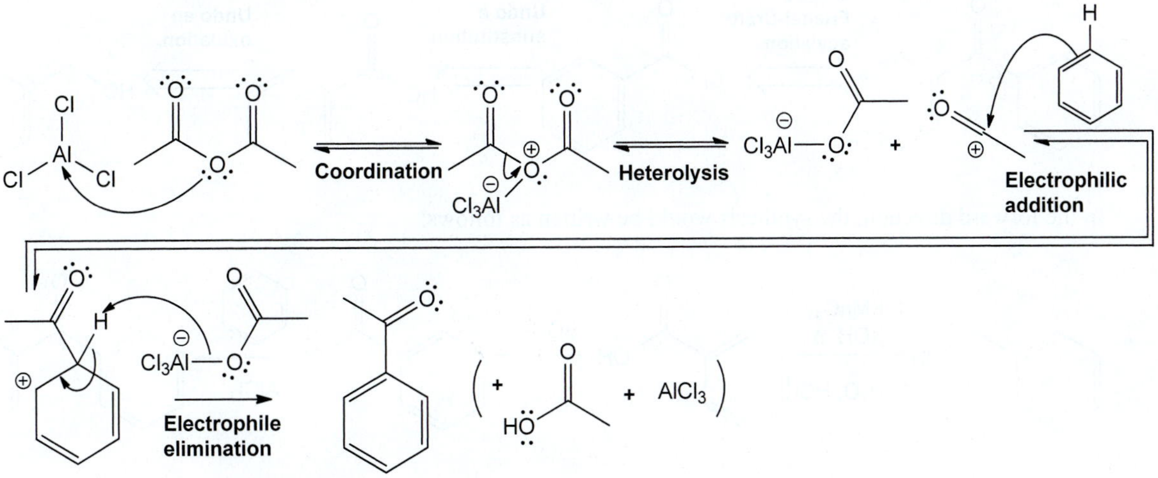

Coordination; Heterolysis; Cl_3Al—O; Electrophilic addition; Electrophile elimination; (+ HO + $AlCl_3$)

Problem 22.12

Think

Which C–C bond was formed from Friedel–Crafts acylation? Which ring came from the benzene ring? What other functional group is present on the other ring? What would the structure of the acid anhydride precursor need to be?

Solve

The C–C bond that formed from the Friedel–Crafts acylation is the one between the C=O and the benzene ring on the right.

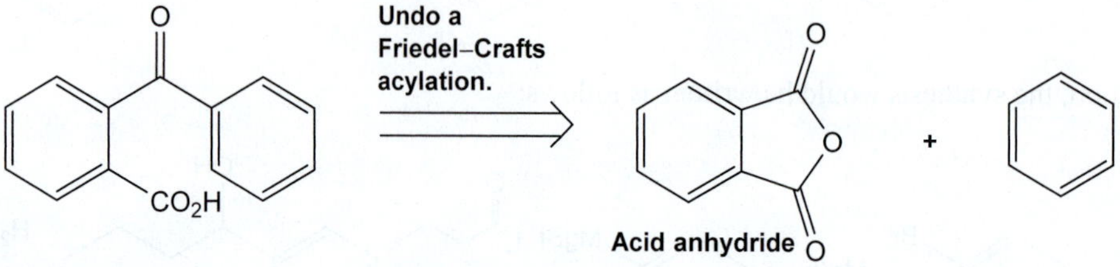

O
CO_2H
Undo a Friedel–Crafts acylation.
O
O
+
Acid anhydride O

The anhydride is shown above. If this anhydride enters into the mechanism shown above for Problem 22-8, we arrive at the product shown. This is because, when the anhydride dissociates in the second step, the two carbon-containing portions are attached to the same ring.

Problem 22.14

Think

The product is an aromatic ketone, so you should consider a Friedel–Crafts acylation. What precursors are necessary to produce the target from a Friedel–Crafts acylation reaction? How can those precursors be produced from the starting material?

Solve

In a Friedel–Crafts acylation, an aromatic carbon forms a bond to an acyl carbon. We begin our retrosynthetic analysis by disconnecting that bond to arrive at the appropriate precursors—an aromatic ring and an acyl chloride. The acyl chloride can be produced from the corresponding carboxylic acid, which can be produced, in turn, by oxidizing a primary alcohol.

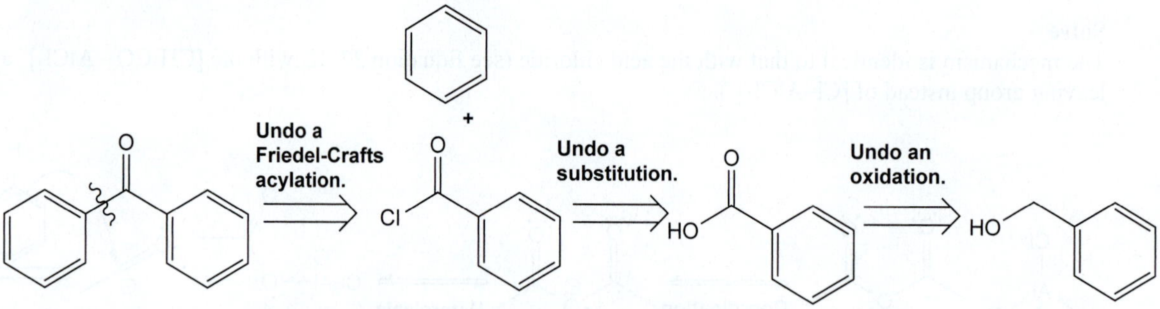

O
Undo a Friedel-Crafts acylation.
+
O
Cl
Undo a substitution.
O
HO
Undo an oxidation.
HO

In the forward direction, the synthesis would be written as follows:

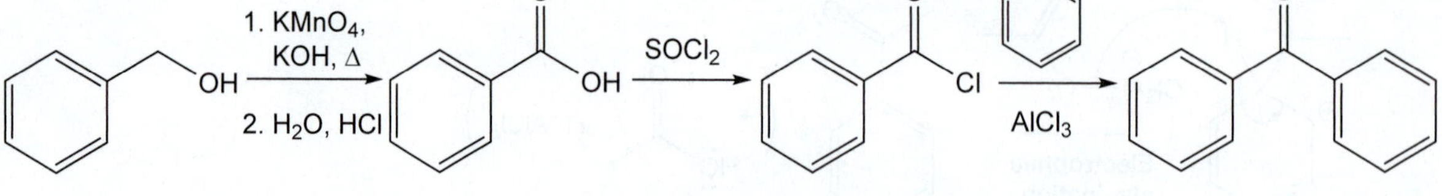

OH
1. $KMnO_4$, KOH, Δ
2. H_2O, HCl
O
OH
$SOCl_2$
O
Cl
$AlCl_3$
O

Problem 22.16

Think

The product is an alkylbenzene. Can a Friedel–Crafts alkylation be used to form the C–C bond to benzene? If not, why not? How can a Friedel–Crafts acylation reaction be incorporated into this synthesis? Which step in Solved Problem 22.15 uses the Clemmensen reduction? Are there other methods to reduce a ketone?

Solve

The reduction could be carried out using a Wolff–Kishner or Raney-nickel reduction. Alternatively, it could be carried out with a hydride reduction, dehydration, and then a hydrogenation, as shown below.

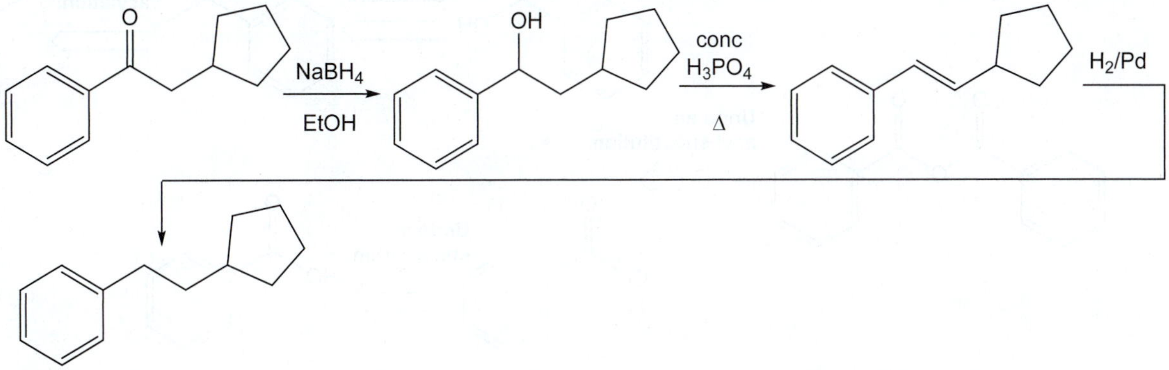

O OH $NaBH_4$ EtOH conc H_3PO_4 Δ H_2/Pd

Problem 22.17

Think

Could Friedel–Crafts alkylation or acylation reactions be used to form the new C–C bonds? How many Friedel–Crafts reactions need to occur? What should be the identity of the Friedel–Crafts starting material? What other reaction must occur to form the target? Are oxidation or reduction reactions necessary?

Solve

Friedel–Crafts alkylation cannot be done because the carbocation formed will be a primary carbocation and will rearrange to a secondary carbocation. The synthesis, however, can be carried out with a Friedel–Crafts acylation. Friedel–Crafts acylation occurs twice, with the following reagent as the starting material:

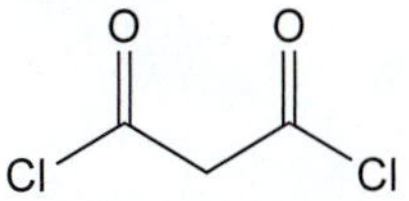

O O Cl Cl

The Friedel–Crafts acylation followed by the Clemmensen (or Wolff–Kishner) reduction is sufficient.

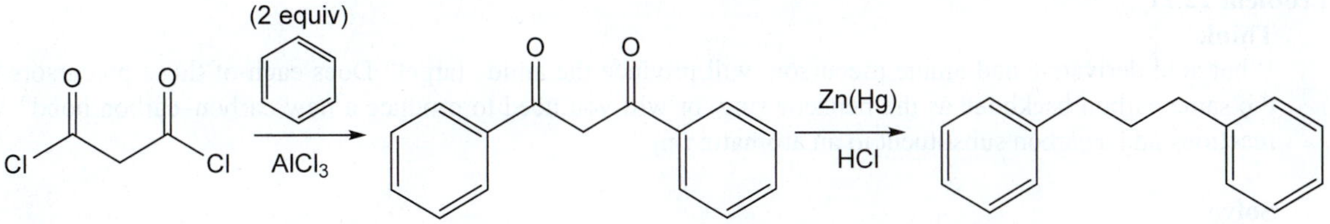

(2 equiv) O O Cl Cl $AlCl_3$ O O Zn(Hg) HCl

Problem 22.19

Think

From what precursors can an acid anhydride be produced? To produce those precursors from the available starting materials, are carbon–carbon bond-forming reactions necessary? Can a Friedel–Crafts alkylation or acylation be used to form those bonds? Are oxidation or reduction reactions necessary?

Solve

In Chapter 21, we learned that an acid anhydride can be synthesized from an acid chloride and a carboxylic acid. Here, the necessary carboxylic acid is benzoic acid, and the necessary acid chloride is benzoyl chloride, which can be produced from benzoic acid. To make benzoic acid using benzene as the only aromatic starting material, a C–C bond must be formed. This can be carried out using a Friedel–Crafts acylation between benzene and acetyl chloride. The resulting ketone can be oxidized to benzoic acid using $KMnO_4$.

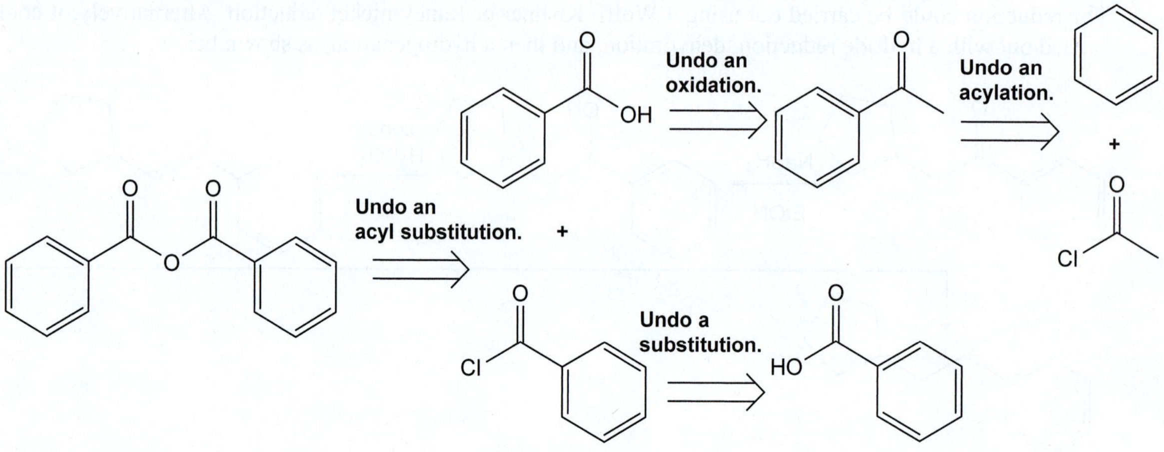

In the forward direction, the synthesis would proceed as follows:

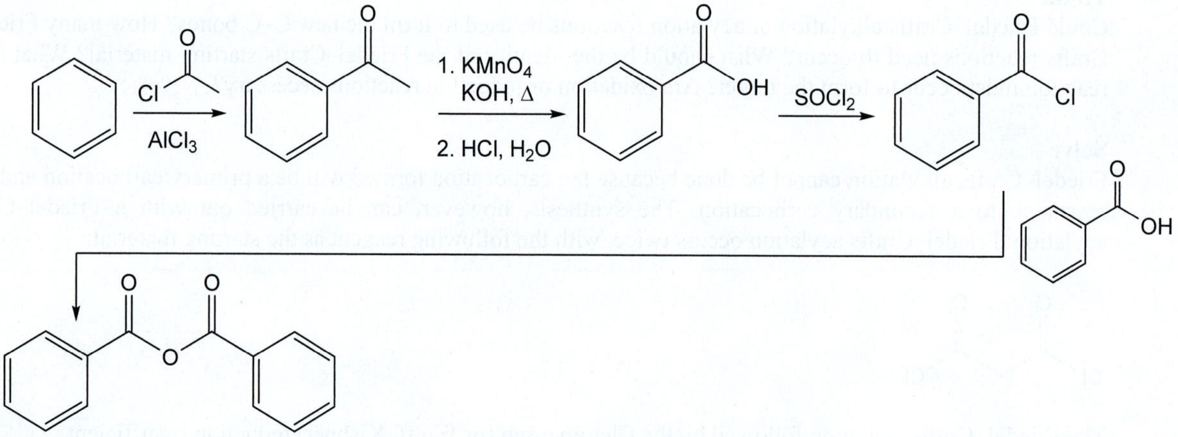

Problem 22.21

Think

What acid derivative and amine precursors will produce the amide target? Does each of those precursors have the same carbon backbone as the benzene ring, or will you need to produce a new carbon–carbon bond? What reactions add a carbon substituent to an aromatic ring?

Solve

(a) An *N,N*-disubstituted amide can be produced from an *N*-substituted amide and an alkyl halide. The *N*-substituted amide can be produced from an amine and an acid chloride—in this case, aniline and ethanoyl chloride. Aniline can be produced by the nitration of benzene, followed by reduction of the nitrobenzene product. The alkyl halide can be produced from Friedel–Crafts alkylation, followed by oxidation, to form benzoic acid. Benzoic acid is then reduced to form the alcohol, which is then brominated. See the figure on the next page.

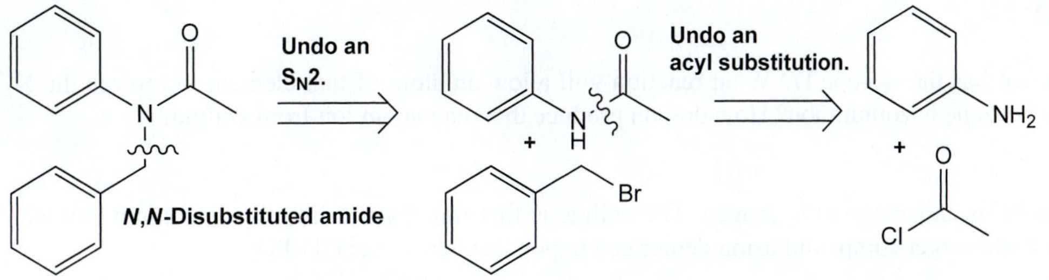

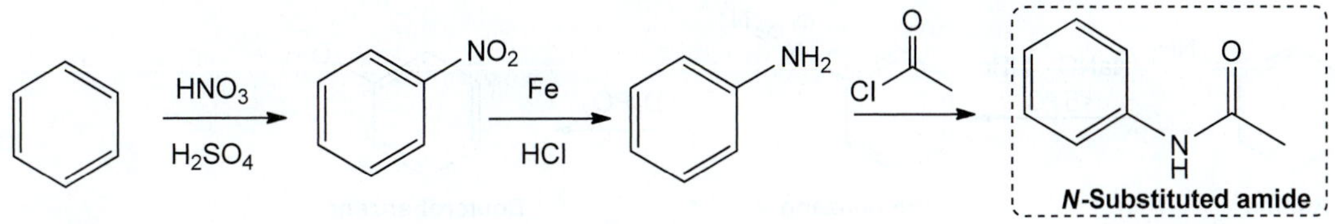

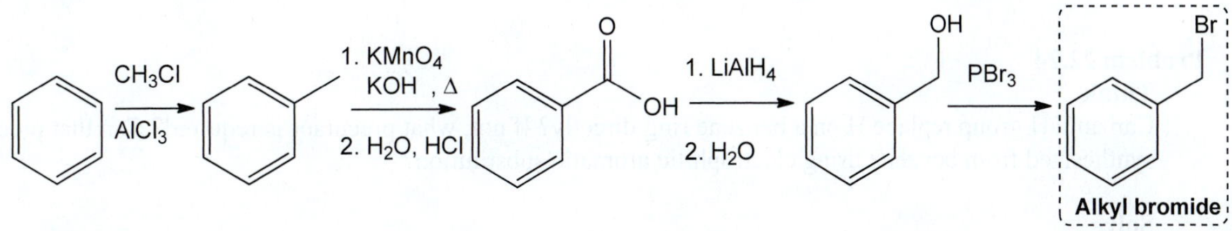

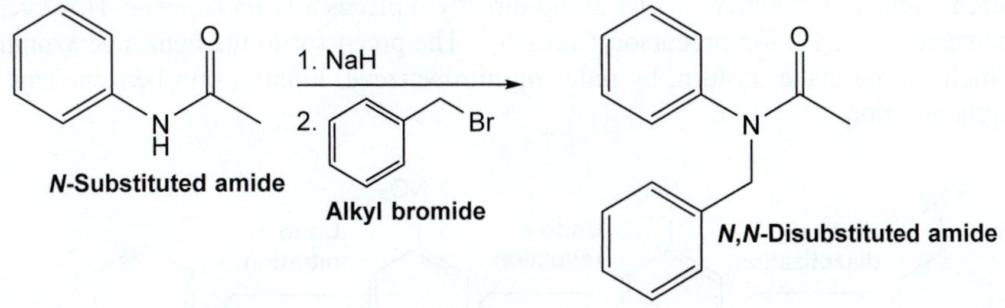

Undo an S_N2.

Undo an acyl substitution.

N,N-Disubstituted amide

+

Br

NH_2

+

O

Cl

HNO_3 / H_2SO_4

NO_2

Fe / HCl

NH_2

Cl

N-Substituted amide

CH_3Cl / $AlCl_3$

1. $KMnO_4$, KOH, Δ
2. H_2O, HCl

OH

1. $LiAlH_4$
2. H_2O

OH

PBr_3

Br

Alkyl bromide

1. NaH
2. Br

N-Substituted amide

Alkyl bromide

N,N-Disubstituted amide

(b) The imine can be produced from the reaction of a ketone and aniline in acidic medium. The aniline was synthesized in **(a)**. The ketone can be synthesized from Friedel–Crafts acylation.

Undo an imine formation reaction.

NH_2

+

Undo a Friedel–Crafts acylation.

Cl

+

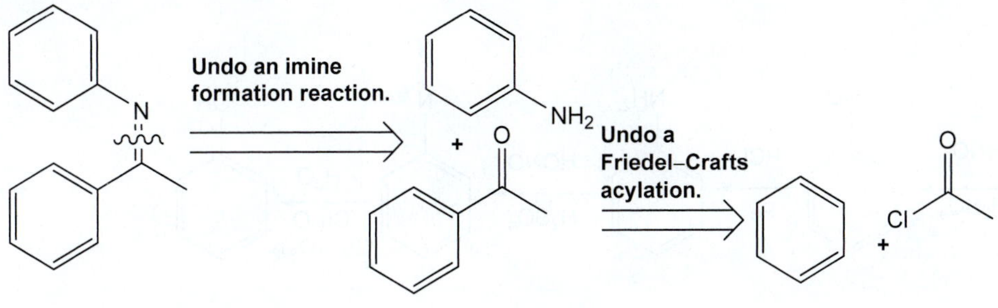

The forward synthesis would proceed as follows:

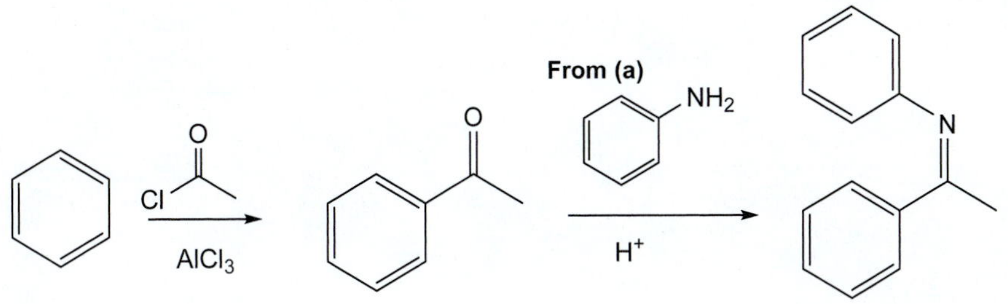

Cl

$AlCl_3$

From (a)

NH_2

H^+

N

Problem 22.22

Think

What element has the isotope D? What reaction will allow an atom of that element to replace the N_2^+ leaving group in a benzenediazonium ion? How do you produce that diazonium ion from aniline?

Solve

Deuterium, D, is an isotope of hydrogen. The aniline is first transformed into the diazonium ion, which can be converted to the target compound using deuterated hypophosphorous acid, D_3PO_2.

NaNO₂, HBr <5 °C; D₃PO₂

Aniline — The benzene diazonium ion — Deuterobenzene

Problem 22.24

Think

Can an OH group replace H on a benzene ring directly? If not, what precursor is required? Can that precursor be synthesized from benzene using electrophilic aromatic substitution?

Solve

We have not encountered a reaction in which an OH group directly replaces a H on benzene. However, phenol can be made from a benzenediazonium ion precursor, $C_6H_5\text{–}N_2^+$. The precursor to the benzenediazonium ion is an aromatic amine, which can be made, in turn, by reducing nitrobenzene. Finally, nitrobenzene can be made directly from benzene via nitration.

Undo phenol formation. **Undo a diazotization.** **Undo a reduction.** **Undo a nitration.**

The synthesis in the forward direction would proceed as follows:

HNO_3, H_2SO_4; HCl, Fe; HONO, H_2SO_4; H_2O, Cu_2O, $Cu(NO_3)_2$

End of Chapter Problems

Sections 22.1–22.5 The General Mechanism, Halogenation, and Friedel–Crafts Reactions

Problem 22.25

Think

Which C–H bonds on the benzene ring are available to undergo electrophilic aromatic substitution? Are those sites chemically distinct?

Solve

Examine the three isomers of trimethylbenzene. **A** has two chemically distinct H atoms, so it can give rise to two distinct bromination products, as shown below. **B** has three distinct H atoms, so it gives rise to three possible bromination products. **C** has just one distinct H, giving rise to one bromination product.

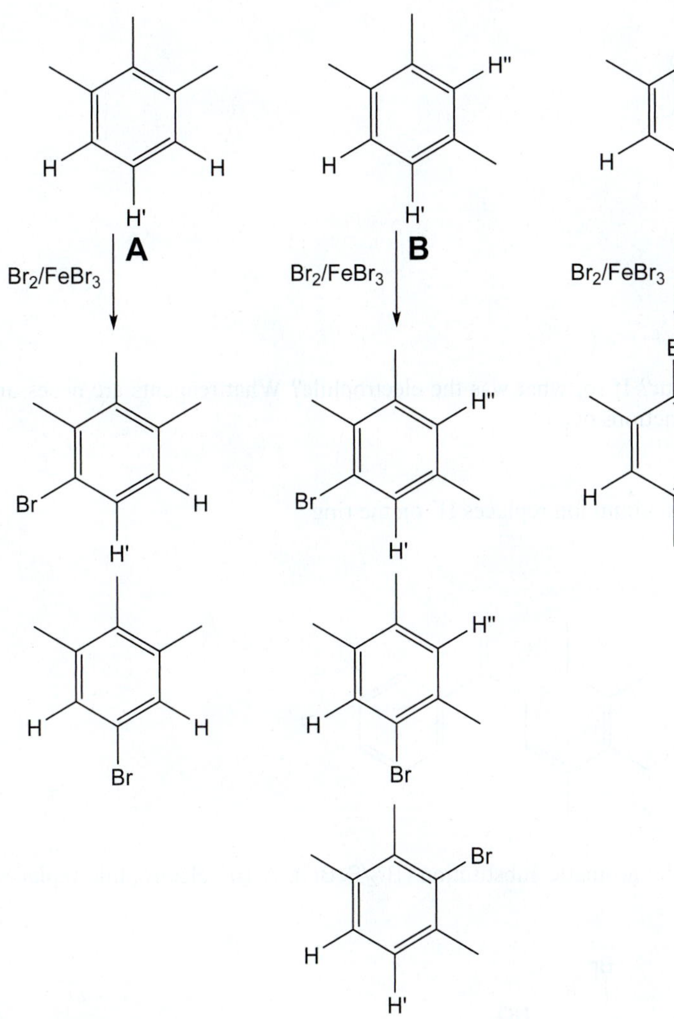

A: $Br_2/FeBr_3$

B: $Br_2/FeBr_3$

C: $Br_2/FeBr_3$

Problem 22.26

Think

What site on the aromatic ring is subject to electrophilic aromatic substitution? What is the electrophile in each reaction?

Solve

Each aromatic C–H can undergo substitution, where H^+ is replaced by an electrophile. The products are given below:

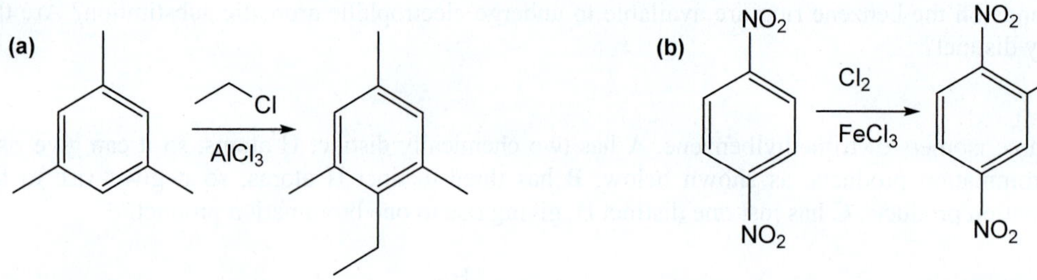

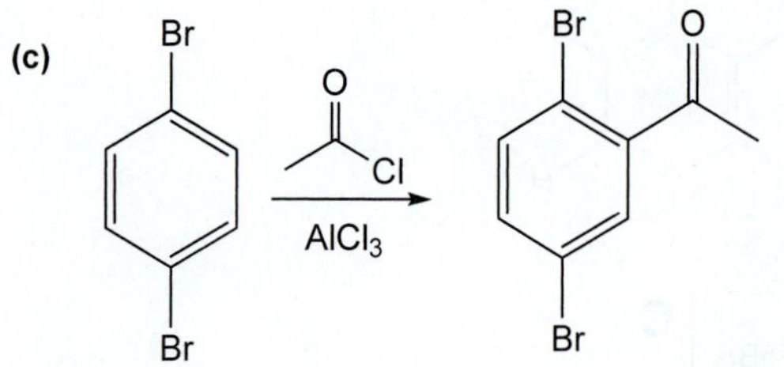

Problem 22.27 (SYN)

Think

Did electrophilic aromatic substitution occur? If so, what was the electrophile? What reagents are necessary to produce that electrophile? Did any other reactions occur?

Solve

(a) This is a Friedel–Crafts acylation. An acylium ion replaces H^+ on the ring.

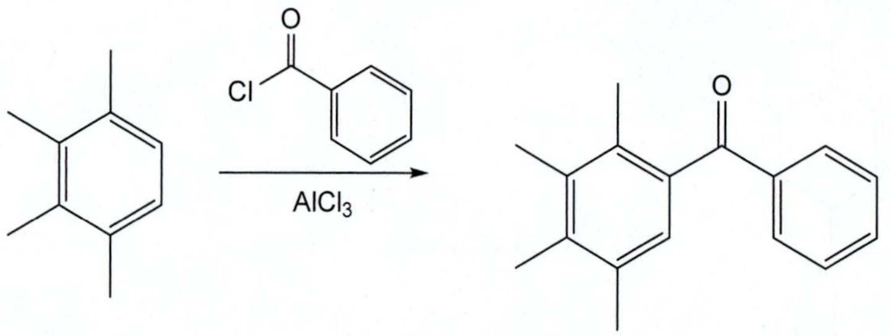

(b) This is a bromination using electrophilic aromatic substitution (Br_2/$FeBr_3$). A Br^+ electrophile replaces H^+ on the ring.

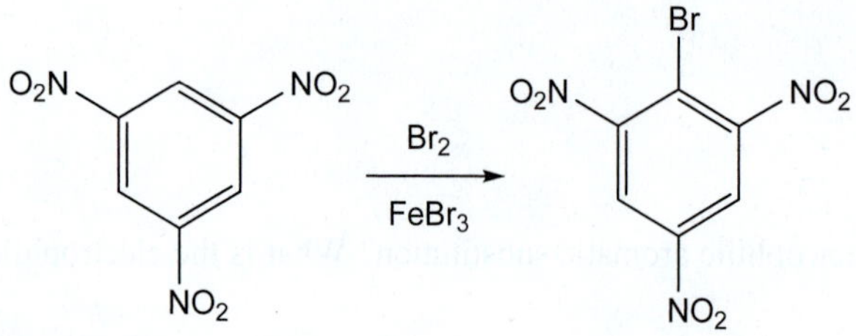

Problem 22.28

Think

Is this reaction an example of electrophilic aromatic substitution? What acts as the electrophile in the electrophilic addition step? What acts as the base to regenerate the aromatic ring?

Solve

Electrophilic addition of H^+ occurs first. In the second step, the first step is simply reversed, but D^+ is eliminated instead of H^+.

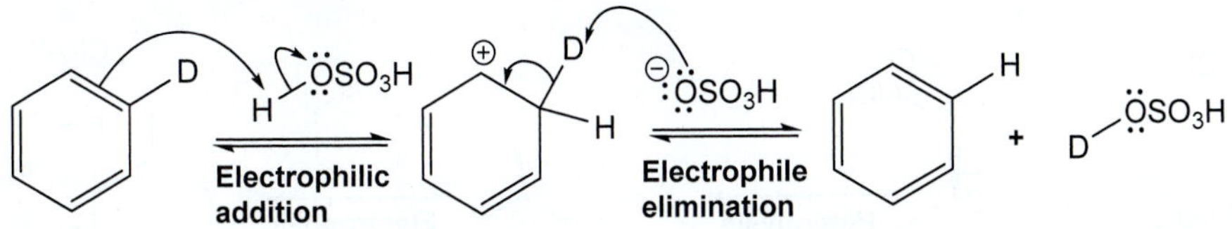

Electrophilic addition

Electrophile elimination

Problem 22.29

Think

How does I–Cl interact with Fe(*s*)? What electrophile must be substituted for H^+? Which atom stabilizes a positive charge better? Is the reaction under kinetic or thermodynamic control? Which electrophile—I^+ or Cl^+—would you expect to be produced more easily?

Solve

These conditions generate a small amount of an iron trihalide catalyst, FeX_3, so the mechanism for each reaction is essentially identical to the bromination reaction we saw earlier in the chapter. The electrophilic aromatic substitution reaction runs under kinetic control, and the reaction rate is governed by the first electrophilic aromatic substitution step, which is electrophilic addition to the benzene ring. The rate of that first step is governed by the concentration of X^+—either Cl^+ or I^+. Formation of X^+ is reversible, and the extent of X^+ that is formed depends on the stability of the ion. I^+ is larger and, therefore, more stable than Cl^+, so there should be more of it available to react with benzene. Thus, the iodobenzene product is favored.

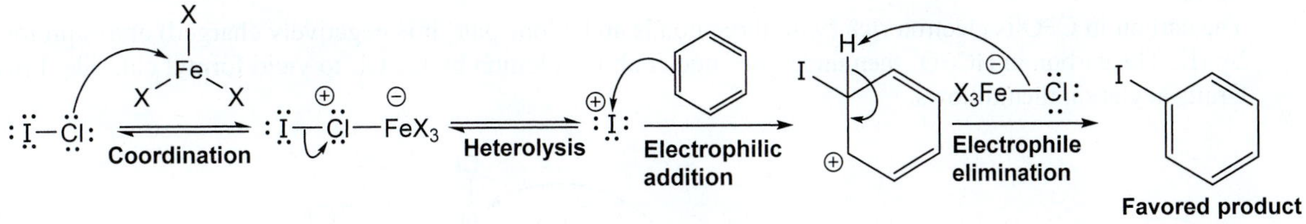

Coordination

Heterolysis

Electrophilic addition

Electrophile elimination

Favored product

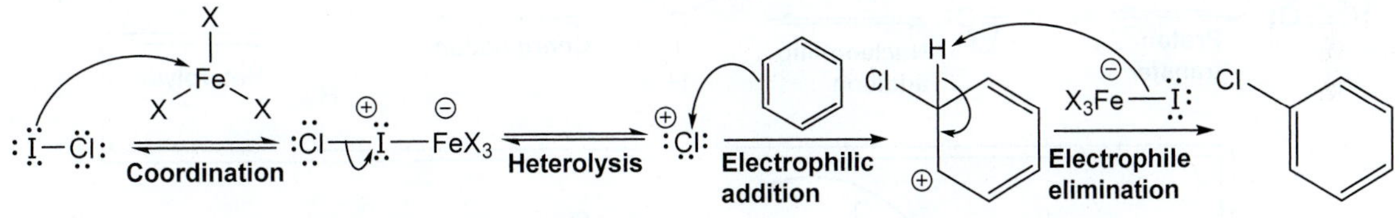

Coordination

Heterolysis

Electrophilic addition

Electrophile elimination

Problem 22.30

Think

What intermediate forms when an alkyl bromide reacts with $AlCl_3$? Is the carbocation chiral? What is the electrophile that must substitute for H^+? How is this mechanism similar to other electrophilic aromatic substitution mechanisms?

Solve

This is a Friedel–Crafts alkylation. There is a stereocenter in the product, but because the halide goes through an achiral carbocation intermediate, the product is formed in a racemic mixture and will not be optically active.

Coordination → Heterolysis → Achiral → Electrophilic addition → Electrophile elimination

Problem 22.31

Think

In the production of the formyl chloride intermediate, is the carbon of C≡O electron rich or poor? Which atom in CO is protonated by HCl? Which atom in CO is attacked by the Cl^-? In the electrophilic aromatic substitution part of the mechanism, what electrophile must substitute for H^+? How is this mechanism similar to other electrophilic aromatic substitution mechanisms?

Solve

The carbon in C≡O is electron rich (with three bonds and a lone pair, it is negatively charged) and is protonated by H^+. The carbon in $HC{\equiv}O^+$ then undergoes nucleophilic addition by the Cl^- to yield formyl chloride. Friedel–Crafts acylation then follows.

Proton transfer → Nucleophilic addition → Coordination → Heterolysis → Electrophilic addition → Electrophile elimination

Problem 22.32

Think

For the given electrophilic aromatic substitution to occur, what should be bonded to the benzene carbon? Which carbon atoms are available to participate in that substitution? Is there symmetry?

Solve

For this electrophilic aromatic substitution to occur, a benzene carbon must possess a H atom. There are aromatic H atoms on only the four outside C atoms. All other aromatic C atoms have no H atoms. The molecule is symmetrical, so all four of those C atoms are equivalent and will lead to the same product. The mechanism is shown on the next page.

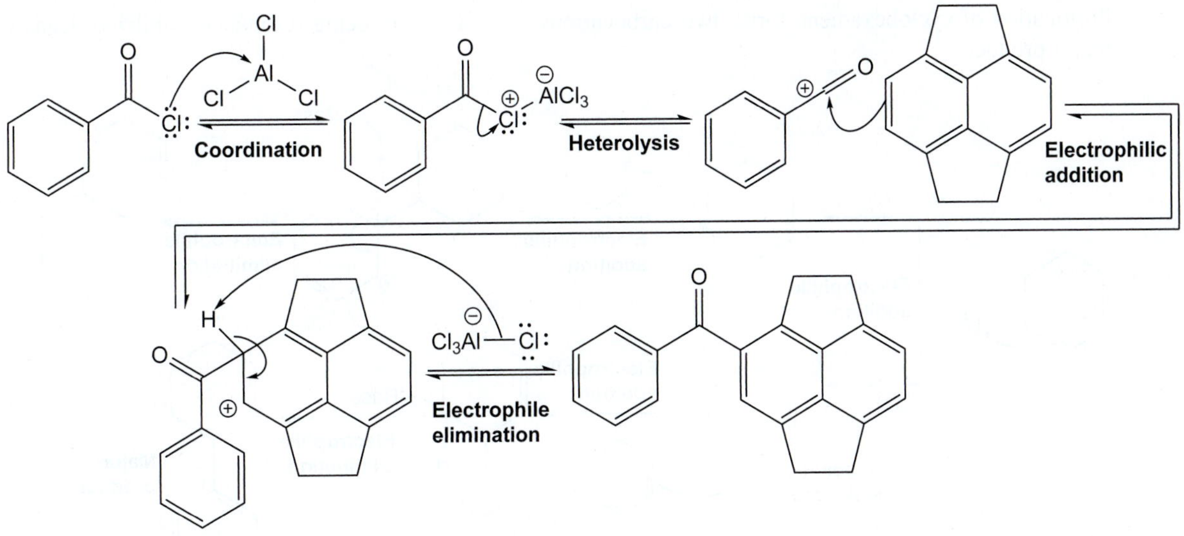

Problem 22.33

Think

What is the electrophile that must substitute for H^+ on the benzene ring? How can that electrophile be produced from the alkene in acid? How many distinct carbocation intermediates are possible? Which one is more stable?

Solve

When a proton adds to 1-methylcyclohexene, two carbocation intermediates are possible, depending on which alkene C gains the proton. The more stable 3° carbocation, however, is formed preferentially. This carbocation is a strong electrophile and alkylates benzene in the same manner as in Problem 22.6.

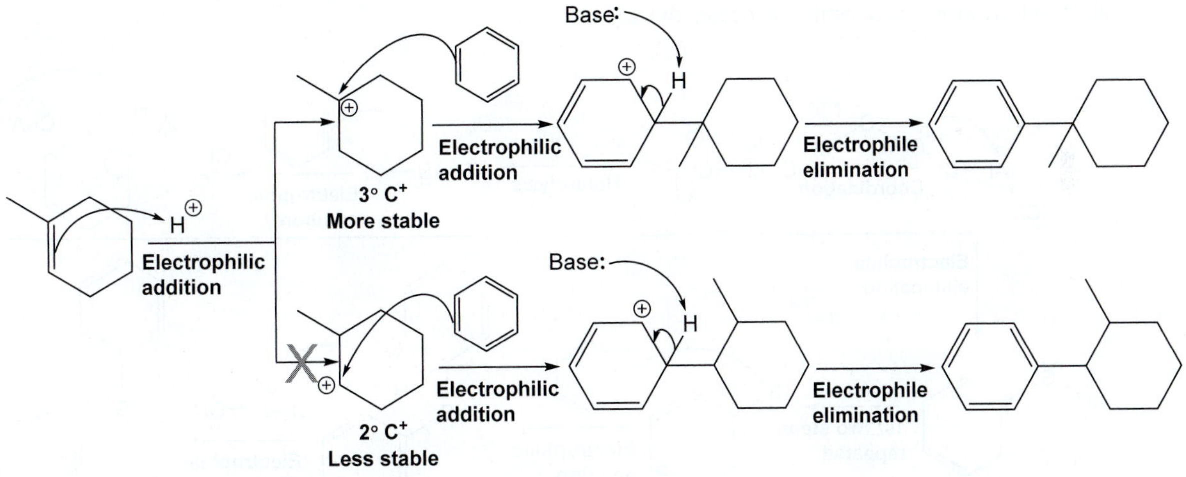

Problem 22.34

Think

How can an electrophile be produced from the diene in acid? How many carbocation intermediates are possible? Which one is more stable? How does that electrophile substitute for H^+?

Solve

Protonation of cyclohexadiene forms two carbocations, one of which, being resonance stabilized, leads to the major product.

Electrophilic addition

Electrophilic addition

Electrophile elimination

:Base

Electrophilic addition

:Base

Electrophile elimination

Resonance stabilized

Major product

Problem 22.35

Think

What electrophile must substitute for H^+? How can that electrophile be produced from the SCl_2 and $AlCl_3$? How is this mechanism similar to other electrophilic aromatic substitution mechanisms? What acts as the base to regenerate the aromatic ring? How many times does the electrophilic aromatic substitution reaction occur?

Solve

The electrophile S^+ is generated in precisely the same way as the R^+ is in Friedel–Crafts alkylation. The electrophilic aromatic substitution occurs twice.

Coordination

Heterolysis

$AlCl_4^-$

Electrophilic addition

Electrophile elimination

1st two steps repeated

Electrophilic addition

Electrophile elimination

Sections 22.6 and 22.7 Nitration and Sulfonation

Problem 22.36

Think

What are the possible isomers of tetramethylbenzene? Which ones have only one distinct H atom?

Solve

Examine the three isomers of tetramethylbenzene. All three isomers, **A**, **B**, and **C**, have just one distinct H, giving rise to only one nitration product for each, so it can be any one of these.

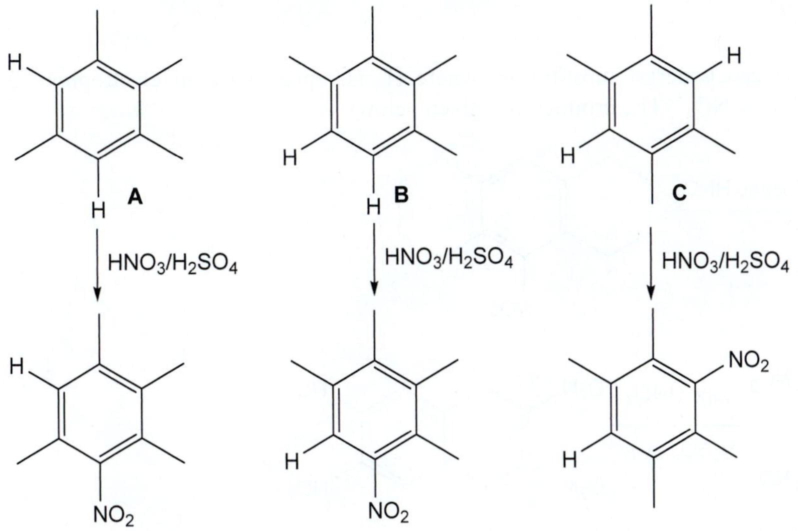

A B C

HNO_3/H_2SO_4 HNO_3/H_2SO_4 HNO_3/H_2SO_4

NO_2 NO_2 NO_2

Problem 22.37

Think

What are the structures of the three isomers of dimethylbenzene? Which one has only one distinct H atom? Which one has two distinct H atoms? Which one has three distinct H atoms?

Solve

Examine the three isomers of dimethylbenzene. If **A** produces only one nitration product, it must have only one distinct H atom and therefore must be the para isomer. If **B** produces only two products, it must only have two distinct H atoms and must be the ortho isomer. If **C** produces three products, it must have three distinct H atoms and must be the meta isomer.

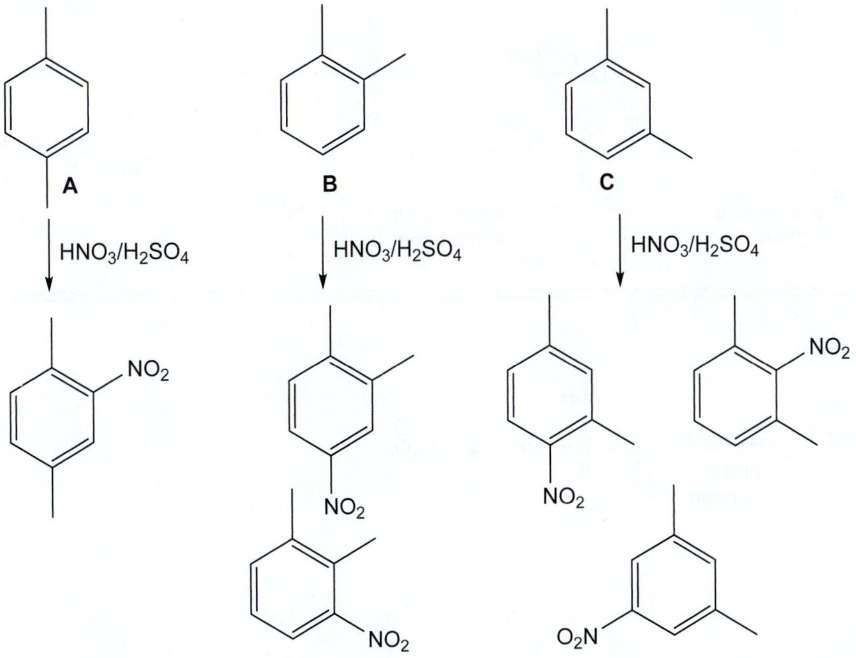

A B C

HNO_3/H_2SO_4 HNO_3/H_2SO_4 HNO_3/H_2SO_4

NO_2 NO_2 NO_2 NO_2 NO_2 O_2N

Problem 22.38

Think

What site on the aromatic ring is subject to electrophilic aromatic substitution? What is the electrophile in each reaction?

Solve

In each case, an aromatic C–H can undergo substitution, where H^+ is replaced by an electrophile, E^+. In aromatic nitration, the electrophile is NO_2^+. The products are given below:

(a)

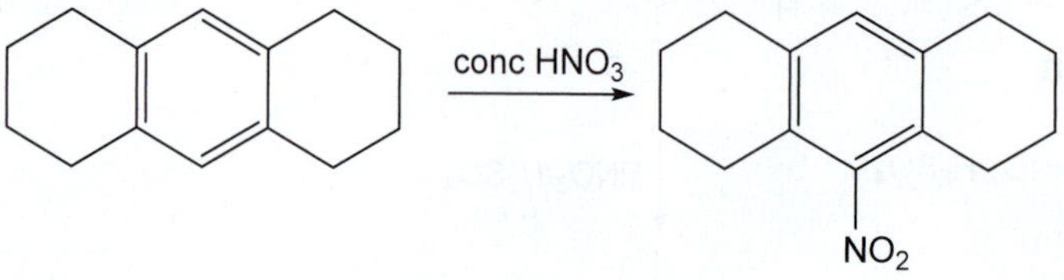

(b)

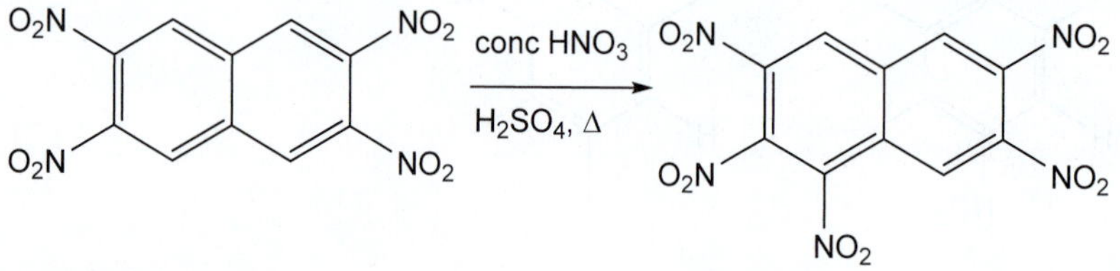

Problem 22.39

Think

What reaction occurs when an arenesulfonic acid, Ar–SO_3H, is treated with aqueous acid? Consult Equation 22-26 and Your Turn 22.10. Desulfonation is the reverse of sulfonation, which is shown in Equation 22-26. How can you arrange the reactants, intermediates, and products in reverse order? Which bonds are broken and formed in each step? Which sites are electron rich and which are electron poor? How can you use curved arrows to show the flow of electrons in each step?

Solve

Desulfonation is the reverse of Steps 1–4 in Equation 22-26.

(a)

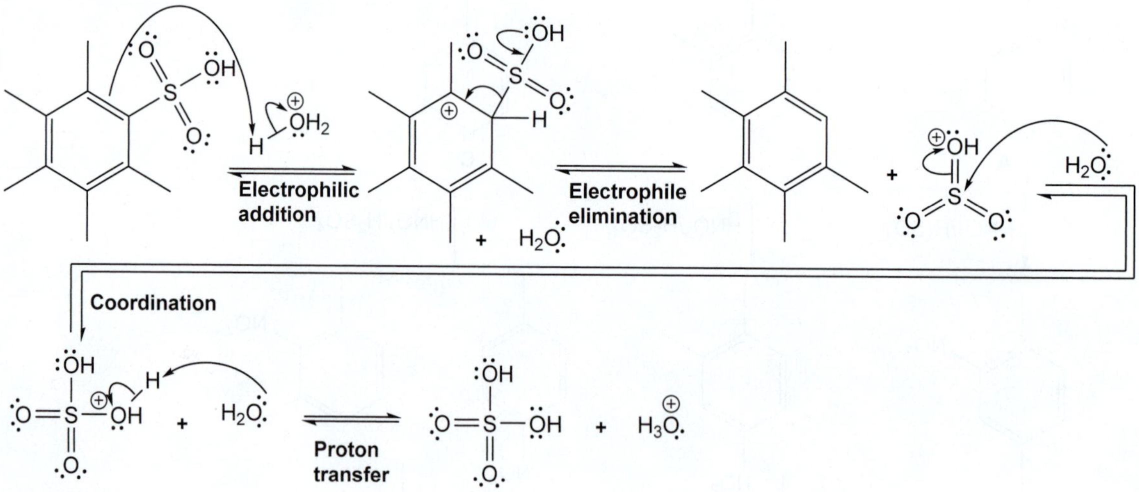

(b) Electrophilic addition; Electrophile elimination; Coordination; Proton transfer

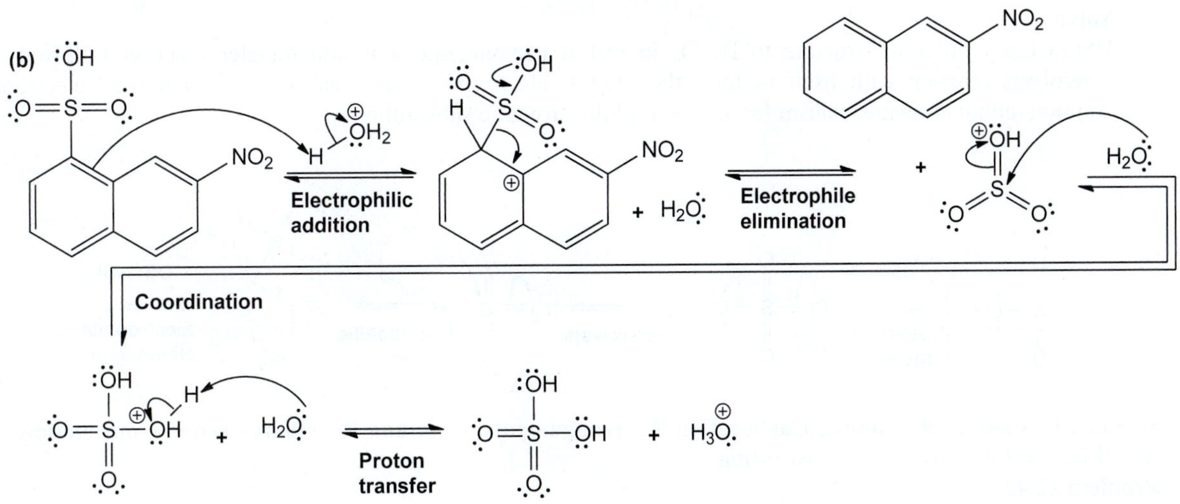

Problem 22.40 (SYN)

Think

Should electrophilic aromatic substitution occur? If so, what should be the electrophile? What reagents are necessary to produce that electrophile? Do any other reactions occur?

Solve

(a) This is a sulfonation using electrophilic aromatic substitution. The electrophile is SO_3H^+, which replaces H^+ on the ring.

SO_3 / H_2SO_4 → SO_3H

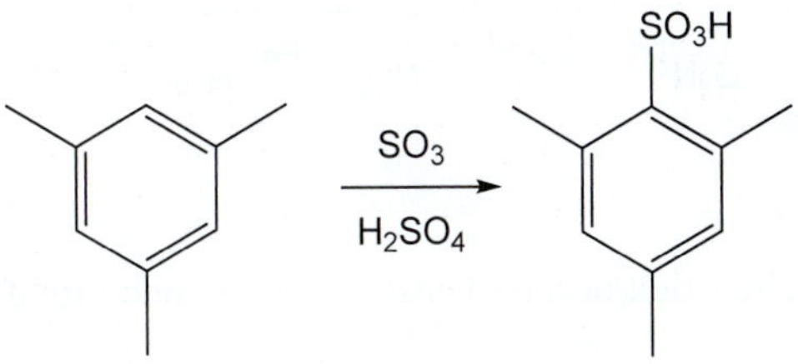

(b) This is a nitration using electrophilic aromatic substitution. The electrophile is NO_2^+, which replaces H^+ on the ring.

HNO_3 / H_2SO_4, Δ → NO_2

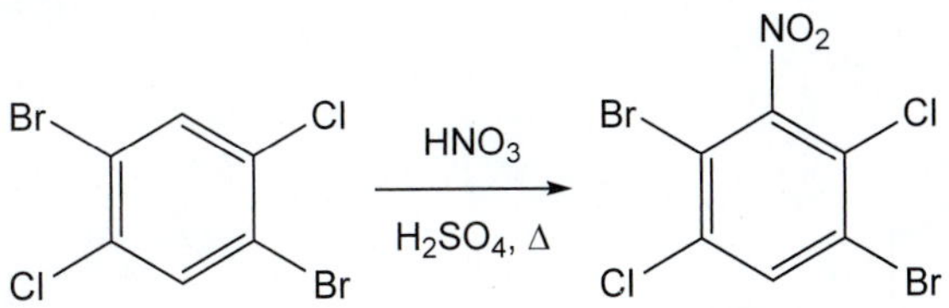

Problem 22.41

Think

How is HSO_3Cl similar in structure to H_2SO_4? How is this mechanism similar to the sulfonation electrophilic aromatic substitution mechanism in Equation 22-26?

Solve

HSO_3Cl is similar in structure to H_2SO_4 in that it can undergo a proton transfer reaction followed by a heterolysis reaction with itself to form the SO_2Cl^+ electrophile. The mechanism is a normal electrophilic addition–elimination mechanism for an electrophilic aromatic substitution.

Proton transfer

Heterolysis

Electrophilic addition

Electrophile elimination

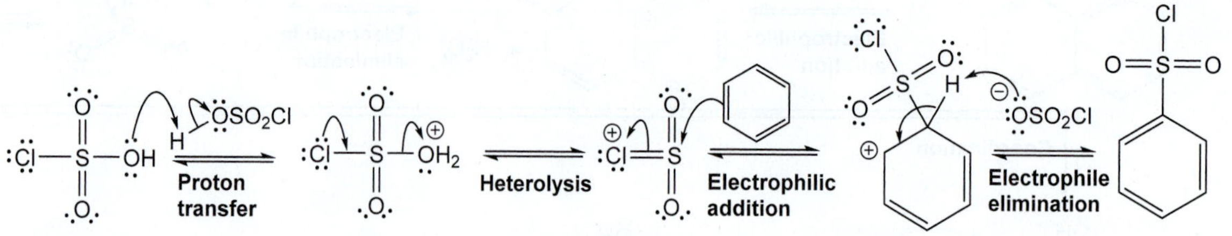

Sections 22.8 and 22.9 Avoiding Carbocation Rearrangements; Common Reactions Used in Conjunction with Electrophilic Aromatic Substitution

Problem 22.42

Think

How does nitrobenzene react with HCl/Fe? Does oxidation or reduction occur? Does the reaction occur more than once?

Solve

Both reactions are a reduction of the Ph–NO_2 group to Ph–NH_2 using HCl/Fe. The reduction occurs twice in **(b)**.

(a) NO_2 → HCl, Fe → NH_2

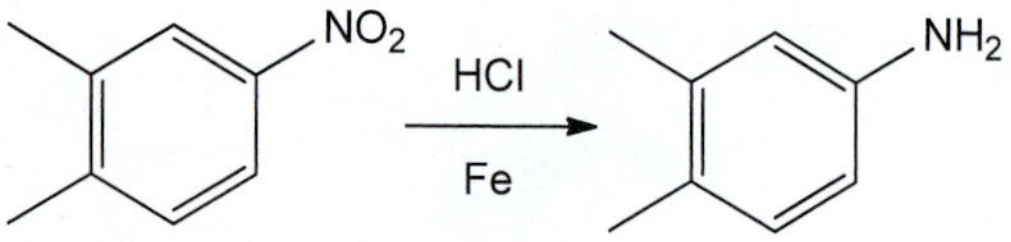

(b) O_2N, NO_2 → HCl, Fe → H_2N, NH_2

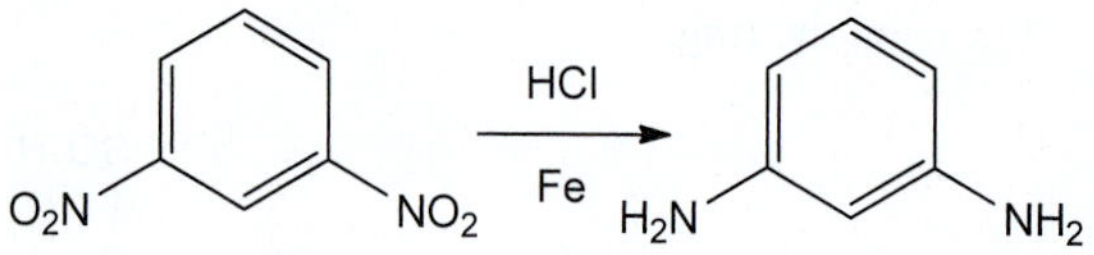

Problem 22.43

Think

Does oxidation or reduction occur? At what site does the reaction occur—benzene ring or side group? At how many sites on the ring can the reaction occur?

Solve

Both reactions are oxidation reactions on the side group to a final Ph–CO_2H product. The reaction in **(a)** occurs at the R group (twice) and the reaction in **(b)** at the C=O group.

(a) 1. $KMnO_4$, KOH, Δ 2. HCl, H_2O → HO, O, OH, O

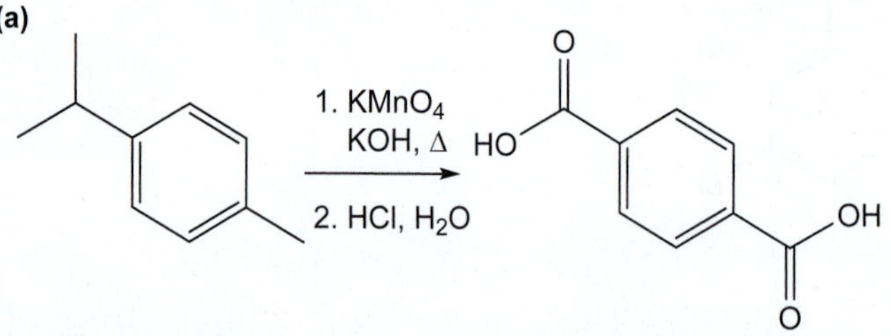

(b) O → 1. $KMnO_4$, KOH, Δ 2. HCl, H_2O → O, OH

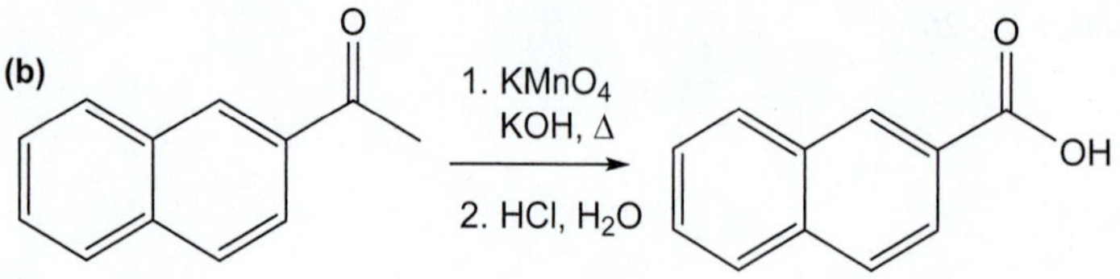

Problem 22.44

Think

If the initial reactant is recovered after the course of the reaction, then did a reaction occur? What type of R group attached to a benzene ring is unreactive in these oxidation reaction conditions?

Solve

If the initial reactant is recovered after the course of the reaction, no reaction occurred. A quaternary R group attached to a benzene ring is unreactive in these oxidation reaction conditions. Potential disubstituted benzene compounds with the formula $C_{14}H_{22}$ are given below.

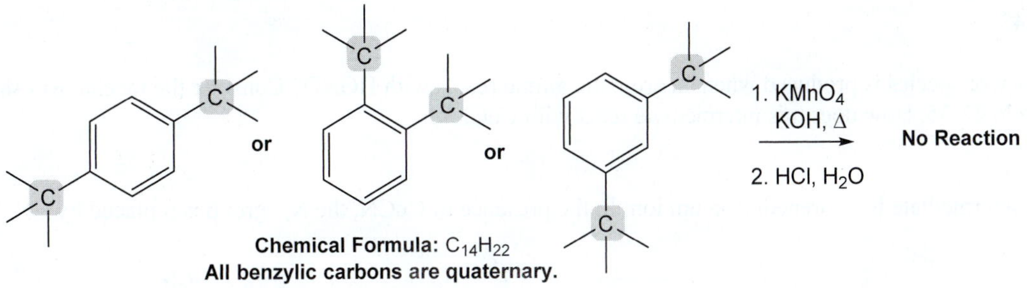

or or

1. $KMnO_4$ KOH, Δ

2. HCl, H_2O

No Reaction

Chemical Formula: $C_{14}H_{22}$

All benzylic carbons are quaternary.

Problem 22.45 (SYN)

Think

Should electrophilic aromatic substitution occur? If so, what should the electrophile be? What reagents are necessary to produce that electrophile? Should any other reactions occur?

Solve

The primary alkyl group can substitute for an aromatic H via a Friedel–Crafts acylation followed by a reduction. A Friedel–Crafts alkylation using 1-chlorobutane would not work, because a rearrangement would occur and the product would have an attached *sec*-butyl group instead.

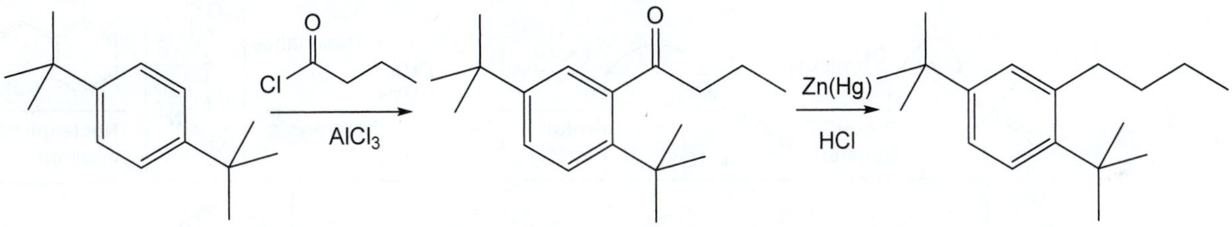

Cl, O, $AlCl_3$, O, Zn(Hg), HCl

Problem 22.46 (SYN)

Think

Should electrophilic aromatic substitution occur? If so, what should the electrophile be? What reagents are necessary to produce that electrophile? Should any other reactions occur?

Solve

(a) The aniline can form from an electrophilic aromatic substitution nitration followed by a reduction.

Br, Cl, Cl, Br; HNO_3, H_2SO_4, Δ; NO_2, Br, Cl, Cl, Br; HCl, Fe; NH_2, Br, Cl, Cl, Br

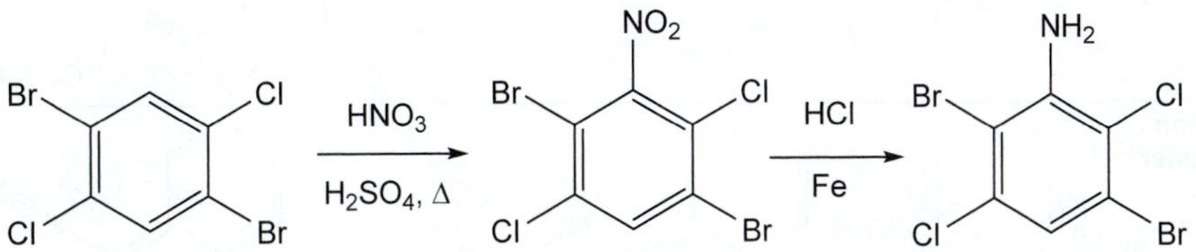

(b) The carboxylic acid can form from a Friedel–Crafts alkylation followed by an oxidation. Alternatively, a Friedel–Crafts acylation could be carried out to produce an aromatic ketone, followed by oxidation of the ketone to produce the carboxylic acid.

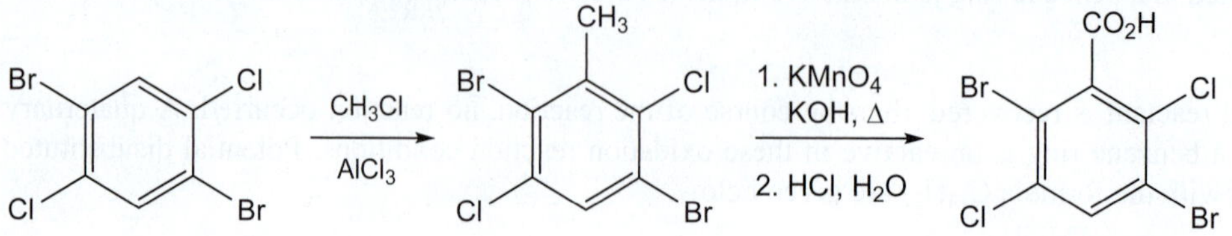

CH_3Cl, $AlCl_3$; 1. $KMnO_4$, KOH, Δ; 2. HCl, H_2O

Problem 22.47

Think

What reactive species is produced when an aromatic amine reacts with HONO? Consider the mechanism shown in Equation 22-35. How does this intermediate react with CuCN?

Solve

(a) The intermediate is an arenediazonium ion. In the presence of CuCN, the N_2^+ group is replaced by CN.

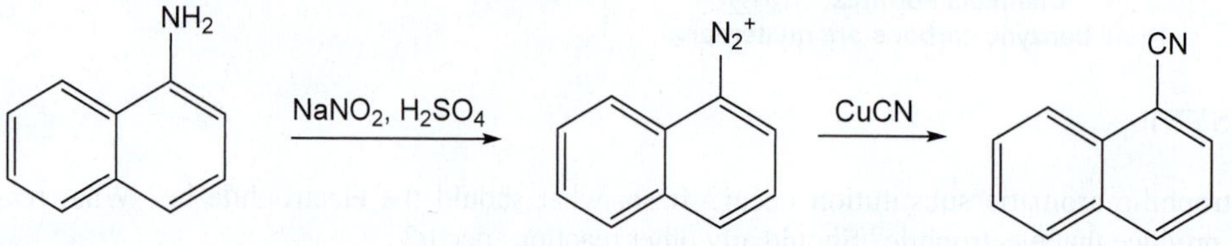

$NaNO_2$, H_2SO_4; CuCN

(b) The mechanism is shown below. First, the *N*-nitrosamine is generated (elementary steps 1–5). Then, three proton transfers (elementary steps 6–8) generate a protonated hydroxyl group that serves as a good leaving group, affording stable molecule of water after heterolysis (step 9) to yield the arenediazonium ion.

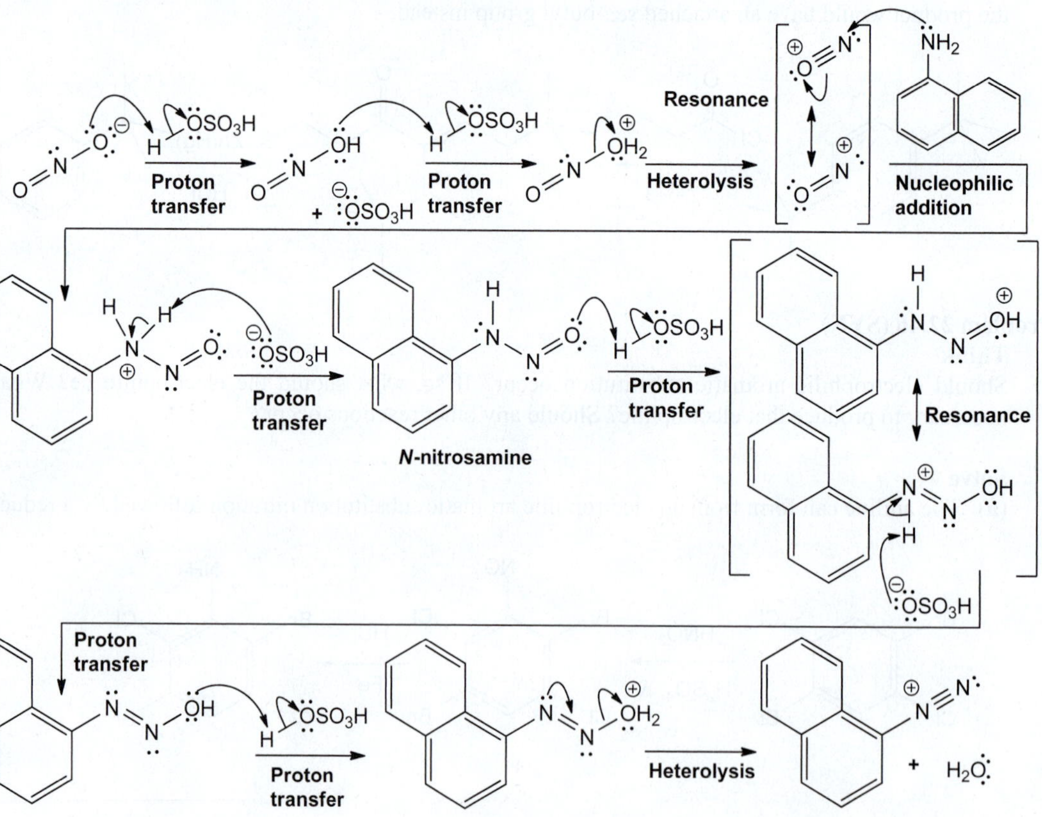

Proton transfer; Proton transfer; Heterolysis; Resonance; Nucleophilic addition; Proton transfer; Proton transfer; Resonance; *N*-nitrosamine; Proton transfer; Proton transfer; Heterolysis

Integrated Problems

Problem 22.48

Think

Is the product aromatic? If Br_2 were to add to just benzene, would the product be aromatic? How does this affect whether the reaction will go forward?

Solve

Addition to the middle ring allows the two outside rings to remain (independently) aromatic. A similar reaction with just benzene isn't feasible, because addition of Br_2 destroys the aromaticity.

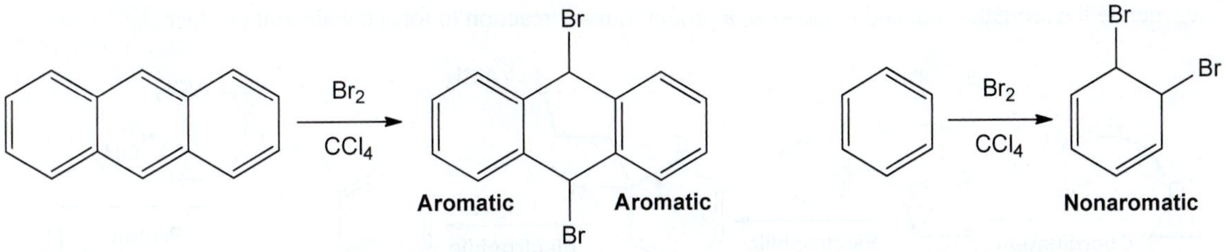

Problem 22.49

Think

What reaction occurs between a halogen and $AlCl_3$? What acts as the electrophile in the electrophilic addition reaction? How can the aromatic ring be regenerated?

Solve

The $AlCl_3$ catalyst undergoes coordination to convert the I substituent into a better cationic leaving group. Then electrophilic aromatic substitution takes place in two steps. Benzene first captures H^+, and then I^- displaces the benzene ring from the I^+. This is analogous to the second step of a typical electrophilic aromatic substitution mechanism, in which a base removes a H^+ from benzene.

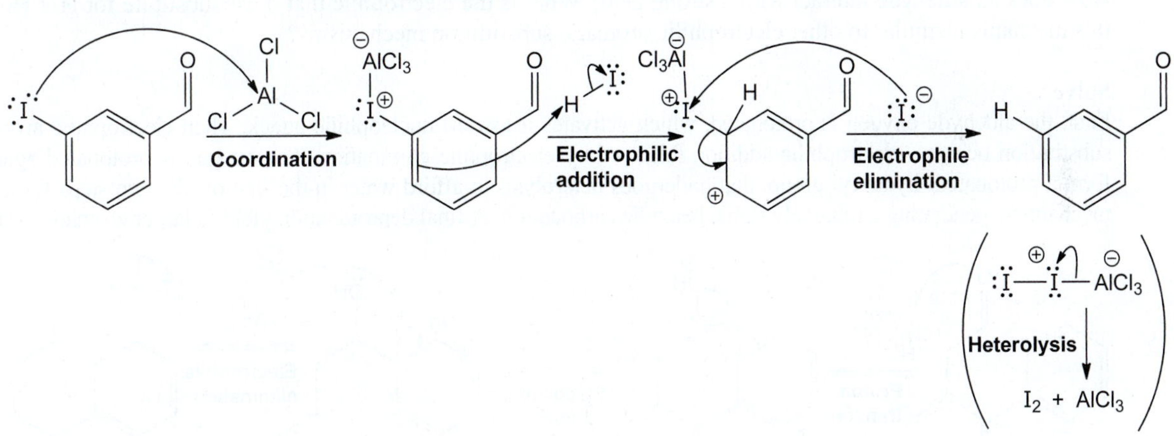

Problem 22.50

Think

In Problem 22.49, what charges appear on the intermediate species? What kind of charge appears on the halogen atom attached to the ring? Which halogen best stabilizes that charge?

Solve

We can see in the mechanism in Problem 22.49 that a positive charge appears on the halogen atom attached to the ring in the intermediate species. Stability decreases in the order $I^+ > Br^+ > Cl^+ > F^+$, because the atom is getting smaller and more electronegative. This increases the reaction's energy barrier and decreases its rate.

Problem 22.51

Think

What is the electrophile that must substitute for H^+? How does the epoxide interact with $AlCl_3$? How can a strong electrophile be produced from the Lewis acid–base complex? How is this mechanism similar to other electrophilic aromatic substitution mechanisms? What acts as the base to regenerate the aromatic ring? Are proton transfer reactions involved?

Solve

The $AlCl_3$ makes the epoxide very electrophilic, so the ring can be opened even by weak nucleophiles like benzene. Note that $AlCl_3$ is regenerated in the final step, so it is a catalyst. The purpose of the water is to serve as the base to regenerate the aromatic ring and to facilitate a proton transfer reaction to form the alcohol product.

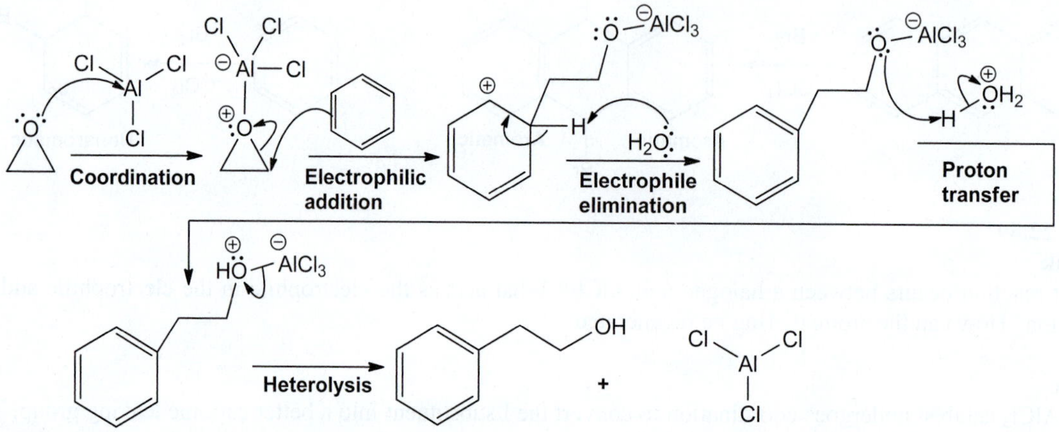

Coordination → Electrophilic addition → Electrophile elimination (H_2O) → Proton transfer → Heterolysis

Problem 22.52

Think

How does an aldehyde interact with a strong acid? What is the electrophile that must substitute for H^+? How is this mechanism similar to other electrophilic aromatic substitution mechanisms?

Solve

First, the aldehyde oxygen is protonated, which activates it toward nucleophilic attack. Then electrophilic aromatic substitution occurs—electrophilic addition followed by electrophile elimination. The oxygen is protonated again to form a protonated hydroxyl group, that undergoes heterolysis to afford water in the first of the two steps for an E1 mechanism, generating a relatively stable benzylic carbocation. A final deprotonation yields a larger aromatic system.

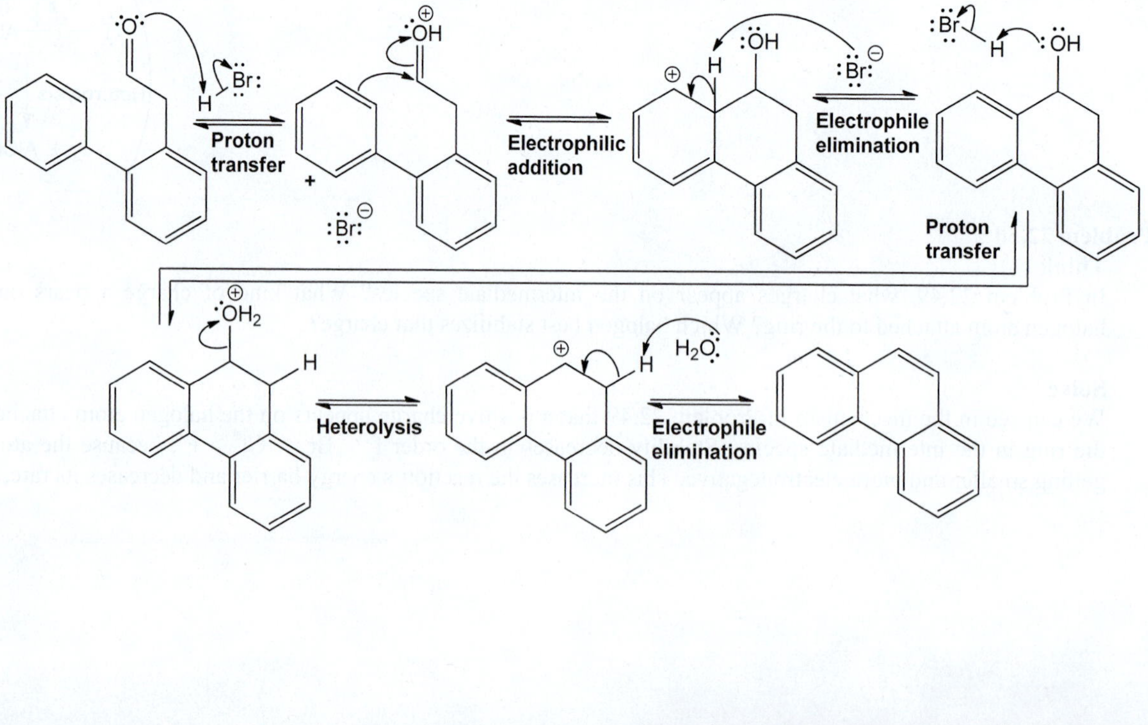

Proton transfer ⇌ Electrophilic addition ⇌ Electrophile elimination ⇌ Proton transfer ⇌ Heterolysis ⇌ Electrophile elimination (H_2O)

Problem 22.53

Think

In each step, consider whether electrophilic aromatic substitution or some other reaction on the side chain (i.e., oxidation, reduction, substitution, elimination, addition) occurs. What acts as the nucleophile and electrophile in each step? Is the carbon skeleton altered? Does functional group transformation occur?

Solve

Compound **A** is formed from an electrophilic aromatic substitution bromination reaction. **A** (bromobenzene) is then transformed into the Grignard reagent **B**, which can open an epoxide ring to form the primary alcohol **C** after acid workup. In the presence of the mild oxidizing agent PCC, the primary alcohol **C** is oxidized to the aldehyde **D**.

Br_2, $FeBr_3$ → **A** (Br); Mg(*s*), Ether → **B** (MgBr); 1. epoxide, 2. H_2O, HCl → **C** (OH); PCC → **D** (O)

Using the primary alcohol **C** as the starting material, **C** reacts with PBr_3 to form the primary alkyl halide. **E** reacts with PPh_3 and Bu–Li to form the Wittig reagent (phosphonium ylide) **F**. The Wittig reagent reacts with the aldehyde **D** to form a new C=C, shown in product **G**.

C (OH) → PBr_3 → **E** (Br); 1. PPh_3, 2. Bu–Li → **F** (PPh_3, ⊕, ⊖); **D** → **G**

Problem 22.54

Think

In each step, consider whether electrophilic aromatic substitution or some other reaction on the side chain (i.e., oxidation, reduction, substitution, elimination, addition) occurs. What acts as the nucleophile and electrophile in each step? Is the carbon skeleton altered? Does functional group transformation occur?

Solve

Compound **A** is formed from a Friedel–Crafts acylation reaction. The next reaction is a haloform reaction. The primary methyl ketone is chlorinated three times at the α carbon to form CCl_3. This intermediate can undergo acyl substitution with HO^- followed by an acid workup to form the carboxylic acid **B**, which is then transformed by $SOCl_2$ into the acyl chloride **C**.

O, Cl, $AlCl_3$ → **A** (O); 1. Cl_2 (excess), NaOH, 2. H_2O, HCl → **B** (O, OH); $SOCl_2$ → **C** (O, Cl)

Nitrobenzene **D** is produced from the electrophilic aromatic substitution reaction of benzene. **D** is reduced to form aniline **E**. The aniline reacts with the acid chloride **C** to form the amide **F**. The amide is reduced to the secondary amine.

$$\text{Benzene} \xrightarrow[H_2SO_4]{HNO_3} \mathbf{D} \xrightarrow[HCl]{Fe} \mathbf{E} \xrightarrow{\mathbf{C}} \mathbf{F} \xrightarrow[2.\ H_2O,\ NH_4Cl]{1.\ LiAlH_4} \mathbf{G}$$

Problem 22.55

Think

In each step, consider whether electrophilic aromatic substitution or some other reaction on the side chain (i.e., oxidation, reduction, substitution, elimination, addition) occurs. What acts as the nucleophile and electrophile in each step? Is the carbon skeleton altered? Does functional group transformation occur?

Solve

Nitrobenzene **A** is produced from the electrophilic aromatic substitution reaction of benzene. **A** is reduced to form aniline **B**. The aniline reacts with the $NaNO_2$/HCl to form **C**, the benzenediazonium ion intermediate, $C_6H_5–N_2^+$. The benzenediazonium ion undergoes a Sandmeyer reaction with CuCN to form the benzonitrile **D**.

$$\text{Benzene} \xrightarrow[H_2SO_4]{HNO_3} \mathbf{A} \xrightarrow[HCl]{Fe} \mathbf{B} \xrightarrow[HCl]{NaNO_2} \mathbf{C} \xrightarrow{CuCN} \mathbf{D}$$

Compound **E** is formed from an electrophilic aromatic bromination reaction. **E** (bromobenzene) is then transformed into the Grignard reagent **F**, which reacts with benzonitrile **D** to form an imine that is hydrolyzed to the ketone **G**. **G** is reduced by $NaBH_4$/EtOH to form the alcohol **H**. The alcohol O–H is deprotonated to form the alkoxide anion, which acts as a nucleophile in an S_N2 reaction (a Williamson ether synthesis) to form the ether **I**.

$$\text{Benzene} \xrightarrow[FeBr_3]{Br_2} \mathbf{E} \xrightarrow[Ether]{Mg(s)} \mathbf{F} \xrightarrow[2.\ H_2O\ (excess),\ H_2SO_4]{1.\ \mathbf{D}} \mathbf{G} \xrightarrow[EtOH]{NaBH_4} \mathbf{H} \xrightarrow[2.\ C_6H_5CH_2Br]{1.\ NaH} \mathbf{I}$$

Problem 22.56

Think

What intermediate forms when the alkene reacts with a strong acid? What acts as the electrophile in the electrophilic addition reaction? What acts as the base to regenerate the aromatic ring?

Solve

This is a Friedel–Crafts alkylation. Protonation occurs first on the nitrogen, and then on the double bond to generate a benzylic carbocation that benzene attacks. Note that under these acidic conditions, the amine N is protonated, but that protonated amine is unreactive.

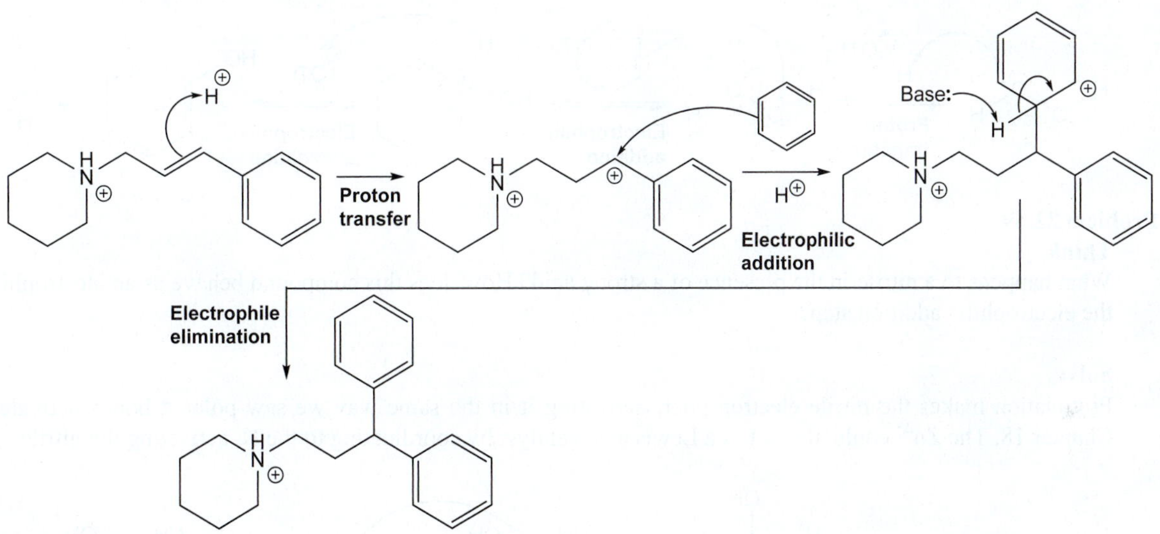

Proton transfer

Electrophilic addition

Electrophile elimination

Problem 22.57

Think

What functional group is present? How does this functional group react with an acid? What acts as the electrophile in the electrophilic addition step? What acts as the base to regenerate the aromatic ring? How does ring strain help drive the reaction?

Solve

The first step is a proton transfer to N to generate a pronated N atom that undergoes heterolysis to open the ring. This yields an acylium cation that undergoes an electrophilic aromatic substitution mechanism.

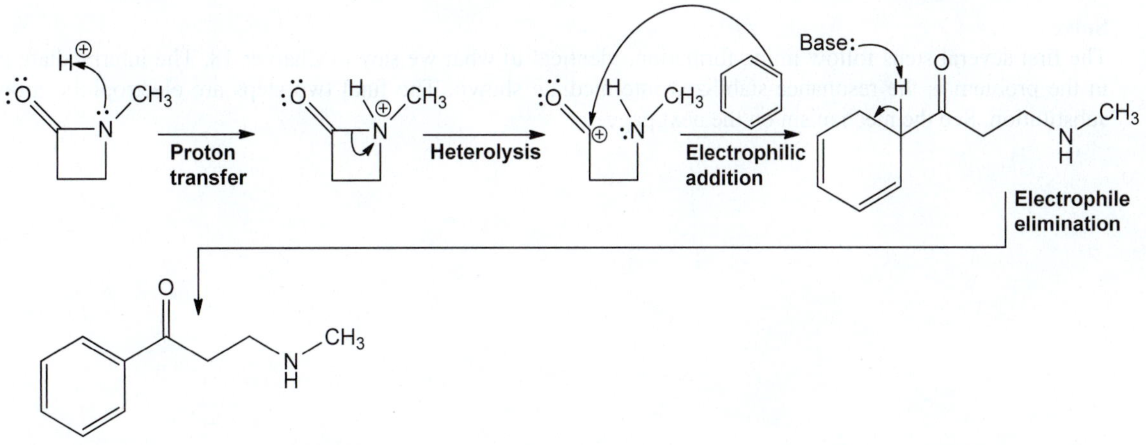

Proton transfer

Heterolysis

Electrophilic addition

Electrophile elimination

Problem 22.58

Think

What is the product of the proton transfer reaction between H_2O_2 and TfOH? What electrophile must substitute for H^+? How is this mechanism similar to other electrophilic aromatic substitution mechanisms? What acts as the base to regenerate the aromatic ring?

Solve

H_2O_2 and TfOH react in a proton transfer reaction to form $HOOH_2^+$, which acts as the electrophile to add HO to the benzene ring. TfO^- acts as the base to regenerate the aromatic ring.

Proton transfer

Electrophilic addition

Electrophile elimination

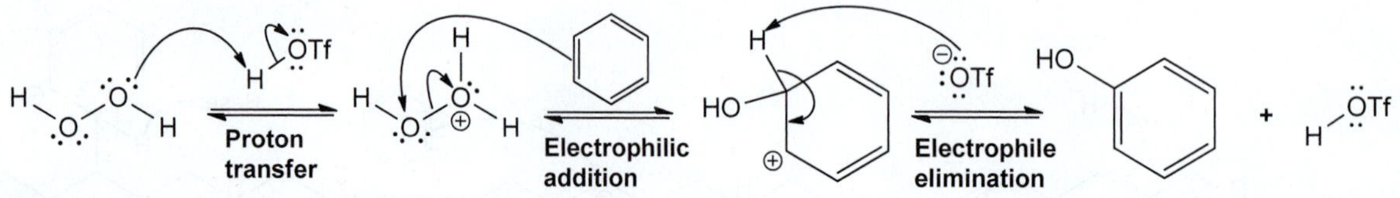

Problem 22.59

Think

What happens to a nitrile in the presence of a strong acid? How does this compound behave as an electrophile in the electrophilic addition step?

Solve

Protonation makes the nitrile electron poor, activating it in the same way we saw polar π bonds activated in Chapter 18. The Zn^{2+} could also act as a Lewis acid catalyst by coordinating to the N, activating the nitrile.

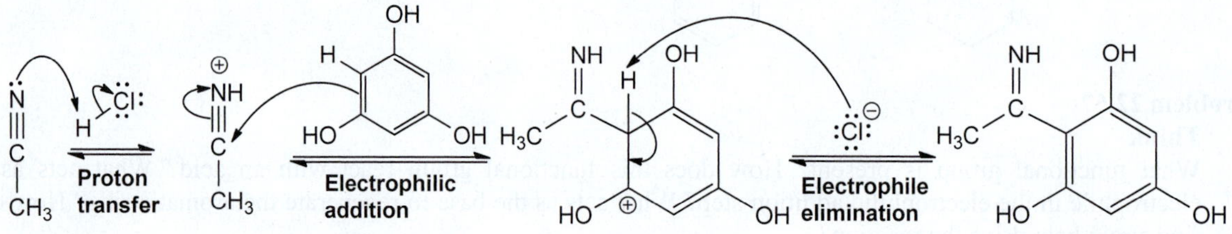

Proton transfer

Electrophilic addition

Electrophile elimination

Problem 22.60

Think

How does an amine react with an aldehyde in the presence of an acid? How is the intermediate shown formed? What is the electrophile in the electrophilic addition step?

Solve

The first several steps follow imine formation, identical to what we saw in Chapter 18. The intermediate given in the problem is the resonance-stabilized intermediate shown. The final two steps are electrophilic aromatic substitution. See the mechanism on the next page.

Nucleophilic addition

Three proton transfers

Heterolysis

Resonance

Electrophilic addition

Electrophile elimination

:Base

Problem 22.61 (SYN)

Think

Recall from Chapter 20 how to form a tertiary alcohol from a carboxylic acid derivative. How can you transform an alcohol into the necessary carboxylic acid derivative? How can you transform benzene into the necessary nucleophile?

Solve

An acid chloride reacts with two equivalents of a Grignard reagent (followed by a proton transfer) to form a tertiary alcohol, where two of the R groups are the same. The precursor to the acid chloride is the carboxylic acid, which could have come from the oxidation of a primary alcohol. The precursor to the Grignard reagent PhMgBr is bromobenzene, which can be formed from benzene.

Undo a Grignard reaction.

Undo acyl chloride formation.

Undo an oxidation.

Undo a Grignard reagent formation.

Undo a bromination.

In the forward direction, the synthesis would proceed as follows (but other synthesis routes are possible):

1. $KMnO_4$, KOH, Δ
2. HCl, H_2O

$SOCl_2$

Then H_3O^+

Br_2, $FeBr_3$

Mg(*s*), Ether

2 equiv.

Problem 22.62 (SYN)

Think

How many carbon atoms in the target compound do not originate from benzene? How many carbon atoms, therefore, would come from the C=O compound? What precursors will make an ether? How can you form those precursors from benzene and that C=O compound?

Solve

Dibenzyl ether can be synthesized from benzyl alcohol and benzyl bromide in an S_N2 reaction (Williamson ether synthesis). The benzyl bromide's precursor is the benzyl alcohol that formed from the Grignard reaction of PhMgBr and formaldehyde.

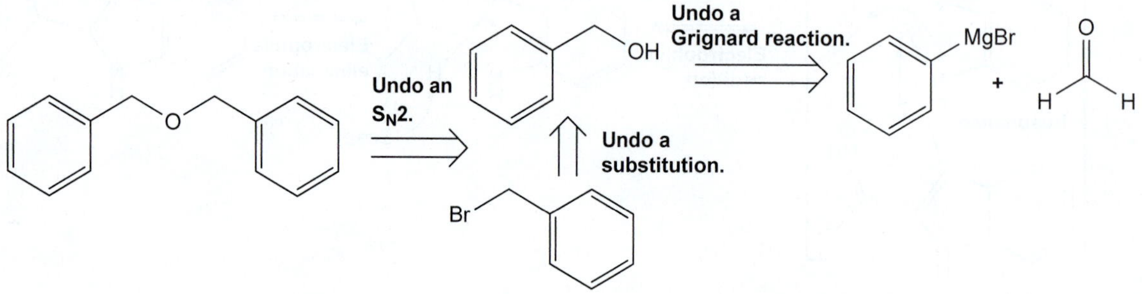

The synthesis in the forward direction would proceed as follows (but other syntheses could be feasible):

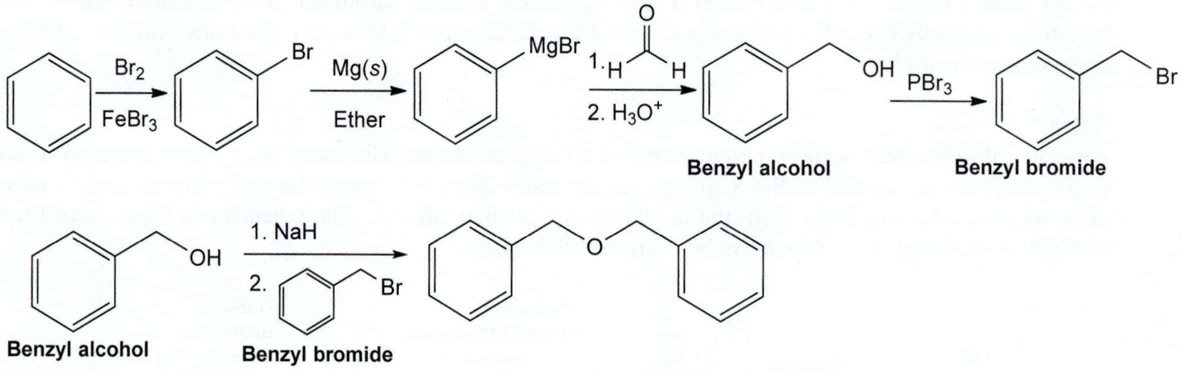

Problem 22.63

Think

How many carbon atoms in the target compound would not originate from benzene? How many carbon atoms, therefore, must come from the carboxylic acid compound? Can the target be produced directly from a simple Friedel–Crafts alkylation? Can another reaction be used to form the appropriate C=C bond?

Solve

The alkene can be formed from the dehydration of an alcohol. This alcohol could have been reduced from a ketone that could come from a Friedel–Crafts acylation. The acid chloride can be synthesized from 2-phenylacetic acid.

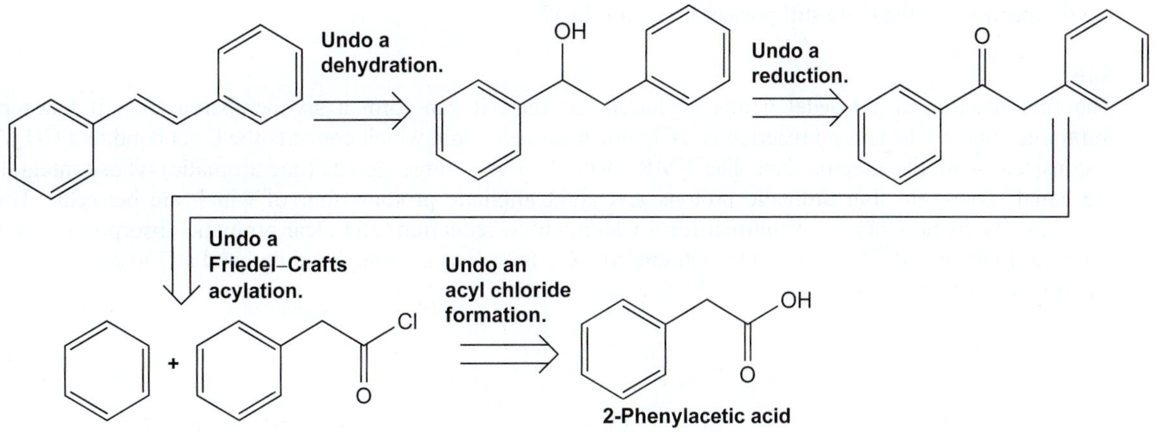

Undo a dehydration.

Undo a reduction.

Undo a Friedel–Crafts acylation.

Undo an acyl chloride formation.

2-Phenylacetic acid

The synthesis in the forward direction would proceed as follows:

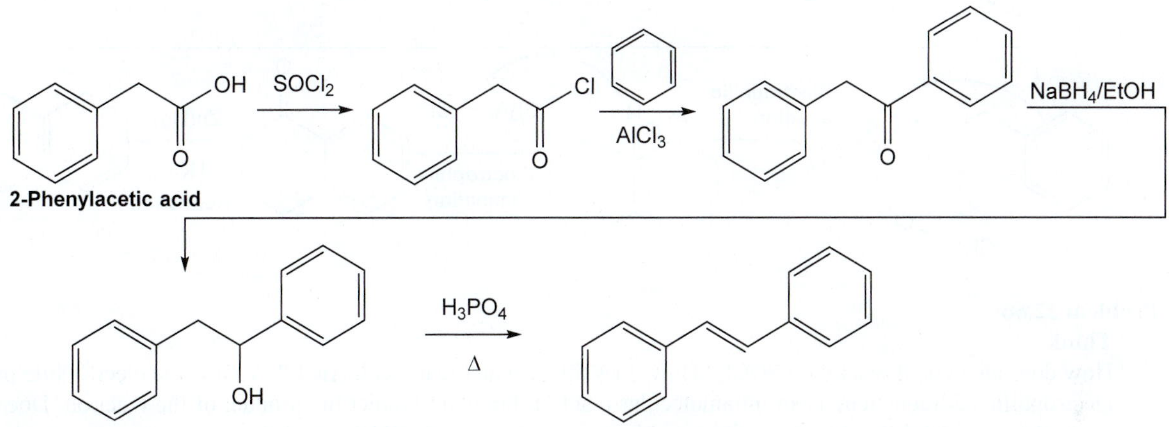

$SOCl_2$

$AlCl_3$

$NaBH_4/EtOH$

2-Phenylacetic acid

H_3PO_4

Δ

Problem 22.64

Think

What are the reaction conditions in Problem 22.35? What sites are available on diphenyl ether for electrophilic addition? To give rise to only six carbon signals, is symmetry present? How can electrophilic aromatic substitution take place without giving rise to more than six signals?

Solve

If a substitution occurs on only one of the two benzene rings in diphenyl ether, the product would have at least seven different carbon signals. The only way for the two separate rings to give rise to fewer signals is to have symmetry. This can be accomplished if substitution takes place on both rings at the position ortho to the oxygen. The following product is consistent with the spectrum:

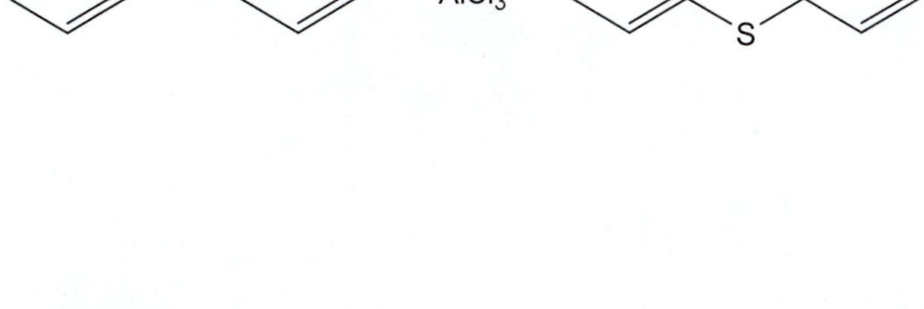

SCl_2

$AlCl_3$

Problem 22.65

Think

What reaction occurs between an aromatic ring and an acid chloride in the presence of $AlCl_3$? Is an intramolecular reaction favored? What type of reagent is HCl, Zn(Hg)? From the ^1H NMR spectrum, how many aliphatic signals are present? What is the relative integration of the three signals? Is symmetry present? From the IR spectrum, is the C=O still present in the product?

Solve

The first reaction is a Friedel–Crafts acylation. Because it can form a six-membered ring, it takes place intramolecularly. The second reaction is a Clemmensen reduction, which converts the C=O bond to a CH_2. This is consistent with the spectral data. The NMR spectrum shows three signals (one aromatic) whose integrations are equal. There are four aromatic protons and eight aliphatic protons, four of which are benzylic. The IR spectrum shows no C=O bond (normal for a Clemmensen reduction) and clear aromatic absorption stretching bands at 3050 cm^{-1} (C–H) and 1600–1450 cm^{-1} (C=C) as well as a strong bending band at 750 cm^{-1}, indicating an ortho-substituted aromatic ring. See the mechanism below.

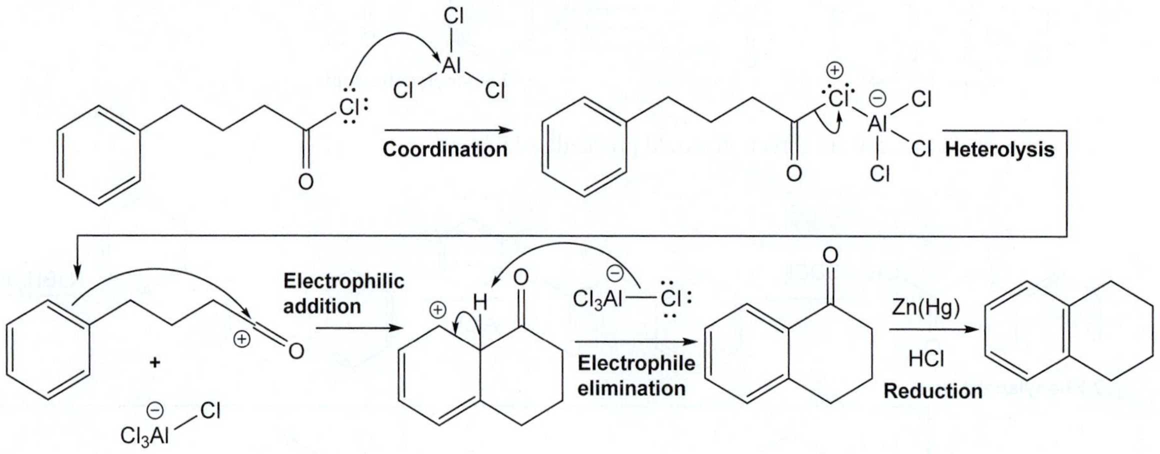

Coordination

Heterolysis

Electrophilic addition

Electrophile elimination

Zn(Hg) HCl Reduction

Problem 22.66

Think

How does an alcohol react with $SOCl_2$? How does this product react with $AlCl_3$? What is the electrophile in the electrophilic addition step? Is an intramolecular reaction favored? Predict the product of the reaction. Does the product have eight distinct C atoms that would give rise to eight ^{13}C NMR signals?

Solve

The first reaction converts the OH group to a Cl. The second reaction is a Friedel–Crafts alkylation. The product has five chemically distinct H atoms. (Note that although a primary carbocation is shown being generated, to be consistent with the mechanism shown in the chapter, it probably does not form, because it would undergo carbocation rearrangement to a more stable secondary carbocation. Rather, the leaving of $AlCl_4^-$ and the electrophilic addition that follows it probably happen in a single, concerted step.) See the mechanism on the next page.

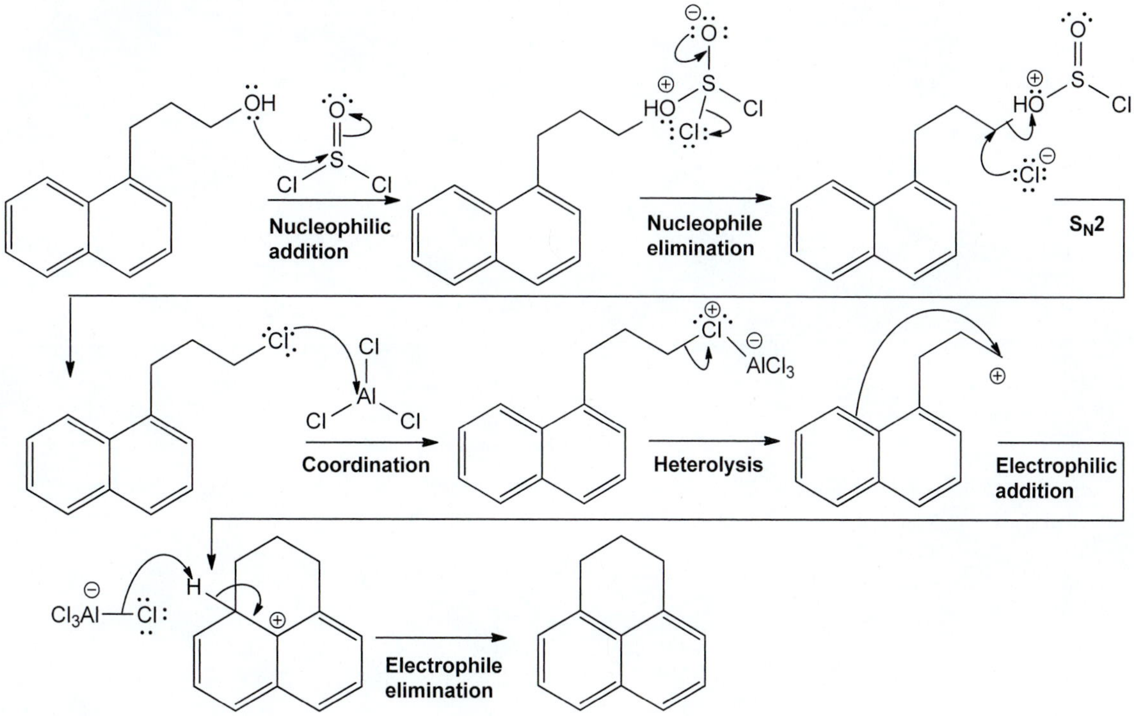

Nucleophilic addition

Nucleophile elimination

S_N2

Coordination

Heterolysis

Electrophilic addition

Electrophile elimination

Problem 22.67

Think

What is the product of the first Friedel–Crafts alkylation? Can that product undergo another Friedel–Crafts alkylation? What are the relative integrations of the aromatic and aliphatic signals? How many signals are in the ^{13}C NMR spectrum? What functional groups are evident from the IR spectrum?

Solve

This comprises two back-to-back Friedel–Crafts alkylation reactions. The integrations of the aromatic and aliphatic peaks give a 10:2 ratio. There are no other functional groups in the compound other than the aromatic ring and *sp*^{3} C–H bonds, and there are five signals in the ^{13}C NMR spectrum (four aromatic and one aliphatic). The product and mechanism are shown below:

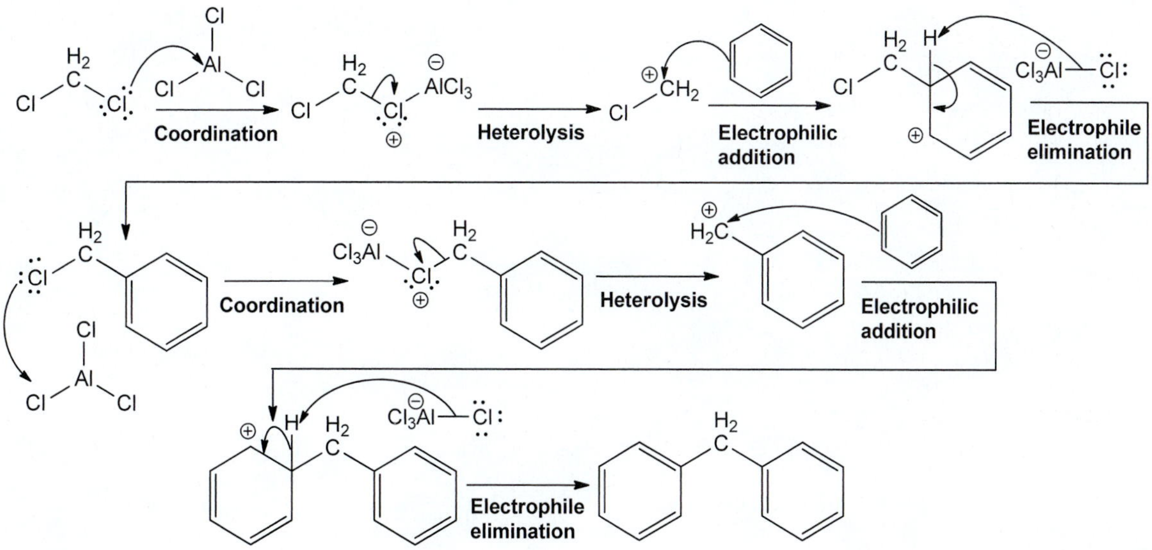

Coordination

Heterolysis

Electrophilic addition

Electrophile elimination

Coordination

Heterolysis

Electrophilic addition

Electrophile elimination

CHAPTER 23 | Aromatic Substitution 2: Reactions of Substituted Benzenes and Other Rings

Your Turn Exercises

Your Turn 23.1

Think

If the reaction mixture contains only ortho, meta, and para products, what percentage of the product mixture is left for the para product? In comparing the relative percentages, which percentage group is the highest: ortho/para or meta?

Solve

The ortho/para percentages are combined, showing that –I is an ortho/para director and –CHO is a meta director.

Sub	*o*	*m*	*p*	*o+p*	Type of Director
–I	45	1	54	99	Ortho/para
–CHO	19	72	9	28	Meta

Your Turn 23.2

Think

Which atom attaches to the benzene ring in each group? Review the rules for Lewis structures and add in lone pairs accordingly. Is there a correlation between the atoms that have a lone pair at the point of attachment and the type of director? Which atoms are electronegative? How does the electronegativity of nearby atoms affect the type of director the substituent is?

Solve

(a) The lone pairs have been added to the atom at the point of attachment for every group in Table 23-1 in the figure below. With the exception of the $-CH_3$ group, all the ortho/para directing groups in the table have a lone pair on the atom that directly attaches to the benzene ring.

(b) The $-CH_3$ group in the ortho/para directing column is attached by an atom that has no lone pairs of electrons. That substituent consists only of C and H atoms, neither of which type is highly electronegative.

(c) Every substituent in the meta-directing column contains an electronegative atom at or near the point of attachment.

Ortho/Para-Directing Substituents					Meta-Directing Substituents				
Substituent	*o*	*m*	*p*	*o+p*	Substituent	*o*	*m*	*p*	*o+p*
$-\ddot{\underset{\cdot\cdot}{O}}H$	**50**	0	**50**	100	$-NO_2$	7	**91**	2	9
$-\ddot{N}HCOCH_3$	**19**	2	**79**	98	$-\overset{\oplus}{N}(CH_3)_3$	2	**87**	11	13
$-CH_3$	**63**	3	**34**	97	$-CO_2H$	22	**76**	2	24
$-\ddot{\underset{\cdot\cdot}{F}}:$	**13**	1	**86**	99	$-CN$	17	**81**	2	19
$-\ddot{\underset{\cdot\cdot}{Cl}}:$	**35**	1	**64**	99	$-CO_2Et$	28	**66**	6	34
$-\ddot{\underset{\cdot\cdot}{Br}}:$	**43**	1	**56**	99	$-COCH_3$	26	**72**	2	28

Your Turn 23.3

Think

Consult Table 23-1 to compare the percentages of meta and para products formed from phenol (C_6H_5–OH). How does the percentage of formation relate to the rate? How is the rate of the reaction determined by the stability of the arenium ion intermediate?

Solve

From Table 23-1, the nitration of phenol (C_6H_5–OH) is 50% para and 0% meta. Because there is much more of the para isomer produced than the meta isomer, we know that the para arenium ion intermediate is lower in energy (more stable) than the meta intermediate.

Your Turn 23.4

Think

How many bonds and lone pairs does a positively charged C have? Does that add up to an octet?

Solve

Each positively charged C (circled) does not possess an octet. Each has three bonds and no lone pairs and, thus, only a share of six electrons. Notice that the ortho intermediate has a resonance structure in which all nonhydrogen atoms have octets, but the meta intermediate does not.

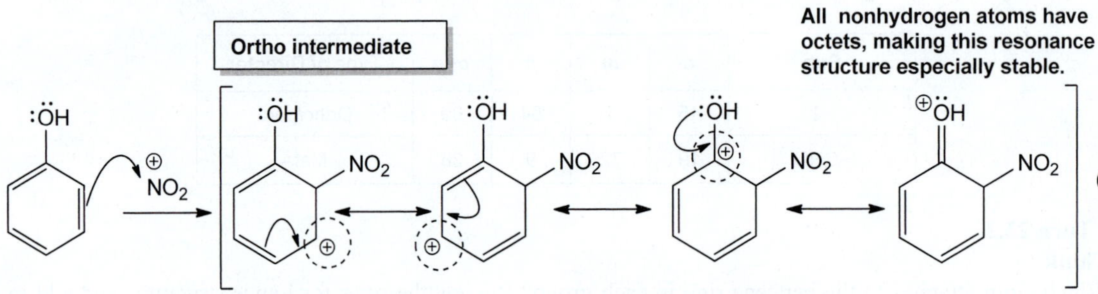

Ortho intermediate

All nonhydrogen atoms have octets, making this resonance structure especially stable.

(23-4)

Meta intermediate

The lone pair on O cannot participate in resonance to stabilize the positive charge.

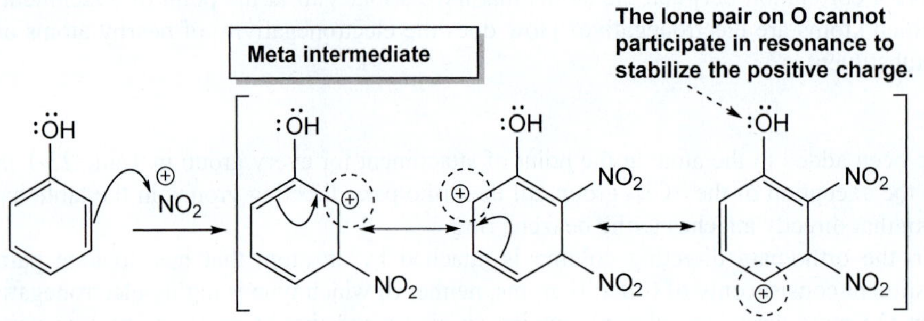

(23-5)

Your Turn 23.5

Think

Are alkyl groups electron-donating or electron-withdrawing groups? In which resonance structure does C^+ have the smallest concentration of positive charge?

Solve

The second resonance structure (boxed) is especially stable owing to the electron-donating CH_3 group attached directly to the C^+. This is because the CH_3 group decreases the concentration of positive charge there.

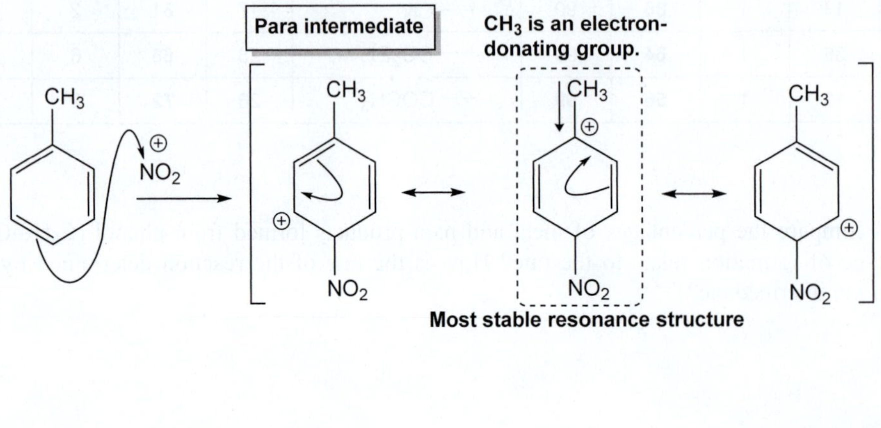

Para intermediate

CH_3 is an electron-donating group.

Most stable resonance structure

Your Turn 23.6

Think

In which resonance structure is the C^+ closest to the NO_2 group? Is the NO_2 group electron donating or withdrawing? What would that do to the concentration of positive charge on C^+? How would that affect the stability of the arenium ion? Alternatively, in comparing the stability of the three resonance structures, what do you notice about the proximity of the positive charge on the C and the positive charge on the N? How do you think this affects the stability of the arenium ion?

Solve

The NO_2 group is electron withdrawing, because the N has no lone pairs, is positively charged, and has two attached electronegative O atoms. Therefore, the C^+ in the middle resonance structure (boxed) would have the greatest concentration of positive charge, making it the least stable resonance structure. Alternatively, when the positive charge on the ring is adjacent to the positive charge on the N, the arenium ion is destabilized owing to electrostatic repulsion of the like charges. The resonance structure, as a result, is much higher in energy than the other two and does not contribute as significantly to the resonance hybrid.

Para intermediate

NO_2 is an electron-withdrawing group.

Least stable resonance structure

Your Turn 23.7

Think

Are there lone-pair electrons on an atom attached to a carbon lacking an octet? In the second resonance structure, what new bond is formed? How many electrons surround each C and the Cl atom? Do any lack an octet? Is Cl more or less electronegative compared to C? How does this group affect the positive charge on the ring?

Solve

The lone pairs on the Cl fold down to the C^+ lacking an octet to form the Cl=C bond. In that resonance structure, all nonhydrogen atoms have an octet. Chlorine is inductively electron withdrawing and pulls electron density *away* from the ring. This increases the positive charge on the ring. These two reasons are why Cl is an ortho/para director (leads to a resonance structure with a full octet when the electrophile has added ortho or para) but is a deactivating group (the Cl is an inductive electron-withdrawing group).

Cl is an electron-withdrawing group.

All nonhydrogen atoms have an octet.

Your Turn 23.8

Think

Which color (red or blue) indicates a greater electron density? In electrophilic aromatic substitution, should the ring be more electron rich or electron poor to be considered activated? Think about the charge needed on the ring if the ring is to react with an electrophile.

Solve

Red indicates greater electron density (electron rich). In electrophilic aromatic substitution, the ring should be sufficiently electron rich compared to the incoming electron-poor electrophile. The ring of the given substituted benzene is deactivated, therefore, because the ring is displaying less negative charge (i.e., less red).

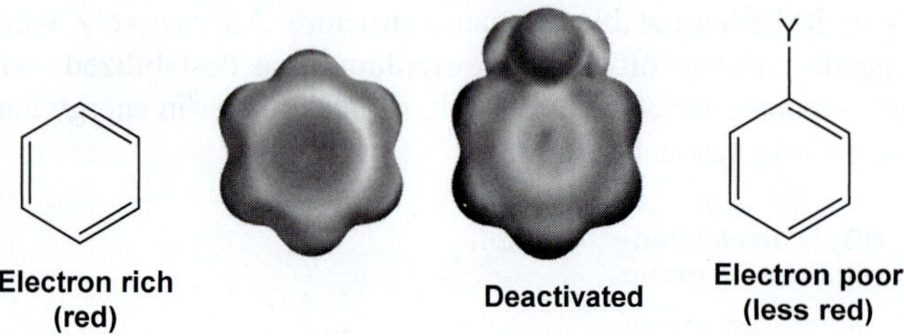

Your Turn 23.9

Think

Are any of the substituted benzene rings in this problem incompatible with Friedel–Crafts reactions? How does the presence of the $AlCl_3$ acid affect the nature of the group attached to the benzene ring?

Solve

The $-N(CH_3)_2$ amino group, although highly activating, is incompatible with Friedel–Crafts reactions, as shown below. The amino group is a relatively strong Lewis base, so it readily coordinates to the $AlCl_3$ Lewis acid. In that complexed form, the N atom possesses a +1 formal charge, making it a highly deactivating group that precludes Friedel–Crafts reactions altogether. The $-C(CH_3)_2$ isopropyl alkyl group is inert to the $AlCl_3$ acid and, therefore, will undergo the indicated Friedel–Crafts reaction.

Cl, O, $AlCl_3$ → $\ominus AlCl_3$, N$\oplus$ — Cl, O → No reaction

Actual product formed

Cl, O, $AlCl_3$ →

Product from compatible Friedel–Crafts reaction

Your Turn 23.10

Think

Are NO_2 groups electron donating or withdrawing? Are CH_3 groups electron donating or withdrawing? How do the electron-donating or electron-withdrawing properties of the substituent affect the stability of an attached C^+? Does a resonance structure contribute more to the resonance hybrid when the resonance structure is more stable or less stable?

Solve

NO_2 groups are electron withdrawing, and the structure where the C^+ is adjacent to the positive charge on the N is the least stable resonance structure. CH_3 groups are electron donating, and the resonance structure that has C^+ adjacent to the CH_3 is the most stable. Overall, this intermediate is more stable owing to the stabilization provided by the third structure; the contribution by the least stable resonance structure is minimized.

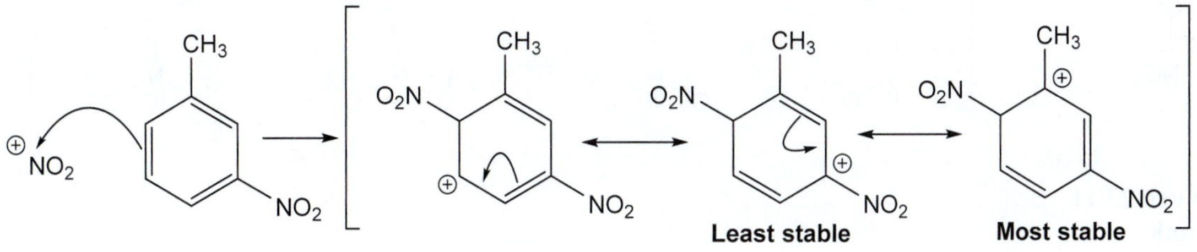

Least stable

Most stable

Your Turn 23.11

Think

Are there any lone pairs of electrons adjacent to π bonds? Is a resonance structure possible involving an sp^3-hybridized C atom?

Solve

Each C^- has a lone pair of electrons and is attached to an atom involved in a separate π bond, so four electrons can be moved to arrive at a new resonance structure. This can be done twice around the ring to arrive at new resonance structures. The uncharged, sp^3-hybridized C atom cannot be involved in resonance. In the final resonance structure, the lone pair of electrons moves outside the ring, and the negative charge moves to the oxygen of the $-NO_2$ group.

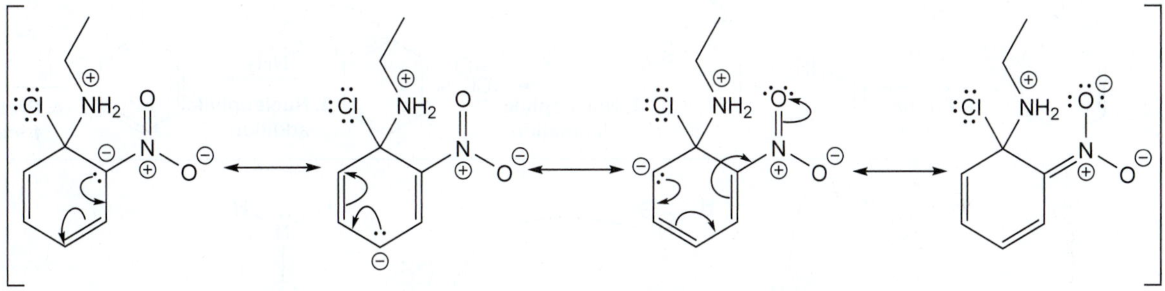

Your Turn 23.12

Think

Are there any lone pairs of electrons adjacent to π bonds? Is a resonance structure possible at an sp^3-hybridized C atom?

Solve

Each C^- has a lone pair of electrons and is attached to an atom that is involved in a separate π bond, so four electrons can be moved to arrive at a new resonance structure. This can be done twice around the ring to arrive at new resonance structures. The uncharged, sp^3-hybridized C atom cannot be involved in resonance. In the second resonance structure, the C^- is attached to NO_2. In the final resonance structure, the lone pair of electrons moves outside the ring, and the negative charge moves to the oxygen of the $-NO_2$ group. The final structure is the most stable, because the negative charge is on an electronegative O atom, whereas the negative charge is on a less electronegative C atom in the other three resonance structures. The mechanism is shown on the following page.

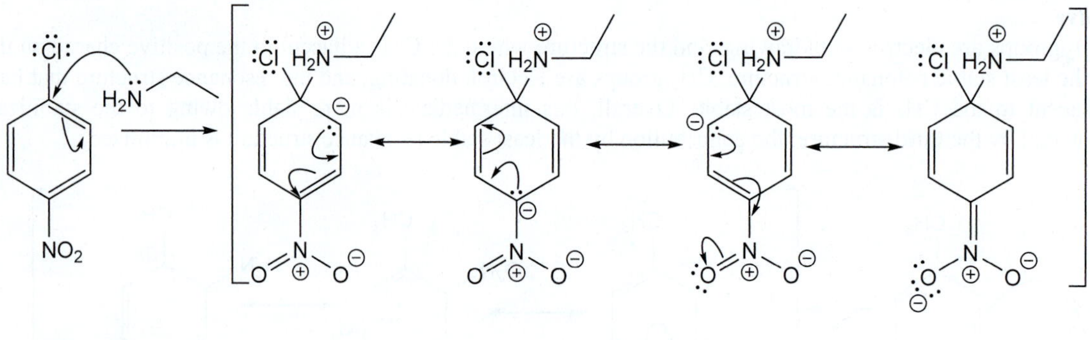

Your Turn 23.11

Think

What bonds are broken and formed in each step? How can you use curved arrows to show the electron movement? Review the 10 elementary steps from Chapters 6 and 7. In which steps are protons involved? Which steps involve the formation of a π bond? Which steps involve breaking a π bond?

Solve

Step 1 is a proton transfer reaction between the very strongly basic NH_2^- anion and the proton ortho to the Cl leaving group. Step 2 is a nucleophile elimination, in which the lone pair on the ring folds over to form the C≡C of the benzyne intermediate and the Cl leaves. Step 3 is nucleophilic addition, in which the NH_2^- anion acts as a nucleophile to add to the C≡C and folds the electrons back to the C, leaving a C=C. Step 4 is a proton transfer step to protonate the C^-. Step 5 is another proton transfer, where the resulting NH_2^- anion acts as a base to deprotonate the aniline N–H. Step 6 is an acid workup, where H_3O^+ is added to yield the final aniline product.

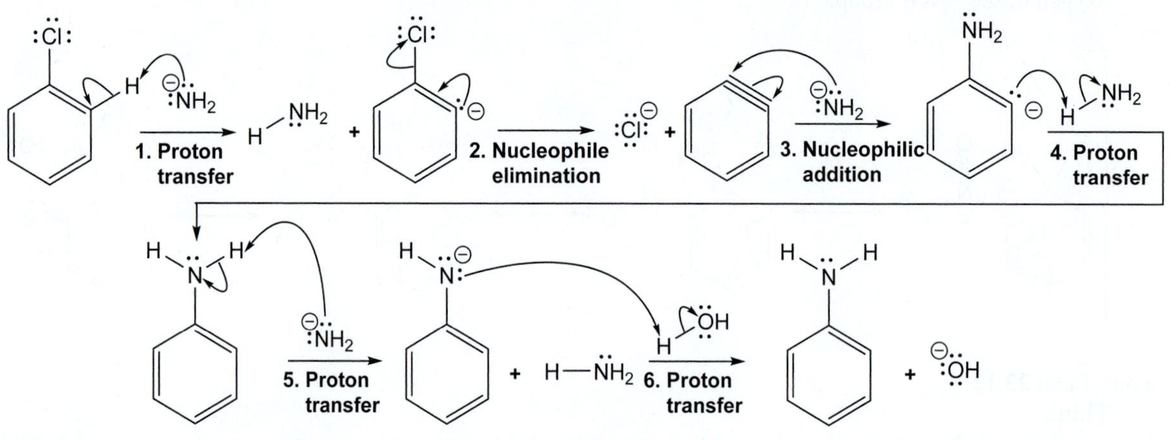

1. Proton transfer
2. Nucleophile elimination
3. Nucleophilic addition
4. Proton transfer
5. Proton transfer
6. Proton transfer

Your Turn 23.14

Think

Is the CO_2H group meta or ortho/para directing? How would the product change if you oxidized the CH_3 into CO_2H prior to the nitration step? Could *m*-nitrobenzoic acid be produced if nitration is the first step? Is more than one route possible?

Solve

The CO_2H group is meta directing, and oxidation of the alkyl group prior to nitration will direct the NO_2 group meta. If nitration were performed first, the ring would be deactivated enough that Friedel–Crafts alkylation would not occur, as shown on the next page. Thus, the first route shown below is the best route to *m*-nitrobenzoic acid.

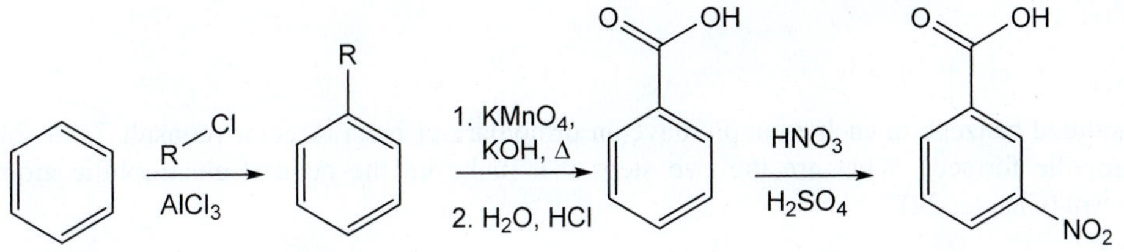

O OH

R

R–Cl, $AlCl_3$ → ; 1. $KMnO_4$, KOH, Δ 2. H_2O, HCl → ; HNO_3, H_2SO_4 →

O OH

NO_2

HNO_3, H_2SO_4 → ; R–Cl, $AlCl_3$ →

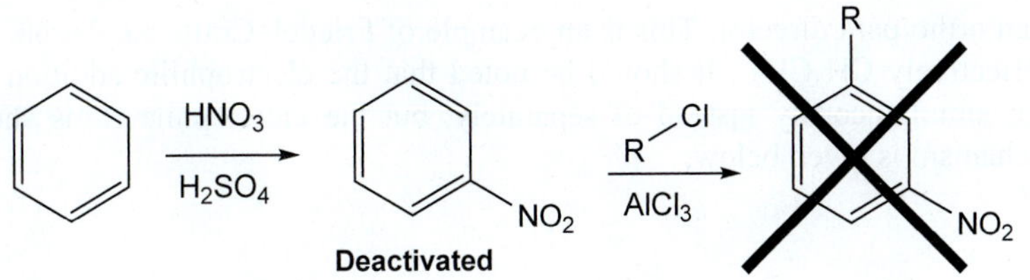

R

NO_2

Deactivated

NO_2

In Chapter Problems

Problem 23.1

Think

Does the monosubstituted benzene in each example have an ortho/para or meta director (consult Table 23-1)? How is each electrophile formed? What are the two steps that make up the general electrophilic aromatic substitution mechanism (Chapter 22)?

Solve

(a) The HO attached to the ring is an ortho/para director. This is an example of Friedel–Crafts alkylation. The electrophile in the reaction is effectively $CH_3CH_2^+$. It should be noted that the electrophilic addition and heterolysis steps probably occur simultaneously instead of separately, but the electrophile R^+ is shown explicitly for emphasis. The mechanism is given below:

Formation of the electrophile

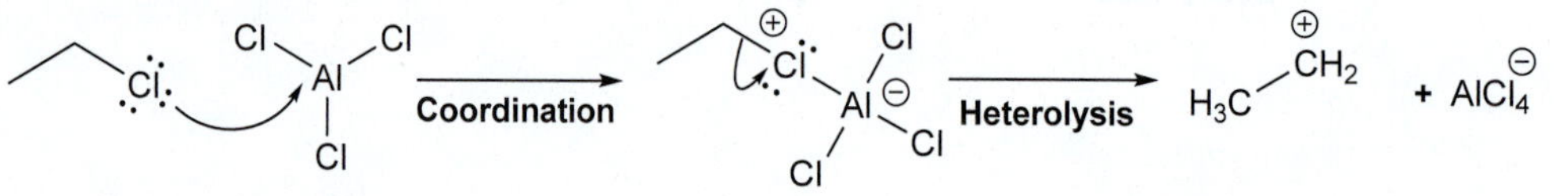

Coordination

Heterolysis

H_3C–CH_2^+ + $AlCl_4^-$

Ortho substitution

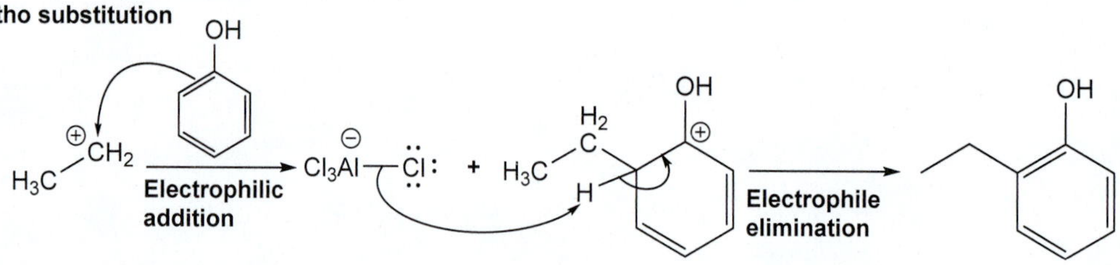

Electrophilic addition

Cl_3Al^-–Cl: + H_3C

Electrophile elimination

Para substitution

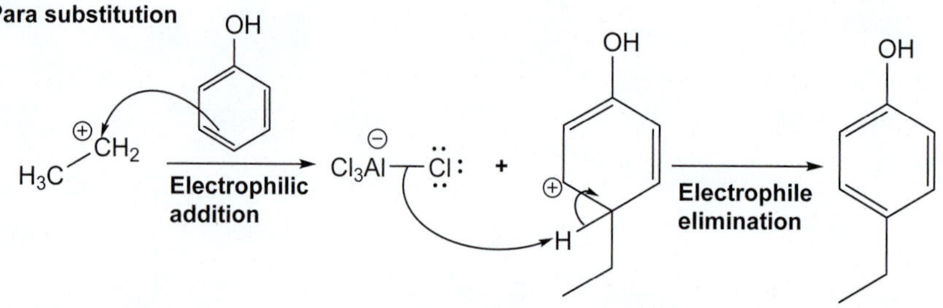

Electrophilic addition

Cl_3Al^-–Cl: +

Electrophile elimination

(b) The NO_2 attached to the ring is a meta director. This is a chlorination reaction, and the electrophile in the reaction is effectively Cl^+. It should be noted that the electrophilic addition and heterolysis steps probably occur simultaneously instead of separately, but the electrophile Cl^+ is shown explicitly for emphasis. The mechanism is given below and on the next page:

Formation of the electrophile

Coordination

Heterolysis

:Cl$^+$ + $FeCl_4^-$

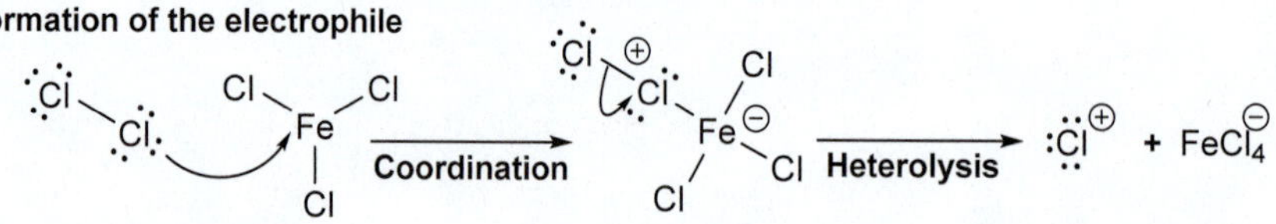

Meta substitution

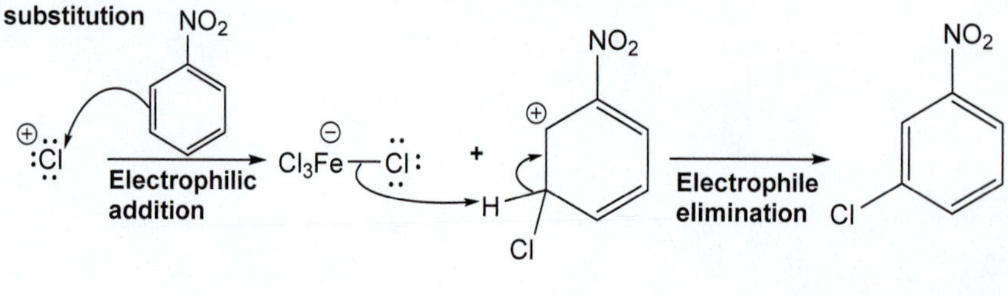

Electrophilic addition

Cl_3Fe^-–Cl: +

Electrophile elimination

(c) The CH_3 attached to the ring is an ortho/para director. This is an example of Friedel–Crafts acylation. The electrophile in the reaction is CH_3CO^+. The mechanism is given below:

Formation of the electrophile

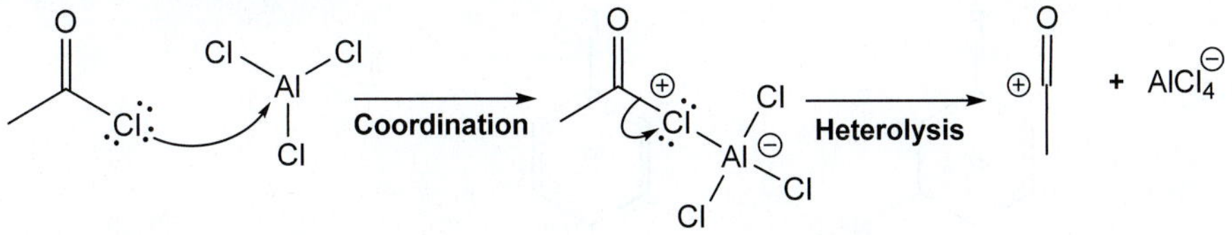

Coordination

Heterolysis

$+ AlCl_4^-$

Ortho substitution

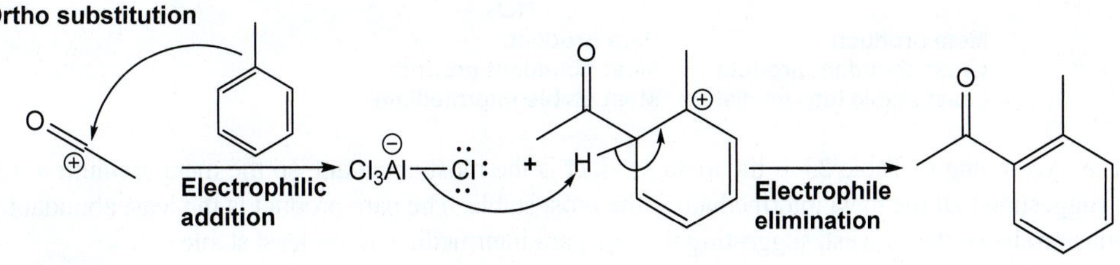

Electrophilic addition

Electrophile elimination

Para substitution

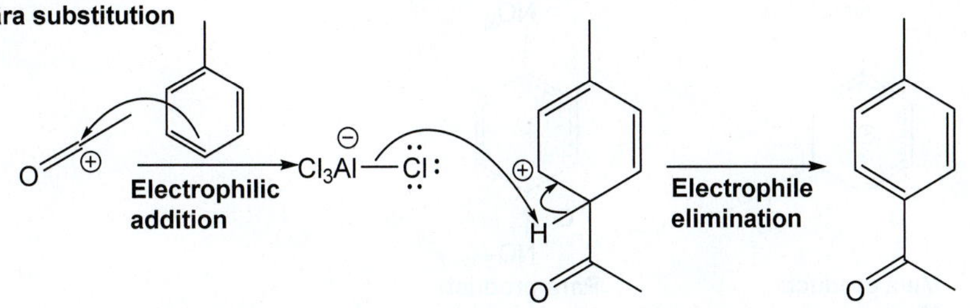

Electrophilic addition

Electrophile elimination

(d) The $CH_3C(O)$ attached to the ring is a meta director. This is a bromination reaction, and the electrophile in the reaction is effectively Br^+. It should be noted that the electrophilic addition and heterolysis steps probably occur simultaneously instead of separately, but the electrophile Br^+ is shown explicitly for emphasis. The mechanism is given below:

Formation of the electrophile

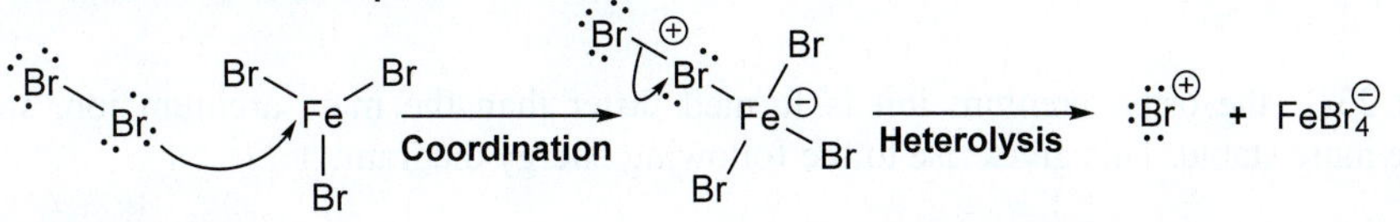

Coordination

Heterolysis

$:Br^+ + FeBr_4^-$

Meta substitution

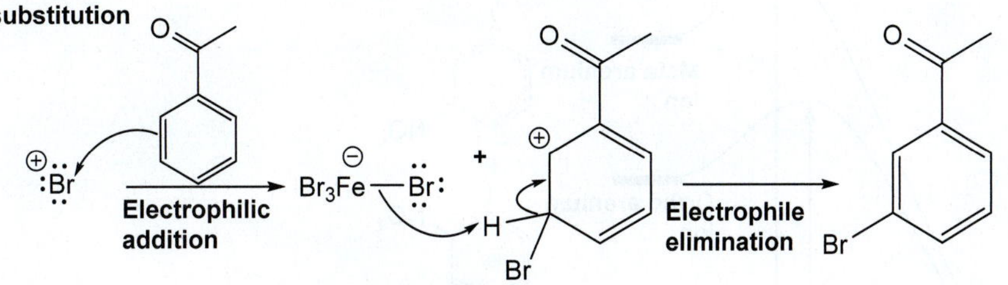

Electrophilic addition

Electrophile elimination

Problem 23.2

Think

Consult Table 23-1. Which product is produced in the greatest abundance: ortho, meta, or para? Which is produced in the least abundance? How does the relative amount of each product correspond to the stability of the arenium ion intermediate from which it was produced?

Solve

Fluorobenzene: According to Table 23-1, the para product is the most abundant, so the para arenium ion is produced the fastest, suggesting that the para arenium ion is the most stable. The meta product is the least abundant, so the meta arenium ion is produced the slowest, suggesting that the meta intermediate is the least stable.

F F F
NO_2 NO_2 NO_2
A B C

Meta product:
Least abundant product
Least stable intermediate

Para product:
Most abundant product
Most stable intermediate

Nitrobenzene: According to Table 23-1, the meta product is the most abundant, so the meta arenium ion is produced the fastest, suggesting that the meta intermediate is the most stable. The para product is the least abundant, so the para arenium ion is produced the slowest, suggesting that the para intermediate is the least stable.

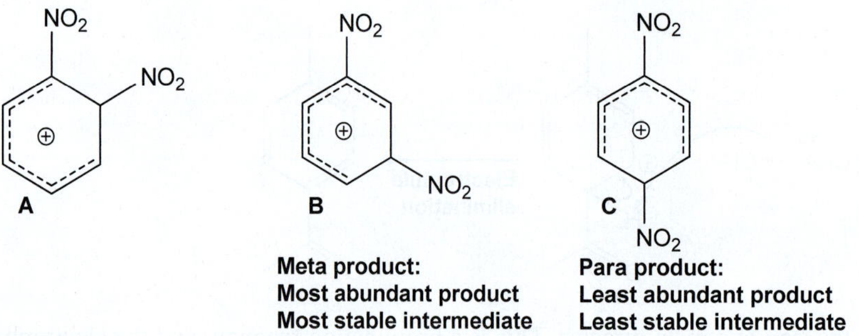

Problem 23.4

Think

Which of the isomeric products—ortho or meta—is formed faster? What does that say about the stability of the respective intermediates? What role is played by resonance delocalization of charge?

Solve

According to Table 23-1, the ortho arenium ion is formed faster than the meta arenium ion, so the ortho intermediate must be more stable. This gives rise to the following energy diagram:

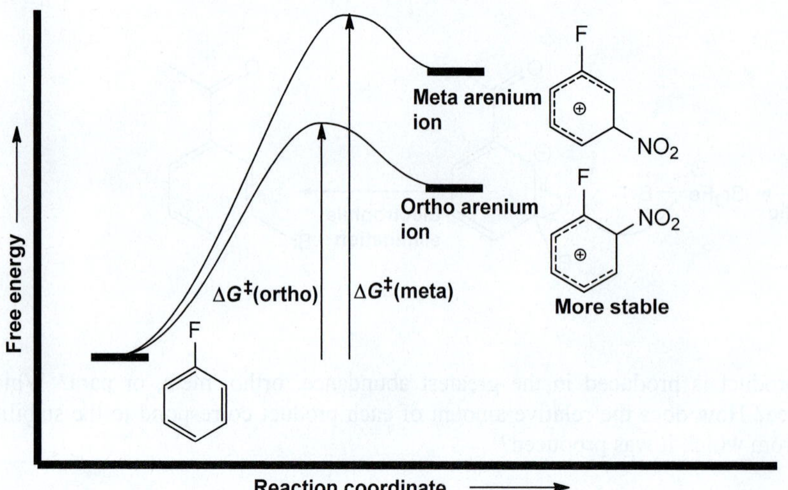

Whereas the meta intermediate only has three resonance structures, the ortho intermediate has *four*, similar to the para intermediate. Furthermore, the one that is boxed below is the most important of the four, because all of the nonhydrogen atoms have complete octets.

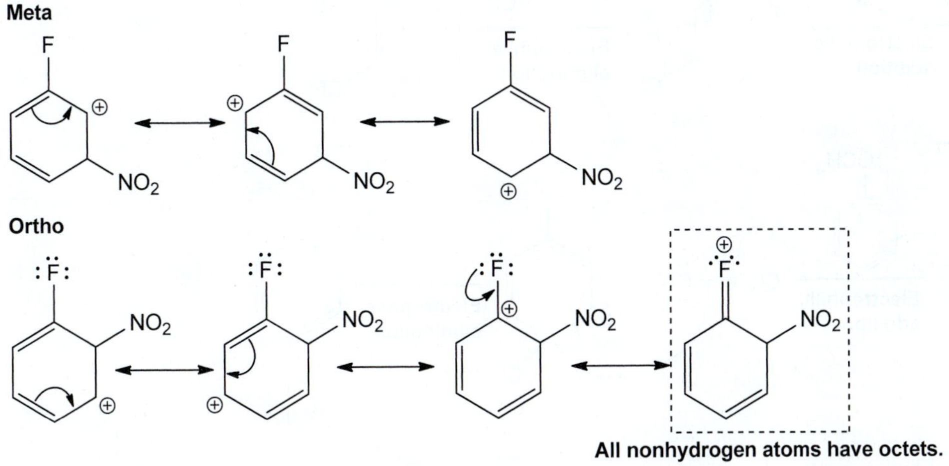

Meta

Ortho

All nonhydrogen atoms have octets.

Problem 23.5

Think

What is the identity of the electrophile in the electrophilic addition–elimination reaction? From which C=C bond should a curved arrow originate to show the π electrons attacking the electrophile in ortho, meta, and para electrophilic aromatic substitution? Is the substituent an ortho/para or meta director?

Solve

(a) This is a chlorination reaction, and the electrophile is effectively the Cl^+ cation. The electrophilic aromatic substitution mechanism is shown for ortho, meta, and para substitution products. OCH_3 is an ortho/para director, because the O atom that attaches it to the ring has a lone pair of electrons, so the major products will be the ortho and para isomers. It should be noted that the electrophilic addition and heterolysis steps probably occur simultaneously instead of separately, but the electrophile Cl^+ is shown explicitly for emphasis.

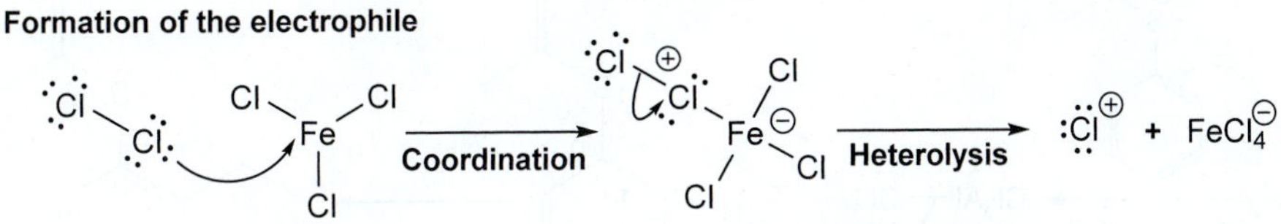

Formation of the electrophile

Coordination

Heterolysis

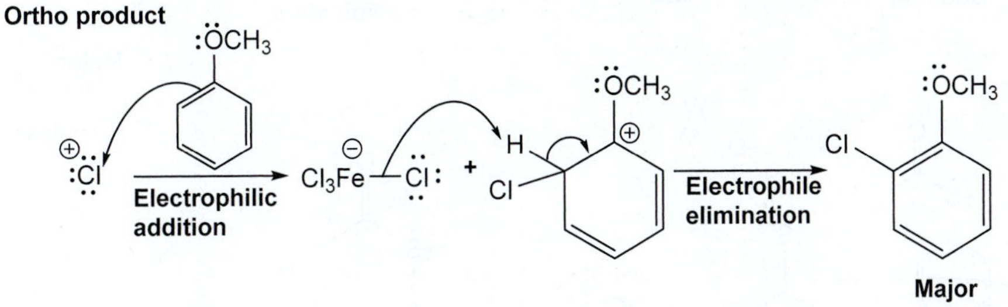

$:Cl^{\oplus}$ + $FeCl_4^{\ominus}$

Ortho product

Electrophilic addition

Electrophile elimination

Major

Meta product

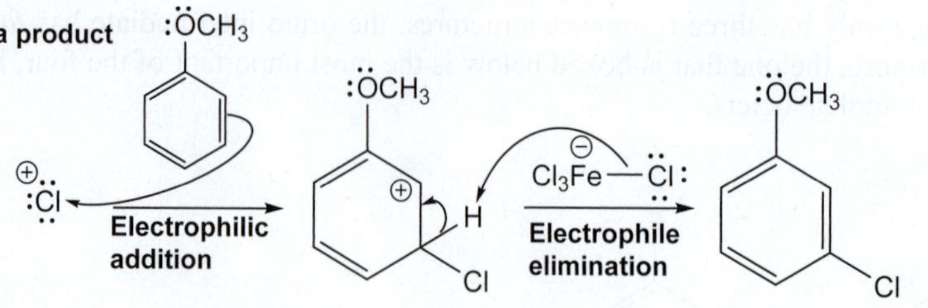

Electrophilic addition

Electrophile elimination

Para product

Electrophilic addition

Electrophile elimination

Major

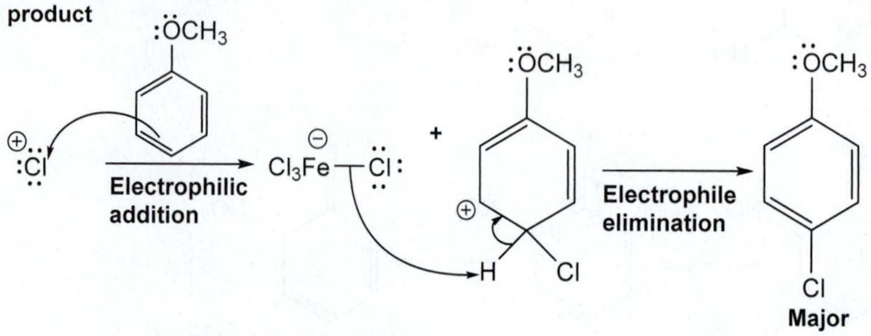

(b) The $CH_3CH_2^+$ cation is effectively the electrophile in this Friedel–Crafts alkylation reaction. The electrophilic aromatic substitution mechanism is shown for ortho, meta, and para substitution products. SC(O)CH_3 is an ortho/para director, because the S atom that attaches it to the ring has a lone pair of electrons, so the major products will be the ortho and para isomers. It should be noted that the electrophilic addition and heterolysis steps probably occur simultaneously instead of separately, but the electrophile R^+ is shown explicitly for emphasis.

Formation of the electrophile

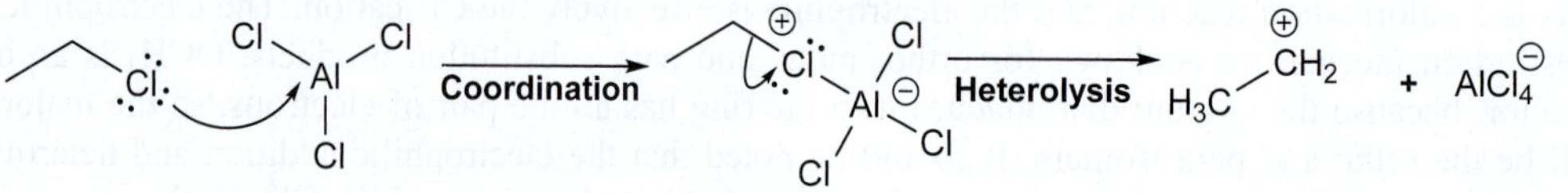

Coordination

Heterolysis

$H_3C\text{–}CH_2^{\oplus}$ + $AlCl_4^{\ominus}$

Ortho product

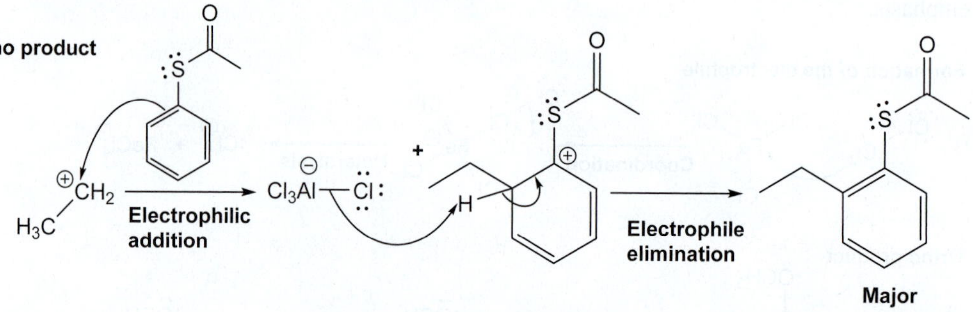

Electrophilic addition

Electrophile elimination

Major

Meta product

Electrophilic addition

Electrophile elimination

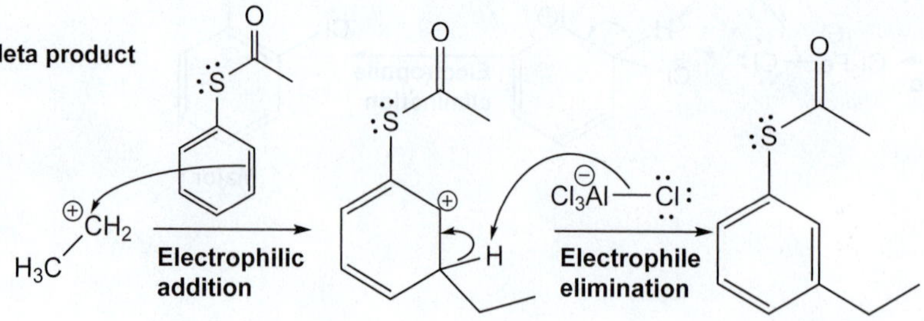

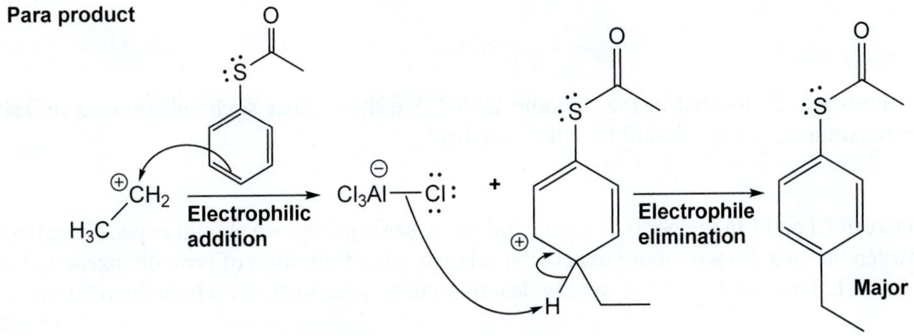

Para product

Electrophilic addition

Electrophile elimination

Major

Problem 23.6

Think

What is the identity of the electrophile in the electrophilic addition–elimination reaction? What does the electrophile replace on the benzene ring? Is CH_3CH_2 an ortho/para or meta director?

Solve

The NO_2^+ cation is the electrophile in this nitration reaction, which replaces an aromatic proton. CH_3CH_2 is an ortho/para director, because it is electron donating, so the major products will be the ortho and para isomers.

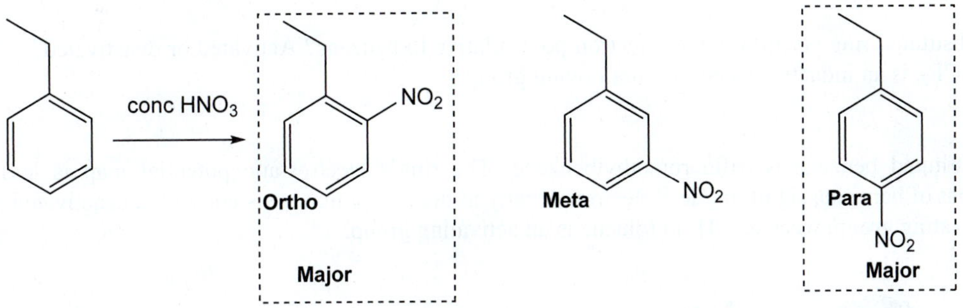

conc HNO_3

Ortho Major

Meta

Para Major

Problem 23.8

Think

Draw the arenium ion intermediate produced when E^+ adds to the para position, and consider all of the resonance structures of that arenium ion intermediate. Do any of the arenium ion resonance structures exhibit the positive charge on the ring directly adjacent to the CF_3 substituent? What effect does the CF_3 have on an adjacent positive charge? Is it stabilizing or destabilizing?

Solve

The resonance structures of the para intermediate are shown below. It is less stable than the meta intermediate, because the second resonance structure has the positive charge on the C directly attached to CF_3. The CF_3 group is electron withdrawing, so it destabilizes that positive charge.

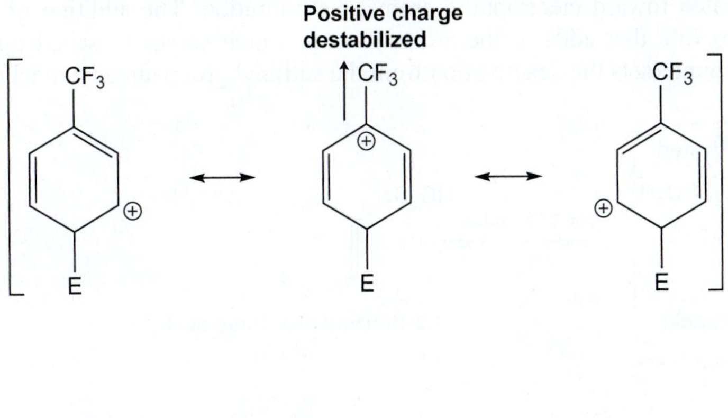

Positive charge destabilized

Problem 23.9

Think

In which group number is Br located in the periodic table? Are there other such substituents in Table 23-2? Relative to those substituents, where should Br fall on the list?

Solve

Because Br is between Cl and I in the periodic table, and in the same group, we should expect its influence to be intermediate between the two. So we should expect the relative rate of nitration of bromobenzene to be between 0.18 and 0.033, or ~0.1. This would make it weakly deactivating relative to H, for which the relative rate is 1.

Problem 23.11

Think

Is CF_3 an inductively electron-donating or electron-withdrawing group? Which type of group activates the benzene ring toward electrophilic aromatic substitution?

Solve

The CF_3 group is inductively electron withdrawing, so it will destabilize the arenium ion intermediate, causing the reaction to be slower. Thus, the ring in trifluoromethylbenzene is deactivated.

Problem 23.12

Think

Is the monosubstituted ring electron rich or electron poor relative to benzene? Activated or deactivated? Which group, CH_3 or CF_3, is an inductive electron-withdrawing group?

Solve

The monosubstituted benzene is trifluoromethylbenzene. The ring's electrostatic potential map is less red compared to that of benzene, signifying less electron density in the π system, consistent with a deactivated ring. CF_3 is a deactivating group, whereas CH_3 in toluene is an activating group.

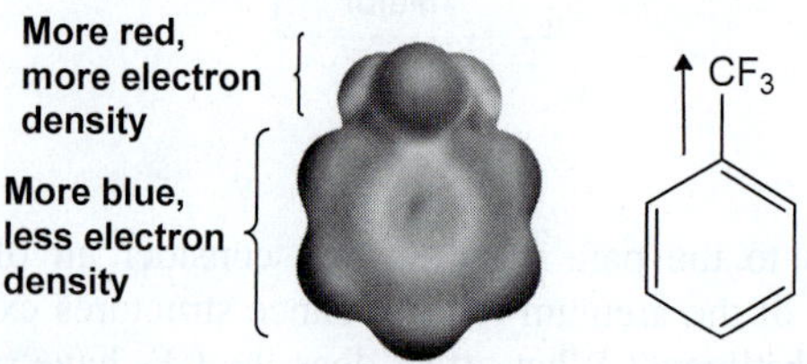

Problem 23.13

Think

Is benzenesulfonic acid activated or deactivated toward electrophilic aromatic substitution? How does this affect the rate of sulfonation? What does the addition of SO_3 do to the rate of sulfonation?

Solve

Benzenesulfonic acid is strongly deactivated toward electrophilic aromatic substitution. The addition of SO_3 increases the concentration of the electrophile that adds to the aromatic ring, which serves to speed up the reaction. The addition of SO_3, therefore, counteracts the deactivation from the sulfonyl group already attached.

conc H_2SO_4, 48 h, 20 °C

Strongly deactivated

SO_3H

Benzenesulfonic acid 66%

conc H_2SO_4, SO_3, 4 h, <90 °C

HO_3S SO_3H

1,3-Benzenedisulfonic acid 90%

Problem 23.14

Think

What type of reaction alkylates a benzene ring to form a new C–C bond? What are the limitations of Friedel–Crafts alkylation? How can Friedel–Crafts acylation yield the target compound? Are additional reactions required?

Solve

Friedel–Crafts alkylation will not yield the target butylbenzene because the carbocation that would be generated from Bu–Cl would be primary and would rearrange prior to reaction with the ring. After the first alkylation, moreover, the benzene ring is activated, and polyalkylated products are possible. To avoid both of these problems, the target can be produced by first performing a Friedel–Crafts acylation, then reducing the C=O to the CH_2.

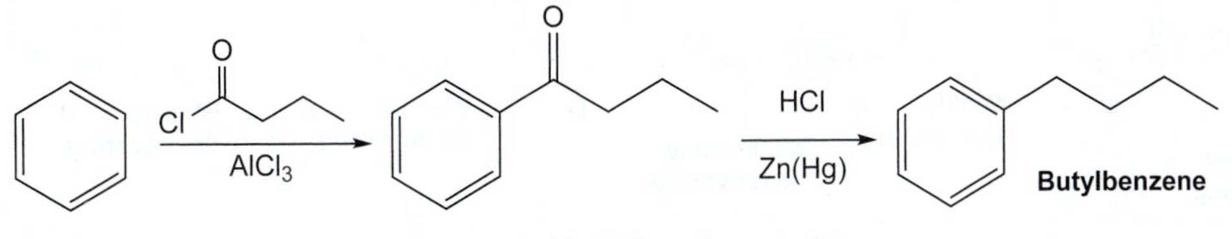

Cl, $AlCl_3$; HCl, Zn(Hg); **Butylbenzene**

Problem 23.15

Think

What type of elementary step yields the phenoxide anion? What is the two-step mechanism by which bromination takes places on the phenoxide anion? How many times does this reaction mechanism occur? How is the uncharged product produced?

Solve

A proton transfer reaction yields the phenoxide anion, which then undergoes three electrophilic addition–elimination mechanisms to give the tribromo phenoxide product. The uncharged phenol product is then produced by a proton transfer reaction.

Proton transfer; **Electrophilic addition**; **Electrophile elimination**; **Electrophilic addition**; **Electrophile elimination**; **Electrophilic addition**; **Electrophile elimination**; **Proton transfer**

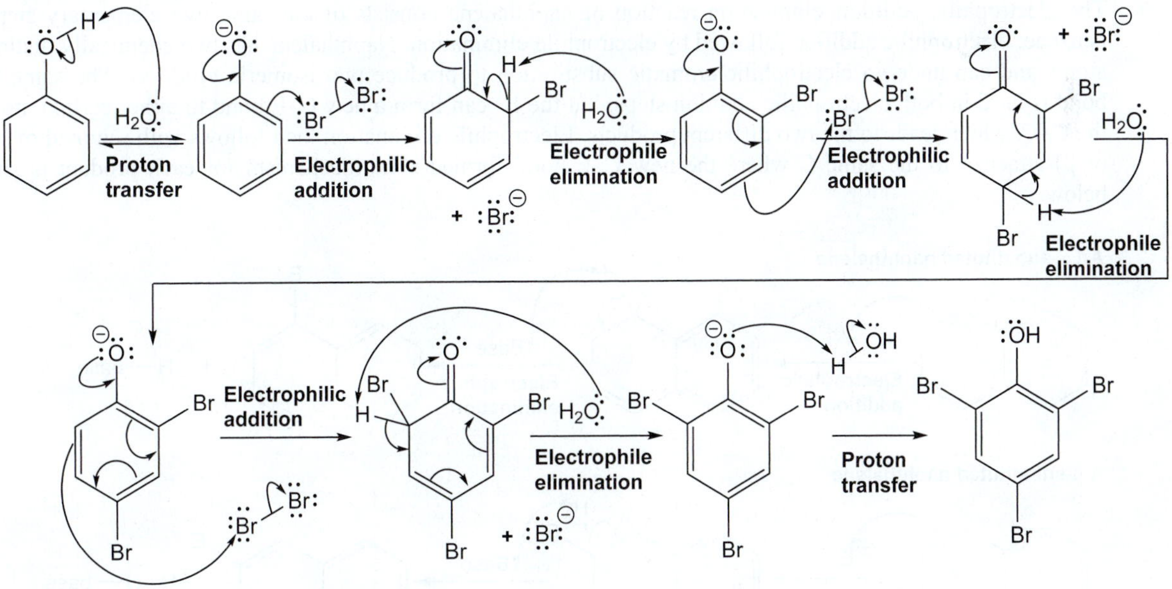

Problem 23.17

Think

Identify each substituent as a weak, moderate, or strong activator or deactivator. How many activating and deactivating groups does each ring have? Are they all the same strength?

Solve

According to Table 23-3, the NO_2 group is strongly deactivating, the $CH_3C{=}O$ group is moderately deactivating, and the Cl and Br groups are weakly deactivating. **C** will react the slowest because it is the only one with three deactivating groups: one that is strongly deactivating, one that is moderately deactivating, and one that is weakly deactivating. **A**, **E**, and **F** have two deactivating groups, and the rate is based on the strength of the deactivating group, **A** < **F** < **E**. Structure **D** only has one deactivating group, so it is next. Benzene has no deactivating groups and reacts the fastest. Therefore, the order of increasing reaction rate is **C** < **A** < **F** < **E**< **D** < **B**.

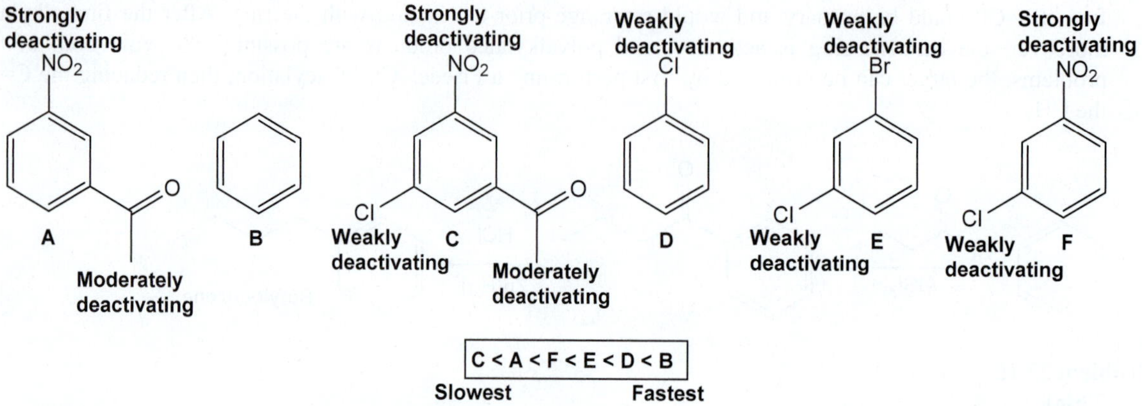

Strongly deactivating NO_2 / Moderately deactivating — A; B; Strongly deactivating NO_2 / Cl Weakly deactivating / Moderately deactivating — C; Weakly deactivating Cl — D; Weakly deactivating Br / Cl Weakly deactivating — E; Strongly deactivating NO_2 / Cl Weakly deactivating — F

C < A < F < E < D < B

Slowest — Fastest

Problem 23.18

Think

How is the electrophilic addition–elimination reaction of naphthalene similar to the electrophilic addition–elimination reaction of benzene? From which C=C bond should a curved arrow originate to show the π electrons attacking the electrophile to yield each of the two products shown? How do you show the elimination of H^+ to restore the aromaticity of the ring?

Solve

The electrophilic addition–elimination reaction of naphthalene consists of the same two elementary steps as benzene: electrophilic addition followed by electrophile elimination. Naphthalene has two chemically distinct H atoms and can undergo electrophilic aromatic substitution to produce two isomeric products. The same C=C bond is used in both electrophilic addition steps, but the E^+ can form a new C–E bond to either carbon atom of the C=C, which leads to the two different products. Electrophile elimination then follows with removal of H^+ (α or β) attached to the same C where the new C–E bond formed. The mechanism for each product is shown below.

An α-substituted naphthalene

$E^{\oplus}$ → Electrophilic addition → (E, H_α, ⊕) :Base → Electrophile elimination → (E) + H—base

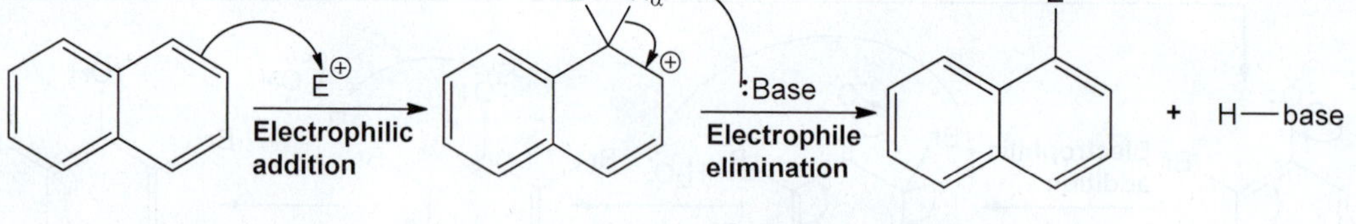

A β-substituted naphthalene

$^{\oplus}E$ → Electrophilic addition → (⊕, H_β, E) :Base → Electrophile elimination → (E) + H—base

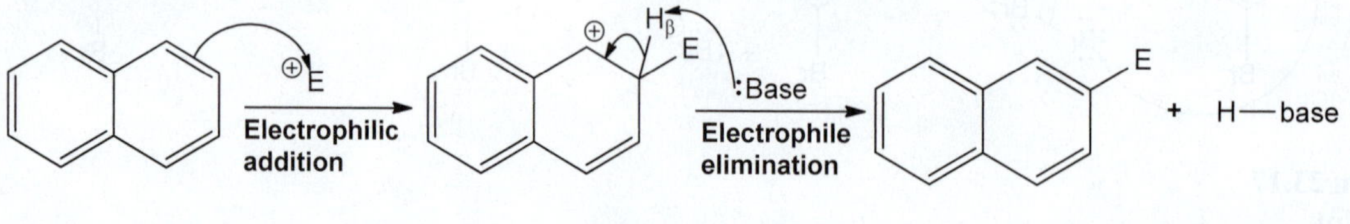

Problem 23.19

Think

How is the electrophilic addition–elimination reaction of pyrrole or pyridine similar to the electrophilic addition–elimination reaction of benzene? From which C=C bond should a curved arrow originate to show the π electrons attacking the electrophile to yield each of the products shown? How do you show the elimination of H^+ to restore the aromaticity of the ring?

Solve

The electrophilic addition–elimination reaction of pyrrole and pyridine consists of the same two elementary steps as benzene: electrophilic addition followed by electrophile elimination.

Pyrrole has two chemically distinct H atoms and can undergo electrophilic aromatic substitution to produce two isomeric products. The same C=C bond is used in both electrophilic addition steps, but the E^+ can form a new C–E bond to either carbon atom of the C=C, which leads to the two different products. Electrophile elimination then follows with removal of H^+ attached to the same C where the new C–E bond formed. The mechanism for each product is shown below.

2 Substitution

Electrophilic addition → Electrophile elimination (:Base) → + H—base

3 Substitution

Electrophilic addition → Electrophile elimination (:Base) → + H—base

Pyridine has three chemically distinct H atoms, so it can undergo electrophilic aromatic substitution to produce three isomeric products. The mechanism for each product is shown below.

2 Substitution

Electrophilic addition → Electrophile elimination (:Base) → + H—base

3 Substitution

Electrophilic addition → Electrophile elimination (:Base) → + H—base

4 Substitution

Electrophilic addition → Electrophile elimination (:Base) → + H—base

Problem 23.20

Think

How is furan similar to pyrrole? Draw all resonance structures for the 2- and the 3-substituted arenium ion intermediates. Do the arenium ion intermediates have the same number of resonance structures? Which substitution forms a more stable intermediate?

Solve

The 2-substituted furan forms an arenium ion intermediate that has three resonance structures, while the 3-substituted arenium ion intermediate has just two resonance structures. Therefore, the 2-substituted arenium ion intermediate is more stable and will lead to the major product.

Addition to 2 position

Electrophilic addition

Addition to 3 position

:Base

Electrophile elimination

Major product

+ H—Base

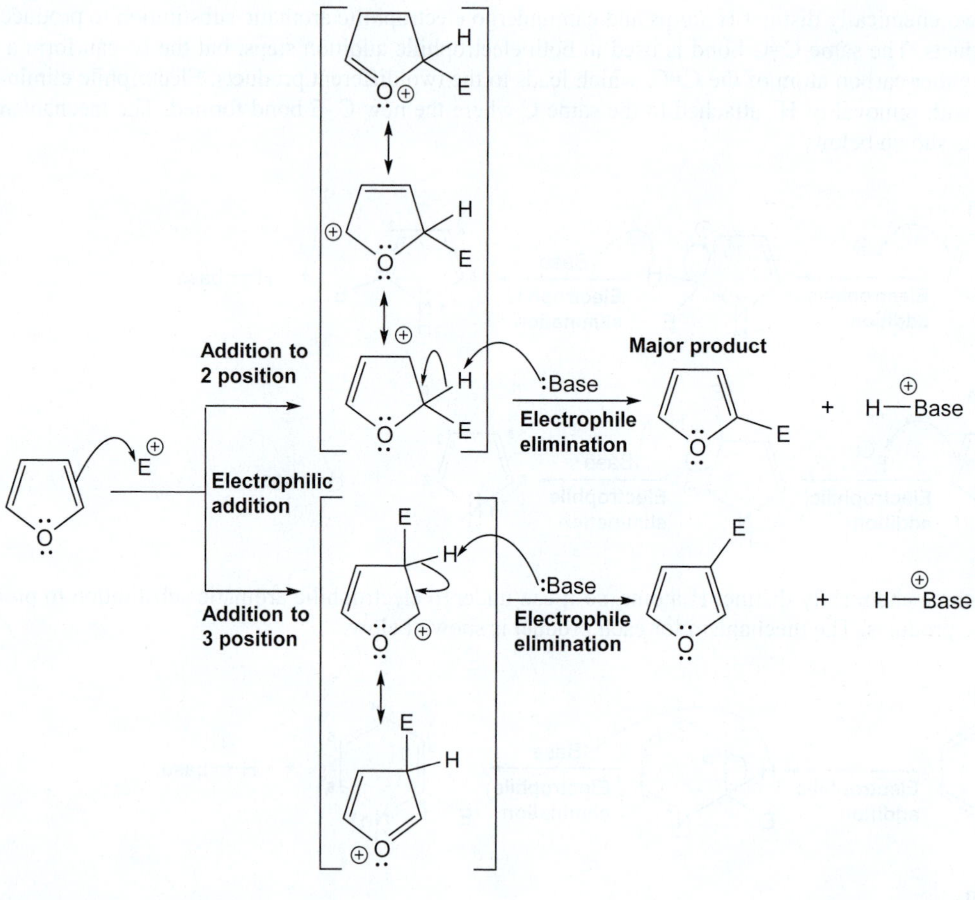

Problem 23.21

Think

Draw all resonance structures for the 4-substituted arenium ion intermediates. Compare this arenium ion intermediate to the 3-substituted intermediate. Do they have the same number of resonance structures? Do all resonance structures contribute equally to the hybrid? Which substitution forms a more stable intermediate?

Solve

The 4-substituted pyridine forms an arenium ion intermediate that has resonance structures (see the mechanism on the next page). Like the arenium ion intermediate produced when E^+ adds to the 2 position (Equation 23-31a), one of the resonance structures of this arenium ion intermediate has N^+ without an octet. When E^+ adds to the 3 position, on the other hand (see Equation 23-31b), all three resonance structures have C^+ without an octet. Therefore, the 3-substituted arenium ion intermediate is the most stable and will lead to the major product.

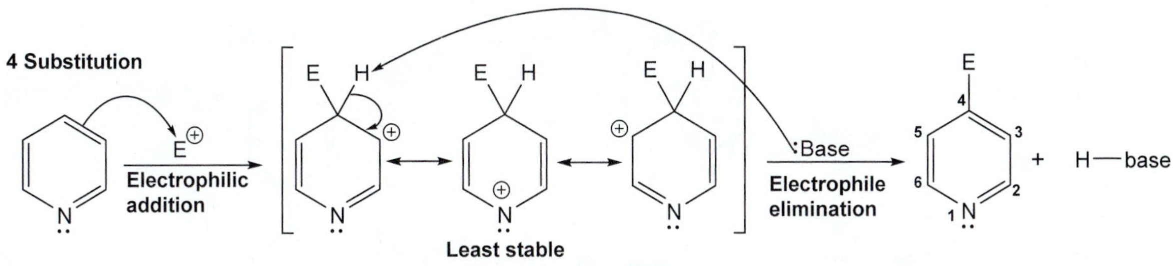

4 Substitution

Electrophilic addition

Least stable

Electrophile elimination

Problem 23.22

Think

What is the identity of the electrophile in the electrophilic addition–elimination reaction? From which C=C bond should a curved arrow originate to show the π electrons attacking the electrophile in ortho, meta, and para electrophilic aromatic substitution? Is the substituent an ortho/para or meta director?

Solve

The arenediazonium ion is electrophilic and can undergo electrophilic aromatic substitution with the activated benzene ring shown. The mechanism is given below.

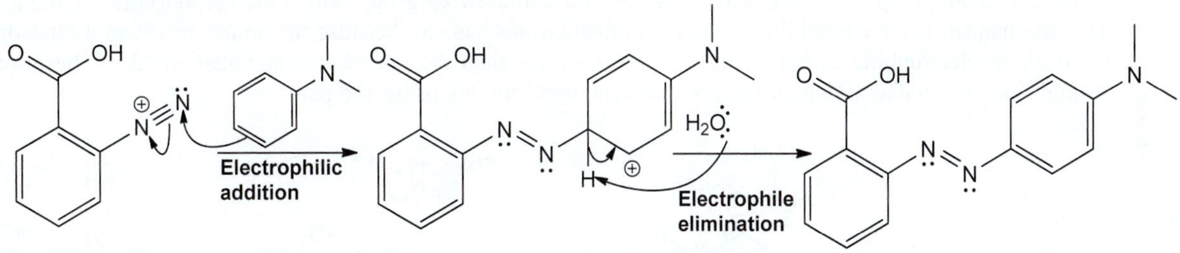

Electrophilic addition

Electrophile elimination

Problem 23.23

Think

How is the structure of methyl orange and acid red 37 similar to the structure of methyl red in Equation 23-33? How will the equation need to be modified to synthesize methyl orange and acid red 37? Does one precursor have an aromatic ring that is activated toward electrophilic aromatic substitution?

Solve

The syntheses of methyl orange and acid red 37 are shown below. In the synthesis of methyl orange, the aniline has an activated aromatic ring and will react with an arenediazonium ion. In the synthesis of acid red 37, there are two rings. The activated ring is the one with the attached NH_2 group and is the ring that will react with the arenediazonium ion.

HCl, H_2O

Methyl orange

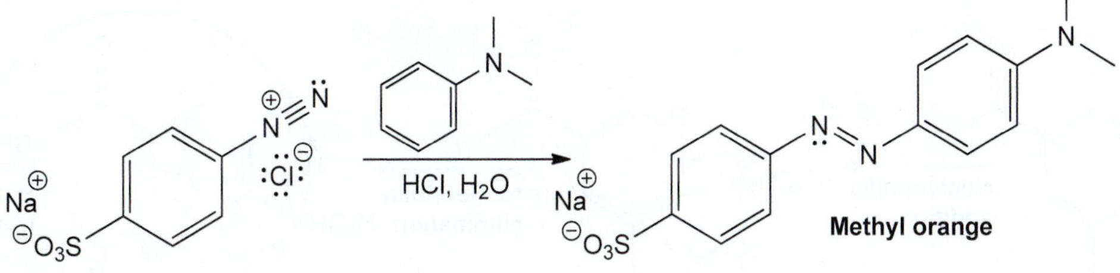

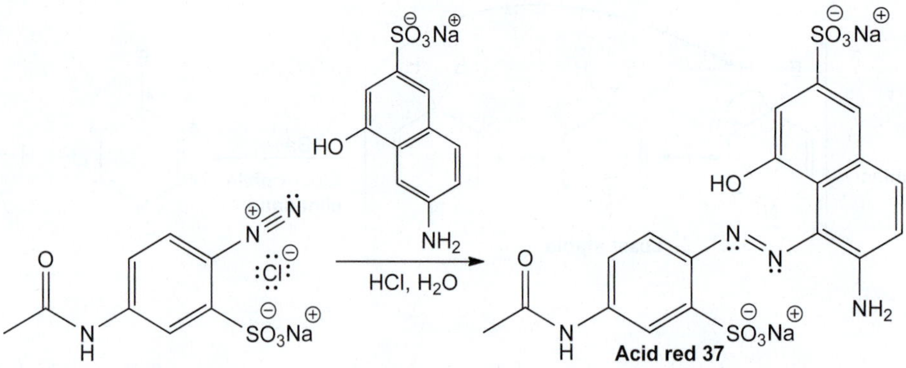

Acid red 37

Problem 23.24

Think

What is the leaving group? Is this mechanism nucleophilic addition–elimination or electrophilic addition–elimination? Is there a strong electron-withdrawing group ortho or para to the leaving group?

Solve

Cl is the leaving group, and there is a strong electron-withdrawing group ortho (and meta) to the leaving group. This mechanism is a nucleophilic addition–elimination mechanism, because the amine serves as a nucleophile to attack an electrophilic carbon. This reaction is *slower* than the one shown in Equation 23-36, because the strongly electron-withdrawing groups are ortho and meta and not ortho and para.

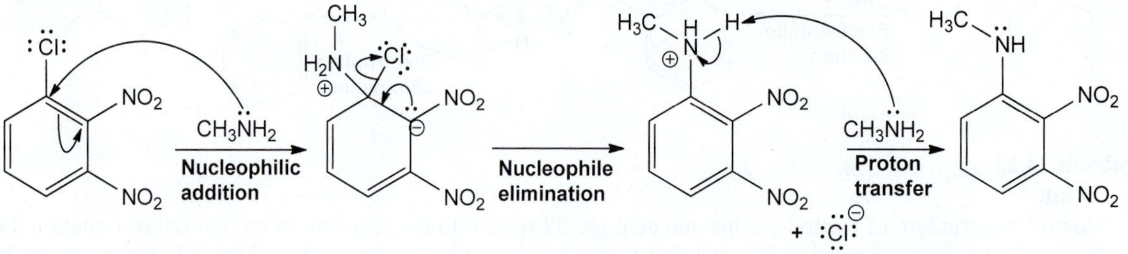

Problem 23.25

Think

What is the leaving group? Is this mechanism nucleophilic addition–elimination or electrophilic addition–elimination? Is there a strong electron-withdrawing group ortho or para to the leaving group?

Solve

(a) Cl is the leaving group, and the amine is the nucleophile in the nucleophilic addition–elimination mechanism. Notice that the leaving group is ortho to one electron-withdrawing group and para to another.

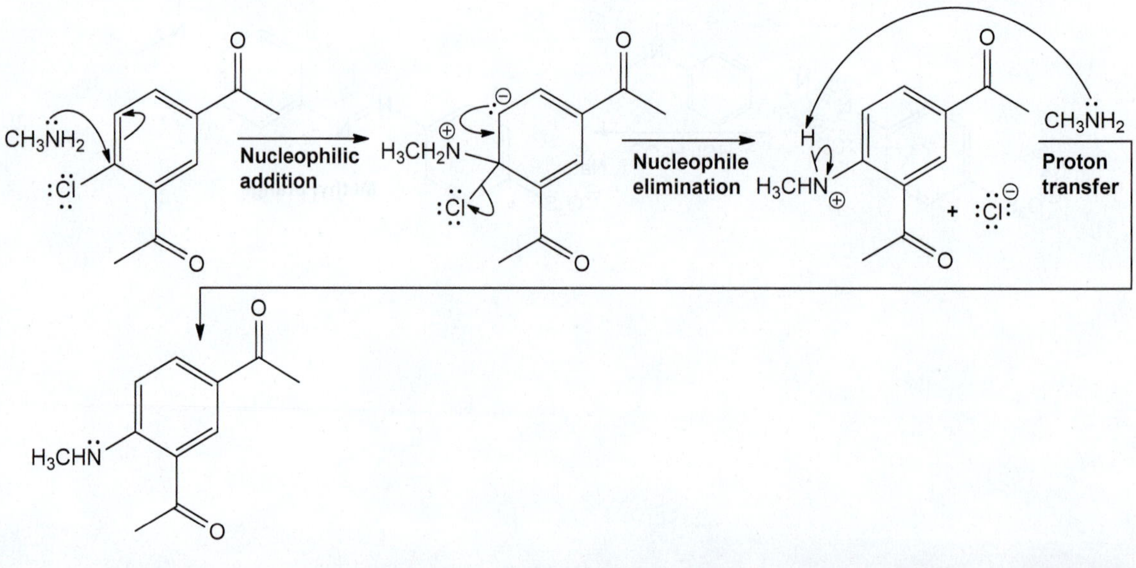

(b) F is the leaving group, and the alkoxide is the nucleophile in the nucleophilic addition–elimination mechanism. Notice that the leaving group is indeed ortho to a strongly electron-withdrawing group.

Nucleophilic addition

Nucleophile elimination

(c) Two F atoms could serve as leaving groups. One F is meta to the NO_2, and the other F is ortho to the NO_2. The Meisenheimer complex is stabilized when the electron-withdrawing group is ortho (or para) to the leaving group.

Nucleophilic addition

Nucleophile elimination

Problem 23.26

Think

How does the bromobenzene interact with a strong base? How does the Br leaving group leave? What intermediate is formed? How does a nucleophile react with that intermediate? In the deuterated compound, which ortho carbon (C–H, C–D, or both) can be deprotonated by the NaOH base? How does this affect the distribution of the D isotope in the product?

Solve

Hydroxide acts as a base to remove a proton from the aromatic ring to produce a carbanion. In Step 2, the leaving group departs in a nucleophile elimination step. The product of Step 2 is a benzyne intermediate, which can undergo nucleophilic addition by hydroxide to produce a deprotonated form of phenol. The carbanion is protonated, and then the phenolic OH is deprotonated, before acid workup replaces the phenolic proton.

Δ Proton transfer

Nucleophile elimination

Nucleophilic addition

Proton transfer

Proton transfer

Add

Proton transfer

H and D can both be eliminated with essentially equal likelihood; therefore, 50% of the product is expected to contain D.

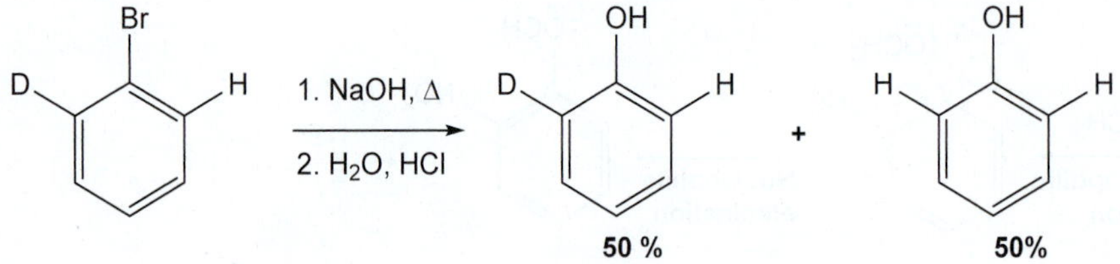

Br D H 1. NaOH, Δ 2. H_2O, HCl OH D H 50 % + OH H H 50%

Problem 23.28

Think

Is the Cl substituent an ortho/para director or a meta director? Is the alkanoyl group an ortho/para director or a meta director? Which group should be on the ring prior to the second substitution?

Solve

The Cl substituent is an ortho/para director, and the alkanoyl group is a meta director. Because our target is a meta isomer, the alkanoyl group should be on the ring before the second substitution is carried out. Therefore, a Friedel–Crafts acylation should take place first, followed by a chlorination, as shown below.

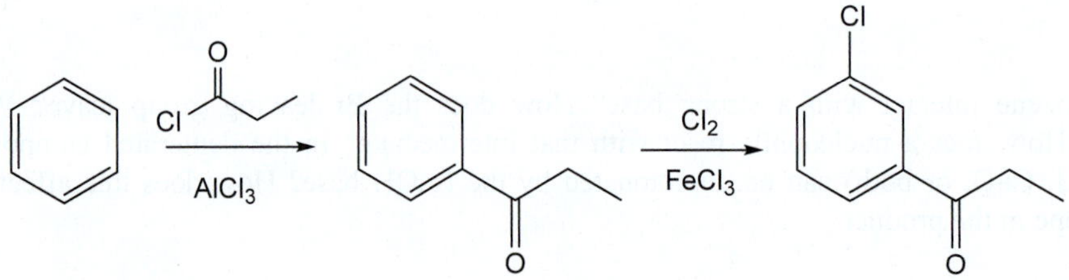

O Cl $AlCl_3$ O Cl_2 $FeCl_3$ Cl O

Problem 23.29

Think

Are the CO_2H groups ortho/para or meta directors? What is the route for adding on the CO_2H group? Does it go through a synthetic intermediate that has an ortho/para director attached to the ring? How can you transform ethanol into the necessary reagent for Friedel–Crafts alkylation?

Solve

A carboxyl group is a meta director, but the target is a para isomer. Adding a carboxyl group to a benzene ring, however, can proceed through a synthetic intermediate in which the substituent is an alkyl group, which is an ortho/para director. Therefore, with the first alkyl group on the ring, the second alkyl group can be added to the para position prior to oxidizing both groups to carboxylic acids. To begin the synthesis, first transform ethanol into the alkyl bromide reagent to conduct a Friedel–Crafts alkylation. The Friedel–Crafts alkylation needs to occur two times to form *p*-diethylbenzene. Then, use $KMnO_4$ followed by acid workup to oxidize the alkyl groups on benzene to the carboxylic acids.

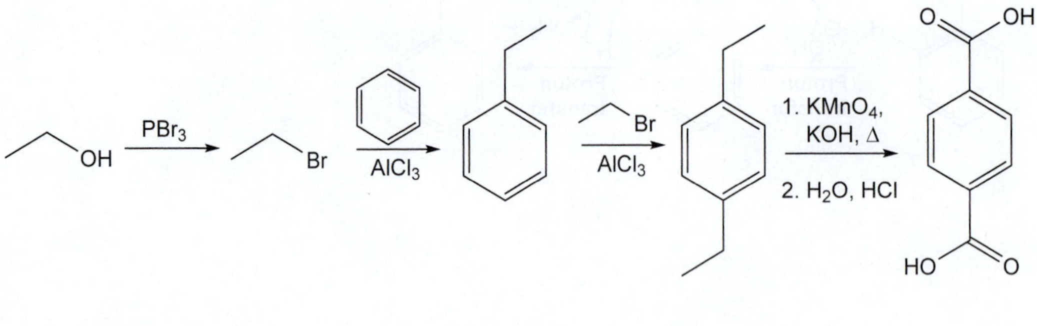

OH PBr_3 Br $AlCl_3$ Br $AlCl_3$ 1. $KMnO_4$, KOH, Δ 2. H_2O, HCl O OH HO O

Problem 23.31

Think

Are the substituents ortho/para or meta directors? What is the route for adding on the NH_2 group? Does it go through a synthetic intermediate that has a meta director attached to the ring?

Solve

To add an NH_2 group to the ring, we would first carry out nitration and then reduce the NO_2 group. Prior to the reduction, we can envision a NO_2 group on the ring to direct the chlorination toward the *meta* position. Afterward, the NO_2 group can be reduced to NH_2.

NO_2 NO_2 NH_2

HNO_3 / H_2SO_4 → ; Cl_2 / $FeCl_3$ → ; Fe / HCl →

Cl Cl

Problem 23.32

Think

Which group is the most activating? Where will it direct incoming electrophiles? What reaction occurs with aniline and a strong acid? Will this interfere with the desired reactions in your synthesis? How can this be avoided by protecting the amino portion of aniline? What type of director does the protected aniline compound have attached to the ring?

Solve

We can envision having an NH_2 group on the ring to direct the alkyl groups ortho and para in a Friedel–Crafts alkylation. The NH_2 group can come from nitration of benzene, followed by reduction of the NO_2. However, we must protect the NH_2 group before the introduction of the strong Lewis acid. Otherwise, the Lewis acid will complex to the NH_2 group, turning it into a strong deactivating group. This would prevent the necessary Friedel–Crafts alkylations from happening. We protect the NH_2 using acetic anhydride and deprotect it in the final step using hydrolysis under basic conditions.

NO_2 NH_2 O O HN O HN O

HNO_3 / H_2SO_4 → ; Fe / HCl → ; O → ; Cl / $AlCl_3$ → ; 1. NaOH, Δ / 2. H_2O, HCl

NH_2

Problem 23.33

Think

How can you block the para position? Are alkyl groups meta or ortho/para directors? Do you add the second alkyl group before or after reduction of the ketone?

Solve

We can envision a Friedel–Crafts acylation taking place with a sulfonyl group temporarily blocking the para position. The sulfonyl group can be placed at the para position after the acetyl group is reduced to an ethyl group to make it an ortho/para director. A final reduction step is necessary to form the propyl group.

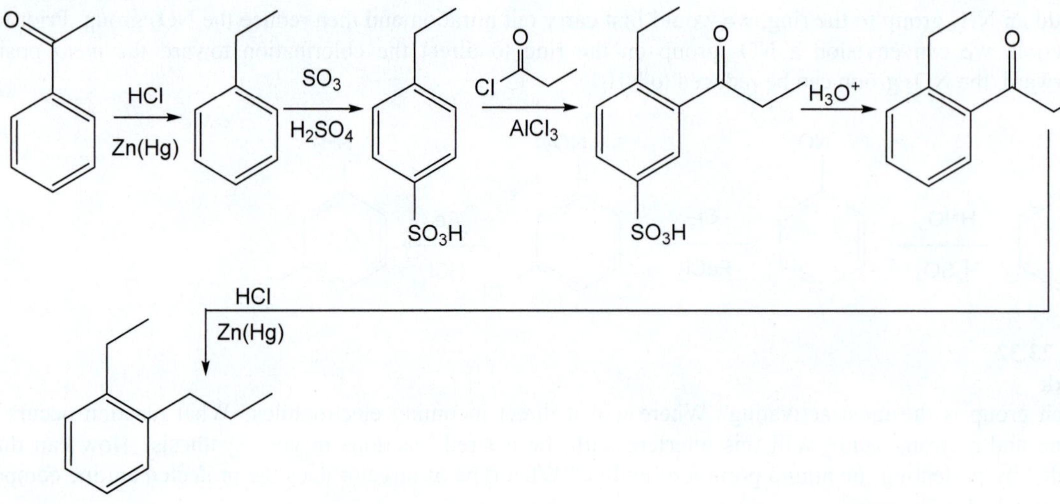

End of Chapter Problems
Sections 23.1 and 23.2 Ortho/Para and Meta Directors in Electrophilic Aromatic Substitution on Monosubstituted Benzenes
Problem 23.34

Think

What is the electrophile in each case? How is the electrophile formed? Is the group on the ring ortho/para or meta directing? Is ortho or para favored? What are the elementary steps for the electrophilic addition–elimination mechanism?

Solve

(a) The electrophile is the NO_2^+ cation. HO_3S is a meta director. The mechanism is shown below:

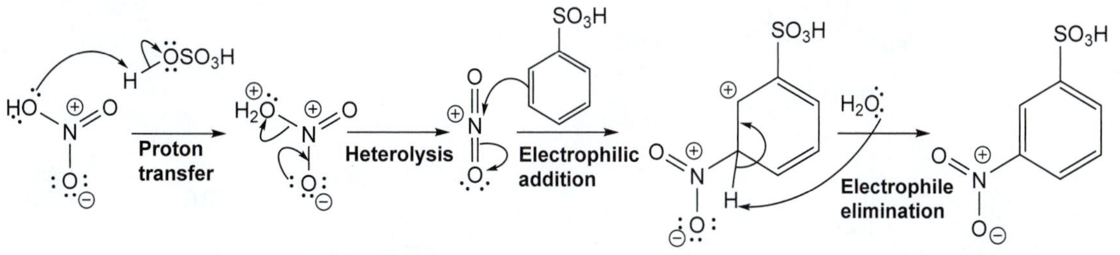

Proton transfer → Heterolysis → Electrophilic addition → Electrophile elimination

(b) The electrophile can be treated as the Cl^+ cation. CH_3CH_2 is an ortho/para director. It should be noted that the electrophilic addition and heterolysis steps probably occur simultaneously instead of separately, but the electrophile Cl^+ is shown explicitly for emphasis. The mechanisms are shown below:

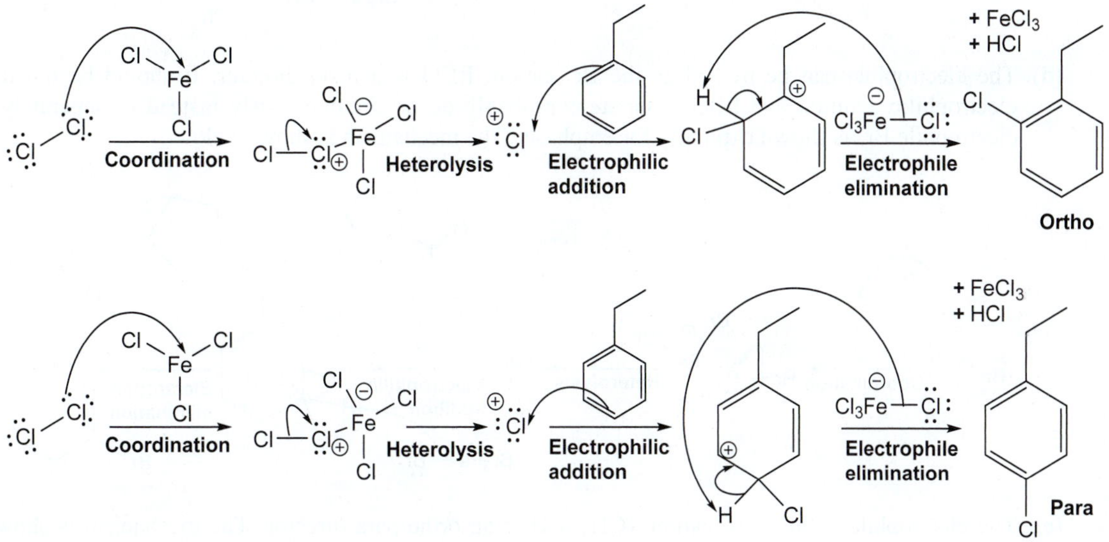

Coordination → Heterolysis → Electrophilic addition → Electrophile elimination (+ $FeCl_3$ + HCl)

Ortho

Coordination → Heterolysis → Electrophilic addition → Electrophile elimination (+ $FeCl_3$ + HCl)

Para

(c) The electrophile is the $PhCO^+$ cation. CH_3O is an ortho/para director. The mechanism is shown below:

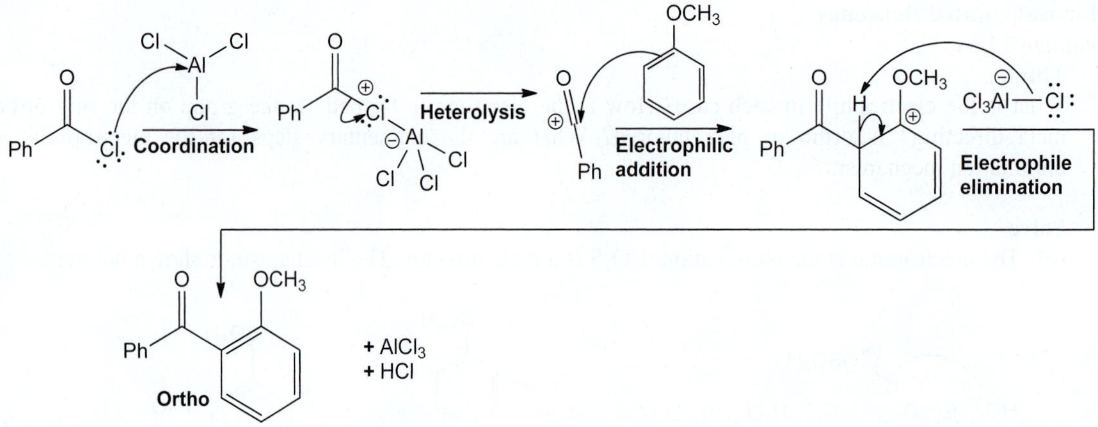

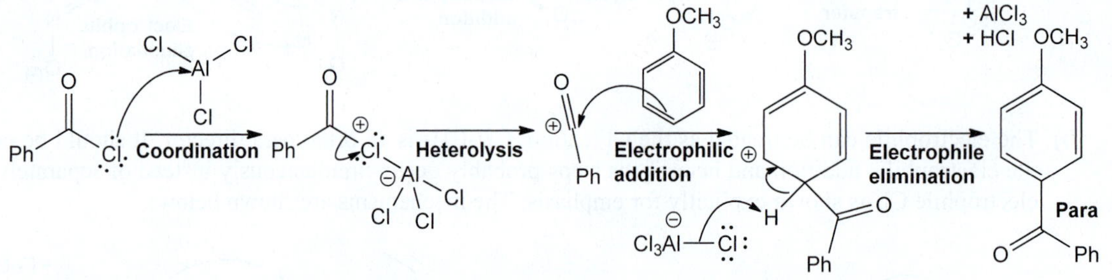

(d) The electrophile can be treated as the Br^+ cation. RCO is a meta director. It should be noted that the electrophilic addition and heterolysis steps probably occur simultaneously instead of separately, but the electrophile Br^+ is shown explicitly for emphasis. The mechanism is shown below:

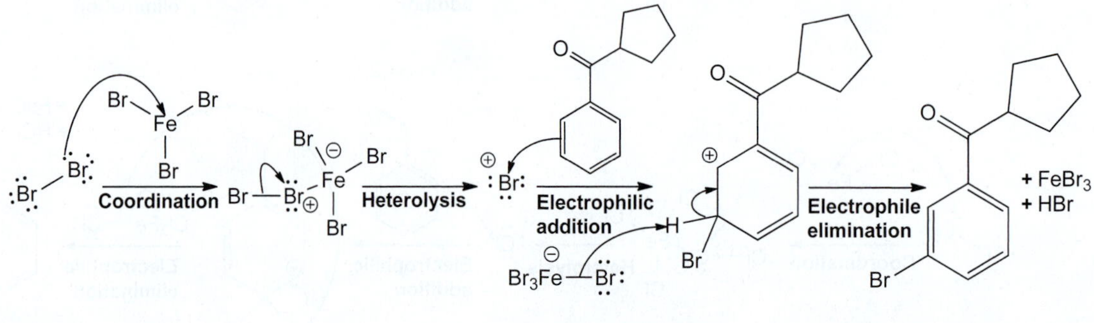

(e) The electrophile is the NO_2^+ cation. $(CH_3)_2CH$ is an ortho/para director. The mechanism is shown below and on the next page:

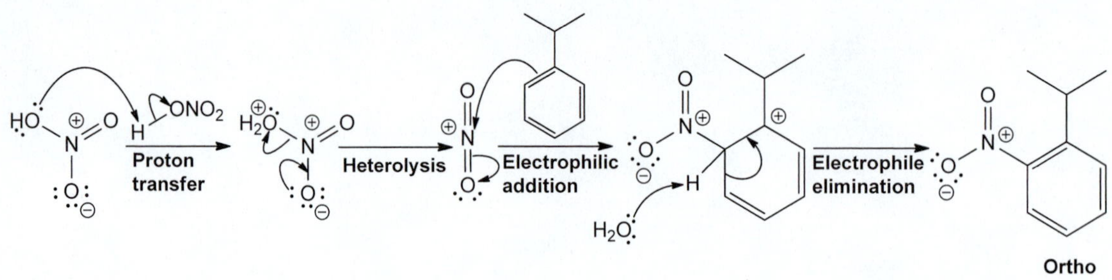

Proton transfer
Heterolysis
Electrophilic addition
Electrophile elimination
Para

Problem 23.35

Think

How many para and ortho sites are present? Statistically, which is the more likely site of reaction? Where is the ortho site located relative to the R group? How does steric crowding affect the product distribution?

Solve

There are two ortho sites and only one para site; statistically, therefore, it would be expected that the ortho site would have twice the product formation, as is observed when CH_3 is the substituent on the ring. However, the R group and electrophile E are next to each other in the ortho-substituted product and are on opposite sides of the ring in the para-substituted product. The electrophile's ability to undergo electrophilic addition is hindered as the sterics involving the R group increases. Therefore, as the steric bulk of the R group increases, the amount of ortho product decreases, and the amount of para product increases.

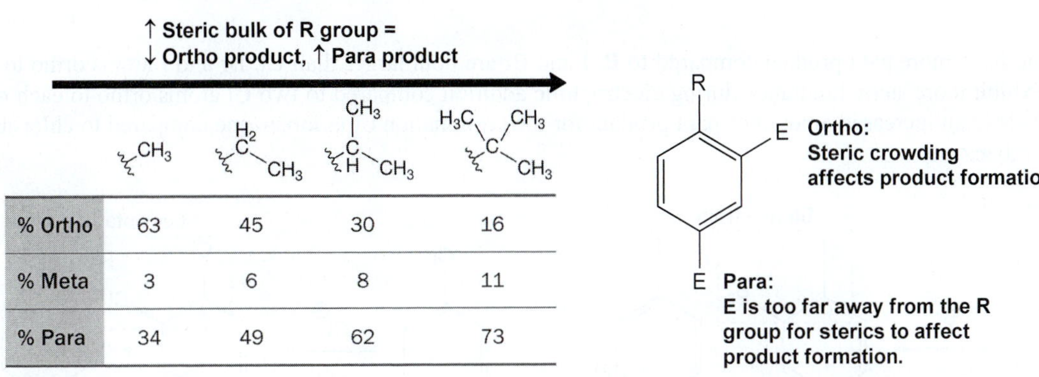
↑ Steric bulk of R group =
↓ Ortho product, ↑ Para product
Ortho:
Steric crowding
affects product formation.
Para:
E is too far away from the R
group for sterics to affect
product formation.

	CH_3	CH_2CH_3	$CH(CH_3)_2$	$C(CH_3)_3$
% Ortho	63	45	30	16
% Meta	3	6	8	11
% Para	34	49	62	73

Problem 23.36

Think

What fraction of the ortho product would come from a reaction involving the C–D bond, and what fraction would come from a reaction involving the C–H bond? How does this affect the product distribution?

Solve

Half of the ortho product would come from a reaction involving the C–D bond, and half would come from a reaction involving the C–H bond. Half of 50% is 25%. Therefore, the product distribution is 50% para, 25% ortho from the reaction of C–D, and 25% ortho from the reaction of C–H.

HNO_3, H_2SO_4

Para 50%

Ortho 25% 25%

Problem 23.37

Think

Which atom is larger, Br or Cl? How does the size of the atom affect the ortho/para product distribution?

Solve

Br is a larger atom compared to Cl. Br atoms ortho to each other exhibit more steric hindrance during electrophilic addition compared to when Cl and Br are ortho to each other. This leads to an increased amount of para product for the bromination of bromobenzene compared to chlorination of bromobenzene.

Steric strain

Less steric strain

13% 85% 42% 53%

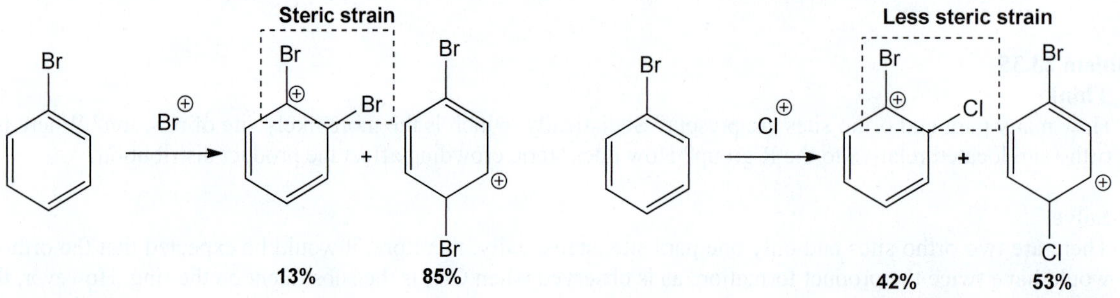

Problem 23.38

Think

Compare the sizes of I, Br, and Cl. How does the size of the atom affect the ortho/para product distribution?

Solve

A would have more para product compared to **B**. I and Br are both larger than Cl. Br and I atoms ortho to each other exhibit more steric hindrance during electrophilic addition compared to two Cl atoms ortho to each other. This leads to an increased amount of para product for the bromination of iodobenzene compared to chlorination of chlorobenzene.

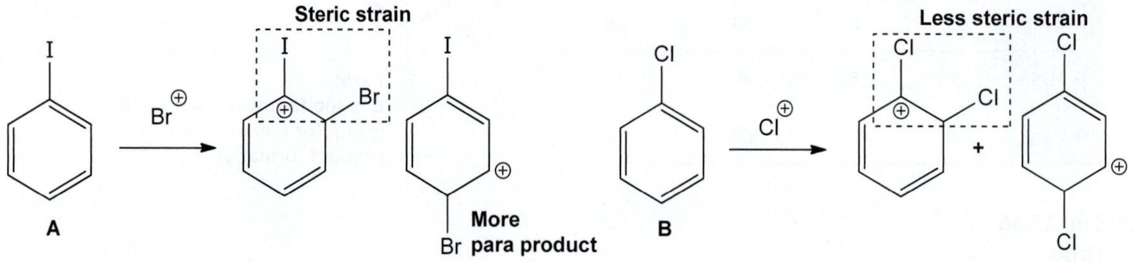

Steric strain

Less steric strain

A

More para product

B

Problem 23.39

Think

Are there any lone pairs of electrons on B? Is the group an inductive electron donator or withdrawer? How would this affect the stability of the arenium ion intermediate produced from ortho, meta, and para attack?

Solve

The substituent is a meta director, and the most likely sites of attack are indicated by asterisks. The B has no lone pairs (which would otherwise make it an ortho/para director). The electronegative O atoms make it such that the substituent would destabilize the arenium ion intermediate if attack were to occur at an ortho or para position.

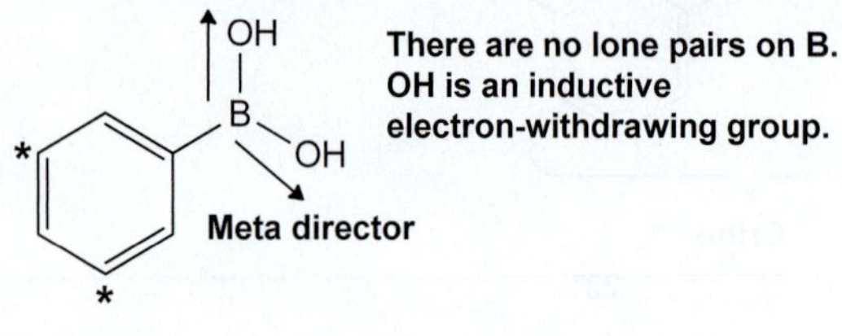

There are no lone pairs on B.
OH is an inductive electron-withdrawing group.

Meta director

Problem 23.40

Think

What is the structure of the −N=O substituent? Are there lone pairs on the N atom? Are additional arenium ion resonance structures possible as a result? Are the additional resonance structures possible when an electrophile adds to the ortho and para positions or the meta position?

Solve

It is an ortho/para-directing group, because the atom at the point of attachment (N) has a lone pair. The lone pair on N participates in resonance in the arenium ion intermediate only when E^+ attacks an ortho or para position. This leads to a resonance structure where all nonhydrogen atoms have a full octet.

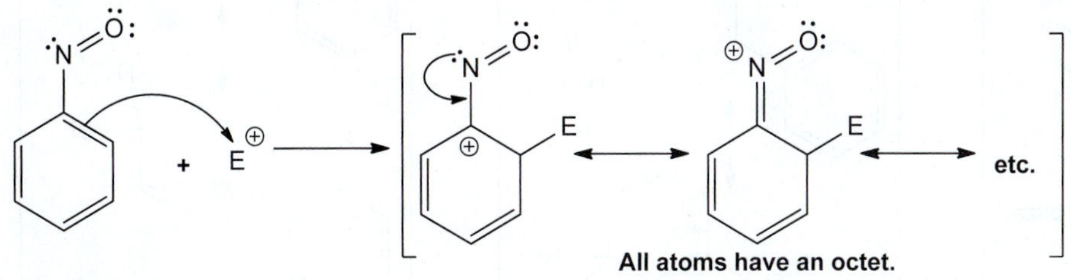

All atoms have an octet.

Problem 23.41

Think

Where are the ortho and para sites on the ring? Which arenium ion intermediate—ortho, meta, or para—allows for the charge delocalization to occur on both rings? Why is this more stable?

Solve

(a) In this bromination reaction, H^+ is replaced by Br^+. Because the phenyl ring is an ortho/para director, this occurs at the ortho and para positions to yield the following:

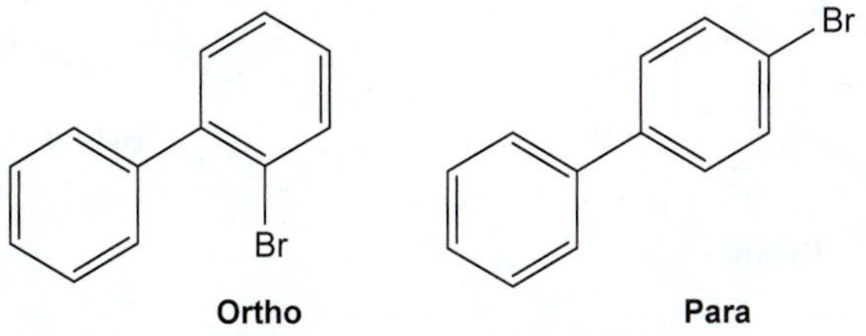

Ortho Para

(b) We can justify why the phenyl ring is an ortho/para director by drawing the three isomeric intermediates, as shown on the next page. When the Br^+ attacks either the ortho or para position, the resulting positive charge is delocalized onto both rings, whereas when the Br^+ attacks at the meta position, it is delocalized only over one ring.

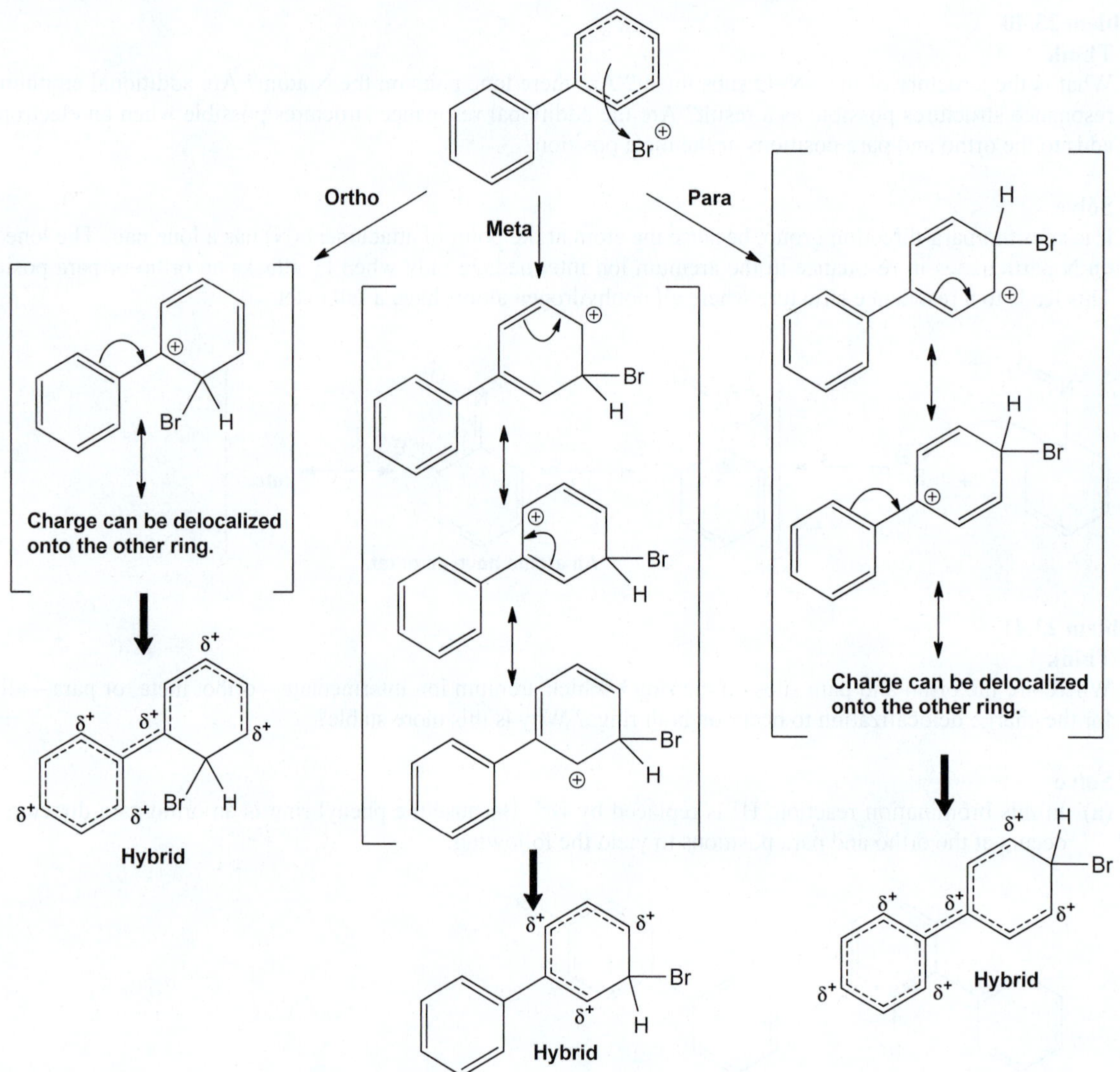

Section 23.3 The Activation and Deactivation of Benzene toward Electrophilic Aromatic Substitution

Problem 23.42

Think

What is the structure of the −NHCOR substituent? Are there lone pairs on the N atom? Are additional arenium ion resonance structures possible as a result? Does the C=O participate in resonance with the lone pair on N? How does that affect the −NHCOR group's activating/deactivating nature?

Solve

(a) It is an activating group, because the participation of the lone pair on N in resonance in the arenium ion intermediate provides more stability to the intermediate than H does.

(b) It is less activating than NH_2, because the C=O group in NHCOR participates in resonance with the lone pair on N. Therefore, the lone pair on N is less available to stabilize the arenium ion intermediate than the lone pair on $-NH_2$. With that lone pair on N less available for resonance with the ring, the arenium ion intermediate is not as stable.

See the figure on the next page.

The lone pair can participate in resonance to stabilize the arenium ion intermediate.

The lone pair is partly tied up in resonance with the C=O.

Problem 23.43

Think

Consult Table 23-3 to identify each substituent as a weak, moderate, or strong activator or deactivator. Which ring is most activated? Which ring is most deactivated? How does the activated/deactivated nature of the ring correspond to the rate of electrophilic aromatic substitution?

Solve

(a) The one on the right; OH is a stronger activator than CH_3.

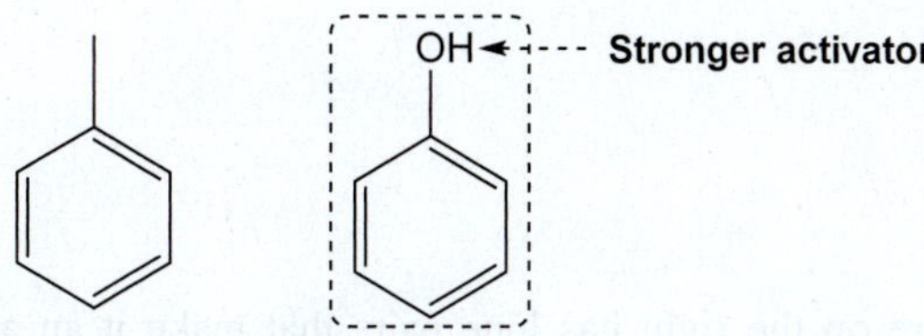

(b) The one on the left; CH_3 is an activating group owing to its electron-donating effect, whereas CF_3 is a deactivating group owing to its electron-withdrawing effect.

Electron donating = activating Electron withdrawing = deactivating

(c) The one on the right; CO_2H is a deactivating group, whereas OCH=O, being connected by an atom with lone pairs, is an activating group. The lone pairs allow for the formation of one resonance structure of the arenium ion intermediate in which all atoms have a full octet.

Electron withdrawing = deactivating Resonance involving lone pair = activating

(d) The one on the left, because, relative to H, NO_2 is an electron withdrawing group and, therefore, a deactivating group.

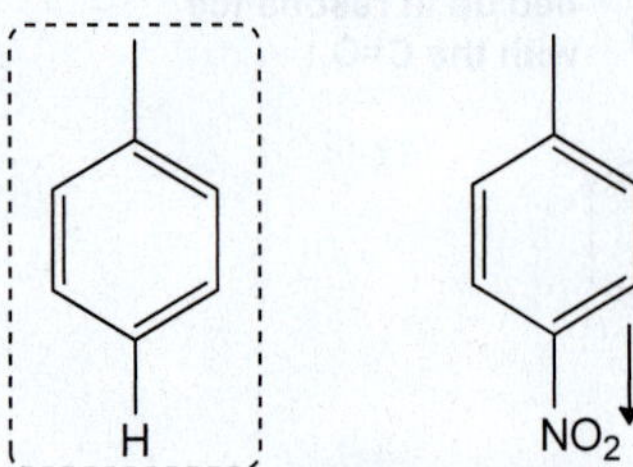

NO_2 is an electron withdrawing group = deactivating

(e) The one on the left, because Cl is a deactivating group, and two deactivating groups slow the reaction more than one.

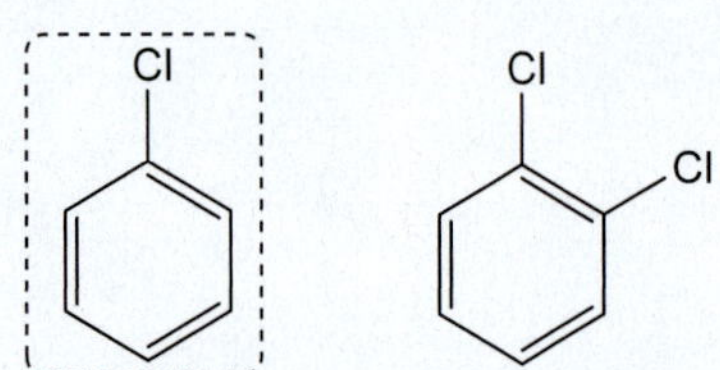

One deactivator Two deactivators

(f) The one on the right. The S atom in the structure on the right has lone pairs that make it an activating group, whereas SO_3H is a deactivating group.

Electron withdrawing = deactivating Resonance involving lone pair = activating

(g) The one on the left, because OH is a stronger activating group than CH_3.

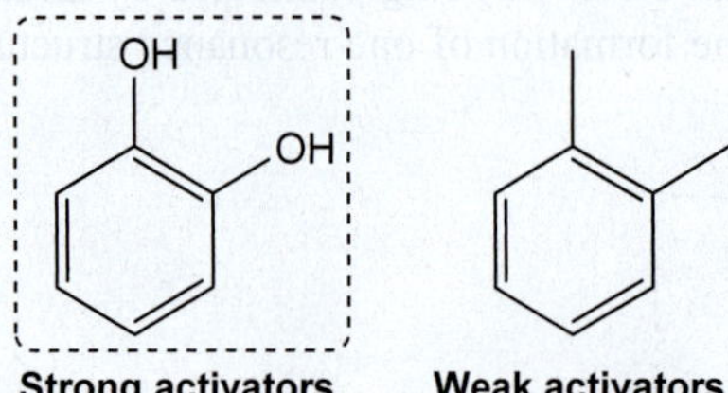

Strong activators Weak activators

(h) The one on the right, because it has two activating groups, whereas the one on the left has just one.

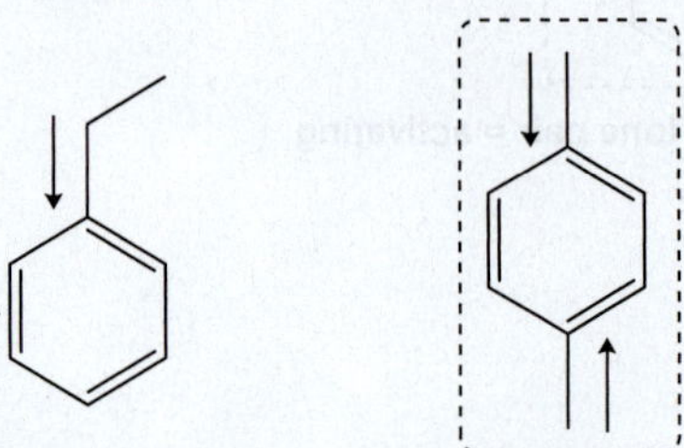

One activating group Two activating groups

Problem 23.44

Think

In phenylmethanol, are there lone pairs on the O atom? What type of atom is in between the HO group and the ring? Are additional arenium ion resonance structures possible? Is the HO inductively electron donating or withdrawing? How does this affect the ring's activation?

Solve

If an OH group is attached *directly* to the ring, as in phenol, the lone pairs on O can participate in resonance to stabilize an arenium ion intermediate. In phenylmethanol, the O is not attached directly to the ring, and an *sp*^{3} C atom is in between. Therefore, the lone pairs cannot participate in resonance to provide stabilization of the arenium ion. Instead, the O atom is highly electronegative and, therefore, is inductively electron withdrawing and destabilizes an arenium ion.

Inductive effects from O destabilize the arenium ion intermediate.

H_2 C OH

Lone pairs cannot participate in resonance with the ring and, therefore, can't stabilize the arenium ion intermediate.

sp^3 C

Problem 23.45

Think

Which sites on the benzene ring are activated from the CH_3 groups? Are any of the sites activated by more than one CH_3? How does steric hindrance come into play?

Solve

Each CH_3 group is an ortho/para director. The symbols on the ring in each molecule below correspond to the CH_3 group that favors that site of attack. In each case, two sites (indicated by boxes) are favored by two separate CH_3 groups. In 1,2,4-trimethylbenzene, one of those sites is sterically hindered, which slows the reaction.

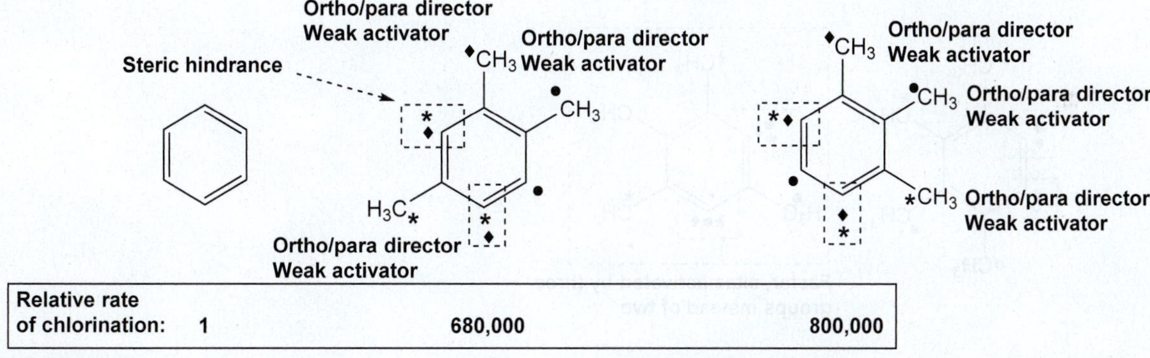

Problem 23.46

Think

Which sites on the benzene ring are activated from the CH_3 groups? Are any of the sites activated by more than one CH_3? How does steric hindrance come into play?

Solve

The symbols on the ring in each molecule below correspond to the CH_3 group that favors that site of attack.

(a) The methyl groups in *o*-dimethylbenzene activate different carbons on the ring, whereas the methyl groups in *m*-dimethylbenzene activate the same carbons in the ring. So *m*-dimethylbenzene will undergo chlorination faster.

Faster, same sites activated

(b) The methyl groups in *p*-dimethylbenzene activate different carbons on the ring, whereas the methyl groups in *m*-dimethylbenzene activate the same carbons on the ring. So *m*-dimethylbenzene will undergo chlorination faster.

Faster, same sites activated

(c) 1,2,3,5-Tetramethylbenzene will undergo chlorination faster, because each position is activated by three methyl groups. In 1,2,3,4-tetramethylbenzene, each position is activated by only two methyl groups.

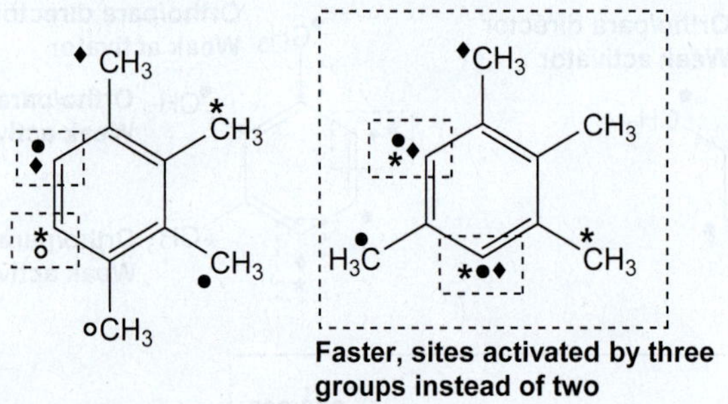

Faster, sites activated by three groups instead of two

Problem 23.47

Think

Is the −N=O group an inductively electron-withdrawing or electron-donating group? How does that affect whether the group is activating or deactivating?

Solve

In Problem 23.40, we learned that the −N=O is ortho/para directing because of the lone pair of electrons on the N. The group is deactivating, however, because the O atom is highly electron withdrawing inductively and is very close to the ring. Its inductive effect is similar to that from a C=O group. See the figure on the next page.

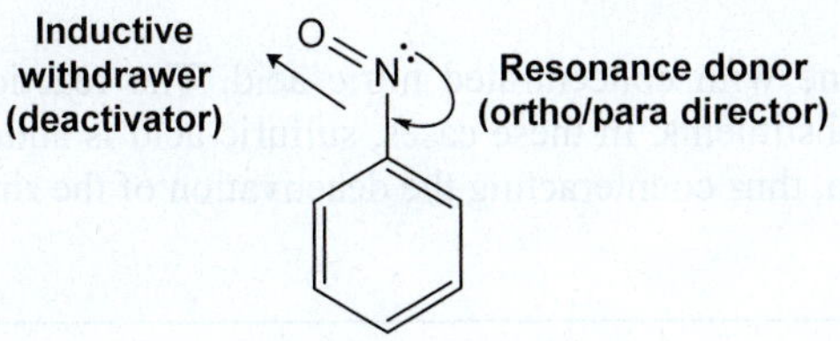

Inductive withdrawer (deactivator)

Resonance donor (ortho/para director)

Problem 23.48

Think

Are CH_3 groups electron donating or withdrawing? How should that effect impact the stability of the arenium ion intermediate and, thus, the rate of electrophilic aromatic substitution when NH_2 is replaced by $N(CH_3)_2$? Are there lone pairs on the N atom? For the lone pair to stabilize the arenium ion intermediate, how must the plane of the N be oriented with respect to the plane of the ring? Can that orientation be achieved when there are two CH_3 groups on the ring in the ortho positions as well as two CH_3 groups attached to N?

Solve

The NH_2 group activates the ring toward electrophilic aromatic substitution because its lone pair can participate in resonance to stabilize the arenium ion intermediate if the electrophile attacks either the ortho or para position. A key resonance structure upon para attack is shown below. In *N,N*-dimethylaniline, the electron-donating capability of the methyl groups further stabilizes this intermediate, which is why the reaction takes place faster.

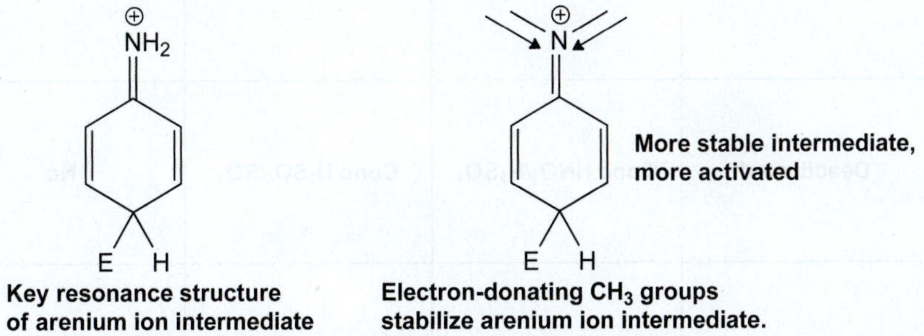

More stable intermediate, more activated

Key resonance structure of arenium ion intermediate

Electron-donating CH_3 groups stabilize arenium ion intermediate.

Notice that this resonance structure can contribute only if the NH_2 or the $N(CH_3)_2$ group is in the same plane as the ring. If, however, two methyl groups are also on the ring at the ortho positions, steric repulsion forces the $N(CH_3)_2$ to be perpendicular to the ring, effectively destroying the resonance stabilization.

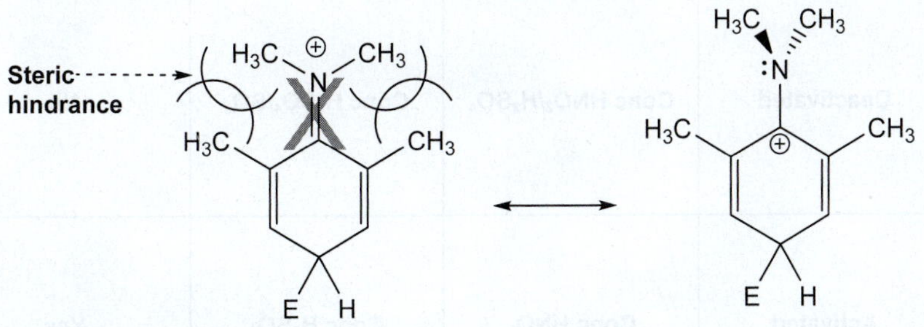

Steric hindrance

Sections 23.4–23.6 Reaction Conditions and Disubstituted Benzenes in Electrophilic Aromatic Substitution Reactions

Problem 23.49 (SYN)

Think

What are the usual conditions to carry out nitration on benzene? What effect does the addition of sulfuric acid have on a nitration reaction? When would that be necessary? In each example, is the ring activated or deactivated?

Solve

Nitration can be carried out on benzene by treating benzene with concentrated nitric acid. The reaction is slowed, however, when the ring has attached deactivating substituents. In these cases, sulfuric acid is added to nitric acid to carry out a nitration to help speed up the reaction, thus counteracting the deactivation of the ring. See the table below for Problems 23.49–23.51.

Example	Activated/ Deactivated	Nitration Conditions	Sulfonation Conditions	Friedel–Crafts Yes/No
(a)	Activated	Conc HNO_3	Conc H_2SO_4	Yes
(b) O	Deactivated	Conc HNO_3/H_2SO_4	Conc H_2SO_4/SO_3	No
(c) NC	Deactivated	Conc HNO_3/H_2SO_4	Conc H_2SO_4/SO_3	No
(d) SO_3H	Deactivated	Conc HNO_3/H_2SO_4	Conc H_2SO_4/SO_3	No
(e) SO_3H Cl	Deactivated	Conc HNO_3/H_2SO_4	Conc H_2SO_4/SO_3	No
(f) OCH_3	Activated	Conc HNO_3	Conc H_2SO_4	Yes

Problem 23.50 (SYN)

Think

What are the usual conditions to carry out sulfonation on benzene? What effect does the addition of SO_3 have on a sulfonation reaction? When would that be necessary? In each example, is the ring activated or deactivated?

Solve

Sulfonation can be carried out on benzene by treating benzene with concentrated nitric acid. The reaction is slowed, however, when the ring has attached deactivating substituents. In these cases, SO_3 is added to concentrated sulfuric acid to carry out a sulfonation to help speed up the reaction, thus counteracting the deactivation of the ring. Answers to **(a)–(f)** are summarized in the table in the solution to Problem 23.49.

Problem 23.51 (SYN)

Think

Are Friedel–Crafts reactions feasible when the aromatic ring is moderately or strongly deactivated? Which rings are moderately or strongly deactivated?

Solve

Friedel–Crafts reactions are unfeasible when the aromatic ring is moderately or strongly deactivated, so only compounds **(a)** and **(f)** would undergo the Friedel–Crafts reaction. Compounds in **(b)–(e)** have moderate and strongly deactivating substituents on the benzene ring. In **(d)**, the weakly activating alkyl substituent is not sufficiently activating to overcome the effects from the strongly deactivating SO_3H group. Answers to **(a)–(f)** are summarized in the table in the solution to Problem 23.49.

Problem 23.52

Think

Consult the partial mechanism in Equation 23-17. For phenol to undergo multiple brominations under neutral conditions, what is the powerfully activating substituent attached to the ring? Under the same conditions, can anisole have the same substituent?

Solve

Under neutral conditions, phenol is deprotonated to a small extent, and the resulting phenoxide anion has the powerfully activating O^- substituent to facilitate multiple brominations.

The O^- substituent is a *very* powerful activator.

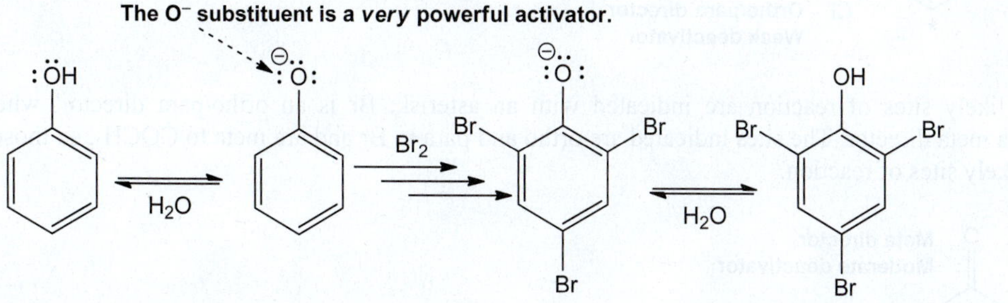

(23-17)

Anisole does not have such an acidic proton, so the activating group under neutral conditions remains OCH_3, which is far less activating than O^-. Thus, anisole undergoes bromination much slower.

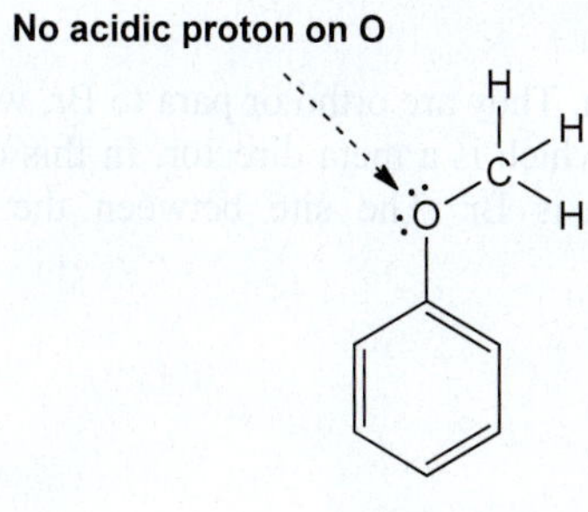

No acidic proton on O

Anisole

Problem 23.53

Think

Identify each substituent as a weak, moderate, or strong activator or deactivator. To what positions (ortho, meta, or para) does each substituent direct an incoming electrophile? Are the two substituents in competition? If so, which one wins? How does steric hindrance come into play?

Solve

(a) The sites indicated with an asterisk are the most likely sites of reaction. OCH_3 is an ortho/para director. The sites indicated with an asterisk are ortho or para to each OCH_3 group; so is the site in between the OCH_3 groups, but steric hindrance prevents attack there.

Ortho/para director
Strong activator
OCH_3
Too sterically crowded
*
*
OCH_3
Ortho/para director
Strong activator

(b) Precisely the same reason as in **(a)**, because both OH and Cl are ortho/para directors.

Ortho/para director
Strong activator
OH
Too sterically crowded
*
*
Cl
Ortho/para director
Weak deactivator

(c) The most likely sites of reaction are indicated with an asterisk. Br is an ortho/para director, whereas $COCH_3$ is a meta director. The sites indicated are ortho and para to Br and are meta to $COCH_3$, so those are the most likely sites of reaction.

O
Meta director
Moderate deactivator
*
*
Br
Ortho/para director
Weak deactivator

(d) The sites indicated with an asterisk are the most likely sites of reaction. They are ortho or para to Br, which is an ortho/para director. But they are also ortho or para to $COCH_3$, which is a meta director. In this case, regiochemistry is governed by the more activating group, which is Br. The site between the two substituents is not attacked significantly, because of steric hindrance.

O
Meta director
Moderate deactivator
*
*
Too sterically crowded
Br
Ortho/para director
Weak deactivator

(e) The most likely sites of reaction are indicated with an asterisk. These sites are ortho or para to the CH_3 group (an ortho/para director) and meta to the NO_2 group (a meta director).

NO_2 **Meta director**
Strong deactivator

Ortho/para director
Weak activator

(f) The most likely sites of reaction are indicated with an asterisk. All four possible sites for attack are activated by an alkyl group. The ones adjacent to the *t*-butyl group, however, are sterically hindered.

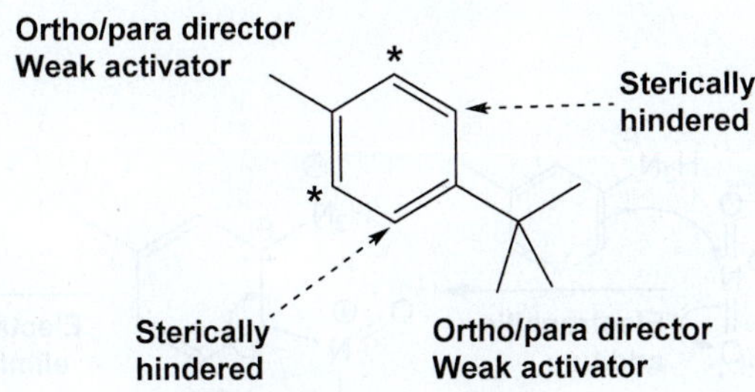

(g) The sites indicated with an asterisk are the most likely sites of reaction. The NH_2 group is the most activating group on the benzene ring, and so attack should be either ortho or para to the NH_2. The position that is ortho to the NH_2 group that is not marked is not favored because of steric hindrance.

Too sterically crowded

Ortho/para director
Strong activator
NH_2

Ortho/para director
Weak activator

Problem 23.54

Think

What is the electrophile in each case? How is the electrophile formed? Is the group on the ring ortho/para or meta directing, and is it activating or deactivating? Do the conditions of the reaction alter the nature of the group attached to the aromatic ring? What are the elementary steps for the electrophilic addition–elimination mechanism?

Solve

CH_3 is an ortho/para director, but the directionality of NH_2 changes depending on the pH of the solution. In both examples, the electrophile is the NO_2^+ cation. The sites indicated with an asterisk are the most likely sites of reaction. In reaction **(b)**, a mildly acidic solution, NH_2 remains unprotonated and is an ortho/para director.

(a) This is a strongly acidic solution; NH_2 is protonated and is a meta director. The CH_3 group is an activator, and the regiochemistry is directed by this substituent.

H_2N — H–OSO_3H — **Proton transfer** → $H_3N^{\oplus}$

Meta director
Deactivator

Too sterically crowded

Ortho/para director
Activator

(b) This is a mildly acidic solution; NH_2 remains unprotonated and is an ortho/para director. The CH_3 is also an ortho/para director, and the regiochemistry of both is in agreement.

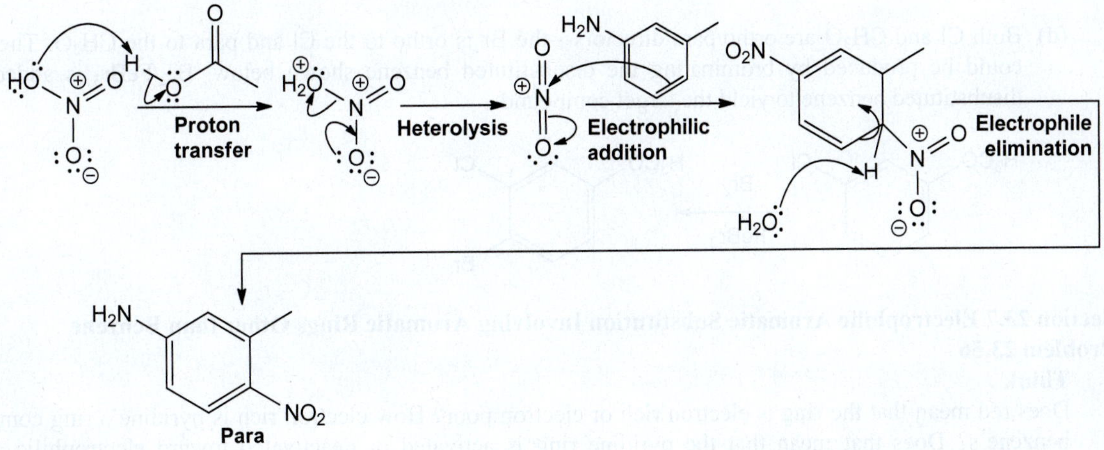

Proton transfer

Heterolysis

Electrophilic addition

Electrophile elimination

Para

Problem 23.55 (SYN)

Think

Which groups in the target are ortho/para directors? Is there a substituent in the target that is ortho or para to one of these groups? Which groups in the target are meta directors? Is there a substituent in the target that is meta to one of these groups? In the disubstituted benzene precursor, are there substituents that would be in disagreement with regard to their ortho/para or meta directing ability? If so, how do you determine which substituent governs the regiochemistry?

Solve

(a) CH_3 is an ortho/para director, and the NO_2 is ortho to one CH_3 and para to the other. The product could be produced by nitrating 1,3-dimethylbenzene, as shown below. Nitric acid is added to this disubstituted benzene to yield the target compound.

HNO_3

NO_2

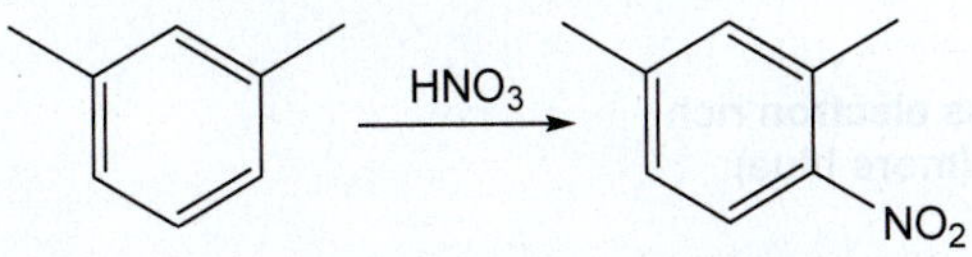

(b) OH is an ortho/para director, and $CH_3C{=}O$ is a meta director—the Cl is ortho to the OH and meta to the $CH_3C{=}O$. The product could be produced by chlorinating the disubstituted benzene shown below. $Cl_2/AlCl_3$ is added to this disubstituted benzene to yield the target compound.

O HO Cl_2 $AlCl_3$ O Cl HO

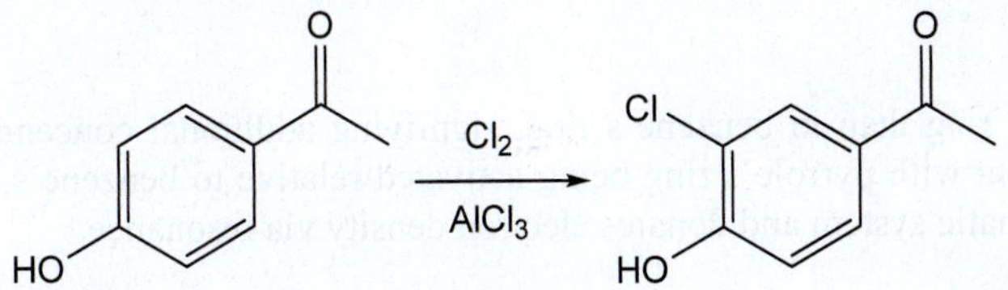

(c) CH_3 is an ortho/para director (activator), and NO_2 is a meta director (deactivator). The SO_3H substituent is ortho to the CH_3 and para to the NO_2. These groups are in disagreement, but the activator CH_3 directs the regiochemistry. The product could be produced by sulfonating the disubstituted benzene shown below. Fuming H_2SO_4 is added to this deactivated disubstituted benzene to yield the target compound.

O_2N H_2SO_4 SO_3 SO_3H O_2N

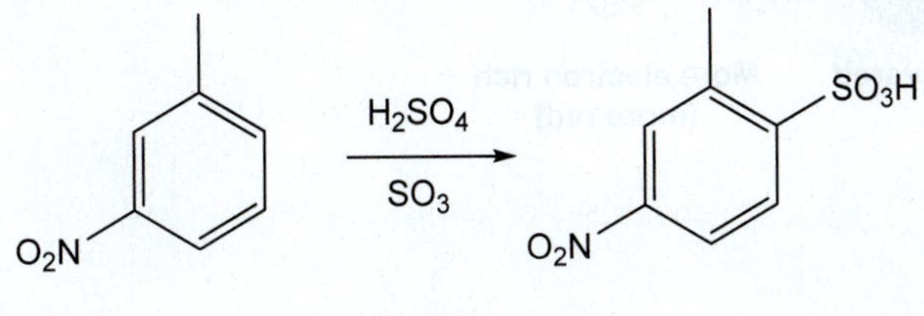

(d) Both Cl and CH_3O are ortho/para directors—the Br is ortho to the Cl and para to the CH_3O. The product could be produced by brominating the disubstituted benzene shown below. $Br_2/FeBr_3$ is added to this disubstituted benzene to yield the target compound.

H_3CO … Cl $\xrightarrow[FeBr_3]{Br_2}$ H_3CO … Cl, Br

Section 23.7 Electrophilic Aromatic Substitution Involving Aromatic Rings Other than Benzene

Problem 23.56

Think

Does red mean that the ring is electron rich or electron poor? How electron rich is pyridine's ring compared to benzene's? Does that mean that the pyridine ring is activated or deactivated toward electrophilic aromatic substitution?

Solve

The pyridine ring is *deactivated* compared to the benzene ring because the nitrogen is an electronegative atom and pulls electron density away from the aromatic ring. In the electrostatic potential maps, red signifies a buildup of negative charge, and blue signifies a buildup of positive charge. Notice the greater intensity of red in the benzene ring than in the pyridine ring.

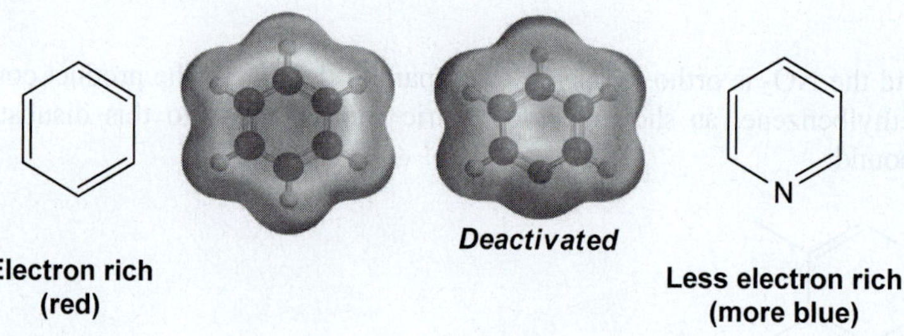

Problem 23.57

Think

Does red mean that the ring is electron rich or electron poor? How electron rich is pyrrole's ring compared to benzene's? Does that mean that the pyrrole ring is activated or deactivated toward electrophilic aromatic substitution?

Solve

There appears to be more red in pyrrole's ring than in benzene's ring, signifying additional concentration of negative charge. Therefore, this is consistent with pyrrole's ring being activated relative to benzene's, because the nitrogen electron pair is part of the aromatic system and donates electron density via resonance.

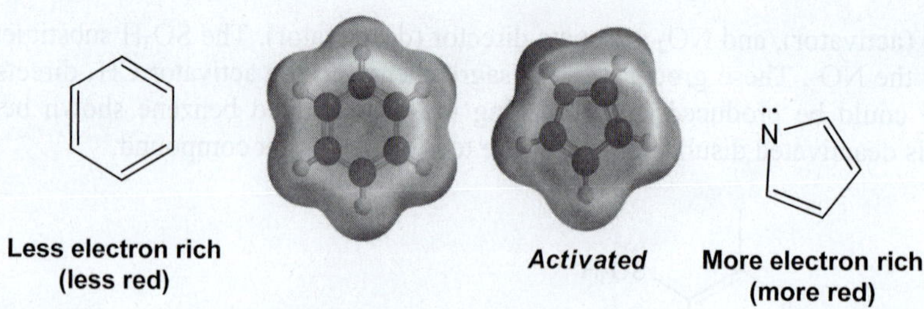

Problem 23.58

Think

What side reaction occurs between the lone pairs on N and a Lewis acid? Are the lone pairs of electrons on the N in pyrrole part of the aromatic system (i.e., tied up in aromaticity)? If so, how does this affect their ability to engage in the side reaction?

Solve

The coordination of N to a Lewis acid catalyst is a problem if the lone pair on N is readily available to form a bond. In aniline, the lone pair on N remains relatively available, because it is only weakly involved in resonance with the benzene ring—any resonance structures involving the lone pair will not exhibit aromaticity. This makes the ring highly deactivated, so a Friedel–Crafts reaction is unfeasible. In pyrrole, the lone pair is not readily available, because it is part of the six π electrons that establish aromaticity in the ring. Therefore the ring remains activated, and the Friedel–Crafts reactions are feasible.

Lone pairs not part of aromatic π system

NH_2 → (O, Cl; Lewis acid catalyst) → Lewis acid, ⊕ NH_2

No Friedel–Crafts acylation reaction

Lone pairs are part of aromatic π system

H, N → (O, Cl; Lewis acid catalyst) → H, N, O

Friedel–Crafts acylation reaction

Problem 23.59

Think

How are the two arenium ion intermediates produced? Draw all resonance structures for each arenium ion intermediate. Do they both have the same number of resonance structures? How many resonance structures of each intermediate preserve the aromaticity? Should a resonance structure that preserves aromaticity have a stronger or weaker contribution to the resonance hybrid?

Solve

The arenium ion intermediates for α and β substitution by Friedel–Crafts acylation of naphthalene are shown below. Both arenium ion intermediates have five total resonance structures. The α-substituted product, however, is the major product, because two of the resonance structures in the arenium ion intermediate maintain aromaticity (six π electrons in the same ring). The arenium ion intermediate produced from β substitution, on the other hand, has just one resonance structure that maintains aromaticity. Thus, the arenium ion intermediate produced from β substitution is more stable.

:O: ⊕ → $AlCl_3$ → [H, O, ⊕ ↔ H, O, ⊕ ↔ **Three other resonance structures destroy aromaticity.**]

α Substitution (major) **Aromatic** **Aromatic**

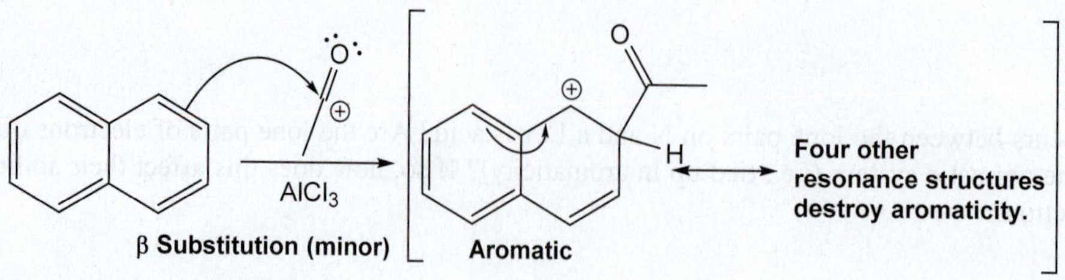

Problem 23.60

Think

What is the structure of the arenium ion intermediate? Is the naphthalene ring more or less activated compared to benzene? Why might this affect the need for an acid catalyst?

Solve

The naphthalene ring is more activated compared to benzene. The arenium ion intermediate that forms in the reaction of naphthalene with Br_2 results in an ion that is still aromatic. Because the aromaticity of one of the rings is not destroyed in the arenium ion intermediate, the energy barrier for the reaction is substantially lower, and milder conditions are permitted.

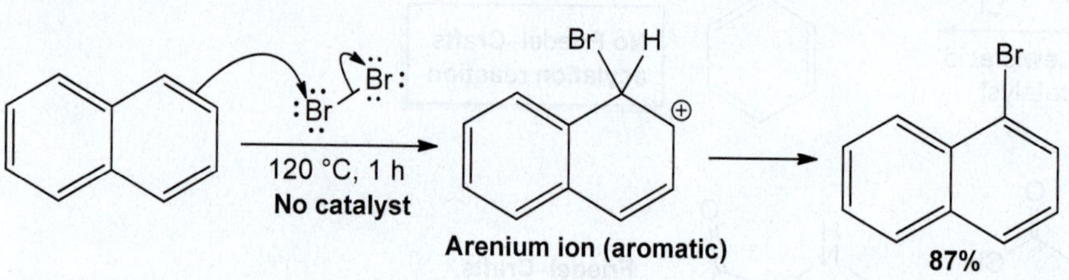

Problem 23.61

Think

Which product has less steric crowding? How does this affect the product formation?

Solve

The β-substituted product has less steric crowding (as shown below) and, therefore, is the more stable product and the one formed as the major product.

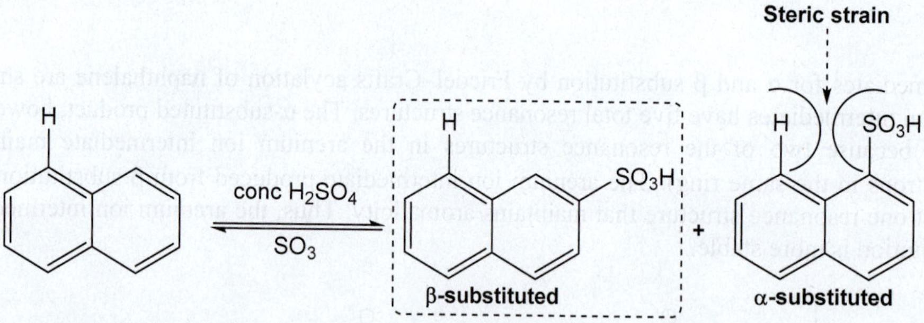

Problem 23.62

Think

When the substituent on naphthalene is an activator like CH_3, which ring is more activated? Which ring is more susceptible to electrophilic aromatic substitution? When the substituent is a deactivator like NO_2, which ring is more activated? Which ring is more susceptible to electrophilic aromatic substitution?

Solve

When the substituent on naphthalene is an activator like CH_3, electrophilic aromatic substitution takes place on the substituted ring because it is the more activated ring. When the substituent is a deactivator like NO_2, electrophilic aromatic substitution takes place on the unsubstituted ring because it is the more activated ring. In electrophilic aromatic substitution, the ring acts as a nucleophile, so when the ring has an electron-donating group, it has more electron density and therefore will be more reactive toward electrophiles. When a ring has an electron-withdrawing group, the other ring is more electron rich.

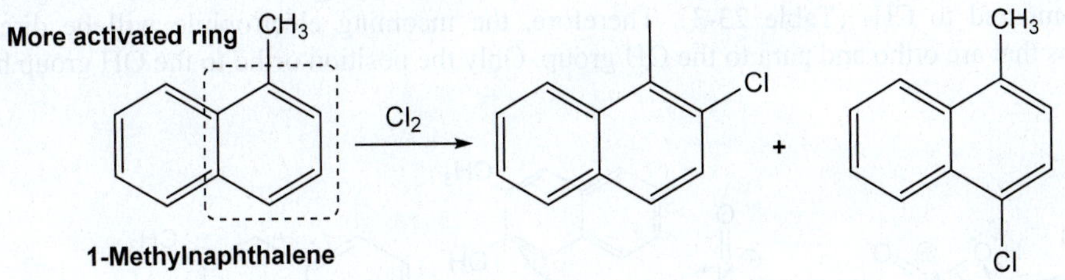

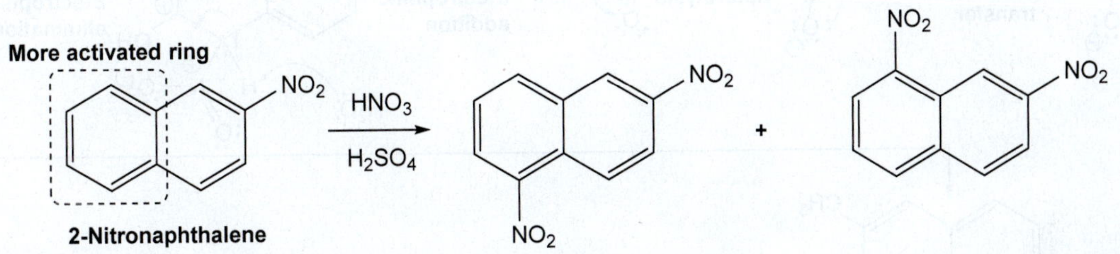

More activated ring CH_3

1-Methylnaphthalene $\xrightarrow{Cl_2}$ (Cl, CH_3 product) + (CH_3, Cl product)

More activated ring

2-Nitronaphthalene $\xrightarrow[H_2SO_4]{HNO_3}$ (NO_2, NO_2 product) + (NO_2, NO_2 product)

Problem 23.63

Think

How are the different arenium ion intermediates produced? Draw all resonance structures for each arenium ion intermediate. Do they both have the same number of resonance structures? How many resonance structures of each intermediate preserve the aromaticity? How does that impact the stability of the resonance hybrid?

Solve

The arenium ions from which the products are formed are shown below. Each has the same number of resonance structures. When an electrophile attacks at the 1 position, however, there are more resonance structures that allow the aromaticity to be preserved in the benzene ring (six π electrons in the same ring). This stabilizes the resonance hybrid and lowers the energy barrier for electrophilic aromatic substitution.

Attack at 1 position: $E^{\oplus}$, CH_3 → [E, ⊕, CH_3 ⟷ E, CH_3, ⊕ ⟷ Three other resonance structures destroy aromaticity.]

Attack at 3 position: CH_3, $E^{\oplus}$ → [CH_3, E, ⊕ ⟷ Four other resonance structures destroy aromaticity.]

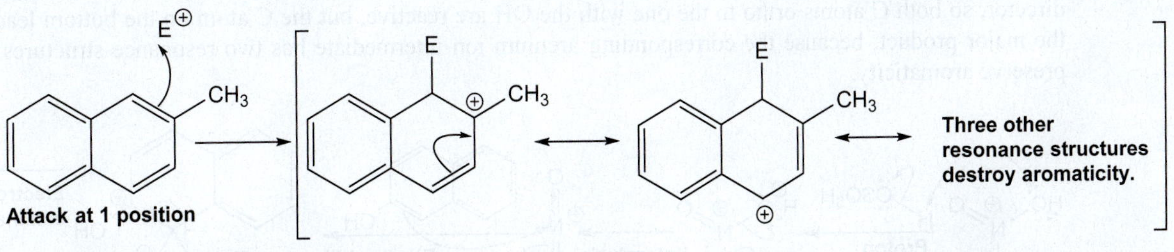

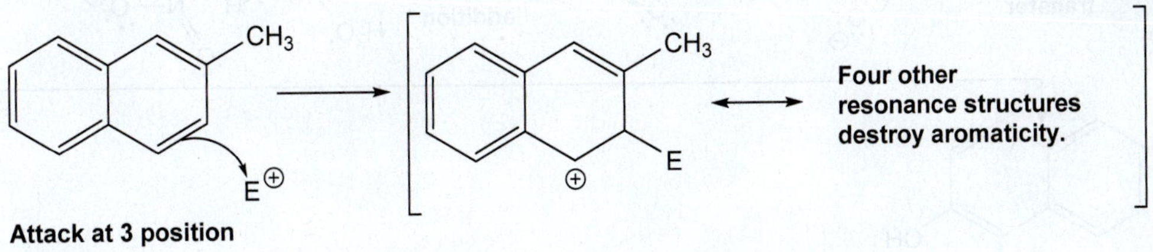

Problem 23.64

Think

Which ring is more activated? On which carbons in that ring can the electrophile be added? Which group, CH_3 or HO, is a stronger activator?

Solve

Electrophilic aromatic substitution will take place in the ring on the right, because both CH_3 and OH are activating groups, making the ring on the right more activated compared to the ring on the left. The HO group is a stronger activator compared to CH_3 (Table 23-3). Therefore, the incoming electrophile will be directed primarily to the positions that are ortho and para to the OH group. Only the position ortho to the OH group has a proton, however.

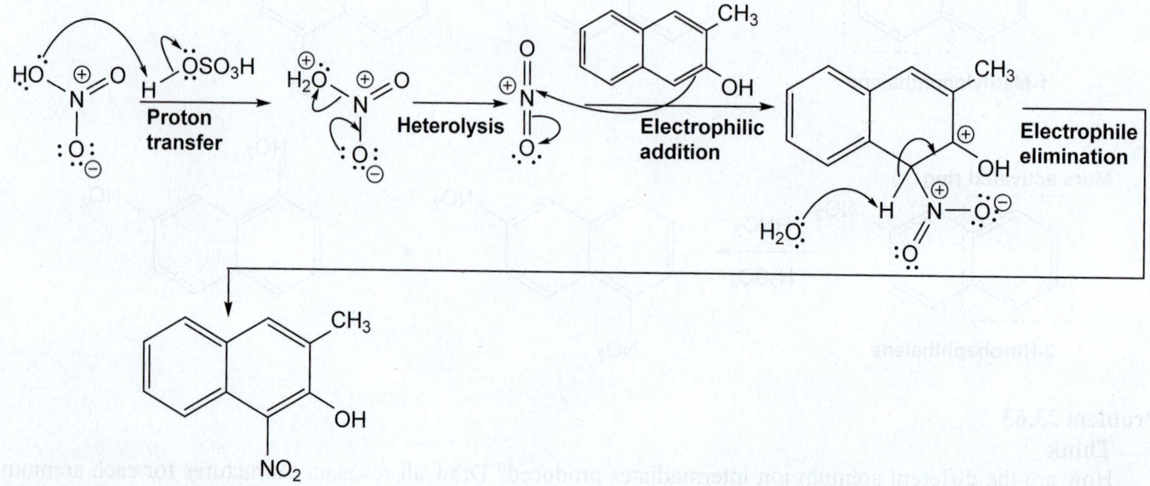

Proton transfer

Heterolysis

Electrophilic addition

Electrophile elimination

Problem 23.65

Think

Which ring is more activated? Which site on the more activated ring leads to the major product? Use Table 23-3 to consider the directing capabilities of the substituent and how many resonance structures of the arenium ion intermediates preserve aromaticity.

Solve

(a) HO is an activating group, so the ring on the right preferentially undergoes reaction. HO is an ortho/para director, so both C atoms ortho to the one with the OH are reactive, but the C atom on the bottom leads to the major product, because the corresponding arenium ion intermediate has two resonance structures that preserve aromaticity.

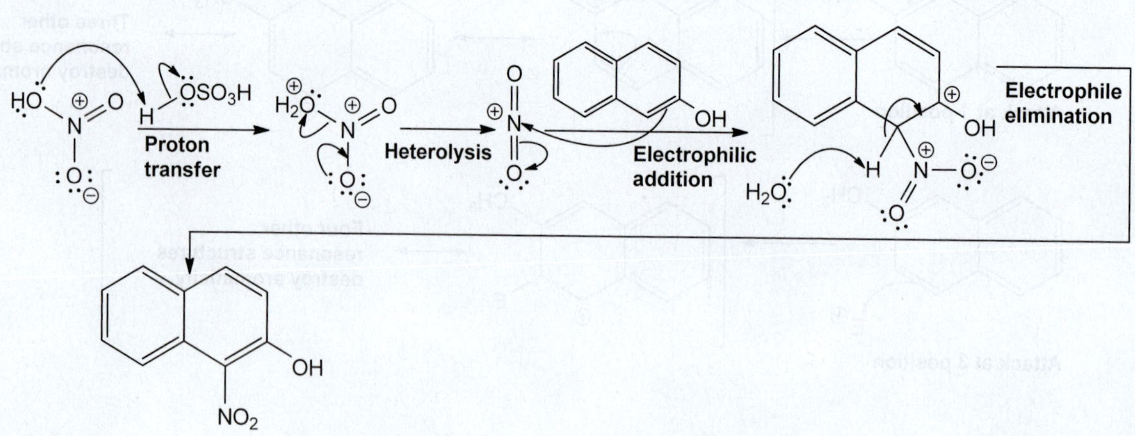

Proton transfer

Heterolysis

Electrophilic addition

Electrophile elimination

(b) CH_3 is an activating group, so the ring with the CH_3 preferentially undergoes reaction. CH_3 is an ortho/para director, so the C atoms that are ortho and para to the one with the CH_3 are reactive, but the para C atom leads to the major product, because the corresponding arenium ion intermediate has two resonance structures that preserve aromaticity.

Coordination → Heterolysis → Electrophilic addition → Electrophile elimination

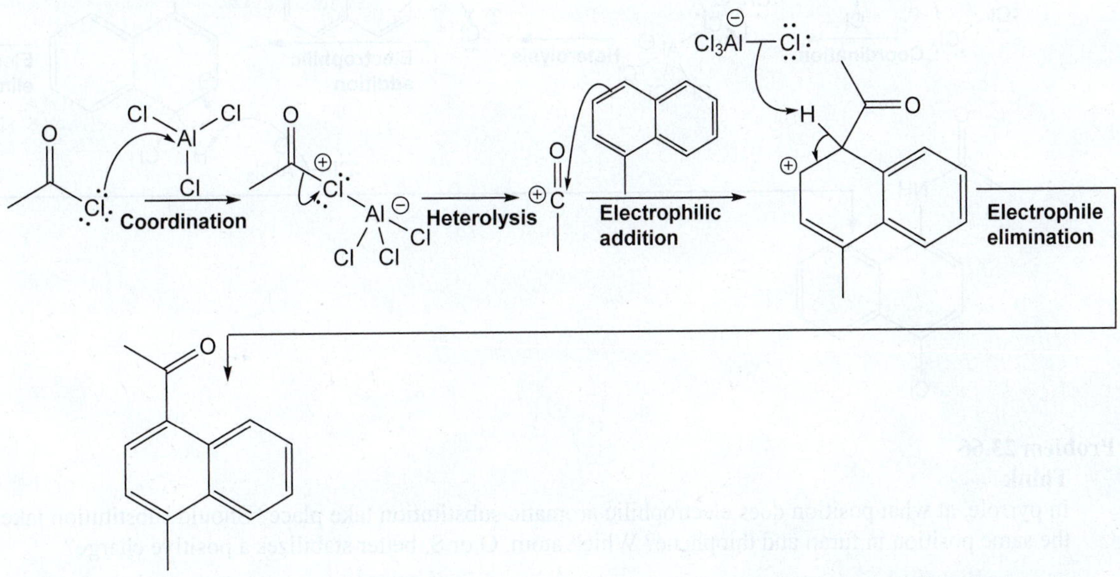

(c) NO_2 is a deactivating group, so the unsubstituted ring is more activated and will undergo reaction. Recall from Problem 23.60 that the α carbon will react preferentially over the β carbon. Of the two α carbons in the unsubstituted ring, the one on the bottom would be more reactive to avoid an arenium ion intermediate that has a resonance structure with the positive charge on the C atom attached to NO_2. It should be noted that the electrophilic addition and heterolysis steps probably occur simultaneously instead of separately, but the electrophile Br^+ is shown explicitly for emphasis.

Coordination → Heterolysis → Electrophilic addition → Electrophile elimination

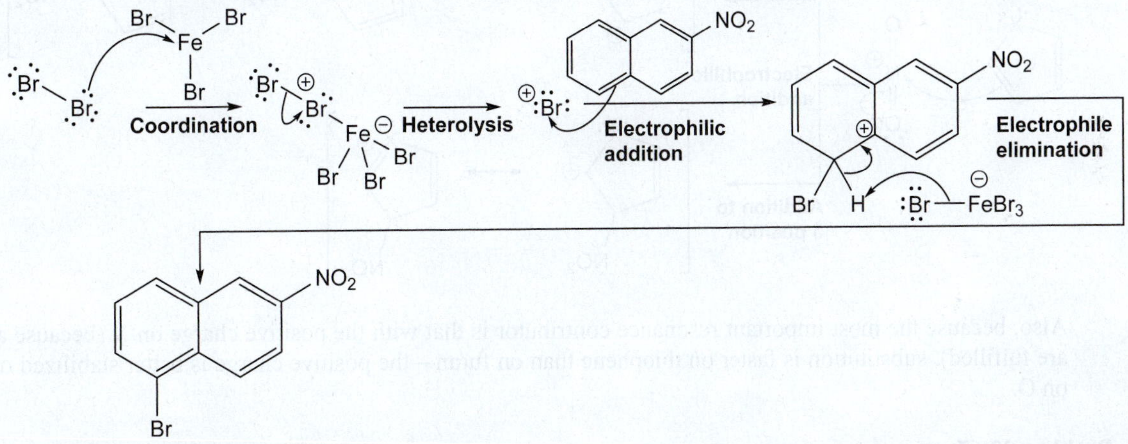

(d) The ring with the CH_3CONH group is more activated, so substitution will occur on the ring with the CH_3CONH group. The CH_3CONH group is an ortho/para director, so substitution will occur primarily at the carbons ortho and para to it. Substitution at the para C will lead to the major product, however, because the corresponding arenium ion intermediate has more resonance structures that preserve aromaticity than the arenium ion produced from reaction at the C atom that is ortho to the CH_3CONH. It should be noted that the electrophilic addition and heterolysis steps probably occur simultaneously instead of separately, but the electrophile Cl^+ is shown explicitly for emphasis. See the mechanism on the next page.

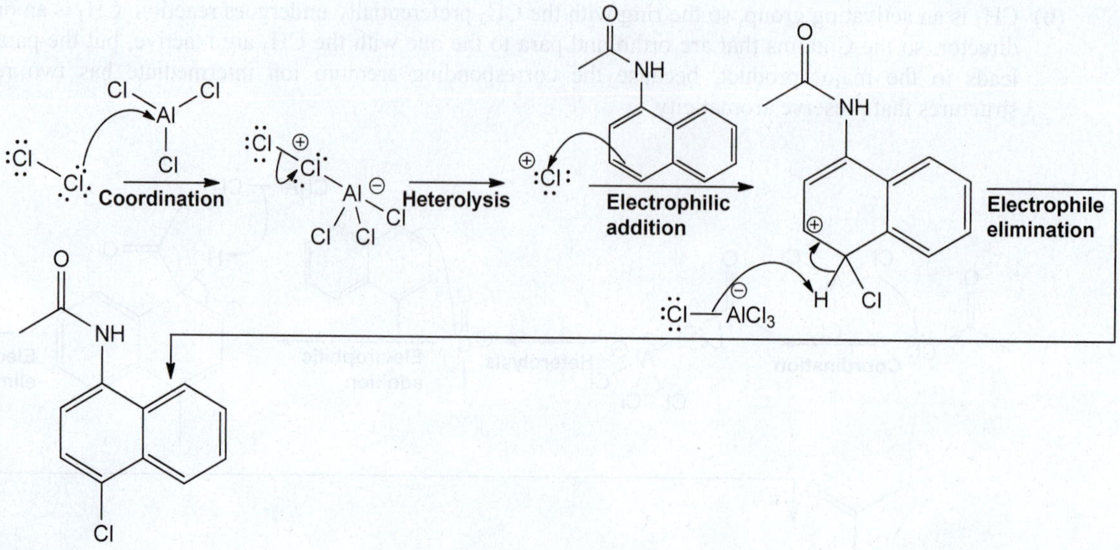

Problem 23.66

Think

In pyrrole, at what position does electrophilic aromatic substitution take place? Should substitution take place at the same position in furan and thiophene? Which atom, O or S, better stabilizes a positive charge?

Solve

As we saw in Equation 23-30, pyrrole undergoes electrophilic aromatic substitution at the 2 position preferentially over the 3 position, because the corresponding arenium ion intermediate has more resonance structures and is more stable. The same is true for furan and thiophene, as shown below, where X is either O or S.

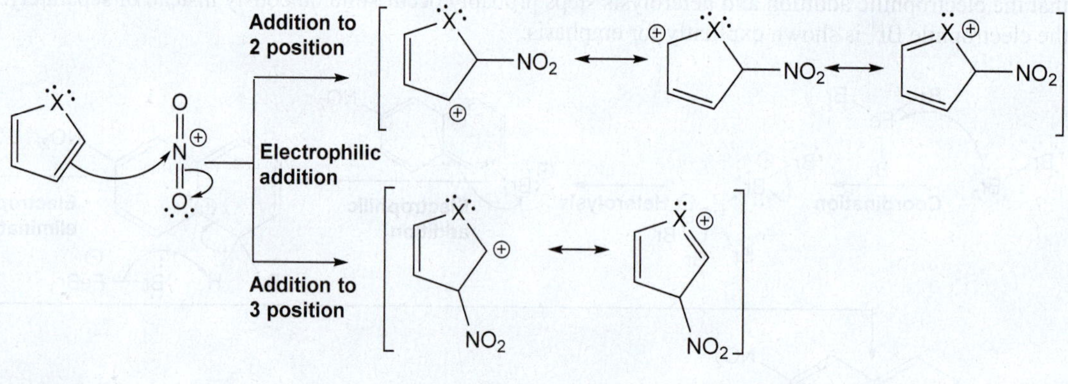

Also, because the most important resonance contributor is that with the positive charge on X (because all octets are fulfilled), substitution is faster on thiophene than on furan—the positive charge is better stabilized on S than on O.

Problem 23.67

Think

In comparing thiophene to pyrrole, at what position does nitration take place? What acts as the electrophile in the electrophilic addition reaction?

Solve

Just as in pyrrole, the arenium ion intermediate is more stable if attack of the electrophile occurs adjacent to the heteroatom (S) than at two carbons away—that is, attack is directed toward the 2 and 5 positions. Because a methyl group already exists at the 2 position, attack is directed toward the 5 position.

Electrophilic addition → Electrophile elimination

Problem 23.68

Think

Which ring is more activated? On the more activated ring, which carbon site most likely reacts with the electrophile? What is the identity of the electrophile in each electrophilic addition reaction?

Solve

(a) The pyrrole ring is more activated than the benzene ring, so substitution will take place on the pyrrole ring, replacing a H^+ for Br^+. The electrophilic aromatic substitution takes place at the 2 position for a pyrrole ring.

Electrophilic addition → Electrophile elimination

(b) The pyridine ring is deactivated relative to the benzene ring, so substitution will take place on the ring on the left. In an irreversible reaction such as this one, similar to naphthalene, H atoms at the α positions are replaced preferentially over those at the β positions. The α position at the bottom undergoes reaction preferentially over the one at the top to avoid an arenium ion intermediate that has a resonance structure in which the positive charge appears on N. It should be noted that the electrophilic addition and heterolysis steps probably occur simultaneously instead of separately, but the electrophile Br^+ is shown explicitly for emphasis.

Coordination → Heterolysis → Electrophilic addition → Electrophile elimination

(c) The C atoms adjacent to S are activated toward substitution. In this nitration reaction, NO_2^+ is the electrophile.

Proton transfer → Heterolysis → Electrophilic addition → :Base Electrophile elimination

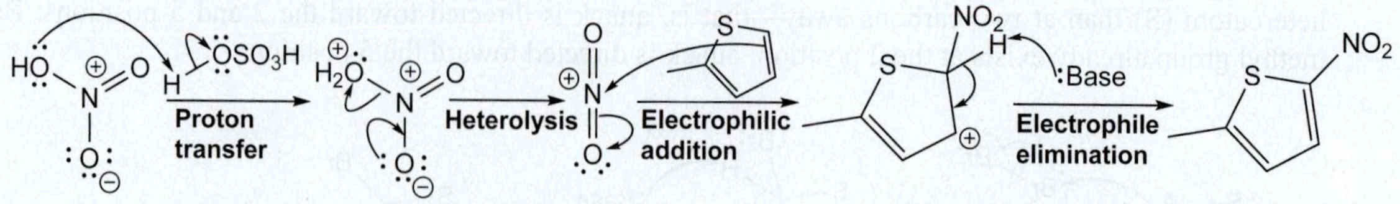

(d) Similar to the pyridine ring, the C atoms adjacent to N are deactivated, directing substitution toward the position two C atoms away. NO_2^+ is the electrophile.

Proton transfer → Heterolysis → Electrophilic addition → :Base Electrophile elimination

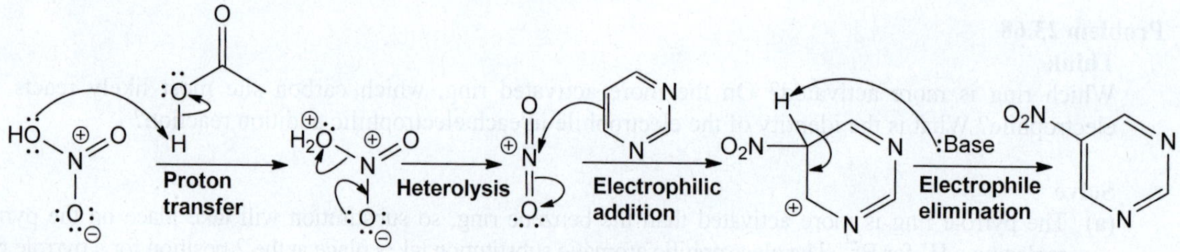

Problem 23.69

Think

Is there a site (or sites) of electrophilic addition on these compounds where the arenium ion intermediate can remain aromatic? If there is more than one of these sites, which arenium ion intermediate exhibits greater stability from aromaticity?

Solve

(a) Substitution is favored at these positions:

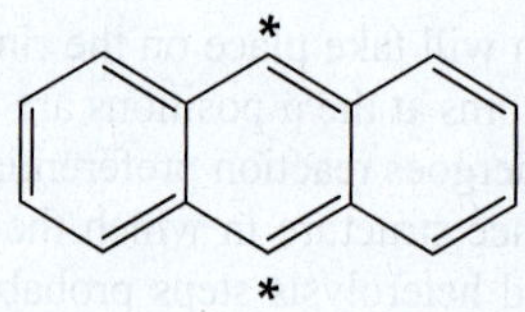

As shown below, addition of an electrophile to one of these carbons produces an arenium ion intermediate in which a resonance structure exhibits two separate benzene rings. Addition of an electrophile to a carbon on one of the outer rings produces an arenium ion intermediate with just a single aromatic ring. The two separate aromatic rings are more stable.

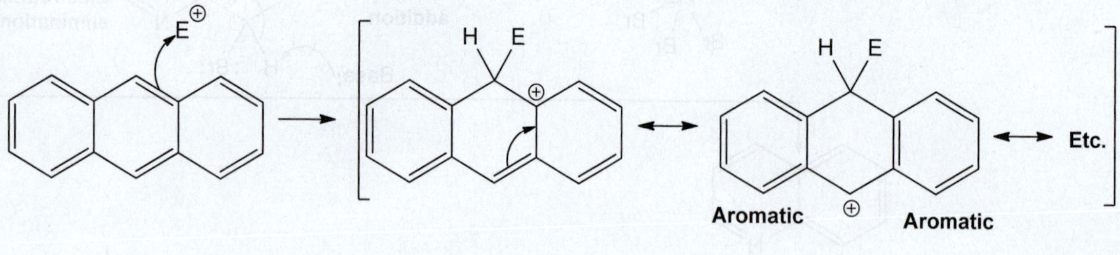

Aromatic Aromatic Etc.

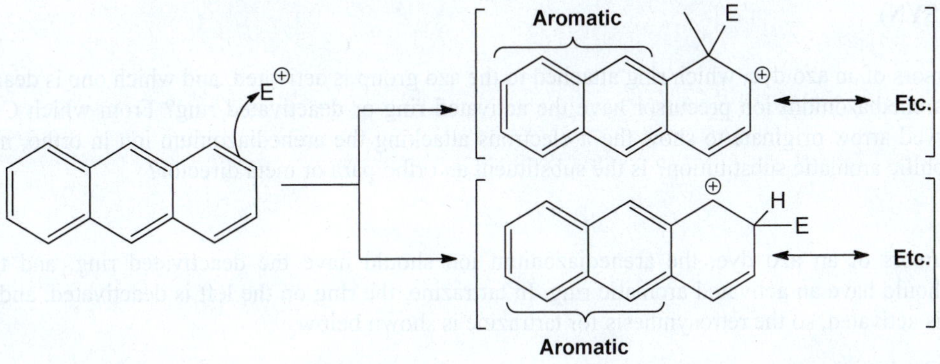

(b) Substitution is favored at these positions.

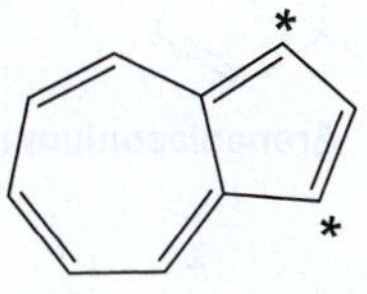

As shown below, electrophilic addition to one of these carbons produces an arenium ion intermediate in which six π electrons are delocalized around the seven-membered ring, making that ring aromatic.

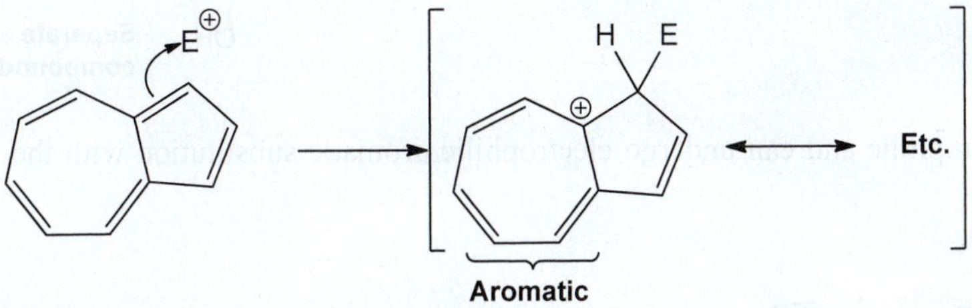

Section 23.8 Azo Coupling and Azo Dyes
Problem 23.70

Think

Is the substituent in the activated ring an ortho/para director or a meta director? What is the identity of the electrophile in the electrophilic addition step? What is the identity of the base in the electrophile elimination step?

Solve

The electrophilic addition–elimination mechanism is shown below:

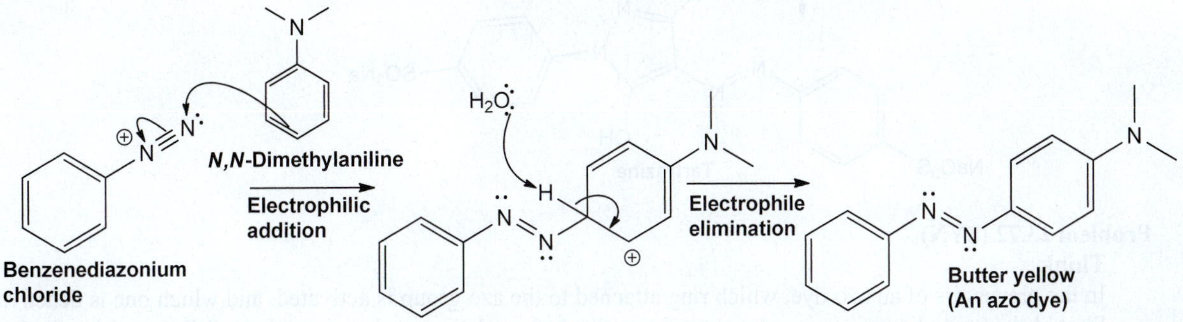

The $N(CH_3)_2$ group is an ortho/para director, but the para product should be favored significantly over the ortho owing to steric hindrance by the CH_3 groups at the ortho positions.

Problem 23.71 (SYN)

Think

In the precursors of an azo dye, which ring attached to the azo group is activated, and which one is deactivated? Should the arenediazonium ion precursor have the activated ring or deactivated ring? From which C=C bond should a curved arrow originate to show the π electrons attacking the arenediazonium ion in ortho, meta, and para electrophilic aromatic substitution? Is the substituent an ortho/para or meta director?

Solve

In the precursors of an azo dye, the arenediazonium ion should have the deactivated ring, and the other compound should have an activated aromatic ring. In tartrazine, the ring on the left is deactivated, and the ring on the right is activated, so the retrosynthesis for tartrazine is shown below.

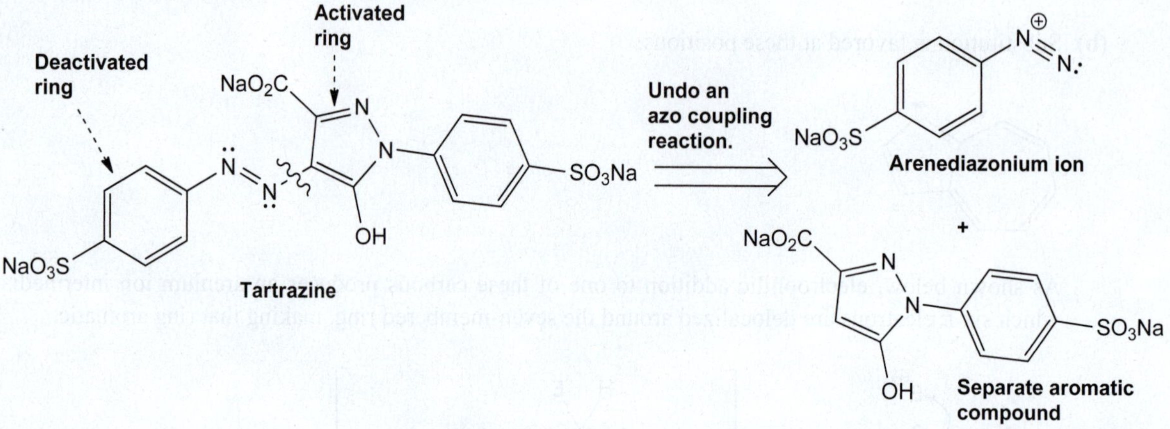

Deactivated ring

Activated ring

NaO_2C

NaO_3S

SO_3Na

OH

Tartrazine

Undo an azo coupling reaction.

Arenediazonium ion

+

Separate aromatic compound

The arenediazonium ion is electrophilic and can undergo electrophilic aromatic substitution with the activated benzene ring shown.

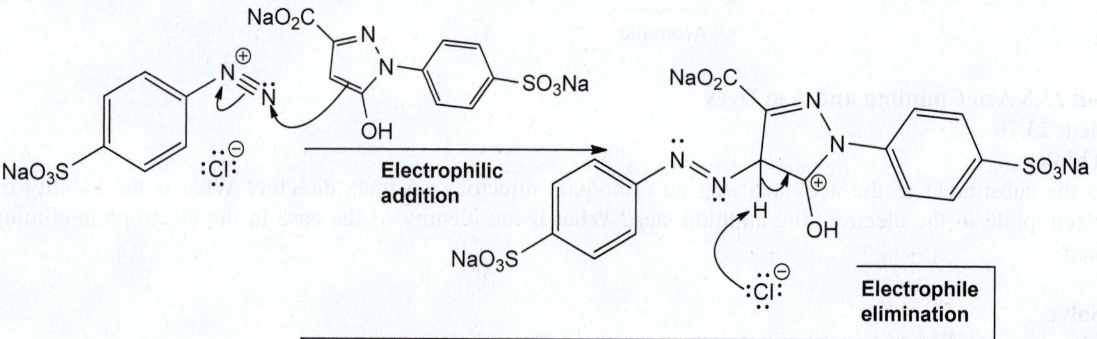

Electrophilic addition

Electrophile elimination

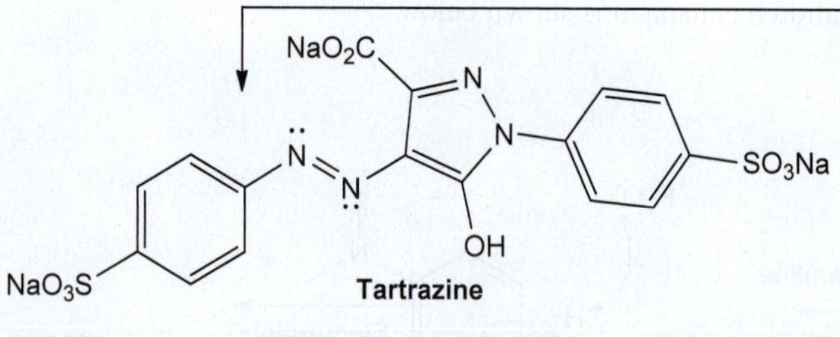

Tartrazine

Problem 23.72 (SYN)

Think

In the precursors of an azo dye, which ring attached to the azo group is activated, and which one is deactivated? Should the arenediazonium ion precursor have the activated ring or deactivated ring? From which C=C bond should a curved arrow originate to show the π electrons attacking the electrophile in ortho, meta, and para electrophilic aromatic substitution? Is the substituent an ortho/para or meta director?

Solve
In the precursors of an azo dye, the arenediazonium ion should have the deactivated ring, and the other precursor should have an activated aromatic ring. In sunset yellow FCF, the ring on the left is deactivated and the ring on the right is activated, so the retrosynthesis for sunset yellow FCF is shown below.

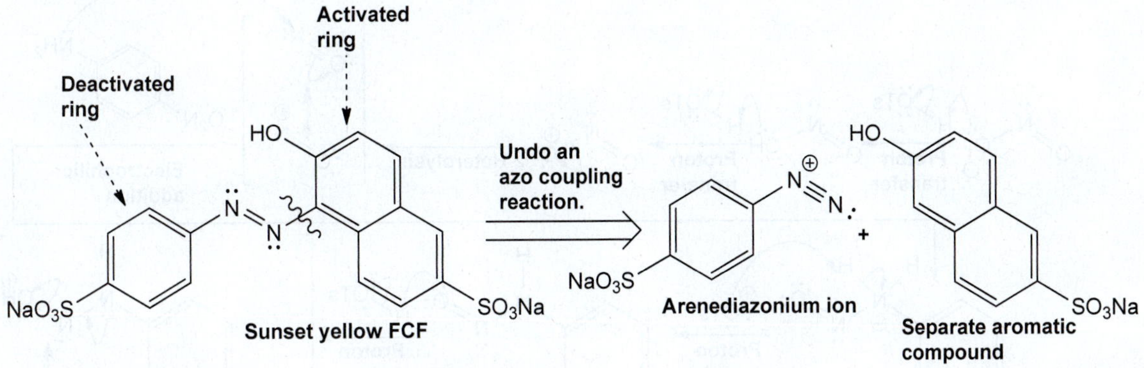

Activated ring
Deactivated ring
HO
NaO$_3$S
SO$_3$Na
Sunset yellow FCF
Undo an azo coupling reaction.
NaO$_3$S
Arenediazonium ion
HO
SO$_3$Na
Separate aromatic compound

The arenediazonium ion is electrophilic and can undergo electrophilic aromatic substitution with the activated benzene ring shown. The mechanism is given below.

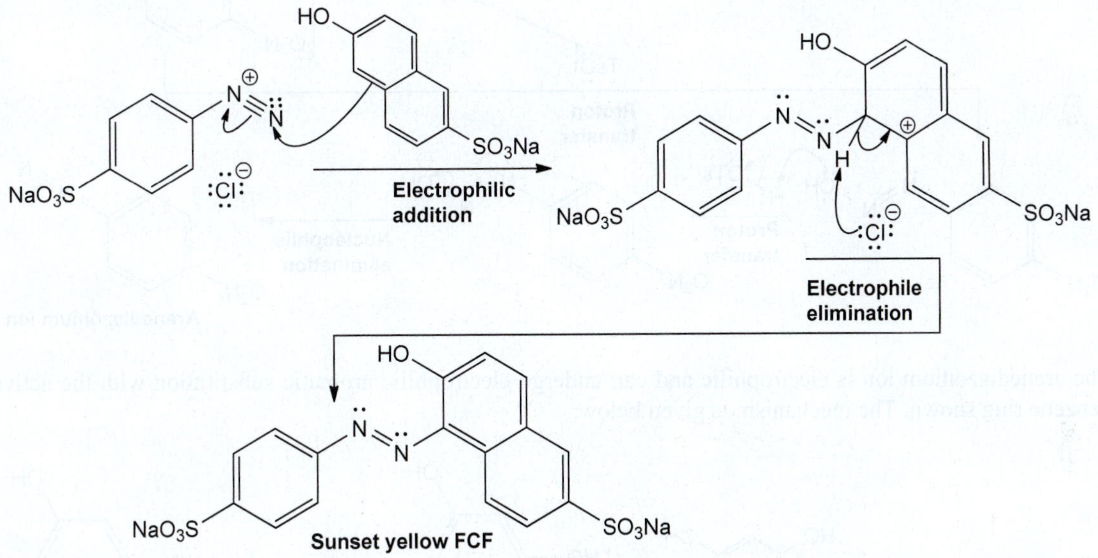

HO
NaO$_3$S
SO$_3$Na
Electrophilic addition
HO
NaO$_3$S
SO$_3$Na
Electrophile elimination
HO
NaO$_3$S
SO$_3$Na
Sunset yellow FCF

Problem 23.73
Think
How does PhNH$_2$ react with NO$_2^-$ in the presence of a strong acid (p-MeC$_6$H$_4$SO$_3$H = TsOH)? What new group is formed? How does this group then react with *m*-hydroxyphenol (resorcinol)? What is the identity of the electrophile in the electrophilic addition–elimination reaction? From which C=C bond should a curved arrow originate to show the π electrons attacking the electrophile in ortho, meta, and para electrophilic aromatic substitution? Is the substituent an ortho/para or meta director?

Solve

$PhNH_2$ reacts with NO_2^- in the presence of a strong acid (p-$MeC_6H_4SO_3H$ = TsOH) to form the arenediazonium ion (review Equation 22-35). The mechanism is shown below.

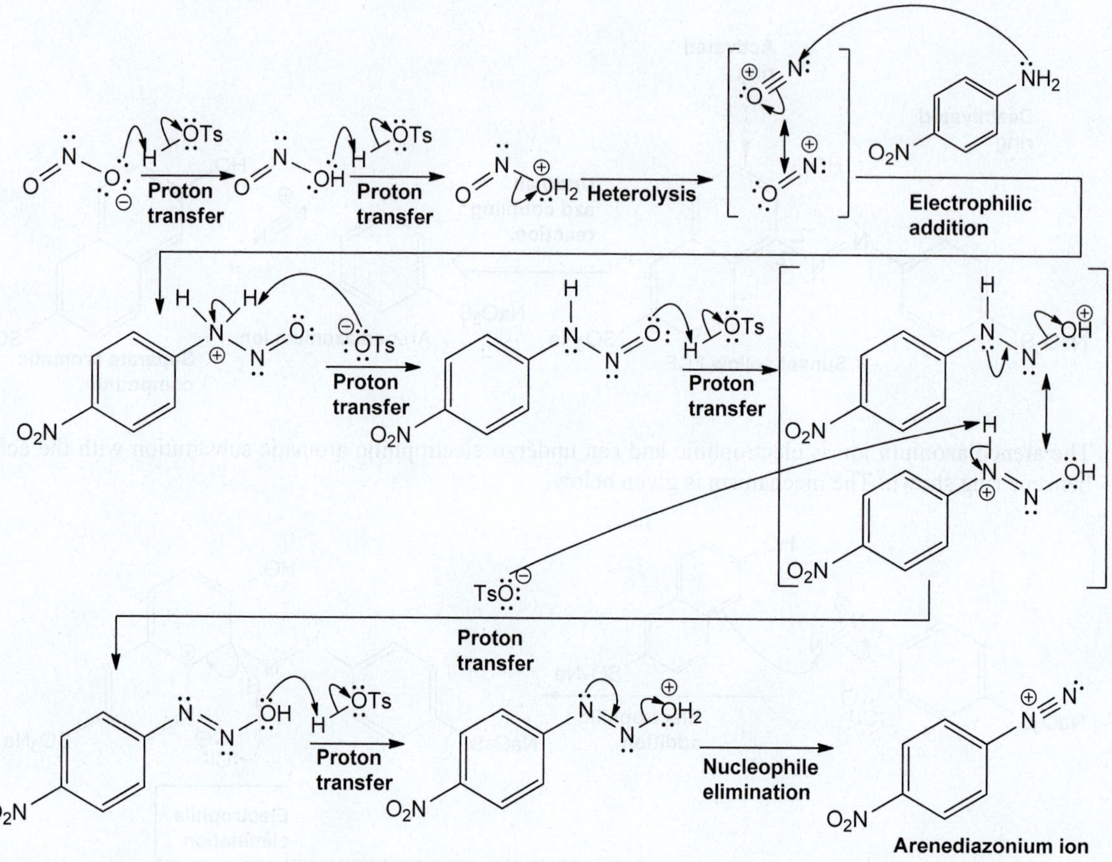

Proton transfer
Proton transfer
Heterolysis
Electrophilic addition
Proton transfer
Proton transfer
Proton transfer
Proton transfer
Nucleophile elimination
Arenediazonium ion

The arenediazonium ion is electrophilic and can undergo electrophilic aromatic substitution with the activated benzene ring shown. The mechanism is given below.

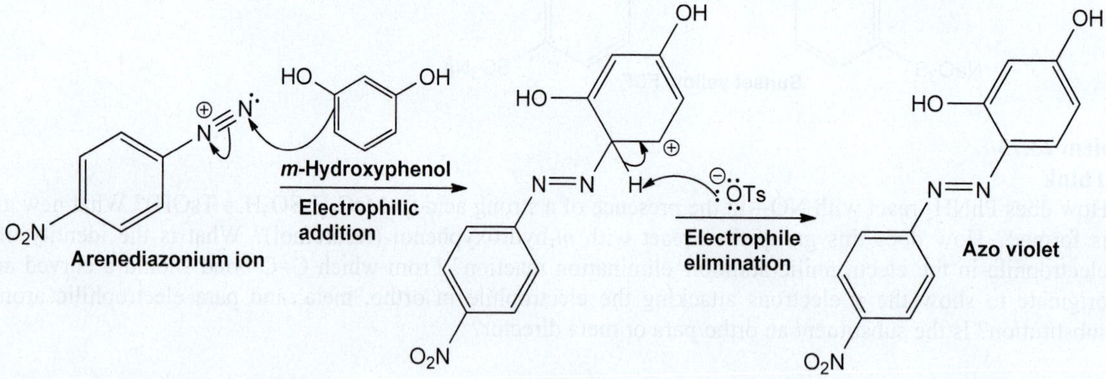

Arenediazonium ion
m-Hydroxyphenol
Electrophilic addition
Electrophilic elimination
Azo violet

Section 23.9 Nucleophilic Aromatic Substitution Mechanisms
Problem 23.74

Think

What is the leaving group? Is this mechanism nucleophilic addition–elimination or electrophilic addition–elimination? Is there a strong electron-withdrawing group ortho or para to the leaving group?

Solve

(a) F^- would be the leaving group, and there is a strong electron-withdrawing group para to the leaving group. This mechanism is a nucleophilic addition–elimination mechanism, because the ethoxide serves as a nucleophile to attack an electrophilic carbon.

Nucleophilic addition

Nucleophile elimination

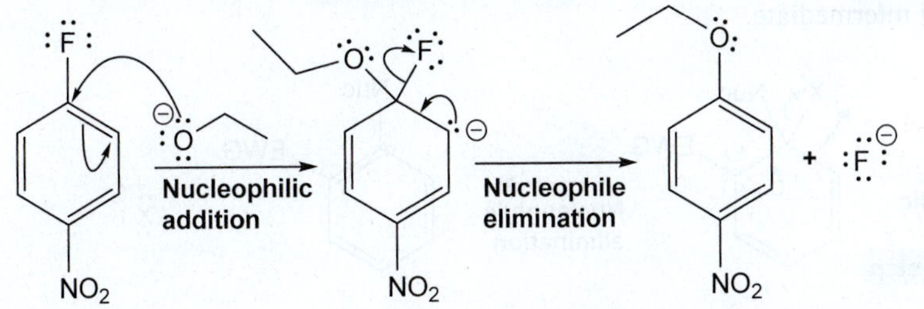

(b) The mechanism below is identical to **(a)** with the addition of a proton transfer step to yield the uncharged product.

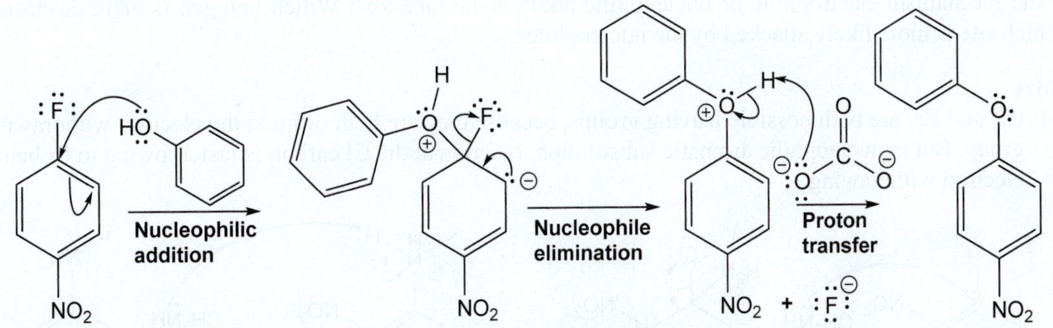

Nucleophilic addition

Nucleophile elimination

Proton transfer

(c) Cl^- would be the leaving group, and there is a strong electron-withdrawing group para and ortho to the leaving group. This mechanism is a nucleophilic addition–elimination mechanism, because the amine serves as a nucleophile to attack an electrophilic carbon.

Nucleophilic addition

Nucleophile elimination

Proton transfer

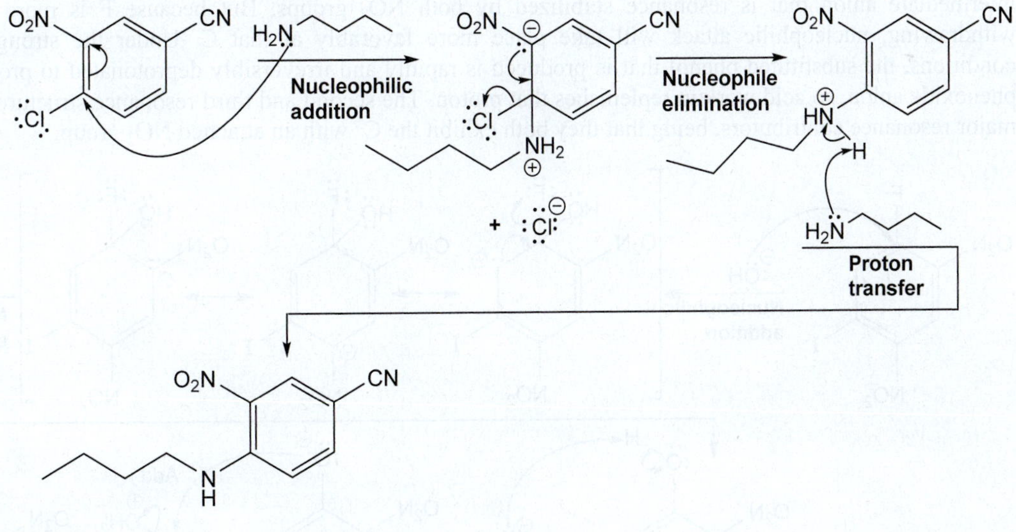

Problem 23.75

Think

What is the mechanism for nucleophilic addition–elimination? Which halogen is most electronegative? Which halogen is the best leaving group?

Solve

Iodine is the best leaving group. If Step 2 were the rate-limiting step, Ar–I would have the fastest rate. This suggests that the first step (nucleophilic addition) is the rate-limiting step. This is in agreement with the relative rates of reaction, because fluorine is the most electronegative atom and makes the carbon the most electrophilic. This draws in the incoming nucleophile more strongly. Moreover, the electron-withdrawing nature of F helps stabilize the negatively charged intermediate.

Nucleophilic addition
Rate-limiting step
Nucleophile elimination

X = F, Cl, Br, I
EWG = Electron-withdrawing group

Problem 23.76

Think

Is the mechanism electrophilic or nucleophilic addition–elimination? Which halogen is more electronegative? Which site is more likely attacked by the nucleophile?

Solve

(a) Cl^- and Br^- are both possible leaving groups, because they are both ortho to the electron-withdrawing NO_2 group. But in nucleophilic aromatic substitution, reaction at the Cl carbon is faster, owing to Cl being more electron withdrawing.

Nucleophilic addition
Nucleophile elimination
Proton transfer

(b) F^- and I^- are both possible leaving groups, and attack at each of their C atoms would lead to an intermediate anion that is resonance stabilized by both NO_2 groups. But because F is more electron withdrawing, nucleophilic attack will take place more favorably at that C. Under the strongly basic conditions, the substituted phenol that is produced is rapidly and irreversibly deprotonated to produce the phenoxide anion, so acid workup replenishes that proton. The second and third resonance structures are the major resonance contributors, being that they both exhibit the C^- with an attached NO_2 group.

Nucleophilic addition
Nucleophile elimination
Proton transfer
Add
Proton transfer

Problem 23.77

Think

How does the substituted chlorobenzene interact with a strong base? How does the Cl^- leaving group leave? What intermediate is formed? How does a nucleophile react with that intermediate?

Solve

(a) Hydroxide acts as a base to remove a proton from the aromatic ring to produce a carbanion. In Step 2, the leaving group departs in a nucleophile elimination step. The product of Step 2 is a benzyne intermediate, which can undergo nucleophilic addition by hydroxide to produce a deprotonated form of phenol. The benzyne intermediate is symmetrical, so only one product is possible. A proton transfer step occurs to yield the uncharged substituted phenol. Under the strongly basic conditions, the substituted phenol that is produced is rapidly and irreversibly deprotonated to produce the corresponding phenoxide anion, so acid workup replenishes that proton.

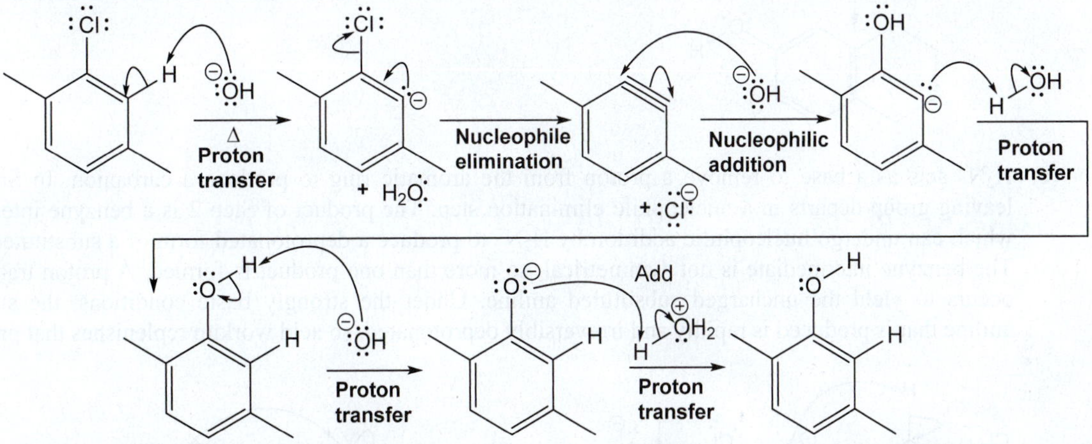

(b) Hydroxide acts as a base to remove a proton from the aromatic ring to produce a carbanion. In Step 2, the leaving group departs in a nucleophile elimination step. The product of Step 2 is a benzyne intermediate, which can undergo nucleophilic addition by hydroxide to produce a deprotonated form of phenol. The benzyne intermediate is not symmetrical, so more than one product is formed. A proton transfer step occurs to yield the uncharged substituted phenol. Under the strongly basic conditions, the substituted phenol that is produced is rapidly and irreversibly deprotonated to produce the corresponding phenoxide anion, so acid workup replenishes that proton. Note that, in the first step, the proton on the other side of the C–Cl could be deprotonated instead, but the product formed from the resulting benzyne intermediate is redundant to the overall product of the second mechanism on the next page.

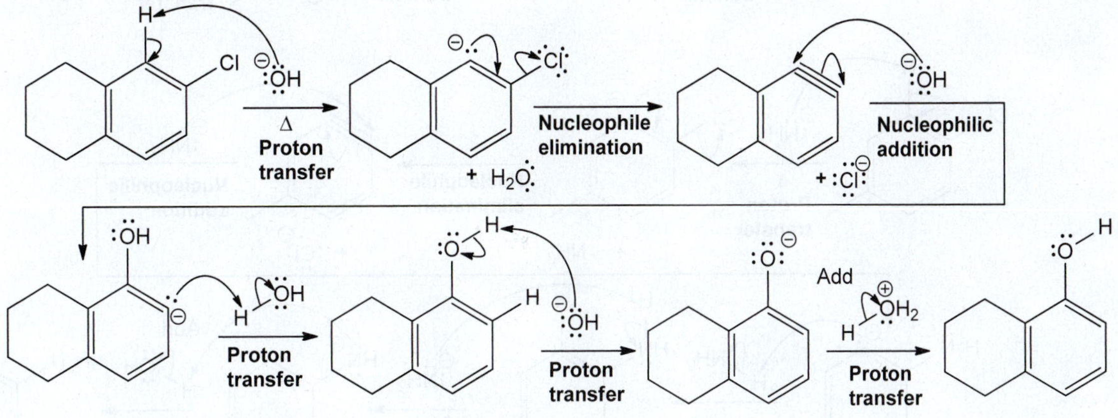

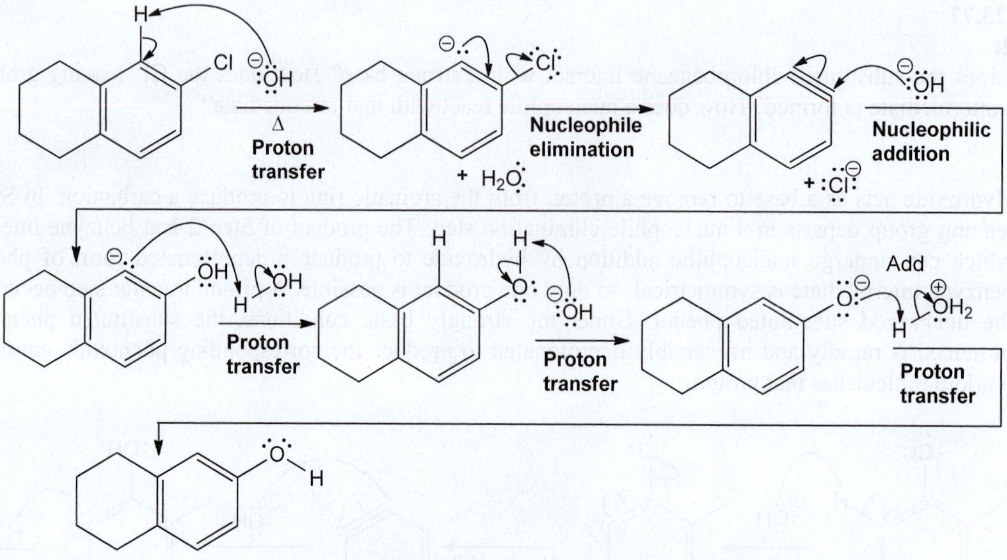

Δ Proton transfer — Nucleophile elimination — Nucleophilic addition — Proton transfer — Proton transfer — Add — Proton transfer

(c) H_2N^- acts as a base to remove a proton from the aromatic ring to produce a carbanion. In Step 2, the leaving group departs in a nucleophile elimination step. The product of Step 2 is a benzyne intermediate, which can undergo nucleophilic addition by H_2N^- to produce a deprotonated form of a substituted aniline. The benzyne intermediate is not symmetrical, so more than one product is formed. A proton transfer step occurs to yield the uncharged substituted aniline. Under the strongly basic conditions, the substituted aniline that is produced is rapidly and irreversibly deprotonated, so acid workup replenishes that proton.

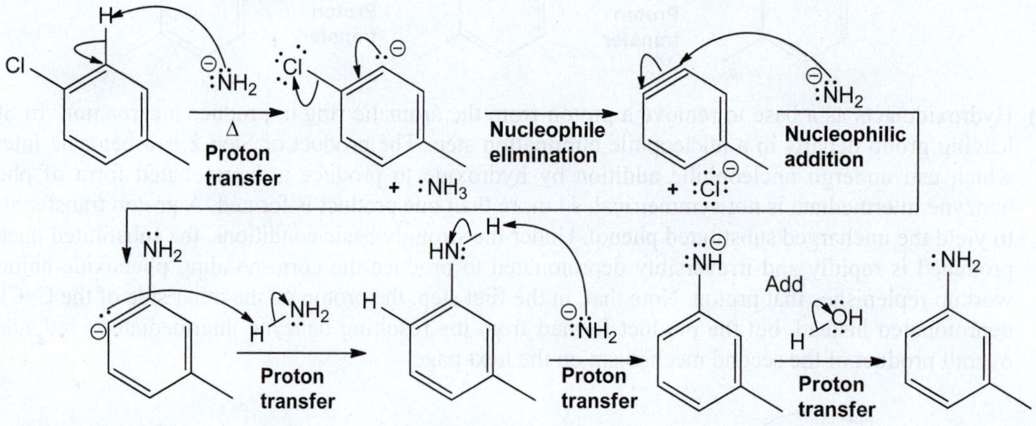

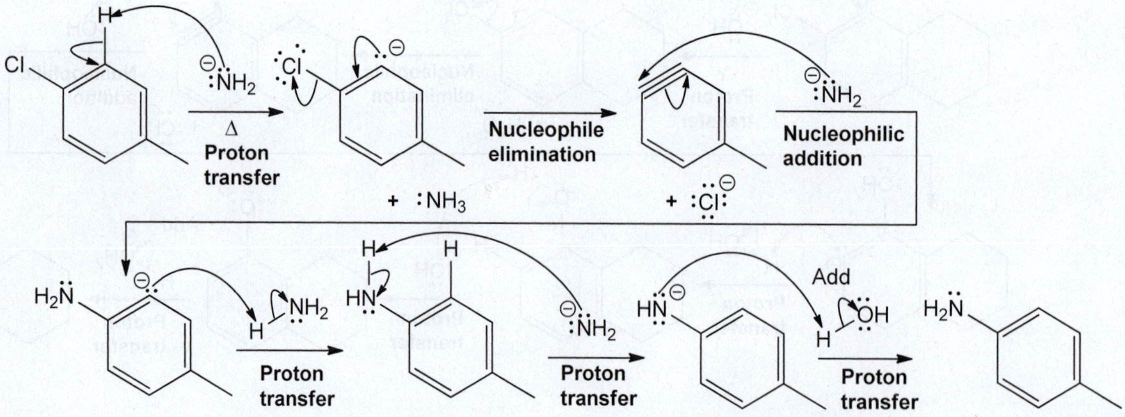

Δ Proton transfer — Nucleophile elimination — Nucleophilic addition — Proton transfer — Proton transfer — Add — Proton transfer

Δ Proton transfer — Nucleophile elimination — Nucleophilic addition — Proton transfer — Proton transfer — Add — Proton transfer

Problem 23.78 (SYN)

Think

What bond is formed in a nucleophilic aromatic substitution? Which groups on the ring activate it toward nucleophilic aromatic substitution? What is the nucleophile in each example? What is the leaving group?

Solve

(a) NO_2 groups activate the ring toward nucleophilic aromatic substitution. A new C–OCH_2CH_3 bond is formed in this reaction. F^- could be the leaving group, and $^-OCH_2CH_3$ is the nucleophile.

NO_2, F, O_2N — NaO⁄ → NO_2, O, O_2N

(b) $CH_3C{=}O$ groups activate the ring toward nucleophilic aromatic substitution. A new C–$NCH(CH_3)_2$ bond is formed in this reaction. F^- could be the leaving group, and $^-NCH(CH_3)_2$ is the nucleophile. An acid workup step is necessary, because the substituted aniline would be deprotonated under the basic conditions of the reaction.

F, O, O — 1. NaN(H); 2. H_2O → H N, O, O

(c) The NO_2 group activates the ring toward nucleophilic aromatic substitution. A new C–OH bond is formed in this reaction. F^- could be the leaving group, and ^-OH is the nucleophile. An acid workup step is necessary, because the substituted phenol would be deprotonated under the basic conditions of the reaction.

NO_2, F — 1. NaOH; 2. H_2O/HCl → NO_2, HO

Problem 23.79 (SYN)

Think

What bond is formed in a nucleophilic aromatic substitution? Are there groups on the ring that make the ring reactive toward nucleophiles? Or is there a strong base that can be used to deprotonate the aromatic ring? What is the nucleophile in each example? What is the leaving group?

Solve

(a) CH_3 groups deactivate the ring toward nucleophilic addition, so substitution should take place via a benzyne intermediate, using HO^- as a strong base and nucleophile. A new C–OH bond is formed in this reaction.

Cl — 1. NaOH, Δ; 2. H_2O/HCl → HO

(b) There are no attached groups to activate the ring toward nucleophilic addition, so substitution should take place via a benzyne intermediate, using H_2N^- as a strong base and nucleophile. A new C–NH_2 bond is formed in this reaction.

1. $NaNH_2$, Δ
2. H_2O

(c) There are no attached groups to activate the ring toward nucleophilic addition, so substitution should take place via a benzyne intermediate, using $CH_3(CH_2)_6O^-$ as a strong base and nucleophile. A new C–$O(CH_2)_6CH_3$ bond is formed in this reaction.

1. $NaO(CH_2)_6CH_3$, Δ
2. H_2O/HCl

Sections 23.10–23.12 Organic Synthesis: Considerations of Regiochemistry; Attaching Groups in the Correct Order; Interconverting Ortho/Para and Meta Directors; Protecting Groups
Problem 23.80 (SYN)

Think

How are the groups positioned relative to one another? Does the positioning match their directing ability? What reactions are a result of electrophilic aromatic substitution? What functional group transformations occur? Are any C–C bond-forming reactions necessary? Oxidation or reduction reactions?

Solve

(a) The acyl group and Br are meta to each other. The acyl group is a meta director and is added to the ring first.

$AlCl_3$; Br_2, $FeBr_3$

(b) The acyl group and Br are para to each other. The Br group is an ortho/para director and is added to the ring first.

Br_2, $FeBr_3$; $AlCl_3$

(c) Using the product in **(b)**, the ketone is transformed into the ester by first converting the ketone to the carboxylic acid and then transforming the carboxylic acid into the ester by Fischer esterification.

1. $KMnO_4$, $HO^{\ominus}$, Δ
2. $H_3O^{\oplus}$

CH_3OH, $H^{\oplus}$

(d) The NH_2 group is not added on the ring by electrophilic aromatic substitution. The ring is nitrated in harsh conditions to get the nitro groups meta. The nitro groups are then reduced to the amino groups.

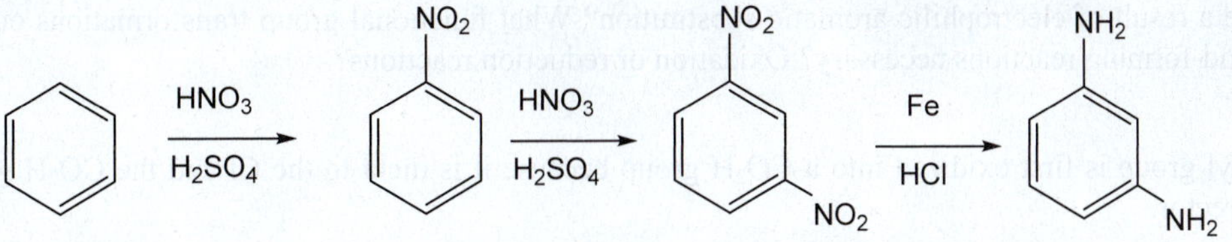

(e) The amino and acyl groups are para to each other. The amino group is an ortho/para director and should be added on first. However, the amine is a Lewis base and would react with the strong Lewis acid, $AlCl_3$, resulting in a highly deactivated ring. This would prevent the necessary Friedel–Crafts acylation. Therefore, the NH_2 group must first be protected before the Friedel–Crafts acylation.

Need to protect amine

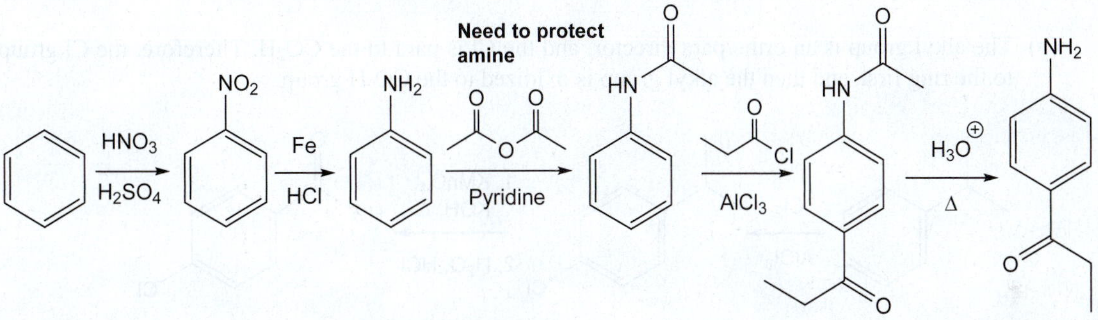

Problem 23.81 (SYN)

Think

How are the groups positioned relative to one another? Does the positioning match their directing ability? What reactions are a result of electrophilic aromatic substitution? What functional group transformations occur? Are any C–C bond-forming reactions necessary? Oxidation or reduction reactions?

Solve

(a) The alkyl group and SO_3H are para to each other. The alkyl group is a para director and is added to the ring first.

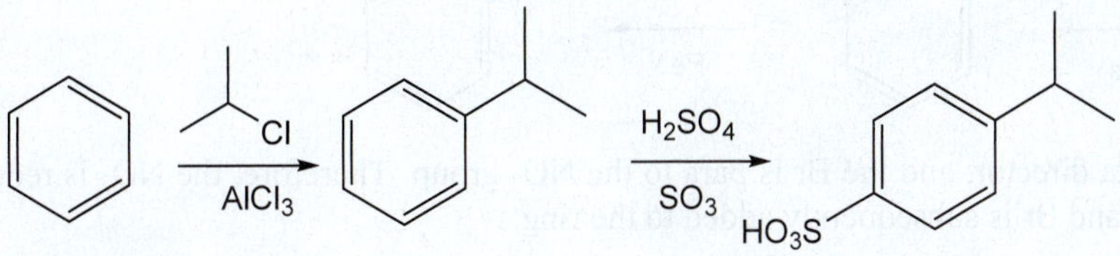

(b) The Br atom and SO_3H are para to each other. The Br atom is a para director and is added to the ring first.

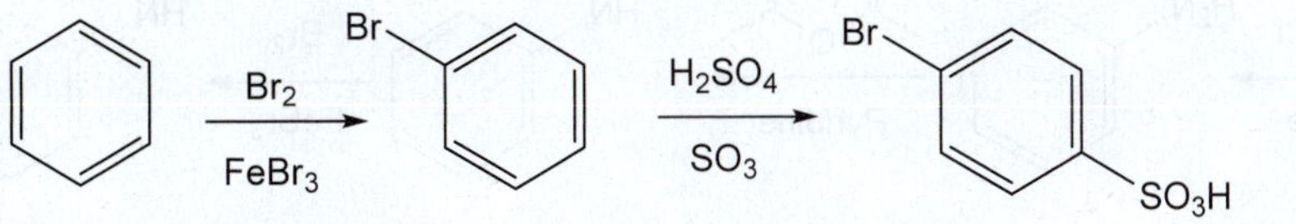

(c) The Cl atom and SO_3H are meta to each other. The SO_3H group is a meta director and is added to the ring first.

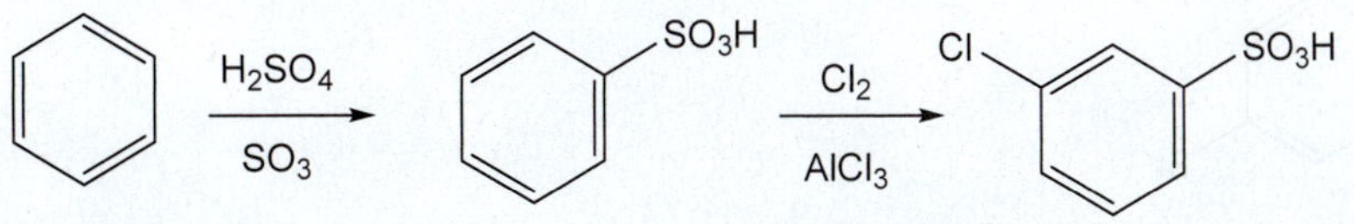

Problem 23.82 (SYN)

Think

How are the groups positioned relative to one another? Does the positioning match their directing ability? What reactions are a result of electrophilic aromatic substitution? What functional group transformations occur? Are any C–C bond-forming reactions necessary? Oxidation or reduction reactions?

Solve

(a) The alkyl group is first oxidized into a CO_2H group because it is meta to the Cl and the CO_2H group is a meta director.

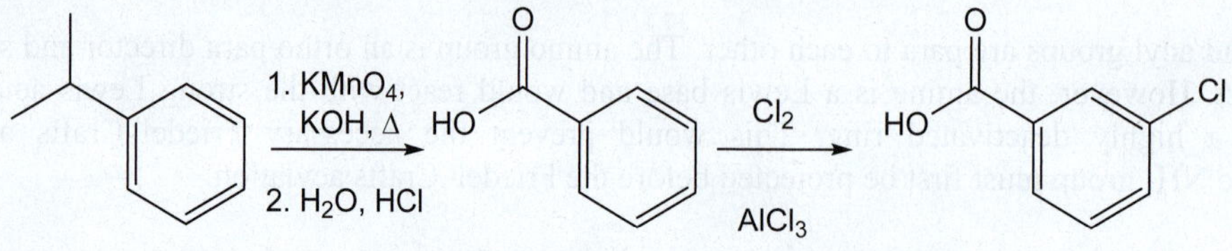

(b) The alkyl group is an ortho/para director, and the Cl is para to the CO_2H. Therefore, the Cl group is added to the ring first, and then the alkyl group is oxidized to the CO_2H group.

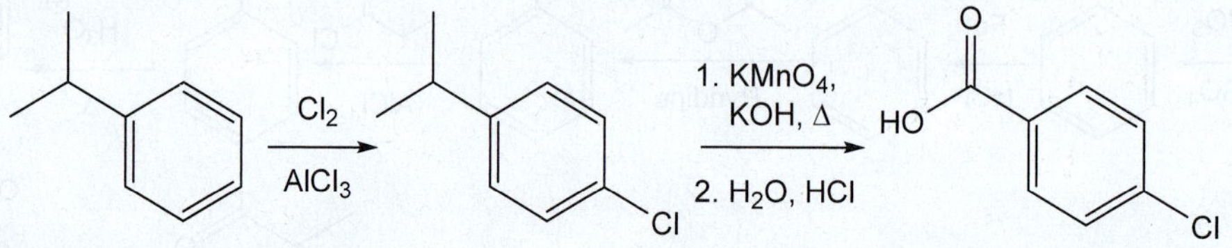

Problem 23.83 (SYN)

Think

How are the groups positioned relative to one another? Does the positioning match their directing ability? What reactions are a result of electrophilic aromatic substitution? What functional group transformations occur? Are any C–C bond-forming reactions necessary? Oxidation or reduction reactions?

Solve

(a) The NO_2 group is a meta director, and the Br is meta to the NO_2 group. Therefore, the Br is added first, and then the NO_2 is reduced to NH_2.

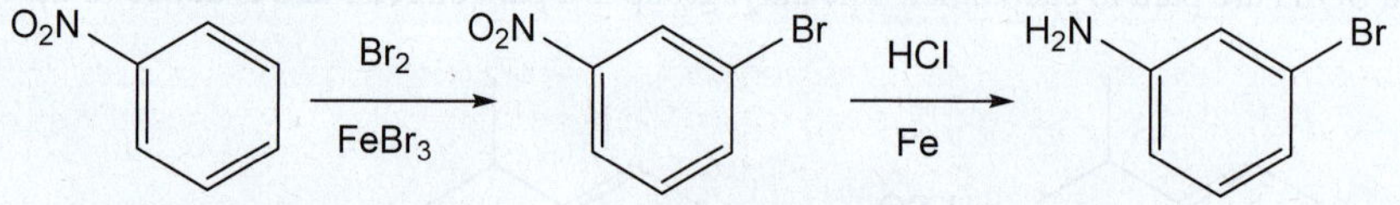

(b) The NO_2 group is a meta director, and the Br is para to the NO_2 group. Therefore, the NO_2 is reduced first to NH_2, a para director, and Br is subsequently added to the ring.

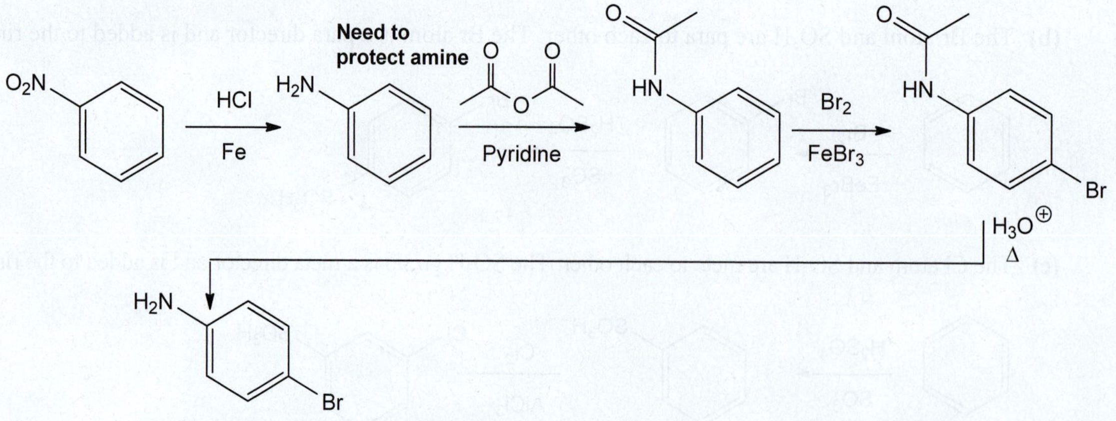

Problem 23.84 (SYN)

Think

How can the NH_2 group be transformed into the Br group? Where are the two Br groups on the ring relative to the NO_2 and NH_2 groups? Does this match the directing nature of either the NO_2 or NH_2 group?

Solve

The retrosynthetic analysis would appear as follows:

Undo a bromination. Undo substitution. Undo diazotization.

The NH_2 group can be replaced by Br by first converting it to a diazonium salt. Bromination can then occur at the positions ortho to the Br group and meta to the NO_2 group. The synthesis in the forward direction would appear as follows:

$NaNO_2$, HCl / H_2O → CuBr → Br_2 / $FeBr_3$

Problem 23.85 (SYN)

Think

How are the groups positioned relative to one another? Does the positioning match their directing ability? What reactions are a result of electrophilic aromatic substitution? What functional group transformations occur? Are any C–C bond-forming reactions necessary? Oxidation or reduction reactions?

Solve

(a) The CH_3 group is added to the ring via Friedel–Crafts acylation using C≡O/HCl (refer to Problem 22.31) followed by a reduction of the HC=O to CH_3. Fuming sulfuric acid is then added to the ring to sulfonate the para position. This blocks the para position and allows addition of two Cl groups ortho to CH_3 in the presence of excess $Cl_2/AlCl_3$. The SO_3H is then removed to yield the target compound.

C≡O / HCl → Zn(Hg) / HCl → H_2SO_4 / SO_3 → 2 equiv Cl_2 / $AlCl_3$ → H_3O^+

(b) The NH_2 group on the ring directs an acyl group para in a Friedel–Crafts acylation, which can then be reduced to convert it to the alkyl group. The NH_2 group can come from nitration of benzene, followed by reduction of the NO_2. However, we must protect the NH_2 group before the introduction of the strong Lewis acid. Otherwise, the Lewis acid will complex to the NH_2 group, turning it into a strong deactivating group. This would prevent the necessary Friedel–Crafts reactions from happening. We protect the NH_2 using acetic anhydride and deprotect it in the final step using hydrolysis under basic conditions. *Note:* Other syntheses are possible, too.

HNO_3, H_2SO_4 → O_2N; Fe, HCl → H_2N; Cl, $AlCl_3$; Zn(Hg), HCl; Cl, $AlCl_3$; 1. NaOH, Δ 2. H_2O, HCl → H_2N

Integrated Problems

Problem 23.86

Think

Identify each substituent as a weak, moderate, or strong activator or deactivator. Which ring is most activated? Which ring is most deactivated? Are the substituents on that ring ortho/para or meta directors?

Solve

(a) The most likely sites of reaction are indicated by an asterisk. The ring on the right is activated, whereas the ring on the left is deactivated. For the ring on the right, the substituent is an ortho/para director.

Meta director
Moderate deactivator

Ortho/para director
Moderate activator

(b) The most likely sites of reaction are indicated by an asterisk. All three rings are activated, but the ring in the middle is activated by two different substituents. Both substituents are ortho/para directors.

Ortho/para director
Strong activator

(c) The most likely sites of reaction are indicated by an asterisk. The ring on the left is activated by the CH_3 group and is deactivated by the C=O group. The ring on the right is deactivated by the C=O group only. The sites indicated are ortho or para to the methyl group (an ortho/para director) and meta to the C=O group (a meta director).

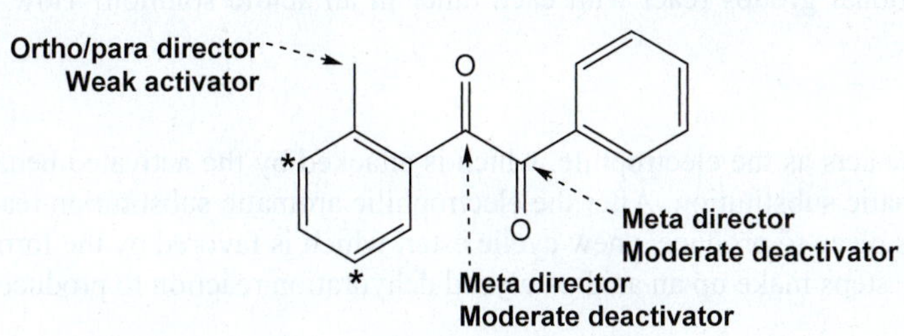

Ortho/para director
Weak activator
Meta director
Moderate deactivator
Meta director
Moderate deactivator

(d) The most likely sites of attack are indicated by an asterisk. Both rings are activated, but the lone pair on N activates the ring on the left more than the lone pair on O activates the ring on the right. The ring on the left has an ortho/para directing substituent.

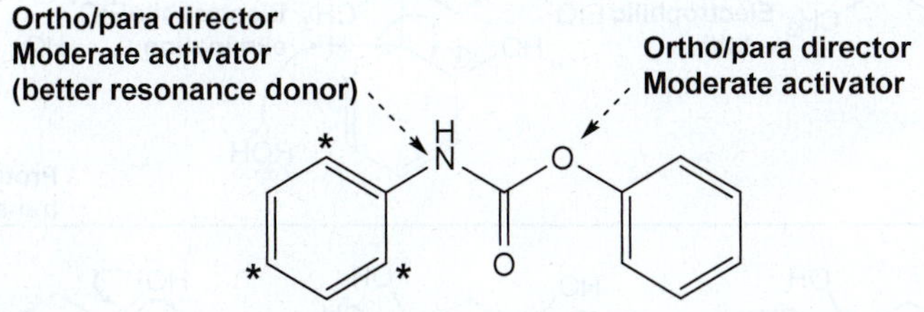

Ortho/para director
Moderate activator
(better resonance donor)
Ortho/para director
Moderate activator

Problem 23.87

Think

How many times does the electrophilic aromatic substitution mechanism need to occur? In the first electrophilic aromatic substitution reaction, what is the identity of the electrophile? How is it produced under acidic conditions? How is the product of that reaction converted to an electrophile under acidic conditions, to participate in the second electrophilic aromatic substitution reaction?

Solve

The ketone O atom is protonated to form the electrophile that the para C of phenol attacks in the electrophilic addition–elimination mechanism. The alcohol is then protonated to allow water to leave and form the tertiary benzylic carbocation. This carbocation acts as the electrophile in the second electrophilic addition–elimination mechanism.

Proton transfer
Electrophilic addition
Electrophile elimination
Proton transfer
Heterolysis
Electrophilic addition
Electrophile elimination

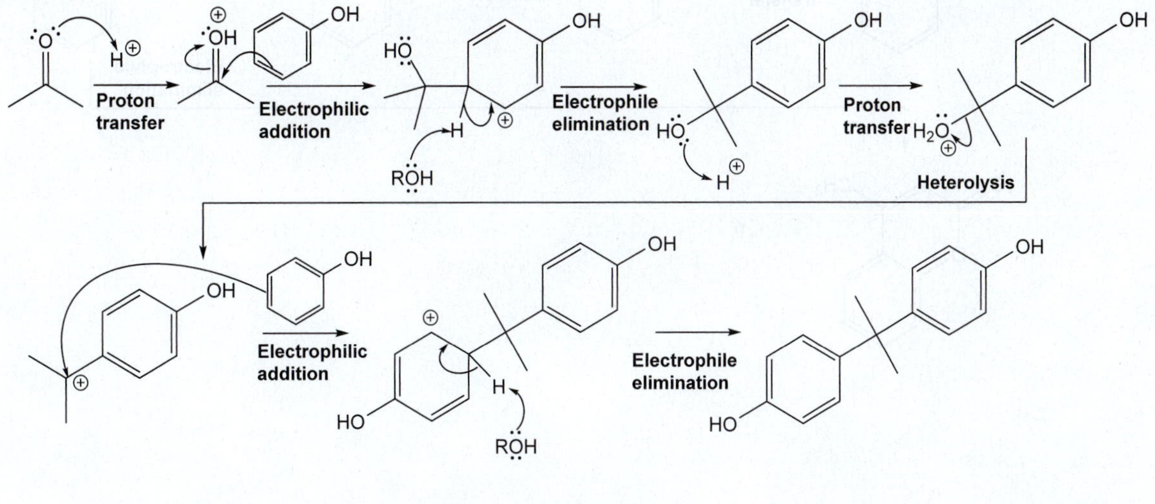

Problem 23.88

Think

What reaction occurs between the β-diketo ester and a strong acid? What is the identity of the electrophile in the electrophilic addition–elimination mechanism? After the electrophilic aromatic substitution, what functional groups are left? How can these functional groups react with each other in an acidic solution? How does the intermediate shown tautomerize?

Solve

A protonated carbonyl from the ketone acts as the electrophile, which is attacked by the activated benzene ring of phenol to initiate electrophilic aromatic substitution. After the electrophilic aromatic substitution reaction, an acid-catalyzed transesterification takes place to produce a new cyclic ester, which is favored by the formation of the six-membered ring. The final three steps make up an acid-catalyzed dehydration reaction to produce the new C=C.

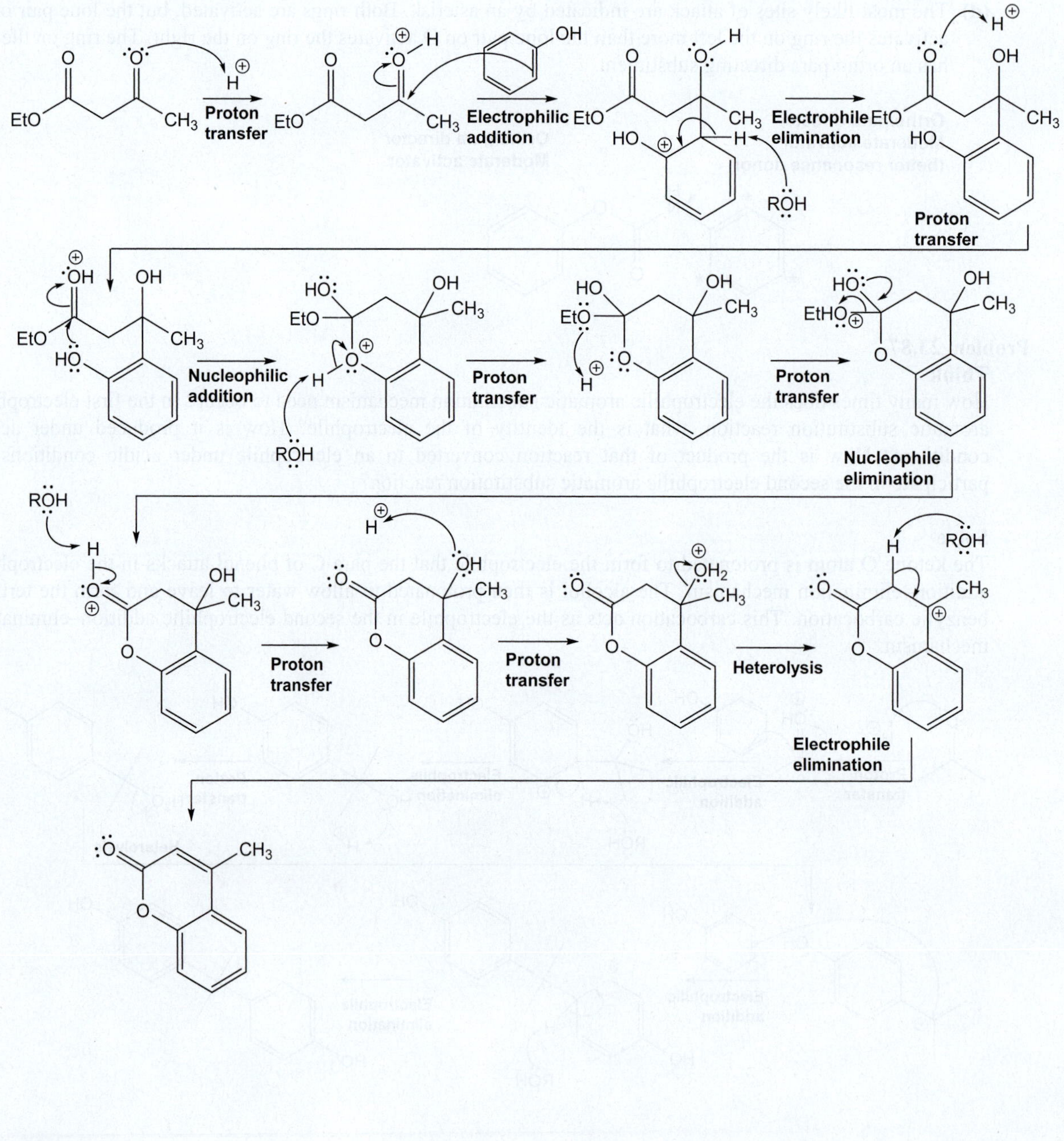

Problem 23.89

Think

What groups are added to the benzene ring? Are these groups ortho/para or meta directors? What reactions occur on functional groups off the ring? Are there any C–C bond-forming reactions? Functional group transformations? Oxidations or reductions?

Solve

Friedel–Crafts acylation forms the ketone product **A**. An acyl group is a meta director, and $Cl_2/FeCl_3$ chlorinates the ring at the meta position to form **B**. A Wittig reaction occurs with the ketone portion of the molecule and $CH_3CH^{-}–^{+}PPh_3$ to form the alkene **C**. The alkene reacts with the peroxyacid MCPBA to form the epoxide **D**. The epoxide is attacked at the *less* crowded carbon by the Grignard reagent C_6H_5MgBr to form the alcohol **E**. The alcohol undergoes a dehydration reaction to form the tetrasubstituted alkene **F**.

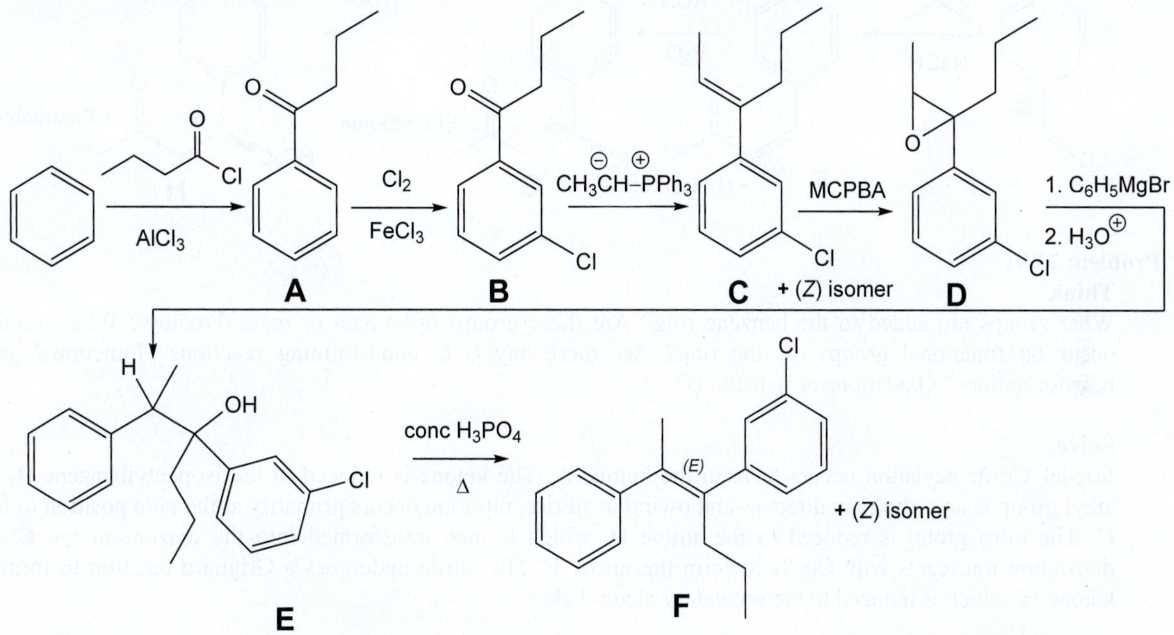

A **B** **C** + (*Z*) isomer **D**

E **F** + (*Z*) isomer

Problem 23.90

Think

What groups are added to the benzene ring? Are these groups ortho/para or meta directors? What reactions occur on functional groups off the ring? Are there any C–C bond-forming reactions? Functional group transformations? Oxidations or reductions?

Solve

Benzene is brominated to form **A**. Bromine is an ortho/para director, but the para isomer is the major product, because the Br and CH_3 have some steric bulk. This leads to **B**. The CH_3 group is oxidized by $KMnO_4$ to the carboxylic acid **C** after acid workup. The carboxylic acid is reduced by $LiAlH_4$ to the alcohol **D**, which is oxidized using PCC to the aldehyde **E**. The aldehyde reacts with acetone in a crossed aldol addition to produce the α,β-unsaturated carbonyl compound **F**, which undergoes conjugate nucleophilic addition to form **G**. The cyano group is hydrolyzed to the carboxylic acid **H**. See the figure on the next page.

Br, Br_2 / $FeBr_3$ → **A**; CH_3Cl / $AlCl_3$ → **B**; 1. $KMnO_4$, KOH, Δ 2. H_2O, HCl → **C** (CO_2H); 1. $LiAlH_4$ 2. $H_3O^{\oplus}$ → **D** (HO); PCC →

E (O) → (O, NaOH, Δ) → **F** (Z) + (E) isomer → (KCN, H_2O) → **G** (R), NC, O + **Enantiomer** → ($H_3O^{\oplus}$, Δ) → **H** (R), HO_2C, O + **Enantiomer**

Problem 23.91

Think

What groups are added to the benzene ring? Are these groups ortho/para or meta directors? What reactions occur on functional groups off the ring? Are there any C–C bond-forming reactions? Functional group transformations? Oxidations or reductions?

Solve

Friedel–Crafts acylation occurs to form the ketone **A**. The ketone is reduced to the isopentylbenzene **B**. The alkyl group is an ortho/para director, and owing to sterics, nitration occurs primarily at the para position to form **C**. The nitro group is reduced to the amine **D**, which is then transformed into the diazonium ion **E**. The diazonium ion reacts with CuCN to form the nitrile **F**. The nitrile undergoes a Grignard reaction to form the ketone **G**, which is reduced to the secondary alcohol **H**.

(O, Cl) / $AlCl_3$ → **A** (O); HCl / Zn/Hg → **B**; HNO_3 / H_2SO_4 → **C** (NO_2); HCl / Fe → **D** (NH_2); $NaNO_2$ / HCl →

E (N_2^+); CuCN → **F** (CN); 1. CH_3CH_2MgBr 2. H_3O^+ → **G** (O); $NaBH_4$ / EtOH → **H** (OH)

Problem 23.92 (SYN)

Think

What is the position of the two groups on the ring? Does that match the ortho/para- or meta-directing capabilities of the substituents? What is the order of electrophilic aromatic substitution for each group? How can a primary alkyl group get added to the ring? Are carbocation rearrangements a concern? From what functional group did the ester originate? Is an oxidation or reduction reaction necessary? Are protecting groups necessary?

Solve

To attach an alkyl group, the alcohol is oxidized to the carboxylic acid, which is transformed into the acid chloride. Friedel–Crafts acylation followed by reduction yields propylbenzene. The propylbenzene is brominated. In that reaction, the propyl group is an ortho/para director, and owing to steric hindrance, the para isomer is the major product. The PhBr is transformed into the Grignard reagent, and reaction with CO_2 followed by an acid workup yields the CO_2H group. A Fischer esterification of the carboxylic acid with propanol yields the target compound.

OH 1. H_2CrO_4 2. $SOCl_2$ O Cl $AlCl_3$ O Zn/Hg HCl

Br_2 $FeBr_3$ Br Mg(s) Ether MgBr 1. CO_2 2. $H_3O^{\oplus}$ O OH

OH $H_3O^{\oplus}$ O O

Problem 23.93 (SYN)

Think

From what type of reaction did the C=C result? How can you prepare those necessary precursors from butan-1-ol? What is the positioning of the Cl and the group containing the C=C? Which one was added first?

Solve

The C=C in both targets could have come from a Wittig reaction between an ylide and a ketone. The ylide is the $CH_3CH_2CH_2CH_2^{-}{-}^{+}PPh_3$.

Ylide prep:

OH PBr_3 Br 1. PPh_3 2. BuLi $\ominus$ $\oplus$ PPh_3

The aromatic ketone with which this ylide would react should be produced by reacting an aromatic ring with butanoyl chloride, which can be prepared as shown below.

Acid chloride prep:

1. H_2CrO_4
2. $SOCl_2$

In **(a)**, the Cl and the group containing the C=C group are para to each other, which suggests that chlorination occurred prior to Friedel–Crafts acylation.

In **(b)**, the Cl and C=C groups are meta to each other, which suggests that chlorination occurred after Friedel–Crafts acylation.

Problem 23.94 (SYN)

Think

How is phenol synthesized from benzene? What type of director is HO? How can the para position be blocked? How can a carboxylic acid be added to a benzene ring? How can an alcohol be transformed into an ester?

Solve

Phenol is formed from the diazonium salt. The carboxylic is added ortho from the Friedel–Crafts alkylation to add CH_3, which is then oxidized. Notice that a reversible sulfonation is used prior to the alkylation step to block the para position, and the sulfonyl group is removed after the alkylation is complete. The Ph–OH is transformed into the ester by reaction with acetic anhydride. *Note:* Other syntheses are feasible, too.

Acetylsalicylic acid (Aspirin)

Problem 23.95

Think

What is the index of hydrogen deficiency (IHD) in the product? How did the carbon skeleton change? What functional group gives rise to the broad peak around 11.7 ppm in the ^1H NMR spectrum? How many unique aromatic H atoms are present? How does NaOH react with the phenol O–H?

Solve

The IHD is 5, 4 of which can come from the aromatic ring and 1 likely from a C=O. Note that the proton signals between about 6 ppm and 8 ppm are the aromatic protons. The broad peak around 11.7 ppm in the ^1H NMR spectrum is due to the overlap of the carboxylic acid OH and the two phenol OH groups. This is indicated by the integration that appears to correspond to the 3H. This peak is significantly broadened owing to H bonding. There are seven carbon signals, indicating that all seven carbons are distinct. The carbonyl carbon signal from the carboxylic acid is around 172 ppm. Sodium hydroxide is a strong enough base to deprotonate a phenolic proton. The O^- is a very activating group, allowing the ring to react with the weak electrophile, CO_2, in an electrophilic aromatic substitution reaction analogous to nitration.

Proton transfer

Electrophilic addition

Add H_3O^+

Electrophile elimination

Proton transfer

Proton transfer

Problem 23.96

Think

What is the IHD? What functional group is evident from the peaks in the ^1H NMR spectrum from 5 to 6 ppm? What is the ratio of integration values? Which C–C bonds formed from dimerization? How does an alkene react with an acid HA? What is the identity of the electrophile in the electrophilic aromatic substitution?

Solve

The IHD is 5, 4 of which come from the aromatic ring and 1 likely from a C=C. The peaks from 5 to 6 ppm in the ^1H NMR spectrum are suggestive of an alkene. The ratio of H atoms is 5 (aromatic):1:1:3. The 3H signal is due to a CH_3. Therefore, the structure is likely prop-1-en-2-ylbenzene.

The alkene reacts with H–A in an electrophilic addition reaction to form the tertiary benzylic cation. This electrophile then reacts with another equivalent of prop-1-en-ylbenzene to form another tertiary benzylic cation. This cation serves as the electrophile in the electrophilic addition–elimination mechanism that takes place intramolecularly. See the mechanism on the next page.

Electrophilic addition

Electrophilic addition

Electrophilic addition

Electrophile elimination

:Base

Problem 23.97

Think

What is the IHD? If the signals are in a ratio of 1:3 and there are 12 H atoms, what is the actual ratio? If there are 14 signals between 120 and 140 in the ^{13}C NMR spectrum, how many C atoms are left in the structure that are not in this range? Is there any symmetry? How does a ketone react with a H–A acid? What is the identity of the electrophile in the electrophilic addition–elimination mechanism?

Solve

The IHD is 10, 8 of which can be from the two initial benzene rings. The actual H ratio is 3:9. The signal representing the three H atoms is likely from a CH_3 group, because all of the C atoms are aromatic, except one. The ketone reacts with the H–A acid via a proton transfer reaction. This protonated ketone is the electrophile in the electrophilic addition–elimination mechanism. The final three steps make up an acid-catalyzed dehydration reaction. The formation of the new ring and the C=C double bond make up the additional IHD of 2.

Proton transfer

Electrophilic addition

Electrophile elimination

Proton transfer

Heterolysis

Electrophile elimination

Chemical Formula: $C_{15}H_{12}$

CHAPTER 24 | The Diels–Alder Reaction and Other Pericyclic Reactions

Your Turn Exercises

Your Turn 24.1

Think

If the arrows move in a counterclockwise direction, what are the differences and similarities between the two mechanisms? How do you use curved arrows to show the formation of a bond? The breaking of a bond? How many arrows are present?

Solve

The difference between the mechanism shown in Equation 24-2 and the new counterclockwise mechanism shown below is the direction of arrow movement. There are still three curved arrows, the same transition state, and, overall, the same product. Three curved arrows are used to show the breaking and formation of bonds.

Same product shown in Equation 24-2

Your Turn 24.2

Think

In how many steps does the Diels–Alder reaction take place? What does it mean for the double bonds in a diene to be conjugated? Does the other molecue have a pair of π electrons that can react with the diene?

Solve

The Diels–Alder is a one-step concerted reaction. The diene has two conjugated C=C bonds, meaning that one C=C is bonded to the other. The dienophile is the C≡C because it has a pair of π electrons that can react with the diene to form a six-membered ring. The π bonds of the triple bond belong to two different π systems.

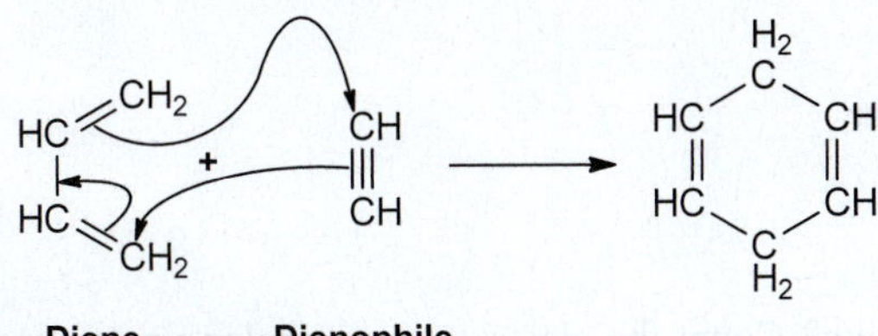

Diene **Dienophile**

Your Turn 24.3

Think

How many π electrons are involved in the formation and breaking of bonds in the cycloaddition reaction? Does this correspond to a Hückel number or an anti-Hückel number?

Solve

Four electrons are involved in the cyclic transition state, indicated by the two curved arrows. Four is an anti-Hückel number (it is an even number of pairs, or $4n$ with $n = 1$), so the transition state is antiaromatic.

Antiaromatic, 4π e^-

Your Turn 24.3

Think

What does the *s* stand for? What groups are being described as cis or trans to each other?

Solve

The *s* stands for *single*, as in single bond. Conformation **B** is *s*-cis, illustrated by the fact that the double bonds are on the same side of the single bond connecting them. Therefore, **B** can undergo a Diels–Alder reaction. Conformation **A** is *s*-trans and cannot undergo a Diels–Alder reaction. Note, however, that conformation **A** can attain conformation **B** via rotation about the single bond, at which point it could undergo a Diels–Alder reaction.

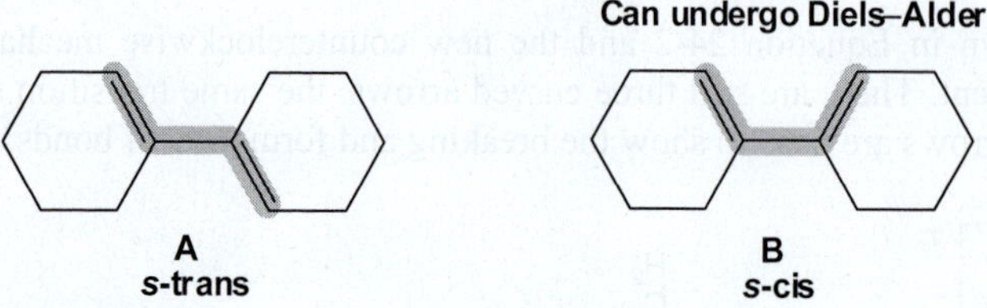

Your Turn 24.5

Think

What does the *s* stand for? What groups are being described as cis or trans to each other? In the current conformation, are the double bonds *s*-cis or *s*-trans? When the single bond connecting the two C=C bonds rotates, do the configurations of the C=C bonds change?

Solve

The current conformation of the diene is *s*-trans because the double bonds are oriented on opposite sides of the single bond that connects them. The single bond can rotate and yield the necessary *s*-cis conformation for Diels–Alder. Both double bonds are in the *E* configuration. This does not change whether the single bond rotates s-trans or s-cis; they are in the *E* configuration in both the *s*-cis and *s*-trans conformations.

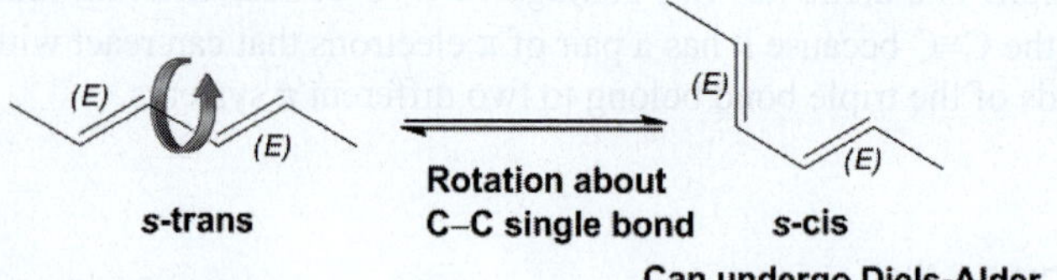

Your Turn 24.6

Think

If a C atom is sp^3 hybridized, can it participate in π bonding? Count the electron groups belonging to each C atom. How many electron groups does an sp^2 atom have? An sp^3 atom?

Solve

If a C atom is sp^3 hybridized, it cannot participate in π bonding; it has four electron groups. A C that is sp^2 hybridized has three electron groups. The two terminal C atoms of the diene and both C atoms of the dienophile are transformed from sp^2 to sp^3 hybridization, as indicated below.

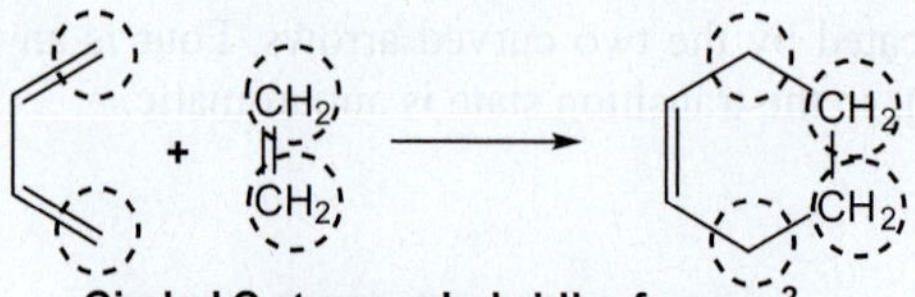

Circled C atoms rehybridize from sp^2 in the reactant to sp^3 in the product.

Your Turn 24.7

Think

Draw in the missing C–H bonds to have a bond drawn for W, X, Y, and Z. Tilt the diene into the paper to have the same orientation as the diene in Equation 24-20. Which groups are wedge and which groups are dash? How do the terminal C atoms in the diene rotate to form the product?

Solve

The H atoms in Equation 24-18 match up with W and Y in Equation 24-20, whereas the CH_3 groups match up with X and Z. Because X and Z end up on opposite sides of the ring in Equation 24-20, so do the CH_3 groups in Equation 24-18, resulting in the trans product.

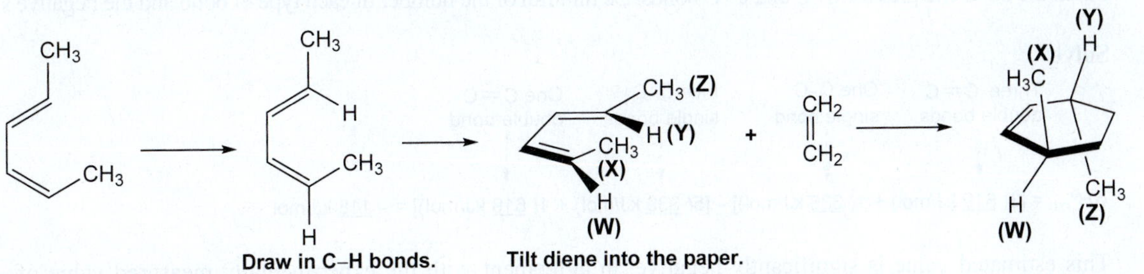

Draw in C–H bonds. Tilt diene into the paper.

Your Turn 24.8

Think

Use Figure 24-4 as a guide. What constitutes a favorable electrostatic interaction? Which orientation, **A** or **B**, shows a favorable electrostatic interaction among atoms undergoing bond formation?

Solve

Attraction between opposite charges constitutes a favorable electrostatic interaction. **B** has a favorable interaction, exhibited by the δ^+ and δ^- on atoms that are undergoing bond formation. **A** does not exhibit this kind of interaction. Therefore, **B** will lead to the major Diels–Alder product.

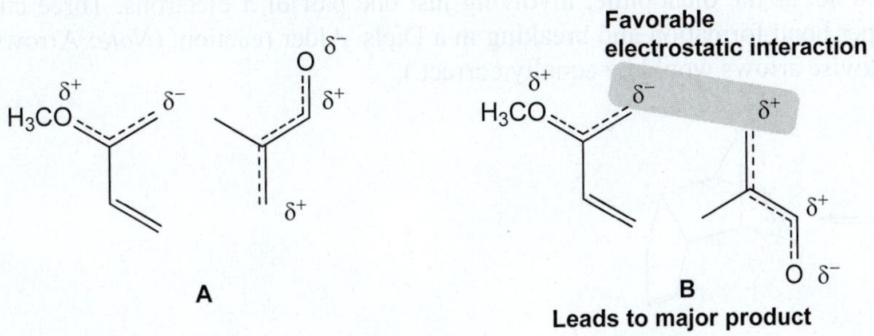

Favorable electrostatic interaction

A

B

Leads to major product

Your Turn 24.9

Think

Refer to Figure 24-4. How do you determine if there is a favorable electrostatic interaction? Draw the reactant molecules approaching without such a favorable interaction. With that orientation of the diene and dienophile, how do you arrive at the product?

Solve

The orientation without a favorable electrostatic interaction is the 180° flip of the orientation of one reactant in Solved Problem 24.15, as shown on the next page. In that orientation, none of the pairs of atoms undergoing bond formation exhibits a favorable interaction between opposite partial charges. The product is obtained by adding the three curved arrows and moving the electron pairs accordingly.

No favorable electrostatic interaction

Minor product

Your Turn 24.10

Think

Fill in the bond energies for C=C and C–C bonds. Be mindful of the number of each type of bond and the negative sign.

Solve

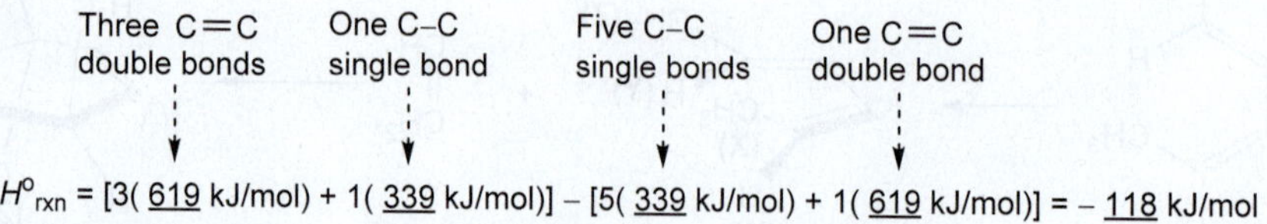

Three C=C double bonds
One C–C single bond
Five C–C single bonds
One C=C double bond
ΔH°rxn = [3(619 kJ/mol) + 1(339 kJ/mol)] – [5(339 kJ/mol) + 1(619 kJ/mol)] = – 118 kJ/mol

This estimated value is significantly negative, in agreement with the experimentally measured value of −168 kJ/mol, but the estimated value is 50 kcal/mol less negative.

Your Turn 24.11

Think

Can you identify one molecule of cyclopentadiene as the diene (two conjugated C=C bonds) and the other as the dienophile (one pair of π electrons)? Can you identify the six-membered ring that has formed in the product? How can you draw the curved arrows to show the formation and breaking of bonds in the concerted Diels–Alder reaction?

Solve

One molecule of cyclopentadiene could act as the diene, involving both pairs of π electrons, and the other molecule of cyclopentadiene could act as the dienophile, involving just one pair of π electrons. Three curved arrows are added to show the proper bond formation and breaking in a Diels–Alder reaction. (*Note:* Arrows are drawn clockwise, but counterclockwise arrows would be equally correct.)

Diene Dienophile

Your Turn 24.12

Think

Compare this Diels–Alder reaction to the one in Equation 24-27, which is irreversible. How do the values of $\Delta H°_{rxn}$ compare? How does that impact the $\Delta G°_{rxn}$? How does the $\Delta G°_{rxn}$ relate to reversibility?

Solve

The $\Delta H°_{rxn}$ for this reaction (−121 kJ/mol) is more negative than the $\Delta H°_{rxn}$ value for the reaction in Equation 24-27 (−113 kJ/mol). With the $\Delta H°_{rxn}$ more negative, this would contribute to a more negative value for $\Delta G°_{rxn}$, too. A significantly negative $\Delta G°_{rxn}$ is associated with an irreversible reaction, so because the reaction in Equation 24-27 is irreversible, this one is irreversible, too.

Your Turn 24.13

Think

Refer to Equation 24-29 as an example of a retro Diels–Alder mechanism. Which bonds are broken and formed to produce the diene and the dienophile?

Solve

The C=C π bond electrons move to form one new C=C bond of the diene product. The adjacent C–C σ bond breaks to form the dienophile C=C. The next C–C σ bond breaks to form the second C=C of the diene product. Note that the curved arrows could have been drawn counterclockwise instead of clockwise.

>800 °C

Your Turn 24.14

Think

Refer to Equation 24-31 as a guide. Which bonds are broken and formed in the concerted cycloaddition reaction to form the manganate ester? How many curved arrows are necessary?

Solve

The [4+2] cycloaddition mechanism of MnO_4^- across the C=C double bond to form a manganate ester is shown below. Three curved arrows are drawn to show the conversion of three π bonds to two new σ bonds and a lone pair.

A manganate ester

Hydrolysis

Your Turn 24.15

Think

In the regions of orbital overlap, do the orbitals have the same or opposite phase? Does that correspond to constructive or destructive interference? Do the two regions of overlap exhibit the same type of interference?

Solve

In both regions of overlap, the orbitals have the same phase, which leads to constructive interference. Because both regions exhibit the same kind of interference, the orbitals have the appropriate symmetries to interact.

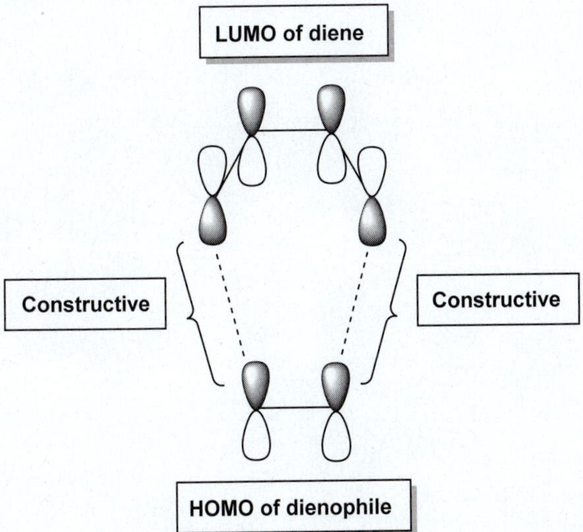

Your Turn 24.16

Think

In the regions of orbital overlap, do the orbitals have the same or opposite phase? Does that correspond to constructive or destructive interference? Do the two regions of overlap exhibit the same type of interference?

Solve

On the left, the orbitals have opposite phase, resulting in destructive interference. On the right, the orbitals have the same phase, resulting in constructive interference. Therefore, there is no net overlap, and the orbitals do not have the appropriate symmetries to interact.

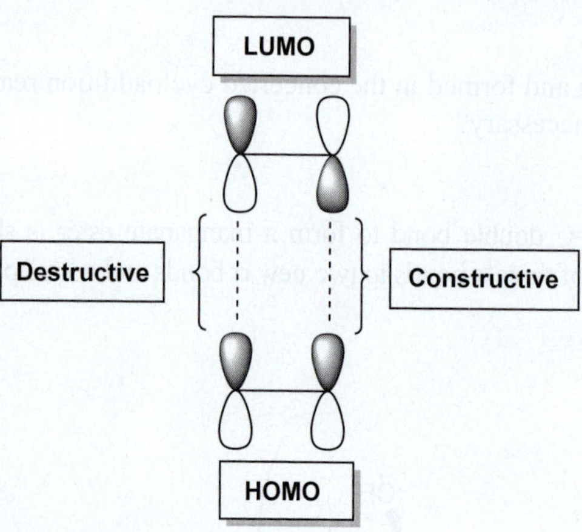

Your Turn 24.17

Think

Referring to Figure 24-11a, see which C atoms of the diene and dienophile are involved in secondary orbital overlap. To which atoms do those correspond in the figure given?

Solve

The *p* orbitals on C2 and C3 of the dienophile are involved in primary orbital overlap in both Figure 24-11a and 24-11b. In Figure 24-11a, the *p* orbital on C1 of the dienophile is involved in secondary orbital overlap with the leftmost *p* orbital of the diene. The same two orbitals in the figure below (circled) are farther away than they are in Figure 24-11a.

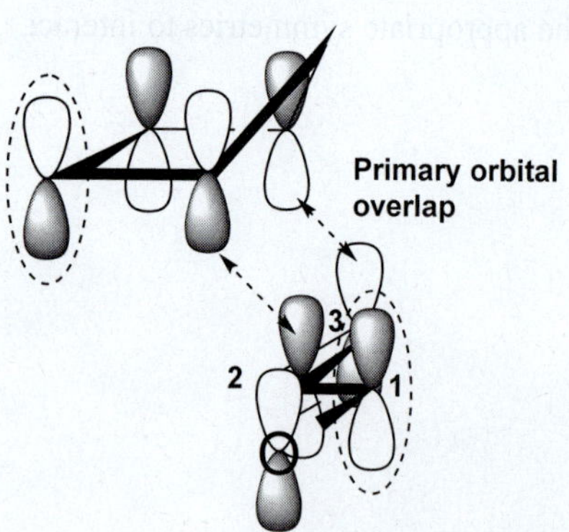

In Chapter Problems

Problem 24.2

Think

In the transition state, are there electrons delocalized over an entire ring? If so, how many electrons? How can you use the number of curved arrows to tell? Is that a Hückel number or an anti-Hückel number? How does that correspond to whether the reaction is allowed or forbidden?

Solve

Eight electrons are delocalized over an entire ring in the transition state (two electrons are represented by each of the four curved arrows). Because eight is an anti-Hückel number of electrons, the transition state is antiaromatic, so the reaction does not take place readily; it is forbidden under normal conditions.

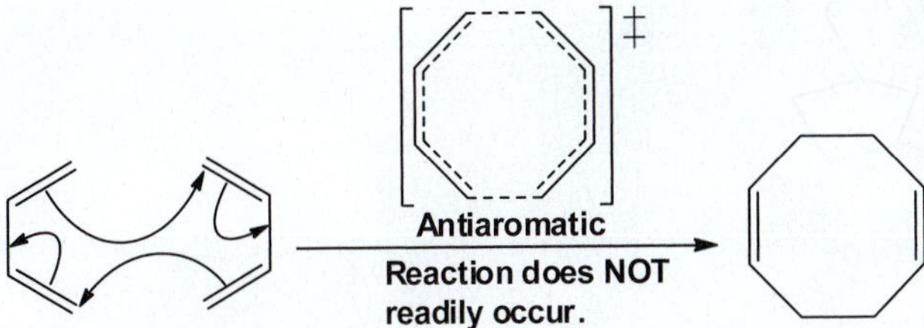

Problem 24.4

Think

Are the dienes in an *s*-cis conformation? If not, can they attain an *s*-cis conformation? Which is more stable in the *s*-cis conformation?

Solve

Neither diene is initially in the *s*-cis conformation; both are in the *s*-trans conformation as drawn. The single bond in the molecule that connects the two double bonds can rotate to attain the *s*-cis conformation, as shown below. **D** will react faster, however, because the steric strain involving the methyl groups helps destabilize the *s*-trans conformation of **D**, which helps favor the *s*-cis conformation.

s-Trans
s-Cis
C
Rotation about the single bond
s-Trans
s-Cis
Too much steric strain
D
Rotation about the single bond

Problem 24.5

Think

If benzene were to react in ethene, what would be the [4+2] cycloaddition product? Evaluate the aromaticity of the reactants and products. Does the product exhibit ring strain? Why might this prohibit the [4+2] cycloaddition reaction?

Solve

First, the aromaticity of benzene would be destroyed if it were to react, because the product would be nonaromatic. Second, there would be significant ring strain in the product; the alkene carbons, being sp^2 hybridized, each have an ideal bond angle of 120°. The bicyclic system would cause significant deviations from this ideal bond angle.

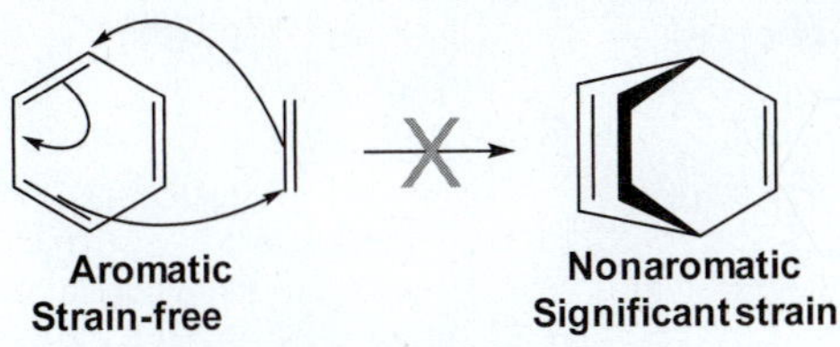

Problem 24.6

Think

What is the mechanism of a [4+2] cycloaddition? What size ring contains the diene? Do you think the diene in this reaction will be locked in the *s*-cis conformation in the same way as it is in Equation 24-10?

Solve

The mechanism for the [4+2] cycloaddition is shown below. The diene in this reaction is part of an eight-membered ring, which is significantly more flexible than the five-membered ring in Equation 24-10. Therefore, the diene will not be locked in the *s*-cis conformation as rigidly as is the dienophile in Equation 24-10, making the reaction slower.

Problem 24.8

Think

Is the dienophile relatively electron rich or electron poor in a standard Diels–Alder reaction? Will electron-donating groups (EDG) or electron-withdrawing groups (EWG) on the dienophile speed up the reaction? What is the relative electron-donating/electron-withdrawing capability of each substituent?

Solve

The dienophile is relatively electron poor in a standard Diels–Alder reaction, so the more electron withdrawing a substituent on the dienophile is, the faster is the reaction. In dienophile **C**, the CH=O substituents are electron withdrawing. In **D**, the alkoxy substituents are electron donating via resonance. Therefore, **C** would react faster.

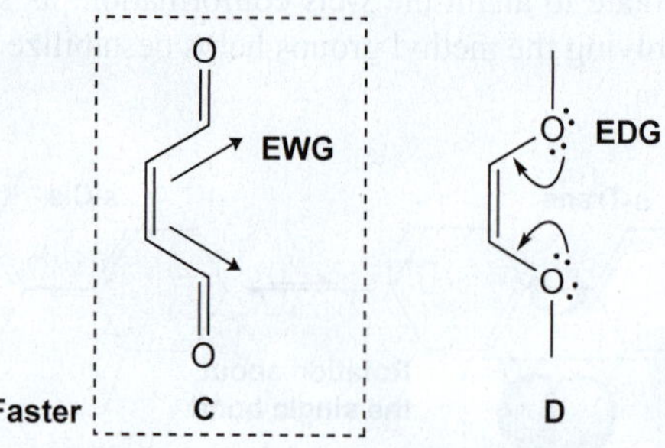

Problem 24.9

Think

What is the configuration involving the NO_2 groups in the alkene dienophile? Is that stereochemistry conserved in a Diels–Alder reaction?

Solve

The NO_2 groups are cis about the C=C bond in the dienophile. The [4+2] cycloaddition reaction is a concerted reaction and conserves this stereochemistry, so the NO_2 groups become cis with respect to the six-membered ring in the product.

Problem 24.10

Think

Which carbons came from the diene? Which came from the dienophile? Do the carbon atoms that came from the dienophile have substituents that are cis or trans to each other in the ring? How does that translate to the stereochemistry of the dienophile?

Solve

(a) The carbon atoms that came from the dienophile are attached to CH=O and methyl groups in the product. These CH=O and methyl groups are cis with respect to the six-membered ring in the Diels–Alder reaction product and, therefore, must have been cis about the C=C bond in the original dienophile reactant.

Undo a Diels–Alder reaction.

CH_3

Cis

(b) The CH=O and methyl group are trans with respect to the six-membered ring in the Diels–Alder reaction product and, therefore, must have been trans about the C=C bond in the original dienophile reactant.

Undo a Diels–Alder reaction.

CH_3

Trans

Problem 24.12

Think

What must the conformation of the diene be to react in a Diels–Alder reaction? Do the carbons at the ends of the diene become chiral centers in the products? What are the configurations about the double bonds in the diene? Do the alkene carbons of the dienophile become chiral centers? What is the configuration about the C=C double bond in the dienophile?

Solve

The dienes must achieve an *s*-cis conformation prior to reaction. This conformational change is shown for each reaction.

(a) If we draw the product without worrying about stereochemistry, notice that each end carbon in the diene becomes a chiral center, and so does each alkene carbon in the dienophile (noted with an asterisk).

CH_3 D H D

Rotation about C–C single bond

D H CH_3 D

H_3C D * * * * H D

The two D–C=C–C configurations are the same (i.e., both trans) along the diene skeleton, so the two D atoms will be cis to each other in the product. In the dienophile, the bonds to the ring are cis about the C=C bond, so they remain cis with respect to the six-membered ring in the product. Overall, there are four results to consider:

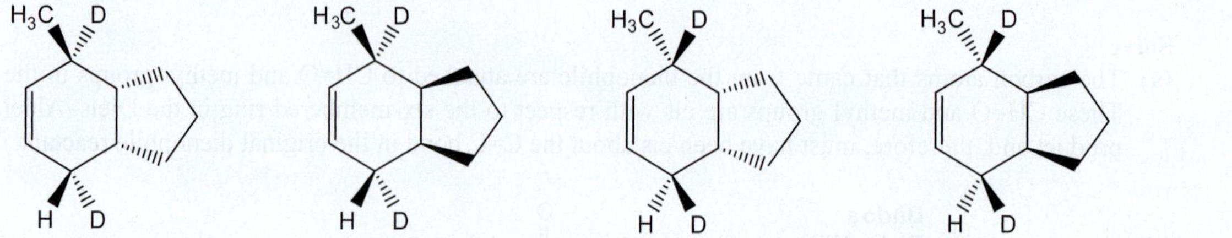

(b) If we draw the product without worrying about stereochemistry, notice that each end carbon in the diene becomes a chiral center, but the two C atoms of the dienophile do not.

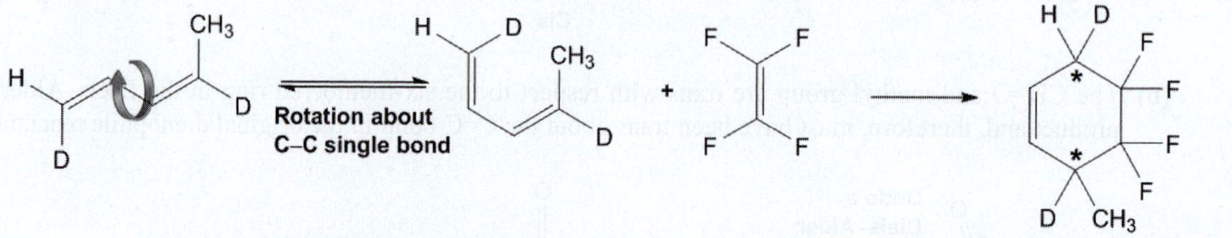

Rotation about C–C single bond

The two D–C=C–C configurations along the diene skeleton are opposite (one is trans and the other is cis), so the two D atoms will be trans to each other with respect to the six-membered ring in the product. Overall, then, there are two results to consider:

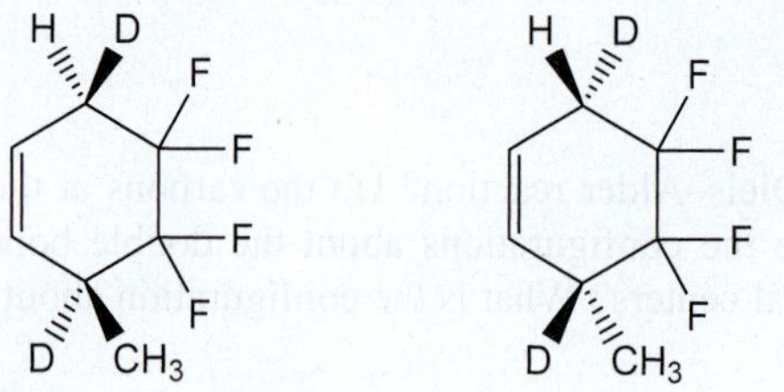

Problem 24.13

Think

Which carbon atoms in the product came from the diene? From the dienophile? Are the OCH_3 groups cis or trans to each other with respect to the plane of the ring? What does this mean for the configurations of the two double bonds in the diene precursor?

Solve

The C atoms bonded to the OCH_3 groups in the product were the carbons at the ends of the conjugated diene in the reactant. The OCH_3 groups in the product are trans to each other. This means that one C=C bond from the diene was cis and the other was trans. The retrosynthesis is shown below:

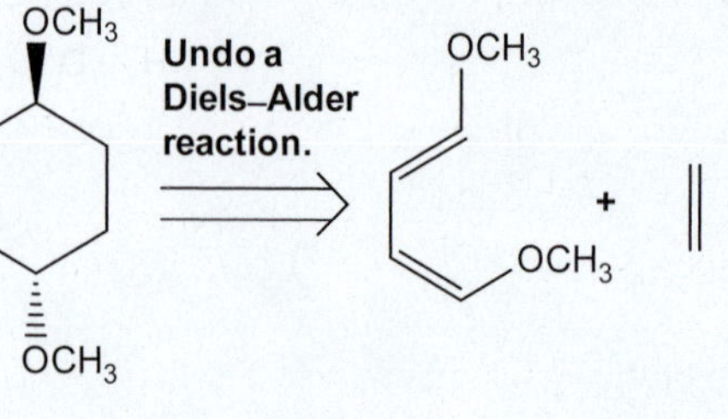

Undo a Diels–Alder reaction.

Problem 24.14

Think

Which reactant acts as the diene and which acts as the dienophile? How do you add curved arrows to show proper bond formation and breaking? Is the major product from the endo approach or the exo approach? In the product, how do you distinguish endo from exo bonds?

Solve

Cyclohexadiene acts as the diene and the diketone acts as the dienophile. The major product is the endo product, which would come from the endo approach, as shown below, where the C=O substituents on the dienophile approach toward the diene. In the endo product, the groups that were attached to the C=C of the dienophile are on the opposite side of the six-membered ring from the groups that were attached to the ends of the conjugated C=C bonds of the diene.

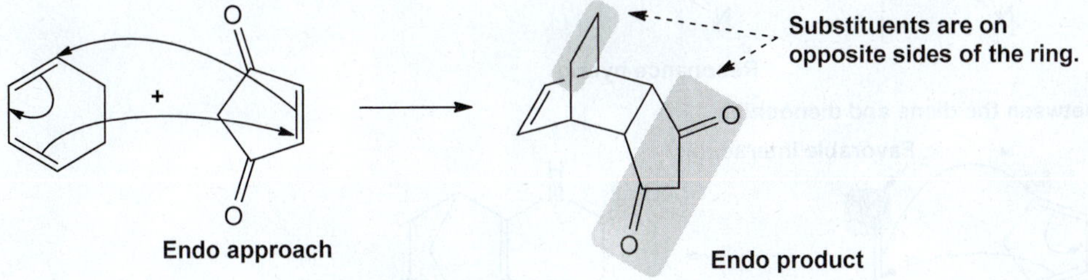

Substituents are on opposite sides of the ring.

Endo approach

Endo product

Problem 24.16

Think

Where are the regions of excess negative charge and excess positive charge in these molecules? How can you use resonance structures to determine this? Which approach leads to favorable electrostatic interactions among the atoms undergoing bond formation?

Solve

The partial charges from resonance hybrids of the diene and dienophile are shown below. The major product comes from the most favorable electrostatic interaction among the atoms undergoing bond formation, which are highlighted in gray screens.

(a) Resonance in the diene and dienophile

Diene

Resonance hybrid

Dienophile

Resonance hybrid

Interactions between the diene and dienophile

Favorable interaction

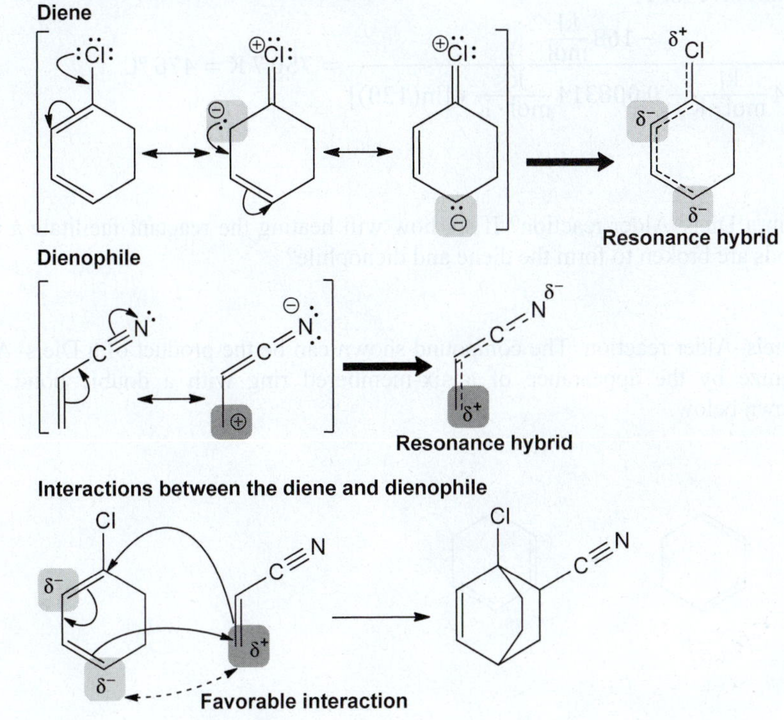

(b) Resonance in the diene and dienophile

Diene

Resonance hybrid

Dienophile

Resonance hybrid

Interactions between the diene and dienophile

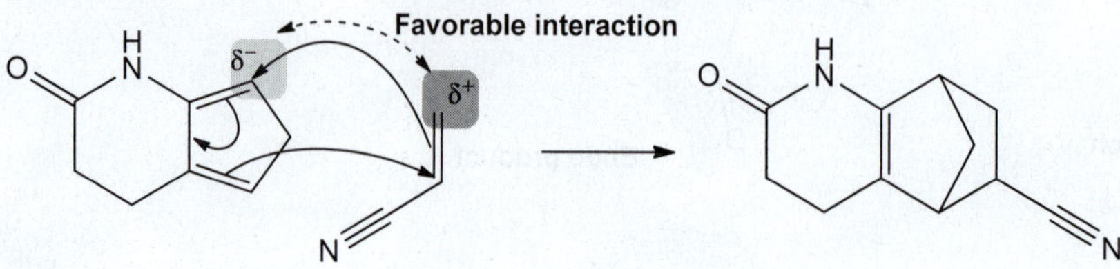

Problem 24.17

Think

What are the values for $\Delta H°$ and $\Delta S°$ for Equation 24-27, and how can you obtain the value for $\Delta G°_{rxn}$? What equation relates temperature, $\Delta G°_{rxn}$, and K_{eq}? When you solve for temperature, what value do you obtain?

Solve

The values for $\Delta H°$ and $\Delta S°$ are −168 kJ/mol and −0.184 kJ/mol•K (from $T\Delta S° = 55$ kJ/mol at 298 K). Therefore, the value for temperature can be solved using the following two $\Delta G°$ relationships:

$$\Delta G°_{rxn} = \Delta H°_{rxn} - T\Delta S°_{rxn} = -RT(\ln K_{eq})$$

Using the second equality, we can solve for T:

$$T = \frac{\Delta H°}{\Delta S° - R[\ln K_{eq}]} = \frac{-168\dfrac{\text{kJ}}{\text{mol}}}{-0.184\dfrac{\text{kJ}}{\text{mol}\cdot\text{K}} - 0.008314\dfrac{\text{kJ}}{\text{mol}\cdot\text{K}} * [\ln(120)]} = 750.7\ \text{K} = 478\ °\text{C}$$

Problem 24.18

Think

Can the reactant be formed from a Diels–Alder reaction? If so, how will heating the reactant facilitate a retro Diels–Alder reaction? What bonds are broken to form the diene and dienophile?

Solve

This is an example of a retro Diels–Alder reaction. The compound shown can be the product of a Diels–Alder reaction, which we can recognize by the appearance of a six-membered ring with a double bond. The mechanism and products are shown below.

Δ

+

Problem 24.19

Think

Which diols appear to be *cis*-1,2 diols in the conformation shown? What about in other conformations about the C–C bond? In the conformation in which the OH groups are syn to each other, what are the relative orientations of the other groups attached to the C–C bond?

Solve

(a) This is an example of a *cis*-1,2 diol. The alkene is cyclohexene.

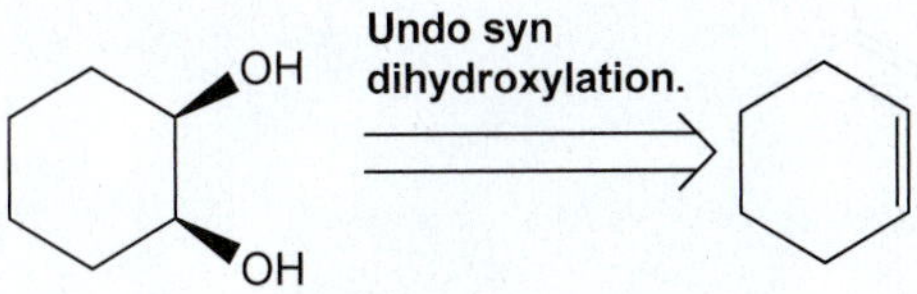

Undo syn dihydroxylation.

(b) This product did not form from syn hydroxylation because the HO groups are anti to each other and a ring does not have free rotation about its single bonds.

(c) This product did not form from syn hydroxylation because the HO groups are in a 1,3 position.

(d) This is an example of a *cis*-1,2 diol. Because the HO groups are already cis, the ethyl groups are trans about the alkene from which it was made.

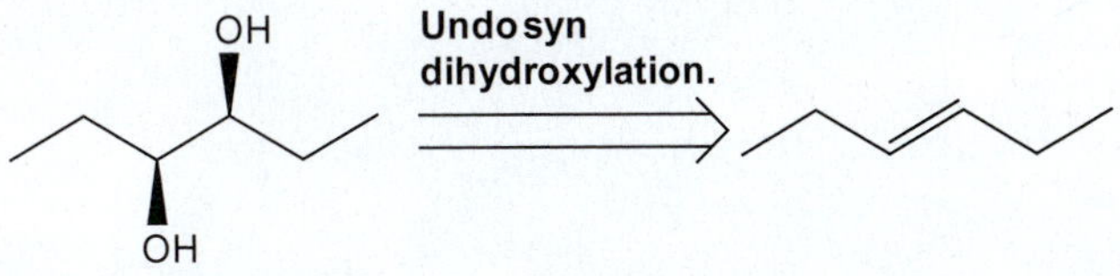

Undo syn dihydroxylation.

(e) The HO groups are trans, but another conformation can be drawn in which the HO groups are *cis*-1,2. Therefore, the ethyl groups are cis in the alkene.

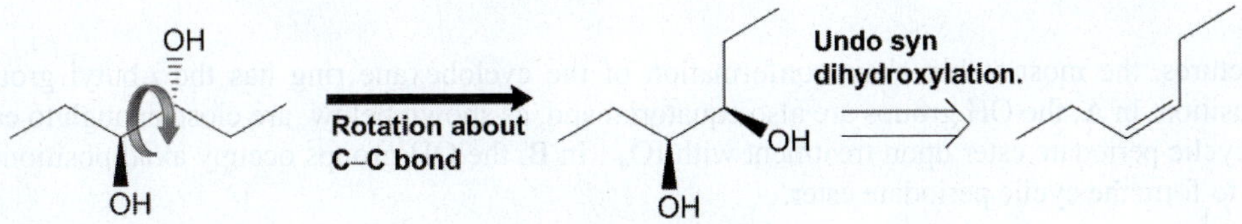

Rotation about C–C bond

Undo syn dihydroxylation.

Problem 24.21

Think

What was the possible original structure given in Solved Problem 24.20? What is the configuration about the C=C bond? Would changing that configuration affect the oxidative cleavage products? How can you connect the C atoms of the ketone C=O and aldehyde C=O groups differently to arrive at the original structure?

Solve

The original structure given in Solved Problem 24.20 is the (*E*) isomer shown below:

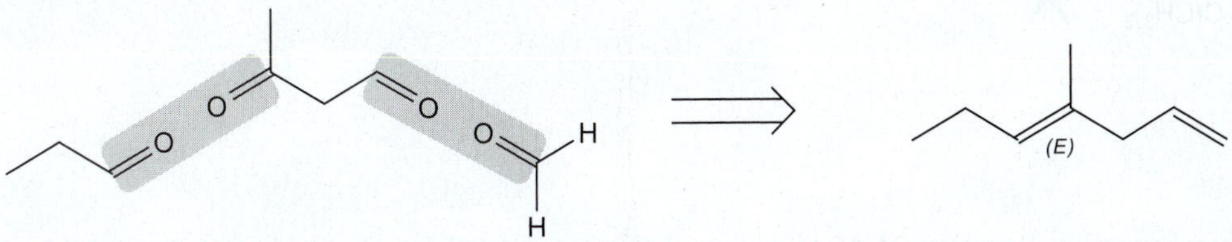

(*E*)

The starting material also could have been the (*Z*) isomer, as shown below. The configuration of the alkene does not impact the identities of the oxidative cleavage products.

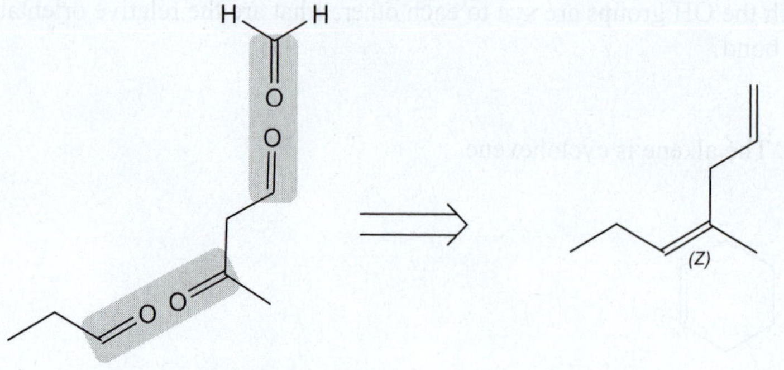

Alternatively, the C=O carbons could have been connected together differently, as shown below.

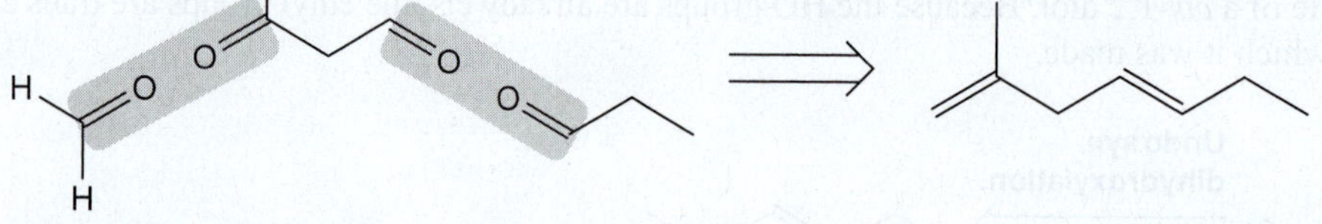

Problem 24.22

Think

What position does the *t*-butyl group occupy on the cyclohexane ring in the most stable chair conformation? What positions do the OH groups occupy in each structure? With the OH groups in those positions, can the cyclic periodate ester form upon treatment with IO_4^-?

Solve

In both structures, the most stable chair conformation of the cyclohexane ring has the *t*-butyl group in the equatorial position. In **A**, the OH groups are also equatorial and, as shown below, are close enough to each other to form the cyclic periodate ester upon treatment with IO_4^-. In **B**, the OH groups occupy axial positions and are too far apart to form the cyclic periodate ester.

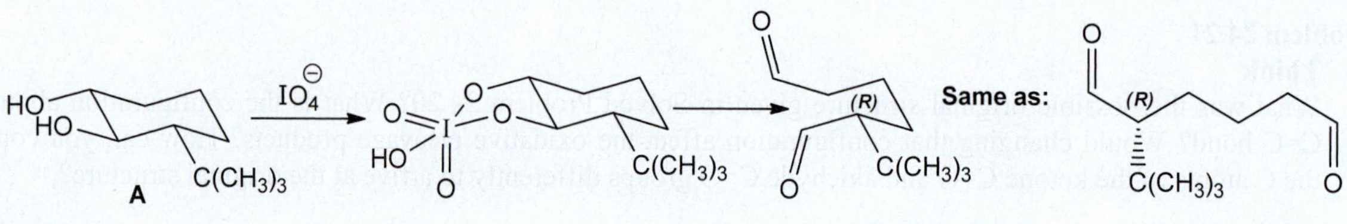

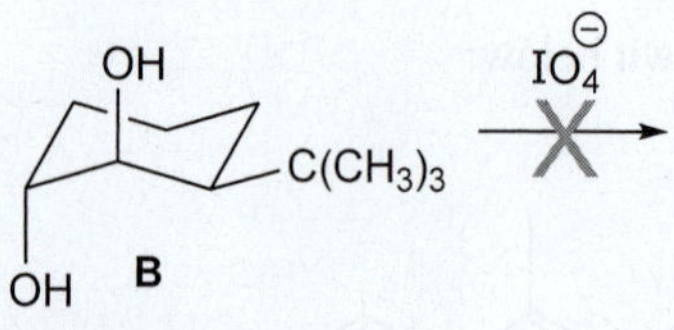

Problem 24.23

Think

Refer to the mechanism in Equation 24-45 to review the structure of the molozonide and the ozonide. What bond is cleaved in the ozonolysis reaction? Is the workup done under oxidative or reducing conditions?

Solve

The structures of the molozonide, ozonide, and final products from the ozonolysis reactions are shown below. In **(a)**, the workup is done under oxidative conditions, so $H_2C{=}O$ is oxidized to CO_2. In **(b)**, the workup is done under reducing conditions, so the aldehyde that is initially formed is not oxidized further.

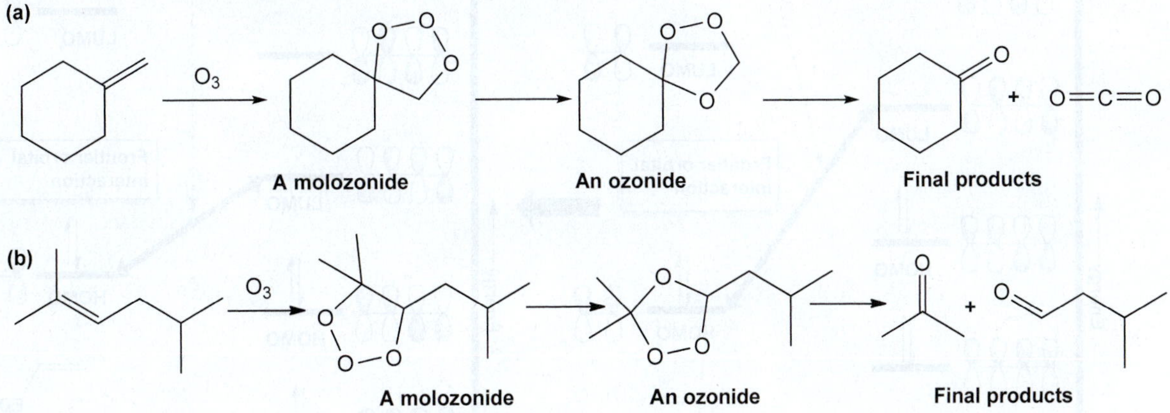

(a) O_3 — A molozonide — An ozonide — Final products (+ O=C=O)

(b) O_3 — A molozonide — An ozonide — Final products

Problem 24.24

Think

Is the standard Diels–Alder reaction favored when the dienophile is electron rich or electron poor? What types of groups are CH_3 and OCH_3? Are they activating or deactivating? Are they equally so?

Solve

The standard Diels–Alder reaction is favored when the dienophile is electron poor—namely, when electron-withdrawing groups are on the dienophile. The OCH_3 and CH_3 are both electron-donating groups (EDGs), but CH_3 is a weaker EDG compared to the OCH_3. Given the choice, the diene will react with the dienophile that has the weaker EDG.

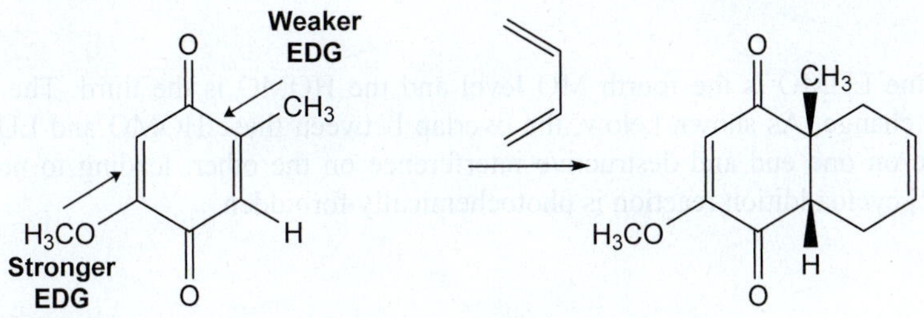

Weaker EDG — CH_3; H_3CO — Stronger EDG; H; O; O → product with CH_3, H, H_3CO

Problem 24.25

Think

Is the energy of an orbital raised or lowered by an attached electron-donating group? Is the energy of an orbital raised or lowered by an attached electron-withdrawing group? Is the HOMO–LUMO gap smaller or larger in this example?

Solve

The energy of an orbital is usually *raised* by an attached electron-donating group, and it is usually *lowered* by an attached electron-withdrawing group. Therefore, the diene's MOs are lowered and the dienophile's MOs are raised. Although this increases the HOMO–LUMO energy gap between the HOMO of the diene and the LUMO of the dienophile, it lowers the HOMO–LUMO energy gap between the HOMO of the dienophile and the LUMO of the diene (as shown on the next page).

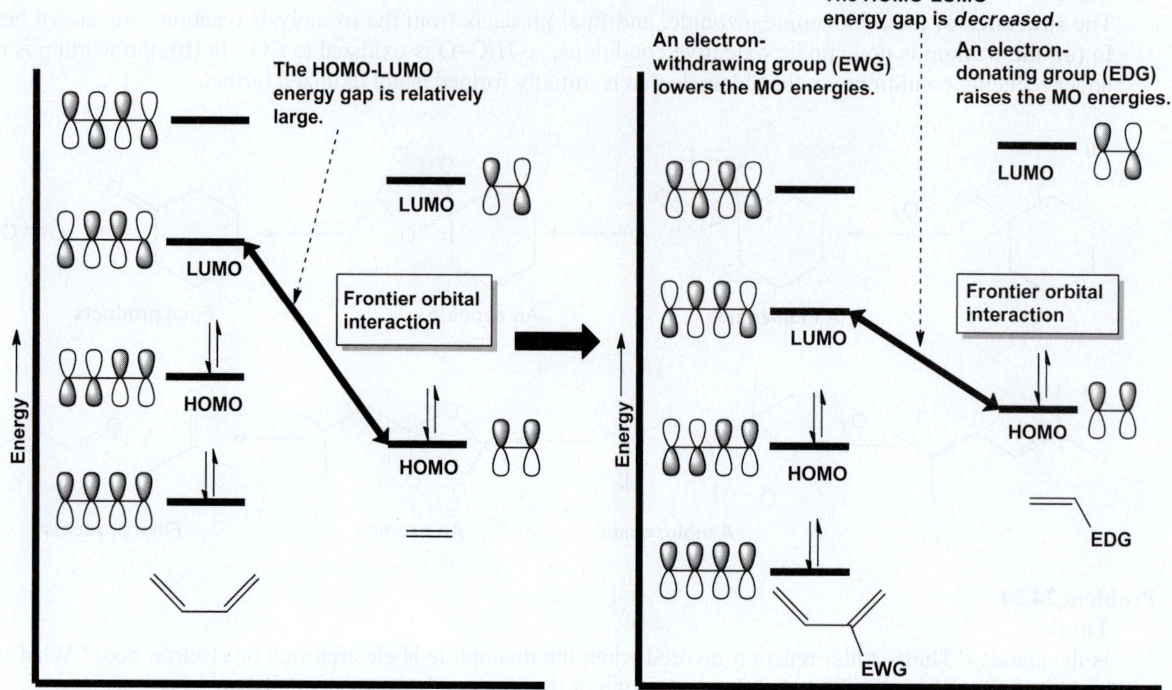

Problem 24.26

Think

If the diene is in the excited state, what are the new HOMO and LUMO? What are the HOMO and LUMO for ethene in its ground state? In this new HOMO–LUMO interaction, are there regions of constructive or destructive interference? Is this forbidden or allowed? How does this all change if the diene is in its ground state and the dienophile is in its excited state?

Solve

If the diene is in the excited state, the LUMO is the fourth MO level and the HOMO is the third. The MO diagram for the dienophile does not change. As shown below, the overlap between these HOMO and LUMO orbitals has constructive interference on one end and destructive interference on the other, leading to no net interaction. This shows why the [4+2] cycloaddition reaction is photochemically forbidden.

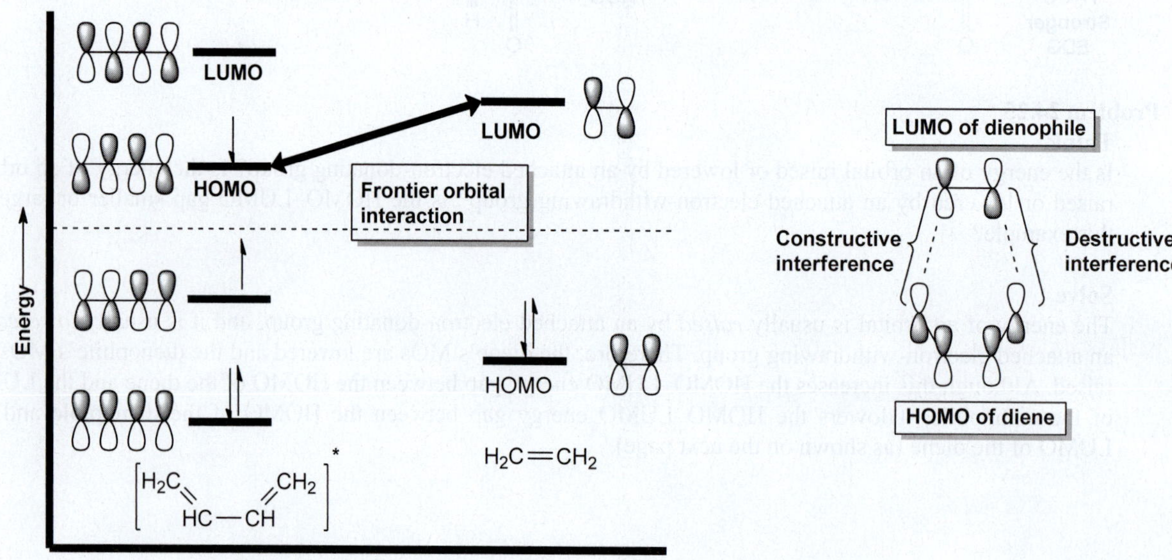

The figure below shows how the energy diagrams change when the dienophile is in its excited state and the diene is in its ground state. Notice, however, that the HOMO–LUMO interaction involves the same two orbitals as above, resulting in no net interaction and a forbidden reaction.

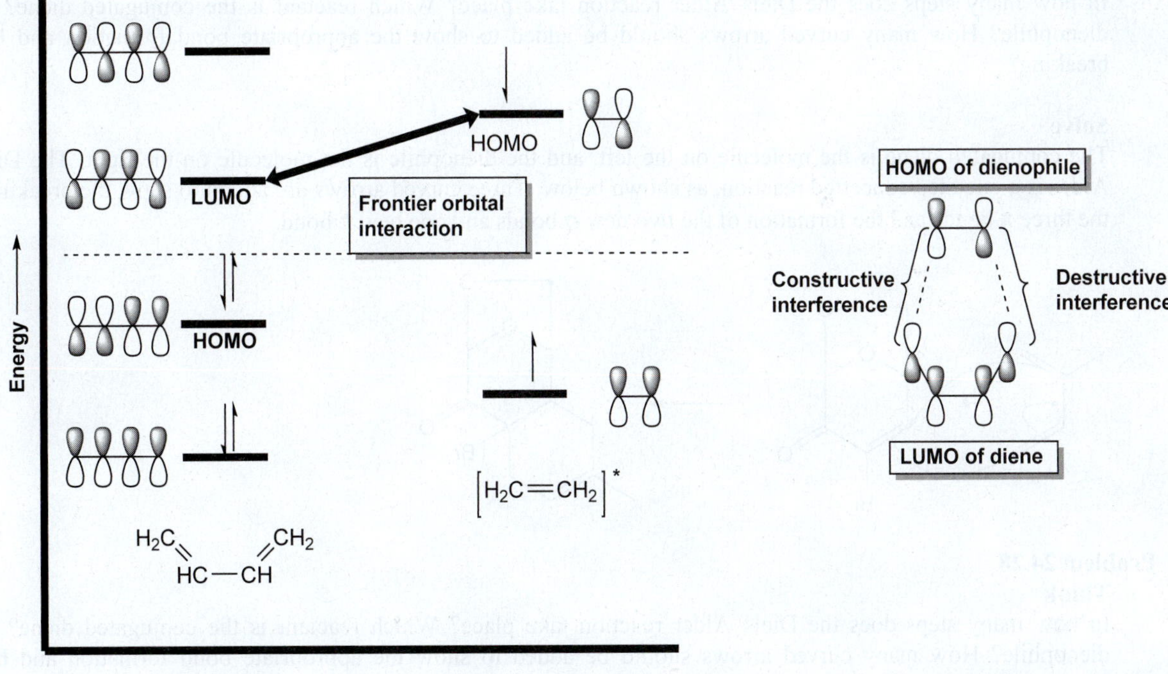

End of Chapter Problems
Sections 24.1 and 24.2 The Diels–Alder Reaction and the *s*-Cis Conformation of the Diene
Problem 24.27

Think

In how many steps does the Diels–Alder reaction take place? Which reactant is the conjugated diene? The dienophile? How many curved arrows should be added to show the appropriate bond formation and bond breaking?

Solve

The conjugated diene is the molecule on the left, and the dienophile is the molecule on the right. The Diels–Alder is a one-step concerted reaction, as shown below. Three curved arrows are added to show the breaking of the three π bonds and the formation of the two new σ bonds and one new π bond.

O
O
+
O
Br
O
O
O
Br

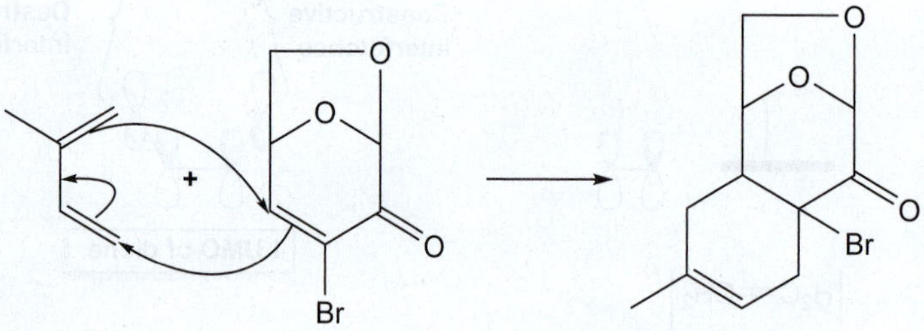

Problem 24.28

Think

In how many steps does the Diels–Alder reaction take place? Which reactant is the conjugated diene? The dienophile? How many curved arrows should be added to show the appropriate bond formation and bond breaking?

Solve

In each case, the conjugated diene is the molecule on the left, and the dienophile is the molecule on the right. The Diels–Alder is a one-step concerted reaction, as shown below. Three curved arrows are added to show the breaking of the three π bonds and the formation of the two new σ bonds and one new π bond.

(a)

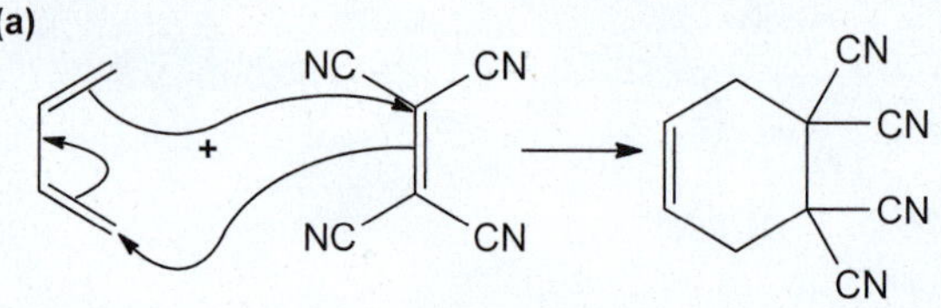

NC CN
+
NC CN
CN
CN
CN
CN

(b)

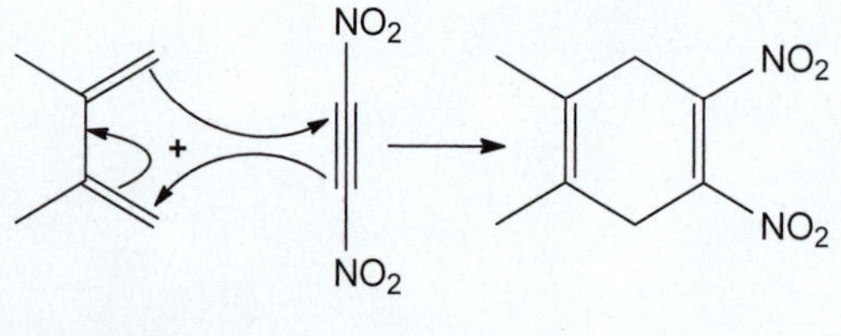

NO_2
+
NO_2
NO_2
NO_2

Problem 24.29 (SYN)

Think

What is the structure of ethene? Which carbons came from the ethene? Which carbons came from the diene? Undo the Diels–Alder reaction to identify the diene that produced the given compounds.

Solve

The carbons that came from the diene have a C=C bond in the six-membered ring product. The carbons that came from the ethene are the other two C atoms, highlighted in each example below. The retrosyntheses show the dienes that would produce compounds **(a)–(d)**.

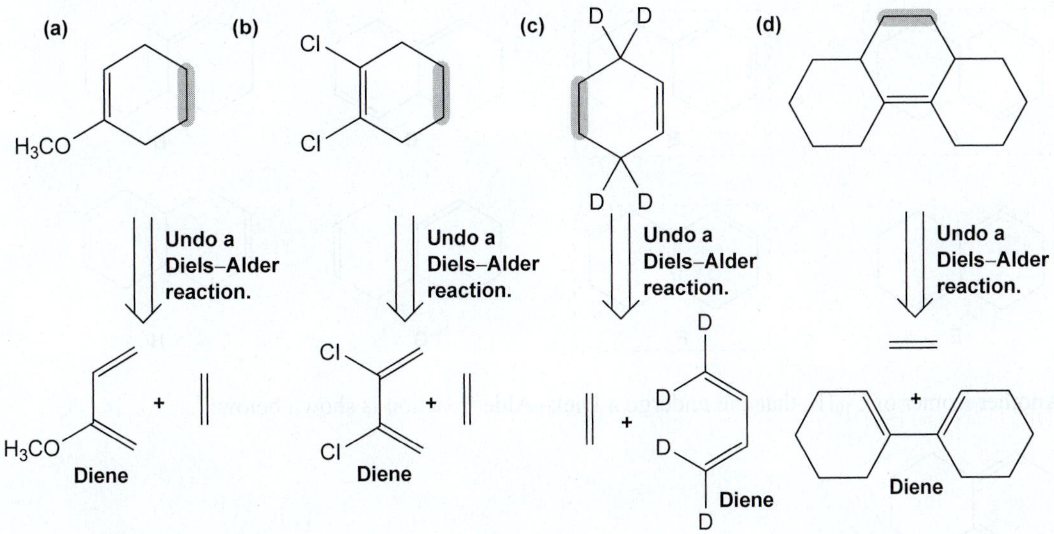

(a) H_3CO (b) Cl Cl (c) D D D D (d)

Undo a Diels–Alder reaction. Undo a Diels–Alder reaction. Undo a Diels–Alder reaction. Undo a Diels–Alder reaction.

H_3CO + Diene; Cl Cl + Diene; + D D D D Diene; + Diene

Problem 24.30 (SYN)

Think

What is the structure of buta-1,3-diene? Which carbons came from buta-1,3-diene? Which carbons came from the dienophile? Undo the Diels–Alder reaction to identify the dienophile that produced the given compounds.

Solve

The carbons that came from the buta-1,3-diene have a C=C in the six-membered ring product and are highlighted in each example below. The other two C atoms in the ring would have come from the dienophile. The retrosyntheses show the dienophiles that would produce compounds **(a)–(c)**.

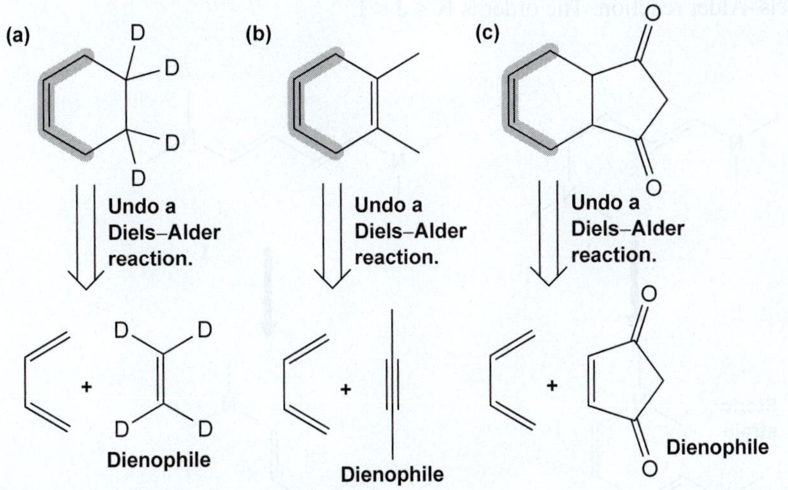

(a) D D D D (b) (c) O O

Undo a Diels–Alder reaction. Undo a Diels–Alder reaction. Undo a Diels–Alder reaction.

+ D D D D Dienophile; + Dienophile; + O O Dienophile

Problem 24.31

Think

Which diene has the C=C bonds conjugated and in an *s*-cis conformation? Are there single bond rotations that would permit an *s*-trans conformation to attain *s*-cis? Is there another isomer that you can draw with the appropriate stipulations?

Solve

(a) Only the three boxed isomers will react. The others have double bonds that are not conjugated dienes or cannot assume the *s*-cis conformation needed to react. The rings prevent an *s*-trans conformation from becoming *s*-cis.

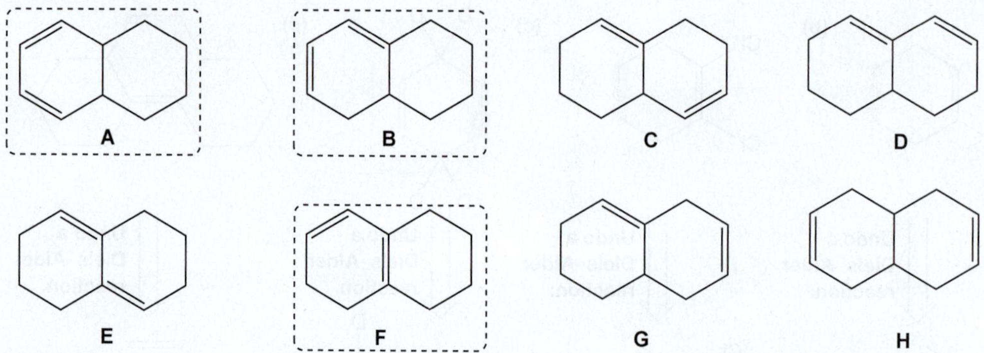

(b) Another isomer of $C_{10}H_{14}$ that can undergo a Diels–Alder reaction is shown below:

Problem 24.32

Think

What is the necessary conformation of the diene in the Diels–Alder reaction? In what conformation is each diene shown? Can rotation about a single bond convert the dienes to the necessary conformation? Which diene is most stable in this conformation? Which is most unstable?

Solve

The diene must be in the *s*-cis conformation to react in a Diels–Alder reaction. Compounds **I**, **J**, and **K** are listed below in order of increasing stability in the *s*-cis conformation (owing to decreasing steric strain) and, thus, also in order of increasing rate of Diels–Alder reaction. The order is **K** < **J** < **I**.

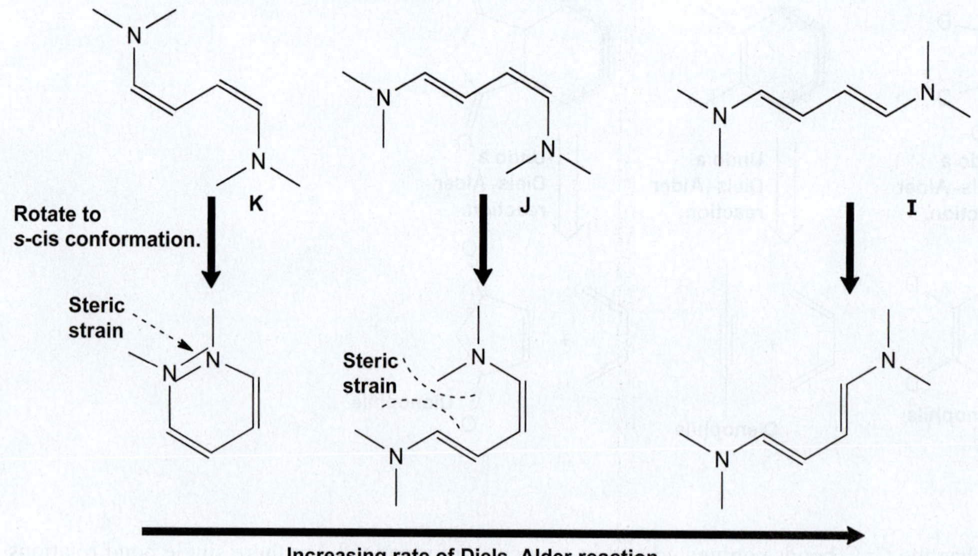

Problem 24.33

Think

What is the necessary conformation of the diene in the Diels–Alder reaction? In what conformation is each diene shown? Can rotation about a single bond convert the dienes to the necessary conformation? Which diene is most stable in this conformation? Which is most unstable?

Solve

The diene must be in the *s*-cis conformation. Compound **M** will react faster as a diene because it has less steric hindrance and can more easily adopt an *s*-cis conformation, as shown below:

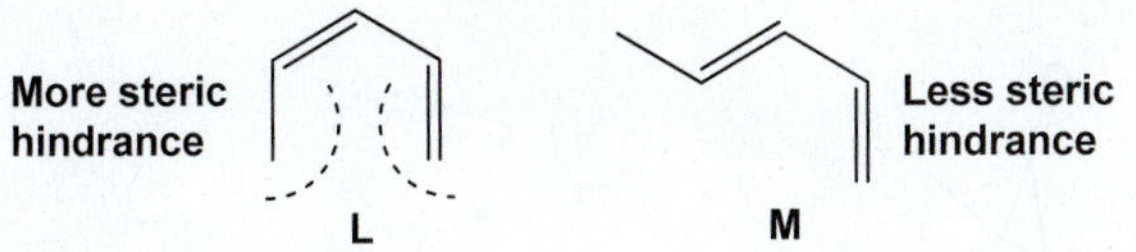

More steric hindrance — **L**

Less steric hindrance — **M**

Problem 24.34

Think

Which diene in anthracene will react with the dienophile and still leave an aromatic compound? Are the dienes in the correct conformation?

Solve

Anthracene can react in one ring as a dienophile, and the aromaticity of the two other rings can be preserved. Either the middle ring or an outer ring can react.

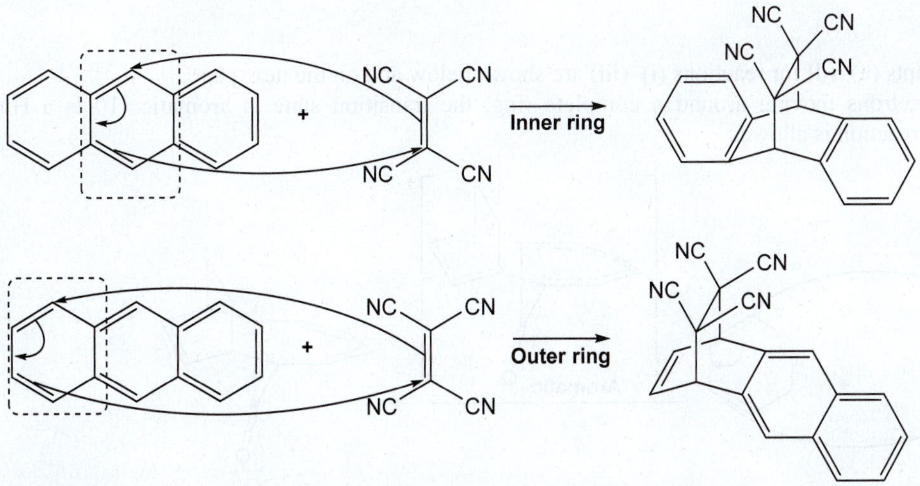

NC, CN, NC, CN — Inner ring

NC, CN, NC, CN — Outer ring

Problem 24.35

Think

Is this a [4+2] cycloaddition? How many curved arrows should be drawn for this type of reaction? What bonds are broken and what bonds are formed? In how many steps does the mechanism occur?

Solve

Three curved arrows are necessary in the Diels–Alder reaction. The mechanism is shown below:

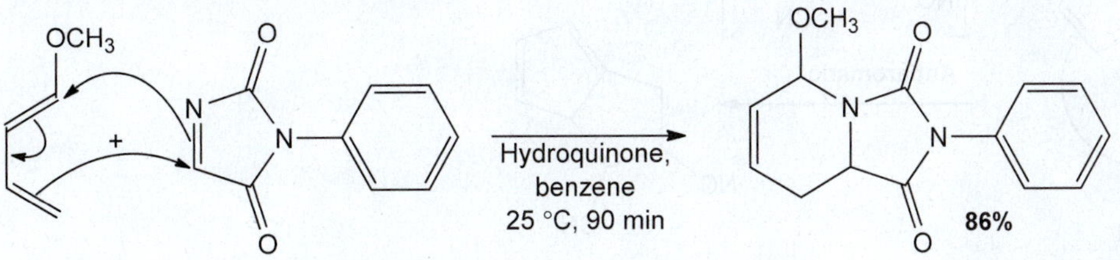

OCH_3 — Hydroquinone, benzene, 25 °C, 90 min — **86%**

Problem 24.36

Think

How many π electrons move in a [6+4] cycloaddition reaction? As a result, how many curved arrows should be drawn? How many π electrons are supplied by each reactant? What bonds are formed and broken when these electrons move in a cyclic fashion?

Solve

The [6+4] cycloaddition involves six electrons from the molecule on the left and four from the molecule on the right, so five curved arrows should be drawn. Five π bonds are broken, and two new σ bonds and three new π bonds are formed. The reaction mechanism is shown below:

Problem 24.37

Think

Can you identify the bonds that were broken and formed? How many π electrons move? In the transition state, are there electrons delocalized over an entire ring? If so, how many electrons? Is that a Hückel number or an anti-Hückel number? How does that correspond to whether the reaction is allowed or forbidden?

Solve

Responses to prompts **(a)–(d)** for reactions **(i)–(iii)** are shown below and on the next page.

(i) With 10 π electrons moving around a complete ring, the transition state is aromatic (10 is a Hückel number). The reaction is allowed.

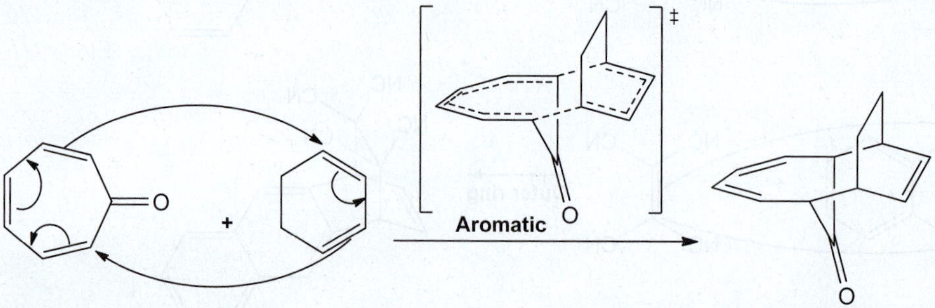

(ii) For this reaction to occur, an eight-electron cyclic transition state would be required, which is an antiaromatic transition state (eight is an anti-Hückel number). This is not allowed under thermal conditions.

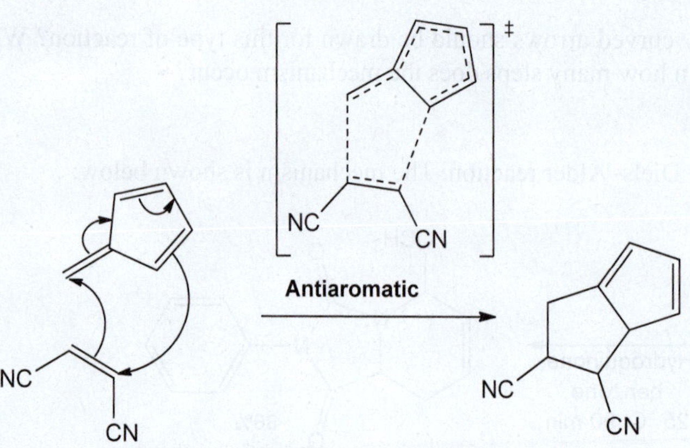

(iii) This reaction proceeds through a 10-electron cyclic transition state, which is an aromatic transition state (10 is a Hückel number) and will be allowed under thermal conditions.

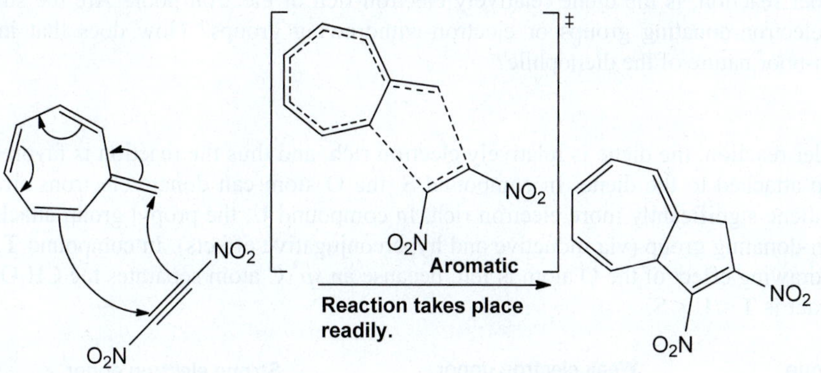

NO_2 O_2N NO_2 O_2N Aromatic NO_2 O_2N

Reaction takes place readily.

Problem 24.38

Think

Follow the flow of electrons. Which bonds are broken, and which new bonds form?

Solve

The products for reactions **(a)–(c)** are given below.

(a) Electrocyclic reaction — Heat

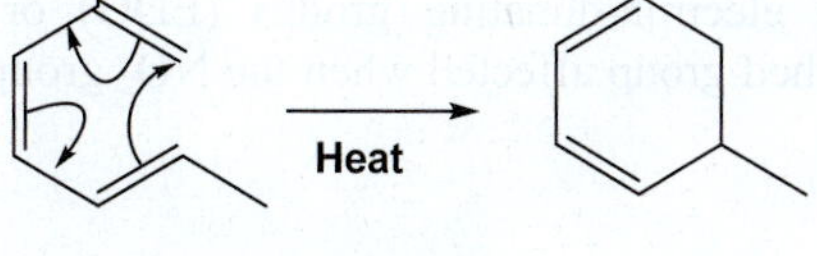

(b) Cope rearrangement — Heat

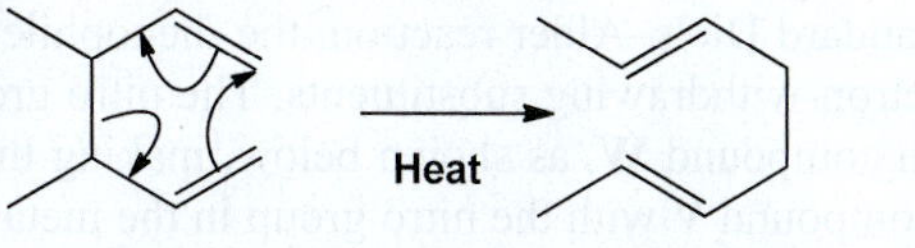

(c) Claisen rearrangement — Heat

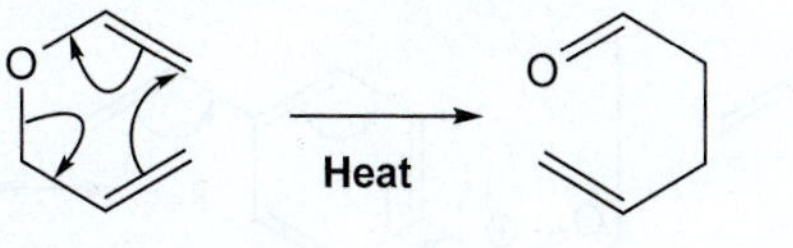

Section 24.3 Substituent Effects on the Reaction Rate

Problem 24.39

Think

In a standard Diels–Alder reaction, is the dienophile relatively electron rich or electron poor? Are the substituents attached to the dienophile electron-donating groups (EDG) or electron-withdrawing groups (EWG)? How does that impact the electron-rich or electron-poor nature of the dienophile?

Solve

In a standard Diels–Alder reaction, the dienophile is relatively electron poor, and thus the reaction is favored with EWG substituents attached to the dienophile. The order of reactivity is **R** < **Q** < **N** < **P** < **O**.

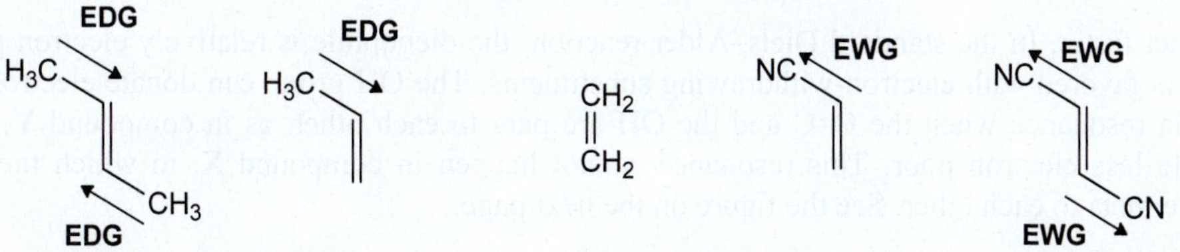

EDG H_3C CH_3 EDG — **R** Least reactive

EDG H_3C — **Q**

CH_2=CH_2 — **N**

NC EWG — **P**

NC EWG CN EWG — **O** Most reactive

Problem 24.40

Think

In a standard Diels–Alder reaction, is the diene relatively electron rich or electron poor? Are the substituents attached to the diene electron-donating groups or electron-withdrawing groups? How does that impact the electron-rich or electron-poor nature of the dienophile?

Solve

In a standard Diels–Alder reaction, the diene is relatively electron rich, and thus the reaction is favored with an electron-donating group attached to the diene. In compound **S**, the O atom can donate electrons strongly via resonance, making the diene significantly more electron rich. In compound **U**, the propyl group attached to the diene is a weak electron-donating group (via inductive and hyperconjugative effects). In compound **T**, only the inductive electron-withdrawing effect of the O atom is felt, because an *sp*^{3} C atom separates the CH_3O from the diene. Therefore, the order is **T** < **U** < **S**.

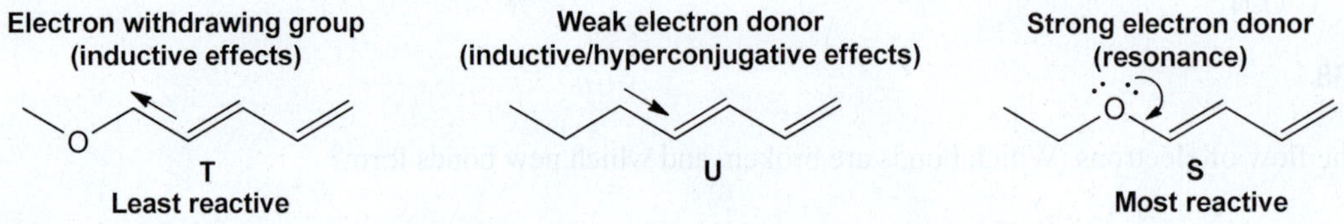

Electron withdrawing group (inductive effects) — **T** — Least reactive

Weak electron donor (inductive/hyperconjugative effects) — **U**

Strong electron donor (resonance) — **S** — Most reactive

Problem 24.41

Think

In a standard Diels–Alder reaction, is the dienophile relatively electron rich or electron poor? Are the substituents attached to the dienophile electron-donating groups (EDG) or electron-withdrawing groups (EWG)? How is this property of the attached group affected when the NO_2 group on the ring occupies the meta versus para position?

Solve

Compound **W** will react faster. In the standard Diels–Alder reaction, the dienophile is relatively electron poor, and thus the reaction is favored with electron-withdrawing substituents. The nitro group can withdraw electrons from the exocyclic C=C via resonance in compound **W**, as shown below, making the dienophile more electron poor. Such resonance cannot happen in compound **V** with the nitro group in the meta position on the ring.

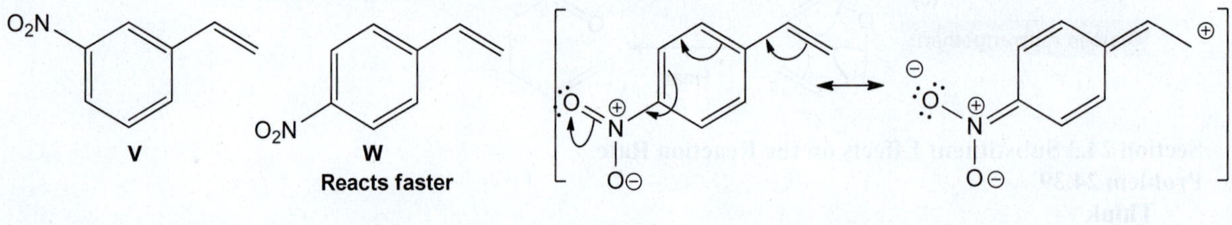

O_2N — **V**

O_2N — **W** — Reacts faster

Problem 24.42

Think

In a standard Diels–Alder reaction, is the dienophile relatively electron rich or electron poor? Are the substituents attached to the dienophile electron-donating groups (EDG) or electron-withdrawing groups (EWG)? How is this property of the attached group affected when the OH group on the ring occupies the meta versus para position?

Solve

Compound **X** will react faster. In the standard Diels–Alder reaction, the dienophile is relatively electron poor, and thus the reaction is favored with electron-withdrawing substituents. The OH group can donate electrons to the exocyclic C=C via resonance when the C=C and the OH are para to each other, as in compound **Y**, thus making the dienophile less electron poor. This resonance cannot happen in compound **X**, in which the OH group and the C=C are meta to each other. See the figure on the next page.

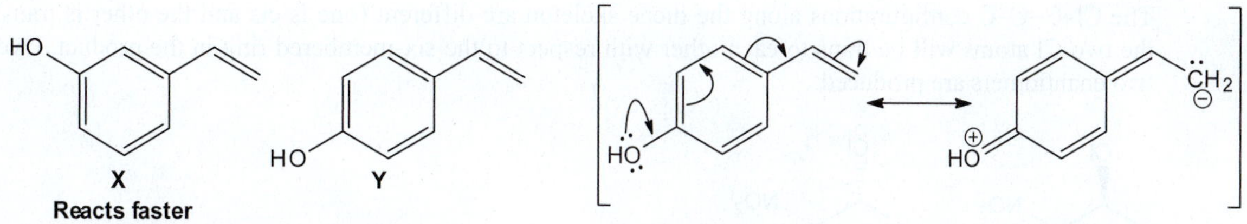

HO

X

Reacts faster

HO

Y

HO

HO

$\ddot{C}H_2$

Problem 24.43 (SYN)

Think

Can the C=C bond in the product be assigned to more than one six-membered ring? Identify the bonds that formed from the Diels–Alder reaction. When you undo the Diels–Alder reaction, what groups become part of the diene and dienophile? Is the reaction more or less favorable when the CO_2H groups are on the diene?

Solve

The C=C bond in the product can be assigned to two different six-membered rings, so we can undo a Diels–Alder reaction two different ways, as shown below. The carboxyl groups can be on either the dienophile or the diene in the precursors. The first reaction will work better, because the carboxyl groups make the dienophile electron poor, and the bonds to CH_2 make the diene electron rich. In the second reaction, the dienophile is neither electron rich nor electron poor, and the diene is made slightly electron poor by the distant CO_2H groups.

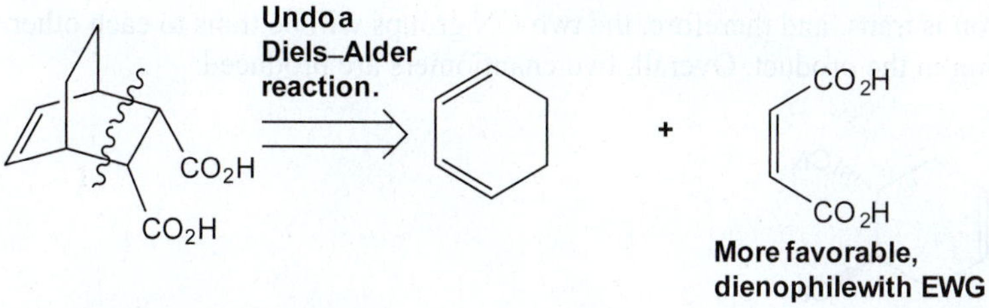

CO_2H

CO_2H

Undo a Diels–Alder reaction.

+

CO_2H

CO_2H

More favorable, dienophile with EWG

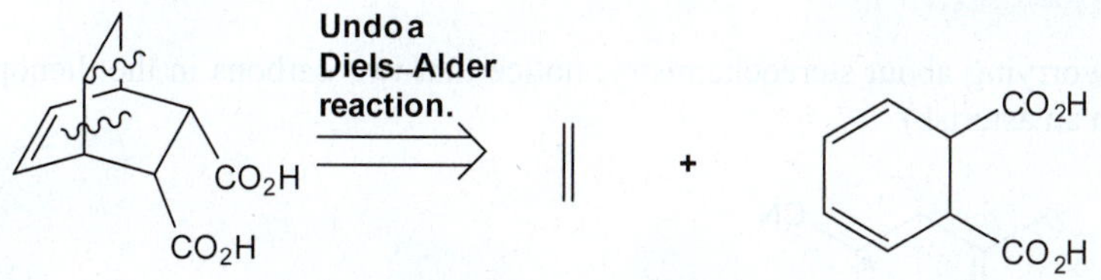

CO_2H

CO_2H

Undo a Diels–Alder reaction.

+

CO_2H

CO_2H

Sections 24.4 and 24.5 Stereochemistry and Regiochemistry of Diels–Alder Reactions

Problem 24.44

Think

What must the conformation of the diene be to react in a Diels–Alder reaction? Do the carbons at the ends of the diene become chiral centers in the products? What are the configurations about the double bonds in the diene? Do the alkene carbons of the dienophile become chiral centers? What is the configuration about the C=C double bond in the dienophile?

Solve

(a) The diene must achieve an *s*-cis conformation prior to reaction, and this conformational change is shown below. If we draw the product without worrying about stereochemistry, notice that each end carbon in the diene becomes a chiral center (noted with an asterisk).

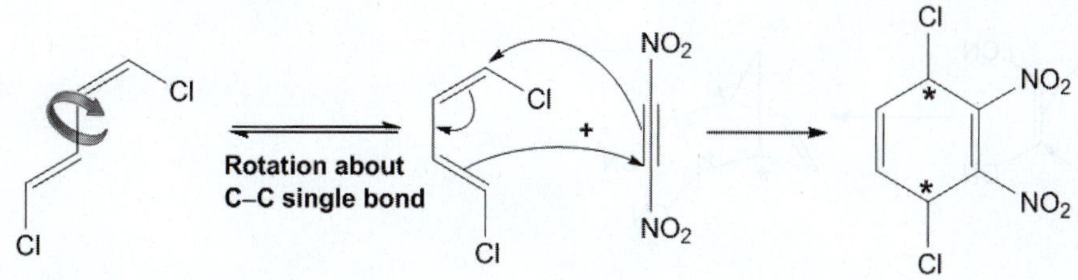

Cl

Cl

Rotation about C–C single bond

Cl

Cl

+

NO_2

NO_2

Cl

NO_2

NO_2

Cl

The Cl–C=C–C configurations along the diene skeleton are different (one is cis and the other is trans), so the two Cl atoms will be trans to each other with respect to the six-membered ring in the product. Overall, two enantiomers are produced:

Cl NO2 NO2 Cl Cl NO2 NO2 Cl

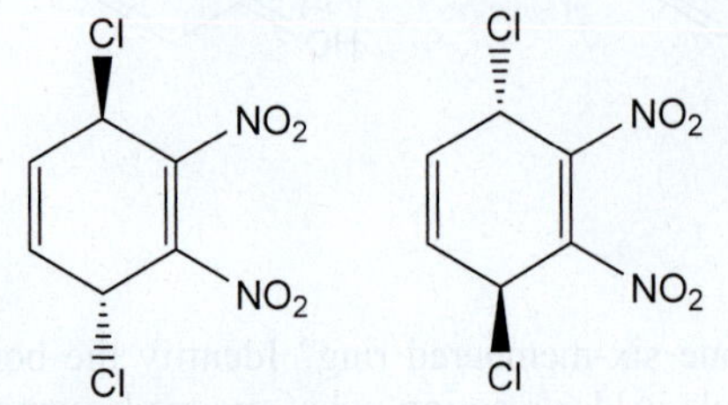

(b) If we draw the product without worrying about stereochemistry, notice that two carbons in the dienophile become chiral centers (noted with an asterisk).

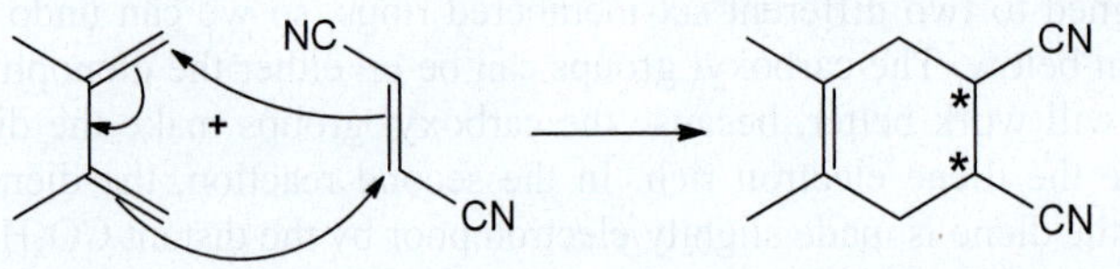

NC + CN CN * * CN

The CN–C=C–CN configuration is trans, and therefore, the two CN groups will be trans to each other with respect to the six-membered ring in the product. Overall, two enantiomers are produced:

CN CN CN CN

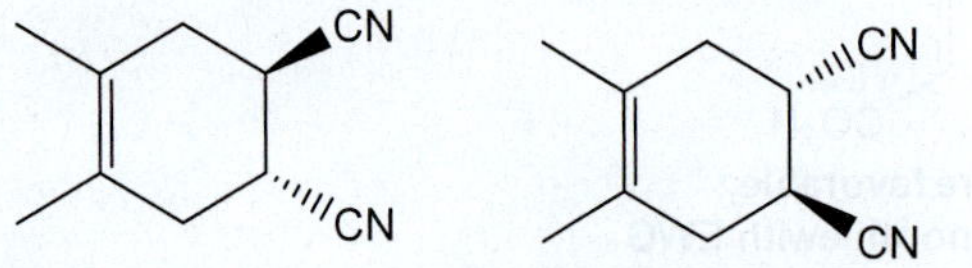

(c) If we draw the product without worrying about stereochemistry, notice that two carbons in the dienophile become chiral centers (noted with an asterisk).

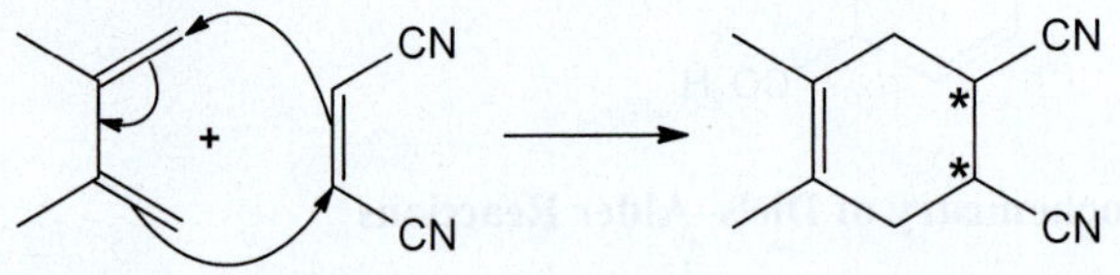

+ CN CN CN * * CN

The CN–C=C–CN configuration is cis, and therefore, the two CN groups will be cis to each other with respect to the six-membered ring in the product. Only the following achiral (meso) product is formed:

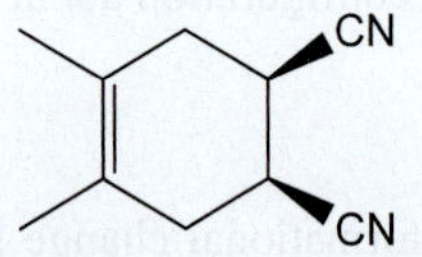

CN CN

(d) If we draw the product without worrying about stereochemistry, notice that two carbons of the diene and two carbons of the dienophile become chiral centers (noted with an asterisk). A bicyclic compound forms.

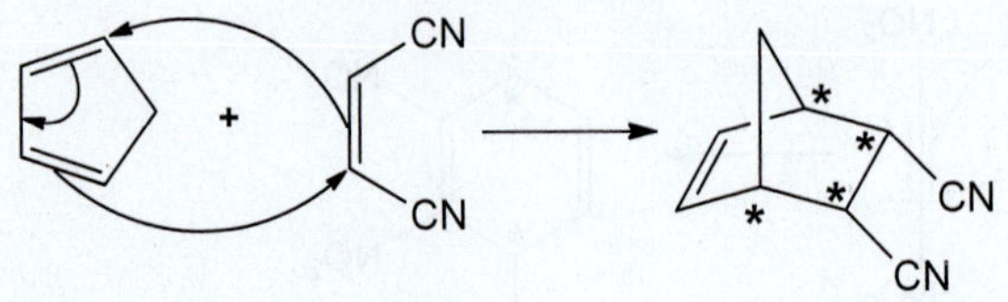

+ CN CN * * * * CN CN

The C–C=C–C configurations along the diene skeleton are the same (both cis), so the two bonds to the CH_2 group will be cis to each other with respect to the six-membered ring in the product. The CN–C=C–CN configuration in the dienophile is cis, and therefore, the two CN groups will be cis to each other with respect to the six-membered ring in the product. Overall, two diastereomers will be produced, and the endo product will be the major product:

CN CN CN CN

Endo
Major product

Exo

Problem 24.45

Think

Identify the diene and dienophile for reach reaction. What orientation of the diene and dienophile in **(a)** leads to a favorable electrostatic interaction among atoms undergoing bond formation? What is the configuration of the C=C bond in the dienophile in **(a)** and of the N=N bond in the dienophile in **(b)**? How do you show that the stereochemistry of the dienophile is retained?

Solve

(a) The partial charges from the resonance hybrids are shown for the diene and dienophile, and the two orient in a way to show favorable electrostatic interaction among the atoms forming bonds. Two enantiomers will be produced:

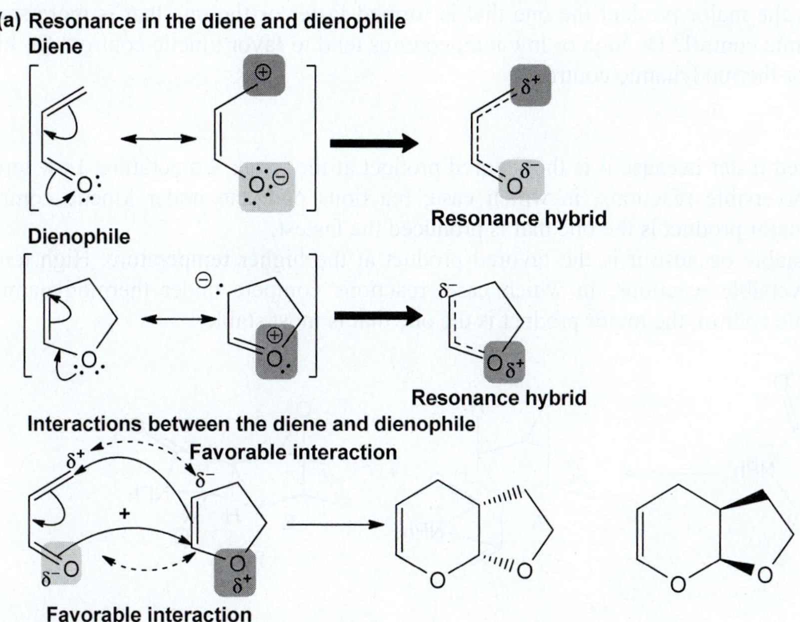

(b) The electron-withdrawing ester groups on the dienophile will be trans in the product, as they were in the reactant. The C–C=C–C configurations along the diene skeleton are the same (both cis), so the two bonds to the CH_2 group will be cis to each other with respect to the six-membered ring in the product. Two enantiomers will be produced:

CO_2CH_3 H_3CO_2C CO_2CH_3 CO_2CH_3 CO_2CH_3 CO_2CH_3

Trans

Problem 24.46 (SYN)

Think

Identify the bonds that formed from the Diels–Alder reaction. When you undo a Diels–Alder reaction, do the OCH_3 groups go to the diene or the dienophile? Can different stereoisomers react to form the same product?

Solve

The bonds that should be disconnected when undoing a Diels–Alder reaction are shown below. The methoxy groups belong to the diene in the precursors. To achieve the cis configuration in the ring product, both C=C bonds in the diene must be trans or both must be cis along the diene skeleton. Both possibilities are shown below. The first is better, because in the second diene, the methoxy groups induce steric strain that would prevent the diene from attaining the *s*-cis conformation.

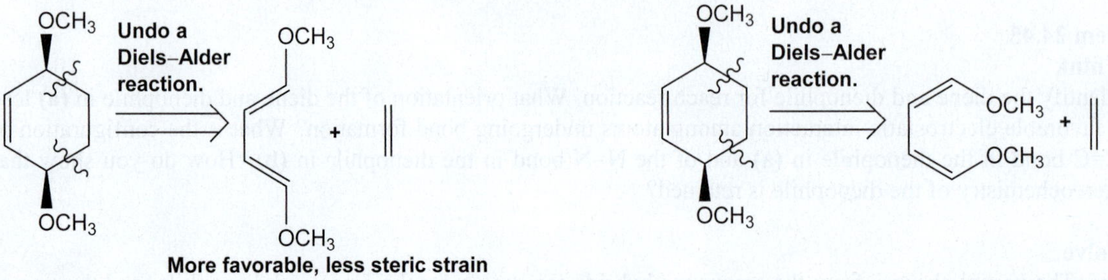

OCH_3 Undo a Diels–Alder reaction. OCH_3 + OCH_3 OCH_3

More favorable, less steric strain

OCH_3 Undo a Diels–Alder reaction. OCH_3 + OCH_3 OCH_3

Problem 24.47

Think

Under kinetic control, is the major product the one that is formed faster or the one that is more stable? How about under thermodynamic control? Do high or low temperatures tend to favor kinetic control? Do high or low temperatures tend to favor thermodynamic control?

Solve

(a) Product **A** is produced faster because it is the favored product at the lower temperature. Low temperatures tend to promote irreversible reactions, in which case, reactions compete under kinetic control. Under kinetic control, the major product is the one that is produced the fastest.

(b) Product **B** is more stable because it is the favored product at the higher temperature. High temperatures tend to promote reversible reactions, in which case, reactions compete under thermodynamic control. Under thermodynamic control, the major product is the one that is most stable.

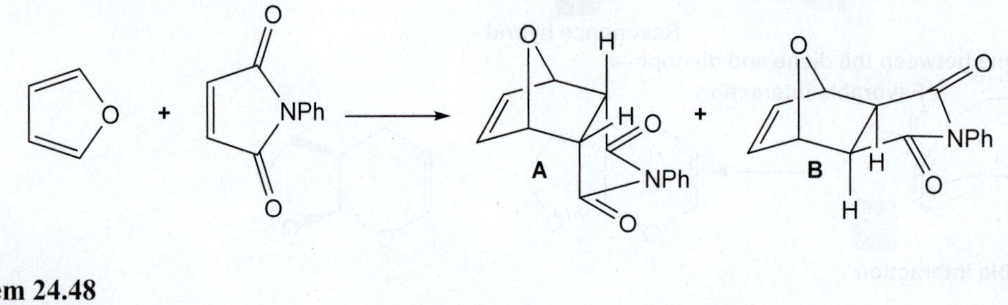

O + O NPh O → O H H O NPh O **A** + O O NPh H O H **B**

Problem 24.48

Think

Where are the regions of excess negative charge and excess positive charge in these molecules? How can you use resonance structures to determine this? Which approach leads to favorable electrostatic interactions among the atoms undergoing bond formation?

Solve

The partial charges from resonance hybrids of the diene and dienophile are shown below. The major product comes from the reaction that exhibits favorable electrostatic interaction among the atoms undergoing bond formation.

(a) Resonance in the diene and dienophile

Diene

Resonance hybrid

Dienophile

Resonance hybrid

Interactions between the diene and dienophile

Favorable interaction

(b) Resonance in the diene and dienophile

Diene

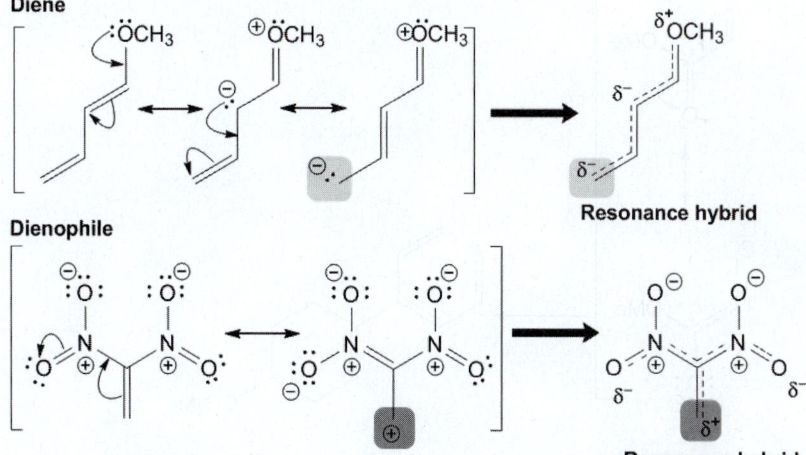

Interactions between the diene and dienophile

Rotation about C–C single bond

Favorable interaction

Problem 24.49

Think

In how many ways can the dienophile approach the diene? Is the phenyl ring electron rich or electron poor? What atoms have partial charges in the resonance hybrid of the diene and dienophile? Which orientation leads to favorable electrostatic interactions among the atoms undergoing bond formation?

Solve

The two possible constitutional isomers are shown below, differing by the approach of the dienophile to the diene.

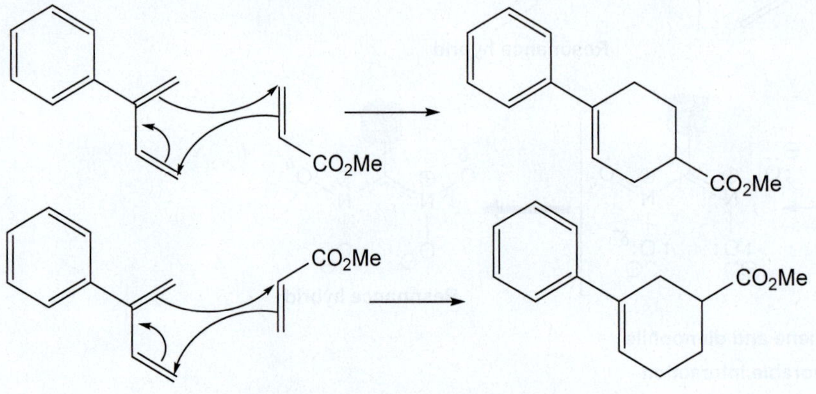

CO_2Me CO_2Me

CO_2Me CO_2Me

The first one is favored because the electron-rich benzene ring induces a partial negative charge on the top right C atom of the diene via resonance, and the CO_2Me group induces a partial positive charge on the terminal carbon of the dienophile. That leads to significant electrostatic attraction among the atoms undergoing bond formation, as shown below.

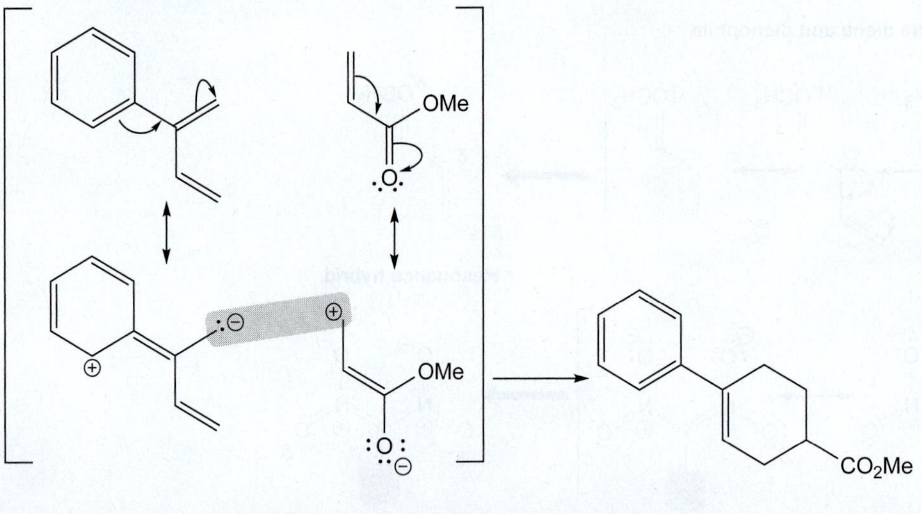

OMe

OMe

CO_2Me

Section 24.6 The Reversibility of Diels–Alder Reactions; the Retro Diels–Alder Reaction

Problem 24.50

Think

How do you calculate $\Delta G°_{rxn}$ from the given $\Delta H°_{rxn}$ and $\Delta S°_{rxn}$ at room temperature? Is $\Delta G°_{rxn}$ positive or negative? How does the $\Delta G°_{rxn}$ relate to reversibility? What equation relates temperature, $\Delta G°_{rxn}$, and K_{eq}? When you solve for temperature, what value do you obtain?

Solve

(a) This Diels–Alder reaction takes place at room temperature (25 °C = 298 K) and $\Delta H^\circ_{rxn} = -54$ kJ/mol and $\Delta S^\circ_{rxn} = -151$ J/mol•K.

$\Delta G^\circ_{rxn} = \Delta H^\circ_{rxn} - T\Delta S^\circ_{rxn} = (-54{,}000 \text{ J/mol}) - (298 \text{ K})(-151 \text{ J/mol•K}) = -9002 \text{ J/mol} = -9.0 \text{ kJ/mol}$

A significantly negative ΔG°_{rxn} is associated with an irreversible reaction.

(b) When $K_{eq} = 1$, $\Delta G^\circ_{rxn} = 0$. Solve for T using $\Delta G^\circ_{rxn} = 0 = \Delta H^\circ_{rxn} - T\Delta S^\circ_{rxn}$, and we get

$$T = \frac{\Delta H^\circ}{\Delta S^\circ} = \frac{-54\dfrac{\text{kJ}}{\text{mol}}}{-0.151\dfrac{\text{kJ}}{\text{mol}\cdot\text{K}}} = 358\text{ K} = 85\text{ °C}$$

The reverse reaction would be spontaneous at temperatures above 85 °C because ΔG° would be positive.

Problem 24.51

Think

Can you identify a six-membered ring that includes a C=C bond? How would such a bond have been formed in a Diels–Alder reaction? What bonds are broken and formed in the retro Diels–Alder mechanism? What are the relative orientations of the substituents (cis or trans)? What does that say about the configurations about the C=C bonds in the diene and dienophile that are produced?

Solve

The mechanism for the retro Diels–Alder reaction is shown below. The C=C in the six-membered ring would have been formed in a Diels–Alder reaction, so that bond is broken in a retro Diels–Alder reaction, and three total curved arrows show cyclic electron movement.

CO_2Et CN CN CO_2Et → **Diene** + EtO_2C CN NC CO_2Et **Dienophile**

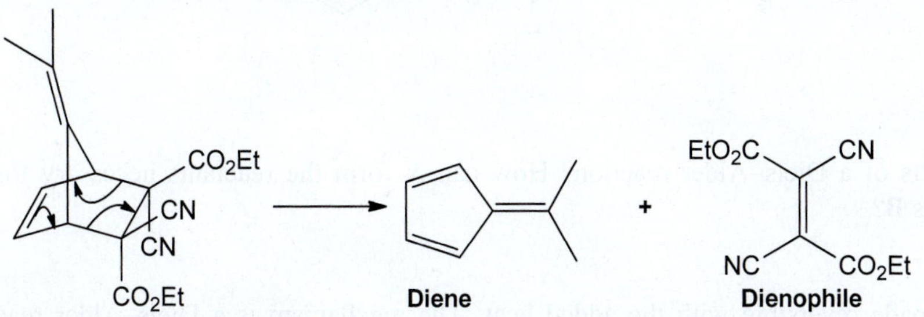

Notice that the CN groups are trans to each other with respect to the six-membered ring, so they are trans to each other about the C=C bond in the dienophile that is produced. The same is true of the CO_2Et groups.

Problem 24.52

Think

Can you identify a six-membered ring that includes a C=C bond? How would such a bond have been formed in a Diels–Alder reaction? What bonds are broken and formed in the retro Diels–Alder mechanism?

Solve

The mechanism for each retro Diels–Alder reaction is shown below. The C=C in the six-membered ring would have been formed in a Diels–Alder reaction, so that bond is broken in a retro Diels–Alder reaction, and three total curved arrows show cyclic electron movement.

(a) → **Diene** + **Dienophile**

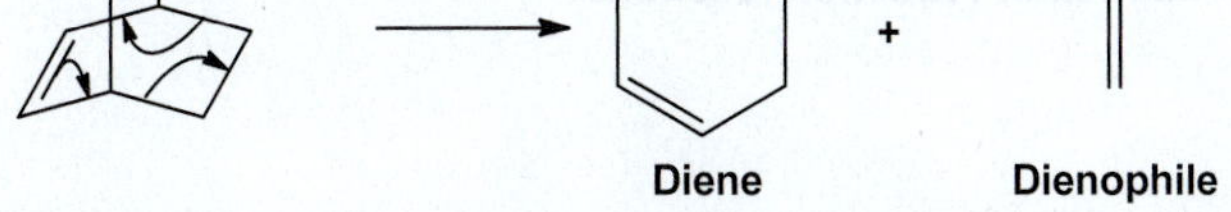

(b)

CO$_2$H CO$_2$H CO$_2$H CO$_2$H → Diene + HO$_2$C CO$_2$H HO$_2$C CO$_2$H Dienophile

Reaction **(b)** would take a place more readily at a lower temperature because, with greater strain in the bicyclic reactant involving the CO_2H and CH_3 groups, the energy barrier for the reaction would be lowered.

Problem 24.53

Think

Are Diels–Alder reactions intrinsically reversible? How does the added heat affect the reversibility of this reaction? What are the products of the Diels–Alder and retro Diels–Alder reactions? Is there symmetry in the Diels–Alder product? How does this account for the isotope scrambling?

Solve

The Diels–Alder reaction is made reversible with the added heat. The mechanism is a Diels–Alder reaction followed by a retro Diels–Alder reaction. Notice that the Diels–Alder product is symmetrical, except for the isotopic labeling.

Diels–Alder ⇌ Retro Diels–Alder

Problem 24.54

Think

Are both **A** and **B** the products of a Diels–Alder reaction? How can **A** form the reactants necessary for the Diels–Alder reaction that forms **B**?

Solve

The Diels–Alder reaction is made reversible with the added heat. The mechanism is a Diels–Alder reaction followed by a retro Diels–Alder reaction.

A ⇌ Retro Diels–Alder ⇌ N—Ph ⇌ Diels–Alder ⇌ B

Sections 24.7 and 24.8 Syn Dihydroxylation and Oxidative Cleavage of Alkenes and Alkynes Using OsO_4 or $KMnO_4$

Problem 24.55

Think

How does a C=C or C≡C bond react when treated with $KMnO_4$/KOH in cold temperatures? With OsO_4? What type of functional group is formed? Are these oxidation or reduction reactions?

Solve

In each of these reactions, the C=C or C≡C reacts, indicated by a squiggly line. A syn 1,2-diol is formed when a C=C is treated with $KMnO_4$/KOH in cold temperatures or with OsO_4. A 1,2-diketone is formed when an internal C≡C is treated with $KMnO_4$/KOH in cold temperatures or with OsO_4. The products for reactions **(a)**–**(d)** are given below.

(a)

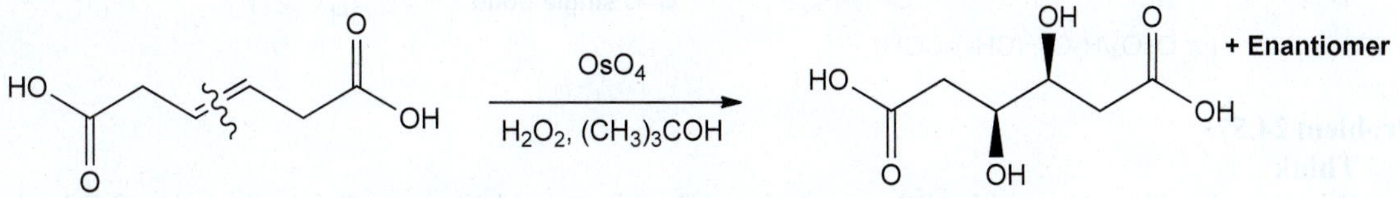
HO
O
OH
OsO4
H2O2, (CH3)3COH
OH
O
+ Enantiomer
HO
OH
O
OH

(b)

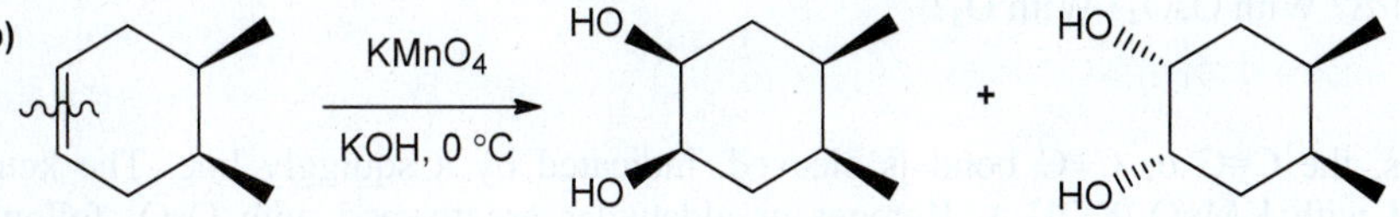
KMnO4
KOH, 0 °C
HO
HO
+
HO
HO

(c)

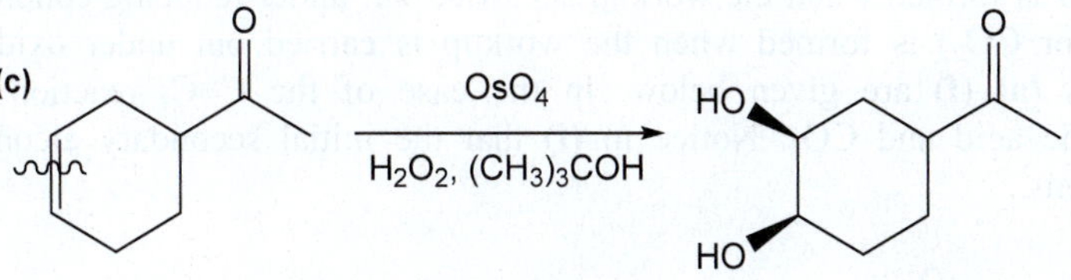
O
OsO4
H2O2, (CH3)3COH
O
HO
HO

(d)

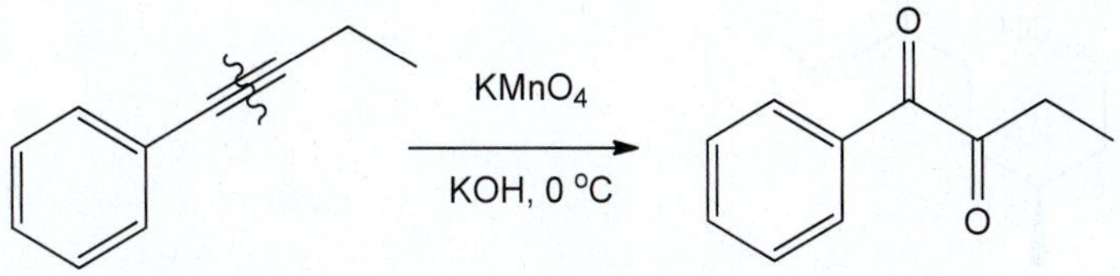
KMnO4
KOH, 0 °C
O
O

Problem 24.56 (SYN)

Think

From what type of bond is a syn 1,2-diol formed? From what type of bond is a 1,2-diketone formed? What are the reaction conditions? Are these oxidation or reduction reactions?

Solve

Products from reactions **(a)**, **(b)**, and **(d)** are syn 1,2-diols. A syn 1,2-diol can be formed when a C=C reacts with $KMnO_4$/KOH in cold temperatures or with OsO_4. Product **(c)** is a 1,2-diketone. A 1,2-diketone can be formed when an internal C≡C reacts with $KMnO_4$/KOH in cold temperatures or with OsO_4. The starting material and reaction conditions for products **(a)**–**(d)** are given below and on the next page.

(a)

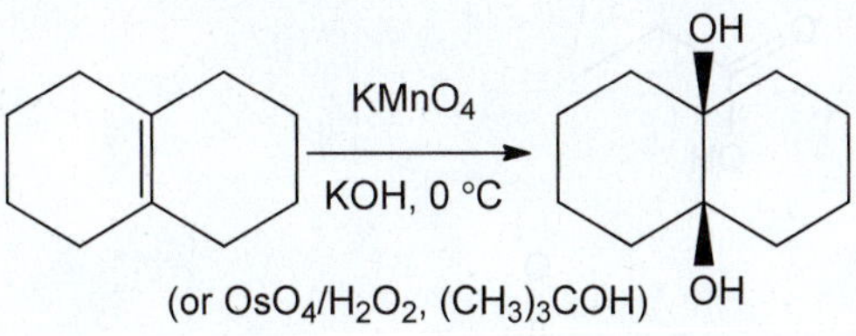
KMnO4
KOH, 0 °C
(or OsO4/H2O2, (CH3)3COH)
OH
OH

(b)

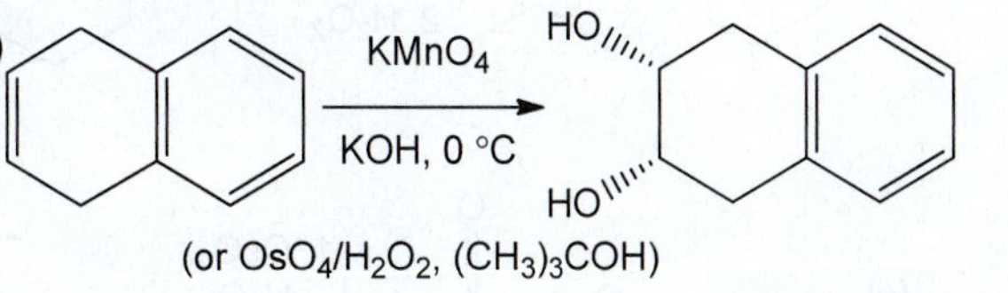
KMnO4
KOH, 0 °C
(or OsO4/H2O2, (CH3)3COH)
HO
HO

(c)

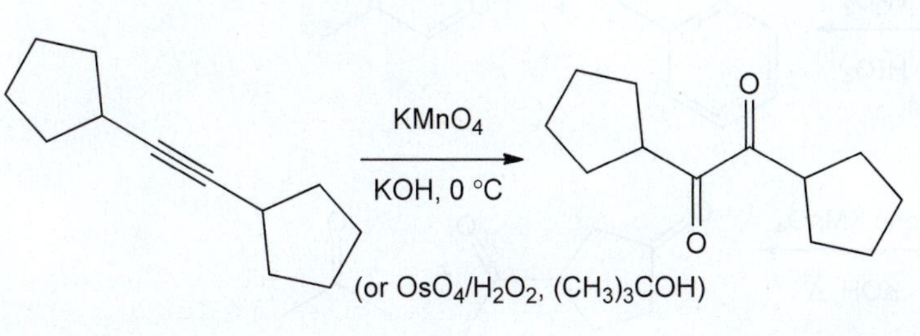
KMnO4
KOH, 0 °C
(or OsO4/H2O2, (CH3)3COH)
O
O

(d) KMnO₄ → KOH, 0 °C (or OsO_4/H_2O_2, $(CH_3)_3COH$); Rotation about C–C single bond

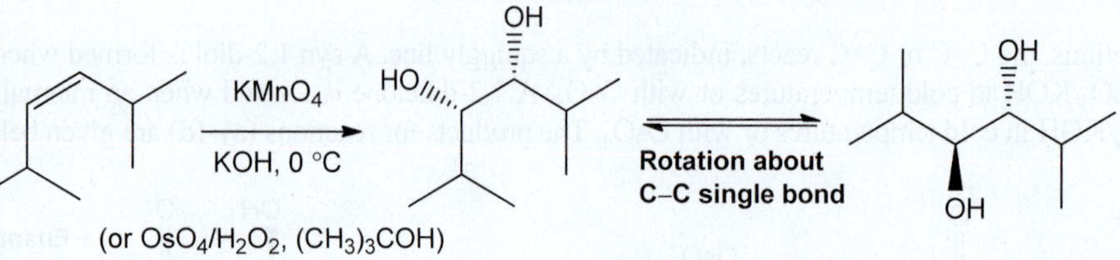

Problem 24.57

Think

What bond is cleaved in each of these reactions? Are these oxidation or reduction reactions? What product is formed with $KMnO_4/KOH/\Delta$? With OsO_4? With O_3?

Solve

In each of these reactions, the C=C or C≡C bond is cleaved, indicated by a squiggly line. The ketone or carboxylic acid is formed with $KMnO_4/KOH/\Delta$. Ketones or aldehydes are formed with OsO_4 followed by periodate. With O_3, a ketone or aldehyde is formed when the workup is carried out under reducing conditions, whereas a ketone or carboxylic acid (or CO_2) is formed when the workup is carried out under oxidizing conditions. The products for reactions **(a)–(f)** are given below. In the case of the C≡C, reaction with $KMnO_4/KOH/\Delta$ produces the carboxylic acid and CO_2. Notice in **(f)** that the initial secondary alcohol is oxidized to a ketone under these conditions.

(a) 1. conc $KMnO_4$, KOH, Δ; 2. $H^{\oplus}$

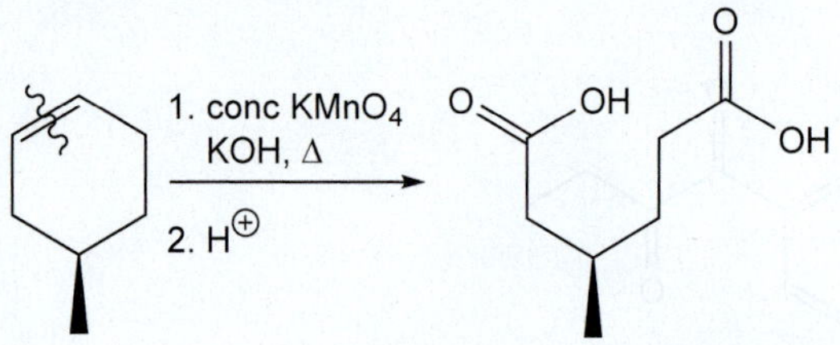

(b) 1. conc $KMnO_4$, KOH, Δ; 2. $H^{\oplus}$ → + $CO_2(g)$

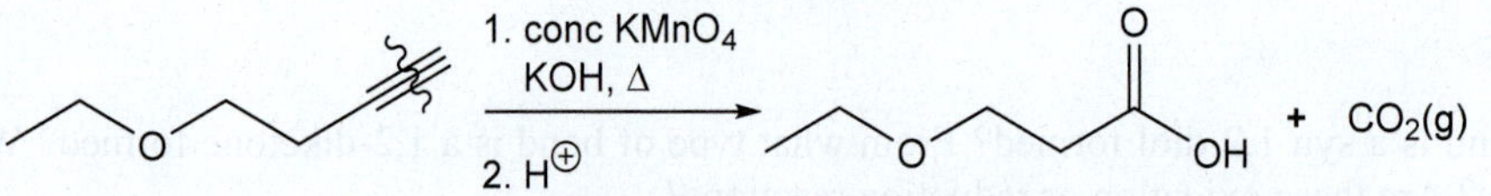

(c) 1. O_3, −78 °C; 2. Zn, HOAc → +

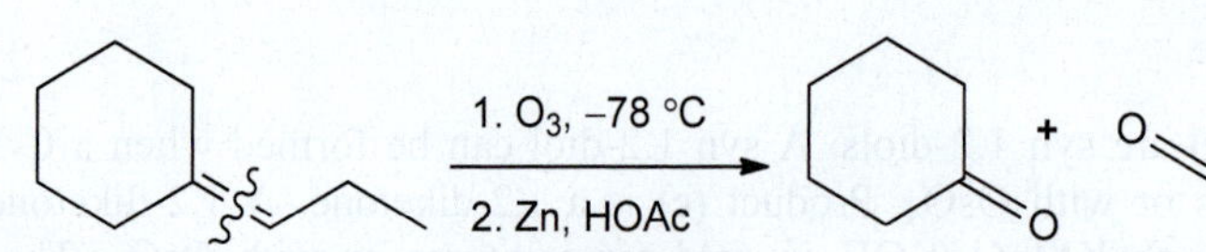

(d) 1. O_3, −78 °C; 2. H_2O_2 → +

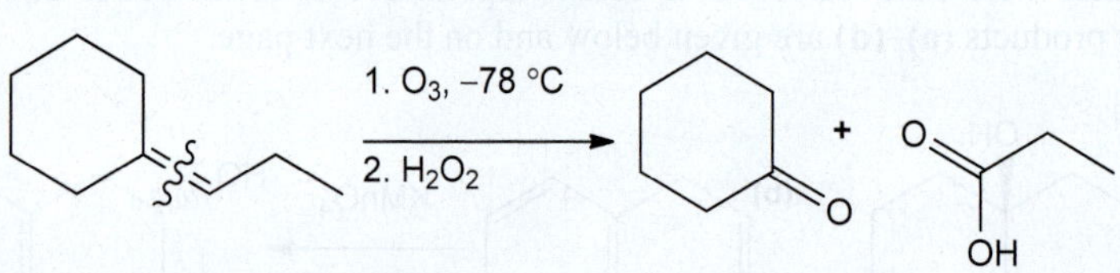

(e) 1. OsO_4, H_2O_2; 2. HIO_4 → +

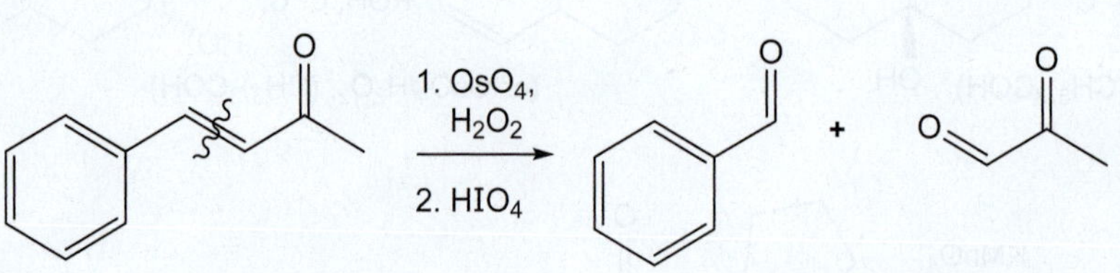

(f) conc $KMnO_4$, KOH, Δ → +

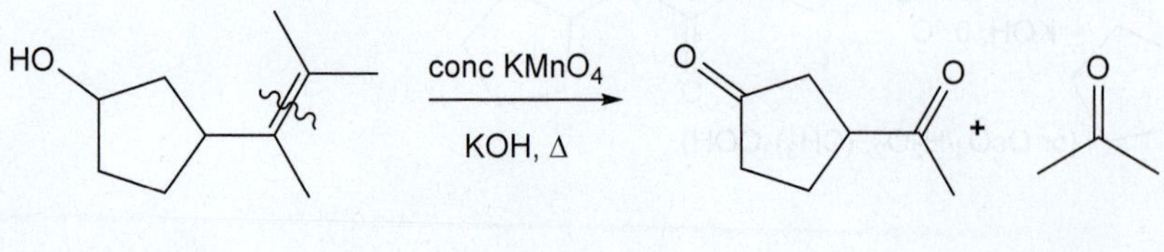

Problem 24.58

Think

What bond is cleaved in each of these reactions? The C=O bonds that are present in the product originated from what bond?

Solve

The C=C bond is cleaved in each of these reactions, and the C atoms of the C=O bonds shown in the products were originally the C=C bond in the reactant. The starting materials for reactions **(a)–(d)** are shown below:

(a) C_9H_{16} — 1. conc $KMnO_4$, KOH, Δ; 2. $H^{\oplus}$ — Originally C=C; OH; O

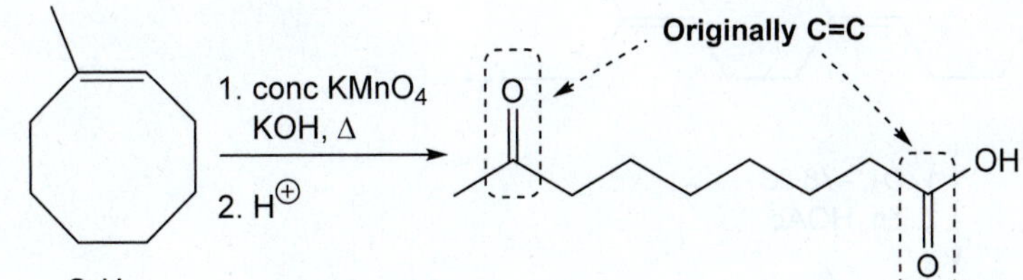

(b) C_6H_{10} — 1. conc $KMnO_4$, KOH, Δ; 2. $H^{\oplus}$ — Originally C=C; O; + CO_2

(c) $C_{10}H_{18}$ — 1. O_3, –78 °C; 2. $(CH_3)_2S$ — Originally C=C; O; + O

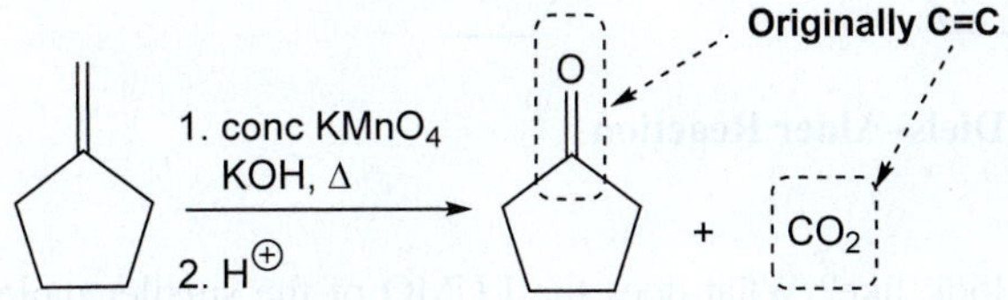

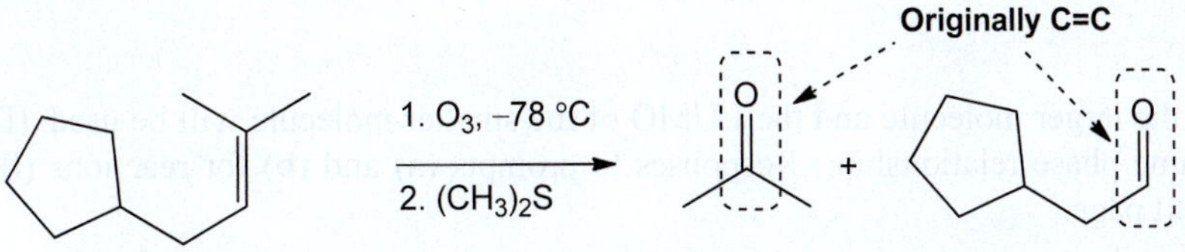

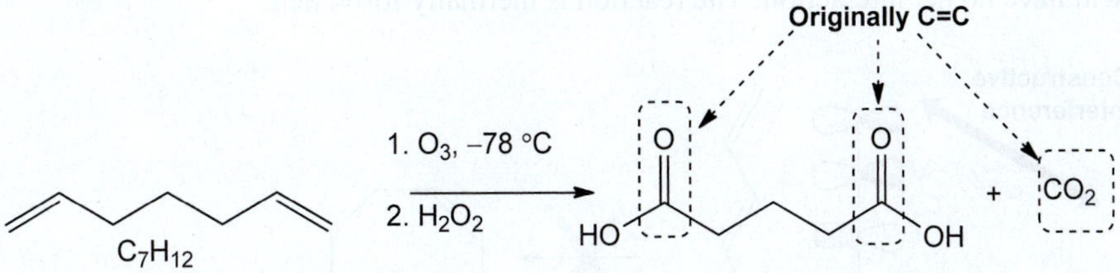

(d) C_7H_{12} — 1. O_3, –78 °C; 2. H_2O_2 — Originally C=C; HO; O; O; OH; + CO_2

Problem 24.59

Think

How does HIO_4 react with a diol compound? What is required of the relative positioning of the OH groups?

Solve

The 1,2-diol OH groups react, and the HO group that is at carbon number 4 does not react. The 1,2-diol reacts with HIO_4 to cleave the C–C bond and form the dialdehyde.

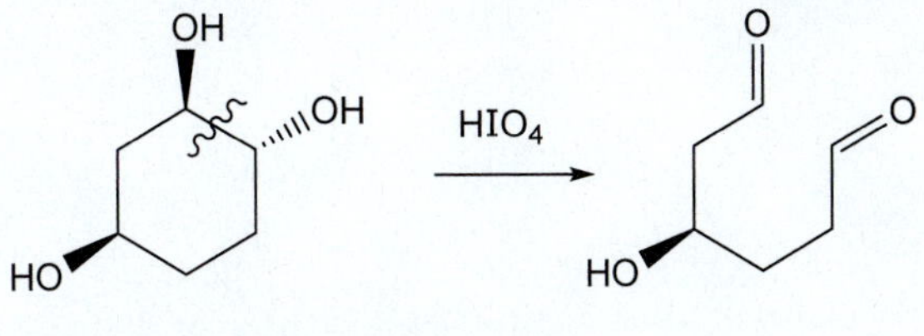

OH; OH; HO — HIO_4 — O; O; HO

Problem 24.60

Think

How does ozone react with an alkene? What product forms? Is the structure still a polymer?

Solve

Ozone, O_3, reacts with an alkene to cleave the C=C bond and form two C=O bonds. When rubber's C=C bonds react with ozone, the polymer breaks down, as shown below:

Rubber

1. O_3, −78 °C
2. Zn, HOAc

Section 24.10 A Molecular Orbital Picture of the Diels–Alder Reaction

Problem 24.61

Think

What does the HOMO of the larger molecule look like? What does the LUMO of the smaller molecule look like? Does constructive or destructive interference occur among the HOMO and LUMO orbitals at the atoms undergoing bond formation? Do the orbitals have the appropriate symmetries to interact?

Solve

In all cases, the HOMO of the larger molecule and the LUMO of the smaller molecule will be used. (Doing the reverse still leads to the same phase relationship.) Responses to prompts **(a)** and **(b)** for reactions **(i)–(iii)** are shown below and on the next page.

(i) The HOMO of the triene and the LUMO of the alkene are in phase at the top but out of phase at the bottom, so the orbitals will have no net interaction. The reaction is thermally forbidden.

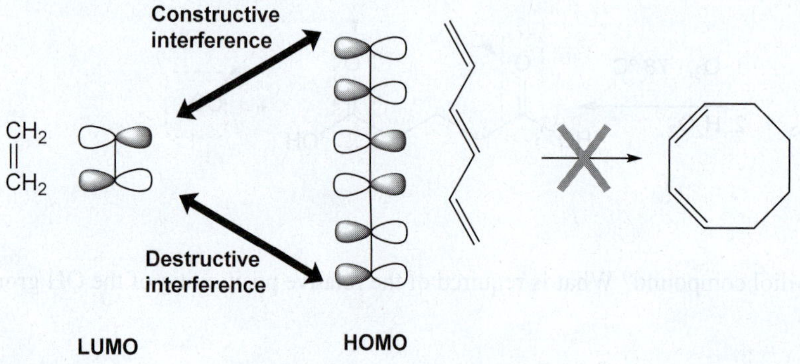

(ii) The HOMO of one diene and the LUMO of the other are in phase at the bottom but out of phase at the top, so the orbitals will have no net interaction. The reaction is thermally forbidden.

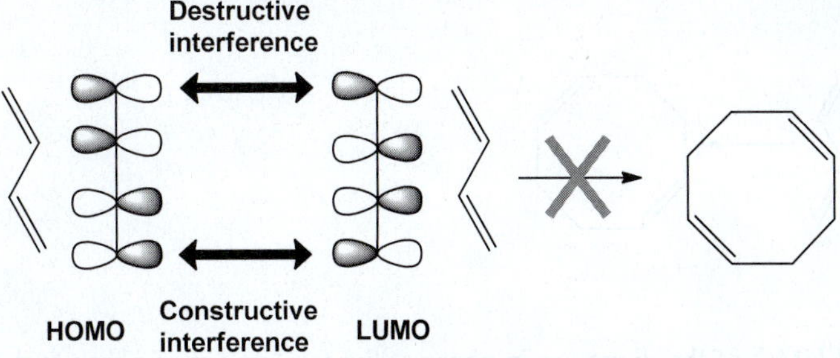

(iii) The HOMO of the triene and the LUMO of the diene are in phase at both the top and the bottom, so the orbitals will interact. This reaction is thermally allowed.

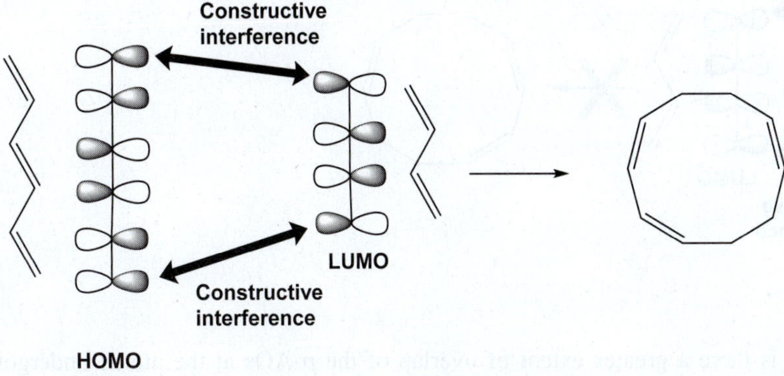

Problem 24.62

Think

Assume that the larger molecule absorbs the photon. How does the HOMO change when the molecule is in the excited state? Does the LUMO of the smaller molecule change if the other molecule didn't absorb a photon? Does constructive or destructive interference occur among the HOMO and LUMO orbitals at the atoms undergoing bond formation? Do the orbitals have the appropriate symmetries to interact?

Solve

As in Problem 24.61, in all cases, the HOMO of the larger molecule and the LUMO of the smaller molecule will be used. (Doing the reverse still leads to the same phase relationship.) For each HOMO, exciting with a photon will promote an electron to the next highest energy MO, which has one additional node.

(i) The HOMO of the triene and the LUMO of the alkene are in phase, so the orbitals will interact. This reaction is photochemically allowed.

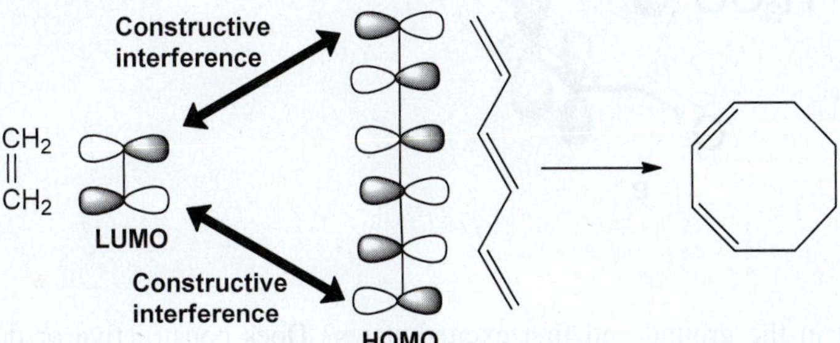

(ii) The HOMO of one diene and the LUMO of the other are in phase, so the orbitals will interact. This reaction is photochemically allowed.

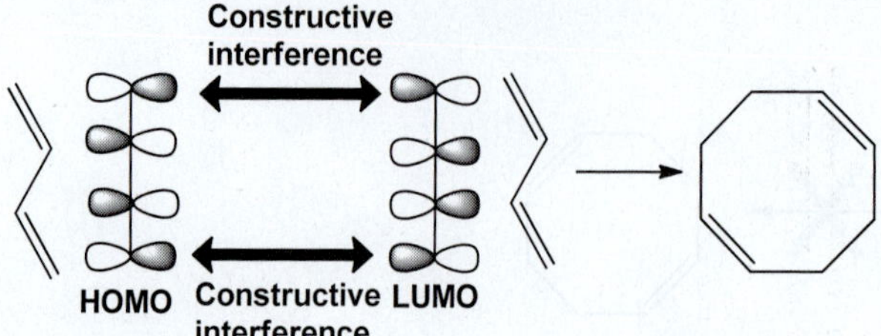

(iii) The HOMO of the triene and the LUMO of the diene are in phase at the top but out of phase at the bottom, so the orbitals will not interact. This reaction is photochemically forbidden.

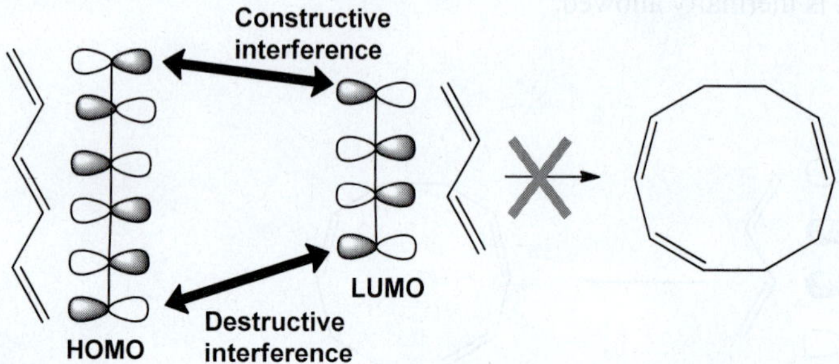

Problem 24.63

Think

In which approach, **A** or **B**, is there a greater extent of overlap of the *p* AOs at the atoms undergoing bond formation? Does this raise or lower the activation energy for the reaction?

Solve

In **A**, the red *p* orbital on C1 interacts with the blue *p* orbital on C3, and the red *p* orbital on C4 interacts with the blue *p* orbital on C2. In both cases, the *p* orbitals are substantially different in size, so the extent of overlap is relatively small. In **B**, the red *p* orbital on C4 is similar in size to the blue *p* orbital on C3, and the red *p* orbital on C1 is similar in size to the blue orbital on C2, so the extent of overlap among these *p* orbitals is greater. With greater overlap among *p* orbitals that are in phase, the transition state is better stabilized, and the energy barrier to the reaction is decreased.

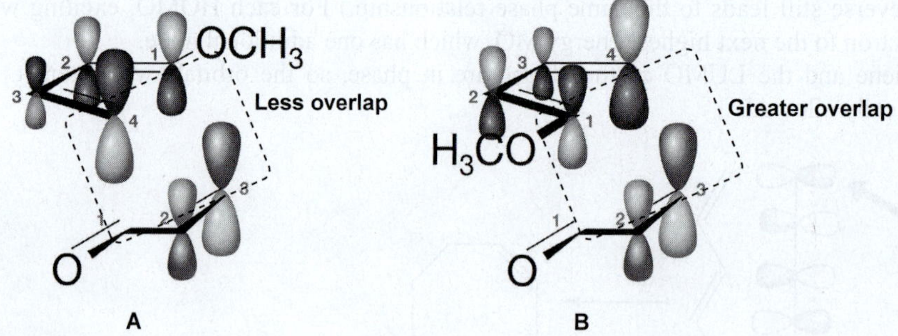

Problem 24.64

Think

What is the HOMO for the reactant in the ground and first excited states? Does constructive or destructive interference occur among the orbitals of the atoms undergoing bond formation when the terminal atoms are rotated in a conrotatory fashion? How about in a disrotatory fashion? Do the orbitals have the appropriate symmetries to interact?

Solve

(a) Thermally. In the ground state, the HOMO of the triene has two nodes, as shown below. When rotation is conrotatory, the *p* orbitals undergo destructive interference, as shown at the left. When rotation is disrotatory, the *p* orbitals undergo constructive interference, as shown on the right. Therefore, the disrotatory reaction is favored, which produces the cis product.

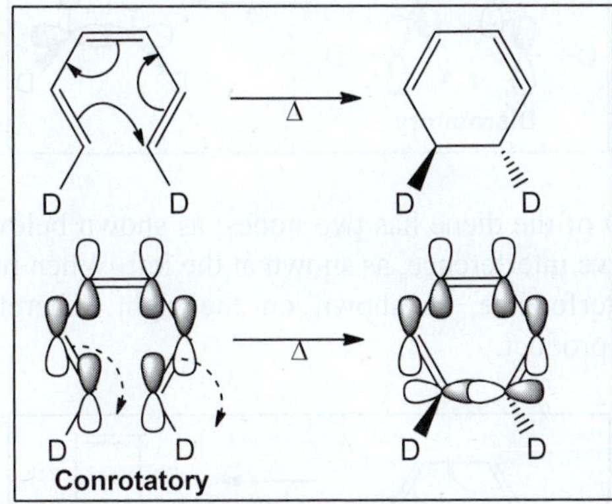

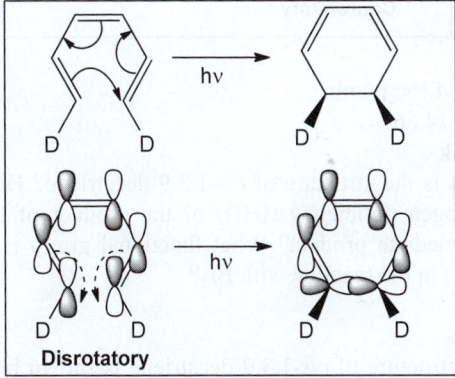

(b) Photochemically. In the first excited state, the HOMO of the triene has three nodes, as shown below. When rotation is conrotatory, the *p* orbitals undergo constructive interference, as shown at the left. When rotation is disrotatory, the *p* orbitals undergo destructive interference, as shown on the right. Therefore, the conrotatory reaction is favored, which produces the trans product.

Problem 24.65

Think

What is the HOMO for the reactant in the ground and first excited states? Does constructive or destructive interference occur among the orbitals of the atoms undergoing bond formation when the terminal atoms are rotated in a conrotatory fashion? How about in a disrotatory fashion? Do the orbitals have the appropriate symmetries to interact?

Solve

(a) Thermally. In the ground state, the HOMO of the diene has one node, as shown below. When rotation is conrotatory, the *p* orbitals undergo constructive interference, as shown at the left. When rotation is disrotatory, the *p* orbitals undergo destructive interference, as shown on the right. Therefore, the conrotatory reaction is favored, which produces the trans product. See the figure on the next page.

(b) Photochemically. In the first excited state, the HOMO of the diene has two nodes, as shown below. When rotation is conrotatory, the *p* orbitals undergo destructive interference, as shown at the left. When rotation is disrotatory, the *p* orbitals undergo constructive interference, as shown on the right. Therefore, the disrotatory reaction is favored, which produces the cis product.

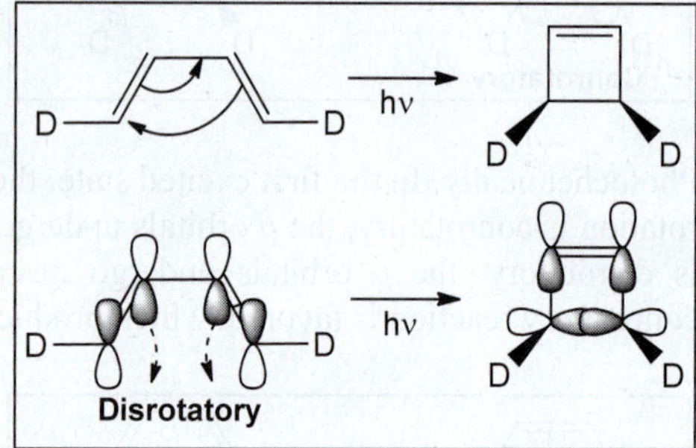

Integrated Problems

Problem 24.66

Think

What is the structure of *cis*-1,3,9-decatriene? How can this compound react with itself? What is the index of hydrogen deficiency (IHD) of the product of the Br_2 reaction? What does this mean for the IHD for the intermediate product? What functional group is likely present in the intermediate product if it gains two Br atoms in the reaction with Br_2?

Solve

The structure of *cis*-1,3,9-decatriene is shown below. The product of the Br_2 reaction has an IHD of 2, which must come from two rings. No C=C bonds should appear in that product, because the reaction with Br_2 would have removed any that existed in the intermediate product. Because the intermediate product reacts with 1 mole of Br_2, it must contain one C=C bond. The intermediate product, therefore, has two rings and one C=C bond. This can be achieved if *cis*-1,3,9-decatriene undergoes an intramolecular Diels–Alder reaction to form a bicyclic alkene.

Chemical formula: $C_{10}H_{16}Br_2$

Problem 24.67

Think

Which pair of C≡C bonds is analogous to a conjugated diene in a regular Diels–Alder reaction? Which C≡C bond is analogous to a dienophile in in a regular Diels–Alder reaction? How do you add the curved arrows? Which bonds form, and which ones break? Does the product of that reaction appear to be a benzyne? Is there another resonance contributor?

Solve

The C≡C bonds that are analogous to a diene and dienophile are indicated below. Like a regular Diels–Alder reaction, this reaction involves the cyclic movement of three pairs of electrons, so three curved arrows are necessary. The product of this curved arrow notation doesn't immediately appear to be a benzyne intermediate, but another resonance contributor of the ring does.

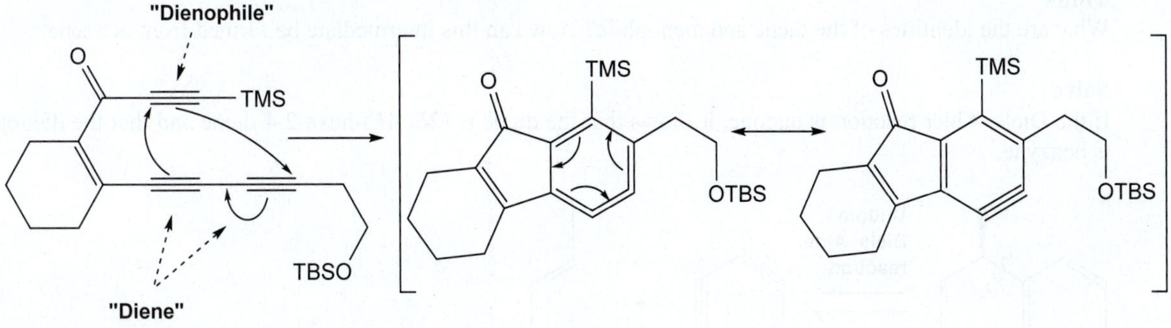

"Dienophile" O TMS TBSO "Diene" O TMS OTBS O TMS OTBS

Problem 24.68

Think

How many π electrons move in an [8+2] cycloaddition reaction? How many π electrons are supplied by each reactant? What bonds are formed and broken when these electrons move in a cyclic fashion?

Solve

An [8+2] cycloaddition reaction involves eight electrons from one π system and two electrons from the other. This can be achieved by the four C=C bonds in the first molecule and the one C=C bond in the second molecule, as shown in the mechanism below:

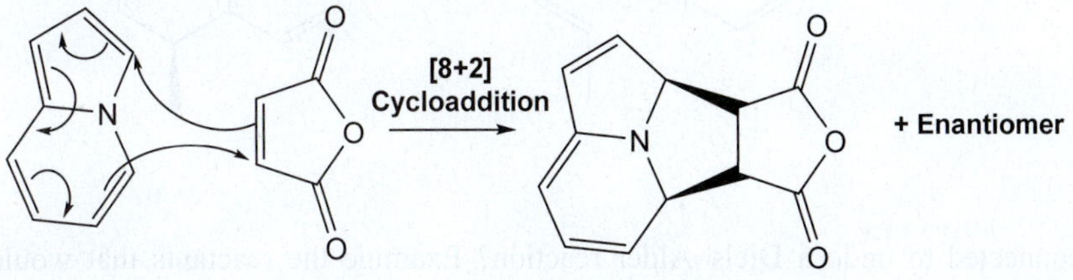

N O O O [8+2] Cycloaddition N O O O + Enantiomer

Notice that eight electrons are supplied by the reactant on the left and two electrons are supplied by the reactant on the right.

Problem 24.69

Think

What is the structure of the benzyne intermediate? How is it formed in this reaction? How might benzyne react with cyclopentadiene to form a product with the formula $C_{11}H_{10}$? Will the benzyne intermediate behave as a diene or dienophile?

Solve

The benzyne intermediate is formed from a proton transfer reaction followed by a nucleophile elimination. The benzyne intermediate will behave as a dienophile in a [4+2] cycloaddition reaction (a Diels–Alder reaction) to form the bicyclic compound shown below:

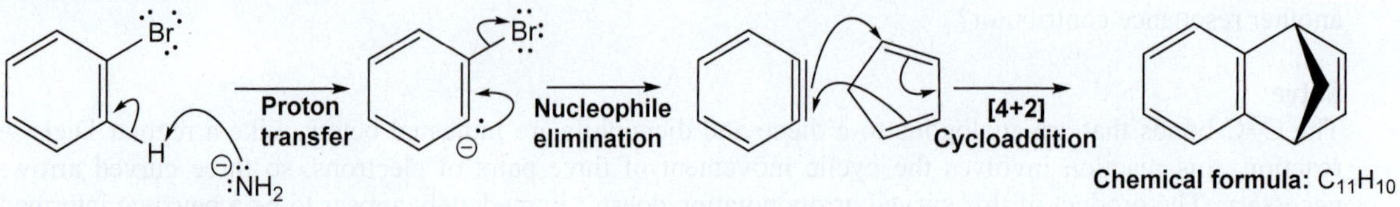

Br: H :NH$_2$ Proton transfer Br: Nucleophile elimination [4+2] Cycloaddition

Chemical formula: $C_{11}H_{10}$

Problem 24.70 (SYN)

Think

What are the identities of the diene and dienophile? How can this intermediate be formed from benzene?

Solve

If the Diels–Alder reaction is undone, it shows that the diene is (2*E*, 4*E*)-hexa-2,4-diene and that the dienophile is benzyne.

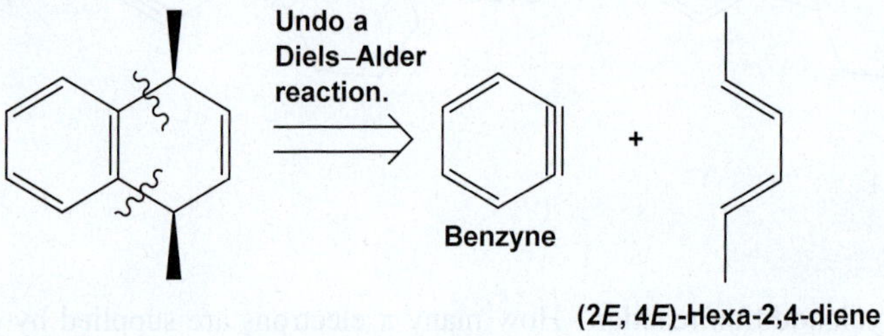

Undo a Diels–Alder reaction. Benzyne + (2*E*, 4*E*)-Hexa-2,4-diene

The synthesis in the forward direction would proceed as follows:

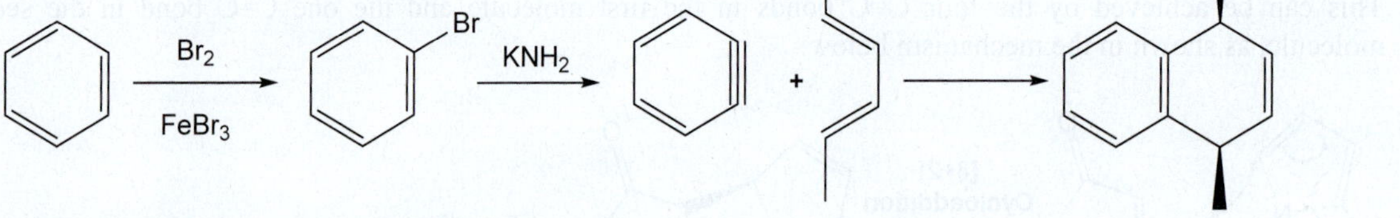

Br_2 / $FeBr_3$ → Br → KNH_2 → benzyne + diene →

Problem 24.71 (SYN)

Think

What bonds need to be disconnected to undo a Diels–Alder reaction? Examine the reactants that would be required. Is this functional group stable? How can you form a similar diene that is stable?

Solve

The diene would have C=C bonds that are each part of an enol and would tautomerize to the more stable keto forms.

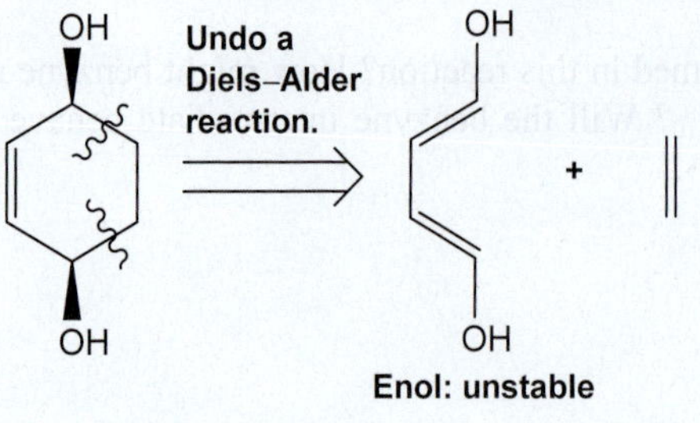

OH OH Undo a Diels–Alder reaction. OH + OH

Enol: unstable

However, if a diether were to be synthesized with alkyl groups that could be removed after the Diels–Alder reaction was over, the synthesis would be feasible. A bis[benzyl ether] is a possibility.

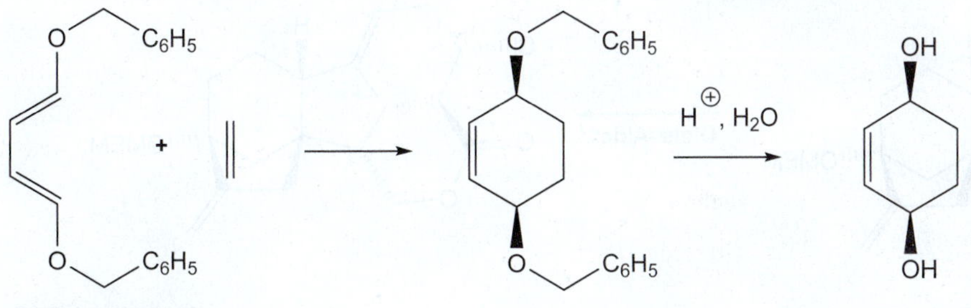

Bis[benzyl ether]

Problem 24.72 (SYN)

Think

What functional group is present in the target compound? How many carbon atoms are in the starting material and target compound? Is the carbon skeleton altered? What reactions do you know that will break a C=C bond? What functional group transformations must occur?

Solve

The target compound is an α,β-unsaturated ketone. This product is a result of an intramolecular aldol condensation reaction. The pentene ring undergoes oxidative cleavage to break the C=C bond, which opens the ring and forms the dicarbonyl compound that can undergo the subsequent aldol condensation. The full synthesis is shown below:

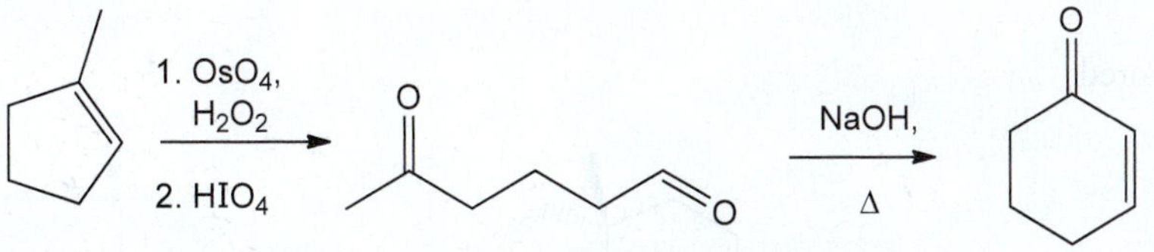

Problem 24.73

Think

Is the standard Diels–Alder reaction favored when the dienophile is electron rich or electron poor? Which alkene C=C is more electron poor? Identify the diene and dienophile in each reaction. What is the mechanism of a [4+2] cycloaddition reaction?

Solve

(a) The Diels–Alder reaction is favored when the dienophile is electron poor, namely, when electron-withdrawing groups are on the dienophile. Therefore, the alkene with the OCH_3 group is more electron rich and will be less likely to react as the dienophile. The mechanism is given below:

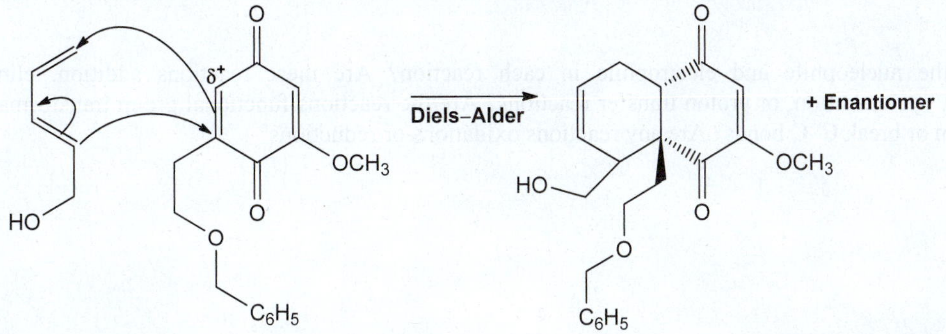

(b) The diene is located in the middle of the structure and is already in the *s*-cis conformation as written. The dienophile is the Cl-substituted alkene, which has the substituents trans.

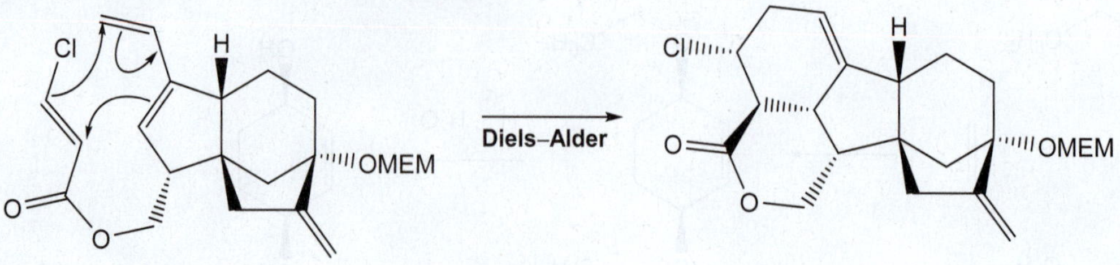

Cl H OMEM O O Diels–Alder Cl H O O OMEM

Problem 24.74

Think

Identify the diene and dienophile in each reaction. What is the mechanism of a [4+2] cycloaddition reaction?

Solve

The mechanisms are given below:

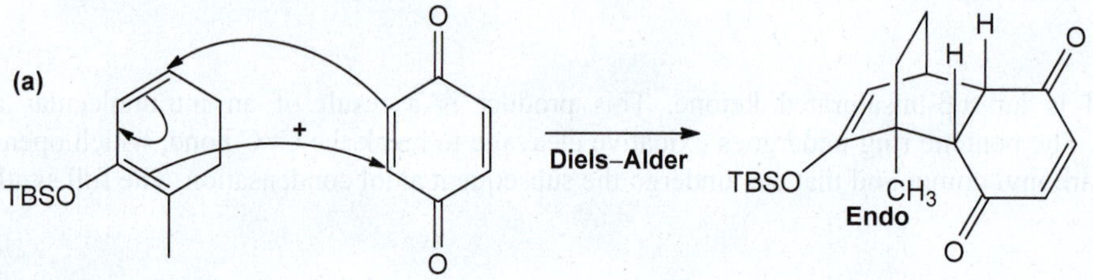

(a) TBSO + O O Diels–Alder H H O TBSO CH_3 Endo O

Notice that the endo product is favored.

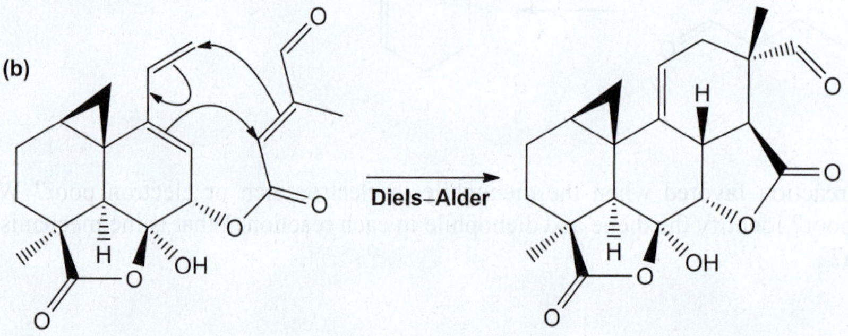

(b) O H OH O O O Diels–Alder H O H O O OH O O

Notice that the aldehyde and ester groups attached to the dienophile are trans about the C=C bond, so those groups are trans with respect to the six-membered ring in the product.

Problem 24.75

Think

What are the nucleophile and electrophile in each reaction? Are these reactions addition, elimination, substitution, cycloaddition, or proton transfer reactions? Are the reactions functional group transformations, or do they form or break C–C bonds? Are any reactions oxidations or reductions?

Solve

The products from each reaction are shown below. **A** is formed from 1,4-addition of HBr to the diene. **B** is formed from an S_N2 substitution reaction, which then undergoes a [4+2] cycloaddition (Diels–Alder) reaction to form **C**. Note that the CH_3 and CH_2CN are trans in the alkene dienophile and trans in the product **C**. The C≡N nitrile is hydrolyzed to the carboxylic acid **D**, which is then transformed to the methyl ester by CH_2N_2. The ester is reduced by DIBAH to form the aldehyde **E**. The aldehyde is protected by ethylene glycol to the acetal **F**, which can now react with furan in a [4+2] cycloaddition to form the product **G**. The alkene is hydrogenated by H_2/Pd to form the alkane **H**. The acetal protecting group is removed to re-form the aldehyde **I**. In the process, the cyclic ether is hydrolyzed to produce the diol.

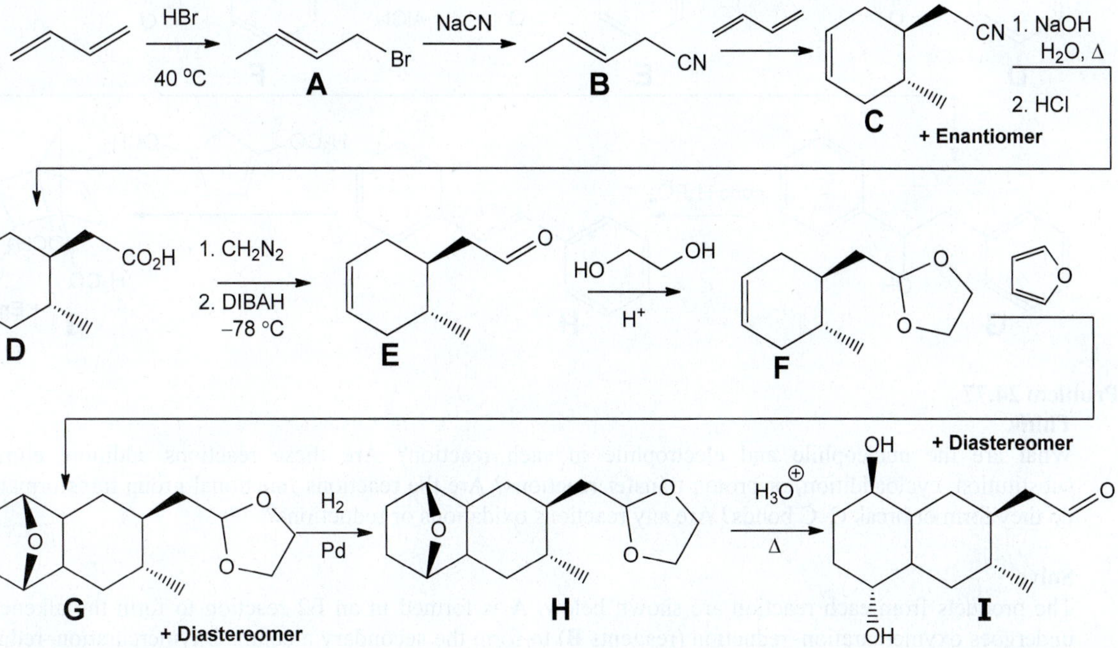

HBr, 40 °C → A; NaCN → B; → C (+ Enantiomer); 1. NaOH, H_2O, Δ; 2. HCl → D; 1. CH_2N_2; 2. DIBAH, −78 °C → E; HO–CH₂CH₂–OH, H^+ → F; furan → G (+ Diastereomer); H_2, Pd → H; H_3O^+, Δ → I (+ Diastereomer)

Problem 24.76

Think

What are the nucleophile and electrophile in each reaction? Are these reactions addition, elimination, substitution, cycloaddition, or proton transfer reactions? Are the reactions functional group transformations, or do they form or break C–C bonds? Are any reactions oxidations or reductions?

Solve

The products from each reaction are shown on the next page. **A** is formed from an electrophilic aromatic bromination reaction. **A** then reacts with Mg(*s*) to form the Grignard reagent **B**. **B** then opens the epoxide ring to form the primary alcohol **C**, which is oxidized to the carboxylic acid **D**. The carboxylic acid is transformed into the acid chloride **E**, which undergoes Friedel–Crafts acylation to form the ketone **F**. The ketone is reduced to the secondary alcohol **G**, which undergoes dehydration to form the conjugated alkene **H**. The alkene **H** serves as the dienophile to undergo a Diels–Alder reaction to form the bicyclic product **I**.

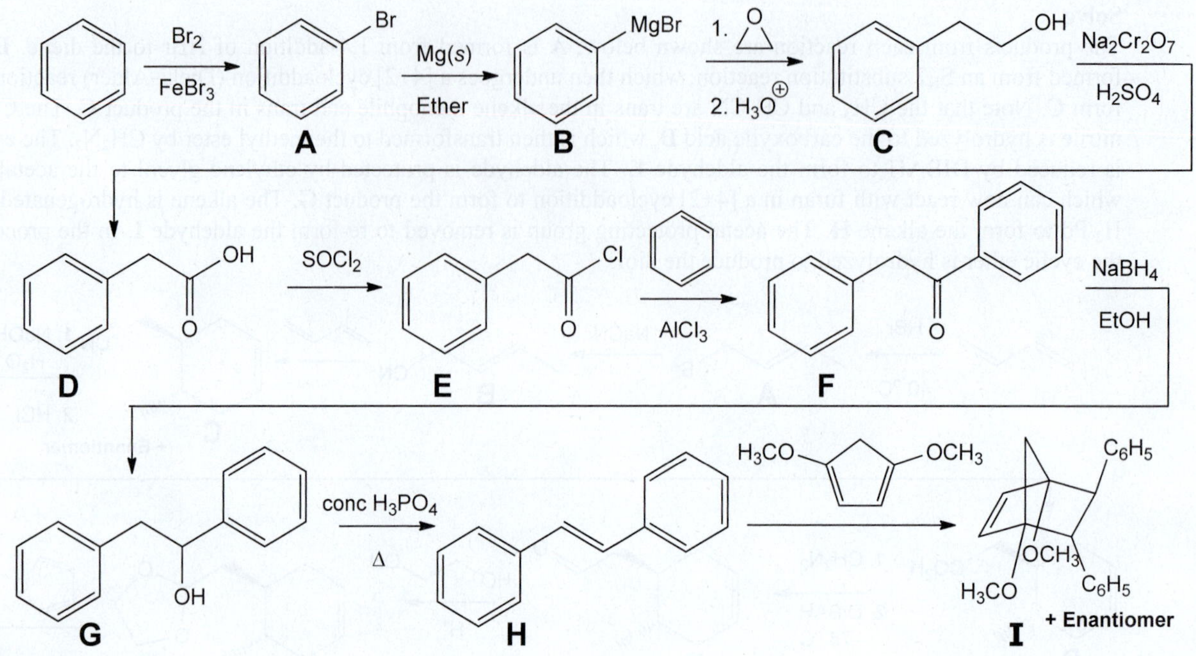

Problem 24.77

Think

What are the nucleophile and electrophile in each reaction? Are these reactions addition, elimination, substitution, cycloaddition, or proton transfer reactions? Are the reactions functional group transformations, or do they form or break C–C bonds? Are any reactions oxidations or reductions?

Solve

The products from each reaction are shown below. **A** is formed in an E2 reaction to form the alkene, which undergoes oxymercuration–reduction (reagents **B**) to form the secondary alcohol. Oxymercuration–reduction is chosen in this reaction to achieve Markovnikov addition of water while avoiding a carbocation rearrangement. The alcohol is oxidized by $KMnO_4$ to the ketone **C**, which reacts with the Grignard reagent to form the tertiary alcohol **D** after acid workup. The alcohol undergoes a dehydration reaction to form alkene **E**. The alkene reacts with OsO_4 to form the syn 1,2-diol **F**.

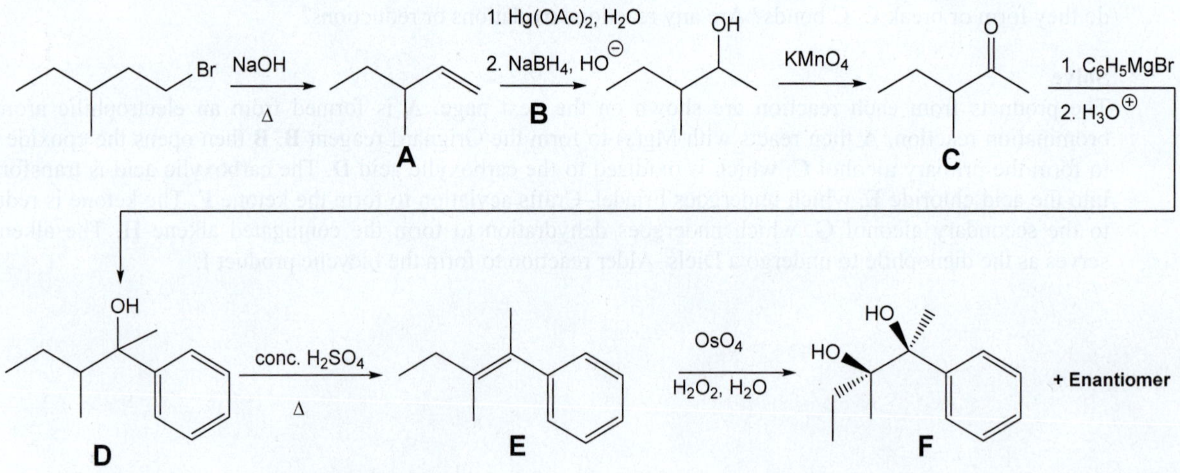

Problem 24.78

Think

What are the nucleophile and electrophile in each reaction? Are these reactions addition, elimination, substitution, cycloaddition, or proton transfer reactions? Are the reactions functional group transformations, or do they form or break C–C bonds? Are any reactions oxidations or reductions?

Solve

The products from each reaction are shown below. **A** is formed from the ozonolysis of the alkene to form a carboxylic acid and ketone. The carboxylic acid is transformed into the acid chloride **B**, which then reacts with the amine $(CH_3)_2NH$ to form the amide **C**. The amide is unreactive with $NaBH_4$, so the ketone is selectively reduced to the alcohol **D**. The alcohol is brominated to the alkyl bromide **E**, which then undergoes an E2 reaction to form the alkene **F**. The alkene serves as the dienophile in the Diels–Alder reaction to form **G**. The trans configuration of the substituents in alkene **F** remain trans in the cyclohexene product. The alkene undergoes reaction with OsO_4 to form the syn diol **H**.

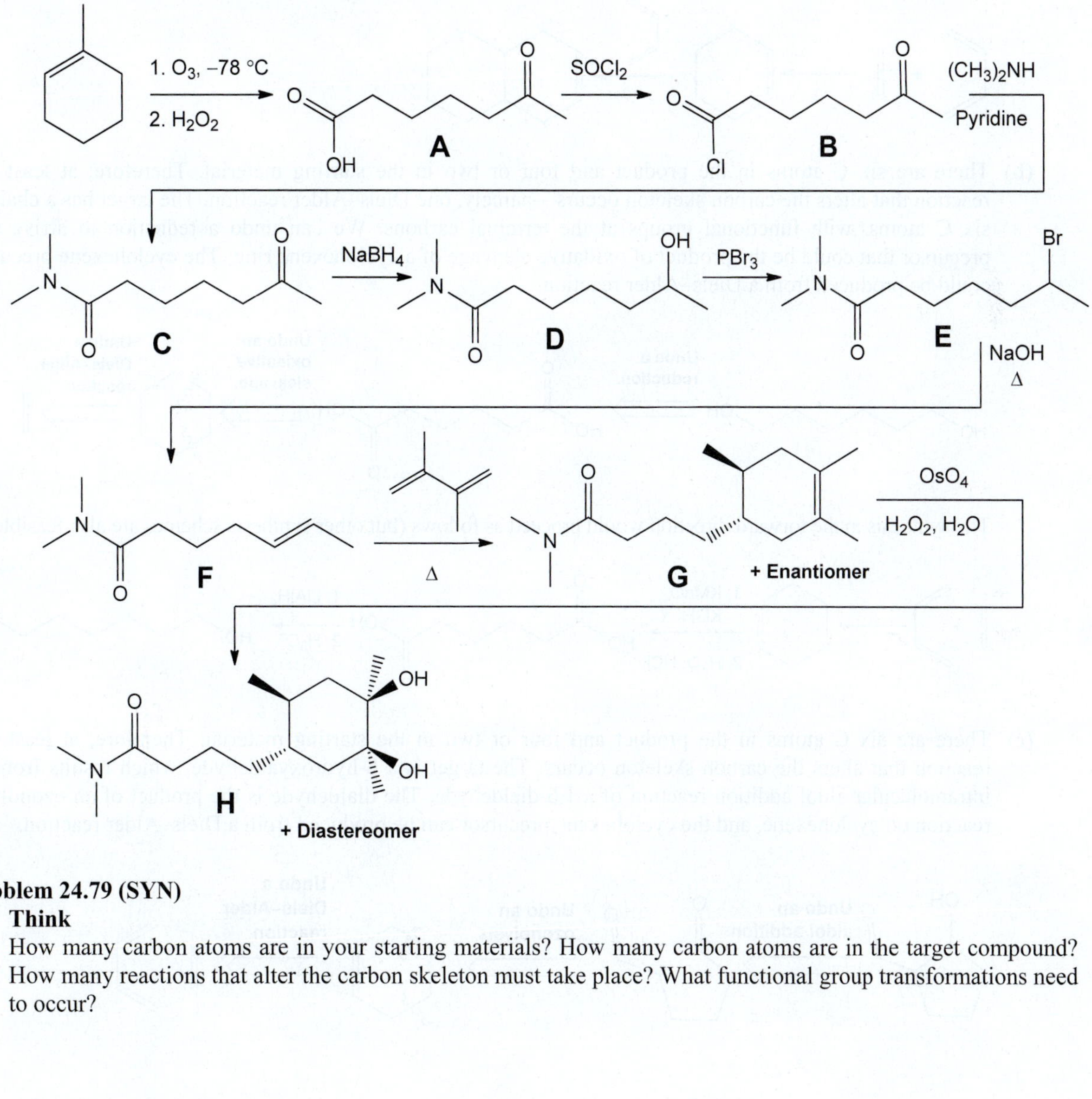

1. O_3, –78 °C
2. H_2O_2

A

$SOCl_2$

B

$(CH_3)_2NH$
Pyridine

C

$NaBH_4$

D

PBr_3

E

NaOH
Δ

F

Δ

G + Enantiomer

OsO_4
H_2O_2, H_2O

H + Diastereomer

Problem 24.79 (SYN)

Think

How many carbon atoms are in your starting materials? How many carbon atoms are in the target compound? How many reactions that alter the carbon skeleton must take place? What functional group transformations need to occur?

Solve

(a) There are 10 C atoms in the product and four or two in the starting material. Therefore, at least two reactions that alter the carbon skeleton must occur, namely, two back-to-back Diels–Alder reactions. The target has a cyclohexene ring, so we can undo a Diels–Alder reaction to arrive at the diene and dienophile precursors. The dienophile precursor is cyclohexene, so we can undo another Diels–Alder reaction.

Undo a Diels–Alder reaction. + Undo a Diels–Alder reaction. +

The synthesis in the forward direction would proceed as follows:

+ Δ Δ

(b) There are six C atoms in the product and four or two in the starting material. Therefore, at least one reaction that alters the carbon skeleton occurs—namely, one Diels–Alder reaction. The target has a chain of six C atoms, with functional groups at the terminal carbons. We can undo a reduction to arrive at a precursor that could be the product of oxidative cleavage of a cyclohexene ring. The cyclohexene precursor could be produced from a Diels–Alder reaction.

HO OH Undo a reduction. HO O OH O Undo an oxidative cleavage. Undo a Diels–Alder reaction.

The synthesis in the forward direction would proceed as follows (but other synthesis schemes are also feasible):

+ 1. $KMnO_4$, KOH, Δ 2. H_2O, HCl HO O OH O 1. $LiAlH_4$ 2. H_3O^+ HO OH

(c) There are six C atoms in the product and four or two in the starting material. Therefore, at least one reaction that alters the carbon skeleton occurs. The target is a β-hydroxyaldehyde, which results from an intramolecular aldol addition reaction of a 1,6-dialdehyde. The dialdehyde is the product of an ozonolysis reaction on cyclohexene, and the cyclohexene precursor can be produced from a Diels–Alder reaction.

OH O Undo an aldol addition. O O Undo an ozonolysis. Undo a Diels–Alder reaction.

The synthesis in the forward direction would proceed as follows (but other synthesis schemes are also feasible):

O_3, then Zn, HOAc (ozonolysis)

NaOH (aldol) EtOH

(d) There are 10 C atoms in the product and four or two in the starting material; therefore, at least two reactions that alter the carbon skeleton must occur. The target has a cyclohexene ring, so we can undo a Diels–Alder reaction to arrive at the diene and dienophile precursors. The dienophile precursor is an □,□-unsaturated aldehyde, which can be produced from an intramolecular aldol condensation. The dialdehyde precursor is the same that was synthesized in **(c)**.

Undo a Diels–Alder reaction.

Undo an aldol condensation.

From (c)

The synthesis direction would proceed as follows:

NaOH, H_2O, Δ (aldol cond)

From (c)

Problem 24.80

Think

What functional group has ^{1}H NMR signals around 7.3 ppm? If there is only one signal, is there a lot of symmetry? Is the starting material strained?

Solve

(a) and **(b)** The proton NMR signal corresponds to an aromatic hydrogen. Because there is a single proton NMR signal, the product has a high degree of symmetry and corresponds to benzene. To break the two bonds of the cyclobutane ring, the reaction must proceed through a transition state in which four π electrons are cyclically delocalized, making it an antiaromatic transition state.

Antiaromatic

(c) The reaction relieves substantial ring strain and also produces two aromatic rings. These two factors combined compensate for the high energy in the antiaromatic transition state.

Problem 24.81

Think

How many carbon atoms are in the reactant? How many are in the product? What functional group has ^{1}H NMR signals between 5 and 7 ppm and ^{13}C NMR signals between 120 and 140 ppm? Is there symmetry in the product?

Solve

The product only has four C atoms; therefore, a retro Diels–Alder reaction occurred to produce buta-1,3-diene. An alkene has ^{1}H NMR signals between 5 and 7 ppm and ^{13}C NMR signals between 120 and 140 ppm. The integration in the ^{1}H NMR spectrum exhibits a 2:4 ratio, which corresponds to the internal:terminal hydrogens in buta-1,3-diene, and the ^{13}C NMR spectrum has two signals, consistent with buta-1,3-diene having two distinct C atoms. The mechanism and product are given below:

700 °C

+

Problem 24.82

Think

How many carbon atoms are in the reactant? How many are in the product? What functional group has ^{1}H NMR signals between 7 and 9 ppm and ^{13}C NMR signals between 120 and 140 ppm? What functional group has a ^{13}C NMR signal around 190 ppm? Is there symmetry in the product?

Solve

The product only has 14 C atoms, and the starting material has 18. It lost four C atoms from two equivalents of ethylene. There are two outer rings that each consist of six C atoms and a C=C bond, so a retro Diels–Alder reaction can occur on both outer rings to produce $H_2C{=}CH_2$. Aromatic ^{1}H NMR signals appear between 7 and 9 ppm, and their ^{13}C NMR signals appear between 120 and 140 ppm. The signal at 190 ppm is due to the C=O.

O O Δ + + O O

Problem 24.83

Think

What is the index of hydrogen deficiency (IHD) of the starting compound? What functional group reacts both with Br_2 and with $KMnO_4/HO^-$? What functional group causes a broad absorption peak in the infrared spectrum (IR) from 2500 to 3300 cm^{-1} and a sharp peak at 1700 cm^{-1}? What information about functional groups and symmetry is given in the NMR spectra?

Solve

The IHD is 2. The functional group that reacts with Br_2 and $KMnO_4/HO^-$ is a C=C bond of an alkene, which accounts for an IHD of 1. Because the compound reacts with just one molar equivalent of Br_2, it should contain one C=C bond, so the other unit of IHD is due to a ring. The IR spectrum suggests a carboxylic acid. The ^{13}C NMR spectrum has six signals, so the product of the reaction has three distinct C atoms. This could occur if there are two carboxyl groups at the ends of a six-carbon chain. That diacid could be the product of oxidative cleavage of cyclohexene:

1. $KMnO_4$, KOH, Δ

2. H_2O, HCl

HO, O, OH, O

Problem 24.84

Think

How does the formula of the product compare to the formula of the reactant? What molecule must have been lost? What is a reasonable [4+2] cycloaddition involving the reactant? What is a reasonable [4+2] cycloelimination involving the product of that reaction, which can result in the loss of N_2? What types of protons have NMR signals between 7 and 9 ppm? What types of protons could give rise to the 2H singlet at 5.46 ppm?

Solve

The signals between 7 and 9 ppm are aromatic protons, and there are 14 of them. The 2H singlet at 5.46 ppm must be from the CH_2 group, which is substantially deshielded. This product has two fewer N atoms than the starting compound ($C_{23}H_{16}N_4O$), so the reaction resulted in the loss of N_2. Notice that the ring containing three N atoms already has two N atoms bonded together, so those two N atoms are probably the ones eliminated in the [4+2] cycloelimination. Prior to that, that same ring can participate in a [4+2] cycloaddition with the C≡N bond serving as the dienophile. This two-step sequence is shown below:

C_6H_5 C_6H_5 N N N N O → [C_6H_5 C_6H_5 N N N N O] → C_6H_5 C_6H_5 N N O + N_2

CHAPTER 25 | Reactions Involving Free Radicals

Your Turn Exercises

Your Turn 25.1

Think

In homolysis, what type of curved arrow is used (single or double barbed) to show bond breaking? How many electrons are in a bond? How many electrons go to each atom connected to the bond? How many curved arrows are required?

Solve

Homolysis uses the single-barbed curved arrow. Two electrons are in each bond, and one electron goes to each atom in the bond, so two arrows are required.

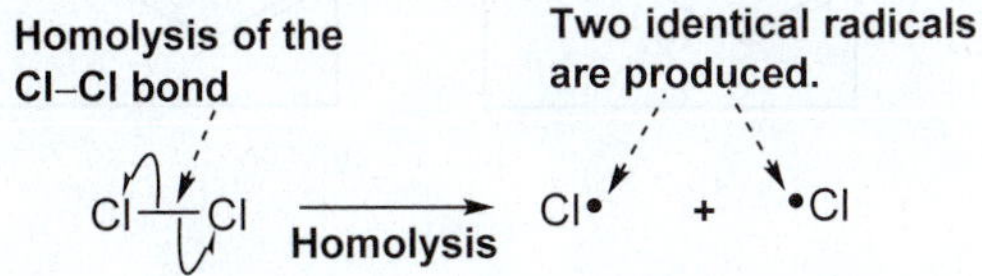

Your Turn 25.2

Think

Refer to Tables 25-1 and 25-2 to determine the bond energies of the bonds indicated. What does bond energy indicate about the strength of the bond?

Solve

Cl–Cl = 243 kJ/mol; Br–Br = 192 kJ/mol; I–I = 151 kJ/mol; HO–OH = 211 kJ/mol; H_3C–H = 439 kJ/mol. The larger the bond energy, the stronger the bond. This shows that the C–H bond is the strongest bond in this list.

Your Turn 25.3

Think

Refer to Table 25-1 to determine the H–Br and H–Cl bond energies. Which bond is more difficult to break? Does that correspond to products that are higher in energy or lower?

Solve

The H–Br and H–Cl bond energies are 368 and 431 kJ/mol, respectively. The H–Cl bond is stronger and requires more energy to cleave homolytically. The product chlorine radical (•Cl), therefore, is higher in energy than the bromine radical (•Br).

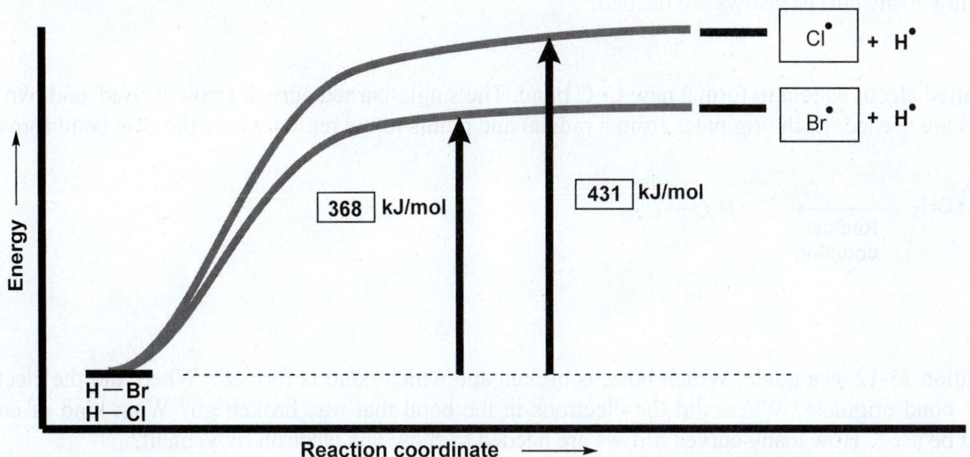

Your Turn 25.4

Think

What is the hybridization of each of the C atoms in the benzyl cation? Are there any atoms with an incomplete valence adjacent to π bonds? How do you use curved arrow notation to show the movement of a pair of electrons? How many times can that electron movement be carried out?

Solve

All C atoms in this structure are sp^2 hybridized. The carbocation C^+ has an incomplete valence and is adjacent to a π bond. One double-barbed arrow is used to show the movement of a pair of electrons to arrive at the next resonance structure. This can be repeated two more times to arrive at a total of five resonance structures.

CH_2 CH_2 CH_2 CH_2 CH_2

Benzyl cation

Your Turn 25.5

Think

Where does the unpaired electron end up in the final resonance structure? Do you need to use single-barbed curved arrows or double-barbed curved arrows to show that? Where do the electrons from the adjacent π bond end up? How many curved arrows are necessary to show that electron movement? Do those arrows point in the same direction?

Solve

The unpaired electron ends up as part of a π bond involving the same C atom, and one single-barbed arrow is needed to show that movement. The two electrons from the adjacent π bond move in opposite directions, requiring two single-barbed arrows. One of those electrons joins the unpaired electron, and the other moves to the CH_2 carbon to become unpaired.

CH_2 Resonance $\dot{C}H_2$

Your Turn 25.6

Think

Where do the unpaired electrons end up in the product? What kind of curved arrows must be used to show that movement? How many curved arrows are needed?

Solve

The two unpaired electrons join to form a new C–C bond. The single-barbed curved arrow is used, and two such curved arrows are needed. Each originates from a radical and points to the region where the new bond forms.

$$H_3C\bullet + \bullet CH_3 \xrightarrow{\text{Radical coupling}} H_3C{-}CH_3$$

Your Turn 25.7

Think

Refer to Equation 25-12 as a guide. Which bond is broken and which bond is formed? Where did the electrons from the new bond originate? Where did the electrons in the bond that was broken go? What kind of curved arrows should be used? How many curved arrows are needed to show this electron movement?

Solve

The C–H bond is broken and the H–Cl bond is formed. The electrons in the C–H bond are cleaved homolytically. One electron from the C–H bond goes to the C to form the radical, and the other electron pairs with the Cl radical to form the H–Cl bond. Three single-barbed curved arrows are needed to show this electron movement.

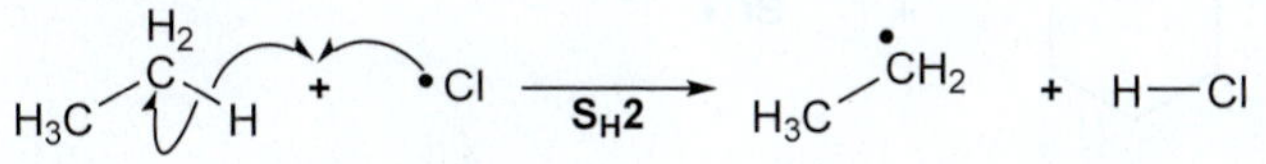

Your Turn 25.8

Think

Refer to Equation 25-14 as a guide. Which bond is broken and which bond is formed? Where did the electrons from the new bond originate? Where did the electrons in the bond that was broken go? What kind of curved arrows should be used? How many curved arrows are needed to show this electron movement?

Solve

The C=C π bond is broken and the C–Cl bond is formed. The electrons in the C=C π bond are homolytically cleaved. One electron from the C=C π bond goes to the C to form the radical, and the other electron joins with the Cl radical to form the C–Cl bond. Three single-barbed curved arrows are needed to show this electron movement.

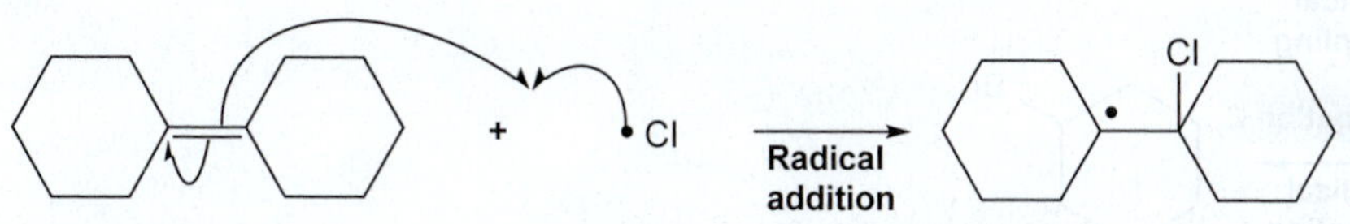

Your Turn 25.9

Think

What does it mean to be an initiation step? What is the weakest bond present? What supplies the energy necessary to carry out the homolysis of that bond? Refer to Equation 25-18 as a guide.

Solve

An initiation step produces radicals from a closed-shell species. The UV light (*hv*) supplies the energy to break the weakest bond homolytically—in this case, the Br–Br bond. The products of the initiation step are two equivalents of Br•.

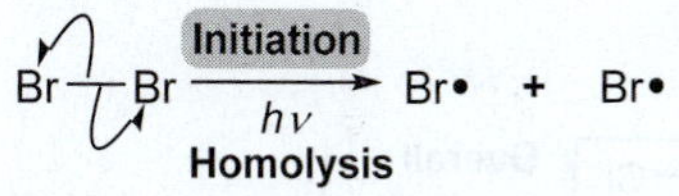

Your Turn 25.10

Think

What is the main difference between the reaction in Equation 25-19 and the one in Your Turn 25.9? What is meant by a propagation step? Are radicals present in the products, in the reactants, or in both? How many propagation steps are necessary? Which bonds are broken and which bonds are formed in each step?

Solve

The difference between this reaction and the one in Equation 25-19 is the presence of Br in place of Cl. A propagation step has radicals present in the reactants and products. The breaking of the C–H bond and formation of the C–Br bond require two propagation steps (as shown on the next page). In the first step, a hydrogen atom is abstracted to form the carbon radical (•C). In the second step, the •C forms the C–Br bond, while the Br–Br bond breaks to form a new bromine radical (•Br).

Propagation 1 S_H2

Propagation 2 S_H2

Your Turn 25.11

Think

What is meant by a termination step? Are radicals present in the products, in the reactants, or in both? Use Equation 25-20 as a guide. Which bonds are formed?

Solve

A termination step decreases the number of free radicals. Two free radicals come together in the reactants and no radicals are present on the product side. The Br–Br bond or the C–Br bond is formed from the two radicals present in solution: •C and •Br.

Termination 1 Radical coupling

Termination 2 Radical coupling

Your Turn 25.12

Think

Review the reactants and products of initiation, termination, and propagation. Which step generates a radical? Which step eliminates a radical? Which step has radicals in both reactant and product? How do the reactions in Figure 25-8 compare to the mechanisms shown in Equations 25-18, 25-19a, 25-19b, 25-20a, and 25-20b?

Solve

The initiation step matches Equation 25-18; the propagation steps match either Equation 25-19a or 25-19b (shown in the figure below). Equations 25-20a and 25-20b are radical coupling or termination steps.

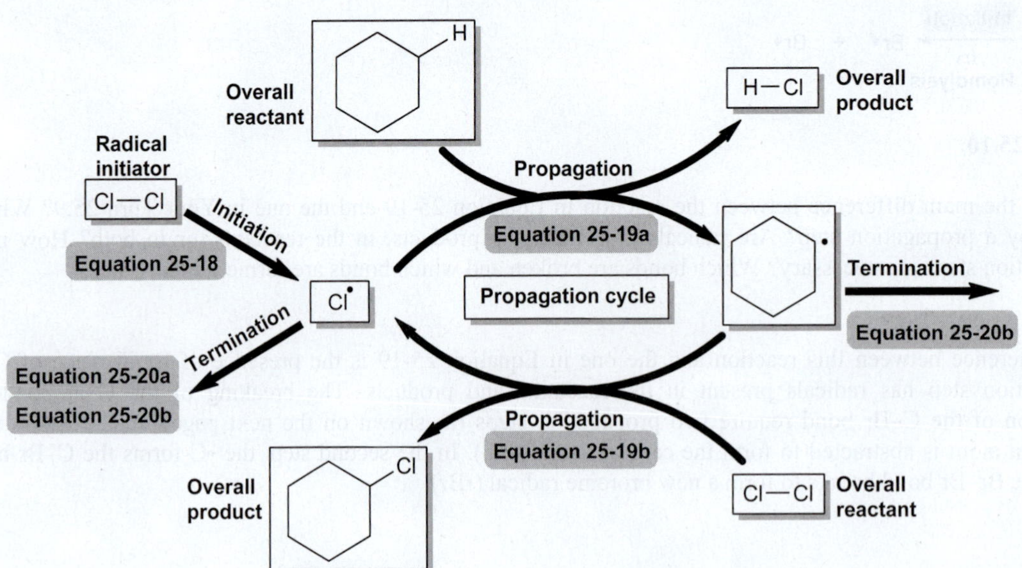

Your Turn 25.13

Think

What are the overall products and reactants for the free radical bromination of methane, CH_4? Which bond in the reactants is the weakest and can undergo homolysis via an initiation step, and what is the product of initiation? What are the products of the first propagation step and of the second propagation step (see the solution to Your Turn 25.10)?

Solve

The overall reaction is

$$CH_4 + Br_2 \xrightarrow{h\nu} H_3C{-}Br + H{-}Br$$

The Br–Br bond is the weakest bond and undergoes homolysis in an initiation step to produce two •Br radicals. The •Br reacts with CH_3–H in the first propagation step to homolytically cleave the C–H bond and form the •C. The •C reacts with Br–Br in the second propagation step to produce CH_3–Br and another •Br. The CH_4/Br• and •CH_3/Br_2 species continue to react through this propagation cycle.

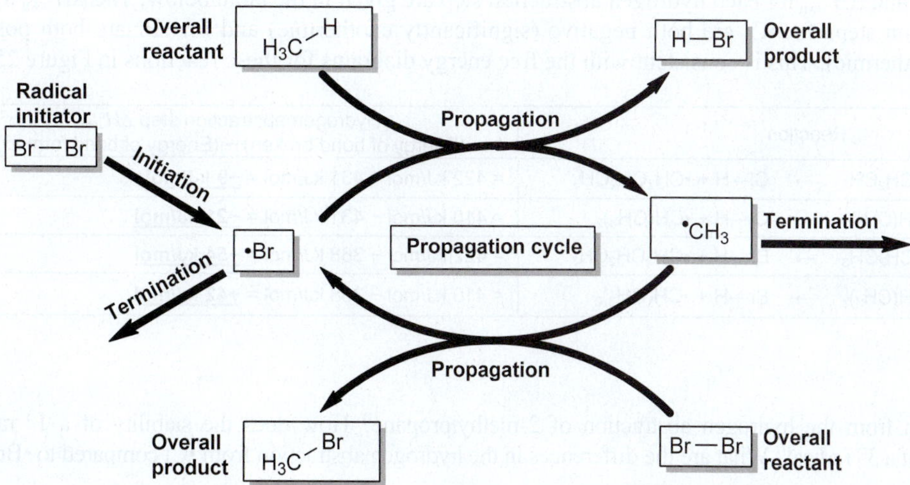

Your Turn 25.14

Think

Refer to Tables 25-1 and 25-2 to fill in the $\Delta H°$ for the two bonds that are broken and the two bonds that are formed. Then use the equation $\Delta H°_{rxn} = \Delta H°_{bond\ broken} - \Delta H°_{bond\ formed}$. Compare your calculated value to the value given in Table 25-3.

Solve

The values for the energy of the bonds broken and formed in the propagation steps of the bromination of CH_4 are given in the table below. $\Delta H°_{rxn}$ for the first step is calculated to be 71 kJ/mol, and $\Delta H°_{rxn}$ for the second step is calculated to be −102 kJ/mol. Both of these values match the ones in Table 1-3.

Reaction	Energy of bond broken	Energy of bond formed	(Energy of bond broken) − (Energy of bond formed)
Br• + CH_4 → HBr + CH_3•	H—CH_3 = **439 kJ/mol**	H—Br = **368 kJ/mol**	= 439 kJ/mol − 368 kJ/mol = **71 kJ/mol**
CH_3• + Br_2 → CH_3Br + Br•	Br—Br = **192 kJ/mol**	Br—CH_3 = **294 kJ/mol**	= 192 kJ/mol − 294 kJ/mol = **−102 kJ/mol**

Your Turn 25.15

Think

Which radical, primary or primary allylic, is more stable? Why? How does the stability of the radical formed contribute to the rate of the reaction?

Solve

The allylic radical produced in **B** is more stable than the propyl radical produced in **A**, owing to resonance delocalization of the unpaired electron (shown below). The rate of free radical halogenation increases with increasing stability of the radical that is formed, so the radical abstraction in **B** will proceed faster.

A
Localized radical

B
Resonance of allylic radical

Your Turn 25.16

Think

Refer to Table 25-1 for the energies of the bonds involved in each hydrogen abstraction step. When you calculate $\Delta H°_{rxn}$ for each step, what do you notice about the signs?

Solve

The bond energies and $\Delta H°_{rxn}$ for each hydrogen abstraction step are given in the table below. The $\Delta H°_{rxn}$ for the hydrogen abstraction steps for •Cl are both negative (significantly exothermic) and for •Br are both positive (significantly endothermic). This is consistent with the free energy diagrams for these reactions in Figure 25-10.

Reaction	Hydrogen abstraction step $\Delta H°_{rxn}$ (Energy of bond broken) − (Energy of bond formed)
Cl• + H—$CH_2CH_2CH_3$ → Cl—H + •$CH_2CH_2CH_3$	= 422 kJ/mol − 431 kJ/mol = **−9 kJ/mol**
Cl• + H—$CH(CH_3)_2$ → Cl—H + •$CH(CH_3)_2$	= 410 kJ/mol − 431 kJ/mol = **−21 kJ/mol**
Br• + H—$CH_2CH_2CH_3$ → Br—H + •$CH_2CH_2CH_3$	= 422 kJ/mol − 368 kJ/mol = **+54 kJ/mol**
Br• + H—$CH(CH_3)_2$ → Br—H + •$CH(CH_3)_2$	= 410 kJ/mol − 368 kJ/mol = **+42 kJ/mol**

Your Turn 25.17

Think

What radicals form from the hydrogen abstraction of 2-methylpropane? How does the stability of a 1° radical compare with that of a 3° radical? What are the differences in the hydrogen abstraction from •Cl compared to •Br?

Solve

The 3° radical is more stable than a 2° or 1° radical. This is reflected in a larger energy difference in the two radicals compared to the 2° and 1° radicals formed from the reactions in Figure 25-10. Hydrogen abstraction by •Cl is exothermic, whereas hydrogen abstraction by •Br is endothermic.

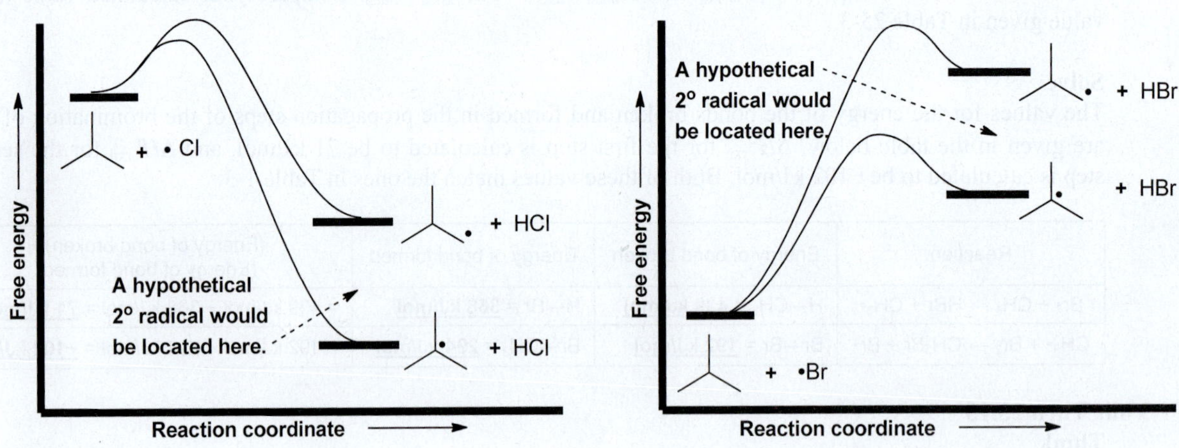

Your Turn 25.18

Think

Is there a π bond adjacent to the N radical? How can you use curved arrow notation to show the unpaired electron on N joining with a single electron from the C=O π bond? Where does the second electron from the C=O π bond end up? How can you use curved arrow notation to undo this electron movement but redo it involving the other C=O π bond? In completing the tautomerization mechanism, are any acids or bases formed in the mechanism? How is this tautomerization similar to the keto–enol tautomerization mechanism?

Solve

The unpaired electron on N joins with a single electron from the C=O π bond to form a C=N, while the other electron from the C=O π bond ends up as an unpaired electron on O•. Three single-barbed curved arrows are necessary to show this electron movement. The third resonance structure is equivalent to the second but involves the other C=O bond. To undo the first electron movement and redo it on the other side, five single-barbed arrows are necessary. In *N*-bromosuccinimide (NBS), the unpaired electron is spread out over two O atoms and one N atom. This stabilizes the radical.

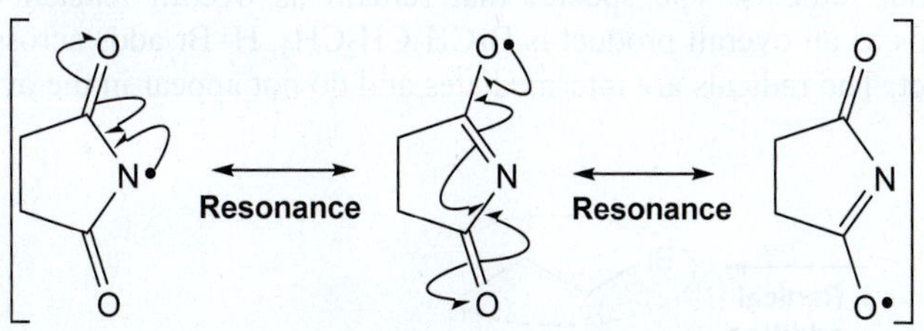

Resonance

Resonance

In Step 2 of the mechanism in Equation 25-28, the strong acid HBr is formed. Therefore, the tautomerization reaction occurs in acidic conditions.

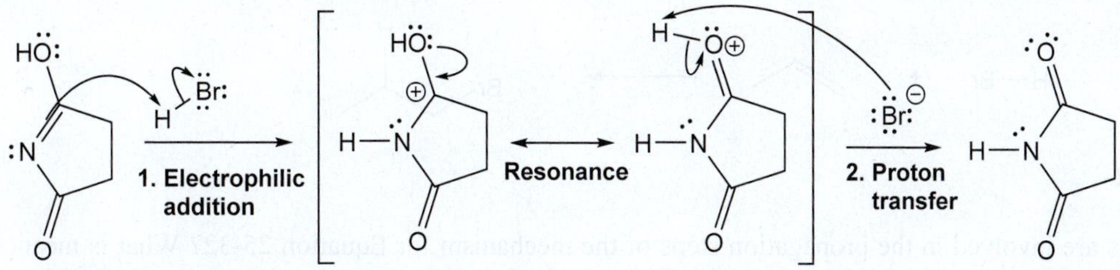

1. Electrophilic addition

Resonance

2. Proton transfer

Your Turn 25.19

Think

Where does the unpaired electron end up in each resonance structure? Do you need to use single-barbed curved arrows or double-barbed curved arrows to show that? Where do the electrons from the adjacent π bond end up? How many curved arrows are necessary to show that electron movement? Do those arrows point in the same direction?

Solve

The unpaired electron from •R abstracts a hydrogen from the phenolic H–O and ends up as an unpaired electron on O•, and three single-barbed curved arrows are needed to show that movement. To show the electron movement that converts the first resonance structure to the second, the electrons in the C=C π bond are homolytically cleaved. One electron from the C=C π bond goes to the C to form the radical, and the other electron joins with the O• radical to form the O=C π bond. Three single-barbed curved arrows are needed to show this electron movement. See the mechanism on the next page.

α-Tocopherol (Vitamin E)

Your Turn 25.20

Think

Which species appear as a reactant in one step but a product in the other? What do you notice about the species that get crossed out? Which species only appear as reactants? Which species only appear as products? Keep in mind that chemical reactions are similar to math equations when you add two equations together.

Solve

The two species that are crossed out are both radicals. The species that remain as overall reactants are $H_2C{=}CHCH_3$ and HBr, and the one that remains as an overall product is $BrCH_2CH_2CH_3$. H–Br adds across the C=C to yield an alkyl bromide as a final product. The radicals are intermediates and do not appear in the overall chemical equation.

Propagation 1 — 3. Radical addition

Propagation 2 — 4. S_H2

Your Turn 25.21

Think

What radicals are involved in the propagation steps of the mechanism for Equation 25-32? What is meant by a termination step? For a termination step, are radicals present in the products, in the reactants, or in both?

Solve

A termination step destroys free radicals that participate in the propagation cycle. Equation 25-33 shows that the radicals involved in the propagation cycle include Br• and the bromoalkyl radical (containing a C•) produced after radical addition. In a termination step, two of these free radicals come together as reactants, and no radicals are present in the product. The Br–Br bond, C–C bond, or C–Br bond is formed from radical coupling of any of those radicals.

Termination 1 — Radical coupling

Termination 2 — Radical coupling

Termination 3 — Radical coupling

Your Turn 25.22

Think

Review the reactants and products of initiation, termination, and propagation. Which step generates a radical? Which step terminates a radical? Which step has radicals in both the reactant and product? How do the reactions in Figure 25-13 compare to the mechanisms shown in Equations 25-33a, 25-33b, 25-33c, and 25-33d?

Solve

The initiation step matches Equations 25-33a and 25-33b, which are responsible for producing the first radical that is part of the propagation cycle. The propagation steps match either Equation 25-33c or Equation 25-33d (shown in the figure below).

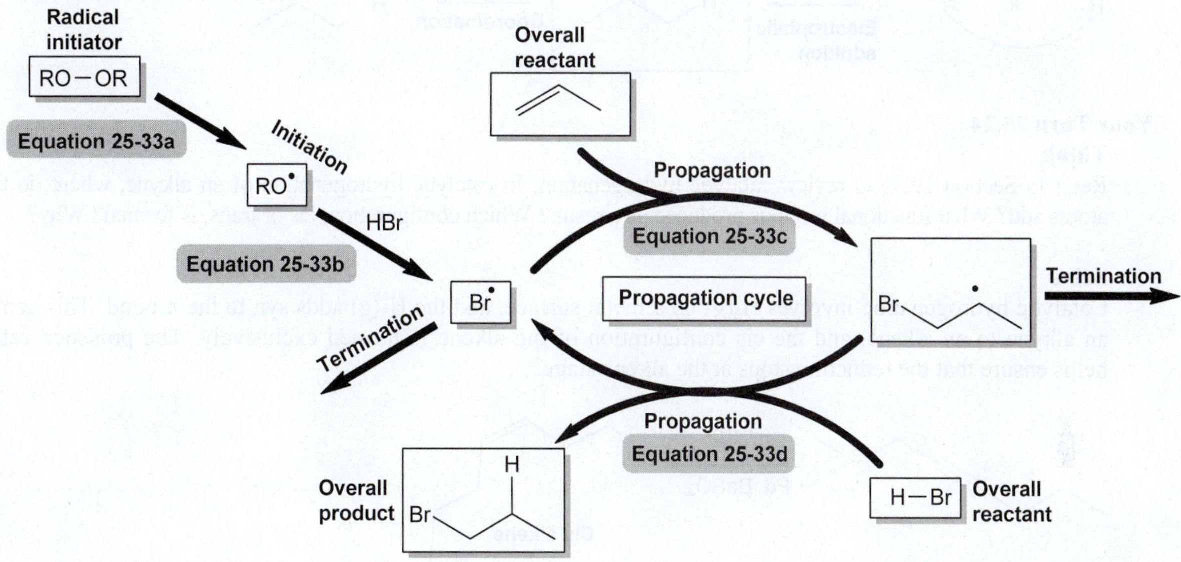

Your Turn 25.23

Think

Which mechanism uses the single-barbed arrows and involves radical intermediates? Which mechanism involves a double-barbed arrow and carbocation intermediates? What type of radical or carbocation intermediate is formed in each reaction? What is the species that adds to the C=C bond in the first step of each reaction? Are they the same or different?

Solve

The radical and closed-shell mechanisms for the addition of HBr to propene are shown on the next page. The regiochemistry of the addition of HBr is different depending on the mechanism. In the radical mechanism, single-barbed curved arrows are used. The Br• radical adds to the alkene first at the 1° C to form the new σ Br–C and yield the more stable 2° C• intermediate. The new σ C–H bond is formed in the second step of the mechanism. In the closed-shell mechanism, double-barbed curved arrows are used. H^+ of HBr adds to the alkene first at the 1° C to form the new σ H–C and yield the more stable 2° C^+ intermediate. The new σ C–Br bond is formed in the second step of the mechanism. In both cases, a 2° reactive intermediate forms in the first step. The difference is the species that adds in the first step: Br• in the radical mechanism and H^+ in the closed-shell mechanism.

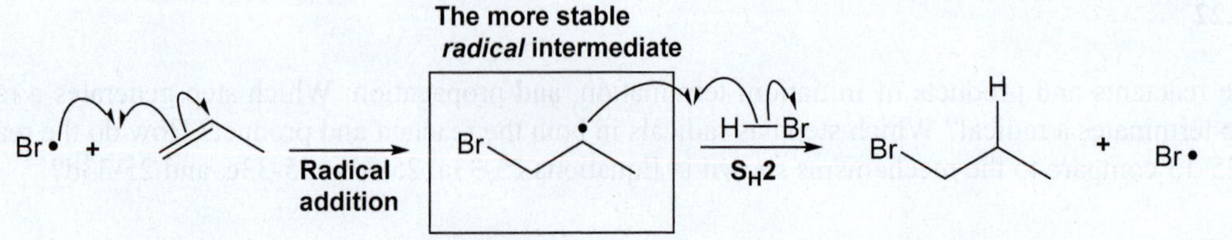

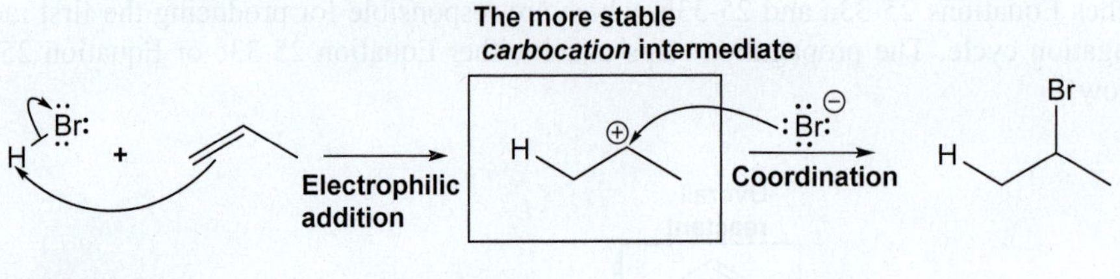

The more stable *radical* intermediate

Br• + [propene] → Radical addition → [Br–CH₂–ĊH–CH₃] + H—Br → S_H2 → Br–CH₂–CH(H)–CH₃ + Br•

The more stable *carbocation* intermediate

H–Br: + [propene] → Electrophilic addition → [H–CH₂–C⁺H–CH₃] + :Br:⁻ → Coordination → H–CH₂–CH(Br)–CH₃

Your Turn 25.24

Think

Refer to Section 19.5b to review catalytic hydrogenation. In catalytic hydrogenation of an alkyne, where do the H atoms add? What functional group is produced as a result? Which configuration, cis or trans, is formed? Why?

Solve

Catalytic hydrogenation involves $H_2(g)$ on a metal surface, and the $H_2(g)$ adds syn to the π bond. This converts an alkyne to an alkene, and the cis configuration of the alkene is formed exclusively. The poisoned catalyst helps ensure that the reduction stops at the alkene stage.

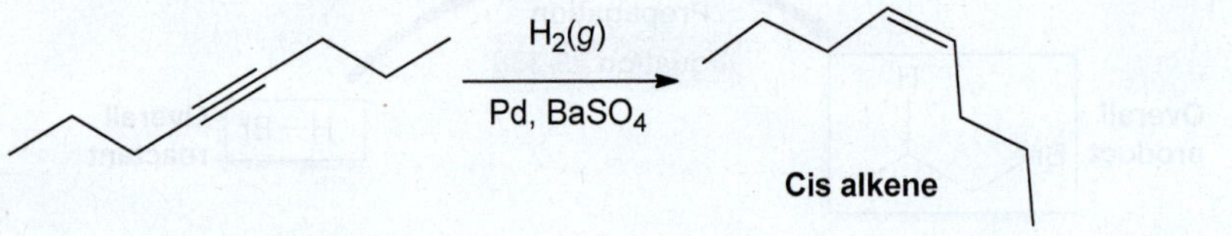

$H_2(g)$; Pd, $BaSO_4$

Cis alkene

In Chapter Problems

Problem 25.1

Think

In homolysis, what type of curved arrow is used (single or double barbed) to show bond breaking? How many electrons are in a bond? How many electrons go to each atom connected to the bond? How many curved arrows are required?

Solve

Homolysis uses the single-barbed arrow. Two electrons are in each bond, and one electron goes to each atom in the bond, so two curved arrows are required.

(a) $H_3C–CH_3$ —Homolysis→ $H_3C\bullet$ + $\bullet CH_3$

(b) $H_3C–CH_2–H$ —Homolysis→ $H_3C–\dot{C}H_2$ + $H\bullet$

(c) $H_3C–CH(Br)–CH_3$ —Homolysis→ $H_3C–\dot{C}H–CH_3$ + $\bullet Br$

(d) $C_6H_5–H$ —Homolysis→ $C_6H_5\bullet$ + $H\bullet$

Problem 25.3

Think

What are the distinct types of bonds in the molecule? Which is the weakest? What are the products upon homolysis of that bond?

Solve

There are three distinct bonds: H–OCH_3, H–CH_2OH, and H_3C–OH. They have the following respective bond dissociation energies: 440 kJ/mol (105 kcal/mol), 402 kJ/mol (96 kcal/mol), and 385 kJ/mol (92 kcal/mol). The weakest bond is the C–O bond, so it is most likely to break in a homolysis reaction.

402 kJ/mol (H–C); 440 kJ/mol (O–H); 385 kJ/mol (C–O)

$H_3C–OH$ —Homolysis→ $H_3C\bullet$ + $\bullet OH$

Problem 25.4

Think

What is the structure of 1,3,5-trimethylcyclohexane? How many distinct H–C bonds are present? What is the type of carbon (methyl, 1°, 2°, 3°)? Which type of C atom stabilizes an unpaired electron the most? The least? Why?

Solve

Three chemically distinct H atoms can be removed. Radical stability decreases in the order 3° > 2° > 1°. This is because a C• is electron poor and each additional alkyl group is electron donating.

Primary Tertiary Secondary

1,3,5-Trimethylcyclohexane

Primary
Least stable radical

Secondary

Tertiary
Most stable radical

Problem 25.5

Think

Use Figure 25-3 and Your Turn 25.3 as guides. What is the identity of the ethyl radical and the allyl radical? From which alkane can each alkyl radical be produced by homolysis? What are the respective bond dissociation energies? Which bond dissociation energy is lower? Why?

Solve

The ethyl radical can be produced by homolysis of a C–H bond in ethane, and the allyl radical can be produced by homolysis of the methyl C–H bond in propene. Bond dissociation energies are from Table 25-1: ethane C–H = 421 kJ/mol (101 kcal/mol) and propene methyl C–H = 369 kJ/mol (88 kcal/mol). Because the allyl radical requires less energy to be formed, it is more stable than the primary ethyl radical. The allyl radical is more stable owing to the presence of resonance structures that delocalize the unpaired electron over two C atoms.

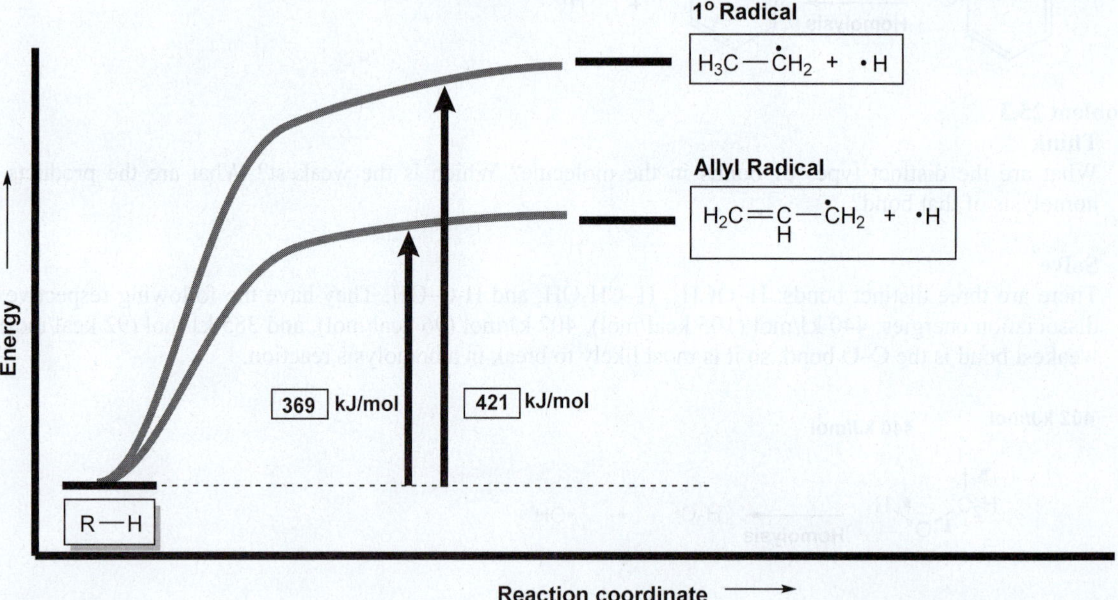

Problem 25.6

Think

Which carbon in Figure 25-5 has the most electron spin density (deepest blue)? Examine the resonance structures for the benzyl radical. What is special about the resonance structure in which the unpaired electron is on that carbon?

Solve

The primary (CH_2) carbon possesses the most spin density (it has the deepest blue color in Fig. 25-5). Examination of the resonance contributors (Equation 25-10) shows that the first radical retains the aromatic character of the benzene ring, whereas the others do not. This contributor is more stable and contributes more to the resonance hybrid.

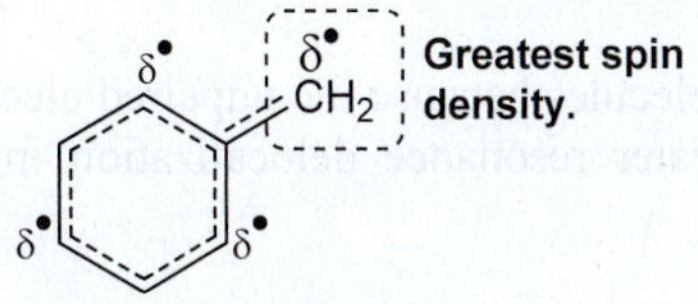

Greatest spin density.

Problem 25.8

Think

What are the distinct types of C–H bonds? What are the products upon homolysis of each of those bonds? Can the stabilities of the product radicals be differentiated based on resonance and/or alkyl substitution?

Solve

There are six distinct C–H single bonds, labeled **A–F** below. The corresponding homolysis products are shown on the right. Only one C–H bond can be broken to produce a resonance-stabilized alkyl radical. Thus, homolysis of C–H bond **D** gives the most stable organic radical as a product, making C–H bond **D** the weakest of the C–H bonds.

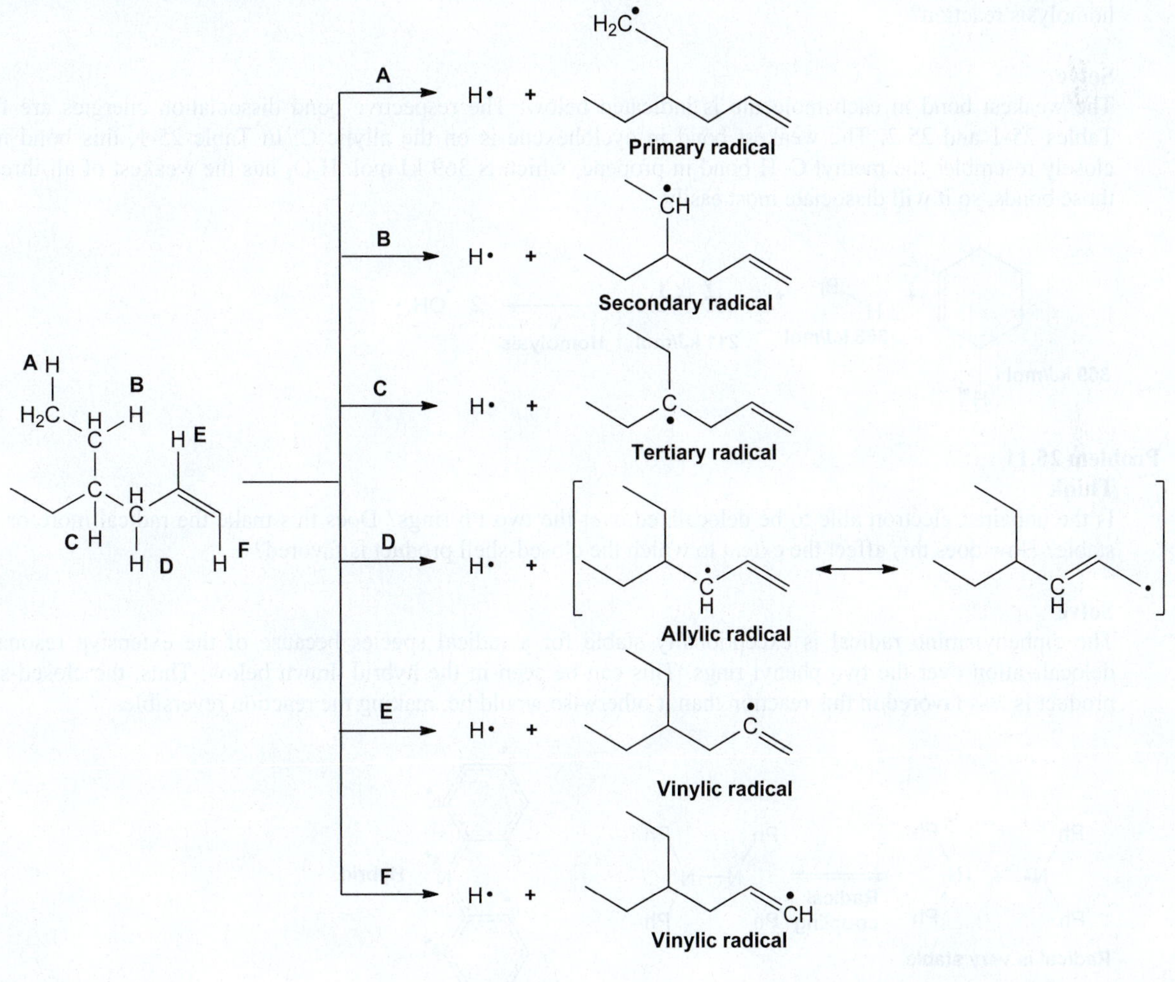

A → H• + Primary radical

B → H• + Secondary radical

C → H• + Tertiary radical

D → H• + Allylic radical

E → H• + Vinylic radical

F → H• + Vinylic radical

Problem 25.9

Think

What are the distinct types of C–H bonds? What are the products upon homolysis of each of those bonds? Can the stabilities of the product radicals be differentiated based on resonance and/or alkyl substitution? Which allylic radical has more resonance structures?

Solve

The allylic hydrogens circled below are the weakest in their respective molecules because the unpaired electron of the resulting radical can be delocalized by resonance. There is greater resonance delocalization in the resulting radical of the second molecule, so its C–H bond energy is lower.

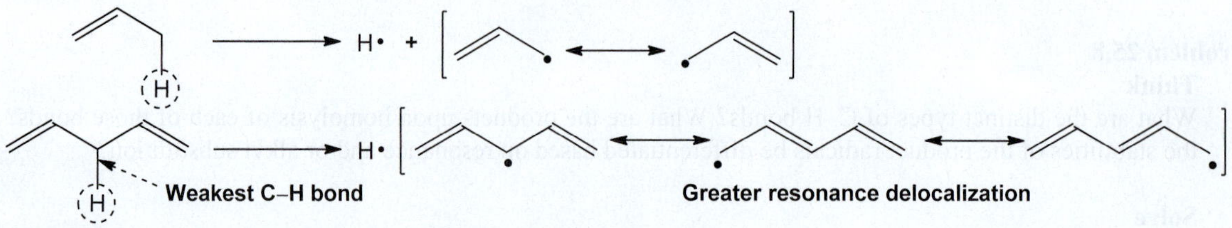

H• +

H•

Weakest C–H bond

Greater resonance delocalization

Problem 25.10

Think

What bond in cyclohexene is the weakest? In HBr? In H_2O_2? Which of those three bonds is the weakest? What is the product of that homolysis reaction? How does bond strength correspond to the likelihood of undergoing a homolysis reaction?

Solve

The weakest bond in each molecule is indicated below. The respective bond dissociation energies are from Tables 25-1 and 25-2. The weakest bond in cyclohexene is on the allylic C. In Table 25-1, this bond most closely resembles the methyl C–H bond in propene, which is 369 kJ/mol. H_2O_2 has the weakest of all three of those bonds, so it will dissociate most easily.

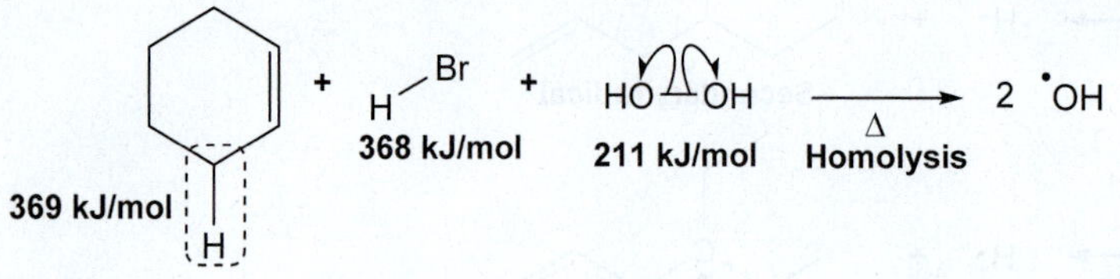

369 kJ/mol + H–Br 368 kJ/mol + HO–OH 211 kJ/mol → (Δ, Homolysis) 2 •OH

Problem 25.11

Think

Is the unpaired electron able to be delocalized over the two Ph rings? Does this make the radical more or less stable? How does this affect the extent to which the closed-shell product is favored?

Solve

The diphenylamino radical is exceptionally stable for a radical species because of the extensive resonance delocalization over the two phenyl rings. This can be seen in the hybrid drawn below. Thus, the closed-shell product is less favored in this reaction than it otherwise would be, making the reaction reversible.

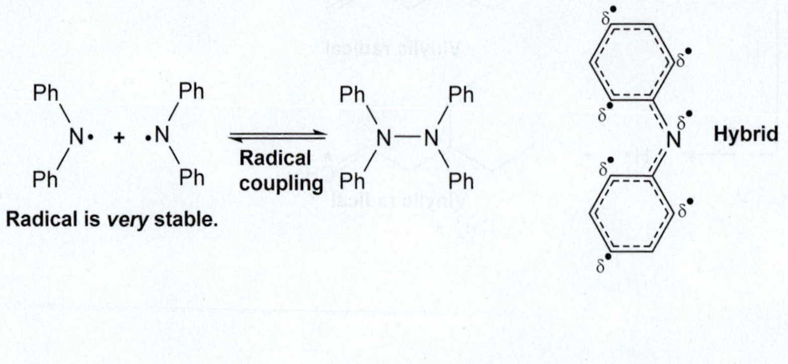

$Ph_2N\bullet$ + $\bullet NPh_2$ ⇌ (Radical coupling) $Ph_2N–NPh_2$

Radical is *very* stable.

Hybrid

Problem 25.12

Think

Which atom(s) can be attacked by Cl•? Which bond(s) can be broken? In an S_H2 step, should you use single-barbed curved arrows or double-barbed ones? How many curved arrows should be drawn?

Solve

We can think of the Cl• attacking either a H atom or a C atom. If it attacks a H atom, then a H atom is abstracted and the ethyl radical is displaced. If it attacks a C atom, either H or CH_3 is displaced. Each S_H2 step incorporates three single-barbed curved arrows.

$$H_3C{-}CH_2{-}H + {\bullet}Cl \xrightarrow{S_H2} H_3C{-}\dot{C}H_2 + H{-}Cl$$

$$H_3C{-}CH_3 + {\bullet}Cl \xrightarrow{S_H2} {\bullet}CH_3 + H_3C{-}Cl$$

$$H_3C{-}CH_2{-}H + {\bullet}Cl \xrightarrow{S_H2} H_3C{-}CH_2Cl + {\bullet}H$$

Problem 25.13

Think

What radicals are possible in Equation 25-17? What radicals couple in the termination steps shown in Equations 25-20a and 25-20b? Which other pairing of radicals is left?

Solve

Two radicals are part of the propagation cycle: the Cl• and the cyclohexyl radical. In Equation 25-20a, two Cl• radicals couple, and in Equation 25-20b, a Cl• radical combines with the cyclohexyl radical. The radical coupling reaction that has yet to be included is the coupling of two cyclohexyl radicals.

cyclohexyl• + •cyclohexyl → (Termination; Radical coupling) → bicyclohexyl

Problem 25.15

Think

For the initiation step, what is the weakest bond in the reactants that can undergo homolysis? For the propagation steps, what S_H2 step can the initial radical undergo to produce the HCl product? What S_H2 step can the resulting radical undergo to produce the CH_3CH_2Cl product and regenerate the initial radical? For the termination steps, what possible radical coupling steps can take place?

Solve

The weakest bond present is the Cl–Cl bond, which undergoes homolysis in an initiation step to produce two Cl• radicals.

$$Cl{-}Cl \xrightarrow[h\nu \text{ Homolysis}]{\text{Initiation}} Cl{\bullet} + {\bullet}Cl$$

In the first of two propagation steps (shown on the next page), a Cl• abstracts a hydrogen from CH_3CH_3 to produce one overall product, HCl, and an ethyl radical, $\bullet CH_2CH_3$. In the second propagation step, the $\bullet CH_2CH_3$ abstracts a Cl atom from Cl_2 to produce the second overall product, $ClCH_2CH_3$, and regenerate the initial Cl•.

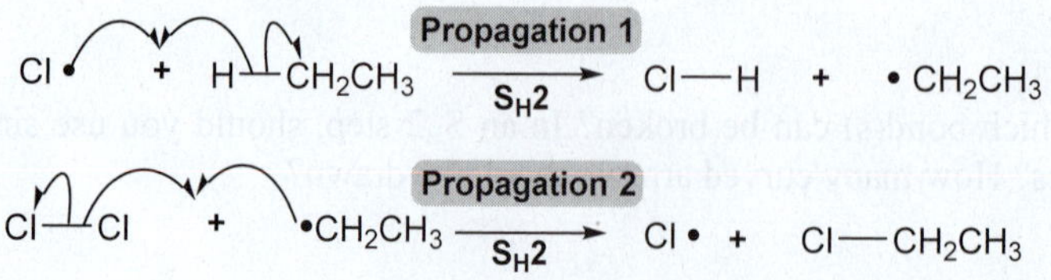

Termination steps can be a radical coupling between any two radicals that appear, such as the following three:

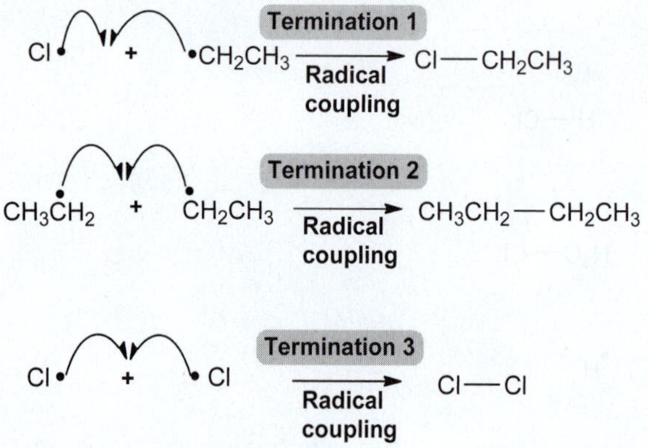

Problem 25.17

Think

What is the initial radical that is formed? By what process? How will that radical interact with the uncharged molecules present? Is regiochemistry a concern? Which hydrogen can be abstracted to produce the most stable alkyl radical?

Solve

Br_2 will undergo homolysis, as suggested by the presence of light, so the initial radical is Br•.

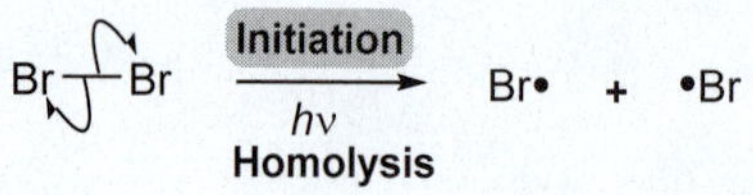

With the alkylbenzene present, radical halogenation will occur via two propagation steps. The Br• abstracts a H atom to yield a benzylic radical, R•. That benzylic radical then abstracts a Br atom from Br_2 to yield $PhCH(Br)CH_2CH_3$ and Br•.

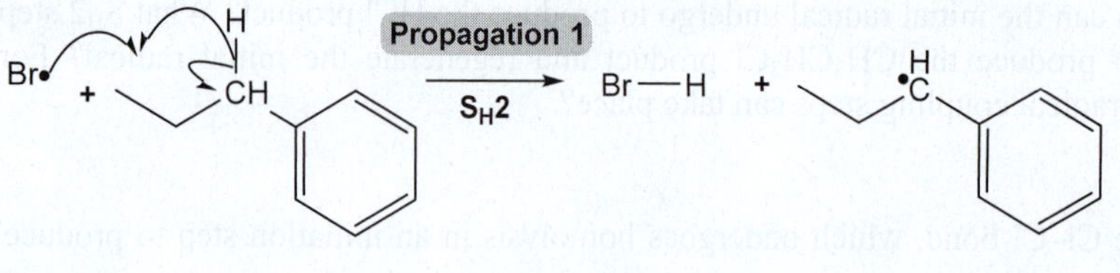

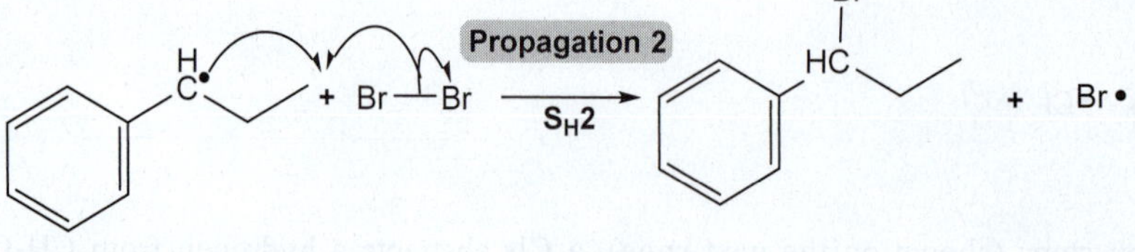

Regiochemistry is a concern because multiple, distinct H atoms can be abstracted by Br•. Recall, however, that Br• is highly selective in hydrogen abstractions, favoring the production of the most stable alkyl radical. In this case, the most stable alkyl radical is produced from abstracting a hydrogen atom attached to the benzylic carbon, as shown above.

Termination steps can be a radical coupling between any two radicals that appear, such as the following:

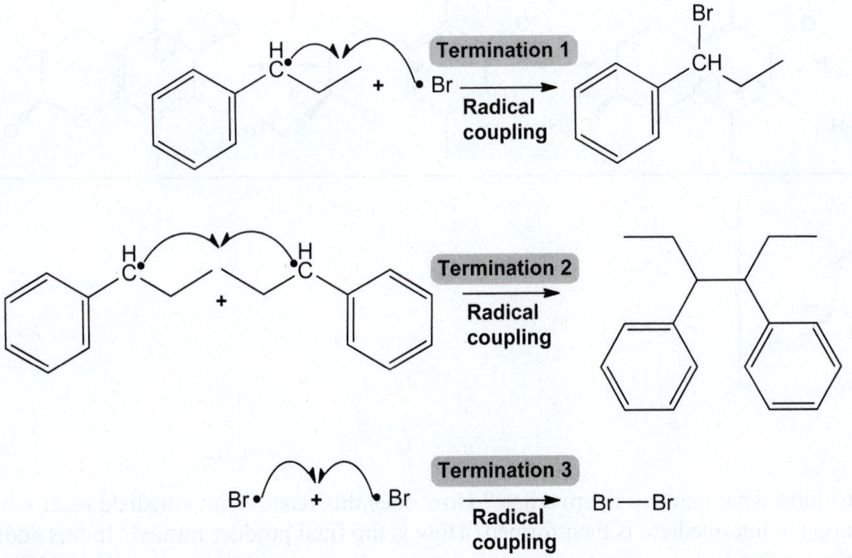

Termination 1
Radical coupling

Termination 2
Radical coupling

Termination 3
Radical coupling

Problem 25.18

Think

What type of reagent is NBS? Which C–H bond in the cycloalkane is most likely to undergo hydrogen atom abstraction? Why? What intermediate is formed? Is resonance possible?

Solve

NBS is a source of a small, steady concentration of Br_2, which produces some Br• upon irradiation with light. The tertiary allylic C–H bond will undergo hydrogen abstration to form the most stable allylic radical. The allylic radical has another resonance structure, and each can abstract a Br atom from Br_2 to form the final product, as shown below.

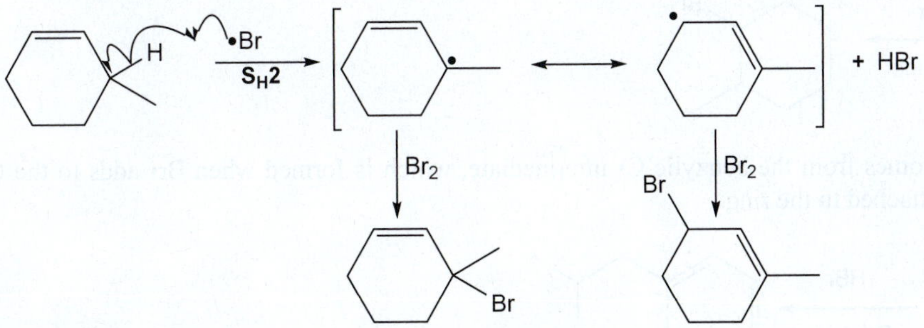

S_H2 + HBr Br_2 Br_2

Problem 25.19

Think

Refer to Your Turn 25.19. Is there an atom with an unpaired electron attached to an atom that is joined to a third by a π bond? Do you need to use single-barbed curved arrows or double-barbed curved arrows to show electron movement? Where do the electrons from the adjacent π bond end up? How many curved arrows are necessary to show that electron movement? Do those arrows point in the same direction?

Solve

In the first structure, O has an unpaired electron and is attached to C=C. Three single-barbed curved arrows are needed to show the electron movement to arrive at the next resonance structures. In the second resonance structure, C has an unpaired electron and is attached to another C=C, so another resonance structure is drawn. This electron movement can continue around the ring.

$C_{16}H_{33}$ $C_{16}H_{33}$ $C_{16}H_{33}$ $C_{16}H_{33}$

$C_{16}H_{33}$

Problem 25.20

Think

HBr and peroxides react to form what reactive intermediate? How does this reactive intermediate react with an alkene C=C bond? What organic intermediate is then formed? How is the final product formed? In this addition reaction, is regiochemistry a concern?

Solve

HBr and peroxides are a source of Br• radical, which reacts with an alkene via radical addition to form the most stable C• intermediate. The C• intermediate reacts with an equivalent of H–Br to form a new C–H bond. This leads to anti-Markovnikov addition of HBr to the C=C. The mechanism goes through the most stable radical intermediate, just as electrophilic addition of HBr goes through the most stable carbocation intermediate. In this reaction, the Br• adds first and the H• adds second. In electrophilic addition of an alkene with H–Br, the H^+ adds first and the Br^- adds second.

(a) The major product comes from the tertiary C• intermediate, which is formed when Br• adds to the secondary C=C carbon.

HBr / Peroxide → Br

(b) The major product comes from the benzylic C• intermediate, which is formed when Br• adds to the C=C carbon not directly attached to the ring.

HBr / Peroxide → Br

Problem 25.21

Think

Are there any chiral centers in the organic reactant? How many chiral centers are in the organic product? Are those stereocenters formed in the same elementary step or in different steps? When each chiral center is formed, is one configuration produced, or is a mixture of configurations produced?

Solve

(a) HBr and peroxides produce a small amount of Br• as a reactive intermediate. Then an organic radical intermediate is formed from radical addition of Br• to the C=C bond. In that step, the carbon that gains the new C–Br bond becomes a chiral center, and both configurations are produced. The C• radical then reacts with H–Br to form a new C–H bond. In that step, a second carbon becomes a chiral center, and both of its configurations are produced. This leads to four possible stereoisomer products.

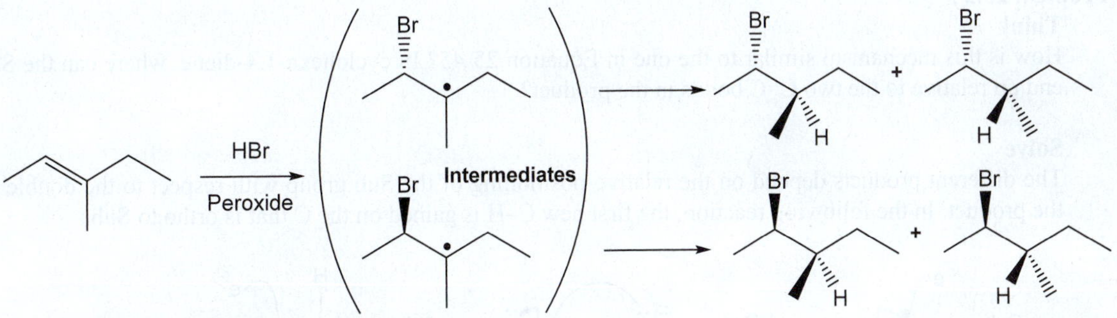

HBr
Peroxide
Intermediates

(b) NBS is a source of low, steady amounts of Br_2, which interacts with light to produce Br•. Then an organic radical intermediate is formed when Br• abstracts a hydrogen to form the secondary allylic C•. This intermediate has two resonance structures, but owing to symmetry, they both lead to the same enantiomeric products. When the C• gains a Br from Br_2, that C becomes a chiral center, and both configurations are produced.

Br_2
$h\nu$

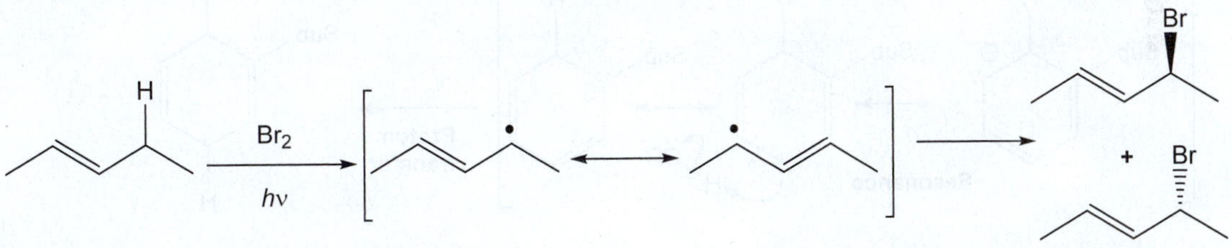

Problem 25.22

Think

Is the cis or trans alkene produced from the reaction of a dissolving metal reduction of an alkyne? Which C=C bond, therefore, was the C≡C in the initial alkyne?

Solve

The trans C=C bond is the one that is produced from dissolving metal reduction of an alkyne, so that was the C≡C in the starting compound.

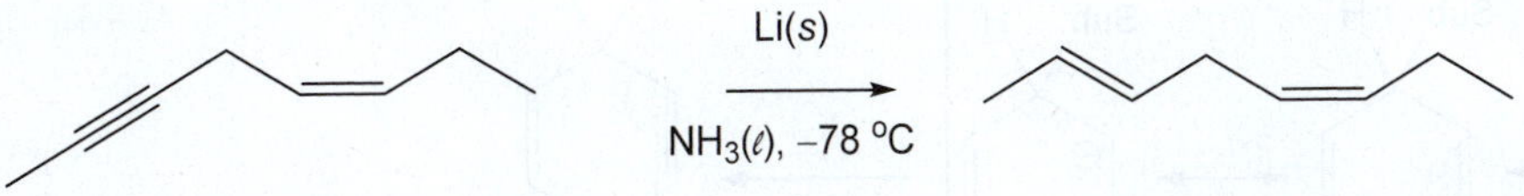

Li(*s*)
$NH_3(\ell)$, –78 °C

Problem 25.23

Think

How is this mechanism similar to the one in Equation 25-40? How does the radical addition reaction occur? How many H atoms are added across the C=O? What is the source of the H atoms?

Solve

This mechanism (shown on the next page) is essentially identical to the one in Equation 25-40. The mechanism begins with a solvated electron adding via radical addition to the C=O. This converts the C=O to an O• and C^-. A proton transfer reaction follows to form the C–H bond. The O• couples with another solvated electron to form the $C:^-$, which is a strong base. Another proton transfer reaction results.

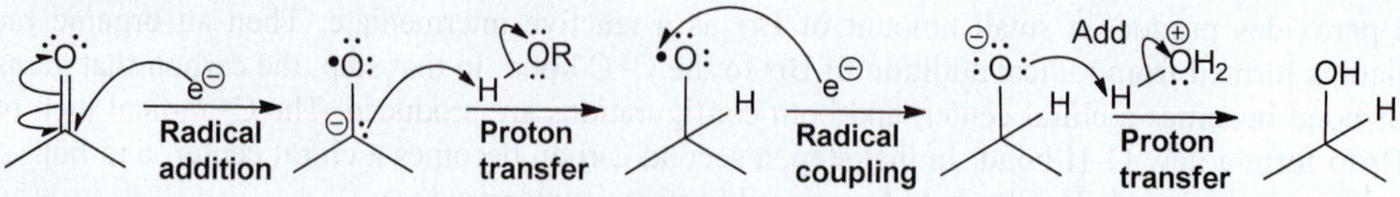

Radical addition
Proton transfer
Radical coupling
Add
Proton transfer

Problem 25.24

Think

How is this mechanism similar to the one in Equation 25-45? In cyclohexa-1,4-diene, where can the Sub group end up relative to the two C=C bonds in the product?

Solve

The different products depend on the relative positioning of the Sub group with respect to the double bonds in the product. In the following reaction, the first new C–H is gained on the C that is ortho to Sub.

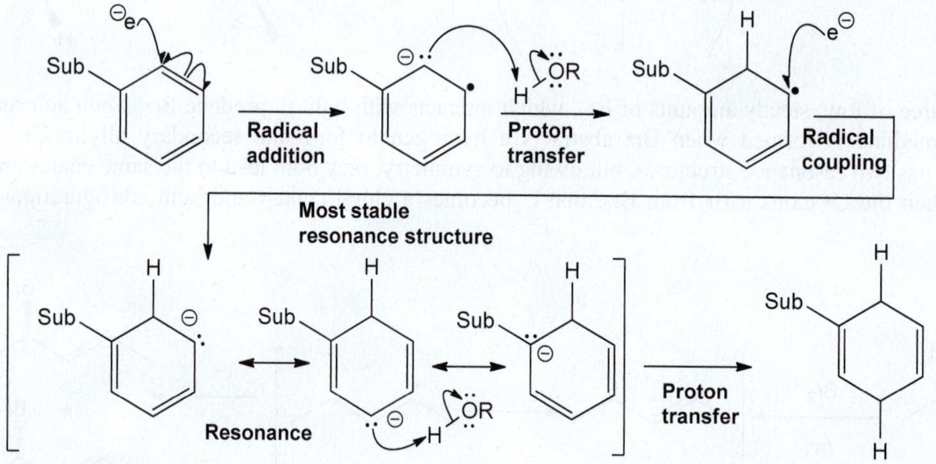

Radical addition
Proton transfer
Radical coupling
Most stable resonance structure
Resonance
Proton transfer

In the following reaction, the first new C–H is gained on the C that is directly attached to Sub.

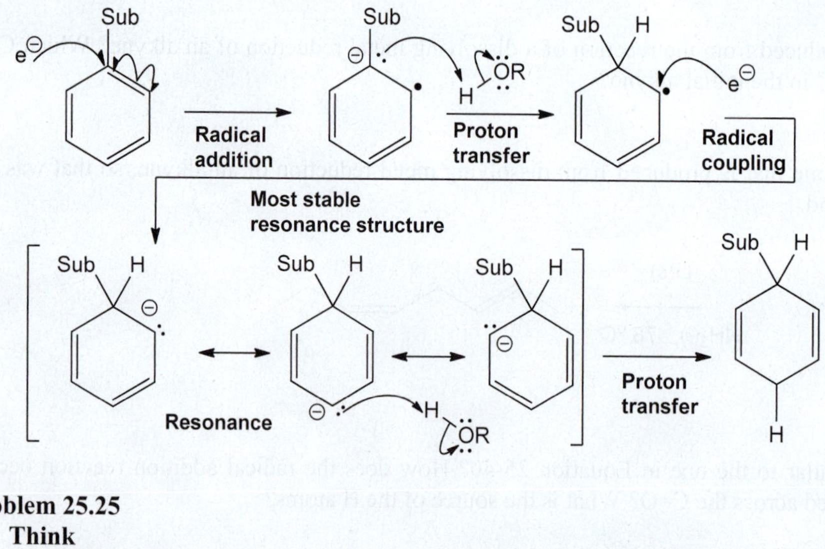

Radical addition
Proton transfer
Radical coupling
Most stable resonance structure
Resonance
Proton transfer

Problem 25.25

Think

Which C–H bond is most likely to undergo radical abstraction? Which C• would be most stable? Where does the C–Br bond form? How many times does this substitution occur? What functional group is present in the product that is not present in the reactant? How is that functional group produced from the synthetic intermediate in the presence of a base?

Solve

(a) Bromination occurs at two benzylic positions. The following are possible isomers that can form:

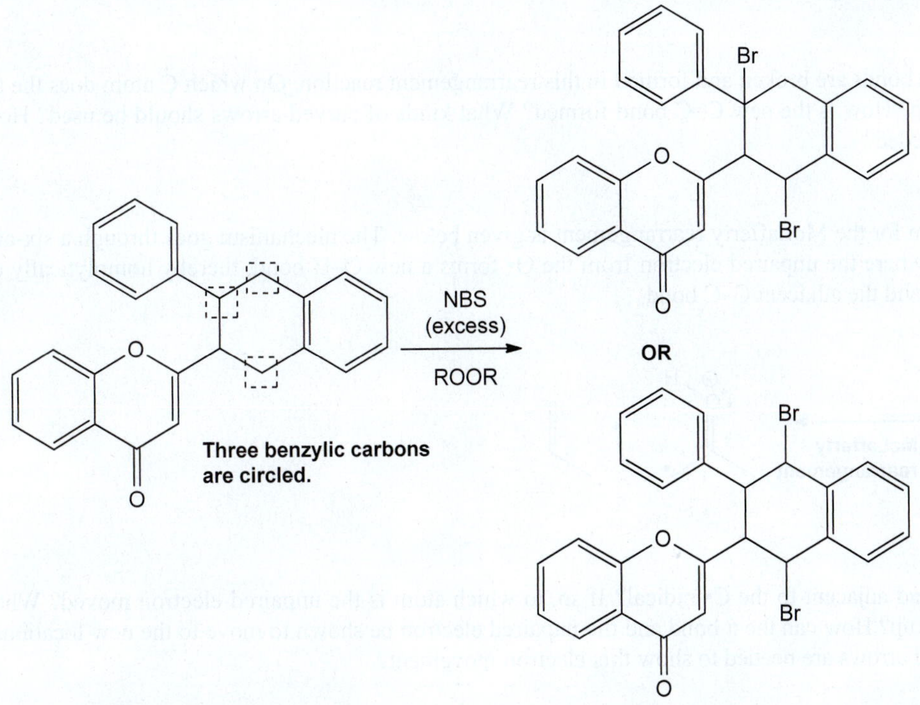

NBS (excess) ROOR

Three benzylic carbons are circled.

OR

(b) Both isomers undergo E2 or E1 reactions twice to form two new C=C bonds. The amine serves as the base for such reactions. The resulting product has a new naphthalene ring.

End of Chapter Problems

Sections 25.1–25.3 Free Radicals: Curved Arrow Notation and Radical Initiators; Structure and Stability; Common Elementary Steps

Problem 25.27

Think

Identify which bonds are broken and formed in this rearrangement reaction. On which C atom does the unpaired electron end up? How is the new C=C bond formed? What kinds of curved arrows should be used? How many of them are needed?

Solve

The mechanism for the McLafferty rearrangement is given below. The mechanism goes through a six-atom ring conformation where the unpaired electron from the O• forms a new O–H bond, thereby homolytically cleaving the H–C bond and the adjacent C–C bond.

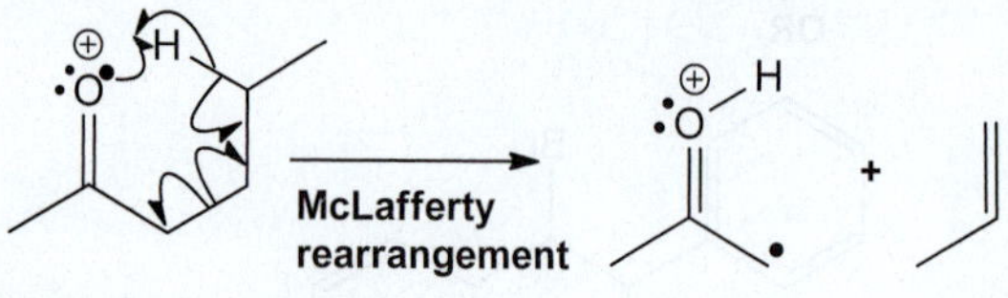

McLafferty rearrangement

Problem 25.28

Think

Is there a π bond adjacent to the C• radical? If so, to which atom is the unpaired electron moved? Where does the π bond end up? How can the π bond and the unpaired electron be shown to move to the new locations? What kinds of curved arrows are needed to show this electron movement?

Solve

There is a π bond adjacent to the C• radical. The resonance structure therefore has a C=C bond, and the unpaired electron then ends up on the O atom. Three single-barbed curved arrows are necessary to show this electron movement.

Resonance

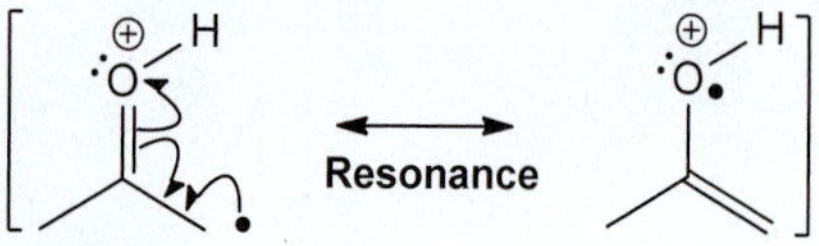

Problem 25.29

Think

What is the weakest bond? Which bond undergoes homolysis in an initiation step? How can this radical react with itself to form CO_2 and R• radical? What kinds of curved arrows should you use?

Solve

The key first step is the breaking of the O–O bond, which is the weakest in the molecule. Two single-barbed curved arrows are used for this step. Generation of the alkyl radical in the second step is facilitated by the production of CO_2, which is very stable and is a gas, so it irreversibly leaves the reaction mixture. Three single-barbed curved arrows are used to show the formation of the new C=O π bond and the breaking of the C–C σ bond.

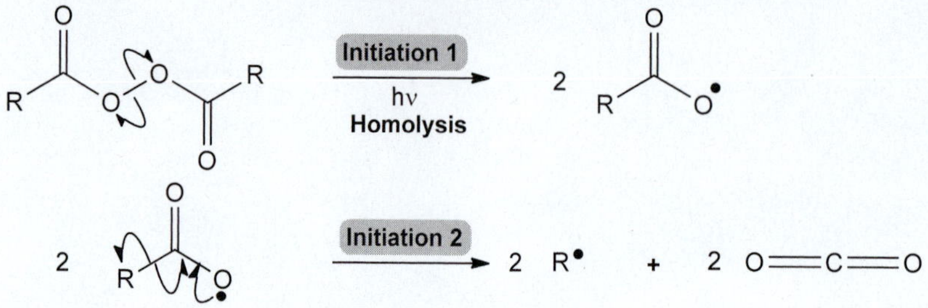

Initiation 1: hν, Homolysis → 2 RCO₂•

Initiation 2: 2 RCO₂• → 2 R• + 2 O=C=O

Problem 25.30

Think

Is a resonance structure possible for the allyl radical ($H_2C{=}CH{-}CH_2\bullet$)? What is the structure of the hybrid? Is the single bond 100% single? What does this mean for the rotation barrier?

Solve

The $CH_3CH_2{-}CH_3$ bond in propane has free rotation because all C atoms are sp^3 hybridized and all bonds are single bonds. The $H_2C{=}CH{-}CH_2\bullet$ bond in the allyl radical is not 100% a single bond, owing to the presence of resonance. The hybrid shows partial double-bond character, which is indicative of a partial π bond. Rotation breaks that partial π bond, which requires additional energy.

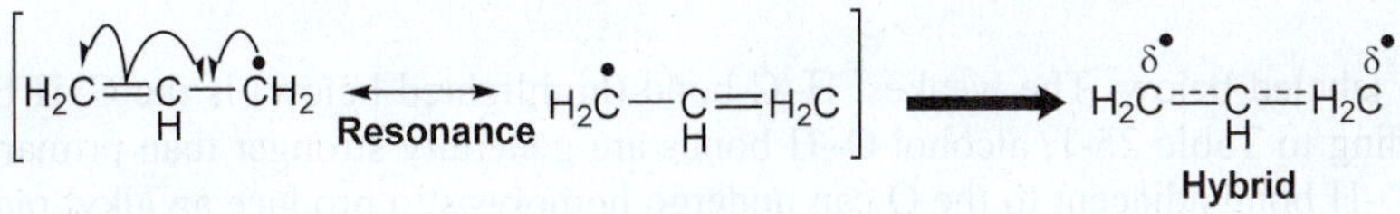

Resonance

Hybrid

Problem 25.31

Think

What bonds are formed and broken in this step? What kinds of curved arrows should be used? How many curved arrows are needed? How is this similar to an S_N2 mechanism?

Solve

The Cl• radical forms a new C–Cl bond to the C atom on the left of the ring, and the bottom C–C bond of the ring breaks. The unpaired electron ends up on the tertiary carbon. Three single-barbed curved arrows are necessary to show the simultaneous formation of the C–Cl bond and the breaking of the C–C bond. This is an S_H2 step because Cl• displaces a C•. The dynamics of an S_H2 reaction are similar to those of an S_N2 reaction, but the reaction arrows differ.

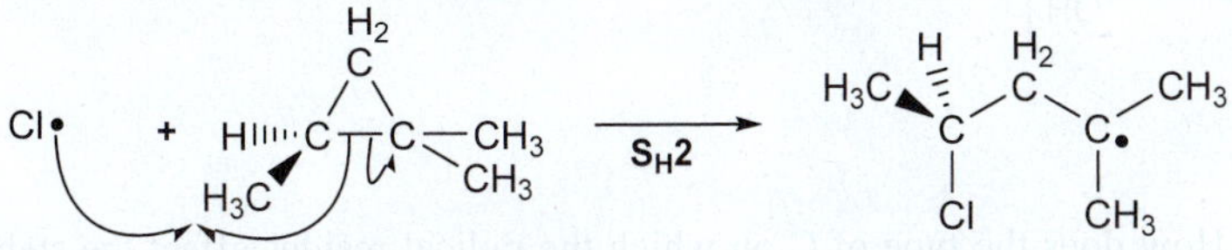

S_H2

Problem 25.32

Think

What type of C–H bonds are present? Refer to Table 25-1 to determine the different H–C bond energies. What is the trend for each type?

Solve

Bonds involving C generally become weaker when the C has more alkyl groups attached (i.e., 1° > 2° > 3°), because the resulting C• is stabilized by additional electron-donating alkyl groups. Bonds involving allylic or benzylic C atoms tend to be weaker because of the resonance stabilization in the resulting alkyl radical. Bonds tend to be stronger when the C is part of a C=C bond (i.e., vinylic or aromatic) owing to the additional *s* character.

(a) The types of H–C bonds are labeled below. The weakest H–C bond (highlighted below) is the tertiary allylic bond.

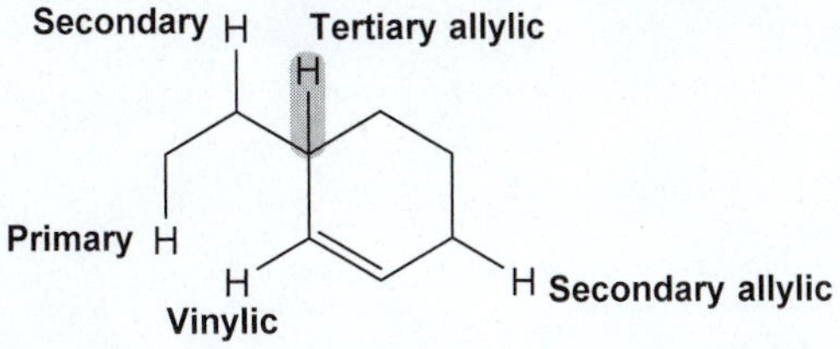

Secondary H

Tertiary allylic

Primary H

Vinylic

Secondary allylic

(b) The types of H–C bonds are labeled below. The weakest H–C bond (highlighted below) is the tertiary benzylic bond.

Primary
H
Tertiary benzylic
H
H Tertiary
H Secondary
H
Aromatic

(c) The types of H–C bonds are labeled below. The weakest H–C bond (highlighted below) is the C–H bond adjacent to the COH. According to Table 25-1, alcohol O–H bonds are generally stronger than primary or secondary C–H bonds. The C–H bond adjacent to the O can undergo homolysis to produce an alkyl radical that is resonance stabilized.

Secondary Primary
H H
H
Primary
OH
H
Secondary

Resulting alkyl radical

OH OH

Problem 25.33

Think

What types of C–C bonds are present? How does the type of C on which the radical resides affect the stability of the radical and, therefore, the bond strength? What is the trend for each type?

Solve

Bonds involving C generally become weaker when the C has more alkyl groups attached (i.e., 1° > 2° > 3°), because the resulting C• is stabilized by additional electron-donating alkyl groups. Bonds involving allylic or benzylic C atoms tend to be weaker because of the resonance stabilization in the resulting alkyl radical. Bonds tend to be stronger when the C is part of a C=C bond (i.e., vinylic or aromatic) owing to the additional *s* character.

(a) The types of C atoms are labeled below. The weakest C–C bond is one between the tertiary allylic C and a secondary C. Two such bonds are highlighted below.

Tertiary allylic
Secondary
Primary
Vinylic
Secondary allylic

(b) The types of C atoms are labeled below. The weakest C–C bond (highlighted below) is the bond between the tertiary benzylic C and the secondary C.

Tertiary benzylic
Primary
Tertiary
Secondary
Aromatic

(c) The types of C atoms are labeled below. The weakest C–C bond is the one on the right (highlighted below). That bond involves a secondary C and primary C atom, and the unpaired electron that develops on the C bonded to O is resonance delocalized on the O.

Secondary
Primary
OH
Primary
Secondary

Resulting alkyl radical

OH
OH

Problem 25.34

Think

What supplies the energy necessary to carry out homolysis? What is the weakest bond present? Do bonds involving Br tend to be stronger or weaker than bonds involving C or H? Do bonds involving an aromatic C tend to be stronger or weaker than bonds involving a benzylic C?

Solve

The UV light ($h\nu$) supplies the energy to break the weakest bond homolytically—in this case, the benzylic C–Br bond (highlighted below). According to Tables 25-1 and 25-2, C–Br bonds tend to be weaker than C–H or C–C bonds. Also, bonds involving benzylic C atoms tend to be weaker than bonds involving aromatic C atoms. The products of this step are two radicals C• and Br•.

Br
Initiation
$h\nu$
Homolysis
+ Br•
Br
Br

Problem 25.35

Think

What types of atoms are abstracted in an S_H2 step? What bonds are formed and broken? What kinds of curved arrows should be used? How many curved arrows are needed? How is this similar to an S_N2 mechanism?

Solve

Typically, H atoms or halogen atoms are abstracted in an S_H2 step. The dynamics of an S_H2 reaction are similar to those of an S_N2 reaction, but the reaction arrows differ—three arrows are required. The mechanisms for reactions **(a)** and **(b)** are shown on the next page.

(a) A H atom is abstracted by Br•. There is only one type of C–H bond. The Br• radical forms a new H–Br bond. The C• then results on the benzylic carbon.

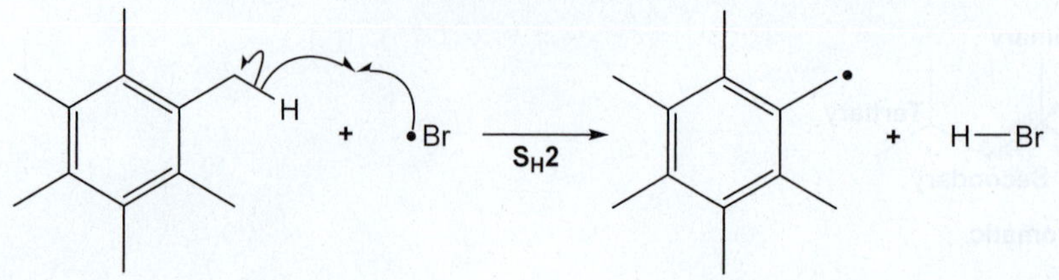

H + •Br $\xrightarrow{S_H2}$ + H—Br

(b) An I atom is abstracted by the C• radical. The C• radical forms a new C–I bond, and the unpaired electron then results on the iodine to form I•.

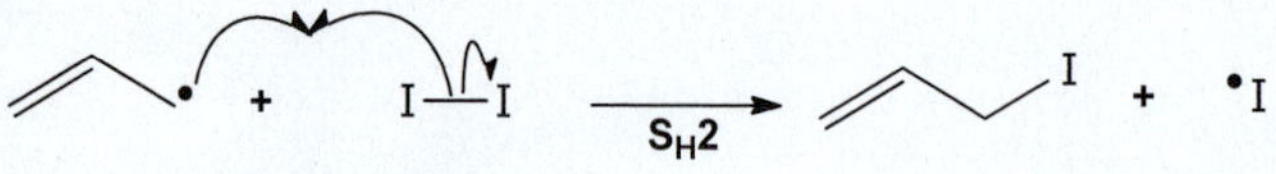

+ I—I $\xrightarrow{S_H2}$ I + •I

Problem 25.36

Think

What bonds break and form in a radical addition step? What types of curved arrows should be used? How many such curved arrows should be drawn? Which C atoms can gain the unpaired electron? Which radical is more stable?

Solve

In a radical addition step, the radical adds to one atom connected by a π bond, and the other atom connected by the π bond gains an unpaired electron.

(a) The two possible radical addition steps are shown below. The more stable product is the tertiary radical C• (boxed below).

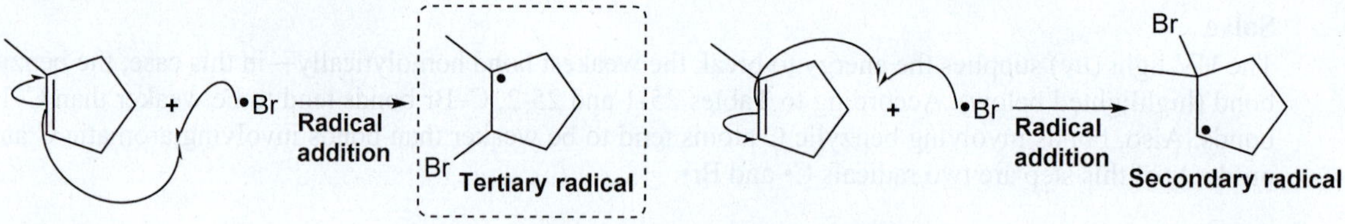

+ •Br → Radical addition → Br Tertiary radical

+ •Br → Radical addition → Br Secondary radical

(b) The two possible radical addition steps are shown below. The more stable product is the secondary allylic radical C• (boxed below).

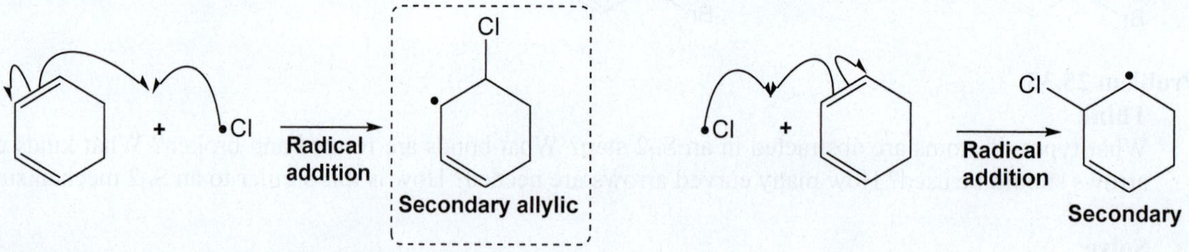

+ •Cl → Radical addition → Cl Secondary allylic

•Cl + → Radical addition → Cl Secondary

Section 25.4 Radical Halogenation of Alkanes: Synthesis of Alkyl Halides

Problem 25.37

Think

What are the products of the various H atom abstractions? What types of carbon atoms are present (methyl, 1°, 2°, 3°, allylic, benzylic)? Which C• would be most stable? How does that affect the likelihood of the H atom being abstracted?

Solve

The most reactive H in each molecule, shown below, is the one with the weakest bond and leads to the formation of the most stable C•. In **(a)** and **(b)**, the H is on a 3° carbon, which forms a 3° carbon radical. A 3° radical is more stable than a 1° or 2° radical. In **(c)**, the H is on a 2° benzylic C, and abstraction of the H forms a benzylic radical. A benzylic radical has the unpaired electron delocalized around the ring, making it more stable than a localized radical. Notice that another benzylic radical could form by abstracting a hydrogen from the benzylic CH_3, but the resulting radical would be less stable, because the radical would be 1°. And in **(d)**, the H is on an allylic C, so the resulting radical would be resonance stabilized.

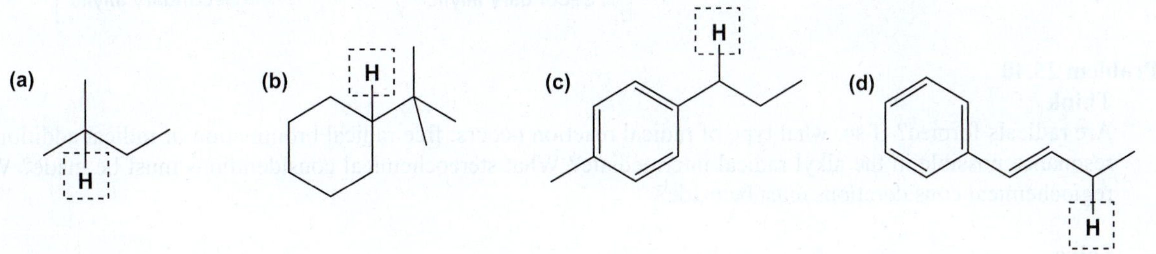

(a) H (b) H (c) H (d) H

Problem 25.38

Think

What is the index of hydrogen deficiency (IHD) of the compound? How many distinct types of reactive C–H bonds are present if only one monochloride isomer is formed? Is there symmetry in the molecule?

Solve

The IHD is 4, so a benzene ring is possible, which would account for six C atoms. Aromatic H atoms are typically not reactive under these conditions, but benzylic H atoms are. All H atoms that are sufficiently reactive must be equivalent, so the remaining three C atoms can be benzylic CH_3 groups, and they must be attached at the 1, 3, and 5 positions.

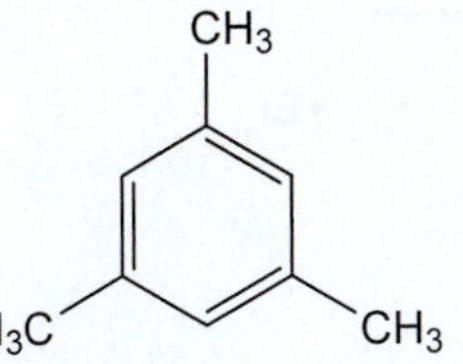

CH_3 H_3C CH_3

Chemical formula: C_9H_{12}

Problem 25.39

Think

What type of elementary step determines where radical substitution occurs? How does the type of C involved in a bond (methyl, 1°, 2°, 3°, allylic) affect the strength of the bond?

Solve

Bonds involving C generally become weaker when the C has more alkyl groups attached (i.e., 1° > 2° > 3°), because the resulting C• is stabilized by additional electron-donating alkyl groups. Bonds involving allylic or benzylic C atoms tend to be weaker because of the resonance stabilization in the resulting alkyl radical. Bonds tend to be stronger when the C is part of a C=C bond (i.e., vinylic or aromatic) owing to the additional *s* character. In this case, a H atom at the allylic position will be most easily abstracted. There are two distinct types of these H atoms, at C1 and C4. Abstraction at C4 leads to a radical in which the unpaired electron is delocalized on two secondary C atoms. Abstraction at C1 leads to a radical in which the unpaired electron is delocalized on one primary and one secondary C. So abstraction at C4 leads to the more stable radical. See the figure on the next page.

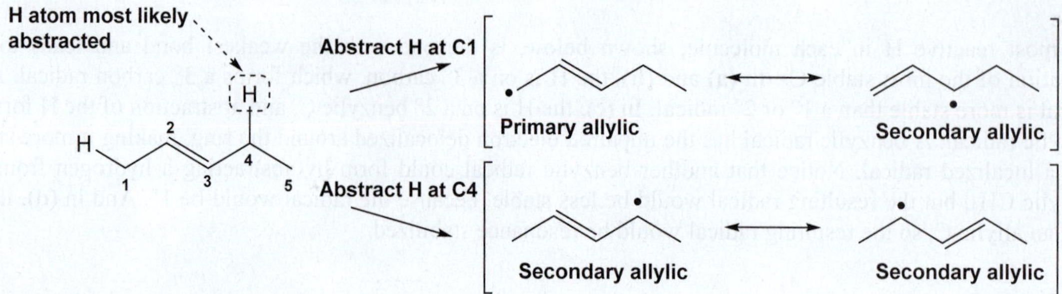

Problem 25.40

Think

Are radicals formed? If so, what type of radical reaction occurs: free radical bromination or radical addition? Is resonance possible in the alkyl radical intermediate? What stereochemical considerations must be made? What regiochemical considerations must be made?

Solve

The reactions in **(a)** and **(b)** are free radical substitution reactions, where H is replaced by Br. The initiation step to form Br• radical is the same for both **(a)** and **(b)**.

Br—Br → (Initiation, $h\nu$, Homolysis) Br• + •Br

The reaction in **(c)** is addition of Br_2 across a C=C bond.

(a) H abstraction takes place most readily at the 3° C. The product is achiral.

(b) H abstraction takes place most easily at the allylic H, and the resulting radical has two resonance contributors. Br can thus add at two different C atoms, giving rise to two constitutional isomers. Each constitutional isomer produced gains a new chiral center, and a mixture of both stereochemical configurations will be produced.

(c) No radicals are produced, so this mechanism involves only closed-shell species. This is an electrophilic addition of Br_2 across the C=C bond, going through a bromonium ion intermediate (Chapter 12).

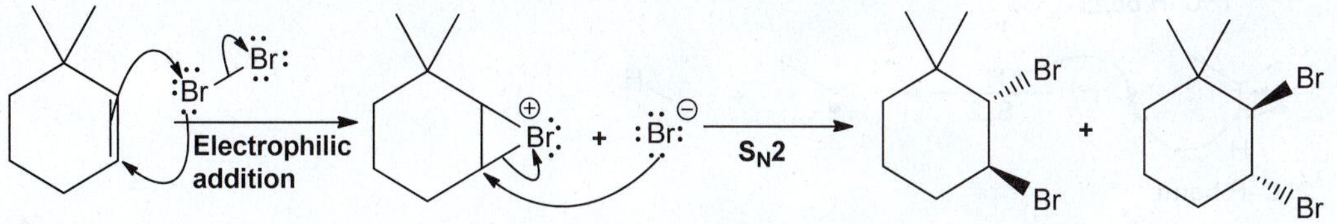

Electrophilic addition

S_N2

Problem 25.41 (SYN)

Think

Compare the formula of the starting reagent to the product shown. What is the difference? What bonds then need to be broken and formed to yield the target shown? What reaction will accomplish that?

Solve

(a) The formula for the starting material is C_9H_{12}, and the formula for the target compound is $C_9H_{11}Br$. Therefore, a halogenation reaction is required, in which the benzylic C–H bond is broken and a new benzylic C–Br bond is formed. NBS is a source of a small, steady concentration of Br_2, which produces some Br• upon irradiation with light. In the propagation cycle, the benzylic C–H bond will undergo hydrogen abstraction to form the most stable radical, which, in this case, is a benzylic radical. Then the C• will abstract Br from Br_2 to form the new C–Br bond.

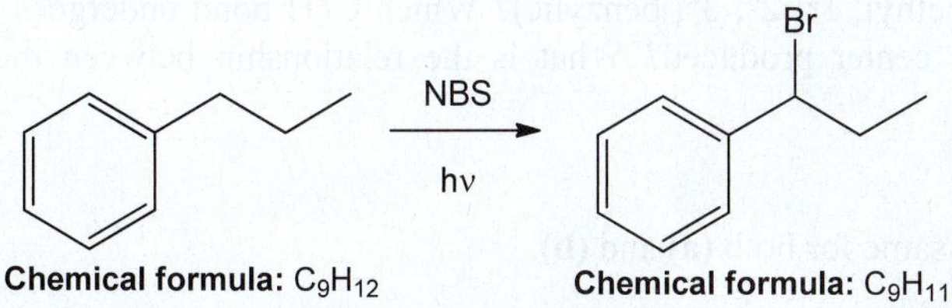

NBS, hν

Chemical formula: C_9H_{12} **Chemical formula:** $C_9H_{11}Br$

(b) The formula for the starting material is C_8H_{18}, and the formula for the target compound is $C_8H_{17}Cl$. Therefore, a halogenation reaction is required, in which one of the primary C–H bonds is broken and a new C–Cl bond is formed. Cl_2 produces Cl• upon irradiation with light. In the propagation cycle, the primary C–H bond will undergo hydrogen abstraction to form a C•, which will then abstract a Cl from Cl_2 to form the new C–Cl bond.

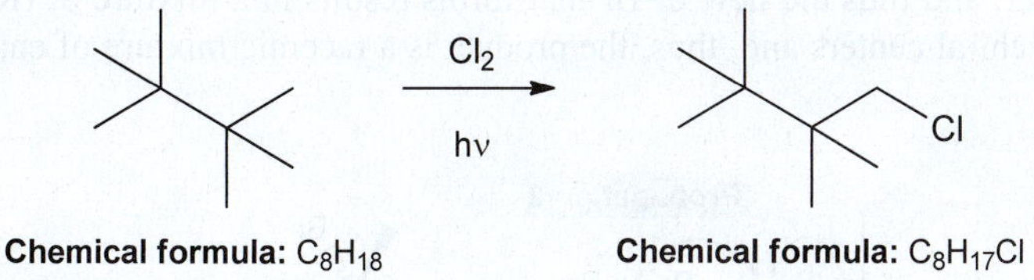

Cl_2, hν

Chemical formula: C_8H_{18} **Chemical formula:** $C_8H_{17}Cl$

Problem 25.42

Think

What types of carbon atoms are present (methyl, 1°, 2°, 3°, benzylic)? How many of each type of C–H bond are present? How does the number of C–H bonds affect the actual product distribution? How does the 5:1 selectivity affect the actual product distribution?

Solve

(a) and **(b)** The radical and curved arrow notation are given below for the two isomeric propyl radicals. One radical comes from hydrogen abstraction at the 1° C atom, and the other comes from hydrogen abstraction at the 2° C atom. See the mechanisms on the next page.

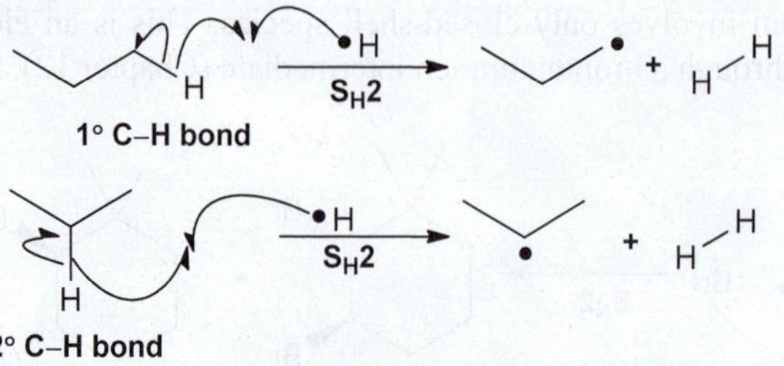

(c) The likelihood of abstracting each type of H increases with an increasing number of H atoms, and it also increases with an increasing selectivity toward abstracting that type of H.

Relative probability of abstracting a 1° H = six 1° H atoms × reactivity of 1 = 6
Relative probability of abstracting a 2° H = two 2° H atoms × reactivity of 5 = 10

Total = 16

Percentage of abstraction of 1° H: (6/16) × 100% = 37.5%
Percentage of abstraction of 2° H: (10/16) × 100% = 62.5%

Problem 25.43

Think

What types of carbon atoms are present (methyl, 1°, 2°, 3°, benzylic)? Which C–H bond undergoes hydrogen abstraction most readily? Is a new chiral center produced? What is the relationship between the isomers produced?

Solve

The initiation step to form Br• radical is the same for both **(a)** and **(b)**.

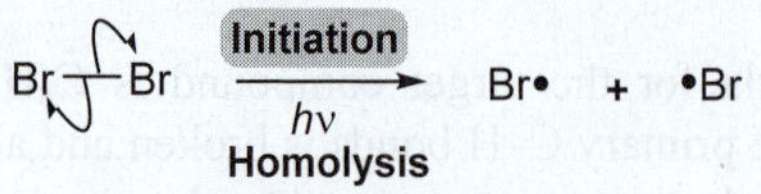

(a) In the propagation cycle, the benzylic C–H bond undergoes hydrogen abstraction, which leaves a C• on the benzylic carbon. Then the C• abstracts a Br from Br_2 to produce the new C–Br bond. In the second step, the C• behaves as if it is trigonal planar, and thus the new C–Br that forms results in a mixture of (R) and (S) configurations. There are no other chiral centers and, thus, the product is a racemic mixture of enantiomers and will not be optically active.

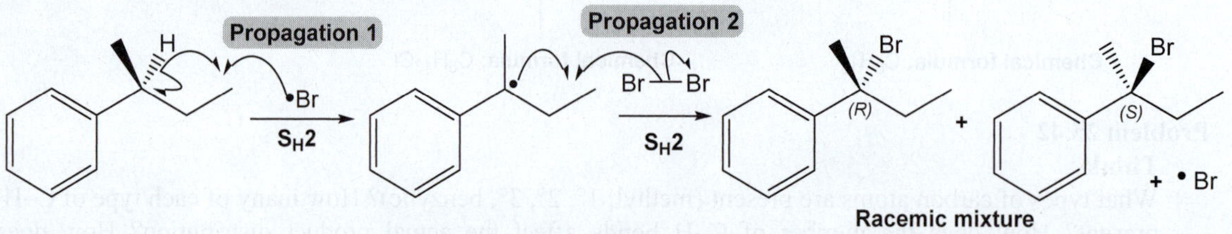

(b) In the propagation cycle (next page), the tertiary C–H bond undergoes hydrogen abstraction, which leaves a C• on the 3° carbon. Then the C• abstracts a Br from Br_2 to produce the new C–Br bond. In the second step, the C• behaves as if it is trigonal planar, and thus the new C–Br that forms results in a mixture of (R) and (S) configurations. The other chiral center remains unchanged during the course of the reaction. Thus, the products are diastereomers, which would form an optically active mixture.

Propagation 1

S_H2

Propagation 2

S_H2

Diastereomers

Problem 25.44

Think

Which bond is the weakest and undergoes homolysis? What radical is formed? How can this radical react with H–H? What radical is then produced, and how can it continue the propagation cycle? For termination steps, what radicals are present in the reaction mixture that can couple?

Solve

The weakest bond present is the X–X bond, which undergoes homolysis in an initiation step to produce two X• radicals.

X–X —Initiation, *hν*, Homolysis→ X• + •X

In the first of two propagation steps, an X• abstracts a hydrogen from H–H to produce the overall product, HX, and a H• radical. In the second propagation step, a H• radical abstracts an X atom from X_2 to produce the second equivalent of HX and regenerate the initial X•.

X• + H–H —Propagation 1, S_H2→ X–H + •H

H• + X–X —Propagation 2, S_H2→ H–X + X•

Termination steps can be a radical coupling between any two radicals that appear in the propagation steps, such as the following:

X• + •X —Termination 1, Radical coupling→ X–X

H• + •H —Termination 2, Radical coupling→ H–H

H• + •X —Termination 3, Radical coupling→ H–X

Problem 25.45

Think

What are the steps in the mechanism of each reaction (see the solution to Problem 25.44 for the propagation steps)? Which step is the rate-determining step? How does the heat of reaction for that step govern the rate of the overall reaction?

Solve

Just as with the radical halogenation of alkanes, the main contribution to the energy barrier for the radical halogenation of H_2 comes from the first propagation step, which is hydrogen atom abstraction. The $\Delta H°_{rxn}$ of this step for each halogenation is shown below and is calculated by subtracting the energy of the X-H bond that is formed from that of the H-H bond that is broken. Because this step is most exothermic with fluorination and most endothermic with iodination, the rate of halogenation increases in the order $I_2 < Br_2 < Cl_2 < F_2$.

First propagation step	$\Delta H°_{rxn}$
F• + H–H → F–H + H•	436 kJ/mol - 569 kJ/mol = -133 kJ/mol
Cl• + H–H → Cl–H + H•	436 kJ/mol - 431 kJ/mol = +5 kJ/mol
Br• + H–H → Br–H + H•	436 kJ/mol - 368 kJ/mol = +68 kJ/mol
I• + H–H → I–H + H•	436 kJ/mol - 297 kJ/mol = +139 kJ/mol

Problem 25.46

Think

Recall that radical chlorination is less selective than radical bromination. Is that because the steps in chlorination are more endothermic or exothermic than in bromination? Does a more stable reactant species lead to a more endothermic or exothermic reaction?

Solve

Chlorination is less selective than bromination because the chlorination steps are more exothermic. The complexation described in this problem is indicated to stabilize chlorine, a reactant, which makes its reaction less exothermic than without the complexation. Thus, the complexation makes the reaction *more* selective.

Problem 25.47

Think

What is the key step that favors bromination at a 3° C? What is the key intermediate that is produced in that step? What is the ideal geometry about the C• radical intermediate? Is this possible in bicyclo[2.2.1]heptane? It may help to build molecular models of each of these compounds.

Solve

The key step that favors reaction at the 3°C is H atom abstraction to produce a 3° carbon radical. Those intermediates are shown below for each reaction. The smaller bicyclic system does not allow the tertiary carbon to become sufficiently planar to allow radical formation. Thus, a tertiary radical that would be formed would be highly strained, making the secondary radical more stable than the tertiary in this case.

Bicyclo[3.3.1]nonane → ($BrCCl_3$, hν) → [3° C• Intermediate] → Br, 100%

Bicyclo[2.2.1]heptane → ($BrCCl_3$, hν) ✗ → [3° C• Intermediate, Too strained to form] → Br, 0%

Problem 25.48

Think

What is the weakest bond in the starting material? Which bond is most likely to undergo homolysis? How does the radical that is produced react with the other molecules of the starting material? Review the mechanism and arrow usage for initiation, propagation, and termination. Do the propagation steps make a cycle? What two radicals could come together in a termination step?

Solve

The Br–C bond is the weakest, so it undergoes homolysis for the initiation step.

Br—CCl3 —Initiation, hν, Homolysis→ Br• + •CCl3

Then, in the first propagation step, the $\bullet CCl_3$ radical abstracts a H atom from the tertiary C–H bond to produce a carbon radical. In the second propagation step, the C• abstracts a Br atom from $BrCCl_3$, producing another $\bullet CCl_3$ that can react further in the propagation cycle.

Propagation 1 (SH2): + •CCl3 → + CHCl3

Propagation 2 (SH2): + Cl3C—Br → Br + Cl3C•

Three possible termination steps:

Termination 1 (Radical coupling): + Br• → Br

Termination 2 (Radical coupling): Cl3C• + •CCl3 → Cl3C—CCl3

Termination 3 (Radical coupling)

Section 25.5 Radical Addition of HBr: Anti-Markovnikov Addition

Problem 25.49

Think

Are radicals formed? If so, what type of reaction occurs between a radical and a C=C bond: free radical bromination or radical addition? Is resonance possible? What stereochemical considerations must be made? What regiochemical considerations must be made?

Solve

(a) This is an example of a radical addition mechanism. In the initiation steps (see Equation 25-33a and 25-33b on p. 1275), Br• is produced. In the propagation cycle, the Br• radical undergoes radical addition to the C=C to form a new C–Br bond so that the C• radical forms on the 3° carbon. The new chiral center is produced as a mixture of both configurations. The C• then abstracts a H atom from HBr to form a new C–H bond. That step also forms a new chiral center in which both configurations are produced and, thus, four stereoisomers are produced.

(b) This not a radical mechanism, since no initiator (heat, light, or peroxides) is present. This mechanism is electrophilic addition of HBr, a strong Brønsted acid, across the C=C bond (Chapter 11). A proton adds in the first step to produce the more stable carbocation intermediate, which then undergoes coordination with Br^-. The product is achiral.

Problem 25.50 (SYN)

Think

Compare the formula of the starting reagent to the product shown. What is the difference? What bonds then need to be broken and formed in order to yield the target shown? What kind of reaction will accomplish that?

Solve

Both reactions are examples of anti-Markovnikov addition of HBr across a C=C bond. HBr and peroxide are a source of Br• radical, which reacts with an alkene via radical addition to form the most stable C• intermediate. The C• intermediate reacts with an equivalent of H–Br to form a new C–Br bond.

(a) The formula for the starting material is C_9H_{10}, and the formula for the target compound is $C_9H_{11}Br$. The starting product is a benzene with a conjugated C=C. HBr adds across the C=C bond and Br is on the carbon one away from the benzylic carbon. This route goes through the more stable benzylic C•.

Starting with a terminal C=C and adding HBr without peroxide will lead to the wrong major product due to C^+ rearrangement from the secondary C^+ to the more stable benzylic C^+.

HBr

Actual major product

Chemical formula: C_9H_{10} Chemical formula: $C_9H_{11}Br$

(b) The formula for the starting material is $C_{10}H_{16}$, and the formula for the target compound is $C_{10}H_{17}Br$. Therefore the HBr is added across the C=C bond, and Br adds to the secondary carbon in order to produce the more stable tertiary carbon radical intermediate. Similar to **(a)**, addition of HBr in a closed-shell mechanism will not produce the desired secondary alkyl bromide because carbocation rearrangement will produce the tertiary alkyl bromide instead.

HBr / Peroxide

Chemical formula: $C_{10}H_{16}$ Chemical formula: $C_{10}H_{17}Br$

Problem 25.51

Think

Which bond homolytically cleaves to form the initial radical? What radical reacts with the C=C? On which C atom does the C• radical form? Why?

Solve

The initiation steps are shown below and will form Br• radicals. If Br• adds to C3 as shown, then the radical on C2 can be stabilized by resonance with the pre-existing Br atom.

RO—OR → (Initiation 1, Homolysis) 2 RO• ; RO• + H—Br → (Initiation 2, S_H2) RO—H + Br•

Propagation 1 (Radical addition); Resonance; Propagation 2 (S_H2) + Br•

Problem 25.52

Think

Which bond homolytically cleaves to form the initial radical? How does a radical react with the C=C? Which atom of the C=C bond gains the unpaired electron? How does the other C=C bond react to form the five-carbon ring?

Solve

The initiation steps are shown on the next page and will form Br• radicals. The Br• radical reacts with the alkene in the first propagation step to form a new C–Br bond and a secondary C•. In a second propagation step, this C• radical then reacts with the other C=C to form a new C–C bond in a cyclopentane ring, leaving another C•. A third propagation step yields the target compound and produces another Br• to complete the propagation cycle.

Termination steps can occur with any two radicals in the propagation cycle undergoing radical coupling, and two examples are shown below.

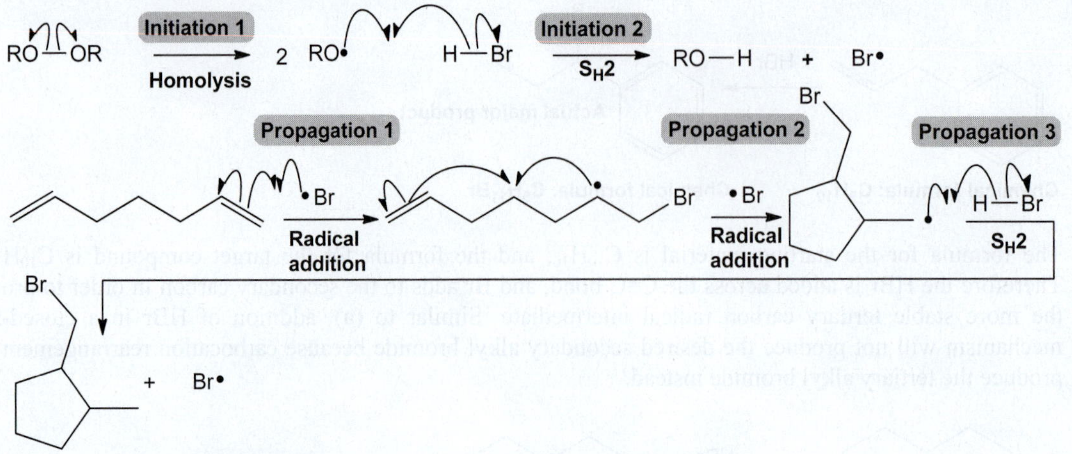

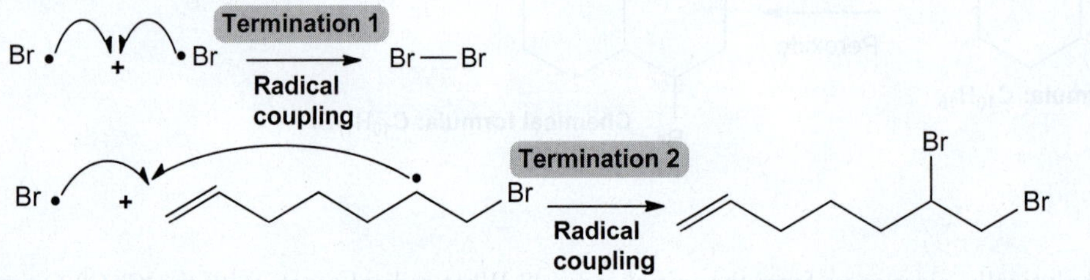

Initiation 1
Homolysis
Initiation 2
S_H2
Propagation 1
Radical addition
Propagation 2
Radical addition
Propagation 3
S_H2
Termination 1
Radical coupling
Termination 2
Radical coupling

Problem 25.53

Think

Which bond homolytically cleaves to form the initial radical? What radical reacts with the C=C? On which C atom does the C• radical form? How do the other two C=C bonds react to form two five-carbon rings?

Solve

The initiation steps are shown above in Problem 25.52 and will form Br• radicals. From there, two radical cyclizations occur; numbering the carbons indicates that C2 joins C6 and C7 joins C11.

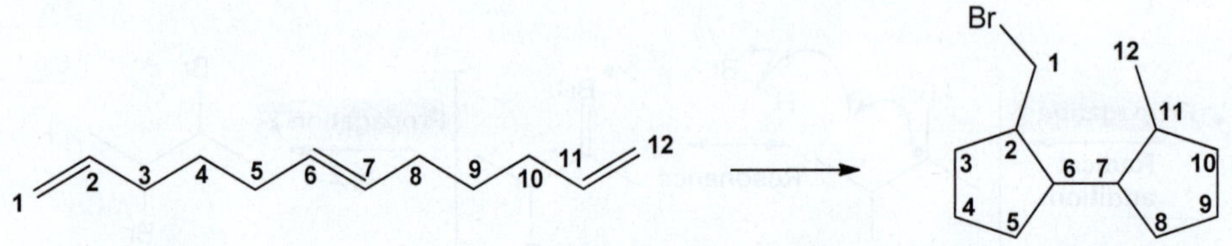

The propagation steps are drawn here. The mechanism involves a 1° radical in the next to last step.

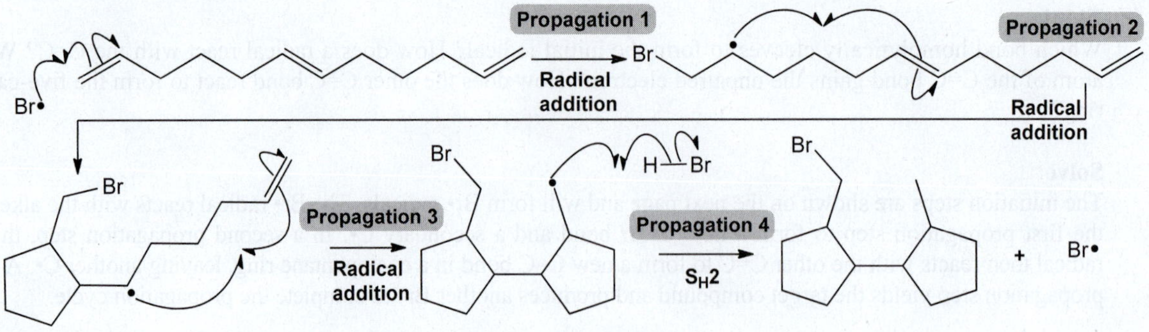

Propagation 1
Radical addition
Propagation 2
Radical addition
Propagation 3
Radical addition
Propagation 4
S_H2

Two possible termination steps are shown below, and others can be drawn involving the coupling of any two radicals that appear in the propagation cycle.

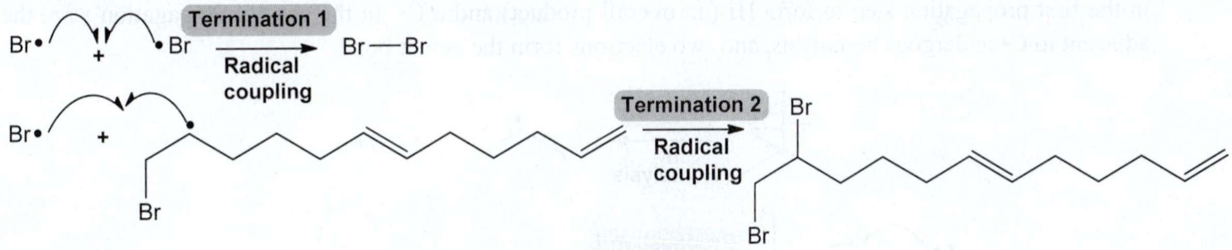

Problem 25.54

Think

Which bond is the weakest? Which bond undergoes homolysis? How does this radical react with the other organic reagent? What propagation and termination steps are possible?

Solve

The weakest bond present is the Cl_3C–Cl bond, which undergoes homolysis in an initiation step to produce a Cl• radical and Cl_3C• radical.

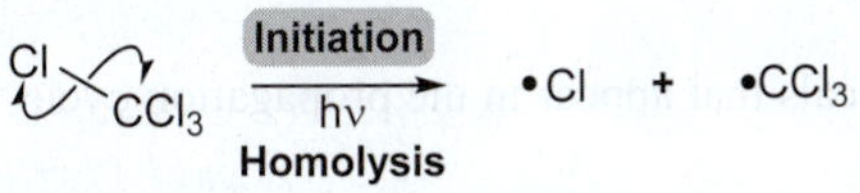

In the first of two propagation steps, a Cl_3C• adds to the C=C via radical addition to form a new C–CCl_3 bond and a C• radical. In the second propagation step, a C• radical reacts with the Cl_3C–Cl bond to form a new C–Cl bond and a Cl_3C• radical.

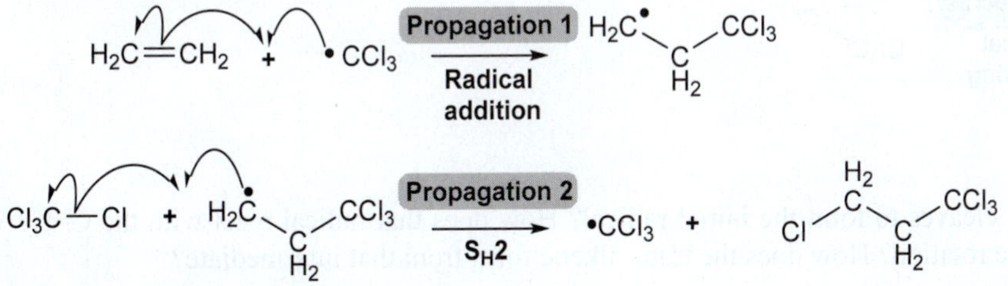

Termination steps can be a radical coupling between any two radicals that appear in the propagation cycle, such as the following two:

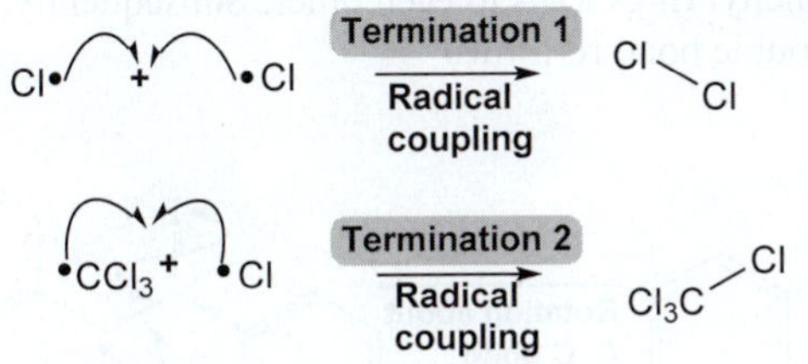

Problem 25.55

Think

Which bond is the weakest? Which bond undergoes homolysis? How does this radical react with the other organic reagent to produce one of the overall products? How does the radical that is produced in that step react to form the other overall product? Which steps make up the propagation cycle? Which radicals can undergo coupling in termination steps? Which steps are summed to obtain the net reaction?

Solve

The chain-reaction mechanism for the elimination of HI from 2-iodopropane to form propene is shown below. The weakest bond is C–I, which undergoes homolysis to form the initial radicals. Then I• abstracts a hydrogen in the first propagation step to form HI (an overall product) and a C•. In the second propagation step, the C–I adjacent to C• undergoes homolysis, and two electrons form the new π bond.

Initiation: hν, Homolysis → + I•

Propagation 1: I• + H– ; S_H2 → H—I +

Propagation 2: Homolysis → + I•

Net reaction → + H—I

Termination steps can be a radical coupling between any two radicals that appear in the propagation cycle, such as the following two:

Termination 1: Cl• + •Cl → Radical coupling → Cl–Cl

Termination 2: $•CCl_3$ + •Cl → Radical coupling → Cl_3C–Cl

Problem 25.56

Think

Which bond homolytically cleaves to form the initial radical? How does that radical react with the C=C? What kinds of bonds undergo free rotation? How does the trans alkene form from that intermediate?

Solve

The I–I bond is the weakest present and is cleaved to form two I• radicals. After I• forms, it adds to the C=C double bond. This converts the double bond to a single bond temporarily, which can freely rotate. Because of the bulkiness of the rings, the favored conformation has the phenyl rings trans to each other. Subsequently, I• is eliminated when the C–I undergoes homolysis and the C=C double bond re-forms.

Initiation: I—I, hν, Homolysis → 2 I•

Propagation 1: Radical addition →

Rotation about C–C bond →

Propagation 2: Homolysis → + I•

Section 25.7 Dissolving Metal Reductions: Hydrogenation of Alkenes and Alkynes
Problem 25.57

Think

What is the result of an alkyne treated with sodium metal dissolved in liquid ammonia? What is the stereochemistry of the product? What is the source of radical? Are proton transfer reactions involved?

Solve

The source of radical is a solvated electron, produced on dissolving solid sodium in liquid ammonia. In Step 1, a solvated electron adds to an alkyne C and converts the triple bond into a double bond. One of the initial alkyne C atoms gains an unpaired electron and the other gains a negative charge. The carbanion is a very strong base and deprotonates NH_3 in Step 2. This produces an uncharged vinylic radical, which, in Step 3, undergoes radical coupling with a second solvated electron to produce a carbanion. In Step 4, the carbanion deprotonates a second molecule of NH_3, yielding the overall product—a trans alkene (less steric strain than cis).

(a)

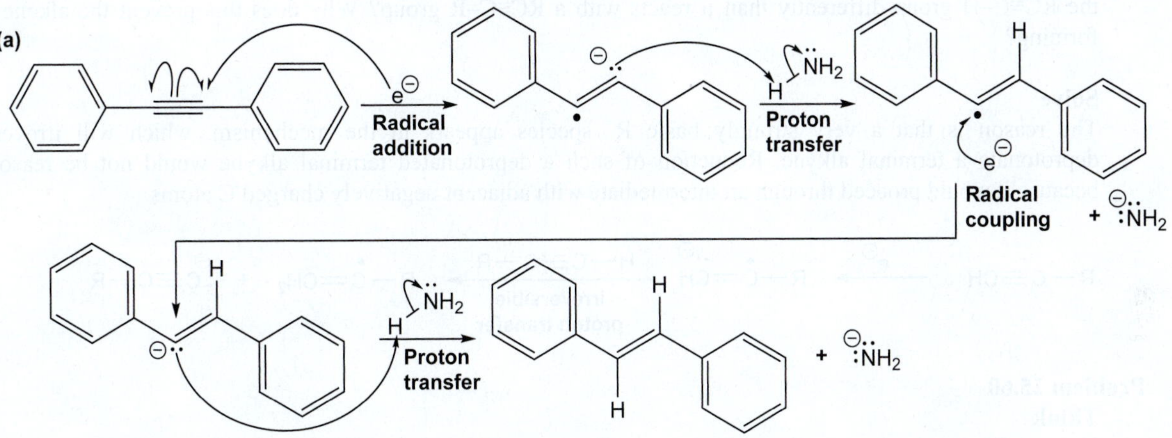

Radical addition

Proton transfer

Radical coupling

$+ :\ddot{N}H_2^{-}$

Proton transfer

$+ :\ddot{N}H_2^{-}$

(b)

Radical addition

Proton transfer

$+ :\ddot{N}H_2^{-}$

Radical coupling

Proton transfer

$+ :\ddot{N}H_2^{-}$

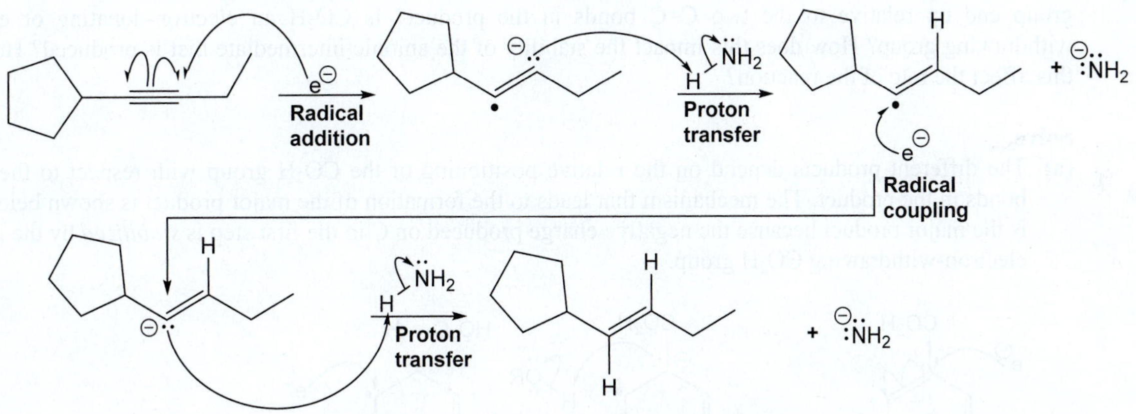

Problem 25.58 (SYN)

Think

What is the structure of 1-phenylprop-1-yne? What reaction conditions yield a cis alkene from an alkyne? What reaction conditions yield a trans alkene from an alkyne?

Solve

The structure of 1-phenylprop-1-yne is shown below. For product **(a)**, catalytic hydrogenation of an alkyne using a poisoned catalyst (Section 19.5b) exclusively forms the cis product. For product **(b)**, a dissolving metal reduction of an alkyne produces the *E* (or trans) alkene almost exclusively. The two reactions are shown on the next page.

H2
Lindlar Catalyst
1-Phenylprop-1-yne
Na(s)
NH3(ℓ), -78°C

Problem 25.59

Think

What type of reactive organic intermediate appears in the mechanism? How does this intermediate react with the RC≡C–H group differently than it reacts with a RC≡C–R group? Why does this prevent the alkene from forming?

Solve

The reason is that a very strongly basic R^- species appears in the mechanism, which will irreversibly deprotonate a terminal alkyne. Reduction of such a deprotonated terminal alkyne would not be reasonable because it would proceed through an intermediate with adjacent negatively charged C atoms.

R—C≡CH → (e⁻) R—Ċ=C̈H⁻ → (H—C≡C—R, Irreversible proton transfer) R—Ċ=CH2 + :C̄≡C—R

Problem 25.60

Think

How is this mechanism similar to the one in Equation 25-45? In cyclohexa-1,4-diene, where can the CO_2H group end up relative to the two C=C bonds in the product? Is CO_2H an electron-donating or electron-withdrawing group? How does this impact the stability of the anionic intermediate that is produced? How does this affect the rate of the reaction?

Solve

(a) The different products depend on the relative positioning of the CO_2H group with respect to the double bonds in the product. The mechanism that leads to the formation of the major product is shown below. This is the major product because the negative charge produced on C in the first step is *stabilized* by the attached electron-withdrawing CO_2H group.

CO2H
Radical addition
Proton transfer
Radical coupling
Most stable resonance structure
HO2C H
Resonance
Proton transfer
H—ÖR

(b) Because the negative charge produced on C in the first step is stabilized by the attached electron-withdrawing CO_2H group, the presence of a CO_2H group should also *speed up* the reaction compared to unsubstituted benzene.

Problem 25.61

Think

How is this mechanism similar to the one in Equation 25-45? In cyclohexa-1,4-diene, where can the CH_3 group end up relative to the two C=C bonds in the product? Is CH_3 an electron-donating or electron-withdrawing group? How does this impact the stability of the anionic intermediate that is produced? How does this affect the rate of the reaction?

Solve

(a) The different products depend on the relative positioning of the CH_3 group with respect to the double bonds in the product. The mechanism that leads to the formation of the major product is shown below. This is the major product because the negative charge produced on C in the first step is *destabilized* by the CH_3 electron-donating group. Therefore the CH_3 should be farther away from the C^-.

H3C
Radical addition
Proton transfer
Radical coupling
Most stable resonance structure
Resonance
Proton transfer

(b) Because the negative charge produced on C in the first step is destabilized by the attached electron-donating CH_3 group, the presence of a CH_3 group should *slow down* the reaction compared to unsubstituted benzene.

Problem 25.62

Think

Which acid, ROH or NH_3, is stronger? In each reduction mechanism, in which step is ROH or NH_3 deprotonated? In that step, which mechanism has the stronger base?

Solve

An alcohol (ROH) with a pK_a of ~16 is a much stronger acid than ammonia with a pK_a of ~36 (see Chapter 6 for pK_a values). When an alkyne is reduced, the species that deprotonates the proton source is a carbanion ($R:^-$) in which the negative charge is localized on an sp^2-hybridized C. That carbanion is a strong enough base to deprotonate NH_3. In a Birch reduction, the carbanion has a resonance-stabilized negative charge, so it is more stable and thus a weaker base. Therefore, it requires a stronger acid, like ROH.

Integrated Problems

Problem 25.63

Think

Oxygen is a diradical. Which C–H bond undergoes H atom abstraction? How does the resulting radical react with another oxygen molecule? Do the propagation steps make a cycle? What termination steps are possible that can reduce the number of radicals?

Solve

Oxygen is a diradical, so it can abstract a H atom to generate the initial radical that enters the propagation cycle. The propagation cycle is defined by the second and third steps below. Notice that the initial ether radical is regenerated.

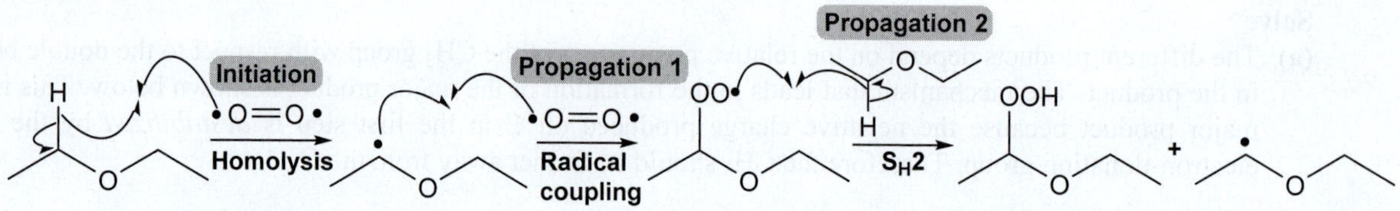

Initiation

Homolysis

Propagation 1

Radical coupling

Propagation 2

S_H2

Two possible termination steps include:

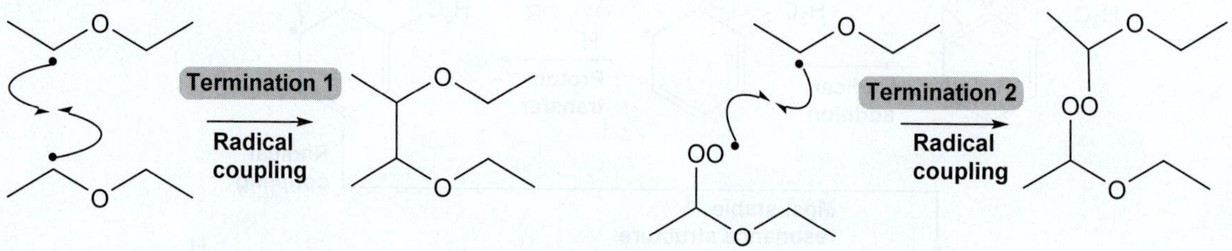

Termination 1

Radical coupling

Termination 2

Radical coupling

Problem 25.64

Think

What is the identity of the two possible radicals that can form from reaction of $\bullet O_2$ and diethyl ether? Are there any inductive and/or resonance factors that stabilize the radical when the unpaired electron is on the C atom that is α to the O?

Solve

The main reason is that the resulting radical formed on abstraction of a hydrogen from the α C is resonance stabilized, as shown below, involving the lone pair of electrons on O. Abstraction of a H atom from the β C results in a localized radical.

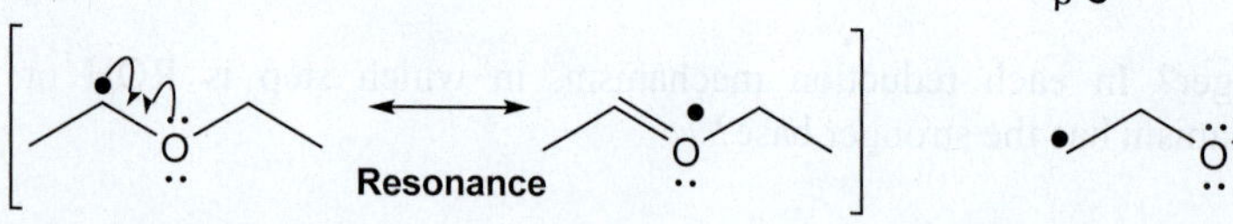

α C•

Resonance

β C•

Problem 25.65

Think

Which steps can be arranged in a way that shows no *net* change in the number of the reactive radicals? How would those steps be classified? What radicals appear in those steps? What other steps are responsible for increasing or decreasing the number of those radicals?

Solve

(a) When steps 2–4 take place together, there is no net change in the number of the radical species. In those steps, each radical appears once on the reactant side and once on the product side, so they all cancel. Steps 2–4, therefore, are the propagation steps. Step 1 is responsible for generating one of the radicals used in the propagation cycle, so it is an initiation step. No termination steps are shown, which would be responsible for decreasing the number of radicals appearing in the propagation cycle.

Initiation: $ArI + e^{\ominus} \longrightarrow ArI^{\bullet\ominus}$

Propagation 1: $ArI^{\bullet\ominus} \longrightarrow Ar^{\bullet} + I^{\ominus}$

Propagation 2: $Ar^{\bullet} + H_2N^{\ominus} \longrightarrow ArNH_2^{\bullet\ominus}$

Propagation 3: $ArNH_2^{\bullet\ominus} + ArI \longrightarrow ArNH_2 + ArI^{\bullet\ominus}$

(b) To arrive at the net reaction, we sum the propagation steps only, crossing out redundant species on either side of the reactions.

$ArI^{\bullet\ominus} \longrightarrow Ar^{\bullet} + I^{\ominus}$

$Ar^{\bullet} + H_2N^{\ominus} \longrightarrow ArNH_2^{\bullet\ominus}$

$ArNH_2^{\bullet\ominus} + ArI \longrightarrow ArNH_2 + ArI^{\bullet\ominus}$

Net reaction: $H_2N^{\ominus} + ArI \quad ArNH_2 + I^{\ominus}$

(c) The only conceivable termination is where two uncharged radicals appearing in the propagation cycle combine or where an uncharged radical combines with a charged radical, as shown below. Combination of two radical anions (like charge) is not reasonable due to charge repulsion.

$Ar\bullet + Ar\bullet \rightarrow Ar\text{–}Ar$
$Ar\bullet + \bullet ArNH_2^{-} \quad Ar\text{–}ArNH_2^{-}$

Problem 25.66

Think

What bonds break and form in each step? What kind of curved arrow should you use? Which steps can be arranged in a way that shows no *net* change in the number of the reactive radicals? How would those steps be classified? What radicals appear in those steps? What other steps are responsible for increasing or decreasing the number of those radicals?

Solve

(a) and **(b)** When the first and fifth steps given in the problem take place together, there is no net change in the number of the radical species. In those steps, each radical appears once on the reactant side and once on the product side, so they all cancel. The first and fifth steps, therefore, are the propagation steps. The second step given is responsible for generating one of the radicals used in the propagation cycle, so it is an initiation step. The third and fourth steps result in decreasing the number of radicals appearing in the propagation cycle, so they are termination steps. See the figure on the next page.

Propagation 1

S_H2

Initiation

hν

Homolysis

Termination 1

Radical coupling

Termination 2

Radical coupling

Propagation 2

S_H2

(c) The net reaction is obtained by summing the propagation steps only.

Propagation #1

Propagation #2

Net reaction

Problem 25.67

Think

What bonds break and form in each step? What kind of curved arrow should you use? Which steps can be arranged in a way that shows no *net* change in the number of the reactive radicals? How would those steps be classified? What radicals appear in those steps? What other steps are responsible for increasing or decreasing the number of those radicals?

Solve

(a) and **(b)** To arrange the steps in a way that shows no net change in the number of radicals, all three steps given must be combined. Only in that way does every radical appear once as a reactant and once as a product. Therefore, all steps given are propagation steps

Propagation 1

S_H2

Propagation 2

S_H2

Propagation 3

S_H2

(c) The net reaction is obtained by summing all three of these propagation steps. The reactants that do not cancel are the RCO_3R and RH species, and the products that do not cancel are the RCO_2R and HO–*t*-Bu species.

Net reaction

R'C(=O)–O–O–C(CH3)3 + R—H ⟶ R'C(=O)–O–R + HO–C(CH3)3

(d) Two termination steps:

R• + •O–C(CH3)3 —(Termination 1; Radical coupling)→ R–O–C(CH3)3

R• + Cu⊕• —(Termination 2; Radical coupling)→ R—Cu⊕

Problem 25.68

Think

What bonds break and form in each step? What kind of curved arrow should you use? Which steps can be arranged in a way that shows no *net* change in the number of the reactive radicals? How would those steps be classified? What radicals appear in those steps? What other steps are responsible for increasing or decreasing the number of those radicals?

Solve

(a) and **(b)** When steps 1–3 take place together, there is no net change in the number of the radical species. In those steps, each radical appears once on the reactant side and once on the product side, so they all cancel. Steps 1–3, therefore, are the propagation steps. Steps 3 and 4, when combined, are responsible for producing the first radical that appears in the propagation cycle, so those are the initiation steps.

Bu3Sn• + Br–(hexenyl) —(Propagation 1; SH2)→ Bu3Sn—Br + •(hexenyl)

(cyclopentylmethyl)• + H—SnBu3 —(Propagation 2; SH2)→ (cyclopentyl)CH2–H + •SnBu3

•(hexenyl) —(Propagation 1; Radical addition)→ (cyclopentylmethyl)•

(H3C)2C(CN)–N=N–C(CN)(CH3)2 —(Initiation; Homolysis)→ 2 H3C–C•(CH3)–CN + N≡N

H3C–C•(CH3)–CN + H—SnBu3 —(Initiation; SH2)→ H3C–CH(CH3)–CN + •SnBu3

(c) The net reaction is obtained by summing all the propagation steps.

Net reaction

H—$SnBu_3$ + Br(CH$_2$)$_4$CH=CH$_2$ ⟶ Bu_3Sn—Br + methylcyclopentane

(d) Two termination steps:

Termination 1 — Radical coupling

Termination 2 — Radical coupling

Problem 25.69

Think

What is the weakest bond in the starting material? Which bond is most likely to undergo homolysis? How does this radical react with the starting material? What radical is formed, and how does it react to produce the overall product? What are the propagation steps? Do they make a cycle? Do the propagation steps sum to the net reaction?

Solve

The C–O bond is weaker than a CH bond, so the C–O bond is the one broken in the initiation step. In the first propagation step, a •CH_3 abstracts a H atom from the ether to produce CH_4 as an overall product and a new radical. In the second step, the new radical undergoes homolysis to produce a molecule of H_2C=O as an overall product and replenish the •CH_3 radical for further reaction.

H_3C–O–CH_3 —(**Initiation**, hν, **Homolysis**)⟶ H_3C–O• + •CH_3

H_3C–O–CH_2–H + •CH_3 —(**Propagation 1**, S_H2)⟶ H_3C–O–•CH_2 + CH_4

H_3C–O–•CH_2 —(**Propagation 2**, **Homolysis**)⟶ H–C(=O)–H + •CH_3

Problem 25.70

Think

What is the weakest bond in the starting material? Which bond is most likely to undergo homolysis? How does the radical that is produced react with the other molecules in the starting material? Review the mechanism and arrow usage for initiation, propagation, and termination. Which steps form a propagation cycle? Do the propagation steps sum to the net reaction? What two radicals could come together in a termination step?

Solve

The mechanism is given below. The O–O single bond is the weakest bond and undergoes homolysis to produce the initial radicals. An initial radical then abstracts CN from ClCN to produce Cl•, which is the first radical that enters the propagation cycle composed of the last two steps.

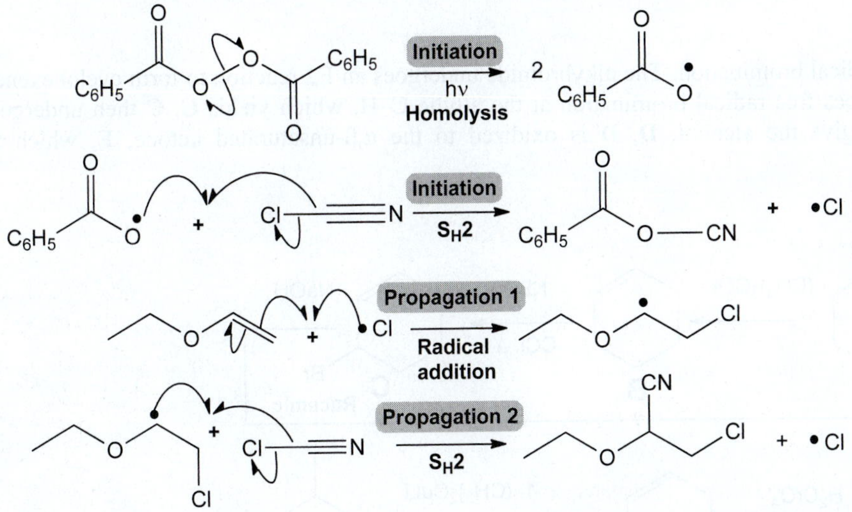

Notice that to obtain the observed regiochemistry, Cl• must add to the C=C first. Specifically, Cl• adds to the terminal C to produce the more stable (resonance stabilized) carbon radical. Then, CN is abstracted from ClCN to complete the overall product and regenerate Cl• for further reaction in the propagation cycle.

Two possible termination steps:

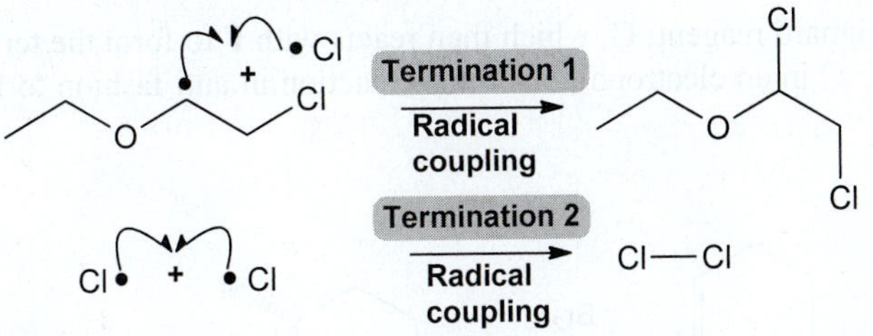

Problem 25.71

Think

What is the initiation step? Which C–H bond is abstracted from the alkyl bromide? What is the identity of the C• radical intermediate? How can the nearby Br atom interact with the C• radical to prevent the production of both configurations when the new chiral center is formed? (*Hint:* Consider the intermediate in the addition of Br_2 across a C=C that is responsible for that reaction's stereoselectivity [Chapter 12].)

Solve

As the H atom is abstracted, the terminal bromine can form a cyclic bromo radical (analogous to the cyclic bromonium ion in Chapter 12). In this step, the stereochemistry at the chiral center is inverted to form the cyclic bromo radical. Subsequent attack of Br_2 can only occur *from the side opposite* the attached bromine, which inverts the stereochemical configuration, just as in an S_N2 reaction. Double inversion at the chiral center leads to a retention of configuration of the C–CH_3

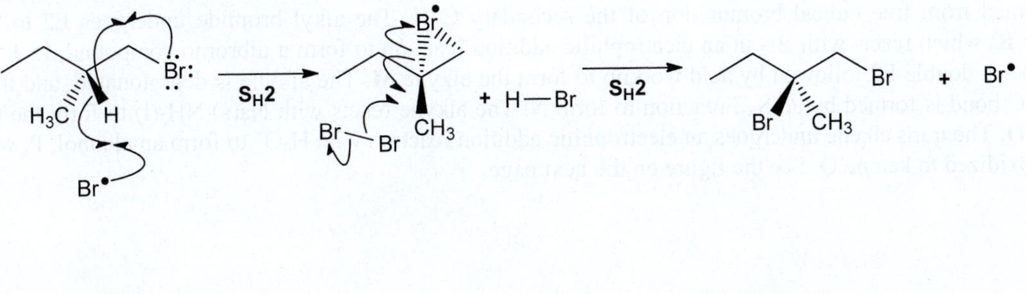

Problem 25.72

Think

Identify the electrophile and nucleophile in each reaction. Are radicals involved? Does a reaction take place that alters the carbon skeleton or functional group? Is the reaction an example of substitution, elimination, addition, or redox? Is regiochemistry or stereochemistry a concern?

Solve

A is formed from free radical bromination. The alkybromide undergoes an E2 reaction to form cyclohexene, **B.** The cyclohexene undergoes free radical bromination at the allylic C–H, which yields **C. C** then undergoes a substitution reaction to give the alcohol, **D. D** is oxidized to the α,β-unsaturated ketone, **E,** which then undergoes 1,4-addition.

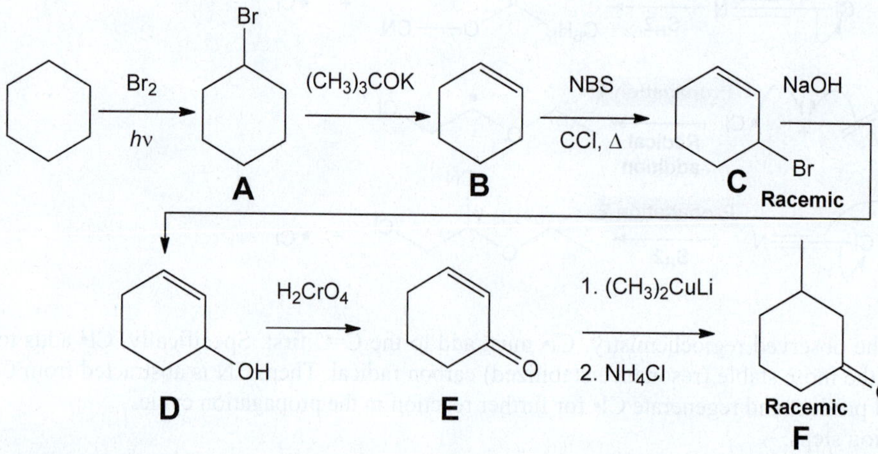

Br_2, $h\nu$ → **A** → $(CH_3)_3COK$ → **B** → NBS, CCl_4, Δ → **C** (Racemic) → NaOH → **D** → H_2CrO_4 → **E** → 1. $(CH_3)_2CuLi$ 2. NH_4Cl → **F** (Racemic)

Alkyl bromide **C** reacts with Mg(*s*) to form a Grignard reagent, **G,** which then reacts with **F** to form the tertiary alcohol, **H,** after acid workup. Br_2 adds to the C=C in an electrophilic addition reaction in anti fashion to form the dibromo compound **I.**

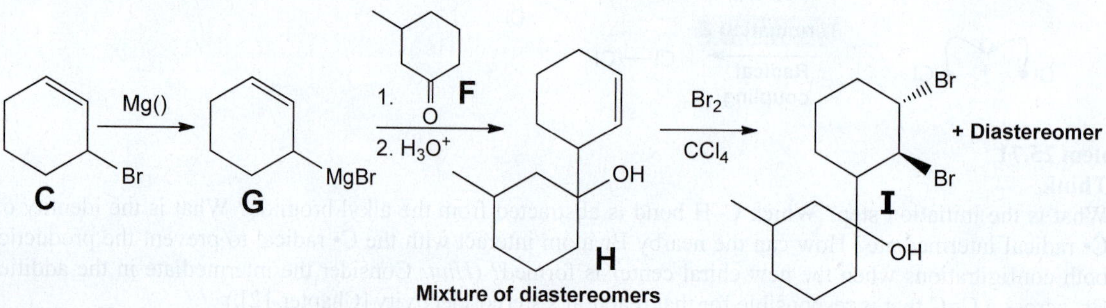

C → Mg() → **G** → 1. **F** 2. H_3O^+ → **H** (Mixture of diastereomers) → Br_2, CCl_4 → **I** + Diastereomer

Problem 25.73

Think

Identify the electrophile and nucleophile in each reaction. Are radicals involved? Does a reaction take place that alters the carbon skeleton or functional group? Is the reaction an example of substitution, elimination, addition, or redox? Is regiochemistry or stereochemistry a concern?

Solve

J is formed from free radical bromination of the secondary C–H. The alkyl bromide undergoes E2 to form propene **K,** which reacts with Br_2 in an electrophilic addition reaction to form a dibromo compound, **L. L** then undergoes a double E2 followed by acid workup to form the alkyne **M.** The alkyne is deprotonated, and then a new C–C bond is formed by an S_N2 reaction to form **N.** The alkyne reacts with Na(s)/NH_3(l) to form the trans alkene, **O.** The trans alkene undergoes an electrophilic addition reaction with H_3O^+ to form an alcohol, **P,** which is then oxidized to ketone **Q.** See the figure on the next page.

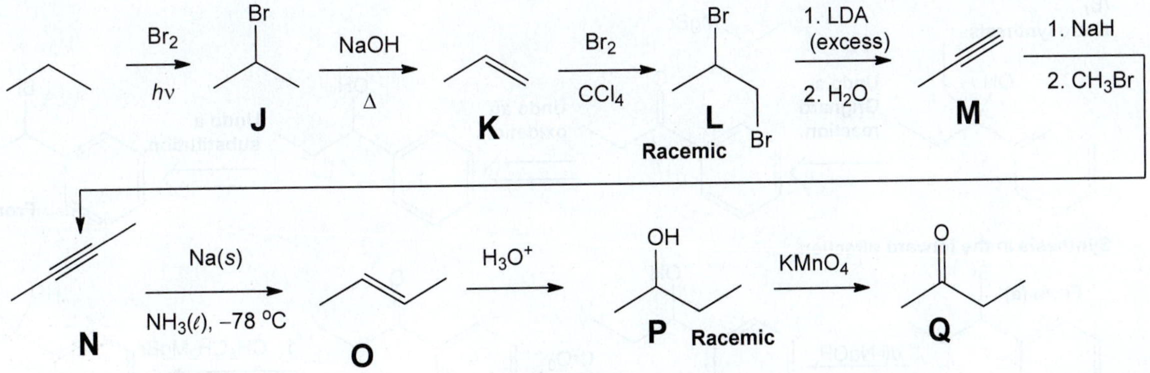

Br$_2$, $h\nu$ → J (Br) → NaOH, Δ → K → Br$_2$, CCl$_4$ → L (Br, Br) Racemic → 1. LDA (excess) 2. H_2O → M → 1. NaH 2. CH_3Br → N → Na(*s*), $NH_3(\ell)$, –78 °C → O → H_3O^+ → P (OH) Racemic → $KMnO_4$ → Q (O)

K undergoes anti-Markovnikov addition of HBr to yield the alkyl halide, **R**, which reacts with PPh$_3$/Bu–Li to form the ylide, **S**. The ylide reacts with ketone **Q** in a Wittig reaction to form an alkene, **T**, which then undergoes a Diels–Alder reaction to form the bicyclic compound **U**.

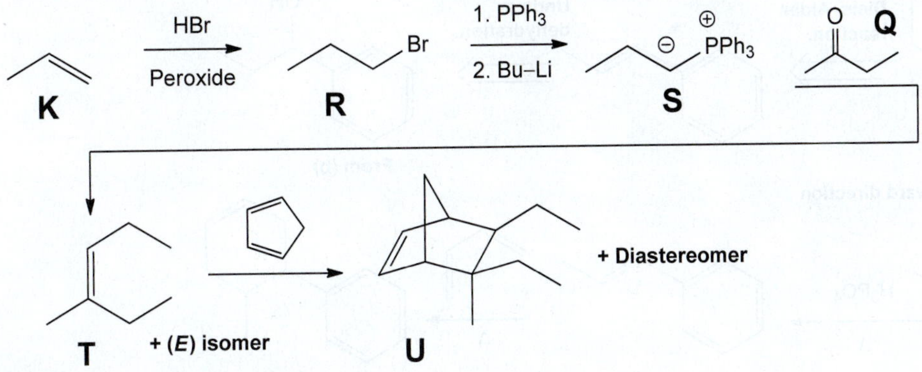

K → HBr, Peroxide → R (Br) → 1. PPh$_3$ 2. Bu–Li → S (⊖, ⊕PPh$_3$) + Q (O) → T + (*E*) isomer → (cyclopentadiene) → U + Diastereomer

Problem 25.74 (SYN)

Think

Which C atom in propylbenzene is most likely to undergo free radical bromination? What functional group transformations can take place on the product? How must the carbon skeleton be altered?

Solve

Retrosynthesis and synthesis: We can functionalize the benzyl radical by bromination, so this is a logical step in each of the syntheses.

(a)
Retrosynthesis

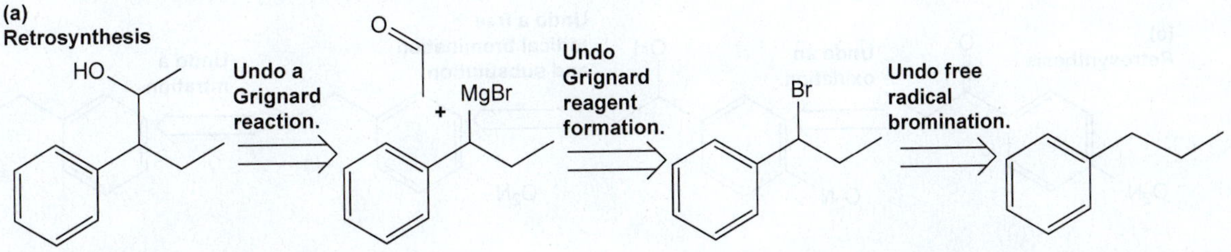

HO ⇒ **Undo a Grignard reaction.** ⇒ O + MgBr ⇒ **Undo Grignard reagent formation.** ⇒ Br ⇒ **Undo free radical bromination.** ⇒

Synthesis in the forward direction

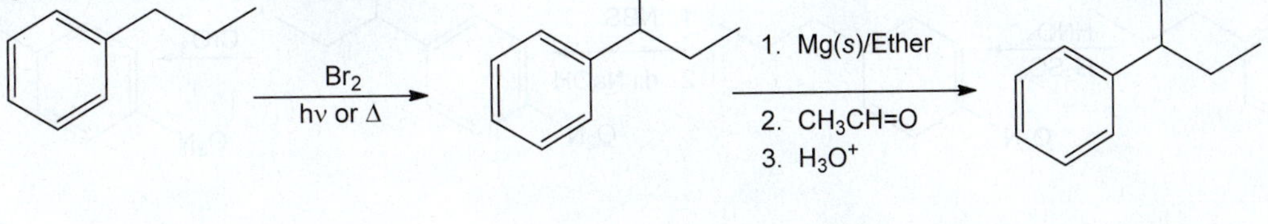

→ Br$_2$, hν or Δ → Br → 1. Mg(*s*)/Ether 2. $CH_3CH{=}O$ 3. H_3O^+ → HO

(b)
Retrosynthesis

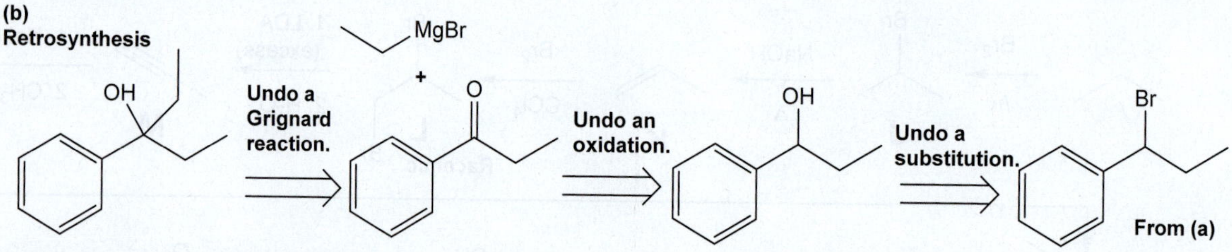

Undo a Grignard reaction. **Undo an oxidation.** **Undo a substitution.**

MgBr + O OH Br

From (a)

Synthesis in the forward direction

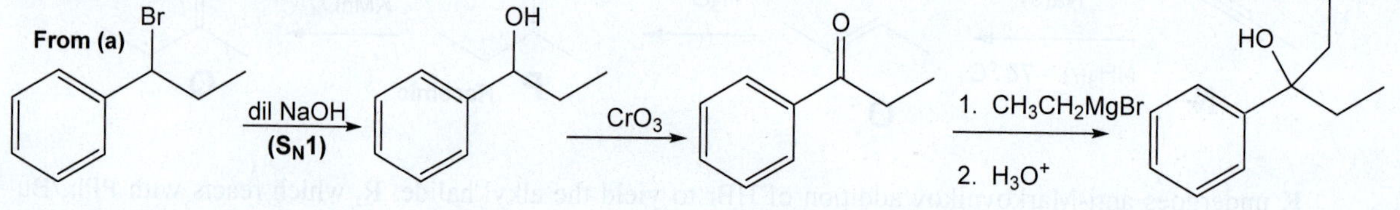

From (a) Br

dil NaOH (S_N1) → OH → CrO_3 → O → 1. CH_3CH_2MgBr 2. H_3O^+ → HO

(c)
Retrosynthesis

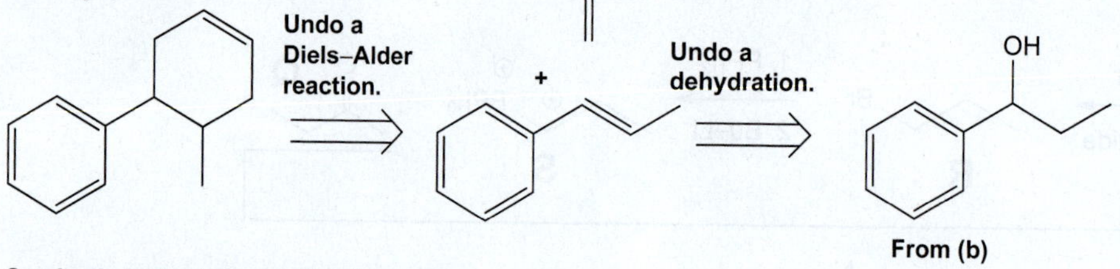

Undo a Diels–Alder reaction. + **Undo a dehydration.** OH

From (b)

Synthesis in the forward direction

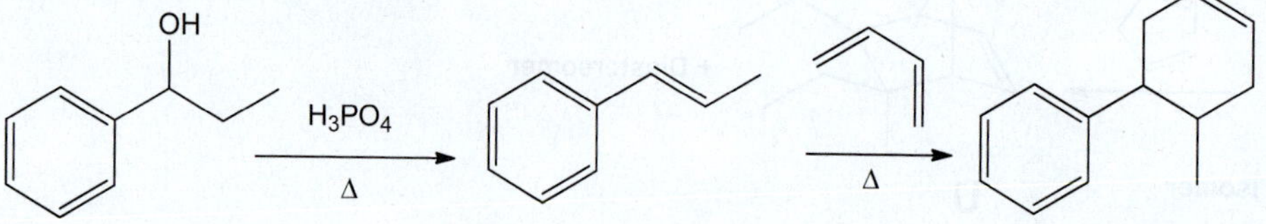

OH → H_3PO_4, Δ → + → Δ →

(d) From (b)

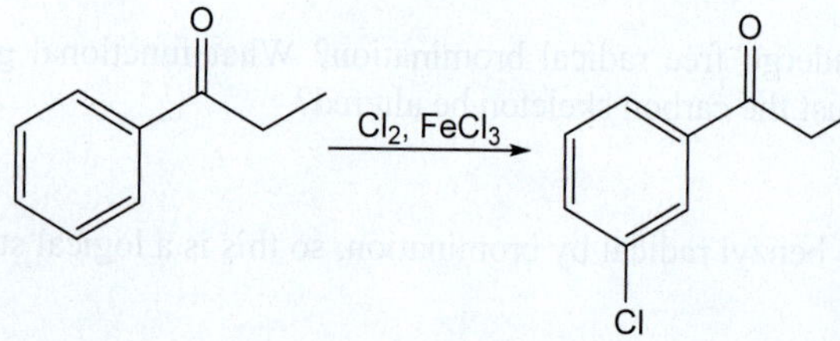

O → Cl_2, $FeCl_3$ → O, Cl

(e)
Retrosynthesis

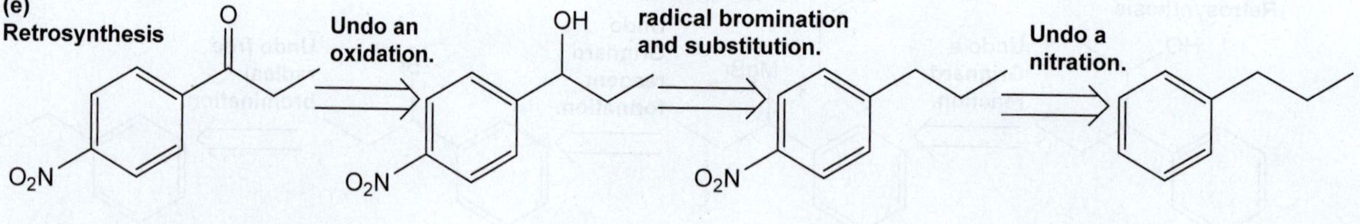

O, O_2N **Undo an oxidation.** OH, O_2N **Undo a free radical bromination and substitution.** O_2N **Undo a nitration.**

Synthesis in the forward direction

HNO_3, H_2SO_4 → O_2N → 1. NBS 2. dil NaOH → OH, O_2N → CrO_3 → O, O_2N

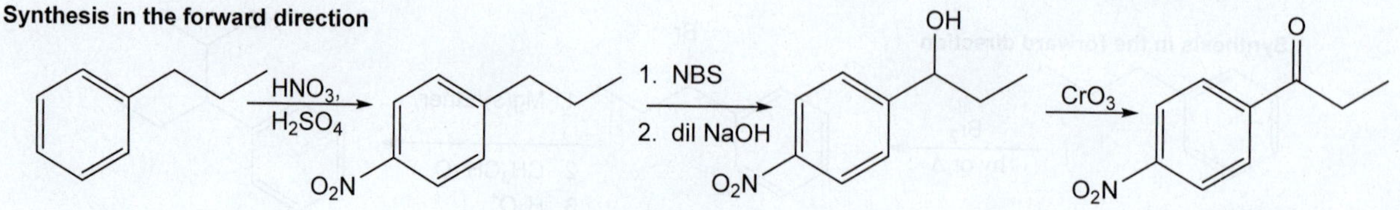

Problem 25.75 (SYN)

Think

Which C atom on 2-methylpropane is most likely to undergo free radical bromination? How can that product react to carry out functional group conversions? How can the carbon skeleton be altered? Are any oxidations or reductions necessary?

Solve

Retrosynthesis and synthesis: We can functionalize the tertiary carbon of 2-methylpropane by bromination, so this is a logical step in each of the syntheses.

(a)
Retrosynthesis

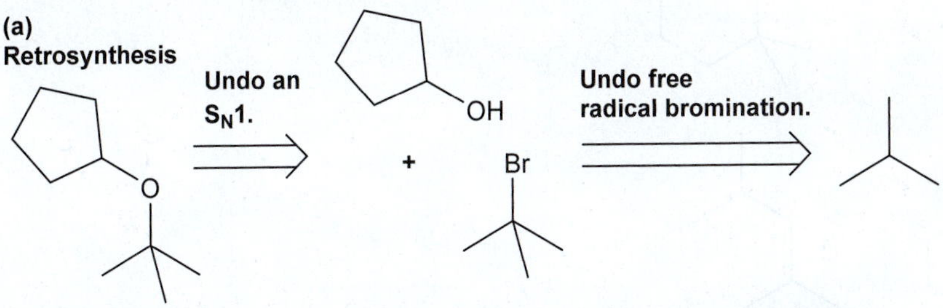

Undo an S_N1. ⟹ OH + Br. Undo free radical bromination. ⟹

Synthesis in the forward direction

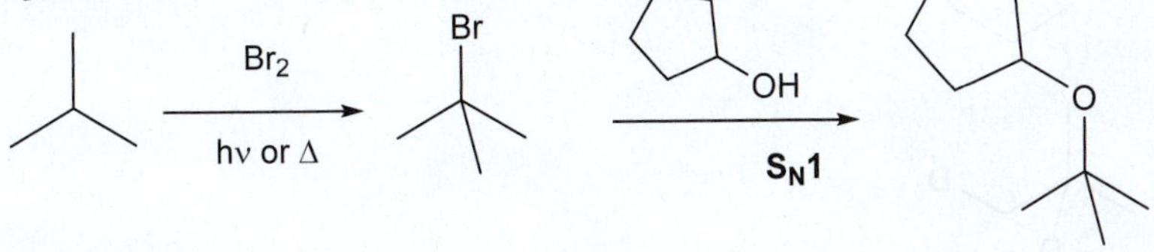

Br_2, hν or Δ → Br; OH, S_N1 → O

(b)
Retrosynthesis

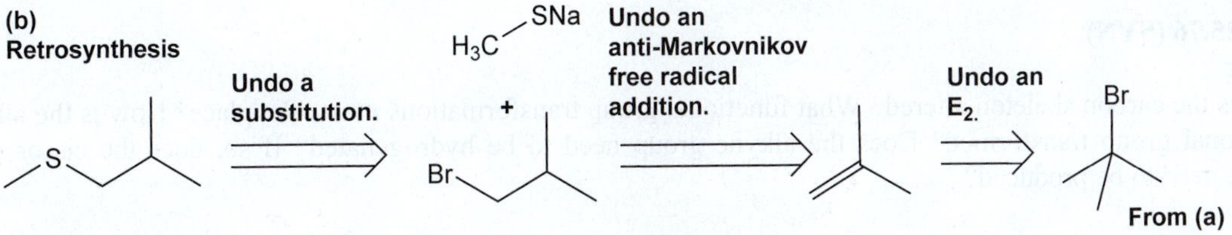

S ⟹ Undo a substitution. H_3C–SNa + Br ⟹ Undo an anti-Markovnikov free radical addition. ⟹ Undo an E_2. ⟹ Br, From (a)

Synthesis in the forward direction

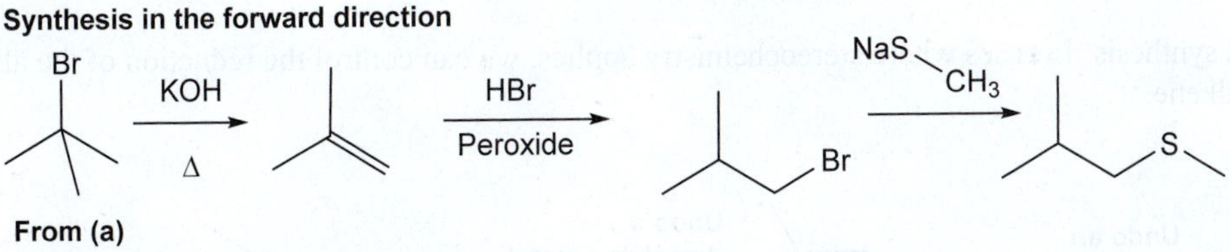

Br (From (a)) → KOH, Δ → HBr, Peroxide → Br → NaS–CH_3 → S

(c)
Retrosynthesis

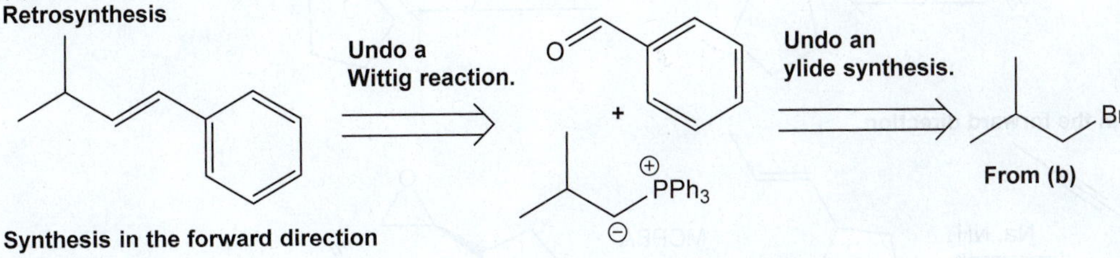

Undo a Wittig reaction. ⟹ O + $\overset{\oplus}{P}Ph_3$ ⊖ ⟹ Undo an ylide synthesis. ⟹ Br, From (b)

Synthesis in the forward direction

Br (From (b)) → 1. PPh_3, 2. BuLi → $\overset{\oplus}{P}Ph_3$ ⊖ → O →

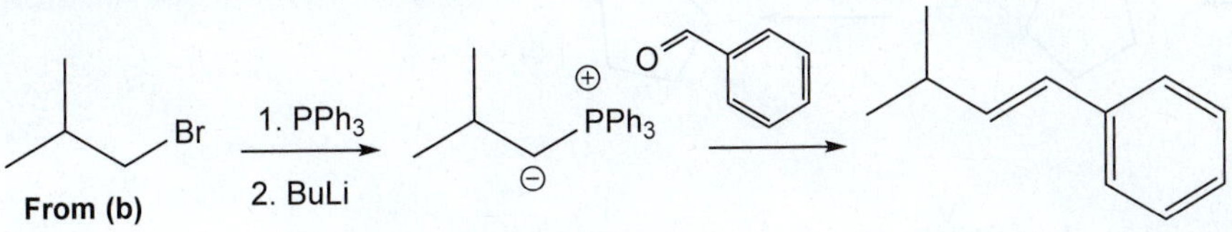

(d)
Retrosynthesis

Undo Diels–Alder

+

From (b)

Synthesis in the forward direction

+ Δ

From (b)

(e)

$\overset{\oplus}{P}Ph_3$ (ylide, $\ominus$ on carbon)

From (c)

(f)

D_2, Pt

D D

From (b)

Problem 25.76 (SYN)

Think

How is the carbon skeleton altered? What functional group transformations must take place? How is the alkyne functional group transformed? Does the alkyne group need to be hydrogenated? If so, does the cis or trans alkene need to be produced?

Solve

Retrosynthesis and synthesis: In cases where stereochemistry applies, we can control the reduction of the alkyne to the cis or trans alkene.

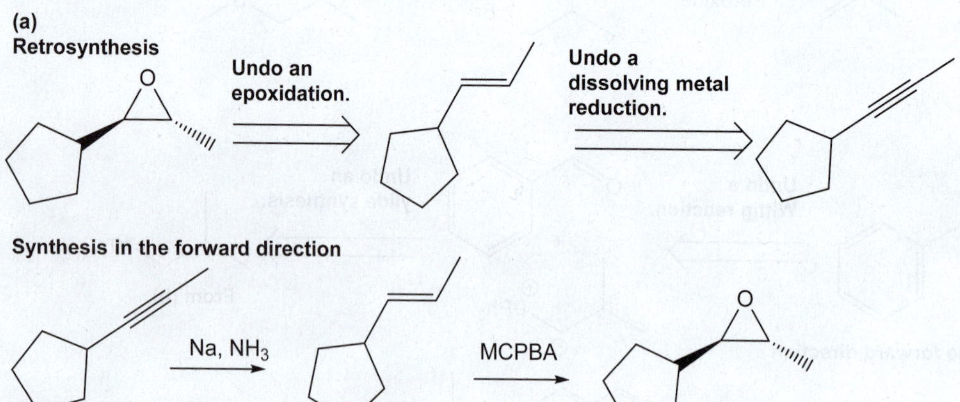

(b)
Retrosynthesis

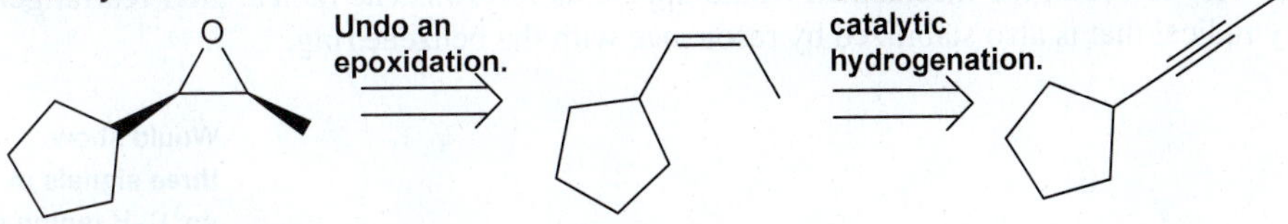

Undo an epoxidation. ⟹ **Undo a catalytic hydrogenation.** ⟹

Synthesis in the forward direction

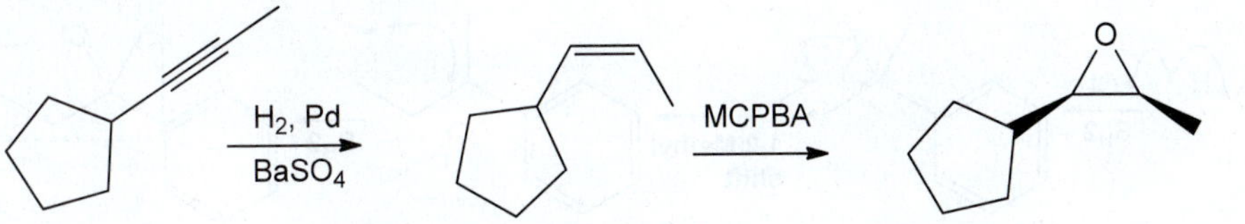

$\xrightarrow[BaSO_4]{H_2,\ Pd}$ $\xrightarrow{MCPBA}$

(c)
Retrosynthesis

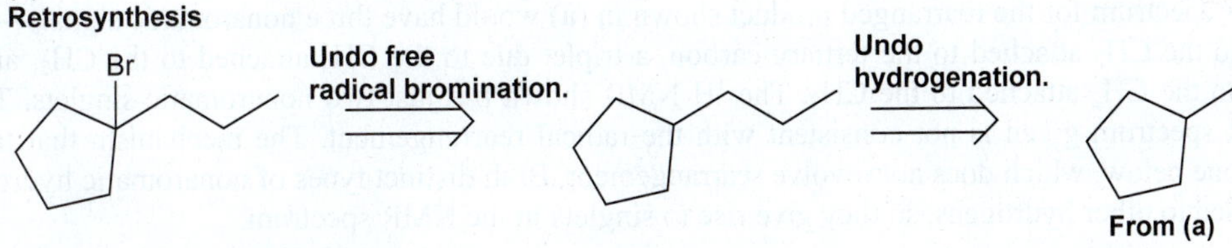

Br

Undo free radical bromination. ⟹ **Undo hydrogenation.** ⟹

From (a)

Synthesis in the forward direction

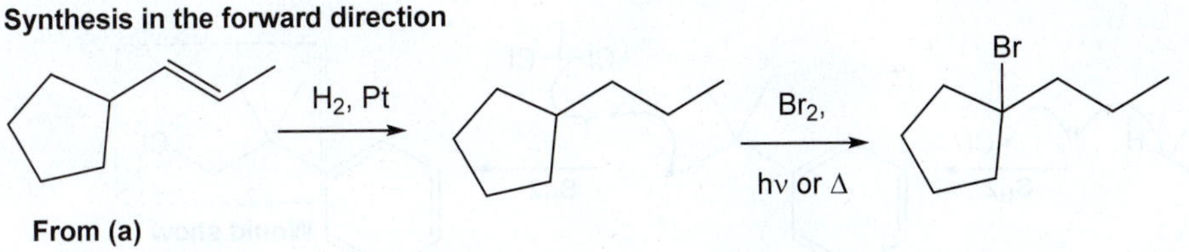

$\xrightarrow{H_2,\ Pt}$ $\xrightarrow[h\nu\ \text{or}\ \Delta]{Br_2,}$ Br

From (a)

(d) $\xrightarrow[\text{(Anti addition)}]{Br_2,}$ Br Br

From (a)

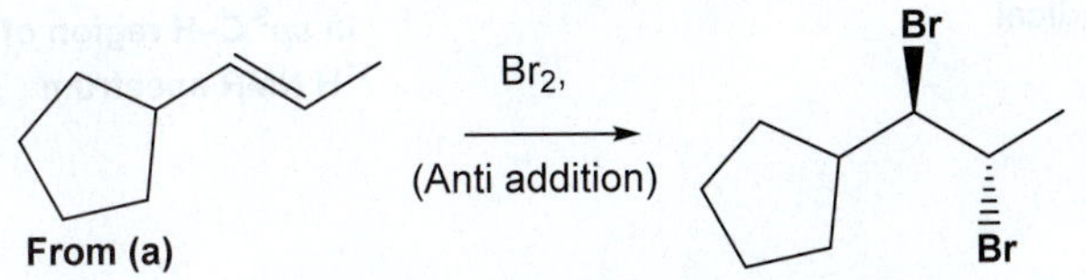

(e) + $\xrightarrow[\text{(Diels–Alder)}]{\Delta}$

From (b)

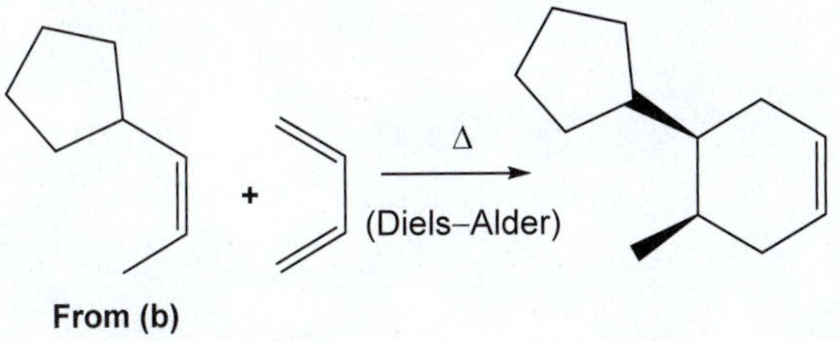

(f) + $\xrightarrow[\text{(Diels–Alder)}]{\Delta}$

From (a)

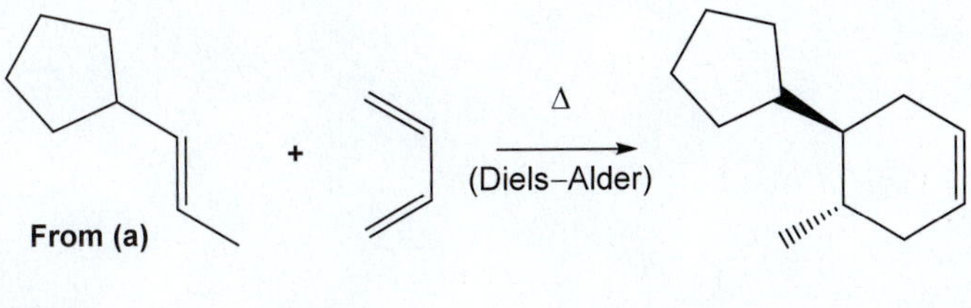

Problem 25.77

Think

What types of carbons are present? What C–H bond undergoes hydrogen atom abstraction? What type of radical forms? Theoretically, is there a more stable radical possible if rearrangement occurs? If so, what product forms? What conclusion do you draw from the ^{1}H NMR spectrum?

Solve

(a) If a rearrangement were to occur, the mechanism would appear as follows. The radical after rearrangement would be a tertiary radical that is also stabilized by resonance with the benzene ring.

S_H2 — 1° Radical — 1,2 Methyl shift — 3° Benzylic radical — S_H2

Would show three signals in *sp*^{3} C–H region of ^{1}H NMR spectrum

(b) The ^{1}H NMR spectrum for the rearranged product shown in **(a)** would have three nonaromatic signals—one singlet due to the CH_3 attached to the tertiary carbon, a triplet due to the CH_3 attached to the CH_2, and a quartet due to the CH_2 attached to the CH_3. The ^{1}H NMR shown has just two nonaromatic singlets. Thus the ^{1}H NMR spectrum given is not consistent with the radical rearrangement. The mechanism that takes place is the one below, which does not involve rearrangement. Both distinct types of nonaromatic hydrogen are not coupled to other hydrogens, so they give rise to singlets in the NMR spectrum.

S_H2 — 1° Radical — S_H2

Would show two singlet signals in *sp*^{3} C–H region of ^{1}H NMR spectrum

INTERCHAPTER G | Fragmentation Pathways in Mass Spectrometry

Your Turn Exercises

Your Turn G.1

Think

What is the identity of the ion that corresponds to the peaks at $m/z = 43$ and $m/z = 57$? Which ion is more stable (Section 6.6e)? Does cation or radical stability win out?

Solve

A mass peak at $m/z = 43$ would correspond to a tertiary isopropyl cation $(CH_3)_2CH^+$, and a mass peak at $m/z = 57$ would correspond to a primary cation $CH_3CH_2CH_2CH_2^+$. The two possible fragmentation pathways are shown below.

Butyl radical + Isopropyl cation $m/z = 43$

Butyl cation $m/z = 57$ + Isopropyl radical

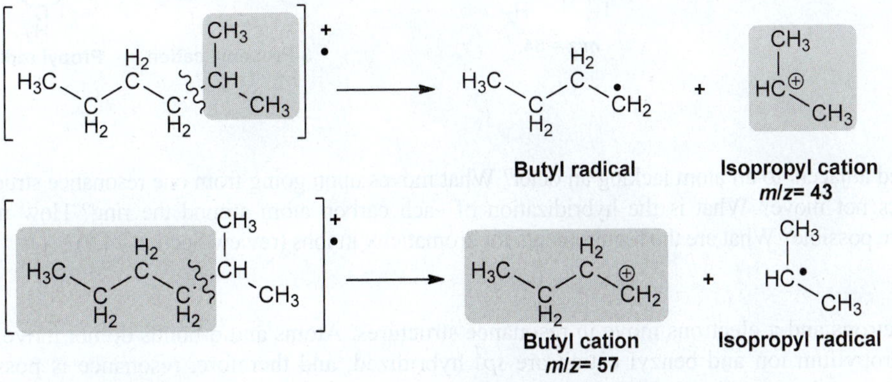

The tertiary isopropyl cation is more stable than the primary butyl cation; therefore, the fragmentation is more likely to produce a mass peak at $m/z = 43$. Cation stability wins out over radical stability, because cations are more electron deficient.

Your Turn G.2

Think

How can the molecular ion fragment produce ions that correspond to the peaks at $m/z = 71$ and $m/z = 29$? Which ion is more stable (Section 6.6e)? Which peak is more intense? How is the intensity of the peak related to the stability of the ion produced?

Solve

Figure G-3 is the mass spectrum of 2-methylpentane. The peak at $m/z = 29$ is from an ethyl cation $CH_3CH_2^+$, and the peak at $m/z = 71$ is from the fragmentation of the methyl group to give $CH_3CH^+CH_2CH_2CH_3$. The fragmentation pathways are shown below.

Isobutyl radical + Ethyl cation $m/z = 29$

2-Pentyl cation $m/z = 71$ + Methyl radical

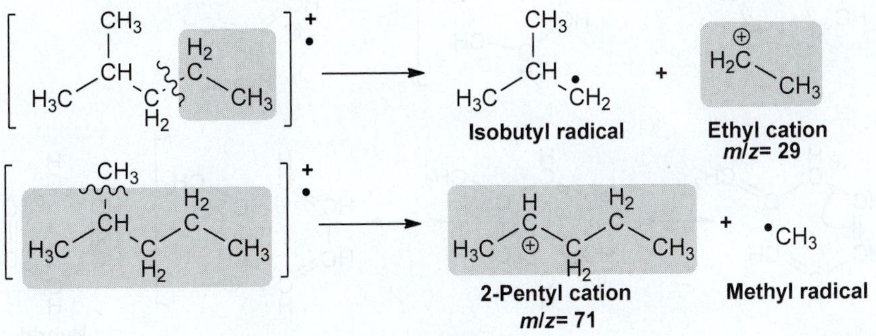

The secondary 2-pentyl cation is more stable than the primary ethyl cation; therefore, the peak is more intense.

Your Turn G.3

Think

What type of fragmentation is common to the molecular ions of alkenes? What is the driving force for this process? What bonds break and form? What kind of curved arrow should be used to show that electron movement?

Solve

An alkene's molecular ion tends to expel an alkyl radical from an allylic carbon, producing a resonance-stabilized allylic cation. This fragmentation of the original allylic C–C bond is common to alkenes, which results in the formation of the allylic 3-propenyl cation. Three single-barbed curved arrows are required to illustrate the electron movement in this fragmentation.

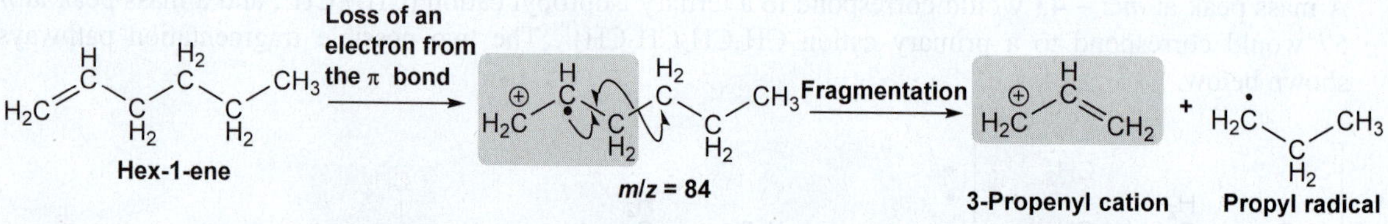

Hex-1-ene —Loss of an electron from the π bond→ $m/z = 84$ —Fragmentation→ 3-Propenyl cation + Propyl radical

Your Turn G.4

Think

Is there a multiple bond adjacent to an atom lacking an octet? What moves upon going from one resonance structure to another? What does not move? What is the hybridization of each carbon atom around the ring? How many resonance structures are possible? What are the requirements for aromaticity in ions (review Section 14.9)?

Solve

Only nonbonding electrons and π electrons move in resonance structures. Atoms and σ bonds do not move. All of the atoms in the tropylium ion and benzyl cation are sp^2 hybridized, and therefore, resonance is possible around the entire ring. For the tropylium ion, there are seven equivalent resonance structures, and all of them exhibit aromaticity (six π electrons completely conjugated in the ring). For the benzylic cation, there are four resonance structures that delocalize the positive charge, but only the first one exhibits aromaticity. The other three do not have a Hückel number of π electrons in the ring.

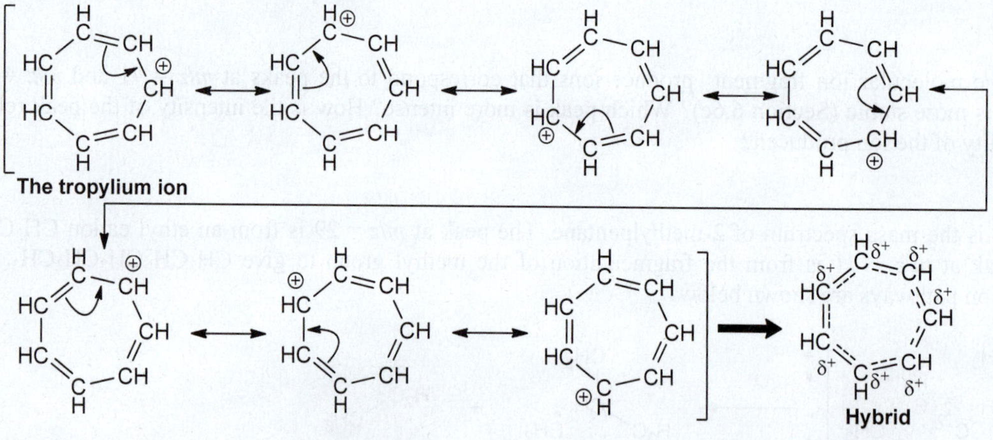

The tropylium ion — Hybrid

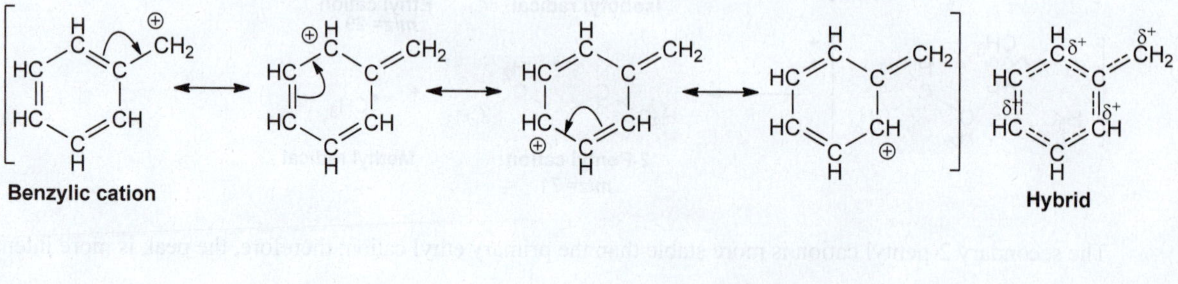

Benzylic cation — Hybrid

Your Turn G.5

Think

Refer to Equation G-11. Which bonds are broken? Which new bonds form? Does the electron movement occur via radicals or two electrons? What types of curved arrows are required?

Solve

The ethyl radical, bonded to the carbon atom that is attached to the Cl, is eliminated. This is α-cleavage and, in this example, requires three single-barbed curved arrows, as shown below.

α-Cleavage → $CH_3CH_2^{\bullet}$ + ; Ethyl radical; $m/z = 77$

Your Turn G.6

Think

What are the masses of diisopropyl ether and pentan-1-ol? What electron is lost to produce the molecular ion? What mass is lost to give the peak at $m/z = 87$ and 31, respectively? What type of fragmentation is common to the molecular ions of ethers and alcohols?

Solve

Diisopropyl ether has a mass of 102 u. The parent ion is produced from the loss of a nonbonding electron on O. The peak at $m/z = 87$ is due to the loss of 15 u from the molecular ion, which is a methyl radical, $CH_3\bullet$. This leaves the cation $^{+}CH(CH_3)–O–CH(CH_3)_2$ at $m/z = 87$. Therefore, the peak at $m/z = 87$ corresponds to loss of a methyl radical from M^+ via α-cleavage.

$m/z = 102$ α-Cleavage → $^{\bullet}CH_3$ + $m/z = 87$

Pentan-1-ol has a mass of 88 u. The parent ion is produced from the loss of a nonbonding electron on O. The peak at $m/z = 31$ is due to the loss of 57 u from the molecular ion, which is a butyl radical, $CH_3CH_2CH_2CH_2\bullet$. This leaves the cation $^{+}CH_2–OH$ at $m/z = 31$. Therefore, the peak at $m/z = 31$ corresponds to loss of a butyl radical from M^+ via α-cleavage.

$m/z = 88$ α-Cleavage → + $m/z = 31$

Your Turn G.7

Think

What is the mass of hexan-2-one? What electron is lost to produce the molecular ion? What mass is lost from the molecular ion to give the peak at $m/z = 85$? What type of fragmentation occurred?

Solve

Hexan-2-one has a mass of 100 u. A nonbonding electron from O is lost to produce the molecular ion. The peak at $m/z = 85$ is due to the loss of 15 u from the molecular ion, which is the loss of a methyl radical, $CH_3\bullet$. This leaves the cation $CH_3CH_2CH_2CH_2C{\equiv}O^+$ at $m/z = 85$. The loss of a methyl radical from M^+ occurs via α-cleavage. See the figure on the next page.

Loss of nonbonding e⁻

α-Cleavage

Hexan-2-one

m/z = 100

An acylium ion
m/z = 85

Your Turn G.8

Think

What is the mass of pentanoic acid? Which electron is lost to produce the molecular ion? What mass is lost to give the peak at $m/z = 60$? What type of conformation does the molecule need to be in for the McLafferty rearrangement to occur?

Solve

The mass of pentanoic acid is 102 u. The peak at $m/z = 60$ is due to the loss of a prop-1-ene, $CH_3CH{=}CH_2$ (mass 42 u), and the formation of the enol radical cation $CH_2{=}C(^{+}OH)OH$. This is a McLafferty rearrangement of a carboxylic acid.

Pentanoic aicd
m/z = 102

An enol radical cation
m/z = 60

End of Chapter Problems

Problem G.1

Think

What are the structures of butylcyclopentane and *tert*-butylcyclopentane? What are likely fragmentation ions for each? What are the corresponding masses of each cation fragment? What peaks are unique to each spectrum?

Solve

The major differences in the two spectra are the peaks at *m*/*z* values 126, 97, and 83 for Spectrum **A** and at the *m*/*z* value of 111 for Spectrum **B**. The structures for butylcyclopentane and *tert*-butylcyclopentane and possible fragmentation ions are shown below.

Butylcyclopentane Spectrum A → **Ethylcyclopentyl cation** *m*/*z* = 97 + **Ethyl radical** ($H_2\dot{C}-CH_3$)

***tert*-Butylcyclopentane Spectrum B** → **Isopropylcyclopentyl cation** *m*/*z* = 111 + **Methyl radical** ($\dot{C}H_3$)

The peak in Spectrum **A** at *m*/*z* = 97 is from the loss of an ethyl radical fragment, and this is only possible for butylcyclopentane. The peak in Spectrum **B** at *m*/*z* = 111 is from the loss of a methyl radical, which is more likely to occur in *tert*-butylcyclopentane owing to the stability of the tertiary cation. Therefore, Spectrum **A** is from butylcyclopentane and Spectrum **B** is from *tert*-butylcyclopentane. Notice that the M^+ peak for **B** is absent, because the 3° carbocation that is produced upon fragmentation is a relatively stable cation.

Problem G.2

Think

What is the significance of the base peak? What is the formula for an ion weighing 57 u? What other atoms would need to be present to account for the molecular formula of the alkane? What, then, is the identity of the ion that corresponds to the peak at *m*/*z* = 57? Why is this ion the base peak?

Solve

The base peak is the tallest peak and corresponds to the most stable ion among the molecular and fragment ions. The peak at *m*/*z* = 57 corresponds to a formula of C_4H_9, which is a *tert*-butyl cation $(CH_3)_3C^+$. If the molecule has a formula of C_7H_{16}, the remaining atoms in the formula are C_3H_7, which could be an isopropyl fragment (in that case, the molecule is 2,2,3-trimethylbutane) or a propyl fragment (in that case, the molecule is 2,2-dimethylpentane). The two possible isomers are shown below and on the next page.

Chemical formula: C_7H_{16} **2,2,3-Trimethylbutane** → ***tert*-Butyl cation** *m*/*z*= 57 + **Isopropyl radical**

Chemical formula: C_7H_{16}
2,2-Dimethylpentane

tert-Butyl cation
m/z = 57

Propyl radical

Problem G.3

Think

What is the mass of hept-3-ene? What mass is lost to give the peak m/z = 83? What common fragmentation pathways are available to the molecular ion of an alkene? What is the identity of the peak at m/z = 69 (refer to Equation G-6)? Which fragment ion is more stable? Which fragment radical is more stable?

Solve

The mass of hept-3-ene is 98 u. The π electron of the C=C is ejected in the ionization process to produce the molecular ion. Hept-3-ene is not symmetrical, and there are two allylic C–C bonds, so fragmentation can happen via two routes, as shown below. The peak m/z = 83 is from the loss of a fragment with 15 u, which is a methyl radical (route 1). The peak m/z = 69 is from the loss of a fragment with 29 u, which is an ethyl radical (route 2). The peak at m/z = 83 is smaller than the peak at m/z = 69, meaning that route 2 is more likely. Both fragment ions are similar in stability, because the positive charge is shared via resonance over a secondary and a primary carbon, so the difference is in the stabilities of the fragment radicals. The ethyl radical is more stable than the methyl radical.

Hept-3-ene

Loss of an electron from the π bond

Route 1 — *m/z* = 98 → 1-Hexenyl cation *m/z* = 83 + Methyl radical

Route 2 — *m/z* = 98 → Ethyl radical + 1-Pentenyl cation *m/z* = 69

Problem G.4

Think

What is the IHD for the formula C_8H_{16}? What is the identity of the ion that corresponds to the peaks at m/z = 41? What is an isomer that is consistent with giving this fragment? Are other isomers possible?

Solve

The IHD for C_8H_{16} is 1, and the molecule has C=C. The mass peak at m/z = 41 corresponds to a propenyl cation, $CH_2{=}CH{-}CH_2^+$. The mechanism for how hex-1-ene expels the propenyl cation is shown on the next page. Many other possible isomers (a few are shown) with the formula C_8H_{16} would give this fragment. The essential part is that the C=C must be terminal to produce the fragment ion $CH_2{=}CH{-}CH_2^+$.

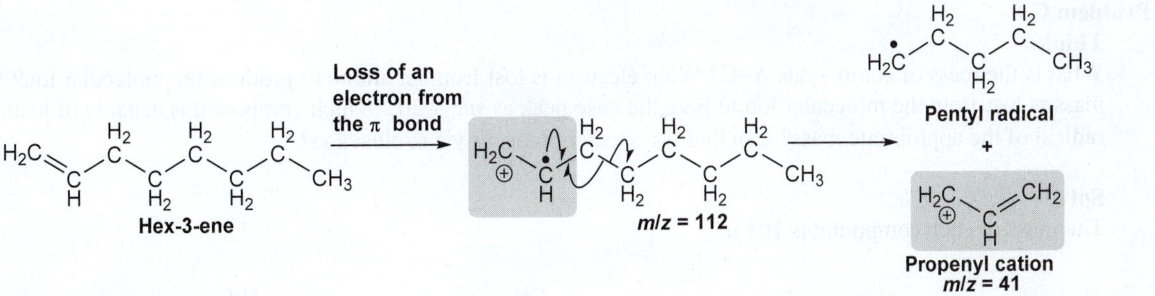

Loss of an electron from the π bond

Hex-3-ene

m/z = 112

Pentyl radical

Propenyl cation
m/z = 41

Other possible isomers, formula C_8H_{16}

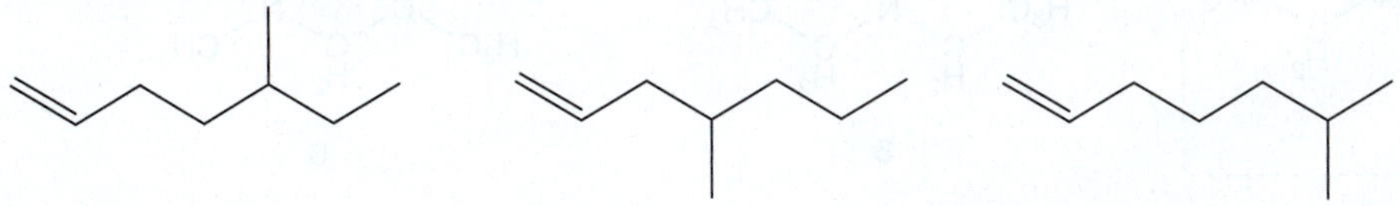

Problem G.5

Think

What are the masses of 1,4-diethylbenzene and 1-methyl-4-propylbenzene? What radical is lost and gives rise to the peak m/z = 105 for Spectrum **A**, and which is lost to give rise to the peak m/z = 119 for Spectrum **B**? Which compound is capable of losing that radical?

Solve

Both 1,4-diethylbenzene and 1-methyl-4-propylbenzene have a mass of 134 u. The base peak in Spectrum **A** at m/z = 105 is due to loss of an ethyl fragment (29 u), and the base peak in Spectrum **B** at m/z = 119 is due to loss of a methyl fragment (15 u). The M^+ of **A** can lose a CH_3 radical via the cleavage of an original benzylic C–C bond, and the M^+ of **B** can lose an ethyl radical via the cleavage of an original benzylic C–C bond. Therefore, 1,4-diethylbenzene produces Spectrum **B**, and 1-methyl-4-propylbenzene produces Spectrum **A**.

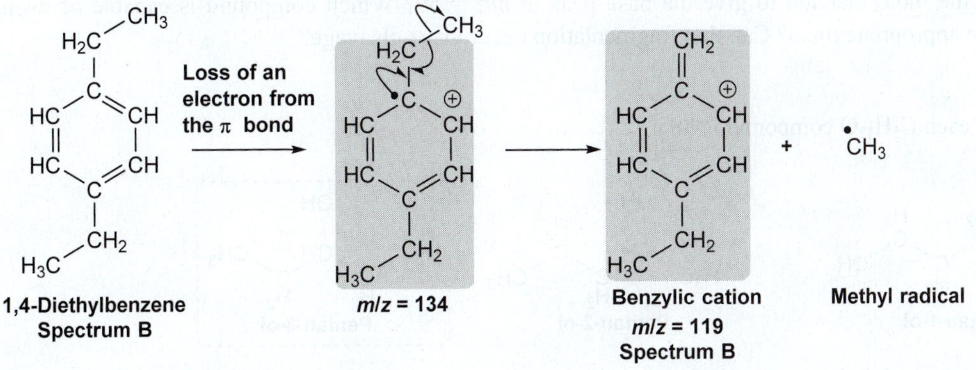

1,4-Diethylbenzene
Spectrum B

Loss of an electron from the π bond

m/z = 134

Benzylic cation
m/z = 119
Spectrum B

Methyl radical

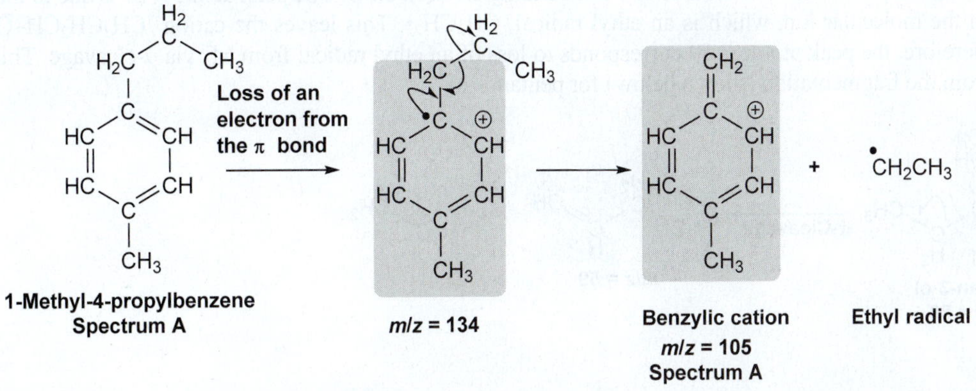

1-Methyl-4-propylbenzene
Spectrum A

Loss of an electron from the π bond

m/z = 134

Benzylic cation
m/z = 105
Spectrum A

Ethyl radical

Problem G.6

Think

What is the mass of compounds **A**–**C**? What electron is lost from an amine to produce the molecular ion? What mass is lost from the molecular ion to give the base peak at $m/z = 30$? Which compound is capable of losing the radical of the appropriate mass? Can that fragmentation occur via α-cleavage?

Solve

The mass of each compound is 101 u.

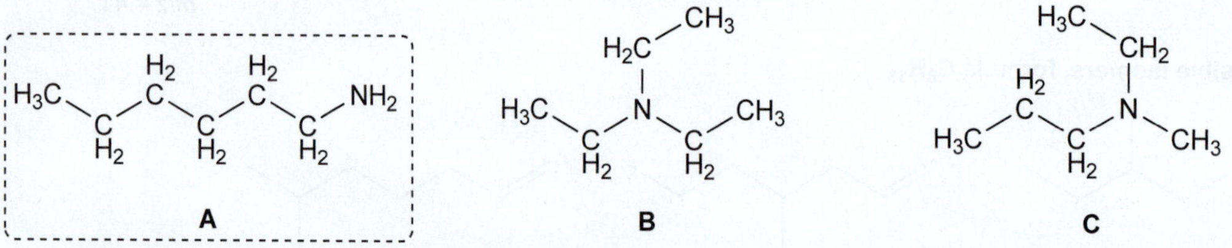

In each case, a nonbonding electron is lost from N to produce the molecular ion. The peak at $m/z = 30$ is due to the loss of a pentyl radical, $CH_3CH_2CH_2CH_2CH_2\bullet$, to form the iminium cation, $H_2C{=}^+NH_2$. Compound **A** is the only compound capable of losing a pentyl radical from α-cleavage.

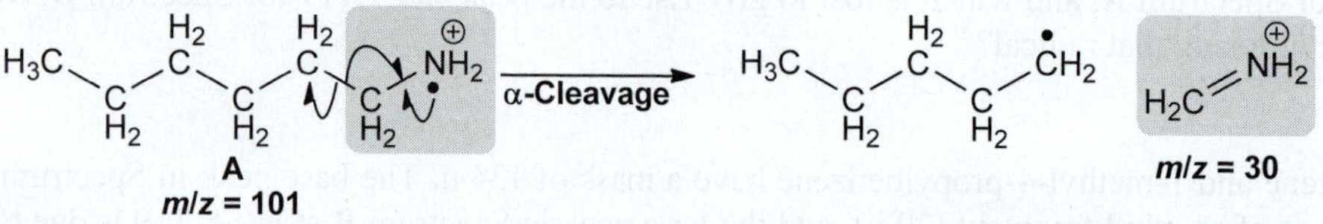

Problem G.7

Think

What is the mass of $C_5H_{12}O$? What are the structures of pentan-1-ol, pentan-2-ol, and pentan-3-ol? What mass is lost from the molecular ion to give the base peak at $m/z = 59$? Which compound is capable of losing the radical of the appropriate mass? Can that fragmentation occur via α-cleavage?

Solve

The mass of each $C_5H_{12}O$ compound is 88 u.

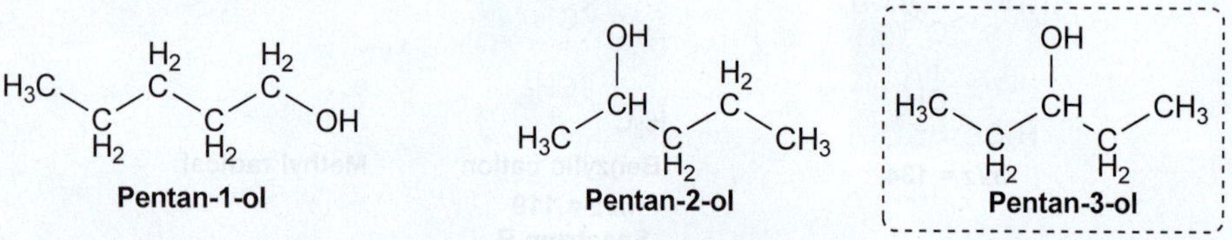

Pentan-1-ol　　Pentan-2-ol　　Pentan-3-ol

The molecular ion is produced from the loss of a nonbonding electron on O. The peak at $m/z = 59$ is due to the loss of 29 u from the molecular ion, which is an ethyl radical, $CH_3CH_2\bullet$. This leaves the cation $^+CH_3CH_2CH{-}OH$ at $m/z = 59$. Therefore, the peak at $m/z = 59$ corresponds to loss of an ethyl radical from M^+ via α-cleavage. This can only come from the fragmentation (shown below) for pentan-3-ol.

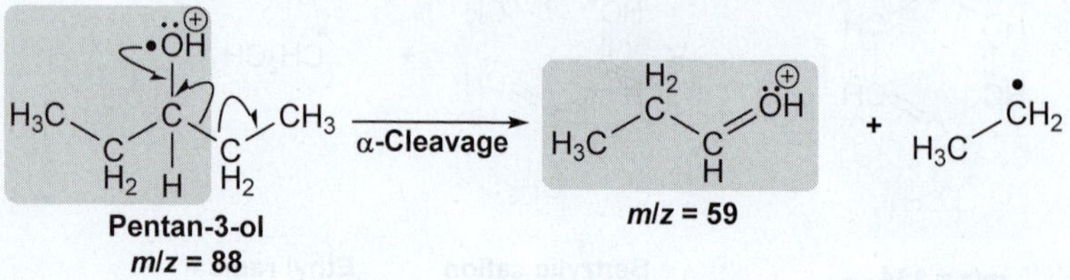

Pentan-3-ol $m/z = 88$ —α-Cleavage→ $m/z = 59$ + $H_3C{-}CH_2\bullet$

Problem G.8

Think

What are the structures of heptan-2-one, heptan-3-one, and heptan-4-one? What is the mass of these three compounds? Which electron is lost to produce the molecular ion? What mass is lost to give the peak at m/z = 57? Which compound is capable of losing the radical of the appropriate mass through a common fragmentation pathway of ketones?

Solve

The mass of each compound is 114 u.

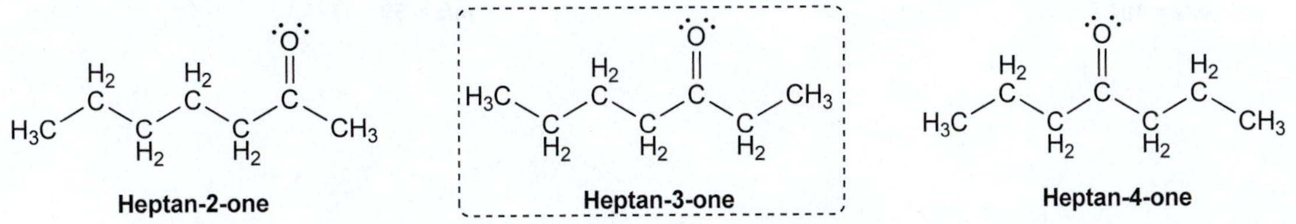

Heptan-2-one **Heptan-3-one** **Heptan-4-one**

In each case, the molecular ion is produced from the loss of a nonbonding electron on O. The peak at m/z = 57 is due to the loss of a butyl radical, $CH_3CH_2CH_2CH_2\bullet$ (57 u), to produce the acylium cation, $CH_3CH_2C{\equiv}O^+$. Heptan-3-one is the only compound capable of losing a butyl radical from α-cleavage.

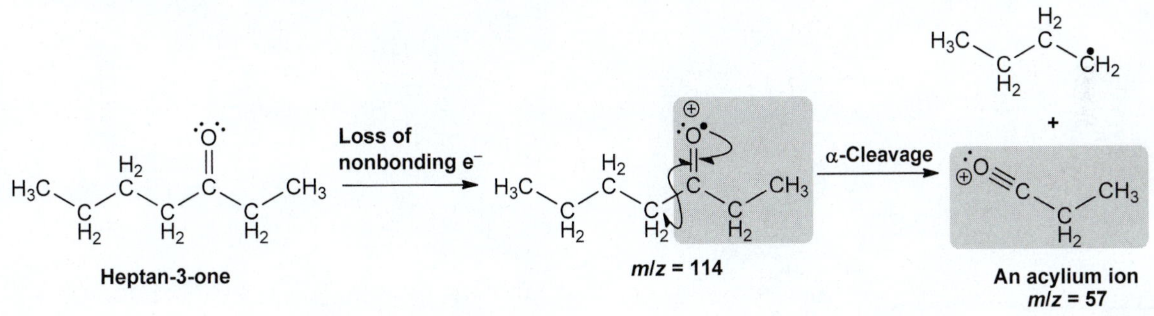

Heptan-3-one —Loss of nonbonding e⁻→ m/z = 114 —α-Cleavage→ **An acylium ion** m/z = 57

Problem G.9

Think

What is the mass of hexanamide? Which electron is lost to produce the molecular ion? What mass is lost to give the peak at m/z = 59? What type of fragmentation occurred?

Solve

The mass of hexanamide is 115 u. The peak at m/z = 59 is due to the loss of a but-1-ene, $CH_3CH_2CH{=}CH_2$, and the formation of the enol radical cation $CH_2{=}C(^+OH)NH_2$. This is a McLafferty rearrangement of the amide's molecular ion.

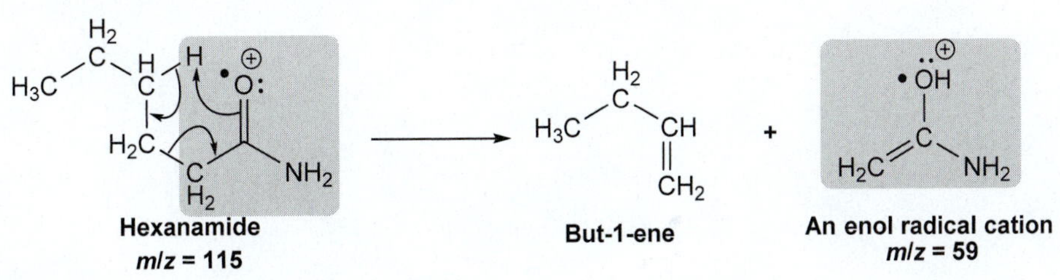

Hexanamide m/z = 115 → **But-1-ene** + **An enol radical cation** m/z = 59

Problem G.10

Think

What is the mass of $C_5H_{11}NO$? Which electron is lost to produce the molecular ion? What mass is lost to give the peak at m/z = 59? What type of fragmentation occurred?

Solve

The mass of $C_5H_{11}NO$ is 101 u. The peak at m/z = 59 is due to the loss of a propene, $CH_3CH{=}CH_2$, and the formation of the enol radical cation $CH_2{=}C(^{+}OH)NH_2$. This is a McLafferty rearrangement of an amide.

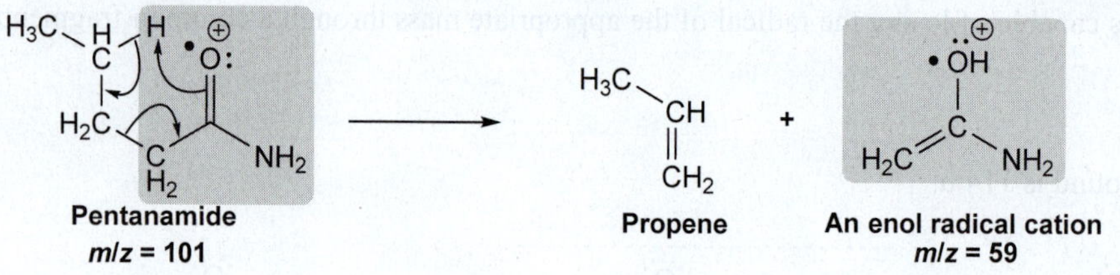

Pentanamide
m/z = 101

Propene

An enol radical cation
m/z = 59

CHAPTER 26 | Polymers

Your Turn Exercises

Your Turn 26.1

Think

Which carbon atoms make up the main chain? Which groups are not part of this main chain?

Solve

The carbon atoms that make up the main chain or backbone are the carbon atoms that are connected to each other, as shown below (boxed). The pendant (or side) groups are the groups "hanging" off the main chain. In this example, the pendant groups are CH_3, or methyl groups.

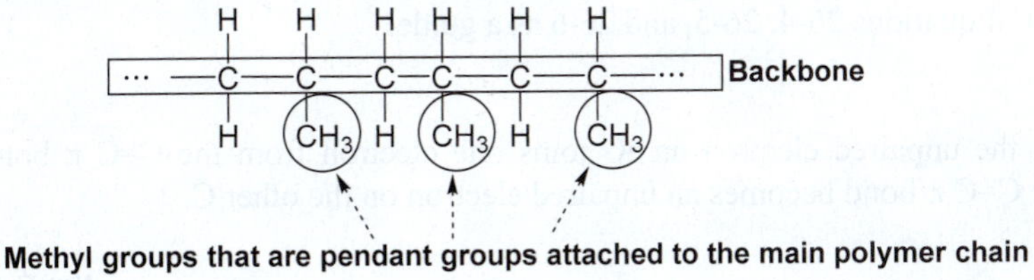

Backbone

Methyl groups that are pendant groups attached to the main polymer chain

Your Turn 26.2

Think

What atoms and bonds make up the main chain? Are the groups attached to the second carbon the same as the ones attached to the first? What about the third carbon? The fourth? What pattern is established, and how many times is it shown to repeat?

Solve

The main chain consists of C–C single bonds only. Two H atoms are attached to the first C shown, whereas the second C has an attached H and CH_3. That pattern is repeated for each of the next pairs of C atoms along the main chain, so the repeating unit involves just two atoms of the main chain. In the structure given, the repeating unit appears three times, so $n = 3$.

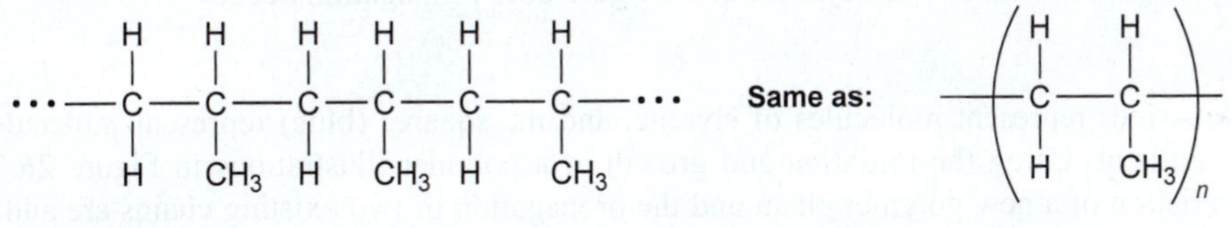

Same as:

Your Turn 26.3

Think

What type of curved arrows is used in forming a radical via homolysis? How many electrons are represented by each curved arrow? Which bonds are broken and formed? To draw a resonance structure of the resulting radical, can you identify a π bond adjacent to the atom that has the unpaired electron? Where do the unpaired electron and the electrons from the π bond end up, and how many curved arrows must be used to show that movement?

Solve

To show electron movement involving radicals, use single-barbed curved arrows; the movement of one electron is represented by each curved arrow. Thus, to move two electrons, two arrows are needed, and to move three electrons, three arrows are needed. In the first equation on the next page, the O–O bond is homolytically cleaved to form the oxygen radical, •O. In that radical, the O that has that unpaired electron is attached to a C=O bond. To arrive at the other resonance structure, the unpaired electron joins with one electron from the C=O π bond, and the other electron from the π bond ends up on the second O. Three single-barbed arrows are used to show this electron movement.

Initiation
Δ
Homolysis
2

Resonance

Your Turn 26.4

Think

Which atoms form the new bond? How is this represented by single-barbed curved arrows? On which atom will an unpaired electron appear? Refer to Equations 26-4, 26-5, and 26-6 as a guide.

Solve

A new C–C bond is formed when the unpaired electron on •C joins one electron from the C=C π bond on styrene. The other electron from the C=C π bond becomes an unpaired electron on the other C.

Propagation
New C–C bond
Radical addition
Reactive polymer chain (n = 3)
Reactive polymer chain (n = 4)

Your Turn 26.5

Think

Refer to Figure 26-7a to 26-7d. How is initiation shown? How is propagation shown? From which symbol in the figure does initiation originate? From which symbol in the figure does propagation occur?

Solve

In Figure 26-7a, the dots (red) represent molecules of styrene, and the squares (blue) represent molecules of benzoyl peroxide—the initiator. Using the initiation and growth of a polymer illustrations in Figure 26-7a to 26-7d as a guide, the initiation of a new polymer chain and the propagation of two existing chains are added to Figure 26-7e in the figure below.

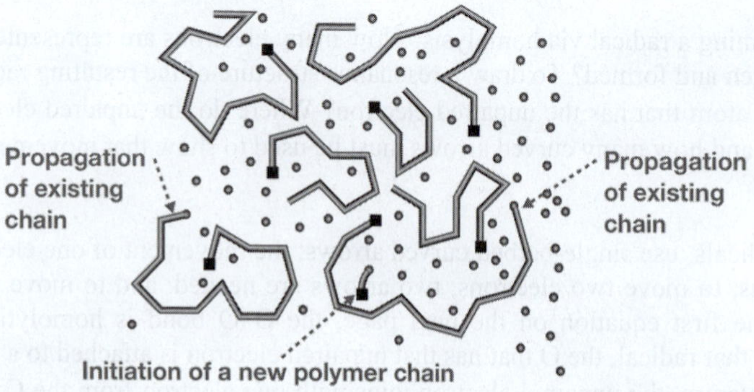

Your Turn 26.6

Think

What are the final radicals in Solved Problem 26.4? What are the reactants and products of a combination step? What are the reactants and products of a disproportionation step? How can you incorporate single-barbed curved arrows to show the appropriate bonds forming and breaking in these steps?

Solve

(a) Two radicals undergo combination to form a new bond in a termination step. Two single-barbed curved arrows are used to show the formation of this bond.

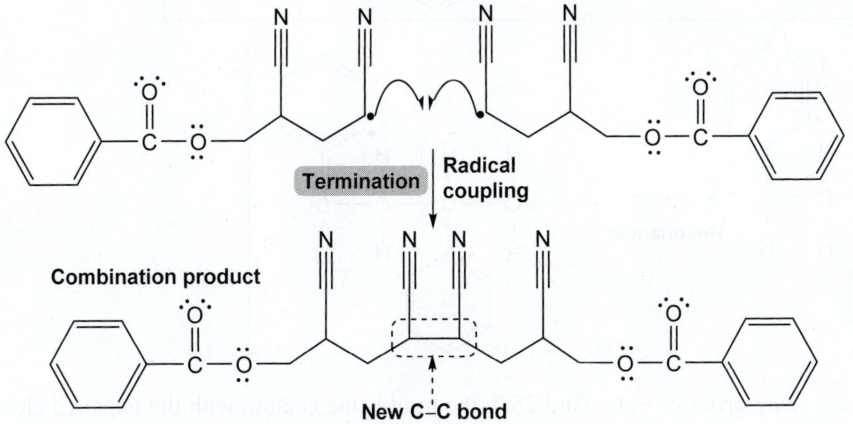

Termination

Radical coupling

Combination product

New C–C bond

(b) Disproportionation generally involves H atom abstraction from the atom adjacent to the C• on one of the growing polymers. Four single-barbed curved arrows are used to show the breaking of the C–H σ bond and the joining of the two resulting unpaired electrons to form the new C=C π bond.

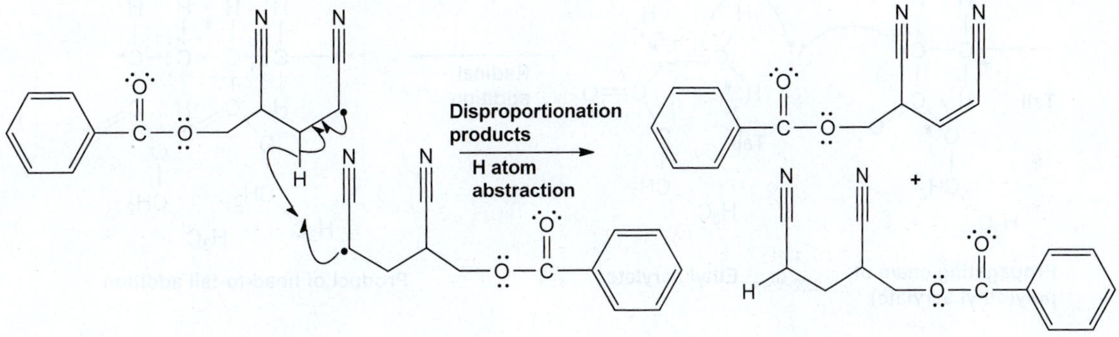

Disproportionation products

H atom abstraction

Your Turn 26.7

Think

Review Section 25.2 for drawing resonance structures of radicals. On the polymer and monomer shown in this problem, which part is the head and which part is the tail? Making sure that the two species are aligned head to tail, draw the curved arrows to show the next radical addition step. How many curved arrows are necessary? In the product, is there an unpaired electron on an atom that is adjacent to a π bond?

Solve

The four resonance structures for the product of head-to-tail addition in Figure 26.11 are shown on the next page.

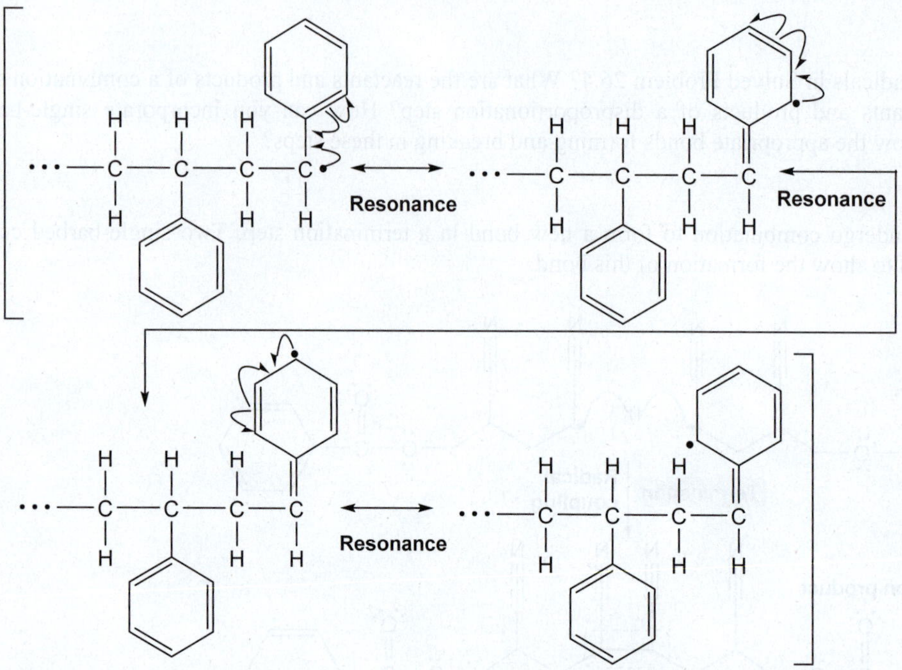

Resonance

Resonance

Resonance

In the growing polymer chain given in Your Turn 26.7, the head is the C atom with the unpaired electron. In a monomer of ethyl acrylate, the tail is the C atom attached to two H atoms. Three single-barbed arrows are required to show radical addition to the C=C bond, as shown below for the head-to-tail bond formation.

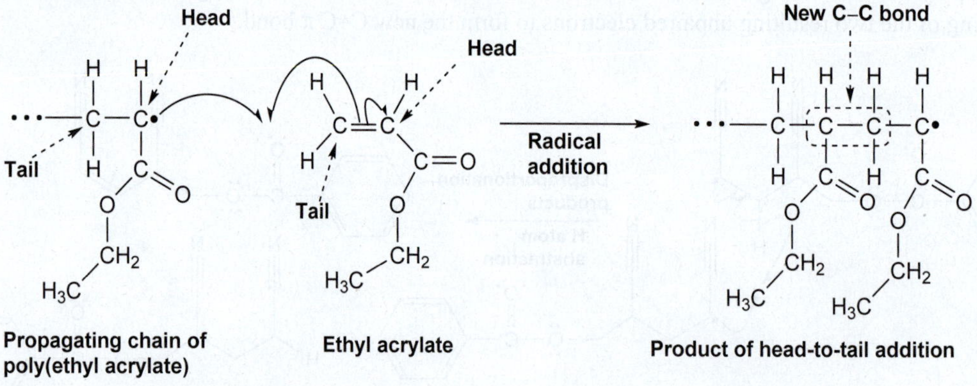

Head

Head

New C–C bond

Tail

Tail

Radical addition

Propagating chain of poly(ethyl acrylate)

Ethyl acrylate

Product of head-to-tail addition

In the product, the unpaired electron is on an atom attached to the C=O π bond, so it has the additional resonance structure shown below:

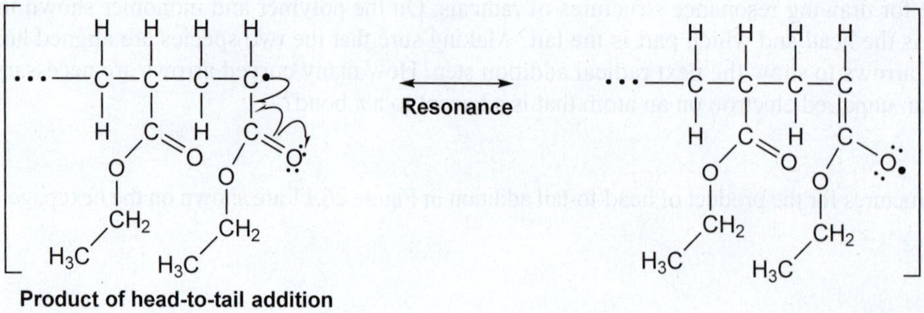

Resonance

Product of head-to-tail addition

In head-to-head addition, shown below, there is no resonance stabilization of the product. Also, the steric interactions between the acrylate groups will hinder head-to-head addition.

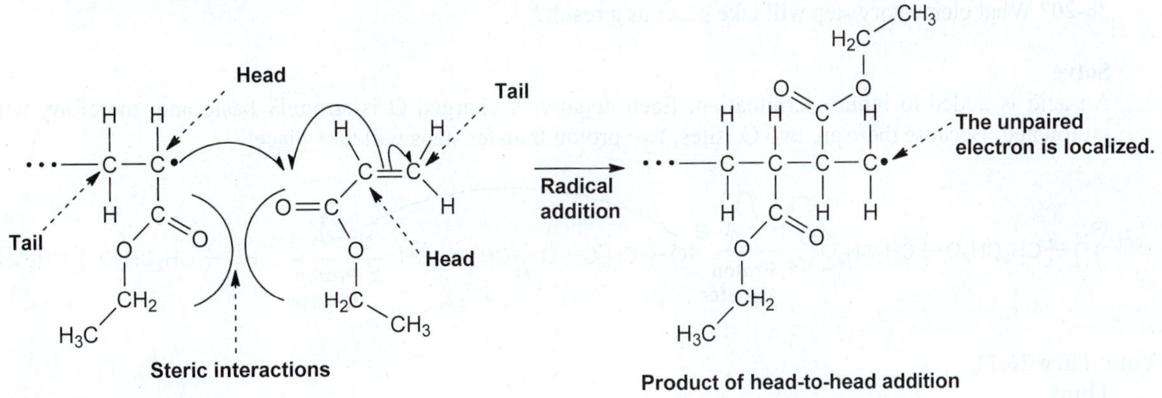

Head

Tail

The unpaired electron is localized.

Radical addition

Tail

Head

Steric interactions

Product of head-to-head addition

Your Turn 26.8

Think

What are stabilizing factors for a negative charge? Are the negative charges on the same atom in the reactant and product? Are the negative charges equally delocalized via resonance? Are inductive effects present?

Solve

The negative charge appears on C in both the reactants and products. The product in Equation 26-12 is resonance stabilized owing to the presence of π bonds next to the C^- electron pair. Resonance is possible around the entire benzene ring. No such resonance exists in the reactants.

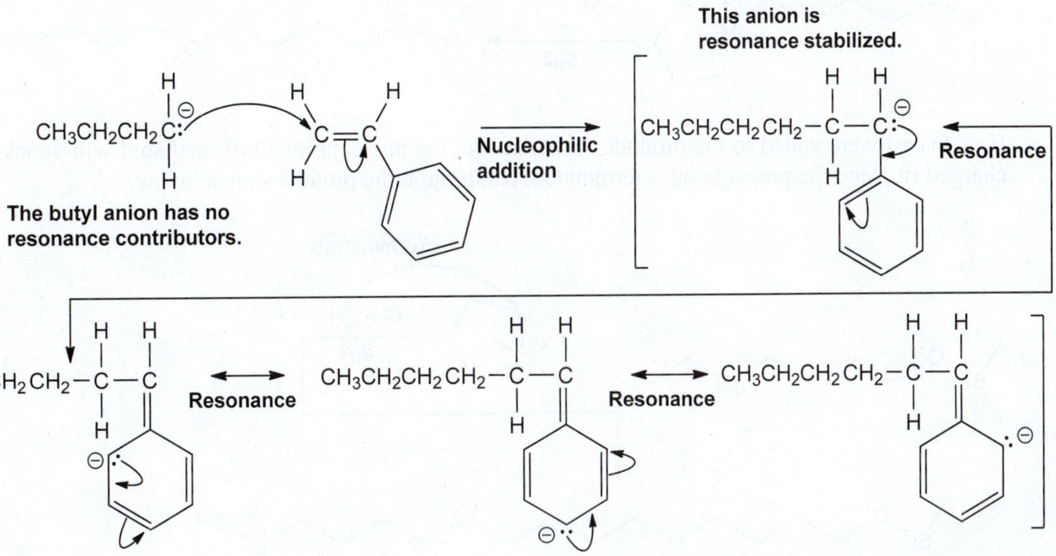

This anion is resonance stabilized.

Nucleophilic addition

Resonance

The butyl anion has no resonance contributors.

Resonance

Resonance

Your Turn 26.9

Think

In a free radical mechanism, what types of species react in a combination or disproportionation step? What types of species would be analogous to these radicals in an anionic mechanism? What factor would prevent these species from coming together?

Solve

In a radical mechanism, combination and disproportionation each require two radicals reacting. The analogous reactive intermediates in an anionic mechanism are anions. Neither of these steps is likely, because the like charges of the two anions would repel each other.

Your Turn 26.10

Think

What type of compound is added to induce termination? What are the reactive parts of the product in Equation 26-20? What elementary step will take place as a result?

Solve

An acid is added to induce termination. Each negatively charged O is strongly basic and, therefore, will be protonated. Because there are two O^- sites, two proton transfer steps will take place.

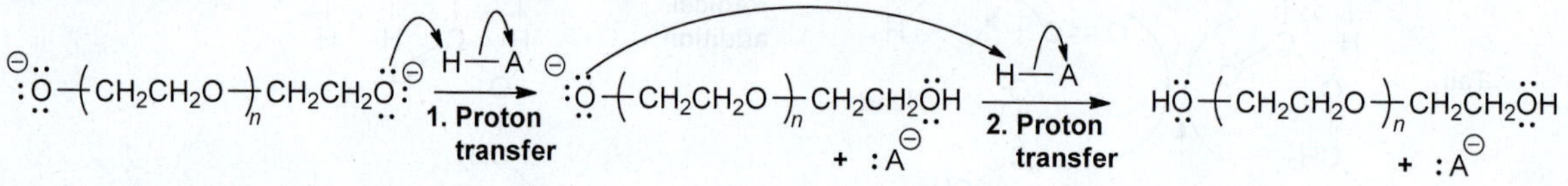

Your Turn 26.11

Think

How is the second propagation step similar to the first propagation step? How is the product different? When methanol is added, how will it react with the growing polymer?

Solve

Propagation occurs when the nucleophilic O on an uncharged molecule of tetrahydrofuran attacks the positively charged ring. The second propagation step occurs via the same mechanism as the first propagation step, as shown below, and a new positively charged ring appears at the end of the growing chain.

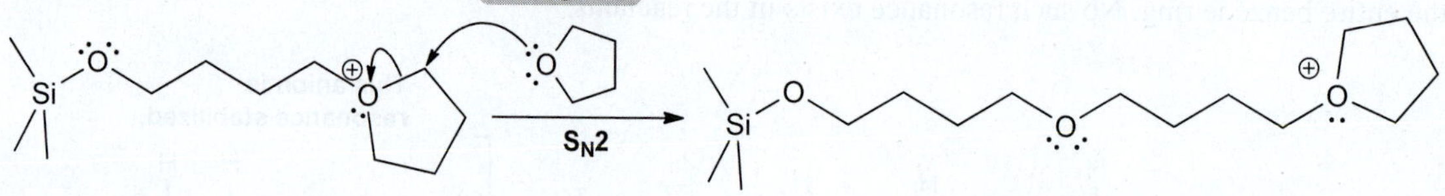

If methanol were added to the product shown above, the nucleophilic O of methanol will attack the positively charged ring and the propagation is terminated, resulting in the product shown below.

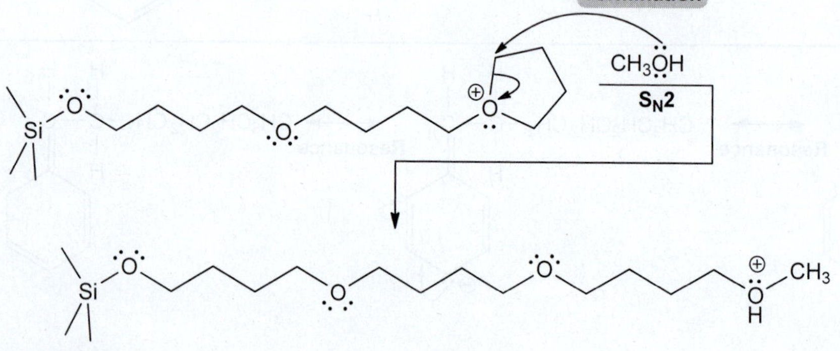

Your Turn 26.12

Think

Review Section 21.7. In an acid-catalyzed reaction, which species is protonated in the first step? What acts as the nucleophile in the nucleophilic addition step? What is the leaving group in the nucleophile elimination step? How are proton transfer reactions incorporated to avoid the appearance of species that are incompatible with the conditions of the reaction?

Solve

The C=O group of the carboxylic acid is protonated to activate the C=O toward nucleophilic addition. The amine acts as the nucleophile in the subsequent nucleophilic addition step. Water is the leaving group (formed after proton transfer) in the nucleophile elimination step. The amide functional group is formed. Notice how proton transfer reactions are incorporated to avoid the appearance of a strongly basic species under the acidic conditions of the reactions.

1. Proton transfer — 2. Nucleophilic addition — 3. Proton transfer — 4. Proton transfer — 5. Nucleophile elimination — 6. Proton transfer

Your Turn 26.13

Think

Refer to Figure 26-14e. Which symbol represents adipic acid? Which symbol represents hexane-1,6-diamine (HD)? Which two symbols connect to link monomers together and link existing chains?

Solve

Adipic acid (AA) is represented by the dots (red), and hexane-1,6-diamine (HD) is represented by the squares (blue). In the figure below, new lines are drawn for

(a) two monomers linking together
(b) an existing polymer chain linking to a monomer
(c) two existing chains linking together

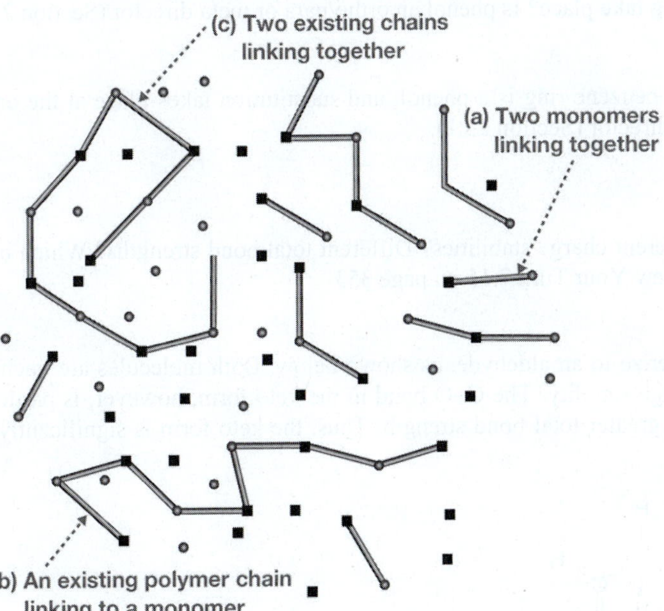

Your Turn 26.14

Think

If radicals are present, what kinds of curved arrows are used to show electron movement? What bonds are formed and broken in the first step? In the second step? What is the name of each elementary step, and how many curved arrows are needed to show the proper electron movement?

Solve

Because radicals are involved, single-barbed curved arrows are used to show electron movement. The first step is a hydrogen atom abstraction, whereby one C–H bond is formed and a second one is broken simultaneously. The second step is a radical addition, whereby a C–C single bond is formed and a C=C π bond is broken. Both steps require three single-barbed curved arrows.

(a)

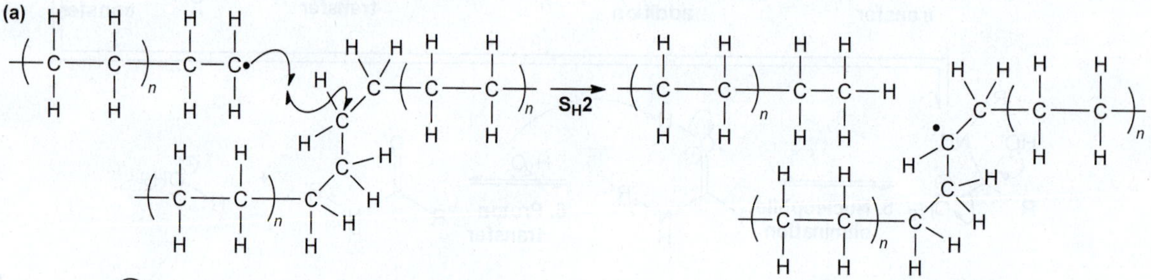

S_H2

(b)

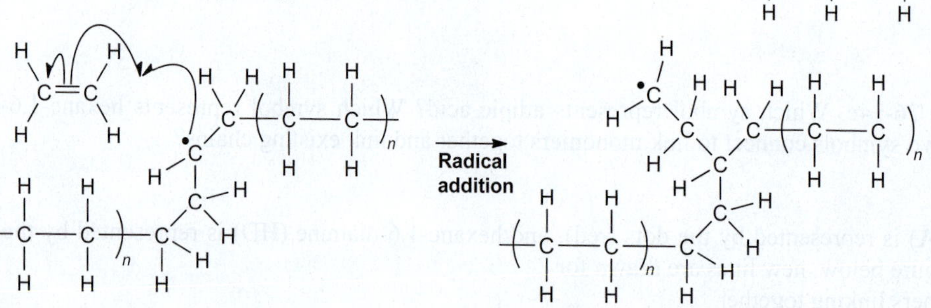

Radical addition

Your Turn 26.15

Think

Where does substitution on the phenol ring take place? Is phenol an ortho/para or meta director (Section 23.1)?

Solve

In Bakelite, the starting monosubstituted benzene ring is a phenol, and substitution takes place at the ortho or para site because phenol is an ortho/para director (Section 23.1).

Your Turn 26.16

Think

Do the keto and enol tautomers have different charge stabilities? Different total bond strengths? Which bond is particularly strong? You may wish to review Your Turn 7.15 on page 353.

Solve

Vinyl alcohol is an enol and will tautomerize to an aldehyde, as shown below. Both molecules are uncharged, so they do not exhibit differences in charge stability. The C=O bond in the keto form, however, is particularly strong, which helps give the keto form a greater total bond strength. Thus, the keto form is significantly more stable than the enol form.

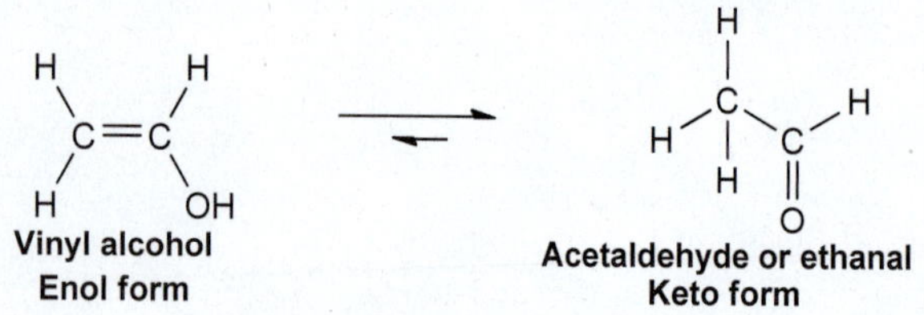

Vinyl alcohol
Enol form

Acetaldehyde or ethanal
Keto form

Your Turn 26.17

Think

What functional groups are present? What type of bond breaks during the course of the reaction? What type of species should CH_3O^- behave like to break that bond? Can that bond break in a single step, or will it require multiple steps?

Solve

An ester group is part of the repeating unit of the polymer, and during the course of the reaction, the bond between O and C=O must break. This can occur when CH_3O^- acts as a nucleophile and RO^- (as part of the polymer) acts as a leaving group from the C=O bond, as shown below. Because the leaving group is attached to a C=O, this reaction takes place in a nucleophilic addition–elimination mechanism. The second product will be methyl acetate.

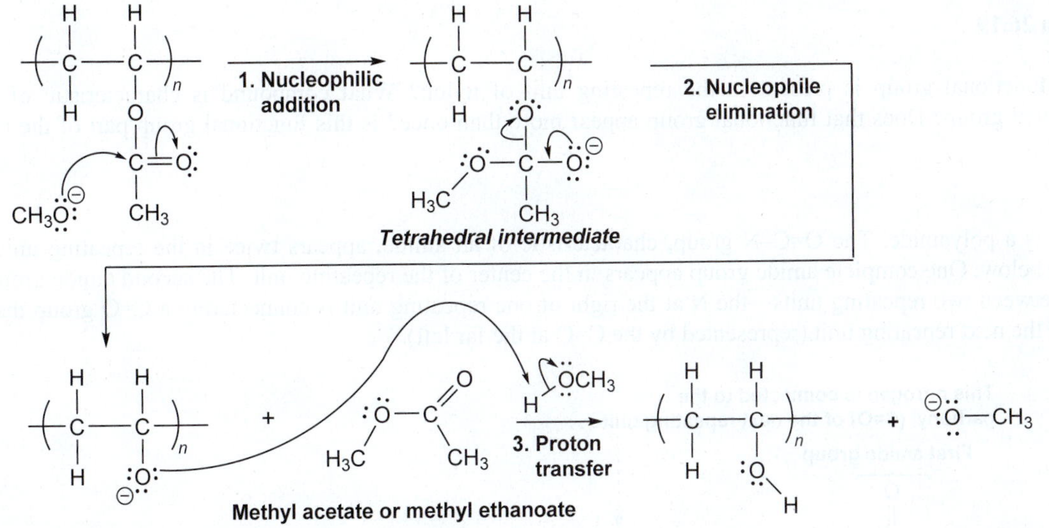

1. Nucleophilic addition

2. Nucleophile elimination

Tetrahedral intermediate

3. Proton transfer

Methyl acetate or methyl ethanoate

Your Turn 26.18

Think

What is the hybridization of the neighboring C atoms? Is resonance possible between the O atom and adjacent benzene rings? If so, what is the hybridization of the O atom in the resonance structure? What is the ideal C–O–C bond angle that corresponds to that hybridization?

Solve

When a heteroatom with a lone pair is adjacent to a benzene ring, the lone pair of electrons is involved in resonance with the ring. For example, a lone pair on the phenoxy O atom can be delocalized into the aromatic ring:

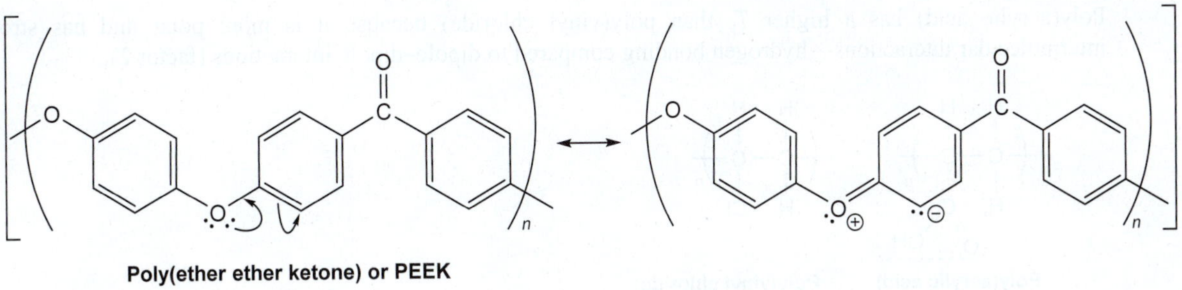

Poly(ether ether ketone) or PEEK

In this resonance structure, the O atom has three electron groups, consistent with an sp^2 hybridization and a ~120° bond angle. Experimental evidence that supports this view includes the bond angles in the following two ethers:

$H_3C-O-CH_3$

Dimethyl ether
C–O–C bond angle: 111°

Diphenyl ether
C–O–C bond angle: 119–129°

Dimethyl ether contains an O that is sp^3 hybridized, and the angle is close to the 109.5° that we expect. Diphenyl ether contains an O that is sp^2 hybridized, so a lone pair on O can be included in the conjugated system and extend the conjugation of the system throughout the whole molecule.

Your Turn 26.19

Think

What functional group is present in the repeating unit of nylon? What compound is characteristic of that functional group? Does that functional group appear more than once? Is this functional group part of the main chain?

Solve

Nylon is a polyamide. The O=C–N group, characteristic of an amide, appears twice in the repeating unit, as shown below. One complete amide group appears in the center of the repeating unit. The second amide group is split between two repeating units—the N at the right of one repeating unit is connected to a C=O group that is part of the next repeating unit (represented by the C=O at the far left).

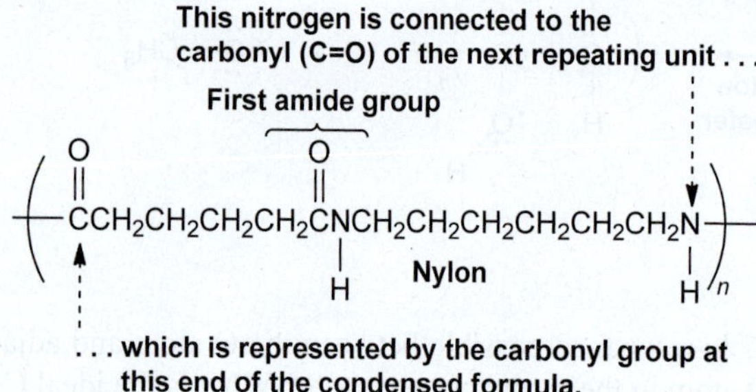
This nitrogen is connected to the carbonyl (C=O) of the next repeating unit . . .
First amide group
Nylon
. . . which is represented by the carbonyl group at this end of the condensed formula.

Your Turn 26.20

Think

What types of intermolecular interactions are present in poly(acrylic acid)? In poly(vinyl chloride)? How do the types of intermolecular interactions affect the T_g?

Solve

Poly(acrylic acid) has a higher T_g than poly(vinyl chloride) because it is more polar and has stronger intermolecular interactions—hydrogen bonding compared to dipole–dipole interactions (factor 2).

Poly(acrylic acid)
Higher T_g
Stronger intermolecular interactions–hydrogen bonding

Poly(vinyl chloride)

Your Turn 26.21

Think

Which type of mechanism is believed to form the double bonds (see Chapters 8 and 9 to review elimination mechanisms)? What is the leaving group? What new functional group is formed? In how many steps does this mechanism occur?

Solve

It is believed that this reaction takes place via an E1 mechanism. The chloride Cl^- is the leaving group, and the new functional group formed is an alkene. The E1 mechanism takes place in two steps—heterolysis followed by elimination of H^+—as shown below:

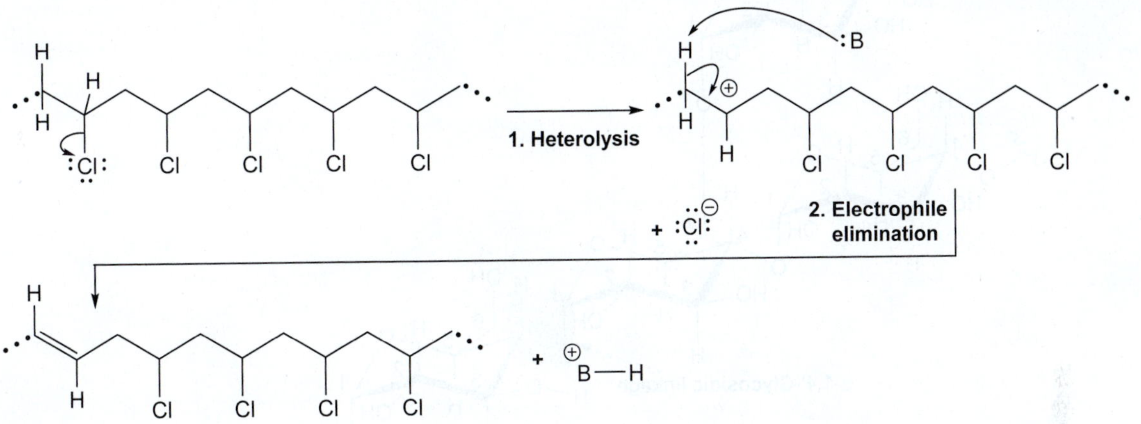

1. Heterolysis

2. Electrophile elimination

Your Turn 26.22

Think

Review the definition of a peptide linkage from Section 21.12. For a hexapeptide, how many peptide linkages are present?

Solve

An O=C–N functional group, characteristic of an amide, is responsible for connecting adjacent amino acids along a protein's main chain, and each of those O=C–N groups is called a peptide linkage. In the hexapeptide in Figure 26-44, there are five peptide linkages, as shown by the gray screens in the figure below.

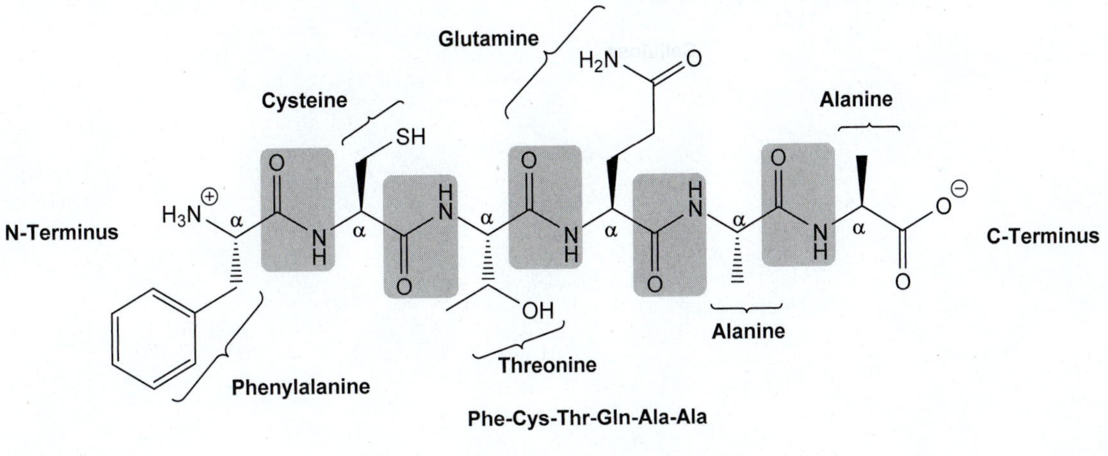

N-Terminus

C-Terminus

Phenylalanine

Cysteine

Threonine

Glutamine

Alanine

Alanine

Phe-Cys-Thr-Gln-Ala-Ala

Your Turn 26.23

Think

What type of functional group characterizes a glycosidic linkage? What locator numbers are assigned to the C atoms in each sugar connected by the glycosidic linkage? How are α and β glycosidic linkages distinguished?

Solve

The missing glycosidic linkages in Figures 26-52 and 26-53 are highlighted (gray screen) and labeled below. The one in Figure 26-52 is an α-1,4′-glycosidic linkage, because the glycosidic linkage connects C1 of one glucose unit to C4 of the next, and the substituent on C1 is axial. The one in Figure 26-53 is a β-1,4′-glycosidic linkage, because the glycosidic linkage connects C1 of one glucose unit to C4 of the next, and the substituent on C1 is equatorial.

Figure 26-52

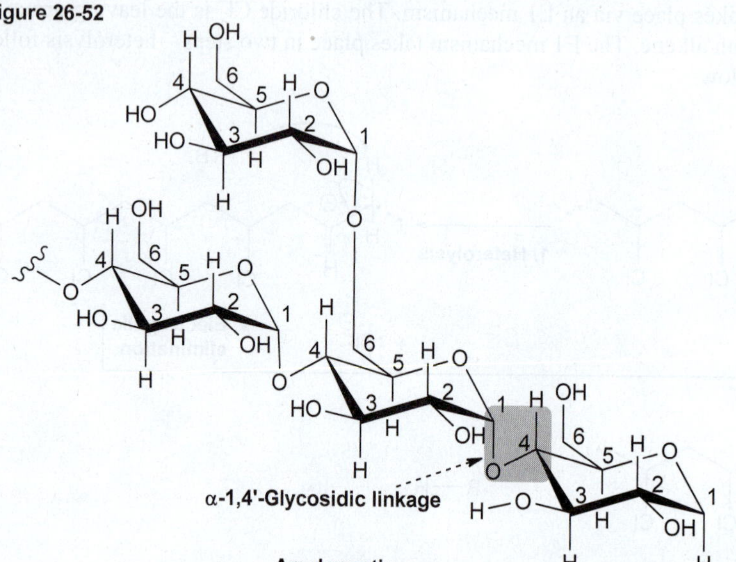

α-1,4′-Glycosidic linkage

Amylopectin

Figure 26-53

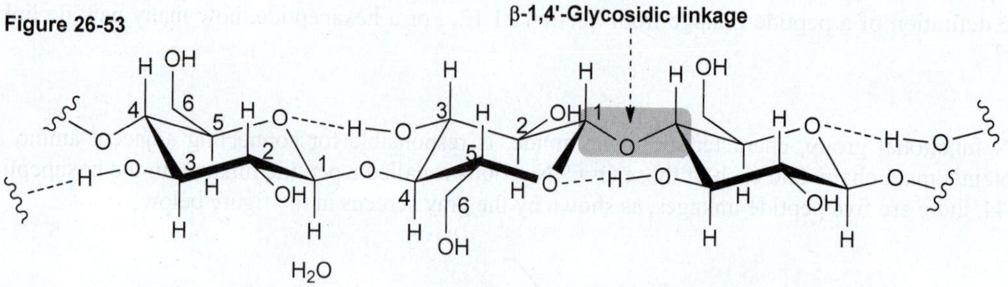

β-1,4′-Glycosidic linkage

H_2O

Cellulose

In Chapter Problems
Problem 26.1

Think

What atoms and bonds should be repeated in the polymer? What bonds in the repeating unit should be used to connect the repeating units together? How many repeating units should be connected?

Solve

The repeating unit is shown in parentheses. In both cases, a new C–C bond is formed using the bonds that are unaccounted for on either side of the repeating unit. In **(a)**, four F substituents are attached to the main chain of the repeating unit, and in **(b)**, there are two Cl substituents. The structures below show the repeating units repeated three times.

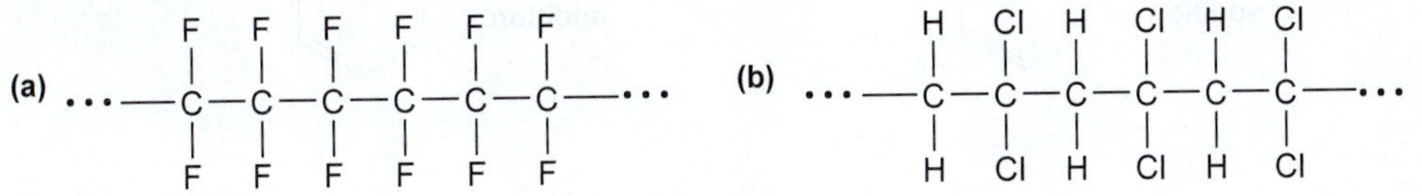

(a) ···—C(F)(F)—C(F)(F)—C(F)(F)—C(F)(F)—C(F)(F)—C(F)(F)—···

(b) ···—C(H)(H)—C(Cl)(Cl)—C(H)(H)—C(Cl)(Cl)—C(H)(H)—C(Cl)(Cl)—···

Problem 26.3

Think

What is the repeating unit of poly(methyl methacrylate)? How is the repeating unit of a vinyl polymer related to the monomer? What bonds exist in the monomer that do not exist in the repeating unit, and vice versa?

Solve

The C atoms in the structure make up the main chain of the molecule. The CH_3 and CO_2CH_3 groups are pendant groups, or side groups, analogous to the phenyl rings in polystyrene. The repeating unit is as follows:

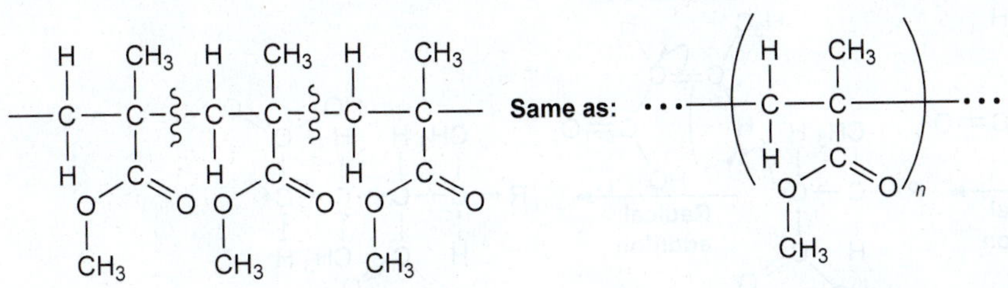

—C(H)(H)—C(CH_3)(CO_2CH_3)—C(H)(H)—C(CH_3)(CO_2CH_3)—C(H)(H)—C(CH_3)(CO_2CH_3)— **Same as:** ···—[C(H)(H)—C(CH_3)(CO_2CH_3)]$_n$—···

To determine the monomer from the repeating unit, remove the single bonds on the outside of the repeating unit and change the C–C single bond to a C=C double bond. The monomer is methyl methacrylate.

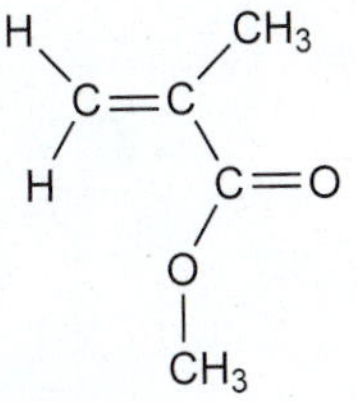

$H_2C=C(CH_3)C(=O)OCH_3$

Problem 26.5

Think

How does benzoyl peroxide produce the initial free radicals? What elementary step occurs between a free radical and a vinyl monomer? Is the product of that step still a free radical?

Solve

Benzoyl peroxide is a radical initiator that undergoes homolysis of the O–O single bond to produce benzoyloxyl radicals. See the figure on the next page.

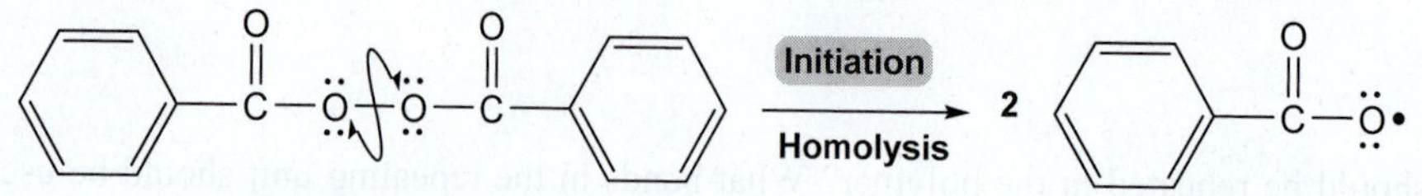

In the first propagation step, a benzoyloxyl radical adds to the C=C bond of a butylene monomer. The product of that step is another radical, which undergoes radical addition to another butylene monomer.

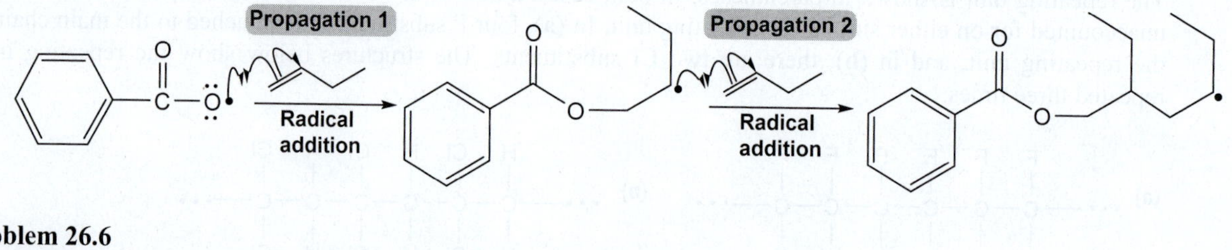

Problem 26.6

Think

What is the difference in structure between the two monomers? Is head-to-tail or tail-to-head polymerization favored? What are the head and tail in each monomer? How does the extra methyl group in crotonic acid affect the ability of the monomer and the growing polymer chain to come together?

Solve

Both monomers will undergo head-to-tail addition. The first two propagation steps for crotonic acid and acrylic acid are shown below.

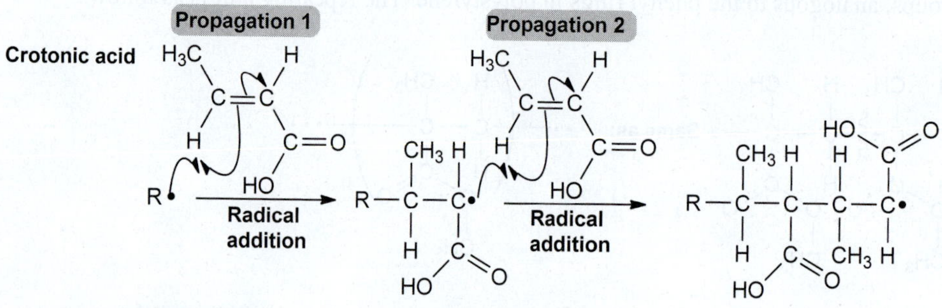

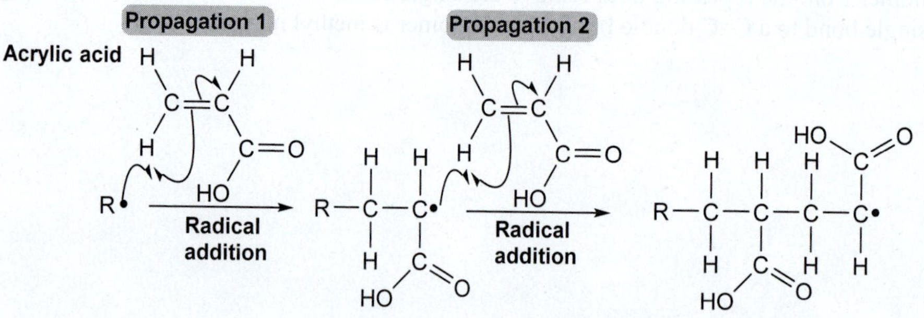

The extra methyl group in crotonic acid creates steric hindrance and slows the rate of polymerization.

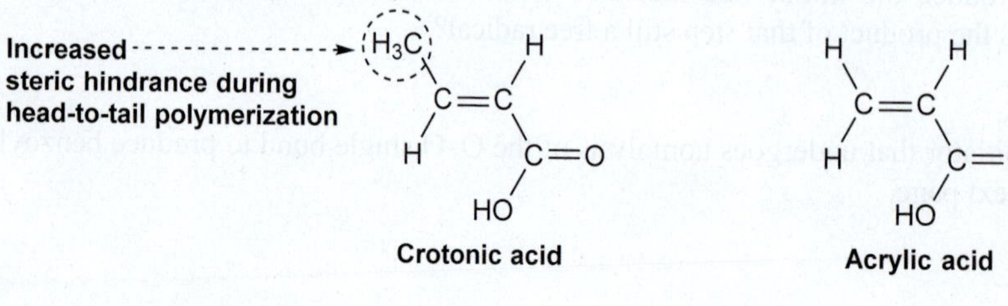

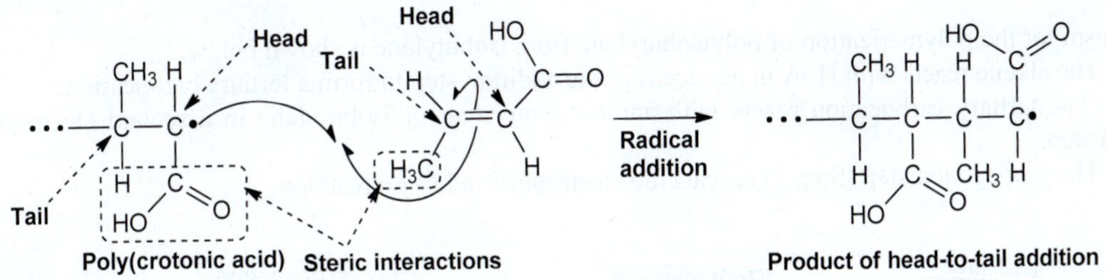

Head
Tail
Head
Tail
Radical addition
Poly(crotonic acid) Steric interactions
Product of head-to-tail addition

Problem 26.7

Think

Where is the new chiral center in polypropylene? What is the repeating unit in polypropylene? What is the pattern of configurations, if any, for an atactic polymer? What is the pattern of configurations, if any, for an isotactic polymer?

Solve

The new chiral center and repeating unit for polypropylene are shown below. In atactic polypropylene, the stereochemical configurations do not establish a regular pattern. In isotactic polypropylene, all stereochemical configurations are the same.

New chiral center
Polypropylene
Atactic
No regular pattern exists for the configurations.
Isotactic
All configurations are the same.

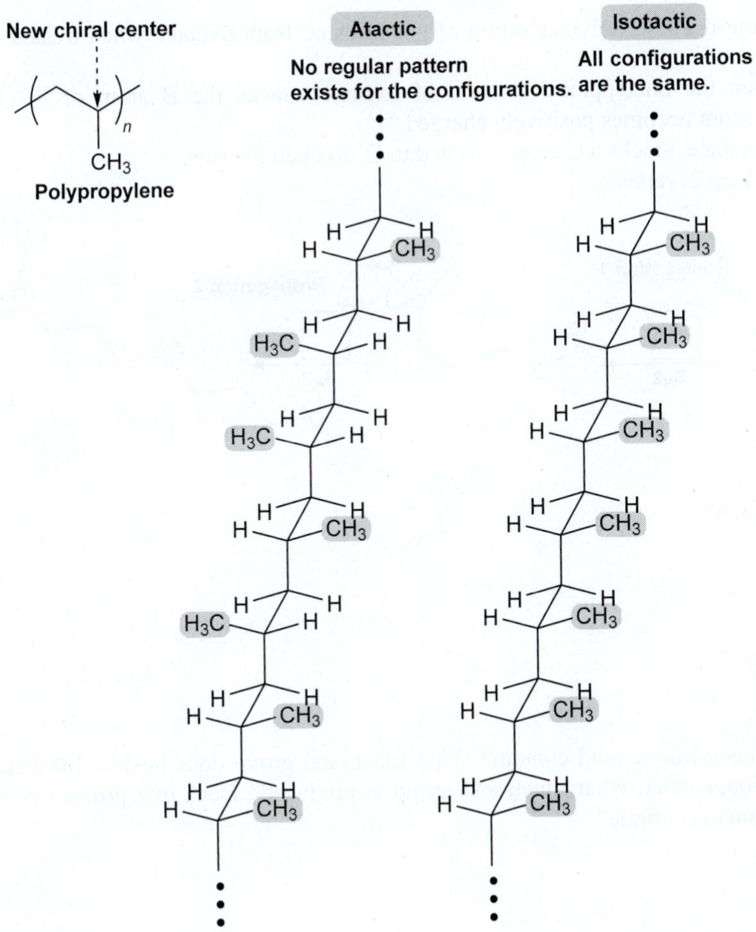

Problem 26.9

Think

How does an alkene π bond react with an acid, HA (Chapter 11)? How does the electrophile formed in the initiation step react with another equivalent of isobutylene?

Solve

The mechanism for the polymerization of polyisobutylene from isobutylene is shown below.

Step 1: The alkene reacts with H–A in an electrophilic addition step to form a tertiary carbocation.

Step 2: The tertiary carbocation reacts with another equivalent of isobutylene in a second electrophilic addition step.

Step 3: The propagation step (Step 2) repeats the electrophilic addition reaction.

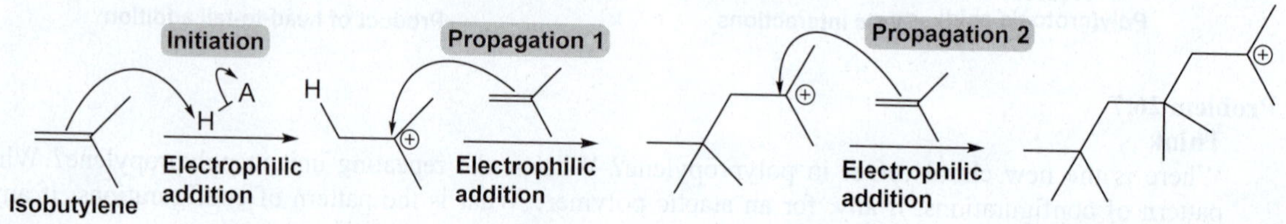

Initiation
Propagation 1
Propagation 2
Isobutylene
Electrophilic addition
Electrophilic addition
Electrophilic addition

Problem 26.10

Think

What is cationic ring-opening polymerization? How can the O atom of oxetane react with BF_3 in an initiation step? How can the resulting species react with another molecule of oxetane in a propagation step?

Solve

The mechanism for the cationic ring-opening polymerization of polyoxetane from oxetane when treated with the Lewis acid, BF_3, is shown below.

Step 1: Initiation occurs when the nucleophilic O atom of oxetane attacks the B atom of BF_3 in a coordination step. Thus, the O atom becomes positively charged.

Step 2: Another molecule of oxetane attacks a C atom attached to O^+ to open the ring.

Step 3: The propagation step (Step 2) repeats.

Initiation
Propagation 1
Propagation 2
Coordination
S_N2
S_N2

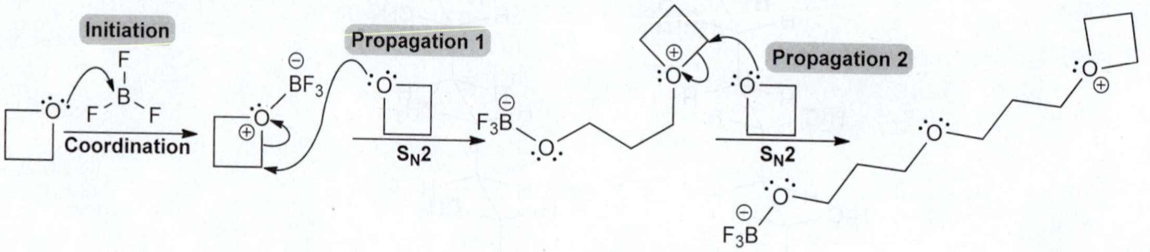

Repeating unit of the resulting polymer:

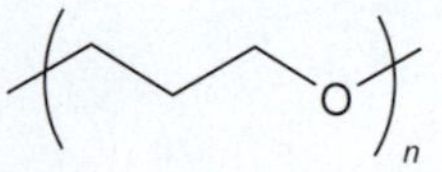

Problem 26.12

Think

What functional group does benzene-1,4-dioic acid contain? What functional group does hexane-1,6-diamine contain? When those functional groups react, what functional group is produced? Does that product contain other reactive sites for polymerization to continue?

Solve

Benzene-1,4-dioic acid has a CO_2H functional group, and hexane-1,6-diamine has an NH_2 functional group. These functional groups react to produce an O=C–N group, characteristic of an amide, that connects the two molecules together, so these two monomers will produce a dimer.

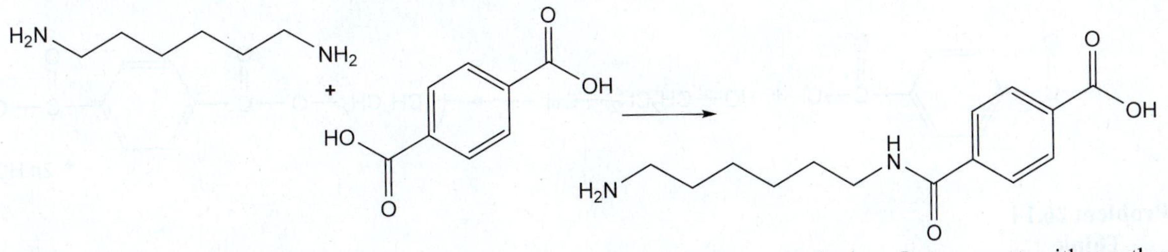

Each end of the dimer has an unreacted functional group that can react further. It can react with another monomer, or, as shown below, two dimers can react to increase the chain length. In the resulting tetramer, the repeating pattern becomes evident, which includes one para disubstituted benzene ring with two attached carbonyl groups—one to the CO_2H and the other to the diamine.

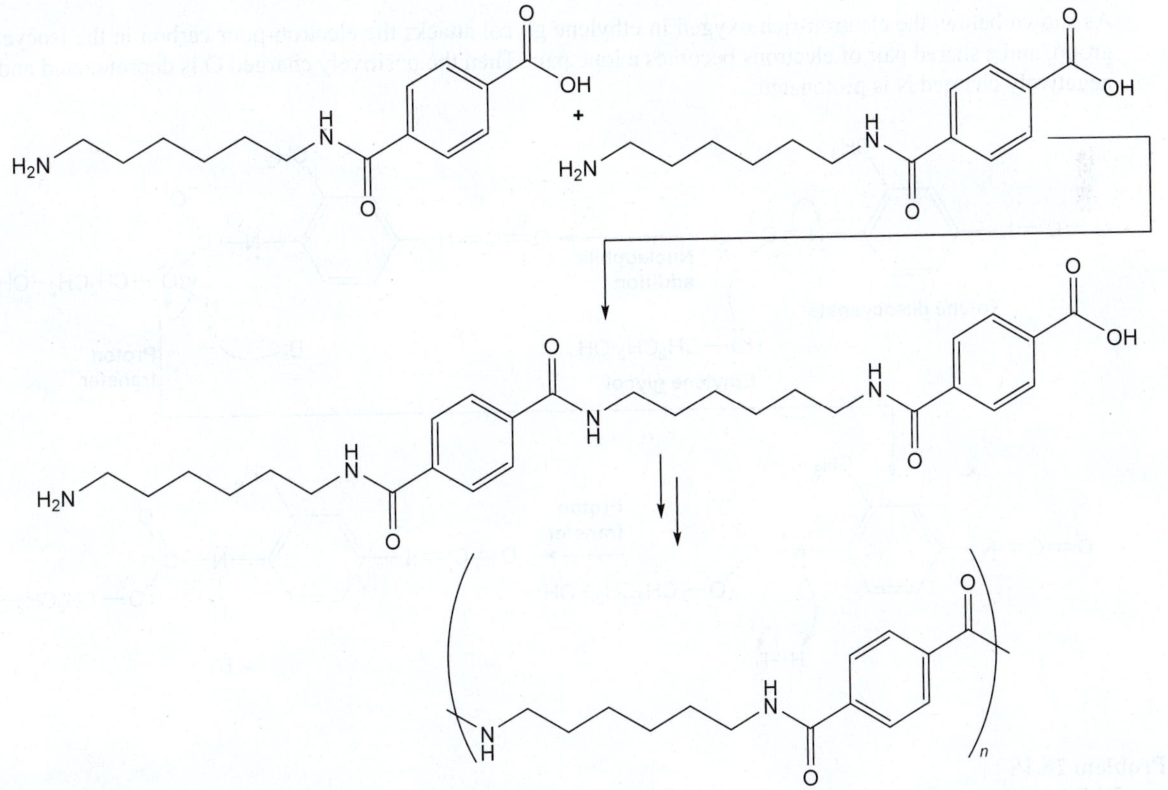

Problem 26.13

Think

What is the structure of poly(ethylene terephthalate) (PET)? What functional group forms from an acid chloride and an alcohol? What is the significance of having a diacid chloride and a diol? How are the starting materials similar/different compared to the ones shown in Equations 26-27 and 26-28?

Solve

One starting material is the same as in Equations 26-27 and 26-28: ethylene glycol. The other material is similar to the one shown in Equations 26-27 and 26-28, with the exception of the carboxylic acid derivative functional group. The new starting material is a diacid chloride rather than a dicarboxylic acid or diester. The reaction is given below:

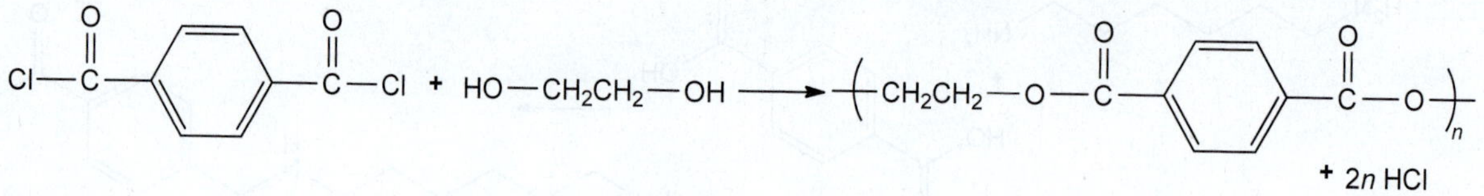

$$\text{Cl–C(=O)–C}_6\text{H}_4\text{–C(=O)–Cl} + \text{HO–CH}_2\text{CH}_2\text{–OH} \longrightarrow \text{–[CH}_2\text{CH}_2\text{–O–C(=O)–C}_6\text{H}_4\text{–C(=O)–O]}_n\text{–} + 2n\,\text{HCl}$$

Problem 26.14

Think

Given that the reaction occurs via one nucleophilic addition to a polar double bond followed by two proton transfers, identify the nucleophile and electrophile in the first step. What species facilitates the proton transfer reactions?

Solve

As shown below, the electron-rich oxygen in ethylene glycol attacks the electron-poor carbon in the isocyanate group, and a shared pair of electrons becomes a lone pair. Then the positively charged O is deprotonated and the negatively charged N is protonated:

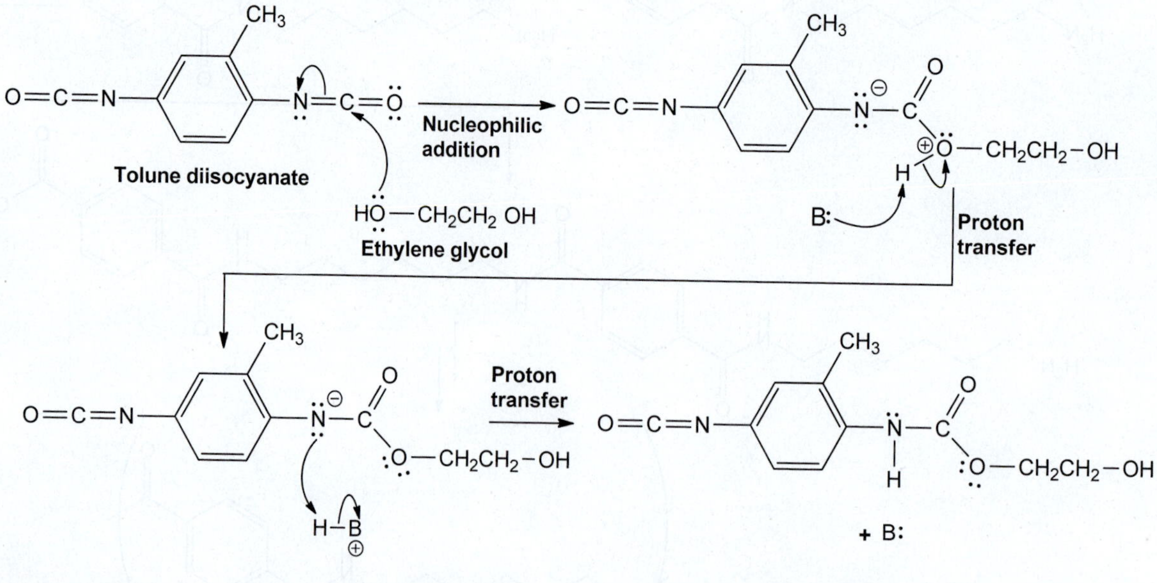

Tolune diisocyanate

$\text{HO–CH}_2\text{CH}_2\text{OH}$

Ethylene glycol

Nucleophilic addition

B:

Proton transfer

$\text{H–B}^{\oplus}$

Proton transfer

+ B:

Problem 26.15

Think

What is the mechanism for electrophilic aromatic substitution? What electrophile is lost from the aromatic ring in PEEK with a *tert*-butyl substituent? What is the stability of this cationic species? Is a linear carbocation as stable?

Solve

Recall the steps for electrophilic aromatic substitution: electrophilic addition followed by electrophile elimination. The mechanism for the trans *tert*-butylation, therefore, is two back-to-back electrophilic aromatic substitution reactions. The electrophile lost in the first electrophile elimination step is the electrophile that reacts with toluene in the second electrophilic aromatic substitution reaction.

CF_3SO_3H

PEEK with a *tert*-butyl substituent

Electrophilic addition

Electrophile elimination

$CF_3SO_3^{\ominus}$

Electrophilic addition

Electrophile elimination

B:

CH_3

$\oplus$ BH

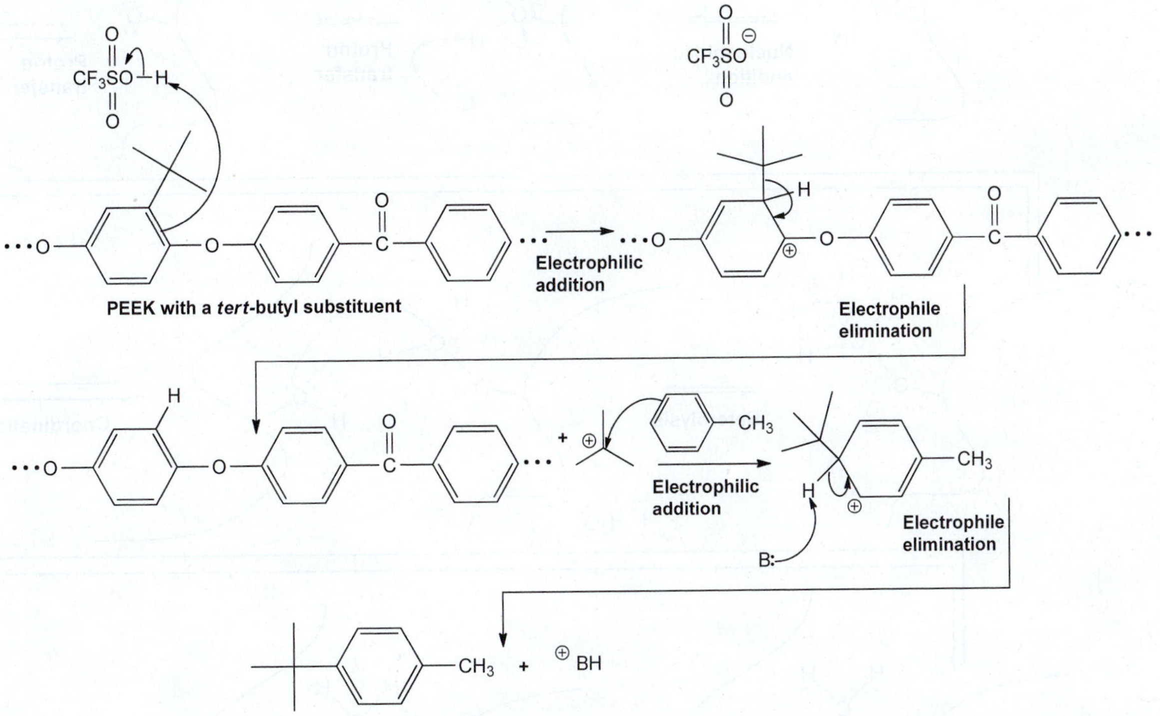

This trans *tert*-butylation would not be as successful if the *tert*-butyl group were a linear butyl group instead because the electrophile that would be eliminated in the second step would be a primary carbocation. A primary carbocation is far less stable than a tertiary carbocation.

Problem 26.16

Think

What functional group is formed in the product? What is the purpose of the acid catalyst? Is the aldehyde or alcohol the electrophile? Which is the nucleophile? Are proton transfer reactions involved? What are reasonable leaving groups in the mechanism?

Solve

The functional group formed is an acetal (Chapter 18), which forms via a nucleophilic addition reaction of the alcohol to the C=O of the aldehyde. The reaction is acid catalyzed, so in the first step, the C=O group is protonated to make the carbon of the C=O more electrophilic. Also, under these acidic conditions, a good leaving group, H_2O, is eventually produced. The full mechanism is given below and on the next page:

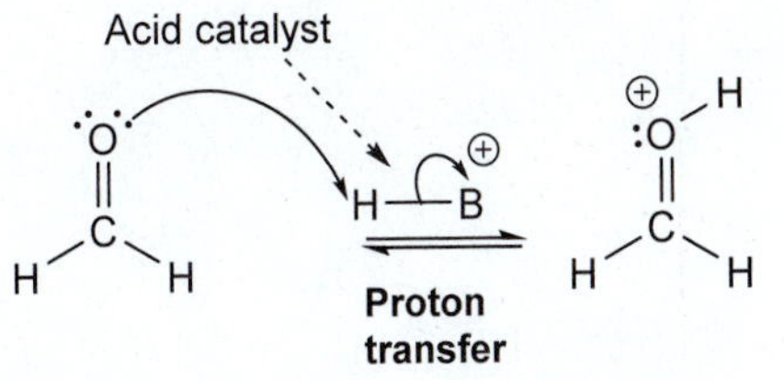

Acid catalyst

Proton transfer

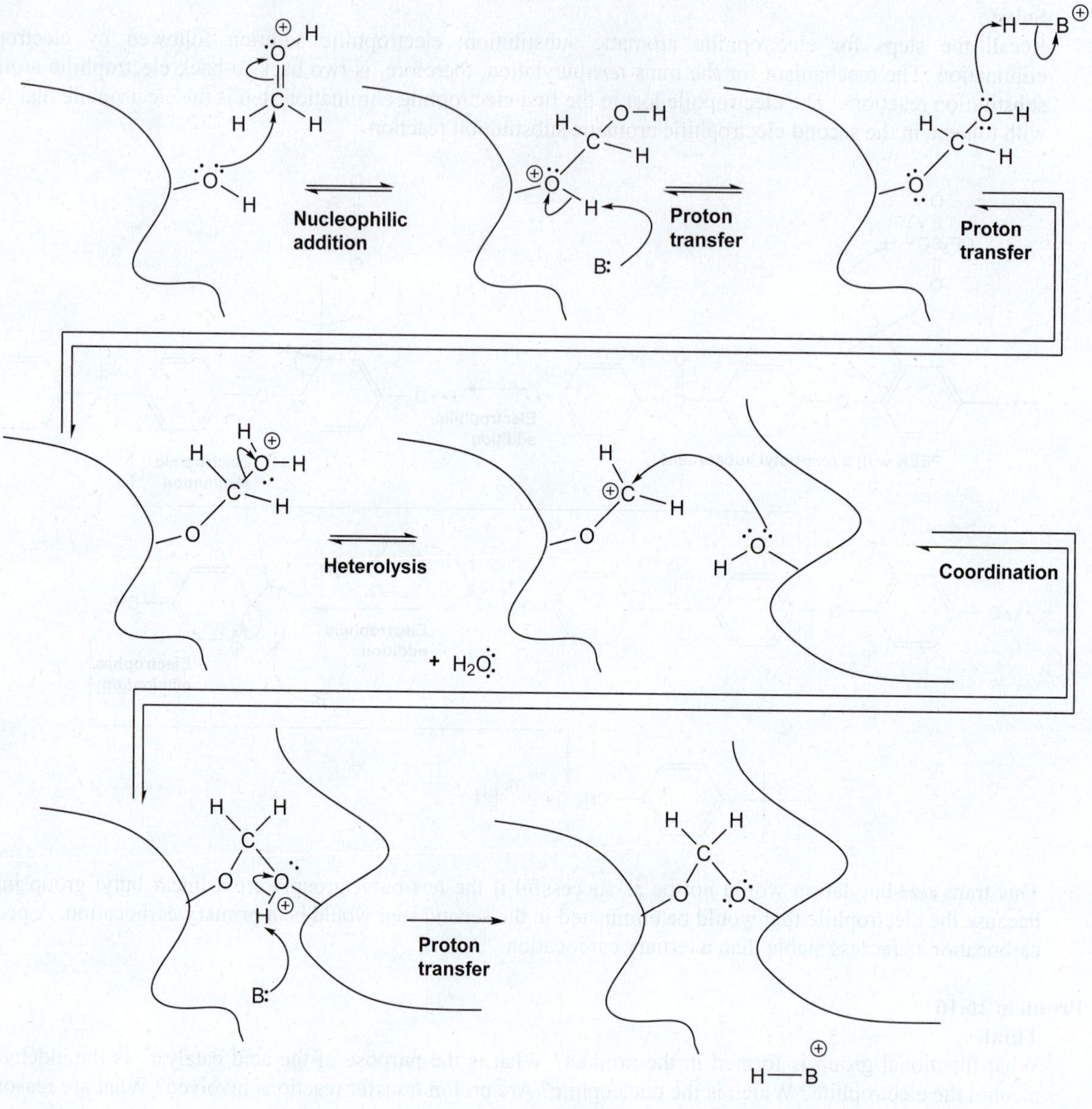

Problem 26.18

Think

What functional group makes up the backbone? What other functional groups are present? Are these functional groups part of the polymer's repeating unit?

Solve

The functional groups present are O=C–N, characteristic of an amide, and OH, characteristic of an alcohol. Therefore, polyamide or polyalcohol could be used to describe polyserine. Because the O=C–N groups are part of the polymer backbone, polyamide refers to the polymeric nature of polyserine.

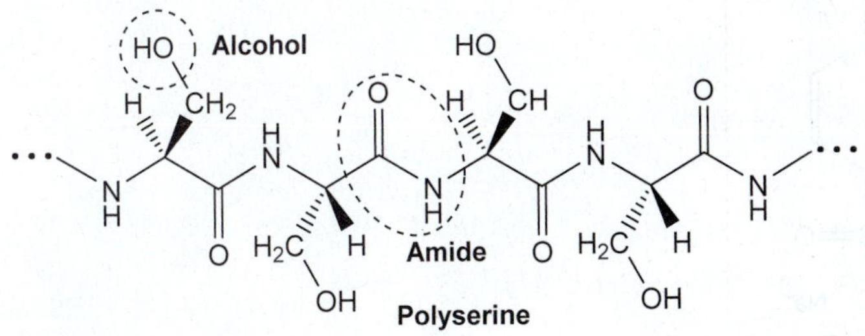

Alcohol

Amide

Polyserine

Problem 26.19

Think

How is the structure of a vinyl monomer related to the structure of the repeating unit? What bonds in the monomer do not appear in the polymer? How many words are in the name of the monomer? Are parentheses necessary in the name of the polymer? What are the initial letters of the different portions of the monomer?

Solve

The condensed formula is given below. The C=C bond in the monomer becomes a C–C bond in the polymer, and the π bond is used to connect the monomers together.

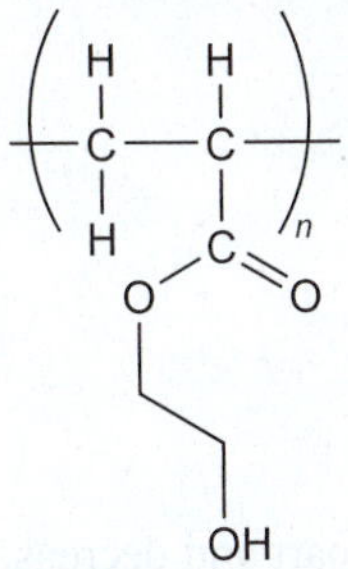

The polymer comes from 2-hydroxyethyl acrylate, which is two words and, thus, requires parentheses. The polymer is, therefore, poly(2-hydroxyethyl acrylate), or PHEA.

Problem 26.21

Think

Review the seven factors that affect the heat required to reach the T_g. What are the structural differences in each monomer? What intermolecular forces will differ?

Solve

(a) **D** will have a higher T_g because it is more polar and, therefore, will have stronger dipole–dipole interactions (factor 2). Even though the first molecule has C–F bonds, which have large bond dipoles, those bond dipoles effectively cancel.

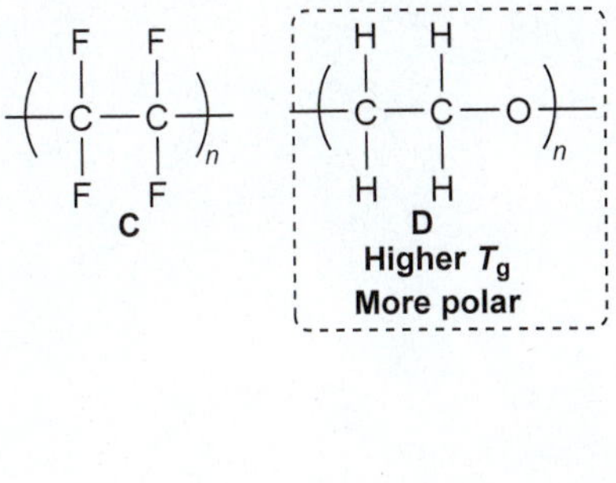

C

D

Higher T_g

More polar

(b) **F** will have a higher T_g because it is has ions and will have ionic interactions (factor 2).

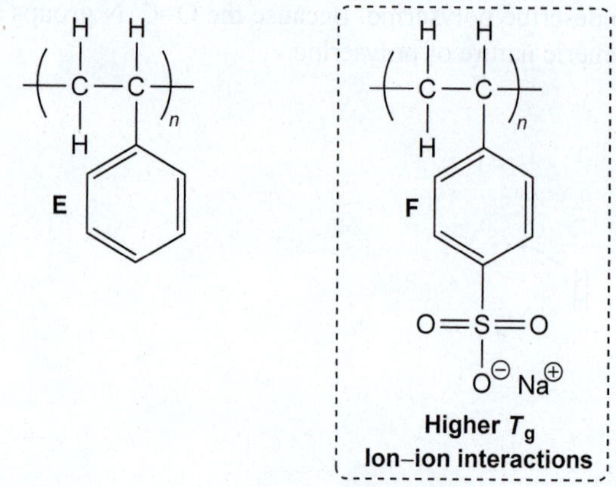

E

F

Higher T_g
Ion–ion interactions

Problem 26.22

Think

Review the seven factors that affect the heat required to reach the T_g. What are the structural differences in each monomer? What intermolecular forces will differ?

Solve

(a) **G** has more branching (factor 4), which contributes to increasing T_g. **H** has more rigidity in its polymer chain owing to the double bond (factor 6), which contributes to increasing its T_g. As a result, the T_g values of the two polymers are not very different: −70° for **G** and −73° for **H**.

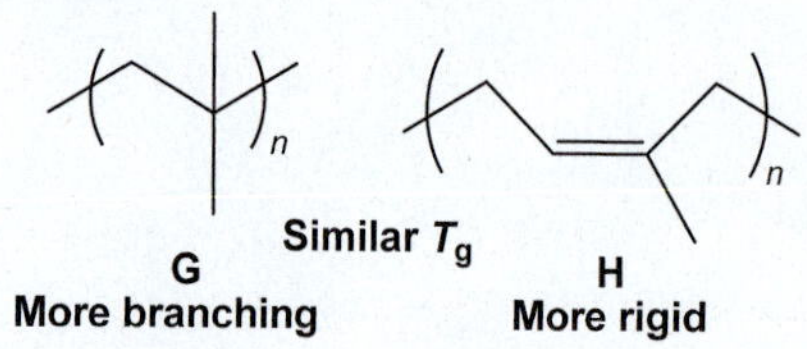

Similar T_g

G
More branching

H
More rigid

(b) **I** will have a higher T_g because the longer side groups of **J** push the polymer chains apart and decrease the strength of the dispersion forces (factor 5).

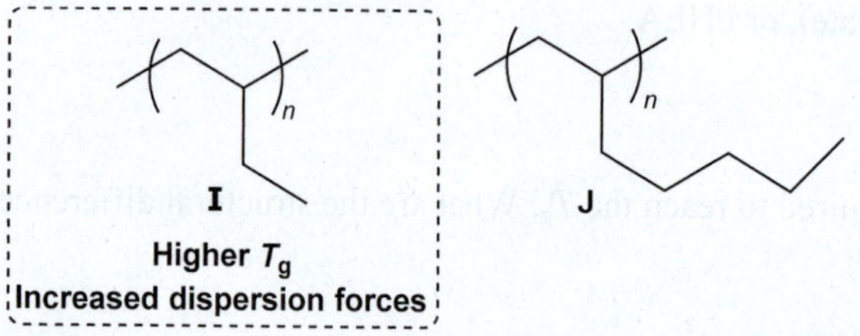

I
Higher T_g
Increased dispersion forces

J

Problem 26.23

Think

Are polar or nonpolar compounds water soluble? Are ionic compounds likely to be water soluble? Which monomers have polar or ionic repeating units?

Solve

The more polar and ionic polymers (**B**, **D**, and **F**) are water soluble and are given below.

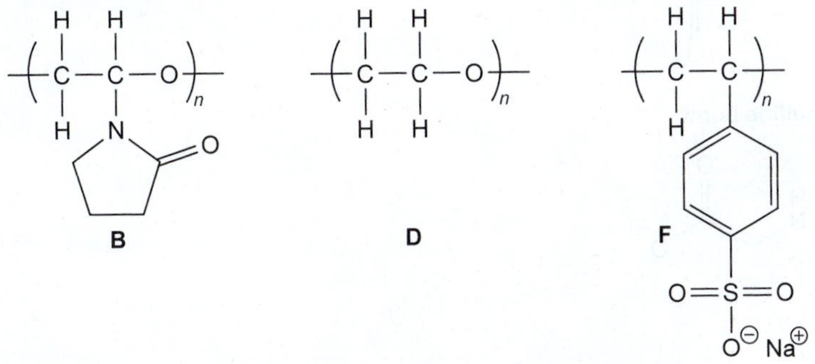

Problem 26.24

Think

Which polymer is rigid? Which polymer is flexible? Which type of polymer is better for each application?

Solve

HDPE is not sufficiently rigid for a disposable razor. PS is too rigid and would crack if dropped, making it inappropriate for a milk container.

Problem 26.25

Think

What do you notice about the arrangement of the double bonds in each subsequent product? Does this lead to a more or less stable product?

Solve

The second double bond is conjugated with the first double bond; subsequent double bonds increase the length of the conjugated system. Conjugated double bonds are more stable than isolated double bonds and will form more readily.

Problem 26.27

Think

How many possibilities are there for the identity of the first amino acid? The second amino acid? The third? How many combinations, therefore, are possible for a dipeptide? By what factor does that number increase for a tripeptide? How can you extend this trend for a protein that is 300 amino acids long?

Solve

In general, there are 20^{Y} combinations for a protein that is Y amino acids long. Therefore, if the protein has 300 amino acids, the number of primary sequences possible is $20^{300} = 2.0 \times 10^{390}$. This is much larger than the number of atoms in Avogadro's number (10^{23}) and the number of atoms in the entire universe, suggested by some estimates to be $\sim 10^{80}$.

Problem 26.28

Think

What atoms make up a disulfide bond? Where is this atom in the monomer? What does the dimer look like when those atoms are connected together?

Solve

A disulfide bond is a S–S bond. The S atom from each monomer forms that bond and loses a bond to H. The dimer is shown on the next page:

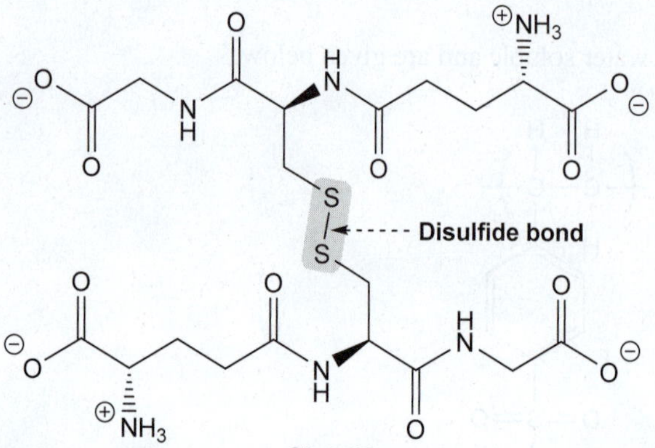

Disulfide bond

Glutathione

Problem 26.29

Think

What are the structures of alanine and isoleucine? Which one has a bulkier side chain? Is the formation of an α-helix favored or disfavored with increased sterics?

Solve

The structures of alanine and isoleucine are shown below. Alanine has only a methyl group as the R group and isoleucine has a *sec*-butyl group. The *sec*-butyl group is bulkier than the methyl group. An α-helix tends to be disrupted when adjacent or nearby amino acids in a protein's sequence are bulky. Therefore, alanine should be found more commonly in an α-helix.

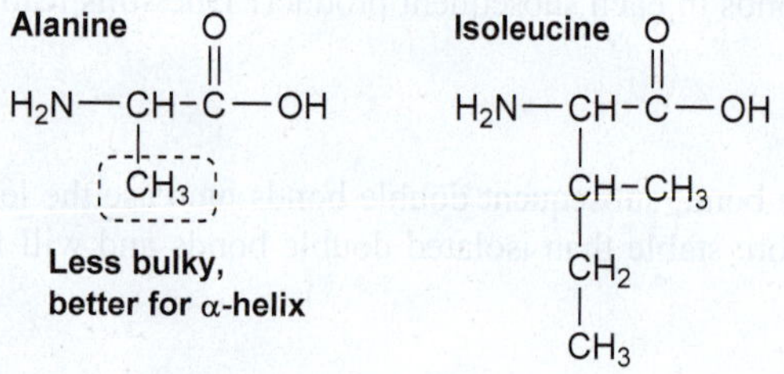

Alanine

Isoleucine

Less bulky, better for α-helix

Problem 26.30

Think

What holds the secondary and tertiary structures in a protein together? How does sodium dodecyl sulfate (SDS) disrupt these interactions?

Solve

The protein's secondary and tertiary structures are held together by intermolecular forces, such as dispersion forces (among the nonpolar regions), hydrogen bonding, and ion–ion interactions. SDS has a long hydrophobic region and an ionic hydrophilic head. The hydrophobic tail can interact favorably with the nonpolar portions of a protein, which tend to be on the interior of the protein when it is properly folded. The ionic head group interacts favorably with the polar regions of a protein and also with the water solvent. Therefore, SDS disrupts the intermolecular forces that a protein normally exhibits between its various portions when it is properly folded, and it also allows the nonpolar interior of a protein to interact more favorably with the water solvent. This makes it much less favorable for the protein to remain folded. See the figure below and on the next page.

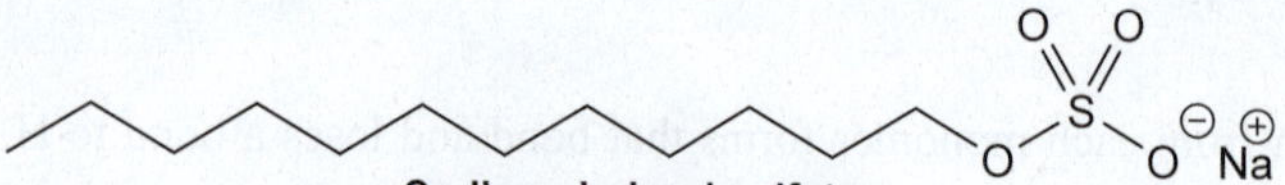

Sodium dodecyl sulfate

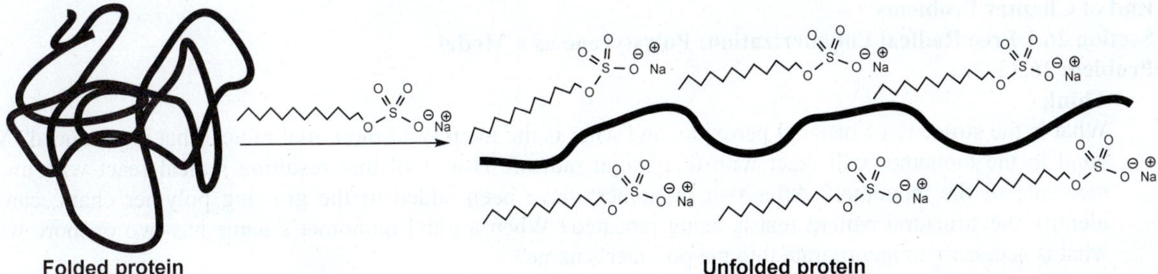

Problem 26.31

Think

Hemoglobin is a biological protein in the blood, and biological environments are generally aqueous. Therefore, would you expect hydrophilic or hydrophobic groups to be on the outside? How does this affect the solubility of hemoglobin in blood? At the surfaces where the subunits bind together, would you expect polar or nonpolar amino acids?

Solve

Hemoglobin is more soluble in blood if hydrophilic or polar amino acids are on the surface. Blood is a water-based liquid, and having a greater number of polar groups on the exterior of the protein makes the protein more soluble in the polar medium. Nonpolar amino acids are found at the surfaces where the subunits bind together, however, because those regions largely exclude water.

Problem 26.32

Think

What is the structure of chitin? Can you draw a portion of its structure by substituting an acetamide group for the OH group at each 2 position of cellulose? What intermolecular force leads to more rigid structures? How does the addition of the acetamide group contribute to the rigidity?

Solve

The partial structure of chitin is shown below. Notice the acetamide groups at the 2 position of each ring. The acetamide group contributes to the rigidity by increasing the possible hydrogen bonding interactions. Whereas the OH group at the 2 position in cellulose has one hydrogen-bond donor and one hydrogen-bond acceptor, the acetamide group has one hydrogen-bond donor (the N–H bond) and two hydrogen-bond acceptors (the N and O atoms). Not only can there be more extensive hydrogen bonding among monomers within a single polymer chain but this also enhances the hydrogen bonding between separate polymer chains.

Acetamide group
More possibilities for H bonding

Chitin

End of Chapter Problems
Section 26.1 Free Radical Polymerization: Polystyrene as a Model
Problem 26.33

Think

What is the structure of benzoyl peroxide, and what is the identity of the initial radical that is produced? What bond in the monomer will react with that initial radical? How will that resulting radical react with another molecule of the monomer? After two monomers have been added to the growing polymer chain, can you identify the structural pattern that is being repeated? When a vinyl monomer's name has two or more words, what is necessary to incorporate into the polymer's name?

Solve

The initiation step is the same for each reaction.

Initiation
Homolysis
Benzoyl peroxide
Radical initiator

In each case, we name the polymer by writing *poly* followed by the monomer's name, and we enclose the monomer's name in parentheses as it consists of two or more words.

(a) Answers for the vinylidene chloride monomer are given below:

(i) The mechanism for the vinylidene chloride monomer's two propagation steps in polymerization is given below. In the first step, the initial radical adds to the C=C bond of the monomer, resulting in a new radical. The new radical reacts with another molecule of the monomer in the same type of radical addition to the C=C.

Propagation 1
Vinylidene chloride
Radical addition
Propagation 2
Radical addition

(ii) Condensed formula:

(iii) Name: Poly(vinylidene chloride)

(b) Answers for the acrylamide monomer are given below:

(i) The mechanism for the acrylamide monomer's two propagation steps in polymerization is given below. In the first step, the initial radical adds to the C=C bond of the monomer, resulting in a new radical. The new radical reacts with another molecule of the monomer in the same type of radical addition to the C=C.

Propagation 1
Acrylamide
Radical addition
Propagation 2
Radical addition

(ii) Condensed formula:

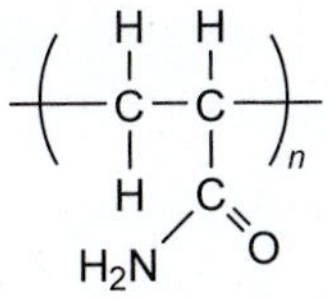

$$\left[\begin{array}{cc} H & H \\ | & | \\ C - & C \\ | & | \\ H & C(=O)NH_2 \end{array} \right]_n$$

(iii) Name: Polyacrylamide

(c) Answers for the methyl vinyl ether monomer are given below:

(i) The mechanism for the methyl vinyl ether monomer's two propagation steps in polymerization is given below. In the first step, the initial radical adds to the C=C bond of the monomer, resulting in a new radical. The new radical reacts with another molecule of the monomer in the same type of radical addition to the C=C.

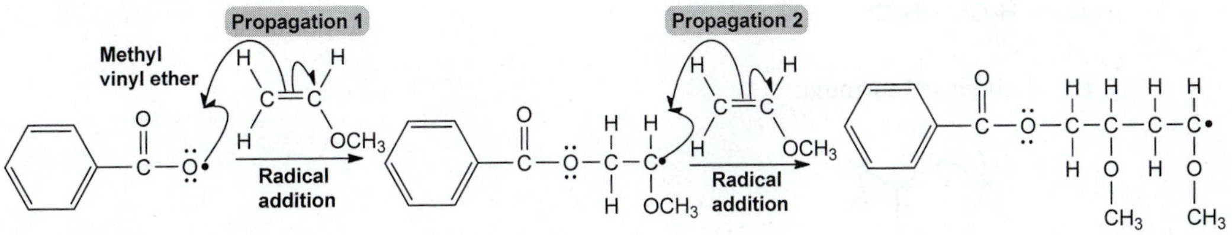

Propagation 1

Methyl vinyl ether

Radical addition

Propagation 2

Radical addition

(ii) Condensed formula:

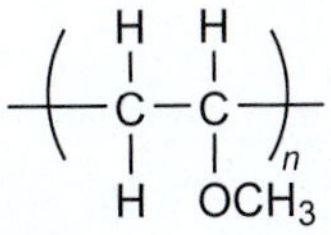

$$\left[\begin{array}{cc} H & H \\ | & | \\ C - & C \\ | & | \\ H & OCH_3 \end{array} \right]_n$$

(iii) Name: Poly(methyl vinyl ether)

Problem 26.34

Think

What is the repeating unit of each polymer? What substituents are attached to the main polymer chain of the polymer? How many C atoms along the main chain must be traversed before the attached substituents repeat? How is a repeating unit of a vinyl polymer related to the structure of the vinyl monomer? What bonds are the same, and what bonds are different?

Solve

In each case, once the repeating unit is determined, the monomer is drawn by removing the single bonds that connect the repeating units together and converting the middle C–C bond to a C=C double bond.

(a) Polymer:

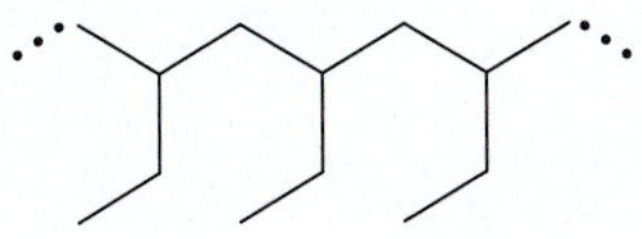

(i) Condensed formula:

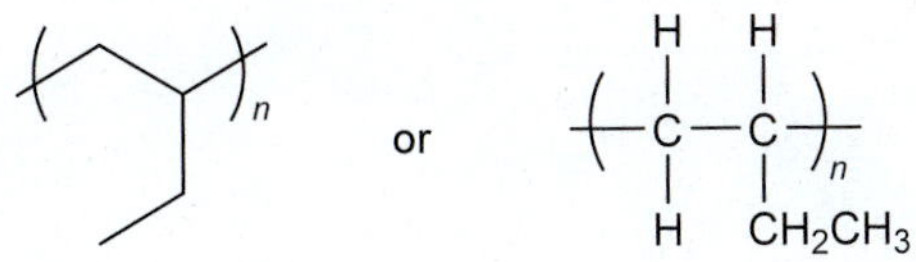

or

$$\left[\begin{array}{cc} H & H \\ | & | \\ C - & C \\ | & | \\ H & CH_2CH_3 \end{array} \right]_n$$

(ii) Monomer structure:

(b) Polymer:

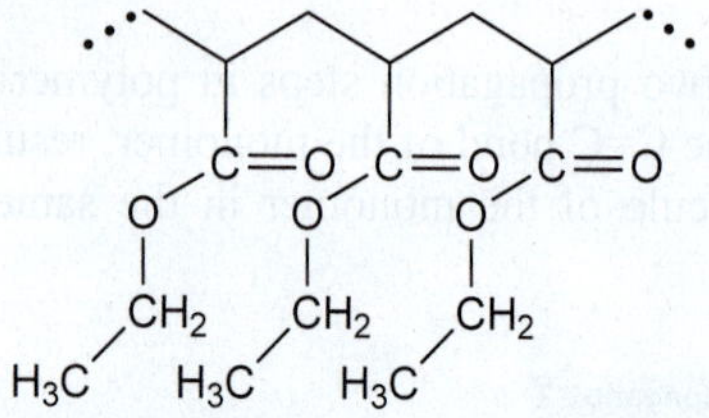

C=O C=O C=O
O O O
CH_2 CH_2 CH_2
H_3C H_3C H_3C

(i) Condensed formula:

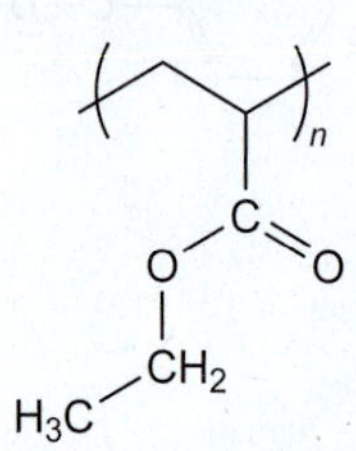

n
C
O O
CH_2
H_3C

(ii) Monomer structure:

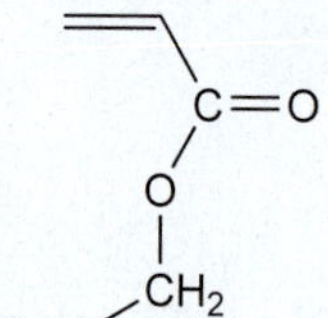

C=O
O
CH_2
H_3C

(c) Polymer:

H H H H H H
···C–C–C–C–C–C···
H H H
N N N

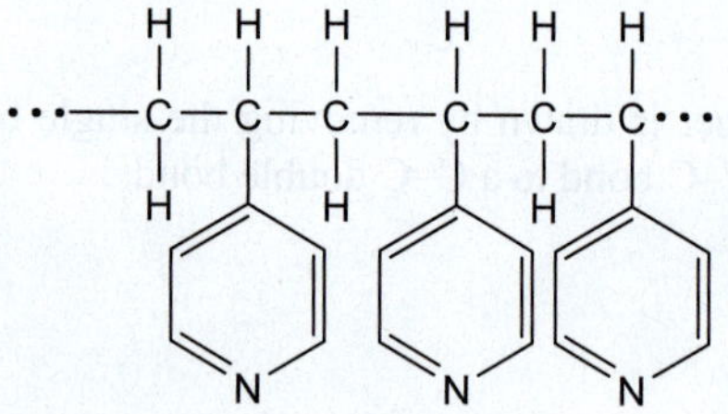

(i) Condensed formula:

H H
C–C
H n
N

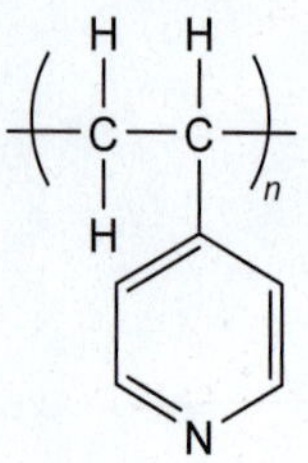

(ii) Monomer structure:

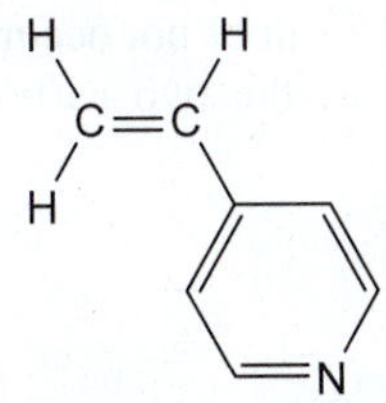

Problem 26.35

Think

What is the structure of the repeating unit? How is a repeating unit of a vinyl polymer related to the structure of the vinyl monomer? What bonds are the same, and what bonds are different? How does H_2O_2 produce the initial free radicals? What elementary step occurs between a free radical and a vinyl monomer? Is the product of that step still a free radical?

Solve

(a) The monomer is derived from the repeating unit by removing the single bonds that connect the repeating units together and converting the middle C–C bond to a C=C double bond. Below is the structure of *N*-vinylpyrrolidone.

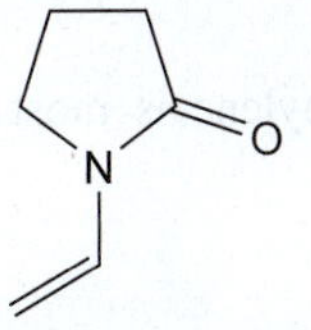

***N*-Vinylpyrrolidone**

(b) Hydrogen peroxide is a radical initiator that undergoes homolysis of the O–O single bond to produce hydroxyl radicals.

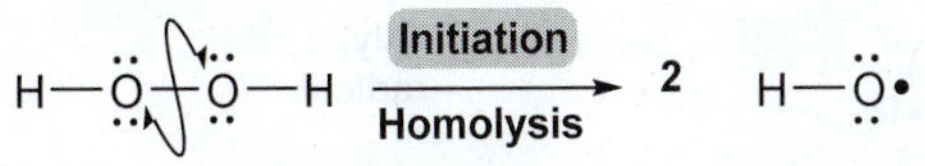

Initiation

Homolysis

In the first propagation step, a hydroxyl radical adds to the C=C bond of an *N*-vinylpyrrolidone monomer. The product of that step is another radical, which undergoes radical addition to another *N*-vinylpyrrolidone monomer.

Propagation 1 **Propagation 2**

Radical addition **Radical addition**

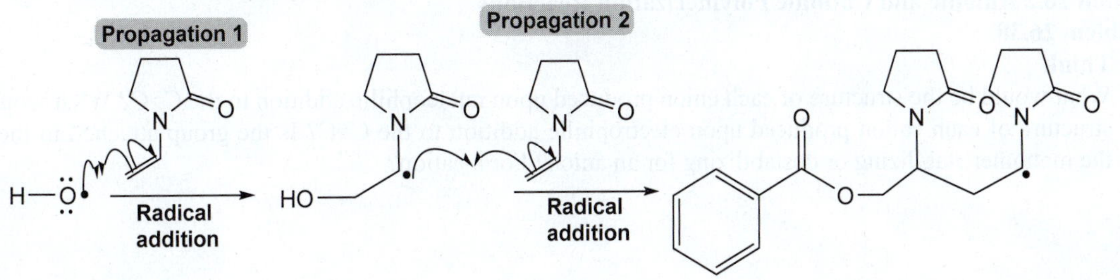

Problem 26.36

Think

What is the structure of Teflon (see Figure 26-25 on page 1343 of the textbook)? What elementary step makes up disproportionation? What atom is abstracted in this step? Is that atom present? What is combination? Can combination still occur between two radicals?

Solve

Two portions of growing Teflon polymers are shown below. Disproportionation generally involves H atom abstraction from the atom adjacent to the C• on one of the growing polymers. This does not occur, because there are no hydrogen atoms for a radical to abstract. Even without such H atoms, the two carbon radicals can undergo combination to form a new C–C in a termination step.

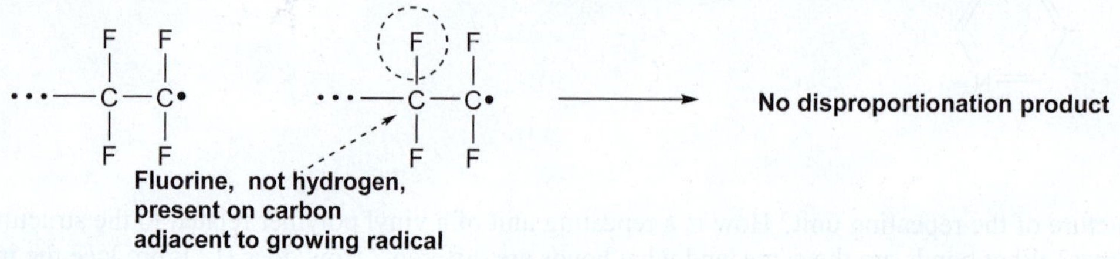

Fluorine, not hydrogen, present on carbon adjacent to growing radical

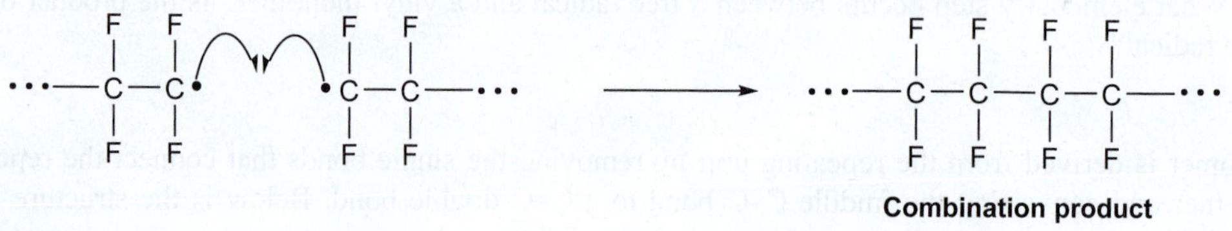

No disproportionation product

Combination product

Problem 26.37

Think

What propagation step is possible for R• and propylene? What H atom in propylene is most likely to be abstracted? Which radical is most stable?

Solve

(a) The propagation and H atom abstraction steps involving R• and propylene are given below:

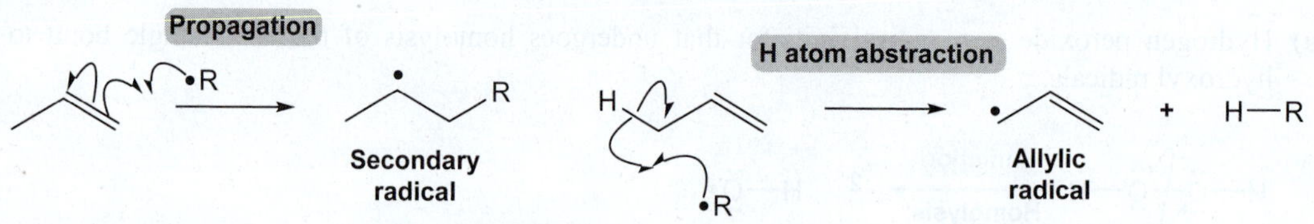

Propagation

Secondary radical

H atom abstraction

Allylic radical

(b) An allylic radical is more stable than the secondary radical owing to resonance.

(c) The step that is necessary for propagation (i.e., radical addition) produces the less stable alkyl radical. H atom abstraction, therefore, tends to take place rapidly and interferes with the formation of the polymer.

Section 26.2 Anionic and Cationic Polymerization Reactions

Problem 26.38

Think

What would be the structure of each anion produced upon nucleophilic addition to the C=C? What would be the structure of each cation produced upon electrophilic addition to the C=C? Is the group attached to the C=C of the monomer stabilizing or destabilizing for an anion? For a cation?

Solve

(a) The anions produced after nucleophilic addition to the C=C for ethyl vinyl ether, but-1-ene, and nitroethylene are shown below. The NO_2 group of nitroethylene is both an inductive and resonance-stabilizing group for the anion, so nitroethylene is most likely to undergo anionic polymerization.

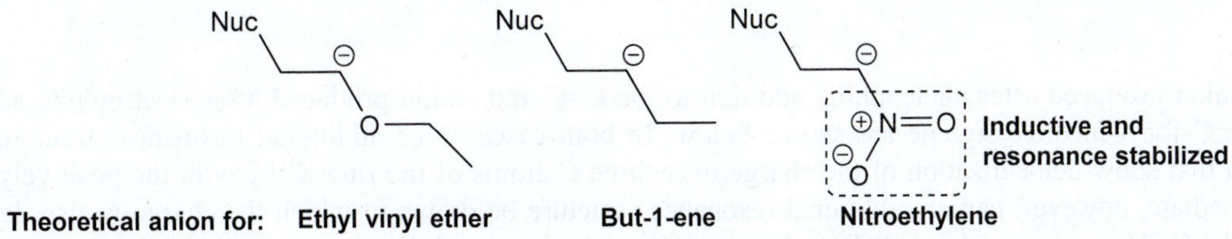

Nuc Nuc Nuc

Inductive and resonance stabilized

Theoretical anion for: Ethyl vinyl ether But-1-ene Nitroethylene

(b) The cations produced after electrophilic addition to the C=C for ethyl vinyl ether, but-1-ene, and nitroethylene are shown below. The ether O of ethyl vinyl ether is a resonance-stabilizing group for the C^+, so ethyl vinyl ether is most likely to undergo cationic polymerization.

E E E

Resonance stabilized

Theoretical cation for: Ethyl vinyl ether But-1-ene Nitroethylene

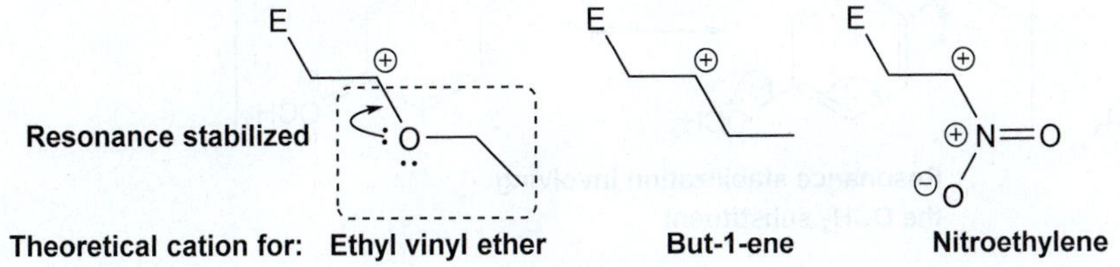

Problem 26.39

Think

What is the structure for the monomer for poly(*N*-vinylcarbazole)? What is cationic polymerization? What would be the structure of each cation produced upon electrophilic addition to the C=C? How is the resulting cation stabilized?

Solve

The structure for the monomer for poly(*N*-vinylcarbazole) is *N*-vinylcarbazole (shown below). The first step of cationic polymerization is production of a cation via electrophilic addition of the C=C bond to an acid. The resulting cation is stabilized via resonance.

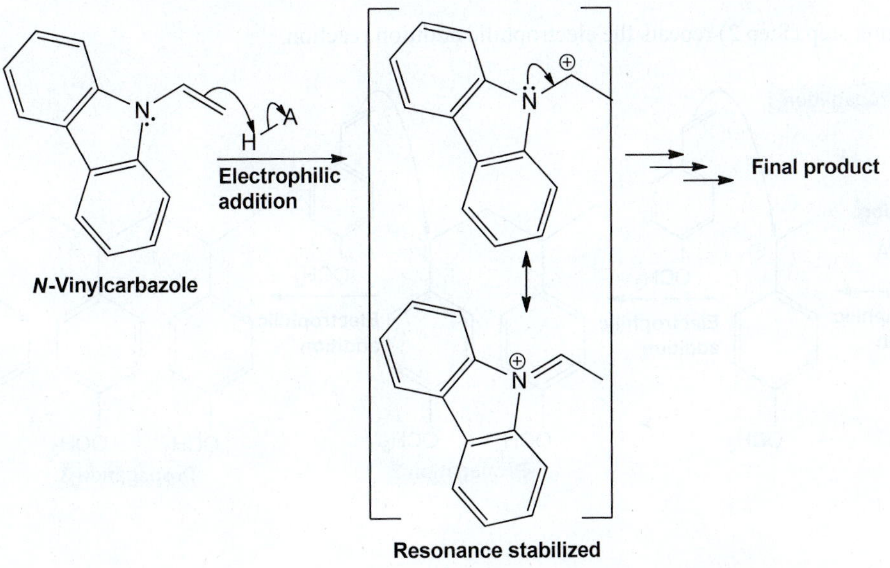

Electrophilic addition

Final product

N-Vinylcarbazole

Resonance stabilized

Problem 26.40

Think

What would be the structure of each anion produced upon nucleophilic addition to the C=C? What would be the structure of each cation produced upon electrophilic addition to the C=C? How are the neighboring groups stabilizing or destabilizing groups for an anion? For a cation?

Solve

The anion produced after nucleophilic addition to the C=C and cation produced after electrophilic addition to the C=C for 4-methoxystyrene are shown below. In both cases, three additional resonance structures can be drawn that show delocalization of the charge over three C atoms of the ring. Only with the positively charged intermediate, however, can an additional resonance structure be drawn in which the charge is also delocalized onto the OCH_3 group, as shown below. This provides additional stabilization to the cationic intermediate, so 4-methoxystyrene would more likely undergo cationic polymerization.

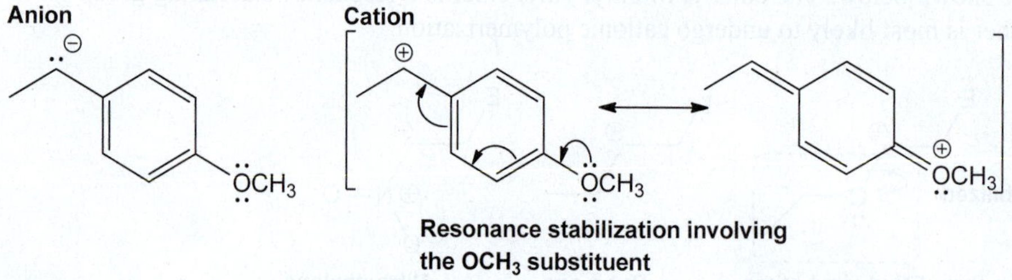

Anion

Cation

OCH_3

OCH_3

OCH_3

Resonance stabilization involving the OCH_3 substituent

Problem 26.41

Think

How does the vinyl C=C π bond react with an acid, HA (Chapter 11)? How does the electrophile formed in the initiation step react with another equivalent of 4-methoxystyrene? What is the repeating unit?

Solve

(a) and **(b)** The mechanism for the polymerization of poly(4-methoxystyrene) from 4-methyoxystyrene is shown below.

Step 1: The alkene reacts with H–A in an electrophilic addition step to form a resonance-stabilized benzylic cation.

Step 2: The benzylic cation reacts with another equivalent of 4-methoxystyrene in a second electrophilic addition step.

Step 3: Propagation step (Step 2) repeats the electrophilic addition reaction.

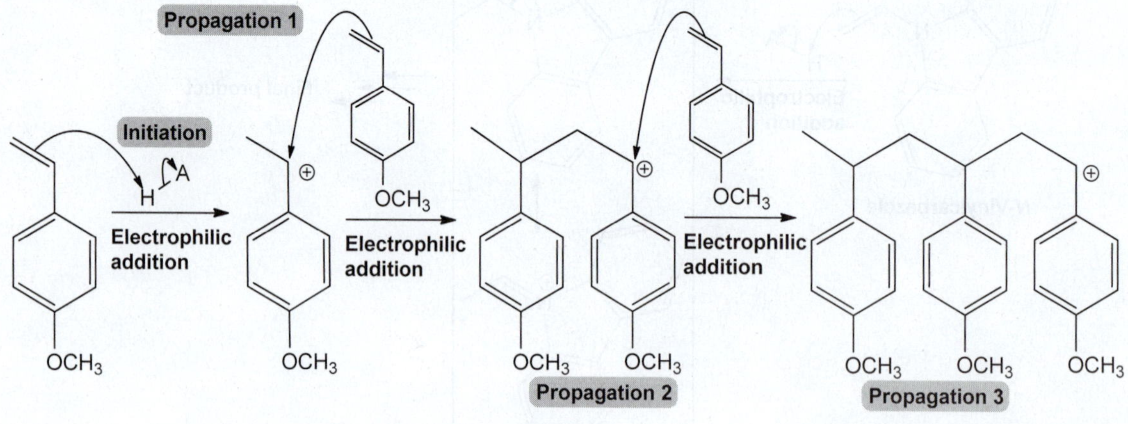

Propagation 1

Initiation

H–A

Electrophilic addition

Electrophilic addition

Electrophilic addition

OCH_3

Propagation 2

Propagation 3

(c) Repeating unit for poly(4-methoxystyrene).

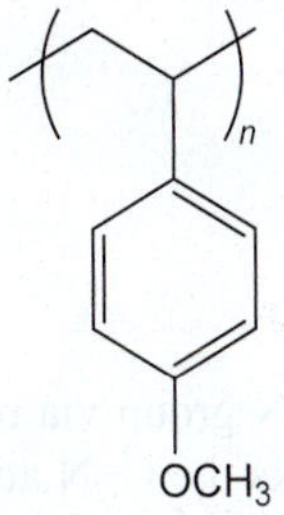

OCH$_3$

Problem 26.42

Think

Which carbon of the C=C in methyl 2-cyanoacrylate would be attacked by HO$^-$ to produce the most stable adduct? What anionic intermediate forms? Can the product of that reaction react in the same way with another equivalent of the monomer? What is the repeating unit? How do the attached CN and CO_2CH_3 groups affect the stability of the charge that develops on the growing polymer chain? Would those attached groups stabilize or destabilize a nearby positive charge?

Solve

(a) and **(b)** The mechanism for the polymerization of poly(methyl 2-cyanoacrylate) from methyl 2-cyanoacrylate is shown below.

Step 1: The alkene reacts with HO$^-$ in a nucleophilic addition step to form a carbanion. The CN and CO_2CH_3 groups stabilize nearby negative charges, so HO$^-$ attacks the C atom without the attached CN and CO_2CH_3 groups to produce the negative charge on the C that has those groups attached.

Step 2: The carbanion reacts with another equivalent of methyl 2-cyanoacrylate in a second nucleophilic addition step.

Step 3: Propagation step (Step 2) repeats the nucleophilic addition reaction.

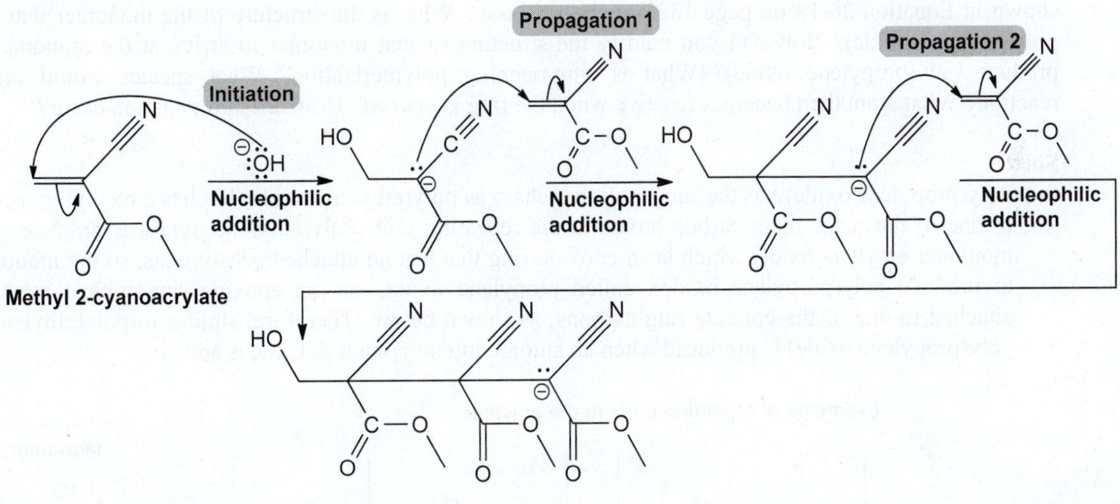

Initiation

Propagation 1

Propagation 2

Nucleophilic addition

Nucleophilic addition

Nucleophilic addition

Methyl 2-cyanoacrylate

(c) Condensed formula for poly(methyl 2-cyanoacrylate):

(d) Methyl 2-cyanoacrylate forms a carbanion intermediate that is stabilized by the C≡N group via resonance and inductive effects. Methyl acrylate has less resonance stabilization because it lacks the C≡N attached to the C^- that develops on the growing polymer chain. Thus, methyl 2-cyanoacrylate polymerizes faster.

Methyl acrylate

Methyl 2-cyanoacrylate
More stable anion

(e) Methyl 2-cyanoacrylate is not likely to polymerize through a cation polymerization mechanism because the cation would be unstable owing to the electron-withdrawing CN and CO_2CH_3 groups. Both of those groups would destabilize the C^+ in the growing polymer chain.

Section 26.3 Ring-Opening Polymerization Reactions

Problem 26.43

Think

How does the structure of poly(propylene oxide) compare to the structure of poly(ethylene oxide), which is shown in Equation 26-18 on page 1323 of the textbook? What is the structure of the monomer that produces poly(ethylene oxide)? How can you modify the structure of that monomer to arrive at the monomer used to produce poly(propylene oxide)? What is ring-opening polymerization? What species would initiate the reaction? What atom then becomes reactive when the ring is opened? How does propagation occur?

Solve

(a) Poly(propylene oxide) has the same polymer chain as poly(ethylene oxide) but has a methyl group attached to one of the main-chain carbon atoms in the repeating unit. Poly(ethylene oxide) is produced from the monomer ethylene oxide, which is an epoxide ring that has no attached substituents, so the monomer used to produce poly(propylene oxide), called propylene oxide, has an epoxide ring with a methyl group attached to one of the epoxide ring carbons, as shown below. Therefore, similar to poly(ethylene oxide), poly(propylene oxide) is produced when an anionic initiator, such as CaO, is added.

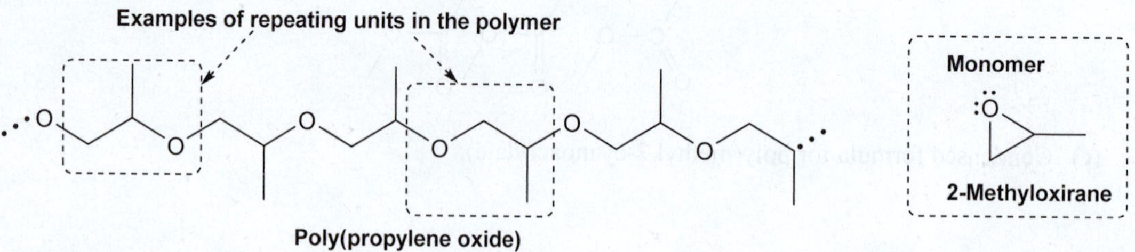

(b) The mechanism for the anionic ring-opening polymerization of poly(propylene oxide) from 2-methyloxirane, when treated with a nucleophile, is shown on the next page.

Step 1: Initiation occurs when the nucleophile attacks the partial positive C atom of the ring at the less substituted C, an S_N2 step. The ring opens, and O^- is the leaving group.

Step 2: This O^- reacts with another equivalence of 2-methyloxirane in an S_N2 step and opens the ring.

Step 3: The propagation step (Step 2) repeats.

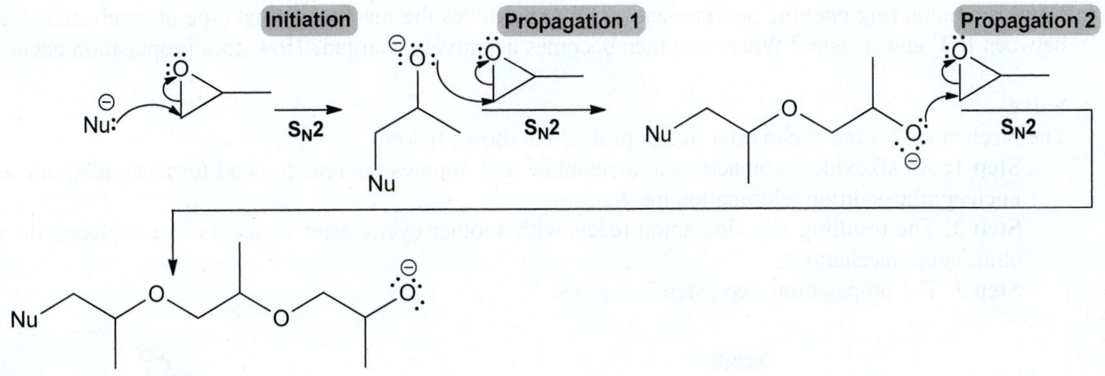

Problem 26.44

Think

Does oxetane have a strained ring? For anionic polymerization, should the initiator be nucleophilic or electrophilic? When the initiator is added, what reaction will take place to open the ring of the oxetane? How will the resulting product of that reaction react with another molecule of the oxetane? Compare the repeating units of poly(ethylene oxide) and polyoxetane. Which one has a greater number of C atoms? How does that affect water solubility? How do the O atoms affect the polarity of the polymer chain, and what is the effect on melting point?

Solve

(a) Oxetane has a strained ring, and anionic polymerization can be initiated with a nucleophile, such as O^{2-} from CaO. The mechanism for the polymerization of polyoxetane from oxetane is shown below.

Step 1: Oxide anion acts as a nucleophile and initiates the reaction, opening the ring and forming an alkoxide anion.

Step 2: The resulting alkoxide anion reacts with another oxetane molecule.

Step 3: The propagation step (Step 2) repeats.

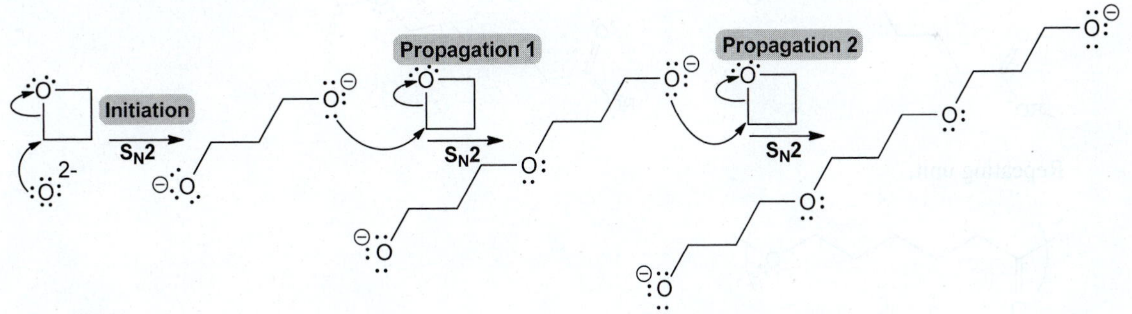

(b) Condensed formula:

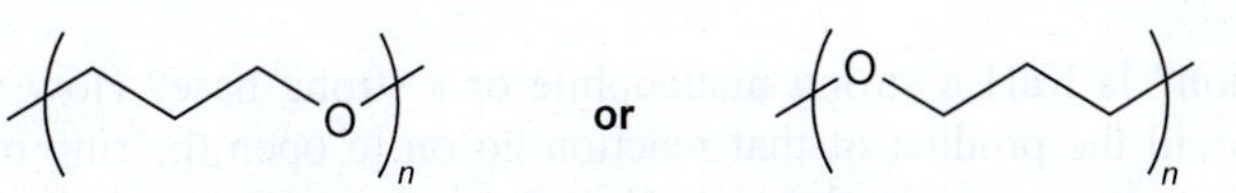

or

(c) Polyoxetane (PO) will be less soluble in water than poly(ethylene oxide) (PEO). The oxygens occur less frequently in the polymer chain in PO, so there will be less hydrogen bonding with water along a chain of PO than along a chain of PEO of the same length.

(d) PEO will have a higher melting point. The bent C–O–C group is somewhat polar. That group occurs less frequently in the polymer chain in PO, making PO slightly less polar.

Problem 26.45

Think

What is anionic ring-opening polymerization? What initiates the reaction? What type of mechanism takes place between RO^- and an ester? What atom then becomes negatively charged? How does propagation occur?

Solve

The mechanism for the cyclic ester in this problem is shown below.

Step 1: An alkoxide anion acts as a nucleophile and initiates the reaction and forms an alkoxide anion in a nucleophilic addition–elimination mechanism.

Step 2: The resulting alkoxide anion reacts with another cyclic ester molecule in a nucleophilic addition–elimination mechanism.

Step 3: The propagation step (Step 2) repeats.

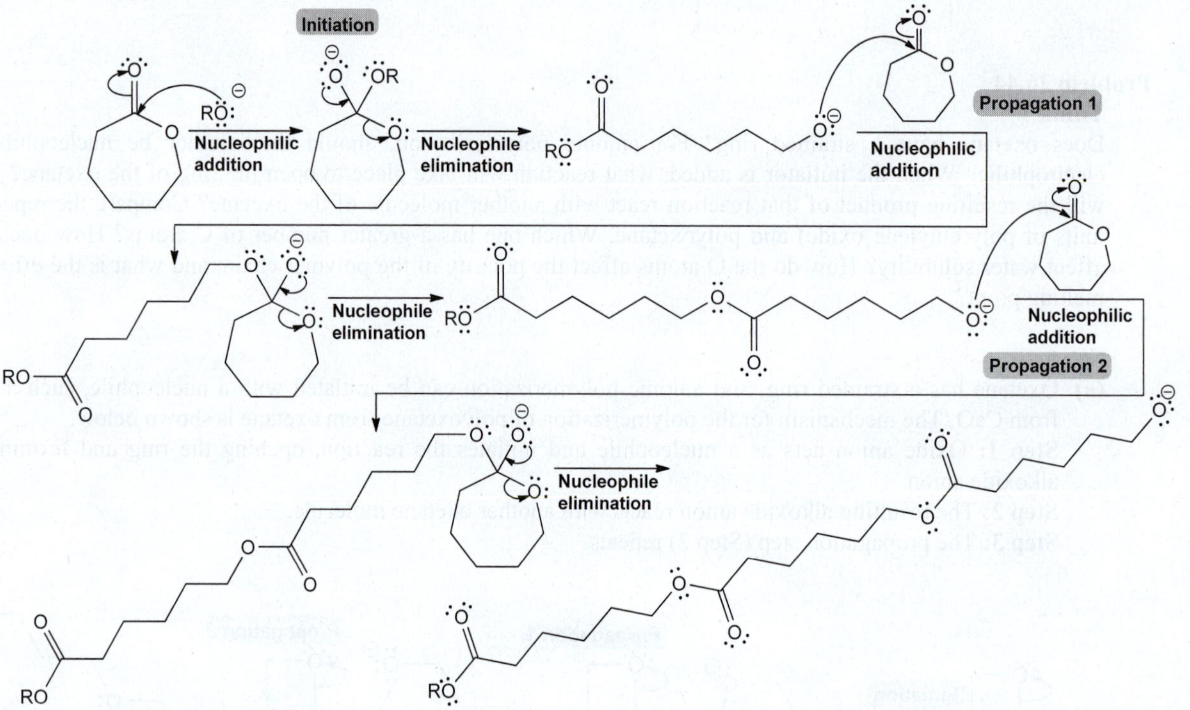

Initiation

RO: Nucleophilic addition → Nucleophile elimination → RO

Propagation 1

Nucleophilic addition → Nucleophile elimination → RO

Propagation 2

Nucleophilic addition → Nucleophile elimination

RO RO

Repeating unit:

n

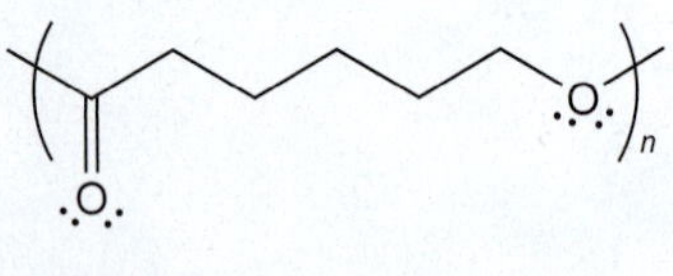

Problem 26.46

Think

What is anionic ring-opening polymerization? Is NaH a strong nucleophile or a strong base? How will NaH react with the cyclic lactam shown? How will the product of that reaction go on to open the ring of another molecule of the lactam? What atom then becomes negatively charged? How does propagation occur?

Solve

The mechanism for the cyclic amide in this problem is shown below.

Step 1: Hydride (H:⁻) from NaH is a poor nucleophile but a very strong base (Section 17.4), so it deprotonates the NH of the lactam to produce a nucleophilic N⁻.

Step 2: The deprotonated lactam acts as a nucleophile and reacts with another molecule of the lactam in a nucleophilic addition–elimination mechanism, opening the ring.

Step 3: The resulting N⁻ anion reacts with another lactam molecule in a nucleophilic addition–elimination mechanism.

Step 4: The propagation step (Step 3) repeats.

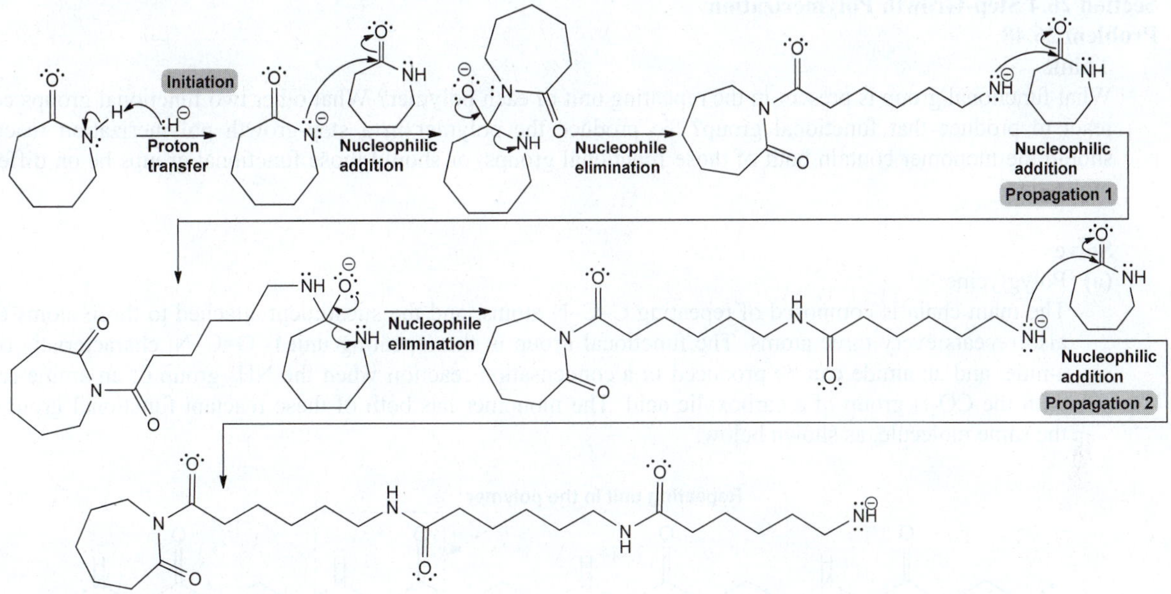

Initiation

Proton transfer

Nucleophilic addition

Nucleophile elimination

Nucleophilic addition

Propagation 1

Nucleophile elimination

Nucleophilic addition

Propagation 2

Repeating unit:

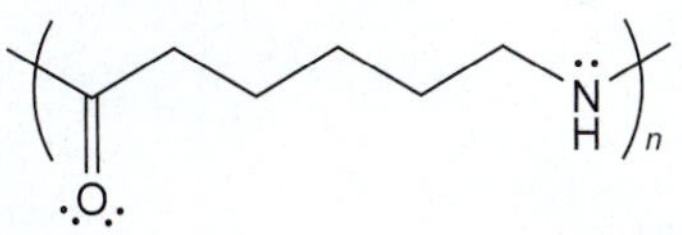

Problem 26.47

Think

What elementary step occurs between a free radical and a vinyl monomer? In the resulting radical, what bond of the ring can break to open the ring? How is the resulting radical stabilized? How does that radical go on to repeat the propagation steps?

Solve

In the first step, the R• radical adds to the C=C bond of the ketene acetal monomer. The product of that step is another radical, in which the unpaired electron is on the C atom between the two O atoms. The next step forms a C=O, opens the ring, and has the radical on the benzylic C. That radical is stabilized by the appearance of a strong C=O bond and resonance delocalization of the unpaired electron in the benzene ring. Propagation continues when that radical reacts with another ketene acetal monomer. See the repeating unit on the next page.

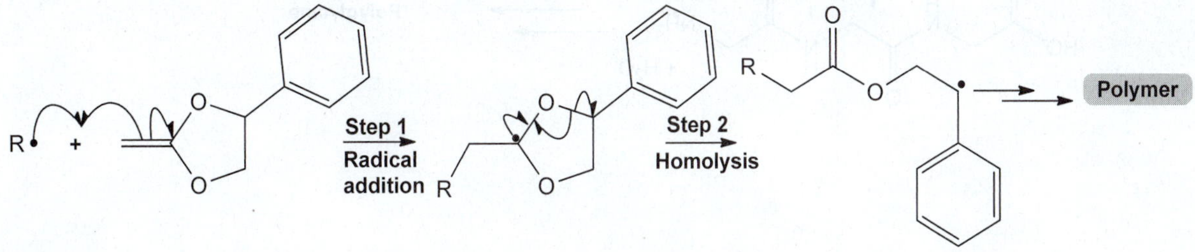

Step 1 Radical addition

Step 2 Homolysis

Polymer

Repeating unit:

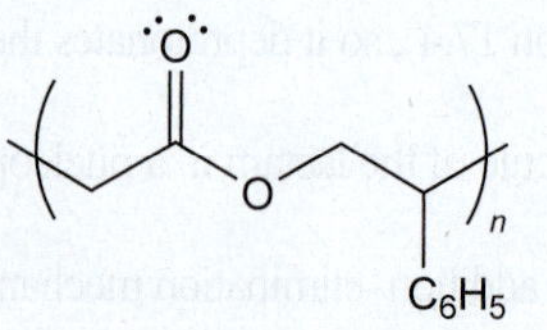

O
O
n
C_6H_5

Section 26.4 Step-Growth Polymerization
Problem 26.48

Think

What functional group is present in the repeating unit of each polymer? What other two functional groups could react to produce that functional group? To produce the polymer in a step-growth polymerization reaction, should the monomer contain both of those functional groups, or should those functional groups be on different molecules?

Solve

(a) Polyglycine

The main chain is composed of repeating C–C–N atoms, and the substituent attached to those atoms (=O) also repeats every three atoms. The functional group in the repeating unit is O=C–N, characteristic of an amide, and an amide can be produced in a condensation reaction when the NH_2 group of an amine reacts with the CO_2H group of a carboxylic acid. The monomer has both of these reactant functional groups on the same molecule, as shown below.

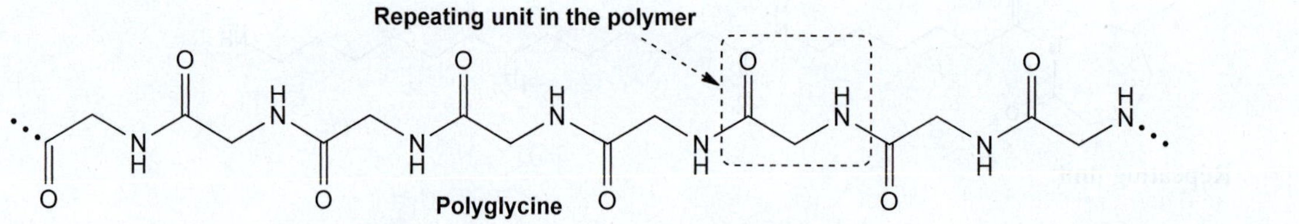

Repeating unit in the polymer

Polyglycine

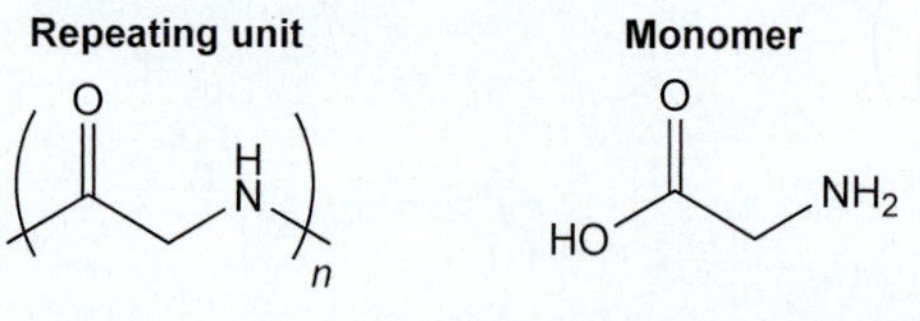

Repeating unit **Monomer**

n
NH_2
HO

When polyglycine is synthesized using this monomer, the polymer can grow one monomer at a time, as shown below, or two growing polymer chains can link together.

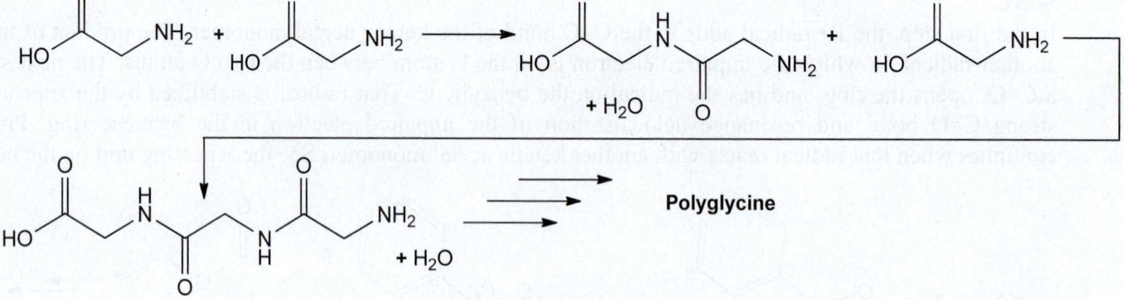

HO NH_2 + HO NH_2 → HO N H NH_2 + HO NH_2
+ H_2O

HO N H N H NH_2 + H_2O → → **Polyglycine**

(b) Poly(trimethylene terephthalate)

The repeating unit and monomer are shown below. In the repeating unit, there are two CO_2C functional groups, characteristic of an ester. One of those groups is completely contained within the repeating unit. The second is formed by the C=O at the left that is attached to another O shown at the right. An ester can be produced by a condensation reaction between the CO_2H group of a carboxylic acid and an OH group of an alcohol, so these groups should appear in the monomers. Notice that the connectivity between the first two ester groups in the polymer unit is different from that between the second two ester groups. Therefore, two different monomers are used. One is a dicarboxylic acid, and the other is a diol, as shown below.

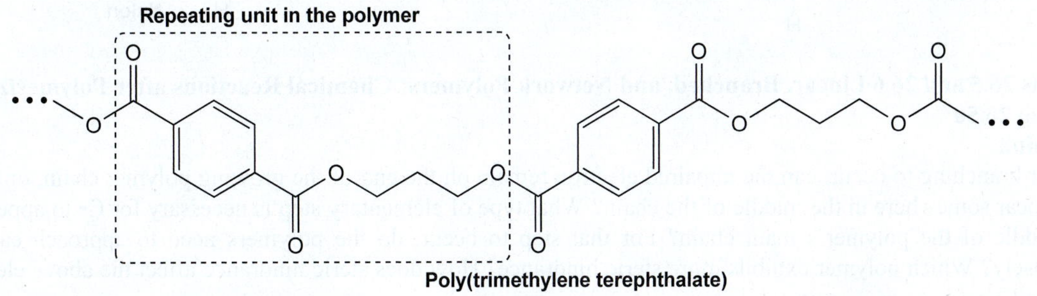

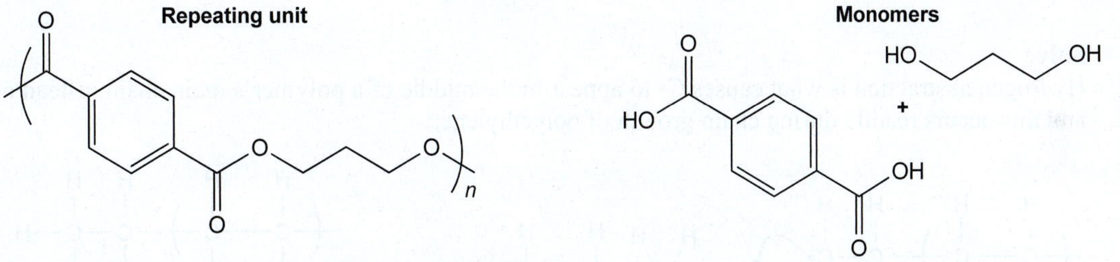

Repeating unit in the polymer

Poly(trimethylene terephthalate)

Repeating unit **Monomers**

HO OH + HO OH

When poly(trimethylene terephthalate) is synthesized using these monomers and an acid catalyst, the polymer can grow one monomer at a time, as shown below, or two growing polymer chains can link together.

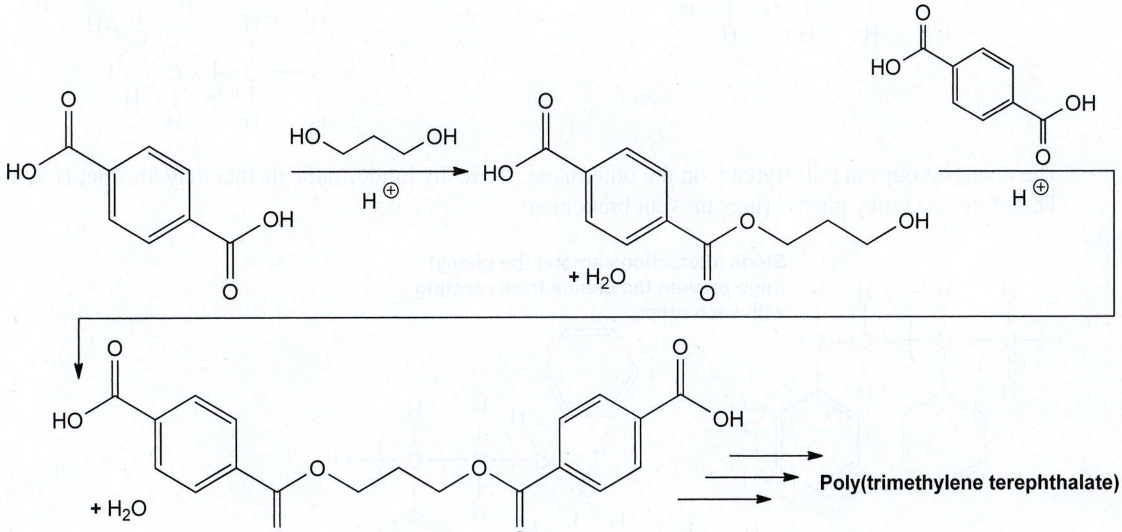

HO OH → (HO OH, $H^{\oplus}$) → HO OH + H_2O → (HO OH, $H^{\oplus}$) → HO OH + H_2O → → → **Poly(trimethylene terephthalate)**

Problem 26.49

Think

What functional group results from the reaction of an acid chloride and an amine? What new bond is formed? What atoms from the acid chloride and amine do not appear in the amide product?

Solve

One product will be the nylon polymer, and the other is an equivalent of hydrochloric acid, HCl. One H is lost from the NH_2 group, and Cl is lost from the O=C–Cl group. However, a base in the system (such as 1,6-hexanediamine) will deprotonate the HCl as it forms, so HCl is not the actual product.

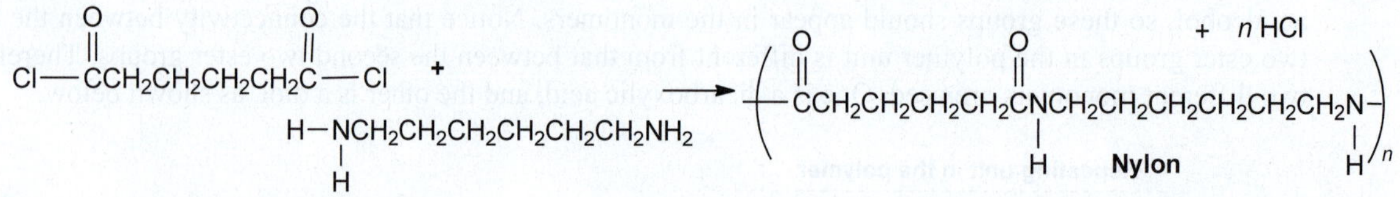

Cl—C(=O)CH₂CH₂CH₂CH₂C(=O)—Cl + H–NHCH₂CH₂CH₂CH₂CH₂CH₂NH₂ ⟶ –(C(=O)CH₂CH₂CH₂CH₂C(=O)N(H)CH₂CH₂CH₂CH₂CH₂CH₂N(H))–ₙ + *n* HCl

Nylon

Sections 26.5 and 26.6 Linear, Branched, and Network Polymers; Chemical Reactions after Polymerization

Problem 26.50

Think

For branching to occur, can the unpaired electron remain on the end of the growing polymer chain, or should it appear somewhere in the middle of the chain? What type of elementary step is necessary for C• to appear in the middle of the polymer's main chain? For that step to occur, do the polymers need to approach each other closely? Which polymer exhibits more steric hindrance? How does steric hindrance affect the above elementary step?

Solve

Hydrogen abstraction is what causes C• to appear in the middle of a polymer's main chain instead of on the end, and this occurs readily during chain growth of polyethylene:

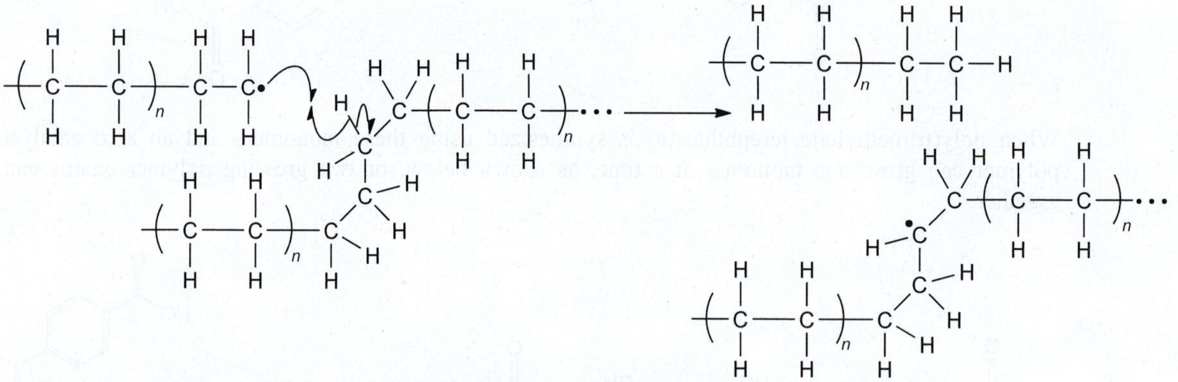

The phenyl groups in polystyrene, on the other hand, sterically hinder radicals that may abstract H atoms. Therefore, the bulky phenyl rings prevent branching:

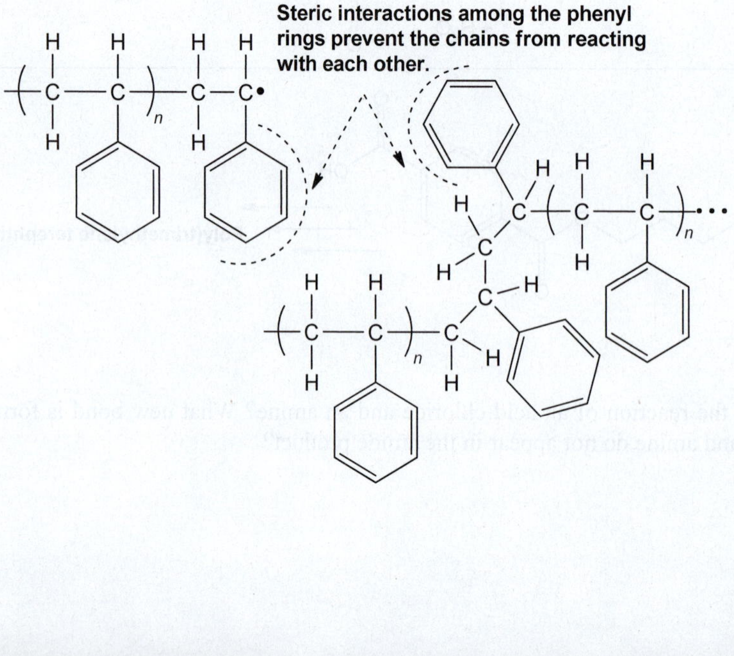

Steric interactions among the phenyl rings prevent the chains from reacting with each other.

Problem 26.51

Think

Will acid promote anionic or cationic ring-opening polymerization? How is the reaction initiated? How is the ring opened? How does propagation occur? Where on the growing polymer chain does the reaction occur to form a linear polymer? Where on the growing polymer chain does the reaction occur to form a branched polymer?

Solve

Acid will promote a cationic polymerization. The mechanism for the cationic ring-opening polymerization of a branched polyethylenimine from aziridine when treated with HCl is shown below. The basic N is protonated first, then the nucleophilic N of another aziridine attacks the C atom of a positively charged ring in an S_N2 step to open the ring. In a growing chain, branching can occur when a nucleophilic N in the middle of the chain attacks a protonated aziridine ring. The mechanism below shows how a tetramer can be produced, in which three monomers form a chain and the fourth forms a branch.

Step 1: Initiation occurs when the nucleophilic N atom of aziridine undergoes a proton transfer reaction with HCl and the N atom becomes positively charged.

Step 2: Another molecule of aziridine acts as a nucleophile to attack the protonated aziridine and open the ring.

Step 3: The propagation step (Step 2) repeats.

Step 4: The propagation step (Step 2) repeats.

Step 5: To have the propagation step form a branch, a N from the chain attacks a protonated aziridine and opens the ring.

Initiation
Proton transfer
Aziridine
Propagation 1
S_N2
Propagation 2
S_N2
Propagation 3
S_N2
Propagation 4
S_N2

Problem 26.52

Think

Review the structures for poly(ethylene terephthalate), terephthalic acid, and ethylene glycol from Equation 26-27. How is the structure for a branched polymer different from the poly(ethylene terephthalate) polymer shown in Equation 26-27? If a linear diol reacts with terephthalic acid, can branching occur?

Solve

The linear alcohol, ethylene glycol, forms a linear polymer. Therefore in order to form a branched polymer, a branched alcohol is required. An example using ethane-1,1,2-triol is used, but other branched alcohols could be used instead.

Linear polymer

Terephthalic acid + Ethylene glycol → $\left(\!-CH_2CH_2-O-\overset{O}{\overset{\|}{C}}-C_6H_4-\overset{O}{\overset{\|}{C}}-O-\right)_n$

Terephthalic acid

Ethylene glycol

Branched polymer

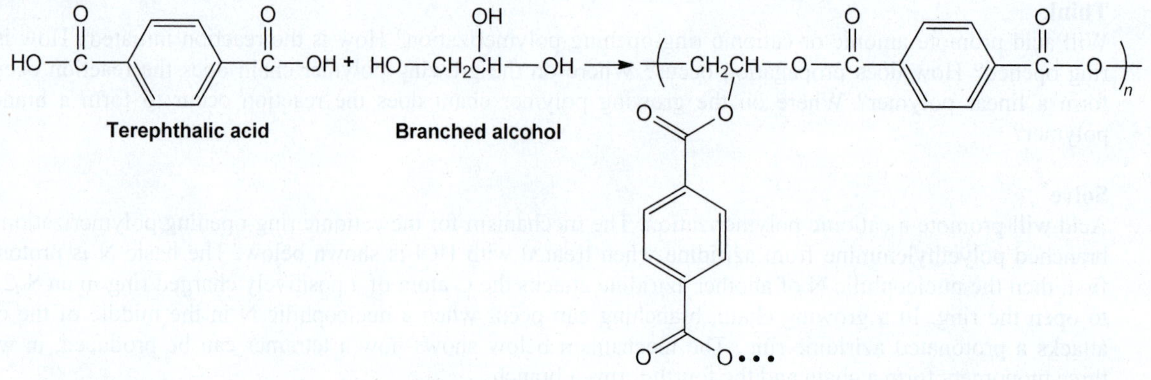

Terephthalic acid

Branched alcohol

Problem 26.53

Think

After polymerization, what is the functional group that appears in the polymer prior to chemical modification? What is the functional group that appears after chemical modification? What reactions do you know that will accomplish this transformation?

Solve

After polymerization, prior to the chemical modification, the polymer's repeating unit contains an O=C–N group, characteristic of an amide. After chemical modification, the functional group is C–N, characteristic of an amine. To carry out this transformation, a nucleophilic addition–elimination is required, which will eliminate the N-containing portion as a leaving group. Hydrolysis under basic conditions could carry out this transformation (Section 20.4), as shown below, but it could also take place under acidic conditions (Section 21.7).

2-Ethyl-2-oxazoline

Linear polyethylenimine

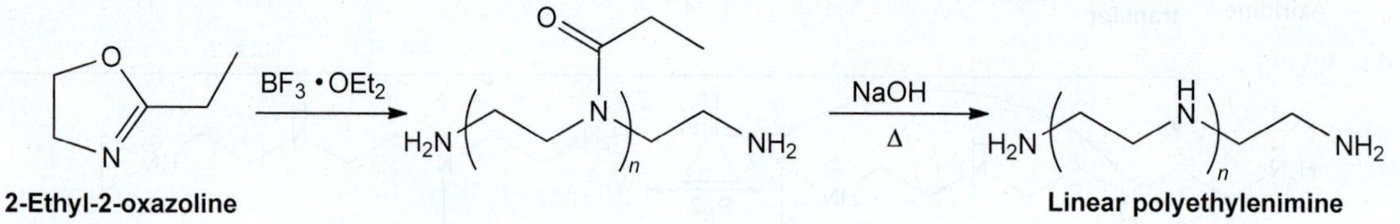

Problem 26.54

Think

How can you initiate the polymerization of a vinyl monomer such as *N*-vinyl formamide? What would be the repeating unit of poly(*N*-vinyl formamide)? What chemical modification of poly(*N*-vinyl formamide) could be carried out to convert it to poly(vinyl amine)?

Solve

N-Vinyl formamide is a vinyl polymer, so it can be polymerized using benzoyl peroxide, as shown below, to produce poly(*N*-vinyl formamide). The repeating unit of poly(*N*-vinyl formamide) has an O=C–N pendant group, characteristic of an amide, but the repeating unit of poly(vinyl amine) has a NH_2 pendant group, which releases the N-containing leaving group as an amine. To carry out this transformation, a nucleophilic addition–elimination is required, which will eliminate the N-containing portion as a leaving group. Hydrolysis under basic conditions could carry out this transformation (Section 20.4), as shown below, but it could also take place under acidic conditions (Section 21.7). See the figure on the next page.

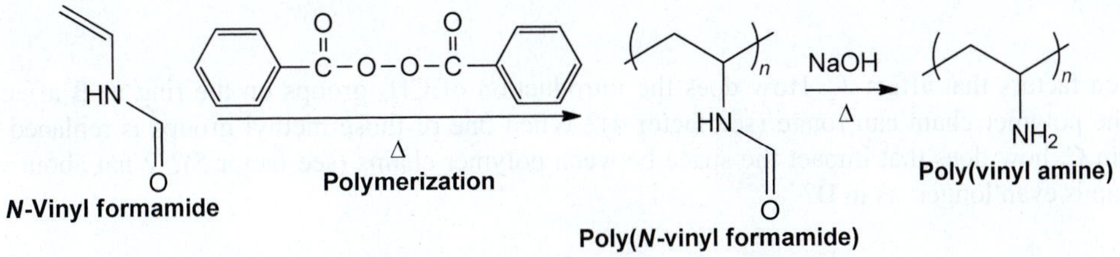

N-Vinyl formamide

Δ

Polymerization

Poly(*N*-vinyl formamide)

NaOH

Δ

Poly(vinyl amine)

Sections 26.7 and 26.8 General Aspects of Polymer Structure; Properties of Polymers

Problem 26.55

Think

How are heterochain polymers distinguished from carbon-chain polymers? What functional group is present in each repeating unit? Is that functional group responsible for connecting the monomers, or is it part of a pendant group?

Solve

In a carbon-chain polymer, the polymer main chain consists of only carbon atoms. In a heterochain polymer, the polymer main chain consists of carbon atoms and heteroatoms. As shown below, all three polymers have heteroatoms in their main chains, so they are heterochain polymers.

Problem	Type of Polymer	Condensed Formula
26.43	Heterochain polymer	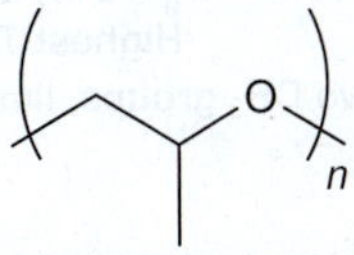
26.48(a)	Heterochain polymer	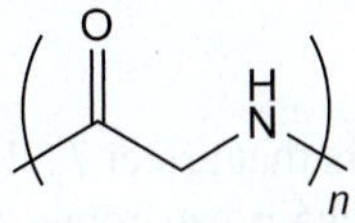
26.48(b)	Heterochain polymer	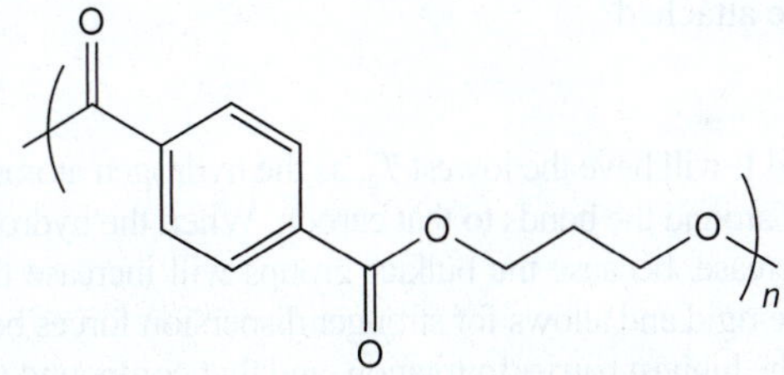

Problem 26.56

Think

What functional groups are present in each condensed formula? Are those functional groups part of the main chain of the polymer? How are the functional groups related to the class of polymer?

Solve

Problem 43: The main chain of poly(propylene oxide) has C–O–C functional groups, characteristic of ethers, so the polymer is a polyether.

Problem 48(a): The main chain of polyglycine has O=C–N functional groups, characteristic of amides, so the polymer is a polyamide.

Problem 48(b): The main chain of poly(trimethylene terephthalate) has CO_2C functional groups, characteristic of esters, so the polymer is a polyester.

Problem 26.57

Think

Review the seven factors that affect T_g. How does the introduction of CH_3 groups on the ring in **B** affect the ease in which the polymer chain can rotate (see factor 4)? When one of those methyl groups is replaced by a phenyl ring, as in **C**, how does that impact the space between polymer chains (see factor 5)? What about when the pendant group is even longer, as in **D**?

Solve

In **A**, the absence of large groups on the benzene ring allows the polymer chains to rotate fairly easily, giving it the lowest T_g. The introduction of the two methyl groups makes rotation of the polymer chains and movement of one past another more difficult and raises T_g. However, when a phenyl ring replaces a methyl group, the phenyl ring pushes neighboring chains farther away, creating space between the polymer molecules, and weakens the intermolecular interactions. This allows for easier movement of polymer molecules and lowers T_g. Lengthening the spacer (a benzyl group instead of a phenyl group in **D**) pushes the molecules even farther apart and lowers T_g.

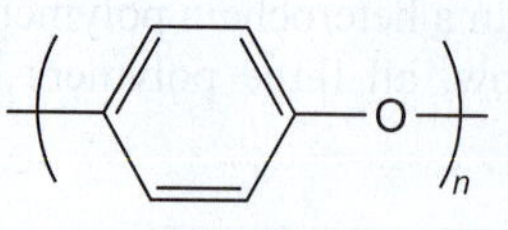

A T_g = 82 °C Lowest T_g

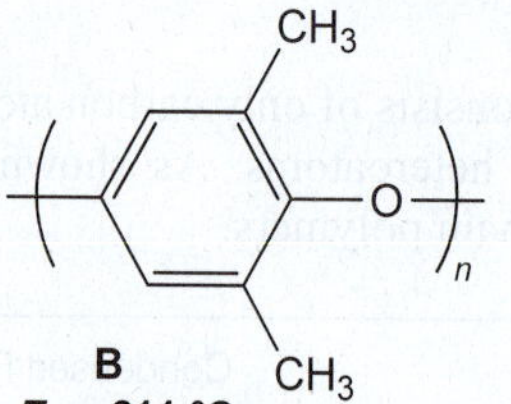

B T_g = 211 °C Highest T_g Two CH_3 groups, limited rotation

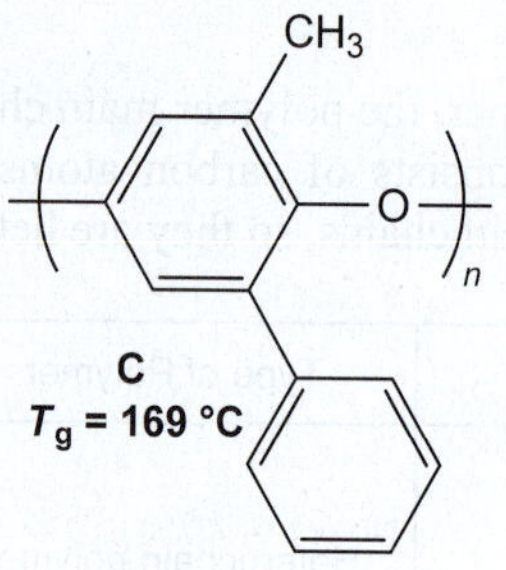

C T_g = 169 °C

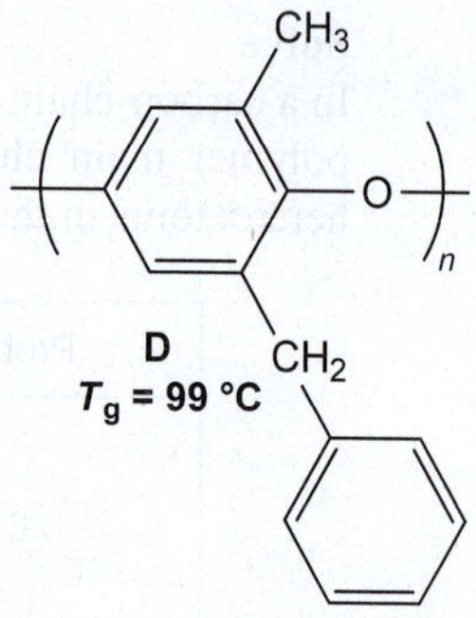

D T_g = 99 °C

Problem 26.58

Think

Review the seven factors that affect T_g. How does the introduction of CH_3 groups along the chain affect the ease in which the polymer chain can rotate (see factor 4)? How is that rotation affected when the bulkier phenyl groups are attached?

Solve

Compound **F** will have the lowest T_g, as the hydrogen atoms attached to the sp^3-hybridized carbon give the lowest barrier to rotation around the bonds to that carbon. When the hydrogens are replaced with methyl groups, as in compound **E**, the T_g will increase, because the bulkier groups will increase the amount of energy required for rotation, which makes the chain more rigid and allows for stronger dispersion forces between the polymers. The two phenyl groups in compound **G** will give the highest barrier to rotation, and that compound will have the highest T_g.

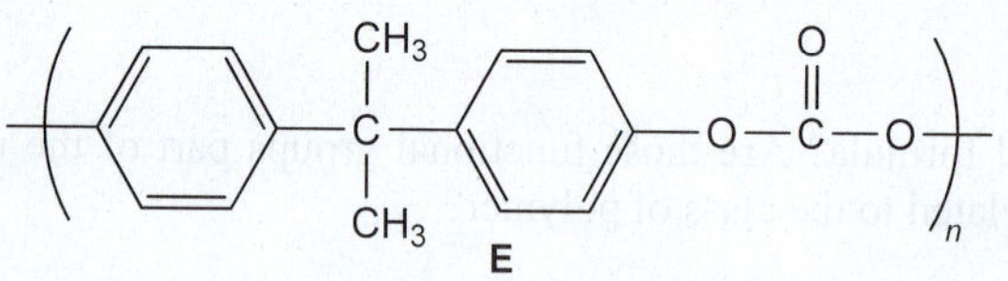

E

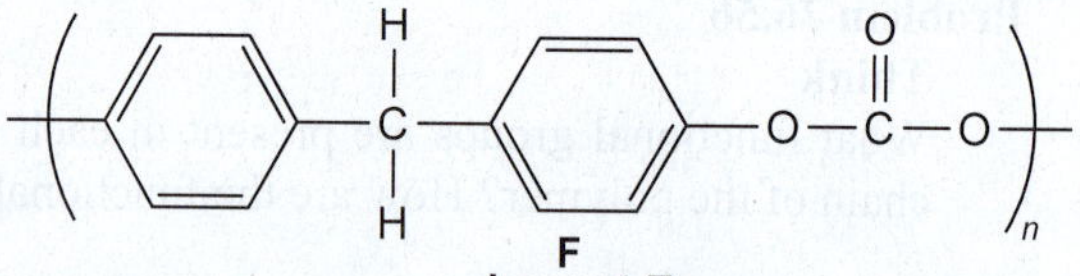

F Lowest T_g Lowest barrier of rotation

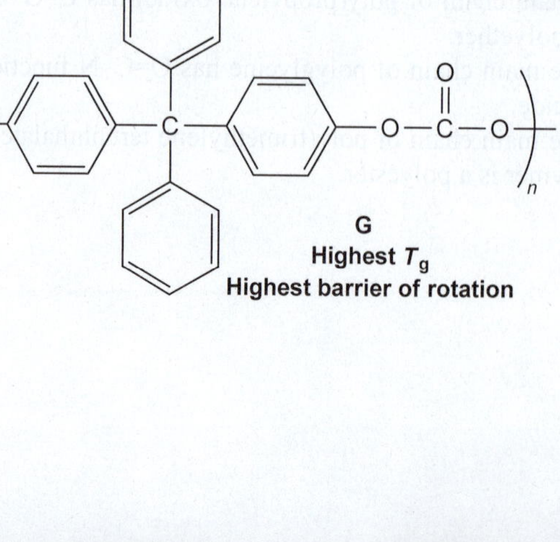

G Highest T_g Highest barrier of rotation

Problem 26.59

Think

Review the structures for Kevlar in Problem 26.11 and the polyphthalamide in Problem 26.12. Review the seven factors that affect T_g. Which factor is affected when the phenyl ring between the two N atoms in the repeating unit is replaced by a hexyl chain? How does that affect the T_g?

Solve

Kevlar will have the higher T_g. In Kevlar, a phenyl ring appears between the two N atoms in the repeating unit, whereas in the polyphthalamide it is a hexyl chain instead. The polymer chain in Kevlar is therefore much more rigid, which, according to factor 6, increases T_g.

Kevlar (Problem 26.11)
Higher T_g, more rigid polymer chain

Polyphthalamide (Problem 26.12)

Problem 26.60

Think

Does the presence of more OH groups usually lead to greater or lesser water solubility? If there are more OH atoms adjacent to each other, is intramolecular hydrogen bonding likely? How does this affect the availability of those OH groups to form hydrogen bonds with water? How does that affect the water solubility of the polymer?

Solve

Usually, the presence of more HO groups increases the water solubility of a molecule (including polymers), owing to increased intermolecular hydrogen bonding capability with water. However, in the case of poly(vinyl alcohol), the presence of more HO groups leads to more *intra*-molecular hydrogen bonding and decreased *inter*-molecular hydrogen bonding with water. The decreased hydrogen bonding with water decreases the polymer's water solubility.

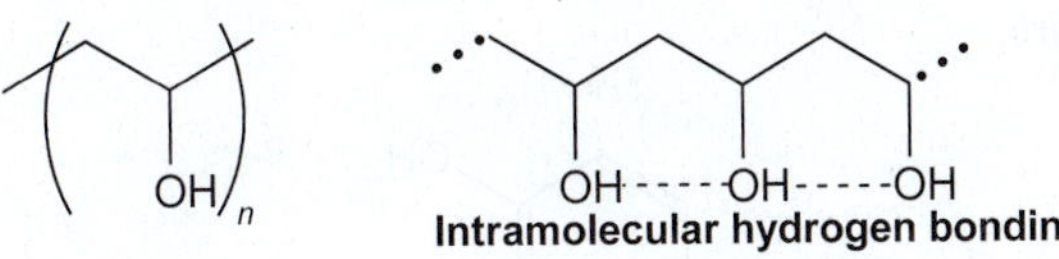

Problem 26.61

Think

Review the structures from Problem 26.34. Which polymers are capable of hydrogen bonding with water? How do you determine the extent of hydrogen bonding that can occur? How does hydrogen bonding assist in a polymer's solubility in ethanol?

Solve

The polymer of ethyl acrylate, poly(ethyl acrylate), **(b)**, will be most soluble in ethanol. The oxygen atoms in the pendant groups are capable of hydrogen bonding with ethanol molecules. The oxygen atoms are hydrogen-bond acceptors. The pendant groups in polymer **(c)** can also undergo hydrogen bonding owing to the presence of N, but each of the ester groups in **(b)** has two hydrogen-bond acceptors, compared to one in the pendant group in **(c)**, which allows for more extensive hydrogen bonding.

Problem 26.62

Think

Review the structures from Problem 26.34. Do the polymers have different polarities? Is hexane polar or nonpolar? Which polymer has intermolecular forces similar to hexane?

Solve

Polybut-1-ene, or **(a)**, will be most soluble in hexane. It is a hydrocarbon, as is hexane, and most similar to hexane in polarity. The polar side chains of **(b)** and **(c)** will make them less likely to dissolve in nonpolar hexane because, upon doing so, more significant intermolecular interactions between polymer molecules would be lost.

Problem 26.63

Think

Is polystyrene polar or nonpolar? Which solvent, benzene or methanol, is nonpolar?

Solve

Benzene is a good solvent for polystyrene (PS), because it is an aromatic hydrocarbon similar to the phenyl groups that are pendant groups (or side groups) on the PS chain. Both benzene and PS are nonpolar. Methanol, however, is a polar solvent, and the addition of methanol to the solution of benzene and PS decreases the solubility of PS until the PS precipitates as a solid.

Problem 26.64

Think

Is the polymer more or less symmetric? How does symmetry affect the ease with which a polymer can form a crystal lattice? Will this favor the polymer existing in a dissolved state or in a crystalline state?

Solve

The addition of a second *tert*-butyl group increases the symmetry of the polymer. This makes it easier for the polymer to form a repeating pattern in its crystal lattice and, therefore, favors the solid form. Just as this increases the T_m of the polymer (factor 4 on p. 1348 of the textbook), it decreases the polymer's solubility.

Section 26.11 The Organic Chemistry of Biomolecules

Problem 26.65

Think

What atoms make up a disulfide bond? Identify this bond in oxytocin. What information does the primary structure of a protein contain? What distinguishes one amino acid from another? Review Figure 26.45a for the abbreviated form of a primary structure.

Solve

(a) A disulfide bond is a S–S bond and is highlighted in light gray in the figure below. The backbone of the protein is highlighted in dark gray in the line structure.

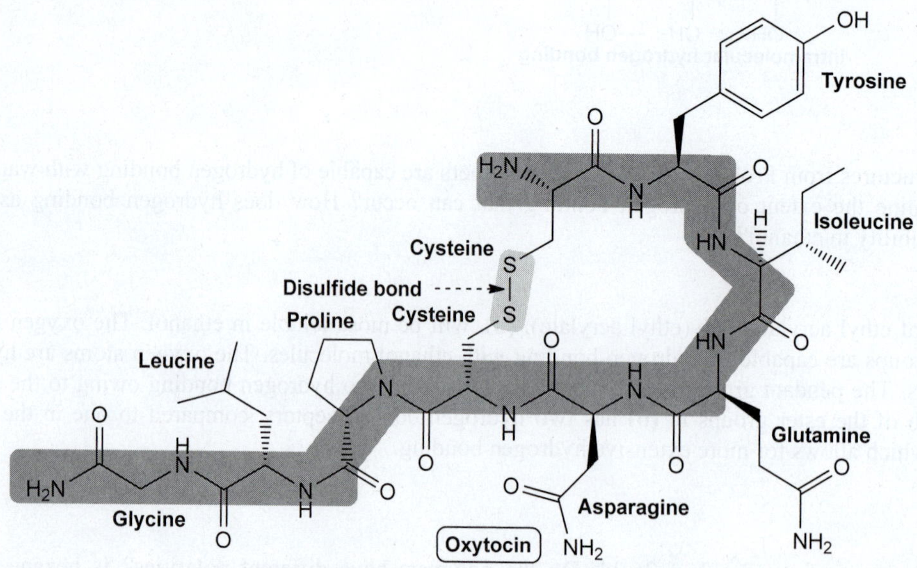

(b) The primary structure of a protein depicts the specific sequence of amino acids. The identities of the amino acids are determined by the side group on each α carbon, as shown in the figure above. For oxytocin, the primary structure is cysteine–tyrosine–isoleucine–glutamine–asparagine–cysteine–proline–leucine–glycine–amide, in which the two cysteine amino acids form the disulfide bond. The abbreviated form is shown below.

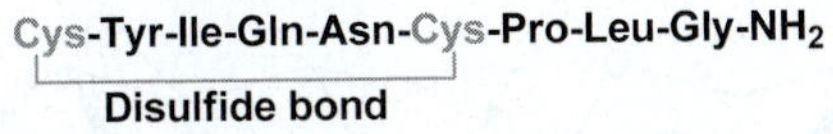

Cys-Tyr-Ile-Gln-Asn-Cys-Pro-Leu-Gly-NH_2

Disulfide bond

Problem 26.66

Think

Is an aqueous environment hydrophobic or hydrophilic? In the aqueous environment, where do the hydrophobic amino acids prefer to reside: exposed to the aqueous environment or hidden from it? Where do the hydrophilic amino acids prefer to reside?

Solve

A coiled coil is stabilized by the hydrophobic effect to hide the hydrophobic side groups in each of the coils from the aqueous environment. Without a second coil present, the hydrophobic side groups of a single coil would be exposed to the aqueous environment. With a second coil, the hydrophobic side groups down the length of one coil aggregate with the hydrophobic side groups down the length of the second coil. In this way, the hydrophobic side groups remain hidden from the aqueous environment.

Problem 26.67

Think

What are the structures of Gly, Ser, and Ala? How bulky are their side groups? Would bulky side groups favor the formation of a β-sheet or disrupt it? Would hydrogen bonding in the side groups favor the formation of a β-sheet or disrupt it?

Solve

The structure for the repeating pattern (Gly-Ser-Gly-Ala-Gly-Ala) within β-pleated sheets for silk is shown below. All of the side groups are rather small, which is important to avoid excessive steric hindrance that would disrupt the formation of a β-sheet. Furthermore, the OH group in serine can form hydrogen bonds with the amide linkage in the protein's backbone, which helps stabilize the β-pleated sheet.

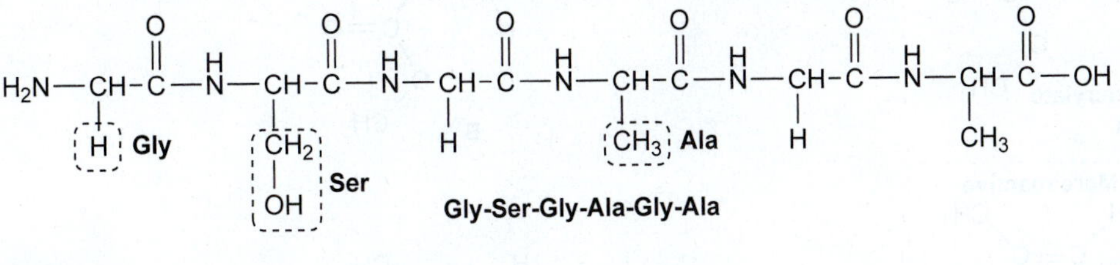

$$H_2N{-}CH(H){-}C(=O){-}NH{-}CH(CH_2OH){-}C(=O){-}NH{-}CH(H){-}C(=O){-}NH{-}CH(CH_3){-}C(=O){-}NH{-}CH(H){-}C(=O){-}NH{-}CH(CH_3){-}C(=O){-}OH$$

Gly **Ser** **Ala**

Gly-Ser-Gly-Ala-Gly-Ala

Problem 26.68

Think

What functional group characterized a glycosidic linkage? What are the types of glycosidic linkages (review Section 9.13)? What is the difference between a storage polysaccharide and a structural polysaccharide? What factors favor each?

Solve

(a) A glycosidic linkage is part of an acetal functional group, in which a C atom is attached to two OR groups. The glycosidic linkages in arabinoxylan are labeled in the figures on the next page (gray screen), and each acetal carbon is circled. The numbers identify the carbons from two different sugar rings that make up the glycosidic linkage. In a six-membered ring sugar, α characterizes the configuration in which the bond between the acetal carbon and OR is axial, and β characterizes the configuration in which that bond is equatorial. In a five-membered ring sugar, the analogous configuration for α has the acetal C–OR bond on the side of the ring opposite the carbon-containing substituent (in this case, CH_2OH); for the β configuration, those groups would be on the same side of the ring.

HOH$_2$C
OH
α-1,3'-Glycosidic linkage
HO
β-1,4'-Glycosidic linkage
β-1,4'-Gylcosidic linkage
HO
HO
OH
HO
HO
OH
HOH$_2$C
HOH$_2$C
OH
OH
HO
CH$_2$OH
α-1,3'-Glycosidic linkage
Arabinoxylan
α-1,3'-Glycosidic linkage

(b) Arabinoxylan is a storage polysaccharide. Storage polysaccharides generally have substantial branching so that the sugar units can be hydrolyzed more rapidly for quick energy.

Integrated Problems

Problem 26.69

Think

What contributes to the radical stability of **B** and **D**? Which radical is more stable? Does an increase in product stability contribute to a faster or slower reaction of the reactant?

Solve

D has an additional CH_3 group attached to the C•, which stabilizes the radical owing to the electron-donating ability of alkyl groups. This greater stability makes **D** less reactive than **B**. Moreover, because **D** is at a lower energy than **B**, **D** is produced more readily from **C** than **B** is produced from **A**. This makes **C** more reactive than **A**.

R• + Methyl acrylate **A** → **B**

R• + Methyl methacrylate **C** (More reactive) → **D**

More stable C• radical product

Problem 26.70

Think

What are the structures of terephthalic acid and butane-1,4-diol? What functional group is present in each compound? What new functional group forms as a result of the reaction under acidic conditions? Review the structure of PET. How does replacing ethane-1,2-diol with butane-1,4-diol change the structure of the polymer? Review the factors for T_m. Which polymer has a longer flexible portion of its repeating unit?

Solve

(a) Structures for terephthalic acid, butane-1,4-diol, the polymer formed, and the condensed formula are given below. The dicarboxylic acid and the diol combine to form a polyester. The reaction that produces each ester is essentially a Fischer esterification reaction (Chapter 21).

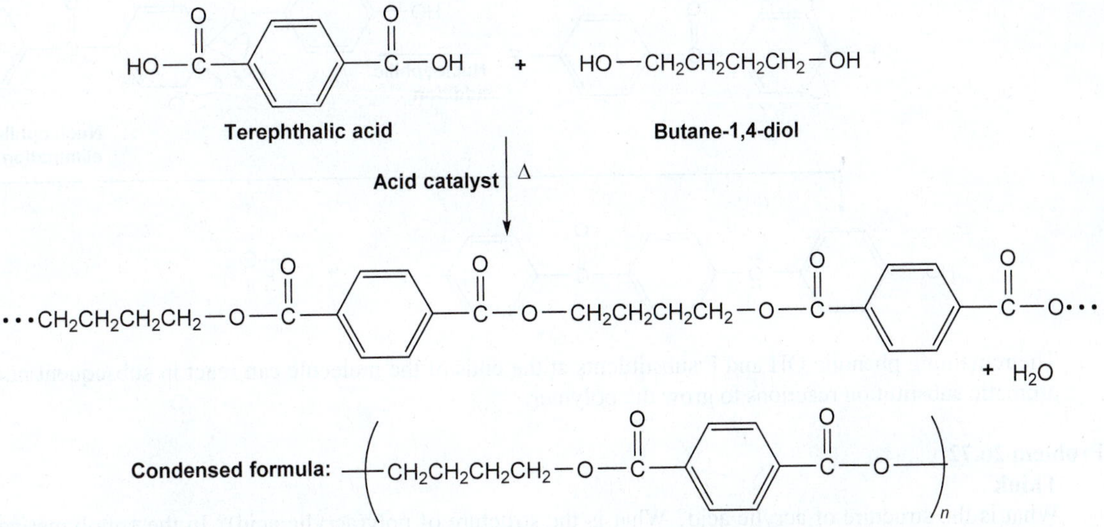

Terephthalic acid

Butane-1,4-diol

Acid catalyst

Δ

$+ H_2O$

Condensed formula:

(b) PBT stands for poly(butylene terephthalate). Recall that PET stands for poly(ethylene terephthalate). In PBT, 1,4-butanediol (butylene glycol) is used instead of 1,2-ethanediol (ethylene glycol), so the *ethylene* in poly(ethylene terephthalate) is replaced by *butylene*.

(c) PBT will have a lower melting point. Increasing the length of the flexible alkyl chains in the polymer (factor 6) will allow easier mobility of the polymer molecules. A lower temperature will supply sufficient kinetic energy for the chains to go from the solid phase to the liquid phase.

(d) The same flexibility described in **(c)** allows the PBT molecules to move into a crystal lattice more quickly than PET molecules.

Problem 26.71

Think

What bonds are broken and formed during the course of the reaction? What is the leaving group? What is the nucleophile? Are the aromatic rings in difluorobenzophenone electron rich or electron poor? Is the reaction acid or base catalyzed?

Solve

This reaction is a nucleophilic aromatic substitution (see Section 23.9a). Each ring of difluorobenzophenone has an electron-poor aromatic ring owing to the electron-withdrawing C=O group and F atom. The rings, therefore, are susceptible to nucleophilic attack and have F^- leaving groups. The first step is an acid–base reaction between the phenolic hydrogen and the carbonate anion, a base:

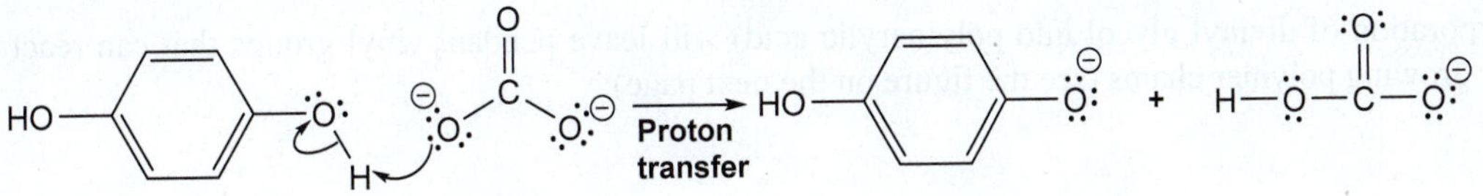

Proton transfer

The phenoxide anion then acts as a nucleophile in the subsequent nucleophilic addition–elimination:

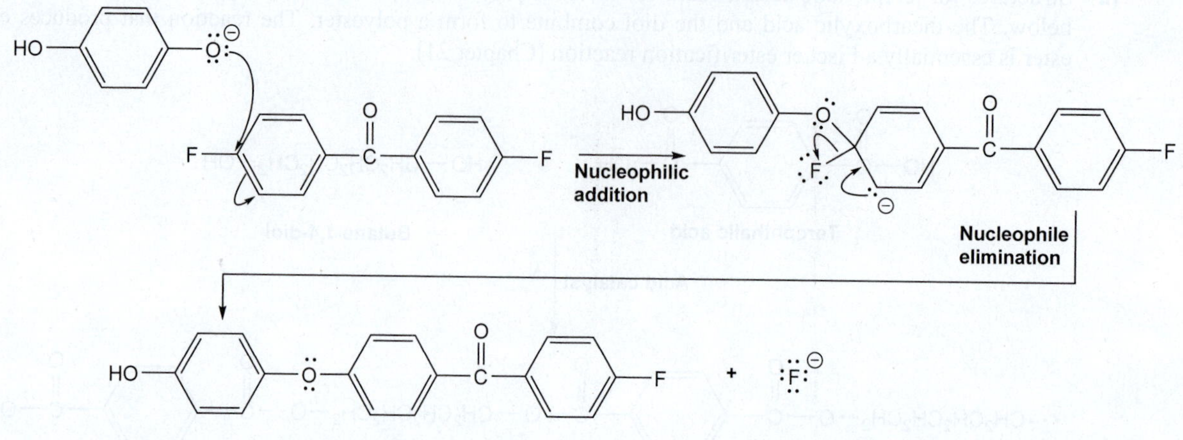

HO F F C O F Nucleophilic addition HO O F C O F Nucleophile elimination HO O C O F + F

The remaining phenolic OH and F substituents at the ends of the molecule can react in subsequent nucleophilic aromatic substitution reactions to grow the polymer.

Problem 26.72

Think

What is the structure of acrylic acid? What is the structure of poly(acrylic acid)? In the copolymerization with divinyl glycol, how does one C=C bond of divinyl glycol react with the growing polymer chain of poly(acrylic acid)? After a molecule of divinyl glycol is incorporated, how can the other C=C on that monomer react with another growing polymer chain? What does the structure of the cross-linked polymer look like? Is the cross-linked polymer more or less water soluble than the linear polymer?

Solve

(a) Acrylic acid has the following structure:

H H C=C H C=O HO

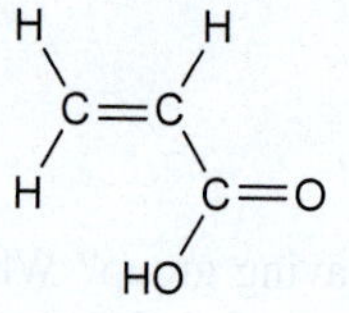

Linear poly(acrylic acid) has the following structure:

C=O C=O C=O C=O C=O C=O C=O C=O C=O HO HO HO HO HO HO HO HO HO

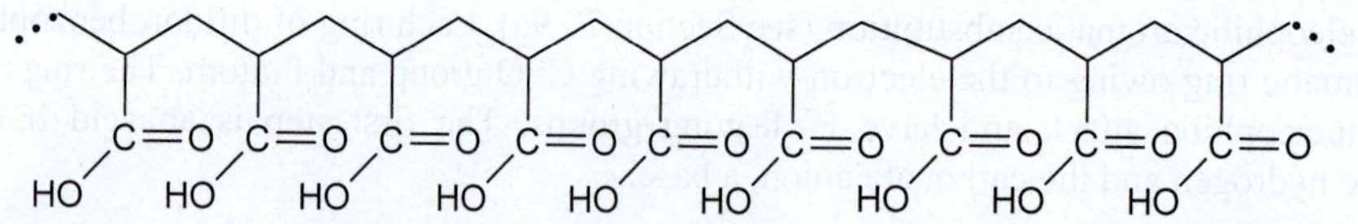

Incorporation of divinyl glycol into poly(acrylic acid) will leave pendant vinyl groups that can react with other growing polymer chains (see the figure on the next page):

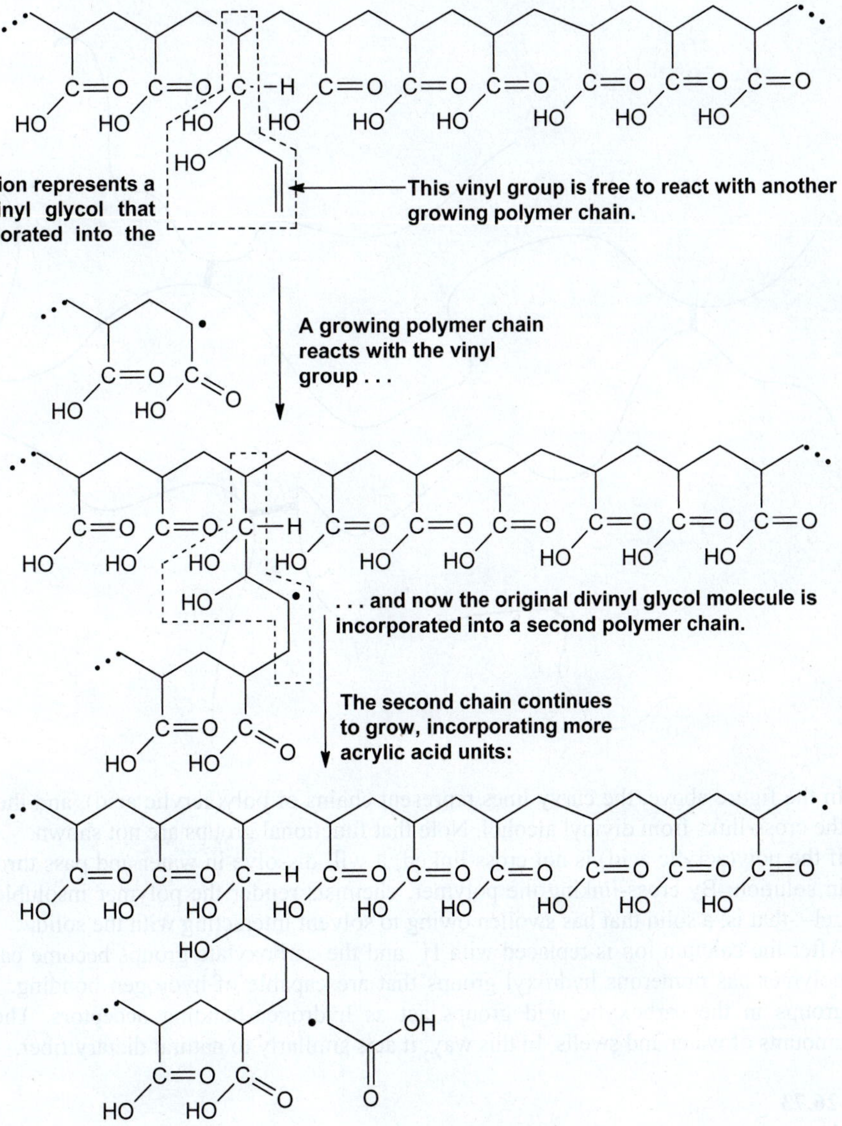

The outlined portion represents a molecule of divinyl glycol that has been incorporated into the polymer chain.

This vinyl group is free to react with another growing polymer chain.

A growing polymer chain reacts with the vinyl group . . .

. . . and now the original divinyl glycol molecule is incorporated into a second polymer chain.

The second chain continues to grow, incorporating more acrylic acid units:

The vinyl glycol acts as a cross-linking agent. Conceptually, the polymer will look like this:

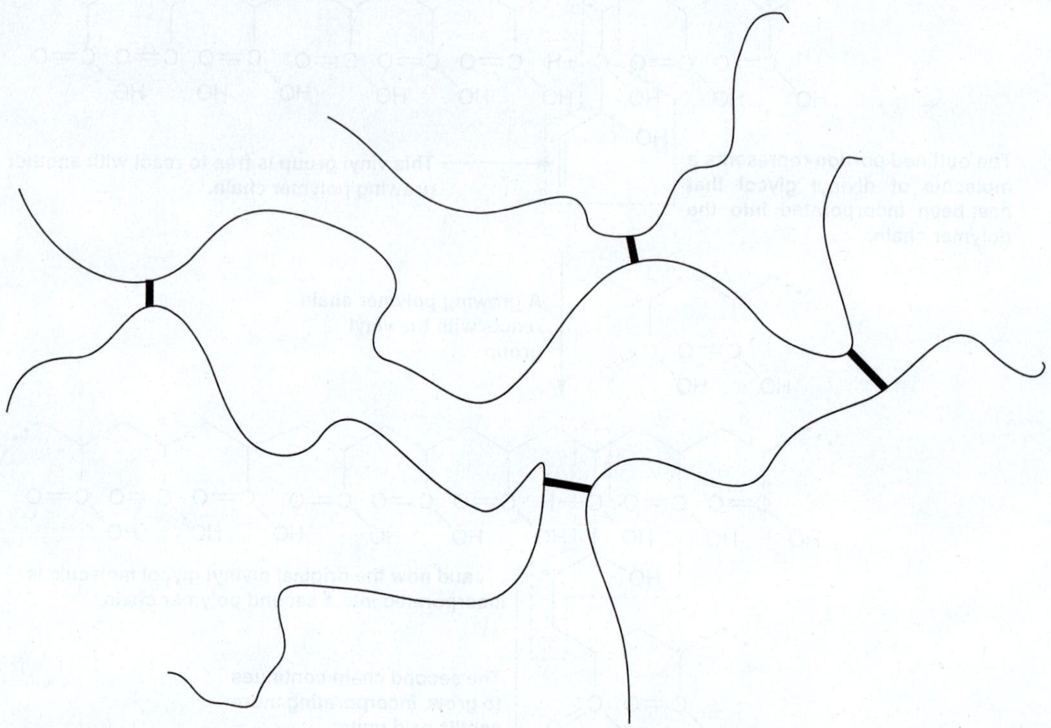

In the figure above, the curvy lines represent chains of poly(acrylic acid), and the bold segments represent the cross-links from divinyl alcohol. Note that functional groups are not shown.

(b) If the poly(acrylic acid) is not cross-linked, it will dissolve in water and pass through the digestive system in solution. By cross-linking the polymer, chemists render the polymer insoluble, so it passes through as gel—that is, a solid that has swollen owing to solvent interacting with the solid.

(c) After the calcium ion is replaced with H^+ and the carboxylate groups become carboxylic acid groups, the polymer has numerous hydroxyl groups that are capable of hydrogen bonding. In addition, the carbonyl groups in the carboxylic acid groups act as hydrogen-bonding acceptors. The polymer absorbs large amounts of water and swells. In this way, it acts similarly to natural dietary fiber.

Problem 26.73

Think

What reaction occurs between formaldehyde and an NH_2 group from melamine? Can the resulting functional group react with an NH_2 group from another melamine molecule? How many amines are there in melamine that can react? In how many directions does polymerization occur with respect to a single monomer? Are network polymers reactive?

Solve

(a) The NH_2 groups on melamine are nucleophilic and can undergo nucleophilic addition to formaldehyde. As shown below, this can occur three times to produce trimethylol melamine.

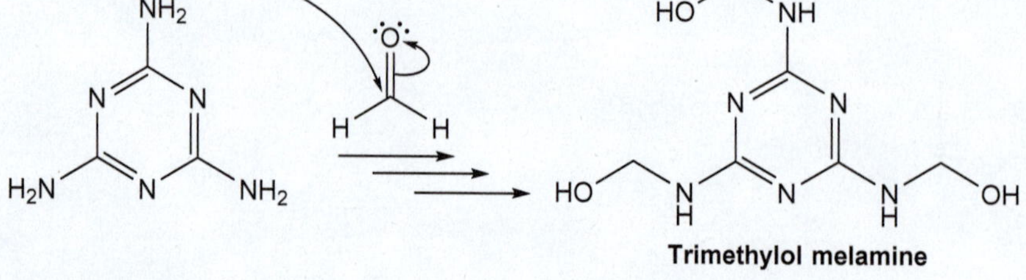

Trimethylol melamine

Under mildly acidic conditions, the OH groups of trimethylol melamine become good leaving groups and can react in a substitution reaction with the NH_2 group of another melamine molecule. This creates a NH–CH_2–NH bridge between two aromatic rings, as shown below.

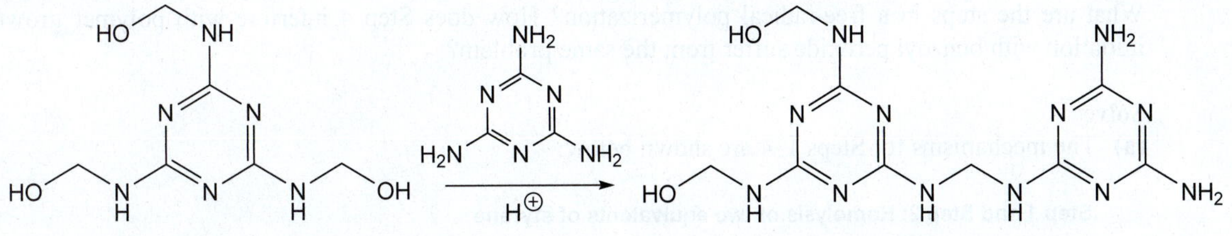

Further reaction results in bridges at three locations on each ring, as shown below.

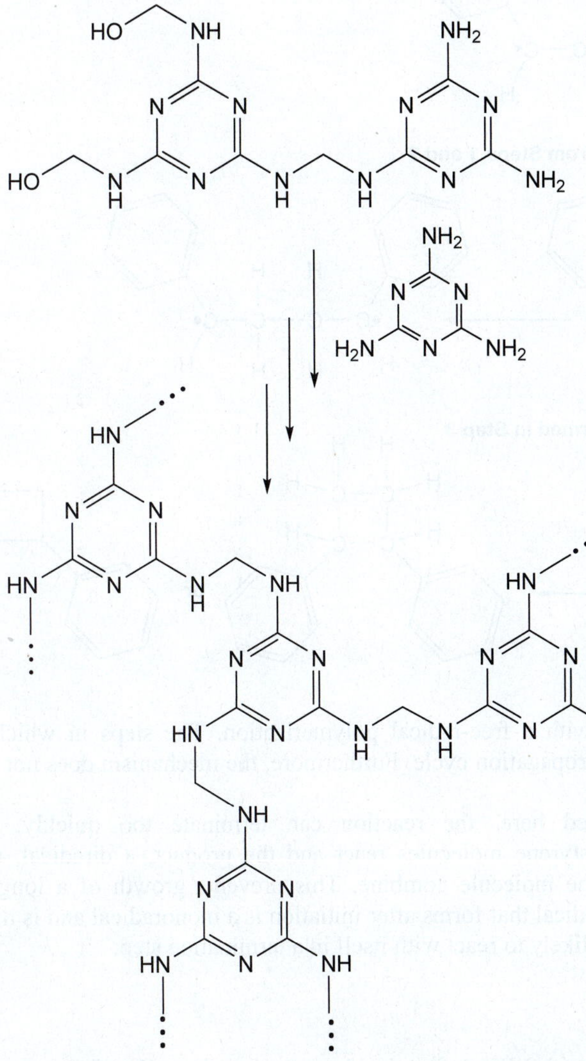

(b) Network polymers are generally insoluble and, therefore, are unaffected by solvents. The polymer will be a good surface to write on because the solvent in the markers will not penetrate the surface, nor will solvents that are used to clean the surface.

Problem 26.74

Think

What is the vinyl π bond? How many single-barbed curved arrows are needed to show a homolysis step? Which part of the radical monomer is the tail, and which part is the head? How can you show tail-to-tail addition? What are the steps in a free-radical polymerization? How does Step 4 interfere with polymer growth? Does initiation with benzoyl peroxide suffer from the same problem?

Solve

(a) The mechanisms for Steps 1–4 are shown below:

Step 1 and Step 2: Homolysis of two equivalents of styrene

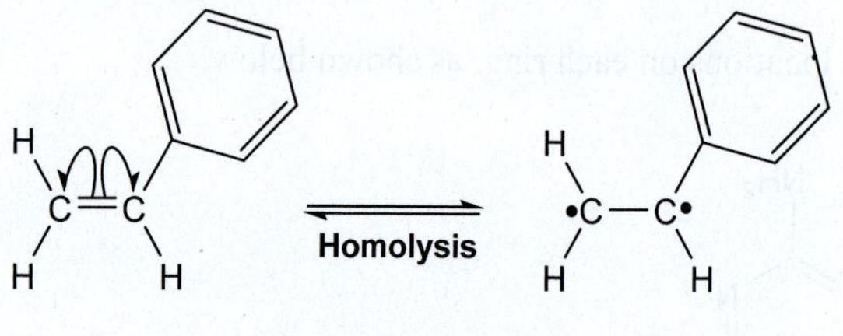

Homolysis

Step 3: Tail-to-tail addition of the radicals from Steps 1 and 2

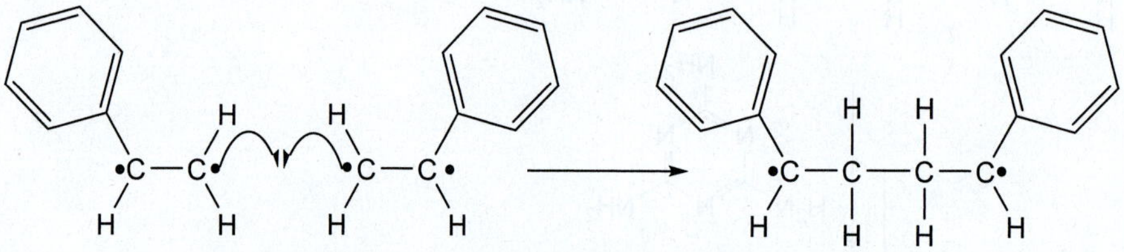

Step 4: Radical coupling of the diradical formed in Step 3

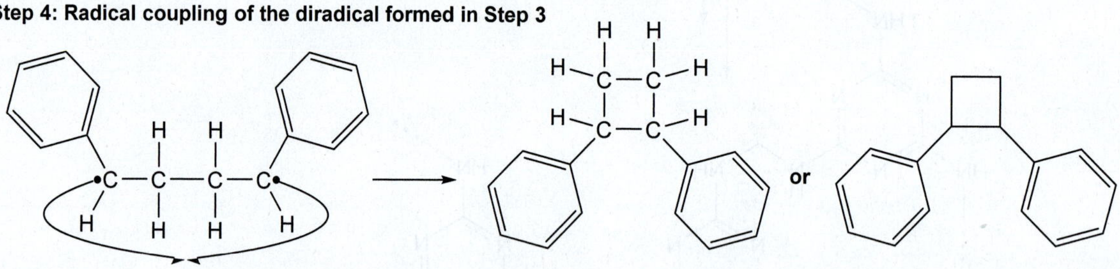

or

(b) No, this mechanism is not consistent with a free-radical polymerization. The steps in which the free radicals are involved do not make up a propagation cycle. Furthermore, the mechanism does not produce a polymer.

(c) In the autoinitiation reaction described here, the reaction can terminate too quickly. The 1,2-phenylcyclobutane forms because two styrene molecules react and the product, a diradical, undergoes termination when the two radicals in the molecule combine. This prevents growth of a long chain of molecules. With benzoyl peroxide, the radical that forms after initiation is a monoradical and is more likely to continue propagation because it is not likely to react with itself in a termination step.

Problem 26.75

Think

How does an alcohol react with an acid chloride? What functional group does it produce? For a polymer to form, should phosgene be able to react with two separate monomers? Which monomer will prevent reaction with phosgene in such a way as to prevent phosgene from being able to react with a second monomer of the diol?

Solve

An alcohol will react with an acid chloride to form an ester. This can happen twice for each diol and twice for phosgene. To form a polymer, however, phosgene should form one ester with one diol monomer and the other ester group with a separate diol monomer. This can happen for Compounds **Y** and **Z**. Compound **X**, on the other hand, will not form a polymer, because the second –OH group is close enough to the first that both react with phosgene, producing a molecule that can no longer react with the monomers.

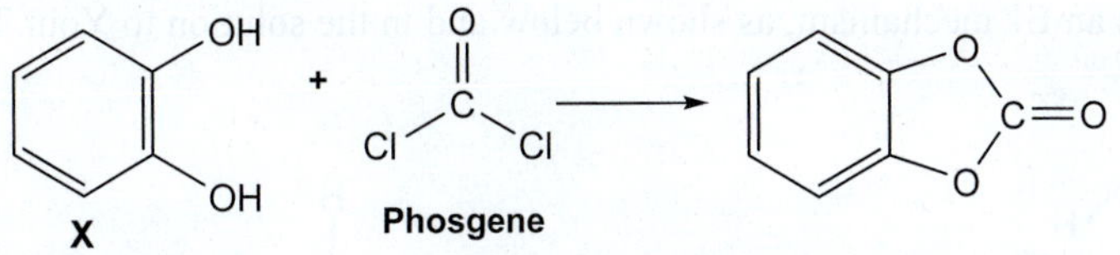

OH, OH, **X** + Cl–C(=O)–Cl **Phosgene** → C=O

Problem 26.76

Think

What functional group repeats in polyurethane? How does that functional group compare to the structure of urea? What atoms are different? What does that suggest needs to be changed in the monomer to produce a polyurea?

Solve

(a) The synthesis of a polyurethane (Equation 26-29) is repeated below. Notice that the O–(C=O)–N motif repeats, and it appears twice in the repeating unit.

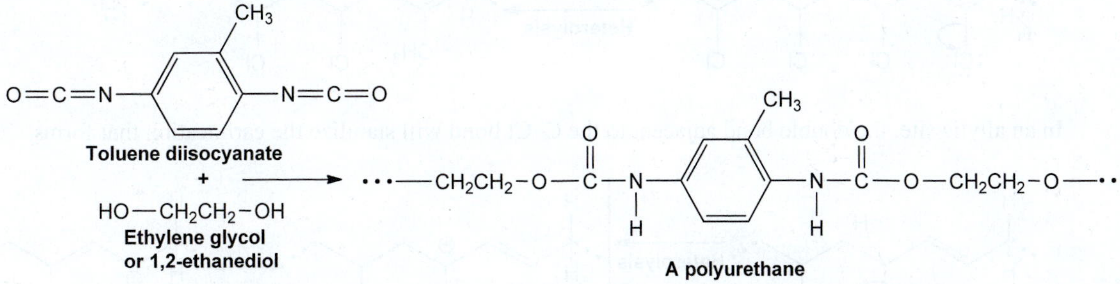

CH_3, O=C=N– –N=C=O **Toluene diisocyanate** + HO—CH_2CH_2–OH **Ethylene glycol or 1,2-ethanediol** → ···—CH_2CH_2–O—C(=O)—N(H)— CH_3 —N(H)—C(=O)—O—CH_2CH_2–O—··· **A polyurethane**

To synthesize a polyurea instead, the O–(C=O)–N motif in the repeating unit needs to be N–(C=O)–N, so you replace the diol in the polyurethane synthesis with a diamine, as shown below:

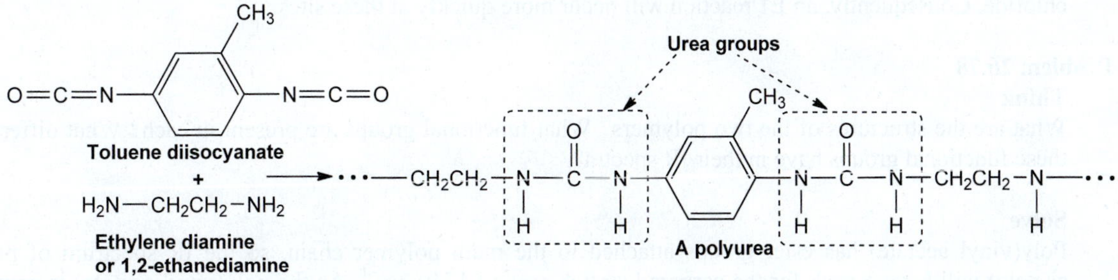

CH_3, O=C=N– –N=C=O **Toluene diisocyanate** + H_2N—CH_2CH_2–NH_2 **Ethylene diamine or 1,2-ethanediamine** → ···—CH_2CH_2–N(H)—C(=O)—N(H)— CH_3 —N(H)—C(=O)—N(H)–CH_2CH_2–N(H)—··· **A polyurea**, **Urea groups**

(b) Condensed formula:

—[C(=O)—N(H)— CH_3 —N(H)—C(=O)—N(H)—CH_2CH_2–N(H)]$_n$—

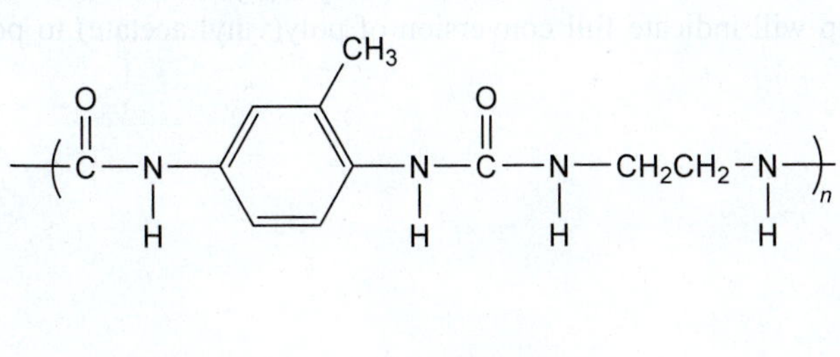

Problem 26.77

Think

Review the mechanism in the solution to Your Turn 26.21 to review how PVC unzips. What is the difference in stability of the intermediate of PVC unzipping compared to PVC with tertiary or allylic chlorides? How will that affect the rate of the reaction?

Solve

The mechanism by which PVC "unzips" is an E1 mechanism, as shown below and in the solution to Your Turn 26.21:

Heterolysis

Electrophile elimination

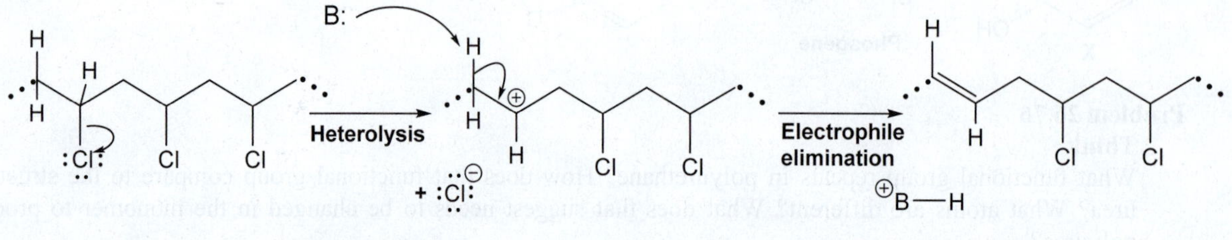

Note that the carbocation formed in the first step is secondary. If the PVC contains sites with tertiary chlorides, the resulting carbocation is tertiary and more stable than a secondary carbocation:

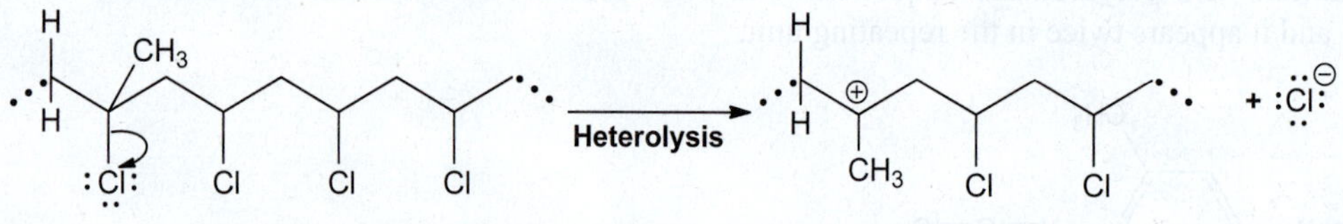

Heterolysis

In an allylic site, the double bond adjacent to the C–Cl bond will stabilize the carbocation that forms:

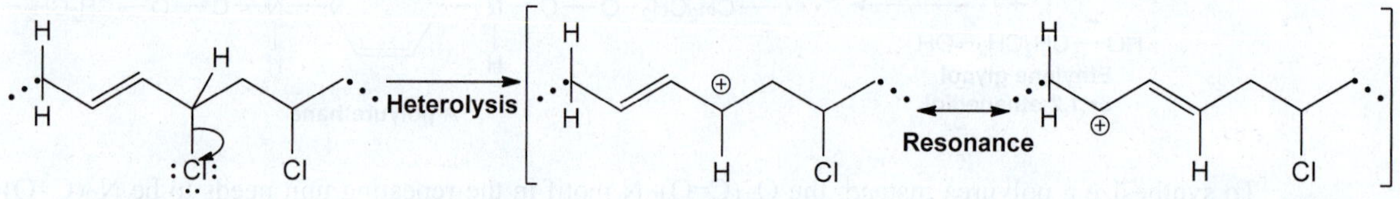

Heterolysis

Resonance

Because of the increased carbocation stability, the Cl^- will leave more readily from either a tertiary or allylic chloride. Consequently, an E1 reaction will occur more quickly at these sites.

Problem 26.78

Think

What are the structures of the two polymers? What functional groups are present in each? What differences do these functional groups have in their IR spectra?

Solve

Poly(vinyl acetate) has ester groups attached to the main polymer chain, so the IR spectrum of poly(vinyl acetate) will have a peak for the carbonyl stretch around 1740 cm^{-1}. As the poly(vinyl acetate) is converted to poly(vinyl alcohol), an absorption stretch of the hydroxyl group will appear around 3400 cm^{-1}. As the conversion proceeds, the carbonyl stretch will decrease in intensity, and the hydroxyl stretch will increase in intensity. The absence of the carbonyl group will indicate full conversion of poly(vinyl acetate) to poly(vinyl alcohol).

CREDITS

p. 698: YT 15.9: © Sigma-Aldrich Co. LLC. Used with permission.
p. 698: YT 15.10: © Sigma-Aldrich Co. LLC. Used with permission.
p. 699: YT 15.11: © Sigma-Aldrich Co. LLC. Used with permission.
p. 699: YT 15.12: © Sigma-Aldrich Co. LLC. Used with permission.
p. 700: YT 15.13: © Sigma-Aldrich Co. LLC. Used with permission.
p. 701: YT 15.15: © Sigma-Aldrich Co. LLC. Used with permission.
p. 702: YT 15.17: © Sigma-Aldrich Co. LLC. Used with permission.
p. 702: YT 15.18: © Sigma-Aldrich Co. LLC. Used with permission.
p. 703: YT 15.19: © Sigma-Aldrich Co. LLC. Used with permission.
p. 704: YT 15.20: © Sigma-Aldrich Co. LLC. Used with permission.
p. 704: YT 15.21–24: © Sigma-Aldrich Co. LLC. Used with permission.
p. 705: YT 15.25–28: © Sigma-Aldrich Co. LLC. Used with permission.
p. 706: YT 15.30: © Sigma-Aldrich Co. LLC. Used with permission.
p. 710: 15.15: © Sigma-Aldrich Co. LLC. Used with permission.
p. 712: 15.20: © Sigma-Aldrich Co. LLC. Used with permission.
p. 712: 15.21: © Sigma-Aldrich Co. LLC. Used with permission.
p. 713: 15.23: © Sigma-Aldrich Co. LLC. Used with permission.
p. 714: 15.24: © Sigma-Aldrich Co. LLC. Used with permission.
p. 714: 15.25: © Sigma-Aldrich Co. LLC. Used with permission.
p. 715: 15.26: © Sigma-Aldrich Co. LLC. Used with permission.
p. 727: 15.53: © Sigma-Aldrich Co. LLC. Used with permission.
p. 729–730: 15.56: © Sigma-Aldrich Co. LLC. Used with permission.
p. 732: 15.59: © Sigma-Aldrich Co. LLC. Used with permission.
p. 732: 15.60: © Sigma-Aldrich Co. LLC. Used with permission.
p. 733: 15.61: © Sigma-Aldrich Co. LLC. Used with permission.
p. 734: 15.62: © Sigma-Aldrich Co. LLC. Used with permission.
p. 734: 15.63: © Sigma-Aldrich Co. LLC. Used with permission.
p. 735: 15.64: © Sigma-Aldrich Co. LLC. Used with permission.
p. 735: 15.65: © Sigma-Aldrich Co. LLC. Used with permission.
p. 736: 15.66: © Sigma-Aldrich Co. LLC. Used with permission.
p. 737: 15.67: © Sigma-Aldrich Co. LLC. Used with permission.
p. 783: 16.64: © Sigma-Aldrich Co. LLC. Used with permission.
p. 783: 16.65: © Sigma-Aldrich Co. LLC. Used with permission.